TO THE STUDENT

As you begin, you may feel anxious about the number of theorems, definitions, procedures, and equations. You may wonder if you can learn it all in time. Don't worry, your concerns are normal. This textbook was written with you in mind. If you attend class, work hard, and read and study this book, you will build the knowledge and skills you need to be successful. Here's how you can use the book to your benefit.

Read Carefully

When you get busy, it's easy to skip reading and go right to the problems. Don't... the book has a large number of examples and clear explanations to help you break down the mathematics into easy-to-understand steps. Reading will provide you with a clearer understanding, beyond simple memorization. Read it before class (not after) so you can ask questions about anything you didn't understand. You'll be amazed at how much more you'll get out of class if you do this.

Use the Features

I use many different methods in the classroom to communicate. Those methods, when incorporated into the book, are called "features". The features serve many purposes, from providing timely review of material you learned before (just when you need it), to providing organized review sessions to help you prepare for quizzes and tests. Take advantage of the features and you will master the material.

To make this easier, I've provided a brief guide to getting the most from this book (just open this insert). Spend fifteen minutes reviewing the guide and familiarizing yourself with the features by flipping to the page numbers provided. Then, as you read, use them. This is the best way to make the most of your textbook.

Please do not hesitate to contact me, through Prentice Hall, with any questions, suggestions, or comments that would improve this text. I look forward to hearing from you, and best of luck with all of your studies.

Best Wishes!
Michael Sullivan

Review
"Study for Quizzes and Tests"

Feature	Description	Benefit	Page
	Chapter Reviews at the end of each chapter contain…		
"Things to Know"	A detailed list of important theorems, formulas, identities, definitions, and functions from the chapter.	Review these and you'll know the most important material in the chapter!	314-15
NEW **"You Should be Able To…"**	Contains a complete list of objectives by section, with corresponding practice exercises	Do the recommended exercises and you'll have mastery over the key material. If you get something wrong, review the suggested page numbers and try again.	315-6
Review Exercises	These provide comprehensive review and practice of key skills, matched to the Learning Objectives for each section.	Practice makes perfect.These problems combine exercises from all sections, giving you a comprehensive review in one place.	316-20
Practice Test	The Review Exercises contain blue-numbered problems. Together, these blue-test numbered problems constitute a Chapter Practice Test.	Be prepared. Take the sample practice test. This will get you ready for your instructor's test.	316-20
NEW **Chapter Projects**	The Chapter Project applies what you've learned in the chapter. Additional projects are available on a website.	The Project gives you an opportunity to apply what you've learned in the chapter to solve a problem related to the opening article. If your instructor allows, these make excellent opportunities to work in a group, which is often the best way of learning math.	320-21
NEW **Cumulative Review**	These problem sets appear at the end of Chapters 2-13. They combine problems from previous chapters, providing an ongoing cumulative review.	These are really important. They will ensure that you are not forgetting anything as you go. These will go a long way toward keeping you constantly primed for quizzes and tests.	321-322

Practice
"Work the Problems"

Feature	Description	Benefit	Page
	"Assess Your Understanding" contains a variety of problems at the end of each section.		
NEW **'Are You Prepared?' Problems**	These assess your retention of the prerequisite material you'll need. Answers are given at the end of the section exercises. This feature is related to the Preparing for This Section feature.	Do you always remember what you've learned? Working these problems is the best way to find out. If you get one wrong, you'll know exactly what you need to review, and where to review it!	255
NEW **Concepts and Vocabulary**	These short-answer problems, namely, Fill-in-the-Blank and True/False items, assess your understanding of key definitions and concepts in the current section.	Learning math is more than memorization – it's about discovering connections. These problems help you understand the 'big ideas' before diving into skill building.	255
Skill Development	Correlated to section examples, these problems provide straightforward practice, organized by difficulty.	It's important to dig in and develop your problem solving skills. These problems provide you with ample practice to do so.	255-7
Graphical	These problems utilize graphs in a variety of ways.	You will supplement your analytical understanding with graphical understanding.	256
Now Work Problems	Many examples refer you to a related homework problem. These related problems are marked by a pencil and yellow numbers.	If you get stuck while working problems, look for the closest Now Work problem and refer back to the related example to see if it helps.	257
Applications	Application problems ("word problems") follow the basic skill development problems.	Math is everywhere, and these problems demonstrate that. You'll learn to approach real problems, and how to break them down into manageable parts. These can be challenging, but are worth the effort.	257-260
Graphing Calculator	These optional problems require the use of a graphing utility, and are marked by a special icon and green numbers.	Your instructor will usually provide guidance on whether or not to do these problems. If so, these problems help to verify and visualize your analytical results.	258
"DWR"	"Discussion, Writing, and Research" problems are marked by a special icon and red numbers. These support class discussion, verbalization of mathematical ideas, and writing and research projects.	To verbalize an idea, or to describe it clearly in writing, shows real understanding. These problems nurture that understanding. They're challenging , but you'll get out what you put in.	260 (#99)

Resources to Help You Study

Student Study-Pak

A single, easy-to-use package, available bundled with your textbook or standalone, for purchase through your bookstore. This invaluable study package contains the following:

➢ **Student Solutions Manual**
A printed manual containing full solutions to odd-numbered textbook exercises.

➢ **Pearson Tutor Center**
Tutors provide one-on-one tutoring for any problem with an answer at the back of the book. You can contact the Tutor Center via a toll-free phone number, fax, or email.

➢ **CD-ROM Lecture Series**
A comprehensive set of textbook-specific CD-ROMS containing short video clips of textbook objectives being reviewed and key examples being solved. These videos provide excellent support for those who require extra assistance, or for those in distance learning and/or self-pace courses.

➢ **Algebra Review**
Four chapters of intermediate algebra review, written in the same style as your book.

Tutorial and Homework Options

➢ **MathXL®**
MathXL® is an online homework, tutorial, and assessment system that accompanies your textbook. Instructors can create and assign online homework and tests using algorithmically generated exercises correlated to the textbook. Your work is tracked in an online gradebook. You can take chapter tests and receive personalized study plans based on your results. The study plan diagnoses weaknesses and links you to tutorial exercises for further study. You can also access video clips from select exercises. For information, visit www.mathxl.com.
(MathXL must be set up and assigned by your instructor)

➢ **MyMathLab (Internet)**
MyMathLab® is a complete online course to help you succeed in learning. MyMathLab contains an online version of your textbook with links to multimedia resource, such as video clips and practice exercises, correlated to the examples and exercises in the text. MyMathLab provides you with online homework and tests and generates a personalized study plan based on your results. The study plan links you to unlimited tutorial exercises for further study, so you can practice until you have mastered the skills. All of the online homework, tests, and tutorial work you do is tracked in your MyMathLab gradebook.
(MyMathLab must be set up and assigned by your instructor)

INSTRUCTOR'S EDITION

SEVENTH EDITION

PRECALCULUS

Michael Sullivan
Chicago State University

PEARSON
Prentice Hall

Upper Saddle River, New Jersey 07458

Editor-in-Chief: Sally Yagan
Senior Acquisitions Editor: Eric Frank
Project Manager/Print Supplements Editor: Dawn Murrin
Vice President/Director of Production and Manufacturing: David W. Riccardi
Executive Managing Editor: Kathleen Schiaparelli
Senior Managing Editor: Linda Mihatov Behrens
Production Editor: Bob Walters, Prepress Management, Inc.
Assistant Manufacturing Manager/Buyer: Michael Bell
Manufacturing Manager: Trudy Pisciotti
Marketing Manager: Halee Dinsey
Marketing Assistant: Rachel Beckman
Editorial Assistant: Tina Magrabi
Art Director: Jonathan Boylan
Interior Designers: Judy Matz-Coniglio, Jonathan Boylan
Cover Designer: Geoffrey Cassar
Creative Director: Carole Anson
Art Editor: Thomas Benfatti
Director of Creative Services: Paul Belfanti
Director, Image Resource Center: Melinda Reo
Manager, Rights and Permissions: Zina Arabia
Interior Image Specialist: Beth Brenzel
Image Permission Coordinator: Cynthia Vincenti
Art Studio: Artworks:
Manager Editor, Audio/Video Assets: Patricia Burns
Production Manager: Ronda Whitson
Manager, Production Technologies: Matt Haas
Illustrators: Stacy Smith, Audrey Simonetti, Mark Landis, Nathan Storck, Ryan Currier, Scott Wieber, Royce Copenheaver, Dan Knopsnyder
Art Quality Assurance: Timothy Nguyen, Stacy Smith, Pamela Taylor

Pearson Prentice Hall
Pearson Education, Inc.
Upper Saddle River, New Jersey 07458

Printed in the United States of America
10 9 8 7 6 5 4 3 2 1

ISBN: 0-13-143133-1
ISBN: 0-13-143120-X (Student Edition)

Pearson Education LTD., *London*
Pearson Education Singapore, Pte. Ltd
Pearson Education Canada, Ltd., *Toronto*
Pearson Education–Japan, *Tokyo*
Pearson Education Australia PTY, Limited, *Sydney*
Pearson Education North Asia Ltd. *Hong Kong*
Pearson Educación de Mexico, S.A. de C.V.

To the Memory of
my Mother and my Father

Contents

CHAPTER 4 Exponential and Logarithmic Functions 223

CHAPTER 5 Trigonometric Functions 323

CHAPTER 6 Analytic Trigonometry 413

Three Distinct Series

Students have different goals, learning styles, and levels of preparation. Instructors have different teaching philosophies, styles, and techniques. Rather than write one series to fit all, the Sullivans have written three distinct series. All share the same goal—to develop a high level of mathematical understanding and an appreciation for the way mathematics can describe the world around us. The manner of reaching that goal, however, differs from series to series.

Contemporary Series, Seventh Edition

The Contemporary Series is the most traditional in approach, yet modern in its treatment of precalculus mathematics. Graphing utility coverage is optional and can be included or excluded based on instructor preference.
College Algebra, College Algebra Essentials, Algebra & Trigonometry, Trigonometry, Precalculus, Precalculus Essentials

Enhanced with Graphing Utilities Series, Third Edition

This series provides a more thorough integration of graphing utilities into topics, allowing students to explore mathematical concepts and foreshadow ideas usually studied in later courses. Using technology, the approach to solving certain problems differs from the Contemporary series, while the emphasis on understanding concepts and building strong skills does not.
College Algebra, Algebra & Trigonometry, Trigonometry, Precalculus

Graphing, Data, and Analysis Series, Third Edition

This series differs from the others, utilizing a *functions approach* which serves as the organizing principle tying concepts together. Each chapter introduces a new type of function, then develops all concepts pertaining to that particular function. The solutions of equations and inequalities, instead of being developed as stand-alone topics, are developed in the context of the underlying functions. Technology is integrated at the same level as the Enhanced with Graphing Utilities series (described above).
College Algebra, Algebra & Trigonometry, Precalculus

Preface to the Instructor

As a professor of mathematics at an urban public university for 35 years, I am aware of the varied needs of precalculus students. As the author of engineering calculus, business calculus, and finite mathematics texts, and, as a teacher, I understand what students must know if they are to be focused and successful in upper level mathematics courses.

Too often, precalculus texts are simply condensed versions of algebra and trigonometry texts. College algebra and algebra & trigonometry courses are different from precalculus courses and their texts should reflect this difference. For example, Chapter 13 A Preview of Calculus: The Limit, Derivative, and Integral of a Function, not only demonstrates to students how the material of Precalculus applies to calculus, but also moves the student into calculus. Throughout this text there are references to calculus, identified in the text by a icon, to further motivate and remind the student that this is mathematics that will be used later. There are other aspects of this text that prepare the student for calculus. For example, many applications that are traditional in calculus are presented as problems in a precalculus setting. These are designed to emphasize the role in calculus of precalculus, and to encourage and motivate students to be successful in precalculus.

A tremendous benefit of authoring a successful series is the broad-based feedback received from teachers and students who have used previous editions. I am sincerely grateful for their support. Virtually every change to this edition is the result of their thoughtful comments and suggestions. My hope is that I have been able to take their ideas, and, building upon a successful foundation of the sixth edition, make this series an even better learning and teaching tool for students and teachers.

New Features in the Seventh Edition

Rather than provide a list of new features here, that information can be found in the foldout entitled "To the Student" on the inside front cover of this book. The new features are easy to spot by the "New" in the left column.

This places the new features in their proper context, as building blocks of an overall learning system that has been carefully crafted over the years to help students get the most out of the time they put into studying. Please take the time to review this, and to discuss it with your students at the beginning of your course. My experience has been that if students utilize these features, they do better in the course.

Organizational Changes in the Seventh Edition

- Composite Functions, formerly in Chapter 2 Functions and Their Graphs now appears as Section 4.1 in Chapter 4 Exponential Functions and Logarithmic Functions.
- The material on Logarithmic Scales, formerly Section 4.8, is now discussed in the Exercises. Exponential and Logarithmic Models, formerly part of Section 4.4, has been expanded to include Logistic Models and is now found in Section 4.9, Fitting Data to Exponential, Logarithmic, and Logistic Functions.
- Systems of Linear Equations: Two Equations Containing Two Variables and Systems of Linear Equations: Three Equations Containing Three Variables have been combined in one section as Systems Linear Equations: Substitution and Elimination.

Using the Seventh Edition Effectively and Efficiently

To meet the varied needs of diverse syllabi, this book contains more content than expected in a *Precalculus* course. As the chart illustrates,this book has been organized with flexibility of use in mind. Within a given chapter, certain sections are optional and can be omitted without loss of continuity.

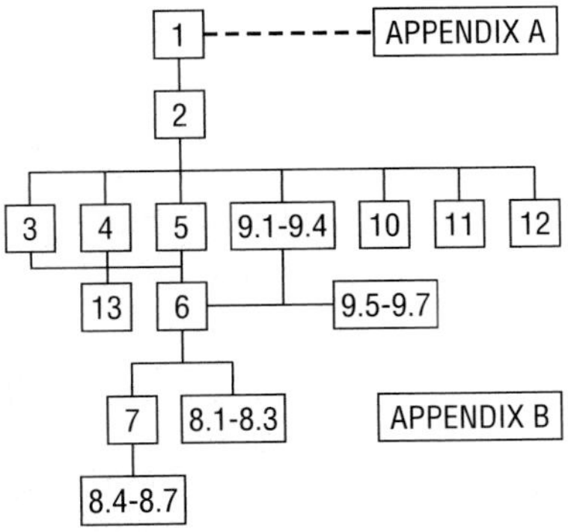

Chapter 1 Graphs
A quick coverage of this chapter, which is mainly review material, will enable you to get to Chapter 2 Functions and Their Graphs earlier. Section 1.4 is optional.

Chapter 2 Functions and Their Graphs
Perhaps the most important chapter. Section 2.6 is optional.

Chapter 3 Polynomial and Rational Functions
Topic selection depends on your syllabus.

Chapter 4 Exponential and Logarithmic Functions
Sections 4.1–4.6 follow in sequence. Sections 4.7, 4.8, and 4.9 are optional.

Chapter 5 Trigonometric Functions
The sections follow in sequence.

Chapter 6 Analytic Trigonometry
Sections 6.2, 6.6, and 6.8 may be omitted in a brief course.

Chapter 7 Applications of Trigonometric Functions
Sections 7.4 and 7.5 may be omitted in a brief course.

Chapter 8 Polar Coordinates; Vectors
Sections 8.1–8.3 and Sections 8.4–8.5 are independent and may be covered separately.

Chapter 9 Analytic Geometry
Sections 9.1–9.4 follow in sequence. Sections 9.5, 9.6, and 9.7 are independent of each other, but each requires Sections 9.1–9.4.

Chapter 10 Systems of Equations and Inequalities
Sections 10.2–10.7 may be covered in any order, but each requires Section 10.1. Section 10.8 requires Section 10.7.

Chapter 11 Sequences; Induction; The Binomial Theorem
There are three independent parts: Sections 11.1–11.3; Section 11.4; and Section 11.5.

Chapter 12 Counting and Probability
The sections follow in sequence.

Chapter 13 A Preview of Calculus: The Limit, Derivative, and Integral of a Function
If time permits, coverage of this chapter will give your students a beneficial head-start in calculus.

Appendix A Review
As the title implies, this is review material. Although it could be used as the first part of a course in Precalculus, its real value lies in its use as a just-in-time review. Specific references to Appendix A occur throughout the text to assist in the review process. Appropriate use of this appendix will allow students to review when they need to and provide more time for instructors to cover the course content.

Acknowledgments

Textbooks are written by authors, but evolve from an idea to final form through the efforts of many people. It was Don Dellen who first suggested this book and series to me. Don is remembered for his extensive contributions to publishing and mathematics.

I would also like to extend thanks to the following individuals for their important assistance and encouragement in the preparation of this edition: From Prentice Hall: Halee Dinsey for her innovative marketing skills; Eric Frank for his substantial contributions, ideas and enthusiasm; Patrice Jones, who remains a huge fan and supporter; Dawn Murrin for her talent and skill in getting supplements out; Bob Walters, who continues to amaze me with his organizational skill as Production Editor; Sally Yagan for her continued support and genuine interest; and the Prentice Hall Sales Team, for their continued confidence and support of my books. I also extend thanks to Tracey Hoy and Anna Maria Mendiola for their thorough accuracy checking of the entire manuscript; Teri Lovelace, Kurt Norlin, and others from Laurel Technical Services, for their dedication to accuracy in checking manuscript and answers. Special thanks to Jill McGowan, Kathleen Miranda, Karla Neal, Philip Piña, and Phoebe Rouse, for their very detailed and helpful reviews in preparation for this edition.

Finally, I offer my grateful thanks to the dedicated users and reviewers of my books, whose collective insights form the backbone of each textbook revision. I apologize for any omissions:

James Africh, *College of DuPage*
Steve Agronsky, *Cal Poly State University*
Grant Alexander, *Joliet Junior College*
Dave Anderson, *South Suburban College*
Joby Milo Anthony, *University of Central Florida*
James E. Arnold, *University of Wisconsin-Milwaukee*
Carolyn Autray, *University of West Georgia*
Agnes Azzolino, *Middlesex County College*
Wilson P Banks, *Illinois State University*

Sudeshna Basu, *Howard University*
Dale R. Bedgood, *East Texas State University*
Beth Beno, *South Suburban College*
Carolyn Bernath, *Tallahassee Community College*
William H. Beyer, *University of Akron*
Annette Blackwelder, *Florida State University*
Richelle Blair, *Lakeland Community College*
Trudy Bratten, *Grossmont College*
Tim Bremer, *Broome Community College*
Joanne Brunner, *Joliet Junior College*
Warren Burch, *Brevard Community College*
Mary Butler, *Lincoln Public Schools*
Jim Butterbach, *Joliet Junior College*
William J. Cable, *University of Wisconsin-Stevens Point*
Lois Calamia, *Brookdale Community College*
Jim Campbell, *Lincoln Public Schools*
Roger Carlsen, *Moraine Valley Community College*
Elena Catoiu, *Joliet Junior College*
Mathews Chakkanakuzhi, *Palomar College*
John Collado, *South Suburban College*
Nelson Collins, *Joliet Junior College*
Jim Cooper, *Joliet Junior College*
Denise Corbett, *East Carolina University*
Theodore C. Coskey, *South Seattle Community College*
Paul Crittenden, *University of Nebraska at Lincoln*
John Davenport, *East Texas State University*
Faye Dang, *Joliet Junior College*
Antonio David, *Del Mar College*
Duane E. Deal, *Ball State University*
Timothy Deis, *University of Wisconsin-Platteville*
Vivian Dennis, *Eastfield College*
Guesna Dohrman, *Tallahassee Community College*
Karen R. Dougan, *University of Florida*
Louise Dyson, *Clark College*
Paul D. East, *Lexington Community College*
Don Edmondson, *University of Texas-Austin*
Erica Egizio, *Joliet Junior College*
Christopher Ennis, *University of Minnesota*
Ralph Esparza, Jr., *Richland College*
Garret J. Etgen, *University of Houston*
Pete Falzone, *Pensacola Junior College*
W.A. Ferguson, *University of Illinois-Urbana/Champaign*
Iris B. Fetta, *Clemson University*
Mason Flake, *student at Edison Community College*
Timothy W. Flood, *Pittsburg State University*
Merle Friel, *Humboldt State University*
Richard A. Fritz, *Moraine Valley Community College*
Carolyn Funk, *South Suburban College*
Dewey Furness, *Ricke College*
Tina Garn, *University of Arizona*
Dawit Getachew, *Chicago State University*
Wayne Gibson, *Rancho Santiago College*
Robert Gill, *University of Minnesota Duluth*
Sudhir Kumar Goel, *Valdosta State University*
Joan Goliday, *Sante Fe Community College*
Frederic Gooding, *Goucher College*
Sue Graupner, *Lincoln Public Schools*
Jennifer L. Grimsley, *University of Charleston*
Ken Gurganus, *University of North Carolina*
James E. Hall, *University of Wisconsin-Madison*
Judy Hall, *West Virginia University*
Edward R. Hancock, *DeVry Institute of Technology*
Julia Hassett, *DeVry Institute-Dupage*
Christopher Hay-Jahans, *University of South Dakota*
Michah Heibel, *Lincoln Public Schools*
LaRae Helliwell, *San Jose City College*
Brother Herron, *Brother Rice High School*
Robert Hoburg, *Western Connecticut State University*
Lynda Hollingsworth, *Northwest Missouri State University*
Lee Hruby, *Naperville North High School*
Kim Hughes, *California State College-San Bernardino*
Ron Jamison, *Brigham Young University*
Richard A. Jensen, *Manatee Community College*
Sandra G. Johnson, *St. Cloud State University*
Tuesday Johnson, *New Mexico State University*
Moana H. Karsteter, *Tallahassee Community College*
Donna Katula, *Joliet Junior College*
Arthur Kaufman, *College of Staten Island*
Thomas Kearns, *North Kentucky University*
Shelia Kellenbarger, *Lincoln Public Schools*
Lynne Kowski, *Raritan Valley Community College*
Keith Kuchar, *Manatee Community College*
Tor Kwembe, *Chicago State University*
Linda J. Kyle, *Tarrant Country Jr. College*
H.E. Lacey, *Texas A & M University*
Harriet Lamm, *Coastal Bend College*
James Lapp, *Fort Lewis College*
Matt Larson, *Lincoln Public Schools*
Christopher Lattin, *Oakton Community College*
Adele LeGere, *Oakton Community College*
Kevin Leith, *University of Houston*
JoAnn Lewin, *Edison College*
Jeff Lewis, *Johnson County Community College*
Stanley Lukawecki, *Clemson University*
Janice C. Lyon, *Tallahassee Community College*
Virginia McCarthy, *Iowa State University*
Jean McArthur, *Joliet Junior College*
Karla McCavit, *Albion College*
Tom McCollow, *DeVry Institute of Technology*
Will McGowant, *Howard University*
Laurence Maher, *North Texas State University*
Jay A. Malmstrom, *Oklahoma City Community College*
Sherry Martina, *Naperville North High School*
Alec Matheson, *Lamar University*
Nancy Matthews, *University of Oklahoma*
James Maxwell, *Oklahoma State University-Stillwater*
Marsha May, *Midwestern State University*
Judy Meckley, *Joliet Junior College*
David Meel, *Bowling Green State University*
Carolyn Meitler, *Concordia University*
Samia Metwali, *Erie Community College*
Rich Meyers, *Joliet Junior College*
Eldon Miller, *University of Mississippi*
James Miller, *West Virginia University*
Michael Miller, *Iowa State University*
Kathleen Miranda, *SUNY at Old Westbury*
Thomas Monaghan, *Naperville North High School*
Craig Morse, *Naperville North High School*
Samad Mortabit, *Metropolitan State University*
Pat Mower, *Washburn University*
A. Muhundan, *Manatee Community College*
Jane Murphy, *Middlesex Community College*
Richard Nadel, *Florida International University*
Gabriel Nagy, *Kansas State University*
Bill Naegele, *South Suburban College*
Lawrence E. Newman, *Holyoke Community College*
James Nymann, *University of Texas-El Paso*
Sharon O'Donnell, *Chicago State University*
Seth F. Oppenheimer, *Mississippi State University*
Linda Padilla, *Joliet Junior College*
E. James Peake, *Iowa State University*
Kelly Pearson, *Murray State University*
Philip Pina, *Florida Atlantic University*
Michael Prophet, *University of Northern Iowa*
Neal C. Raber, *University of Akron*
Thomas Radin, *San Joaquin Delta College*
Ken A. Rager, *Metropolitan State College*
Kenneth D. Reeves, *San Antonio College*
Elsi Reinhardt, *Truckee Meadows Community College*
Jane Ringwald, *Iowa State University*
Stephen Rodi, *Austin Community College*
William Rogge, *Lincoln Northeast High School*
Howard L. Rolf, *Baylor University*
Phoebe Rouse, *Lousiana State University*
Edward Rozema, *University of Tennessee at Chattanooga*
Dennis C. Runde, *Manatee Community College*
Alan Saleski, *Loyola University of Chicago*
John Sanders, *Chicago State University*
Susan Sandmeyer, *Jamestown Community College*
Linda Schmidt, *Greenville Technical College*
A.K. Shamma, *University of West Florida*
Martin Sherry, *Lower Columbia College*
Tatrana Shubin, *San Jose State University*
Anita Sikes, *Delgado Community College*
Timothy Sipka, *Alma College*
Lori Smellegar, *Manatee Community College*
John Spellman, *Southwest Texas State University*
Rajalakshmi Sriram, *Okaloosa-Walton Community College*
Becky Stamper, *Western Kentucky University*
Judy Staver, *Florida Community College-South*
Neil Stephens, *Hinsdale South High School*
Patrick Stevens, *Joliet Junior College*
Christopher Terry, *Augusta State University*

Diane Tesar, *South Suburban College*
Tommy Thompson, *Brookhaven College*
Richard J. Tondra, *Iowa State University*
Marvel Townsend, *University of Florida*
Jim Trudnowski, *Carroll College*
Robert Tuskey, *Joliet Junior College*
Richard G. Vinson, *University of South Alabama*
Mary Voxman, *University of Idaho*
Jennifer Walsh, *Daytona Beach Community College*
Donna Wandke, *Naperville North High School*
Darlene Whitkenack, *Northern Illinois University*
Christine Wilson, *West Virginia University*
Brad Wind, *Florida International University*
Mary Wolyniak, *Broome Community College*
Canton Woods, *Auburn University*
Tamara S. Worner, *Wayne State College*
Terri Wright, *New Hampshire Community Technical College, Manchester*
George Zazi, *Chicago State University*

Michael Sullivan

STUDENT RESOURCES	INSTRUCTOR RESOURCES
Student Study Pack Everything a student needs to succeed in one place. Free packaged with the book, or available for purchase stand-alone. Study-Pack contains: • ***Student Solutions Manual*** Fully worked solutions to odd-numbered exercises. • ***Pearson Tutor Center*** Tutors provide one-on-one tutoring for any problem with an answer at the back of the book. Students access the Tutor Center via toll-free phone, fax, or email. • ***CD Lecture Series*** A comprehensive set of CD-ROMs, tied to the textbook, containing short video clips of an instructor working key book examples. • ***Algebra Review*** Four chapters of intermediate algebra review. Perfect for a slower-paced course or for individual review.	**Instructor Resource Distribution** All instructor resources can be downloaded from a single website (URL and password can be obtained from your PH Representative), or ordered individually: • ***TestGen*** Easily create tests from textbook section objectives. Questions are algorithmically generated allowing for unlimited versions. Edit problems or create your own. • ***Test Item File*** A printed test bank derived from TestGen. • ***PowerPoint Lecture Slides*** Fully editable slides that follow the textbook. Project in class or post to a website in an online course. • ***Instructor Solutions Manual*** Fully worked solutions to all textbook exercises. **Instructor's Edition** Provides answers to *all* exercises in the back of the text

MathXL®

MathXL® is a powerful online homework, tutorial, and assessment system that accompanies your textbook. Instructors can create, edit, and assign online homework and tests using algorithmically generated exercises correlated at the objective level to the textbook. Student work is tracked in an online gradebook. Students can take chapter tests and receive personalized study plans based on their results. The study plan diagnoses weaknesses and links students to tutorial exercises for objectives they need to study. Students can also access video clips from selected exercises. MathXL is available to qualified adopters. For more information, visit our website at **www.mathxl.com**, or contact your Prentice Hall sales representative for a demonstration.

MyMathLab®

MyMathLab® is a text-specific, customizable online course for your textbooks. MyMathLab is powered by CourseCompass™—Pearson Education's online teaching and learning environment—and by MathXL®—our online homework, tutorial, and assessment system. MyMathLab gives you the tools you need to deliver all or a portion of your course online, whether your students are in a lab setting or working from home.

MyMathLab provides a rich and flexible set of course materials, featuring free-response exercises that are algorithmically generated for unlimited practice. Students can use online tools such as video lectures and a multimedia textbook to improve their performance. Instructors can use MyMathLab's homework and test managers to select and assign online exercises correlated to the textbook, and can import TestGen tests for added flexibility. The online gradebook—designed specifically for mathematics—automatically tracks students' homework and test results and gives the instructor control over how to calculate final grades. MyMathLab is available to qualified adopters. For more information, visit our website at **www.mymathlab.com** or contact your Prentice Hall sales representative for a product demonstration.

WebCT or BlackBoard Premium

A suite of book-specific resources available for use within the WebCT or BlackBoard systems. Resources include a multimedia textbook, book-specific section videos, student and instructor solutions manuals, practice assignments with immediate feedback, question banks for building homework assignments, quizzes or tests, online graphing calculator manuals, PowerPoint lecture slides, and more.

List of Applications

Photo Credits

Chapter 1	Pages 1 and 48, Najlah Feanny/Stock Boston; Page 36, ©Kathy McLaughlin/The Image Works; Page 38, Tony Freeman/PhotoEdit.
Chapter 2	Pages 49 and 121, Doug Menuez/Getty Images; Page 63, Courtesy of NASA/JPL/Caltech; Page 70, Jamie Squire/Getty Images; Page 106, Tony Freeman/PhotoEdit.
Chapter 3	Pages 123 and 220, John Whalen/Northrop Grumman Newport News; Page 142, Erin Garvey/Index Stock Imagery, Inc.; Page 187, Oliver Drum Band/Getty Images.
Chapter 4	Pages 223 and 320, Amy C. Etra/PhotoEdit; Page 280, The Granger Collection; Page 292, Jim Pickerell/The Image Works; Page 306, ©Theo Allofs/CORBIS.
Chapter 5	Pages 323 and 411, Ned Haines/Photo Researchers, Inc.; Page 336, Doug Pensinger/Getty Images; Page 386, P. Berndt/Custom Medical Stock Photo, Inc.
Chapter 6	Pages 413 and 479, Steve Starr/Stock Boston.
Chapter 7	Pages 481 and 533, Fred Maroon/Photo Researchers, Inc.; Page 519, ©1989 Richard Megna/Fundamental Photographs.
Chapter 8	Pages 535 and 611, Art Matrix/Visuals Unlimited; Pages 559 and 568, CORBIS BETTMANN; Page 579, Library of Congress.
Chapter 9	Pages 613 and 681, CORBIS BETTMANN.
Chapter 10	Pages 683 and 781, Elena Rooraid/PhotoEdit; Page 740, CORBIS BETTMAN.
Chapter 11	Pages 783 and 825, NASA/GFSC/Tom Stack & Associates, Inc.; Page 808, The Granger Collection; Page 821, CORBIS BETTMANN.
Chapter 12	Pages 827 and 859, Myrleen Ferguson/PhotoEdit; Page 853, The Granger Collection.
Chapter 13	Pages 861 and 902, NASA/GFSC/Tom Stack & Associates, Inc.

1 Graphs

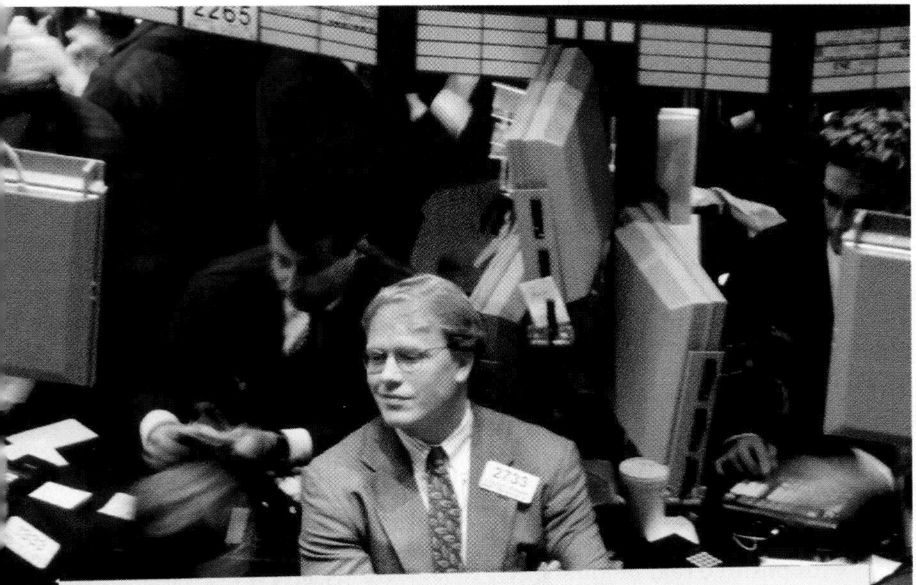

OUTLINE

The Use of Statistics in Making Investment Decisions: Using Beta

One of the most common statistics used by investors is beta, which measures a security or portfolio's movement relative to the S&P 500. When determining beta graphically, returns are plotted on a graph, with the horizontal axis representing the S&P returns and the vertical axis representing the security or portfolio returns; the points that are plotted represent the return of the security versus the return of the S&P 500 over the same time period. A regression line is then drawn, which is a straight line that most closely approximates the points in the graph. Beta represents the slope of that line; a steep slope (45 degrees or greater) indicates that when the stock market moved up or down the security or portfolio moved up or down on average to the same degree or greater, and a flatter slope indicates that when the stock market moved up or down the security or portfolio moved to a lesser degree. Since beta measures variability, it is used as a measure of risk; the greater the beta, the greater the risk.

SOURCE: Paul E. Hoffman, *Journal of the American Association of Individual Investors*, September, 1991; http://www.aaii.com

—SEE CHAPTER PROJECT 1.

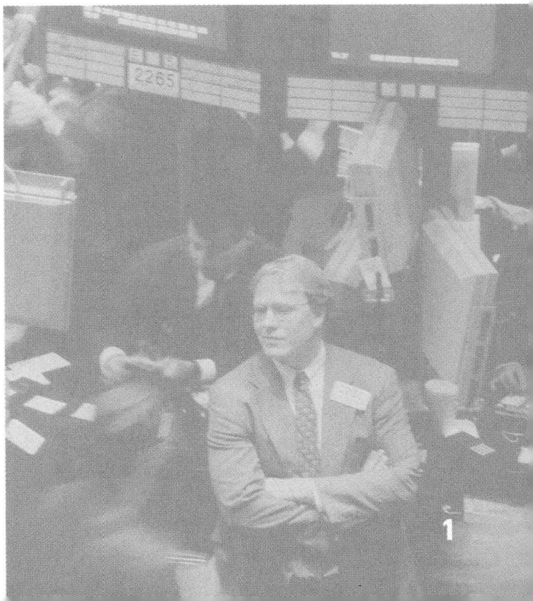

1.1 Rectangular Coordinates

PREPARING FOR THIS BOOK *Before getting started, read "To the Student" at the front of this book*

PREPARING FOR THIS SECTION *Before getting started, review the following:*

- Algebra Review (Appendix A, Section A.1, pp. 903–911)
- Geometry Review (Appendix A, Section A.2, pp. 913–916)

Now work the 'Are You Prepared?' problems on page 6.

OBJECTIVES
1 Use the Distance Formula
2 Use the Midpoint Formula

We locate a point on the real number line by assigning it a single real number, called the *coordinate of the point*. For work in a two-dimensional plane, we locate points by using two numbers.

We begin with two real number lines located in the same plane: one horizontal and the other vertical. We call the horizontal line the ***x*-axis;** the vertical line, the ***y*-axis;** and the point of intersection, the **origin *O*.** We assign coordinates to every point on these number lines as shown in Figure 1, using a convenient scale. In mathematics, we usually use the same scale on each axis; in applications, a different scale is often used on each axis.

Figure 1

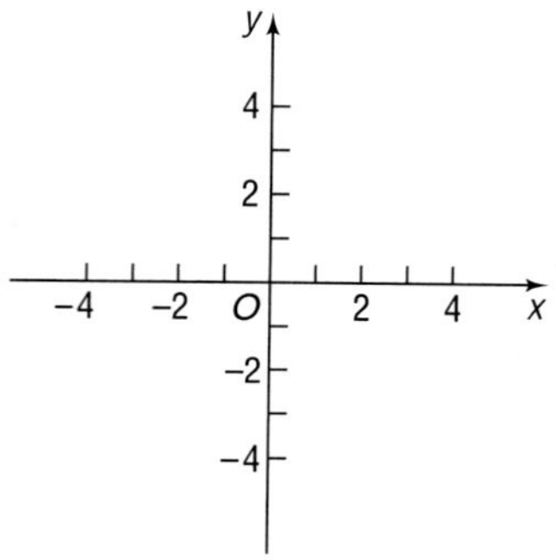

The origin O has a value of 0 on both the x-axis and the y-axis. We follow the usual convention that points on the x-axis to the right of O are associated with positive real numbers, and those to the left of O are associated with negative real numbers. Points on the y-axis above O are associated with positive real numbers, and those below O are associated with negative real numbers. In Figure 1, the x-axis and y-axis are labeled as x and y, respectively, and we have used an arrow at the end of each axis to denote the positive direction.

The coordinate system described here is called a **rectangular** or **Cartesian*** coordinate system. The plane formed by the x-axis and y-axis is sometimes called the ***xy*-plane,** and the x-axis and y-axis are referred to as the **coordinate axes.**

Any point P in the xy-plane can then be located by using an **ordered pair** (x, y) of real numbers. Let x denote the signed distance of P from the y-axis (*signed* in the sense that, if P is to the right of the y-axis, then $x > 0$, and if P is to the left of the y-axis, then $x < 0$); and let y denote the signed distance of P from the x-axis. The ordered pair (x, y), also called the **coordinates** of P, then gives us enough information to locate the point P in the plane.

Figure 2

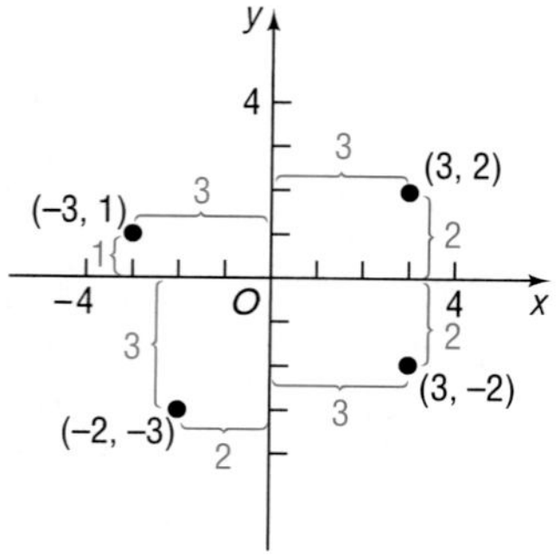

For example, to locate the point whose coordinates are $(-3, 1)$, go 3 units along the x-axis to the left of O and then go straight up 1 unit. We **plot** this point by placing a dot at this location. See Figure 2, in which the points with coordinates $(-3, 1)$, $(-2, -3)$, $(3, -2)$, and $(3, 2)$ are plotted.

The origin has coordinates $(0, 0)$. Any point on the x-axis has coordinates of the form $(x, 0)$, and any point on the y-axis has coordinates of the form $(0, y)$.

*Named after René Descartes (1596–1650), a French mathematician, philosopher, and theologian.

Figure 3

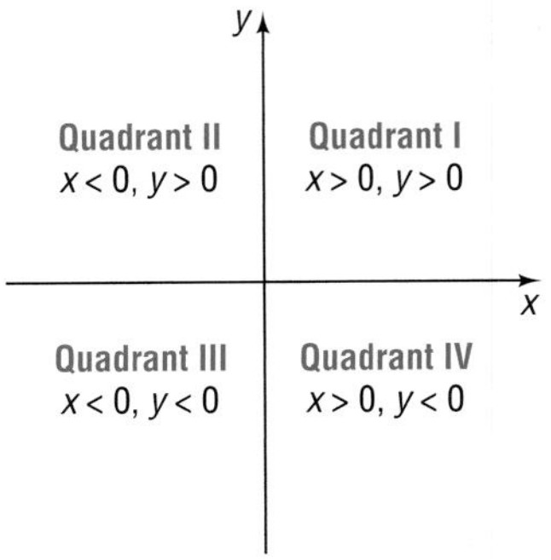

If (x, y) are the coordinates of a point P, then x is called the ***x*-coordinate,** or **abscissa,** of P and y is the ***y*-coordinate,** or **ordinate,** of P. We identify the point P by its coordinates (x, y) by writing $P = (x, y)$, referring to it as "the point (x, y)," rather than "the point whose coordinates are (x, y)."

The coordinate axes divide the xy-plane into four sections, called **quadrants,** as shown in Figure 3. In quadrant I, both the x-coordinate and the y-coordinate of all points are positive; in quadrant II, x is negative and y is positive; in quadrant III, both x and y are negative; and in quadrant IV, x is positive and y is negative. Points on the coordinate axes belong to no quadrant.

NOW WORK PROBLEM 11.

Figure 4

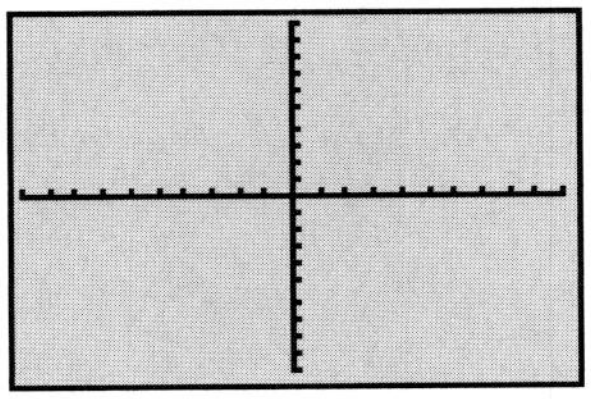

Comment: On a graphing calculator, you can set the scale on each axis. Once this has been done, you obtain the **viewing rectangle.** See Figure 4 for a typical viewing rectangle. You should now read Section B.1, *The Viewing Rectangle*, in Appendix B. ■

Distance between Points

1 If the same unit of measurement, such as inches or centimeters, is used for both the x-axis and the y-axis, then all distances in the xy-plane can be measured using this unit of measurement.

The **distance formula** provides a straightforward method for computing the distance between two points.

Theorem

Distance Formula

The distance between two points $P_1 = (x_1, y_1)$ and $P_2 = (x_2, y_2)$, denoted by $d(P_1, P_2)$, is

$$d(P_1, P_2) = \sqrt{(x_2 - x_1)^2 + (y_2 - y_1)^2} \qquad \textbf{(1)}$$

EXAMPLE 1 **Finding the Distance between Two Points**

Find the distance d between the points $(-4, 5)$ and $(3, 2)$.

Solution Using the distance formula (1), the solution is obtained as follows:

$$d = \sqrt{[3 - (-4)]^2 + (2 - 5)^2} = \sqrt{7^2 + (-3)^2}$$
$$= \sqrt{49 + 9} = \sqrt{58} \approx 7.62$$ ◀

NOW WORK PROBLEMS 15 AND 19.

Figure 5

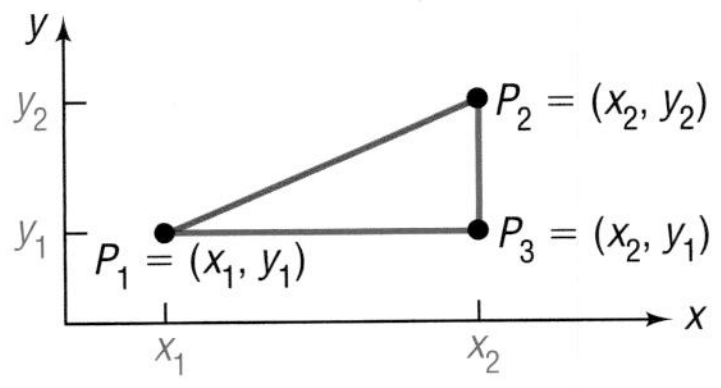

(a)

Proof of the Distance Formula Let (x_1, y_1) denote the coordinates of point P_1, and let (x_2, y_2) denote the coordinates of point P_2. Assume that the line joining P_1 and P_2 is neither horizontal nor vertical. Refer to Figure 5(a). The coordinates of P_3 are (x_2, y_1). The horizontal distance from P_1 to P_3 is the absolute value of the difference of the x-coordinates, $|x_2 - x_1|$. The vertical distance from P_3 to P_2 is the absolute value of the

Figure 5

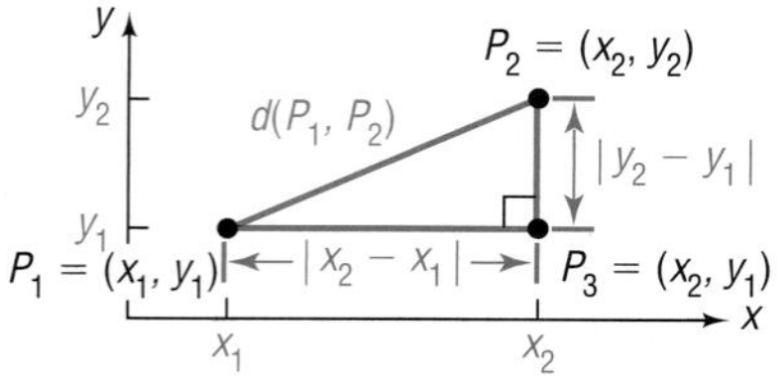

(b)

difference of the y-coordinates, $|y_2 - y_1|$. See Figure 5(b). The distance $d(P_1, P_2)$ that we seek is the length of the hypotenuse of the right triangle, so, by the Pythagorean Theorem, it follows that

$$[d(P_1, P_2)]^2 = |x_2 - x_1|^2 + |y_2 - y_1|^2$$
$$= (x_2 - x_1)^2 + (y_2 - y_1)^2$$
$$d(P_1, P_2) = \sqrt{(x_2 - x_1)^2 + (y_2 - y_1)^2}$$

Now, if the line joining P_1 and P_2 is horizontal, then the y-coordinate of P_1 equals the y-coordinate of P_2; that is, $y_1 = y_2$. Refer to Figure 6(a). In this case, the distance formula (1) still works, because, for $y_1 = y_2$, it reduces to

$$d(P_1, P_2) = \sqrt{(x_2 - x_1)^2 + 0^2} = \sqrt{(x_2 - x_1)^2} = |x_2 - x_1|$$

A similar argument holds if the line joining P_1 and P_2 is vertical. See Figure 6(b).

Figure 6

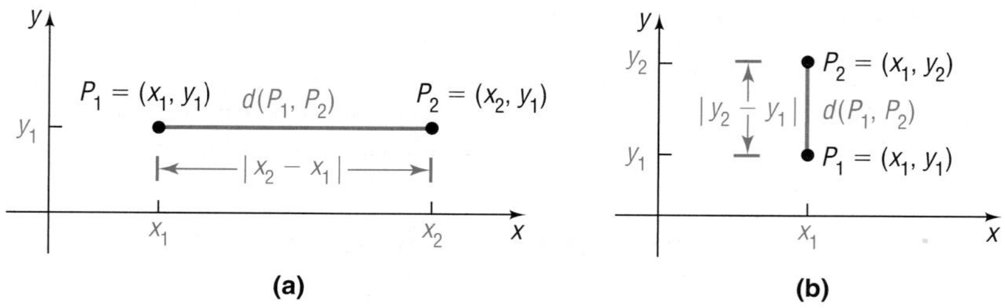

The distance formula is valid in all cases. ■

The distance between two points $P_1 = (x_1, y_1)$ and $P_2 = (x_2, y_2)$ is never a negative number. Furthermore, the distance between two points is 0 only when the points are identical, that is, when $x_1 = x_2$ and $y_1 = y_2$. Also, because $(x_2 - x_1)^2 = (x_1 - x_2)^2$ and $(y_2 - y_1)^2 = (y_1 - y_2)^2$, it makes no difference whether the distance is computed from P_1 to P_2 or from P_2 to P_1; that is, $d(P_1, P_2) = d(P_2, P_1)$.

Rectangular coordinates enable us to translate geometry problems into algebra problems, and vice versa. The next example shows how algebra (the distance formula) can be used to solve geometry problems.

EXAMPLE 2 Using Algebra to Solve Geometry Problems

Consider the three points $A = (-2, 1)$, $B = (2, 3)$, and $C = (3, 1)$.

(a) Plot each point and form the triangle ABC.

(b) Find the length of each side of the triangle.

(c) Verify that the triangle is a right triangle.

(d) Find the area of the triangle.

Figure 7

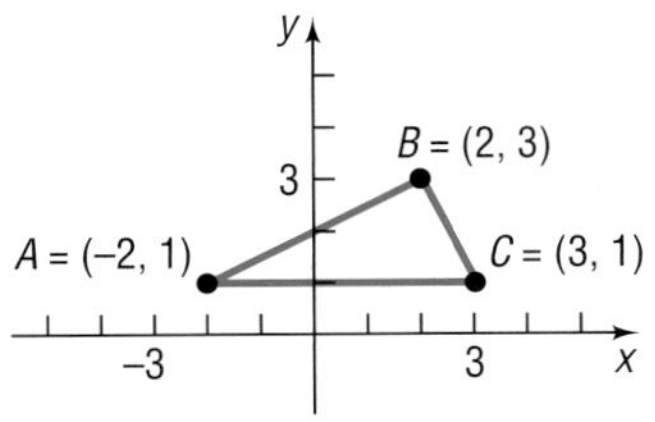

Solution (a) Points A, B, and C and triangle ABC are plotted in Figure 7.

(b) $d(A, B) = \sqrt{[2 - (-2)]^2 + (3 - 1)^2} = \sqrt{16 + 4} = \sqrt{20} = 2\sqrt{5}$

$$d(B, C) = \sqrt{(3 - 2)^2 + (1 - 3)^2} = \sqrt{1 + 4} = \sqrt{5}$$

$$d(A, C) = \sqrt{[3 - (-2)]^2 + (1 - 1)^2} = \sqrt{25 + 0} = 5$$

(c) To show that the triangle is a right triangle, we need to show that the sum of the squares of the lengths of two of the sides equals the square of the length of the third side. (Why is this sufficient?) Looking at Figure 7, it seems reasonable to conjecture that the right angle is at vertex B. To verify, we check to see whether

$$[d(A, B)]^2 + [d(B, C)]^2 = [d(A, C)]^2$$

We find that

$$\begin{aligned} [d(A, B)]^2 + [d(B, C)]^2 &= \left(2\sqrt{5}\right)^2 + \left(\sqrt{5}\right)^2 \\ &= 20 + 5 = 25 = [d(A, C)]^2 \end{aligned}$$

so it follows from the converse of the Pythagorean Theorem that triangle ABC is a right triangle.

(d) Because the right angle is at B, the sides AB and BC form the base and altitude of the triangle. Its area is therefore

$$\text{Area} = \frac{1}{2}(\text{Base})(\text{Altitude}) = \frac{1}{2}\left(2\sqrt{5}\right)\left(\sqrt{5}\right) = 5 \text{ square units}$$

◀

NOW WORK PROBLEM 29.

Midpoint Formula

2 We now derive a formula for the coordinates of the **midpoint of a line segment.** Let $P_1 = (x_1, y_1)$ and $P_2 = (x_2, y_2)$ be the endpoints of a line segment, and let $M = (x, y)$ be the point on the line segment that is the same distance from P_1 as it is from P_2. See Figure 8. The triangles P_1AM and MBP_2 are congruent.* [Do you see why? Angle AP_1M = Angle BMP_2.† Angle P_1MA = Angle MP_2B, and $d(P_1, M) = d(M, P_2)$ is given. We have Angle–Side–Angle.] Because triangles P_1AM and MBP_2 are congruent, corresponding sides are equal in length. That is,

Figure 8

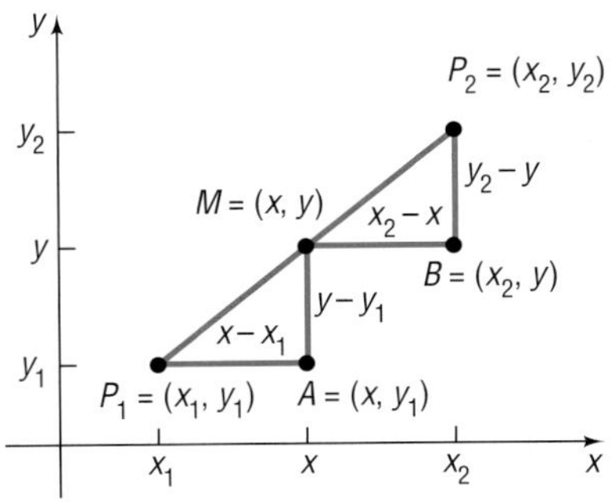

$$\begin{aligned} x - x_1 &= x_2 - x & \quad \text{and} \quad y - y_1 &= y_2 - y \\ 2x &= x_1 + x_2 & 2y &= y_1 + y_2 \\ x &= \frac{x_1 + x_2}{2} & y &= \frac{y_1 + y_2}{2} \end{aligned}$$

*The following statement is a postulate from geometry. Two triangles are congruent if their sides are the same length (SSS), or if two sides and the included angle are the same (SAS), or if two angles and the included side are the same (ASA).

†Another postulate from geometry states that the transversal $\overline{P_1P_2}$ forms equal corresponding angles with the parallel line segments $\overline{P_1A}$ and $\overline{MB}$.

Theorem

In Words
To find the midpoint of a line segment, average the x-coordinates and average the y-coordinates of the endpoints.

Midpoint Formula

The midpoint $M = (x, y)$ of the line segment from $P_1 = (x_1, y_1)$ to $P_2 = (x_2, y_2)$ is

$$M = (x, y) = \left(\frac{x_1 + x_2}{2}, \frac{y_1 + y_2}{2}\right) \qquad \textbf{(2)}$$

EXAMPLE 3 **Finding the Midpoint of a Line Segment**

Find the midpoint of the line segment from $P_1 = (-5, 5)$ to $P_2 = (3, 1)$. Plot the points P_1 and P_2 and their midpoint. Check your answer.

Solution We apply the midpoint formula (2) using $x_1 = -5$, $y_1 = 5$, $x_2 = 3$, and $y_2 = 1$. Then the coordinates (x, y) of the midpoint M are

$$x = \frac{x_1 + x_2}{2} = \frac{-5 + 3}{2} = -1 \quad \text{and} \quad y = \frac{y_1 + y_2}{2} = \frac{5 + 1}{2} = 3$$

Figure 9

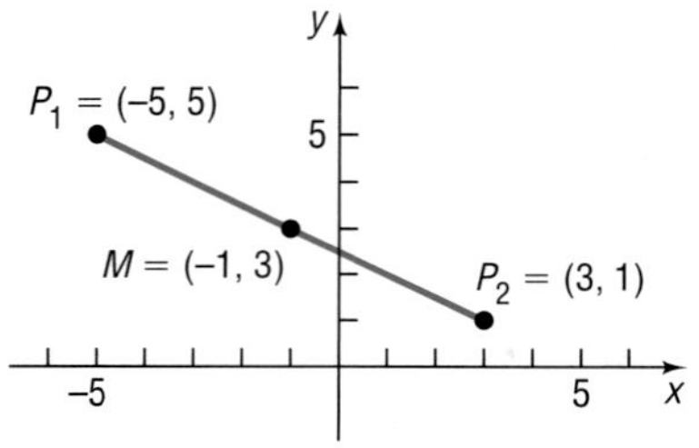

That is, $M = (-1, 3)$. See Figure 9.

✔ **CHECK:** Because M is the midpoint, we check the answer by verifying that $d(P_1, M) = d(M, P_2)$:

$$d(P_1, M) = \sqrt{[-1 - (-5)]^2 + (3 - 5)^2} = \sqrt{16 + 4} = \sqrt{20}$$

$$d(M, P_2) = \sqrt{[3 - (-1)]^2 + (1 - 3)^2} = \sqrt{16 + 4} = \sqrt{20}$$ ◀

NOW WORK PROBLEM 39.

1.1 Assess Your Understanding

'Are You Prepared?' *Answers are given at the end of these exercises. If you get to wrong answer, read the pages listed in red.*

1. On the real number line the origin is assigned the number ________. (pp. 903–911)

2. If 3 and 4 are the legs of a right triangle, the hypotenuse is ________. (pp. 913–916)

3. If -3 and 5 are the coordinates of two points on the real number line, the distance between these points is ________. (pp. 903–911)

Concepts and Vocabulary

4. If (x, y) are the coordinates of a point P in the xy-plane, then x is called the ________ of P and y is the ________ of P.

5. The coordinate axes divide the xy-plane into four sections called ________.

6. If three distinct points P, Q, and R all lie on a line and if $d(P, Q) = d(Q, R)$, then Q is called the ________ of the line segment from P to R.

7. The point $(a, 0)$ lies on the ________ axis.

8. *True or False:* The distance between two points is sometimes a negative number.

9. *True or False:* The point $(-1, 4)$ lies in quadrant IV of the Cartesian plane.

10. *True or False:* The midpoint of a line segment is found by averaging the x-coordinates and averaging the y-coordinates of the endpoints.

Exercises

In Problems 11 and 12, plot each point in the xy-plane. Tell in which quadrant or on what coordinate axis each point lies.

11. (a) $A = (-3, 2)$
(b) $B = (6, 0)$
(c) $C = (-2, -2)$
(d) $D = (6, 5)$
(e) $E = (0, -3)$
(f) $F = (6, -3)$

12. (a) $A = (1, 4)$
(b) $B = (-3, -4)$
(c) $C = (-3, 4)$
(d) $D = (4, 1)$
(e) $E = (0, 1)$
(f) $F = (-3, 0)$

13. Plot the points $(2, 0)$, $(2, -3)$, $(2, 4)$, $(2, 1)$, and $(2, -1)$. Describe the set of all points of the form $(2, y)$, where y is a real number.

14. Plot the points $(0, 3)$, $(1, 3)$, $(-2, 3)$, $(5, 3)$, and $(-4, 3)$. Describe the set of all points of the form $(x, 3)$, where x is a real number.

In Problems 15–28, find the distance $d(P_1, P_2)$ between the points P_1 and P_2.

15.
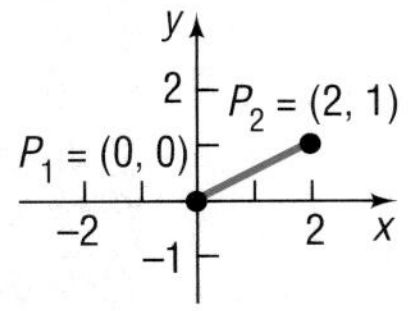

16.
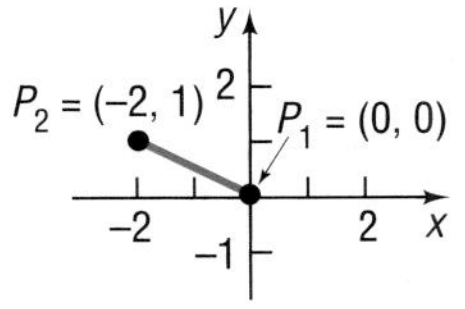

17.
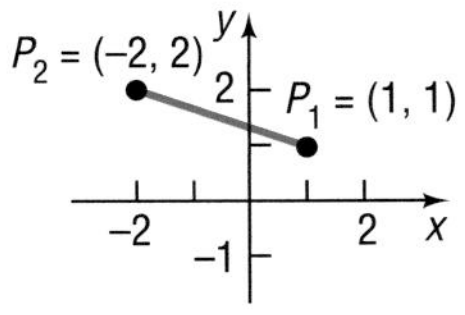

18.
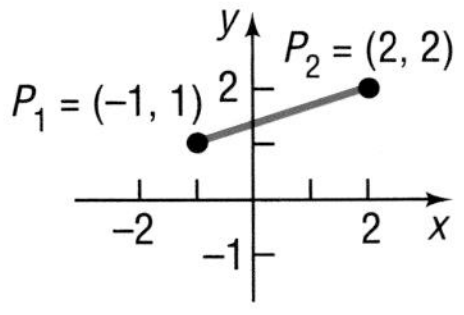

19. $P_1 = (3, -4)$; $P_2 = (5, 4)$

20. $P_1 = (-1, 0)$; $P_2 = (2, 4)$

21. $P_1 = (-3, 2)$; $P_2 = (6, 0)$

22. $P_1 = (2, -3)$; $P_2 = (4, 2)$

23. $P_1 = (4, -3)$; $P_2 = (6, 4)$

24. $P_1 = (-4, -3)$; $P_2 = (6, 2)$

25. $P_1 = (-0.2, 0.3)$; $P_2 = (2.3, 1.1)$

26. $P_1 = (1.2, 2.3)$; $P_2 = (-0.3, 1.1)$

27. $P_1 = (a, b)$; $P_2 = (0, 0)$

28. $P_1 = (a, a)$; $P_2 = (0, 0)$

In Problems 29–34, plot each point and form the triangle ABC. Verify that the triangle is a right triangle. Find its area.

29. $A = (-2, 5)$; $B = (1, 3)$; $C = (-1, 0)$

30. $A = (-2, 5)$; $B = (12, 3)$; $C = (10, -11)$

31. $A = (-5, 3)$; $B = (6, 0)$; $C = (5, 5)$

32. $A = (-6, 3)$; $B = (3, -5)$; $C = (-1, 5)$

33. $A = (4, -3)$; $B = (0, -3)$; $C = (4, 2)$

34. $A = (4, -3)$; $B = (4, 1)$; $C = (2, 1)$

35. Find all points having an x-coordinate of 2 whose distance from the point $(-2, -1)$ is 5.

36. Find all points having a y-coordinate of -3 whose distance from the point $(1, 2)$ is 13.

37. Find all points on the x-axis that are 5 units from the point $(4, -3)$.

38. Find all points on the y-axis that are 5 units from the point $(4, 4)$.

In Problems 39–48, find the midpoint of the line segment joining the points P_1 and P_2.

39. $P_1 = (5, -4)$; $P_2 = (3, 2)$

40. $P_1 = (-1, 0)$; $P_2 = (2, 4)$

41. $P_1 = (-3, 2)$; $P_2 = (6, 0)$

42. $P_1 = (2, -3)$; $P_2 = (4, 2)$

43. $P_1 = (4, -3)$; $P_2 = (6, 1)$

44. $P_1 = (-4, -3)$; $P_2 = (2, 2)$

45. $P_1 = (-0.2, 0.3)$; $P_2 = (2.3, 1.1)$

46. $P_1 = (1.2, 2.3)$; $P_2 = (-0.3, 1.1)$

47. $P_1 = (a, b)$; $P_2 = (0, 0)$

48. $P_1 = (a, a)$; $P_2 = (0, 0)$

49. The **medians** of a triangle are the line segments from each vertex to the midpoint of the opposite side (see the figure). Find the lengths of the medians of the triangle with vertices at $A = (0, 0)$, $B = (0, 6)$, and $C = (4, 4)$.

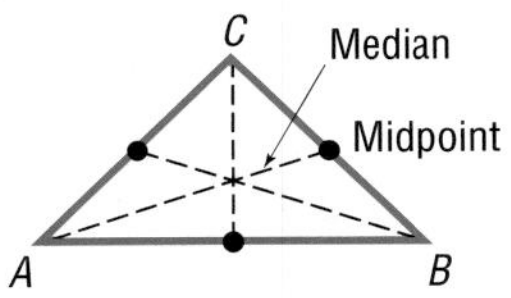

50. An **equilateral triangle** is one in which all three sides are of equal length. If two vertices of an equilateral triangle are $(0, 4)$ and $(0, 0)$, find the third vertex. How many of these triangles are possible?

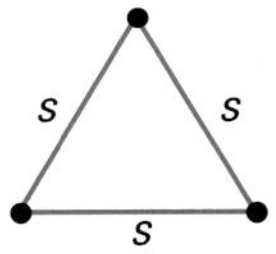

In Problems 51–54, find the length of each side of the triangle determined by the three points P_1, P_2, and P_3. State whether the triangle is an isosceles triangle, a right triangle, neither of these, or both. (An **isosceles triangle** *is one in which at least two of the sides are of equal length.)*

51. $P_1 = (2, 1)$; $P_2 = (-4, 1)$; $P_3 = (-4, -3)$

52. $P_1 = (-1, 4)$; $P_2 = (6, 2)$; $P_3 = (4, -5)$

53. $P_1 = (-2, -1)$; $P_2 = (0, 7)$; $P_3 = (3, 2)$

54. $P_1 = (7, 2)$; $P_2 = (-4, 0)$; $P_3 = (4, 6)$

In Problems 55–58, find the length of the line segment. Assume that the endpoints of each line segment have integer coordinates.

55.

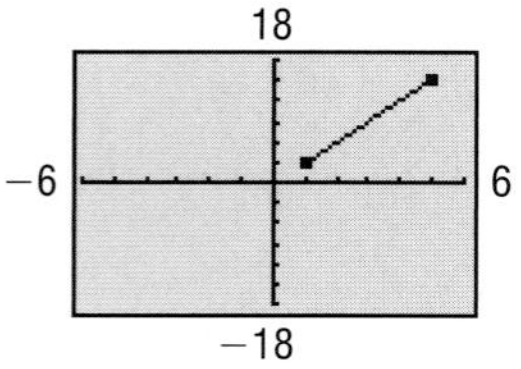

56.

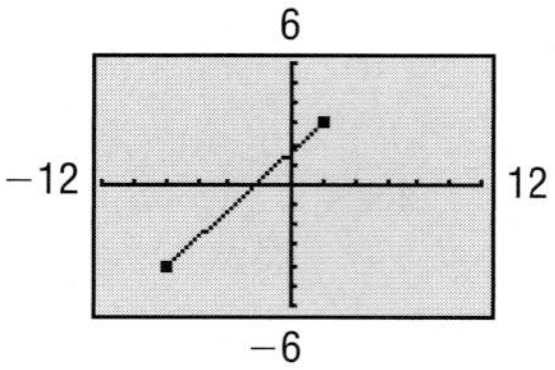

57.

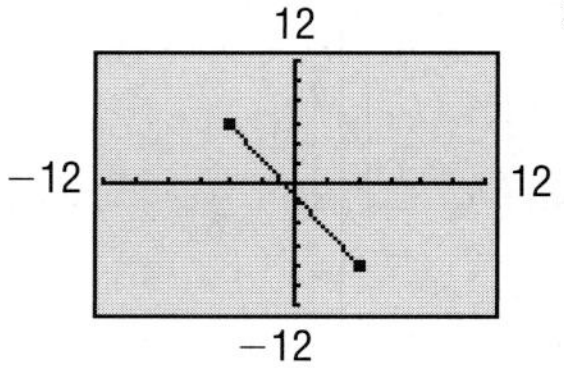

58.

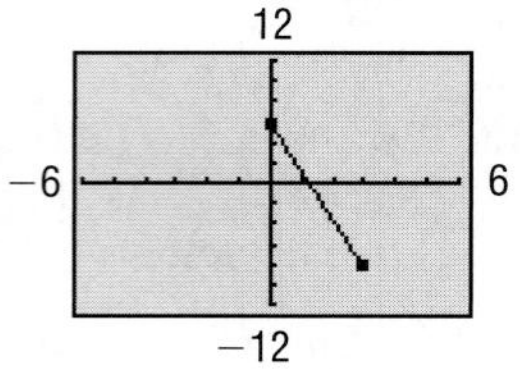

59. Geometry Find the midpoint of each diagonal of a square with side of length s. Draw the conclusion that the diagonals of a square intersect at their midpoints.
[**Hint:** Use $(0, 0)$, $(0, s)$, $(s, 0)$, and (s, s) as the vertices of the square.]

60. Geometry Verify that the points $(0, 0)$, $(a, 0)$, and $\left(\frac{a}{2}, \frac{\sqrt{3}a}{2}\right)$ are the vertices of an equilateral triangle. Then show that the midpoints of the three sides are the vertices of a second equilateral triangle (refer to Problem 50).

61. Baseball A major league baseball "diamond" is actually a square, 90 feet on a side (see the figure). What is the distance directly from home plate to second base (the diagonal of the square)?

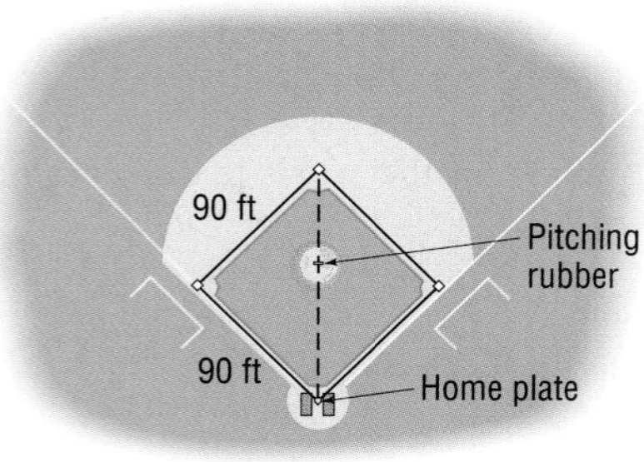

62. Little League Baseball The layout of a Little League playing field is a square, 60 feet on a side. How far is it directly from home plate to second base (the diagonal of the square)? SOURCE: *Little League Baseball, Official Regulations and Playing Rules*, 2000.

63. Baseball Refer to Problem 61. Overlay a rectangular coordinate system on a major league baseball diamond so that the origin is at home plate, the positive x-axis lies in the direction from home plate to first base, and the positive y-axis lies in the direction from home plate to third base.

(a) What are the coordinates of first base, second base, and third base? Use feet as the unit of measurement.

(b) If the right fielder is located at $(310, 15)$, how far is it from the right fielder to second base?

(c) If the center fielder is located at $(300, 300)$, how far is it from the center fielder to third base?

64. Little League Baseball Refer to Problem 62. Overlay a rectangular coordinate system on a Little League baseball diamond so that the origin is at home plate, the positive x-axis lies in the direction from home plate to first base, and the positive y-axis lies in the direction from home plate to third base.

(a) What are the coordinates of first base, second base, and third base? Use feet as the unit of measurement.

(b) If the right fielder is located at $(180, 20)$, how far is it from the right fielder to second base?

(c) If the center fielder is located at $(220, 220)$, how far is it from the center fielder to third base?

65. A Dodge Intrepid and a Mack truck leave an intersection at the same time. The Intrepid heads east at an average speed of 30 miles per hour, while the truck heads south at an average speed of 40 miles per hour. Find an expression for their distance d apart (in miles) at the end of t hours.

66. A hot-air balloon, headed due east at an average speed of 15 miles per hour and at a constant altitude of 100 feet, passes over an intersection (see the figure). Find an expression for the distance d (measured in feet) from the balloon to the intersection t seconds later.

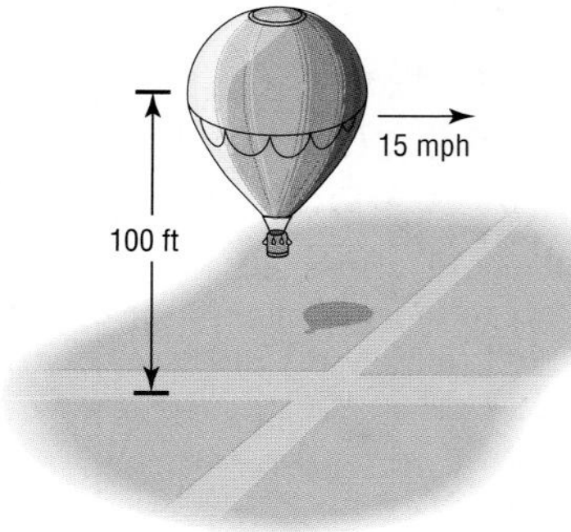

'Are You Prepared?' Answers

1. 0 **2.** 5 **3.** 8

1.2 Graphs of Equations; Circles

PREPARING FOR THIS SECTION *Before getting started, review the following:*

- Square Root Method (Appendix A, Section A.5, pp. 940–941)
- Completing the Square (Appendix A, Section A.5, pp. 940–941)

Now work the 'Are You Prepared?' problems on page 19.

OBJECTIVES
1. Graph Equations by Plotting Points
2. Find Intercepts from a Graph
3. Find Intercepts from an Equation
4. Test an Equation for Symmetry with Respect to the *x*-Axis, the *y*-Axis, and the Origin
5. Write the Standard Form of the Equation of a Circle
6. Graph a Circle
7. Find the Center and Radius of a Circle in General Form and Graph It

An **equation in two variables,** say x and y, is a statement in which two expressions involving x and y are equal. The expressions are called the **sides** of the equation. Since an equation is a statement, it may be true or false, depending on the value of the variables. Any values of x and y that result in a true statement are said to **satisfy** the equation.

For example, the following are all equations in two variables x and y:

$$x^2 + y^2 = 5 \qquad 2x - y = 6 \qquad y = 2x + 5 \qquad x^2 = y$$

The first of these, $x^2 + y^2 = 5$, is satisfied for $x = 1$, $y = 2$, since $1^2 + 2^2 = 1 + 4 = 5$. Other choices of x and y also satisfy this equation. It is not satisfied for $x = 2$ and $y = 3$, since $2^2 + 3^2 = 4 + 9 = 13 \neq 5$.

The **graph of an equation in two variables** x and y consists of the set of points in the xy-plane whose coordinates (x, y) satisfy the equation.

EXAMPLE 1 Determining Whether a Point Is on the Graph of an Equation

Determine if the following points are on the graph of the equation $2x - y = 6$.

(a) $(2, 3)$ (b) $(2, -2)$

Solution (a) For the point $(2, 3)$, we check to see if $x = 2$, $y = 3$ satisfies the equation $2x - y = 6$.

$$2x - y = 2(2) - 3 = 4 - 3 = 1 \neq 6$$

The equation is not satisfied, so the point $(2, 3)$ is not on the graph.

(b) For the point $(2, -2)$, we have

$$2x - y = 2(2) - (-2) = 4 + 2 = 6$$

The equation is satisfied, so the point $(2, -2)$ is on the graph. ◄

NOW WORK PROBLEM 31.

1 **EXAMPLE 2**

Graphing an Equation by Plotting Points

Graph the equation: $y = 2x + 5$

Solution

We want to find all points (x, y) that satisfy the equation. To locate some of these points (and to get an idea of the pattern of the graph), we assign some numbers to x and find corresponding values for y.

If	Then	Point on Graph
$x = 0$	$y = 2(0) + 5 = 5$	$(0, 5)$
$x = 1$	$y = 2(1) + 5 = 7$	$(1, 7)$
$x = -5$	$y = 2(-5) + 5 = -5$	$(-5, -5)$
$x = 10$	$y = 2(10) + 5 = 25$	$(10, 25)$

By plotting these points and then connecting them, we obtain the graph of the equation (a *line*), as shown in Figure 10. ◀

Figure 10

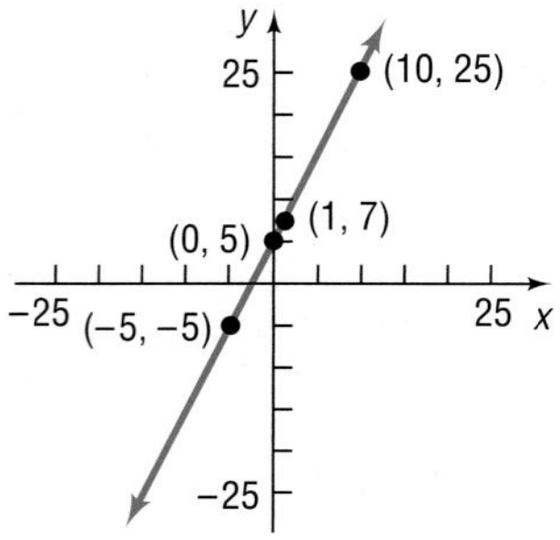

EXAMPLE 3

Graphing an Equation by Plotting Points

Graph the equation: $y = x^2$

Solution

Table 1 provides several points on the graph. In Figure 11 we plot these points and connect them with a smooth curve to obtain the graph (a *parabola*). ◀

Table 1

x	$y = x^2$	(x, y)
−4	16	(−4, 16)
−3	9	(−3, 9)
−2	4	(−2, 4)
−1	1	(−1, 1)
0	0	(0, 0)
1	1	(1, 1)
2	4	(2, 4)
3	9	(3, 9)
4	16	(4, 16)

Figure 11

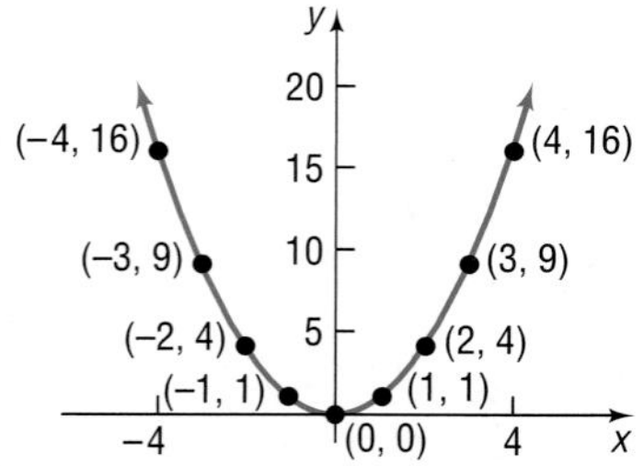

The graphs of the equations shown in Figures 10 and 11 do not show all points. For example, in Figure 10, the point $(20, 45)$ is a part of the graph of $y = 2x + 5$, but it is not shown. Since the graph of $y = 2x + 5$ could be extended out as far as we please, we use arrows to indicate that the pattern shown continues. It is important when illustrating a graph to present enough of the graph so that any viewer of the illustration will "see" the rest of it as an obvious continuation of what is actually there. This is referred to as a **complete graph.**

So, one way to obtain a complete graph of an equation is to plot a sufficient number of points on the graph until a pattern becomes evident. Then these points are connected with a smooth curve following the suggested pattern. But how many points are sufficient? Sometimes knowledge about the equation tells us. For example, we will learn in the next section that, if an equation is of the form $y = mx + b$, then its graph is a line. In this case, only two points are needed to obtain the graph.

One purpose of this book is to investigate the properties of equations in order to decide whether a graph is complete. At first we shall graph equations by plotting a sufficient number of points. Shortly, we shall investigate various techniques that will enable us to graph an equation without plotting so many points. Other times we shall graph equations based solely on properties of the equation.

COMMENT: Another way to obtain the graph of an equation is to use a graphing utility. Read Section B.2, *Using a Graphing Utility to Graph Equations*, in Appendix B. ■

Two techniques that reduce the number of points required to graph an equation involve finding *intercepts* and checking for *symmetry.*

Intercepts

2 The points, if any, at which a graph crosses or touches the coordinate axes are called the **intercepts.** See Figure 12. The x-coordinate of a point at which the graph crosses or touches the x-axis is an ***x*-intercept,** and the y-coordinate of a point at which the graph crosses or touches the y-axis is a ***y*-intercept.**

Figure 12

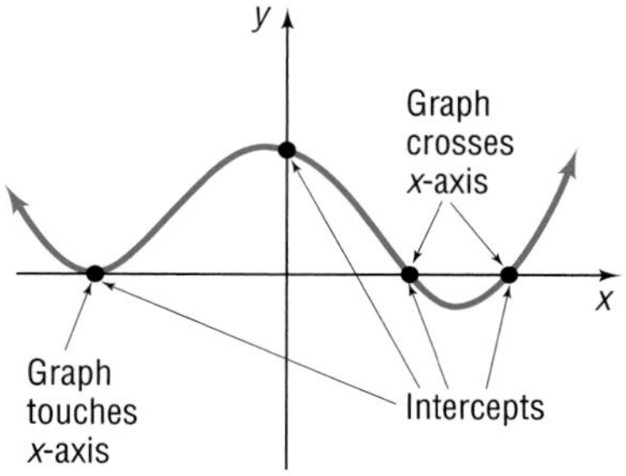

EXAMPLE 4 **Finding Intercepts from a Graph**

Find the intercepts of the graph in Figure 13. What are its x-intercepts? What are its y-intercepts?

Figure 13

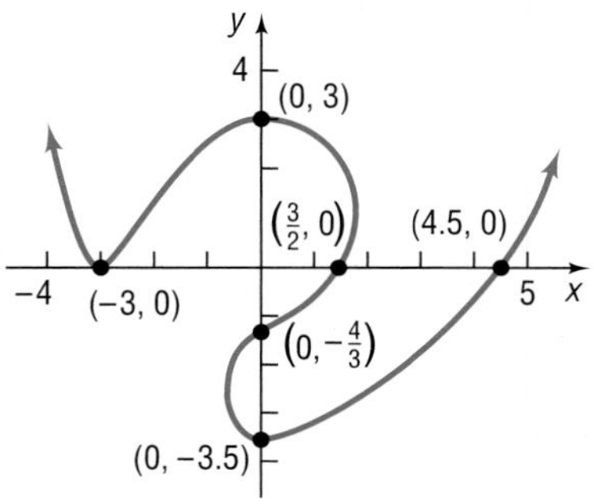

Solution The intercepts of the graph are the points

$$(-3, 0), \quad (0, 3), \quad \left(\frac{3}{2}, 0\right), \quad \left(0, -\frac{4}{3}\right), \quad (0, -3.5), \quad (4.5, 0)$$

The x-intercepts are $-3, \frac{3}{2}$, and 4.5; the y-intercepts are $-3.5, -\frac{4}{3}$, and 3. ◀

NOW WORK PROBLEM 19(a).

3 The intercepts of the graph of an equation can be found by using the fact that points on the x-axis have y-coordinates equal to 0, and points on the y-axis have x-coordinates equal to 0.

Procedure for Finding Intercepts

1. To find the x-intercept(s), if any, of the graph of an equation, let $y = 0$ in the equation and solve for x.
2. To find the y-intercept(s), if any, of the graph of an equation, let $x = 0$ in the equation and solve for y.

Because the x-intercepts of the graph of an equation are those x-values for which $y = 0$, they are also called the **zeros** (or **roots**) of the equation.

EXAMPLE 5 Finding Intercepts from an Equation

Find the x-intercept(s) and the y-intercept(s) of the graph of $y = x^2 - 4$.

Solution To find the x-intercept(s), we let $y = 0$ and obtain the equation

$$x^2 - 4 = 0$$

$$(x + 2)(x - 2) = 0 \quad \text{Factor.}$$

$$x + 2 = 0 \quad \text{or} \quad x - 2 = 0 \quad \text{Zero-Product Property}$$

$$x = -2 \quad \text{or} \quad x = 2 \quad \text{Solve}$$

The equation has the solution set $\{-2, 2\}$. The x-intercepts are -2 and 2.

To find the y-intercept(s), we let $x = 0$ and obtain the equation

$$y = -4$$

The y-intercept is -4. ◄

NOW WORK PROBLEM 41 **(list the intercepts).**

COMMENT: For many equations, finding intercepts may not be so easy. In such cases, a graphing utility can be used. Read Section B.3, *Using a Graphing Utility to Locate Intercepts and Check for Symmetry* in Appendix B to find out how a graphing utility locates intercepts. ■

Symmetry

We have just seen the role that intercepts play in obtaining key points on the graph of an equation. Another helpful tool for graphing equations involves *symmetry*, particularly symmetry with respect to the x-axis, the y-axis, and the origin.

Figure 14
Symmetry with respect to the x-axis

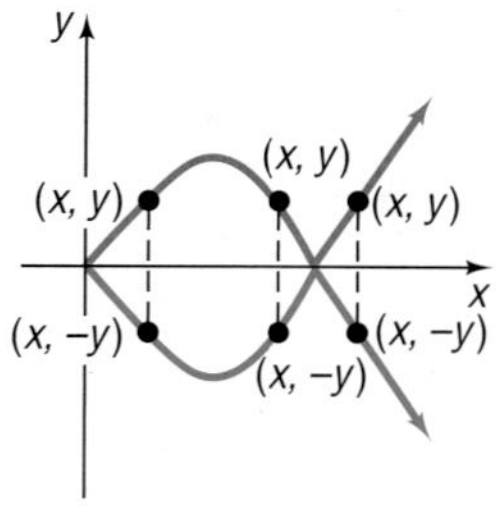

A graph is said to be **symmetric with respect to the x-axis** if, for every point (x, y) on the graph, the point $(x, -y)$ is also on the graph.

Figure 14 illustrates the definition. When a graph is symmetric with respect to the x-axis, notice that the part of the graph above the x-axis is a reflection or mirror image of the part below it, and vice versa.

EXAMPLE 6 Points Symmetric with Respect to the x-Axis

If a graph is symmetric with respect to the x-axis and the point $(3, 2)$ is on the graph, then the point $(3, -2)$ is also on the graph. ◄

Figure 15
Symmetry with respect to the y-axis

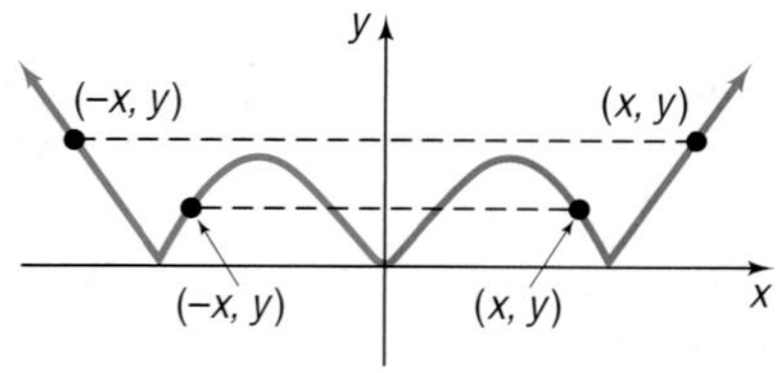

A graph is said to be **symmetric with respect to the y-axis** if, for every point (x, y) on the graph, the point $(-x, y)$ is also on the graph.

Figure 15 illustrates the definition. When a graph is symmetric with respect to the y-axis, notice that the part of the graph to the right of the y-axis is a reflection of the part to the left of it, and vice versa.

EXAMPLE 7 Points Symmetric with Respect to the y-Axis

If a graph is symmetric with respect to the y-axis and the point $(5, 8)$ is on the graph, then the point $(-5, 8)$ is also on the graph. ◄

A graph is said to be **symmetric with respect to the origin** if, for every point (x, y) on the graph, the point $(-x, -y)$ is also on the graph.

Figure 16
Symmetry with respect to the origin.

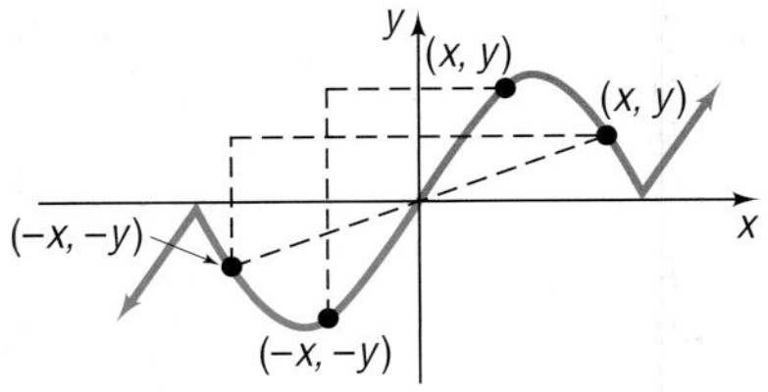

Figure 16 illustrates the definition. Notice that symmetry with respect to the origin may be viewed in two ways:

1. As a reflection about the y-axis, followed by a reflection about the x-axis.
2. As a projection along a line through the origin so that the distances from the origin are equal.

EXAMPLE 8 Points Symmetric with Respect to the Origin

If a graph is symmetric with respect to the origin and the point $(4, 2)$ is on the graph, then the point $(-4, -2)$ is also on the graph. ◄

NOW WORK PROBLEMS 9 AND 19(b).

4 When the graph of an equation is symmetric with respect to a coordinate axis or the origin, the number of points that you need to plot in order to see the pattern is reduced. For example, if the graph of an equation is symmetric with respect to the y-axis, then, once points to the right of the y-axis are plotted, an equal number of points on the graph can be obtained by reflecting them about the y-axis. Because of this, before we graph an equation, we first want to determine whether it has any symmetry. The following tests are used for this purpose.

Tests for Symmetry

To test the graph of an equation for symmetry with respect to the

x-Axis Replace y by $-y$ in the equation. If an equivalent equation results, the graph of the equation is symmetric with respect to the x-axis.

y-Axis Replace x by $-x$ in the equation. If an equivalent equation results, the graph of the equation is symmetric with respect to the y-axis.

Origin Replace x by $-x$ and y by $-y$ in the equation. If an equivalent equation results, the graph of the equation is symmetric with respect to the origin.

EXAMPLE 9 Graphing an Equation by Finding Intercepts and Checking for Symmetry

Graph the equation $y = x^3$. Find any intercepts and check for symmetry first.

Solution First, we seek the intercepts. When $x = 0$, then $y = 0$; and when $y = 0$, then $x = 0$. The origin $(0, 0)$ is the only intercept. Now we test for symmetry.

x-Axis Replace y by $-y$. Since the result, $-y = x^3$, is not equivalent to $y = x^3$, the graph is not symmetric with respect to the x-axis.

y-Axis Replace x by $-x$. Since the result, $y = (-x)^3 = -x^3$, is not equivalent to $y = x^3$, the graph is not symmetric with respect to the x-axis.

Origin Replace x by $-x$ and y by $-y$. Since the result, $-y = -x^3$, is equivalent to $y = x^3$, the graph is symmetric with respect to the origin.

To graph $y = x^3$, we use the equation to obtain several points on the graph. Because of the symmetry, we only need to locate points on the graph for which $x \geq 0$. See Table 2. Figure 17 shows the graph.

Table 2

x	$y = x^3$	(x, y)
0	0	(0, 0)
1	1	(1, 1)
2	8	(2, 8)
3	27	(3, 27)

Figure 17

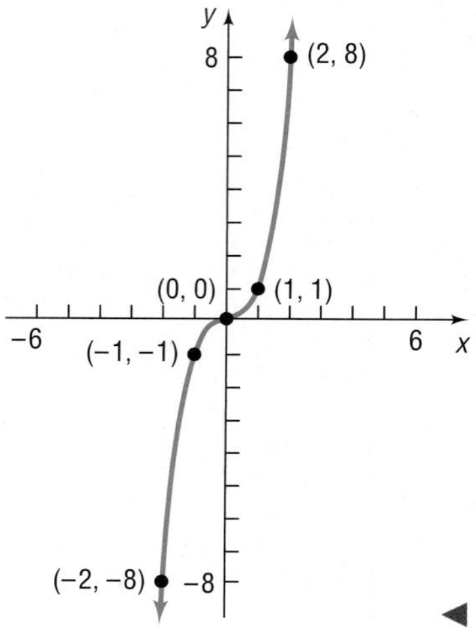

NOW WORK PROBLEM 41 (test for symmetry).

EXAMPLE 10 Graphing an Equation

Graph the equation $x = y^2$. Find any intercepts and check for symmetry first.

Solution The lone intercept is $(0, 0)$. The graph is symmetric with respect to the x-axis. (Do you see why? Replace y by $-y$.) Figure 18 shows the graph.

Figure 18

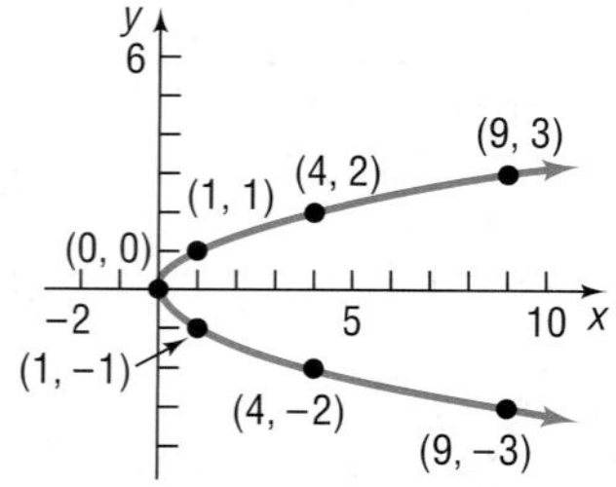

If we restrict y so that $y \geq 0$, the equation $x = y^2$, $y \geq 0$, may be written equivalently as $y = \sqrt{x}$. The portion of the graph of $x = y^2$ in quadrant I is therefore the graph of $y = \sqrt{x}$. See Figure 19.

COMMENT: To see the graph of the equation $x = y^2$ on a graphing calculator, you will need to graph two equations: $Y_1 = \sqrt{x}$ and $Y_2 = -\sqrt{x}$. We discuss why in the next chapter. See Figure 20.

Figure 19

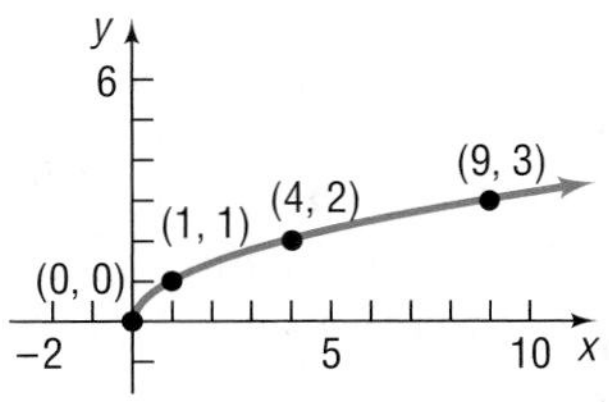

Figure 20

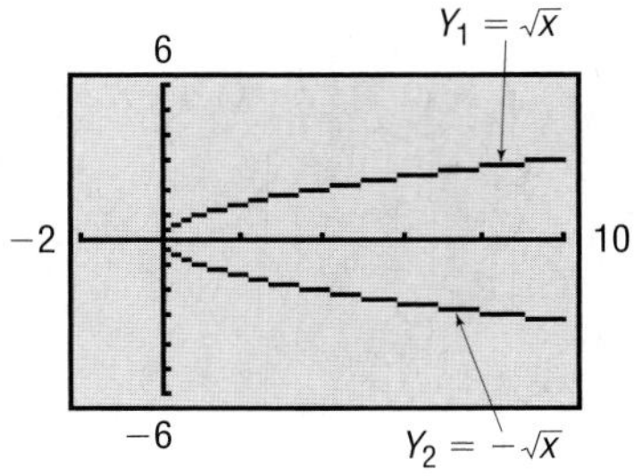

EXAMPLE 11

Graphing the Equation $y = \frac{1}{x}$

Graph the equation: $y = \frac{1}{x}$

Find any intercepts and check for symmetry first.

Solution We check for intercepts first. If we let $x = 0$, we obtain a 0 denominator, which is not defined. We conclude that there is no y-intercept. If we let $y = 0$, we get the equation $\frac{1}{x} = 0$, which has no solution. We conclude that there is no x-intercept. The graph of $y = \frac{1}{x}$ does not cross or touch the coordinate axes.

Next we check for symmetry:

x-Axis Replacing y by $-y$ yields $-y = \frac{1}{x}$, which is not equivalent to $y = \frac{1}{x}$.

y-Axis Replacing x by $-x$ yields $y = \frac{1}{-x}$, which is not equivalent to $y = \frac{1}{x}$.

Origin Replacing x by $-x$ and y by $-y$ yields $-y = -\frac{1}{x}$, which is equivalent to $y = \frac{1}{x}$.

The graph is symmetric with respect to the origin.

Table 3

x	$y = \frac{1}{x}$	(x, y)
$\frac{1}{10}$	10	$\left(\frac{1}{10}, 10\right)$
$\frac{1}{3}$	3	$\left(\frac{1}{3}, 3\right)$
$\frac{1}{2}$	2	$\left(\frac{1}{2}, 2\right)$
1	1	$(1, 1)$
2	$\frac{1}{2}$	$\left(2, \frac{1}{2}\right)$
3	$\frac{1}{3}$	$\left(3, \frac{1}{3}\right)$
10	$\frac{1}{10}$	$\left(10, \frac{1}{10}\right)$

Finally, we set up Table 3, listing several points on the graph. Because of the symmetry with respect to the origin, we use only positive values of x. From Table 3 we infer that if x is a large and positive number then $y = \frac{1}{x}$ is a positive number close to 0. We also infer that if x is a positive number close to 0, then $y = \frac{1}{x}$ is a large and positive number. Armed with this information, we can graph the equation. Figure 21 illustrates some of these points and the graph of $y = \frac{1}{x}$. Observe how the absence of intercepts and the existence of symmetry with respect to the origin were utilized. ◀

Figure 21

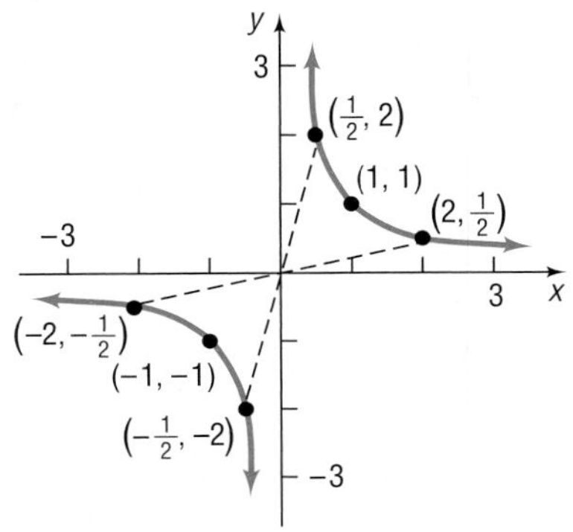

COMMENT: Refer to Example 3 in Appendix B, Section B.3, for the graph of $y = \frac{1}{x}$ using a graphing utility. ■

Circles

5 One advantage of a coordinate system is that it enables us to translate a geometric statement into an algebraic statement, and vice versa. Consider, for example, the following geometric statement that defines a circle.

> A **circle** is a set of points in the xy-plane that are a fixed distance r from a fixed point (h, k). The fixed distance r is called the **radius,** and the fixed point (h, k) is called the **center** of the circle.

Figure 22

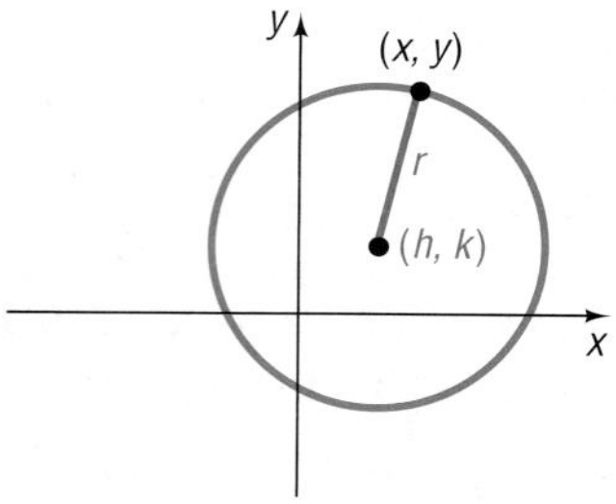

Figure 22 shows the graph of a circle. Is there an equation having this graph? If so, what is the equation? To find the equation, we let (x, y) represent the coordinates of any point on a circle with radius r and center (h, k). Then the distance between the points (x, y) and (h, k) must always equal r. That is, by the distance formula

$$\sqrt{(x - h)^2 + (y - k)^2} = r$$

or, equivalently,

$$(x - h)^2 + (y - k)^2 = r^2$$

> The **standard form of an equation of a circle** with radius r and center (h, k) is
>
> $$(x - h)^2 + (y - k)^2 = r^2 \qquad \textbf{(1)}$$

> The standard form of an equation of a circle of radius r with center at the origin $(0, 0)$ is
>
> $$x^2 + y^2 = r^2$$

> If the radius $r = 1$, the circle whose center is at the origin is called the **unit circle** and has the equation
>
> $$x^2 + y^2 = 1$$

Figure 23
Unit circle $x^2 + y^2 = 1$

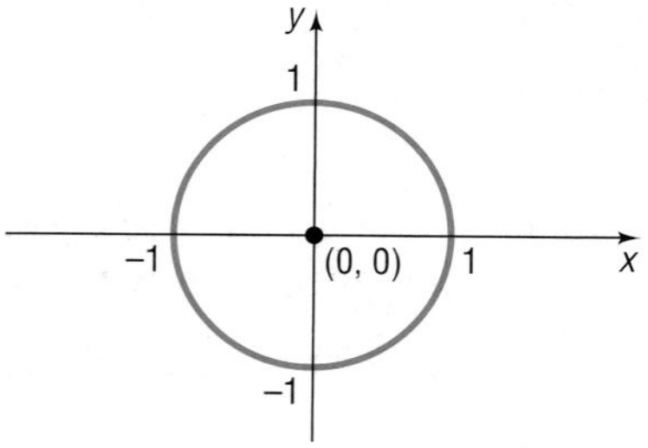

See Figure 23. Notice that the graph of the unit circle is symmetric with respect to the x-axis, the y-axis, and the origin.

EXAMPLE 12 Writing the Standard Form of the Equation of a Circle

Write the standard form of the equation of the circle with radius 5 and center $(-3, 6)$.

Solution Using the form of equation (1) and substituting the values $r = 5$, $h = -3$, and $k = 6$, we have

$$(x - h)^2 + (y - k)^2 = r^2$$
$$(x + 3)^2 + (y - 6)^2 = 25$$ ◀

NOW WORK PROBLEM 61.

6 The graph of any equation of the form of equation (1) is that of a circle with radius r and center (h, k).

EXAMPLE 13 Graphing a Circle

Graph the equation: $(x + 3)^2 + (y - 2)^2 = 16$

Solution The equation is of the form of equation (1), so its graph is a circle. To graph the equation, we first compare the given equation to the standard form of the equation of a circle. The comparison yields information about the circle.

$$(x + 3)^2 + (y - 2)^2 = 16$$
$$(x - (-3))^2 + (y - 2)^2 = 4^2$$
$$\uparrow \qquad \uparrow \qquad \uparrow$$
$$(x - h)^2 + (y - k)^2 = r^2$$

Figure 24

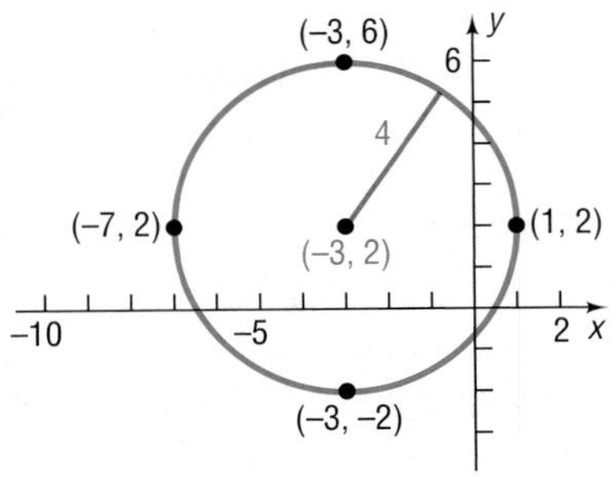

We see that $h = -3$, $k = 2$, and $r = 4$. The circle has center $(-3, 2)$ and a radius of 4 units. To graph this circle, we first plot the center $(-3, 2)$. Since the radius is 4, we can locate four points on the circle by plotting points 4 units to the left, to the right, and up and down from the center. These four points can then be used as guides to obtain the graph. See Figure 24. ◀

NOW WORK PROBLEM 79(a) AND (b).

EXAMPLE 14 Finding the Intercepts of a Circle

For the circle $(x + 3)^2 + (y - 2)^2 = 16$, find the intercepts, if any, of its graph.

Solution This is the equation discussed and graphed in Example 13. To find the x-intercepts, if any, let $y = 0$. Then

$$(x + 3)^2 + (y - 2)^2 = 16$$
$$(x + 3)^2 + (0 - 2)^2 = 16 \quad y = 0$$
$$(x + 3)^2 + 4 = 16 \quad \text{Simplify.}$$
$$(x + 3)^2 = 12 \quad \text{Simplify.}$$
$$x + 3 = \pm\sqrt{12} \quad \text{Apply the Square Root Method.}$$
$$x = -3 \pm 2\sqrt{3} \quad \text{Solve for x.}$$

The x-intercepts are $-3 - 2\sqrt{3} \approx -6.46$ and $-3 + 2\sqrt{3} \approx 0.46$.

To find the y-intercepts, if any, we let $x = 0$. Then

$$\begin{aligned} (x+3)^2 + (y-2)^2 &= 16 \\ 9 + (y-2)^2 &= 16 && x = 0 \\ (y-2)^2 &= 7 && \text{Simplify.} \\ y - 2 &= \pm\sqrt{7} && \text{Apply the Square Root Method.} \\ y &= 2 \pm \sqrt{7} && \text{Solve for } y. \end{aligned}$$

The y-intercepts are $2 - \sqrt{7} \approx -0.65$ and $2 + \sqrt{7} \approx 4.65$.

Look back at Figure 24 to verify the approximate location of the intercepts. ◀

NOW WORK PROBLEM 79(c).

If we eliminate the parentheses from the equation of the circle $(x+3)^2 + (y-2)^2 = 6$, we get

$$\begin{aligned} (x+3)^2 + (y-2)^2 &= 16 \\ x^2 + 6x + 9 + y^2 - 4y + 4 &= 16 \end{aligned}$$

which we find, upon simplifying, is equivalent to

$$x^2 + y^2 + 6x - 4y - 3 = 0 \tag{2}$$

COMMENT: We could have used the equivalent form of the equation of the circle given in (2) to find its intercepts. Try it and compare the different steps required to obtain the same solutions. ■

It can be shown that any equation of the form

$$x^2 + y^2 + ax + by + c = 0$$

has a graph that is a circle or a point, or it has no graph at all. For example, the graph of the equation $x^2 + y^2 = 0$ is the single point $(0, 0)$. The equation $x^2 + y^2 + 5 = 0$, or $x^2 + y^2 = -5$, has no graph, because sums of squares of real numbers are never negative.

When its graph is a circle, the equation

$$x^2 + y^2 + ax + by + c = 0$$

is referred to as the **general form of the equation of a circle.**

NOW WORK PROBLEM 67.

7 If an equation of a circle is in the general form, we use the method of completing the square to put the equation in standard form so that we can identify its center and radius.

EXAMPLE 15 Graphing a Circle Whose Equation Is in General Form

Graph the equation: $x^2 + y^2 + 4x - 6y + 12 = 0$

Solution We complete the square in both x and y to put the equation in standard form. Group the expressions involving x, group the expressions involving y, and put the constant on the right side of the equation. The result is

$$(x^2 + 4x) + (y^2 - 6y) = -12$$

Next, complete the square of each expression in parentheses. Remember that any number added on the left side of the equation must be added on the right.

$$(x^2 + 4x + 4) + (y^2 - 6y + 9) = -12 + 4 + 9$$

$$\left(\frac{4}{2}\right)^2 = 4 \qquad \left(\frac{-6}{2}\right)^2 = 9$$

$$(x + 2)^2 + (y - 3)^2 = 1 \qquad \text{Factor.}$$

We recognize this equation as the standard form of the equation of a circle with radius 1 and center $(-2, 3)$.

To graph the equation, we use the center $(-2, 3)$ and the radius 1. See Figure 25. ◀

Figure 25

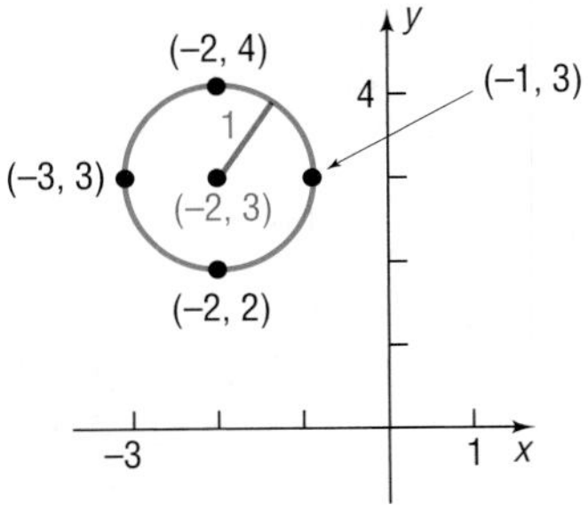

NOW WORK PROBLEM 81.

COMMENT: Now read Section B.5, *Square Screens*, in Appendix B. ■

EXAMPLE 16 Using a Graphing Utility to Graph a Circle

Graph the equation: $x^2 + y^2 = 4$

Solution This is the equation of a circle with center at the origin and radius 2. To graph this equation, we must first solve for y.

$$x^2 + y^2 = 4$$

$$y^2 = 4 - x^2 \qquad \text{Subtract } x^2 \text{ from each side.}$$

$$y = \pm\sqrt{4 - x^2} \qquad \text{Apply the Square Root Method to solve for } y.$$

There are two equations to graph: first, we graph $Y_1 = \sqrt{4 - x^2}$ and then $Y_2 = -\sqrt{4 - x^2}$ on the same square screen. (Your circle will appear oval if you do not use a square screen.) See Figure 26. ◀

Figure 26

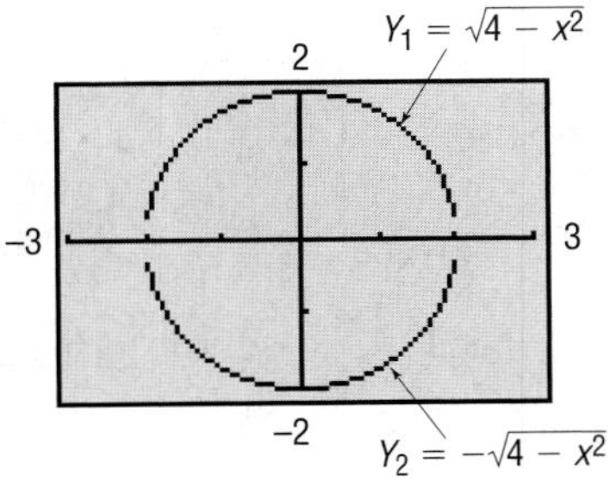

1.2 Assess Your Understanding

'Are You Prepared?' *Answers are given at the end of these exercises. If you get a wrong answer, read the pages listed in red.*

1. To complete the square of the expression $x + 4x$, you would __________ the number __________. (pp. 941–942)

2. Use the Square Root Method to find the solution set of the equation $(x - 2)^2 = 4$. (pp. 940–941)

Concepts and Vocabulary

3. The points, if any, at which a graph crosses or touches the coordinate axes are called __________.

4. *True or False:* To find the x-intercepts of the graph of an equation, let $y = 0$ and solve for x.

5. If a graph is symmetric with respect to the origin and if $(-3, 4)$ is a point on the graph, then the point __________ is also on the graph.

6. For a circle, the __________ is the distance from the center to any point on the circle.

7. *True or False:* A circle with center at the origin is symmetric with respect to the x-axis, y-axis, and origin.

8. The center and radius of the circle $(x - 2)^2 + (y + 5)^2 = 36$ are __________ and __________.

Exercises

In Problems 9–18, plot each point. Then plot the point that is symmetric to it with respect to (a) the x-axis; (b) the y-axis; (c) the origin.

9. $(3, 4)$ **10.** $(5, 3)$ **11.** $(-2, 1)$ **12.** $(4, -2)$ **13.** $(1, 1)$

14. $(-1, -1)$ **15.** $(-3, -4)$ **16.** $(4, 0)$ **17.** $(0, -3)$ **18.** $(-3, 0)$

In Problems 19–30, the graph of an equation is given.
(a) *List the intercepts of the graph.*
(b) *Based on the graph, tell whether the graph is symmetric with respect to the x-axis, the y-axis, and/or the origin.*

19.

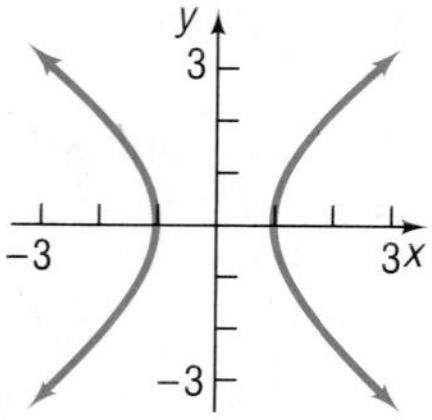

20.

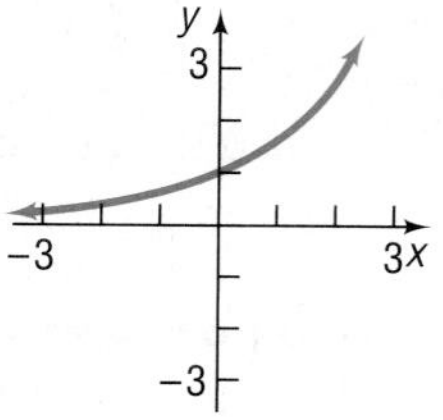

21.

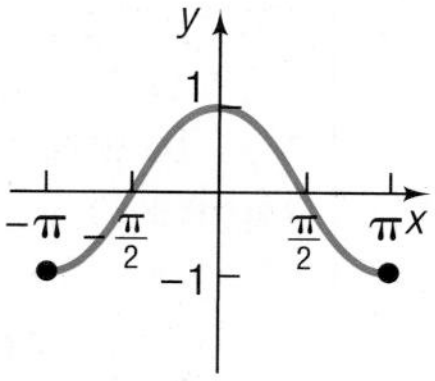

22.

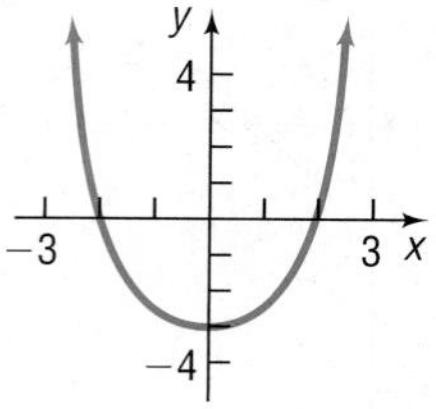

23.

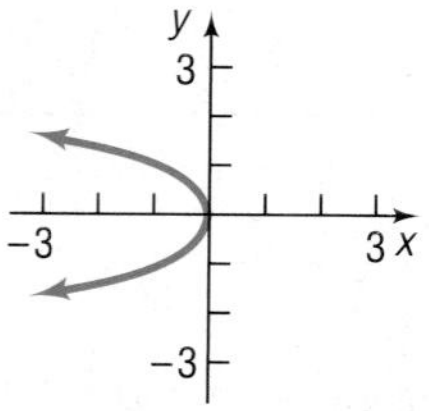

24.

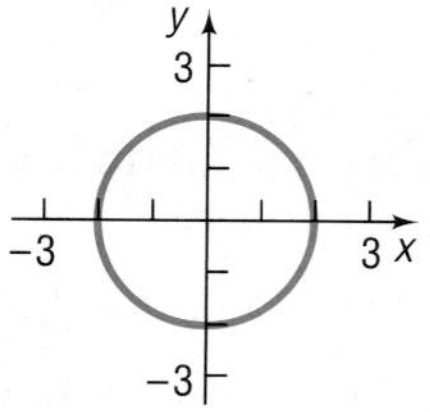

25.

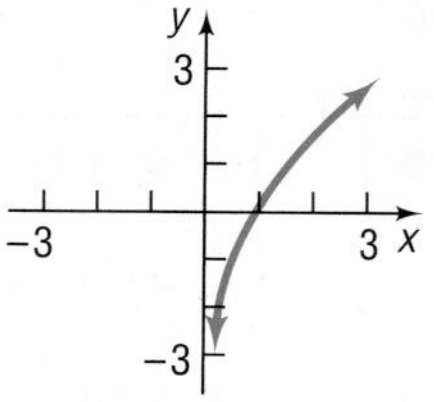

26.

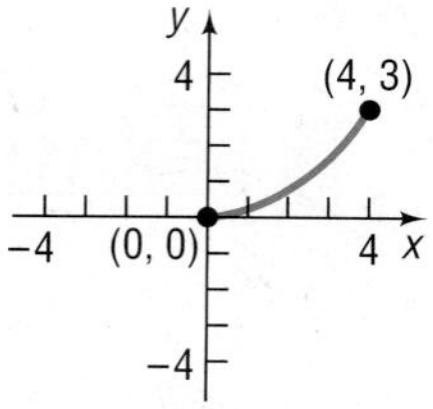

27.

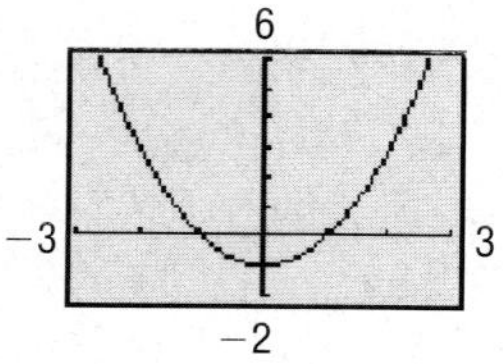

28.

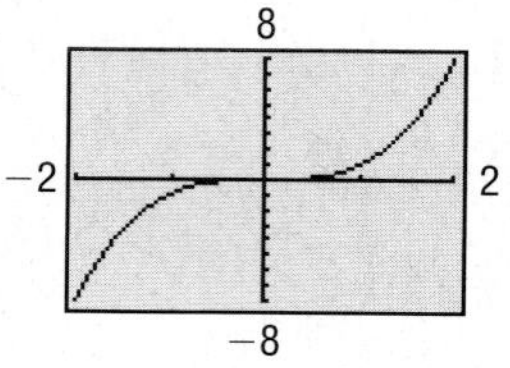

29.

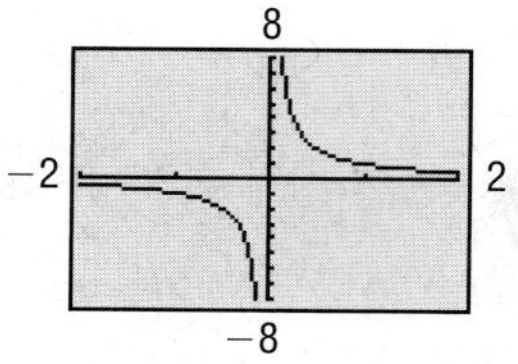

30.

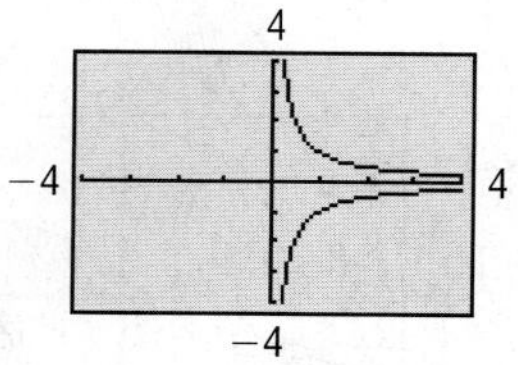

In Problems 31–36, determine whether the given points are on the graph of the equation.

31. Equation: $y = x^4 - \sqrt{x}$
Points: $(0, 0)$; $(1, 1)$; $(-1, 0)$

32. Equation: $y = x^3 - 2\sqrt{x}$
Points: $(0, 0)$; $(1, 1)$; $(1, -1)$

33. Equation: $y^2 = x^2 + 9$
Points: $(0, 3)$; $(3, 0)$; $(-3, 0)$

34. Equation: $y^3 = x + 1$
Points: $(1, 2)$; $(0, 1)$; $(-1, 0)$

35. Equation: $x^2 + y^2 = 4$
Points: $(0, 2)$; $(-2, 2)$; $\left(\sqrt{2}, \sqrt{2}\right)$

36. Equation: $x^2 + 4y^2 = 4$
Points: $(0, 1)$; $(2, 0)$; $\left(2, \frac{1}{2}\right)$

In Problems 37–52, list the intercepts and test for symmetry.

37. $x^2 = y$ **38.** $y^2 = x$ **39.** $y = 3x$ **40.** $y = -5x$

41. $x^2 + y - 9 = 0$ **42.** $y^2 - x - 4 = 0$ **43.** $9x^2 + 4y^2 = 36$ **44.** $4x^2 + y^2 = 4$

45. $y = x^3 - 27$ **46.** $y = x^4 - 1$ **47.** $y = x^2 - 3x - 4$ **48.** $y = x^2 + 4$

49. $y = \dfrac{3x}{x^2 + 9}$ **50.** $y = \dfrac{x^2 - 4}{2x}$ **51.** $y = \dfrac{-x^3}{x^2 - 9}$ **52.** $y = \dfrac{x^4 + 1}{2x^5}$

In Problems 53–56, draw a quick sketch of each equation.

53. $y = x^3$ **54.** $x = y^2$ **55.** $y = \sqrt{x}$ **56.** $y = \dfrac{1}{x}$

57. If $(a, 2)$ is a point on the graph of $y = 3x + 5$, what is a?

58. If $(2, b)$ is a point on the graph of $y = x^2 + 4x$, what is b?

59. If (a, b) is a point on the graph of $2x + 3y = 6$, write an equation that relates a to b.

60. If $(2, 0)$ and $(0, 5)$ are points on the graph of $y = mx + b$, what are m and b?

In Problems 61–64, find the center and radius of each circle. Write the standard form of the equation.

61.
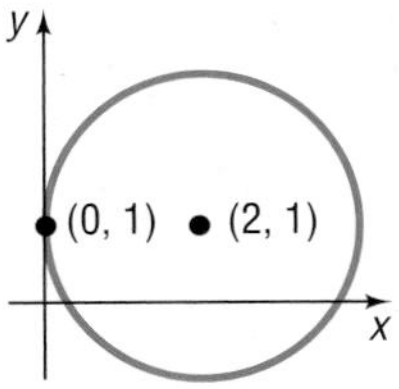

62.
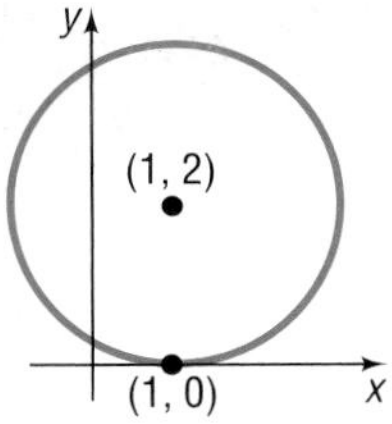

63.
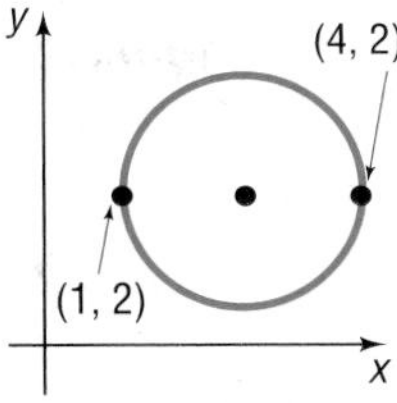

64.
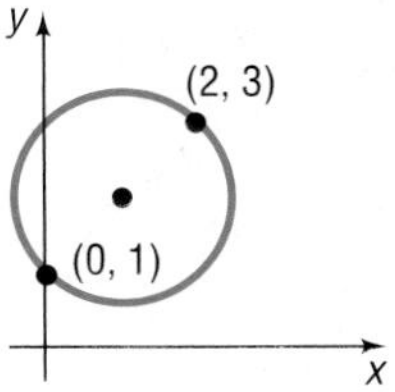

In Problems 65–76, write the standard form of the equation and the general form of the equation of each circle of radius r and center (h, k). Graph each circle.

65. $r = 2$; $(h, k) = (0, 0)$
66. $r = 3$; $(h, k) = (0, 0)$
67. $r = 1$; $(h, k) = (1, -1)$
68. $r = 2$; $(h, k) = (-2, 1)$
69. $r = 2$; $(h, k) = (0, 2)$
70. $r = 3$; $(h, k) = (1, 0)$
71. $r = 5$; $(h, k) = (4, -3)$
72. $r = 4$; $(h, k) = (2, -3)$
73. $r = 6$; $(h, k) = (-3, -6)$
74. $r = 5$; $(h, k) = (-5, 2)$
75. $r = 3$; $(h, k) = (0, -3)$
76. $r = 2$; $(h, k) = (-2, 0)$

In Problems 77–88, (a) find the center (h, k) and radius r of each circle; (b) graph each circle; (c) find the intercepts, if any, of the graph.

77. $x^2 + y^2 = 4$
78. $x^2 + (y - 1)^2 = 1$
79. $2(x - 3)^2 + 2y^2 = 8$
80. $3(x + 1)^2 + 3(y - 1)^2 = 6$
81. $x^2 + y^2 + 4x - 4y - 1 = 0$
82. $x^2 + y^2 - 6x + 2y + 9 = 0$
83. $x^2 + y^2 - 2x + 4y - 4 = 0$
84. $x^2 + y^2 + 4x + 2y - 20 = 0$
85. $x^2 + y^2 - x + 2y + 1 = 0$
86. $x^2 + y^2 + x + y - \frac{1}{2} = 0$
87. $2x^2 + 2y^2 - 12x + 8y - 24 = 0$
88. $2x^2 + 2y^2 + 8x + 7 = 0$

In Problems 89–94, find the general form of the equation of each circle.

89. Center at the origin and containing the point $(-3, 2)$

90. Center at the point $(1, 0)$ and containing the point $(-2, 3)$

91. Center at the point $(2, 3)$ and tangent to the x-axis

92. Center at the point $(-3, 1)$ and tangent to the y-axis

93. With endpoints of a diameter at the points $(1, 4)$ and $(-3, 2)$

94. With endpoints of a diameter at the points $(4, 3)$ and $(0, 1)$

In Problems 95–98, match each graph with the correct equation.

(a) $(x - 3)^2 + (y + 3)^2 = 9$
(b) $(x + 1)^2 + (y - 2)^2 = 4$
(c) $(x - 1)^2 + (y + 2)^2 = 4$
(d) $(x + 3)^2 + (y - 3)^2 = 9$

95.
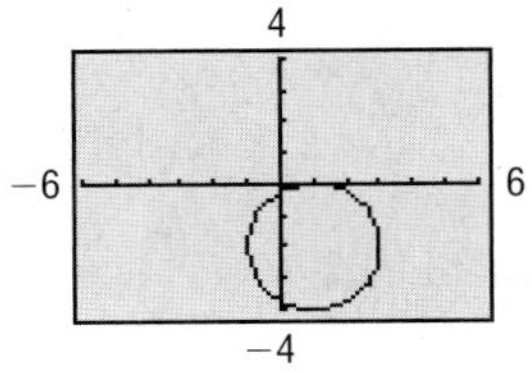

96.
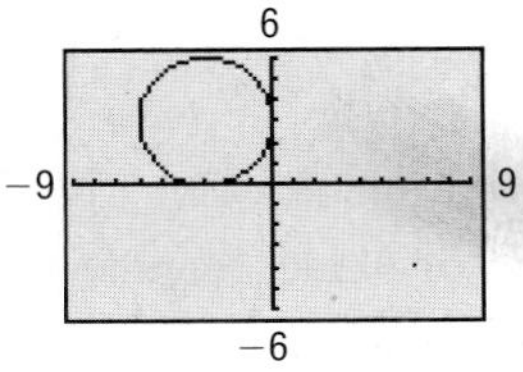

97.
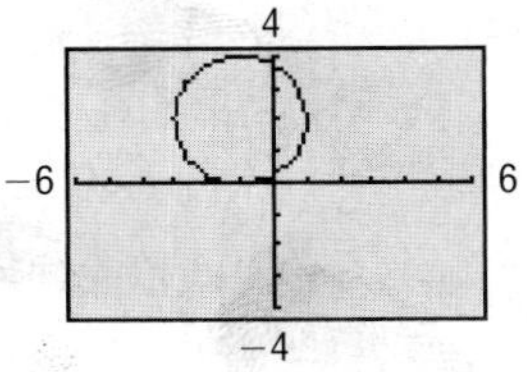

98.
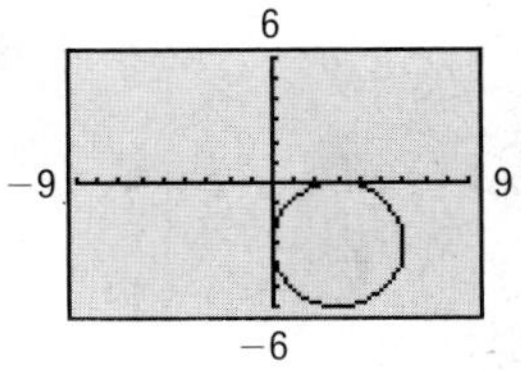

99. Which of the following equations might have the graph shown? (More than one answer is possible.)
(a) $(x - 2)^2 + (y + 3)^2 = 13$
(b) $(x - 2)^2 + (y - 2)^2 = 8$
(c) $(x - 2)^2 + (y - 3)^2 = 13$
(d) $(x + 2) + (y - 2)^2 = 8$
(e) $x^2 + y^2 - 4x - 9y = 0$
(f) $x^2 + y^2 + 4x - 2y = 0$
(g) $x^2 + y^2 - 9x - 4y = 0$
(h) $x^2 + y^2 - 4x - 4y = 4$

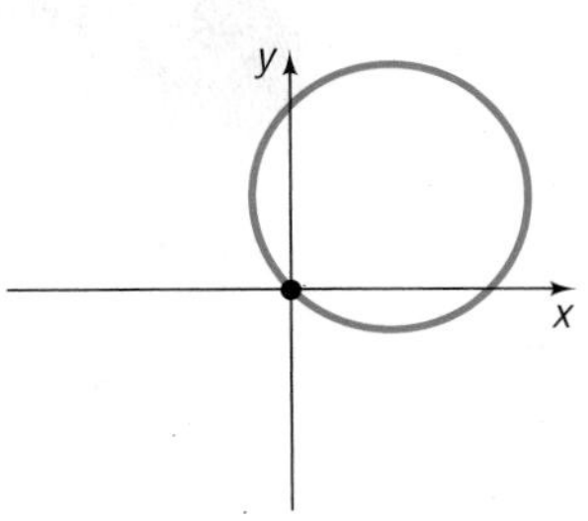

100. Which of the following equations might have the graph shown? (More than one answer is possible.)
(a) $(x - 2)^2 + y^2 = 3$
(b) $(x + 2)^2 + y^2 = 3$
(c) $x^2 + (y - 2)^2 = 3$
(d) $(x + 2)^2 + y^2 = 4$
(e) $x^2 + y^2 + 10x + 16 = 0$
(f) $x^2 + y^2 + 10x - 2y = 1$
(g) $x^2 + y^2 + 9x + 10 = 0$
(h) $x^2 + y^2 - 9x - 10 = 0$

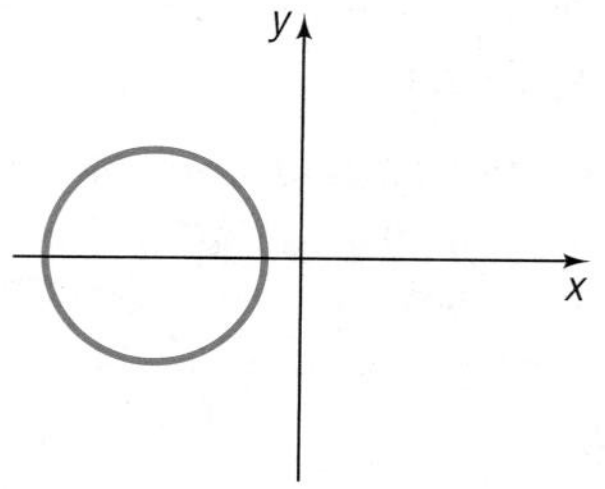

101. Weather Satellites Earth is represented on a map of a portion of the solar system so that its surface is the circle with equation $x^2 + y^2 + 2x + 4y - 4091 = 0$. A weather satellite circles 0.6 unit above Earth with the center of its circular orbit at the center of Earth. Find the equation for the orbit of the satellite on this map.

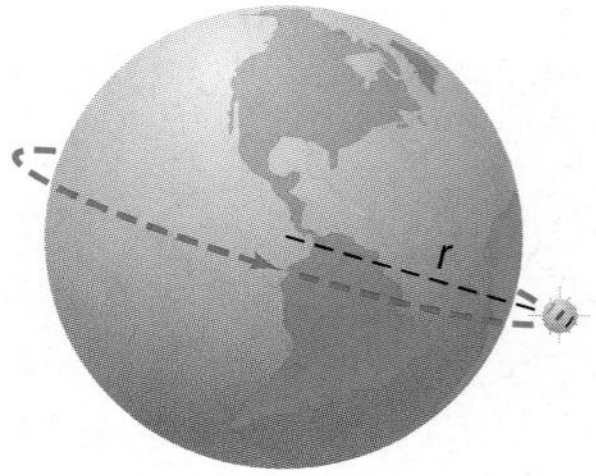

In Problem 102, you may use a graphing utility, but it is not required.

102. (a) Graph $y = \sqrt{x^2}$, $y = x$, $y = |x|$, and $y = (\sqrt{x})^2$, noting which graphs are the same.
(b) Explain why the graphs of $y = \sqrt{x^2}$ and $y = |x|$ are the same.
(c) Explain why the graphs of $y = x$ and $y = (\sqrt{x})^2$ are not the same.
(d) Explain why the graphs of $y = \sqrt{x^2}$ and $y = x$ are not the same.

103. Find an equation with the intercepts $(2, 0)$, $(4, 0)$, and $(0, 1)$. Compare your equation with a friend's equation. Comment on any similarities.

104. An equation is being tested for symmetry with respect to the x-axis, the y-axis, and the origin. Explain why, if two of these symmetries are present, the remaining one must also be present.

105. Draw a graph that contains the points $(-2, -1)$, $(0, 1)$, $(1, 3)$, and $(3, 5)$. Compare your graph with those of other students. Are most of the graphs almost straight lines? How many are "curved"? Discuss the various ways that these points might be connected.

'Are You Prepared?' Answers

1. add; 4 **2.** $\{0, 4\}$

1.3 Lines

OBJECTIVES
1 Calculate and Interpret the Slope of a Line
2 Graph Lines Given a Point and the Slope
3 Find the Equation of a Vertical Line
4 Use the Point–Slope Form of a Line; Identify Horizontal Lines
5 Find the Equation of a Line Given Two Points
6 Write the Equation of a Line in Slope–Intercept Form
7 Identify the Slope and y-Intercept of a Line from Its Equation
8 Write the Equation of a Line in General Form
9 Find Equations of Parallel Lines
10 Find Equations of Perpendicular Lines

In this section we study a certain type of equation that contains two variables, called a *linear equation,* and its graph, a *line.*

Slope of a Line

Figure 27

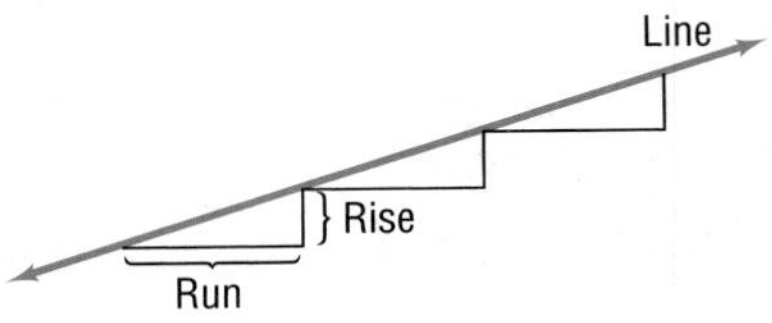

1 Consider the staircase illustrated in Figure 27. Each step contains exactly the same horizontal **run** and the same vertical **rise.** The ratio of the rise to the run, called the *slope*, is a numerical measure of the steepness of the staircase. For example, if the run is increased and the rise remains the same, the staircase becomes less steep. If the run is kept the same, but the rise is increased, the staircase becomes more steep. The slope of a line is best defined using rectangular coordinates.

Let $P = (x_1, y_1)$ and $Q = (x_2, y_2)$ be two distinct points. If $x_1 \neq x_2$, the **slope *m*** of the nonvertical line L containing P and Q is defined by the formula

$$m = \frac{y_2 - y_1}{x_2 - x_1} \qquad x_1 \neq x_2 \qquad \textbf{(1)}$$

If $x_1 = x_2$, L is a **vertical line** and the slope m of L is **undefined** (since this results in division by 0).

Figure 28(a) provides an illustration of the slope of a nonvertical line, and Figure 28(b) illustrates a vertical line.

Figure 28

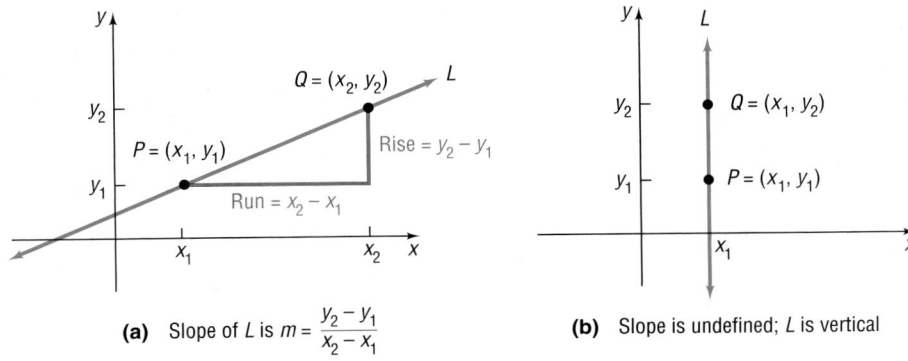

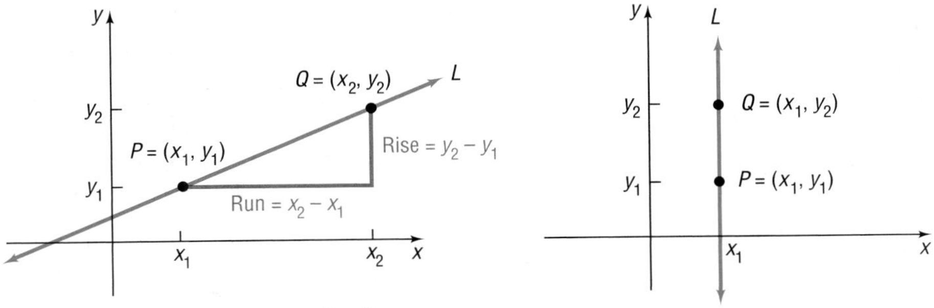

(a) Slope of L is $m = \frac{y_2 - y_1}{x_2 - x_1}$ **(b)** Slope is undefined; L is vertical

As Figure 28(a) illustrates, the slope m of a nonvertical line may be viewed as

$$m = \frac{y_2 - y_1}{x_2 - x_1} = \frac{\text{Rise}}{\text{Run}}$$

We can also express the slope m of a nonvertical line as

$$m = \frac{y_2 - y_1}{x_2 - x_1} = \frac{\text{Change in } y}{\text{Change in } x} = \frac{\Delta y}{\Delta x}$$

That is, the slope m of a nonvertical line L measures how y changes as x changes from x_1 to x_2. This is called the **average rate of change** of y with respect to x.

Two comments about computing the slope of a nonvertical line may prove helpful.

1. Any two distinct points on the line can be used to compute the slope of the line. (See Figure 29 for justification.)

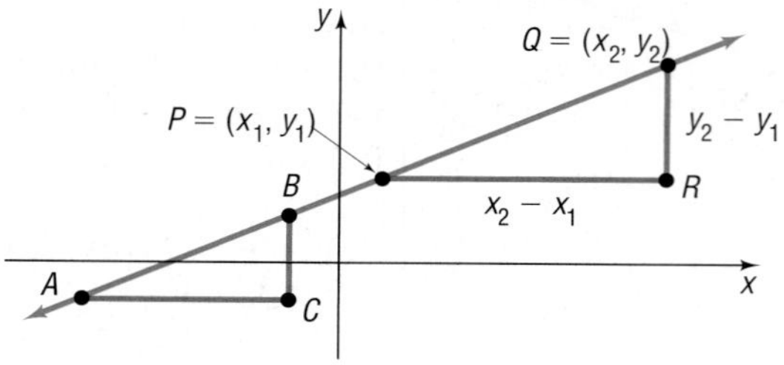

Figure 29
Triangles ABC and PQR are similar (equal angles). Then, ratios of corresponding sides are proportional so that

$$\text{Slope using } P \text{ and } Q = \frac{y_2 - y_1}{x_2 - x_1}$$

$$= \text{Slope using } A \text{ and } B = \frac{d(B, C)}{d(A, C)}$$

2. The slope of a line may be computed from $P = (x_1, y_1)$ to $Q = (x_2, y_2)$ or from Q to P, because

$$\frac{y_2 - y_1}{x_2 - x_1} = \frac{y_1 - y_2}{x_1 - x_2}$$

EXAMPLE 1 Finding and Interpreting the Slope of a Line Containing Two Points

The slope m of the line containing the points $(1, 2)$ and $(5, -3)$ may be computed as

$$m = \frac{-3 - 2}{5 - 1} = \frac{-5}{4} = -\frac{5}{4} \quad \text{or as} \quad m = \frac{2 - (-3)}{1 - 5} = \frac{5}{-4} = -\frac{5}{4}$$

For every 4-unit change in x, y will change by -5 units. That is, if x increases by 4 units, then y decreases by 5 units. The average rate of change of y with respect to x is $-\frac{5}{4}$. ◀

NOW WORK PROBLEMS 11 AND 17.

To get a better idea of the meaning of the slope m of a line L, consider the following example.

EXAMPLE 2 Finding the Slopes of Various Lines Containing the Same Point (2, 3)

Compute the slopes of the lines L_1, L_2, L_3, and L_4 containing the following pairs of points. Graph all four lines on the same set of coordinate axes.

L_1: $P = (2, 3)$ $Q_1 = (-1, -2)$

L_2: $P = (2, 3)$ $Q_2 = (3, -1)$

L_3: $P = (2, 3)$ $Q_3 = (5, 3)$

L_4: $P = (2, 3)$ $Q_4 = (2, 5)$

Figure 30

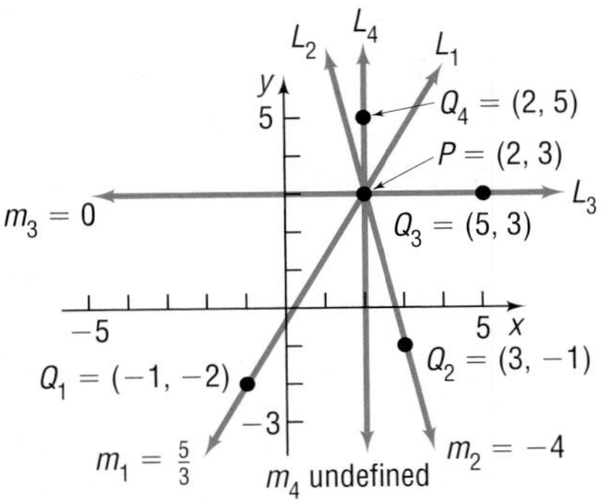

Solution Let m_1, m_2, m_3, and m_4 denote the slopes of the lines L_1, L_2, L_3, and L_4, respectively. Then

$$m_1 = \frac{-2-3}{-1-2} = \frac{-5}{-3} = \frac{5}{3} \quad \text{A rise of 5 divided by a run of 3}$$

$$m_2 = \frac{-1-3}{3-2} = \frac{-4}{1} = -4$$

$$m_3 = \frac{3-3}{5-2} = \frac{0}{3} = 0$$

m_4 is undefined

The graphs of these lines are given in Figure 30. ◄

Figure 30 illustrates the following facts:

1. When the slope of a line is positive, the line slants upward from left to right (L_1).
2. When the slope of a line is negative, the line slants downward from left to right (L_2).
3. When the slope is 0, the line is horizontal (L_3).
4. When the slope is undefined, the line is vertical (L_4).

Seeing the Concept

On the same square screen, graph the following equations:

$Y_1 = 0$ Slope of line is 0.

$Y_2 = \frac{1}{4}x$ Slope of line is $\frac{1}{4}$.

$Y_3 = \frac{1}{2}x$ Slope of line is $\frac{1}{2}$.

$Y_4 = x$ Slope of line is 1.

$Y_5 = 2x$ Slope of line is 2.

$Y_6 = 6x$ Slope of line is 6.

See Figure 31.

Figure 31

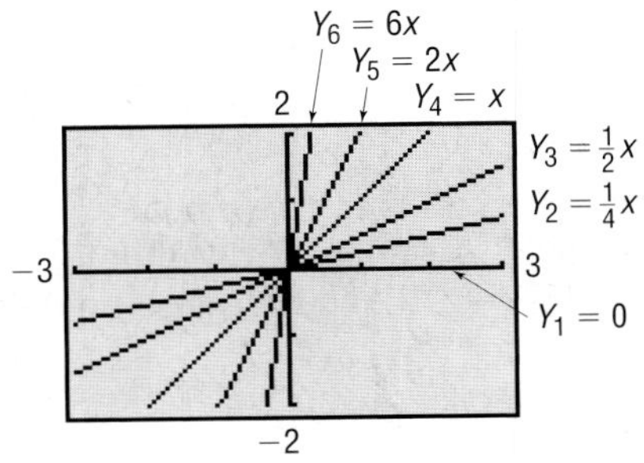

Seeing the Concept

On the same square screen, graph the following equations:

$Y_1 = 0$ Slope of line is 0.

$Y_2 = -\frac{1}{4}x$ Slope of line is $-\frac{1}{4}$.

$Y_3 = -\frac{1}{2}x$ Slope of line is $-\frac{1}{2}$.

$Y_4 = -x$ Slope of line is −1.

$Y_5 = -2x$ Slope of line is −2.

$Y_6 = -6x$ Slope of line is −6.

See Figure 32.

Figure 32

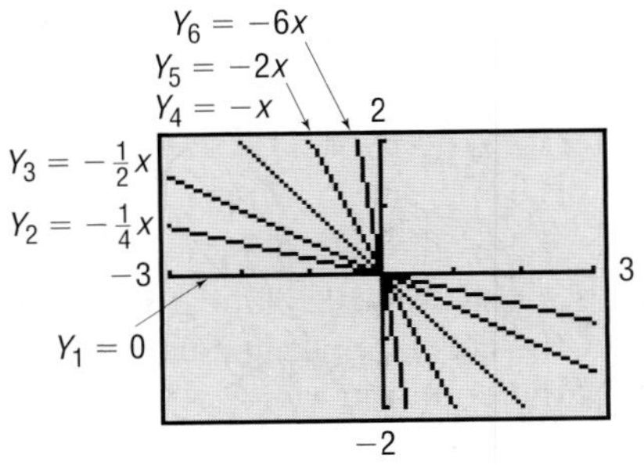

Figures 31 and 32 illustrate that the closer the line is to the vertical position, the greater the magnitude of the slope.

2 The next example illustrates how the slope of a line can be used to graph the line.

EXAMPLE 3 Graphing a Line Given a Point and a Slope

Draw a graph of the line that contains the point (3, 2) and has a slope of

(a) $\frac{3}{4}$ (b) $-\frac{4}{5}$

Figure 33

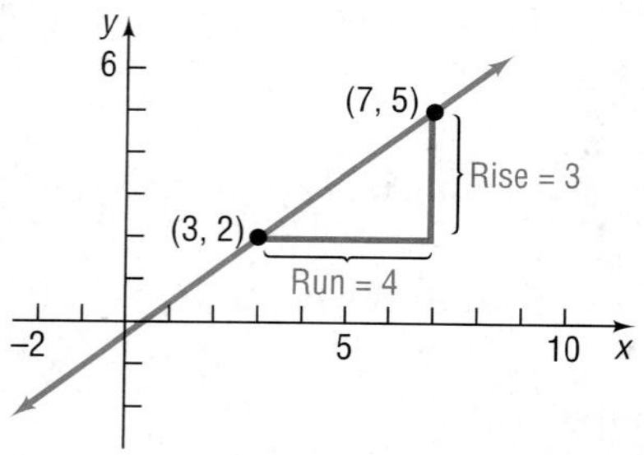

Solution (a) Slope $= \frac{\text{Rise}}{\text{Run}}$. The fact that the slope is $\frac{3}{4}$ means that for every horizontal movement (run) of 4 units to the right there will be a vertical movement (rise) of 3 units. If we start at the given point (3, 2) and move 4 units to the right and 3 units up, we reach the point (7, 5). By drawing the line through this point and the point (3, 2), we have the graph. See Figure 33.

(b) The fact that the slope is

$$-\frac{4}{5} = \frac{-4}{5} = \frac{\text{Rise}}{\text{Run}}$$

means that for every horizontal movement of 5 units (run = 5) to the right there will be a corresponding vertical movement of −4 units (rise = −4, a downward movement of 4 units). If we start at the given point (3, 2) and move 5 units to the right and then 4 units down, we arrive at the point (8, −2). By drawing the line through these points, we have the graph. See Figure 34.

Figure 34

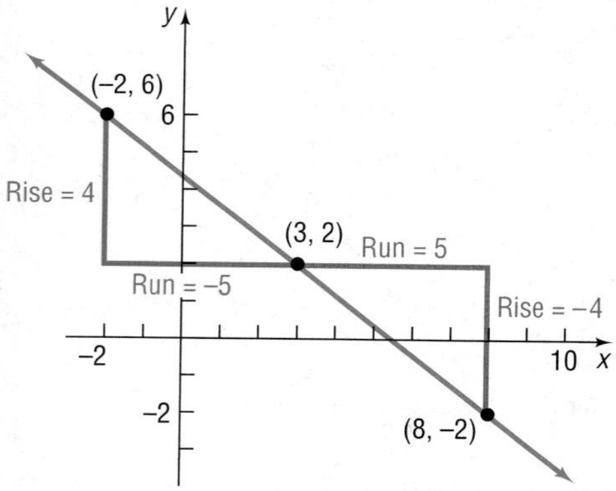

Alternatively, we can set

$$-\frac{4}{5} = \frac{4}{-5} = \frac{\text{Rise}}{\text{Run}}$$

so that for every horizontal movement of −5 units (a movement to the left) there will be a corresponding vertical movement of 4 units (upward). This approach brings us to the point (−2, 6), which is also on the graph shown in Figure 34. ◀

NOW WORK PROBLEM 25.

Equations of Lines

3 Now that we have discussed the slope of a line, we are ready to derive equations of lines. As we shall see, there are several forms of the equation of a line. Let's start with an example.

EXAMPLE 4 Graphing a Line

Figure 35

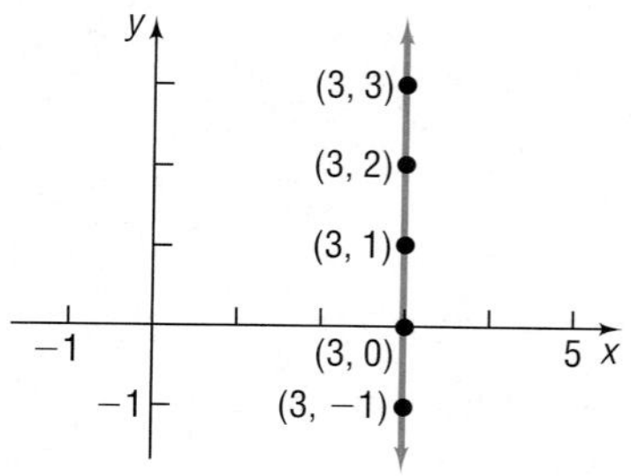

Graph the equation: $x = 3$

Solution We are looking for all points (x, y) in the plane for which $x = 3$. No matter what y-coordinate is used, the corresponding x-coordinate always equals 3. Consequently, the graph of the equation $x = 3$ is a vertical line with x-intercept 3 and undefined slope. See Figure 35. ◀

As suggested by Example 4, we have the following result:

Theorem

Equation of a Vertical Line

A vertical line is given by an equation of the form

$$x = a$$

where a is the x-intercept.

COMMENT: To graph an equation using a graphing utility, we need to express the equation in the form y = expression in x. But $x = 3$ cannot be put in this form. To overcome this, most graphing utilities have special commands for drawing vertical lines. LINE, PLOT, and VERT are among the more common ones. Consult your manual to determine the correct methodology for your graphing utility. ■

Figure 36

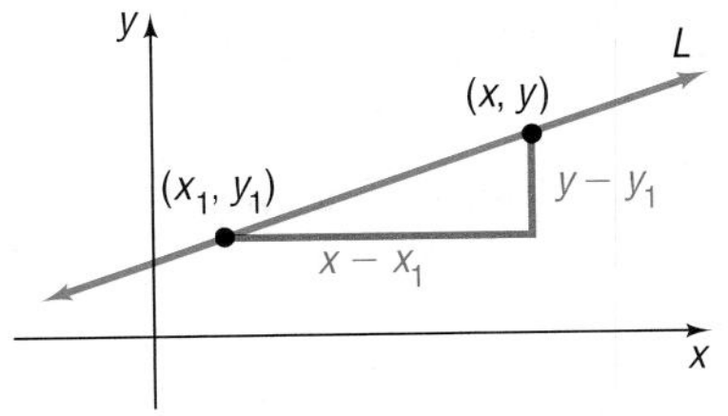

4 Now let L be a nonvertical line with slope m containing the point (x_1, y_1). See Figure 36. For any other point (x, y) on L, we have

$$m = \frac{y - y_1}{x - x_1} \quad \text{or} \quad y - y_1 = m(x - x_1)$$

Theorem

Point–Slope Form of an Equation of a Line

An equation of a nonvertical line of slope m that contains the point (x_1, y_1) is

$$y - y_1 = m(x - x_1) \qquad \textbf{(2)}$$

Figure 37

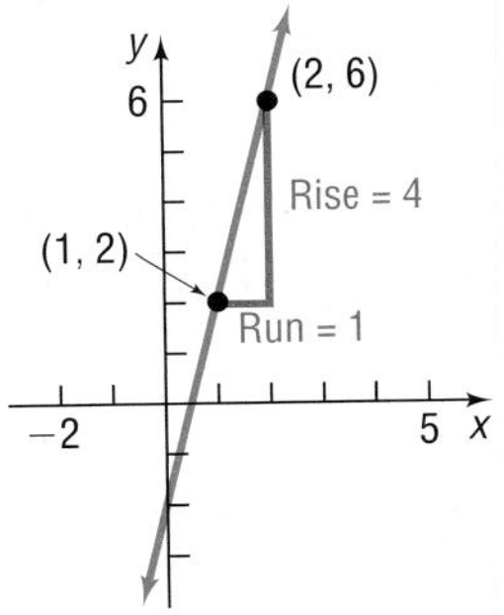

EXAMPLE 5 **Using the Point–Slope Form of a Line**

An equation of the line with slope 4 and containing the point $(1, 2)$ can be found by using the point–slope form with $m = 4$, $x_1 = 1$, and $y_1 = 2$.

$$y - y_1 = m(x - x_1) \qquad \text{Point-slope form}$$
$$y - 2 = 4(x - 1) \qquad m = 4, x_1 = 1, y_1 = 2$$

See Figure 37. ◄

EXAMPLE 6 **Finding the Equation of a Horizontal Line**

Find an equation of the horizontal line containing the point $(3, 2)$.

Figure 38

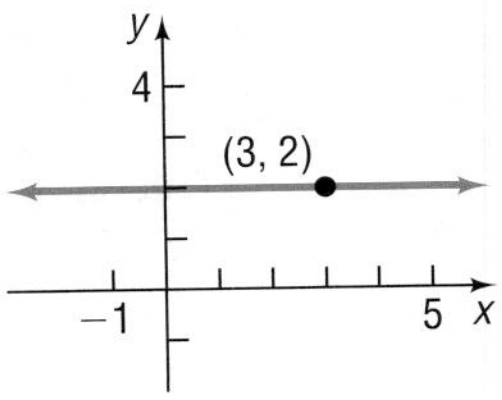

Solution The slope of a horizontal line is 0. To get an equation, we use the point–slope form with $m = 0$, $x_1 = 3$, and $y_1 = 2$.

$$y - y_1 = m(x - x_1) \qquad \text{Point-slope form}$$
$$y - 2 = 0 \cdot (x - 3) \qquad m = 0, x_1 = 3, y_1 = 2$$
$$y - 2 = 0$$
$$y = 2$$

See Figure 38 for the graph. ◄

As suggested by Example 6, we have the following result:

Theorem

Equation of a Horizontal Line

A horizontal line is given by an equation of the form

$$y = b$$

where b is the y-intercept.

5 **EXAMPLE 7** **Finding an Equation of a Line Given Two Points**

Find an equation of the line L containing the points $(2, 3)$ and $(-4, 5)$. Graph the line L.

Solution Since two points are given, we first compute the slope of the line.

$$m = \frac{5 - 3}{-4 - 2} = \frac{2}{-6} = -\frac{1}{3}$$

Figure 39

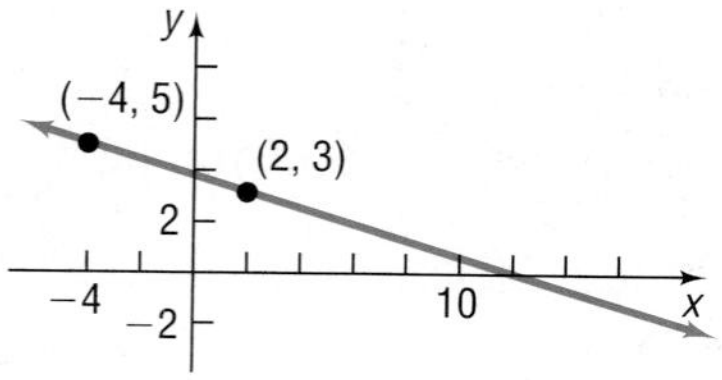

We use the point $(2, 3)$ and the fact that the slope $m = -\frac{1}{3}$ to get the point–slope form of the equation of the line.

$$y - y_1 = m(x - x_1) \quad \text{Point–slope form}$$

$$y - 3 = -\frac{1}{3}(x - 2) \quad m = -\frac{1}{3}, x_1 = 2, y_1 = 3$$

See Figure 39 for the graph. ◀

In the solution to Example 7, we could have used the other point, $(-4, 5)$, instead of the point $(2, 3)$. The equation that results, although it looks different, is equivalent to the equation that we obtained in the example. (Try it for yourself.)

6 Another useful equation of a line is obtained when the slope m and y-intercept b are known. In this event, we know both the slope m of the line and a point $(0, b)$ on the line; we use the point–slope form, equation (2), to obtain the following equation:

$$y - y_1 = m(x - x_1) \quad \text{Point–slope form}$$

$$y - b = m(x - 0) \quad x_1 = 0; y_1 = b$$

$$y - b = mx \quad \text{Simplify.}$$

$$y = mx + b \quad \text{Solve for } y.$$

Theorem

Slope–Intercept Form of an Equation of a Line

An equation of a line L with slope m and y-intercept b is

$$y = mx + b \qquad \textbf{(3)}$$

NOW WORK PROBLEM 31 (express answer in slope-intercept form).

Figure 40
$y = mx + 2$

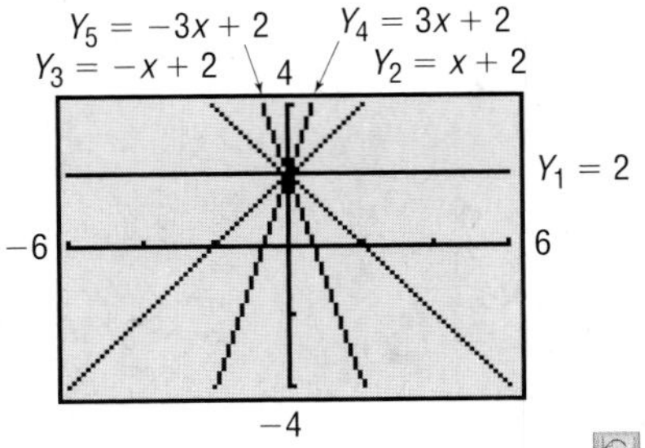

Seeing the Concept

To see the role that the slope m plays, graph the following lines on the same square screen.

$$Y_1 = 2$$
$$Y_2 = x + 2$$
$$Y_3 = -x + 2$$
$$Y_4 = 3x + 2$$
$$Y_5 = -3x + 2$$

See Figure 40. What do you conclude about the lines $y = mx + 2$?

Figure 41
$y = 2x + b$

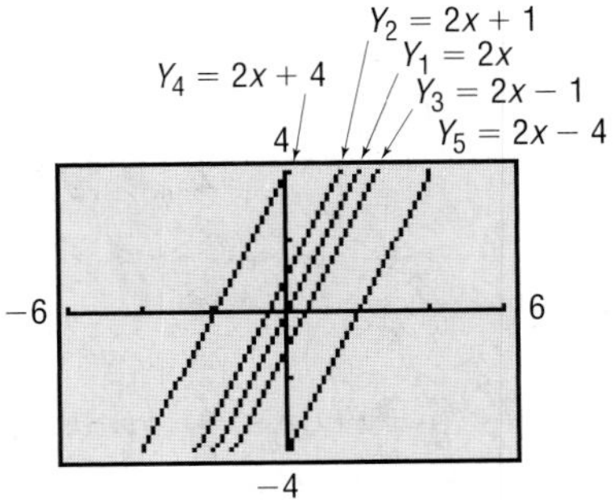

Seeing the Concept

To see the role of the y-intercept b, graph the following lines on the same square screen.

$$Y_1 = 2x$$
$$Y_2 = 2x + 1$$
$$Y_3 = 2x - 1$$
$$Y_4 = 2x + 4$$
$$Y_5 = 2x - 4$$

See Figure 41. What do you conclude about the lines $y = 2x + b$?

7 When the equation of a line is written in slope–intercept form, it is easy to find the slope m and y-intercept b of the line. For example, suppose that the equation of a line is

$$y = -2x + 3$$

Compare it to $y = mx + b$:

$$\begin{array}{c} y = -2x + 3 \\ \quad\uparrow \quad\ \uparrow \\ y = mx + b \end{array}$$

The slope of this line is -2 and its y-intercept is 3.

EXAMPLE 8 Finding the Slope and y-Intercept

Find the slope m and y-intercept b of the line with equation $2x + 4y = 8$. Graph the equation.

Solution To obtain the slope and y-intercept, we transform the equation into its slope–intercept form by solving the equation for y.

$$2x + 4y = 8$$
$$4y = -2x + 8$$
$$y = -\frac{1}{2}x + 2 \qquad y = mx + b$$

The coefficient of x, $-\frac{1}{2}$, is the slope, and the y-intercept is 2.

We can graph the line in two ways:

1. Use the facts that the y-intercept is 2 and the slope is $-\frac{1}{2}$. Then, starting at the point $(0, 2)$, go to the right 2 units and then down 1 unit to the point $(2, 1)$. See Figure 42.

Figure 42

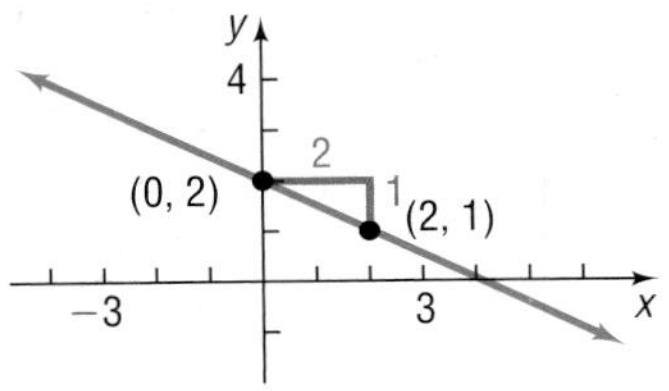

Figure 43

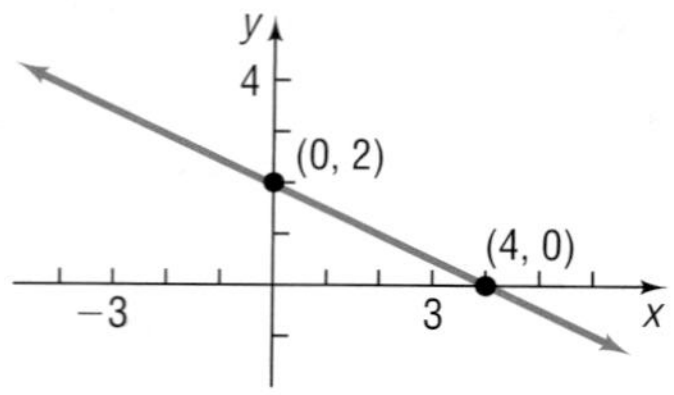

2. Locate the intercepts. Because the y-intercept is 2, we know that one intercept is $(0, 2)$. To obtain the x-intercept, let $y = 0$ and solve for x. When $y = 0$, we have

$$2x + 4y = 8$$
$$2x + 4 \cdot 0 = 8 \qquad y = 0$$
$$2x = 8$$
$$x = 4$$

The intercepts are $(4, 0)$ and $(0, 2)$. See Figure 43. ◀

NOW WORK PROBLEM 71.

8 The form of the equation of the line in Example 8, $2x + 4y = 8$, is called the *general form.*

The equation of a line L is in **general form*** when it is written as

$$Ax + By = C \qquad \textbf{(4)}$$

where A, B, and C are real numbers and A and B are not both 0.

Every line has an equation that is equivalent to an equation written in general form. For example, a vertical line whose equation is

$$x = a$$

can be written in the general form

$$1 \cdot x + 0 \cdot y = a \qquad A = 1, B = 0, C = a$$

A horizontal line whose equation is

$$y = b$$

can be written in the general form

$$0 \cdot x + 1 \cdot y = b \qquad A = 0, B = 1, C = b$$

Lines that are neither vertical nor horizontal have general equations of the form

$$Ax + By = C \qquad A \neq 0 \text{ and } B \neq 0$$

Because the equation of every line can be written in general form, any equation equivalent to (4) is called a **linear equation.**

NOW WORK PROBLEM 43.

Figure 44

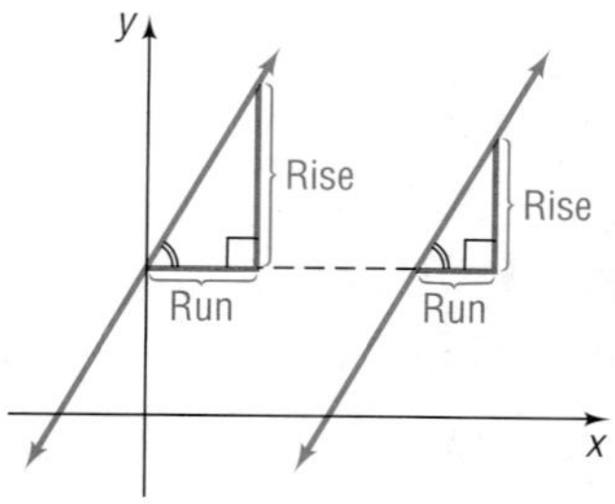

Parallel and Perpendicular Lines

9 When two lines (in the plane) do not intersect (that is, they have no points in common), they are said to be **parallel.** Look at Figure 44. There we have drawn two lines and have constructed two right triangles by drawing sides parallel to the coordinate axes. These lines are parallel if and only if the right triangles are similar. (Do you see why? Two angles are equal.) The triangles are similar if and only if the ratios of corresponding sides are equal.

*Some books use the term **standard form.**

This suggests the following result:

Theorem

Criterion for Parallel Lines

Two nonvertical lines are parallel if and only if their slopes are equal and they have different y-intercepts.

The use of the words "if and only if" in the preceding theorem means that actually two statements are being made, one the converse of the other.

If two nonvertical lines are parallel, then their slopes are equal and they have different y-intercepts.

If two nonvertical lines have equal slopes and different y-intercepts, then they are parallel.

Refer to *Seeing the Concept* and Figure 41, $y = 2x + b$, on page 29.

EXAMPLE 9 Showing That Two Lines Are Parallel

Show that the lines given by the following equations are parallel:

$$L_1: \quad 2x + 3y = 6 \qquad L_2: \quad 4x + 6y = 0$$

Solution To determine whether these lines have equal slopes, we write each equation in slope–intercept form:

$$\begin{aligned} L_1: \quad 2x + 3y &= 6 \\ 3y &= -2x + 6 \\ y &= -\frac{2}{3}x + 2 \\ \text{Slope} &= -\frac{2}{3} \\ y\text{-intercept} &= 2 \end{aligned} \qquad \begin{aligned} L_2: \quad 4x + 6y &= 0 \\ 6y &= -4x \\ y &= -\frac{2}{3}x \\ \text{Slope} &= -\frac{2}{3} \\ y\text{-intercept} &= 0 \end{aligned}$$

Figure 45 Parallel lines

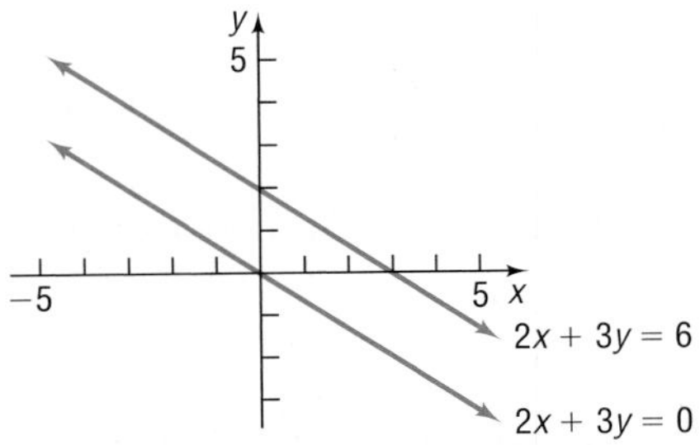

Because these lines have the same slope, $-\frac{2}{3}$, but different y-intercepts, the lines are parallel. See Figure 45. ◀

EXAMPLE 10 Finding a Line That Is Parallel to a Given Line

Find an equation for the line that contains the point $(2, -3)$ and is parallel to the line $2x + y = 6$.

Solution Since the two lines are to be parallel, the slope of the line that we seek equals the slope of the line $2x + y = 6$. We begin by writing the equation of the line $2x + y = 6$ in slope–intercept form.

$$\begin{aligned} 2x + y &= 6 \\ y &= -2x + 6 \end{aligned}$$

The slope is -2. Since the line that we seek contains the point $(2, -3)$, we use the point–slope form to obtain

$$\begin{aligned} y - y_1 &= m(x - x_1) && \text{Point–slope form} \\ y + 3 &= -2(x - 2) && m = -2,\ x_1 = 2,\ y_1 = -3 \\ y + 3 &= -2x + 4 && \\ y &= -2x + 1 && \text{Slope–intercept form} \\ 2x + y &= 1 && \text{General form} \end{aligned}$$

This line is parallel to the line $2x + y = 6$ and contains the point $(2, -3)$. See Figure 46. ◀

Figure 46

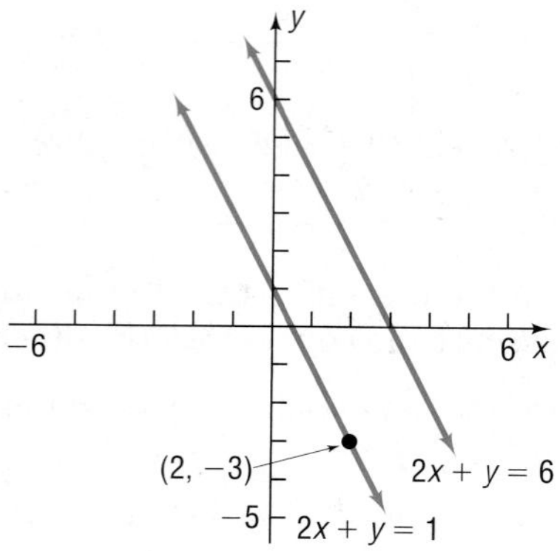

NOW WORK PROBLEM 53.

10 When two lines intersect at a right angle (90°), they are said to be **perpendicular.** See Figure 47.

The following result gives an algebraic condition, in terms of their slopes, for two lines to be perpendicular.

Figure 47 Perpendicular lines

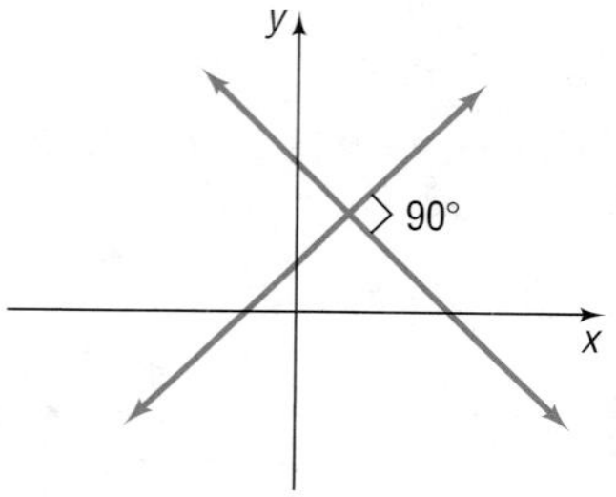

Theorem

Criterion for Perpendicular Lines

Two nonvertical lines are perpendicular if and only if the product of their slopes is -1.

You may find it easier to remember the condition for two nonvertical lines to be perpendicular by observing that the equality $m_1 m_2 = -1$ means that m_1 and m_2 are negative reciprocals of each other; that is, $m_1 = -\dfrac{1}{m_2}$ and $m_2 = -\dfrac{1}{m_1}$.

Here, we shall prove the "only if" part of the statement:

> If two nonvertical lines are perpendicular, then the product of their slopes is -1.

Figure 48

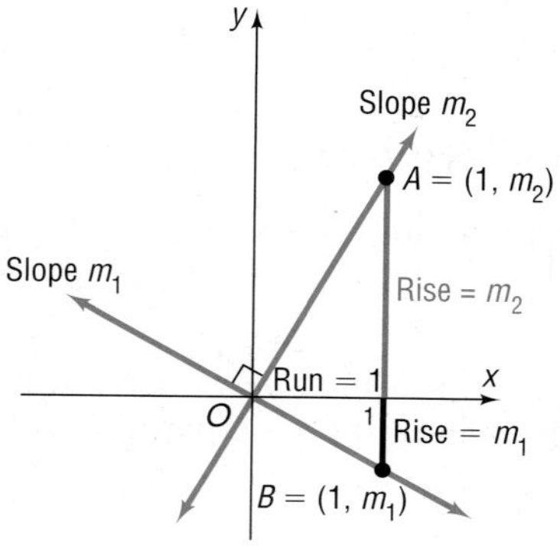

Proof Let m_1 and m_2 denote the slopes of the two lines. There is no loss in generality (that is, neither the angle nor the slopes are affected) if we situate the lines so that they meet at the origin. See Figure 48. The point $A = (1, m_2)$ is on the line having slope m_2, and the point $B = (1, m_1)$ is on the line having slope m_1. (Do you see why this must be true?)

Suppose that the lines are perpendicular. Then triangle OAB is a right triangle. As a result of the Pythagorean Theorem, it follows that

$$[d(O, A)]^2 + [d(O, B)]^2 = [d(A, B)]^2 \tag{5}$$

By the distance formula, we can write each of these distances as

$$\begin{aligned} [d(O, A)]^2 &= (1 - 0)^2 + (m_2 - 0)^2 = 1 + (m_2)^2 \\ [d(O, B)]^2 &= (1 - 0)^2 + (m_1 - 0)^2 = 1 + (m_1)^2 \\ [d(A, B)]^2 &= (1 - 1)^2 + (m_2 - m_1)^2 = (m_2)^2 - 2m_1 m_2 + (m_1)^2 \end{aligned}$$

Using these facts in equation (5), we get

$$1 + (m_2)^2 + 1 + (m_1)^2 = (m_2)^2 - 2m_1m_2 + (m_1)^2$$

which, upon simplification, can be written as

$$m_1m_2 = -1$$

If the lines are perpendicular, the product of their slopes is -1. ■

In Problem 122, you are asked to prove the "if" part of the theorem; that is:

> If two nonvertical lines have slopes whose product is -1, then the lines are perpendicular.

EXAMPLE 11 Finding the Slope of a Line Perpendicular to a Given Line

If a line has slope $\frac{3}{2}$, any line having slope $-\frac{2}{3}$ is perpendicular to it. ◀

EXAMPLE 12 Finding the Equation of a Line Perpendicular to a Given Line

Find an equation of the line containing the point $(1, -2)$ that is perpendicular to the line $x + 3y = 6$. Graph the two lines.

Solution We first write the equation of the given line in slope–intercept form to find its slope.

$$x + 3y = 6$$

$$3y = -x + 6 \qquad \text{Proceed to solve for } y.$$

$$y = -\frac{1}{3}x + 2 \qquad \text{Place in the form } y = mx + b.$$

The given line has slope $-\frac{1}{3}$. Any line perpendicular to this line will have slope 3. Because we require the point $(1, -2)$ to be on this line with slope 3, we use the point–slope form of the equation of a line.

$$y - y_1 = m(x - x_1) \qquad \text{Point–slope form}$$

$$y - (-2) = 3(x - 1) \qquad m = 3, x_1 = 1, y_1 = -2$$

To obtain other forms of the equation, we proceed as follows:

$$y + 2 = 3(x - 1)$$

$$y + 2 = 3x - 3$$

$$y = 3x - 5 \qquad \text{Slope–intercept form}$$

$$3x - y = 5 \qquad \text{General form}$$

Figure 49 shows the graphs. ◀

Figure 49

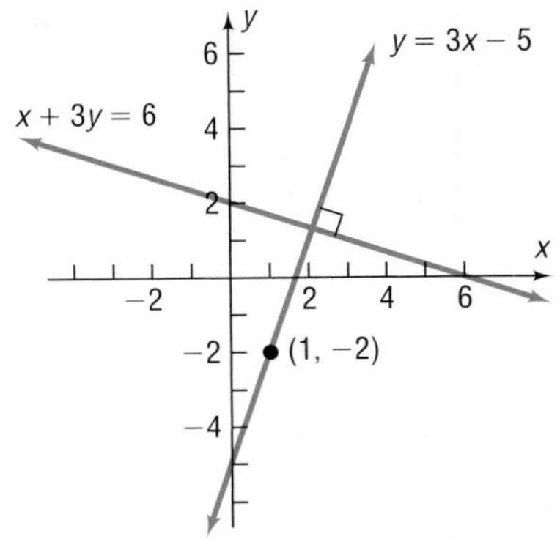

NOW WORK PROBLEM 59.

WARNING: Be sure to use a square screen when you graph perpendicular lines. Otherwise, the angle between the two lines will appear distorted. ■

1.3 Assess Your Understanding

Concepts and Vocabulary

1. The slope of a vertical line is __________; the slope of a horizontal line is __________.

2. For the line $2x + 3y = 6$, the x-intercept is __________ and the y-intercept is __________.

3. A horizontal line is given by an equation of the form __________, where b is the __________.

4. *True or False:* Vertical lines have an undefined slope.

5. *True or False:* The slope of the line $2y = 3x + 5$ is 3.

6. *True or False:* The point $(1, 2)$ is on the line $2x + y = 4$.

7. Two nonvertical lines have slopes m_1 and m_2, respectively. The lines are parallel if __________ and the __________ are unequal; the lines are perpendicular if __________.

8. The lines $y = 2x + 3$ and $y = ax + 5$ are parallel if $a =$ __________.

9. The lines $y = 2x - 1$ and $y = ax + 2$ are perpendicular if $a =$ __________.

10. *True or False:* Perpendicular lines have slopes that are reciprocals of one another.

Exercises

In Problems 11–14, (a) find the slope of the line and (b) interpret the slope.

11.
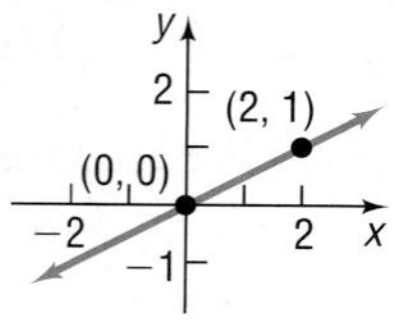

12.
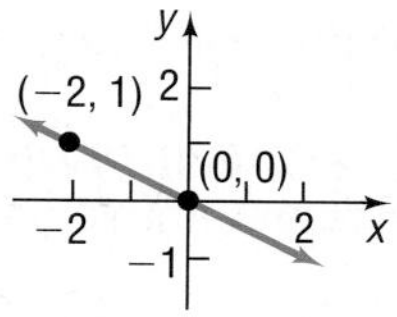

13.
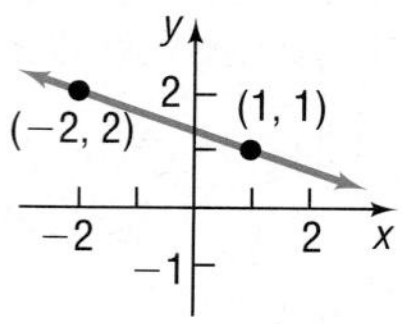

14.
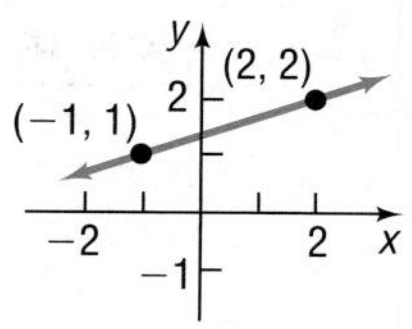

In Problems 15–22, plot each pair of points and determine the slope of the line containing them. Graph the line.

15. $(2, 3)$; $(4, 0)$
16. $(4, 2)$; $(3, 4)$
17. $(-2, 3)$; $(2, 1)$
18. $(-1, 1)$; $(2, 3)$
19. $(-3, -1)$; $(2, -1)$
20. $(4, 2)$; $(-5, 2)$
21. $(-1, 2)$; $(-1, -2)$
22. $(2, 0)$; $(2, 2)$

In Problems 23–30, graph the line containing the point P and having slope m.

23. $P = (1, 2)$; $m = 3$
24. $P = (2, 1)$; $m = 4$
25. $P = (2, 4)$; $m = -\frac{3}{4}$
26. $P = (1, 3)$; $m = -\frac{2}{5}$
27. $P = (-1, 3)$; $m = 0$
28. $P = (2, -4)$; $m = 0$
29. $P = (0, 3)$; slope undefined
30. $P = (-2, 0)$; slope undefined

In Problems 31–38, find an equation of the line L. Express your answer using either the general form or the slope–intercept form of the equation of a line, whichever you prefer.

31.
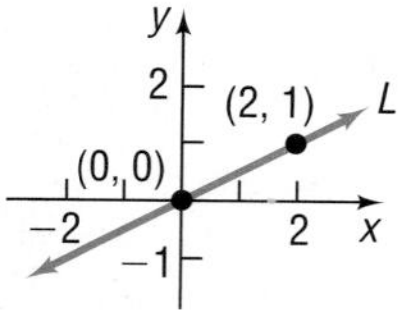

32.
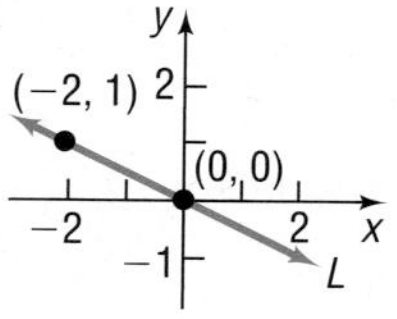

33.
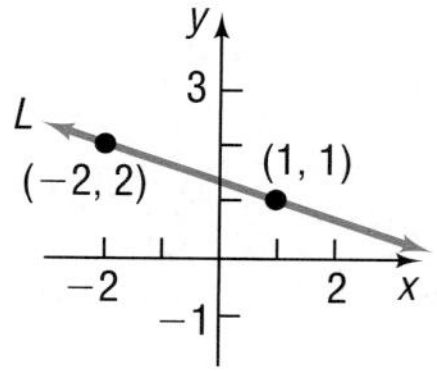

34.
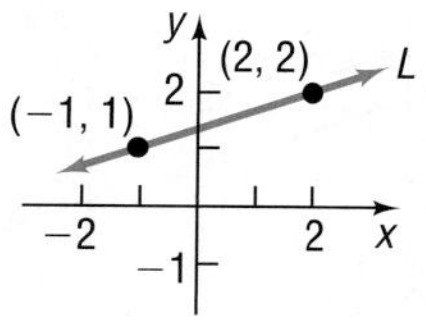

35.
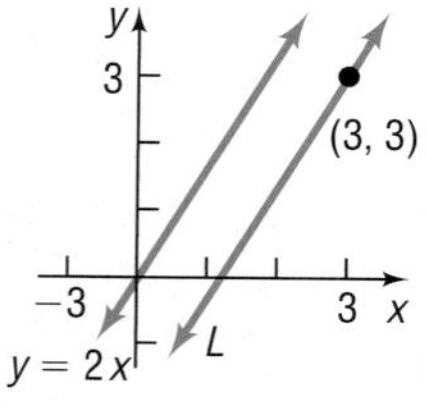

L is parallel to $y = 2x$

36.
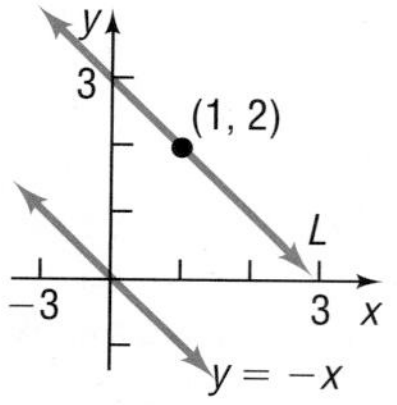

L is parallel to $y = -x$

37.
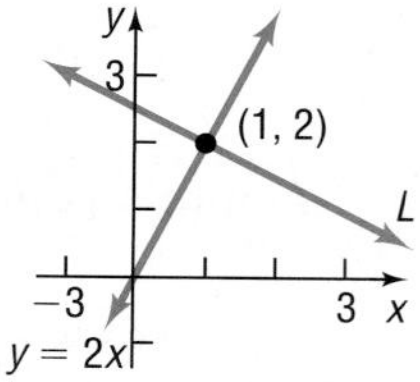

L is perpendicular to $y = 2x$

38.
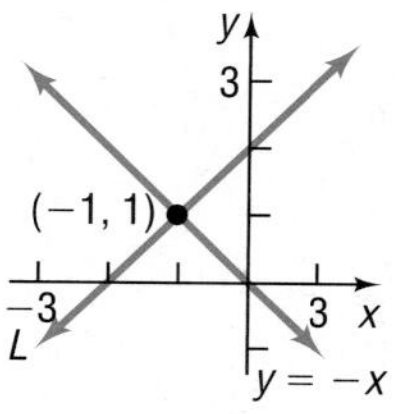

L is perpendicular to $y = -x$

In Problems 39–64, find an equation for the line with the given properties. Express your answer using either the general form or the slope–intercept form of the equation of a line, whichever you prefer.

39. Slope = 3; containing the point $(-2, 3)$

40. Slope = 2; containing the point $(4, -3)$

41. Slope $= -\frac{2}{3}$; containing the point $(1, -1)$

42. Slope $= \frac{1}{2}$; containing the point $(3, 1)$

43. Containing the points $(1, 3)$ and $(-1, 2)$

44. Containing the points $(-3, 4)$ and $(2, 5)$

45. Slope $= -3$; y-intercept $= 3$

46. Slope $= -2$; y-intercept $= -2$

47. x-intercept $= 2$; y-intercept $= -1$

48. x-intercept $= -4$; y-intercept $= 4$

49. Slope undefined; containing the point $(2, 4)$

50. Slope undefined; containing the point $(3, 8)$

51. Horizontal; containing the point $(-3, 2)$

52. Vertical; containing the point $(4, -5)$

53. Parallel to the line $y = 2x$; containing the point $(-1, 2)$

54. Parallel to the line $y = -3x$; containing the point $(-1, 2)$

55. Parallel to the line $2x - y = -2$; containing the point $(0, 0)$

56. Parallel to the line $x - 2y = -5$; containing the point $(0, 0)$

57. Parallel to the line $x = 5$; containing the point $(4, 2)$

58. Parallel to the line $y = 5$; containing the point $(4, 2)$

59. Perpendicular to the line $y = \frac{1}{2}x + 4$; containing the point $(1, -2)$

60. Perpendicular to the line $y = 2x - 3$; containing the point $(1, -2)$

61. Perpendicular to the line $2x + y = 2$; containing the point $(-3, 0)$

62. Perpendicular to the line $x - 2y = -5$; containing the point $(0, 4)$

63. Perpendicular to the line $x = 8$; containing the point $(3, 4)$

64. Perpendicular to the line $y = 8$; containing the point $(3, 4)$

In Problems 65–84, find the slope and y-intercept of each line. Graph the line.

65. $y = 2x + 3$

66. $y = -3x + 4$

67. $\frac{1}{2}y = x - 1$

68. $\frac{1}{3}x + y = 2$

69. $y = \frac{1}{2}x + 2$

70. $y = 2x + \frac{1}{2}$

71. $x + 2y = 4$

72. $-x + 3y = 6$

73. $2x - 3y = 6$

74. $3x + 2y = 6$

75. $x + y = 1$

76. $x - y = 2$

77. $x = -4$

78. $y = -1$

79. $y = 5$

80. $x = 2$

81. $y - x = 0$

82. $x + y = 0$

83. $2y - 3x = 0$

84. $3x + 2y = 0$

In Problems 85 to 88, the equations of two lines are given. Determine if the lines are parallel, perpendicular, or neither.

85. $y = 2x - 3$
$y = 2x + 4$

86. $y = \frac{1}{2}x - 3$
$y = -2x + 4$

87. $y = 4x + 5$
$y = -4x + 2$

88. $y = -2x + 3$
$y = -\frac{1}{2}x + 2$

89. Geometry Use slopes to show that the triangle whose vertices are $(-2, 5)$, $(1, 3)$, and $(-1, 0)$ is a right triangle.

90. Geometry Use slopes to show that the quadrilateral whose vertices are $(1, -1)$, $(4, 1)$, $(2, 2)$, and $(5, 4)$ is a parallelogram.

91. Geometry Use slopes to show that the quadrilateral whose vertices are $(-1, 0)$, $(2, 3)$, $(1, -2)$, and $(4, 1)$ is a rectangle.

92. Geometry Use slopes and the distance formula to show that the quadrilateral whose vertices are $(0, 0)$, $(1, 3)$, $(4, 2)$, and $(3, -1)$ is a square.

In Problems 93–96, write an equation of each line. Express your answer using either the general form or the slope–intercept form of the equation of a line, whichever you prefer.

93.

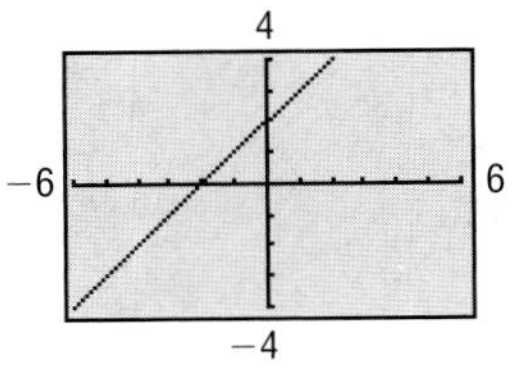

94.

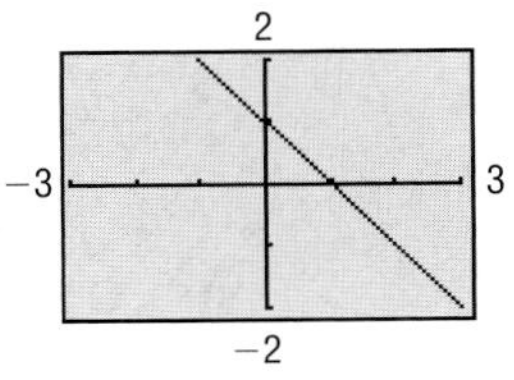

95.

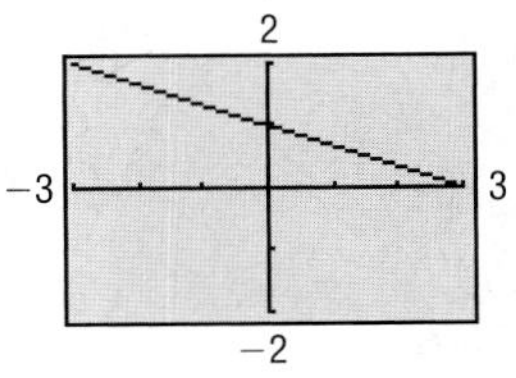

96.

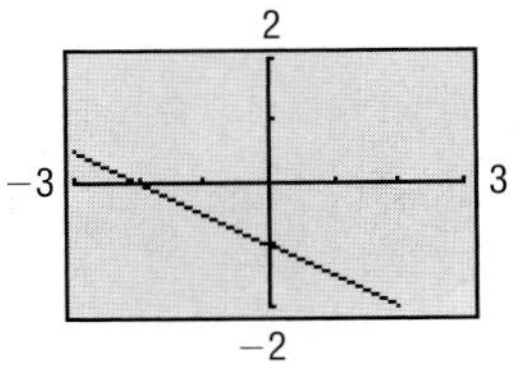

97. Find an equation of the x-axis.

98. Find an equation of the y-axis.

99. Truck Rentals A truck rental company rents a moving truck for one day by charging \$29 plus \$0.07 per mile. Write a linear equation that relates the cost C, in dollars, of renting the truck to the number x of miles driven. What is the cost of renting the truck if the truck is driven 110 miles? 230 miles?

100. Cost Equation The **fixed costs** of operating a business are the costs incurred regardless of the level of production. Fixed costs include rent, fixed salaries, and costs of buying machinery. The **variable costs** of operating a business are the costs that change with the level of output. Variable costs include raw materials, hourly wages, and electricity. Suppose that a manufacturer of jeans has fixed costs of \$500 and variable costs of \$8 for each pair of jeans manufactured. Write a linear equation that relates the cost C, in dollars, of manufacturing the jeans to the number x of pairs of jeans manufactured. What is the cost of manufacturing 400 pairs of jeans? 740 pairs?

101. Cost of Sunday Home Delivery The cost to the *Chicago Tribune* for Sunday home delivery is approximately \$0.53 per newspaper with fixed costs of \$1,070,000. Write an equation that relates the cost C and the number x of copies delivered.
SOURCE: Chicago Tribune, 2002.

102. Wages of a Car Salesperson Dan receives \$375 per week for selling new and used cars at a car dealership in Oak Lawn, Illinois. In addition, he receives 5% of the profit on any sales that he generates. Write an equation that relates Dan's weekly salary S when he has sales that generate a profit of x dollars.

103. Electricity Rates in Illinois Commonwealth Edison Company supplies electricity to residential customers for a monthly customer charge of \$7.58 plus 8.275 cents per kilowatt-hour for up to 400 kilowatt-hours.
(a) Write an equation that relates the monthly charge C, in dollars, to the number x of kilowatt-hours used in a month, $0 \le x \le 400$.
(b) Graph this equation.
(c) What is the monthly charge for using 100 kilowatt-hours?
(d) What is the monthly charge for using 300 kilowatt-hours?
(e) Interpret the slope of the line.
SOURCE: Commonwealth Edison Company, December, 2002.

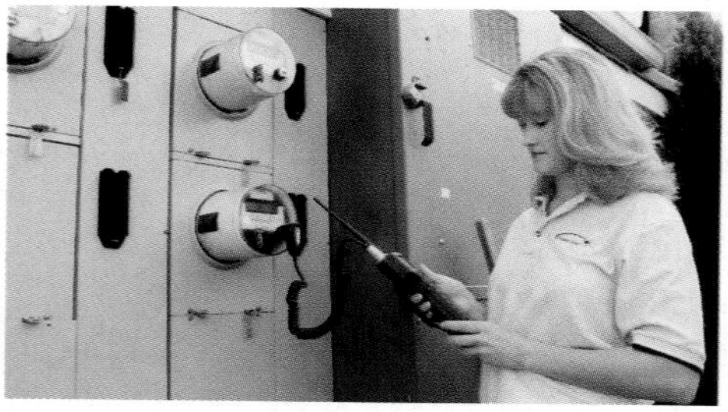

104. Electricity Rates in Florida Florida Power & Light Company supplies electricity to residential customers for a monthly customer charge of \$5.25 plus 6.787 cents per kilowatt-hour for up to 750 kilowatt-hours.
(a) Write an equation that relates the monthly charge C, in dollars, to the number x of kilowatt-hours used in a month, $0 \le x \le 750$.
(b) Graph this equation.
(c) What is the monthly charge for using 200 kilowatt-hours?
(d) What is the monthly charge for using 500 kilowatt-hours?
(e) Interpret the slope of the line.
SOURCE: Florida Power & Light Company, January 2003.

105. Measuring Temperature The relationship between Celsius (°C) and Fahrenheit (°F) degrees for measuring temperature is linear. Find an equation relating °C and °F if 0°C corresponds to 32°F and 100°C corresponds to 212°F. Use the equation to find the Celsius measure of 70°F.

106. Measuring Temperature The Kelvin scale for measuring temperature is obtained by adding 273 kelvins (K) to the Celsius temperature.
(a) Write an equation relating K and °C.
(b) Write an equation relating K and °F (see Problem 105).

107. Product Promotion A cereal company finds that the number of people who will buy one of its products the first month it is introduced is linearly related to the amount of money it spends on advertising. If it spends \$40,000 on advertising, then 100,000 boxes of cereal will be sold, and if it spends \$60,000, then 200,000 boxes will be sold.
(a) Write an equation describing the relation between the amount A spent on advertising and the number x of boxes sold.
(b) How much advertising is needed to sell 300,000 boxes of cereal?
(c) Interpret the slope.

108. Show that an equation for a line with nonzero x- and y-intercepts can be written as

$$\frac{x}{a} + \frac{y}{b} = 1$$

where a is the x-intercept and b is the y-intercept. This is called the **intercept form** of the equation of a line.

109. Which of the following equations might have the graph shown? (More than one answer is possible.)

(a) $2x + 3y = 6$	(e) $x - y = -1$
(b) $-2x + 3y = 6$	(f) $y = 3x - 5$
(c) $3x - 4y = -12$	(g) $y = 2x + 3$
(d) $x - y = 1$	(h) $y = -3x + 3$

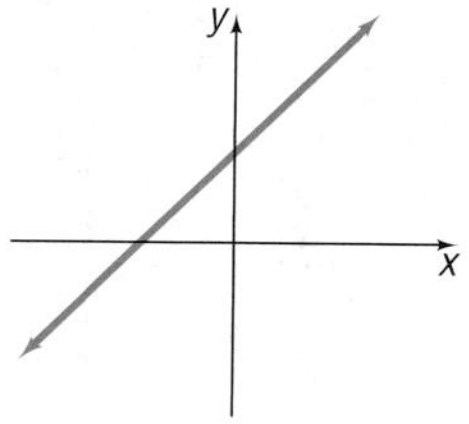

110. Which of the following equations might have the graph shown? (More than one answer is possible.)

(a) $2x + 3y = 6$ (e) $x - y = -1$
(b) $2x - 3y = 6$ (f) $y = -2x + 1$
(c) $3x + 4y = 12$ (g) $y = -\frac{1}{2}x + 10$
(d) $x - y = 1$ (h) $y = x + 4$

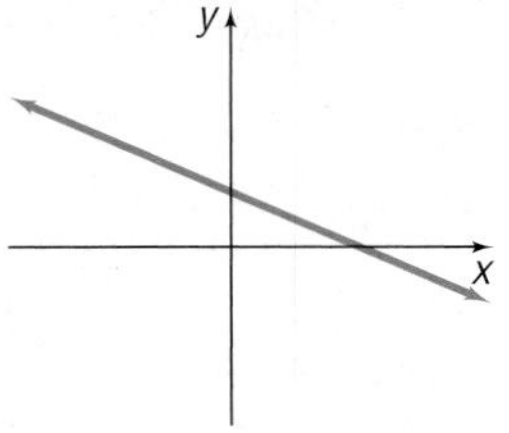

111. The figure shows the graph of two parallel lines. Which of the following pairs of equations might have such a graph?

(a) $x - 2y = 3$
$x + 2y = 7$
(b) $x + y = 2$
$x + y = -1$
(c) $x - y = -2$
$x - y = 1$
(d) $x - y = -2$
$2x - 2y = -4$
(e) $x + 2y = 2$
$x + 2y = -1$

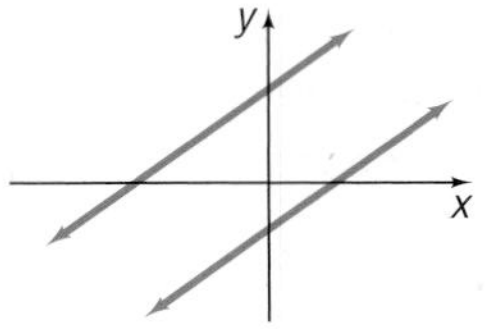

112. The figure shows the graph of two perpendicular lines. Which of the following pairs of equations might have such a graph?

(a) $y - 2x = 2$
$y + 2x = -1$
(b) $y - 2x = 0$
$2y + x = 0$
(c) $2y - x = 2$
$2y + x = -2$
(d) $y - 2x = 2$
$x + 2y = -1$
(e) $2x + y = -2$
$2y + x = -2$

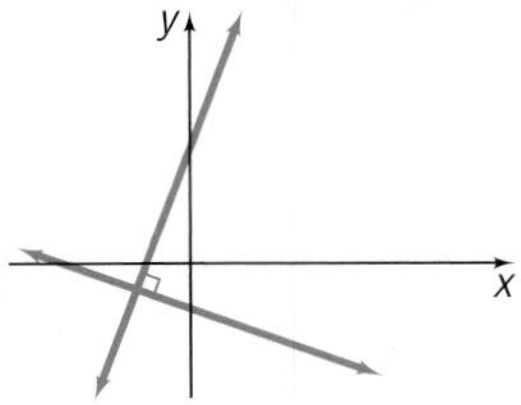

113. Geometry The **tangent line** to a circle may be defined as the line that intersects the circle in a single point, called the **point of tangency** (see the figure). If the equation of the circle is $x^2 + y^2 = r^2$ and the equation of the tangent line is $y = mx + b$, show that:

(a) $r^2(1 + m^2) = b^2$
[**Hint:** The quadratic equation $x^2 + (mx + b)^2 = r^2$ has exactly one solution.]

(b) The point of tangency is $\left(-\frac{r^2 m}{b}, \frac{r^2}{b}\right)$.

(c) The tangent line is perpendicular to the line containing the center of the circle and the point of tangency.

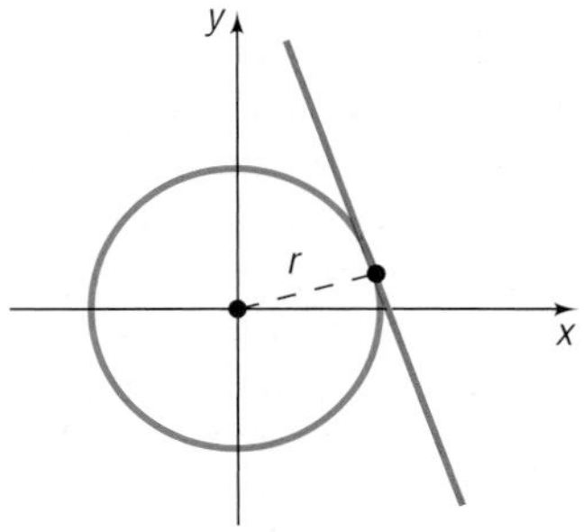

114. The **Greek method** for finding the equation of the tangent line to a circle used the fact that at any point on a circle the line containing the radius and the tangent line are perpendicular (see Problem 113). Use this method to find an equation of the tangent line to the circle $x^2 + y^2 = 9$ at the point $(1, 2\sqrt{2})$.

115. Use the Greek method described in Problem 114 to find an equation of the tangent line to the circle $x^2 + y^2 - 4x + 6y + 4 = 0$ at the point $(3, 2\sqrt{2} - 3)$.

116. Refer to Problem 113. The line $x - 2y = -4$ is tangent to a circle at $(0, 2)$. The line $y = 2x - 7$ is tangent to the same circle at $(3, -1)$. Find the center of the circle.

117. Find an equation of the line containing the centers of the two circles

$$x^2 + y^2 - 4x + 6y + 4 = 0$$

and

$$x^2 + y^2 + 6x + 4y + 9 = 0$$

118. Show that the line containing the points (a, b) and (b, a), $a \neq b$, is perpendicular to the line $y = x$. Also show that the midpoint of (a, b) and (b, a) lies on the line $y = x$.

119. The equation $2x - y = C$ defines a **family of lines,** one line for each value of C. On one set of coordinate axes, graph the members of the family when $C = -4$, $C = 0$, and $C = 2$. Can you draw a conclusion from the graph about each member of the family?

120. Rework Problem 119 for the family of lines $Cx + y = -4$.

121. If a circle of radius 2 is made to roll along the x-axis, what is an equation for the path of the center of the circle?

122. Prove that if two nonvertical lines have slopes whose product is -1, then the lines are perpendicular. [**Hint:** Refer to Figure 48 and use the converse of the Pythagorean Theorem].

123. Which form of the equation of a line do you prefer to use? Justify your position with an example that shows that your choice is better than another. Have reasons.

124. Can every line be written in slope–intercept form? Explain.

125. Does every line have two distinct intercepts? Explain. Are there lines that have no intercepts? Explain.

126. What can you say about two lines that have equal slopes and equal y-intercepts?

127. What can you say about two lines with the same x-intercept and the same y-intercept? Assume that the x-intercept is not 0.

128. If two lines have the same slope, but different x-intercepts, can they have the same y-intercept?

129. If two lines have the same y-intercept, but different slopes, can they have the same x-intercept? What is the only way that this can happen?

130. The accepted symbol used to denote the slope of a line is the letter m. Investigate the origin of this symbolism. Begin by consulting a French dictionary and looking up the French word *monter*. Write a brief essay on your findings.

131. **Grade of a Road** The term *grade* is used to describe the inclination of a road. How does this term relate to the notion of slope of a line? Is a 4% grade very steep? Investigate the grades of some mountainous roads and determine their slopes. Write a brief essay on your findings.

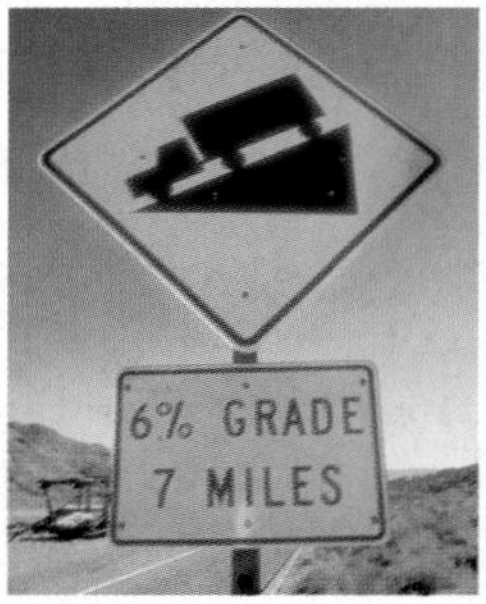

132. **Carpentry** Carpenters use the term *pitch* to describe the steepness of staircases and roofs. How does pitch relate to slope? Investigate typical pitches used for stairs and for roofs. Write a brief essay on your findings.

1.4 Scatter Diagrams; Linear Curve Fitting

OBJECTIVES
1 Draw and Interpret Scatter Diagrams
2 Distinguish between Linear and Nonlinear Relations
3 Use a Graphing Utility to Find the Line of Best Fit

Scatter Diagrams

1 A **relation** is a correspondence between two sets. If x and y are two elements in these sets and if a relation exists between x and y, then we say that x **corresponds to** y or that y **depends on** x and write $x \rightarrow y$. We may also write $x \rightarrow y$ as the ordered pair (x, y). Here, y is referred to as the **dependent** variable and x is called the **independent** variable.

Often we are interested in specifying the type of relation (such as an equation) that might exist between two variables. The first step in finding this relation is to plot the ordered pairs using rectangular coordinates. The resulting graph is called a **scatter diagram.**

EXAMPLE 1 **Drawing a Scatter Diagram**

The data listed in Table 4 represent the apparent temperature versus the relative humidity in a room whose actual temperature is 72° Fahrenheit.

(a) Draw a scatter diagram by hand.

(b) Use a graphing utility to draw a scatter diagram.*

(c) Describe what happens to the apparent temperature as the relative humidity increases.

*Consult your owner's manual for the proper keystrokes.

Table 4

Relative Humidity (%), x	Apparent Temperature (°F), y	(x, y)	Relative Humidity (%), x	Apparent Temperature (°F), y	(x, y)
0	64	(0, 64)	60	72	(60, 72)
10	65	(10, 65)	70	73	(70, 73)
20	67	(20, 67)	80	74	(80, 74)
30	68	(30, 68)	90	75	(90, 75)
40	70	(40, 70)	100	76	(100, 76)
50	71	(50, 71)			

Solution (a) To draw a scatter diagram by hand, we plot the ordered pairs listed in Table 4, with the relative humidity as the x-coordinate and the apparent temperature as the y-coordinate. See Figure 50(a). Notice that the points in a scatter diagram are not connected.

(b) Figure 50(b) shows a scatter diagram using a graphing utility.

(c) We see from the scatter diagrams that, as the relative humidity increases, the apparent temperature increases.

Figure 50

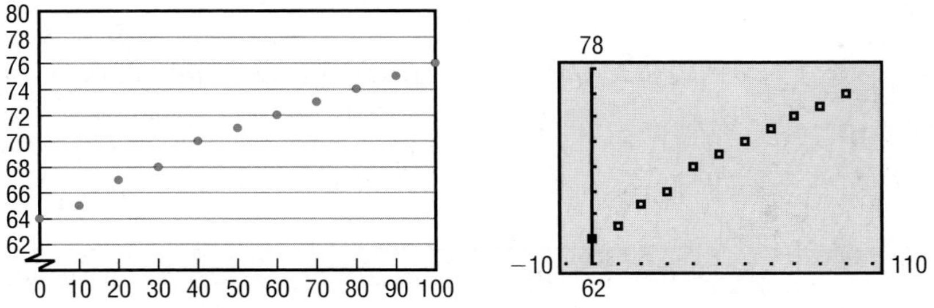

(a) **(b)**

NOW WORK PROBLEM 9(a).

Curve Fitting

2 Scatter diagrams are used to help us to see the type of relation that may exist between two variables. In this text, we will discuss a variety of different relations that may exist between two variables. For now, we concentrate on distinguishing between linear and nonlinear relations. See Figure 51.

Figure 51

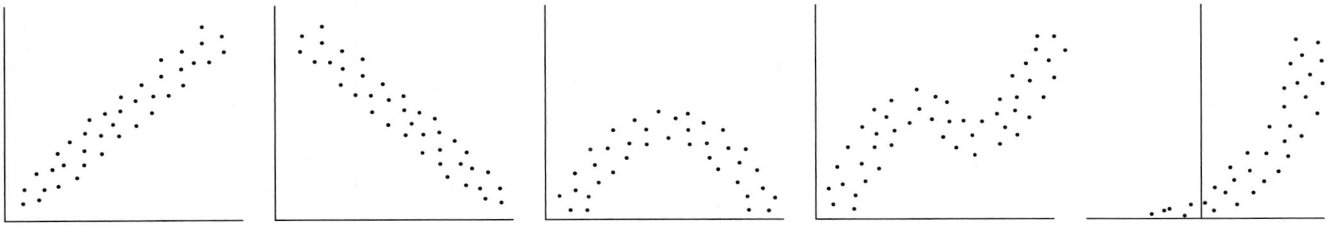

(a) Linear $y = mx + b, m > 0$ **(b)** Linear $y = mx + b, m < 0$ **(c)** Nonlinear **(d)** Nonlinear **(e)** Nonlinear

EXAMPLE 2 Distinguishing between Linear and Nonlinear Relations

Determine whether the relation between the two variables in each graph of Figure 52 is linear or nonlinear.

Figure 52

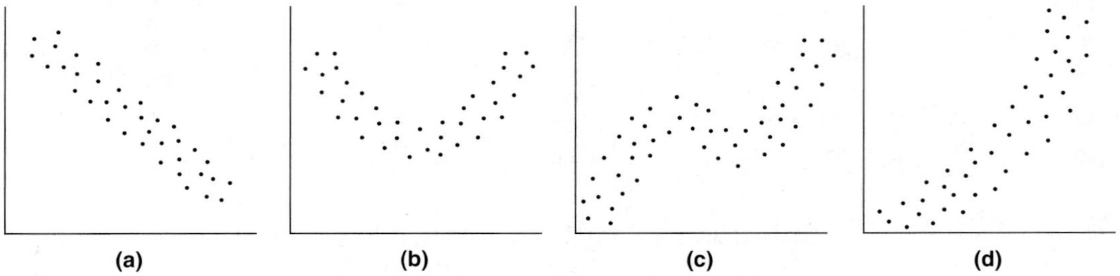

Solution (a) Linear (b) Nonlinear (c) Nonlinear (d) Nonlinear ◀

NOW WORK PROBLEM 3.

In this section we will study data whose scatter diagrams imply that a linear relation exists between the two variables. Nonlinear data will be discussed in later chapters.

Suppose that the scatter diagram of a set of data appears to be linearly related as in Figure 51(a) or (b). We might wish to find an equation of a line that relates the two variables. One way to obtain an equation for such data is to draw a line through two points on the scatter diagram and estimate the equation of the line.

EXAMPLE 3 Find an Equation for Linearly Related Data

Using the data in Table 4 from Example 1, select two points from the data and find an equation of the line containing the points.

(a) Graph the line on the scatter diagram obtained in Example 1(a).

(b) Graph the line on the scatter diagram obtained in Example 1(b).

Solution Select two points, say (10, 65) and (70, 73). (You should select your own two points and complete the solution.) The slope of the line joining the points (10, 65) and (70, 73) is

$$m = \frac{73 - 65}{70 - 10} = \frac{8}{60} = \frac{2}{15}$$

The equation of the line with slope $\frac{2}{15}$ and passing through (10, 65) is found using the point–slope form with $m = \frac{2}{15}$, $x_1 = 10$, and $y_1 = 65$.

$$y - y_1 = m(x - x_1) \quad \text{Point–slope form}$$

$$y - 65 = \frac{2}{15}(x - 10) \quad m = \frac{2}{15};\ x_1 = 10;\ y_1 = 65$$

$$y = \frac{2}{15}x + \frac{191}{3} \quad y = mx + b$$

(a) Figure 53(a) shows the scatter diagram with the graph of the line drawn by hand.

(b) Figure 53(b) shows the scatter diagram with the graph of the line using a graphing utility.

Figure 53

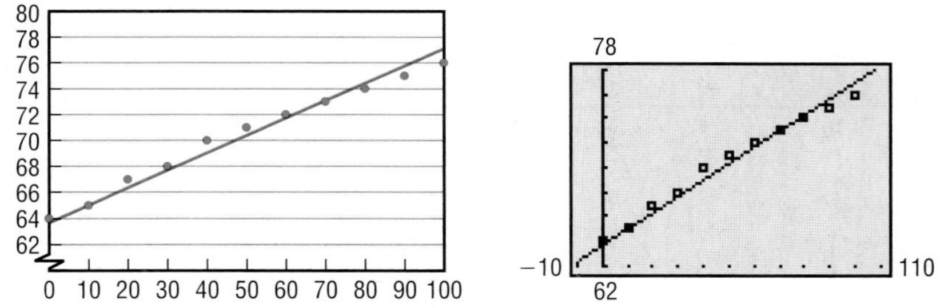

(a) (b) ◀

NOW WORK PROBLEMS 9(b) AND (c).

Line of Best Fit

3 The line obtained in Example 3 depends on the selection of points, which will vary from person to person. So the line that we found might be different from the line that you found. Although the line that we found in Example 3 appears to "fit" the data well, there may be a line that "fits it better." Do you think your line fits the data better? Is there a line of *best fit*? As it turns out, there is a method for finding the line that best fits linearly related data (called the *line of best fit*).*

EXAMPLE 4 Finding the Line of Best Fit

Using the data in Table 4 from Example 1:

(a) Find the line of best fit using a graphing utility.

(b) Graph the line of best fit on the scatter diagram obtained in Example 1(b).

(c) Interpret the slope of the line of best fit.

(d) Use the line of best fit to predict the apparent temperature of a room whose actual temperature is 72° F and relative humidity is 45%.

Solution (a) Graphing utilities contain built-in programs that find the line of best fit for a collection of points in a scatter diagram. (Look in your owner's manual under Linear Regression or Line of Best Fit for details on how to execute the program.) Upon executing the LINear REGression program, we obtain the results shown in Figure 54. The output that the utility provides shows us the equation $y = ax + b$, where a is the slope of the line and b is the y-intercept. The line of best fit that relates relative humidity to apparent temperature may be expressed as the line $y = 0.121x + 64.409$.

Figure 54

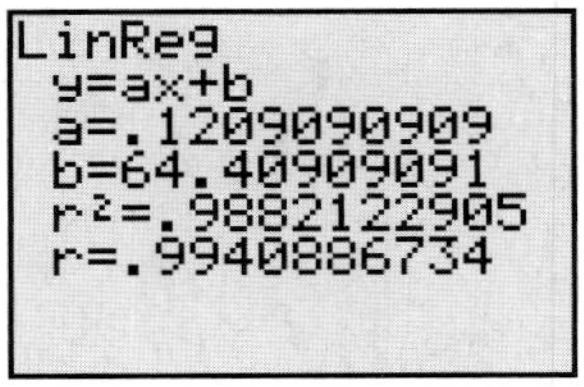

*We shall not discuss in this book the underlying mathematics of lines of best fit. Most books in statistics and many in linear algebra discuss this topic.

Figure 55

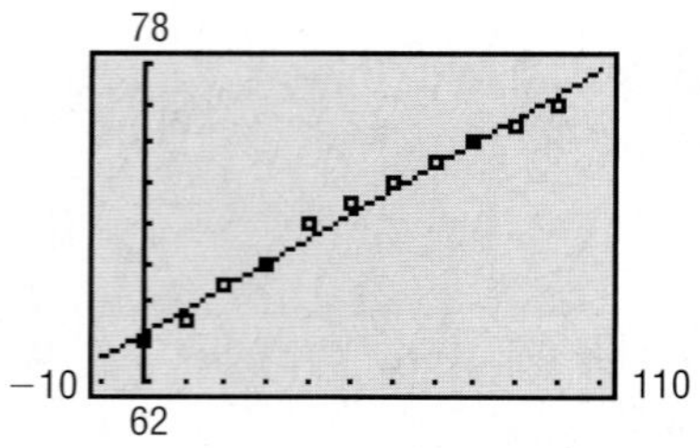

(b) Figure 55 shows the graph of the line of best fit, along with the scatter diagram.

(c) The slope of the line of best fit is approximately 0.121, which means that, for every 1% increase in the relative humidity, apparent room temperature increases 0.121°F.

(d) Letting $x = 45$ in the equation of the line of best fit, we obtain $y = 0.121(45) + 64.409 \approx 70°F$, which is the apparent temperature in the room. ◀

NOW WORK PROBLEMS 9(d) AND (e).

Does the line of best fit appear to be a good fit? In other words, does the line appear to accurately describe the relation between temperature and relative humidity?

And just how "good" is this line of best fit? The answers are given by what is called the *correlation coefficient*. Look again at Figure 54. The last line of output is $r = 0.994$. This number, called the **correlation coefficient, *r***, $-1 \le r \le 1$, is a measure of the strength of the *linear relation* that exists between two variables. The closer that $|r|$ is to 1, the more perfect the linear relationship is. If r is close to 0, there is little or no *linear* relationship between the variables. A negative value of r, $r < 0$, indicates that as x increases y decreases; a positive value of r, $r > 0$, indicates that as x increases y does also. The data given in Example 1, having a correlation coefficient of 0.994, are indicative of a strong linear relationship with positive slope.

1.4 Assess Your Understanding

Concepts and Vocabulary

1. A _______ _______ is used to help us to see the type of relation, if any, that may exist between two variables.

2. *True or False:* The correlation coefficient is a measure of the strength of a linear relation between two variables and must lie between −1 and 1, inclusive.

Exercises

In Problems 3–8, examine the scatter diagram and determine whether the type of relation, if any, that may exist is linear or nonlinear.

3.

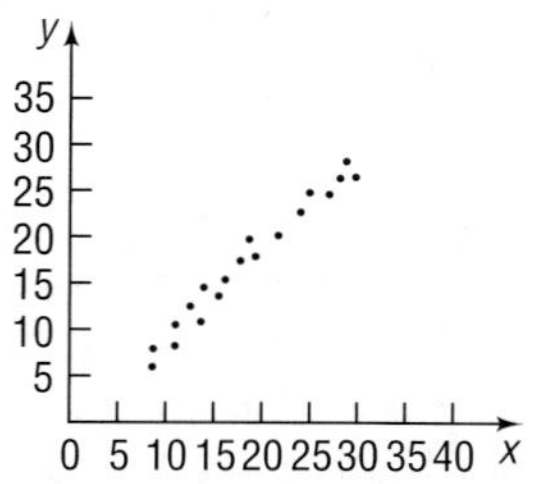

4.

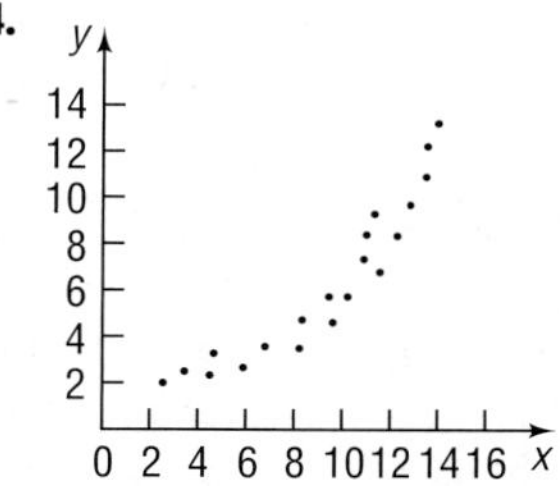

5.

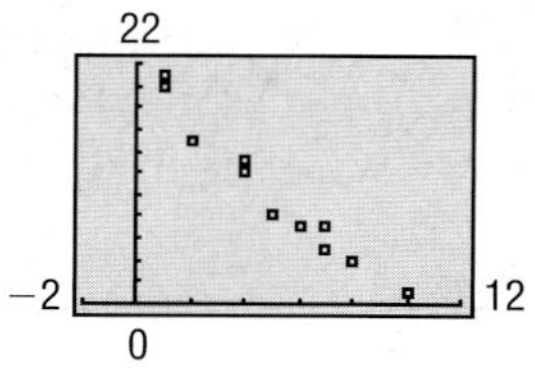

6.

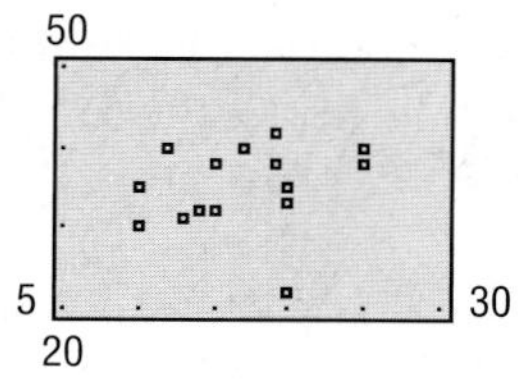

7.

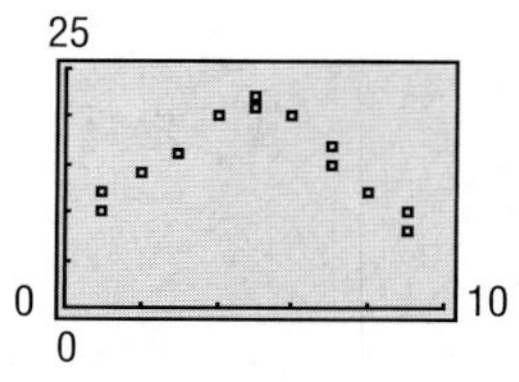

8.

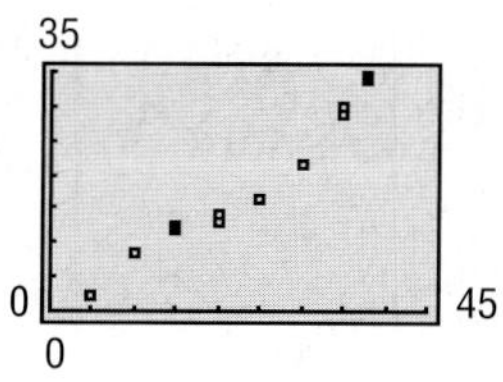

In Problems 9–14:

(a) *Draw a scatter diagram by hand.*
(b) *Select two points from the scatter diagram and find the equation of the line containing the points selected.**
(c) *Graph the line found in part (b) on the scatter diagram.*
(d) *Use a graphing utility to find the line of best fit.*
(e) *Use a graphing utility to graph the line of best fit on the scatter diagram.*

9.

x	3	4	5	6	7	8	9
y	4	6	7	10	12	14	16

10.

x	3	5	7	9	11	13
y	0	2	3	6	9	11

11.

x	−2	−1	0	1	2
y	−4	0	1	4	5

12.

x	−2	−1	0	1	2
y	7	6	3	2	0

13.

x	−20	−17	−15	−14	−10
y	100	120	118	130	140

14.

x	−30	−27	−25	−20	−14
y	10	12	13	13	18

15. Consumption and Disposable Income An economist wants to estimate a line that relates personal consumption expenditures C and disposable income I. Both C and I are in thousands of dollars. She interviews eight heads of households for families of size 3 and obtains the following data. Let I represent the independent variable and C the dependent variable.

(a) Draw a scatter diagram by hand.
(b) Find a line that fits the data.*
(c) Interpret the slope. The slope of this line is called the **marginal propensity to consume.**
(d) Predict the consumption of a family whose disposable income is $42,000.
(e) Use a graphing utility to find the line of best fit to the data.

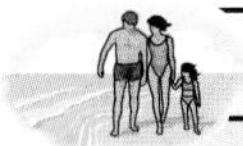

I (000)	C (000)
20	16
20	18
18	13
27	21
36	27
37	26
45	36
50	39

16. Marginal Propensity to Save The same economist as in Problem 15 wants to estimate a line that relates savings S and disposable income I. Let $S = I - C$ be the dependent variable and I the independent variable.

(a) Draw a scatter diagram by hand.
(b) Find a line that fits the data.
(c) Interpret the slope. The slope of this line is called the **marginal propensity to save.**
(d) Predict the savings of a family whose income is $42,000.
(e) Use a graphing utility to find the line of best fit.

17. Mortgage Qualification The amount of money that a lending institution will allow you to borrow mainly depends on the interest rate and your annual income. The following data represent the annual income I required by a bank in order to lend L dollars at an interest rate of 7.5% for 30 years.

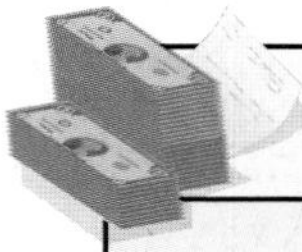

Annual Income, I ($)	Loan Amount, L ($)
15,000	44,600
20,000	59,500
25,000	74,500
30,000	89,400
35,000	104,300
40,000	119,200
45,000	134,100
50,000	149,000
55,000	163,900
60,000	178,800
65,000	193,700
70,000	208,600

SOURCE: *Information Please Almanac*, 1999

Let I represent the independent variable and L the dependent variable.

(a) Use a graphing utility to draw a scatter diagram of the data.
(b) Use a graphing utility to find the line of best fit to the data.
(c) Graph the line of best fit on the scatter diagram drawn in part (a).
(d) Interpret the slope of the line of best fit.
(e) Determine the loan amount that an individual will qualify for if her income is $42,000.

18. Mortgage Qualification The amount of money that a lending institution will allow you to borrow mainly depends on the interest rate and your annual income.

*Answers will vary. We will use the first and last data points in the answer section.

The following data represent the annual income I required by a bank in order to lend L dollars at an interest rate of 8.5% for 30 years.

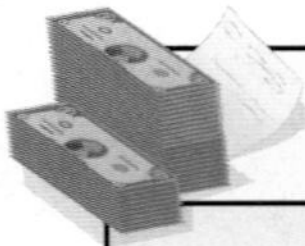

Annual Income, I ($)	Loan Amount, L ($)
15,000	40,600
20,000	54,100
25,000	67,700
30,000	81,200
35,000	94,800
40,000	108,300
45,000	121,900
50,000	135,400
55,000	149,000
60,000	162,500
65,000	176,100
70,000	189,600

SOURCE: *Information Please Almanac*, 1999

Let I represent the independent variable and L the dependent variable.

(a) Use a graphing utility to draw a scatter diagram of the data.
(b) Use a graphing utility to find the line of best fit to the data.
(c) Graph the line of best fit on the scatter diagram drawn in part (a).
(d) Interpret the slope of the line of best fit.
(e) Determine the loan amount that an individual will qualify for if her income is $42,000.

19. Apparent Room Temperature The following data represent the apparent temperature versus the relative humidity in a room whose actual temperature is 65° Fahrenheit.

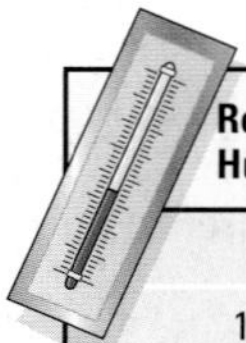

Relative Humidity, h (%)	Apparent Temperature, T (°F)
0	59
10	60
20	61
30	61
40	62
50	63
60	64
70	65
80	65
90	66
100	67

SOURCE: National Oceanic and Atmospheric Administration

Let h represent the independent variable and T the dependent variable.

(a) Use a graphing utility to draw a scatter diagram of the data.
(b) Use a graphing utility to find the line of best fit to the data.
(c) Graph the line of best fit on the scatter diagram drawn in part (a).
(d) Interpret the slope of the line of best fit.
(e) Determine the apparent temperature of a room whose actual temperature is 65° F if the relative humidity is 75%.

20. Gestation Period versus Life Expectancy A researcher would like to estimate the linear function relating the gestation period of an animal, G, and its life expectancy, L. She collects the following data.

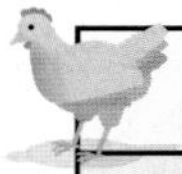

Animal	Gestation (or incubation) Period, G (days)	Life Expectancy, L (years)
Cat	63	11
Chicken	22	7.5
Dog	63	11
Duck	28	10
Goat	151	12
Lion	108	10
Parakeet	18	8
Pig	115	10
Rabbit	31	7
Squirrel	44	9

SOURCE: *Time Almanac* 2000.

Let G represent the independent variable and L the dependent variable.

(a) Use a graphing utility to draw a scatter diagram of the data.
(b) Use a graphing utility to find the line of best fit to the data.
(c) Interpret the slope of the line of best fit.
(d) Graph the line of best fit on the scatter diagram.
(e) Predict the life expectancy of an animal whose gestation period is 89 days.

Chapter Review

Things to Know

Formulas

Distance formula (p. 3)	$d = \sqrt{(x_2 - x_1)^2 + (y_2 - y_1)^2}$
Midpoint formula (p. 6)	$M = (x, y) = \left(\frac{x_1 + x_2}{2}, \frac{y_1 + y_2}{2}\right)$
Slope (p. 23)	$m = \frac{y_2 - y_1}{x_2 - x_1}$, if $x_1 \neq x_2$; undefined if $x_1 = x_2$
Parallel lines (p. 31)	Equal slopes ($m_1 = m_2$) and different y-intercepts ($b_1 \neq b_2$)
Perpendicular lines (p. 32)	Product of slopes is -1 ($m_1 \cdot m_2 = -1$)

Equations of Lines and Circles

Vertical line (p. 27)	$x = a$
Horizontal line (p. 28)	$y = b$
Point–slope form of the equation of a line (p. 27)	$y - y_1 = m(x - x_1)$; m is the slope of the line, (x_1, y_1) is a point on the line.
Slope–intercept form of the equation of a line (p. 28)	$y = mx + b$; m is the slope of the line, b is the y-intercept.
General form of the equation of a line (p. 30)	$Ax + By = C$; A, B not both 0
Standard form of the equation of a circle (p. 16)	$(x - h)^2 + (y - k)^2 = r^2$; r is the radius of the circle, (h, k) is the center of the circle.
Equation of the unit circle (p. 16)	$x^2 + y^2 = 1$
General form of the equation of a circle (p. 18)	$x^2 + y^2 + ax + by + c = 0$

Objectives

Section		You should be able to . . .	Review Exercises
1.1	1	Use the distance formula (p. 3)	1(a)–6(a), 42, 43(a), 44
	2	Use the midpoint formula (p. 5)	1(b)–6(b), 44
1.2	1	Graph equations by plotting points (p. 10)	7
	2	Find intercepts from a graph (p. 11)	8
	3	Find intercepts from an equation (p. 11)	9–16
	4	Test an equation for symmetry with respect to the x-axis, the y-axis, and the origin (p. 13)	9–16
	5	Write the standard form of the equation of a circle (p. 16)	17–20
	6	Graph a circle (p. 17)	21–26
	7	Find the center and radius of a circle from an equation in general form and graph it (p. 18)	23–26
1.3	1	Calculate and interpret the slope of a line (p. 23)	1(c)–6(c); 1(d)–6(d)
	2	Graph lines given a point and the slope (p. 26)	41
	3	Find the equation of a vertical line (p. 26)	29
	4	Use the point–slope form of a line; identify horizontal lines (p. 27)	27, 28
	5	Find the equation of a line given two points (p. 28)	30–32
	6	Write the equation of a line in slope–intercept form (p. 28)	27–36

	7	Identify the slope and y-intercept of a line from its equation (p. 29)	37–40
	8	Write the equation of a line in general form (p. 30)	27–36
	9	Find equations of parallel lines (p. 30)	33–34
	10	Find equations of perpendicular lines (p. 32)	35, 36, 43(b)
1.4	1	Draw and interpret scatter diagrams (p. 38)	46(a)
	2	Distinguish between linear and nonlinear relations (p. 39)	46(b)
	3	Use a graphing utility to find the line of best fit (p. 41)	46(c)

Review Exercises

(Blue problem numbers indicate the author's suggestions for a Practice Test.)

In Problems 1–6, find the following for each pair of points:
(a) *The distance between the points.*
(b) *The midpoint of the line segment connecting the points.*
(c) *The slope of the line containing the points.*
(d) *Interpret the slope found in part (c).*

1. $(0, 0); (4, 2)$ **2.** $(0, 0); (-4, 6)$ 3. $(1, -1); (-2, 3)$ **4.** $(-2, 2); (1, 4)$ **5.** $(4, -4); (4, 8)$ **6.** $(-3, 4); (2, 4)$

7. Graph $y = x^2 + 4$ by plotting points.

8. List the intercepts of the following graph.

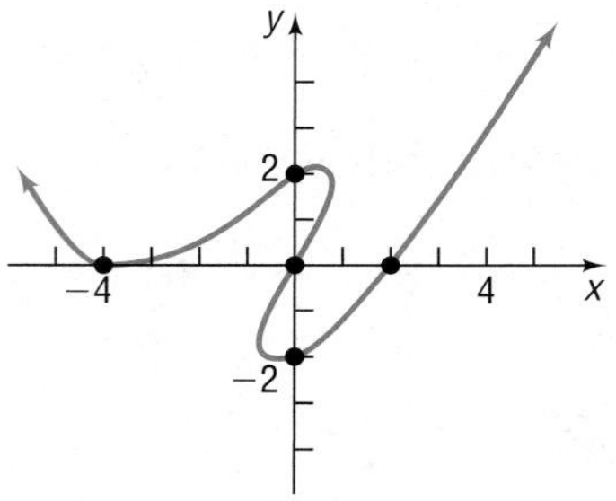

In Problems 9–16, list the intercepts and test for symmetry with respect to the x-axis, the y-axis, and origin.

9. $2x = 3y^2$ **10.** $y = 5x$ **11.** $x^2 + 4y^2 = 16$ **12.** $9x^2 - y^2 = 9$

13. $y = x^4 + 2x^2 + 1$ **14.** $y = x^3 - x$ **15.** $x^2 + x + y^2 + 2y = 0$ **16.** $x^2 + 4x + y^2 - 2y = 0$

In Problems 17–20, find the standard form of the equation of the circle whose center and radius are given.

17. $(h, k) = (-2, 3); r = 4$ **18.** $(h, k) = (3, 4); r = 4$ 19. $(h, k) = (-1, -2); r = 1$ **20.** $(h, k) = (2, -4); r = 3$

In Problems 21–26, find the center and radius of each circle. Graph each circle.

21. $x^2 + (y - 1)^2 = 4$ **22.** $(x + 2)^2 + y^2 = 9$ 23. $x^2 + y^2 - 2x + 4y - 4 = 0$

24. $x^2 + y^2 + 4x - 4y - 1 = 0$ **25.** $3x^2 + 3y^2 - 6x + 12y = 0$ **26.** $2x^2 + 2y^2 - 4x = 0$

In Problems 27–36, find an equation of the line having the given properties. Express your answer using either the general form or the slope–intercept form of the equation of a line, whichever you prefer.

27. Slope $= -2$; containing the point $(3, -1)$

28. Slope $= 0$; containing the point $(-5, 4)$

29. Vertical; containing the point $(-3, 4)$

30. x-intercept $= 2$; containing the point $(4, -5)$

31. y-intercept $= -2$; containing the point $(5, -3)$

32. Containing the points $(3, -4)$ and $(2, 1)$

33. Parallel to the line $2x - 3y = -4$; containing the point $(-5, 3)$

34. Parallel to the line $x + y = 2$; containing the point $(1, -3)$

35. Perpendicular to the line $x + y = 2$; containing the point $(4, -3)$

36. Perpendicular to the line $3x - y = -4$; containing the point $(-2, 4)$

In Problems 37–40, find the slope and y-intercept of each line.

37. $4x - 5y = -20$

38. $3x + 4y = 12$

39. $\frac{1}{2}x - \frac{1}{3}y = -\frac{1}{6}$

40. $-\frac{3}{4}x + \frac{1}{2}y = 0$

41. Graph the line with slope $\frac{2}{3}$ containing the point $(1, 2)$.

42. Show that the points $A = (3, 4)$, $B = (1, 1)$, and $C = (-2, 3)$ are the vertices of an isosceles triangle.

43. Show that the points $A = (-2, 0)$, $B = (-4, 4)$, and $C = (8, 5)$ are the vertices of a right triangle in two ways:
(a) By using the converse of the Pythagorean Theorem
(b) By using the slopes of the lines joining the vertices

44. The endpoints of the diameter of a circle are $(-3, 2)$ and $(5, -6)$. Find the center and radius of the circle. Write the general equation of this circle.

45. Show that the points $A = (2, 5)$, $B = (6, 1)$, and $C = (8, -1)$ lie on a line by using slopes.

46. Bone Length Research performed at NASA, led by Dr. Emily R. Morey-Holton, measured the lengths of the right humerus and right tibia in 11 rats that were sent to space on Spacelab Life Sciences 2. The following data were collected.

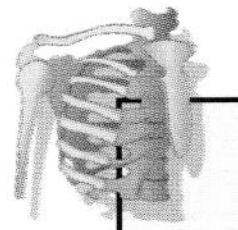

Right Humerus (mm), *x*	Right Tibia (mm), *y*
24.80	36.05
24.59	35.57
24.59	35.57
24.29	34.58
23.81	34.20
24.87	34.73
25.90	37.38
26.11	37.96
26.63	37.46
26.31	37.75
26.84	38.50

SOURCE: NASA Life Sciences Data Archive

(a) Draw a scatter diagram of the data treating length of the right humerus as the independent variable.
(b) Based on the scatter diagram, do you think that there is a linear relation between the length of the right humerus and the length of the right tibia?
(c) If the two variables appear to be linearly related, use a graphing utility to find the line of best fit relating the length of the right humerus and the length of the right tibia.
(d) Predict the length of the right tibia on a rat whose right humerus is 26.5 mm.

47. Create four problems that you might be asked to do given the two points $(-3, 4)$ and $(6, 1)$. Each problem should involve a different concept. Be sure that your directions are clearly stated.

48. Describe each of the following graphs in the xy-plane. Give justification.
(a) $x = 0$
(b) $y = 0$
(c) $x + y = 0$
(d) $xy = 0$
(e) $x^2 + y^2 = 0$

49. Suppose that you have a rectangular field that requires watering. Your watering system consists of an arm of variable length that rotates so that the watering pattern is a circle. Decide where to position the arm and what length it should be so that the entire field is watered most efficiently. When does it become desirable to use more than one arm?

[**Hint:** Use a rectangular coordinate system positioned as shown in the figures below. Write equations for the circle(s) swept out by the watering arm(s).]

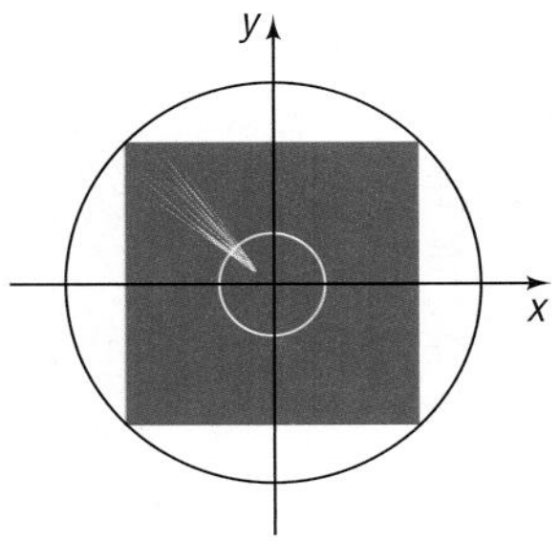

Square field

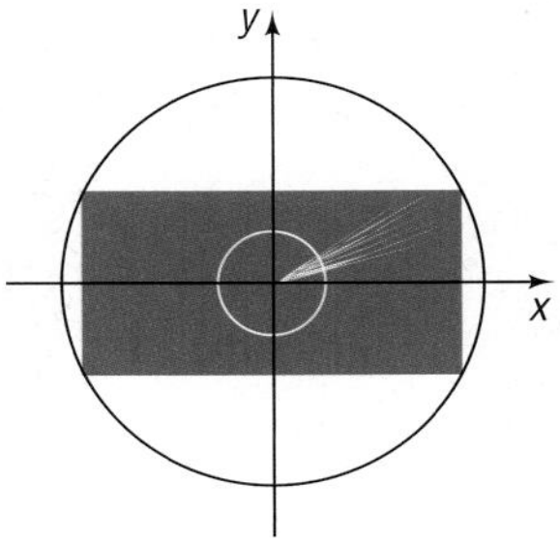

Rectangular field, one arm

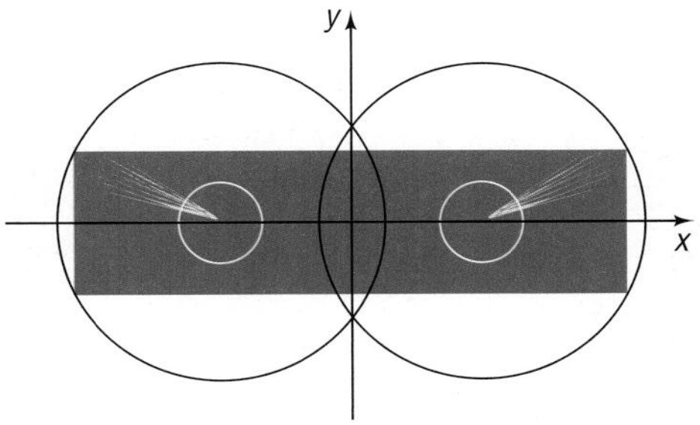

Rectangular field, two arms

Chapter Projects

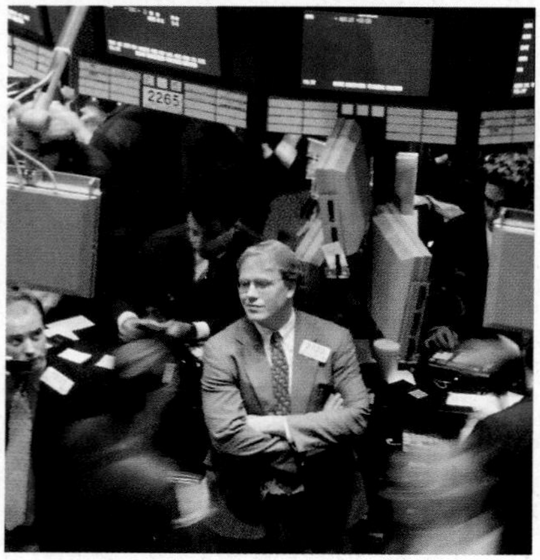

1. **Analyzing a Stock** The beta, β, of a stock represents the relative risk of a stock compared with a market basket of stocks, such as Standard and Poor's 500 index of stocks. Beta is computed by finding the slope of the line of best fit between the rate of return of the stock and rate of return of the S&P 500. The rates of return are computed on a weekly basis.
   (a) Find the weekly closing price of your favorite stock and the S&P 500 for 20 weeks. One good source is on the Internet at *http://finance.yahoo.com.*
   (b) Compute the rate of return by computing the weekly percentage change in the closing price of your stock and the weekly percentage change in the S&P 500 using the following formula:

$$\text{Weekly \% change} = \frac{P_2 - P_1}{P_1}$$

   where P_1 is last week's price and P_2 is this week's price.
   (c) Using a graphing utility, find the line of best fit, treating the weekly rate of return of the S&P 500 as the independent variable and the weekly rate of return of your stock as the dependent variable.
   (d) What is the beta of your stock?
   (e) Compare your result with that of the Value Line Investment Survey found in your library. What might account for any differences?

The following projects are available at www.prenhall.com/sullivan7e.

2. **Project at Motorola** ***Mobile Phone Usage***
3. **Economics** ***Isocost Lines***

2 Functions and Their Graphs

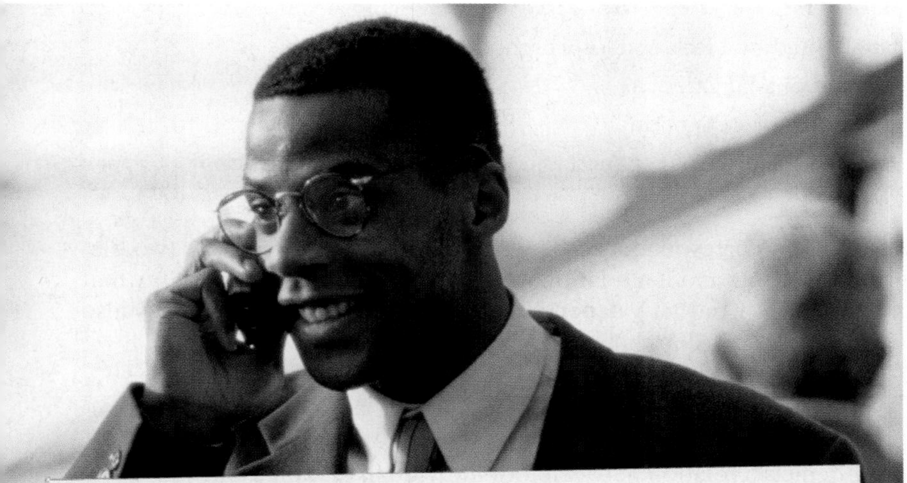

OUTLINE

Portable Phones Power Up

2000 Census Bureau Data Show Rise in Cell Phones

Jan. 24—Cell phone ownership jumped from just over 5.2 million in 1990 to nearly 110 million in 2000—a more than twenty-fold increase in just 10 years, new figures show.

Driving the phenomenal growth are several factors—one of which is the decreasing costs. The data note that the average monthly cell phone bill has been almost cut in half—from \$81 to just over \$45—over the past decade.

Survey data, compiled by the Cellular Telecommunications and Internet Association (CTIA) in Washington, D.C., and published in the U.S. Census Bureau's *Statistical Abstract of the United States: 2001* report, point to the rapid adoption for cell phone technology as a key reason behind the growth.

"The cell phone industry has shown remarkable growth over the decade," says Glenn King, chief of the statistical compendia branch of the Commerce Department that produces the abstract. Noting that there were only 55 million users in 1997, "It's doubled in the last three years alone," he says.

"With cost of service coming down, cell phones service becomes accessible to everyone," says Charles Golvin, a senior analyst with Forrester Research. "People can be in touch anytime and anywhere they want to."

By Paul Eng, abcNEWS.com

—SEE CHAPTER PROJECT 1.

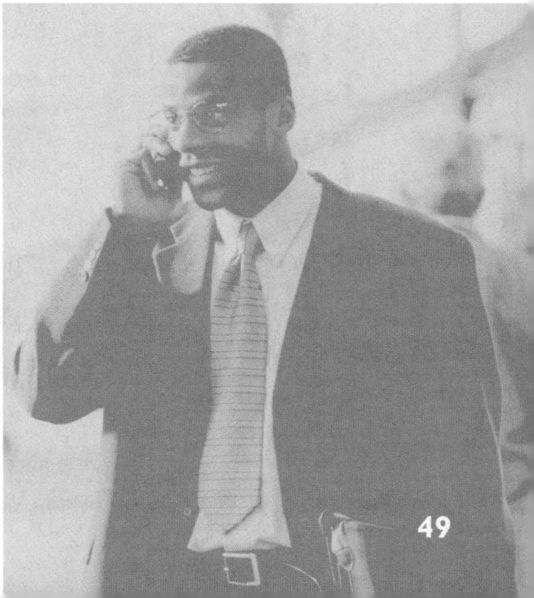

2.1 Functions

PREPARING FOR THIS SECTION *Before getting started, review the following:*

- Intervals (Appendix A, Section A.8, pp. 968–969)
- Evaluating Algebraic Expressions, Domain of a Variable (Appendix A, Section A.1, pp. 905–906)
- Solving Inequalities (Appendix A, Section A.8, pp. 971–974)

Now work the 'Are You Prepared?' problems on page 60.

OBJECTIVES
1. Determine Whether a Relation Represents a Function
2. Find the Value of a Function
3. Find the Domain of a Function
4. Form the Sum, Difference, Product, and Quotient of Two Functions

1 A **relation** is a correspondence between two sets. If x and y are two elements in these sets and if a relation exists between x and y, then we say that x **corresponds** to y or that y **depends on** x, and we write $x \rightarrow y$. We may also write $x \rightarrow y$ as the ordered pair (x, y).

EXAMPLE 1 An Example of a Relation

Figure 1 depicts a relation between four individuals and their birthdays. The relation might be named "was born on." Then Katy corresponds to June 20, Dan corresponds to Sept. 4, and so on. Using ordered pairs, this relation would be expressed as

$$\{(\text{Katy, June 20}), (\text{Dan, Sept. 4}), (\text{Patrick, Dec. 31}), (\text{Phoebe, Dec. 31})\}$$

Figure 1

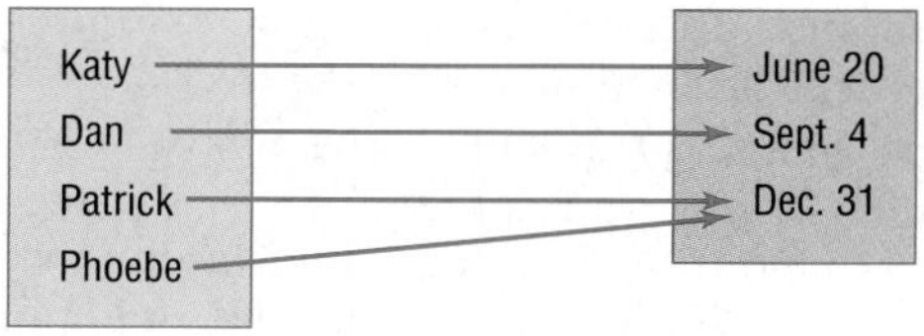

Often, we are interested in specifying the type of relation (such as an equation) that exists between the two variables. For example, the relation between the revenue R resulting from the sale of x items selling for \$10 each may be expressed by the equation $R = 10x$. If we know how many items have been sold, then we can calculate the revenue by using the equation $R = 10x$. This equation is an example of a *function*.

As another example, suppose that an icicle falls off a building from a height of 64 feet above the ground. According to a law of physics, the distance s (in feet) of the icicle from the ground after t seconds is given (approximately) by the formula $s = 64 - 16t^2$. When $t = 0$ seconds, the icicle is $s = 64$ feet above the ground. After 1 second, the icicle is $s = 64 - 16(1)^2 = 48$ feet above the ground. After 2 seconds, the icicle strikes the ground. The formula $s = 64 - 16t^2$ provides a way of finding the distance s for any time t $(0 \le t \le 2)$. There is a correspondence between each time t in the interval $0 \le t \le 2$ and the distance s. We say that the distance s is a function of the time t because:

1. There is a correspondence between the set of times and the set of distances.
2. There is exactly one distance s obtained for any time t in the interval $0 \le t \le 2$.

Let's now look at the definition of a function.

Definition of Function

Let X and Y be two nonempty sets.* A **function** from X into Y is a relation that associates with each element of X exactly one element of Y.

Figure 2

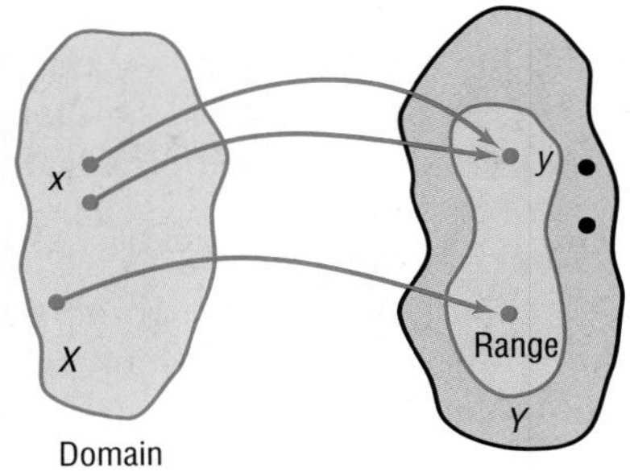

The set X is called the **domain** of the function. For each element x in X, the corresponding element y in Y is called the **value** of the function at x, or the image of x. The set of all images of the elements of the domain is called the **range** of the function. See Figure 2.

Since there may be some elements in Y that are not the image of any x in X, it follows that the range of a function may be a subset of Y, as shown in Figure 2.

Not all relations between two sets are functions. The next example shows how to determine whether a relation is a function or not.

EXAMPLE 2 Determining Whether a Relation Represents a Function

Determine whether the following relations represent functions.

(a) See Figure 3. For this relation, the domain represents four individuals and the range represents their birthdays.

Figure 3

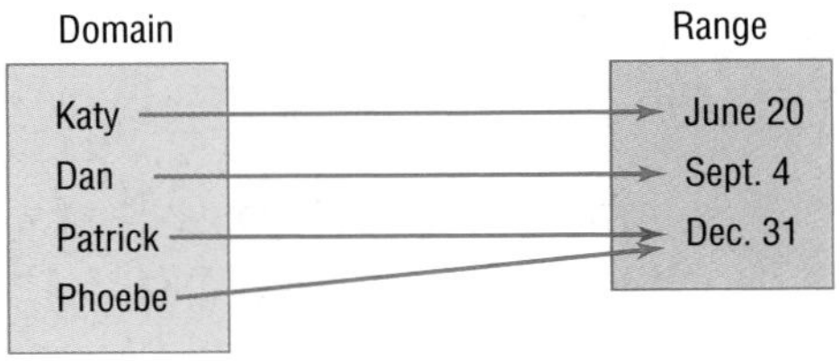

(b) See Figure 4. For this relation, the domain represents the employees of Sara's Pre-Owned Car Mart and the range represents their phone number(s).

Figure 4

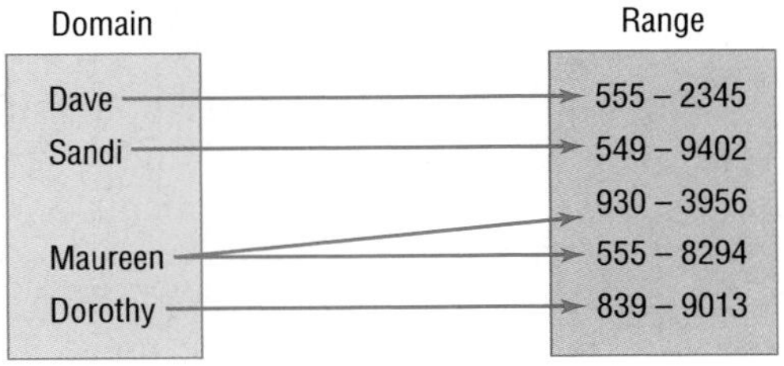

*The sets X and Y will usually be sets of real numbers. The sets X and Y can also be sets of complex numbers and then we have defined a complex function. In the broad definition (due to Lejeune Dirichlet), X and Y can be any two sets.

Solution (a) The relation in Figure 3 is a function because each element in the domain corresponds to exactly one element in the range. Notice that more than one element in the domain can correspond to the same element in the range (Phoebe and Patrick were born on the same day of the year).

(b) The relation in Figure 4 is not a function because each element in the domain does not correspond to exactly one element in the range. Maureen has two telephone numbers; therefore, if Maureen is chosen from the domain, a single telephone number cannot be assigned to her. ◀

NOW WORK PROBLEM 15.

In Words
For a function, no input has more than one output.

In Words
For a function, the domain is the set of inputs, and the range is the set of outputs.

The idea behind a function is its predictability. If the input is known, we can use the function to determine the output. In "nonfunctions," we don't have this predictability. Look back at Figure 3. The inputs are {Katy, Dan, Patrick, Phoebe}. The correspondence is "was born on" and the outputs are {June 20, Sept. 4, Dec. 31}. If asked "What is Katy's birthday?", we can use the correspondence to answer "June 20." Now consider Figure 4. If asked "What is Maureen's phone number?", we could not give a single response because there are two outputs that result from the single input "Maureen." For this reason, the relation in Figure 4 is not a function.

We may think of a function as a set of ordered pairs (x, y) in which no two ordered pairs have the same first element, but different second elements. The set of all first elements x is the domain of the function, and the set of all second elements y is its range. Each element x in the domain corresponds to exactly one element y in the range.

EXAMPLE 3 Determining Whether a Relation Represents a Function

Determine whether each relation represents a function. If it is a function, state the domain and range.

(a) $\{(1, 4), (2, 5), (3, 6), (4, 7)\}$

(b) $\{(1, 4), (2, 4), (3, 5), (6, 10)\}$

(c) $\{(-3, 9), (-2, 4), (0, 0), (1, 1), (-3, 8)\}$

Solution (a) This relation is a function because there are no ordered pairs with the same first element and different second elements. The domain of this function is $\{1, 2, 3, 4\}$, and its range is $\{4, 5, 6, 7\}$.

(b) This relation is a function because there are no ordered pairs with the same first element and different second elements. The domain of this function is $\{1, 2, 3, 6\}$, and its range is $\{4, 5, 10\}$.

(c) This relation is not a function because there are two ordered pairs, $(-3, 9)$ and $(-3, 8)$, that have the same first element, but different second elements. ◀

In Example 3(b), notice that 1 and 2 in the domain each have the same image in the range. This does not violate the definition of a function; two different first elements can have the same second element. A violation of the definition occurs when two ordered pairs have the same first element and different second elements, as in Example 3(c).

NOW WORK PROBLEM 19.

Example 2(a) demonstrates that a function may be defined by some correspondence between two sets. Examples 3(a) and 3(b) demonstrate that a function may be defined by a set of ordered pairs. A function may also be defined by an equation in two variables, usually denoted x and y.

EXAMPLE 4 **Example of a Function**

Consider the function defined by the equation

$$y = 2x - 5, \qquad 1 \le x \le 6$$

Notice that for each input x there corresponds exactly one output y. For example, if $x = 1$, then $y = 2(1) - 5 = -3$. If $x = 3$, then $y = 2(3) - 5 = 1$. For this reason, the equation is a function. Since we restrict the inputs to the real numbers between 1 and 6, inclusive, the domain of the function is $\{x | 1 \le x \le 6\}$. The function specifies that in order to get the image of x we multiply x by 2 and then subtract 5 from this product. ◀

Function Notation

2 Functions are often denoted by letters such as f, F, g, G, and others. If f is a function, then for each number x in its domain the corresponding image in the range is designated by the symbol $f(x)$, read as "f of x" or as "f at x." We refer to $f(x)$ as the **value of f at the number x;** $f(x)$ is the number that results when x is given and the function f is applied; $f(x)$ does *not* mean "f times x." For example, the function given in Example 4 may be written as $y = f(x) = 2x - 5, 1 \le x \le 6$. Then $f\left(\frac{3}{2}\right) = -2$.

Figure 5 illustrates some other functions. Notice that, in every function illustrated, for each x in the domain there is one value in the range.

Figure 5

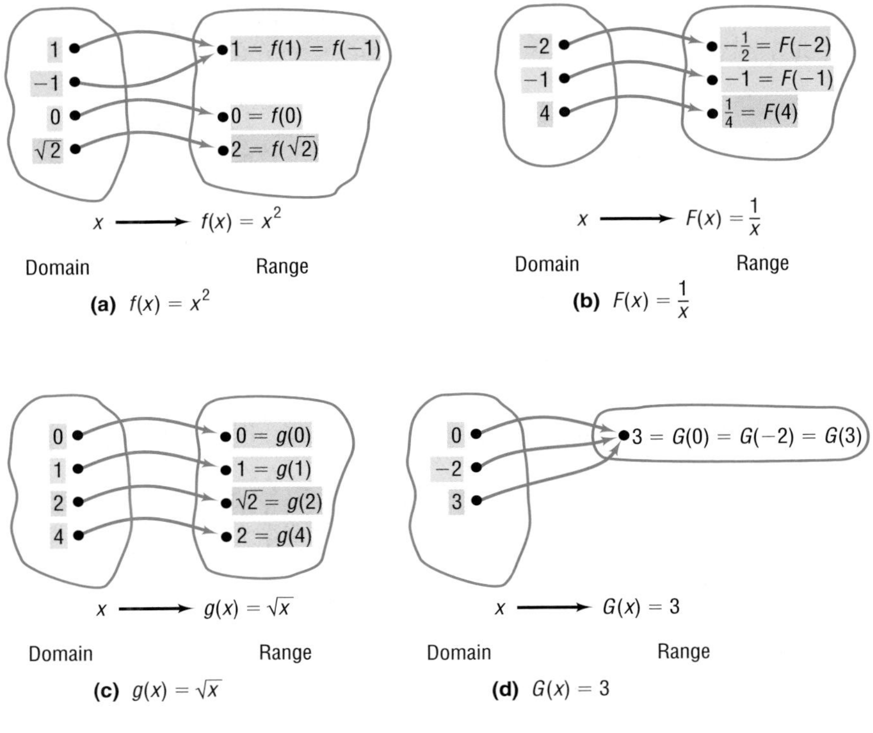

(a) $f(x) = x^2$ **(b)** $F(x) = \frac{1}{x}$ **(c)** $g(x) = \sqrt{x}$ **(d)** $G(x) = 3$

Figure 6

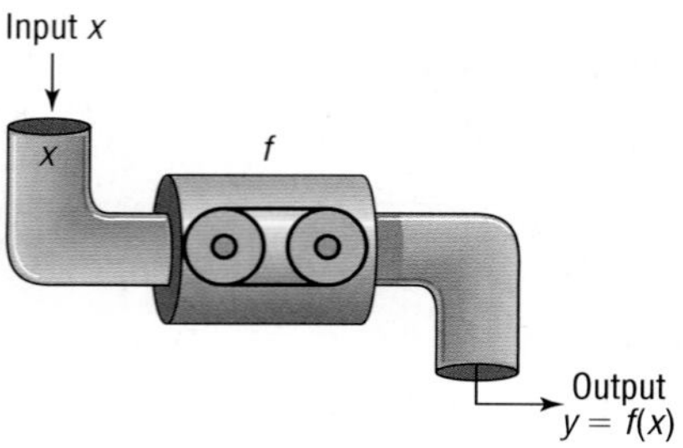

Sometimes it is helpful to think of a function f as a machine that receives as input a number from the domain, manipulates it, and outputs the value. See Figure 6.

The restrictions on this input/output machine are as follows:

1. It only accepts numbers from the domain of the function.
2. For each input, there is exactly one output (which may be repeated for different inputs).

For a function $y = f(x)$, the variable x is called the **independent variable,** because it can be assigned any of the permissible numbers from the domain. The variable y is called the **dependent variable,** because its value depends on x.

Any symbol can be used to represent the independent and dependent variables. For example, if f is the *cube function*, then f can be given by $f(x) = x^3$ or $f(t) = t^3$ or $f(z) = z^3$. All three functions are the same. Each tells us to cube the independent variable. In practice, the symbols used for the independent and dependent variables are based on common usage, such as using C for cost in business.

The independent variable is also called the **argument** of the function. Thinking of the independent variable as an argument can sometimes make it easier to find the value of a function. For example, if f is the function defined by $f(x) = x^3$, then f tells us to cube the argument. Thus, $f(2)$ means to cube 2, $f(a)$ means to cube the number a, and $f(x + h)$ means to cube the quantity $x + h$.

EXAMPLE 5 Finding Values of a Function

For the function f defined by $f(x) = 2x^2 - 3x$, evaluate

(a) $f(3)$ (b) $f(x) + f(3)$ (c) $f(-x)$

(d) $-f(x)$ (e) $f(x + 3)$ (f) $\dfrac{f(x + h) - f(x)}{h}, \quad h \neq 0$

Solution (a) We substitute 3 for x in the equation for f to get

$$f(3) = 2(3)^2 - 3(3) = 18 - 9 = 9$$

(b) $f(x) + f(3) = (2x^2 - 3x) + (9) = 2x^2 - 3x + 9$

(c) We substitute $-x$ for x in the equation for f.

$$f(-x) = 2(-x)^2 - 3(-x) = 2x^2 + 3x$$

(d) $-f(x) = -(2x^2 - 3x) = -2x^2 + 3x$

(e) $f(x + 3) = 2(x + 3)^2 - 3(x + 3)$ Notice the use of parentheses here.

$$= 2(x^2 + 6x + 9) - 3x - 9$$
$$= 2x^2 + 12x + 18 - 3x - 9$$
$$= 2x^2 + 9x + 9$$

$$\text{(f)}\quad \frac{f(x+h)-f(x)}{h} = \frac{[2(x+h)^2 - 3(x+h)] - [2x^2 - 3x]}{h}$$

$\uparrow$ $f(x + h) = 2(x + h)^2 - 3(x + h)$

$$= \frac{2(x^2 + 2xh + h^2) - 3x - 3h - 2x^2 + 3x}{h} \qquad \text{Simplify.}$$

$$= \frac{2x^2 + 4xh + 2h^2 - 3h - 2x^2}{h}$$

$$= \frac{4xh + 2h^2 - 3h}{h}$$

$$= \frac{h(4x + 2h - 3)}{h} \qquad \text{Factor out } h.$$

$$= 4x + 2h - 3 \qquad \text{Cancel } h.$$

◀

Notice in this example that $f(x + 3) \neq f(x) + f(3)$ and $f(-x) \neq -f(x)$.

The expression in part (f) is called the **difference quotient** of f, an important expression in calculus.

NOW WORK PROBLEMS 27 AND 73.

Most calculators have special keys that enable you to find the value of certain commonly used functions. For example, you should be able to find the square function $f(x) = x^2$, the square root function $f(x) = \sqrt{x}$, the reciprocal function $f(x) = \frac{1}{x} = x^{-1}$, and many others that will be discussed later in this book (such as $\ln x$ and $\log x$). Verify the results of Example 6, which follows, on your calculator.

EXAMPLE 6 Finding Values of a Function on a Calculator

(a) $f(x) = x^2$; $f(1.234) = 1.522756$

(b) $F(x) = \frac{1}{x}$; $F(1.234) = 0.8103727715$

(c) $g(x) = \sqrt{x}$; $g(1.234) = 1.110855526$ ◀

COMMENT: Graphing calculators can be used to evaluate any function that you wish. Figure 7 shows the result obtained in Example 5(a) on a TI-83 graphing calculator with the function to be evaluated, $f(x) = 2x^2 - 3x$, in Y_1.*

Figure 7

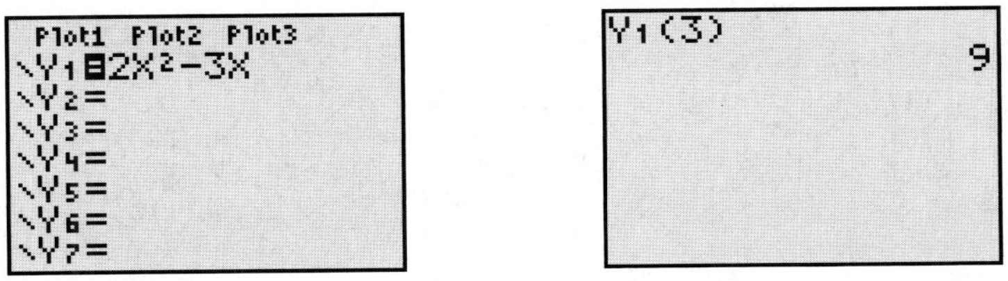

■

*Consult your owner's manual for the required keystrokes.

Implicit Form of a Function

In general, when a function f is defined by an equation in x and y, we say that the function f is given **implicitly.** If it is possible to solve the equation for y in terms of x, then we write $y = f(x)$ and say that the function is given **explicitly.** For example,

Implicit Form	Explicit Form
$3x + y = 5$	$y = f(x) = -3x + 5$
$x^2 - y = 6$	$y = f(x) = x^2 - 6$
$xy = 4$	$y = f(x) = \dfrac{4}{x}$

Not all equations in x and y define a function $y = f(x)$. If an equation is solved for y and two or more values of y can be obtained for a given x, then the equation does not define a function.

EXAMPLE 7 Determining Whether an Equation Is a Function

Determine if the equation $x^2 + y^2 = 1$ is a function.

Solution To determine whether the equation $x^2 + y^2 = 1$, which defines the unit circle, is a function, we need to solve the equation for y.

$$x^2 + y^2 = 1$$
$$y^2 = 1 - x^2$$
$$y = \pm\sqrt{1 - x^2} \quad \text{Square Root Method}$$

For values of x between -1 and 1, two values of y result. This means that the equation $x^2 + y^2 = 1$ does not define a function. ◀

NOW WORK PROBLEM 41.

COMMENT: The explicit form of a function is the form required by a graphing calculator. Now do you see why it is necessary to graph a circle in two "pieces"? ■

We list next a summary of some important facts to remember about a function f.

Summary

Important Facts About Functions

(a) For each x in the domain of f, there is exactly one image $f(x)$ in the range; however, an element in the range can result from more than one x in the domain.
(b) f is the symbol that we use to denote the function. It is symbolic of the equation that we use to get from an x in the domain to $f(x)$ in the range.
(c) If $y = f(x)$, then x is called the independent variable or argument of f, and y is called the dependent variable or the value of f at x.

Domain of a Function

3 Often the domain of a function f is not specified; instead, only the equation defining the function is given. In such cases, we agree that the domain of f is the largest set of real numbers for which the value $f(x)$ is a real number. The domain of a function f is the same as the domain of the variable x in the expression $f(x)$.

EXAMPLE 8 **Finding the Domain of a Function**

Find the domain of each of the following functions:

(a) $f(x) = x^2 + 5x$ (b) $g(x) = \dfrac{3x}{x^2 - 4}$ (c) $h(t) = \sqrt{4 - 3t}$

Solution (a) The function tells us to square a number and then add five times the number. Since these operations can be performed on any real number, we conclude that the domain of f is all real numbers.

(b) The function g tells us to divide $3x$ by $x^2 - 4$. Since division by 0 is not defined, the denominator $x^2 - 4$ can never be 0, so x can never equal -2 or 2. The domain of the function g is $\{x | x \neq -2, x \neq 2\}$.

(c) The function h tells us to take the square root of $4 - 3t$. But only nonnegative numbers have real square roots, so the expression under the square root must be nonnegative. This requires that

$$\begin{aligned} 4 - 3t &\geq 0 \\ -3t &\geq -4 \\ t &\leq \frac{4}{3} \end{aligned}$$

The domain of h is $\left\{t \middle| t \leq \frac{4}{3}\right\}$ or the interval $\left(-\infty, \frac{4}{3}\right]$. ◀

NOW WORK PROBLEM 51.

If x is in the domain of a function f, we shall say that **f is defined at x**, or **$f(x)$ exists**. If x is not in the domain of f, we say that **f is not defined at x**, or **$f(x)$ does not exist**. For example, if $f(x) = \dfrac{x}{x^2 - 1}$, then $f(0)$ exists, but $f(1)$ and $f(-1)$ do not exist. (Do you see why?)

We have not said much about finding the range of a function. The reason is that when a function is defined by an equation it is often difficult to find the range.* Therefore, we shall usually be content to find just the domain of a function when only the rule for the function is given. We shall express the domain of a function using inequalities, interval notation, set notation, or words, whichever is most convenient.

Applications

When we use functions in applications, the domain may be restricted by physical or geometric considerations. For example, the domain of the function f defined by $f(x) = x^2$ is the set of all real numbers. However, if f is

*In Section 4.2 we discuss a way to find the range for a special class of functions.

used to obtain the area of a square when the length x of a side is known, then we must restrict the domain of f to the positive real numbers, since the length of a side can never be 0 or negative.

EXAMPLE 9 Area of a Circle

Express the area of a circle as a function of its radius.

Figure 8

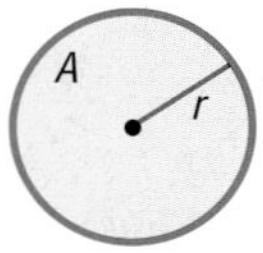

Solution See Figure 8. We know that the formula for the area A of a circle of radius r is $A = \pi r^2$. If we use r to represent the independent variable and A to represent the dependent variable, the function expressing this relationship is

$$A(r) = \pi r^2$$

In this setting, the domain is $\{r | r > 0\}$. (Do you see why?) ◀

Observe in the solution to Example 9 that we used the symbol A in two ways: It is used to name the function, and it is used to symbolize the dependent variable. This double use is common in applications and should not cause any difficulty.

NOW WORK PROBLEM 85.

Operations on Functions

4 Next we introduce some operations on functions. We shall see that functions, like numbers, can be added, subtracted, multiplied, and divided. For example, if $f(x) = x^2 + 9$ and $g(x) = 3x + 5$, then

$$f(x) + g(x) = (x^2 + 9) + (3x + 5) = x^2 + 3x + 14$$

The new function $y = x^2 + 3x + 14$ is called the *sum function* $f + g$. Similarly,

$$f(x) \cdot g(x) = (x^2 + 9)(3x + 5) = 3x^3 + 5x^2 + 27x + 45$$

The new function $y = 3x^3 + 5x^2 + 27x + 45$ is called the *product function* $f \cdot g$.

The general definitions are given next.

If f and g are functions:

The **sum** $\boldsymbol{f + g}$ is the function defined by

$$(f + g)(x) = f(x) + g(x)$$

The domain of $f + g$ consists of the numbers x that are in the domains of both f and g.

The **difference** $\boldsymbol{f - g}$ is the function defined by

$$(f - g)(x) = f(x) - g(x)$$

The domain of $f - g$ consists of the numbers x that are in the domains of both f and g.

The product $\boldsymbol{f \cdot g}$ is the function defined by

$$(f \cdot g)(x) = f(x) \cdot g(x)$$

The domain of $f \cdot g$ consists of the numbers x that are in the domains of both f and g.

The **quotient** $\dfrac{\boldsymbol{f}}{\boldsymbol{g}}$ is the function defined by

$$\left(\frac{f}{g}\right)(x) = \frac{f(x)}{g(x)}, \qquad g(x) \neq 0$$

The domain of $\dfrac{f}{g}$ consists of the numbers x for which $g(x) \neq 0$ that are in the domains of both f and g.

EXAMPLE 10 Operations on Functions

Let f and g be two functions defined as

$$f(x) = \frac{1}{x+2} \quad \text{and} \quad g(x) = \frac{x}{x-1}$$

Find the following, and determine the domain in each case.

(a) $(f + g)(x)$ (b) $(f - g)(x)$ (c) $(f \cdot g)(x)$ (d) $\left(\dfrac{f}{g}\right)(x)$

Solution The domain of f is $\{x \mid x \neq -2\}$ and the domain of g is $\{x \mid x \neq 1\}$.

(a) $$(f + g)(x) = f(x) + g(x) = \frac{1}{x+2} + \frac{x}{x-1} = \frac{x-1}{(x+2)(x-1)} + \frac{x(x+2)}{(x+2)(x-1)} = \frac{x^2 + 3x - 1}{(x+2)(x-1)}$$

The domain of $f + g$ consists of those numbers x that are in the domains of both f and g. Therefore, the domain of $f + g$ is $\{x \mid x \neq -2, x \neq 1\}$.

(b) $$(f - g)(x) = f(x) - g(x) = \frac{1}{x+2} - \frac{x}{x-1} = \frac{x-1}{(x+2)(x-1)} - \frac{x(x+2)}{(x+2)(x-1)} = \frac{-(x^2 + x + 1)}{(x+2)(x-1)}$$

The domain of $f - g$ consists of those numbers x that are in the domains of both f and g. Therefore, the domain of $f - g$ is $\{x \mid x \neq -2, x \neq 1\}$.

(c) $$(f \cdot g)(x) = f(x) \cdot g(x) = \frac{1}{x+2} \cdot \frac{x}{x-1} = \frac{x}{(x+2)(x-1)}$$

The domain of $f \cdot g$ consists of those numbers x that are in the domains of both f and g. Therefore, the domain of $f \cdot g$ is $\{x \mid x \neq -2, x \neq 1\}$.

(d) $\left(\dfrac{f}{g}\right)(x) = \dfrac{f(x)}{g(x)} = \dfrac{\dfrac{1}{x+2}}{\dfrac{x}{x-1}} = \dfrac{1}{x+2} \cdot \dfrac{x-1}{x} = \dfrac{x-1}{x(x+2)}$

The domain of $\dfrac{f}{g}$ consists of the numbers x for which $g(x) \neq 0$ that are in the domains of both f and g. Since $g(x) = 0$ when $x = 0$, we exclude 0 as well as -2 and 1. The domain of $\dfrac{f}{g}$ is $\{x | x \neq -2, x \neq 0, x \neq 1\}$. ◀

NOW WORK PROBLEM 61.

In calculus, it is sometimes helpful to view a complicated function as the sum, difference, product, or quotient of simpler functions. For example,

$F(x) = x^2 + \sqrt{x}$ is the sum of $f(x) = x^2$ and $g(x) = \sqrt{x}$.

$H(x) = \dfrac{x^2 - 1}{x^2 + 1}$ is the quotient of $f(x) = x^2 - 1$ and $g(x) = x^2 + 1$.

Summary

We list here some of the important vocabulary introduced in this section, with a brief description of each term.

Function A relation between two sets of real numbers so that each number x in the first set, the domain, has corresponding to it exactly one number y in the second set.
A set of ordered pairs (x, y) or $(x, f(x))$ in which no first element is paired with two different second elements.
The range is the set of y values of the function for the x values in the domain.
A function f may be defined implicitly by an equation involving x and y or explicitly by writing $y = f(x)$.

Unspecified domain If a function f is defined by an equation and no domain is specified, then the domain will be taken to be the largest set of real numbers for which the equation defines a real number.

Function notation $y = f(x)$
f is a symbol for the function.
x is the independent variable or argument.
y is the dependent variable.
$f(x)$ is the value of the function at x, or the image of x.

2.1 Assess Your Understanding

'Are You Prepared?' *Answers are given at the end of these exercises. If you get the wrong answer, read the pages listed in red.*

1. The inequality $-1 < x < 3$ can be written in interval notation as __________. (pp. 968–969)

2. If $x = -2$, the value of the expression $3x^2 - 5x + \dfrac{1}{x}$ is __________. (pp. 905–906)

3. The domain of the variable in the expression $\dfrac{x-3}{x+4}$ is __________. (pp. 905–906)

4. Solve the inequality: $3 - 2x > 5$. Graph the solution set. (pp. 971–974)

Concepts and Vocabulary

5. If f is a function defined by the equation $y = f(x)$, then x is called the __________ variable and y is the __________ variable.

6. The set of all images of the elements in the domain of a function is called the __________.

7. If the domain of f is all real numbers in the interval $[0, 7]$ and the domain of g is all real numbers in the interval $[-2, 5]$, the domain of $f + g$ is all real numbers in the interval __________.

8. The domain of $\frac{f}{g}$ consists of numbers x for which $g(x)$ __________ 0 that are in the domains of both __________ and __________.

9. If $f(x) = x + 1$ and $g(x) = x^3$, then __________ $= x^3 - (x + 1)$.

10. *True or False:* Every relation is a function.

11. *True or False:* The domain of $(f \cdot g)(x)$ consists of the numbers x that are in the domains of both f and g.

12. *True or False:* The independent variable is sometimes referred to as the argument of the function.

13. *True or False:* If no domain is specified for a function f, then the domain of f is taken to be the set of real numbers.

14. *True or False:* The domain of the function $f(x) = \frac{x^2 - 4}{x}$ is $\{x \mid x \neq \pm 2\}$.

Exercises

In Problems 15–26, determine whether each relation represents a function. For each function, state the domain and range.

15.

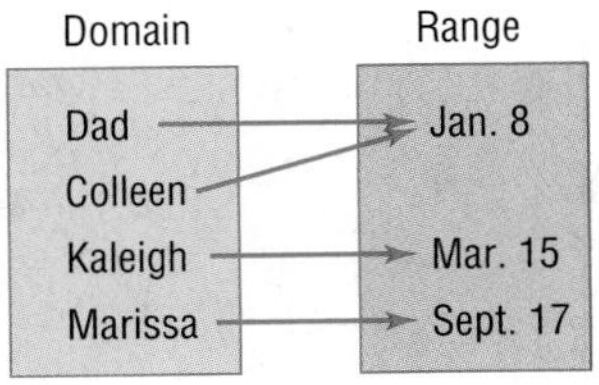

16.

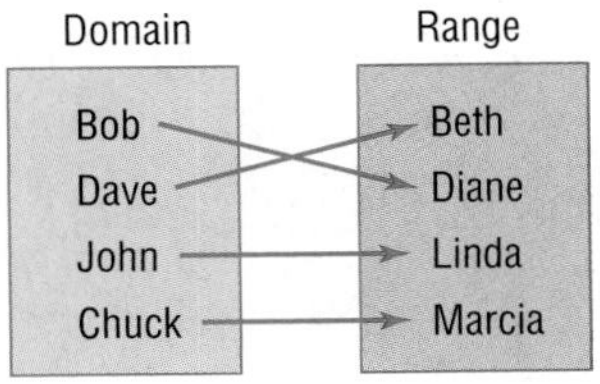

17.

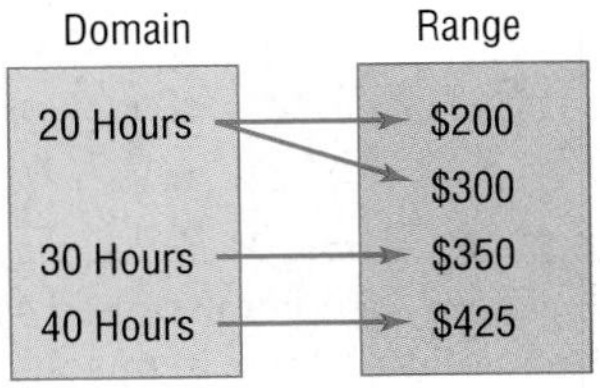

18.

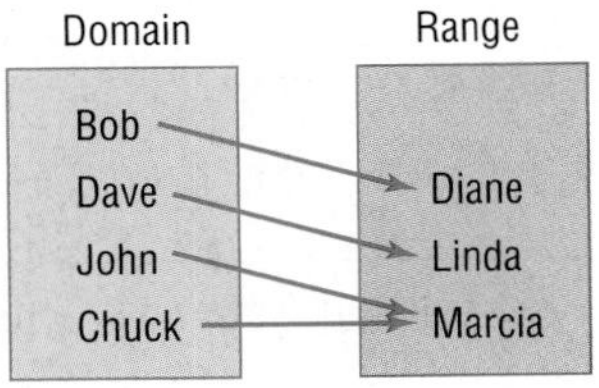

19. $\{(2, 6), (-3, 6), (4, 9), (2, 10)\}$

20. $\{(-2, 5), (-1, 3), (3, 7), (4, 12)\}$

21. $\{(1, 3), (2, 3), (3, 3), (4, 3)\}$

22. $\{(0, -2), (1, 3), (2, 3), (3, 7)\}$

23. $\{(-2, 4), (-2, 6), (0, 3), (3, 7)\}$

24. $\{(-4, 4), (-3, 3), (-2, 2), (-1, 1), (-4, 0)\}$

25. $\{(-2, 4), (-1, 1), (0, 0), (1, 1)\}$

26. $\{(-2, 16), (-1, 4), (0, 3), (1, 4)\}$

In Problems 27–34, find the following values for each function:

(a) $f(0)$ (b) $f(1)$ (c) $f(-1)$ (d) $f(-x)$ (e) $-f(x)$ (f) $f(x + 1)$ (g) $f(2x)$ (h) $f(x + h)$

27. $f(x) = 3x^2 + 2x - 4$

28. $f(x) = -2x^2 + x - 1$

29. $f(x) = \dfrac{x}{x^2 + 1}$

30. $f(x) = \dfrac{x^2 - 1}{x + 4}$

31. $f(x) = |x| + 4$

32. $f(x) = \sqrt{x^2 + x}$

33. $f(x) = \dfrac{2x + 1}{3x - 5}$

34. $f(x) = 1 - \dfrac{1}{(x + 2)^2}$

In Problems 35–46, determine whether the equation is a function.

35. $y = x^2$

36. $y = x^3$

37. $y = \dfrac{1}{x}$

38. $y = |x|$

39. $y^2 = 4 - x^2$

40. $y = \pm\sqrt{1 - 2x}$

41. $x = y^2$

42. $x + y^2 = 1$

43. $y = 2x^2 - 3x + 4$

44. $y = \dfrac{3x - 1}{x + 2}$

45. $2x^2 + 3y^2 = 1$

46. $x^2 - 4y^2 = 1$

In Problems 47–60, find the domain of each function.

47. $f(x) = -5x + 4$

48. $f(x) = x^2 + 2$

49. $f(x) = \dfrac{x}{x^2 + 1}$

50. $f(x) = \dfrac{x^2}{x^2 + 1}$

51. $g(x) = \dfrac{x}{x^2 - 16}$

52. $h(x) = \dfrac{2x}{x^2 - 4}$

53. $F(x) = \dfrac{x - 2}{x^3 + x}$

54. $G(x) = \dfrac{x + 4}{x^3 - 4x}$

55. $h(x) = \sqrt{3x - 12}$

56. $G(x) = \sqrt{1 - x}$

57. $f(x) = \dfrac{4}{\sqrt{x - 9}}$

58. $f(x) = \dfrac{x}{\sqrt{x - 4}}$

59. $p(x) = \sqrt{\dfrac{2}{x - 1}}$

60. $q(x) = \sqrt{-x - 2}$

In Problems 61–70, for the given functions f and g, find the following functions and state the domain of each.

(a) $f + g$ (b) $f - g$ (c) $f \cdot g$ (d) $\dfrac{f}{g}$

61. $f(x) = 3x + 4;\quad g(x) = 2x - 3$

62. $f(x) = 2x + 1;\quad g(x) = 3x - 2$

63. $f(x) = x - 1;\quad g(x) = 2x^2$

64. $f(x) = 2x^2 + 3;\quad g(x) = 4x^3 + 1$

65. $f(x) = \sqrt{x};\quad g(x) = 3x - 5$

66. $f(x) = |x|;\quad g(x) = x$

67. $f(x) = 1 + \dfrac{1}{x};\quad g(x) = \dfrac{1}{x}$

68. $f(x) = \sqrt{x - 2};\quad g(x) = \sqrt{4 - x}$

69. $f(x) = \dfrac{2x + 3}{3x - 2};\quad g(x) = \dfrac{4x}{3x - 2}$

70. $f(x) = \sqrt{x + 1};\quad g(x) = \dfrac{2}{x}$

71. Given $f(x) = 3x + 1$ and $(f + g)(x) = 6 - \dfrac{1}{2}x$, find the function g.

72. Given $f(x) = \dfrac{1}{x}$ and $\left(\dfrac{f}{g}\right)(x) = \dfrac{x + 1}{x^2 - x}$, find the function g.

In Problems 73–78, find the difference quotient of f, that is, find $\dfrac{f(x + h) - f(x)}{h}$, $h \neq 0$, *for each function. Be sure to simplify.*

73. $f(x) = 4x + 3$

74. $f(x) = -3x + 1$

75. $f(x) = x^2 - x + 4$

76. $f(x) = x^2 + 5x - 1$

77. $f(x) = x^3 - 2$

78. $f(x) = \dfrac{1}{x + 3}$

---

79. If $f(x) = 2x^3 + Ax^2 + 4x - 5$ and $f(2) = 5$, what is the value of A?

80. If $f(x) = 3x^2 - Bx + 4$ and $f(-1) = 12$, what is the value of B?

81. If $f(x) = \dfrac{3x + 8}{2x - A}$ and $f(0) = 2$, what is the value of A?

82. If $f(x) = \dfrac{2x - B}{3x + 4}$ and $f(2) = \dfrac{1}{2}$, what is the value of B?

83. If $f(x) = \dfrac{2x - A}{x - 3}$ and $f(4) = 0$, what is the value of A? Where is f not defined?

84. If $f(x) = \dfrac{x - B}{x - A}$, $f(2) = 0$, and $f(1)$ is undefined, what are the values of A and B?

85. Geometry Express the area A of a rectangle as a function of the length x if the length of the rectangle is twice its width.

86. Geometry Express the area A of an isosceles right triangle as a function of the length x of one of the two equal sides.

87. Constructing Functions Express the gross salary G of a person who earns \$10 per hour as a function of the number x of hours worked.

88. Constructing Functions Tiffany, a commissioned salesperson, earns \$100 base pay plus \$10 per item sold. Express her gross salary G as a function of the number x of items sold.

89. Effect of Gravity on Earth If a rock falls from a height of 20 meters on Earth, the height H (in meters) after x seconds is approximately

$$H(x) = 20 - 4.9x^2$$

(a) What is the height of the rock when $x = 1$ second? $x = 1.1$ seconds? $x = 1.2$ seconds? $x = 1.3$ seconds?

(b) When is the height of the rock 15 meters? When is it 10 meters? When is it 5 meters?

(c) When does the rock strike the ground?

90. Effect of Gravity on Jupiter If a rock falls from a height of 20 meters on the planet Jupiter, its height H (in meters) after x seconds is approximately

$$H(x) = 20 - 13x^2$$

(a) What is the height of the rock when $x = 1$ second? $x = 1.1$ seconds? $x = 1.2$ seconds?

(b) When is the height of the rock 15 meters? When is it 10 meters? When is it 5 meters?

(c) When does the rock strike the ground?

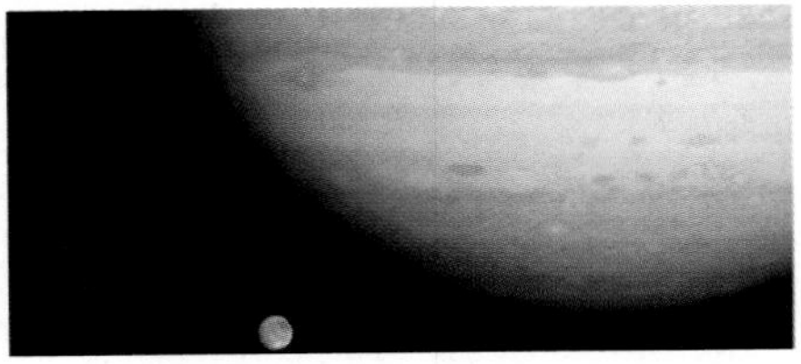

91. Cost of Trans-Atlantic Travel A Boeing 747 crosses the Atlantic Ocean (3000 miles) with an airspeed of 500 miles per hour. The cost C (in dollars) per passenger is given by

$$C(x) = 100 + \frac{x}{10} + \frac{36{,}000}{x}$$

where x is the ground speed (airspeed $\pm$ wind).
(a) What is the cost per passenger for quiescent (no wind) conditions?
(b) What is the cost per passenger with a head wind of 50 miles per hour?
(c) What is the cost per passenger with a tail wind of 100 miles per hour?
(d) What is the cost per passenger with a head wind of 100 miles per hour?

92. Cross-sectional Area The cross-sectional area of a beam cut from a log with radius 1 foot is given by the function $A(x) = 4x\sqrt{1 - x^2}$, where x represents the length of half the base of the beam. See the figure. Determine the cross-sectional area of the beam if the length of half the base of the beam is as follows:
(a) One-third of a foot
(b) One-half of a foot
(c) Two-thirds of a foot

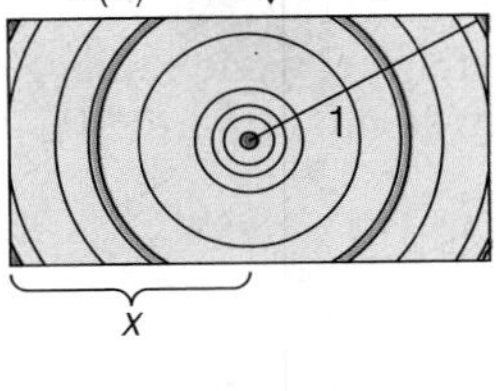

93. Economics The **participation rate** is the number of people in the labor force divided by the civilian population (excludes military). Let $L(x)$ represent the size of the labor force in year x and $P(x)$ represent the civilian population in year x. Determine a function that represents the participation rate R as a function of x.

94. Crimes Suppose that $V(x)$ represents the number of violent crimes committed in year x and $P(x)$ represents the number of property crimes committed in year x. Determine a function T that represents the combined total of violent crimes and property crimes in year x.

95. Health Care Suppose that $P(x)$ represents the percentage of income spent on health care in year x and $I(x)$ represents income in year x. Determine a function H that represents total health care expenditures in year x.

96. Income Tax Suppose that $I(x)$ represents the income of an individual in year x before taxes and $T(x)$ represents the individual's tax bill in year x. Determine a function N that represents the individual's net income (income after taxes) in year x.

97. Some functions f have the property that $f(a + b) = f(a) + f(b)$ for all real numbers a and b. Which of the following functions have this property?
(a) $h(x) = 2x$ (b) $g(x) = x^2$
(c) $F(x) = 5x - 2$ (d) $G(x) = \frac{1}{x}$

98. Are the functions $f(x) = x - 1$ and $g(x) = \frac{x^2 - 1}{x + 1}$ the same? Explain.

99. Investigate when, historically, the use of the function notation $y = f(x)$ first appeared.

'Are You Prepared?' Answers

1. $(-1, 3)$ **2.** 21.5
3. $\{x \mid x \neq -4\}$ **4.** $\{x \mid x < -1\}$

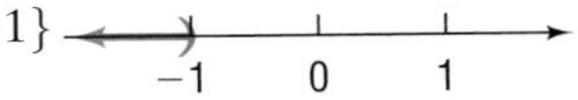

2.2 The Graph of a Function

PREPARING FOR THIS SECTION *Before getting started, review the following:*

- Graphs of Equations (Section 1.2, pp. 9–10)
- Intercepts (Section 1.2, pp. 11–12)

Now work the 'Are You Prepared?' problems on page 68.

OBJECTIVES
1 Identify the Graph of a Function
2 Obtain Information from or about the Graph of a Function

In applications, a graph often demonstrates more clearly the relationship between two variables than, say, an equation or table would. For example, Table 1 shows the price per share of IBM stock at the end of each month

Table 1

Date	Closing Price ($)
6/30/02	72.00
7/31/02	66.40
8/31/02	75.38
9/30/02	60.36
10/31/02	74.56
11/30/02	87.10
12/31/02	77.36
1/31/03	78.20
2/28/03	77.95
3/31/03	78.43
4/30/03	84.90
5/31/03	88.04
6/30/03	82.50

from 6/30/02 through 6/30/03. If we plot these data using the date as the x-coordinate and the price as the y-coordinate and then connect the points, we obtain Figure 9.

Figure 9
Monthly closing prices of IBM stock 6/30/02 through 6/30/03

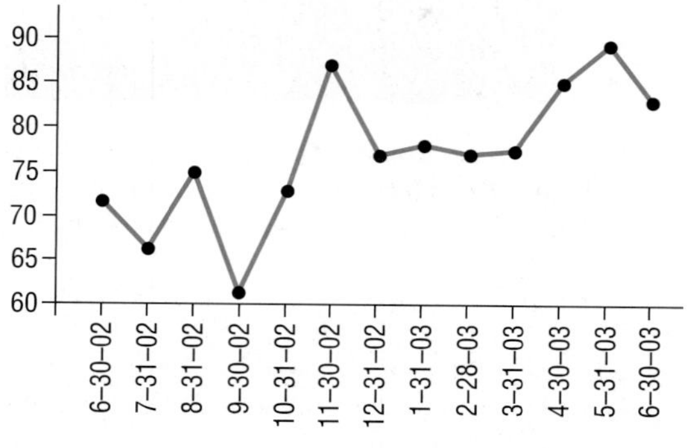

We can see from the graph that the price of the stock was rising rapidly from 9/30/02 through 11/30/02 and was falling from 8/31/02 through 9/30/02. The graph also shows that the lowest price occurred at the end of September 2002, whereas the highest occurred at the end of May 2003. Equations and tables, on the other hand, usually require some calculations and interpretation before this kind of information can be "seen."

Look again at Figure 9. The graph shows that for each date on the horizontal axis there is only one price on the vertical axis. The graph represents a function, although the exact rule for getting from date to price is not given.

When a function is defined by an equation in x and y, the **graph of the function** is the graph of the equation, that is, the set of points (x, y) in the xy-plane that satisfies the equation.

COMMENT: When we select a viewing rectangle to graph a function, the values of Xmin, Xmax give the domain that we wish to view, while Ymin, Ymax give the range that we wish to view. These settings usually do not represent the actual domain and range of the function. ■

1 Not every collection of points in the xy-plane represents the graph of a function. Remember, for a function, each number x in the domain has exactly one image y in the range. This means that the graph of a function cannot contain two points with the same x-coordinate and different y-coordinates. Therefore, the graph of a function must satisfy the following **vertical-line test.**

Theorem

Vertical-line Test

A set of points in the xy-plane is the graph of a function if and only if every vertical line intersects the graph in at most one point.

In other words, if any vertical line intersects a graph at more than one point, the graph is not the graph of a function.

EXAMPLE 1 Identifying the Graph of a Function

Which of the graphs in Figure 10 are graphs of functions?

Figure 10

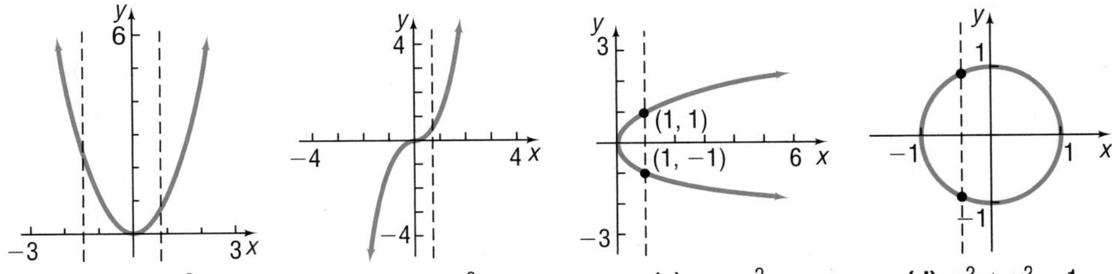

(a) $y = x^2$ **(b)** $y = x^3$ **(c)** $x = y^2$ **(d)** $x^2 + y^2 = 1$

Solution The graphs in Figures 10(a) and 10(b) are graphs of functions, because every vertical line intersects each graph in at most one point. The graphs in Figures 10(c) and 10(d) are not graphs of functions, because there is a vertical line that intersects each graph in more than one point. ◀

NOW WORK PROBLEM 15.

2 If (x, y) is a point on the graph of a function f, then y is the value of f at x; that is, $y = f(x)$. The next example illustrates how to obtain information about a function if its graph is given.

EXAMPLE 2 Obtaining Information from the Graph of a Function

Figure 11

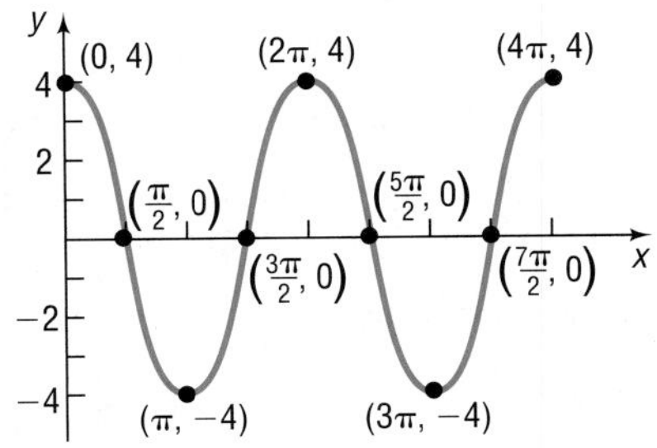

Let f be the function whose graph is given in Figure 11. (The graph of f might represent the distance that the bob of a pendulum is from its *at-rest* position. Negative values of y mean that the pendulum is to the left of the at-rest position, and positive values of y mean that the pendulum is to the right of the at-rest position.)

(a) What are $f(0)$, $f\left(\frac{3\pi}{2}\right)$, and $f(3\pi)$?

(b) What is the domain of f?

(c) What is the range of f?

(d) List the intercepts. (Recall that these are the points, if any, where the graph crosses or touches the coordinate axes.)

(e) How often does the line $y = 2$ intersect the graph?

(f) For what values of x does $f(x) = -4$?

(g) For what values of x is $f(x) > 0$?

Solution (a) Since $(0, 4)$ is on the graph of f, the y-coordinate 4 is the value of f at the x-coordinate 0; that is, $f(0) = 4$. In a similar way, we find that when $x = \frac{3\pi}{2}$ then $y = 0$, so $f\left(\frac{3\pi}{2}\right) = 0$. When $x = 3\pi$, then $y = -4$, so $f(3\pi) = -4$.

(b) To determine the domain of f, we notice that the points on the graph of f will have x-coordinates between 0 and 4π, inclusive; and for each number x between 0 and 4π, there is a point $(x, f(x))$ on the graph. The domain of f is $\{x|0 \le x \le 4\pi\}$ or the interval $[0, 4\pi]$.

(c) The points on the graph all have y-coordinates between -4 and 4, inclusive; and for each such number y, there is at least one number x in the domain. The range of f is $\{y|-4 \le y \le 4\}$ or the interval $[-4, 4]$.

(d) The intercepts are

$$(0, 4), \left(\frac{\pi}{2}, 0\right), \left(\frac{3\pi}{2}, 0\right), \left(\frac{5\pi}{2}, 0\right), \text{ and } \left(\frac{7\pi}{2}, 0\right).$$

(e) If we draw the horizontal line $y = 2$ on the graph in Figure 11, then we find that it intersects the graph four times.

(f) Since $(\pi, -4)$ and $(3\pi, -4)$ are the only points on the graph for which $y = f(x) = -4$, we have $f(x) = -4$ when $x = \pi$ and $x = 3\pi$.

(g) To determine where $f(x) > 0$, we look at Figure 11 and determine the x-values for which the y-coordinate is positive. This occurs on the intervals $\left[0, \frac{\pi}{2}\right), \left(\frac{3\pi}{2}, \frac{5\pi}{2}\right)$, and $\left(\frac{7\pi}{2}, 4\pi\right]$. Using inequality notation, $f(x) > 0$ for $0 \le x < \frac{\pi}{2}, \frac{3\pi}{2} < x < \frac{5\pi}{2}$, and $\frac{7\pi}{2} < x \le 4\pi$. ◀

When the graph of a function is given, its domain may be viewed as the shadow created by the graph on the x-axis by vertical beams of light. Its range can be viewed as the shadow created by the graph on the y-axis by horizontal beams of light. Try this technique with the graph given in Figure 11.

NOW WORK PROBLEMS 9 AND 13.

EXAMPLE 3 Obtaining Information about the Graph of a Function

Consider the function: $f(x) = \dfrac{x}{x + 2}$

(a) Is the point $\left(1, \frac{1}{2}\right)$ on the graph of f?

(b) If $x = -1$, what is $f(x)$? What point is on the graph of f?

(c) If $f(x) = 2$, what is x? What point is on the graph of f?

Solution (a) When $x = 1$, then

$$f(x) = \frac{x}{x + 2}$$

$$f(1) = \frac{1}{1 + 2} = \frac{1}{3}$$

The point $\left(1, \frac{1}{3}\right)$ is on the graph of f; the point $\left(1, \frac{1}{2}\right)$ is not.

(b) If $x = -1$, then

$$f(x) = \frac{x}{x + 2}$$

$$f(-1) = \frac{-1}{-1 + 2} = -1$$

The point $(-1, -1)$ is on the graph of f.

(c) If $f(x) = 2$, then

$$\begin{aligned} f(x) &= 2 \\ \frac{x}{x+2} &= 2 \\ x &= 2(x+2) && \text{Multiply both sides by } x + 2. \\ x &= 2x + 4 && \text{Remove parentheses.} \\ x &= -4 && \text{Solve for } x. \end{aligned}$$

If $f(x) = 2$, then $x = -4$. The point $(-4, 2)$ is on the graph of f. ◄

NOW WORK PROBLEM 25.

EXAMPLE 4 Average Cost Function

The average cost $\overline{C}$ of manufacturing x computers per day is given by the function

$$\overline{C}(x) = 0.56x^2 - 34.39x + 1212.57 + \frac{20{,}000}{x}$$

Determine the average cost of manufacturing:

(a) 30 computers in a day

(b) 40 computers in a day

(c) 50 computers in a day

(d) Graph the function $\overline{C} = \overline{C}(x), 0 < x \le 80$.

(e) Create a TABLE with TblStart = 1 and ΔTbl = 1.* Which value of x minimizes the average cost?

Solution (a) The average cost of manufacturing $x = 30$ computers is

$$\overline{C}(30) = 0.56(30)^2 - 34.39(30) + 1212.57 + \frac{20{,}000}{30} = \$1351.54$$

(b) The average cost of manufacturing $x = 40$ computers is

$$\overline{C}(40) = 0.56(40)^2 - 34.39(40) + 1212.57 + \frac{20{,}000}{40} = \$1232.97$$

(c) The average cost of manufacturing $x = 50$ computers is

$$\overline{C}(50) = 0.56(50)^2 - 34.39(50) + 1212.57 + \frac{20{,}000}{50} = \$1293.07$$

(d) See Figure 12 for the graph of $\overline{C} = \overline{C}(x)$.

Figure 12

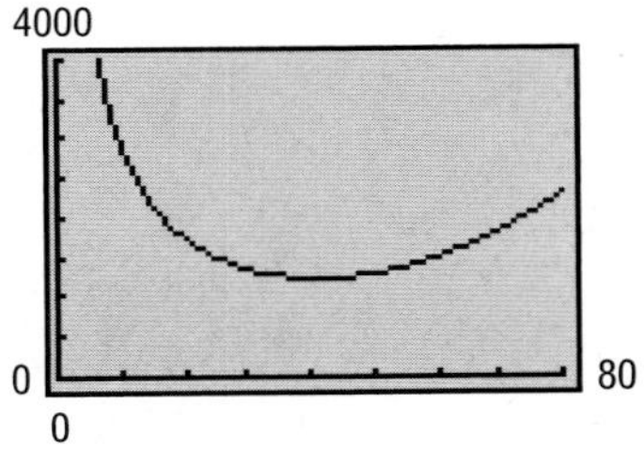

(e) With the function $\overline{C} = \overline{C}(x)$ in Y_1, we create Table 2. We scroll down until we find a value of x for which Y_1 is smallest. Table 3 shows that

*Check your owner's manual for the keystrokes to use.

manufacturing $x = 41$ computers minimizes the average cost at \$1231.74 per computer.

Table 2

X	Y_1
1	21179
2	11146
3	7781.1
4	6084
5	5054.6
6	4359.7
7	3856.4

Y_1▮.56X²−34.39X...

Table 3

X	Y_1
38	1240.7
39	1235.9
40	1233
41	1231.7
42	1232.2
43	1234.4
44	1238.1

Y_1=1231.74487805

◀

NOW WORK PROBLEM 29.

Summary

Graph of a function The collection of points (x, y) that satisfies the equation $y = f(x)$.
A collection of points is the graph of a function provided that every vertical line intersects the graph in at most one point (vertical-line test).

2.2 Assess Your Understanding

'Are You Prepared?' *Answers are given at the end of these exercises. If you get the wrong answer, read the pages listed in red.*

1. The intercepts of the equation $x^2 + 4y^2 = 16$ are ________. (pp. 11–12)

2. *True or False:* The point $(-2, 3)$ is on the graph of the equation $x = 2y - 2$. (pp. 9–10)

Concepts and Vocabulary

3. A set of points in the xy-plane is the graph of a function if and only if every ________ line intersects the graph in at most one point.

4. If the point $(5, -3)$ is a point on the graph of f, then $f(\_\_\_\_\_\_) = \_\_\_\_\_\_$.

5. Find a so that the point $(-1, 2)$ is on the graph of $f(x) = ax^2 + 4$.

6. *True or False:* A function can have more than one y-intercept.

7. *True or False:* The graph of a function $y = f(x)$ always crosses the y-axis.

8. *True or False:* The y-intercept of the graph of the function $y = f(x)$, whose domain is all real numbers, is $f(0)$.

Exercises

9. Use the graph of the function f given below to answer parts (a)–(n).

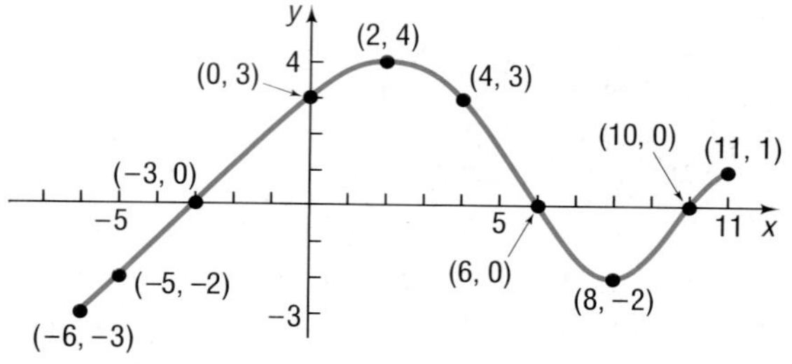

(a) Find $f(0)$ and $f(-6)$.
(b) Find $f(6)$ and $f(11)$.
(c) Is $f(3)$ positive or negative?
(d) Is $f(-4)$ positive or negative?
(e) For what numbers x is $f(x) = 0$?
(f) For what numbers x is $f(x) > 0$?
(g) What is the domain of f?
(h) What is the range of f?
(i) What are the x-intercepts?
(j) What is the y-intercept?
(k) How often does the line $y = \frac{1}{2}$ intersect the graph?
(l) How often does the line $x = 5$ intersect the graph?
(m) For what values of x does $f(x) = 3$?
(n) For what values of x does $f(x) = -2$?

10. Use the graph of the function f given below to answer parts (a)–(n).

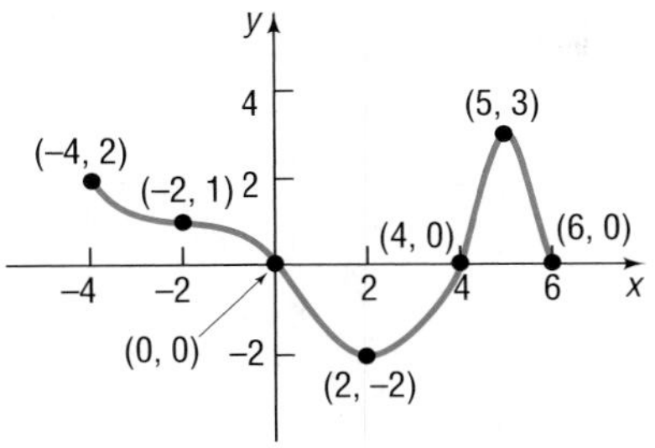

(a) Find $f(0)$ and $f(6)$.
(b) Find $f(2)$ and $f(-2)$.
(c) Is $f(3)$ positive or negative?
(d) Is $f(-1)$ positive or negative?
(e) For what numbers is $f(x) = 0$?
(f) For what numbers is $f(x) < 0$?
(g) What is the domain of f?
(h) What is the range of f?
(i) What are the x-intercepts?
(j) What is the y-intercept?
(k) How often does the line $y = -1$ intersect the graph?
(l) How often does the line $x = 1$ intersect the graph?
(m) For what value of x does $f(x) = 3$?
(n) For what value of x does $f(x) = -2$?

In Problems 11–22, determine whether the graph is that of a function by using the vertical-line test. If it is, use the graph to find:
(a) *Its domain and range*
(b) *The intercepts, if any*
(c) *Any symmetry with respect to the x-axis, the y-axis, or the origin*

11.

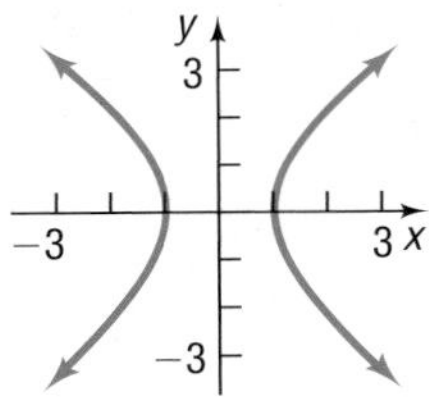

12.

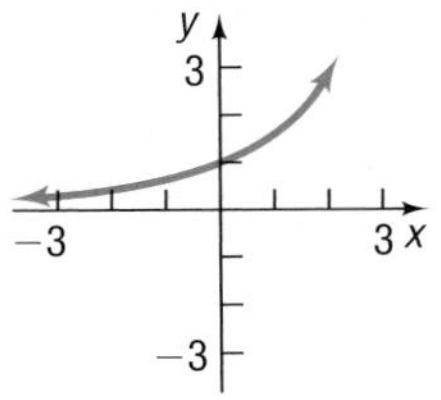

13.

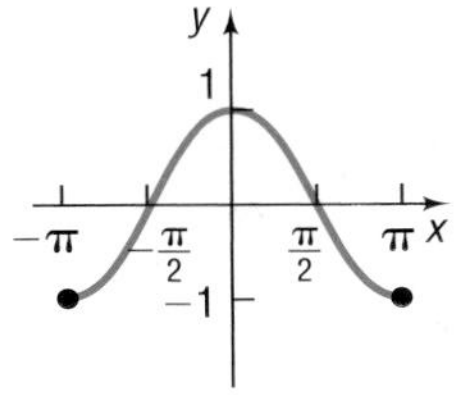

14.

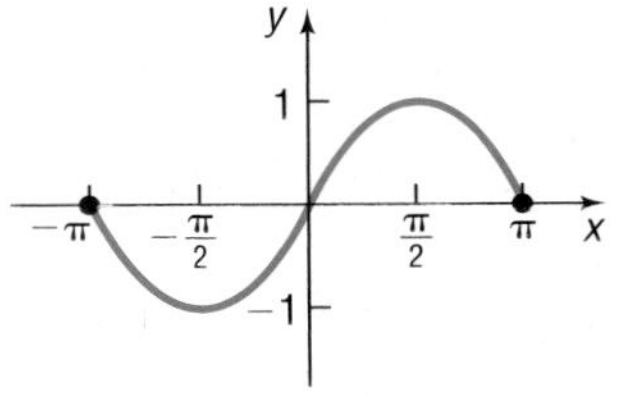

15.

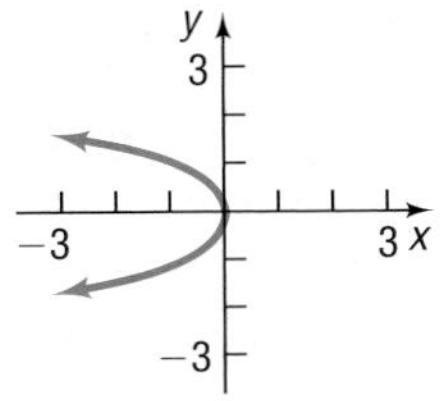

16.

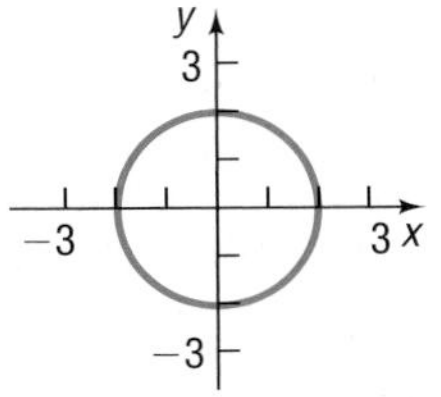

17.

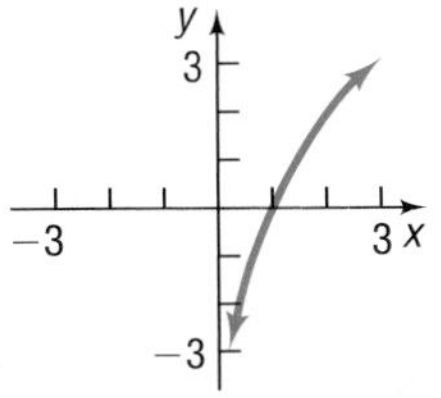

18.

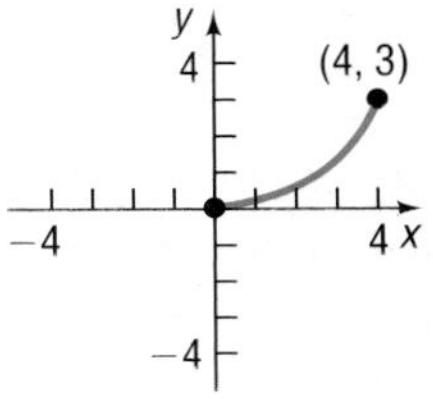

19.

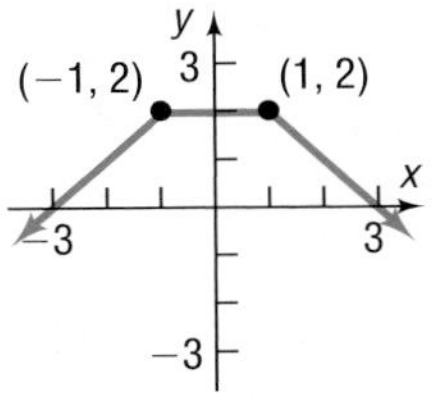

20.

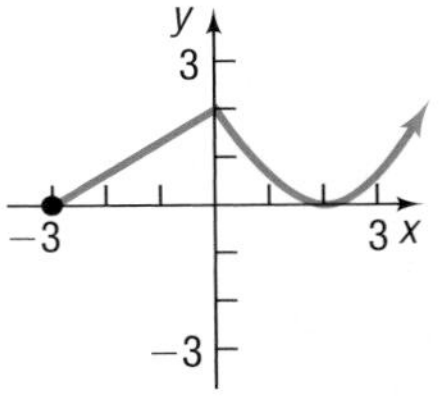

21.

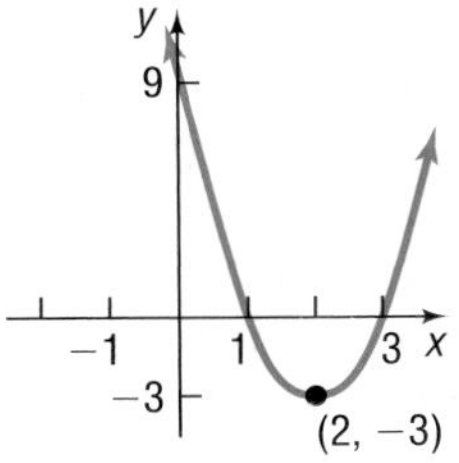

22.

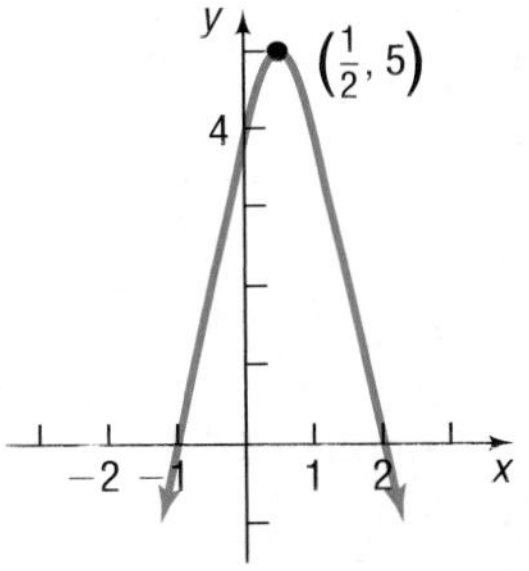

In Problems 23–28, answer the questions about the given function.

23. $f(x) = 2x^2 - x - 1$
(a) Is the point $(-1, 2)$ on the graph of f?
(b) If $x = -2$, what is $f(x)$? What point is on the graph of f?
(c) If $f(x) = -1$, what is x? What point(s) are on the graph of f?
(d) What is the domain of f?
(e) List the x-intercepts, if any, of the graph of f.
(f) List the y-intercept, if there is one, of the graph of f.

24. $f(x) = -3x^2 + 5x$
(a) Is the point $(-1, 2)$ on the graph of f?
(b) If $x = -2$, what is $f(x)$? What point is on the graph of f?
(c) If $f(x) = -2$, what is x? What point(s) are on the graph of f?
(d) What is the domain of f?
(e) List the x-intercepts, if any, of the graph of f.
(f) List the y-intercept, if there is one, of the graph of f.

25. $f(x) = \frac{x+2}{x-6}$
   (a) Is the point (3, 14) on the graph of f?
   (b) If $x = 4$, what is $f(x)$? What point is on the graph of f?
   (c) If $f(x) = 2$, what is x? What point(s) are on the graph of f?
   (d) What is the domain of f?
   (e) List the x-intercepts, if any, of the graph of f.
   (f) List the y-intercept, if there is one, of the graph of f.

26. $f(x) = \frac{x^2+2}{x+4}$
   (a) Is the point $\left(1, \frac{3}{5}\right)$ on the graph of f?
   (b) If $x = 0$, what is $f(x)$? What point is on the graph of f?
   (c) If $f(x) = \frac{1}{2}$, what is x? What point(s) are on the graph of f?
   (d) What is the domain of f?
   (e) List the x-intercepts, if any, of the graph of f.
   (f) List the y-intercept, if there is one, of the graph of f.

27. $f(x) = \frac{2x^2}{x^4+1}$
   (a) Is the point $(-1, 1)$ on the graph of f?
   (b) If $x = 2$, what is $f(x)$? What point is on the graph of f?
   (c) If $f(x) = 1$, what is x? What point(s) are on the graph of f?
   (d) What is the domain of f?
   (e) List the x-intercepts, if any, of the graph of f.
   (f) List the y-intercept, if there is one, of the graph of f.

28. $f(x) = \frac{2x}{x-2}$
   (a) Is the point $\left(\frac{1}{2}, -\frac{2}{3}\right)$ on the graph of f?
   (b) If $x = 4$, what is $f(x)$? What point is on the graph of f?
   (c) If $f(x) = 1$, what is x? What point(s) are on the graph of f?
   (d) What is the domain of f?
   (e) List the x-intercepts, if any, of the graph of f.
   (f) List the y-intercept, if there is one, of the graph of f.

29. **Motion of a Golf Ball** A golf ball is hit with an initial velocity of 130 feet per second at an inclination of 45° to the horizontal. In physics, it is established that the height h of the golf ball is given by the function

$$h(x) = \frac{-32x^2}{130^2} + x$$

where x is the horizontal distance that the golf ball has traveled.

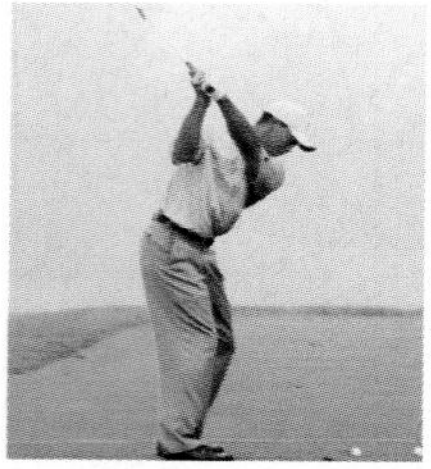

   (a) Determine the height of the golf ball after it has traveled 100 feet.
   (b) What is the height after it has traveled 300 feet?
   (c) What is the height after it has traveled 500 feet?
   (d) How far was the golf ball hit?
   (e) Graph the function $h = h(x)$.
   (f) Use a graphing utility to determine the distance that the ball has traveled when the height of the ball is 90 feet.
   (g) Create a TABLE with TblStart = 0 and ΔTbl = 25. To the nearest 25 feet, how far does the ball travel before it reaches a maximum height? What is the maximum height?
   (h) Adjust the value of ΔTbl until you determine the distance, to within 1 foot, that the ball travels before it reaches a maximum height.

30. **Effect of Elevation on Weight** If an object weighs m pounds at sea level, then its weight W (in pounds) at a height of h miles above sea level is given approximately by

$$W(h) = m\left(\frac{4000}{4000 + h}\right)^2$$

   (a) If Amy weighs 120 pounds at sea level, how much will she weigh on Pike's Peak, which is 14,110 feet above sea level?
   (b) Use a graphing utility to graph the function $W = W(h)$. Use $m = 120$ pounds.
   (c) Create a Table with TblStart = 0 and ΔTbl = 0.5 to see how the weight W varies as h changes from 0 to 5 miles.
   (d) At what height will Amy weigh 119.95 pounds?
   (e) Does your answer to part (d) seem reasonable?

31. Match each function with the graph that best describes the situation. Discuss the reason for your choice.
   (a) The cost of building a house as a function of its square footage
   (b) The height of an egg dropped from a 300-foot building as a function of time
   (c) The height of a human as a function of time
   (d) The demand for Big Macs as a function of price
   (e) The height of a child on a swing as a function of time

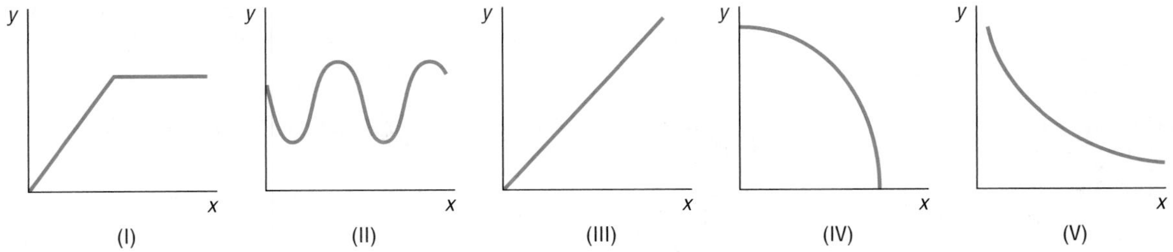

32. Match each function with the graph that best describes the situation. Discuss the reason for your choice.

(a) The temperature of a bowl of soup as a function of time
(b) The number of hours of daylight per day over a 2-year period
(c) The population of Florida as a function of time
(d) The distance of a car traveling at a constant velocity as a function of time
(e) The height of a golf ball hit with a 7-iron as a function of time

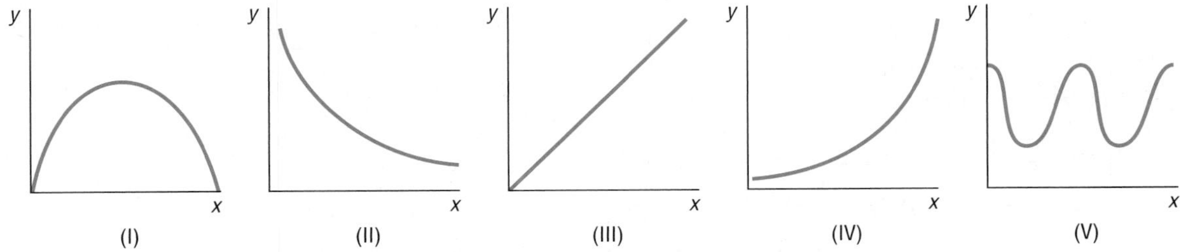

33. Consider the following scenario: Barbara decides to take a walk. She leaves home, walks 2 blocks in 5 minutes at a constant speed, and realizes that she forgot to lock the door. So Barbara runs home in 1 minute. While at her doorstep, it takes her 1 minute to find her keys and lock the door. Barbara walks 5 blocks in 15 minutes and then decides to jog home. It takes her 7 minutes to get home. Draw a graph of Barbara's distance from home (in blocks) as a function of time.

34. Consider the following scenario: Jayne enjoys riding her bicycle through the woods. At the forest preserve, she gets on her bicycle and rides up a 2000-foot incline in 10 minutes. She then travels down the incline in 3 minutes. The next 5000 feet is level terrain and she covers the distance in 20 minutes. She rests for 15 minutes. Jayne then travels 10,000 feet in 30 minutes. Draw a graph of Jayne's distance traveled (in feet) as a function of time.

35. The following sketch represents the distance d (in miles) that Kevin is from home as a function of time t (in hours). Answer the questions based on the graph. In parts (a)–(g), how many hours elapsed and how far was Kevin from home during this time?

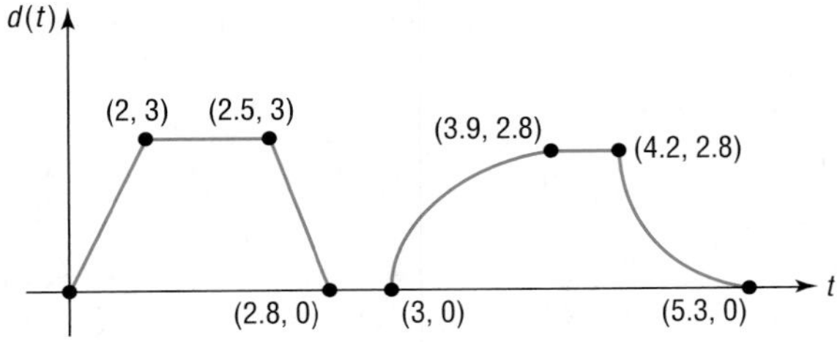

(a) From $t = 0$ to $t = 2$
(b) From $t = 2$ to $t = 2.5$
(c) From $t = 2.5$ to $t = 2.8$
(d) From $t = 2.8$ to $t = 3$
(e) From $t = 3$ to $t = 3.9$
(f) From $t = 3.9$ to $t = 4.2$
(g) From $t = 4.2$ to $t = 5.3$
(h) What is the farthest distance that Kevin is from home?
(i) How many times did Kevin return home?

36. The following sketch represents the speed v (in miles per hour) of Michael's car as a function of time t (in minutes).

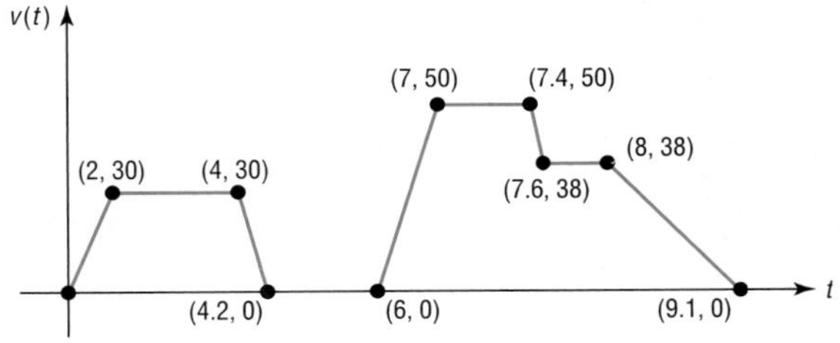

(a) Over what interval of time is Michael traveling fastest?
(b) Over what interval(s) of time is Michael's speed zero?
(c) What is Michael's speed between 0 and 2 minutes?
(d) What is Michael's speed between 4.2 and 6 minutes?
(e) What is Michael's speed between 7 and 7.4 minutes?
(f) When is Michael's speed constant?

37. Draw the graph of a function whose domain is $\{x \mid -3 \le x \le 8, \ x \ne 5\}$ and whose range is $\{y \mid -1 \le y \le 2, \ y \ne 0\}$. What point(s) in the rectangle $-3 \le x \le 8$, $-1 \le y \le 2$ cannot be on the graph? Compare your graph with those of other students. What differences do you see?

38. Describe how you would proceed to find the domain and range of a function if you were given its graph. How would your strategy change if, instead, you were given the equation defining the function?

39. How many x-intercepts can the graph of a function have? How many y-intercepts can it have?

40. Is a graph that consists of a single point the graph of a function? If so, can you write the equation of such a function?

41. Is there a function whose graph is symmetric with respect to the x-axis? Explain.

'Are You Prepared?' Answers

1. $(-4, 0), (4, 0), (0, -2), (0, 2)$

2. False

2.3 Properties of Functions

PREPARING FOR THIS SECTION *Before getting started, review the following:*

- Intervals (Appendix A, Section A.8, pp. 968–969)
- Slope of a Line (Section 1.3, pp. 23–26)
- Point–slope Form of a Line (Section 1.3, p. 27)
- Tests for Symmetry of an Equation (Section 1.2, pp. 12–14)
- Intercepts (Section 1.2, pp. 11–12)

Now work the 'Are You Prepared?' problems on page 80.

OBJECTIVES
1. Determine Even and Odd Functions from a Graph
2. Identify Even and Odd Functions from the Equation
3. Use a Graph to Determine Where a Function Is Increasing, Is Decreasing, or Is Constant
4. Use a Graph to Locate Local Maxima and Minima
5. Use a Graphing Utility to Approximate Local Maxima and Minima and to Determine Where a Function Is Increasing or Decreasing
6. Find the Average Rate of Change of a Function

It is easiest to obtain the graph of a function $y = f(x)$ by knowing certain properties that the function has and the impact of these properties on the way that the graph will look. In this section we describe some properties of functions that we will use in subsequent chapters.

We begin with the familiar notions of intercepts and symmetry.

Intercepts

If $x = 0$ is in the domain of a function $y = f(x)$, then the y-intercept of the graph of f is the value of f at 0, which is $f(0)$. The x-intercepts of the graph of f, if there are any, are the solutions of the equation $f(x) = 0$.

The x-intercepts of the graph of a function f are called the **zeros of f**.

Even and Odd Functions

1 The words *even* and *odd*, when applied to a function f, describe the symmetry that exists for the graph of the function.

A function f is even if and only if, whenever the point (x, y) is on the graph of f, then the point $(-x, y)$ is also on the graph. Using function notation, we define an even function as follows:

A function f is **even** if, for every number x in its domain, the number $-x$ is also in the domain and

$$f(-x) = f(x)$$

A function f is odd if and only if, whenever the point (x, y) is on the graph of f, then the point $(-x, -y)$ is also on the graph. Using function notation, we define an odd function as follows:

A function f is **odd** if, for every number x in its domain, the number $-x$ is also in the domain and

$$f(-x) = -f(x)$$

Refer to Section 1.2, where the tests for symmetry are listed. The following results are then evident.

Theorem A function is even if and only if its graph is symmetric with respect to the y-axis. A function is odd if and only if its graph is symmetric with respect to the origin.

EXAMPLE 1 Determining Even and Odd Functions from the Graph

Determine whether each graph given in Figure 13 is the graph of an even function, an odd function, or a function that is neither even nor odd.

Figure 13

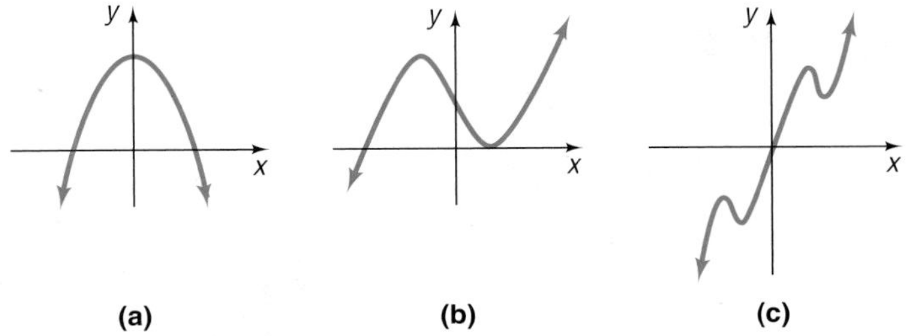

Solution The graph in Figure 13(a) is that of an even function, because the graph is symmetric with respect to the y-axis. The function whose graph is given in Figure 13(b) is neither even nor odd, because the graph is neither symmetric with respect to the y-axis nor symmetric with respect to the origin. The function whose graph is given in Figure 13(c) is odd, because its graph is symmetric with respect to the origin. ◀

NOW WORK PROBLEMS 21(a), (b), and (d).

2 In the next example, we use algebraic techniques to verify whether a given function is even, odd, or neither.

EXAMPLE 2 Identifying Even and Odd Functions Algebraically

Determine whether each of the following functions is even, odd, or neither. Then determine whether the graph is symmetric with respect to the y-axis or with respect to the origin.

(a) $f(x) = x^2 - 5$ (b) $g(x) = x^3 - 1$

(c) $h(x) = 5x^3 - x$ (d) $F(x) = |x|$

Solution (a) To determine whether f is even, odd, or neither, we replace x by $-x$ in $f(x) = x^2 - 5$. Then

$$f(-x) = (-x)^2 - 5 = x^2 - 5 = f(x)$$

Since $f(-x) = f(x)$, we conclude that f is an even function, and the graph is symmetric with respect to the y-axis.

(b) We replace x by $-x$ in $g(x) = x^3 - 1$. Then

$$g(-x) = (-x)^3 - 1 = -x^3 - 1$$

Since $g(-x) \neq g(x)$ and $g(-x) \neq -g(x) = -(x^3 - 1) = -x^3 + 1$, we conclude that g is neither even nor odd. The graph is not symmetric with respect to the y-axis nor is it symmetric with respect to the origin.

(c) We replace x by $-x$ in $h(x) = 5x^3 - x$. Then

$$h(-x) = 5(-x)^3 - (-x) = -5x^3 + x = -(5x^3 - x) = -h(x)$$

Since $h(-x) = -h(x)$, h is an odd function, and the graph of h is symmetric with respect to the origin.

(d) We replace x by $-x$ in $F(x) = |x|$. Then

$$F(-x) = |-x| = |-1| \cdot |x| = |x| = F(x)$$

Since $F(-x) = F(x)$, F is an even function, and the graph of F is symmetric with respect to the y-axis. ◀

NOW WORK PROBLEM 47.

Increasing and Decreasing Functions

3 Consider the graph given in Figure 14. If you look from left to right along the graph of the function, you will notice that parts of the graph are rising, parts are falling, and parts are horizontal. In such cases, the function is described as *increasing*, *decreasing*, or *constant*, respectively.

Figure 14

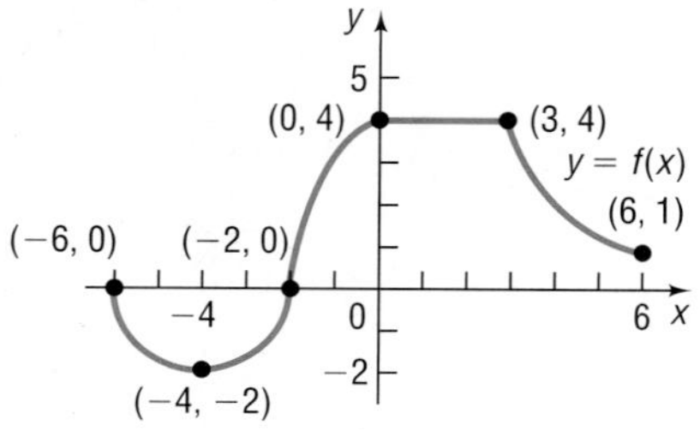

EXAMPLE 3 **Determining Where a Function Is Increasing, Decreasing, or Constant from Its Graph**

Where is the function in Figure 14 increasing? Where is it decreasing? Where is it constant?

Solution To answer the question of where a function is increasing, where it is decreasing, and where it is constant, we use strict inequalities involving the independent variable x, or we use open intervals* of x-coordinates. The graph in Figure 14 is rising (increasing) from the point $(-4, -2)$ to the point $(0, 4)$, so we conclude that it is increasing on the open interval $(-4, 0)$ or for $-4 < x < 0$. The graph is falling (decreasing) from the point $(-6, 0)$ to the point $(-4, -2)$ and from the point $(3, 4)$ to the point $(6, 1)$. We conclude that the graph is decreasing on the open intervals $(-6, -4)$ and $(3, 6)$ or for $-6 < x < -4$ and $3 < x < 6$. The graph is constant on the open interval $(0, 3)$ or for $0 < x < 3$. ◀

More precise definitions follow:

A function f is **increasing** on an open interval I if, for any choice of x_1 and x_2 in I, with $x_1 < x_2$, we have $f(x_1) < f(x_2)$.

A function f is **decreasing** on an open interval I if, for any choice of x_1 and x_2 in I, with $x_1 < x_2$, we have $f(x_1) > f(x_2)$.

A function f is **constant** on an interval I if, for all choices of x in I, the values $f(x)$ are equal.

Figure 15 illustrates the definitions. The graph of an increasing function goes up from left to right, the graph of a decreasing function goes down from left to right, and the graph of a constant function remains at a fixed height.

Figure 15

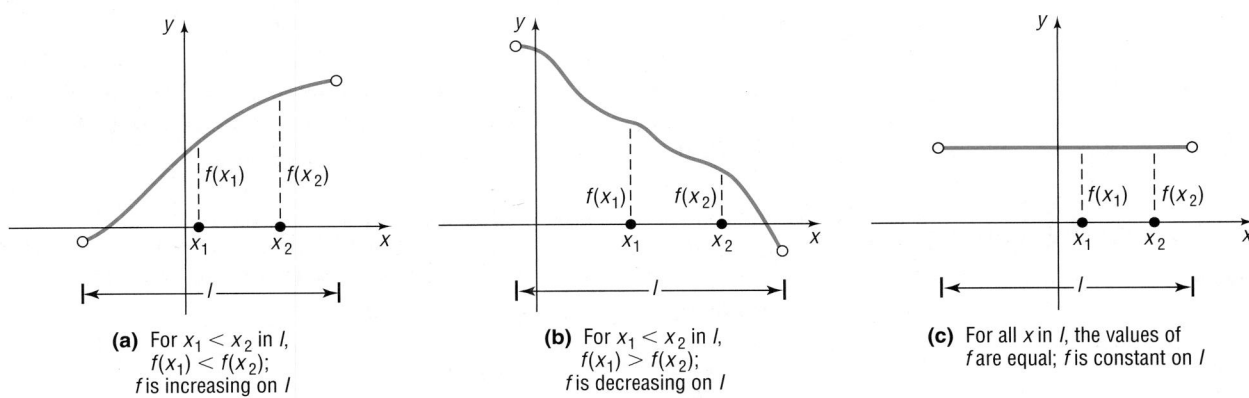

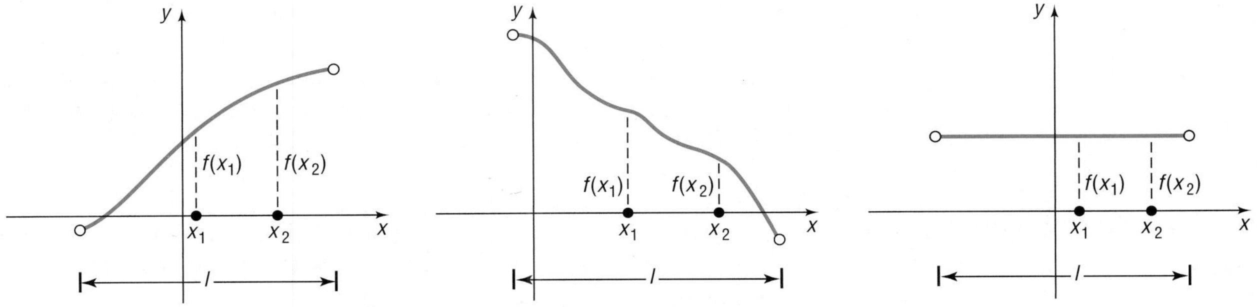

(a) For $x_1 < x_2$ in I, $f(x_1) < f(x_2)$; f is increasing on I

(b) For $x_1 < x_2$ in I, $f(x_1) > f(x_2)$; f is decreasing on I

(c) For all x in I, the values of f are equal; f is constant on I

NOW WORK PROBLEMS **11, 13, 15, AND 21(c).**

*The open interval (a, b) consists of all real numbers x for which $a < x < b$. Refer to Appendix A, Section A.8, if necessary.

Local Maximum; Local Minimum

4 When the graph of a function is increasing to the left of $x = c$ and decreasing to the right of $x = c$, then at c the value of f is largest. This value is called a *local maximum* of f. See Figure 16(a).

Figure 16

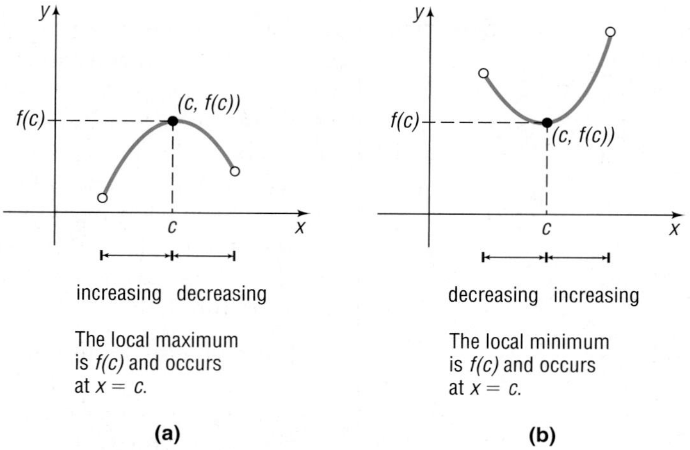

When the graph of a function is decreasing to the left of $x = c$ and is increasing to the right of $x = c$, then at c the value of f is the smallest. This value is called a *local minimum* of f. See Figure 16(b).

A function f has a **local maximum at** c if there is an open interval I containing c so that, for all $x \neq c$ in I, $f(x) < f(c)$. We call $f(c)$ a **local maximum of *f*.**

A function f has a **local minimum at** c if there is an open interval I containing c so that, for all $x \neq c$ in I, $f(x) > f(c)$. We call $f(c)$ a **local minimum of *f*.**

If f has a local maximum at c, then the value of f at c is greater than the values of f near c. If f has a local minimum at c, then the value of f at c is less than the values of f near c. The word *local* is used to suggest that it is only near c that the value $f(c)$ is largest or smallest.

EXAMPLE 4

Finding Local Maxima and Local Minima from the Graph of a Function and Determining Where the Function Is Increasing, Decreasing, or Constant

Figure 17

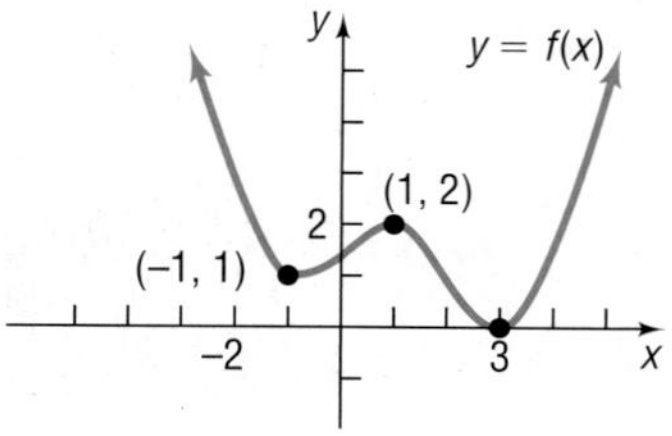

Figure 17 shows the graph of a function f.

(a) At what number(s), if any, does f have a local maximum?
(b) What are the local maxima?
(c) At what number(s), if any, does f have a local minimum?
(d) What are the local minima?

(e) List the intervals on which f is increasing. List the intervals on which f is decreasing.

Solution The domain of f is the set of real numbers.

(a) f has a local maximum at 1, since for all x close to 1, $x \neq 1$, we have $f(x) < f(1)$.

(b) The local maximum is $f(1) = 2$.

(c) f has a local minimum at -1 and at 3.

(d) The local minima are $f(-1) = 1$ and $f(3) = 0$.

(e) The function whose graph is given in Figure 17 is increasing on the interval $(-1, 1)$. The function is also increasing for all values of x greater than 3. That is, the function is increasing on the intervals $(-1, 1)$ and $(3, \infty)$ or for $-1 < x < 1$ and $x > 3$. The function is decreasing for all values of x less than -1. The function is also decreasing on the interval $(1, 3)$. That is, the function is decreasing on the intervals $(-\infty, -1)$ and $(1, 3)$ or for $x < -1$ and $1 < x < 3$. ◀

NOW WORK PROBLEMS 17 AND 19.

5 To locate the exact value at which a function f has a local maximum or a local minimum usually requires calculus. However, a graphing utility may be used to approximate these values by using the MAXIMUM and MINIMUM features.*

EXAMPLE 5 Using a Graphing Utility to Approximate Local Maxima and Minima and to Determine Where a Function Is Increasing or Decreasing

(a) Use a graphing utility to graph $f(x) = 6x^3 - 12x + 5$ for $-2 < x < 2$. Approximate where f has a local maximum and where f has a local minimum.

(b) Determine where f is increasing and where it is decreasing.

Solution (a) Graphing utilities have a feature that finds the maximum or minimum point of a graph within a given interval. Graph the function f for $-2 < x < 2$. Using MAXIMUM, we find that the local maximum is 11.53 and it occurs at $x = -0.82$, rounded to two decimal places. See Figure 18(a). Using MINIMUM, we find that the local minimum is -1.53 and it occurs at $x = 0.82$, rounded to two decimal places. See Figure 18(b).

Figure 18

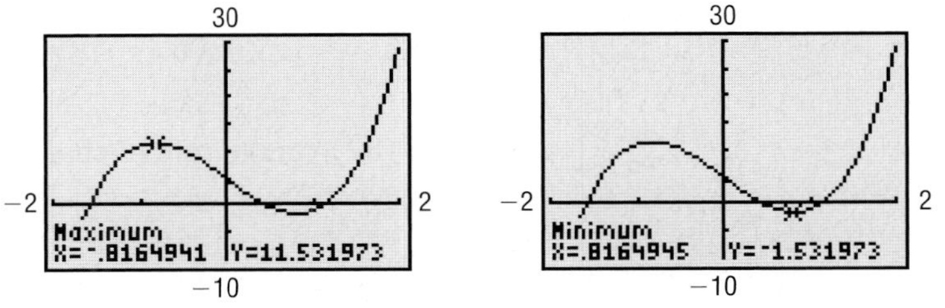

(a) (b)

*Consult your owner's manual for the appropriate keystrokes.

(b) Looking at Figures 18(a) and (b), we see that the graph of f is rising (increasing) from $x = -2$ to $x = -0.82$ and from $x = 0.82$ to $x = 2$, so f is increasing on the intervals $(-2, -0.82)$ and $(0.82, 2)$ or for $-2 < x < -0.82$ and $0.82 < x < 2$. The graph is falling (decreasing) from $x = -0.82$ to $x = 0.82$, so f is decreasing on the interval $(-0.82, 0.82)$ or for $-0.82 < x < 0.82$. ◀

NOW WORK PROBLEM 59.

Average Rate of Change

6 In Section 1.3 we said that the slope of a line could be interpreted as the average rate of change. Often we are interested in the rate at which functions change. To find the average rate of change of a function between any two points on its graph, we calculate the slope of the line containing the two points.

If c is in the domain of a function $y = f(x)$, the **average rate of change of f** from c to x is defined as

$$\text{Average rate of change} = \frac{\Delta y}{\Delta x} = \frac{f(x) - f(c)}{x - c}, \qquad x \neq c \qquad \textbf{(1)}$$

In calculus, this expression is called the **difference quotient** of f at c.

Recall that the symbol Δy in (1) is the "change in y" and Δx is the "change in x." The average rate of change of f is the change in y divided by the change in x.

EXAMPLE 6 Finding the Average Rate of Change

Find the average rate of change of $f(x) = 3x^2$:

(a) From 1 to 3 (b) From 1 to 5 (c) From 1 to 7

Solution (a) The average rate of change of $f(x) = 3x^2$ from 1 to 3 is

$$\frac{\Delta y}{\Delta x} = \frac{f(3) - f(1)}{3 - 1} = \frac{27 - 3}{3 - 1} = \frac{24}{2} = 12$$

(b) The average rate of change of $f(x) = 3x^2$ from 1 to 5 is

$$\frac{\Delta y}{\Delta x} = \frac{f(5) - f(1)}{5 - 1} = \frac{75 - 3}{5 - 1} = \frac{72}{4} = 18$$

(c) The average rate of change of $f(x) = 3x^2$ from 1 to 7 is

$$\frac{\Delta y}{\Delta x} = \frac{f(7) - f(1)}{7 - 1} = \frac{147 - 3}{7 - 1} = \frac{144}{6} = 24$$ ◀

The average rate of change of a function has an important geometric interpretation. Look at the graph of $y = f(x)$ in Figure 19. We have labeled

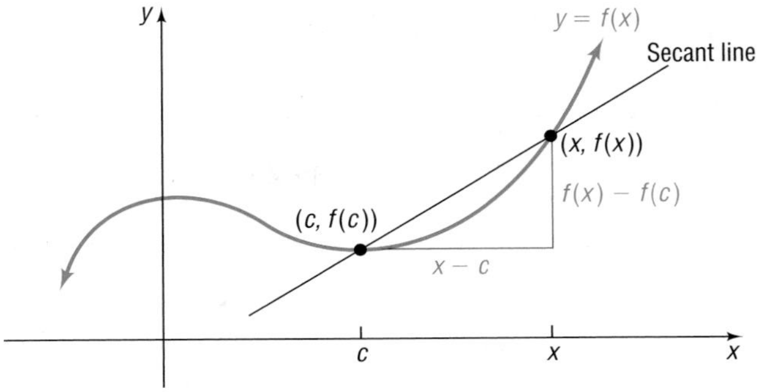

Figure 19

two points on the graph: $(c, f(c))$ and $(x, f(x))$. The line containing these two points is called a **secant line;** its slope is

$$m_{\text{sec}} = \frac{f(x) - f(c)}{x - c}$$

Theorem

Slope of the Secant Line

The average rate of change of a function equals the slope of the secant line containing two points on its graph.

EXAMPLE 7 Finding the Average Rate of Change of a Function

(a) Find the average rate of change of $f(x) = 2x^2 - 3x$ from 1 to x.

(b) Use this result to find the slope of the secant line containing $(1, f(1))$ and $(2, f(2))$.

(c) Find an equation of this secant line.

Solution (a) The average rate of change of f from 1 to x is

$$\begin{aligned}
\frac{\Delta y}{\Delta x} &= \frac{f(x) - f(1)}{x - 1} && x \neq 1 \\
&= \frac{(2x^2 - 3x) - (-1)}{x - 1} && f(x) = 2x^2 - 3x;\ f(1) = 2 \cdot 1^2 - 3(1) = -1 \\
&= \frac{2x^2 - 3x + 1}{x - 1} && \text{Simplify.} \\
&= \frac{(2x - 1)(x - 1)}{x - 1} && \text{Factor the numerator.} \\
&= 2x - 1 && x \neq 1;\ \text{cancel } x - 1
\end{aligned}$$

(b) The slope of the secant line containing $(1, f(1))$ and $(2, f(2))$ is the average rate of change of f from 1 to 2. Using $x = 2$ in part (a), we obtain $m_{\text{sec}} = 2(2) - 1 = 3$.

(c) Use the point–slope form to find the equation of the secant line.

$$\begin{aligned}
y - y_1 &= m_{\text{sec}}(x - x_1) && \text{Point–slope form of the secant line} \\
y + 1 &= 3(x - 1) && x_1 = 1,\ y_1 = f(1) = -1;\ m_{\text{sec}} = 3 \\
y + 1 &= 3x - 3 \\
y &= 3x - 4 && \text{Slope–intercept form of the secant line}
\end{aligned}$$

◀

NOW WORK PROBLEM 39.

2.3 Assess Your Understanding

'Are You Prepared?' *Answers are given at the end of these exercises. If you get the wrong answer, read the pages listed in red.*

1. The interval $[2, 5]$ can be written as the inequality __________. (pp. 968–969)

2. The slope of the line containing the points $(-2, 3)$ and $(3, 8)$ is __________. (pp. 23–26)

3. Test the equation $y = 2x^2 + 3$ for symmetry with respect to the x-axis, the y-axis, and the origin. (pp. 12–14)

4. Write the point–slope form of the line with slope 5 containing the point $(3, -2)$. (p. 27)

5. The intercepts of the equation $y = x^2 - 9$ are __________. (pp. 11–12)

Concepts and Vocabulary

6. A function f is __________ on an open interval I if for any choice of x_1 and x_2 in I, with $x_1 < x_2$, we have $f(x_1) < f(x_2)$.

7. A(n) __________ function f is one for which $f(-x) = f(x)$ for every x in the domain of f; a(n) __________ function f is one for which $f(-x) = -f(x)$ for every x in the domain of f.

8. *True or False:* A function f is decreasing on an open interval I if, for any choice of x_1 and x_2 in I, with $x_1 < x_2$, we have $f(x_1) > f(x_2)$.

9. *True or False:* A function f has a local maximum at c if there is an open interval I containing c so that, for all $x \neq c$ in I, $f(x) < f(c)$.

10. *True or False:* Even functions have graphs that are symmetric with respect to the origin.

Exercises

In Problems 11–20, use the graph of the function f given below.

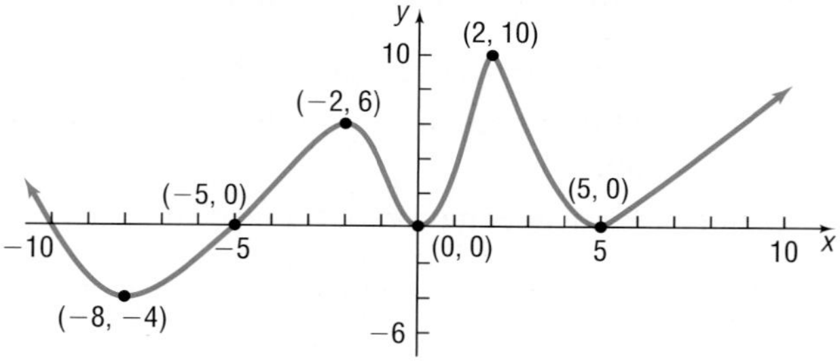

11. Is f increasing on the interval $(-8, -2)$?

12. Is f decreasing on the interval $(-8, -4)$?

13. Is f increasing on the interval $(2, 10)$?

14. Is f decreasing on the interval $(2, 5)$?

15. List the interval(s) on which f is increasing.

16. List the interval(s) on which f is decreasing.

17. Is there a local maximum at 2? If yes, what is it?

18. Is there a local maximum at 5? If yes, what is it?

19. List the numbers at which f has a local maximum. What are these local maxima?

20. List the numbers at which f has a local minimum. What are these local minima?

In Problems 21–28, the graph of a function is given. Use the graph to find:

(a) *The intercepts, if any*
(b) *Its domain and range*
(c) *The intervals on which it is increasing, decreasing, or constant*
(d) *Whether it is even, odd, or neither*

21.

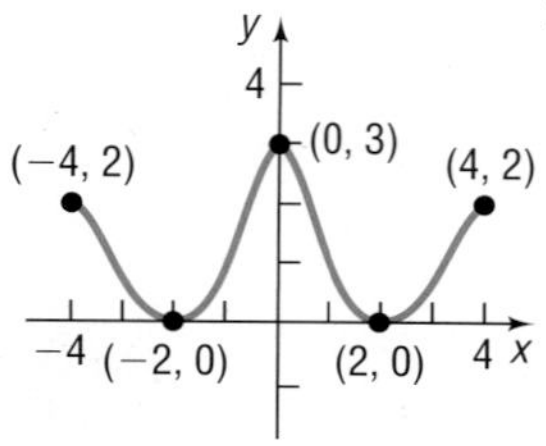

22.

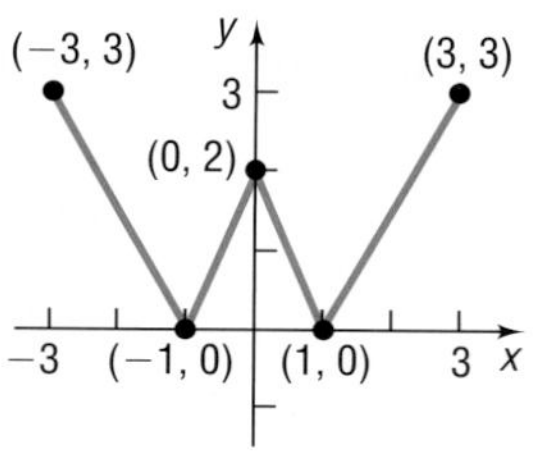

23.

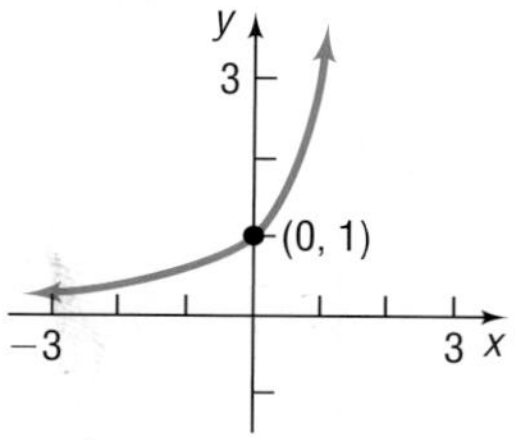

24.

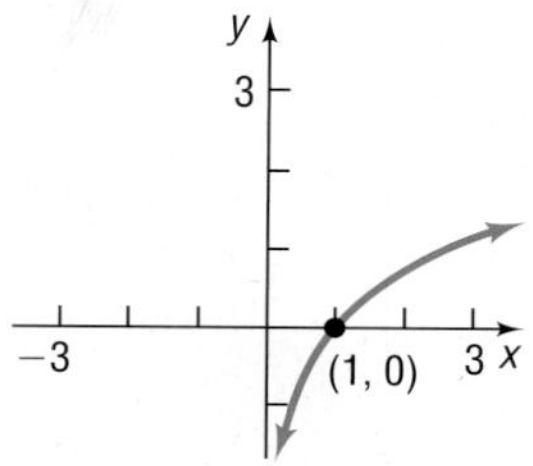

25.

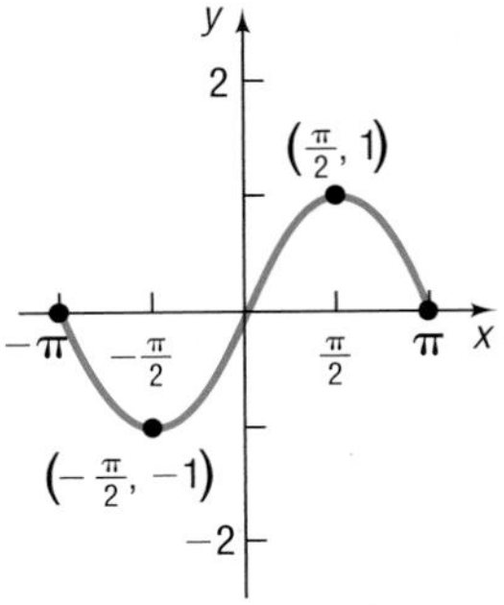

26.

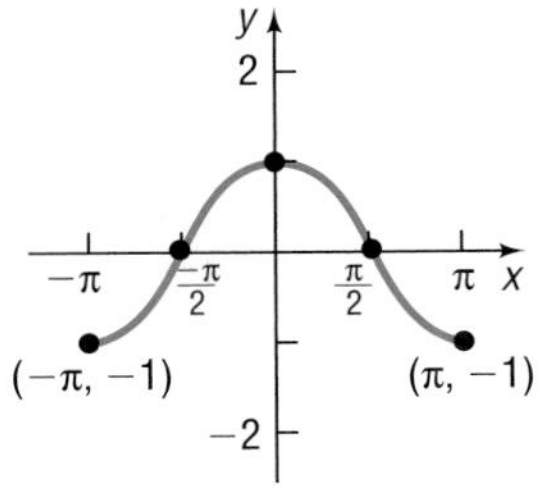

27.

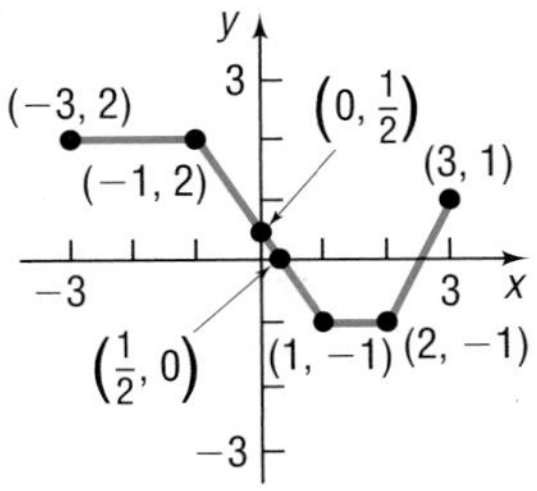

28.

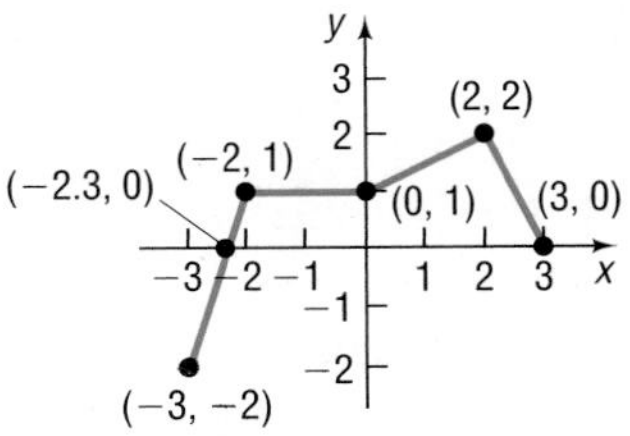

In Problems 29–32, the graph of a function f is given. Use the graph to find:

(a) *The numbers, if any, at which f has a local maximum. What are these local maxima?*
(b) *The numbers, if any, at which f has a local minimum. What are these local minima?*

29.

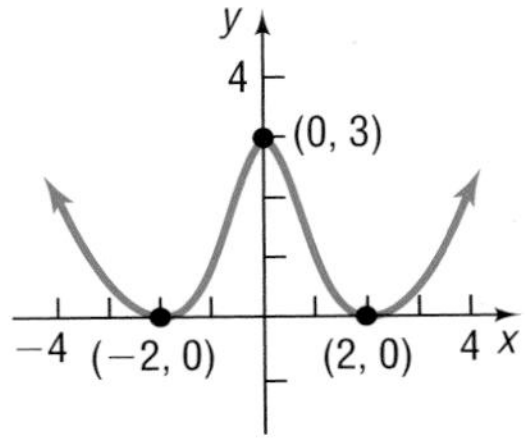

30.

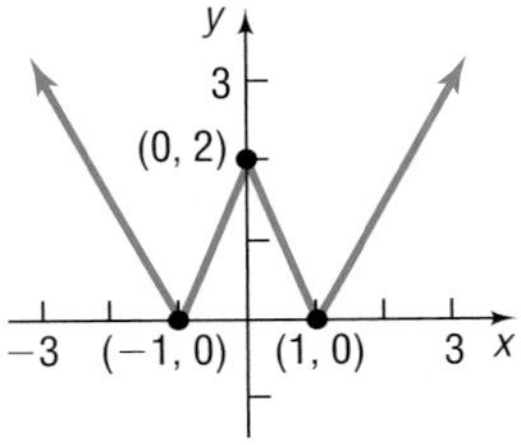

31.

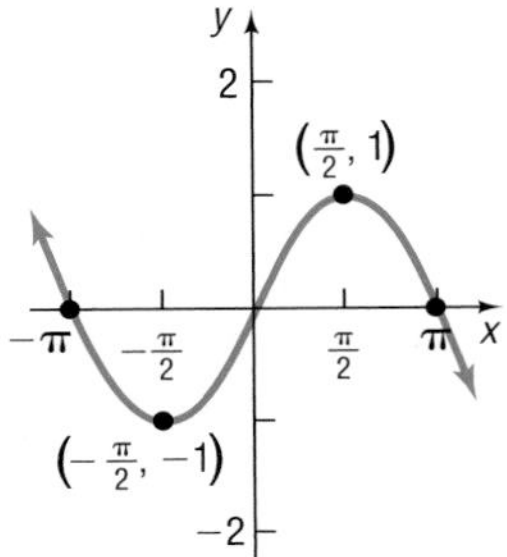

32.

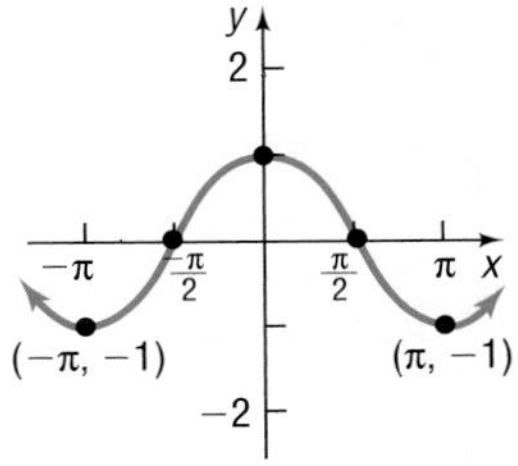

33. Find the average rate of change of $f(x) = -2x^2 + 4$:
(a) From 0 to 2
(b) From 1 to 3
(c) From 1 to 4

34. Find the average rate of change of $f(x) = -x^3 + 1$:
(a) From 0 to 2
(b) From 1 to 3
(c) From -1 to 1

In Problems 35–46,
(a) *For each function find the average rate of change of f from 1 to x:*

$$\frac{f(x) - f(1)}{x - 1}, \quad x \neq 1$$

(b) *Use the result from part (a) to compute the average rate of change from $x = 1$ to $x = 2$. Be sure to simplify.*
(c) *Find an equation of the secant line containing $(1, f(1))$ and $(2, f(2))$.*

35. $f(x) = 5x$ **36.** $f(x) = -4x$ **37.** $f(x) = 1 - 3x$ **38.** $f(x) = x^2 + 1$

39. $f(x) = x^2 - 2x$ **40.** $f(x) = x - 2x^2$ **41.** $f(x) = x^3 - x$ **42.** $f(x) = x^3 + x$

43. $f(x) = \dfrac{2}{x + 1}$ **44.** $f(x) = \dfrac{1}{x^2}$ **45.** $f(x) = \sqrt{x}$ **46.** $f(x) = \sqrt{x + 3}$

In Problems 47–58, determine algebraically whether each function is even, odd, or neither.

47. $f(x) = 4x^3$ **48.** $f(x) = 2x^4 - x^2$ **49.** $g(x) = -3x^2 - 5$ **50.** $h(x) = 3x^3 + 5$

51. $F(x) = \sqrt[3]{x}$ **52.** $G(x) = \sqrt{x}$ **53.** $f(x) = x + |x|$ **54.** $f(x) = \sqrt[3]{2x^2 + 1}$

55. $g(x) = \dfrac{1}{x^2}$ **56.** $h(x) = \dfrac{x}{x^2 - 1}$ **57.** $h(x) = \dfrac{-x^3}{3x^2 - 9}$ **58.** $F(x) = \dfrac{2x}{|x|}$

In Problems 59–66, use a graphing utility to graph each function over the indicated interval and approximate any local maxima and local minima. Determine where the function is increasing and where it is decreasing. Round answers to two decimal places.

59. $f(x) = x^3 - 3x + 2 \quad (-2, 2)$ **60.** $f(x) = x^3 - 3x^2 + 5 \quad (-1, 3)$

61. $f(x) = x^5 - x^3 \quad (-2, 2)$

62. $f(x) = x^4 - x^2 \quad (-2, 2)$

63. $f(x) = -0.2x^3 - 0.6x^2 + 4x - 6 \quad (-6, 4)$

64. $f(x) = -0.4x^3 + 0.6x^2 + 3x - 2 \quad (-4, 5)$

65. $f(x) = 0.25x^4 + 0.3x^3 - 0.9x^2 + 3 \quad (-3, 2)$

66. $f(x) = -0.4x^4 - 0.5x^3 + 0.8x^2 - 2 \quad (-3, 2)$

---

67. Constructing an Open Box An open box with a square base is to be made from a square piece of cardboard 24 inches on a side by cutting out a square from each corner and turning up the sides (see the figure).

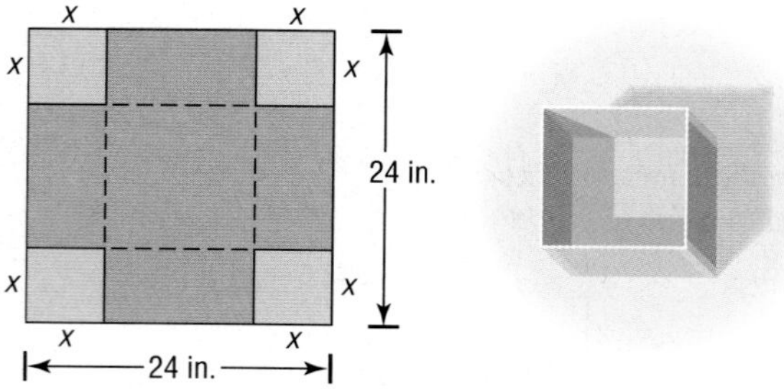

(a) Express the volume V of the box as a function of the length x of the side of the square cut from each corner.
(b) What is the volume if a 3-inch square is cut out?
(c) What is the volume if a 10-inch square is cut out?
(d) Graph $V = V(x)$. For what value of x is V largest?

68. Constructing an Open Box A open box with a square base is required to have a volume of 10 cubic feet.
(a) Express the amount A of material used to make such a box as a function of the length x of a side of the square base.
(b) How much material is required for a base 1 foot by 1 foot?
(c) How much material is required for a base 2 feet by 2 feet?
(d) Graph $A = A(x)$. For what value of x is A smallest?

69. Maximum Height of a Ball The height s of a ball (in feet) thrown with an initial velocity of 80 feet per second from an initial height of 6 feet is given as a function of the time t (in seconds) by

$$s(t) = -16t^2 + 80t + 6$$

(a) Graph s.
(b) Determine the time at which height is maximum.
(c) What is the maximum height?

70. Minimum Average Cost The average cost of producing x riding lawn mowers per hour is given by

$$\overline{C}(x) = 0.3x^2 + 21x - 251 + \frac{2500}{x}$$

(a) Graph $\overline{C}$.
(b) Determine the number of riding lawn mowers to produce in order to minimize average cost.
(c) What is the minimum average cost?

71. For the function $f(x) = x^2$, compute each average rate of change:
(a) From 0 to 1
(b) From 0 to 0.5
(c) From 0 to 0.1
(d) From 0 to 0.01
(e) From 0 to 0.001
(f) Graph each of the secant lines.
(g) What do you think is happening to the secant lines?
(h) What is happening to the slopes of the secant lines? Is there some number that they are getting closer to? What is that number?

72. For the function $f(x) = x^2$, compute each average rate of change:
(a) From 1 to 2
(b) From 1 to 1.5
(c) From 1 to 1.1
(d) From 1 to 1.01
(e) From 1 to 1.001
(f) Graph each of the secant lines.
(g) What do you think is happening to the secant lines?
(h) What is happening to the slopes of the secant lines? Is there some number that they are getting closer to? What is that number?

73. Draw the graph of a function that has the following characteristics: Domain: all real numbers; Range: all real numbers; Intercepts: $(0, -3)$ and $(2, 0)$; A local maximum of -2 is at -1; a local minimum of -6 is at 2. Compare your graph with others. Comment on any differences.

74. Redo Problem 73 with the following additional information: Increasing on $(-\infty, -1)$, $(2, \infty)$; decreasing on $(-1, 2)$. Again compare your graph with others and comment on any differences.

75. How many x-intercepts can a function defined on an interval have if it is increasing on that interval? Explain.

76. Suppose that a friend of yours does not understand the idea of increasing and decreasing functions. Provide an explanation complete with graphs that clarifies the idea.

77. Can a function be both even and odd? Explain.

'Are You Prepared?' Answers

1. $2 \le x \le 5$ **2.** 1 **3.** y-axis **4.** $y + 2 = 5(x - 3)$ **5.** $(-3, 0)$, $(3, 0)$, $(0, -9)$

2.4 Library of Functions; Piecewise-defined Functions

PREPARING FOR THIS SECTION *Before getting started, review the following:*

- Intercepts (Section 1.2, pp. 11–12)
- Graphs of Key Equations (Section 1.2: Example 3, p. 10; Example 9, p. 13; Example 10, p. 14; and Example 11, p. 15)

Now work the 'Are You Prepared?' problems on page 90.

OBJECTIVES 1 Graph the Functions Listed in the Library of Functions
2 Graph Piecewise-defined Functions

We now introduce a few more functions to add to our list of important functions. We begin with the *square root function*.

In Section 1.2 we graphed the equation $x = y^2$. If we solve the equation for y and restrict y so that $y \geq 0$, the equation $x = y^2, y \geq 0$, is written as $y = f(x) = \sqrt{x}$. Figure 20 shows a graph of $f(x) = \sqrt{x}$.

Figure 20
$f(x) = \sqrt{x}$

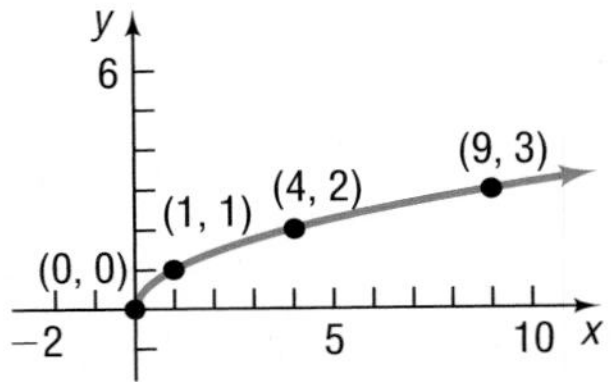

Based on the graph of $f(x) = \sqrt{x}$, we have the following properties:

Properties of $f(x) = \sqrt{x}$

1. The x-intercept of the graph of $f(x) = \sqrt{x}$ is 0. The y-intercept of the graph of $f(x) = \sqrt{x}$ is also 0.
2. The function is neither even nor odd.
3. It is increasing on the interval $(0, \infty)$.
4. It has a local minimum of 0 at $x = 0$.

EXAMPLE 1 **Graphing the Cube Root Function**

(a) Determine whether $f(x) = \sqrt[3]{x}$ is even, odd, or neither. State whether the graph of f is symmetric with respect to the y-axis or symmetric with respect to the origin.

(b) Determine the intercepts, if any, of the graph of $f(x) = \sqrt[3]{x}$.

(c) Graph $f(x) = \sqrt[3]{x}$.

Solution (a) Because

$$f(-x) = \sqrt[3]{-x} = -\sqrt[3]{x} = -f(x)$$

the function is odd. The graph of f is symmetric with respect to the origin.

(b) The y-intercept is $f(0) = \sqrt[3]{0} = 0$. The x-intercept is found by solving the equation $f(x) = 0$.

$$f(x) = 0$$
$$\sqrt[3]{x} = 0 \quad f(x) = \sqrt[3]{x}$$
$$x = 0 \quad \text{Cube both sides of the equation.}$$

The x-intercept is also 0.

(c) We use the function to form Table 4 and obtain some points on the graph. Because of the symmetry with respect to the origin, we only need to find points (x, y) for which $x \geq 0$. Figure 21 shows the graph of $f(x) = \sqrt[3]{x}$.

Table 4

x	$y = f(x) = \sqrt[3]{x}$	(x, y)
0	0	(0, 0)
$\frac{1}{8}$	$\frac{1}{2}$	$\left(\frac{1}{8}, \frac{1}{2}\right)$
1	1	(1, 1)
2	$\sqrt[3]{2} \approx 1.26$	$(2, \sqrt[3]{2})$
8	2	(8, 2)

Figure 21

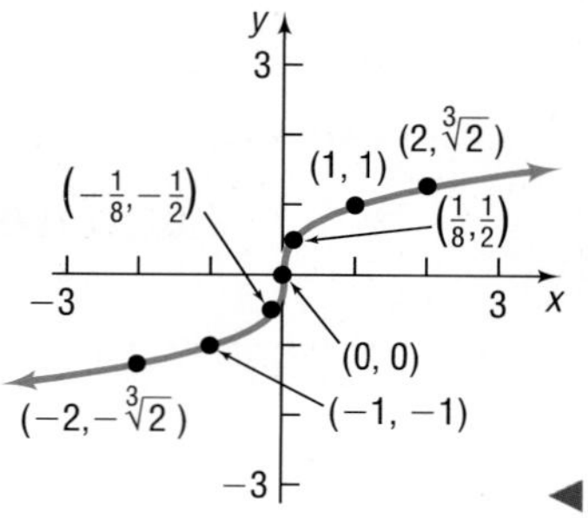

From the results of Example 1 and Figure 21, we have the following properties of the cube root function.

Properties of $f(x) = \sqrt[3]{x}$

1. The x-intercept of the graph of $f(x) = \sqrt[3]{x}$ is 0. The y-intercept of the graph of $f(x) = \sqrt[3]{x}$ is also 0.
2. The function is odd.
3. It is increasing on the interval $(-\infty, \infty)$.
4. It does not have a local minimum or a local maximum.

EXAMPLE 2 Graphing the Absolute Value Function

(a) Determine whether $f(x) = |x|$ is even, odd, or neither. State whether the graph of f is symmetric with respect to the y-axis or symmetric with respect to the origin.

(b) Determine the intercepts, if any, of the graph of $f(x) = |x|$.

(c) Graph $f(x) = |x|$.

Solution (a) Because

$$f(-x) = |-x| = |x| = f(x)$$

the function is even. The graph of f is symmetric with respect to the y-axis.

(b) The y-intercept is $f(0) = |0| = 0$. The x-intercept is found by solving the equation $f(x) = |x| = 0$. So the x-intercept is 0.

(c) We use the function to form Table 5 and obtain some points on the graph. Because of the symmetry with respect to the y-axis, we only need to find points (x, y) for which $x \geq 0$. Figure 22 shows the graph of $f(x) = |x|$.

Table 5

x	$y = f(x) = |x|$	(x, y)
0	0	(0, 0)
1	1	(1, 1)
2	2	(2, 2)
3	3	(3, 3)

Figure 22

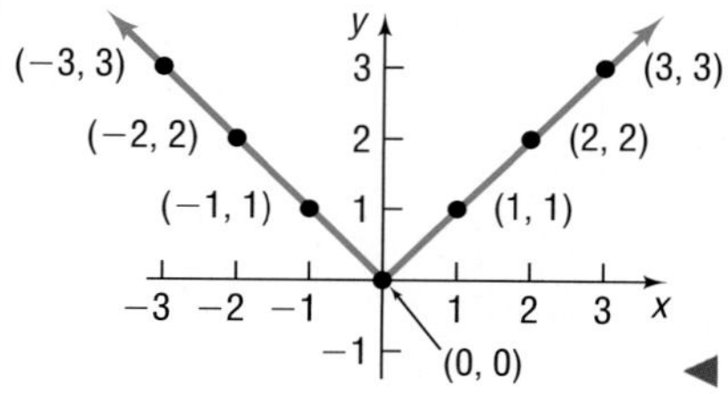

From the results of Example 2 and Figure 22, we have the following properties of the absolute value function.

Properties of $f(x) = |x|$

1. The x-intercept of the graph of $f(x) = |x|$ is 0. The y-intercept of the graph of $f(x) = |x|$ is also 0.
2. The function is even.
3. It is decreasing on the interval $(-\infty, 0)$. It is increasing on the interval $(0, \infty)$.
4. It has a local minimum of 0 at $x = 0$.

Seeing the Concept

Graph $y = |x|$ on a square screen and compare what you see with Figure 22. Note that some graphing calculators use the symbols abs(x) for absolute value. If your utility has no built-in absolute value function, you can still graph $y = |x|$ by using the fact that $|x| = \sqrt{x^2}$.

Library of Functions

1 We now provide a summary of the key functions that we have encountered. In going through this list, pay special attention to the properties of each function, particularly to the shape of each graph. Knowing these graphs will lay the foundation for later graphing techniques.

Figure 23

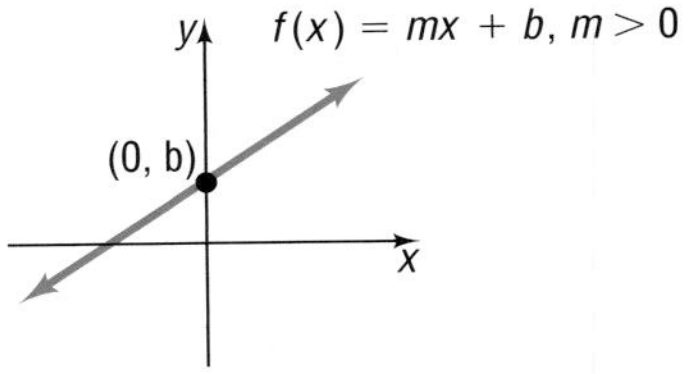

Linear Function

$$f(x) = mx + b, \quad m \text{ and } b \text{ are real numbers}$$

See Figure 23.

The domain of a **linear function** is the set of all real numbers. The graph of this function is a nonvertical line with slope m and y-intercept b. A linear function is increasing if $m > 0$, decreasing if $m < 0$, and constant if $m = 0$.

Figure 24

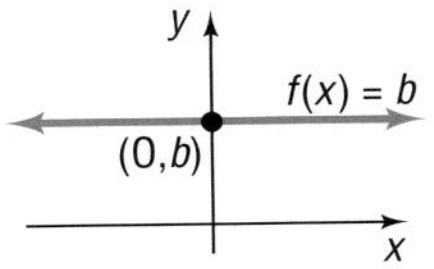

Constant Function

$$f(x) = b, \quad b \text{ is a real number}$$

See Figure 24.

A **constant function** is a special linear function ($m = 0$). Its domain is the set of all real numbers; its range is the set consisting of a single number b. Its graph is a horizontal line whose y-intercept is b. The constant function is an even function whose graph is constant over its domain.

Figure 25
Identity Function

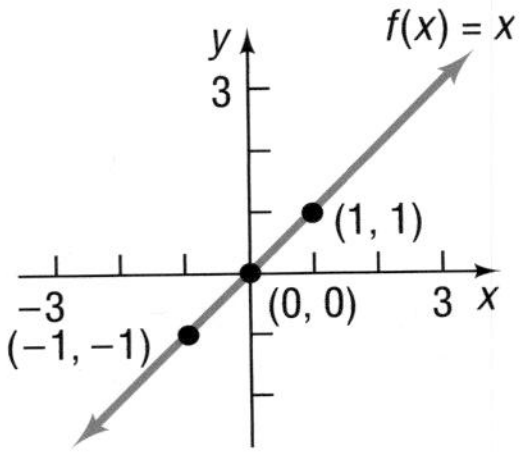

Identity Function

$$f(x) = x$$

See Figure 25.

The **identity function** is also a special linear function. Its domain and range are the set of all real numbers. Its graph is a line whose slope is $m = 1$ and whose y-intercept is 0. The line consists of all points for which the x-coordinate equals the y-coordinate. The identity function is an odd function that is increasing over its domain. Note that the graph bisects quadrants I and III.

Figure 26
Square Function

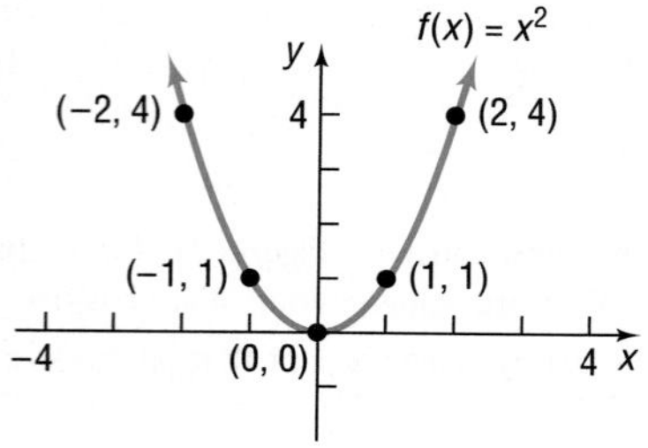

Square Function

$$f(x) = x^2$$

See Figure 26.

The domain of the **square function** f is the set of all real numbers; its range is the set of nonnegative real numbers. The graph of this function is a parabola whose intercept is at $(0, 0)$. The square function is an even function that is decreasing on the interval $(-\infty, 0)$ and increasing on the interval $(0, \infty)$.

Figure 27
Cube Function

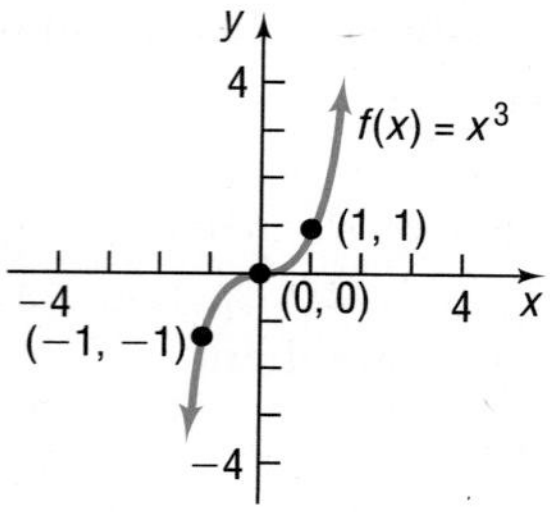

Cube Function

$$f(x) = x^3$$

See Figure 27.

The domain and the range of the **cube function** is the set of all real numbers. The intercept of the graph is at $(0, 0)$. The cube function is odd and is increasing on the interval $(-\infty, \infty)$.

Figure 28
Square Root Function

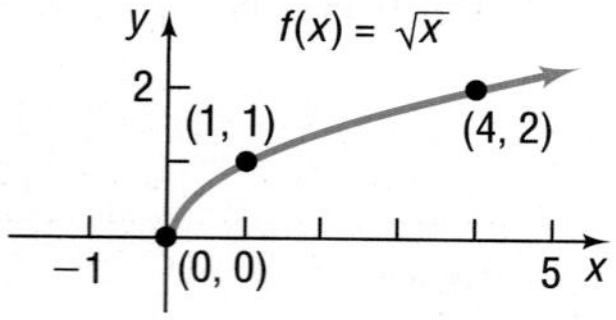

Square Root Function

$$f(x) = \sqrt{x}$$

See Figure 28.

The domain and the range of the **square root function** is the set of nonnegative real numbers. The intercept of the graph is at $(0, 0)$. The square root function is neither even nor odd and is increasing on the interval $(0, \infty)$.

Figure 29
Cube Root Function

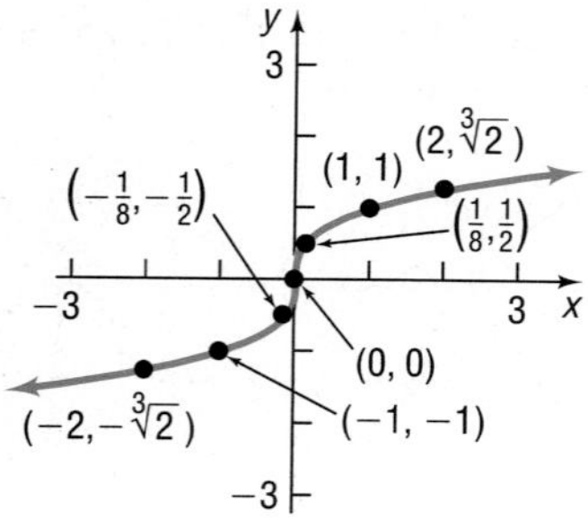

Cube Root Function

$$f(x) = \sqrt[3]{x}$$

See Figure 29.

The domain and the range of the **cube root function** is the set of all real numbers. The intercept of the graph is at $(0, 0)$. The cube root function is an odd function that is increasing on the interval $(-\infty, \infty)$.

Figure 30
Reciprocal Function

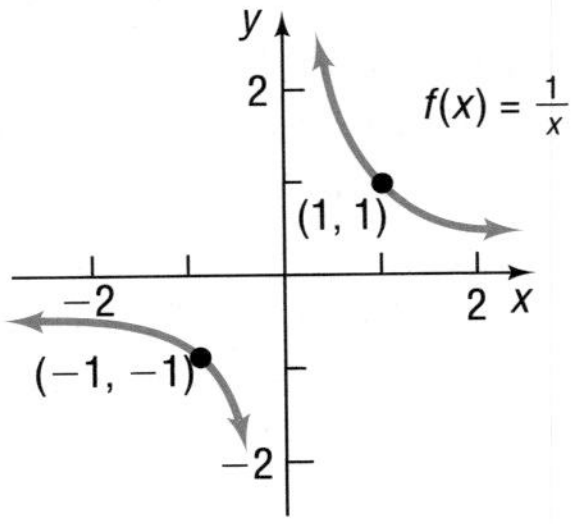

Reciprocal Function

$$f(x) = \frac{1}{x}$$

Refer to Example 11, p. 15, for a discussion of the equation $y = \frac{1}{x}$. See Figure 30.

The domain and the range of the **reciprocal function** is the set of all nonzero real numbers. The graph has no intercepts. The reciprocal function is decreasing on the intervals $(-\infty, 0)$ and $(0, \infty)$ and is an odd function.

Figure 31
Absolute Value Function

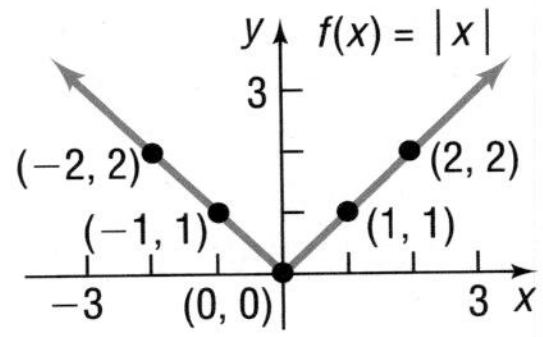

Absolute Value Function

$$f(x) = |x|$$

See Figure 31.

The domain of the **absolute value function** is the set of all real numbers; its range is the set of nonnegative real numbers. The intercept of the graph is at $(0, 0)$. If $x \geq 0$, then $f(x) = x$, and the graph of f is part of the line $y = x$; if $x < 0$, then $f(x) = -x$, and the graph of f is part of the line $y = -x$. The absolute value function is an even function; it is decreasing on the interval $(-\infty, 0)$ and increasing on the interval $(0, \infty)$.

The notation $\text{int}(x)$ stands for the largest integer less than or equal to x. For example,

$$\text{int}(1) = 1 \quad \text{int}(2.5) = 2 \quad \text{int}\left(\frac{1}{2}\right) = 0 \quad \text{int}\left(-\frac{3}{4}\right) = -1 \quad \text{int}(\pi) = 3$$

This type of correspondence occurs frequently enough in mathematics that we give it a name.

Table 6

x	$y = f(x) = \text{int}(x)$	(x, y)
-1	-1	$(-1, -1)$
$-\frac{1}{2}$	-1	$\left(-\frac{1}{2}, -1\right)$
$-\frac{1}{4}$	-1	$\left(-\frac{1}{4}, -1\right)$
0	0	$(0, 0)$
$\frac{1}{4}$	0	$\left(\frac{1}{4}, 0\right)$
$\frac{1}{2}$	0	$\left(\frac{1}{2}, 0\right)$
$\frac{3}{4}$	0	$\left(\frac{3}{4}, 0\right)$

Greatest Integer Function

$$f(x) = \text{int}(x) = \text{Greatest integer less than or equal to } x$$

Note: Some books use the notation $f(x) = [\![x]\!]$ instead of $\text{int}(x)$.

We obtain the graph of $f(x) = \text{int}(x)$ by plotting several points. See Table 6. For values of x, $-1 \leq x < 0$, the value of $f(x) = \text{int}(x)$ is -1; for values of x, $0 \leq x < 1$, the value of f is 0. See Figure 32 for the graph.

Figure 32
Greatest Integer Function

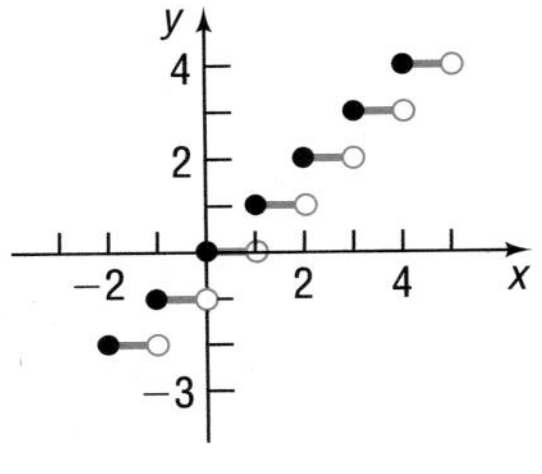

The domain of the **greatest integer function** is the set of all real numbers; its range is the set of integers. The y-intercept of the graph is 0. The x-intercepts lie in the interval $[0, 1)$. The greatest integer function is neither even nor odd. It is constant on every interval of the form $[k, k + 1)$, for k an integer. In Figure 32, we use a solid dot to indicate, for example, that at $x = 1$ the value of f is $f(1) = 1$; we use an open circle to illustrate that the function does not assume the value of 0 at $x = 1$.

From the graph of the greatest integer function, we can see why it also called a **step function.** At $x = 0, x = \pm 1, x = \pm 2$, and so on, this function exhibits what is called a *discontinuity*; that is, at integer values, the graph suddenly "steps" from one value to another without taking on any of the intermediate values. For example, to the immediate left of $x = 3$, the y-coordinates are 2, and to the immediate right of $x = 3$, the y-coordinates are 3.

COMMENT: When graphing a function, you can choose either the **connected mode,** in which points plotted on the screen are connected, making the graph appear without any breaks, or the **dot mode,** in which only the points plotted appear. When graphing the greatest integer function with a graphing utility, it is necessary to be in the **dot mode.** This is to prevent the utility from "connecting the dots" when $f(x)$ changes from one integer value to the next. See Figure 33.

Figure 33
$f(x) = \text{int}(x)$

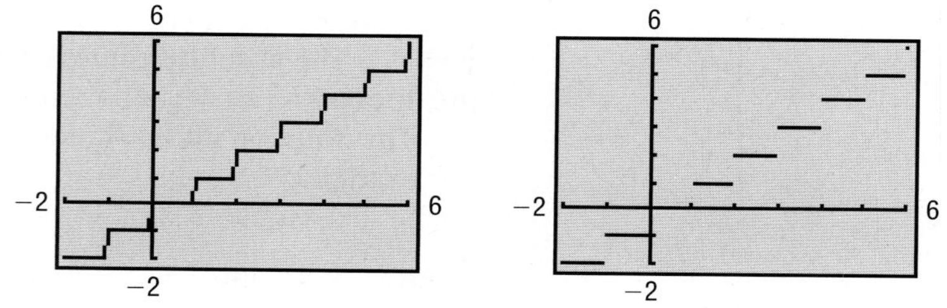

(a) Connected mode **(b)** Dot mode

The functions that we have discussed so far are basic. Whenever you encounter one of them, you should see a mental picture of its graph. For example, if you encounter the function $f(x) = x^2$, you should see in your mind's eye a picture like Figure 26.

NOW WORK PROBLEMS 9 THROUGH 16.

Piecewise-defined Functions

2 Sometimes a function is defined differently on different parts of its domain. For example, the absolute value function $f(x) = |x|$ is actually defined by two equations: $f(x) = x$ if $x \geq 0$ and $f(x) = -x$ if $x < 0$. For convenience, we generally combine these equations into one expression as

$$f(x) = |x| = \begin{cases} x & \text{if } x \geq 0 \\ -x & \text{if } x < 0 \end{cases}$$

When functions are defined by more than one equation, they are called **piecewise-defined** functions.

Let's look at another example of a piecewise-defined function.

EXAMPLE 3 Analyzing a Piecewise-defined Function

The function f is defined as

$$f(x) = \begin{cases} -x + 1 & \text{if } -1 \le x < 1 \\ 2 & \text{if } x = 1 \\ x^2 & \text{if } x > 1 \end{cases}$$

(a) Find $f(0)$, $f(1)$, and $f(2)$.

(b) Determine the domain of f.

(c) Graph f.

(d) Use the graph to find the range of f.

Solution (a) To find $f(0)$, we observe that when $x = 0$ the equation for f is given by $f(x) = -x + 1$. So we have

$$f(0) = -0 + 1 = 1$$

When $x = 1$, the equation for f is $f(x) = 2$. Thus,

$$f(1) = 2$$

When $x = 2$, the equation for f is $f(x) = x^2$. So

$$f(2) = 2^2 = 4$$

(b) To find the domain of f, we look at its definition. We conclude that the domain of f is $\{x | x \ge -1\}$, or the interval $[-1, \infty)$.

(c) To graph f, we graph "each piece." First we graph the line $y = -x + 1$ and keep only the part for which $-1 \le x < 1$. Then we plot the point $(1, 2)$ because, when $x = 1$, $f(x) = 2$. Finally, we graph the parabola $y = x^2$ and keep only the part for which $x > 1$. See Figure 34.

Figure 34

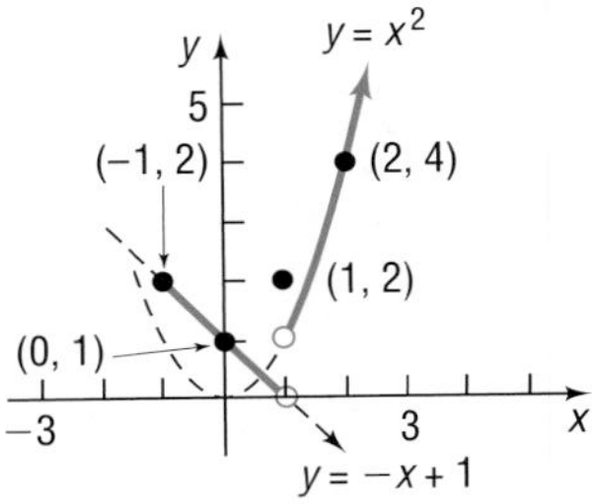

(d) From the graph, we conclude that the range of f is $\{y | y > 0\}$, or the interval $(0, \infty)$. ◀

NOW WORK PROBLEM 29.

EXAMPLE 4 Cost of Electricity

In May 2003, Commonwealth Edison Company supplied electricity to residences for a monthly customer charge of \$7.58 plus 8.275¢ per kilowatt-hour (kWhr) for the first 400 kWhr supplied in the month and 6.574¢ per kWhr for all usage over 400 kWhr in the month.

(a) What is the charge for using 300 kWhr in a month?

(b) What is the charge for using 700 kWhr in a month?

(c) If C is the monthly charge for x kWhr, express C as a function of x.

SOURCE: Commonwealth Edison Co., Chicago, Illinois, 2003.

Solution (a) For 300 kWhr, the charge is \$7.58 plus 8.275¢ = \$0.08275 per kWhr. That is,

$$\text{Charge} = \$7.58 + \$0.08275(300) = \$32.41$$

(b) For 700 kWhr, the charge is \$7.58 plus 8.275¢ per kWhr for the first 400 kWhr plus 6.574¢ per kWhr for the 300 kWhr in excess of 400. That is,

$$\text{Charge} = \$7.58 + \$0.08275(400) + \$0.06574(300) = \$60.40$$

(c) If $0 \le x \le 400$, the monthly charge C (in dollars) can be found by multiplying x times \$0.08275 and adding the monthly customer charge of \$7.58. So, if $0 \le x \le 400$, then $C(x) = 0.08275x + 7.58$. For $x > 400$, the charge is $0.08275(400) + 7.58 + 0.06574(x - 400)$, since $x - 400$ equals the usage in excess of 400 kWhr, which costs \$0.06574 per kWhr. That is, if $x > 400$, then

$$\begin{aligned} C(x) &= 0.08275(400) + 7.58 + 0.06574(x - 400) \\ &= 40.68 + 0.06574(x - 400) \\ &= 0.06574x + 14.38 \end{aligned}$$

The rule for computing C follows two equations:

$$C(x) = \begin{cases} 0.08275x + 7.58 & \text{if } 0 \le x \le 400 \\ 0.06574x + 14.38 & \text{if } x > 400 \end{cases}$$

See Figure 35 for the graph. ◀

Figure 35

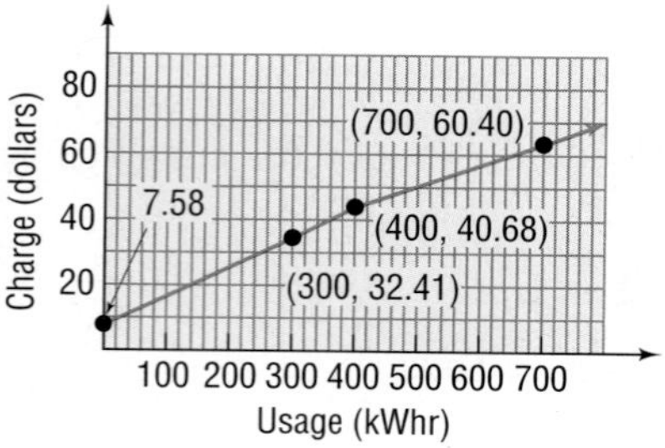

2.4 Assess Your Understanding

'Are You Prepared?' *Answers are given at the end of these exercises. If you get the wrong answer, read the pages listed in red.*

1. Sketch the graph of $y = \sqrt{x}$. (p. 14)

2. Sketch the graph of $y = \dfrac{1}{x}$. (p. 15)

3. List the intercepts of the equation $y = x^3 - 8$. (pp. 11–12)

Concepts and Vocabulary

4. The graph of $f(x) = mx + b$ is decreasing if m is ________ than zero.

5. When functions are defined by more than one equation they are called ________ functions.

6. *True or False:* The cube function is odd and is increasing on the interval $(-\infty, \infty)$.

7. *True or False:* The cube root function is odd and is decreasing on the interval $(-\infty, \infty)$.

8. *True or False:* The domain and the range of the reciprocal function is the set of all real numbers.

Exercises

In Problems 9–16, match each graph to the function listed whose graph most resembles the one given.

A. *Constant function* B. *Linear function* C. *Square function*
D. *Cube function* E. *Square root function* F. *Reciprocal function*
G. *Absolute value function* H. *Cube root function*

9. 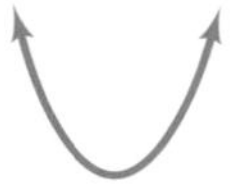

10. 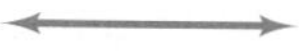

11. 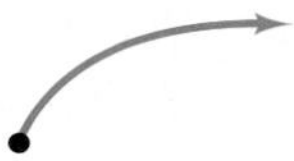

12. 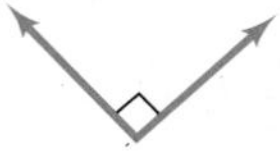

13. 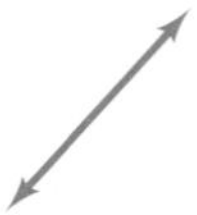

14. 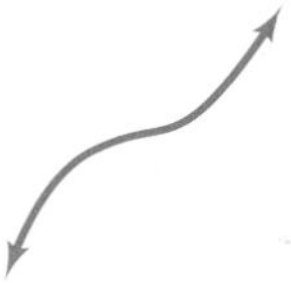

15. 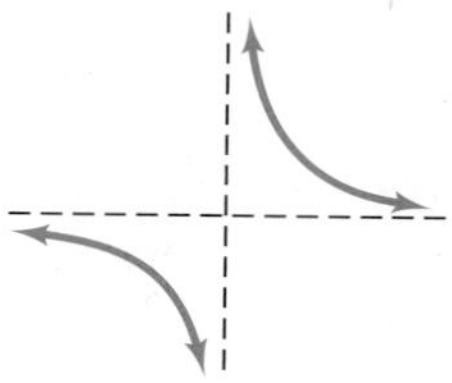

16. 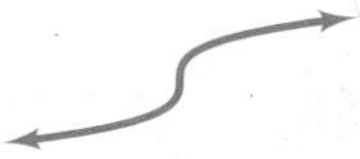

In Problems 17–24, sketch the graph of each function. Be sure to label at least three points on the graph.

17. $f(x) = x$ **18.** $f(x) = x^2$ **19.** $f(x) = x^3$ **20.** $f(x) = \sqrt{x}$

21. $f(x) = \dfrac{1}{x}$ **22.** $f(x) = |x|$ **23.** $f(x) = \sqrt[3]{x}$ **24.** $f(x) = 3$

25. If $f(x) = \begin{cases} x^2 & \text{if } x < 0 \\ 2 & \text{if } x = 0 \\ 2x + 1 & \text{if } x > 0 \end{cases}$

find: (a) $f(-2)$ (b) $f(0)$ (c) $f(2)$

26. If $f(x) = \begin{cases} x^3 & \text{if } x < 0 \\ 3x + 2 & \text{if } x \geq 0 \end{cases}$

find: (a) $f(-1)$ (b) $f(0)$ (c) $f(1)$

27. If $f(x) = \text{int}(2x)$, find: (a) $f(1.2)$ (b) $f(1.6)$ (c) $f(-1.8)$

28. If $f(x) = \text{int}\left(\dfrac{x}{2}\right)$, find: (a) $f(1.2)$ (b) $f(1.6)$ (c) $f(-1.8)$

In Problems 29–40:

(a) *Find the domain of each function.*
(b) *Locate any intercepts.*
(c) *Graph each function.*
(d) *Based on the graph, find the range.*

29. $f(x) = \begin{cases} 2x & \text{if } x \neq 0 \\ 1 & \text{if } x = 0 \end{cases}$

30. $f(x) = \begin{cases} 3x & \text{if } x \neq 0 \\ 4 & \text{if } x = 0 \end{cases}$

31. $f(x) = \begin{cases} -2x + 3 & x < 1 \\ 3x - 2 & x \geq 1 \end{cases}$

32. $f(x) = \begin{cases} x + 3 & x < -2 \\ -2x - 3 & x \geq -2 \end{cases}$

33. $f(x) = \begin{cases} x + 3 & -2 \leq x < 1 \\ 5 & x = 1 \\ -x + 2 & x > 1 \end{cases}$

34. $f(x) = \begin{cases} 2x + 5 & -3 \leq x < 0 \\ -3 & x = 0 \\ -5x & x > 0 \end{cases}$

35. $f(x) = \begin{cases} 1 + x & \text{if } x < 0 \\ x^2 & \text{if } x \geq 0 \end{cases}$

36. $f(x) = \begin{cases} \dfrac{1}{x} & \text{if } x < 0 \\ \sqrt[3]{x} & \text{if } x \geq 0 \end{cases}$

37. $f(x) = \begin{cases} |x| & \text{if } -2 \leq x < 0 \\ 1 & \text{if } x = 0 \\ x^3 & \text{if } x > 0 \end{cases}$

38. $f(x) = \begin{cases} 3 + x & \text{if } -3 \leq x < 0 \\ 3 & \text{if } x = 0 \\ \sqrt{x} & \text{if } x > 0 \end{cases}$

39. $f(x) = 2\,\text{int}(x)$

40. $f(x) = \text{int}(2x)$

In Problems 41–44, the graph of a piecewise-defined function is given. Write a definition for each function.

41.

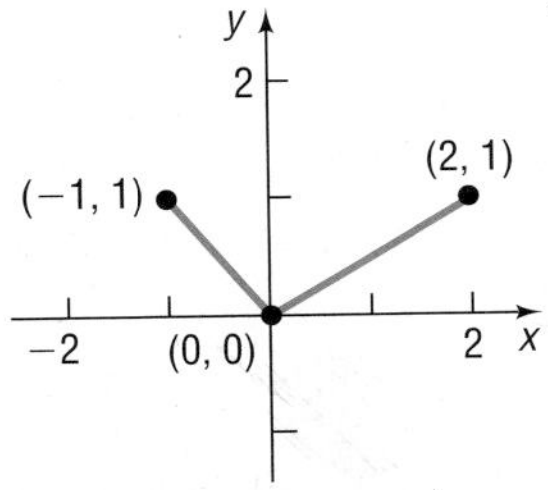

42.

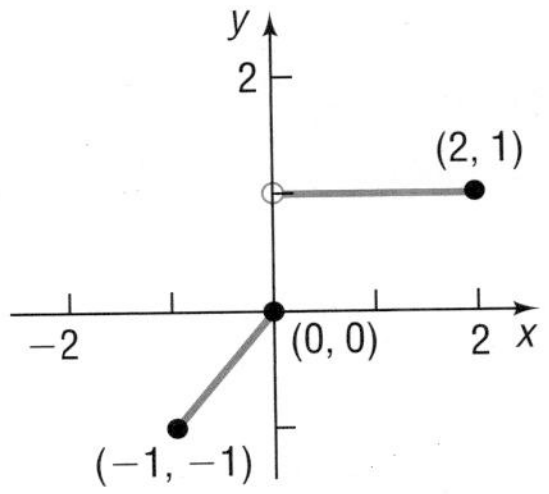

43.

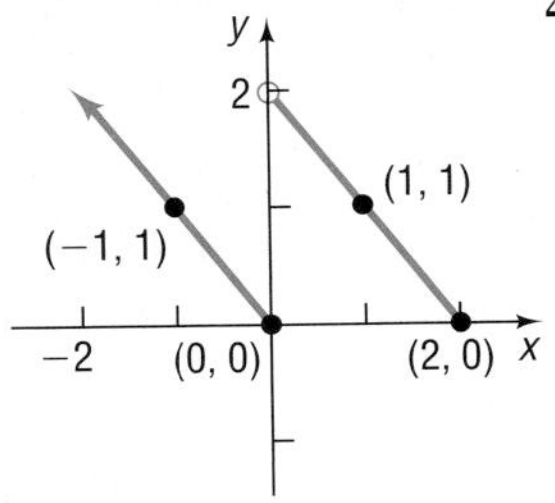

44.

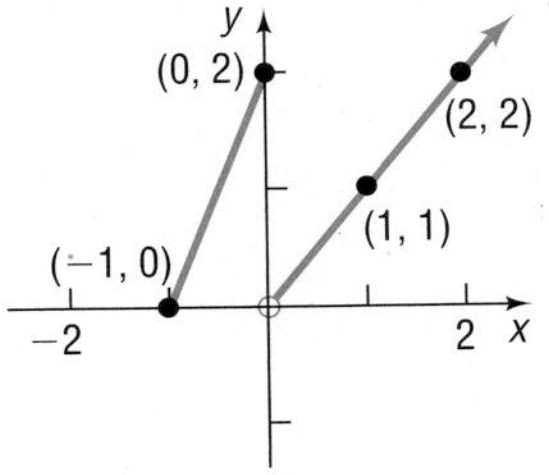

45. Cell Phone Service Sprint PCS offers a monthly cellular phone plan for \$39.99. It includes 350 anytime minutes plus \$0.25 per minute for additional minutes. The following function is used to compute the monthly cost for a subscriber:

$$C(x) = \begin{cases} 39.99 & \text{if } 0 < x \leq 350 \\ 0.25x - 47.51 & \text{if } x > 350 \end{cases}$$

where x is the number of anytime minutes used. Compute the monthly cost of the cellular phone for use of the following anytime minutes:

(a) 200 (b) 365 (c) 351

46. First-class Letter According to the U.S. Postal Service, first-class mail is used for personal and business correspondence. Any mailable item may be sent as first-class mail. It includes postcards, letters, large envelopes, and small packages. The maximum weight is 13 ounces. The following function is used to compute the cost of mailing a first-class letter.

$$C(x) = \begin{cases} 0.37 & \text{if } 0 < x \leq 1 \\ 0.23\,\text{int}(x) + 0.37 & \text{if } 1 < x \leq 13 \end{cases}$$

where x is the weight of the package in ounces. Compute the cost of mailing a package via first class for the following weights:

(a) A letter weighing 4.3 ounces
(b) A postcard weighing 0.4 ounce
(c) A package weighing 12.2 ounces

47. Cost of Natural Gas In May 2003, the Peoples Gas Company had the following rate schedule for natural gas usage in single-family residences:

Monthly service charge	\$9.45
Per therm service charge	
1st 50 therms	\$0.36375/therm
Over 50 therms	\$0.11445/therm
Gas charge	\$0.6338/therm

(a) What is the charge for using 50 therms in a month?
(b) What is the charge for using 500 therms in a month?
(c) Construct a function that relates the monthly charge C for x therms of gas.
(d) Graph this function.

SOURCE: The Peoples Gas Company, Chicago, Illinois, 2003.

48. Cost of Natural Gas In May 2003, Nicor Gas had the following rate schedule for natural gas usage in single-family residences:

Monthly customer charge	\$6.45
Distribution charge	
1st 20 therms	\$0.2012/therm
Next 30 therms	\$0.1117/therm
Over 50 therms	\$0.0374/therm
Gas supply charge	\$0.7268/therm

(a) What is the charge for using 40 therms in a month?
(b) What is the charge for using 202 therms in a month?
(c) Construct a function that gives the monthly charge C for x therms of gas.
(d) Graph this function.

SOURCE: Nicor Gas, Aurora, Illinois, 2003.

49. Federal Income Tax Two 2003 Tax Rate Schedules are given in the accompanying table. If x equals taxable income and y equals the tax due, construct a function $y = f(x)$ for Schedule X.

REVISED 2003 TAX RATE SCHEDULES

Schedule X—

	If Taxable Income Is Over	But Not Over	Then The Tax Is This Amount	Plus This %	Of the Excess Over
Single	\$0	\$7,000	\$0.00	10%	\$0.00
	\$7,000	\$28,400	\$700.00	15%	\$7,000
	\$28,400	\$68,800	\$3,910.00	25%	\$28,400
	\$68,800	\$143,500	\$14,010.00	28%	\$68,800
	\$143,500	\$311,950	\$34,926.00	33%	\$143,500
	\$311,950	—	\$90,514.50	35%	\$311,950

Schedule Y-1—

	If Taxable Income Is Over	But Not Over	Then The Tax Is This Amount	Plus This %	Of the Excess Over
Married Filing Jointly or Qualifying Widow(er)	\$0	\$14,000	\$0.00	10%	\$0.00
	\$14,000	\$56,800	\$1,400.00	15%	\$14,000
	\$56,800	\$114,650	\$7,820.00	25%	\$56,800
	\$114,650	\$174,700	\$22,282.50	28%	\$114,650
	\$174,700	\$311,950	\$39,096.50	33%	\$174,700
	\$311,950	—	\$84,389.00	35%	\$311,950

SOURCE: Internal Revenue Service

50. Federal Income Tax Refer to the revised 2003 tax rate schedules. If x equals taxable income and y equals the tax due, construct a function $y = f(x)$ for Schedule Y-1.

51. Cost of Transporting Goods A trucking company transports goods between Chicago and New York, a distance of 960 miles. The company's policy is to charge, for each pound, \$0.50 per mile for the first 100 miles, \$0.40 per mile for the next 300 miles, \$0.25 per mile for the next 400 miles, and no charge for the remaining 160 miles.

(a) Graph the relationship between the cost of transportation in dollars and mileage over the entire 960-mile route.
(b) Find the cost as a function of mileage for hauls between 100 and 400 miles from Chicago.
(c) Find the cost as a function of mileage for hauls between 400 and 800 miles from Chicago.

52. Car Rental Costs An economy car rented in Florida from National Car Rental® on a weekly basis costs \$95 per week. Extra days cost \$24 per day until the day rate exceeds the weekly rate, in which case the weekly rate applies. Find the cost C of renting an economy car as a piecewise-defined function of the number x of days used, where $7 \le x \le 14$. Graph this function.

Note: Any part of a day counts as a full day.

53. Minimum Payments for Credit Cards Holders of credit cards issued by banks, department stores, oil companies, and so on, receive bills each month that state minimum amounts that must be paid by a certain due date. The minimum due depends on the total amount owed. One such credit card company uses the following rules: For a bill of less than \$10, the entire amount is due. For a bill of at least \$10 but less than \$500, the minimum due is \$10. There is a minimum of \$30 due on a bill of at least \$500 but less than \$1000, a minimum of \$50 due on a bill of at least \$1000 but less than \$1500, and a minimum of \$70 due on bills of \$1500 or more. Find the function f that describes the minimum payment due on a bill of x dollars. Graph f.

54. Interest Payments for Credit Cards Refer to Problem 53. The card holder may pay any amount between the minimum due and the total owed. The organization issuing the card charges the card holder interest of 1.5% per month for the first \$1000 owed and 1% per month on any unpaid balance over \$1000. Find the function g that gives the amount of interest charged per month on a balance of x dollars. Graph g.

55. Wind Chill The wind chill factor represents the equivalent air temperature at a standard wind speed that would produce the same heat loss as the given temperature and wind speed. One formula for computing the equivalent temperature is

$$W = \begin{cases} t & 0 \le v < 1.79 \\ 33 - \dfrac{(10.45 + 10\sqrt{v} - v)(33 - t)}{22.04} & 1.79 \le v \le 20 \\ 33 - 1.5958(33 - t) & v > 20 \end{cases}$$

where v represents the wind speed (in meters per second) and t represents the air temperature (°C). Compute the wind chill for the following:

(a) An air temperature of 10°C and a wind speed of 1 meter per second (m/sec)
(b) An air temperature of 10°C and a wind speed of 5 m/sec
(c) An air temperature of 10°C and a wind speed of 15 m/sec
(d) An air temperature of 10°C and a wind speed of 25 m/sec
(e) Explain the physical meaning of the equation corresponding to $0 \le v < 1.79$.
(f) Explain the physical meaning of the equation corresponding to $v > 20$.

56. Wind Chill Redo Problem 55(a)–(d) for an air temperature of -10°C.

57. Exploration Graph $y = x^2$. Then on the same screen graph $y = x^2 + 2$, followed by $y = x^2 + 4$, followed by $y = x^2 - 2$. What pattern do you observe? Can you predict the graph of $y = x^2 - 4$? Of $y = x^2 + 5$.

58. Exploration Graph $y = x^2$. Then on the same screen graph $y = (x - 2)^2$, followed by $y = (x - 4)^2$, followed by $y = (x + 2)^2$. What pattern do you observe? Can you predict the graph of $y = (x + 4)^2$? Of $y = (x - 5)^2$?

59. Exploration Graph $y = |x|$. Then on the same screen graph $y = 2|x|$, followed by $y = 4|x|$, followed by $y = \frac{1}{2}|x|$. What pattern do you observe? Can you predict the graph of $y = \frac{1}{4}|x|$? Of $y = 5|x|$?

60. Exploration Graph $y = x^2$. Then on the same screen graph $y = -x^2$. What pattern do you observe? Now try $y = |x|$ and $y = -|x|$. What do you conclude?

61. Exploration Graph $y = \sqrt{x}$. Then on the same screen graph $y = \sqrt{-x}$. What pattern do you observe? Now try $y = 2x + 1$ and $y = 2(-x) + 1$. What do you conclude?

62. Exploration Graph $y = x^3$. Then on the same screen graph $y = (x - 1)^3 + 2$. Could you have predicted the result?

63. Exploration Graph $y = x^2$, $y = x^4$, and $y = x^6$ on the same screen. What do you notice is the same about each graph? What do you notice that is different?

64. Exploration Graph $y = x^3$, $y = x^5$, and $y = x^7$ on the same screen. What do you notice is the same about each graph? What do you notice that is different?

65. Consider the equation

$$y = \begin{cases} 1 & \text{if } x \text{ is rational} \\ 0 & \text{if } x \text{ is irrational} \end{cases}$$

Is this a function? What is its domain? What is its range? What is its y-intercept, if any? What are its x-intercepts, if any? Is it even, odd, or neither? How would you describe its graph?

66. Define some functions that pass through $(0, 0)$ and $(1, 1)$ and are increasing for $x \ge 0$. Begin your list with $y = \sqrt{x}$, $y = x$, and $y = x^2$. Can you propose a general result about such functions?

'Are You Prepared?' Answers

1.

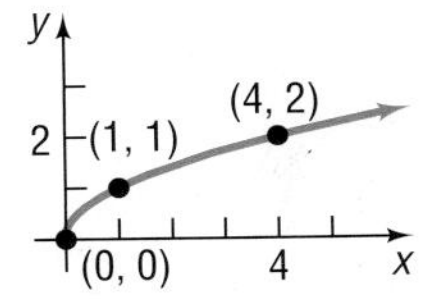

2.

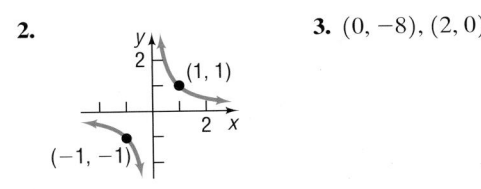

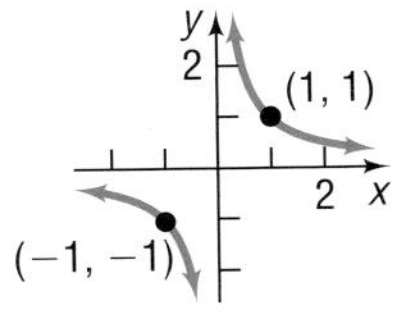

3. $(0, -8)$, $(2, 0)$

2.5 Graphing Techniques: Transformations

OBJECTIVES
1. Graph Functions Using Horizontal and Vertical Shifts
2. Graph Functions Using Compressions and Stretches
3. Graph Functions Using Reflections about the x-Axis or y-Axis

At this stage, if you were asked to graph any of the functions defined by $y = x$, $y = x^2$, $y = x^3$, $y = \sqrt{x}$, $y = \sqrt[3]{x}$, $y = |x|$, or $y = \frac{1}{x}$, your response should be, "Yes, I recognize these functions and know the general shapes of their graphs." (If this is not your answer, review the previous section, Figures 25 through 31.)

Sometimes we are asked to graph a function that is "almost" like one that we already know how to graph. In this section, we look at some of these functions and develop techniques for graphing them. Collectively, these techniques are referred to as **transformations.**

1 Vertical Shifts

EXAMPLE 1 Vertical Shift Up

Use the graph of $f(x) = x^2$ to obtain the graph of $g(x) = x^2 + 3$.

Solution We begin by obtaining some points on the graphs of f and g. For example, when $x = 0$, then $y = f(0) = 0$ and $y = g(0) = 3$. When $x = 1$, then $y = f(1) = 1$ and $y = g(1) = 4$. Table 7 lists these and a few other points on each graph. We conclude that the graph of g is identical to that of f, except that it is shifted vertically up 3 units. See Figure 36.

Table 7

x	$y = f(x) = x^2$	$y = g(x) = x^2 + 3$
−2	4	7
−1	1	4
0	0	3
1	1	4
2	4	7

Figure 36

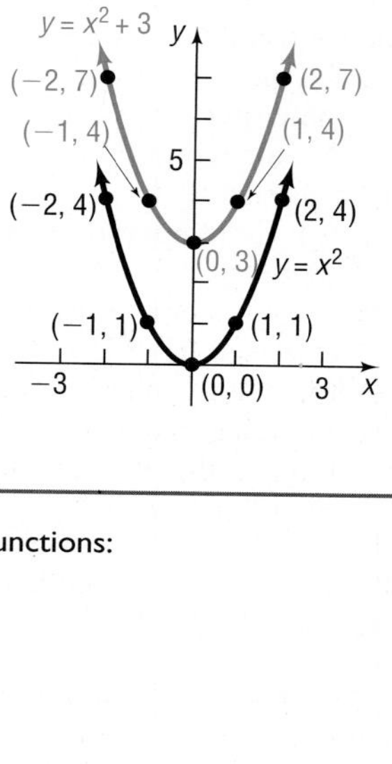

◀

Seeing the Concept

On the same screen, graph each of the following functions:

$$Y_1 = x^2$$
$$Y_2 = x^2 + 1$$
$$Y_3 = x^2 + 2$$
$$Y_4 = x^2 - 1$$
$$Y_5 = x^2 - 2$$

Figure 37

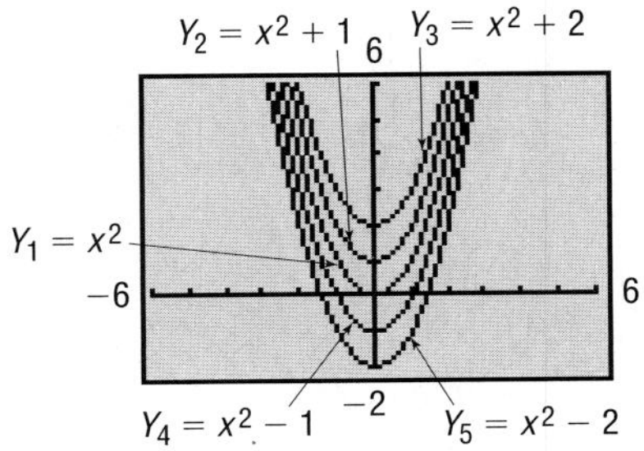

Figure 37 illustrates the graphs. You should have observed a general pattern. With $Y_1 = x^2$ on the screen, the graph of $Y_2 = x^2 + 1$ is identical to that of $Y_1 = x^2$, except that it is shifted vertically up 1 unit. Similarly, $Y_3 = x^2 + 2$ is identical to that of $Y_1 = x^2$, except that it is shifted vertically up 2 units. The graph of $Y_4 = x^2 - 1$ is identical to that of $Y_1 = x^2$, except that it is shifted vertically down 1 unit.

We are led to the following conclusion:

> If a real number k is added to the right side of a function $y = f(x)$, the graph of the new function $y = f(x) + k$ is the graph of f **shifted vertically up** (if $k > 0$) or **down** (if $k < 0$).

Let's look at another example.

EXAMPLE 2 Vertical Shift Down

Use the graph of $f(x) = x^2$ to obtain the graph of $g(x) = x^2 - 4$.

Solution Table 8 lists some points on the graphs of f and g. Notice that each y-coordinate of g is 4 units less than the corresponding y-coordinate of f. The graph of g is identical to that of f, except that it is shifted down 4 units. See Figure 38.

Table 8

x	$y = f(x) = x^2$	$y = g(x) = x^2 - 4$
−2	4	0
−1	1	−3
0	0	−4
1	1	−3
2	4	0

Figure 38

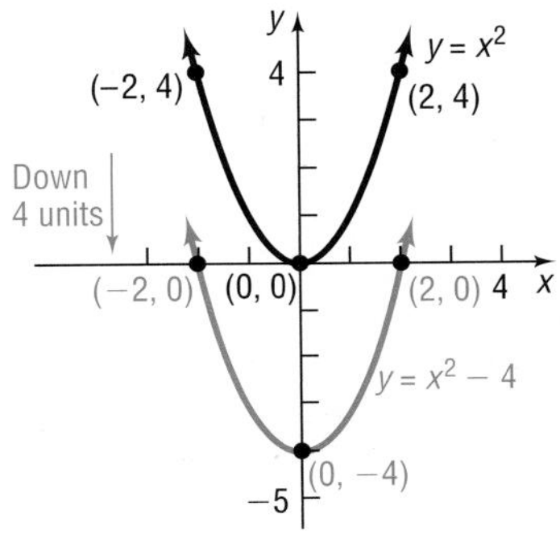

NOW WORK PROBLEM 35.

Horizontal Shifts

EXAMPLE 3 Horizontal Shift to the Right

Use the graph of $f(x) = x^2$ to obtain the graph of $g(x) = (x - 2)^2$.

Solution The function $g(x) = (x - 2)^2$ is basically a square function. Table 9 lists some points on the graphs of f and g. Note that when $f(x) = 0$ then $x = 0$, and when $g(x) = 0$, then $x = 2$. Also, when $f(x) = 4$, then $x = -2$ or 2, and when $g(x) = 4$, then $x = 0$ or 4. We conclude that the graph of g is identical to that of f, except that it is shifted 2 units to the right. See Figure 39.

Table 9

x	$y = f(x) = x^2$	$y = g(x) = (x - 2)^2$
−2	4	16
0	0	4
2	4	0
4	16	4

Figure 39

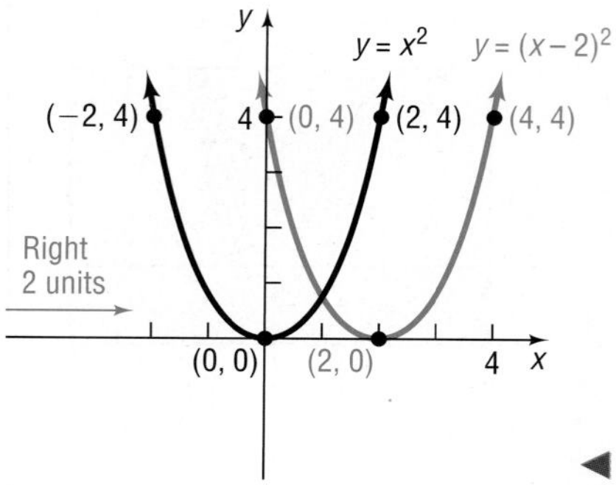

Seeing the Concept

On the same screen, graph each of the following functions:

$$Y_1 = x^2$$
$$Y_2 = (x - 1)^2$$
$$Y_3 = (x - 3)^2$$
$$Y_4 = (x + 2)^2$$

Figure 40

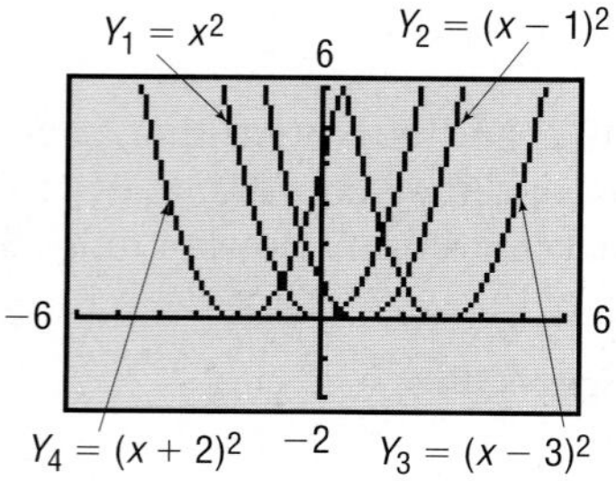

Figure 40 illustrates the graphs.

You should have observed the following pattern. With the graph of $Y_1 = x^2$ on the screen, the graph of $Y_2 = (x - 1)^2$ is identical to that of $Y_1 = x^2$, except that it is shifted horizontally to the right 1 unit. Similarly, the graph of $Y_3 = (x - 3)^2$ is identical to that of $Y_1 = x^2$, except that it is shifted horizontally to the right 3 units. Finally, the graph of $Y_4 = (x + 2)^2$ is identical to that of $Y_1 = x^2$, except that it is shifted horizontally to the left 2 units.

We are led to the following conclusion.

> If the argument x of a function f is replaced by $x - h$, h a real number, the graph of the new function $y = f(x - h)$ is the graph of f **shifted horizontally left** (if $h < 0$) or **right** (if $h > 0$).

EXAMPLE 4 Horizontal Shift to the Left

Use the graph of $f(x) = x^2$ to obtain the graph of $g(x) = (x + 4)^2$.

Solution Again, the function $g(x) = (x + 4)^2$ is basically a square function. Its graph is the same as that of f, except that it is shifted horizontally 4 units to the left. (Do you see why? $(x + 4)^2 = [x - (-4)]^2$.) See Figure 41.

Figure 41

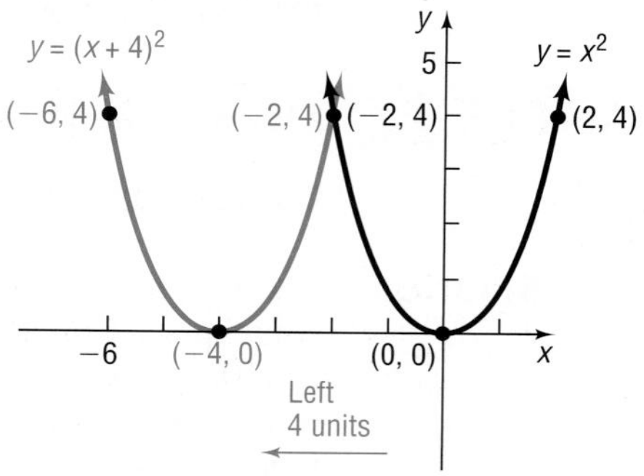

NOW WORK PROBLEM 39.

Vertical and horizontal shifts are sometimes combined.

EXAMPLE 5 Combining Vertical and Horizontal Shifts

Graph the function: $f(x) = (x + 3)^2 - 5$

Solution We graph f in steps. First, we note that the rule for f is basically a square function, so we begin with the graph of $y = x^2$ as shown in Figure 42(a). Next, to get the graph of $y = (x + 3)^2$, we shift the graph of $y = x^2$ horizontally 3 units to the left. See Figure 42(b). Finally, to get the graph of $y = (x + 3)^2 - 5$, we shift the graph of $y = (x + 3)^2$ vertically down 5 units. See Figure 42(c). Note the points plotted on each graph. Using key points can be helpful in keeping track of the transformation that has taken place.

Figure 42

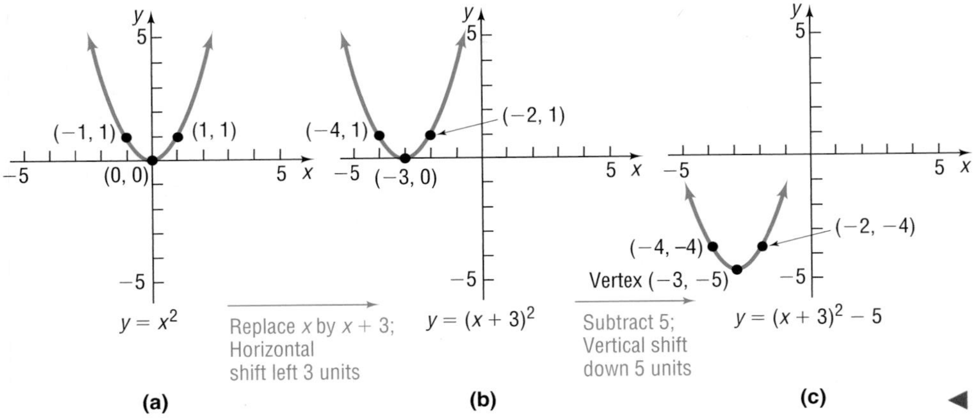

CHECK: Graph $Y_1 = f(x) = (x + 3)^2 - 5$ and compare the graph to Figure 42(c).

In Example 5, if the vertical shift had been done first, followed by the horizontal shift, the final graph would have been the same. Try it for yourself.

NOW WORK PROBLEM 41.

2 Compressions and Stretches

EXAMPLE 6 Vertical Stretch

Use the graph of $f(x) = |x|$ to obtain the graph of $g(x) = 2|x|$.

Solution To see the relationship between the graphs of f and g, we form Table 10, listing points on each graph. For each x, the y-coordinate of a point on the graph of g is 2 times as large as the corresponding y-coordinate on the graph of f. The graph of $f(x) = |x|$ is vertically stretched by a factor of 2 [for example, from (1, 1) to (1, 2)] to obtain the graph of $g(x) = 2|x|$. See Figure 43.

Table 10

x	$y = f(x) = \|x\|$	$y = g(x) = 2\|x\|$
−2	2	4
−1	1	2
0	0	0
1	1	2
2	2	4

Figure 43

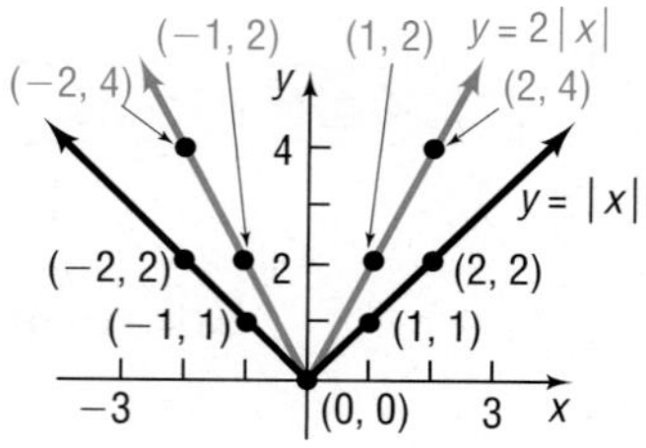

EXAMPLE 7 Vertical Compression

Use the graph of $f(x) = |x|$ to obtain the graph of $g(x) = \frac{1}{2}|x|$.

Solution For each x, the y-coordinate of a point on the graph of g is $\frac{1}{2}$ as large as the corresponding y-coordinate on the graph of f. The graph of $f(x) = |x|$ is vertically compressed by a factor of $\frac{1}{2}$ [for example, from (2, 2) to (2, 1)] to obtain the graph of $g(x) = \frac{1}{2}|x|$. See Table 11 and Figure 44.

Table 11

x	$y = f(x) = \|x\|$	$y = g(x) = \frac{1}{2}\|x\|$
−2	2	1
−1	1	$\frac{1}{2}$
0	0	0
1	1	$\frac{1}{2}$
2	2	1

Figure 44

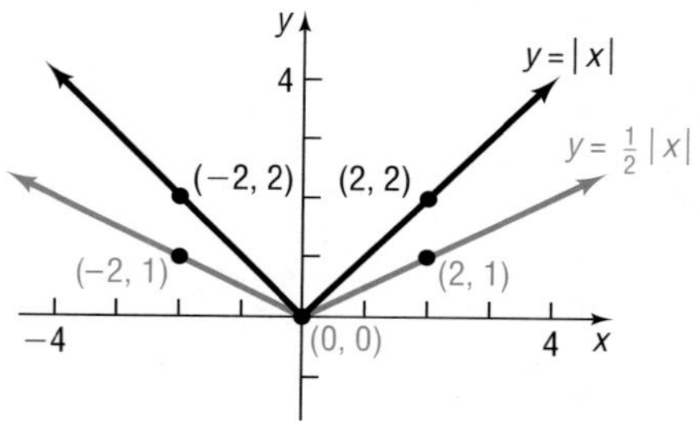

When the right side of a function $y = f(x)$ is multiplied by a positive number a, the graph of the new function $y = af(x)$ is obtained by multiplying each y-coordinate of $y = f(x)$ by a. A **vertical compression** results if $0 < a < 1$ and a **vertical stretch** occurs if $a > 1$.

NOW WORK PROBLEM 43.

What happens if the argument x of a function $y = f(x)$ is multiplied by a positive number a, creating a new function $y = f(ax)$? To find the answer, we look at the following Exploration.

Exploration

On the same screen, graph each of the following functions:

$$Y_1 = f(x) = x^2$$
$$Y_2 = f(2x) = (2x)^2$$
$$Y_3 = f\left(\frac{1}{2}x\right) = \left(\frac{1}{2}x\right)^2$$

Figure 45

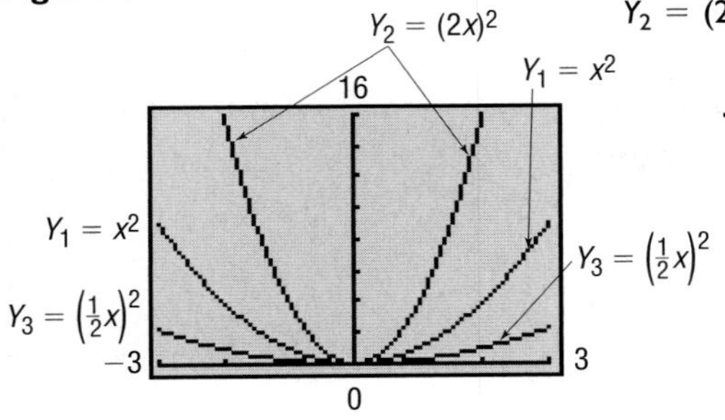

RESULT You should have obtained the graphs shown in Figure 45. The graph of $Y_2 = (2x)^2$ is the graph of $Y_1 = x^2$ compressed horizontally. Look at Table 12.

Table 12

X	Y1	Y2
0	0	0
.5	.25	1
1	1	4
2	4	16
4	16	64
8	64	256
16	256	1024

Y2=(2X)²

(a)

X	Y1	Y3
0	0	0
.5	.25	.0625
1	1	.25
2	4	1
4	16	4
8	64	16
16	256	64

Y3=(X/2)²

(b)

Notice that (1, 1), (2, 4), (4, 16), and (16, 256) are points on the graph of $Y_1 = x^2$. Also, (0.5, 1), (1, 4), (2, 16), and (8, 256) are points on the graph of $Y_2 = (2x)^2$. For each y-coordinate, the x-coordinate on the graph of Y_2 is $\frac{1}{2}$ the x-coordinate on Y_1. The graph of $Y_2 = (2x)^2$ is obtained by multiplying the x-coordinate of each point on the graph of $Y_1 = x^2$ by $\frac{1}{2}$.

The graph of $Y_3 = \left(\frac{1}{2}x\right)^2$ is the graph of $Y_1 = x^2$ stretched horizontally. Look at Table 12(b). Notice that (0.5, 0.25), (1, 1), (2, 4), and (4, 16) are points on the graph of $Y_1 = x^2$. Also, (1, 0.25), (2, 1), (4, 4), and (8, 16) are points on the graph of $Y_3 = \left(\frac{1}{2}x\right)^2$. For each y-coordinate, the x-coordinate on the graph of Y_3 is 2 times the x-coordinate on Y_1. The graph of $Y_3 = \left(\frac{1}{2}x\right)^2$ is obtained by multiplying the x-coordinate of each point on the graph of $Y_1 = x^2$ by a factor of 2.

> If the argument x of a function $y = f(x)$ is multiplied by a positive number a, the graph of the new function $y = f(ax)$ is obtained by multiplying each x-coordinate of $y = f(x)$ by $\frac{1}{a}$. A **horizontal compression** results if $a > 1$, and a **horizontal stretch** occurs if $0 < a < 1$.

Let's look at an example.

EXAMPLE 8 Graphing Using Stretches and Compressions

The graph of $y = f(x)$ is given in Figure 46. Use this graph to find the graphs of:

(a) $y = 3f(x)$ (b) $y = f(3x)$

Figure 46

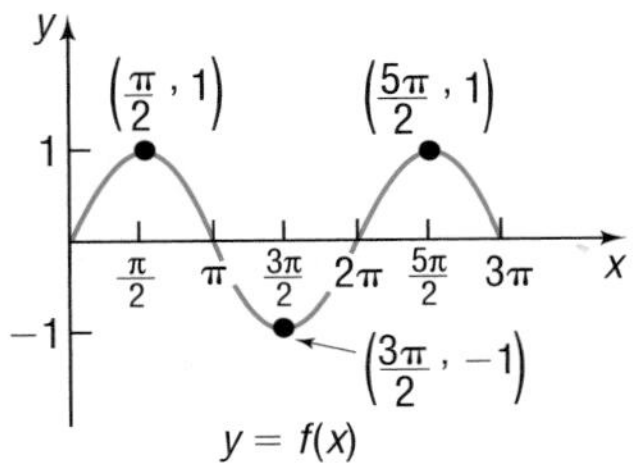

Solution (a) The graph of $y = 3f(x)$ is obtained by multiplying each y-coordinate of $y = f(x)$ by a factor of 3. See Figure 47(a).

(b) The graph of $y = f(3x)$ is obtained from the graph of $y = f(x)$ by multiplying each x-coordinate of $y = f(x)$ by a factor of $\frac{1}{3}$. See Figure 47(b).

Figure 47

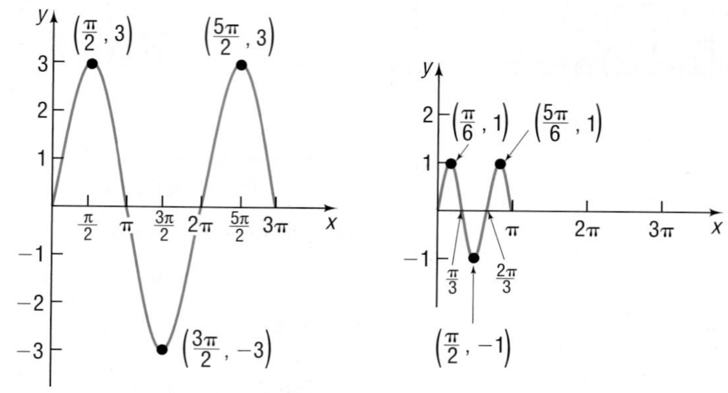

(a) $y = 3f(x)$ (b) $y = f(3x)$

NOW WORK PROBLEMS 65(e) AND (g).

3 Reflections about the x-Axis and the y-Axis

EXAMPLE 9 Reflection about the x-Axis

Graph the function: $f(x) = -x^2$

Solution We begin with the graph of $y = x^2$, as shown in Figure 48. For each point (x, y) on the graph of $y = x^2$, the point $(x, -y)$ is on the graph of $y = -x^2$, as indicated in Table 13. We can draw the graph of $y = -x^2$ by reflecting the graph of $y = x^2$ about the x-axis. See Figure 48.

Figure 48

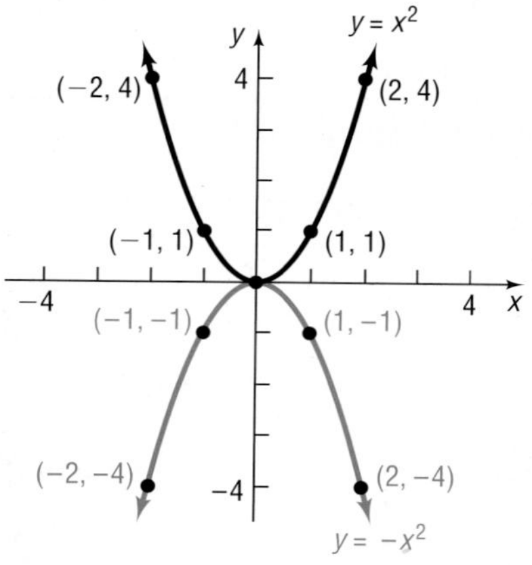

Table 13

x	$y = x^2$	$y = -x^2$
−2	4	−4
−1	1	−1
0	0	0
1	1	−1
2	4	−4

When the right side of the function $y = f(x)$ is multiplied by -1, the graph of the new function $y = -f(x)$ is the **reflection about the x-axis** of the graph of the function $y = f(x)$.

NOW WORK PROBLEM 47.

EXAMPLE 10 Reflection about the y-Axis

Graph the function: $f(x) = \sqrt{-x}$

Solution First, notice that the domain of f consists of all real numbers x for which $-x \geq 0$ or, equivalently, $x \leq 0$. To get the graph of $f(x) = \sqrt{-x}$, we begin with the graph of $y = \sqrt{x}$, as shown in Figure 49. For each point (x, y) on the graph of $y = \sqrt{x}$, the point $(-x, y)$ is on the graph of $y = \sqrt{-x}$. We obtain the graph of $y = \sqrt{-x}$ by reflecting the graph of $y = \sqrt{x}$ about the y-axis. See Figure 49.

Figure 49

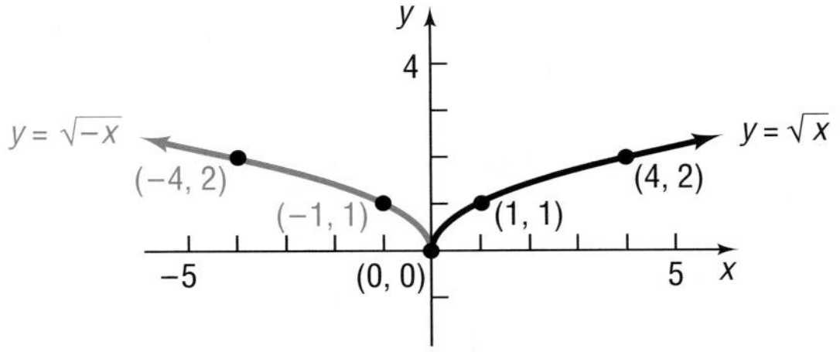

When the graph of the function $y = f(x)$ is known, the graph of the new function $y = f(-x)$ is the **reflection about the y-axis** of the graph of the function $y = f(x)$.

Summary

Summary of Graphing Techniques

Table 14 summarizes the graphing procedures that we have just discussed.

Table 14

To Graph:	Draw the Graph of f and:	Functional Change to $f(x)$
Vertical shifts		
$y = f(x) + k, \quad k > 0$	Raise the graph of f by k units.	Add k to $f(x)$.
$y = f(x) - k, \quad k > 0$	Lower the graph of f by k units.	Subtract k from $f(x)$.
Horizontal shifts		
$y = f(x + h), \quad h > 0$	Shift the graph of f to the left h units.	Replace x by $x + h$.
$y = f(x - h), \quad h > 0$	Shift the graph of f to the right h units.	Replace x by $x - h$.
Compressing or stretching		
$y = af(x), \quad a > 0$	Multiply each y-coordinate of $y = f(x)$ by a. Stretch the graph of f vertically if $a > 1$. Compress the graph of f vertically if $0 < a < 1$.	Multiply $f(x)$ by a.
$y = f(ax), \quad a > 0$	Multiply each x-coordinate of $y = f(x)$ by $\frac{1}{a}$. Stretch the graph of f horizontally if $0 < a < 1$. Compress the graph of f horizontally if $a > 1$.	Replace x by ax.
Reflection about the x-axis		
$y = -f(x)$	Reflect the graph of f about the x-axis.	Multiply $f(x)$ by -1.
Reflection about the y-axis		
$y = f(-x)$	Reflect the graph of f about the y-axis.	Replace x by $-x$.

The examples that follow combine some of the procedures outlined in this section to get the required graph.

EXAMPLE 11 Determining the Function Obtained from a Series of Transformations

Find the function that is finally graphed after the following three transformations are applied to the graph of $y = |x|$.

1. Shift left 2 units.
2. Shift up 3 units.
3. Reflect about the y-axis.

Solution

1. Shift left 2 units: Replace x by $x + 2$. $y = |x + 2|$
2. Shift up 3 units: Add 3. $y = |x + 2| + 3$
3. Reflect about the y-axis: Replace x by $-x$. $y = |-x + 2| + 3$ ◀

NOW WORK PROBLEM 27.

EXAMPLE 12 Combining Graphing Procedures

Graph the function: $f(x) = \dfrac{3}{x-2} + 1$

Solution We use the following steps to obtain the graph of f:

Step 1: $y = \dfrac{1}{x}$ Reciprocal function.

Step 2: $y = \dfrac{3}{x}$ Multiply by 3; vertical stretch of the graph of $y = \dfrac{1}{x}$ by a factor of 3.

Step 3: $y = \dfrac{3}{x-2}$ Replace x by $x - 2$; horizontal shift to the right 2 units.

Step 4: $y = \dfrac{3}{x-2} + 1$ Add 1; vertical shift up 1 unit.

See Figure 50.

Figure 50

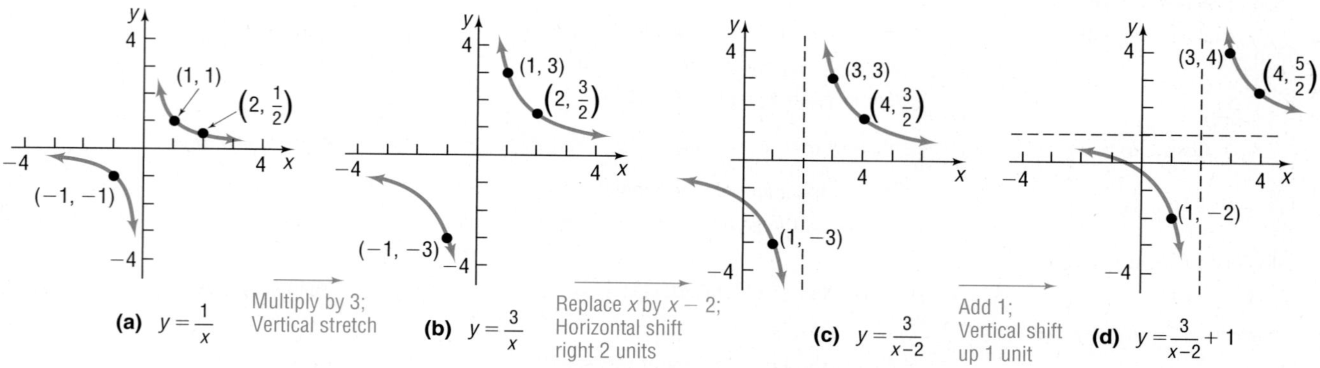

(a) $y = \dfrac{1}{x}$ (b) $y = \dfrac{3}{x}$ (c) $y = \dfrac{3}{x-2}$ (d) $y = \dfrac{3}{x-2} + 1$

◀

Other ordering of the steps shown in Example 12 would also result in the graph of f. For example, try this one:

STEP 1: $y = \dfrac{1}{x}$ — Reciprocal function

STEP 2: $y = \dfrac{1}{x - 2}$ — Replace x by $x - 2$; horizontal shift to the right 2 units.

STEP 3: $y = \dfrac{3}{x - 2}$ — Multiply by 3; vertical stretch of the graph of $y = \dfrac{1}{x - 2}$ by a factor of 3.

STEP 4: $y = \dfrac{3}{x - 2} + 1$ — Add 1; vertical shift up 1 unit.

NOW WORK PROBLEM 57.

EXAMPLE 13 **Combining Graphing Procedures**

Graph the function: $f(x) = \sqrt{1 - x} + 2$

Solution We use the following steps to get the graph of $y = \sqrt{1 - x} + 2$:

STEP 1: $y = \sqrt{x}$ — Square root function

STEP 2: $y = \sqrt{x + 1}$ — Replace x by $x + 1$; horizontal shift left 1 unit.

STEP 3: $y = \sqrt{-x + 1} = \sqrt{1 - x}$ — Replace x by $-x$; reflect about y-axis.

STEP 4: $y = \sqrt{1 - x} + 2$ — Add 2; vertical shift up 2 units.

See Figure 51.

Figure 51

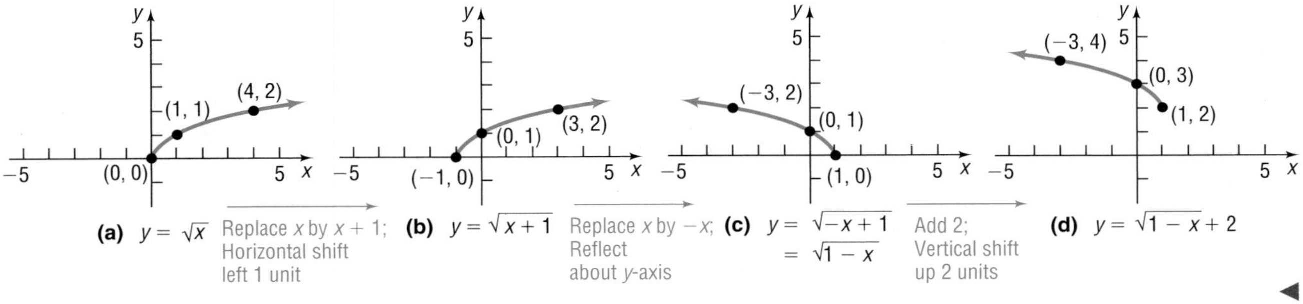

2.5 Assess Your Understanding

Concepts and Vocabulary

1. Suppose that the graph of a function f is known. Then the graph of $y = f(x - 2)$ may be obtained by a(n) __________ shift of the graph of f to the __________ a distance of 2 units.

2. Suppose that the graph of a function f is known. Then the graph of $y = f(-x)$ may be obtained by a reflection about the __________-axis of the graph of the function $y = f(x)$.

3. Suppose that the x-intercepts of the graph of $y = f(x)$ are -2, 1, and 5. The x-intercepts of $y = f(x + 3)$ are __________, __________, and __________.

4. *True or False:* The graph of $y = -f(x)$ is the reflection about the x-axis of the graph of $y = f(x)$.

5. *True or False:* To obtain the graph of $y = f(x + 2) - 3$, shift the graph of $y = f(x)$ horizontally to the right 2 units and vertically down 3 units.

6. *True or False:* Suppose that the x-intercepts of the graph of $y = f(x)$ are -3 and 2. Then the x-intercepts of the graph of $y = 2f(x)$ are -3 and 2.

Exercises

In Problems 7–18, match each graph to one of the following functions.

A. $y = x^2 + 2$	*B.* $y = -x^2 + 2$	*C.* $y = \lvert x\rvert + 2$	*D.* $y = -\lvert x\rvert + 2$
E. $y = (x - 2)^2$	*F.* $y = -(x + 2)^2$	*G.* $y = \lvert x - 2\rvert$	*H.* $y = -\lvert x + 2\rvert$
I. $y = 2x^2$	*J.* $y = -2x^2$	*K.* $y = 2\lvert x\rvert$	*L.* $y = -2\lvert x\rvert$

7.

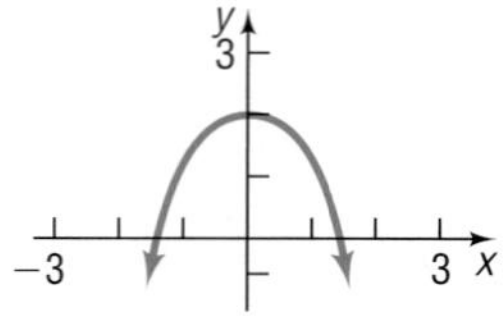

8.

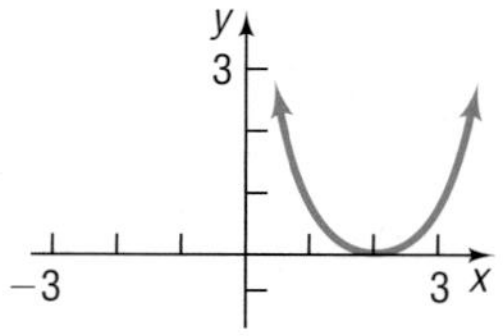

9.

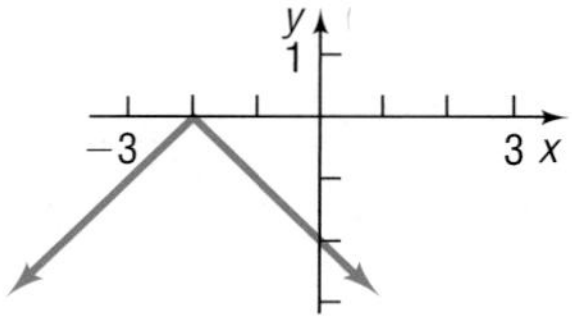

10.

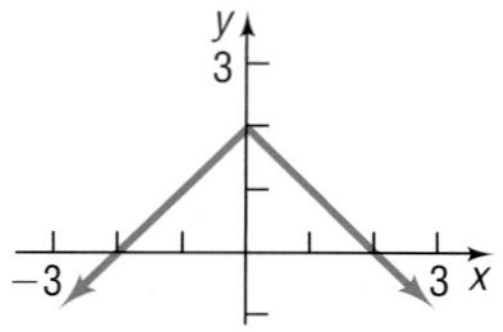

11.

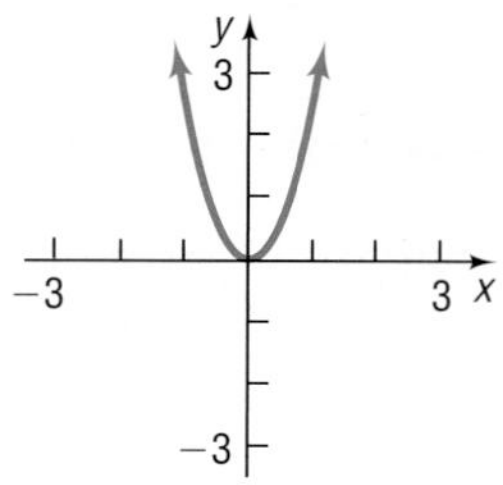

12.

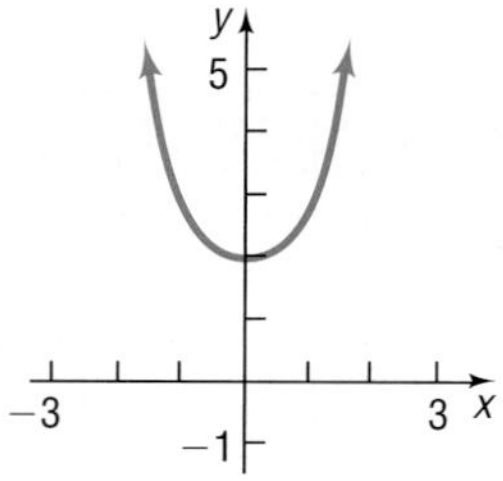

13.

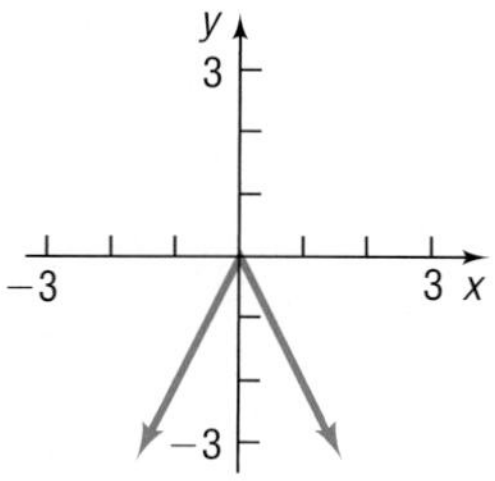

14.

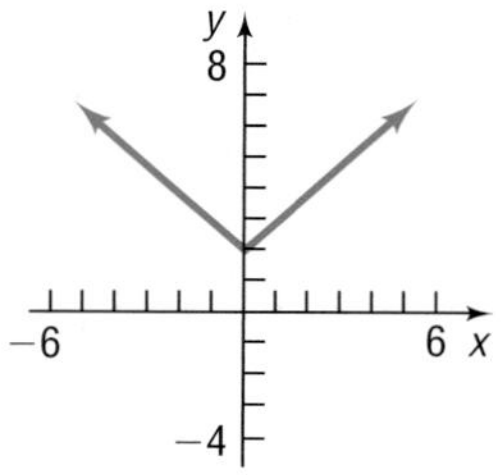

15.

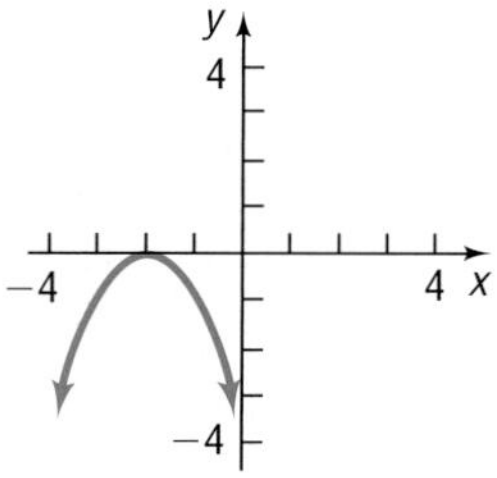

16.

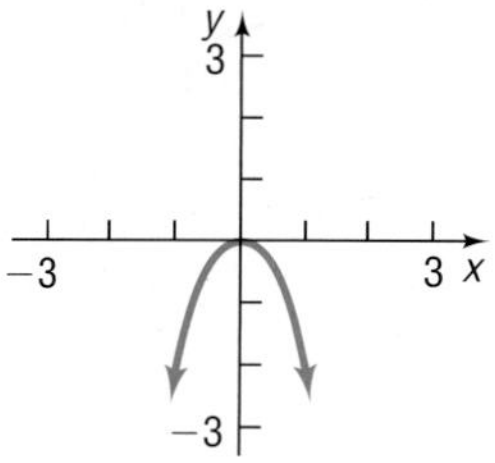

17.

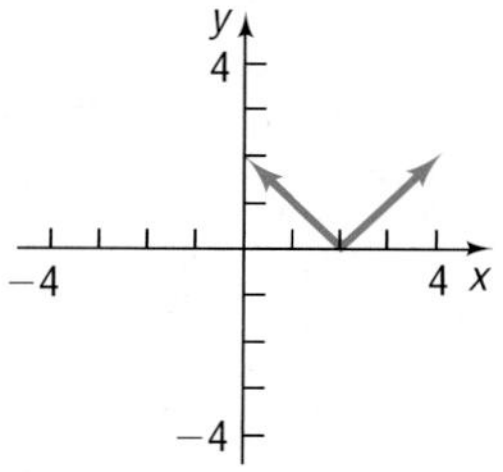

18.

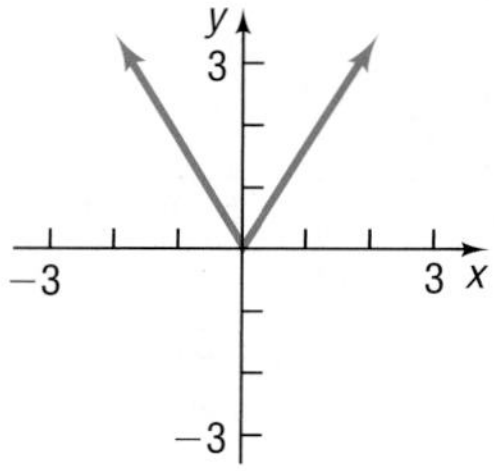

In Problems 19–26, write the function whose graph is the graph of $y = x^3$, *but is:*

19. Shifted to the right 4 units

20. Shifted to the left 4 units

21. Shifted up 4 units

22. Shifted down 4 units

23. Reflected about the *y*-axis

24. Reflected about the *x*-axis

25. Vertically stretched by a factor of 4

26. Horizontally stretched by a factor of 4

In Problems 27–30, find the function that is finally graphed after the following transformations are applied to the graph of $y = \sqrt{x}$.

27. (1) Shift up 2 units
(2) Reflect about the *x*-axis
(3) Reflect about the *y*-axis

28. (1) Reflect about the *x*-axis
(2) Shift right 3 units
(3) Shift down 2 units

29. (1) Reflect about the *x*-axis
(2) Shift up 2 units
(3) Shift left 3 units

30. (1) Shift up 2 units
(2) Reflect about the *y*-axis
(3) Shift left 3 units

31. If $(3, 0)$ is a point on the graph of $y = f(x)$, which of the following must be on the graph of $y = -f(x)$?
(a) $(0, 3)$ (b) $(0, -3)$
(c) $(3, 0)$ (d) $(-3, 0)$

32. If $(3, 0)$ is a point on the graph of $y = f(x)$, which of the following must be on the graph of $y = f(-x)$?
(a) $(0, 3)$ (b) $(0, -3)$
(c) $(3, 0)$ (d) $(-3, 0)$

33. If $(0, 3)$ is a point on the graph of $y = f(x)$, which of the following must be on the graph of $y = 2f(x)$?
(a) $(0, 3)$ (b) $(0, 2)$
(c) $(0, 6)$ (d) $(6, 0)$

34. If $(3, 0)$ is a point on the graph of $y = f(x)$, which of the following must be on the graph of $y = \frac{1}{2}f(x)$?
(a) $(3, 0)$ (b) $\left(\frac{3}{2}, 0\right)$
(c) $\left(0, \frac{3}{2}\right)$ (d) $\left(\frac{1}{2}, 0\right)$

In Problems 35–64, graph each function using the techniques of shifting, compressing, stretching, and/or reflecting. Start with the graph of the basic function (for example, $y = x^2$) and show all stages.

35. $f(x) = x^2 - 1$
36. $f(x) = x^2 + 4$
37. $g(x) = x^3 + 1$
38. $g(x) = x^3 - 1$
39. $h(x) = \sqrt{x - 2}$
40. $h(x) = \sqrt{x + 1}$
41. $f(x) = (x - 1)^3 + 2$
42. $f(x) = (x + 2)^3 - 3$
43. $g(x) = 4\sqrt{x}$
44. $g(x) = \frac{1}{2}\sqrt{x}$
45. $h(x) = \frac{1}{2x}$
46. $h(x) = 3\sqrt[3]{x}$
47. $f(x) = -\sqrt[3]{x}$
48. $f(x) = -\sqrt{x}$
49. $g(x) = |-x|$
50. $g(x) = \sqrt[3]{-x}$
51. $h(x) = -x^3 + 2$
52. $h(x) = \frac{1}{-x} + 2$
53. $f(x) = 2(x + 1)^2 - 3$
54. $f(x) = 3(x - 2)^2 + 1$
55. $g(x) = \sqrt{x - 2} + 1$
56. $g(x) = |x + 1| - 3$
57. $h(x) = \sqrt{-x} - 2$
58. $h(x) = \frac{4}{x} + 2$
59. $f(x) = -(x + 1)^3 - 1$
60. $f(x) = -4\sqrt{x - 1}$
61. $g(x) = 2|1 - x|$
62. $g(x) = 4\sqrt{2 - x}$
63. $h(x) = 2\,\text{int}(x - 1)$
64. $h(x) = \text{int}(-x)$

In Problems 65–70, the graph of a function f is illustrated. Use the graph of f as the first step toward graphing each of the following functions:

(a) $F(x) = f(x) + 3$ (b) $G(x) = f(x + 2)$ (c) $P(x) = -f(x)$ (d) $H(x) = f(x + 1) - 2$
(e) $Q(x) = \frac{1}{2}f(x)$ (f) $g(x) = f(-x)$ (g) $h(x) = f(2x)$

65.

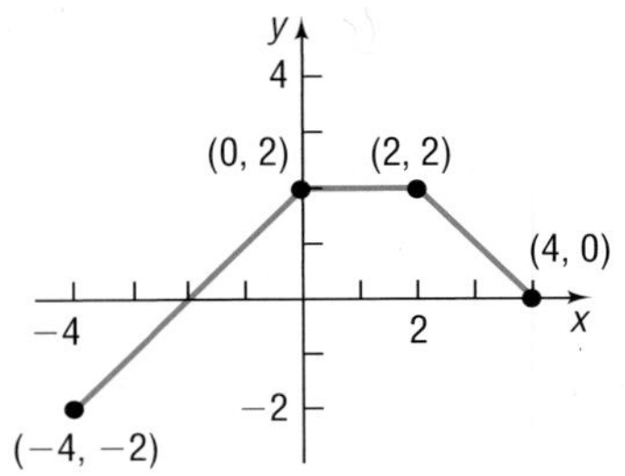

66.

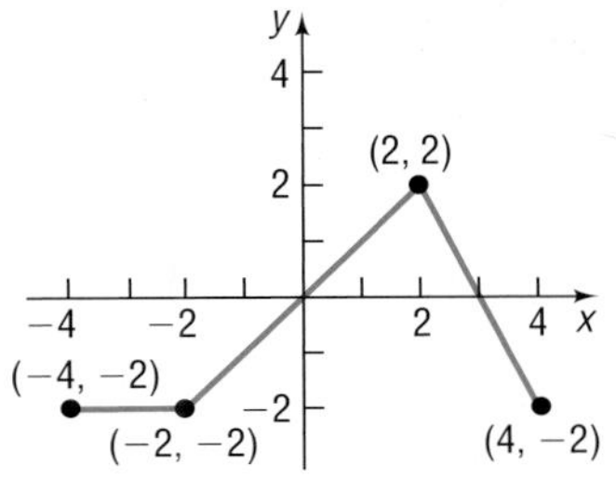

67.

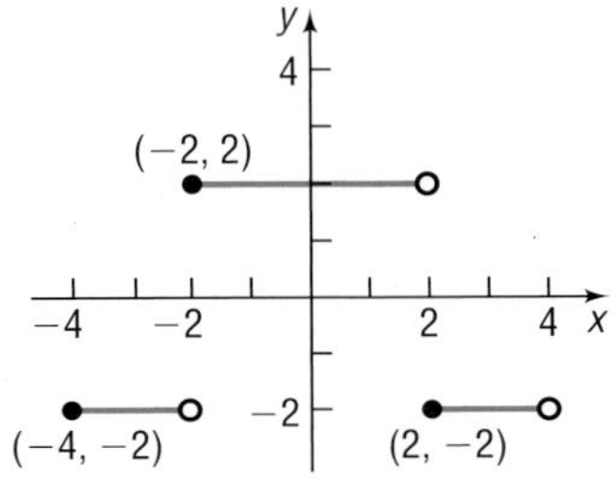

68.

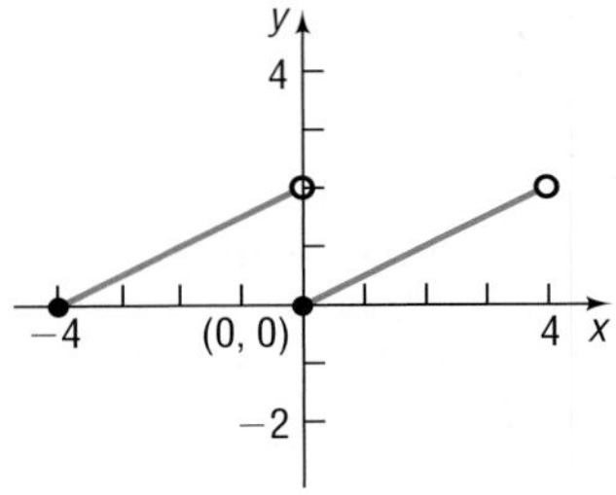

69.

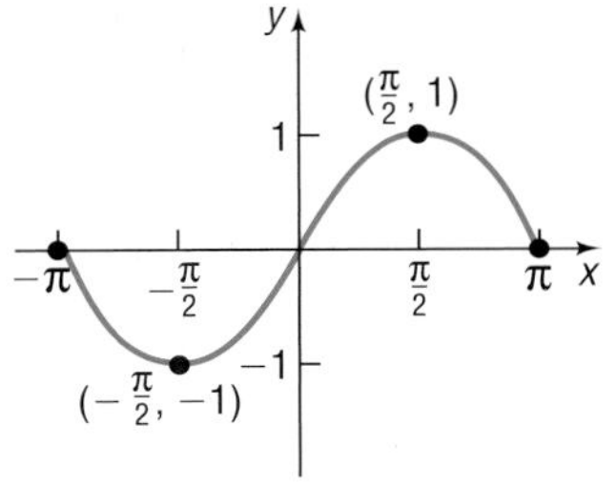

70.

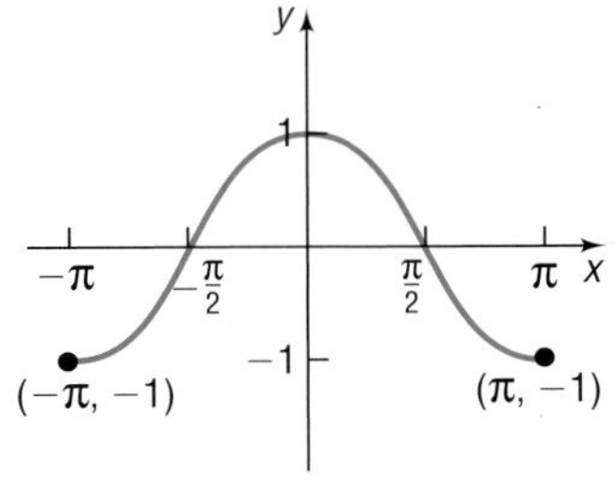

71. Exploration
(a) Use a graphing utility to graph $y = x + 1$ and $y = |x + 1|$.
(b) Graph $y = 4 - x^2$ and $y = |4 - x^2|$.
(c) Graph $y = x^3 + x$ and $y = |x^3 + x|$.
(d) What do you conclude about the relationship between the graphs of $y = f(x)$ and $y = |f(x)|$?

72. Exploration
(a) Use a graphing utility to graph $y = x + 1$ and $y = |x| + 1$.
(b) Graph $y = 4 - x^2$ and $y = 4 - |x|^2$.
(c) Graph $y = x^3 + x$ and $y = |x|^3 + |x|$.
(d) What do you conclude about the relationship between the graphs of $y = f(x)$ and $y = f(|x|)$?

73. The graph of a function f is illustrated in the figure.

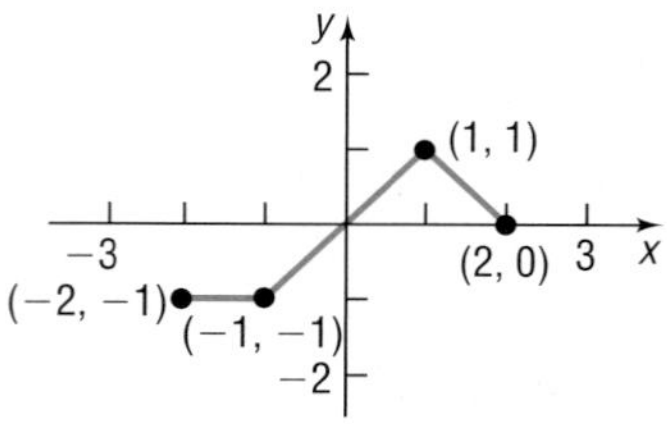

(a) Draw the graph of $y = |f(x)|$.
(b) Draw the graph of $y = f(|x|)$.

74. The graph of a function f is illustrated in the figure.

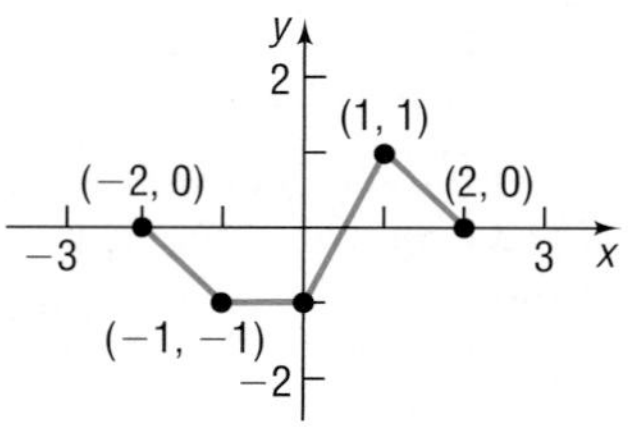

(a) Draw the graph of $y = |f(x)|$.
(b) Draw the graph of $y = f(|x|)$.

In Problems 75–80, complete the square of each quadratic expression. Then graph each function using the technique of shifting. (If necessary, refer to Appendix A, Section A.5 to review completing the square.)

75. $f(x) = x^2 + 2x$
76. $f(x) = x^2 - 6x$
77. $f(x) = x^2 - 8x + 1$
78. $f(x) = x^2 + 4x + 2$
79. $f(x) = x^2 + x + 1$
80. $f(x) = x^2 - x + 1$

81. The equation $y = (x - c)^2$ defines a *family of parabolas*, one parabola for each value of c. On one set of coordinate axes, graph the members of the family for $c = 0$, $c = 3$, and $c = -2$.

82. Repeat Problem 81 for the family of parabolas $y = x^2 + c$.

83. Temperature Measurements The relationship between the Celsius (°C) and Fahrenheit (°F) scales for measuring temperature is given by the equation

$$F = \frac{9}{5}C + 32$$

The relationship between the Celsius (°C) and Kelvin (K) scales is $K = C + 273$. Graph the equation $F = \frac{9}{5}C + 32$ using degrees Fahrenheit on the y-axis and degrees Celsius on the x-axis. Use the techniques introduced in this section to obtain the graph showing the relationship between Kelvin and Fahrenheit temperatures.

84. Period of a Pendulum The period T (in seconds) of a simple pendulum is a function of its length l (in feet) defined by the equation

$$T = 2\pi\sqrt{\frac{l}{g}}$$

where $g \approx 32.2$ feet per second per second is the acceleration of gravity.

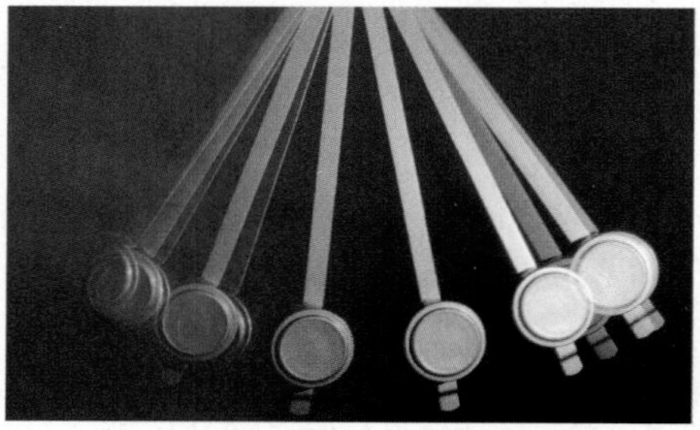

(a) Use a graphing utility to graph the function $T = T(l)$.
(b) Now graph the functions $T = T(l + 1)$, $T = T(l + 2)$, and $T = T(l + 3)$.
(c) Discuss how adding to the length l changes the period T.
(d) Now graph the functions $T = T(2l)$, $T = T(3l)$, and $T = T(4l)$.
(e) Discuss how multiplying the length l by factors of 2, 3, and 4 changes the period T.

85. Cigar Company Profits The daily profits of a cigar company from selling x cigars are given by

$$p(x) = -0.05x^2 + 100x - 2000$$

The government wishes to impose a tax on cigars (sometimes called a *sin tax*) that gives the company the option of either paying a flat tax of \$10,000 per day or a tax of 10% on profits. As chief financial officer of the company, you need to decide which tax is the better option for the company.

(a) On the same screen, graph $Y_1 = p(x) - 10{,}000$ and $Y_2 = (1 - 0.10)p(x)$.

(b) Based on the graph, which option would you select? Why?

(c) Using the terminology learned in this section, describe each graph in terms of the graph of $p(x)$.

(d) Suppose that the government offered the options of a flat tax of \$4800 or a tax of 10% on profits. Which would you select? Why?

86. Suppose that the graph of a function f is known. Explain how the graph of $y = 4f(x)$ differs from the graph of $y = f(4x)$.

2.6 Mathematical Models: Constructing Functions

OBJECTIVE 1 Construct and Analyze Functions

1 Real-world problems often result in mathematical models that involve functions. These functions need to be constructed or built based on the information given. In constructing functions, we must be able to translate the verbal description into the language of mathematics. We do this by assigning symbols to represent the independent and dependent variables and then finding the function or rule that relates these variables.

EXAMPLE 1 Area of a Rectangle with Fixed Perimeter

The perimeter of a rectangle is 50 feet. Express its area A as a function of the length l of a side.

Figure 52

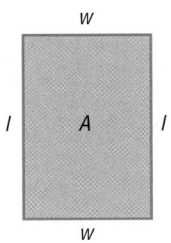

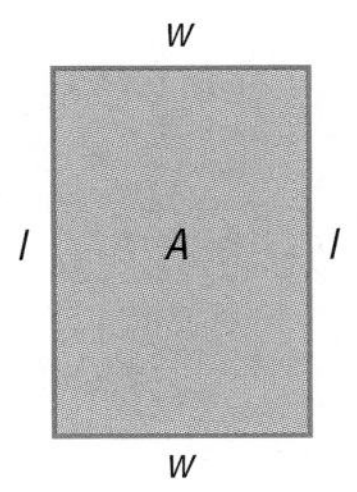

Solution Consult Figure 52. If the length of the rectangle is l and if w is its width, then the sum of the lengths of the sides is the perimeter, 50.

$$l + w + l + w = 50$$
$$2l + 2w = 50$$
$$l + w = 25$$
$$w = 25 - l$$

The area A is length times width, so

$$A = lw = l(25 - l)$$

The area A as a function of l is

$$A(l) = l(25 - l)$$

◄

Note that we use the symbol A as the dependent variable and also as the name of the function that relates the length l to the area A. As we mentioned earlier, this double usage is common in applications and should cause no difficulties.

EXAMPLE 2 Economics: Demand Equations

In economics, revenue R is defined as the amount of money received from the sale of a product and is equal to the unit selling price p of the product times the number x of units actually sold. That is,

$$R = xp$$

In economics, the Law of Demand states that p and x are related: As one increases, the other decreases. Suppose that p and x are related by the following **demand equation:**

$$p = -\frac{1}{10}x + 20, \qquad 0 \le x \le 200$$

Express the revenue R as a function of the number x of units sold.

Solution Since $R = xp$ and $p = -\frac{1}{10}x + 20$, it follows that

$$R(x) = xp = x\left(-\frac{1}{10}x + 20\right) = -\frac{1}{10}x^2 + 20x$$

NOW WORK PROBLEM 3.

EXAMPLE 3 Finding the Distance from the Origin to a Point on a Graph

Let $P = (x, y)$ be a point on the graph of $y = x^2 - 1$.

(a) Express the distance d from P to the origin O as a function of x.

(b) What is d if $x = 0$?

(c) What is d if $x = 1$?

(d) What is d if $x = \frac{\sqrt{2}}{2}$?

(e) Use a graphing utility to graph the function $d = d(x)$, $x \ge 0$. Rounded to two decimal places, find the value(s) of x at which d has a local minimum. [This gives the point(s) on the graph of $y = x^2 - 1$ closest to the origin.]

Figure 53

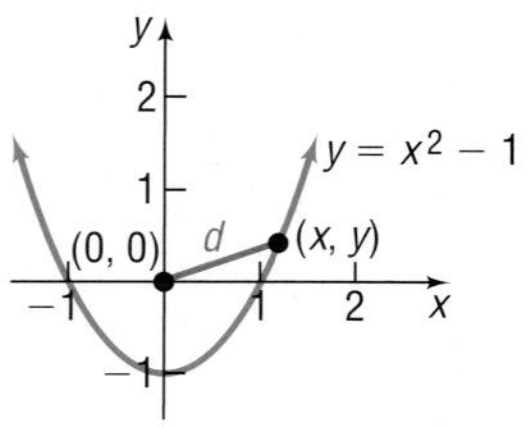

Solution (a) Figure 53 illustrates the graph. The distance d from P to O is

$$d = \sqrt{(x - 0)^2 + (y - 0)^2} = \sqrt{x^2 + y^2}$$

Since P is a point on the graph of $y = x^2 - 1$, we have

$$d(x) = \sqrt{x^2 + (x^2 - 1)^2} = \sqrt{x^4 - x^2 + 1}$$

We have expressed the distance d as a function of x.

(b) If $x = 0$, the distance d is

$$d(0) = \sqrt{1} = 1$$

(c) If $x = 1$, the distance d is

$$d(1) = \sqrt{1 - 1 + 1} = 1$$

(d) If $x = \frac{\sqrt{2}}{2}$, the distance d is

$$d\left(\frac{\sqrt{2}}{2}\right) = \sqrt{\left(\frac{\sqrt{2}}{2}\right)^4 - \left(\frac{\sqrt{2}}{2}\right)^2 + 1} = \sqrt{\frac{1}{4} - \frac{1}{2} + 1} = \frac{\sqrt{3}}{2}$$

Figure 54

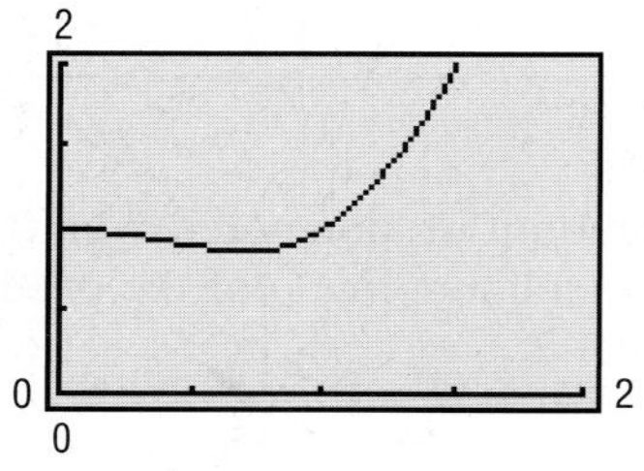

(e) Figure 54 shows the graph of $Y_1 = \sqrt{x^4 - x^2 + 1}$. Using the MINIMUM feature on a graphing utility, we find that when $x \approx 0.71$ the value of d is smallest ($d \approx 0.87$ rounded to two decimal places is a local minimum). By symmetry, it follows that when $x \approx -0.71$, the value of d is also a local minimum.

NOW WORK PROBLEM 9.

EXAMPLE 4 Filling a Swimming Pool

A rectangular swimming pool 20 meters long and 10 meters wide is 4 meters deep at one end and 1 meter deep at the other. Figure 55 illustrates a cross-sectional view of the pool. Water is being pumped into the pool to a height of 3 meters at the deep end.

Figure 55

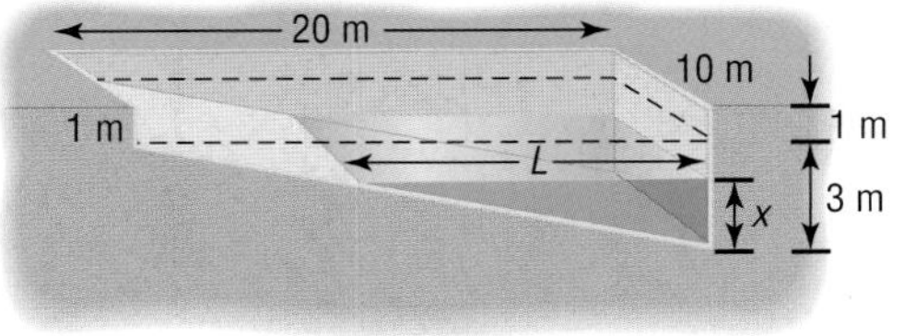

(a) Find a function that expresses the volume V of water in the pool as a function of the height x of the water at the deep end.

(b) Find the volume when the height is 1 meter.

(c) Find the volume when the height is 2 meters.

(d) Use a graphing utility to graph the function $V = V(x)$. At what height is the volume 20 cubic meters? 100 cubic meters?

Solution (a) Let L denote the distance (in meters) measured at water level from the deep end to the short end. Notice that L and x form the sides of a triangle that is similar to the triangle whose sides are 20 meters by 3 meters. Thus, L and x are related by the equation

$$\frac{L}{x} = \frac{20}{3} \quad \text{or} \quad L = \frac{20x}{3}, \qquad 0 \le x \le 3$$

The volume V of water in the pool at any time is

$$V = \begin{pmatrix}\text{cross-sectional}\\ \text{triangular area}\end{pmatrix}(\text{width}) = \left(\frac{1}{2}Lx\right)(10) \quad \text{cubic meters}$$

Since $L = \dfrac{20x}{3}$, we have

$$V(x) = \left(\frac{1}{2}\cdot\frac{20x}{3}\cdot x\right)(10) = \frac{100}{3}x^2 \quad \text{cubic meters}$$

(b) When the height x of the water is 1 meter, the volume $V = V(x)$ is

$$V(1) = \frac{100}{3}\cdot 1^2 = 33\frac{1}{3} \quad \text{cubic meters}$$

(c) When the height x of the water is 2 meters, the volume $V = V(x)$ is

$$V(2) = \frac{100}{3}\cdot 2^2 = \frac{400}{3} = 133\frac{1}{3} \quad \text{cubic meters}$$

Figure 56

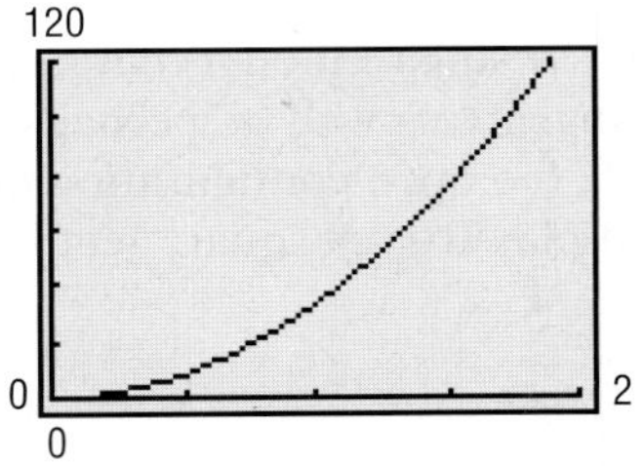

(d) See Figure 56. Use TRACE. When $x \approx 0.77$ meter, the volume is 20 cubic meters. When $x \approx 1.73$ meters, the volume is 100 cubic meters. ◀

EXAMPLE 5 Area of a Rectangle

A rectangle has one corner on the graph of $y = 25 - x^2$, another at the origin, a third on the positive y-axis, and the fourth on the positive x-axis. See Figure 57.

Figure 57

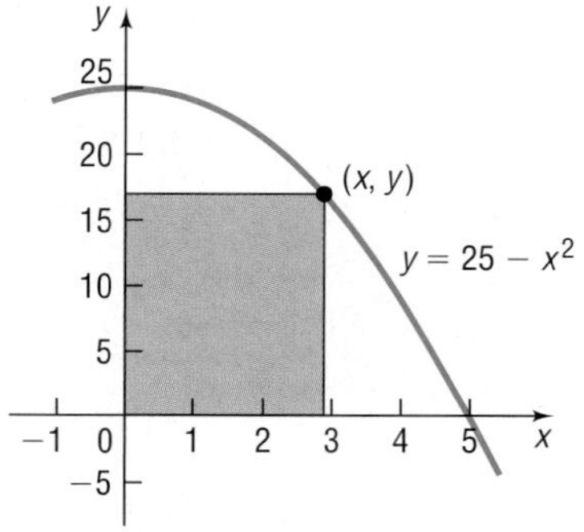

(a) Express the area A of the rectangle as a function of x.

(b) What is the domain of A?

(c) Graph $A = A(x)$.

(d) For what value of x is the area largest?

Solution (a) The area A of the rectangle is $A = xy$, where $y = 25 - x^2$. Substituting this expression for y, we obtain $A(x) = x(25 - x^2) = 25x - x^3$.

(b) Since x represents a side of the rectangle, we have $x > 0$. In addition, the area must be positive, so $y = 25 - x^2 > 0$, which implies that $x^2 < 25$ so $-5 < x < 5$. Combining these restrictions, we have the domain of A as $\{x \mid 0 < x < 5\}$ or $(0, 5)$ using interval notation.

(c) See Figure 58 for the graph of $A(x)$.

(d) Using MAXIMUM, we find that the area is a maximum of 48.11 at $x = 2.89$, each rounded to two decimal places. See Figure 59.

Figure 58

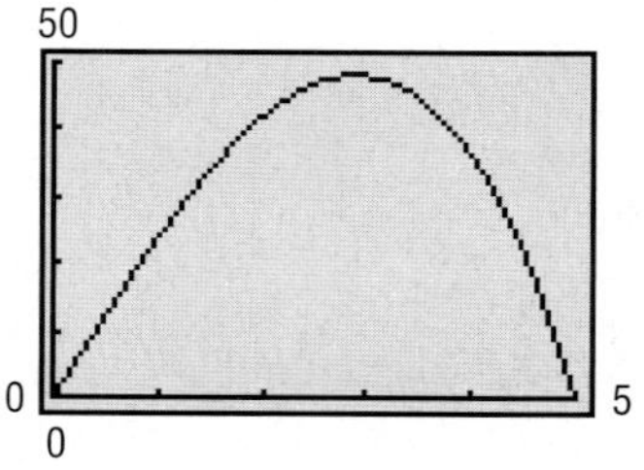

Figure 59

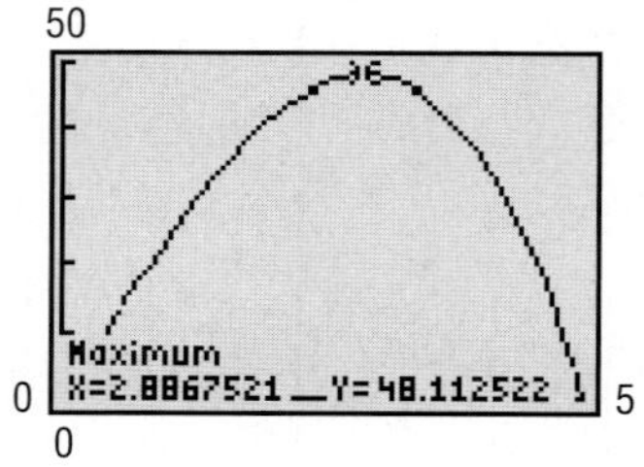

NOW WORK PROBLEM 15.

EXAMPLE 6 Making a Playpen*

A manufacturer of children's playpens makes a square model that can be opened at one corner and attached at right angles to a wall or, perhaps, the side of a house. If each side is 3 feet in length, the open configuration doubles the available area in which the child can play from 9 square feet to 18 square feet. See Figure 60.

*Adapted from Proceedings, *Summer Conference for College Teachers on Applied Mathematics* (University of Missouri, Rolla), 1971.

Now suppose that we place hinges at the outer corners to allow for a configuration like the one shown in Figure 61.

Figure 60

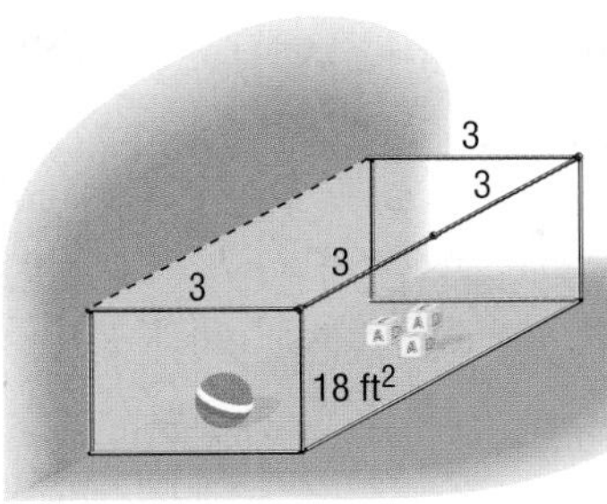

Figure 61

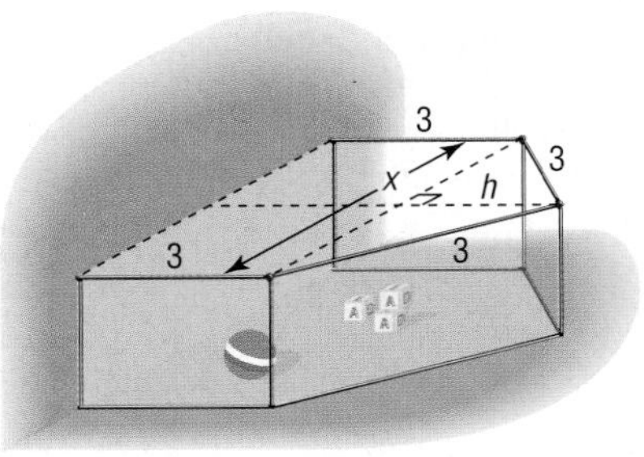

(a) Express the area A of this configuration as a function of the distance x between the two parallel sides.

(b) Find the domain of A.

(c) Find A if $x = 5$.

(d) Graph $A = A(x)$. For what value of x is the area largest? What is the maximum area?

Solution (a) Refer to Figure 61. The area A that we seek consists of the area of a rectangle (with width 3 and length x) and the area of an isosceles triangle (with base x and two equal sides of length 3). The height h of the triangle may be found using the Pythagorean Theorem.

$$h^2 = 3^2 - \left(\frac{x}{2}\right)^2 = 9 - \frac{x^2}{4} = \frac{36 - x^2}{4}$$

$$h = \frac{1}{2}\sqrt{36 - x^2}$$

The area A enclosed by the playpen is

$$A = \text{area of rectangle} + \text{area of triangle} = 3x + \frac{1}{2}x\left(\frac{1}{2}\sqrt{36 - x^2}\right)$$

$$A(x) = 3x + \frac{x\sqrt{36 - x^2}}{4}$$

Now the area A is expressed as a function of x.

(b) To find the domain of A, we note first that $x > 0$, since x is a length. Also, the expression under the radical must be positive, so

$$36 - x^2 > 0$$
$$x^2 < 36$$
$$-6 < x < 6$$

Combining these restrictions, we find that the domain of A is $0 < x < 6$, or $(0, 6)$ using interval notation.

Figure 62

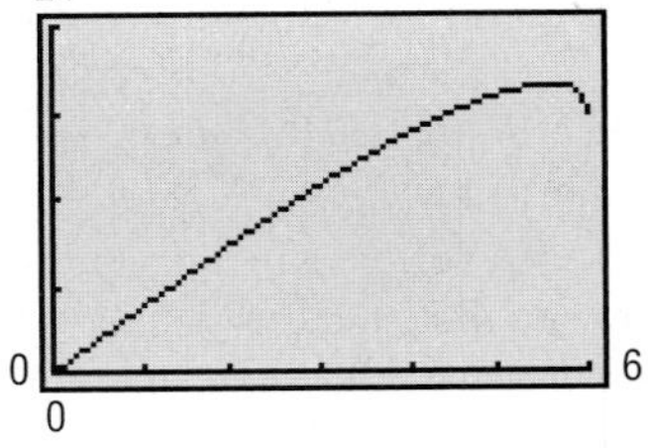

(c) If $x = 5$, the area is

$$A(5) = 3(5) + \frac{5}{4}\sqrt{36 - (5)^2} \approx 19.15 \text{ square feet}$$

If the width of the playpen is 5 feet, its area is 19.15 square feet.

(d) See Figure 62. The maximum area is about 19.82 square feet, obtained when x is about 5.58 feet. ◀

2.6 Assess Your Understanding

Exercises

1. **Volume of a Cylinder** The volume V of a right circular cylinder of height h and radius r is $V = \pi r^2 h$. If the height is twice the radius, express the volume V as a function of r.

2. **Volume of a Cone** The volume V of a right circular cone is $V = \frac{1}{3}\pi r^2 h$. If the height is twice the radius, express the volume V as a function of r.

3. **Demand Equation** The price p and the quantity x sold of a certain product obey the demand equation

$$p = -\frac{1}{6}x + 100, \qquad 0 \le x \le 600$$

(a) Express the revenue R as a function of x. (Remember, $R = xp$.)
(b) What is the revenue if 200 units are sold?
(c) Graph the revenue function using a graphing utility.
(d) What quantity x maximizes revenue? What is the maximum revenue?
(e) What price should the company charge to maximize revenue?

4. **Demand Equation** The price p and the quantity x sold of a certain product obey the demand equation

$$p = -\frac{1}{3}x + 100, \qquad 0 \le x \le 300$$

(a) Express the revenue R as a function of x.
(b) What is the revenue if 100 units are sold?
(c) Graph the revenue function using a graphing utility.
(d) What quantity x maximizes revenue? What is the maximum revenue?
(e) What price should the company charge to maximize revenue?

5. **Demand Equation** The price p and the quantity x sold of a certain product obey the demand equation

$$x = -5p + 100, \qquad 0 \le p \le 20$$

(a) Express the revenue R as a function of x.
(b) What is the revenue if 15 units are sold?
(c) Graph the revenue function using a graphing utility.
(d) What quantity x maximizes revenue? What is the maximum revenue?
(e) What price should the company charge to maximize revenue?

6. **Demand Equation** The price p and the quantity x sold of a certain product obey the demand equation

$$x = -20p + 500, \qquad 0 \le p \le 25$$

(a) Express the revenue R as a function of x.
(b) What is the revenue if 20 units are sold?
(c) Graph the revenue function using a graphing utility.
(d) What quantity x maximizes revenue? What is the maximum revenue?
(e) What price should the company charge to maximize revenue?

7. **Enclosing a Rectangular Field** David has available 400 yards of fencing and wishes to enclose a rectangular area.
(a) Express the area A of the rectangle as a function of the width x of the rectangle.
(b) What is the domain of A?
(c) Graph $A = A(x)$ using a graphing utility. For what value of x is the area largest?

8. **Enclosing a Rectangular Field along a River** Beth has 3000 feet of fencing available to enclose a rectangular field. One side of the field lies along a river, so only three sides require fencing.
(a) Express the area A of the rectangle as a function of x, where x is the length of the side parallel to the river.
(b) Graph $A = A(x)$ using a graphing utility. For what value of x is the area largest?

9. Let $P = (x, y)$ be a point on the graph of $y = x^2 - 8$.
(a) Express the distance d from P to the origin as a function of x.
(b) What is d if $x = 0$?
(c) What is d if $x = 1$?
(d) Use a graphing utility to graph $d = d(x)$.
(e) For what values of x is d smallest?

10. Let $P = (x, y)$ be a point on the graph of $y = x^2 - 8$.
(a) Express the distance d from P to the point $(0, -1)$ as a function of x.
(b) What is d if $x = 0$?
(c) What is d if $x = -1$?
(d) Use a graphing utility to graph $d = d(x)$.
(e) For what values of x is d smallest?

11. Let $P = (x, y)$ be a point on the graph of $y = \sqrt{x}$.
(a) Express the distance d from P to the point $(1, 0)$ as a function of x.
(b) Use a graphing utility to graph $d = d(x)$.
(c) For what values of x is d smallest?

12. Let $P = (x, y)$ be a point on the graph of $y = \frac{1}{x}$.
(a) Express the distance d from P to the origin as a function of x.
(b) Use a graphing utility to graph $d = d(x)$.
(c) For what values of x is d smallest?

13. A right triangle has one vertex on the graph of $y = x^3, x > 0$, at (x, y), another at the origin, and the third on the positive y-axis at $(0, y)$, as shown in the figure. Express the area A of the triangle as a function of x.

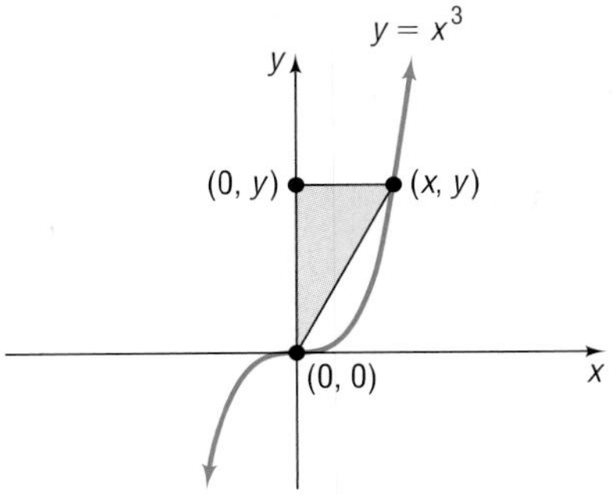

14. A right triangle has one vertex on the graph of $y = 9 - x^2$, $x > 0$, at (x, y), another at the origin, and the third on the positive x-axis at $(x, 0)$. Express the area A of the triangle as a function of x.

15. A rectangle has one corner on the graph of $y = 16 - x^2$, another at the origin, a third on the positive y-axis, and the fourth on the positive x-axis (see the figure).

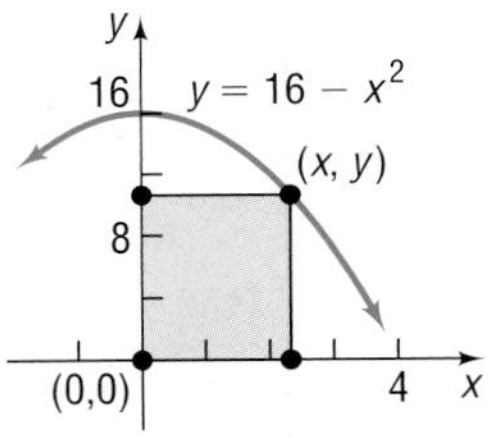

(a) Express the area A of the rectangle as a function of x.
(b) What is the domain of A?
(c) Graph $A = A(x)$. For what value of x is A largest?

16. A rectangle is inscribed in a semicircle of radius 2 (see the figure). Let $P = (x, y)$ be the point in quadrant I that is a vertex of the rectangle and is on the circle.

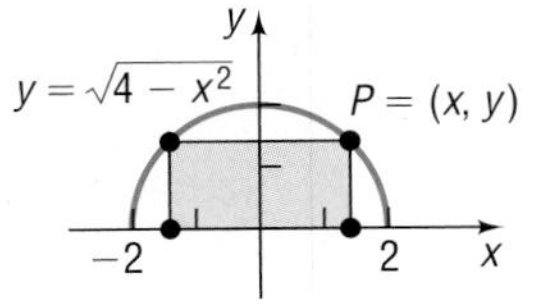

(a) Express the area A of the rectangle as a function of x.
(b) Express the perimeter p of the rectangle as a function of x.
(c) Graph $A = A(x)$. For what value of x is A largest?
(d) Graph $p = p(x)$. For what value of x is p largest?

17. A rectangle is inscribed in a circle of radius 2 (see the figure). Let $P = (x, y)$ be the point in quadrant I that is a vertex of the rectangle and is on the circle.

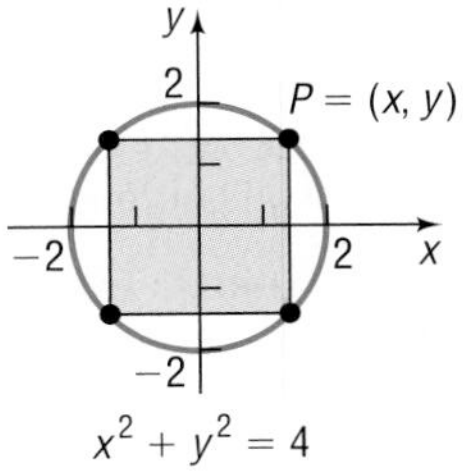

(a) Express the area A of the rectangle as a function of x.
(b) Express the perimeter p of the rectangle as a function of x.
(c) Graph $A = A(x)$. For what value of x is A largest?
(d) Graph $p = p(x)$. For what value of x is p largest?

18. A circle of radius r is inscribed in a square (see the figure).

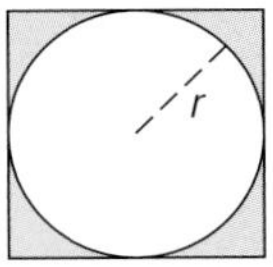

(a) Express the area A of the square as a function of the radius r of the circle.
(b) Express the perimeter p of the square as a function of r.

19. A wire 10 meters long is to be cut into two pieces. One piece will be shaped as a square, and the other piece will be shaped as a circle (see the figure).

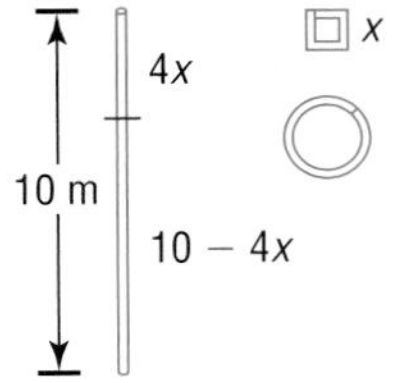

(a) Express the total area A enclosed by the pieces of wire as a function of the length x of a side of the square.
(b) What is the domain of A?
(c) Graph $A = A(x)$. For what value of x is A smallest?

20. A wire 10 meters long is to be cut into two pieces. One piece will be shaped as an equilateral triangle, and the other piece will be shaped as a circle.
(a) Express the total area A enclosed by the pieces of wire as a function of the length x of a side of the equilateral triangle.
(b) What is the domain of A?
(c) Graph $A = A(x)$. For what value of x is A smallest?

21. A wire of length x is bent into the shape of a circle.
(a) Express the circumference of the circle as a function of x.
(b) Express the area of the circle as a function of x.

22. A wire of length x is bent into the shape of a square.
(a) Express the perimeter of the square as a function of x.
(b) Express the area of the square as a function of x.

23. A semicircle of radius r is inscribed in a rectangle so that the diameter of the semicircle is the length of the rectangle (see the figure).

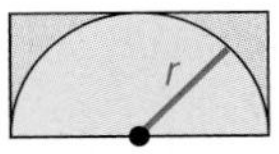

(a) Express the area A of the rectangle as a function of the radius r of the semicircle.
(b) Express the perimeter p of the rectangle as a function of r.

24. An equilateral triangle is inscribed in a circle of radius r. See the figure. Express the circumference C of the circle as a function of the length x of a side of the triangle.

[**Hint:** First show that $r^2 = \frac{x^2}{3}$.]

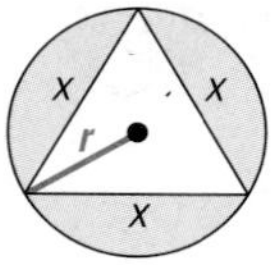

25. An equilateral triangle is inscribed in a circle of radius r. See the figure. Express the area A within the circle, but outside the triangle, as a function of the length x of a side of the triangle.

26. Two cars leave an intersection at the same time. One is headed south at a constant speed of 30 miles per hour, and the other is headed west at a constant speed of 40 miles per hour (see the figure). Express the distance d between the cars as a function of the time t.

[**Hint:** At $t = 0$, the cars leave the intersection.]

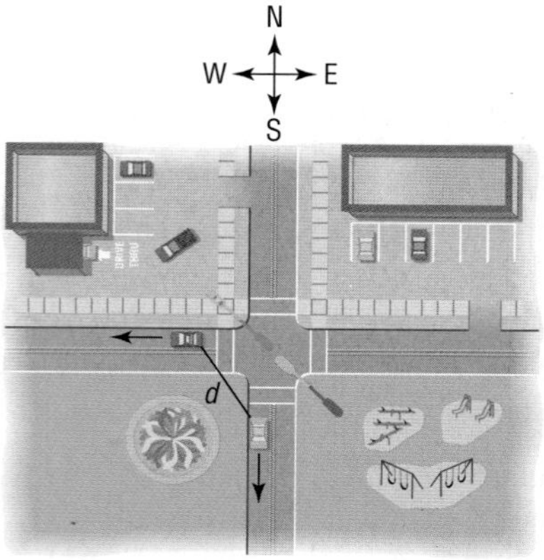

27. Two cars are approaching an intersection. One is 2 miles south of the intersection and is moving at a constant speed of 30 miles per hour. At the same time, the other car is 3 miles east of the intersection and is moving at a constant speed of 40 miles per hour.

(a) Express the distance d between the cars as a function of time t.

[**Hint:** At $t = 0$, the cars are 2 miles south and 3 miles east of the intersection, respectively.]

(b) Use a graphing utility to graph $d = d(t)$. For what value of t is d smallest?

28. Inscribing a Cylinder in a Sphere Inscribe a right circular cylinder of height h and radius r in a sphere of fixed radius R. See the illustration. Express the volume V of the cylinder as a function of h.

[**Hint:** $V = \pi r^2 h$. Note also the right triangle.]

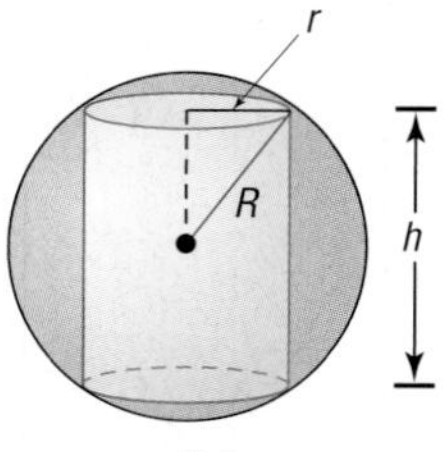

Sphere

29. Inscribing a Cylinder in a Cone Inscribe a right circular cylinder of height h and radius r in a cone of fixed radius R and fixed height H. See the illustration. Express the volume V of the cylinder as a function of r.

[**Hint:** $V = \pi r^2 h$. Note also the similar triangles.]

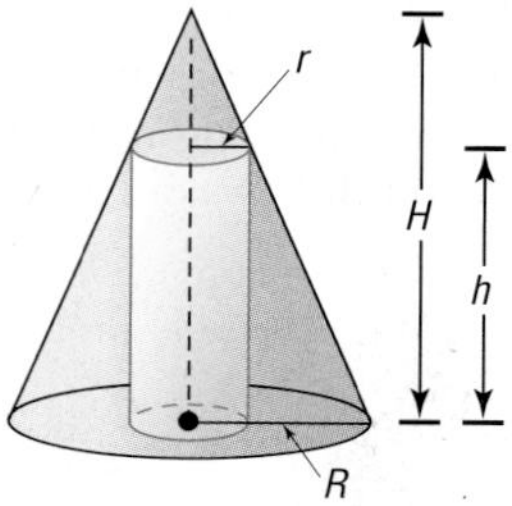

Cone

30. Installing Cable TV MetroMedia Cable is asked to provide service to a customer whose house is located 2 miles from the road along which the cable is buried. The nearest connection box for the cable is located 5 miles down the road (see the figure).

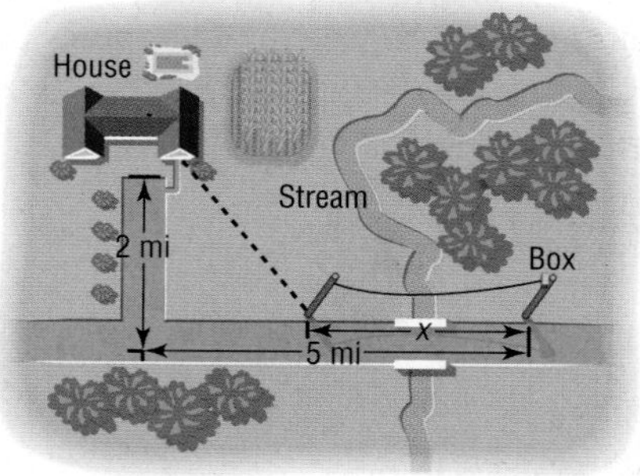

(a) If the installation cost is \$10 per mile along the road and \$14 per mile off the road, express the total cost C of installation as a function of the distance x (in miles) from the connection box to the point where the cable installation turns off the road. Give the domain.

(b) Compute the cost if $x = 1$ mile.

(c) Compute the cost if $x = 3$ miles.

(d) Graph the function $C = C(x)$. Use TRACE to see how the cost C varies as x changes from 0 to 5.

(e) What value of x results in the least cost?

31. Time Required to Go from an Island to a Town An island is 2 miles from the nearest point P on a straight shoreline. A town is 12 miles down the shore from P. See the illustration.

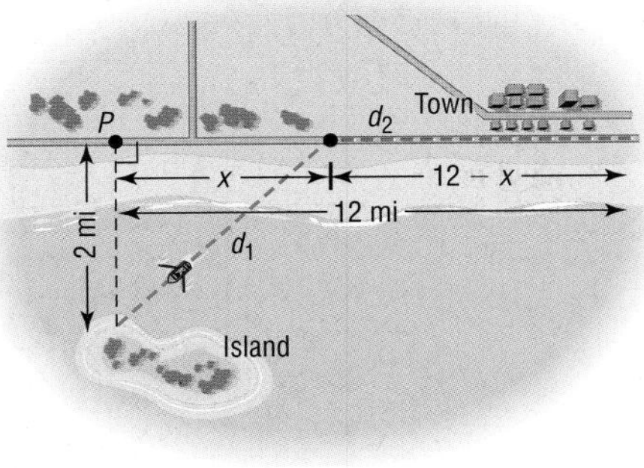

(a) If a person can row a boat at an average speed of 3 miles per hour and the same person can walk 5 miles per hour, express the time T that it takes to go from the island to town as a function of the distance x from P to where the person lands the boat.

(b) What is the domain of T?

(c) How long will it take to travel from the island to town if the person lands the boat 4 miles from P?

(d) How long will it take if the person lands the boat 8 miles from P?

32. Water is poured into a container in the shape of a right circular cone with radius 4 feet and height 16 feet (see the figure). Express the volume V of the water in the cone as a function of the height h of the water.
[**Hint:** The volume V of a cone of radius r and height h is $V = \frac{1}{3}\pi r^2 h$.]

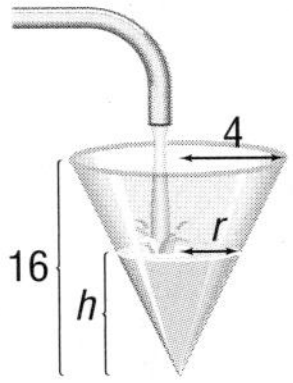

Chapter Review

Library of Functions

Linear function (p. 85)

$f(x) = mx + b$

Graph is a line with slope m and y-intercept b.

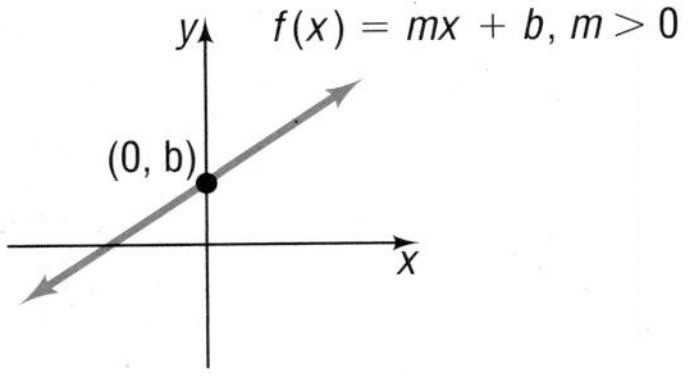

Constant function (p. 85)

$f(x) = b$

Graph is a horizontal line with y-intercept b.

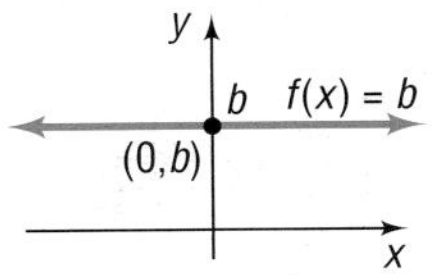

Identity function (p. 85)

$f(x) = x$

Graph is a line with slope 1 and y-intercept 0.

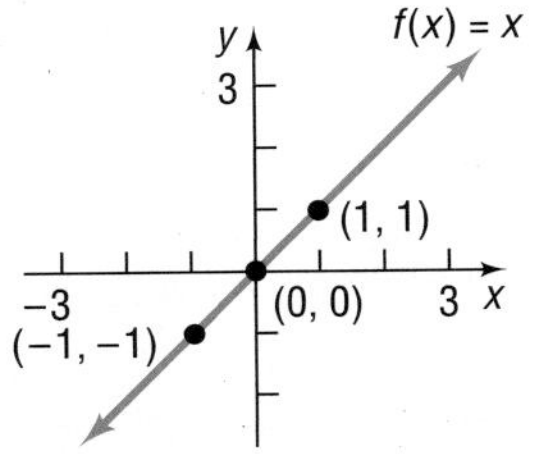

Square function (p. 86)

$f(x) = x^2$

Graph is a parabola with intercept at $(0, 0)$.

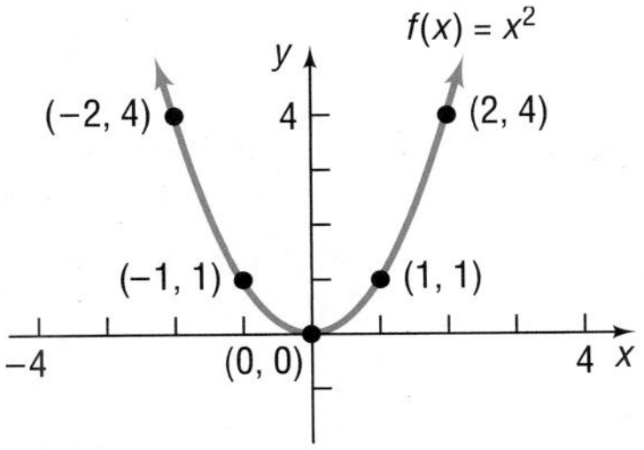

Cube function (p. 86)

$f(x) = x^3$

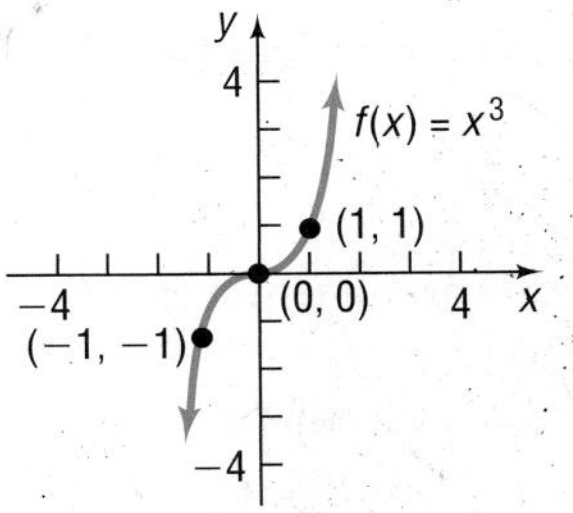

Square root function (p. 86)

$f(x) = \sqrt{x}$

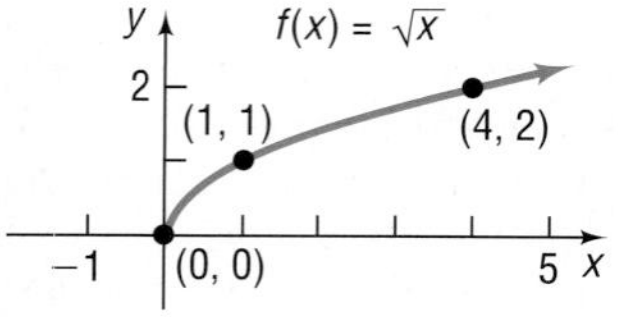

Cube root function (p. 86)

$f(x) = \sqrt[3]{x}$

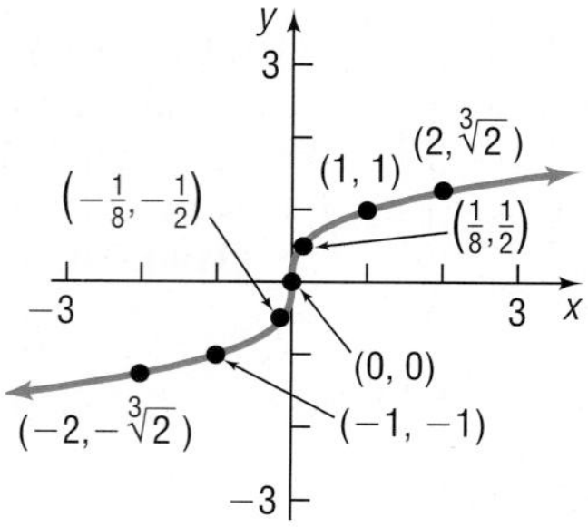

Reciprocal function (p. 87)

$f(x) = \dfrac{1}{x}$

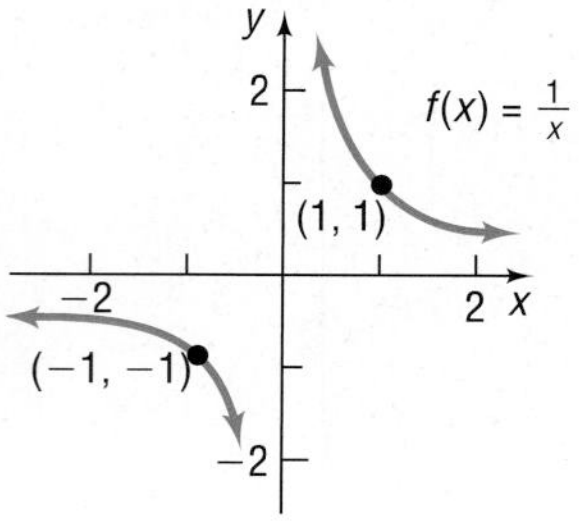

Absolute value function (p. 87)

$f(x) = |x|$

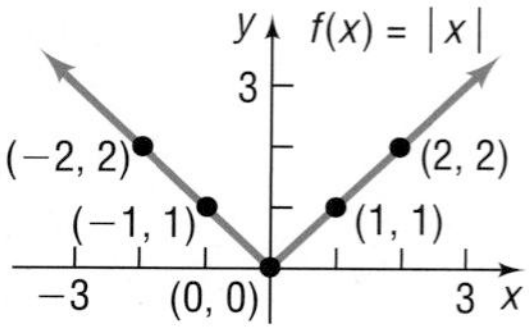

Greatest integer function (p. 87)

$f(x) = \text{int}(x)$

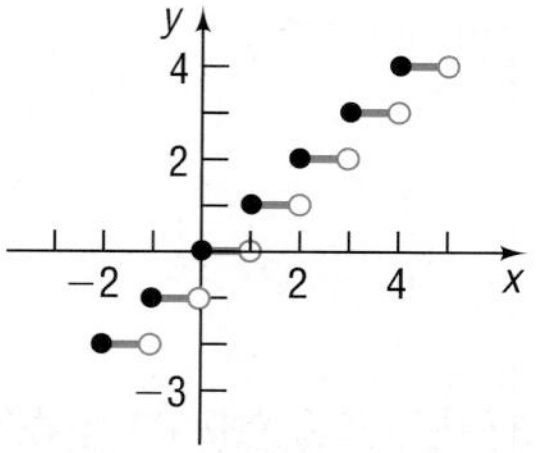

Things to Know

Function (pp. 51–54)

A relation between two sets of real numbers so that each number x in the first set, the domain, has corresponding to it exactly one number y in the second set. The range is the set of y values of the function for the x values in the domain.

x is the independent variable; y is the dependent variable.

A function can also be characterized as a set of ordered pairs (x, y) or $(x, f(x))$ in which no first element is paired with two different second elements.

Function notation (pp. 53–55)

$y = f(x)$

f is a symbol for the function.

x is the argument, or independent variable.

y is the dependent variable.

$f(x)$ is the value of the function at x, or the image of x.

A function f may be defined implicitly by an equation involving x and y or explicitly by writing $y = f(x)$.

Difference quotient of *f* (p. 55)

$\dfrac{f(x + h) - f(x)}{h}, \quad h \neq 0$

Domain (p. 57)

If unspecified, the domain of a function f is the largest set of real numbers for which $f(x)$ is a real number.

Vertical-line test (p. 64)

A set of points in the plane is the graph of a function if and only if every vertical line intersects the graph in at most one point.

Even function *f* (p. 73)

$f(-x) = f(x)$ for every x in the domain ($-x$ must also be in the domain).

Odd function *f* (p. 73)

$f(-x) = -f(x)$ for every x in the domain ($-x$ must also be in the domain).

Increasing function (p. 75)	A function f is increasing on an open interval I if, for any choice of x_1 and x_2 in I, with $x_1 < x_2$, we have $f(x_1) < f(x_2)$.
Decreasing function (p. 75)	A function f is decreasing on an open interval I if, for any choice of x_1 and x_2 in I, with $x_1 < x_2$, we have $f(x_1) > f(x_2)$.
Constant function (p. 75)	A function f is constant on an interval I if, for all choices of x in I, the values of $f(x)$ are equal.
Local maximum (p. 76)	A function f has a local maximum at c if there is an open interval I containing c so that, for all $x \neq c$ in I, $f(x) < f(c)$.
Local minimum (p. 76)	A function f has a local minimum at c if there is an open interval I containing c so that, for all $x \neq c$ in I, $f(x) > f(c)$.
Average rate of change of a function (p. 78)	The average rate of change of f from c to x is $\frac{\Delta y}{\Delta x} = \frac{f(x) - f(c)}{x - c}, \quad x \neq c$

Objectives

Section		You should be able to . . .	Review Exercises
2.1	1	Determine whether a relation represents a function (p. 50)	1, 2
	2	Find the value of a function (p. 53)	3–8, 23–24, 67–70
	3	Find the domain of a function (p. 57)	1, 9–16, 17–22, 63(a)–66(a)
	4	Form the sum, difference, product, and quotient of two functions (p. 58)	17–22
2.2	1	Identify the graph of a function (p. 64)	47
	2	Obtain information from or about the graph of a function (p. 65)	25(a)–(e); 26(a)–(e); 27(a), (d), (f); 28(a), (d), (f)
2.3	1	Determine even and odd functions from a graph (p. 72)	27(e); 28(e)
	2	Identify even or odd functions from the equation (p. 73)	29–36
	3	Use a graph to determine where a function is increasing, is decreasing, or is constant (p. 74)	27(b), 28(b)
	4	Use a graph to locate local maxima and minima (p. 76)	27(c), 28(c)
	5	Use a graphing utility to approximate local maxima and minima and to determine where a function is increasing or decreasing (p. 77)	37–40
	6	Find the average rate of a change of a function (p. 78)	41–46
2.4	1	Graph the functions listed in the library of functions (p. 85)	48–50
	2	Graph piecewise-defined functions (p. 88)	63(c)–66(c)
2.5	1	Graph functions using horizontal and vertical shifts (p. 94)	25(f); 26(f), (g); 51, 52, 55–62
	2	Graph functions using compressions and stretches (p. 97)	25(g), 26(h), 53, 54, 61, 62
	3	Graph functions using reflections about the x-axis or y-axis (p. 100)	25(h), 53, 57, 58, 62
2.6	1	Construct and analyze functions (p. 107)	67–68, 71–78

Review Exercises *(Blue problem numbers indicate the author's suggestions for a Practice Test.)*

In Problems 1 and 2, determine whether each relation represents a function. For each function, state the domain and range.

1. $\{(-1, 0), (2, 3), (4, 0)\}$ **2.** $\{(4, -1), (2, 1), (4, 2)\}$

In Problems 3–8, find the following for each function:

(a) $f(2)$ (b) $f(-2)$ (c) $f(-x)$ (d) $-f(x)$ (e) $f(x-2)$ (f) $f(2x)$

3. $f(x) = \frac{3x}{x^2 - 1}$ **4.** $f(x) = \frac{x^2}{x+1}$ **5.** $f(x) = \sqrt{x^2 - 4}$

6. $f(x) = |x^2 - 4|$ **7.** $f(x) = \frac{x^2 - 4}{x^2}$ **8.** $f(x) = \frac{x^3}{x^2 - 9}$

In Problems 9–16, find the domain of each function.

9. $f(x) = \frac{x}{x^2 - 9}$ **10.** $f(x) = \frac{3x^2}{x - 2}$ **11.** $f(x) = \sqrt{2 - x}$ **12.** $f(x) = \sqrt{x + 2}$

13. $h(x) = \frac{\sqrt{x}}{|x|}$ **14.** $g(x) = \frac{|x|}{x}$ 15. $f(x) = \frac{x}{x^2 + 2x - 3}$ **16.** $F(x) = \frac{1}{x^2 - 3x - 4}$

In Problems 17–22, find $f + g$, $f - g$, $f \cdot g$, *and* $\frac{f}{g}$ *for each pair of functions. State the domain of each.*

17. $f(x) = 2 - x$; $g(x) = 3x + 1$ **18.** $f(x) = 2x - 1$; $g(x) = 2x + 1$ **19.** $f(x) = 3x^2 + x + 1$; $g(x) = 3x$

20. $f(x) = 3x$; $g(x) = 1 + x + x^2$ 21. $f(x) = \frac{x+1}{x-1}$; $g(x) = \frac{1}{x}$ **22.** $f(x) = \frac{1}{x-3}$; $g(x) = \frac{3}{x}$

In Problems 23 and 24, find the difference quotient of each function f; *that is, find*

$$\frac{f(x+h) - f(x)}{h}, \quad h \neq 0$$

23. $f(x) = -2x^2 + x + 1$ **24.** $f(x) = 3x^2 - 2x + 4$

25. Using the graph of the function f shown:
(a) Find the domain and range of f.
(b) List the intercepts.
(c) Find $f(-2)$.
(d) For what values of x does $f(x) = -3$?
(e) Solve $f(x) > 0$.
(f) Graph $y = f(x - 3)$.
(g) Graph $y = f\left(\frac{1}{2}x\right)$.
(h) Graph $y = -f(x)$.

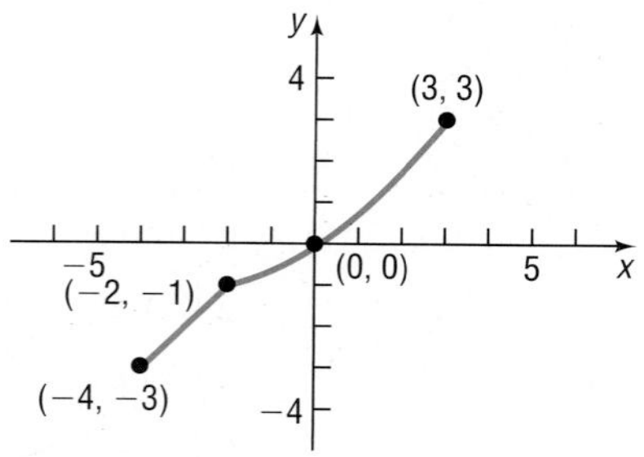

26. Using the graph of the function g shown:
(a) Find the domain and range of g.
(b) Find $g(-1)$.
(c) List the intercepts.
(d) For what value of x does $g(x) = -3$?
(e) Solve $g(x) > 0$.
(f) Graph $y = g(x - 2)$.
(g) Graph $y = g(x) + 1$.
(h) Graph $y = 2g(x)$.

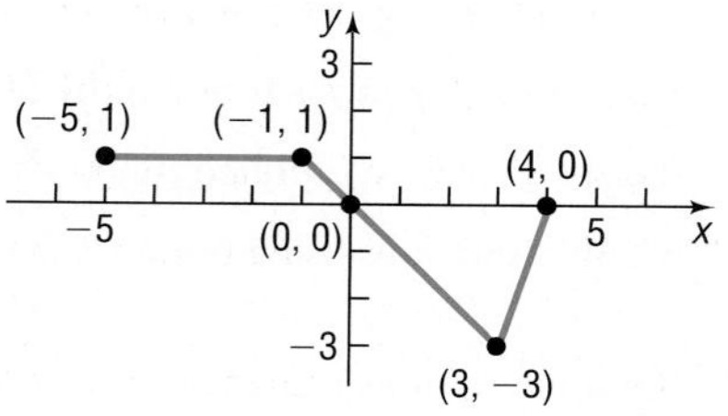

In Problems 27 and 28, use the graph of the function f to find:
(a) *The domain and the range of f*
(b) *The intervals on which f is increasing, decreasing, or constant*
(c) *The local minima and local maxima*
(d) *Whether the graph is symmetric with respect to the x-axis, the y-axis, or the origin*
(e) *Whether the function is even, odd, or neither*
(f) *The intercepts, if any*

27.

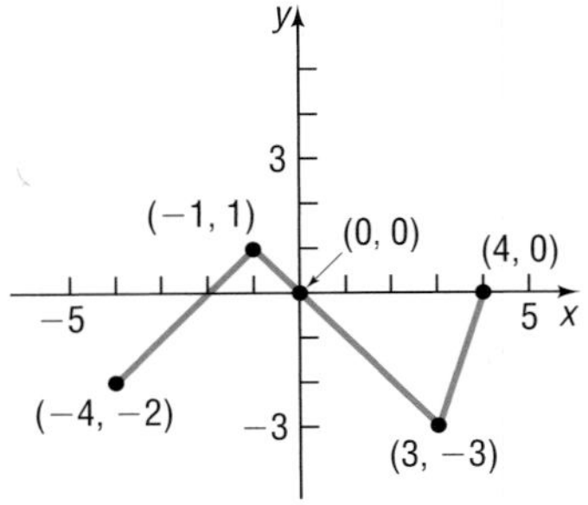

28.

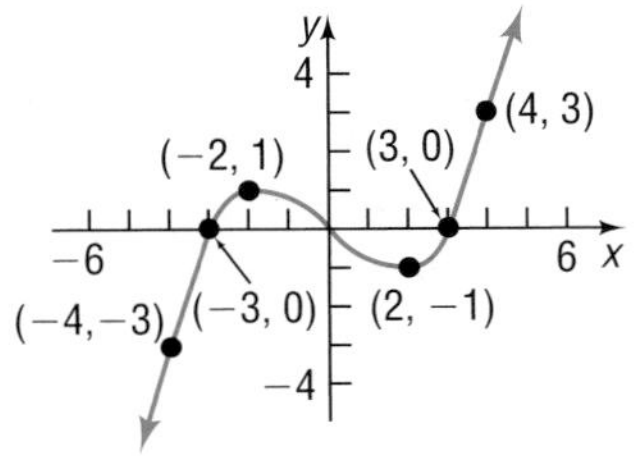

In Problems 29–36, determine (algebraically) whether the given function is even, odd, or neither.

29. $f(x) = x^3 - 4x$ **30.** $g(x) = \dfrac{4 + x^2}{1 + x^4}$ **31.** $h(x) = \dfrac{1}{x^4} + \dfrac{1}{x^2} + 1$ **32.** $F(x) = \sqrt{1 - x^3}$

33. $G(x) = 1 - x + x^3$ **34.** $H(x) = 1 + x + x^2$ 35. $f(x) = \dfrac{x}{1 + x^2}$ **36.** $g(x) = \dfrac{1 + x^2}{x^3}$

In Problems 37–40, use a graphing utility to graph each function over the indicated interval. Approximate any local maxima and local minima. Determine where the function is increasing and where it is decreasing.

37. $f(x) = 2x^3 - 5x + 1 \quad (-3, 3)$ **38.** $f(x) = -x^3 + 3x - 5 \quad (-3, 3)$

39. $f(x) = 2x^4 - 5x^3 + 2x + 1 \quad (-2, 3)$ **40.** $f(x) = -x^4 + 3x^3 - 4x + 3 \quad (-2, 3)$

In Problems 41 and 42, find the average rate of change of f:
(a) *From 1 to 2* (b) *From 0 to 1* (c) *From 2 to 4*

41. $f(x) = 8x^2 - x$ **42.** $f(x) = 2x^3 + x$

In Problems 43–46, find the average rate of change from 2 to x for each function f. Be sure to simplify.

43. $f(x) = 2 - 5x$ **44.** $f(x) = 2x^2 + 7$ 45. $f(x) = 3x - 4x^2$ **46.** $f(x) = x^2 - 3x + 2$

47. Tell which of the following are graphs of functions.

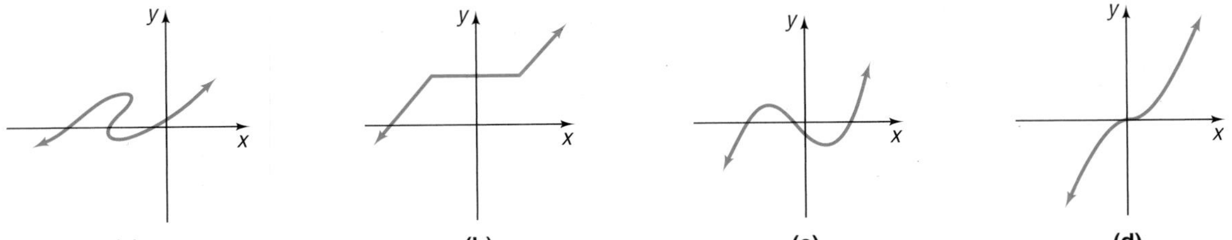

(a) (b) (c) (d)

In Problems 48–50, sketch the graph of each function. Be sure to label at least three points.

48. $f(x) = |x|$ **49.** $f(x) = \sqrt[3]{x}$ **50.** $f(x) = \sqrt{x}$

In Problems 51–62, graph each function using the techniques of shifting, compressing or stretching, and reflections. Identify any intercepts on the graph. State the domain and, based on the graph, find the range.

51. $F(x) = |x| - 4$ **52.** $f(x) = |x| + 4$ **53.** $g(x) = -2|x|$ **54.** $g(x) = \dfrac{1}{2}|x|$

55. $h(x) = \sqrt{x - 1}$ **56.** $h(x) = \sqrt{x} - 1$ 57. $f(x) = \sqrt{1 - x}$ **58.** $f(x) = -\sqrt{x + 3}$

59. $h(x) = (x - 1)^2 + 2$ **60.** $h(x) = (x + 2)^2 - 3$ **61.** $g(x) = 3(x - 1)^3 + 1$ **62.** $g(x) = -2(x + 2)^3 - 8$

In Problems 63–66:

(a) *Find the domain of each function.*
(b) *Locate any intercepts.*
(c) *Graph each function.*
(d) *Based on the graph, find the range.*

63. $f(x) = \begin{cases} 3x & -2 < x \le 1 \\ x + 1 & x > 1 \end{cases}$

64. $f(x) = \begin{cases} x - 1 & -3 < x < 0 \\ 3x - 1 & x \ge 0 \end{cases}$

65. $f(x) = \begin{cases} x & -4 \le x < 0 \\ 1 & x = 0 \\ 3x & x > 0 \end{cases}$

66. $f(x) = \begin{cases} x^2 & -2 \le x \le 2 \\ 2x - 1 & x > 2 \end{cases}$

67. Given that f is a linear function, $f(4) = -5$ and $f(0) = 3$, write the equation that defines f.

68. Given that g is a linear function with slope $= -4$ and $g(-2) = 2$, write the equation that defines g.

69. A function f is defined by

$$f(x) = \frac{Ax + 5}{6x - 2}$$

If $f(1) = 4$, find A.

70. A function g is defined by

$$g(x) = \frac{A}{x} + \frac{8}{x^2}$$

If $g(-1) = 0$, find A.

---

71. Temperature Conversion The temperature T of the air is approximately a linear function of the altitude h for altitudes within 10,000 meters of the surface of Earth. If the surface temperature is 30°C and the temperature at 10,000 meters is 5°C, find the function $T = T(h)$.

72. Speed as a Function of Time The speed v (in feet per second) of a car is a linear function of the time t (in seconds) for $10 \le t \le 30$. If after each second the speed of the car has increased by 5 feet per second and if after 20 seconds the speed is 80 feet per second, how fast is the car going after 30 seconds? Find the function $v = v(t)$.

73. Spheres The volume V of a sphere of radius r is $V = \frac{4}{3}\pi r^3$; the surface area S of this sphere is $S = 4\pi r^2$. Express the volume V as a function of the surface area S. If the surface area doubles, how does the volume change?

74. Page Design A page with dimensions of $8\frac{1}{2}$ inches by 11 inches has a border of uniform width x surrounding the printed matter of the page, as shown in the figure.

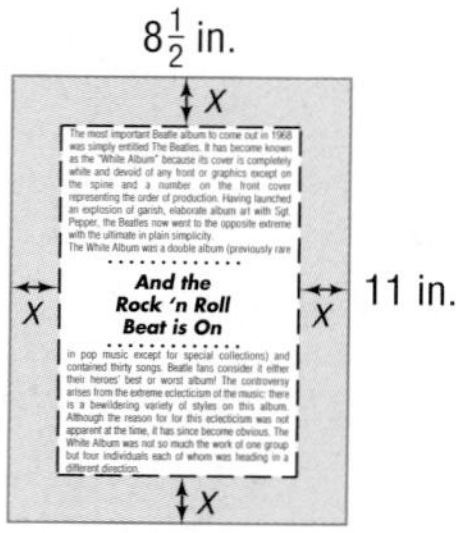

(a) Write a formula for the area A of the printed part of the page as a function of the width x of the border.
(b) Give the domain and range of A.
(c) Find the area of the printed page for borders of widths 1 inch, 1.2 inches, and 1.5 inches.
(d) Graph the function $A = A(x)$.
(e) Use TRACE to determine what margin should be used to obtain an area of 70 square inches and of 50 square inches.

75. Strength of a Beam The strength of a rectangular wooden beam is proportional to the product of the width and the cube of its depth (see the figure). If the beam is to be cut from a log in the shape of a cylinder of radius 3 feet, express the strength S of the beam as a function of the width x. What is the domain of S?

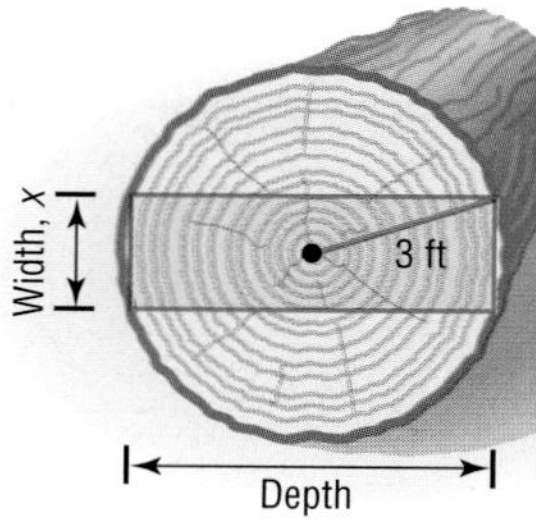

76. Material Needed to Make a Drum A steel drum in the shape of a right circular cylinder is required to have a volume of 100 cubic feet.
(a) Express the amount A of material required to make the drum as a function of the radius r of the cylinder.
(b) How much material is required if the drum is of radius 3 feet?
(c) How much material is required if the drum is of radius 4 feet?
(d) How much material is required if the drum is of radius 5 feet?
(e) Graph $A = A(r)$. For what value of r is A smallest?

77. Cost of a Drum A drum in the shape of a right circular cylinder is required to have a volume of 500 cubic centimeters. The top and bottom are made of material that costs 6¢ per square centimeter; the sides are made of material that costs 4¢ per square centimeter.

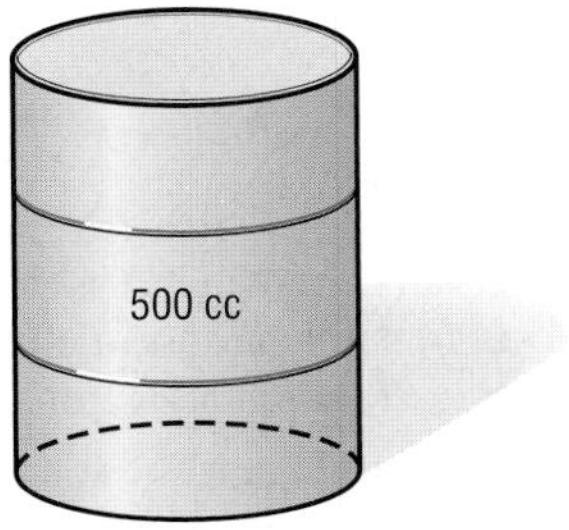

(a) Express the total cost C of the material as a function of the radius r of the cylinder.
(b) What is the cost if the radius is 4 cm?
(c) What is the cost if the radius is 8 cm?
(d) Graph $C = C(r)$. For what value of r is the cost C least?

78. Constructing a Closed Box A closed box with a square base is required to have a volume of 10 cubic feet.
(a) Express the amount A of material used to make such a box as a function of the length x of a side of the square base.
(b) How much material is required for a base 1 foot by 1 foot?
(c) How much material is required for a base 2 feet by 2 feet?
(d) Graph $A = A(x)$. For what value of x is A smallest?

Chapter Projects

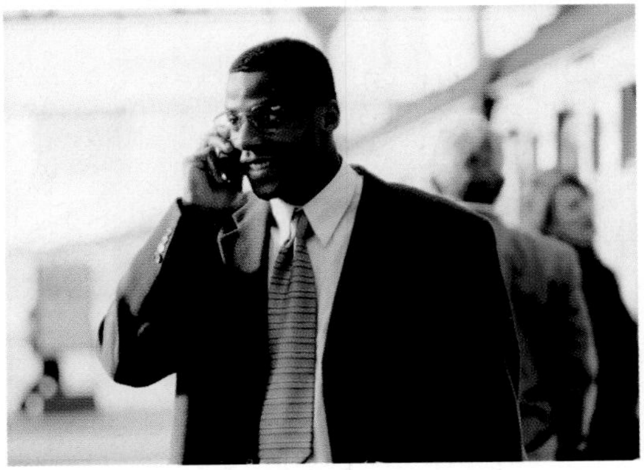

1. Cell Phone Service In purchasing cell phone service, you must shop around to determine the best rates and services that apply to your situation. Suppose that you want to upgrade your old-model cell phone. You have gathered the following information on cell phone services:

Nokia 5165	\$19.99 for the phone
Ericsson A1228di	Free (after \$20 rebate)

Option 1: \$29.99 for 250 peak time minutes, unlimited nights and weekends; extra peak time minutes cost \$0.45 per minute
Option 2: \$39.99 for 400 peak time minutes, unlimited nights and weekends; extra peak time minutes cost \$0.45 per minute
Option 3: \$49.99 for 600 peak time minutes, unlimited nights and weekends; extra peak time minutes cost \$0.35 per minute

To qualify for these phone prices, you must sign up for a 2-year contract.
(a) List all the different combinations of phone and services.
(b) Determine the total cost of each combination described in part (a) for the life of the contract (24-months), assuming that you stay within the allotted peak time minutes provided by each contract.
(c) If you expect to use 260 peak time minutes per month, which option provides the best deal? If you expect to use 320 peak time minutes per month, which option provides the best deal?
(d) If you expect to use 410 peak time minutes per month, which option provides the best deal? If you expect to use 450 peak time minutes per month, which option provides the best deal?
(e) Each monthly charge includes a specific number of peak time minutes included in the monthly fee. Write a function for each available option, where C is the monthly cost and x is the number of peak time minutes used.
(f) Graph the functions corresponding to each option.
(g) At what point does Option 2 become a better deal than Option 1?
(h) At what point does Option 3 become a better deal than Option 2?

This project is based on information given in a "Cingular Wireless" advertisement in the *Denton Record Chronicle* on September 11, 2001.

The following projects are available at www.prenhall.com/sullivan7e.

2. **Project at Motorola** *Pricing Wireless Services*
3. **Cost of Cable**
4. **Oil Spill**

Cumulative Review

In Problems 1–8, find the real solutions of each equation.

1. $-5x + 4 = 0$

2. $x^2 - 7x + 12 = 0$

3. $3x^2 - 5x - 2 = 0$

4. $4x^2 + 4x + 1 = 0$

5. $4x^2 - 2x + 4 = 0$

6. $\sqrt[3]{1 - x} = 2$

7. $\sqrt[5]{1 - x} = 2$

8. $|2 - 3x| = 1$

9. In the complex number system, solve $4x^2 - 2x + 4 = 0$.

10. Solve the inequality $-2 < 3x - 5 < 7$. Graph the solution set.

In Problems 11–14, graph each equation.

11. $-3x + 4y = 12$

12. $f(x) = 3x + 12$

13. $x^2 + y^2 + 2x - 4y + 4 = 0$

14. $f(x) = (x + 1)^2 - 3$

15. For the graph of the function f shown:
(a) Find the domain and the range of f.
(b) Find the intercepts.
(c) Is the graph of f symmetric with respect to the x-axis, the y-axis, or the origin?

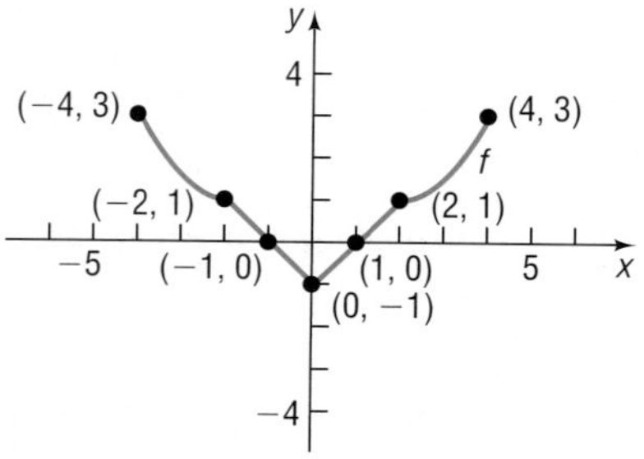

(d) Find $f(2)$.
(e) For what value(s) of x is $f(x) = 3$?
(f) Solve $f(x) < 0$.
(g) Graph $y = f(x + 2)$
(h) Graph $y = f(-x)$.
(i) Graph $y = 2f(x)$.
(j) Is f even, odd, or neither?
(k) Find the interval(s) on which f is increasing.
(l) Find the interval(s) on which f is decreasing.
(m) Find the local maxima and local minima.
(n) Find the average rate of change of f from 1 to 4.

16. Productivity versus Earnings The following data represent the average hourly earnings and productivity (output per hour) of production workers for the years 1986–1995. Let productivity x be the independent variable and average hourly earnings y be the dependent variable.

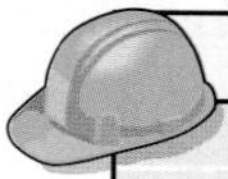

Productivity	Average Hourly Earnings
94.2	8.76
94.1	8.98
94.6	9.28
95.3	9.66
96.1	10.01
96.7	10.32
100	10.57
100.2	10.83
100.7	11.12
100.8	11.44

SOURCE: Bureau of Labor Statistics.

(a) Draw a scatter diagram of the data.
(b) Draw a line from the point (94.2, 8.76) to (96.7, 10.32) on the scatter diagram found in part (a).
(c) Find the average rate of change of hourly earnings for productivity from 94.2 to 96.7.
(d) Interpret the average rate of change found in part (c).
(e) Draw a line from the point (96.7, 10.32) to (100.8, 11.44) on the scatter diagram found in part (a).
(f) Find the average rate of change of hourly earnings for productivity from 96.7 to 100.8.
(g) Interpret the average rate of change found in part (f).
(h) What is happening to the average rate of change of hourly earnings as productivity increases?

3 Polynomial and Rational Functions

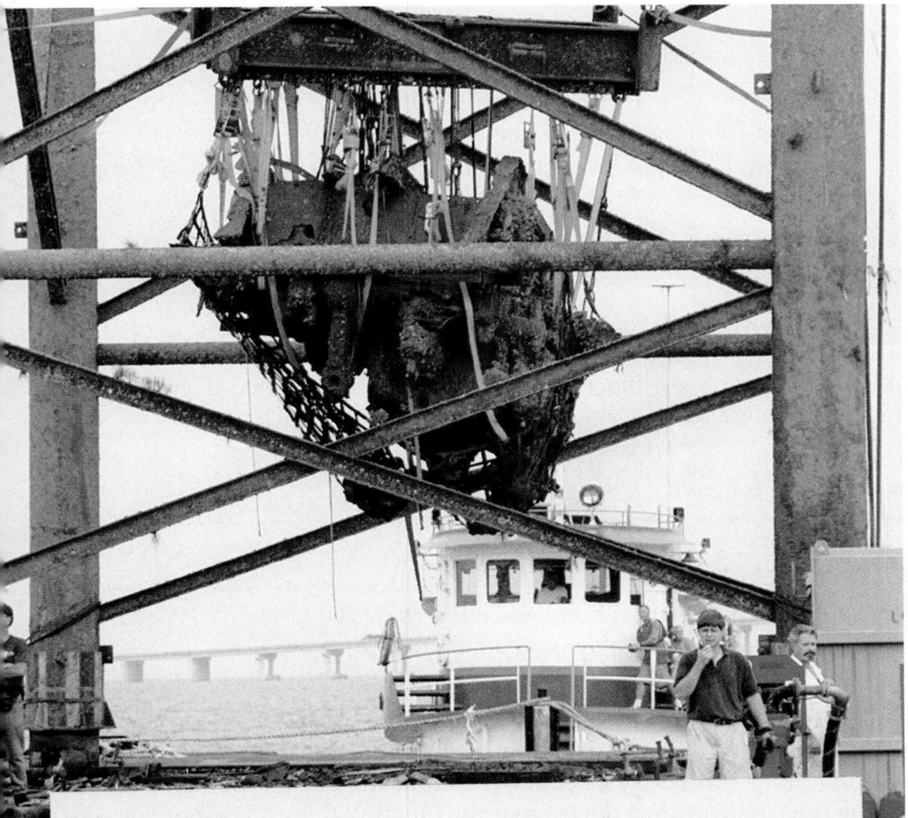

OUTLINE

Clash of Iron

One hundred and forty years after its guns fell silent, the turret of the Union ironclad *Monitor* was pulled from the ocean last week, bearing secrets. Now scientists hope to unlock more mysteries from the ship that, in a single battle, ended the era of the wooden navy.

The Turret

Ericsson obtained dozens of patents for the features of the turret alone. It weighed more than 120 tons without the cannon, and was held in place by its sheer weight. The turret was driven by gears below the floor, and could make 2.5 revolutions a minute. A crew of 17 men and two officers worked the guns in the 20-ft. diameter space.

(From *Clash of Iron*, pages 54–55, of *TIME*, August 19, 2002.)

—SEE CHAPTER PROJECT 1.

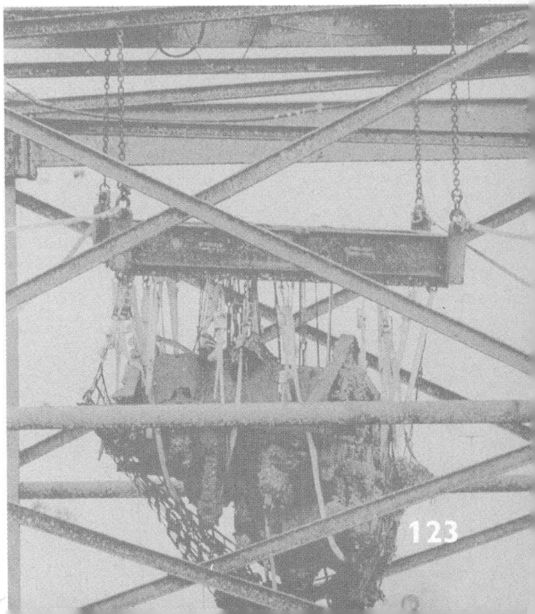

3.1 Quadratic Functions and Models

PREPARING FOR THIS SECTION *Before getting started, review the following:*

- Intercepts (Section 1.2, pp. 11–12)
- Solving Quadratic Equations (Appendix A, Section A.5, pp. 942–945)
- Completing the Square (Appendix A, Section A.5, p. 941)
- Graphing Techniques: Transformations (Section 2.5, pp. 94–103)

Now work the 'Are You Prepared?' problems on page 138.

OBJECTIVES
1. Graph a Quadratic Function Using Transformations
2. Identify the Vertex and Axis of Symmetry of a Quadratic Function
3. Graph a Quadratic Function Using Its Vertex, Axis, and Intercepts
4. Use the Maximum or Minimum Value of a Quadratic Function to Solve Applied Problems
5. Use a Graphing Utility to Find the Quadratic Function of Best Fit to Data

Quadratic Functions

A *quadratic function* is a function that is defined by a second-degree polynomial in one variable.

A **quadratic function** is a function of the form

$$f(x) = ax^2 + bx + c \qquad \textbf{(1)}$$

where a, b, and c are real numbers and $a \neq 0$. The domain of a quadratic function is the set of all real numbers.

Many applications require a knowledge of quadratic functions. For example, suppose that Texas Instruments collects the data shown in Table 1, which relate the number of calculators sold at the price p per calculator. Since the price of a product determines the quantity that will be purchased, we treat price as the independent variable. The relationship between the number x of calculators sold and the price p per calculator may be approximated by the linear equation

$$x = 21{,}000 - 150p$$

Table 1

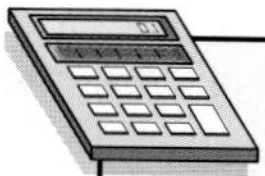

Price per Calculator, p (Dollars)	Number of Calculators, x
60	11,100
65	10,115
70	9,652
75	8,731
80	8,087
85	7,205
90	6,439

Then the revenue R derived from selling x calculators at the price p per calculator is

$$\begin{aligned} R &= xp \\ R(p) &= (21{,}000 - 150p)p \\ &= -150p^2 + 21{,}000p \end{aligned}$$

So the revenue R is a quadratic function of the price p. Figure 1 illustrates the graph of this revenue function, whose domain is $0 \le p \le 140$, since both x and p must be nonnegative. Later in this section we shall determine the price p that maximizes revenue.

Figure 1
Graph of a revenue function: $R = -150p^2 + 21{,}000p$

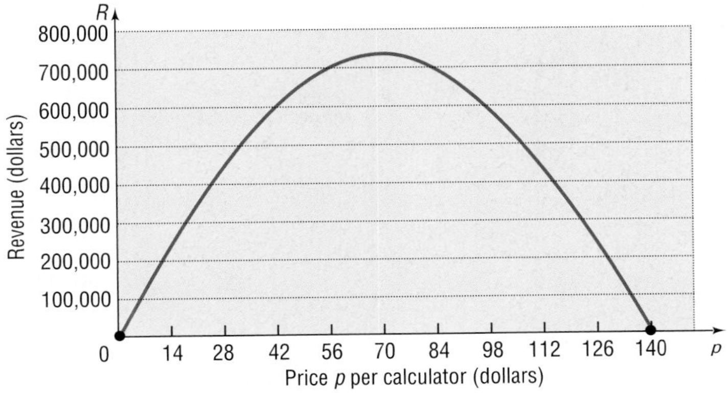

A second situation in which a quadratic function appears involves the motion of a projectile. Based on Newton's second law of motion (force equals mass times acceleration, $F = ma$), it can be shown that, ignoring air resistance, the path of a projectile propelled upward at an inclination to the horizontal is the graph of a quadratic function. See Figure 2 for an illustration. Later in this section we shall analyze the path of a projectile.

Figure 2
Path of a cannonball

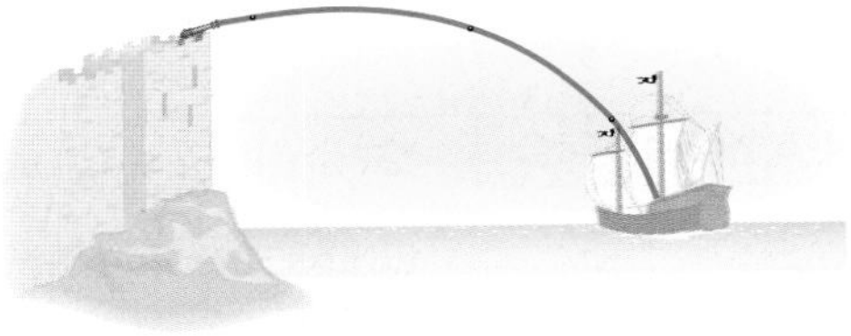

Graphing Quadratic Functions

Figure 3

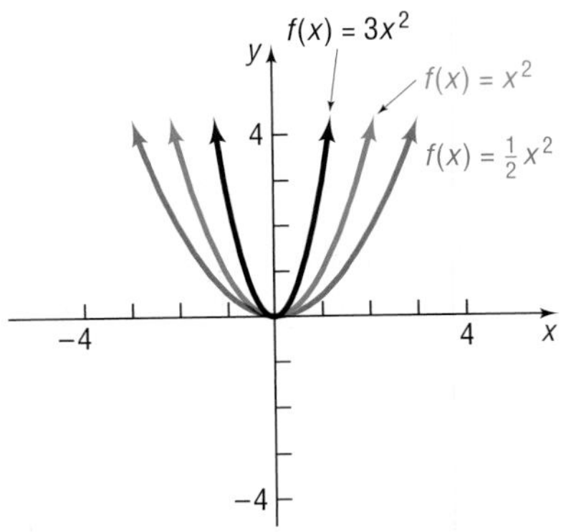

We know how to graph the quadratic function $f(x) = x^2$. Figure 3 shows the graph of three functions of the form $f(x) = ax^2$, $a > 0$, for $a = 1$, $a = \frac{1}{2}$, and $a = 3$. Notice that the larger the value of a, the "narrower" the graph, and the smaller the value of a, the "wider" the graph.

Figure 4 on page 126 shows the graphs of $f(x) = ax^2$ for $a < 0$. Notice that these graphs are reflections about the x-axis of the graphs in Figure 3. Based on the results of these two figures, we can draw some general conclusions about the graph of $f(x) = ax^2$. First, as $|a|$ increases, the graph becomes *narrower* (a vertical stretch), and as $|a|$ gets closer to zero, the graph gets *wider* (a vertical compression). Second, if a is positive, then the graph opens *up*, and if a is negative, the graph opens *down*.

Figure 4

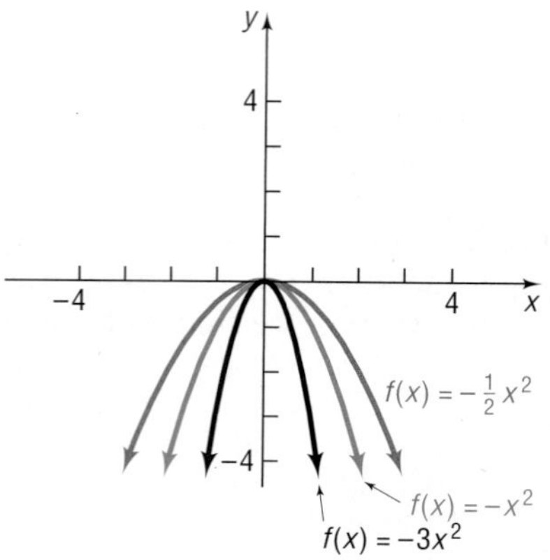

The graphs in Figures 3 and 4 are typical of the graphs of all quadratic functions, which we call **parabolas.*** Refer to Figure 5, where two parabolas are pictured. The one on the left **opens up** and has a lowest point; the one on the right **opens down** and has a highest point. The lowest or highest point of a parabola is called the **vertex.** The vertical line passing through the vertex in each parabola in Figure 5 is called the **axis of symmetry** (sometimes abbreviated to **axis**) of the parabola. Because the parabola is symmetric about its axis, the axis of symmetry of a parabola can be used to find additional points on the parabola.

Figure 5
Graphs of a quadratic function, $f(x) = ax^2 + bx + c, a \neq 0$

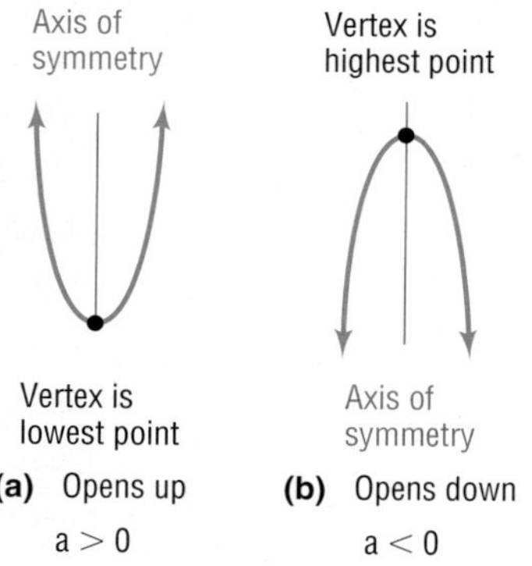

The parabolas shown in Figure 5 are the graphs of a quadratic function $f(x) = ax^2 + bx + c$, $a \neq 0$. Notice that the coordinate axes are not included in the figure. Depending on the values of a, b, and c, the axes could be anywhere. The important fact is that, except possibly for compression or stretching, the shape of the graph of a quadratic function will look like one of the parabolas in Figure 5.

1 In the following example, we use techniques from Section 2.5 to graph a quadratic function $f(x) = ax^2 + bx + c$, $a \neq 0$. In so doing, we shall complete the square and write the function f in the form $f(x) = a(x - h)^2 + k$.

EXAMPLE 1 **Graphing a Quadratic Function Using Transformations**

Graph the function $f(x) = 2x^2 + 8x + 5$. Find the vertex and axis of symmetry.

Solution We begin by completing the square on the right side.

$$\begin{aligned} f(x) &= 2x^2 + 8x + 5 \\ &= 2(x^2 + 4x) + 5 && \text{Factor out the 2 from } 2x^2 + 8x. \\ &= 2(x^2 + 4x + 4) + 5 - 8 && \text{Complete the square of } 2(x^2 + 4x). \\ &= 2(x + 2)^2 - 3 && \text{Notice that the factor of 2 requires that 8 be added and subtracted.} \end{aligned} \quad (2)$$

The graph of f can be obtained in three stages, as shown in Figure 6. Now compare this graph to the graph in Figure 5(a). The graph of $f(x) = 2x^2 + 8x + 5$ is a parabola that opens up and has its vertex (lowest point) at $(-2, -3)$. Its axis of symmetry is the line $x = -2$.

*We shall study parabolas using a geometric definition later in this book.

Figure 6

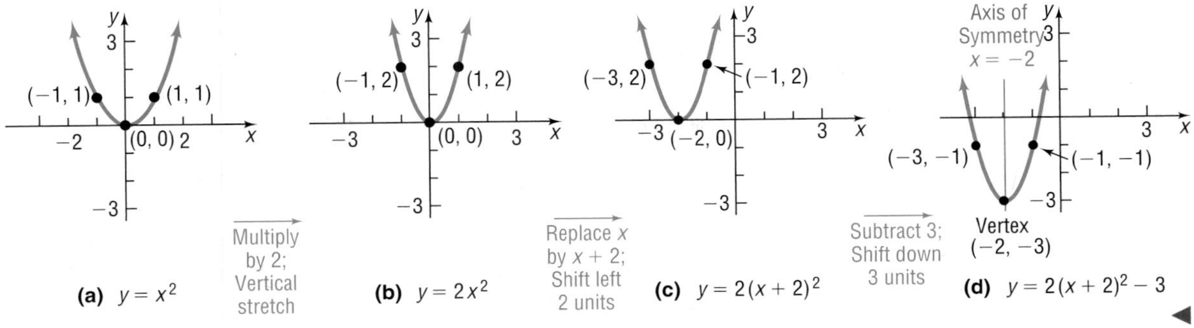

CHECK: Graph $f(x) = 2x^2 + 8x + 5$ and use the MINIMUM command to locate its vertex.

NOW WORK PROBLEM 27.

The method used in Example 1 can be used to graph any quadratic function $f(x) = ax^2 + bx + c$, $a \neq 0$, as follows:

$$f(x) = ax^2 + bx + c$$

$$= a\left(x^2 + \frac{b}{a}x\right) + c \qquad \text{Factor out } a \text{ from } ax^2 + bx.$$

$$= a\left(x^2 + \frac{b}{a}x + \frac{b^2}{4a^2}\right) + c - a\left(\frac{b^2}{4a^2}\right) \qquad \text{Complete the square by adding and subtracting } a\left(\frac{b^2}{4a^2}\right)\text{. Look closely at this step!}$$

$$= a\left(x + \frac{b}{2a}\right)^2 + c - \frac{b^2}{4a}$$

$$= a\left(x + \frac{b}{2a}\right)^2 + \frac{4ac - b^2}{4a} \qquad c - \frac{b^2}{4a} = c \cdot \frac{4a}{4a} - \frac{b^2}{4a} = \frac{4ac - b^2}{4a}$$

Based on these results, we conclude the following:

If $h = -\dfrac{b}{2a}$ and $k = \dfrac{4ac - b^2}{4a}$, then

$$f(x) = ax^2 + bx + c = a(x - h)^2 + k \qquad (3)$$

The graph of f is the parabola $y = ax^2$ shifted horizontally h units and vertically k units. As a result, the vertex is at (h, k), and the graph opens up if $a > 0$ and down if $a < 0$. The axis of symmetry is the vertical line $x = h$.

For example, compare equation (3) with equation (2) of Example 1.

$$f(x) = 2(x + 2)^2 - 3$$
$$= a(x - h)^2 + k$$

We conclude that $a = 2$, so the graph opens up. Also, we find that $h = -2$ and $k = -3$, so its vertex is at $(-2, -3)$.

2 It is not required to complete the square to obtain the vertex. In almost every case, it is easier to obtain the vertex of a quadratic function f by remembering that its x-coordinate is $h = -\dfrac{b}{2a}$. The y-coordinate can then be found by evaluating f at $-\dfrac{b}{2a}$.

We summarize these remarks as follows:

Properties of the Graph of a Quadratic Function

$$f(x) = ax^2 + bx + c, \quad a \neq 0$$

$$\text{Vertex} = \left(-\frac{b}{2a}, f\left(-\frac{b}{2a}\right)\right) \quad \text{Axis of symmetry: the line } x = -\frac{b}{2a} \quad \textbf{(4)}$$

Parabola opens up if $a > 0$; the vertex is a minimum point.
Parabola opens down if $a < 0$; the vertex is a maximum point.

EXAMPLE 2 Locating the Vertex without Graphing

Without graphing, locate the vertex and axis of symmetry of the parabola defined by $f(x) = -3x^2 + 6x + 1$. Does it open up or down?

Solution For this quadratic function, $a = -3$, $b = 6$, and $c = 1$. The x-coordinate of the vertex is

$$h = -\frac{b}{2a} = -\frac{6}{-6} = 1$$

The y-coordinate of the vertex is therefore

$$k = f\left(-\frac{b}{2a}\right) = f(1) = -3 + 6 + 1 = 4$$

The vertex is located at the point $(1, 4)$. The axis of symmetry is the line $x = 1$. Finally, because $a = -3 < 0$, the parabola opens down. ◀

3 The information we gathered in Example 2, together with the location of the intercepts, usually provides enough information to graph the quadratic function $f(x) = ax^2 + bx + c$, $a \neq 0$. The y-intercept is the value of f at $x = 0$, that is, $f(0) = c$.

The x-intercepts, if there are any, are found by solving the quadratic equation

$$f(x) = ax^2 + bx + c = 0$$

This equation has two, one, or no real solutions, depending on whether the discriminant $b^2 - 4ac$ is positive, 0, or negative. Depending on the value of the discriminant, the graph of f has x-intercepts, as follows:

The x-Intercepts of a Quadratic Function

1. If the discriminant $b^2 - 4ac > 0$, the graph of $f(x) = ax^2 + bx + c$ has two distinct x-intercepts and so will cross the x-axis in two places.
2. If the discriminant $b^2 - 4ac = 0$, the graph of $f(x) = ax^2 + bx + c$ has one x-intercept and touches the x-axis at its vertex.
3. If the discriminant $b^2 - 4ac < 0$, the graph of $f(x) = ax^2 + bx + c$ has no x-intercept and so will not cross or touch the x-axis.

Figure 7 illustrates these possibilities for parabolas that open up.

Figure 7
$f(x) = ax^2 + bx + c, \quad a > 0$

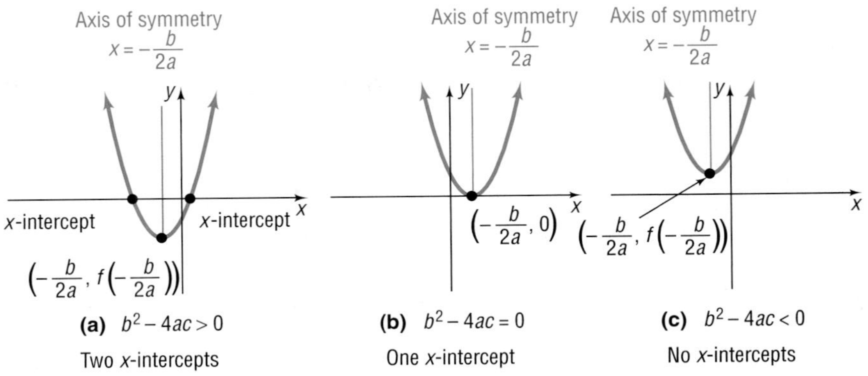

EXAMPLE 3 Graphing a Quadratic Function Using Its Vertex, Axis, and Intercepts

Use the information from Example 2 and the locations of the intercepts to graph $f(x) = -3x^2 + 6x + 1$.

Solution In Example 2, we found the vertex to be at $(1, 4)$ and the axis of symmetry to be $x = 1$. The y-intercept is found by letting $x = 0$. The y-intercept is $f(0) = 1$. The x-intercepts are found by solving the equation $f(x) = 0$. This results in the equation

$$-3x^2 + 6x + 1 = 0 \qquad a = -3, b = 6, c = 1$$

The discriminant $b^2 - 4ac = (6)^2 - 4(-3)(1) = 36 + 12 = 48 > 0$, so the equation has two real solutions and the graph has two x-intercepts. Using the quadratic formula, we find that

$$x = \frac{-b + \sqrt{b^2 - 4ac}}{2a} = \frac{-6 + \sqrt{48}}{-6} = \frac{-6 + 4\sqrt{3}}{-6} \approx -0.15$$

and

$$x = \frac{-b - \sqrt{b^2 - 4ac}}{2a} = \frac{-6 - \sqrt{48}}{-6} = \frac{-6 - 4\sqrt{3}}{-6} \approx 2.15$$

Figure 8

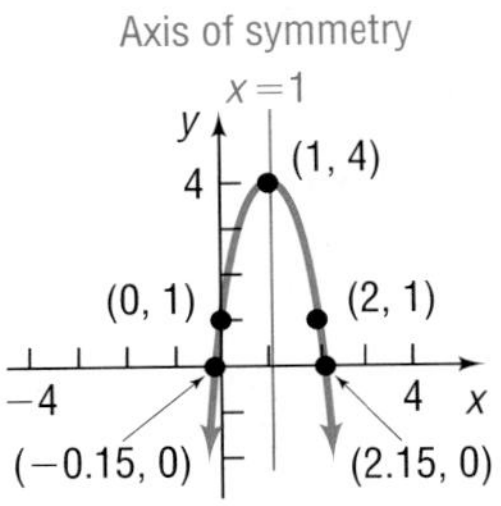

The x-intercepts are approximately -0.15 and 2.15.

The graph is illustrated in Figure 8. Notice how we used the y-intercept and the axis of symmetry, $x = 1$, to obtain the additional point $(2, 1)$ on the graph. ◀

GRAPH THE FUNCTION IN EXAMPLE 3 USING THE METHOD PRESENTED IN EXAMPLE 1. WHICH OF THE TWO METHODS DO YOU PREFER? GIVE REASONS.

CHECK: Graph $f(x) = -3x^2 + 6x + 1$. Use ROOT or ZERO to locate the two x-intercepts and use MAXIMUM to locate the vertex.

NOW WORK PROBLEM 35.

If the graph of a quadratic function has only one x-intercept or none, it is usually necessary to plot an additional point to obtain the graph.

EXAMPLE 4 Graphing a Quadratic Function Using Its Vertex, Axis, and Intercepts

Graph $f(x) = x^2 - 6x + 9$ by determining whether the graph opens up or down. Find its vertex, axis of symmetry, y-intercept, and x-intercepts, if any.

Solution For $f(x) = x^2 - 6x + 9$, we have $a = 1$, $b = -6$, and $c = 9$. Since $a = 1 > 0$, the parabola opens up. The x-coordinate of the vertex is

$$h = -\frac{b}{2a} = -\frac{-6}{2(1)} = 3$$

The y-coordinate of the vertex is

$$k = f(3) = (3)^2 - 6(3) + 9 = 0$$

So the vertex is at $(3, 0)$. The axis of symmetry is the line $x = 3$. The y-intercept is $f(0) = 9$. Since the vertex $(3, 0)$ lies on the x-axis, the graph touches the x-axis at the x-intercept. By using the axis of symmetry and the y-intercept at $(0, 9)$, we can locate the additional point $(6, 9)$ on the graph. See Figure 9. ◀

Figure 9

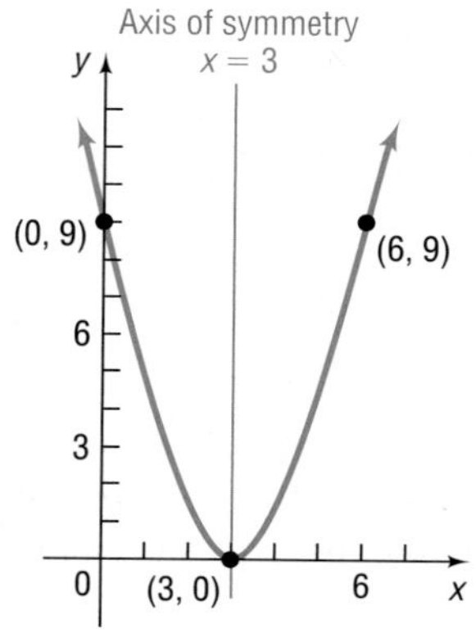

GRAPH THE FUNCTION IN EXAMPLE 4 USING THE METHOD PRESENTED IN EXAMPLE 1. WHICH OF THE TWO METHODS DO YOU PREFER? GIVE REASONS.

NOW WORK PROBLEM 43.

EXAMPLE 5 Graphing a Quadratic Function Using Its Vertex, Axis, and Intercepts

Graph $f(x) = 2x^2 + x + 1$ by determining whether the graph opens up or down. Find its vertex, axis of symmetry, y-intercept, and x-intercepts, if any.

Solution For $f(x) = 2x^2 + x + 1$, we have $a = 2, b = 1$, and $c = 1$. Since $a = 2 > 0$, the parabola opens up. The x-coordinate of the vertex is

$$h = -\frac{b}{2a} = -\frac{1}{4}$$

The y-coordinate of the vertex is

$$k = f\left(-\frac{1}{4}\right) = 2\left(\frac{1}{16}\right) + \left(-\frac{1}{4}\right) + 1 = \frac{7}{8}$$

So the vertex is at $\left(-\frac{1}{4}, \frac{7}{8}\right)$. The axis of symmetry is the line $x = -\frac{1}{4}$. The y-intercept is $f(0) = 1$. The x-intercept(s), if any, obey the equation

$2x^2 + x + 1 = 0$. Since the discriminant $b^2 - 4ac = (1)^2 - 4(2)(1) = -7 < 0$, this equation has no real solutions, and therefore the graph has no x-intercepts. We use the point $(0, 1)$ and the axis of symmetry $x = -\frac{1}{4}$ to locate the additional point $\left(-\frac{1}{2}, 1\right)$ on the graph. See Figure 10. ◀

Figure 10

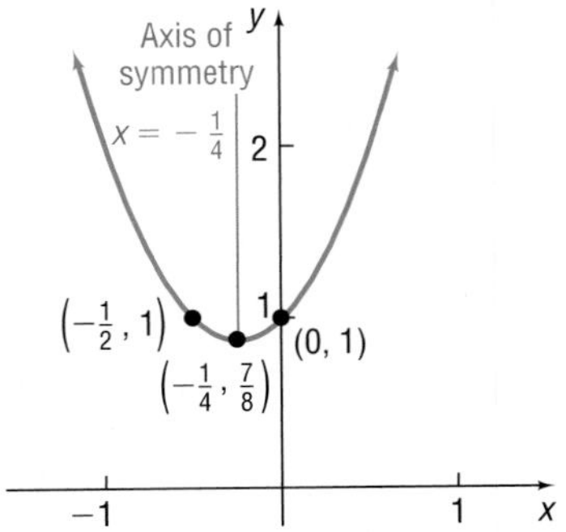

NOW WORK PROBLEM 47.

Given the vertex and one additional point on the graph of a quadratic function $f(x) = ax^2 + bx + c$, $a \neq 0$, we can use

$$f(x) = a(x - h)^2 + k \tag{5}$$

where (h, k) is the vertex, to obtain the quadratic function.

EXAMPLE 6 Finding the Quadratic Function Given Its Vertex and One Other Point

Determine the quadratic function whose vertex is $(1, -5)$ and whose y-intercept is -3. The graph of the parabola is shown in Figure 11.

Figure 11

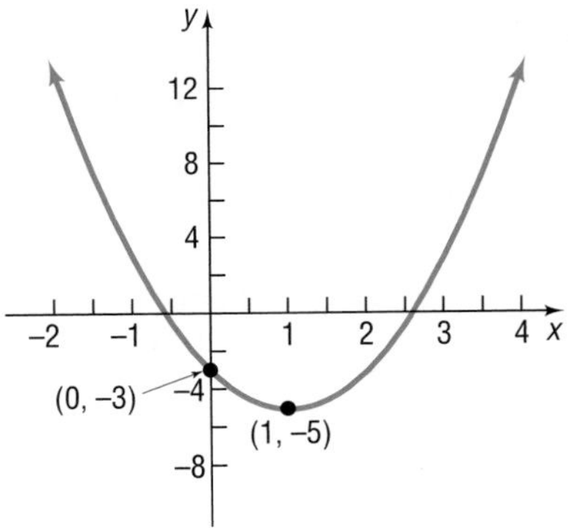

Solution The vertex is $(1, -5)$, so $h = 1$ and $k = -5$. Substitute these values into equation (5).

$$f(x) = a(x - h)^2 + k \quad \text{Equation (5)}$$
$$f(x) = a(x - 1)^2 - 5 \quad h = 1, k = -5$$

To determine the value of a, we use the fact that $f(0) = -3$ (the y-intercept).

$$f(x) = a(x - 1)^2 - 5$$
$$-3 = a(0 - 1)^2 - 5 \quad x = 0, y = f(0) = -3$$
$$-3 = a - 5$$
$$a = 2$$

The quadratic function whose graph is shown in Figure 11 is

$$f(x) = 2(x - 1)^2 - 5 = 2x^2 - 4x - 3$$ ◀

NOW WORK PROBLEM 53.

Summary

Steps for Graphing a Quadratic Function $f(x) = ax^2 + bx + c, \quad a \neq 0$

Option 1

STEP 1: Complete the square in x to write the quadratic function in the form $f(x) = a(x - h)^2 + k$.
STEP 2: Graph the function in stages using transformations.

Option 2

STEP 1: Determine the vertex $\left(-\frac{b}{2a}, f\left(-\frac{b}{2a}\right)\right)$.
STEP 2: Determine the axis of symmetry, $x = -\frac{b}{2a}$.
STEP 3: Determine the y-intercept, $f(0)$.
STEP 4: (a) If $b^2 - 4ac > 0$, then the graph of the quadratic function has two x-intercepts, which are found by solving the equation $ax^2 + bx + c = 0$.
(b) If $b^2 - 4ac = 0$, the vertex is the x-intercept.
(c) If $b^2 - 4ac < 0$, there are no x-intercepts.
STEP 5: Determine an additional point if $b^2 - 4ac \leq 0$ by using the y-intercept and the axis of symmetry.
STEP 6: Plot the points and draw the graph.

Quadratic Models

4 When a mathematical model leads to a quadratic function, the properties of this quadratic function can provide important information about the model. For example, for a quadratic revenue function, we can find the maximum revenue; for a quadratic cost function, we can find the minimum cost.

To see why, recall that the graph of a quadratic function

$$f(x) = ax^2 + bx + c, \quad a \neq 0$$

is a parabola with vertex at $\left(-\frac{b}{2a}, f\left(-\frac{b}{2a}\right)\right)$. This vertex is the highest point on the graph if $a < 0$ and the lowest point on the graph if $a > 0$. If the vertex is the highest point ($a < 0$), then $f\left(-\frac{b}{2a}\right)$ is the **maximum value** of f. If the vertex is the lowest point ($a > 0$), then $f\left(-\frac{b}{2a}\right)$ is the **minimum value** of f.

This property of the graph of a quadratic function enables us to answer questions involving optimization (finding maximum or minimum values) in models involving quadratic functions.

EXAMPLE 7 Finding the Maximum or Minimum Value of a Quadratic Function

Determine whether the quadratic function

$$f(x) = x^2 - 4x + 7$$

has a maximum or minimum value. Then find the maximum or minimum value.

Solution We compare $f(x) = x^2 - 4x + 7$ to $f(x) = ax^2 + bx + c$. We conclude that $a = 1, b = -4$, and $c = 7$. Since $a > 0$, the graph of f opens up, so the vertex is a minimum point. The minimum value occurs at

$$x = -\frac{b}{2a} = -\frac{-4}{2(1)} = \frac{4}{2} = 2$$

$a = 1, b = -4$

The minimum value is

$$f\left(-\frac{b}{2a}\right) = f(2) = 2^2 - 4(2) + 7 = 4 - 8 + 7 = 3$$

NOW WORK PROBLEM 61.

EXAMPLE 8 Maximizing Revenue

The marketing department at Texas Instruments has found that, when certain calculators are sold at a price of p dollars per unit, the revenue R (in dollars) as a function of the price p is

$$R(p) = -150p^2 + 21{,}000p$$

What unit price should be established in order to maximize revenue? If this price is charged, what is the maximum revenue?

Solution The revenue R is

$$R(p) = -150p^2 + 21{,}000p \qquad R(p) = ap^2 + bp + c$$

The function R is a quadratic function with $a = -150$, $b = 21{,}000$, and $c = 0$. Because $a < 0$, the vertex is the highest point on the parabola. The revenue R is therefore a maximum when the price p is

$$p = -\frac{b}{2a} = -\frac{21{,}000}{2(-150)} = \frac{-21{,}000}{-300} = \$70.00$$

The maximum revenue R is

$$R(70) = -150(70)^2 + 21{,}000(70) = \$735{,}000$$

See Figure 12 for an illustration.

Figure 12

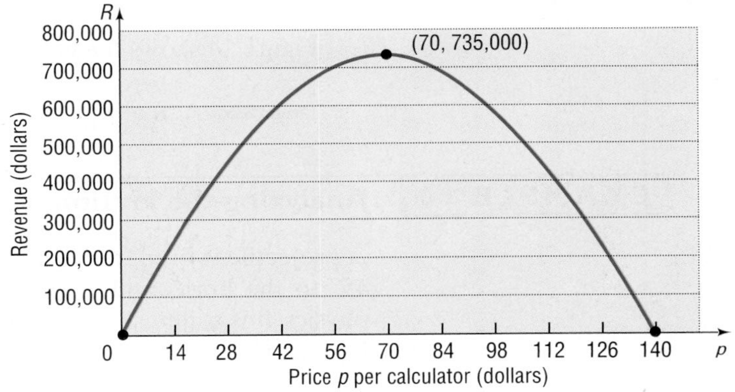

NOW WORK PROBLEM 69.

EXAMPLE 9 Maximizing the Area Enclosed by a Fence

A farmer has 2000 yards of fence to enclose a rectangular field. What is the largest area that can be enclosed?

Figure 13

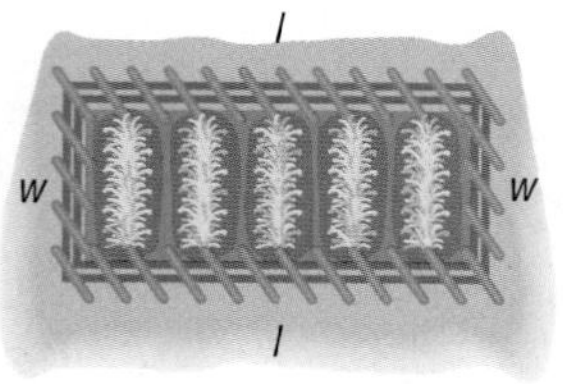

Solution Figure 13 illustrates the situation. The available fence represents the perimeter of the rectangle. If l is the length and w is the width, then

$$\text{perimeter} = 2l + 2w = 2000$$

$$2l + 2w = 2000 \qquad \textbf{(6)}$$

The area A of the rectangle is

$$A = lw$$

To express A in terms of a single variable, we solve equation (6) for w and substitute the result in $A = lw$. Then A involves only the variable l. [You could also solve equation (6) for l and express A in terms of w alone. Try it!]

$$2l + 2w = 2000 \qquad \text{Equation (6)}$$

$$2w = 2000 - 2l \qquad \text{Solve for } w.$$

$$w = \frac{2000 - 2l}{2} = 1000 - l$$

Then the area A is

$$A = lw = l(1000 - l) = -l^2 + 1000l$$

Now, A is a quadratic function of l.

$$A(l) = -l^2 + 1000l \qquad a = -1, b = 1000, c = 0$$

Since $a < 0$, the vertex is a maximum point on the graph of A. The maximum value occurs at

$$l = -\frac{b}{2a} = -\frac{1000}{2(-1)} = 500$$

Figure 14

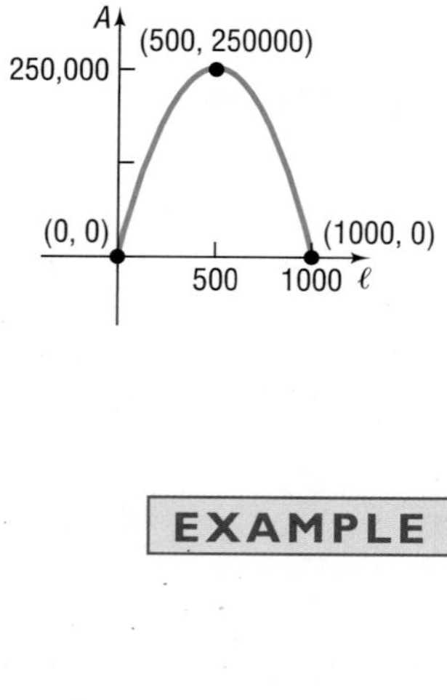

The maximum value of A is

$$A\left(-\frac{b}{2a}\right) = A(500) = -500^2 + 1000(500)$$

$$= -250{,}000 + 500{,}000 = 250{,}000$$

The largest area that can be enclosed by 2000 yards of fence in the shape of a rectangle is 250,000 square yards. ◀

Figure 14 shows the graph of $A(l) = -l^2 + 1000l$.

NOW WORK PROBLEM 75.

EXAMPLE 10 Analyzing the Motion of a Projectile

A projectile is fired from a cliff 500 feet above the water at an inclination of 45° to the horizontal, with a muzzle velocity of 400 feet per second. In physics, it is established that the height h of the projectile above the water is given by

$$h(x) = \frac{-32x^2}{(400)^2} + x + 500$$

where x is the horizontal distance of the projectile from the base of the cliff. See Figure 15.

Figure 15

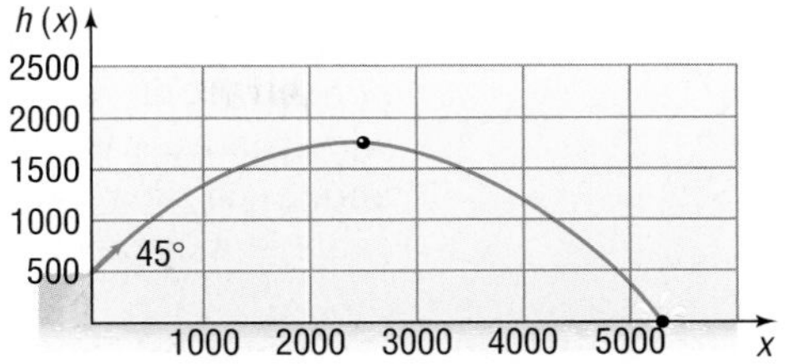

(a) Find the maximum height of the projectile.

(b) How far from the base of the cliff will the projectile strike the water?

Solution (a) The height of the projectile is given by a quadratic function.

$$h(x) = \frac{-32x^2}{(400)^2} + x + 500 = \frac{-1}{5000}x^2 + x + 500$$

We are looking for the maximum value of h. Since the maximum value is obtained at the vertex, we compute

$$x = -\frac{b}{2a} = -\frac{1}{2\left(\frac{-1}{5000}\right)} = \frac{5000}{2} = 2500$$

The maximum height of the projectile is

$$h(2500) = \frac{-1}{5000}(2500)^2 + 2500 + 500$$
$$= -1250 + 2500 + 500 = 1750 \text{ ft}$$

(b) The projectile will strike the water when the height is zero. To find the distance x traveled, we need to solve the equation

$$h(x) = \frac{-1}{5000}x^2 + x + 500 = 0$$

We use the quadratic formula with

$$b^2 - 4ac = 1 - 4\left(\frac{-1}{5000}\right)(500) = 1.4$$

$$x = \frac{-1 \pm \sqrt{1.4}}{2\left(\frac{-1}{5000}\right)} \approx \begin{cases} -458 \\ 5458 \end{cases}$$

We discard the negative solution and find that the projectile will strike the water at a distance of about 5458 feet from the base of the cliff. ◀

Seeing the Concept

Graph

$$h(x) = \frac{-1}{5000}x^2 + x + 500, \qquad 0 \le x \le 5500$$

Use MAXIMUM to find the maximum height of the projectile, and use ROOT or ZERO to find the distance from the base of the cliff to where the projectile strikes the water. Compare your results with those obtained in the text. TRACE the path of the projectile. How far from the base of the cliff is the projectile when its height is 1000 ft? 1500 ft?

NOW WORK PROBLEM 79.

EXAMPLE 11 The Golden Gate Bridge

The Golden Gate Bridge, a suspension bridge, spans the entrance to San Francisco Bay. Its 746-foot-tall towers are 4200 feet apart. The bridge is suspended from two huge cables more than 3 feet in diameter; the 90-foot-wide roadway is 220 feet above the water. The cables are parabolic in shape and touch the road surface at the center of the bridge. Find the height of the cable at a distance of 1000 feet from the center.

Solution We begin by choosing the placement of the coordinate axes so that the x-axis coincides with the road surface and the origin coincides with the center of the bridge. As a result, the twin towers will be vertical (height $746 - 220 = 526$ feet above the road) and located 2100 feet from the center. Also, the cable, which has the shape of a parabola, will extend from the towers, open up, and have its vertex at $(0, 0)$. As illustrated in Figure 16, the choice of placement of the axes enables us to identify the equation of the parabola as $y = ax^2$, $a > 0$. We can also see that the points $(-2100, 526)$ and $(2100, 526)$ are on the graph.

Figure 16

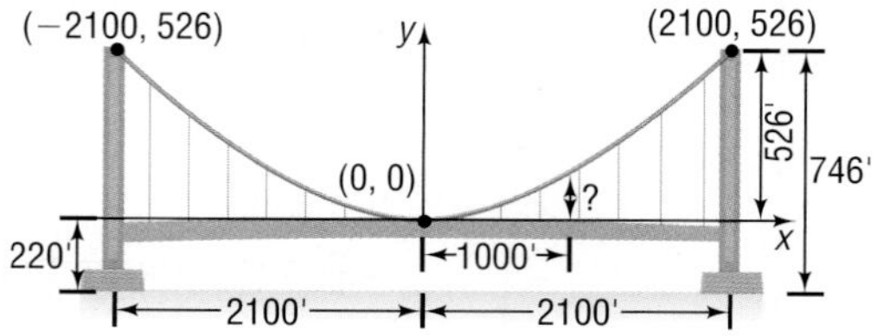

Based on these facts, we can find the value of a in $y = ax^2$.

$$y = ax^2$$
$$526 = a(2100)^2 \quad y = 526;\ x = 2100$$
$$a = \frac{526}{(2100)^2}$$

The equation of the parabola is therefore

$$y = \frac{526}{(2100)^2}x^2$$

The height of the cable when $x = 1000$ is

$$y = \frac{526}{(2100)^2}(1000)^2 \approx 119.3 \text{ feet}$$

The cable is 119.3 feet high at a distance of 1000 feet from the center of the bridge. ◀

NOW WORK PROBLEM 81.

Fitting a Quadratic Function to Data

5 In Section 1.4 we found the line of best fit for data that appeared to be linearly related. It was noted that data may also follow a nonlinear relation. Figures 17(a) and (b) show scatter diagrams of data that follow a quadratic relation.

Figure 17

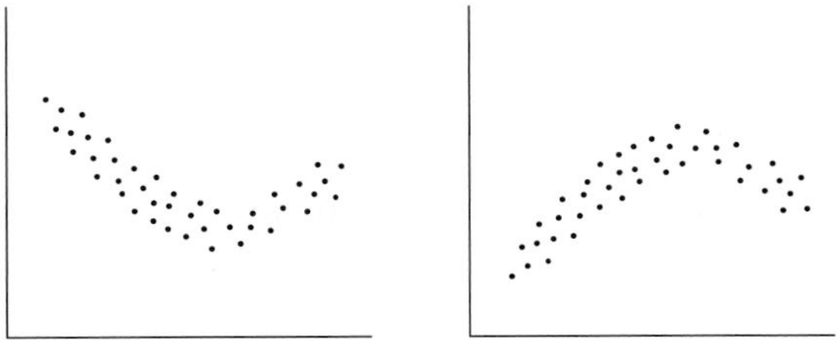

$y = ax^2 + bx + c, a > 0$
(a)

$y = ax^2 + bx + c, a < 0$
(b)

EXAMPLE 12 Fitting a Quadratic Function to Data

A farmer collected the data given in Table 2, which shows crop yields Y for various amounts of fertilizer used, x.

Table 2

Plot	Fertilizer, x (Pounds/100 ft^2)	Yield (Bushels)
1	0	4
2	0	6
3	5	10
4	5	7
5	10	12
6	10	10
7	15	15
8	15	17
9	20	18
10	20	21
11	25	20
12	25	21
13	30	21
14	30	22
15	35	21
16	35	20
17	40	19
18	40	19

(a) Draw a scatter diagram of the data. Comment on the type of relation that may exist between the two variables.

(b) The quadratic function of best fit to these data is

$$Y(x) = -0.0171x^2 + 1.0765x + 3.8939$$

Use this function to determine the optimal amount of fertilizer to apply.

(c) Use the function to predict crop yield when the optimal amount of fertilizer is applied.

(d) Use a graphing utility to verify that the function given in part (b) is the quadratic function of best fit.

(e) With a graphing utility, draw a scatter diagram of the data and then graph the quadratic function of best fit on the scatter diagram.

Solution (a) Figure 18 shows the scatter diagram. It appears that the data follow a quadratic relation, with $a < 0$.

Figure 18

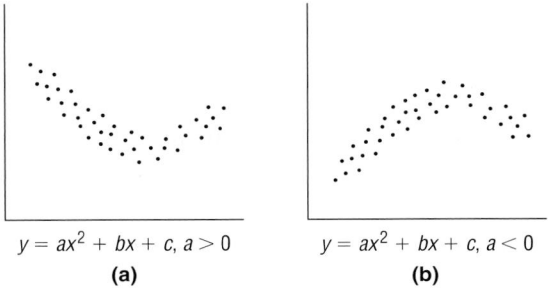

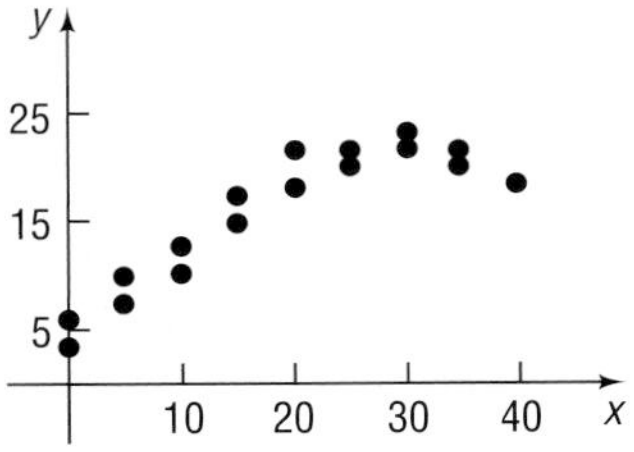

(b) Based on the quadratic function of best fit, the optimal amount of fertilizer to apply is

$$h = -\frac{b}{2a} = -\frac{1.0765}{2(-0.0171)} \approx 31.5 \text{ pounds of fertilizer per 100 square feet}$$

(c) We evaluate the function $Y(x)$ for $x = 31.5$.

$$Y(31.5) = -0.0171(31.5)^2 + 1.0765(31.5) + 3.8939 \approx 20.8 \text{ bushels}$$

If we apply 31.5 pounds of fertilizer per 100 square feet, the crop yield will be 20.8 bushels according to the quadratic function of best fit.

(d) Upon executing the QUADratic REGression program, we obtain the results shown in Figure 19. The output of the utility shows us the equation $y = ax^2 + bx + c$. The quadratic function of best fit is $Y(x) = -0.0171x^2 + 1.0765x + 3.8939$, where x represents the amount of fertilizer used and Y represents crop yield.

(e) Figure 20 shows the graph of the quadratic function found in part (d) drawn on the scatter diagram.

Figure 19

Figure 20

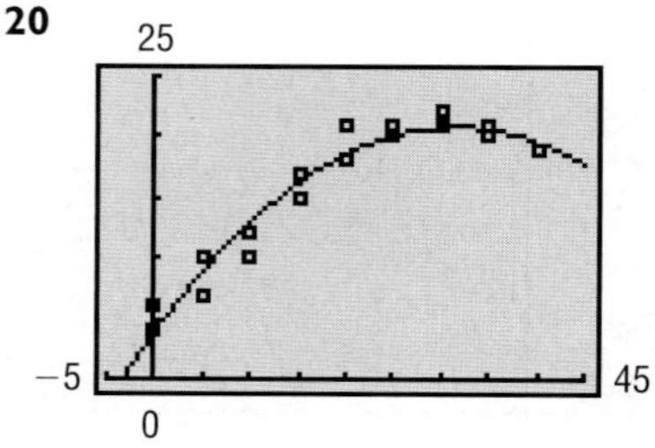

◀

Look again at Figure 19. Notice that the output given by the graphing calculator does not include r, the correlation coefficient. Recall that the correlation coefficient is a measure of the strength of a **linear** relation that exists between two variables. The graphing calculator does not provide an indication of how well the function fits the data in terms of r since a quadratic function cannot be expressed as a linear function.

NOW WORK PROBLEM 91.

3.1 Assess Your Understanding

'Are You Prepared?' *Answers are given at the end of these exercises. If you get a wrong answer, read the pages listed in red.*

1. List the intercepts of the equation $y = x^2 - 9$. (pp. 11–12)
2. Solve the equation: $2x^2 + 7x - 4 = 0$. (pp. 942–945)
3. To complete the square of $x^2 - 5x$, you would add the number __________. (pp. 941)
4. To graph $y = (x - 4)^2$, you would shift the graph of $y = x^2$ to the __________ a distance of __________ units. (pp. 94–103)

Concepts and Vocabulary

5. The graph of a quadratic function is called a(n) __________.
6. The vertical line passing through the vertex of a parabola is called the __________.
7. The x-coordinate of the vertex of $f(x) = ax^2 + bx + c$, $a \neq 0$, is __________.
8. *True or False:* The graph of $f(x) = 2x^2 + 3x - 4$ opens up.
9. *True or False:* The x-coordinate of the vertex of $f(x) = -x^2 + 4x + 5$ is $f(2)$.
10. *True or False:* If the discriminant $b^2 - 4ac = 0$, the graph of $f(x) = ax^2 + bx + c$, $a \neq 0$, will touch the x-axis at its vertex.

Exercises

In Problems 11–18, match each graph to one the following functions without using a graphing utility.

11. $f(x) = x^2 - 1$ **12.** $f(x) = -x^2 - 1$ **13.** $f(x) = x^2 - 2x + 1$ **14.** $f(x) = x^2 + 2x + 1$

15. $f(x) = x^2 - 2x + 2$

16. $f(x) = x^2 + 2x$

17. $f(x) = x^2 - 2x$

18. $f(x) = x^2 + 2x + 2$

A.

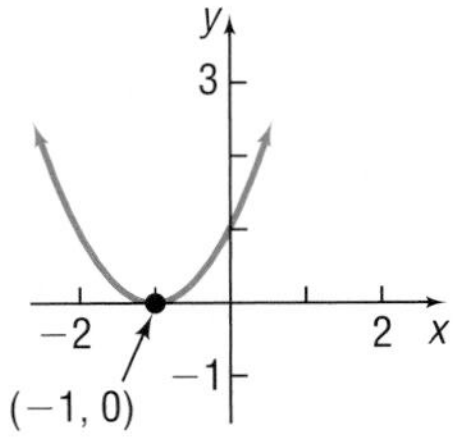

B.

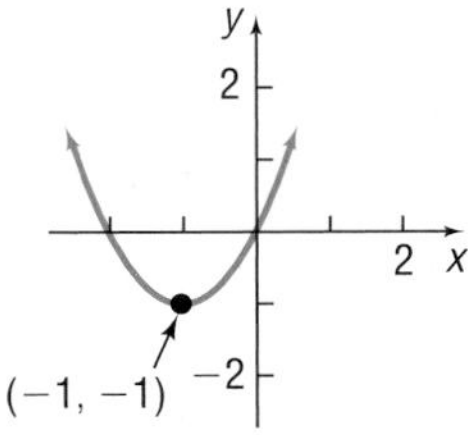

C.

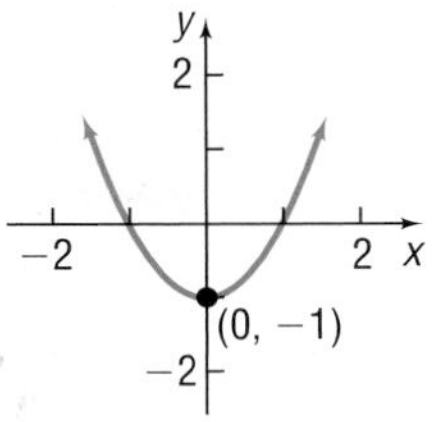

D.

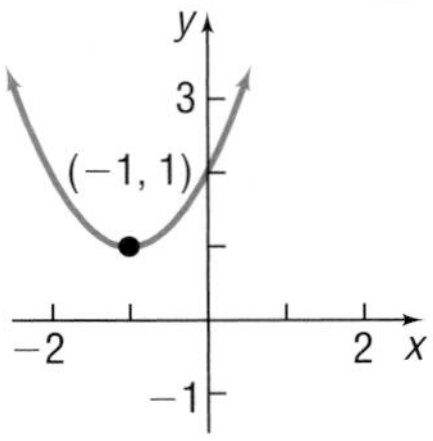

E.

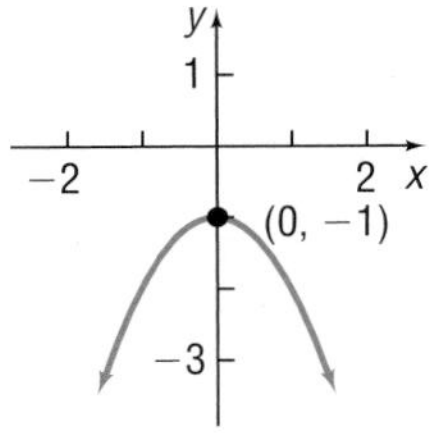

F.

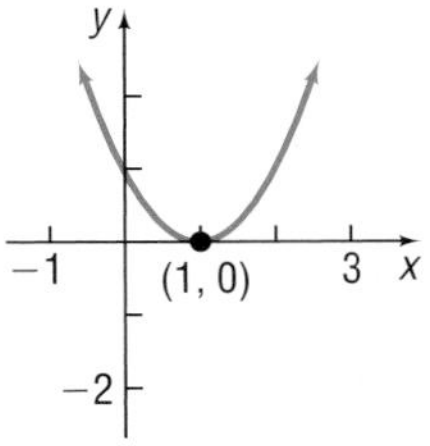

G.

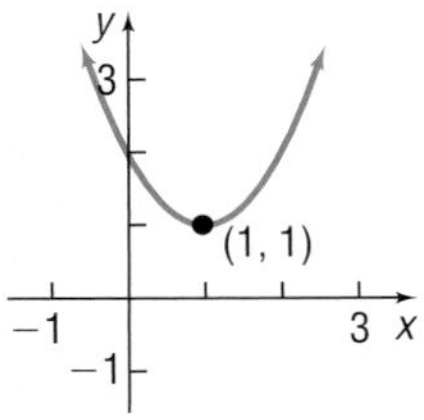

H.

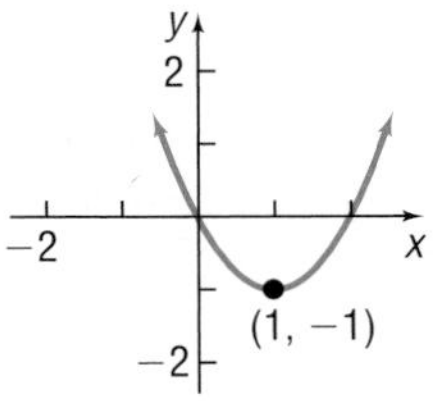

In Problems 19–34, graph the function f by starting with the graph of $y = x^2$ and using transformations (shifting, compressing, stretching, and/or reflection).

[**Hint:** If necessary, write f in the form $f(x) = a(x - h)^2 + k$.]

19. $f(x) = \frac{1}{4}x^2$

20. $f(x) = 2x^2$

21. $f(x) = \frac{1}{4}x^2 - 2$

22. $f(x) = 2x^2 - 3$

23. $f(x) = \frac{1}{4}x^2 + 2$

24. $f(x) = 2x^2 + 4$

25. $f(x) = \frac{1}{4}x^2 + 1$

26. $f(x) = -2x^2 - 2$

27. $f(x) = x^2 + 4x + 2$

28. $f(x) = x^2 - 6x - 1$

29. $f(x) = 2x^2 - 4x + 1$

30. $f(x) = 3x^2 + 6x$

31. $f(x) = -x^2 - 2x$

32. $f(x) = -2x^2 + 6x + 2$

33. $f(x) = \frac{1}{2}x^2 + x - 1$

34. $f(x) = \frac{2}{3}x^2 + \frac{4}{3}x - 1$

In Problems 35–52, graph each quadratic function by determining whether its graph opens up or down and by finding its vertex, axis of symmetry, y-intercept, and x-intercepts, if any. Determine the domain and the range of the function. Determine where the function is increasing and where it is decreasing.

35. $f(x) = x^2 + 2x$

36. $f(x) = x^2 - 4x$

37. $f(x) = -x^2 - 6x$

38. $f(x) = -x^2 + 4x$

39. $f(x) = 2x^2 - 8x$

40. $f(x) = 3x^2 + 18x$

41. $f(x) = x^2 + 2x - 8$

42. $f(x) = x^2 - 2x - 3$

43. $f(x) = x^2 + 2x + 1$

44. $f(x) = x^2 + 6x + 9$

45. $f(x) = 2x^2 - x + 2$

46. $f(x) = 4x^2 - 2x + 1$

47. $f(x) = -2x^2 + 2x - 3$

48. $f(x) = -3x^2 + 3x - 2$

49. $f(x) = 3x^2 + 6x + 2$

50. $f(x) = 2x^2 + 5x + 3$

51. $f(x) = -4x^2 - 6x + 2$

52. $f(x) = 3x^2 - 8x + 2$

In Problems 53–58, determine the quadratic function whose graph is given.

53.

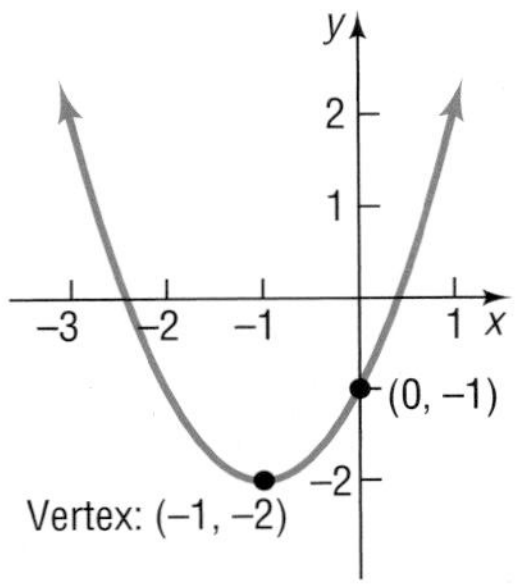

54.

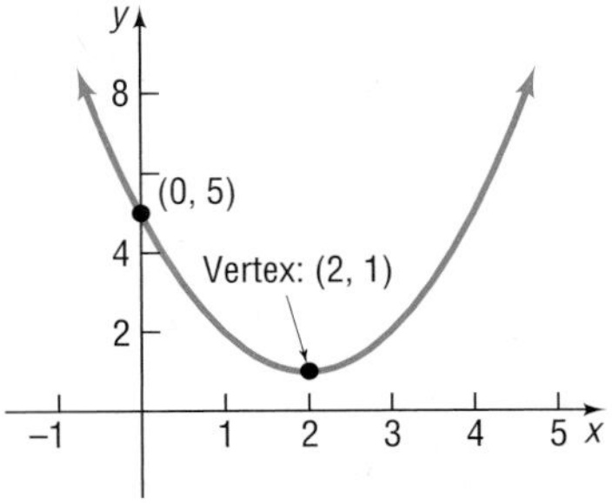

55.

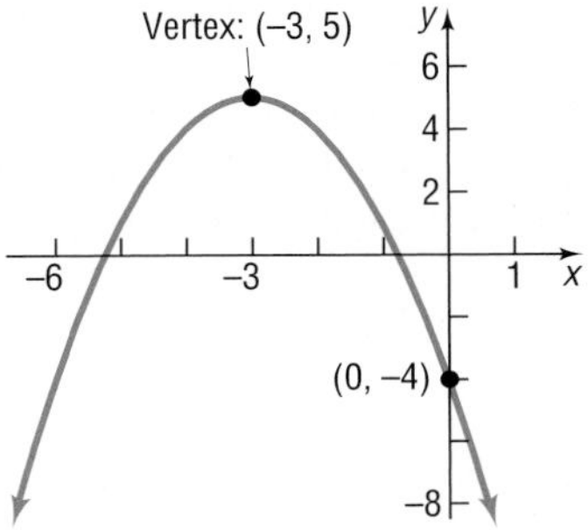

56.

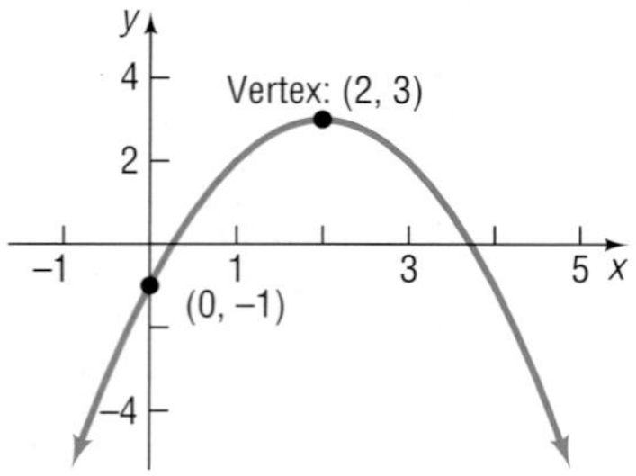

57.

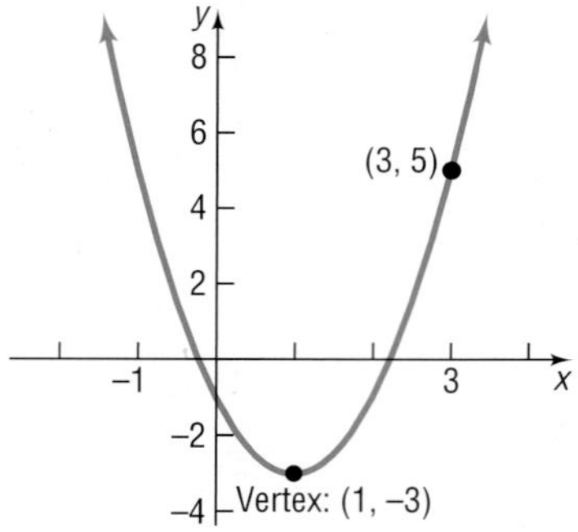

58.

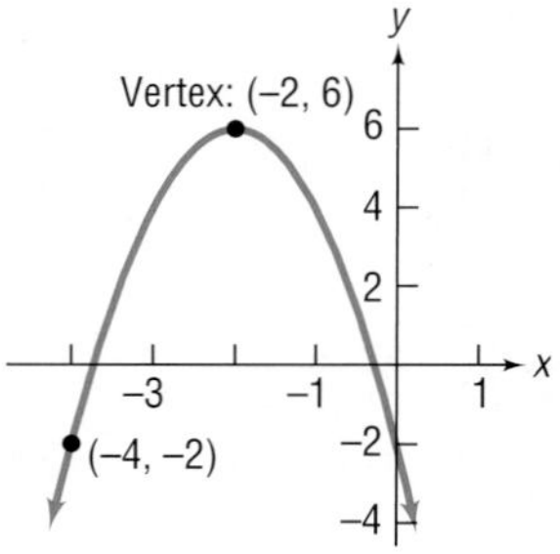

In Problems 59–66, determine, without graphing, whether the given quadratic function has a maximum value or a minimum value and then find the value.

59. $f(x) = 2x^2 + 12x$ **60.** $f(x) = -2x^2 + 12x$ **61.** $f(x) = 2x^2 + 12x - 3$ **62.** $f(x) = 4x^2 - 8x + 3$

63. $f(x) = -x^2 + 10x - 4$ **64.** $f(x) = -2x^2 + 8x + 3$ **65.** $f(x) = -3x^2 + 12x + 1$ **66.** $f(x) = 4x^2 - 4x$

Answer Problems 67 and 68 using the following: A quadratic function of the form $f(x) = ax^2 + bx + c$ with $b^2 - 4ac > 0$ may also be written in the form $f(x) = a(x - r_1)(x - r_2)$, where r_1 and r_2 are the x-intercepts of the graph of the quadratic function.

67. (a) Find a quadratic function whose x-intercepts are -3 and 1 with $a = 1$; $a = 2$; $a = -2$; $a = 5$.
(b) How does the value of a affect the intercepts?
(c) How does the value of a affect the axis of symmetry?
(d) How does the value of a affect the vertex?
(e) Compare the x-coordinate of the vertex with the midpoint of the x-intercepts. What might you conclude?

68. (a) Find a quadratic function whose x-intercepts are -5 and 3 with $a = 1$; $a = 2$; $a = -2$; $a = 5$.
(b) How does the value of a affect the intercepts?
(c) How does the value of a affect the axis of symmetry?
(d) How does the value of a affect the vertex?
(e) Compare the x-coordinate of the vertex with the midpoint of the x-intercepts. What might you conclude?

69. Maximizing Revenue Suppose that the manufacturer of a gas clothes dryer has found that, when the unit price is p dollars, the revenue R (in dollars) is

$$R(p) = -4p^2 + 4000p$$

What unit price should be established for the dryer to maximize revenue? What is the maximum revenue?

70. Maximizing Revenue The John Deere company has found that the revenue from sales of heavy-duty tractors is a function of the unit price p that it charges. If the revenue R is

$$R(p) = -\frac{1}{2}p^2 + 1900p$$

what unit price p should be charged to maximize revenue? What is the maximum revenue?

71. Demand Equation The price p and the quantity x sold of a certain product obey the demand equation

$$p = -\frac{1}{6}x + 100, \qquad 0 \le x \le 600$$

(a) Express the revenue R as a function of x. (Remember, $R = xp$.)
(b) What is the revenue if 200 units are sold?
(c) What quantity x maximizes revenue? What is the maximum revenue?
(d) What price should the company charge to maximize revenue?

72. Demand Equation The price p and the quantity x sold of a certain product obey the demand equation

$$p = -\frac{1}{3}x + 100, \qquad 0 \le x \le 300$$

(a) Express the revenue R as a function of x.
(b) What is the revenue if 100 units are sold?
(c) What quantity x maximizes revenue? What is the maximum revenue?
(d) What price should the company charge to maximize revenue?

73. Demand Equation The price p and the quantity x sold of a certain product obey the demand equation

$$x = -5p + 100, \qquad 0 \le p \le 20$$

(a) Express the revenue R as a function of x.
(b) What is the revenue if 15 units are sold?
(c) What quantity x maximizes revenue? What is the maximum revenue?
(d) What price should the company charge to maximize revenue?

74. Demand Equation The price p and the quantity x sold of a certain product obey the demand equation

$$x = -20p + 500, \qquad 0 \le p \le 25$$

(a) Express the revenue R as a function of x.
(b) What is the revenue if 20 units are sold?
(c) What quantity x maximizes revenue? What is the maximum revenue?
(d) What price should the company charge to maximize revenue?

75. Enclosing a Rectangular Field David has available 400 yards of fencing and wishes to enclose a rectangular area.
(a) Express the area A of the rectangle as a function of the width w of the rectangle.
(b) For what value of w is the area largest?
(c) What is the maximum area?

76. Enclosing a Rectangular Field Beth has 3000 feet of fencing available to enclose a rectangular field.
(a) Express the area A of the rectangle as a function of x where x is the length of the rectangle.
(b) For what value of x is the area largest?
(c) What is the maximum area?

77. Enclosing the Most Area with a Fence A farmer with 4000 meters of fencing wants to enclose a rectangular plot that borders on a river. If the farmer does not fence the side along the river, what is the largest area that can be enclosed? (See the figure.)

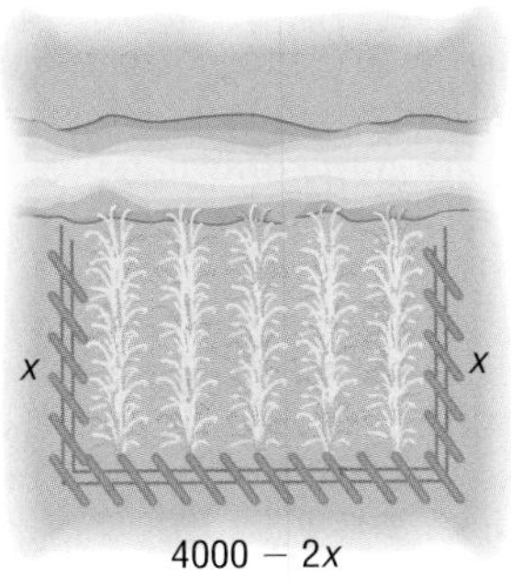

78. Enclosing the Most Area with a Fence A farmer with 2000 meters of fencing wants to enclose a rectangular plot that borders on a straight highway. If the farmer does not fence the side along the highway, what is the largest area that can be enclosed?

79. Analyzing the Motion of a Projectile A projectile is fired from a cliff 200 feet above the water at an inclination of 45° to the horizontal, with a muzzle velocity of 50 feet per second. The height h of the projectile above the water is given by

$$h(x) = \frac{-32x^2}{(50)^2} + x + 200$$

where x is the horizontal distance of the projectile from the base of the cliff.
(a) How far from the base of the cliff is the height of the projectile a maximum?
(b) Find the maximum height of the projectile.
(c) How far from the base of the cliff will the projectile strike the water?
(d) Using a graphing utility, graph the function h, $0 \le x \le 200$.
(e) When the height of the projectile is 100 feet above the water, how far is it from the cliff?

80. Analyzing the Motion of a Projectile A projectile is fired at an inclination of 45° to the horizontal, with a muzzle velocity of 100 feet per second. The height h of the projectile is given by

$$h(x) = \frac{-32x^2}{(100)^2} + x$$

where x is the horizontal distance of the projectile from the firing point.
(a) How far from the firing point is the height of the projectile a maximum?
(b) Find the maximum height of the projectile.
(c) How far from the firing point will the projectile strike the ground?
(d) Using a graphing utility, graph the function h, $0 \le x \le 350$.
(e) When the height of the projectile is 50 feet above the ground, how far has it traveled horizontally?

81. Suspension Bridge A suspension bridge with weight uniformly distributed along its length has twin towers that extend 75 meters above the road surface and are 400 meters apart. The cables are parabolic in shape and are suspended from the tops of the towers. The cables touch the road surface at the center of the bridge. Find the height of the cables at a point 100 meters from the center. (Assume that the road is level.)

82. Architecture A parabolic arch has a span of 120 feet and a maximum height of 25 feet. Choose suitable rectangular coordinate axes and find the equation of the parabola. Then calculate the height of the arch at points 10 feet, 20 feet, and 40 feet from the center.

83. Constructing Rain Gutters A rain gutter is to be made of aluminum sheets that are 12 inches wide by turning up the edges 90°. What depth will provide maximum cross-sectional area and hence allow the most water to flow?

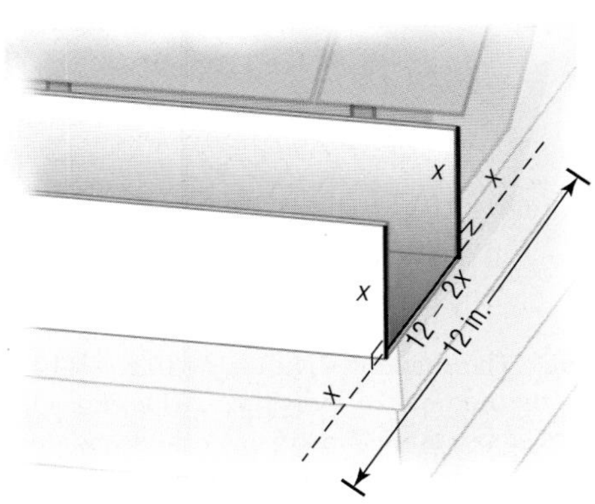

84. Norman Windows A Norman window has the shape of a rectangle surmounted by a semicircle of diameter equal to the width of the rectangle (see the figure). If the perimeter of the window is 20 feet, what dimensions will admit the most light (maximize the area)?

[**Hint:** Circumference of a circle $= 2\pi r$; area of a circle $= \pi r^2$, where r is the radius of the circle.]

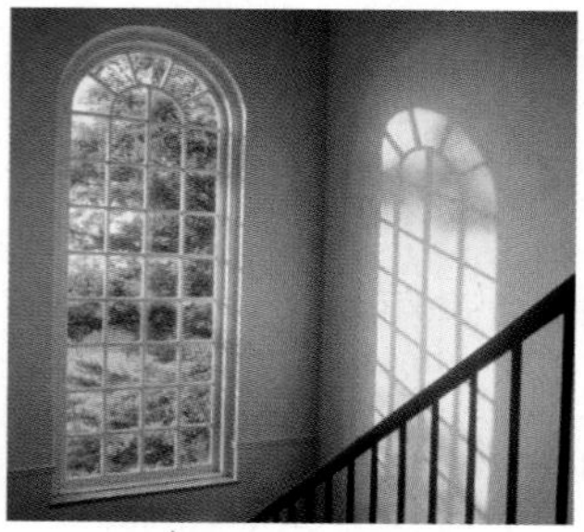

85. Constructing a Stadium A track and field playing area is in the shape of a rectangle with semicircles at each end (see the figure). The inside perimeter of the track is to be 1500 meters. What should the dimensions of the rectangle be so that the area of the rectangle is a maximum?

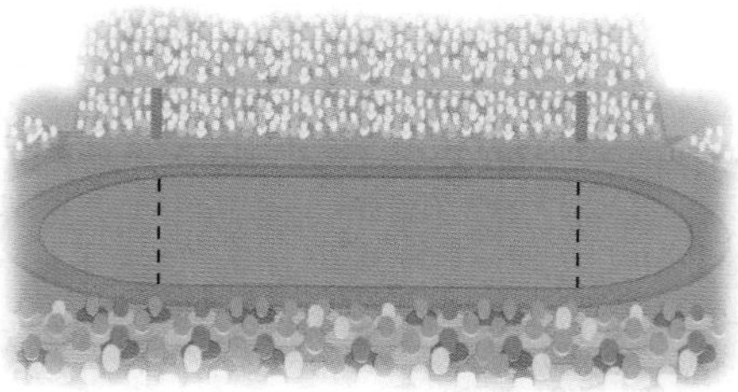

86. Architecture A special window has the shape of a rectangle surmounted by an equilateral triangle (see the figure). If the perimeter of the window is 16 feet, what dimensions will admit the most light?

[**Hint:** Area of an equilateral triangle $= \left(\frac{\sqrt{3}}{4}\right)x^2$, where x is the length of a side of the triangle.]

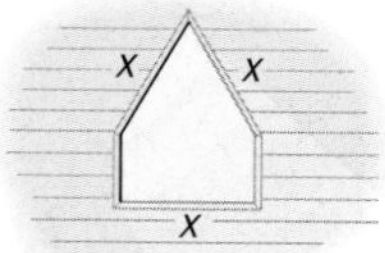

87. Hunting The function $H(x) = -1.01x^2 + 114.3x + 451.0$ models the number of individuals who engage in hunting activities whose annual income is x thousand dollars.
SOURCE: Based on data obtained from the National Sporting Goods Association.

(a) What is the income level for which there are the most hunters? Approximately how many hunters earn this amount?

(b) Using a graphing utility, graph $H = H(x)$. Are the number of hunters increasing or decreasing for individuals earning between \$20,000 and \$40,000?

88. Advanced Degrees The function $P(x) = -0.008x^2 + 0.868x - 11.884$ models the percentage of the U.S. population whose age is given by x that has earned an advanced degree (more than a bachelor's degree) in March 2000.
SOURCE: Based on data obtained from the U.S. Census Bureau.

(a) What is the age for which the highest percentage of Americans have earned an advanced degree? What is the highest percentage?

(b) Using a graphing utility, graph $P = P(x)$. Is the percentage of Americans that have earned an advanced degree increasing or decreasing for individuals between the ages of 40 and 50?

89. Male Murder Victims The function $M(x) = 0.76x^2 - 107.00x + 3854.18$ models the number of male murder victims who are x years of age $(20 \le x < 90)$.
SOURCE: Based on data obtained from the Federal Bureau of Investigation.

(a) Use the model to approximate the number of male murder victims who are $x = 23$ years of age.

(b) At what age is the number of male murder victims 1456?

(c) Using a graphing utility, graph $M = M(x)$.

(d) Based on the graph drawn in part (c), describe what happens to the number of male murder victims as age increases.

90. Health Care Expenditures The function $H(x) = 0.004x^2 - 0.197x + 5.406$ models the percentage of total income that an individual that is x years of age spends on health care.
SOURCE: Based on data obtained from the Bureau of Labor Statistics.

(a) Use the model to approximate the percentage of total income that an individual that is $x = 45$ years of age spends on health care.

(b) At what age is the percentage of income spent on health care 10%?

(c) Using a graphing utility, graph $H = H(x)$.

(d) Based on the graph drawn in part (c), describe what happens to the percentage of income spent on health care as individuals age.

91. Life Cycle Hypothesis An individual's income varies with his or her age. The following table shows the median income I of individuals of different age groups within the United States for 1995. For each age group, let the class midpoint represent the independent variable, x. For the class "65 years and older," we will assume that the class midpoint is 69.5.

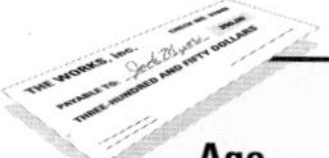

Age	Class Midpoint, x	Median Income, I
15–24 years	19.5	\$20,979
25–34 years	29.5	\$34,701
35–44 years	39.5	\$43,465
45–54 years	49.5	\$48,058
55–64 years	59.5	\$38,077
65 years and older	69.5	\$19,096

SOURCE: U.S. Census Bureau

(a) Draw a scatter diagram of the data. Comment on the type of relation that may exist between the two variables.
(b) The quadratic function of best fit to these data is

$$I(x) = -42.6x^2 + 3806x - 38{,}526$$

Use this function to determine the age at which an individual can expect to earn the most income.
(c) Use the function to predict the peak income earned.
(d) Use a graphing utility to verify that the function given in part (b) is the quadratic function of best fit.
(e) With a graphing utility, draw a scatter diagram of the data and then graph the quadratic function of best fit on the scatter diagram.

92. Life Cycle Hypothesis An individual's income varies with his or her age. The following table shows the median income I of individuals of different age groups within the United States for 1996. For each age group, the class midpoint represents the independent variable, x. For the age group "65 years and older," we will assume that the class midpoint is 69.5.

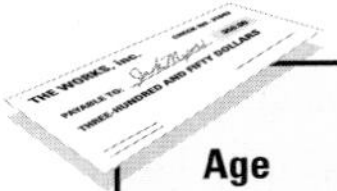

Age	Class Midpoint, x	Median Income, I
15–24 years	19.5	\$21,438
25–34 years	29.5	\$35,888
35–44 years	39.5	\$44,420
45–54 years	49.5	\$50,472
55–64 years	59.5	\$39,815
65 years and older	69.5	\$19,448

SOURCE: U.S. Census Bureau

(a) Draw a scatter diagram of the data. Comment on the type of relation that may exist between the two variables.
(b) The quadratic function of best fit to these data is

$$I(x) = -44.8x^2 + 4009x - 41392$$

Use this function to determine the age at which an individual can expect to earn the most income.
(c) Use the function to predict the peak income earned.
(d) Use a graphing utility to verify that the function given in part (b) is the quadratic function of best fit.
(e) With a graphing utility, draw a scatter diagram of the data and then graph the quadratic function of best fit on the scatter diagram.

93. Height of a Ball A shot-putter throws a ball at an inclination of 45° to the horizontal. The following data represent the height of the ball h at the instant that it has traveled x feet horizontally.

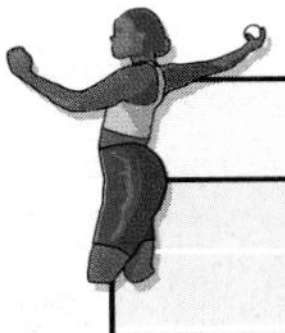

Distance, x	Height, h
20	25
40	40
60	55
80	65
100	71
120	77
140	77
160	75
180	71
200	64

(a) Draw a scatter diagram of the data. Comment on the type of relation that may exist between the two variables.
(b) The quadratic function of best fit to these data is

$$h(x) = -0.0037x^2 + 1.03x + 5.7$$

Use this function to determine how far the ball will travel before it reaches its maximum height.
(c) Use the function to find the maximum height of the ball.
(d) Use a graphing utility to verify that the function given in part (b) is the quadratic function of best fit.
(e) With a graphing utility, draw a scatter diagram of the data and then graph the quadratic function of best fit on the scatter diagram.

94. Miles per Gallon An engineer collects data showing the speed s of a Ford Taurus and its average miles per gallon, M. See the table.

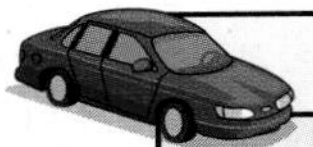

Speed, s	Miles per Gallon, M
30	18
35	20
40	23
40	25
45	25
50	28
55	30
60	29
65	26
65	25
70	25

(a) Draw a scatter diagram of the data. Comment on the type of relation that may exist between the two variables.

(b) The quadratic function of best fit to these data is

$$M(s) = -0.018s^2 + 1.93s - 25.34$$

Use this function to determine the speed that maximizes miles per gallon.

(c) Use the function to predict miles per gallon for a speed of 63 miles per hour.

(d) Use a graphing utility to verify that the function given in part (b) is the quadratic function of best fit.

(e) With a graphing utility, draw a scatter diagram of the data and then graph the quadratic function of best fit on the scatter diagram.

95. Chemical Reactions A self-catalytic chemical reaction results in the formation of a compound that causes the formation ratio to increase. If the reaction rate V is given by

$$V(x) = kx(a - x), \qquad 0 \le x \le a$$

where k is a positive constant, a is the initial amount of the compound, and x is the variable amount of the compound, for what value of x is the reaction rate a maximum?

96. Calculus: Simpson's Rule The figure shows the graph of $y = ax^2 + bx + c$. Suppose that the points $(-h, y_0)$, $(0, y_1)$, and (h, y_2) are on the graph. It can be shown that the area enclosed by the parabola, the x-axis, and the lines $x = -h$ and $x = h$ is

$$\text{Area} = \frac{h}{3}(2ah^2 + 6c)$$

Show that this area may also be given by

$$\text{Area} = \frac{h}{3}(y_0 + 4y_1 + y_2)$$

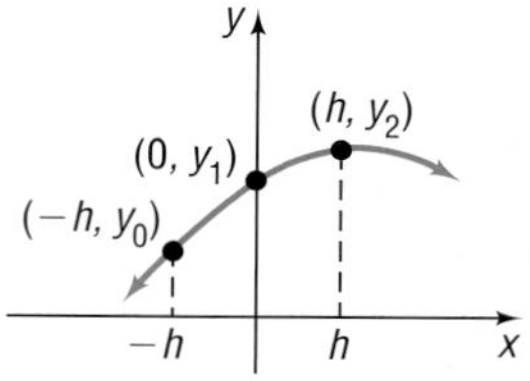

97. Use the result obtained in Problem 96 to find the area enclosed by $f(x) = -5x^2 + 8$, the x-axis, and the lines $x = -1$ and $x = 1$.

98. Use the result obtained in Problem 96 to find the area enclosed by $f(x) = 2x^2 + 8$, the x-axis, and the lines $x = -2$ and $x = 2$.

99. Use the result obtained in Problem 96 to find the area enclosed by $f(x) = x^2 + 3x + 5$, the x-axis, and the lines $x = -4$ and $x = 4$.

100. Use the result obtained in Problem 96 to find the area enclosed by $f(x) = -x^2 + x + 4$, the x-axis, and the lines $x = -1$ and $x = 1$.

101. A rectangle has one vertex on the line $y = 10 - x$, $x > 0$, another at the origin, one on the positive x-axis, and one on the positive y-axis. Find the largest area A that can be enclosed by the rectangle.

102. Let $f(x) = ax^2 + bx + c$, where a, b, and c are odd integers. If x is an integer, show that $f(x)$ must be an odd integer.
[**Hint:** x is either an even integer or an odd integer.]

103. Make up a quadratic function that opens down and has only one x-intercept. Compare yours with others in the class. What are the similarities? What are the differences?

104. On one set of coordinate axes, graph the family of parabolas $f(x) = x^2 + 2x + c$ for $c = -3$, $c = 0$, and $c = 1$. Describe the characteristics of a member of this family.

105. On one set of coordinate axes, graph the family of parabolas $f(x) = x^2 + bx + 1$ for $b = -4$, $b = 0$, and $b = 4$. Describe the general characteristics of this family.

106. State the circumstances that cause the graph of a quadratic function $f(x) = ax^2 + bx + c$ to have no x-intercepts.

107. Why does the graph of a quadratic function open up if $a > 0$ and down if $a < 0$?

108. Refer to Example 8 on page 133. Notice that if the price charged for the calculators is \$0 or \$140 the revenue is \$0. It is easy to explain why revenue would be \$0 if the priced charged is \$0, but how can revenue be \$0 if the price charged is \$140?

'Are You Prepared?' Answers

1. $(0, -9), (-3, 0), (3, 0)$ **2.** $\left\{-4, \frac{1}{2}\right\}$

3. $\frac{25}{4}$ **4.** right; 4

3.2 Polynomial Functions

PREPARING FOR THIS SECTION *Before getting started, review the following:*

- Polynomials (Appendix A, Section A.3, pp. 918–923)
- Intercepts (Section 1.2, pp. 11–12)
- Graphing Techniques: Transformations (Section 2.5, pp. 94–103)
- Intercepts of a Function (Section 2.3, p. 72)

Now work the 'Are You Prepared?' problems on page 158.

OBJECTIVES
1 Identify Polynomial Functions and Their Degree
2 Graph Polynomial Functions Using Transformations
3 Identify the Zeros of a Polynomial Function and Their Multiplicity
4 Analyze the Graph of a Polynomial Function

1 *Polynomial functions* are among the simplest expressions in algebra. They are easy to evaluate: only addition and repeated multiplication are required. Because of this, they are often used to approximate other, more complicated functions. In this section, we investigate properties of this important class of functions.

A **polynomial function** is a function of the form

$$f(x) = a_n x^n + a_{n-1}x^{n-1} + \cdots + a_1 x + a_0 \tag{1}$$

where $a_n, a_{n-1}, \ldots, a_1, a_0$ are real numbers and n is a nonnegative integer. The domain is the set of all real numbers.

A polynomial function is a function whose rule is given by a polynomial in one variable. The **degree** of a polynomial function is the degree of the polynomial in one variable, that is, the largest power of x that appears.

EXAMPLE 1 Identifying Polynomial Functions

Determine which of the following are polynomial functions. For those that are, state the degree; for those that are not, tell why not.

(a) $f(x) = 2 - 3x^4$ (b) $g(x) = \sqrt{x}$ (c) $h(x) = \dfrac{x^2 - 2}{x^3 - 1}$

(d) $F(x) = 0$ (e) $G(x) = 8$ (f) $H(x) = -2x^3(x - 1)^2$

Solution

(a) f is a polynomial function of degree 4.

(b) g is not a polynomial function. The variable x is raised to the $\frac{1}{2}$ power, which is not a nonnegative integer.

(c) h is not a polynomial function. It is the ratio of two polynomials, and the polynomial in the denominator is of positive degree.

(d) F is the zero polynomial function; it is not assigned a degree.

(e) G is a nonzero constant function, a polynomial function of degree 0 since $G(x) = 8 = 8x^0$.

(f) $H(x) = -2x^3(x - 1)^2 = -2x^3(x^2 - 2x + 1) = -2x^5 + 4x^4 - 2x^3$. So H is a polynomial function of degree 5. Do you see how to find the degree of H without multiplying out? ◀

NOW WORK PROBLEMS 11 AND 15.

We have already discussed in detail polynomial functions of degrees 0, 1, and 2. See Table 3 for a summary of the properties of the graphs of these polynomial functions.

Table 3

Degree	Form	Name	Graph
No degree	$f(x) = 0$	Zero function	The x-axis
0	$f(x) = a_0, \quad a_0 \neq 0$	Constant function	Horizontal line with y-intercept a_0
1	$f(x) = a_1x + a_0, \quad a_1 \neq 0$	Linear function	Nonvertical, nonhorizontal line with slope a_1 and y-intercept a_0
2	$f(x) = a_2x^2 + a_1x + a_0, \quad a_2 \neq 0$	Quadratic function	Parabola: Graph opens up if $a_2 > 0$; graph opens down if $a_2 < 0$

One objective of this section is to analyze the graph of a polynomial function. If you take a course in calculus, you will learn that the graph of every polynomial function is both smooth and continuous. By **smooth,** we mean that the graph contains no sharp corners or cusps; by **continuous,** we mean that the graph has no gaps or holes and can be drawn without lifting pencil from paper. See Figures 21(a) and (b).

Figure 21

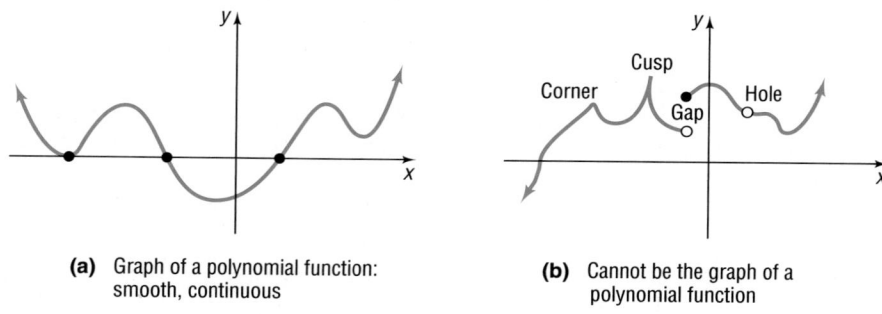

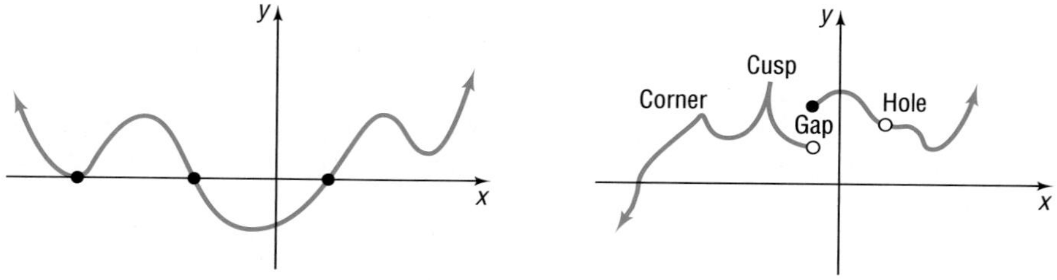

(a) Graph of a polynomial function: smooth, continuous

(b) Cannot be the graph of a polynomial function

We begin the analysis of the graph of a polynomial function by discussing *power functions*, a special kind of polynomial function.

In Words

A power function is a function that is defined by a single monomial.

Power Function

A **power function of degree** n is a function of the form

$$f(x) = ax^n \qquad \textbf{(2)}$$

where a is a real number, $a \neq 0$, and $n > 0$ is an integer.

The graph of a power function of degree 1, $f(x) = ax$, is a straight line, with slope a, that passes through the origin. The graph of a power function of degree 2, $f(x) = ax^2$, is a parabola, with vertex at the origin, that opens up if $a > 0$ and down if $a < 0$.

If we know how to graph a power function of the form $f(x) = x^n$, then a compression or stretch and, perhaps, a reflection about the x-axis will enable us to obtain the graph of $g(x) = ax^n$. Consequently, we shall concentrate on graphing power functions of the form $f(x) = x^n$.

We begin with power functions of even degree of the form $f(x) = x^n$, $n \geq 2$ and n even. The domain of f is the set of all real numbers, and the range is the set of nonnegative real numbers. Such a power function is an even function (do you see why?), so its graph is symmetric with respect to the y-axis. Its graph always contains the origin and the points $(-1, 1)$ and $(1, 1)$.

If $n = 2$, the graph is the familiar parabola $y = x^2$ that opens up, with vertex at the origin. If $n \geq 4$, the graph of $f(x) = x^n$, n even, will be closer to the x-axis than the parabola $y = x^2$ if $-1 < x < 1$ and farther from the x-axis than the parabola $y = x^2$ if $x < -1$ or if $x > 1$. Figure 22(a) illustrates this conclusion. Figure 22(b) shows the graphs of $y = x^4$ and $y = x^8$ for comparison.

From Figure 22, we can see that as n increases the graph of $f(x) = x^n$, $n \geq 2$ and n even, tends to flatten out near the origin and to increase very rapidly when x is far from 0. For large n, it may appear that the graph coincides with the x-axis near the origin, but it does not; the graph actually

Figure 22

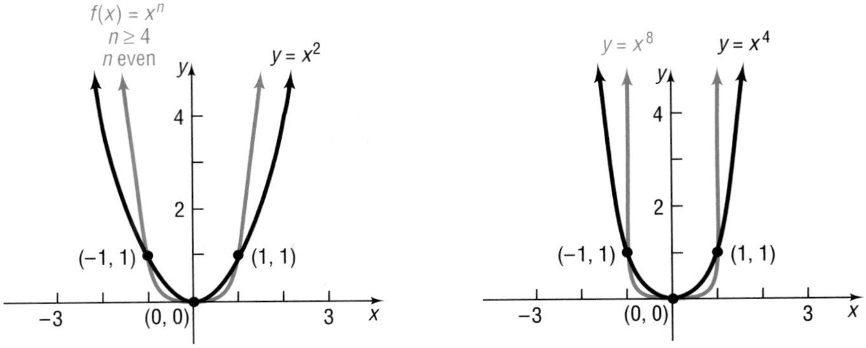

(a) (b)

touches the x-axis only at the origin (see Table 4). Also, for large n, it may appear that for $x < -1$ or for $x > 1$ the graph is vertical, but it is not; it is only increasing very rapidly in these intervals. If the graphs were enlarged many times, these distinctions would be clear.

Table 4

	$x = 0.1$	$x = 0.3$	$x = 0.5$
$f(x) = x^8$	10^{-8}	0.0000656	0.0039063
$f(x) = x^{20}$	10^{-20}	$3.487 \cdot 10^{-11}$	0.000001
$f(x) = x^{40}$	10^{-40}	$1.216 \cdot 10^{-21}$	$9.095 \cdot 10^{-13}$

Seeing the Concept

Graph $Y_1 = x^4$, $Y_2 = x^8$, and $Y_3 = x^{12}$ using the viewing rectangle $-2 \leq x \leq 2$, $-4 \leq y \leq 16$. Then graph each again using the viewing rectangle $-1 \leq x \leq 1$, $0 \leq y \leq 1$. See Figure 23. TRACE along one of the graphs to confirm that for x close to 0 the graph is above the x-axis and that for $x > 0$ the graph is increasing.

Figure 23

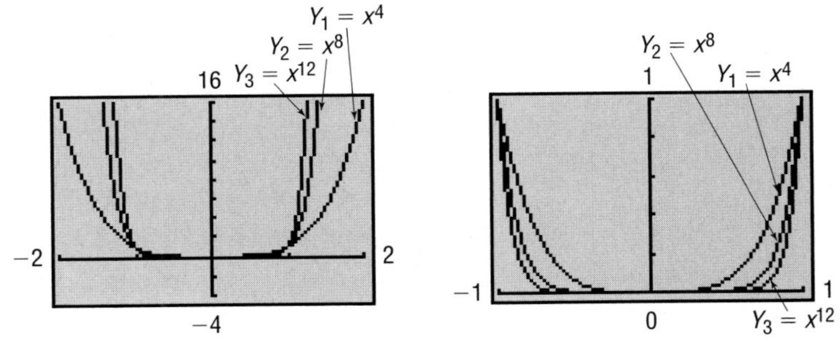

(a) (b)

Properties of Power Functions, $y = x^n$, n is an Even Integer

1. The graph is symmetric with respect to the y-axis.
2. The domain is the set of all real numbers. The range is the set of nonnegative real numbers.
3. The graph always contains the points $(0, 0)$, $(1, 1)$, and $(-1, 1)$.
4. As the exponent n increases in magnitude, the graph becomes more vertical when $x < -1$ or $x > 1$; but for x near the origin, the graph tends to flatten out and lie closer to the x-axis.

Now we consider power functions of odd degree of the form $f(x) = x^n$, $n \geq 3$ and n odd. The domain and the range of f is the set of real numbers. Such a power function is an odd function (do you see why?), so its graph is symmetric with respect to the origin. Its graph always contains the origin and the points $(-1, -1)$ and $(1, 1)$.

The graph of $f(x) = x^n$ when $n = 3$ has been shown several times and is repeated in Figure 24. If $n \geq 5$, the graph of $f(x) = x^n$, n odd, will be closer to the x-axis than that of $y = x^3$ if $-1 < x < 1$ and farther from the x-axis than that of $y = x^3$ if $x < -1$ or if $x > 1$. Figure 24 also illustrates this conclusion.

Figure 25 shows the graph of $y = x^5$ and the graph of $y = x^9$ for further comparison.

Figure 24

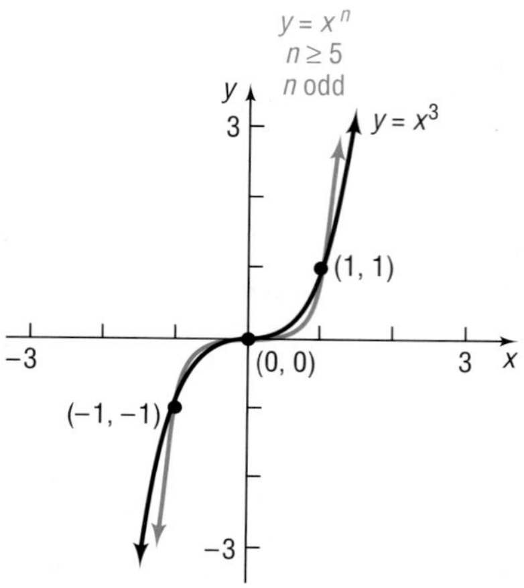

Figure 25

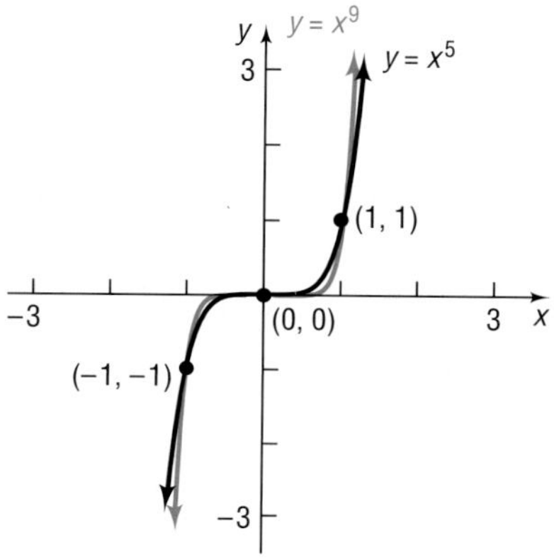

It appears that each graph coincides with the x-axis near the origin, but it does not; each graph actually touches the x-axis only at the origin. Also, it appears that as x increases the graph becomes vertical, but it does not; each graph is increasing very rapidly.

Seeing the Concept

Graph $Y_1 = x^3$, $Y_2 = x^7$, and $Y_3 = x^{11}$ using the viewing rectangle $-2 \leq x \leq 2$, $-16 \leq y \leq 16$. Then graph each again using the viewing rectangle $-1 \leq x \leq 1$, $0 \leq y \leq 1$. See Figure 26. TRACE along one of the graphs to confirm that the graph is increasing and only touches the x-axis at the origin.

Figure 26

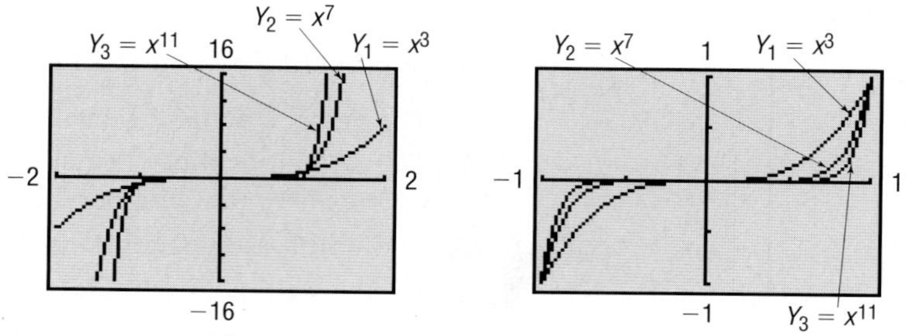

(a) (b)

To summarize:

> **Properties of Power Functions, $y = x^n$, n Is an Odd Integer**
>
> 1. The graph is symmetric with respect to the origin.
> 2. The domain and the range is the set of all real numbers.
> 3. The graph always contains the points $(0, 0)$, $(1, 1)$, and $(-1, -1)$.
> 4. As the exponent n increases in magnitude, the graph becomes more vertical when $x < -1$ or $x > 1$; but for x near the origin, the graph tends to flatten out and lie closer to the x-axis.

2 The methods of shifting, compression, stretching, and reflection studied in Section 2.5, when used with the facts just presented, will enable us to graph polynomial functions that are transformations of power functions.

EXAMPLE 2 **Graphing Polynomial Functions Using Transformations**

Graph: $f(x) = 1 - x^5$

Solution Figure 27 shows the required stages.

Figure 27

(a) $y = x^5$ (b) $y = -x^5$ (c) $y = -x^5 + 1 = 1 - x^5$

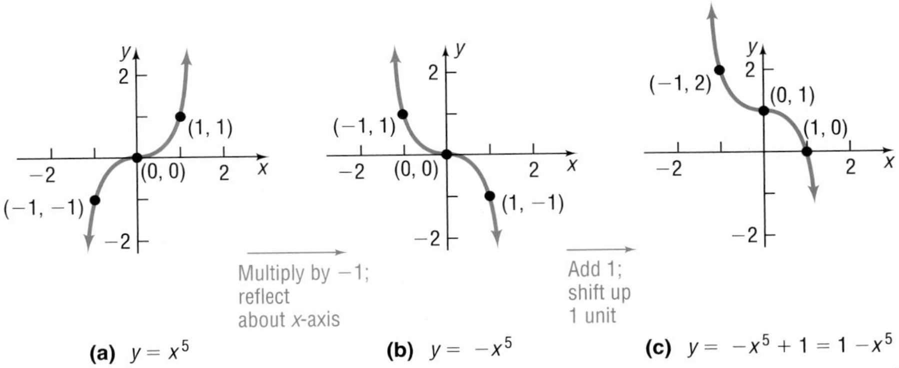

EXAMPLE 3 **Graphing Polynomial Functions Using Transformations**

Graph: $f(x) = \dfrac{1}{2}(x - 1)^4$

Solution Figure 28 shows the required stages.

Figure 28

(a) $y = x^4$ (b) $y = (x - 1)^4$ (c) $y = \frac{1}{2}(x - 1)^4$

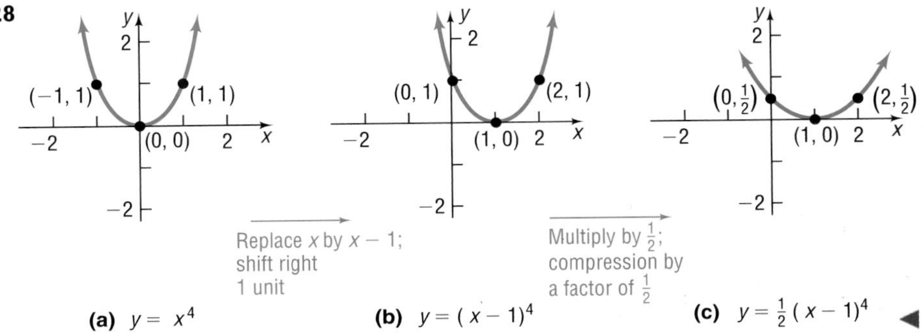

NOW WORK PROBLEMS 23 AND 27.

Graphing Other Polynomials

To graph most polynomials of degree 3 or higher requires advanced techniques. However, if we can locate the x-intercepts of the graph, then algebraic techniques can be used to obtain the graph.

Figure 29 shows the graph of a polynomial function with four x-intercepts. Notice that at the x-intercepts the graph must either cross the x-axis or touch the x-axis. Consequently, between consecutive x-intercepts the graph is either above the x-axis or below the x-axis. We will make use of this property of the graph of a polynomial shortly.

Figure 29

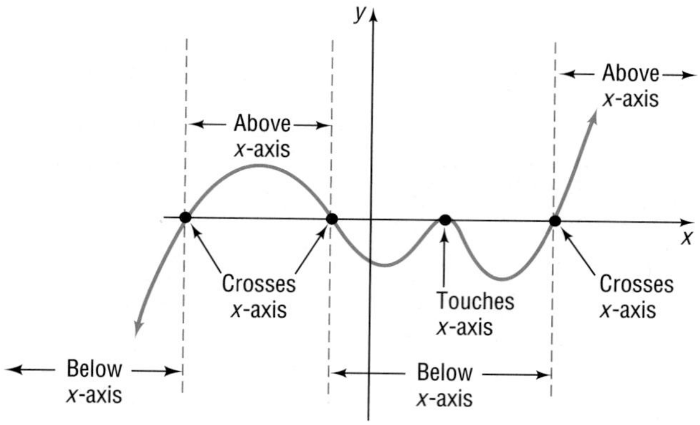

3 If a polynomial function f is factored completely, it is easy to solve the equation $f(x) = 0$ and locate the x-intercepts of the graph. For example, if $f(x) = (x - 1)^2(x + 3)$, then the solutions of the equation

$$f(x) = (x - 1)^2(x + 3) = 0$$

are identified as 1 and -3. Based on this result, we make the following observations:

> If f is a polynomial function and r is a real number for which $f(r) = 0$, then r is called a (real) **zero of f**, or **root of f.** If r is a (real) zero of f, then
>
> (a) r is an x-intercept of the graph of f.
>
> (b) $(x - r)$ is a factor of f.

So the real zeros of a function are the x-intercepts of its graph and they are found by solving the equation $f(x) = 0$.

EXAMPLE 4 Finding a Polynomial from Its Zeros

(a) Find a polynomial of degree 3 whose zeros are -3, 2, and 5.

(b) Graph the polynomial found in part (a) to verify your result.

Solution (a) If r is a zero of a polynomial f, then $x - r$ is a factor of f. This means that $x - (-3) = x + 3$, $x - 2$, and $x - 5$ are factors of f. As a result, any polynomial of the form

$$f(x) = a(x + 3)(x - 2)(x - 5)$$

where a is any nonzero real number, qualifies.

Figure 30

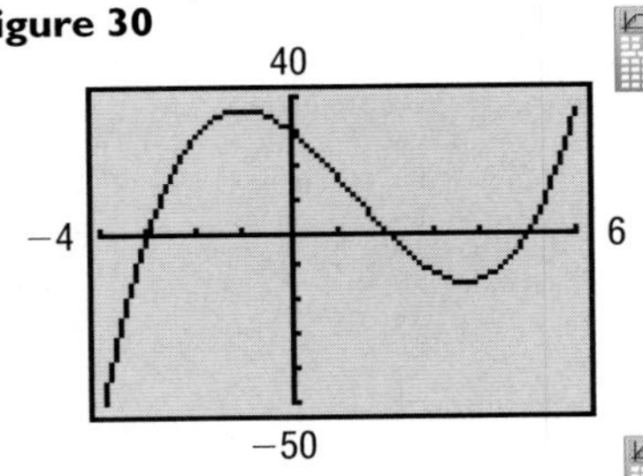

(b) The value of a causes a stretch, compression, or reflection, but does not affect the x-intercepts. We choose to graph f with $a = 1$.

$$f(x) = (x + 3)(x - 2)(x - 5) = x^3 - 4x^2 - 11x + 30$$

Figure 30 shows the graph of f. Notice that the x-intercepts are -3, 2, and 5. ◀

Seeing the Concept

Graph the function found in Example 4 for $a = 2$ and $a = -1$. Does the value of a affect the zeros of f? How does the value of a affect the graph of f?

NOW WORK PROBLEM 37.

If the same factor $x - r$ occurs more than once, then r is called a **repeated,** or **multiple, zero of *f*.** More precisely, we have the following definition.

> If $(x - r)^m$ is a factor of a polynomial f and $(x - r)^{m+1}$ is not a factor of f, then r is called a **zero of multiplicity *m* of *f*.**

EXAMPLE 5 Identifying Zeros and Their Multiplicities

For the polynomial

$$f(x) = 5(x - 2)(x + 3)^2\left(x - \frac{1}{2}\right)^4$$

2 is a zero of multiplicity 1.

-3 is a zero of multiplicity 2.

$\frac{1}{2}$ is a zero of multiplicity 4. ◀

NOW WORK PROBLEM 45(a).

In Example 5 notice that, if you add the multiplicities $(1 + 2 + 4 = 7)$, you obtain the degree of the polynomial.

4 Suppose that it is possible to factor completely a polynomial function and, as a result, locate all the x-intercepts of its graph (the real zeros of the function). As mentioned earlier, these x-intercepts then divide the x-axis into open intervals and, on each such interval, the graph of the polynomial will be either above or below the x-axis. Let's look at an example.

EXAMPLE 6 Graphing a Polynomial Using Its x-Intercepts

For the polynomial: $f(x) = x^2(x - 2)$

(a) Find the x- and y-intercepts of the graph of f.

(b) Use the x-intercepts to find the intervals on which the graph of f is above the x-axis and the intervals on which the graph of f is below the x-axis.

(c) Locate other points on the graph and connect all the points plotted with a smooth, continuous curve.

Solution (a) The y-intercept is $f(0) = 0^2(0 - 2) = 0$. The x-intercepts satisfy the equation

$$f(x) = x^2(x - 2) = 0$$

from which we find

$$x^2 = 0 \quad \text{or} \quad x - 2 = 0$$
$$x = 0 \qquad\qquad x = 2$$

The x-intercepts are 0 and 2.

(b) The two x-intercepts divide the x-axis into three intervals:

$$(-\infty, 0) \qquad (0, 2) \qquad (2, \infty)$$

Since the graph of f crosses or touches the x-axis only at $x = 0$ and $x = 2$, it follows that the graph of f is either above the x-axis $[f(x) > 0]$ or below the x-axis $[f(x) < 0]$ on each of these three intervals. To see where the graph lies, we only need to pick a number in each interval, evaluate f there, and see whether the value is positive (above the x-axis) or negative (below the x-axis). See Table 5.

(c) In constructing Table 5, we obtained three additional points on the graph: $(-1, -3)$, $(1, -1)$, and $(3, 9)$. Figure 31 illustrates these points, the intercepts, and a smooth, continuous curve (the graph of f) connecting them.

Table 5

	0	2	x
Interval	$(-\infty, 0)$	$(0, 2)$	$(2, \infty)$
Number Chosen	-1	1	3
Value of f	$f(-1) = -3$	$f(1) = -1$	$f(3) = 9$
Location of Graph	Below x-axis	Below x-axis	Above x-axis
Point on Graph	$(-1, -3)$	$(1, -1)$	$(3, 9)$

Figure 31

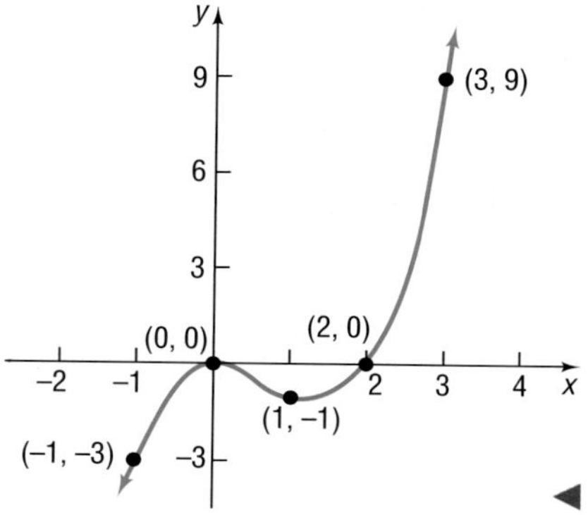

Look again at Table 5. Since the graph of f is below the x-axis on both sides of 0, the graph of f *touches* the x-axis at $x = 0$, a *zero of multiplicity 2*. Since the graph of f is below the x-axis for $x < 2$ and above the x-axis for $x > 2$, the graph of f *crosses* the x-axis at $x = 2$, a *zero of multiplicity 1*.

This suggests the following results:

If *r* Is a Zero of Even Multiplicity

Sign of $f(x)$ does not change from one side to the other side of r. — Graph **touches** x-axis at r.

If *r* Is a Zero of Odd Multiplicity

Sign of $f(x)$ changes from one side to the other side of r. — Graph **crosses** x-axis at r.

NOW WORK PROBLEM 45(b).

Look again at Figure 31. We cannot be sure just how low the graph actually goes between $x = 0$ and $x = 2$. But we do know that somewhere in the interval $(0, 2)$ the graph of f must change direction (from decreasing to increasing). The points at which a graph changes direction are called **turning points.** In calculus, such points are called **local maxima** or **local minima,** and techniques for locating them are given. So we shall not ask for the location of turning points in our graphs. Instead, we will use the following result from calculus, which tells us the maximum number of turning points that the graph of a polynomial function can have.

Theorem

If f is a polynomial function of degree n, then f has at most $n - 1$ turning points.

For example, the graph of $f(x) = x^2(x - 2)$ shown in Figure 31 is the graph of a polynomial of degree 3 and has $3 - 1 = 2$ turning points: one at $(0, 0)$ and the other somewhere between $x = 0$ and $x = 2$.

Exploration

A graphing utility can be used to locate the turning points of a graph. Graph $y = x^2(x - 2)$. Use MINIMUM to find the location of the turning point for $0 < x < 2$. See Figure 32.

Figure 32

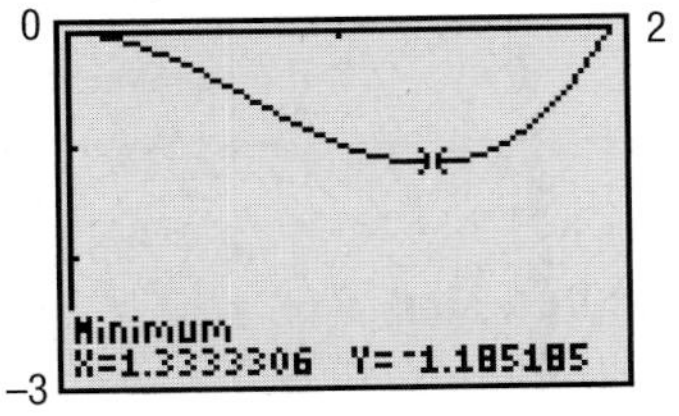

One last remark about Figure 31. Notice that the graph of $f(x) = x^2(x - 2)$ looks somewhat like the graph of $y = x^3$. In fact, for very large values of x, either positive or negative, there is little difference. To see for yourself, use your calculator to compare the values of $f(x) = x^2(x - 2)$ and $y = x^3$ for $x = -100{,}000$ and $x = 100{,}000$. The behavior of the graph of a function for large values of x, either positive or negative, is referred to as its **end behavior.**

Theorem

End Behavior

For large values of x, either positive or negative, the graph of the polynomial

$$f(x) = a_n x^n + a_{n-1}x^{n-1} + \cdots + a_1 x + a_0$$

resembles the graph of the power function

$$y = a_n x^n$$

Look back at Figures 22 and 24. Based on the above theorem and the previous discussion on power functions, the end behavior of a polynomial can only be of four types. See Figure 33.

Figure 33
End Behavior

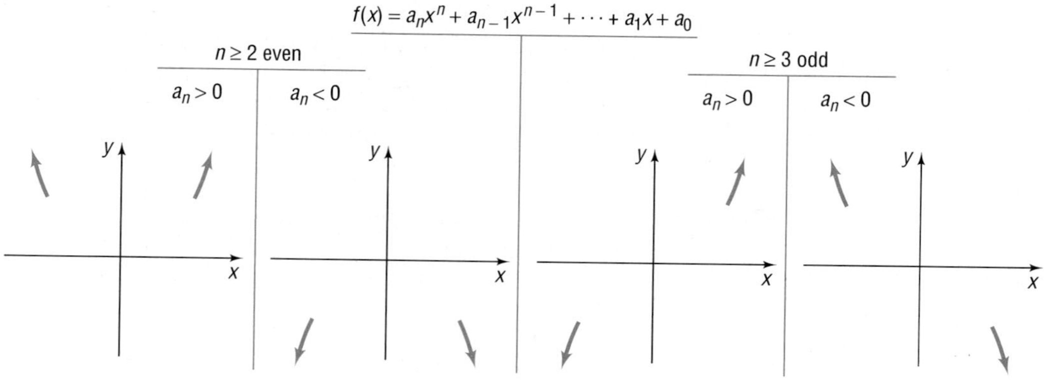

(a) $n \geq 2$ even; $a_n > 0$

(b) $n \geq 2$ even; $a_n < 0$

(c) $n \geq 3$ odd; $a_n > 0$

(d) $n \geq 3$ odd; $a_n < 0$

For example, consider the polynomial function $f(x) = -2x^4 + x^3 + 4x^2 - 7x + 1$. The graph of f will resemble the graph of the power function $y = -2x^4$ for large $|x|$. The graph of f will look like Figure 33(b) for large $|x|$.

NOW WORK PROBLEM 45(c).

Summary

Graph of a Polynomial Function $f(x) = a_n x^n + a_{n-1}x^{n-1} + \cdots + a_1 x + a_0, \quad a_n \neq 0$

Degree of the polynomial f: n

Maximum number of turning points: $n - 1$

At a zero of even multiplicity: The graph of f touches the x-axis.

At a zero of odd multiplicity: The graph of f crosses the x-axis.

Between zeros, the graph of f is either above or below the x-axis.

End behavior: For large $|x|$, the graph of f behaves like the graph of $y = a_n x^n$.

EXAMPLE 7 Analyzing the Graph of a Polynomial Function

For the polynomial $f(x) = x^3 + x^2 - 12x$:

(a) Find the x- and y-intercepts of the graph of f.

(b) Determine whether the graph crosses or touches the x-axis at each x-intercept.

(c) End behavior: find the power function that the graph of f resembles for large values of x.

(d) Determine the maximum number of turning points on the graph of f.

(e) Use the x-intercepts to find the intervals on which the graph of f is above the x-axis and the intervals on which the graph is below the x-axis.

(f) Put all the information together and connect the points with a smooth, continuous curve to obtain the graph of f.

Solution (a) The y-intercept is $f(0) = 0$. To find the x-intercepts, if any, we factor f.

$$f(x) = x^3 + x^2 - 12x = x(x^2 + x - 12) = x(x + 4)(x - 3)$$

Solving the equation $f(x) = x(x + 4)(x - 3) = 0$, we find that the x-intercepts, or zeros of f, are -4, 0, and 3.

(b) Since each zero of f is of multiplicity 1, the graph of f will cross the x-axis at each x-intercept.

(c) End behavior: the graph of f resembles that of the power function $y = x^3$ for large values of $|x|$.

(d) The graph of f will contain at most two turning points.

(e) The three x-intercepts divide the x-axis into four intervals:

$$(-\infty, -4) \qquad (-4, 0) \qquad (0, 3) \qquad (3, \infty)$$

To determine the location of the graph of $f(x)$ in each interval, we create Table 6.

(f) The graph of f is given in Figure 34.

Table 6

	-4	0	3	x
Interval	$(-\infty, -4)$	$(-4, 0)$	$(0, 3)$	$(3, \infty)$
Number Chosen	-5	-2	1	4
Value of f	$f(-5) = -40$	$f(-2) = 20$	$f(1) = -10$	$f(4) = 32$
Location of Graph	Below x-axis	Above x-axis	Below x-axis	Above x-axis
Point on Graph	$(-5, -40)$	$(-2, 20)$	$(1, -10)$	$(4, 32)$

Figure 34

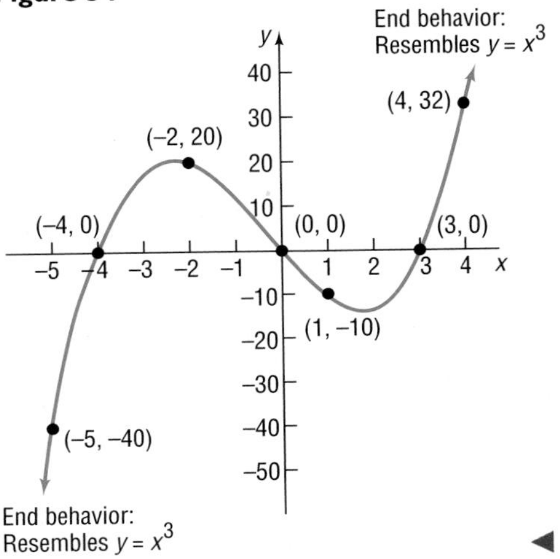

Exploration

Graph $y = x^3 + x^2 - 12x$. Compare what you see with Figure 34. Use MAXIMUM/MINIMUM to locate the two turning points.

NOW WORK PROBLEM 59.

EXAMPLE 8 Analyzing the Graph of a Polynomial Function

Follow the instructions of Example 7 for the following polynomial:

$$f(x) = x^2(x - 4)(x + 1)$$

Solution (a) The y-intercept is $f(0) = 0$. The x-intercepts satisfy the equation

$$f(x) = x^2(x - 4)(x + 1) = 0$$

So

$$x^2 = 0 \quad \text{or} \quad x - 4 = 0 \quad \text{or} \quad x + 1 = 0$$
$$x = 0 \qquad\qquad x = 4 \qquad\qquad x = -1$$

The x-intercepts are -1, 0, and 4.

(b) The intercept 0 is a zero of multiplicity 2, so the graph of f will touch the x-axis at 0; 4 and -1 are zeros of multiplicity 1, so the graph of f will cross the x-axis at 4 and -1.

(c) End behavior: the graph of f resembles that of the power function $y = x^4$ for large values of $|x|$.

(d) The graph of f will contain at most three turning points.

(e) The three x-intercepts divide the x-axis into four intervals:

$$(-\infty, -1) \qquad (-1, 0) \qquad (0, 4) \qquad (4, \infty)$$

To determine the location of the graph of $f(x)$ in each interval, we create Table 7.

Table 7

	-1	0	4	x
Interval	$(-\infty, -1)$	$(-1, 0)$	$(0, 4)$	$(4, \infty)$
Number Chosen	-2	$-\frac{1}{2}$	2	5
Value of f	$f(-2) = 24$	$f\left(-\frac{1}{2}\right) = -\frac{9}{16}$	$f(2) = -24$	$f(5) = 150$
Location of Graph	Above x-axis	Below x-axis	Below x-axis	Above x-axis
Point on Graph	$(-2, 24)$	$\left(-\frac{1}{2}, -\frac{9}{16}\right)$	$(2, -24)$	$(5, 150)$

(f) The graph of f is given in Figure 35.

Figure 35

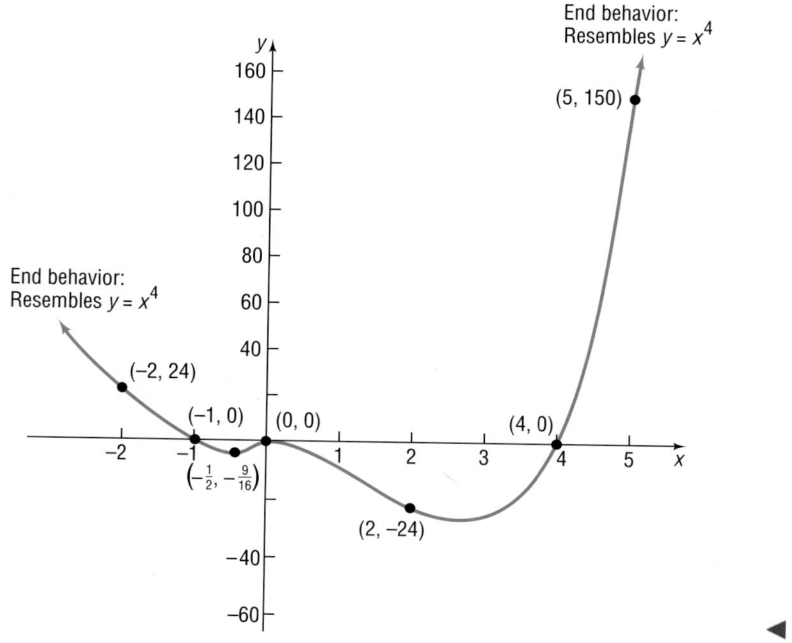

Exploration

Graph $y = x^2(x - 4)(x + 1)$. Compare what you see with Figure 35. Use MAXIMUM/MINIMUM to locate the two turning points besides (0, 0).

NOW WORK PROBLEM 69.

For polynomial functions that have noninteger coefficients and for polynomials that are not easily factored, we utilize the graphing utility early in the analysis of the graph. This is because the amount of information that can be obtained from algebraic analysis is limited.

EXAMPLE 9 Using a Graphing Utility to Analyze the Graph of a Polynomial Function

For the polynomial $f(x) = x^3 + 2.48x^2 - 4.3155x + 2.484406$:

(a) Find the degree of the polynomial. Determine the end behavior: that is, find the power function that the graph of f resembles for large values of $|x|$.

(b) Graph f using a graphing utility.

(c) Find the x- and y-intercepts of the graph.

(d) Use a TABLE to find points on the graph around each x-intercept. Determine on which intervals the graph is above and below the x-axis.

(e) Determine the local maxima and local minima, if any exist, rounded to two decimal places. That is, locate any turning points.

(f) Use the information obtained in parts (a)–(e) to draw a complete graph of f by hand. Be sure to label the intercepts, turning points, and the points obtained in part (d).

(g) Find the domain of f. Use the graph to find the range of f.

(h) Use the graph to determine where f is increasing and decreasing.

Figure 36

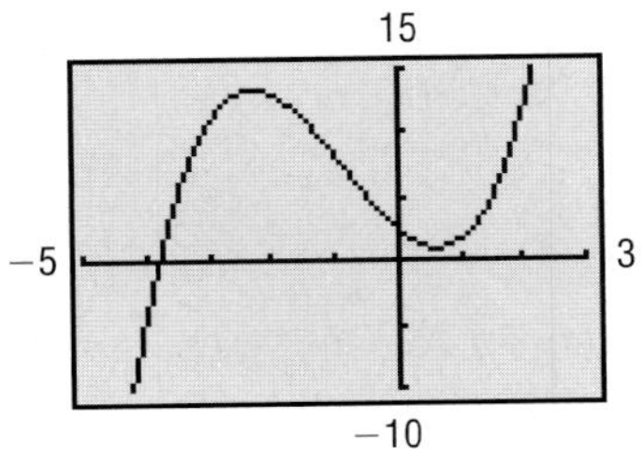

Table 8

X	Y1
-4	-4.574
-2	13.035

Y1=X^3+2.48X²−4...

Solution

(a) The degree of the polynomial is 3. End behavior: the graph of f resembles that of the power function $y = x^3$ for large values of $|x|$.

(b) See Figure 36 for the graph of f.

(c) The y-intercept is $f(0) = 2.484406$. In Example 7 we could easily factor $f(x)$ to find the x-intercepts. However, it is not readily apparent how $f(x)$ factors in this example. Therefore, we use a graphing utility's ZERO (or ROOT) feature and find the lone x-intercept to be -3.79, rounded to two decimal places.

(d) Table 8 shows values of x around the x-intercept. The points $(-4, -4.57)$ and $(-2, 13.04)$ are on the graph. The graph is below the x-axis on the interval $(-\infty, -3.79)$ and above the x-axis on the interval $(-3.79, \infty)$.

(e) From the graph we see that it has two turning points: one between -3 and -2, the other between 0 and 1. Rounded to two decimal places, the local maximum is 13.36 and occurs at $x = -2.28$; the local minimum is 1 and occurs at $x = 0.63$. The turning points are $(-2.28, 13.36)$ and $(0.63, 1)$.

(f) Figure 37 shows a graph of f drawn by hand using the information obtained in parts (a) to (e).

Figure 37

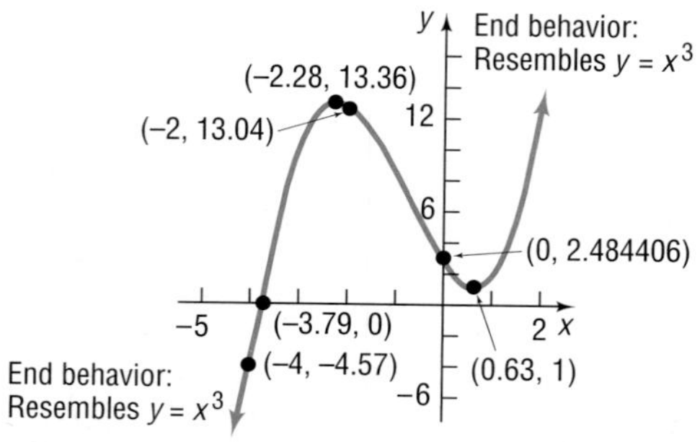

(g) The domain and the range of f is the set of all real numbers.

(h) Based on the graph, f is decreasing on the interval $(-2.28, 0.63)$ and is increasing on the intervals $(-\infty, -2.28)$ and $(0.63, \infty)$. ◀

NOW WORK PROBLEM 85.

Summary Steps for Analyzing the Graph of a Polynomial

To analyze the graph of a polynomial function $y = f(x)$, follow these steps:

STEP 1: (a) Find the x-intercepts, if any, by solving the equation $f(x) = 0$.
(b) Find the y-intercept by letting $x = 0$ and finding the value of $f(0)$.

STEP 2: Determine whether the graph of f crosses or touches the x-axis at each x-intercept.

STEP 3: End behavior: find the power function that the graph of f resembles for large values of x.

STEP 4: Determine the maximum number of turning points on the graph of f.

STEP 5: Use the x-intercept(s) to find the intervals on which the graph of f is above the x-axis and the intervals on which the graph is below the x-axis.

STEP 6: Plot the points obtained in Steps 1 and 5, and use the remaining information to connect them with a smooth, continuous curve.

3.2 Assess Your Understanding

'Are You Prepared?' *Answers are given at the end of these exercises. If you get a wrong answer, read the pages listed in red.*

1. The intercepts of the equation $9x^2 + 4y = 36$ are ________. (pp. 11–12)

2. *True or False:* The expression $4x^3 - 3.6x^2 - \sqrt{2}$ is a polynomial. (pp. 918–923)

3. To graph $y = x^2 - 4$, you would shift the graph of $y = x^2$ ________ a distance of ________ units. (pp. 94–103)

4. *True or False:* The x-intercepts of the graph of a function $y = f(x)$ are the real solutions of the equation $f(x) = 0$. (p. 72)

Concepts and Vocabulary

5. The graph of every polynomial function is both ________ and ________.

6. A number r for which $f(r) = 0$ is called a(n) ________ of the function f.

7. If r is a zero of even multiplicity for a function f, the graph of f _________ the x-axis at r.

8. *True or False:* The graph of $f(x) = x^2(x - 3)(x + 4)$ has exactly three x-intercepts.

9. *True or False:* The x-intercepts of the graph of a polynomial function are called turning points.

10. *True or False:* End behavior: the graph of the function $f(x) = 3x^4 - 6x^2 + 2x + 5$ resembles $y = x^4$ for large values of $|x|$.

Exercises

In Problems 11–22, determine which functions are polynomial functions. For those that are, state the degree. For those that are not, tell why not.

11. $f(x) = 4x + x^3$ **12.** $f(x) = 5x^2 + 4x^4$ **13.** $g(x) = \dfrac{1 - x^2}{2}$

14. $h(x) = 3 - \dfrac{1}{2}x$ **15.** $f(x) = 1 - \dfrac{1}{x}$ **16.** $f(x) = x(x - 1)$

17. $g(x) = x^{3/2} - x^2 + 2$ **18.** $h(x) = \sqrt{x}(\sqrt{x} - 1)$ **19.** $F(x) = 5x^4 - \pi x^3 + \dfrac{1}{2}$

20. $F(x) = \dfrac{x^2 - 5}{x^3}$ **21.** $G(x) = 2(x - 1)^2(x^2 + 1)$ **22.** $G(x) = -3x^2(x + 2)^3$

In Problems 23–36, use transformations of the graph of $y = x^4$ or $y = x^5$ to graph each function.

23. $f(x) = (x + 1)^4$ **24.** $f(x) = (x - 2)^5$ **25.** $f(x) = x^5 - 3$ **26.** $f(x) = x^4 + 2$

27. $f(x) = \dfrac{1}{2}x^4$ **28.** $f(x) = 3x^5$ **29.** $f(x) = -x^5$ **30.** $f(x) = -x^4$

31. $f(x) = (x - 1)^5 + 2$ **32.** $f(x) = (x + 2)^4 - 3$ **33.** $f(x) = 2(x + 1)^4 + 1$ **34.** $f(x) = \dfrac{1}{2}(x - 1)^5 - 2$

35. $f(x) = 4 - (x - 2)^5$ **36.** $f(x) = 3 - (x + 2)^4$

In Problems 37–44, form a polynomial whose zeros and degree are given.

37. Zeros: $-1, 1, 3$; degree 3 **38.** Zeros: $-2, 2, 3$; degree 3 **39.** Zeros: $-3, 0, 4$; degree 3

40. Zeros: $-4, 0, 2$; degree 3 **41.** Zeros: $-4, -1, 2, 3$; degree 4 **42.** Zeros: $-3, -1, 2, 5$; degree 4

43. Zeros: -1, multiplicity 1; 3, multiplicity 2; degree 3 **44.** Zeros: -2, multiplicity 2; 4, multiplicity 1; degree 3

In Problems 45–56, for each polynomial function: (a) List each real zero and its multiplicity; (b) Determine whether the graph crosses or touches the x-axis at each x-intercept. (c) Find the power function that the graph of f resembles for large values of $|x|$.

45. $f(x) = 3(x - 7)(x + 3)^2$ **46.** $f(x) = 4(x + 4)(x + 3)^3$ **47.** $f(x) = 4(x^2 + 1)(x - 2)^3$

48. $f(x) = 2(x - 3)(x + 4)^3$ **49.** $f(x) = -2\left(x + \dfrac{1}{2}\right)^2(x^2 + 4)^2$ **50.** $f(x) = \left(x - \dfrac{1}{3}\right)^2(x - 1)^3$

51. $f(x) = (x - 5)^3(x + 4)^2$ **52.** $f(x) = (x + \sqrt{3})^2(x - 2)^4$ **53.** $f(x) = 3(x^2 + 8)(x^2 + 9)^2$

54. $f(x) = -2(x^2 + 3)^3$ **55.** $f(x) = -2x^2(x^2 - 2)$ **56.** $f(x) = 4x(x^2 - 3)$

In Problems 57–80, for each polynomial function f:

(a) *Find the x- and y-intercepts of f.*
(b) *Determine whether the graph of f crosses or touches the x-axis at each x-intercept.*
(c) *End behavior: find the power function that the graph of f resembles for large values of $|x|$.*
(d) *Determine the maximum number of turning points on the graph of f.*
(e) *Use the x-intercept(s) to find the intervals on which the graph of f is above and below the x-axis.*
(f) *Plot the points obtained in parts (a) and (e), and use the remaining information to connect them with a smooth, continuous curve.*

57. $f(x) = (x - 1)^2$ **58.** $f(x) = (x - 2)^3$ **59.** $f(x) = x^2(x - 3)$

60. $f(x) = x(x + 2)^2$ **61.** $f(x) = 6x^3(x + 4)$ **62.** $f(x) = 5x(x - 1)^3$

63. $f(x) = -4x^2(x + 2)$ **64.** $f(x) = -\dfrac{1}{2}x^3(x + 4)$ **65.** $f(x) = (x - 1)(x - 2)(x + 4)$

66. $f(x) = (x + 1)(x + 4)(x - 3)$ **67.** $f(x) = 4x - x^3$ **68.** $f(x) = x - x^3$

69. $f(x) = x^2(x - 2)(x + 2)$
70. $f(x) = x^2(x - 3)(x + 4)$
71. $f(x) = (x + 2)^2(x - 2)^2$
72. $f(x) = (x + 1)^3(x - 3)$
73. $f(x) = (x - 1)^2(x - 3)(x + 1)$
74. $f(x) = (x + 1)^2(x - 3)(x - 1)$
75. $f(x) = (x + 2)^2(x - 4)^2$
76. $f(x) = (x - 2)^2(x + 2)(x + 4)$
77. $f(x) = x^2(x - 2)(x^2 + 3)$
78. $f(x) = x^2(x^2 + 1)(x + 4)$
79. $f(x) = -x^2(x^2 - 1)(x + 1)$
80. $f(x) = -x^2(x^2 - 4)(x - 5)$

In Problems 81–84, decide which of the polynomial functions in the list might have the given graph. (More than one answer may be possible.)

81.

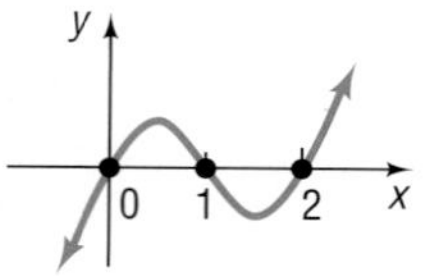

(a) $y = -4x(x - 1)(x - 2)$
(b) $y = x^2(x - 1)^2(x - 2)$
(c) $y = 3x(x - 1)(x - 2)$
(d) $y = x(x - 1)^2(x - 2)^2$
(e) $y = x^3(x - 1)(x - 2)$
(f) $y = -x(1 - x)(x - 2)$

82.

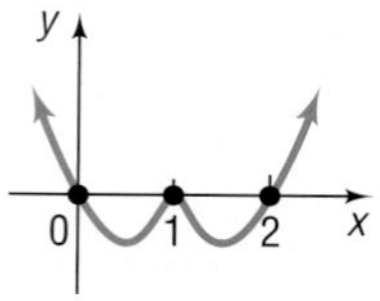

(a) $y = 2x^3(x - 1)(x - 2)^2$
(b) $y = x^2(x - 1)(x - 2)$
(c) $y = x^3(x - 1)^2(x - 2)$
(d) $y = x^2(x - 1)^2(x - 2)^2$
(e) $y = 5x(x - 1)^2(x - 2)$
(f) $y = -2x(x - 1)^2(2 - x)$

83.

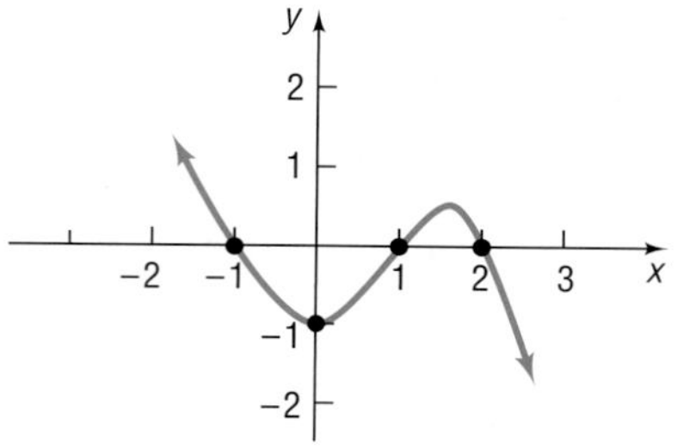

(a) $y = \frac{1}{2}(x^2 - 1)(x - 2)$
(b) $y = -\frac{1}{2}(x^2 + 1)(x - 2)$
(c) $y = (x^2 - 1)\left(1 - \frac{x}{2}\right)$
(d) $y = -\frac{1}{2}(x^2 - 1)^2(x - 2)$
(e) $y = \left(x^2 + \frac{1}{2}\right)(x^2 - 1)(2 - x)$
(f) $y = -(x - 1)(x - 2)(x + 1)$

84.

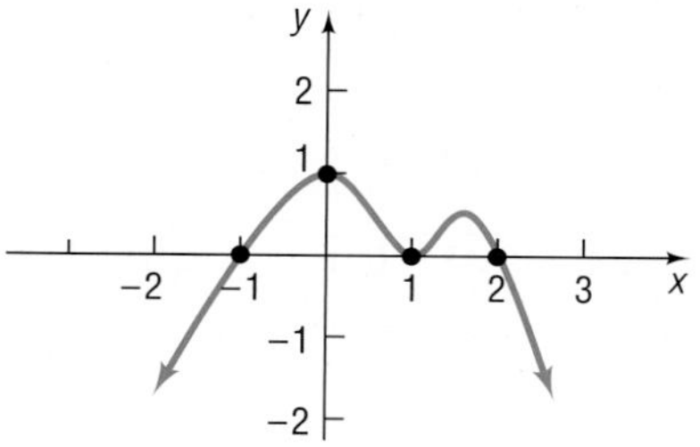

(a) $y = -\frac{1}{2}(x^2 - 1)(x - 2)(x + 1)$
(b) $y = -\frac{1}{2}(x^2 + 1)(x - 2)(x + 1)$
(c) $y = -\frac{1}{2}(x + 1)^2(x - 1)(x - 2)$
(d) $y = (x - 1)^2(x + 1)\left(1 - \frac{x}{2}\right)$
(e) $y = -(x - 1)^2(x - 2)(x + 1)$
(f) $y = -\left(x^2 + \frac{1}{2}\right)(x - 1)^2(x + 1)(x - 2)$

In Problems 85–94, for each polynomial function f:
(a) *Find the degree of the polynomial. Determine the end behavior: that is, find the power function that the graph of f resembles for large values of $|x|$.*
(b) *Graph f using a graphing utility.*
(c) *Find the x- and y-intercepts of the graph.*
(d) *Use TABLE to find points on the graph around each x-intercept. Determine on which intervals the graph is above and below the x-axis.*
(e) *Determine the local maxima and local minima, if any exist, rounded to two decimal places. That is, locate any turning points.*
(f) *Use the information obtained in parts (a)–(e) to draw a complete graph of f by hand. Be sure to label the intercepts, turning points, and the points obtained in part (d).*
(g) *Find the domain of f. Use the graph to find the range of f.*
(h) *Use the graph to determine where f is increasing and where f is decreasing.*

85. $f(x) = x^3 + 0.2x^2 - 1.5876x - 0.31752$
86. $f(x) = x^3 - 0.8x^2 - 4.6656x + 3.73248$
87. $f(x) = x^3 + 2.56x^2 - 3.31x + 0.89$
88. $f(x) = x^3 - 2.91x^2 - 7.668x - 3.8151$

89. $f(x) = x^4 - 2.5x^2 + 0.5625$

90. $f(x) = x^4 - 18.5x^2 + 50.2619$

91. $f(x) = 2x^4 - \pi x^3 + \sqrt{5}x - 4$

92. $f(x) = -1.2x^4 + 0.5x^2 - \sqrt{3}x + 2$

93. $f(x) = -2x^5 - \sqrt{2}x^2 - x - \sqrt{2}$

94. $f(x) = \pi x^5 + \pi x^4 + \sqrt{3}x + 1$

95. Motor Vehicle Thefts The following data represent the number of motor vehicle thefts (in thousands) in the United States for the years 1987–1997, where 1 represents 1987, 2 represents 1988, and so on.

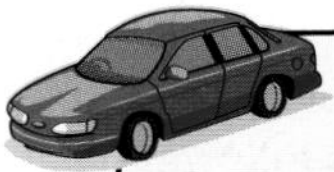

Year, x	Motor Vehicle Thefts, T
1987, 1	1289
1988, 2	1433
1989, 3	1565
1990, 4	1636
1991, 5	1662
1992, 6	1611
1993, 7	1563
1994, 8	1539
1995, 9	1472
1996, 10	1394
1997, 11	1354

SOURCE: U.S. Federal Bureau of Investigation

(a) Draw a scatter diagram of the data. Comment on the type of relation that may exist between the two variables.

(b) The cubic function of best fit to these data is

$$T(x) = 1.52x^3 - 39.81x^2 + 282.29x + 1035.5$$

Use this function to predict the number of motor vehicle thefts in 1994.

(c) Use a graphing utility to verify that the function given in part (b) is the cubic function of best fit.

(d) With a graphing utility, draw a scatter diagram of the data and then graph the cubic function of best fit on the scatter diagram.

(e) Do you think that the function given in part (b) will be useful in predicting the number of motor vehicle thefts in 1999?

96. Cost of Manufacturing The following data represent the cost C (in thousands of dollars) of manufacturing Chevy Cavaliers and the number x of Cavaliers produced.

(a) Draw a scatter diagram of the data. Comment on the type of relation that may exist between the two variables.

(b) Find the average rate of change in cost from four to five Cavaliers.

(c) What is the average rate of change in cost from eight to nine Cavaliers?

(d) The cubic function of best fit to these data is

$$C(x) = 0.2x^3 - 2.3x^2 + 14.3x + 10.2$$

Use this function to predict the cost of manufacturing eleven Cavaliers.

(e) Use a graphing utility to verify that the function given in part (d) is the cubic function of best fit.

(f) With a graphing utility, draw a scatter diagram of the data and then graph the cubic function of best fit on the scatter diagram.

(g) Interpret the y-intercept.

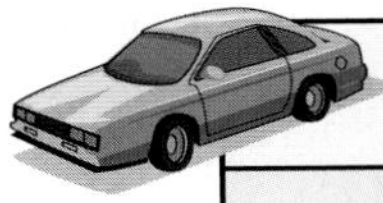

Number of Cavaliers Produced, x	Cost, C
0	10
1	23
2	31
3	38
4	43
5	50
6	59
7	70
8	85
9	105
10	135

97. Cost of Printing The following data represent the weekly cost C (in thousands of dollars) of printing textbooks and the number x (in thousands of units) of texts printed.

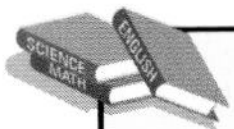

Number of Textbooks, x	Cost, C
0	100
5	128.1
10	144
13	153.5
17	161.2
18	162.6
20	166.3
23	178.9
25	190.2
27	221.8

(a) Draw a scatter diagram of the data. Comment on the type of relation that may exist between the two variables.

(b) Find the average rate of change in cost from 10,000 to 13,000 textbooks.

(c) What is the average rate of change in cost from 18,000 to 20,000 textbooks?

(d) The cubic function of best fit to these data is

$$C(x) = 0.015x^3 - 0.595x^2 + 9.15x + 98.43$$

Use this function to predict the cost of printing 22,000 texts per week.

(e) Use a graphing utility to verify that the function given in part (d) is the cubic function of best fit.

(f) With a graphing utility, draw a scatter diagram of the data and then graph the cubic function of best fit on the scatter diagram.

(g) Interpret the y-intercept.

98. Total Car Sales The following data represent the total car sales S (used plus new car sales) in thousands of cars in the United States for the years 1990–1998, where $x = 1$ represents 1990, $x = 2$ represents 1991, and so on.

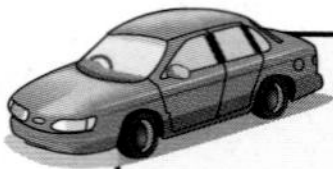

Year, x	Total Car Sales, S (in thousands)
1990, 1	46,830
1991, 2	45,465
1992, 3	45,163
1993, 4	46,575
1994, 5	49,132
1995, 6	50,353
1996, 7	49,355
1997, 8	48,542
1998, 9	48,372

SOURCE: Statistical Abstract of the United States, 2000

(a) Draw a scatter diagram of the data using x as the independent variable and S as the dependent variable. Comment on the type of relation that may exist between the two variables S and x.

(b) Use a graphing utility to find the cubic function of best fit $S = S(x)$.

(c) Graph the cubic function of best fit on the scatter diagram.

(d) Use the function found in part (b) to predict total car sales in 1999.

(e) Check the prediction in part (d) against actual data. Do you think that the function found in part (b) will be useful in predicting total car sales in 2004?

99. Can the graph of a polynomial function have no y-intercept? Can it have no x-intercepts? Explain.

100. Write a few paragraphs that provide a general strategy for graphing a polynomial function. Be sure to mention the following: degree, intercepts, end behavior, and turning points.

101. Make up a polynomial that has the following characteristics: crosses the x-axis at -1 and 4, touches the x-axis at 0 and 2, and is above the x-axis between 0 and 2. Give your polynomial to a fellow classmate and ask for a written critique of your polynomial.

102. Make up two polynomials, not of the same degree, with the following characteristics: crosses the x-axis at -2, touches the x-axis at 1, and is above the x-axis between -2 and 1. Give your polynomials to a fellow classmate and ask for a written critique of your polynomials.

103. The graph of a polynomial function is always smooth and continuous. Name a function studied earlier that is smooth and not continuous. Name one that is continuous, but not smooth.

104. Which of the following statements are true regarding the graph of the cubic polynomial $f(x) = x^3 + bx^2 + cx + d$? (Give reasons for your conclusions.)

(a) It intersects the y-axis in one and only one point.
(b) It intersects the x-axis in at most three points.
(c) It intersects the x-axis at least once.
(d) For x very large, it behaves like the graph of $y = x^3$.
(e) It is symmetric with respect to the origin.
(f) It passes through the origin.

'Are You Prepared?' Answers

1. $(-2, 0), (2, 0), (0, 9)$ **2.** True
3. down; 4 **4.** True

3.3 Rational Functions I

PREPARING FOR THIS SECTION *Before getting started, review the following:*

- Rational Expressions (Appendix A, Section A.3, pp. 923–927)
- Polynomial Division; Synthetic Division (Appendix A, Section A.4, pp. 929–934)
- Graph of $f(x) = \frac{1}{x}$ (Section 1.2, Example 11, p. 15)
- Graphing Techniques: Transformations (Section 2.5, pp. 94–103)

Now work the 'Are You Prepared?' problems on page 171.

OBJECTIVES
1 Find the Domain of a Rational Function
2 Determine the Vertical Asymptotes of a Rational Function
3 Determine the Horizontal or Oblique Asymptotes of a Rational Function

Ratios of integers are called *rational numbers*. Similarly, ratios of polynomial functions are called *rational functions*.

A **rational function** is a function of the form

$$R(x) = \frac{p(x)}{q(x)}$$

where p and q are polynomial functions and q is not the zero polynomial. The domain is the set of all real numbers except those for which the denominator q is 0.

1 **EXAMPLE 1** **Finding the Domain of a Rational Function**

(a) The domain of $R(x) = \frac{2x^2 - 4}{x + 5}$ is the set of all real numbers x except -5, that is, $\{x|x \neq -5\}$.

(b) The domain of $R(x) = \frac{1}{x^2 - 4}$ is the set of all real numbers x except -2 and 2, that is, $\{x|x \neq -2, x \neq 2\}$.

(c) The domain of $R(x) = \frac{x^3}{x^2 + 1}$ is the set of all real numbers.

(d) The domain of $R(x) = \frac{-x^2 + 2}{3}$ is the set of all real numbers.

(e) The domain of $R(x) = \frac{x^2 - 1}{x - 1}$ is the set of all real numbers x except 1, that is, $\{x|x \neq 1\}$. ◀

It is important to observe that the functions

$$R(x) = \frac{x^2 - 1}{x - 1} \quad \text{and} \quad f(x) = x + 1$$

are not equal, since the domain of R is $\{x|x \neq 1\}$ and the domain of f is the set of all real numbers.

NOW WORK PROBLEM 13.

If $R(x) = \frac{p(x)}{q(x)}$ is a rational function and if p and q have no common factors, then the rational function R is said to be in **lowest terms.** For a rational function $R(x) = \frac{p(x)}{q(x)}$ in lowest terms, the zeros, if any, of the numerator are the x-intercepts of the graph of R and so will play a major role in the graph of R. The zeros of the denominator of R [that is, the numbers x, if any, for which $q(x) = 0$], although not in the domain of R, also play a major role in the graph of R. We will discuss this role shortly.

We have already discussed the properties of the rational function $f(x) = \frac{1}{x}$. (Refer to Example 11, page 15.) The next rational function that we take up is $H(x) = \frac{1}{x^2}$.

EXAMPLE 2 Graphing $y = \frac{1}{x^2}$

Analyze the graph of: $H(x) = \frac{1}{x^2}$

Solution The domain of $H(x) = \frac{1}{x^2}$ is the set of all real numbers x except 0. The graph has no y-intercept, because x can never equal 0. The graph has no x-intercept because the equation $H(x) = 0$ has no solution. Therefore, the graph of H will not cross either of the coordinate axes. Because

$$H(-x) = \frac{1}{(-x)^2} = \frac{1}{x^2} = H(x)$$

H is an even function, so its graph is symmetric with respect to the y-axis.

Table 9

x	$H(x) = \frac{1}{x^2}$
$\frac{1}{2}$	4
$\frac{1}{10}$	100
$\frac{1}{100}$	10,000
1	1
2	$\frac{1}{4}$
10	$\frac{1}{100}$
100	$\frac{1}{10,000}$

Table 9 shows the behavior of $H(x) = \frac{1}{x^2}$ for selected positive numbers x (we will use symmetry to obtain the graph of H when $x < 0$). From the first three rows of Table 9, we see that, as the values of x approach (get closer to) 0, the values of $H(x)$ become larger and larger positive numbers. When this happens, we say that H is **unbounded in the positive direction.** We symbolize this by writing $H \to \infty$ (read as "H **approaches infinity**"). In calculus, **limits** are used to convey these ideas. There we use the symbolism $\lim_{x \to 0} H(x) = \infty$, read "the limit of $H(x)$ as x approaches zero equals infinity," to mean that $H(x) \to \infty$ as $x \to 0$.

Look at the last four rows of Table 9. As $x \to \infty$, the values of $H(x)$ approach 0 (the end behavior of the graph). In calculus, this is symbolized by writing $\lim_{x \to \infty} H(x) = 0$. Figure 38 shows the graph. Notice the use of red dashed lines to convey the ideas discussed above.

Figure 38
$H(x) = \frac{1}{x^2}$

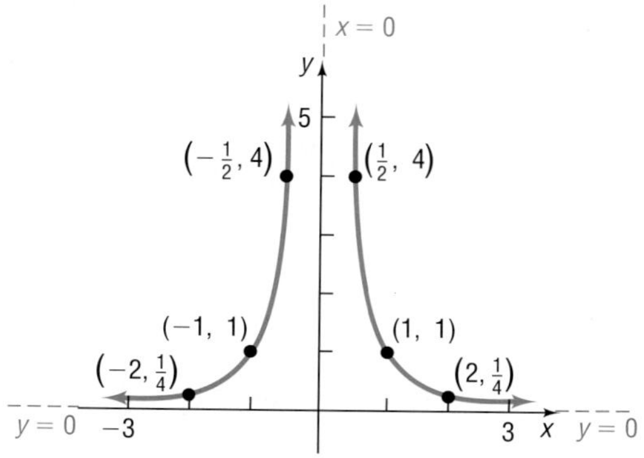

Sometimes transformations (shifting, compressing, stretching, and reflection) can be used to graph a rational function.

EXAMPLE 3 Using Transformations to Graph a Rational Function

Graph the rational function: $R(x) = \frac{1}{(x-2)^2} + 1$

Solution First, we take note of the fact that the domain of R is the set of all real numbers except $x = 2$. To graph R, we start with the graph of $y = \frac{1}{x^2}$. See Figure 39 for the steps.

Figure 39

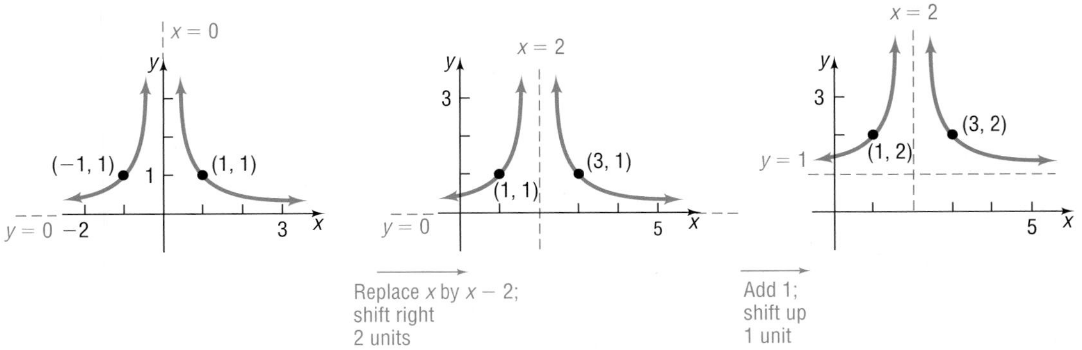

(a) $y = \frac{1}{x^2}$ (b) $y = \frac{1}{(x-2)^2}$ (c) $y = \frac{1}{(x-2)^2} + 1$

NOW WORK PROBLEM 31.

Asymptotes

In Figure 39(c), notice that as the values of x become more negative, that is, as x becomes **unbounded in the negative direction** ($x \to -\infty$, read as "x **approaches negative infinity**"), the values $R(x)$ approach 1. In fact, we can conclude the following from Figure 39(c):

1. As $x \to -\infty$, the values $R(x)$ approach 1. $[\lim_{x \to -\infty} R(x) = 1]$
2. As x approaches 2, the values $R(x) \to \infty$. $[\lim_{x \to 2} R(x) = \infty]$
3. As $x \to \infty$, the values $R(x)$ approach 1. $[\lim_{x \to \infty} R(x) = 1]$

This behavior of the graph is depicted by the vertical line $x = 2$ and the horizontal line $y = 1$. These lines are called *asymptotes* of the graph, which we define as follows:

Let R denote a function.

If, as $x \to -\infty$ or as $x \to \infty$, the values of $R(x)$ approach some fixed number, L, then the line $y = L$ is a **horizontal asymptote** of the graph of R.

If, as x approaches some number c, the values $|R(x)| \to \infty$, then the line $x = c$ is a **vertical asymptote** of the graph of R. The graph of R never intersects a vertical asymptote.

Even though the asymptotes of a function are not part of the graph of the function, they provide information about how the graph looks. Figure 40 illustrates some of the possibilities.

Figure 40

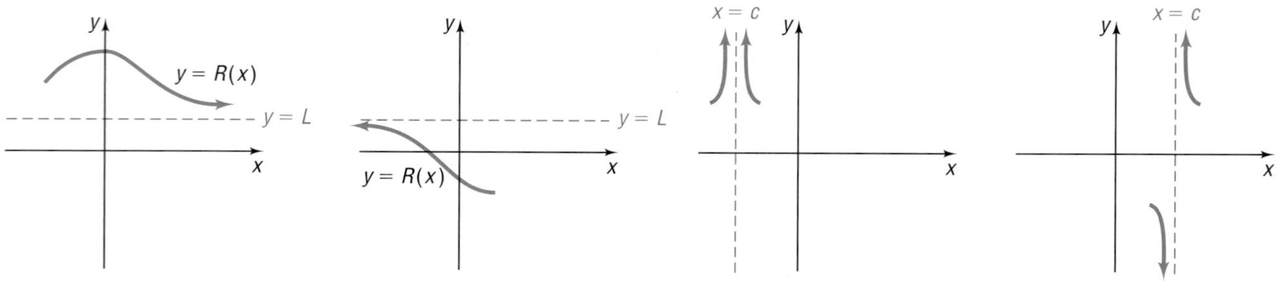

(a) End behavior: As $x \to \infty$, the values of $R(x)$ approach L. $[\lim_{x\to\infty} R(x) = L]$. That is, the points on the graph of R are getting closer to the line $y = L$; $y = L$ is a horizontal asymptote.

(b) End behavior: As $x \to -\infty$, the values of $R(x)$ approach L. $[\lim_{x\to-\infty} R(x) = L]$. That is, the points on the graph of R are getting closer to the line $y = L$; $y = L$ is a horizontal asymptote.

(c) As x approaches c, the values of $|R(x)| \to \infty$, $[\lim_{x\to c^-} R(x) = \infty$, $\lim_{x\to c^+} R(x) = \infty]$. That is, the points on the graph of R are getting closer to the line $x = c$; $x = c$ is a vertical asymptote.

(d) As x approaches c, the values of $|R(x)| \to \infty$, $[\lim_{x\to c^-} R(x) = -\infty$; $\lim_{x\to c^+} R(x) = \infty]$. That is, the points on the graph of R are getting closer to the line $x = c$; $x = c$ is a vertical asymptote.

Figure 41
Oblique asymptote

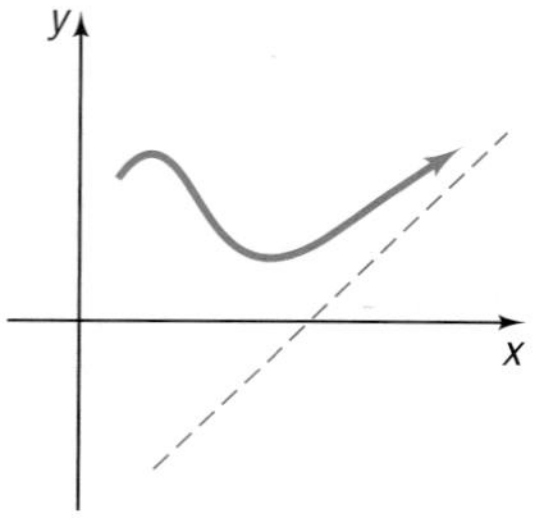

A horizontal asymptote, when it occurs, describes a certain behavior of the graph as $x \to \infty$ or as $x \to -\infty$, that is, its end behavior. The graph of a function may intersect a horizontal asymptote.

A vertical asymptote, when it occurs, describes a certain behavior of the graph when x is close to some number c. The graph of the function will never intersect a vertical asymptote.

If an asymptote is neither horizontal nor vertical, it is called **oblique.** Figure 41 shows an oblique asymptote. An oblique asymptote, when it occurs, describes the end behavior of the graph. The graph of a function may intersect an oblique asymptote.

Finding Asymptotes

2 The vertical asymptotes, if any, of a rational function $R(x) = \dfrac{p(x)}{q(x)}$, in lowest terms, are located at the zeros of the denominator $q(x)$. Suppose that r is a zero so that $x - r$ is a factor. Now, as x approaches r, symbolized as $x \to r$, the values of $x - r$ approach 0, causing the ratio to become unbounded, that is, causing $|R(x)| \to \infty$. Based on the definition, we conclude that the line $x = r$ is a vertical asymptote.

Theorem

Locating Vertical Asymptotes

A rational function $R(x) = \dfrac{p(x)}{q(x)}$, *in lowest terms*, will have a vertical asymptote $x = r$ if r is a real zero of the *denominator* q. That is, if $x - r$ is a factor of the denominator q of a rational function $R(x) = \dfrac{p(x)}{q(x)}$, in lowest terms, then R will have the vertical asymptote $x = r$.

WARNING: If a rational function is not in lowest terms, an application of this theorem may result in an incorrect listing of vertical asymptotes. ■

EXAMPLE 4 Finding Vertical Asymptotes

Find the vertical asymptotes, if any, of the graph of each rational function.

(a) $R(x) = \dfrac{x}{x^2 - 4}$ (b) $F(x) = \dfrac{x + 3}{x - 1}$

(c) $H(x) = \dfrac{x^2}{x^2 + 1}$ (d) $G(x) = \dfrac{x^2 - 9}{x^2 + 4x - 21}$

Solution (a) R is in lowest terms and the zeros of the denominator $x^2 - 4$ are -2 and 2. Hence, the lines $x = -2$ and $x = 2$ are the vertical asymptotes of the graph of R.

(b) F is in lowest terms and the only zero of the denominator is 1. Hence, the line $x = 1$ is the only vertical asymptote of the graph of F.

(c) H is in lowest terms and the denominator has no real zeros. Hence, the graph of H has no vertical asymptotes.

(d) Factor $G(x)$ to determine if it is in lowest terms.

$$G(x) = \frac{x^2 - 9}{x^2 + 4x - 21} = \frac{(x + 3)(x - 3)}{(x + 7)(x - 3)} = \frac{x + 3}{x + 7}, \qquad x \neq 3$$

The only zero of the denominator of $G(x)$ in lowest terms is -7. Hence, the line $x = -7$ is the only vertical asymptote of the graph of G. ◀

As Example 4 points out, rational functions can have no vertical asymptotes, one vertical asymptote, or more than one vertical asymptote. However, the graph of a rational function will never intersect any of its vertical asymptotes. (Do you know why?)

Exploration

Graph each of the following rational functions:

$$R(x) = \frac{1}{x - 1} \quad R(x) = \frac{1}{(x - 1)^2} \quad R(x) = \frac{1}{(x - 1)^3} \quad R(x) = \frac{1}{(x - 1)^4}$$

Each has the vertical asymptote $x = 1$. What happens to the value of $R(x)$ as x approaches 1 from the right side of the vertical asymptote; that is, what is $\lim_{x \to 1^+} R(x)$? What happens to the value of $R(x)$ as x approaches 1 from the left side of the vertical asymptote; that is, what is $\lim_{x \to 1^-} R(x)$? How does the multiplicity of the zero in the denominator affect the graph of R?

NOW WORK PROBLEM 45. (find the vertical asymptotes, if any.)

3 The procedure for finding horizontal and oblique asymptotes is somewhat more involved. To find such asymptotes, we need to know how the values of a function behave as $x \to -\infty$ or as $x \to \infty$.

If a rational function $R(x)$ is **proper,** that is, if the degree of the numerator is less than the degree of the denominator, then as $x \to -\infty$ or as $x \to \infty$ the value of $R(x)$ approaches 0. Consequently, the line $y = 0$ (the x-axis) is a horizontal asymptote of the graph.

Theorem If a rational function is proper, the line $y = 0$ is a horizontal asymptote of its graph.

EXAMPLE 5 Finding Horizontal Asymptotes

Find the horizontal asymptotes, if any, of the graph of

$$R(x) = \frac{x - 12}{4x^2 + x + 1}$$

Solution The rational function R is proper, since the degree of the numerator, 1, is less than the degree of the denominator, 2. We conclude that the line $y = 0$ is a horizontal asymptote of the graph of R. ◀

To see why $y = 0$ is a horizontal asymptote of the function R in Example 5, we need to investigate the behavior of R as $x \to -\infty$ and $x \to \infty$. When $|x|$ is unbounded, the numerator of R, which is $x - 12$, can be approximated by the power function $y = x$, while the denominator of R, which is $4x^2 + x + 1$, can be approximated by the power function $y = 4x^2$. Applying these ideas to $R(x)$, we find that

$$R(x) = \frac{x - 12}{4x^2 + x + 1} \underset{\uparrow}{\approx} \frac{x}{4x^2} = \frac{1}{4x} \underset{\uparrow}{\to} 0$$

For $|x|$ unbounded   As $x \to -\infty$ or $x \to \infty$

This shows that the line $y = 0$ is a horizontal asymptote of the graph of R.

If a rational function $R(x) = \frac{p(x)}{q(x)}$ is **improper,** that is, if the degree of the numerator is greater than or equal to the degree of the denominator, we must use long division to write the rational function as the sum of a polynomial $f(x)$ plus a proper rational function $\frac{r(x)}{q(x)}$. That is, we write

$$R(x) = \frac{p(x)}{q(x)} = f(x) + \frac{r(x)}{q(x)}$$

where $f(x)$ is a polynomial and $\frac{r(x)}{q(x)}$ is a proper rational function. Since $\frac{r(x)}{q(x)}$ is proper, then $\frac{r(x)}{q(x)} \to 0$ as $x \to -\infty$ or as $x \to \infty$. As a result,

$$R(x) = \frac{p(x)}{q(x)} \to f(x), \qquad \text{as } x \to -\infty \text{ or as } x \to \infty$$

The possibilities are listed next.

1. If $f(x) = b$, a constant, then the line $y = b$ is a horizontal asymptote of the graph of R.
2. If $f(x) = ax + b$, $a \neq 0$, then the line $y = ax + b$ is an oblique asymptote of the graph of R.
3. In all other cases, the graph of R approaches the graph of f, and there are no horizontal or oblique asymptotes.

The following examples demonstrate these conclusions.

EXAMPLE 6 Finding Horizontal or Oblique Asymptotes

Find the horizontal or oblique asymptotes, if any, of the graph of

$$H(x) = \frac{3x^4 - x^2}{x^3 - x^2 + 1}$$

Solution The rational function H is improper, since the degree of the numerator, 4, is larger than the degree of the denominator, 3. To find any horizontal or oblique asymptotes, we use long division.

$$\begin{array}{r|l} & 3x + 3 \\ x^3 - x^2 + 1 & 3x^4 \qquad\qquad - x^2 \\ & \underline{3x^4 - 3x^3 \qquad\quad + 3x} \\ & 3x^3 - x^2 - 3x \\ & \underline{3x^3 - 3x^2 \qquad + 3} \\ & 2x^2 - 3x - 3 \end{array}$$

As a result,

$$H(x) = \frac{3x^4 - x^2}{x^3 - x^2 + 1} = 3x + 3 + \frac{2x^2 - 3x - 3}{x^3 - x^2 + 1}$$

Then, as $x \to -\infty$ or as $x \to \infty$,

$$\frac{2x^2 - 3x - 3}{x^3 - x^2 + 1} \approx \frac{2x^2}{x^3} = \frac{2}{x} \to 0$$

So, as $x \to -\infty$ or as $x \to \infty$, we have $H(x) \to 3x + 3$. We conclude that the graph of the rational function H has an oblique asymptote $y = 3x + 3$. ◀

EXAMPLE 7 Finding Horizontal or Oblique Asymptotes

Find the horizontal or oblique asymptotes, if any, of the graph of

$$R(x) = \frac{8x^2 - x + 2}{4x^2 - 1}$$

Solution The rational function R is improper, since the degree of the numerator, 2, equals the degree of the denominator, 2. To find any horizontal or oblique asymptotes, we use long division.

$$\begin{array}{r|l} & 2 \\ 4x^2 - 1 & 8x^2 - x + 2 \\ & \underline{8x^2 \qquad\quad -2} \\ & -x + 4 \end{array}$$

As a result,

$$R(x) = \frac{8x^2 - x + 2}{4x^2 - 1} = 2 + \frac{-x + 4}{4x^2 - 1}$$

Then, as $x \to -\infty$ or as $x \to \infty$,

$$\frac{-x + 4}{4x^2 - 1} \approx \frac{-x}{4x^2} = \frac{-1}{4x} \to 0$$

So, as $x \to -\infty$ or as $x \to \infty$, we have $R(x) \to 2$. We conclude that $y = 2$ is a horizontal asymptote of the graph. ◀

In Example 7, we note that the quotient 2 obtained by long division is the quotient of the leading coefficients of the numerator polynomial and the denominator polynomial $\left(\frac{8}{4}\right)$. This means that we can avoid the long division process for rational functions whose numerator and denominator *are of the same degree* and conclude that the quotient of the leading coefficients will give us the horizontal asymptote.

NOW WORK PROBLEM 41.

EXAMPLE 8 Finding Horizontal or Oblique Asymptotes

Find the horizontal or oblique asymptotes, if any, of the graph of

$$G(x) = \frac{2x^5 - x^3 + 2}{x^3 - 1}$$

Solution The rational function G is improper, since the degree of the numerator, 5, is larger than the degree of the denominator, 3. To find any horizontal or oblique asymptotes, we use long division.

$$\begin{array}{r} 2x^2 - 1 \\ x^3 - 1\overline{)2x^5 - x^3 + 2} \\ \underline{2x^5 - 2x^2 } \\ -x^3 + 2x^2 + 2 \\ \underline{-x^3 + 1} \\ 2x^2 + 1 \end{array}$$

As a result,

$$G(x) = \frac{2x^5 - x^3 + 2}{x^3 - 1} = 2x^2 - 1 + \frac{2x^2 + 1}{x^3 - 1}$$

Then, as $x \to -\infty$ or as $x \to \infty$,

$$\frac{2x^2 + 1}{x^3 - 1} \approx \frac{2x^2}{x^3} = \frac{2}{x} \to 0$$

So, as $x \to -\infty$ or as $x \to \infty$, we have $G(x) \to 2x^2 - 1$. We conclude that, for large values of $|x|$, the graph of G approaches the graph of $y = 2x^2 - 1$. That is, the graph of G will look like the graph of $y = 2x^2 - 1$ as $x \to -\infty$ or $x \to \infty$. Since $y = 2x^2 - 1$ is not a linear function, G has no horizontal or oblique asymptotes. ◄

We now summarize the procedure for finding horizontal and oblique asymptotes.

Summary

Finding Horizontal and Oblique Asymptotes of a Rational Function *R*

Consider the rational function

$$R(x) = \frac{p(x)}{q(x)} = \frac{a_n x^n + a_{n-1}x^{n-1} + \cdots + a_1 x + a_0}{b_m x^m + b_{m-1}x^{m-1} + \cdots + b_1 x + b_0}$$

in which the degree of the numerator is n and the degree of the denominator is m.

1. If $n < m$, then R is a proper rational function, and the graph of R will have the horizontal asymptote $y = 0$ (the x-axis).
2. If $n \geq m$, then R is improper. Here long division is used.
   (a) If $n = m$, the quotient obtained will be a number $L\left(= \frac{a_n}{b_m}\right)$, and the line $y = L\left(= \frac{a_n}{b_m}\right)$ is a horizontal asymptote.
   (b) If $n = m + 1$, the quotient obtained is of the form $ax + b$ (a polynomial of degree 1), and the line $y = ax + b$ is an oblique asymptote.
   (c) If $n > m + 1$, the quotient obtained is a polynomial of degree 2 or higher, and R has neither a horizontal nor an oblique asymptote. In this case, for x unbounded, the graph of R will behave like the graph of the quotient.

Note: The graph of a rational function either has one horizontal or one oblique asymptote or else has no horizontal and no oblique asymptote.

3.3 Assess Your Understanding

'Are You Prepared?' *Answers are given at the end of these exercises. If you get a wrong answer, read the pages listed in red.*

1. *True or False:* The quotient of two polynomial expressions is a rational expression. (pp. 923–927)

2. What is the quotient and remainder when $3x^3 - 6x^2 + 3x - 4$ is divided by $x^2 - 1$? (pp. 929–934)

3. Graph $y = \frac{1}{x}$. (p. 15)

4. *True or False:* To graph $y = -x^2$, you would reflect the graph of $y = x^2$ about the x-axis. (pp. 94–103)

Concepts and Vocabulary

5. The line ________ is a horizontal asymptote of $R(x) = \frac{x^3 - 1}{x^3 + 1}$.

6. The line ________ is a vertical asymptote of $R(x) = \frac{x^3 - 1}{x^3 + 1}$.

7. For a rational function R, if the degree of the numerator is less than the degree of the denominator, then R is ________.

8. *True or False:* The domain of every rational function is the set of all real numbers.

9. *True or False:* If an asymptote is neither horizontal nor vertical, it is called oblique.

10. *True or False:* If the degree of the numerator of a rational function equals the degree of the denominator, then the ratio of the leading coefficients gives rise to the horizontal asymptote.

Exercises

In Problems 11–22, find the domain of each rational function.

11. $R(x) = \frac{4x}{x - 3}$

12. $R(x) = \frac{5x^2}{3 + x}$

13. $H(x) = \frac{-4x^2}{(x - 2)(x + 4)}$

14. $G(x) = \frac{6}{(x + 3)(4 - x)}$

15. $F(x) = \frac{3x(x - 1)}{2x^2 - 5x - 3}$

16. $Q(x) = \frac{-x(1 - x)}{3x^2 + 5x - 2}$

17. $R(x) = \frac{x}{x^3 - 8}$

18. $R(x) = \frac{x}{x^4 - 1}$

19. $H(x) = \frac{3x^2 + x}{x^2 + 4}$

20. $G(x) = \frac{x - 3}{x^4 + 1}$

21. $R(x) = \frac{3(x^2 - x - 6)}{4(x^2 - 9)}$

22. $F(x) = \frac{-2(x^2 - 4)}{3(x^2 + 4x + 4)}$

In Problems 23–28, use the graph shown to find:

(a) *The domain and range of each function* (b) *The intercepts, if any* (c) *Horizontal asymptotes, if any*
(d) *Vertical asymptotes, if any* (e) *Oblique asymptotes, if any*

23.

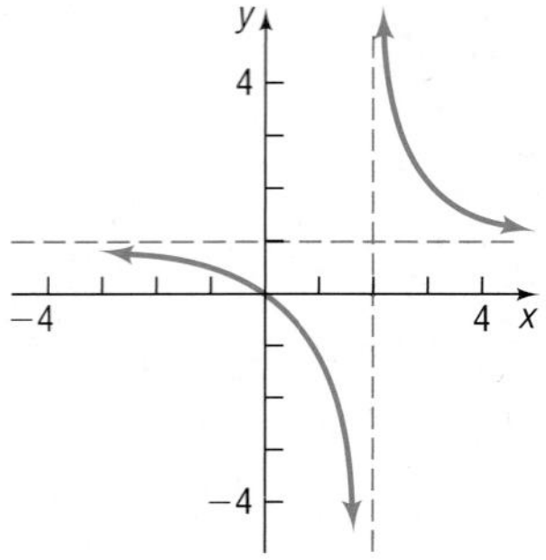

24.

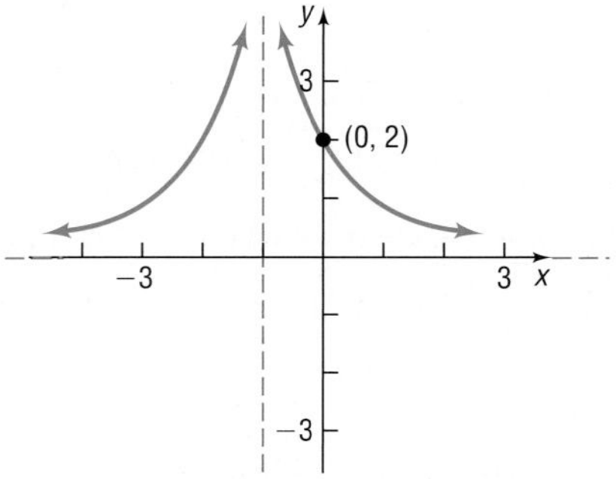

25.

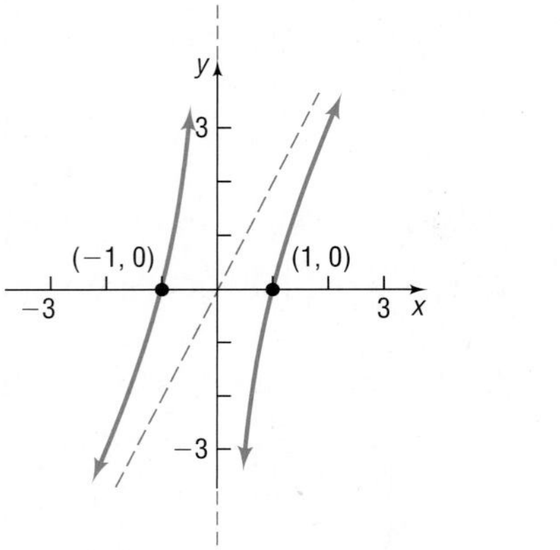

26.

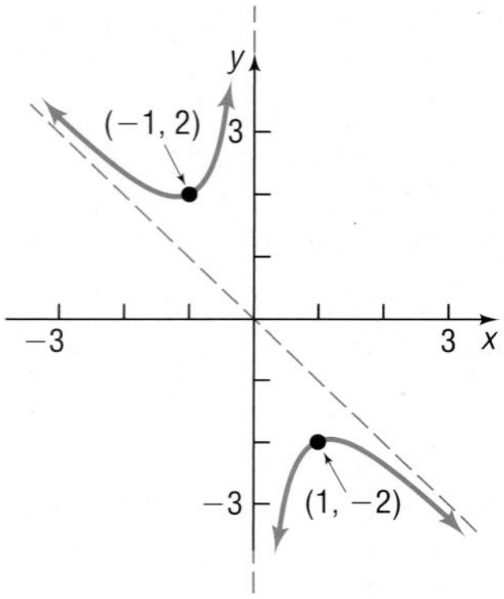

27.

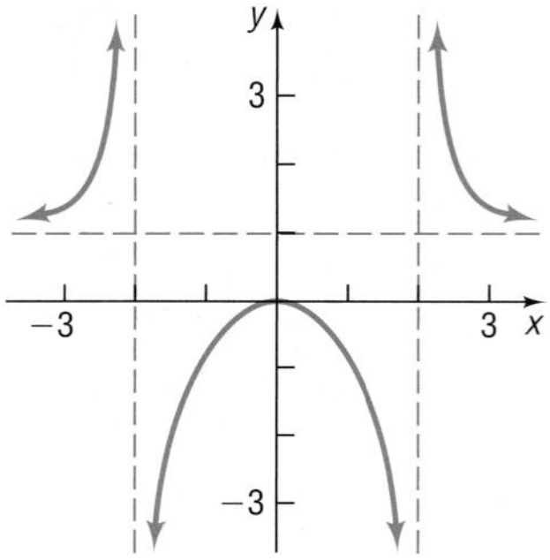

28.

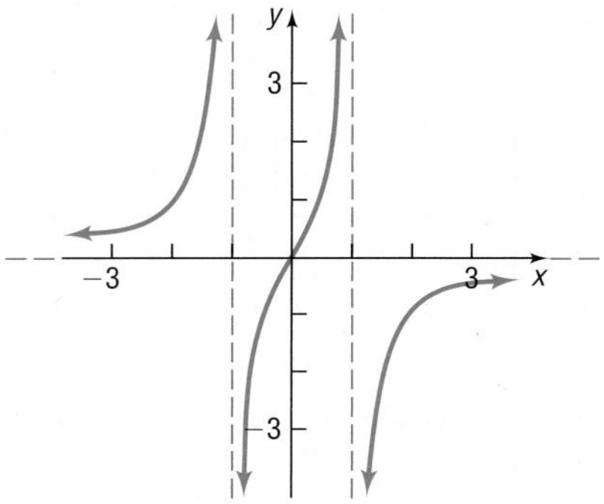

In Problems 29–40, graph each rational function using transformations.

29. $F(x) = 2 + \dfrac{1}{x}$ **30.** $Q(x) = 3 + \dfrac{1}{x^2}$ **31.** $R(x) = \dfrac{1}{(x-1)^2}$ **32.** $R(x) = \dfrac{3}{x}$

33. $H(x) = \dfrac{-2}{x+1}$ **34.** $G(x) = \dfrac{2}{(x+2)^2}$ **35.** $R(x) = \dfrac{-1}{x^2 + 4x + 4}$ **36.** $R(x) = \dfrac{1}{x-1} + 1$

37. $G(x) = 1 + \dfrac{2}{(x-3)^2}$ **38.** $F(x) = 2 - \dfrac{1}{x+1}$ **39.** $R(x) = \dfrac{x^2 - 4}{x^2}$ **40.** $R(x) = \dfrac{x-4}{x}$

In Problems 41–52, find the vertical, horizontal, and oblique asymptotes, if any, of each rational function.

41. $R(x) = \dfrac{3x}{x+4}$ **42.** $R(x) = \dfrac{3x+5}{x-6}$ **43.** $H(x) = \dfrac{x^4 + 2x^2 + 1}{x^2 - x + 1}$ **44.** $G(x) = \dfrac{-x^2 + 1}{x+5}$

45. $T(x) = \dfrac{x^3}{x^4 - 1}$ **46.** $P(x) = \dfrac{4x^5}{x^3 - 1}$ **47.** $Q(x) = \dfrac{5 - x^2}{3x^4}$ **48.** $F(x) = \dfrac{-2x^2 + 1}{2x^3 + 4x^2}$

49. $R(x) = \dfrac{3x^4 + 4}{x^3 + 3x}$ **50.** $R(x) = \dfrac{6x^2 + x + 12}{3x^2 - 5x - 2}$ **51.** $G(x) = \dfrac{x^3 - 1}{x - x^2}$ **52.** $F(x) = \dfrac{x-1}{x - x^3}$

53. Gravity In physics, it is established that the acceleration due to gravity, g, (in meters/sec^2), at a height h meters above sea level is given by

$$g(h) = \frac{3.99 \times 10^{14}}{(6.374 \times 10^6 + h)^2}$$

where 6.374×10^6 is the radius of Earth in meters.

(a) What is the acceleration due to gravity at sea level?
(b) The Sears Tower in Chicago, Illinois, is 443 meters tall. What is the acceleration due to gravity at the top of the Sears Tower?
(c) The peak of Mount Everest is 8848 meters above sea level. What is the acceleration due to gravity on the peak of Mount Everest?
(d) Find the horizontal asymptote of $g(h)$.
(e) Solve $g(h) = 0$. How do you interpret your answer?

54. Population Model A rare species of insect was discovered in the Amazon Rain Forest. To protect the species, environmentalists declare the insect endangered and transplant the insects into a protected area. The population of the insect t months after being transplanted is given by P.

$$P(t) = \frac{50(1 + 0.5t)}{(2 + 0.01t)}$$

(a) How many insects were discovered? In other words, what was the population when $t = 0$?
(b) What will the population be after 5 years?
(c) Determine the horizontal asymptote of $P(t)$. What is the largest population that the protected area can sustain?

55. If the graph of a rational function R has the vertical asymptote $x = 4$, then the factor $x - 4$ must be present in the denominator of R. Explain why.

56. If the graph of a rational function R has the horizontal asymptote $y = 2$, then the degree of the numerator of R equals the degree of the denominator of R. Explain why.

57. Can the graph of a rational function have both a horizontal and an oblique asymptote? Explain.

58. Make up a rational function that has $y = 2x + 1$ as an oblique asymptote. Explain the methodology that you used.

'Are You Prepared?' Answers

1. True

2. Quotient: $3x - 6$; Remainder: $6x - 10$

3.

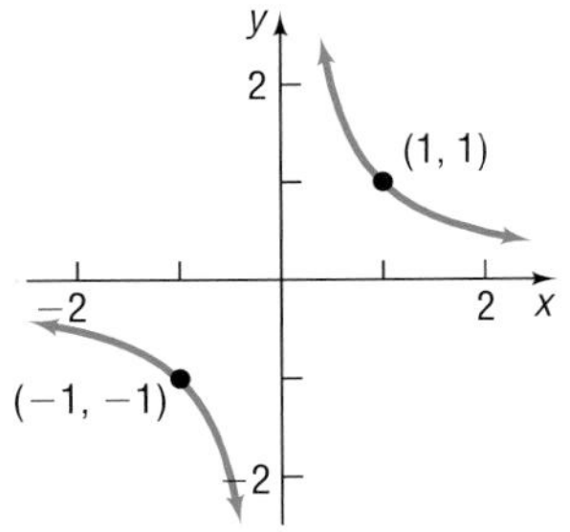

4. True

3.4 Rational Functions II: Analyzing Graphs

PREPARING FOR THIS SECTION *Before getting started, review the following:*

- Intercepts of a Function (Section 2.3, p. 72)
- Even and Odd Functions (Section 2.3, pp. 72–74)

Now work the 'Are You Prepared?' problems on page 185.

OBJECTIVES 1 Analyze the Graph of a Rational Function
2 Solve Applied Problems Involving Rational Functions

Graphing Rational Functions

1 We commented earlier that calculus provides the tools required to graph a polynomial function accurately. The same holds true for rational functions. However, we can gather together quite a bit of information about their graphs to get an idea of the general shape and position of the graph.

In the examples that follow, we will analyze the graph of a rational function $R(x) = \frac{p(x)}{q(x)}$ by applying the following steps:

Analyzing the Graph of a Rational Function

STEP 1: Find the domain of the rational function.

STEP 2: Locate the intercepts, if any, of the graph. The x-intercepts, if any, of $R(x) = \dfrac{p(x)}{q(x)}$ in lowest terms, are numbers in the domain that satisfy the equation $p(x) = 0$. The y-intercept, if there is one, is $R(0)$.

STEP 3: Test for symmetry. Replace x by $-x$ in $R(x)$. If $R(-x) = R(x)$, there is symmetry with respect to the y-axis; if $R(-x) = -R(x)$, there is symmetry with respect to the origin.

STEP 4: Write R in lowest terms and find the real zeros of the denominator. With R in lowest terms, each zero will give rise to a vertical asymptote.

STEP 5: Locate the horizontal or oblique asymptotes, if any, using the procedure given earlier. Determine points, if any, at which the graph of R intersects these asymptotes.

STEP 6: Determine where the graph is above the x-axis and where the graph is below the x-axis, using the zeros of the numerator and the denominator to divide the x-axis into intervals.

STEP 7: Graph the asymptotes, if any, found in Steps 4 and 5. Plot the points found in Steps 2, 5, and 6. Use all the information to connect the points and graph R.

EXAMPLE 1 Analyzing the Graph of a Rational Function

Analyze the graph of the rational function: $R(x) = \dfrac{x - 1}{x^2 - 4}$

Solution First, we factor both the numerator and the denominator of R.

$$R(x) = \frac{x - 1}{(x + 2)(x - 2)}$$

R is in lowest terms.

STEP 1: The domain of R is $\{x \mid x \neq -2, x \neq 2\}$.

STEP 2: We locate the x-intercepts by finding the zeros of the numerator. By inspection, 1 is the only x-intercept. The y-intercept is $R(0) = \dfrac{1}{4}$.

STEP 3: Because

$$R(-x) = \frac{-x - 1}{x^2 - 4}$$

we conclude that R is neither even nor odd. There is no symmetry with respect to the y-axis or the origin.

STEP 4: We locate the vertical asymptotes by factoring the denominator: $x^2 - 4 = (x + 2)(x - 2)$. Since R is in lowest terms, the graph of R has two vertical asymptotes: the lines $x = -2$ and $x = 2$.

STEP 5: The degree of the numerator is less than the degree of the denominator, so R is proper and the line $y = 0$ (the x-axis) is a horizontal asymptote of the graph. To determine if the graph of R intersects the horizontal asymptote, we solve the equation $R(x) = 0$.

$$\frac{x-1}{x^2-4} = 0$$
$$x - 1 = 0$$
$$x = 1$$

The only solution is $x = 1$, so the graph of R intersects the horizontal asymptote at $(1, 0)$.

STEP 6: The zero of the numerator, 1, and the zeros of the denominator, -2 and 2, divide the x-axis into four intervals:

$$(-\infty, -2) \qquad (-2, 1) \qquad (1, 2) \qquad (2, \infty)$$

Now construct Table 10.

Table 10

	-2	1	2	x
Interval	$(-\infty, -2)$	$(-2, 1)$	$(1, 2)$	$(2, \infty)$
Number Chosen	-3	0	$\frac{3}{2}$	3
Value of R	$R(-3) = -0.8$	$R(0) = \frac{1}{4}$	$R\left(\frac{3}{2}\right) = -\frac{2}{7}$	$R(3) = 0.4$
Location of Graph	Below x-axis	Above x-axis	Below x-axis	Above x-axis
Point on Graph	$(-3, -0.8)$	$\left(0, \frac{1}{4}\right)$	$\left(\frac{3}{2}, -\frac{2}{7}\right)$	$(3, 0.4)$

STEP 7: We begin by graphing the asymptotes and plotting the points found in Steps 2, 5, and 6. See Figure 42(a). Next, we determine the behavior of the graph near the asymptotes. Since the x-axis is a horizontal asymptote and the graph lies below the x-axis for $-\infty < x < -2$, we can sketch a portion of the graph by placing a small arrow to the far left and under the x-axis. Since the line $x = -2$ is a vertical asymptote and the graph lies below the x-axis for $-\infty < x < -2$, we

Figure 42

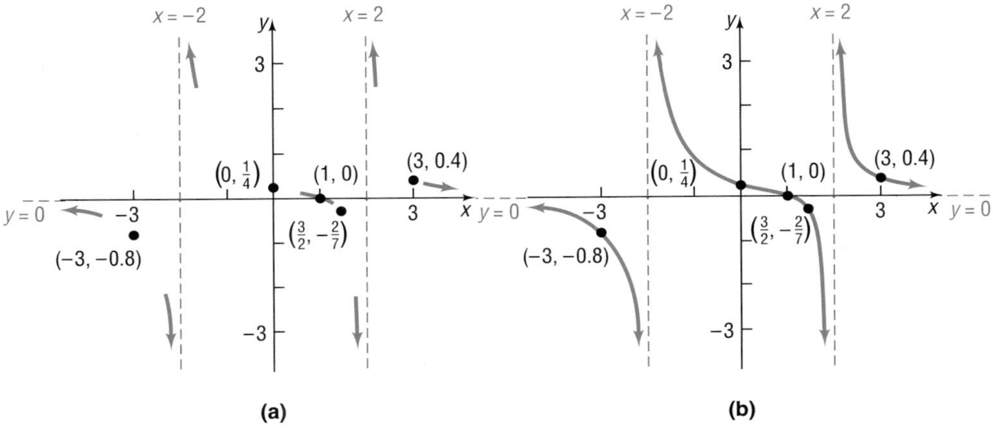

continue the sketch with an arrow placed well below the x-axis and approaching the line $x = -2$ on the left. Similar explanations account for the positions of the other portions of the graph. In particular, note how we use the facts that the graph lies above the x-axis for $-2 < x < 1$ and below the x-axis for $1 < x < 2$ to draw the conclusion that the graph crosses the x-axis at $(1, 0)$. Figure 42(b) shows the complete graph. ◀

Exploration

Graph $R(x) = \dfrac{x - 1}{x^2 - 4}$.

SOLUTION The analysis just completed in Example 1 helps us to set the viewing rectangle to obtain a complete graph. Figure 43(a) shows the graph of $R(x) = \dfrac{x - 1}{x^2 - 4}$ in connected mode, and Figure 43(b) shows it in dot mode. Notice in Figure 43(a) that the graph has vertical lines at $x = -2$ and $x = 2$. This is due to the fact that, when the graphing utility is in connected mode, it will "connect the dots" between consecutive pixels. We know that the graph of R does not cross the lines $x = -2$ and $x = 2$, since R is not defined at $x = -2$ or $x = 2$. So, when graphing rational functions, dot mode should be used to avoid extraneous vertical lines that are not part of the graph.

Figure 43

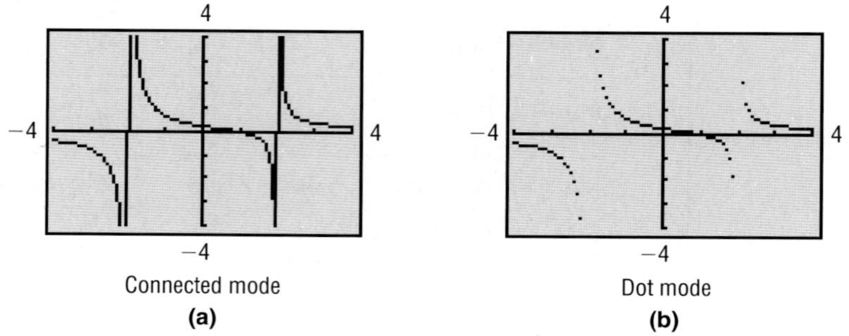

NOW WORK PROBLEM 7.

EXAMPLE 2 Analyzing the Graph of a Rational Function

Analyze the graph of the rational function: $R(x) = \dfrac{x^2 - 1}{x}$

Solution

STEP 1: The domain of R is $\{x \mid x \neq 0\}$.

STEP 2: The graph has two x-intercepts: -1 and 1. There is no y-intercept, since x cannot equal 0.

STEP 3: Since $R(-x) = -R(x)$, the function is odd and the graph is symmetric with respect to the origin.

STEP 4: R is in lowest terms, so the graph of R has the line $x = 0$ (the y-axis) as a vertical asymptote.

STEP 5: The rational function R is improper, since the degree of the numerator, 2, is larger than the degree of the denominator, 1. To find any horizontal or oblique asymptotes, we use long division.

$$\begin{array}{r} x \\ x\overline{)x^2 - 1} \\ \underline{x^2 } \\ -1 \end{array}$$

The quotient is x, so the line $y = x$ is an oblique asymptote of the graph. To determine whether the graph of R intersects the asymptote $y = x$, we solve the equation $R(x) = x$.

$$R(x) = \frac{x^2 - 1}{x} = x$$
$$x^2 - 1 = x^2$$
$$-1 = 0 \quad \text{Impossible}$$

We conclude that the equation $\frac{x^2 - 1}{x} = x$ has no solution, so the graph of R does not intersect the line $y = x$.

Step 6: The zeros of the numerator are -1 and 1; the denominator has the zero 0. We divide the x-axis into four intervals:

$$(-\infty, -1) \qquad (-1, 0) \qquad (0, 1) \qquad (1, \infty)$$

Now we construct Table 11.

Table 11

	-1	0	1	x
Interval	$(-\infty, -1)$	$(-1, 0)$	$(0, 1)$	$(1, \infty)$
Number Chosen	-2	$-\frac{1}{2}$	$\frac{1}{2}$	2
Value of R	$R(-2) = -\frac{3}{2}$	$R\left(-\frac{1}{2}\right) = \frac{3}{2}$	$R\left(\frac{1}{2}\right) = -\frac{3}{2}$	$R(2) = \frac{3}{2}$
Location of Graph	Below x-axis	Above x-axis	Below x-axis	Above x-axis
Point on Graph	$\left(-2, -\frac{3}{2}\right)$	$\left(-\frac{1}{2}, \frac{3}{2}\right)$	$\left(\frac{1}{2}, -\frac{3}{2}\right)$	$\left(2, \frac{3}{2}\right)$

Step 7: Figure 44(a) shows a partial graph using the facts that we have gathered. The complete graph is given in Figure 44(b).

Figure 44

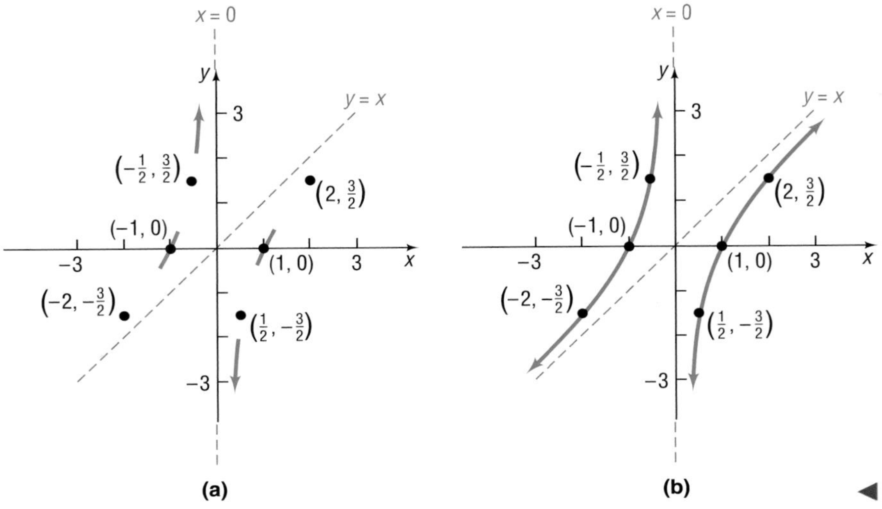

◀

Seeing the Concept

Graph $R(x) = \dfrac{x^2 - 1}{x}$ and compare what you see with Figure 44(b). Could you have predicted from the graph that $y = x$ is an oblique asymptote? Graph $y = x$ and ZOOM-OUT. What do you observe?

NOW WORK PROBLEM 15.

EXAMPLE 3 Analyzing the Graph of a Rational Function

Analyze the graph of the rational function: $R(x) = \dfrac{x^4 + 1}{x^2}$

Solution

STEP 1: The domain of R is $\{x \mid x \neq 0\}$.

STEP 2: The graph has no x-intercepts and no y-intercepts.

STEP 3: Since $R(-x) = R(x)$, the function is even and the graph is symmetric with respect to the y-axis.

STEP 4: R is in lowest terms, so the graph of R has the line $x = 0$ (the y-axis) as a vertical asymptote.

STEP 5: The rational function R is improper. To find any horizontal or oblique asymptotes, we use long division.

$$\begin{array}{r} x^2 \\ x^2 \overline{)\, x^4 + 1} \\ \underline{x^4 } \\ 1 \end{array}$$

The quotient is x^2, so the graph has no horizontal or oblique asymptotes. However, the graph of R will approach the graph of $y = x^2$ as $x \to -\infty$ and as $x \to \infty$.

STEP 6: The numerator has no zeros, and the denominator has one zero at 0. We divide the x-axis into the two intervals

$$(-\infty, 0) \qquad (0, \infty)$$

Now we construct Table 12.

Figure 45

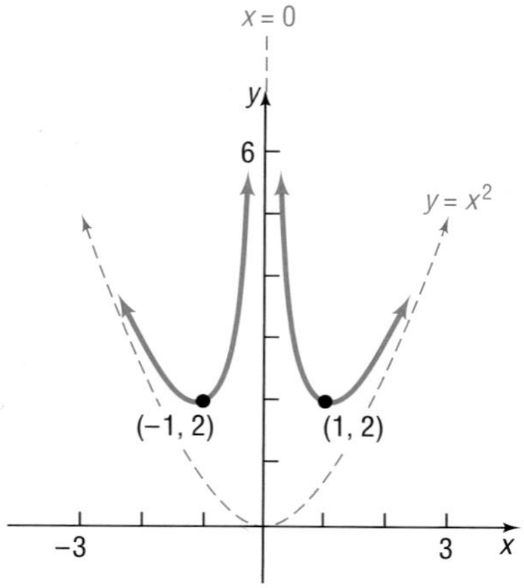

Table 12

	$(-\infty, 0)$	$(0, \infty)$
Interval	$(-\infty, 0)$	$(0, \infty)$
Number Chosen	-1	1
Value of R	$R(-1) = 2$	$R(1) = 2$
Location of Graph	Above x-axis	Above x-axis
Point on Graph	$(-1, 2)$	$(1, 2)$

STEP 7: Figure 45 shows the graph. ◀

Seeing the Concept

Graph $R(x) = \dfrac{x^4 + 1}{x^2}$ and compare what you see with Figure 45. Use MINIMUM to find the two turning points. Enter $y = x^2$ and ZOOM-OUT. What do you see?

NOW WORK PROBLEM 13.

EXAMPLE 4 Analyzing the Graph of a Rational Function

Analyze the graph of the rational function: $R(x) = \dfrac{3x^2 - 3x}{x^2 + x - 12}$

Solution We factor R to get

$$R(x) = \frac{3x(x - 1)}{(x + 4)(x - 3)}$$

R is in lowest terms.

STEP 1: The domain of R is $\{x | x \neq -4, x \neq 3\}$.

STEP 2: The graph has two x-intercepts: 0 and 1. The y-intercept is $R(0) = 0$.

STEP 3: There is no symmetry with respect to the y-axis or the origin.

STEP 4: Since R is in lowest terms, the graph of R has two vertical asymptotes: $x = -4$ and $x = 3$.

STEP 5: Since the degree of the numerator equals the degree of the denominator, the graph has a horizontal asymptote. To find it, we either use long division or form the quotient of the leading coefficient of the numerator, 3, and the leading coefficient of the denominator, 1. The graph of R has the horizontal asymptote $y = 3$. To find out whether the graph of R intersects the asymptote, we solve the equation $R(x) = 3$.

$$R(x) = \frac{3x^2 - 3x}{x^2 + x - 12} = 3$$
$$3x^2 - 3x = 3x^2 + 3x - 36$$
$$-6x = -36$$
$$x = 6$$

The graph intersects the line $y = 3$ only at $x = 6$, and $(6, 3)$ is a point on the graph of R.

STEP 6: The zeros of the numerator, 0 and 1, and the zeros of the denominator, -4 and 3, divide the x-axis into five intervals:

$$(-\infty, -4) \qquad (-4, 0) \qquad (0, 1) \qquad (1, 3) \qquad (3, \infty)$$

Now we construct Table 13.

Table 13

	-4	0	1	3	x
Interval	$(-\infty, -4)$	$(-4, 0)$	$(0, 1)$	$(1, 3)$	$(3, \infty)$
Number Chosen	-5	-2	$\frac{1}{2}$	2	4
Value of R	$R(-5) = 11.25$	$R(-2) = -1.8$	$R\left(\frac{1}{2}\right) = \frac{1}{15}$	$R(2) = -1$	$R(4) = 4.5$
Location of Graph	Above x-axis	Below x-axis	Above x-axis	Below x-axis	Above x-axis
Point on Graph	$(-5, 11.25)$	$(-2, -1.8)$	$\left(\frac{1}{2}, \frac{1}{15}\right)$	$(2, -1)$	$(4, 4.5)$

STEP 7: Figure 46(a) shows a partial graph. Notice that we have not yet used the fact that the line $y = 3$ is a horizontal asymptote, because we do not know yet whether the graph of R crosses or touches the line $y = 3$ at $(6, 3)$. To see whether the graph, in fact, crosses or touches the line $y = 3$, we plot an additional point to the right of $(6, 3)$. We use $x = 7$ to find $R(7) = \frac{63}{22} < 3$. The graph crosses $y = 3$ at $x = 6$. Because $(6, 3)$ is the only point where the graph of R intersects the asymptote $y = 3$, the graph must approach the line $y = 3$ from above as $x \to -\infty$ and approach the line $y = 3$ from below as $x \to \infty$. See Figure 46(b). The completed graph is shown in Figure 46(c).

Figure 46

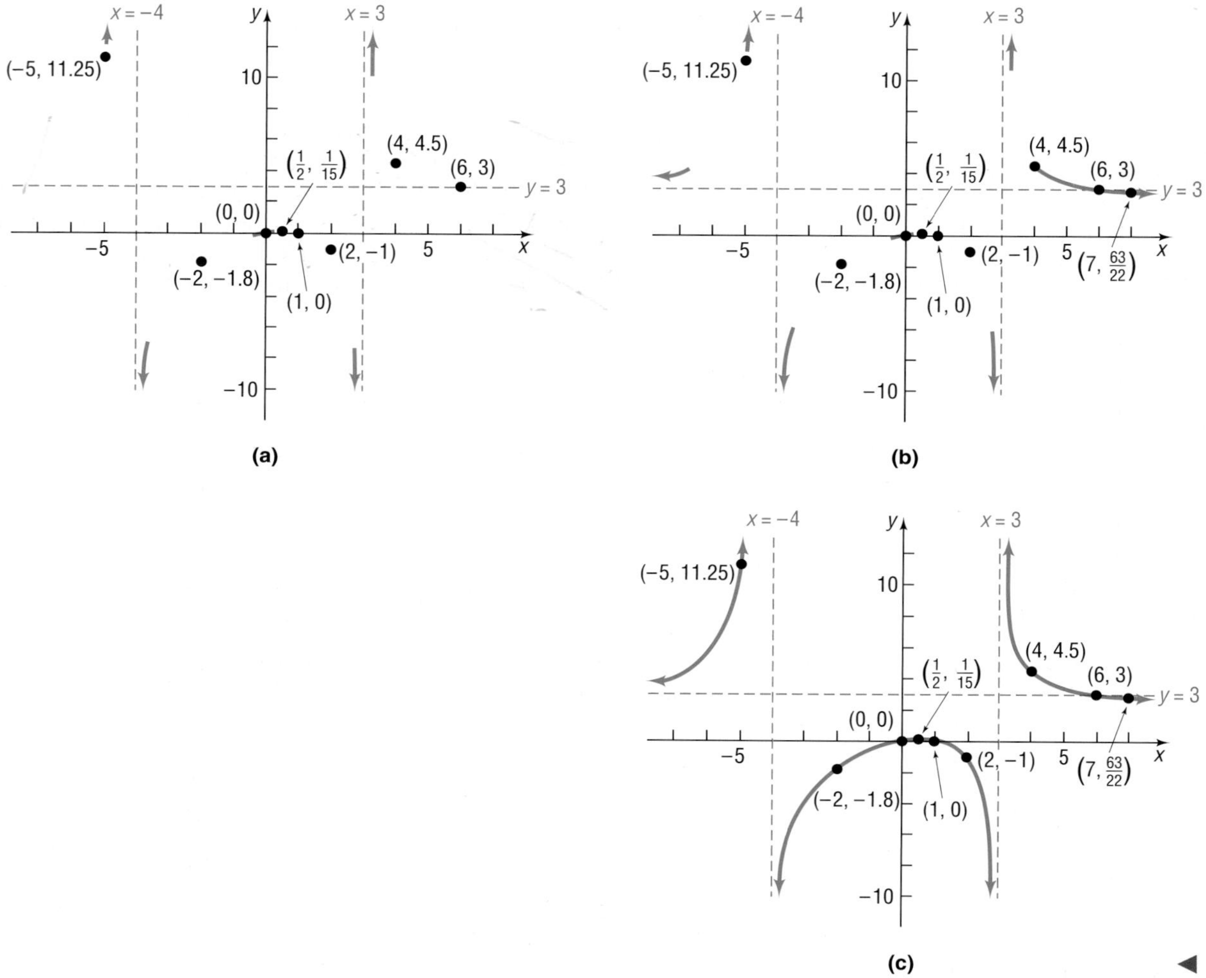

◀

Exploration

Graph $R(x) = \dfrac{3x^2 - 3x}{x^2 + x - 12}$.

SOLUTION Figure 47 shows the graph in connected mode, and Figure 48(a) shows it in dot mode. Neither graph displays clearly the behavior between the two x-intercepts, 0 and 1.

Nor do they clearly display the fact that the graph crosses the horizontal asymptote at (6, 3). To see these parts better, we graph R for $-1 \leq x \leq 2$, Figure 48(b), and for $4 \leq x \leq 60$, Figure 49(b).

Figure 47

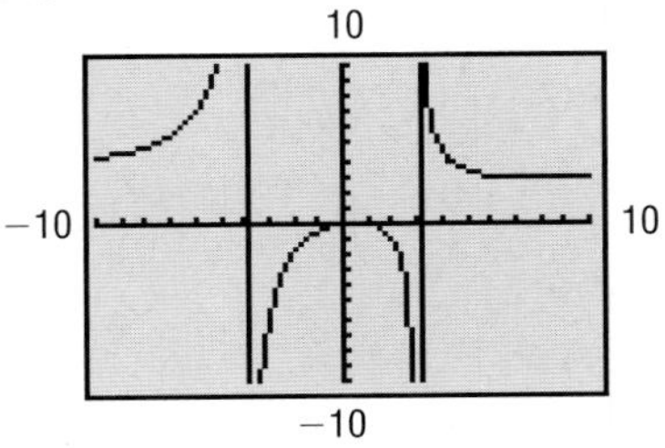

Connected mode

Figure 48

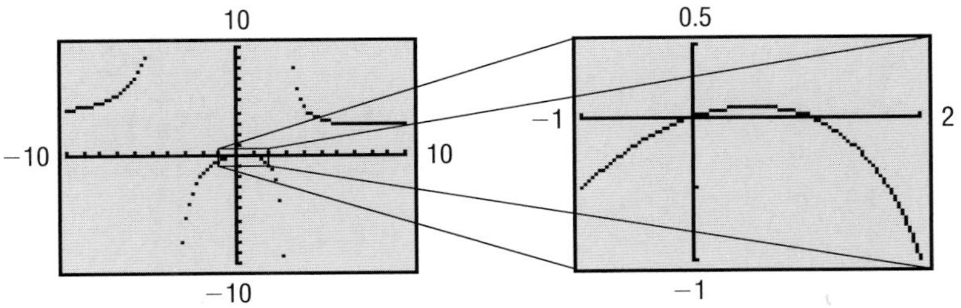

Dot mode
(a) **(b)**

Figure 49

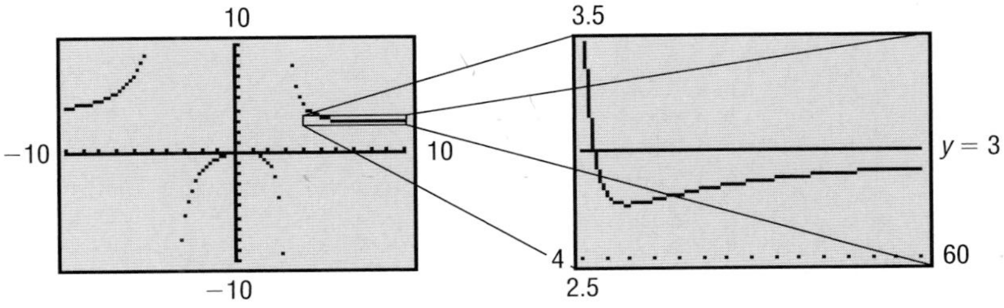

Dot mode
(a) **(b)**

The new graphs reflect the behavior produced by the analysis. Furthermore, we observe two turning points, one between 0 and 1 and the other to the right of 4. Rounded to two decimal places, these turning points are (0.52, 0.07) and (11.48, 2.75).

EXAMPLE 5 Analyzing the Graph of a Rational Function with a Hole

Analyze the graph of the rational function: $R(x) = \dfrac{2x^2 - 5x + 2}{x^2 - 4}$

Solution We factor R and obtain

$$R(x) = \frac{(2x - 1)(x - 2)}{(x + 2)(x - 2)}$$

In lowest terms,

$$R(x) = \frac{2x - 1}{x + 2}, \qquad x \neq -2$$

Step 1: The domain of R is $\{x | x \neq -2, x \neq 2\}$.

Step 2: The graph has one x-intercept: $\frac{1}{2}$. The y-intercept is $R(0) = -\frac{1}{2}$.

Step 3: There is no symmetry with respect to the y-axis or the origin.

Step 4: The graph has one vertical asymptote, $x = -2$, since $x + 2$ is the only factor of the denominator of $R(x)$ *in lowest terms*. However, the rational function is undefined at both $x = 2$ and $x = -2$.

STEP 5: Since the degree of the numerator equals the degree of the denominator, the graph has a horizontal asymptote. To find it, we either use long division or form the quotient of the leading coefficient of the numerator, 2, and the leading coefficient of the denominator, 1. The graph of R has the horizontal asymptote $y = 2$. To find out whether the graph of R intersects the asymptote, we solve the equation $R(x) = 2$.

$$R(x) = \frac{2x - 1}{x + 2} = 2$$
$$2x - 1 = 2(x + 2)$$
$$2x - 1 = 2x + 4$$
$$-1 = 4 \quad \text{Impossible}$$

The graph does not intersect the line $y = 2$.

STEP 6: The zeros of the numerator and denominator, $-2, \frac{1}{2}$, and 2, divide the x-axis into four intervals:

$$(-\infty, -2) \quad \left(-2, \frac{1}{2}\right) \quad \left(\frac{1}{2}, 2\right) \quad (2, \infty)$$

Now we construct Table 14.

Table 14

	-2	$1/2$	2	x
Interval	$(-\infty, -2)$	$\left(-2, \frac{1}{2}\right)$	$\left(\frac{1}{2}, 2\right)$	$(2, \infty)$
Number Chosen	-3	-1	1	3
Value of R	$R(-3) = 7$	$R(-1) = -3$	$R(1) = \frac{1}{3}$	$R(3) = 1$
Location of Graph	Above x-axis	Below x-axis	Above x-axis	Above x-axis
Point on Graph	$(-3, 7)$	$(-1, -3)$	$\left(1, \frac{1}{3}\right)$	$(3, 1)$

STEP 7: See Figure 50. Notice the vertical asymptote $x = -2$ and the hole at the point $\left(2, \frac{3}{4}\right)$. R is not defined at -2 and 2.

Figure 50

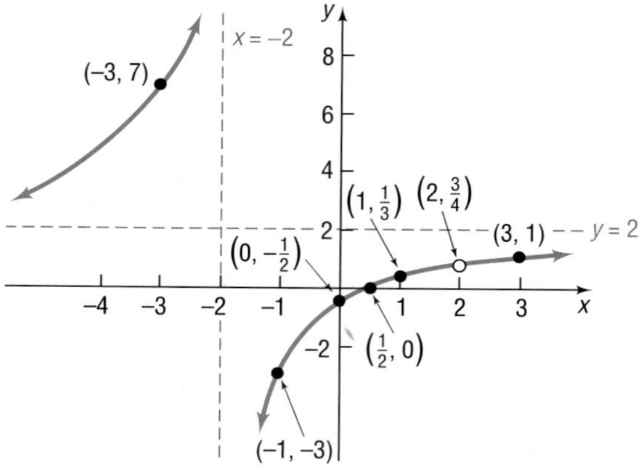

Note: The coordinates of the hole were obtained by evaluating R *in lowest terms* at 2. R in lowest terms is $\frac{2x-1}{x+2}$, which, at $x = 2$, is $\frac{2(2)-1}{2+2} = \frac{3}{4}$. ◀

As Example 5 shows, the zeros of the denominator of a rational function give rise to either vertical asymptotes or holes on the graph.

Exploration

Graph $R(x) = \frac{2x^2 - 5x + 2}{x^2 - 4}$. Do you see the hole at $\left(2, \frac{3}{4}\right)$? TRACE along the graph. Did you obtain an ERROR at $x = 2$? Are you convinced that an algebraic analysis of a rational function is required in order to accurately interpret the graph obtained with a graphing utility?

NOW WORK PROBLEM 33.

We now discuss the problem of finding a rational function from its graph.

EXAMPLE 6 Constructing a Rational Function from Its Graph

Find a rational function that might have the graph shown in Figure 51.

Figure 51

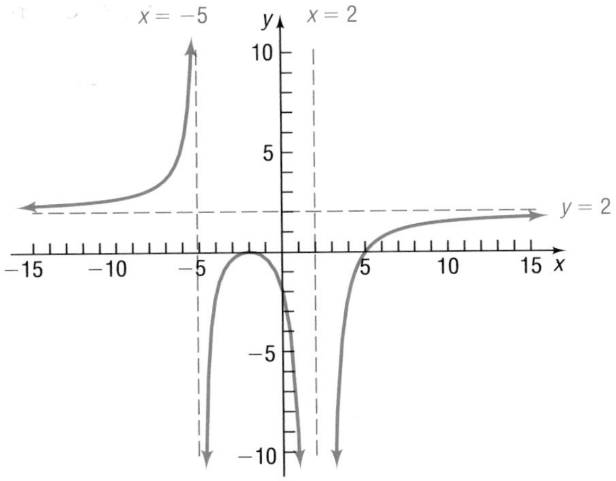

Solution The numerator of a rational function $R(x) = \frac{p(x)}{q(x)}$ in lowest terms determines the x-intercepts of its graph. The graph shown in Figure 51 has x-intercepts -2 (even multiplicity; graph touches the x-axis) and 5 (odd multiplicity; graph crosses the x-axis). So one possibility for the numerator is $p(x) = (x + 2)^2(x - 5)$. The denominator of a rational function in lowest terms determines the vertical asymptotes of its graph. The vertical asymptotes of the graph are $x = -5$ and $x = 2$. Since $R(x)$ approaches ∞ from the left of $x = -5$ and $R(x)$ approaches $-\infty$ from the right of $x = -5$, we know that $(x + 5)$ is a factor of odd multiplicity in $q(x)$. Also, $R(x)$ approaches $-\infty$ from both sides of $x = 2$, so $(x - 2)$ is a factor of even multiplicity in $q(x)$. A possibility for the denominator is $q(x) = (x + 5)(x - 2)^2$. So far we have $R(x) = \frac{(x + 2)^2(x - 5)}{(x + 5)(x - 2)^2}$.

However, the horizontal asymptote of the graph given in Figure 51, is

$y = 2$, so we know that the degree of the numerator must equal the degree in the denominator and the quotient of leading coefficients must be $\frac{2}{1}$. This leads to

$$R(x) = \frac{2(x+2)^2(x-5)}{(x+5)(x-2)^2}$$

Figure 52

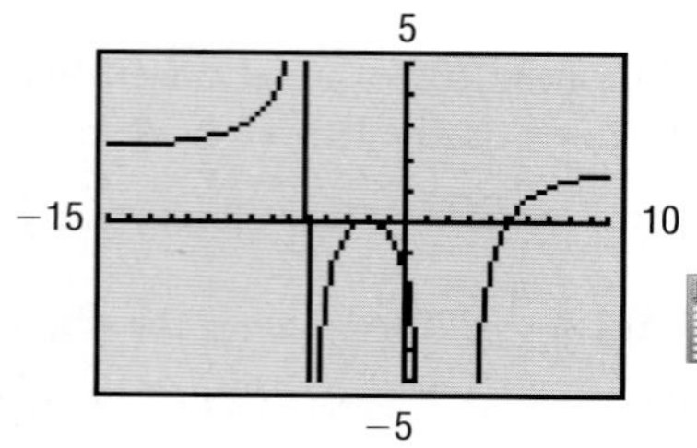

CHECK: Figure 52 shows the graph of R on a graphing utility. Since Figure 52 looks similar to Figure 51, we have found a rational function R for the graph in Figure 51.

NOW WORK PROBLEM 45.

2 Application

EXAMPLE 7 Finding the Least Cost of a Can

Reynolds Metal Company manufactures aluminum cans in the shape of a cylinder with a capacity of 500 cubic centimeters $\left(\frac{1}{2}\text{ liter}\right)$. The top and bottom of the can are made of a special aluminum alloy that costs 0.05¢ per square centimeter. The sides of the can are made of material that costs 0.02¢ per square centimeter.

(a) Express the cost of material for the can as a function of the radius r of the can.

(b) Use a graphing utility to graph the function $C = C(r)$.

(c) What value of r will result in the least cost?

(d) What is this least cost?

Solution (a) Figure 53 illustrates the components of a can in the shape of a right circular cylinder. Notice that the material required to produce a cylindrical can of height h and radius r consists of a rectangle of area $2\pi rh$ and two circles, each of area πr^2. The total cost C (in cents) of manufacturing the can is therefore

Figure 53

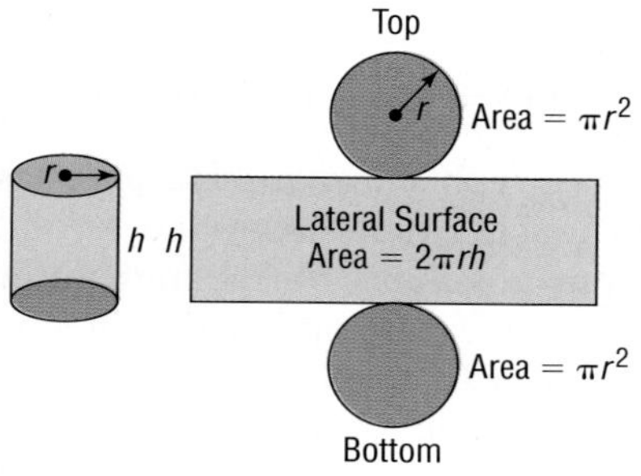

$$\begin{aligned} C &= \text{Cost of top and bottom} + \text{Cost of side} \\ &= \underbrace{2(\pi r^2)}_{\substack{\text{Total area}\\ \text{of top and}\\ \text{bottom}}} \underbrace{(0.05)}_{\substack{\text{Cost/unit}\\ \text{area}}} + \underbrace{(2\pi rh)}_{\substack{\text{Total}\\ \text{area of}\\ \text{side}}} \underbrace{(0.02)}_{\substack{\text{Cost/unit}\\ \text{area}}} \\ &= 0.10\pi r^2 + 0.04\pi rh \end{aligned}$$

But we have the additional restriction that the height h and radius r must be chosen so that the volume V of the can is 500 cubic centimeters. Since $V = \pi r^2 h$, we have

$$500 = \pi r^2 h \quad \text{so} \quad h = \frac{500}{\pi r^2}$$

Figure 54

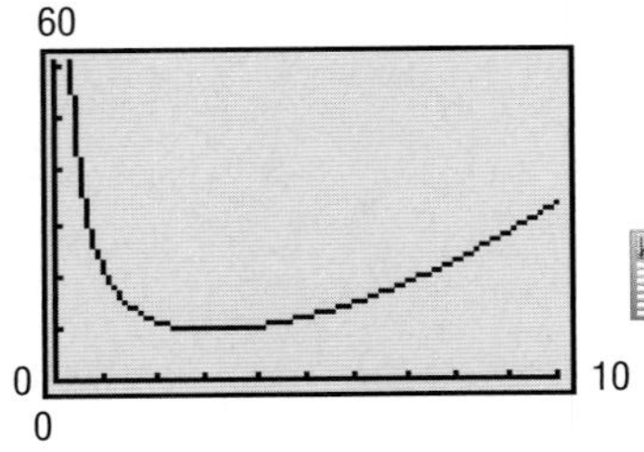

Substituting this expression for h, the cost C, in cents, as a function of the radius r is

$$C(r) = 0.10\pi r^2 + 0.04\pi r \frac{500}{\pi r^2} = 0.10\pi r^2 + \frac{20}{r} = \frac{0.10\pi r^3 + 20}{r}$$

(b) See Figure 54 for the graph of $C(r)$.

(c) Using the MINIMUM command, the cost is least for a radius of about 3.17 centimeters.

(d) The least cost is $C(3.17) \approx 9.47¢$.

3.4 Assess Your Understanding

'Are You Prepared?' *Answers are given at the end of these exercises. If you get a wrong answer, read the pages listed in red.*

1. *True or False:* The graph of a function will have at least one intercept. (p. 72)

2. If the graph of $y = f(x)$ is symmetric with respect to the origin and if $f(4) = 2$, then the points __________ and __________ are on the graph of f. (pp. 72–74)

Concepts and Vocabulary

3. If the numerator and the denominator of a rational function have no common factors, the rational function is __________.

4. *True or False:* The graph of a polynomial function sometimes has a hole.

5. *True or False:* The graph of a rational function never intersects a horizontal asymptote.

6. *True or False:* The graph of a rational function sometimes has a hole.

Exercises

In Problems 7–44, follow Steps 1 through 7 on page 174 to analyze the graph of each function.

7. $R(x) = \dfrac{x+1}{x(x+4)}$

8. $R(x) = \dfrac{x}{(x-1)(x+2)}$

9. $R(x) = \dfrac{3x+3}{2x+4}$

10. $R(x) = \dfrac{2x+4}{x-1}$

11. $R(x) = \dfrac{3}{x^2-4}$

12. $R(x) = \dfrac{6}{x^2-x-6}$

13. $P(x) = \dfrac{x^4+x^2+1}{x^2-1}$

14. $Q(x) = \dfrac{x^4-1}{x^2-4}$

15. $H(x) = \dfrac{x^3-1}{x^2-9}$

16. $G(x) = \dfrac{x^3+1}{x^2+2x}$

17. $R(x) = \dfrac{x^2}{x^2+x-6}$

18. $R(x) = \dfrac{x^2+x-12}{x^2-4}$

19. $G(x) = \dfrac{x}{x^2-4}$

20. $G(x) = \dfrac{3x}{x^2-1}$

21. $R(x) = \dfrac{3}{(x-1)(x^2-4)}$

22. $R(x) = \dfrac{-4}{(x+1)(x^2-9)}$

23. $H(x) = 4\dfrac{x^2-1}{x^4-16}$

24. $H(x) = \dfrac{x^2+4}{x^4-1}$

25. $F(x) = \dfrac{x^2-3x-4}{x+2}$

26. $F(x) = \dfrac{x^2+3x+2}{x-1}$

27. $R(x) = \dfrac{x^2+x-12}{x-4}$

28. $R(x) = \dfrac{x^2-x-12}{x+5}$

29. $F(x) = \dfrac{x^2+x-12}{x+2}$

30. $G(x) = \dfrac{x^2-x-12}{x+1}$

31. $R(x) = \dfrac{x(x-1)^2}{(x+3)^3}$

32. $R(x) = \dfrac{(x-1)(x+2)(x-3)}{x(x-4)^2}$

33. $R(x) = \dfrac{x^2+x-12}{x^2-x-6}$

34. $R(x) = \dfrac{x^2 + 3x - 10}{x^2 + 8x + 15}$

35. $R(x) = \dfrac{6x^2 - 7x - 3}{2x^2 - 7x + 6}$

36. $R(x) = \dfrac{8x^2 + 26x + 15}{2x^2 - x - 15}$

37. $R(x) = \dfrac{x^2 + 5x + 6}{x + 3}$

38. $R(x) = \dfrac{x^2 + x - 30}{x + 6}$

39. $f(x) = x + \dfrac{1}{x}$

40. $f(x) = 2x + \dfrac{9}{x}$

41. $f(x) = x^2 + \dfrac{1}{x}$

42. $f(x) = 2x^2 + \dfrac{9}{x}$

43. $f(x) = x + \dfrac{1}{x^3}$

44. $f(x) = 2x + \dfrac{9}{x^3}$

In Problems 45–48, find a rational function that might have the given graph. (More than one answer might be possible.)

45.

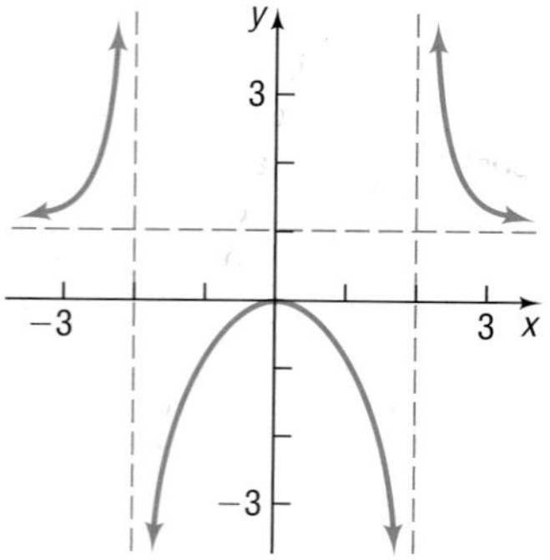

46.

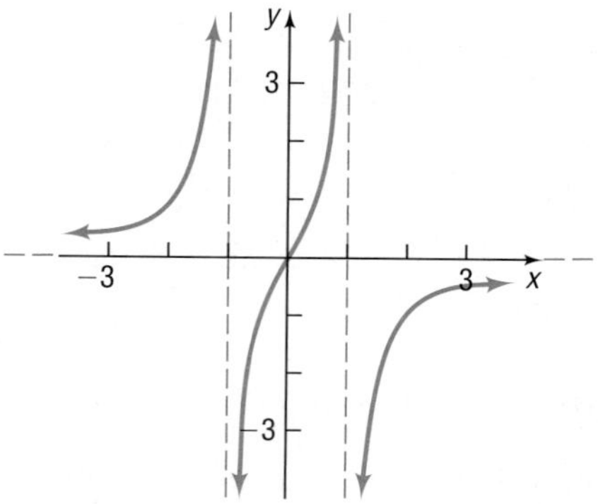

47.

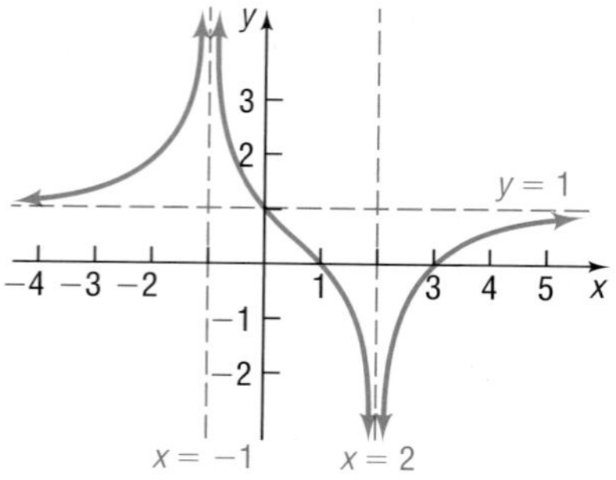

48.

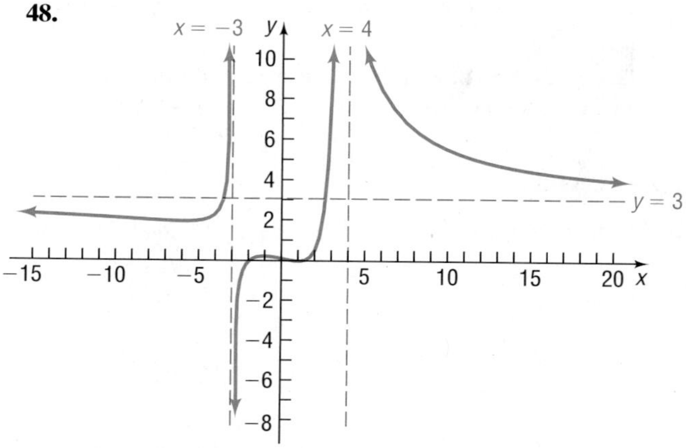

49. Drug Concentration The concentration C of a certain drug in a patient's bloodstream t hours after injection is given by

$$C(t) = \frac{t}{2t^2 + 1}$$

(a) Find the horizontal asymptote of $C(t)$. What happens to the concentration of the drug as t increases?
(b) Using your graphing utility, graph $C(t)$.
(c) Determine the time at which the concentration is highest.

50. Drug Concentration The concentration C of a certain drug in a patient's bloodstream t minutes after injection is given by

$$C(t) = \frac{50t}{t^2 + 25}$$

(a) Find the horizontal asymptote of $C(t)$. What happens to the concentration of the drug as t increases?
(b) Using your graphing utility, graph $C(t)$.
(c) Determine the time at which the concentration is highest.

51. Average Cost In Problem 96, Exercise 3.2, the cost function C (in thousands of dollars) for manufacturing x Chevy Cavaliers was given as

$$C(x) = 0.2x^3 - 2.3x^2 + 14.3x + 10.2$$

Economists define the **average cost function** as

$$\overline{C}(x) = \frac{C(x)}{x}$$

(a) Find the average cost function.

(b) What is the average cost of producing six Cavaliers?
(c) What is the average cost of producing nine Cavaliers?
(d) Using your graphing utility, graph the average cost function.
(e) Using your graphing utility, find the number of Cavaliers that should be produced to minimize the average cost.
(f) What is the minimum average cost?

52. Average Cost In Problem 97, Exercise 3.2, the cost function C (in thousands of dollars) for printing x textbooks (in thousands of units) was given as

$$C(x) = 0.015x^3 - 0.595x^2 + 9.15x + 98.43$$

(a) Find the average cost function (refer to Problem 51).
(b) What is the average cost of printing 13,000 textbooks per week?
(c) What is the average cost of printing 25,000 textbooks per week?
(d) Using your graphing utility, graph the average cost function.
(e) Using your graphing utility, find the number of textbooks that should be printed to minimize the average cost.
(f) What is the minimum average cost?

53. Minimizing Surface Area United Parcel Service has contracted you to design a closed box with a square base that has a volume of 10,000 cubic inches. See the illustration.

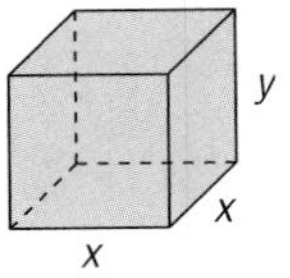

(a) Find a function for the surface area of the box.
(b) Using a graphing utility, graph the function found in part (a).
(c) What is the minimum amount of cardboard that can be used to construct the box?
(d) What are the dimensions of the box that minimize the surface area?
(e) Why might UPS be interested in designing a box that minimizes the surface area?

54. Minimizing Surface Area United Parcel Service has contracted you to design a closed box with a square base that has a volume of 5000 cubic inches. See the illustration.

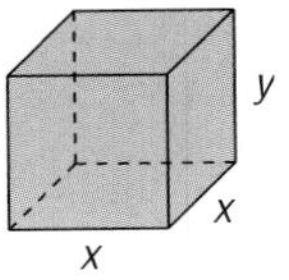

(a) Find a function for the surface area of the box.
(b) Using a graphing utility, graph the function found in part (a).
(c) What is the minimum amount of cardboard that can be used to construct the box?
(d) What are the dimensions of the box that minimize the surface area?
(e) Why might UPS be interested in designing a box that minimizes the surface area?

55. Cost of a Can A can in the shape of a right circular cylinder is required to have a volume of 500 cubic centimeters. The top and bottom are made of material that costs 6¢ per square centimeter, while the sides are made of material that costs 4¢ per square centimeter.
(a) Express the total cost C of the material as a function of the radius r of the cylinder. (Refer to Figure 53.)
(b) Graph $C = C(r)$. For what value of r is the cost C a minimum?

56. Material Needed to Make a Drum A steel drum in the shape of a right circular cylinder is required to have a volume of 100 cubic feet.

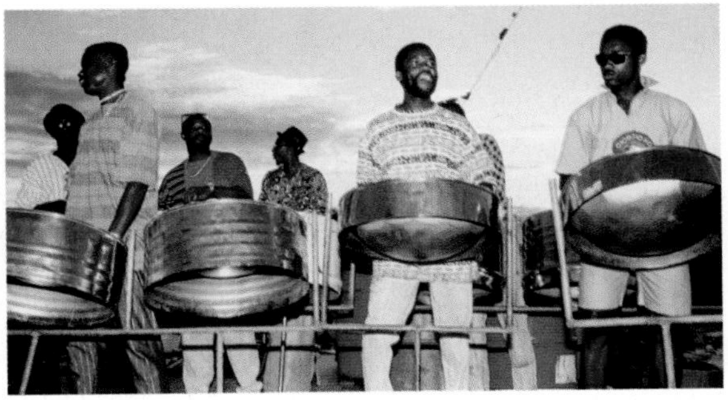

(a) Express the amount A of material required to make the drum as a function of the radius r of the cylinder.
(b) How much material is required if the drum's radius is 3 feet?
(c) How much material is required if the drum's radius is 4 feet?
(d) How much material is required if the drum's radius is 5 feet?
(e) Graph $A = A(r)$. For what value of r is A smallest?

57. Graph each of the following functions:

$$y = \frac{x^2 - 1}{x - 1} \qquad y = \frac{x^3 - 1}{x - 1}$$

$$y = \frac{x^4 - 1}{x - 1} \qquad y = \frac{x^5 - 1}{x - 1}$$

Is $x = 1$ a vertical asymptote? Why not? What is happening for $x = 1$? What do you conjecture about $y = \frac{x^n - 1}{x - 1}$, $n \geq 1$ an integer, for $x = 1$?

58. Graph each of the following functions:

$$y = \frac{x^2}{x-1} \qquad y = \frac{x^4}{x-1} \qquad y = \frac{x^6}{x-1} \qquad y = \frac{x^8}{x-1}$$

What similarities do you see? What differences?

In Problems 59–64, graph each function and use MINIMUM to obtain the minimum value, rounded to two decimal places.

59. $f(x) = x + \frac{1}{x}, \quad x > 0$

60. $f(x) = 2x + \frac{9}{x}, \quad x > 0$

61. $f(x) = x^2 + \frac{1}{x}, \quad x > 0$

62. $f(x) = 2x^2 + \frac{9}{x}, \quad x > 0$

63. $f(x) = x + \frac{1}{x^3}, \quad x > 0$

64. $f(x) = 2x + \frac{9}{x^3}, \quad x > 0$

---

65. Write a few paragraphs that provide a general strategy for graphing a rational function. Be sure to mention the following: proper, improper, intercepts, and asymptotes.

66. Create a rational function that has the following characteristics: crosses the x-axis at 2; touches the x-axis at -1; one vertical asymptote at $x = -5$ and another at $x = 6$; and one horizontal asymptote, $y = 3$. Compare yours to a fellow classmate's. How do they differ? What are the similarities?

67. Create a rational function that has the following characteristics: crosses the x-axis at 3; touches the x-axis at -2; one vertical asymptote, $x = 1$; and one horizontal asymptote, $y = 2$. Give your rational function to a fellow classmate and ask for a written critique of your rational function.

'Are You Prepared?' Answers

1. False **2.** $(4, 2)$; $(-4, -2)$

3.5 Polynomial and Rational Inequalities

PREPARING FOR THIS SECTION *Before getting started, review the following:*

- Solving Inequalities (Appendix A, Section A.8, pp. 971–974)

Now work the 'Are You Prepared?' problem on page 192.

OBJECTIVES 1 Solve Polynomial Inequalities
2 Solve Rational Inequalities

1 In this section we solve inequalities that involve polynomials of degree 2 and higher, as well as some that involve rational expressions. To solve such inequalities, we use the information obtained in the previous three sections about the graph of polynomial and rational functions. The general idea follows:

Suppose that the polynomial or rational inequality is in one of the forms

$$f(x) < 0 \qquad f(x) > 0 \qquad f(x) \le 0 \qquad f(x) \ge 0$$

Locate the zeros of f if f is a polynomial function, and locate the zeros of the numerator and the denominator if f is a rational function. If we use these zeros to divide the real number line into intervals, then we know that on each interval the graph of f is either above the x-axis $[f(x) > 0]$ or below the x-axis $[f(x) < 0]$. In other words, we have found the solution of the inequality.

The following steps provide more detail.

Steps for Solving Polynomial and Rational Inequalities

STEP 1: Write the inequality so that a polynomial or rational expression f is on the left side and zero is on the right side in one of the following forms:

$$f(x) > 0 \qquad f(x) \geq 0 \qquad f(x) < 0 \qquad f(x) \leq 0$$

For rational expressions, be sure that the left side is written as a single quotient.

STEP 2: Determine the numbers at which the expression f on the left side equals zero and, if the expression is rational, the numbers at which the expression f on the left side is undefined.

STEP 3: Use the numbers found in Step 2 to separate the real number line into intervals.

STEP 4: Select a number in each interval and evaluate f at the number.

(a) If the value of f is positive, then $f(x) > 0$ for all numbers x in the interval.

(b) If the value of f is negative, then $f(x) < 0$ for all numbers x in the interval.

If the inequality is not strict, include the solutions of $f(x) = 0$ in the solution set.

EXAMPLE 1 Solving a Polynomial Inequality

Solve the inequality $x^2 \leq 4x + 12$, and graph the solution set.

Solution **STEP 1:** Rearrange the inequality so that 0 is on the right side.

$$x^2 \leq 4x + 12$$
$$x^2 - 4x - 12 \leq 0 \quad \text{Subtract } 4x + 12 \text{ from both sides of the inequality.}$$

This inequality is equivalent to the one that we wish to solve.

STEP 2: Find the zeros of $f(x) = x^2 - 4x - 12$ by solving the equation $x^2 - 4x - 12 = 0$.

$$x^2 - 4x - 12 = 0$$
$$(x + 2)(x - 6) = 0 \quad \text{Factor.}$$
$$x = -2 \quad \text{or} \quad x = 6$$

STEP 3: We use the zeros of f to separate the real number line into three intervals:

$$(-\infty, -2) \qquad (-2, 6) \qquad (6, \infty)$$

STEP 4: Select a number in each interval and evaluate $f(x) = x^2 - 4x - 12$ to determine if $f(x)$ is positive or negative. See Table 15.

Table 15

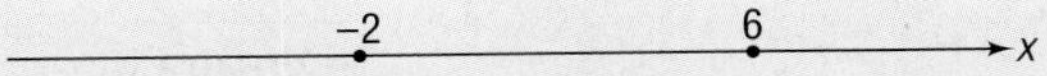

Interval	$(-\infty, -2)$	$(-2, 6)$	$(6, \infty)$
Number Chosen	-3	0	7
Value of f	$f(-3) = 9$	$f(0) = -12$	$f(7) = 9$
Conclusion	Positive	Negative	Positive

Based on Table 15, we know that $f(x) < 0$ for all x in the interval $(-2, 6)$, that is, for all x such that $-2 < x < 6$. However, because the original inequality is not strict, numbers x that satisfy the equation $f(x) = x^2 - 4x - 12 = 0$ are also solutions of the inequality $x^2 \leq 4x + 12$. Thus, we include -2 and 6. The solution set of the given inequality is $\{x | -2 \leq x \leq 6\}$ or, using interval notation, $[-2, 6]$.

Figure 55

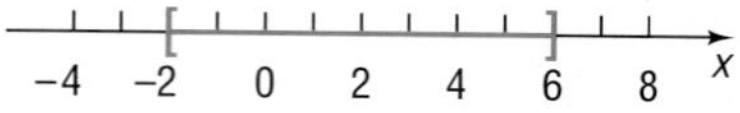

Figure 55 shows the graph of the solution set. ◀

NOW WORK PROBLEMS 5 AND 9.

EXAMPLE 2 Solving a Polynomial Inequality

Solve the inequality $x^4 > x$, and graph the solution set.

Solution **STEP 1:** Rearrange the inequality so that 0 is on the right side.

$$x^4 > x$$
$$x^4 - x > 0 \quad \text{Subtract } x \text{ from both sides of the inequality.}$$

This inequality is equivalent to the one that we wish to solve.

STEP 2: Find the zeros of $f(x) = x^4 - x$ by solving $x^4 - x = 0$.

$$x^4 - x = 0$$
$$x(x^3 - 1) = 0 \quad \text{Factor out } x.$$
$$x(x - 1)(x^2 + x + 1) = 0 \quad \text{Factor the difference of two cubes.}$$
$$x = 0 \text{ or } x - 1 = 0 \text{ or } x^2 + x + 1 = 0 \quad \text{Set each factor equal to zero and solve.}$$
$$x = 0 \text{ or } x = 1$$

The equation $x^2 + x + 1 = 0$ has no real solutions. (Do you see why?)

STEP 3: We use the zeros to separate the real number line into three intervals:

$$(-\infty, 0) \qquad (0, 1) \qquad (1, \infty)$$

STEP 4: We choose a number in each interval and evaluate $f(x) = x^4 - x$ to determine if $f(x)$ is positive or negative. See Table 16.

Table 16

	0	1	x
Interval	$(-\infty, 0)$	$(0, 1)$	$(1, \infty)$
Number Chosen	-1	$\frac{1}{2}$	2
Value of f	$f(-1) = 2$	$f\left(\frac{1}{2}\right) = -\frac{7}{16}$	$f(2) = 14$
Conclusion	Positive	Negative	Positive

Based on Table 16, we know that $f(x) > 0$ for all x in the intervals $(-\infty, 0)$ or $(1, \infty)$, that is, for all numbers x for which $x < 0$ or $x > 1$. Because the original inequality is strict, the solution set of the given inequality is $\{x | x < 0 \text{ or } x > 1\}$ or, using interval notation, $(-\infty, 0)$ or $(1, \infty)$.

Figure 56

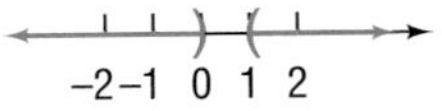

Figure 56 shows the graph of the solution set. ◀

NOW WORK PROBLEM 21.

2 Let's solve a rational inequality.

EXAMPLE 3 Solving a Rational Inequality

Solve the inequality $\dfrac{(x+3)(2-x)}{(x-1)^2} > 0$, and graph the solution set.

Solution **Step 1:** The domain of the variable x is $\{x \mid x \neq 1\}$. The inequality is already in a form with 0 on the right side.

Step 2: Let $f(x) = \dfrac{(x+3)(2-x)}{(x-1)^2}$. The zeros of the numerator of f are -3 and 2; the zero of the denominator is 1.

Step 3: We use the zeros found in Step 2 to separate the real number line into four intervals:

$$(-\infty, -3) \qquad (-3, 1) \qquad (1, 2) \qquad (2, \infty)$$

Step 4: Select a number in each interval and evaluate $f(x) = \dfrac{(x+3)(2-x)}{(x-1)^2}$ to determine if $f(x)$ is positive or negative. See Table 17.

Table 17

	−3	1	2	→ x
Interval	$(-\infty, -3)$	$(-3, 1)$	$(1, 2)$	$(2, \infty)$
Number Chosen	-4	0	$\frac{3}{2}$	3
Value of f	$f(-4) = -\frac{6}{25}$	$f(0) = 6$	$f\left(\frac{3}{2}\right) = 9$	$f(3) = -\frac{3}{2}$
Conclusion	Negative	Positive	Positive	Negative

Based on Table 17, we know that $f(x) > 0$ for all x in the intervals $(-3, 1)$ or $(1, 2)$, that is, for all x such that $-3 < x < 1$ or $1 < x < 2$. Because the original inequality is strict, the solution set of the given inequality is $\{x \mid -3 < x < 2, x \neq 1\}$ or, using interval notation, $(-3, 1)$ or $(1, 2)$. Figure 57 shows the graph of the solution set. Notice the hole at $x = 1$ to indicate that 1 is to be excluded. ◀

Figure 57

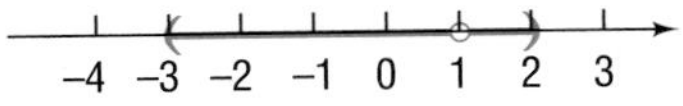

NOW WORK PROBLEM 31.

EXAMPLE 4 Solving a Rational Inequality

Solve the inequality $\dfrac{4x+5}{x+2} \geq 3$, and graph the solution set.

Solution **Step 1:** The domain of the variable x is $\{x \mid x \neq -2\}$. We rearrange terms so that 0 is on the right side.

$$\frac{4x+5}{x+2} - 3 \geq 0 \qquad \text{Subtract 3 from both sides of the inequality.}$$

STEP 2: Let $f(x) = \dfrac{4x + 5}{x + 2} - 3$. To find the zeros of the numerator and the denominator, we must express f as a quotient.

$$\begin{aligned} f(x) &= \frac{4x+5}{x+2} - 3 && \text{Least Common Denominator: } x + 2 \\ &= \frac{4x+5}{x+2} - 3\left(\frac{x+2}{x+2}\right) && \text{Multiply } -3 \text{ by } \frac{x+2}{x+2}. \\ &= \frac{4x + 5 - 3x - 6}{x+2} && \text{Write as a single quotient.} \\ &= \frac{x-1}{x+2} && \text{Combine like terms.} \end{aligned}$$

The zero of the numerator of f is 1, and the zero of the denominator is -2.

STEP 3: We use the zeros found in Step 2 to separate the real number line into three intervals:

$$(-\infty, -2) \qquad (-2, 1) \qquad (1, \infty)$$

STEP 4: Select a number in each interval and evaluate $f(x) = \dfrac{4x + 5}{x + 2} - 3$ to determine if $f(x)$ is positive or negative. See Table 18.

Table 18

	-2	1	x
Interval	$(-\infty, -2)$	$(-2, 1)$	$(1, \infty)$
Number Chosen	-3	0	2
Value of *f*	$f(-3) = 4$	$f(0) = -\frac{1}{2}$	$f(2) = \frac{1}{4}$
Conclusion	Positive	Negative	Positive

Based on Table 18, we know that $f(x) > 0$ for all x in the intervals $(-\infty, -2)$ or $(1, \infty)$, that is, for all x such that $x < -2$ or $x > 1$. Because the original inequality is not strict, numbers x that satisfy the equation $f(x) = \dfrac{x-1}{x+2} = 0$ are also solutions of the inequality. Since $\dfrac{x-1}{x+2} = 0$ only if $x = 1$, we conclude that the solution set is $\{x \mid x < -2 \text{ or } x \geq 1\}$ or, using interval notation, $(-\infty, -2)$ or $[1, \infty)$.

Figure 58

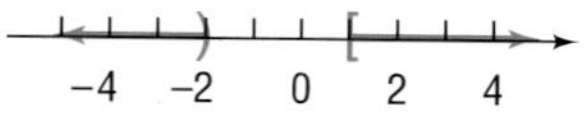

Figure 58 shows the graph of the solution set. ◀

NOW WORK PROBLEM 39.

3.5 Assess Your Understanding

'Are You Prepared?' *Answer is given at the end of these exercises. If you get a wrong answer, read the pages listed in red.*

1. Solve the inequality: $3 - 4x > 5$. Graph the solution set. (pp. 971–974)

Concepts and Vocabulary

2. *True or False:* A test number for the interval $-5 < x < 1$ is 0.

Exercises

In Problems 3–50, solve each inequality.

3. $(x-5)(x+2)<0$ **4.** $(x-5)(x+2)>0$ 5. $x^2-4x\geq 0$ **6.** $x^2+8x\geq 0$

7. $x^2-9<0$ **8.** $x^2-1<0$ 9. $x^2+x\geq 2$ **10.** $x^2+7x\leq -12$

11. $2x^2\leq 5x+3$ **12.** $6x^2\leq 6+5x$ **13.** $x(x-7)>8$ **14.** $x(x+1)>20$

15. $4x^2+9<6x$ **16.** $25x^2+16<40x$ **17.** $6(x^2-1)>5x$ **18.** $2(2x^2-3x)>-9$

19. $(x-1)(x^2+x+4)\geq 0$ **20.** $(x+2)(x^2-x+1)\geq 0$

21. $(x-1)(x-2)(x-3)\leq 0$ **22.** $(x+1)(x+2)(x+3)\leq 0$

23. $x^3-2x^2-3x>0$ **24.** $x^3+2x^2-3x>0$ **25.** $x^4>x^2$ **26.** $x^4<4x^2$

27. $x^3\geq 4x^2$ **28.** $x^3\leq 9x^2$ **29.** $x^4>1$ **30.** $x^3>1$

31. $\dfrac{x+1}{x-1}>0$ **32.** $\dfrac{x-3}{x+1}>0$ **33.** $\dfrac{(x-1)(x+1)}{x}\leq 0$ **34.** $\dfrac{(x-3)(x+2)}{x-1}\leq 0$

35. $\dfrac{(x-2)^2}{x^2-1}\geq 0$ **36.** $\dfrac{(x+5)^2}{x^2-4}\geq 0$ **37.** $6x-5<\dfrac{6}{x}$ **38.** $x+\dfrac{12}{x}<7$

39. $\dfrac{x+4}{x-2}\leq 1$ **40.** $\dfrac{x+2}{x-4}\geq 1$ **41.** $\dfrac{3x-5}{x+2}\leq 2$ **42.** $\dfrac{x-4}{2x+4}\geq 1$

43. $\dfrac{1}{x-2}<\dfrac{2}{3x-9}$ **44.** $\dfrac{5}{x-3}>\dfrac{3}{x+1}$ **45.** $\dfrac{2x+5}{x+1}>\dfrac{x+1}{x-1}$ **46.** $\dfrac{1}{x+2}>\dfrac{3}{x+1}$

47. $\dfrac{x^2(3+x)(x+4)}{(x+5)(x-1)}\geq 0$ **48.** $\dfrac{x(x^2+1)(x-2)}{(x-1)(x+1)}\geq 0$ **49.** $\dfrac{(3-x)^3(2x+1)}{x^3-1}<0$ **50.** $\dfrac{(2-x)^3(3x-2)}{x^3+1}<0$

51. For what positive numbers will the cube of a number exceed four times its square?

52. For what positive numbers will the square of a number exceed twice the number?

53. What is the domain of the function $f(x)=\sqrt{x^2-16}$?

54. What is the domain of the function $f(x)=\sqrt{x^3-3x^2}$?

55. What is the domain of the function $f(x)=\sqrt{\dfrac{x-2}{x+4}}$?

56. What is the domain of the function $f(x)=\sqrt{\dfrac{x-1}{x+4}}$?

57. Physics A ball is thrown vertically upward with an initial velocity of 80 feet per second. The distance s (in feet) of the ball from the ground after t seconds is $s=80t-16t^2$. For what time interval is the ball more than 96 feet above the ground? (See the figure.)

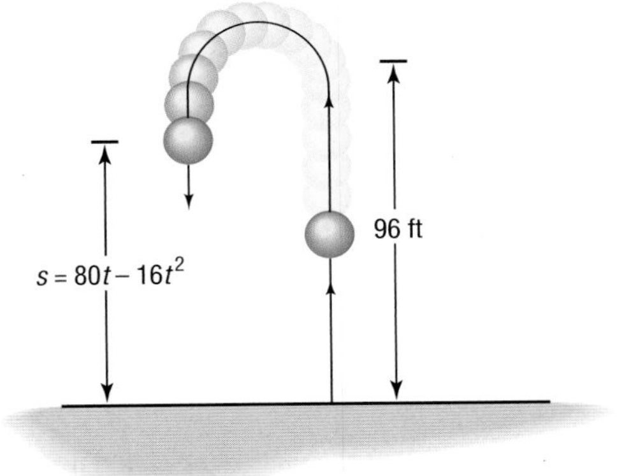

58. Physics A ball is thrown vertically upward with an initial velocity of 96 feet per second. The distance s (in feet) of the ball from the ground after t seconds is $s=96t-16t^2$. For what time interval is the ball more than 112 feet above the ground?

59. Business The monthly revenue achieved by selling x wristwatches is calculated to be $x(40-0.2x)$ dollars. The wholesale cost of each watch is \$32. How many watches must be sold each month to achieve a profit (revenue − cost) of at least \$50?

60. Business The monthly revenue achieved by selling x boxes of candy is calculated to be $x(5-0.05x)$ dollars. The wholesale cost of each box of candy is \$1.50. How many boxes must be sold each month to achieve a profit of at least \$60?

61. Find k such that the equation $x^2+kx+1=0$ has no real solution.

62. Find k such that the equation $kx^2+2x+1=0$ has two distinct real solutions.

63. Make up an inequality that has no solution. Make up one that has exactly one solution.

64. The inequality $x^2+1<-5$ has no solution. Explain why.

'Are You Prepared?' Answer

1. $\{x|x<-\frac{1}{2}\}$

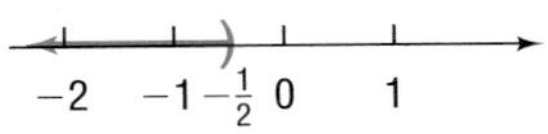

3.6 The Real Zeros of a Polynomial Function

PREPARING FOR THIS SECTION *Before getting started, review the following:*

- Evaluating Functions (Section 2.1, pp. 53–55)
- Factoring Polynomials (Appendix A, Section A.3, pp. 918–923)
- Polynomial Division; Synthetic Division (Appendix A, Section A.4, pp. 929–934)
- Quadratic Formula (Appendix A, Section A.5, pp. 942–945)

Now work the 'Are You Prepared?' problems on page 206.

OBJECTIVES
1. Use the Remainder and Factor Theorems
2. Use Descartes' Rule of Signs to Determine the Number of Positive and the Number of Negative Real Zeros of a Polynomial Function
3. Use the Rational Zeros Theorem to List the Potential Rational Zeros of a Polynomial Function
4. Find the Real Zeros of a Polynomial Function
5. Solve Polynomial Equations
6. Use the Theorem for Bounds on Zeros
7. Use the Intermediate Value Theorem

In this section, we discuss techniques that can be used to find the real zeros of a polynomial function. Recall that if r is a real zero of a polynomial function f then $f(r) = 0$, r is an x-intercept of the graph of f, and r is a solution of the equation $f(x) = 0$. For polynomial and rational functions, we have seen the importance of the zeros for graphing. In most cases, however, the zeros of a polynomial function are difficult to find using algebraic methods. No nice formulas like the quadratic formula are available to help us find zeros for polynomials of degree 3 or higher. Formulas do exist for solving any third- or fourth-degree polynomial equation, but they are somewhat complicated. No general formulas exist for polynomial equations of degree 5 or higher. Refer to the Historical Feature at the end of this section for more information.

Remainder and Factor Theorems

1 When we divide one polynomial (the dividend) by another (the divisor), we obtain a quotient polynomial and a remainder, the remainder being either the zero polynomial or a polynomial whose degree is less than the degree of the divisor. To check our work, we verify that

$$(\text{Quotient})(\text{Divisor}) + \text{Remainder} = \text{Dividend}$$

This checking routine is the basis for a famous theorem called the **division algorithm*** **for polynomials**, which we now state without proof.

*A systematic process in which certain steps are repeated a finite number of times is called an **algorithm.** For example, long division is an algorithm.

Theorem

Division Algorithm for Polynomials

If $f(x)$ and $g(x)$ denote polynomial functions and if $g(x)$ is not the zero polynomial, then there are unique polynomial functions $q(x)$ and $r(x)$ such that

$$\frac{f(x)}{g(x)} = q(x) + \frac{r(x)}{g(x)} \quad \text{or} \quad \underset{\text{dividend}}{f(x)} = \underset{\text{quotient}}{q(x)}\,\underset{\text{divisor}}{g(x)} + \underset{\text{remainder}}{r(x)} \qquad \textbf{(1)}$$

where $r(x)$ is either the zero polynomial or a polynomial of degree less than that of $g(x)$.

In equation (1), $f(x)$ is the **dividend,** $g(x)$ is the **divisor,** $q(x)$ is the **quotient,** and $r(x)$ is the **remainder.**

If the divisor $g(x)$ is a first-degree polynomial of the form

$$g(x) = x - c, \qquad c \text{ a real number}$$

then the remainder $r(x)$ is either the zero polynomial or a polynomial of degree 0. As a result, for such divisors, the remainder is some number, say R, and we may write

$$f(x) = (x - c)q(x) + R \qquad \textbf{(2)}$$

This equation is an identity in x and is true for all real numbers x. Suppose that $x = c$. Then equation (2) becomes

$$f(c) = (c - c)q(c) + R$$
$$f(c) = R$$

Substitute $f(c)$ for R in equation (2) to obtain

$$f(x) = (x - c)q(x) + f(c) \qquad \textbf{(3)}$$

We have now proved the **Remainder Theorem.**

Remainder Theorem

Let f be a polynomial function. If $f(x)$ is divided by $x - c$, then the remainder is $f(c)$.

EXAMPLE 1 **Using the Remainder Theorem**

Find the remainder if $f(x) = x^3 - 4x^2 - 5$ is divided by

(a) $x - 3$ (b) $x + 2$

Solution (a) We could use long division or synthetic division, but it is easier to use the Remainder Theorem, which says that the remainder is $f(3)$.

$$f(3) = (3)^3 - 4(3)^2 - 5 = 27 - 36 - 5 = -14$$

The remainder is -14.

(b) To find the remainder when $f(x)$ is divided by $x + 2 = x - (-2)$, we evaluate $f(-2)$.

$$f(-2) = (-2)^3 - 4(-2)^2 - 5 = -8 - 16 - 5 = -29$$

The remainder is -29. ◀

Compare the method used in Example 1(a) above with the method used in Example 4 of Appendix A, Section A.4. Which method do you prefer? Give reasons.

COMMENT: A graphing utility provides another way to find the value of a function, using the eVALUEate feature. Consult your manual for details. Then check the results of Example 1. ■

An important and useful consequence of the Remainder Theorem is the **Factor Theorem.**

Factor Theorem

Let f be a polynomial function. Then $x - c$ is a factor of $f(x)$ if and only if $f(c) = 0$.

The Factor Theorem actually consists of two separate statements:

1. If $f(c) = 0$, then $x - c$ is a factor of $f(x)$.
2. If $x - c$ is a factor of $f(x)$, then $f(c) = 0$.

The proof requires two parts.

Proof

1. Suppose that $f(c) = 0$. Then, by equation (3), we have
$$f(x) = (x - c)q(x)$$
for some polynomial $q(x)$. That is, $x - c$ is a factor of $f(x)$.
2. Suppose that $x - c$ is a factor of $f(x)$. Then there is a polynomial function q such that
$$f(x) = (x - c)q(x)$$
Replacing x by c, we find that
$$f(c) = (c - c)q(c) = 0 \cdot q(c) = 0$$
This completes the proof. ■

One use of the Factor Theorem is to determine whether a polynomial has a particular factor.

EXAMPLE 2 Using the Factor Theorem

Use the Factor Theorem to determine whether the function
$$f(x) = 2x^3 - x^2 + 2x - 3$$
has the factor: (a) $x - 1$ (b) $x + 3$

Solution The Factor Theorem states that if $f(c) = 0$ then $x - c$ is a factor.

(a) Because $x - 1$ is of the form $x - c$ with $c = 1$, we find the value of $f(1)$. We choose to use substitution.
$$f(1) = 2(1)^3 - (1)^2 + 2(1) - 3 = 2 - 1 + 2 - 3 = 0$$
By the Factor Theorem, $x - 1$ is a factor of $f(x)$.

(b) To test the factor $x + 3$, we first need to write it in the form $x - c$. Since $x + 3 = x - (-3)$, we find the value of $f(-3)$. We choose to use synthetic division.

$$\begin{array}{r|rrrr} -3 & 2 & -1 & 2 & -3 \\ & & -6 & 21 & -69 \\ \hline & 2 & -7 & 23 & -72 \end{array}$$

Because $f(-3) = -72 \neq 0$, we conclude from the Factor Theorem that $x - (-3) = x + 3$ is not a factor of $f(x)$. ◀

NOW WORK PROBLEM 11.

In Example 2(a), we found that $x - 1$ was a factor of f. To write f in factored form, we use long division or synthetic division. Using synthetic division, we find that

$$\begin{array}{r|rrrr} 1 & 2 & -1 & 2 & -3 \\ & & 2 & 1 & 3 \\ \hline & 2 & 1 & 3 & 0 \end{array}$$

The quotient is $q(x) = 2x^2 + x + 3$ with a remainder of 0, as expected. We can write f in factored form as

$$f(x) = 2x^3 - x^2 + 2x - 3 = (x - 1)(2x^2 + x + 3)$$

The Number and Location of Real Zeros

The next theorem concerns the number of real zeros that a polynomial function may have. In counting the zeros of a polynomial, we count each zero as many times as its multiplicity.

Theorem

Number of Real Zeros

A polynomial function cannot have more real zeros than its degree.

Proof The proof is based on the Factor Theorem. If r is a zero of a polynomial function f, then $f(r) = 0$ and, hence, $x - r$ is a factor of $f(x)$. Each zero corresponds to a factor of degree 1. Because f cannot have more first-degree factors than its degree, the result follows. ■

2 **Descartes' Rule of Signs** provides information about the number and location of the real zeros of a polynomial function written in standard form (descending powers of x). It requires that we count the number of variations in the sign of the coefficients of $f(x)$ and $f(-x)$.

For example, the following polynomial function has two variations in the signs of coefficients.

$$\begin{aligned} f(x) &= -3x^7 + 4x^4 + 3x^2 - 2x - 1 \\ &= \underbrace{-3x^7 + 0x^6 + 0x^5 + 4x^4}_{-\text{ to }+} + 0x^3 + \underbrace{3x^2 - 2x}_{+\text{ to }-} - 1 \end{aligned}$$

Notice that we ignored the zero coefficients in $0x^6$, $0x^5$ and $0x^3$ in counting the number of variations in the sign of $f(x)$. Replacing x by $-x$, we get

$$\begin{aligned} f(-x) &= -3(-x)^7 + 4(-x)^4 + 3(-x)^2 - 2(-x) - 1 \\ &= 3x^7 + 4x^4 + 3x^2 + 2x - 1 \end{aligned}$$

+ to −

which has one variation in sign.

Theorem

Descartes' Rule of Signs

Let f denote a polynomial function written in standard form.

The number of positive real zeros of f either equals the number of variations in the sign of the nonzero coefficients of $f(x)$ or else equals that number less an even integer.

The number of negative real zeros of f either equals the number of variations in the sign of the nonzero coefficients of $f(-x)$ or else equals that number less an even integer.

We shall not prove Descartes' Rule of Signs. Let's see how it is used.

EXAMPLE 3 Using the Number of Real Zeros Theorem and Descartes' Rule of Signs

Discuss the real zeros of $f(x) = 3x^6 - 4x^4 + 3x^3 + 2x^2 - x - 3$.

Solution Because the polynomial is of degree 6, by the Number of Real Zeros Theorem there are at most six real zeros. Since there are three variations in the sign of the nonzero coefficients of $f(x)$, by Descartes' Rule of Signs we expect either three or one positive real zero. To continue, we look at $f(-x)$.

$$f(-x) = 3x^6 - 4x^4 - 3x^3 + 2x^2 + x - 3$$

There are three variations in sign, so we expect either three (or one) negative real zeros. Equivalently, we now know that the graph of f has either three or one positive x-intercept and three or one negative x-intercept. ◀

NOW WORK PROBLEM 21.

Rational Zeros Theorem

3 The next result, called the **Rational Zeros Theorem,** provides information about the rational zeros of a polynomial *with integer coefficients.*

Theorem

Rational Zeros Theorem

Let f be a polynomial function of degree 1 or higher of the form

$$f(x) = a_n x^n + a_{n-1} x^{n-1} + \cdots + a_1 x + a_0, \qquad a_n \neq 0, \quad a_0 \neq 0$$

where each coefficient is an integer. If $\frac{p}{q}$, in lowest terms, is a rational zero of f, then p must be a factor of a_0, and q must be a factor of a_n.

EXAMPLE 4 **Listing Potential Rational Zeros**

List the potential rational zeros of

$$f(x) = 2x^3 + 11x^2 - 7x - 6$$

Solution Because f has integer coefficients, we may use the Rational Zeros Theorem. First, we list all the integers p that are factors of the constant term $a_0 = -6$ and all the integers q that are factors of the leading coefficient $a_3 = 2$.

$$p: \quad \pm 1, \pm 2, \pm 3, \pm 6 \qquad \text{Factors of } -6$$
$$q: \quad \pm 1, \pm 2 \qquad \text{Factors of } 2$$

Now we form all possible ratios $\frac{p}{q}$.

$$\frac{p}{q}: \quad \pm 1, \pm 2, \pm 3, \pm 6, \pm\frac{1}{2}, \pm\frac{3}{2}$$

If f has a rational zero, it will be found in this list, which contains 12 possibilities. ◀

NOW WORK PROBLEM 33.

Be sure that you understand what the Rational Zeros Theorem says: For a polynomial with integer coefficients, *if* there is a rational zero, it is one of those listed. It may be the case that the function does not have any rational zeros.

4 Long division, synthetic division, or substitution can be used to test each potential rational zero to determine whether it is indeed a zero. To make the work easier, the integers are usually tested first. Let's continue this example.

EXAMPLE 5 **Finding the Rational Zeros of a Polynomial Function**

Continue working with Example 4 to find the rational zeros of

$$f(x) = 2x^3 + 11x^2 - 7x - 6$$

Write f in factored form.

Solution We gather all the information that we can about the zeros.

STEP 1: There are at most three real zeros.

STEP 2: By Descartes' Rule of Signs, there is one positive real zero. Also, because

$$f(-x) = -2x^3 + 11x^2 + 7x - 6$$

there are two negative zeros or no negative zeros.

STEP 3: Now we use the list of potential rational zeros obtained in Example 4: $\pm 1, \pm 2, \pm 3, \pm 6, \pm\frac{1}{2}, \pm\frac{3}{2}$. We choose to test the potential rational zero 1 using substitution.

$$f(1) = 2(1)^3 + 11(1)^2 - 7(1) - 6 = 2 + 11 - 7 - 6 = 0$$

Since $f(1) = 0$, 1 is a zero and $x - 1$ is a factor of f. We can use long division or synthetic division to factor f.

$$\begin{aligned} f(x) &= 2x^3 + 11x^2 - 7x - 6 \\ &= (x - 1)(2x^2 + 13x + 6) \end{aligned}$$

Now any solution of the equation $2x^2 + 13x + 6 = 0$ will be a zero of f. Because of this, we call the equation $2x^2 + 13x + 6 = 0$ a **depressed equation** of f. Since the degree of the depressed equation of f is less than that of the original polynomial, we work with the depressed equation to find the zeros of f.

Step 4: The depressed equation $2x^2 + 13x + 6 = 0$ is a quadratic equation with discriminant $b^2 - 4ac = 169 - 48 = 121 > 0$. The equation has two real solutions, which can be found by factoring.

$$2x^2 + 13x + 6 = (2x + 1)(x + 6) = 0$$

$$2x + 1 = 0 \quad \text{or} \quad x + 6 = 0$$

$$x = -\frac{1}{2} \qquad\qquad x = -6$$

The zeros of f are -6, $-\frac{1}{2}$, and 1.

We use the Factor Theorem to factor f. Each zero gives rise to a factor, so $x - (-6) = x + 6$, $x - \left(-\frac{1}{2}\right) = x + \frac{1}{2}$, and $x - 1$ are factors of f. Since the leading coefficient of f is 2 and f is of degree 3, we have

$$f(x) = 2x^3 + 11x^2 - 7x - 6 = 2(x + 6)\left(x + \frac{1}{2}\right)(x - 1)$$

Notice that the three zeros of f found in this example are among those given in the list of potential rational zeros in Example 4. ◀

To obtain information about the real zeros of a polynomial function, follow these steps:

Steps for Finding the Real Zeros of a Polynomial Function

Step 1: Use the degree of the polynomial to determine the maximum number of zeros.

Step 2: Use Descartes' Rule of Signs to determine the possible number of positive zeros and negative zeros.

Step 3: (a) If the polynomial has integer coefficients, use the Rational Zeros Theorem to identify those rational numbers that potentially can be zeros.

(b) Use substitution, synthetic division, or long division to test each potential rational zero.

(c) Each time that a zero (and thus a factor) is found, repeat Step 3 on the depressed equation.

Step 4: In attempting to find the zeros, remember to use (if possible) the factoring techniques that you already know (special products, factoring by grouping, and so on).

EXAMPLE 6 Finding the Real Zeros of a Polynomial Function

Find the real zeros of $f(x) = x^5 - 5x^4 + 12x^3 - 24x^2 + 32x - 16$. Write f in factored form.

Solution We gather all the information that we can about the zeros.

STEP 1: There are at most five real zeros.

STEP 2: By Descartes' Rule of Signs, there are five, three, or one positive zero. Because

$$f(-x) = -x^5 - 5x^4 - 12x^3 - 24x^2 - 32x - 16$$

there are no negative zeros.

STEP 3: Because the leading coefficient $a_5 = 1$ and there are no negative zeros, the potential rational zeros are the integers 1, 2, 4, 8, and 16, the positive factors of the constant term, 16. We test the potential rational zero 1 first, using synthetic division.

$$\begin{array}{r|rrrrrr} 1) & 1 & -5 & 12 & -24 & 32 & -16 \\ & & 1 & -4 & 8 & -16 & 16 \\ \hline & 1 & -4 & 8 & -16 & 16 & 0 \end{array}$$

The remainder is $f(1) = 0$, so 1 is a zero and $x - 1$ is a factor of f. Using the entries in the bottom row of the synthetic division, we can begin to factor f.

$$\begin{aligned} f(x) &= x^5 - 5x^4 + 12x^3 - 24x^2 + 32x - 16 \\ &= (x - 1)(x^4 - 4x^3 + 8x^2 - 16x + 16) \end{aligned}$$

We now work with the first depressed equation:

$$q_1(x) = x^4 - 4x^3 + 8x^2 - 16x + 16 = 0$$

REPEAT STEP 3: The potential rational zeros of q_1 are still 1, 2, 4, 8, and 16. We test 1 first, since it may be a repeated zero.

$$\begin{array}{r|rrrrr} 1) & 1 & -4 & 8 & -16 & 16 \\ & & 1 & -3 & 5 & -11 \\ \hline & 1 & -3 & 5 & -11 & 5 \end{array}$$

Since the remainder is 5, 1 is not a repeated zero. We try 2 next.

$$\begin{array}{r|rrrrr} 2) & 1 & -4 & 8 & -16 & 16 \\ & & 2 & -4 & 8 & -16 \\ \hline & 1 & -2 & 4 & -8 & 0 \end{array}$$

The remainder is $f(2) = 0$, so 2 is a zero and $x - 2$ is a factor of f. Again using the bottom row, we find

$$\begin{aligned} f(x) &= x^5 - 5x^4 + 12x^3 - 24x^2 + 32x - 16 \\ &= (x - 1)(x - 2)(x^3 - 2x^2 + 4x - 8) \end{aligned}$$

The remaining zeros satisfy the new depressed equation

$$q_2(x) = x^3 - 2x^2 + 4x - 8 = 0$$

Notice that $q_2(x)$ can be factored using grouping. (Alternatively, you could repeat Step 3 and check the potential rational zero 2.) Then

$$\begin{aligned} x^3 - 2x^2 + 4x - 8 &= 0 \\ x^2(x-2) + 4(x-2) &= 0 \\ (x^2+4)(x-2) &= 0 \\ x^2 + 4 = 0 \quad \text{or} \quad x - 2 &= 0 \\ x &= 2 \end{aligned}$$

Since $x^2 + 4 = 0$ has no real solutions, the real zeros of f are 1 and 2, the latter being a zero of multiplicity 2. The factored form of f is

$$\begin{aligned} f(x) &= x^5 - 5x^4 + 12x^3 - 24x^2 + 32x - 16 \\ &= (x-1)(x-2)^2(x^2+4) \end{aligned}$$ ◀

NOW WORK PROBLEM 45.

5 **EXAMPLE 7** **Solving a Polynomial Equation**

Solve the equation: $x^5 - 5x^4 + 12x^3 - 24x^2 + 32x - 16 = 0$

Solution The solutions of this equation are the zeros of the polynomial function

$$f(x) = x^5 - 5x^4 + 12x^3 - 24x^2 + 32x - 16$$

Using the result of Example 6, the real zeros of f are 1 and 2. These are the real solutions of the equation

$$x^5 - 5x^4 + 12x^3 - 24x^2 + 32x - 16 = 0$$ ◀

NOW WORK PROBLEM 57.

In Example 6, the quadratic factor $x^2 + 4$ that appears in the factored form of f is called *irreducible*, because the polynomial $x^2 + 4$ cannot be factored over the real numbers. In general, we say that a quadratic factor $ax^2 + bx + c$ is **irreducible** if it cannot be factored over the real numbers, that is, if it is prime over the real numbers.

Refer to Examples 5 and 6. The polynomial function of Example 5 has three real zeros, and its factored form contains three linear factors. The polynomial function of Example 6 has two distinct real zeros, and its factored form contains two distinct linear factors and one irreducible quadratic factor.

Theorem

Every polynomial function (with real coefficients) can be uniquely factored into a product of linear factors and/or irreducible quadratic factors.

We shall prove this result in Section 3.7, and, in fact, we shall draw several additional conclusions about the zeros of a polynomial function. One conclusion is worth noting now. If a polynomial (with real coefficients) is of odd degree, then it must contain at least one linear factor. (Do you see why?) This means that it must have at least one real zero.

Corollary

A polynomial function (with real coefficients) of odd degree has at least one real zero.

Bounds on Zeros

6 The search for the real zeros of a polynomial function can be reduced somewhat if *bounds* on the zeros are found. A number M is a **bound** on the zeros of a polynomial if every zero lies between $-M$ and M, inclusive. That is, M is a bound to the zeros of a polynomial f if

$$-M \leq \text{any zero of } f \leq M$$

Theorem

Bounds on Zeros

Let f denote a polynomial function whose leading coefficient is 1.

$$f(x) = x^n + a_{n-1}x^{n-1} + \cdots + a_1 x + a_0$$

A bound M on the zeros of f is the smaller of the two numbers

$$\text{Max}\{1, |a_0| + |a_1| + \cdots + |a_{n-1}|\},\ 1 + \text{Max}\{|a_0|, |a_1|, \ldots, |a_{n-1}|\} \quad \textbf{(4)}$$

where Max{ } means "choose the largest entry in { }."

An example will help to make the theorem clear.

EXAMPLE 8 Using the Theorem for Finding Bounds on Zeros

Find a bound to the zeros of each polynomial.

(a) $f(x) = x^5 + 3x^3 - 9x^2 + 5$ (b) $g(x) = 4x^5 - 2x^3 + 2x^2 + 1$

Solution (a) The leading coefficient of f is 1.

$$f(x) = x^5 + 3x^3 - 9x^2 + 5 \qquad a_4 = 0, a_3 = 3, a_2 = -9, a_1 = 0, a_0 = 5$$

We evaluate the two expressions in (4).

$$\begin{aligned} \text{Max}\{1, |a_0| + |a_1| + \cdots + |a_{n-1}|\} &= \text{Max}\{1, |5| + |0| + |-9| + |3| + |0|\} \\ &= \text{Max}\{1, 17\} = 17 \\ 1 + \text{Max}\{|a_0|, |a_1|, \ldots, |a_{n-1}|\} &= 1 + \text{Max}\{|5|, |0|, |-9|, |3|, |0|\} \\ &= 1 + 9 = 10 \end{aligned}$$

The smaller of the two numbers, 10, is the bound. Every zero of f lies between -10 and 10.

(b) First we write g so that it is the product of a constant times a polynomial whose leading coefficient is 1.

$$g(x) = 4x^5 - 2x^3 + 2x^2 + 1 = 4\left(x^5 - \frac{1}{2}x^3 + \frac{1}{2}x^2 + \frac{1}{4}\right)$$

Next we evaluate the two expressions in (4) with $a_4 = 0$, $a_3 = -\frac{1}{2}$, $a_2 = \frac{1}{2}$, $a_1 = 0$, and $a_0 = \frac{1}{4}$.

$$\text{Max}\{1, |a_0| + |a_1| + \cdots + |a_{n-1}|\} = \text{Max}\left\{1, \left|\frac{1}{4}\right| + |0| + \left|\frac{1}{2}\right| + \left|-\frac{1}{2}\right| + |0|\right\}$$
$$= \text{Max}\left\{1, \frac{5}{4}\right\} = \frac{5}{4}$$
$$1 + \text{Max}\{|a_0|, |a_1|, \ldots, |a_{n-1}|\} = 1 + \text{Max}\left\{\left|\frac{1}{4}\right|, |0|, \left|\frac{1}{2}\right|, \left|-\frac{1}{2}\right|, |0|\right\}$$
$$= 1 + \frac{1}{2} = \frac{3}{2}$$

The smaller of the two numbers, $\frac{5}{4}$, is the bound. Every zero of g lies between $-\frac{5}{4}$ and $\frac{5}{4}$. ◀

COMMENT: The bounds on the zeros of a polynomial provide good choices for setting Xmin and Xmax of the viewing rectangle. With these choices, all the x-intercepts of the graph can be seen. ■

NOW WORK PROBLEM 81.

Intermediate Value Theorem

7 The next result, called the **Intermediate Value Theorem,** is based on the fact that the graph of a polynomial function is continuous; that is, it contains no "holes" or "gaps."

Theorem

Intermediate Value Theorem

Let f denote a polynomial function. If $a < b$ and if $f(a)$ and $f(b)$ are of opposite sign, then there is at least one zero of f between a and b.

Although the proof of this result requires advanced methods in calculus, it is easy to "see" why the result is true. Look at Figure 59.

Figure 59
If $f(a) < 0$ and $f(b) > 0$, there is a zero between a and b.

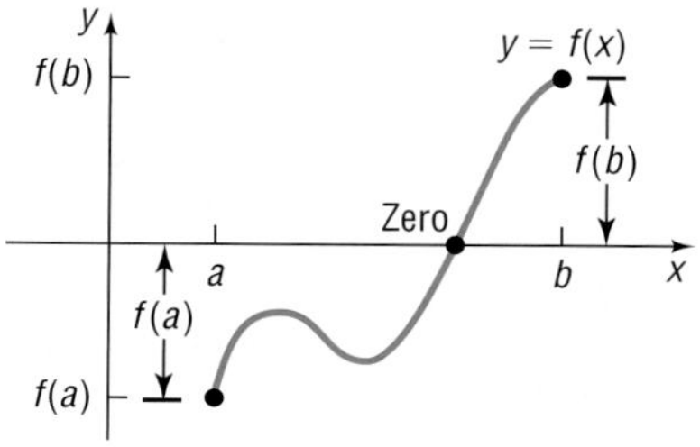

EXAMPLE 9 Using the Intermediate Value Theorem to Locate Zeros

Show that $f(x) = x^5 - x^3 - 1$ has a zero between 1 and 2.

Solution We evaluate f at 1 and at 2.

$$f(1) = -1 \quad \text{and} \quad f(2) = 23$$

Because $f(1) < 0$ and $f(2) > 0$, it follows from the Intermediate Value Theorem that f has a zero between 1 and 2. ◀

NOW WORK PROBLEM 89.

Let's look at the polynomial f of Example 9 more closely. Based on Descartes' Rule of Signs, f has exactly one positive real zero. Based on the Rational Zeros Theorem, 1 is the only potential positive rational zero. Since $f(1) \neq 0$, we conclude that the zero between 1 and 2 is irrational. We can use the Intermediate Value Theorem to approximate it. The steps to use follow:

Approximating the Zeros of a Polynomial Function

STEP 1: Find two consecutive integers a and $a + 1$ such that f has a zero between them.

STEP 2: Divide the interval $[a, a + 1]$ into 10 equal subintervals.

STEP 3: Evaluate f at each endpoint of the subintervals until the Intermediate Value Theorem applies; this interval then contains a zero.

STEP 4: Repeat the process starting at Step 2 until the desired accuracy is achieved.

EXAMPLE 10 Approximating the Zeros of a Polynomial Function

Find the positive zero of $f(x) = x^5 - x^3 - 1$ correct to two decimal places.

Solution From Example 9 we know that the positive zero is between 1 and 2. We divide the interval $[1, 2]$ into 10 equal subintervals: $[1, 1.1]$, $[1.1, 1.2]$, $[1.2, 1.3]$, $[1.3, 1.4]$, $[1.4, 1.5]$, $[1.5, 1.6]$, $[1.6, 1.7]$, $[1.7, 1.8]$, $[1.8, 1.9]$, $[1.9, 2]$. Now we find the value of f at each endpoint until the Intermediate Value Theorem applies.

$$f(x) = x^5 - x^3 - 1$$
$$f(1.0) = -1 \qquad f(1.2) = -0.23968$$
$$f(1.1) = -0.72049 \qquad f(1.3) = 0.51593$$

We can stop here and conclude that the zero is between 1.2 and 1.3. Now we divide the interval $[1.2, 1.3]$ into 10 equal subintervals and proceed to evaluate f at each endpoint.

$$f(1.20) = -0.23968 \qquad f(1.23) \approx -0.0455613$$
$$f(1.21) \approx -0.1778185 \qquad f(1.24) \approx 0.025001$$
$$f(1.22) \approx -0.1131398$$

We conclude that the zero lies between 1.23 and 1.24, and so, correct to two decimal places, the zero is 1.23. ◀

Figure 60

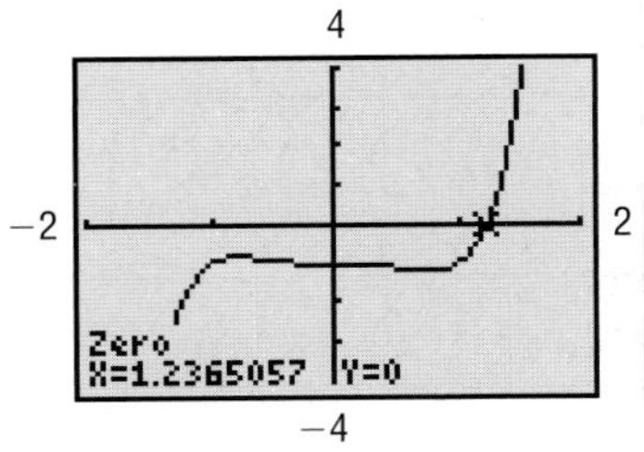

Exploration

We examine the polynomial f given in Example 10. The Theorem on Bounds of Zeros tells us that every zero is between -2 and 2. If we graph f using $-2 \leq x \leq 2$, we see that f has exactly one x-intercept. See Figure 60. Using ZERO or ROOT, we find this zero to be 1.24 rounded to two decimal places. Correct to two decimal places, the zero is 1.23.

There are many other numerical techniques for approximating the zeros of a polynomial. The one outlined in Example 10 (a variation of the *bisection method*) has the advantages that it will always work, that it can be programmed rather easily on a computer, and each time it is used another decimal place of accuracy is achieved. See Problem 119 for the bisection method, which places the zero in a succession of intervals, with each new interval being half the length of the preceding one.

Historical Feature

Formulas for the solution of third- and fourth-degree polynomial equations exist, and, while not very practical, they do have an interesting history.

In the 1500s in Italy, mathematical contests were a popular pastime, and persons possessing methods for solving problems kept them secret. (Solutions that were published were already common knowledge.) Niccolo of Brescia (1499–1557), commonly referred to as Tartaglia ("the stammerer"), had the secret for solving cubic (third-degree) equations, which gave him a decided advantage in the contests. Girolamo Cardano (1501–1576) found out that Tartaglia had the secret, and, being interested in cubics, he requested it from Tartaglia. The reluctant Tartaglia hesitated for some time, but finally, swearing Cardano to secrecy with midnight oaths by candlelight, told him the secret. Cardano then published the solution in his book *Ars Magna* (1545), giving Tartaglia the credit but rather compromising the secrecy. Tartaglia exploded into bitter recriminations, and each wrote pamphlets that reflected on the other's mathematics, moral character, and ancestry.

The quartic (fourth-degree) equation was solved by Cardano's student Lodovico Ferrari, and this solution also was included, with credit and this time with permission, in the *Ars Magna*.

Attempts were made to solve the fifth-degree equation in similar ways, all of which failed. In the early 1800s, P. Ruffini, Niels Abel, and Evariste Galois all found ways to show that it is not possible to solve fifth-degree equations by formula, but the proofs required the introduction of new methods. Galois's methods eventually developed into a large part of modern algebra.

Historical Problems

Problems 1–8 develop the Tartaglia–Cardano solution of the cubic equation and show why it is not altogether practical.

1. Show that the general cubic equation $y^3 + by^2 + cy + d = 0$ can be transformed into an equation of the form $x^3 + px + q = 0$ by using the substitution $y = x - \frac{b}{3}$.
2. In the equation $x^3 + px + q = 0$, replace x by $H + K$. Let $3HK = -p$, and show that $H^3 + K^3 = -q$.
3. Based on Problem 2, we have the two equations

$$3HK = -p \quad \text{and} \quad H^3 + K^3 = -q$$

Solve for K in $3HK = -p$ and substitute into $H^3 + K^3 = -q$. Then show that

$$H = \sqrt[3]{\frac{-q}{2} + \sqrt{\frac{q^2}{4} + \frac{p^3}{27}}}$$

[**Hint:** Look for an equation that is quadratic in form.]

4. Use the solution for H from Problem 3 and the equation $H^3 + K^3 = -q$ to show that

$$K = \sqrt[3]{\frac{-q}{2} - \sqrt{\frac{q^2}{4} + \frac{p^3}{27}}}$$

5. Use the results from Problems 2–4 to show that the solution of $x^3 + px + q = 0$ is

$$x = \sqrt[3]{\frac{-q}{2} + \sqrt{\frac{q^2}{4} + \frac{p^3}{27}}} + \sqrt[3]{\frac{-q}{2} - \sqrt{\frac{q^2}{4} + \frac{p^3}{27}}}$$

6. Use the result of Problem 5 to solve the equation $x^3 - 6x - 9 = 0$.
7. Use a calculator and the result of Problem 5 to solve the equation $x^3 + 3x - 14 = 0$.
8. Use the methods of this chapter to solve the equation $x^3 + 3x - 14 = 0$.

3.6 Assess Your Understanding

'Are You Prepared?' *Answers are given at the end of these exercises. If you get a wrong answer, read the pages listed in red.*

1. Find $f(-1)$ if $f(x) = 2x^2 - x$. (pp. 53–55)
2. Factor the expression $6x^2 + x - 2$. (pp. 918–923)
3. Find the quotient and remainder if $3x^4 - 5x^3 + 7x - 4$ is divided by $x - 3$. (pp. 929–934)
4. Solve the equation $x^2 + x - 3 = 0$. (pp. 942–945)

Concepts and Vocabulary

5. In the process of polynomial division, (Divisor)(Quotient) + __________ = __________.

6. When a polynomial function f is divided by $x - c$, the remainder is __________.

7. If a function f, whose domain is all real numbers, is even and if 4 is a zero of f, then __________ is also a zero.

8. *True or False:* Every polynomial function of degree 3 with real coefficients has exactly three real zeros.

9. *True or False:* The only potential rational zeros of $f(x) = 2x^5 - x^3 + x^2 - x + 1$ are $\pm 1, \pm 2$.

10. *True or False:* If f is a polynomial function of degree 4 and if $f(2) = 5$, then

$$\frac{f(x)}{x-2} = p(x) + \frac{5}{x-2}$$

where $p(x)$ is a polynomial of degree 3.

Exercises

In Problems 11–20, use the Factor Theorem to determine whether $x - c$ is a factor of $f(x)$.

11. $f(x) = 4x^3 - 3x^2 - 8x + 4;\quad x - 2$
12. $f(x) = -4x^3 + 5x^2 + 8;\quad x + 3$
13. $f(x) = 3x^4 - 6x^3 - 5x + 10;\quad x - 2$
14. $f(x) = 4x^4 - 15x^2 - 4;\quad x - 2$
15. $f(x) = 3x^6 + 82x^3 + 27;\quad x + 3$
16. $f(x) = 2x^6 - 18x^4 + x^2 - 9;\quad x + 3$
17. $f(x) = 4x^6 - 64x^4 + x^2 - 15;\quad x + 4$
18. $f(x) = x^6 - 16x^4 + x^2 - 16;\quad x + 4$
19. $f(x) = 2x^4 - x^3 + 2x - 1;\quad x - \frac{1}{2}$
20. $f(x) = 3x^4 + x^3 - 3x + 1;\quad x + \frac{1}{3}$

In Problems 21–32, tell the maximum number of zeros that each polynomial function may have. Then use Descartes' Rule of Signs to determine how many positive and how many negative zeros each polynomial function may have. Do not attempt to find the zeros.

21. $f(x) = -4x^7 + x^3 - x^2 + 2$
22. $f(x) = 5x^4 + 2x^2 - 6x - 5$
23. $f(x) = 2x^6 - 3x^2 - x + 1$
24. $f(x) = -3x^5 + 4x^4 + 2$
25. $f(x) = 3x^3 - 2x^2 + x + 2$
26. $f(x) = -x^3 - x^2 + x + 1$
27. $f(x) = -x^4 + x^2 - 1$
28. $f(x) = x^4 + 5x^3 - 2$
29. $f(x) = x^5 + x^4 + x^2 + x + 1$
30. $f(x) = x^5 - x^4 + x^3 - x^2 + x - 1$
31. $f(x) = x^6 - 1$
32. $f(x) = x^6 + 1$

In Problems 33–44, list the potential rational zeros of each polynomial function. Do not attempt to find the zeros.

33. $f(x) = 3x^4 - 3x^3 + x^2 - x + 1$
34. $f(x) = x^5 - x^4 + 2x^2 + 3$
35. $f(x) = x^5 - 6x^2 + 9x - 3$
36. $f(x) = 2x^5 - x^4 - x^2 + 1$
37. $f(x) = -4x^3 - x^2 + x + 2$
38. $f(x) = 6x^4 - x^2 + 2$
39. $f(x) = 6x^4 - x^2 + 9$
40. $f(x) = -4x^3 + x^2 + x + 6$
41. $f(x) = 2x^5 - x^3 + 2x^2 + 12$
42. $f(x) = 3x^5 - x^2 + 2x + 18$
43. $f(x) = 6x^4 + 2x^3 - x^2 + 20$
44. $f(x) = -6x^3 - x^2 + x + 10$

In Problems 45–56, use Descartes' Rule of Signs and the Rational Zeros Theorem to find all the real zeros of each polynomial function. Use the zeros to factor f over the real numbers.

45. $f(x) = x^3 + 2x^2 - 5x - 6$
46. $f(x) = x^3 + 8x^2 + 11x - 20$
47. $f(x) = 2x^3 - x^2 + 2x - 1$
48. $f(x) = 2x^3 + x^2 + 2x + 1$
49. $f(x) = x^4 + x^2 - 2$
50. $f(x) = x^4 - 3x^2 - 4$
51. $f(x) = 4x^4 + 7x^2 - 2$
52. $f(x) = 4x^4 + 15x^2 - 4$
53. $f(x) = x^4 + x^3 - 3x^2 - x + 2$
54. $f(x) = x^4 - x^3 - 6x^2 + 4x + 8$
55. $f(x) = 4x^5 - 8x^4 - x + 2$
56. $f(x) = 4x^5 + 12x^4 - x - 3$

In Problems 57–68, solve each equation in the real number system.

57. $x^4 - x^3 + 2x^2 - 4x - 8 = 0$
58. $2x^3 + 3x^2 + 2x + 3 = 0$
59. $3x^3 + 4x^2 - 7x + 2 = 0$
60. $2x^3 - 3x^2 - 3x - 5 = 0$
61. $3x^3 - x^2 - 15x + 5 = 0$
62. $2x^3 - 11x^2 + 10x + 8 = 0$
63. $x^4 + 4x^3 + 2x^2 - x + 6 = 0$
64. $x^4 - 2x^3 + 10x^2 - 18x + 9 = 0$
65. $x^3 - \frac{2}{3}x^2 + \frac{8}{3}x + 1 = 0$
66. $x^3 + \frac{3}{2}x^2 + 3x - 2 = 0$
67. $2x^4 - 19x^3 + 57x^2 - 64x + 20 = 0$
68. $2x^4 + x^3 - 24x^2 + 20x + 16 = 0$

In Problems 69–80, find the intercepts of each polynomial function $f(x)$. Find the intervals x for which the graph of f is above the x-axis and below the x-axis. Obtain several other points on the graph, and connect them with a smooth curve. [**Hint:** Use the factored form of f (see Problems 45–56).]

69. $f(x) = x^3 + 2x^2 - 5x - 6$
70. $f(x) = x^3 + 8x^2 + 11x - 20$
71. $f(x) = 2x^3 - x^2 + 2x - 1$
72. $f(x) = 2x^3 + x^2 + 2x + 1$
73. $f(x) = x^4 + x^2 - 2$
74. $f(x) = x^4 - 3x^2 - 4$
75. $f(x) = 4x^4 + 7x^2 - 2$
76. $f(x) = 4x^4 + 15x^2 - 4$
77. $f(x) = x^4 + x^3 - 3x^2 - x + 2$
78. $f(x) = x^4 - x^3 - 6x^2 + 4x + 8$
79. $f(x) = 4x^5 - 8x^4 - x + 2$
80. $f(x) = 4x^5 + 12x^4 - x - 3$

In Problems 81–88, find a bound on the real zeros of each polynomial function.

81. $f(x) = x^4 - 3x^2 - 4$
82. $f(x) = x^4 - 5x^2 - 36$
83. $f(x) = x^4 + x^3 - x - 1$
84. $f(x) = x^4 - x^3 + x - 1$
85. $f(x) = 3x^4 + 3x^3 - x^2 - 12x - 12$
86. $f(x) = 3x^4 - 3x^3 - 5x^2 + 27x - 36$
87. $f(x) = 4x^5 - x^4 + 2x^3 - 2x^2 + x - 1$
88. $f(x) = 4x^5 + x^4 + x^3 + x^2 - 2x - 2$

In Problems 89–94, use the Intermediate Value Theorem to show that each polynomial function has a zero in the given interval.

89. $f(x) = 8x^4 - 2x^2 + 5x - 1;\quad [0, 1]$
90. $f(x) = x^4 + 8x^3 - x^2 + 2;\quad [-1, 0]$
91. $f(x) = 2x^3 + 6x^2 - 8x + 2;\quad [-5, -4]$
92. $f(x) = 3x^3 - 10x + 9;\quad [-3, -2]$
93. $f(x) = x^5 - x^4 + 7x^3 - 7x^2 - 18x + 18;\quad [1.4, 1.5]$
94. $f(x) = x^5 - 3x^4 - 2x^3 + 6x^2 + x + 2;\quad [1.7, 1.8]$

In Problems 95–98, each equation has a solution r in the interval indicated. Use the method of Example 10 to approximate this solution correct to two decimal places.

95. $8x^4 - 2x^2 + 5x - 1 = 0;\quad 0 \le r \le 1$
96. $x^4 + 8x^3 - x^2 + 2 = 0;\quad -1 \le r \le 0$
97. $2x^3 + 6x^2 - 8x + 2 = 0;\quad -5 \le r \le -4$
98. $3x^3 - 10x + 9 = 0;\quad -3 \le r \le -2$

In Problems 99–102, each polynomial function has exactly one positive zero. Use the method of Example 10 to approximate the zero correct to two decimal places.

99. $f(x) = x^3 + x^2 + x - 4$
100. $f(x) = 2x^4 + x^2 - 1$
101. $f(x) = 2x^4 - 3x^3 - 4x^2 - 8$
102. $f(x) = 3x^3 - 2x^2 - 20$

---

103. Find k such that $f(x) = x^3 - kx^2 + kx + 2$ has the factor $x - 2$.

104. Find k such that $f(x) = x^4 - kx^3 + kx^2 + 1$ has the factor $x + 2$.

105. What is the remainder when $f(x) = 2x^{20} - 8x^{10} + x - 2$ is divided by $x - 1$?

106. What is the remainder when $f(x) = -3x^{17} + x^9 - x^5 + 2x$ is divided by $x + 1$?

107. Use the Factor Theorem to prove that $x - c$ is a factor of $x^n - c^n$ for any positive integer n.

108. Use the Factor Theorem to prove that $x + c$ is a factor of $x^n + c^n$ if $n \ge 1$ is an odd integer.

109. One solution of the equation $x^3 - 8x^2 + 16x - 3 = 0$ is 3. Find the sum of the remaining solutions.

110. One solution of the equation $x^3 + 5x^2 + 5x - 2 = 0$ is -2. Find the sum of the remaining solutions.

111. Is $\frac{1}{3}$ a zero of $f(x) = 2x^3 + 3x^2 - 6x + 7$? Explain.

112. Is $\frac{1}{3}$ a zero of $f(x) = 4x^3 - 5x^2 - 3x + 1$? Explain.

113. Is $\frac{3}{5}$ a zero of $f(x) = 2x^6 - 5x^4 + x^3 - x + 1$? Explain.

114. Is $\frac{2}{3}$ a zero of $f(x) = x^7 + 6x^5 - x^4 + x + 2$? Explain.

115. What is the length of the edge of a cube if, after a slice 1 inch thick is cut from one side, the volume remaining is 294 cubic inches?

116. What is the length of the edge of a cube if its volume could be doubled by an increase of 6 centimeters in one edge, an increase of 12 centimeters in a second edge, and a decrease of 4 centimeters in the third edge?

117. Let $f(x)$ be a polynomial function whose coefficients are integers. Suppose that r is a real zero of f and that the leading coefficient of f is 1. Use the Rational Zeros Theorem to show that r is either an integer or an irrational number.

118. Prove the Rational Zeros Theorem.

[**Hint:** Let $\frac{p}{q}$, where p and q have no common factors except 1 and -1, be a zero of the polynomial function $f(x) = a_n x^n + a_{n-1}x^{n-1} + \cdots + a_1 x + a_0$, whose coefficients are all integers. Show that $a_n p^n + a_{n-1}p^{n-1}q + \cdots + a_1 pq^{n-1} + a_0 q^n = 0$. Now, because p is a factor of the first n terms of this equation, p must also be a factor of the term $a_0 q^n$. Since p is not a factor of q (why?), p must be a factor of a_0. Similarly, q must be a factor of a_n.]

119. Bisection Method for Approximating Zeros of a Function f We begin with two consecutive integers, a and $a + 1$, such that $f(a)$ and $f(a + 1)$ are of opposite sign. Evaluate f at the midpoint m_1 of a and $a + 1$. If $f(m_1) = 0$, then m_1 is the zero of f, and we are finished. Otherwise, $f(m_1)$ is of opposite sign to either $f(a)$ or $f(a + 1)$. Suppose that it is $f(a)$ and $f(m_1)$ that are of opposite sign. Now evaluate f at the midpoint m_2 of a and m_1. Repeat this process until the desired degree of accuracy is obtained. Note that each iteration places the zero in an interval whose length is half that of the previous interval. Use the bisection method to solve Problems 95–102. [**Hint:** The process ends when both endpoints agree to the desired number of decimal places.]

'Are You Prepared?' Answers

1. 3

2. $(3x + 2)(2x - 1)$

3. Quotient: $3x^3 + 4x^2 + 12x + 43$; Remainder: 125

4. $\left\{\frac{-1 - \sqrt{13}}{2}, \frac{-1 + \sqrt{13}}{2}\right\}$

3.7 Complex Zeros; Fundamental Theorem of Algebra

PREPARING FOR THIS SECTION *Before getting started, review the following:*

- Complex Numbers (Appendix A, Section A.6, pp. 947–952)
- Quadratic Equations with a Negative Discriminant (Appendix A, Section A.6, pp. 952–954)

Now work the 'Are You Prepared?' problems on page 214.

OBJECTIVES
1. Use the Conjugate Pairs Theorem to Find the Complex Zeros of a Polynomial
2. Find a Polynomial Function with Specified Zeros
3. Find the Complex Zeros of a Polynomial

In Section 3.6 we found the **real** zeros of a polynomial function. In this section we will find the **complex** zeros of a polynomial function. Finding the complex zeros of a function requires finding all zeros of the form $a + bi$. These zeros will be real if $b = 0$.

A variable in the complex number system is referred to as a **complex variable**. A **complex polynomial function f** of degree n is a function of the form

$$f(x) = a_n x^n + a_{n-1}x^{n-1} + \cdots + a_1 x + a_0 \quad (1)$$

where $a_n, a_{n-1}, \ldots, a_1, a_0$ are complex numbers, $a_n \neq 0$, n is a nonnegative integer, and x is a complex variable. As before, a_n is called the **leading coefficient** of f. A complex number r is called a **(complex) zero** of f if $f(r) = 0$.

We have learned that some quadratic equations have no real solutions, but that in the complex number system every quadratic equation has a solution, either real or complex. The next result, proved by Karl Friedrich Gauss (1777–1855) when he was 22 years old,* gives an extension to complex polynomials. In fact, this result is so important and useful that it has become known as the **Fundamental Theorem of Algebra**.

Fundamental Theorem of Algebra

Every complex polynomial function $f(x)$ of degree $n \geq 1$ has at least one complex zero.

We shall not prove this result, as the proof is beyond the scope of this book. However, using the Fundamental Theorem of Algebra and the Factor Theorem, we can prove the following result:

*In all, Gauss gave four different proofs of this theorem, the first one in 1799 being the subject of his doctoral dissertation.

Theorem

Every complex polynomial function $f(x)$ of degree $n \geq 1$ can be factored into n linear factors (not necessarily distinct) of the form

$$f(x) = a_n(x - r_1)(x - r_2) \cdot \ldots \cdot (x - r_n) \quad \textbf{(2)}$$

where $a_n, r_1, r_2, \ldots, r_n$ are complex numbers. That is, every complex polynomial function of degree $n \geq 1$ has exactly n (not necessarily distinct) zeros.

Proof Let

$$f(x) = a_n x^n + a_{n-1}x^{n-1} + \cdots + a_1 x + a_0$$

By the Fundamental Theorem of Algebra, f has at least one zero, say r_1. Then, by the Factor Theorem, $x - r_1$ is a factor, and

$$f(x) = (x - r_1)q_1(x)$$

where $q_1(x)$ is a complex polynomial of degree $n - 1$ whose leading coefficient is a_n. Again by the Fundamental Theorem of Algebra, the complex polynomial $q_1(x)$ has at least one zero, say r_2. By the Factor Theorem, $q_1(x)$ has the factor $x - r_2$, so

$$q_1(x) = (x - r_2)q_2(x)$$

where $q_2(x)$ is a complex polynomial of degree $n - 2$ whose leading coefficient is a_n. Consequently,

$$f(x) = (x - r_1)(x - r_2)q_2(x)$$

Repeating this argument n times, we finally arrive at

$$f(x) = (x - r_1)(x - r_2) \cdot \ldots \cdot (x - r_n)q_n(x)$$

where $q_n(x)$ is a complex polynomial of degree $n - n = 0$ whose leading coefficient is a_n. Thus, $q_n(x) = a_n x^0 = a_n$, and so

$$f(x) = a_n(x - r_1)(x - r_2) \cdot \ldots \cdot (x - r_n)$$

We conclude that every complex polynomial function $f(x)$ of degree $n \geq 1$ has exactly n (not necessarily distinct) zeros. ■

Complex Zeros of Polynomials with Real Coefficients

1 We can use the Fundamental Theorem of Algebra to obtain valuable information about the complex zeros of polynomials whose coefficients are real numbers.

Conjugate Pairs Theorem

Let $f(x)$ be a polynomial whose coefficients are real numbers. If $r = a + bi$ is a zero of f, then the complex conjugate $\bar{r} = a - bi$ is also a zero of f.

In other words, for polynomials whose coefficients are real numbers, the zeros occur in conjugate pairs.

Proof Let

$$f(x) = a_n x^n + a_{n-1} x^{n-1} + \cdots + a_1 x + a_0$$

where $a_n, a_{n-1}, \ldots, a_1, a_0$ are real numbers and $a_n \neq 0$. If $r = a + bi$ is a zero of f, then $f(r) = f(a + bi) = 0$, so

$$a_n r^n + a_{n-1} r^{n-1} + \cdots + a_1 r + a_0 = 0$$

We take the conjugate of both sides to get

$$\overline{a_n r^n + a_{n-1} r^{n-1} + \cdots + a_1 r + a_0} = \overline{0}$$

$$\overline{a_n r^n} + \overline{a_{n-1} r^{n-1}} + \cdots + \overline{a_1 r} + \overline{a_0} = \overline{0}$$

The conjugate of a sum equals the sum of the conjugates (see Appendix A, Section A.6).

$$\overline{a_n}(\overline{r})^n + \overline{a_{n-1}}(\overline{r})^{n-1} + \cdots + \overline{a_1}\,\overline{r} + \overline{a_0} = \overline{0}$$

The conjugate of a product equals the product of the conjugates.

$$a_n(\overline{r})^n + a_{n-1}(\overline{r})^{n-1} + \cdots + a_1 \overline{r} + a_0 = 0$$

The conjugate of a real number equals the real number.

This last equation states that $f(\overline{r}) = 0$; that is, $\overline{r} = a - bi$ is a zero of f. ■

The importance of this result should be clear. Once we know that, say, $3 + 4i$ is a zero of a polynomial with real coefficients, then we know that $3 - 4i$ is also a zero. This result has an important corollary.

Corollary

A polynomial f of odd degree with real coefficients has at least one real zero.

Proof Because complex zeros occur as conjugate pairs in a polynomial with real coefficients, there will always be an even number of zeros that are not real numbers. Consequently, since f is of odd degree, one of its zeros has to be a real number. ■

For example, the polynomial $f(x) = x^5 - 3x^4 + 4x^3 - 5$ has at least one zero that is a real number, since f is of degree 5 (odd) and has real coefficients.

EXAMPLE 1 Using the Conjugate Pairs Theorem

A polynomial f of degree 5 whose coefficients are real numbers has the zeros $1, 5i$, and $1 + i$. Find the remaining two zeros.

Solution Since f has coefficients that are real numbers, complex zeros appear as conjugate pairs. It follows that $-5i$, the conjugate of $5i$, and $1 - i$, the conjugate of $1 + i$, are the two remaining zeros. ◀

NOW WORK PROBLEM 7.

2 **EXAMPLE 2** **Finding a Polynomial Function Whose Zeros Are Given**

Find a polynomial f of degree 4 whose coefficients are real numbers and that has the zeros 1, 1, and $-4 + i$.

Solution Since $-4 + i$ is a zero, by the Conjugate Pairs Theorem, $-4 - i$ must also be a zero of f. Because of the Factor Theorem, if $f(c) = 0$, then $x - c$ is a factor of $f(x)$. So we can now write f as

$$f(x) = a(x - 1)(x - 1)[x - (-4 + i)][x - (-4 - i)]$$

where a is any real number. Then

$$\begin{aligned} f(x) &= a(x - 1)(x - 1)[x - (-4 + i)][x - (-4 - i)] \\ &= a(x^2 - 2x + 1)[x^2 - (-4 + i)x - (-4 - i)x + (-4 + i)(-4 - i)] \\ &= a(x^2 - 2x + 1)(x^2 + 4x - ix + 4x + ix + 16 + 4i - 4i - i^2) \\ &= a(x^2 - 2x + 1)(x^2 + 8x + 17) \\ &= a(x^4 + 8x^3 + 17x^2 - 2x^3 - 16x^2 - 34x + x^2 + 8x + 17) \\ &= a(x^4 + 6x^3 + 2x^2 - 26x + 17) \end{aligned}$$

◀

Exploration

Graph the function f found in Example 2 for $a = 1$. Does the value of a affect the zeros of f? How does the value of a affect the graph of f?

SOLUTION A quick analysis of the polynomial f tells us what to expect:

At most three turning points.

For large $|x|$, the graph will behave like $y = x^4$.

A repeated real zero at 1 so that the graph will touch the x-axis at 1.

The only x-intercept is at 1; the y-intercept is 17.

Figure 61

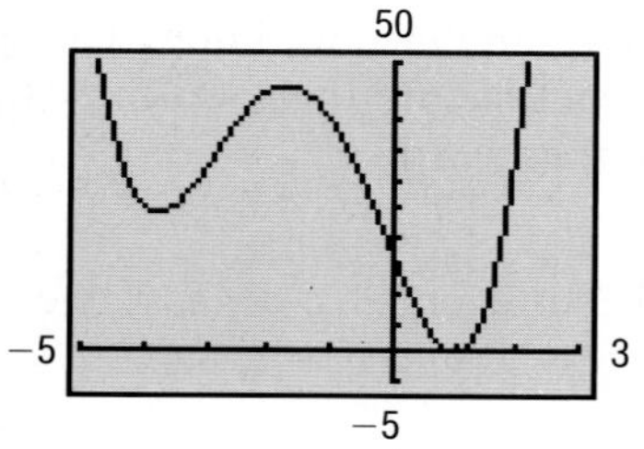

Figure 61 shows the complete graph. (Do you see why? The graph has exactly three turning points.) The value of a causes a stretch or compression; a reflection also occurs if $a < 0$. The zeros are not affected.

Now we can prove the theorem we conjectured in Section 3.6.

Theorem Every polynomial function with real coefficients can be uniquely factored over the real numbers into a product of linear factors and/or irreducible quadratic factors.

Proof Every complex polynomial f of degree n has exactly n zeros and can be factored into a product of n linear factors. If its coefficients are real, then those zeros that are complex numbers will always occur as conjugate pairs. As a result, if $r = a + bi$ is a complex zero, then so is $\bar{r} = a - bi$. Consequently, when the linear factors $x - r$ and $x - \bar{r}$ of f are multiplied, we have

$$(x - r)(x - \bar{r}) = x^2 - (r + \bar{r})x + r\bar{r} = x^2 - 2ax + a^2 + b^2$$

This second-degree polynomial has real coefficients and is irreducible (over the real numbers). Thus, the factors of f are either linear or irreducible quadratic factors. ■

3 **EXAMPLE 3** **Finding the Complex Zeros of a Polynomial**

Find the complex zeros of the polynomial function

$$f(x) = 3x^4 + 5x^3 + 25x^2 + 45x - 18$$

Write f in factored form.

Solution **STEP 1:** The degree of f is 4. So f will have four complex zeros.

STEP 2: Descartes' Rule of Signs provides information about the real zeros. For this polynomial, there is one positive real zero. Because

$$f(-x) = 3x^4 - 5x^3 + 25x^2 - 45x - 18$$

there are three or one negative real zero.

STEP 3: The Rational Zeros Theorem provides information about the potential rational zeros of polynomials with integer coefficients. For this polynomial (which has integer coefficients), the potential rational zeros are

$$\pm\frac{1}{3}, \quad \pm\frac{2}{3}, \quad \pm 1, \quad \pm 2, \quad \pm 3, \quad \pm 6, \quad \pm 9, \quad \pm 18$$

We test 1 first:

$$\begin{array}{r|rrrrr} 1 & 3 & 5 & 25 & 45 & -18 \\ & & 3 & 8 & 33 & 78 \\ \hline & 3 & 8 & 33 & 78 & 60 \end{array}$$

We test -1:

$$\begin{array}{r|rrrrr} -1 & 3 & 5 & 25 & 45 & -18 \\ & & -3 & -2 & -23 & -22 \\ \hline & 3 & 2 & 23 & 22 & -40 \end{array}$$

We test 2:

$$\begin{array}{r|rrrrr} 2 & 3 & 5 & 25 & 45 & -18 \\ & & 6 & 22 & 94 & 278 \\ \hline & 3 & 11 & 47 & 139 & 260 \end{array}$$

We test -2:

$$\begin{array}{r|rrrrr} -2 & 3 & 5 & 25 & 45 & -18 \\ & & -6 & 2 & -54 & 18 \\ \hline & 3 & -1 & 27 & -9 & 0 \end{array}$$

Since $f(-2) = 0$, then -2 is a zero and $x + 2$ is a factor of f. The depressed equation is

$$3x^3 - x^2 + 27x - 9 = 0$$

STEP 4: We factor by grouping.

$$3x^3 - x^2 + 27x - 9 = 0$$

$$x^2(3x - 1) + 9(3x - 1) = 0 \qquad \text{Factor } x^2 \text{ from } 3x^3 - x^2 \text{ and 9 from } 27x - 9.$$

$$(x^2 + 9)(3x - 1) = 0 \qquad \text{Factor out the common factor } 3x - 1.$$

$$x^2 + 9 = 0 \quad \text{or} \quad 3x - 1 = 0 \qquad \text{Apply Zero-Product Property.}$$

$$x^2 = -9 \quad \text{or} \quad x = \frac{1}{3}$$

$$x = -3i, \quad x = 3i, \quad \text{or} \quad x = \frac{1}{3}$$

The four complex zeros of f are $\left\{-3i, 3i, -2, \frac{1}{3}\right\}$.

The factored form of f is

$$f(x) = 3x^4 + 5x^3 + 25x^2 + 45x - 18$$
$$= 3(x + 3i)(x - 3i)(x + 2)\left(x - \frac{1}{3}\right)$$

NOW WORK PROBLEM 33.

3.7 Assess Your Understanding

'Are You Prepared?' *Answers are given at the end of these exercises. If you get the wrong answer, read the pages listed in red.*

1. Find the sum and the product of the complex numbers $3 - 2i$ and $-3 + 5i$. (pp. 947–952)

2. In the complex number system, solve the equation $x^2 + 2x + 2 = 0$. (pp. 952–954)

Concepts and Vocabulary

3. Every polynomial function of odd degree with real coefficients will have at least __________ real zero(s).

4. If $3 + 4i$ is a zero of a polynomial function of degree 5 with real coefficients, then so is __________.

5. *True or False:* A polynomial function of degree n with real coefficients has exactly n complex zeros. At most n of them are real zeros.

6. *True or False:* A polynomial function of degree 4 with real coefficients could have $-3, 2 + i, 2 - i$, and $-3 + 5i$ as its zeros.

Exercises

In Problems 7–16, information is given about a polynomial $f(x)$ whose coefficients are real numbers. Find the remaining zeros of f.

7. Degree 3; zeros: $3, 4 - i$

8. Degree 3; zeros: $4, 3 + i$

9. Degree 4; zeros: $i, 1 + i$

10. Degree 4; zeros: $1, 2, 2 + i$

11. Degree 5; zeros: $1, i, 2i$

12. Degree 5; zeros: $0, 1, 2, i$

13. Degree 4; zeros: $i, 2, -2$

14. Degree 4; zeros: $2 - i, -i$

15. Degree 6; zeros: $2, 2 + i, -3 - i, 0$

16. Degree 6; zeros: $i, 3 - 2i, -2 + i$

In Problems 17–22, form a polynomial $f(x)$ with real coefficients having the given degree and zeros.

17. Degree 4; zeros: $3 + 2i$; 4, multiplicity 2

18. Degree 4; zeros: $i, 1 + 2i$

19. Degree 5; zeros: $2; -i; 1 + i$

20. Degree 6; zeros: $i, 4 - i; 2 + i$

21. Degree 4; zeros: 3, multiplicity 2; $-i$

22. Degree 5; zeros: 1, multiplicity 3; $1 + i$

In Problems 23–30, use the given zero to find the remaining zeros of each function.

23. $f(x) = x^3 - 4x^2 + 4x - 16$; zero: $2i$

24. $g(x) = x^3 + 3x^2 + 25x + 75$; zero: $-5i$

25. $f(x) = 2x^4 + 5x^3 + 5x^2 + 20x - 12$; zero: $-2i$

26. $h(x) = 3x^4 + 5x^3 + 25x^2 + 45x - 18$; zero: $3i$

27. $h(x) = x^4 - 9x^3 + 21x^2 + 21x - 130$; zero: $3 - 2i$

28. $f(x) = x^4 - 7x^3 + 14x^2 - 38x - 60$; zero: $1 + 3i$

29. $h(x) = 3x^5 + 2x^4 + 15x^3 + 10x^2 - 528x - 352$; zero: $-4i$

30. $g(x) = 2x^5 - 3x^4 - 5x^3 - 15x^2 - 207x + 108$; zero: $3i$

In Problems 31–40, find the complex zeros of each polynomial function. Write f in factored form.

31. $f(x) = x^3 - 1$

32. $f(x) = x^4 - 1$

33. $f(x) = x^3 - 8x^2 + 25x - 26$

34. $f(x) = x^3 + 13x^2 + 57x + 85$

35. $f(x) = x^4 + 5x^2 + 4$

36. $f(x) = x^4 + 13x^2 + 36$

37. $f(x) = x^4 + 2x^3 + 22x^2 + 50x - 75$

38. $f(x) = x^4 + 3x^3 - 19x^2 + 27x - 252$

39. $f(x) = 3x^4 - x^3 - 9x^2 + 159x - 52$

40. $f(x) = 2x^4 + x^3 - 35x^2 - 113x + 65$

In Problems 41 and 42, explain why the facts given are contradictory.

41. $f(x)$ is a polynomial of degree 3 whose coefficients are real numbers; its zeros are $4 + i, 4 - i$, and $2 + i$.

42. $f(x)$ is a polynomial of degree 3 whose coefficients are real numbers; its zeros are $2, i$, and $3 + i$.

43. $f(x)$ is a polynomial of degree 4 whose coefficients are real numbers; three of its zeros are $2, 1 + 2i$, and $1 - 2i$. Explain why the remaining zero must be a real number.

44. $f(x)$ is a polynomial of degree 4 whose coefficients are real numbers; two of its zeros are -3 and $4 - i$. Explain why one of the remaining zeros must be a real number. Write down one of the missing zeros.

'Are You Prepared?' Answers

1. Sum: $3i$; product: $1 + 21i$ **2.** $\{-1 - i, -1 + i\}$

Chapter Review

Things to Know

Quadratic function (pp. 124–132)

$f(x) = ax^2 + bx + c$

Graph is a parabola that opens up if $a > 0$ and opens down if $a < 0$.

Vertex: $\left(-\dfrac{b}{2a}, f\left(-\dfrac{b}{2a}\right)\right)$

Axis of symmetry: $x = -\dfrac{b}{2a}$

y-intercept: $f(0)$

x-intercept(s): If any, found by finding the real solutions of the equation $ax^2 + bx + c = 0$.

Power function (p. 146)

$f(x) = x^n$, $n \geq 2$ even (p. 147)

Even function
Passes through $(-1, 1), (0, 0), (1, 1)$
Opens up

$f(x) = x^n$, $n \geq 3$ odd (p. 149)

Odd function
Passes through $(-1, -1), (0, 0), (1, 1)$
Increasing

Polynomial function (pp. 145, 149–154)

$f(x) = a_n x^n + a_{n-1}x^{n-1} + \cdots + a_1 x + a_0$, $a_n \neq 0$

Domain: all real numbers
At most $n - 1$ turning points
End behavior: Behaves like $y = a_n x^n$ for large $|x|$

Zeros of a polynomial f (p. 150)

Numbers for which $f(x) = 0$; the real zeros of f are the x-intercepts of the graph of f.

Rational function (p. 163)

$R(x) = \dfrac{p(x)}{q(x)}$

p, q are polynomial functions. Domain: $\{x | q(x) \neq 0\}$

Remainder Theorem (p. 195)

If a polynomial $f(x)$ is divided by $x - c$, then the remainder is $f(c)$.

Factor Theorem (p. 196)

$x - c$ is a factor of a polynomial $f(x)$ if and only if $f(c) = 0$.

Descartes' Rule of Signs (p. 198)

Let f denote a polynomial function. The number of positive zeros of f either equals the number of variations in sign of the nonzero coefficients of $f(x)$ or else equals that number less some even integer. The number of negative zeros of f either equals the number of variations in sign of the nonzero coefficients of $f(-x)$ or else equals that number less some even integer.

Rational Zeros Theorem (p. 198)

Let f be a polynomial function of degree 1 or higher of the form

$$f(x) = a_n x^n + a_{n-1}x^{n-1} + \cdots + a_1 x + a_0, \quad a_n \neq 0, a_0 \neq 0$$

where each coefficient is an integer. If $\dfrac{p}{q}$, in lowest terms, is a rational zero of f, then p must be a factor of a_0 and q must be a factor of a_n.

Intermediate Value Theorem (p. 204)

Let f be a polynomial function. If $a < b$ and $f(a)$ and $f(b)$ are of opposite sign, then there is at least one real zero of f between a and b.

Fundamental Theorem of Algebra (p. 209)	Every complex polynomial function $f(x)$ of degree $n \geq 1$ has at least one complex zero.
Conjugate Pairs Theorem (p. 210)	Let $f(x)$ be a polynomial whose coefficients are real numbers. If $r = a + bi$ is a zero of f, then its complex conjugate $\bar{r} = a - bi$ is also a zero of f.

Objectives

Section		You should be able to . . .	Review Exercises
3.1	1	Graph a quadratic function using transformations (p. 126)	1–6
	2	Identify the vertex and axis of symmetry of a quadratic function (p. 127)	7–16
	3	Graph a quadratic function using its vertex, axis, and intercepts (p. 128)	7–16
	4	Use the maximum or minimum value of a quadratic function to solve applied problems (p. 132)	111–118
	5	Use a graphing utility to find the quadratic function of best fit to data (p. 136)	117–118
3.2	1	Identify polynomial functions and their degree (p. 145)	23–26
	2	Graph polynomial functions using transformations (p. 149)	1–6, 27–32
	3	Identify the zeros of a polynomial function and their multiplicity (p. 150)	33–40
	4	Analyze the graph of a polynomial function (p. 151)	33–40
3.3	1	Find the domain of a rational function (p. 163)	41–44
	2	Determine the vertical asymptotes of a rational function (p. 166)	41–44
	3	Determine the horizontal or oblique asymptotes of a rational function (p. 167)	41–44
3.4	1	Analyze the graph of a rational function (p. 173)	45–56
	2	Solve applied problems involving rational functions (p. 184)	120
3.5	1	Solve polynomial inequalities (p. 188)	57–58
	2	Solve rational inequalities (p. 191)	59–66
3.6	1	Use the Remainder and Factor Theorems (p. 194)	67–72
	2	Use Descartes' Rule of Signs (p. 197)	73–74
	3	Use the Rational Zeros Theorem (p. 198)	75–76
	4	Find the real zeros of a polynomial function (p. 199)	77–82
	5	Solve polynomial equations (p. 202)	83–86
	6	Use the Theorem for Bounds on Zeros (p. 203)	95–98
	7	Use the Intermediate Value Theorem (p. 204)	99–106
3.7	1	Use the Conjugate Pairs Theorem (p. 210)	107–110
	2	Find a polynomial function with specified zeros (p. 212)	107–110
	3	Find the complex zeros of a polynomial (p. 213)	87–94

Review Exercises

(Blue problem numbers indicate the author's suggestions for use in a Practice Test.)

In Problems 1–6, graph each function using transformations (shifting, compressing, stretching, and reflection).

1. $f(x) = (x - 2)^2 + 2$ **2.** $f(x) = (x + 1)^2 - 4$ **3.** $f(x) = -(x - 1)^2$

4. $f(x) = (x - 1)^2 - 2$ **5.** $f(x) = (x - 1)^2 + 2$ **6.** $f(x) = (1 - x)^2$

In Problems 7–16, graph each quadratic function by determining whether its graph opens up or down and by finding its vertex, axis of symmetry, y-intercept, and x-intercepts, if any.

7. $f(x) = (x - 2)^2 + 2$ **8.** $f(x) = (x + 1)^2 - 4$ **9.** $f(x) = \frac{1}{4}x^2 - 16$ **10.** $f(x) = -\frac{1}{2}x^2 + 2$

11. $f(x) = -4x^2 + 4x$ **12.** $f(x) = 9x^2 - 6x + 3$ **13.** $f(x) = \frac{9}{2}x^2 + 3x + 1$ **14.** $f(x) = -x^2 + x + \frac{1}{2}$

15. $f(x) = 3x^2 + 4x - 1$ **16.** $f(x) = -2x^2 - x + 4$

In Problems 17–22, determine whether the given quadratic function has a maximum value or a minimum value, and then find the value.

17. $f(x) = 3x^2 - 6x + 4$
18. $f(x) = 2x^2 + 8x + 5$
19. $f(x) = -x^2 + 8x - 4$
20. $f(x) = -x^2 - 10x - 3$
21. $f(x) = -3x^2 + 12x + 4$
22. $f(x) = -2x^2 + 4$

In Problems 23–26, determine which functions are polynomial functions. For those that are, state the degree. For those that are not, tell why not.

23. $f(x) = 4x^5 - 3x^2 + 5x - 2$
24. $f(x) = \dfrac{3x^5}{2x + 1}$
25. $f(x) = 3x^2 + 5x^{1/2} - 1$
26. $f(x) = 3$

In Problems 27–32, graph each function using transformations (shifting, compressing, stretching, and reflection). Show all the stages.

27. $f(x) = (x + 2)^3$
28. $f(x) = -x^3 + 3$
29. $f(x) = -(x - 1)^4$
30. $f(x) = (x - 1)^4 - 2$
31. $f(x) = (x - 1)^4 + 2$
32. $f(x) = (1 - x)^3$

In Problems 33–40:

(a) Find the x- and y-intercepts of each polynomial function f.
(b) Determine whether the graph of f touches or crosses the x-axis at each x-intercept.
(c) End behavior: find the power function that the graph of f resembles for large values of $|x|$.
(d) Determine the maximum number of turning points of the graph of f.
(e) Use the x-intercept(s) to find the intervals on which the graph of f is above and below the x-axis.
(f) Plot the points obtained in parts (a) and (e), and use the remaining information to connect them with a smooth curve.

33. $f(x) = x(x + 2)(x + 4)$
34. $f(x) = x(x - 2)(x - 4)$
35. $f(x) = (x - 2)^2(x + 4)$
36. $f(x) = (x - 2)(x + 4)^2$
37. $f(x) = -2x^3 + 4x^2$
38. $f(x) = -4x^3 + 4x$
39. $f(x) = (x - 1)^2(x + 3)(x + 1)$
40. $f(x) = (x - 4)(x + 2)^2(x - 2)$

In Problems 41–44, find the domain of each rational function. Find any horizontal, vertical, or oblique asymptotes.

41. $R(x) = \dfrac{x + 2}{x^2 - 9}$
42. $R(x) = \dfrac{x^2 + 4}{x - 2}$
43. $R(x) = \dfrac{x^2 + 3x + 2}{(x + 2)^2}$
44. $R(x) = \dfrac{x^3}{x^3 - 1}$

In Problems 45–56, discuss each rational function following the seven steps outlined in Section 3.4.

45. $R(x) = \dfrac{2x - 6}{x}$
46. $R(x) = \dfrac{4 - x}{x}$
47. $H(x) = \dfrac{x + 2}{x(x - 2)}$
48. $H(x) = \dfrac{x}{x^2 - 1}$
49. $R(x) = \dfrac{x^2 + x - 6}{x^2 - x - 6}$
50. $R(x) = \dfrac{x^2 - 6x + 9}{x^2}$
51. $F(x) = \dfrac{x^3}{x^2 - 4}$
52. $F(x) = \dfrac{3x^3}{(x - 1)^2}$
53. $R(x) = \dfrac{2x^4}{(x - 1)^2}$
54. $R(x) = \dfrac{x^4}{x^2 - 9}$
55. $G(x) = \dfrac{x^2 - 4}{x^2 - x - 2}$
56. $F(x) = \dfrac{(x - 1)^2}{x^2 - 1}$

In Problems 57–66, solve each inequality.

57. $2x^2 + 5x - 12 < 0$
58. $3x^2 - 2x - 1 \geq 0$
59. $\dfrac{6}{x + 3} \geq 1$
60. $\dfrac{-2}{1 - 3x} < 1$
61. $\dfrac{2x - 6}{1 - x} < 2$
62. $\dfrac{3 - 2x}{2x + 5} \geq 2$
63. $\dfrac{(x - 2)(x - 1)}{x - 3} \geq 0$
64. $\dfrac{x + 1}{x(x - 5)} \leq 0$
65. $\dfrac{x^2 - 8x + 12}{x^2 - 16} > 0$
66. $\dfrac{x(x^2 + x - 2)}{x^2 + 9x + 20} \leq 0$

In Problems 67–70, find the quotient $q(x)$ and remainder R when $f(x)$ is divided by $g(x)$. Is g a factor of f?

67. $f(x) = 8x^3 - 3x^2 + x + 4;\quad g(x) = x - 1$
68. $f(x) = 2x^3 + 8x^2 - 5x + 5;\quad g(x) = x - 2$
69. $f(x) = x^4 - 2x^3 + 15x - 2;\quad g(x) = x + 2$
70. $f(x) = x^4 - x^2 + 2x + 2;\quad g(x) = x + 1$

71. Find the value of $f(x) = 12x^6 - 8x^4 + 1$ at $x = 4$.

72. Find the value of $f(x) = -16x^3 + 18x^2 - x + 2$ at $x = -2$.

In Problems 73 and 74, use Descartes' Rule of Signs to determine how many positive and negative zeros each polynomial function may have. Do not attempt to find the zeros.

73. $f(x) = 12x^8 - x^7 + 8x^4 - 2x^3 + x + 3$
74. $f(x) = -6x^5 + x^4 + 5x^3 + x + 1$

75. List all the potential rational zeros of $f(x) = 12x^8 - x^7 + 6x^4 - x^3 + x - 3$.

76. List all the potential rational zeros of $f(x) = -6x^5 + x^4 + 2x^3 - x + 1$.

In Problems 77–82, use Descartes' Rule of Signs and the Rational Zeros Theorem to find all the real zeros of each polynomial function. Use the zeros to factor f over the real numbers.

77. $f(x) = x^3 - 3x^2 - 6x + 8$

78. $f(x) = x^3 - x^2 - 10x - 8$

79. $f(x) = 4x^3 + 4x^2 - 7x + 2$

80. $f(x) = 4x^3 - 4x^2 - 7x - 2$

81. $f(x) = x^4 - 4x^3 + 9x^2 - 20x + 20$

82. $f(x) = x^4 + 6x^3 + 11x^2 + 12x + 18$

In Problems 83–86, solve each equation in the real number system.

83. $2x^4 + 2x^3 - 11x^2 + x - 6 = 0$

84. $3x^4 + 3x^3 - 17x^2 + x - 6 = 0$

85. $2x^4 + 7x^3 + x^2 - 7x - 3 = 0$

86. $2x^4 + 7x^3 - 5x^2 - 28x - 12 = 0$

In Problems 87–94, find the complex zeros of each polynomial function $f(x)$. Write f in factored form.

87. $f(x) = x^3 - 3x^2 - 6x + 8$

88. $f(x) = x^3 - x^2 - 10x - 8$

89. $f(x) = 4x^3 + 4x^2 - 7x + 2$

90. $f(x) = 4x^3 - 4x^2 - 7x - 2$

91. $f(x) = x^4 - 4x^3 + 9x^2 - 20x + 20$

92. $f(x) = x^4 + 6x^3 + 11x^2 + 12x + 18$

93. $f(x) = 2x^4 + 2x^3 - 11x^2 + x - 6$

94. $f(x) = 3x^4 + 3x^3 - 17x^2 + x - 6$

In Problems 95–98, find a bound to the zeros of each polynomial function.

95. $f(x) = x^3 - x^2 - 4x + 2$

96. $f(x) = x^3 + x^2 - 10x - 5$

97. $f(x) = 2x^3 - 7x^2 - 10x + 35$

98. $f(x) = 3x^3 - 7x^2 - 6x + 14$

In Problems 99–102, use the Intermediate Value Theorem to show that each polynomial has a zero in the given interval.

99. $f(x) = 3x^3 - x - 1$; $[0, 1]$

100. $f(x) = 2x^3 - x^2 - 3$; $[1, 2]$

101. $f(x) = 8x^4 - 4x^3 - 2x - 1$; $[0, 1]$

102. $f(x) = 3x^4 + 4x^3 - 8x - 2$; $[1, 2]$

In Problems 103–106, each polynomial has exactly one positive zero. Approximate the zero correct to two decimal places.

103. $f(x) = x^3 - x - 2$

104. $f(x) = 2x^3 - x^2 - 3$

105. $f(x) = 8x^4 - 4x^3 - 2x - 1$

106. $f(x) = 3x^4 + 4x^3 - 8x - 2$

In Problems 107–110, information is given about a complex polynomial $f(x)$ whose coefficients are real numbers. Find the remaining zeros of f.

107. Degree 3; zeros: $4 + i, 6$

108. Degree 3; zeros: $3 + 4i, 5$

109. Degree 4; zeros: $i, 1 + i$

110. Degree 4; zeros: $1, 2, 1 + i$

111. Find the point on the line $y = x$ that is closest to the point $(3, 1)$.

[**Hint:** Find the minimum value of the function $f(x) = d^2$, where d is the distance from $(3, 1)$ to a point on the line.]

112. **Landscaping** A landscape engineer has 200 feet of border to enclose a rectangular pond. What dimensions will result in the largest pond?

113. **Enclosing the Most Area with a Fence** A farmer with 10,000 meters of fencing wants to enclose a rectangular field and then divide it into two plots with a fence parallel to one of the sides (see the figure). What is the largest area that can be enclosed?

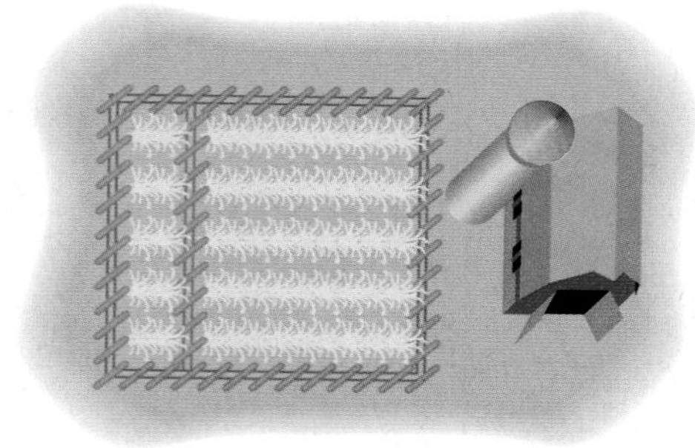

114. A rectangle has one vertex on the line $y = 8 - 2x$, $x > 0$, another at the origin, one on the positive x-axis, and one on the positive y-axis. Find the largest area A that can be enclosed by the rectangle.

115. Architecture A special window in the shape of a rectangle with semicircles at each end is to be constructed so that the outside dimensions are 100 feet in length. See the illustration. Find the dimensions of the rectangle that maximizes its area.

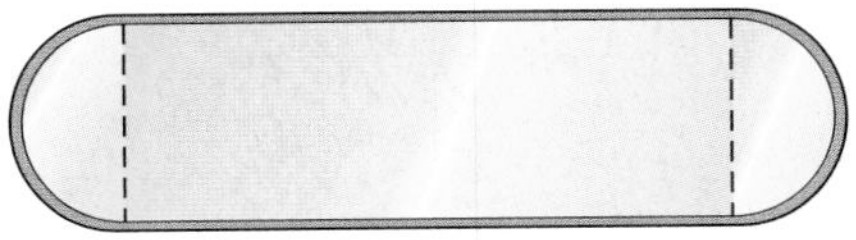

116. Parabolic Arch Bridges A horizontal bridge is in the shape of a parabolic arch. Given the information shown in the figure, what is the height h of the arch 2 feet from shore?

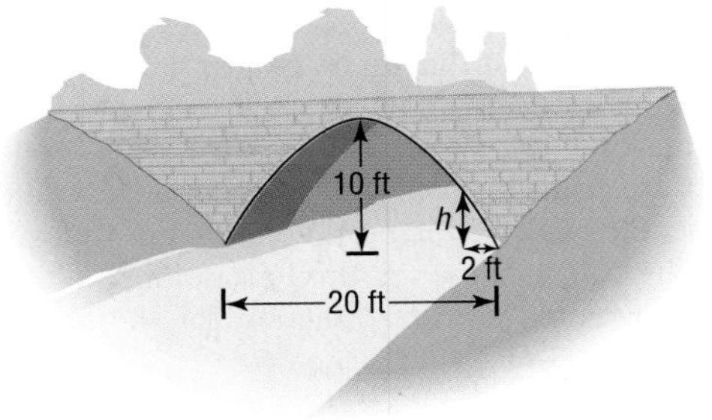

117. Minimizing Marginal Cost The marginal cost of a product can be thought of as the cost of producing one additional unit of output. For example, if the marginal cost of producing the 50th product is \$6.20, then it cost \$6.20 to increase production from 49 to 50 units of output. Callaway Golf Company has determined that the marginal cost C of manufacturing x Big Bertha golf clubs may be expressed by the quadratic function

$$C(x) = 4.9x^2 - 617.4x + 19{,}600$$

(a) How many clubs should be manufactured to minimize the marginal cost?
(b) At this level of production, what is the marginal cost?

118. Violent Crimes The function

$$V(t) = -10.0t^2 + 39.2t + 1862.6$$

models the number V (in thousands) of violent crimes committed in the United States t years after 1990. So $t = 0$ represents 1990, $t = 1$ represents 1991, and so on.
(a) Determine the year in which the most violent crimes were committed.
(b) Approximately how many violent crimes were committed during this year?
(c) Using a graphing utility, graph $V = V(t)$. Were the number of violent crimes increasing or decreasing during the years 1994 to 1998?

SOURCE: Based on data obtained from the Federal Bureau of Investigation.

119. AIDS Cases in the United States The following data represent the cumulative number of reported AIDS cases in the United States for 1990–1997.

Year, t	Number of AIDS Cases, A
1990, 1	193,878
1991, 2	251,638
1992, 3	326,648
1993, 4	399,613
1994, 5	457,280
1995, 6	528,215
1996, 7	594,760
1997, 8	653,253

SOURCE: U.S. Center for Disease Control and Prevention.

(a) Draw a scatter diagram of the data.
(b) The cubic function of best fit to these data is

$$A(t) = -212t^3 + 2429t^2 + 59{,}569t + 130{,}003$$

Use this function to predict the cumulative number of AIDS cases reported in the United States in 2000.
(c) Use a graphing utility to verify that the function given in part (b) is the cubic function of best fit.
(d) With a graphing utility, draw a scatter diagram of the data and then graph the cubic function of best fit on the scatter diagram.
(e) Do you think the function found in part (b) will be useful in predicting the number of AIDS cases in 2005?

120. Making a Can A can in the shape of a right circular cylinder is required to have a volume of 250 cubic centimeters.
(a) Express the amount A of material to make the can as a function of the radius r of the cylinder.
(b) How much material is required if the can is of radius 3 centimeters?
(c) How much material is required if the can is of radius 5 centimeters?
(d) Graph $A = A(r)$. For what value of r is A smallest?

121. Design a polynomial function with the following characteristics: degree 6; four real zeros, one of multiplicity 3; y-intercept 3; behaves like $y = -5x^6$ for large values of $|x|$. Is this polynomial unique? Compare your polynomial with those of other students. What terms will be the same as everyone else's? Add some more characteristics, such as symmetry or naming the real zeros. How does this modify the polynomial?

122. Design a rational function with the following characteristics: three real zeros, one of multiplicity 2; y-intercept 1; vertical asymptotes $x = -2$ and $x = 3$; oblique asymptote $y = 2x + 1$. Is this rational function unique? Compare yours with those of other students. What will be the same as everyone else's? Add some more characteristics, such as symmetry or naming the real zeros. How does this modify the rational function?

123. The illustration on the right shows the graph of a polynomial function.
(a) Is the degree of the polynomial even or odd?
(b) Is the leading coefficient positive or negative?
(c) Is the function even, odd, or neither?
(d) Why is x^2 necessarily a factor of the polynomial?
(e) What is the minimum degree of the polynomial?
(f) Formulate five different polynomials whose graphs could look like the one shown. Compare yours to those of other students. What similarities do you see? What differences?

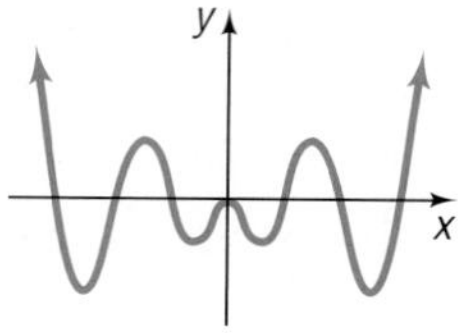

Chapter Projects

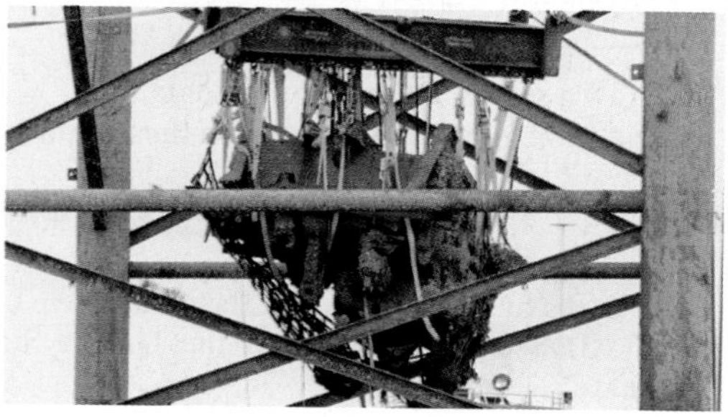

1. Cannons The velocity of a projectile depends on many factors, in particular, the weight of the ammunition.
(a) Plot a scatter diagram of the data in the table below. Let x be the weight in kilograms and let y be the velocity in meters per second.

Type	Weight (kg)	Initial Velocity (m/sec)
MG 17	10.2	905
MG 131	19.7	710
MG 151	41.5	850
MG 151/20	42.3	695
MG/FF	35.7	575
MK 103	145	860
MK 108	58	520
WGr 21	111	315

SOURCE: Data and information taken from "Flugzeug-Handbuch, Ausgabe Dezember 1996: Guns and Cannons of the Jagdwaffe" at www.xs4all.nl/~rhorta/jgguns.htm.

(b) Determine which type of function would fit these data the best: linear or quadratic. Use a graphing utility to find the function of best fit. Are the results reasonable?
(c) Based on velocity, we can determine how high a projectile will travel before it begins to come back down. If a cannon is fired at an angle of 45° to the horizontal, then the function for the height of the projectile is given by $s(t) = -4.9t^2 + \frac{\sqrt{2}}{2}v_0t + s_0$, where v_0 is the velocity at which the shell leaves the cannon (initial velocity), and s_0 is the initial height of the nose of the cannon (because cannons are not very long, we may assume that the nose and the firing pin at the back are at the same height for simplicity). Graph the function $s = s(t)$ for each of the guns described in the table. Which gun would be the best for antiaircraft if the gun were sitting on the ground? Which would be the best to have mounted on a hilltop or on the top of a tall building? If the guns were on the turret of a ship, which would be the most effective?

The following projects are available at www.prenhall.com/sullivan7e.

2. **Project at Motorola** ***How Many Cellphones Can I Make?***
3. **First and Second Differences**
4. **Weed Pollen**
5. **Maclaurin Series**
6. **Theory of Equations**
7. **CBL Experiment**

Cumulative Review

1. Find the distance between the points $P = (1, 3)$ and $Q = (-4, 2)$.

2. Solve the inequality $x^2 \geq x$ and graph the solution set.

3. Solve the inequality $x^2 - 3x < 4$ and graph the solution set.

4. Find a linear function with slope -3 that contains the point $(-1, 4)$. Graph the function.

5. Find the equation of the line parallel to the line $y = 2x + 1$ and containing the point $(3, 5)$. Express your answer in slope–intercept form and graph the line.

6. Graph the equation $y = x^3$.

7. Does the following relation represent a function? $\{(3, 6), (1, 3), (2, 5), (3, 8)\}$. Why or why not?

8. Solve the equation $x^3 - 6x^2 + 8x = 0$.

9. Solve the inequality $3x + 2 \leq 5x - 1$ and graph the solution set.

10. Find the center and radius of the circle $x^2 + 4x + y^2 - 2y - 4 = 0$. Graph the circle

11. For the equation $y = x^3 - 9x$, determine the intercepts and test for symmetry.

12. Find an equation of the line perpendicular to $3x - 2y = 7$ that contains the point $(1, 5)$.

13. Is the following graph the graph of a function? Why or why not?

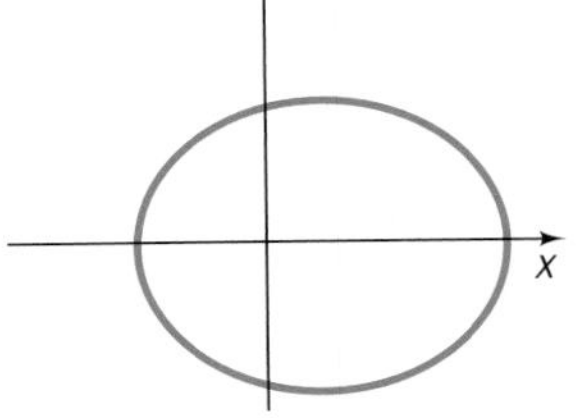

14. For the function $f(x) = x^2 + 5x - 2$, find

(a) $f(3)$ (b) $f(-x)$

(c) $-f(x)$ (d) $f(3x)$

(e) $\dfrac{f(x + h) - f(x)}{h}, h \neq 0$

15. Answer the following questions regarding the function $f(x) = \dfrac{x + 5}{x - 1}$.

(a) What is the domain of f?

(b) Is the point $(2, 6)$ on the graph of f?

(c) If $x = 3$, what is $f(x)$? What point is on the graph of f?

(d) If $f(x) = 9$, what is x? What point is on the graph of f?

16. Graph the function $f(x) = -3x + 7$.

17. Graph $f(x) = 2x^2 - 4x + 1$ by determining whether its graph opens up or down and by finding its vertex, axis of symmetry, y-intercept, and x-intercepts, if any.

18. Find the average rate of change of $f(x) = x^2 + 3x + 1$ from 1 to x. Use this result to find the slope of the secant line containing $(1, f(1))$ and $(2, f(2))$.

19. In parts (a) to (f) use the following graph.

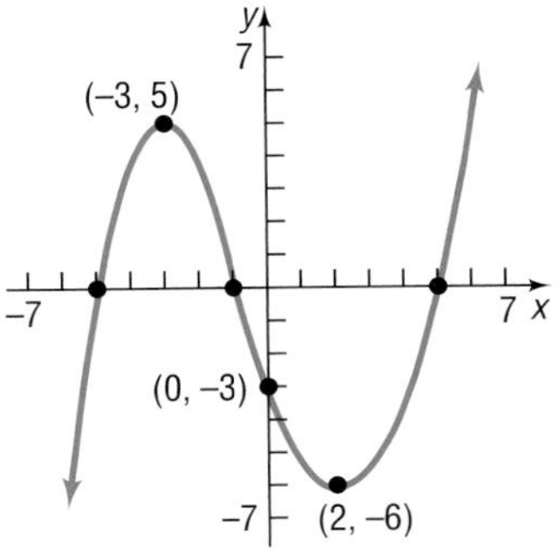

(a) Determine the intercepts.

(b) Based on the graph, tell whether the graph is symmetric with respect to the x-axis, the y-axis, and/or the origin.

(c) Based on the graph, tell whether the function is even, odd, or neither.

(d) List the intervals on which f is increasing. List the intervals on which f is decreasing.

(e) List the numbers, if any, at which f has a local maximum. What are these local maxima?

(f) List the numbers, if any, at which f has a local minimum. What are these local minima?

20. Determine algebraically whether the function $f(x) = \dfrac{5x}{x^2 - 9}$ is even, odd, or neither.

21. For the function $f(x) = \begin{cases} 2x + 1 & \text{if } -3 < x < 2 \\ -3x + 4 & \text{if } x \geq 2 \end{cases}$

(a) Find the domain of f.
(b) Locate any intercepts.
(c) Graph the function.
(d) Based on the graph, find the range.

22. Graph the function $f(x) = -3(x + 1)^2 + 5$ using transformations.

23. Suppose that $f(x) = x^2 - 5x + 1$ and $g(x) = -4x - 7$.

(a) Find $f + g$ and state its domain.
(b) Find $\dfrac{f}{g}$ and state its domain.

24. Demand Equation The price p (in dollars) and the quantity x sold of a certain product obey the demand equation

$$p = -\frac{1}{10}x + 150, \qquad 0 \leq x \leq 1500$$

(a) Express the revenue R as a function of x.
(b) What is the revenue if 100 units are sold?
(c) What quantity x maximizes revenue? What is the maximum revenue?
(d) What price should the company charge to maximize revenue?

4 Exponential and Logarithmic Functions

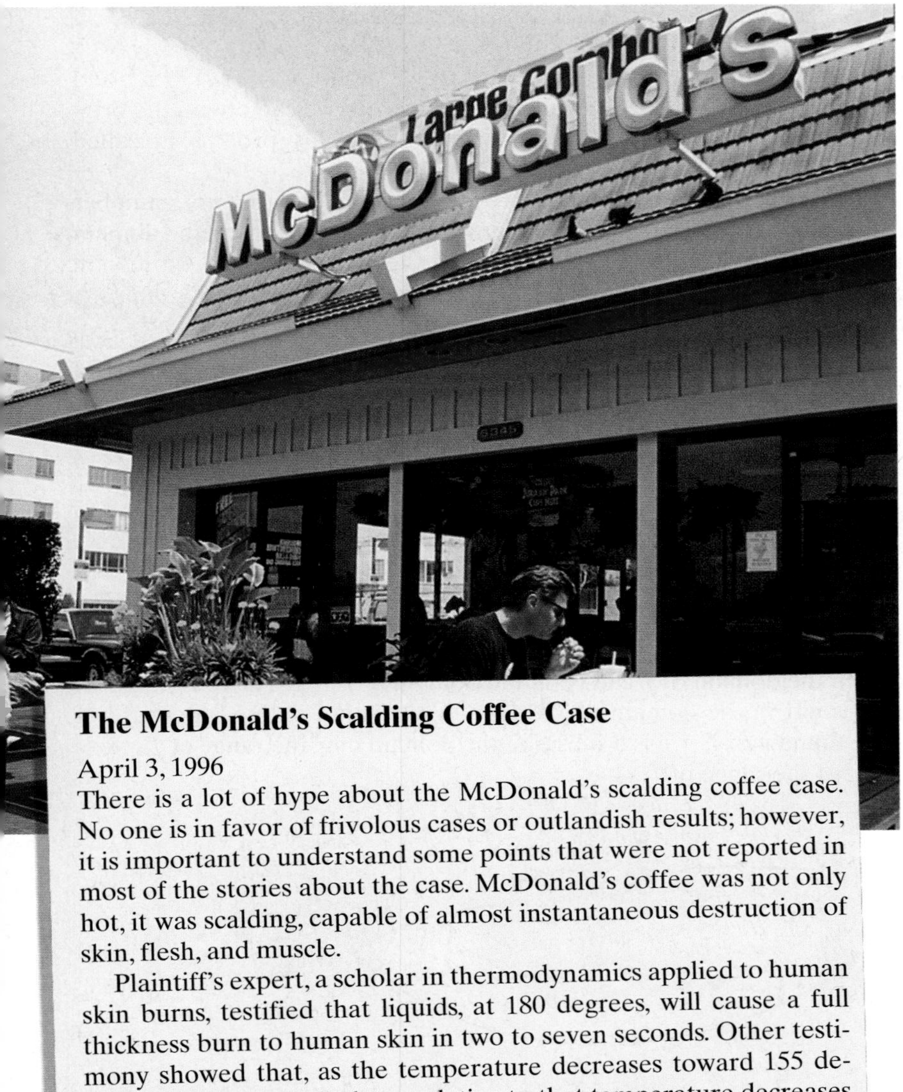

The McDonald's Scalding Coffee Case

April 3, 1996

There is a lot of hype about the McDonald's scalding coffee case. No one is in favor of frivolous cases or outlandish results; however, it is important to understand some points that were not reported in most of the stories about the case. McDonald's coffee was not only hot, it was scalding, capable of almost instantaneous destruction of skin, flesh, and muscle.

Plaintiff's expert, a scholar in thermodynamics applied to human skin burns, testified that liquids, at 180 degrees, will cause a full thickness burn to human skin in two to seven seconds. Other testimony showed that, as the temperature decreases toward 155 degrees, the extent of the burn relative to that temperature decreases exponentially. Thus, if (the) spill had involved coffee at 155 degrees, the liquid would have cooled and given her time to avoid a serious burn.

Miller, Norman, & Associates, Ltd., Attorney Moorhead, MN.

—SEE CHAPTER PROJECT 1.

OUTLINE

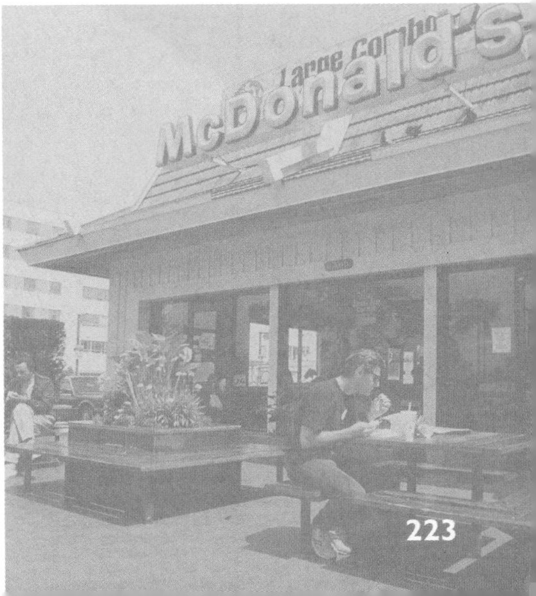

4.1 Composite Functions

PREPARING FOR THIS SECTION *Before getting started, review the following:*

- Finding Values of a Function (Section 2.1, pp. 53–55)
- Domain of a Function (Section 2.1, pp. 57–58)

Now work the 'Are You Prepared?' problems on page 229.

OBJECTIVE 1 Form a Composite Function and Find Its Domain

1 Consider the function $y = (2x + 3)^2$. If we write $y = f(u) = u^2$ and $u = g(x) = 2x + 3$, then, by a substitution process, we can obtain the original function: $y = f(u) = f(g(x)) = (2x + 3)^2$. This process is called **composition.**

In general, suppose that f and g are two functions and that x is a number in the domain of g. By evaluating g at x, we get $g(x)$. If $g(x)$ is in the domain of f, then we may evaluate f at $g(x)$ and thereby obtain the expression $f(g(x))$. The correspondence from x to $f(g(x))$ is called a *composite function* $f \circ g$.

Given two functions f and g, the **composite function,** denoted by $f \circ g$ (read as "f composed with g"), is defined by

$$(f \circ g)(x) = f(g(x))$$

The domain of $f \circ g$ is the set of all numbers x in the domain of g such that $g(x)$ is in the domain of f.

Look carefully at Figure 1. Only those x's in the domain of g for which $g(x)$ is in the domain of f can be in the domain of $f \circ g$. The reason is that if $g(x)$ is not in the domain of f then $f(g(x))$ is not defined. Because of this, the domain of $f \circ g$ is a subset of the domain of g; the range of $f \circ g$ is a subset of the range of f.

Figure 1

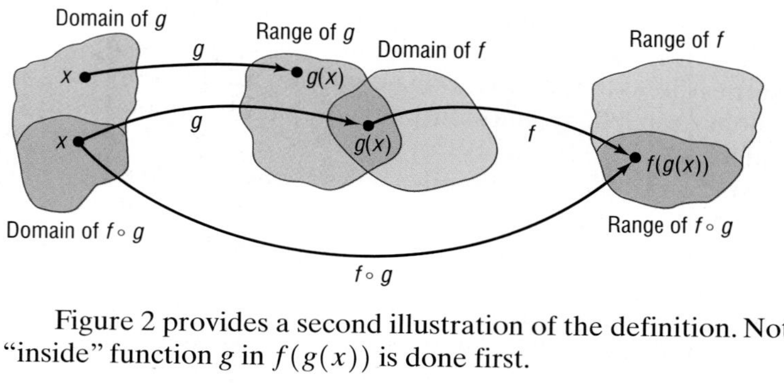

Figure 2 provides a second illustration of the definition. Notice that the "inside" function g in $f(g(x))$ is done first.

Figure 2

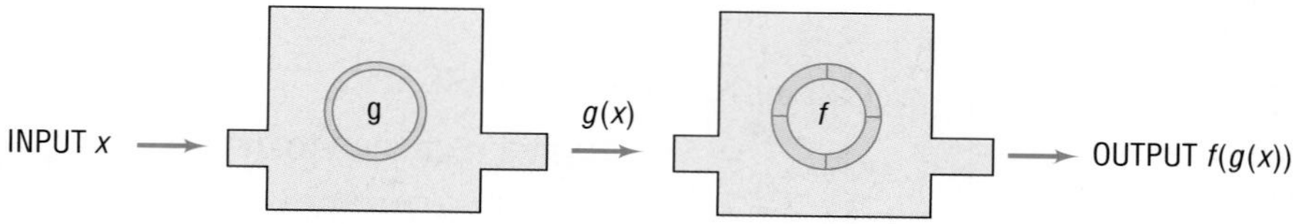

Let's look at some examples.

EXAMPLE 1 Evaluating a Composite Function

Suppose that $f(x) = 2x^2 - 3$ and $g(x) = 4x$. Find:

(a) $(f \circ g)(1)$ (b) $(g \circ f)(1)$ (c) $(f \circ f)(-2)$ (d) $(g \circ g)(-1)$

Solution (a) $(f \circ g)(1) = f(g(1)) = f(4) = 2 \cdot 16 - 3 = 29$

$g(x) = 4x$, $g(1) = 4$; $f(x) = 2x^2 - 3$

(b) $(g \circ f)(1) = g(f(1)) = g(-1) = 4 \cdot (-1) = -4$

$f(x) = 2x^2 - 3$, $f(1) = -1$; $g(x) = 4x$

(c) $(f \circ f)(-2) = f(f(-2)) = f(5) = 2 \cdot 25 - 3 = 47$

$f(-2) = 5$

(d) $(g \circ g)(-1) = g(g(-1)) = g(-4) = 4 \cdot (-4) = -16$

$g(-1) = -4$

◀

NOW WORK PROBLEM 9.

EXAMPLE 2 Finding a Composite Function

Suppose that $f(x) = x^2 + 3x - 1$ and $g(x) = 2x + 3$. Find:

(a) $f \circ g$ (b) $g \circ f$

State the domain of each composite function.

Solution The domain of f and the domain of g are all real numbers.

(a) $f \circ g = f(g(x)) = f(2x + 3) = (2x + 3)^2 + 3(2x + 3) - 1$

$f(x) = x^2 + 3x - 1$

$= 4x^2 + 12x + 9 + 6x + 9 - 1 = 4x^2 + 18x + 17$

Since the domains of both f and g are all real numbers, the domain of $f \circ g$ is all real numbers.

(b) $g \circ f = g(f(x)) = g(x^2 + 3x - 1) = 2(x^2 + 3x - 1) + 3$

$g(x) = 2x + 3$

$= 2x^2 + 6x - 2 + 3 = 2x^2 + 6x + 1$

Since the domains of both f and g are all real numbers, the domain of $f \circ g$ is all real numbers. ◀

Look back at Figure 1. In determining the domain of the composite function $(f \circ g)(x) = f(g(x))$, keep the following two thoughts in mind about the input x.

1. $g(x)$ must be defined so that any x not in the domain of g must be excluded.

2. $f(g(x))$ must be defined so that any x for which $g(x)$ is not in the domain of f is excluded.

EXAMPLE 3 Finding the Domain of $f \circ g$

Find the domain of $(f \circ g)(x)$ if $f(x) = \dfrac{1}{x+2}$ and $g(x) = \dfrac{4}{x-1}$.

Solution For $(f \circ g)(x) = f(g(x))$, we first note that the domain of g is $\{x|x \neq 1\}$, so we exclude 1 from the domain of $f \circ g$. Next, we note that the domain of f is $\{x|x \neq -2\}$, which means that $g(x)$ cannot equal -2. We solve the equation $g(x) = -2$ to determine what values of x to exclude.

$$\begin{aligned} \frac{4}{x-1} &= -2 \qquad g(x) = -2 \\ 4 &= -2(x-1) \\ 4 &= -2x + 2 \\ 2x &= -2 \\ x &= -1 \end{aligned}$$

We also exclude -1 from the domain of $f \circ g$. The domain of $f \circ g$ is $\{x|x \neq -1, x \neq 1\}$.

CHECK: For $x = 1$, $g(x) = \dfrac{4}{x-1}$ is not defined, so $(f \circ g)(x) = f(g(x))$ is not defined.

For $x = -1$, $g(-1) = \dfrac{4}{-2} = -2$, and $(f \circ g)(-1) = f(g(-1)) = f(-2)$ is not defined. ◀

NOW WORK PROBLEM 19.

EXAMPLE 4 Finding a Composite Function

Suppose that $f(x) = \dfrac{1}{x+2}$ and $g(x) = \dfrac{4}{x-1}$

Find: (a) $f \circ g$ (b) $f \circ f$

Then find the domain of each composite function.

Solution The domain of f is $\{x|x \neq -2\}$ and the domain of g is $\{x|x \neq 1\}$.

(a) $$(f \circ g)(x) = f(g(x)) = f\left(\frac{4}{x-1}\right) = \frac{1}{\dfrac{4}{x-1} + 2} = \frac{x-1}{4 + 2(x-1)} = \frac{x-1}{2x+2} = \frac{x-1}{2(x+1)}$$

$\uparrow$ $f(x) = \dfrac{1}{x+2}$ $\qquad$ $\uparrow$ Multiply by $\dfrac{x-1}{x-1}$.

In Example 3 we found the domain of $f \circ g$ to be $\{x|x \neq -1, x \neq 1\}$. We could also find the domain of $f \circ g$ by first looking at the domain of g: $\{x|x \neq 1\}$. We exclude 1 from the domain of $f \circ g$ as a result. Then we look at $(f \circ g)(x) = \dfrac{x-1}{2(x+1)}$ and notice that x cannot equal -1, since $x = -1$ results in division by zero. So we also exclude -1 from the domain of $f \circ g$. The domain of $f \circ g$ is $\{x|x \neq -1, x \neq 1\}$.

(b) $$(f \circ f)(x) = f(f(x)) = f\left(\frac{1}{x+2}\right) = \frac{1}{\dfrac{1}{x+2} + 2} = \frac{x+2}{1 + 2(x+2)} = \frac{x+2}{2x+5}$$

$\uparrow$ $f(x) = \dfrac{1}{x+2}$ $\qquad$ $\uparrow$ Multiply by $\dfrac{x+2}{x+2}$.

The domain of $f \circ f$ consists of those x in the domain of f, $\{x|x \neq -2\}$, for which

$$f(x) = \frac{1}{x+2} \neq -2$$

$$\frac{1}{x+2} = -2$$
$$1 = -2(x+2)$$
$$1 = -2x - 4$$
$$2x = -5$$
$$x = -\frac{5}{2}$$

or, equivalently,

$$x \neq -\frac{5}{2}$$

The domain of $f \circ f$ is $\left\{x \middle| x \neq -\dfrac{5}{2}, x \neq -2\right\}$.

We could also find the domain of $f \circ f$ by recognizing that -2 is not in the domain of f and so should be excluded from the domain of $f \circ f$. Then looking at $f \circ f$, we see that x cannot equal $-\dfrac{5}{2}$. (Do you see why?) Therefore, the domain of $f \circ f$ is $\left\{x \middle| x \neq -\dfrac{5}{2}, x \neq -2\right\}$. ◀

NOW WORK PROBLEMS 31 AND 33.

Examples 1(a), 1(b), and 2 illustrate that, in general, $f \circ g \neq g \circ f$. However, sometimes $f \circ g$ does equal $g \circ f$, as shown in the next example.

EXAMPLE 5 Showing That Two Composite Functions Are Equal

If $f(x) = 3x - 4$ and $g(x) = \dfrac{1}{3}(x + 4)$, show that

$$(f \circ g)(x) = (g \circ f)(x) = x$$

for every x.

Solution

$$\begin{aligned}(f \circ g)(x) &= f(g(x))\\ &= f\left(\frac{x+4}{3}\right) && g(x) = \frac{1}{3}(x+4) = \frac{x+4}{3}\\ &= 3 \cdot \frac{x+4}{3} - 4 && \text{Substitute } g(x) \text{ into the rule for } f,\ f(x) = 3x - 4.\\ &= x + 4 - 4 = x\end{aligned}$$

$$\begin{aligned}(g \circ f)(x) &= g(f(x))\\ &= g(3x - 4) && f(x) = 3x - 4\\ &= \frac{1}{3}[(3x - 4) + 4] && \text{Substitute } f(x) \text{ into the rule for } g,\ g(x) = \frac{1}{3}(x+4).\\ &= \frac{1}{3}(3x) = x\end{aligned}$$

We conclude that $(f \circ g)(x) = (g \circ f)(x) = x$. ◀

— Seeing the Concept —

Using a graphing utility, let

$Y_1 = f(x) = 3x - 4$

$Y_2 = g(x) = \frac{1}{3}(x + 4)$

$Y_3 = f \circ g$ and $Y_4 = g \circ f$

Using the viewing window $-3 \le x \le 3$, $-2 \le y \le 2$, graph only Y_3 and Y_4. What do you see? TRACE or create a TABLE to verify that $Y_3 = Y_4 = x$.

In the next section we shall see that there is an important relationship between functions f and g for which $(f \circ g)(x) = (g \circ f)(x) = x$.

NOW WORK PROBLEM 43.

Calculus Application

Some techniques in calculus require that we be able to determine the components of a composite function. For example, the function $H(x) = \sqrt{x+1}$ is the composition of the functions f and g, where $f(x) = \sqrt{x}$ and $g(x) = x + 1$, because $H(x) = (f \circ g)(x) = f(g(x)) = f(x+1) = \sqrt{x+1}$.

EXAMPLE 6 Finding the Components of a Composite Function

Find functions f and g such that $f \circ g = H$ if $H(x) = (x^2 + 1)^{50}$.

Solution The function H takes $x^2 + 1$ and raises it to the power 50. A natural way to decompose H is to raise the function $g(x) = x^2 + 1$ to the power 50. If we let $f(x) = x^{50}$ and $g(x) = x^2 + 1$, then

$$\begin{aligned}(f \circ g)(x) &= f(g(x))\\ &= f(x^2 + 1)\\ &= (x^2 + 1)^{50} = H(x)\end{aligned}$$ ◀

Other functions f and g may be found for which $f \circ g = H$ in Example 6. For example, if $f(x) = x^2$ and $g(x) = (x^2 + 1)^{25}$, then

$$(f \circ g)(x) = f(g(x)) = f((x^2+1)^{25}) = [(x^2+1)^{25}]^2 = (x^2+1)^{50}$$

Although the functions f and g found as a solution to Example 6 are not unique, there is usually a "natural" selection for f and g that comes to mind first.

EXAMPLE 7 **Finding the Components of a Composite Function**

Find functions f and g such that $f \circ g = H$ if $H(x) = \dfrac{1}{x+1}$.

Solution Here H is the reciprocal of $g(x) = x + 1$. If we let $f(x) = \dfrac{1}{x}$ and $g(x) = x + 1$, we find that

$$(f \circ g)(x) = f(g(x)) = f(x+1) = \frac{1}{x+1} = H(x)$$

◀

4.1 Assess Your Understanding

'Are You Prepared?' *Answers are given at the end of these exercises. If you get a wrong answer, read the pages listed in red.*

1. Find $f(3)$ if $f(x) = -4x^2 + 5x$. (pp. 53–55)
2. Find $f(3x)$ if $f(x) = 4 - 2x^2$. (pp. 53–55)
3. Find the domain of the function $f(x) = \dfrac{x^2 - 1}{x^2 - 4}$. (pp. 57–58)

Concepts and Vocabulary

4. If $f(x) = x + 1$ and $g(x) = x^3$, then __________ $= (x+1)^3$.
5. *True or False:* $f(g(x)) = f(x) \cdot g(x)$.
6. *True or False:* The domain of the composite function $(f \circ g)(x)$ is the same as the domain of $g(x)$.

Exercises

In Problems 7 and 8, evaluate each expression using the values given in the table.

7.

x	−3	−2	−1	0	1	2	3
f(x)	−7	−5	−3	−1	3	5	5
g(x)	8	3	0	−1	0	3	8

(a) $(f \circ g)(1)$ (b) $(f \circ g)(-1)$ (c) $(g \circ f)(-1)$
(d) $(g \circ f)(0)$ (e) $(g \circ g)(-2)$ (f) $(f \circ f)(-1)$

8.

x	−3	−2	−1	0	1	2	3
f(x)	11	9	7	5	3	1	−1
g(x)	−8	−3	0	1	0	−3	−8

(a) $(f \circ g)(1)$ (b) $(f \circ g)(2)$ (c) $(g \circ f)(2)$
(d) $(g \circ f)(3)$ (e) $(g \circ g)(1)$ (f) $(f \circ f)(3)$

In Problems 9–18, for the given functions f and g, find

(a) $(f \circ g)(4)$ (b) $(g \circ f)(2)$ (c) $(f \circ f)(1)$ (d) $(g \circ g)(0)$

9. $f(x) = 2x;\quad g(x) = 3x^2 + 1$
10. $f(x) = 3x + 2;\quad g(x) = 2x^2 - 1$
11. $f(x) = 4x^2 - 3;\quad g(x) = 3 - \frac{1}{2}x^2$
12. $f(x) = 2x^2;\quad g(x) = 1 - 3x^2$
13. $f(x) = \sqrt{x};\quad g(x) = 2x$
14. $f(x) = \sqrt{x+1};\quad g(x) = 3x$

15. $f(x) = |x|;\ g(x) = \dfrac{1}{x^2 + 1}$

16. $f(x) = |x - 2|;\ g(x) = \dfrac{3}{x^2 + 2}$

17. $f(x) = \dfrac{3}{x + 1};\ g(x) = \sqrt[3]{x}$

18. $f(x) = x^{3/2};\ g(x) = \dfrac{2}{x + 1}$

In Problems 19–26, find the domain of the composite function $f \circ g$.

19. $f(x) = \dfrac{3}{x - 1};\ g(x) = \dfrac{2}{x}$

20. $f(x) = \dfrac{1}{x + 3};\ g(x) = \dfrac{-2}{x}$

21. $f(x) = \dfrac{x}{x - 1};\ g(x) = \dfrac{-4}{x}$

22. $f(x) = \dfrac{x}{x + 3};\ g(x) = \dfrac{2}{x}$

23. $f(x) = \sqrt{x};\ g(x) = 2x + 3$

24. $f(x) = x - 2;\ g(x) = \sqrt{1 - x}$

25. $f(x) = x^2 + 1;\ g(x) = \sqrt{x - 1}$

26. $f(x) = x^2 + 4;\ g(x) = \sqrt{x - 2}$

In Problems 27–42, for the given functions f and g, find

(a) $f \circ g$ (b) $g \circ f$ (c) $f \circ f$ (d) $g \circ g$

State the domain of each composite function.

27. $f(x) = 2x + 3;\ g(x) = 3x$

28. $f(x) = -x;\ g(x) = 2x - 4$

29. $f(x) = 3x + 1;\ g(x) = x^2$

30. $f(x) = x + 1;\ g(x) = x^2 + 4$

31. $f(x) = x^2;\ g(x) = x^2 + 4$

32. $f(x) = x^2 + 1;\ g(x) = 2x^2 + 3$

33. $f(x) = \dfrac{3}{x - 1};\ g(x) = \dfrac{2}{x}$

34. $f(x) = \dfrac{1}{x + 3};\ g(x) = \dfrac{-2}{x}$

35. $f(x) = \dfrac{x}{x - 1};\ g(x) = \dfrac{-4}{x}$

36. $f(x) = \dfrac{x}{x + 3};\ g(x) = \dfrac{2}{x}$

37. $f(x) = \sqrt{x};\ g(x) = 2x + 3$

38. $f(x) = \sqrt{x - 2};\ g(x) = 1 - 2x$

39. $f(x) = x^2 + 1;\ g(x) = \sqrt{x - 1}$

40. $f(x) = x^2 + 4;\ g(x) = \sqrt{x - 2}$

41. $f(x) = ax + b;\ g(x) = cx + d$

42. $f(x) = \dfrac{ax + b}{cx + d};\ g(x) = mx$

In Problems 43–50, show that $(f \circ g)(x) = (g \circ f)(x) = x$.

43. $f(x) = 2x;\ g(x) = \dfrac{1}{2}x$

44. $f(x) = 4x;\ g(x) = \dfrac{1}{4}x$

45. $f(x) = x^3;\ g(x) = \sqrt[3]{x}$

46. $f(x) = x + 5;\ g(x) = x - 5$

47. $f(x) = 2x - 6;\ g(x) = \dfrac{1}{2}(x + 6)$

48. $f(x) = 4 - 3x;\ g(x) = \dfrac{1}{3}(4 - x)$

49. $f(x) = ax + b;\ g(x) = \dfrac{1}{a}(x - b),\ \ a \neq 0$

50. $f(x) = \dfrac{1}{x};\ g(x) = \dfrac{1}{x}$

In Problems 51–56, find functions f and g so that $f \circ g = H$.

51. $H(x) = (2x + 3)^4$

52. $H(x) = (1 + x^2)^3$

53. $H(x) = \sqrt{x^2 + 1}$

54. $H(x) = \sqrt{1 - x^2}$

55. $H(x) = |2x + 1|$

56. $H(x) = |2x^2 + 3|$

57. If $f(x) = 2x^3 - 3x^2 + 4x - 1$ and $g(x) = 2$, find $(f \circ g)(x)$ and $(g \circ f)(x)$.

58. If $f(x) = \dfrac{x}{x - 1}$, find $(f \circ f)(x)$.

59. If $f(x) = 2x^2 + 5$ and $g(x) = 3x + a$, find a so that the graph of $f \circ g$ crosses the y-axis at 23.

60. If $f(x) = 3x^2 - 7$ and $g(x) = 2x + a$, find a so that the graph of $f \circ g$ crosses the y-axis at 68.

61. Surface Area of a Balloon The surface area S (in square meters) of a hot-air balloon is given by

$$S(r) = 4\pi r^2$$

where r is the radius of the balloon (in meters). If the radius r is increasing with time t (in seconds) according to the formula $r(t) = \dfrac{2}{3}t^3$, $t \geq 0$, find the surface area S of the balloon as a function of the time t.

62. Volume of a Balloon The volume V (in cubic meters) of the hot-air balloon described in Problem 61 is given by $V(r) = \frac{4}{3}\pi r^3$. If the radius r is the same function of t as in Problem 61, find the volume V as a function of the time t.

63. **Automobile Production** The number N of cars produced at a certain factory in 1 day after t hours of operation is given by $N(t) = 100t - 5t^2, 0 \le t \le 10$. If the cost C (in dollars) of producing N cars is $C(N) = 15{,}000 + 8000N$, find the cost C as a function of the time t of operation of the factory.

64. **Environmental Concerns** The spread of oil leaking from a tanker is in the shape of a circle. If the radius r (in feet) of the spread after t hours is $r(t) = 200\sqrt{t}$, find the area A of the oil slick as a function of the time t.

65. **Production Cost** The price p of a certain product and the quantity x sold obey the demand equation

$$p = -\frac{1}{4}x + 100, \qquad 0 \le x \le 400$$

Suppose that the cost C of producing x units is

$$C = \frac{\sqrt{x}}{25} + 600$$

Assuming that all items produced are sold, find the cost C as a function of the price p.

[**Hint:** Solve for x in the demand equation and then form the composite.]

66. **Cost of a Commodity** The price p of a certain commodity and the quantity x sold obey the demand equation

$$p = -\frac{1}{5}x + 200, \qquad 0 \le x \le 1000$$

Suppose that the cost C of producing x units is

$$C = \frac{\sqrt{x}}{10} + 400$$

Assuming that all items produced are sold, find the cost C as a function of the price p.

67. **Volume of a Cylinder** The volume V of a right circular cylinder of height h and radius r is $V = \pi r^2 h$. If the height is twice the radius, express the volume V as a function of r.

68. **Volume of a Cone** The volume V of a right circular cone is $V = \frac{1}{3}\pi r^2 h$. If the height is twice the radius, express the volume V as a function of r.

69. **Foreign Exchange** Traders often buy foreign currency in hope of making money when the currency's value changes. For example, on October 20, 2003, one U.S. dollar could purchase 0.857118 Euros and one Euro could purchase 128.6054 yen. Let $f(x)$ represent the number of Euros one can buy with x dollars and let $g(x)$ represent the number of yen one can buy with x Euros.
   (a) Find a function that relates dollars to Euros.
   (b) Find a function that relates Euros to yen.
   (c) Use the results of parts (a) and (b) to find a function that relates dollars to yen. That is, find $g(f(x))$.
   (d) What is $g(f(1000))$?

70. If f and g are odd functions, show that the composite function $f \circ g$ is also odd.

71. If f is an odd function and g is an even function, show that the composite functions $f \circ g$ and $g \circ f$ are also even.

'Are You Prepared?' Answers

1. -21 **2.** $4 - 18x^2$ **3.** $\{x \mid x \ne -2, x \ne 2\}$

4.2 Inverse Functions

PREPARING FOR THIS SECTION *Before getting started, review the following:*

- Functions (Section 2.1, pp. 53–58)
- Increasing/Decreasing Functions (Section 2.3, pp. 74–75)

Now work the 'Are You Prepared?' problems on page 241.

OBJECTIVES
1 Determine the Inverse of a Function
2 Obtain the Graph of the Inverse Function from the Graph of the Function
3 Find the Inverse Function f^{-1}

1 In Section 2.1 we said that a function f can be thought of as a machine that receives as input a number, say x, from the domain, manipulates it, and outputs the value $f(x)$. The **inverse of f** receives as input a number $f(x)$, manipulates it, and outputs the value x.

EXAMPLE 1 Finding the Inverse of a Function

Find the inverse of the following functions.

(a) Let the domain of the function represent the employees of Yolanda's Preowned Car Mart and let the range represent their base salaries.

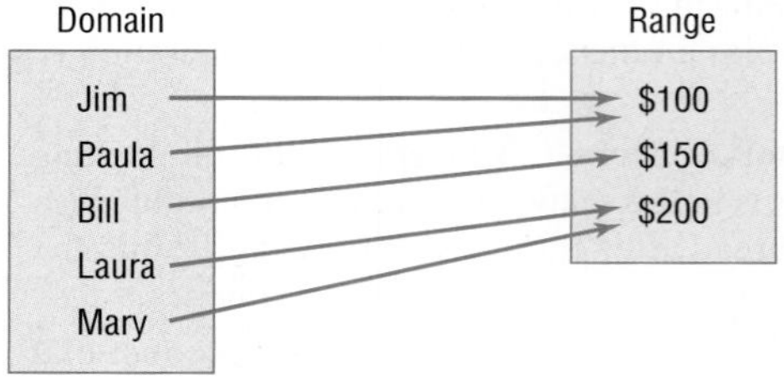

(b) Let the domain of the function represent the employees of Yolanda's Preowned Car Mart and let the range represent their spouse's names.

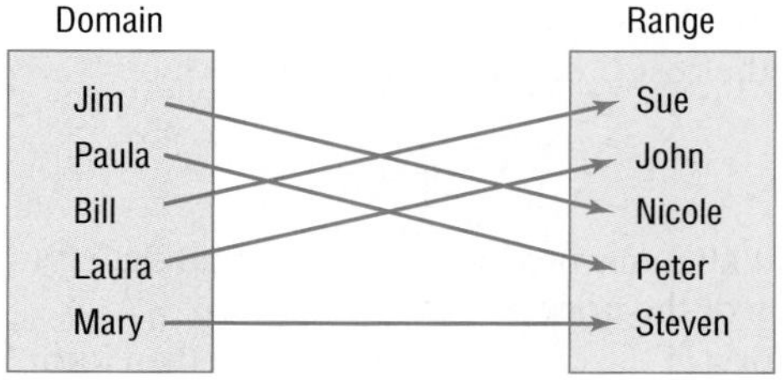

Solution (a) The elements in the domain represent inputs to the function, and the elements in the range represent the outputs. To find the inverse, interchange the elements in the domain with the elements in the range. For example, the function receives as input Bill and outputs $150. So the inverse receives an input $150 and outputs Bill. The inverse of the given function takes the form

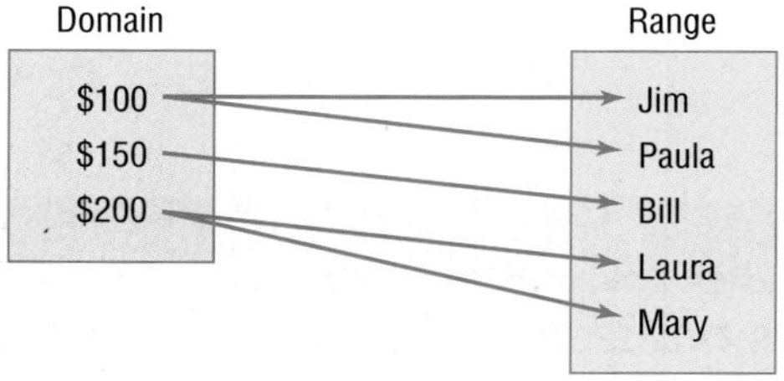

(b) The inverse of the given function is

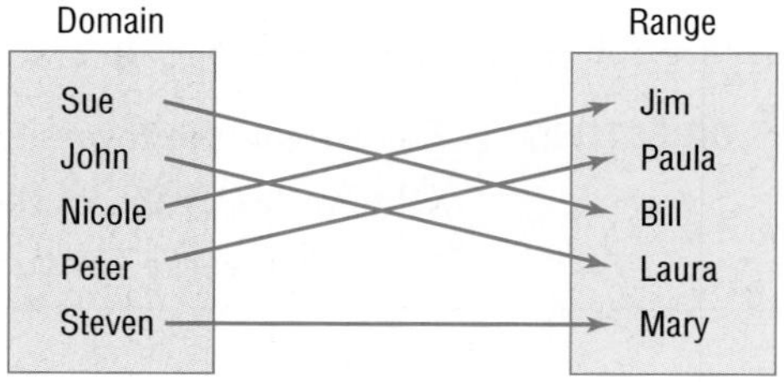

◀

NOW WORK PROBLEM 9(a).

If the function f is a set of ordered pairs (x, y), then the inverse of f is the set of ordered pairs (y, x).

EXAMPLE 2 Finding the Inverse of a Function

Find the inverse of the following functions:

(a) $\{(-3, 9), (-2, 4), (-1, 1), (0, 0), (1, 1), (2, 4), (3, 9)\}$
(b) $\{(-3, -27), (-2, -8), (-1, -1), (0, 0), (1, 1), (2, 8), (3, 27)\}$

Solution (a) The inverse of the given function is found by interchanging the entries in each ordered pair and so is given by

$$\{(9, -3), (4, -2), (1, -1), (0, 0), (1, 1), (4, 2), (9, 3)\}$$

(b) The inverse of the given function is

$$\{(-27, -3), (-8, -2), (-1, -1), (0, 0), (1, 1), (8, 2), (27, 3)\}$$ ◄

NOW WORK PROBLEM 13(a).

Exploration

Look back at the functions from Examples 1 and 2. Notice that the functions in Examples 1(b) and 2(b) have inverses that are also functions. However, the functions given in Examples 1(a) and 2(a) have inverses that are not functions. How are the functions in Examples 1(b) and 2(b) similar? How are the functions in Examples 1(a) and 2(a) similar?

RESULT Examples 1(b) and 2(b) are similar in that each element in the domain corresponds to one element in the range. Consider Example 1(b), where Jim corresponds to Nicole, Paula corresponds to Peter, and so on. However, in Examples 1(a) and 2(a), we notice that two different elements in the domain correspond to the same element in the range. In Example 2(a), we see that -3 corresponds to 9 and 3 also corresponds to 9.

The results of the Exploration lead to the following definition.

In Words
A function is one-to-one if any two different inputs never correspond to the same output.

When the inverse of a function f is itself a function, then f is said to be a **one-to-one function.** That is, f is **one-to-one** if, for any choice of elements x_1 and x_2 in the domain of f, with $x_1 \neq x_2$, the corresponding values $f(x_1)$ and $f(x_2)$ are unequal, $f(x_1) \neq f(x_2)$.

In other words, a function f is one-to-one if no y in the range is the image of more than one x in the domain. A function is not one-to-one if two different elements in the domain correspond to the same element in the range. In Example 2(a), the elements -3 and 3 both correspond to 9, so the function is not one-to-one. Figure 3 illustrates the definition.

Figure 3

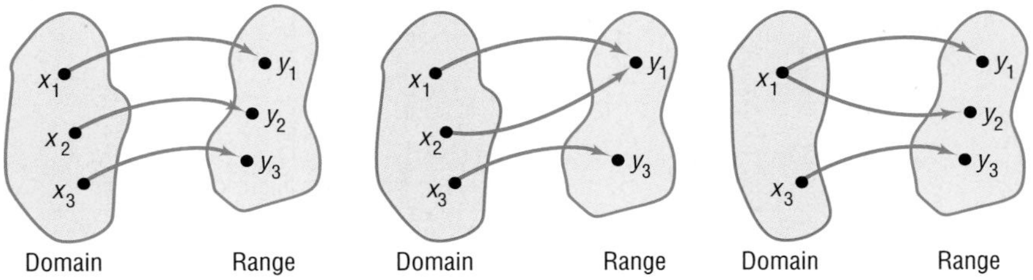

(a) One-to-one function: Each x in the domain has one and only one image in the range. No y in the range is the image of more than one x.

(b) Not a one-to-one function: y_1 is the image of both x_1 and x_2

(c) Not a function: x_1 has two images, y_1 and y_2

NOW WORK PROBLEMS 9(b) AND 13(b).

If the graph of a function f is known, there is a simple test, called the **horizontal-line test,** to determine whether f is one-to-one.

Theorem

Horizontal-line Test

If every horizontal line intersects the graph of a function f in at most one point, then f is one-to-one.

Figure 4
$f(x_1) = f(x_2) = h$, and $x_1 \neq x_2$; f is not a one-to-one function.

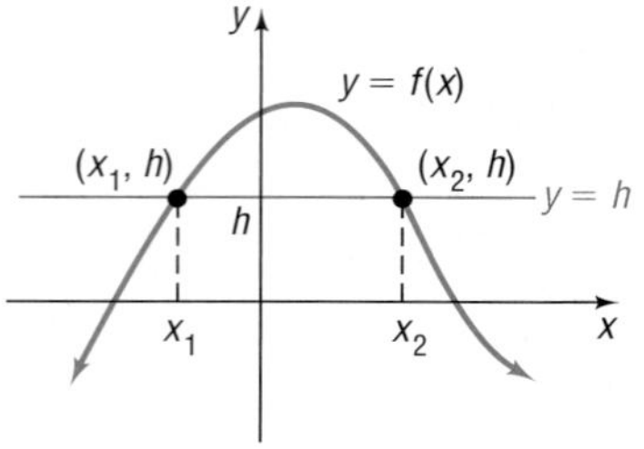

The reason that this test works can be seen in Figure 4, where the horizontal line $y = h$ intersects the graph at two distinct points, (x_1, h) and (x_2, h). Since h is the image of both x_1 and x_2, $x_1 \neq x_2$, f is not one-to-one. Based on Figure 4, we can state the horizontal-line test in another way: If the graph of any horizontal line intersects the graph of a function f at more than one point, then f is not one-to-one.

EXAMPLE 3 Using the Horizontal-line Test

For each function, use the graph to determine whether the function is one-to-one.

(a) $f(x) = x^2$ (b) $g(x) = x^3$

Solution

(a) Figure 5(a) illustrates the horizontal-line test for $f(x) = x^2$. The horizontal line $y = 1$ intersects the graph of f twice, at $(1, 1)$ and at $(-1, 1)$, so f is not one-to-one.

(b) Figure 5(b) illustrates the horizontal-line test for $g(x) = x^3$. Because every horizontal line will intersect the graph of g exactly once, it follows that g is one-to-one.

Figure 5

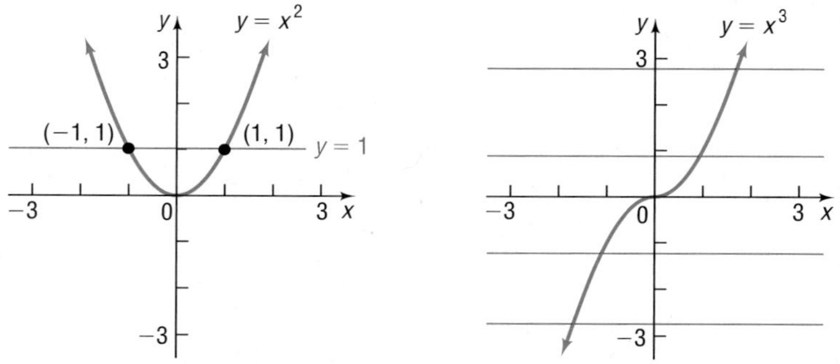

(a) A horizontal line intersects the graph twice; thus, f is not one-to-one

(b) Every horizontal line intersects the graph exactly once; thus, g is one-to-one

NOW WORK PROBLEM 17.

Let's look more closely at the one-to-one function $g(x) = x^3$. This function is an increasing function on the interval $(-\infty, \infty)$, its domain. Because an increasing (or decreasing) function will always have different

y-values for unequal x-values, it follows that a function that is increasing (or decreasing) on an interval is also a one-to-one function on the interval.

Theorem

> A function that is increasing on an interval I is a one-to-one function on I.
>
> A function that is decreasing on an interval I is a one-to-one function on I.

Inverse Function of $y = f(x)$

If f is a one-to-one function, its inverse is a function. Then, to each x in the domain of f, there is exactly one y in the range (because f is a function); and to each y in the range of f, there is exactly one x in the domain (because f is one-to-one). The correspondence from the range of f back to the domain of f is called the **inverse function of f** and is denoted by the symbol f^{-1}. Figure 6 illustrates this definition.

Figure 6

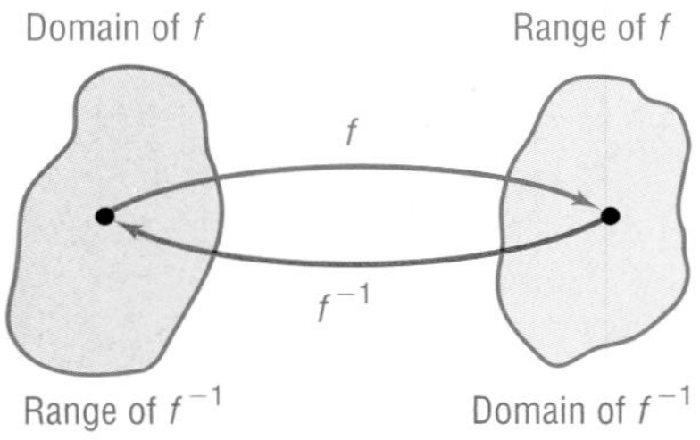

WARNING: Be careful! f^{-1} is a symbol for the inverse function of f. The -1 used in f^{-1} is not an exponent. That is, f^{-1} does *not* mean the reciprocal of f; $f^{-1}(x)$ is not equal to $\dfrac{1}{f(x)}$. ■

Two facts are now apparent about a function f and its inverse f^{-1}.

Domain of f = Range of f^{-1}	Range of f = Domain of f^{-1}

Look again at Figure 6 to visualize the relationship. If we start with x, apply f, and then apply f^{-1}, we get x back again. If we start with x, apply f^{-1}, and then apply f, we get the number x back again. To put it simply, what f does, f^{-1} undoes, and vice versa.

Input x $\xrightarrow{\text{Apply } f}$ $f(x)$ $\xrightarrow{\text{Apply } f^{-1}}$ $f^{-1}(f(x)) = x$

Input x $\xrightarrow{\text{Apply } f^{-1}}$ $f^{-1}(x)$ $\xrightarrow{\text{Apply } f}$ $f(f^{-1}(x)) = x$

In other words,

$$f^{-1}(f(x)) = x \quad \text{and} \quad f(f^{-1}(x)) = x$$

For example, the function $f(x) = 2x$ multiplies the argument x by 2. The inverse function f^{-1} undoes whatever f does. So the inverse function of f is $f^{-1}(x) = \dfrac{1}{2}x$, which divides the argument by 2. For $x = 3$, we have

$$f(3) = 2 \cdot 3 = 6 \quad \text{and} \quad f^{-1}(6) = \frac{1}{2} \cdot 6 = 3,$$

so f^{-1} undoes what f did. We can verify this by showing that

$$f^{-1}(f(x)) = f^{-1}(2x) = \frac{1}{2}(2x) = x \quad \text{and} \quad f(f^{-1}(x)) = f\left(\frac{1}{2}x\right) = 2\left(\frac{1}{2}x\right) = x$$

See Figure 7.

Figure 7

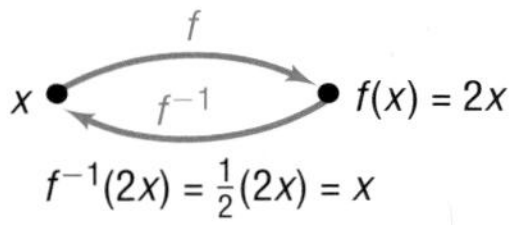

EXAMPLE 4 **Verifying Inverse Functions**

(a) We verify that the inverse of $g(x) = x^3$ is $g^{-1}(x) = \sqrt[3]{x}$ by showing that

$$g^{-1}(g(x)) = g^{-1}(x^3) = \sqrt[3]{x^3} = x$$

and

$$g(g^{-1}(x)) = g(\sqrt[3]{x}) = (\sqrt[3]{x})^3 = x$$

(b) We verify that the inverse of $h(x) = 3x$ is $h^{-1}(x) = \frac{1}{3}x$ by showing that

$$h^{-1}(h(x)) = h^{-1}(3x) = \frac{1}{3}(3x) = x$$

and

$$h(h^{-1}(x)) = h\left(\frac{1}{3}x\right) = 3\left(\frac{1}{3}x\right) = x$$

(c) We verify that the inverse of $f(x) = 2x + 3$ is $f^{-1}(x) = \frac{1}{2}(x - 3)$ by showing that

$$f^{-1}(f(x)) = f^{-1}(2x + 3) = \frac{1}{2}[(2x + 3) - 3] = \frac{1}{2}(2x) = x$$

and

$$f(f^{-1}(x)) = f\left(\frac{1}{2}(x - 3)\right) = 2\left[\frac{1}{2}(x - 3)\right] + 3 = (x - 3) + 3 = x$$ ◀

NOW WORK PROBLEM 29.

Exploration

Simultaneously graph $Y_1 = x$, $Y_2 = x^3$, and $Y_3 = \sqrt[3]{x}$ on a square screen, using the viewing rectangle $-3 \le x \le 3$, $-2 \le y \le 2$. What do you observe about the graphs of $Y_2 = x^3$, its inverse $Y_3 = \sqrt[3]{x}$, and the line $Y_1 = x$?

Repeat this experiment by simultaneously graphing $Y_1 = x$, $Y_2 = 2x + 3$, and $Y_3 = \frac{1}{2}(x - 3)$, using the viewing rectangle $-6 \le x \le 3$, $-8 \le y \le 4$. Do you see the symmetry of the graph of Y_2 and its inverse Y_3 with respect to the line $Y_1 = x$?

Geometric Interpretation

Figure 8

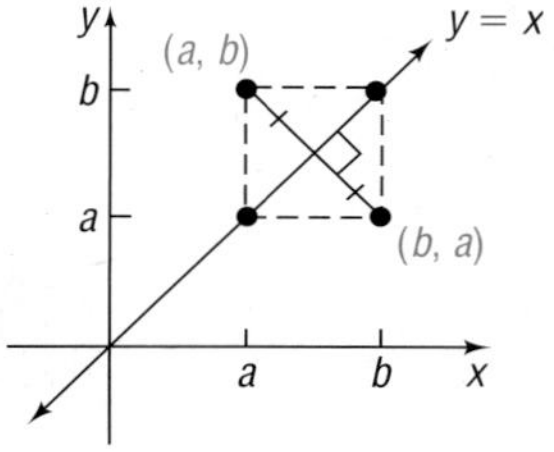

2 Suppose that (a, b) is a point on the graph of the one-to-one function f defined by $y = f(x)$. Then $b = f(a)$. This means that $a = f^{-1}(b)$, so (b, a) is a point on the graph of the inverse function f^{-1}. The relationship between the point (a, b) on f and the point (b, a) on f^{-1} is shown in Figure 8. The line segment containing (a, b) and (b, a) is perpendicular to the line $y = x$ and is bisected by the line $y = x$. (Do you see why?) It follows that the point (b, a) on f^{-1} is the reflection about the line $y = x$ of the point (a, b) on f.

Theorem The graph of a function f and the graph of its inverse f^{-1} are symmetric with respect to the line $y = x$.

Figure 9 illustrates this result. Notice that, once the graph of f is known, the graph of f^{-1} may be obtained by reflecting the graph of f about the line $y = x$.

Figure 9

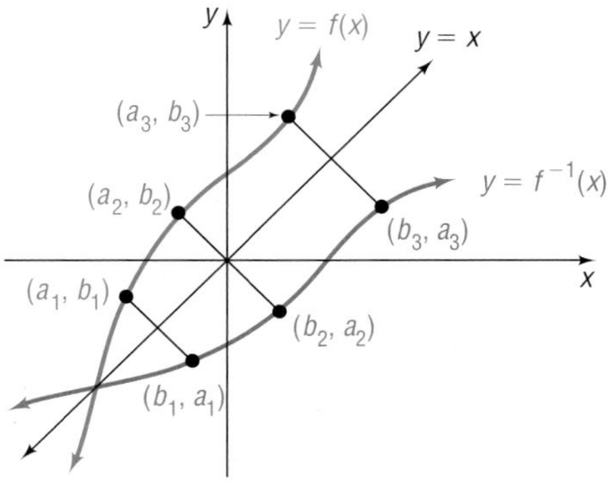

EXAMPLE 5 Graphing the Inverse Function

The graph in Figure 10(a) is that of a one-to-one function $y = f(x)$. Draw the graph of its inverse.

Solution We begin by adding the graph of $y = x$ to Figure 10(a). Since the points $(-2, -1)$, $(-1, 0)$, and $(2, 1)$ are on the graph of f, we know that the points $(-1, -2)$, $(0, -1)$, and $(1, 2)$ must be on the graph of f^{-1}. Keeping in mind that the graph of f^{-1} is the reflection about the line $y = x$ of the graph of f, we can draw f^{-1}. See Figure 10(b).

Figure 10

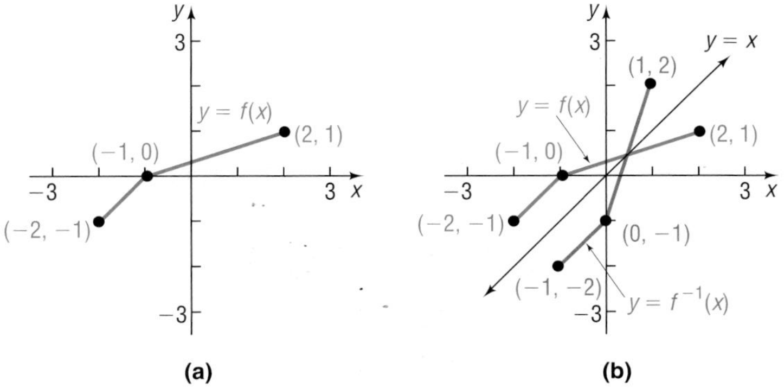

NOW WORK PROBLEM 23.

3 Finding the Inverse Function f^{-1}

The fact that the graphs of a one-to-one function f and its inverse function f^{-1} are symmetric with respect to the line $y = x$ tells us more. It says that we can obtain f^{-1} by interchanging the roles of x and y in f. Look again at Figure 9. If f is defined by the equation

$$y = f(x)$$

then f^{-1} is defined by the equation

$$x = f(y)$$

The equation $x = f(y)$ defines f^{-1} *implicitly*. If we can solve this equation for y, we will have the *explicit* form of f^{-1}, that is,

$$y = f^{-1}(x)$$

Let's use this procedure to find the inverse of $f(x) = 2x + 3$. (Since f is a linear function and is increasing, we know that f is one-to-one and so has an inverse function.)

EXAMPLE 6 Finding the Inverse Function f^{-1}

Find the inverse of $f(x) = 2x + 3$. Also find the domain and range of f and f^{-1}. Graph f and f^{-1} on the same coordinate axes.

Solution In the equation $y = 2x + 3$, interchange the variables x and y. The result,

$$x = 2y + 3$$

is an equation that defines the inverse f^{-1} implicitly. To find the explicit form, we solve for y.

$$2y + 3 = x$$
$$2y = x - 3$$
$$y = \frac{1}{2}(x - 3)$$

Figure 11

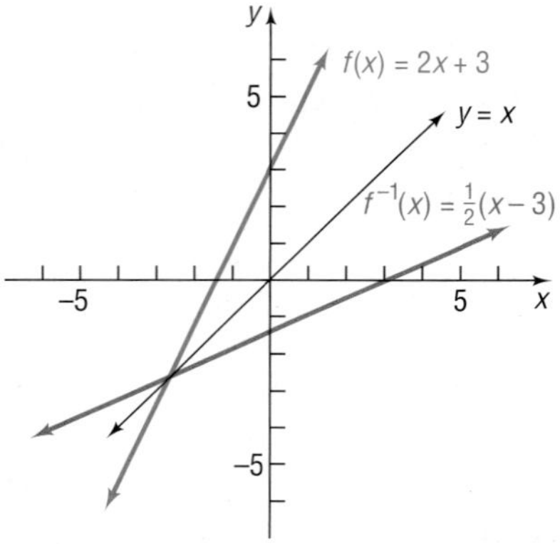

The explicit form of the inverse f^{-1} is therefore

$$f^{-1}(x) = \frac{1}{2}(x - 3)$$

which we verified in Example 4(c).

Next we find

$$\text{Domain of } f = \text{Range of } f^{-1} = (-\infty, \infty)$$
$$\text{Range of } f = \text{Domain of } f^{-1} = (-\infty, \infty)$$

The graphs of $f(x) = 2x + 3$ and its inverse $f^{-1}(x) = \frac{1}{2}(x - 3)$ are shown in Figure 11. Note the symmetry of the graphs with respect to the line $y = x$. ◀

We outline next the steps to follow for finding the inverse of a one-to-one function.

Procedure for Finding the Inverse of a One-to-One Function

STEP 1: In $y = f(x)$, interchange the variables x and y to obtain

$$x = f(y)$$

This equation defines the inverse function f^{-1} implicitly.

STEP 2: If possible, solve the implicit equation for y in terms of x to obtain the explicit form of f^{-1}.

$$y = f^{-1}(x)$$

STEP 3: Check the result by showing that

$$f^{-1}(f(x)) = x \quad \text{and} \quad f(f^{-1}(x)) = x$$

EXAMPLE 7 Finding the Inverse Function

The function

$$f(x) = \frac{2x + 1}{x - 1}, \qquad x \neq 1$$

is one-to-one. Find its inverse and check the result.

Solution **STEP 1:** Interchange the variables x and y in

$$y = \frac{2x + 1}{x - 1}$$

to obtain

$$x = \frac{2y + 1}{y - 1}$$

STEP 2: Solve for y.

$$x = \frac{2y + 1}{y - 1}$$

$$x(y - 1) = 2y + 1 \qquad \text{Multiply both sides by } y - 1.$$

$$xy - x = 2y + 1 \qquad \text{Apply the Distributive Property.}$$

$$xy - 2y = x + 1 \qquad \text{Subtract } 2y \text{ from both sides; add } x \text{ to both sides.}$$

$$(x - 2)y = x + 1 \qquad \text{Factor the left side.}$$

$$y = \frac{x + 1}{x - 2} \qquad \text{Divide by } x - 2.$$

The inverse is

$$f^{-1}(x) = \frac{x + 1}{x - 2}, \qquad x \neq 2 \qquad \text{Replace } y \text{ by } f^{-1}(x).$$

STEP 3: CHECK:

$$f^{-1}(f(x)) = f^{-1}\left(\frac{2x + 1}{x - 1}\right) = \frac{\frac{2x + 1}{x - 1} + 1}{\frac{2x + 1}{x - 1} - 2} = \frac{2x + 1 + x - 1}{2x + 1 - 2(x - 1)} = \frac{3x}{3} = x$$

$$f(f^{-1}(x)) = f\left(\frac{x + 1}{x - 2}\right) = \frac{2\left(\frac{x + 1}{x - 2}\right) + 1}{\frac{x + 1}{x - 2} - 1} = \frac{2(x + 1) + x - 2}{x + 1 - (x - 2)} = \frac{3x}{3} = x$$

◀

Exploration

In Example 7, we found that, if $f(x) = \frac{2x + 1}{x - 1}$, then $f^{-1}(x) = \frac{x + 1}{x - 2}$. Compare the vertical and horizontal asymptotes of f and f^{-1}. What did you find? Are you surprised?

RESULT You should have determined that the vertical asymptote of f is $x = 1$ and the horizontal asymptote is $y = 2$. The vertical asymptote of f^{-1} is $x = 2$, and the horizontal asymptote is $y = 1$.

NOW WORK PROBLEM 41.

We said in Chapter 2 that finding the range of a function f is not easy. However, if f is one-to-one, we can find its range by finding the domain of the inverse function f^{-1}.

EXAMPLE 8 Finding the Range of a Function

Find the domain and range of

$$f(x) = \frac{2x + 1}{x - 1}$$

Solution The domain of f is $\{x | x \neq 1\}$. To find the range of f, we first find the inverse f^{-1}. Based on Example 7, we have

$$f^{-1}(x) = \frac{x + 1}{x - 2}$$

The domain of f^{-1} is $\{x | x \neq 2\}$, so the range of f is $\{y | y \neq 2\}$. ◀

NOW WORK PROBLEM 55.

If a function is not one-to-one, then its inverse is not a function. Sometimes, though, an appropriate restriction on the domain of such a function will yield a new function that is one-to-one. Then its inverse is a function. Let's look at an example of this common practice.

EXAMPLE 9 Finding the Inverse of a Domain-restricted Function

Find the inverse of $y = f(x) = x^2$ if $x \geq 0$. Graph f and its inverse, f^{-1}.

Solution The function $y = x^2$ is not one-to-one. [Refer to Example 3(a).] However, if we restrict this function to only that part of its domain for which $x \geq 0$, as indicated, we have a new function that is increasing on the interval $[0, \infty)$ and therefore is one-to-one on $[0, \infty)$. As a result, the function defined by $y = f(x) = x^2$, $x \geq 0$, has an inverse function, f^{-1}.

We follow the steps given previously to find f^{-1}.

STEP 1: In the equation $y = x^2$, $x \geq 0$, interchange the variables x and y. The result is

$$x = y^2, \qquad y \geq 0$$

This equation defines (implicitly) the inverse function.

STEP 2: We solve for y to get the explicit form of the inverse. Since $y \geq 0$, only one solution for y is obtained, namely,

$$y = \sqrt{x}$$

So

$$f^{-1}(x) = \sqrt{x}$$

STEP 3: **CHECK:** $f^{-1}(f(x)) = f^{-1}(x^2) = \sqrt{x^2} = |x| = x$, since $x \geq 0$

$$f(f^{-1}(x)) = f(\sqrt{x}) = (\sqrt{x})^2 = x.$$

Figure 12 illustrates the graphs of $f(x) = x^2$, $x \geq 0$, and $f^{-1}(x) = \sqrt{x}$. ◀

Figure 12

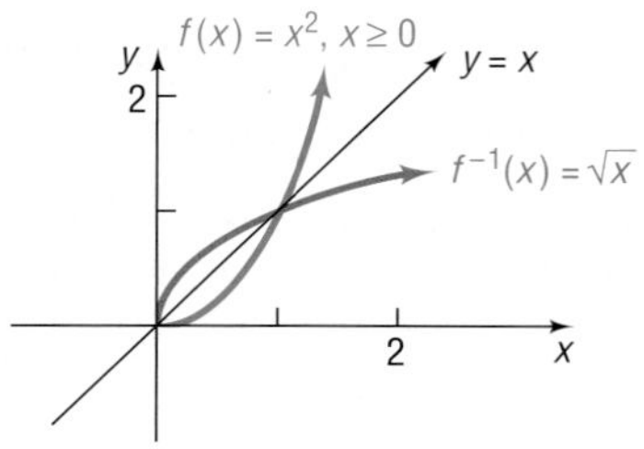

Summary

1. If a function f is one-to-one, then it has an inverse function f^{-1}.
2. Domain of f = Range of f^{-1}; Range of f = Domain of f^{-1}.
3. To verify that f^{-1} is the inverse of f, show that $f^{-1}(f(x)) = x$ and $f(f^{-1}(x)) = x$.
4. The graphs of f and f^{-1} are symmetric with respect to the line $y = x$.
5. To find the range of a one-to-one function f, find the domain of the inverse function f^{-1}.

4.2 Assess Your Understanding

'Are You Prepared?' *Answers are given at the end of these exercises. If you get a wrong answer, read the pages listed in red.*

1. Is the set of ordered pairs $\{(1, 3), (2, 3), (-1, 2)\}$ a function? Why or why not? (pp. 53–58)
2. Where is the function $f(x) = x^2$ increasing? Where is it decreasing? (pp. 74–75)
3. Where is the function $f(x) = x^3$ increasing? Where is it decreasing? (pp. 74–75)

Concepts and Vocabulary

4. If every horizontal line intersects the graph of a function f at no more than one point, then f is a(n) __________ function.
5. If f^{-1} denotes the inverse of a function f, then the graphs of f and f^{-1} are symmetric with respect to the line __________.
6. If the domain of a one-to-one function f is $[4, \infty)$, then the range of its inverse, f^{-1}, is __________.
7. *True or False:* If f and g are inverse functions, then the domain of f is the same as the domain of g.
8. *True or False:* If f and g are inverse functions, then their graphs are symmetric with respect to the line $y = x$.

Exercises

In Problems 9–16, (a) find the inverse and (b) determine whether the inverse is a function.

9.
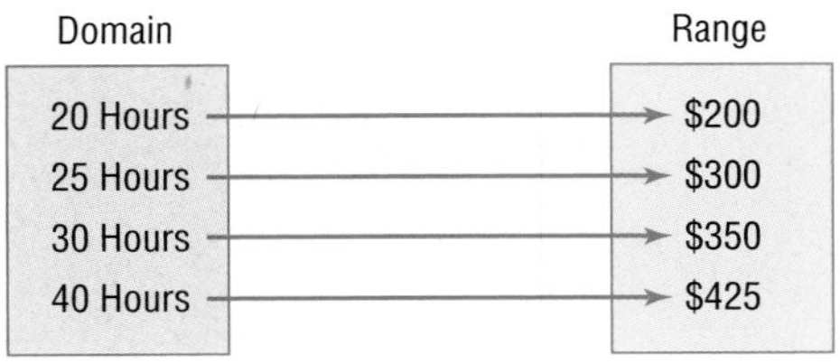

10.
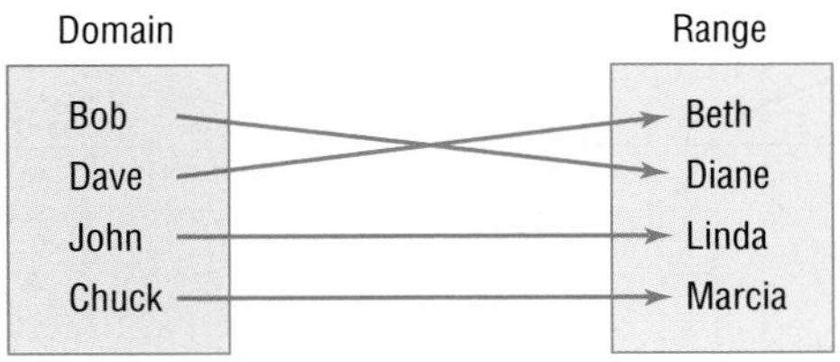

11.
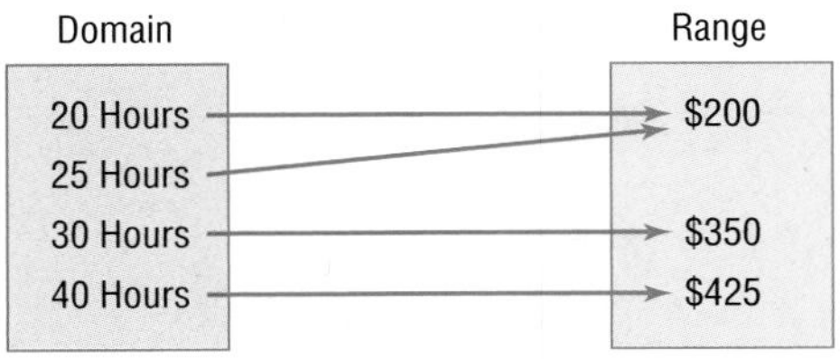

12.
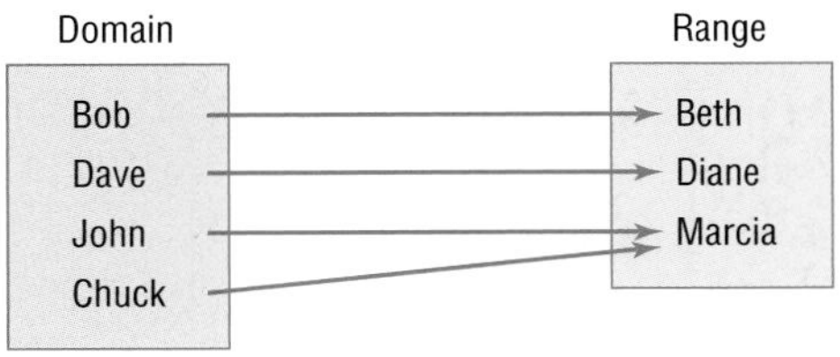

13. $\{(2, 6), (-3, 6), (4, 9), (1, 10)\}$

14. $\{(-2, 5), (-1, 3), (3, 7), (4, 12)\}$

15. $\{(0, 0), (1, 1), (2, 16), (3, 81)\}$

16. $\{(1, 2), (2, 8), (3, 18), (4, 32)\}$

In Problems 17–22, the graph of a function f is given. Use the horizontal-line test to determine whether f is one-to-one.

17.
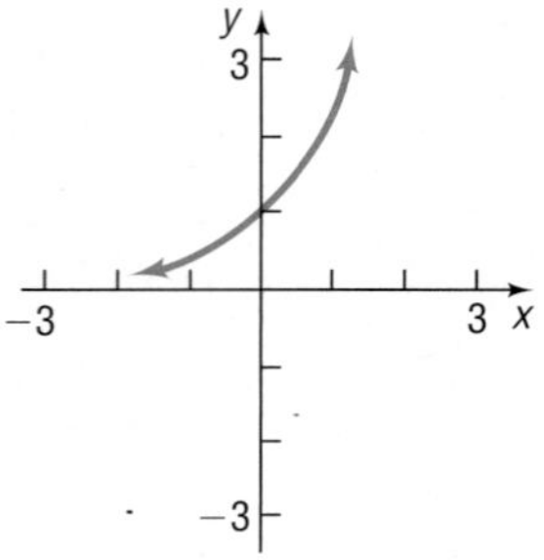

18.
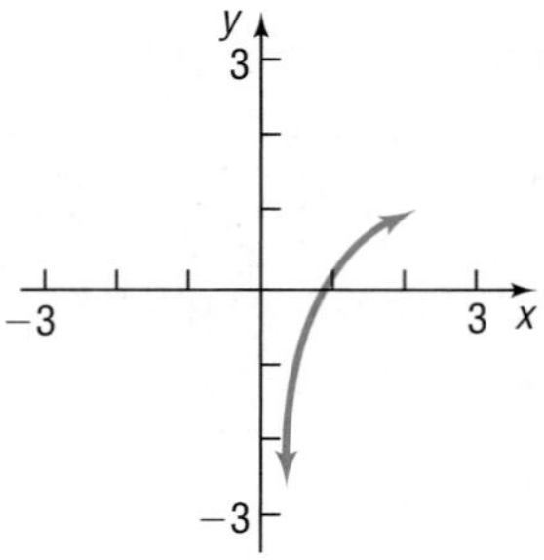

19.
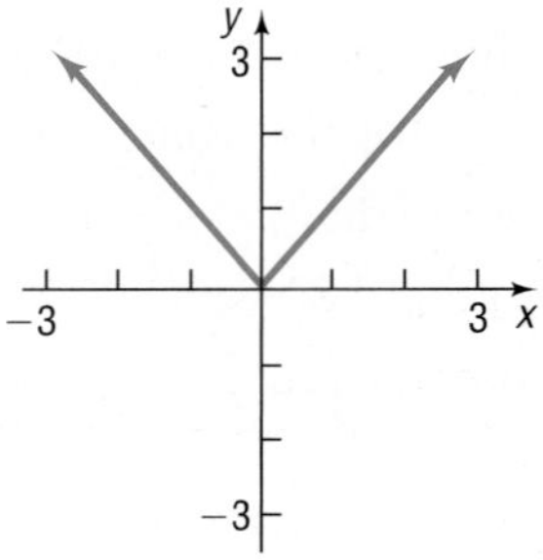

20.
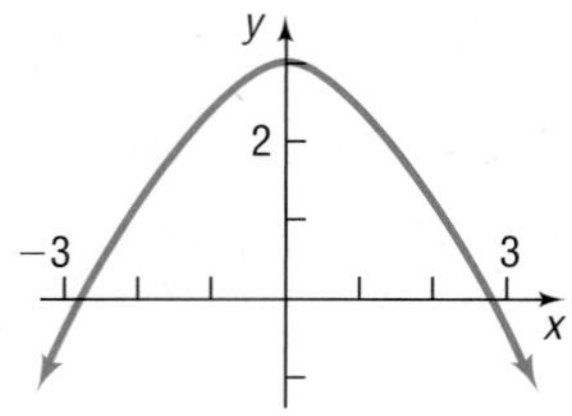

21.
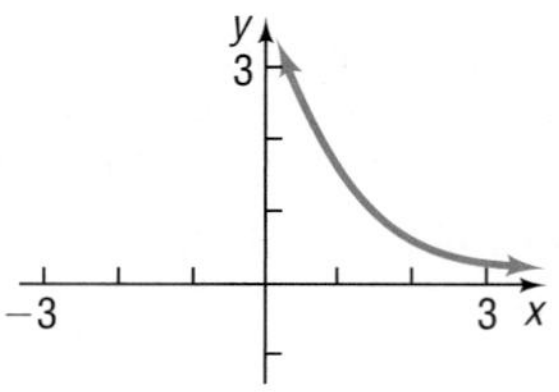

22.
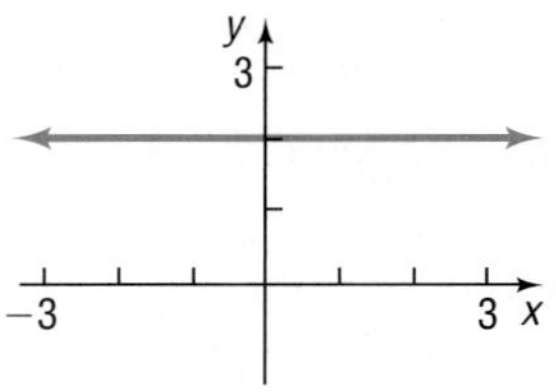

In Problems 23–28, the graph of a one-to-one function f is given. Draw the graph of the inverse function f^{-1}. For convenience (and as a hint), the graph of $y = x$ is also given.

23.
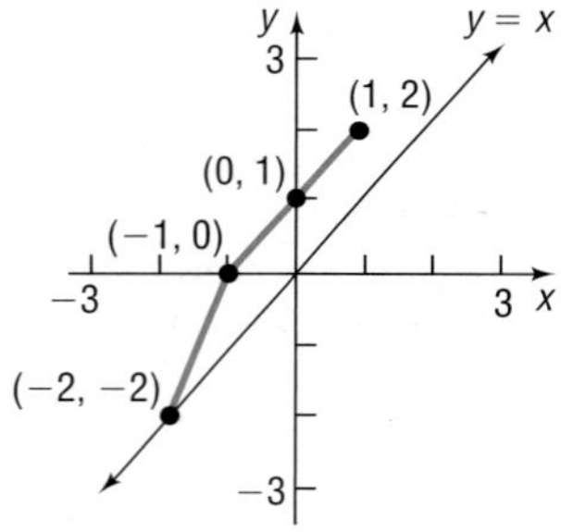

24.
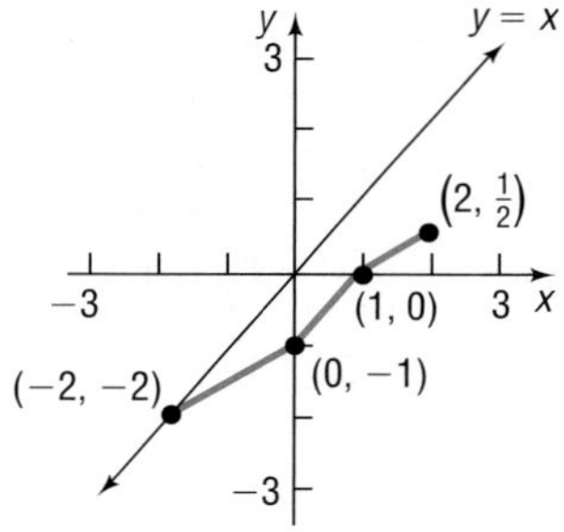

25.
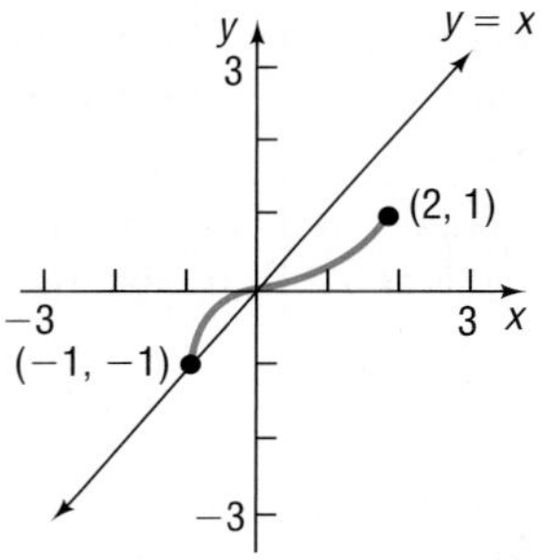

26.
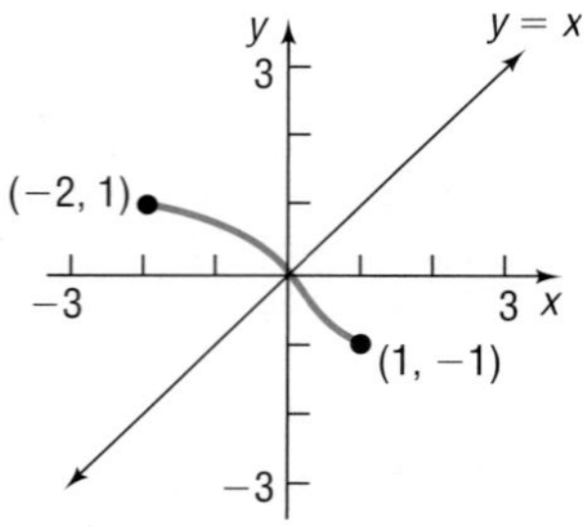

27.
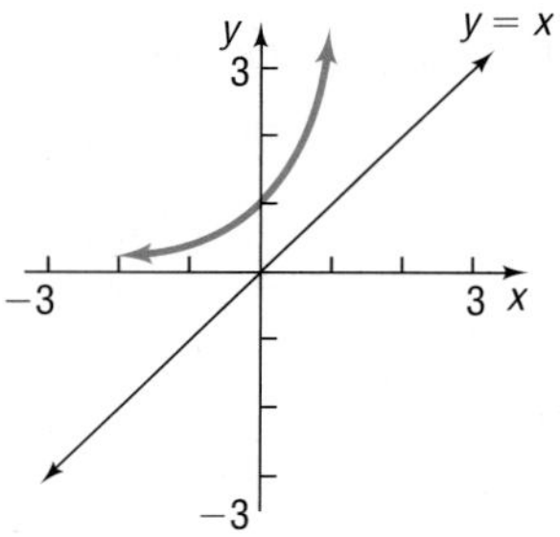

28.
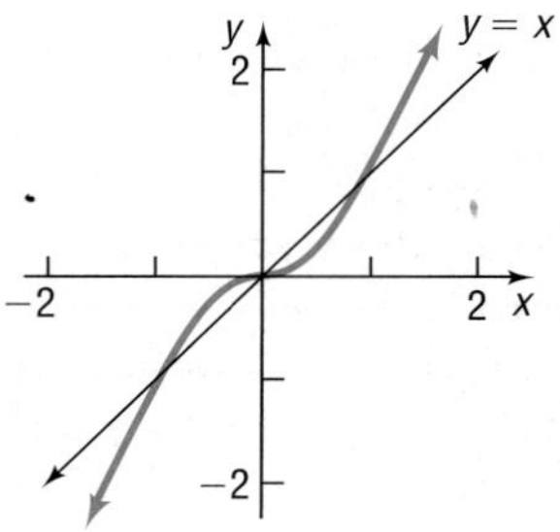

In Problems 29–38, verify that the functions f and g are inverses of each other by showing that $f(g(x)) = x$ and $g(f(x)) = x$.

29. $f(x) = 3x + 4; \quad g(x) = \frac{1}{3}(x - 4)$

30. $f(x) = 3 - 2x; \quad g(x) = -\frac{1}{2}(x - 3)$

31. $f(x) = 4x - 8; \quad g(x) = \frac{x}{4} + 2$

32. $f(x) = 2x + 6; \quad g(x) = \frac{1}{2}x - 3$

33. $f(x) = x^3 - 8; \quad g(x) = \sqrt[3]{x + 8}$

34. $f(x) = (x - 2)^2, x \geq 2; \quad g(x) = \sqrt{x} + 2$

35. $f(x) = \frac{1}{x}; \quad g(x) = \frac{1}{x}$

36. $f(x) = x; \quad g(x) = x$

37. $f(x) = \frac{2x + 3}{x + 4}; \quad g(x) = \frac{4x - 3}{2 - x}$

38. $f(x) = \frac{x - 5}{2x + 3}; \quad g(x) = \frac{3x + 5}{1 - 2x}$

In Problems 39–50, the function f is one-to-one. Find its inverse and check your answer. State the domain and range of f and f^{-1}. Graph f, f^{-1}, and $y = x$ on the same coordinate axes.

39. $f(x) = 3x$ **40.** $f(x) = -4x$ **41.** $f(x) = 4x + 2$

42. $f(x) = 1 - 3x$ **43.** $f(x) = x^3 - 1$ **44.** $f(x) = x^3 + 1$

45. $f(x) = x^2 + 4, \quad x \geq 0$ **46.** $f(x) = x^2 + 9, \quad x \geq 0$ **47.** $f(x) = \dfrac{4}{x}$

48. $f(x) = -\dfrac{3}{x}$ **49.** $f(x) = \dfrac{1}{x - 2}$ **50.** $f(x) = \dfrac{4}{x + 2}$

In Problems 51–62, the function f is one-to-one. Find its inverse and check your answer. State the domain of f and find its range using f^{-1}.

51. $f(x) = \dfrac{2}{3 + x}$ **52.** $f(x) = \dfrac{4}{2 - x}$ **53.** $f(x) = \dfrac{3x}{x + 2}$

54. $f(x) = \dfrac{-2x}{x - 1}$ **55.** $f(x) = \dfrac{2x}{3x - 1}$ **56.** $f(x) = \dfrac{3x + 1}{-x}$

57. $f(x) = \dfrac{3x + 4}{2x - 3}$ **58.** $f(x) = \dfrac{2x - 3}{x + 4}$ **59.** $f(x) = \dfrac{2x + 3}{x + 2}$

60. $f(x) = \dfrac{-3x - 4}{x - 2}$ **61.** $f(x) = \dfrac{x^2 - 4}{2x^2}, \quad x > 0$ **62.** $f(x) = \dfrac{x^2 + 3}{3x^2}, \quad x > 0$

---

63. Use the graph of $y = f(x)$ given in Problem 23 to evaluate the following:
(a) $f(-1)$ (b) $f(1)$ (c) $f^{-1}(1)$ (d) $f^{-1}(2)$

64. Use the graph of $y = f(x)$ given in Problem 24 to evaluate the following:
(a) $f(2)$ (b) $f(1)$ (c) $f^{-1}(0)$ (d) $f^{-1}(-1)$

65. Find the inverse of the linear function

$$f(x) = mx + b, \quad m \neq 0$$

66. Find the inverse of the function

$$f(x) = \sqrt{r^2 - x^2}, \quad 0 \leq x \leq r$$

67. A function f has an inverse function. If the graph of f lies in quadrant I, in which quadrant does the graph of f^{-1} lie?

68. A function f has an inverse function. If the graph of f lies in quadrant II, in which quadrant does the graph of f^{-1} lie?

69. The function $f(x) = |x|$ is not one-to-one. Find a suitable restriction on the domain of f so that the new function that results is one-to-one. Then find the inverse of f.

70. The function $f(x) = x^4$ is not one-to-one. Find a suitable restriction on the domain of f so that the new function that results is one-to-one. Then find the inverse of f.

71. Height versus Head Circumference The head circumference C of a child is related to the height H of the child (both in inches) through the function

$$H(C) = 2.15C - 10.53.$$

(a) Express the head circumference C as a function of height H.
(b) Predict the head circumference of a child who is 26 inches tall.

72. Temperature Conversion To convert from x degrees Celsius to y degrees Fahrenheit, we use the formula $y = f(x) = \dfrac{9}{5}x + 32$. To convert from x degrees Fahrenheit to y degrees Celsius, we use the formula $y = g(x) = \dfrac{5}{9}(x - 32)$. Show that f and g are inverse functions.

73. Demand for Corn The demand for corn obeys the equation $p(x) = 300 - 50x$, where p is the price per bushel (in dollars) and x is the number of bushels produced, in millions. Express the production amount x as a function of the price p.

74. Period of a Pendulum The period T (in seconds) of a simple pendulum is a function of its length l (in feet), given by $T(l) = 2\pi\sqrt{\dfrac{l}{g}}$, where $g \approx 32.2$ feet per second per second is the acceleration of gravity. Express the length l as a function of the period T.

75. Given

$$f(x) = \frac{ax + b}{cx + d}$$

find $f^{-1}(x)$. If $c \neq 0$, under what conditions on a, b, c, and d is $f = f^{-1}$?

76. Can a one-to-one function and its inverse be equal? What must be true about the graph of f for this to happen? Give some examples to support your conclusion.

77. Draw the graph of a one-to-one function that contains the points $(-2, -3)$, $(0, 0)$, and $(1, 5)$. Now draw the graph of its inverse. Compare your graph to those of other students. Discuss any similarities. What differences do you see?

78. Give an example of a function whose domain is the set of real numbers and that is neither increasing nor decreasing on its domain, but is one-to-one.
[Hint: Use a piecewise-defined function.]

79. If a function f is even, can it be one-to-one? Explain.

80. Is every odd function one-to-one? Explain.

81. If the graph of a function and its inverse intersect, where must this necessarily occur? Can they intersect anywhere else? Must they intersect?

'Are You Prepared?' Answers

1. Yes; for each input x there is one output y.
2. On $(0, \infty)$; on $(-\infty, 0)$.
3. It is increasing on its domain $(-\infty, \infty)$.

4.3 Exponential Functions

PREPARING FOR THIS SECTION *Before getting started, review the following:*

- Exponents (Appendix A, Section A.1, pp. 908–910 and Section A.9, pp. 977-982)
- Graphing Techniques: Transformations (Section 2.5, pp. 94–103)
- Average Rate of Change (Section 2.3, pp. 78–79)
- Solving Equations (Appendix A, Section A.5, pp. 936–945)
- Horizontal Asymptotes (Section 3.3, pp. 165–166)

Now work the 'Are You Prepared?' problems on page 255.

OBJECTIVES
1 Evaluate Exponential Functions
2 Graph Exponential Functions
3 Define the Number e
4 Solve Exponential Equations

1 In Appendix A, Section A.9, we give a definition for raising a real number a to a rational power. Based on that discussion, we gave meaning to expressions of the form

$$a^r$$

where the base a is a positive real number and the exponent r is a rational number.

But what is the meaning of a^x, where the base a is a positive real number and the exponent x is an irrational number? Although a rigorous definition requires methods discussed in calculus, the basis for the definition is easy to follow: Select a rational number r that is formed by truncating (removing) all but a finite number of digits from the irrational number x. Then it is reasonable to expect that

$$a^x \approx a^r$$

For example, take the irrational number $\pi = 3.14159\ldots$. Then an approximation to a^π is

$$a^\pi \approx a^{3.14}$$

where the digits after the hundredths position have been removed from the value for π. A better approximation would be

$$a^\pi \approx a^{3.14159}$$

where the digits after the hundred-thousandths position have been removed. Continuing in this way, we can obtain approximations to a^π to any desired degree of accuracy.

Most calculators have an $\boxed{x^y}$ key or a caret key $\boxed{\wedge}$ for working with exponents. To evaluate expressions of the form a^x, enter the base a, then press the $\boxed{x^y}$ key (or the $\boxed{\wedge}$ key), enter the exponent x, and press $\boxed{=}$ (or $\boxed{\text{enter}}$).

EXAMPLE 1 Using a Calculator to Evaluate Powers of 2

Using a calculator, evaluate:

(a) $2^{1.4}$ (b) $2^{1.41}$ (c) $2^{1.414}$ (d) $2^{1.4142}$ (e) $2^{\sqrt{2}}$

Solution (a) $2^{1.4} \approx 2.639015822$ (b) $2^{1.41} \approx 2.657371628$
(c) $2^{1.414} \approx 2.66474965$ (d) $2^{1.4142} \approx 2.665119089$
(e) $2^{\sqrt{2}} \approx 2.665144143$ ◀

NOW WORK PROBLEM 11.

It can be shown that the Laws of Exponents hold for real exponents.

Theorem

Laws of Exponents

If s, t, a, and b are real numbers, with $a > 0$ and $b > 0$, then

$$a^s \cdot a^t = a^{s+t} \qquad (a^s)^t = a^{st} \qquad (ab)^s = a^s \cdot b^s$$
$$1^s = 1 \qquad a^{-s} = \frac{1}{a^s} = \left(\frac{1}{a}\right)^s \qquad a^0 = 1 \qquad \textbf{(1)}$$

We are now ready for the following definition:

An **exponential function** is a function of the form

$$f(x) = a^x$$

where a is a positive real number $(a > 0)$ and $a \neq 1$. The domain of f is the set of all real numbers.

We exclude the base $a = 1$, because this function is simply the constant function $f(x) = 1^x = 1$. We also need to exclude bases that are negative, because, otherwise, we would have to exclude many values of x from the domain, such as $x = \frac{1}{2}$ and $x = \frac{3}{4}$.
[Recall that $(-2)^{1/2} = \sqrt{-2}$, $(-3)^{3/4} = \sqrt[4]{(-3)^3} = \sqrt[4]{-27}$, and so on, are not defined in the system of real numbers.]

CAUTION: It is important to distinguish a power function $g(x) = x^n$, $n \geq 2$ an integer, from an exponential function $f(x) = a^x$, $a > 0$, $a \neq 1$, a real. In a power function the base is a variable and the exponent is a constant. In an exponential function the base is a constant and the exponent is a variable. ■

Some examples of exponential functions are

$$f(x) = 2^x, \qquad F(x) = \left(\frac{1}{3}\right)^x$$

Notice that in each example the base is a constant and the exponent is a variable.

You may wonder what role the base a plays in the exponential function $f(x) = a^x$. We use the following Exploration to find out.

Exploration

(a) Evaluate $f(x) = 2^x$ at $x = -2, -1, 0, 1, 2$, and 3.

(b) Evaluate $g(x) = 3x + 2$ at $x = -2, -1, 0, 1, 2$, and 3.

(c) Comment on the pattern that exists in the values of f and g.

RESULT

(a) Table 1 shows the values of $f(x) = 2^x$ for $x = -2, -1, 0, 1, 2$, and 3.

(b) Table 2 shows the values of $g(x) = 3x + 2$ for $x = -2, -1, 0, 1, 2$, and 3.

Table 1

x	$f(x) = 2^x$
-2	$f(-2) = 2^{-2} = \frac{1}{2^2} = \frac{1}{4}$
-1	$\frac{1}{2}$
0	1
1	2
2	4
3	8

Table 2

x	$g(x) = 3x + 2$
-2	$g(-2) = 3(-2) + 2 = -4$
-1	-1
0	2
1	5
2	8
3	11

(c) In Table 1 we notice that each value of the exponential function $f(x) = a^x = 2^x$ could be found by multiplying the previous value of the function by the base, $a = 2$. For example,

$$f(-1) = 2 \cdot f(-2) = 2 \cdot \frac{1}{4} = \frac{1}{2}, \quad f(0) = 2 \cdot f(-1) = 2 \cdot \frac{1}{2} = 1, \quad f(1) = 2 \cdot f(0) = 2 \cdot 1 = 2$$

and so on.

In Words
For an exponential function $f(x) = a^x$, for 1-unit changes in the input x, the ratio of consecutive outputs is the constant a.

Put another way, we see that the ratio of consecutive outputs is constant for unit increases in the input. The constant equals the value of the base of the exponential function a. For example, for the function $f(x) = 2^x$, we notice that

$$\frac{f(-1)}{f(-2)} = \frac{\frac{1}{2}}{\frac{1}{4}} = 2, \quad \frac{f(1)}{f(0)} = \frac{2}{1} = 2, \quad \frac{f(x+1)}{f(x)} = \frac{2^{x+1}}{2^x} = 2$$

and so on.

From Table 2 we see that the function $g(x) = 3x + 2$ does not have the ratio of consecutive outputs that are constant because it is not exponential. For example,

$$\frac{g(-1)}{g(-2)} = \frac{-1}{-4} = \frac{1}{4} \neq \frac{g(1)}{g(0)} = \frac{5}{2}$$

Instead, because $g(x) = 3x + 2$ is a linear function, for unit increases in the input, the outputs increase by a fixed amount equal to the value of the slope, 3.

The results of the Exploration lead to the following result.

Theorem

For an exponential function $f(x) = a^x, a > 0, a \neq 1$, if x is any real number, then

$$\frac{f(x+1)}{f(x)} = a$$

Proof $$\frac{f(x+1)}{f(x)} = \frac{a^{x+1}}{a^x} = a^{x+1-x} = a^1 = a$$ ■

NOW WORK PROBLEM 21.

Graphs of Exponential Functions

2 First, we graph the exponential function $f(x) = 2^x$.

EXAMPLE 2 **Graphing an Exponential Function**

Graph the exponential function: $f(x) = 2^x$

Solution The domain of $f(x) = 2^x$ consists of all real numbers. We begin by locating some points on the graph of $f(x) = 2^x$, as listed in Table 3.

Since $2^x > 0$ for all x, the range of f is the interval $(0, \infty)$. From this, we conclude that the graph has no x-intercepts, and, in fact, the graph will lie above the x-axis. As Table 3 indicates, the y-intercept is 1. Table 3 also indicates that as $x \to -\infty$ the value of $f(x) = 2^x$ gets closer and closer to 0. We conclude that the x-axis is a horizontal asymptote to the graph as $x \to -\infty$. This gives us the end behavior of the graph for x large and negative.

To determine the end behavior for x large and positive, look again at Table 3. As $x \to \infty$, $f(x) = 2^x$ grows very quickly, causing the graph of $f(x) = 2^x$ to rise very rapidly. It is apparent that f is an increasing function and hence is one-to-one.

Using all this information, we plot some of the points from Table 3 and connect them with a smooth, continuous curve, as shown in Figure 13.

Table 3

x	$f(x) = 2^x$
-10	$2^{-10} \approx 0.00098$
-3	$2^{-3} = \frac{1}{8}$
-2	$2^{-2} = \frac{1}{4}$
-1	$2^{-1} = \frac{1}{2}$
0	$2^0 = 1$
1	$2^1 = 2$
2	$2^2 = 4$
3	$2^3 = 8$
10	$2^{10} = 1024$

Figure 13

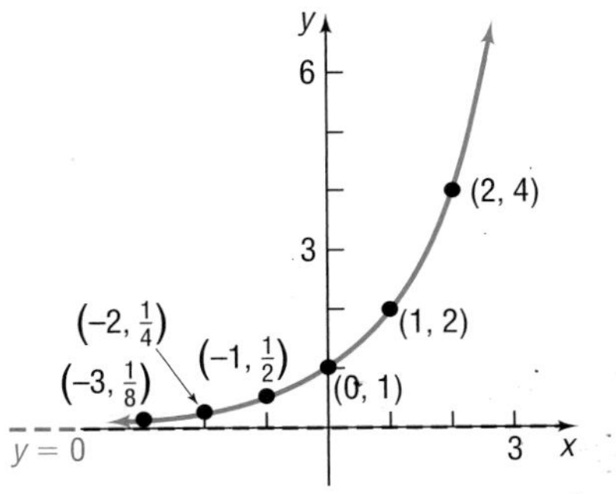

◄

As we shall see, graphs that look like the one in Figure 13 occur very frequently in a variety of situations. For example, look at the graph in Figure 14, which illustrates the closing price of a share of Harley Davidson stock. Investors might conclude from this graph that the price of Harley Davidson stock is *behaving exponentially*; that is, the graph exhibits rapid, or exponential, growth.

Figure 14

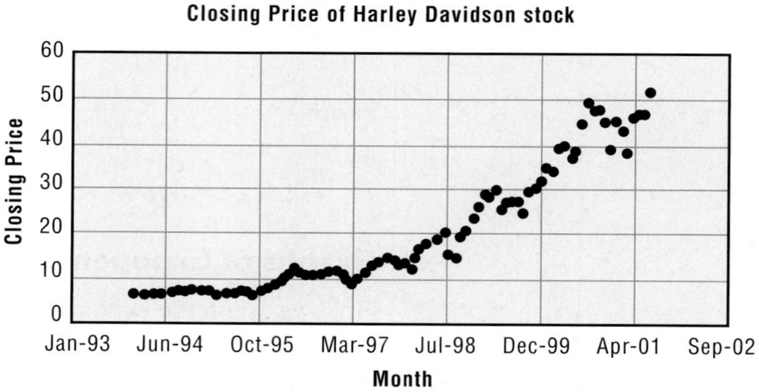

Figure 15

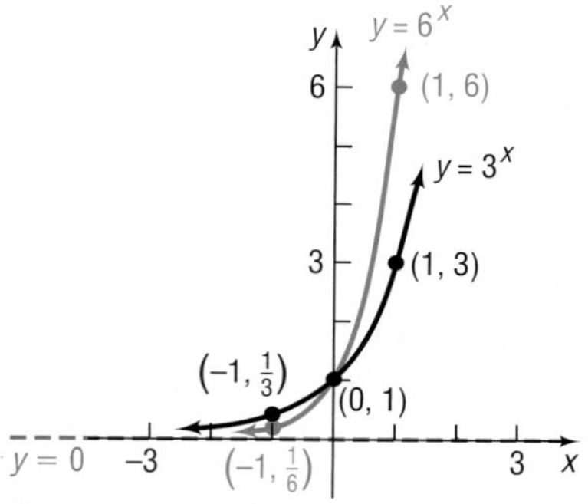

We shall have more to say about situations that lead to exponential growth later in this chapter. For now, we continue to seek properties of the exponential functions.

The graph of $f(x) = 2^x$ in Figure 13 is typical of all exponential functions that have a base larger than 1. Such functions are increasing functions and hence are one-to-one. Their graphs lie above the x-axis, pass through the point $(0, 1)$, and thereafter rise rapidly as $x \to \infty$. As $x \to -\infty$, the x-axis $(y = 0)$ is a horizontal asymptote. There are no vertical asymptotes. Finally, the graphs are smooth and continuous, with no corners or gaps.

Figure 15 illustrates the graphs of two more exponential functions whose bases are larger than 1. Notice that for the larger base the graph is steeper when $x > 0$ and is closer to the x-axis when $x < 0$.

Seeing the Concept

Graph $y = 2^x$ and compare what you see to Figure 13. Clear the screen and graph $y = 3^x$ and $y = 6^x$ and compare what you see to Figure 15. Clear the screen and graph $y = 10^x$ and $y = 100^x$. What viewing rectangle seems to work best?

The following display summarizes the information that we have about $f(x) = a^x, a > 1$.

Properties of the Exponential Function $f(x) = a^x$, $a > 1$

1. The domain is the set of all real numbers; the range is the set of positive real numbers.
2. There are no x-intercepts; the y-intercept is 1.
3. The x-axis $(y = 0)$ is a horizontal asymptote as $x \to -\infty$.
4. $f(x) = a^x, a > 1$, is an increasing function and is one-to-one.
5. The graph of f contains the points $(0, 1)$, $(1, a)$, and $\left(-1, \frac{1}{a}\right)$.
6. The graph of f is smooth and continuous, with no corners or gaps. See Figure 16.

Figure 16
$f(x) = a^x, a > 1$

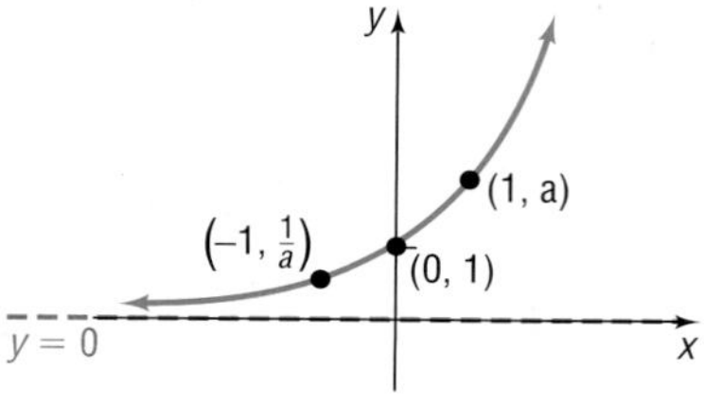

Now we consider $f(x) = a^x$ when $0 < a < 1$.

EXAMPLE 3 Graphing an Exponential Function

Graph the exponential function: $f(x) = \left(\frac{1}{2}\right)^x$

Solution The domain of $f(x) = \left(\frac{1}{2}\right)^x$ consists of all real numbers. As before, we locate some points on the graph by creating Table 4. Since $\left(\frac{1}{2}\right)^x > 0$ for all x, the range of f is the interval $(0, \infty)$. The graph lies above the x-axis and so has no x-intercepts. The y-intercept is 1. As $x \to -\infty$, $f(x) = \left(\frac{1}{2}\right)^x$ grows very quickly. As $x \to \infty$, the values of $f(x)$ approach 0. The x-axis $(y = 0)$ is a horizontal asymptote as $x \to \infty$. It is apparent that f is a decreasing function and hence is one-to-one. Figure 17 illustrates the graph.

Table 4

x	$f(x) = \left(\frac{1}{2}\right)^x$
-10	$\left(\frac{1}{2}\right)^{-10} = 1024$
-3	$\left(\frac{1}{2}\right)^{-3} = 8$
-2	$\left(\frac{1}{2}\right)^{-2} = 4$
-1	$\left(\frac{1}{2}\right)^{-1} = 2$
0	$\left(\frac{1}{2}\right)^{0} = 1$
1	$\left(\frac{1}{2}\right)^{1} = \frac{1}{2}$
2	$\left(\frac{1}{2}\right)^{2} = \frac{1}{4}$
3	$\left(\frac{1}{2}\right)^{3} = \frac{1}{8}$
10	$\left(\frac{1}{2}\right)^{10} \approx 0.00098$

Figure 17

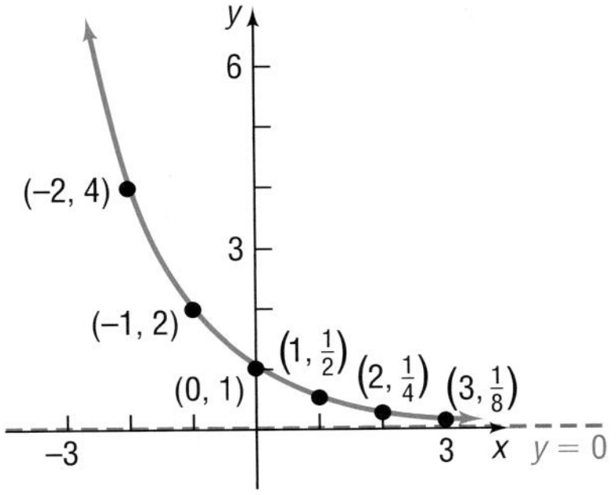

◄

We could have obtained the graph of $y = \left(\frac{1}{2}\right)^x$ from the graph of $y = 2^x$ using transformations. If $f(x) = 2^x$, then $f(-x) = 2^{-x} = \frac{1}{2^x} = \left(\frac{1}{2}\right)^x$. The graph of $y = \left(\frac{1}{2}\right)^x = 2^{-x}$ is a reflection about the y-axis of the graph of $y = 2^x$. See Figures 18(a) and (b).

Figure 18

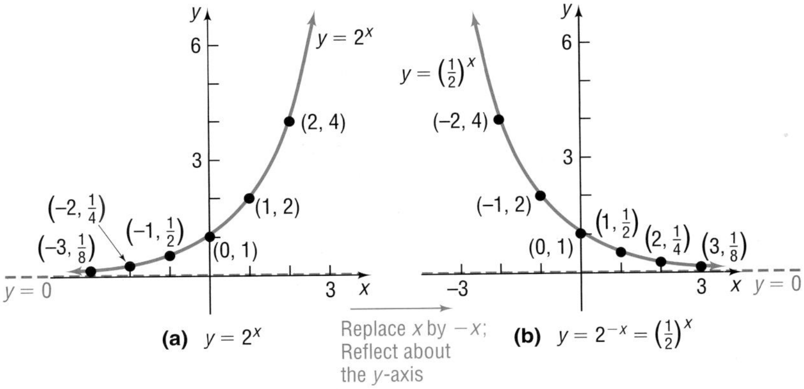

Seeing the Concept

Using a graphing utility, simultaneously graph

(a) $Y_1 = 3^x$, $Y_2 = \left(\frac{1}{3}\right)^x$ (b) $Y_1 = 6^x$, $Y_2 = \left(\frac{1}{6}\right)^x$

Conclude that the graph of $Y_2 = \left(\frac{1}{a}\right)^x$, for $a > 0$, is the reflection about the y-axis of the graph of $Y_1 = a^x$.

Figure 19

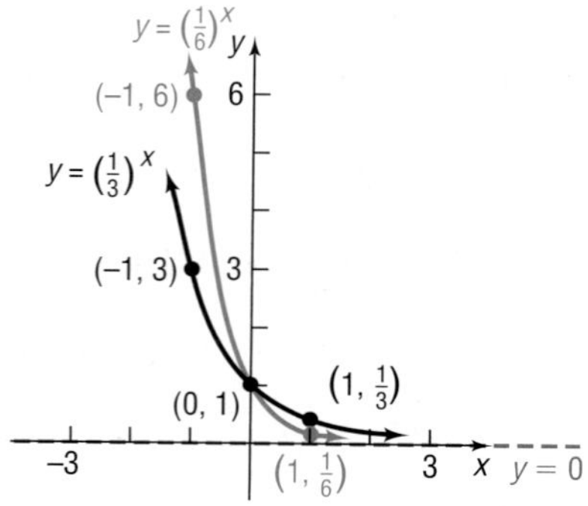

The graph of $f(x) = \left(\frac{1}{2}\right)^x$ in Figure 17 is typical of all exponential functions that have a base between 0 and 1. Such functions are decreasing and one-to-one. Their graphs lie above the x-axis and pass through the point $(0, 1)$. The graphs rise rapidly as $x \to -\infty$. As $x \to \infty$, the x-axis is a horizontal asymptote. There are no vertical asymptotes. Finally, the graphs are smooth and continuous, with no corners or gaps.

Figure 19 illustrates the graphs of two more exponential functions whose bases are between 0 and 1. Notice that the choice of a base closer to 0 results in a graph that is steeper when $x < 0$ and closer to the x-axis when $x > 0$.

Seeing the Concept

Graph $y = \left(\frac{1}{2}\right)^x$ and compare what you see to Figure 17. Clear the screen and graph $y = \left(\frac{1}{3}\right)^x$ and $y = \left(\frac{1}{6}\right)^x$ and compare what you see to Figure 19. Clear the screen and graph $y = \left(\frac{1}{10}\right)^x$ and $y = \left(\frac{1}{100}\right)^x$. What viewing rectangle seems to work best?

The following display summarizes the information that we have about the function $f(x) = a^x$, $0 < a < 1$.

Properties of the Graph of an Exponential Function $f(x) = a^x$, $0 < a < 1$

1. The domain is the set of all real numbers; the range is the set of positive real numbers.
2. There are no x-intercepts; the y-intercept is 1.
3. The x-axis ($y = 0$) is a horizontal asymptote as $x \to \infty$.
4. $f(x) = a^x$, $0 < a < 1$, is a decreasing function and is one-to-one.
5. The graph of f contains the points $(0, 1)$, $(1, a)$, and $\left(-1, \frac{1}{a}\right)$.
6. The graph of f is smooth and continuous, with no corners or gaps. See Figure 20.

Figure 20
$f(x) = a^x$, $0 < a < 1$

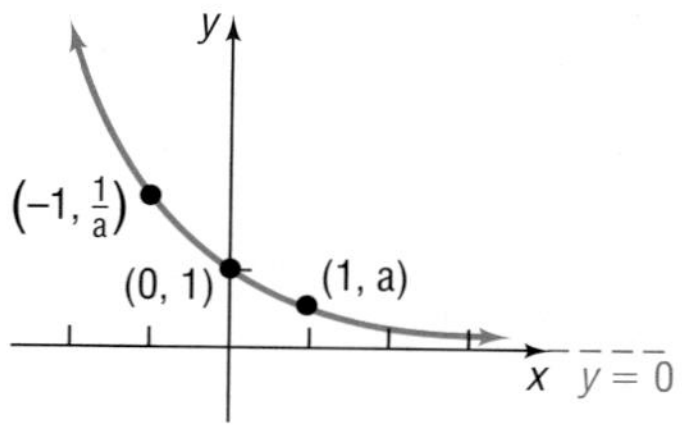

EXAMPLE 4 Graphing Exponential Functions Using Transformations

Graph $f(x) = 2^{-x} - 3$ and determine the domain, range, and horizontal asymptote of f.

Solution We begin with the graph of $y = 2^x$. Figure 21 shows the various steps. As Figure 21(c) illustrates, the domain of $f(x) = 2^{-x} - 3$ is the interval $(-\infty, \infty)$ and the range is the interval $(-3, \infty)$. The horizontal asymptote of f is the line $y = -3$.

Figure 21

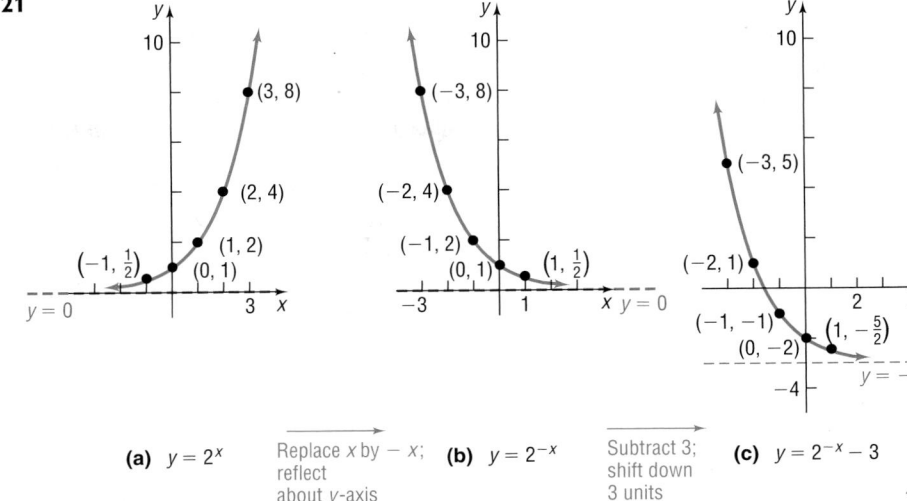

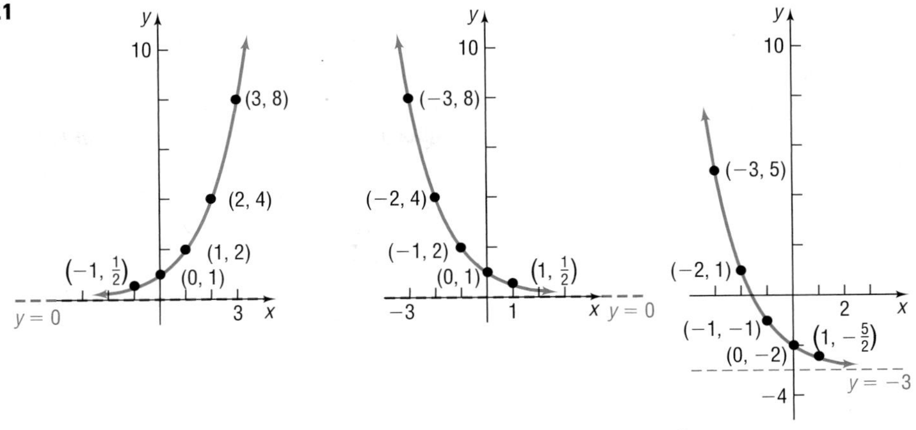

(a) $y = 2^x$ → Replace x by $-x$; reflect about y-axis → (b) $y = 2^{-x}$ → Subtract 3; shift down 3 units → (c) $y = 2^{-x} - 3$ ◀

NOW WORK PROBLEM 37.

The Base e

3 As we shall see shortly, many problems that occur in nature require the use of an exponential function whose base is a certain irrational number, symbolized by the letter e.

Let's look now at one way of arriving at this important number e.

The **number e** is defined as the number that the expression

$$\left(1 + \frac{1}{n}\right)^n \qquad (2)$$

approaches as $n \to \infty$. In calculus, this is expressed using limit notation as

$$e = \lim_{n\to\infty}\left(1 + \frac{1}{n}\right)^n$$

Table 5 illustrates what happens to the defining expression (2) as n takes on increasingly large values. The last number in the last column in the

Table 5

n	$\frac{1}{n}$	$1 + \frac{1}{n}$	$\left(1 + \frac{1}{n}\right)^n$
1	1	2	2
2	0.5	1.5	2.25
5	0.2	1.2	2.48832
10	0.1	1.1	2.59374246
100	0.01	1.01	2.704813829
1,000	0.001	1.001	2.716923932
10,000	0.0001	1.0001	2.718145927
100,000	0.00001	1.00001	2.718268237
1,000,000	0.000001	1.000001	2.718280469
1,000,000,000	10^{-9}	$1 + 10^{-9}$	2.718281828

table is correct to nine decimal places and is the same as the entry given for e on your calculator (if expressed correctly to nine decimal places).

The exponential function $f(x) = e^x$, whose base is the number e, occurs with such frequency in applications that it is usually referred to as *the* exponential function. Indeed, most calculators have the key* $\boxed{e^x}$ or $\boxed{\exp(x)}$, which may be used to evaluate the exponential function for a given value of x.

Now use your calculator to approximate e^x for $x = -2$, $x = -1$, $x = 0$, $x = 1$, and $x = 2$, as we have done to create Table 6. The graph of the exponential function $f(x) = e^x$ is given in Figure 22. Since $2 < e < 3$, the graph of $y = e^x$ lies between the graphs of $y = 2^x$ and $y = 3^x$. Do you see why? (Refer to Figures 13 and 15.)

Table 6

x	e^x
−2	0.14
−1	0.37
0	1
1	2.72
2	7.39

Figure 22
$y = e^x$

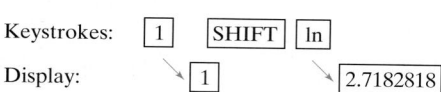

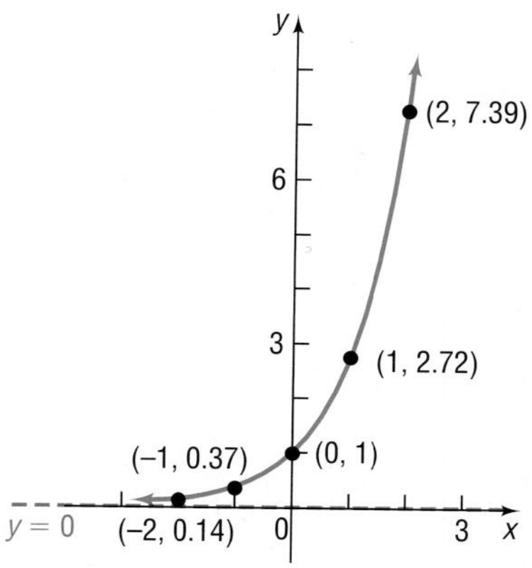

Seeing the Concept

Graph $Y_1 = e^x$ and compare what you see to Figure 22. Use eVALUEate or TABLE to verify the points on the graph shown in Figure 22. Now graph $Y_2 = 2^x$ and $Y_3 = 3^x$ on the same screen as $Y_1 = e^x$. Notice that the graph of $Y_1 = e^x$ lies between these two graphs.

EXAMPLE 5 Graphing Exponential Functions Using Transformations

Graph $f(x) = -e^{x-3}$ and determine the domain, range, and horizontal asymptote of f.

Solution We begin with the graph of $y = e^x$. Figure 23 shows the various steps.

As Figure 23(c) illustrates, the domain of $f(x) = -e^{x-3}$ is the interval $(-\infty, \infty)$ and the range is the interval $(-\infty, 0)$. The horizontal asymptote is the line $y = 0$.

*If your calculator does not have this key but does have a $\boxed{\text{SHIFT}}$ key (or $\boxed{2^{\text{nd}}}$ key) and an $\boxed{\ln}$ key, you can display the number e as follows:

Keystrokes:	$\boxed{1}$	$\boxed{\text{SHIFT}}$	$\boxed{\ln}$
Display:	$\boxed{1}$		$\boxed{2.7182818}$

The reason this works will become clear in Section 4.4.

Figure 23

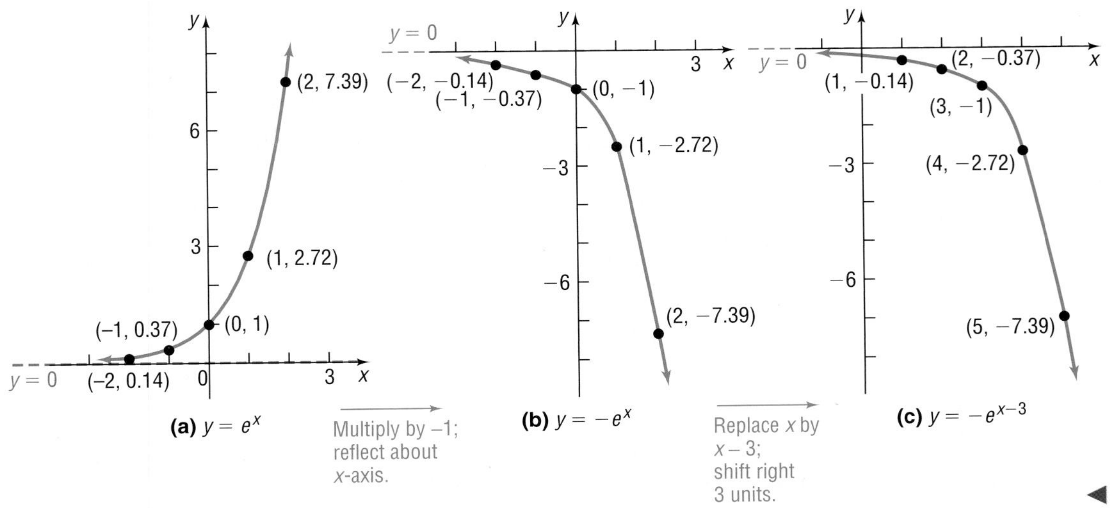

NOW WORK PROBLEM 45.

Exponential Equations

4 Equations that involve terms of the form a^x, $a > 0$, $a \neq 1$, are often referred to as **exponential equations.** Such equations can sometimes be solved by appropriately applying the Laws of Exponents and property (3).

$$\text{If } a^u = a^v, \quad \text{then } u = v \qquad (3)$$

Property (3) is a consequence of the fact that exponential functions are one-to-one. To use property (3), each side of the equality must be written with the same base.

EXAMPLE 6 Solving an Exponential Equation

Solve: $3^{x+1} = 81$

Solution Since $81 = 3^4$, we can write the equation as

$$3^{x+1} = 81 = 3^4$$

Now we have the same base, 3, on each side, so we can apply property (3) to obtain

$$x + 1 = 4$$
$$x = 3$$

NOW WORK PROBLEM 53.

EXAMPLE 7 Solving an Exponential Equation

Solve: $e^{-x^2} = (e^x)^2 \cdot \dfrac{1}{e^3}$

Solution We use the Laws of Exponents first to get the base e on the right side.

$$(e^x)^2 \cdot \frac{1}{e^3} = e^{2x} \cdot e^{-3} = e^{2x-3}$$

As a result,

$$e^{-x^2} = e^{2x-3}$$

$$-x^2 = 2x - 3 \quad \text{Apply property (3).}$$

$$x^2 + 2x - 3 = 0 \quad \text{Place the quadratic equation in standard form.}$$

$$(x + 3)(x - 1) = 0 \quad \text{Factor.}$$

$$x = -3 \quad \text{or} \quad x = 1 \quad \text{Use the Zero-Product Property.}$$

The solution set is $\{-3, 1\}$. ◀

Application

Many applications involve exponential functions. Let's look at one.

EXAMPLE 8 Exponential Probability

Between 9:00 PM and 10:00 PM cars arrive at Burger King's drive-thru at the rate of 12 cars per hour (0.2 car per minute). The following formula from the field of probability can be used to determine the probability that a car will arrive within t minutes of 9:00 PM.

$$F(t) = 1 - e^{-0.2t}$$

(a) Determine the probability that a car will arrive within 5 minutes of 9 PM (that is, before 9:05 PM).

(b) Determine the probability that a car will arrive within 30 minutes of 9 PM (before 9:30 PM).

(c) What value does F approach as t becomes unbounded in the positive direction?

(d) Graph $F(t) = 1 - e^{-0.2t}, t > 0$. Use eVALUEate or TABLE to compare the values of F at $t = 5$ [part (a)] and at $t = 30$ [part (b)].

(e) Within how many minutes of 9 PM will the probability of a car arriving equal 50%? [**Hint**: Use TRACE or TABLE.]

Solution (a) The probability that a car will arrive within 5 minutes is found by evaluating $F(t)$ at $t = 5$.

$$F(5) = 1 - e^{-0.2(5)} \approx 0.63212$$

↑ Use a calculator.

We conclude that there is a 63% probability that a car will arrive within 5 minutes.

(b) The probability that a car will arrive within 30 minutes is found by evaluating $F(t)$ at $t = 30$.

$$F(30) = 1 - e^{-0.2(30)} \approx 0.9975$$

↑ Use a calculator.

There is a 99.75% probability that a car will arrive within 30 minutes.

Figure 24

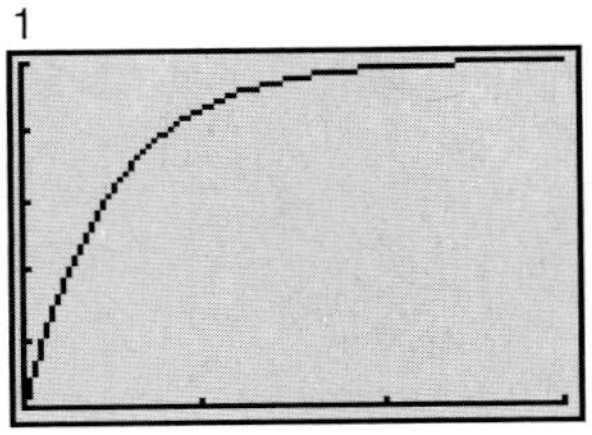

(c) As time passes, the probability that a car will arrive increases. The value that F approaches can be found by letting $t \to \infty$. Since $e^{-0.2t} = \frac{1}{e^{0.2t}}$, it follows that $e^{-0.2t} \to 0$ as $t \to \infty$. Thus, F approaches 1 as t gets large.

(d) See Figure 24 for the graph of F.

(e) Within 3.5 minutes of 9 PM, the probability of a car arriving equals 50%. ◀

Summary

Properties of the Exponential Function

$f(x) = a^x, \quad a > 1$ — Domain: the interval $(-\infty, \infty)$; Range: the interval $(0, \infty)$; x-intercepts: none; y-intercept: 1; horizontal asymptote: x-axis as $x \to -\infty$; increasing; one-to-one; smooth; continuous. See Figure 16 for a typical graph.

$f(x) = a^x, \quad 0 < a < 1$ — Domain: the interval $(-\infty, \infty)$; Range: the interval $(0, \infty)$; x-intercepts: none; y-intercept: 1; horizontal asymptote: x-axis as $x \to \infty$; decreasing; one-to-one; smooth; continuous. See Figure 20 for a typical graph.

If $a^u = a^v$, then $u = v$.

4.3 Assess Your Understanding

'Are You Prepared?' *Answers are given at the end of these exercises. If you get a wrong answer, read the pages listed in red.*

1. $4^3 =$ ______; $8^{2/3} =$ ______; $3^{-2} =$ ______. (pp. 908–910 and 977–982)

2. Solve: $3x^2 + 5x - 2 = 0$. (pp. 936–945)

3. Find the average rate of change of $f(x) = 3x - 5$ from $x = 0$ to $x = c$. (pp. 78–79)

4. *True or False*: The function $f(x) = \frac{2x}{x-3}$ has a horizontal asymptote. (pp. 165–166)

5. *True or False:* To graph $y = (x-2)^3$, shift the graph of $y = x^3$ to the left 2 units. (pp. 94–103)

Concepts and Vocabulary

6. The graph of every exponential function $f(x) = a^x, a > 0, a \neq 1$, passes through three points: ______, ______, and ______.

7. If the graph of the exponential function $f(x) = a^x$, $a > 0, a \neq 1$, is decreasing, then a must be less than ______.

8. If $3^x = 3^4$, then $x =$ ______.

9. *True or False:* The graphs of $y = 3^x$ and $y = \left(\frac{1}{3}\right)^x$ are identical.

10. *True or False:* The range of the exponential function $f(x) = a^x, a > 0, a \neq 1$, is the set of all real numbers.

Exercises

In Problems 11–20, approximate each number using a calculator. Express your answer rounded to three decimal places.

11. (a) $3^{2.2}$ (b) $3^{2.23}$ (c) $3^{2.236}$ (d) $3^{\sqrt{5}}$

12. (a) $5^{1.7}$ (b) $5^{1.73}$ (c) $5^{1.732}$ (d) $5^{\sqrt{3}}$

13. (a) $2^{3.14}$ (b) $2^{3.141}$ (c) $2^{3.1415}$ (d) 2^{π}

14. (a) $2^{2.7}$ (b) $2^{2.71}$ (c) $2^{2.718}$ (d) 2^{e}

15. (a) $3.1^{2.7}$ (b) $3.14^{2.71}$ (c) $3.141^{2.718}$ (d) π^{e}

16. (a) $2.7^{3.1}$ (b) $2.71^{3.14}$ (c) $2.718^{3.141}$ (d) e^{π}

17. $e^{1.2}$

18. $e^{-1.3}$

19. $e^{-0.85}$

20. $e^{2.1}$

In Problems 21–28, determine whether the given function is exponential or not. For those that are exponential functions, identify the value of the base a. [**Hint:** Look at the ratio of consecutive values.]

21.

x	f(x)
−1	3
0	6
1	12
2	18
3	30

22.

x	g(x)
−1	2
0	5
1	8
2	11
3	14

23.

x	H(x)
−1	$\frac{1}{4}$
0	1
1	4
2	16
3	64

24.

x	F(x)
−1	$\frac{2}{3}$
0	1
1	$\frac{3}{2}$
2	$\frac{9}{4}$
3	$\frac{27}{8}$

25.

x	f(x)
−1	$\frac{3}{2}$
0	3
1	6
2	12
3	24

26.

x	g(x)
−1	6
0	1
1	0
2	3
3	10

27.

x	H(x)
−1	2
0	4
1	6
2	8
3	10

28.

x	F(x)
−1	$\frac{1}{2}$
0	$\frac{1}{4}$
1	$\frac{1}{8}$
2	$\frac{1}{16}$
3	$\frac{1}{32}$

In Problems 29–36, the graph of an exponential function is given. Match each graph to one of the following functions.

A. $y = 3^x$ B. $y = 3^{-x}$ C. $y = -3^x$ D. $y = -3^{-x}$
E. $y = 3^x - 1$ F. $y = 3^{x-1}$ G. $y = 3^{1-x}$ H. $y = 1 - 3^x$

29.

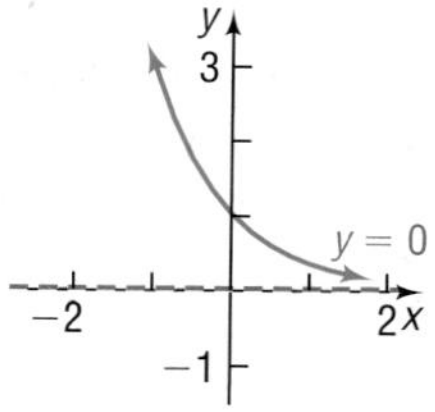

30.

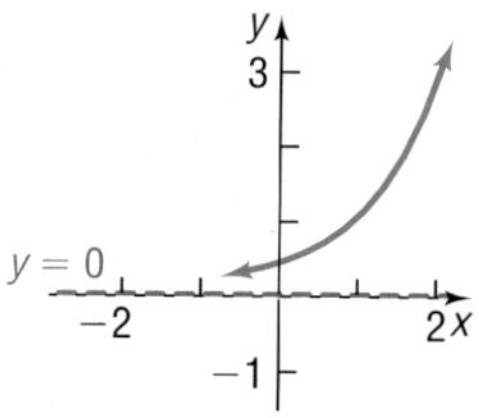

31.

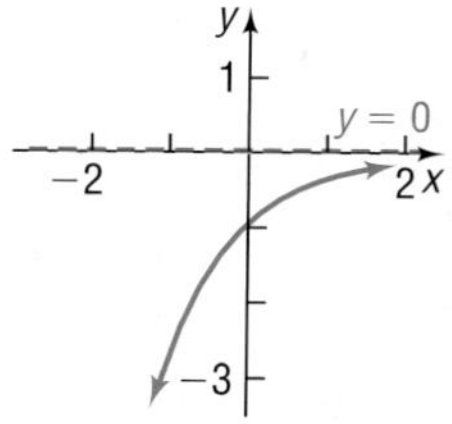

32.

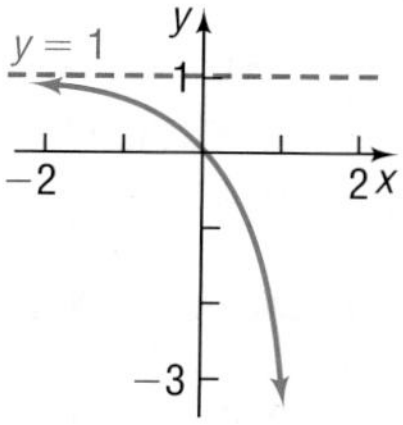

33.

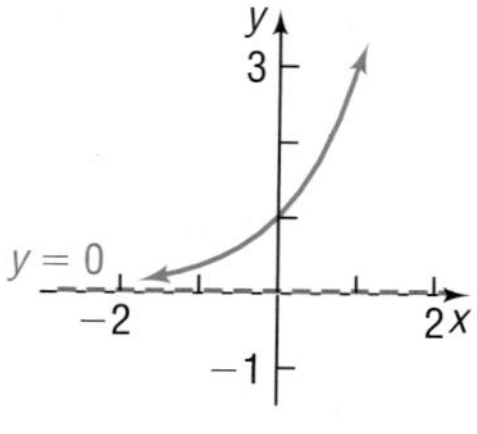

34.

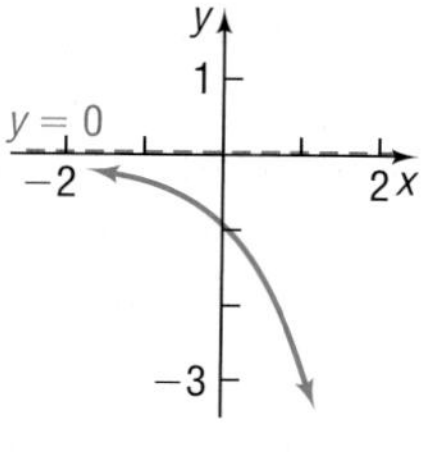

35.

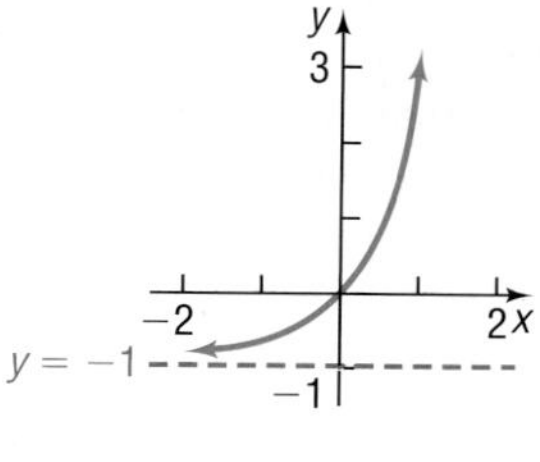

36.

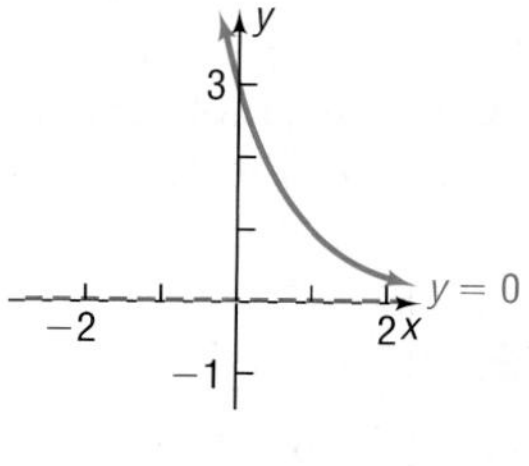

In Problems 37–44, use transformations to graph each function. Determine the domain, range, and horizontal asymptote of each function.

37. $f(x) = 2^x + 1$
38. $f(x) = 2^{x+2}$
39. $f(x) = 3^{-x} - 2$
40. $f(x) = -3^x + 1$
41. $f(x) = 2 + 3(4^x)$
42. $f(x) = 1 - 3(2^x)$
43. $f(x) = 2 + 3^{x/2}$
44. $f(x) = 1 - 2^{-x/3}$

In Problems 45–52, begin with the graph of $y = e^x$ (Figure 22) and use transformations to graph each function. Determine the domain, range, and horizontal asymptote of each function.

45. $f(x) = e^{-x}$
46. $f(x) = -e^x$
47. $f(x) = e^{x+2}$
48. $f(x) = e^x - 1$
49. $f(x) = 5 - e^{-x}$
50. $f(x) = 9 - 3e^{-x}$
51. $f(x) = 2 - e^{-x/2}$
52. $f(x) = 7 - 3e^{2x}$

In Problems 53–66, solve each equation.

53. $2^{2x+1} = 4$
54. $5^{1-2x} = \frac{1}{5}$
55. $3^{x^3} = 9^x$
56. $4^{x^2} = 2^x$
57. $8^{x^2-2x} = \frac{1}{2}$
58. $9^{-x} = \frac{1}{3}$
59. $2^x \cdot 8^{-x} = 4^x$
60. $\left(\frac{1}{2}\right)^{1-x} = 4$
61. $\left(\frac{1}{5}\right)^{2-x} = 25$
62. $4^x - 2^x = 0$
63. $4^x = 8$
64. $9^{2x} = 27$
65. $e^{x^2} = (e^{3x}) \cdot \frac{1}{e^2}$
66. $(e^4)^x \cdot e^{x^2} = e^{12}$

67. If $4^x = 7$, what does 4^{-2x} equal?

68. If $2^x = 3$, what does 4^{-x} equal?

69. If $3^{-x} = 2$, what does 3^{2x} equal?

70. If $5^{-x} = 3$, what does 5^{3x} equal?

In Problems 71–74, determine the exponential function whose graph is given.

71.

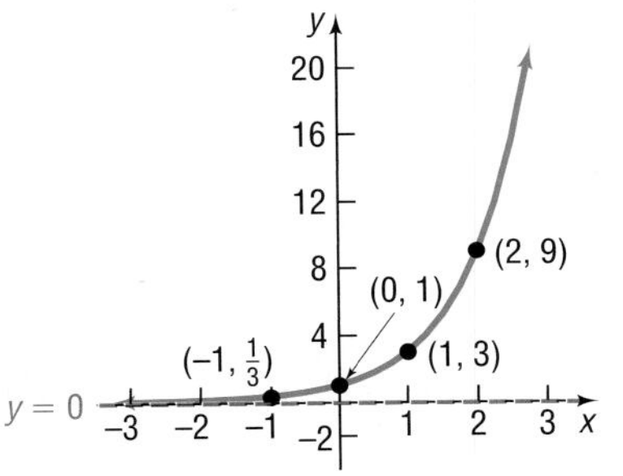

72.

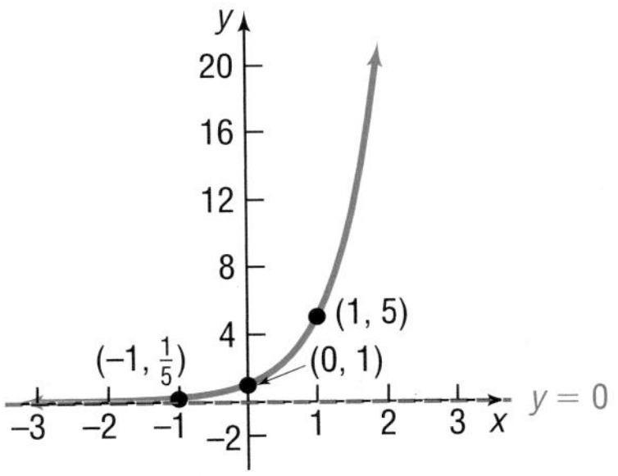

73.

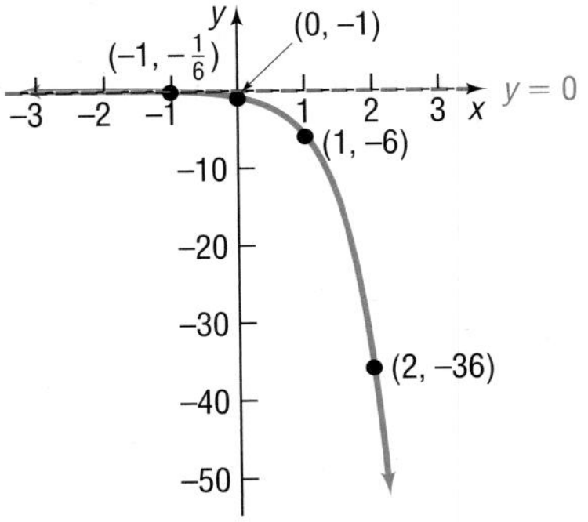

74.

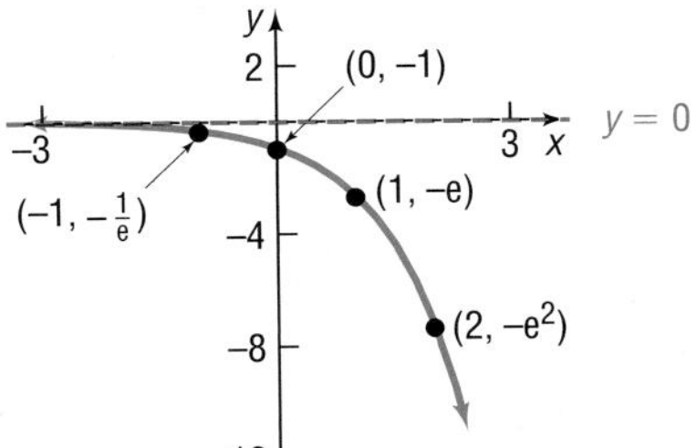

75. Optics If a single pane of glass obliterates 3% of the light passing through it, then the percent p of light that passes through n successive panes is given approximately by the function

$$p(n) = 100e^{-0.03n}$$

(a) What percent of light will pass through 10 panes?
(b) What percent of light will pass through 25 panes?

76. Atmospheric Pressure The atmospheric pressure p on a balloon or plane decreases with increasing height. This pressure, measured in millimeters of mercury, is related to the number of kilometers h above sea level by the function

$$p(h) = 760e^{-0.145h}$$

(a) Find the atmospheric pressure at a height of 2 kilometers (over 1 mile).
(b) What is it at a height of 10 kilometers (over 30,000 feet)?

77. Space Satellites The number of watts w provided by a space satellite's power supply after a period of d days is given by the function

$$w(d) = 50e^{-0.004d}$$

(a) How much power will be available after 30 days?
(b) How much power will be available after 1 year (365 days)?

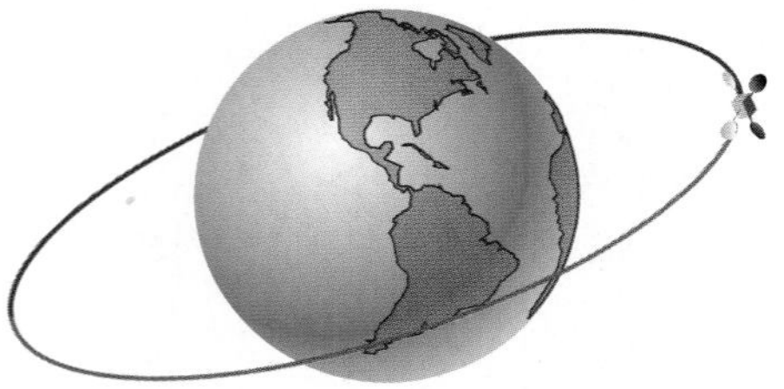

78. Healing of Wounds The normal healing of wounds can be modeled by an exponential function. If A_0 represents the original area of the wound and if A equals the area of the wound after n days, then the function

$$A(n) = A_0 e^{-0.35n}$$

describes the area of a wound on the nth day following an injury when no infection is present to retard the healing. Suppose that a wound initially had an area of 100 square millimeters.
(a) If healing is taking place, how large will the area of the wound be after 3 days?
(b) How large will it be after 10 days?

79. Drug Medication The function

$$D(h) = 5e^{-0.4h}$$

can be used to find the number of milligrams D of a certain drug that is in a patient's bloodstream h hours after the drug has been administered. How many milligrams will be present after 1 hour? After 6 hours?

80. Spreading of Rumors A model for the number of people N in a college community who have heard a certain rumor is

$$N = P(1 - e^{-0.15d})$$

where P is the total population of the community and d is the number of days that have elapsed since the rumor began. In a community of 1000 students, how many students will have heard the rumor after 3 days?

81. Exponential Probability Between 12:00 PM and 1:00 PM, cars arrive at Citibank's drive-thru at the rate of 6 cars per hour (0.1 car per minute). The following formula from probability can be used to determine the probability that a car will arrive within t minutes of 12:00 PM:

$$F(t) = 1 - e^{-0.1t}$$

(a) Determine the probability that a car will arrive within 10 minutes of 12:00 PM (that is, before 12:10 PM).
(b) Determine the probability that a car will arrive within 40 minutes of 12:00 PM (before 12:40 PM).
(c) What value does F approach as t becomes unbounded in the positive direction?
(d) Graph F using a graphing utility.
(e) Using TRACE, determine how many minutes are needed for the probability to reach 50%.

82. Exponential Probability Between 5:00 PM and 6:00 PM, cars arrive at Jiffy Lube at the rate of 9 cars per hour (0.15 car per minute). The following formula from the field of probability can be used to determine the probability that a car will arrive within t minutes of 5:00 PM:

$$F(t) = 1 - e^{-0.15t}$$

(a) Determine the probability that a car will arrive within 15 minutes of 5:00 PM (that is, before 5:15 PM).
(b) Determine the probability that a car will arrive within 30 minutes of 5:00 PM (before 5:30 PM).
(c) What value does F approach as t becomes unbounded in the positive direction?
(d) Graph F using a graphing utility.
(e) Using TRACE, determine how many minutes are needed for the probability to reach 60%.

83. Poisson Probability Between 5:00 PM and 6:00 PM, cars arrive at McDonald's drive-thru at the rate of 20 cars per hour. The following formula from the field of probability can be used to determine the probability that x cars will arrive between 5:00 PM and 6:00 PM.

$$P(x) = \frac{20^x e^{-20}}{x!}$$

where

$$x! = x \cdot (x-1) \cdot (x-2) \cdot \cdots \cdot 3 \cdot 2 \cdot 1$$

(a) Determine the probability that $x = 15$ cars will arrive between 5:00 PM and 6:00 PM.
(b) Determine the probability that $x = 20$ cars will arrive between 5:00 PM and 6:00 PM.

84. Poisson Probability People enter a line for the *Demon Roller Coaster* at the rate of 4 per minute. The following formula from probability can be used to determine the probability that x people will arrive within the next minute.

$$P(x) = \frac{4^x e^{-4}}{x!}$$

where

$$x! = x \cdot (x-1) \cdot (x-2) \cdot \cdots \cdot 3 \cdot 2 \cdot 1$$

(a) Determine the probability that $x = 5$ people will arrive within the next minute.
(b) Determine the probability that $x = 8$ people will arrive within the next minute.

85. Depreciation The price p of a Honda Civic DX Sedan that is x years old is given by

$$p(x) = 16{,}630(0.90)^x$$

(a) How much does a 3-year-old Civic DX Sedan cost?
(b) How much does a 9-year-old Civic DX Sedan cost?

86. Learning Curve Suppose that a student has 500 vocabulary words to learn. If the student learns 15 words after 5 minutes, the function

$$L(t) = 500(1 - e^{-0.0061t})$$

approximates the number of words L that the student will learn after t minutes.

(a) How many words will the student learn after 30 minutes?

(b) How many words will the student learn after 60 minutes?

87. Current in a *RL* Circuit The equation governing the amount of current I (in amperes) after time t (in seconds) in a single RL circuit consisting of a resistance R (in ohms), an inductance L (in henrys), and an electromotive force E (in volts) is

$$I = \frac{E}{R}[1 - e^{-(R/L)t}]$$

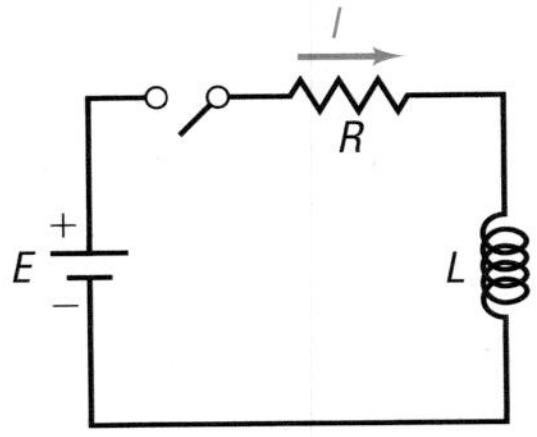

(a) If $E = 120$ volts, $R = 10$ ohms, and $L = 5$ henrys, how much current I_1 is flowing after 0.3 second? After 0.5 second? After 1 second?

(b) What is the maximum current?

(c) Graph this function $I = I_1(t)$, measuring I along the y-axis and t along the x-axis.

(d) If $E = 120$ volts, $R = 5$ ohms, and $L = 10$ henrys, how much current I_2 is flowing after 0.3 second? After 0.5 second? After 1 second?

(e) What is the maximum current?

(f) Graph this function $I = I_2(t)$ on the same coordinate axes as $I_1(t)$.

88. Current in a *RC* Circuit The equation governing the amount of current I (in amperes) after time t (in microseconds) in a single RC circuit consisting of a resistance R (in ohms), a capacitance C (in microfarads), and an electromotive force E (in volts) is

$$I = \frac{E}{R}e^{-t/(RC)}$$

(a) If $E = 120$ volts, $R = 2000$ ohms, and $C = 1.0$ microfarad, how much current I_1 is flowing initially ($t = 0$)? After 1000 microseconds? After 3000 microseconds?

(b) What is the maximum current?

(c) Graph this function $I = I_1(t)$, measuring I along the y-axis and t along the x-axis.

(d) If $E = 120$ volts, $R = 1000$ ohms, and $C = 2.0$ microfarads, how much current I_2 is flowing initially? After 1000 microseconds? After 3000 microseconds?

(e) What is the maximum current?

(f) Graph this function $I = I_2(t)$ on the same coordinate axes as $I_1(t)$.

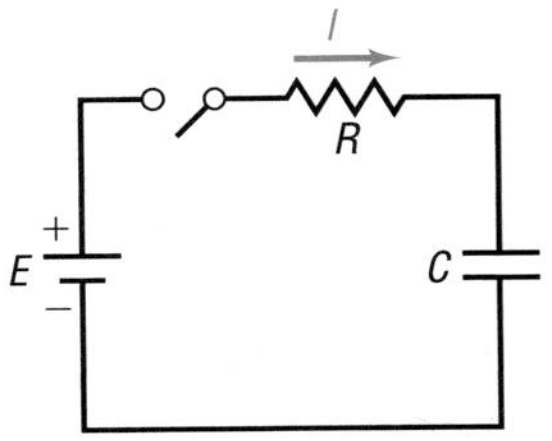

89. Another Formula for *e* Use a calculator to compute the values of

$$2 + \frac{1}{2!} + \frac{1}{3!} + \cdots + \frac{1}{n!}$$

for $n = 4, 6, 8$, and 10. Compare each result with e.

[**Hint:** $1! = 1, 2! = 2 \cdot 1, 3! = 3 \cdot 2 \cdot 1$,
$n! = n(n-1) \cdot \cdots \cdot (3)(2)(1)$.]

90. Another Formula for *e* Use a calculator to compute the various values of the expression. Compare the values to e.

$$2 + \cfrac{1}{1 + \cfrac{1}{2 + \cfrac{2}{3 + \cfrac{3}{4 + \cfrac{4}{\text{etc.}}}}}}$$

91. Difference Quotient If $f(x) = a^x$, show that

$$\frac{f(x+h) - f(x)}{h} = a^x \cdot \frac{a^h - 1}{h}$$

92. If $f(x) = a^x$, show that $f(A + B) = f(A) \cdot f(B)$.

93. If $f(x) = a^x$, show that $f(-x) = \dfrac{1}{f(x)}$.

94. If $f(x) = a^x$, show that $f(\alpha x) = [f(x)]^\alpha$.

95. Relative Humidity The relative humidity is the ratio (expressed as a percent) of the amount of water vapor in the air to the maximum amount that it can hold at a specific temperature. The relative humidity, R, is found using the following formula:

$$R = 10^{\left(\frac{4221}{T+459.4} - \frac{4221}{D+459.4} + 2\right)}$$

where T is the air temperature (in °F) and D is the dew point temperature (in °F).

(a) Determine the relative humidity if the air temperature is 50° Fahrenheit and the dew point temperature is 41° Fahrenheit.

(b) Determine the relative humidity if the air temperature is 68° Fahrenheit and the dew point temperature is 59° Fahrenheit.

(c) What is the relative humidity if the air temperature and the dew point temperature are the same?

96. Historical Problem Pierre de Fermat (1601–1665) conjectured that the function

$$f(x) = 2^{(2^x)} + 1$$

for $x = 1, 2, 3, \ldots$, would always have a value equal to a prime number. But Leonhard Euler (1707–1783) showed that this formula fails for $x = 5$. Use a calculator to determine the prime numbers produced by f for $x = 1, 2, 3, 4$. Then show that $f(5) = 641 \times 6{,}700{,}417$, which is not prime.

Problems 97 and 98 provide definitions for two other transcendental functions.

97. The **hyperbolic sine function,** designated by sinh x, is defined as

$$\sinh x = \frac{1}{2}(e^x - e^{-x})$$

(a) Show that $f(x) = \sinh x$ is an odd function.

(b) Graph $f(x) = \sinh x$ using a graphing utility.

98. The **hyperbolic cosine function,** designated by cosh x, is defined as

$$\cosh x = \frac{1}{2}(e^x + e^{-x})$$

(a) Show that $f(x) = \cosh x$ is an even function.

(b) Graph $f(x) = \cosh x$ using a graphing utility.

(c) Refer to Problem 97. Show that, for every x,

$$(\cosh x)^2 - (\sinh x)^2 = 1$$

99. The bacteria in a 4-liter container double every minute. After 60 minutes the container is full. How long did it take to fill half the container?

100. Explain in your own words what the number e is. Provide at least two applications that require the use of this number.

101. Do you think that there is a power function that increases more rapidly than an exponential function whose base is greater than 1? Explain.

102. As the base a of an exponential function $f(x) = a^x$, $a > 1$, increases, what happens to the behavior of its graph for $x > 0$? What happens to the behavior of its graph for $x < 0$?

103. The graphs of $y = a^{-x}$ and $y = \left(\frac{1}{a}\right)^x$ are identical. Why?

'Are You Prepared?' Answers

1. $64; 4; \frac{1}{9}$ **2.** $\left\{-2, \frac{1}{3}\right\}$ **3.** False **4.** 3 **5.** True

4.4 Logarithmic Functions

PREPARING FOR THIS SECTION *Before getting started, review the following:*

- Solving Inequalities (Appendix A, Section A.8, pp. 971–974)
- Polynomial and Rational Inequalities (Section 3.5, pp. 188–192)

Now work the 'Are You Prepared?' problems on page 269.

OBJECTIVES
1. Change Exponential Expressions to Logarithmic Expressions
2. Change Logarithmic Expressions to Exponential Expressions
3. Evaluate Logarithmic Functions
4. Determine the Domain of a Logarithmic Function
5. Graph Logarithmic Functions
6. Solve Logarithmic Equations

Recall that a one-to-one function $y = f(x)$ has an inverse function that is defined implicitly by the equation $x = f(y)$. In particular, the exponential function $y = f(x) = a^x$, $a > 0$, $a \neq 1$, is one-to-one and hence has an inverse function that is defined implicitly by the equation

$$x = a^y, \quad a > 0, \quad a \neq 1$$

This inverse function is so important that it is given a name, the *logarithmic function*.

The **logarithmic function to the base *a*,** where $a > 0$ and $a \neq 1$, is denoted by $y = \log_a x$ (read as "y is the logarithm to the base a of x") and is defined by

$$y = \log_a x \quad \text{if and only if} \quad x = a^y$$

The domain of the logarithmic function $y = \log_a x$ is $x > 0$.

A *logarithm* is merely a name for a certain exponent.

EXAMPLE 1 Relating Logarithms to Exponents

(a) If $y = \log_3 x$, then $x = 3^y$. For example, $4 = \log_3 81$ is equivalent to $81 = 3^4$.

(b) If $y = \log_5 x$, then $x = 5^y$. For example, $-1 = \log_5\left(\frac{1}{5}\right)$ is equivalent to $\frac{1}{5} = 5^{-1}$.

◄

1 ### EXAMPLE 2 Changing Exponential Expressions to Logarithmic Expressions

Change each exponential expression to an equivalent expression involving a logarithm.

(a) $1.2^3 = m$ (b) $e^b = 9$ (c) $a^4 = 24$

Solution We use the fact that $y = \log_a x$ and $x = a^y$, $a > 0$, $a \neq 1$, are equivalent.

(a) If $1.2^3 = m$, then $3 = \log_{1.2} m$.
(b) If $e^b = 9$, then $b = \log_e 9$.
(c) If $a^4 = 24$, then $4 = \log_a 24$.

◄

NOW WORK PROBLEM 9.

2 ### EXAMPLE 3 Changing Logarithmic Expressions to Exponential Expressions

Change each logarithmic expression to an equivalent expression involving an exponent.

(a) $\log_a 4 = 5$ (b) $\log_e b = -3$ (c) $\log_3 5 = c$

Solution
(a) If $\log_a 4 = 5$, then $a^5 = 4$.
(b) If $\log_e b = -3$, then $e^{-3} = b$.
(c) If $\log_3 5 = c$, then $3^c = 5$.

◄

NOW WORK PROBLEM 21.

3 To find the exact value of a logarithm, we write the logarithm in exponential notation and use the fact that if $a^u = a^v$ then $u = v$.

EXAMPLE 4 **Finding the Exact Value of a Logarithmic Expression**

Find the exact value of

(a) $\log_2 16$ (b) $\log_3 \frac{1}{27}$

Solution (a)

$$\begin{aligned} y &= \log_2 16 \\ 2^y &= 16 && \text{Change to exponential form.} \\ 2^y &= 2^4 && 16 = 2^4 \\ y &= 4 && \text{Equate exponents.} \end{aligned}$$

Therefore, $\log_2 16 = 4$.

(b)

$$\begin{aligned} y &= \log_3 \frac{1}{27} \\ 3^y &= \frac{1}{27} && \text{Change to exponential form.} \\ 3^y &= 3^{-3} && \frac{1}{27} = \frac{1}{3^3} = 3^{-3} \\ y &= -3 && \text{Equate exponents.} \end{aligned}$$

Therefore, $\log_3 \frac{1}{27} = -3$. ◀

NOW WORK PROBLEM 33.

Domain of a Logarithmic Function

4 The logarithmic function $y = \log_a x$ has been defined as the inverse of the exponential function $y = a^x$. That is, if $f(x) = a^x$, then $f^{-1}(x) = \log_a x$. Based on the discussion given in Section 4.2 on inverse functions, we know that, for a function f and its inverse f^{-1},

$$\text{Domain of } f^{-1} = \text{Range of } f \quad \text{and} \quad \text{Range of } f^{-1} = \text{Domain of } f$$

Consequently, it follows that

> Domain of the logarithmic function = Range of the exponential function = $(0, \infty)$
> Range of the logarithmic function = Domain of the exponential function = $(-\infty, \infty)$

In the next box, we summarize some properties of the logarithmic function:

> $y = \log_a x$ (defining equation: $x = a^y$)
> Domain: $0 < x < \infty$ Range: $-\infty < y < \infty$

The domain of a logarithmic function consists of the *positive* real numbers, so the argument of a logarithmic function must be greater than zero.

EXAMPLE 5 **Finding the Domain of a Logarithmic Function**

Find the domain of each logarithmic function.

(a) $F(x) = \log_2(x - 5)$ (b) $g(x) = \log_5\left(\frac{1 + x}{1 - x}\right)$ (c) $h(x) = \log_{1/2}|x|$

Solution (a) The domain of F consists of all x for which $x - 5 > 0$, that is, all $x > 5$, or, using interval notation, $(5, \infty)$.

(b) The domain of g is restricted to

$$\frac{1 + x}{1 - x} > 0$$

Solving this inequality, we find that the domain of g consists of all x between -1 and 1, that is, $-1 < x < 1$, or, using interval notation, $(-1, 1)$.

(c) Since $|x| > 0$, provided that $x \neq 0$, the domain of h consists of all nonzero real numbers or, using interval notation, $(-\infty, 0)$ or $(0, \infty)$. ◀

NOW WORK PROBLEMS 47 AND 53.

Graphs of Logarithmic Functions

5 Since exponential functions and logarithmic functions are inverses of each other, the graph of the logarithmic function $y = \log_a x$ is the reflection about the line $y = x$ of the graph of the exponential function $y = a^x$, as shown in Figure 25.

Figure 25

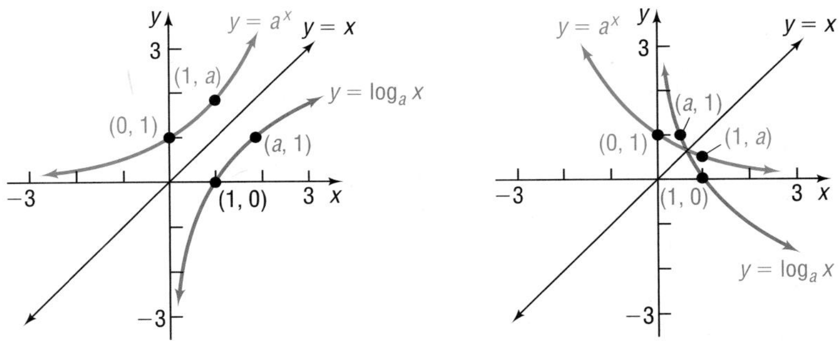

(a) $a > 1$ (b) $0 < a < 1$

For example, to graph $y = \log_2 x$, graph $y = 2^x$ and reflect it about the line $y = x$. See Figure 26. To graph $y = \log_{1/3} x$, graph $y = \left(\frac{1}{3}\right)^x$ and reflect it about the line $y = x$. See Figure 27.

Figure 26

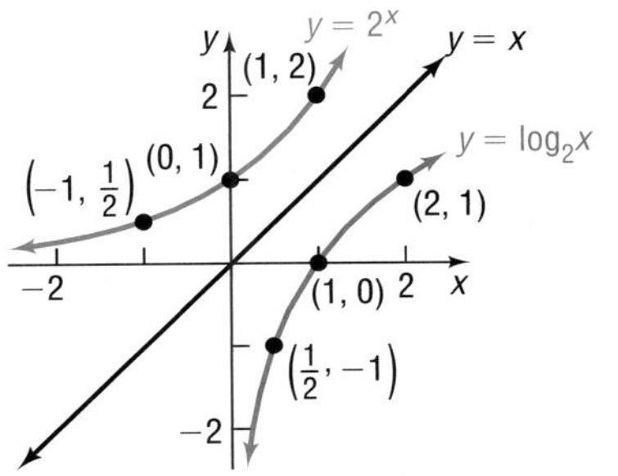

Figure 27

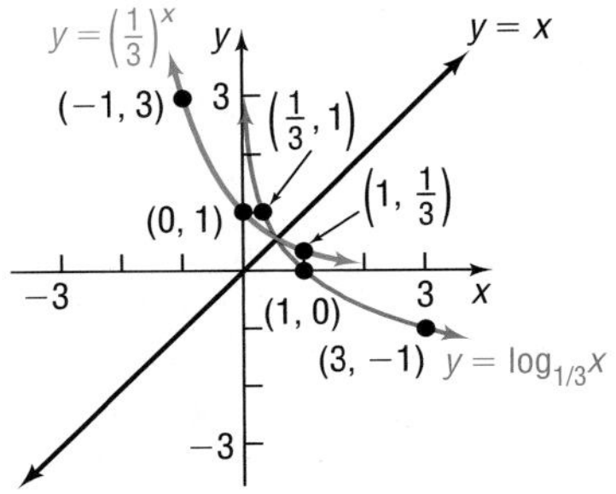

NOW WORK PROBLEM 63.

Properties of the Graph of a Logarithmic Function $f(x) = \log_a x$

1. The domain is the set of positive real numbers; the range is the set of all real numbers.
2. The x-intercept of the graph is 1. There is no y-intercept.
3. The y-axis ($x = 0$) is a vertical asymptote of the graph.
4. A logarithmic function is decreasing if $0 < a < 1$ and increasing if $a > 1$.
5. The graph of f contains the points $(1, 0)$, $(a, 1)$, and $\left(\frac{1}{a}, -1\right)$.
6. The graph is smooth and continuous, with no corners or gaps.

If the base of a logarithmic function is the number e, then we have the **natural logarithm function.** This function occurs so frequently in applications that it is given a special symbol, **ln** (from the Latin, *logarithmus naturalis*). Thus,

$$y = \log_e x = \ln x \quad \text{if and only if} \quad x = e^y \tag{1}$$

Since $y = \ln x$ and the exponential function $y = e^x$ are inverse functions, we can obtain the graph of $y = \ln x$ by reflecting the graph of $y = e^x$ about the line $y = x$. See Figure 28.

Using a calculator with an [ln] key, we can obtain other points on the graph of $f(x) = \ln x$. See Table 7.

Figure 28

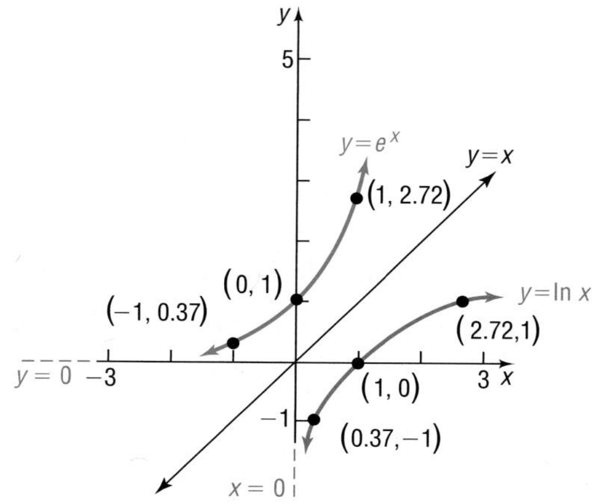

Table 7

x	$\ln x$
$\frac{1}{2}$	−0.69
2	0.69
3	1.10

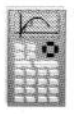

Seeing the Concept

Graph $Y_1 = e^x$ and $Y_2 = \ln x$ on the same square screen. Use eVALUEate to verify the points on the graph given in Figure 28. Do you see the symmetry of the two graphs with respect to the line $y = x$?

EXAMPLE 6 Graphing Logarithmic Functions Using Transformations

Graph $f(x) = -\ln(x + 2)$ by starting with the graph of $y = \ln x$. Determine the domain, range, and vertical asymptote of f.

Solution The domain of f consists of all x for which

$$x + 2 > 0 \quad \text{or} \quad x > -2$$

To obtain the graph of $y = -\ln(x + 2)$, we use the steps illustrated in Figure 29.

Figure 29

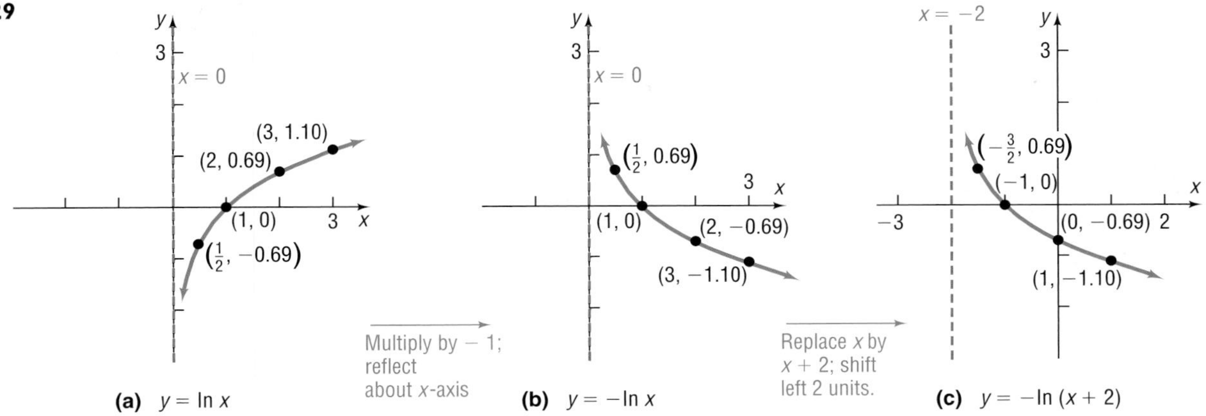

(a) $y = \ln x$ **(b)** $y = -\ln x$ **(c)** $y = -\ln(x + 2)$

The range of $f(x) = -\ln(x + 2)$ is the interval $(-\infty, \infty)$, and the vertical asymptote is $x = -2$. [Do you see why? The original asymptote $(x = 0)$ is shifted to the left 2 units.] ◀

NOW WORK PROBLEM 75.

If the base of a logarithmic function is the number 10, then we have the **common logarithm function.** If the base a of the logarithmic function is not indicated, it is understood to be 10. Thus,

$$y = \log x \quad \text{if and only if} \quad x = 10^y$$

Since $y = \log x$ and the exponential function $y = 10^x$ are inverse functions, we can obtain the graph of $y = \log x$ by reflecting the graph of $y = 10^x$ about the line $y = x$. See Figure 30.

Figure 30

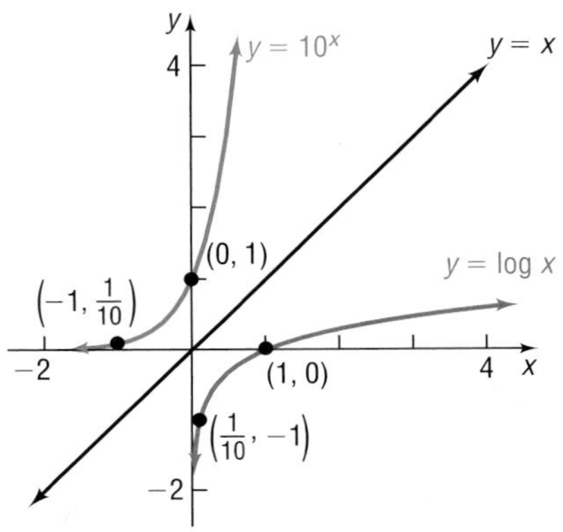

EXAMPLE 7 Graphing Logarithmic Functions Using Transformations

Graph $f(x) = 3\log(x - 1)$. Determine the domain, range, and vertical asymptote of f.

Solution The domain consists of all x for which

$$x - 1 > 0 \quad \text{or} \quad x > 1$$

To obtain the graph of $y = 3\log(x - 1)$, we use the steps illustrated in Figure 31. The range of $f(x) = 3\log(x - 1)$ is the interval $(-\infty, \infty)$, and the vertical asymptote is $x = 1$.

Figure 31

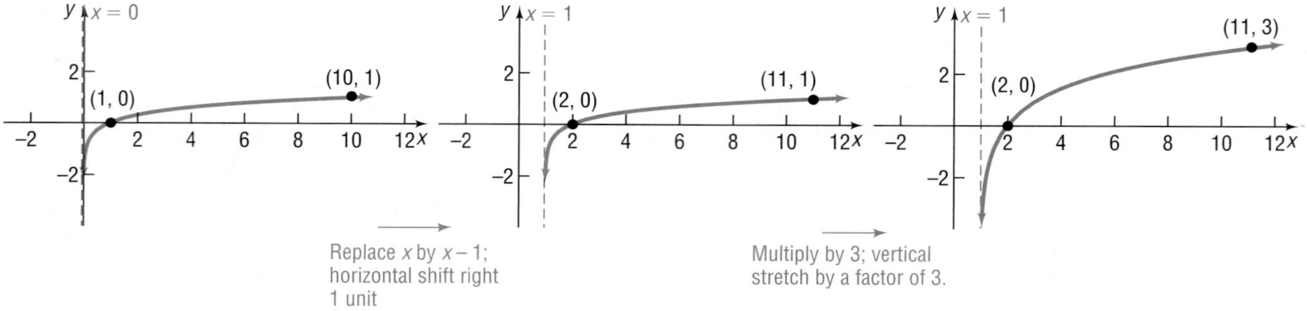

(a) $y = \log x$ **(b)** $y = \log(x - 1)$ **(c)** $y = 3\log(x - 1)$ ◀

NOW WORK PROBLEM 85.

Logarithmic Equations

6 Equations that contain logarithms are called **logarithmic equations.** Care must be taken when solving logarithmic equations algebraically. Be sure to check each apparent solution in the original equation and discard any that are extraneous. In the expression $\log_a M$, remember that a and M are positive and $a \neq 1$.

Some logarithmic equations can be solved by changing from a logarithmic expression to an exponential expression.

EXAMPLE 8 Solving a Logarithmic Equation

Solve: (a) $\log_3(4x - 7) = 2$ (b) $\log_x 64 = 2$

Solution (a) We can obtain an exact solution by changing the logarithm to exponential form.

$$\begin{aligned} \log_3(4x - 7) &= 2 \\ 4x - 7 &= 3^2 \quad \text{Change to exponential form.} \\ 4x - 7 &= 9 \\ 4x &= 16 \\ x &= 4 \end{aligned}$$

CHECK: $\log_3(4x - 7) = \log_3(16 - 7) = \log_3 9 = 2$ $\quad 3^2 = 9$

(b) We can obtain an exact solution by changing the logarithm to exponential form.

$$\log_x 64 = 2$$
$$x^2 = 64 \qquad \text{Change to exponential form.}$$
$$x = \pm\sqrt{64} = \pm 8$$

The base of a logarithm is always positive. As a result, we discard -8; the only solution is 8.

CHECK: $\log_8 64 = 2 \quad (8^2 = 64)$. ◀

EXAMPLE 9 Using Logarithms to Solve Exponential Equations

Solve: $e^{2x} = 5$

Solution We can obtain an exact solution by changing the exponential equation to logarithmic form.

$$e^{2x} = 5$$
$$\ln 5 = 2x \qquad \text{Change to a logarithmic expression using (1).}$$
$$x = \frac{\ln 5}{2} \qquad \text{Exact solution.}$$
$$\approx 0.805 \qquad \text{Approximate solution.}$$
◀

NOW WORK PROBLEMS 91 AND 103.

EXAMPLE 10 Alcohol and Driving

The concentration of alcohol in a person's blood is measurable. Recent medical research suggests that the risk R (given as a percent) of having an accident while driving a car can be modeled by the equation

$$R = 6e^{kx}$$

where x is the variable concentration of alcohol in the blood and k is a constant.

(a) Suppose that a concentration of alcohol in the blood of 0.04 results in a 10% risk ($R = 10$) of an accident. Find the constant k in the equation.

(b) Using this value of k, what is the risk if the concentration is 0.17?

(c) Using the same value of k, what concentration of alcohol corresponds to a risk of 100%?

(d) If the law asserts that anyone with a risk of having an accident of 20% or more should not have driving privileges, at what concentration of alcohol in the blood should a driver be arrested and charged with a DUI (Driving Under the Influence)?

Solution (a) For a concentration of alcohol in the blood of 0.04 and a risk of 10%, we let $x = 0.04$ and $R = 10$ in the equation and solve for k.

$$R = 6e^{kx}$$

$$10 = 6e^{k(0.04)} \qquad R = 10;\ x = 0.04$$

$$\frac{10}{6} = e^{0.04k} \qquad \text{Divide both sides by 6.}$$

$$0.04k = \ln\frac{10}{6} \approx 0.510826 \qquad \text{Change to a logarithmic expression.}$$

$$k \approx 12.77 \qquad \text{Solve for } k.$$

(b) Using $k = 12.77$ and $x = 0.17$ in the equation, we find the risk R to be

$$R = 6e^{kx} = 6e^{(12.77)(0.17)} \approx 52.6$$

For a concentration of alcohol in the blood of 0.17, the risk of an accident is about 52.6%.

(c) Using $k = 12.77$ and $R = 100$ in the equation, we find the concentration x of alcohol in the blood to be

$$R = 6e^{kx}$$

$$100 = 6e^{12.77x} \qquad R = 100;\ k = 12.77$$

$$\frac{100}{6} = e^{12.77x} \qquad \text{Divide both sides by 6.}$$

$$12.77x = \ln\frac{100}{6} \approx 2.8134 \qquad \text{Change to a logarithmic expression.}$$

$$x \approx 0.22 \qquad \text{Solve for } x.$$

For a concentration of alcohol in the blood of 0.22, the risk of an accident is 100%.

(d) Using $k = 12.77$ and $R = 20$ in the equation, we find the concentration x of alcohol in the blood to be

$$R = 6e^{kx}$$

$$20 = 6e^{12.77x}$$

$$\frac{20}{6} = e^{12.77x}$$

$$12.77x = \ln\frac{20}{6} \approx 1.204$$

$$x \approx 0.094$$

A driver with a concentration of alcohol in the blood of 0.094 or more (9.4%) should be arrested and charged with DUI. ◀

[*Note:* Most states use 0.08 or 0.10 as the blood alcohol content at which a DUI citation is given.]

Summary

Properties of the Logarithmic Function

$f(x) = \log_a x, \quad a > 1$
($y = \log_a x$ means $x = a^y$)

Domain: the interval $(0, \infty)$; Range: the interval $(-\infty, \infty)$; x-intercept: 1; y-intercept: none; vertical asymptote: $x = 0$ (y-axis); increasing; one-to-one
See Figure 32(a) for a typical graph.

$f(x) = \log_a x, \quad 0 < a < 1$
($y = \log_a x$ means $x = a^y$)

Domain: the interval $(0, \infty)$; Range: the interval $(-\infty, \infty)$; x-intercept: 1; y-intercept: none; vertical asymptote: $x = 0$ (y-axis); decreasing; one-to-one
See Figure 32(b) for a typical graph.

Figure 32

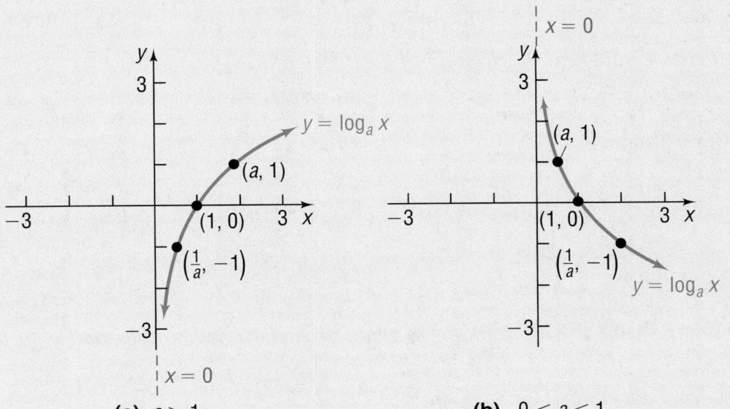

(a) $a > 1$ (b) $0 < a < 1$

4.4 Assess Your Understanding

'Are You Prepared?' *Answers are given at the end of these exercises. If you get a wrong answer, read the pages listed in red.*

1. Solve the inequality: $3x - 7 \leq 8 - 2x$ (pp. 971–974)
2. Solve the inequality: $x^2 - x - 6 > 0$ (pp. 188–192)
3. Solve the inequality: $\dfrac{x-1}{x+4} > 0$ (pp. 188–192)

Concepts and Vocabulary

4. The domain of the logarithmic function $f(x) = \log_a x$ is __________.
5. The graph of every logarithmic function $f(x) = \log_a x$, $a > 0$, $a \neq 1$, passes through three points: __________, __________, and __________.
6. If the graph of a logarithmic function $f(x) = \log_a x$, $a > 0$, $a \neq 1$, is increasing, then its base must be larger than __________.
7. *True or False:* If $y = \log_a x$, then $y = a^x$.
8. *True or False:* The graph of every logarithmic function $f(x) = \log_a x$, $a > 0$, $a \neq 1$, will contain the points $(1, 0)$, $(a, 1)$ and $\left(\dfrac{1}{a}, -1\right)$.

Exercises

In Problems 9–20, change each exponential expression to an equivalent expression involving a logarithm.

9. $9 = 3^2$
10. $16 = 4^2$
11. $a^2 = 1.6$
12. $a^3 = 2.1$

13. $1.1^2 = M$ **14.** $2.2^3 = N$ **15.** $2^x = 7.2$ **16.** $3^x = 4.6$

17. $x^{\sqrt{2}} = \pi$ **18.** $x^\pi = e$ **19.** $e^x = 8$ **20.** $e^{2.2} = M$

In Problems 21–32, change each logarithmic expression to an equivalent expression involving an exponent.

21. $\log_2 8 = 3$ **22.** $\log_3\left(\frac{1}{9}\right) = -2$ **23.** $\log_a 3 = 6$ **24.** $\log_b 4 = 2$

25. $\log_3 2 = x$ **26.** $\log_2 6 = x$ **27.** $\log_2 M = 1.3$ **28.** $\log_3 N = 2.1$

29. $\log_{\sqrt{2}} \pi = x$ **30.** $\log_\pi x = \frac{1}{2}$ **31.** $\ln 4 = x$ **32.** $\ln x = 4$

In Problems 33–44, find the exact value of each logarithm without using a calculator.

33. $\log_2 1$ **34.** $\log_8 8$ **35.** $\log_5 25$ **36.** $\log_3\left(\frac{1}{9}\right)$

37. $\log_{1/2} 16$ **38.** $\log_{1/3} 9$ **39.** $\log_{10} \sqrt{10}$ **40.** $\log_5 \sqrt[3]{25}$

41. $\log_{\sqrt{2}} 4$ **42.** $\log_{\sqrt{3}} 9$ **43.** $\ln \sqrt{e}$ **44.** $\ln e^3$

In Problems 45–56, find the domain of each function.

45. $f(x) = \ln(x - 3)$ **46.** $g(x) = \ln(x - 1)$ **47.** $F(x) = \log_2 x^2$

48. $H(x) = \log_5 x^3$ **49.** $f(x) = 3 - 2\log_4 \frac{x}{2}$ **50.** $g(x) = 8 + 5\ln(2x)$

51. $f(x) = \ln\left(\frac{1}{x + 1}\right)$ **52.** $g(x) = \ln\left(\frac{1}{x - 5}\right)$ **53.** $g(x) = \log_5\left(\frac{x + 1}{x}\right)$

54. $h(x) = \log_3\left(\frac{x}{x - 1}\right)$ **55.** $f(x) = \sqrt{\ln x}$ **56.** $g(x) = \frac{1}{\ln x}$

In Problems 57–60, use a calculator to evaluate each expression. Round your answer to three decimal places.

57. $\ln \frac{5}{3}$ **58.** $\frac{\ln 5}{3}$ **59.** $\frac{\ln \frac{10}{3}}{0.04}$ **60.** $\frac{\ln \frac{2}{3}}{-0.1}$

61. Find a so that the graph of $f(x) = \log_a x$ contains the point $(2, 2)$.

62. Find a so that the graph of $f(x) = \log_a x$ contains the point $\left(\frac{1}{2}, -4\right)$.

In Problems 63–66, graph each logarithmic function.

63. $y = \log_3 x$ **64.** $y = \log_{1/2} x$ **65.** $y = \log_{1/4} x$ **66.** $y = \log_5 x$

In Problems 67–74, the graph of a logarithmic function is given. Match each graph to one of the following functions:

A. $y = \log_3 x$ *B.* $y = \log_3(-x)$ *C.* $y = -\log_3 x$ *D.* $y = -\log_3(-x)$

E. $y = \log_3 x - 1$ *F.* $y = \log_3(x - 1)$ *G.* $y = \log_3(1 - x)$ *H.* $y = 1 - \log_3 x$

67.

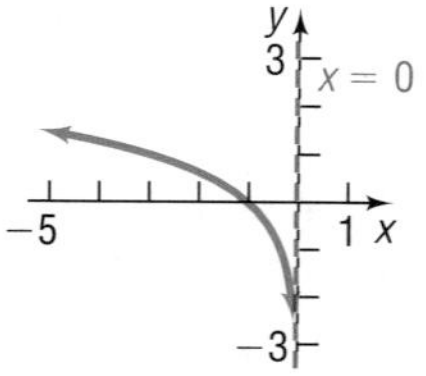

68.

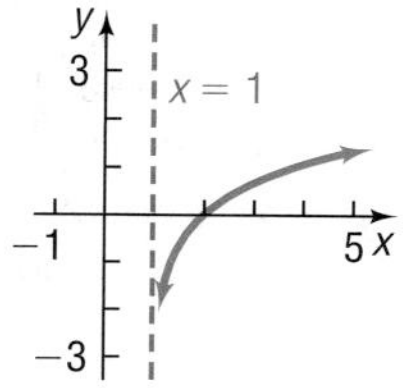

69.

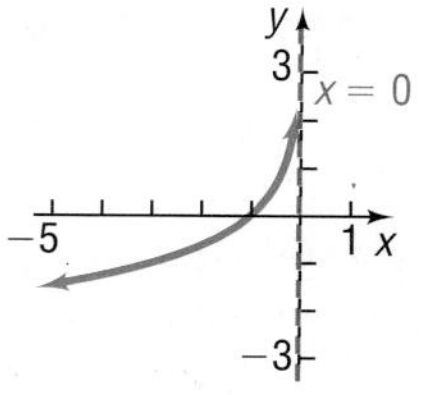

70.

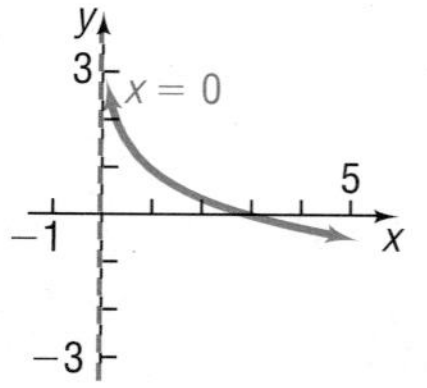

71.

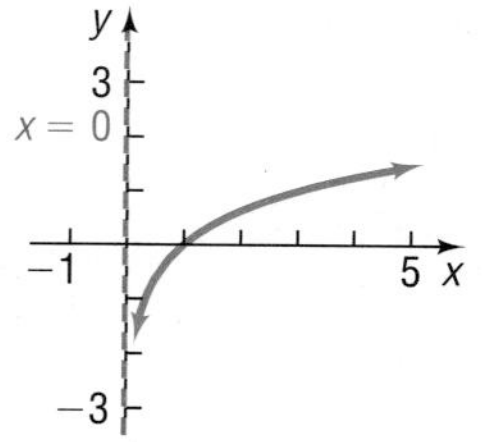

72.

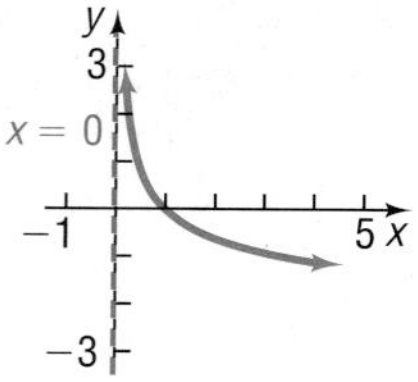

73.

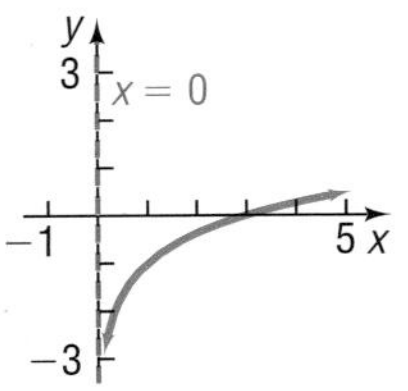

74.

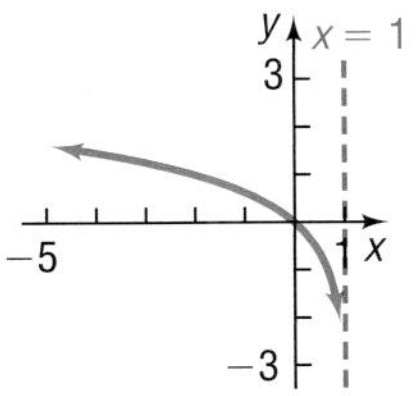

In Problems 75–90, use transformations to graph each function. Determine the domain, range, and vertical asymptote of each function.

75. $f(x) = \ln(x + 4)$ **76.** $f(x) = \ln(x - 3)$ **77.** $f(x) = 2 + \ln x$ **78.** $f(x) = -\ln(-x)$

79. $g(x) = \ln(2x)$ **80.** $h(x) = \ln\left(\frac{1}{2}x\right)$ **81.** $f(x) = 3 \ln x$ **82.** $f(x) = -2 \ln x$

83. $f(x) = \log(x - 4)$ **84.** $f(x) = \log(x + 5)$ **85.** $h(x) = 4 \log x$ **86.** $g(x) = -3 \log x$

87. $F(x) = \log(2x)$ **88.** $G(x) = \log(5x)$ **89.** $h(x) = 3 + \log(x + 2)$ **90.** $g(x) = 2 - \log(x + 1)$

In Problems 91–110, solve each equation.

91. $\log_3 x = 2$ **92.** $\log_5 x = 3$ **93.** $\log_2(2x + 1) = 3$ **94.** $\log_3(3x - 2) = 2$

95. $\log_x 4 = 2$ **96.** $\log_x\left(\frac{1}{8}\right) = 3$ **97.** $\ln e^x = 5$ **98.** $\ln e^{-2x} = 8$

99. $\log_4 64 = x$ **100.** $\log_5 625 = x$ **101.** $\log_3 243 = 2x + 1$ **102.** $\log_6 36 = 5x + 3$

103. $e^{3x} = 10$ **104.** $e^{-2x} = \frac{1}{3}$ **105.** $e^{2x+5} = 8$ **106.** $e^{-2x+1} = 13$

107. $\log_3(x^2 + 1) = 2$ **108.** $\log_5(x^2 + x + 4) = 2$ **109.** $\log_2 8^x = -3$ **110.** $\log_3 3^x = -1$

111. Chemistry The pH of a chemical solution is given by the formula

$$\text{pH} = -\log_{10}[\text{H}^+]$$

where $[\text{H}^+]$ is the concentration of hydrogen ions in moles per liter. Values of pH range from 0 (acidic) to 14 (alkaline).

(a) What is the pH of a solution for which $[\text{H}^+]$ is 0.1?
(b) What is the pH of a solution for which $[\text{H}^+]$ is 0.01?
(c) What is the pH of a solution for which $[\text{H}^+]$ is 0.001?
(d) What happens to pH as the hydrogen ion concentration decreases?
(e) Determine the hydrogen ion concentration of an orange (pH = 3.5).
(f) Determine the hydrogen ion concentration of human blood (pH = 7.4).

112. Diversity Index Shannon's diversity index is a measure of the diversity of a population. The diversity index is given by the formula

$$H = -(p_1 \log p_1 + p_2 \log p_2 + \cdots + p_n \log p_n)$$

where p_1 is the proportion of the population that is species 1, p_2 is the proportion of the population that is species 2, and so on.

(a) According to the U.S. Census Bureau, the distribution of race in the United States in 2000 was as follows:

Race	Proportion
American Indian or Native Alaskan	0.014
Asian	0.041
Black or African American	0.128
Hispanic	0.124
Native Hawaiian or Pacific Islander	0.003
White	0.690

SOURCE: U.S. Census Bureau

Compute the diversity index of the United States in 2000.

(b) The largest value of the diversity index is given by $H_{\max} = \log(S)$, where S is the number of categories of race. Compute $H_{\max}$.

(c) The evenness ratio is given by $E_H = \frac{H}{H_{\max}}$, where $0 \le E_H \le 1$. If $E_H = 1$, there is complete evenness. Compute the evenness ratio for the United States.

(d) Obtain the distribution of race for the United States in 1990 from the Census Bureau. Compute Shannon's diversity index. Is the United States becoming more diverse? Why?

113. Atmospheric Pressure The atmospheric pressure p on a balloon or an aircraft decreases with increasing height. This pressure, measured in millimeters of mercury, is related to the height h (in kilometers) above sea level by the formula

$$p = 760e^{-0.145h}$$

(a) Find the height of an aircraft if the atmospheric pressure is 320 millimeters of mercury.
(b) Find the height of a mountain if the atmospheric pressure is 667 millimeters of mercury.

114. Healing of Wounds The normal healing of wounds can be modeled by an exponential function. If A_0 represents the original area of the wound and if A equals the area of the wound after n days, then the formula

$$A = A_0 e^{-0.35n}$$

describes the area of a wound on the nth day following an injury when no infection is present to retard the healing. Suppose that a wound initially had an area of 100 square millimeters.
(a) If healing is taking place, how many days will pass before the wound is one-half its original size?
(b) How long before the wound is 10% of its original size?

115. Exponential Probability Between 12:00 PM and 1:00 PM, cars arrive at Citibank's drive-thru at the rate of 6 cars per hour (0.1 car per minute). The following formula from statistics can be used to determine the probability that a car will arrive within t minutes of 12:00 PM.

$$F(t) = 1 - e^{-0.1t}$$

(a) Determine how many minutes are needed for the probability to reach 50%.
(b) Determine how many minutes are needed for the probability to reach 80%.
(c) Is it possible for the probability to equal 100%? Explain.

116. Exponential Probability Between 5:00 PM and 6:00 PM, cars arrive at Jiffy Lube at the rate of 9 cars per hour (0.15 car per minute). The following formula from statistics can be used to determine the probability that a car will arrive within t minutes of 5:00 PM.

$$F(t) = 1 - e^{-0.15t}$$

(a) Determine how many minutes are needed for the probability to reach 50%.
(b) Determine how many minutes are needed for the probability to reach 80%.

117. Drug Medication The formula

$$D = 5e^{-0.4h}$$

can be used to find the number of milligrams D of a certain drug that is in a patient's bloodstream h hours after the drug has been administered. When the number of milligrams reaches 2, the drug is to be administered again. What is the time between injections?

118. Spreading of Rumors A model for the number of people N in a college community who have heard a certain rumor is

$$N = P(1 - e^{-0.15d})$$

where P is the total population of the community and d is the number of days that have elapsed since the rumor began. In a community of 1000 students, how many days will elapse before 450 students have heard the rumor?

119. Current in a *RL* Circuit The equation governing the amount of current I (in amperes) after time t (in seconds) in a simple RL circuit consisting of a resistance R (in ohms), an inductance L (in henrys), and an electromotive force E (in volts) is

$$I = \frac{E}{R}[1 - e^{-(R/L)t}]$$

If $E = 12$ volts, $R = 10$ ohms, and $L = 5$ henrys, how long does it take to obtain a current of 0.5 ampere? Of 1.0 ampere? Graph the equation.

120. Learning Curve Psychologists sometimes use the function

$$L(t) = A(1 - e^{-kt})$$

to measure the amount L learned at time t. The number A represents the amount to be learned, and the number k measures the rate of learning. Suppose that a student has an amount A of 200 vocabulary words to learn. A psychologist determines that the student learned 20 vocabulary words after 5 minutes.
(a) Determine the rate of learning k.
(b) Approximately how many words will the student have learned after 10 minutes?
(c) After 15 minutes?
(d) How long does it take for the student to learn 180 words?

Loudness of Sound *Problems 121–124 use the following discussion: The* **loudness** $L(x)$, *measured in decibels, of a sound of intensity x, measured in watts per square meter, is defined as* $L(x) = 10 \log \frac{x}{I_0}$, *where* $I_0 = 10^{-12}$ *watt per square meter is the least intense sound that a human ear can detect. Determine the loudness, in decibels, of each of the following sounds.*

121. Normal conversation: intensity of $x = 10^{-7}$ watt per square meter.

122. Heavy city traffic: intensity of $x = 10^{-3}$ watt per square meter.

123. Amplified rock music: intensity of 10^{-1} watt per square meter.

124. Diesel truck traveling 40 miles per hour 50 feet away: intensity 10 times that of a passenger car traveling 50 miles per hour 50 feet away whose loudness is 70 decibels.

Problems 125 and 126 use the following discussion: The **Richter scale** *is one way of converting seismographic readings into numbers that provide an easy reference for measuring the magnitude M of an earthquake. All earthquakes are compared to a* **zero-level earthquake** *whose seismographic reading measures 0.001 millimeter at a distance of 100 kilometers from the epicenter. An earthquake whose seismographic reading measures x millimeters has* **magnitude** *$M(x)$, given by*

$$M(x) = \log\left(\frac{x}{x_0}\right)$$

where $x_0 = 10^{-3}$ is the reading of a zero-level earthquake the same distance from its epicenter. Determine the magnitude of the following earthquakes.

125. Magnitude of an Earthquake Mexico City in 1985: seismographic reading of 125,892 millimeters 100 kilometers from the center.

126. Magnitude of an Earthquake San Francisco in 1906: seismographic reading of 7943 millimeters 100 kilometers from the center.

127. Alcohol and Driving The concentration of alcohol in a person's blood is measurable. Suppose that the risk R (given as a percent) of having an accident while driving a car can be modeled by the equation

$$R = 3e^{kx}$$

where x is the variable concentration of alcohol in the blood and k is a constant.

(a) Suppose that a concentration of alcohol in the blood of 0.06 results in a 10% risk ($R = 10$) of an accident. Find the constant k in the equation.

(b) Using this value of k, what is the risk if the concentration is 0.17?

(c) Using the same value of k, what concentration of alcohol corresponds to a risk of 100%?

(d) If the law asserts that anyone with a risk of having an accident of 15% or more should not have driving privileges, at what concentration of alcohol in the blood should a driver be arrested and charged with a DUI?

(e) Compare this situation with that of Example 10. If you were a lawmaker, which situation would you support? Give your reasons.

128. Is there any function of the form $y = x^{\alpha}, 0 < \alpha < 1$, that increases more slowly than a logarithmic function whose base is greater than 1? Explain.

129. In the definition of the logarithmic function, the base a is not allowed to equal 1. Why?

130. Critical Thinking In buying a new car, one consideration might be how well the price of the car holds up over time. Different makes of cars have different depreciation rates. One way to compute a depreciation rate for a car is given here. Suppose that the current prices of a certain Mercedes automobile are as follows:

	Age in Years				
New	1	2	3	4	5
$38,000	$36,600	$32,400	$28,750	$25,400	$21,200

Use the formula New $=$ Old(e^{Rt}) to find R, the annual depreciation rate, for a specific time t. When might be the best time to trade in the car? Consult the NADA ("blue") book and compare two like models that you are interested in. Which has the better depreciation rate?

'Are You Prepared?' Answers

1. $x \le 3$ **2.** $x < -2$ or $x > 3$ **3.** $x < -4$ or $x > 1$

4.5 Properties of Logarithms

OBJECTIVES
1. Work with the Properties of Logarithms
2. Write a Logarithmic Expression as a Sum or Difference of Logarithms
3. Write a Logarithmic Expression as a Single Logarithm
4. Evaluate Logarithms Whose Base Is Neither 10 nor e

1 Logarithms have some very useful properties that can be derived directly from the definition and the laws of exponents.

EXAMPLE 1 Establishing Properties of Logarithms

(a) Show that $\log_a 1 = 0$. (b) Show that $\log_a a = 1$.

Solution (a) This fact was established when we graphed $y = \log_a x$ (see Figure 25). To show the result algebraically, let $y = \log_a 1$. Then

$$\begin{aligned} y &= \log_a 1 & \\ a^y &= 1 & \text{Change to an exponent.} \\ a^y &= a^0 & a^0 = 1 \\ y &= 0 & \text{Solve for } y. \\ \log_a 1 &= 0 & y = \log_a 1 \end{aligned}$$

(b) Let $y = \log_a a$. Then

$$\begin{aligned} y &= \log_a a & \\ a^y &= a & \text{Change to an exponent.} \\ a^y &= a^1 & a^1 = a \\ y &= 1 & \text{Solve for } y. \\ \log_a a &= 1 & y = \log_a a \end{aligned}$$

◀

To summarize:

$$\log_a 1 = 0 \qquad \log_a a = 1$$

Theorem

Properties of Logarithms

In the properties given next, M and a are positive real numbers, with $a \neq 1$, and r is any real number.
The number $\log_a M$ is the exponent to which a must be raised to obtain M. That is,

$$a^{\log_a M} = M \qquad \textbf{(1)}$$

The logarithm to the base a of a raised to a power equals that power. That is,

$$\log_a a^r = r \qquad \textbf{(2)}$$

The proof uses the fact that $y = a^x$ and $y = \log_a x$ are inverses.

Proof of Property (1) For inverse functions,

$$f(f^{-1}(x)) = x$$

Using $f(x) = a^x$ and $f^{-1}(x) = \log_a x$, we find

$$f(f^{-1}(x)) = a^{\log_a x} = x$$

Now let $x = M$ to obtain $a^{\log_a M} = M$. ■

Proof of Property (2) For inverse functions,

$$f^{-1}(f(x)) = x$$

Using $f(x) = a^x$ and $f^{-1}(x) = \log_a x$, we find

$$f^{-1}(f(x)) = \log_a a^x = x$$

Now let $x = r$ to obtain $\log_a a^r = r$. ■

EXAMPLE 2 **Using Properties (1) and (2)**

(a) $2^{\log_2 \pi} = \pi$ (b) $\log_{0.2} 0.2^{-\sqrt{2}} = -\sqrt{2}$ (c) $\ln e^{kt} = kt$ ◄

NOW WORK PROBLEM 9.

Other useful properties of logarithms are given next.

Theorem

Properties of Logarithms

In the following properties, M, N, and a are positive real numbers, with $a \neq 1$, and r is any real number.

The Log of a Product Equals the Sum of the Logs

$$\log_a(MN) = \log_a M + \log_a N \quad \textbf{(3)}$$

The Log of a Quotient Equals the Difference of the Logs

$$\log_a\left(\frac{M}{N}\right) = \log_a M - \log_a N \quad \textbf{(4)}$$

The Log of a Power Equals the Product of the Power and the Log

$$\log_a M^r = r \log_a M \quad \textbf{(5)}$$

We shall derive properties (3) and (5) leave the derivation of property (4) as an exercise (see Problem 101).

Proof of Property (3) Let $A = \log_a M$ and let $B = \log_a N$. These expressions are equivalent to the exponential expressions

$$a^A = M \quad \text{and} \quad a^B = N$$

Now

$$\begin{aligned} \log_a(MN) = \log_a(a^A a^B) &= \log_a a^{A+B} && \text{Law of Exponents} \\ &= A + B && \text{Property (2) of logarithms} \\ &= \log_a M + \log_a N \end{aligned}$$

■

Proof of Property (5) Let $A = \log_a M$. This expression is equivalent to

$$a^A = M$$

Now

$$\log_a M^r = \log_a(a^A)^r = \log_a a^{rA} \quad \text{Law of Exponents}$$
$$= rA \quad \text{Property (2) of logarithms}$$
$$= r \log_a M$$

■

NOW WORK PROBLEM 13.

2 Logarithms can be used to transform products into sums, quotients into differences, and powers into factors. Such transformations prove useful in certain types of calculus problems.

EXAMPLE 3 Writing a Logarithmic Expression as a Sum of Logarithms

Write $\log_a\left(x\sqrt{x^2+1}\right)$, $x > 0$, as a sum of logarithms. Express all powers as factors.

Solution

$$\log_a\left(x\sqrt{x^2+1}\right) = \log_a x + \log_a \sqrt{x^2+1} \quad \text{Property (3)}$$
$$= \log_a x + \log_a(x^2+1)^{1/2}$$
$$= \log_a x + \frac{1}{2}\log_a(x^2+1) \quad \text{Property (5)}$$

◀

EXAMPLE 4 Writing a Logarithmic Expression as a Difference of Logarithms

Write

$$\ln \frac{x^2}{(x-1)^3}, \qquad x > 1$$

as a difference of logarithms. Express all powers as factors.

Solution

$$\ln \frac{x^2}{(x-1)^3} \underset{\text{Property (4)}}{=} \ln x^2 - \ln(x-1)^3 \underset{\text{Property (5)}}{=} 2\ln x - 3\ln(x-1)$$

◀

EXAMPLE 5 Writing a Logarithmic Expression as a Sum and Difference of Logarithms

Write

$$\log_a \frac{\sqrt{x^2+1}}{x^3(x+1)^4}, \qquad x > 0$$

as a sum and difference of logarithms. Express all powers as factors.

Solution

$$\log_a \frac{\sqrt{x^2+1}}{x^3(x+1)^4} = \log_a \sqrt{x^2+1} - \log_a[x^3(x+1)^4] \quad \text{Property (4)}$$
$$= \log_a \sqrt{x^2+1} - [\log_a x^3 + \log_a(x+1)^4] \quad \text{Property (3)}$$
$$= \log_a(x^2+1)^{1/2} - \log_a x^3 - \log_a(x+1)^4$$
$$= \frac{1}{2}\log_a(x^2+1) - 3\log_a x - 4\log_a(x+1) \quad \text{Property (5)}$$

◀

CAUTION: In using properties (3) through (5), be careful about the values that the variable may assume. For example, the domain of the variable for $\log_a x$ is $x > 0$ and for $\log_a(x - 1)$ it is $x > 1$. If we add these functions, the domain is $x > 1$. That is, the equality

$$\log_a x + \log_a(x - 1) = \log_a[x(x - 1)]$$

is true only for $x > 1$. ■

NOW WORK PROBLEM 45.

3 Another use of properties (3) through (5) is to write sums and/or differences of logarithms with the same base as a single logarithm. This skill will be needed to solve certain logarithmic equations discussed in the next section.

EXAMPLE 6 Writing Expressions as a Single Logarithm

Write each of the following as a single logarithm.

(a) $\log_a 7 + 4\log_a 3$ (b) $\frac{2}{3}\ln 8 - \ln(3^4 - 8)$

(c) $\log_a x + \log_a 9 + \log_a(x^2 + 1) - \log_a 5$

Solution (a)

$$\begin{aligned} \log_a 7 + 4\log_a 3 &= \log_a 7 + \log_a 3^4 && \text{Property (5)} \\ &= \log_a 7 + \log_a 81 \\ &= \log_a(7 \cdot 81) && \text{Property (3)} \\ &= \log_a 567 \end{aligned}$$

(b)

$$\begin{aligned} \frac{2}{3}\ln 8 - \ln(3^4 - 8) &= \ln 8^{2/3} - \ln(81 - 8) && \text{Property (5)} \\ &= \ln 4 - \ln 73 \\ &= \ln\left(\frac{4}{73}\right) && \text{Property (4)} \end{aligned}$$

(c)

$$\begin{aligned} \log_a x + \log_a 9 + \log_a(x^2 + 1) - \log_a 5 &= \log_a(9x) + \log_a(x^2 + 1) - \log_a 5 \\ &= \log_a[9x(x^2 + 1)] - \log_a 5 \\ &= \log_a\left[\frac{9x(x^2 + 1)}{5}\right] \end{aligned}$$

◀

WARNING: A common error made by some students is to express the logarithm of a sum as the sum of logarithms.

$$\log_a(M + N) \quad \text{is not equal to} \quad \log_a M + \log_a N$$

Correct statement $\log_a(MN) = \log_a M + \log_a N$ Property (3)

Another common error is to express the difference of logarithms as the quotient of logarithms.

$$\log_a M - \log_a N \quad \text{is not equal to} \quad \frac{\log_a M}{\log_a N}$$

Correct statement $\log_a M - \log_a N = \log_a\left(\frac{M}{N}\right)$ Property (4)

A third common error is to express a logarithm raised to a power as the product of the power times the logarithm.

$$(\log_a M)^r \quad \text{is not equal to} \quad r \log_a M$$

Correct statement $\log_a M^r = r \log_a M$ Property (5) ■

NOW WORK PROBLEM 51.

Two other properties of logarithms that we need to know are consequences of the fact that the logarithmic function $y = \log_a x$ is one-to-one.

Theorem

Properties of Logarithms

In the following properties, M, N, and a are positive real numbers, with $a \neq 1$.

If $M = N$, then $\log_a M = \log_a N$. **(6)**

If $\log_a M = \log_a N$, then $M = N$. **(7)**

When property (6) is used, we start with the equation $M = N$ and say "take the logarithm of both sides" to obtain $\log_a M = \log_a N$.

Properties (6) and (7) are useful for solving *exponential and logarithmic equations*, a topic discussed in the next section.

Using a Calculator to Evaluate Logarithms with Bases Other Than 10 or e

4 Logarithms to the base 10, common logarithms, were used to facilitate arithmetic computations before the widespread use of calculators. (See the Historical Feature at the end of this section.) Natural logarithms, that is, logarithms whose base is the number e, remain very important because they arise frequently in the study of natural phenomena.

Common logarithms are usually abbreviated by writing **log,** with the base understood to be 10, just as natural logarithms are abbreviated by **ln,** with the base understood to be e.

Most calculators have both [log] and [ln] keys to calculate the common logarithm and natural logarithm of a number. Let's look at an example to see how to approximate logarithms having a base other than 10 or e.

EXAMPLE 7 Approximating Logarithms Whose Base Is Neither 10 nor e

Approximate $\log_2 7$. Round the answer to four decimal places.

Solution Let $y = \log_2 7$. Then $2^y = 7$, so

$$2^y = 7$$

$$\ln 2^y = \ln 7 \quad \text{Property (6)}$$

$$y \ln 2 = \ln 7 \quad \text{Property (5)}$$

$$y = \frac{\ln 7}{\ln 2} \quad \text{Exact solution}$$

$$y \approx 2.8074 \quad \text{Approximate solution rounded to four decimal places}$$

◀

Example 7 shows how to approximate a logarithm whose base is 2 by changing to logarithms involving the base e. In general, we use the **Change-of-Base Formula**.

Theorem

Change-of-Base Formula

If $a \neq 1, b \neq 1$, and M are positive real numbers, then

$$\log_a M = \frac{\log_b M}{\log_b a} \quad \textbf{(8)}$$

Proof We derive this formula as follows: Let $y = \log_a M$. Then

$$\begin{aligned} a^y &= M \\ \log_b a^y &= \log_b M && \text{Property (6)} \\ y \log_b a &= \log_b M && \text{Property (5)} \\ y &= \frac{\log_b M}{\log_b a} && \text{Solve for } y. \\ \log_a M &= \frac{\log_b M}{\log_b a} && y = \log_a M \end{aligned}$$

■

Since calculators have keys only for [log] and [ln,] in practice, the Change-of-Base Formula uses either $b = 10$ or $b = e$. Thus,

$$\log_a M = \frac{\log M}{\log a} \quad \text{and} \quad \log_a M = \frac{\ln M}{\ln a} \quad \textbf{(9)}$$

EXAMPLE 8 Using the Change-of-Base Formula

Approximate: (a) $\log_5 89$ (b) $\log_{\sqrt{2}} \sqrt{5}$
Round answers to four decimal places.

Solution (a) $\log_5 89 = \dfrac{\log 89}{\log 5} \approx \dfrac{1.949390007}{0.6989700043} \approx 2.7889$

or

$$\log_5 89 = \frac{\ln 89}{\ln 5} \approx \frac{4.48863637}{1.609437912} \approx 2.7889$$

(b) $\log_{\sqrt{2}} \sqrt{5} = \dfrac{\log \sqrt{5}}{\log \sqrt{2}} = \dfrac{\frac{1}{2}\log 5}{\frac{1}{2}\log 2} \approx 2.3219$

or

$$\log_{\sqrt{2}} \sqrt{5} = \frac{\ln \sqrt{5}}{\ln \sqrt{2}} = \frac{\frac{1}{2}\ln 5}{\frac{1}{2}\ln 2} \approx 2.3219$$

◀

COMMENT: To graph logarithmic functions when the base is different from e or 10 requires the Change-of-Base Formula. For example, to graph $y = \log_2 x$, we would instead graph $y = \dfrac{\ln x}{\ln 2}$. Try it. ■

NOW WORK PROBLEMS 17 AND 65.

Summary

Properties of Logarithms

In the list that follows, $a > 0$, $a \neq 1$, and $b > 0$, $b \neq 1$; also, $M > 0$ and $N > 0$.

Definition	$y = \log_a x$ means $x = a^y$	
Properties of logarithms	$\log_a 1 = 0$; $\log_a a = 1$	$\log_a M^r = r \log_a M$
	$a^{\log_a M} = M$; $\log_a a^r = r$	If $M = N$, then $\log_a M = \log_a N$.
	$\log_a(MN) = \log_a M + \log_a N$	If $\log_a M = \log_a N$, then $M = N$.
	$\log_a\left(\dfrac{M}{N}\right) = \log_a M - \log_a N$	
Change-of-Base Formula	$\log_a M = \dfrac{\log_b M}{\log_b a}$	

HISTORICAL FEATURE

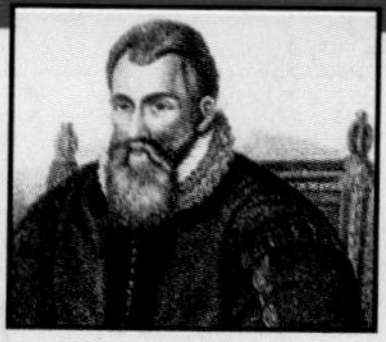

John Napier (1550–1617)

Logarithms were invented about 1590 by John Napier (1550–1617) and Joost Bürgi (1552–1632), working independently. Napier, whose work had the greater influence, was a Scottish lord, a secretive man whose neighbors were inclined to believe him to be in league with the devil. His approach to logarithms was very different from ours; it was based on the relationship between arithmetic and geometric sequences, discussed in a later chapter, and not on the inverse function relationship of logarithms to exponential functions (described in Section 4.4). Napier's tables, published in 1614, listed what would now be called *natural logarithms* of sines and were rather difficult to use. A London professor, Henry Briggs, became interested in the tables and visited Napier. In their conversations, they developed the idea of common logarithms, which were published in 1617. Their importance for calculation was immediately recognized, and by 1650 they were being printed as far away as China. They remained an important calculation tool until the advent of the inexpensive handheld calculator about 1972, which has decreased their calculational, but not their theoretical, importance.

A side effect of the invention of logarithms was the popularization of the decimal system of notation for real numbers.

4.5 Assess Your Understanding

Concepts and Vocabulary

1. The logarithm of a product equals the __________ of the logarithms.

2. If $\log_8 M = \dfrac{\log_5 7}{\log_5 8}$, then $M =$ __________.

3. $\log_a M^r =$ __________.

4. *True or False:* $\ln(x + 3) - \ln(2x) = \dfrac{\ln(x + 3)}{\ln(2x)}$

5. *True or False:* $\log_2(3x^4) = 4 \log_2(3x)$

6. *True or False:* $\log_2 16 = \dfrac{\ln 16}{\ln 2}$

Exercises

In Problems 7–22, use properties of logarithms to find the exact value of each expression. Do not use a calculator.

7. $\log_3 3^{71}$ **8.** $\log_2 2^{-13}$ **9.** $\ln e^{-4}$ **10.** $\ln e^{\sqrt{2}}$

11. $2^{\log_2 7}$ **12.** $e^{\ln 8}$ **13.** $\log_8 2 + \log_8 4$ **14.** $\log_6 9 + \log_6 4$

15. $\log_6 18 - \log_6 3$ **16.** $\log_8 16 - \log_8 2$ **17.** $\log_2 6 \cdot \log_6 4$ **18.** $\log_3 8 \cdot \log_8 9$

19. $3^{\log_3 5 - \log_3 4}$ **20.** $5^{\log_5 6 + \log_5 7}$ **21.** $e^{\log_{e^2} 16}$ **22.** $e^{\log_{e^2} 9}$

In Problems 23–30, suppose that $\ln 2 = a$ *and* $\ln 3 = b$*. Use properties of logarithms to write each logarithm in terms of a and b.*

23. $\ln 6$ **24.** $\ln \dfrac{2}{3}$ **25.** $\ln 1.5$ **26.** $\ln 0.5$

27. $\ln 8$ **28.** $\ln 27$ **29.** $\ln \sqrt[5]{6}$ **30.** $\ln \sqrt[4]{\dfrac{2}{3}}$

In Problems 31–50, write each expression as a sum and/or difference of logarithms. Express powers as factors.

31. $\log_5(25x)$ **32.** $\log_3 \dfrac{x}{9}$ **33.** $\log_2 z^3$ **34.** $\log_7(x^5)$

35. $\ln(ex)$ **36.** $\ln \dfrac{e}{x}$ **37.** $\ln(xe^x)$ **38.** $\ln \dfrac{x}{e^x}$

39. $\log_a(u^2v^3)$, $u > 0, v > 0$ **40.** $\log_2\left(\dfrac{a}{b^2}\right)$, $a > 0, b > 0$ **41.** $\ln\left(x^2\sqrt{1-x}\right)$, $0 < x < 1$ **42.** $\ln\left(x\sqrt{1+x^2}\right)$, $x > 0$

43. $\log_2\left(\dfrac{x^3}{x-3}\right)$, $x > 3$ **44.** $\log_5\left(\dfrac{\sqrt[3]{x^2+1}}{x^2-1}\right)$, $x > 1$ **45.** $\log\left[\dfrac{x(x+2)}{(x+3)^2}\right]$, $x > 0$ **46.** $\log\left[\dfrac{x^3\sqrt{x+1}}{(x-2)^2}\right]$, $x > 2$

47. $\ln\left[\dfrac{x^2-x-2}{(x+4)^2}\right]^{1/3}$, $x > 2$ **48.** $\ln\left[\dfrac{(x-4)^2}{x^2-1}\right]^{2/3}$, $x > 4$

49. $\ln \dfrac{5x\sqrt{1+3x}}{(x-4)^3}$, $x > 4$ **50.** $\ln\left[\dfrac{5x^2\sqrt[3]{1-x}}{4(x+1)^2}\right]$, $0 < x < 1$

In Problems 51–64, write each expression as a single logarithm.

51. $3\log_5 u + 4\log_5 v$ **52.** $2\log_3 u - \log_3 v$

53. $\log_3 \sqrt{x} - \log_3 x^3$ **54.** $\log_2\left(\dfrac{1}{x}\right) + \log_2\left(\dfrac{1}{x^2}\right)$

55. $\log_4(x^2-1) - 5\log_4(x+1)$ **56.** $\log(x^2+3x+2) - 2\log(x+1)$

57. $\ln\left(\dfrac{x}{x-1}\right) + \ln\left(\dfrac{x+1}{x}\right) - \ln(x^2-1)$ **58.** $\log\left(\dfrac{x^2+2x-3}{x^2-4}\right) - \log\left(\dfrac{x^2+7x+6}{x+2}\right)$

59. $8\log_2\sqrt{3x-2} - \log_2\left(\dfrac{4}{x}\right) + \log_2 4$ **60.** $21\log_3\sqrt[3]{x} + \log_3(9x^2) - \log_3 9$

61. $2\log_a(5x^3) - \dfrac{1}{2}\log_a(2x+3)$ **62.** $\dfrac{1}{3}\log(x^3+1) + \dfrac{1}{2}\log(x^2+1)$

63. $2\log_2(x+1) - \log_2(x+3) - \log_2(x-1)$ **64.** $3\log_5(3x+1) - 2\log_5(2x-1) - \log_5 x$

In Problems 65–72, use the Change-of-Base Formula and a calculator to evaluate each logarithm. Round your answer to three decimal places.

65. $\log_3 21$ **66.** $\log_5 18$ **67.** $\log_{1/3} 71$ **68.** $\log_{1/2} 15$

69. $\log_{\sqrt{2}} 7$ **70.** $\log_{\sqrt{5}} 8$ **71.** $\log_\pi e$ **72.** $\log_\pi \sqrt{2}$

In Problems 73–78, graph each function using a graphing utility and the Change-of-Base Formula.

73. $y = \log_4 x$ **74.** $y = \log_5 x$ **75.** $y = \log_2(x+2)$ **76.** $y = \log_4(x-3)$

77. $y = \log_{x-1}(x+1)$ **78.** $y = \log_{x+2}(x-2)$

In Problems 79–88, express y as a function of x. The constant C is a positive number.

79. $\ln y = \ln x + \ln C$

80. $\ln y = \ln(x + C)$

81. $\ln y = \ln x + \ln(x + 1) + \ln C$

82. $\ln y = 2 \ln x - \ln(x + 1) + \ln C$

83. $\ln y = 3x + \ln C$

84. $\ln y = -2x + \ln C$

85. $\ln(y - 3) = -4x + \ln C$

86. $\ln(y + 4) = 5x + \ln C$

87. $3 \ln y = \frac{1}{2}\ln(2x + 1) - \frac{1}{3}\ln(x + 4) + \ln C$

88. $2 \ln y = -\frac{1}{2}\ln x + \frac{1}{3}\ln(x^2 + 1) + \ln C$

89. Find the value of $\log_2 3 \cdot \log_3 4 \cdot \log_4 5 \cdot \log_5 6 \cdot \log_6 7 \cdot \log_7 8$.

90. Find the value of $\log_2 4 \cdot \log_4 6 \cdot \log_6 8$.

91. Find the value of $\log_2 3 \cdot \log_3 4 \cdot \cdots \cdot \log_n(n + 1) \cdot \log_{n+1} 2$.

92. Find the value of $\log_2 2 \cdot \log_2 4 \cdot \cdots \cdot \log_2 2^n$.

93. Show that $\log_a\left(x + \sqrt{x^2 - 1}\right) + \log_a\left(x - \sqrt{x^2 - 1}\right) = 0$.

94. Show that $\log_a\left(\sqrt{x} + \sqrt{x - 1}\right) + \log_a\left(\sqrt{x} - \sqrt{x - 1}\right) = 0$.

95. Show that $\ln(1 + e^{2x}) = 2x + \ln(1 + e^{-2x})$.

96. Difference Quotient If $f(x) = \log_a x$, show that $\frac{f(x + h) - f(x)}{h} = \log_a\left(1 + \frac{h}{x}\right)^{1/h}, \quad h \neq 0$.

97. If $f(x) = \log_a x$, show that $-f(x) = \log_{1/a} x$.

98. If $f(x) = \log_a x$, show that $f(AB) = f(A) + f(B)$.

99. If $f(x) = \log_a x$, show that $f\left(\frac{1}{x}\right) = -f(x)$.

100. If $f(x) = \log_a x$, show that $f(x^\alpha) = \alpha f(x)$.

101. Show that $\log_a\left(\frac{M}{N}\right) = \log_a M - \log_a N$, where a, M, and N are positive real numbers, with $a \neq 1$.

102. Show that $\log_a\left(\frac{1}{N}\right) = -\log_a N$, where a and N are positive real numbers, with $a \neq 1$.

103. Graph $Y_1 = \log(x^2)$ and $Y_2 = 2\log(x)$ using a graphing utility. Are they equivalent? What might account for any differences in the two functions?

4.6 Logarithmic and Exponential Equations

OBJECTIVES
1 Solve Logarithmic Equations Using the Properties of Logarithms
2 Solve Exponential Equations
3 Solve Logarithmic and Exponential Equations Using a Graphing Utility

Logarithmic Equations

1 In Section 4.4 we solved logarithmic equations by changing a logarithm to exponential form. Often, however, some manipulation of the equation (usually using the properties of logarithms) is required before we can change to exponential form.

Our practice will be to solve equations, whenever possible, by finding exact solutions using algebraic methods. When algebraic methods cannot be used, approximate solutions will be obtained using a graphing utility. The reader is encouraged to pay particular attention to the form of equations for which exact solutions are possible.

EXAMPLE 1 **Solving a Logarithmic Equation**

Solve: $2 \log_5 x = \log_5 9$

Solution The domain of the variable in this equation is $x > 0$. Because each logarithm has the same base, 5, we can obtain an exact solution as follows:

$$2\log_5 x = \log_5 9$$

$$\log_5 x^2 = \log_5 9 \qquad \log_a M^r = r\log_a M$$

$$x^2 = 9 \qquad \text{If } \log_a M = \log_a N, \text{ then } M = N.$$

$$x = 3 \quad \text{or} \quad \cancel{x = -3} \qquad \text{Recall that } x \text{ must be positive. Therefore, } -3 \text{ is extraneous and we discard it.}$$

The equation has only one solution, 3. ◀

NOW WORK PROBLEM 5.

EXAMPLE 2 Solving a Logarithmic Equation

Solve: $\log_4(x + 3) + \log_4(2 - x) = 1$

Solution The domain of the variable in this equation requires that $x + 3 > 0$ and $2 - x > 0$, so $x > -3$ and $x < 2$. That is, any solution must satisfy $-3 < x < 2$. To obtain an exact solution, we need to express the left side as a single logarithm. Then we will change the expression to exponential form.

$$\log_4(x + 3) + \log_4(2 - x) = 1$$

$$\log_4[(x + 3)(2 - x)] = 1 \qquad \log_a M + \log_a N = \log_a(MN)$$

$$(x + 3)(2 - x) = 4^1 = 4 \qquad \text{Change to an exponential expression.}$$

$$-x^2 - x + 6 = 4 \qquad \text{Simplify.}$$

$$x^2 + x - 2 = 0 \qquad \text{Place the quadratic equation in standard form.}$$

$$(x + 2)(x - 1) = 0 \qquad \text{Factor.}$$

$$x = -2 \quad \text{or} \quad x = 1 \qquad \text{Zero-Product Property}$$

Since both $x = -2$ and $x = 1$ satisfy $-3 < x < 2$, neither is extraneous. The solution set is $\{-2, 1\}$. ◀

NOW WORK PROBLEM 9.

Exponential Equations

2 In Sections 4.3 and 4.4, we solved certain exponential equations by expressing each side of the equation with the same base. However, many exponential equations cannot be rewritten so that each side has the same base. In such cases, properties of logarithms along with algebraic techniques can sometimes be used to obtain a solution.

EXAMPLE 3 Solving an Exponential Equation

Solve: $4^x - 2^x - 12 = 0$

Solution We note that $4^x = (2^2)^x = 2^{2x} = (2^x)^2$, so the equation is actually quadratic in form, and we can rewrite it as

$$(2^x)^2 - 2^x - 12 = 0 \qquad \text{Let } u = 2^x\text{; then } u^2 - u - 12 = 0.$$

Now we can factor as usual.

$$(2^x - 4)(2^x + 3) = 0 \qquad (u - 4)(u + 3) = 0$$

$$2^x - 4 = 0 \quad \text{or} \quad 2^x + 3 = 0 \qquad u - 4 = 0 \text{ or } u + 3 = 0$$

$$2^x = 4 \qquad 2^x = -3 \qquad u = 2^x = 4 \qquad u = 2^x = -3$$

The equation on the left has the solution $x = 2$, since $2^x = 4 = 2^2$; the equation on the right has no solution, since $2^x > 0$ for all x. The only solution is 2. ◀

In Example 3, we were able to write the exponential expression using the same base after utilizing some algebra, obtaining an exact solution to the equation. When this is not possible, logarithms can sometimes be used to obtain the solution.

EXAMPLE 4 Solving an Exponential Equation

Solve: $2^x = 5$

Solution A We write the exponential equation as the equivalent logarithmic equation.

$$2^x = 5$$

$$x = \log_2 5 \underset{\uparrow}{=} \frac{\ln 5}{\ln 2} \quad \text{Exact solution}$$

Change-of-Base Formula (9), Section 4.5

Solution B Alternatively, we can solve the equation $2^x = 5$ by taking the logarithm of both sides [refer to Property (6), Section 4.5]. Taking the natural logarithm,

$$\begin{aligned} 2^x &= 5 \\ \ln 2^x &= \ln 5 && \text{If } M = N, \log_a M = \log_a N. \\ x \ln 2 &= \ln 5 && \log_a M^r = r \log_a M \\ x &= \frac{\ln 5}{\ln 2} && \text{Exact solution} \end{aligned}$$

Using a calculator, the solution, rounded to three decimal places, is

$$x = \frac{\ln 5}{\ln 2} \approx 2.322 \quad \text{Approximate solution}$$

◀

NOW WORK PROBLEM 17.

EXAMPLE 5 Solving an Exponential Equation

Solve: $8 \cdot 3^x = 5$

Solution A Solve for 3^x.

$$\begin{aligned} 8 \cdot 3^x &= 5 \\ 3^x &= \frac{5}{8} && \text{Solve for } 3^x. \\ x &= \log_3\left(\frac{5}{8}\right) = \frac{\ln \frac{5}{8}}{\ln 3} && \text{Exact solution} \end{aligned}$$

The solution, rounded to three decimal places, is

$$x = \frac{\ln\left(\frac{5}{8}\right)}{\ln 3} \approx -0.428 \quad \text{Approximate solution}$$

Solution B Take logarithms of both sides.

$$
\begin{aligned}
8 \cdot 3^x &= 5 \\
\ln(8 \cdot 3^x) &= \ln 5 && \text{If } M = N, \text{ then } \ln M = \ln N \\
\ln 8 + \ln 3^x &= \ln 5 && \ln(MN) = \ln M + \ln N \\
\ln 8 + x \ln 3 &= \ln 5 && \ln M^r = r \ln M \\
x \ln 3 &= \ln 5 - \ln 8 \\
x &= \frac{\ln 5 - \ln 8}{\ln 3} && \text{Divide by } \ln 3. \\
&\approx -0.428
\end{aligned}
$$

◀

EXAMPLE 6 Solving an Exponential Equation

Solve: $5^{x-2} = 3^{3x+2}$

Solution Because the bases are different, we first apply Property (6), Section 4.5 (taking the natural logarithm of both sides), and then use appropriate properties of logarithms. The result is an equation in x that we can solve.

$$
\begin{aligned}
5^{x-2} &= 3^{3x+2} \\
\ln 5^{x-2} &= \ln 3^{3x+2} && \text{If } M = N, \log_a M = \log_a N. \\
(x - 2) \ln 5 &= (3x + 2) \ln 3 && \log_a M^r = r \log_a M \\
x \ln 5 - 2 \ln 5 &= 3x \ln 3 + 2 \ln 3 && \text{Distribute.} \\
x \ln 5 - 3x \ln 3 &= 2 \ln 3 + 2 \ln 5 && \text{Place terms involving } x \text{ on the left.} \\
(\ln 5 - 3 \ln 3)x &= 2 (\ln 3 + \ln 5) && \text{Factor.} \\
x &= \frac{2(\ln 3 + \ln 5)}{\ln 5 - 3 \ln 3} && \text{Exact solution} \\
&\approx -3.212 && \text{Approximate solution}
\end{aligned}
$$

◀

NOW WORK PROBLEM 25.

Graphing Utility Solutions

3 The techniques introduced in this section apply only to certain types of logarithmic and exponential equations. Solutions for other types are usually studied in calculus, using numerical methods. However, we can use a graphing utility to approximate the solution.

EXAMPLE 7 Solving Equations Using a Graphing Utility

Solve: $\log_3 x + \log_4 x = 4$
Express the solution(s) rounded to two decimal places.

Figure 33

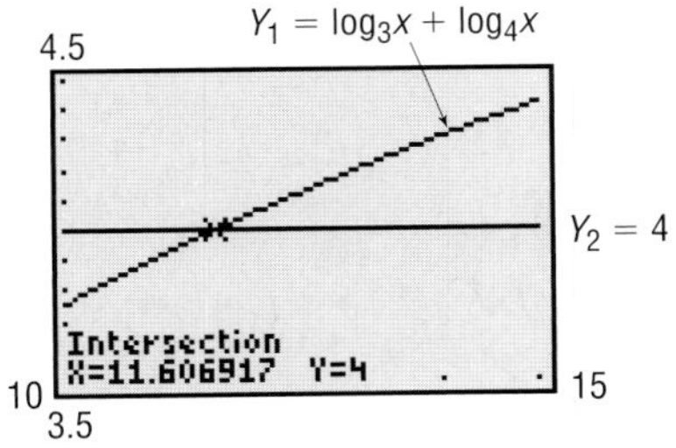

Solution The solution is found by graphing

$$Y_1 = \log_3 x + \log_4 x = \frac{\log x}{\log 3} + \frac{\log x}{\log 4} \quad \text{and} \quad Y_2 = 4$$

(Remember that you must use the Change-of-Base Formula to graph Y_1.) Y_1 is an increasing function (do you know why?), and so there is only one point of intersection for Y_1 and Y_2. Figure 33 shows the graphs of Y_1 and Y_2. Using the INTERSECT command, the solution is 11.61, rounded to two decimal places. ◀

Exploration

Can you discover an algebraic solution to Example 7?

[**Hint:** Factor $\log x$ from Y_1.]

EXAMPLE 8 **Solving Equations Using a Graphing Utility**

Figure 34

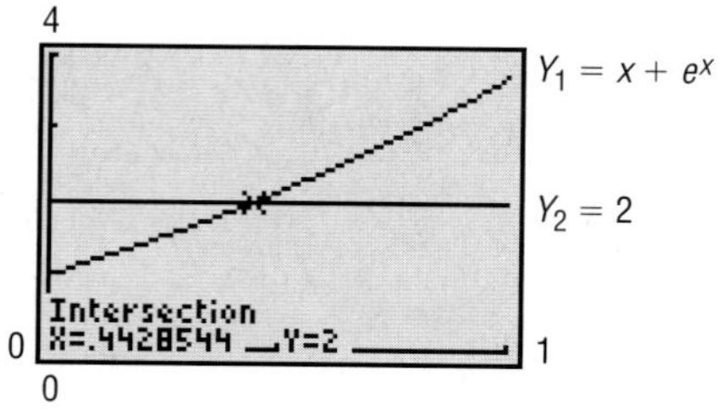

Solve: $x + e^x = 2$

Express the solution(s) rounded to two decimal places.

Solution The solution is found by graphing $Y_1 = x + e^x$ and $Y_2 = 2$. Y_1 is an increasing function (do you know why?), and so there is only one point of intersection for Y_1 and Y_2. Figure 34 shows the graphs of Y_1 and Y_2. Using the INTERSECT command, the solution is 0.44 rounded to two decimal places. ◀

4.6 Assess Your Understanding

Exercises

In Problems 1–44, solve each equation. Express irrational solutions in exact form and as a decimal rounded to 3 decimal places.

1. $\log_4(x + 2) = \log_4 8$

2. $\log_5(2x + 3) = \log_5 3$

3. $\frac{1}{2}\log_3 x = 2\log_3 2$

4. $-2\log_4 x = \log_4 9$

5. $2\log_5 x = 3\log_5 4$

6. $3\log_2 x = -\log_2 27$

7. $3\log_2(x - 1) + \log_2 4 = 5$

8. $2\log_3(x + 4) - \log_3 9 = 2$

9. $\log x + \log(x + 15) = 2$

10. $\log_4 x + \log_4(x - 3) = 1$

11. $\ln x + \ln(x + 2) = 4$

12. $\ln(x + 1) - \ln x = 2$

13. $2^{2x} + 2^x - 12 = 0$

14. $3^{2x} + 3^x - 2 = 0$

15. $3^{2x} + 3^{x+1} - 4 = 0$

16. $2^{2x} + 2^{x+2} - 12 = 0$

17. $2^x = 10$

18. $3^x = 14$

19. $8^{-x} = 1.2$

20. $2^{-x} = 1.5$

21. $3^{1-2x} = 4^x$

22. $2^{x+1} = 5^{1-2x}$

23. $\left(\frac{3}{5}\right)^x = 7^{1-x}$

24. $\left(\frac{4}{3}\right)^{1-x} = 5^x$

25. $1.2^x = (0.5)^{-x}$

26. $(0.3)^{1+x} = 1.7^{2x-1}$

27. $\pi^{1-x} = e^x$

28. $e^{x+3} = \pi^x$

29. $5(2^{3x}) = 8$

30. $0.3(4^{0.2x}) = 0.2$

31. $\log_a(x - 1) - \log_a(x + 6) = \log_a(x - 2) - \log_a(x + 3)$

32. $\log_a x + \log_a(x - 2) = \log_a(x + 4)$

33. $\log_{1/3}(x^2 + x) - \log_{1/3}(x^2 - x) = -1$

34. $\log_4(x^2 - 9) - \log_4(x + 3) = 3$

35. $\log_2(x + 1) - \log_4 x = 1$
[**Hint:** Change $\log_4 x$ to base 2.]

36. $\log_2(3x + 2) - \log_4 x = 3$

37. $\log_{16} x + \log_4 x + \log_2 x = 7$

38. $\log_9 x + 3\log_3 x = 14$

39. $\left(\sqrt[3]{2}\right)^{2-x} = 2^{x^2}$

40. $\log_2 x^{\log_2 x} = 4$

41. $\dfrac{e^x + e^{-x}}{2} = 1$
[**Hint:** Multiply each side by e^x.]

42. $\dfrac{e^x + e^{-x}}{2} = 3$

43. $\dfrac{e^x - e^{-x}}{2} = 2$

44. $\dfrac{e^x - e^{-x}}{2} = -2$

In Problems 45–60, use a graphing utility to solve each equation. Express your answer rounded to two decimal places.

45. $\log_5 x + \log_3 x = 1$

46. $\log_2 x + \log_6 x = 3$

47. $\log_5(x + 1) - \log_4(x - 2) = 1$

48. $\log_2(x - 1) - \log_6(x + 2) = 2$

49. $e^x = -x$ **50.** $e^{2x} = x + 2$ **51.** $e^x = x^2$ **52.** $e^x = x^3$

53. $\ln x = -x$ **54.** $\ln(2x) = -x + 2$ **55.** $\ln x = x^3 - 1$ **56.** $\ln x = -x^2$

57. $e^x + \ln x = 4$ **58.** $e^x - \ln x = 4$ **59.** $e^{-x} = \ln x$ **60.** $e^{-x} = -\ln x$

61. Fill in reasons for each step in the following two solutions.

Solve: $\log_3(x-1)^2 = 2$

Solution A	**Solution B**
$\log_3(x-1)^2 = 2$	$\log_3(x-1)^2 = 2$
$(x-1)^2 = 3^2 = 9$ ________	$2\log_3(x-1) = 2$ ________
$(x-1) = \pm 3$ ________	$\log_3(x-1) = 1$ ________
$x - 1 = -3$ or $x - 1 = 3$ ________	$x - 1 = 3^1 = 3$ ________
$x = -2$ or $x = 4$ ________	$x = 4$ ________

Both solutions given in Solution A check. Explain what caused the solution $x = -2$ to be lost in Solution B.

4.7 Compound Interest

PREPARING FOR THIS SECTION *Before getting started, review the following:*

- Simple Interest (Appendix A, Section A.7, pp. 957–958)

Now work the 'Are You Prepared?' problems on page 294.

OBJECTIVES
1. Determine the Future Value of a Lump Sum of Money
2. Calculate Effective Rates of Return
3. Determine the Present Value of a Lump Sum of Money
4. Determine the Time Required to Double or Triple a Lump Sum of Money

1 Interest is money paid for the use of money. The total amount borrowed (whether by an individual from a bank in the form of a loan or by a bank from an individual in the form of a savings account) is called the **principal.** The **rate of interest,** expressed as a percent, is the amount charged for the use of the principal for a given period of time, usually on a yearly (that is, per annum) basis.

Simple Interest Formula

If a principal of P dollars is borrowed for a period of t years at a per annum interest rate r, expressed as a decimal, the interest I charged is

$$I = Prt \qquad \textbf{(1)}$$

Interest charged according to formula (1) is called **simple interest.**

In working with problems involving interest, we define the term **payment period** as follows:

Annually	Once per year	Monthly	12 times per year
Semiannually	Twice per year	Daily	365 times per year*
Quarterly	Four times per year		

*Most banks use a 360-day "year." Why do you think they do?

When the interest due at the end of a payment period is added to the principal so that the interest computed at the end of the next payment period is based on this new principal amount (old principal + interest), the interest is said to have been **compounded. Compound interest** is interest paid on principal and previously earned interest.

EXAMPLE 1 Computing Compound Interest

A credit union pays interest of 8% per annum compounded quarterly on a certain savings plan. If \$1000 is deposited in such a plan and the interest is left to accumulate, how much is in the account after 1 year?

Solution We use the simple interest formula, $I = Prt$. The principal P is \$1000 and the rate of interest is $8\% = 0.08$. After the first quarter of a year, the time t is $\frac{1}{4}$ year, so the interest earned is

$$I = Prt = (\$1000)(0.08)\left(\frac{1}{4}\right) = \$20$$

The new principal is $P + I = \$1000 + \$20 = \$1020$. At the end of the second quarter, the interest on this principal is

$$I = (\$1020)(0.08)\left(\frac{1}{4}\right) = \$20.40$$

At the end of the third quarter, the interest on the new principal of $\$1020 + \$20.40 = \$1040.40$ is

$$I = (\$1040.40)(0.08)\left(\frac{1}{4}\right) = \$20.81$$

Finally, after the fourth quarter, the interest is

$$I = (\$1061.21)(0.08)\left(\frac{1}{4}\right) = \$21.22$$

After 1 year the account contains $\$1061.21 + \$21.22 = \$1082.43$. ◀

The pattern of the calculations performed in Example 1 leads to a general formula for compound interest. To fix our ideas, let P represent the principal to be invested at a per annum interest rate r that is compounded n times per year, so the time of each compounding period is $\frac{1}{n}$ years. (For computing purposes, r is expressed as a decimal.) The interest earned after each compounding period is given by formula (1).

$$\text{Interest} = \text{principal} \times \text{rate} \times \text{time} = P \cdot r \cdot \frac{1}{n} = P \cdot \left(\frac{r}{n}\right)$$

The amount A after one compounding period is

$$A = P + I = P + P \cdot \left(\frac{r}{n}\right) = P \cdot \left(1 + \frac{r}{n}\right)$$

After two compounding periods, the amount A, based on the new principal $P \cdot \left(1 + \frac{r}{n}\right)$, is

$$\underbrace{P \cdot \left(1 + \frac{r}{n}\right)}_{\text{New principal}} + \underbrace{P \cdot \left(1 + \frac{r}{n}\right)\left(\frac{r}{n}\right)}_{\text{Interest on new principal}} = P \cdot \left(1 + \frac{r}{n}\right)\left(1 + \frac{r}{n}\right) = P \cdot \left(1 + \frac{r}{n}\right)^2$$

After three compounding periods, the amount A is

$$A = P \cdot \left(1 + \frac{r}{n}\right)^2 + P \cdot \left(1 + \frac{r}{n}\right)^2\left(\frac{r}{n}\right) = P \cdot \left(1 + \frac{r}{n}\right)^2 \cdot \left(1 + \frac{r}{n}\right) = P \cdot \left(1 + \frac{r}{n}\right)^3$$

Continuing this way, after n compounding periods (1 year), the amount A is

$$A = P \cdot \left(1 + \frac{r}{n}\right)^n$$

Because t years will contain $n \cdot t$ compounding periods, after t years we have

$$A = P \cdot \left(1 + \frac{r}{n}\right)^{nt}$$

Theorem

Compound Interest Formula

The amount A after t years due to a principal P invested at an annual interest rate r compounded n times per year is

$$A = P \cdot \left(1 + \frac{r}{n}\right)^{nt} \qquad \textbf{(2)}$$

For example, to rework Example 1, we would use $P = \$1000$, $r = 0.08$, $n = 4$ (quarterly compounding), and $t = 1$ year to obtain

$$A = P \cdot \left(1 + \frac{r}{n}\right)^{nt} = 1000\left(1 + \frac{0.08}{4}\right)^4 = \$1082.43$$

In equation (2), the amount A is typically referred to as the **accumulated value or future value** of the account, while P is called the **present value.**

Exploration

To see the effects of compounding interest monthly on an initial deposit of \$1, graph $Y_1 = \left(1 + \frac{r}{12}\right)^{12x}$ with $r = 0.06$ and $r = 0.12$ for $0 \leq x \leq 30$. What is the future value of \$1 in 30 years when the interest rate per annum is $r = 0.06$ (6%)? What is the future value of \$1 in 30 years when the interest rate per annum is $r = 0.12$ (12%)? Does doubling the interest rate double the future value?

Note: In using your calculator, be sure to use stored values, rather than approximations, in order to avoid round-off errors. At the final step, round money to the nearest cent.

NOW WORK PROBLEM 3.

EXAMPLE 2 Comparing Investments Using Different Compounding Periods

Investing \$1000 at an annual rate of 10% compounded annually, semiannually, quarterly, monthly, and daily will yield the following amounts after 1 year:

Annual compounding: $A = P\cdot(1 + r)$
$= (\$1000)(1 + 0.10) = \$1100.00$

Semiannual compounding: $A = P\cdot\left(1 + \frac{r}{2}\right)^2$
$= (\$1000)(1 + 0.05)^2 = \$1102.50$

Quarterly compounding: $A = P\cdot\left(1 + \frac{r}{4}\right)^4$
$= (\$1000)(1 + 0.025)^4 = \$1103.81$

Monthly compounding: $A = P\cdot\left(1 + \frac{r}{12}\right)^{12}$
$= (\$1000)(1 + 0.00833)^{12} = \$1104.71$

Daily compounding: $A = P\cdot\left(1 + \frac{r}{365}\right)^{365}$
$= (\$1000)(1 + 0.000274)^{365} = \$1105.16$

◀

From Example 2 we can see that the effect of compounding more frequently is that the amount after 1 year is higher: \$1000 compounded 4 times a year at 10% results in \$1103.81; \$1000 compounded 12 times a year at 10% results in \$1104.71; and \$1000 compounded 365 times a year at 10% results in \$1105.16. This leads to the following question: What would happen to the amount after 1 year if the number of times that the interest is compounded were increased without bound?

Let's find the answer. Suppose that P is the principal, r is the per annum interest rate, and n is the number of times that the interest is compounded each year. The amount after 1 year is

$$A = P\cdot\left(1 + \frac{r}{n}\right)^n$$

Rewrite this expression as follows:

$$A = P\cdot\left(1 + \frac{r}{n}\right)^n = P\cdot\left(1 + \frac{1}{\frac{n}{r}}\right)^n = P\cdot\left[\left(1 + \frac{1}{\frac{n}{r}}\right)^{n/r}\right]^r \underset{\substack{\uparrow \\ h = \frac{n}{r}}}{=} P\cdot\left[\left(1 + \frac{1}{h}\right)^h\right]^r \qquad (3)$$

Now suppose that the number n of times that the interest is compounded per year gets larger and larger; that is, suppose that $n \to \infty$. Then $h = \frac{n}{r} \to \infty$, and the expression in brackets equals e. [Refer to equation (2), page 251.] That is, $A \to Pe^r$.

Table 8 compares $\left(1 + \frac{r}{n}\right)^n$, for large values of n, to e^r for $r = 0.05$, $r = 0.10$, $r = 0.15$, and $r = 1$. The larger that n gets, the closer $\left(1 + \frac{r}{n}\right)^n$

Table 8

	$\left(1+\frac{r}{n}\right)^n$			
	$n = 100$	$n = 1000$	$n = 10{,}000$	e^r
$r = 0.05$	1.0512580	1.0512698	1.051271	1.0512711
$r = 0.10$	1.1051157	1.1051654	1.1051704	1.1051709
$r = 0.15$	1.1617037	1.1618212	1.1618329	1.1618342
$r = 1$	2.7048138	2.7169239	2.7181459	2.7182818

gets to e^r. No matter how frequent the compounding, the amount after 1 year has the definite ceiling Pe^r.

When interest is compounded so that the amount after 1 year is Pe^r, we say the interest is **compounded continuously.**

Theorem

Continuous Compounding

The amount A after t years due to a principal P invested at an annual interest rate r compounded continuously is

$$A = Pe^{rt} \tag{4}$$

EXAMPLE 3 **Using Continuous Compounding**

The amount A that results from investing a principal P of \$1000 at an annual rate r of 10% compounded continuously for a time t of 1 year is

$$A = \$1000e^{0.10} = (\$1000)(1.10517) = \$1105.17$$

◀

NOW WORK PROBLEM 11.

2 The **effective rate of interest** is the equivalent annual simple rate of interest that would yield the same amount as compounding after 1 year. For example, based on Example 3, a principal of \$1000 will result in \$1105.17 at a rate of 10% compounded continuously. To get this same amount using a simple rate of interest would require that interest of \$1105.17 − \$1000.00 = \$105.17 be earned on the principal. Since \$105.17 is 10.517% of \$1000, a simple rate of interest of 10.517% is needed to equal 10% compounded continuously. The effective rate of interest of 10% compounded continuously is 10.517%.

Based on the results of Examples 2 and 3, we find the following comparisons:

	Annual Rate	Effective Rate
Annual compounding	10%	10%
Semiannual compounding	10%	10.25%
Quarterly compounding	10%	10.381%
Monthly compounding	10%	10.471%
Daily compounding	10%	10.516%
Continuous compounding	10%	10.517%

NOW WORK PROBLEM 23.

EXAMPLE 4 Computing the Value of an IRA

On January 2, 2004, \$2000 is placed in an Individual Retirement Account (IRA) that will pay interest of 10% per annum compounded continuously.

(a) What will the IRA be worth on January 1, 2024?

(b) What is the effective rate of interest?

Solution (a) The amount A after 20 years is

$$A = Pe^{rt} = \$2000e^{(0.10)(20)} = \$14{,}778.11$$

(b) First, we compute the interest earned on \$2000 at $r = 10\%$ compounded continuously for 1 year.

$$\begin{aligned} A &= \$2000e^{0.10(1)} \\ &= \$2210.34 \end{aligned}$$

So the interest earned is $\$2210.34 - \$2000.00 = \$210.34$. Use the simple interest formula $I = Prt$, with $I = \$210.34$, $P = \$2000$, and $t = 1$, and solve for r, the effective rate of interest.

$$\begin{aligned} \$210.34 &= \$2000 \cdot r \cdot 1 \\ r &= \frac{\$210.34}{\$2000} = 0.10517 \end{aligned}$$

The effective rate of interest is 10.517%. ◀

Exploration

For the IRA described in Example 4, how long will it be until $A = \$4000$? \$6000?

[**Hint:** Graph $Y_1 = 2000e^{0.1x}$ and $Y_2 = 4000$. Use INTERSECT to find x.]

3 When people engaged in finance speak of the "time value of money," they are usually referring to the *present value* of money. The **present value** of A dollars to be received at a future date is the principal that you would need to invest now so that it would grow to A dollars in the specified time period. The present value of money to be received at a future date is always less than the amount to be received, since the amount to be received will equal the present value (money invested now) *plus* the interest accrued over the time period.

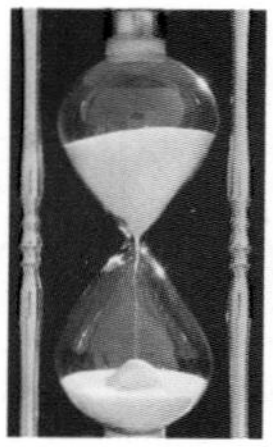

We use the compound interest formula (2) to get a formula for present value. If P is the present value of A dollars to be received after t years at a per annum interest rate r compounded n times per year, then, by formula (2),

$$A = P \cdot \left(1 + \frac{r}{n}\right)^{nt}$$

To solve for P, we divide both sides by $\left(1 + \frac{r}{n}\right)^{nt}$. The result is

$$\frac{A}{\left(1 + \frac{r}{n}\right)^{nt}} = P \quad \text{or} \quad P = A \cdot \left(1 + \frac{r}{n}\right)^{-nt}$$

Theorem

Present Value Formulas

The present value P of A dollars to be received after t years, assuming a per annum interest rate r compounded n times per year, is

$$P = A \cdot \left(1 + \frac{r}{n}\right)^{-nt} \tag{5}$$

If the interest is compounded continuously, then

$$P = Ae^{-rt} \tag{6}$$

To prove (6), solve formula (4) for P.

EXAMPLE 5 Computing the Value of a Zero-Coupon Bond

A zero-coupon (noninterest-bearing) bond can be redeemed in 10 years for \$1000. How much should you be willing to pay for it now if you want a return of

(a) 8% compounded monthly?

(b) 7% compounded continuously?

Solution (a) We are seeking the present value of \$1000. We use formula (5) with $A = \$1000$, $n = 12$, $r = 0.08$, and $t = 10$.

$$P = A \cdot \left(1 + \frac{r}{n}\right)^{-nt} = \$1000\left(1 + \frac{0.08}{12}\right)^{-12(10)} = \$450.52$$

For a return of 8% compounded monthly, you should pay \$450.52 for the bond.

(b) Here we use formula (6) with $A = \$1000$, $r = 0.07$, and $t = 10$.

$$P = Ae^{-rt} = \$1000e^{-(0.07)(10)} = \$496.59$$

For a return of 7% compounded continuously, you should pay \$496.59 for the bond. ◀

NOW WORK PROBLEM 13.

EXAMPLE 6 Rate of Interest Required to Double an Investment

What annual rate of interest compounded annually should you seek if you want to double your investment in 5 years?

Solution If P is the principal and we want P to double, the amount A will be $2P$. We use the compound interest formula with $n = 1$ and $t = 5$ to find r.

$$A = P \cdot \left(1 + \frac{r}{n}\right)^{nt}$$

$$2P = P \cdot (1 + r)^5 \qquad A = 2P, n = 1, t = 5$$

$$2 = (1 + r)^5 \qquad \text{Cancel the } P\text{'s.}$$

$$1 + r = \sqrt[5]{2} \qquad \text{Take the fifth root of each side.}$$

$$r = \sqrt[5]{2} - 1 \approx 1.148698 - 1 = 0.148698$$

The annual rate of interest needed to double the principal in 5 years is 14.87% ◀

NOW WORK PROBLEM 25.

4 **EXAMPLE 7** **Doubling and Tripling Time for an Investment**

(a) How long will it take for an investment to double in value if it earns 5% compounded continuously?

(b) How long will it take to triple at this rate?

Solution (a) If P is the initial investment and we want P to double, the amount A will be $2P$. We use formula (4) for continuously compounded interest with $r = 0.05$. Then

$$
\begin{aligned}
A &= Pe^{rt} \\
2P &= Pe^{0.05t} && A = 2P,\ r = 0.05 \\
2 &= e^{0.05t} && \text{Cancel the } P\text{'s.} \\
0.05t &= \ln 2 && \text{Rewrite as a logarithm.} \\
t &= \frac{\ln 2}{0.05} \approx 13.86 && \text{Solve for } t.
\end{aligned}
$$

It will take about 14 years to double the investment.

(b) To triple the investment, we set $A = 3P$ in formula (4).

$$
\begin{aligned}
A &= Pe^{rt} \\
3P &= Pe^{0.05t} && A = 3P,\ n = 0.05 \\
3 &= e^{0.05t} && \text{Cancel the } P\text{'s.} \\
0.05t &= \ln 3 && \text{Rewrite as a logarithm.} \\
t &= \frac{\ln 3}{0.05} \approx 21.97 && \text{Solve for } t.
\end{aligned}
$$

It will take about 22 years to triple the investment. ◀

NOW WORK PROBLEM 31.

4.7 Assess Your Understanding

'Are You Prepared?' *Answers are given at the end of these exercises. If you get a wrong answer, read the pages listed in red.*

1. What is the interest due if \$500 is borrowed for 6 months at a simple interest rate of 6% per annum? (pp. 957–958)

2. If you borrow \$5000 and, after 9 months, pay off the loan in the amount of \$5500, what per annum rate of interest was charged? (pp. 957–958)

Exercises

In Problems 3–12, find the amount that results from each investment.

3. \$100 invested at 4% compounded quarterly after a period of 2 years

4. \$50 invested at 6% compounded monthly after a period of 3 years

5. \$500 invested at 8% compounded quarterly after a period of $2\frac{1}{2}$ years

6. \$300 invested at 12% compounded monthly after a period of $1\frac{1}{2}$ years

7. \$600 invested at 5% compounded daily after a period of 3 years

8. \$700 invested at 6% compounded daily after a period of 2 years

9. \$10 invested at 11% compounded continuously after a period of 2 years

10. \$40 invested at 7% compounded continuously after a period of 3 years

11. \$100 invested at 10% compounded continuously after a period of $2\frac{1}{4}$ years

12. \$100 invested at 12% compounded continuously after a period of $3\frac{3}{4}$ years

In Problems 13–22, find the principal needed now to get each amount; that is, find the present value.

13. To get \$100 after 2 years at 6% compounded monthly

14. To get \$75 after 3 years at 8% compounded quarterly

15. To get \$1000 after $2\frac{1}{2}$ years at 6% compounded daily

16. To get \$800 after $3\frac{1}{2}$ years at 7% compounded monthly

17. To get \$600 after 2 years at 4% compounded quarterly

18. To get \$300 after 4 years at 3% compounded daily

19. To get \$80 after $3\frac{1}{4}$ years at 9% compounded continuously

20. To get \$800 after $2\frac{1}{2}$ years at 8% compounded continuously

21. To get \$400 after 1 year at 10% compounded continuously

22. To get \$1000 after 1 year at 12% compounded continuously

23. Find the effective rate of interest for $5\frac{1}{4}$% compounded quarterly.

24. What interest rate compounded quarterly will give an effective interest rate of 7%?

25. What rate of interest compounded annually is required to double an investment in 3 years?

26. What rate of interest compounded annually is required to double an investment in 10 years?

In Problems 27–30, which of the two rates would yield the larger amount in 1 year?

[**Hint:** Start with a principal of \$10,000 in each instance.]

27. 6% compounded quarterly or $6\frac{1}{4}$% compounded annually

28. 9% compounded quarterly or $9\frac{1}{4}$% compounded annually

29. 9% compounded monthly or 8.8% compounded daily

30. 8% compounded semiannually or 7.9% compounded daily

---

31. How long does it take for an investment to double in value if it is invested at 8% per annum compounded monthly? Compounded continuously?

32. How long does it take for an investment to double in value if it is invested at 10% per annum compounded monthly? Compounded continuously?

33. If Tanisha has \$100 to invest at 8% per annum compounded monthly, how long will it be before she has \$150? If the compounding is continuous, how long will it be?

34. If Angela has \$100 to invest at 10% per annum compounded monthly, how long will it be before she has \$175? If the compounding is continuous, how long will it be?

35. How many years will it take for an initial investment of \$10,000 to grow to \$25,000? Assume a rate of interest of 6% compounded continuously.

36. How many years will it take for an initial investment of \$25,000 to grow to \$80,000? Assume a rate of interest of 7% compounded continuously.

37. What will a \$90,000 house cost 5 years from now if the inflation rate over that period averages 3% compounded annually?

38. Sears charges 1.25% per month on the unpaid balance for customers with charge accounts (interest is compounded monthly). A customer charges \$200 and does not pay her bill for 6 months. What is the bill at that time?

39. Jerome will be buying a used car for \$15,000 in 3 years. How much money should he ask his parents for now so that, if he invests it at 5% compounded continuously, he will have enough to buy the car?

40. John will require \$3000 in 6 months to pay off a loan that has no prepayment privileges. If he has the \$3000 now, how much of it should he save in an account paying 3% compounded monthly so that in 6 months he will have exactly \$3000?

41. George is contemplating the purchase of 100 shares of a stock selling for \$15 per share. The stock pays no dividends. The history of the stock indicates that it should grow at an annual rate of 15% per year. How much will the 100 shares of stock be worth in 5 years?

42. Tracy is contemplating the purchase of 100 shares of a stock selling for \$15 per share. The stock pays no dividends. Her broker says that the stock will be worth \$20 per share in 2 years. What is the annual rate of return on this investment?

43. A business purchased for \$650,000 in 1994 is sold in 1997 for \$850,000. What is the annual rate of return for this investment?

44. Tanya has just inherited a diamond ring appraised at \$5000. If diamonds have appreciated in value at an annual rate of 8%, what was the value of the ring 10 years ago when the ring was purchased?

45. Jim places \$1000 in a bank account that pays 5.6% compounded continuously. After 1 year, will he have enough money to buy a computer system that costs \$1060? If another bank will pay Jim 5.9% compounded monthly, is this a better deal?

46. On January 1, Kim places \$1000 in a certificate of deposit that pays 6.8% compounded continuously and matures in 3 months. Then Kim places the \$1000 and the

interest in a passbook account that pays 5.25% compounded monthly. How much does Kim have in the passbook account on May 1?

47. Will invests \$2000 in a bond trust that pays 9% interest compounded semiannually. His friend Henry invests \$2000 in a certificate of deposit that pays $8\frac{1}{2}\%$ compounded continuously. Who has more money after 20 years, Will or Henry?

48. Suppose that April has access to an investment that will pay 10% interest compounded continuously. Which is better: To be given \$1000 now so that she can take advantage of this investment opportunity or to be given \$1325 after 3 years?

49. Colleen and Bill have just purchased a house for \$150,000, with the seller holding a second mortgage of \$50,000. They promise to pay the seller \$50,000 plus all accrued interest 5 years from now. The seller offers them three interest options on the second mortgage:
(a) Simple interest at 12% per annum
(b) $11\frac{1}{2}\%$ interest compounded monthly
(c) $11\frac{1}{4}\%$ interest compounded continuously

Which option is best; that is, which results in the least interest on the loan?

50. The First National Bank advertises that it pays interest on savings accounts at the rate of 4.25% compounded daily. Find the effective rate if the bank uses (a) 360 days or (b) 365 days in determining the daily rate.

Problems 51–54 involve zero-coupon bonds. A zero-coupon bond is a bond that is sold now at a discount and will pay its face value at the time when it matures; no interest payments are made.

51. A zero-coupon bond can be redeemed in 20 years for \$10,000. How much should you be willing to pay for it now if you want a return of:
(a) 10% compounded monthly?
(b) 10% compounded continuously?

52. A child's grandparents are considering buying a \$40,000 face value zero-coupon bond at birth so that she will have enough money for her college education 17 years later. If they want a rate of return of 8% compounded annually, what should they pay for the bond?

53. How much should a \$10,000 face value zero-coupon bond, maturing in 10 years, be sold for now if its rate of return is to be 8% compounded annually?

54. If Pat pays \$12,485.52 for a \$25,000 face value zero-coupon bond that matures in 8 years, what is his annual rate of return?

55. Time to Double or Triple an Investment The formula

$$t = \frac{\ln m}{n \ln\left(1 + \frac{r}{n}\right)}$$

can be used to find the number of years t required to multiply an investment m times when r is the per annum interest rate compounded n times a year.
(a) How many years will it take to double the value of an IRA that compounds annually at the rate of 12%?
(b) How many years will it take to triple the value of a savings account that compounds quarterly at an annual rate of 6%?
(c) Give a derivation of this formula.

56. Time to Reach an Investment Goal The formula

$$t = \frac{\ln A - \ln P}{r}$$

can be used to find the number of years t required for an investment P to grow to a value A when compounded continuously at an annual rate r.
(a) How long will it take to increase an initial investment of \$1000 to \$8000 at an annual rate of 10%?
(b) What annual rate is required to increase the value of a \$2000 IRA to \$30,000 in 35 years?
(c) Give a derivation of this formula.

57. Explain in your own words what the term *compound interest* means. What does *continuous compounding* mean?

58. Explain in your own words the meaning of *present value*.

59. Critical Thinking You have just contracted to buy a house and will seek financing in the amount of \$100,000. You go to several banks. Bank 1 will lend you \$100,000 at the rate of 8.75% amortized over 30 years with a loan origination fee of 1.75%. Bank 2 will lend you \$100,000 at the rate of 8.375% amortized over 15 years with a loan origination fee of 1.5%. Bank 3 will lend you \$100,000 at the rate of 9.125% amortized over 30 years with no loan origination fee. Bank 4 will lend you \$100,000 at the rate of 8.625% amortized over 15 years with no loan origination fee. Which loan would you take? Why? Be sure to have sound reasons for your choice. Use the information in the table to assist you. If the amount of the monthly payment does not matter to you, which loan would you take? Again, have sound reasons for your choice. Compare your final decision with others in the class. Discuss.

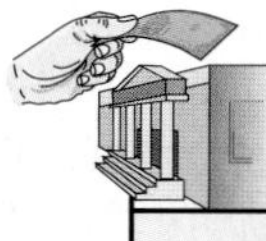

	Monthly Payment	Loan Origination Fee
Bank 1	\$786.70	\$1,750.00
Bank 2	\$977.42	\$1,500.00
Bank 3	\$813.63	\$0.00
Bank 4	\$990.68	\$0.00

'Are You Prepared?' Answers

1. \$15 **2.** 13.33%

4.8 Exponential Growth and Decay; Newton's Law; Logistic Models

OBJECTIVES
1. Find Equations of Populations That Obey the Law of Uninhibited Growth
2. Find Equations of Populations That Obey the Law of Decay
3. Use Newton's Law of Cooling
4. Use Logistic Models

1 Many natural phenomena have been found to follow the law that an amount A varies with time t according to

$$A(t) = A_0 e^{kt} \tag{1}$$

where $A_0 = A(0)$ is the original amount ($t = 0$) and $k \neq 0$ is a constant.

If $k > 0$, then equation (1) states that the amount A is increasing over time; if $k < 0$, the amount A is decreasing over time. In either case, when an amount A varies over time according to equation (1), it is said to follow the **exponential law** or the **law of uninhibited growth** ($k > 0$) **or decay** ($k < 0$). See Figure 35.

Figure 35

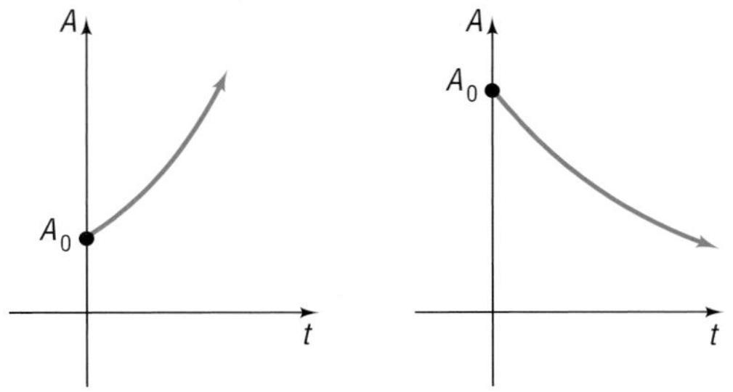

(a) $A(t) = A_0 e^{kt}, k > 0$ **(b)** $A(t) = A_0 e^{kt}, k < 0$

For example, we saw in Section 4.7 that continuously compounded interest follows the law of uninhibited growth. In this section we shall look at three additional phenomena that follow the exponential law.

Uninhibited Growth

Cell division is a process in the growth of many living organisms, such as amoebas, plants, and human skin cells. Based on an ideal situation in which no cells die and no by-products are produced, the number of cells present at a given time follows the law of uninhibited growth. Actually, however, after enough time has passed, growth at an exponential rate will cease due to the influence of factors such as lack of living space and dwindling food supply. The law of uninhibited growth accurately reflects only the early stages of the cell division process.

The cell division process begins with a culture containing N_0 cells. Each cell in the culture grows for a certain period of time and then divides into two identical cells. We assume that the time needed for each cell to divide in two is constant and does not change as the number of cells increases. These new cells then grow, and eventually each divides in two, and so on.

Uninhibited Growth of Cells

A model that gives the number N of cells in the culture after a time t has passed (in the early stages of growth) is

$$N(t) = N_0 e^{kt}, \qquad k > 0 \qquad \textbf{(2)}$$

where $N_0 = N(0)$ is the initial number of cells and k is a positive constant that represents the growth rate of the cells.

In using formula (2) to model the growth of cells, we are using a function that yields positive real numbers, even though we are counting the number of cells, which must be an integer. This is a common practice in many applications.

EXAMPLE 1 **Bacterial Growth**

A colony of bacteria grows according to the law of uninhibited growth according to the function $N(t) = 100e^{0.045t}$, where N is measured in grams and t is measured in days.

(a) Determine the initial amount of bacteria.
(b) What is the growth rate of the bacteria?
(c) What is the population after 5 days?
(d) How long will it take for the population to reach 140 grams?
(e) What is the doubling time for the population?

Solution (a) The initial amount of bacteria, N_0, is obtained when $t = 0$, so

$$N_0 = N(0) = 100e^{0.045(0)} = 100 \text{ grams.}$$

(b) Compare $N(t) = 100e^{0.045t}$ to $N(t) = 100e^{kt}$. The value of k, 0.045, indicates a growth rate of 4.5%.

(c) The population after 5 days is $N(5) = 100e^{0.045(5)} \approx 125.2$ grams.

(d) To find how long it takes for the population to reach 140 grams, we solve the equation $N(t) = 140$.

$$\begin{aligned} 100e^{0.045t} &= 140 \\ e^{0.045t} &= 1.4 && \text{Divide both sides of the equation by 100.} \\ 0.045t &= \ln 1.4 && \text{Rewrite as a logarithm.} \\ t &= \frac{\ln 1.4}{0.045} && \text{Divide both sides of the equation by 0.045.} \\ &\approx 7.5 \text{ days} \end{aligned}$$

(e) The population doubles when $N(t) = 200$ grams, so we find the doubling time by solving the equation $200 = 100e^{0.045t}$ for t.

$$\begin{aligned} 200 &= 100e^{0.045t} \\ 2 &= e^{0.045t} && \text{Divide both sides of the equation by 100.} \\ \ln 2 &= 0.045t && \text{Rewrite as a logarithm.} \\ t &= \frac{\ln 2}{0.045} && \text{Divide both sides of the equation by 0.045.} \\ &\approx 15.4 \text{ days} \end{aligned}$$

The population doubles approximately every 15.4 days. ◄

NOW WORK PROBLEM 1.

EXAMPLE 2 Bacterial Growth

A colony of bacteria increases according to the law of uninhibited growth.

(a) If the number of bacteria doubles in 3 hours, find the function that gives the number of cells in the culture.

(b) How long will it take for the size of the colony to triple?

(c) How long will it take for the population to double a second time (that is, increase four times)?

Solution (a) Using formula (2), the number N of cells at a time t is

$$N(t) = N_0 e^{kt}$$

where N_0 is the initial number of bacteria present and k is a positive number. We first seek the number k. The number of cells doubles in 3 hours, so we have

$$N(3) = 2N_0$$

But $N(3) = N_0 e^{k(3)}$, so

$$N_0 e^{k(3)} = 2N_0$$

$$e^{3k} = 2 \quad \text{Divide both sides by } N_0.$$

$$3k = \ln 2 \quad \text{Write the exponential equation as a logarithm.}$$

$$k = \frac{1}{3}\ln 2 \approx \frac{1}{3}(0.6931) \approx 0.2310$$

Formula (2) for this growth process is therefore

$$N(t) = N_0 e^{\left(\frac{1}{3}\ln 2\right)t}$$

(b) The time t needed for the size of the colony to triple requires that $N = 3N_0$. We substitute $3N_0$ for N to get

$$3N_0 = N_0 e^{\left(\frac{1}{3}\ln 2\right)t}$$

$$3 = e^{\left(\frac{1}{3}\ln 2\right)t}$$

$$\left(\frac{1}{3}\ln 2\right)t = \ln 3$$

$$t = \frac{3\ln 3}{\ln 2} \approx 4.755 \text{ hours}$$

It will take about 4.755 hours or 4 hours, 45 minutes for the size of the colony to triple.

(c) If a population doubles in 3 hours, it will double a second time in 3 more hours, for a total time of 6 hours. ◀

Radioactive Decay

2 Radioactive materials follow the law of uninhibited decay.

Uninhibited Radioactive Decay

The amount A of a radioactive material present at time t is given by

$$A(t) = A_0 e^{kt}, \qquad k < 0 \tag{3}$$

where A_0 is the original amount of radioactive material and k is a negative number that represents the rate of decay.

All radioactive substances have a specific **half-life,** which is the time required for half of the radioactive substance to decay. In **carbon dating,** we use the fact that all living organisms contain two kinds of carbon, carbon 12 (a stable carbon) and carbon 14 (a radioactive carbon, with a half-life of 5600 years). While an organism is living, the ratio of carbon 12 to carbon 14 is constant. But when an organism dies, the original amount of carbon 12 present remains unchanged, whereas the amount of carbon 14 begins to decrease. This change in the amount of carbon 14 present relative to the amount of carbon 12 present makes it possible to calculate when an organism died.

EXAMPLE 3 Estimating the Age of Ancient Tools

Traces of burned wood along with ancient stone tools in an archeological dig in Chile were found to contain approximately 1.67% of the original amount of carbon 14. If the half-life of carbon 14 is 5600 years, approximately when was the tree cut and burned?

Solution Using formula (3), the amount A of carbon 14 present at time t is

$$A(t) = A_0 e^{kt}$$

where A_0 is the original amount of carbon 14 present and k is a negative number. We first seek the number k. To find it, we use the fact that after 5600 years half of the original amount of carbon 14 remains, so $A(5600) = \frac{1}{2}A_0$. Then,

$$\begin{aligned}
\frac{1}{2}A_0 &= A_0 e^{k(5600)} \\
\frac{1}{2} &= e^{5600k} \\
5600k &= \ln\frac{1}{2} \\
k &= \frac{\ln\frac{1}{2}}{5600} \approx -0.000124
\end{aligned}$$

Formula (3) therefore becomes

$$A(t) = A_0 e^{-0.000124t}$$

If the amount A of carbon 14 now present is 1.67% of the original amount, it follows that

$$\begin{aligned} 0.0167A_0 &= A_0e^{-0.000124t} \\ 0.0167 &= e^{-0.000124t} \\ -0.000124t &= \ln 0.0167 \\ t &= \frac{\ln 0.0167}{-0.000124} \approx 33{,}000 \text{ years} \end{aligned}$$

The tree was cut and burned about 33,000 years ago. Some archeologists use this conclusion to argue that humans lived in the Americas 33,000 years ago, much earlier than is generally accepted. ◀

NOW WORK PROBLEM 3.

Newton's Law of Cooling

3 **Newton's Law of Cooling*** states that the temperature of a heated object decreases exponentially over time toward the temperature of the surrounding medium.

Newton's Law of Cooling

The temperature u of a heated object at a given time t can be modeled by the following function:

$$u(t) = T + (u_0 - T)e^{kt}, \qquad k < 0 \qquad \textbf{(4)}$$

where T is the constant temperature of the surrounding medium, u_0 is the initial temperature of the heated object, and k is a negative constant.

EXAMPLE 4 Using Newton's Law of Cooling

An object is heated to 100°C (degrees Celsius) and is then allowed to cool in a room whose air temperature is 30°C.

(a) If the temperature of the object is 80°C after 5 minutes, when will its temperature be 50°C?

(b) Determine the elapsed time before the temperature of the object is 35°C.

(c) What do you notice about $u(t)$, the temperature, as t, time, passes?

Solution (a) Using formula (4) with $T = 30$ and $u_0 = 100$, the temperature (in degrees Celsius) of the object at time t (in minutes) is

$$u(t) = 30 + (100 - 30)e^{kt} = 30 + 70e^{kt} \qquad \textbf{(5)}$$

*Named after Sir Isaac Newton (1642–1727), one of the cofounders of calculus.

where k is a negative constant. To find k, we use the fact that $u = 80$ when $t = 5$. Then

$$\begin{aligned} u(t) &= 30 + 70e^{kt} \\ 80 &= 30 + 70e^{k(5)} && t = 5;\ u(5) = 80 \\ 50 &= 70e^{5k} \\ e^{5k} &= \frac{50}{70} \\ 5k &= \ln\frac{5}{7} \\ k &= \frac{\ln\frac{5}{7}}{5} \approx -0.0673 \end{aligned}$$

Formula (5) therefore becomes

$$u(t) = 30 + 70e^{-0.0673t} \tag{6}$$

We want to find t when $u = 50°\text{C}$, so

$$\begin{aligned} 50 &= 30 + 70e^{-0.0673t} \\ 20 &= 70e^{-0.0673t} \\ e^{-0.0673t} &= \frac{20}{70} \\ -0.0673t &= \ln\frac{2}{7} \\ t &= \frac{\ln\frac{2}{7}}{-0.0673} \approx 18.6 \text{ minutes} \end{aligned}$$

The temperature of the object will be 50°C after about 18.6 minutes.

(b) If $u = 35°\text{C}$, then, based on equation (6), we have

$$\begin{aligned} 35 &= 30 + 70e^{-0.0673t} \\ 5 &= 70e^{-0.0673t} \\ e^{-0.0673t} &= \frac{5}{70} \\ -0.0673t &= \ln\frac{5}{70} \\ t &= \frac{\ln\frac{5}{70}}{-0.0673} \approx 39.2 \text{ minutes} \end{aligned}$$

The object will reach a temperature of 35°C after about 39.2 minutes.

(c) Refer to equation (6). As time passes, the value of t increases, the value of $e^{-0.0673t}$ approaches zero, and the value of $u(t)$ approaches 30°C. ◀

NOW WORK PROBLEM 13.

Logistic Models

4 The exponential growth model $A(t) = A_0e^{kt}$, $k > 0$, assumes uninhibited growth, meaning that the value of the function grows without limit. Earlier we stated that cell division could be modeled using this function, assuming that no cells die and no by-products are produced. However, cell division would eventually be limited by factors such as living space and food supply. The **logistic growth model** is an exponential function that can model situations where the growth of the dependent variable is limited.

Other situations that lead to a logistic growth model include population growth and the sales of a product due to advertising. See Problems 21 through 24. The logistic growth model is given next.

Logistic Growth Model

In a logistic growth model, the population P after time t obeys the equation

$$P(t) = \frac{c}{1 + ae^{-bt}}$$

where a, b, and c are constants with $c > 0$ and $b > 0$.

The number c is called the **carrying capacity** because the value $P(t)$ approaches c as t approaches infinity; that is, $\lim_{t\to\infty} P(t) = c$.

EXAMPLE 5 Fruit Fly Population

Fruit flies are placed in a half-pint milk bottle with a banana (for food) and yeast plants (for food and to provide a stimulus to lay eggs). Suppose that the fruit fly population P after t days is given by

$$P(t) = \frac{230}{1 + 56.5e^{-0.37t}}$$

(a) What is the carrying capacity of the half-pint bottle? That is, what is $P(t)$ as $t \to \infty$?

(b) How many fruit flies were initially placed in the half-pint bottle?

(c) When will the population of fruit flies be 180?

(d) Using a graphing utility, graph $P(t)$.

Solution (a) As $t \to \infty$, $e^{-0.37t} \to 0$ and $P(t) \to \frac{230}{1}$. The carrying capacity of the half-pint bottle is 230 fruit flies.

(b) To find the initial number of fruit flies in the half-pint bottle, we evaluate $P(0)$.

$$\begin{aligned} P(0) &= \frac{230}{1 + 56.5e^{-0.37(0)}} \\ &= \frac{230}{1 + 56.5} \\ &= 4 \end{aligned}$$

So initially there were four fruit flies in the half-pint bottle.

(c) To determine when the population of fruit flies will be 180, we solve the equation

$$
\begin{aligned}
\frac{230}{1 + 56.5e^{-0.37t}} &= 180 \\
230 &= 180(1 + 56.5e^{-0.37t}) \\
1.2778 &= 1 + 56.5e^{-0.37t} && \text{Divide both sides by 180.} \\
0.2778 &= 56.5e^{-0.37t} && \text{Subtract 1 from both sides.} \\
0.0049 &= e^{-0.37t} && \text{Divide both sides by 56.5.} \\
\ln(0.0049) &= -0.37t && \text{Rewrite as a logarithmic expression.} \\
t &\approx 14.4 \text{ days} && \text{Divide both sides by } -0.37.
\end{aligned}
$$

It will take approximately 14.4 days for the population to reach 180 fruit flies.

(d) See Figure 36 for the graph of $P(t)$. ◀

Figure 36

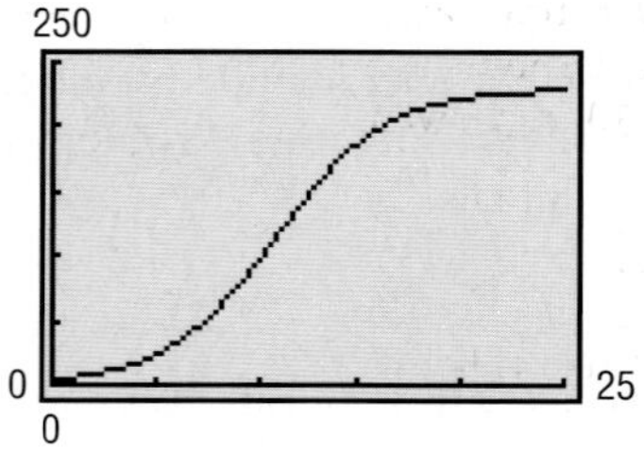

Exploration

On the same viewing rectangle, graph $Y_1 = \frac{500}{1 + 24e^{-0.03t}}$ and $Y_2 = \frac{500}{1 + 24e^{-0.08t}}$. What effect does b have on the logistic growth function?

4.8 Assess Your Understanding

Exercises

1. **Growth of an Insect Population** The size P of a certain insect population at time t (in days) obeys the function $P(t) = 500e^{0.02t}$.
(a) Determine the number of insects at $t = 0$ days.
(b) What is the growth rate of the insect population?
(c) What is the population after 10 days?
(d) When will the insect population reach 800?
(e) When will the insect population double?

2. **Growth of Bacteria** The number N of bacteria present in a culture at time t (in hours) obeys the function $N(t) = 1000e^{0.01t}$.
(a) Determine the number of bacteria at $t = 0$ hours.
(b) What is the growth rate of the bacteria?
(c) What is the population after 4 hours?
(d) When will the number of bacteria reach 1700?
(e) When will the number of bacteria double?

3. **Radioactive Decay** Strontium 90 is a radioactive material that decays according to the function $A(t) = A_0e^{-0.0244t}$, where A_0 is the initial amount present and A is the amount present at time t (in years). Assume that a scientist has a sample of 500 grams of strontium 90.
(a) What is the decay rate of strontium 90?
(b) How much strontium 90 is left after 10 years?
(c) When will 400 grams of strontium 90 be left?
(d) What is the half-life of strontium 90?

4. **Radioactive Decay** Iodine 131 is a radioactive material that decays according to the function $A(t) = A_0e^{-0.087t}$, where A_0 is the initial amount present and A is the amount present at time t (in days). Assume that a scientist has a sample of 100 grams of iodine 131.
(a) What is the decay rate of iodine 131?
(b) How much iodine 131 is left after 9 days?
(c) When will 70 grams of iodine 131 be left?
(d) What is the half-life of iodine 131?

5. **Growth of a Colony of Mosquitoes** The population of a colony of mosquitoes obeys the law of uninhibited growth. If there are 1000 mosquitoes initially and there are 1800 after 1 day, what is the size of the colony after 3 days? How long is it until there are 10,000 mosquitoes?

6. **Bacterial Growth** A culture of bacteria obeys the law of uninhibited growth. If 500 bacteria are present initially and there are 800 after 1 hour, how many will be present in the culture after 5 hours? How long is it until there are 20,000 bacteria?

7. **Population Growth** The population of a southern city follows the exponential law. If the population doubled in size over an 18-month period and the current population is 10,000, what will the population be 2 years from now?

8. **Population Growth** The population of a midwestern city follows the exponential law. If the population decreased from 900,000 to 800,000 from 1993 to 1995, what will the population be in 1997?

9. **Radioactive Decay** The half-life of radium is 1590 years. If 10 grams is present now, how much will be present in 50 years?

10. **Radioactive Decay** The half-life of radioactive potassium is 1.3 billion years. If 10 grams is present now, how much will be present in 100 years? In 1000 years?

11. **Estimating the Age of a Tree** A piece of charcoal is found to contain 30% of the carbon 14 that it originally had. When did the tree from which the charcoal came die? Use 5600 years as the half-life of carbon 14.

12. **Estimating the Age of a Fossil** A fossilized leaf contains 70% of its normal amount of carbon 14. How old is the fossil?

13. **Cooling Time of a Pizza** A pizza baked at 450°F is removed from the oven at 5:00 PM into a room that is a constant 70°F. After 5 minutes, the pizza is at 300°F.
   (a) At what time can you begin eating the pizza if you want its temperature to be 135°F?
   (b) Determine the time that needs to elapse before the pizza is 160°F.
   (c) What do you notice about the temperature as time passes?

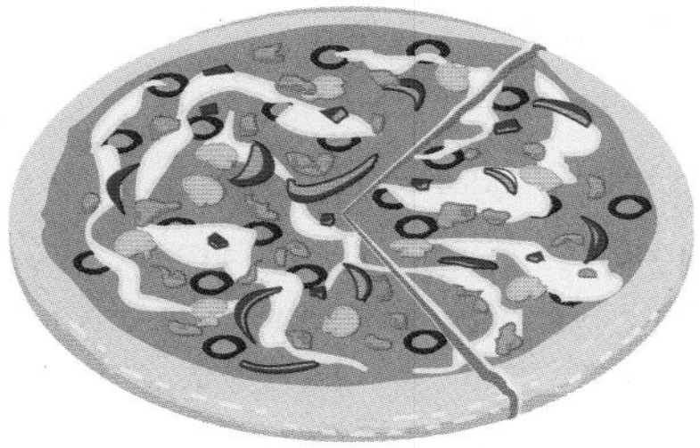

14. **Newton's Law of Cooling** A thermometer reading 72°F is placed in a refrigerator where the temperature is a constant 38°F.
   (a) If the thermometer reads 60°F after 2 minutes, what will it read after 7 minutes?
   (b) How long will it take before the thermometer reads 39°F?
   (c) Determine the time needed to elapse before the thermometer reads 45°F.
   (d) What do you notice about the temperature as time passes?

15. **Newton's Law of Heating** A thermometer reading 8°C is brought into a room with a constant temperature of 35°C. If the thermometer reads 15°C after 3 minutes, what will it read after being in the room for 5 minutes? For 10 minutes?
   [**Hint:** You need to construct a formula similar to equation (4).]

16. **Thawing Time of a Steak** A frozen steak has a temperature of 28°F. It is placed in a room with a constant temperature of 70°F. After 10 minutes, the temperature of the steak has risen to 35°F. What will the temperature of the steak be after 30 minutes? How long will it take the steak to thaw to a temperature of 45°F? [See the hint given for Problem 15.]

17. **Decomposition of Salt in Water** Salt (NaCl) decomposes in water into sodium (NA^+) and chloride (Cl^-) ions according to the law of uninhibited decay. If the initial amount of salt is 25 kilograms and, after 10 hours, 15 kilograms of salt is left, how much salt is left after 1 day? How long does it take until $\frac{1}{2}$ kilogram of salt is left?

18. **Voltage of a Conductor** The voltage of a certain conductor decreases over time according to the law of uninhibited decay. If the initial voltage is 40 volts, and 2 seconds later it is 10 volts, what is the voltage after 5 seconds?

19. **Radioactivity from Chernobyl** After the release of radioactive material into the atmosphere from a nuclear power plant at Chernobyl (Ukraine) in 1986, the hay in Austria was contaminated by iodine 131 (half-life 8 days). If it is all right to feed the hay to cows when 10% of the iodine 131 remains, how long did the farmers need to wait to use this hay?

20. **Pig Roasts** The hotel Bora-Bora is having a pig roast. At noon, the chef put the pig in a large earthen oven. The pig's original temperature was 75°F. At 2:00 PM the chef checked the pig's temperature and was upset because it had reached only 100°F. If the oven's temperature remains a constant 325°F, at what time may the hotel serve its guests, assuming that pork is done when it reaches 175°F?

21. **Proportion of the Population That Owns a DVD** The logistic growth model

$$P(t) = \frac{0.9}{1 + 6e^{-0.32t}}$$

relates the proportion of U.S. households that own a DVD to the year. Let $t = 0$ represent 2004, $t = 1$ represent 2005, and so on.
   (a) What proportion of the U.S. households owned a DVD in 2004?
   (b) Determine the maximum proportion of households that will own a DVD.
   (c) When will 0.8 (80%) of U.S. households own a DVD?

22. **Market Penetration of Intel's Coprocessor** The logistic growth model

$$P(t) = \frac{0.90}{1 + 3.5e^{-0.339t}}$$

relates the proportion of new personal computers sold at Best Buy that have Intel's latest coprocessor t months after it has been introduced.
   (a) What proportion of new personal computers sold at Best Buy will have Intel's latest coprocessor when it is first introduced (that is, at $t = 0$)?
   (b) Determine the maximum proportion of new personal computers sold at Best Buy that will have Intel's latest coprocessor.
   (c) When will 0.75 (75%) of new personal computers sold at Best Buy have Intel's latest coprocessor?

23. Population of a Bacteria Culture The logistic growth model

$$P(t) = \frac{1000}{1 + 32.33e^{-0.439t}}$$

represents the population of a bacterium after t hours.
(a) What is the carrying capacity of the environment?
(b) What was the initial amount of bacteria in the population?
(c) When will the amount of bacteria be 800?

24. Population of a Endangered Species Often environmentalists will capture an endangered species and transport the species to a controlled environment where the species can produce offspring and regenerate its population. Suppose that six American bald eagles are captured, transported to Montana, and set free. Based on experience, the environmentalists expect the population to grow according to the model

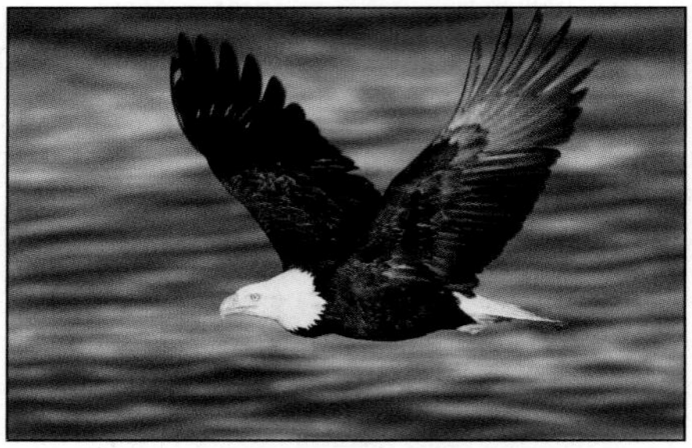

$$P(t) = \frac{500}{1 + 83.33e^{-0.162t}}$$

where $P(t)$ is the population after t years.
(a) What is the carrying capacity of the environment?
(b) What is the predicted population of this species of American bald eagle in 20 years?
(c) When will the population be 300?

4.9 Fitting Data to Exponential, Logarithmic, and Logistic Functions

PREPARING FOR THIS SECTION *Before getting started, review the following:*

- Scatter Diagrams; Linear Curve Fitting (Section 1.4, pp. 38–42)
- Quadratic Models (Section 3.1, pp. 136–138)

OBJECTIVES
1 Use a Graphing Utility to Fit an Exponential Function to Data
2 Use a Graphing Utility to Fit a Logarithmic Function to Data
3 Use a Graphing Utility to Fit a Logistic Function to Data

In Section 1.4 we discussed how to find the linear function of best fit $(y = ax + b)$ and in Section 3.1 we discussed how to find the quadratic function of best fit $(y = ax^2 + bx + c)$.

In this section we will discuss how to use a graphing utility to find equations of best fit that describe the relation between two variables when the relation is thought to be exponential $(y = ab^x)$, logarithmic $(y = a + b \ln x)$, or logistic $\left(y = \dfrac{c}{1 + ae^{-bx}}\right)$. As before, we draw a scatter diagram of the data to help to determine the appropriate model to use.

Figure 37 shows scatter diagrams that will typically be observed for the three models. Below each scatter diagram are any restrictions on the values of the parameters.

Figure 37

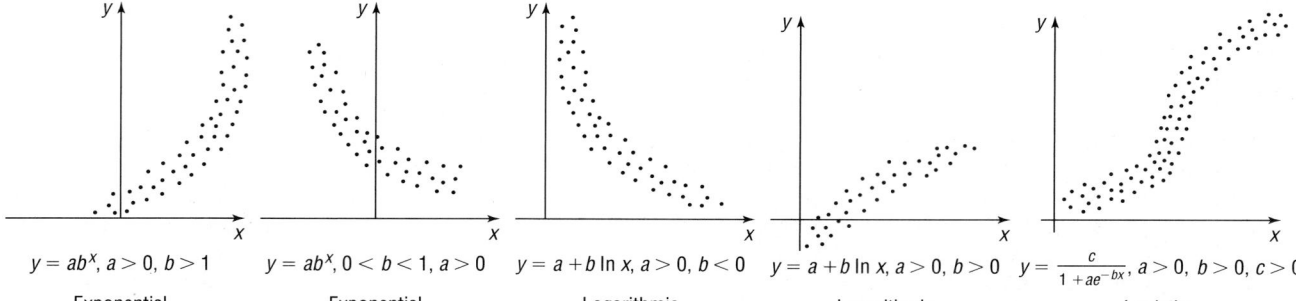

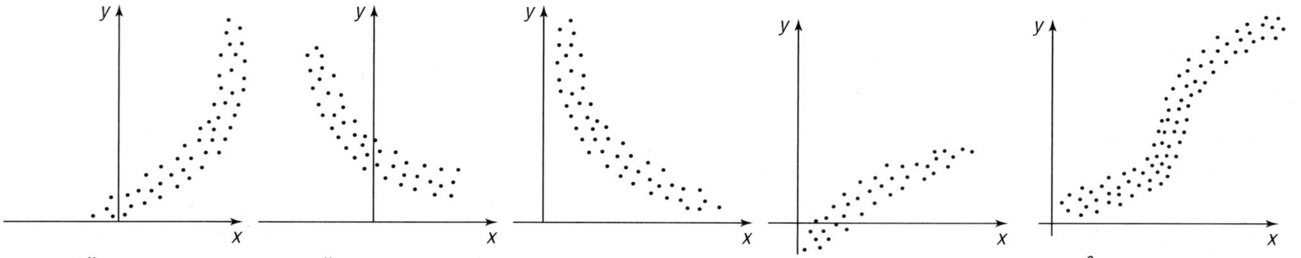

$y = ab^x$, $a > 0$, $b > 1$ — Exponential

$y = ab^x$, $0 < b < 1$, $a > 0$ — Exponential

$y = a + b \ln x$, $a > 0$, $b < 0$ — Logarithmic

$y = a + b \ln x$, $a > 0$, $b > 0$ — Logarithmic

$y = \dfrac{c}{1 + ae^{-bx}}$, $a > 0$, $b > 0$, $c > 0$ — Logistic

Most graphing utilities have REGression options that fit data to a specific type of curve. Once the data have been entered and a scatter diagram obtained, the type of curve that you want to fit to the data is selected. Then that REGression option is used to obtain the curve of "best fit" of the type selected.

The correlation coefficient r will appear only if the model can be written as a linear expression. As it turns out, r will appear for the linear, power, exponential, and logarithmic models, since these models can be written as a linear expression. Remember, the closer $|r|$ is to 1, the better the fit.

Let's look at some examples.

Exponential Models

1 We saw in Section 4.7 that the future value of money behaves exponentially, and we saw in Section 4.8 that growth and decay models also behave exponentially. The next example shows how data can lead to an exponential model.

EXAMPLE 1 Fitting an Exponential Function to Data

Beth is interested in finding a function that explains the closing price of Harley Davidson stock at the end of each year. She obtains the data shown in Table 9.

(a) Using a graphing utility, draw a scatter diagram with year as the independent variable.

(b) Using a graphing utility, fit an exponential function to the data.

(c) Express the function found in part (b) in the form $A = A_0 e^{kt}$.

(d) Graph the exponential function found in part (b) or (c) on the scatter diagram.

(e) Using the solution to part (b) or (c), predict the closing price of Harley Davidson stock at the end of 2002.

(f) Interpret the value of k found in part (c).

Table 9

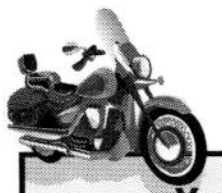

Year, x	Closing Price, y
1987 ($x = 1$)	0.392
1988 ($x = 2$)	0.7652
1989 ($x = 3$)	1.1835
1990 ($x = 4$)	1.1609
1991 ($x = 5$)	2.6988
1992 ($x = 6$)	4.5381
1993 ($x = 7$)	5.3379
1994 ($x = 8$)	6.8032
1995 ($x = 9$)	7.0328
1996 ($x = 10$)	11.5585
1997 ($x = 11$)	13.4799
1998 ($x = 12$)	23.5424
1999 ($x = 13$)	31.9342
2000 ($x = 14$)	39.7277
2001 ($x = 15$)	54.31

SOURCE: http://finance.yahoo.com

Solution (a) Enter the data into the graphing utility, letting 1 represent 1987, 2 represent 1998, and so on. We obtain the scatter diagram shown in Figure 38.

(b) A graphing utility fits the data in Figure 38 to an exponential function of the form $y = ab^x$ by using the EXPonential REGression option. See Figure 39. Thus, $y = ab^x = 0.40257(1.39745)^x$.

Figure 38

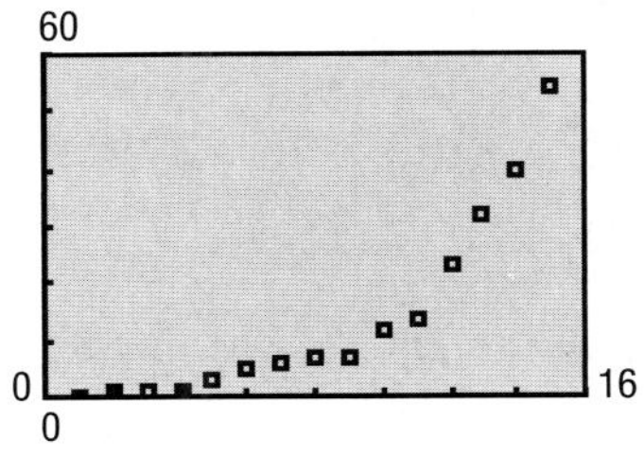

Figure 39

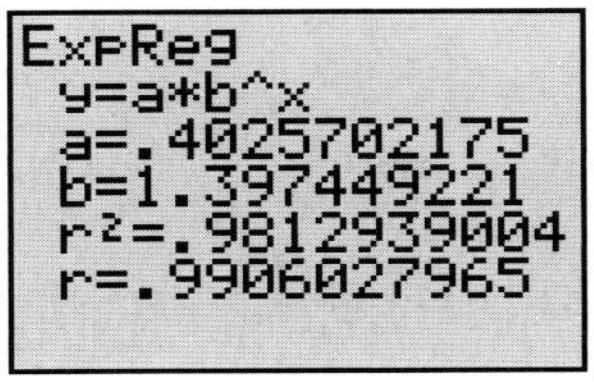

(c) To express $y = ab^x$ in the form $A = A_0e^{kt}$, where $x = t$ and $y = A$, we proceed as follows:

$$ab^x = A_0e^{kt}, \quad x = t$$

When $x = t = 0$, we find $a = A_0$. This leads to

$$a = A_0, \qquad b^x = e^{kt}$$
$$b^x = (e^k)^t$$
$$b = e^k \qquad x = t$$

Since $y = ab^x = 0.40257(1.39745)^x$, we find that $a = 0.40257$ and $b = 1.39745$,

$$a = A_0 = 0.40257 \quad \text{and} \quad b = 1.39745 = e^k$$

We want to find k, so we rewrite $1.39745 = e^k$ as a logarithm and obtain

$$k = \ln(1.39745) \approx 0.3346$$

As a result, $A = A_0e^{kt} = 0.40257e^{0.3346t}$.

(d) See Figure 40 for the graph of the exponential function of best fit.

Figure 40

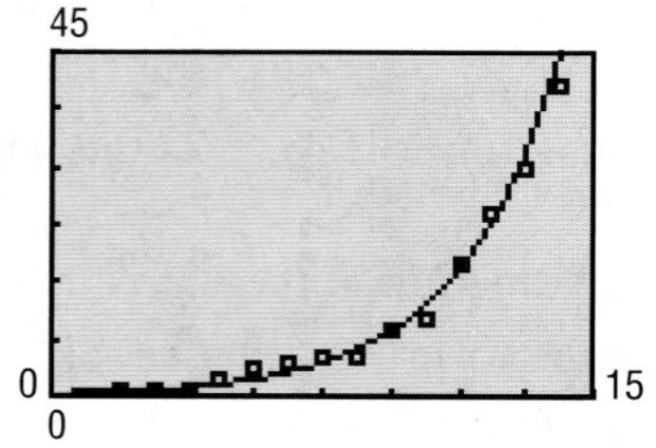

(e) Let $t = 16$ (end of 2002) in the function found in part (c). The predicted closing price of Harley Davidson stock at the end of 2002 is

$$A = 0.40257e^{0.3346(16)} = \$85.09$$

(f) The value of k represents the annual interest rate compounded continuously.

$$A = A_0e^{kt} = 0.40257e^{0.3346t}$$
$$= Pe^{rt} \qquad \text{Equation (4), Section 4.7}$$

The price of Harley Davidson stock has grown at an annual rate of 33.46% (compounded continuously) between 1987 and 2001. ◀

NOW WORK PROBLEM 1.

Logarithmic Models

2 Many relations between variables do not follow an exponential model, but, instead, the independent variable is related to the dependent variable using a logarithmic model.

EXAMPLE 2 Fitting a Logarithmic Function to Data

Jodi, a meteorologist, is interested in finding a function that explains the relation between the height of a weather balloon (in kilometers) and the atmospheric pressure (measured in millimeters of mercury) on the balloon. She collects the data shown in Table 10.

(a) Using a graphing utility, draw a scatter diagram of the data with atmospheric pressure as the independent variable.

Table 10

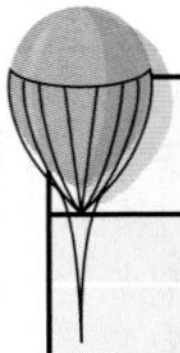

Atmospheric Pressure, *p*	Height, *h*
760	0
740	0.184
725	0.328
700	0.565
650	1.079
630	1.291
600	1.634
580	1.862
550	2.235

(b) It is known that the relation between atmospheric pressure and height follows a logarithmic model. Using a graphing utility, fit a logarithmic function to the data.

(c) Draw the logarithmic function found in part (b) on the scatter diagram.

(d) Use the function found in part (b) to predict the height of the weather balloon if the atmospheric pressure is 560 millimeters of mercury.

Solution (a) After entering the data into the graphing utility, we obtain the scatter diagram shown in Figure 41.

(b) A graphing utility fits the data in Figure 41 to a logarithmic function of the form $y = a + b \ln x$ by using the Logarithm REGression option. See Figure 42. The logarithmic function of best fit to the data is

$$h(p) = 45.7863 - 6.9025 \ln p$$

where h is the height of the weather balloon and p is the atmospheric pressure. Notice that $|r|$ is close to 1, indicating a good fit.

(c) Figure 43 shows the graph of $h(p) = 45.7863 - 6.9025 \ln p$ on the scatter diagram.

Figure 41

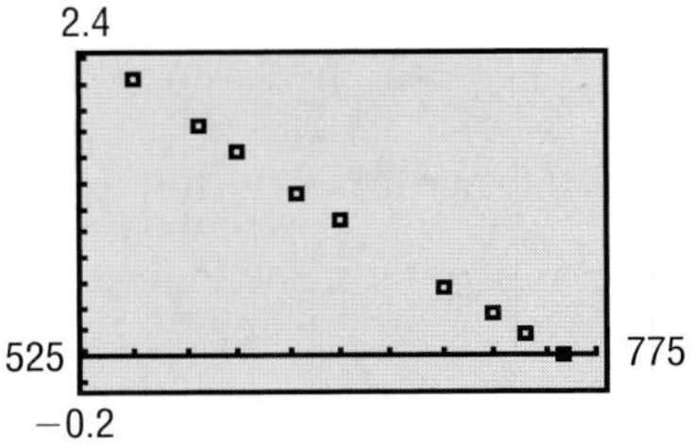

Figure 42

Figure 43

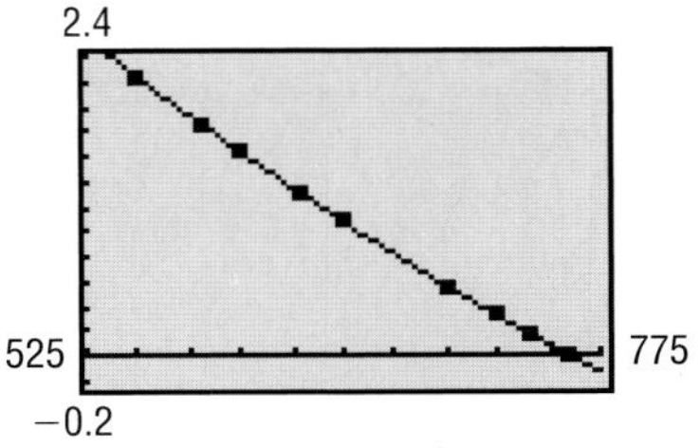

(d) Using the function found in part (b), Jodi predicts the height of the weather balloon when the atmospheric pressure is 560 to be

$$h(560) = 45.7863 - 6.9025 \ln 560$$
$$\approx 2.108 \text{ kilometers}$$

◀

NOW WORK PROBLEM 7.

Logistic Models

3 Logistic growth models can be used to model situations for which the value of the dependent variable is limited. Many real-world situations conform to this scenario. For example, the population of the human race is limited by the availability of natural resources such as food and shelter. When the value of the dependent variable is limited, a logistic growth model is often appropriate.

EXAMPLE 3 **Fitting a Logistic Function to Data**

The data in Table 11, obtained from Tor Carlson (Über Geschwindigkeit und Grösse der Hefevermehrung in Würze, *Biochemische Zeitschrift*, Bd. 57, pp. 313–334, 1913), represent the amount of yeast biomass after t hours in a culture.

Table 11

Time (in hours)	Yeast Biomass	Time (in hours)	Yeast Biomass
0	9.6	10	513.3
1	18.3	11	559.7
2	29.0	12	594.8
3	47.2	13	629.4
4	71.1	14	640.8
5	119.1	15	651.1
6	174.6	16	655.9
7	257.3	17	659.6
8	350.7	18	661.8
9	441.0		

(a) Using a graphing utility, draw a scatter diagram of the data with time as the independent variable.
(b) Using a graphing utility, fit a logistic function to the data.
(c) Using a graphing utility, graph the function found in part (b) on the scatter diagram.
(d) What is the predicted carrying capacity of the culture?
(e) Use the function found in part (b) to predict the population of the culture at $t = 19$ hours.

Solution (a) See Figure 44 for a scatter diagram of the data.

(b) A graphing utility fits a logistic growth model of the form $y = \dfrac{c}{1 + ae^{-bx}}$ by using the LOGISTIC regression option. See Figure 45. The logistic function of best fit to the data is

$$y = \frac{663.0}{1 + 71.6e^{-0.5470x}}$$

where y is the amount of yeast biomass in the culture and x is the time.

(c) See Figure 46 for the graph of the logistic function of best fit.

Figure 44

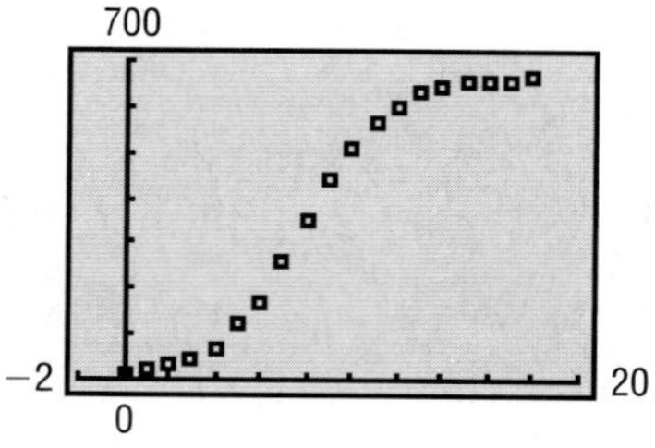

Figure 45

Figure 46

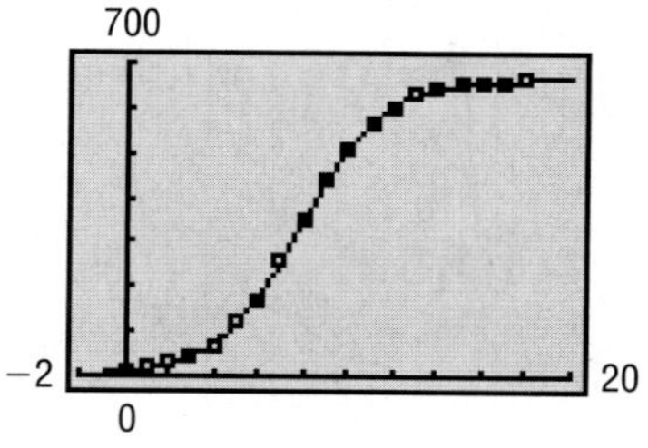

(d) Based on the logistic growth function found in part (b), the carrying capacity of the culture is 663.

(e) Using the logistic growth function found in part (b), the predicted amount of yeast biomass at $t = 19$ hours is

$$y = \frac{663.0}{1 + 71.6e^{-0.5470(19)}} = 661.5$$

◀

NOW WORK PROBLEM 9.

4.9 Assess Your Understanding

Exercises

1. **Biology** A strain of E-coli Beu 397-recA441 is placed into a petri dish at 30°Celsius and allowed to grow. The following data are collected. Theory states that the number of bacteria in the petri dish will initially grow according to the law of uninhibited growth. The population is measured using an optical device in which the amount of light that passes through the petri dish is measured.

Time (hours), x	Population, y
0	0.09
2.5	0.18
3.5	0.26
4.5	0.35
6	0.50

SOURCE: Dr. Polly Lavery, Joliet Junior College

(a) Draw a scatter diagram treating time as the predictor variable.
(b) Using a graphing utility, fit an exponential function to the data.
(c) Express the function found in part (b) in the form $N(t) = N_0e^{kt}$.
(d) Graph the exponential function found in part (b) or (c) on the scatter diagram.
(e) Use the exponential function from part (b) or (c) to predict the population at $x = 7$ hours.
(f) Use the exponential function from part (b) or (c) to predict when the population will reach 0.75.

2. **Biology** A strain of E-coli SC18del-recA718 is placed into a petri dish at 30°Celsius and allowed to grow. The following data are collected. Theory states that the number of bacteria in the petri dish will initially grow according to the law of uninhibited growth. The population is measured using an optical device in which the amount of light that passes through the petri dish is measured.

Time (hours), x	Population, y
2.5	0.175
3.5	0.38
4.5	0.63
4.75	0.76
5.25	1.20

SOURCE: Dr. Polly Lavery, Joliet Junior College

(a) Draw a scatter diagram treating time as the predictor variable.
(b) Using a graphing utility, fit an exponential function to the data.
(c) Express the function found in part (b) in the form $N(t) = N_0e^{kt}$.
(d) Graph the exponential function found in part (b) or (c) on the scatter diagram.
(e) Use the exponential function from part (b) or (c) to predict the population at $x = 6$ hours.
(f) Use the exponential function from part (b) or (c) to predict when the population will reach 2.1.

3. **Chemistry** A chemist has a 100-gram sample of a radioactive material. He records the amount of radioactive material every week for 6 weeks and obtains the following data:

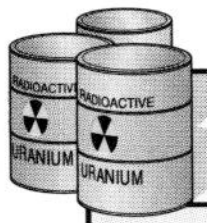

Week	Weight (in Grams)
0	100.0
1	88.3
2	75.9
3	69.4
4	59.1
5	51.8
6	45.5

(a) Using a graphing utility, draw a scatter diagram with week as the independent variable.
(b) Using a graphing utility, fit an exponential function to the data.
(c) Express the function found in part (b) in the form $A(t) = A_0e^{kt}$.
(d) Graph the exponential function found in part (b) or (c) on the scatter diagram.
(e) From the result found in part (b), determine the half-life of the radioactive material.
(f) How much radioactive material will be left after 50 weeks?
(g) When will there be 20 grams of radioactive material?

4. **Chemistry** A chemist has a 1000-gram sample of a radioactive material. She records the amount of radioactive material remaining in the sample every day for a week and obtains the following data:

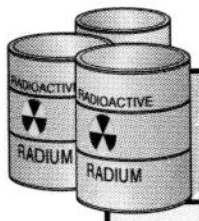

Day	Weight (in Grams)
0	1000.0
1	897.1
2	802.5
3	719.8
4	651.1
5	583.4
6	521.7
7	468.3

(a) Using a graphing utility, draw a scatter diagram with day as the independent variable.

(b) Using a graphing utility, fit an exponential function to the data.

(c) Express the function found in part (b) in the form $A(t) = A_0e^{kt}$.

(d) Graph the exponential function found in part (b) or (c) on the scatter diagram.

(e) From the result found in part (b), find the half-life of the radioactive material.

(f) How much radioactive material will be left after 20 days?

(g) When will there be 200 grams of radioactive material?

5. **Finance** The following data represent the amount of money an investor has in an investment account each year for 10 years. She wishes to determine the average annual rate of return on her investment.

Year	Value of Account
1994	\$10,000
1995	\$10,573
1996	\$11,260
1997	\$11,733
1998	\$12,424
1999	\$13,269
2000	\$13,968
2001	\$14,823
2002	\$15,297
2003	\$16,539

(a) Using a graphing utility, draw a scatter diagram with time as the independent variable and the value of the account as the dependent variable.

(b) Using a graphing utility, fit an exponential function to the data.

(c) Based on the answer in part (b), what was the average annual rate of return from this account over the past 10 years?

(d) If the investor plans on retiring in 2021, what will the predicted value of this account be?

(e) When will the account be worth \$50,000?

6. **Finance** The following data show the amount of money an investor has in an investment account each year for 7 years. He wishes to determine the average annual rate of return on his investment.

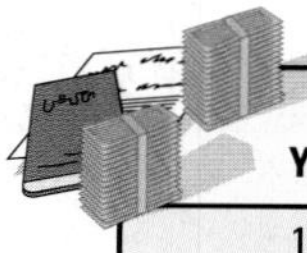

Year	Value of Account
1997	\$20,000
1998	\$21,516
1999	\$23,355
2000	\$24,885
2001	\$27,434
2002	\$30,053
2003	\$32,622

(a) Using a graphing utility, draw a scatter diagram with time as the independent variable and the value of the account as the dependent variable.

(b) Using a graphing utility, fit an exponential function to the data.

(c) Based on the answer to part (b), what was the average annual rate of return from this account over the past 7 years?

(d) If the investor plans on retiring in 2020, what will the predicted value of this account be?

(e) When will the account be worth \$80,000?

7. **Economics and Marketing** The following data represent the price and quantity demanded in 2004 for IBM personal computers.

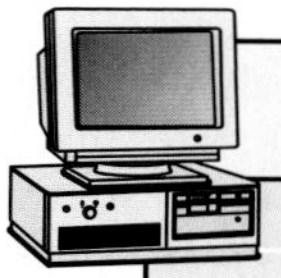

Price (\$/Computer)	Quantity Demanded
2300	152
2000	159
1700	164
1500	171
1300	176
1200	180
1000	189

(a) Using a graphing utility, draw a scatter diagram of the data with price as the dependent variable.

(b) Using a graphing utility, fit a logarithmic function to the data.

(c) Using a graphing utility, draw the logarithmic function found in part (b) on the scatter diagram.

(d) Use the function found in part (b) to predict the number of IBM personal computers that would be demanded if the price were \$1650.

8. Economics and Marketing The following data represent the price and quantity supplied in 2004 for IBM personal computers.

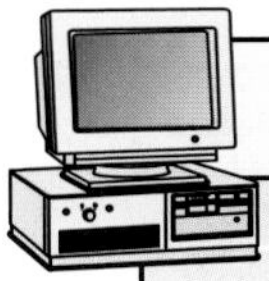

Price ($/Computer)	Quantity Supplied
2300	180
2000	173
1700	160
1500	150
1300	137
1200	130
1000	113

(a) Using a graphing utility, draw a scatter diagram of the data with price as the dependent variable.
(b) Using a graphing utility, fit a logarithmic function to the data.
(c) Using a graphing utility, draw the logarithmic function found in part (b) on the scatter diagram.
(d) Use the function found in part (b) to predict the number of IBM personal computers that would be supplied if the price were $1650.

9. Population Model The following data represent the population of the United States. An ecologist is interested in finding a function that describes the population of the United States.

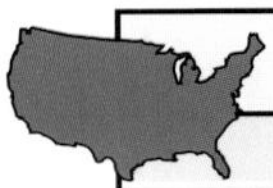

Year	Population
1900	76,212,168
1910	92,228,496
1920	106,021,537
1930	123,202,624
1940	132,164,569
1950	151,325,798
1960	179,323,175
1970	203,302,031
1980	226,542,203
1990	248,709,873
2000	281,421,906

SOURCE: U.S. Census Bureau

(a) Using a graphing utility, draw a scatter diagram of the data using the year as the independent variable and population as the dependent variable.
(b) Using a graphing utility, fit a logistic function to the data.
(c) Using a graphing utility, draw the function found in part (b) on the scatter diagram.
(d) Based on the function found in part (b), what is the carrying capacity of the United States?
(e) Use the function found in part (b) to predict the population of the United States in 2004.
(f) When will the United States population be 300,000,000?

10. Population Model The following data represent the world population. An ecologist is interested in finding a function that describes the world population.

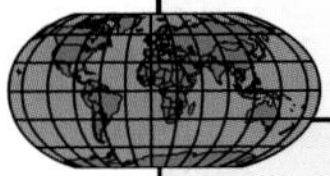

Year	Population (in Billions)
1993	5.531
1994	5.611
1995	5.691
1996	5.769
1997	5.847
1998	5.925
1999	6.003
2000	6.080
2001	6.157

SOURCE: U.S. Census Bureau

(a) Using a graphing utility, draw a scatter diagram of the data using year as the independent variable and population as the dependent variable.
(b) Using a graphing utility, fit a logistic function to the data.
(c) Using a graphing utility, draw the function found in part (b) on the scatter diagram.
(d) Based on the function found in part (b), what is the carrying capacity of the world?
(e) Use the function found in part (b) to predict the population of the world in 2004.
(f) When will world population be 7 billion?

11. Population Model The following data represent the population of Illinois. An urban economist is interested in finding a model that describes the population of Illinois.

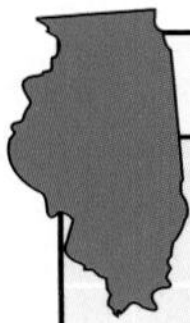

Year	Population
1900	4,821,550
1910	5,638,591
1920	6,485,280
1930	7,630,654
1940	7,897,241
1950	8,712,176
1960	10,081,158
1970	11,110,285
1980	11,427,409
1990	11,430,602
2000	12,419,293

SOURCE: U.S. Census Bureau

(a) Using a graphing utility, draw a scatter diagram of the data using year as the independent variable and population as the dependent variable.
(b) Using a graphing utility, fit a logistic function to the data.
(c) Using a graphing utility, draw the function found in part (b) on the scatter diagram.
(d) Based on the function found in part (b), what is the carrying capacity of Illinois?
(e) Use the function found in part (b) to predict the population of Illinois in 2010.

12. Population Model The data to the right represent the population of Pennsylvania. An urban economist is interested in finding a model that describes the population of Pennsylvania.
(a) Using a graphing utility, draw a scatter diagram of the data using year as the independent variable and population as the dependent variable.
(b) Using a graphing utility, fit a logistic function to the data.
(c) Using a graphing utility, draw the function found in part (b) on the scatter diagram.
(d) Based on the function found in part (b), what is the carrying capacity of Pennsylvania?
(e) Use the function found in part (b) to predict the population of Pennsylvania in 2010.

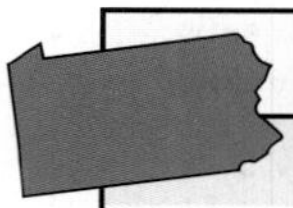

Year	Population
1900	6,302,115
1910	7,665,111
1920	8,720,017
1930	9,631,350
1940	9,900,180
1950	10,498,012
1960	11,319,366
1970	11,800,766
1980	11,864,720
1990	11,881,643
2000	12,281,054

SOURCE: U.S. Census Bureau

Chapter Review

Things to Know

Composite function (p. 224) $(f \circ g)(x) = f(g(x))$.

One-to-one function f (p. 233) A function whose inverse is also a function
For any choice of elements x_1, x_2 in the domain of f, if $x_1 \neq x_2$, then $f(x_1) \neq f(x_2)$.

Horizontal-line test (p. 234) If every horizontal line intersects the graph of a function f in at most one point, then f is one-to-one.

Inverse function f^{-1} of f (pp. 235–237) Domain of f = Range of f^{-1}; Range of f = Domain of f^{-1}
$f^{-1}(f(x)) = x$ and $f(f^{-1}(x)) = x$
Graphs of f and f^{-1} are symmetric with respect to the line $y = x$.

Properties of the exponential function (pp. 248 and 250)

$f(x) = a^x, \quad a > 1$ Domain: the interval $(-\infty, \infty)$;
Range: the interval $(0, \infty)$;
x-intercepts: none; y-intercept: 1;
horizontal asymptote: x-axis ($y = 0$) as $x \to -\infty$;
increasing; one-to-one; smooth; continuous
See Figure 16 for a typical graph.

$f(x) = a^x, \quad 0 < a < 1$ Domain: the interval $(-\infty, \infty)$;
Range: the interval $(0, \infty)$;
x-intercepts: none; y-intercept: 1;
horizontal asymptote: x-axis ($y = 0$) as $x \to \infty$;
decreasing; one-to-one; smooth; continuous
See Figure 20 for a typical graph.

Number e (p. 251) Value approached by the expression $\left(1 + \frac{1}{n}\right)^n$ as $n \to \infty$; that is, $\lim_{n \to \infty} \left(1 + \frac{1}{n}\right)^n = e$.

Property of exponents (p. 253)	If $a^u = a^v$, then $u = v$.	
Properties of the logarithmic functions (pp. 264–265)	$f(x) = \log_a x, \quad a > 1$ $(y = \log_a x \text{ means } x = a^y)$	Domain: the interval $(0, \infty)$; Range: the interval $(-\infty, \infty)$; x-intercept: 1; y-intercept: none; vertical asymptote: $x = 0$ (y-axis); increasing; one-to-one; smooth; continuous See Figure 25(a) for a typical graph.
	$f(x) = \log_a x, \quad 0 < a < 1$ $(y = \log_a x \text{ means } x = a^y)$	Domain: the interval $(0, \infty)$; Range: the interval $(-\infty, \infty)$; x-intercept: 1; y-intercept: none; vertical asymptote: $x = 0$ (y-axis); decreasing; one-to-one; smooth; continuous See Figure 25(b) for a typical graph.
Natural logarithm (p. 264)	$y = \ln x$ means $x = e^y$.	

Properties of logarithms (pp. 274–275, 278)

$$\log_a 1 = 0 \qquad \log_a a = 1 \qquad a^{\log_a M} = M \qquad \log_a a^r = r$$

$$\log_a(MN) = \log_a M + \log_a N \qquad \log_a\left(\frac{M}{N}\right) = \log_a M - \log_a N$$

$$\log_a M^r = r \log_a M$$

If $M = N$, then $\log_a M = \log_a N$.

If $\log_a M = \log_a N$, then $M = N$.

Formulas

Change-of-Base Formula (p. 279)	$\log_a M = \dfrac{\log_b M}{\log_b a}$
Compound Interest Formula (p. 289)	$A = P \cdot \left(1 + \dfrac{r}{n}\right)^{nt}$
Continuous compounding (p. 291)	$A = Pe^{rt}$
Present Value Formulas (p. 293)	$P = A \cdot \left(1 + \dfrac{r}{n}\right)^{-nt}$ or $P = Ae^{-rt}$
Growth and decay models (p. 297)	$A(t) = A_0 e^{kt}$
Newton's Law of Cooling (p. 301)	$u(t) = T + (u_0 - T)e^{kt}, \quad k < 0$
Logistic growth model (p. 303)	$P(t) = \dfrac{c}{1 + ae^{-bt}}$

Objectives

Section		You should be able to . . .	Review Exercises
4.1	1	Form a composite function and find its domain (p. 224)	1–12
4.2	1	Determine the inverse of a function (p. 231)	13, 14
	2	Obtain the graph of the inverse function from the graph of the function (p. 236)	15, 16
	3	Find the inverse function f^{-1} (p. 237)	17–22
4.3	1	Evaluate exponential functions (p. 244)	23(a), (c); 24(a), (c)
	2	Graph exponential functions (p. 247)	55–59, 62, 63
	3	Define the number e (p. 251)	59, 62, 63
	4	Solve exponential equations (p. 253)	65–68, 73, 74, 76, 77, 78

4.4	1	Change exponential expressions to logarithmic expressions (p. 261)	25, 26
	2	Change logarithmic expressions to exponential expressions (p. 261)	27, 28
	3	Evaluate logarithmic functions (p. 261)	23(b), (d), 24(b), (d), 33–34, 85, 86
	4	Determine the domain of a logarithmic function (p. 262)	29–32
	5	Graph logarithmic functions (p. 263)	60, 61, 64
	6	Solve logarithmic equations (p. 266)	69, 70, 75
4.5	1	Work with the properties of logarithms (p. 273)	35–38
	2	Write a logarithmic expression as a sum or difference of logarithms (p. 276)	39–44
	3	Write a logarithmic expression as a single logarithm (p. 277)	45–50
	4	Evaluate logarithms whose base is neither 10 nor *e* (p. 278)	51, 52
4.6	1	Solve logarithmic equations using the properties of logarithms (p. 282)	79, 80
	2	Solve exponential equations (p. 283)	71, 72, 81–84
	3	Solve logarithmic and exponential equations using a graphing utility (p. 285)	65–84
4.7	1	Determine the future value of a lump sum of money (p. 287)	90
	2	Calculate effective rates of return (p. 291)	90
	3	Determine the present value of a lump sum of money (p. 292)	91
	4	Determine the time required to double or triple a lump sum of money (p. 294)	90
4.8	1	Find equations of populations that obey the law of uninhibited growth (p. 297)	95
	2	Find equations of populations that obey the law of decay (p. 300)	93, 96
	3	Use Newton's Law of Cooling (p. 301)	94
	4	Use logistic models (p. 303)	97
4.9	1	Use a graphing utility to fit an exponential function to data (p. 307)	98
	2	Use a graphing utility to fit a logarithmic function to data (p. 308)	99
	3	Use a graphing utility to fit a logistic function to data (p. 309)	100

Review Exercises

(Blue problem numbers indicate the author's suggestions for use in a Practice Test.)

In Problems 1–6, for the given functions f and g find:

(a) $(f \circ g)(2)$ (b) $(g \circ f)(-2)$ (c) $(f \circ f)(4)$ (d) $(g \circ g)(-1)$

1. $f(x) = 3x - 5;\quad g(x) = 1 - 2x^2$
2. $f(x) = 4 - x;\quad g(x) = 1 + x^2$
3. $f(x) = \sqrt{x + 2};\quad g(x) = 2x^2 + 1$
4. $f(x) = 1 - 3x^2;\quad g(x) = \sqrt{4 - x}$
5. $f(x) = e^x;\quad g(x) = 3x - 2$
6. $f(x) = \dfrac{2}{1 + 2x^2};\quad g(x) = 3x$

In Problems 7–12, find $f \circ g$, $g \circ f$, $f \circ f$, and $g \circ g$ for each pair of functions. State the domain of each composite function.

7. $f(x) = 2 - x;\quad g(x) = 3x + 1$
8. $f(x) = 2x - 1;\quad g(x) = 2x + 1$
9. $f(x) = 3x^2 + x + 1;\quad g(x) = |3x|$
10. $f(x) = \sqrt{3x};\quad g(x) = 1 + x + x^2$
11. $f(x) = \dfrac{x + 1}{x - 1};\quad g(x) = \dfrac{1}{x}$
12. $f(x) = \sqrt{x - 3};\quad g(x) = \dfrac{3}{x}$

In Problems 13 and 14, (a) find the inverse of the given function, and (b) determine whether the inverse represents a function.

13. $\{(1, 2), (3, 5), (5, 8), (6, 10)\}$
14. $\{(-1, 4), (0, 2), (1, 4), (3, 7)\}$

In Problems 15 and 16, the graph of a one-to-one function is given. Draw the graph of the inverse function f^{-1}. For convenience (and as a hint), the graph of $y = x$ is also given.

15.

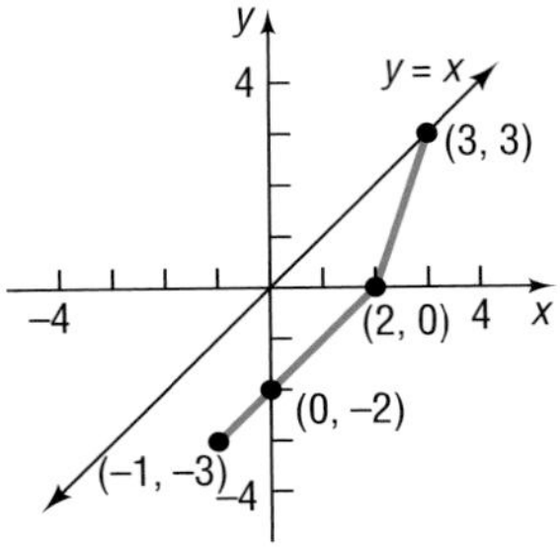

16.

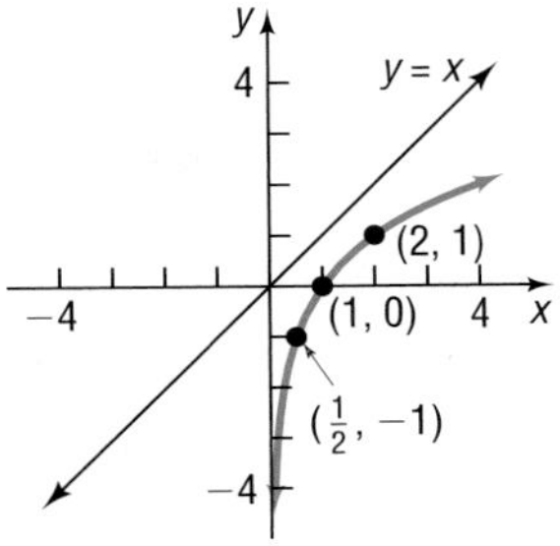

In Problems 17–22, the function f is one-to-one. Find the inverse of each function and check your answer. Find the domain and the range of f and f^{-1}.

17. $f(x) = \dfrac{2x + 3}{5x - 2}$

18. $f(x) = \dfrac{2 - x}{3 + x}$

19. $f(x) = \dfrac{1}{x - 1}$

20. $f(x) = \sqrt{x - 2}$

21. $f(x) = \dfrac{3}{x^{1/3}}$

22. $f(x) = x^{1/3} + 1$

In Problems 23 and 24, $f(x) = 3^x$ and $g(x) = \log_3 x$.

23. Evaluate:
(a) $f(4)$ (b) $g(9)$ (c) $f(-2)$ (d) $g\left(\dfrac{1}{27}\right)$

24. Evaluate:
(a) $f(1)$ (b) $g(81)$ (c) $f(-4)$ (d) $g\left(\dfrac{1}{243}\right)$

In Problems 25 and 26, convert each exponential expression to an equivalent expression involving a logarithm. In Problems 27 and 28, convert each logarithmic expression to an equivalent expression involving an exponent.

25. $5^2 = z$

26. $a^5 = m$

27. $\log_5 u = 13$

28. $\log_a 4 = 3$

In Problems 29–32, find the domain of each logarithmic function.

29. $f(x) = \log(3x - 2)$

30. $F(x) = \log_5(2x + 1)$

31. $H(x) = \log_2(x^2 - 3x + 2)$

32. $F(x) = \ln(x^2 - 9)$

In Problems 33–38, evaluate each expression. Do not use a calculator.

33. $\log_2\left(\dfrac{1}{8}\right)$

34. $\log_3 81$

35. $\ln e^{\sqrt{2}}$

36. $e^{\ln 0.1}$

37. $2^{\log_2 0.4}$

38. $\log_2 2^{\sqrt{3}}$

In Problems 39–44, write each expression as the sum and/or difference of logarithms. Express powers as factors.

39. $\log_3\left(\dfrac{uv^2}{w}\right), u > 0, v > 0, w > 0$

40. $\log_2(a^2\sqrt{b})^4, a > 0, b > 0$

41. $\log\left(x^2\sqrt{x^3 + 1}\right), x > 0$

42. $\log_5\left(\dfrac{x^2 + 2x + 1}{x^2}\right), x > 0$

43. $\ln\left(\dfrac{x\sqrt[3]{x^2 + 1}}{x - 3}\right), x > 3$

44. $\ln\left(\dfrac{2x + 3}{x^2 - 3x + 2}\right)^2, x > 2$

In Problems 45–50, write each expression as a single logarithm.

45. $3\log_4 x^2 + \dfrac{1}{2}\log_4\sqrt{x}$

46. $-2\log_3\left(\dfrac{1}{x}\right) + \dfrac{1}{3}\log_3\sqrt{x}$

47. $\ln\left(\dfrac{x - 1}{x}\right) + \ln\left(\dfrac{x}{x + 1}\right) - \ln(x^2 - 1)$

48. $\log(x^2 - 9) - \log(x^2 + 7x + 12)$

49. $2\log 2 + 3\log x - \dfrac{1}{2}[\log(x + 3) + \log(x - 2)]$

50. $\dfrac{1}{2}\ln(x^2 + 1) - 4\ln\dfrac{1}{2} - \dfrac{1}{2}[\ln(x - 4) + \ln x]$

In Problems 51 and 52, use the Change-of-Base Formula and a calculator to evaluate each logarithm. Round your answer to three decimal places.

51. $\log_4 19$

52. $\log_2 21$

In Problems 53 and 54, graph each function using a graphing utility and the Change-of-Base Formula.

53. $y = \log_3 x$

54. $y = \log_7 x$

In Problems 55–64, use transformations to graph each function. Determine the domain, range, and any asymptotes.

55. $f(x) = 2^{x-3}$

56. $f(x) = -2^x + 3$

57. $f(x) = \frac{1}{2}(3^{-x})$

58. $f(x) = 1 + 3^{2x}$

59. $f(x) = 1 - e^x$

60. $f(x) = 3 + \ln x$

61. $f(x) = \frac{1}{2}\ln x$

62. $f(x) = 3e^x$

63. $f(x) = 3 - e^{-x}$

64. $f(x) = 4 - \ln(-x)$

In Problems 65–84, solve each equation.

65. $4^{1-2x} = 2$

66. $8^{6+3x} = 4$

67. $3^{x^2+x} = \sqrt{3}$

68. $4^{x-x^2} = \frac{1}{2}$

69. $\log_x 64 = -3$

70. $\log_{\sqrt{2}} x = -6$

71. $5^x = 3^{x+2}$

72. $5^{x+2} = 7^{x-2}$

73. $9^{2x} = 27^{3x-4}$

74. $25^{2x} = 5^{x^2-12}$

75. $\log_3 \sqrt{x-2} = 2$

76. $2^{x+1} \cdot 8^{-x} = 4$

77. $8 = 4^{x^2} \cdot 2^{5x}$

78. $2^x \cdot 5 = 10^x$

79. $\log_6(x+3) + \log_6(x+4) = 1$

80. $\log(7x - 12) = 2\log x$

81. $e^{1-x} = 5$

82. $e^{1-2x} = 4$

83. $2^{3x} = 3^{2x+1}$

84. $2^{x^3} = 3^{x^2}$

In Problems 85 and 86, use the following result: If x is the atmospheric pressure (measured in millimeters of mercury), then the formula for the altitude h(x) (measured in meters above sea level) is

$$h(x) = (30T + 8000)\log\left(\frac{P_0}{x}\right)$$

where T is the temperature (in degrees Celsius) and P_0 is the atmospheric pressure at sea level, which is approximately 760 millimeters of mercury.

85. **Finding the Altitude of an Airplane** At what height is a Piper Cub whose instruments record an outside temperature of 0°C and a barometric pressure of 300 millimeters of mercury?

86. **Finding the Height of a Mountain** How high is a mountain if instruments placed on its peak record a temperature of 5°C and a barometric pressure of 500 millimeters of mercury?

---

87. **Amplifying Sound** An amplifier's power output P (in watts) is related to its decibel voltage gain d by the formula $P = 25e^{0.1d}$.

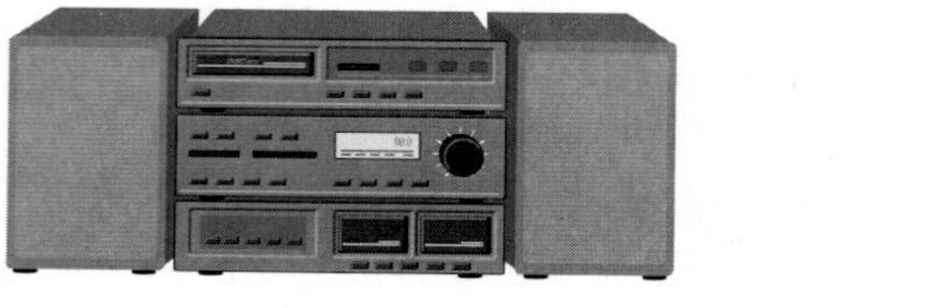

(a) Find the power output for a decibel voltage gain of 4 decibels.

(b) For a power output of 50 watts, what is the decibel voltage gain?

88. **Limiting Magnitude of a Telescope** A telescope is limited in its usefulness by the brightness of the star that it is aimed at and by the diameter of its lens. One measure of a star's brightness is its *magnitude*; the dimmer the star, the larger its magnitude. A formula for the limiting magnitude L of a telescope, that is, the magnitude of the dimmest star that it can be used to view, is given by

$$L = 9 + 5.1\log d$$

where d is the diameter (in inches) of the lens.

(a) What is the limiting magnitude of a 3.5-inch telescope?

(b) What diameter is required to view a star of magnitude 14?

89. Salvage Value The number of years n for a piece of machinery to depreciate to a known salvage value can be found using the formula

$$n = \frac{\log s - \log i}{\log(1 - d)}$$

where s is the salvage value of the machinery, i is its initial value, and d is the annual rate of depreciation.

(a) How many years will it take for a piece of machinery to decline in value from \$90,000 to \$10,000 if the annual rate of depreciation is 0.20 (20%)?

(b) How many years will it take for a piece of machinery to lose half of its value if the annual rate of depreciation is 15%?

90. Funding a College Education A child's grandparents purchase a \$10,000 bond fund that matures in 18 years to be used for her college education. The bond fund pays 4% interest compounded semiannually. How much will the bond fund be worth at maturity? What is the effective rate of interest? How long would it take the bond to double in value under these terms?

91. Funding a College Education A child's grandparents wish to purchase a bond that matures in 18 years to be used for her college education. The bond pays 4% interest compounded semiannually. How much should they pay so that the bond will be worth \$85,000 at maturity?

92. Funding an IRA First Colonial Bankshares Corporation advertised the following IRA investment plans.

Target IRA Plans

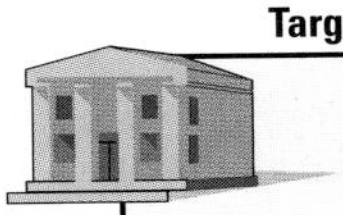

For each \$5000 Maturity Value Desired

Deposit:	At a Term of:
\$620.17	20 Years
\$1045.02	15 Years
\$1760.92	10 Years
\$2967.26	5 Years

(a) Assuming continuous compounding, what was the annual rate of interest that they offered?

(b) First Colonial Bankshares claims that \$4000 invested today will have a value of over \$32,000 in 20 years. Use the answer found in part (a) to find the actual value of \$4000 in 20 years. Assume continuous compounding.

93. Estimating the Date That a Prehistoric Man Died The bones of a prehistoric man found in the desert of New Mexico contain approximately 5% of the original amount of carbon 14. If the half-life of carbon 14 is 5600 years, approximately how long ago did the man die?

94. Temperature of a Skillet A skillet is removed from an oven whose temperature is 450°F and placed in a room whose temperature is 70°F. After 5 minutes, the temperature of the skillet is 400°F. How long will it be until its temperature is 150°F?

95. World Population The growth rate of the world's population in 2003 was $k = 1.16\% = 0.0116$. The population of the world in 2003 was 6,302,486,693. Letting $t = 0$ represent 2003, use the uninhibited growth model to predict the world's population in the year 2010.

SOURCE: U.S. Census Bureau.

96. Radioactive Decay The half-life of radioactive cobalt is 5.27 years. If 100 grams of radioactive cobalt is present now, how much will be present in 20 years? In 40 years?

97. Logistic Growth The logistic growth model

$$P(t) = \frac{0.8}{1 + 1.67e^{-0.16t}}$$

represents the proportion of new cars with a Global Positioning System (GPS). Let $t = 0$ represent 2003, $t = 1$ represent 2004, and so on.

(a) What proportion of new cars in 2003 had a GPS?

(b) Determine the maximum proportion of new cars that have a GPS.

(c) Using a graphing utility, graph $P(t)$.

(d) When will 75% of new cars have a GPS?

98. CBL Experiment The following data were collected by placing a temperature probe in a portable heater, removing the probe, and then recording temperature over time.

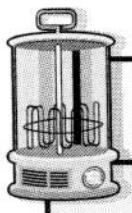

Time (sec.)	Temperature (°F)
0	165.07
1	164.77
2	163.99
3	163.22
4	162.82
5	161.96
6	161.20
7	160.45
8	159.35
9	158.61
10	157.89
11	156.83
12	156.11
13	155.08
14	154.40
15	153.72

According to Newton's Law of Cooling, these data should follow an exponential model.

(a) Using a graphing utility, draw a scatter diagram for the data.
(b) Using a graphing utility, fit an exponential function to the data.
(c) Graph the exponential function found in part (b) on the scatter diagram.
(d) Predict how long it will take for the probe to reach a temperature of 110°F.

99. Wind Chill Factor The following data represent the wind speed (mph) and wind chill factor at an air temperature of 15°F.

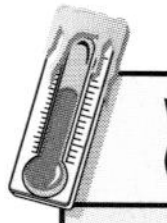

Wind Speed (mph)	Wind Chill Factor (°F)
5	7
10	3
15	0
20	−2
25	−4
30	−5
35	−7

SOURCE: U.S. National Weather Service

(a) Using a graphing utility, draw a scatter diagram with wind speed as the independent variable.
(b) Using a graphing utility, fit a logarithmic function to the data.
(c) Using a graphing utility, draw the logarithmic function found in part (b) on the scatter diagram.
(d) Use the function found in part (b) to predict the wind chill factor if the air temperature is 15°F and the wind speed is 23 mph.

100. Spreading of a Disease Jack and Diane live in a small town of 50 people. Unfortunately, Jack and Diane both have a cold. Those who come in contact with someone who has this cold will themselves catch the cold. The following data represent the number of people in the small town who have caught the cold after t days.

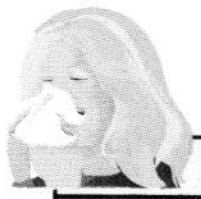

Days, t	Number of People with Cold, C
0	2
1	4
2	8
3	14
4	22
5	30
6	37
7	42
8	44

(a) Using a graphing utility, draw a scatter diagram of the data. Comment on the type of relation that appears to exist between the days and number of people with a cold.
(b) Using a graphing utility, fit a logistic function to the data.
(c) Graph the function found in part (b) on the scatter diagram.
(d) According to the function found in part (b), what is the maximum number of people who will catch the cold? In reality, what is the maximum number of people who could catch the cold?
(e) Sometime between the second and third day, 10 people in the town had a cold. According to the model found in part (b), when did 10 people have a cold?
(f) How long will it take for 46 people to catch the cold?

Chapter Projects

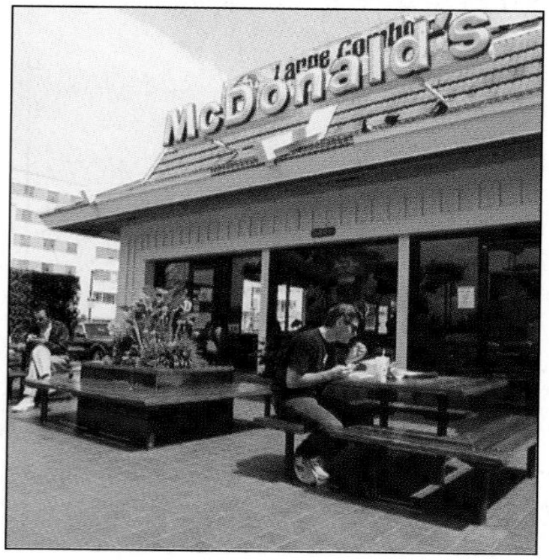

1. Hot Coffee A fast-food restaurant wants a special container to hold coffee. The restaurant wishes the container to quickly cool the coffee from 200° to 130°F and keep the liquid between 110° and 130°F as long as possible. The restaurant has three containers to select from.

1. The CentiKeeper Company has a container that reduces the temperature of a liquid from 200°F to 100°F in 30 minutes by maintaining a constant temperature of 70°F.

2. The TempControl Company has a container that reduces the temperature of a liquid from 200°F to 110°F in 25 minutes by maintaining a constant temperature of 60°F.

3. The Hot'n'Cold Company has a container that reduces the temperature of a liquid from 200°F to 120°F in 20 minutes by maintaining a constant temperature of 65°F.

You need to recommend which container the restaurant should purchase.

(a) Use Newton's Law of Cooling to find a function relating the temperature of the liquid over time for each container.

(b) How long does it take each container to lower the coffee temperature from 200° to 130°F?

(c) How long will the coffee temperature remain between 110° and 130°F? This temperature is considered the optimal drinking temperature.

(d) Graph each function using a graphing utility.

(e) Which company would you recommend to the restaurant? Why?

(f) How might the cost of the container affect your decision?

The following projects are available at www.prenhall.com/sullivan7e.

2. **Project at Motorola** ***Thermal Fatigue of Solder Connections***
3. **Depreciation of a New Car**
4. **CBL Experiment**

Cumulative Review

1. Is the following graph the graph of a function? If it is, is the function one-to-one?

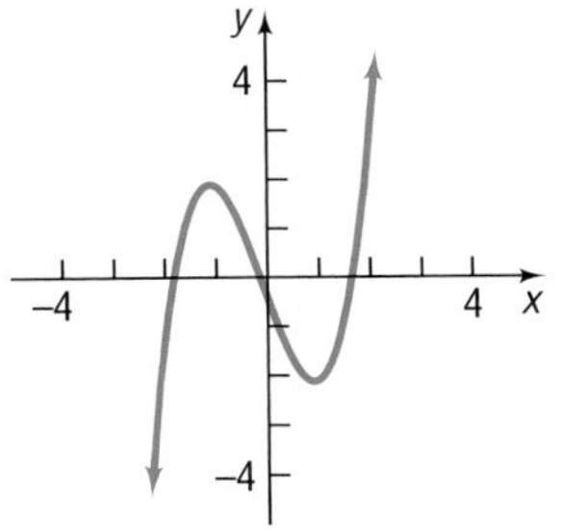

2. For the function $f(x) = 2x^2 - 3x + 1$, find the following:

(a) $f(3)$ (b) $f(-x)$ (c) $f(x + h)$

3. Determine which of the following points are on the graph of $x^2 + y^2 = 1$.

(a) $\left(\frac{1}{2}, \frac{1}{2}\right)$ (b) $\left(\frac{1}{2}, \frac{\sqrt{3}}{2}\right)$

4. Solve the equation $3(x - 2) = 4(x + 5)$.

5. Graph the line $2x - 4y = 16$.

6. (a) Graph the quadratic function $f(x) = -x^2 + 2x - 3$ by determining whether its graph opens up or down and by finding its vertex, axis of symmetry, y-intercept, and x-intercept(s), if any.

(b) Solve $f(x) \le 0$.

7. Determine the quadratic function whose graph is given in the figure.

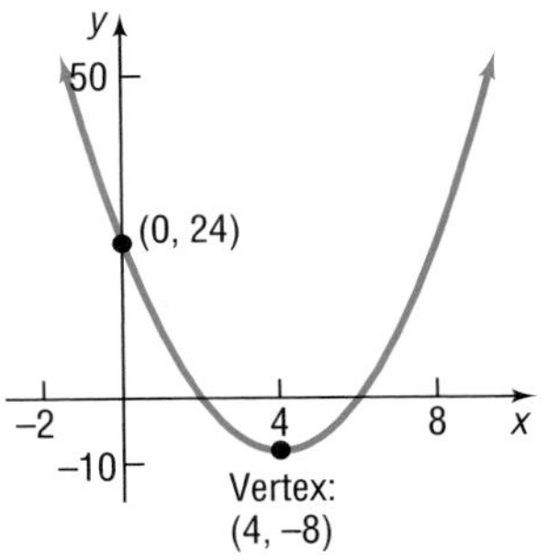

8. Graph $f(x) = 3(x + 1)^3 - 2$ using transformations.

9. Given that $f(x) = x^2 + 2$ and $g(x) = \frac{2}{x - 3}$, find $f(g(x))$ and state its domain. What is $f(g(5))$?

10. For the polynomial function $f(x) = 4x^3 + 9x^2 - 30x - 8$

(a) Find the real zeros of f.

(b) Determine the intercepts of the graph of f.

(c) Use a graphing utility to approximate the local maxima and local minima.

(d) By hand, draw a complete graph of f. Be sure to label the intercepts and turning points.

11. For the function $g(x) = 3^x + 2$:

(a) Graph g using transformations. State the domain, range, and horizontal asymptote of g.

(b) Determine the inverse of g. State the domain, range, and vertical asymptote of g^{-1}.

(c) On the same graph as g, graph g^{-1}.

12. Solve the equation $4^{x-3} = 8^{2x}$.

13. Solve the equation
$\log_3(x + 1) + \log_3(2x - 3) = \log_9 9$

14. Suppose that $f(x) = \log_3(x + 2)$.

(a) Solve $f(x) = 0$.

(b) Solve $f(x) > 0$.

(c) Solve $f(x) = 3$.

15. Data Analysis The following data represent the percent of all drivers that have been stopped by the police for any reason within the past year by age. The median age represents the midpoint of the upper and lower limit for the age range.

Age Range	Median Age, x	Percentage Stopped, y
16–19	17.5	18.2
20–29	24.5	16.8
30–39	34.5	11.3
40–49	44.5	9.4
50–59	54.5	7.7
≥ 60	69.5	3.8

(a) Using your graphing utility, draw a scatter diagram of the data treating median age, x, as the independent variable.

(b) Determine a model that you feel best describes the relation between median age and percentage stopped. You may choose from among linear, quadratic, cubic, power, exponential, logarithmic, or logistic models.

(c) Provide a justification for the model that you selected in part (b).

5 Trigonometric Functions

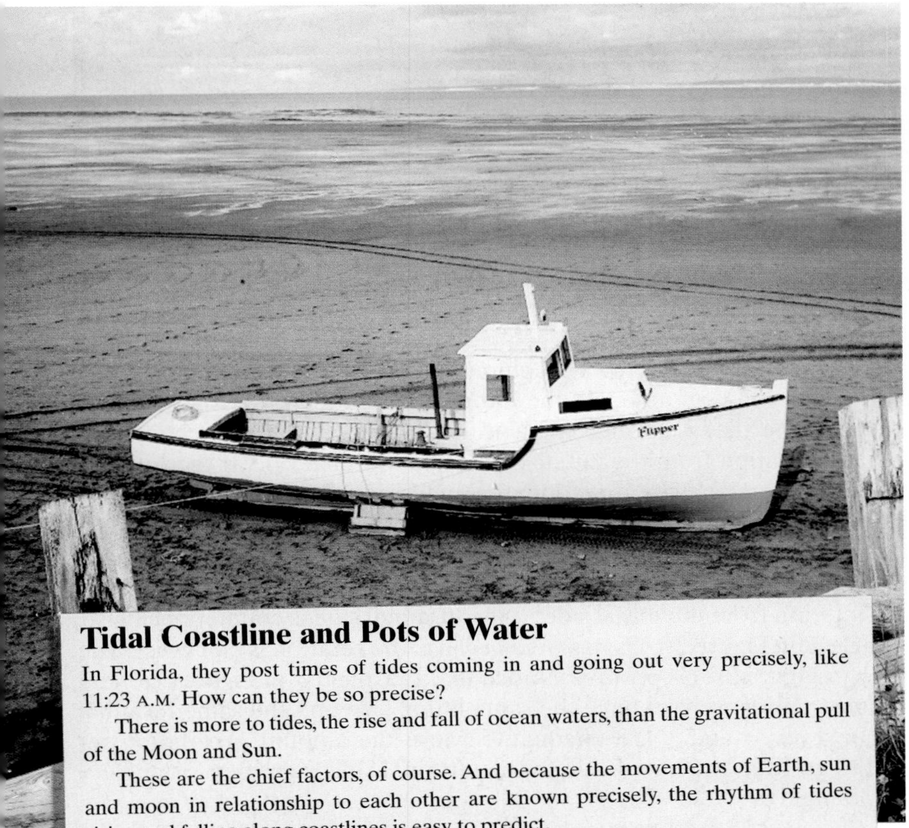

Tidal Coastline and Pots of Water

In Florida, they post times of tides coming in and going out very precisely, like 11:23 A.M. How can they be so precise?

There is more to tides, the rise and fall of ocean waters, than the gravitational pull of the Moon and Sun.

These are the chief factors, of course. And because the movements of Earth, sun and moon in relationship to each other are known precisely, the rhythm of tides rising and falling along coastlines is easy to predict.

Yet the time and heights of high and low tide may vary along different stretches of the same coast, though they are reacting to similar driving forces and pressures.

Historic observation makes possible the exact timing of high tides and low tides along a particular section of coast for a month, a year, or far into the future.

The reason for the difference is oscillation. Think of pots and pans filled with varying levels of water on a table, says Charles O'Reilly, Chief of Tidal Analysis for the Geological Survey of Canada's hydrographic service in Dartmouth, Nova Scotia. Then kick the table.

"You'll notice the water in the pots and pans will slosh differently. That's their natural oscillation," he said. "If you kick the table rhythmically, you'll find each pot continues to slosh differently because it has its own rhythm.

"Now, if you join those pots and pans together, that's sort of like a coastal ocean. They're all feeling the same 'kick,' but they are all responding differently. In order to predict a tide, you have to measure them for some period of time."

SOURCE: *Toronto Star*, June 13, 2001, p. GT02. Reprinted with permission—Torstar Syndication Services.

—SEE CHAPTER PROJECT 1.

OUTLINE

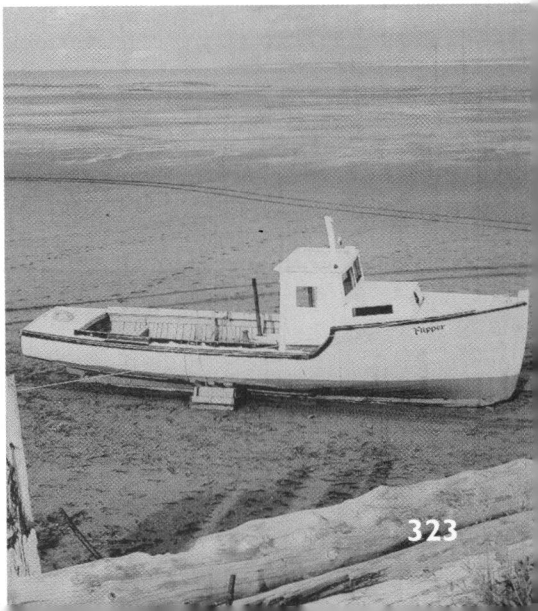

5.1 Angles and Their Measure

PREPARING FOR THIS SECTION *Before getting started, review the following:*

- Circumference and Area of a Circle (Appendix A, Section A.2, p. 915)

Now work the 'Are You Prepared?' problems on page 334.

OBJECTIVES
1. Convert between Degrees, Minutes, Seconds, and Decimal Forms for Angles
2. Find the Arc Length of a Circle
3. Convert from Degrees to Radians
4. Convert from Radians to Degrees
5. Find the Area of a Sector of a Circle
6. Find the Linear Speed of an Object Traveling in Circular Motion

A **ray,** or **half-line,** is that portion of a line that starts at a point V on the line and extends indefinitely in one direction. The starting point V of a ray is called its **vertex.** See Figure 1.

Figure 1

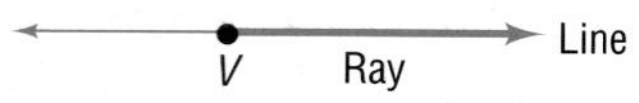

If two rays are drawn with a common vertex, they form an **angle.** We call one of the rays of an angle the **initial side** and the other the **terminal side.** The angle that is formed is identified by showing the direction and amount of rotation from the initial side to the terminal side. If the rotation is in the counterclockwise direction, the angle is **positive;** if the rotation is clockwise, the angle is **negative.** See Figure 2. Lowercase Greek letters, such as α (alpha), β (beta), γ (gamma), and θ (theta), will be used to denote angles. Notice in Figure 2(a) that the angle α is positive because the direction of the rotation from the initial side to the terminal side is counterclockwise. The angle β in Figure 2(b) is negative because the rotation is clockwise. The angle γ in Figure 2(c) is positive. Notice that the angle α in Figure 2(a) and the angle γ in Figure 2(c) have the same initial side and the same terminal side. However, α and γ are unequal, because the amount of rotation required to go from the initial side to the terminal side is greater for angle γ than for angle α.

Figure 2

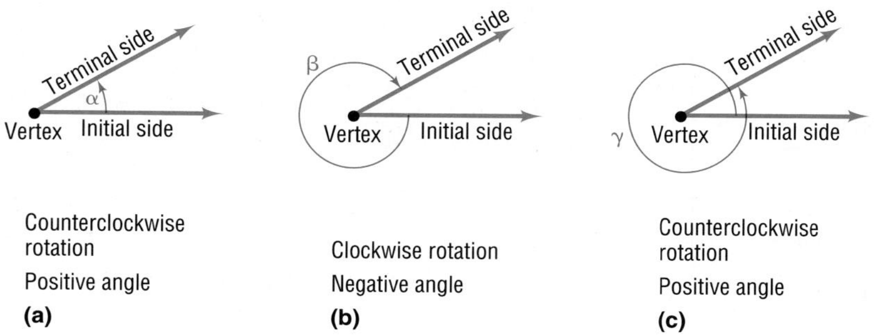

An angle θ is said to be in **standard position** if its vertex is at the origin of a rectangular coordinate system and its initial side coincides with the positive x-axis. See Figure 3.

Figure 3

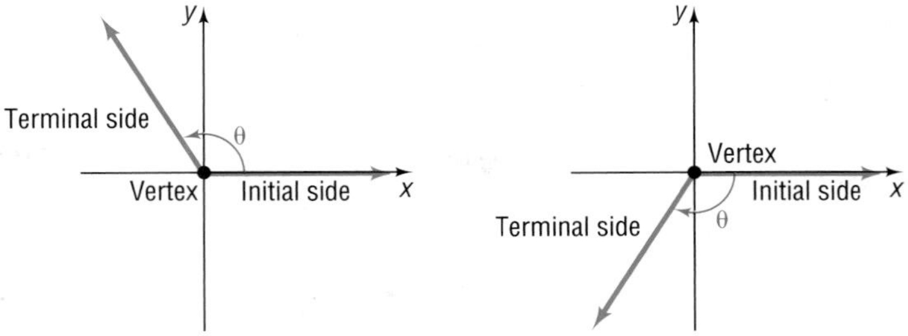

(a) θ is in standard position; θ is positive

(b) θ is in standard position; θ is negative

When an angle θ is in standard position, the terminal side will lie either in a quadrant, in which case we say that θ **lies in that quadrant**, or θ will lie on the x-axis or the y-axis, in which case we say that θ is a **quadrantal angle**. For example, the angle θ in Figure 4(a) lies in quadrant II, the angle θ in Figure 4(b) lies in quadrant IV, and the angle θ in Figure 4(c) is a quadrantal angle.

Figure 4

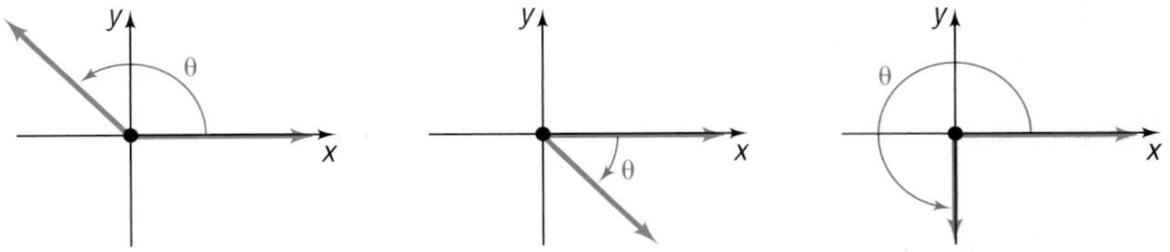

(a) θ lies in quadrant II

(b) θ lies in quadrant IV

(c) θ is a quadrantal angle

We measure angles by determining the amount of rotation needed for the initial side to become coincident with the terminal side. The two commonly used measures for angles are *degrees* and *radians*.

Degrees

The angle formed by rotating the initial side exactly once in the counterclockwise direction until it coincides with itself (1 revolution) is said to measure 360 degrees, abbreviated 360°. **One degree, 1°**, is $\frac{1}{360}$ revolution. A **right angle** is an angle that measures 90°, or $\frac{1}{4}$ revolution; a **straight angle** is an angle that measures 180°, or $\frac{1}{2}$ revolution. See Figure 5. As Figure 5(b) shows, it is customary to indicate a right angle by using the symbol ∟.

Figure 5

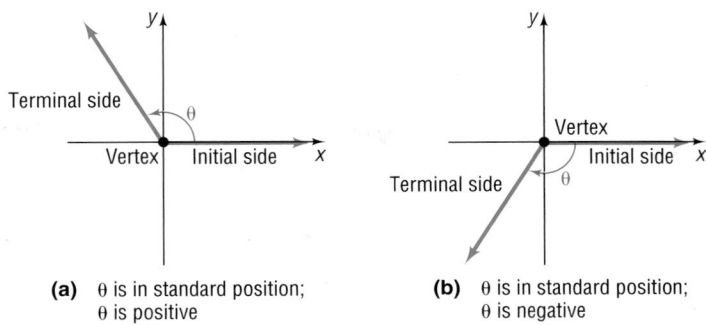

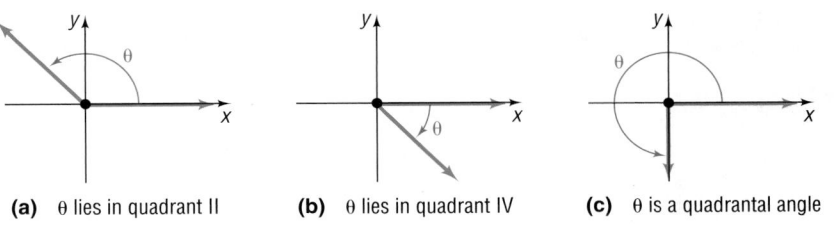

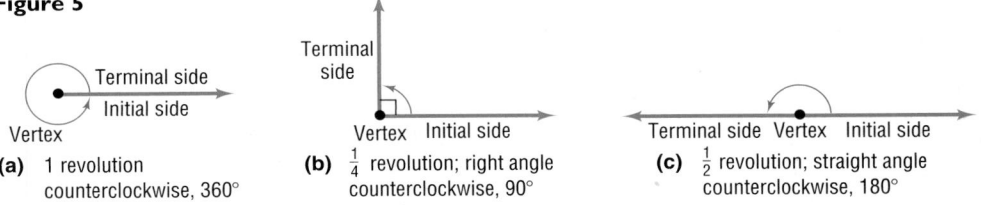

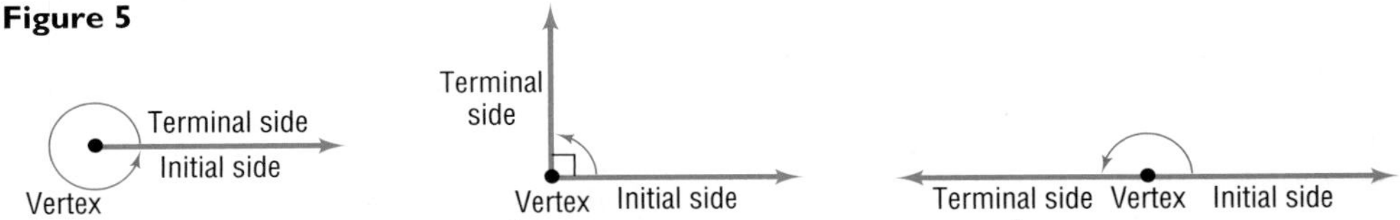

(a) 1 revolution counterclockwise, 360°

(b) $\frac{1}{4}$ revolution; right angle counterclockwise, 90°

(c) $\frac{1}{2}$ revolution; straight angle counterclockwise, 180°

It is also customary to refer to an angle that measures θ degrees as an angle of θ degrees.

EXAMPLE 1 Drawing an Angle

Draw each angle.

(a) $45°$ (b) $-90°$ (c) $225°$ (d) $405°$

Solution

(a) An angle of $45°$ is $\frac{1}{2}$ of a right angle. See Figure 6.

(b) An angle of $-90°$ is $\frac{1}{4}$ revolution in the clockwise direction. See Figure 7.

Figure 6

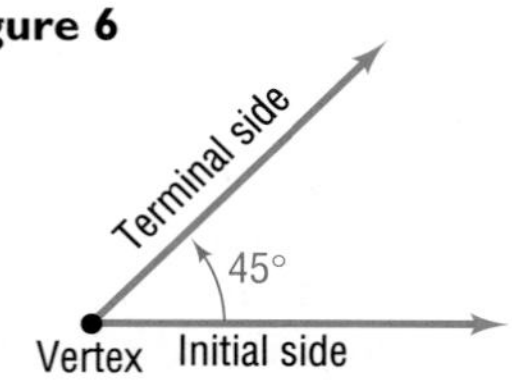

Figure 7

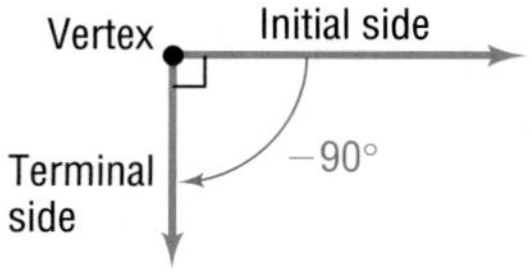

(c) An angle of $225°$ consists of a rotation through $180°$ followed by a rotation through $45°$. See Figure 8.

(d) An angle of $405°$ consists of 1 revolution ($360°$) followed by a rotation through $45°$. See Figure 9.

Figure 8

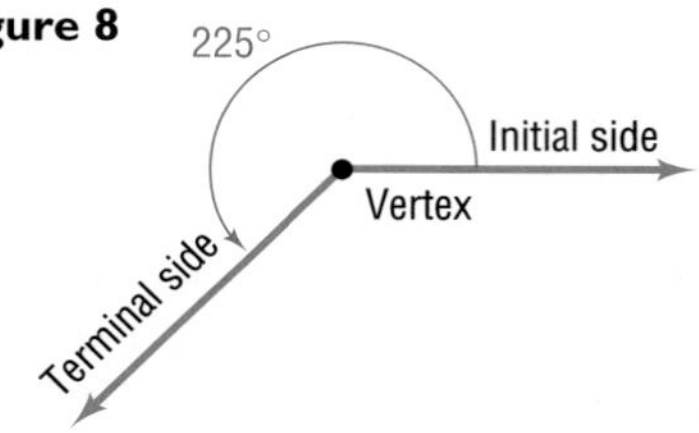

Figure 9

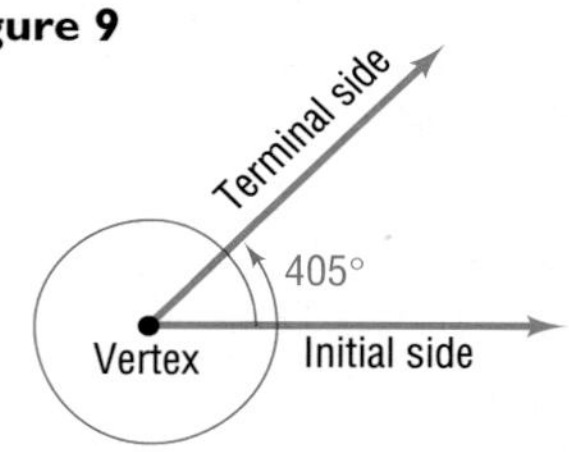

◀

NOW WORK PROBLEM 11.

1 Although subdivisions of a degree may be obtained by using decimals, we also may use the notion of *minutes* and *seconds*. **One minute**, denoted by **1′**, is defined as $\frac{1}{60}$ degree. **One second**, denoted by **1″**, is defined as $\frac{1}{60}$ minute, or equivalently, $\frac{1}{3600}$ degree. An angle of, say, 30 degrees, 40 minutes, 10 seconds is written compactly as $30°40'10''$. To summarize:

$$1 \text{ counterclockwise revolution} = 360°$$
$$1° = 60' \qquad 1' = 60'' \tag{1}$$

It is sometimes necessary to convert from the degree, minute, second notation (D°M′S″) to a decimal form, and vice versa. Check your calculator; it should be capable of doing the conversion for you.

Before getting started, though, you must set the mode to degrees, because there are two common ways to measure angles: degree mode and radian mode. (We will define radians shortly.) Usually, a menu is used to change from one mode to another. Check your owner's manual to find out how your particular calculator works.

Now let's see how to convert by hand from the degree, minute, second notation (D°M′S″) to a decimal form, and vice versa, by looking at some examples:

$$15°30' = 15.5° \quad \text{because} \quad 30' \underset{\uparrow}{=} 30 \cdot \left(\frac{1}{60}\right)^{\circ} = 0.5°$$

$$1' = \left(\frac{1}{60}\right)^{\circ}$$

$$32.25° = 32°15' \quad \text{because} \quad 0.25° = \left(\frac{1}{4}\right)^{\circ} \underset{\uparrow}{=} \frac{1}{4}(60') = 15'.$$

$$1° = 60'$$

EXAMPLE 2 Converting between Degrees, Minutes, Seconds, and Decimal Forms by Hand

(a) Convert 50°6′21″ to a decimal in degrees.

(b) Convert 21.256° to the D°M′S″ form.

Solution (a) Because $1' = \left(\frac{1}{60}\right)^{\circ}$ and $1'' = \left(\frac{1}{60}\right)' = \left(\frac{1}{60} \cdot \frac{1}{60}\right)^{\circ}$, we convert as follows:

$$\begin{aligned} 50°6'21'' &= 50° + 6' + 21'' \\ &= 50° + 6 \cdot \left(\frac{1}{60}\right)^{\circ} + 21 \cdot \left(\frac{1}{60} \cdot \frac{1}{60}\right)^{\circ} \\ &\approx 50° + 0.1° + 0.005833° \\ &= 50.105833° \end{aligned}$$

(b) We proceed as follows:

$$\begin{aligned} 21.256° &= 21° + 0.256° \\ &= 21° + (0.256)(60') && \text{Convert fraction of degree to minutes; } 1° = 60' \\ &= 21° + 15.36' \\ &= 21° + 15' + 0.36' \\ &= 21° + 15' + (0.36)(60'') && \text{Convert fraction of minute to seconds; } 1' = 60'' \\ &= 21° + 15' + 21.6'' \\ &\approx 21°15'22'' \end{aligned}$$

◀

NOW WORK PROBLEMS 23 AND 29.

In many applications, such as describing the exact location of a star or the precise position of a boat at sea, angles measured in degrees, minutes, and even seconds are used. For calculation purposes, these are transformed to decimal form. In other applications, especially those in calculus, angles are measured using *radians*.

Radians

A **central angle** is an angle whose vertex is at the center of a circle. The rays of a central angle subtend (intersect) an arc on the circle. If the radius of the circle is r and the length of the arc subtended by the central angle is also r, then the measure of the angle is **1 radian**. See Figure 10(a).

For a circle of radius 1, the rays of a central angle with measure 1 radian would subtend an arc of length 1. For a circle of radius 3, the rays of a central angle with measure 1 radian would subtend an arc of length 3. See Figure 10(b).

Figure 10

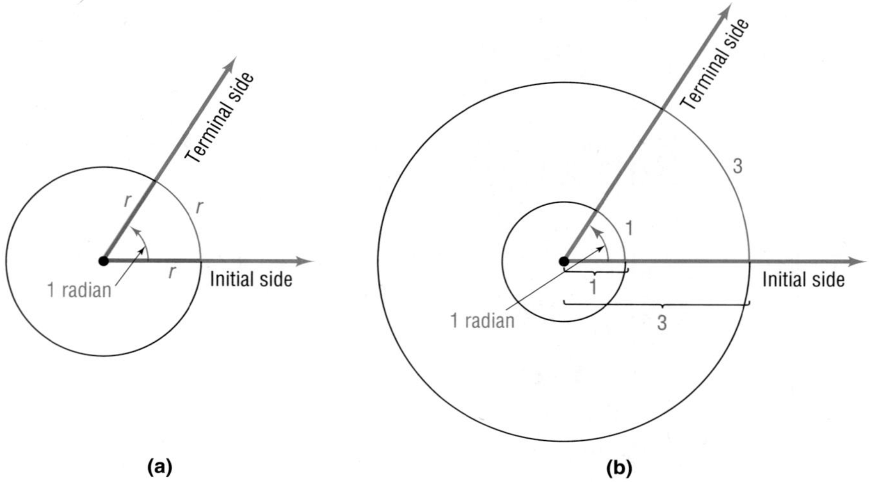

Figure 11

$\frac{\theta}{\theta_1} = \frac{s}{s_1}$

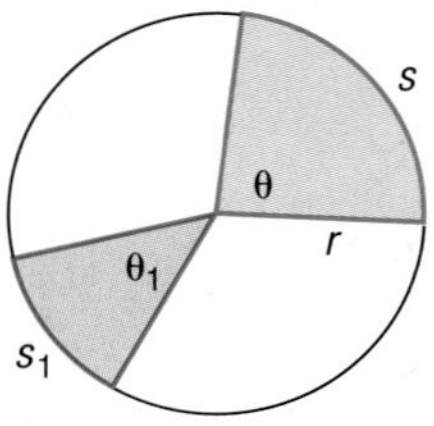

2 Now consider a circle of radius r and two central angles, θ and θ_1, measured in radians. Suppose that these central angles subtend arcs of lengths s and s_1, respectively, as shown in Figure 11. From geometry, we know that the ratio of the measures of the angles equals the ratio of the corresponding lengths of the arcs subtended by these angles; that is,

$$\frac{\theta}{\theta_1} = \frac{s}{s_1} \tag{2}$$

Suppose that $\theta_1 = 1$ radian. Refer again to Figure 10(a). The amount of arc s_1 subtended by the central angle $\theta_1 = 1$ radian equals the radius r of the circle. Then $s_1 = r$, so equation (2) reduces to

$$\frac{\theta}{1} = \frac{s}{r} \quad \text{or} \quad s = r\theta \tag{3}$$

Theorem

Arc Length

For a circle of radius r, a central angle of θ radians subtends an arc whose length s is

$$s = r\theta \tag{4}$$

Note: Formulas must be consistent with regard to the units used. In equation (4), we write

$$s = r\theta$$

To see the units, however, we must go back to equation (3) and write

$$\frac{\theta \text{ radians}}{1 \text{ radian}} = \frac{s \text{ length units}}{r \text{ length units}}$$

$$s \text{ length units} = r \text{ length units} \frac{\theta \text{ radians}}{1 \text{ radian}}$$

Since the radians cancel, we are left with

$$s \text{ length units} = (r \text{ length units})\theta \qquad s = r\theta$$

where θ appears to be "dimensionless" but, in fact, is measured in radians. So, in using the formula $s = r\theta$, the dimension for θ is radians, and any convenient unit of length (such as inches or meters) may be used for s and r. ■

EXAMPLE 3 Finding the Length of an Arc of a Circle

Find the length of the arc of a circle of radius 2 meters subtended by a central angle of 0.25 radian.

Solution We use equation (4) with $r = 2$ meters and $\theta = 0.25$. The length s of the arc is

$$s = r\theta = 2(0.25) = 0.5 \text{ meter}$$ ◀

NOW WORK PROBLEM 71.

Relationship between Degrees and Radians

Figure 12
1 revolution $= 2\pi$ radians

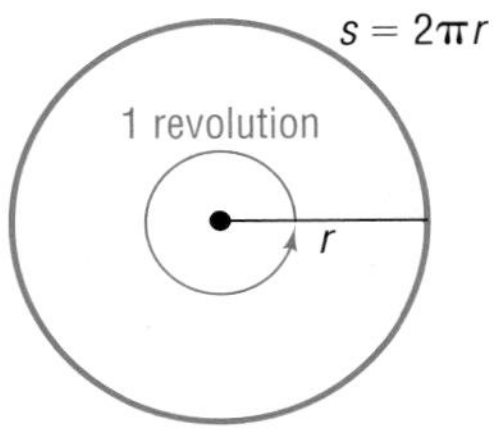

Consider a circle of radius r. A central angle of 1 revolution will subtend an arc equal to the circumference of the circle (Figure 12). Because the circumference of a circle equals $2\pi r$, we use $s = 2\pi r$ in equation (4) to find that, for an angle θ of 1 revolution,

$$s = r\theta$$

$$2\pi r = r\theta \qquad \theta = 1 \text{ revolution; } s = 2\pi r.$$

$$\theta = 2\pi \text{ radians} \qquad \text{Solve for } \theta.$$

From this we have,

$$1 \text{ revolution} = 2\pi \text{ radians} \tag{5}$$

so

$$360° = 2\pi \text{ radians}$$

or

$$180° = \pi \text{ radians} \tag{6}$$

Divide both sides of equation (6) by 180. Then

$$1 \text{ degree} = \frac{\pi}{180} \text{ radian}$$

Divide both sides of (6) by π. Then

$$\frac{180}{\pi} \text{ degrees} = 1 \text{ radian}$$

We have the following two conversion formulas:

$$1 \text{ degree} = \frac{\pi}{180} \text{ radian} \qquad 1 \text{ radian} = \frac{180}{\pi} \text{ degrees} \qquad \textbf{(7)}$$

3 **EXAMPLE 4** **Converting from Degrees to Radians**

Convert each angle in degrees to radians.

(a) 60° (b) 150° (c) −45° (d) 90° (e) 107°

Solution (a) $60° = 60 \cdot 1 \text{ degree} = 60 \cdot \frac{\pi}{180} \text{ radian} = \frac{\pi}{3} \text{ radians}$

(b) $150° = 150 \cdot \frac{\pi}{180} \text{ radian} = \frac{5\pi}{6} \text{ radians}$

(c) $-45° = -45 \cdot \frac{\pi}{180} \text{ radian} = -\frac{\pi}{4} \text{ radian}$

(d) $90° = 90 \cdot \frac{\pi}{180} \text{ radian} = \frac{\pi}{2} \text{ radians}$

(e) $107° = 107 \cdot \frac{\pi}{180} \text{ radian} \approx 1.868 \text{ radians}$ ◄

Example 4 illustrates that angles that are fractions of a revolution, (a)–(d), are expressed in radian measure as fractional multiples of π, rather than as decimals. For example, a right angle, as in Example 4(d), is left in the form $\frac{\pi}{2}$ radians, which is exact, rather than using the approximation $\frac{\pi}{2} \approx \frac{3.1416}{2} = 1.5708$ radians.

NOW WORK PROBLEMS 35 AND 61.

4 **EXAMPLE 5** **Converting Radians to Degrees**

Convert each angle in radians to degrees.

(a) $\frac{\pi}{6}$ radian (b) $\frac{3\pi}{2}$ radians (c) $-\frac{3\pi}{4}$ radians

(d) $\frac{7\pi}{3}$ radians (e) 3 radians

Solution (a) $\frac{\pi}{6} \text{ radian} = \frac{\pi}{6} \cdot 1 \text{ radian} = \frac{\pi}{6} \cdot \frac{180}{\pi} \text{ degrees} = 30°$

(b) $\frac{3\pi}{2} \text{ radians} = \frac{3\pi}{2} \cdot \frac{180}{\pi} \text{ degrees} = 270°$

(c) $-\frac{3\pi}{4} \text{ radians} = -\frac{3\pi}{4} \cdot \frac{180}{\pi} \text{ degrees} = -135°$

(d) $\frac{7\pi}{3}$ radians $= \frac{7\pi}{3} \cdot \frac{180}{\pi}$ degrees $= 420°$

(e) 3 radians $= 3 \cdot \frac{180}{\pi}$ degrees $\approx 171.89°$

NOW WORK PROBLEM 47.

Table 1 lists the degree and radian measures of some commonly encountered angles. You should learn to feel equally comfortable using degree or radian measure for these angles.

Table 1

Degrees	0°	30°	45°	60°	90°	120°	135°	150°	180°
Radians	0	$\frac{\pi}{6}$	$\frac{\pi}{4}$	$\frac{\pi}{3}$	$\frac{\pi}{2}$	$\frac{2\pi}{3}$	$\frac{3\pi}{4}$	$\frac{5\pi}{6}$	π
Degrees		210°	225°	240°	270°	300°	315°	330°	360°
Radians		$\frac{7\pi}{6}$	$\frac{5\pi}{4}$	$\frac{4\pi}{3}$	$\frac{3\pi}{2}$	$\frac{5\pi}{3}$	$\frac{7\pi}{4}$	$\frac{11\pi}{6}$	2π

EXAMPLE 6 Finding the Distance between Two Cities

See Figure 13(a). The latitude of a location L is the angle formed by a ray drawn from the center of Earth to the Equator and a ray drawn from the center of Earth to L. See Figure 13(b). Glasgow, Montana, is due north of Albuquerque, New Mexico. Find the distance between Glasgow (48°9′ north latitude) and Albuquerque (35°5′ north latitude). Assume that the radius of Earth is 3960 miles.

Figure 13

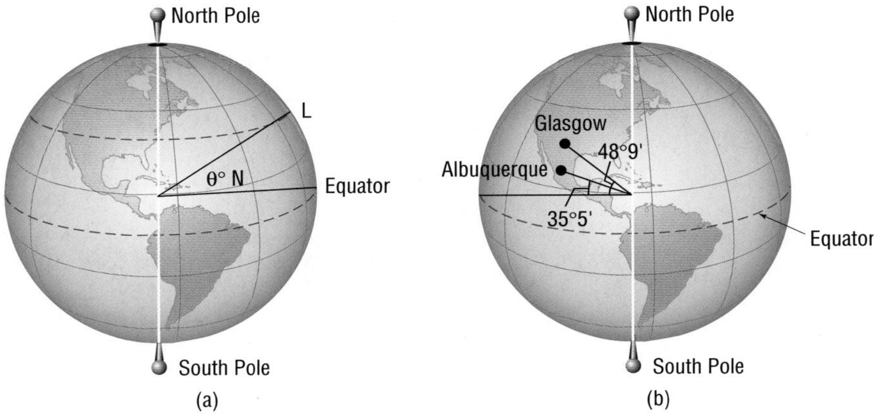

Solution The measure of the central angle between the two cities is $48°9' - 35°5' = 13°4'$. We use equation (4), $s = r\theta$, but first we must convert the angle of 13°4′ to radians.

$$\theta = 13°4' \approx 13.0667° = 13.0667 \cdot \frac{\pi}{180} \text{ radian} \approx 0.228 \text{ radian}$$

We use $\theta = 0.228$ radian and $r = 3960$ miles in equation (4). The distance between the two cities is

$$s = r\theta = 3960 \cdot 0.228 \approx 903 \text{ miles}$$ ◄

When an angle is measured in degrees, the degree symbol will always be shown. However, when an angle is measured in radians, we will follow the usual practice and omit the word *radians*. So, if the measure of an angle is given as $\frac{\pi}{6}$, it is understood to mean $\frac{\pi}{6}$ radian.

NOW WORK PROBLEM 101.

Figure 14

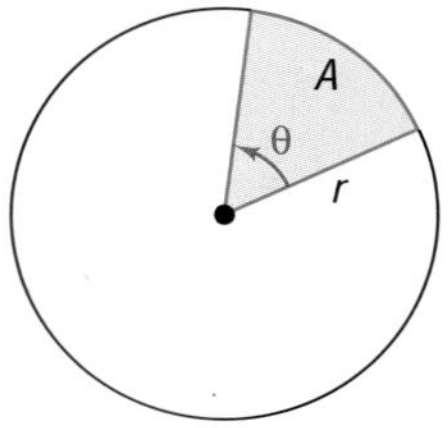

Figure 15

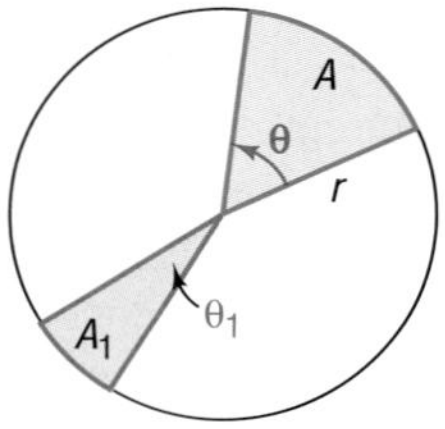

Area of a Sector

5 Consider a circle of radius r. Suppose that θ, measured in radians, is a central angle of this circle. See Figure 14. We seek a formula for the area A of the sector formed by the angle θ (shown in blue).

Now consider a circle of radius r and two central angles θ and θ_1, both measured in radians. See Figure 15. From geometry, we know the ratio of the measures of the angles equals the ratio of the corresponding areas of the sectors formed by these angles. That is,

$$\frac{\theta}{\theta_1} = \frac{A}{A_1}$$

Suppose that $\theta_1 = 2\pi$ radians. Then A_1 = area of the circle $= \pi r^2$. Solving for A, we find

$$A = A_1\frac{\theta}{\theta_1} = \pi r^2\frac{\theta}{2\pi} = \frac{1}{2}r^2\theta$$

Theorem

Area of a Sector

The area A of the sector of a circle of radius r formed by a central angle of θ radians is

$$A = \frac{1}{2}r^2\theta \quad \textbf{(8)}$$

EXAMPLE 7 **Finding the Area of a Sector of a Circle**

Find the area of the sector of a circle of radius 2 feet formed by an angle of 30°. Round the answer to two decimal places.

Solution We use equation (8) with $r = 2$ feet and $\theta = 30° = \frac{\pi}{6}$ radians. [Remember, in equation (8), θ must be in radians.] The area A of the sector is

$$A = \frac{1}{2}r^2\theta = \frac{1}{2}(2)^2\frac{\pi}{6} = \frac{\pi}{3} \text{ square feet} \approx 1.05 \text{ square feet}$$

rounded to two decimal places. ◄

NOW WORK PROBLEM 79.

Circular Motion

6 We have already defined the average speed of an object as the distance traveled divided by the elapsed time. Suppose that an object moves around a circle of radius r at a constant speed. If s is the distance traveled in time t around this circle, then the **linear speed** v of the object is defined as

$$v = \frac{s}{t} \tag{9}$$

Figure 16

$v = \frac{s}{t} \quad \omega = \frac{\theta}{t}$

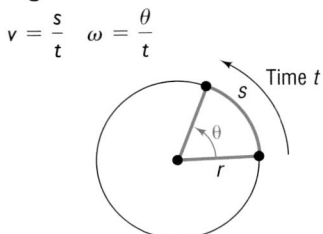

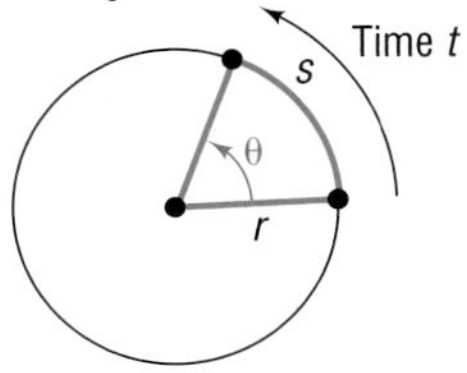

As this object travels around the circle, suppose that θ (measured in radians) is the central angle swept out in time t. See Figure 16. Then the **angular speed** ω (the Greek letter omega) of this object is the angle (measured in radians) swept out divided by the elapsed time; that is,

$$\omega = \frac{\theta}{t} \tag{10}$$

Angular speed is the way the turning rate of an engine is described. For example, an engine idling at 900 rpm (revolutions per minute) is one that rotates at an angular speed of

$$900\frac{\text{revolutions}}{\text{minute}} = 900\frac{\cancel{\text{revolutions}}}{\text{minute}} \cdot 2\pi\frac{\text{radians}}{\cancel{\text{revolution}}} = 1800\pi\frac{\text{radians}}{\text{minute}}$$

There is an important relationship between linear speed and angular speed:

$$\text{linear speed} = v = \underset{\substack{\uparrow \\ (9)}}{\frac{s}{t}} \underset{\substack{\uparrow \\ s = r\theta}}{=} \frac{r\theta}{t} = r\left(\frac{\theta}{t}\right)$$

Then, using equation (10), we obtain

$$v = r\omega \tag{11}$$

where ω is measured in radians per unit time.

When using equation (11), remember that $v = \frac{s}{t}$ (the linear speed) has the dimensions of length per unit of time (such as feet per second or miles per hour), r (the radius of the circular motion) has the same length dimension as s, and ω (the angular speed) has the dimensions of radians per unit of time. If the angular speed is given in terms of *revolutions* per unit of time (as is often the case), be sure to convert it to *radians* per unit of time before attempting to use equation (11).

EXAMPLE 8 Finding Linear Speed

A child is spinning a rock at the end of a 2-foot rope at the rate of 180 revolutions per minute (rpm). Find the linear speed of the rock when it is released.

Figure 17

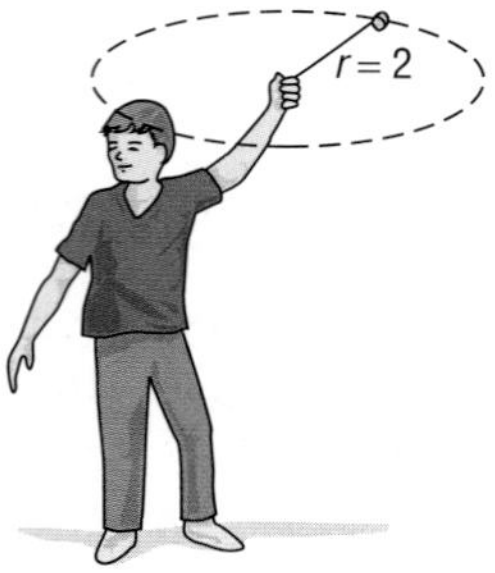

Solution Look at Figure 17. The rock is moving around a circle of radius $r = 2$ feet. The angular speed ω of the rock is

$$\omega = 180\frac{\text{revolutions}}{\text{minute}} = 180\frac{\cancel{\text{revolutions}}}{\text{minute}}\cdot 2\pi\frac{\text{radians}}{\cancel{\text{revolution}}} = 360\pi\frac{\text{radians}}{\text{minute}}$$

From equation (11), the linear speed v of the rock is

$$v = r\omega = 2\text{ feet}\cdot 360\pi\frac{\text{radians}}{\text{minute}} = 720\pi\frac{\text{feet}}{\text{minute}} \approx 2262\frac{\text{feet}}{\text{minute}}$$

The linear speed of the rock when it is released is $2262\text{ ft/min} \approx 25.7\text{ mi/hr}$. ◀

NOW WORK PROBLEM 97.

HISTORICAL FEATURE

Trigonometry was developed by Greek astronomers, who regarded the sky as the inside of a sphere, so it was natural that triangles on a sphere were investigated early (by Menelaus of Alexandria about AD 100) and that triangles in the plane were studied much later. The first book containing a systematic treatment of plane and spherical trigonometry was written by the Persian astronomer Nasîr Eddîn (about AD 1250).

Regiomontanus (1436–1476) is the person most responsible for moving trigonometry from astronomy into mathematics. His work was improved by Copernicus (1473–1543) and Copernicus's student Rhaeticus (1514–1576). Rhaeticus's book was the first to define the six trigonometric functions as ratios of sides of triangles, although he did not give the functions their present names. Credit for this is due to Thomas Finck (1583), but Finck's notation was by no means universally accepted at the time. The notation was finally stabilized by the textbooks of Leonhard Euler (1707–1783).

Trigonometry has since evolved from its use by surveyors, navigators, and engineers to present applications involving ocean tides, the rise and fall of food supplies in certain ecologies, brain wave patterns, and many other phenomena.

5.1 Assess Your Understanding

'Are You Prepared?' *Answers are given at the end of these exercises. If you get a wrong answer, read the pages listed in red.*

1. What is the formula for the circumference C of a circle of radius r? (p. 915)

2. What is the formula for the area A of a circle of radius r? (p. 915)

Concepts and Vocabulary

3. An angle θ is in __________ __________ if its vertex is at the origin of a rectangular coordinate system and its initial side coincides with the positive x-axis.

4. On a circle of radius r, a central angle of θ radians subtends an arc of length $s =$ __________; the area of the sector formed by this angle θ is $A =$ __________.

5. An object travels around a circle of radius r with constant speed. If s is the distance traveled in time t around the circle and θ is the central angle (in radians) swept out in time t, then the linear speed of the object is $v =$ __________ and the angular speed of the object is $w =$ __________.

6. *True or False:* $\pi = 180$.

7. *True or False:* $180° = \pi$ radians.

8. *True or False:* On the unit circle, if s is the length of the arc subtended by a central angle θ, measured in radians, then $s = \theta$.

9. *True or False:* The area A of the sector of a circle of radius r formed by a central angle of θ degrees is $A = \frac{1}{2}r^2\theta$.

10. *True or False:* For circular motion on a circle of radius r, linear speed equals angular speed divided by r.

Exercises

In Problems 11–22, draw each angle.

11. 30° **12.** 60° **13.** 135° **14.** −120° **15.** 450° **16.** 540°

17. $\frac{3\pi}{4}$ **18.** $\frac{4\pi}{3}$ **19.** $-\frac{\pi}{6}$ **20.** $-\frac{2\pi}{3}$ **21.** $\frac{16\pi}{3}$ **22.** $\frac{21\pi}{4}$

In Problems 23–28, convert each angle to a decimal in degrees. Round your answer to two decimal places.

23. 40°10′25″ **24.** 61°42′21″ **25.** 1°2′3″ **26.** 73°40′40″ **27.** 9°9′9″ **28.** 98°22′45″

In Problems 29–34, convert each angle to D°M′S″ form. Round your answer to the nearest second.

29. 40.32° **30.** 61.24° **31.** 18.255° **32.** 29.411° **33.** 19.99° **34.** 44.01°

In Problems 35–46, convert each angle in degrees to radians. Express your answer as a multiple of π.

35. 30° **36.** 120° **37.** 240° **38.** 330° **39.** −60° **40.** −30°

41. 180° **42.** 270° **43.** −135° **44.** −225° **45.** −90° **46.** −180°

In Problems 47–58, convert each angle in radians to degrees.

47. $\frac{\pi}{3}$ **48.** $\frac{5\pi}{6}$ **49.** $-\frac{5\pi}{4}$ **50.** $-\frac{2\pi}{3}$ **51.** $\frac{\pi}{2}$ **52.** 4π

53. $\frac{\pi}{12}$ **54.** $\frac{5\pi}{12}$ **55.** $-\frac{\pi}{2}$ **56.** $-\pi$ **57.** $-\frac{\pi}{6}$ **58.** $-\frac{3\pi}{4}$

In Problems 59–64, convert each angle in degrees to radians. Express your answer in decimal form, rounded to two decimal places.

59. 17° **60.** 73° **61.** −40° **62.** −51° **63.** 125° **64.** 350°

In Problems 65–70, convert each angle in radians to degrees. Express your answer in decimal form, rounded to two decimal places.

65. 3.14 **66.** 0.75 **67.** 2 **68.** 3 **69.** 6.32 **70.** $\sqrt{2}$

In Problems 71–78, s denotes the length of the arc of a circle of radius r subtended by the central angle θ. Find the missing quantity. Round answers to three decimal places.

71. $r = 10$ meters, $\theta = \frac{1}{2}$ radian, $s = ?$

72. $r = 6$ feet, $\theta = 2$ radians, $s = ?$

73. $\theta = \frac{1}{3}$ radian, $s = 2$ feet, $r = ?$

74. $\theta = \frac{1}{4}$ radian, $s = 6$ centimeters, $r = ?$

75. $r = 5$ miles, $s = 3$ miles, $\theta = ?$

76. $r = 6$ meters, $s = 8$ meters, $\theta = ?$

77. $r = 2$ inches, $\theta = 30°$, $s = ?$

78. $r = 3$ meters, $\theta = 120°$, $s = ?$

In Problems 79–86, A denotes the area of the sector of a circle of radius r formed by the central angle θ. Find the missing quantity. Round answers to three decimal places.

79. $r = 10$ meters, $\theta = \frac{1}{2}$ radian, $A = ?$

80. $r = 6$ feet, $\theta = 2$ radians, $A = ?$

81. $\theta = \frac{1}{3}$ radian, $A = 2$ square feet, $r = ?$

82. $\theta = \frac{1}{4}$ radian, $A = 6$ square centimeters, $r = ?$

83. $r = 5$ miles, $A = 3$ square miles, $\theta = ?$

84. $r = 6$ meters, $A = 8$ square meters, $\theta = ?$

85. $r = 2$ inches, $\theta = 30°$, $A = ?$

86. $r = 3$ meters, $\theta = 120°$, $A = ?$

In Problems 87–90, find the length s and area A. Round answers to three decimal places.

87.

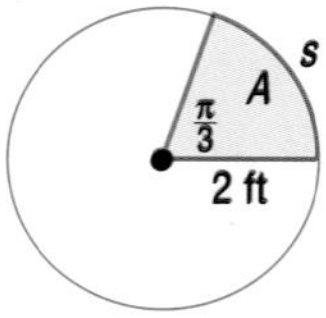

88.

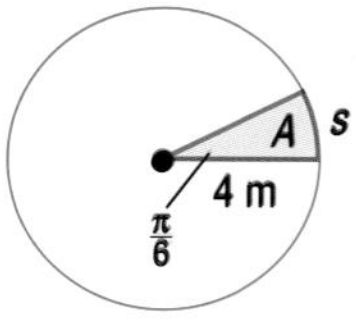

89.

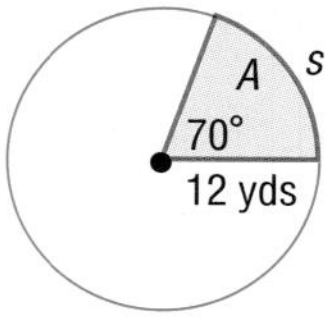

90.

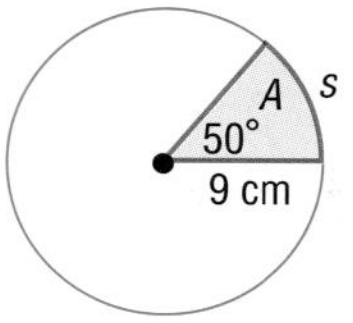

91. Minute Hand of a Clock The minute hand of a clock is 6 inches long. How far does the tip of the minute hand move in 15 minutes? How far does it move in 25 minutes?

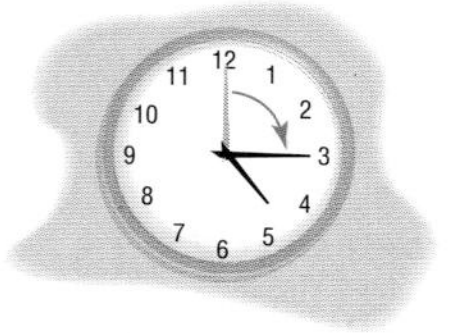

92. Movement of a Pendulum A pendulum swings through an angle of 20° each second. If the pendulum is 40 inches long, how far does its tip move each second?

93. Area of a Sector Find the area of the sector of a circle of radius 4 meters formed by an angle of 45°. Round the answer to two decimal places.

94. Area of a Sector Find the area of the sector of a circle of radius 3 centimeters formed by an angle of 60°. Round the answer to two decimal places.

95. Watering a Lawn A water sprinkler sprays water over a distance of 30 feet while rotating through an angle of 135°. What area of lawn receives water?

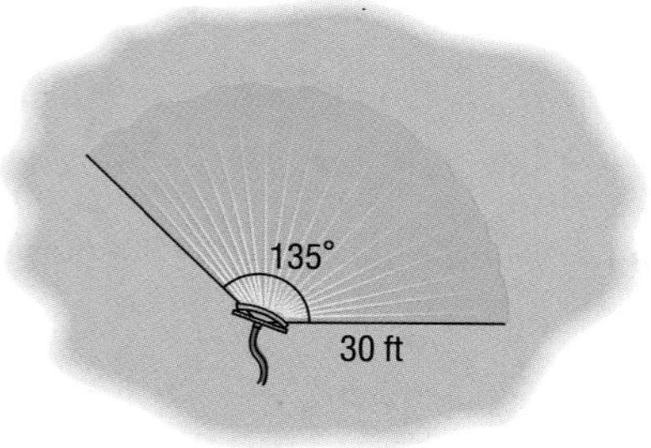

96. Designing a Water Sprinkler An engineer is asked to design a water sprinkler that will cover a field of 100 square yards that is in the shape of a sector of a circle of radius 50 yards. Through what angle should the sprinkler rotate?

97. Motion on a Circle An object is traveling around a circle with a radius of 5 centimeters. If in 20 seconds a central angle of $\frac{1}{3}$ radian is swept out, what is the angular speed of the object? What is its linear speed?

98. Motion on a Circle An object is traveling around a circle with a radius of 2 meters. If in 20 seconds the object travels 5 meters, what is its angular speed? What is its linear speed?

99. Bicycle Wheels The diameter of each wheel of a bicycle is 26 inches. If you are traveling at a speed of 35 miles per hour on this bicycle, through how many revolutions per minute are the wheels turning?

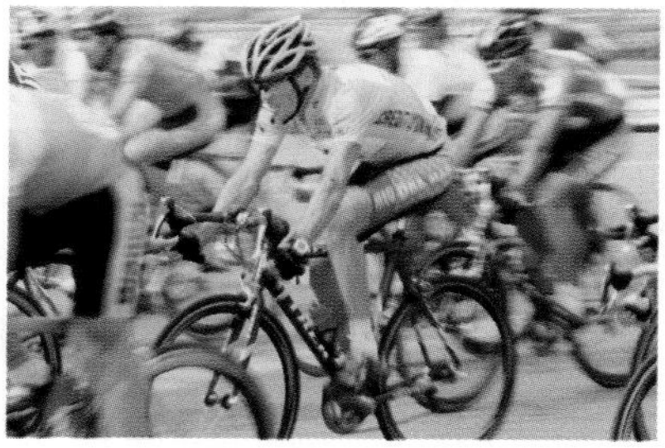

100. Car Wheels The radius of each wheel of a car is 15 inches. If the wheels are turning at the rate of 3 revolutions per second, how fast is the car moving? Express your answer in inches per second and in miles per hour.

In Problems 101–104, the latitude of a location L is the angle formed by a ray drawn from the center of Earth to the Equator and a ray drawn from the center of Earth to L. See the figure.

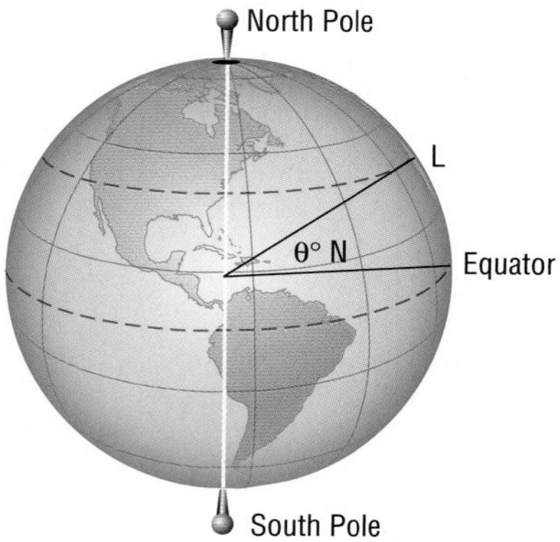

101. Distance between Cities Memphis, Tennessee, is due north of New Orleans, Louisiana. Find the distance between Memphis (35°9′ north latitude) and New Orleans (29°57′ north latitude). Assume that the radius of Earth is 3960 miles.

102. Distance between Cities Charleston, West Virginia, is due north of Jacksonville, Florida. Find the distance between Charleston (38°21′ north latitude) and Jacksonville (30°20′ north latitude). Assume that the radius of Earth is 3960 miles.

103. Linear Speed on Earth Earth rotates on an axis through its poles. The distance from the axis to a location on Earth 30° north latitude is about 3429.5 miles. Therefore, a location on Earth at 30° north latitude is spinning on a circle of radius 3429.5 miles. Compute the linear speed on the surface of Earth at 30° north latitude.

104. Linear Speed on Earth Earth rotates on an axis through its poles. The distance from the axis to a location on Earth 40° north latitude is about 3033.5 miles. Therefore, a location on Earth at 40° north latitude is spinning on a circle of radius 3033.5 miles. Compute the linear speed on the surface of Earth at 40° north latitude.

105. Speed of the Moon The mean distance of the Moon from Earth is 2.39×10^5 miles. Assuming that the orbit of the Moon around Earth is circular and that 1 revolution takes 27.3 days, find the linear speed of the Moon. Express your answer in miles per hour.

106. Speed of Earth The mean distance of Earth from the Sun is 9.29×10^7 miles. Assuming that the orbit of Earth around the Sun is circular and that 1 revolution takes 365 days, find the linear speed of Earth. Express your answer in miles per hour.

107. Pulleys Two pulleys, one with radius 2 inches and the other with radius 8 inches, are connected by a belt. (See the figure.) If the 2-inch pulley is caused to rotate at 3 revolutions per minute, determine the revolutions per minute of the 8-inch pulley.

[**Hint:** The linear speeds of the pulleys are the same, both equal the speed of the belt.]

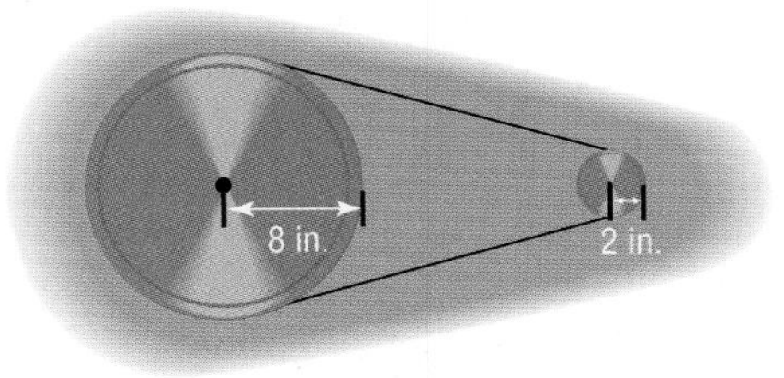

108. Ferris Wheels A neighborhood carnival has a Ferris wheel whose radius is 30 feet. You measure the time it takes for one revolution to be 70 seconds. What is the linear speed (in feet per second) of this Ferris wheel? What is the angular speed in radians per second?

109. Computing the Speed of a River Current To approximate the speed of the current of a river, a circular paddle wheel with radius 4 feet is lowered into the water. If the current causes the wheel to rotate at a speed of 10 revolutions per minute, what is the speed of the current? Express your answer in miles per hour.

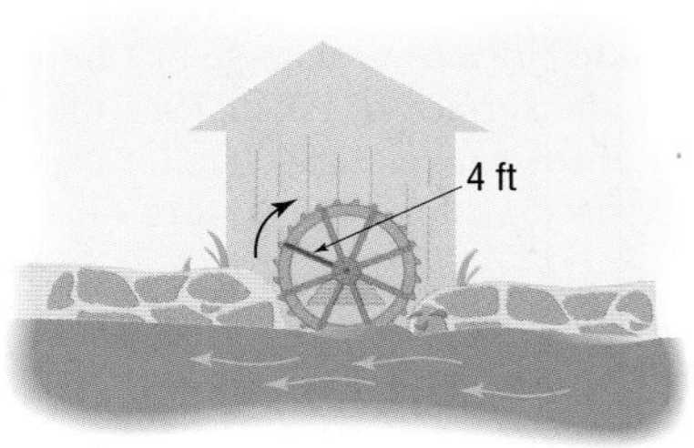

110. Spin Balancing Tires A spin balancer rotates the wheel of a car at 480 revolutions per minute. If the diameter of the wheel is 26 inches, what road speed is being tested? Express your answer in miles per hour. At how many revolutions per minute should the balancer be set to test a road speed of 80 miles per hour?

111. The Cable Cars of San Francisco At the Cable Car Museum you can see the four cable lines that are used to pull cable cars up and down the hills of San Francisco. Each cable travels at a speed of 9.55 miles per hour, caused by a rotating wheel whose diameter is 8.5 feet. How fast is the wheel rotating? Express your answer in revolutions per minute.

112. Difference in Time of Sunrise Naples, Florida, is approximately 90 miles due west of Ft. Lauderdale. How much sooner would a person in Ft. Lauderdale first see the rising Sun than a person in Naples?

[**Hint:** Consult the figure. When a person at Q sees the first rays of the Sun, a person at P is still in the dark. The person at P sees the first rays after Earth has rotated so that P is at the location Q. Now use the fact that at the latitude of Ft. Lauderdale in 24 hours a length of arc of $2\pi(3559)$ miles is subtended.]

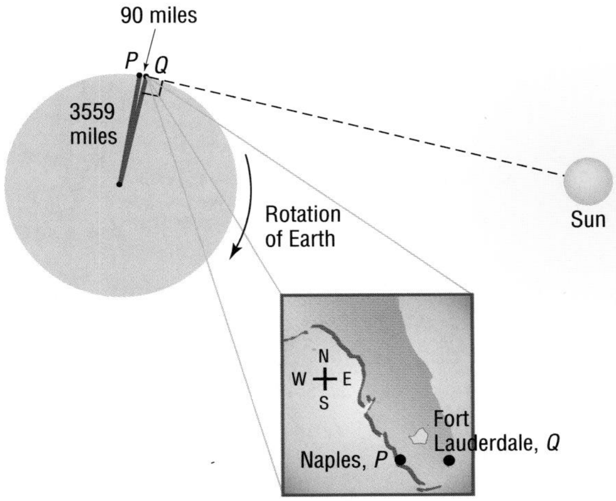

113. Keeping Up with the Sun How fast would you have to travel on the surface of Earth at the equator to keep up with the Sun (that is, so that the Sun would appear to remain in the same position in the sky)?

114. **Nautical Miles** A **nautical mile** equals the length of arc subtended by a central angle of 1 minute on a great circle* on the surface of Earth. (See the figure.) If the radius of Earth is taken as 3960 miles, express 1 nautical mile in terms of ordinary, or **statute,** miles.

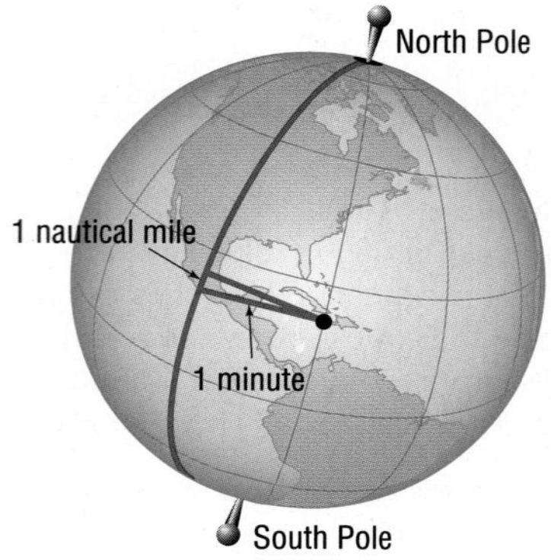

115. **Pulleys** Two pulleys, one with radius r_1 and the other with radius r_2, are connected by a belt. The pulley with radius r_1 rotates at ω_1 revolutions per minute, whereas the pulley with radius r_2 rotates at ω_2 revolutions per minute. Show that $\dfrac{r_1}{r_2} = \dfrac{\omega_2}{\omega_1}$.

116. Do you prefer to measure angles using degrees or radians? Provide justification and a rationale for your choice.

117. What is 1 radian?

118. Which angle has the larger measure: 1 degree or 1 radian? Or are they equal?

119. Explain the difference between linear speed and angular speed.

120. For a circle of radius r, a central angle of θ degrees subtends an arc whose length s is $s = \dfrac{\pi}{180} r\theta$. Discuss whether this is a true or false statement. Give reasons to defend your position.

121. Discuss why ships and airplanes use nautical miles to measure distance. Explain the difference between a nautical mile and a statute mile.

122. Investigate the way that speed bicycles work. In particular, explain the differences and similarities between 5-speed and 9-speed derailleurs. Be sure to include a discussion of linear speed and angular speed.

'Are You Prepared?' Answers

1. $C = 2\pi r$ **2.** $A = \pi r^2$

*Any circle drawn on the surface of Earth that divides Earth into two equal hemispheres.

5.2 Trigonometric Functions: Unit Circle Approach

PREPARING FOR THIS SECTION *Before getting started, review the following:*

- Pythagorean Theorem (Appendix A, Section A.2, pp. 913–915)
- Unit Circle (Section 1.2, p. 16)
- Symmetry (Section 1.2, pp. 12–14)
- Functions (Section 2.1, pp. 50–58)

Now work the 'Are You Prepared?' problems on page 352.

OBJECTIVES
1 Find the Exact Values of the Trigonometric Functions Using a Point on the Unit Circle
2 Find the Exact Values of the Trigonometric Functions of Quadrantal Angles
3 Find the Exact Values of the Trigonometric Functions of $\dfrac{\pi}{4} = 45°$
4 Find the Exact Values of the Trigonometric Functions of $\dfrac{\pi}{6} = 30°$ and $\dfrac{\pi}{3} = 60°$
5 Find the Exact Values for Certain Integral Multiples of $\dfrac{\pi}{6} = 30°$, $\dfrac{\pi}{4} = 45°$, and $\dfrac{\pi}{3} = 60°$
6 Use a Calculator to Approximate the Value of a Trigonometric Function

We are now ready to introduce trigonometric functions. The approach that we take uses the unit circle.

The Unit Circle

Recall that the unit circle is a circle whose radius is 1 and whose center is at the origin of a rectangular coordinate system. Also recall that any circle of radius r has circumference of length $2\pi r$. Therefore, the unit circle (radius $= 1$) has a circumference of length 2π. In other words, for 1 revolution around the unit circle the length of the arc is 2π units.

The following discussion sets the stage for defining the trigonometric functions.

Let t be any real number. We position the t-axis so it is vertical with positive direction up. We place this t-axis in the xy-plane, so that $t = 0$ is located at the point $(1, 0)$ in the xy-plane.

If $t \geq 0$, let s be the distance from the origin to t on the t-axis. See the red portion of Figure 18(a).

Now look at the unit circle in Figure 18(a). Beginning at the point $(1, 0)$ on the unit circle, travel $s = t$ units in the counterclockwise direction along the circle, to arrive at the point $P = (x, y)$. In this sense, the length $s = t$ units is being **wrapped** around the unit circle.

If $t < 0$, we begin at the point $(1, 0)$ on the unit circle and travel $s = |t|$ units in the clockwise direction to arrive at the point $P = (x, y)$. See Figure 18(b).

Figure 18

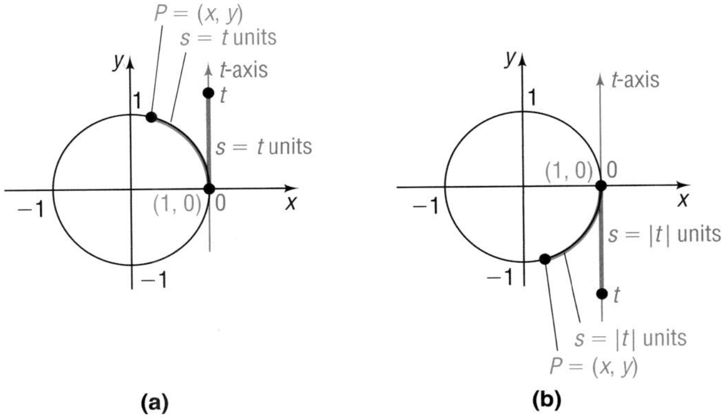

If $t > 2\pi$ or if $t < -2\pi$, it will be necessary to travel around the unit circle more than once before arriving at the point P. Do you see why?

Let's describe this process another way. Picture a string of length $s = |t|$ units being wrapped around a circle of radius 1 unit. We start wrapping the string around the circle at the point $(1, 0)$. If $t \geq 0$, we wrap the string in the counterclockwise direction; if $t < 0$, we wrap the string in the clockwise direction. The point $P = (x, y)$ is the point where the string ends.

This discussion tells us that, for any real number t, we can locate a unique point $P = (x, y)$ on the unit circle. We call P **the point on the unit circle that corresponds to *t*.** This is the important idea here. No matter what real number t is chosen, there is a unique point P on the unit circle corresponding to it. We use the coordinates of the point $P = (x, y)$ on the unit circle corresponding to the real number t to define the **six trigonometric functions of *t*.**

Let t be a real number and let $P = (x, y)$ be the point on the unit circle that corresponds to t.

The **sine function** associates with t the y-coordinate of P and is denoted by

$$\sin t = y$$

The **cosine function** associates with t the x-coordinate of P and is denoted by

$$\cos t = x$$

If $x \neq 0$, the **tangent function** is defined as

$$\tan t = \frac{y}{x}$$

If $y \neq 0$, the **cosecant function** is defined as

$$\csc t = \frac{1}{y}$$

If $x \neq 0$, the **secant function** is defined as

$$\sec t = \frac{1}{x}$$

If $y \neq 0$, the **cotangent function** is defined as

$$\cot t = \frac{x}{y}$$

Notice in these definitions that if $x = 0$, that is, if the point $P = (0, y)$ is on the y-axis, then the tangent function and the secant function are undefined. Also, if $y = 0$, that is, if the point $P = (x, 0)$ is on the x-axis, then the cosecant function and the cotangent function are undefined.

Because we use the unit circle in these definitions of the trigonometric functions, they are also sometimes referred to as **circular functions.**

1 **EXAMPLE 1**

Finding the Values of the Six Trigonometric Functions Using a Point on the Unit Circle

Let t be a real number and let $P = \left(-\frac{1}{2}, \frac{\sqrt{3}}{2}\right)$ be the point on the unit circle that corresponds to t. Find the values of $\sin t$, $\cos t$, $\tan t$, $\csc t$, $\sec t$, and $\cot t$.

Solution See Figure 19. We follow the definition of the six trigonometric functions, using $P = \left(-\frac{1}{2}, \frac{\sqrt{3}}{2}\right) = (x, y)$. Then, with $x = -\frac{1}{2}$, $y = \frac{\sqrt{3}}{2}$, we have

Figure 19

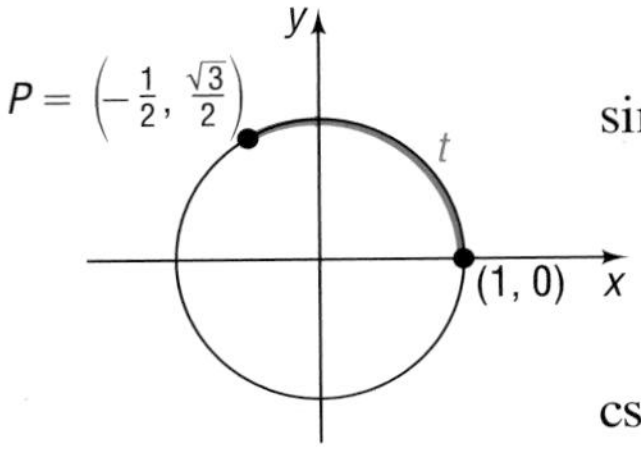

$$\sin t = y = \frac{\sqrt{3}}{2} \qquad \cos t = x = -\frac{1}{2} \qquad \tan t = \frac{y}{x} = \frac{\frac{\sqrt{3}}{2}}{-\frac{1}{2}} = -\sqrt{3}$$

$$\csc t = \frac{1}{y} = \frac{1}{\frac{\sqrt{3}}{2}} = \frac{2\sqrt{3}}{3} \qquad \sec t = \frac{1}{x} = \frac{1}{-\frac{1}{2}} = -2 \qquad \cot t = \frac{x}{y} = \frac{-\frac{1}{2}}{\frac{\sqrt{3}}{2}} = -\frac{\sqrt{3}}{3}$$

◄

NOW WORK PROBLEM 11.

Trigonometric Functions of Angles

Let $P = (x, y)$ be the point on the unit circle corresponding to the real number t. See Figure 20(a). Let θ be the angle in standard position, measured in radians, whose terminal side is the ray from the origin through P. See Figure 20(b). Since the unit circle has radius 1 unit, from the formula for arc length, $s = r\theta$, we find that

$$s = r\theta = \theta$$
$$\uparrow$$
$$r = 1$$

So, if $s = |t|$ units, then $\theta = t$ radians. See Figures 20(c) and (d).

Figure 20

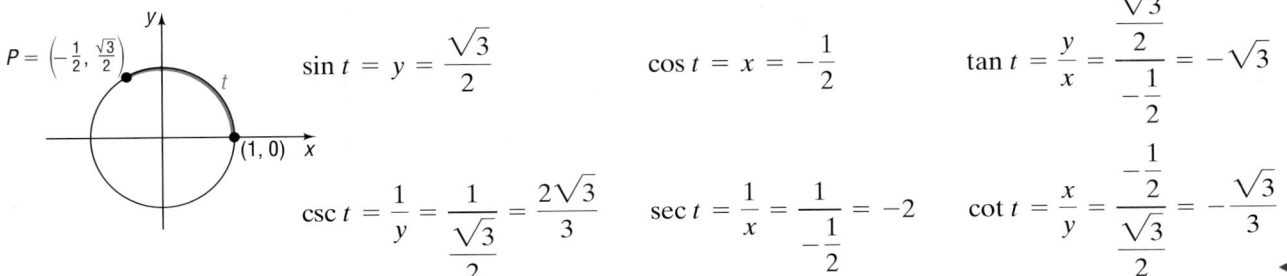

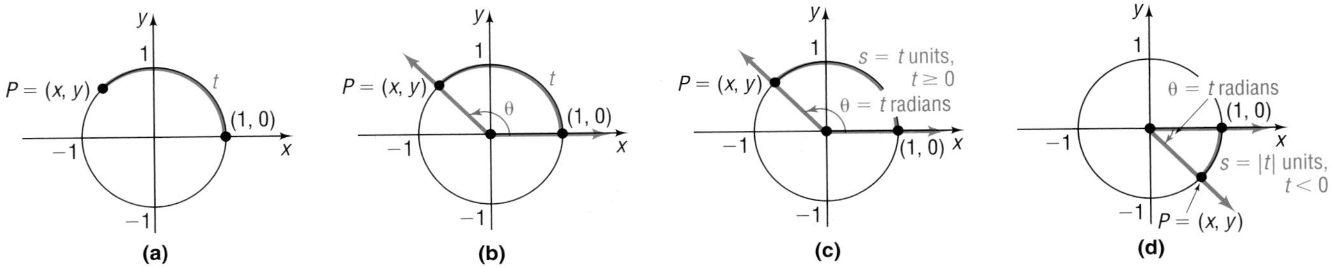

The point $P = (x, y)$ on the unit circle that corresponds to the real number t is the point P on the terminal side of the angle $\theta = t$ radians. As a result, we can say that

$$\underset{\text{Real number}}{\sin t} = \underset{\theta = t \text{ radians}}{\sin \theta}$$

and so on. We can now define the trigonometric functions of the angle θ.

If $\theta = t$ radians, the **six trigonometric functions of the angle θ** are defined as

$\sin \theta = \sin t$	$\cos \theta = \cos t$	$\tan \theta = \tan t$
$\csc \theta = \csc t$	$\sec \theta = \sec t$	$\cot \theta = \cot t$

Even though the distinction between trigonometric functions of real numbers and trigonometric functions of angles is important, it is customary to refer to trigonometric functions of real numbers and trigonometric functions of angles collectively as the *trigonometric functions*. We shall follow this practice from now on.

If an angle θ is measured in degrees, we shall use the degree symbol when writing a trigonometric function of θ, as, for example, in $\sin 30°$ and $\tan 45°$. If an angle θ is measured in radians, then no symbol is used when writing a trigonometric function of θ, as, for example, in $\cos \pi$ and $\sec \frac{\pi}{3}$.

Finally, since the values of the trigonometric functions of an angle θ are determined by the coordinates of the point $P = (x, y)$ on the unit circle corresponding to θ, the units used to measure the angle θ are irrelevant. For example, it does not matter whether we write $\theta = \frac{\pi}{2}$ radians or $\theta = 90°$. The point on the unit circle corresponding to this angle is $P = (0, 1)$. Hence,

$$\sin \frac{\pi}{2} = \sin 90° = 1 \quad \text{and} \quad \cos \frac{\pi}{2} = \cos 90° = 0$$

Evaluating the Trigonometric Functions

2 To find the exact value of a trigonometric function of an angle θ or a real number t requires that we locate the point $P = (x, y)$ on the unit circle that corresponds to t. This is not always easy to do. In the examples that follow, we will evaluate the trigonometric functions of certain angles or real numbers for which this process is relatively easy. A calculator will be used to evaluate the trigonometric functions of most other angles.

EXAMPLE 2 **Finding the Exact Values of the Six Trigonometric Functions of Quadrantal Angles**

Find the exact values of the six trigonometric functions of:

(a) $\theta = 0 = 0°$ (b) $\theta = \frac{\pi}{2} = 90°$

(c) $\theta = \pi = 180°$ (d) $\theta = \frac{3\pi}{2} = 270°$

Figure 21(a)

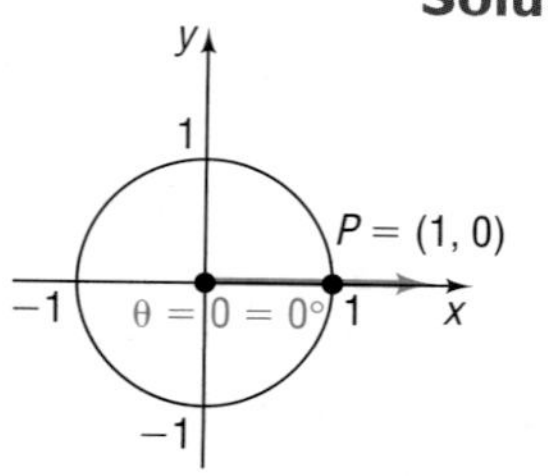

(a)

Solution (a) The point on the unit circle that corresponds to $\theta = 0 = 0°$ is $P = (1, 0)$. See Figure 21(a). Then

$$\sin 0 = \sin 0° = y = 0 \qquad \cos 0 = \cos 0° = x = 1$$

$$\tan 0 = \tan 0° = \frac{y}{x} = 0 \qquad \sec 0 = \sec 0° = \frac{1}{x} = 1$$

Since the y-coordinate of P is 0, csc 0 and cot 0 are not defined.

(b) The point on the unit circle that corresponds to $\theta = \frac{\pi}{2} = 90°$ is $P = (0, 1)$. See Figure 21(b). Then

$$\sin\frac{\pi}{2} = \sin 90° = y = 1 \qquad \cos\frac{\pi}{2} = \cos 90° = x = 0$$

$$\csc\frac{\pi}{2} = \csc 90° = \frac{1}{y} = 1 \qquad \cot\frac{\pi}{2} = \cot 90° = \frac{x}{y} = 0$$

Since the x-coordinate of P is 0, $\tan\frac{\pi}{2}$ and $\sec\frac{\pi}{2}$ are not defined.

Figure 21(b)

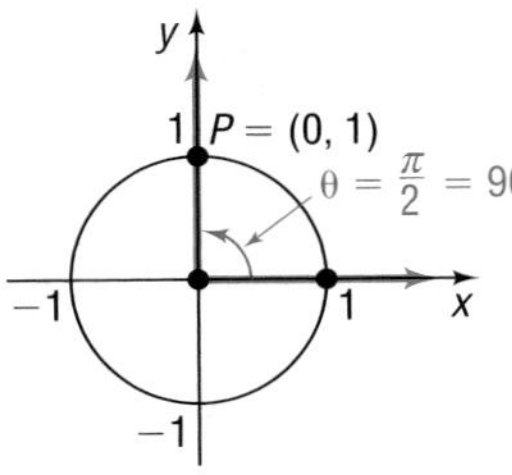

(b)

(c) The point on the unit circle that corresponds to $\theta = \pi = 180°$ is $P = (-1, 0)$. See Figure 21(c). Then

$$\sin\pi = \sin 180° = y = 0 \qquad \cos\pi = \cos 180° = x = -1$$

$$\tan\pi = \tan 180° = \frac{y}{x} = 0 \qquad \sec\pi = \sec 180° = \frac{1}{x} = -1$$

Since the y-coordinate of P is 0, $\csc\pi$ and $\cot\pi$ are not defined.

Figure 21(c)

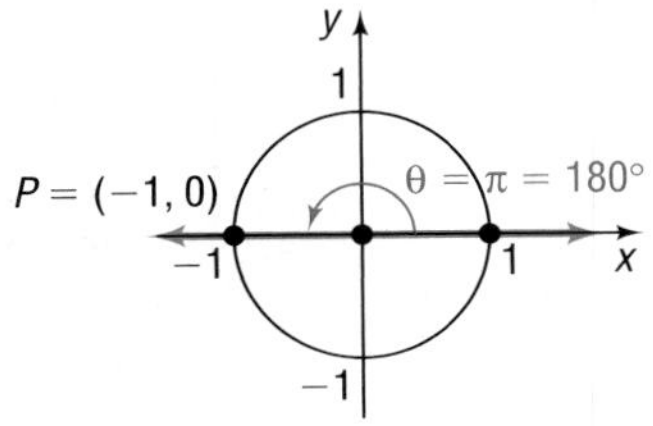

(c)

(d) The point on the unit circle that corresponds to $\theta = \frac{3\pi}{2} = 270°$ is $P = (0, -1)$. See Figure 21(d). Then

$$\sin\frac{3\pi}{2} = \sin 270° = y = -1 \qquad \cos\frac{3\pi}{2} = \cos 270° = x = 0$$

$$\csc\frac{3\pi}{2} = \csc 270° = \frac{1}{y} = -1 \qquad \cot\frac{3\pi}{2} = \cot 270° = \frac{x}{y} = 0$$

Since the x-coordinate of P is 0, $\tan\frac{3\pi}{2}$ and $\sec\frac{3\pi}{2}$ are not defined. ◀

Figure 21(d)

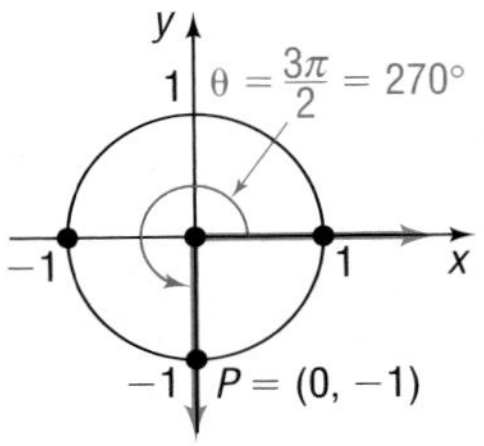

(d)

Table 2 summarizes the values of the trigonometric functions found in Example 2.

Table 2

Quadrantal Angles

θ (Radians)	θ (Degrees)	$\sin\theta$	$\cos\theta$	$\tan\theta$	$\csc\theta$	$\sec\theta$	$\cot\theta$
0	0°	0	1	0	Not defined	1	Not defined
$\frac{\pi}{2}$	90°	1	0	Not defined	1	Not defined	0
π	180°	0	−1	0	Not defined	−1	Not defined
$\frac{3\pi}{2}$	270°	−1	0	Not defined	−1	Not defined	0

There is no need to memorize Table 2. To find the value of a trigonometric function of a quadrantal angle, draw the angle and apply the definition, as we did in Example 2.

EXAMPLE 3 **Finding Exact Values of the Trigonometric Functions of Angles That Are Integral Multiples of Quadrantal Angles**

Find the exact value of:

(a) $\sin(3\pi)$ (b) $\cos(-270^\circ)$

Solution (a) See Figure 22. The point P on the unit circle that corresponds to $\theta = 3\pi$ is $P = (-1, 0)$, so $\sin(3\pi) = 0$.

(b) See Figure 23. The point P on the unit circle that corresponds to $\theta = -270^\circ$ is $P = (0, 1)$, so $\cos(-270^\circ) = 0$.

Figure 22

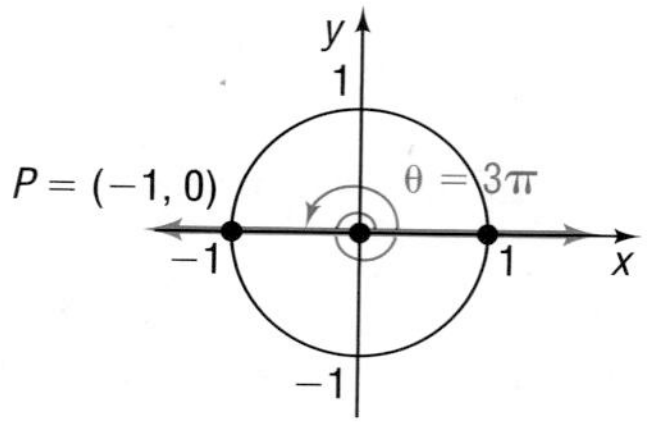

Figure 23

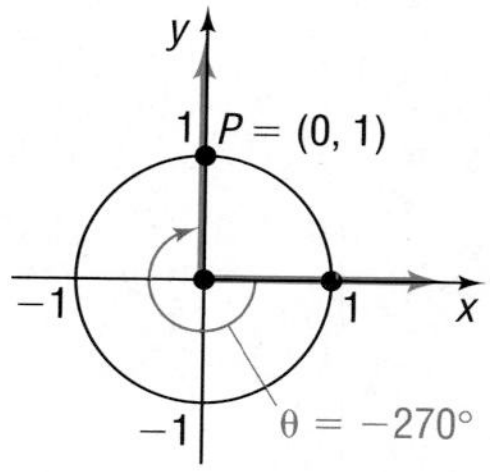

◀

NOW WORK PROBLEMS 19 AND 63.

3 Trigonometric Functions of $\frac{\pi}{4} = 45^\circ$

EXAMPLE 4 **Finding the Exact Values of the Trigonometric Functions of $\frac{\pi}{4} = 45^\circ$**

Find the exact values of the six trigonometric functions of $\frac{\pi}{4} = 45^\circ$.

Solution We seek the coordinates of the point $P = (x, y)$ on the unit circle that corresponds to $\theta = \frac{\pi}{4} = 45^\circ$. See Figure 24. First, we observe that P lies on the line $y = x$. (Do you see why? Since $\theta = 45^\circ = \frac{1}{2} \cdot 90^\circ$, P must lie on the line that bisects quadrant I.) Since $P = (x, y)$ also lies on the unit circle, $x^2 + y^2 = 1$, it follows that

Figure 24

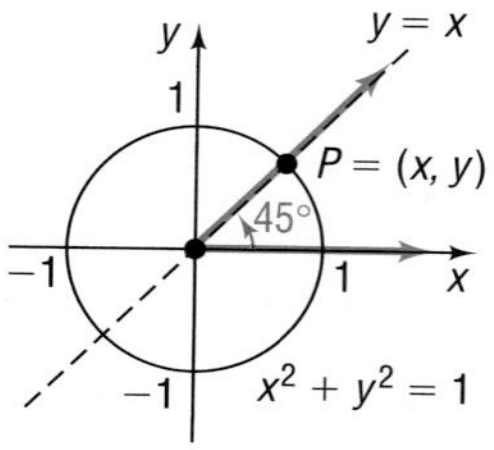

$$x^2 + y^2 = 1$$
$$x^2 + x^2 = 1 \quad y = x, x > 0, y > 0$$
$$2x^2 = 1$$
$$x = \frac{1}{\sqrt{2}} = \frac{\sqrt{2}}{2}, \qquad y = \frac{\sqrt{2}}{2}$$

Then

$$\sin\frac{\pi}{4} = \sin 45° = \frac{\sqrt{2}}{2} \qquad \cos\frac{\pi}{4} = \cos 45° = \frac{\sqrt{2}}{2} \qquad \tan\frac{\pi}{4} = \tan 45° = \frac{\frac{\sqrt{2}}{2}}{\frac{\sqrt{2}}{2}} = 1$$

$$\csc\frac{\pi}{4} = \csc 45° = \frac{1}{\frac{\sqrt{2}}{2}} = \sqrt{2} \qquad \sec\frac{\pi}{4} = \sec 45° = \frac{1}{\frac{\sqrt{2}}{2}} = \sqrt{2} \qquad \cot\frac{\pi}{4} = \cot 45° = \frac{\frac{\sqrt{2}}{2}}{\frac{\sqrt{2}}{2}} = 1$$

◀

EXAMPLE 5 Finding the Exact Value of a Trigonometric Expression

Find the exact value of each expression.

(a) $\sin 45° \cos 180°$ (b) $\tan\frac{\pi}{4} - \sin\frac{3\pi}{2}$ (c) $\left(\sec\frac{\pi}{4}\right)^2 + \csc\frac{\pi}{2}$

Solution (a) $\sin 45° \cos 180° = \frac{\sqrt{2}}{2} \cdot (-1) = -\frac{\sqrt{2}}{2}$

↑ From Example 4 ↑ From Table 2

(b) $\tan\frac{\pi}{4} - \sin\frac{3\pi}{2} = 1 - (-1) = 2$

↑ From Example 4 ↑ From Table 2

(c) $\left(\sec\frac{\pi}{4}\right)^2 + \csc\frac{\pi}{2} = \left(\sqrt{2}\right)^2 + 1 = 2 + 1 = 3$ ◀

NOW WORK PROBLEM 33.

Trigonometric Functions of $\frac{\pi}{6} = 30°$ and $\frac{\pi}{3} = 60°$

4 Consider a right triangle in which one of the angles is $\frac{\pi}{6} = 30°$. It then follows that the third angle is $\frac{\pi}{3} = 60°$. Figure 25(a) illustrates such a triangle with hypotenuse of length 1. Our problem is to determine a and b.

We begin by placing next to this triangle another triangle congruent to the first, as shown in Figure 25(b). Notice that we now have a triangle whose angles are each 60°. This triangle is therefore equilateral, so each side is of length 1. In particular, the base is $2a = 1$, and so $a = \frac{1}{2}$. By the Pythagorean Theorem, b satisfies the equation $a^2 + b^2 = c^2$, so we have

Figure 25

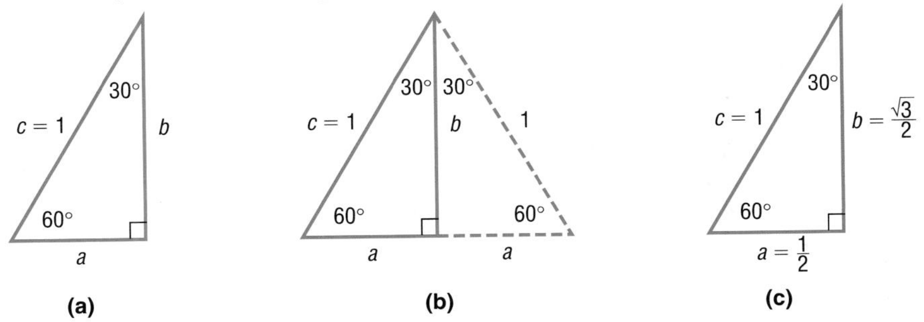

$$a^2 + b^2 = c^2 \qquad a = \frac{1}{2}, c = 1$$

$$\frac{1}{4} + b^2 = 1$$

$$b^2 = 1 - \frac{1}{4} = \frac{3}{4}$$

$$b = \frac{\sqrt{3}}{2}$$

This results in Figure 25(c).

EXAMPLE 6 **Finding the Exact Values of the Trigonometric Functions of $\frac{\pi}{3} = 60°$**

Find the exact values of the six trigonometric functions of $\frac{\pi}{3} = 60°$.

Solution Position the triangle in Figure 25(c) so that the 60° angle is in the standard position. See Figure 26.

Figure 26

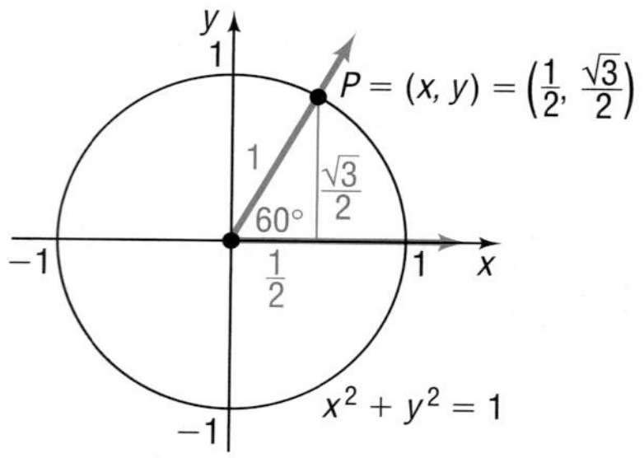

The point on the unit circle that corresponds to $\theta = \frac{\pi}{3} = 60°$ is $P = \left(\frac{1}{2}, \frac{\sqrt{3}}{2}\right)$. Then

$$\sin\frac{\pi}{3} = \sin 60° = \frac{\sqrt{3}}{2} \qquad \cos\frac{\pi}{3} = \cos 60° = \frac{1}{2}$$

$$\csc\frac{\pi}{3} = \csc 60° = \frac{1}{\frac{\sqrt{3}}{2}} = \frac{2}{\sqrt{3}} = \frac{2\sqrt{3}}{3} \qquad \sec\frac{\pi}{3} = \sec 60° = \frac{1}{\frac{1}{2}} = 2$$

$$\tan\frac{\pi}{3} = \tan 60° = \frac{\frac{\sqrt{3}}{2}}{\frac{1}{2}} = \sqrt{3} \qquad \cot\frac{\pi}{3} = \cot 60° = \frac{\frac{1}{2}}{\frac{\sqrt{3}}{2}} = \frac{1}{\sqrt{3}} = \frac{\sqrt{3}}{3}$$

◀

EXAMPLE 7 **Finding the Exact Values of the Trigonometric Functions of $\frac{\pi}{6} = 30°$**

Find the exact values of the trigonometric functions of $\frac{\pi}{6} = 30°$.

Figure 27

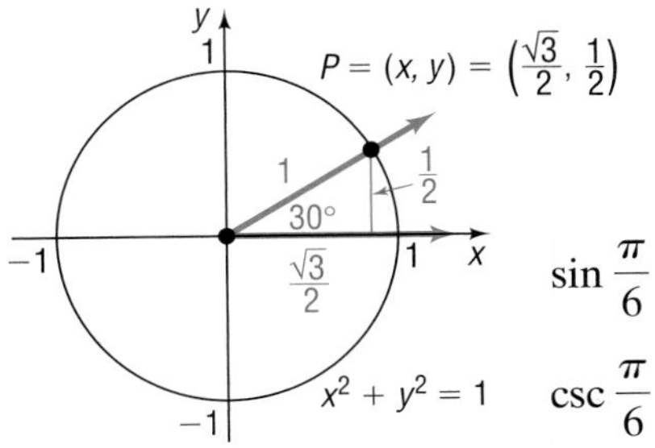

Solution Position the triangle in Figure 25(c) so that the 30° angle is in the standard position. See Figure 27.

The point on the unit circle that corresponds to $\theta = \frac{\pi}{6} = 30°$ is $P = \left(\frac{\sqrt{3}}{2}, \frac{1}{2}\right)$. Then

$$\sin\frac{\pi}{6} = \sin 30° = \frac{1}{2} \qquad \cos\frac{\pi}{6} = \cos 30° = \frac{\sqrt{3}}{2}$$

$$\csc\frac{\pi}{6} = \csc 30° = \frac{1}{\frac{1}{2}} = 2 \qquad \sec\frac{\pi}{6} = \sec 30° = \frac{1}{\frac{\sqrt{3}}{2}} = \frac{2}{\sqrt{3}} = \frac{2\sqrt{3}}{3}$$

$$\tan\frac{\pi}{6} = \tan 30° = \frac{\frac{1}{2}}{\frac{\sqrt{3}}{2}} = \frac{1}{\sqrt{3}} = \frac{\sqrt{3}}{3} \qquad \cot\frac{\pi}{6} = \cot 30° = \frac{\frac{\sqrt{3}}{2}}{\frac{1}{2}} = \sqrt{3}$$

◀

Table 3 summarizes the information just derived for $\frac{\pi}{6} = 30°$, $\frac{\pi}{4} = 45°$, and $\frac{\pi}{3} = 60°$. Until you memorize the entries in Table 3, you should draw an appropriate diagram to determine the values given in the table.

Table 3

θ (Radians)	θ (Degrees)	$\sin\theta$	$\cos\theta$	$\tan\theta$	$\csc\theta$	$\sec\theta$	$\cot\theta$
$\frac{\pi}{6}$	30°	$\frac{1}{2}$	$\frac{\sqrt{3}}{2}$	$\frac{\sqrt{3}}{3}$	2	$\frac{2\sqrt{3}}{3}$	$\sqrt{3}$
$\frac{\pi}{4}$	45°	$\frac{\sqrt{2}}{2}$	$\frac{\sqrt{2}}{2}$	1	$\sqrt{2}$	$\sqrt{2}$	1
$\frac{\pi}{3}$	60°	$\frac{\sqrt{3}}{2}$	$\frac{1}{2}$	$\sqrt{3}$	$\frac{2\sqrt{3}}{3}$	2	$\frac{\sqrt{3}}{3}$

NOW WORK PROBLEM 39.

EXAMPLE 8 Constructing a Rain Gutter

A rain gutter is to be constructed of aluminum sheets 12 inches wide. After marking off a length of 4 inches from each edge, this length is bent up at an angle θ. See Figure 28. The area A of the opening may be expressed as a function of θ as

$$A(\theta) = 16\sin\theta(\cos\theta + 1)$$

Find the area A of the opening for $\theta = 30°$, $\theta = 45°$, and $\theta = 60°$.

Figure 28

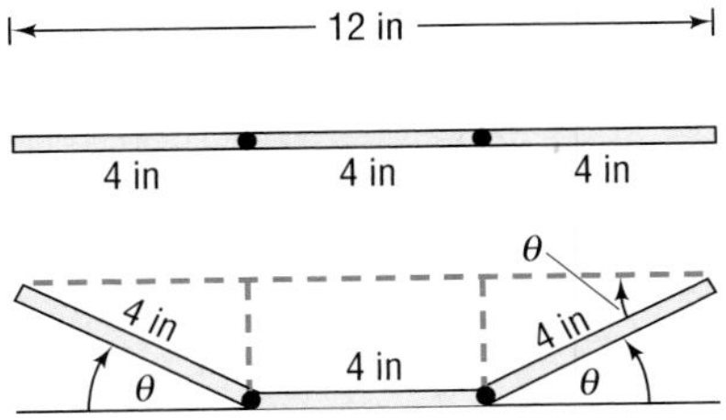

Solution For $\theta = 30°$: $A(30°) = 16\sin 30°(\cos 30° + 1)$

$$= 16\left(\frac{1}{2}\right)\left(\frac{\sqrt{3}}{2} + 1\right) = 4\sqrt{3} + 8$$

The area of the opening for $\theta = 30°$ is about 14.9 square inches.

For $\theta = 45°$: $A(45°) = 16 \sin 45°(\cos 45° + 1)$

$$= 16\left(\frac{\sqrt{2}}{2}\right)\left(\frac{\sqrt{2}}{2} + 1\right) = 8 + 8\sqrt{2}$$

The area of the opening for $\theta = 45°$ is about 19.3 square inches.

For $\theta = 60°$: $A(60°) = 16 \sin 60°(\cos 60° + 1)$

$$= 16\left(\frac{\sqrt{3}}{2}\right)\left(\frac{1}{2} + 1\right) = 12\sqrt{3}$$

The area of the opening for $\theta = 60°$ is about 20.8 square inches. ◀

Exact Values for Certain Integral Multiples of $\frac{\pi}{6} = 30°$, $\frac{\pi}{4} = 45°$, and $\frac{\pi}{3} = 60°$

5 We know the exact values of the trigonometric functions of $\frac{\pi}{4} = 45°$. Using symmetry, we can find the exact values of the trigonometric functions of $\frac{3\pi}{4} = 135°$, $\frac{5\pi}{4} = 225°$, and $\frac{7\pi}{4} = 315°$. Figure 29 shows how.

Figure 29

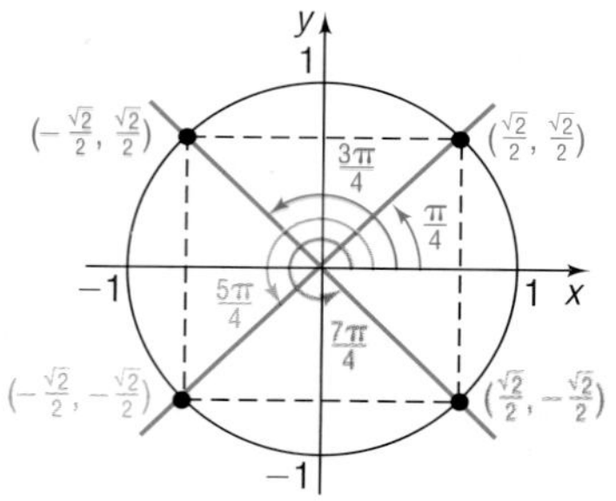

As Figure 29 shows, using symmetry with respect to the y-axis, the point $\left(-\frac{\sqrt{2}}{2}, \frac{\sqrt{2}}{2}\right)$ is the point on the unit circle that corresponds to the angle $\frac{3\pi}{4} = 135°$. Similarly, using symmetry with respect to the origin, the point $\left(-\frac{\sqrt{2}}{2}, -\frac{\sqrt{2}}{2}\right)$ is the point on the unit circle that corresponds to the angle $\frac{5\pi}{4} = 225°$. Finally, using symmetry with respect to the x-axis, the point $\left(\frac{\sqrt{2}}{2}, -\frac{\sqrt{2}}{2}\right)$ is the point on the unit circle that corresponds to the angle $\frac{7\pi}{4} = 315°$.

EXAMPLE 9 Finding Exact Values for Multiples of $\frac{\pi}{4} = 45°$

Based on Figure 29, we see that

(a) $\sin 135° = \frac{\sqrt{2}}{2}$ (b) $\cos \frac{5\pi}{4} = -\frac{\sqrt{2}}{2}$ (c) $\tan 315° = \frac{-\frac{\sqrt{2}}{2}}{\frac{\sqrt{2}}{2}} = -1$ ◀

Figure 29 can also be used to find exact values for other multiples of $\frac{\pi}{4} = 45°$. For example, the point $\left(\frac{\sqrt{2}}{2}, -\frac{\sqrt{2}}{2}\right)$ is the point on the unit circle that corresponds to the angle $-\frac{\pi}{4} = -45°$; the point $\left(\frac{\sqrt{2}}{2}, \frac{\sqrt{2}}{2}\right)$ is the point on the unit circle that corresponds to the angle $\frac{9\pi}{4} = 405°$.

NOW WORK PROBLEMS 53 AND 57.

The use of symmetry also provides information about certain integral multiples of the angles $\frac{\pi}{6} = 30°$ and $\frac{\pi}{3} = 60°$. See Figures 30 and 31.

Figure 30

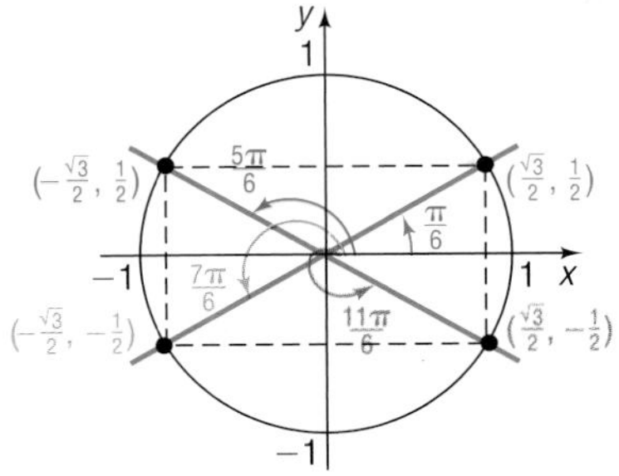

Figure 31

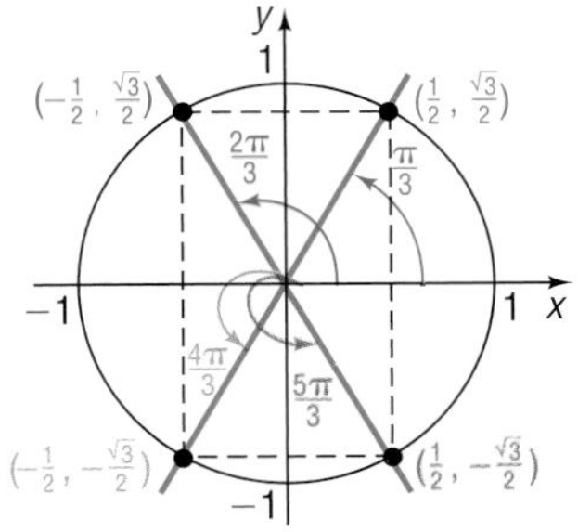

EXAMPLE 10 Using Figures 30 and 31

Based on Figures 30 and 31, we see that

$$\text{(a) } \cos 210° = -\frac{\sqrt{3}}{2} \qquad \text{(b) } \sin(-60°) = -\frac{\sqrt{3}}{2} \qquad \text{(c) } \tan\frac{5\pi}{3} = \frac{-\frac{\sqrt{3}}{2}}{\frac{1}{2}} = -\sqrt{3}$$

◀

NOW WORK PROBLEM 49.

Using a Calculator to Find Values of Trigonometric Functions

6 Before getting started, you must first decide whether to enter the angle in the calculator using radians or degrees and then set the calculator to the correct MODE.* Check your instruction manual to find out how your calculator handles degrees and radians. Your calculator has keys marked [sin], [cos], and [tan]. To find the values of the remaining three trigonometric functions, secant, cosecant, and cotangent, we use the fact that, if $P = (x, y)$ is a point on the unit circle on the terminal side of θ, then

$$\sec\theta = \frac{1}{x} = \frac{1}{\cos\theta} \qquad \csc\theta = \frac{1}{y} = \frac{1}{\sin\theta} \qquad \cot\theta = \frac{x}{y} = \frac{1}{\frac{y}{x}} = \frac{1}{\tan\theta}$$

EXAMPLE 11 Using a Calculator to Approximate the Value of a Trigonometric Function

Use a calculator to find the approximate value of:

(a) $\cos 48°$ (b) $\csc 21°$ (c) $\tan\frac{\pi}{12}$

Express your answers rounded to two decimal places.

*If your calculator does not display the MODE, you can determine the current mode by evaluating [sin] [30]. If you are in the degree mode, the display will show [0.5] ($\sin 30° = 0.5$). If you are in the radian mode, the display will show [−0.9880316].

Solution (a) Set the mode to receive degrees. Rounded to two decimal places,

$$\cos 48^\circ = 0.66991306 \approx 0.67$$

(b) Most calculators do not have a csc key. The manufacturers assume that the user knows some trigonometry. To find the value of csc 21°, use the fact that $\csc 21^\circ = \dfrac{1}{\sin 21^\circ}$. Rounded to two decimal places,

$$\csc 21^\circ \approx 2.79$$

Figure 32

(c) Set the MODE to receive radians. Figure 32 shows the solution using a TI-83 Plus graphing calculator. Rounded to two decimal places,

$$\tan \frac{\pi}{12} \approx 0.27$$

◄

NOW WORK PROBLEM 67.

Using a Circle of Radius r to Evaluate the Trigonometric Functions

Figure 33

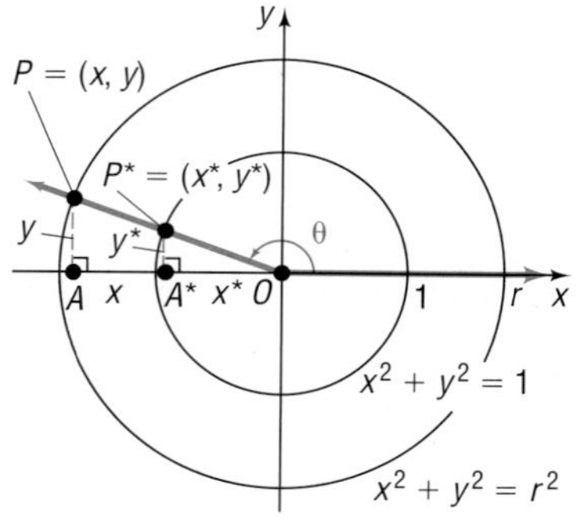

Until now, to find the exact value of a trigonometric function of an angle θ required that we locate the corresponding point $P = (x, y)$ on the unit circle. In fact, though, any circle whose center is at the origin can be used.

Let θ be any nonquadrantal angle placed in standard position. Let $P = (x, y)$ be the point on the circle $x^2 + y^2 = r^2$ that corresponds to θ, and let $P^* = (x^*, y^*)$ be the point on the unit circle that corresponds to θ. See Figure 33.

Notice that the triangles OA^*P^* and OAP are similar; as a result, the ratios of corresponding sides are equal.

$$\frac{y^*}{1} = \frac{y}{r} \qquad \frac{x^*}{1} = \frac{x}{r} \qquad \frac{y^*}{x^*} = \frac{y}{x}$$

$$\frac{1}{y^*} = \frac{r}{y} \qquad \frac{1}{x^*} = \frac{r}{x} \qquad \frac{x^*}{y^*} = \frac{x}{y}$$

These results lead us to formulate the following theorem:

Theorem For an angle θ in standard position, let $P = (x, y)$ be the point on the terminal side of θ that is also on the circle $x^2 + y^2 = r^2$. Then

$$\sin \theta = \frac{y}{r} \qquad \cos \theta = \frac{x}{r} \qquad \tan \theta = \frac{y}{x}, \quad x \neq 0$$

$$\csc \theta = \frac{r}{y}, \quad y \neq 0 \qquad \sec \theta = \frac{r}{x}, \quad x \neq 0 \qquad \cot \theta = \frac{x}{y}, \quad y \neq 0$$

EXAMPLE 12 Finding the Exact Values of the Six Trigonometric Functions

Find the exact values of each of the six trigonometric functions of an angle θ if $(4, -3)$ is a point on its terminal side.

Figure 34

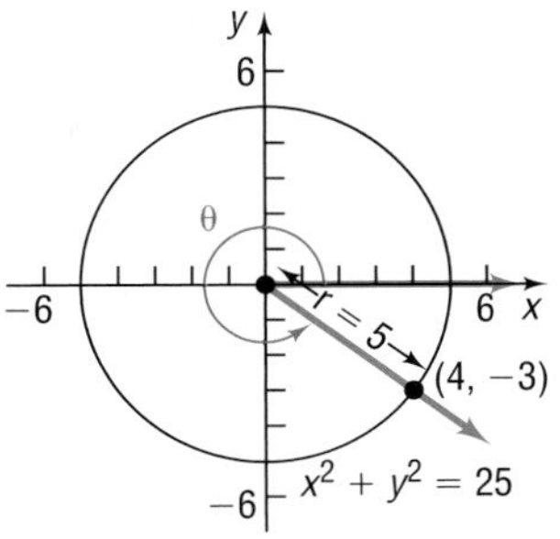

Solution The point $(4, -3)$ is on a circle of radius $r = \sqrt{4^2 + (-3)^2} = \sqrt{16 + 9} = \sqrt{25} = 5$ with the center at the origin. For the point $(x, y) = (4, -3)$, we have $x = 4$ and $y = -3$. Since $r = 5$, we find

$$\sin\theta = \frac{y}{r} = -\frac{3}{5} \qquad \cos\theta = \frac{x}{r} = \frac{4}{5} \qquad \tan\theta = \frac{y}{x} = -\frac{3}{4}$$

$$\csc\theta = \frac{r}{y} = -\frac{5}{3} \qquad \sec\theta = \frac{r}{x} = \frac{5}{4} \qquad \cot\theta = \frac{x}{y} = -\frac{4}{3}$$

◀

NOW WORK PROBLEM 83.

HISTORICAL FEATURE

The name *sine* for the sine function is due to a medieval confusion. The name comes from the Sanskrit word *jīva* (meaning chord), first used in India by Araybhata the Elder (AD 510). He really meant half-chord, but abbreviated it. This was brought into Arabic as *jība*, which was meaningless. Because the proper Arabic word *jaib* would be written the same way (short vowels are not written out in Arabic), *jība* was pronounced as *jaib*, which meant bosom or hollow, and *jība* remains as the Arabic word for sine to this day. Scholars translating the Arabic works into Latin found that the word *sinus* also meant bosom or hollow, and from *sinus* we get the word *sine*.

The name *tangent*, due to Thomas Finck (1583), can be understood by looking at Figure 35. The line segment $\overline{DC}$ is tangent to the circle at C. If $d(O, B) = d(O, C) = 1$, then the length of the line segment $\overline{DC}$ is

$$d(D, C) = \frac{d(D, C)}{1} = \frac{d(D, C)}{d(O, C)} = \tan\alpha$$

The old name for the tangent is *umbra versa* (meaning turned shadow), referring to the use of the tangent in solving height problems with shadows.

The names of the remaining functions came about as follows. If α and β are complementary angles, then $\cos\alpha = \sin\beta$. Because β is the complement of α, it was natural to write the cosine of α as *sin co* α. Probably for reasons involving ease of pronunciation, the *co* migrated to the front, and then cosine received a three-letter abbreviation to match sin, sec, and tan. The two other cofunctions were similarly treated, except that the long forms *cotan* and *cosec* survive to this day in some countries.

Figure 35

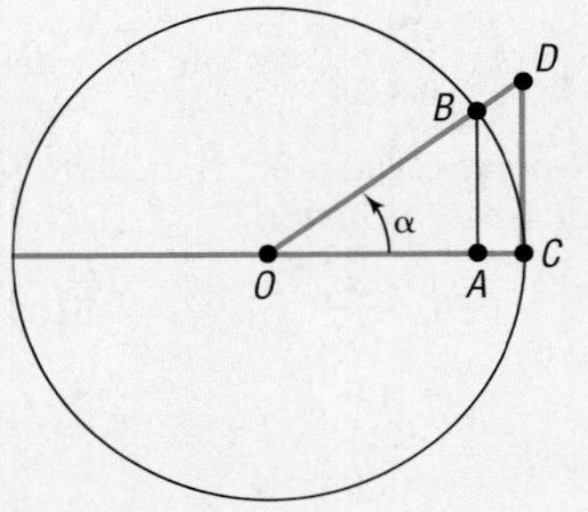

5.2 Assess Your Understanding

'Are You Prepared?' *Answers are given at the end of these exercises. If you get a wrong answer, read the pages in red.*

1. In a right triangle, with legs a and b and hypotenuse c, the Pythagorean Theorem states that ________. (pp. 913–915)

2. The value of the function $f(x) = 3x - 7$ at 5 is ________. (pp. 53–56)

3. *True or False:* For a function $y = f(x)$, for each x in the domain, there is exactly one element y in the range. (pp. 53–56)

4. What is the equation of the unit circle? (p. 16)

5. What point is symmetric with respect to the y-axis to the point $\left(\frac{1}{2}, \frac{\sqrt{3}}{2}\right)$? (pp. 12–16)

6. If (x, y) is a point on the unit circle in quadrant IV and if $x = \frac{\sqrt{3}}{2}$, what is y? (p. 16)

Concepts and Vocabulary

7. $\tan\frac{\pi}{4} + \sin 30° =$ ________.

8. Using a calculator, $\sin 2 =$ ________, rounded to two decimal places.

9. *True or False:* Exact values can be found for the trigonometric functions of 60°.

10. *True or False:* Exact values can be found for the sine of any angle.

Exercises

In Problems 11–18, t is a real number and $P = (x, y)$ is the point on the unit circle that corresponds to t. Find the exact values of the six trigonometric functions of t.

11. $\left(\frac{\sqrt{3}}{2}, \frac{1}{2}\right)$ **12.** $\left(\frac{1}{2}, -\frac{\sqrt{3}}{2}\right)$ **13.** $\left(-\frac{2}{5}, \frac{\sqrt{21}}{5}\right)$ **14.** $\left(-\frac{1}{5}, \frac{2\sqrt{6}}{5}\right)$

15. $\left(-\frac{\sqrt{2}}{2}, \frac{\sqrt{2}}{2}\right)$ **16.** $\left(\frac{\sqrt{2}}{2}, \frac{\sqrt{2}}{2}\right)$ **17.** $\left(\frac{2\sqrt{2}}{3}, -\frac{1}{3}\right)$ **18.** $\left(-\frac{\sqrt{5}}{3}, -\frac{2}{3}\right)$

In Problems 19–28, find the exact value. Do not use a calculator.

19. $\sin\frac{11\pi}{2}$ **20.** $\cos(7\pi)$ **21.** $\tan(6\pi)$ **22.** $\cot\frac{7\pi}{2}$ **23.** $\csc\frac{11\pi}{2}$

24. $\sec(8\pi)$ **25.** $\cos\left(-\frac{3\pi}{2}\right)$ **26.** $\sin(-3\pi)$ **27.** $\sec(-\pi)$ **28.** $\tan(-3\pi)$

In Problems 29–48, find the exact value of each expression. Do not use a calculator.

29. $\sin 45° + \cos 60°$ **30.** $\sin 30° - \cos 45°$ **31.** $\sin 90° + \tan 45°$ **32.** $\cos 180° - \sin 180°$

33. $\sin 45° \cos 45°$ **34.** $\tan 45° \cos 30°$ **35.** $\csc 45° \tan 60°$ **36.** $\sec 30° \cot 45°$

37. $4\sin 90° - 3\tan 180°$ **38.** $5\cos 90° - 8\sin 270°$ **39.** $2\sin\frac{\pi}{3} - 3\tan\frac{\pi}{6}$ **40.** $2\sin\frac{\pi}{4} + 3\tan\frac{\pi}{4}$

41. $\sin\frac{\pi}{4} - \cos\frac{\pi}{4}$ **42.** $\tan\frac{\pi}{3} + \cos\frac{\pi}{3}$ **43.** $2\sec\frac{\pi}{4} + 4\cot\frac{\pi}{3}$ **44.** $3\csc\frac{\pi}{3} + \cot\frac{\pi}{4}$

45. $\tan\pi - \cos 0$ **46.** $\sin\frac{3\pi}{2} + \tan\pi$ **47.** $\csc\frac{\pi}{2} + \cot\frac{\pi}{2}$ **48.** $\sec\pi - \csc\frac{\pi}{2}$

In Problems 49–66, find the exact values of the six trigonometric functions of the given angle. If any are not defined, say "not defined." Do not use a calculator.

49. $\frac{2\pi}{3}$ **50.** $\frac{5\pi}{6}$ **51.** 210° **52.** 240° **53.** $\frac{3\pi}{4}$ **54.** $\frac{11\pi}{4}$

55. $\frac{8\pi}{3}$ **56.** $\frac{13\pi}{6}$ **57.** 405° **58.** 390° **59.** $-\frac{\pi}{6}$ **60.** $-\frac{\pi}{3}$

61. $-45°$ **62.** $-60°$ **63.** $\frac{5\pi}{2}$ **64.** 5π **65.** 720° **66.** 630°

In Problems 67–82, use a calculator to find the approximate value of each expression rounded to two decimal places.

67. $\sin 28^\circ$ **68.** $\cos 14^\circ$ **69.** $\tan 21^\circ$ **70.** $\cot 70^\circ$

71. $\sec 41^\circ$ **72.** $\csc 55^\circ$ **73.** $\sin \frac{\pi}{10}$ **74.** $\cos \frac{\pi}{8}$

75. $\tan \frac{5\pi}{12}$ **76.** $\cot \frac{\pi}{18}$ **77.** $\sec \frac{\pi}{12}$ **78.** $\csc \frac{5\pi}{13}$

79. $\sin 1$ **80.** $\tan 1$ **81.** $\sin 1^\circ$ **82.** $\tan 1^\circ$

In Problems 83–92, a point on the terminal side of an angle θ is given. Find the exact values of the six trigonometric functions of θ.

83. $(-3, 4)$ **84.** $(5, -12)$ **85.** $(2, -3)$ **86.** $(-1, -2)$ **87.** $(-2, -2)$

88. $(1, -1)$ **89.** $(-3, -2)$ **90.** $(2, 2)$ **91.** $\left(\frac{1}{3}, -\frac{1}{4}\right)$ **92.** $(-0.3, -0.4)$

93. Find the exact value of $\sin 45^\circ + \sin 135^\circ + \sin 225^\circ + \sin 315^\circ$.

94. Find the exact value of $\tan 60^\circ + \tan 150^\circ$.

95. If $\sin\theta = 0.1$, find $\sin(\theta + \pi)$.

96. If $\cos\theta = 0.3$, find $\cos(\theta + \pi)$.

97. If $\tan\theta = 3$, find $\tan(\theta + \pi)$.

98. If $\cot\theta = -2$, find $\cot(\theta + \pi)$.

99. If $\sin\theta = \frac{1}{5}$, find $\csc\theta$.

100. If $\cos\theta = \frac{2}{3}$, find $\sec\theta$.

In Problems 101–112, $f(\theta) = \sin\theta$ and $g(\theta) = \cos\theta$. Find the exact value of each function below if $\theta = 60^\circ$. Do not use a calculator.

101. $f(\theta)$ **102.** $g(\theta)$ **103.** $f\left(\frac{\theta}{2}\right)$ **104.** $g\left(\frac{\theta}{2}\right)$ **105.** $[f(\theta)]^2$ **106.** $[g(\theta)]^2$

107. $f(2\theta)$ **108.** $g(2\theta)$ **109.** $2f(\theta)$ **110.** $2g(\theta)$ **111.** $f(-\theta)$ **112.** $g(-\theta)$

113. Use a calculator in radian mode to complete the following table.

What can you conclude about the ratio $\frac{\sin\theta}{\theta}$ as θ approaches 0?

θ	0.5	0.4	0.2	0.1	0.01	0.001	0.0001	0.00001
$\sin\theta$								
$\frac{\sin\theta}{\theta}$								

114. Use a calculator in radian mode to complete the following table.

What can you conclude about the ratio $\frac{\cos\theta - 1}{\theta}$ as θ approaches 0?

θ	0.5	0.4	0.2	0.1	0.01	0.001	0.0001	0.00001
$\cos\theta - 1$								
$\frac{\cos\theta - 1}{\theta}$								

Projectile Motion *The path of a projectile fired at an inclination θ to the horizontal with initial speed v_0 is a parabola (see the figure).*

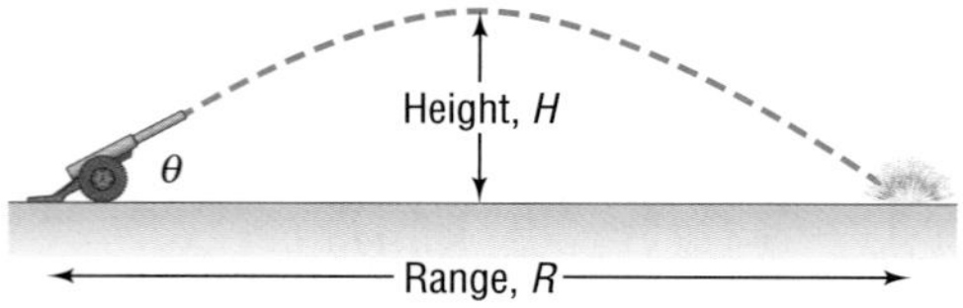

The range R of the projectile, that is, the horizontal distance that the projectile travels, is found by using the formula

$$R = \frac{v_0^2 \sin(2\theta)}{g}$$

where $g \approx 32.2$ feet per second per second ≈ 9.8 meters per second per second is the acceleration due to gravity. The maximum height H of the projectile is

$$H = \frac{v_0^2 \sin^2 \theta}{2g}$$

In Problems 115–118, find the range R and maximum height H.

115. The projectile is fired at an angle of 45° to the horizontal with an initial speed of 100 feet per second.

116. The projectile is fired at an angle of 30° to the horizontal with an initial speed of 150 meters per second.

117. The projectile is fired at an angle of 25° to the horizontal with an initial speed of 500 meters per second.

118. The projectile is fired at an angle of 50° to the horizontal with an initial speed of 200 feet per second.

119. Inclined Plane If friction is ignored, the time t (in seconds) required for a block to slide down an inclined plane (see the figure) is given by the formula

$$t = \sqrt{\frac{2a}{g \sin \theta \cos \theta}}$$

where a is the length (in feet) of the base and $g \approx 32$ feet per second per second is the acceleration of gravity. How long does it take a block to slide down an inclined plane with base $a = 10$ feet when:
(a) $\theta = 30°$? (b) $\theta = 45°$? (c) $\theta = 60°$?

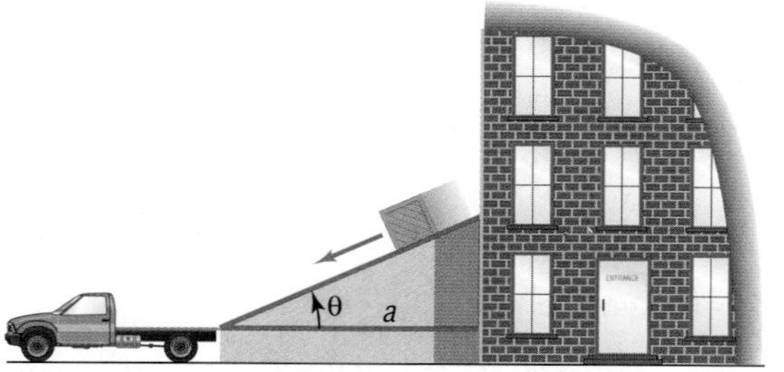

120. Piston Engines In a certain piston engine, the distance x (in centimeters) from the center of the drive shaft to the head of the piston is given by

$$x = \cos \theta + \sqrt{16 + 0.5 \cos(2\theta)}$$

where θ is the angle between the crank and the path of the piston head (see the figure). Find x when $\theta = 30°$ and when $\theta = 45°$.

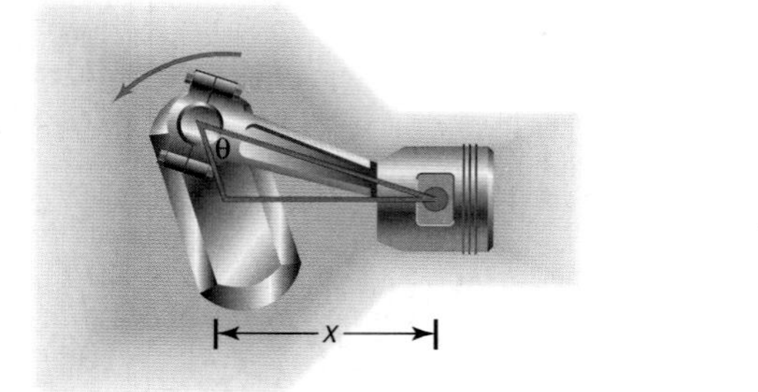

121. Calculating the Time of a Trip Two oceanfront homes are located 8 miles apart on a straight stretch of beach, each a distance of 1 mile from a paved road that parallels the ocean. Sally can jog 8 miles per hour along the paved road, but only 3 miles per hour in the sand on the beach. Because of a river directly between the two houses, it is necessary to jog in the sand to the road, continue on the road, and then jog directly back in the sand to get from one house to the other. See the illustration. The time T to get from one house to the other as a function of the angle θ shown in the illustration is

$$T(\theta) = 1 + \frac{2}{3 \sin \theta} - \frac{1}{4 \tan \theta}, \qquad 0° < \theta < 90°$$

(a) Calculate the time T for $\theta = 30°$. How long is Sally on the paved road?
(b) Calculate the time T for $\theta = 45°$. How long is Sally on the paved road?
(c) Calculate the time T for $\theta = 60°$. How long is Sally on the paved road?
(d) Calculate the time T for $\theta = 90°$. Describe the path taken. Why can't the formula for T be used?

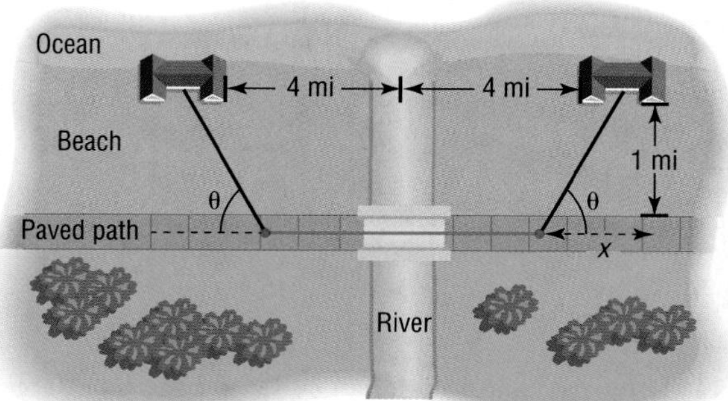

122. Designing Fine Decorative Pieces A designer of decorative art plans to market solid gold spheres encased in clear crystal cones. Each sphere is of fixed radius R and will be enclosed in a cone of height h and radius r. See the illustration. Many cones can be used to enclose the sphere, each having a different slant angle θ. The volume V of the cone can be expressed as a function of the slant angle θ of the cone as

$$V(\theta) = \frac{1}{3}\pi R^3 \frac{(1 + \sec\theta)^3}{\tan^2\theta}, \qquad 0° < \theta < 90°$$

What volume V is required to enclose a sphere of radius 2 centimeters in a cone whose slant angle θ is 30°? 45°? 60°?

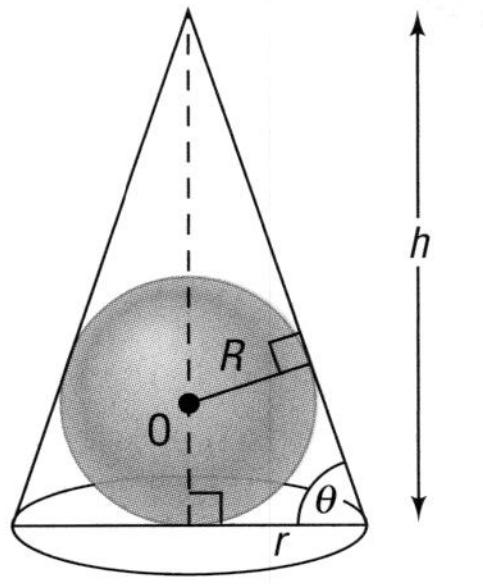

123. Projectile Motion An object is propelled upward at an angle θ, $45° < \theta < 90°$, to the horizontal with an initial velocity of v_0 feet per second from the base of an inclined plane that makes an angle of 45° with the horizontal. See the illustration. If air resistance is ignored, the distance R that it travels up the inclined plane is given by

$$R = \frac{v_0^2\sqrt{2}}{32}[\sin(2\theta) - \cos(2\theta) - 1]$$

(a) Find the distance R that the object travels along the inclined plane if the initial velocity is 32 feet per second and $\theta = 60°$.

(b) Graph $R = R(\theta)$ if the initial velocity is 32 feet per second.

(c) What value of θ makes R largest?

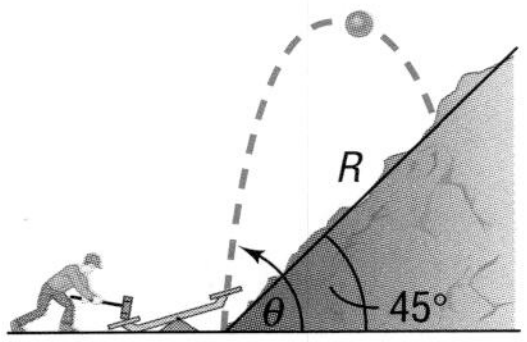

124. If θ $(0 < \theta < \pi)$ is the angle between a horizontal ray directed to the right (say, the positive x-axis) and a nonhorizontal, nonvertical line L, show that the slope m of L equals $\tan\theta$. The angle θ is called the **inclination** of L.

[**Hint:** See the illustration, where we have drawn the line L^* parallel to L and passing through the origin. Use the fact that L^* intersects the unit circle at the point $(\cos\theta, \sin\theta)$.]

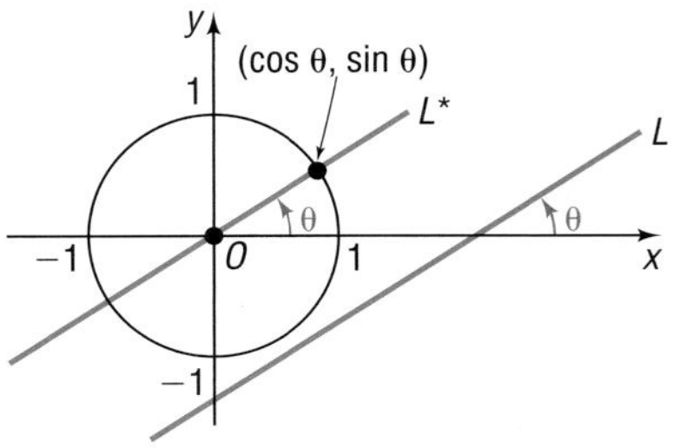

125. Write a brief paragraph that explains how to quickly compute the trigonometric functions of 30°, 45°, and 60°.

126. Write a brief paragraph that explains how to quickly compute the trigonometric functions of 0°, 90°, 180°, and 270°.

127. How would you explain the meaning of the sine function to a fellow student who has just completed college algebra?

'Are You Prepared?' Answers

1. $c^2 = a^2 + b^2$ **2.** 8

3. True **4.** $x^2 + y^2 = 1$

5. $\left(-\frac{1}{2}, \frac{\sqrt{3}}{2}\right)$ **6.** $-\frac{1}{2}$

5.3 Properties of the Trigonometric Functions

PREPARING FOR THIS SECTION *Before getting started, review the following:*

- Functions (Section 2.1, pp. 50-58)
- Identity (Appendix A, Section A.5, p. 936)
- Even and Odd Functions (Section 2.3, pp. 72–74)

Now work the 'Are You Prepared?' problems on page 367.

OBJECTIVES

1 Determine the Domain and Range of the Trigonometric Functions
2 Determine the Period of the Trigonometric Functions
3 Determine the Signs of the Trigonometric Functions in a Given Quadrant
4 Find the Values of the Trigonometric Functions Utilizing Fundamental Identities
5 Find the Exact Values of the Trigonometric Functions of an Angle Given One of the Functions and the Quadrant of the Angle
6 Use Even–Odd Properties to Find the Exact Values of the Trigonometric Functions

Domain and Range of the Trigonometric Functions

Figure 36

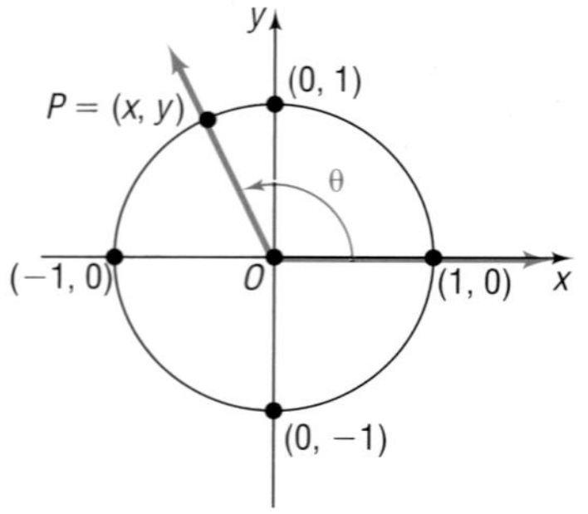

1 Let θ be an angle in standard position, and let $P = (x, y)$ be the point on the unit circle that corresponds to θ. See Figure 36. Then, by definition,

$\sin\theta = y$	$\cos\theta = x$	$\tan\theta = \dfrac{y}{x}, \quad x \neq 0$
$\csc\theta = \dfrac{1}{y}, \quad y \neq 0$	$\sec\theta = \dfrac{1}{x}, \quad x \neq 0$	$\cot\theta = \dfrac{x}{y}, \quad y \neq 0$

For $\sin\theta$ and $\cos\theta$, θ can be any angle, so it follows that the domain of the sine function and cosine function is the set of all real numbers.

> The domain of the sine function is the set of all real numbers.
> The domain of the cosine function is the set of all real numbers.

If $x = 0$, then the tangent function and the secant function are not defined. That is, for the tangent function and secant function, the x-coordinate of $P = (x, y)$ cannot be 0. On the unit circle, there are two such points $(0, 1)$ and $(0, -1)$. These two points correspond to the angles $\frac{\pi}{2}(90°)$ and $\frac{3\pi}{2}(270°)$ or, more generally, to any angle that is an odd multiple of $\frac{\pi}{2}(90°)$, such as $\frac{\pi}{2}(90°)$, $\frac{3\pi}{2}(270°)$, $\frac{5\pi}{2}(450°)$, $-\frac{\pi}{2}(-90°)$, $-\frac{3\pi}{2}(-270°)$, and so on. Such angles must therefore be excluded from the domain of the tangent function and secant function.

> The domain of the tangent function is the set of all real numbers, except odd multiples of $\frac{\pi}{2}(90°)$.
>
> The domain of the secant function is the set of all real numbers, except odd multiples of $\frac{\pi}{2}(90°)$.

If $y = 0$, then the cotangent function and the cosecant function are not defined. For the cotangent function and cosecant function, the y-coordinate of $P = (x, y)$ cannot be 0. On the unit circle, there are two such points, $(1, 0)$ and $(-1, 0)$. These two points correspond to the angles $0(0°)$ and $\pi(180°)$ or, more generally, to any angle that is an integral multiple of $\pi(180°)$, such as $0(0°)$, $\pi(180°)$, $2\pi(360°)$, $3\pi(540°)$, $-\pi(-180°)$, and so on. Such angles must therefore be excluded from the domain of the cotangent function and cosecant function.

> The domain of the cotangent function is the set of all real numbers, except integral multiples of $\pi(180°)$.
>
> The domain of the cosecant function is the set of all real numbers, except integral multiples of $\pi(180°)$.

Next we determine the range of each of the six trigonometric functions. Refer again to Figure 36. Let $P = (x, y)$ be the point on the unit circle that corresponds to the angle θ. It follows that $-1 \le x \le 1$ and $-1 \le y \le 1$. Consequently, since $\sin\theta = y$ and $\cos\theta = x$, we have

$$-1 \le \sin\theta \le 1 \qquad -1 \le \cos\theta \le 1$$

The range of both the sine function and the cosine function consists of all real numbers between -1 and 1, inclusive. Using absolute value notation, we have $|\sin\theta| \le 1$ and $|\cos\theta| \le 1$.

If θ is not a multiple of $\pi(180°)$, then $\csc\theta = \frac{1}{y}$. Since $y = \sin\theta$ and $|y| = |\sin\theta| \le 1$, it follows that $|\csc\theta| = \frac{1}{|\sin\theta|} = \frac{1}{|y|} \ge 1$. The range of the cosecant function consists of all real numbers less than or equal to -1 or greater than or equal to 1. That is,

$$\csc\theta \le -1 \quad \text{or} \quad \csc\theta \ge 1$$

Using absolute value notation, we have $|\csc\theta| \ge 1$.

If θ is not an odd multiple of $\frac{\pi}{2}(90°)$, then, by definition, $\sec\theta = \frac{1}{x}$. Since $x = \cos\theta$ and $|x| = |\cos\theta| \le 1$, it follows that $|\sec\theta| = \frac{1}{|\cos\theta|} = \frac{1}{|x|} \ge 1$. The range of the secant function consists of all real numbers less than or equal to -1 or greater than or equal to 1. That is,

$$\sec\theta \le -1 \quad \text{or} \quad \sec\theta \ge 1$$

Using absolute value notation, we have $|\sec\theta| \ge 1$.

The range of both the tangent function and the cotangent function is the set of all real numbers. You are asked to prove this in Problems 121 and 122.

$$-\infty < \tan\theta < \infty \qquad -\infty < \cot\theta < \infty$$

Table 4 summarizes these results.

Table 4

Function	Symbol	Domain	Range
sine	$f(\theta) = \sin\theta$	All real numbers	All real numbers from -1 to 1, inclusive
cosine	$f(\theta) = \cos\theta$	All real numbers	All real numbers from -1 to 1, inclusive
tangent	$f(\theta) = \tan\theta$	All real numbers, except odd multiples of $\frac{\pi}{2}(90°)$	All real numbers
cosecant	$f(\theta) = \csc\theta$	All real numbers, except integral multiples of $\pi(180°)$	All real numbers greater than or equal to 1 or less than or equal to -1
secant	$f(\theta) = \sec\theta$	All real numbers, except odd multiples of $\frac{\pi}{2}(90°)$	All real numbers greater than or equal to 1 or less than or equal to -1
cotangent	$f(\theta) = \cot\theta$	All real numbers, except integral multiples of $\pi(180°)$	All real numbers

NOW WORK PROBLEM 97.

Period of the Trigonometric Functions

Figure 37

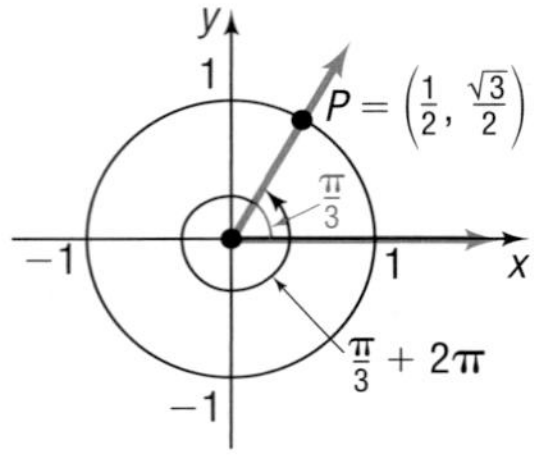

2 Look at Figure 37. This figure shows that for an angle of $\frac{\pi}{3}$ radians the corresponding point P on the unit circle is $\left(\frac{1}{2}, \frac{\sqrt{3}}{2}\right)$. Notice that, for an angle of $\frac{\pi}{3} + 2\pi$ radians, the corresponding point P on the unit circle is also $\left(\frac{1}{2}, \frac{\sqrt{3}}{2}\right)$. Then

$$\sin\frac{\pi}{3} = \frac{\sqrt{3}}{2} \quad \text{and} \quad \sin\left(\frac{\pi}{3} + 2\pi\right) = \frac{\sqrt{3}}{2}$$

$$\cos\frac{\pi}{3} = \frac{1}{2} \quad \text{and} \quad \cos\left(\frac{\pi}{3} + 2\pi\right) = \frac{1}{2}$$

Figure 38

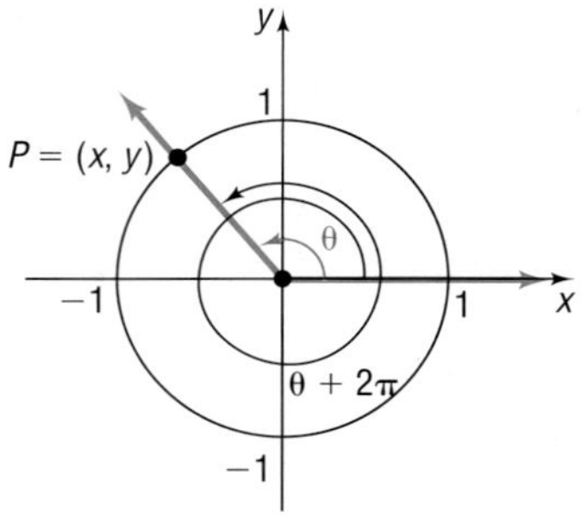

This example illustrates a more general situation. For a given angle θ, measured in radians, suppose that we know the corresponding point $P = (x, y)$ on the unit circle. Now add 2π to θ. The point on the unit circle corresponding to $\theta + 2\pi$ is identical to the point P on the unit circle corresponding to θ. See Figure 38. The values of the trigonometric functions of $\theta + 2\pi$ are equal to the values of the corresponding trigonometric functions of θ.

If we add (or subtract) integral multiples of 2π to θ, the trigonometric values remain unchanged. That is, for all θ.

$$\sin(\theta + 2\pi k) = \sin\theta \qquad \cos(\theta + 2\pi k) = \cos\theta$$
where k is any integer. **(1)**

Functions that exhibit this kind of behavior are called *periodic functions*.

A function f is called **periodic** if there is a positive number p such that, whenever θ is in the domain of f, so is $\theta + p$, and

$$f(\theta + p) = f(\theta)$$

If there is a smallest such number p, this smallest value is called the **(fundamental) period** of f.

Based on equation (1), the sine and cosine functions are periodic. In fact, the sine and cosine functions have period 2π. You are asked to prove this fact in Problems 123 and 124. The secant and cosecant functions are also periodic with period 2π, and the tangent and cotangent functions are periodic with period π. You are asked to prove these statements in Problems 125 through 128.

These facts are summarized as follows:

In Words
Tangent and cotangent have period π; the others have period 2π.

Periodic Properties

$$\sin(\theta + 2\pi) = \sin\theta \quad \cos(\theta + 2\pi) = \cos\theta \quad \tan(\theta + \pi) = \tan\theta$$
$$\csc(\theta + 2\pi) = \csc\theta \quad \sec(\theta + 2\pi) = \sec\theta \quad \cot(\theta + \pi) = \cot\theta$$

Because the sine, cosine, secant, and cosecant functions have period 2π, once we know their values for $0 \le \theta < 2\pi$, we know all their values; similarly, since the tangent and cotangent functions have period π, once we know their values for $0 \le \theta < \pi$, we know all their values.

EXAMPLE 1 Finding Exact Values Using Periodic Properties

Find the exact value of:

(a) $\sin \frac{17\pi}{4}$ (b) $\cos(5\pi)$ (c) $\tan \frac{5\pi}{4}$

Solution (a) It is best to sketch the angle first, as shown in Figure 39(a). Since the period of the sine function is 2π, each full revolution can be ignored. This leaves the angle $\frac{\pi}{4}$. Then

$$\sin \frac{17\pi}{4} = \sin\left(\frac{\pi}{4} + 4\pi\right) = \sin \frac{\pi}{4} = \frac{\sqrt{2}}{2}$$

(b) See Figure 39(b). Since the period of the cosine function is 2π, each full revolution can be ignored. This leaves the angle π. Then

$$\cos(5\pi) = \cos(\pi + 4\pi) = \cos \pi = -1$$

(c) See Figure 39(c). Since the period of the tangent function is π, each half-revolution can be ignored. This leaves the angle $\frac{\pi}{4}$. Then

$$\tan \frac{5\pi}{4} = \tan\left(\frac{\pi}{4} + \pi\right) = \tan \frac{\pi}{4} = 1$$

Figure 39

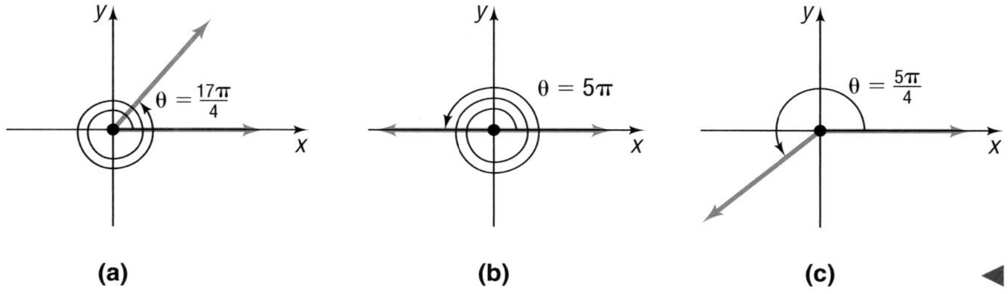

◀

The periodic properties of the trigonometric functions will be very helpful to us when we study their graphs later in the chapter.

NOW WORK PROBLEM 11.

The Signs of the Trigonometric Functions

3 Let $P = (x, y)$ be the point on the unit circle that corresponds to the angle θ. If we know in which quadrant the point P lies, then we can determine the signs of the trigonometric functions of θ. For example, if $P = (x, y)$ lies in

Figure 40

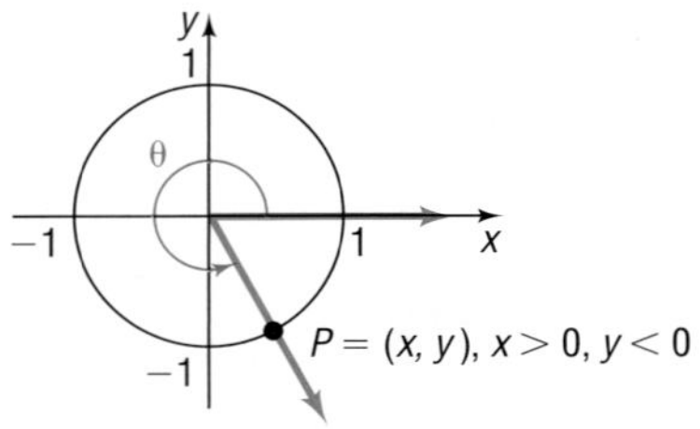

quadrant IV, as shown in Figure 40, then we know that $x > 0$ and $y < 0$. Consequently,

$$\sin\theta = y < 0 \qquad \cos\theta = x > 0 \qquad \tan\theta = \frac{y}{x} < 0$$

$$\csc\theta = \frac{1}{y} < 0 \qquad \sec\theta = \frac{1}{x} > 0 \qquad \cot\theta = \frac{x}{y} < 0$$

Table 5 lists the signs of the six trigonometric functions for each quadrant. See also Figure 41.

Table 5

Quadrant of *P*	sin θ, csc θ	cos θ, sec θ	tan θ, cot θ
I	Positive	Positive	Positive
II	Positive	Negative	Negative
III	Negative	Negative	Positive
IV	Negative	Positive	Negative

Figure 41

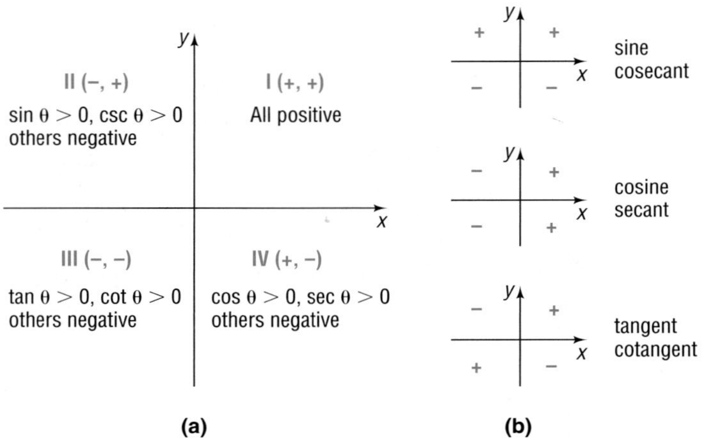

EXAMPLE 2 Finding the Quadrant in Which an Angle θ Lies

If $\sin\theta < 0$ and $\cos\theta < 0$, name the quadrant in which the angle θ lies.

Solution Let $P = (x, y)$ be the point on the unit circle corresponding to θ. Then $\sin\theta = y < 0$ and $\cos\theta = x < 0$. The point $P = (x, y)$ must be in quadrant III, so θ lies in quadrant III. ◀

NOW WORK PROBLEM 27.

Fundamental Identities

If $P = (x, y)$ is the point on the unit circle corresponding to θ, then

$$\sin\theta = y \qquad \cos\theta = x \qquad \tan\theta = \frac{y}{x}, \quad \text{if } x \neq 0$$

$$\csc\theta = \frac{1}{y}, \quad \text{if } y \neq 0 \qquad \sec\theta = \frac{1}{x}, \quad \text{if } x \neq 0 \qquad \cot\theta = \frac{x}{y}, \quad \text{if } y \neq 0$$

Based on these definitions, we have the **reciprocal identities:**

Reciprocal Identities

$$\csc \theta = \frac{1}{\sin \theta} \qquad \sec \theta = \frac{1}{\cos \theta} \qquad \cot \theta = \frac{1}{\tan \theta} \qquad \textbf{(2)}$$

Two other fundamental identities are the **quotient identities.**

Quotient Identities

$$\tan \theta = \frac{\sin \theta}{\cos \theta} \qquad \cot \theta = \frac{\cos \theta}{\sin \theta} \qquad \textbf{(3)}$$

The proofs of identities (2) and (3) follow from the definitions of the trigonometric functions. (See Problems 129 and 130.)

If $\sin \theta$ and $\cos \theta$ are known, identities (2) and (3) make it easy to find the values of the remaining trigonometric functions.

EXAMPLE 3 Finding Exact Values Using Identities When Sine and Cosine Are Given

Given $\sin \theta = \frac{\sqrt{5}}{5}$ and $\cos \theta = \frac{2\sqrt{5}}{5}$, find the exact values of the four remaining trigonometric functions of θ using identities.

Solution Based on a quotient identity from (3), we have

$$\tan \theta = \frac{\sin \theta}{\cos \theta} = \frac{\frac{\sqrt{5}}{5}}{\frac{2\sqrt{5}}{5}} = \frac{1}{2}$$

Then we use the reciprocal identities from (2) to get

$$\csc \theta = \frac{1}{\sin \theta} = \frac{1}{\frac{\sqrt{5}}{5}} = \frac{5}{\sqrt{5}} = \sqrt{5} \qquad \sec \theta = \frac{1}{\cos \theta} = \frac{1}{\frac{2\sqrt{5}}{5}} = \frac{5}{2\sqrt{5}} = \frac{\sqrt{5}}{2} \qquad \cot \theta = \frac{1}{\tan \theta} = \frac{1}{\frac{1}{2}} = 2$$

◄

NOW WORK PROBLEM 35.

The equation of the unit circle is $x^2 + y^2 = 1$, or equivalently, $y^2 + x^2 = 1$. If $P = (x, y)$ is the point on the unit circle that corresponds to the angle θ, then

$$y^2 + x^2 = 1$$

But $y = \sin \theta$ and $x = \cos \theta$, so

$$(\sin \theta)^2 + (\cos \theta)^2 = 1 \qquad \textbf{(4)}$$

It is customary to write $\sin^2 \theta$ instead of $(\sin \theta)^2$, $\cos^2 \theta$ instead of $(\cos \theta)^2$, and so on. With this notation, we can rewrite equation (4) as

$$\sin^2 \theta + \cos^2 \theta = 1 \qquad \textbf{(5)}$$

If $\cos\theta \neq 0$, we can divide each side of equation (5) by $\cos^2\theta$.

$$\frac{\sin^2\theta}{\cos^2\theta} + 1 = \frac{1}{\cos^2\theta}$$

$$\left(\frac{\sin\theta}{\cos\theta}\right)^2 + 1 = \left(\frac{1}{\cos\theta}\right)^2$$

Now use identities (2) and (3) to get

$$\tan^2\theta + 1 = \sec^2\theta \qquad \textbf{(6)}$$

Similarly, if $\sin\theta \neq 0$, we can divide equation (5) by $\sin^2\theta$ and use identities (2) and (3) to get $1 + \cot^2\theta = \csc^2\theta$, which we write as

$$\cot^2\theta + 1 = \csc^2\theta \qquad \textbf{(7)}$$

Collectively, the identities in (5), (6), and (7) are referred to as the **Pythagorean identities.**

Let's pause here to summarize the fundamental identities.

Fundamental Identities

$$\tan\theta = \frac{\sin\theta}{\cos\theta} \qquad \cot\theta = \frac{\cos\theta}{\sin\theta}$$

$$\csc\theta = \frac{1}{\sin\theta} \qquad \sec\theta = \frac{1}{\cos\theta} \qquad \cot\theta = \frac{1}{\tan\theta}$$

$$\sin^2\theta + \cos^2\theta = 1 \qquad \tan^2\theta + 1 = \sec^2\theta \qquad \cot^2\theta + 1 = \csc^2\theta$$

The Pythagorean identity

$$\sin^2\theta + \cos^2\theta = 1$$

can be solved for $\sin\theta$ in terms of $\cos\theta$ (or vice versa) as follows:

$$\sin^2\theta = 1 - \cos^2\theta$$

$$\sin\theta = \pm\sqrt{1 - \cos^2\theta}$$

where the $+$ sign is used if $\sin\theta > 0$ and the $-$ sign is used if $\sin\theta < 0$. Similarly, in $\tan^2\theta + 1 = \sec^2\theta$, we can solve for $\tan\theta$ (or $\sec\theta$), and in $\cot^2\theta + 1 = \csc^2\theta$, we can solve for $\cot\theta$ (or $\csc\theta$).

4 **EXAMPLE 4**

Finding the Exact Value of a Trigonometric Expression Using Identities

Find the exact value of each expression. Do not use a calculator.

(a) $\tan 20° - \dfrac{\sin 20°}{\cos 20°}$ (b) $\sin^2\dfrac{\pi}{12} + \dfrac{1}{\sec^2\dfrac{\pi}{12}}$

Solution (a) $\tan 20° - \dfrac{\sin 20°}{\cos 20°} = \tan 20° - \tan 20° = 0$

$\uparrow$ $\dfrac{\sin\theta}{\cos\theta} = \tan\theta$

(b) $\sin^2\dfrac{\pi}{12} + \dfrac{1}{\sec^2\dfrac{\pi}{12}} = \sin^2\dfrac{\pi}{12} + \cos^2\dfrac{\pi}{12} = 1$

$\uparrow$ $\cos\theta = \dfrac{1}{\sec\theta}$ $\uparrow$ $\sin^2\theta + \cos^2\theta = 1$

NOW WORK PROBLEM 79.

5 Many problems require finding the exact values of the remaining trigonometric functions when the value of one of them is known and the quadrant in which θ lies can be found.

EXAMPLE 5 Finding Exact Values Given One Value and the Sign of Another

Given that $\sin\theta = \dfrac{1}{3}$ and $\cos\theta < 0$, find the exact values of each of the remaining five trigonometric functions.

Solution 1 Using the Definition Suppose that $P = (x, y)$ is the point on the unit circle that corresponds to θ. Since $\sin\theta = \dfrac{1}{3} = y$ and $\cos\theta = x < 0$, the point $P = (x, y) = \left(x, \dfrac{1}{3}\right)$ is in quadrant II. See Figure 42. Then

Figure 42

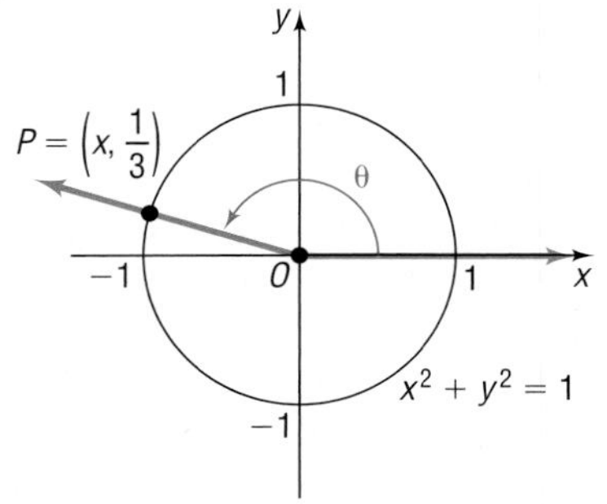

$$x^2 + y^2 = 1$$
$$x^2 + \left(\frac{1}{3}\right)^2 = 1 \qquad y = \frac{1}{3}$$
$$x^2 = \frac{8}{9}$$
$$x = -\frac{2\sqrt{2}}{3} \qquad x < 0$$

Since $x = -\dfrac{2\sqrt{2}}{3}$ and $y = \dfrac{1}{3}$, we find that

$$\cos\theta = x = -\frac{2\sqrt{2}}{3} \qquad \tan\theta = \frac{y}{x} = \frac{\frac{1}{3}}{\frac{-2\sqrt{2}}{3}} = -\frac{1}{2\sqrt{2}} = -\frac{\sqrt{2}}{4}$$

$$\csc\theta = \frac{1}{y} = \frac{1}{\frac{1}{3}} = 3 \qquad \sec\theta = \frac{1}{x} = \frac{1}{\frac{-2\sqrt{2}}{3}} = -\frac{3}{2\sqrt{2}} = -\frac{3\sqrt{2}}{4} \qquad \cot\theta = \frac{x}{y} = \frac{\frac{-2\sqrt{2}}{3}}{\frac{1}{3}} = -2\sqrt{2}$$

Solution 2 Using Identities First, we solve equation (5) for $\cos\theta$.

$$\sin^2\theta + \cos^2\theta = 1$$
$$\cos^2\theta = 1 - \sin^2\theta$$
$$\cos\theta = \pm\sqrt{1 - \sin^2\theta}$$

Because $\cos\theta < 0$, we choose the minus sign.

$$\cos\theta = -\sqrt{1 - \sin^2\theta} = -\sqrt{1 - \frac{1}{9}} = -\sqrt{\frac{8}{9}} = -\frac{2\sqrt{2}}{3}$$

$\uparrow$ $\sin\theta = \frac{1}{3}$

Now we know the values of $\sin\theta$ and $\cos\theta$, so we can use identities (2) and (3) to get

$$\tan\theta = \frac{\sin\theta}{\cos\theta} = \frac{\frac{1}{3}}{\frac{-2\sqrt{2}}{3}} = \frac{1}{-2\sqrt{2}} = -\frac{\sqrt{2}}{4} \qquad \cot\theta = \frac{1}{\tan\theta} = -2\sqrt{2}$$

$$\sec\theta = \frac{1}{\cos\theta} = \frac{1}{\frac{-2\sqrt{2}}{3}} = \frac{-3}{2\sqrt{2}} = -\frac{3\sqrt{2}}{4} \qquad \csc\theta = \frac{1}{\sin\theta} = \frac{1}{\frac{1}{3}} = 3$$

◀

Finding the Values of the Trigonometric Functions When One Is Known

Given the value of one trigonometric function and the quadrant in which θ lies, the exact value of each of the remaining five trigonometric functions can be found in either of two ways.

Method 1 Using the Definition

STEP 1: Draw a circle showing the location of the angle θ and the point $P = (x, y)$ that corresponds to θ. The radius of the circle is $r = \sqrt{x^2 + y^2}$.

STEP 2: Assign a value to two of the three variables x, y, r based on the value of the given trigonometric fuction.

STEP 3: Use the fact that P lies on the circle $x^2 + y^2 = r^2$ to find the value of the missing variable.

STEP 4: Apply the theorem on page 350 to find the values of the remaining trigonometric functions.

Method 2 Using Identities

Use appropriately selected identities to find the value of each of the remaining trigonometric functions.

EXAMPLE 6 **Given One Value of a Trigonometric Function, Find the Remaining Ones**

Given that $\tan\theta = \frac{1}{2}$ and $\sin\theta < 0$, find the exact value of each of the remaining five trigonometric functions of θ.

Solution 1 Using the Definition

Figure 43

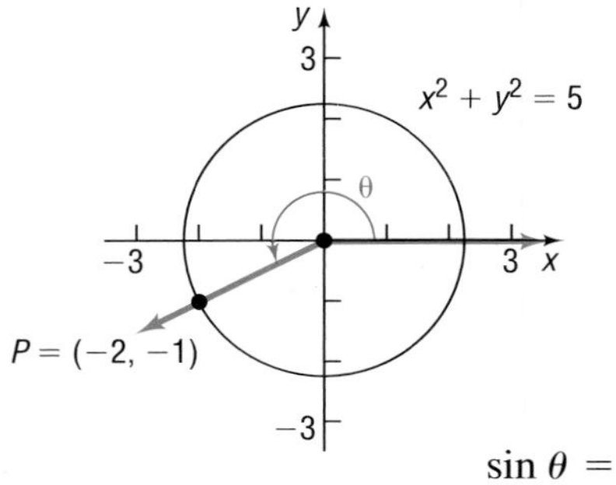

STEP 1: Since $\tan\theta = \dfrac{1}{2} > 0$ and $\sin\theta < 0$, the point $P = (x, y)$ that corresponds to θ lies in quadrant III. See Figure 43.

STEP 2: Since $\tan\theta = \dfrac{1}{2} = \dfrac{y}{x}$ and θ lies in quadrant III, we let $x = -2$ and $y = -1$.

STEP 3: Then $r = \sqrt{x^2 + y^2} = \sqrt{(-2)^2 + (-1)^2} = \sqrt{5}$, and P lies on the circle $x^2 + y^2 = 5$.

STEP 4: Now apply the theorem on p. 350 using $x = -2$, $y = -1$, $r = \sqrt{5}$.

$$\sin\theta = \frac{y}{r} = \frac{-1}{\sqrt{5}} = -\frac{\sqrt{5}}{5} \qquad \cos\theta = \frac{x}{r} = \frac{-2}{\sqrt{5}} = -\frac{2\sqrt{5}}{5}$$

$$\csc\theta = \frac{r}{y} = \frac{\sqrt{5}}{-1} = -\sqrt{5} \qquad \sec\theta = \frac{r}{x} = \frac{\sqrt{5}}{-2} = -\frac{\sqrt{5}}{2} \qquad \cot\theta = \frac{x}{y} = \frac{-2}{-1} = 2$$

Solution 2 Using Identities

We use the Pythagorean identity that involves $\tan\theta$, that is, $\tan^2\theta + 1 = \sec^2\theta$. Since $\tan\theta = \dfrac{1}{2} > 0$ and $\sin\theta < 0$, then θ lies in quadrant III, where $\sec\theta < 0$. Then

$$\tan^2\theta + 1 = \sec^2\theta \qquad \text{Pythagorean identity}$$

$$\left(\frac{1}{2}\right)^2 + 1 = \sec^2\theta \qquad \tan\theta = \frac{1}{2}$$

$$\sec^2\theta = \frac{1}{4} + 1 = \frac{5}{4} \qquad \text{Proceed to solve for } \sec\theta.$$

$$\sec\theta = -\frac{\sqrt{5}}{2} \qquad \sec\theta < 0$$

Now

$$\cos\theta = \frac{1}{\sec\theta} = \frac{1}{-\frac{\sqrt{5}}{2}} = -\frac{2}{\sqrt{5}} = -\frac{2\sqrt{5}}{5}$$

$$\tan\theta = \frac{\sin\theta}{\cos\theta} \quad \text{so} \quad \sin\theta = \tan\theta \cdot \cos\theta = \left(\frac{1}{2}\right)\cdot\left(-\frac{2\sqrt{5}}{5}\right) = -\frac{\sqrt{5}}{5}$$

$$\csc\theta = \frac{1}{\sin\theta} = \frac{1}{-\frac{\sqrt{5}}{5}} = -\frac{5}{\sqrt{5}} = -\sqrt{5}$$

$$\cot\theta = \frac{1}{\tan\theta} = \frac{1}{\frac{1}{2}} = 2$$

◀

NOW WORK PROBLEM 43.

Even–Odd Properties

6 Recall that a function f is even if $f(-\theta) = f(\theta)$ for all θ in the domain of f; a function f is odd if $f(-\theta) = -f(\theta)$ for all θ in the domain of f. We will now show that the trigonometric functions sine, tangent, cotangent, and cosecant are odd functions, whereas the functions cosine and secant are even functions.

Theorem

In Words

Cosine and secant are even functions; the others are odd functions.

Even–Odd Properties

$$\sin(-\theta) = -\sin\theta \qquad \cos(-\theta) = \cos\theta \qquad \tan(-\theta) = -\tan\theta$$
$$\csc(-\theta) = -\csc\theta \qquad \sec(-\theta) = \sec\theta \qquad \cot(-\theta) = -\cot\theta$$

Figure 44

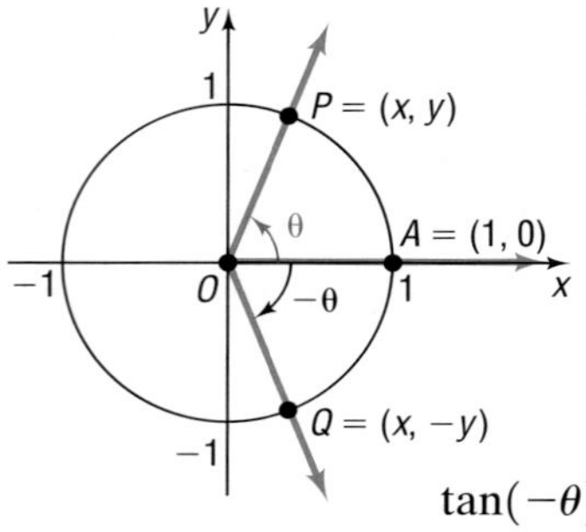

Proof Let $P = (x, y)$ be the point on the unit circle that corresponds to the angle θ. (See Figure 44.) The point Q on the unit circle that corresponds to the angle $-\theta$ will have coordinates $(x, -y)$. Using the definition of the trigonometric functions, we have

$$\sin\theta = y \qquad \cos\theta = x \qquad \sin(-\theta) = -y \qquad \cos(-\theta) = x$$

so

$$\sin(-\theta) = -y = -\sin\theta \qquad \cos(-\theta) = x = \cos\theta$$

Now, using these results and some of the fundamental identities, we have

$$\tan(-\theta) = \frac{\sin(-\theta)}{\cos(-\theta)} = \frac{-\sin\theta}{\cos\theta} = -\tan\theta \qquad \cot(-\theta) = \frac{1}{\tan(-\theta)} = \frac{1}{-\tan\theta} = -\cot\theta$$

$$\sec(-\theta) = \frac{1}{\cos(-\theta)} = \frac{1}{\cos\theta} = \sec\theta \qquad \csc(-\theta) = \frac{1}{\sin(-\theta)} = \frac{1}{-\sin\theta} = -\csc\theta$$ ■

EXAMPLE 7 **Finding Exact Values Using Even–Odd Properties**

Find the exact value of:

(a) $\sin(-45°)$ (b) $\cos(-\pi)$ (c) $\cot\left(-\frac{3\pi}{2}\right)$ (d) $\tan\left(-\frac{37\pi}{4}\right)$

Solution (a) $\sin(-45°) = -\sin 45° = -\frac{\sqrt{2}}{2}$ (Odd function) (b) $\cos(-\pi) = \cos\pi = -1$ (Even function)

(c) $\cot\left(-\frac{3\pi}{2}\right) = -\cot\frac{3\pi}{2} = 0$ (Odd function)

(d) $\tan\left(-\frac{37\pi}{4}\right) = -\tan\frac{37\pi}{4} = -\tan\left(\frac{\pi}{4} + 9\pi\right) = -\tan\frac{\pi}{4} = -1$ (Odd function; Period is π.) ◄

NOW WORK PROBLEM 59.

5.3 Assess Your Understanding

'Are You Prepared?' *Answers are given at the end of these exercises. If you get a wrong answer, read the pages in red.*

1. The domain of the function $f(x) = 3x - 6$ is ________. (pp. 50–58)

2. A function for which $f(x) = f(-x)$ for all x in the domain of f is called a(n) ________ function. (pp. 72–74)

3. The range of the function $f(x) = \sqrt{x}$ is ________. (pp. 50–58)

4. *True or False:* The equation $x^2 + 2x = (x + 1)^2 - 1$ is an identity. (p. 936)

Concepts and Vocabulary

5. The sine, cosine, cosecant, and secant functions have period ________; the tangent and cotangent functions have period ________.

6. The domain of the tangent function is ________.

7. The range of the sine function is ________.

8. *True or False:* The only even trigonometric functions are the cosine and secant functions.

9. *True or False:* All the trigonometric functions are periodic, with period 2π.

10. *True or False:* The range of the secant function is the set of positive real numbers.

Exercises

In Problems 11–26, use the fact that the trigonometric functions are periodic to find the exact value of each expression. Do not use a calculator.

11. $\sin 405°$ **12.** $\cos 420°$ **13.** $\tan 405°$ **14.** $\sin 390°$ **15.** $\csc 450°$ **16.** $\sec 540°$

17. $\cot 390°$ **18.** $\sec 420°$ **19.** $\cos \dfrac{33\pi}{4}$ **20.** $\sin \dfrac{9\pi}{4}$ **21.** $\tan(21\pi)$ **22.** $\csc \dfrac{9\pi}{2}$

23. $\sec \dfrac{17\pi}{4}$ **24.** $\cot \dfrac{17\pi}{4}$ **25.** $\tan \dfrac{19\pi}{6}$ **26.** $\sec \dfrac{25\pi}{6}$

In Problems 27–34, name the quadrant in which the angle θ lies.

27. $\sin\theta > 0, \quad \cos\theta < 0$ **28.** $\sin\theta < 0, \quad \cos\theta > 0$ **29.** $\sin\theta < 0, \quad \tan\theta < 0$

30. $\cos\theta > 0, \quad \tan\theta > 0$ **31.** $\cos\theta > 0, \quad \tan\theta < 0$ **32.** $\cos\theta < 0, \quad \tan\theta > 0$

33. $\sec\theta < 0, \quad \sin\theta > 0$ **34.** $\csc\theta > 0, \quad \cos\theta < 0$

In Problems 35–42, $\sin\theta$ and $\cos\theta$ are given. Find the exact value of each of the four remaining trigonometric functions.

35. $\sin\theta = -\dfrac{3}{5}, \quad \cos\theta = \dfrac{4}{5}$ **36.** $\sin\theta = \dfrac{4}{5}, \quad \cos\theta = -\dfrac{3}{5}$ **37.** $\sin\theta = \dfrac{2\sqrt{5}}{5}, \quad \cos\theta = \dfrac{\sqrt{5}}{5}$

38. $\sin\theta = -\dfrac{\sqrt{5}}{5}, \quad \cos\theta = -\dfrac{2\sqrt{5}}{5}$ **39.** $\sin\theta = \dfrac{1}{2}, \quad \cos\theta = \dfrac{\sqrt{3}}{2}$ **40.** $\sin\theta = \dfrac{\sqrt{3}}{2}, \quad \cos\theta = \dfrac{1}{2}$

41. $\sin\theta = -\dfrac{1}{3}, \quad \cos\theta = \dfrac{2\sqrt{2}}{3}$ **42.** $\sin\theta = \dfrac{2\sqrt{2}}{3}, \quad \cos\theta = -\dfrac{1}{3}$

In Problems 43–58, find the exact value of each of the remaining trigonometric functions of θ.

43. $\sin\theta = \dfrac{12}{13}$, θ in quadrant II **44.** $\cos\theta = \dfrac{3}{5}$, θ in quadrant IV **45.** $\cos\theta = -\dfrac{4}{5}$, θ in quadrant III

46. $\sin\theta = -\dfrac{5}{13}$, θ in quadrant III **47.** $\sin\theta = \dfrac{5}{13}, \quad 90° < \theta < 180°$ **48.** $\cos\theta = \dfrac{4}{5}, \quad 270° < \theta < 360°$

49. $\cos\theta = -\dfrac{1}{3}, \quad \dfrac{\pi}{2} < \theta < \pi$ **50.** $\sin\theta = -\dfrac{2}{3}, \quad \pi < \theta < \dfrac{3\pi}{2}$ **51.** $\sin\theta = \dfrac{2}{3}, \quad \tan\theta < 0$

52. $\cos\theta = -\dfrac{1}{4}, \quad \tan\theta > 0$ **53.** $\sec\theta = 2, \quad \sin\theta < 0$ **54.** $\csc\theta = 3, \quad \cot\theta < 0$

55. $\tan\theta = \dfrac{3}{4}, \quad \sin\theta < 0$ **56.** $\cot\theta = \dfrac{4}{3}, \quad \cos\theta < 0$ **57.** $\tan\theta = -\dfrac{1}{3}, \quad \sin\theta > 0$

58. $\sec\theta = -2, \quad \tan\theta > 0$

In Problems 59–76, use the even–odd properties to find the exact value of each expression. Do not use a calculator.

59. $\sin(-60°)$ **60.** $\cos(-30°)$ **61.** $\tan(-30°)$ **62.** $\sin(-135°)$ **63.** $\sec(-60°)$ **64.** $\csc(-30°)$

65. $\sin(-90°)$ **66.** $\cos(-270°)$ **67.** $\tan\left(-\frac{\pi}{4}\right)$ **68.** $\sin(-\pi)$ **69.** $\cos\left(-\frac{\pi}{4}\right)$ **70.** $\sin\left(-\frac{\pi}{3}\right)$

71. $\tan(-\pi)$ **72.** $\sin\left(-\frac{3\pi}{2}\right)$ **73.** $\csc\left(-\frac{\pi}{4}\right)$ **74.** $\sec(-\pi)$ **75.** $\sec\left(-\frac{\pi}{6}\right)$ **76.** $\csc\left(-\frac{\pi}{3}\right)$

In Problems 77–88, use properties of the trigonometric functions to find the exact value of each expression. Do not use a calculator.

77. $\sin^2 40° + \cos^2 40°$ **78.** $\sec^2 18° - \tan^2 18°$ **79.** $\sin 80° \csc 80°$ **80.** $\tan 10° \cot 10°$

81. $\tan 40° - \frac{\sin 40°}{\cos 40°}$ **82.** $\cot 20° - \frac{\cos 20°}{\sin 20°}$ **83.** $\cos 400° \cdot \sec 40°$ **84.** $\tan 200° \cdot \cot 20°$

85. $\sin\left(-\frac{\pi}{12}\right)\csc\frac{25\pi}{12}$ **86.** $\sec\left(-\frac{\pi}{18}\right)\cdot\cos\frac{37\pi}{18}$ **87.** $\frac{\sin(-20°)}{\cos 380°} + \tan 200°$ **88.** $\frac{\sin 70°}{\cos(-430°)} + \tan(-70°)$

89. If $\sin\theta = 0.3$, find the value of:
$\sin\theta + \sin(\theta + 2\pi) + \sin(\theta + 4\pi)$.

90. If $\cos\theta = 0.2$, find the value of:
$\cos\theta + \cos(\theta + 2\pi) + \cos(\theta + 4\pi)$.

91. If $\tan\theta = 3$, find the value of:
$\tan\theta + \tan(\theta + \pi) + \tan(\theta + 2\pi)$.

92. If $\cot\theta = -2$, find the value of:
$\cot\theta + \cot(\theta - \pi) + \cot(\theta - 2\pi)$.

93. Find the exact value of
$\sin 1° + \sin 2° + \sin 3° + \cdots + \sin 358° + \sin 359°$.

94. Find the exact value of
$\cos 1° + \cos 2° + \cos 3° + \cdots + \cos 358° + \cos 359°$.

95. What is the domain of the sine function?

96. What is the domain of the cosine function?

97. For what numbers θ is $f(\theta) = \tan\theta$ not defined?

98. For what numbers θ is $f(\theta) = \cot\theta$ not defined?

99. For what numbers θ is $f(\theta) = \sec\theta$ not defined?

100. For what numbers θ is $f(\theta) = \csc\theta$ not defined?

101. What is the range of the sine function?

102. What is the range of the cosine function?

103. What is the range of the tangent function?

104. What is the range of the cotangent function?

105. What is the range of the secant function?

106. What is the range of the cosecant function?

107. Is the sine function even, odd, or neither? Is its graph symmetric? With respect to what?

108. Is the cosine function even, odd, or neither? Is its graph symmetric? With respect to what?

109. Is the tangent function even, odd, or neither? Is its graph symmetric? With respect to what?

110. Is the cotangent function even, odd, or neither? Is its graph symmetric? With respect to what?

111. Is the secant function even, odd, or neither? Is its graph symmetric? With respect to what?

112. Is the cosecant function even, odd, or neither? Is its graph symmetric? With respect to what?

In Problems 113–118, use the periodic and even–odd properties.

113. If $f(\theta) = \sin\theta$ and $f(a) = \frac{1}{3}$, find the exact value of:
(a) $f(-a)$ (b) $f(a) + f(a + 2\pi) + f(a + 4\pi)$

114. If $f(\theta) = \cos\theta$ and $f(a) = \frac{1}{4}$, find the exact value of:
(a) $f(-a)$ (b) $f(a) + f(a + 2\pi) + f(a - 2\pi)$

115. If $f(\theta) = \tan\theta$ and $f(a) = 2$, find the exact value of:
(a) $f(-a)$ (b) $f(a) + f(a + \pi) + f(a + 2\pi)$

116. If $f(\theta) = \cot\theta$ and $f(a) = -3$, find the exact value of:
(a) $f(-a)$ (b) $f(a) + f(a + \pi) + f(a + 4\pi)$

117. If $f(\theta) = \sec\theta$ and $f(a) = -4$, find the exact value of:
(a) $f(-a)$ (b) $f(a) + f(a + 2\pi) + f(a + 4\pi)$

118. If $f(\theta) = \csc\theta$ and $f(a) = 2$, find the exact value of:
(a) $f(-a)$ (b) $f(a) + f(a + 2\pi) + f(a + 4\pi)$

119. Calculating the Time of a Trip From a parking lot, you want to walk to a house on the ocean. The house is located 1500 feet down a paved path that parallels the ocean, which is 500 feet away. See the illustration. Along the path you can walk 300 feet per minute, but in the sand on the beach you can only walk 100 feet per minute.

The time T to get from the parking lot to the beachhouse can be expressed as a function of the angle θ shown in the illustration and is

$$T(\theta) = 5 - \frac{5}{3\tan\theta} + \frac{5}{\sin\theta}, \quad 0 < \theta < \frac{\pi}{2}$$

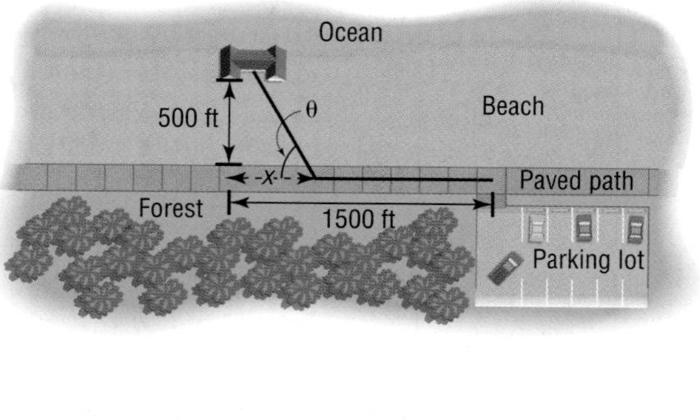

Calculate the time T if you walk directly from the parking lot to the house.

[**Hint:** $\tan \theta = \frac{500}{1500}$.]

120. Calculating the Time of a Trip Two oceanfront homes are located 8 miles apart on a straight stretch of beach, each a distance of 1 mile from a paved road that parallels the ocean. Sally can jog 8 miles per hour along the paved road, but only 3 miles per hour in the sand on the beach. Because of a river directly between the two houses, it is necessary to jog in the sand to the road, continue on the road, and then jog directly back in the sand to get from one house to the other. See the illustration. The time T to get from one house to the other as a function of the angle θ shown in the illustration is

$$T(\theta) = 1 + \frac{2}{3 \sin \theta} - \frac{1}{4 \tan \theta} \qquad 0 < \theta < \frac{\pi}{2}$$

(a) Calculate the time T for $\tan \theta = \frac{1}{4}$.
(b) Describe the path taken.
(c) Explain why θ must be larger than 14°.

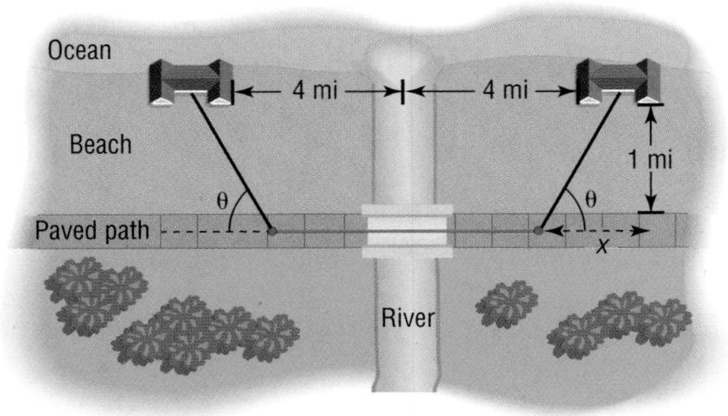

121. Show that the range of the tangent function is the set of all real numbers.

122. Show that the range of the cotangent function is the set of all real numbers.

123. Show that the period of $f(\theta) = \sin \theta$ is 2π.
[**Hint:** Assume that $0 < p < 2\pi$ exists so that $\sin(\theta + p) = \sin \theta$ for all θ. Let $\theta = 0$ to find p. Then let $\theta = \frac{\pi}{2}$ to obtain a contradiction.]

124. Show that the period of $f(\theta) = \cos \theta$ is 2π.

125. Show that the period of $f(\theta) = \sec \theta$ is 2π.

126. Show that the period of $f(\theta) = \csc \theta$ is 2π.

127. Show that the period of $f(\theta) = \tan \theta$ is π.

128. Show that the period of $f(\theta) = \cot \theta$ is π.

129. Prove the reciprocal identities given in formula (2).

130. Prove the quotient identities given in formula (3).

131. Establish the identity:
$$(\sin \theta \cos \phi)^2 + (\sin \theta \sin \phi)^2 + \cos^2 \theta = 1$$

132. Write down five properties of the tangent function. Explain the meaning of each.

133. Describe your understanding of the meaning of a periodic function.

134. Explain how to find the value of sin 390° using periodic properties.

135. Explain how to find the value of cos(−45°) using even–odd properties.

'Are You Prepared?' Answers

1. The set of all real numbers **2.** Even
3. The set of nonnegative real numbers **4.** True

5.4 Graphs of the Sine and Cosine Functions

PREPARING FOR THIS SECTION *Before getting started, review the following:*

- Graphing Techniques: Transformations (Section 2.5, pp. 94–103)

Now work the 'Are You Prepared?' problems on page 381.

OBJECTIVES
1 Graph Transformations of the Sine Function
2 Graph Transformations of the Cosine Function
3 Determine the Amplitude and Period of Sinusoidal Functions
4 Graph Sinusoidal Functions: $y = A \sin(\omega x)$
5 Find an Equation for a Sinusoidal Graph

Since we want to graph the trigonometric functions in the xy-plane, we shall use the traditional symbols x for the independent variable (or argument) and y for the dependent variable (or value at x) for each function. So we write the six trigonometric functions as

$y = f(x) = \sin x$	$y = f(x) = \cos x$	$y = f(x) = \tan x$
$y = f(x) = \csc x$	$y = f(x) = \sec x$	$y = f(x) = \cot x$

Here the independent variable x represents an angle, measured in radians. In calculus, x will usually be treated as a real number. As we said earlier, these are equivalent ways of viewing x.

The Graph of y = sin x

Table 6

x	$y = \sin x$	(x, y)
0	0	$(0, 0)$
$\frac{\pi}{6}$	$\frac{1}{2}$	$\left(\frac{\pi}{6}, \frac{1}{2}\right)$
$\frac{\pi}{2}$	1	$\left(\frac{\pi}{2}, 1\right)$
$\frac{5\pi}{6}$	$\frac{1}{2}$	$\left(\frac{5\pi}{6}, \frac{1}{2}\right)$
π	0	$(\pi, 0)$
$\frac{7\pi}{6}$	$-\frac{1}{2}$	$\left(\frac{7\pi}{6}, -\frac{1}{2}\right)$
$\frac{3\pi}{2}$	-1	$\left(\frac{3\pi}{2}, -1\right)$
$\frac{11\pi}{6}$	$-\frac{1}{2}$	$\left(\frac{11\pi}{6}, -\frac{1}{2}\right)$
2π	0	$(2\pi, 0)$

Since the sine function has period 2π, we need to graph $y = \sin x$ only on the interval $[0, 2\pi]$. The remainder of the graph will consist of repetitions of this portion of the graph.

We begin by constructing Table 6, which lists some points on the graph of $y = \sin x$, $0 \le x \le 2\pi$. As the table shows, the graph of $y = \sin x$, $0 \le x \le 2\pi$, begins at the origin. As x increases from 0 to $\frac{\pi}{2}$, the value of $y = \sin x$ increases from 0 to 1; as x increases from $\frac{\pi}{2}$ to π to $\frac{3\pi}{2}$, the value of y decreases from 1 to 0 to -1; as x increases from $\frac{3\pi}{2}$ to 2π, the value of y increases from -1 to 0. If we plot the points listed in Table 6 and connect them with a smooth curve, we obtain the graph shown in Figure 45.

Figure 45
$y = \sin x$, $0 \le x \le 2\pi$

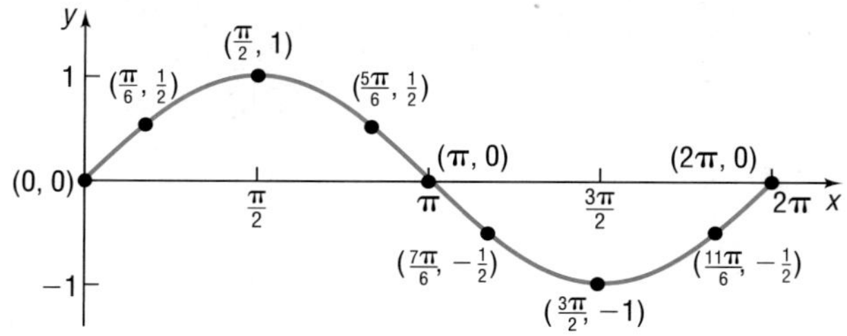

The graph in Figure 45 is one period, or **cycle,** of the graph of $y = \sin x$. To obtain a more complete graph of $y = \sin x$, we repeat this period in each direction, as shown in Figure 46.

Figure 46
$y = \sin x$, $-\infty < x < \infty$

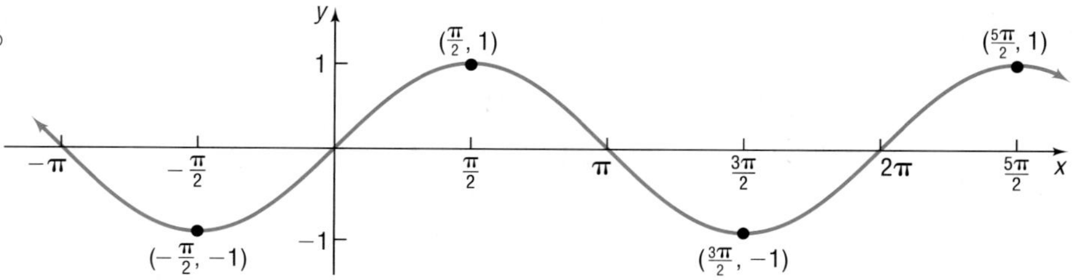

The graph of $y = \sin x$ illustrates some of the facts that we already know about the sine function.

Properties of the Sine Function

1. The domain is the set of all real numbers.
2. The range consists of all real numbers from -1 to 1, inclusive.
3. The sine function is an odd function, as the symmetry of the graph with respect to the origin indicates.
4. The sine function is periodic, with period 2π.
5. The x-intercepts are $\ldots, -2\pi, -\pi, 0, \pi, 2\pi, 3\pi, \ldots$; the y-intercept is 0.
6. The maximum value is 1 and occurs at $x = \ldots, -\frac{3\pi}{2}, \frac{\pi}{2}, \frac{5\pi}{2}, \frac{9\pi}{2}, \ldots$; the minimum value is -1 and occurs at $x = \ldots, -\frac{\pi}{2}, \frac{3\pi}{2}, \frac{7\pi}{2}, \frac{11\pi}{2}, \ldots$.

NOW WORK PROBLEMS 9, 11, AND 13.

1 The graphing techniques introduced in Chapter 2 may be used to graph functions that are transformations of the sine function (refer to Section 2.5).

EXAMPLE 1 Graphing Variations of $y = \sin x$ Using Transformations

Use the graph of $y = \sin x$ to graph $y = \sin\left(x - \frac{\pi}{2}\right)$.

Solution Figure 47 illustrates the steps.

Figure 47

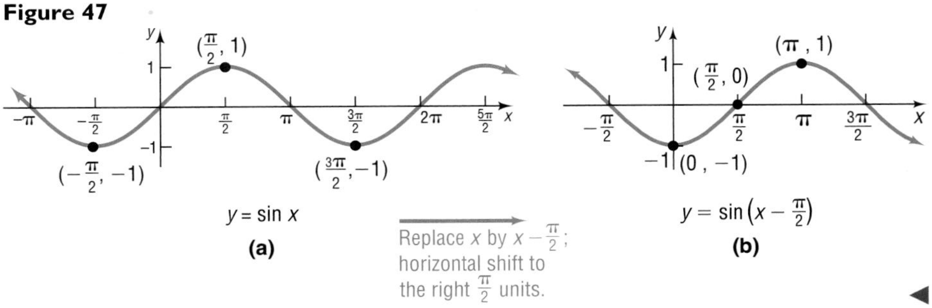

✔ **CHECK:** Graph $Y_1 = \sin\left(x - \frac{\pi}{2}\right)$ and compare the result with Figure 47(b).

EXAMPLE 2 Graphing Variations of $y = \sin x$ Using Transformations

Use the graph of $y = \sin x$ to graph $y = -\sin x + 2$.

Solution Figure 48 illustrates the steps.

Figure 48

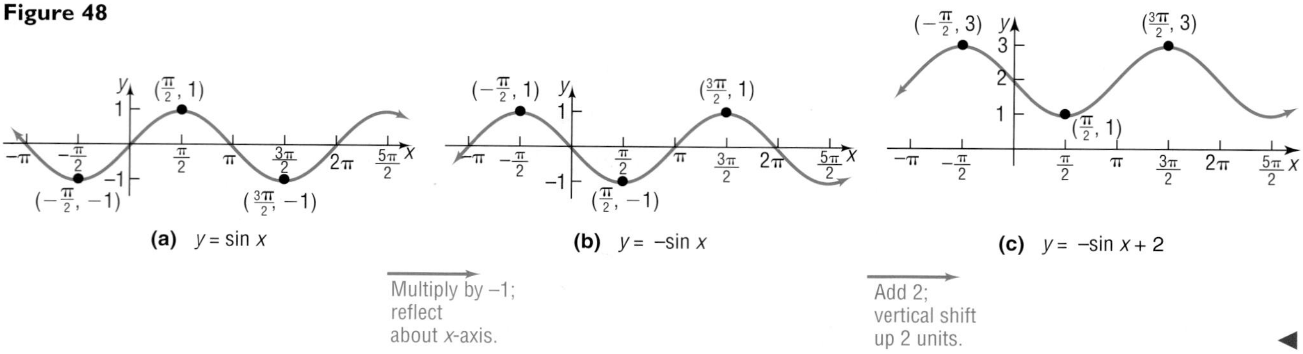

✔ **CHECK:** Graph $Y_1 = -\sin x + 2$ and compare the result with Figure 48(c).

NOW WORK PROBLEM 25.

The Graph of $y = \cos x$

The cosine function also has period 2π. We proceed as we did with the sine function by constructing Table 7, which lists some points on the graph of $y = \cos x$, $0 \le x \le 2\pi$. As the table shows, the graph of $y = \cos x$, $0 \le x \le 2\pi$, begins at the point $(0, 1)$. As x increases from 0 to $\frac{\pi}{2}$ to π, the value of y decreases from 1 to 0 to -1; as x increases from π to $\frac{3\pi}{2}$ to 2π, the value of y increases from -1 to 0 to 1. As before, we plot the points in Table 7 to get one period or cycle of the graph. See Figure 49.

Table 7

x	$y = \cos x$	(x, y)
0	1	$(0, 1)$
$\frac{\pi}{3}$	$\frac{1}{2}$	$\left(\frac{\pi}{3}, \frac{1}{2}\right)$
$\frac{\pi}{2}$	0	$\left(\frac{\pi}{2}, 0\right)$
$\frac{2\pi}{3}$	$-\frac{1}{2}$	$\left(\frac{2\pi}{3}, -\frac{1}{2}\right)$
π	-1	$(\pi, -1)$
$\frac{4\pi}{3}$	$-\frac{1}{2}$	$\left(\frac{4\pi}{3}, -\frac{1}{2}\right)$
$\frac{3\pi}{2}$	0	$\left(\frac{3\pi}{2}, 0\right)$
$\frac{5\pi}{3}$	$\frac{1}{2}$	$\left(\frac{5\pi}{3}, \frac{1}{2}\right)$
2π	1	$(2\pi, 1)$

Figure 49
$y = \cos x$, $0 \le x \le 2\pi$

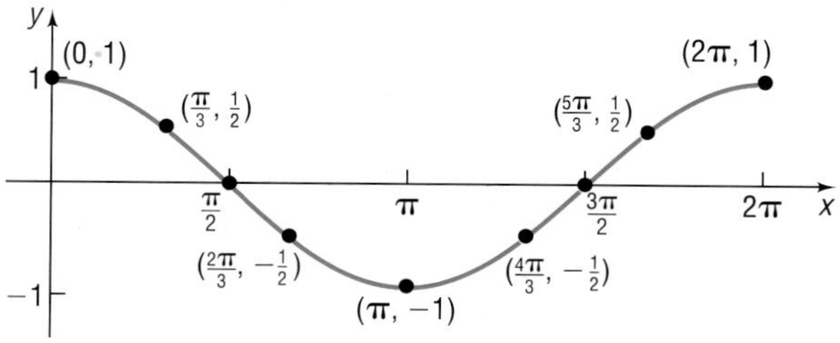

A more complete graph of $y = \cos x$ is obtained by repeating this period in each direction, as shown in Figure 50.

Figure 50
$y = \cos x$, $-\infty < x < \infty$

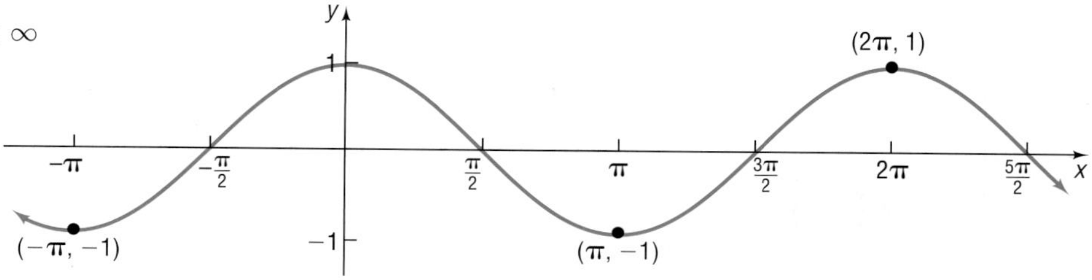

The graph of $y = \cos x$ illustrates some of the facts that we already know about the cosine function.

Properties of the Cosine Function

1. The domain is the set of all real numbers.
2. The range consists of all real numbers from -1 to 1, inclusive.
3. The cosine function is an even function, as the symmetry of the graph with respect to the y-axis indicates.
4. The cosine function is periodic, with period 2π.
5. The x-intercepts are $\dots, -\frac{3\pi}{2}, -\frac{\pi}{2}, \frac{\pi}{2}, \frac{3\pi}{2}, \frac{5\pi}{2}, \dots$; the y-intercept is 1.
6. The maximum value is 1 and occurs at $x = \dots, -2\pi, 0, 2\pi, 4\pi, 6\pi, \dots$; the minimum value is -1 and occurs at $x = \dots, -\pi, \pi, 3\pi, 5\pi, \dots$.

2 Again, the graphing techniques from Chapter 2 may be used to graph transformations of the cosine function.

EXAMPLE 3 Graphing Variations of $y = \cos x$ Using Transformations

Use the graph of $y = \cos x$ to graph $y = 3 \cos x$.

Solution Figure 51 illustrates the steps.

Figure 51

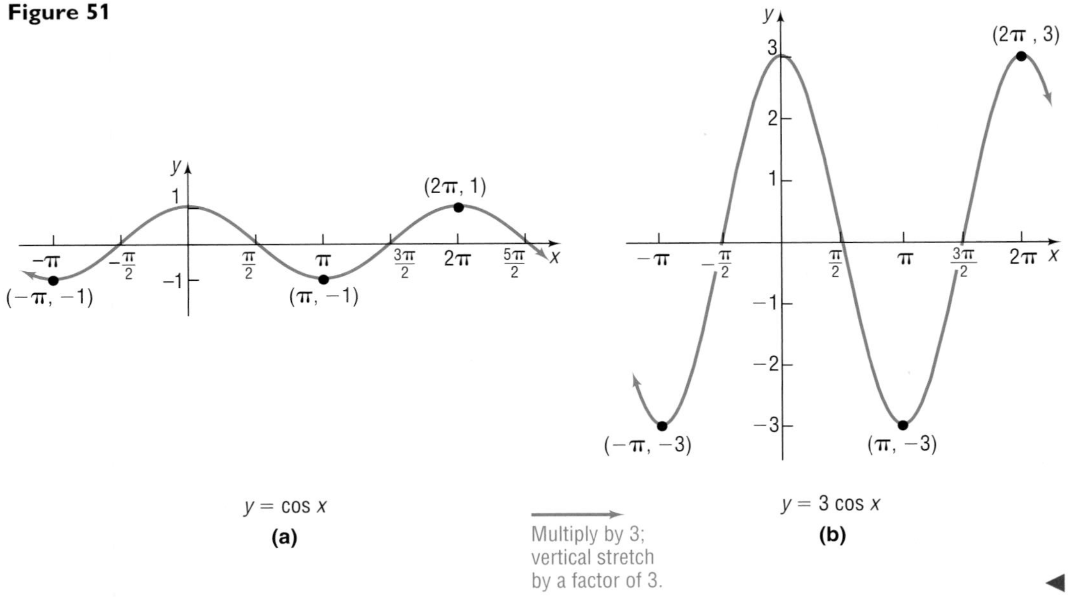

✔ **CHECK:** Graph $Y_1 = 3 \cos x$ and compare the result with Figure 51(b).

EXAMPLE 4 Graphing Variations of $y = \cos x$ Using Transformations

Use the graph of $y = \cos x$ to graph $y = \cos(2x)$.

Solution Figure 52 illustrates the graph, which is a horizontal compression of the graph of $y = \cos x$. $\left(\text{Multiply each } x\text{-coordinate by } \frac{1}{2}.\right)$ Notice that, due to this compression, the period of $y = \cos(2x)$ is $\frac{1}{2}(2\pi) = \pi$, whereas the period of $y = \cos x$ is 2π.

Figure 52

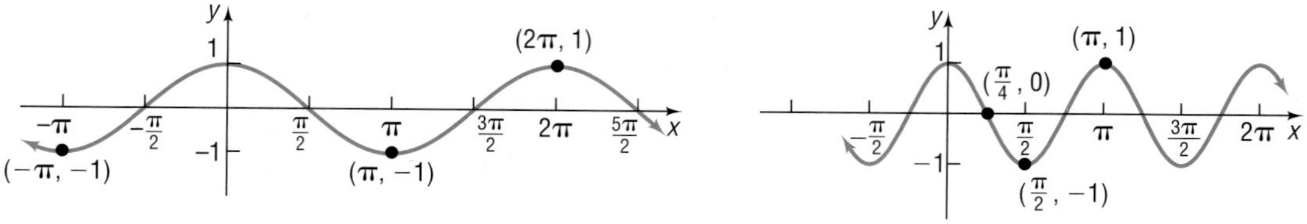

$y = \cos x$
(a)

Replace x by $2x$; horizontal compression by a factor of $\frac{1}{2}$.

$y = \cos(2x)$
(b)

✔ **CHECK:** Graph $Y_1 = \cos(2x)$. Use TRACE to verify that the period is π.

NOW WORK PROBLEM 33.

Sinusoidal Graphs

Shift the graph of $y = \cos x$ to the right $\frac{\pi}{2}$ units to obtain the graph of $y = \cos\left(x - \frac{\pi}{2}\right)$. See Figure 53(a). Now look at the graph of $y = \sin x$ in Figure 53(b). We see that the graph of $y = \sin x$ is the same as the graph of $y = \cos\left(x - \frac{\pi}{2}\right)$.

Figure 53

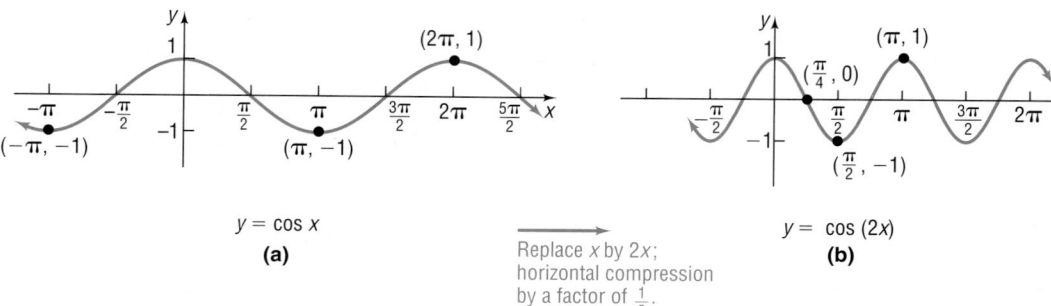

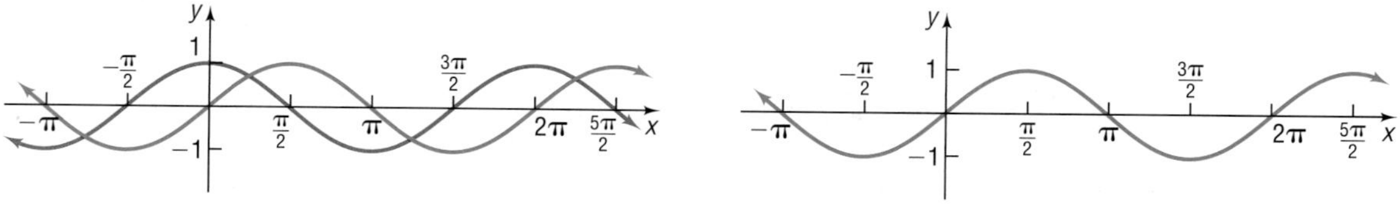

(a) $y = \cos x$ $y = \cos\left(x - \frac{\pi}{2}\right)$ **(b)** $y = \sin x$

Based on Figure 53, we conjecture that

$$\sin x = \cos\left(x - \frac{\pi}{2}\right)$$

(We shall prove this fact in Chapter 6.) Because of this similarity, the graphs of sine functions and cosine functions are referred to as **sinusoidal graphs.**

Seeing the Concept

Graph $Y_1 = \sin x$ and $Y_2 = \cos\left(x - \frac{\pi}{2}\right)$. How many graphs do you see?

Let's look at some general properties of sinusoidal graphs.

3 In Example 3 we obtained the graph of $y = 3\cos x$, which we reproduce in Figure 54. Notice that the values of $y = 3\cos x$ lie between -3 and 3, inclusive.

Figure 54
$y = 3\cos x$

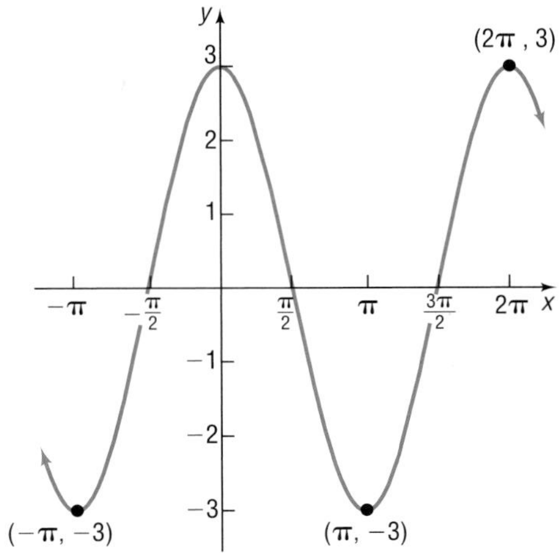

In general, the values of the functions $y = A\sin x$ and $y = A\cos x$, where $A \neq 0$, will always satisfy the inequalities

$$-|A| \leq A\sin x \leq |A| \quad \text{and} \quad -|A| \leq A\cos x \leq |A|$$

respectively. The number $|A|$ is called the **amplitude** of $y = A\sin x$ or $y = A\cos x$. See Figure 55.

Figure 55

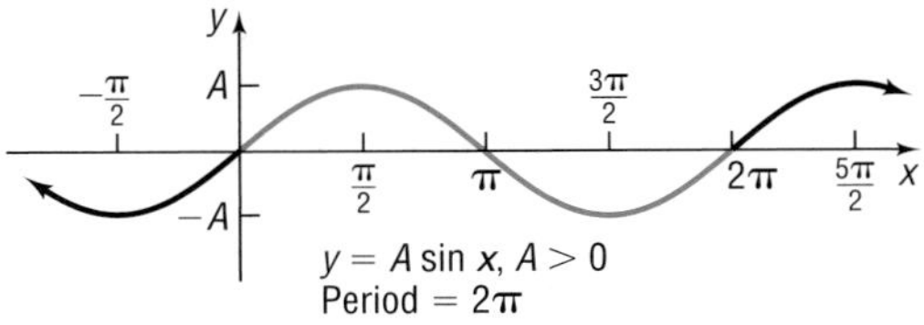

In Example 4, we obtained the graph of $y = \cos(2x)$, which we reproduce in Figure 56. Notice that the period of this function is π.

Figure 56
$y = \cos(2x)$

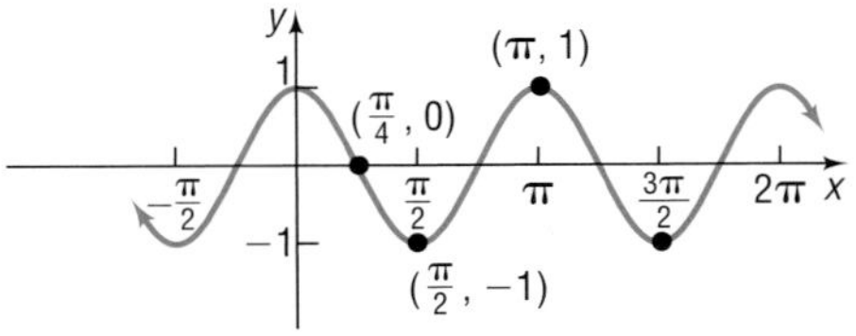

In general, if $\omega > 0$, the functions $y = \sin(\omega x)$ and $y = \cos(\omega x)$ will have period $T = \dfrac{2\pi}{\omega}$. To see why, recall that the graph of $y = \sin(\omega x)$ is obtained from the graph of $y = \sin x$ by performing a horizontal compression or stretch by a factor $\frac{1}{\omega}$. This horizontal compression replaces the interval $[0, 2\pi]$, which contains one period of the graph of $y = \sin x$, by the interval $\left[0, \frac{2\pi}{\omega}\right]$, which contains one period of the graph of $y = \sin(\omega x)$. The period of the functions $y = \sin(\omega x)$ and $y = \cos(\omega x)$, $\omega > 0$, is $\frac{2\pi}{\omega}$.

For example, for the function $y = \cos(2x)$ graphed in Figure 56, $\omega = 2$, so the period is $\frac{2\pi}{\omega} = \frac{2\pi}{2} = \pi$.

One period of the graph of $y = \sin(\omega x)$ or $y = \cos(\omega x)$ is called a **cycle.** Figure 57 illustrates the general situation. The blue portion of the graph is one cycle.

Figure 57

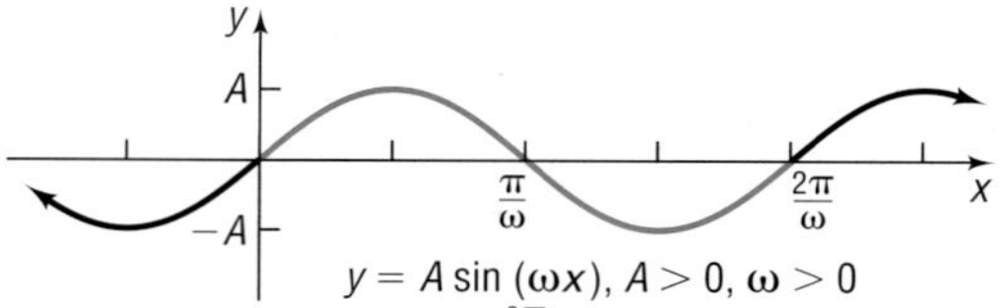

If $\omega < 0$ in $y = \sin(\omega x)$ or $y = \cos(\omega x)$, we use the even–odd properties of the sine and cosine functions as follows:

$$\sin(-\omega x) = -\sin(\omega x) \quad \text{and} \quad \cos(-\omega x) = \cos(\omega x)$$

This gives us an equivalent form in which the coefficient of x is positive. For example,

$$\sin(-2x) = -\sin(2x) \quad \text{and} \quad \cos(-\pi x) = \cos(\pi x)$$

Theorem

If $\omega > 0$, the amplitude and period of $y = A\sin(\omega x)$ and $y = A\cos(\omega x)$ are given by

$$\text{Amplitude} = |A| \qquad \text{Period} = T = \frac{2\pi}{\omega} \tag{1}$$

EXAMPLE 5 Finding the Amplitude and Period of a Sinusoidal Function

Determine the amplitude and period of $y = 3\sin(4x)$.

Solution Comparing $y = 3\sin(4x)$ to $y = A\sin(\omega x)$, we find that $A = 3$ and $\omega = 4$. From equation (1),

$$\text{Amplitude} = |A| = 3 \qquad \text{Period} = T = \frac{2\pi}{\omega} = \frac{2\pi}{4} = \frac{\pi}{2}$$

◀

NOW WORK PROBLEM 41.

Earlier, we graphed sine and cosine functions using tranformations. We now introduce another method that can be used to graph these functions.

Figure 58 shows one cycle of the graphs of $y = \sin x$ and $y = \cos x$ on the interval $[0, 2\pi]$. Notice that each graph consists of four parts corresponding to the four subintervals:

$$\left[0, \frac{\pi}{2}\right], \quad \left[\frac{\pi}{2}, \pi\right], \quad \left[\pi, \frac{3\pi}{2}\right], \quad \left[\frac{3\pi}{2}, 2\pi\right]$$

Each subinterval is of length $\frac{\pi}{2}$ (the period 2π divided by 4), and the endpoints of these intervals give rise to five key points, as shown in Figure 58.

Figure 58

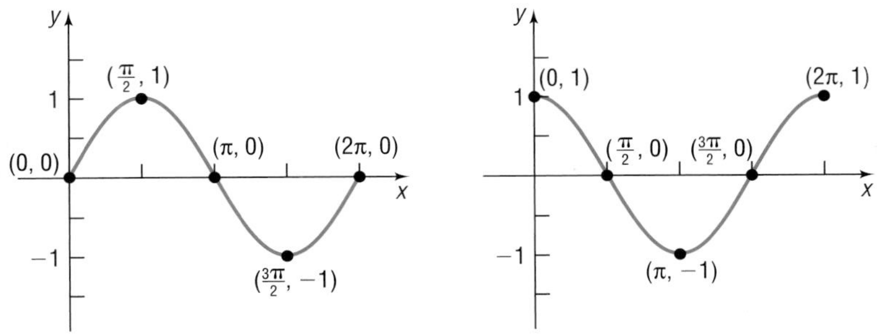

4 When graphing a sinusoidal function of the form $y = A\sin(\omega x)$ or $y = A\cos(\omega x)$, we use the amplitude to determine the maximum and minimum values of the function. The period is used to divide the x-axis into four subintervals. The endpoints of the subintervals give rise to five key points on the graph, which are used to sketch one cycle. Finally, extend the graph in either direction to make it complete. Let's look at an example.

EXAMPLE 6 Graphing a Sinusoidal Function

Graph: $y = 3\sin(4x)$

Solution From Example 5, the amplitude is 3 and the period is $\frac{\pi}{2}$. The graph of $y = 3\sin(4x)$ will lie between -3 and 3 on the y-axis. One cycle will begin at $x = 0$ and end at $x = \frac{\pi}{2}$.

We divide the interval $\left[0, \frac{\pi}{2}\right]$ into four subintervals, each of length $\frac{\pi}{2} \div 4 = \frac{\pi}{8}$:

$$\left[0, \frac{\pi}{8}\right], \quad \left[\frac{\pi}{8}, \frac{\pi}{4}\right], \quad \left[\frac{\pi}{4}, \frac{3\pi}{8}\right], \quad \left[\frac{3\pi}{8}, \frac{\pi}{2}\right]$$

The endpoints of these intervals give rise to five key points on the graph:

$$(0, 0), \quad \left(\frac{\pi}{8}, 3\right), \quad \left(\frac{\pi}{4}, 0\right), \quad \left(\frac{3\pi}{8}, -3\right), \quad \left(\frac{\pi}{2}, 0\right)$$

We plot these five points and fill in the graph of the sine curve as shown in Figure 59(a). If we extend the graph in either direction, we obtain the complete graph shown in Figure 59(b).

Figure 59

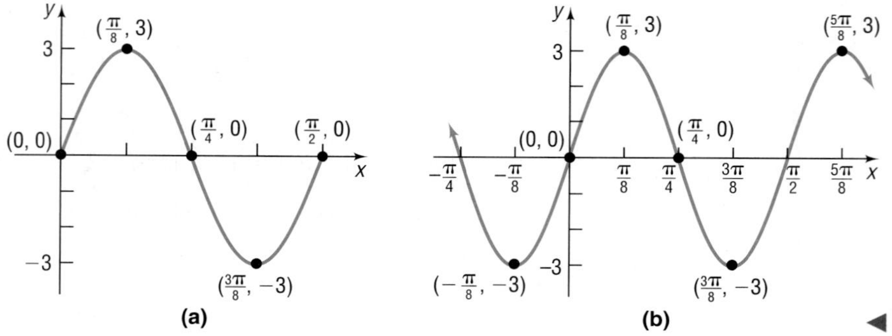

✔ **CHECK:** Graph $y = 3\sin(4x)$ using transformations.

Which graphing method do you prefer?

✔ **CHECK:** Graph $Y_1 = 3\sin(4x)$ using a graphing utility.

[**Hint**: Use the amplitude to set Ymin, Ymax. Use the period to set Xmin, Xmax.]

NOW WORK PROBLEM 47.

EXAMPLE 7 Finding the Amplitude and Period of a Sinusoidal Function and Graphing It

Determine the amplitude and period of $y = -4\cos(\pi x)$, and graph the function.

Solution Comparing $y = -4\cos(\pi x)$ with $y = A\cos(\omega x)$, we find that $A = -4$ and $\omega = \pi$. The amplitude is $|A| = |-4| = 4$, and the period is $T = \frac{2\pi}{\omega} = \frac{2\pi}{\pi} = 2$.

The graph of $y = -4\cos(\pi x)$ will lie between -4 and 4 on the y-axis. One cycle will begin at $x = 0$ and end at $x = 2$. We divide the interval $[0, 2]$ into four subintervals, each of length $2 \div 4 = \frac{1}{2}$:

$$\left[0, \frac{1}{2}\right], \quad \left[\frac{1}{2}, 1\right], \quad \left[1, \frac{3}{2}\right], \quad \left[\frac{3}{2}, 2\right]$$

The five key points on the graph are

$$(0, -4), \quad \left(\frac{1}{2}, 0\right), \quad (1, 4), \quad \left(\frac{3}{2}, 0\right), \quad (2, -4)$$

We plot these five points and fill in the graph of the cosine function as shown in Figure 60(a). Extending the graph in either direction, we obtain Figure 60(b).

Figure 60

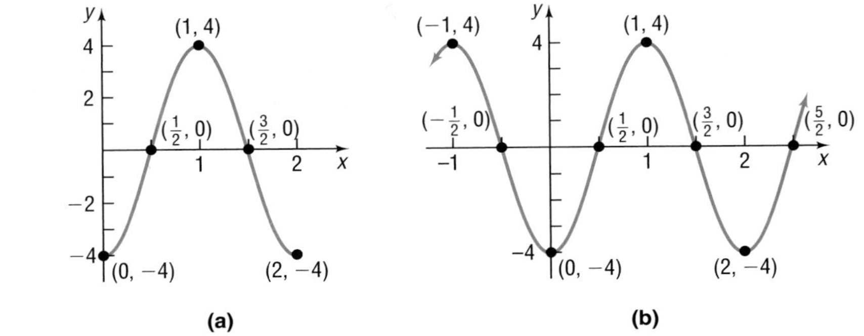

✔ **CHECK:** Graph $y = -4\cos(\pi x)$ using transformations.

Which graphing method do you prefer?

✔ **CHECK:** Graph $Y_1 = -4\cos(\pi x)$ using a graphing utility.

EXAMPLE 8 Finding the Amplitude and Period of a Sinusoidal Function and Graphing It

Determine the amplitude and period of $y = 2\sin\left(-\frac{\pi}{2}x\right)$, and graph the function.

Solution Since the sine function is odd, we can use the equivalent form:

$$y = -2\sin\left(\frac{\pi}{2}x\right)$$

Comparing $y = -2\sin\left(\frac{\pi}{2}x\right)$ to $y = A\sin(\omega x)$, we find that $A = -2$ and $\omega = \frac{\pi}{2}$. The amplitude is $|A| = 2$, and the period is $T = \frac{2\pi}{\omega} = \frac{2\pi}{\frac{\pi}{2}} = 4$.

The graph of $y = -2\sin\left(\frac{\pi}{2}x\right)$ will lie between -2 and 2 on the y-axis. One cycle will begin at $x = 0$ and end at $x = 4$. We divide the interval $[0, 4]$ into four subintervals, each of length $4 \div 4 = 1$:

$$[0, 1], \quad [1, 2], \quad [2, 3], \quad [3, 4]$$

The five key points on the graph are

$$(0, 0), \quad (1, -2), \quad (2, 0), \quad (3, 2), \quad (4, 0)$$

We plot these five points and fill in the graph of the sine function as shown in Figure 61(a). Extending the graph in either direction, we obtain Figure 61(b).

Figure 61

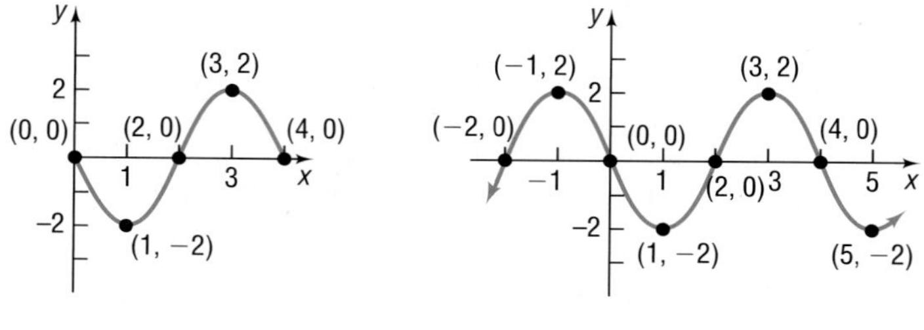

✔ **CHECK:** Graph $y = 2\sin\left(-\frac{\pi}{2}x\right)$ using transformations.

Which graphing method do you prefer?

✔ **CHECK:** Graph $Y_1 = 2\sin\left(-\frac{\pi}{2}x\right)$ using a graphing utility.

NOW WORK PROBLEM 61.

5 We can also use the ideas of amplitude and period to identify a sinusoidal function when its graph is given.

EXAMPLE 9 Finding an Equation for a Sinusoidal Graph

Find an equation for the graph shown in Figure 62.

Figure 62

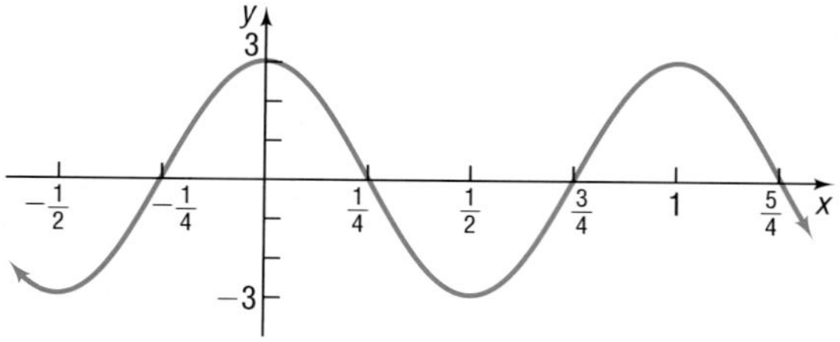

Solution The graph has the characteristics of a cosine function. Do you see why? So we view the equation as a cosine function $y = A\cos(\omega x)$ with $A = 3$ and period $T = 1$. Then $\frac{2\pi}{\omega} = 1$, so $\omega = 2\pi$. The cosine function whose graph is given in Figure 62 is

$$y = A\cos(\omega x) = 3\cos(2\pi x)$$

✔ **CHECK:** Graph $Y_1 = 3\cos(2\pi x)$ and compare the result with Figure 62.

EXAMPLE 10 Finding an Equation for a Sinusoidal Graph

Find an equation for the graph shown in Figure 63.

Figure 63

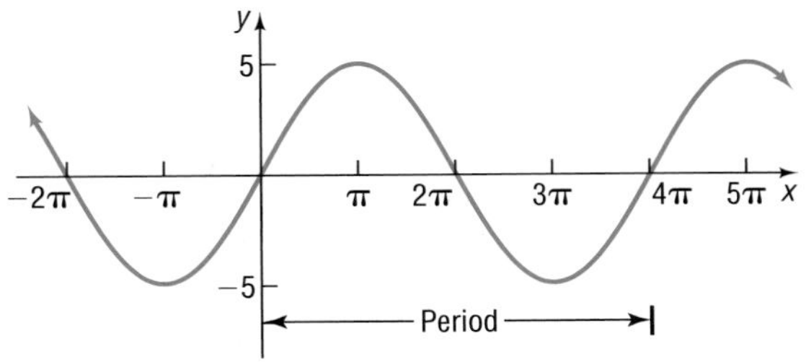

Solution This graph has the characteristics of a sine function* $y = A\sin(\omega x)$ with $A = 5$. The period T is observed to be 4π. By equation (1),

$$T = \frac{2\pi}{\omega}$$

$$4\pi = \frac{2\pi}{\omega}$$

$$\omega = \frac{2\pi}{4\pi} = \frac{1}{2}$$

The sine function whose graph is given in Figure 63 is

$$y = A\sin(\omega x) = 5\sin\left(\frac{1}{2}x\right)$$

◀

✔ **CHECK:** Graph $Y_1 = 5\sin\left(\frac{1}{2}x\right)$ and compare the result with Figure 63.

NOW WORK PROBLEMS 71 AND 75.

*The graph could also be viewed as a cosine function with a horizontal shift, but viewing it as a sine function is easier because the graph passes through the origin.

5.4 Assess Your Understanding

'Are You Prepared?' *Answers are given at the end of these exercises. If you get a wrong answer, read the pages listed in red.*

1. Use transformations to graph $y = 3x^2$ (pp. 94–103)

2. Use transformations to graph $y = -x^2$. (pp. 94–103)

Concepts and Vocabulary

3. The maximum value of $y = \sin x$ is ________ and occurs at $x =$ ________.

4. The function $y = A\sin(\omega x)$, $A > 0$, has amplitude 3 and period 2; then $A =$ ________ and $\omega =$ ________.

5. The function $y = 3\cos(6x)$ has amplitude ________ and period ________.

6. *True or False:* The graphs of $y = \sin x$ and $y = \cos x$ are identical except for a horizontal shift.

7. *True or False:* For $y = 2\sin(\pi x)$, the amplitude is 2 and the period is $\frac{\pi}{2}$.

8. *True or False:* The graph of the sine function has infinitely many x-intercepts.

Exercises

In Problems 9–18, if necessary, refer to the graphs to answer each question.

9. What is the y-intercept of $y = \sin x$?

10. What is the y-intercept of $y = \cos x$?

11. For what numbers x, $-\pi \le x \le \pi$, is the graph of $y = \sin x$ increasing?

12. For what numbers x, $-\pi \le x \le \pi$, is the graph of $y = \cos x$ decreasing?

13. What is the largest value of $y = \sin x$?

14. What is the smallest value of $y = \cos x$?

15. For what numbers x, $0 \le x \le 2\pi$, does $\sin x = 0$?

16. For what numbers x, $0 \le x \le 2\pi$, does $\cos x = 0$?

17. For what numbers x, $-2\pi \le x \le 2\pi$, does $\sin x = 1$? What about $\sin x = -1$?

18. For what numbers x, $-2\pi \le x \le 2\pi$, does $\cos x = 1$? What about $\cos x = -1$?

In Problems 19 and 20, match the graph to a function. Three answers are possible.

A. $y = -\sin x$

B. $y = -\cos x$

C. $y = \sin\left(x - \frac{\pi}{2}\right)$

D. $y = -\cos\left(x - \frac{\pi}{2}\right)$

E. $y = \sin(x + \pi)$

F. $y = \cos(x + \pi)$

19.

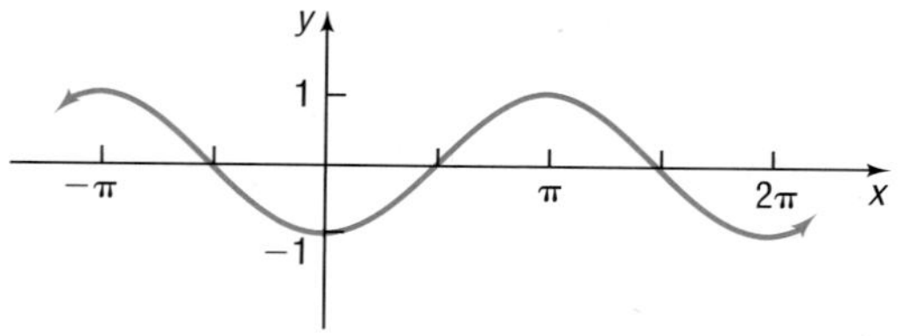

20.

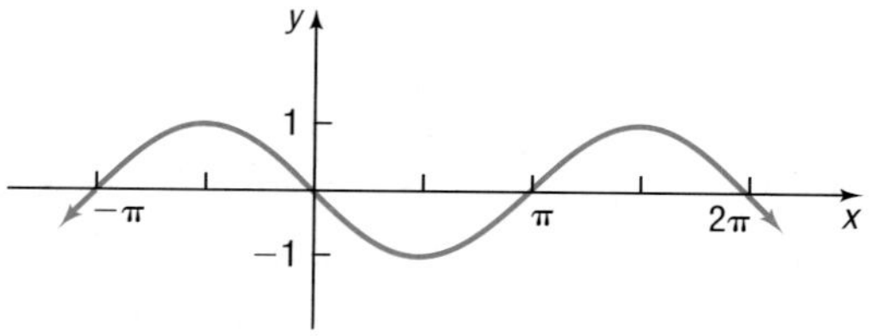

In Problems 21–36, use transformations to graph each function.

21. $y = 3\sin x$

22. $y = 4\cos x$

23. $y = -\cos x$

24. $y = -\sin x$

25. $y = \sin x - 1$

26. $y = \cos x + 1$

27. $y = \sin(x - \pi)$

28. $y = \cos(x + \pi)$

29. $y = \sin(\pi x)$

30. $y = \cos\left(\frac{\pi}{2}x\right)$

31. $y = 2\sin x + 2$

32. $y = 3\cos x + 3$

33. $y = 4\cos(2x)$

34. $y = 3\sin\left(\frac{1}{2}x\right)$

35. $y = -2\sin x + 2$

36. $y = -3\cos x - 2$

In Problems 37–46, determine the amplitude and period of each function without graphing.

37. $y = 2\sin x$

38. $y = 3\cos x$

39. $y = -4\cos(2x)$

40. $y = -\sin\left(\frac{1}{2}x\right)$

41. $y = 6\sin(\pi x)$

42. $y = -3\cos(3x)$

43. $y = -\frac{1}{2}\cos\left(\frac{3}{2}x\right)$

44. $y = \frac{4}{3}\sin\left(\frac{2}{3}x\right)$

45. $y = \frac{5}{3}\sin\left(-\frac{2\pi}{3}x\right)$

46. $y = \frac{9}{5}\cos\left(-\frac{3\pi}{2}x\right)$

In Problems 47–56, match the given function to one of the graphs (A)–(J).

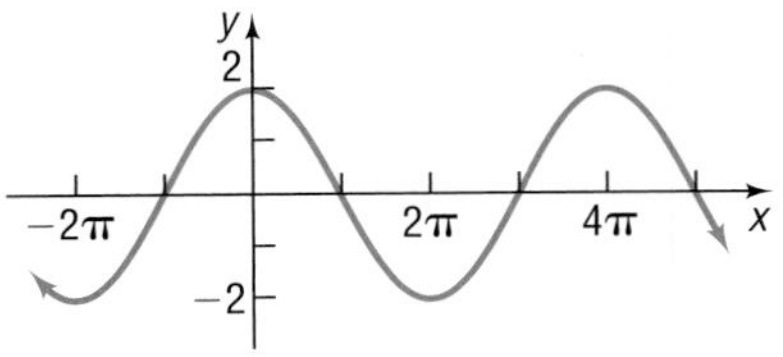

(A)

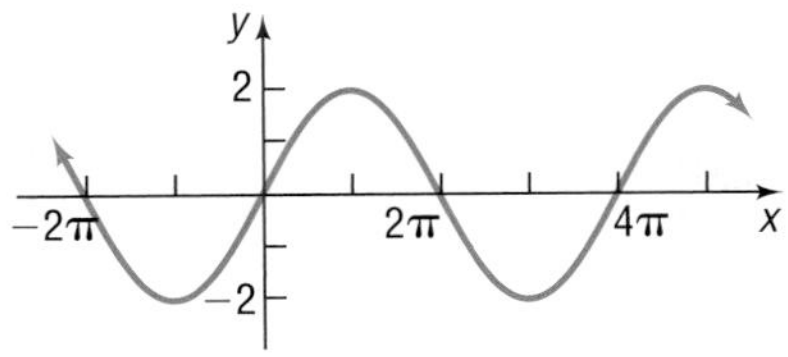

(B)

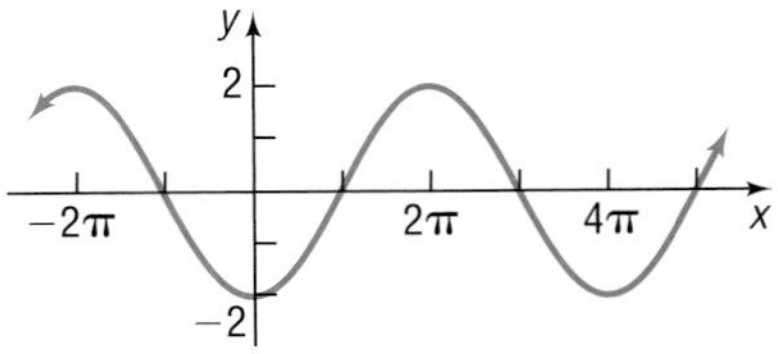

(C)

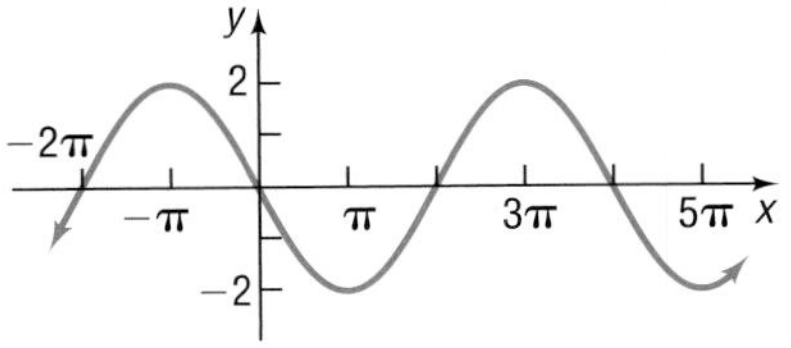

(D)

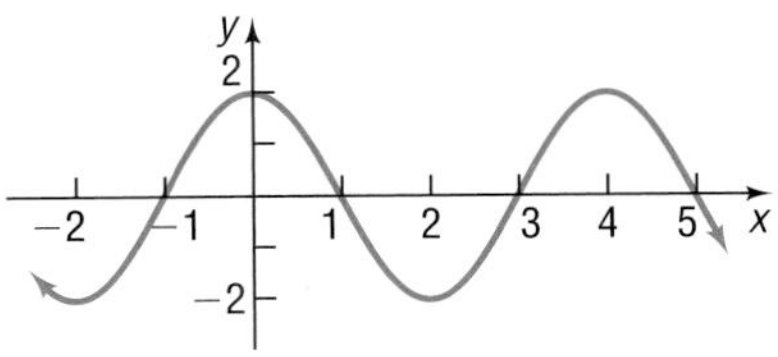

(E)

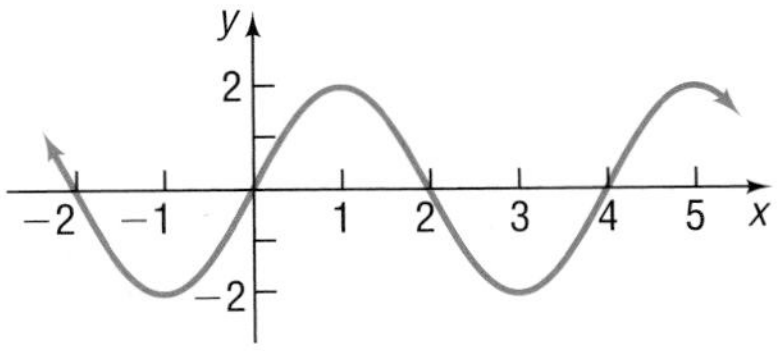

(F)

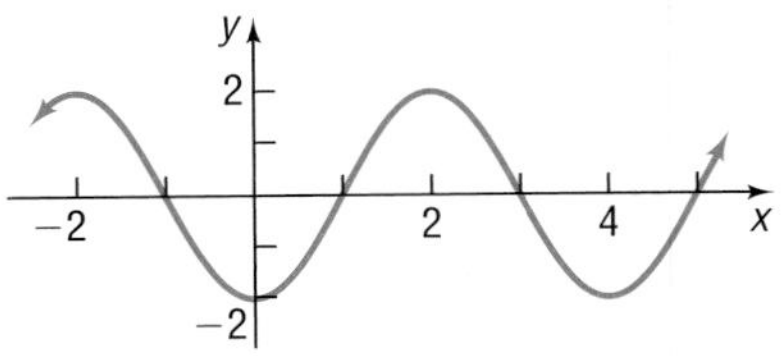

(G)

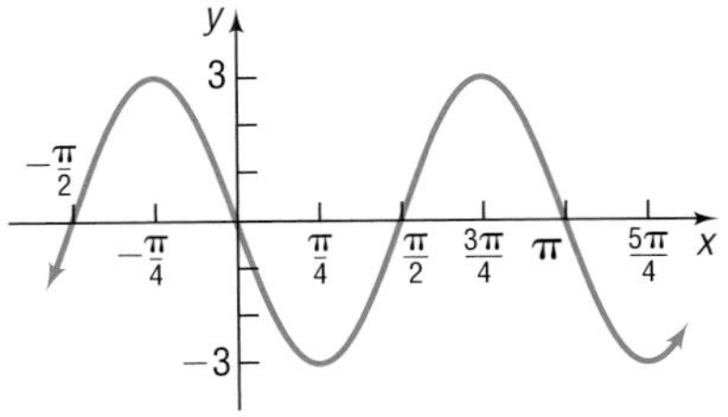

(H)

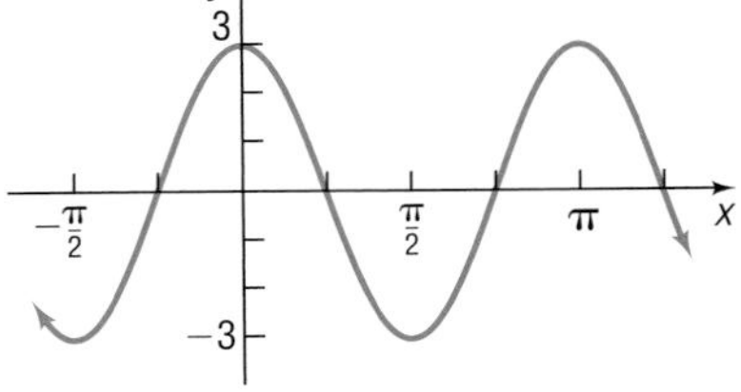

(I)

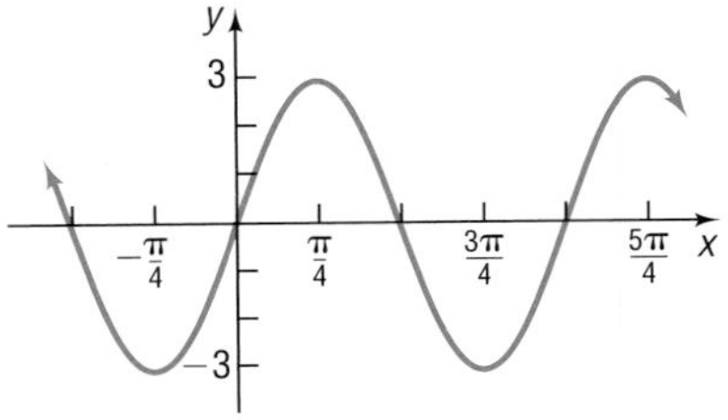

(J)

47. $y = 2\sin\left(\frac{\pi}{2}x\right)$

48. $y = 2\cos\left(\frac{\pi}{2}x\right)$

49. $y = 2\cos\left(\frac{1}{2}x\right)$

50. $y = 3\cos(2x)$

51. $y = -3\sin(2x)$

52. $y = 2\sin\left(\frac{1}{2}x\right)$

53. $y = -2\cos\left(\frac{1}{2}x\right)$

54. $y = -2\cos\left(\frac{\pi}{2}x\right)$

55. $y = 3\sin(2x)$

56. $y = -2\sin\left(\frac{1}{2}x\right)$

In Problems 57–60, match the given function to one of the graphs (A)–(D).

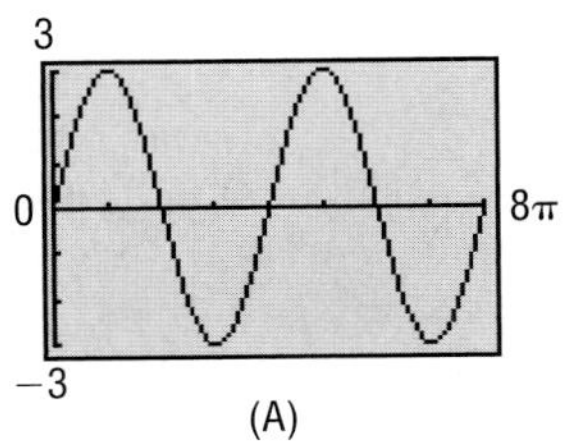

(A)

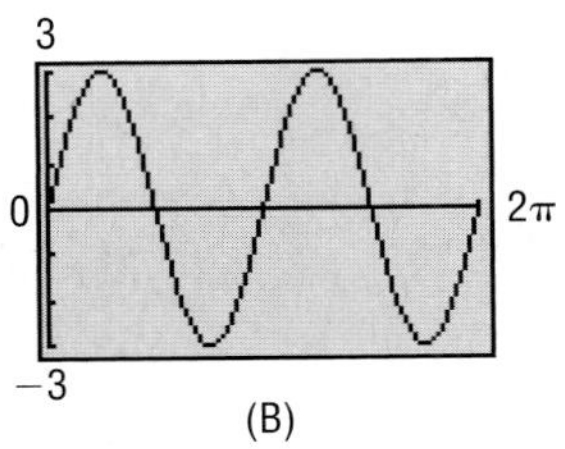

(B)

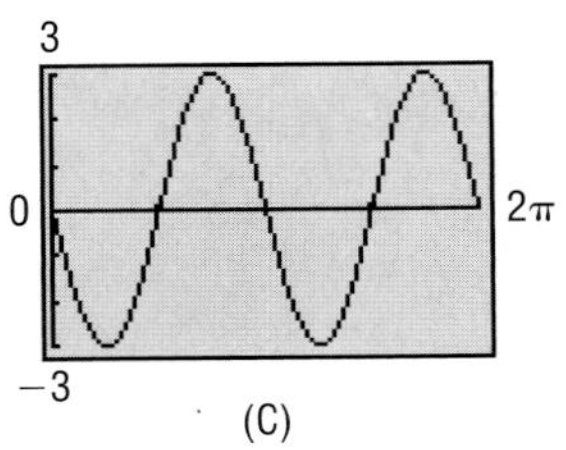

(C)

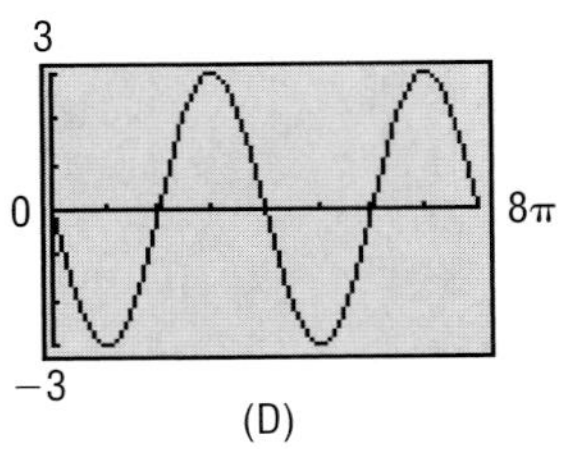

(D)

57. $y = 3\sin\left(\frac{1}{2}x\right)$

58. $y = -3\sin(2x)$

59. $y = 3\sin(2x)$

60. $y = -3\sin\left(\frac{1}{2}x\right)$

In Problems 61–70, graph each sinusoidal function.

61. $y = 5\sin(4x)$ **62.** $y = 4\cos(6x)$ **63.** $y = 5\cos(\pi x)$ **64.** $y = 2\sin(\pi x)$

65. $y = -2\cos(2\pi x)$ **66.** $y = -5\cos(2\pi x)$ **67.** $y = -4\sin\left(\frac{1}{2}x\right)$ **68.** $y = -2\cos\left(\frac{1}{2}x\right)$

69. $y = \frac{3}{2}\sin\left(-\frac{2}{3}x\right)$ **70.** $y = \frac{4}{3}\cos\left(-\frac{1}{3}x\right)$

In Problems 71–74, write the equation of a sine function that has the given characteristics.

71. Amplitude: 3
Period: π

72. Amplitude: 2
Period: 4π

73. Amplitude: 3
Period: 2

74. Amplitude: 4
Period: 1

In Problems 75–88, find an equation for each graph.

75.

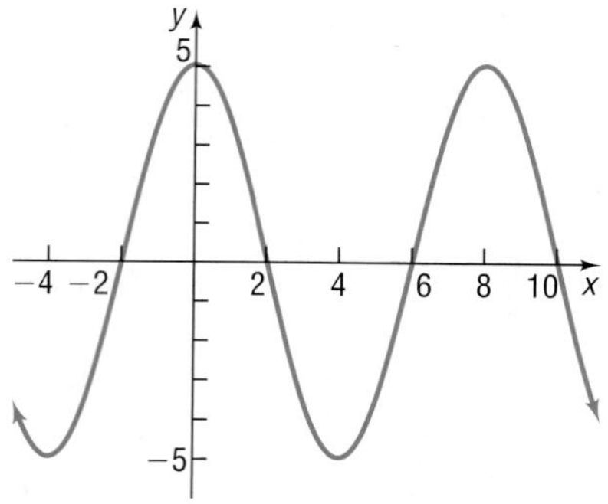

76.

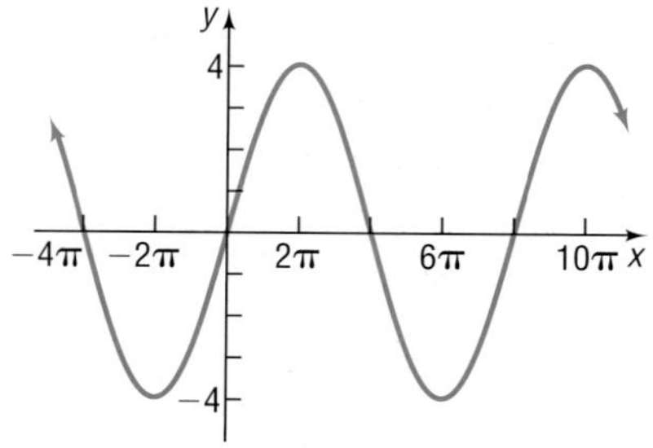

77.

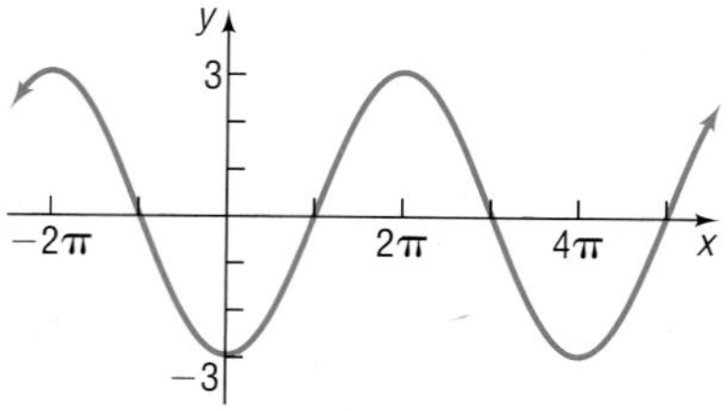

78.

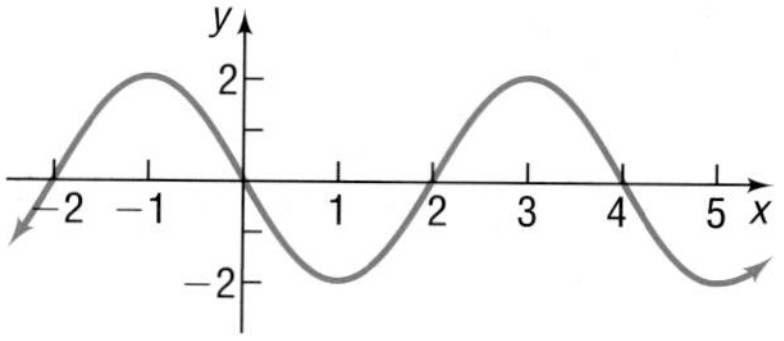

79.

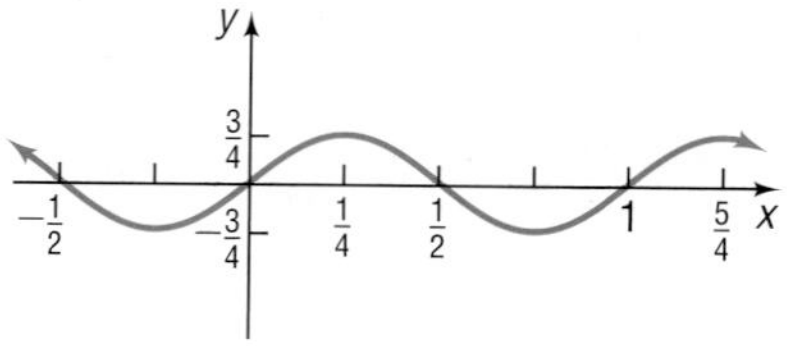

80.

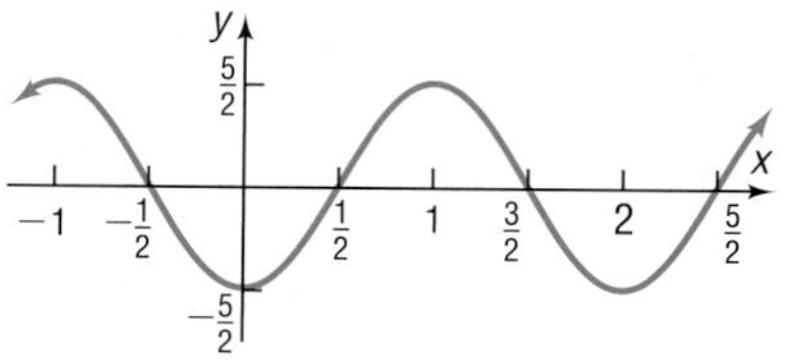

81.

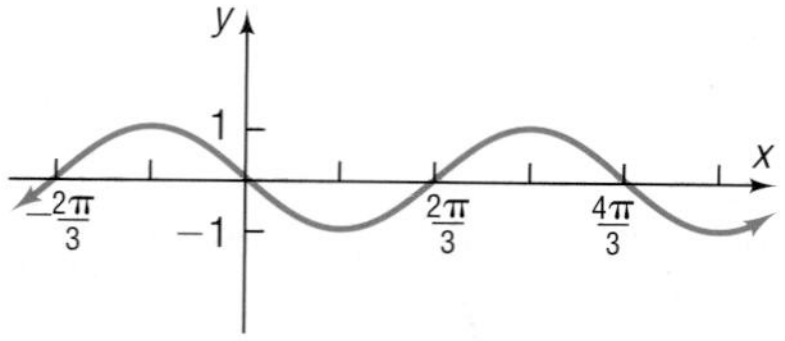

82.

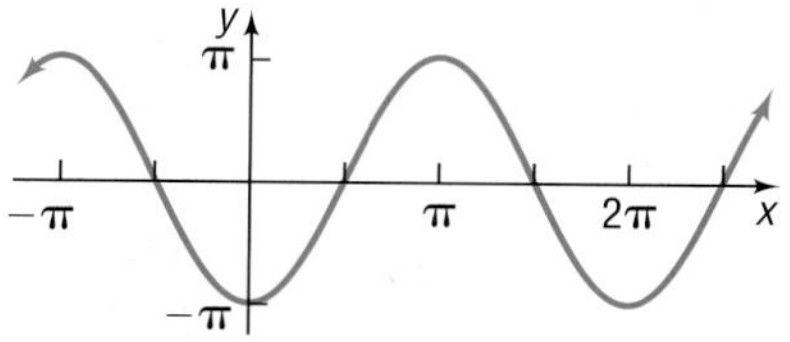

83.

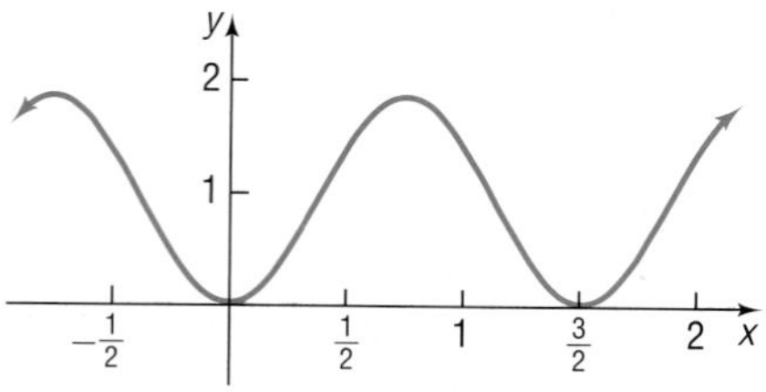

84.

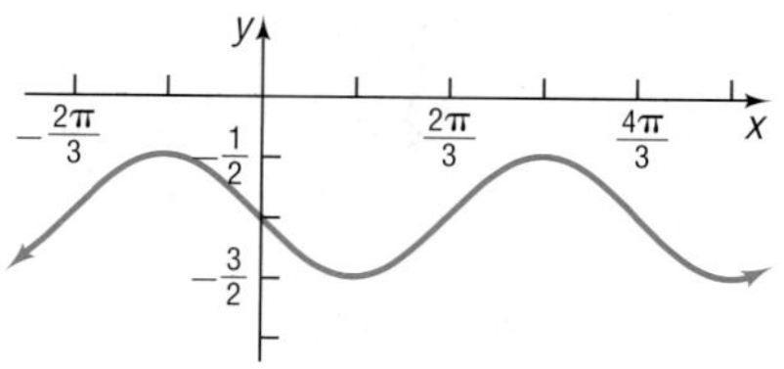

85.

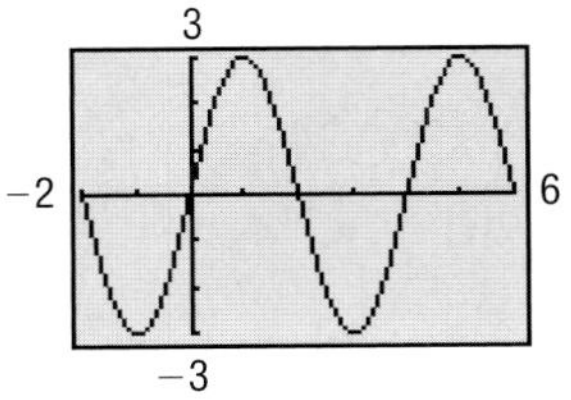

86.

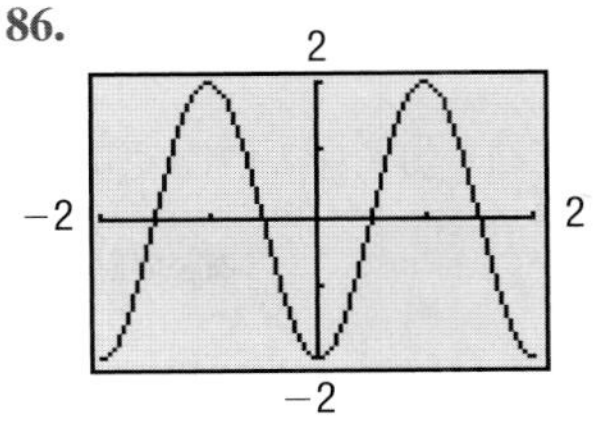

87.

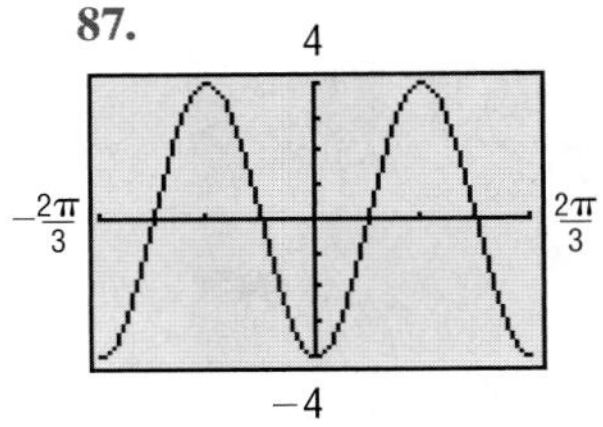

88.

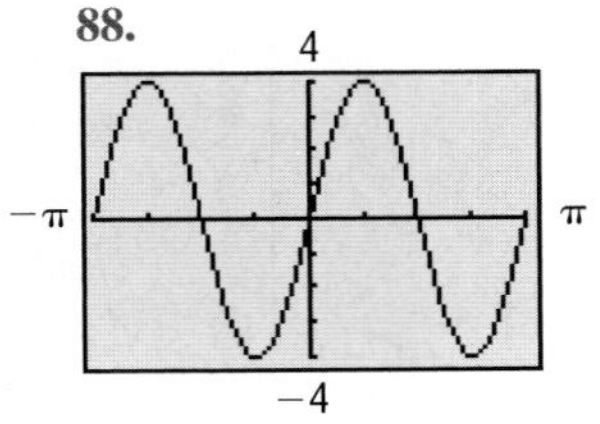

89. Alternating Current (ac) Circuits The current I, in amperes, flowing through an ac (alternating current) circuit at time t is

$$I = 220 \sin(60\pi t), \qquad t \geq 0$$

What is the period? What is the amplitude? Graph this function over two periods.

90. Alternating Current (ac) Circuits The current I, in amperes, flowing through an ac (alternating current) circuit at time t is

$$I = 120 \sin(30\pi t), \qquad t \geq 0$$

What is the period? What is the amplitude? Graph this function over two periods.

91. Alternating Current (ac) Generators The voltage V produced by an ac generator is

$$V = 220 \sin(120\pi t)$$

(a) What is the amplitude? What is the period?
(b) Graph V over two periods, beginning at $t = 0$.
(c) If a resistance of $R = 10$ ohms is present, what is the current I?
[**Hint:** Use Ohm's Law, $V = IR$.]
(d) What is the amplitude and period of the current I?
(e) Graph I over two periods, beginning at $t = 0$.

92. Alternating Current (ac) Generators The voltage V produced by an ac generator is

$$V = 120 \sin(120\pi t)$$

(a) What is the amplitude? What is the period?
(b) Graph V over two periods, beginning at $t = 0$.
(c) If a resistance of $R = 20$ ohms is present, what is the current I?
[**Hint:** Use Ohm's Law, $V = IR$.]
(d) What is the amplitude and period of the current I?
(e) Graph I over two periods, beginning at $t = 0$.

93. Alternating Current (ac) Generators The voltage V produced by an ac generator is sinusoidal. As a function of time, the voltage V is

$$V = V_0 \sin(2\pi f t)$$

where f is the **frequency,** the number of complete oscillations (cycles) per second. [In the United States and Canada, f is 60 hertz (Hz).] The **power** P delivered to a resistance R at any time t is defined as

$$P = \frac{V^2}{R}$$

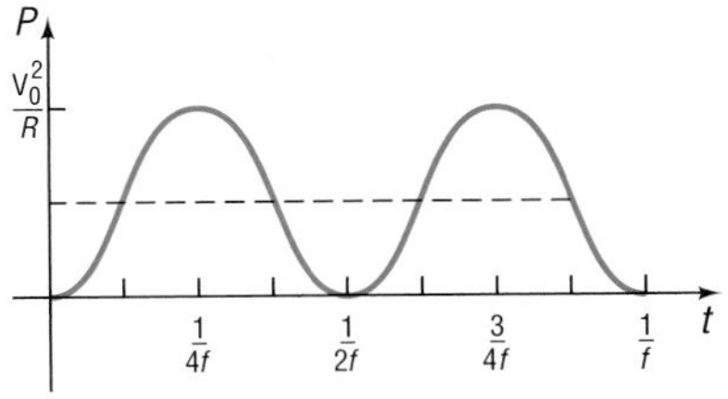

Power in an ac generator

(a) Show that $P = \frac{V_0^2}{R} \sin^2(2\pi f t)$.
(b) The graph of P is shown in the figure. Express P as a sinusoidal function.
(c) Deduce that

$$\sin^2(2\pi f t) = \frac{1}{2}[1 - \cos(4\pi f t)]$$

94. Biorhythms In the theory of biorhythms, a sine function of the form

$$P = 50 \sin(\omega t) + 50$$

is used to measure the percent P of a person's potential at time t, where t is measured in days and $t = 0$ is the person's birthday. Three characteristics are commonly measured:

Physical potential: period of 23 days
Emotional potential: period of 28 days
Intellectual potential: period of 33 days

(a) Find ω for each characteristic.
(b) Graph all three functions.
(c) Is there a time t when all three characteristics have 100% potential? When is it?
(d) Suppose that you are 20 years old today ($t = 7305$ days). Describe your physical, emotional, and intellectual potential for the next 30 days.

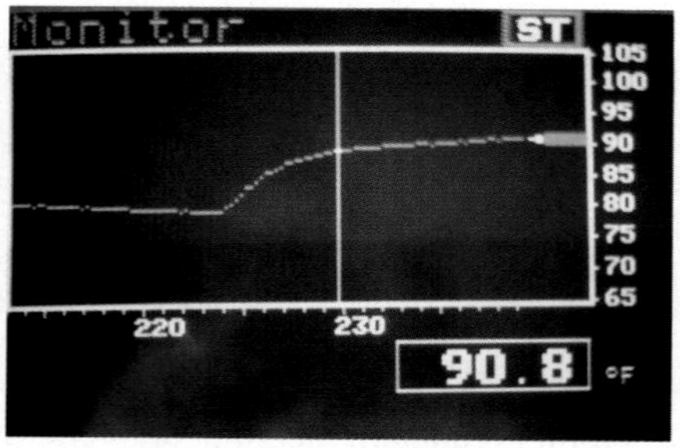

95. Graph $y = |\cos x|, -2\pi \le x \le 2\pi$.

96. Graph $y = |\sin x|, -2\pi \le x \le 2\pi$.

97. Draw a quick sketch of $y = \sin x$. Be sure to label at least five points.

98. Explain how you would scale the x-axis and y-axis before graphing $y = 3\cos(\pi x)$.

99. Explain the term *amplitude* as it relates to the graph of a sinusoidal function.

100. Explain how the amplitude and period of a sinusoidal graph are used to establish the scale on each coordinate axis.

101. Find an application in your major field that leads to a sinusoidal graph. Write a paper about your findings.

'Are You Prepared?' Answers

1. Vertical stretch by a factor of 3

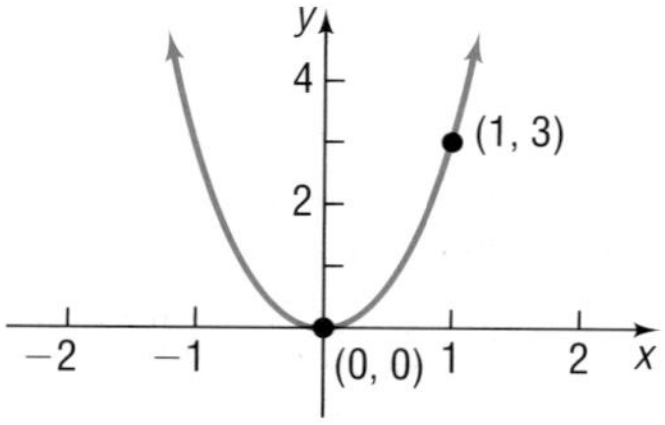

2. Reflection about the x-axis

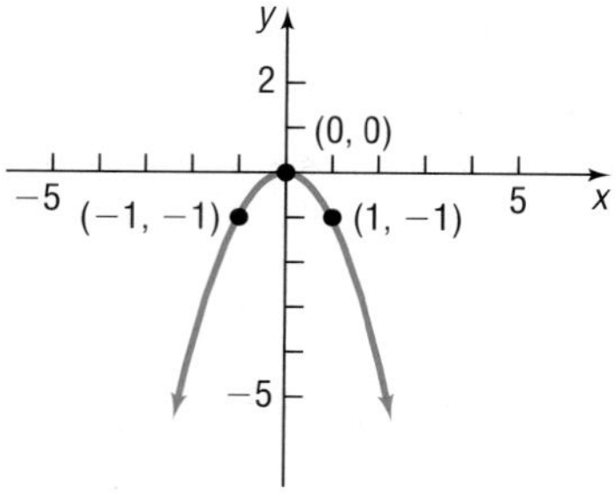

5.5 Graphs of the Tangent, Cotangent, Cosecant, and Secant Functions

PREPARING FOR THIS SECTION *Before getting started, review the following:*

- Vertical Asymptotes (Section 3.3, pp. 165–167)

Now work the 'Are You Prepared?' problems on page 392.

OBJECTIVES
1 Graph Transformations of the Tangent Function and Cotangent Function
2 Graph Transformations of the Cosecant Function and Secant Function

The Graphs of $y = \tan x$ and $y = \cot x$

1 Because the tangent function has period π, we only need to determine the graph over some interval of length π. The rest of the graph will consist of repetitions of that graph. Because the tangent function is not defined at $\ldots, -\frac{3\pi}{2}, -\frac{\pi}{2}, \frac{\pi}{2}, \frac{3\pi}{2}, \ldots$, we will concentrate on the interval $\left(-\frac{\pi}{2}, \frac{\pi}{2}\right)$, of length π, and construct Table 8, which lists some points on the graph of $y = \tan x$, $-\frac{\pi}{2} < x < \frac{\pi}{2}$. We plot the points in the table and connect them with a smooth curve. See Figure 64 for a partial graph of $y = \tan x$, where $-\frac{\pi}{3} \le x \le \frac{\pi}{3}$.

Table 8

x	$y = \tan x$	(x, y)
$-\frac{\pi}{3}$	$-\sqrt{3} \approx -1.73$	$\left(-\frac{\pi}{3}, -\sqrt{3}\right)$
$-\frac{\pi}{4}$	-1	$\left(-\frac{\pi}{4}, -1\right)$
$-\frac{\pi}{6}$	$-\frac{\sqrt{3}}{3} \approx -0.58$	$\left(-\frac{\pi}{6}, -\frac{\sqrt{3}}{3}\right)$
0	0	(0, 0)
$\frac{\pi}{6}$	$\frac{\sqrt{3}}{3} \approx 0.58$	$\left(\frac{\pi}{6}, \frac{\sqrt{3}}{3}\right)$
$\frac{\pi}{4}$	1	$\left(\frac{\pi}{4}, 1\right)$
$\frac{\pi}{3}$	$\sqrt{3} \approx 1.73$	$\left(\frac{\pi}{3}, \sqrt{3}\right)$

Figure 64

$y = \tan x, -\frac{\pi}{3} \leq x \leq \frac{\pi}{3}$

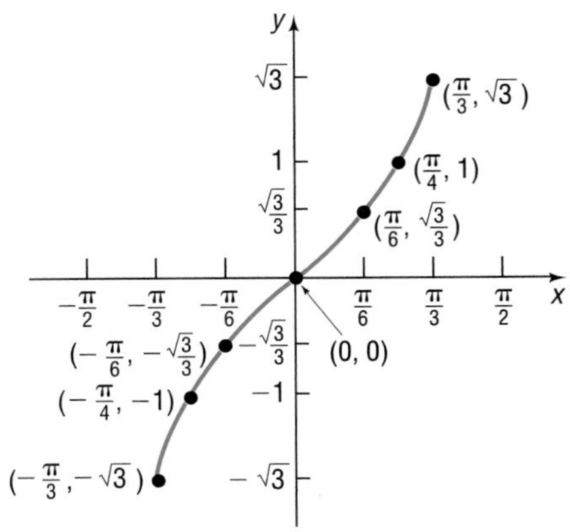

To complete one period of the graph of $y = \tan x$, we need to investigate the behavior of the function as x approaches $-\frac{\pi}{2}$ and $\frac{\pi}{2}$. We must be careful, though, because $y = \tan x$ is not defined at these numbers. To determine this behavior, we use the identity

$$\tan x = \frac{\sin x}{\cos x}$$

See Table 9. If x is close to $\frac{\pi}{2} \approx 1.5708$, but remains less than $\frac{\pi}{2}$, then $\sin x$ will be close to 1 and $\cos x$ will be positive and close to 0. (Refer back to the graphs of the sine function and the cosine function.) Hence, the ratio $\frac{\sin x}{\cos x}$ will be positive and large. In fact, the closer x gets to $\frac{\pi}{2}$, the closer

$\sin x$ gets to 1 and $\cos x$ gets to 0, so $\tan x$ approaches ∞ $\left(\lim_{x \to \frac{\pi}{2}^-} \tan x = \infty\right)$.

In other words, the vertical line $x = \frac{\pi}{2}$ is a vertical asymptote to the graph of $y = \tan x$.

Table 9

x	$\sin x$	$\cos x$	$y = \tan x$
$\frac{\pi}{3} \approx 1.05$	$\frac{\sqrt{3}}{2}$	$\frac{1}{2}$	$\sqrt{3} \approx 1.73$
1.5	0.9975	0.0707	14.1
1.57	0.9999	$7.96E^{-4}$	1255.8
1.5707	0.9999	$9.6E^{-5}$	10381
$\frac{\pi}{2} \approx 1.5708$	1	0	Undefined

If x is close to $-\frac{\pi}{2}$, but remains greater than $-\frac{\pi}{2}$, then $\sin x$ will be close to -1 and $\cos x$ will be positive and close to 0. Hence, the ratio $\frac{\sin x}{\cos x}$ approaches $-\infty$ ($\lim_{x \to -\frac{\pi}{2}^+} \tan x = -\infty$). In other words, the vertical line $x = -\frac{\pi}{2}$ is also a vertical asymptote to the graph.

With these observations, we can complete one period of the graph. We obtain the complete graph of $y = \tan x$ by repeating this period, as shown in Figure 65.

Figure 65

$y = \tan x$, $-\infty < x < \infty$, x not equal to odd multiples of $\frac{\pi}{2}$, $-\infty < y < \infty$

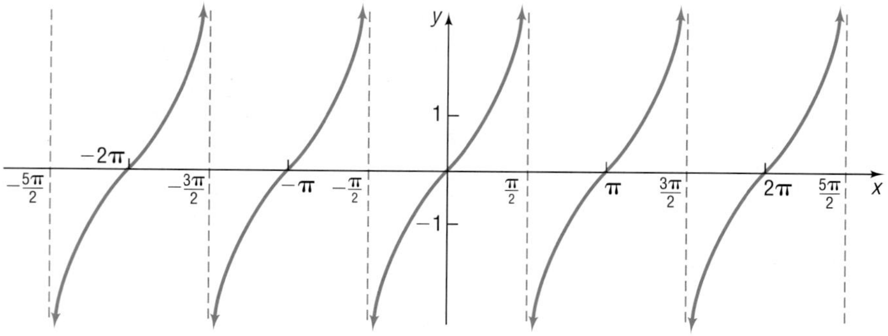

✔ **CHECK:** Graph $Y_1 = \tan x$ and compare the result with Figure 65. Use TRACE to see what happens as x gets close to $\frac{\pi}{2}$, but is less than $\frac{\pi}{2}$. Be sure to set the WINDOW accordingly and use DOT mode.

The graph of $y = \tan x$ illustrates some facts that we already know about the tangent function.

Properties of the Tangent Function

1. The domain is the set of all real numbers, except odd multiples of $\frac{\pi}{2}$.
2. The range is the set of all real numbers.
3. The tangent function is an odd function, as the symmetry of the graph with respect to the origin indicates.
4. The tangent function is periodic, with period π.
5. The x-intercepts are $\ldots, -2\pi, -\pi, 0, \pi, 2\pi, 3\pi, \ldots$; the y-intercept is 0.
6. Vertical asymptotes occur at $x = \ldots, -\frac{3\pi}{2}, -\frac{\pi}{2}, \frac{\pi}{2}, \frac{3\pi}{2}, \ldots$.

NOW WORK PROBLEMS 7 AND 15.

EXAMPLE 1 Graphing Variations of $y = \tan x$ Using Transformations

Graph: $y = 2 \tan x$

Solution We start with the graph of $y = \tan x$ and vertically stretch it by a factor of 2. See Figure 66.

Figure 66

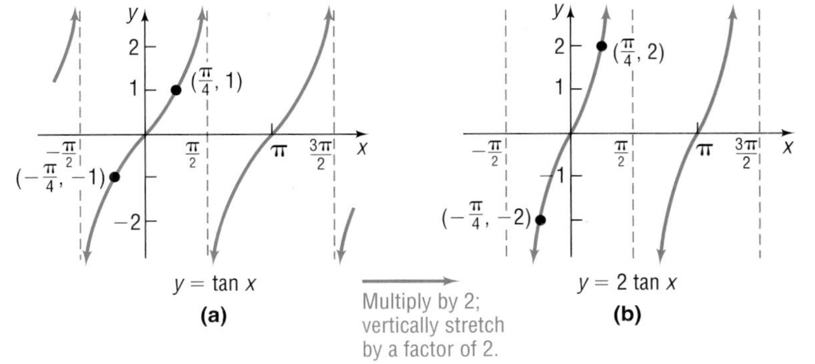

◀

✔ **CHECK:** Graph $Y_1 = 2 \tan x$ and compare the result to Figure 66(b).

EXAMPLE 2 Graphing Variations of $y = \tan x$ Using Transformations

Graph: $y = -\tan\left(x + \dfrac{\pi}{4}\right)$

Solution We start with the graph of $y = \tan x$. See Figure 67.

Figure 67

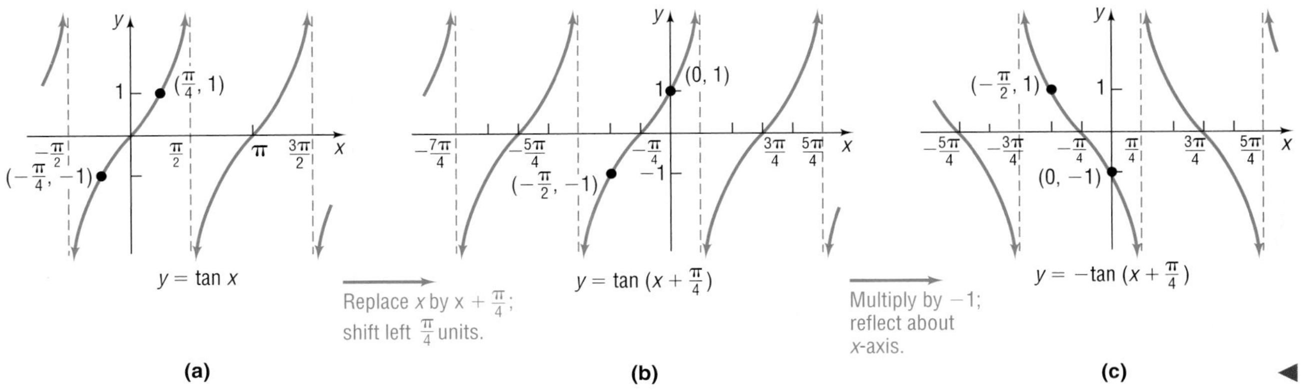

◀

✔ **CHECK:** Graph $Y_1 = -\tan\left(x + \dfrac{\pi}{4}\right)$ and compare the result to Figure 67(c).

NOW WORK PROBLEM 25.

We obtain the graph of $y = \cot x$ as we did the graph of $y = \tan x$. The period of $y = \cot x$ is π. Because the cotangent function is not defined for

Table 10

x	$y = \cot x$	(x, y)
$\frac{\pi}{6}$	$\sqrt{3}$	$\left(\frac{\pi}{6}, \sqrt{3}\right)$
$\frac{\pi}{4}$	1	$\left(\frac{\pi}{4}, 1\right)$
$\frac{\pi}{3}$	$\frac{\sqrt{3}}{3}$	$\left(\frac{\pi}{3}, \frac{\sqrt{3}}{3}\right)$
$\frac{\pi}{2}$	0	$\left(\frac{\pi}{2}, 0\right)$
$\frac{2\pi}{3}$	$-\frac{\sqrt{3}}{3}$	$\left(\frac{2\pi}{3}, -\frac{\sqrt{3}}{3}\right)$
$\frac{3\pi}{4}$	-1	$\left(\frac{3\pi}{4}, -1\right)$
$\frac{5\pi}{6}$	$-\sqrt{3}$	$\left(\frac{5\pi}{6}, -\sqrt{3}\right)$

integral multiples of π, we will concentrate on the interval $(0, \pi)$. Table 10 lists some points on the graph of $y = \cot x$, $0 < x < \pi$. As x approaches 0, but remains greater than 0, the value of $\cos x$ will be close to 1 and the value of $\sin x$ will be positive and close to 0. Hence, the ratio $\frac{\cos x}{\sin x} = \cot x$ will be positive and large; so as x approaches 0, with $x > 0$, $\cot x$ approaches $\infty\left(\lim_{x\to 0^+} \cot x = \infty\right)$. Similarly, as x approaches π, but remains less than π, the value of $\cos x$ will be close to -1, and the value of $\sin x$ will be positive and close to 0. Hence, the ratio $\frac{\cos x}{\sin x} = \cot x$ will be negative and will approach $-\infty$ as x approaches $\pi\left(\lim_{x\to\pi^-} \cot x = -\infty\right)$. Figure 68 shows the graph.

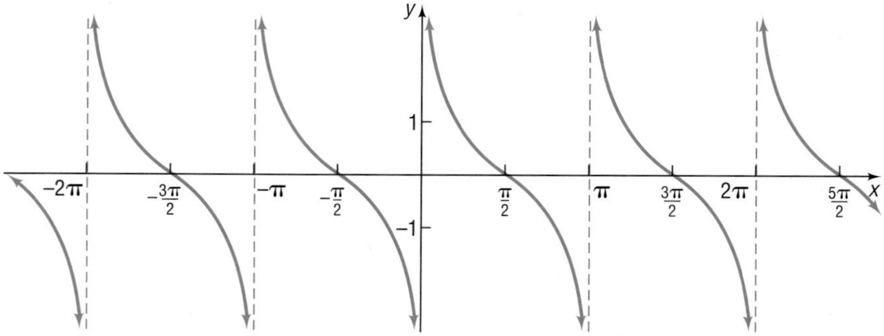

Figure 68
$y = \cot x$, $-\infty < x < \infty$, x not equal to integral multiples of π, $-\infty < y < \infty$

✔ **CHECK:** Graph $Y_1 = \cot x$ and compare the result with Figure 68. Use TRACE to see what happens when x is close to 0.

NOW WORK PROBLEM 31.

The Graphs of $y = \csc x$ and $y = \sec x$

2 The cosecant and secant functions, sometimes referred to as **reciprocal functions,** are graphed by making use of the reciprocal identities

$$\csc x = \frac{1}{\sin x} \quad \text{and} \quad \sec x = \frac{1}{\cos x}$$

For example, the value of the cosecant function $y = \csc x$ at a given number x equals the reciprocal of the corresponding value of the sine function, provided that the value of the sine function is not 0. If the value of $\sin x$ is 0, then, at such numbers x, the cosecant function is not defined. In fact, the graph of the cosecant function has vertical asymptotes at integral multiples of π. Figure 69 shows the graph.

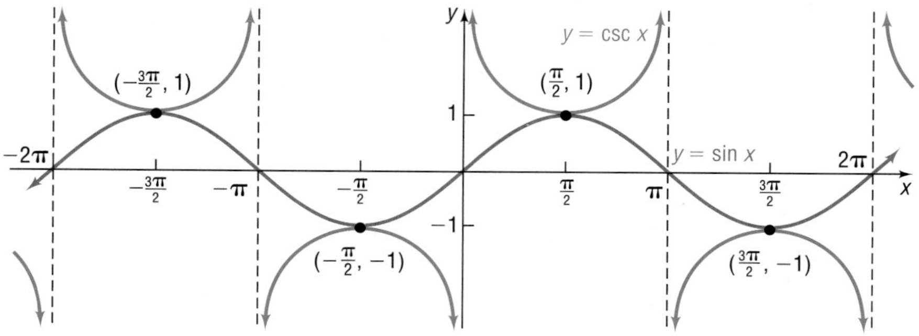

Figure 69
$y = \csc x$, $-\infty < x < \infty$, x not equal to integral multiples of π, $|y| \geq 1$

✔ **CHECK:** Graph $Y_1 = \csc x$ and compare the result with Figure 69. Use TRACE to see what happens when x is close to 0.

EXAMPLE 3 Graphing Variations of $y = \csc x$ Using Transformations

Graph: $y = 2 \csc x - 1$

Solution Figure 70 shows the required steps.

Figure 70

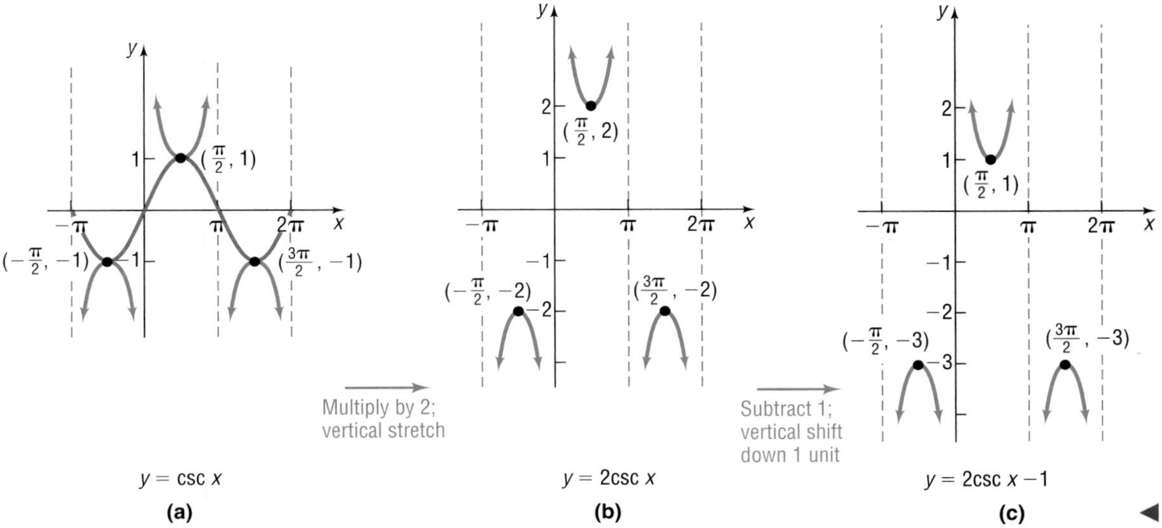

✔ **CHECK:** Graph $Y_1 = 2 \csc x - 1$ and compare the result with Figure 70(c).

NOW WORK PROBLEM 37.

Using the idea of reciprocals, we can similarly obtain the graph of $y = \sec x$. See Figure 71.

Figure 71
$y = \sec x$, $-\infty < x < \infty$, x not equal to odd multiples of $\frac{\pi}{2}$, $|y| \geq 1$

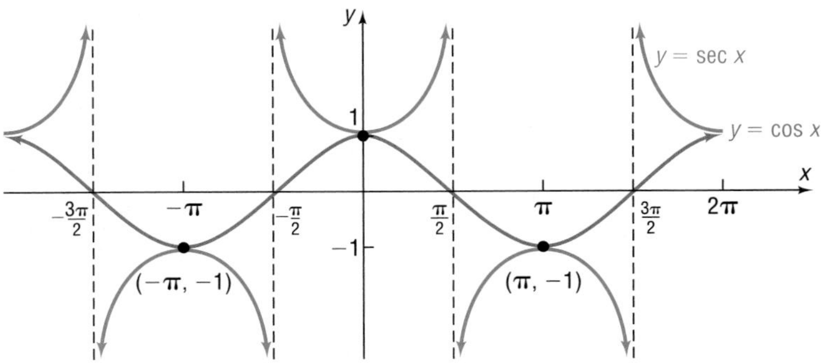

5.5 Assess Your Understanding

'Are You Prepared?' *Answers are given at the end of these exercises. If you get a wrong answer, read the pages listed in red.*

1. The graph of $y = \dfrac{3x - 6}{x - 4}$ has a vertical asymptote. What is it? (pp. 165–167)

2. *True or False:* If a function f has the vertical asymptote $x = c$, then $f(c)$ is not defined. (pp. 165–167)

Concepts and Vocabulary

3. The graph of $y = \tan x$ is symmetric with respect to the __________ and has vertical asymptotes at __________.

4. The graph of $y = \sec x$ is symmetric with respect to the __________ and has vertical asymptotes at __________.

5. It is easiest to graph $y = \sec x$ by first sketching the graph of __________.

6. *True or False:* The graphs of $y = \tan x$, $y = \cot x$, $y = \sec x$, and $y = \csc x$ each have infinitely many vertical asymptotes.

Exercises

In Problems 7–16, if necessary, refer to the graphs to answer each question.

7. What is the y-intercept of $y = \tan x$?

8. What is the y-intercept of $y = \cot x$?

9. What is the y-intercept of $y = \sec x$?

10. What is the y-intercept of $y = \csc x$?

11. For what numbers x, $-2\pi \le x \le 2\pi$, does $\sec x = 1$? What about $\sec x = -1$?

12. For what numbers x, $-2\pi \le x \le 2\pi$, does $\csc x = 1$? What about $\csc x = -1$?

13. For what numbers x, $-2\pi \le x \le 2\pi$, does the graph of $y = \sec x$ have vertical asymptotes?

14. For what numbers x, $-2\pi \le x \le 2\pi$, does the graph of $y = \csc x$ have vertical asymptotes?

15. For what numbers x, $-2\pi \le x \le 2\pi$, does the graph of $y = \tan x$ have vertical asymptotes?

16. For what numbers x, $-2\pi \le x \le 2\pi$, does the graph of $y = \cot x$ have vertical asymptotes?

In Problems 17–20, match each function to its graph.

A. $y = -\tan x$ B. $y = \tan\left(x + \dfrac{\pi}{2}\right)$ C. $y = \tan(x + \pi)$ D. $y = -\tan\left(x - \dfrac{\pi}{2}\right)$

17.

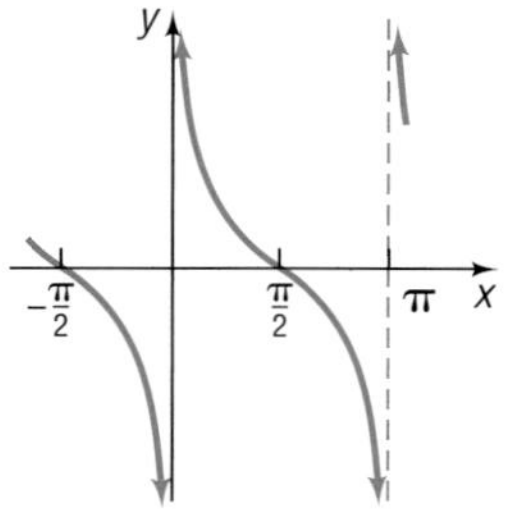

18.

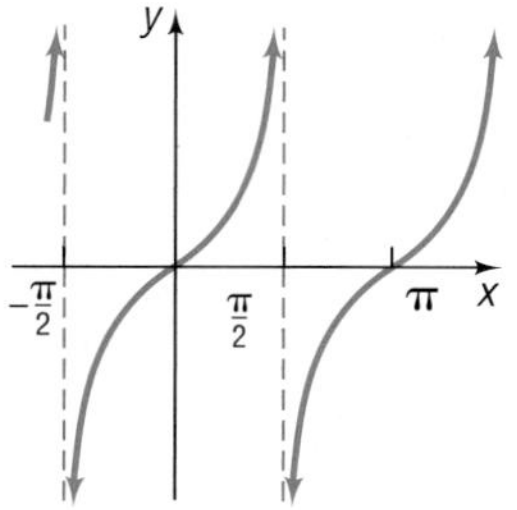

19.

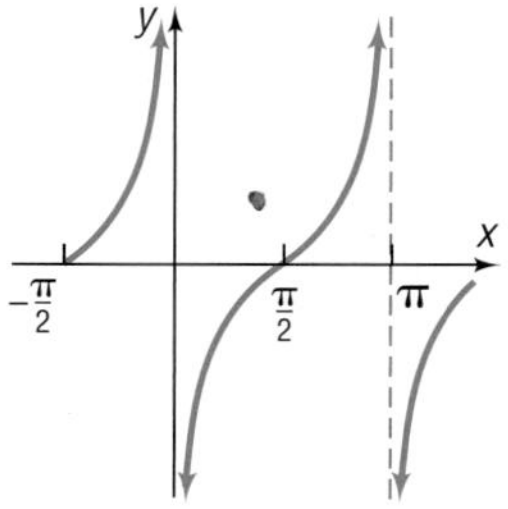

20.

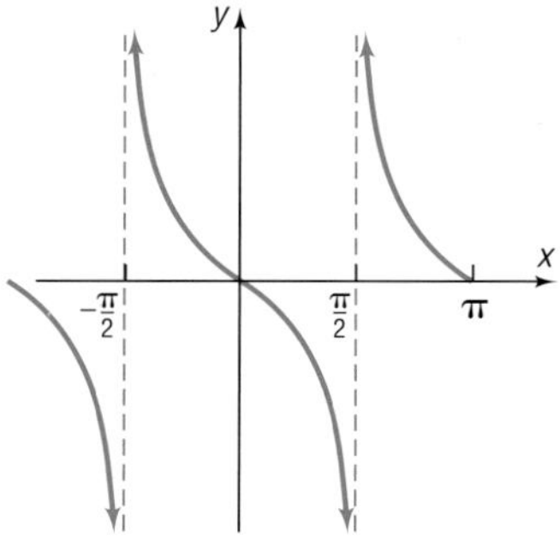

In Problems 21–40, use tranformations to graph each function.

21. $y = -\sec x$

22. $y = -\cot x$

23. $y = \sec\left(x - \dfrac{\pi}{2}\right)$

24. $y = \csc(x - \pi)$

25. $y = \tan(x - \pi)$

26. $y = \cot(x - \pi)$

27. $y = 3\tan(2x)$

28. $y = 4\tan\left(\dfrac{1}{2}x\right)$

29. $y = \sec(2x)$

30. $y = \csc\left(\dfrac{1}{2}x\right)$

31. $y = \cot(\pi x)$

32. $y = \cot(2x)$

33. $y = -3\tan(4x)$

34. $y = -3\tan(2x)$

35. $y = 2\sec\left(\dfrac{1}{2}x\right)$

36. $y = 2\sec(3x)$

37. $y = -3\csc\left(x + \dfrac{\pi}{4}\right)$

38. $y = -2\tan\left(x + \dfrac{\pi}{4}\right)$

39. $y = \dfrac{1}{2}\cot\left(x - \dfrac{\pi}{4}\right)$

40. $y = 3\sec\left(x + \dfrac{\pi}{2}\right)$

41. Carrying a Ladder around a Corner Two hallways, one of width 3 feet, the other of width 4 feet, meet at a right angle. See the illustration.

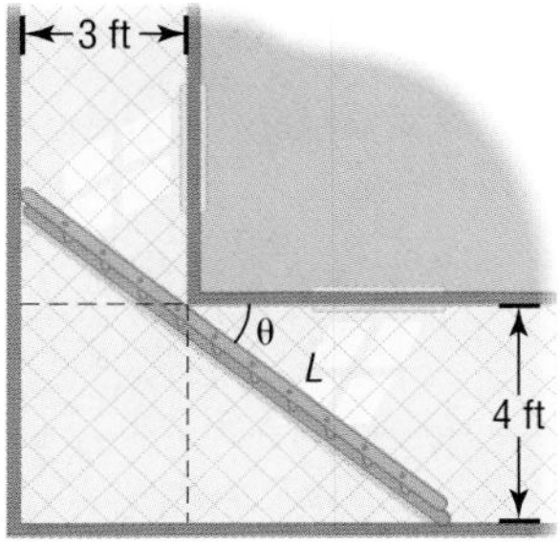

(a) Show that the length L of the line segment shown as a function of the angle θ is

$$L(\theta) = 3\sec\theta + 4\csc\theta$$

(b) Graph L, $0 < \theta < \frac{\pi}{2}$.

(c) For what value of θ is L the least?

(d) What is the length of the longest ladder that can be carried around the corner? Why is this also the least value of L?

42. Exploration Graph

$$y = \tan x \quad \text{and} \quad y = -\cot\left(x + \frac{\pi}{2}\right)$$

Do you think that $\tan x = -\cot\left(x + \frac{\pi}{2}\right)$?

'Are You Prepared?' Answers

1. $x = 4$ **2.** True

5.6 Phase Shift; Sinusoidal Curve Fitting

OBJECTIVES
1 Determine the Phase Shift of a Sinusoidal Function
2 Graph Sinusoidal Functions: $y = A\sin(\omega x - \phi)$
3 Find a Sinusoidal Function from Data

Phase Shift

1 We have seen that the graph of $y = A\sin(\omega x)$, $\omega > 0$, has amplitude $|A|$ and period $T = \frac{2\pi}{\omega}$. One cycle can be drawn as x varies from 0 to $\frac{2\pi}{\omega}$ or, equivalently, as ωx varies from 0 to 2π. See Figure 72.

Figure 72
One cycle $y = A\sin(\omega x)$, $A > 0$, $\omega > 0$

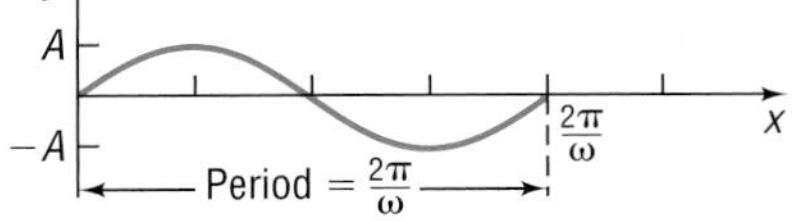

We now want to discuss the graph of

$$y = A\sin(\omega x - \phi)$$

which may also be written as

$$y = A\sin\left[\omega\left(x - \frac{\phi}{\omega}\right)\right]$$

where $\omega > 0$ and ϕ (the Greek letter phi) are real numbers. The graph will be a sine curve of amplitude $|A|$. As $\omega x - \phi$ varies from 0 to 2π, one period will be traced out. This period will begin when

$$\omega x - \phi = 0 \quad \text{or} \quad x = \frac{\phi}{\omega}$$

and will end when

$$\omega x - \phi = 2\pi \quad \text{or} \quad x = \frac{2\pi}{\omega} + \frac{\phi}{\omega}$$

See Figure 73.

Figure 73
One cycle $y = A\sin(\omega x - \phi)$, $A > 0$, $\omega > 0$, $\phi > 0$

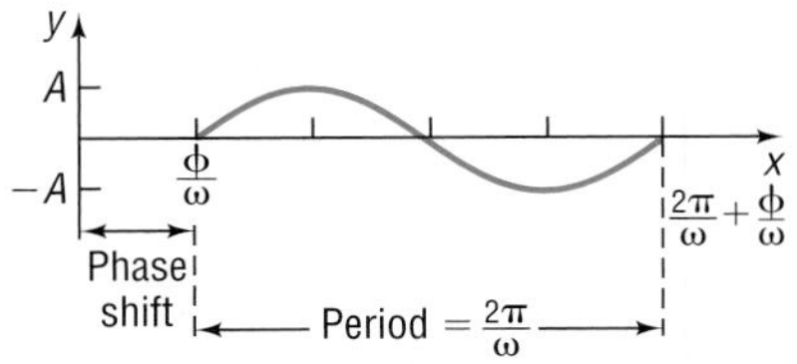

We see that the graph of $y = A\sin(\omega x - \phi) = A\sin\left[\omega\left(x - \frac{\phi}{\omega}\right)\right]$ is the same as the graph of $y = A\sin(\omega x)$, except that it has been shifted $\frac{\phi}{\omega}$ units (to the right if $\phi > 0$ and to the left if $\phi < 0$). This number $\frac{\phi}{\omega}$ is called the **phase shift** of the graph of $y = A\sin(\omega x - \phi)$.

For the graphs of $y = A\sin(\omega x - \phi)$ or $y = A\cos(\omega x - \phi)$, $\omega > 0$,

Amplitude $= \lvert A \rvert$	Period $= T = \frac{2\pi}{\omega}$	Phase shift $= \frac{\phi}{\omega}$

The phase shift is to the left if $\phi < 0$ and to the right if $\phi > 0$.

2 **EXAMPLE 1**

Finding the Amplitude, Period, and Phase Shift of a Sinusoidal Function and Graphing It

Find the amplitude, period, and phase shift of $y = 3\sin(2x - \pi)$, and graph the function.

Solution Comparing

$$y = 3\sin(2x - \pi) = 3\sin\left[2\left(x - \frac{\pi}{2}\right)\right]$$

to

$$y = A\sin(\omega x - \phi) = A\sin\left[\omega\left(x - \frac{\phi}{\omega}\right)\right]$$

we find that $A = 3$, $\omega = 2$, and $\phi = \pi$. The graph is a sine curve with amplitude $|A| = 3$, period $T = \frac{2\pi}{\omega} = \frac{2\pi}{2} = \pi$, and phase shift $= \frac{\phi}{\omega} = \frac{\pi}{2}$.

The graph of $y = 3\sin(2x - \pi)$ will lie between -3 and 3 on the y-axis. One cycle will begin at $x = \frac{\phi}{\omega} = \frac{\pi}{2}$ and end at $x = \frac{2\pi}{\omega} + \frac{\phi}{\omega} = \pi + \frac{\pi}{2} = \frac{3\pi}{2}$. We divide the interval $\left[\frac{\pi}{2}, \frac{3\pi}{2}\right]$ into four subintervals, each of length $\pi \div 4 = \frac{\pi}{4}$:

$$\left[\frac{\pi}{2}, \frac{3\pi}{4}\right], \quad \left[\frac{3\pi}{4}, \pi\right], \quad \left[\pi, \frac{5\pi}{4}\right], \quad \left[\frac{5\pi}{4}, \frac{3\pi}{2}\right]$$

The five key points on the graph are

$$\left(\frac{\pi}{2}, 0\right), \quad \left(\frac{3\pi}{4}, 3\right), \quad (\pi, 0), \quad \left(\frac{5\pi}{4}, -3\right), \quad \left(\frac{3\pi}{2}, 0\right)$$

We plot these five points and fill in the graph of the sine function as shown in Figure 74(a). Extending the graph in either direction, we obtain Figure 74(b).

Figure 74

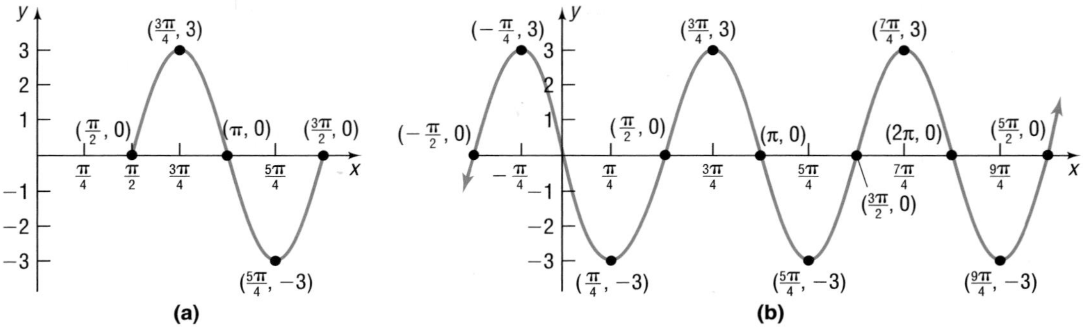

The graph of $y = 3\sin(2x - \pi) = 3\sin\left[2\left(x - \frac{\pi}{2}\right)\right]$ may also be obtained using transformations. See Figure 75.

Figure 75

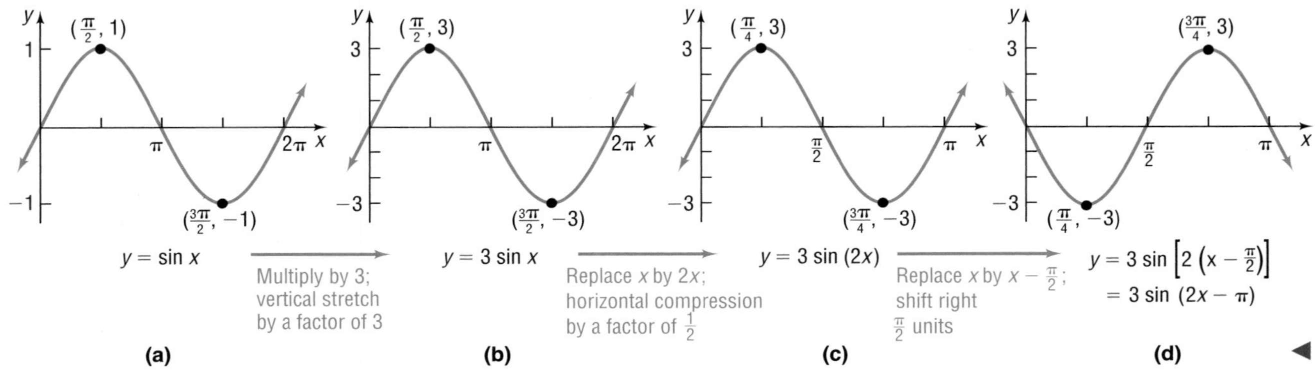

✔ **CHECK:** Graph $Y_1 = 3\sin(2x - \pi)$ using a graphing utility.

EXAMPLE 2 Finding the Amplitude, Period, and Phase Shift of a Sinusoidal Function and Graphing It

Find the amplitude, period, and phase shift of $y = 2\cos(4x + 3\pi)$, and graph the function.

Solution Comparing

$$y = 2\cos(4x + 3\pi) = 2\cos\left[4\left(x + \frac{3\pi}{4}\right)\right]$$

to

$$y = A\cos(\omega x - \phi) = A\cos\left[\omega\left(x - \frac{\phi}{\omega}\right)\right]$$

we see that $A = 2$, $\omega = 4$, and $\phi = -3\pi$. The graph is a cosine curve with amplitude $|A| = 2$, period $T = \frac{2\pi}{\omega} = \frac{2\pi}{4} = \frac{\pi}{2}$, and phase shift $= \frac{\phi}{\omega} = -\frac{3\pi}{4}$.

The graph of $y = 2\cos(4x + 3\pi)$ will lie between -2 and 2 on the y-axis. One cycle will begin at $x = \frac{\phi}{\omega} = -\frac{3\pi}{4}$ and end at $x = \frac{2\pi}{\omega} + \frac{\phi}{\omega} = \frac{\pi}{2} + \left(-\frac{3\pi}{4}\right) = -\frac{\pi}{4}$. We divide the interval $\left[-\frac{3\pi}{4}, -\frac{\pi}{4}\right]$ into four subintervals, each of length $\frac{\pi}{2} \div 4 = \frac{\pi}{8}$:

$$\left[-\frac{3\pi}{4}, -\frac{5\pi}{8}\right], \quad \left[-\frac{5\pi}{8}, -\frac{\pi}{2}\right], \quad \left[-\frac{\pi}{2}, -\frac{3\pi}{8}\right], \quad \left[-\frac{3\pi}{8}, -\frac{\pi}{4}\right]$$

The five key points on the graph are

$$\left(-\frac{3\pi}{4}, 2\right), \quad \left(-\frac{5\pi}{8}, 0\right), \quad \left(-\frac{\pi}{2}, -2\right), \quad \left(-\frac{3\pi}{8}, 0\right), \quad \left(-\frac{\pi}{4}, 2\right)$$

We plot these five points and fill in the graph of the cosine function as shown in Figure 76(a). Extending the graph in either direction, we obtain Figure 76(b).

Figure 76

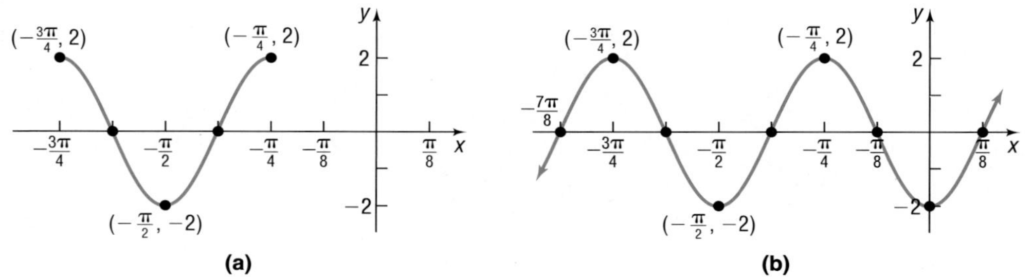

The graph of $y = 2\cos(4x + 3\pi) = 2\cos\left[4\left(x + \frac{3\pi}{4}\right)\right]$ may also be obtained using transformations. See Figure 77.

Figure 77

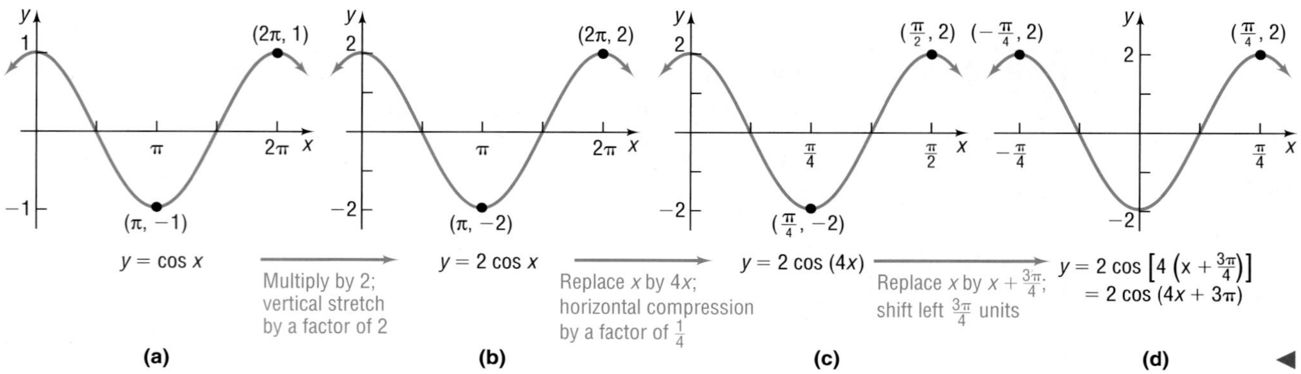

✔ **CHECK:** Graph $Y_1 = 2\cos(4x + 3\pi)$ using a graphing utility.

NOW WORK PROBLEM 3.

Summary

Steps for Graphing Sinusoidal Functions

To graph sinusoidal functions of the form $y = A\sin(\omega x - \phi)$ or $y = A\cos(\omega x - \phi)$:

STEP 1: Determine the amplitude $|A|$ and period $T = \frac{2\pi}{\omega}$.

STEP 2: Determine the starting point of one cycle of the graph, $\frac{\phi}{\omega}$.

STEP 3: Determine the ending point of one cycle of the graph, $\frac{2\pi}{\omega} + \frac{\phi}{\omega}$.

STEP 4: Divide the interval $\left[\frac{\phi}{\omega}, \frac{2\pi}{\omega} + \frac{\phi}{\omega}\right]$ into four subintervals, each of length $\frac{2\pi}{\omega} \div 4$.

STEP 5: Use the endpoints of the subintervals to find the five key points on the graph.

STEP 6: Fill in one cycle of the graph.

STEP 7: Extend the graph in each direction to make it complete.

Finding Sinusoidal Functions from Data

3 Scatter diagrams of data sometimes take the form of a sinusoidal function. Let's look at an example.

The data given in Table 11 represent the average monthly temperatures in Denver, Colorado. Since the data represent *average* monthly temperatures collected over many years, the data will not vary much from year to year and so will essentially repeat each year. In other words, the data are periodic. Figure 78 shows the scatter diagram of these data repeated over two years, where $x = 1$ represents January, $x = 2$ represents February, and so on.

Table 11

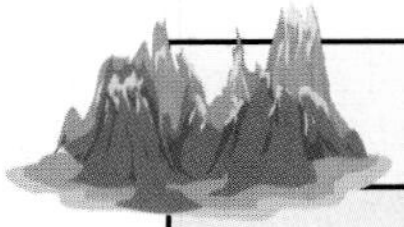

Month, x	Average Monthly Temperature, °F
January, 1	29.7
February, 2	33.4
March, 3	39.0
April, 4	48.2
May, 5	57.2
June, 6	66.9
July, 7	73.5
August, 8	71.4
September, 9	62.3
October, 10	51.4
November, 11	39.0
December, 12	31.0

SOURCE: U.S. National Oceanic and Atmospheric Administration

Figure 78

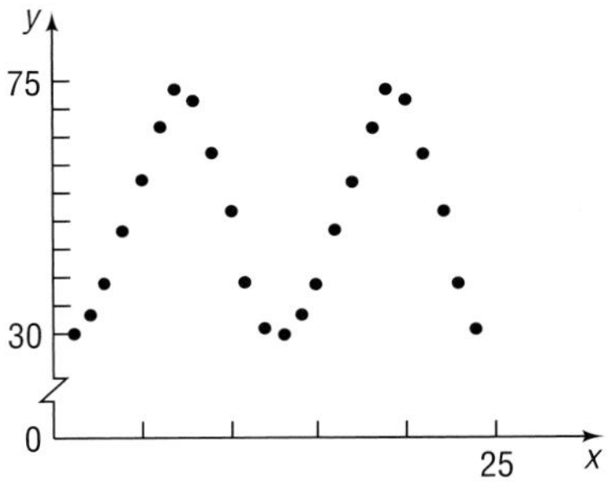

Notice that the scatter diagram looks like the graph of a sinusoidal function. We choose to fit the data to a sine function of the form

$$y = A\sin(\omega x - \phi) + B$$

where A, B, ω, and ϕ are constants.

EXAMPLE 3 Finding a Sinusoidal Function from Temperature Data

Fit a sine function to the data in Table 11.

Solution We begin with a scatter diagram of the data for one year. See Figure 79. The data will be fitted to a sine function of the form

$$y = A\sin(\omega x - \phi) + B$$

Figure 79

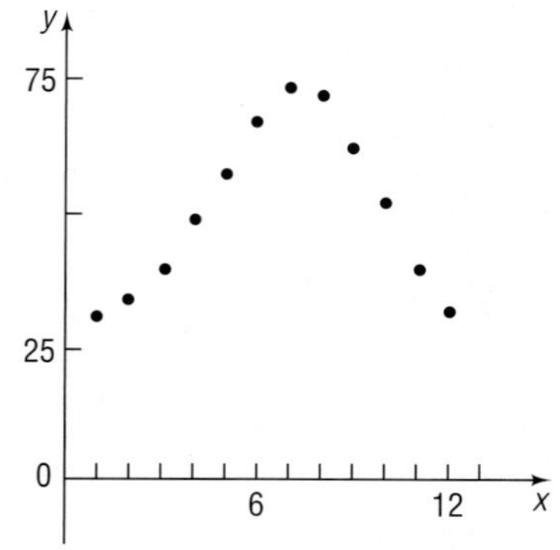

Step 1: To find the amplitude A, we compute

$$\text{Amplitude} = \frac{\text{largest data value} - \text{smallest data value}}{2}$$
$$= \frac{73.5 - 29.7}{2} = 21.9$$

To see the remaining steps in this process, we superimpose the graph of the function $y = 21.9\sin x$, where x represents months, on the scatter diagram. Figure 80 shows the two graphs.

To fit the data, the graph needs to be shifted vertically, shifted horizontally, and stretched horizontally.

Figure 80

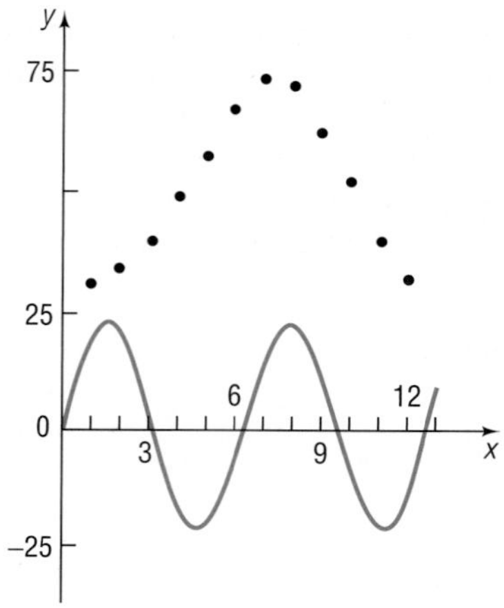

Step 2: We determine the vertical shift by finding the average of the highest and lowest data value.

$$\text{Vertical shift} = \frac{73.5 + 29.7}{2} = 51.6$$

Now we superimpose the graph of $y = 21.9\sin x + 51.6$ on the scatter diagram. See Figure 81.

We see that the graph needs to be shifted horizontally and stretched horizontally.

Step 3: It is easier to find the horizontal stretch factor first. Since the temperatures repeat every 12 months, the period of the function is $T = 12$. Since $T = \frac{2\pi}{\omega} = 12$, we find

$$\omega = \frac{2\pi}{12} = \frac{\pi}{6}$$

Now we superimpose the graph of $y = 21.9\sin\left(\frac{\pi}{6}x\right) + 51.6$ on the scatter diagram. See Figure 82. We see that the graph still needs to be shifted horizontally.

Figure 81

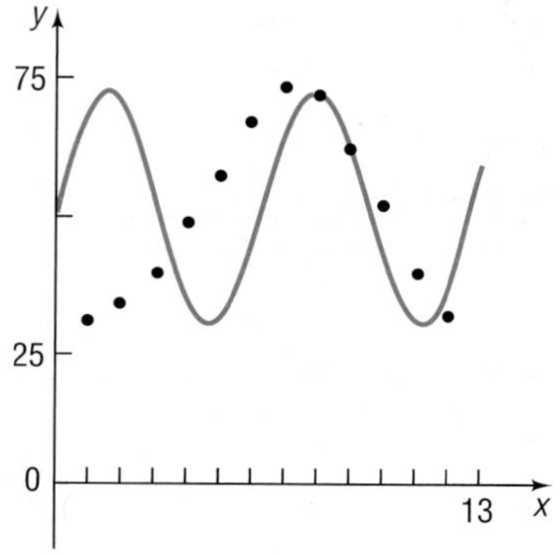

Figure 82

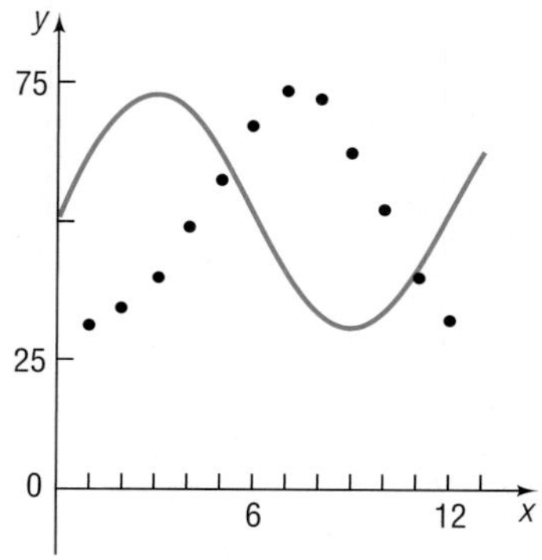

STEP 4: To determine the horizontal shift, we solve the equation

$$y = 21.9 \sin\left(\frac{\pi}{6}x - \phi\right) + 51.6$$

for ϕ by letting $y = 29.7$ and $x = 1$ (the average temperature in Denver in January).*

$$29.7 = 21.9 \sin\left(\frac{\pi}{6} \cdot 1 - \phi\right) + 51.6$$

$$-21.9 = 21.9 \sin\left(\frac{\pi}{6} - \phi\right) \quad \text{Subtract 51.6 from both sides of the equation.}$$

$$-1 = \sin\left(\frac{\pi}{6} - \phi\right) \quad \text{Divide both sides of the equation by 21.9.}$$

$$-\frac{\pi}{2} = \frac{\pi}{6} - \phi \quad \sin\theta = -1 \text{ when } \theta = -\frac{\pi}{2}.$$

$$\phi = \frac{2\pi}{3} \quad \text{Solve for } \phi.$$

The sine function that fits the data is

$$y = 21.9 \sin\left(\frac{\pi}{6}x - \frac{2\pi}{3}\right) + 51.6$$

The graph of $y = 21.9 \sin\left(\frac{\pi}{6}x - \frac{2\pi}{3}\right) + 51.6$ and the scatter diagram of the data are shown in Figure 83. ◀

Figure 83

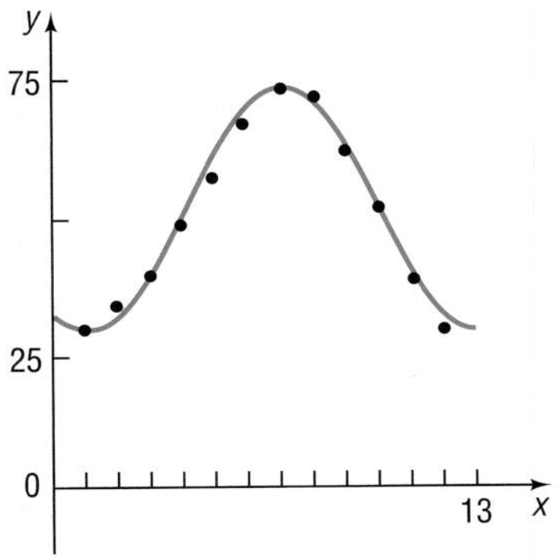

The steps to fit a sine function

$$y = A \sin(\omega x - \phi) + B$$

to sinusoidal data follow:

Steps for Fitting Data to a Sine Function $y = A \sin(\omega x - \phi) + B$

STEP 1: Determine A, the amplitude of the function.

$$\text{Amplitude} = \frac{\text{largest data value} - \text{smallest data value}}{2}$$

STEP 2: Determine B, the vertical shift of the function.

$$\text{Vertical shift} = \frac{\text{largest data value} + \text{smallest data value}}{2}$$

STEP 3: Determine ω. Since the period T, the time it takes for the data to repeat, is $T = \frac{2\pi}{\omega}$, we have

$$\omega = \frac{2\pi}{T}$$

Continues on next page

*The data point selected here to find ϕ is arbitrary. Selecting a different data point will usually result in a different value for ϕ. To maintain consistency, we will always choose the data point for which y is smallest (in this case, January gives the lowest temperature).

STEP 4: Determine the horizontal shift of the function by solving the equation

$$y = A\sin(\omega x - \phi) + B$$

for ϕ by choosing an ordered pair (x, y) from the data. Since answers will vary depending on the ordered pair selected, we will always choose the ordered pair for which y is smallest in order to maintain consistency.

NOW WORK PROBLEM **21(a)–(c).**

Let's look at another example. Since the number of hours of sunlight in a day cycles annually, the number of hours of sunlight in a day for a given location can be modeled by a sinusoidal function.

The longest day of the year (in terms of hours of sunlight) occurs on the day of the summer solstice, usually June 21, the 172nd day of the year. The summer solstice is the time when the Sun is farthest north (for locations in the northern hemisphere). The shortest day of the year occurs on the day of the winter solstice, usually December 21, the 355th day of the year. The winter solstice is the time when the Sun is farthest south (again, for locations in the northern hemisphere).

EXAMPLE 4 Finding a Sinusoidal Function for Hours of Daylight

According to the *Old Farmer's Almanac*, the number of hours of sunlight in Boston on the summer solstice is 15.283 and the number of hours of sunlight on the winter solstice is 9.067.

(a) Find a sinusoidal function of the form $y = A\sin(\omega x - \phi) + B$ that fits the data.

(b) Use the function found in part (a) to predict the number of hours of sunlight on April 1, the 91st day of the year.

(c) Draw a graph of the function found in part (a).

(d) Look up the number of hours of sunlight for April 1 in the *Old Farmer's Almanac* and compare the actual hours of daylight to the results found in part (b).

Solution (a) **STEP 1:**

$$\text{Amplitude} = \frac{\text{largest data value} - \text{smallest data value}}{2} = \frac{15.283 - 9.067}{2} = 3.108$$

STEP 2:

$$\text{Vertical shift} = \frac{\text{largest data value} + \text{smallest data value}}{2} = \frac{15.283 + 9.067}{2} = 12.175$$

STEP 3: The data repeat every 365 days. Since $T = \frac{2\pi}{\omega} = 365$, we find

$$\omega = \frac{2\pi}{365}$$

So far, we have $y = 3.108 \sin\left(\frac{2\pi}{365}x - \phi\right) + 12.175$.

STEP 4: To determine the horizontal shift, we solve the equation

$$y = 3.108 \sin\left(\frac{2\pi}{365}x - \phi\right) + 12.175$$

for ϕ by letting $y = 9.067$ and $x = 355$ (the number of hours of daylight in Boston on December 21).

$$9.067 = 3.108 \sin\left(\frac{2\pi}{365} \cdot 355 - \phi\right) + 12.175$$

$$-3.108 = 3.108 \sin\left(\frac{2\pi}{365} \cdot 355 - \phi\right)$$ Subtract 12.175 from both sides of the equation.

$$-1 = \sin\left(\frac{2\pi}{365} \cdot 355 - \phi\right)$$ Divide both sides of the equation by 3.108

$$-\frac{\pi}{2} = \frac{2\pi}{365} \cdot 355 - \phi$$ $\sin\theta = -1$ when $\theta = -\frac{\pi}{2}$.

$$\phi \approx 2.45\pi$$ Solve for ϕ.

The function that provides the number of hours of daylight in Boston for any day, x, is given by

$$y = 3.108 \sin\left(\frac{2\pi}{365}x - 2.45\pi\right) + 12.175$$

(b) To predict the number of hours of daylight on April 1, we let $x = 91$ in the function found in part (a) and obtain

$$y = 3.108 \sin\left(\frac{2\pi}{365} \cdot 91 - 2.45\pi\right) + 12.175$$

$$\approx 3.108 \sin(-1.95\pi) + 12.175$$

$$\approx 12.66$$

Figure 84

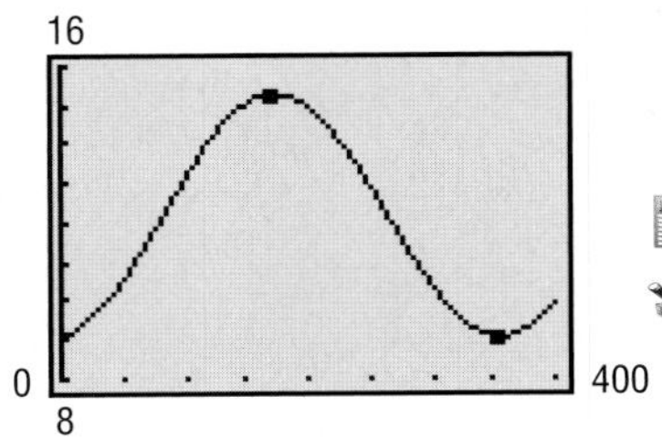

So we predict that there will be about 12.66 hours of sunlight on April 1 in Boston.

(c) The graph of the function found in part (a) is given in Figure 84.

(d) According to the *Old Farmer's Almanac*, there will be 12 hours 43 minutes of sunlight on April 1 in Boston. Our prediction of 12.66 hours converts to 12 hours 40 minutes. ◀

NOW WORK PROBLEM **27**.

Certain graphing utilities (such as a TI-83Plus and TI-86) have the capability of finding the sine function of best fit for sinusoidal data. At least four data points are required for this process.

EXAMPLE 5 Finding the Sine Function of Best Fit

Figure 85

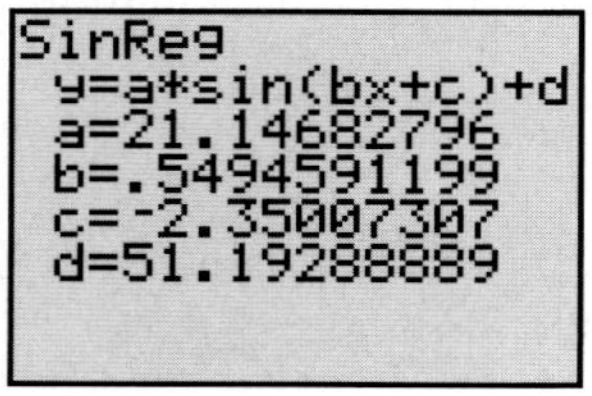

Figure 86

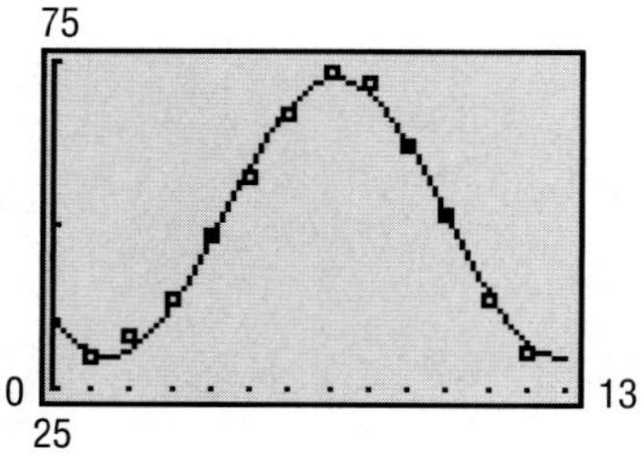

Use a graphing utility to find the sine function of best fit for the data in Table 11. Graph this function with the scatter diagram of the data.

Solution Enter the data from Table 11 and execute the SINe REGression program. The result is shown in Figure 85.

The output that the utility provides shows us the equation

$$y = a\sin(bx + c) + d$$

The sinusoidal function of best fit is

$$y = 21.15\sin(0.55x - 2.35) + 51.19$$

where x represents the month and y represents the average temperature. Figure 86 shows the graph of the sinusoidal function of best fit on the scatter diagram. ◀

NOW WORK PROBLEM 21(d)–(e).

5.6 Assess Your Understanding

Concepts and Vocabulary

1. For the graph of $y = A\sin(\omega x - \phi)$, the number $\dfrac{\phi}{\omega}$ is called the __________.

2. *True or False:* Only two data points are required by a graphing utility to find the sine function of best fit.

Exercises

In Problems 3–14, find the amplitude, period, and phase shift of each function. Graph each function. Show at least one period.

3. $y = 4\sin(2x - \pi)$

4. $y = 3\sin(3x - \pi)$

5. $y = 2\cos\left(3x + \dfrac{\pi}{2}\right)$

6. $y = 3\cos(2x + \pi)$

7. $y = -3\sin\left(2x + \dfrac{\pi}{2}\right)$

8. $y = -2\cos\left(2x - \dfrac{\pi}{2}\right)$

9. $y = 4\sin(\pi x + 2)$

10. $y = 2\cos(2\pi x + 4)$

11. $y = 3\cos(\pi x - 2)$

12. $y = 2\cos(2\pi x - 4)$

13. $y = 3\sin\left(-2x + \dfrac{\pi}{2}\right)$

14. $y = 3\cos\left(-2x + \dfrac{\pi}{2}\right)$

In Problems 15–18, write the equation of a sine function that has the given characteristics.

15. Amplitude: 2
Period: π
Phase shift: $\dfrac{1}{2}$

16. Amplitude: 3
Period: $\dfrac{\pi}{2}$
Phase shift: 2

17. Amplitude: 3
Period: 3π
Phase shift: $-\dfrac{1}{3}$

18. Amplitude: 2
Period: π
Phase shift: -2

19. Alternating Current (ac) Circuits The current I, in amperes, flowing through an ac (alternating current) circuit at time t is

$$I = 120 \sin\left(30\pi t - \frac{\pi}{3}\right), \qquad t \geq 0$$

What is the period? What is the amplitude? What is the phase shift? Graph this function over two periods.

20. Alternating Current (ac) Circuits The current I, in amperes, flowing through an ac (alternating current) circuit at time t is

$$I = 220 \sin\left(60\pi t - \frac{\pi}{6}\right), \qquad t \geq 0$$

What is the period? What is the amplitude? What is the phase shift? Graph this function over two periods.

21. Monthly Temperature The following data represent the average monthly temperatures for Juneau, Alaska.

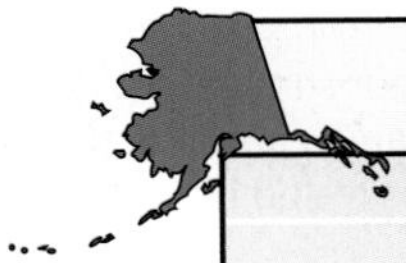

Month, x	Average Monthly Temperature, °F
January, 1	24.2
February, 2	28.4
March, 3	32.7
April, 4	39.7
May, 5	47.0
June, 6	53.0
July, 7	56.0
August, 8	55.0
September, 9	49.4
October, 10	42.2
November, 11	32.0
December, 12	27.1

SOURCE: U.S. National Oceanic and Atmospheric Administration

(a) Draw a scatter diagram of the data for one period.

(b) Find a sinusoidal function of the form $y = A \sin(\omega x - \phi) + B$ that fits the data.

(c) Draw the sinusoidal function found in part (b) on the scatter diagram.

(d) Use a graphing utility to find the sinusoidal function of best fit.

(e) Draw the sinusoidal function of best fit on the scatter diagram.

22. Monthly Temperature The following data represent the average monthly temperatures for Washington, D.C.

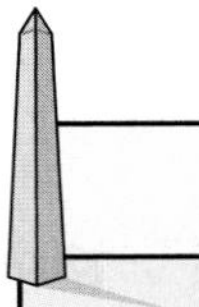

Month, x	Average Monthly Temperature, °F
January, 1	34.6
February, 2	37.5
March, 3	47.2
April, 4	56.5
May, 5	66.4
June, 6	75.6
July, 7	80.0
August, 8	78.5
September, 9	71.3
October, 10	59.7
November, 11	49.8
December, 12	39.4

SOURCE: U.S. National Oceanic and Atmospheric Administration

(a) Draw a scatter diagram of the data for one period.

(b) Find a sinusoidal function of the form $y = A \sin(\omega x - \phi) + B$ that fits the data.

(c) Draw the sinusoidal function found in part (b) on the scatter diagram.

(d) Use a graphing utility to find the sinusoidal function of best fit.

(e) Graph the sinusoidal function of best fit on the scatter diagram.

23. Monthly Temperature The following data represent the average monthly temperatures for Indianapolis, Indiana.

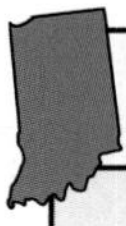

Month, x	Average Monthly Temperature, °F
January, 1	25.5
February, 2	29.6
March, 3	41.4
April, 4	52.4
May, 5	62.8
June, 6	71.9
July, 7	75.4
August, 8	73.2
September, 9	66.6
October, 10	54.7
November, 11	43.0
December, 12	30.9

SOURCE: U.S. National Oceanic and Atmospheric Administration

(a) Draw a scatter diagram of the data for one period.

(b) Find a sinusoidal function of the form $y = A \sin(\omega x - \phi) + B$ that fits the data.

(c) Draw the sinusoidal function found in part (b) on the scatter diagram.

(d) Use a graphing utility to find the sinusoidal function of best fit.

(e) Graph the sinusoidal function of best fit on the scatter diagram.

24. **Monthly Temperature** The following data represent the average monthly temperatures for Baltimore, Maryland.

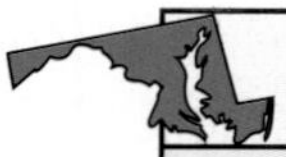

Month, x	Average Monthly Temperature, °F
January, 1	31.8
February, 2	34.8
March, 3	44.1
April, 4	53.4
May, 5	63.4
June, 6	72.5
July, 7	77.0
August, 8	75.6
September, 9	68.5
October, 10	56.6
November, 11	46.8
December, 12	36.7

SOURCE: U.S. National Oceanic and Atmospheric Administration

(a) Draw a scatter diagram of the data for one period.

(b) Find a sinusoidal function of the form $y = A\sin(\omega x - \phi) + B$ that fits the data.

(c) Draw the sinusoidal function found in part (b) on the scatter diagram.

(d) Use a graphing utility to find the sinusoidal function of best fit.

(e) Graph the sinusoidal function of best fit on the scatter diagram.

25. **Tides** Suppose that the length of time between consecutive high tides is approximately 12.5 hours. According to the National Oceanic and Atmospheric Administration, on Saturday, June 28, 1997, in Savannah, Georgia, high tide occurred at 3:38 AM (3.6333 hours) and low tide occurred at 10:08 AM (10.1333 hours). Water heights are measured as the amounts above or below the mean lower low water. The height of the water at high tide was 8.2 feet and the height of the water at low tide was −0.6 foot.

(a) Approximately when will the next high tide occur?

(b) Find a sinusoidal function of the form $y = A\sin(\omega x - \phi) + B$ that fits the data.

(c) Draw a graph of the function found in part (b).

(d) Use the function found in part (b) to predict the height of the water at the next high tide.

26. **Tides** Suppose that the length of time between consecutive high tides is approximately 12.5 hours. According to the National Oceanic and Atmospheric Administration, on Saturday, June 28, 1997, in Juneau, Alaska, high tide occurred at 8:11 AM (8.1833 hours) and low tide occurred at 2:14 PM (14.2333 hours). Water heights are measured as the amounts above or below the mean lower low water. The height of the water at high tide was 13.2 feet and the height of the water at low tide was 2.2 feet.

(a) Approximately when will the next high tide occur?

(b) Find a sinusoidal function of the form $y = A\sin(\omega x - \phi) + B$ that fits the data.

(c) Draw a graph of the function found in part (b).

(d) Use the function found in part (b) to predict the height of the water at the next high tide.

27. **Hours of Daylight** According to the *Old Farmer's Almanac*, in Miami, Florida, the number of hours of sunlight on the summer solstice is 12.75 and the number of hours of sunlight on the winter solstice is 10.583.

(a) Find a sinusoidal function of the form $y = A\sin(\omega x - \phi) + B$ that fits the data.

(b) Use the function found in part (a) to predict the number of hours of sunlight on April 1, the 91st day of the year.

(c) Draw a graph of the function found in part (a).

(d) Look up the number of hours of sunlight for April 1 in the *Old Farmer's Almanac*, and compare the actual hours of daylight to the results found in part (b).

28. **Hours of Daylight** According to the *Old Farmer's Almanac*, in Detroit, Michigan, the number of hours of sunlight on the summer solstice is 13.65 and the number of hours of sunlight on the winter solstice is 9.067.

(a) Find a sinusoidal function of the form $y = A\sin(\omega x - \phi) + B$ that fits the data.

(b) Use the function found in part (a) to predict the number of hours of sunlight on April 1, the 91st day of the year.

(c) Draw a graph of the function found in part (a).

(d) Look up the number of hours of sunlight for April 1 in the *Old Farmer's Almanac*, and compare the actual hours of daylight to the results found in part (b).

29. **Hours of Daylight** According to the *Old Farmer's Almanac*, in Anchorage, Alaska, the number of hours of sunlight on the summer solstice is 16.233 and the number of hours of sunlight on the winter solstice is 5.45.

(a) Find a sinusoidal function of the form $y = A\sin(\omega x - \phi) + B$ that fits the data.

(b) Use the function found in part (a) to predict the number of hours of sunlight on April 1, the 91st day of the year.

(c) Draw a graph of the function found in part (a).

(d) Look up the number of hours of sunlight for April 1 in the *Old Farmer's Almanac*, and compare the actual hours of daylight to the results found in part (b).

30. Hours of Daylight According to the *Old Farmer's Almanac*, in Honolulu, Hawaii, the number of hours of sunlight on the summer solstice is 12.767 and the number of hours of sunlight on the winter solstice is 10.783.

(a) Find a sinusoidal function of the form $y = A\sin(\omega x - \phi) + B$ that fits the data.

(b) Use the function found in part (a) to predict the number of hours of sunlight on April 1, the 91st day of the year.

(c) Draw a graph of the function found in part (a).

(d) Look up the number of hours of sunlight for April 1 in the *Old Farmer's Almanac*, and compare the actual hours of daylight to the results found in part (b).

31. Explain how the amplitude and period of a sinusoidal graph are used to establish the scale on each coordinate axis.

32. Find an application in your major field that leads to a sinusoidal graph. Write a paper about your findings.

Chapter Review

Things to Know

Definitions

Angle in standard position (p. 324) — Vertex is at the origin; initial side is along the positive x-axis

1 Degree (1°) (p. 325) — $1° = \frac{1}{360}$ revolution

1 Radian (p. 328) — The measure of a central angle of a circle whose rays subtend an arc whose length is the radius of the circle

Trigonometric functions (p. 339–341) — $P = (x, y)$ is the point on the unit circle corresponding to $\theta = t$ radians.

$$\sin t = \sin\theta = y \qquad \cos t = \cos\theta = x \qquad \tan t = \tan\theta = \frac{y}{x}, \quad x \neq 0$$

$$\csc t = \csc\theta = \frac{1}{y}, \quad y \neq 0 \qquad \sec t = \sec\theta = \frac{1}{x}, \quad x \neq 0 \qquad \cot t = \cot\theta = \frac{x}{y}, \quad y \neq 0$$

Periodic function (p. 358) — $f(\theta + p) = f(\theta)$, for all θ, $p > 0$, where the smallest such p is the fundamental period

Formulas

1 revolution $= 360°$ (p. 326)
$= 2\pi$ radians (p. 329)

$s = r\theta$ (p. 328) — θ is measured in radians; s is the length of arc subtended by the central angle θ of the circle of radius r; A is the area of the sector.

$A = \frac{1}{2}r^2\theta$ (p. 332)

$v = r\omega$ (p. 333) — v is the linear speed along the circle of radius r; ω is the angular speed (measured in radians per unit time).

Table of Values

θ (Radians)	θ (Degrees)	$\sin\theta$	$\cos\theta$	$\tan\theta$	$\csc\theta$	$\sec\theta$	$\cot\theta$
0	0°	0	1	0	Not defined	1	Not defined
$\frac{\pi}{6}$	30°	$\frac{1}{2}$	$\frac{\sqrt{3}}{2}$	$\frac{\sqrt{3}}{3}$	2	$\frac{2\sqrt{3}}{3}$	$\sqrt{3}$
$\frac{\pi}{4}$	45°	$\frac{\sqrt{2}}{2}$	$\frac{\sqrt{2}}{2}$	1	$\sqrt{2}$	$\sqrt{2}$	1
$\frac{\pi}{3}$	60°	$\frac{\sqrt{3}}{2}$	$\frac{1}{2}$	$\sqrt{3}$	$\frac{2\sqrt{3}}{3}$	2	$\frac{\sqrt{3}}{3}$
$\frac{\pi}{2}$	90°	1	0	Not defined	1	Not defined	0
π	180°	0	-1	0	Not defined	-1	Not defined
$\frac{3\pi}{2}$	270°	-1	0	Not defined	-1	Not defined	0

Fundamental Identities (p. 362)

$$\tan\theta = \frac{\sin\theta}{\cos\theta}, \quad \cot\theta = \frac{\cos\theta}{\sin\theta}$$

$$\csc\theta = \frac{1}{\sin\theta}, \quad \sec\theta = \frac{1}{\cos\theta}, \quad \cot\theta = \frac{1}{\tan\theta}$$

$$\sin^2\theta + \cos^2\theta = 1, \quad \tan^2\theta + 1 = \sec^2\theta, \quad 1 + \cot^2\theta = \csc^2\theta$$

Properties of the Trigonometric Functions

$y = \sin x$ (p. 371)
Domain: $-\infty < x < \infty$
Range: $-1 \le y \le 1$
Periodic: period $= 2\pi$ (360°)
Odd function

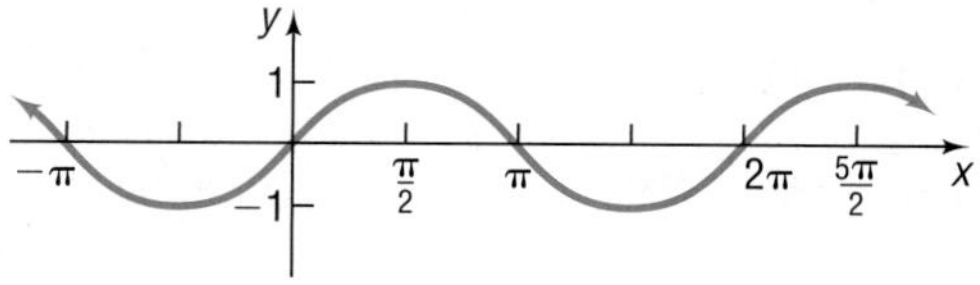

$y = \cos x$ (p. 373)
Domain: $-\infty < x < \infty$
Range: $-1 \le y \le 1$
Periodic: period $= 2\pi$ (360°)
Even function

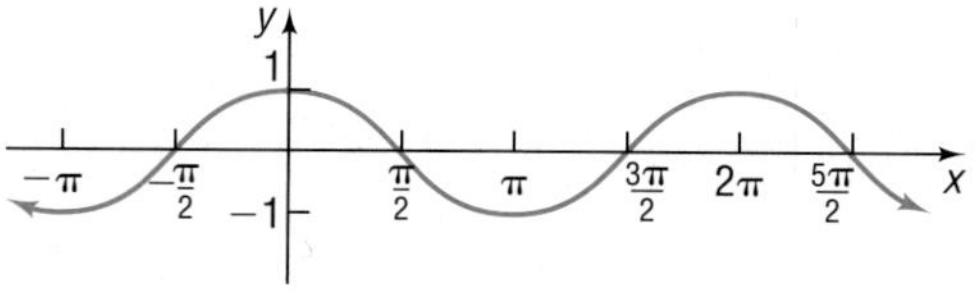

$y = \tan x$ (p. 388)
Domain: $-\infty < x < \infty$, except odd multiples of $\frac{\pi}{2}$ (90°)
Range: $-\infty < y < \infty$
Periodic: period $= \pi$ (180°)
Odd function

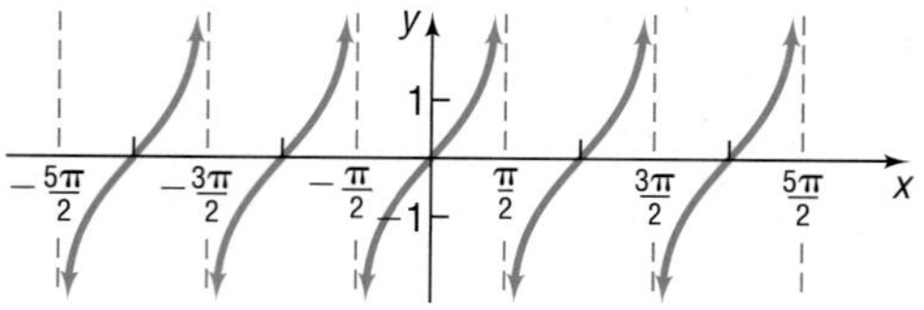

$y = \cot x$ (p. 390)
Domain: $-\infty < x < \infty$, except integral multiples of π (180°)
Range: $-\infty < y < \infty$
Periodic: period $= \pi$ (180°)
Odd function

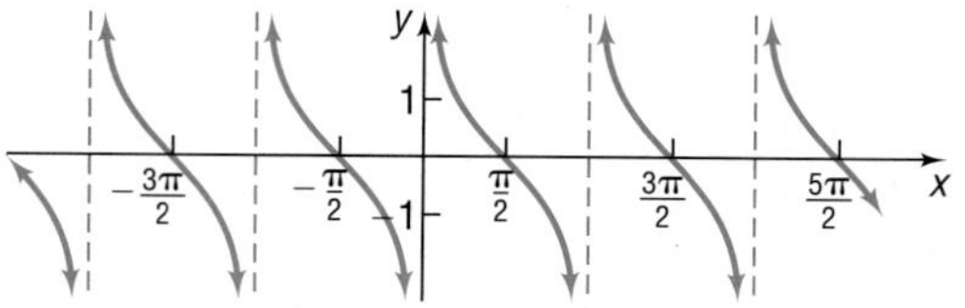

$y = \csc x$ (p. 390) Domain: $-\infty < x < \infty$, except integral multiples of π (180°)
Range: $|y| \geq 1$
Periodic: period $= 2\pi$ (360°)
Odd function

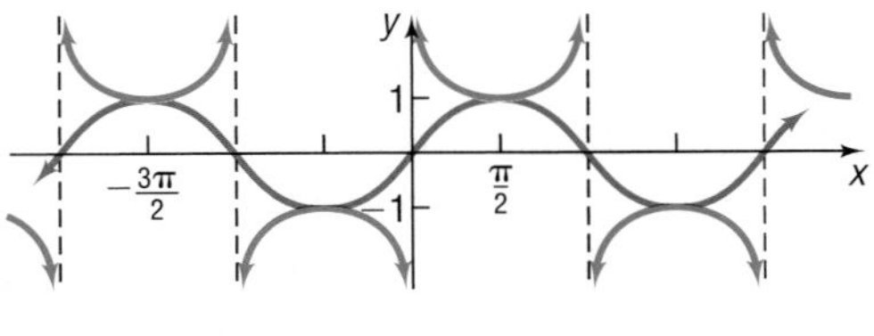

$y = \sec x$ (p. 391) Domain: $-\infty < x < \infty$, except odd multiples of $\frac{\pi}{2}$ (90°)
Range: $|y| \geq 1$
Periodic: period $= 2\pi$ (360°)
Even function

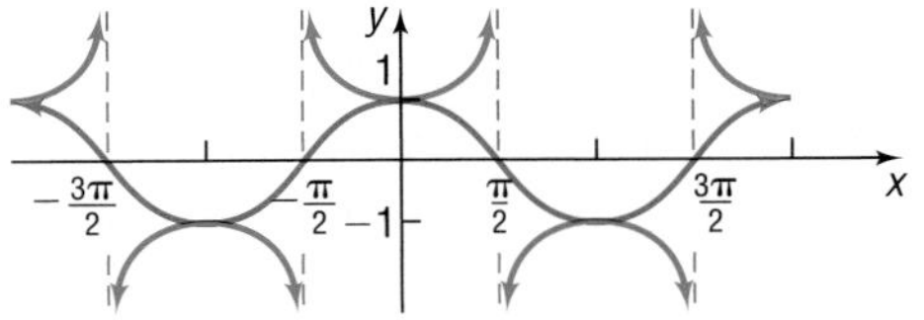

Sinusoidal graphs (p. 376, 394)

$y = A\sin(\omega x), \quad \omega > 0$ Period $= \dfrac{2\pi}{\omega}$

$y = A\cos(\omega x), \quad \omega > 0$ Amplitude $= |A|$

$y = A\sin(\omega x - \phi) = A\sin\left[\omega\left(x - \dfrac{\phi}{\omega}\right)\right]$ Phase shift $= \dfrac{\phi}{\omega}$

$y = A\cos(\omega x - \phi) = A\cos\left[\omega\left(x - \dfrac{\phi}{\omega}\right)\right]$

Objectives

Section		You should be able to . . .	Review Exercises
5.1	1	Convert between degrees, minutes, seconds, and decimal forms for angles (p. 326)	82
	2	Find the arc length of a circle (p. 328)	83, 84
	3	Convert from degrees to radians (p. 330)	1–4
	4	Convert from radians to degrees (p. 330)	5–8
	5	Find the area of a sector of a circle (p. 332)	83
	6	Find the linear speed of an object traveling in circular motion (p. 333)	85–88
5.2	1	Find the exact values of the trigonometric functions using a point on the unit circle (p. 340)	79, 80
	2	Find the exact values of the trigonometric functions of quadrantal angles (p. 342)	17, 18, 20
	3	Find the exact values of the trigonometric functions of $\frac{\pi}{4} = 45°$ (p. 344)	9, 11, 13, 15, 16, 19
	4	Find the exact values of the trigonometric functions of $\frac{\pi}{6} = 30°$ and $\frac{\pi}{3} = 60°$ (p. 345)	9–15
	5	Find the exact values for certain integral multiples of $\frac{\pi}{6} = 30°$, $\frac{\pi}{4} = 45°$, and $\frac{\pi}{3} = 60°$ (p. 348)	13–16
	6	Use a calculator to approximate the value of a trigonometric function (p. 349)	75, 76
5.3	1	Determine the domain and the range of the trigonometric functions (p. 356)	81
	2	Determine the period of the trigonometric functions (p. 358)	81
	3	Determine the signs of the trigonometric functions in a given quadrant (p. 359)	77–78
	4	Find the values of the trigonometric functions utilizing fundamental identities (p. 362)	21–30
	5	Find the exact values of the trigonometric functions of an angle given one of the functions and the quadrant of the angle (p. 363)	31–46

	6	Use even–odd properties to find the exact values of the trigonometric functions (p. 366)	27–30
5.4	1	Graph transformations of the sine function (p. 371)	47, 50
	2	Graph transformations of the cosine function (p. 373)	48, 49
	3	Determine the amplitude and period of sinusoidal functions (p. 375)	59–62, 89
	4	Graph sinusoidal functions: $y = A\sin(\omega x)$ (p. 377)	47, 48, 63-65, 89
	5	Find an equation for a sinusoidal graph (p. 380)	71–74
5.5	1	Graph transformations of the tangent function and cotangent function (p. 386)	51–56
	2	Graph transformations of the cosecant function and secant function (p. 390)	57, 58
5.6	1	Determine the phase shift of a sinusoidal function (p. 393)	65–70, 90
	2	Graph sinusoidal functions: $y = A\sin(\omega x - \phi)$ (p. 394)	65–70, 90
	3	Find a sinusoidal function from data (p. 397)	91–94

Review Exercises *(Blue problem numbers indicate the author's suggestions for use in a Practice Test.)*

In Problems 1–4, convert each angle in degrees to radians. Express your answer as a multiple of π.

1. $135°$ **2.** $210°$ **3.** $18°$ **4.** $15°$

In Problems 5–8, convert each angle in radians to degrees.

5. $\frac{3\pi}{4}$ **6.** $\frac{2\pi}{3}$ **7.** $-\frac{5\pi}{2}$ **8.** $-\frac{3\pi}{2}$

In Problems 9–30, find the exact value of each expression. Do not use a calculator.

9. $\tan\frac{\pi}{4} - \sin\frac{\pi}{6}$ **10.** $\cos\frac{\pi}{3} + \sin\frac{\pi}{2}$ 11. $3\sin 45° - 4\tan\frac{\pi}{6}$

12. $4\cos 60° + 3\tan\frac{\pi}{3}$ **13.** $6\cos\frac{3\pi}{4} + 2\tan\left(-\frac{\pi}{3}\right)$ **14.** $3\sin\frac{2\pi}{3} - 4\cos\frac{5\pi}{2}$

15. $\sec\left(-\frac{\pi}{3}\right) - \cot\left(-\frac{5\pi}{4}\right)$ **16.** $4\csc\frac{3\pi}{4} - \cot\left(-\frac{\pi}{4}\right)$ **17.** $\tan\pi + \sin\pi$

18. $\cos\frac{\pi}{2} - \csc\left(-\frac{\pi}{2}\right)$ **19.** $\cos 540° - \tan(-45°)$ **20.** $\sin 630° + \cos(-180°)$

21. $\sin^2 20° + \frac{1}{\sec^2 20°}$ **22.** $\frac{1}{\cos^2 40°} - \frac{1}{\cot^2 40°}$ **23.** $\sec 50° \cos 50°$

24. $\tan 10° \cot 10°$ 25. $\sec^2 20° - \tan^2 20°$ **26.** $\frac{1}{\sec^2 40°} + \frac{1}{\csc^2 40°}$

27. $\sin(-40°)\csc 40°$ **28.** $\tan(-20°)\cot 20°$ **29.** $\cos 410° \sec(-50°)$ **30.** $\cot 200° \cot(-70°)$

In Problems 31–46, find the exact value of each of the remaining trigonometric functions.

31. $\sin\theta = \frac{4}{5}$, θ is acute **32.** $\cos\theta = \frac{3}{5}$, θ is acute **33.** $\tan\theta = \frac{12}{5}$, $\sin\theta < 0$

34. $\cot\theta = \frac{12}{5}$, $\cos\theta < 0$ **35.** $\sec\theta = -\frac{5}{4}$, $\tan\theta < 0$ **36.** $\csc\theta = -\frac{5}{3}$, $\cot\theta < 0$

37. $\sin\theta = \frac{12}{13}$, θ in quadrant II

38. $\cos\theta = -\frac{3}{5}$, θ in quadrant III

39. $\sin\theta = -\frac{5}{13}$, $\frac{3\pi}{2} < \theta < 2\pi$

40. $\cos\theta = \frac{12}{13}$, $\frac{3\pi}{2} < \theta < 2\pi$

41. $\tan\theta = \frac{1}{3}$, $180° < \theta < 270°$

42. $\tan\theta = -\frac{2}{3}$, $90° < \theta < 180°$

43. $\sec\theta = 3$, $\frac{3\pi}{2} < \theta < 2\pi$

44. $\csc\theta = -4$, $\pi < \theta < \frac{3\pi}{2}$

45. $\cot\theta = -2$, $\frac{\pi}{2} < \theta < \pi$

46. $\tan\theta = -2$, $\frac{3\pi}{2} < \theta < 2\pi$

In Problems 47–58, graph each function. Each graph should contain at least one period.

47. $y = 2\sin(4x)$

48. $y = -3\cos(2x)$

49. $y = -2\cos\left(x + \frac{\pi}{2}\right)$

50. $y = 3\sin(x - \pi)$

51. $y = \tan(x + \pi)$

52. $y = -\tan\left(x - \frac{\pi}{2}\right)$

53. $y = -2\tan(3x)$

54. $y = 4\tan(2x)$

55. $y = \cot\left(x + \frac{\pi}{4}\right)$

56. $y = -4\cot(2x)$

57. $y = \sec\left(x - \frac{\pi}{4}\right)$

58. $y = \csc\left(x + \frac{\pi}{4}\right)$

In Problems 59–62, determine the amplitude and period of each function without graphing.

59. $y = 4\cos x$

60. $y = \sin(2x)$

61. $y = -8\sin\left(\frac{\pi}{2}x\right)$

62. $y = -2\cos(3\pi x)$

In Problems 63–70, find the amplitude, period, and phase shift of each function. Graph each function. Show at least one period.

63. $y = 4\sin(3x)$

64. $y = 2\cos\left(\frac{1}{3}x\right)$

65. $y = 2\sin(2x - \pi)$

66. $y = -\cos\left(\frac{1}{2}x + \frac{\pi}{2}\right)$

67. $y = \frac{1}{2}\sin\left(\frac{3}{2}x - \pi\right)$

68. $y = \frac{3}{2}\cos(6x + 3\pi)$

69. $y = -\frac{2}{3}\cos(\pi x - 6)$

70. $y = -7\sin\left(\frac{\pi}{3}x + \frac{4}{3}\right)$

In Problems 71–74, find a function whose graph is given.

71.

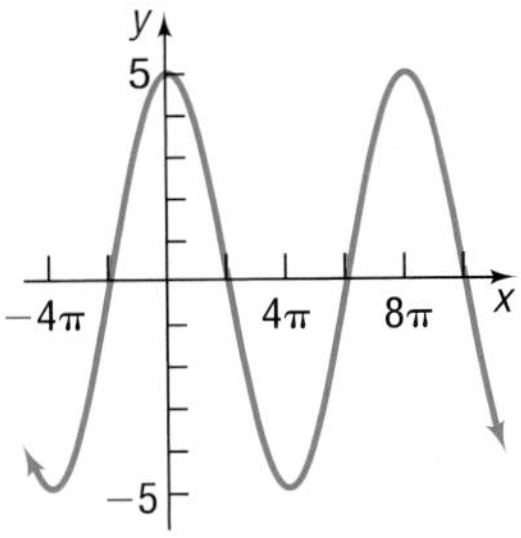

72.

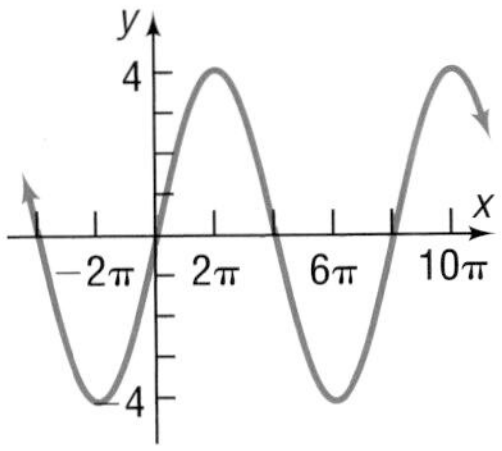

73.

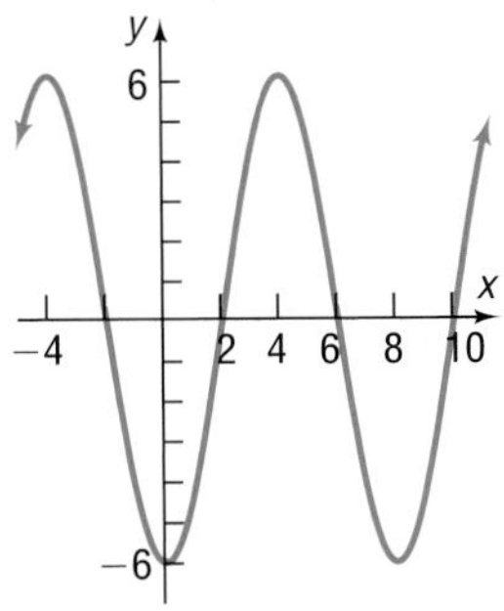

74.

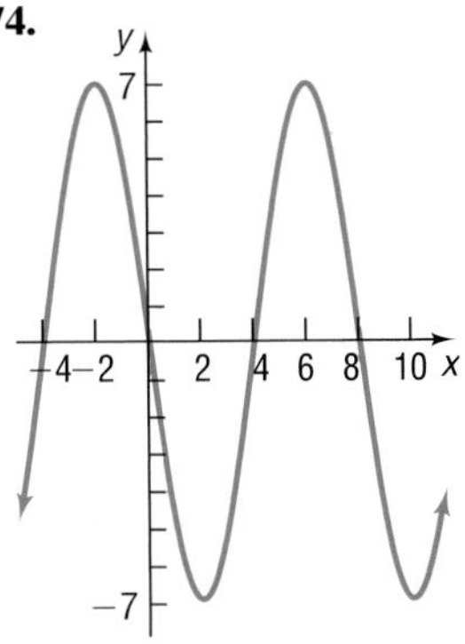

75. Use a calculator to approximate $\sin\frac{\pi}{8}$. Round the answer to two decimal places.

76. Use a calculator to approximate sec 10°. Round the answer to two decimal places.

77. Determine the signs of the six trigonometric functions of an angle θ whose terminal side is in quadrant III.

78. Name the quadrant θ lies in if $\cos\theta > 0$ and $\tan\theta < 0$.

79. Find the exact values of the six trigonometric functions if $P = \left(-\frac{1}{3}, \frac{2\sqrt{2}}{3}\right)$ is the point on the unit circle that corresponds to t.

80. Find the exact value of $\sin t$, $\cos t$, and $\tan t$ if $P = \left(-\frac{3}{5}, \frac{4}{5}\right)$ is the point on the unit circle that corresponds to t.

81. What is the domain and the range of the secant function? What is the period?

82. (a) Convert the angle 32°20′35″ to a decimal in degrees. Round the answer to two decimal places.
(b) Convert the angle 63.18° to D°M′S″ form. Express the answer to the nearest second.

83. Find the length of the arc subtended by a central angle of 30° on a circle of radius 2 feet. What is the area of the sector?

84. The minute hand of a clock is 8 inches long. How far does the tip of the minute hand move in 30 minutes? How far does it move in 20 minutes?

85. Angular Speed of a Race Car A race car is driven around a circular track at a constant speed of 180 miles per hour. If the diameter of the track is $\frac{1}{2}$ mile, what is the angular speed of the car? Express your answer in revolutions per hour (which is equivalent to laps per hour).

86. Merry-Go-Rounds A neighborhood carnival has a merry-go-round whose radius is 25 feet. If the time for one revolution is 30 seconds, how fast is the merry-go-round going?

87. Lighthouse Beacons The Montauk Point Lighthouse on Long Island has dual beams (two light sources opposite each other). Ships at sea observe a blinking light every 5 seconds. What rotation speed is required to do this?

88. Spin Balancing Tires The radius of each wheel of a car is 16 inches. At how many revolutions per minute should a spin balancer be set to balance the tires at a speed of 90 miles per hour? Is the setting different for a wheel of radius 14 inches? If so, what is this setting?

89. Alternating Voltage The electromotive force E, in volts, in a certain ac circuit obeys the equation

$$E = 120\sin(120\pi t), \qquad t \geq 0$$

where t is measured in seconds.
(a) What is the maximum value of E?
(b) What is the period?
(c) Graph this function over two periods.

90. Alternating Current The current I, in amperes, flowing through an ac (alternating current) circuit at time t is

$$I = 220\sin\left(30\pi t + \frac{\pi}{6}\right), \qquad t \geq 0$$

(a) What is the period?
(b) What is the amplitude?
(c) What is the phase shift?
(d) Graph this function over two periods.

91. Monthly Temperature The following data represent the average monthly temperatures for Phoenix, Arizona.

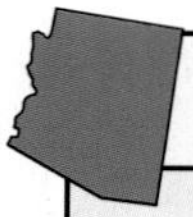

Month, m	Average Monthly Temperature, T
January, 1	51
February, 2	55
March, 3	63
April, 4	67
May, 5	77
June, 6	86
July, 7	90
August, 8	90
September, 9	84
October, 10	71
November, 11	59
December, 12	52

SOURCE: U.S. National Oceanic and Atmospheric Administration

(a) Use a graphing utility to draw a scatter diagram of the data for one period.
(b) By hand, find a sinusoidal function of the form $y = A\sin(\omega x - \phi) + B$ that fits the data.
(c) Draw the sinusoidal function found in part (b) on the scatter diagram.
(d) Use a graphing utility to find the sinusoidal function of best fit.
(e) Graph the sinusoidal function of best fit on the scatter diagram.

92. Monthly Temperature The following data represent the average monthly temperatures for Chicago, Illinois.

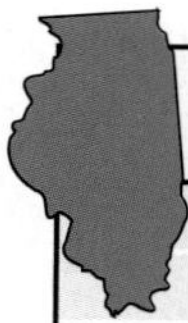

Month, m	Average Monthly Temperature, T
January, 1	25
February, 2	28
March, 3	36
April, 4	48
May, 5	61
June, 6	72
July, 7	74
August, 8	75
September, 9	66
October, 10	55
November, 11	39
December, 12	28

SOURCE: U.S. National Oceanic and Atmospheric Administration

(a) Use a graphing utility to draw a scatter diagram of the data for one period.
(b) By hand, find a sinusoidal function of the form $y = A\sin(\omega x - \phi) + B$ that fits the data.

(c) Draw the sinusoidal function found in part (b) on the scatter diagram.
(d) Use a graphing utility to find the sinusoidal function of best fit.
(e) Graph the sinusoidal function of best fit on the scatter diagram.

93. Hours of Daylight According to the *Old Farmer's Almanac,* in Las Vegas, Nevada, the number of hours of sunlight on the summer solstice is 13.367 and the number of hours of sunlight on the winter solstice is 9.667.
(a) Find a sinusoidal function of the form

$$y = A\sin(\omega x - \phi) + B$$

that fits the data.
(b) Draw a graph of the function found in part (a).
(c) Use the function found in part (a) to predict the number of hours of sunlight on April 1, the 91st day of the year.
(d) Look up the number of hours of sunlight for April 1 in the *Old Farmer's Almanac* and compare the actual hours of daylight to the results found in part (c).

94. Hours of Daylight According to the *Old Farmer's Almanac,* in Seattle, Washington, the number of hours of sunlight on the summer solstice is 13.967 and the number of hours of sunlight on the winter solstice is 8.417.
(a) Find a sinusoidal function of the form

$$y = A\sin(\omega x - \phi) + B$$

that fits the data.
(b) Draw a graph of the function found in part (a).
(c) Use the function found in part (a) to predict the number of hours of sunlight on April 1, the 91st day of the year.
(d) Look up the number of hours of sunlight for April 1 in the *Old Farmer's Almanac* and compare the actual hours of daylight to the results found in part (c).

Chapter Projects

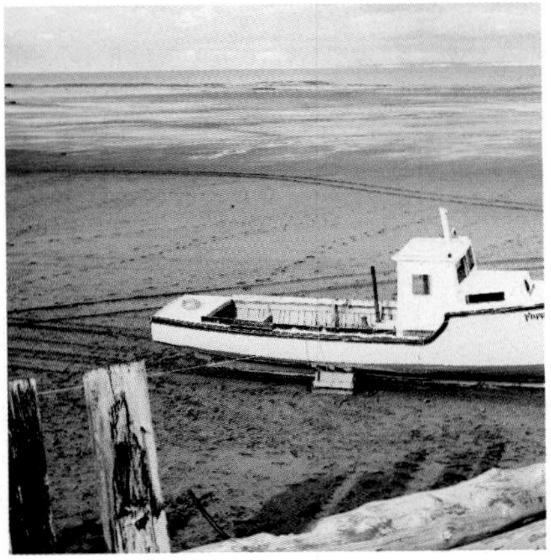

1. Tides A partial tide table for September 2001 for Sabine Pass along the Texas Gulf Coast is given in the table.

(a) On September 15, when was the tide high? This is called *high tide*. On September 19, when was the tide low? This is called *low tide*. Most days will have two low tides and two high tides.

(b) Why do you think there is a negative height for the low tide on September 14? What is the tide height measured against?

(c) On your graphing utility, draw a scatter diagram for the data in the table. Let T (time) be the independent variable, with $T = 0$ being 12:00 AM on September 1, $T = 24$ being 12:00 AM on September 2, and so on. Remember that there are 60 minutes in an hour. Let H be the height in feet when converting the times. Also, make sure that your graphing utility is in radian mode.

(d) What shape does the data take? What is the period of the data? What is the amplitude? Is the amplitude constant? Explain.

	High Tide		High Tide		Low Tide		Low Tide		Sun/Moon phase
Sept	Time	Ht (ft)	Time	Ht (ft)	Time	Ht (ft)	Time	Ht (ft)	Rise/Set
F 14	03:08a	2.4	11:12a	2.2	08:14a	2.0	07:19p	−0.1	7:00a/7:23p
S 15	03:33a	2.4	12:56p	2.2	08:15a	1.9	08:13p	0.0	7:00a/7:22p
S 16	03:57a	2.3	02:17p	2.3	08:45a	1.6	09:05p	0.3	7:01a/7:20p
M17	04:20a	2.2	03:33p	2.3	09:24a	1.4	09:54p	0.5	7:01a/7:19p
T18	04:41a	2.2	04:47p	2.3	10:08a	1.0	10:43p	1.0	7:02a/7:08p
W19	05:01a	2.0	06:04p	2.3	10:54a	0.7	11:32p	1.4	7:02a/7:17p
T20	05:20a	2.0	07:27p	2.3	11:44a	0.4			7:03a/7:15p

SOURCE: www.harbortides.com

(e) Using Steps 1–4 given on pages 399–400, fit a sine curve to the data. Let the amplitude be the average of the amplitudes that you found in part (c), unless the amplitude was constant. Is there a vertical shift? Is there a phase shift?

(f) Using your graphing utility, find the sinusoidal function of best fit. How does it compare to your equation?

(g) Using the equation found in part (e) and the sinusoidal equation of best fit found in part (f), predict the high tides and the low tides on September 21.

(h) Looking at the times of day that the low tides occur, what do you think causes the low tides to vary so much each day? Explain. Does this seem to have the same type of effect on the high tides? Explain.

The following projects are available at www.prenhall.com/sullivan7e

2. **Project at Motorola** ***Digital Transmission over the Air***
3. **Identifying Mountain Peaks in Hawaii**
4. **CBL Experiment**

Cumulative Review

1. Find the real solutions, if any, of the equation $2x^2 + x - 1 = 0$.

2. Find an equation for the line with slope -3 containing the point $(-2, 5)$.

3. Find an equation for the circle of radius 4 and center at the point $(0, -2)$.

4. Discuss the equation $2x - 3y = 12$. Graph it.

5. Discuss the equation $x^2 + y^2 - 2x + 4y - 4 = 0$. Graph it.

6. Use transformations to graph the function $y = (x - 3)^2 + 2$.

7. Sketch a graph of each of the following functions. Label at least three points on each graph.
   (a) $y = x^2$ (b) $y = x^3$
   (c) $y = e^x$ (d) $y = \ln x$
   (e) $y = \sin x$ (f) $y = \tan x$

8. Find the inverse function of $f(x) = 3x - 2$.

9. Find the exact value of $(\sin 14°)^2 + (\cos 14°)^2 - 3$.

10. Graph $y = 3\sin(2x)$.

11. Find the exact value of $\tan\frac{\pi}{4} - 3\cos\frac{\pi}{6} + \csc\frac{\pi}{6}$.

12. Find an exponential function for the following graph. Express your answer in the form $y = Ab^x$.

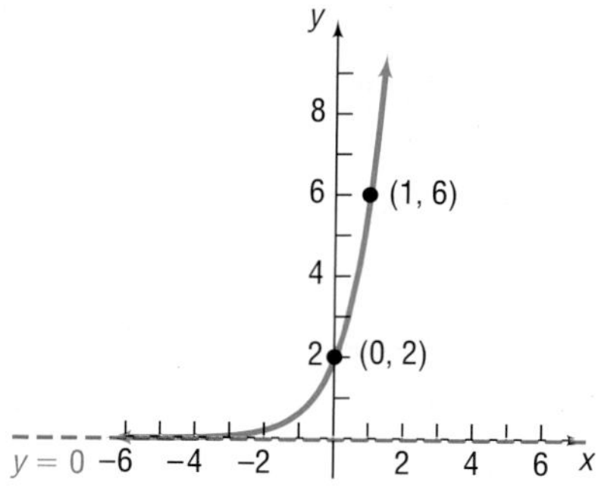

13. Find a sinusoidal function for the following graph.

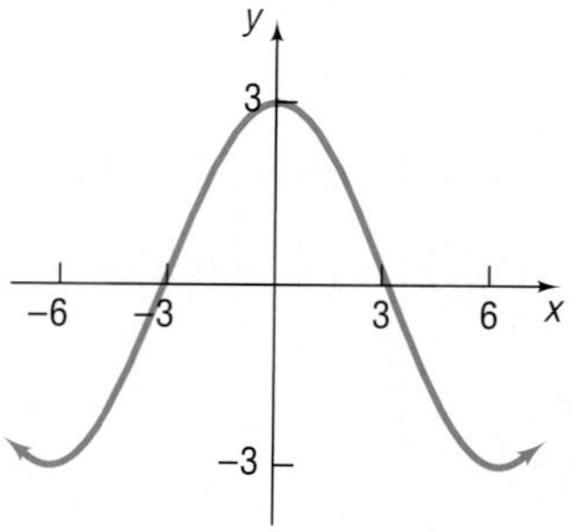

14. (a) Find a linear function that contains the points $(-2, 3)$ and $(1, -6)$. What is the slope? What are the intercepts of the function? Graph the function. Be sure to label the intercepts.
    (b) Find a quadratic function that contains the point $(-2, 3)$ with vertex $(1, -6)$. What are the intercepts of the function? Graph the function.
    (c) Show that there is no exponential function of the form $f(x) = ae^x$ that contains the points $(-2, 3)$ and $(1, -6)$.

15. (a) Find a polynomial function of degree 3 whose y-intercept is 5 and whose x-intercepts are -2, 3, and 5. Graph the function. Label the local minima and local maxima.
    (b) Find a rational function whose y-intercept is 5 and whose x-intercepts are -2, 3, and 5 that has the line $x = 2$ as a vertical asymptote. Graph the function. (Answers may vary.) Label any local maxima or minima.

6 Analytic Trigonometry

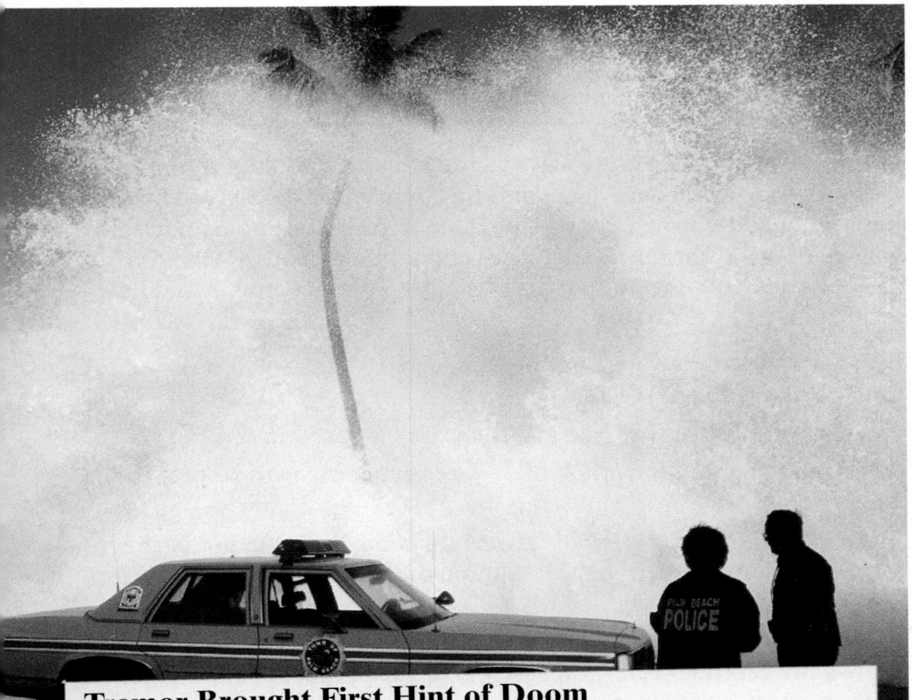

Tremor Brought First Hint of Doom

Underwater earthquakes that triggered the devastating tsunami in PNG would have been felt by villagers about 30 minutes before the waves struck, scientists said yesterday. "The tremor was felt by coastal residents who may not have realized its significance or did not have time to retreat," said associate Professor Ted Bryant, a geoscientist at the University of Wollongong.

The tsunami would have sounded like a fleet of bombers as it crashed into a 30-kilometer stretch of coast. "The tsunami would have been caused by a rapid uplift or drop of the sea floor," said an applied mathematician and cosmologist from Monash University, Professor Joe Monaghan, who is one of Australia's leading experts on tsunamis....

Tsunamis, ridges of water hundreds of kilometers long and stretching from front to back for several kilometers, line up parallel to the beach. "We are talking about a huge volume of water moving very fast—300 kilometers per hour would be typical," Professor Monaghan said.

(Source: Peter Spinks, *The Age*, Tuesday, July 21, 1998. Peter Spinks runs science-writing and media-skills workshops, on request worldwide [www.dreamwater.org/workshop])

—SEE CHAPTER PROJECT 1.

OUTLINE

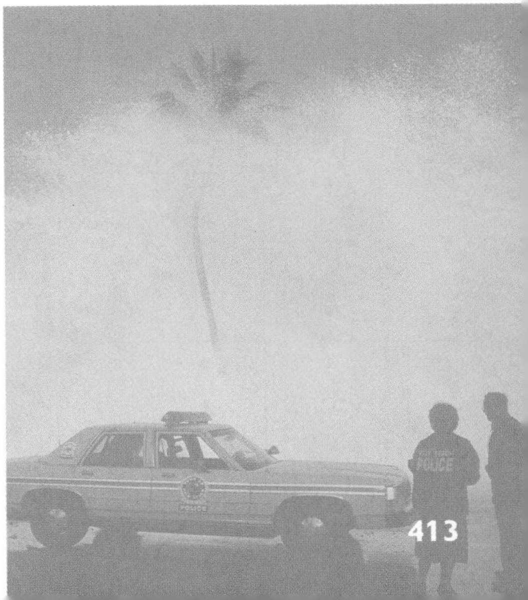

6.1 The Inverse Sine, Cosine, and Tangent Functions

PREPARING FOR THIS SECTION *Before getting started, review the following:*

- Inverse Functions (Section 4.2, pp. 231–241)
- Values of the Trigonometric Functions (Section 5.2, pp. 342–351)
- Domain and Range of the Sine, Cosine, and Tangent Functions (Section 5.3, pp. 356–357)
- Graphs of the Sine, Cosine, and Tangent Functions (Section 5.4, pp. 370–374, and Section 5.5, pp. 386–388)

Now work the 'Are You Prepared?' problems on page 423.

OBJECTIVES 1 Find the Exact Value of the Inverse Sine, Cosine, and Tangent Functions
2 Find an Approximate Value of the Inverse Sine, Cosine, and Tangent Functions

In Section 4.2 we discussed inverse functions, and we noted that if a function is one-to-one it will have an inverse function. We also observed that if a function is not one-to-one it may be possible to restrict its domain in some suitable manner so that the restricted function is one-to-one.

Next, we review some properties of a one-to-one function f and its inverse function f^{-1}.

1. $f^{-1}(f(x)) = x$ for every x in the domain of f and $f(f^{-1}(x)) = x$ for every x in the domain of f^{-1}.
2. Domain of f = range of f^{-1}, and range of f = domain of f^{-1}.
3. The graph of f and the graph of f^{-1} are symmetric with respect to the line $y = x$.
4. If a function $y = f(x)$ has an inverse function, the equation of the inverse function is $x = f(y)$. The solution of this equation is $y = f^{-1}(x)$.

The Inverse Sine Function

In Figure 1, we reproduce the graph of $y = \sin x$. Because every horizontal line $y = b$, where b is between -1 and 1, intersects the graph of $y = \sin x$ infinitely many times, it follows from the horizontal-line test that the function $y = \sin x$ is not one-to-one.

Figure 1
$y = \sin x, -\infty < x < \infty, -1 \le y \le 1$

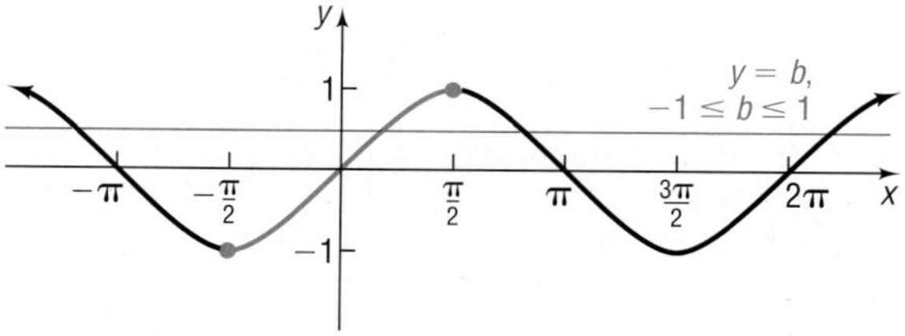

However, if we restrict the domain of $y = \sin x$ to the interval $\left[-\frac{\pi}{2}, \frac{\pi}{2}\right]$, the restricted function

$$y = \sin x, \qquad -\frac{\pi}{2} \le x \le \frac{\pi}{2}$$

is one-to-one and hence will have an inverse function.* See Figure 2.

Figure 2
$y = \sin x, -\frac{\pi}{2} \le x \le \frac{\pi}{2}, -1 \le y \le 1$

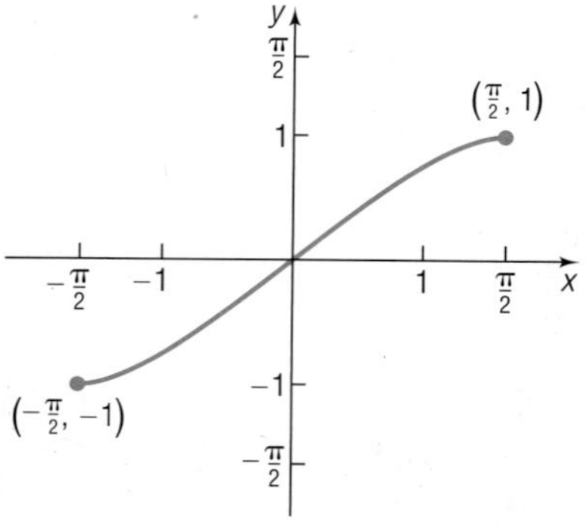

*Although there are many other ways to restrict the domain and obtain a one-to-one function, mathematicians have agreed on a consistent use of the interval $\left[-\frac{\pi}{2}, \frac{\pi}{2}\right]$, in order to define the inverse of $y = \sin x$.

An equation for the inverse of $y = f(x) = \sin x$ is obtained by interchanging x and y. The implicit form of the inverse function is $x = \sin y$, $-\frac{\pi}{2} \le y \le \frac{\pi}{2}$. The explicit form is called the **inverse sine** of x and is symbolized by $y = f^{-1}(x) = \sin^{-1} x$.

$$y = \sin^{-1} x \quad \text{means} \quad x = \sin y$$
$$\text{where} \quad -1 \le x \le 1 \quad \text{and} \quad -\frac{\pi}{2} \le y \le \frac{\pi}{2} \qquad \textbf{(1)}$$

Because $y = \sin^{-1} x$ means $x = \sin y$, we read $y = \sin^{-1} x$ as "y is the angle or real number whose sine equals x." Alternatively, we can say that "y is the inverse sine of x." Be careful about the notation used. The superscript -1 that appears in $y = \sin^{-1} x$ is not an exponent, but is reminiscent of the symbolism f^{-1} used to denote the inverse function of f. (To avoid this notation, some books use the notation $y = \arcsin x$ instead of $y = \sin^{-1} x$.)

The inverse of a function f receives as input an element from the range of f and returns as output an element in the domain of f. The restricted sine function, $y = f(x) = \sin x$, receives as input an angle or real number x in the interval $\left[-\frac{\pi}{2}, \frac{\pi}{2}\right]$ and outputs a real number in the interval $[-1, 1]$. Therefore, the inverse sine function $y = \sin^{-1} x$ receives as input a real number in the interval $[-1, 1]$ or $-1 \le x \le 1$, its domain, and outputs an angle or real number in the interval $\left[-\frac{\pi}{2}, \frac{\pi}{2}\right]$ or $-\frac{\pi}{2} \le y \le \frac{\pi}{2}$, its range.

Figure 3

$y = \sin^{-1} x,\ -1 \le x \le 1,\ -\frac{\pi}{2} \le y \le \frac{\pi}{2}$

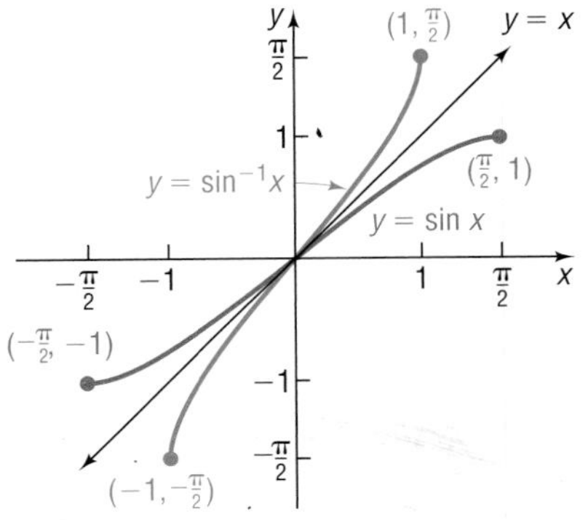

The graph of the inverse sine function can be obtained by reflecting the restricted portion of the graph of $y = f(x) = \sin x$ about the line $y = x$, as shown in Figure 3.

CHECK: Graph $Y = \sin^{-1} x$ and compare the result with Figure 3.

1 For some numbers x, it is possible to find the exact value of $y = \sin^{-1} x$.

EXAMPLE 1 Finding the Exact Value of an Inverse Sine Function

Find the exact value of: $\sin^{-1} 1$

Solution Let $\theta = \sin^{-1} 1$. We seek the angle, θ, $-\frac{\pi}{2} \le \theta \le \frac{\pi}{2}$, whose sine equals 1.

$$\theta = \sin^{-1} 1, \qquad -\frac{\pi}{2} \le \theta \le \frac{\pi}{2}$$

$$\sin \theta = 1, \qquad -\frac{\pi}{2} \le \theta \le \frac{\pi}{2} \qquad \text{By definition of } y = \sin^{-1} x$$

Now look at Table 1 and Figure 4.

Table 1

θ	$\sin\theta$
$-\frac{\pi}{2}$	-1
$-\frac{\pi}{3}$	$-\frac{\sqrt{3}}{2}$
$-\frac{\pi}{4}$	$-\frac{\sqrt{2}}{2}$
$-\frac{\pi}{6}$	$-\frac{1}{2}$
0	0
$\frac{\pi}{6}$	$\frac{1}{2}$
$\frac{\pi}{4}$	$\frac{\sqrt{2}}{2}$
$\frac{\pi}{3}$	$\frac{\sqrt{3}}{2}$
$\frac{\pi}{2}$	1

Figure 4

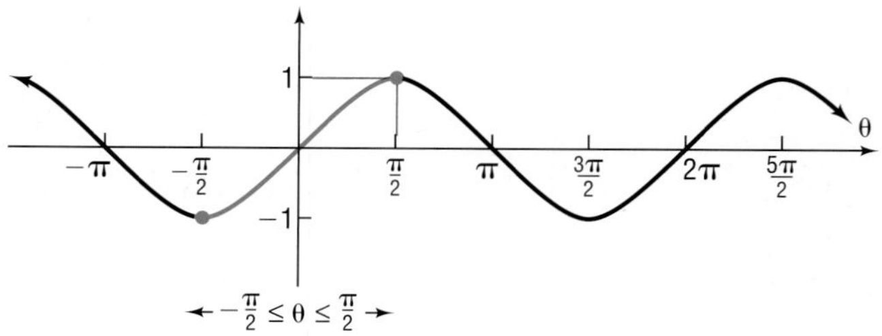

We see that the only angle θ within the interval $\left[-\frac{\pi}{2}, \frac{\pi}{2}\right]$ whose sine is 1 is $\frac{\pi}{2}$. (Note that $\sin\frac{5\pi}{2}$ also equals 1, but $\frac{5\pi}{2}$ lies outside the interval $\left[-\frac{\pi}{2}, \frac{\pi}{2}\right]$ and hence is not admissible.) So, since $\sin\frac{\pi}{2} = 1$ and $\frac{\pi}{2}$ is in $\left[-\frac{\pi}{2}, \frac{\pi}{2}\right]$, we conclude that

$$\sin^{-1} 1 = \frac{\pi}{2}$$

◀

NOW WORK PROBLEM 13.

EXAMPLE 2 Finding the Exact Value of an Inverse Sine Function

Find the exact value of: $\sin^{-1}\left(-\frac{1}{2}\right)$

Solution Let $\theta = \sin^{-1}\left(-\frac{1}{2}\right)$. We seek the angle θ, $-\frac{\pi}{2} \le \theta \le \frac{\pi}{2}$, whose sine equals $-\frac{1}{2}$.

$$\theta = \sin^{-1}\left(-\frac{1}{2}\right), \qquad -\frac{\pi}{2} \le \theta \le \frac{\pi}{2}$$

$$\sin\theta = -\frac{1}{2}, \qquad -\frac{\pi}{2} \le \theta \le \frac{\pi}{2}$$

(Refer to Table 1 and Figure 4, if necessary.) The only angle within the interval $\left[-\frac{\pi}{2}, \frac{\pi}{2}\right]$ whose sine is $-\frac{1}{2}$ is $-\frac{\pi}{6}$. So, since $\sin\left(-\frac{\pi}{6}\right) = -\frac{1}{2}$ and $-\frac{\pi}{6}$ is in the interval $\left[-\frac{\pi}{2}, \frac{\pi}{2}\right]$, we conclude that

$$\sin^{-1}\left(-\frac{1}{2}\right) = -\frac{\pi}{6}$$

◀

NOW WORK PROBLEM 19.

2 For most numbers x, the value $y = \sin^{-1} x$ must be approximated.

EXAMPLE 3 Finding an Approximate Value of an Inverse Sine Function

Find an approximate value of:

(a) $\sin^{-1}\frac{1}{3}$ (b) $\sin^{-1}\left(-\frac{1}{4}\right)$

Express the answer in radians rounded to two decimal places.

Solution Because we want the angle measured in radians, we first set the mode to radians.

(a) $\sin^{-1}\dfrac{1}{3} = 0.34$, rounded to two decimal places.*

(b) Figure 5 shows the solution using a TI-83 graphing calculator. Then

$$\sin^{-1}\left(-\frac{1}{4}\right) = -0.25$$

rounded to two decimal places. ◀

Figure 5

NOW WORK PROBLEM 25.

When we discussed functions and their inverses in Section 4.2, we found that $f^{-1}(f(x)) = x$ and $f(f^{-1}(x)) = x$. In terms of the sine function and its inverse, these properties are of the form

$$f^{-1}(f(x)) = \sin^{-1}(\sin x) = x, \quad \text{where } -\frac{\pi}{2} \le x \le \frac{\pi}{2} \qquad \textbf{(2a)}$$

$$f(f^{-1}(x)) = \sin(\sin^{-1} x) = x, \quad \text{where } -1 \le x \le 1 \qquad \textbf{(2b)}$$

For example, because $\dfrac{\pi}{8}$ lies in the interval $\left[-\dfrac{\pi}{2}, \dfrac{\pi}{2}\right]$, the restricted domain of the sine function, we can apply (2a) to get

$$\sin^{-1}\left[\sin\left(\frac{\pi}{8}\right)\right] = \frac{\pi}{8}$$

Also, because 0.8 lies in the interval $[-1, 1]$, the domain of the inverse sine function, we can apply (2b) to get

$$\sin[\sin^{-1}(0.8)] = 0.8$$

See Figure 6 for these calculations on a graphing calculator.

Figure 6

See Figure 7. Because $\dfrac{5\pi}{8}$ is not in the interval $\left[-\dfrac{\pi}{2}, \dfrac{\pi}{2}\right]$,

$$\sin^{-1}\left[\sin\left(\frac{5\pi}{8}\right)\right] \ne \frac{5\pi}{8}$$

To find $\sin^{-1}\left(\sin\dfrac{5\pi}{8}\right)$, we use the fact that $\sin\dfrac{5\pi}{8} = \sin\dfrac{3\pi}{8}$.

Figure 7

Since $\dfrac{3\pi}{8}$ lies in the interval $\left[-\dfrac{\pi}{2}, \dfrac{\pi}{2}\right]$, we can apply (2a) to get

$$\sin^{-1}\left(\sin\frac{5\pi}{8}\right) = \sin^{-1}\left(\sin\frac{3\pi}{8}\right) = \frac{3\pi}{8} \approx 1.178097245$$

*On most calculators, the inverse sine is obtained by pressing SHIFT or 2nd, followed by sin. On some calculators, $\sin^{-1}$ is pressed first, then 1/3 is entered; on others, this sequence is reversed. Consult your owner's manual for the correct sequence.

Also, because 1.8 is not in the interval $[-1, 1]$,

$$\sin[\sin^{-1}(1.8)] \neq 1.8$$

See Figure 8. Can you explain why the error appears?

Figure 8

NOW WORK PROBLEM 37.

The Inverse Cosine Function

In Figure 9 we reproduce the graph of $y = \cos x$. Because every horizontal line $y = b$, where b is between -1 and 1, intersects the graph of $y = \cos x$ infinitely many times, it follows that the cosine function is not one-to-one.

Figure 9
$y = \cos x$, $-\infty < x < \infty$, $-1 \leq y \leq 1$

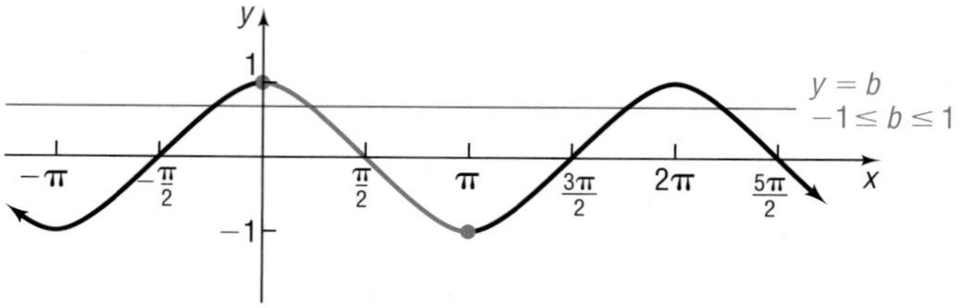

However, if we restrict the domain of $y = \cos x$ to the interval $[0, \pi]$, the restricted function

$$y = \cos x, \qquad 0 \leq x \leq \pi$$

is one-to-one and hence will have an inverse function.* See Figure 10.

Figure 10
$y = \cos x$, $0 \leq x \leq \pi$, $-1 \leq y \leq 1$

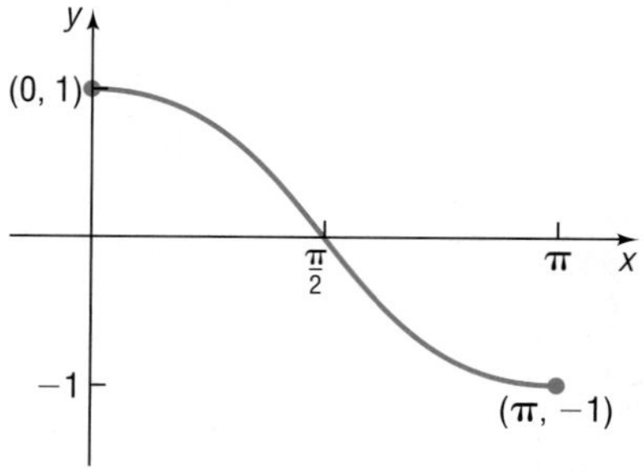

An equation for the inverse of $y = f(x) = \cos x$ is obtained by interchanging x and y. The implicit form of the inverse function is $x = \cos y$, $0 \leq y \leq \pi$. The explicit form is called the **inverse cosine** of x and is symbolized by $y = f^{-1}(x) = \cos^{-1} x$ (or by $y = \arccos x$).

$$y = \cos^{-1} x \quad \text{means} \quad x = \cos y \qquad (3)$$
$$\text{where} \quad -1 \leq x \leq 1 \quad \text{and} \quad 0 \leq y \leq \pi$$

Here y is the angle whose cosine is x. The domain of the function $y = \cos^{-1} x$ is $-1 \leq x \leq 1$, and its range is $0 \leq y \leq \pi$. (Do you know why?)

*This is the generally accepted restriction to define the inverse.

The graph of $y = \cos^{-1} x$ can be obtained by reflecting the restricted portion of the graph of $y = \cos x$ about the line $y = x$, as shown in Figure 11.

Figure 11
$y = \cos^{-1} x, -1 \le x \le 1, 0 \le y \le \pi$

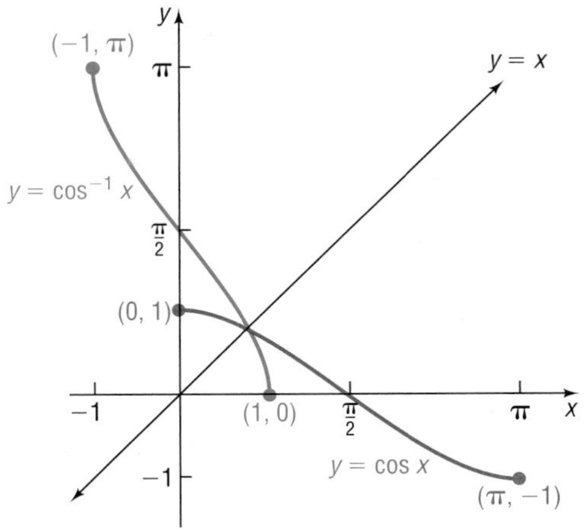

CHECK: Graph $Y = \cos^{-1} x$ and compare the result with Figure 11.

EXAMPLE 4 Finding the Exact Value of an Inverse Cosine Function

Find the exact value of: $\cos^{-1} 0$

Solution Let $\theta = \cos^{-1} 0$. We seek the angle θ, $0 \le \theta \le \pi$, whose cosine equals 0.

$$\theta = \cos^{-1} 0, \qquad 0 \le \theta \le \pi$$

$$\cos\theta = 0, \qquad 0 \le \theta \le \pi$$

Look at Table 2 and Figure 12.

Table 2

θ	$\cos\theta$
0	1
$\frac{\pi}{6}$	$\frac{\sqrt{3}}{2}$
$\frac{\pi}{4}$	$\frac{\sqrt{2}}{2}$
$\frac{\pi}{3}$	$\frac{1}{2}$
$\frac{\pi}{2}$	0
$\frac{2\pi}{3}$	$-\frac{1}{2}$
$\frac{3\pi}{4}$	$-\frac{\sqrt{2}}{2}$
$\frac{5\pi}{6}$	$-\frac{\sqrt{3}}{2}$
π	-1

Figure 12

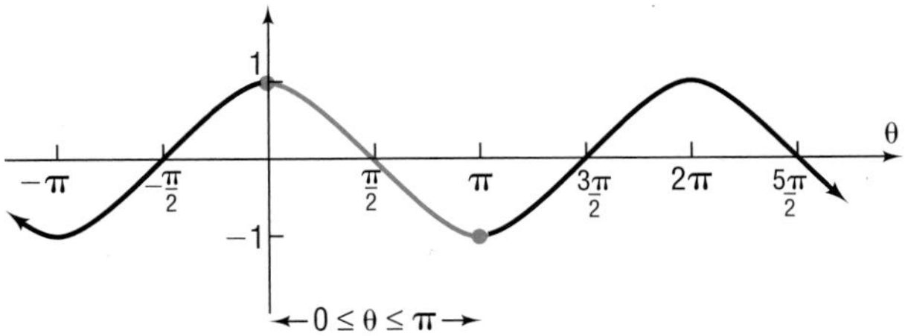

We see that the only angle θ within the interval $[0, \pi]$ whose cosine is 0 is $\frac{\pi}{2}$. (Note that $\cos\frac{3\pi}{2}$ also equals 0, but $\frac{3\pi}{2}$ lies outside the interval $[0, \pi]$ and hence is not admissible.) So, since $\cos\frac{\pi}{2} = 0$ and $\frac{\pi}{2}$ is in the interval $[0, \pi]$, we conclude that

$$\cos^{-1} 0 = \frac{\pi}{2}$$

◀

EXAMPLE 5 Finding the Exact Value of an Inverse Cosine Function

Find the exact value of: $\cos^{-1}\left(-\frac{\sqrt{2}}{2}\right)$

Solution Let $\theta = \cos^{-1}\left(-\frac{\sqrt{2}}{2}\right)$. We seek the angle θ, $0 \le \theta \le \pi$, whose cosine equals $-\frac{\sqrt{2}}{2}$.

$$\theta = \cos^{-1}\left(-\frac{\sqrt{2}}{2}\right), \qquad 0 \le \theta \le \pi$$

$$\cos\theta = -\frac{\sqrt{2}}{2}, \qquad 0 \le \theta \le \pi$$

Look at Table 2 and Figure 13.

Figure 13

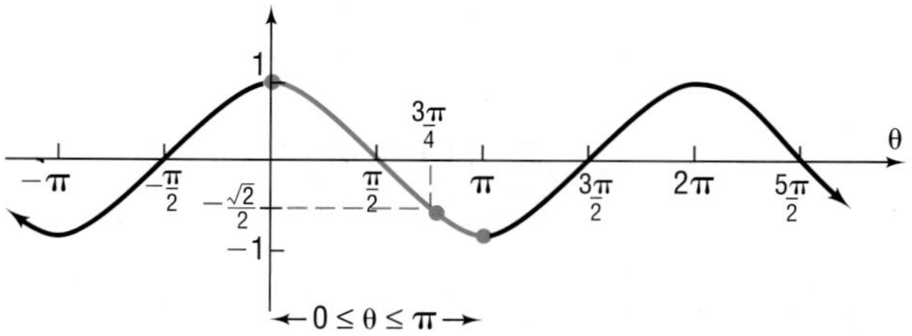

We see that the only angle θ within the interval $[0, \pi]$, whose cosine is $-\frac{\sqrt{2}}{2}$ is $\frac{3\pi}{4}$. So, since $\cos\frac{3\pi}{4} = -\frac{\sqrt{2}}{2}$ and $\frac{3\pi}{4}$ is in the interval $[0, \pi]$, we conclude that

$$\cos^{-1}\left(\frac{\sqrt{2}}{2}\right) = \frac{3\pi}{4}$$

◀

NOW WORK PROBLEM 23.

For the cosine function and its inverse, the following properties hold:

$$f^{-1}(f(x)) = \cos^{-1}(\cos x) = x, \qquad \text{where } 0 \le x \le \pi \qquad \textbf{(4a)}$$
$$f(f^{-1}(x)) = \cos(\cos^{-1} x) = x, \qquad \text{where } -1 \le x \le 1 \qquad \textbf{(4b)}$$

EXAMPLE 6 Finding the Exact Value of a Composite Function

Find the exact value of: (a) $\cos^{-1}\left[\cos\left(\frac{\pi}{12}\right)\right]$ (b) $\cos[\cos^{-1}(-0.4)]$

Solution (a) $\cos^{-1}\left[\cos\left(\frac{\pi}{12}\right)\right] = \frac{\pi}{12}$ By Property (4a)

(b) $\cos[\cos^{-1}(-0.4)] = -0.4$ By Property (4b) ◀

NOW WORK PROBLEM 39.

The Inverse Tangent Function

In Figure 14 we reproduce the graph of $y = \tan x$. Because every horizontal line intersects the graph infinitely many times, it follows that the tangent function is not one-to-one.

Figure 14

$y = \tan x, -\infty < x < \infty$, x not equal to odd multiples of $\frac{\pi}{2}$, $-\infty < y < \infty$

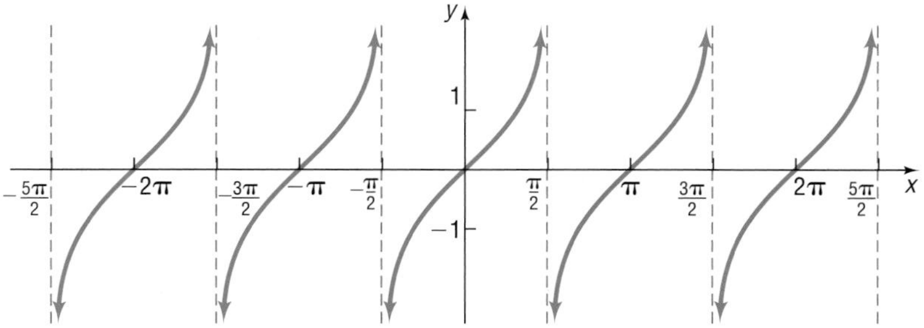

However, if we restrict the domain of $y = \tan x$ to the interval $\left(-\frac{\pi}{2}, \frac{\pi}{2}\right)$,* the restricted function

$$y = \tan x, \qquad -\frac{\pi}{2} < x < \frac{\pi}{2}$$

is one-to-one and hence has an inverse function. See Figure 15.

Figure 15

$y = \tan x, -\frac{\pi}{2} < x < \frac{\pi}{2}$, $-\infty < y < \infty$

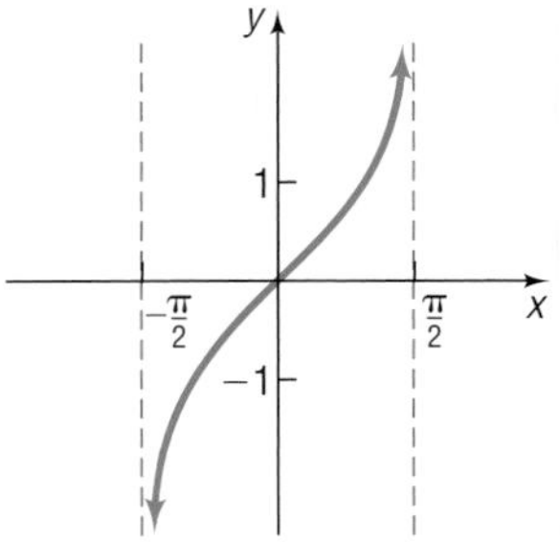

An equation for the inverse of $y = f(x) = \tan x$ is obtained by interchanging x and y. The implicit form of the inverse function is $x = \tan y$, $-\frac{\pi}{2} < y < \frac{\pi}{2}$. The explicit form is called the **inverse tangent** of x and is symbolized by $y = f^{-1}(x) = \tan^{-1} x$ (or by $y = \arctan x$).

$$y = \tan^{-1} x \quad \text{means} \quad x = \tan y$$
$$\text{where} \quad -\infty < x < \infty \quad \text{and} \quad -\frac{\pi}{2} < y < \frac{\pi}{2} \qquad \textbf{(5)}$$

Here y is the angle whose tangent is x. The domain of the function $y = \tan^{-1} x$ is $-\infty < x < \infty$, and its range is $-\frac{\pi}{2} < y < \frac{\pi}{2}$. The graph of $y = \tan^{-1} x$ can be obtained by reflecting the restricted portion of the graph of $y = \tan x$ about the line $y = x$, as shown in Figure 16.

Figure 16

$y = \tan^{-1} x, -\infty < x < \infty$, $-\frac{\pi}{2} < y < \frac{\pi}{2}$

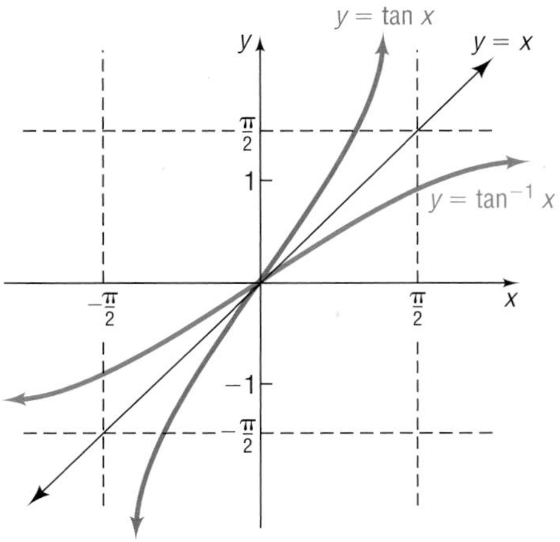

*This is the generally accepted restriction.

CHECK: Graph $Y = \tan^{-1} x$ and compare the result with Figure 16.

EXAMPLE 7 Finding the Exact Value of an Inverse Tangent Function

Find the exact value of: $\tan^{-1} 1$

Solution Let $\theta = \tan^{-1} 1$. We seek the angle θ, $-\frac{\pi}{2} < \theta < \frac{\pi}{2}$, whose tangent equals 1.

$$\theta = \tan^{-1} 1, \qquad -\frac{\pi}{2} < \theta < \frac{\pi}{2}$$

$$\tan \theta = 1, \qquad -\frac{\pi}{2} < \theta < \frac{\pi}{2}$$

Look at Table 3. The only angle θ within the interval $\left(-\frac{\pi}{2}, \frac{\pi}{2}\right)$ whose tangent is 1 is $\frac{\pi}{4}$. So, since $\tan \frac{\pi}{4} = 1$ and $\frac{\pi}{4}$ is in the interval $\left(-\frac{\pi}{2}, \frac{\pi}{2}\right)$, we conclude that

$$\tan^{-1} 1 = \frac{\pi}{4}$$

◀

Table 3

θ	$\tan \theta$
$-\frac{\pi}{2}$	Undefined
$-\frac{\pi}{3}$	$-\sqrt{3}$
$-\frac{\pi}{4}$	-1
$-\frac{\pi}{6}$	$-\frac{\sqrt{3}}{3}$
0	0
$\frac{\pi}{6}$	$\frac{\sqrt{3}}{3}$
$\frac{\pi}{4}$	1
$\frac{\pi}{3}$	$\sqrt{3}$
$\frac{\pi}{2}$	Undefined

NOW WORK PROBLEM 17.

EXAMPLE 8 Finding the Exact Value of an Inverse Tangent Function

Find the exact value of: $\tan^{-1}(-\sqrt{3})$

Solution Let $\theta = \tan^{-1}(-\sqrt{3})$. We seek the angle θ, $-\frac{\pi}{2} < \theta < \frac{\pi}{2}$, whose tangent equals $-\sqrt{3}$.

$$\theta = \tan^{-1}(-\sqrt{3}), \qquad -\frac{\pi}{2} < \theta < \frac{\pi}{2}$$

$$\tan \theta = -\sqrt{3}, \qquad -\frac{\pi}{2} < \theta < \frac{\pi}{2}$$

Look at Table 3 or Figure 15 if necessary. The only angle θ within the interval $\left(-\frac{\pi}{2}, \frac{\pi}{2}\right)$ whose tangent is $-\sqrt{3}$ is $-\frac{\pi}{3}$. So, since $\tan\left(-\frac{\pi}{3}\right) = -\sqrt{3}$ and $-\frac{\pi}{3}$ is in the interval $\left(-\frac{\pi}{2}, \frac{\pi}{2}\right)$, we conclude that

$$\tan^{-1}(-\sqrt{3}) = -\frac{\pi}{3}$$

◀

For the tangent function and its inverse, the following properties hold:

$$f^{-1}(f(x)) = \tan^{-1}(\tan x) = x, \qquad \text{where } -\frac{\pi}{2} < x < \frac{\pi}{2}$$

$$f(f^{-1}(x)) = \tan(\tan^{-1} x) = x, \qquad \text{where } -\infty < x < \infty$$

6.1 Assess Your Understanding

'Are You Prepared?' *Answers are given at the end of these exercises. If you get a wrong answer, read the pages listed in red.*

1. What is the domain and the range of $y = \sin x$? (pp. 356–357)

2. For a function f and its inverse f^{-1}, $f(f^{-1}(x))$ ________ = ________. (pp. 231–241)

3. *True or False:* If $y = f(x)$ is a one-to-one function, then $x = f(y)$. (pp. 231–241)

4. *True or False:* The graph of $y = \cos x$ is decreasing on the interval $[0, \pi]$. (pp. 370–374)

5. $\tan \frac{\pi}{4} =$ ________; $\sin \frac{\pi}{3} =$ ________ (pp. 342–351)

6. $\sin\left(-\frac{\pi}{6}\right) =$ ________; $\cos \pi =$ ________ (pp. 342–351)

Concepts and Vocabulary

7. $y = \sin^{-1} x$ means ________, where $-1 \le x \le 1$ and $-\frac{\pi}{2} \le y \le \frac{\pi}{2}$.

8. The value of $\sin^{-1}\left[\cos \frac{\pi}{2}\right]$ is ________.

9. $\cos^{-1}\left[\cos \frac{\pi}{5}\right] =$ ________.

10. *True or False:* The domain of $y = \sin^{-1} x$ is $-\frac{\pi}{2} \le x \le \frac{\pi}{2}$.

11. *True or False:* $\cos(\sin^{-1} 0) = 1$ and $\sin(\cos^{-1} 0) = 1$.

12. *True or False:* $y = \tan^{-1} x$ means $x = \tan y$, where $-\infty < x < \infty$ and $-\frac{\pi}{2} < y < \frac{\pi}{2}$.

Exercises

In Problems 13–24, find the exact value of each expression.

13. $\sin^{-1} 0$ **14.** $\cos^{-1} 1$ **15.** $\sin^{-1}(-1)$ **16.** $\cos^{-1}(-1)$

17. $\tan^{-1} 0$ **18.** $\tan^{-1}(-1)$ **19.** $\sin^{-1} \frac{\sqrt{2}}{2}$ **20.** $\tan^{-1} \frac{\sqrt{3}}{3}$

21. $\tan^{-1} \sqrt{3}$ **22.** $\sin^{-1}\left(-\frac{\sqrt{3}}{2}\right)$ **23.** $\cos^{-1}\left(-\frac{\sqrt{3}}{2}\right)$ **24.** $\sin^{-1}\left(-\frac{\sqrt{2}}{2}\right)$

In Problems 25–36, use a calculator to find the value of each expression rounded to two decimal places.

25. $\sin^{-1} 0.1$ **26.** $\cos^{-1} 0.6$ **27.** $\tan^{-1} 5$ **28.** $\tan^{-1} 0.2$

29. $\cos^{-1} \frac{7}{8}$ **30.** $\sin^{-1} \frac{1}{8}$ **31.** $\tan^{-1}(-0.4)$ **32.** $\tan^{-1}(-3)$

33. $\sin^{-1}(-0.12)$ **34.** $\cos^{-1}(-0.44)$ **35.** $\cos^{-1} \frac{\sqrt{2}}{3}$ **36.** $\sin^{-1} \frac{\sqrt{3}}{5}$

In Problems 37–44, find the exact value of each expression. Do not use a calculator.

37. $\sin[\sin^{-1}(0.54)]$ **38.** $\tan[\tan^{-1}(7.4)]$ **39.** $\cos^{-1}\left[\cos\left(\frac{4\pi}{5}\right)\right]$ **40.** $\sin^{-1}\left[\sin\left(-\frac{\pi}{10}\right)\right]$

41. $\tan[\tan^{-1}(-3.5)]$ **42.** $\cos[\cos^{-1}(-0.05)]$ **43.** $\sin^{-1}\left[\sin\left(-\frac{3\pi}{7}\right)\right]$ **44.** $\tan^{-1}\left[\tan\left(\frac{2\pi}{5}\right)\right]$

In Problems 45–56, do not use a calculator. For your answer, also say why or why not.

45. Does $\sin^{-1}\left[\sin\left(-\frac{\pi}{6}\right)\right] = -\frac{\pi}{6}$? **46.** Does $\sin^{-1}\left[\sin\left(\frac{2\pi}{3}\right)\right] = \frac{2\pi}{3}$? **47.** Does $\sin[\sin^{-1}(2)] = 2$?

48. Does $\sin\left[\sin^{-1}\left(-\frac{1}{2}\right)\right] = -\frac{1}{2}$? **49.** Does $\cos^{-1}\left[\cos\left(-\frac{\pi}{6}\right)\right] = -\frac{\pi}{6}$? **50.** Does $\cos^{-1}\left[\cos\left(\frac{2\pi}{3}\right)\right] = \frac{2\pi}{3}$?

51. Does $\cos\left[\cos^{-1}\left(-\frac{1}{2}\right)\right] = -\frac{1}{2}$? **52.** Does $\cos[\cos^{-1}(2)] = 2$? **53.** Does $\tan^{-1}\left[\tan\left(-\frac{\pi}{3}\right)\right] = -\frac{\pi}{3}$?

54. Does $\tan^{-1}\left[\tan\left(\frac{2\pi}{3}\right)\right] = \frac{2\pi}{3}$? **55.** Does $\tan[\tan^{-1}(2)] = 2$? **56.** Does $\tan\left[\tan^{-1}\left(-\frac{1}{2}\right)\right] = -\frac{1}{2}$?

In Problems 57–62, use the following: The formula

$$D = 24\left[1 - \frac{\cos^{-1}(\tan i \tan \theta)}{\pi}\right]$$

can be used to approximate the number of hours of daylight when the declination of the Sun is $i°$ *at a location* $\theta°$ *north latitude for any date between the vernal equinox and autumnal equinox. The declination of the Sun is defined as the angle i between the equatorial plane and any ray of light from the Sun. The latitude of a location is the angle* θ *between the Equator and the location on the surface of Earth, with the vertex of the angle located at the center of Earth. See the figure. To use the formula,* $\cos^{-1}(\tan i \tan \theta)$ *must be expressed in radians.*

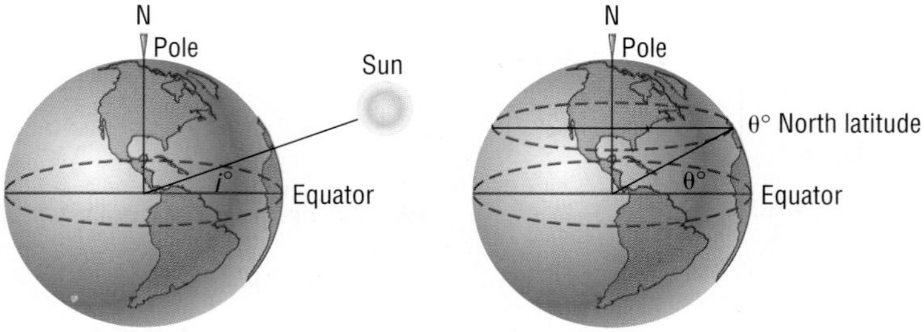

57. Approximate the number of hours of daylight in Houston, Texas (29°45′ north latitude), for the following dates:
(a) Summer solstice ($i = 23.5°$)
(b) Vernal equinox ($i = 0°$)
(c) July 4 ($i = 22°48'$)

58. Approximate the number of hours of daylight in New York, New York (40°45′ north latitude), for the following dates:
(a) Summer solstice ($i = 23.5°$)
(b) Vernal equinox ($i = 0°$)
(c) July 4 ($i = 22°48'$)

59. Approximate the number of hours of daylight in Honolulu, Hawaii (21°18′ north latitude), for the following dates:
(a) Summer solstice ($i = 23.5°$)
(b) Vernal equinox ($i = 0°$)
(c) July 4 ($i = 22°48'$)

60. Approximate the number of hours of daylight in Anchorage, Alaska (61°10′ north latitude), for the following dates:
(a) Summer solstice ($i = 23.5°$)
(b) Vernal equinox ($i = 0°$)
(c) July 4 ($i = 22°48'$)

61. Approximate the number of hours of daylight at the Equator (0° north latitude) for the following dates:
(a) Summer solstice ($i = 23.5°$)
(b) Vernal equinox ($i = 0°$)
(c) July 4 ($i = 22°48'$)
(d) What do you conclude about the number of hours of daylight throughout the year for a location at the Equator?

62. Approximate the number of hours of daylight for any location that is 66°30′ north latitude for the following dates:
(a) Summer solstice ($i = 23.5°$)
(b) Vernal equinox ($i = 0°$)
(c) July 4 ($i = 22°48'$)
(d) The number of hours of daylight on the winter solstice may be found by computing the number of hours of daylight on the summer solstice and subtracting this result from 24 hours, due to the symmetry of the orbital path of Earth around the Sun. Compute the number of hours of daylight for this location on the winter solstice. What do you conclude about daylight for a location at 66°30′ north latitude?

63. Being the First to See the Rising Sun Cadillac Mountain, elevation 1530 feet, is located in Acadia National Park, Maine, and is the highest peak on the east coast of the United States. It is said that a person standing on the summit will be the first person in the United States to see the rays of the rising Sun. How much sooner would a person atop Cadillac Mountain see the first rays than a person standing below, at sea level?

[**Hint:** Consult the figure. When the person at D sees the first rays of the Sun, the person at P does not. The person at P sees the first rays of the Sun only after Earth has rotated so that P is at location Q. Compute the length of the arc subtended by the central angle θ. Then use the fact that, at the latitude of Cadillac Mountain, in 24 hours a length of 2π (2710) miles is subtended, and find the time that it takes to subtend this length.]

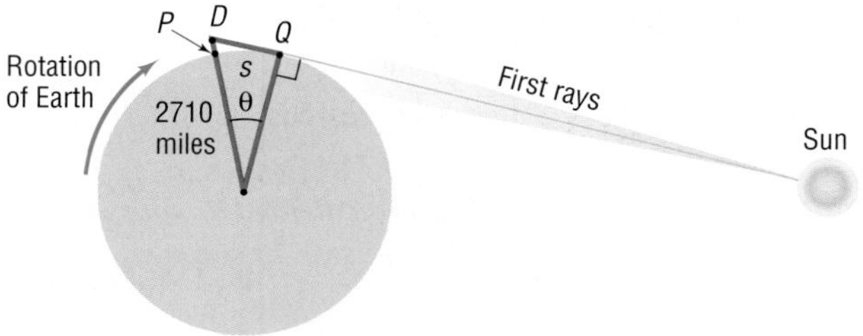

'Are You Prepared?' Answers

1. domain: $-\infty < x < \infty$; range: $-1 \le y \le 1$
2. $f^{-1}(f(x)) = x$ **3.** False **4.** True
5. $1; \frac{\sqrt{3}}{2}$ **6.** $-\frac{1}{2}; -1$

6.2 The Inverse Trigonometric Functions (Continued)

PREPARING FOR THIS SECTION *Before getting started, review the following concepts:*

- Finding Exact Values Given the Value of a Trigonometric Function and the Quadrant of the Angle (Section 5.2, pp. 363–365)
- Graphs of the Secant, Cosecant, and Cotangent Functions (Section 5.5, pp. 386–391)
- Domain and Range of the Secant, Cosecant, and Cotangent Functions (Section 5.3, pp. 356–357)

Now work the 'Are You Prepared?' problems on page 429.

OBJECTIVES
1. Find the Exact Value of Expressions Involving the Inverse Sine, Cosine, and Tangent Functions
2. Know the Definition of the Inverse Secant, Cosecant, and Cotangent Functions
3. Use a Calculator to Evaluate $\sec^{-1} x$, $\csc^{-1} x$, and $\cot^{-1} x$

1 In this section we continue our discussion of the inverse trigonometric functions.

EXAMPLE 1 **Finding the Exact Value of Expressions Involving Inverse Trigonometric Functions**

Find the exact value of: $\sin^{-1}\left(\sin \frac{5\pi}{4}\right)$

Solution

$$\sin^{-1}\left(\sin \frac{5\pi}{4}\right) = \sin^{-1}\left(-\frac{\sqrt{2}}{2}\right) = -\frac{\pi}{4}$$

Notice in the solution to Example 1 that we did not use Property (2a), page 417. This is because the argument of the sine function is not in the interval $\left[-\frac{\pi}{2}, \frac{\pi}{2}\right]$, as required. If we use the fact that

$$\sin \frac{5\pi}{4} = -\sin \frac{\pi}{4} \underset{y = \sin x \text{ is odd}}{=} \sin\left(-\frac{\pi}{4}\right)$$

then we can use Property (2a):

$$\sin^{-1}\left(\sin \frac{5\pi}{4}\right) = \sin^{-1}\left[\sin\left(-\frac{\pi}{4}\right)\right] \underset{\text{Property (2a)}}{=} -\frac{\pi}{4}$$

NOW WORK PROBLEM 21.

EXAMPLE 2 Finding the Exact Value of Expressions Involving Inverse Trigonometric Functions

Find the exact value of: $\sin\left(\tan^{-1}\frac{1}{2}\right)$

Figure 17

$\tan\theta = \frac{1}{2}$

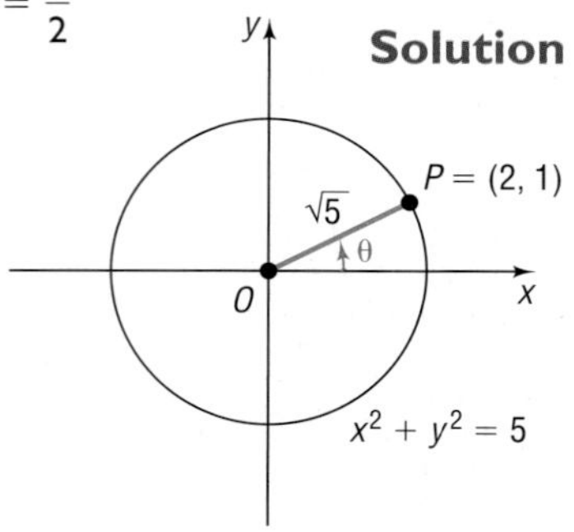

Solution Let $\theta = \tan^{-1}\frac{1}{2}$. Then $\tan\theta = \frac{1}{2}$, where $-\frac{\pi}{2} < \theta < \frac{\pi}{2}$. Because $\tan\theta > 0$, it follows that $0 < \theta < \frac{\pi}{2}$, so θ lies in quadrant I. Since $\tan\theta = \frac{1}{2} = \frac{y}{x}$, we let $x = 2$ and $y = 1$. Since $r = d(O, P) = \sqrt{2^2 + 1^2} = \sqrt{5}$, the point $P = (x, y) = (2, 1)$ is on the circle $x^2 + y^2 = 5$. See Figure 17. Then, with $x = 2$, $y = 1$, and $r = \sqrt{5}$, we have

$$\sin\left(\tan^{-1}\frac{1}{2}\right) = \sin\theta \underset{\sin\theta = \frac{y}{r}}{=} \frac{1}{\sqrt{5}} = \frac{\sqrt{5}}{5}$$

◀

EXAMPLE 3 Finding the Exact Value of Expressions Involving Inverse Trigonometric Functions

Find the exact value of: $\cos\left[\sin^{-1}\left(-\frac{1}{3}\right)\right]$

Figure 18

$\sin\theta = -\frac{1}{3}$

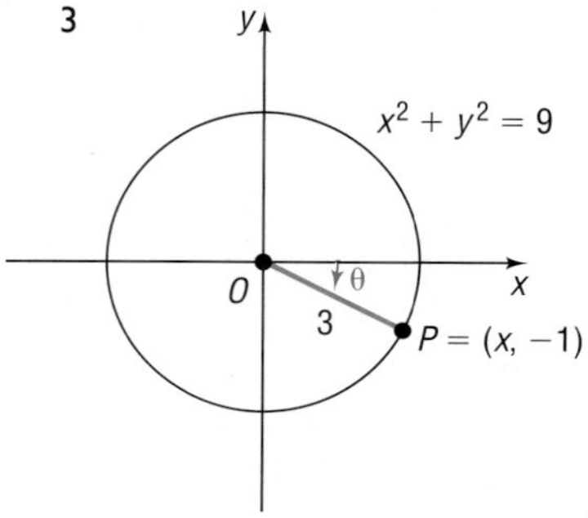

Solution Let $\theta = \sin^{-1}\left(-\frac{1}{3}\right)$. Then $\sin\theta = -\frac{1}{3}$ and $-\frac{\pi}{2} \le \theta \le \frac{\pi}{2}$. Because $\sin\theta < 0$, it follows that $-\frac{\pi}{2} \le \theta < 0$, so θ lies in quadrant IV. Since $\sin\theta = \frac{-1}{3} = \frac{y}{r}$, we let $y = -1$ and $r = 3$. The point $P = (x, y) = (x, -1)$, $x > 0$, is on a circle of radius 3, namely, $x^2 + y^2 = 9$. See Figure 18. Then,

$$x^2 + y^2 = 9$$
$$x^2 + (-1)^2 = 9 \qquad y = -1$$
$$x^2 = 8$$
$$x = 2\sqrt{2} \qquad x > 0$$

Then we have $x = 2\sqrt{2}$, $y = -1$, $r = 3$, so that

$$\cos\left[\sin^{-1}\left(-\frac{1}{3}\right)\right] = \cos\theta \underset{\cos\theta = \frac{x}{r}}{=} \frac{2\sqrt{2}}{3}$$

◀

EXAMPLE 4 Finding the Exact Value of Expressions Involving Inverse Trigonometric Functions

Find the exact value of: $\tan\left[\cos^{-1}\left(-\frac{1}{3}\right)\right]$

Solution Let $\theta = \cos^{-1}\left(-\frac{1}{3}\right)$. Then $\cos\theta = -\frac{1}{3}$ and $0 \le \theta \le \pi$. Because $\cos\theta < 0$, it follows that $\frac{\pi}{2} < \theta \le \pi$, so θ lies in quadrant II. Since $\cos\theta = \frac{-1}{3} = \frac{x}{r}$,

Figure 19
$\cos\theta = -\frac{1}{3}$

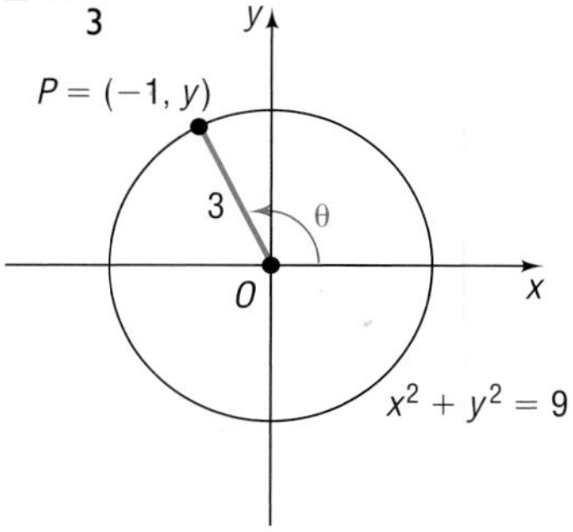

we let $x = -1$ and $r = 3$. The point $P = (x, y) = (-1, y)$, $y > 0$, is on a circle of radius $r = 3$, namely, $x^2 + y^2 = 9$. See Figure 19. Then,

$$\begin{aligned} x^2 + y^2 &= 9 \\ (-1)^2 + y^2 &= 9 \qquad x = -1 \\ y^2 &= 8 \\ y &= 2\sqrt{2} \qquad y > 0 \end{aligned}$$

Then, we have $x = -1$, $y = 2\sqrt{2}$, and $r = 3$, so that

$$\tan\left[\cos^{-1}\left(-\frac{1}{3}\right)\right] = \tan\theta = \frac{2\sqrt{2}}{-1} = -2\sqrt{2}$$

$\uparrow$ $\tan\theta = \frac{y}{x}$

◀

NOW WORK PROBLEMS 9 AND 27.

The Remaining Inverse Trigonometric Functions

2 The inverse secant, inverse cosecant, and inverse cotangent functions are defined as follows:

$$y = \sec^{-1} x \quad \text{means} \quad x = \sec y$$
$$\text{where} \quad |x| \geq 1 \quad \text{and} \quad 0 \leq y \leq \pi, \quad y \neq \frac{\pi}{2}^{*} \qquad \textbf{(1)}$$

$$y = \csc^{-1} x \quad \text{means} \quad x = \csc y$$
$$\text{where} \quad |x| \geq 1 \quad \text{and} \quad -\frac{\pi}{2} \leq y \leq \frac{\pi}{2}, \quad y \neq 0^{\dagger} \qquad \textbf{(2)}$$

$$y = \cot^{-1} x \quad \text{means} \quad x = \cot y$$
$$\text{where} \quad -\infty < x < \infty \quad \text{and} \quad 0 < y < \pi \qquad \textbf{(3)}$$

You are encouraged to review the graphs of the cotangent, cosecant, and secant functions in Figures 68, 69, and 71 in Section 5.5 to help you to see the basis for these definitions.

EXAMPLE 5 **Finding the Exact Value of an Inverse Cosecant Function**

Find the exact value of: $\csc^{-1} 2$

Solution Let $\theta = \csc^{-1} 2$. We seek the angle θ, $-\frac{\pi}{2} \leq \theta \leq \frac{\pi}{2}$, $\theta \neq 0$, whose cosecant equals 2 $\left(\text{or, equivalently, whose sine equals } \frac{1}{2}\right)$.

$$\theta = \csc^{-1} 2, \qquad -\frac{\pi}{2} \leq \theta \leq \frac{\pi}{2}, \quad \theta \neq 0$$
$$\csc\theta = 2, \qquad -\frac{\pi}{2} \leq \theta \leq \frac{\pi}{2}, \quad \theta \neq 0$$

The only angle θ in the interval $-\frac{\pi}{2} \leq \theta \leq \frac{\pi}{2}$, $\theta \neq 0$, whose cosecant is 2 is $\frac{\pi}{6}$, so $\csc^{-1} 2 = \frac{\pi}{6}$.

◀

NOW WORK PROBLEM 39.

*Most books use this definition. A few use the restriction $0 \leq y < \frac{\pi}{2}$, $\pi \leq y < \frac{3\pi}{2}$.

†Most books use this definition. A few use the restriction $-\pi < y \leq -\frac{\pi}{2}$, $0 < y \leq \frac{\pi}{2}$.

3 Most calculators do not have keys for evaluating the inverse cotangent, cosecant, and secant functions. The easiest way to evaluate them is to convert to an inverse trigonometric function whose range is the same as the one to be evaluated. In this regard, notice that $y = \cot^{-1} x$ and $y = \sec^{-1} x$ (except where undefined) each have the same range as $y = \cos^{-1} x$; $y = \csc^{-1} x$, except where undefined, has the same range as $y = \sin^{-1} x$.

EXAMPLE 6 **Approximating the Value of Inverse Trigonometric Functions**

Use a calculator to approximate each expression in radians rounded to two decimal places.

(a) $\sec^{-1} 3$ (b) $\csc^{-1}(-4)$ (c) $\cot^{-1} \dfrac{1}{2}$ (d) $\cot^{-1}(-2)$

Solution First, set your calculator to radian mode.

(a) Let $\theta = \sec^{-1} 3$. Then $\sec \theta = 3$ and $0 \le \theta \le \pi$, $\theta \ne \dfrac{\pi}{2}$. Since $\cos \theta = \dfrac{1}{3}$ and $\theta = \cos^{-1} \dfrac{1}{3}$, we have

$$\sec^{-1} 3 = \theta = \cos^{-1} \frac{1}{3} \approx 1.23$$

↑ Use a calculator.

(b) Let $\theta = \csc^{-1}(-4)$. Then $\csc \theta = -4$, $-\dfrac{\pi}{2} \le \theta \le \dfrac{\pi}{2}$, $\theta \ne 0$. Since $\sin \theta = -\dfrac{1}{4}$ and $\theta = \sin^{-1}\left(-\dfrac{1}{4}\right)$, we have

$$\csc^{-1}(-4) = \theta = \sin^{-1}\left(-\frac{1}{4}\right) \approx -0.25$$

Figure 20 $\cot \theta = \dfrac{1}{2}, 0 < \theta < \pi$

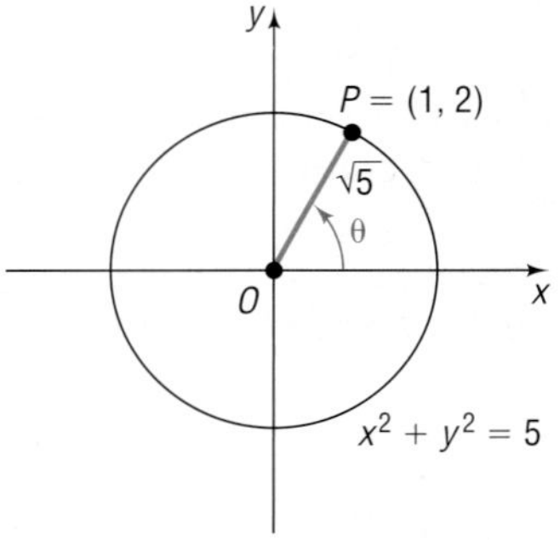

(c) Let $\theta = \cot^{-1} \dfrac{1}{2}$. Then $\cot \theta = \dfrac{1}{2}, 0 < \theta < \pi$. From these facts we know that θ lies in quadrant I. Since $\cot \theta = \dfrac{1}{2} = \dfrac{x}{y}$, we let $x = 1$ and $y = 2$. Since $r = d(O, P) = \sqrt{1^2 + 2^2} = \sqrt{5}$, the point $P = (x, y) = (1, 2)$ is on the circle $x^2 + y^2 = 5$. See Figure 20. Then $\cos \theta = \dfrac{x}{r} = \dfrac{1}{\sqrt{5}}$, so, $\theta = \cos^{-1}\left(\dfrac{1}{\sqrt{5}}\right)$. As a result,

$$\cot^{-1} \frac{1}{2} = \theta = \cos^{-1}\left(\frac{1}{\sqrt{5}}\right) \approx 1.11$$

Figure 21 $\cot \theta = -2, 0 < \theta < \pi$

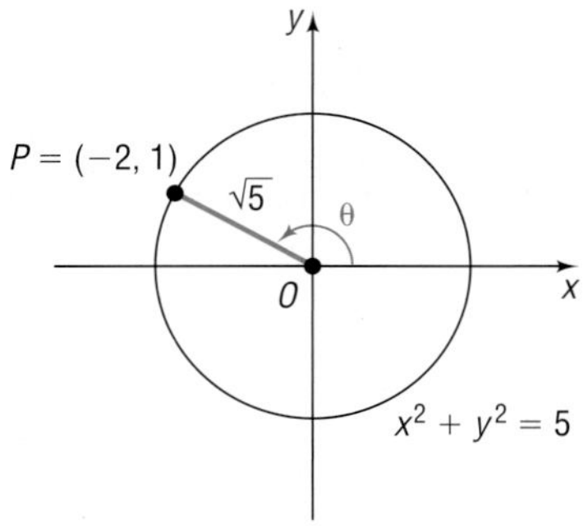

(d) Let $\theta = \cot^{-1}(-2)$. Then $\cot \theta = -2, 0 < \theta < \pi$. From these facts we know that θ lies in quadrant II. Since $\cot \theta = -2 = \dfrac{x}{y}$, $x < 0$, $y > 0$, we let $x = -2$ and $y = 1$. Since $r = d(O, P) = \sqrt{(-2)^2 + 1^2} = \sqrt{5}$, the point $P = (x, y) = (-2, 1)$ is on the circle $x^2 + y^2 = 5$. See Figure 21. Then $\cos \theta = \dfrac{x}{r} = \dfrac{-2}{\sqrt{5}}$, so $\theta = \cos^{-1}\left(\dfrac{-2}{\sqrt{5}}\right)$, and

$$\cot^{-1}(-2) = \theta = \cos^{-1}\left(-\frac{2}{\sqrt{5}}\right) \approx 2.68$$

◀

NOW WORK PROBLEM 45.

6.2 Assess Your Understanding

'Are You Prepared?' *Answers are given at the end of these exercises. If you get a wrong answer, read the pages listed in red*

1. What is the domain and the range of $y = \sec x$? (pp. 356–357)

2. *True or False:* The graph of $y = \sec x$ is increasing on the interval $\left[0, \frac{\pi}{2}\right)$ and on the interval $\left(\frac{\pi}{2}, \pi\right]$. (pp. 386–391)

3. If $\cot \theta = -2$ and $0 < \theta < \pi$, then $\cos \theta =$ _________. (pp. 363–365)

Concepts and Vocabulary

4. $y = \sec^{-1} x$ means _________, where $|x|$ _________ and _________ $\le y \le$ _________, $y \ne \frac{\pi}{2}$.

5. $\cos(\tan^{-1} 1) =$ _________.

6. *True or False:* It is impossible to obtain exact values for the inverse secant function.

7. *True or False:* $\csc^{-1} 0.5$ is not defined.

8. *True or False:* The domain of the inverse cotangent function is the set of real numbers.

Exercises

In Problems 9–36, find the exact value of each expression.

9. $\cos\left(\sin^{-1} \frac{\sqrt{2}}{2}\right)$ **10.** $\sin\left(\cos^{-1} \frac{1}{2}\right)$ **11.** $\tan\left[\cos^{-1}\left(-\frac{\sqrt{3}}{2}\right)\right]$ **12.** $\tan\left[\sin^{-1}\left(-\frac{1}{2}\right)\right]$

13. $\sec\left(\cos^{-1} \frac{1}{2}\right)$ **14.** $\cot\left[\sin^{-1}\left(-\frac{1}{2}\right)\right]$ **15.** $\csc(\tan^{-1} 1)$ **16.** $\sec(\tan^{-1} \sqrt{3})$

17. $\sin[\tan^{-1}(-1)]$ **18.** $\cos\left[\sin^{-1}\left(-\frac{\sqrt{3}}{2}\right)\right]$ **19.** $\sec\left[\sin^{-1}\left(-\frac{1}{2}\right)\right]$ **20.** $\csc\left[\cos^{-1}\left(-\frac{\sqrt{3}}{2}\right)\right]$

21. $\cos^{-1}\left(\cos \frac{5\pi}{4}\right)$ **22.** $\tan^{-1}\left(\tan \frac{2\pi}{3}\right)$ **23.** $\sin^{-1}\left[\sin\left(-\frac{7\pi}{6}\right)\right]$ **24.** $\cos^{-1}\left[\cos\left(-\frac{\pi}{3}\right)\right]$

25. $\tan\left(\sin^{-1} \frac{1}{3}\right)$ **26.** $\tan\left(\cos^{-1} \frac{1}{3}\right)$ 27. $\sec\left(\tan^{-1} \frac{1}{2}\right)$ **28.** $\cos\left(\sin^{-1} \frac{\sqrt{2}}{3}\right)$

29. $\cot\left[\sin^{-1}\left(-\frac{\sqrt{2}}{3}\right)\right]$ **30.** $\csc[\tan^{-1}(-2)]$ **31.** $\sin[\tan^{-1}(-3)]$ **32.** $\cot\left[\cos^{-1}\left(-\frac{\sqrt{3}}{3}\right)\right]$

33. $\sec\left(\sin^{-1} \frac{2\sqrt{5}}{5}\right)$ **34.** $\csc\left(\tan^{-1} \frac{1}{2}\right)$ **35.** $\sin^{-1}\left(\cos \frac{3\pi}{4}\right)$ **36.** $\cos^{-1}\left(\sin \frac{7\pi}{6}\right)$

In Problems 37–44, find the exact value of each expression.

37. $\cot^{-1} \sqrt{3}$ **38.** $\cot^{-1} 1$ 39. $\csc^{-1}(-1)$ **40.** $\csc^{-1} \sqrt{2}$

41. $\sec^{-1} \frac{2\sqrt{3}}{3}$ **42.** $\sec^{-1}(-2)$ **43.** $\cot^{-1}\left(-\frac{\sqrt{3}}{3}\right)$ **44.** $\csc^{-1}\left(-\frac{2\sqrt{3}}{3}\right)$

In Problems 45–56, use a calculator to find the value of each expression rounded to two decimal places.

45. $\sec^{-1} 4$ **46.** $\csc^{-1} 5$ **47.** $\cot^{-1} 2$ **48.** $\sec^{-1}(-3)$

49. $\csc^{-1}(-3)$ **50.** $\cot^{-1}\left(-\frac{1}{2}\right)$ **51.** $\cot^{-1}(-\sqrt{5})$ **52.** $\cot^{-1}(-8.1)$

53. $\csc^{-1}\left(-\frac{3}{2}\right)$ **54.** $\sec^{-1}\left(-\frac{4}{3}\right)$ **55.** $\cot^{-1}\left(-\frac{3}{2}\right)$ **56.** $\cot^{-1}(-\sqrt{10})$

57. Using a graphing utility, graph $y = \cot^{-1} x$.

58. Using a graphing utility, graph $y = \sec^{-1} x$.

59. Using a graphing utility, graph $y = \csc^{-1} x$.

60. Explain in your own words how you would use your calculator to find the value of $\cot^{-1} 10$.

61. Consult three books on calculus and write down the definition in each of $y = \sec^{-1} x$ and $y = \csc^{-1} x$. Compare these with the definition given in this book.

'Are You Prepared?' Answers

1. Domain: $\{x \mid x \ne (2k + 1)\frac{\pi}{2}\}$; Range: $\{y \mid |y| \ge 1\}$

2. True **3.** $-\frac{2}{\sqrt{5}}$

6.3 Trigonometric Identities

PREPARING FOR THIS SECTION *Before getting started, review the following:*

- Fundamental Identities (Section 5.2, p. 362)

Now work the 'Are You Prepared?' problems on page 435.

OBJECTIVES 1 Use Algebra to Simplify Trigonometric Expressions
2 Establish Identities

We saw in the previous chapter that the trigonometric functions lend themselves to a wide variety of identities. Before establishing some additional identities, let's review the definition of an *identity*.

Two functions f and g are said to be **identically equal** if

$$f(x) = g(x)$$

for every value of x for which both functions are defined. Such an equation is referred to as an **identity.** An equation that is not an identity is called a **conditional equation.**

For example, the following are identities:

$$(x + 1)^2 = x^2 + 2x + 1 \qquad \sin^2 x + \cos^2 x = 1 \qquad \csc x = \frac{1}{\sin x}$$

The following are conditional equations:

$2x + 5 = 0$ — True only if $x = -\frac{5}{2}$.

$\sin x = 0$ — True only if $x = k\pi$, k an integer.

$\sin x = \cos x$ — True only if $x = \frac{\pi}{4} + 2k\pi$ or $x = \frac{5\pi}{4} + 2k\pi$, k an integer.

The following boxes summarize the trigonometric identities that we have established thus far.

Quotient Identities

$$\tan \theta = \frac{\sin \theta}{\cos \theta} \qquad \cot \theta = \frac{\cos \theta}{\sin \theta}$$

Reciprocal Identities

$$\csc \theta = \frac{1}{\sin \theta} \qquad \sec \theta = \frac{1}{\cos \theta} \qquad \cot \theta = \frac{1}{\tan \theta}$$

Pythagorean Identities

$$\sin^2\theta + \cos^2\theta = 1 \qquad \tan^2\theta + 1 = \sec^2\theta$$
$$\cot^2\theta + 1 = \csc^2\theta$$

Even–Odd Identities

$$\sin(-\theta) = -\sin\theta \qquad \cos(-\theta) = \cos\theta \qquad \tan(-\theta) = -\tan\theta$$
$$\csc(-\theta) = -\csc\theta \qquad \sec(-\theta) = \sec\theta \qquad \cot(-\theta) = -\cot\theta$$

This list of identities comprises what we shall refer to as the **basic trigonometric identities.** These identities should not merely be memorized, but should be *known* (just as you know your name rather than have it memorized). In fact, minor variations of a basic identity are often used. For example, we might want to use

$$\sin^2\theta = 1 - \cos^2\theta \quad \text{or} \quad \cos^2\theta = 1 - \sin^2\theta$$

instead of $\sin^2\theta + \cos^2\theta = 1$. For this reason, among others, you need to know these relationships and be quite comfortable with variations of them.

1 The ability to use algebra to manipulate trigonometric expressions is a key skill that one must have in order to establish identities. Some of the techniques that are used in establishing identities are multiplying by a "well-chosen 1," writing a trigonometric expression over a common denominator, rewriting a trigonometric expression in terms of sine and cosine only, and factoring.

EXAMPLE 1 **Using Algebraic Techniques to Simplify Trigonometric Expressions**

(a) Simplify $\dfrac{\cot\theta}{\csc\theta}$ by rewriting each trigonometric function in terms of sine and cosine.

(b) Simplify $\dfrac{\cos\theta}{1 + \sin\theta}$ by multiplying the numerator and denominator by $1 - \sin\theta$.

(c) Simplify $\dfrac{1 + \sin\theta}{\sin\theta} + \dfrac{\cot\theta - \cos\theta}{\cos\theta}$ by rewriting the expression over a common denominator.

(d) Simplify $\dfrac{\sin^2\theta - 1}{\tan\theta\sin\theta - \tan\theta}$ by factoring.

Solution (a) $$\frac{\cot\theta}{\csc\theta} = \frac{\dfrac{\cos\theta}{\sin\theta}}{\dfrac{1}{\sin\theta}} = \frac{\cos\theta}{\sin\theta}\cdot\frac{\sin\theta}{1} = \cos\theta$$

(b) $$\frac{\cos\theta}{1+\sin\theta} = \frac{\cos\theta}{1+\sin\theta}\cdot\frac{1-\sin\theta}{1-\sin\theta} = \frac{\cos\theta(1-\sin\theta)}{1-\sin^2\theta} = \frac{\cos\theta(1-\sin\theta)}{\cos^2\theta} = \frac{1-\sin\theta}{\cos\theta}$$

↑ Well-chosen 1: $\frac{1-\sin\theta}{1-\sin\theta}$

(c) $$\frac{1+\sin\theta}{\sin\theta} + \frac{\cot\theta-\cos\theta}{\cos\theta} = \frac{1+\sin\theta}{\sin\theta}\cdot\frac{\cos\theta}{\cos\theta} + \frac{\cot\theta-\cos\theta}{\cos\theta}\cdot\frac{\sin\theta}{\sin\theta}$$

$$= \frac{\cos\theta+\sin\theta\cos\theta+\cot\theta\sin\theta-\cos\theta\sin\theta}{\sin\theta\cos\theta} = \frac{\cos\theta+\frac{\cos\theta}{\sin\theta}\cdot\sin\theta}{\sin\theta\cos\theta}$$

↑ $\cot\theta = \frac{\cos\theta}{\sin\theta}$

$$= \frac{\cos\theta+\cos\theta}{\sin\theta\cos\theta} = \frac{2\cos\theta}{\sin\theta\cos\theta} = \frac{2}{\sin\theta}$$

(d) $$\frac{\sin^2\theta-1}{\tan\theta\sin\theta-\tan\theta} = \frac{(\sin\theta+1)(\sin\theta-1)}{\tan\theta(\sin\theta-1)} = \frac{\sin\theta+1}{\tan\theta}$$ ◀

NOW WORK PROBLEMS 9, 11, AND 13.

2 In the examples that follow, the directions will read "Establish the identity...." As you will see, this is accomplished by starting with one side of the given equation (usually the one containing the more complicated expression) and, using appropriate basic identities and algebraic manipulations, arriving at the other side. The selection of appropriate basic identities to obtain the desired result is learned only through experience and lots of practice.

EXAMPLE 2 Establishing an Identity

Establish the identity: $\csc\theta\cdot\tan\theta = \sec\theta$

Solution We start with the left side, because it contains the more complicated expression, and apply a reciprocal identity and a quotient identity.

$$\csc\theta\cdot\tan\theta = \frac{1}{\cancel{\sin\theta}}\cdot\frac{\cancel{\sin\theta}}{\cos\theta} = \frac{1}{\cos\theta} = \sec\theta$$

Having arrived at the right side, the identity is established. ◀

COMMENT: A graphing utility can be used to provide evidence of an identity. For example, if we graph $Y_1 = \csc\theta\cdot\tan\theta$ and $Y_2 = \sec\theta$, the graphs appear to be the same. This provides evidence that $Y_1 = Y_2$. However, it does not prove their equality. A graphing utility *cannot be used to establish an identity*—identities must be established algebraically. ■

NOW WORK PROBLEM 19.

EXAMPLE 3 Establishing an Identity

Establish the identity: $\sin^2(-\theta) + \cos^2(-\theta) = 1$

Solution We begin with the left side and apply Even–Odd Identities.

$$\begin{aligned}\sin^2(-\theta)+\cos^2(-\theta) &= [\sin(-\theta)]^2 + [\cos(-\theta)]^2\\ &= (-\sin\theta)^2 + (\cos\theta)^2 && \text{Even–Odd Identities}\\ &= (\sin\theta)^2 + (\cos\theta)^2\\ &= 1 && \text{Pythagorean Identity}\end{aligned}$$ ◀

EXAMPLE 4

Establishing an Identity

Establish the identity: $\dfrac{\sin^2(-\theta) - \cos^2(-\theta)}{\sin(-\theta) - \cos(-\theta)} = \cos\theta - \sin\theta$

Solution We begin with two observations: The left side appears to contain the more complicated expression. Also, the left side contains expressions with the argument $-\theta$, whereas the right side contains expressions with the argument θ. We decide, therefore, to start with the left side and apply Even–Odd Identities.

$$\begin{aligned}\frac{\sin^2(-\theta) - \cos^2(-\theta)}{\sin(-\theta) - \cos(-\theta)} &= \frac{[\sin(-\theta)]^2 - [\cos(-\theta)]^2}{\sin(-\theta) - \cos(-\theta)} \\ &= \frac{(-\sin\theta)^2 - (\cos\theta)^2}{-\sin\theta - \cos\theta} && \text{Even–Odd Identities} \\ &= \frac{(\sin\theta)^2 - (\cos\theta)^2}{-\sin\theta - \cos\theta} && \text{Simplify.} \\ &= \frac{(\sin\theta - \cos\theta)\cancel{(\sin\theta + \cos\theta)}}{-\cancel{(\sin\theta + \cos\theta)}} && \text{Factor.} \\ &= \cos\theta - \sin\theta && \text{Cancel and simplify.}\end{aligned}$$

◀

EXAMPLE 5

Establishing an Identity

Establish the identity: $\dfrac{1 + \tan\theta}{1 + \cot\theta} = \tan\theta$

Solution

$$\frac{1 + \tan\theta}{1 + \cot\theta} = \frac{1 + \tan\theta}{1 + \dfrac{1}{\tan\theta}} = \frac{1 + \tan\theta}{\dfrac{\tan\theta + 1}{\tan\theta}} = \frac{\tan\theta\cancel{(1 + \tan\theta)}}{\cancel{\tan\theta + 1}} = \tan\theta$$

◀

NOW WORK PROBLEM 27.

When sums or differences of quotients appear, it is usually best to rewrite them as a single quotient, especially if the other side of the identity consists of only one term.

EXAMPLE 6

Establishing an Identity

Establish the identity: $\dfrac{\sin\theta}{1 + \cos\theta} + \dfrac{1 + \cos\theta}{\sin\theta} = 2\csc\theta$

Solution The left side is more complicated, so we start with it and proceed to add.

$$\begin{aligned}\frac{\sin\theta}{1 + \cos\theta} + \frac{1 + \cos\theta}{\sin\theta} &= \frac{\sin^2\theta + (1 + \cos\theta)^2}{(1 + \cos\theta)(\sin\theta)} && \text{Add the quotients.} \\ &= \frac{\sin^2\theta + 1 + 2\cos\theta + \cos^2\theta}{(1 + \cos\theta)(\sin\theta)} && \text{Remove parentheses in the numerator.} \\ &= \frac{(\sin^2\theta + \cos^2\theta) + 1 + 2\cos\theta}{(1 + \cos\theta)(\sin\theta)} && \text{Regroup.} \\ &= \frac{2 + 2\cos\theta}{(1 + \cos\theta)(\sin\theta)} && \text{Pythagorean Identity} \\ &= \frac{2\cancel{(1 + \cos\theta)}}{\cancel{(1 + \cos\theta)}(\sin\theta)} && \text{Factor and cancel.} \\ &= \frac{2}{\sin\theta} \\ &= 2\csc\theta && \text{Reciprocal Identity}\end{aligned}$$

◀

Sometimes it helps to write one side in terms of sines and cosines only.

EXAMPLE 7 Establishing an Identity

Establish the identity: $\dfrac{\tan\theta + \cot\theta}{\sec\theta\csc\theta} = 1$

Solution

$$\frac{\tan\theta + \cot\theta}{\sec\theta\csc\theta} = \frac{\dfrac{\sin\theta}{\cos\theta} + \dfrac{\cos\theta}{\sin\theta}}{\dfrac{1}{\cos\theta}\cdot\dfrac{1}{\sin\theta}} = \frac{\dfrac{\sin^2\theta + \cos^2\theta}{\cos\theta\sin\theta}}{\dfrac{1}{\cos\theta\sin\theta}}$$

Change to sines and cosines. Add the quotients in the numerator.

$$= \frac{1}{\cos\theta\sin\theta}\cdot\frac{\cos\theta\sin\theta}{1} = 1$$

Divide quotient; $\sin^2\theta + \cos^2\theta = 1$ ◀

NOW WORK PROBLEM 69.

Sometimes, multiplying the numerator and denominator by an appropriate factor will result in a simplification.

EXAMPLE 8 Establishing an Identity

Establish the identity: $\dfrac{1 - \sin\theta}{\cos\theta} = \dfrac{\cos\theta}{1 + \sin\theta}$

Solution We start with the left side and multiply the numerator and the denominator by $1 + \sin\theta$. (Alternatively, we could multiply the numerator and denominator of the right side by $1 - \sin\theta$.)

$$\frac{1 - \sin\theta}{\cos\theta} = \frac{1 - \sin\theta}{\cos\theta}\cdot\frac{1 + \sin\theta}{1 + \sin\theta} \quad \text{Multiply the numerator and denominator by } 1 + \sin\theta.$$

$$= \frac{1 - \sin^2\theta}{\cos\theta(1 + \sin\theta)}$$

$$= \frac{\cos^2\theta}{\cos\theta(1 + \sin\theta)} \quad 1 - \sin^2\theta = \cos^2\theta$$

$$= \frac{\cos\theta}{1 + \sin\theta} \quad \text{Cancel.}$$

◀

NOW WORK PROBLEM 53.

EXAMPLE 9 Establishing an Identity Involving Inverse Trigonometric Functions

Show that $\sin(\tan^{-1} v) = \dfrac{v}{\sqrt{1 + v^2}}$.

Solution Let $\theta = \tan^{-1} v$ so that $\tan \theta = v$, $-\frac{\pi}{2} < \theta < \frac{\pi}{2}$. As a result, we know that $\sec \theta > 0$.

$$\sin(\tan^{-1} v) = \sin\theta = \sin\theta \cdot \frac{\cos\theta}{\cos\theta} = \tan\theta\cos\theta = \frac{\tan\theta}{\sec\theta} = \frac{\tan\theta}{\sqrt{1+\tan^2\theta}} = \frac{v}{\sqrt{1+v^2}}$$

Multiply by 1: $\frac{\cos\theta}{\cos\theta}$; $\frac{\sin\theta}{\cos\theta} = \tan\theta$; $\sec^2\theta = 1 + \tan^2\theta$, $\sec\theta > 0$ ◀

NOW WORK PROBLEM 99.

Although a lot of practice is the only real way to learn how to establish identities, the following guidelines should prove helpful.

Guidelines for Establishing Identities

1. It is almost always preferable to start with the side containing the more complicated expression.
2. Rewrite sums or differences of quotients as a single quotient.
3. Sometimes rewriting one side in terms of sines and cosines only will help.
4. Always keep your goal in mind. As you manipulate one side of the expression, you must keep in mind the form of the expression on the other side.

WARNING: Be careful not to handle identities to be established as if they were conditional equations. You *cannot* establish an identity by such methods as adding the same expression to each side and obtaining a true statement. This practice is not allowed, because the original statement is precisely the one that you are trying to establish. You do not know until it has been established that it is, in fact, true. ■

6.3 Assess Your Understanding

'Are You Prepared?' *Answers are given at the end of these exercises. If you get a wrong answer, read the pages listed in red.*

1. *True or False:* $\sin^2\theta = 1 - \cos^2\theta$. (p. 362)
2. The quotient identity states that $\tan\theta =$ ________. (p. 362)

Concepts and Vocabulary

3. Suppose that f and g are two functions with the same domain. If $f(x) = g(x)$ for every x in the domain, the equation is called a(n) ________. Otherwise, it is called a(n) ________ equation.
4. $\tan^2\theta - \sec^2\theta =$ ________.
5. $\cos(-\theta) - \cos\theta =$ ________.
6. *True or False:* $\sin(-\theta) + \sin\theta = 0$ for any value of θ.
7. *True or False:* In establishing an identity, it is often easiest to just multiply both sides by a well-chosen nonzero expression involving the variable.
8. *True or False:* $\tan\theta \cdot \cos\theta = \sin\theta$ for any $\theta \neq (2k+1)\frac{\pi}{2}$.

Exercises

In Problems 9–18, simplify each trigonometric expression by following the indicated direction.

9. Rewrite in terms of sine and cosine: $\tan\theta \cdot \csc\theta$.
10. Rewrite in terms of sine and cosine: $\cot\theta \cdot \sec\theta$.
11. Multiply $\frac{\cos\theta}{1 - \sin\theta}$ by $\frac{1 + \sin\theta}{1 + \sin\theta}$.
12. Multiply $\frac{\sin\theta}{1 + \cos\theta}$ by $\frac{1 - \cos\theta}{1 - \cos\theta}$.

13. Rewrite over a common denominator:

$$\frac{\sin\theta + \cos\theta}{\cos\theta} + \frac{\cos\theta - \sin\theta}{\sin\theta}$$

14. Rewrite over a common denominator:

$$\frac{1}{1 - \cos\theta} + \frac{1}{1 + \cos\theta}$$

15. Multiply and simplify: $\dfrac{(\sin\theta + \cos\theta)(\sin\theta + \cos\theta) - 1}{\sin\theta\cos\theta}$

16. Multiply and simplify: $\dfrac{(\tan\theta + 1)(\tan\theta + 1) - \sec^2\theta}{\tan\theta}$

17. Factor and simplify: $\dfrac{3\sin^2\theta + 4\sin\theta + 1}{\sin^2\theta + 2\sin\theta + 1}$

18. Factor and simplify: $\dfrac{\cos^2\theta - 1}{\cos^2\theta - \cos\theta}$

In Problems 19–98, establish each identity.

19. $\csc\theta \cdot \cos\theta = \cot\theta$

20. $\sec\theta \cdot \sin\theta = \tan\theta$

21. $1 + \tan^2(-\theta) = \sec^2\theta$

22. $1 + \cot^2(-\theta) = \csc^2\theta$

23. $\cos\theta(\tan\theta + \cot\theta) = \csc\theta$

24. $\sin\theta(\cot\theta + \tan\theta) = \sec\theta$

25. $\tan\theta\cot\theta - \cos^2\theta = \sin^2\theta$

26. $\sin\theta\csc\theta - \cos^2\theta = \sin^2\theta$

27. $(\sec\theta - 1)(\sec\theta + 1) = \tan^2\theta$

28. $(\csc\theta - 1)(\csc\theta + 1) = \cot^2\theta$

29. $(\sec\theta + \tan\theta)(\sec\theta - \tan\theta) = 1$

30. $(\csc\theta + \cot\theta)(\csc\theta - \cot\theta) = 1$

31. $\cos^2\theta(1 + \tan^2\theta) = 1$

32. $(1 - \cos^2\theta)(1 + \cot^2\theta) = 1$

33. $(\sin\theta + \cos\theta)^2 + (\sin\theta - \cos\theta)^2 = 2$

34. $\tan^2\theta\cos^2\theta + \cot^2\theta\sin^2\theta = 1$

35. $\sec^4\theta - \sec^2\theta = \tan^4\theta + \tan^2\theta$

36. $\csc^4\theta - \csc^2\theta = \cot^4\theta + \cot^2\theta$

37. $\sec\theta - \tan\theta = \dfrac{\cos\theta}{1 + \sin\theta}$

38. $\csc\theta - \cot\theta = \dfrac{\sin\theta}{1 + \cos\theta}$

39. $3\sin^2\theta + 4\cos^2\theta = 3 + \cos^2\theta$

40. $9\sec^2\theta - 5\tan^2\theta = 5 + 4\sec^2\theta$

41. $1 - \dfrac{\cos^2\theta}{1 + \sin\theta} = \sin\theta$

42. $1 - \dfrac{\sin^2\theta}{1 - \cos\theta} = -\cos\theta$

43. $\dfrac{1 + \tan\theta}{1 - \tan\theta} = \dfrac{\cot\theta + 1}{\cot\theta - 1}$

44. $\dfrac{\csc\theta - 1}{\csc\theta + 1} = \dfrac{1 - \sin\theta}{1 + \sin\theta}$

45. $\dfrac{\sec\theta}{\csc\theta} + \dfrac{\sin\theta}{\cos\theta} = 2\tan\theta$

46. $\dfrac{\csc\theta - 1}{\cot\theta} = \dfrac{\cot\theta}{\csc\theta + 1}$

47. $\dfrac{1 + \sin\theta}{1 - \sin\theta} = \dfrac{\csc\theta + 1}{\csc\theta - 1}$

48. $\dfrac{\cos\theta + 1}{\cos\theta - 1} = \dfrac{1 + \sec\theta}{1 - \sec\theta}$

49. $\dfrac{1 - \sin\theta}{\cos\theta} + \dfrac{\cos\theta}{1 - \sin\theta} = 2\sec\theta$

50. $\dfrac{\cos\theta}{1 + \sin\theta} + \dfrac{1 + \sin\theta}{\cos\theta} = 2\sec\theta$

51. $\dfrac{\sin\theta}{\sin\theta - \cos\theta} = \dfrac{1}{1 - \cot\theta}$

52. $1 - \dfrac{\sin^2\theta}{1 + \cos\theta} = \cos\theta$

53. $\dfrac{1 - \sin\theta}{1 + \sin\theta} = (\sec\theta - \tan\theta)^2$

54. $\dfrac{1 - \cos\theta}{1 + \cos\theta} = (\csc\theta - \cot\theta)^2$

55. $\dfrac{\cos\theta}{1 - \tan\theta} + \dfrac{\sin\theta}{1 - \cot\theta} = \sin\theta + \cos\theta$

56. $\dfrac{\cot\theta}{1 - \tan\theta} + \dfrac{\tan\theta}{1 - \cot\theta} = 1 + \tan\theta + \cot\theta$

57. $\tan\theta + \dfrac{\cos\theta}{1 + \sin\theta} = \sec\theta$

58. $\dfrac{\sin\theta\cos\theta}{\cos^2\theta - \sin^2\theta} = \dfrac{\tan\theta}{1 - \tan^2\theta}$

59. $\dfrac{\tan\theta + \sec\theta - 1}{\tan\theta - \sec\theta + 1} = \tan\theta + \sec\theta$

60. $\dfrac{\sin\theta - \cos\theta + 1}{\sin\theta + \cos\theta - 1} = \dfrac{\sin\theta + 1}{\cos\theta}$

61. $\dfrac{\tan\theta - \cot\theta}{\tan\theta + \cot\theta} = \sin^2\theta - \cos^2\theta$

62. $\dfrac{\sec\theta - \cos\theta}{\sec\theta + \cos\theta} = \dfrac{\sin^2\theta}{1 + \cos^2\theta}$

63. $\dfrac{\tan\theta - \cot\theta}{\tan\theta + \cot\theta} + 1 = 2\sin^2\theta$

64. $\dfrac{\tan\theta - \cot\theta}{\tan\theta + \cot\theta} + 2\cos^2\theta = 1$

65. $\dfrac{\sec\theta + \tan\theta}{\cot\theta + \cos\theta} = \tan\theta\sec\theta$

66. $\dfrac{\sec\theta}{1 + \sec\theta} = \dfrac{1 - \cos\theta}{\sin^2\theta}$

67. $\dfrac{1 - \tan^2\theta}{1 + \tan^2\theta} + 1 = 2\cos^2\theta$

68. $\dfrac{1 - \cot^2\theta}{1 + \cot^2\theta} + 2\cos^2\theta = 1$

69. $\dfrac{\sec\theta - \csc\theta}{\sec\theta\csc\theta} = \sin\theta - \cos\theta$

70. $\dfrac{\sin^2\theta - \tan\theta}{\cos^2\theta - \cot\theta} = \tan^2\theta$

71. $\sec\theta - \cos\theta = \sin\theta\tan\theta$

72. $\tan\theta + \cot\theta = \sec\theta\csc\theta$

73. $\dfrac{1}{1 - \sin\theta} + \dfrac{1}{1 + \sin\theta} = 2\sec^2\theta$

74. $\dfrac{1 + \sin\theta}{1 - \sin\theta} - \dfrac{1 - \sin\theta}{1 + \sin\theta} = 4\tan\theta\sec\theta$

75. $\dfrac{\sec\theta}{1 - \sin\theta} = \dfrac{1 + \sin\theta}{\cos^3\theta}$

76. $\dfrac{1 - \sin\theta}{1 + \sin\theta} = (\sec\theta - \tan\theta)^2$

77. $\dfrac{(\sec\theta - \tan\theta)^2 + 1}{\csc\theta(\sec\theta - \tan\theta)} = 2\tan\theta$

78. $\dfrac{\sec^2\theta - \tan^2\theta + \tan\theta}{\sec\theta} = \sin\theta + \cos\theta$

79. $\dfrac{\sin\theta + \cos\theta}{\cos\theta} - \dfrac{\sin\theta - \cos\theta}{\sin\theta} = \sec\theta\csc\theta$

80. $\dfrac{\sin\theta + \cos\theta}{\sin\theta} - \dfrac{\cos\theta - \sin\theta}{\cos\theta} = \sec\theta\csc\theta$

81. $\dfrac{\sin^3\theta + \cos^3\theta}{\sin\theta + \cos\theta} = 1 - \sin\theta\cos\theta$

82. $\dfrac{\sin^3\theta + \cos^3\theta}{1 - 2\cos^2\theta} = \dfrac{\sec\theta - \sin\theta}{\tan\theta - 1}$

83. $\dfrac{\cos^2\theta - \sin^2\theta}{1 - \tan^2\theta} = \cos^2\theta$

84. $\dfrac{\cos\theta + \sin\theta - \sin^3\theta}{\sin\theta} = \cot\theta + \cos^2\theta$

85. $\dfrac{(2\cos^2\theta - 1)^2}{\cos^4\theta - \sin^4\theta} = 1 - 2\sin^2\theta$

86. $\dfrac{1 - 2\cos^2\theta}{\sin\theta\cos\theta} = \tan\theta - \cot\theta$

87. $\dfrac{1 + \sin\theta + \cos\theta}{1 + \sin\theta - \cos\theta} = \dfrac{1 + \cos\theta}{\sin\theta}$

88. $\dfrac{1 + \cos\theta + \sin\theta}{1 + \cos\theta - \sin\theta} = \sec\theta + \tan\theta$

89. $(a\sin\theta + b\cos\theta)^2 + (a\cos\theta - b\sin\theta)^2 = a^2 + b^2$

90. $(2a\sin\theta\cos\theta)^2 + a^2(\cos^2\theta - \sin^2\theta)^2 = a^2$

91. $\dfrac{\tan\alpha + \tan\beta}{\cot\alpha + \cot\beta} = \tan\alpha\tan\beta$

92. $(\tan\alpha + \tan\beta)(1 - \cot\alpha\cot\beta) + (\cot\alpha + \cot\beta)(1 - \tan\alpha\tan\beta) = 0$

93. $(\sin\alpha + \cos\beta)^2 + (\cos\beta + \sin\alpha)(\cos\beta - \sin\alpha) = 2\cos\beta(\sin\alpha + \cos\beta)$

94. $(\sin\alpha - \cos\beta)^2 + (\cos\beta + \sin\alpha)(\cos\beta - \sin\alpha) = -2\cos\beta(\sin\alpha - \cos\beta)$

95. $\ln|\sec\theta| = -\ln|\cos\theta|$

96. $\ln|\tan\theta| = \ln|\sin\theta| - \ln|\cos\theta|$

97. $\ln|1 + \cos\theta| + \ln|1 - \cos\theta| = 2\ln|\sin\theta|$

98. $\ln|\sec\theta + \tan\theta| + \ln|\sec\theta - \tan\theta| = 0$

99. Show that $\sec(\tan^{-1} v) = \sqrt{1 + v^2}$.

100. Show that $\tan(\sin^{-1} v) = \dfrac{v}{\sqrt{1 - v^2}}$.

101. Show that $\tan(\cos^{-1} v) = \dfrac{\sqrt{1 - v^2}}{v}$.

102. Show that $\sin(\cos^{-1} v) = \sqrt{1 - v^2}$.

103. Show that $\cos(\sin^{-1} v) = \sqrt{1 - v^2}$.

104. Show that $\cos(\tan^{-1} v) = \dfrac{1}{\sqrt{1 + v^2}}$.

---

105. Write a few paragraphs outlining your strategy for establishing identities.

106. Write down the three Pythagorean Identities.

107. Why do you think it is usually preferable to start with the side containing the more complicated expression when establishing an identity?

108. Make up an identity that is not a Fundamental Identity.

'Are You Prepared?' Answers

1. True **2.** $\dfrac{\sin\theta}{\cos\theta}$

6.4 Sum and Difference Formulas

PREPARING FOR THIS SECTION *Before getting started, review the following:*

- Distance Formula (Section 1.1, p. 3)
- Values of the Trigonometric Functions (Section 5.2, pp. 342–351)
- Finding Exact Values Given the Value of a Trigonometric Function and the Quadrant of the Angle (Section 5.2, pp. 363–365)

Now work the 'Are You Prepared?' problems on page 445.

OBJECTIVES
1 Use Sum and Difference Formulas to Find Exact Values
2 Use Sum and Difference Formulas to Establish Identities
3 Use Sum and Difference Formulas Involving Inverse Trigonometric Functions

In this section, we continue our derivation of trigonometric identities by obtaining formulas that involve the sum or difference of two angles, such as $\cos(\alpha + \beta)$, $\cos(\alpha - \beta)$, or $\sin(\alpha + \beta)$. These formulas are referred to as the **sum and difference formulas.** We begin with the formulas for $\cos(\alpha + \beta)$ and $\cos(\alpha - \beta)$.

Theorem

Sum and Difference Formulas for Cosines

$$\cos(\alpha + \beta) = \cos \alpha \cos \beta - \sin \alpha \sin \beta \quad (1)$$

$$\cos(\alpha - \beta) = \cos \alpha \cos \beta + \sin \alpha \sin \beta \quad (2)$$

In Words

Formula (1) states that the cosine of the sum of two angles equals the cosine of the first angle times the cosine of the second angle minus the sine of the first angle times the sine of the second angle.

Proof We will prove formula (2) first. Although this formula is true for all numbers α and β, we shall assume in our proof that $0 < \beta < \alpha < 2\pi$. We begin with the unit circle and place the angles α and β in standard position, as shown in Figure 22(a). The point P_1 lies on the terminal side of β, so its coordinates are $(\cos \beta, \sin \beta)$; and the point P_2 lies on the terminal side of α, so its coordinates are $(\cos \alpha, \sin \alpha)$.

Figure 22

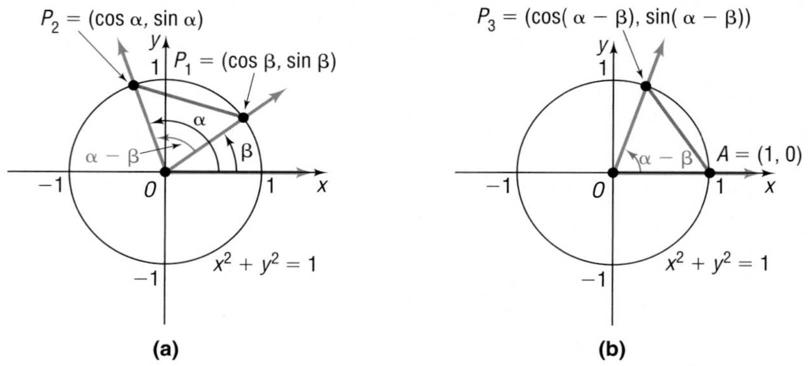

Now, place the angle $\alpha - \beta$ in standard position, as shown in Figure 22(b). The point A has coordinates $(1, 0)$, and the point P_3 is on the terminal side of the angle $\alpha - \beta$, so its coordinates are $(\cos(\alpha - \beta), \sin(\alpha - \beta))$.

Looking at triangle OP_1P_2 in Figure 22(a) and triangle OAP_3 in Figure 22(b), we see that these triangles are congruent. (Do you see why? Two sides and the included angle, $\alpha - \beta$, are equal.) As a result, the unknown side of each triangle must be equal; that is,

$$d(A, P_3) = d(P_1, P_2)$$

Using the distance formula, we find that

$$\sqrt{[\cos(\alpha - \beta) - 1]^2 + [\sin(\alpha - \beta) - 0]^2} = \sqrt{(\cos \alpha - \cos \beta)^2 + (\sin \alpha - \sin \beta)^2} \qquad d(A, P_3) = d(P_1, P_2)$$

$$[\cos(\alpha - \beta) - 1]^2 + \sin^2(\alpha - \beta) = (\cos \alpha - \cos \beta)^2 + (\sin \alpha - \sin \beta)^2 \qquad \text{Square both sides.}$$

$$\cos^2(\alpha - \beta) - 2\cos(\alpha - \beta) + 1 + \sin^2(\alpha - \beta) = \cos^2 \alpha - 2\cos \alpha \cos \beta + \cos^2 \beta + \sin^2 \alpha - 2\sin \alpha \sin \beta + \sin^2 \beta \qquad \text{Multiply out the squared terms.}$$

$$2 - 2\cos(\alpha - \beta) = 2 - 2\cos \alpha \cos \beta - 2\sin \alpha \sin \beta \qquad \text{Apply a Pythagorean Identity (3 times).}$$

$$-2\cos(\alpha - \beta) = -2\cos \alpha \cos \beta - 2\sin \alpha \sin \beta \qquad \text{Subtract 2 from each side.}$$

$$\cos(\alpha - \beta) = \cos \alpha \cos \beta + \sin \alpha \sin \beta \qquad \text{Divide each side by } -2.$$

which is formula (2).

The proof of formula (1) follows from formula (2) and the Even–Odd Identities. We use the fact that $\alpha + \beta = \alpha - (-\beta)$. Then

$$\begin{aligned}\cos(\alpha + \beta) &= \cos[\alpha - (-\beta)] \\ &= \cos\alpha\cos(-\beta) + \sin\alpha\sin(-\beta) && \text{Use formula (2).} \\ &= \cos\alpha\cos\beta - \sin\alpha\sin\beta && \text{Even–Odd Identities}\end{aligned}$$

■

1 One use of formulas (1) and (2) is to obtain the exact value of the cosine of an angle that can be expressed as the sum or difference of angles whose sine and cosine are known exactly.

EXAMPLE 1 Using the Sum Formula to Find Exact Values

Find the exact value of $\cos 75°$.

Solution Since $75° = 45° + 30°$, we use formula (1) to obtain

$$\cos 75° = \cos(45° + 30°) = \cos 45° \cos 30° - \sin 45° \sin 30°$$

↑ Formula (1)

$$= \frac{\sqrt{2}}{2} \cdot \frac{\sqrt{3}}{2} - \frac{\sqrt{2}}{2} \cdot \frac{1}{2} = \frac{1}{4}\left(\sqrt{6} - \sqrt{2}\right)$$

◀

EXAMPLE 2 Using the Difference Formula to Find Exact Values

Find the exact value of $\cos \frac{\pi}{12}$.

Solution

$$\cos\frac{\pi}{12} = \cos\left(\frac{3\pi}{12} - \frac{2\pi}{12}\right) = \cos\left(\frac{\pi}{4} - \frac{\pi}{6}\right)$$

$$= \cos\frac{\pi}{4}\cos\frac{\pi}{6} + \sin\frac{\pi}{4}\sin\frac{\pi}{6} \qquad \text{Use formula (2).}$$

$$= \frac{\sqrt{2}}{2} \cdot \frac{\sqrt{3}}{2} + \frac{\sqrt{2}}{2} \cdot \frac{1}{2} = \frac{1}{4}\left(\sqrt{6} + \sqrt{2}\right)$$

◀

NOW WORK PROBLEM 11.

2 Another use of formulas (1) and (2) is to establish other identities. One important pair of identities is given next.

$$\cos\left(\frac{\pi}{2} - \theta\right) = \sin\theta \qquad \textbf{(3a)}$$

$$\sin\left(\frac{\pi}{2} - \theta\right) = \cos\theta \qquad \textbf{(3b)}$$

Seeing the Concept

Graph $Y_1 = \cos\left(\frac{\pi}{2} - \theta\right)$ and $Y_2 = \sin\theta$ on the same screen. Does this demonstrate the result 3(a)? How would you demonstrate the result 3(b)?

Proof To prove formula (3a), we use the formula for $\cos(\alpha - \beta)$ with $\alpha = \frac{\pi}{2}$ and $\beta = \theta$.

$$\cos\left(\frac{\pi}{2} - \theta\right) = \cos\frac{\pi}{2}\cos\theta + \sin\frac{\pi}{2}\sin\theta = 0 \cdot \cos\theta + 1 \cdot \sin\theta = \sin\theta$$

To prove formula (3b), we make use of the identity (3a) just established.

$$\sin\left(\frac{\pi}{2} - \theta\right) \underset{\uparrow \text{ Use (3a)}}{=} \cos\left[\frac{\pi}{2} - \left(\frac{\pi}{2} - \theta\right)\right] = \cos\theta$$

■

Also, since

$$\cos\left(\frac{\pi}{2} - \theta\right) = \cos\left[-\left(\theta - \frac{\pi}{2}\right)\right] \underset{\uparrow \text{ Even Property of Cosine}}{=} \cos\left(\theta - \frac{\pi}{2}\right)$$

and

$$\cos\left(\frac{\pi}{2} - \theta\right) \underset{\uparrow \text{ 3(a)}}{=} \sin\theta$$

it follows that $\cos\left(\theta - \frac{\pi}{2}\right) = \sin\theta$. The graphs of $y = \cos\left(\theta - \frac{\pi}{2}\right)$ and $y = \sin\theta$ are identical, a fact that we conjectured earlier in Section 5.4.

Formulas for $\sin(\alpha + \beta)$ and $\sin(\alpha - \beta)$

Having established the identities in formulas (3a) and (3b), we now can derive the sum and difference formulas for $\sin(\alpha + \beta)$ and $\sin(\alpha - \beta)$.

Proof

$$\begin{aligned}
\sin(\alpha + \beta) &= \cos\left[\frac{\pi}{2} - (\alpha + \beta)\right] && \text{Formula (3a)} \\
&= \cos\left[\left(\frac{\pi}{2} - \alpha\right) - \beta\right] \\
&= \cos\left(\frac{\pi}{2} - \alpha\right)\cos\beta + \sin\left(\frac{\pi}{2} - \alpha\right)\sin\beta && \text{Formula (2)} \\
&= \sin\alpha\cos\beta + \cos\alpha\sin\beta && \text{Formulas (3a) and (3b)} \\
\sin(\alpha - \beta) &= \sin[\alpha + (-\beta)] \\
&= \sin\alpha\cos(-\beta) + \cos\alpha\sin(-\beta) && \text{Use the sum formula for sine just obtained.} \\
&= \sin\alpha\cos\beta + \cos\alpha(-\sin\beta) && \text{Even–Odd Identities.} \\
&= \sin\alpha\cos\beta - \cos\alpha\sin\beta
\end{aligned}$$

■

Theorem

Sum and Difference Formulas for Sines

$$\sin(\alpha + \beta) = \sin\alpha\cos\beta + \cos\alpha\sin\beta \quad \textbf{(4)}$$
$$\sin(\alpha - \beta) = \sin\alpha\cos\beta - \cos\alpha\sin\beta \quad \textbf{(5)}$$

In Words

Formula (4) states that the sine of the sum of two angles equals the sine of the first angle times the cosine of the second angle plus the cosine of the first angle times the sine of the second angle.

EXAMPLE 3 Using the Sum Formula to Find Exact Values

Find the exact value of $\sin\frac{7\pi}{12}$.

Solution

$$\sin\frac{7\pi}{12} = \sin\left(\frac{3\pi}{12} + \frac{4\pi}{12}\right) = \sin\left(\frac{\pi}{4} + \frac{\pi}{3}\right)$$
$$= \sin\frac{\pi}{4}\cos\frac{\pi}{3} + \cos\frac{\pi}{4}\sin\frac{\pi}{3} \qquad \text{Formula (4)}$$
$$= \frac{\sqrt{2}}{2}\cdot\frac{1}{2} + \frac{\sqrt{2}}{2}\cdot\frac{\sqrt{3}}{2} = \frac{1}{4}\left(\sqrt{2} + \sqrt{6}\right)$$

◀

NOW WORK PROBLEM 17.

EXAMPLE 4 Using the Difference Formula to Find Exact Values

Find the exact value of $\sin 80° \cos 20° - \cos 80° \sin 20°$.

Solution The form of the expression $\sin 80° \cos 20° - \cos 80° \sin 20°$ is that of the right side of the formula (5) for $\sin(\alpha - \beta)$ with $\alpha = 80°$ and $\beta = 20°$. Thus,

$$\sin 80° \cos 20° - \cos 80° \sin 20° = \sin(80° - 20°) = \sin 60° = \frac{\sqrt{3}}{2}$$

◀

NOW WORK PROBLEMS 23 AND 27.

EXAMPLE 5 Finding Exact Values

If it is known that $\sin\alpha = \frac{4}{5}, \frac{\pi}{2} < \alpha < \pi$, and that $\sin\beta = -\frac{2}{\sqrt{5}} = -\frac{2\sqrt{5}}{5}$, $\pi < \beta < \frac{3\pi}{2}$, find the exact value of

(a) $\cos\alpha$ (b) $\cos\beta$ (c) $\cos(\alpha + \beta)$ (d) $\sin(\alpha + \beta)$

Figure 23

$\sin\alpha = \frac{4}{5}, \frac{\pi}{2} < \alpha < \pi$

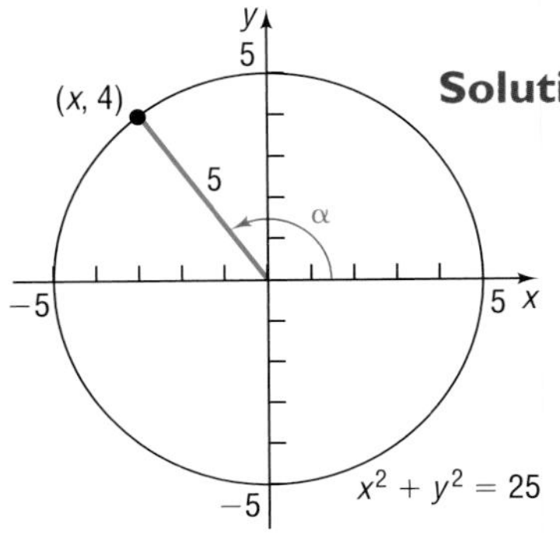

Solution (a) Since $\sin\alpha = \frac{4}{5} = \frac{y}{r}$ and $\frac{\pi}{2} < \alpha < \pi$, we let $y = 4$ and $r = 5$ and place α in quadrant II. The point $P = (x, y) = (x, 4)$, $x < 0$, is on a circle of radius 5, namely, $x^2 + y^2 = 25$. See Figure 23. Then,

$$x^2 + y^2 = 25, \qquad x < 0,\ y = 4$$
$$x^2 + 16 = 25$$
$$x^2 = 25 - 16 = 9$$
$$x = -3$$

Then,

$$\cos\alpha = \frac{x}{r} = -\frac{3}{5}$$

Alternatively, we can find $\cos\alpha$ using identities, as follows:

$$\cos\alpha = -\sqrt{1-\sin^2\alpha} = -\sqrt{1-\frac{16}{25}} = -\sqrt{\frac{9}{25}} = -\frac{3}{5}$$

↑ α in quadrant II, $\cos\alpha < 0$

(b) Since $\sin\beta = \frac{-2}{\sqrt{5}} = \frac{y}{r}$ and $\pi < \beta < \frac{3\pi}{2}$, we let $y = -2$ and $r = \sqrt{5}$ and place β in quadrant III. The point $P = (x, y) = (x, -2)$, $x < 0$, is on a circle of radius $\sqrt{5}$, namely, $x^2 + y^2 = 5$. See Figure 24. Then,

Figure 24

Given $\sin\beta = \frac{-2}{\sqrt{5}}, \pi < \beta < \frac{3\pi}{2}$

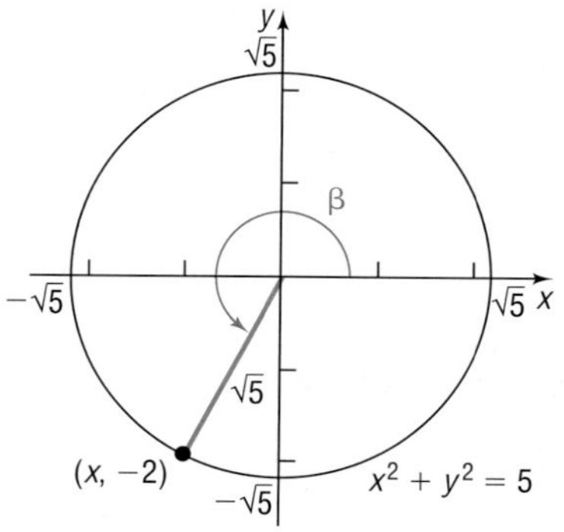

$$\begin{aligned} x^2 + y^2 &= 5, \qquad x < 0,\ y = -2 \\ x^2 + 4 &= 5 \\ x^2 &= 1 \\ x &= -1 \end{aligned}$$

Then,

$$\cos\beta = \frac{x}{r} = \frac{-1}{\sqrt{5}} = -\frac{\sqrt{5}}{5}$$

Alternatively, we can find $\cos\beta$ using identities, as follows:

$$\cos\beta = -\sqrt{1-\sin^2\beta} = -\sqrt{1-\frac{4}{5}} = -\sqrt{\frac{1}{5}} = -\frac{\sqrt{5}}{5}$$

(c) Using the results found in parts (a) and (b) and formula (1), we have

$$\begin{aligned} \cos(\alpha+\beta) &= \cos\alpha\cos\beta - \sin\alpha\sin\beta \\ &= -\frac{3}{5}\left(-\frac{\sqrt{5}}{5}\right) - \frac{4}{5}\left(-\frac{2\sqrt{5}}{5}\right) = \frac{11\sqrt{5}}{25} \end{aligned}$$

(d)

$$\begin{aligned} \sin(\alpha+\beta) &= \sin\alpha\cos\beta + \cos\alpha\sin\beta \\ &= \frac{4}{5}\left(-\frac{\sqrt{5}}{5}\right) + \left(-\frac{3}{5}\right)\left(-\frac{2\sqrt{5}}{5}\right) = \frac{2\sqrt{5}}{25} \end{aligned}$$

◀

NOW WORK PROBLEMS 31(a), (b), AND (c).

EXAMPLE 6 Establishing an Identity

Establish the identity: $\frac{\cos(\alpha-\beta)}{\sin\alpha\sin\beta} = \cot\alpha\cot\beta + 1$

Solution

$$\begin{aligned} \frac{\cos(\alpha-\beta)}{\sin\alpha\sin\beta} &= \frac{\cos\alpha\cos\beta + \sin\alpha\sin\beta}{\sin\alpha\sin\beta} \\ &= \frac{\cos\alpha\cos\beta}{\sin\alpha\sin\beta} + \frac{\sin\alpha\sin\beta}{\sin\alpha\sin\beta} \\ &= \frac{\cos\alpha}{\sin\alpha}\cdot\frac{\cos\beta}{\sin\beta} + 1 \\ &= \cot\alpha\cot\beta + 1 \end{aligned}$$

◀

NOW WORK PROBLEMS 39 AND 51.

Formulas for tan(α + β) and tan(α − β)

We use the identity $\tan\theta = \dfrac{\sin\theta}{\cos\theta}$ and the sum formulas for $\sin(\alpha+\beta)$ and $\cos(\alpha+\beta)$ to derive a formula for $\tan(\alpha+\beta)$.

Proof $\tan(\alpha+\beta) = \dfrac{\sin(\alpha+\beta)}{\cos(\alpha+\beta)} = \dfrac{\sin\alpha\cos\beta + \cos\alpha\sin\beta}{\cos\alpha\cos\beta - \sin\alpha\sin\beta}$

Now we divide the numerator and denominator by $\cos\alpha\cos\beta$.

$$\tan(\alpha+\beta) = \frac{\dfrac{\sin\alpha\cos\beta + \cos\alpha\sin\beta}{\cos\alpha\cos\beta}}{\dfrac{\cos\alpha\cos\beta - \sin\alpha\sin\beta}{\cos\alpha\cos\beta}} = \frac{\dfrac{\sin\alpha\cos\beta}{\cos\alpha\cos\beta} + \dfrac{\cos\alpha\sin\beta}{\cos\alpha\cos\beta}}{\dfrac{\cos\alpha\cos\beta}{\cos\alpha\cos\beta} - \dfrac{\sin\alpha\sin\beta}{\cos\alpha\cos\beta}}$$

$$= \frac{\dfrac{\sin\alpha}{\cos\alpha} + \dfrac{\sin\beta}{\cos\beta}}{1 - \dfrac{\sin\alpha\sin\beta}{\cos\alpha\cos\beta}} = \frac{\tan\alpha + \tan\beta}{1 - \tan\alpha\tan\beta}$$

We use the sum formula for $\tan(\alpha+\beta)$ and even–odd properties to get the difference formula.

$$\tan(\alpha-\beta) = \tan[\alpha + (-\beta)] = \frac{\tan\alpha + \tan(-\beta)}{1 - \tan\alpha\tan(-\beta)} = \frac{\tan\alpha - \tan\beta}{1 + \tan\alpha\tan\beta} \quad ■$$

We have proved the following results:

Theorem

Sum and Difference Formulas for Tangents

$$\tan(\alpha+\beta) = \frac{\tan\alpha + \tan\beta}{1 - \tan\alpha\tan\beta} \qquad (6)$$

$$\tan(\alpha-\beta) = \frac{\tan\alpha - \tan\beta}{1 + \tan\alpha\tan\beta} \qquad (7)$$

In Words
Formula (6) states that the tangent of the sum of two angles equals the tangent of the first angle plus the tangent of the second angle, all divided by 1 minus their product.

NOW WORK PROBLEM **31(d).**

EXAMPLE 7 Establishing an Identity

Prove the identity: $\tan(\theta+\pi) = \tan\theta$

Solution $\tan(\theta+\pi) = \dfrac{\tan\theta + \tan\pi}{1 - \tan\theta\tan\pi} = \dfrac{\tan\theta + 0}{1 - \tan\theta\cdot 0} = \tan\theta$ ◀

The result obtained in Example 7 verifies that the tangent function is periodic with period π, a fact that we mentioned earlier.

WARNING: Be careful when using formulas (6) and (7). These formulas can be used only for angles α and β for which $\tan\alpha$ and $\tan\beta$ are defined, that is, all angles except odd multiples of $\dfrac{\pi}{2}$. ■

EXAMPLE 8 **Establishing an Identity**

Prove the identity: $\tan\left(\theta + \frac{\pi}{2}\right) = -\cot\theta$

Solution We cannot use formula (6), since $\tan\frac{\pi}{2}$ is not defined. Instead, we proceed as follows:

$$\tan\left(\theta + \frac{\pi}{2}\right) = \frac{\sin\left(\theta + \frac{\pi}{2}\right)}{\cos\left(\theta + \frac{\pi}{2}\right)} = \frac{\sin\theta\cos\frac{\pi}{2} + \cos\theta\sin\frac{\pi}{2}}{\cos\theta\cos\frac{\pi}{2} - \sin\theta\sin\frac{\pi}{2}}$$

$$= \frac{(\sin\theta)(0) + (\cos\theta)(1)}{(\cos\theta)(0) - (\sin\theta)(1)} = \frac{\cos\theta}{-\sin\theta} = -\cot\theta$$

3 **EXAMPLE 9** **Finding the Exact Value of an Expression Involving Inverse Trigonometric Functions**

Find the exact value of: $\sin\left(\cos^{-1}\frac{1}{2} + \sin^{-1}\frac{3}{5}\right)$

Solution We seek the sine of the sum of two angles, $\alpha = \cos^{-1}\frac{1}{2}$ and $\beta = \sin^{-1}\frac{3}{5}$. Then

$$\cos\alpha = \frac{1}{2}, \quad 0 \le \alpha \le \pi, \quad \text{and} \quad \sin\beta = \frac{3}{5}, \quad -\frac{\pi}{2} \le \beta \le \frac{\pi}{2}$$

We use Pythagorean Identities to obtain $\sin\alpha$ and $\cos\beta$. Since $\sin\alpha \ge 0$ and $\cos\beta \ge 0$ (do you know why?), we find

$$\sin\alpha = \sqrt{1 - \cos^2\alpha} = \sqrt{1 - \frac{1}{4}} = \sqrt{\frac{3}{4}} = \frac{\sqrt{3}}{2}$$

$$\cos\beta = \sqrt{1 - \sin^2\beta} = \sqrt{1 - \frac{9}{25}} = \sqrt{\frac{16}{25}} = \frac{4}{5}$$

As a result,

$$\sin\left(\cos^{-1}\frac{1}{2} + \sin^{-1}\frac{3}{5}\right) = \sin(\alpha + \beta) = \sin\alpha\cos\beta + \cos\alpha\sin\beta$$

$$= \frac{\sqrt{3}}{2}\cdot\frac{4}{5} + \frac{1}{2}\cdot\frac{3}{5} = \frac{4\sqrt{3} + 3}{10}$$

NOW WORK PROBLEM 67.

EXAMPLE 10 **Writing a Trigonometric Expression as an Algebraic Expression**

Write $\sin(\sin^{-1} u + \cos^{-1} v)$ as an algebraic expression containing u and v (that is, without any trigonometric functions).

Solution Let $\alpha = \sin^{-1} u$ and $\beta = \cos^{-1} v$. Then

$$\sin\alpha = u, -\frac{\pi}{2} \le \alpha \le \frac{\pi}{2}, \quad \text{and} \quad \cos\beta = v, 0 \le \beta \le \pi$$

Since $-\frac{\pi}{2} \le \alpha \le \frac{\pi}{2}$, we know that $\cos\alpha \ge 0$. As a result,

$$\cos\alpha = \sqrt{1 - \sin^2\alpha} = \sqrt{1 - u^2}$$

Similarly, since $0 \le \beta \le \pi$, we know that $\sin\beta \ge 0$. Then,

$$\sin\beta = \sqrt{1 - \cos^2\beta} = \sqrt{1 - v^2}$$

Now

$$\sin(\sin^{-1} u + \cos^{-1} v) = \sin(\alpha + \beta) = \sin\alpha\cos\beta + \cos\alpha\sin\beta$$

$$= uv + \sqrt{1 - u^2}\sqrt{1 - v^2}$$ ◄

NOW WORK PROBLEM 77.

Summary

Sum and Difference Formulas

$$\cos(\alpha + \beta) = \cos\alpha\cos\beta - \sin\alpha\sin\beta \qquad \cos(\alpha - \beta) = \cos\alpha\cos\beta + \sin\alpha\sin\beta$$

$$\sin(\alpha + \beta) = \sin\alpha\cos\beta + \cos\alpha\sin\beta \qquad \sin(\alpha - \beta) = \sin\alpha\cos\beta - \cos\alpha\sin\beta$$

$$\tan(\alpha + \beta) = \frac{\tan\alpha + \tan\beta}{1 - \tan\alpha\tan\beta} \qquad \tan(\alpha - \beta) = \frac{\tan\alpha - \tan\beta}{1 + \tan\alpha\tan\beta}$$

6.4 Assess Your Understanding

'Are You Prepared?' *Answers are given at the end of these exercises. If you get a wrong answer, read the pages listed in red.*

1. The distance d from the point $(2, -3)$ to the point $(5, 1)$ is ________. (p. 3)

2. If $\sin\theta = \frac{4}{5}$ and θ is in quadrant II, then $\cos\theta =$ ________. (pp. 363–365)

3. (a) $\sin\frac{\pi}{4} \cdot \cos\frac{\pi}{3} =$ ________. (pp. 342–351)

(b) $\tan\frac{\pi}{4} - \sin\frac{\pi}{6} =$ ________. (pp. 342–351)

Concepts and Vocabulary

4. $\cos(\alpha + \beta) = \cos\alpha\cos\beta$ ________ $\sin\alpha\sin\beta$.

5. $\sin(\alpha - \beta) = \sin\alpha\cos\beta$ ________ $\cos\alpha\sin\beta$.

6. *True or False:* $\sin(\alpha + \beta) = \sin\alpha + \sin\beta + 2\sin\alpha\sin\beta$

7. *True or False:* $\tan 75° = \tan 30° + \tan 45°$

8. *True or False:* $\cos\left(\frac{\pi}{2} - \theta\right) = \cos\theta$

Exercises

In Problems 9–20, find the exact value of each trigonometric function.

9. $\sin\frac{5\pi}{12}$ **10.** $\sin\frac{\pi}{12}$ **11.** $\cos\frac{7\pi}{12}$ **12.** $\tan\frac{7\pi}{12}$ **13.** $\cos 165°$ **14.** $\sin 105°$

15. $\tan 15°$ **16.** $\tan 195°$ **17.** $\sin\frac{17\pi}{12}$ **18.** $\tan\frac{19\pi}{12}$ **19.** $\sec\left(-\frac{\pi}{12}\right)$ **20.** $\cot\left(-\frac{5\pi}{12}\right)$

In Problems 21–30, find the exact value of each expression.

21. $\sin 20°\cos 10° + \cos 20°\sin 10°$

22. $\sin 20°\cos 80° - \cos 20°\sin 80°$

23. $\cos 70°\cos 20° - \sin 70°\sin 20°$

24. $\cos 40°\cos 10° + \sin 40°\sin 10°$

25. $\dfrac{\tan 20° + \tan 25°}{1 - \tan 20° \tan 25°}$

26. $\dfrac{\tan 40° - \tan 10°}{1 + \tan 40° \tan 10°}$

27. $\sin\dfrac{\pi}{12}\cos\dfrac{7\pi}{12} - \cos\dfrac{\pi}{12}\sin\dfrac{7\pi}{12}$

28. $\cos\dfrac{5\pi}{12}\cos\dfrac{7\pi}{12} - \sin\dfrac{5\pi}{12}\sin\dfrac{7\pi}{12}$

29. $\cos\dfrac{\pi}{12}\cos\dfrac{5\pi}{12} + \sin\dfrac{5\pi}{12}\sin\dfrac{\pi}{12}$

30. $\sin\dfrac{\pi}{18}\cos\dfrac{5\pi}{18} + \cos\dfrac{\pi}{18}\sin\dfrac{5\pi}{18}$

In Problems 31–36, find the exact value of each of the following under the given conditions:

(a) $\sin(\alpha + \beta)$ (b) $\cos(\alpha + \beta)$ (c) $\sin(\alpha - \beta)$ (d) $\tan(\alpha - \beta)$

31. $\sin\alpha = \dfrac{3}{5}, 0 < \alpha < \dfrac{\pi}{2};\quad \cos\beta = \dfrac{2\sqrt{5}}{5}, -\dfrac{\pi}{2} < \beta < 0$

32. $\cos\alpha = \dfrac{\sqrt{5}}{5}, 0 < \alpha < \dfrac{\pi}{2};\quad \sin\beta = -\dfrac{4}{5}, -\dfrac{\pi}{2} < \beta < 0$

33. $\tan\alpha = -\dfrac{4}{3}, \dfrac{\pi}{2} < \alpha < \pi;\quad \cos\beta = \dfrac{1}{2}, 0 < \beta < \dfrac{\pi}{2}$

34. $\tan\alpha = \dfrac{5}{12}, \pi < \alpha < \dfrac{3\pi}{2};\quad \sin\beta = -\dfrac{1}{2}, \pi < \beta < \dfrac{3\pi}{2}$

35. $\sin\alpha = \dfrac{5}{13}, -\dfrac{3\pi}{2} < \alpha < -\pi;\quad \tan\beta = -\sqrt{3}, \dfrac{\pi}{2} < \beta < \pi$

36. $\cos\alpha = \dfrac{1}{2}, -\dfrac{\pi}{2} < \alpha < 0;\quad \sin\beta = \dfrac{1}{3}, 0 < \beta < \dfrac{\pi}{2}$

37. If $\sin\theta = \dfrac{1}{3}$, θ in quadrant II, find the exact value of:

(a) $\cos\theta$ (b) $\sin\left(\theta + \dfrac{\pi}{6}\right)$ (c) $\cos\left(\theta - \dfrac{\pi}{3}\right)$ (d) $\tan\left(\theta + \dfrac{\pi}{4}\right)$

38. If $\cos\theta = \dfrac{1}{4}$, θ in quadrant IV, find the exact value of:

(a) $\sin\theta$ (b) $\sin\left(\theta - \dfrac{\pi}{6}\right)$ (c) $\cos\left(\theta + \dfrac{\pi}{3}\right)$ (d) $\tan\left(\theta - \dfrac{\pi}{4}\right)$

In Problems 39–64, establish each identity.

39. $\sin\left(\dfrac{\pi}{2} + \theta\right) = \cos\theta$

40. $\cos\left(\dfrac{\pi}{2} + \theta\right) = -\sin\theta$

41. $\sin(\pi - \theta) = \sin\theta$

42. $\cos(\pi - \theta) = -\cos\theta$

43. $\sin(\pi + \theta) = -\sin\theta$

44. $\cos(\pi + \theta) = -\cos\theta$

45. $\tan(\pi - \theta) = -\tan\theta$

46. $\tan(2\pi - \theta) = -\tan\theta$

47. $\sin\left(\dfrac{3\pi}{2} + \theta\right) = -\cos\theta$

48. $\cos\left(\dfrac{3\pi}{2} + \theta\right) = \sin\theta$

49. $\sin(\alpha + \beta) + \sin(\alpha - \beta) = 2\sin\alpha\cos\beta$

50. $\cos(\alpha + \beta) + \cos(\alpha - \beta) = 2\cos\alpha\cos\beta$

51. $\dfrac{\sin(\alpha + \beta)}{\sin\alpha\cos\beta} = 1 + \cot\alpha\tan\beta$

52. $\dfrac{\sin(\alpha + \beta)}{\cos\alpha\cos\beta} = \tan\alpha + \tan\beta$

53. $\dfrac{\cos(\alpha + \beta)}{\cos\alpha\cos\beta} = 1 - \tan\alpha\tan\beta$

54. $\dfrac{\cos(\alpha - \beta)}{\sin\alpha\cos\beta} = \cot\alpha + \tan\beta$

55. $\dfrac{\sin(\alpha + \beta)}{\sin(\alpha - \beta)} = \dfrac{\tan\alpha + \tan\beta}{\tan\alpha - \tan\beta}$

56. $\dfrac{\cos(\alpha + \beta)}{\cos(\alpha - \beta)} = \dfrac{1 - \tan\alpha\tan\beta}{1 + \tan\alpha\tan\beta}$

57. $\cot(\alpha + \beta) = \dfrac{\cot\alpha\cot\beta - 1}{\cot\beta + \cot\alpha}$

58. $\cot(\alpha - \beta) = \dfrac{\cot\alpha\cot\beta + 1}{\cot\beta - \cot\alpha}$

59. $\sec(\alpha + \beta) = \dfrac{\csc\alpha\csc\beta}{\cot\alpha\cot\beta - 1}$

60. $\sec(\alpha - \beta) = \dfrac{\sec\alpha\sec\beta}{1 + \tan\alpha\tan\beta}$

61. $\sin(\alpha - \beta)\sin(\alpha + \beta) = \sin^2\alpha - \sin^2\beta$

62. $\cos(\alpha - \beta)\cos(\alpha + \beta) = \cos^2\alpha - \sin^2\beta$

63. $\sin(\theta + k\pi) = (-1)^k\sin\theta$, k any integer

64. $\cos(\theta + k\pi) = (-1)^k\cos\theta$, k any integer

In Problems 65–76, find the exact value of each expression.

65. $\sin\left(\sin^{-1}\frac{1}{2} + \cos^{-1} 0\right)$ **66.** $\sin\left(\sin^{-1}\frac{\sqrt{3}}{2} + \cos^{-1} 1\right)$ **67.** $\sin\left[\sin^{-1}\frac{3}{5} - \cos^{-1}\left(-\frac{4}{5}\right)\right]$ **68.** $\sin\left[\sin^{-1}\left(-\frac{4}{5}\right) - \tan^{-1}\frac{3}{4}\right]$

69. $\cos\left(\tan^{-1}\frac{4}{3} + \cos^{-1}\frac{5}{13}\right)$ **70.** $\cos\left[\tan^{-1}\frac{5}{12} - \sin^{-1}\left(-\frac{3}{5}\right)\right]$ **71.** $\cos\left(\sin^{-1}\frac{5}{13} - \tan^{-1}\frac{3}{4}\right)$ **72.** $\cos\left(\tan^{-1}\frac{4}{3} + \cos^{-1}\frac{12}{13}\right)$

73. $\tan\left(\sin^{-1}\frac{3}{5} + \frac{\pi}{6}\right)$ **74.** $\tan\left(\frac{\pi}{4} - \cos^{-1}\frac{3}{5}\right)$ **75.** $\tan\left(\sin^{-1}\frac{4}{5} + \cos^{-1} 1\right)$ **76.** $\tan\left(\cos^{-1}\frac{4}{5} + \sin^{-1} 1\right)$

In Problems 77–82, write each trigonometric expression as an algebraic expression containing u and v.

77. $\cos(\cos^{-1} u + \sin^{-1} v)$ **78.** $\sin(\sin^{-1} u - \cos^{-1} v)$ **79.** $\sin(\tan^{-1} u - \sin^{-1} v)$

80. $\cos(\tan^{-1} u + \tan^{-1} v)$ **81.** $\tan(\sin^{-1} u - \cos^{-1} v)$ **82.** $\sec(\tan^{-1} u + \cos^{-1} v)$

83. Show that $\sin^{-1} v + \cos^{-1} v = \frac{\pi}{2}$.

84. Show that $\tan^{-1} v + \cot^{-1} v = \frac{\pi}{2}$.

85. Show that $\tan^{-1}\left(\frac{1}{v}\right) = \frac{\pi}{2} - \tan^{-1} v$, if $v > 0$.

86. Show that $\cot^{-1} e^v = \tan^{-1} e^{-v}$.

87. Show that $\sin(\sin^{-1} v + \cos^{-1} v) = 1$.

88. Show that $\cos(\sin^{-1} v + \cos^{-1} v) = 0$.

---

89. Calculus Show that the difference quotient for $f(x) = \sin x$ is given by

$$\frac{f(x+h) - f(x)}{h} = \frac{\sin(x+h) - \sin x}{h} = \cos x \cdot \frac{\sin h}{h} - \sin x \cdot \frac{1 - \cos h}{h}$$

90. Calculus Show that the difference quotient for $f(x) = \cos x$ is given by

$$\frac{f(x+h) - f(x)}{h} = \frac{\cos(x+h) - \cos x}{h} = -\sin x \cdot \frac{\sin h}{h} - \cos x \cdot \frac{1 - \cos h}{h}$$

91. Explain why formula (7) cannot be used to show that

$$\tan\left(\frac{\pi}{2} - \theta\right) = \cot\theta$$

Establish this identity by using formulas (3a) and (3b).

92. If $\tan\alpha = x + 1$ and $\tan\beta = x - 1$, show that $2\cot(\alpha - \beta) = x^2$.

93. Geometry: Angle Between Two Lines Let L_1 and L_2 denote two nonvertical intersecting lines, and let θ denote the acute angle between L_1 and L_2 (see the figure). Show that

$$\tan\theta = \frac{m_2 - m_1}{1 + m_1 m_2}$$

where m_1 and m_2 are the slopes of L_1 and L_2, respectively.

[**Hint:** Use the facts that $\tan\theta_1 = m_1$ and $\tan\theta_2 = m_2$.]

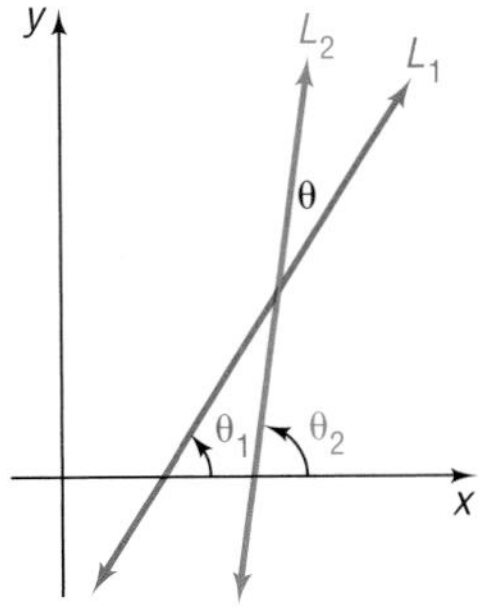

94. If $\alpha + \beta + \gamma = 180°$ and

$$\cot\theta = \cot\alpha + \cot\beta + \cot\gamma, \quad 0 < \theta < 90°$$

show that

$$\sin^3\theta = \sin(\alpha - \theta)\sin(\beta - \theta)\sin(\gamma - \theta)$$

95. Discuss the following derivation:

$$\tan\left(\theta + \frac{\pi}{2}\right) = \frac{\tan\theta + \tan\frac{\pi}{2}}{1 - \tan\theta\tan\frac{\pi}{2}} = \frac{\dfrac{\tan\theta}{\tan\frac{\pi}{2}} + 1}{\dfrac{1}{\tan\frac{\pi}{2}} - \tan\theta} = \frac{0 + 1}{0 - \tan\theta} = \frac{1}{-\tan\theta} = -\cot\theta$$

Can you justify each step?

'Are You Prepared?' Answers

1. 5 **2.** $-\frac{3}{5}$ **3.** (a) $\frac{\sqrt{2}}{4}$ (b) $\frac{1}{2}$

6.5 Double-Angle and Half-Angle Formulas

OBJECTIVES
1. Use Double-Angle Formulas to Find Exact Values
2. Use Double-Angle and Half-Angle Formulas to Establish Identities
3. Use Half-Angle Formulas to Find Exact Values

In this section we derive formulas for $\sin(2\theta)$, $\cos(2\theta)$, $\sin\left(\frac{1}{2}\theta\right)$, and $\cos\left(\frac{1}{2}\theta\right)$ in terms of $\sin\theta$ and $\cos\theta$. They are easily derived using the sum formulas.

Double-Angle Formulas

In the sum formulas for $\sin(\alpha+\beta)$ and $\cos(\alpha+\beta)$, let $\alpha=\beta=\theta$. Then

$$\begin{aligned}\sin(\alpha+\beta) &= \sin\alpha\cos\beta+\cos\alpha\sin\beta\\ \sin(\theta+\theta) &= \sin\theta\cos\theta+\cos\theta\sin\theta\\ \sin(2\theta) &= 2\sin\theta\cos\theta\end{aligned}$$

and

$$\begin{aligned}\cos(\alpha+\beta) &= \cos\alpha\cos\beta-\sin\alpha\sin\beta\\ \cos(\theta+\theta) &= \cos\theta\cos\theta-\sin\theta\sin\theta\\ \cos(2\theta) &= \cos^2\theta-\sin^2\theta\end{aligned}$$

An application of the Pythagorean Identity $\sin^2\theta+\cos^2\theta=1$ results in two other ways to express $\cos(2\theta)$.

$$\cos(2\theta)=\cos^2\theta-\sin^2\theta=(1-\sin^2\theta)-\sin^2\theta=1-2\sin^2\theta$$

and

$$\cos(2\theta)=\cos^2\theta-\sin^2\theta=\cos^2\theta-(1-\cos^2\theta)=2\cos^2\theta-1$$

We have established the following **Double-Angle Formulas:**

Theorem

Double-Angle Formulas

$$\sin(2\theta)=2\sin\theta\cos\theta \qquad \textbf{(1)}$$

$$\cos(2\theta)=\cos^2\theta-\sin^2\theta \qquad \textbf{(2)}$$

$$\cos(2\theta)=1-2\sin^2\theta \qquad \textbf{(3)}$$

$$\cos(2\theta)=2\cos^2\theta-1 \qquad \textbf{(4)}$$

1 **EXAMPLE 1** **Finding Exact Values Using the Double-Angle Formula**

If $\sin\theta=\frac{3}{5}$, $\frac{\pi}{2}<\theta<\pi$, find the exact value of:

(a) $\sin(2\theta)$ (b) $\cos(2\theta)$

Solution (a) Because $\sin(2\theta) = 2\sin\theta\cos\theta$ and we already know that $\sin\theta = \frac{3}{5}$, we only need to find $\cos\theta$. Since $\sin\theta = \frac{3}{5} = \frac{y}{r}$, $\frac{\pi}{2} < \theta < \pi$, we let $y = 3$ and $r = 5$ and place θ in quadrant II. The point $P = (x, y) = (x, 3)$, $x < 0$, is on a circle of radius 5, namely, $x^2 + y^2 = 25$. See Figure 25. Then

Figure 25

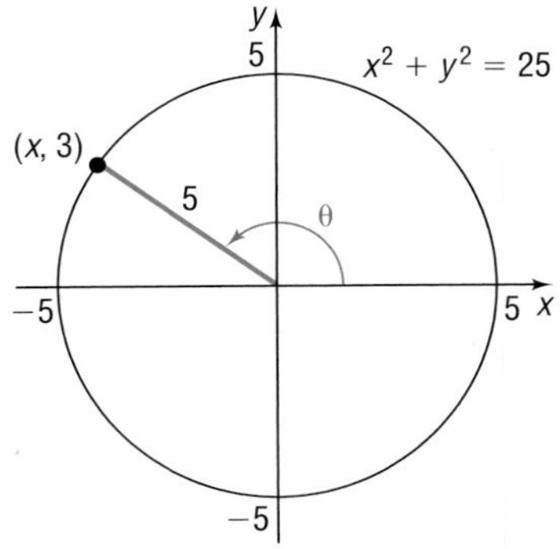

$$x^2 + y^2 = 25, \quad x < 0, y = 3$$
$$x^2 + 9 = 25$$
$$x^2 = 25 - 9 = 16$$
$$x = -4$$

We find that $\cos\theta = \frac{x}{r} = \frac{-4}{5}$. Now we use formula (1) to obtain

$$\sin(2\theta) = 2\sin\theta\cos\theta = 2\left(\frac{3}{5}\right)\left(-\frac{4}{5}\right) = -\frac{24}{25}$$

(b) Because we are given $\sin\theta = \frac{3}{5}$, it is easiest to use formula (4b) to get $\cos(2\theta)$.

$$\cos(2\theta) = 1 - 2\sin^2\theta = 1 - 2\left(\frac{9}{25}\right) = 1 - \frac{18}{25} = \frac{7}{25}$$ ■

WARNING: In finding $\cos(2\theta)$ in Example 1(b), we chose to use a version of the Double-Angle Formula, formula (3). Note that we are unable to use the Pythagorean Identity $\cos(2\theta) = \pm\sqrt{1 - \sin^2(2\theta)}$, with $\sin(2\theta) = -\frac{24}{25}$, because we have no way of knowing which sign to choose. ■

NOW WORK PROBLEMS 7(a) AND (b).

2 **EXAMPLE 2** **Establishing Identities**

(a) Develop a formula for $\tan(2\theta)$ in terms of $\tan\theta$.
(b) Develop a formula for $\sin(3\theta)$ in terms of $\sin\theta$ and $\cos\theta$.

Solution (a) In the sum formula for $\tan(\alpha + \beta)$, let $\alpha = \beta = \theta$. Then

$$\tan(\alpha + \beta) = \frac{\tan\alpha + \tan\beta}{1 - \tan\alpha\tan\beta}$$
$$\tan(\theta + \theta) = \frac{\tan\theta + \tan\theta}{1 - \tan\theta\tan\theta}$$

$$\tan(2\theta) = \frac{2\tan\theta}{1 - \tan^2\theta} \qquad (5)$$

(b) To get a formula for $\sin(3\theta)$, we use the sum formula and write 3θ as $2\theta + \theta$.

$$\sin(3\theta) = \sin(2\theta + \theta) = \sin(2\theta)\cos\theta + \cos(2\theta)\sin\theta$$

Now use the Double-Angle Formulas to get

$$\begin{aligned}\sin(3\theta) &= (2\sin\theta\cos\theta)(\cos\theta) + (\cos^2\theta - \sin^2\theta)(\sin\theta)\\ &= 2\sin\theta\cos^2\theta + \sin\theta\cos^2\theta - \sin^3\theta\\ &= 3\sin\theta\cos^2\theta - \sin^3\theta\end{aligned}$$ ◀

The formula obtained in Example 2(b) can also be written as

$$\begin{aligned}\sin(3\theta) &= 3\sin\theta\cos^2\theta - \sin^3\theta = 3\sin\theta\,(1 - \sin^2\theta) - \sin^3\theta \\ &= 3\sin\theta - 4\sin^3\theta\end{aligned}$$

That is, $\sin(3\theta)$ is a third-degree polynomial in the variable $\sin\theta$. In fact, $\sin(n\theta)$, n a positive odd integer, can always be written as a polynomial of degree n in the variable $\sin\theta$.[*]

NOW WORK PROBLEM 53.

Other Variations of the Double-Angle Formulas

By rearranging the Double-Angle Formulas (3) and (4), we obtain other formulas that we will use later in this section.

We begin with formula (3) and proceed to solve for $\sin^2\theta$.

$$\cos(2\theta) = 1 - 2\sin^2\theta$$

$$2\sin^2\theta = 1 - \cos(2\theta)$$

$$\sin^2\theta = \frac{1 - \cos(2\theta)}{2} \qquad \textbf{(6)}$$

Similarly, using formula (4), we proceed to solve for $\cos^2\theta$.

$$\cos(2\theta) = 2\cos^2\theta - 1$$

$$2\cos^2\theta = 1 + \cos(2\theta)$$

$$\cos^2\theta = \frac{1 + \cos(2\theta)}{2} \qquad \textbf{(7)}$$

Formulas (6) and (7) can be used to develop a formula for $\tan^2\theta$.

$$\tan^2\theta = \frac{\sin^2\theta}{\cos^2\theta} = \frac{\dfrac{1 - \cos(2\theta)}{2}}{\dfrac{1 + \cos(2\theta)}{2}}$$

$$\tan^2\theta = \frac{1 - \cos(2\theta)}{1 + \cos(2\theta)} \qquad \textbf{(8)}$$

Formulas (6) through (8) do not have to be memorized since their derivations are so straightforward.

Formulas (6) and (7) are important in calculus. The next example illustrates a problem that arises in calculus requiring the use of formula (7).

[*]Due to the work done by P. L. Chebyshëv, these polynomials are sometimes called *Chebyshëv polynomials.*

EXAMPLE 3 Establishing an Identity

Write an equivalent expression for $\cos^4\theta$ that does not involve any powers of sine or cosine greater than 1.

Solution The idea here is to apply formula (7) twice.

$$\begin{aligned}
\cos^4\theta = (\cos^2\theta)^2 &= \left(\frac{1+\cos(2\theta)}{2}\right)^2 && \text{Formula (7)}\\
&= \frac{1}{4}[1 + 2\cos(2\theta) + \cos^2(2\theta)]\\
&= \frac{1}{4} + \frac{1}{2}\cos(2\theta) + \frac{1}{4}\cos^2(2\theta)\\
&= \frac{1}{4} + \frac{1}{2}\cos(2\theta) + \frac{1}{4}\left\{\frac{1+\cos[2(2\theta)]}{2}\right\} && \text{Formula (7)}\\
&= \frac{1}{4} + \frac{1}{2}\cos(2\theta) + \frac{1}{8}[1+\cos(4\theta)]\\
&= \frac{3}{8} + \frac{1}{2}\cos(2\theta) + \frac{1}{8}\cos(4\theta)
\end{aligned}$$

◀

NOW WORK PROBLEM 29.

Identities, such as the Double-Angle Formulas, can sometimes be used to rewrite expressions in a more suitable form. Let's look at an example.

EXAMPLE 4 Projectile Motion

Figure 26

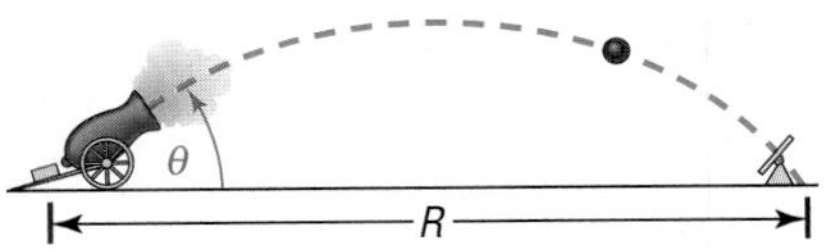

An object is propelled upward at an angle θ to the horizontal with an initial velocity of v_0 feet per second. See Figure 26. If air resistance is ignored, the **range** R, the horizontal distance that the object travels, is given by

$$R = \frac{1}{16}v_0^2\sin\theta\cos\theta$$

(a) Show that $R = \frac{1}{32}v_0^2\sin(2\theta)$.

(b) Find the angle θ for which R is a maximum.

Solution (a) We rewrite the given expression for the range using the Double-Angle Formula $\sin(2\theta) = 2\sin\theta\cos\theta$. Then

$$R = \frac{1}{16}v_0^2\sin\theta\cos\theta = \frac{1}{16}v_0^2\frac{2\sin\theta\cos\theta}{2} = \frac{1}{32}v_0^2\sin(2\theta)$$

(b) In this form, the largest value for the range R can be found. For a fixed initial speed v_0, the angle θ of inclination to the horizontal determines the value of R. Since the largest value of a sine function is 1, occurring when the argument 2θ is 90°, it follows that for maximum R we must have

$$\begin{aligned}
2\theta &= 90^\circ\\
\theta &= 45^\circ
\end{aligned}$$

An inclination to the horizontal of 45° results in maximum range. ◀

Half-Angle Formulas

Another important use of formulas (6) through (8) is to prove the **Half-Angle Formulas.** In formulas (6) through (8), let $\theta = \frac{\alpha}{2}$. Then

$$\sin^2\frac{\alpha}{2} = \frac{1-\cos\alpha}{2} \qquad \cos^2\frac{\alpha}{2} = \frac{1+\cos\alpha}{2} \qquad \tan^2\frac{\alpha}{2} = \frac{1-\cos\alpha}{1+\cos\alpha} \quad \textbf{(9)}$$

The identities in box (9) will prove useful in integral calculus.

If we solve for the trigonometric functions on the left sides of equations (9), we obtain the Half-angle Formulas.

Theorem

Half-Angle Formulas

$$\sin\frac{\alpha}{2} = \pm\sqrt{\frac{1-\cos\alpha}{2}} \quad \textbf{(10a)}$$

$$\cos\frac{\alpha}{2} = \pm\sqrt{\frac{1+\cos\alpha}{2}} \quad \textbf{(10b)}$$

$$\tan\frac{\alpha}{2} = \pm\sqrt{\frac{1-\cos\alpha}{1+\cos\alpha}} \quad \textbf{(10c)}$$

where the + or − sign is determined by the quadrant of the angle $\frac{\alpha}{2}$.

We use the Half-Angle Formulas in the next example.

3 **EXAMPLE 5** **Finding Exact Values Using Half-Angle Formulas**

Use a Half-Angle Formula find the exact value of:

(a) $\cos 15°$ (b) $\sin(-15°)$

Solution (a) Because $15° = \frac{30°}{2}$, we can use the Half-Angle Formula for $\cos\frac{\alpha}{2}$ with $\alpha = 30°$. Also, because $15°$ is in quadrant I, $\cos 15° > 0$, we choose the + sign in using formula (10b):

$$\cos 15° = \cos\frac{30°}{2} = \sqrt{\frac{1+\cos 30°}{2}}$$

$$= \sqrt{\frac{1+\sqrt{3}/2}{2}} = \sqrt{\frac{2+\sqrt{3}}{4}} = \frac{\sqrt{2+\sqrt{3}}}{2}$$

(b) We use the fact that $\sin(-15°) = -\sin 15°$ and then apply formula (10a).

$$\sin(-15°) = -\sin\frac{30°}{2} = -\sqrt{\frac{1-\cos 30°}{2}}$$

$$= -\sqrt{\frac{1-\sqrt{3}/2}{2}} = -\sqrt{\frac{2-\sqrt{3}}{4}} = -\frac{\sqrt{2-\sqrt{3}}}{2}$$

◀

It is interesting to compare the answer found in Example 5(a) with the answer to Example 2 of Section 6.4. There we calculated

$$\cos\frac{\pi}{12} = \cos 15° = \frac{1}{4}\left(\sqrt{6} + \sqrt{2}\right)$$

Based on this and the result of Example 5(a), we conclude that

$$\frac{1}{4}\left(\sqrt{6} + \sqrt{2}\right) \quad \text{and} \quad \frac{\sqrt{2 + \sqrt{3}}}{2}$$

are equal. (Since each expression is positive, you can verify this equality by squaring each expression.) Two very different looking, yet correct, answers can be obtained, depending on the approach taken to solve a problem.

NOW WORK PROBLEM 19.

EXAMPLE 6 Finding Exact Values Using Half-Angle Formulas

If $\cos\alpha = -\frac{3}{5}$, $\pi < \alpha < \frac{3\pi}{2}$, find the exact value of:

(a) $\sin\frac{\alpha}{2}$ (b) $\cos\frac{\alpha}{2}$ (c) $\tan\frac{\alpha}{2}$

Solution First, we observe that if $\pi < \alpha < \frac{3\pi}{2}$ then $\frac{\pi}{2} < \frac{\alpha}{2} < \frac{3\pi}{4}$. As a result, $\frac{\alpha}{2}$ lies in quadrant II.

(a) Because $\frac{\alpha}{2}$ lies in quadrant II, $\sin\frac{\alpha}{2} > 0$, so we use the + sign in formula (10a) to get

$$\sin\frac{\alpha}{2} = \sqrt{\frac{1 - \cos\alpha}{2}} = \sqrt{\frac{1 - \left(-\frac{3}{5}\right)}{2}}$$

$$= \sqrt{\frac{\frac{8}{5}}{2}} = \sqrt{\frac{4}{5}} = \frac{2}{\sqrt{5}} = \frac{2\sqrt{5}}{5}$$

(b) Because $\frac{\alpha}{2}$ lies in quadrant II, $\cos\frac{\alpha}{2} < 0$, so we use the − sign in formula (10b) to get

$$\cos\frac{\alpha}{2} = -\sqrt{\frac{1 + \cos\alpha}{2}} = -\sqrt{\frac{1 + \left(-\frac{3}{5}\right)}{2}}$$

$$= -\sqrt{\frac{\frac{2}{5}}{2}} = -\frac{1}{\sqrt{5}} = -\frac{\sqrt{5}}{5}$$

(c) Because $\frac{\alpha}{2}$ lies in quadrant II, $\tan\frac{\alpha}{2} < 0$, so we use the $-$ sign in formula (10c) to get

$$\tan\frac{\alpha}{2} = -\sqrt{\frac{1-\cos\alpha}{1+\cos\alpha}} = -\sqrt{\frac{1-\left(-\frac{3}{5}\right)}{1+\left(-\frac{3}{5}\right)}} = -\sqrt{\frac{\frac{8}{5}}{\frac{2}{5}}} = -2$$

Another way to solve Example 6(c) is to use the solutions found in parts (a) and (b).

$$\tan\frac{\alpha}{2} = \frac{\sin\frac{\alpha}{2}}{\cos\frac{\alpha}{2}} = \frac{\frac{2\sqrt{5}}{5}}{-\frac{\sqrt{5}}{5}} = -2$$

NOW WORK PROBLEMS 7(c) AND (d).

There is a formula for $\tan\frac{\alpha}{2}$ that does not contain $+$ and $-$ signs, making it more useful than Formula 10(c). To derive it, we use the formulas:

$$1 - \cos\alpha = 2\sin^2\frac{\alpha}{2} \qquad \text{Formula (9)}$$

and

$$\sin\alpha = \sin\left[2\left(\frac{\alpha}{2}\right)\right] = 2\sin\frac{\alpha}{2}\cos\frac{\alpha}{2} \qquad \text{Double-Angle formula}$$

Then

$$\frac{1-\cos\alpha}{\sin\alpha} = \frac{2\sin^2\frac{\alpha}{2}}{2\sin\frac{\alpha}{2}\cos\frac{\alpha}{2}} = \frac{\sin\frac{\alpha}{2}}{\cos\frac{\alpha}{2}} = \tan\frac{\alpha}{2}$$

Since it also can be shown that

$$\frac{1-\cos\alpha}{\sin\alpha} = \frac{\sin\alpha}{1+\cos\alpha}$$

we have the following two Half-Angle Formulas:

Half-Angle Formulas for $\tan\frac{\alpha}{2}$

$$\tan\frac{\alpha}{2} = \frac{1-\cos\alpha}{\sin\alpha} = \frac{\sin\alpha}{1+\cos\alpha} \qquad \textbf{(11)}$$

With this formula, the solution to Example 6(c) can be given as

$$\cos\alpha = -\frac{3}{5}$$

$$\sin\alpha = -\sqrt{1-\cos^2\alpha} = -\sqrt{1-\frac{9}{25}} = -\sqrt{\frac{16}{25}} = -\frac{4}{5}$$

Then, by equation (11),

$$\tan\frac{\alpha}{2} = \frac{1-\cos\alpha}{\sin\alpha} = \frac{1-\left(-\frac{3}{5}\right)}{-\frac{4}{5}} = \frac{\frac{8}{5}}{-\frac{4}{5}} = -2$$

6.5 Assess Your Understanding

Concepts and Vocabulary

1. $\cos(2\theta) = \cos^2\theta -$ __________ $=$ __________ $- 1 = 1 -$ __________.

2. $\sin^2\frac{\theta}{2} = \frac{\_\_\_}{2}$.

3. $\tan\frac{\theta}{2} = \frac{1-\cos\theta}{\_\_\_}$

4. *True or False:* $\cos(2\theta)$ has three equivalent forms:

$\cos^2\theta - \sin^2\theta$, $1 - 2\sin^2\theta$, and $2\cos^2\theta - 1$

5. *True or False:* $\sin(2\theta)$ has two equivalent forms:

$2\sin\theta\cos\theta$ and $\sin^2\theta - \cos^2\theta$

6. *True or False:* $\tan(2\theta) + \tan(2\theta) = \tan(4\theta)$

Exercises

In Problems 7–18, use the information given about the angle θ, $0 \le \theta \le 2\pi$, to find the exact value of

(a) $\sin(2\theta)$ (b) $\cos(2\theta)$ (c) $\sin\frac{\theta}{2}$ (d) $\cos\frac{\theta}{2}$

7. $\sin\theta = \frac{3}{5}$, $0 < \theta < \frac{\pi}{2}$

8. $\cos\theta = \frac{3}{5}$, $0 < \theta < \frac{\pi}{2}$

9. $\tan\theta = \frac{4}{3}$, $\pi < \theta < \frac{3\pi}{2}$

10. $\tan\theta = \frac{1}{2}$, $\pi < \theta < \frac{3\pi}{2}$

11. $\cos\theta = -\frac{\sqrt{6}}{3}$, $\frac{\pi}{2} < \theta < \pi$

12. $\sin\theta = -\frac{\sqrt{3}}{3}$, $\frac{3\pi}{2} < \theta < 2\pi$

13. $\sec\theta = 3$, $\sin\theta > 0$

14. $\csc\theta = -\sqrt{5}$, $\cos\theta < 0$

15. $\cot\theta = -2$, $\sec\theta < 0$

16. $\sec\theta = 2$, $\csc\theta < 0$

17. $\tan\theta = -3$, $\sin\theta < 0$

18. $\cot\theta = 3$, $\cos\theta < 0$

In Problems 19–28, use the Half-Angle Formulas to find the exact value of each trigonometric function.

19. $\sin 22.5°$

20. $\cos 22.5°$

21. $\tan\frac{7\pi}{8}$

22. $\tan\frac{9\pi}{8}$

23. $\cos 165°$

24. $\sin 195°$

25. $\sec\frac{15\pi}{8}$

26. $\csc\frac{7\pi}{8}$

27. $\sin\left(-\frac{\pi}{8}\right)$

28. $\cos\left(-\frac{3\pi}{8}\right)$

29. Show that $\sin^4\theta = \frac{3}{8} - \frac{1}{2}\cos(2\theta) + \frac{1}{8}\cos(4\theta)$.

30. Develop a formula for $\cos(3\theta)$ as a third-degree polynomial in the variable $\cos\theta$.

31. Show that $\sin(4\theta) = (\cos\theta)(4\sin\theta - 8\sin^3\theta)$.

32. Develop a formula for $\cos(4\theta)$ as a fourth-degree polynomial in the variable $\cos\theta$.

33. Find an expression for $\sin(5\theta)$ as a fifth-degree polynomial in the variable $\sin\theta$.

34. Find an expression for $\cos(5\theta)$ as a fifth-degree polynomial in the variable $\cos\theta$.

In Problems 35–56, establish each identity.

35. $\cos^4\theta - \sin^4\theta = \cos(2\theta)$

36. $\frac{\cot\theta - \tan\theta}{\cot\theta + \tan\theta} = \cos(2\theta)$

37. $\cot(2\theta) = \frac{\cot^2\theta - 1}{2\cot\theta}$

38. $\cot(2\theta) = \frac{1}{2}(\cot\theta - \tan\theta)$

39. $\sec(2\theta) = \frac{\sec^2\theta}{2 - \sec^2\theta}$

40. $\csc(2\theta) = \frac{1}{2}\sec\theta\csc\theta$

41. $\cos^2(2\theta) - \sin^2(2\theta) = \cos(4\theta)$

42. $(4 \sin\theta \cos\theta)(1 - 2\sin^2\theta) = \sin(4\theta)$

43. $\dfrac{\cos(2\theta)}{1 + \sin(2\theta)} = \dfrac{\cot\theta - 1}{\cot\theta + 1}$

44. $\sin^2\theta \cos^2\theta = \dfrac{1}{8}[1 - \cos(4\theta)]$

45. $\sec^2\dfrac{\theta}{2} = \dfrac{2}{1 + \cos\theta}$

46. $\csc^2\dfrac{\theta}{2} = \dfrac{2}{1 - \cos\theta}$

47. $\cot^2\dfrac{\theta}{2} = \dfrac{\sec\theta + 1}{\sec\theta - 1}$

48. $\tan\dfrac{\theta}{2} = \csc\theta - \cot\theta$

49. $\cos\theta = \dfrac{1 - \tan^2\dfrac{\theta}{2}}{1 + \tan^2\dfrac{\theta}{2}}$

50. $1 - \dfrac{1}{2}\sin(2\theta) = \dfrac{\sin^3\theta + \cos^3\theta}{\sin\theta + \cos\theta}$

51. $\dfrac{\sin(3\theta)}{\sin\theta} - \dfrac{\cos(3\theta)}{\cos\theta} = 2$

52. $\dfrac{\cos\theta + \sin\theta}{\cos\theta - \sin\theta} - \dfrac{\cos\theta - \sin\theta}{\cos\theta + \sin\theta} = 2\tan(2\theta)$

53. $\tan(3\theta) = \dfrac{3\tan\theta - \tan^3\theta}{1 - 3\tan^2\theta}$

54. $\tan\theta + \tan(\theta + 120°) + \tan(\theta + 240°) = 3\tan(3\theta)$

55. $\ln|\sin\theta| = \dfrac{1}{2}(\ln|1 - \cos(2\theta)| - \ln 2)$

56. $\ln|\cos\theta| = \dfrac{1}{2}(\ln|1 + \cos(2\theta)| - \ln 2)$

In Problems 57–68, find the exact value of each expression.

57. $\sin\left(2\sin^{-1}\dfrac{1}{2}\right)$

58. $\sin\left[2\sin^{-1}\dfrac{\sqrt{3}}{2}\right]$

59. $\cos\left(2\sin^{-1}\dfrac{3}{5}\right)$

60. $\cos\left(2\cos^{-1}\dfrac{4}{5}\right)$

61. $\tan\left[2\cos^{-1}\left(-\dfrac{3}{5}\right)\right]$

62. $\tan\left(2\tan^{-1}\dfrac{3}{4}\right)$

63. $\sin\left(2\cos^{-1}\dfrac{4}{5}\right)$

64. $\cos\left[2\tan^{-1}\left(-\dfrac{4}{3}\right)\right]$

65. $\sin^2\left(\dfrac{1}{2}\cos^{-1}\dfrac{3}{5}\right)$

66. $\cos^2\left(\dfrac{1}{2}\sin^{-1}\dfrac{3}{5}\right)$

67. $\sec\left(2\tan^{-1}\dfrac{3}{4}\right)$

68. $\csc\left[2\sin^{-1}\left(-\dfrac{3}{5}\right)\right]$

69. If $x = 2\tan\theta$, express $\sin(2\theta)$ as a function of x.

70. If $x = 2\tan\theta$, express $\cos(2\theta)$ as a function of x.

71. Find the value of the number C:

$$\frac{1}{2}\sin^2 x + C = -\frac{1}{4}\cos(2x)$$

72. Find the value of the number C:

$$\frac{1}{2}\cos^2 x + C = \frac{1}{4}\cos(2x)$$

73. If $z = \tan\dfrac{\alpha}{2}$, show that $\sin\alpha = \dfrac{2z}{1 + z^2}$.

74. If $z = \tan\dfrac{\alpha}{2}$, show that $\cos\alpha = \dfrac{1 - z^2}{1 + z^2}$.

---

75. Area of an Isosceles Triangle Show that the area A of an isosceles triangle whose equal sides are of length s and θ is the angle between them is

$$\frac{1}{2}s^2\sin\theta$$

[**Hint:** See the illustration. The height h bisects the angle θ and is the perpendicular bisector of the base.]

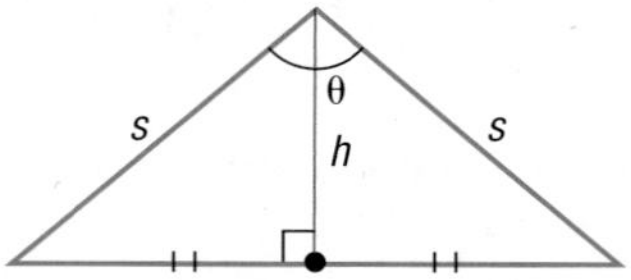

76. Geometry A rectangle is inscribed in a semicircle of radius 1. See the illustration.

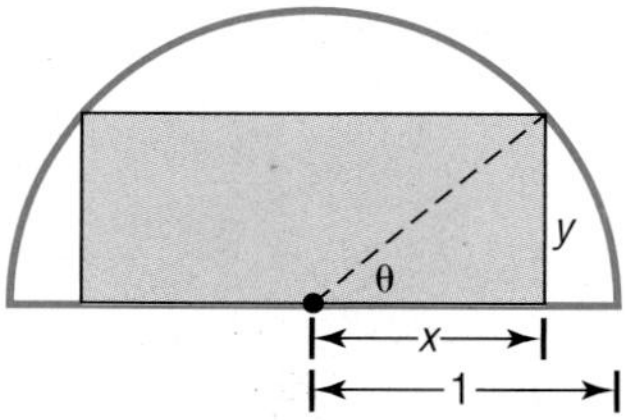

(a) Express the area A of the rectangle as a function of the angle θ shown in the illustration.
(b) Show that $A = \sin(2\theta)$.
(c) Find the angle θ that results in the largest area A.
(d) Find the dimensions of this largest rectangle.

77. Graph $f(x) = \sin^2 x = \dfrac{1 - \cos(2x)}{2}$ for $0 \le x \le 2\pi$ by using transformations.

78. Repeat Problem 77 for $g(x) = \cos^2 x$.

79. Use the fact that

$$\cos\frac{\pi}{12} = \frac{1}{4}(\sqrt{6} + \sqrt{2})$$

to find $\sin\dfrac{\pi}{24}$ and $\cos\dfrac{\pi}{24}$.

80. Show that

$$\cos\frac{\pi}{8} = \frac{\sqrt{2 + \sqrt{2}}}{2}$$

and use it to find $\sin\dfrac{\pi}{16}$ and $\cos\dfrac{\pi}{16}$.

81. Show that

$$\sin^3\theta + \sin^3(\theta + 120°) + \sin^3(\theta + 240°) = -\frac{3}{4}\sin(3\theta)$$

82. If $\tan\theta = a\tan\frac{\theta}{3}$, express $\tan\frac{\theta}{3}$ in terms of a.

83. Projectile Motion An object is propelled upward at an angle θ, $45° < \theta < 90°$, to the horizontal with an initial velocity of v_0 feet per second from the base of a plane that makes an angle of 45° with the horizontal. See the illustration. If air resistance is ignored, the distance R that it travels up the inclined plane is given by

$$R = \frac{v_0^2\sqrt{2}}{16}\cos\theta(\sin\theta - \cos\theta)$$

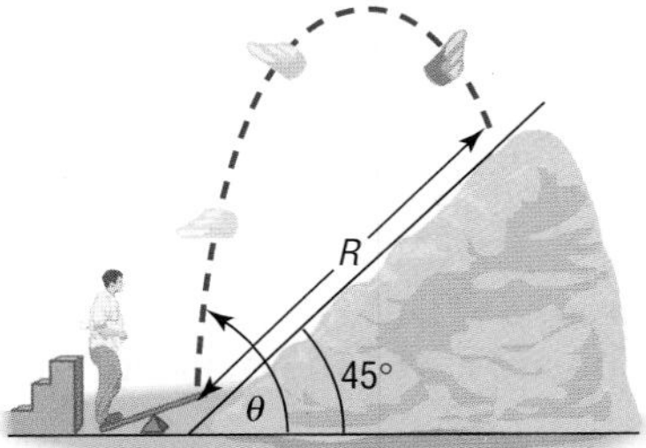

(a) Show that

$$R = \frac{v_0^2\sqrt{2}}{32}[\sin(2\theta) - \cos(2\theta) - 1]$$

(b) Graph $R = R(\theta)$. (Use $v_0 = 32$ feet per second.)
(c) What value of θ makes R the largest? (Use $v_0 = 32$ feet per second.)

84. Sawtooth Curve An oscilloscope often displays a sawtooth curve. This curve can be approximated by sinusoidal curves of varying periods and amplitudes. A first approximation to the sawtooth curve is given by

$$y = \frac{1}{2}\sin(2\pi x) + \frac{1}{4}\sin(4\pi x)$$

Show that $y = \sin(2\pi x)\cos^2(\pi x)$.

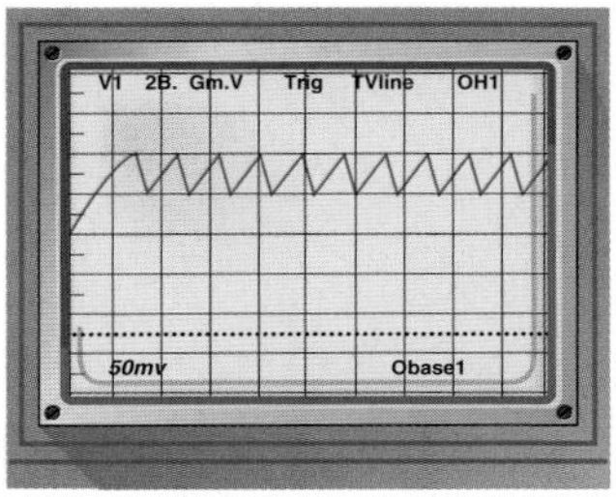

85. Go to the library and research Chebyshëv polynomials. Write a report on your findings.

6.6 Product-to-Sum and Sum-to-Product Formulas

OBJECTIVES 1 Express Products as Sums
2 Express Sums as Products

1 Sum and difference formulas can be used to derive formulas for writing the products of sines and/or cosines as sums or differences. These identities are usually called the **Product-to-Sum Formulas.**

Theorem **Product-to-Sum Formulas**

$$\sin\alpha\sin\beta = \frac{1}{2}[\cos(\alpha - \beta) - \cos(\alpha + \beta)] \quad (1)$$

$$\cos\alpha\cos\beta = \frac{1}{2}[\cos(\alpha - \beta) + \cos(\alpha + \beta)] \quad (2)$$

$$\sin\alpha\cos\beta = \frac{1}{2}[\sin(\alpha + \beta) + \sin(\alpha - \beta)] \quad (3)$$

These formulas do not have to be memorized. Instead, you should remember how they are derived. Then, when you want to use them, either look them up or derive them, as needed.

To derive formulas (1) and (2), write down the sum and difference formulas for the cosine:

$$\cos(\alpha - \beta) = \cos\alpha\cos\beta + \sin\alpha\sin\beta \tag{4}$$

$$\cos(\alpha + \beta) = \cos\alpha\cos\beta - \sin\alpha\sin\beta \tag{5}$$

Subtract equation (5) from equation (4) to get

$$\cos(\alpha - \beta) - \cos(\alpha + \beta) = 2\sin\alpha\sin\beta$$

from which

$$\sin\alpha\sin\beta = \frac{1}{2}[\cos(\alpha - \beta) - \cos(\alpha + \beta)]$$

Now add equations (4) and (5) to get

$$\cos(\alpha - \beta) + \cos(\alpha + \beta) = 2\cos\alpha\cos\beta$$

from which

$$\cos\alpha\cos\beta = \frac{1}{2}[\cos(\alpha - \beta) + \cos(\alpha + \beta)]$$

To derive Product-to-Sum Formula (3), use the sum and difference formulas for sine in a similar way. (You are asked to do this in Problem 41.)

EXAMPLE 1 Expressing Products as Sums

Express each of the following products as a sum containing only sines or cosines.

(a) $\sin(6\theta)\sin(4\theta)$ (b) $\cos(3\theta)\cos\theta$ (c) $\sin(3\theta)\cos(5\theta)$

Solution (a) We use formula (1) to get

$$\begin{aligned}\sin(6\theta)\sin(4\theta) &= \frac{1}{2}[\cos(6\theta - 4\theta) - \cos(6\theta + 4\theta)]\\ &= \frac{1}{2}[\cos(2\theta) - \cos(10\theta)]\end{aligned}$$

(b) We use formula (2) to get

$$\begin{aligned}\cos(3\theta)\cos\theta &= \frac{1}{2}[\cos(3\theta - \theta) + \cos(3\theta + \theta)]\\ &= \frac{1}{2}[\cos(2\theta) + \cos(4\theta)]\end{aligned}$$

(c) We use formula (3) to get

$$\begin{aligned}\sin(3\theta)\cos(5\theta) &= \frac{1}{2}[\sin(3\theta + 5\theta) + \sin(3\theta - 5\theta)]\\ &= \frac{1}{2}[\sin(8\theta) + \sin(-2\theta)] = \frac{1}{2}[\sin(8\theta) - \sin(2\theta)]\end{aligned}$$ ◄

NOW WORK PROBLEM 1.

2 The **Sum-to-Product Formulas** are given next.

Theorem

Sum-to-Product Formulas

$$\sin\alpha + \sin\beta = 2\sin\frac{\alpha+\beta}{2}\cos\frac{\alpha-\beta}{2} \quad (6)$$

$$\sin\alpha - \sin\beta = 2\sin\frac{\alpha-\beta}{2}\cos\frac{\alpha+\beta}{2} \quad (7)$$

$$\cos\alpha + \cos\beta = 2\cos\frac{\alpha+\beta}{2}\cos\frac{\alpha-\beta}{2} \quad (8)$$

$$\cos\alpha - \cos\beta = -2\sin\frac{\alpha+\beta}{2}\sin\frac{\alpha-\beta}{2} \quad (9)$$

We will derive formula (6) and leave the derivations of formulas (7) through (9) as exercises (see Problems 42 through 44).

Proof

$$2\sin\frac{\alpha+\beta}{2}\cos\frac{\alpha-\beta}{2} \underset{\uparrow}{=} 2\cdot\frac{1}{2}\left[\sin\left(\frac{\alpha+\beta}{2}+\frac{\alpha-\beta}{2}\right)+\sin\left(\frac{\alpha+\beta}{2}-\frac{\alpha-\beta}{2}\right)\right]$$

Product-to-Sum Formula (3)

$$= \sin\frac{2\alpha}{2} + \sin\frac{2\beta}{2} = \sin\alpha + \sin\beta$$

■

EXAMPLE 2 Expressing Sums (or Differences) as a Product

Express each sum or difference as a product of sines and/or cosines.

(a) $\sin(5\theta) - \sin(3\theta)$ (b) $\cos(3\theta) + \cos(2\theta)$

Solution (a) We use formula (7) to get

$$\sin(5\theta) - \sin(3\theta) = 2\sin\frac{5\theta-3\theta}{2}\cos\frac{5\theta+3\theta}{2}$$
$$= 2\sin\theta\cos(4\theta)$$

(b) $$\cos(3\theta) + \cos(2\theta) = 2\cos\frac{3\theta+2\theta}{2}\cos\frac{3\theta-2\theta}{2} \quad \text{Formula (8)}$$
$$= 2\cos\frac{5\theta}{2}\cos\frac{\theta}{2}$$

◄

NOW WORK PROBLEM 11.

6.6 Assess Your Understanding

Exercises

In Problems 1–10, express each product as a sum containing only sines or cosines.

1. $\sin(4\theta)\sin(2\theta)$
2. $\cos(4\theta)\cos(2\theta)$
3. $\sin(4\theta)\cos(2\theta)$
4. $\sin(3\theta)\sin(5\theta)$
5. $\cos(3\theta)\cos(5\theta)$
6. $\sin(4\theta)\cos(6\theta)$
7. $\sin\theta\sin(2\theta)$
8. $\cos(3\theta)\cos(4\theta)$
9. $\sin\frac{3\theta}{2}\cos\frac{\theta}{2}$
10. $\sin\frac{\theta}{2}\cos\frac{5\theta}{2}$

In Problems 11–18, express each sum or difference as a product of sines and/or cosines.

11. $\sin(4\theta) - \sin(2\theta)$ **12.** $\sin(4\theta) + \sin(2\theta)$ **13.** $\cos(2\theta) + \cos(4\theta)$ **14.** $\cos(5\theta) - \cos(3\theta)$

15. $\sin\theta + \sin(3\theta)$ **16.** $\cos\theta + \cos(3\theta)$ **17.** $\cos\frac{\theta}{2} - \cos\frac{3\theta}{2}$ **18.** $\sin\frac{\theta}{2} - \sin\frac{3\theta}{2}$

In Problems 19–36, establish each identity.

19. $\dfrac{\sin\theta + \sin(3\theta)}{2\sin(2\theta)} = \cos\theta$ **20.** $\dfrac{\cos\theta + \cos(3\theta)}{2\cos(2\theta)} = \cos\theta$ **21.** $\dfrac{\sin(4\theta) + \sin(2\theta)}{\cos(4\theta) + \cos(2\theta)} = \tan(3\theta)$

22. $\dfrac{\cos\theta - \cos(3\theta)}{\sin(3\theta) - \sin\theta} = \tan(2\theta)$ **23.** $\dfrac{\cos\theta - \cos(3\theta)}{\sin\theta + \sin(3\theta)} = \tan\theta$ **24.** $\dfrac{\cos\theta - \cos(5\theta)}{\sin\theta + \sin(5\theta)} = \tan(2\theta)$

25. $\sin\theta[\sin\theta + \sin(3\theta)] = \cos\theta[\cos\theta - \cos(3\theta)]$

26. $\sin\theta[\sin(3\theta) + \sin(5\theta)] = \cos\theta[\cos(3\theta) - \cos(5\theta)]$

27. $\dfrac{\sin(4\theta) + \sin(8\theta)}{\cos(4\theta) + \cos(8\theta)} = \tan(6\theta)$

28. $\dfrac{\sin(4\theta) - \sin(8\theta)}{\cos(4\theta) - \cos(8\theta)} = -\cot(6\theta)$

29. $\dfrac{\sin(4\theta) + \sin(8\theta)}{\sin(4\theta) - \sin(8\theta)} = -\dfrac{\tan(6\theta)}{\tan(2\theta)}$

30. $\dfrac{\cos(4\theta) - \cos(8\theta)}{\cos(4\theta) + \cos(8\theta)} = \tan(2\theta)\tan(6\theta)$

31. $\dfrac{\sin\alpha + \sin\beta}{\sin\alpha - \sin\beta} = \tan\dfrac{\alpha+\beta}{2}\cot\dfrac{\alpha-\beta}{2}$

32. $\dfrac{\cos\alpha + \cos\beta}{\cos\alpha - \cos\beta} = -\cot\dfrac{\alpha+\beta}{2}\cot\dfrac{\alpha-\beta}{2}$

33. $\dfrac{\sin\alpha + \sin\beta}{\cos\alpha + \cos\beta} = \tan\dfrac{\alpha+\beta}{2}$

34. $\dfrac{\sin\alpha - \sin\beta}{\cos\alpha - \cos\beta} = -\cot\dfrac{\alpha+\beta}{2}$

35. $1 + \cos(2\theta) + \cos(4\theta) + \cos(6\theta) = 4\cos\theta\cos(2\theta)\cos(3\theta)$

36. $1 - \cos(2\theta) + \cos(4\theta) - \cos(6\theta) = 4\sin\theta\cos(2\theta)\sin(3\theta)$

---

37. Touch-Tone Phones On a Touch-Tone phone, each button produces a unique sound. The sound produced is the sum of two tones, given by

$$y = \sin(2\pi lt) \quad \text{and} \quad y = \sin(2\pi ht)$$

where l and h are the low and high frequencies (cycles per second) shown on the illustration. For example, if you touch 7, the low frequency is $l = 852$ cycles per second and the high frequency is $h = 1209$ cycles per second. The sound emitted by touching 7 is

$$y = \sin[2\pi(852)t] + \sin[2\pi(1209)t]$$

Touch-Tone phone

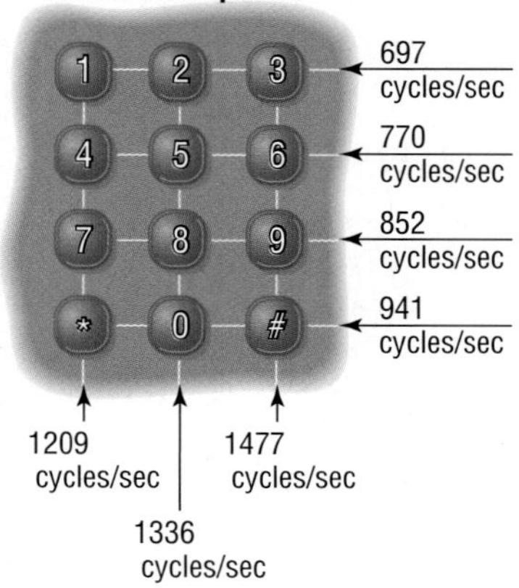

(a) Write this sound as a product of sines and/or cosines.
(b) Determine the maximum value of y.
(c) Graph the sound emitted by touching 7.

38. Touch-Tone Phones
(a) Write the sound emitted by touching the # key as a product of sines and/or cosines.
(b) Determine the maximum value of y.
(c) Graph the sound emitted by touching the # key.

39. If $\alpha + \beta + \gamma = \pi$, show that

$$\sin(2\alpha) + \sin(2\beta) + \sin(2\gamma) = 4\sin\alpha\sin\beta\sin\gamma$$

40. If $\alpha + \beta + \gamma = \pi$, show that

$$\tan\alpha + \tan\beta + \tan\gamma = \tan\alpha\tan\beta\tan\gamma$$

41. Derive formula (3).

42. Derive formula (7).

43. Derive formula (8).

44. Derive formula (9).

6.7 Trigonometric Equations (I)

PREPARING FOR THIS SECTION *Before getting started, review the following:*

- Solving Equations (Appendix A, Section A.5, pp. 936–945)
- Values of the Trigonometric Functions (Section 5.2, pp. 342–351)

Now work the 'Are You Prepared?' problems on page 465.

OBJECTIVE 1 Solve Equations Involving a Single Trigonometric Function

1 The previous four sections of this chapter were devoted to trigonometric identities, that is, equations involving trigonometric functions that are satisfied by every value in the domain of the variable. In the remaining two sections, we discuss **trigonometric equations,** that is, equations involving trigonometric functions that are satisfied only by some values of the variable (or, possibly, are not satisfied by any values of the variable). The values that satisfy the equation are called **solutions** of the equation.

EXAMPLE 1 **Checking Whether a Given Number Is a Solution of a Trigonometric Equation**

Determine whether $\theta = \frac{\pi}{4}$ is a solution of the equation $\sin\theta = \frac{1}{2}$. Is $\theta = \frac{\pi}{6}$ a solution?

Solution Replace θ by $\frac{\pi}{4}$ in the given equation. The result is

$$\sin\frac{\pi}{4} = \frac{\sqrt{2}}{2} \neq \frac{1}{2}$$

We conclude that $\frac{\pi}{4}$ is not a solution.

Next replace θ by $\frac{\pi}{6}$ in the equation. The result is

$$\sin\frac{\pi}{6} = \frac{1}{2}$$

We conclude that $\frac{\pi}{6}$ is a solution of the given equation. ◀

The equation given in Example 1 has other solutions besides $\theta = \frac{\pi}{6}$. For example, $\theta = \frac{5\pi}{6}$ is also a solution, as is $\theta = \frac{13\pi}{6}$. (You should check this for yourself.) In fact, the equation has an infinite number of solutions due to the periodicity of the sine function, as can be seen in Figure 27.

Figure 27

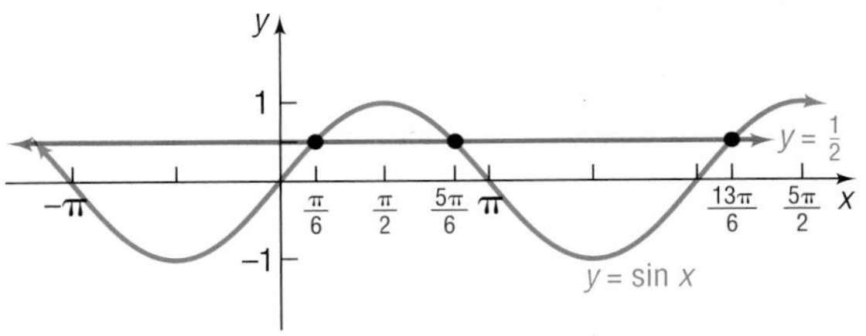

Unless the domain of the variable is restricted, we need to find *all* the solutions of a trigonometric equation. As the next example illustrates, finding all the solutions can be accomplished by first finding solutions over an interval whose length equals the period of the function and then adding multiples of that period to the solutions found. Let's look at some examples.

EXAMPLE 2 Finding the Solutions of a Trigonometric Equation

Solve the equation: $\cos\theta = \frac{1}{2}$

Give a general formula for all the solutions. List six solutions.

Solution The period of the cosine function is 2π. In the interval $[0, 2\pi)$, there are two angles θ for which $\cos\theta = \frac{1}{2}$: $\theta = \frac{\pi}{3}$ and $\theta = \frac{5\pi}{3}$. See Figure 28.

Figure 28

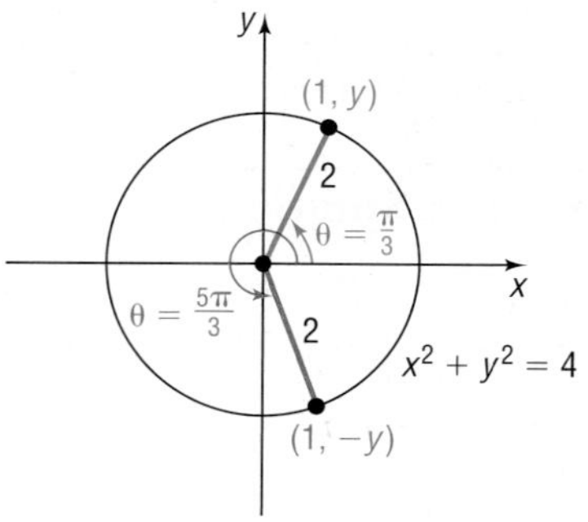

Because the cosine function has period 2π, all the solutions of $\cos\theta = \frac{1}{2}$ may be given by the general formula

$$\theta = \frac{\pi}{3} + 2k\pi \quad \text{or} \quad \theta = \frac{5\pi}{3} + 2k\pi \qquad k \text{ any integer}$$

Some of the solutions are

$$\underbrace{-\frac{5\pi}{3}, -\frac{\pi}{3}}_{k=-1}, \underbrace{\frac{\pi}{3}, \frac{5\pi}{3}}_{k=0}, \underbrace{\frac{7\pi}{3}, \frac{11\pi}{3}}_{k=1}, \underbrace{\frac{13\pi}{3}, \frac{17\pi}{3}}_{k=2}, \text{ and so on}$$

◀

Figure 29

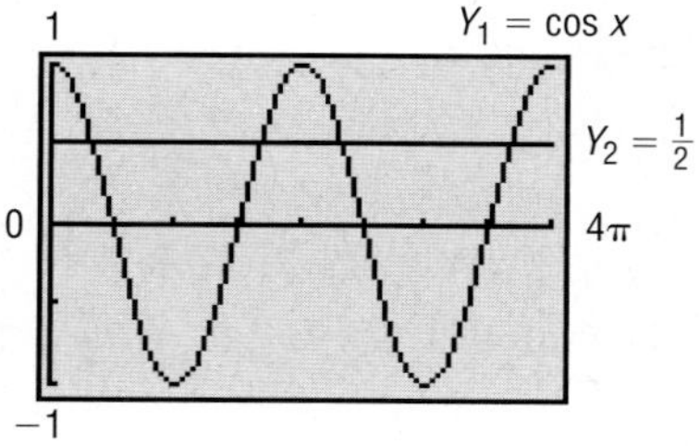

CHECK: We can verify the solutions by graphing $Y_1 = \cos x$ and $Y_2 = \frac{1}{2}$ to determine where the graphs intersect. (Be sure to graph in radian mode.) See Figure 29. The graph of Y_1 intersects the graph of Y_2 at $x = 1.05\left(\approx \frac{\pi}{3}\right)$, $5.24\left(\approx \frac{5\pi}{3}\right)$, $7.33\left(\approx \frac{7\pi}{3}\right)$, and $11.52\left(\approx \frac{11\pi}{3}\right)$, rounded to two decimal places.

NOW WORK PROBLEM 31.

In most of our work, we shall be interested only in finding solutions of trigonometric equations for $0 \le \theta < 2\pi$.

EXAMPLE 3 Solving a Linear Trigonometric Equation

Solve the equation: $2\sin\theta + \sqrt{3} = 0, \quad 0 \le \theta < 2\pi$

Solution We solve the equation for $\sin\theta$.

$$2\sin\theta + \sqrt{3} = 0$$
$$2\sin\theta = -\sqrt{3} \qquad \text{Subtract } \sqrt{3} \text{ from both sides.}$$
$$\sin\theta = -\frac{\sqrt{3}}{2} \qquad \text{Divide both sides by 2.}$$

The period of the sine function is 2π. In the interval $[0, 2\pi)$, there are two angles θ for which $\sin\theta = -\frac{\sqrt{3}}{2}$: $\theta = \frac{4\pi}{3}$ and $\theta = \frac{5\pi}{3}$. ◀

NOW WORK PROBLEM 7.

EXAMPLE 4 Solving a Trigonometric Equation

Solve the equation: $\sin(2\theta) = \frac{1}{2}, \quad 0 \le \theta < 2\pi$

Solution The period of the sine function is 2π. In the interval $[0, 2\pi)$, the sine function has a value $\frac{1}{2}$ at $\frac{\pi}{6}$ and $\frac{5\pi}{6}$. See Figure 30. Consequently, because the argument is 2θ in the equation $\sin(2\theta) = \frac{1}{2}$, we have

Figure 30

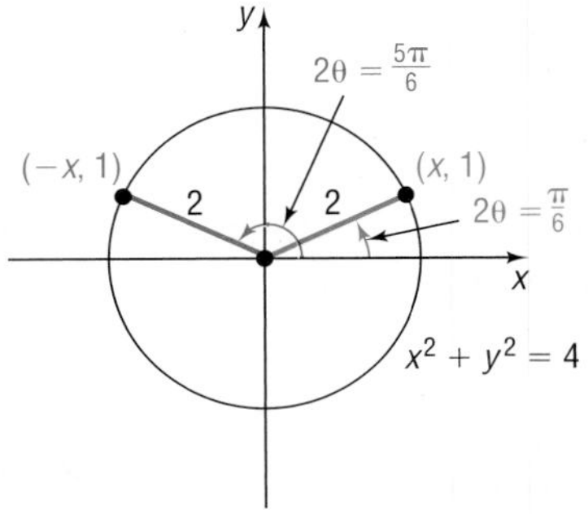

$$2\theta = \frac{\pi}{6} + 2k\pi \quad \text{or} \quad 2\theta = \frac{5\pi}{6} + 2k\pi \qquad k \text{ any integer}$$

$$\theta = \frac{\pi}{12} + k\pi \qquad\qquad \theta = \frac{5\pi}{12} + k\pi \qquad \text{Divide by 2.}$$

Then

$$\theta = \frac{\pi}{12} + (-1)\pi = \frac{-11\pi}{12} \qquad k = -1 \qquad \theta = \frac{5\pi}{12} + (-1)\pi = \frac{-7\pi}{12}$$

$$\theta = \frac{\pi}{12} + (0)\pi = \frac{\pi}{12} \qquad k = 0 \qquad \theta = \frac{5\pi}{12} + (0)\pi = \frac{5\pi}{12}$$

$$\theta = \frac{\pi}{12} + (1)\pi = \frac{13\pi}{12} \qquad k = 1 \qquad \theta = \frac{5\pi}{12} + (1)\pi = \frac{17\pi}{12}$$

$$\theta = \frac{\pi}{12} + (2)\pi = \frac{25\pi}{12} \qquad k = 2 \qquad \theta = \frac{5\pi}{12} + (2)\pi = \frac{29\pi}{12}$$

In the interval $[0, 2\pi)$, the solutions of $\sin(2\theta) = \frac{1}{2}$ are $\theta = \frac{\pi}{12}$, $\theta = \frac{5\pi}{12}$, $\theta = \frac{13\pi}{12}$, and $\theta = \frac{17\pi}{12}$. ◀

CHECK: Verify these solutions by graphing $Y_1 = \sin(2x)$ and $Y_2 = \frac{1}{2}$ for $0 \le x \le 2\pi$.

WARNING: In solving a trigonometric equation for θ, $0 \le \theta < 2\pi$, in which the argument is not θ (as in Example 4), you must write down all the solutions first and then list those that are in the interval $[0, 2\pi)$. Otherwise, solutions may be lost. For example, in solving $\sin(2\theta) = \frac{1}{2}$, if you merely write the solutions $2\theta = \frac{\pi}{6}$ and $2\theta = \frac{5\pi}{6}$, you will find only $\theta = \frac{\pi}{12}$ and $\theta = \frac{5\pi}{12}$ and miss the other solutions. ■

NOW WORK PROBLEM 13.

EXAMPLE 5 Solving a Trigonometric Equation

Solve the equation: $\tan\left(\theta - \frac{\pi}{2}\right) = 1, \quad 0 \le \theta < 2\pi$

Solution The period of the tangent function is π. In the interval $[0, \pi)$, the tangent function has the value 1 when the argument is $\frac{\pi}{4}$. Because the argument is $\theta - \frac{\pi}{2}$ in the given equation, we have

$$\theta - \frac{\pi}{2} = \frac{\pi}{4} + k\pi \qquad k \text{ any integer}$$

$$\theta = \frac{3\pi}{4} + k\pi$$

In the interval $[0, 2\pi)$, $\theta = \frac{3\pi}{4}$ and $\theta = \frac{3\pi}{4} + \pi = \frac{7\pi}{4}$ are the only solutions. ◀

CHECK: Verify these solutions using a graphing utility.

The next example illustrates how to solve trigonometric equations using a calculator. Remember that the function keys on a calculator will only give values consistent with the definition of the function.

EXAMPLE 6 Solving a Trigonometric Equation with a Calculator

Use a calculator to solve the equation: $\sin\theta = 0.3, \quad 0 \le \theta < 2\pi$
Express any solutions in radians, rounded to two decimal places.

Solution To solve $\sin\theta = 0.3$ on a calculator, first set the mode to radians. Then use the $\boxed{\sin^{-1}}$ key to obtain

$$\theta = \sin^{-1}(0.3) \approx 0.3046927$$

Figure 31

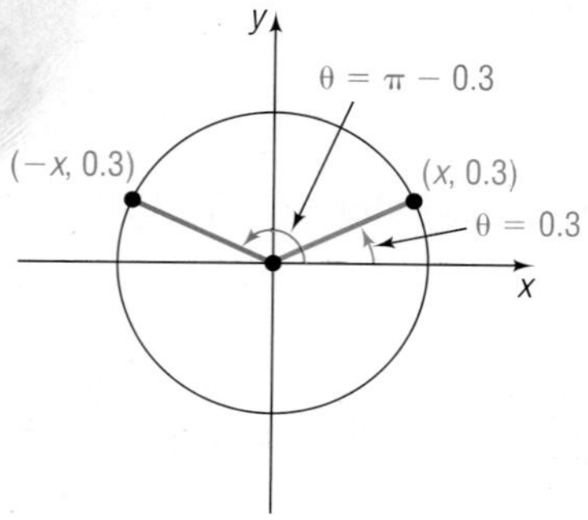

Rounded to two decimal places, $\theta = \sin^{-1}(0.3) = 0.30$ radian. Because of the definition of $y = \sin^{-1} x$, the angle θ that we obtain is the angle $-\frac{\pi}{2} \le \theta \le \frac{\pi}{2}$ for which $\sin\theta = 0.3$. Another angle for which $\sin\theta = 0.3$ is $\pi - 0.30$. See Figure 31. The angle $\pi - 0.30$ is the angle in quadrant II, where $\sin\theta = 0.3$. The solutions for $\sin\theta = 0.3$, $0 \le \theta < 2\pi$, are

$$\theta = 0.30 \text{ radian} \quad \text{and} \quad \theta = \pi - 0.30 \approx 2.84 \text{ radians}$$ ◀

WARNING: Example 6 illustrates that caution must be exercised when solving trigonometric equations on a calculator. Remember that the calculator supplies an angle only within the restrictions of the definition of the inverse trigonometric function. To find the remaining solutions, you must identify other quadrants, if any, in which a solution may be located. ■

NOW WORK PROBLEM 41.

6.7 Assess Your Understanding

'Are You Prepared?' *Answers are given at the end of these exercises. If you get a wrong answer, read the pages listed in red.*

1. Solve: $3x - 5 = -x + 1$. (pp. 936–945)

2. $\sin\left(\frac{\pi}{4}\right) =$ ________; $\cos\left(\frac{8\pi}{3}\right) =$ ________. (pp. 342–351)

Concepts and Vocabulary

3. Two solutions of the equation $\sin\theta = \frac{1}{2}$ are ________ and ________.

4. All the solutions of the equation $\sin\theta = \frac{1}{2}$ are ________.

5. *True or False:* Most trigonometric equations have unique solutions.

6. *True or False:* The equation $\sin\theta = 2$ has a real solution that can be found using a graphing calculator.

Exercises

In Problems 7–30, solve each equation on the interval $0 \le \theta < 2\pi$.

7. $2\sin\theta + 3 = 2$
8. $1 - \cos\theta = \frac{1}{2}$
9. $4\cos^2\theta = 1$
10. $\tan^2\theta = \frac{1}{3}$
11. $2\sin^2\theta - 1 = 0$
12. $4\cos^2\theta - 3 = 0$
13. $\sin(3\theta) = -1$
14. $\tan\frac{\theta}{2} = \sqrt{3}$
15. $\cos(2\theta) = -\frac{1}{2}$
16. $\tan(2\theta) = -1$
17. $\sec\frac{3\theta}{2} = -2$
18. $\cot\frac{2\theta}{3} = -\sqrt{3}$
19. $2\sin\theta + 1 = 0$
20. $\cos\theta + 1 = 0$
21. $\tan\theta + 1 = 0$
22. $\sqrt{3}\cot\theta + 1 = 0$
23. $4\sec\theta + 6 = -2$
24. $5\csc\theta - 3 = 2$
25. $3\sqrt{2}\cos\theta + 2 = -1$
26. $4\sin\theta + 3\sqrt{3} = \sqrt{3}$
27. $\cos\left(2\theta - \frac{\pi}{2}\right) = -1$
28. $\sin\left(3\theta + \frac{\pi}{18}\right) = 1$
29. $\tan\left(\frac{\theta}{2} + \frac{\pi}{3}\right) = 1$
30. $\cos\left(\frac{\theta}{3} - \frac{\pi}{4}\right) = \frac{1}{2}$

In Problems 31–40, solve each equation. Give a general formula for all the solutions. List six solutions.

31. $\sin\theta = \frac{1}{2}$
32. $\tan\theta = 1$
33. $\tan\theta = -\frac{\sqrt{3}}{3}$
34. $\cos\theta = -\frac{\sqrt{3}}{2}$
35. $\cos\theta = 0$
36. $\sin\theta = \frac{\sqrt{2}}{2}$
37. $\cos(2\theta) = -\frac{1}{2}$
38. $\sin(2\theta) = -1$
39. $\sin\frac{\theta}{2} = -\frac{\sqrt{3}}{2}$
40. $\tan\frac{\theta}{2} = -1$

In Problems 41–52, use a calculator to solve each equation on the interval $0 \le \theta < 2\pi$. Round answers to two decimal places.

41. $\sin\theta = 0.4$
42. $\cos\theta = 0.6$
43. $\tan\theta = 5$
44. $\cot\theta = 2$
45. $\cos\theta = -0.9$
46. $\sin\theta = -0.2$
47. $\sec\theta = -4$
48. $\csc\theta = -3$
49. $5\tan\theta + 9 = 0$
50. $4\cot\theta = -5$
51. $3\sin\theta - 2 = 0$
52. $4\cos\theta + 3 = 0$

53. Suppose that $f(x) = 3\sin x$.
(a) Solve $f(x) = \frac{3}{2}$.
(b) For what values of x is $f(x) > \frac{3}{2}$ on the interval $[0, 2\pi)$?

54. Suppose that $f(x) = 2\cos x$.
(a) Solve $f(x) = -\sqrt{3}$.
(b) For what values of x is $f(x) < -\sqrt{3}$ on the interval $[0, 2\pi)$?

55. Suppose that $f(x) = 4\tan x$.
(a) Solve $f(x) = -4$.
(b) For what values of x is $f(x) < -4$ on the interval $\left(-\frac{\pi}{2}, \frac{\pi}{2}\right)$?

56. Suppose that $f(x) = \cot x$.
(a) Solve $f(x) = -\sqrt{3}$.
(b) For what values of x is $f(x) > -\sqrt{3}$ on the interval $(0, \pi)$?

57. The Ferris Wheel In 1893, George Ferris engineered the Ferris Wheel. It was 250 feet in diameter. If the wheel makes 1 revolution every 40 seconds, then

$$h(t) = 125\sin\left(0.157t - \frac{\pi}{2}\right) + 125$$

represents the height h, in feet, of a seat on the wheel as a function of time t, where t is measured in seconds. The ride begins when $t = 0$.
(a) During the first 40 seconds of the ride, at what time t is an individual on the Ferris Wheel exactly 125 feet above the ground?

(b) During the first 80 seconds of the ride, at what time t is an individual on the Ferris Wheel exactly 250 feet above the ground?
(c) During the first 40 seconds of the ride, over what interval of time t is an individual on the Ferris Wheel more than 125 feet above the ground?

58. Tire Rotation The P215/65R15 Cobra Radial G/T tire has a diameter of exactly 26 inches. Suppose that a car's wheel is making 2 revolutions per second (the car is traveling a little less than 5 miles per hour). Then

$$h(t) = 13 \sin\left(4\pi t - \frac{\pi}{2}\right) + 13$$

represents the height h (in inches) of a point on the tire as a function of time t (in seconds). The car starts to move when $t = 0$.
(a) During the first second that the car is moving, at what time t is the point on the tire exactly 13 inches above the ground?
(b) During the first second that the car is moving, at what time t is the point on the tire exactly 6.5 inches above the ground?
(c) During the first second that the car is moving, at what time t is the point on the tire more than 13 inches above the ground?

SOURCE: Cobra Tire

59. Holding Pattern Suppose that an airplane is asked to stay within a holding pattern near Chicago's O'Hare International Airport. The function $d(x) = 70 \sin(0.65x) + 150$ represents the distance d, in miles, that the airplane is from the airport at time x, in minutes.
(a) When the plane enters the holding pattern, $x = 0$, how far is it from O'Hare?
(b) During the first 20 minutes after the plane enters the holding pattern, at what time x is the plane exactly 100 miles from the airport?
(c) During the first 20 minutes after the plane enters the holding pattern, at what time x is the plane more than 100 miles from the airport?
(d) While the plane is in the holding pattern, will it ever be within 70 miles of the airport? Why?

60. Projectile Motion A golfer hits a golf ball with an initial velocity of 100 miles per hour. The range R of the ball as a function of the angle θ to the horizontal is given by $R(\theta) = 672 \sin(2\theta)$, where R is measured in feet.
(a) At what angle θ should the ball be hit if the golfer wants the ball to travel 450 feet (150 yards)?
(b) At what angle θ should the ball be hit if the golfer wants the ball to travel 540 feet (180 yards)?
(c) At what angle θ should the ball be hit if the golfer wants the ball to travel at least 480 feet (160 yards)?
(d) Can the golfer hit the ball 720 feet (240 yards)?

The following discussion of Snell's Law of Refraction (named after Willebrord Snell, 1580–1626) is needed for Problems 61–67. Light, sound, and other waves travel at different speeds, depending on the media (air, water, wood, and so on) through which they pass. Suppose that light travels from a point A in one medium, where its speed is v_1, to a point B in another medium, where its speed is v_2. Refer to the figure, where the angle θ_1 is called the **angle of incidence** *and the angle θ_2 is the* **angle of refraction.** *Snell's Law,* which can be proved using calculus, states that*

$$\frac{\sin \theta_1}{\sin \theta_2} = \frac{v_1}{v_2}$$

The ratio $\frac{v_1}{v_2}$ is called the **index of refraction.** *Some values are given in the following table.*

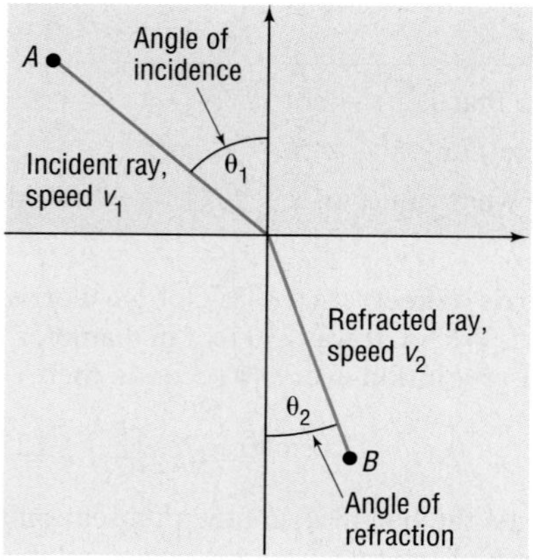

Some Indexes of Refraction

Medium	Index of Refraction†
Water	1.33
Ethyl alcohol	1.36
Carbon disulfide	1.63
Air (1 atm and 20°C)	1.0003
Methylene iodide	1.74
Fused quartz	1.46
Glass, crown	1.52
Glass, dense flint	1.66
Sodium chloride	1.54

†For light of wavelength 589 nanometers, measured with respect to a vacuum. The index with respect to air is negligibly different in most cases.

*Because this law was also deduced by René Descartes in France, it is also known as Descartes' Law.

61. The index of refraction of light in passing from a vacuum into water is 1.33. If the angle of incidence is 40°, determine the angle of refraction.

62. The index of refraction of light in passing from a vacuum into dense glass is 1.66. If the angle of incidence is 50°, determine the angle of refraction.

63. Ptolemy, who lived in the city of Alexandria in Egypt during the second century AD, gave the measured values in the table below for the angle of incidence θ_1 and the angle of refraction θ_2 for a light beam passing from air into water. Do these values agree with Snell's Law? If so, what index of refraction results? (These data are interesting as the oldest recorded physical measurements.)*

θ_1	θ_2	θ_1	θ_2
10°	7°45′	50°	35°0′
20°	15°30′	60°	40°30′
30°	22°30′	70°	45°30′
40°	29°0′	80°	50°0′

64. The speed of yellow sodium light (wavelength of 589 nanometers) in a certain liquid is measured to be 1.92×10^8 meters per second. What is the index of refraction of this liquid, with respect to air, for sodium light?†

[**Hint:** The speed of light in air is approximately 2.997×10^8 meters per second.]

65. A beam of light with a wavelength of 589 nanometers traveling in air makes an angle of incidence of 40° on a slab of transparent material, and the refracted beam makes an angle of refraction of 26°. Find the index of refraction of the material.*

66. A light ray with a wavelength of 589 nanometers (produced by a sodium lamp) traveling through air makes an angle of incidence of 30° on a smooth, flat slab of crown glass. Find the angle of refraction.*

67. A light beam passes through a thick slab of material whose index of refraction is n_2. Show that the emerging beam is parallel to the incident beam.†

68. Explain in your own words how you would use your calculator to solve the equation $\sin x = 0.3$, $0 \le x < 2\pi$. How would you modify your approach in order to solve the equation $\cot x = 5$, $0 < x < 2\pi$?

'Are You Prepared?' Answers

1. $\left\{\frac{3}{2}\right\}$ **2.** $\frac{\sqrt{2}}{2}; -\frac{1}{2}$

*Adapted from Halliday and Resnick, *Physics, Parts 1 & 2*, 3rd ed. New York: Wiley.

†*Physics for Scientists & Engineers* 3/E by Serway. ©1990. Reprinted with permission of Brooks/Cole, a division of Thomson Learning: www.thomsonrights.com. Fax 800-730-2215.

6.8 Trigonometric Equations (II)

PREPARING FOR THIS SECTION *Before getting started, review the following:*

- Solving Quadratic Equations by Factoring (Appendix A, Section A.5, pp. 939–940)
- The Quadratic Formula (Appendix A, Section A.5, pp. 943–945)

Now work the 'Are You Prepared?' problems on page 473.

OBJECTIVES
1 Solve Trigonometric Equations Quadratic in Form
2 Solve Trigonometric Equations Using Identities
3 Solve Trigonometric Equations Linear in Sine and Cosine
4 Solve Trigonometric Equations Using a Graphing Utility

1 In this section we continue our study of trigonometric equations. Many trigonometric equations can be solved by applying techniques that we already know, such as applying the quadratic formula (if the equation is a second-degree polynomial) or factoring.

EXAMPLE 1 Solving a Trigonometric Equation Quadratic in Form

Solve the equation: $2\sin^2\theta - 3\sin\theta + 1 = 0, \quad 0 \le \theta < 2\pi$

Solution The equation that we wish to solve is a quadratic equation (in $\sin\theta$) that can be factored.

$$\begin{aligned} 2\sin^2\theta - 3\sin\theta + 1 &= 0 \qquad 2x^2 - 3x + 1 = 0,\quad x = \sin\theta \\ (2\sin\theta - 1)(\sin\theta - 1) &= 0 \qquad (2x - 1)(x - 1) = 0 \\ 2\sin\theta - 1 &= 0 \quad \text{or} \quad \sin\theta - 1 = 0 \\ \sin\theta &= \frac{1}{2} \qquad\qquad \sin\theta = 1 \end{aligned}$$

Solving each equation in the interval $[0, 2\pi)$, we obtain

$$\theta = \frac{\pi}{6}, \qquad \theta = \frac{5\pi}{6}, \qquad \theta = \frac{\pi}{2}$$

NOW WORK PROBLEM 5.

2 When a trigonometric equation contains more than one trigonometric function, identities sometimes can be used to obtain an equivalent equation that contains only one trigonometric function.

EXAMPLE 2 Solving a Trigonometric Equation Using Identities

Solve the equation: $3\cos\theta + 3 = 2\sin^2\theta,\quad 0 \le \theta < 2\pi$

Solution The equation in its present form contains sines and cosines. However, a form of the Pythagorean Identity can be used to transform the equation into an equivalent expression containing only cosines.

$$\begin{aligned} 3\cos\theta + 3 &= 2\sin^2\theta \\ 3\cos\theta + 3 &= 2(1 - \cos^2\theta) \qquad \sin^2\theta = 1 - \cos^2\theta \\ 3\cos\theta + 3 &= 2 - 2\cos^2\theta \\ 2\cos^2\theta + 3\cos\theta + 1 &= 0 \qquad \text{Quadratic in } \cos\theta \\ (2\cos\theta + 1)(\cos\theta + 1) &= 0 \qquad \text{Factor.} \\ 2\cos\theta + 1 &= 0 \quad \text{or} \quad \cos\theta + 1 = 0 \\ \cos\theta &= -\frac{1}{2} \qquad\qquad \cos\theta = -1 \end{aligned}$$

Solving each equation in the interval $[0, 2\pi)$, we obtain

$$\theta = \frac{2\pi}{3}, \qquad \theta = \frac{4\pi}{3}, \qquad \theta = \pi$$

CHECK: Graph $Y_1 = 3\cos x + 3$ and $Y_2 = 2\sin^2 x$, $0 \le x \le 2\pi$, and find the points of intersection. How close are your approximate solutions to the exact ones found in this example?

EXAMPLE 3 Solving a Trigonometric Equation Using Identities

Solve the equation: $\cos(2\theta) + 3 = 5\cos\theta,\quad 0 \le \theta < 2\pi$

Solution First, we observe that the given equation contains two cosine functions, but with different arguments, θ and 2θ. We use the Double-Angle Formula

$\cos(2\theta) = 2\cos^2\theta - 1$ to obtain an equivalent equation containing only $\cos\theta$.

$$\begin{aligned} \cos(2\theta) + 3 &= 5\cos\theta \\ (2\cos^2\theta - 1) + 3 &= 5\cos\theta \quad && \cos(2\theta) = 2\cos^2\theta - 1 \\ 2\cos^2\theta - 5\cos\theta + 2 &= 0 && \text{Place in standard form.} \\ (\cos\theta - 2)(2\cos\theta - 1) &= 0 && \text{Factor.} \\ \cos\theta = 2 \quad \text{or} \quad \cos\theta &= \frac{1}{2} \end{aligned}$$

For any angle θ, $-1 \le \cos\theta \le 1$; therefore, the equation $\cos\theta = 2$ has no solution. The solutions of $\cos\theta = \frac{1}{2}$, $0 \le \theta < 2\pi$, are

$$\theta = \frac{\pi}{3}, \qquad \theta = \frac{5\pi}{3}$$ ◀

CHECK: Graph $Y_1 = \cos(2x) + 3$ and $Y_2 = 5\cos x$, $0 \le x \le 2\pi$, and find the points of intersection. Compare your results with those of Example 3.

NOW WORK PROBLEM 21.

EXAMPLE 4 Solving a Trigonometric Equation Using Identities

Solve the equation: $\cos^2\theta + \sin\theta = 2, \quad 0 \le \theta < 2\pi$

Solution This equation involves two trigonometric functions, sine and cosine. Since it is easier to work with only one, we use a form of the Pythagorean Identity, $\sin^2\theta + \cos^2\theta = 1$ to rewrite the equation.

$$\begin{aligned} \cos^2\theta + \sin\theta &= 2 \\ (1 - \sin^2\theta) + \sin\theta &= 2 \quad && \cos^2\theta = 1 - \sin^2\theta \\ \sin^2\theta - \sin\theta + 1 &= 0 \end{aligned}$$

This is a quadratic equation in $\sin\theta$. The discriminant is $b^2 - 4ac = 1 - 4 = -3 < 0$. Therefore, the equation has no real solution. ◀

CHECK: Graph $Y_1 = \cos^2 x + \sin x$ and $Y_2 = 2$ to see that the two graphs do not intersect anywhere.

EXAMPLE 5 Solving a Trigonometric Equation Using Identities

Solve the equation: $\sin\theta\cos\theta = -\frac{1}{2}, \quad 0 \le \theta < 2\pi$

Solution The left side of the given equation is in the form of the Double-Angle Formula $2\sin\theta\cos\theta = \sin(2\theta)$, except for a factor of 2. We multiply each side by 2.

$$\begin{aligned} \sin\theta\cos\theta &= -\frac{1}{2} \\ 2\sin\theta\cos\theta &= -1 \quad && \text{Multiply each side by 2.} \\ \sin(2\theta) &= -1 && \text{Double-Angle Formula} \end{aligned}$$

The argument here is 2θ. So we need to write all the solutions of this equation and then list those that are in the interval $[0, 2\pi)$.

$$2\theta = \frac{3\pi}{2} + 2k\pi \quad k \text{ any integer}$$
$$\theta = \frac{3\pi}{4} + k\pi$$

$$\underset{k=-1}{\theta = \frac{3\pi}{4} + (-1)\pi = -\frac{\pi}{4}}, \quad \underset{k=0}{\theta = \frac{3\pi}{4} + (0)\pi = \frac{3\pi}{4}}, \quad \underset{k=1}{\theta = \frac{3\pi}{4} + (1)\pi = \frac{7\pi}{4}}, \quad \underset{k=2}{\theta = \frac{3\pi}{4} + (2)\pi = \frac{11\pi}{4}}$$

The solutions in the interval $[0, 2\pi)$ are

$$\theta = \frac{3\pi}{4}, \qquad \theta = \frac{7\pi}{4}$$

◀

3 Sometimes it is necessary to square both sides of an equation in order to obtain expressions that allow the use of identities. Remember, however, that when squaring both sides extraneous solutions may be introduced. As a result, apparent solutions must be checked.

EXAMPLE 6 Other Methods for Solving a Trigonometric Equation

Solve the equation: $\sin\theta + \cos\theta = 1, \quad 0 \le \theta < 2\pi$

Solution A Attempts to use available identities do not lead to equations that are easy to solve. (Try it yourself.) Given the form of this equation, we decide to square each side.

$$\sin\theta + \cos\theta = 1$$
$$(\sin\theta + \cos\theta)^2 = 1 \quad \text{Square each side.}$$
$$\sin^2\theta + 2\sin\theta\cos\theta + \cos^2\theta = 1 \quad \text{Remove parentheses.}$$
$$2\sin\theta\cos\theta = 0 \quad \sin^2\theta + \cos^2\theta = 1$$
$$\sin\theta\cos\theta = 0$$

Setting each factor equal to zero, we obtain

$$\sin\theta = 0 \quad \text{or} \quad \cos\theta = 0$$

The apparent solutions are

$$\theta = 0, \qquad \theta = \pi, \qquad \theta = \frac{\pi}{2}, \qquad \theta = \frac{3\pi}{2}$$

Because we squared both sides of the original equation, we must check these apparent solutions to see if any are extraneous.

$$\theta = 0: \quad \sin 0 + \cos 0 = 0 + 1 = 1 \quad \text{A solution}$$
$$\theta = \pi: \quad \sin\pi + \cos\pi = 0 + (-1) = -1 \quad \text{Not a solution}$$
$$\theta = \frac{\pi}{2}: \quad \sin\frac{\pi}{2} + \cos\frac{\pi}{2} = 1 + 0 = 1 \quad \text{A solution}$$
$$\theta = \frac{3\pi}{2}: \quad \sin\frac{3\pi}{2} + \cos\frac{3\pi}{2} = -1 + 0 = -1 \quad \text{Not a solution}$$

Therefore, $\theta = \pi$ and $\theta = \frac{3\pi}{2}$ are extraneous. The only solutions are $\theta = 0$ and $\theta = \frac{\pi}{2}$.

Solution B We start with the equation

$$\sin\theta + \cos\theta = 1$$

and divide each side by $\sqrt{2}$. (The reason for this choice will become apparent shortly.) Then

$$\frac{1}{\sqrt{2}}\sin\theta + \frac{1}{\sqrt{2}}\cos\theta = \frac{1}{\sqrt{2}}$$

The left side now resembles the formula for the sine of the sum of two angles, one of which is θ. The other angle is unknown (call it ϕ.) Then

$$\sin(\theta + \phi) = \sin\theta\cos\phi + \cos\theta\sin\phi = \frac{1}{\sqrt{2}} = \frac{\sqrt{2}}{2} \quad \textbf{(1)}$$

where

$$\cos\phi = \frac{1}{\sqrt{2}} = \frac{\sqrt{2}}{2}, \qquad \sin\phi = \frac{1}{\sqrt{2}} = \frac{\sqrt{2}}{2}, \qquad 0 \le \phi < 2\pi$$

Figure 32

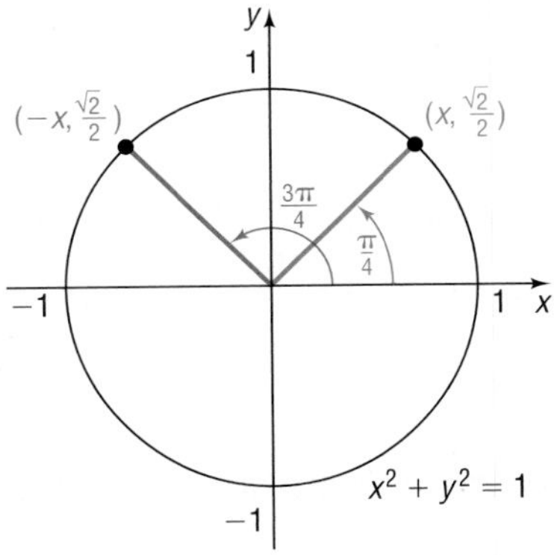

The angle ϕ is therefore $\frac{\pi}{4}$. As a result, equation (1) becomes

$$\sin\left(\theta + \frac{\pi}{4}\right) = \frac{\sqrt{2}}{2}$$

There are two angles whose sine is $\frac{\sqrt{2}}{2}$: $\frac{\pi}{4}$ and $\frac{3\pi}{4}$. See Figure 32. As a result,

$$\theta + \frac{\pi}{4} = \frac{\pi}{4} \quad \text{or} \quad \theta + \frac{\pi}{4} = \frac{3\pi}{4}$$

$$\theta = 0 \qquad\qquad \theta = \frac{\pi}{2}$$

These solutions agree with the solutions found earlier. ◀

This second method of solution can be used to solve any linear equation in the variables $\sin\theta$ and $\cos\theta$.

EXAMPLE 7 Solving a Trigonometric Equation Linear in $\sin\theta$ and $\cos\theta$

Solve:

$$a\sin\theta + b\cos\theta = c, \qquad 0 \le \theta < 2\pi \quad \textbf{(2)}$$

where a, b, and c are constants and either $a \neq 0$ or $b \neq 0$.

Solution We divide each side of equation (2) by $\sqrt{a^2 + b^2}$. Then

$$\frac{a}{\sqrt{a^2 + b^2}}\sin\theta + \frac{b}{\sqrt{a^2 + b^2}}\cos\theta = \frac{c}{\sqrt{a^2 + b^2}} \quad \textbf{(3)}$$

There is a unique angle ϕ, $0 \le \phi < 2\pi$, for which

$$\cos\phi = \frac{a}{\sqrt{a^2 + b^2}} \quad \text{and} \quad \sin\phi = \frac{b}{\sqrt{a^2 + b^2}} \quad \textbf{(4)}$$

Figure 33

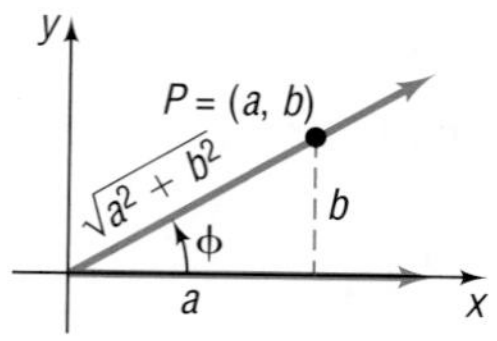

(see Figure 33). Equation (3) may be written as

$$\sin\theta\cos\phi + \cos\theta\sin\phi = \frac{c}{\sqrt{a^2+b^2}}$$

or, equivalently,

$$\sin(\theta+\phi) = \frac{c}{\sqrt{a^2+b^2}} \qquad (5)$$

where ϕ satisfies equation (4).

If $|c| > \sqrt{a^2+b^2}$, then $\sin(\theta+\phi) > 1$ or $\sin(\theta+\phi) < -1$, and equation (5) has no solution.

If $|c| \le \sqrt{a^2+b^2}$, then the solutions of equation (5) are

$$\theta+\phi = \sin^{-1}\frac{c}{\sqrt{a^2+b^2}} \quad \text{or} \quad \theta+\phi = \pi - \sin^{-1}\frac{c}{\sqrt{a^2+b^2}}$$

Because the angle ϕ is determined by equations (4), these are the solutions to equation (2). ◀

NOW WORK PROBLEM 39.

4 Graphing Utility Solutions

The techniques introduced in this section apply only to certain types of trigonometric equations. Solutions for other types are usually studied in calculus, using numerical methods. In the next example, we show how a graphing utility may be used to obtain solutions.

EXAMPLE 8 Solving Trigonometric Equations Using a Graphing Utility

Solve: $5\sin x + x = 3$

Express the solution(s) rounded to two decimal places.

Solution This type of trigonometric equation cannot be solved by previous methods. A graphing utility, though, can be used here. The solution(s) of this equation is the same as the points of intersection of the graphs of $Y_1 = 5\sin x + x$ and $Y_2 = 3$. See Figure 34.

Figure 34

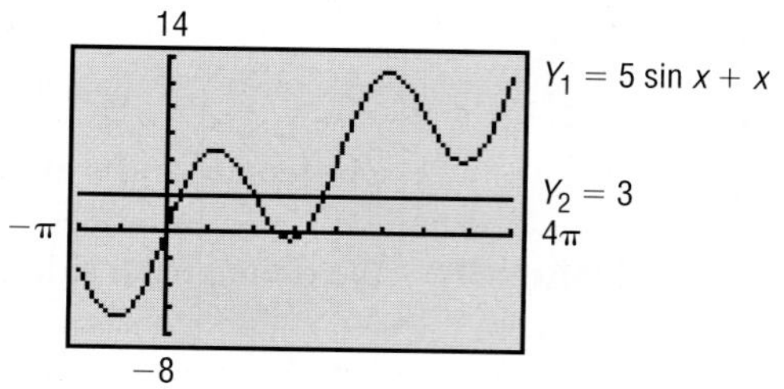

There are three points of intersection; the x-coordinates are the solutions that we seek. Using INTERSECT, we find

$$x = 0.52, \qquad x = 3.18, \qquad x = 5.71$$ ◀

NOW WORK PROBLEM 51.

6.8 Assess Your Understanding

'Are You Prepared?' *Answers are given at the end of these exercises. If you get a wrong answer, read the pages listed in red.*

1. Find the real solutions of $4x^2 - x - 5 = 0$. (pp. 939–940)

2. Find the real solutions of $x^2 - x - 1 = 0$. (pp. 943–945)

Exercises

In Problems 3–44, solve each equation on the interval $0 \le \theta < 2\pi$.

3. $2\cos^2\theta + \cos\theta = 0$

4. $\sin^2\theta - 1 = 0$

5. $2\sin^2\theta - \sin\theta - 1 = 0$

6. $2\cos^2\theta + \cos\theta - 1 = 0$

7. $(\tan\theta - 1)(\sec\theta - 1) = 0$

8. $(\cot\theta + 1)\left(\csc\theta - \frac{1}{2}\right) = 0$

9. $\sin^2\theta - \cos^2\theta = 1 + \cos\theta$

10. $\cos^2\theta - \sin^2\theta + \sin\theta = 0$

11. $\sin^2\theta = 6(\cos\theta + 1)$

12. $2\sin^2\theta = 3(1 - \cos\theta)$

13. $\cos(2\theta) + 6\sin^2\theta = 4$

14. $\cos(2\theta) = 2 - 2\sin^2\theta$

15. $\cos\theta = \sin\theta$

16. $\cos\theta + \sin\theta = 0$

17. $\tan\theta = 2\sin\theta$

18. $\sin(2\theta) = \cos\theta$

19. $\sin\theta = \csc\theta$

20. $\tan\theta = \cot\theta$

21. $\cos(2\theta) = \cos\theta$

22. $\sin(2\theta)\sin\theta = \cos\theta$

23. $\sin(2\theta) + \sin(4\theta) = 0$

24. $\cos(2\theta) + \cos(4\theta) = 0$

25. $\cos(4\theta) - \cos(6\theta) = 0$

26. $\sin(4\theta) - \sin(6\theta) = 0$

27. $1 + \sin\theta = 2\cos^2\theta$

28. $\sin^2\theta = 2\cos\theta + 2$

29. $2\sin^2\theta - 5\sin\theta + 3 = 0$

30. $2\cos^2\theta - 7\cos\theta - 4 = 0$

31. $3(1 - \cos\theta) = \sin^2\theta$

32. $4(1 + \sin\theta) = \cos^2\theta$

33. $\tan^2\theta = \frac{3}{2}\sec\theta$

34. $\csc^2\theta = \cot\theta + 1$

35. $3 - \sin\theta = \cos(2\theta)$

36. $\cos(2\theta) + 5\cos\theta + 3 = 0$

37. $\sec^2\theta + \tan\theta = 0$

38. $\sec\theta = \tan\theta + \cot\theta$

39. $\sin\theta - \sqrt{3}\cos\theta = 1$

40. $\sqrt{3}\sin\theta + \cos\theta = 1$

41. $\tan(2\theta) + 2\sin\theta = 0$

42. $\tan(2\theta) + 2\cos\theta = 0$

43. $\sin\theta + \cos\theta = \sqrt{2}$

44. $\sin\theta + \cos\theta = -\sqrt{2}$

In Problems 45–50, solve each equation for x, $-\pi \le x \le \pi$. *Express the solution(s) rounded to two decimal places.*

45. Solve the equation $\cos x = e^x$ by graphing $Y_1 = \cos x$ and $Y_2 = e^x$ and finding their point(s) of intersection.

46. Solve the equation $\cos x = e^x$ by graphing $Y_1 = \cos x - e^x$ and finding the x-intercept(s).

47. Solve the equation $2\sin x = 0.7x$ by graphing $Y_1 = 2\sin x$ and $Y_2 = 0.7x$ and finding their point(s) of intersection.

48. Solve the equation $2\sin x = 0.7x$ by graphing $Y_1 = 2\sin x - 0.7x$ and finding the x-intercept(s).

49. Solve the equation $\cos x = x^2$ by graphing $Y_1 = \cos x$ and $Y_2 = x^2$ and finding their point(s) of intersection.

50. Solve the equation $\cos x = x^2$ by graphing $Y_1 = \cos x - x^2$ and finding the x-intercept(s).

In Problems 51–62, use a graphing utility to solve each equation. Express the solution(s) rounded to two decimal places.

51. $x + 5\cos x = 0$

52. $x - 4\sin x = 0$

53. $22x - 17\sin x = 3$

54. $19x + 8\cos x = 2$

55. $\sin x + \cos x = x$

56. $\sin x - \cos x = x$

57. $x^2 - 2\cos x = 0$

58. $x^2 + 3\sin x = 0$

59. $x^2 - 2\sin(2x) = 3x$

60. $x^2 = x + 3\cos(2x)$

61. $6\sin x - e^x = 2, \quad x > 0$

62. $4\cos(3x) - e^x = 1, \quad x > 0$

63. Constructing a Rain Gutter A rain gutter is to be constructed of aluminum sheets 12 inches wide. After marking off a length of 4 inches from each edge, this length is bent up at an angle θ. See the illustration. The area A of the opening as a function of θ is given by

$$A(\theta) = 16 \sin\theta(\cos\theta + 1), \quad 0° < \theta < 90°$$

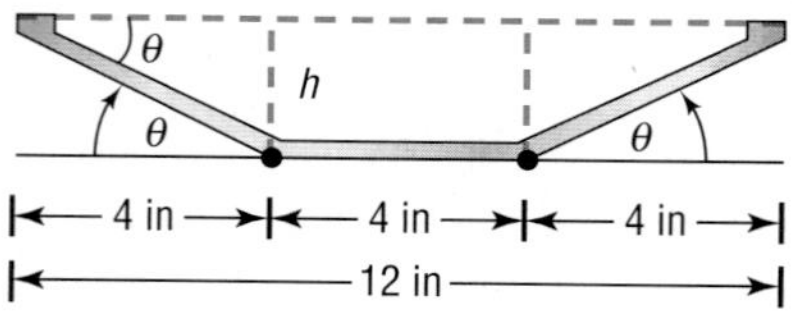

(a) In calculus, you will be asked to find the angle θ that maximizes A by solving the equation

$$\cos(2\theta) + \cos\theta = 0, \quad 0° < \theta < 90°$$

Solve this equation for θ by using the Double-Angle Formula.

(b) Solve the equation for θ by writing the sum of the two cosines as a product.

(c) What is the maximum area A of the opening?

(d) Graph A, $0° \le \theta \le 90°$, and find the angle θ that maximizes the area A. Also find the maximum area. Compare the results to the answers found earlier.

64. Projectile Motion An object is propelled upward at an angle θ, $45° < \theta < 90°$, to the horizontal with an initial velocity of v_0 feet per second from the base of a plane that makes an angle of 45° with the horizontal. See the illustration. If air resistance is ignored, the distance R that it travels up the inclined plane is given by

$$R = \frac{v_0^2\sqrt{2}}{32}[\sin(2\theta) - \cos(2\theta) - 1]$$

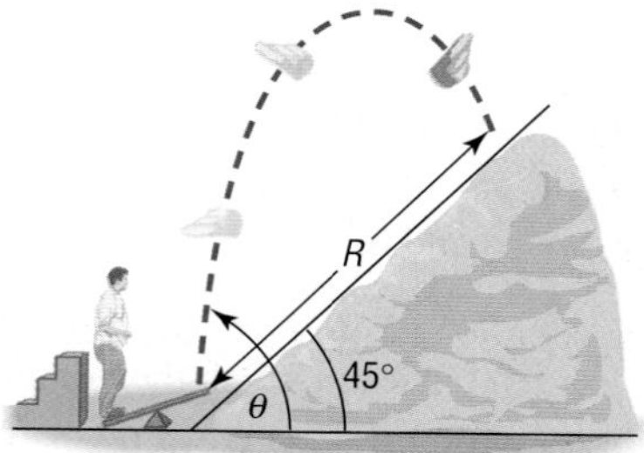

(a) In calculus, you will be asked to find the angle θ that maximizes R by solving the equation

$$\sin(2\theta) + \cos(2\theta) = 0$$

Solve this equation for θ using the method of Example 7.

(b) Solve this equation for θ by dividing each side by $\cos(2\theta)$.

(c) What is the maximum distance R if $v_0 = 32$ feet per second?

(d) Graph R, $45° \le \theta \le 90°$, and find the angle θ that maximizes the distance R. Also find the maximum distance. Use $v_0 = 32$ feet per second. Compare the results with the answers found earlier.

65. Heat Transfer In the study of heat transfer, the equation $x + \tan x = 0$ occurs. Graph $Y_1 = -x$ and $Y_2 = \tan x$ for $x \ge 0$. Conclude that there are an infinite number of points of intersection of these two graphs. Now find the first two positive solutions of $x + \tan x = 0$ rounded to two decimal places.

66. Carrying a Ladder Around a Corner Two hallways, one of width 3 feet, the other of width 4 feet, meet at a right angle. See the illustration.

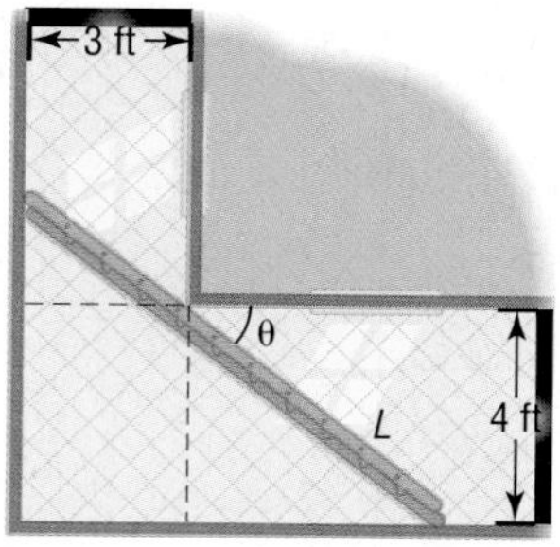

(a) Express the length L of the line segment shown as a function of θ.

(b) In calculus, you will be asked to find the length of the longest ladder that can turn the corner by solving the equation

$$3\sec\theta\tan\theta - 4\csc\theta\cot\theta = 0, \quad 0° < \theta < 90°$$

Solve this equation for θ.

(c) What is the length of the longest ladder that can be carried around the corner?

(d) Graph L, $0° \le \theta \le 90°$, and find the angle θ that minimizes the length L.

(e) Compare the result with the one found in part (b). Explain why the two answers are the same.

67. Projectile Motion The horizontal distance that a projectile will travel in the air is given by the equation

$$R = \frac{v_0^2 \sin(2\theta)}{g}$$

where v_0 is the initial velocity of the projectile, θ is the angle of elevation, and g is acceleration due to gravity (9.8 meters per second squared).

(a) If you can throw a baseball with an initial speed of 34.8 meters per second, at what angle of elevation θ should you direct the throw so that the ball travels a distance of 107 meters before striking the ground?

(b) Determine the maximum distance that you can throw the ball.

(c) Graph R, with $v_0 = 34.8$ meters per second.

(d) Verify the results obtained in parts (a) and (b) using ZERO or ROOT.

68. Projectile Motion Refer to Problem 67.

(a) If you can throw a baseball with an initial speed of 40 meters per second, at what angle of elevation θ should you direct the throw so that the ball travels a distance of 110 meters before striking the ground?

(b) Determine the maximum distance that you can throw the ball.

(c) Graph R, with $v_0 = 40$ meters per second.

(d) Verify the results obtained in parts (a) and (b) using ZERO or ROOT.

'Are You Prepared?' Answers

1. $\left\{-1, \frac{5}{4}\right\}$ **2.** $\left\{\frac{1-\sqrt{5}}{2}, \frac{1+\sqrt{5}}{2}\right\}$

Chapter Review

Things to Know

Definitions of the six inverse trigonometric functions

$y = \sin^{-1} x$ means $x = \sin y$ where $-1 \le x \le 1, \quad -\frac{\pi}{2} \le y \le \frac{\pi}{2}$ (p. 415)

$y = \cos^{-1} x$ means $x = \cos y$ where $-1 \le x \le 1, \quad 0 \le y \le \pi$ (p. 418)

$y = \tan^{-1} x$ means $x = \tan y$ where $-\infty < x < \infty, \quad -\frac{\pi}{2} < y < \frac{\pi}{2}$ (p. 421)

$y = \sec^{-1} x$ means $x = \sec y$ where $|x| \ge 1, \quad 0 \le y \le \pi, \quad y \ne \frac{\pi}{2}$ (p. 427)

$y = \csc^{-1} x$ means $x = \csc y$ where $|x| \ge 1, \quad -\frac{\pi}{2} \le y \le \frac{\pi}{2}, \quad y \ne 0$ (p. 427)

$y = \cot^{-1} x$ means $x = \cot y$ where $-\infty < x < \infty, \quad 0 < y < \pi$ (p. 427)

Sum and Difference Formulas (pp. 438, 441, and 443)

$$\cos(\alpha + \beta) = \cos\alpha\cos\beta - \sin\alpha\sin\beta$$

$$\cos(\alpha - \beta) = \cos\alpha\cos\beta + \sin\alpha\sin\beta$$

$$\sin(\alpha + \beta) = \sin\alpha\cos\beta + \cos\alpha\sin\beta$$

$$\sin(\alpha - \beta) = \sin\alpha\cos\beta - \cos\alpha\sin\beta$$

$$\tan(\alpha + \beta) = \frac{\tan\alpha + \tan\beta}{1 - \tan\alpha\tan\beta}$$

$$\tan(\alpha - \beta) = \frac{\tan\alpha - \tan\beta}{1 + \tan\alpha\tan\beta}$$

Double-Angle Formulas (pp. 448 and 449)

$$\sin(2\theta) = 2\sin\theta\cos\theta \qquad \cos(2\theta) = \cos^2\theta - \sin^2\theta \qquad \cos(2\theta) = 1 - 2\sin^2\theta$$

$$\cos(2\theta) = 2\cos^2\theta - 1 \qquad \tan(2\theta) = \frac{2\tan\theta}{1 - \tan^2\theta}$$

Half-Angle Formulas (pp. 452 and 454)

$$\sin^2\frac{\alpha}{2} = \frac{1-\cos\alpha}{2} \qquad \cos^2\frac{\alpha}{2} = \frac{1+\cos\alpha}{2} \qquad \tan^2\frac{\alpha}{2} = \frac{1-\cos\alpha}{1+\cos\alpha}$$

$$\sin\frac{\alpha}{2} = \pm\sqrt{\frac{1-\cos\alpha}{2}} \qquad \cos\frac{\alpha}{2} = \pm\sqrt{\frac{1+\cos\alpha}{2}} \qquad \tan\frac{\alpha}{2} = \pm\sqrt{\frac{1-\cos\alpha}{1+\cos\alpha}} = \frac{1-\cos\alpha}{\sin\alpha} = \frac{\sin\alpha}{1+\cos\alpha}$$

where the + or − is determined by the quadrant of $\frac{\alpha}{2}$

Product-to-Sum Formulas (p. 457)

$$\sin\alpha\sin\beta = \frac{1}{2}[\cos(\alpha - \beta) - \cos(\alpha + \beta)]$$

$$\cos\alpha\cos\beta = \frac{1}{2}[\cos(\alpha - \beta) + \cos(\alpha + \beta)]$$

$$\sin\alpha\cos\beta = \frac{1}{2}[\sin(\alpha + \beta) + \sin(\alpha - \beta)]$$

Sum-to-Product Formulas (p. 459)

$$\sin\alpha + \sin\beta = 2\sin\frac{\alpha + \beta}{2}\cos\frac{\alpha - \beta}{2}$$

$$\sin\alpha - \sin\beta = 2\sin\frac{\alpha - \beta}{2}\cos\frac{\alpha + \beta}{2}$$

$$\cos\alpha + \cos\beta = 2\cos\frac{\alpha + \beta}{2}\cos\frac{\alpha - \beta}{2}$$

$$\cos\alpha - \cos\beta = -2\sin\frac{\alpha + \beta}{2}\sin\frac{\alpha - \beta}{2}$$

Objectives

Section		You should be able to . . .	Review Exercises
6.1	1	Find the exact value of the inverse sine, cosine, and tangent functions (p. 415)	1–6
	2	Find an approximate value of the inverse sine, cosine, and tangent functions (p. 416)	101–104
6.2	1	Find the exact value of expressions involving the inverse sine, cosine, and tangent functions (p. 425)	9–20
	2	Find the exact value of the inverse secant, cosecant, and cotangent functions (p. 427)	7–8
	3	Use a calculator to evaluate $\sec^{-1} x$, $\csc^{-1} x$, and $\cot^{-1} x$ (p. 428)	105–106
6.3	1	Use algebra to simplify trigonometric expressions (p. 431)	21–52
	2	Establish identities (p. 432)	21–38
6.4	1	Use sum and difference formulas to find exact values (p. 439)	53–60, 61–70(a)–(d)
	2	Use sum and difference formulas to establish identities (p. 439)	39–42
	3	Use sum and difference formulas involving inverse trigonometric functions (p. 444)	71–74
6.5	1	Use Double-Angle Formulas to find exact values (p. 448)	61–70(e)–(f), 75, 76
	2	Use Double-Angle and Half-Angle Formulas to establish identities (p. 449)	43–47
	3	Use Half-Angle Formulas to find exact values (p. 452)	61–70(g)–(h), 113
6.6	1	Express products as sums (p. 457)	48
	2	Express sums as products (p. 459)	49–52
6.7	1	Solve equations involving a single trigonometric function (p. 461)	77–86
6.8	1	Solve trigonometric equations quadratic in form (p. 467)	93–94
	2	Solve trigonometric equations using identities (p. 468)	87–92, 95–98
	3	Solve trigonometric equations linear in sine and cosine (p. 470)	99–100
	4	Solve trigonometric equations using a graphing utility (p. 472)	107–112

Review Exercises

(Blue problem numbers indicate the author's suggestions for use in a Practice Test.)

In Problems 1–20, find the exact value of each expression. Do not use a calculator.

1. $\sin^{-1} 1$
2. $\cos^{-1} 0$
3. $\tan^{-1} 1$
4. $\sin^{-1}\left(-\frac{1}{2}\right)$
5. $\cos^{-1}\left(-\frac{\sqrt{3}}{2}\right)$
6. $\tan^{-1}(-\sqrt{3})$
7. $\sec^{-1}\sqrt{2}$
8. $\cot^{-1}(-1)$
9. $\tan\left[\sin^{-1}\left(-\frac{\sqrt{3}}{2}\right)\right]$
10. $\tan\left[\cos^{-1}\left(-\frac{1}{2}\right)\right]$
11. $\sec\left(\tan^{-1}\frac{\sqrt{3}}{3}\right)$
12. $\csc\left(\sin^{-1}\frac{\sqrt{3}}{2}\right)$
13. $\sin\left(\tan^{-1}\frac{3}{4}\right)$
14. $\cos\left(\sin^{-1}\frac{3}{5}\right)$
15. $\tan\left[\sin^{-1}\left(-\frac{4}{5}\right)\right]$
16. $\tan\left[\cos^{-1}\left(-\frac{3}{5}\right)\right]$
17. $\sin^{-1}\left(\cos\frac{2\pi}{3}\right)$
18. $\cos^{-1}\left(\tan\frac{3\pi}{4}\right)$
19. $\tan^{-1}\left(\tan\frac{7\pi}{4}\right)$
20. $\cos^{-1}\left(\cos\frac{7\pi}{6}\right)$

In Problems 21–52, establish each identity.

21. $\tan\theta\cot\theta - \sin^2\theta = \cos^2\theta$
22. $\sin\theta\csc\theta - \sin^2\theta = \cos^2\theta$
23. $\cos^2\theta(1 + \tan^2\theta) = 1$
24. $(1 - \cos^2\theta)(1 + \cot^2\theta) = 1$
25. $4\cos^2\theta + 3\sin^2\theta = 3 + \cos^2\theta$
26. $4\sin^2\theta + 2\cos^2\theta = 4 - 2\cos^2\theta$
27. $\frac{1 - \cos\theta}{\sin\theta} + \frac{\sin\theta}{1 - \cos\theta} = 2\csc\theta$
28. $\frac{\sin\theta}{1 + \cos\theta} + \frac{1 + \cos\theta}{\sin\theta} = 2\csc\theta$
29. $\frac{\cos\theta}{\cos\theta - \sin\theta} = \frac{1}{1 - \tan\theta}$
30. $1 - \frac{\cos^2\theta}{1 + \sin\theta} = \sin\theta$
31. $\frac{\csc\theta}{1 + \csc\theta} = \frac{1 - \sin\theta}{\cos^2\theta}$
32. $\frac{1 + \sec\theta}{\sec\theta} = \frac{\sin^2\theta}{1 - \cos\theta}$
33. $\csc\theta - \sin\theta = \cos\theta\cot\theta$
34. $\frac{\csc\theta}{1 - \cos\theta} = \frac{1 + \cos\theta}{\sin^3\theta}$
35. $\frac{1 - \sin\theta}{\sec\theta} = \frac{\cos^3\theta}{1 + \sin\theta}$
36. $\frac{1 - \cos\theta}{1 + \cos\theta} = (\csc\theta - \cot\theta)^2$
37. $\frac{1 - 2\sin^2\theta}{\sin\theta\cos\theta} = \cot\theta - \tan\theta$
38. $\frac{(2\sin^2\theta - 1)^2}{\sin^4\theta - \cos^4\theta} = 1 - 2\cos^2\theta$
39. $\frac{\cos(\alpha + \beta)}{\cos\alpha\sin\beta} = \cot\beta - \tan\alpha$
40. $\frac{\sin(\alpha - \beta)}{\sin\alpha\cos\beta} = 1 - \cot\alpha\tan\beta$
41. $\frac{\cos(\alpha - \beta)}{\cos\alpha\cos\beta} = 1 + \tan\alpha\tan\beta$
42. $\frac{\cos(\alpha + \beta)}{\sin\alpha\cos\beta} = \cot\alpha - \tan\beta$
43. $(1 + \cos\theta)\tan\frac{\theta}{2} = \sin\theta$
44. $\sin\theta\tan\frac{\theta}{2} = 1 - \cos\theta$
45. $2\cot\theta\cot(2\theta) = \cot^2\theta - 1$
46. $2\sin(2\theta)(1 - 2\sin^2\theta) = \sin(4\theta)$
47. $1 - 8\sin^2\theta\cos^2\theta = \cos(4\theta)$
48. $\frac{\sin(3\theta)\cos\theta - \sin\theta\cos(3\theta)}{\sin(2\theta)} = 1$
49. $\frac{\sin(2\theta) + \sin(4\theta)}{\cos(2\theta) + \cos(4\theta)} = \tan(3\theta)$
50. $\frac{\sin(2\theta) + \sin(4\theta)}{\sin(2\theta) - \sin(4\theta)} + \frac{\tan(3\theta)}{\tan\theta} = 0$
51. $\frac{\cos(2\theta) - \cos(4\theta)}{\cos(2\theta) + \cos(4\theta)} - \tan\theta\tan(3\theta) = 0$
52. $\cos(2\theta) - \cos(10\theta) = \tan(4\theta)[\sin(2\theta) + \sin(10\theta)]$

In Problems 53–60, find the exact value of each expression.

53. $\sin 165°$
54. $\tan 105°$
55. $\cos\frac{5\pi}{12}$
56. $\sin\left(-\frac{\pi}{12}\right)$
57. $\cos 80°\cos 20° + \sin 80°\sin 20°$
58. $\sin 70°\cos 40° - \cos 70°\sin 40°$
59. $\tan\frac{\pi}{8}$
60. $\sin\frac{5\pi}{8}$

In Problems 61–70, use the information given about the angles α and β to find the exact value of:

(a) $\sin(\alpha + \beta)$ (b) $\cos(\alpha + \beta)$ (c) $\sin(\alpha - \beta)$ (d) $\tan(\alpha + \beta)$

(e) $\sin(2\alpha)$ (f) $\cos(2\beta)$ (g) $\sin \dfrac{\beta}{2}$ (h) $\cos \dfrac{\alpha}{2}$

61. $\sin \alpha = \frac{4}{5}, 0 < \alpha < \frac{\pi}{2}; \sin \beta = \frac{5}{13}, \frac{\pi}{2} < \beta < \pi$

62. $\cos \alpha = \frac{4}{5}, 0 < \alpha < \frac{\pi}{2}; \cos \beta = \frac{5}{13}, -\frac{\pi}{2} < \beta < 0$

63. $\sin \alpha = -\frac{3}{5}, \pi < \alpha < \frac{3\pi}{2}; \cos \beta = \frac{12}{13}, \frac{3\pi}{2} < \beta < 2\pi$

64. $\sin \alpha = -\frac{4}{5}, -\frac{\pi}{2} < \alpha < 0; \cos \beta = -\frac{5}{13}, \frac{\pi}{2} < \beta < \pi$

65. $\tan \alpha = \frac{3}{4}, \pi < \alpha < \frac{3\pi}{2}; \tan \beta = \frac{12}{5}, 0 < \beta < \frac{\pi}{2}$

66. $\tan \alpha = -\frac{4}{3}, \frac{\pi}{2} < \alpha < \pi; \cot \beta = \frac{12}{5}, \pi < \beta < \frac{3\pi}{2}$

67. $\sec \alpha = 2, -\frac{\pi}{2} < \alpha < 0; \sec \beta = 3, \frac{3\pi}{2} < \beta < 2\pi$

68. $\csc \alpha = 2, \frac{\pi}{2} < \alpha < \pi; \sec \beta = -3, \frac{\pi}{2} < \beta < \pi$

69. $\sin \alpha = -\frac{2}{3}, \pi < \alpha < \frac{3\pi}{2}; \cos \beta = -\frac{2}{3}, \pi < \beta < \frac{3\pi}{2}$

70. $\tan \alpha = -2, \frac{\pi}{2} < \alpha < \pi; \cot \beta = -2, \frac{\pi}{2} < \beta < \pi$

In Problems 71–76, find the exact value of each expression.

71. $\cos\left(\sin^{-1} \frac{3}{5} - \cos^{-1} \frac{1}{2}\right)$

72. $\sin\left(\cos^{-1} \frac{5}{13} - \cos^{-1} \frac{4}{5}\right)$

73. $\tan\left[\sin^{-1}\left(-\frac{1}{2}\right) - \tan^{-1} \frac{3}{4}\right]$

74. $\cos\left[\tan^{-1}(-1) + \cos^{-1}\left(-\frac{4}{5}\right)\right]$

75. $\sin\left[2 \cos^{-1}\left(-\frac{3}{5}\right)\right]$

76. $\cos\left(2 \tan^{-1} \frac{4}{3}\right)$

In Problems 77–100, solve each equation on the interval $0 \le \theta < 2\pi$.

77. $\cos \theta = \frac{1}{2}$

78. $\sin \theta = -\frac{\sqrt{3}}{2}$

79. $2 \cos \theta + \sqrt{2} = 0$

80. $\tan \theta + \sqrt{3} = 0$

81. $\sin(2\theta) + 1 = 0$

82. $\cos(2\theta) = 0$

83. $\tan(2\theta) = 0$

84. $\sin(3\theta) = 1$

85. $\sec^2 \theta = 4$

86. $\csc^2 \theta = 1$

87. $\sin \theta = \tan \theta$

88. $\cos \theta = \sec \theta$

89. $\sin \theta + \sin(2\theta) = 0$

90. $\cos(2\theta) = \sin \theta$

91. $\sin(2\theta) - \cos \theta - 2 \sin \theta + 1 = 0$

92. $\sin(2\theta) - \sin \theta - 2 \cos \theta + 1 = 0$

93. $2 \sin^2 \theta - 3 \sin \theta + 1 = 0$

94. $2 \cos^2 \theta + \cos \theta - 1 = 0$

95. $4 \sin^2 \theta = 1 + 4 \cos \theta$

96. $8 - 12 \sin^2 \theta = 4 \cos^2 \theta$

97. $\sin(2\theta) = \sqrt{2} \cos \theta$

98. $1 + \sqrt{3} \cos \theta + \cos(2\theta) = 0$

99. $\sin \theta - \cos \theta = 1$

100. $\sin \theta - \sqrt{3} \cos \theta = 2$

In Problems 101–106, use a calculator to find an approximate value for each expression, rounded to two decimal places.

101. $\sin^{-1} 0.7$

102. $\cos^{-1} \frac{4}{5}$

103. $\tan^{-1}(-2)$

104. $\cos^{-1}(-0.2)$

105. $\sec^{-1} 3$

106. $\cot^{-1}(-4)$

In Problems 107–112, use a graphing utility to solve each equation on the interval $0 \le x \le 2\pi$. Approximate any solutions rounded to two decimal places.

107. $2x = 5 \cos x$

108. $2x = 5 \sin x$

109. $2 \sin x + 3 \cos x = 4x$

110. $3 \cos x + x = \sin x$

111. $\sin x = \ln x$

112. $\sin x = e^{-x}$

113. Use a half-angle formula to find the exact value of $\sin 15°$. Then use a difference formula to find the exact value of $\sin 15°$. Show that the answers found are the same.

114. If you are given the value of $\cos \theta$ and want the exact value of $\cos(2\theta)$, what form of the Double-angle Formula for $\cos(2\theta)$ is most efficient to use?

Chapter Projects

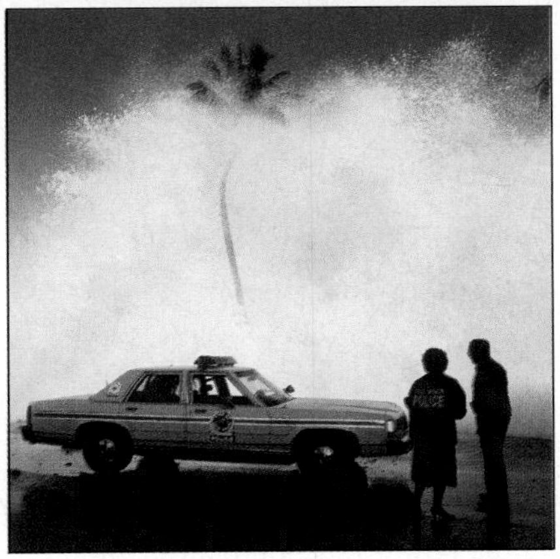

1. **Waves** A stretched string that is attached at both ends, pulled in a direction perpendicular to the string, and released has motion that is described as wave motion. If we assume no friction and a length such that there are no "echoes" (that is, the wave doesn't bounce back), the transverse motion (motion perpendicular to the string) can be described by the equation

$$y = y_m \sin(kx - \omega t)$$

where y_m is the amplitude measured in meters and k and ω are constants. The height of the sound wave depends on the distance x from one endpoint of the string and on time t. Thus, a typical wave has horizontal and vertical motion over time.

(a) What is the amplitude of the wave $y = 0.00421 \sin(68.3x - 2.68t)$?

(b) The value of ω is the angular frequency measured in radians per second. What is the angular frequency of the wave given in part (a)?

(c) The frequency f is the number of vibrations per second (hertz) made by the wave as it passes a certain point. Its value is found using the formula $f = \dfrac{\omega}{2\pi}$. What is the frequency of the wave given in part (a)?

(d) The wavelength, λ, of a wave is the shortest distance at which the wave pattern repeats itself for a constant t. Thus, $\lambda = \dfrac{2\pi}{k}$. What is the wavelength of the wave given in part (a)?

(e) Graph the height of the string a distance $x = 1$ meter from an endpoint.

(f) If two waves travel simultaneously along the same stretched string, the vertical displacement of the string when both waves act is $y = y_1 + y_2$, where y_1 is the vertical displacement of the first wave and y_2 is the vertical displacement of the second wave. This result is called the Principle of Superposition and was analyzed by the French mathematician Jean Baptiste Fourier (1768–1830). When two waves travel along the same string, one wave will differ from the other wave by a phase constant ϕ.

$$y_1 = y_m \sin(kx - \omega t)$$
$$y_2 = y_m \sin(kx - \omega t + \phi)$$

assuming that each wave has the same amplitude. Write $y_1 + y_2$ as a product using the Sum-to-Product Formulas.

(g) Suppose that two waves are moving in the same direction along a stretched string. The amplitude of each wave is 0.0045 meter, and the phase difference between them is 2.5 radians. The wavelength, λ, of each wave is 0.09 meter and the frequency, f, is 2.3 hertz. Find y_1, y_2, and $y_1 + y_2$.

(h) Using a graphing utility, graph y_1, y_2, and $y_1 + y_2$ on the same viewing window.

(i) Redo parts (g) and (h) with the phase difference between the waves being 0.4 radian.

(j) What effect does the phase difference have on the amplitude of $y_1 + y_2$?

The following chapter projects may be found at www.prenhall.com/sullivan7e.

2. **Project at Motorola** *Sending Pictures Wirelessly*
3. **Jacob's Field**
4. **Calculus of Differences**

Cumulative Review

1. Find the real solutions, if any, of the equation $3x^2 + x - 1 = 0$.

2. Find an equation for the line containing the points $(-2, 5)$ and $(4, -1)$. What is the distance between these points? What is their midpoint?

3. Test the equation $3x + y^2 = 9$ for symmetry with respect to the x-axis, y-axis, and origin. List the intercepts.

4. Use transformations to graph the equation $y = |x - 3| + 2$.

5. Use transformations to graph the equation $y = 3e^x - 2$.

6. Use transformations to graph the equation

$$y = \cos\left(x - \frac{\pi}{2}\right) - 1.$$

7. Sketch a graph of each of the following functions. Label at least three points on each graph. Name the inverse function of each and show its graph.

(a) $y = x^3$

(b) $y = e^x$

(c) $y = \sin x, \quad -\frac{\pi}{2} \le x \le \frac{\pi}{2}$

(d) $y = \cos x, \quad 0 \le x \le \pi$

8. If $\sin\theta = -\frac{1}{3}$ and $\pi < \theta < \frac{3\pi}{2}$, find the exact value of:

(a) $\cos\theta$ (b) $\tan\theta$ (c) $\sin(2\theta)$

(d) $\cos(2\theta)$ (e) $\sin\left(\frac{1}{2}\theta\right)$ (f) $\cos\left(\frac{1}{2}\theta\right)$

9. Find the exact value of $\cos(\tan^{-1} 2)$.

10. If $\sin\alpha = \frac{1}{3}, \frac{\pi}{2} < \alpha < \pi$, and $\cos\beta = -\frac{1}{3}, \pi < \beta < \frac{3\pi}{2}$, find the exact value of:

(a) $\cos\alpha$ (b) $\sin\beta$ (c) $\cos(2\alpha)$

(d) $\cos(\alpha + \beta)$ (e) $\sin\frac{\beta}{2}$

11. For the function

$$f(x) = 2x^5 - x^4 - 4x^3 + 2x^2 + 2x - 1:$$

(a) Find the real zeros and their multiplicity.

(b) Find the intercepts.

(c) Find the power function that the graph of f resembles for large $|x|$.

(d) Graph f using a graphing utility.

(e) Approximate the turning points, if any exist.

(f) Use the information obtained in parts (a)–(e) to sketch a graph of f by hand.

(g) Identify the intervals on which f is increasing, decreasing, or constant.

12. If $f(x) = 2x^2 + 3x + 1$ and $g(x) = x^2 + 3x + 2$, solve:

(a) $f(x) = 0$

(b) $f(x) = g(x)$

(c) $f(x) > 0$

(d) $f(x) \ge g(x)$

7 Applications of Trigonometric Functions

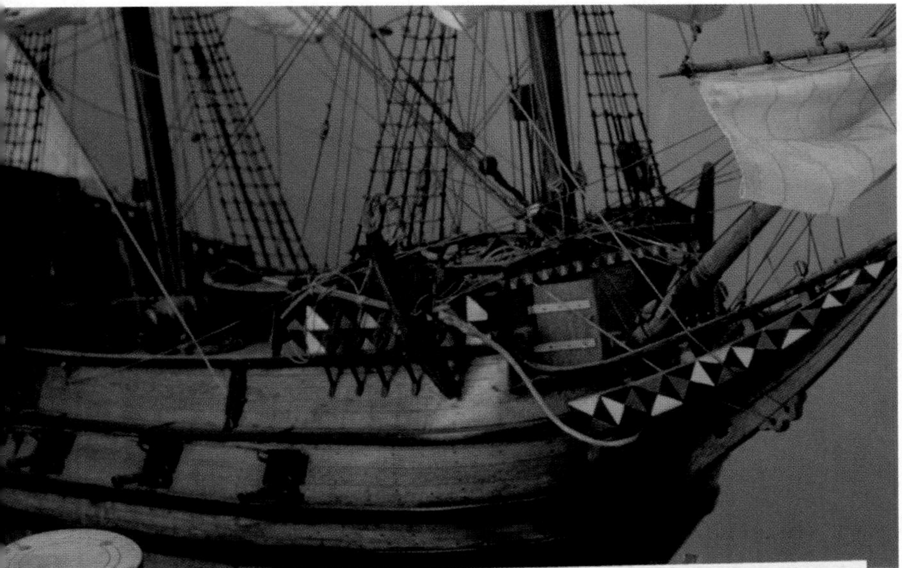

New maps pinpoint Lewis and Clark's journey through Missouri

KANSAS CITY, Mo.—Nearly two centuries ago, Congress commissioned Meriwether Lewis and William Clark to explore trade routes to the West.

Their long-documented journey took them through Missouri, but their exact route has been up for debate.

Now, modern computer technology combined with 19th century land surveys may provide a precise picture. The latest computer-generated maps of Lewis and Clark's expedition were unveiled here this week by Missouri Secretary of State Matt Blunt.

The maps combine the terrain of the early 19th century with contemporary geographical markers. They are the most precise to date of Lewis and Clark's journeys through Missouri in 1804 and 1806, said the project's lead researcher, Jim Harlan.

"We've known what they've done, but not with this much certainty," said Harlan, geographic resource project director at the University of Missouri-Columbia. "I think we need more information and less speculation. That's what this is all about."

SOURCE: Sophia Maines, "New Maps Pinpoint Lewis and Clark's Journey Through Missouri," *The Kansas City Star*, August 1, 2001. Distributed by Knight Ridder/Tribune Information Services.

—SEE CHAPTER PROJECT 1.

OUTLINE

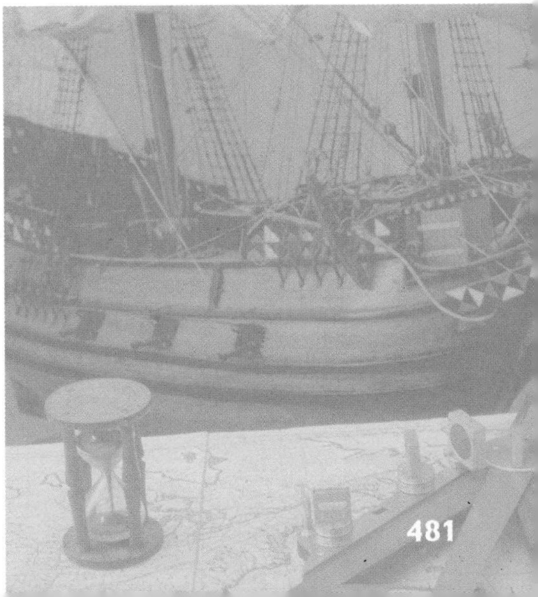

7.1 Right Triangle Trigonometry; Applications

PREPARING FOR THIS SECTION *Before getting started, review the following:*

- Pythagorean Theorem (Appendix A, Section A.2, pp. 913–915)
- Trigonometric Equations (I) (Section 6.7, pp. 461–464)

Now work the 'Are You Prepared?' problems on page 490.

OBJECTIVES
1. Find the Value of Trigonometric Functions of Acute Angles
2. Use the Complementary Angle Theorem
3. Solve Right Triangles
4. Solve Applied Problems

1 A triangle in which one angle is a right angle (90°) is called a **right triangle.** Recall that the side opposite the right angle is called the **hypotenuse,** and the remaining two sides are called the **legs** of the triangle. In Figure 1(a), we have labeled the hypotenuse as c to indicate that its length is c units, and, in a like manner, we have labeled the legs as a and b. Because the triangle is a right triangle, the Pythagorean Theorem tells us that

$$a^2 + b^2 = c^2$$

In Figure 1(a), we also show the angle θ. The angle θ is an **acute angle:** that is, $0° < \theta < 90°$, if θ is measured in degrees, or $0 < \theta < \frac{\pi}{2}$, if θ is measured in radians. Place θ in standard position, as shown in Figure 1(b). Then the coordinates of the point P are (a, b). Also, P is a point on the terminal side of θ that is also on the circle $x^2 + y^2 = c^2$. (Do you see why?)

Figure 1

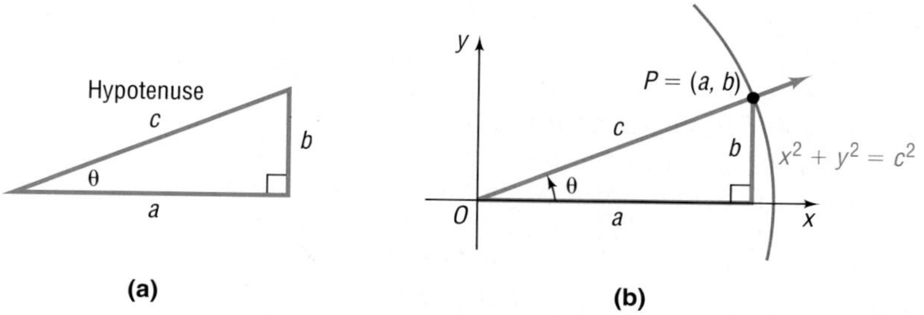

Now use the theorem on page 350. By referring to the lengths of the sides of the triangle by the names hypotenuse (c), opposite (b), and adjacent (a), as indicated in Figure 2, we can express the trigonometric functions of θ as ratios of the sides of a right triangle.

Figure 2

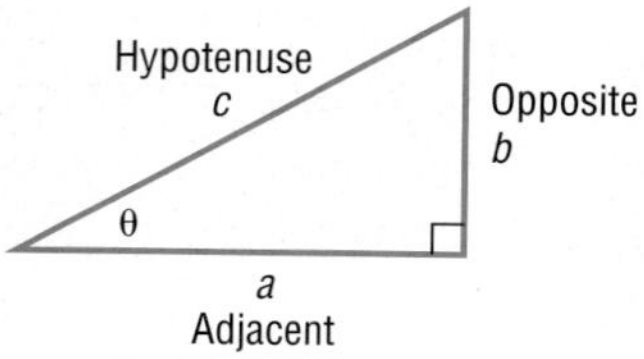

$$\sin\theta = \frac{\text{Opposite}}{\text{Hypotenuse}} = \frac{b}{c} \qquad \cos\theta = \frac{\text{Adjacent}}{\text{Hypotenuse}} = \frac{a}{c}$$
$$\tan\theta = \frac{\text{Opposite}}{\text{Adjacent}} = \frac{b}{a} \qquad \csc\theta = \frac{\text{Hypotenuse}}{\text{Opposite}} = \frac{c}{b} \qquad (1)$$
$$\sec\theta = \frac{\text{Hypotenuse}}{\text{Adjacent}} = \frac{c}{a} \qquad \cot\theta = \frac{\text{Adjacent}}{\text{Opposite}} = \frac{a}{b}$$

Notice that each of the trigonometric functions of the acute angle θ is positive.

EXAMPLE 1 Finding the Value of Trigonometric Functions from a Right Triangle

Find the exact value of the six trigonometric functions of the angle θ in Figure 3.

Figure 3

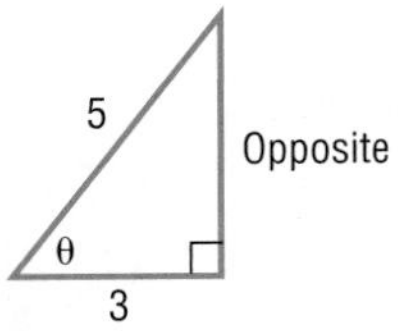

Solution We see in Figure 3 that the two given sides of the triangle are

$$c = \text{Hypotenuse} = 5, \quad a = \text{Adjacent} = 3$$

To find the length of the opposite side, we use the Pythagorean Theorem.

$$\begin{aligned}(\text{Adjacent})^2 + (\text{Opposite})^2 &= (\text{Hypotenuse})^2\\ 3^2 + (\text{Opposite})^2 &= 5^2\\ (\text{Opposite})^2 &= 25 - 9 = 16\\ \text{Opposite} &= 4\end{aligned}$$

Now that we know the lengths of the three sides, we use the ratios in equations (1) to find the value of each of the six trigonometric functions.

$$\sin\theta = \frac{\text{Opposite}}{\text{Hypotenuse}} = \frac{4}{5} \qquad \cos\theta = \frac{\text{Adjacent}}{\text{Hypotenuse}} = \frac{3}{5} \qquad \tan\theta = \frac{\text{Opposite}}{\text{Adjacent}} = \frac{4}{3}$$

$$\csc\theta = \frac{\text{Hypotenuse}}{\text{Opposite}} = \frac{5}{4} \qquad \sec\theta = \frac{\text{Hypotenuse}}{\text{Adjacent}} = \frac{5}{3} \qquad \cot\theta = \frac{\text{Adjacent}}{\text{Opposite}} = \frac{3}{4}$$ ◀

NOW WORK PROBLEM 9.

The values of the trigonometric functions of an acute angle are ratios of the lengths of the sides of a right triangle. This way of viewing the trigonometric functions leads to many applications and, in fact, was the point of view used by early mathematicians (before calculus) in studying the subject of trigonometry.

We look at one such application next.

EXAMPLE 2 Constructing a Rain Gutter

A rain gutter is to be constructed of aluminum sheets 12 inches wide. After marking off a length of 4 inches from each edge, this length is bent up at an angle θ. See Figure 4(a).

Figure 4(a)

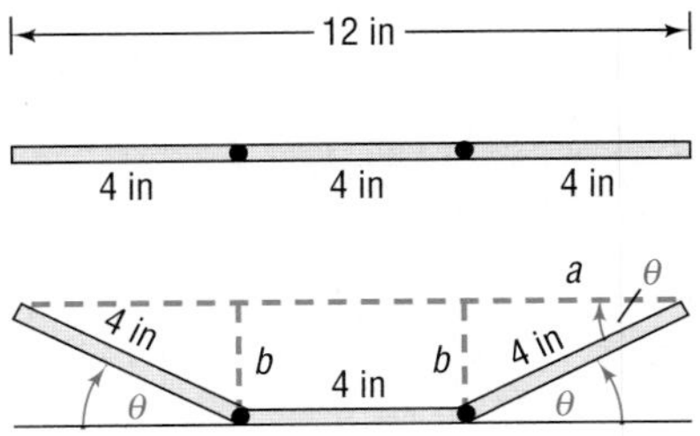

(a) Express the area A of the opening as a function of θ.
[**Hint:** Let b denote the vertical height of the bend.]

(b) Graph $A = A(\theta)$. Find the angle θ that makes A largest. (This bend will allow the most water to flow through the gutter.)

Solution (a) Look again at Figure 4(a). The area A of the opening is the sum of the areas of two congruent right triangles and one rectangle. Look at Figure 4(b), showing the triangle in Figure 4(a) redrawn. We see that

Figure 4(b)

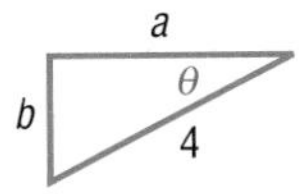

$$\cos\theta = \frac{a}{4} \quad \text{so} \quad a = 4\cos\theta \qquad \sin\theta = \frac{b}{4} \quad \text{so} \quad b = 4\sin\theta$$

The area of the triangle is

$$\text{area} = \frac{1}{2}(\text{base})(\text{height}) = \frac{1}{2}ab = \frac{1}{2}(4\cos\theta)(4\sin\theta) = 8\sin\theta\cos\theta$$

So the area of the two triangles is $16\sin\theta\cos\theta$.

The rectangle has length 4 and height b, so its area is

$$4b = 4(4\sin\theta) = 16\sin\theta$$

The area A of the opening is

$$A = \text{area of the two triangles} + \text{area of the rectangle}$$
$$A(\theta) = 16\sin\theta\cos\theta + 16\sin\theta = 16\sin\theta(\cos\theta + 1)$$

Figure 5

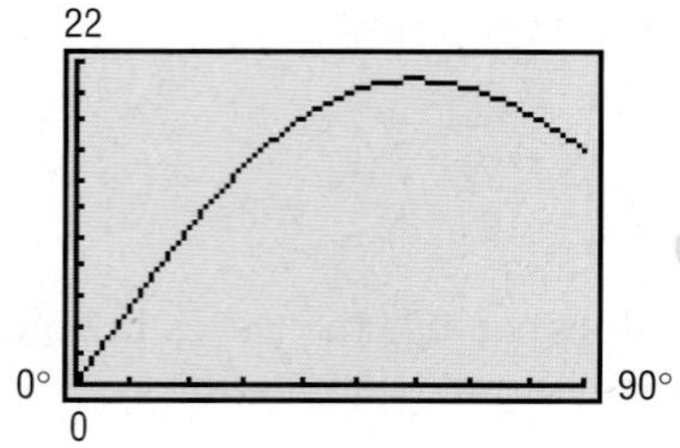

(b) Figure 5 shows the graph of $A = A(\theta)$. Using MAXIMUM, the angle θ that makes A largest is 60°. ◀

Complementary Angles: Cofunctions

2 Two acute angles are called **complementary** if their sum is a right angle. Because the sum of the angles of any triangle is 180°, it follows that, for a right triangle, the two acute angles are complementary.

Refer now to Figure 6; we have labeled the angle opposite side b as β and the angle opposite side a as α. Notice that side b is adjacent to angle α and side a is adjacent to angle β. As a result,

Figure 6

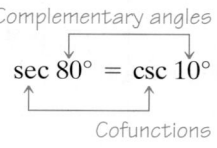

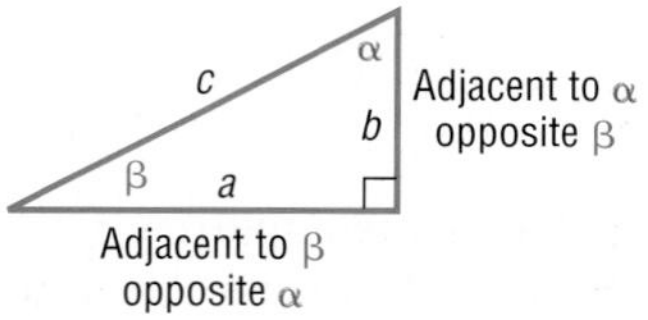

$$\sin\beta = \frac{b}{c} = \cos\alpha \qquad \cos\beta = \frac{a}{c} = \sin\alpha \qquad \tan\beta = \frac{b}{a} = \cot\alpha$$
$$\csc\beta = \frac{c}{b} = \sec\alpha \qquad \sec\beta = \frac{c}{a} = \csc\alpha \qquad \cot\beta = \frac{a}{b} = \tan\alpha \tag{2}$$

Because of these relationships, the functions sine and cosine, tangent and cotangent, and secant and cosecant are called **cofunctions** of each other. The identities (2) may be expressed in words as follows:

Complementary Angle Theorem

Cofunctions of complementary angles are equal.

Examples of this theorem are given next:

Complementary angles: $\sin 30° = \cos 60°$ (Cofunctions) Complementary angles: $\tan 40° = \cot 50°$ (Cofunctions) Complementary angles: $\sec 80° = \csc 10°$ (Cofunctions)

EXAMPLE 3 **Using the Complementary Angle Theorem**

(a) $\sin 62° = \cos(90° - 62°) = \cos 28°$

(b) $\tan\frac{\pi}{12} = \cot\left(\frac{\pi}{2} - \frac{\pi}{12}\right) = \cot\frac{5\pi}{12}$

(c) $\sin^2 40° + \sin^2 50° = \sin^2 40° + \cos^2 40° = 1$

$\sin 50° = \cos 40°$ ◀

NOW WORK PROBLEM 19.

Solving Right Triangles

Figure 7

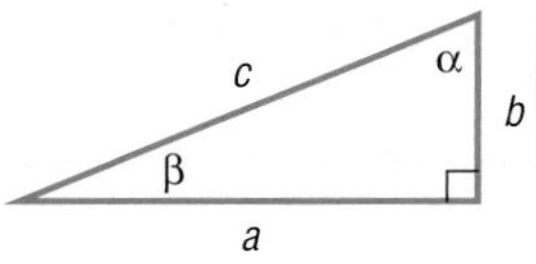

3 In the discussion that follows, we will always label a right triangle so that side a is opposite angle α, side b is opposite angle β, and side c is the hypotenuse, as shown in Figure 7. **To solve a right triangle** means to find the missing lengths of its sides and the measurements of its angles. We shall follow the practice of expressing the lengths of the sides rounded to two decimal places and of expressing angles in degrees rounded to one decimal place. (Be sure that your calculator is in degree mode.)

To solve a right triangle, we need to know one of the acute angles α or β and a side, or else two sides. Then we make use of the Pythagorean Theorem and the fact that the sum of the angles of a triangle is 180°. The sum of the angles α and β in a right triangle is therefore 90°.

For the right triangle shown in Figure 1, we have

$$c^2 = a^2 + b^2, \qquad \alpha + \beta = 90°$$

EXAMPLE 4 **Solving a Right Triangle**

Figure 8

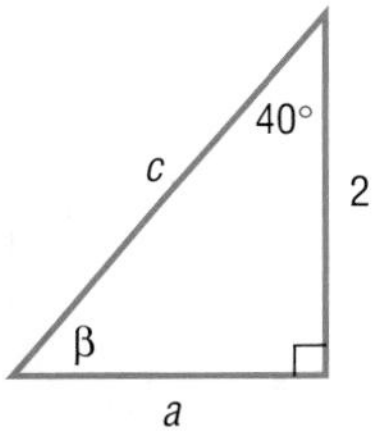

Use Figure 8. If $b = 2$ and $\alpha = 40°$, find $a, c,$ and β.

Solution Since $\alpha = 40°$ and $\alpha + \beta = 90°$, we find that $\beta = 50°$. To find the sides a and c, we use the facts that

$$\tan 40° = \frac{a}{2} \quad \text{and} \quad \cos 40° = \frac{2}{c}$$

Now solve for a and c.

$$a = 2 \tan 40° \approx 1.68 \quad \text{and} \quad c = \frac{2}{\cos 40°} \approx 2.61$$

◀

NOW WORK PROBLEM 29.

EXAMPLE 5 **Solving a Right Triangle**

Use Figure 9. If $a = 3$ and $b = 2$, find $c, \alpha,$ and β.

Figure 9

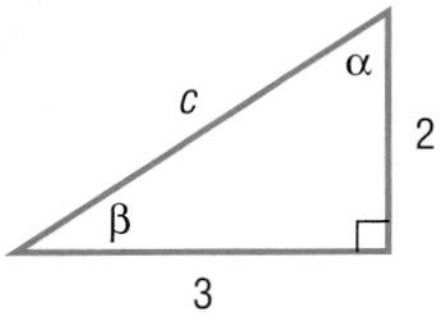

Solution Since $a = 3$ and $b = 2$, then, by the Pythagorean Theorem, we have

$$c^2 = a^2 + b^2 = 3^2 + 2^2 = 9 + 4 = 13$$
$$c = \sqrt{13} \approx 3.61$$

To find angle α, we use the fact that

$$\tan \alpha = \frac{3}{2} \quad \text{so} \quad \alpha = \tan^{-1}\frac{3}{2}$$

Set the mode on your calculator to degrees. Then, rounded to one decimal place, we find that $\alpha = 56.3°$. Since $\alpha + \beta = 90°$, we find that $\beta = 33.7°$. ◄

Note: To avoid round-off errors when using a calculator, we will store unrounded values in memory for use in subsequent calculations.

NOW WORK PROBLEM 39.

Applications

4 One common use for trigonometry is to measure heights and distances that are either awkward or impossible to measure by ordinary means.

EXAMPLE 6 Finding the Width of a River

A surveyor can measure the width of a river by setting up a transit* at a point C on one side of the river and taking a sighting of a point A on the other side. Refer to Figure 10. After turning through an angle of 90° at C, the surveyor walks a distance of 200 meters to point B. Using the transit at B, the angle β is measured and found to be 20°. What is the width of the river rounded to the nearest meter?

Figure 10

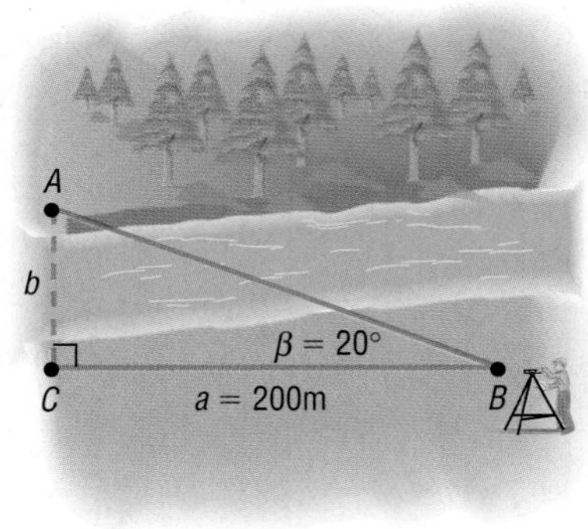

Solution We seek the length of side b. We know a and β, so we use the fact that

$$\tan \beta = \frac{b}{a}$$

to get

$$\tan 20° = \frac{b}{200}$$

$$b = 200 \tan 20° \approx 72.79 \text{ meters}$$

The width of the river is 73 meters, rounded to the nearest meter. ◄

NOW WORK PROBLEM 49.

*An instrument used in surveying to measure angles.

EXAMPLE 7 Finding the Inclination of a Mountain Trail

A straight trail leads from the Alpine Hotel, elevation 8000 feet, to a scenic overlook, elevation 11,100 feet. The length of the trail is 14,100 feet. What is the inclination (grade) of the trail? That is, what is the angle β in Figure 11?

Figure 11

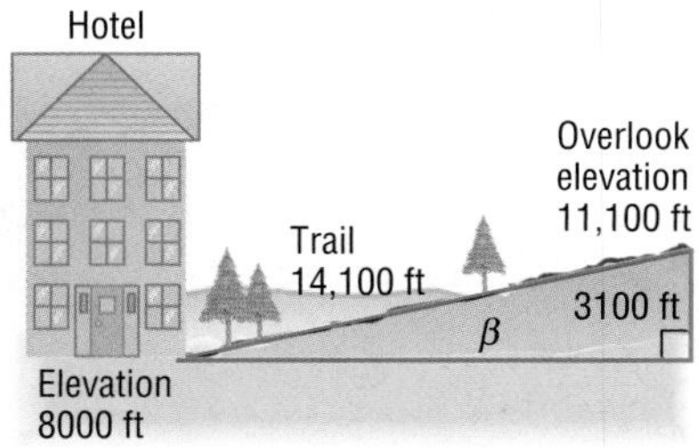

Solution As Figure 11 illustrates, the angle β obeys the equation

$$\sin \beta = \frac{3100}{14{,}100}$$

Using a calculator,

$$\beta = \sin^{-1} \frac{3100}{14{,}100} \approx 12.7^\circ$$

The inclination (grade) of the trail is approximately 12.7°. ◀

Vertical heights can sometimes be measured using either the *angle of elevation* or the *angle of depression*. If a person is looking up at an object, the acute angle measured from the horizontal to a line-of-sight observation of the object is called the **angle of elevation.** See Figure 12(a).

If a person is standing on a cliff looking down at an object, the acute angle made by the line-of-sight observation of the object and the horizontal is called the **angle of depression.** See Figure 12(b).

Figure 12

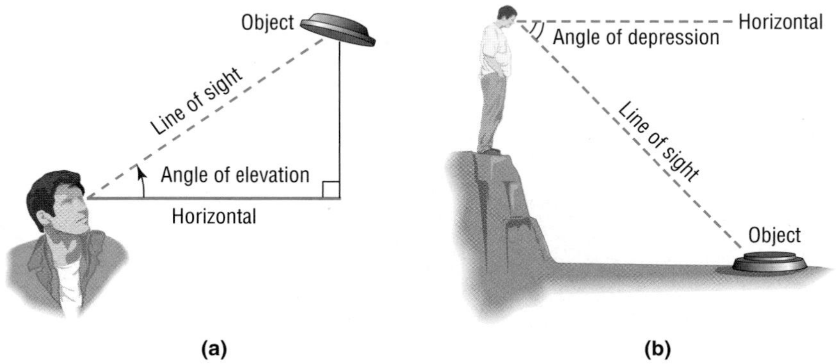

EXAMPLE 8 Finding the Height of a Cloud

Meteorologists find the height of a cloud using an instrument called a **ceilometer.** A ceilometer consists of a **light projector** that directs a vertical light beam up to the cloud base and a **light detector** that scans the cloud to detect the light beam. See Figure 13(a). On December 1, 2003, at Midway Airport in Chicago, a ceilometer with a base of 300 feet was employed to find the height of the cloud cover. If the angle of elevation of the light detector is 75°, what is the height of the cloud cover? Round the answer to the nearest foot.

Figure 13

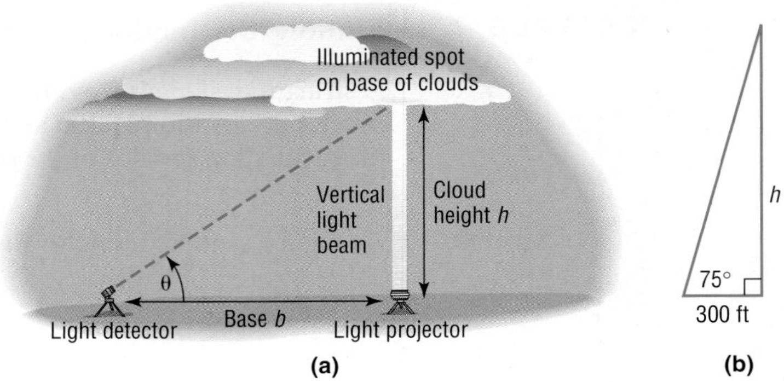

Solution Figure 13(b) illustrates the situation. To find the height h, we use the fact that $\tan 75° = \frac{h}{300}$, so

$$h = 300 \tan 75° \approx 1120 \text{ feet}$$

The ceiling (height to the base of the cloud cover) is approximately 1120 feet. ◀

NOW WORK PROBLEM 51.

The idea behind Example 8 can also be used to find the height of an object with a base that is not accessible to the horizontal.

EXAMPLE 9 Finding the Height of a Statue on a Building

Adorning the top of the Board of Trade building in Chicago is a statue of Ceres, the Roman goddess of wheat. From street level, two observations are taken 400 feet from the center of the building. The angle of elevation to the base of the statue is found to be 55.1°; the angle of elevation to the top of the statue is 56.5°. See Figure 14(a). What is the height of the statue? Round the answer to the nearest foot.

Figure 14

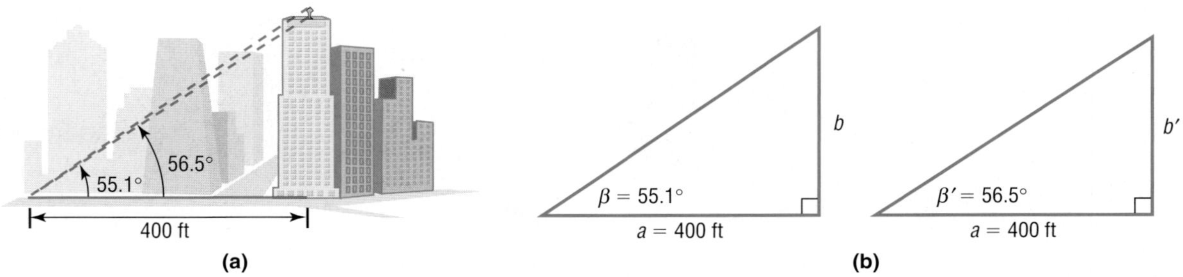

Solution Figure 14(b) shows two triangles that replicate Figure 14(a). The height of the statue of Ceres will be $b' - b$. To find b and b', we refer to Figure 14(b).

$$\tan 55.1° = \frac{b}{400} \qquad \tan 56.5° = \frac{b'}{400}$$

$$b = 400 \tan 55.1° \approx 573 \qquad b' = 400 \tan 56.5° \approx 604$$

The height of the statue is approximately $604 - 573 = 31$ feet. ◀

NOW WORK PROBLEM 59.

EXAMPLE 10 The Gibb's Hill Lighthouse, Southampton, Bermuda

In operation since 1846, the Gibb's Hill Lighthouse stands 117 feet high on a hill 245 feet high, so its beam of light is 362 feet above sea level. A brochure states that the light can be seen on the horizon about 26 miles distant. Verify the accuracy of this statement.

Figure 15

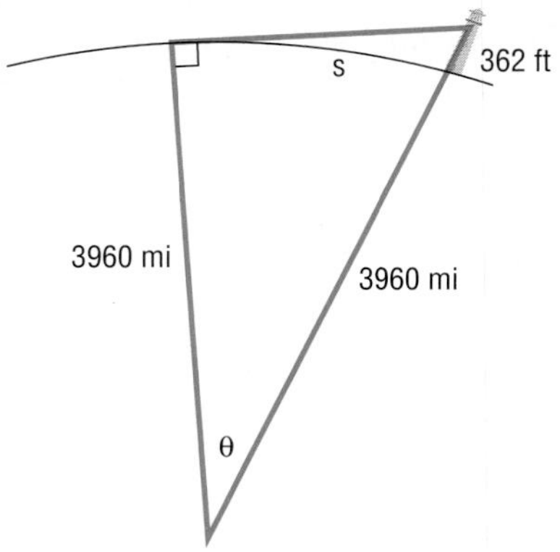

Solution Figure 15 illustrates the situation. The central angle θ, positioned at the center of Earth, radius 3960 miles, obeys the equation

$$\cos\theta = \frac{3960}{3960 + \frac{362}{5280}} \approx 0.999982687 \qquad \text{1 mile} = \text{5280 feet}$$

Solving for θ, we find

$$\theta \approx 0.33715° \approx 20.23'$$

The brochure does not indicate whether the distance is measured in nautical miles or statute miles. Let's calculate both distances.

The distance s in nautical miles (refer to Problem 114, p. 338) is the measure of angle θ in minutes, so $s = 20.23$ nautical miles.

The distance s in statute miles is given by the formula $s = r\theta$, where θ is measured in radians. Then, since

$$\theta = 20.23' \underset{1' = \frac{1}{60}°}{\approx} 0.33715° \underset{1° = \frac{\pi}{180} \text{ radian}}{\approx} 0.00588 \text{ radian}$$

we find that

$$s = r\theta = (3960)(0.00588) = 23.3 \text{ miles}$$

In either case, it would seem that the brochure overstated the distance somewhat. ◀

Figure 16

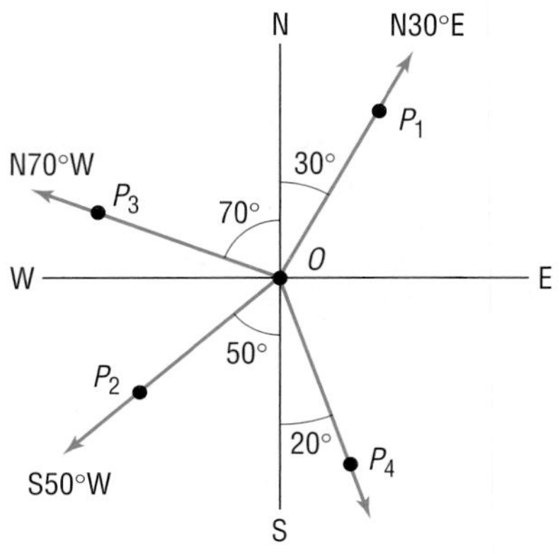

In navigation and surveying, the **direction** or **bearing** from a point O to a point P equals the acute angle θ between the ray OP and the vertical line through O, the north–south line.

Figure 16 illustrates some bearings. Notice that the bearing from O to P_1 is denoted by the symbolism N30°E, indicating that the bearing is 30° east of north. In writing the bearing from O to P, the direction north or south always appears first, followed by an acute angle, followed by east or west. In Figure 16, the bearing from O to P_2 is S50°W, and from O to P_3 it is N70°W.

EXAMPLE 11 Finding the Bearing of an Object

In Figure 16, what is the bearing from O to an object at P_4?

Solution The acute angle between the ray OP_4 and the north–south line through O is given as 20°. The bearing from O to P_4 is S20°E. ◀

EXAMPLE 12 Finding the Bearing of an Airplane

A Boeing 777 aircraft takes off from O'Hare Airport on runway 2 LEFT, which has a bearing of N20°E.* After flying for 1 mile, the pilot of the aircraft requests permission to turn 90° and head toward the northwest. The request is granted. After the plane goes 2 miles in this direction, what bearing should the control tower use to locate the aircraft?

Figure 17

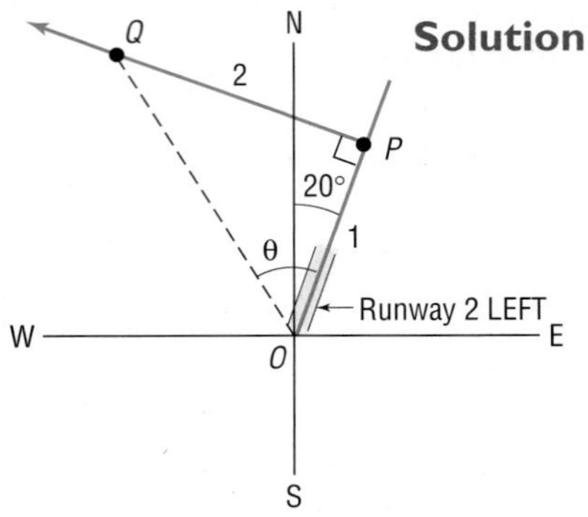

Solution Figure 17 illustrates the situation. After flying 1 mile from the airport O (the control tower), the aircraft is at P. After turning 90° toward the northwest and flying 2 miles, the aircraft is at the point Q. In triangle OPQ, the angle θ obeys the equation

$$\tan\theta = \frac{2}{1} = 2 \quad \text{so} \quad \theta = \tan^{-1} 2 \approx 63.4^\circ$$

The acute angle between north and the ray OQ is $63.4^\circ - 20^\circ = 43.4^\circ$. The bearing of the aircraft from O to Q is N43.4°W. ◀

NOW WORK PROBLEM 67.

*In air navigation, the term **azimuth** is employed to denote the positive angle measured clockwise from the north (N) to a ray OP. In Figure 16, the azimuth from O to P_1 is 30°; the azimuth from O to P_2 is 230°; the azimuth from O to P_3 is 290°. In naming runways, the units digit is left off the azimuth. Runway 2 LEFT means the left runway with a direction of azimuth 20° (bearing N20°E). Runway 23 is the runway with azimuth 230° and bearing S50°W.

7.1 Assess Your Understanding

'Are You Prepared?' *Answers are given at the end of these exercises. If you get a wrong answer, read the pages listed in red.*

1. In a right triangle, if the length of the hypotenuse is 5 and the length of one of the other sides is 3, what is the length of the third side? (pp. 913–915)

2. If θ is an acute angle, solve the equation $\tan\theta = \frac{1}{2}$. (pp. 461–464)

Concepts and Vocabulary

3. *True or False:* The angles 52° and 48° are complementary.

4. *True or False:* In a right triangle, one of the angles is 90° and the sum of the other two angles is 90°.

5. When you look up at an object, the acute angle measured from the horizontal to a line-of-sight observation of the object is called the ________ ________ ________.

6. When you look down at an object, the acute angle described in Problem 5 is called the ________ ________ ________.

7. *True or False:* In a right triangle, if two sides are known, we can solve the triangle.

8. *True or False:* In a right triangle, if we know the two acute angles, we can solve the triangle.

Exercises

In Problems 9–18, find the exact value of the six trigonometric functions of the angle θ in each figure.

9.

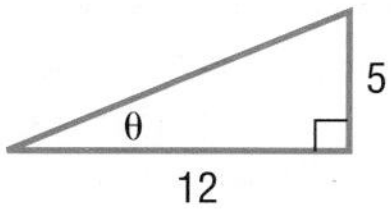

10.

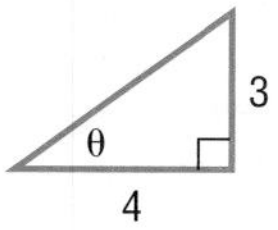

11.

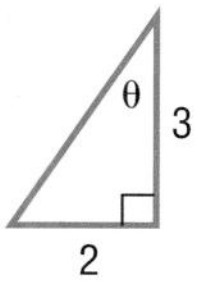

12.

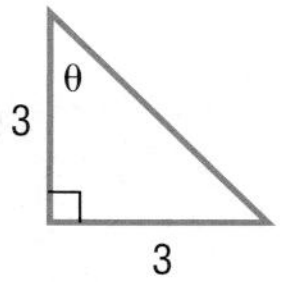

13.

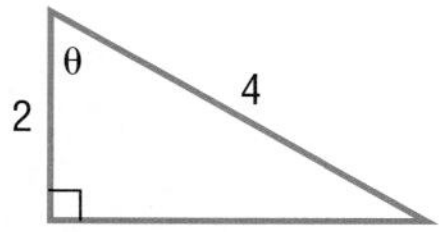

14.

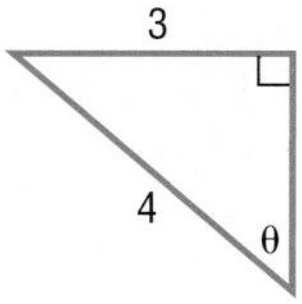

15.

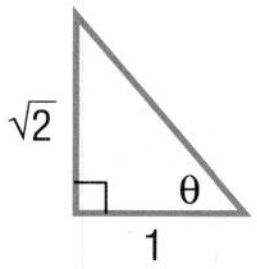

16.

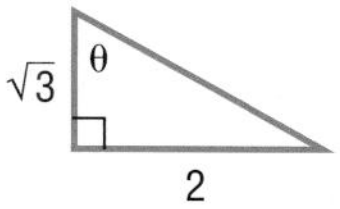

17.

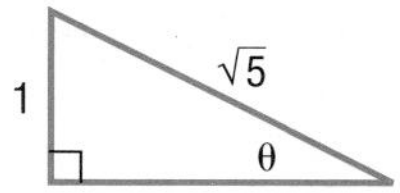

18.

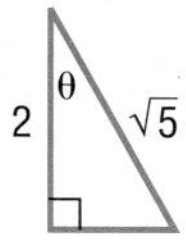

In Problems 19–28, find the exact value of each expression. Do not use a calculator.

19. $\sin 38^\circ - \cos 52^\circ$

20. $\tan 12^\circ - \cot 78^\circ$

21. $\dfrac{\cos 10^\circ}{\sin 80^\circ}$

22. $\dfrac{\cos 40^\circ}{\sin 50^\circ}$

23. $1 - \cos^2 20^\circ - \cos^2 70^\circ$

24. $1 + \tan^2 5^\circ - \csc^2 85^\circ$

25. $\tan 20^\circ - \dfrac{\cos 70^\circ}{\cos 20^\circ}$

26. $\cot 40^\circ - \dfrac{\sin 50^\circ}{\sin 40^\circ}$

27. $\cos 35^\circ \sin 55^\circ + \sin 35^\circ \cos 55^\circ$

28. $\sec 35^\circ \csc 55^\circ - \tan 35^\circ \cot 55^\circ$

In Problems 29–42, use the right triangle shown in the margin. Then, using the given information, solve the triangle.

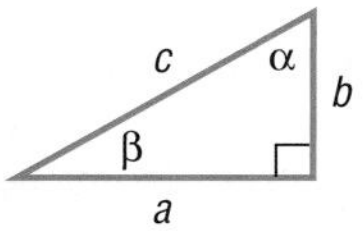

29. $b = 5, \quad \beta = 20^\circ;$ find $a, c,$ and α

30. $b = 4, \quad \beta = 10^\circ;$ find $a, c,$ and α

31. $a = 6, \quad \beta = 40^\circ;$ find $b, c,$ and α

32. $a = 7, \quad \beta = 50^\circ;$ find $b, c,$ and α

33. $b = 4, \quad \alpha = 10^\circ;$ find $a, c,$ and β

34. $b = 6, \quad \alpha = 20^\circ;$ find $a, c,$ and β

35. $a = 5, \quad \alpha = 25^\circ;$ find $b, c,$ and β

36. $a = 6, \quad \alpha = 40^\circ;$ find $b, c,$ and β

37. $c = 9, \quad \beta = 20^\circ;$ find $b, a,$ and α

38. $c = 10, \quad \alpha = 40^\circ;$ find $b, a,$ and β

39. $a = 5, \quad b = 3;$ find $c, \alpha,$ and β

40. $a = 2, \quad b = 8;$ find $c, \alpha,$ and β

41. $a = 2, \quad c = 5;$ find $b, \alpha,$ and β

42. $b = 4, \quad c = 6;$ find $a, \alpha,$ and β

43. Geometry A right triangle has a hypotenuse of length 8 inches. If one angle is 35°, find the length of each leg.

44. Geometry A right triangle has a hypotenuse of length 10 centimeters. If one angle is 40°, find the length of each leg.

45. Geometry A right triangle contains a 25° angle. If one leg is of length 5 inches, what is the length of the hypotenuse?

[**Hint:** Two answers are possible.]

46. Geometry A right triangle contains an angle of $\dfrac{\pi}{8}$ radian. If one leg is of length 3 meters, what is the length of the hypotenuse?

[**Hint:** Two answers are possible.]

47. Geometry The hypotenuse of a right triangle is 5 inches. If one leg is 2 inches, find the degree measure of each angle.

48. Geometry The hypotenuse of a right triangle is 3 feet. If one leg is 1 foot, find the degree measure of each angle.

49. Finding the Width of a Gorge Find the distance from A to C across the gorge illustrated in the figure.

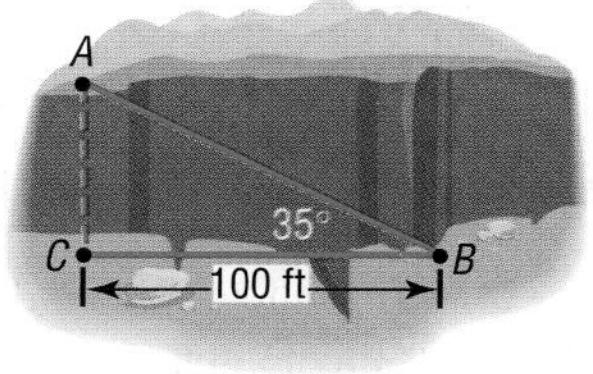

50. Finding the Distance across a Pond Find the distance from A to C across the pond illustrated in the figure.

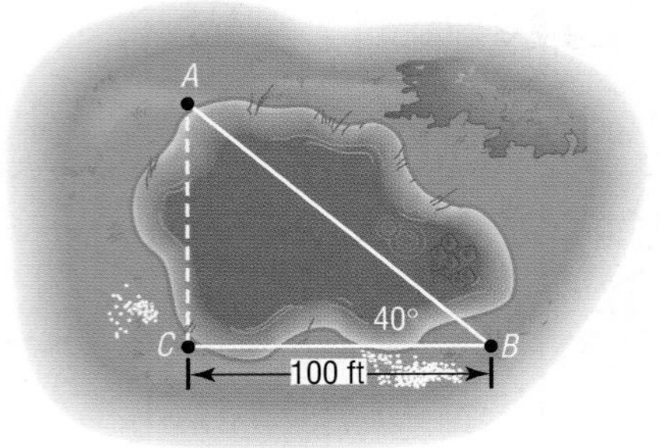

51. The Eiffel Tower The tallest tower built before the era of television masts, the Eiffel Tower was completed on March 31, 1889. Find the height of the Eiffel Tower (before a television mast was added to the top) using the information given in the illustration.

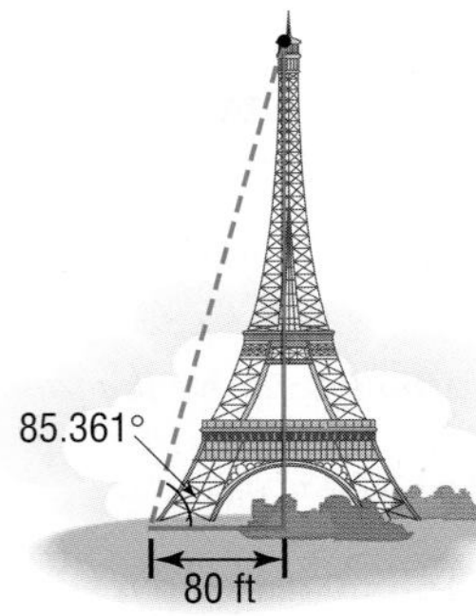

52. Finding the Distance of a Ship from Shore A ship, offshore from a vertical cliff known to be 100 feet in height, takes a sighting of the top of the cliff. If the angle of elevation is found to be 25°, how far offshore is the ship?

53. Finding the Distance to a Plateau Suppose that you are headed toward a plateau 50 meters high. If the angle of elevation to the top of the plateau is 20°, how far are you from the base of the plateau?

54. Statue of Liberty A ship is just offshore of New York City. A sighting is taken of the Statue of Liberty, which is about 305 feet tall. If the angle of elevation to the top of the statue is 20°, how far is the ship from the base of the statue?

55. Finding the Reach of a Ladder A 22-foot extension ladder leaning against a building makes a 70° angle with the ground. How far up the building does the ladder touch?

56. Finding the Height of a Building To measure the height of a building, two sightings are taken a distance of 50 feet apart. If the first angle of elevation is 40° and the second is 32°, what is the height of the building?

57. Finding the Distance between Two Objects A blimp, suspended in the air at a height of 500 feet, lies directly over a line from Soldier Field to the Adler Planetarium on Lake Michigan (see the figure). If the angle of depression from the blimp to the stadium is 32° and from the blimp to the planetarium is 23°, find the distance between Soldier Field and the Adler Planetarium.

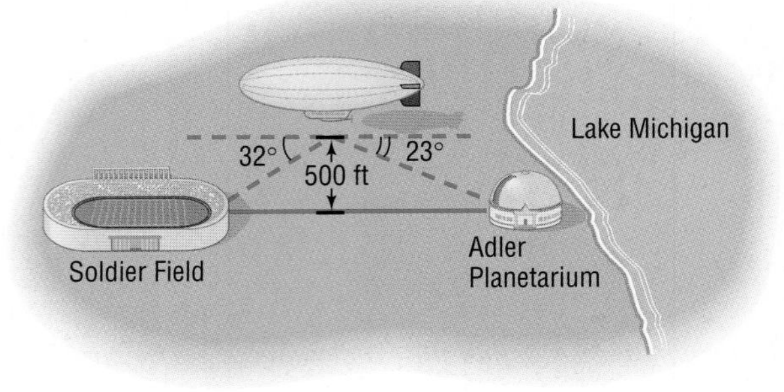

58. Finding the Angle of Elevation of the Sun At 10 AM on April 26, 2004, a building 300 feet high casts a shadow 50 feet long. What is the angle of elevation of the Sun?

59. Mt. Rushmore To measure the height of Lincoln's caricature on Mt. Rushmore, two sightings 800 feet from the base of the mountain are taken. If the angle of elevation to the bottom of Lincoln's face is 32° and the angle of elevation to the top is 35°, what is the height of Lincoln's face?

60. Directing a Laser Beam A laser beam is to be directed through a small hole in the center of a circle of radius 10 feet. The origin of the beam is 35 feet from the circle (see the figure). At what angle of elevation should the beam be aimed to ensure that it goes through the hole?

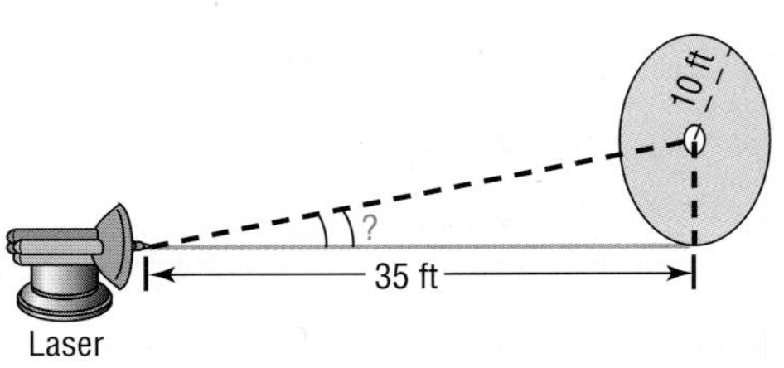

61. Finding the Length of a Guy Wire A radio transmission tower is 200 feet high. How long should a guy wire be if it is to be attached to the tower 10 feet from the top and is to make an angle of 21° with the ground?

62. Finding the Height of a Tower A guy wire 80 feet long is attached to the top of a radio transmission tower, making an angle of 25° with the ground. How high is the tower?

63. Washington Monument The angle of elevation of the Sun is 35.1° at the instant it casts a shadow 789 feet long of the Washington Monument. Use this information to calculate the height of the monument.

64. Finding the Length of a Mountain Trail A straight trail with an inclination of 17° leads from a hotel at an elevation of 9000 feet to a mountain lake at an elevation of 11,200 feet. What is the length of the trail?

65. Finding the Speed of a Truck A state trooper is hidden 30 feet from a highway. One second after a truck passes, the angle θ between the highway and the line of observation from the patrol car to the truck is measured. See the illustration.

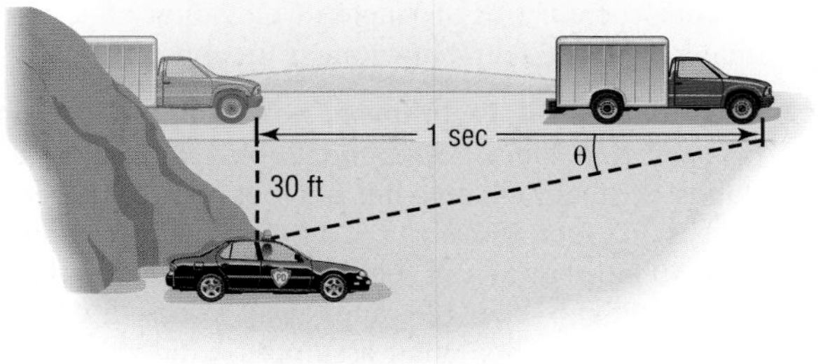

(a) If the angle measures 15°, how fast is the truck traveling? Express the answer in feet per second and in miles per hour.
(b) If the angle measures 20°, how fast is the truck traveling? Express the answer in feet per second and in miles per hour.
(c) If the speed limit is 55 miles per hour and a speeding ticket is issued for speeds of 5 miles per hour or more over the limit, for what angles should the trooper issue a ticket?

66. Security A security camera in a neighborhood bank is mounted on a wall 9 feet above the floor. What angle of depression should be used if the camera is to be directed to a spot 6 feet above the floor and 12 feet from the wall?

67. Finding the Bearing of an Aircraft A DC-9 aircraft leaves Midway Airport from runway 4 RIGHT, whose bearing is N40°E. After flying for $\frac{1}{2}$ mile, the pilot requests permission to turn 90° and head toward the southeast. The permission is granted. After the airplane goes 1 mile in this direction, what bearing should the control tower use to locate the aircraft?

68. Finding the Bearing of a Ship A ship leaves the port of Miami with a bearing of S80°E and a speed of 15 knots. After 1 hour, the ship turns 90° toward the south. After 2 hours, maintaining the same speed, what is the bearing to the ship from port?

69. Finding the Pitch of a Roof A carpenter is preparing to put a roof on a garage that is 20 feet by 40 feet by 20 feet. A steel support beam 46 feet in length is positioned in the center of the garage. To support the roof, another beam will be attached to the top of the center beam (see the figure). At what angle of elevation is the new beam? In other words, what is the pitch of the roof?

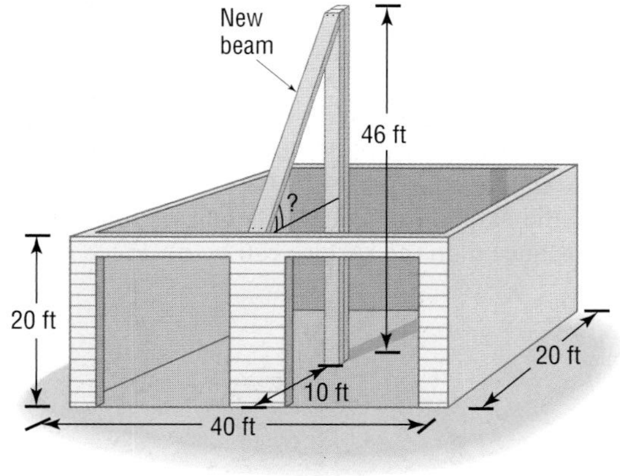

70. Shooting Free Throws in Basketball The eyes of a basketball player are 6 feet above the floor. The player is at the free-throw line, which is 15 feet from the center of the basket rim (see the figure). What is the angle of elevation from the player's eyes to the center of the rim? [**Hint:** The rim is 10 feet above the floor.]

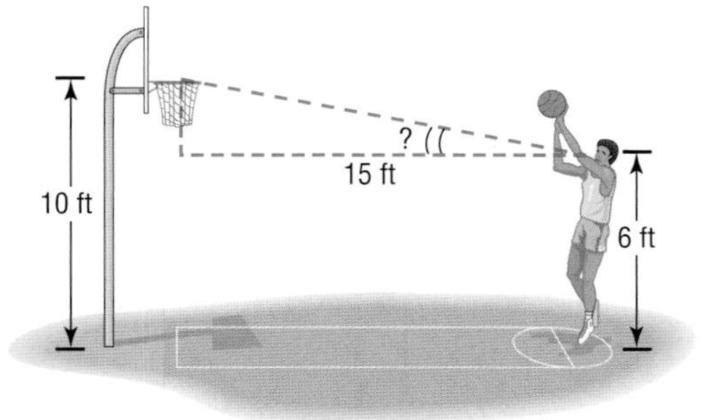

71. Constructing a Highway A highway whose primary directions are north–south is being constructed along the west coast of Florida. Near Naples, a bay obstructs the straight path of the road. Since the cost of a bridge is prohibitive, engineers decide to go around the bay. The illustration shows the path that they decide on and the measurements taken. What is the length of highway needed to go around the bay?

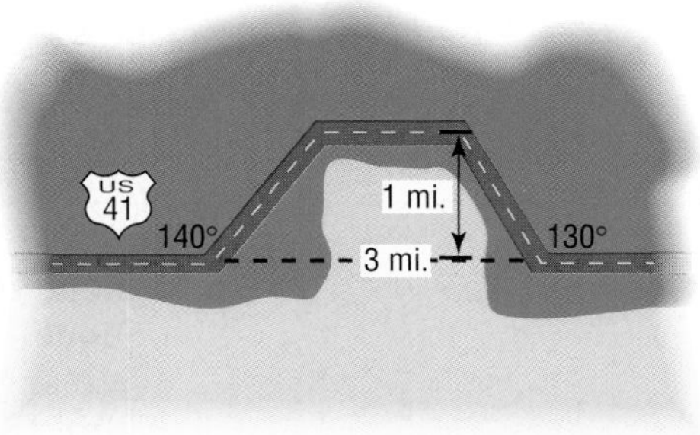

72. Surveillance Satellites A surveillance satellite circles Earth at a height of h miles above the surface. Suppose that d is the distance, in miles, on the surface of Earth that can be observed from the satellite. See the illustration.

(a) Find an equation that relates the central angle θ to the height h.

(b) Find an equation that relates the observable distance d and θ.

(c) Find an equation that relates d and h.

(d) If d is to be 2500 miles, how high must the satellite orbit above Earth?

(e) If the satellite orbits at a height of 300 miles, what distance d on the surface can be observed?

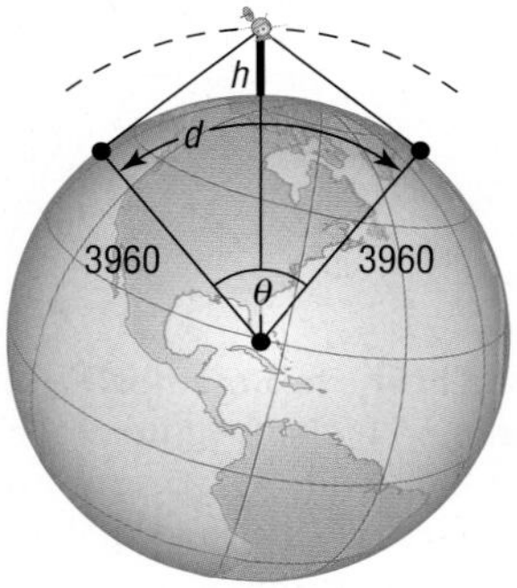

73. Photography A camera is mounted on a tripod 4 feet high at a distance of 10 feet from George, who is 6 feet tall. See the illustration. If the camera lens has angles of depression and elevation of 20°, will George's feet and head be seen by the lens? If not, how far back will the camera need to be moved to include George's feet and head?

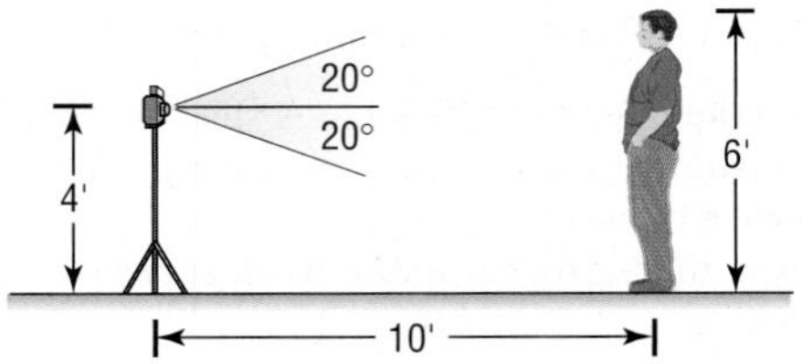

74. Construction A ramp for wheelchair accessibility is to be constructed with an angle of elevation of 15° and a final height of 5 feet. How long is the ramp?

75. The Gibb's Hill Lighthouse, Southampton, Bermuda In operation since 1846, the Gibb's Hill Lighthouse stands 117 feet high on a hill 245 feet high, so its beam of light is 362 feet above sea level. A brochure states that ships 40 miles away can see the light and planes flying at 10,000 feet can see it 120 miles away. Verify the accuracy of these statements. What assumption did the brochure make about the height of the ship?

76. Explain how you would measure the width of the Grand Canyon from a point on its ridge.

77. Explain how you would measure the height of a TV tower that is on the roof of a tall building.

'Are You Prepared?' Answers

1. 4 **2.** 26.6°

7.2 The Law of Sines

PREPARING FOR THIS SECTION *Before getting started, review the following:*

- Trigonometric Equations (I) (Section 6.7, pp. 461–464)
- Difference Formula for Sine (Section 6.4, p. 441)

Now work the 'Are You Prepared?' problems on page 502.

OBJECTIVES 1 Solve SAA or ASA Triangles
2 Solve SSA Triangles
3 Solve Applied Problems

If none of the angles of a triangle is a right angle, the triangle is called **oblique.** An oblique triangle will have either three acute angles or two acute angles and one obtuse angle (an angle between 90° and 180°). See Figure 18.

Figure 18

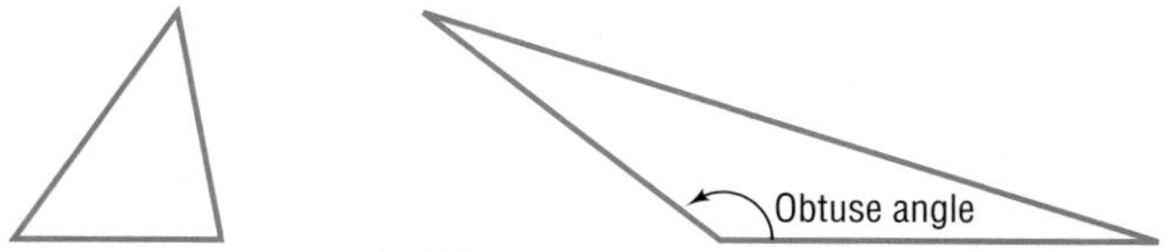

(a) All angles are acute (b) Two acute angles and one obtuse angle

Figure 19

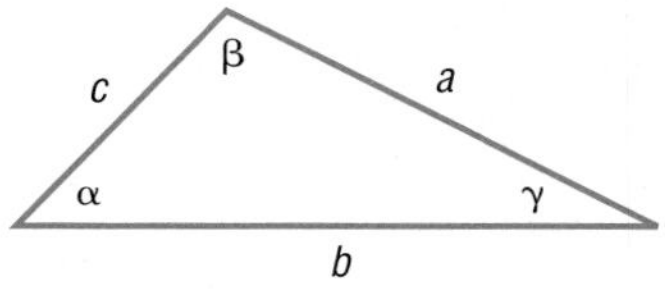

In the discussion that follows, we will always label an oblique triangle so that side a is opposite angle α, side b is opposite angle β, and side c is opposite angle γ, as shown in Figure 19.

To **solve an oblique triangle** means to find the lengths of its sides and the measurements of its angles. To do this, we shall need to know the length of one side* along with: (i) two angles; (ii) one angle and one other side; or (iii) the other two sides. There are four possibilities to consider:

CASE 1: One side and two angles are known (ASA or SAA).
CASE 2: Two sides and the angle opposite one of them are known (SSA).
CASE 3: Two sides and the included angle are known (SAS).
CASE 4: Three sides are known (SSS).

Figure 20 illustrates the four cases.

Figure 20

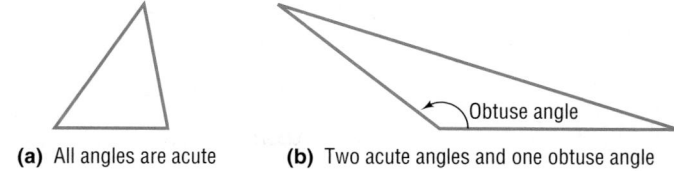

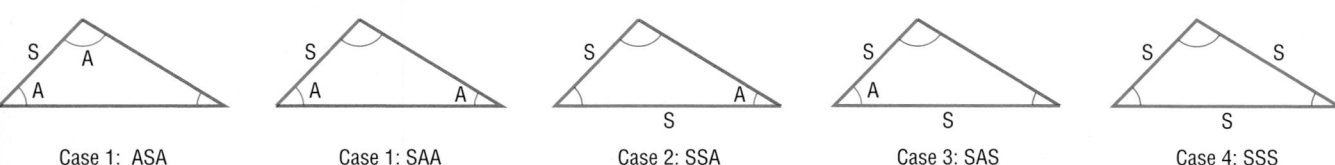

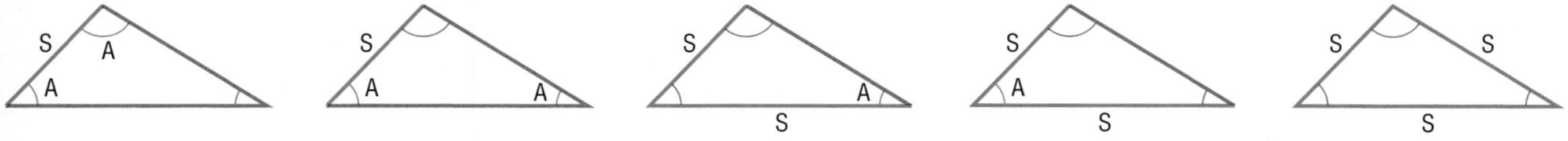

Case 1: ASA Case 1: SAA Case 2: SSA Case 3: SAS Case 4: SSS

The **Law of Sines** is used to solve triangles for which Case 1 or 2 holds. Cases 3 and 4 are considered when we study the Law of Cosines in the next section.

Theorem

Law of Sines

For a triangle with sides a, b, c and opposite angles α, β, γ, respectively,

$$\frac{\sin \alpha}{a} = \frac{\sin \beta}{b} = \frac{\sin \gamma}{c} \qquad (1)$$

A proof of the Law of Sines is given at the end of this section.

*The reason we need to know the length of one side is that, if we only know the angles, this will result in a family of *similar triangles*.

In applying the Law of Sines to solve triangles, we use the fact that the sum of the angles of any triangle equals 180°; that is,

$$\alpha + \beta + \gamma = 180^\circ \qquad (2)$$

1 Our first two examples show how to solve a triangle when one side and two angles are known (Case 1: SAA or ASA).

EXAMPLE 1 Using the Law of Sines to Solve a SAA Triangle

Solve the triangle: $\alpha = 40^\circ, \beta = 60^\circ, a = 4$

Figure 21

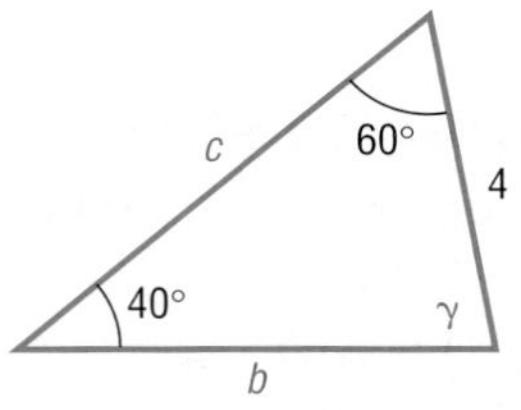

Solution Figure 21 shows the triangle that we want to solve. The third angle γ is found using equation (2).

$$\alpha + \beta + \gamma = 180^\circ$$
$$40^\circ + 60^\circ + \gamma = 180^\circ$$
$$\gamma = 80^\circ$$

Now we use the Law of Sines (twice) to find the unknown sides b and c.

$$\frac{\sin\alpha}{a} = \frac{\sin\beta}{b} \qquad \frac{\sin\alpha}{a} = \frac{\sin\gamma}{c}$$

Because $a = 4, \alpha = 40^\circ, \beta = 60^\circ$, and $\gamma = 80^\circ$, we have

$$\frac{\sin 40^\circ}{4} = \frac{\sin 60^\circ}{b} \qquad \frac{\sin 40^\circ}{4} = \frac{\sin 80^\circ}{c}$$

Solving for b and c, we find that

$$b = \frac{4\sin 60^\circ}{\sin 40^\circ} \approx 5.39 \qquad c = \frac{4\sin 80^\circ}{\sin 40^\circ} \approx 6.13$$

◀

Notice in Example 1 that we found b and c by working with the given side a. This is better than finding b first and working with a rounded value of b to find c.

NOW WORK PROBLEM 9.

EXAMPLE 2 Using the Law of Sines to Solve an ASA Triangle

Figure 22

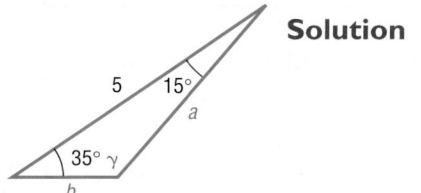

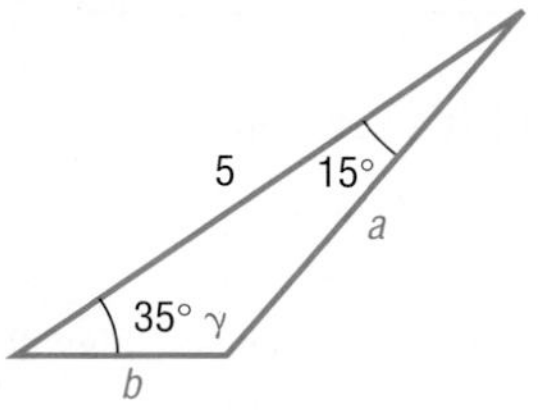

Solve the triangle: $\alpha = 35^\circ, \beta = 15^\circ, c = 5$

Solution Figure 22 illustrates the triangle that we want to solve. Because we know two angles ($\alpha = 35^\circ$ and $\beta = 15^\circ$), we find the third angle using equation (2).

$$\alpha + \beta + \gamma = 180^\circ$$
$$35^\circ + 15^\circ + \gamma = 180^\circ$$
$$\gamma = 130^\circ$$

Now we know the three angles and one side ($c = 5$) of the triangle. To find the remaining two sides a and b, we use the Law of Sines (twice).

$$\frac{\sin \alpha}{a} = \frac{\sin \gamma}{c} \qquad \frac{\sin \beta}{b} = \frac{\sin \gamma}{c}$$

$$\frac{\sin 35^\circ}{a} = \frac{\sin 130^\circ}{5} \qquad \frac{\sin 15^\circ}{b} = \frac{\sin 130^\circ}{5}$$

$$a = \frac{5 \sin 35^\circ}{\sin 130^\circ} \approx 3.74 \qquad b = \frac{5 \sin 15^\circ}{\sin 130^\circ} \approx 1.69$$

◀

NOW WORK PROBLEM 23.

The Ambiguous Case

Figure 23

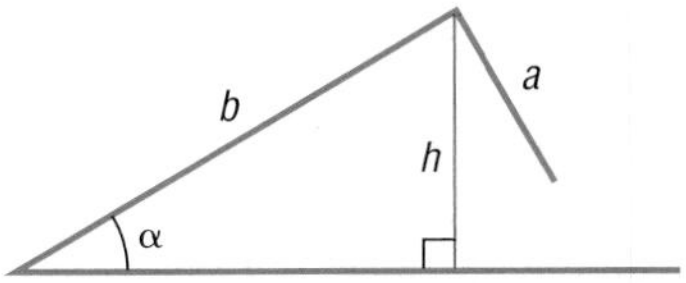

2 Case 2 (SSA), which applies to triangles for which two sides and the angle opposite one of them are known, is referred to as the **ambiguous case,** because the known information may result in one triangle, two triangles, or no triangle at all. Suppose that we are given sides a and b and angle α, as illustrated in Figure 23. The key to determining the possible triangles, if any, that may be formed from the given information lies primarily with the height h and the fact that $h = b \sin \alpha$.

No Triangle If $a < h = b \sin \alpha$, then side a is not sufficiently long to form a triangle. See Figure 24.

Figure 24
$a < h = b \sin \alpha$

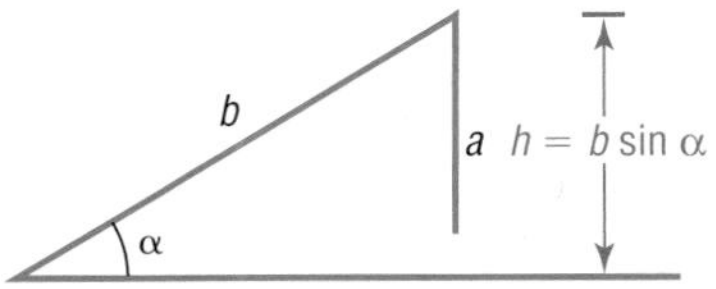

One Right Triangle If $a = h = b \sin \alpha$, then side a is just long enough to form a right triangle. See Figure 25.

Figure 25
$a = b \sin \alpha$

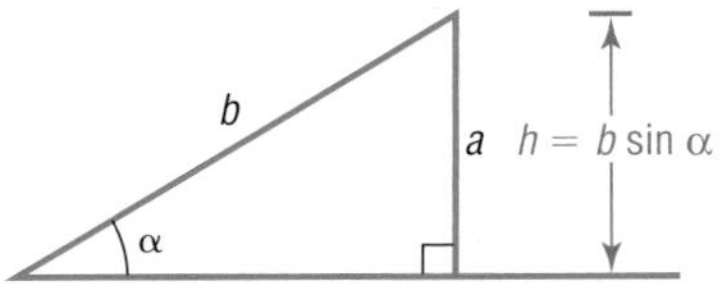

Two Triangles If $a < b$ and $h = b \sin \alpha < a$, then two distinct triangles can be formed from the given information. See Figure 26.

Figure 26
$b \sin \alpha < a$ and $a < b$

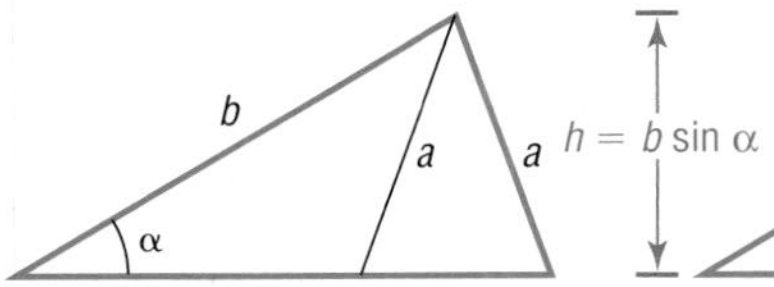

One Triangle If $a \geq b$, then only one triangle can be formed. See Figure 27.

Figure 27
$a \geq b$

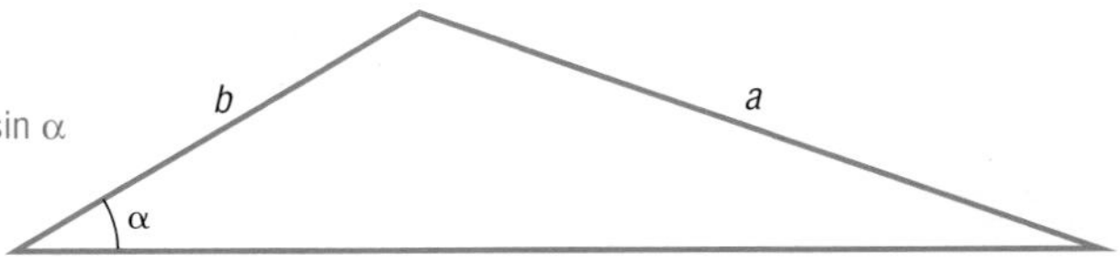

Fortunately, we do not have to rely on an illustration to draw the correct conclusion in the ambiguous case. The Law of Sines will lead us to the correct determination. Let's see how.

EXAMPLE 3 Using the Law of Sines to Solve a SSA Triangle (One Solution)

Solve the triangle: $a = 3, b = 2, \alpha = 40°$

Figure 28(a)

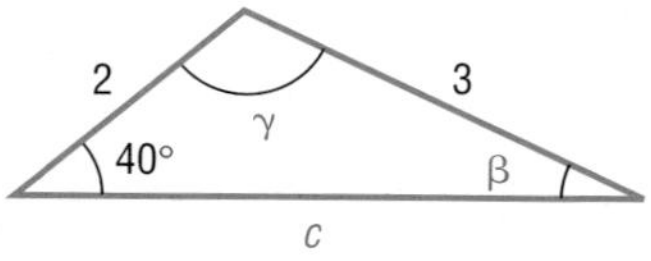

Solution See Figure 28(a). Because $a = 3, b = 2$, and $\alpha = 40°$ are known, we use the Law of Sines to find the angle β.

$$\frac{\sin \alpha}{a} = \frac{\sin \beta}{b}$$

Then

$$\frac{\sin 40°}{3} = \frac{\sin \beta}{2}$$

$$\sin \beta = \frac{2 \sin 40°}{3} \approx 0.43$$

There are two angles β, $0° < \beta < 180°$, for which $\sin \beta \approx 0.43$.

$$\beta_1 \approx 25.4° \quad \text{and} \quad \beta_2 \approx 154.6°$$

Note: Here we computed β using the stored value of $\sin \beta$. If you use the rounded value, $\sin \beta \approx 0.43$, you will obtain slightly different results.

The second possibility, $\beta_2 \approx 154.6°$, is ruled out, because $\alpha = 40°$, making $\alpha + \beta_2 \approx 194.6° > 180°$. Now, using $\beta_1 \approx 25.4°$, we find that

$$\gamma = 180° - \alpha - \beta_1 \approx 180° - 40° - 25.4° = 114.6°$$

The third side c may now be determined using the Law of Sines.

$$\frac{\sin \alpha}{a} = \frac{\sin \gamma}{c}$$

$$\frac{\sin 40°}{3} = \frac{\sin 114.6°}{c}$$

$$c = \frac{3 \sin 114.6°}{\sin 40°} \approx 4.24$$

Figure 28(b)

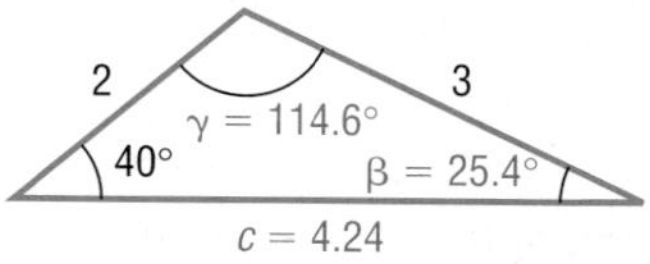

Figure 28(b) illustrates the solved triangle. ◀

EXAMPLE 4 Using the Law of Sines to Solve a SSA Triangle (Two Solutions)

Solve the triangle: $a = 6, b = 8, \alpha = 35°$

Figure 29(a)

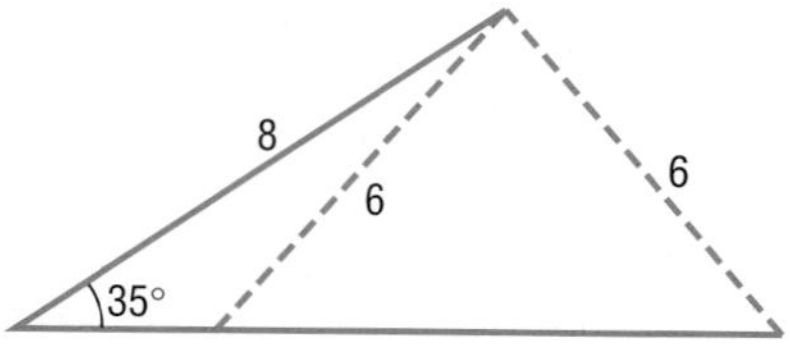

Solution See Figure 29(a). Because $a = 6, b = 8$, and $\alpha = 35°$ are known, we use the Law of Sines to find the angle β.

$$\frac{\sin \alpha}{a} = \frac{\sin \beta}{b}$$

Then

$$\frac{\sin 35°}{6} = \frac{\sin \beta}{8}$$

$$\sin \beta = \frac{8 \sin 35°}{6} \approx 0.76$$

$$\beta_1 \approx 49.9° \quad \text{or} \quad \beta_2 \approx 130.1°$$

For both choices of β, we have $\alpha + \beta < 180°$. There are two triangles, one containing the angle $\beta_1 \approx 49.9°$ and the other containing the angle $\beta_2 \approx 130.1°$. The third angle γ is either

$$\gamma_1 = 180° - \alpha - \beta_1 \approx 95.1° \quad \text{or} \quad \gamma_2 = 180° - \alpha - \beta_2 \approx 14.9°$$

$$\uparrow \; \alpha = 35°, \; \beta_1 = 49.9° \qquad \uparrow \; \alpha = 35°, \; \beta_2 = 130.1°$$

The third side c obeys the Law of Sines, so we have

Figure 29(b)

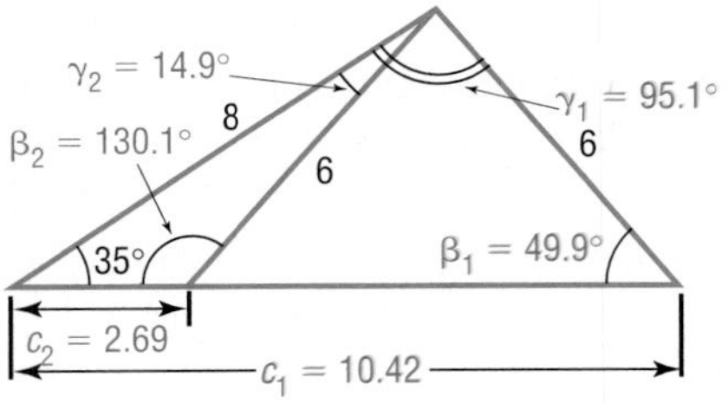

$$\frac{\sin \alpha}{a} = \frac{\sin \gamma_1}{c_1} \qquad \frac{\sin \alpha}{a} = \frac{\sin \gamma_2}{c_2}$$

$$\frac{\sin 35°}{6} = \frac{\sin 95.1°}{c_1} \qquad \frac{\sin 35°}{6} = \frac{\sin 14.9°}{c_2}$$

$$c_1 = \frac{6 \sin 95.1°}{\sin 35°} \approx 10.42 \qquad c_2 = \frac{6 \sin 14.9°}{\sin 35°} \approx 2.69$$

The two solved triangles are illustrated in Figure 29(b). ◀

EXAMPLE 5

Using the Law of Sines to Solve a SSA Triangle (No Solution)

Solve the triangle: $a = 2, c = 1, \gamma = 50°$

Solution Because $a = 2$, $c = 1$, and $\gamma = 50°$ are known, we use the Law of Sines to find the angle α.

$$\frac{\sin \alpha}{a} = \frac{\sin \gamma}{c}$$

$$\frac{\sin \alpha}{2} = \frac{\sin 50°}{1}$$

$$\sin \alpha = 2 \sin 50° \approx 1.53$$

Figure 30

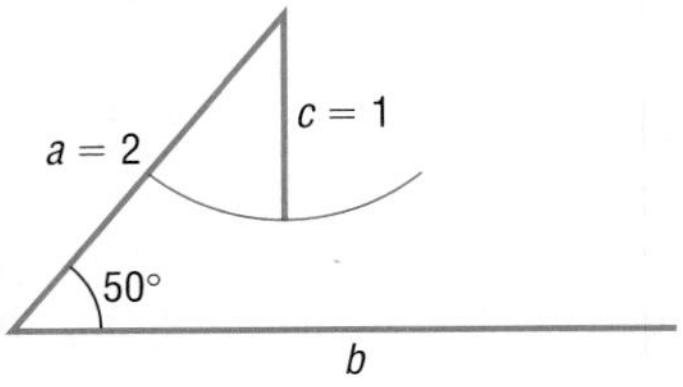

Since there is no angle α for which $\sin \alpha > 1$, there can be no triangle with the given measurements. Figure 30 illustrates the measurements given. Notice that, no matter how we attempt to position side c, it will never touch side b to form a triangle. ◀

NOW WORK PROBLEMS 25 AND 31.

Applications

3 The Law of Sines is particularly useful for solving certain applied problems.

EXAMPLE 6 Finding the Height of a Mountain

To measure the height of a mountain, a surveyor takes two sightings of the peak at a distance 900 meters apart on a direct line to the mountain.* See Figure 31(a). The first observation results in an angle of elevation of 47°, whereas the second results in an angle of elevation of 35°. If the transit is 2 meters high, what is the height h of the mountain?

Figure 31

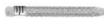

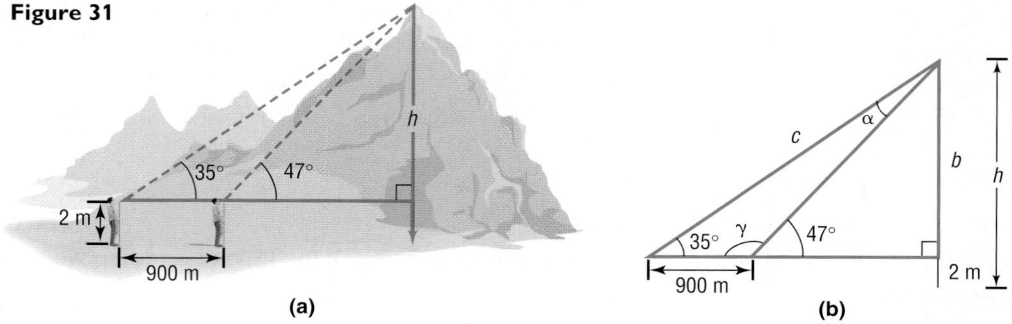

Solution Figure 31(b) shows the triangles that replicate the illustration in Figure 31(a). Since $\gamma + 47° = 180°$, we find that $\gamma = 133°$. Also, since $\alpha + \gamma + 35° = 180°$, we find that $\alpha = 180° - 35° - \gamma = 145° - 133° = 12°$. We use the Law of Sines to find c.

$$\frac{\sin \alpha}{a} = \frac{\sin \gamma}{c} \qquad \alpha = 12°, \gamma = 133°, a = 900$$

$$c = \frac{900 \sin 133°}{\sin 12°} \approx 3165.86$$

Using the larger right triangle, we have

$$\sin 35° = \frac{b}{c} \qquad c = 3165.86$$

$$b = 3165.86 \sin 35° \approx 1815.86 \approx 1816 \text{ meters}$$

The height of the peak from ground level is approximately $1816 + 2 = 1818$ meters. ◀

NOW WORK PROBLEM 39.

EXAMPLE 7 Rescue at Sea

Coast Guard Station Zulu is located 120 miles due west of Station X-ray. A ship at sea sends an SOS call that is received by each station. The call to Station Zulu indicates that the bearing of the ship from Zulu is N40°E (40° east of north). The call to Station X-ray indicates that the bearing of the ship from X-ray is N30°W (30° west of north).

(a) How far is each station from the ship?

(b) If a helicopter capable of flying 200 miles per hour is dispatched from the nearest station to the ship, how long will it take to reach the ship?

*For simplicity, we assume that these sightings are at the same level.

Figure 32

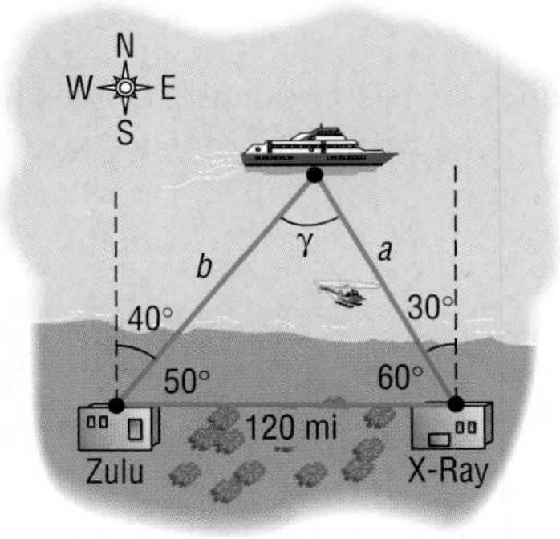

Solution (a) Figure 32 illustrates the situation. The angle γ is found to be

$$\gamma = 180^\circ - 50^\circ - 60^\circ = 70^\circ$$

The Law of Sines can now be used to find the two distances a and b that we seek.

$$\frac{\sin 50^\circ}{a} = \frac{\sin 70^\circ}{120}$$

$$a = \frac{120 \sin 50^\circ}{\sin 70^\circ} \approx 97.82 \text{ miles}$$

$$\frac{\sin 60^\circ}{b} = \frac{\sin 70^\circ}{120}$$

$$b = \frac{120 \sin 60^\circ}{\sin 70^\circ} \approx 110.59 \text{ miles}$$

Station Zulu is about 111 miles from the ship, and Station X-ray is about 98 miles from the ship.

(b) The time t needed for the helicopter to reach the ship from Station X-ray is found by using the formula

$$(\text{Velocity}, v)(\text{Time}, t) = \text{Distance}, a$$

Then

$$t = \frac{a}{v} = \frac{97.82}{200} \approx 0.49 \text{ hour} \approx 29 \text{ minutes}$$

It will take about 29 minutes for the helicopter to reach the ship. ◀

NOW WORK PROBLEM 37.

Figure 33

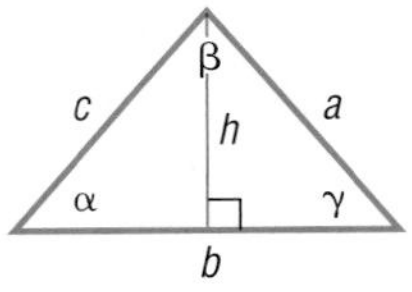

(a)

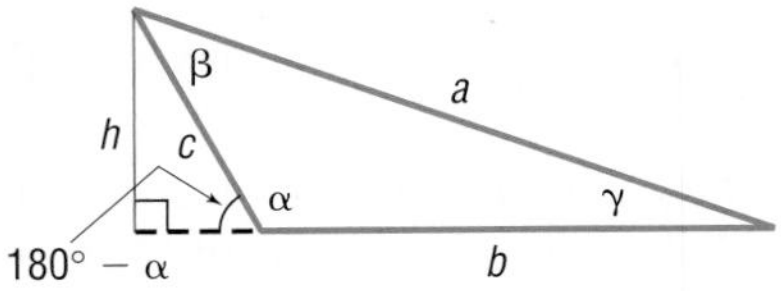

(b)

Proof of the Law of Sines To prove the Law of Sines, we construct an altitude of length h from one of the vertices of a triangle. Figure 33(a) shows h for a triangle with three acute angles, and Figure 33(b) shows h for a triangle with an obtuse angle. In each case, the altitude is drawn from the vertex at β. Using either illustration, we have

$$\sin \gamma = \frac{h}{a}$$

from which

$$h = a \sin \gamma \qquad (3)$$

From Figure 33(a), it also follows that

$$\sin \alpha = \frac{h}{c}$$

from which

$$h = c \sin \alpha \qquad (4)$$

From Figure 33(b), it follows that

$$\sin(180^\circ - \alpha) \underset{\uparrow}{=} \sin \alpha = \frac{h}{c}$$

Difference formula

which again gives

$$h = c \sin \alpha$$

So, whether the triangle has three acute angles or has two acute angles and one obtuse angle, equations (3) and (4) hold. As a result, we may equate the expressions for h in equations (3) and (4) to get

$$a \sin \gamma = c \sin \alpha$$

from which

$$\frac{\sin \alpha}{a} = \frac{\sin \gamma}{c} \qquad \textbf{(5)}$$

Figure 34

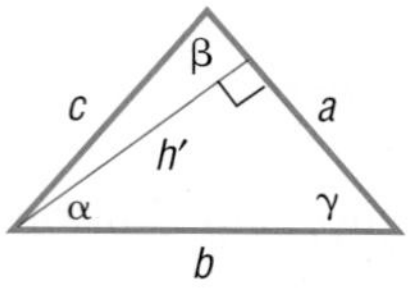

(a)

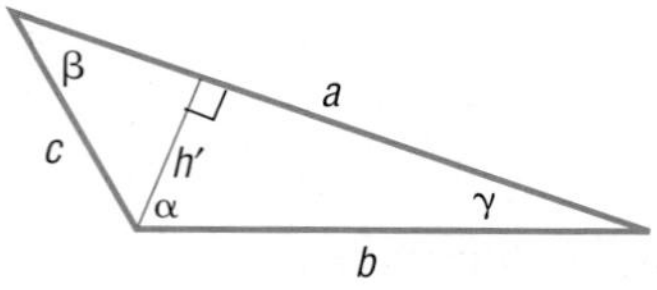

(b)

In a similar manner, by constructing the altitude h' from the vertex of angle α as shown in Figure 34, we can show that

$$\sin \beta = \frac{h'}{c} \quad \text{and} \quad \sin \gamma = \frac{h'}{b}$$

Equating the expressions for h', we find that

$$h' = c \sin \beta = b \sin \gamma$$

from which

$$\frac{\sin \beta}{b} = \frac{\sin \gamma}{c} \qquad \textbf{(6)}$$

When equations (5) and (6) are combined, we have equation (1), the Law of Sines. ■

7.2 Assess Your Understanding

'Are You Prepared?' *Answers are given at the end of these exercises. If you get a wrong answer, read the pages listed in red.*

1. The difference formula for sine is $\sin(\alpha - \beta) =$ ________. (p. 441)

2. If θ is an acute angle, solve the equation $\sin \theta = \frac{1}{2}$. (pp. 461–464)

3. If θ is an acute angle, solve the equation $\sin \theta = 2$. (pp. 461–464)

Concepts and Vocabulary

4. If none of the angles of a triangle is a right angle, the triangle is called ________.

5. For a triangle with sides a, b, c and opposite angles α, β, γ, the Law of Sines states that ________.

6. *True or False:* An oblique triangle in which two sides and an angle are given always results in at least one triangle.

7. *True or False:* The sum of the angles of any triangle equals 180°.

8. *True or False:* The ambiguous case refers to the fact that, when two sides and the angle opposite one of them are given, sometimes the Law of Sines cannot be used.

Exercises

In Problems 9–16, solve each triangle.

9.

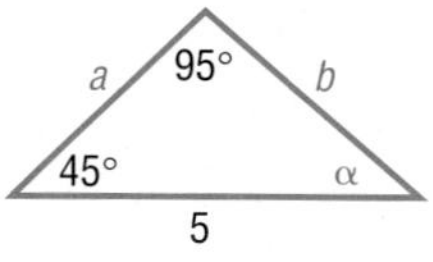

10.

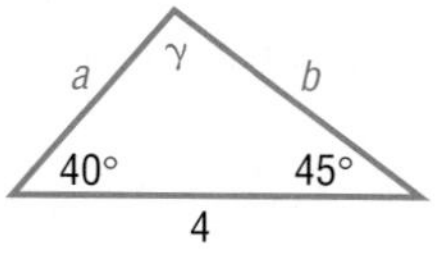

11.

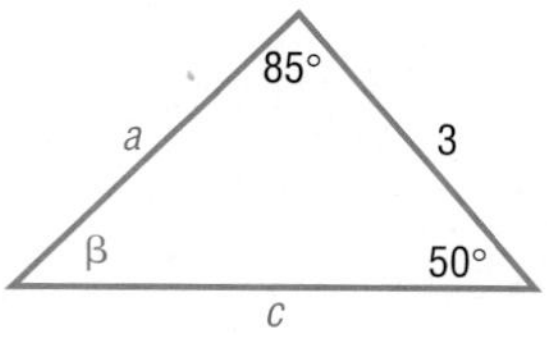

12.

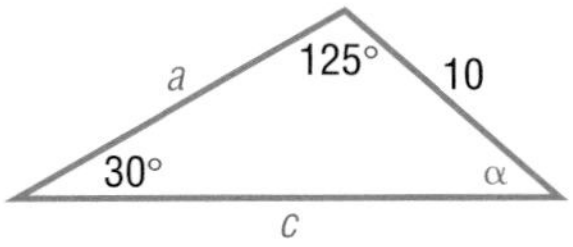

13.

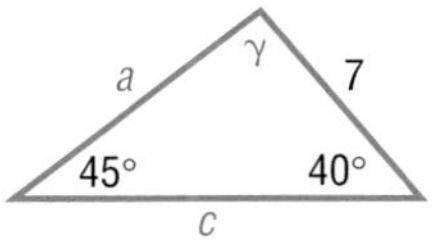

14.

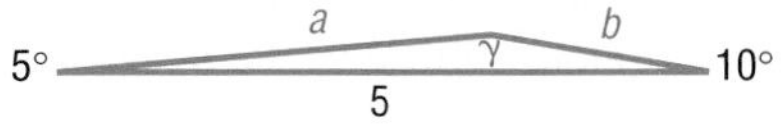

15.

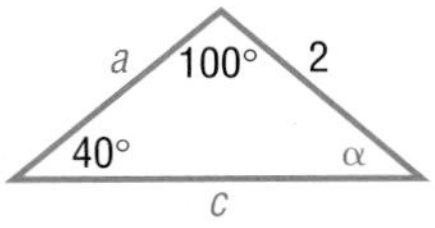

16.

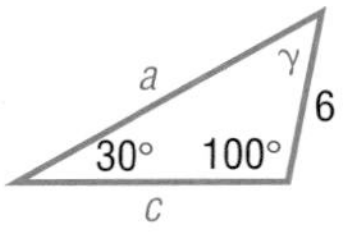

In Problems 17–24, solve each triangle.

17. $\alpha = 40°, \quad \beta = 20°, \quad a = 2$

18. $\alpha = 50°, \quad \gamma = 20°, \quad a = 3$

19. $\beta = 70°, \quad \gamma = 10°, \quad b = 5$

20. $\alpha = 70°, \quad \beta = 60°, \quad c = 4$

21. $\alpha = 110°, \quad \gamma = 30°, \quad c = 3$

22. $\beta = 10°, \quad \gamma = 100°, \quad b = 2$

23. $\alpha = 40°, \quad \beta = 40°, \quad c = 2$

24. $\beta = 20°, \quad \gamma = 70°, \quad a = 1$

In Problems 25–36, two sides and an angle are given. Determine whether the given information results in one triangle, two triangles, or no triangle at all. Solve any triangle(s) that results.

25. $a = 3, \quad b = 2, \quad \alpha = 50°$

26. $b = 4, \quad c = 3, \quad \beta = 40°$

27. $b = 5, \quad c = 3, \quad \beta = 100°$

28. $a = 2, \quad c = 1, \quad \alpha = 120°$

29. $a = 4, \quad b = 5, \quad \alpha = 60°$

30. $b = 2, \quad c = 3, \quad \beta = 40°$

31. $b = 4, \quad c = 6, \quad \beta = 20°$

32. $a = 3, \quad b = 7, \quad \alpha = 70°$

33. $a = 2, \quad c = 1, \quad \gamma = 100°$

34. $b = 4, \quad c = 5, \quad \beta = 95°$

35. $a = 2, \quad c = 1, \quad \gamma = 25°$

36. $b = 4, \quad c = 5, \quad \beta = 40°$

---

37. Rescue at Sea Coast Guard Station Able is located 150 miles due south of Station Baker. A ship at sea sends an SOS call that is received by each station. The call to Station Able indicates that the ship is located N55°E; the call to Station Baker indicates that the ship is located S60°E.

(a) How far is each station from the ship?

(b) If a helicopter capable of flying 200 miles per hour is dispatched from the nearest station to the ship, how long will it take to reach the ship?

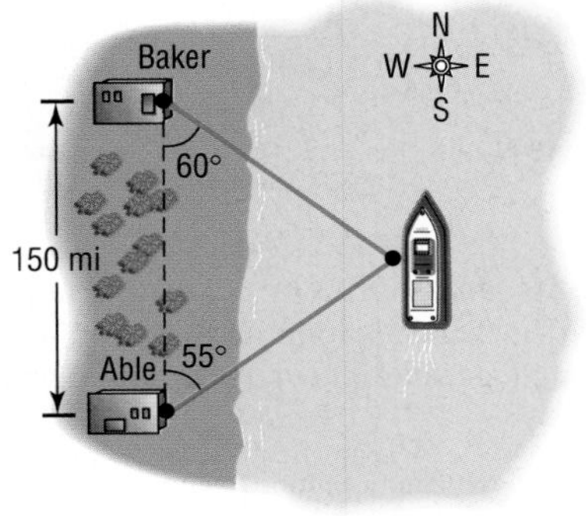

38. Surveying Consult the figure. To find the distance from the house at A to the house at B, a surveyor measures the angle BAC to be 40° and then walks off a distance of 100 feet to C and measures the angle ACB to be 50°. What is the distance from A to B?

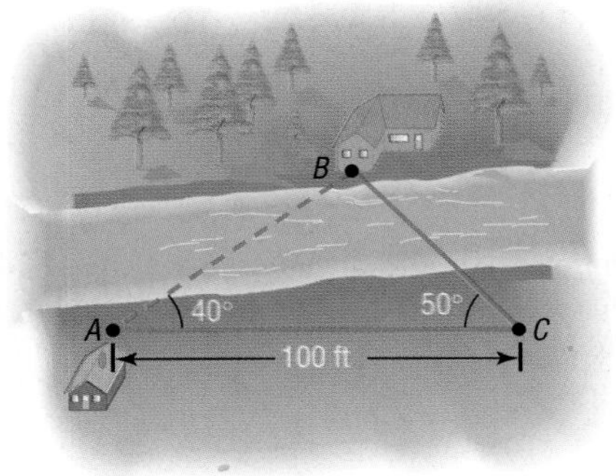

39. Finding the Length of a Ski Lift Consult the figure. To find the length of the span of a proposed ski lift from A to B, a surveyor measures the angle DAB to be 25° and then walks off a distance of 1000 feet to C and measures the angle ACB to be 15°. What is the distance from A to B?

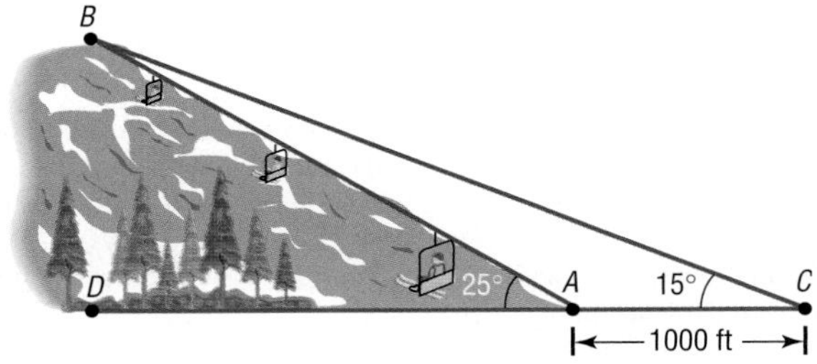

40. Finding the Height of a Mountain Use the illustration in Problem 39 to find the height BD of the mountain at B.

41. Finding the Height of an Airplane An aircraft is spotted by two observers who are 1000 feet apart. As the airplane passes over the line joining them, each observer takes a sighting of the angle of elevation to the plane, as indicated in the figure. How high is the airplane?

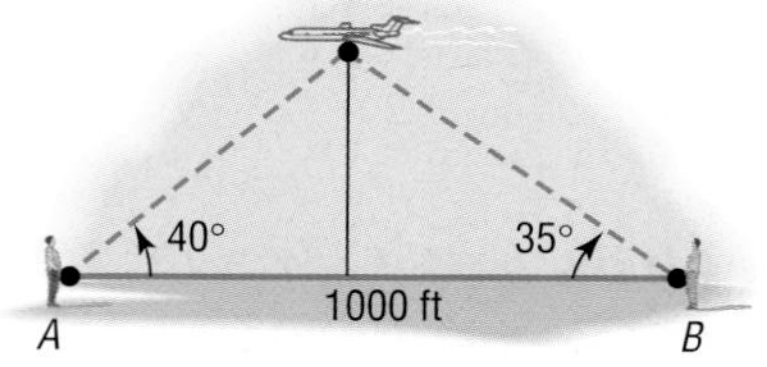

42. Finding the Height of the Bridge over the Royal Gorge The highest bridge in the world is the bridge over the Royal Gorge of the Arkansas River in Colorado. Sightings to the same point at water level directly under the bridge are taken from each side of the 880-foot-long bridge, as indicated in the figure. How high is the bridge?

SOURCE: *Guinness Book of World Records*.

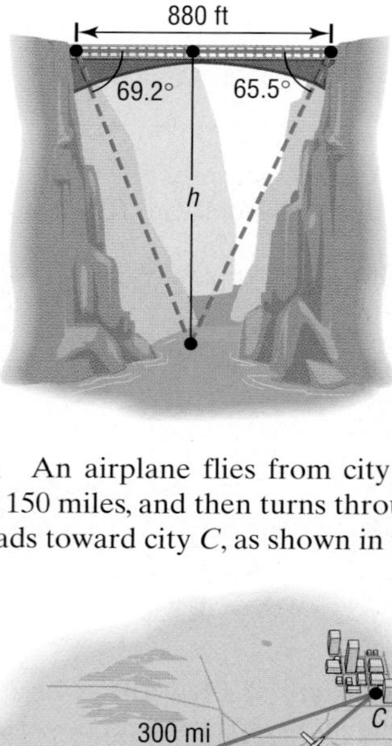

43. Navigation An airplane flies from city A to city B, a distance of 150 miles, and then turns through an angle of 40° and heads toward city C, as shown in the figure.

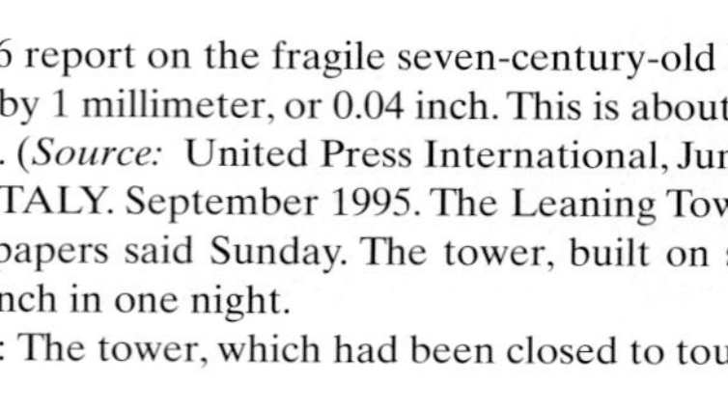

(a) If the distance between cities A and C is 300 miles, how far is it from city B to city C?

(b) Through what angle should the pilot turn at city C to return to city A?

44. Time Lost due to a Navigation Error In attempting to fly from city A to city B, an aircraft followed a course that was 10° in error, as indicated in the figure. After flying a distance of 50 miles, the pilot corrected the course by turning at point C and flying 70 miles farther. If the constant speed of the aircraft was 250 miles per hour, how much time was lost due to the error?

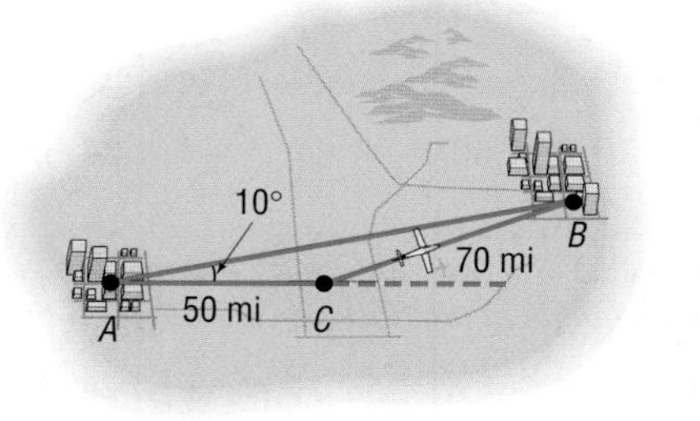

45. Finding the Lean of the Leaning Tower of Pisa The famous Leaning Tower of Pisa was originally 184.5 feet high.† At a distance of 123 feet from the base of the tower, the angle of elevation to the top of the tower is found to be 60°. Find the angle CAB indicated in the figure. Also, find the perpendicular distance from C to AB.

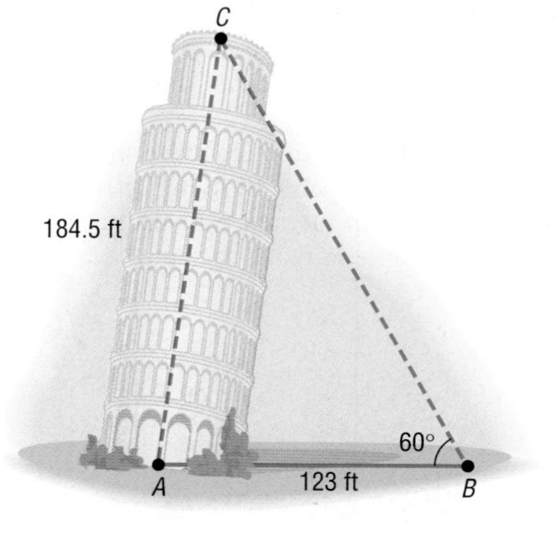

46. Crankshafts on Cars On a certain automobile, the crankshaft is 3 inches long and the connecting rod is 9 inches long (see the figure). At the time when the angle

†In their 1986 report on the fragile seven-century-old bell tower, scientists in Pisa, Italy, said that the Leaning Tower of Pisa had increased its famous lean by 1 millimeter, or 0.04 inch. This is about the annual average, although the tilting had slowed to about half that much in the previous 2 years. (*Source:* United Press International, June 29, 1986.)

PISA, ITALY. September 1995. The Leaning Tower of Pisa has suddenly shifted, jeopardizing years of preservation work to stabilize it, Italian newspapers said Sunday. The tower, built on shifting subsoil between 1174 and 1350 as a belfry for the nearby cathedral, recently moved 0.07 inch in one night.

Update: The tower, which had been closed to tourists since 1990, was reopened in December, 2001 after reinforcement of its base.

OPA is 15°, how far is the piston (P) from the center (O) of the crankshaft?

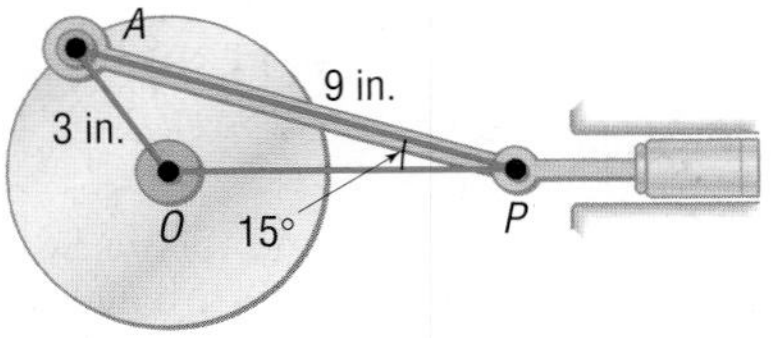

47. Constructing a Highway U.S. 41, a highway whose primary directions are north–south, is being constructed along the west coast of Florida. Near Naples, a bay obstructs the straight path of the road. Since the cost of a bridge is prohibitive, engineers decide to go around the bay. The illustration shows the path that they decide on and the measurements taken. What is the length of highway needed to go around the bay?

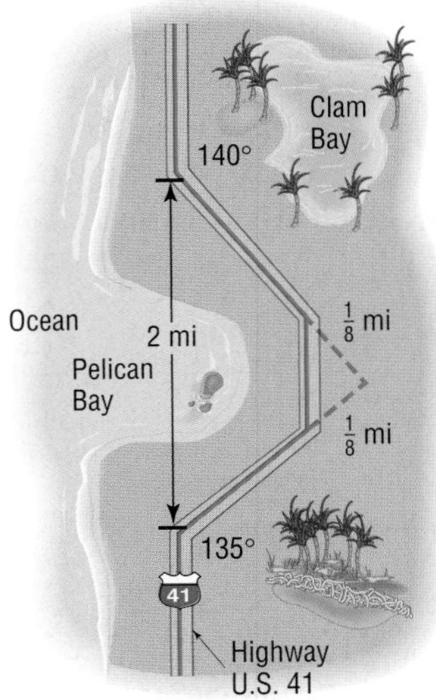

48. Calculating Distances at Sea The navigator of a ship at sea spots two lighthouses that she knows to be 3 miles apart along a straight seashore. She determines that the angles formed between two line-of-sight observations of the lighthouses and the line from the ship directly to shore are 15° and 35°. See the illustration.

(a) How far is the ship from lighthouse A?
(b) How far is the ship from lighthouse B?
(c) How far is the ship from shore?

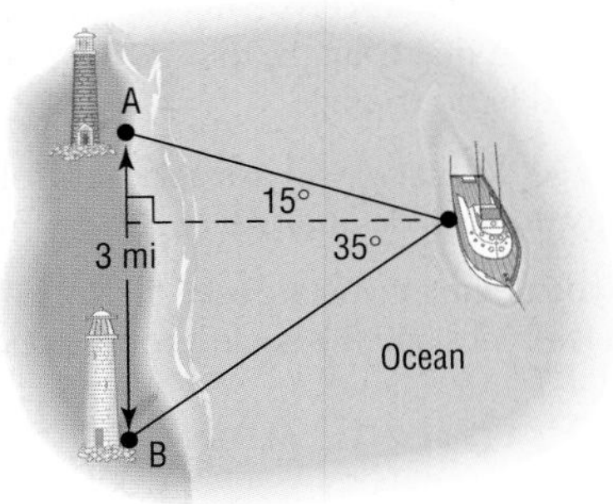

49. Designing an Awning An awning that covers a sliding glass door that is 88 inches tall forms an angle of 50° with the wall. The purpose of the awning is to prevent sunlight from entering the house when the angle of elevation of the sun is more than 65°. See the figure. Find the length L of the awning.

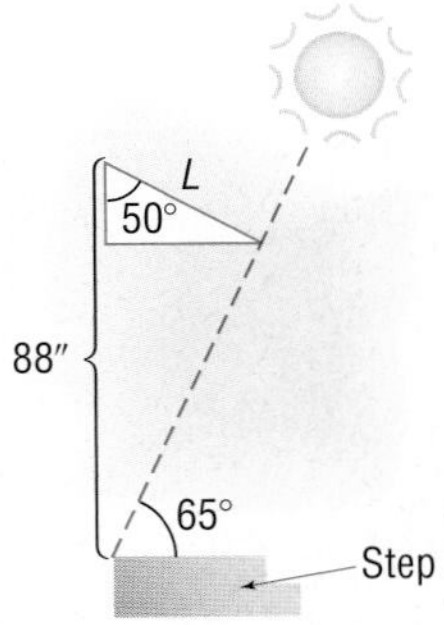

50. Finding Distances A forest ranger is walking on a path inclined at 5° to the horizontal directly toward a 100-foot-tall fire observation tower. The angle of elevation from the path to the top of the tower is 40°. How far is the ranger from the tower at this time?

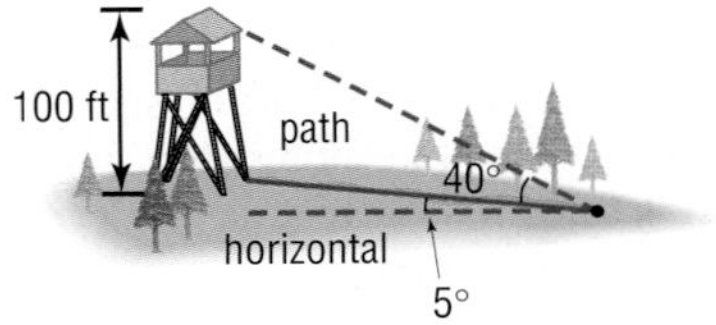

51. Great Pyramid of Cheops One of the original Seven Wonders of the World, the Great Pyramid of Cheops was built about 2580 BC. Its original height was 480 feet 11 inches, but due to the loss of its topmost stones, it is now shorter. Find the current height of the Great Pyramid, using the information given in the illustration.

SOURCE: *Guinness Book of World Records*.

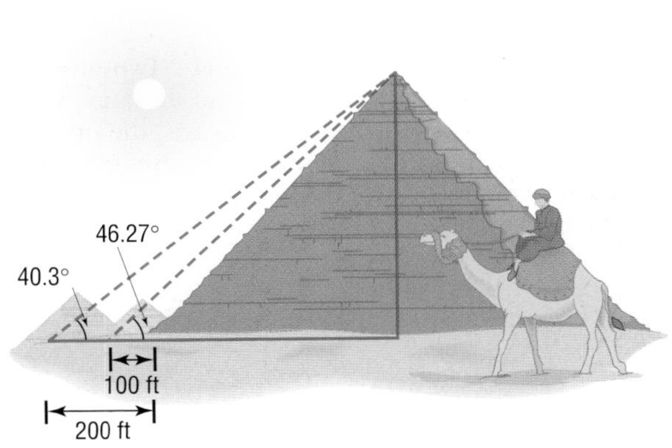

52. Determining the Height of an Aircraft Two sensors are spaced 700 feet apart along the approach to a small airport. When an aircraft is nearing the airport, the angle of elevation from the first sensor to the aircraft is 20°, and from the second sensor to the aircraft it is 15°. Determine how high the aircraft is at this time.

53. Mercury The distance from the Sun to Earth is approximately 149,600,000 kilometers (km). The distance from the Sun to Mercury is approximately 57,910,000 km. The **elongation angle** α is the angle formed between the line of sight from Earth to the Sun and the line of sight from Earth to Mercury. See the figure. Suppose that the elongation angle for Mercury is 15°. Use this information to find the possible distances between Earth and Mercury.

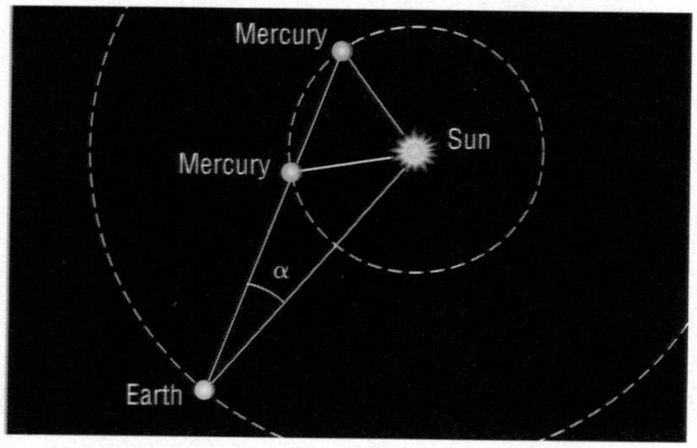

54. Venus The distance from the Sun to Earth is approximately 149,600,000 km. The distance from the Sun to Venus is approximately 108,200,000 km. The elongation angle is the angle formed between the line of sight from Earth to the Sun and the line of sight from Earth to Venus. Suppose that the elongation angle for Venus is 10°. Use this information to find the possible distances between Earth and Venus.

55. Landscaping Pat needs to determine the height of a tree before cutting it down to be sure that it will not fall on a nearby fence. The angle of elevation of the tree from one position on a flat path from the tree is 30°, and from a second position 40 feet farther along this path it is 20°. What is the height of the tree?

56. Construction A loading ramp 10 feet long that makes an angle of 18° with the horizontal is to be replaced by one that makes an angle of 12° with the horizontal. How long is the new ramp?

57. Finding the Height of a Helicopter Two observers simultaneously measure the angle of elevation of a helicopter. One angle is measured as 25°, the other as 40° (see the figure). If the observers are 100 feet apart and the helicopter lies over the line joining them, how high is the helicopter?

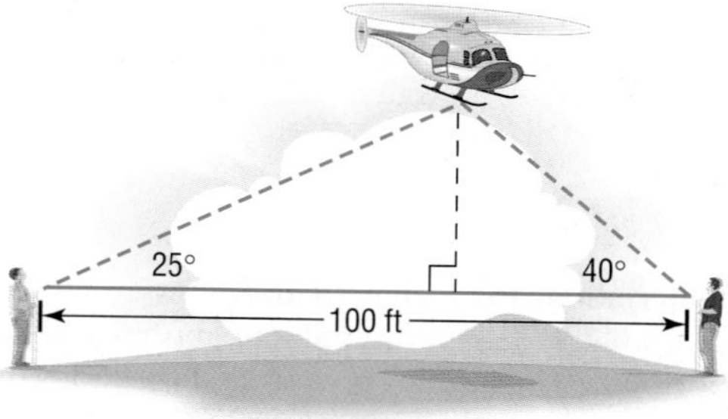

58. Mollweide's Formula For any triangle, Mollweide's Formula (named after Karl Mollweide, 1774–1825) states that

$$\frac{a+b}{c} = \frac{\cos\left[\frac{1}{2}(\alpha - \beta)\right]}{\sin\left(\frac{1}{2}\gamma\right)}$$

Derive it.

[**Hint:** Use the Law of Sines and then a Sum-to-Product Formula. Notice that this formula involves all six parts of a triangle. As a result, it is sometimes used to check the solution of a triangle.]

59. Mollweide's Formula Another form of Mollweide's Formula is

$$\frac{a-b}{c} = \frac{\sin\left[\frac{1}{2}(\alpha - \beta)\right]}{\cos\left(\frac{1}{2}\gamma\right)}$$

Derive it.

60. For any triangle, derive the formula

$$a = b\cos\gamma + c\cos\beta$$

[**Hint:** Use the fact that $\sin\alpha = \sin(180° - \beta - \gamma)$.]

61. Law of Tangents For any triangle, derive the Law of Tangents.

$$\frac{a-b}{a+b} = \frac{\tan\left[\frac{1}{2}(\alpha - \beta)\right]}{\tan\left[\frac{1}{2}(\alpha + \beta)\right]}$$

[**Hint:** Use Mollweide's Formula.]

62. Circumscribing a Triangle Show that

$$\frac{\sin\alpha}{a} = \frac{\sin\beta}{b} = \frac{\sin\gamma}{c} = \frac{1}{2r}$$

where r is the radius of the circle circumscribing the triangle ABC whose sides are a, b, and c, as shown in the figure. [**Hint:** Draw the diameter AB'. Then β = angle ABC = angle $AB'C$, and angle $ACB' = 90°$.]

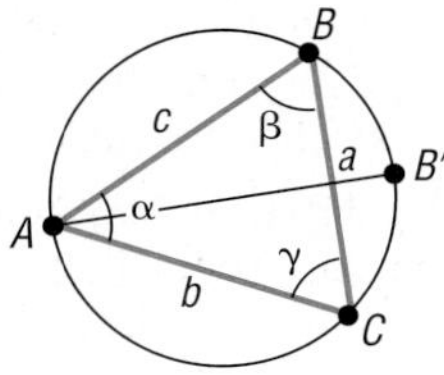

63. Create three problems involving oblique triangles. One should result in one triangle, the second in two triangles, and the third in no triangle.

64. What do you do first if you are asked to solve a triangle and are given one side and two angles?

65. What do you do first if you are asked to solve a triangle and are given two sides and the angle opposite one of them?

'Are You Prepared?' Answers

1. $\sin\alpha\cos\beta - \cos\alpha\sin\beta$ **2.** $\left\{\frac{\pi}{6}\right\}$ **3.** No solution

7.3 The Law of Cosines

PREPARING FOR THIS SECTION *Before getting started, review the following:*

- Trigonometric Equations (I) (Section 6.7, pp. 461–464)
- Distance Formula (Section 1.1, p. 3)

Now work the 'Are You Prepared?' problems on page 510.

OBJECTIVES
1. Solve SAS Triangles
2. Solve SSS Triangles
3. Solve Applied Problems

In the previous section, we used the Law of Sines to solve Case 1 (SAA or ASA) and Case 2 (SSA) of an oblique triangle. In this section, we derive the Law of Cosines and use it to solve the remaining cases, 3 and 4.

CASE 3: Two sides and the included angle are known (SAS).

CASE 4: Three sides are known (SSS).

Theorem

Law of Cosines

For a triangle with sides a, b, c and opposite angles α, β, γ, respectively,

$$c^2 = a^2 + b^2 - 2ab\cos\gamma \qquad (1)$$

$$b^2 = a^2 + c^2 - 2ac\cos\beta \qquad (2)$$

$$a^2 = b^2 + c^2 - 2bc\cos\alpha \qquad (3)$$

Figure 35

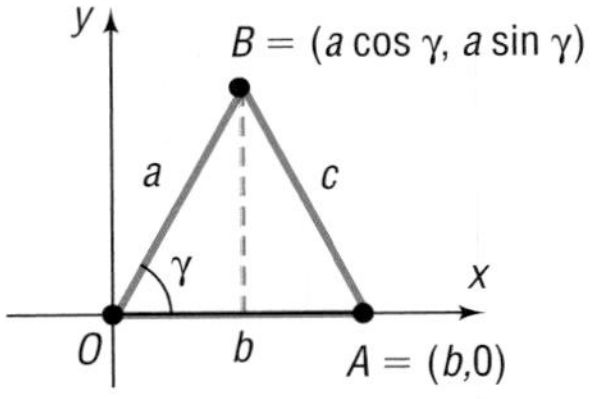

(a) Angle γ is acute

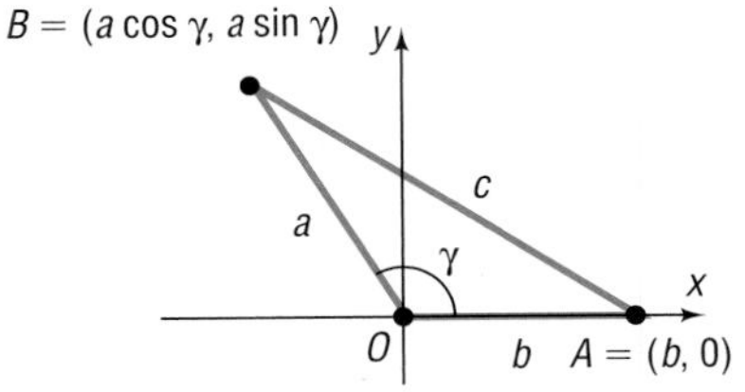

(b) Angle γ is obtuse

Proof We will prove only formula (1) here. Formulas (2) and (3) may be proved using the same argument.

We begin by strategically placing a triangle on a rectangular coordinate system so that the vertex of angle γ is at the origin and side b lies along the positive x-axis. Regardless of whether γ is acute, as in Figure 35(a), or obtuse, as in Figure 35(b), the vertex B has coordinates $(a\cos\gamma, a\sin\gamma)$. Vertex A has coordinates $(b, 0)$.

We can now use the distance formula to compute c^2.

$$\begin{aligned} c^2 &= (b - a\cos\gamma)^2 + (0 - a\sin\gamma)^2 \\ &= b^2 - 2ab\cos\gamma + a^2\cos^2\gamma + a^2\sin^2\gamma \\ &= b^2 - 2ab\cos\gamma + a^2(\cos^2\gamma + \sin^2\gamma) \\ &= a^2 + b^2 - 2ab\cos\gamma \end{aligned}$$

■

Each of formulas (1), (2), and (3) may be stated in words as follows:

Theorem

Law of Cosines

The square of one side of a triangle equals the sum of the squares of the other two sides minus twice their product times the cosine of their included angle.

Observe that if the triangle is a right triangle (so that, say, $\gamma = 90°$) then formula (1) becomes the familiar Pythagorean Theorem: $c^2 = a^2 + b^2$. Thus, the Pythagorean Theorem is a special case of the Law of Cosines.

1 Let's see how to use the Law of Cosines to solve Case 3 (SAS), which applies to triangles for which two sides and the included angle are known.

EXAMPLE 1 Using the Law of Cosines to Solve a SAS Triangle

Solve the triangle: $a = 2, b = 3, \gamma = 60°$

Solution See Figure 36. The Law of Cosines makes it easy to find the third side, c.

Figure 36

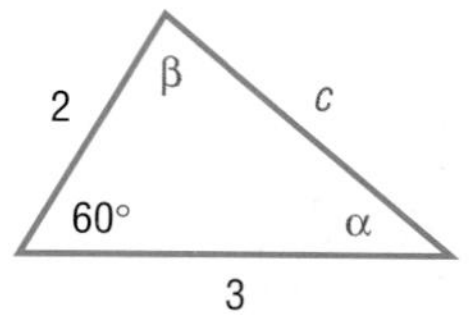

$$\begin{aligned} c^2 &= a^2 + b^2 - 2ab\cos\gamma \\ &= 4 + 9 - 2\cdot 2\cdot 3\cdot \cos 60° \\ &= 13 - \left(12\cdot\frac{1}{2}\right) = 7 \\ c &= \sqrt{7} \end{aligned}$$

Side c is of length $\sqrt{7}$. To find the angles α and β, we may use either the Law of Sines or the Law of Cosines. It is preferable to use the Law of Cosines, since it will lead to an equation with one solution. Using the Law of Sines would lead to an equation with two solutions that would need to be checked to determine which solution fits the given data. We choose to use formulas (2) and (3) of the Law of Cosines to find α and β.

For α:

$$\begin{aligned} a^2 &= b^2 + c^2 - 2bc\cos\alpha \\ 2bc\cos\alpha &= b^2 + c^2 - a^2 \\ \cos\alpha &= \frac{b^2 + c^2 - a^2}{2bc} = \frac{9 + 7 - 4}{2\cdot 3\sqrt{7}} = \frac{12}{6\sqrt{7}} = \frac{2\sqrt{7}}{7} \\ \alpha &\approx 40.9° \end{aligned}$$

For β:

$$\begin{aligned} b^2 &= a^2 + c^2 - 2ac\cos\beta \\ \cos\beta &= \frac{a^2 + c^2 - b^2}{2ac} = \frac{4 + 7 - 9}{4\sqrt{7}} = \frac{1}{2\sqrt{7}} = \frac{\sqrt{7}}{14} \\ \beta &\approx 79.1° \end{aligned}$$

Notice that $\alpha + \beta + \gamma = 40.9° + 79.1° + 60° = 180°$, as required. ◀

NOW WORK PROBLEM 9.

2 The next example illustrates how the Law of Cosines is used when three sides of a triangle are known, Case 4 (SSS).

EXAMPLE 2 Using the Law of Cosines to Solve a SSS Triangle

Solve the triangle: $a = 4, b = 3, c = 6$

Figure 37

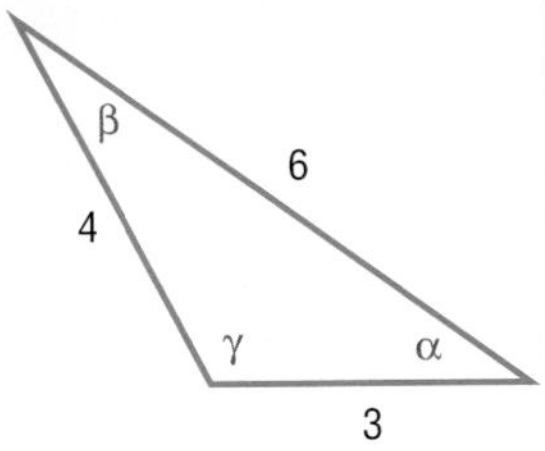

Solution See Figure 37. To find the angles α, β, and γ, we proceed as we did in the latter part of the solution to Example 1.

For α:

$$\cos\alpha = \frac{b^2 + c^2 - a^2}{2bc} = \frac{9 + 36 - 16}{2\cdot3\cdot6} = \frac{29}{36}$$
$$\alpha \approx 36.3^\circ$$

For β:

$$\cos\beta = \frac{a^2 + c^2 - b^2}{2ac} = \frac{16 + 36 - 9}{2\cdot4\cdot6} = \frac{43}{48}$$
$$\beta \approx 26.4^\circ$$

Since we know α and β,

$$\gamma = 180^\circ - \alpha - \beta \approx 180^\circ - 36.3^\circ - 26.4^\circ = 117.3^\circ$$ ◀

NOW WORK PROBLEM 15.

3 **EXAMPLE 3** **Correcting a Navigational Error**

A motorized sailboat leaves Naples, Florida, bound for Key West, 150 miles away. Maintaining a constant speed of 15 miles per hour, but encountering heavy crosswinds and strong currents, the crew finds, after 4 hours, that the sailboat is off course by 20°.

(a) How far is the sailboat from Key West at this time?

(b) Through what angle should the sailboat turn to correct its course?

(c) How much time has been added to the trip because of this? (Assume that the speed remains at 15 miles per hour.)

Solution See Figure 38. With a speed of 15 miles per hour, the sailboat has gone 60 miles after 4 hours. We seek the distance x of the sailboat from Key West. We also seek the angle θ that the sailboat should turn through to correct its course.

Figure 38

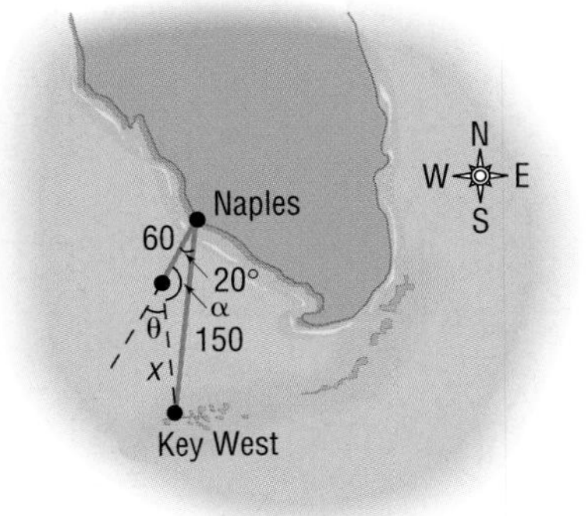

(a) To find x, we use the Law of Cosines, since we know two sides and the included angle.

$$x^2 = 150^2 + 60^2 - 2(150)(60)\cos 20^\circ \approx 9186$$
$$x \approx 95.8$$

The sailboat is about 96 miles from Key West.

(b) We now know three sides of the triangle, so we can use the Law of Cosines again to find the angle α opposite the side of length 150 miles.

$$150^2 = 96^2 + 60^2 - 2(96)(60)\cos\alpha$$
$$9684 = -11{,}520\cos\alpha$$
$$\cos\alpha \approx -0.8406$$
$$\alpha \approx 147.2^\circ$$

The sailboat should turn through an angle of

$$\theta = 180^\circ - \alpha \approx 180^\circ - 147.2^\circ = 32.8^\circ$$

The sailboat should turn through an angle of about 33° to correct its course.

(c) The total length of the trip is now $60 + 96 = 156$ miles. The extra 6 miles will only require about 0.4 hour or 24 minutes more if the speed of 15 miles per hour is maintained. ◀

NOW WORK PROBLEM 35.

Historical Feature

The Law of Sines was known vaguely long before it was explicitly stated by Nasîr Eddîn (about AD 1250). Ptolemy (about AD 150) was aware of it in a form using a chord function instead of the sine function. But it was first clearly stated in Europe by Regiomontanus, writing in 1464.

The Law of Cosines appears first in Euclid's *Elements* (Book II), but in a well-disguised form in which squares built on the sides of triangles are added and a rectangle representing the cosine term is subtracted. It was thus known to all mathematicians because of their familiarity with Euclid's work.

An early modern form of the Law of Cosines, that for finding the angle when the sides are known, was stated by François Viète (in 1593).

The Law of Tangents (see Problem 61 of Exercise 7.2) has become obsolete. In the past it was used in place of the Law of Cosines, because the Law of Cosines was very inconvenient for calculation with logarithms or slide rules. Mixing of addition and multiplication is now very easy on a calculator, however, and the Law of Tangents has been shelved along with the slide rule.

7.3 Assess Your Understanding

'Are You Prepared?' *Answers are given at the end of these exercises. If you get a wrong answer, read the pages listed in red.*

1. Write down the formula for the distance d from $P_1 = (x_1, y_1)$ to $P_2 = (x_2, y_2)$. (p. 3)

2. If θ is an acute angle, solve the equation $\cos\theta = \frac{\sqrt{2}}{2}$. (pp. 461–464)

Concepts and Vocabulary

3. If three sides of a triangle are given, the Law of __________ is used to solve the triangle.

4. If one side and two angles of a triangle are given, the Law of __________ is used to solve the triangle.

5. If two sides and the included angle of a triangle are given, the Law of __________ is used to solve the triangle.

6. *True or False:* Given only the three sides of a triangle, there is insufficient information to solve the triangle.

7. *True or False:* Given two sides and the included angle, the first thing to do to solve the triangle is to use the Law of Sines.

8. *True or False:* A special case of the Law of Cosines is the Pythagorean Theorem.

Exercises

In Problems 9–16, solve each triangle.

9.

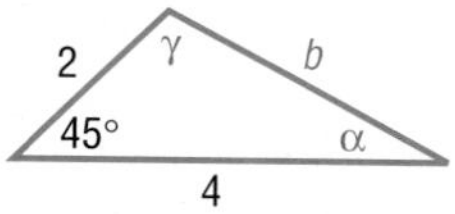

10.

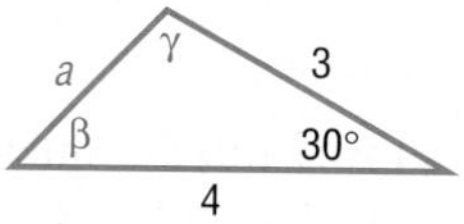

11.

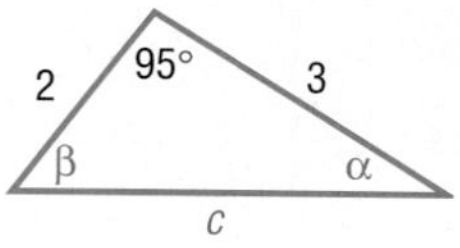

12.

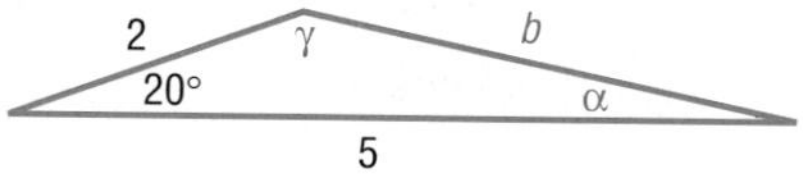

13.

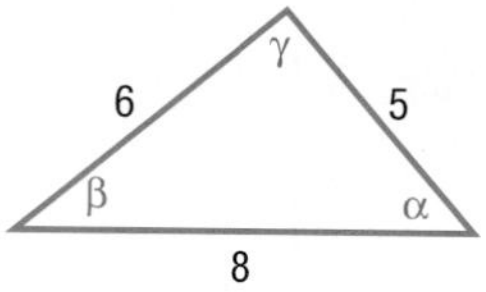

14.

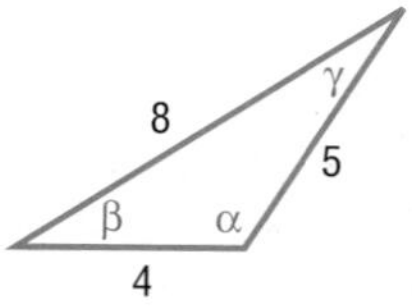

15.

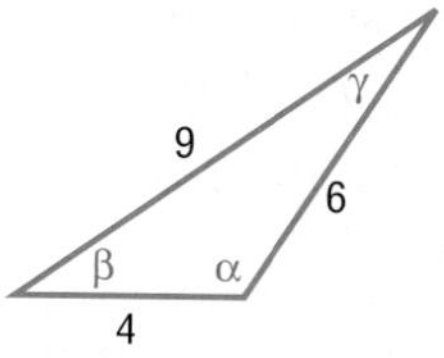

16.

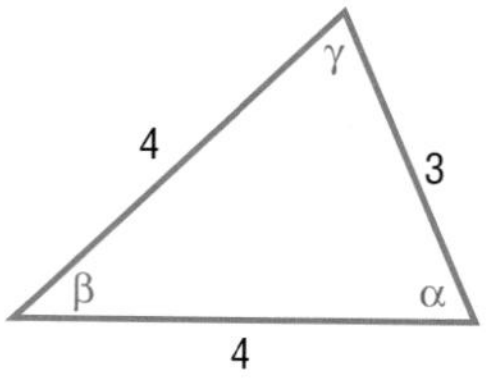

In Problems 17–32, solve each triangle.

17. $a = 3, \quad b = 4, \quad \gamma = 40°$

18. $a = 2, \quad c = 1, \quad \beta = 10°$

19. $b = 1, \quad c = 3, \quad \alpha = 80°$

20. $a = 6, \quad b = 4, \quad \gamma = 60°$
21. $a = 3, \quad c = 2, \quad \beta = 110°$
22. $b = 4, \quad c = 1, \quad \alpha = 120°$
23. $a = 2, \quad b = 2, \quad \gamma = 50°$
24. $a = 3, \quad c = 2, \quad \beta = 90°$
25. $a = 12, \quad b = 13, \quad c = 5$
26. $a = 4, \quad b = 5, \quad c = 3$
27. $a = 2, \quad b = 2, \quad c = 2$
28. $a = 3, \quad b = 3, \quad c = 2$
29. $a = 5, \quad b = 8, \quad c = 9$
30. $a = 4, \quad b = 3, \quad c = 6$
31. $a = 10, \quad b = 8, \quad c = 5$
32. $a = 9, \quad b = 7, \quad c = 10$

33. Surveying Consult the figure. To find the distance from the house at A to the house at B, a surveyor measures the angle ACB, which is found to be $70°$, and then walks off the distance to each house, 50 feet and 70 feet, respectively. How far apart are the houses?

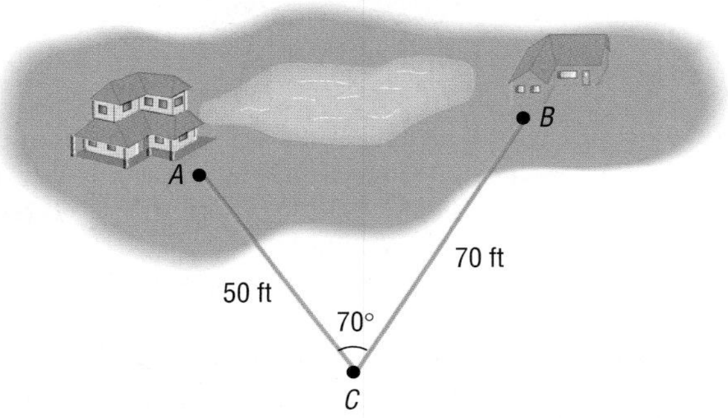

34. Navigation An airplane flies from Ft. Myers to Sarasota, a distance of 150 miles, and then turns through an angle of $50°$ and flies to Orlando, a distance of 100 miles (see the figure).

(a) How far is it from Ft. Myers to Orlando?
(b) Through what angle should the pilot turn at Orlando to return to Ft. Myers?

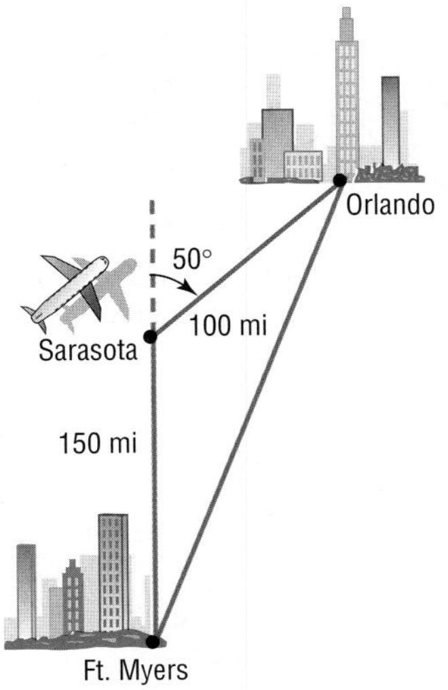

35. Avoiding a Tropical Storm A cruise ship maintains an average speed of 15 knots in going from San Juan, Puerto Rico, to Barbados, West Indies, a distance of 600 nautical miles. To avoid a tropical storm, the captain heads out of San Juan in a direction of $20°$ off a direct heading to Barbados. The captain maintains the 15-knot speed for 10 hours, after which time the path to Barbados becomes clear of storms.

(a) Through what angle should the captain turn to head directly to Barbados?
(b) Once the turn is made, how long will it be before the ship reaches Barbados if the same 15-knot speed is maintained?

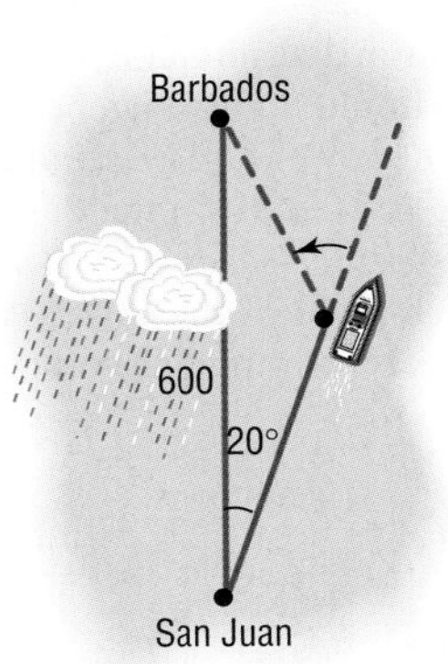

36. Revising a Flight Plan In attempting to fly from Chicago to Louisville, a distance of 330 miles, a pilot inadvertently took a course that was $10°$ in error, as indicated in the figure.

(a) If the aircraft maintains an average speed of 220 miles per hour and if the error in direction is discovered after 15 minutes, through what angle should the pilot turn to head toward Louisville?
(b) What new average speed should the pilot maintain so that the total time of the trip is 90 minutes?

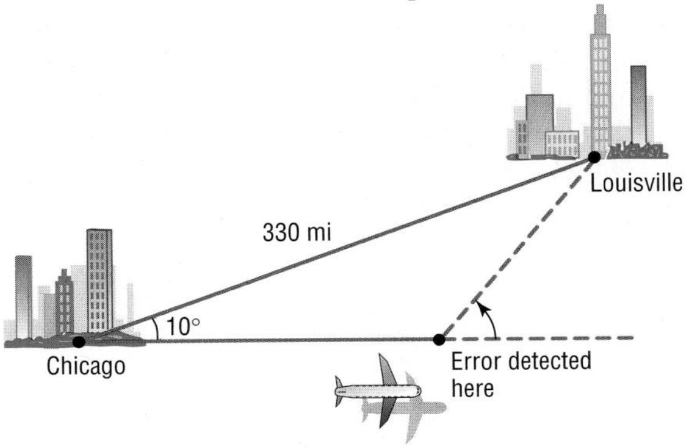

37. Major League Baseball Field A Major League baseball diamond is actually a square 90 feet on a side. The pitching rubber is located 60.5 feet from home plate on a line joining home plate and second base.

(a) How far is it from the pitching rubber to first base?
(b) How far is it from the pitching rubber to second base?
(c) If a pitcher faces home plate, through what angle does he need to turn to face first base?

38. Little League Baseball Field According to Little League baseball official regulations, the diamond is a square 60 feet on a side. The pitching rubber is located 46 feet from home plate on a line joining home plate and second base.

(a) How far is it from the pitching rubber to first base?
(b) How far is it from the pitching rubber to second base?
(c) If a pitcher faces home plate, through what angle does he need to turn to face first base?

39. Finding the Length of a Guy Wire The height of a radio tower is 500 feet, and the ground on one side of the tower slopes upward at an angle of 10° (see the figure).

(a) How long should a guy wire be if it is to connect to the top of the tower and be secured at a point on the sloped side 100 feet from the base of the tower?
(b) How long should a second guy wire be if it is to connect to the middle of the tower and be secured at a point 100 feet from the base on the flat side?

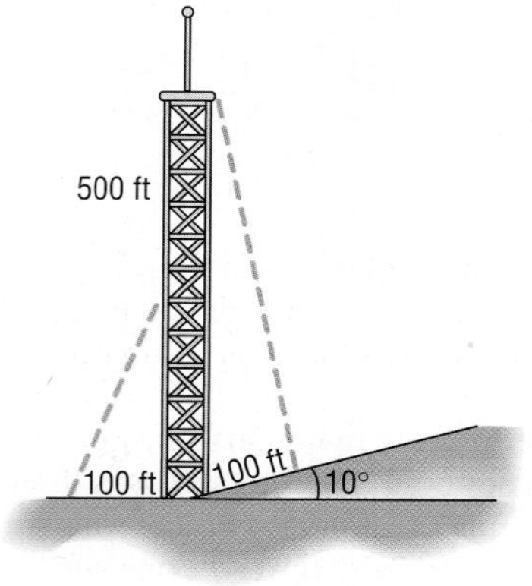

40. Finding the Length of a Guy Wire A radio tower 500 feet high is located on the side of a hill with an inclination to the horizontal of 5° (see the figure). How long should two guy wires be if they are to connect to the top of the tower and be secured at two points 100 feet directly above and directly below the base of the tower?

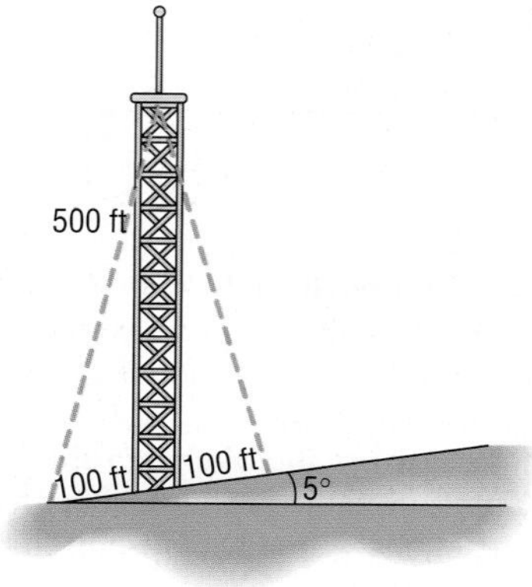

41. Wrigley Field, Home of the Chicago Cubs The distance from home plate to the fence in dead center in Wrigley Field is 400 feet (see the figure). How far is it from the fence in dead center to third base?

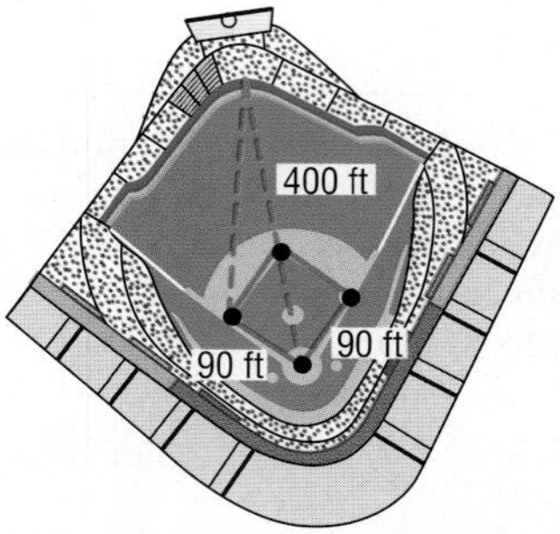

42. Little League Baseball The distance from home plate to the fence in dead center at the Oak Lawn Little League field is 280 feet. How far is it from the fence in dead center to third base?
[**Hint:** The distance between the bases in Little League is 60 feet.]

43. Rods and Pistons Rod OA (see the figure) rotates about the fixed point O so that point A travels on a circle of radius r. Connected to point A is another rod AB of length $L > 2r$, and point B is connected to a piston. Show that the distance x between point O and point B is given by

$$x = r\cos\theta + \sqrt{r^2\cos^2\theta + L^2 - r^2}$$

where θ is the angle of rotation of rod OA.

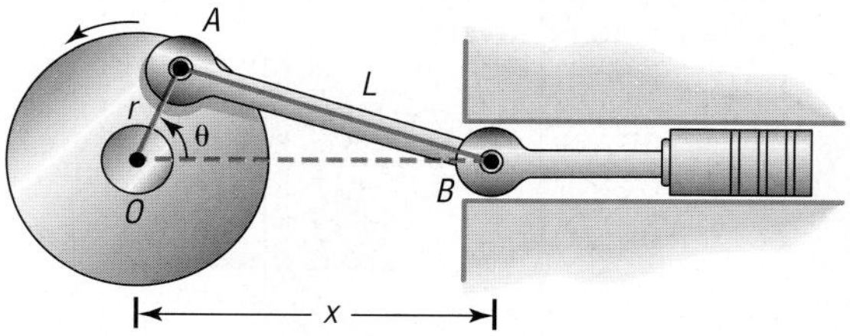

44. Geometry Show that the length d of a chord of a circle of radius r is given by the formula

$$d = 2r\sin\frac{\theta}{2}$$

where θ is the central angle formed by the radii to the ends of the chord (see the figure). Use this result to derive the fact that $\sin\theta < \theta$, where $\theta > 0$ is measured in radians.

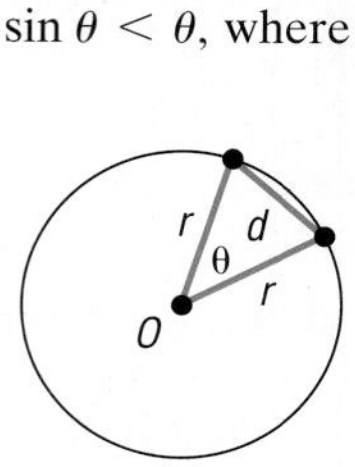

45. For any triangle, show that

$$\cos\frac{\gamma}{2} = \sqrt{\frac{s(s-c)}{ab}}$$

where $s = \frac{1}{2}(a + b + c)$.

[**Hint:** Use a Half-angle Formula and the Law of Cosines.]

46. For any triangle show that

$$\sin \frac{\gamma}{2} = \sqrt{\frac{(s - a)(s - b)}{ab}}$$

where $s = \frac{1}{2}(a + b + c)$.

47. Use the Law of Cosines to prove the identity

$$\frac{\cos \alpha}{a} + \frac{\cos \beta}{b} + \frac{\cos \gamma}{c} = \frac{a^2 + b^2 + c^2}{2abc}$$

48. What do you do first if you are asked to solve a triangle and are given two sides and the included angle?

49. What do you do first if you are asked to solve a triangle and are given three sides?

50. Make up an applied problem that requires using the Law of Cosines.

51. Write down your strategy for solving an oblique triangle.

'Are You Prepared?' Answers

1. $d = \sqrt{(x_2 - x_1)^2 + (y_2 - y_1)^2}$ **2.** $\theta = 45°$

7.4 Area of a Triangle

PREPARING FOR THIS SECTION *Before getting started, review the following:*

- Geometry Review (Appendix A, Section A.2, pp. 915–916)

Now work the 'Are You Prepared?' problem on page 516.

OBJECTIVES 1 Find the Area of SAS Triangles
2 Find the Area of SSS Triangles

In this section, we will derive several formulas for calculating the area A of a triangle. The most familiar of these is the following:

Theorem The area A of a triangle is

$$A = \frac{1}{2}bh \quad (1)$$

where b is the base and h is an altitude drawn to that base.

Figure 39

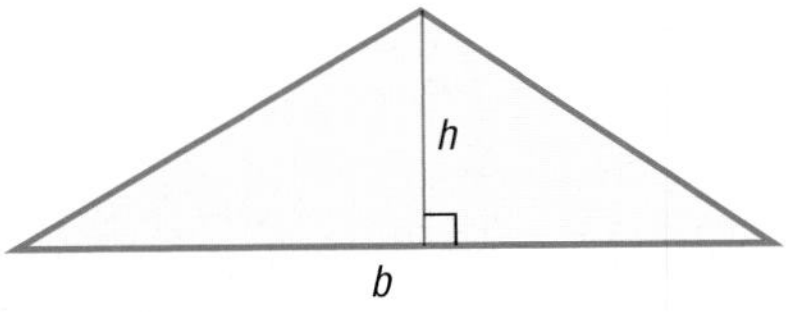

Figure 40

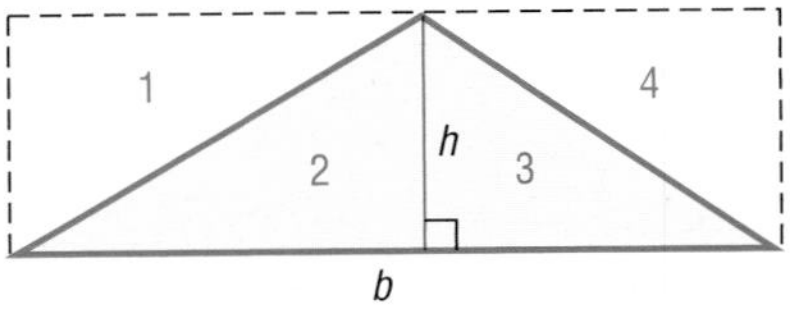

Figure 41

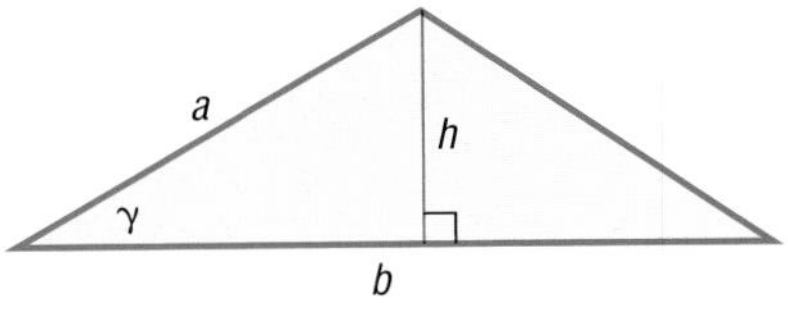

Proof The derivation of this formula is rather easy once a rectangle of base b and height h is constructed around the triangle. See Figures 39 and 40.

Triangles 1 and 2 in Figure 40 are equal in area, as are triangles 3 and 4. Consequently, the area of the triangle with base b and altitude h is exactly half the area of the rectangle, which is bh. ■

If the base b and altitude h to that base are known, then we can find the area of such a triangle using formula (1). Usually, though, the information required to use formula (1) is not given. Suppose, for example, that we know two sides a and b and the included angle γ (see Figure 41). Then the altitude h can be found by noting that

$$\frac{h}{a} = \sin \gamma$$

so that

$$h = a \sin \gamma$$

Using this fact in formula (1) produces

$$A = \frac{1}{2}bh = \frac{1}{2}b(a \sin \gamma) = \frac{1}{2}ab \sin \gamma$$

We now have the formula

$$A = \frac{1}{2}ab \sin \gamma \qquad \textbf{(2)}$$

By dropping altitudes from the other two vertices of the triangle, we obtain the following corresponding formulas:

$$A = \frac{1}{2}bc \sin \alpha \qquad \textbf{(3)}$$

$$A = \frac{1}{2}ac \sin \beta \qquad \textbf{(4)}$$

It is easiest to remember these formulas using the following wording:

Theorem The area A of a triangle equals one-half the product of two of its sides times the sine of their included angle.

1 **EXAMPLE 1** **Finding the Area of a SAS Triangle**

Find the area A of the triangle for which $a = 8$, $b = 6$, and $\gamma = 30°$.

Solution See Figure 42. We use formula (2) to get

Figure 42

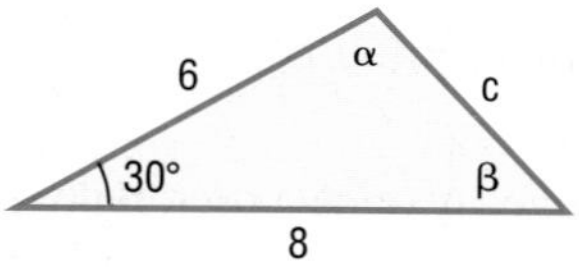

$$A = \frac{1}{2}ab \sin \gamma = \frac{1}{2} \cdot 8 \cdot 6 \sin 30° = 12$$

◀

NOW WORK PROBLEM 5.

2 If the three sides of a triangle are known, another formula, called **Heron's Formula** (named after Heron of Alexandria), can be used to find the area of a triangle.

Theorem **Heron's Formula**

The area A of a triangle with sides a, b, and c is

$$A = \sqrt{s(s-a)(s-b)(s-c)} \qquad \textbf{(5)}$$

where $s = \frac{1}{2}(a + b + c)$.

A proof of Heron's Formula is given at the end of this section.

EXAMPLE 2 **Finding the Area of a SSS Triangle**

Find the area of a triangle whose sides are 4, 5, and 7.

Solution We let $a = 4$, $b = 5$, and $c = 7$. Then

$$s = \frac{1}{2}(a + b + c) = \frac{1}{2}(4 + 5 + 7) = 8$$

Heron's Formula then gives the area A as

$$A = \sqrt{s(s - a)(s - b)(s - c)} = \sqrt{8 \cdot 4 \cdot 3 \cdot 1} = \sqrt{96} = 4\sqrt{6}$$ ◀

NOW WORK PROBLEM 11.

Proof of Heron's Formula The proof that we shall give uses the Law of Cosines and is quite different from the proof given by Heron.

From the Law of Cosines,

$$c^2 = a^2 + b^2 - 2ab \cos \gamma$$

and the half-angle formula

$$\cos^2 \frac{\gamma}{2} = \frac{1 + \cos \gamma}{2}$$

we find that

$$\cos^2 \frac{\gamma}{2} = \frac{1 + \cos \gamma}{2} = \frac{1 + \dfrac{a^2 + b^2 - c^2}{2ab}}{2}$$

$$= \frac{a^2 + 2ab + b^2 - c^2}{4ab} = \frac{(a + b)^2 - c^2}{4ab}$$

$$= \frac{(a + b - c)(a + b + c)}{4ab} = \frac{2(s - c) \cdot 2s}{4ab} = \frac{s(s - c)}{ab} \quad \textbf{(6)}$$

↑ $a + b - c = a + b + c - 2c = 2s - 2c = 2(s - c)$

Similarly,

$$\sin^2 \frac{\gamma}{2} = \frac{(s - a)(s - b)}{ab} \quad \textbf{(7)}$$

Now we use formula (2) for the area.

$$A = \frac{1}{2}ab \sin \gamma$$

$$= \frac{1}{2}ab \cdot 2 \sin \frac{\gamma}{2} \cos \frac{\gamma}{2}$$ $\quad \sin \gamma = \sin\left[2\left(\frac{\gamma}{2}\right)\right] = 2 \sin \frac{\gamma}{2} \cos \frac{\gamma}{2}$

$$= ab\sqrt{\frac{(s - a)(s - b)}{ab}}\sqrt{\frac{s(s - c)}{ab}}$$ Use equations (6) and (7).

$$= \sqrt{s(s - a)(s - b)(s - c)}$$ ■

Historical Feature

Heron's Formula (also known as *Hero's Formula*) is due to Heron of Alexandria (first century AD), who had, besides his mathematical talents, a good deal of engineering skills. In various temples his mechanical devices produced effects that seemed supernatural, and visitors presumably were thus influenced to generosity. Heron's book *Metrica*, on making such devices, has survived and was discovered in 1896 in the city of Constantinople.

Heron's Formulas for the area of a triangle caused some mild discomfort in Greek mathematics, because a product with two factors was an area, while one with three factors was a volume, but four factors seemed contradictory in Heron's time.

7.4 Assess Your Understanding

'Are You Prepared?' *Answers are given at the end of these exercises. If you get a wrong answer, read the pages listed in red.*

1. The area A of a triangle whose base is b and whose height is h is ________. (pp. 915–916)

Concepts and Vocabulary

2. If three sides of a triangle are given, ________ Formula is used to find the area of the triangle.
3. *True or False:* No formula exists for finding the area of a triangle when only three sides are given.
4. *True or False:* Given two sides and the included angle, there is a formula that can be used to find the area of the triangle.

Exercises

In Problems 5–12, find the area of each triangle. Round answers to two decimal places.

5.
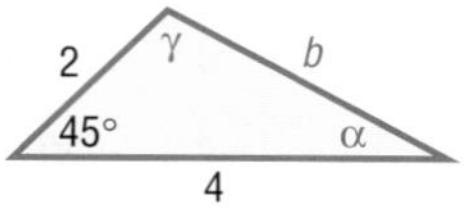

6.
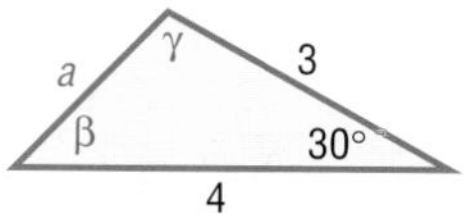

7.
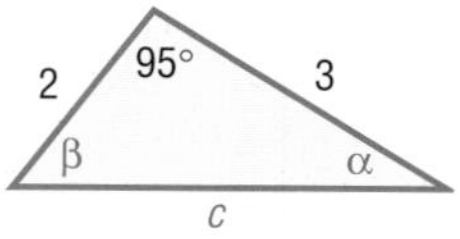

8.
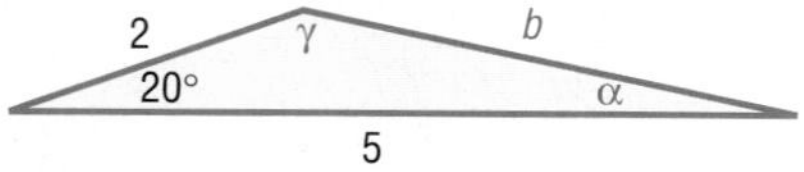

9.
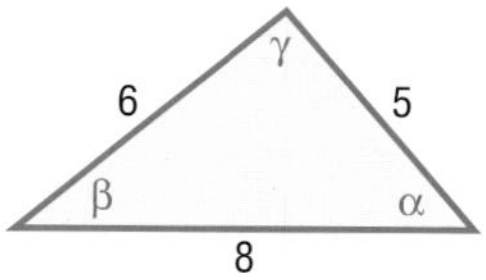

10.
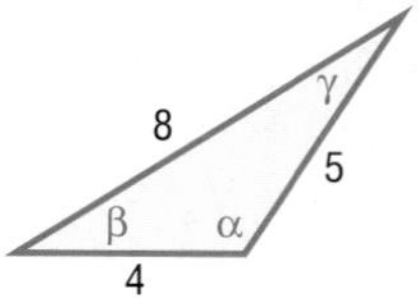

11.
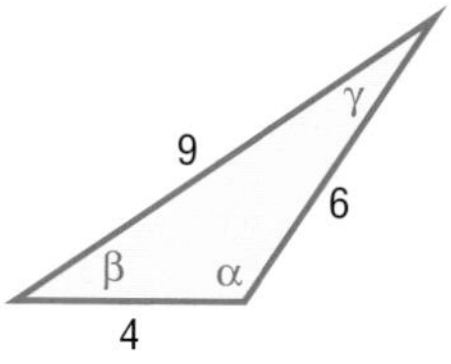

12.
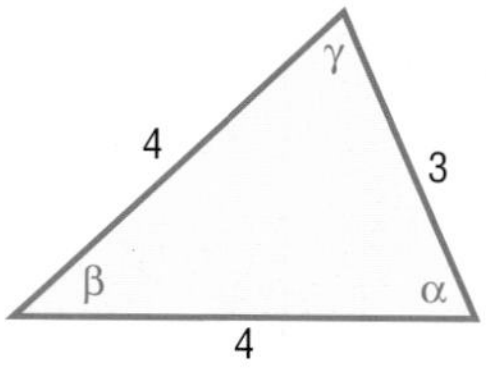

In Problems 13–24, find the area of each triangle. Round answers to two decimal places.

13. $a = 3, \quad b = 4, \quad \gamma = 40°$
14. $a = 2, \quad c = 1, \quad \beta = 10°$
15. $b = 1, \quad c = 3, \quad \alpha = 80°$
16. $a = 6, \quad b = 4, \quad \gamma = 60°$
17. $a = 3, \quad c = 2, \quad \beta = 110°$
18. $b = 4, \quad c = 1, \quad \alpha = 120°$
19. $a = 12, \quad b = 13, \quad c = 5$
20. $a = 4, \quad b = 5, \quad c = 3$
21. $a = 2, \quad b = 2, \quad c = 2$
22. $a = 3, \quad b = 3, \quad c = 2$
23. $a = 5, \quad b = 8, \quad c = 9$
24. $a = 4, \quad b = 3, \quad c = 6$

25. **Area of a Triangle** Prove that the area A of a triangle is given by the formula

$$A = \frac{a^2 \sin \beta \sin \gamma}{2 \sin \alpha}$$

26. **Area of a Triangle** Prove the two other forms of the formula given in Problem 25.

$$A = \frac{b^2 \sin \alpha \sin \gamma}{2 \sin \beta} \quad \text{and} \quad A = \frac{c^2 \sin \alpha \sin \beta}{2 \sin \gamma}$$

In Problems 27–32, use the results of Problem 25 or 26 to find the area of each triangle. Round answers to two decimal places.

27. $\alpha = 40°, \quad \beta = 20°, \quad a = 2$
28. $\alpha = 50°, \quad \gamma = 20°, \quad a = 3$
29. $\beta = 70°, \quad \gamma = 10°, \quad b = 5$
30. $\alpha = 70°, \quad \beta = 60°, \quad c = 4$
31. $\alpha = 110°, \quad \gamma = 30°, \quad c = 3$
32. $\beta = 10°, \quad \gamma = 100°, \quad b = 2$

33. Area of a Segment Find the area of the segment (shaded in blue in the figure) of a circle whose radius is 8 feet, formed by a central angle of 70°.

[**Hint:** Subtract the area of the triangle from the area of the sector to obtain the area of the segment.]

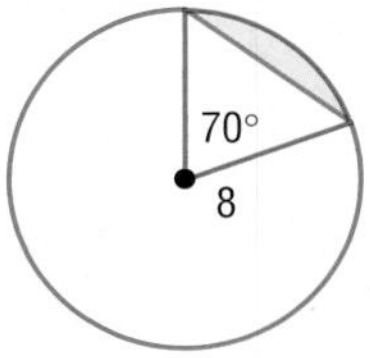

34. Area of a Segment Find the area of the segment of a circle whose radius is 5 inches, formed by a central angle of 40°.

35. Cost of a Triangular Lot The dimensions of a triangular lot are 100 feet by 50 feet by 75 feet. If the price of such land is $3 per square foot, how much does the lot cost?

36. Amount of Materials to Make a Tent A cone-shaped tent is made from a circular piece of canvas 24 feet in diameter by removing a sector with central angle 100° and connecting the ends. What is the surface area of the tent?

37. Computing Areas Find the area of the shaded region enclosed in a semicircle of diameter 8 centimeters. The length of the chord AB is 6 centimeters.

[**Hint:** Triangle ABC is a right triangle.]

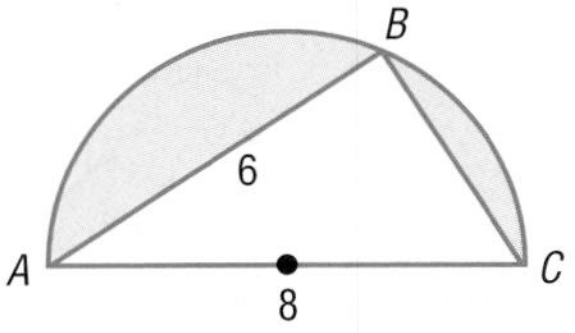

38. Computing Areas Find the area of the shaded region enclosed in a semicircle of diameter 10 inches. The length of the chord AB is 8 inches.

[**Hint:** Triangle ABC is a right triangle.]

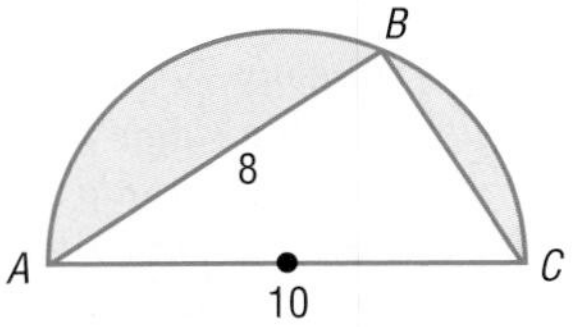

39. Geometry Consult the figure, which shows a circle of radius r with center at O. Find the area A of the shaded region as a function of the central angle θ.

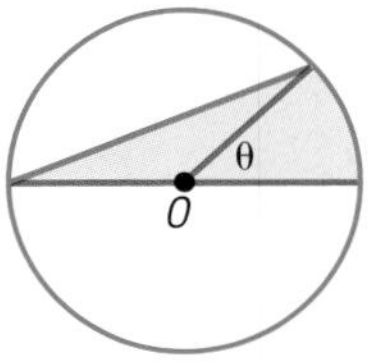

40. Approximating the Area of a Lake To approximate the area of a lake, a surveyor walks around the perimeter of the lake, taking the measurements shown in the illustration. Using this technique, what is the approximate area of the lake?

[**Hint:** Use the Law of Cosines on the three triangles shown and then find the sum of their areas.]

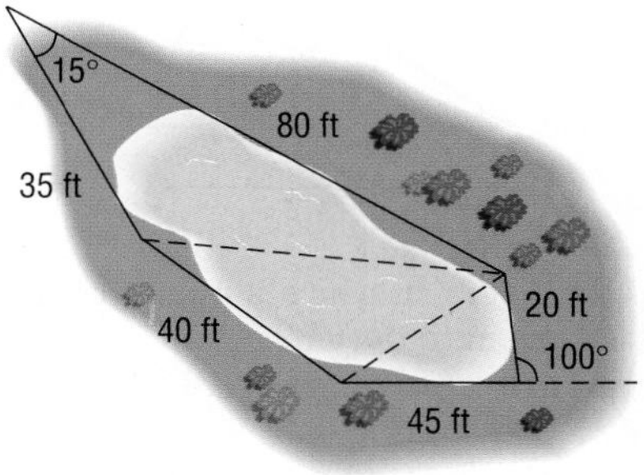

41. Geometry A rectangle is inscribed in a semicircle of radius 1. See the illustration.

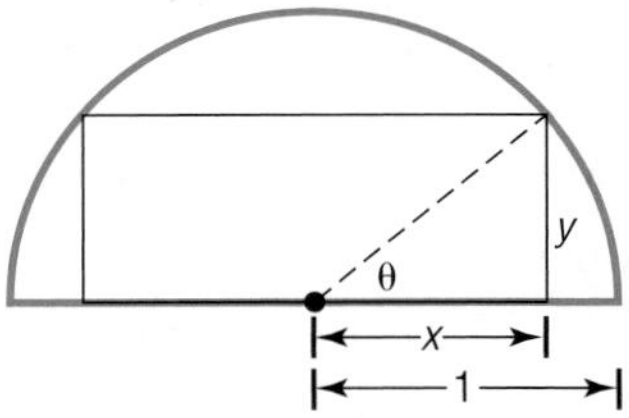

(a) Express the area A of the rectangle as a function of the angle θ shown in the illustration.
(b) Show that $A = \sin(2\theta)$.
(c) Find the angle θ that results in the largest area A.
(d) Find the dimensions of this largest rectangle.

42. Area of an Isosceles Triangle Show that the area A of an isosceles triangle whose equal sides are of length s and the angle between them is θ is

$$A = \frac{1}{2}s^2 \sin \theta$$

[**Hint:** See the illustration. The height h bisects the angle θ and is the perpendicular bisector of the base.]

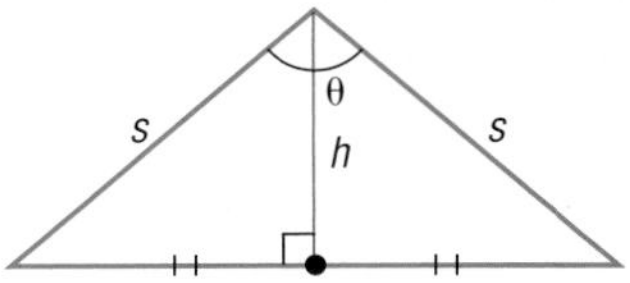

43. Refer to the figure on page 518. If $|OA| = 1$, show that:

(a) Area $\Delta OAC = \frac{1}{2} \sin \alpha \cos \alpha$

(b) Area $\Delta OCB = \dfrac{1}{2}|OB|^2 \sin \beta \cos \beta$

(c) Area $\Delta OAB = \dfrac{1}{2}|OB| \sin(\alpha + \beta)$

(d) $|OB| = \dfrac{\cos \alpha}{\cos \beta}$

(e) $\sin(\alpha + \beta) = \sin \alpha \cos \beta + \cos \alpha \sin \beta$

[**Hint:** Area ΔOAB = Area ΔOAC + Area ΔOCB.]

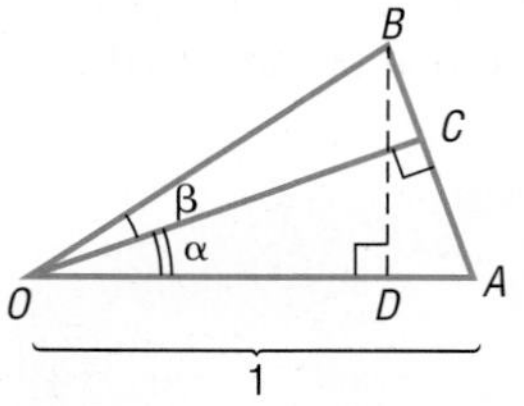

44. Refer to the figure, in which a unit circle is drawn. The line DB is tangent to the circle.

(a) Express the area of ΔOBC in terms of $\sin \theta$ and $\cos \theta$.

(b) Express the area of ΔOBD in terms of $\sin \theta$ and $\cos \theta$.

(c) The area of the sector $\widehat{OBC}$ of the circle is $\dfrac{1}{2}\theta$, where θ is measured in radians. Use the results of parts (a) and (b) and the fact that

$$\text{Area } \Delta OBC < \text{Area } \widehat{OBC} < \text{Area } \Delta OBD$$

to show that

$$1 < \frac{\theta}{\sin \theta} < \frac{1}{\cos \theta}$$

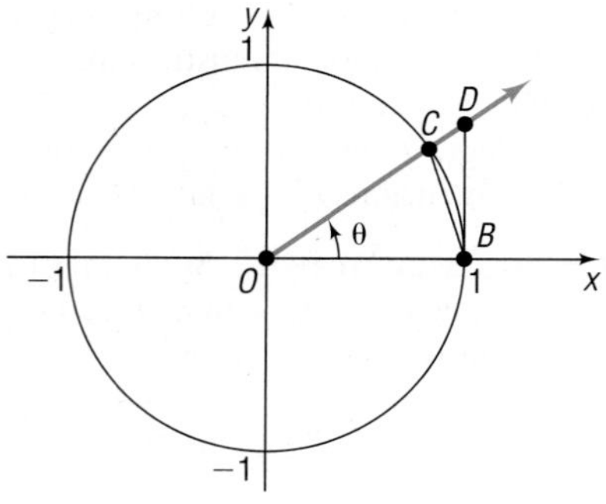

45. The Cow Problem[*] A cow is tethered to one corner of a square barn, 10 feet by 10 feet, with a rope 100 feet long. What is the maximum grazing area for the cow? [**Hint:** See the illustration.]

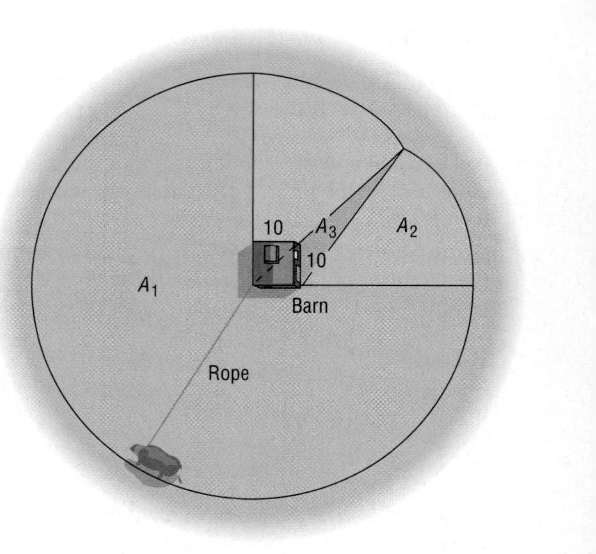

46. Another Cow Problem If the barn in Problem 45 is rectangular, 10 feet by 20 feet, what is the maximum grazing area for the cow?

47. If h_1, h_2, and h_3 are the altitudes dropped from A, B, and C, respectively, in a triangle (see the figure), show that

$$\frac{1}{h_1} + \frac{1}{h_2} + \frac{1}{h_3} = \frac{s}{K}$$

where K is the area of the triangle and $s = \dfrac{1}{2}(a + b + c)$. [**Hint:** $h_1 = \dfrac{2K}{a}$.]

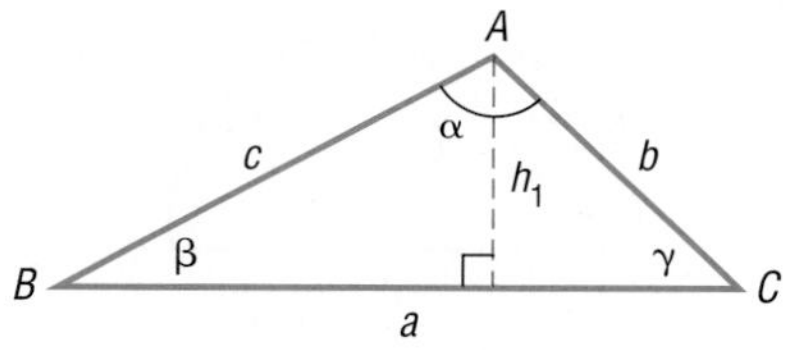

48. Show that a formula for the altitude h from a vertex to the opposite side a of a triangle is

$$h = \frac{a \sin \beta \sin \gamma}{\sin \alpha}$$

Inscribed Circle *For Problems 49–52, the lines that bisect each angle of a triangle meet in a single point O, and the perpendicular distance r from O to each side of the triangle is the same. The circle with center at O and radius r is called the* **inscribed circle** *of the triangle (see the figure).*

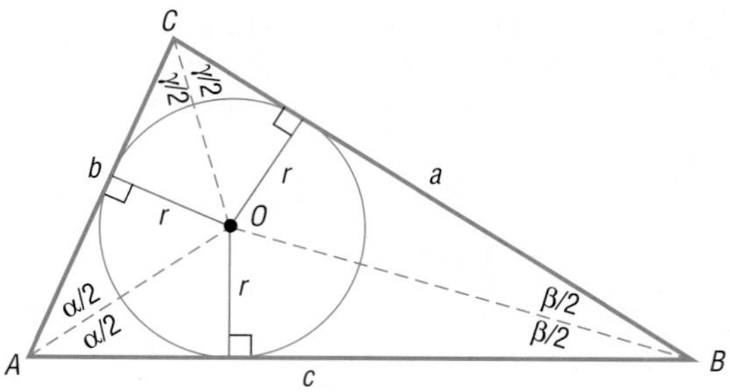

[*] Suggested by Professor Teddy Koukounas of Suffolk Community College, who learned of it from an old farmer in Virginia. Solution provided by Professor Kathleen Miranda of SUNY at Old Westbury.

49. Apply Problem 48 to triangle OAB to show that

$$r = \frac{c \sin \frac{\alpha}{2} \sin \frac{\beta}{2}}{\cos \frac{\gamma}{2}}$$

50. Use the results of Problems 45 and 46 in Section 7.3 to show that

$$\cot \frac{\gamma}{2} = \frac{s - c}{r}$$

51. Show that

$$\cot \frac{\alpha}{2} + \cot \frac{\beta}{2} + \cot \frac{\gamma}{2} = \frac{s}{r}$$

52. Show that the area K of triangle ABC is $K = rs$. Then show that

$$r = \sqrt{\frac{(s - a)(s - b)(s - c)}{s}}$$

where $s = \frac{1}{2}(a + b + c)$.

53. What do you do first if you are asked to find the area of a triangle and are given two sides and the included angle?

54. What do you do first if you are asked to find the area of a triangle and are given three sides?

'Are You Prepared?' Answer

1. $A = \frac{1}{2}bh$

7.5 Simple Harmonic Motion; Damped Motion; Combining Waves

PREPARING FOR THIS SECTION *Before getting started, review the following:*

- Sinusoidal Graphs (Section 5.4, pp. 374–381)

Now work the 'Are You Prepared?' problem on page 526.

OBJECTIVES
1. Find an Equation for an Object in Simple Harmonic Motion
2. Analyze Simple Harmonic Motion
3. Analyze an Object in Damped Motion
4. Graph the Sum of Two Functions

Simple Harmonic Motion

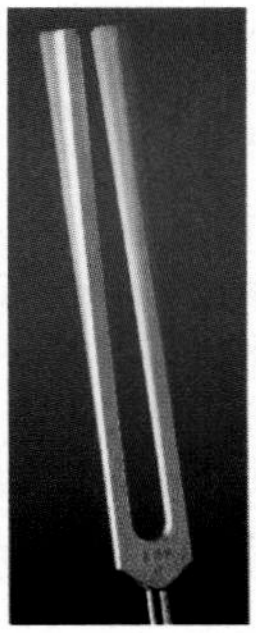

Vibrating tuning fork

Many physical phenomena can be described as simple harmonic motion. Radio and television waves, light waves, sound waves, and water waves exhibit motion that is simple harmonic.

The swinging of a pendulum, the vibrations of a tuning fork, and the bobbing of a weight attached to a coiled spring are examples of vibrational motion. In this type of motion, an object swings back and forth over the same path. In Figure 43, the point B is the **equilibrium (rest) position** of the vibrating object. The **amplitude** is the distance from the ob-

Figure 43

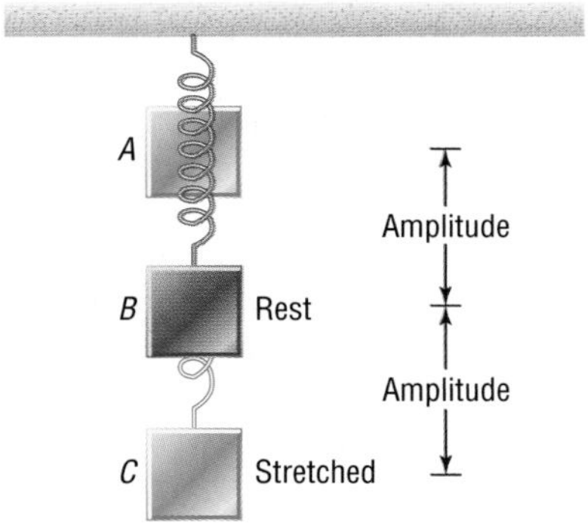

Coiled spring

ject's rest position to its point of greatest displacement (either point A or point C in Figure 43). The **period** is the time required to complete one vibration, that is, the time it takes to go from, say, point A through B to C and back to A.

Simple harmonic motion is a special kind of vibrational motion in which the acceleration a of the object is directly proportional to the negative of its displacement d from its rest position. That is, $a = -kd, k > 0$.

For example, when the mass hanging from the spring in Figure 43 is pulled down from its rest position B to the point C, the force of the spring tries to restore the mass to its rest position. Assuming that there is no frictional force* to retard the motion, the amplitude will remain constant. The force increases in direct proportion to the distance that the mass is pulled from its rest position. Since the force increases directly, the acceleration of the mass of the object must do likewise, because (by Newton's Second Law of Motion) force is directly proportional to acceleration. Thus, the acceleration of the object varies directly with its displacement, and the motion is an example of simple harmonic motion.

Simple harmonic motion is related to circular motion. To see this relationship, consider a circle of radius a, with center at $(0, 0)$. See Figure 44. Suppose that an object initially placed at $(a, 0)$ moves counterclockwise around the circle at a constant angular speed ω. Suppose further that after time t has elapsed the object is at the point $P = (x, y)$ on the circle. The angle θ, in radians, swept out by the ray $\overrightarrow{OP}$ in this time t is

Figure 44

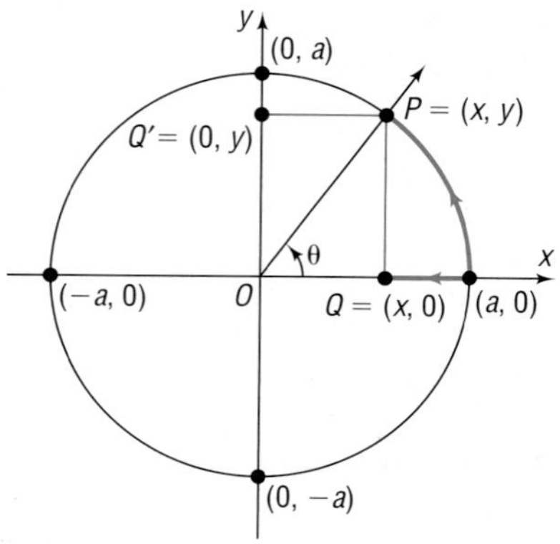

$$\theta = \omega t$$

The coordinates of the point P at time t are

$$x = a \cos \theta = a \cos(\omega t)$$

$$y = a \sin \theta = a \sin(\omega t)$$

Corresponding to each position $P = (x, y)$ of the object moving about the circle, there is the point $Q = (x, 0)$, called the **projection of P on the x-axis.** As P moves around the circle at a constant rate, the point Q moves back and forth between the points $(a, 0)$ and $(-a, 0)$ along the x-axis with a motion that is simple harmonic. Similarly, for each point P there is a point $Q' = (0, y)$, called the **projection of P on the y-axis.** As P moves around the circle, the point Q' moves back and forth between the points $(0, a)$ and $(0, -a)$ on the y-axis with a motion that is simple harmonic. Simple harmonic motion can be described as the projection of constant circular motion on a coordinate axis.

To put it another way, again consider a mass hanging from a spring where the mass is pulled down from its rest position to the point C and then released. See Figure 45(a). The graph shown in Figure 45(b) describes the displacement d of the object from its rest position as a function of time t, assuming that no frictional force is present.

*If friction is present, the amplitude will decrease with time to 0. This type of motion is an example of **damped motion,** which is discussed later in this section.

Figure 45

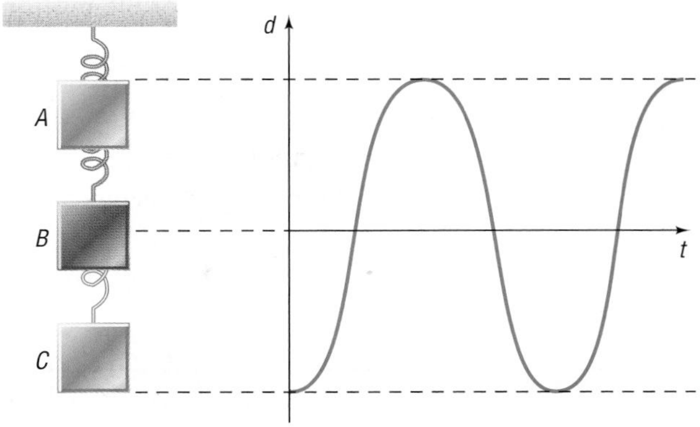

Theorem

Simple Harmonic Motion

An object that moves on a coordinate axis so that its distance d from its rest position at time t is given by either

$$d = a\cos(\omega t) \quad \text{or} \quad d = a\sin(\omega t)$$

where a and $\omega > 0$ are constants, moves with simple harmonic motion. The motion has amplitude $|a|$ and period $\dfrac{2\pi}{\omega}$.

The **frequency** f of an object in simple harmonic motion is the number of oscillations per unit time. Since the period is the time required for one oscillation, it follows that the frequency is the reciprocal of the period; that is,

$$f = \frac{\omega}{2\pi}, \qquad \omega > 0$$

Figure 46

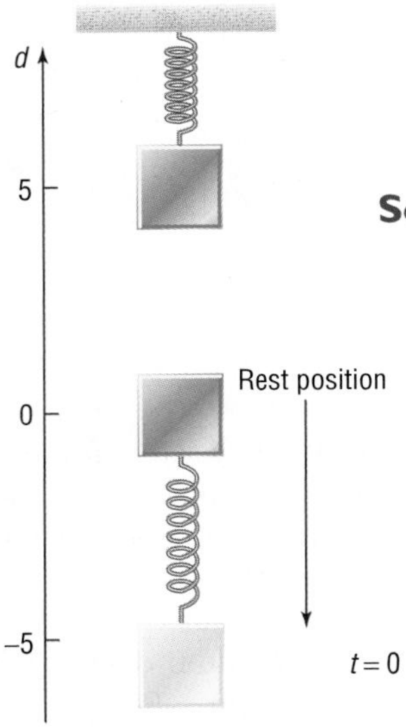

EXAMPLE 1 Finding an Equation for an Object in Harmonic Motion

1 Suppose that an object attached to a coiled spring is pulled down a distance of 5 inches from its rest position and then released. If the time for one oscillation is 3 seconds, write an equation that relates the displacement d of the object from its rest position after time t (in seconds). Assume no friction.

Solution The motion of the object is simple harmonic. See Figure 46. When the object is released ($t = 0$), the displacement of the object from the rest position is -5 units (since the object was pulled down). Because $d = -5$ when $t = 0$, it is easier to use the cosine function*

$$d = a\cos(\omega t)$$

to describe the motion. Now the amplitude is $|-5| = 5$ and the period is 3, so

$$a = -5 \quad \text{and} \quad \frac{2\pi}{\omega} = \text{period} = 3, \quad \omega = \frac{2\pi}{3}$$

*No phase shift is required if a cosine function is used.

An equation of the motion of the object is

$$d = -5\cos\left[\frac{2\pi}{3}t\right]$$

◀

Note: In the solution to Example 1, we let $a = -5$, since the initial motion is down. If the initial direction were up, we would let $a = 5$.

NOW WORK PROBLEM 5.

2 **EXAMPLE 2** **Analyzing the Motion of an Object**

Suppose that the displacement d (in meters) of an object at time t (in seconds) satisfies the equation

$$d = 10\sin(5t)$$

(a) Describe the motion of the object.

(b) What is the maximum displacement from its resting position?

(c) What is the time required for one oscillation?

(d) What is the frequency?

Solution We observe that the given equation is of the form

$$d = a\sin(\omega t) \qquad d = 10\sin(5t)$$

where $a = 10$ and $\omega = 5$.

(a) The motion is simple harmonic.

(b) The maximum displacement of the object from its resting position is the amplitude: $|a| = 10$ meters.

(c) The time required for one oscillation is the period:

$$\text{Period} = \frac{2\pi}{\omega} = \frac{2\pi}{5} \text{ seconds}$$

(d) The frequency is the reciprocal of the period. Thus,

$$\text{Frequency} = f = \frac{5}{2\pi} \text{ oscillations per second}$$

◀

NOW WORK PROBLEM 13.

Damped Motion

3 Most physical phenomena are affected by friction or other resistive forces. These forces remove energy from a moving system and thereby damp its motion. For example, when a mass hanging from a spring is pulled down a distance a and released, the friction in the spring causes the distance that the mass moves from its at-rest position to decrease over time. Thus, the amplitude of any real oscillating spring or swinging pendulum decreases with time due to air resistance, friction, and so forth. See Figure 47.

Figure 47

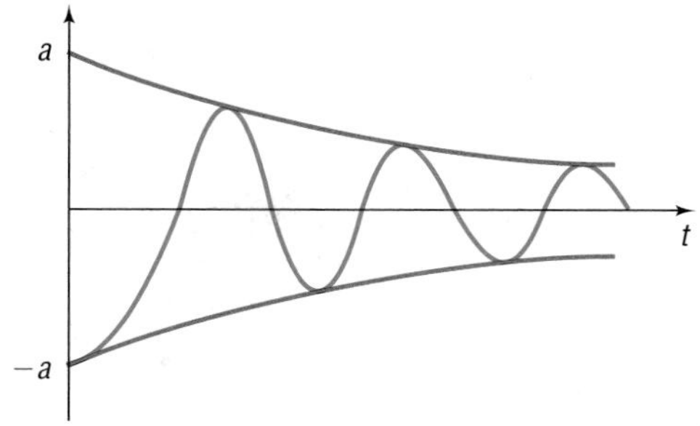

A function that describes this phenomenon maintains a sinusoidal component, but the amplitude of this component will decrease with time in order to account for the damping effect. In addition, the period of the oscillating component will be affected by the damping. The next result, from physics, describes damped motion.

Theorem

Damped Motion

The displacement d of an oscillating object from its rest position at time t is given by

$$d(t) = ae^{-bt/2m}\cos\left(\sqrt{\omega^2 - \frac{b^2}{4m^2}}\,t\right)$$

where b is a **damping factor** (most physics texts call this a **damping coefficient**) and m is the mass of the oscillating object.

Notice for $b = 0$ (zero damping) that we have the formula for simple harmonic motion with amplitude $|a|$ and period $\frac{2\pi}{\omega}$.

EXAMPLE 3 Analyzing a Damped Vibration Curve

Analyze the damped vibration curve

$$d(t) = e^{-t/\pi}\cos t, \quad t \geq 0$$

Solution The displacement d is the product of $y = e^{-t/\pi}$ and $y = \cos t$. Using properties of absolute value and the fact that $|\cos t| \leq 1$, we find that

$$|d(t)| = |e^{-t/\pi}\cos t| = |e^{-t/\pi}||\cos t| \leq |e^{-t/\pi}| = e^{-t/\pi}$$

↑ $e^{-t/\pi} > 0$

As a result,

$$-e^{-t/\pi} \leq d(t) \leq e^{-t/\pi}$$

This means that the graph of d will lie between the graphs of $y = e^{-t/\pi}$ and $y = -e^{-t/\pi}$, the **bounding curves** of d.

Also, the graph of d will touch these graphs when $|\cos t| = 1$, that is, when $t = 0, \pi, 2\pi$, and so on. The x-intercepts of the graph of d occur when $\cos t = 0$, that is, at $\frac{\pi}{2}, \frac{3\pi}{2}, \frac{5\pi}{2}$, and so on. See Table 1.

Table 1

t	0	$\frac{\pi}{2}$	π	$\frac{3\pi}{2}$	2π
$e^{-t/\pi}$	1	$e^{-1/2}$	e^{-1}	$e^{-3/2}$	e^{-2}
$\cos t$	1	0	-1	0	1
$d(t) = e^{-t/\pi}\cos t$	1	0	$-e^{-1}$	0	e^{-2}
Point on graph of d	$(0, 1)$	$\left(\frac{\pi}{2}, 0\right)$	$(\pi, -e^{-1})$	$\left(\frac{3\pi}{2}, 0\right)$	$(2\pi, e^{-2})$

We graph $y = \cos t$, $y = e^{-t/\pi}$, $y = -e^{-t/\pi}$, and $d(t) = e^{-t/\pi}\cos t$ in Figure 48.

Figure 48

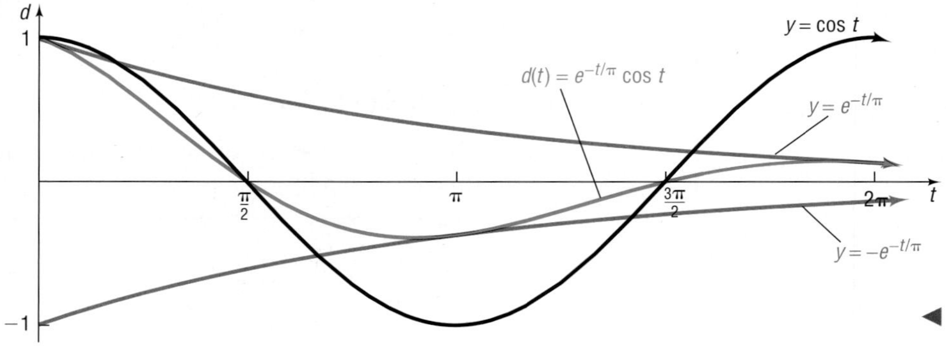

Exploration

Graph $Y_1 = e^{-x/\pi}\cos x$, along with $Y_2 = e^{-x/\pi}$, and $Y_3 = -e^{-x/\pi}$, for $0 \le x \le \pi$. Determine where Y_1 has its first turning point (local minimum). Compare this to where Y_1 intersects Y_3.

SOLUTION Figure 49 shows the graphs of $Y_1 = e^{-x/\pi}\cos x$, $Y_2 = e^{-x/\pi}$, and $Y_3 = -e^{-x/\pi}$. Using MINIMUM, the first turning point occurs at $x \approx 2.83$; Y_1 INTERSECTS Y_3 at $x = \pi \approx 3.14$.

Figure 49

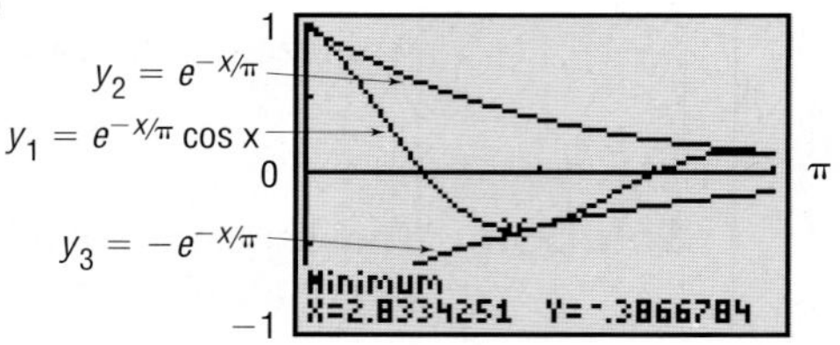

NOW WORK PROBLEM 21.

Combining Waves

4 Many physical and biological applications require the graph of the sum of two functions, such as

$$f(x) = x + \sin x \quad \text{or} \quad g(x) = \sin x + \cos 2x$$

For example, on a Touch-Tone phone, two tones are emitted and the sound produced is the sum of the waves produced by the two tones. See Problem 35 for an explanation of Touch-Tone phones.

To graph the sum of two (or more) functions, we can use the method of adding y-coordinates described next.

EXAMPLE 4 Graphing the Sum of Two Functions

Use the method of adding y-coordinates to graph $f(x) = x + \sin x$.

Solution First, we graph the component functions,

$$y = f_1(x) = x \qquad y = f_2(x) = \sin x$$

using the same coordinate system. See Figure 50(a). Now, select several values of x, say, $x = 0$, $x = \frac{\pi}{2}$, $x = \pi$, $x = \frac{3\pi}{2}$, and $x = 2\pi$, at which we compute $f(x) = f_1(x) + f_2(x)$. Table 2 shows the computation. We plot these points and connect them to get the graph, as shown in Figure 50(b).

Table 2

x	0	$\frac{\pi}{2}$	π	$\frac{3\pi}{2}$	2π
$y = f_1(x) = x$	0	$\frac{\pi}{2}$	π	$\frac{3\pi}{2}$	2π
$y = f_2(x) = \sin x$	0	1	0	-1	0
$f(x) = x + \sin x$	0	$\frac{\pi}{2} + 1 \approx 2.57$	π	$\frac{3\pi}{2} - 1 \approx 3.71$	2π
Point on graph of f	$(0, 0)$	$\left(\frac{\pi}{2}, 2.57\right)$	(π, π)	$\left(\frac{3\pi}{2}, 3.71\right)$	$(2\pi, 2\pi)$

Figure 50

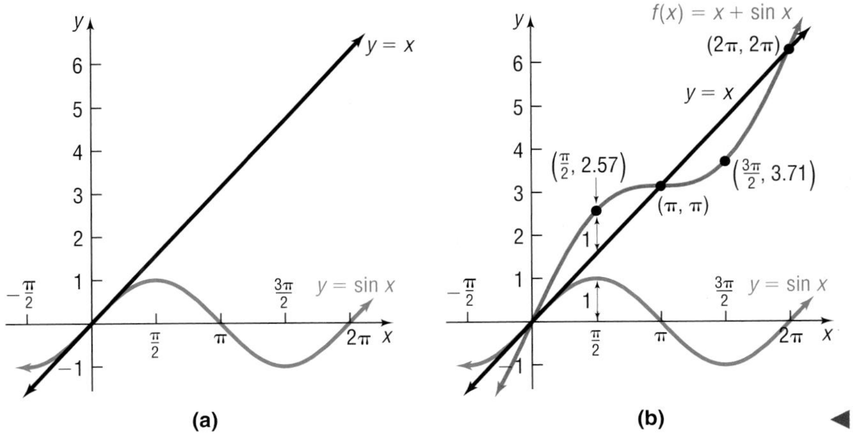

In Example 4, note that the graph of $f(x) = x + \sin x$ intersects the line $y = x$ whenever $\sin x = 0$. Also, notice that the graph of f is not periodic.

CHECK: Graph $y = x + \sin x$ and compare the result with Figure 50(b). Use INTERSECT to verify that the graphs intersect when $\sin x = 0$.

NOW WORK PROBLEM 25.

The next example shows a periodic graph of the sum of two functions.

EXAMPLE 5 Graphing the Sum of Two Sinusoidal Functions

Use the method of adding y-coordinates to graph

$$f(x) = \sin x + \cos(2x)$$

Solution Table 3 shows the steps for computing several points on the graph of f. Figure 51 illustrates the graphs of the component functions, $y = f_1(x) = \sin x$ and $y = f_2(x) = \cos(2x)$, and the graph of $f(x) = \sin x + \cos(2x)$, which is shown in red.

Table 3

x	$-\frac{\pi}{2}$	0	$\frac{\pi}{2}$	π	$\frac{3\pi}{2}$	2π
$y = f_1(x) = \sin x$	-1	0	1	0	-1	0
$y = f_2(x) = \cos(2x)$	-1	1	-1	1	-1	1
$f(x) = \sin x + \cos(2x)$	-2	1	0	1	-2	1
Point on graph of f	$\left(-\frac{\pi}{2}, -2\right)$	$(0, 1)$	$\left(-\frac{\pi}{2}, 0\right)$	$(\pi, 1)$	$\left(\frac{3\pi}{2}, -2\right)$,	$(2\pi, 1)$

Figure 51
$f(x) = \sin x + \cos(2x)$

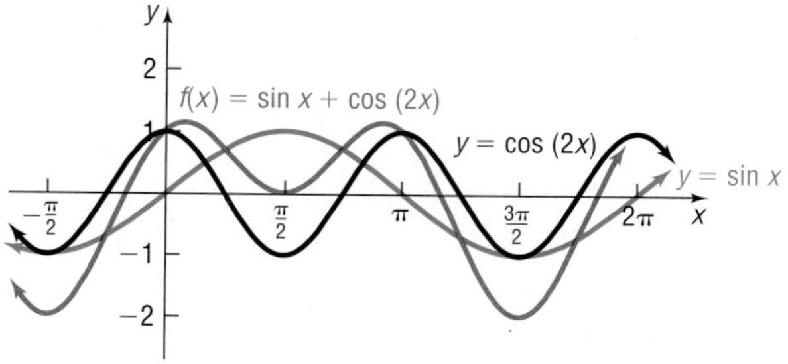

CHECK: Graph $y = \sin x + \cos(2x)$ and compare the result with Figure 51. ◀

7.5 Assess Your Understanding

'Are You Prepared?' *Answers are given at the end of these exercises. If you get a wrong answer, read the pages listed in red.*

1. The amplitude A and period T of $f(x) = 5\sin(4x)$ are ________ and ________. (pp. 374–381)

Concepts and Vocabulary

2. The motion of an object obeys the equation $d = 4\cos(6t)$. Such motion is described as ________ ________. The number 4 is called the ________.

3. When a mass hanging from a spring is pulled down and then released, the motion is called ________ ________ ________ if there is no frictional force to retard the motion, and the motion is called ________ ________ if there is friction.

4. *True or False:* If the distance d of an object from its rest position at time t is given by a sinusoidal graph, the motion of the object is simple harmonic motion.

Exercises

In Problems 5–8, an object attached to a coiled spring is pulled down a distance a from its rest position and then released. Assuming that the motion is simple harmonic with period T, write an equation that relates the displacement d of the object from its rest position after t seconds. Also assume that the positive direction of the motion is up.

5. $a = 5$; $T = 2$ seconds

6. $a = 10$; $T = 3$ seconds

7. $a = 6$; $T = \pi$ seconds

8. $a = 4$; $T = \frac{\pi}{2}$ seconds

9. Rework Problem 5 under the same conditions except that, at time $t = 0$, the object is at its rest position and moving down.

10. Rework Problem 6 under the same conditions except that, at time $t = 0$, the object is at its rest position and moving down.

11. Rework Problem 7 under the same conditions except that, at time $t = 0$, the object is at its rest position and moving down.

12. Rework Problem 8 under the same conditions except that, at time $t = 0$, the object is at its rest position and moving down.

In Problems 13–20, the displacement d (in meters) of an object at time t (in seconds) is given.
(a) *Describe the motion of the object.*
(b) *What is the maximum displacement from its resting position?*
(c) *What is the time required for one oscillation?*
(d) *What is the frequency?*

13. $d = 5\sin(3t)$ **14.** $d = 4\sin(2t)$ **15.** $d = 6\cos(\pi t)$ **16.** $d = 5\cos\frac{\pi}{2}t$

17. $d = -3\sin\left(\frac{1}{2}t\right)$ **18.** $d = -2\cos(2t)$ **19.** $d = 6 + 2\cos(2\pi t)$ **20.** $d = 4 + 3\sin(\pi t)$

In Problems 21–24, graph each damped vibration curve for $0 \le t \le 2\pi$.

21. $d(t) = e^{-t/\pi}\cos(2t)$ **22.** $d(t) = e^{-t/2\pi}\cos(2t)$ **23.** $d(t) = e^{-t/2\pi}\cos t$ **24.** $d(t) = e^{-t/4\pi}\cos t$

In Problems 25–32, use the method of adding y-coordinates to graph each function.

25. $f(x) = x + \cos x$ **26.** $f(x) = x + \cos(2x)$ **27.** $f(x) = x - \sin x$

28. $f(x) = x - \cos x$ **29.** $f(x) = \sin x + \cos x$ **30.** $f(x) = \sin(2x) + \cos x$

31. $g(x) = \sin x + \sin(2x)$ **32.** $g(x) = \cos(2x) + \cos x$

33. Charging a Capacitor If a charged capacitor is connected to a coil by closing a switch (see the figure), energy is transferred to the coil and then back to the capacitor in an oscillatory motion. The voltage V (in volts) across the capacitor will gradually diminish to 0 with time t (in seconds).

(a) Graph the equation relating V and t:

$$V(t) = e^{-t/3}\cos(\pi t), \qquad 0 \le t \le 3$$

(b) At what times t will the graph of V touch the graph of $y = e^{-t/3}$? When does V touch the graph of $y = -e^{-t/3}$?

(c) When will the voltage V be between -0.4 and 0.4 volt?

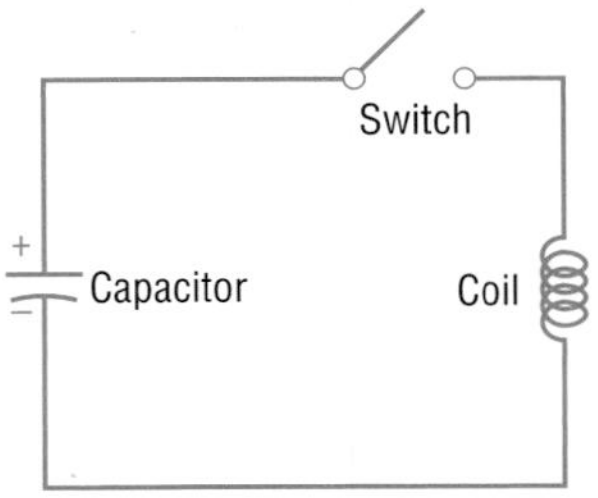

34. The Sawtooth Curve An oscilloscope often displays a *sawtooth curve.* This curve can be approximated by sinusoidal curves of varying periods and amplitudes.

(a) Graph the following function, which can be used to approximate the sawtooth curve.

$$f(x) = \frac{1}{2}\sin(2\pi x) + \frac{1}{4}\sin(4\pi x), \qquad 0 \le x \le 2$$

(b) A better approximation to the sawtooth curve is given by

$$f(x) = \frac{1}{2}\sin(2\pi x) + \frac{1}{4}\sin(4\pi x) + \frac{1}{8}\sin(8\pi x)$$

Graph this function for $0 \le x \le 4$ and compare the result to the graph obtained in part (a).

(c) A third and even better approximation to the sawtooth curve is given by

$$f(x) = \frac{1}{2}\sin(2\pi x) + \frac{1}{4}\sin(4\pi x) + \frac{1}{8}\sin(8\pi x) + \frac{1}{16}\sin(16\pi x)$$

Graph this function for $0 \le x \le 4$ and compare the result to the graphs obtained in parts (a) and (b).

(d) What do you think the next approximation to the sawtooth curve is?

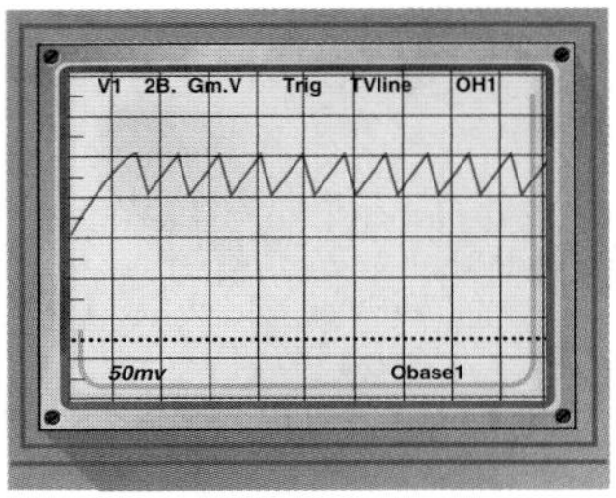

35. Touch-Tone Phones On a Touch-Tone phone, each button produces a unique sound. The sound produced is the sum of two tones, given by

$$y = \sin(2\pi l t) \quad \text{and} \quad y = \sin(2\pi h t)$$

where l and h are the low and high frequencies (cycles per second) shown on the illustration. For example, if you touch 7, the low frequency is $l = 852$ cycles per second and the high frequency is $h = 1209$ cycles per second. The sound emitted by touching 7 is

$$y = \sin[2\pi(852)t] + \sin[2\pi(1209)t]$$

Graph the sound emitted by touching 7.

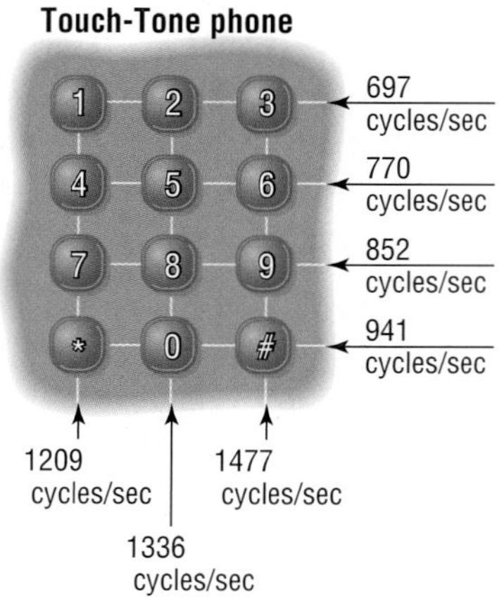

36. Graph the sound emitted by the * key on a Touch-Tone phone. See Problem 35.

37. Graph the function $f(x) = \dfrac{\sin x}{x}$, $x > 0$. Based on the graph, what do you conjecture about the value of $\dfrac{\sin x}{x}$ for x close to 0?

38. Graph $y = x \sin x$, $y = x^2 \sin x$, and $y = x^3 \sin x$ for $x > 0$. What patterns do you observe?

39. Graph $y = \dfrac{1}{x} \sin x$, $y = \dfrac{1}{x^2} \sin x$, and $y = \dfrac{1}{x^3} \sin x$ for $x > 0$. What patterns do you observe?

40. CBL Experiment Pendulum motion is analyzed to estimate simple harmonic motion. A plot is generated with the position of the pendulum over time. The graph is used to find a sinusoidal curve of the form $y = A\cos[B(x - C)] + D$. Determine the amplitude, period, and frequency. (Activity 16, Real-World Math with the CBL System.)

41. CBL Experiment The sound from a tuning fork is collected over time. The amplitude, frequency, and period of the graph are determined. A model of the form $y = A\cos B(x - C)$ is fitted to the data. (Activity 23, Real-World Math with the CBL System.)

42. How would you explain to a friend what simple harmonic motion is? How would you explain damped motion?

'Are You Prepared?' Answers

1. $A = 5$; $T = \dfrac{\pi}{2}$

Chapter Review

Things to Know

Formulas

Law of Sines (p. 495) $\quad \dfrac{\sin\alpha}{a} = \dfrac{\sin\beta}{b} = \dfrac{\sin\gamma}{c}$

Law of Cosines (p. 507)

$$c^2 = a^2 + b^2 - 2ab\cos\gamma$$
$$b^2 = a^2 + c^2 - 2ac\cos\beta$$
$$a^2 = b^2 + c^2 - 2bc\cos\alpha$$

Area of a triangle (pp. 513–514)

$$A = \frac{1}{2}bh$$
$$A = \frac{1}{2}ab\sin\gamma$$
$$A = \frac{1}{2}bc\sin\alpha$$
$$A = \frac{1}{2}ac\sin\beta$$
$$A = \sqrt{s(s-a)(s-b)(s-c)}, \quad \text{where} \quad s = \frac{1}{2}(a + b + c)$$

Objectives

Section		You should be able to . . .	Review Exercises
7.1	1	Find the value of trigonometric functions of acute angles (p. 482)	1–4
	2	Use the Complementary Angle Theorem (p. 484)	5–10
	3	Solve right triangles (p. 485)	11–14
	4	Solve applied problems using right triangle trigonometry (p. 486)	45–50, 60
7.2	1	Solve SAA or ASA triangles (p. 496)	15–16, 32
	2	Solve SSA triangles (p. 497)	17–20, 22, 27–28, 31
	3	Solve applied problems using the Law of Sines (p. 499)	51, 53–54
7.3	1	Solve SAS triangles (p. 508)	21, 25–26, 33–34
	2	Solve SSS triangles (p. 508)	23–24, 29–30
	3	Solve applied problems using the Law of Cosines (p. 509)	52, 55, 56–57
7.4	1	Find the area of SAS triangles (p. 514)	35–38, 57–59
	2	Find the area of SSS triangles (p. 514)	39–42
7.5	1	Find an equation for an object in simple harmonic motion (p. 521)	63–64
	2	Analyze simple harmonic motion (p. 522)	65–68
	3	Analyze an object in damped motion (p. 522)	69–72
	4	Graph the sum of two functions (p. 524)	73–74

Review Exercises *(Blue problem numbers indicate the author's suggestions for use in a Practice Test.)*

In Problems 1–4, find the exact value of the six trigonometric functions of the angle θ in each figure.

1.
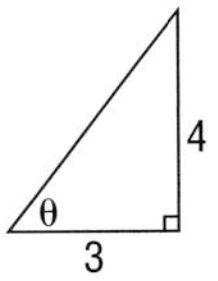

2.
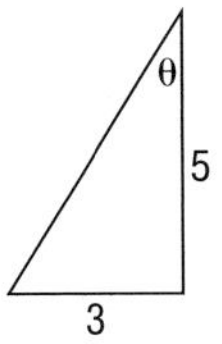

3.
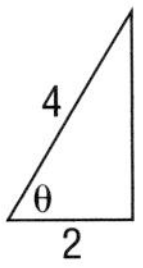

4.
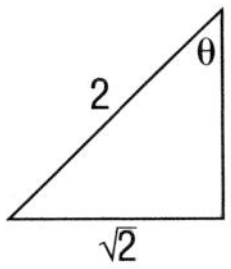

In Problems 5–10, find the exact value of each expression. Do not use a calculator.

5. $\cos 62° - \sin 28°$

6. $\tan 15° - \cot 75°$

7. $\dfrac{\sec 55°}{\csc 35°}$

8. $\dfrac{\tan 40°}{\cot 50°}$

9. $\cos^2 40° + \cos^2 50°$

10. $\tan^2 40° - \csc^2 50°$

In Problems 11–14, solve each triangle.

11.
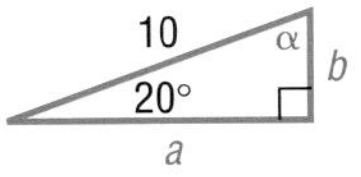

12.
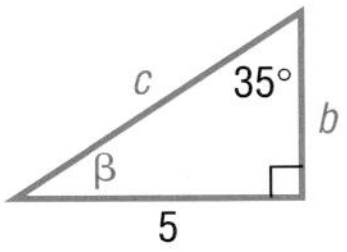

13.
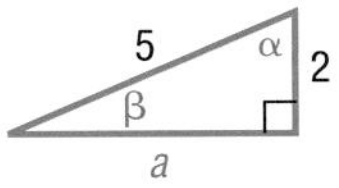

14.
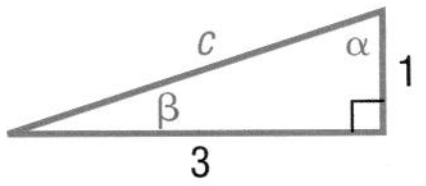

In Problems 15–34, find the remaining angle(s) and side(s) of each triangle, if it (they) exists. If no triangle exists, say "No triangle."

15. $\alpha = 50°, \quad \beta = 30°, \quad a = 1$

16. $\alpha = 10°, \quad \gamma = 40°, \quad c = 2$

17. $\alpha = 100°, \quad a = 5, \quad c = 2$

18. $a = 2, \quad c = 5, \quad \alpha = 60°$

19. $a = 3, \quad c = 1, \quad \gamma = 110°$

20. $a = 3, \quad c = 1, \quad \gamma = 20°$

21. $a = 3, \quad c = 1, \quad \beta = 100°$

22. $a = 3, \quad b = 5, \quad \beta = 80°$

23. $a = 2, \quad b = 3, \quad c = 1$

24. $a = 10, \quad b = 7, \quad c = 8$

25. $a = 1, \quad b = 3, \quad \gamma = 40°$

26. $a = 4, \quad b = 1, \quad \gamma = 100°$

27. $a = 5, \quad b = 3, \quad \alpha = 80°$
28. $a = 2, \quad b = 3, \quad \alpha = 20°$
29. $a = 1, \quad b = \frac{1}{2}, \quad c = \frac{4}{3}$
30. $a = 3, \quad b = 2, \quad c = 2$
31. $a = 3, \quad \alpha = 10°, \quad b = 4$
32. $a = 4, \quad \alpha = 20°, \quad \beta = 100°$
33. $c = 5, \quad b = 4, \quad \alpha = 70°$
34. $a = 1, \quad b = 2, \quad \gamma = 60°$

In Problems 35–44, find the area of each triangle.

35. $a = 2, \quad b = 3, \quad \gamma = 40°$
36. $b = 5, \quad c = 5, \quad \alpha = 20°$
37. $b = 4, \quad c = 10, \quad \alpha = 70°$
38. $a = 2, \quad b = 1, \quad \gamma = 100°$
39. $a = 4, \quad b = 3, \quad c = 5$
40. $a = 10, \quad b = 7, \quad c = 8$
41. $a = 4, \quad b = 2, \quad c = 5$
42. $a = 3, \quad b = 2, \quad c = 2$
43. $\alpha = 50°, \quad \beta = 30°, \quad a = 1$
44. $\alpha = 10°, \quad \gamma = 40°, \quad c = 3$

45. Measuring the Length of a Lake From a stationary hot-air balloon 500 feet above the ground, two sightings of a lake are made (see the figure). How long is the lake?

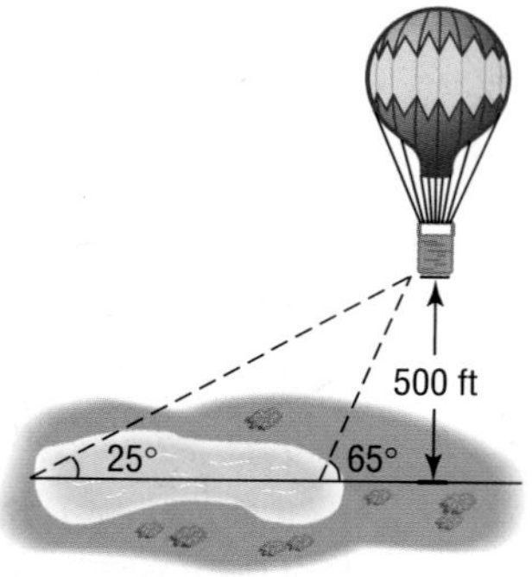

46. Finding the Speed of a Glider From a glider 200 feet above the ground, two sightings of a stationary object directly in front are taken 1 minute apart (see the figure). What is the speed of the glider?

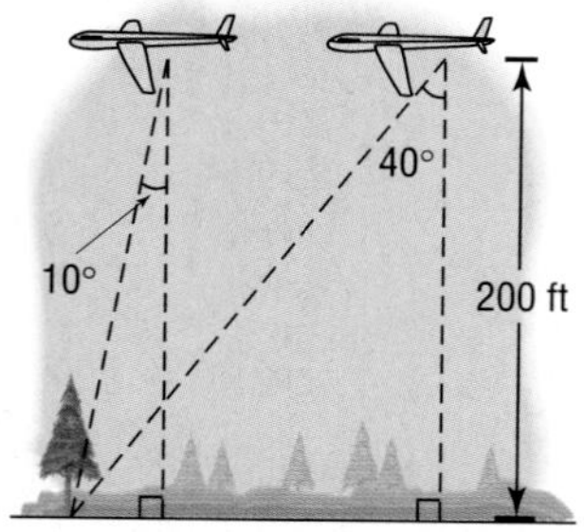

47. Finding the Width of a River Find the distance from A to C across the river illustrated in the figure.

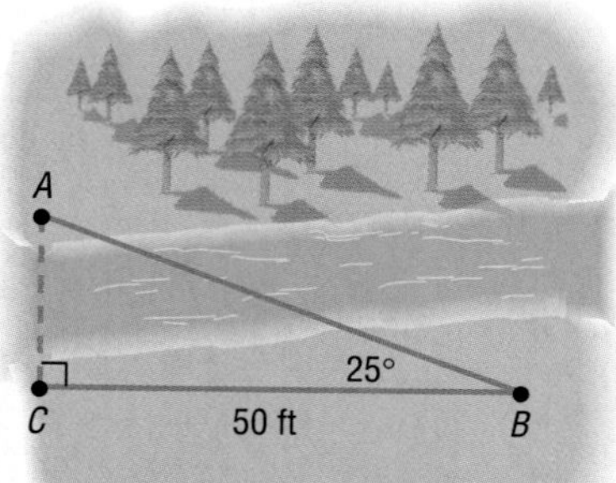

48. Finding the Height of a Building Find the height of the building shown in the figure.

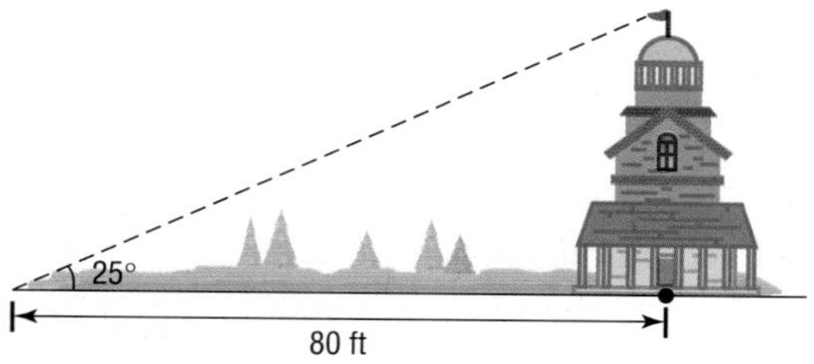

49. Finding the Distance to Shore The Sears Tower in Chicago is 1454 feet tall and is situated about 1 mile inland from the shore of Lake Michigan, as indicated in the figure. An observer in a pleasure boat on the lake directly in front of the Sears Tower looks at the top of the tower and measures the angle of elevation as 5°. How far offshore is the boat?

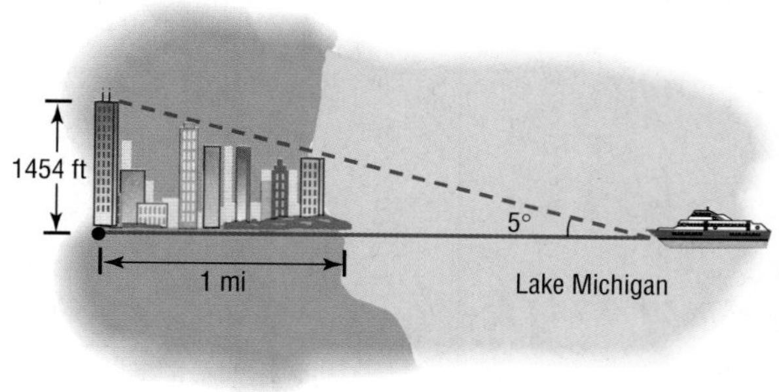

50. Finding the Grade of a Mountain Trail A straight trail with a uniform inclination leads from a hotel, elevation 5000 feet, to a lake in a valley, elevation 4100 feet. The length of the trail is 4100 feet. What is the inclination (grade) of the trail?

51. Navigation An airplane flies from city A to city B, a distance of 100 miles, and then turns through an angle of 20° and heads toward city C, as indicated in the figure. If the distance from A to C is 300 miles, how far is it from city B to city C?

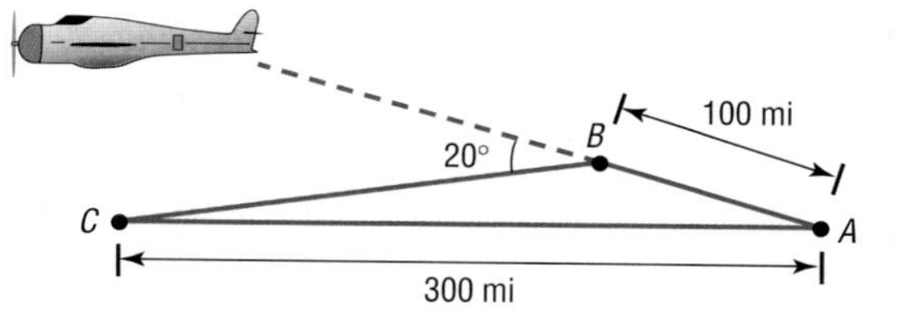

52. Correcting a Navigation Error Two cities A and B are 300 miles apart. In flying from city A to city B, a pilot inadvertently took a course that was 5° in error.

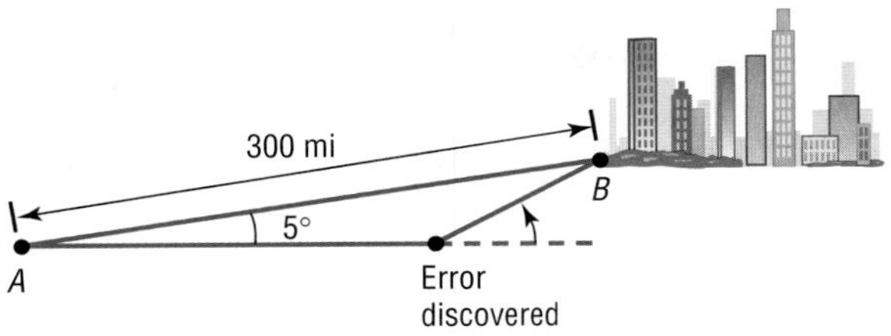

(a) If the error was discovered after flying 10 minutes at a constant speed of 420 miles per hour, through what angle should the pilot turn to correct the course? (Consult the figure.)
(b) What new constant speed should be maintained so that no time is lost due to the error? (Assume that the speed would have been a constant 420 miles per hour if no error had occurred.)

53. Determining Distances at Sea Rebecca, the navigator of a ship at sea, spots two lighthouses that she knows to be 2 miles apart along a straight shoreline. She determines that the angles formed between two line-of-sight observations of the lighthouses and the line from the ship directly to shore are 12° and 30°. See the illustration.
(a) How far is the ship from lighthouse A?
(b) How far is the ship from lighthouse B?
(c) How far is the ship from shore?

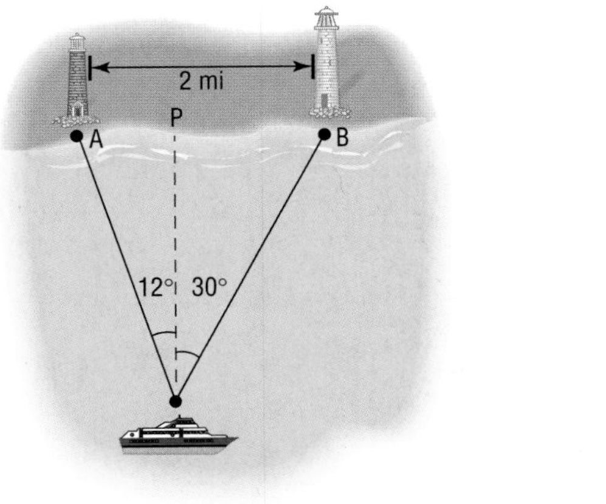

54. Constructing a Highway A highway whose primary directions are north–south is being constructed along the west coast of Florida. Near Naples, a bay obstructs the straight path of the road. Since the cost of a bridge is prohibitive, engineers decide to go around the bay. The illustration shows the path that they decide on and the measurements taken. What is the length of highway needed to go around the bay?

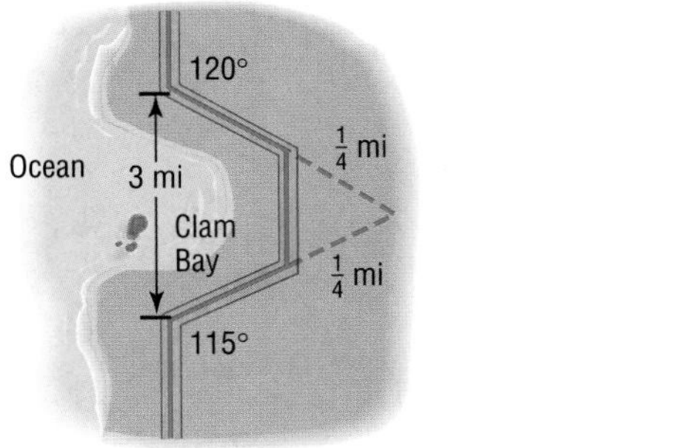

55. Correcting a Navigational Error A sailboat leaves St. Thomas bound for an island in the British West Indies, 200 miles away. Maintaining a constant speed of 18 miles per hour, but encountering heavy crosswinds and strong currents, the crew finds after 4 hours that the sailboat is off course by 15°.
(a) How far is the sailboat from the island at this time?
(b) Through what angle should the sailboat turn to correct its course?
(c) How much time has been added to the trip because of this? (Assume that the speed remains at 18 miles per hour.)

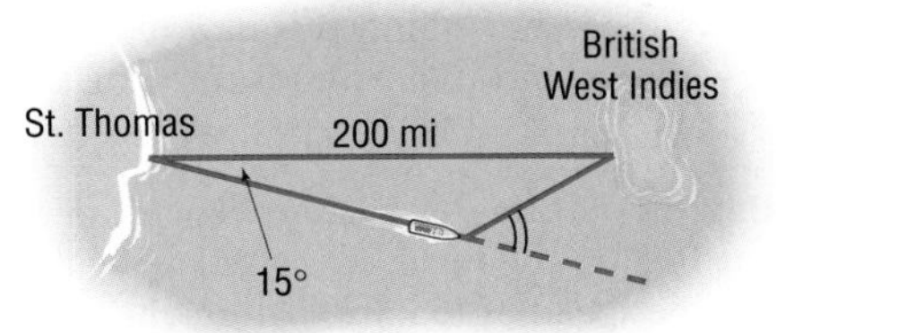

56. Surveying Two homes are located on opposite sides of a small hill. See the illustration. To measure the distance between them, a surveyor walks a distance of 50 feet from house A to point C, uses a transit to measure the angle ACB, which is found to be 80°, and then walks to house B, a distance of 60 feet. How far apart are the houses?

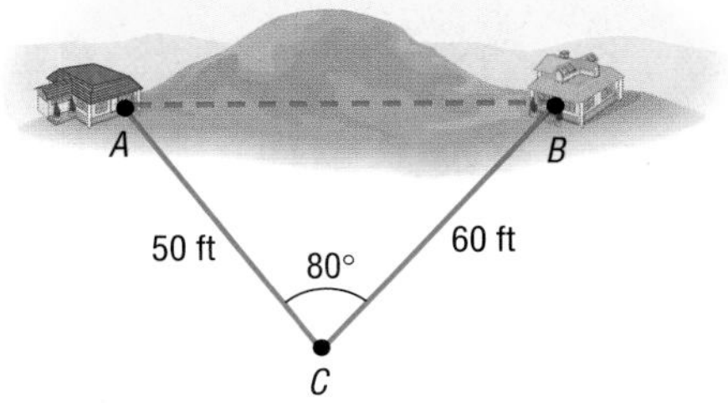

57. Approximating the Area of a Lake To approximate the area of a lake, Cindy walks around the perimeter of the lake, taking the measurements shown in the illustration. Using this technique, what is the approximate area of the lake?

[**Hint:** Use the Law of Cosines on the three triangles shown and then find the sum of their areas.]

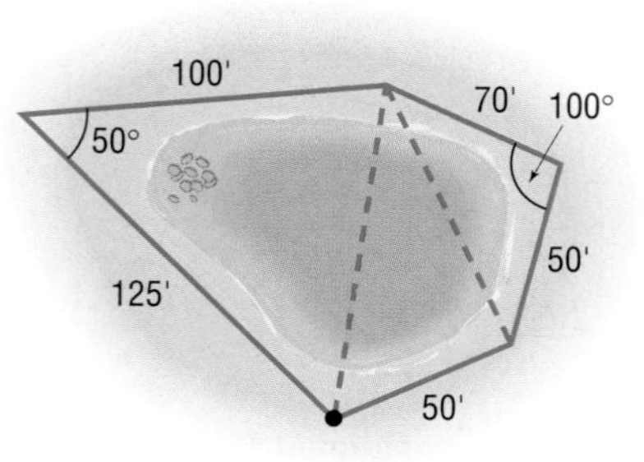

58. Calculating the Cost of Land The irregular parcel of land shown in the figure is being sold for \$100 per square foot. What is the cost of this parcel?

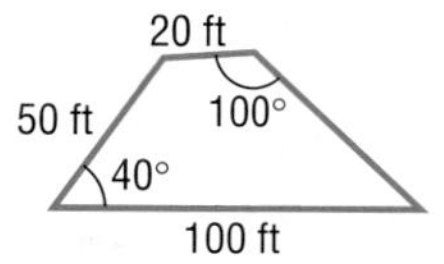

59. Area of a Segment Find the area of the segment of a circle whose radius is 6 inches formed by a central angle of 50°.

60. Finding the Bearing of a Ship The *Majesty* leaves the Port at Boston for Bermuda with a bearing of S80°E at an average speed of 10 knots. After 1 hour, the ship turns 90° toward the southwest. After 2 hours at an average speed of 20 knots, what is the bearing of the ship from Boston?

61. The drive wheel of an engine is 13 inches in diameter, and the pulley on the rotary pump is 5 inches in diameter. If the shafts of the drive wheel and the pulley are 2 feet apart, what length of belt is required to join them as shown in the figure?

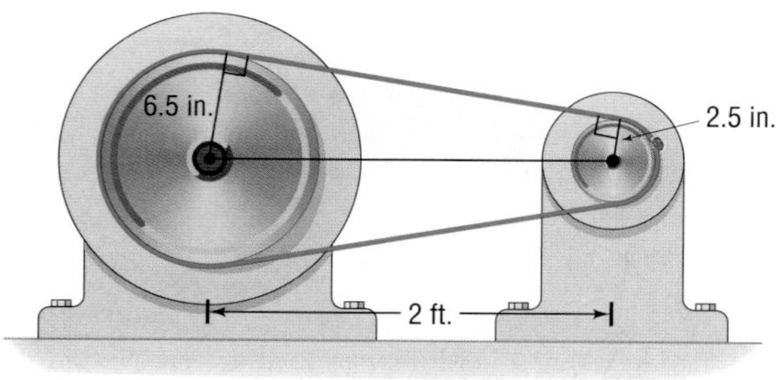

62. Rework Problem 61 if the belt is crossed, as shown in the figure.

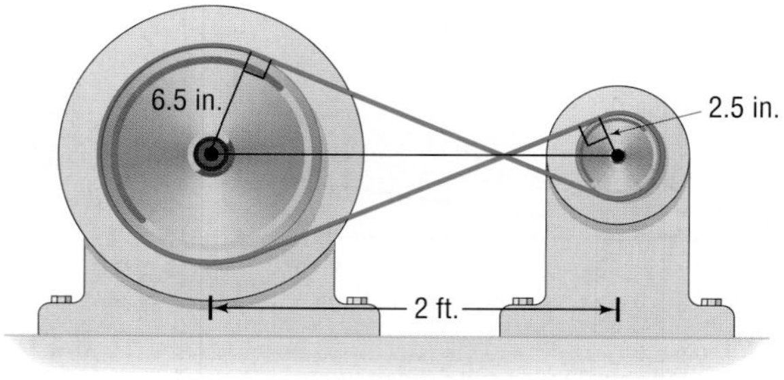

In Problems 63–64, an object attached to a coiled spring is pulled down a distance a from its rest position and then released. Assuming that the motion is simple harmonic with period T, write an equation that relates the displacement d of the object from its rest position after t seconds. Also assume that the positive direction of the motion is up.

63. $a = 3; T = 4$ seconds

64. $a = 5; T = 6$ seconds

In Problems 65–68, the distance d (in feet) that an object travels in time t (in seconds) is given.

(a) *Describe the motion of the object.*
(b) *What is the maximum displacement from its rest position?*
(c) *What is the time required for one oscillation?*
(d) *What is the frequency?*

65. $d = 6\sin(2t)$

66. $d = 2\cos(4t)$

67. $d = -2\cos(\pi t)$

68. $d = -3\sin\left[\frac{\pi}{2}t\right]$

In Problems 69–74, graph each function.

69. $y = e^{-x/2\pi}\sin(2x), \quad 0 \le x \le 2\pi$

70. $y = e^{-x/3\pi}\cos(4x), \quad 0 \le x \le 2\pi$

71. $y = x\cos x, \quad 0 \le x \le 2\pi$

72. $y = x\sin(2x), \quad 0 \le x \le 2\pi$

73. $y = 2\sin x + \cos(2x), \quad 0 \le x \le 2\pi$

74. $y = 2\cos(2x) + \sin\frac{x}{2}, \quad 0 \le x \le 2\pi$

Chapter Projects

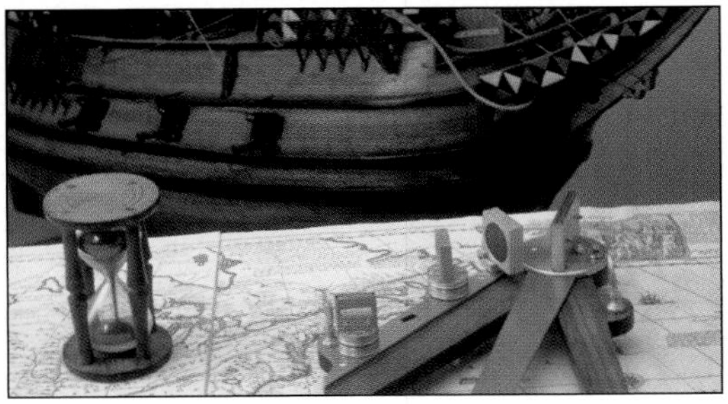

1. **A. Spherical Trigonometry** When the distance between two locations on the surface of Earth is small, we can compute the distance in statutory miles. Using this assumption, we can use the Law of Sines and the Law of Cosines to approximate distances and angles. However, if you look at a globe, you notice that Earth is a sphere, so, as the distance between two points on its surface increases, the linear distance is less accurate because of curvature. Under this circumstance, we need to take into account the curvature of Earth when using the Law of Sines and the Law of Cosines.

(a) Draw a spherical triangle and label each vertex by A, B, and C. Then connect each vertex by a radius to the center O of the sphere. Now, draw tangent lines to the sides a and b of the triangle that go through C. Extend the lines OA and OB to intersect the tangent lines at P and Q, respectively. See the diagram. List the plane right triangles. Determine the measures of the central angles.

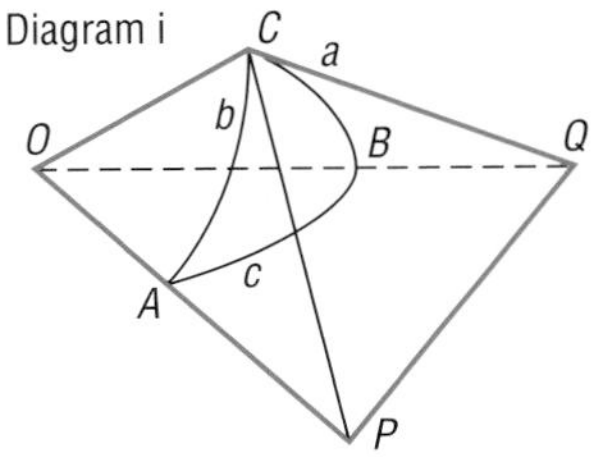

(b) Apply the Law of Cosines to triangles OPQ and CPQ to find two expressions for the length of PQ.

(c) Subtract the expressions in part (b) from each other. Solve for the term containing $\cos C$.

(d) Use the Pythagorean Theorem to find another value for $OQ^2 - CQ^2$ and $OP^2 - CP^2$. Now solve for $\cos C$.

(e) Replacing the ratios in part (d) by the cosines of the sides of the spherical triangle, you should now have the Law of Cosines for spherical triangles:

$$\cos C = \cos A \cos B + \sin A \sin B \cos C$$

SOURCE: For spherical Law of Cosines: *Mathematics from the Birth of Numbers* by Jan Gullberg. W.W. Norton & Co., Publishers. 1996, pp. 491–494.

B. The Lewis and Clark Expedition Lewis and Clark followed several rivers in their trek from what is now Great Falls, Montana to the Pacific coast. First, they went down the Missouri and Jefferson Rivers from Great Falls to Lemhi, Idaho. Because the two cities are on different longitudes and different latitudes, we must account for the curvature of Earth when computing the distance that they traveled. Assume that the radius of Earth is 3960 miles.

(a) Great Falls is at approximately 47.5°N and 111.3°W. Lemhi is at approximately 45.0°N and 113.5°W. (We will assume that the rivers flow straight from Great Falls to Lemhi on the surface of Earth.) This line is called a geodesic line. Apply the Law of Cosines for a spherical triangle to find the angle between Great Falls and Lemhi. (The central angles are found by using the differences in the latitudes and longitudes of the towns. See the diagram.) Then find the length of the arc joining the two towns. (Recall $s = r\theta$.)

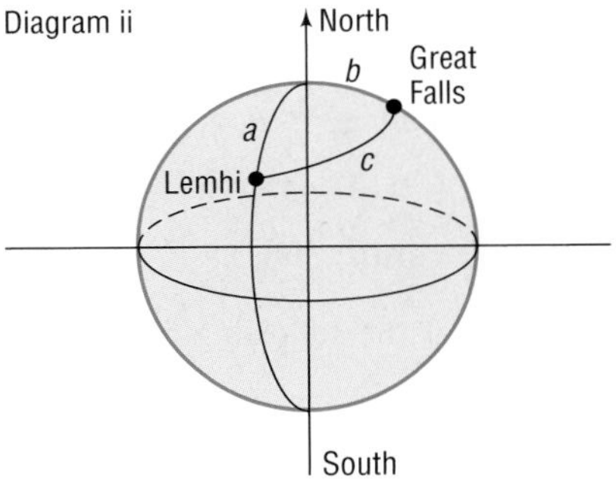

(b) From Lemhi, they went up the Bitteroot River and the Snake River to what is now Lewiston and Clarkston on the border of Idaho and Washington. Although this is not really a side to a triangle, we will make a side that goes from Lemhi to Lewiston and Clarkston. If Lewiston and Clarkston are at about 46.5°N 117.0°W, find the distance from Lemhi using the Law of Cosines for a spherical triangle and the arc length.

(c) How far did the explorers travel just to get that far?

(d) Draw a plane triangle connecting the three towns. If the distance from Lewiston to Great Falls is 282 miles and the angle at Great Falls is 42° and the angle at Lewiston is 48.5°, find the distance from Great Falls to Lemhi and from Lemhi to Lewiston. How do these distances compare with the ones computed in parts (a) and (b)?

SOURCE: For Lewis and Clark Expedition: *American Journey: The Quest for Liberty to 1877, Texas Edition*. Prentice Hall, 1992, p. 345.

SOURCE: For map coordinates: *National Geographic Atlas of the World*, published by National Geographic Society, 1981, pp. 74–75.

The following projects may be found at www.prenhall.com/sullivan7e.

2. **Project at Motorola** ***How Can You Build or Analyze a Vibration Profile?***
3. **Leaning Tower of Pisa**
4. **Locating Lost Treasure**

Cumulative Review

1. Find the real solutions, if any, of the equation $3x^2 + 1 = 4x$.

2. Find an equation for the circle with center at the point $(-5, 1)$ and radius 3. Graph this circle.

3. What is the domain of the function

$$f(x) = \sqrt{x^2 - 3x - 4}\ ?$$

4. Graph the function $y = 3\sin(\pi x)$.

5. Graph the function $y = -2\cos(2x - \pi)$.

6. If $\tan\theta = -2$ and $\frac{3\pi}{2} < \theta < 2\pi$, find the exact value of:

(a) $\sin\theta$ (b) $\cos\theta$ (c) $\sin(2\theta)$

(d) $\cos(2\theta)$ (e) $\sin\left(\frac{1}{2}\theta\right)$ (f) $\cos\left(\frac{1}{2}\theta\right)$

7. Graph each of the following functions on the interval $[0, 4]$:

(a) $y = e^x$ (b) $y = \sin x$

(c) $y = e^x \sin x$ (d) $y = 2x + \sin x$

8. Sketch the graph of each of the following functions:

(a) $y = x$ (b) $y = x^2$ (c) $y = \sqrt{x}$

(d) $y = x^3$ (e) $y = e^x$ (f) $y = \ln x$

(g) $y = \sin x$ (h) $y = \cos x$ (i) $y = \tan x$

9. Solve the triangle:

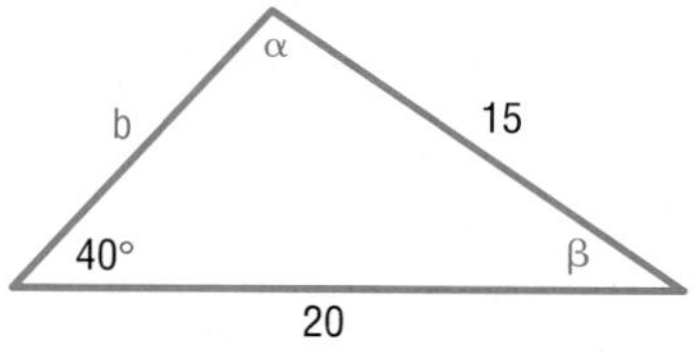

10. In the complex number system, solve the equation

$$3x^5 - 10x^4 + 21x^3 - 42x^2 + 36x - 8 = 0$$

11. Analyze the graph of the rational function

$$R(x) = \frac{2x^2 - 7x - 4}{x^2 + 2x - 15}$$

12. Solve $3^x = 12$. Round your answer to two decimal places.

13. Solve $\log_3(x + 8) + \log_3 x = 2$.

14. Suppose that $f(x) = 4x + 5$ and $g(x) = x^2 + 5x - 24$.

(a) Solve $f(x) = 0$. (b) Solve $f(x) = 13$.

(c) Solve $f(x) = g(x)$. (d) Solve $f(x) > 0$.

(e) Solve $g(x) \le 0$. (f) Graph $y = f(x)$.

(g) Graph $y = g(x)$.

8 Polar Coordinates; Vectors

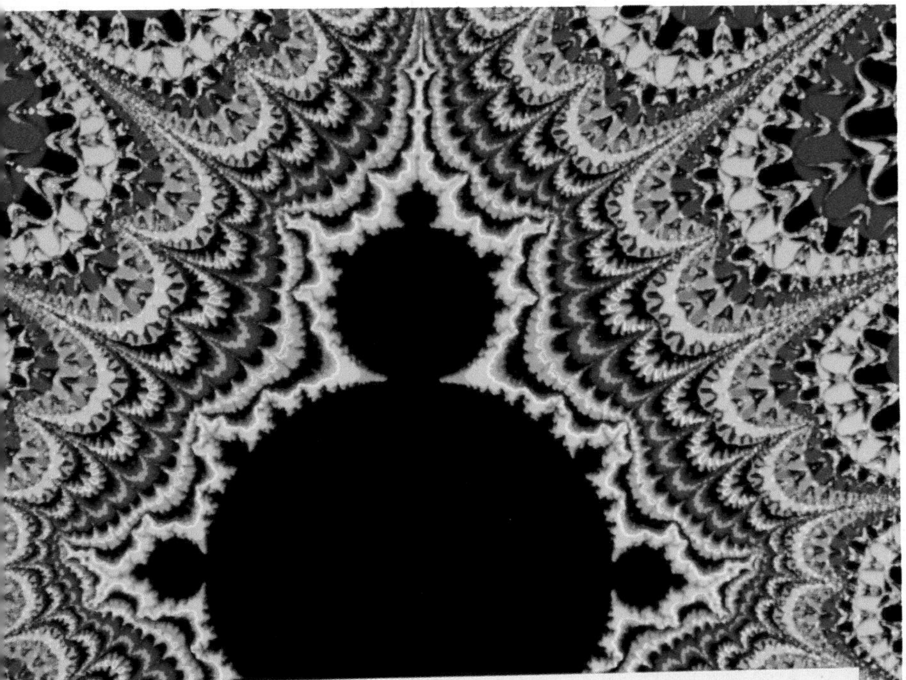

OUTLINE

Multifractals and the Market

An extensive mathematical basis already exists for fractals and multifractals. Fractal patterns appear not just in the price changes of securities, but in the distribution of galaxies throughout the cosmos, in the shape of coastlines, and in the decorative designs generated by innumerable computer programs.

A fractal is a geometric shape that can be separated into parts, each of which is a reduced scale version of the whole. In finance, this concept is not a rootless abstraction but a theoretical reformulation of a down-to-earth bit of market folklore—namely, that movements of stock or currency all look alike when a market chart is enlarged or reduced so that it fits the same time and price scale. An observer then cannot tell which of the data concern prices that change from week to week, day to day, or hour to hour. This quality defines charts as fractal curves and makes available many powerful tools of mathematical and computer analysis.

SOURCE: Benoit Mandelbrot, *Scientific American*, February 1999.

—SEE CHAPTER PROJECT 1.

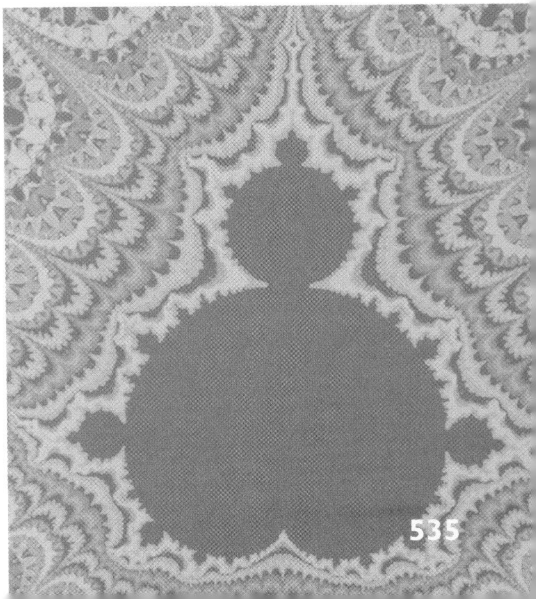

8.1 Polar Coordinates

PREPARING FOR THIS SECTION *Before getting started, review the following:*

- Rectangular Coordinates (Section 1.1, pp. 2–6)
- Definitions of the Sine and Cosine Functions (Section 5.2, pp. 339–341)
- Inverse Tangent Function (Section 6.1, pp. 421–422)
- Completing the Square (Appendix A, Section A.5, p. 941)

Now work the 'Are You Prepared?' problems on page 543.

OBJECTIVES
1. Plot Points Using Polar Coordinates
2. Convert from Polar Coordinates to Rectangular Coordinates
3. Convert from Rectangular Coordinates to Polar Coordinates

So far, we have always used a system of rectangular coordinates to plot points in the plane. Now we are ready to describe another system called *polar coordinates*. As we shall soon see, in many instances polar coordinates offer certain advantages over rectangular coordinates.

Figure 1

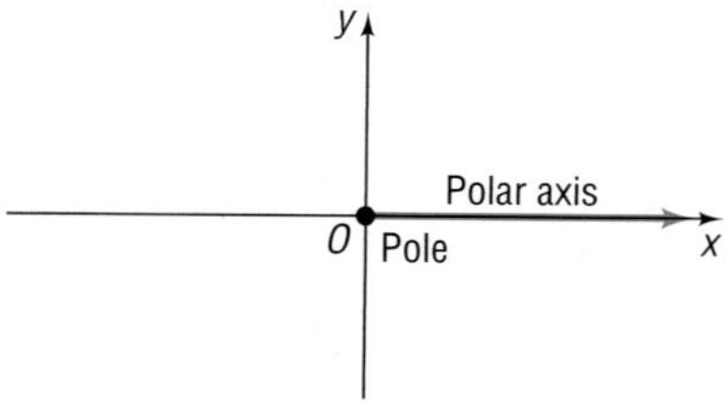

In a rectangular coordinate system, you will recall, a point in the plane is represented by an ordered pair of numbers (x, y), where x and y equal the signed distance of the point from the y-axis and x-axis, respectively. In a polar coordinate system, we select a point, called the **pole,** and then a ray with vertex at the pole, called the **polar axis.** Comparing the rectangular and polar coordinate systems, we see (in Figure 1) that the origin in rectangular coordinates coincides with the pole in polar coordinates, and the positive x-axis in rectangular coordinates coincides with the polar axis in polar coordinates.

1 A point P in a polar coordinate system is represented by an ordered pair of numbers (r, θ). If $r > 0$, then r is the distance of the point from the pole; θ is an angle (in degrees or radians) formed by the polar axis and a ray from the pole through the point. We call the ordered pair (r, θ) the **polar coordinates** of the point. See Figure 2.

As an example, suppose that the polar coordinates of a point P are $\left(2, \frac{\pi}{4}\right)$. We locate P by first drawing an angle of $\frac{\pi}{4}$ radian, placing its vertex at the pole and its initial side along the polar axis. Then we go out a distance of 2 units along the terminal side of the angle to reach the point P. See Figure 3.

Figure 2

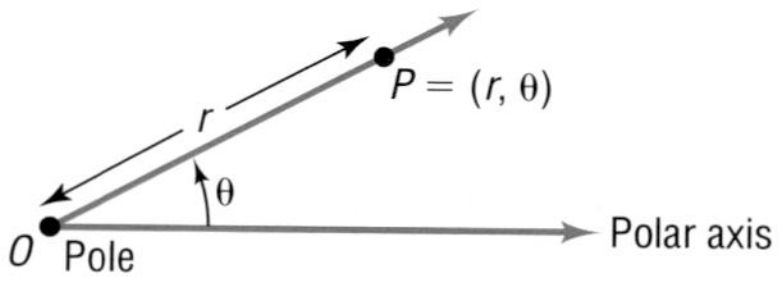

Figure 3

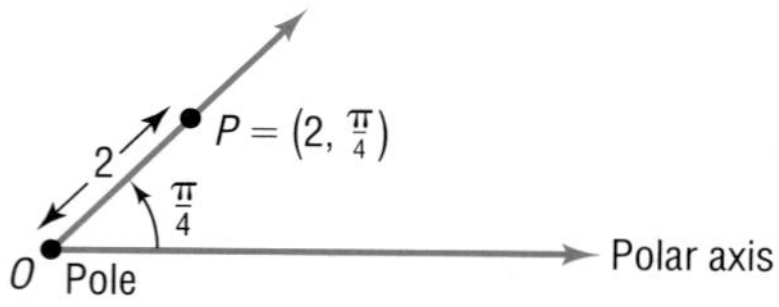

NOW WORK PROBLEM 19.

In using polar coordinates (r, θ), it is possible for the first entry r to be negative. When this happens, instead of the point being on the terminal side of θ, it is on the ray from the pole extending in the direction *opposite* the terminal side of θ at a distance $|r|$ units from the pole. See Figure 4 for an illustration.

For example, to plot the point $\left(-3, \frac{2\pi}{3}\right)$, we use the ray in the opposite direction of $\frac{2\pi}{3}$ and go out $|-3| = 3$ units along that ray. See Figure 5.

Figure 4

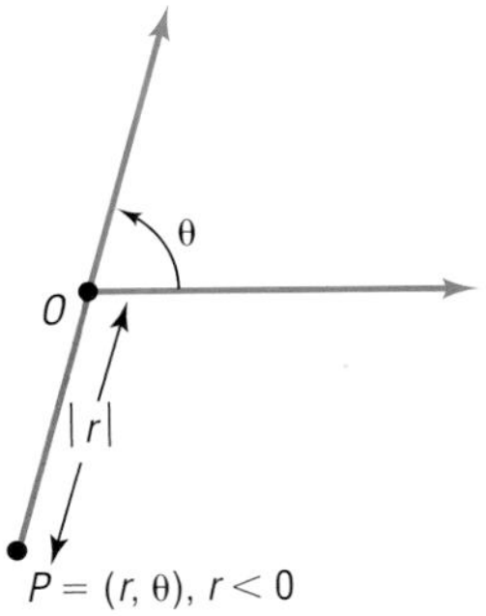

Figure 5

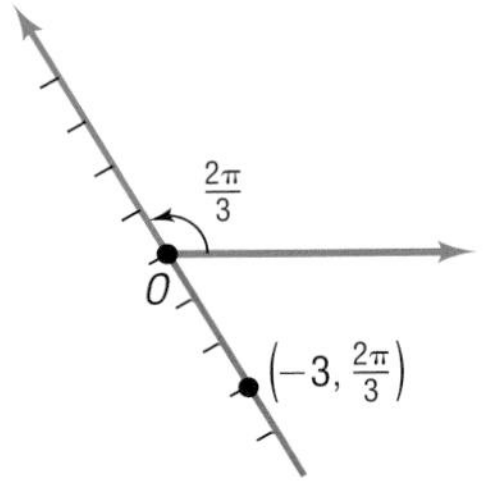

EXAMPLE 1 Plotting Points Using Polar Coordinates

Plot the points with the following polar coordinates:

(a) $\left(3, \frac{5\pi}{3}\right)$ (b) $\left(2, -\frac{\pi}{4}\right)$ (c) $(3, 0)$ (d) $\left(-2, \frac{\pi}{4}\right)$

Solution Figure 6 shows the points.

Figure 6

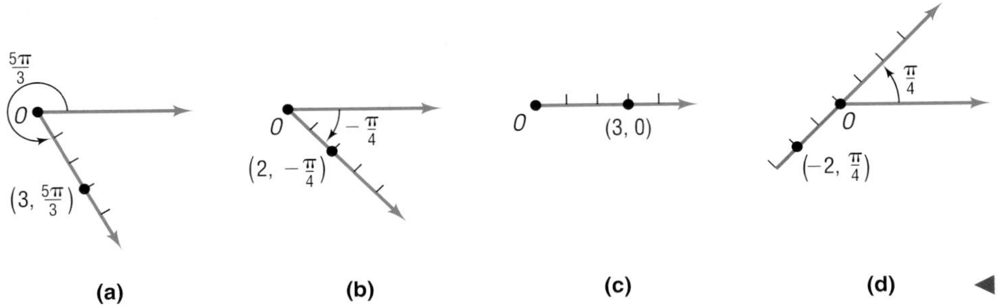

NOW WORK PROBLEMS 11 AND 27.

Recall that an angle measured counterclockwise is positive, whereas an angle measured clockwise is negative. This convention has some interesting consequences relating to polar coordinates. Let's see what these consequences are.

EXAMPLE 2 Finding Several Polar Coordinates of a Single Point

Consider again the point P with polar coordinates $\left(2, \frac{\pi}{4}\right)$, as shown in Figure 7(a). Because $\frac{\pi}{4}, \frac{9\pi}{4}$, and $-\frac{7\pi}{4}$ all have the same terminal side, we also could have located this point P by using the polar coordinates $\left(2, \frac{9\pi}{4}\right)$ or $\left(2, -\frac{7\pi}{4}\right)$, as shown in Figures 7(b) and (c). The point $\left(2, \frac{\pi}{4}\right)$ can also be represented by the polar coordinates $\left(-2, \frac{5\pi}{4}\right)$. See Figure 7(d).

Figure 7

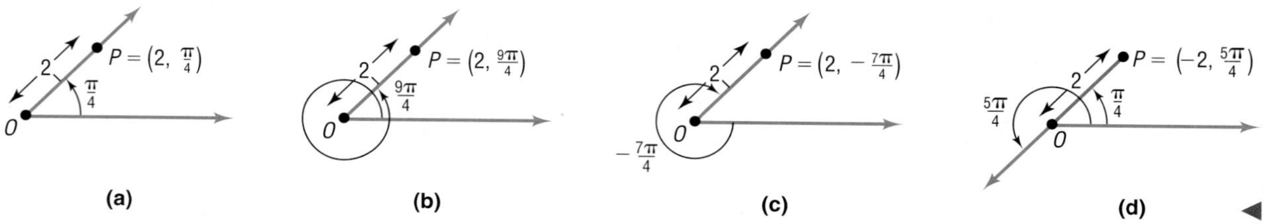

EXAMPLE 3 Finding Other Polar Coordinates of a Given Point

Plot the point P with polar coordinates $\left(3, \frac{\pi}{6}\right)$, and find other polar coordinates (r, θ) of this same point for which:

(a) $r > 0, \quad 2\pi \le \theta < 4\pi$ (b) $r < 0, \quad 0 \le \theta < 2\pi$

(c) $r > 0, \quad -2\pi \le \theta < 0$

Figure 8

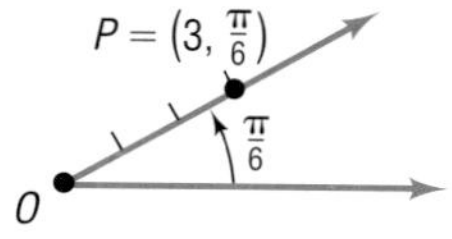

Solution The point $\left(3, \frac{\pi}{6}\right)$ is plotted in Figure 8.

(a) We add 1 revolution (2π radians) to the angle $\frac{\pi}{6}$ to get $P = \left(3, \frac{\pi}{6} + 2\pi\right) = \left(3, \frac{13\pi}{6}\right)$. See Figure 9.

(b) We add $\frac{1}{2}$ revolution (π radians) to the angle $\frac{\pi}{6}$ and replace 3 by -3 to get $P = \left(-3, \frac{\pi}{6} + \pi\right) = \left(-3, \frac{7\pi}{6}\right)$. See Figure 10.

(c) We subtract 2π from the angle $\frac{\pi}{6}$ to get $P = \left(3, \frac{\pi}{6} - 2\pi\right) = \left(3, -\frac{11\pi}{6}\right)$. See Figure 11.

Figure 9

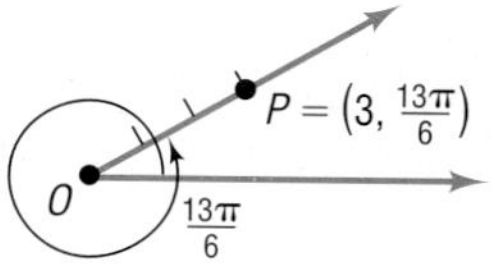

Figure 10

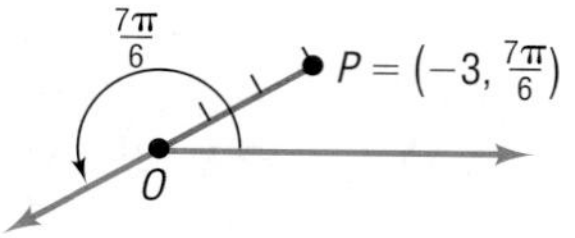

Figure 11

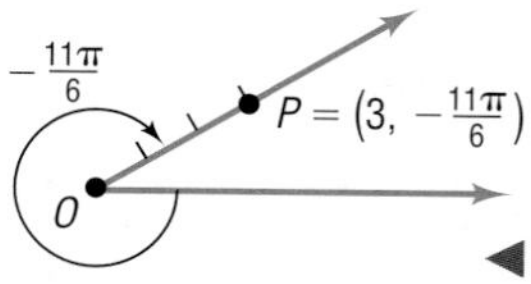

NOW WORK PROBLEM 31.

These examples show a major difference between rectangular coordinates and polar coordinates. In the former, each point has exactly one pair of rectangular coordinates; in the latter, a point can have infinitely many pairs of polar coordinates.

Summary

A point with polar coordinates (r, θ) also can be represented by either of the following:

$$(r, \theta + 2k\pi) \quad \text{or} \quad (-r, \theta + \pi + 2k\pi), \qquad k \text{ any integer}$$

The polar coordinates of the pole are $(0, \theta)$, where θ can be any angle.

Conversion from Polar Coordinates to Rectangular Coordinates, and Vice Versa

2 It is sometimes convenient and, indeed, necessary to be able to convert coordinates or equations in rectangular form to polar form, and vice versa. To do this, we recall that the origin in rectangular coordinates is the pole in polar coordinates and that the positive x-axis in rectangular coordinates is the polar axis in polar coordinates.

Theorem

Conversion from Polar Coordinates to Rectangular Coordinates

If P is a point with polar coordinates (r, θ), the rectangular coordinates (x, y) of P are given by

$$x = r\cos\theta \qquad y = r\sin\theta \tag{1}$$

Figure 12

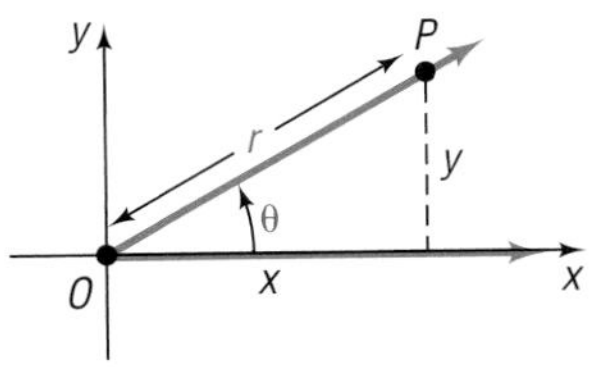

Proof Suppose that P has the polar coordinates (r, θ). We seek the rectangular coordinates (x, y) of P. Refer to Figure 12.

If $r = 0$, then, regardless of θ, the point P is the pole, for which the rectangular coordinates are $(0, 0)$. Formula (1) is valid for $r = 0$.

If $r > 0$, the point P is on the terminal side of θ, and $r = d(O, P) = \sqrt{x^2 + y^2}$. Since

$$\cos\theta = \frac{x}{r} \qquad \sin\theta = \frac{y}{r}$$

we have

$$x = r\cos\theta \qquad y = r\sin\theta$$

If $r < 0$, then the point $P = (r, \theta)$ can be represented as $(-r, \pi + \theta)$, where $-r > 0$. Since

$$\cos(\pi + \theta) = -\cos\theta = \frac{x}{-r} \qquad \sin(\pi + \theta) = -\sin\theta = \frac{y}{-r}$$

we have

$$x = r\cos\theta \qquad y = r\sin\theta$$

■

EXAMPLE 4 Converting from Polar Coordinates to Rectangular Coordinates

Find the rectangular coordinates of the points with the following polar coordinates:

(a) $\left(6, \dfrac{\pi}{6}\right)$ (b) $\left(-4, -\dfrac{\pi}{4}\right)$

Solution We use formula (1): $x = r\cos\theta$ and $y = r\sin\theta$.

Figure 13

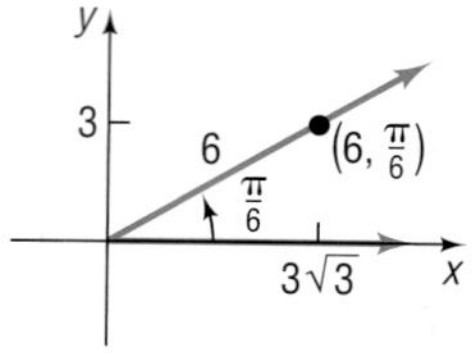

(a)

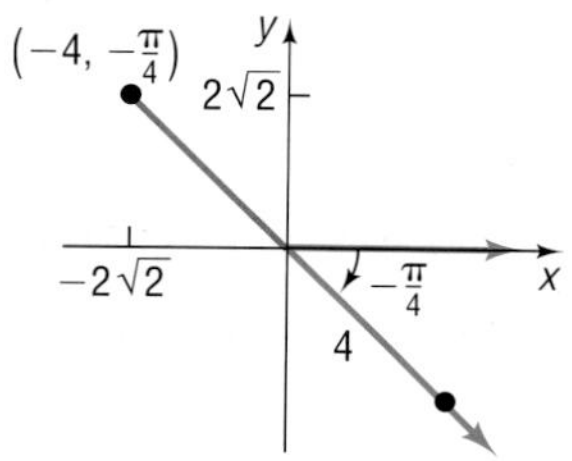

(b)

(a) Figure 13(a) shows $\left(6, \dfrac{\pi}{6}\right)$ plotted. With $r = 6$ and $\theta = \dfrac{\pi}{6}$, we have

$$x = r\cos\theta = 6\cos\frac{\pi}{6} = 6 \cdot \frac{\sqrt{3}}{2} = 3\sqrt{3}$$

$$y = r\sin\theta = 6\sin\frac{\pi}{6} = 6 \cdot \frac{1}{2} = 3$$

The rectangular coordinates of the point $\left(6, \dfrac{\pi}{6}\right)$ are $(3\sqrt{3}, 3)$.

(b) Figure 13(b) shows $\left(-4, -\dfrac{\pi}{4}\right)$ plotted. With $r = -4$ and $\theta = -\dfrac{\pi}{4}$, we have

$$x = r\cos\theta = -4\cos\left(-\frac{\pi}{4}\right) = -4 \cdot \frac{\sqrt{2}}{2} = -2\sqrt{2}$$

$$y = r\sin\theta = -4\sin\left(-\frac{\pi}{4}\right) = -4\left(-\frac{\sqrt{2}}{2}\right) = 2\sqrt{2}$$

The rectangular coordinates of the point $\left(-4, -\dfrac{\pi}{4}\right)$ are $(-2\sqrt{2}, 2\sqrt{2})$.

◀

Note: Most calculators have the capability of converting from polar coordinates to rectangular coordinates. Consult your owner's manual for the proper key strokes. Since in most cases this procedure is tedious, you will find that using formula (1) is faster.

NOW WORK PROBLEMS **39** AND **51**.

3 Converting from rectangular coordinates (x, y) to polar coordinates (r, θ) is a little more complicated. Notice that we begin each example by plotting the given rectangular coordinates.

EXAMPLE 5 Converting from Rectangular Coordinates to Polar Coordinates

Find polar coordinates of a point whose rectangular coordinates are $(0, 3)$.

Figure 14

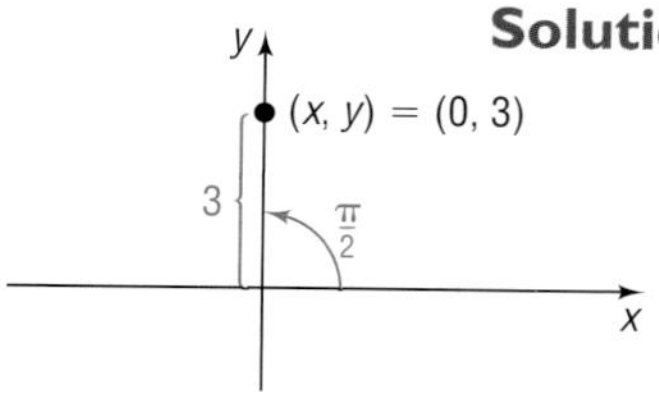

Solution See Figure 14. The point $(0, 3)$ lies on the y-axis a distance of 3 units from the origin (pole), so $r = 3$. A ray with vertex at the pole through $(0, 3)$ forms an angle $\theta = \dfrac{\pi}{2}$ with the polar axis. Polar coordinates for this point can be given by $\left(3, \dfrac{\pi}{2}\right)$.

◀

Figure 15 shows polar coordinates of points that lie on either the x-axis or the y-axis. In each illustration, $a > 0$.

Figure 15

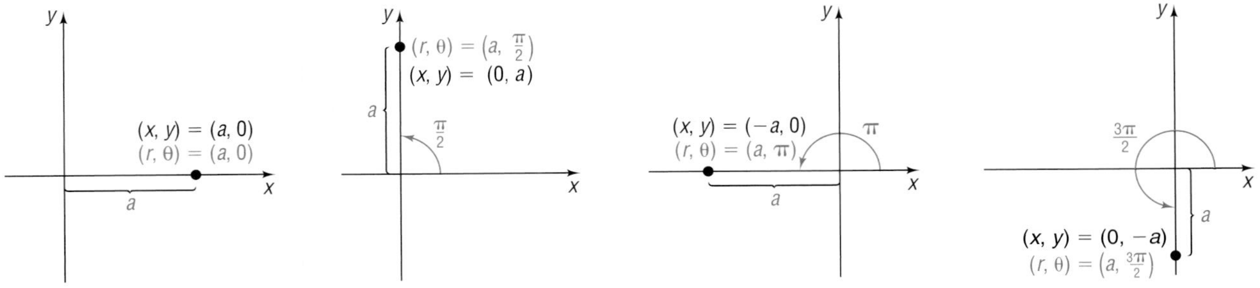

(a) $(x, y) = (a, 0), a > 0$ **(b)** $(x, y) = (0, a), a > 0$ **(c)** $(x, y) = (-a, 0), a > 0$ **(d)** $(x, y) = (0, -a), a > 0$

NOW WORK PROBLEM 55.

EXAMPLE 6 Converting from Rectangular Coordinates to Polar Coordinates

Find polar coordinates of a point whose rectangular coordinates are:

(a) $(2, -2)$ (b) $(-1, -\sqrt{3})$

Figure 16

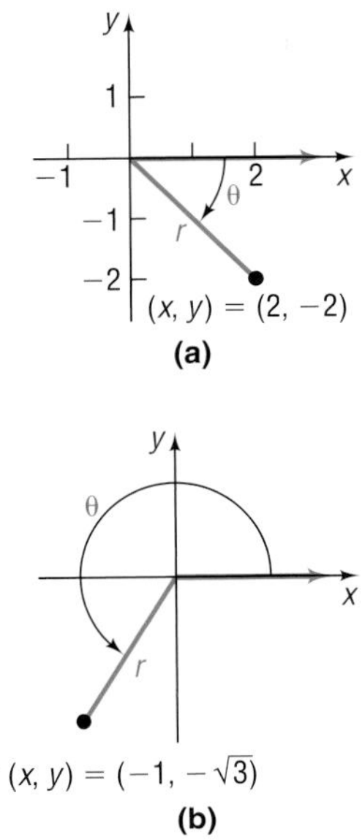

Solution (a) See Figure 16(a). The distance r from the origin to the point $(2, -2)$ is

$$r = \sqrt{x^2 + y^2} = \sqrt{(2)^2 + (-2)^2} = \sqrt{8} = 2\sqrt{2}$$

We find θ by recalling that $\tan\theta = \frac{y}{x}$, so $\theta = \tan^{-1}\frac{y}{x}, -\frac{\pi}{2} < \theta < \frac{\pi}{2}$.

Since $(2, -2)$ lies in quadrant IV, we know that $-\frac{\pi}{2} < \theta < 0$. As a result,

$$\theta = \tan^{-1}\frac{y}{x} = \tan^{-1}\left(\frac{-2}{2}\right) = \tan^{-1}(-1) = -\frac{\pi}{4}$$

A set of polar coordinates for this point is $\left(2\sqrt{2}, -\frac{\pi}{4}\right)$. Other possible representations include $\left(2\sqrt{2}, \frac{7\pi}{4}\right)$ and $\left(-2\sqrt{2}, \frac{3\pi}{4}\right)$.

(b) See Figure 16(b). The distance r from the origin to the point $(-1, -\sqrt{3})$ is

$$r = \sqrt{(-1)^2 + (-\sqrt{3})^2} = \sqrt{4} = 2$$

To find θ, we use $\theta = \tan^{-1}\frac{y}{x}, -\frac{\pi}{2} < \theta < \frac{\pi}{2}$. Since the point $(-1, -\sqrt{3})$ lies in quadrant III and the inverse tangent function gives an angle in quadrant I, we add π to the result to obtain an angle in quadrant III.

$$\theta = \pi + \tan^{-1}\left(\frac{-\sqrt{3}}{-1}\right) = \pi + \tan^{-1}\sqrt{3} = \pi + \frac{\pi}{3} = \frac{4\pi}{3}$$

A set of polar coordinates for this point is $\left(2, \frac{4\pi}{3}\right)$. Other possible representations include $\left(-2, \frac{\pi}{3}\right)$ and $\left(2, -\frac{2\pi}{3}\right)$. ◀

Figure 17 shows how to find polar coordinates of a point that lies in a quadrant when its rectangular coordinates (x, y) are given.

Figure 17

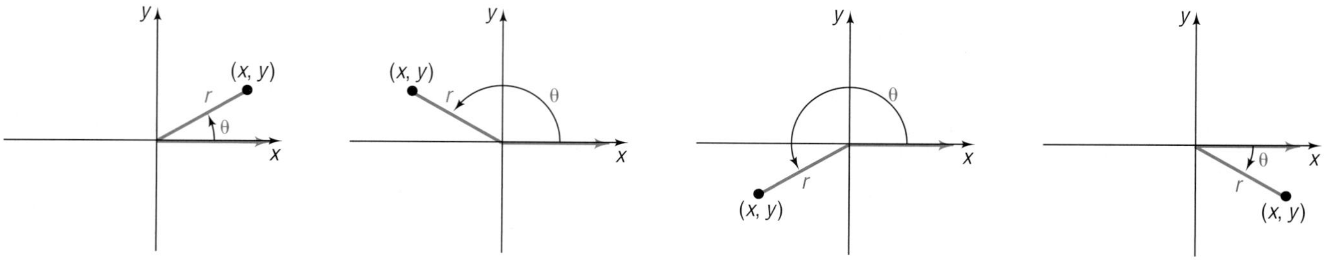

(a) $r = \sqrt{x^2 + y^2}$, $\theta = \tan^{-1}\frac{y}{x}$

(b) $r = \sqrt{x^2 + y^2}$, $\theta = \pi + \tan^{-1}\frac{y}{x}$

(c) $r = \sqrt{x^2 + y^2}$, $\theta = \pi + \tan^{-1}\frac{y}{x}$

(d) $r = \sqrt{x^2 + y^2}$, $\theta = \tan^{-1}\frac{y}{x}$

Based on the preceding discussion, we have the formulas

$$r^2 = x^2 + y^2 \qquad \tan\theta = \frac{y}{x} \qquad \text{if } x \neq 0 \tag{2}$$

To use formula (2) effectively, follow these steps:

Steps for Converting from Rectangular to Polar Coordinates

STEP 1: Always plot the point (x, y) first, as we did in Examples 5 and 6.

STEP 2: To find r, compute the distance from the origin to (x, y).

STEP 3: To find θ, first determine the quadrant that the point lies in.

Quadrant I: $\theta = \tan^{-1}\frac{y}{x}$ Quadrant II: $\theta = \pi + \tan^{-1}\frac{y}{x}$

Quadrant III: $\theta = \pi + \tan^{-1}\frac{y}{x}$ Quadrant IV: $\theta = \tan^{-1}\frac{y}{x}$

Look again at Figure 17 and Example 6.

NOW WORK PROBLEM 59.

Formulas (1) and (2) may also be used to transform equations.

EXAMPLE 7 Transforming an Equation from Polar to Rectangular Form

Transform the equation $r = 4\sin\theta$ from polar coordinates to rectangular coordinates, and identify the graph.

Solution If we multiply each side by r, it will be easier to apply formulas (1) and (2).

$$r = 4\sin\theta$$

$$r^2 = 4r\sin\theta \qquad \text{Multiply each side by } r.$$

$$x^2 + y^2 = 4y \qquad r^2 = x^2 + y^2;\ y = r\sin\theta$$

This is the equation of a circle; we proceed to complete the square to obtain the standard form of the equation.

$$x^2 + y^2 = 4y$$
$$x^2 + (y^2 - 4y) = 0 \quad \text{General form}$$
$$x^2 + (y^2 - 4y + 4) = 4 \quad \text{Complete the square in } y.$$
$$x^2 + (y - 2)^2 = 4 \quad \text{Standard form}$$

The center of the circle is at $(0, 2)$, and its radius is 2. ◄

NOW WORK PROBLEM 75.

EXAMPLE 8 Transforming an Equation from Rectangular to Polar Form

Transform the equation $4xy = 9$ from rectangular coordinates to polar coordinates.

Solution We use formula (1).

$$4xy = 9$$
$$4(r\cos\theta)(r\sin\theta) = 9 \quad x = r\cos\theta,\ y = r\sin\theta$$
$$4r^2\cos\theta\sin\theta = 9$$
$$2r^2(2\sin\theta\cos\theta) = 9 \quad \text{Factor out } 2r^2.$$
$$2r^2\sin(2\theta) = 9 \quad \text{Double-angle Formula}$$
◄

8.1 Assess Your Understanding

'Are You Prepared?' *Answers are given at the end of these exercises. If you get a wrong answer, read the pages listed in red.*

1. The rectangular coordinates of a point are $(3, -1)$. Plot it. (pp. 2–6)
2. To complete the square of $x^2 + 6x$, add ________. (p. 941)
3. If $P = (x, y)$ is a point on the unit circle that is also on the terminal side of the angle θ, then $\sin\theta =$ ________. (pp. 339–341)
4. $\tan^{-1}(-1) =$ ________ (pp. 421–422)

Concepts and Vocabulary

5. In polar coordinates, the origin is called the ________ and the positive x-axis is referred to as the ________ ________.
6. Another representation in polar coordinates for the point $\left(2, \frac{\pi}{3}\right)$ is $\left(\_\_\_\_\_\_,\ \frac{4\pi}{3}\right)$.
7. The polar coordinates $\left(-2, \frac{\pi}{6}\right)$ are represented in rectangular coordinates by (________, ________).
8. *True or False:* The polar coordinates of a point are unique.
9. *True or False:* The rectangular coordinates of a point are unique.
10. *True or False:* In (r, θ), the number r can be negative.

Exercises

In Problems 11–18, match each point in polar coordinates with either A, B, C, or D on the graph.

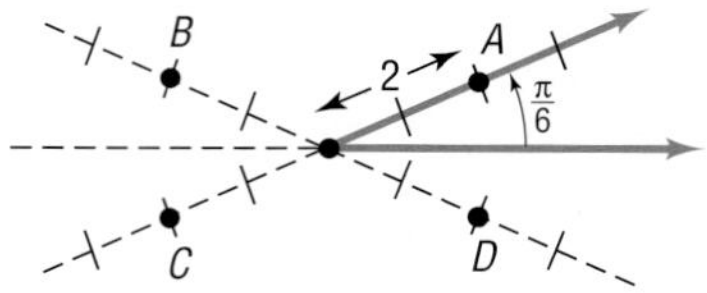

11. $\left(2, -\frac{11\pi}{6}\right)$ **12.** $\left(-2, -\frac{\pi}{6}\right)$ **13.** $\left(-2, \frac{\pi}{6}\right)$ **14.** $\left(2, \frac{7\pi}{6}\right)$

15. $\left(2, \frac{5\pi}{6}\right)$ **16.** $\left(-2, \frac{5\pi}{6}\right)$ **17.** $\left(-2, \frac{7\pi}{6}\right)$ **18.** $\left(2, \frac{11\pi}{6}\right)$

In Problems 19–30, plot each point given in polar coordinates.

19. $(3, 90^\circ)$ **20.** $(4, 270^\circ)$ **21.** $(-2, 0)$ **22.** $(-3, \pi)$

23. $\left(6, \frac{\pi}{6}\right)$ **24.** $\left(5, \frac{5\pi}{3}\right)$ **25.** $(-2, 135^\circ)$ **26.** $(-3, 120^\circ)$

27. $\left(-1, -\frac{\pi}{3}\right)$ **28.** $\left(-3, -\frac{3\pi}{4}\right)$ **29.** $(-2, -\pi)$ **30.** $\left(-3, -\frac{\pi}{2}\right)$

In Problems 31–38, plot each point given in polar coordinates, and find other polar coordinates (r, θ) of the point for which:
(a) $r > 0, \quad -2\pi \le \theta < 0$ (b) $r < 0, \quad 0 \le \theta < 2\pi$ (c) $r > 0, \quad 2\pi \le \theta < 4\pi$

31. $\left(5, \frac{2\pi}{3}\right)$ **32.** $\left(4, \frac{3\pi}{4}\right)$ **33.** $(-2, 3\pi)$ **34.** $(-3, 4\pi)$

35. $\left(1, \frac{\pi}{2}\right)$ **36.** $(2, \pi)$ **37.** $\left(-3, -\frac{\pi}{4}\right)$ **38.** $\left(-2, -\frac{2\pi}{3}\right)$

In Problems 39–54, the polar coordinates of a point are given. Find the rectangular coordinates of each point.

39. $\left(3, \frac{\pi}{2}\right)$ **40.** $\left(4, \frac{3\pi}{2}\right)$ **41.** $(-2, 0)$ **42.** $(-3, \pi)$

43. $(6, 150^\circ)$ **44.** $(5, 300^\circ)$ **45.** $\left(-2, \frac{3\pi}{4}\right)$ **46.** $\left(-2, \frac{2\pi}{3}\right)$

47. $\left(-1, -\frac{\pi}{3}\right)$ **48.** $\left(-3, -\frac{3\pi}{4}\right)$ **49.** $(-2, -180^\circ)$ **50.** $(-3, -90^\circ)$

51. $(7.5, 110^\circ)$ **52.** $(-3.1, 182^\circ)$ **53.** $(6.3, 3.8)$ **54.** $(8.1, 5.2)$

In Problems 55–66, the rectangular coordinates of a point are given. Find polar coordinates for each point.

55. $(3, 0)$ **56.** $(0, 2)$ **57.** $(-1, 0)$ **58.** $(0, -2)$

59. $(1, -1)$ **60.** $(-3, 3)$ **61.** $\left(\sqrt{3}, 1\right)$ **62.** $\left(-2, -2\sqrt{3}\right)$

63. $(1.3, -2.1)$ **64.** $(-0.8, -2.1)$ **65.** $(8.3, 4.2)$ **66.** $(-2.3, 0.2)$

In Problems 67–74, the letters x and y represent rectangular coordinates. Write each equation using polar coordinates (r, θ).

67. $2x^2 + 2y^2 = 3$ **68.** $x^2 + y^2 = x$ **69.** $x^2 = 4y$ **70.** $y^2 = 2x$

71. $2xy = 1$ **72.** $4x^2y = 1$ **73.** $x = 4$ **74.** $y = -3$

In Problems 75–82, the letters r and θ represent polar coordinates. Write each equation using rectangular coordinates (x, y).

75. $r = \cos\theta$ **76.** $r = \sin\theta + 1$ **77.** $r^2 = \cos\theta$ **78.** $r = \sin\theta - \cos\theta$

79. $r = 2$ **80.** $r = 4$ **81.** $r = \frac{4}{1 - \cos\theta}$ **82.** $r = \frac{3}{3 - \cos\theta}$

83. Show that the formula for the distance d between two points $P_1 = (r_1, \theta_1)$ and $P_2 = (r_2, \theta_2)$ is

$$d = \sqrt{r_1^2 + r_2^2 - 2r_1r_2\cos(\theta_2 - \theta_1)}$$

84. In converting from polar coordinates to rectangular coordinates, what formulas will you use?

85. Explain how you proceed to convert from rectangular coordinates to polar coordinates.

86. Is the street system in your town based on a rectangular coordinate system, a polar coordinate system, or some other system? Explain.

'Are You Prepared?' Answers

1.

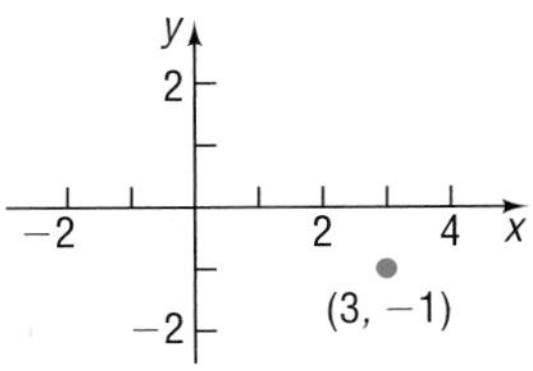

2. 9 **3.** y **4.** $-\frac{\pi}{4}$

8.2 Polar Equations and Graphs

PREPARING FOR THIS SECTION *Before getting started, review the following:*

- Graphs of Equations (Section 1.2, pp. 9–16)
- Even–Odd Properties of Trigonometric Functions (Section 5.3, p. 366)
- Circles (Section 1.2, pp. 16–19)
- Difference Formulas for Sine and Cosine (Section 6.4, pp. 438 and 441)
- Value of the Sine and Cosine Functions at Certain Angles (Section 5.2, pp. 342–351)

Now work the 'Are You Prepared?' problems on page 559.

OBJECTIVES
1 Graph and Identify Polar Equations by Converting to Rectangular Equations
2 Test Polar Equations for Symmetry
3 Graph Polar Equations by Plotting Points

Just as a rectangular grid may be used to plot points given by rectangular coordinates, as in Figure 18(a), we can use a grid consisting of concentric circles (with centers at the pole) and rays (with vertices at the pole) to plot points given by polar coordinates, as shown in Figure 18(b). We shall use such **polar grids** to graph *polar equations*.

Figure 18

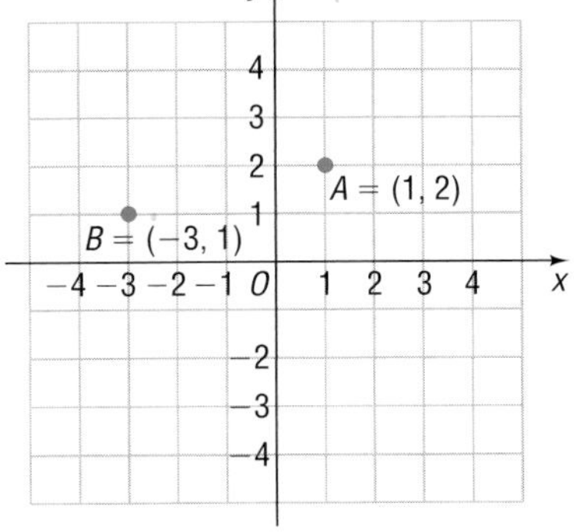

(a) Rectangular grid

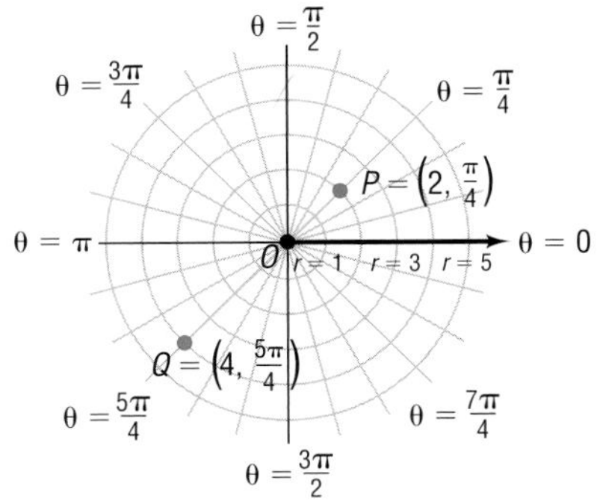

(b) Polar grid

An equation whose variables are polar coordinates is called a **polar equation.** The **graph of a polar equation** consists of all points whose polar coordinates satisfy the equation.

1 One method that we can use to graph a polar equation is to convert the equation to rectangular coordinates. In the discussion that follows, (x, y) represent the rectangular coordinates of a point P, and (r, θ) represent polar coordinates of the point P.

EXAMPLE 1 Identifying and Graphing a Polar Equation (Circle)

Identify and graph the equation: $r = 3$

Solution We convert the polar equation to a rectangular equation.

$$r = 3$$
$$r^2 = 9 \quad \text{Square both sides.}$$
$$x^2 + y^2 = 9 \quad r^2 = x^2 + y^2$$

The graph of $r = 3$ is a circle, with center at the pole and radius 3. See Figure 19.

Figure 19
$r = 3$ or $x^2 + y^2 = 9$

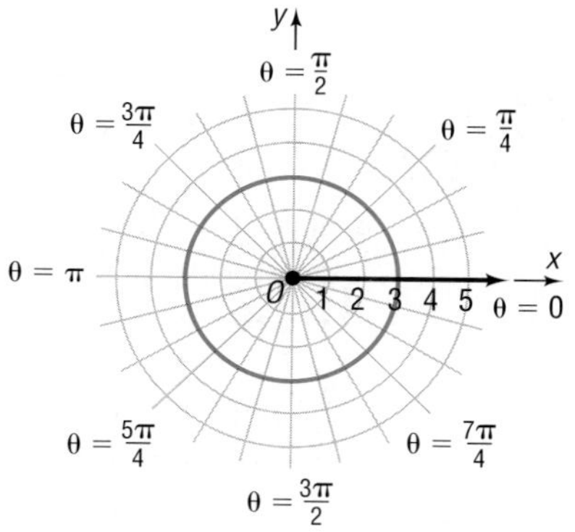

NOW WORK PROBLEM 13.

EXAMPLE 2 Identifying and Graphing a Polar Equation (Line)

Identify and graph the equation: $\theta = \dfrac{\pi}{4}$

Solution We convert the polar equation to a rectangular equation.

$$\theta = \frac{\pi}{4}$$
$$\tan\theta = \tan\frac{\pi}{4} = 1$$
$$\frac{y}{x} = 1 \quad \tan\theta = \frac{y}{x}$$
$$y = x$$

The graph of $\theta = \frac{\pi}{4}$ is a line passing through the pole making an angle of $\frac{\pi}{4}$ with the polar axis. See Figure 20.

Figure 20
$\theta = \frac{\pi}{4}$ or $y = x$

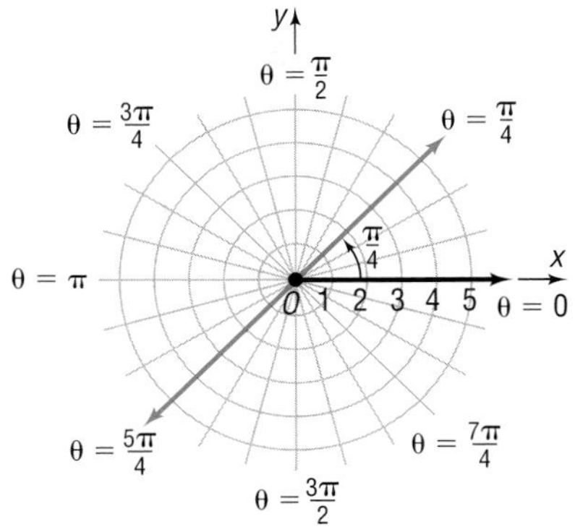

◀

NOW WORK PROBLEM 15.

EXAMPLE 3 Identifying and Graphing a Polar Equation (Horizontal Line)

Identify and graph the equation: $r \sin \theta = 2$

Solution Since $y = r \sin \theta$, we can write the equation as

$$y = 2$$

We conclude that the graph of $r \sin \theta = 2$ is a horizontal line 2 units above the pole. See Figure 21.

Figure 21
$r \sin \theta = 2$ or $y = 2$

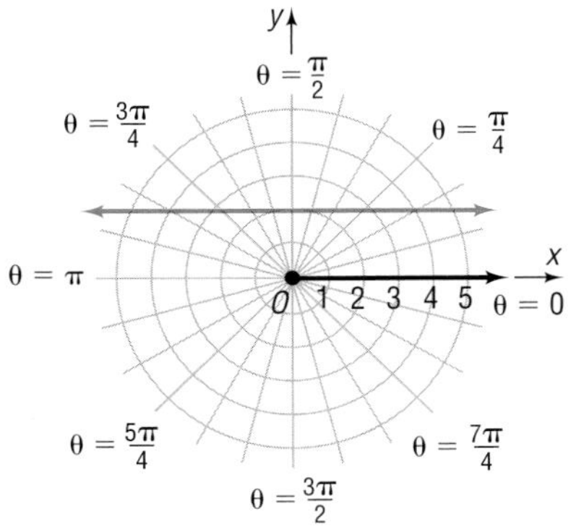

◀

COMMENT: A graphing utility can be used to graph polar equations. Read *Using a Graphing Utility to Graph a Polar Equation,* Appendix, Section B.8. ■

EXAMPLE 4 Identifying and Graphing a Polar Equation (Vertical Line)

Identify and graph the equation: $r \cos \theta = -3$

Solution Since $x = r \cos \theta$, we can write the equation as

$$x = -3$$

We conclude that the graph of $r \cos \theta = -3$ is a vertical line 3 units to the left of the pole. Figure 22 shows the graph.

Figure 22
$r \cos \theta = -3$ or $x = -3$

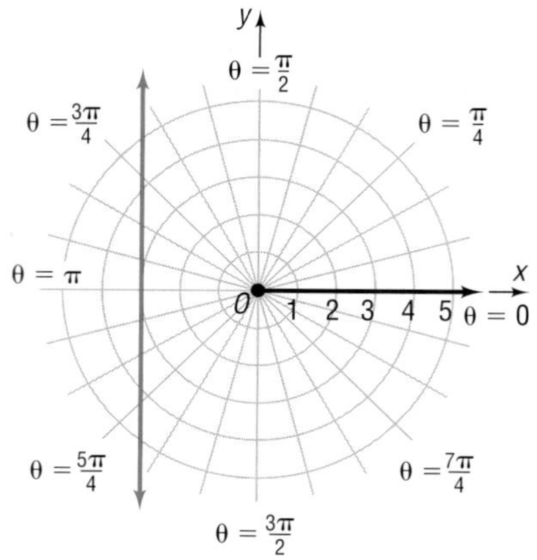

CHECK: Graph $r = -\dfrac{3}{\cos \theta}$ using $\theta \min = 0, \theta \max = 2\pi$, and $\theta \text{ step} = \dfrac{\pi}{24}$. Compare the result to Figure 22.

Based on Examples 3 and 4, we are led to the following results. (The proofs are left as exercises.)

Theorem

Let a be a nonzero real number. Then the graph of the equation

$$r \sin \theta = a$$

is a horizontal line a units above the pole if $a > 0$ and $|a|$ units below the pole if $a < 0$.

The graph of the equation

$$r \cos \theta = a$$

is a vertical line a units to the right of the pole if $a > 0$ and $|a|$ units to the left of the pole if $a < 0$.

NOW WORK PROBLEM 19.

EXAMPLE 5 Identifying and Graphing a Polar Equation (Circle)

Identify and graph the equation: $r = 4 \sin \theta$

Solution To transform the equation to rectangular coordinates, we multiply each side by r.

$$r^2 = 4r \sin \theta$$

Now we use the facts that $r^2 = x^2 + y^2$ and $y = r \sin \theta$. Then

$$x^2 + y^2 = 4y$$
$$x^2 + (y^2 - 4y) = 0$$
$$x^2 + (y^2 - 4y + 4) = 4 \quad \text{Complete the square in } y.$$
$$x^2 + (y - 2)^2 = 4 \quad \text{Standard equation of a circle}$$

This is the equation of a circle with center at $(0, 2)$ in rectangular coordinates and radius 2. Figure 23 shows the graph.

Figure 23
$r = 4 \sin \theta$ or $x^2 + (y - 2)^2 = 4$

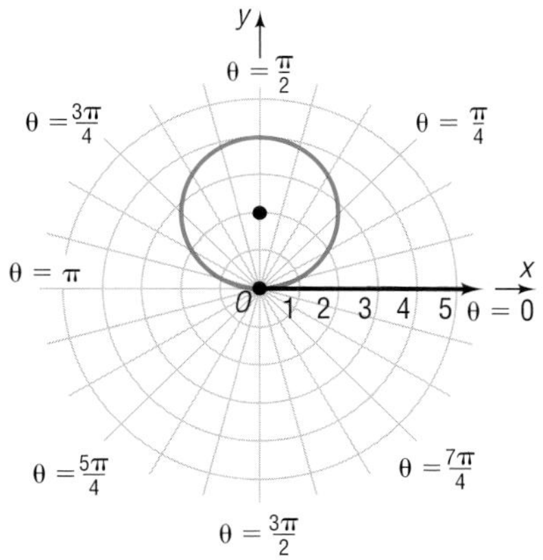

◀

EXAMPLE 6 Identifying and Graphing a Polar Equation (Circle)

Identify and graph the equation: $r = -2 \cos \theta$

Solution We proceed as in Example 5.

$$r^2 = -2r \cos \theta \quad \text{Multiply both sides by } r.$$
$$x^2 + y^2 = -2x \quad r^2 = x^2 + y^2;\ x = r\cos\theta$$
$$x^2 + 2x + y^2 = 0$$
$$(x^2 + 2x + 1) + y^2 = 1 \quad \text{Complete the square in } x.$$
$$(x + 1)^2 + y^2 = 1 \quad \text{Standard equation of a circle}$$

This is the equation of a circle with center at $(-1, 0)$ in rectangular coordinates and radius 1. Figure 24 shows the graph.

Figure 24
$r = -2 \cos \theta$ or $(x + 1)^2 + y^2 = 1$

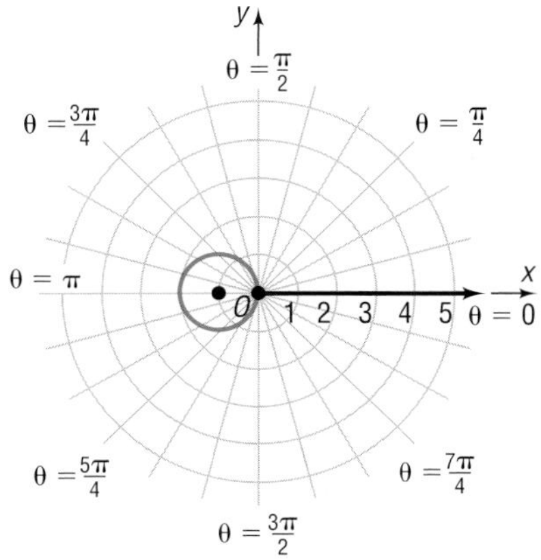

◀

CHECK: Graph $r = 4 \sin \theta$ and compare the result with Figure 23. Clear the screen and do the same for $r = -2 \cos \theta$ and compare with Figure 24. Be sure to use a square screen.

Exploration

Using a square screen, graph $r_1 = \sin\theta$, $r_2 = 2\sin\theta$, and $r_3 = 3\sin\theta$. Do you see the pattern? Clear the screen and graph $r_1 = -\sin\theta$, $r_2 = -2\sin\theta$, and $r_3 = -3\sin\theta$. Do you see the pattern? Clear the screen and graph $r_1 = \cos\theta$, $r_2 = 2\cos\theta$, and $r_3 = 3\cos\theta$. Do you see the pattern? Clear the screen and graph $r_1 = -\cos\theta$, $r_2 = -2\cos\theta$, and $r_3 = -3\cos\theta$. Do you see the pattern?

Based on Examples 5 and 6 and the preceding Exploration, we are led to the following results. (The proofs are left as exercises.)

Theorem

Let a be a positive real number. Then

Equation	**Description**
(a) $r = 2a\sin\theta$	Circle: radius a; center at $(0, a)$ in rectangular coordinates
(b) $r = -2a\sin\theta$	Circle: radius a; center at $(0, -a)$ in rectangular coordinates
(c) $r = 2a\cos\theta$	Circle: radius a; center at $(a, 0)$ in rectangular coordinates
(d) $r = -2a\cos\theta$	Circle: radius a; center at $(-a, 0)$ in rectangular coordinates

Each circle passes through the pole.

NOW WORK PROBLEM 21.

The method of converting a polar equation to an identifiable rectangular equation in order to obtain the graph is not always helpful, nor is it always necessary. Usually, we set up a table that lists several points on the graph. By checking for symmetry, it may be possible to reduce the number of points needed to draw the graph.

Symmetry

2 In polar coordinates, the points (r, θ) and $(r, -\theta)$ are symmetric with respect to the polar axis (and to the x-axis). See Figure 25(a). The points

Figure 25

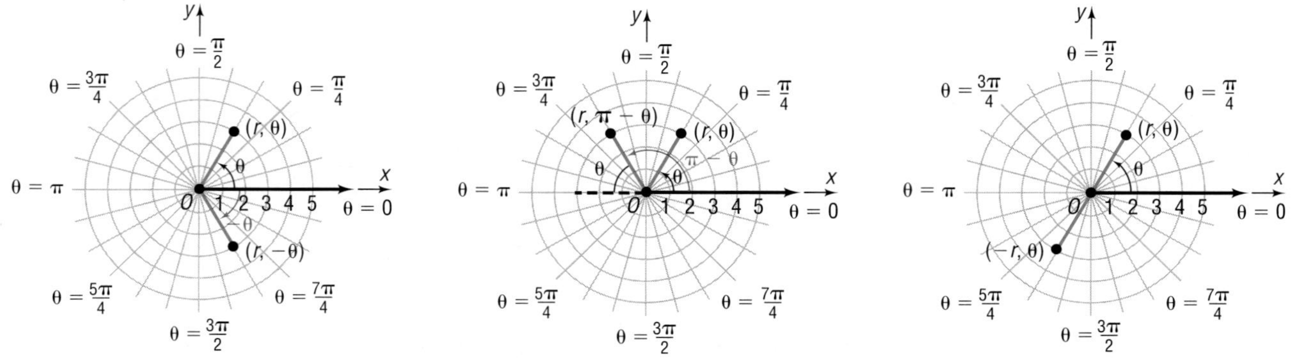

(a) Points symmetric with respect to the polar axis

(b) Points symmetric with respect to the line $\theta = \frac{\pi}{2}$

(c) Points symmetric with respect to the pole

(r, θ) and $(r, \pi - \theta)$ are symmetric with respect to the line $\theta = \frac{\pi}{2}$ (the y-axis). See Figure 25(b). The points (r, θ) and $(-r, \theta)$ are symmetric with respect to the pole (the origin). See Figure 25(c).

The following tests are consequences of these observations.

Theorem

Tests for Symmetry

Symmetry with Respect to the Polar Axis (x-Axis)

In a polar equation, replace θ by $-\theta$. If an equivalent equation results, the graph is symmetric with respect to the polar axis.

Symmetry with Respect to the Line $\theta = \frac{\pi}{2}$ (y-Axis)

In a polar equation, replace θ by $\pi - \theta$. If an equivalent equation results, the graph is symmetric with respect to the line $\theta = \frac{\pi}{2}$.

Symmetry with Respect to the Pole (Origin)

In a polar equation, replace r by $-r$. If an equivalent equation results, the graph is symmetric with respect to the pole.

The three tests for symmetry given here are *sufficient* conditions for symmetry, but they are not *necessary* conditions. That is, an equation may fail these tests and still have a graph that is symmetric with respect to the polar axis, the line $\theta = \frac{\pi}{2}$, or the pole. For example, the graph of $r = \sin(2\theta)$ turns out to be symmetric with respect to the polar axis, the line $\theta = \frac{\pi}{2}$, and the pole, but all three tests given here fail. See also Problems 81, 82, and 83.

3 **EXAMPLE 7** **Graphing a Polar Equation (Cardioid)**

Graph the equation: $r = 1 - \sin\theta$

Solution We check for symmetry first.

Polar Axis: Replace θ by $-\theta$. The result is

$$r = 1 - \sin(-\theta) = 1 + \sin\theta$$

The test fails, so the graph may or may not be symmetric with respect to the polar axis.

The Line $\theta = \frac{\pi}{2}$: Replace θ by $\pi - \theta$. The result is

$$\begin{aligned} r = 1 - \sin(\pi - \theta) &= 1 - (\sin\pi\cos\theta - \cos\pi\sin\theta) \\ &= 1 - [0 \cdot \cos\theta - (-1)\sin\theta] = 1 - \sin\theta \end{aligned}$$

The test is satisfied, so the graph is symmetric with respect to the line $\theta = \frac{\pi}{2}$.

The Pole: Replace r by $-r$. Then the result is $-r = 1 - \sin\theta$, so $r = -1 + \sin\theta$. The test fails, so the graph may or may not be symmetric with respect to the pole.

Table 1

θ	$r = 1 - \sin\theta$
$-\frac{\pi}{2}$	$1 - (-1) = 2$
$-\frac{\pi}{3}$	$1 - \left(-\frac{\sqrt{3}}{2}\right) \approx 1.87$
$-\frac{\pi}{6}$	$1 - \left(-\frac{1}{2}\right) = \frac{3}{2}$
0	$1 - 0 = 1$
$\frac{\pi}{6}$	$1 - \frac{1}{2} = \frac{1}{2}$
$\frac{\pi}{3}$	$1 - \frac{\sqrt{3}}{2} \approx 0.13$
$\frac{\pi}{2}$	$1 - 1 = 0$

Next, we identify points on the graph by assigning values to the angle θ and calculating the corresponding values of r. Due to the symmetry with respect to the line $\theta = \frac{\pi}{2}$, we only need to assign values to θ from $-\frac{\pi}{2}$ to $\frac{\pi}{2}$, as given in Table 1.

Now we plot the points (r, θ) from Table 1 and trace out the graph, beginning at the point $\left(2, -\frac{\pi}{2}\right)$ and ending at the point $\left(0, \frac{\pi}{2}\right)$. Then we reflect this portion of the graph about the line $\theta = \frac{\pi}{2}$ (the y-axis) to obtain the complete graph. Figure 26 shows the graph.

Figure 26
$r = 1 - \sin\theta$

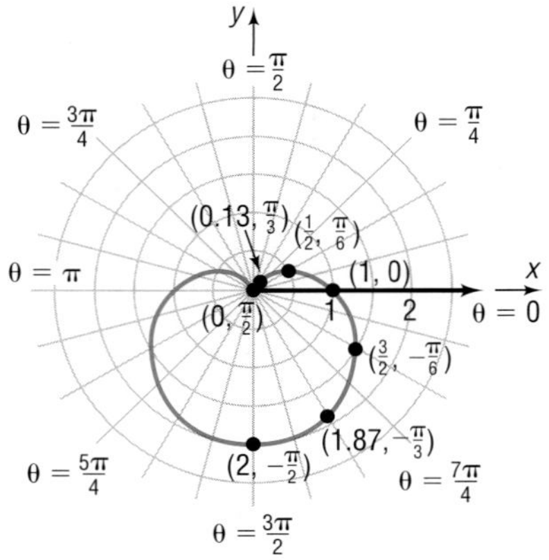

◀

Exploration

Graph $r_1 = 1 + \sin\theta$. Clear the screen and graph $r_1 = 1 - \cos\theta$. Clear the screen and graph $r_1 = 1 + \cos\theta$. Do you see a pattern?

The curve in Figure 26 is an example of a *cardioid* (a heart-shaped curve).

Cardioids are characterized by equations of the form

$$r = a(1 + \cos\theta) \qquad r = a(1 + \sin\theta)$$
$$r = a(1 - \cos\theta) \qquad r = a(1 - \sin\theta)$$

where $a > 0$. The graph of a cardioid passes through the pole.

NOW WORK PROBLEM 37.

EXAMPLE 8 Graphing a Polar Equation (Limaçon without Inner Loop)

Graph the equation: $r = 3 + 2\cos\theta$

Solution We check for symmetry first.

Polar Axis: Replace θ by $-\theta$. The result is

$$r = 3 + 2\cos(-\theta) = 3 + 2\cos\theta$$

The test is satisfied, so the graph is symmetric with respect to the polar axis.

The Line $\theta = \frac{\pi}{2}$: Replace θ by $\pi - \theta$. The result is

$$r = 3 + 2\cos(\pi - \theta) = 3 + 2(\cos\pi\cos\theta + \sin\pi\sin\theta)$$
$$= 3 - 2\cos\theta$$

The test fails, so the graph may or may not be symmetric with respect to the line $\theta = \frac{\pi}{2}$.

The Pole: Replace r by $-r$. The test fails, so the graph may or may not be symmetric with respect to the pole.

Next, we identify points on the graph by assigning values to the angle θ and calculating the corresponding values of r. Due to the symmetry with respect to the polar axis, we only need to assign values to θ from 0 to π, as given in Table 2.

Now we plot the points (r, θ) from Table 2 and trace out the graph, beginning at the point $(5, 0)$ and ending at the point $(1, \pi)$. Then we reflect this portion of the graph about the polar axis (the x-axis) to obtain the complete graph. Figure 27 shows the graph.

Table 2

θ	$r = 3 + 2\cos\theta$
0	$3 + 2(1) = 5$
$\frac{\pi}{6}$	$3 + 2\left(\frac{\sqrt{3}}{2}\right) \approx 4.73$
$\frac{\pi}{3}$	$3 + 2\left(\frac{1}{2}\right) = 4$
$\frac{\pi}{2}$	$3 + 2(0) = 3$
$\frac{2\pi}{3}$	$3 + 2\left(-\frac{1}{2}\right) = 2$
$\frac{5\pi}{6}$	$3 + 2\left(-\frac{\sqrt{3}}{2}\right) \approx 1.27$
π	$3 + 2(-1) = 1$

Figure 27
$r = 3 + 2\cos\theta$

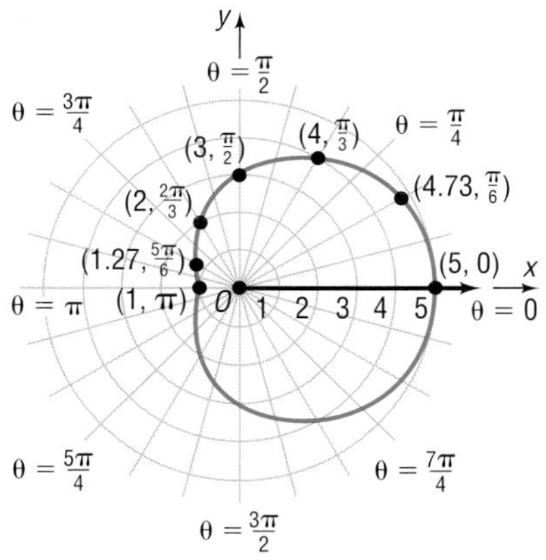

◀

Exploration

Graph $r_1 = 3 - 2\cos\theta$. Clear the screen and graph $r_1 = 3 + 2\sin\theta$. Clear the screen and graph $r_1 = 3 - 2\sin\theta$. Do you see a pattern?

The curve in Figure 27 is an example of a limaçon (the French word for *snail*) without an inner loop.

Limaçons without an inner loop are characterized by equations of the form

$$r = a + b\cos\theta \qquad r = a + b\sin\theta$$
$$r = a - b\cos\theta \qquad r = a - b\sin\theta$$

where $a > 0, b > 0$, and $a > b$. The graph of a limaçon without an inner loop does not pass through the pole.

NOW WORK PROBLEM 43.

EXAMPLE 9 Graphing a Polar Equation (Limaçon with Inner Loop)

Graph the equation: $r = 1 + 2\cos\theta$

Solution First, we check for symmetry.

Polar Axis: Replace θ by $-\theta$. The result is

$$r = 1 + 2\cos(-\theta) = 1 + 2\cos\theta$$

The test is satisfied, so the graph is symmetric with respect to the polar axis.

The Line $\theta = \frac{\pi}{2}$: Replace θ by $\pi - \theta$. The result is

$$r = 1 + 2\cos(\pi - \theta) = 1 + 2(\cos\pi\cos\theta + \sin\pi\sin\theta)$$
$$= 1 - 2\cos\theta$$

The test fails, so the graph may or may not be symmetric with respect to the line $\theta = \frac{\pi}{2}$.

The Pole: Replace r by $-r$. The test fails, so the graph may or may not be symmetric with respect to the pole.

Next, we identify points on the graph of $r = 1 + 2\cos\theta$ by assigning values to the angle θ and calculating the corresponding values of r. Due to the symmetry with respect to the polar axis, we only need to assign values to θ from 0 to π, as given in Table 3.

Now we plot the points (r, θ) from Table 3, beginning at $(3, 0)$ and ending at $(-1, \pi)$. See Figure 28(a). Finally, we reflect this portion of the graph about the polar axis (the x-axis) to obtain the complete graph. See Figure 28(b).

Table 3

θ	$r = 1 + 2\cos\theta$
0	$1 + 2(1) = 3$
$\frac{\pi}{6}$	$1 + 2\left(\frac{\sqrt{3}}{2}\right) \approx 2.73$
$\frac{\pi}{3}$	$1 + 2\left(\frac{1}{2}\right) = 2$
$\frac{\pi}{2}$	$1 + 2(0) = 1$
$\frac{2\pi}{3}$	$1 + 2\left(-\frac{1}{2}\right) = 0$
$\frac{5\pi}{6}$	$1 + 2\left(-\frac{\sqrt{3}}{2}\right) \approx -0.73$
π	$1 + 2(-1) = -1$

Figure 28

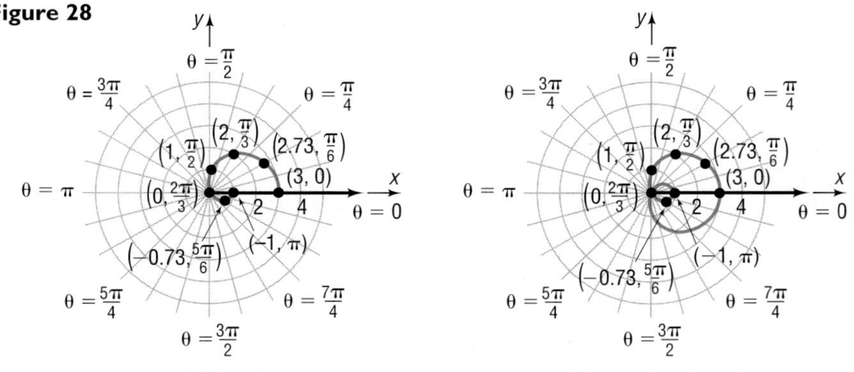

(a)

(b) $r = 1 + 2\cos\theta$

Exploration

Graph $r_1 = 1 - 2\cos\theta$. Clear the screen and graph $r_1 = 1 + 2\sin\theta$. Clear the screen and graph $r_1 = 1 - 2\sin\theta$. Do you see a pattern?

The curve in Figure 28(b) is an example of a limaçon with an inner loop.

Limaçons with an inner loop are characterized by equations of the form

$$r = a + b\cos\theta \qquad r = a + b\sin\theta$$
$$r = a - b\cos\theta \qquad r = a - b\sin\theta$$

where $a > 0$, $b > 0$, and $a < b$. The graph of a limaçon with an inner loop will pass through the pole twice.

NOW WORK PROBLEM 45.

EXAMPLE 10 Graphing a Polar Equation (Rose)

Graph the equation: $r = 2\cos(2\theta)$

Solution We check for symmetry.

Polar Axis: If we replace θ by $-\theta$, the result is

$$r = 2\cos[2(-\theta)] = 2\cos(2\theta)$$

The test is satisfied, so the graph is symmetric with respect to the polar axis.

The Line $\theta = \frac{\pi}{2}$: If we replace θ by $\pi - \theta$, we obtain

$$r = 2\cos[2(\pi - \theta)] = 2\cos(2\pi - 2\theta) = 2\cos(2\theta)$$

The test is satisfied, so the graph is symmetric with respect to the line $\theta = \frac{\pi}{2}$.

The Pole: Since the graph is symmetric with respect to both the polar axis and the line $\theta = \frac{\pi}{2}$, it must be symmetric with respect to the pole.

Next, we construct Table 4. Due to the symmetry with respect to the polar axis, the line $\theta = \frac{\pi}{2}$, and the pole, we consider only values of θ from 0 to $\frac{\pi}{2}$.

Table 4

θ	$r = 2\cos(2\theta)$
0	$2(1) = 2$
$\frac{\pi}{6}$	$2\left(\frac{1}{2}\right) = 1$
$\frac{\pi}{4}$	$2(0) = 0$
$\frac{\pi}{3}$	$2\left(-\frac{1}{2}\right) = -1$
$\frac{\pi}{2}$	$2(-1) = -2$

We plot and connect these points in Figure 29(a). Finally, because of symmetry, we reflect this portion of the graph first about the polar axis (the x-axis) and then about the line $\theta = \frac{\pi}{2}$ (the y-axis) to obtain the complete graph. See Figure 29(b).

Figure 29

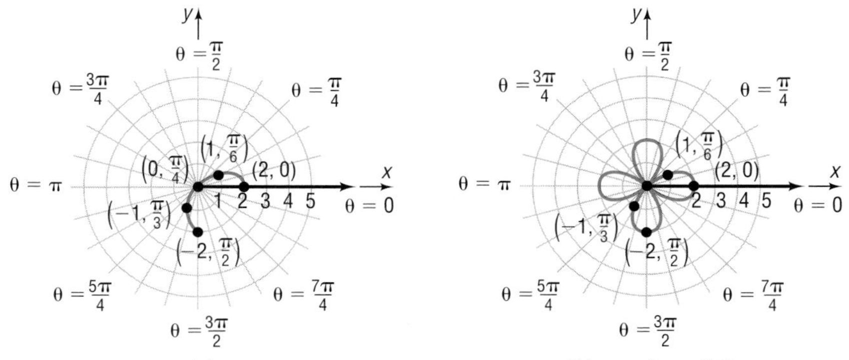

(a) **(b)** $r = 2\cos(2\theta)$

Exploration

Graph $r = 2\cos(4\theta)$; clear the screen and graph $r = 2\cos(6\theta)$. How many petals did each of these graphs have?

Clear the screen and graph, in order, each on a clear screen, $r = 2\cos(3\theta)$, $r = 2\cos(5\theta)$, and $r = 2\cos(7\theta)$. What do you notice about the number of petals?

The curve in Figure 29(b) is called a *rose* with four petals.

Rose curves are characterized by equations of the form

$$r = a\cos(n\theta), \qquad r = a\sin(n\theta), \qquad a \neq 0$$

and have graphs that are rose shaped. If $n \neq 0$ is even, the rose has $2n$ petals; if $n \neq \pm 1$ is odd, the rose has n petals.

NOW WORK PROBLEM 49.

EXAMPLE 11 Graphing a Polar Equation (Lemniscate)

Graph the equation: $r^2 = 4\sin(2\theta)$

Solution We leave it to you to verify that the graph is symmetric with respect to the pole. Table 5 lists points on the graph for values of $\theta = 0$ through $\theta = \frac{\pi}{2}$. Note that there are no points on the graph for $\frac{\pi}{2} < \theta < \pi$ (quadrant II), since $\sin(2\theta) < 0$ for such values. The points from Table 5 where $r \geq 0$ are plotted in Figure 30(a). The remaining points on the graph may be obtained by using symmetry. Figure 30(b) shows the final graph.

Table 5

θ	$r^2 = 4\sin(2\theta)$	r
0	$4(0) = 0$	0
$\frac{\pi}{6}$	$4\left(\frac{\sqrt{3}}{2}\right) = 2\sqrt{3}$	± 1.9
$\frac{\pi}{4}$	$4(1) = 4$	± 2
$\frac{\pi}{3}$	$4\left(\frac{\sqrt{3}}{2}\right) = 2\sqrt{3}$	± 1.9
$\frac{\pi}{2}$	$4(0) = 4$	0

Figure 30

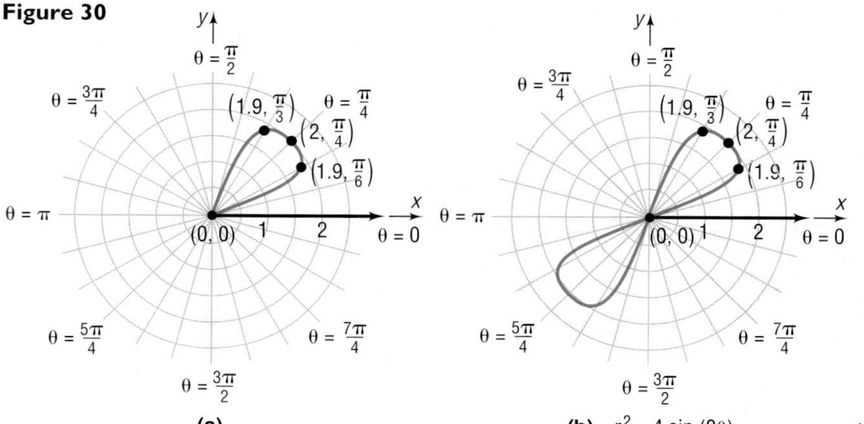

(a) (b) $r^2 = 4\sin(2\theta)$

The curve in Figure 30(b) is an example of a *lemniscate* (from the Greek word *ribbon*).

Lemniscates are characterized by equations of the form

$$r^2 = a^2\sin(2\theta) \qquad r^2 = a^2\cos(2\theta)$$

where $a > 0$, and have graphs that are propeller shaped.

NOW WORK PROBLEM 53.

EXAMPLE 12 Graphing a Polar Equation (Spiral)

Graph the equation: $r = e^{\theta/5}$

Solution The tests for symmetry with respect to the pole, the polar axis, and the line $\theta = \frac{\pi}{2}$ fail. Furthermore, there is no number θ for which $r = 0$, so the graph does not pass through the pole. We observe that r is positive for all θ, r increases as θ increases, $r \to 0$ as $\theta \to -\infty$, and $r \to \infty$ as $\theta \to \infty$. With the help of a calculator, we obtain the values in Table 6. See Figure 31 for the graph.

Table 6

θ	$r = e^{\theta/5}$
$-\frac{3\pi}{2}$	0.39
$-\pi$	0.53
$-\frac{\pi}{2}$	0.73
$-\frac{\pi}{4}$	0.85
0	1
$\frac{\pi}{4}$	1.17
$\frac{\pi}{2}$	1.37
π	1.87
$\frac{3\pi}{2}$	2.57
2π	3.51

Figure 31
$r = e^{\theta/5}$

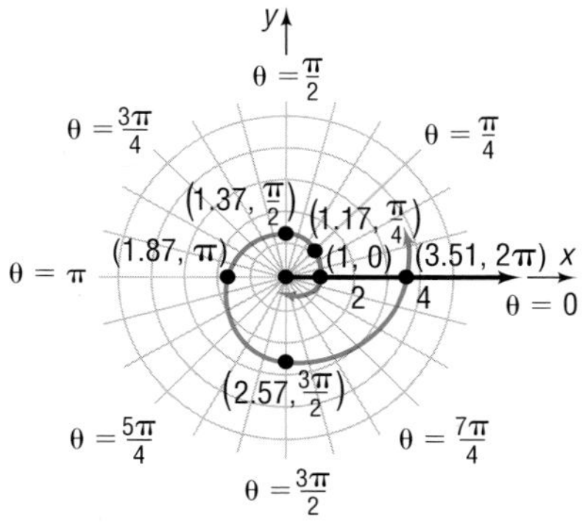

◀

The curve in Figure 31 is called a **logarithmic spiral,** since its equation may be written as $\theta = 5 \ln r$ and it spirals infinitely both toward the pole and away from it.

Classification of Polar Equations

The equations of some lines and circles in polar coordinates and their corresponding equations in rectangular coordinates are given in Table 7 on page 558. Also included are the names and graphs of a few of the more frequently encountered polar equations.

Sketching Quickly

If a polar equation only involves a sine (or cosine) function, you can quickly obtain a sketch of its graph by making use of Table 7, periodicity, and a short table.

EXAMPLE 13 Sketching the Graph of a Polar Equation Quickly by Hand

Graph the equation: $r = 2 + 2\sin\theta$

Solution We recognize the polar equation: Its graph is a cardioid. The period of $\sin\theta$ is 2π, so we form a table using $0 \le \theta \le 2\pi$, compute r, plot the points (r, θ), and sketch the graph of a cardioid as θ varies from 0 to 2π. See Table 8 and Figure 32 on page 559.

Table 7

Lines			
Description	Line passing through the pole making an angle α with the polar axis	Vertical line	Horizontal line
Rectangular equation	$y = (\tan \alpha)x$	$x = a$	$y = b$
Polar equation	$\theta = \alpha$	$r \cos \theta = a$	$r \sin \theta = b$
Typical graph	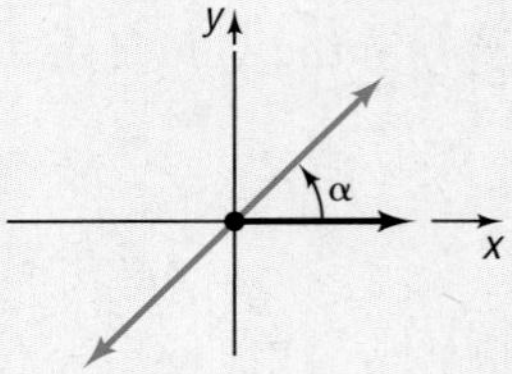	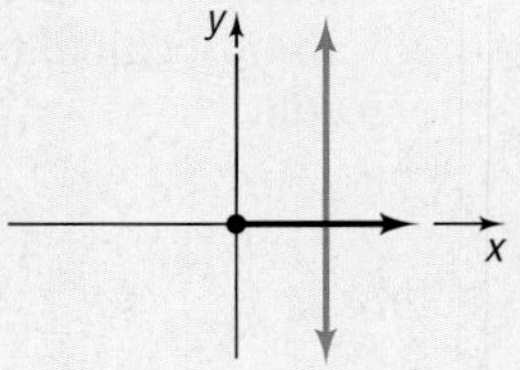	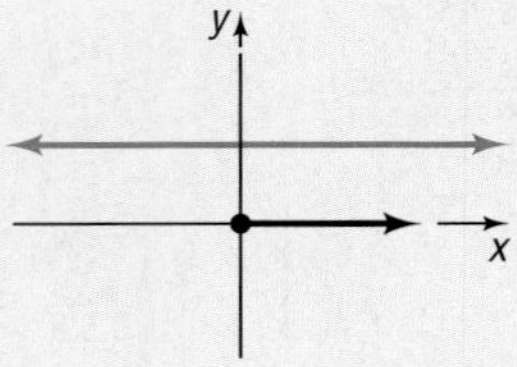

Circles			
Description	Center at the pole, radius a	Passing through the pole, tangent to the line $\theta = \frac{\pi}{2}$, center on the polar axis, radius a	Passing through the pole, tangent to the polar axis, center on the line $\theta = \frac{\pi}{2}$, radius a
Rectangular equation	$x^2 + y^2 = a^2, \quad a > 0$	$x^2 + y^2 = \pm 2ax, \quad a > 0$	$x^2 + y^2 = \pm 2ay, \quad a > 0$
Polar equation	$r = a, \quad a > 0$	$r = \pm 2a \cos \theta, \quad a > 0$	$r = \pm 2a \sin \theta, \quad a > 0$
Typical graph	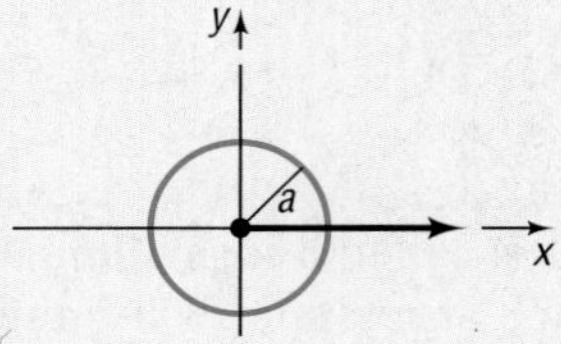	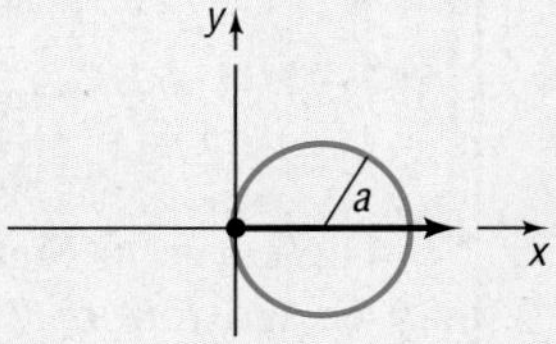	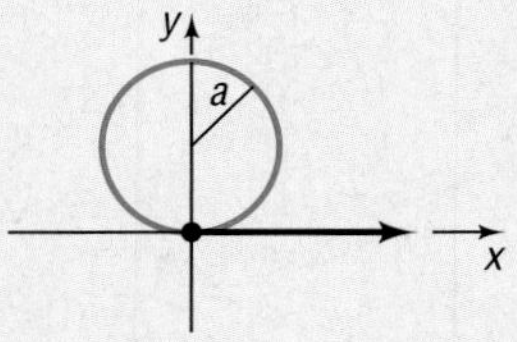

Other Equations			
Name	Cardioid	Limaçon without inner loop	Limaçon with inner loop
Polar equations	$r = a \pm a \cos \theta, \quad a > 0$ $r = a \pm a \sin \theta, \quad a > 0$	$r = a \pm b \cos \theta, \quad 0 < b < a$ $r = a \pm b \sin \theta, \quad 0 < b < a$	$r = a \pm b \cos \theta, \quad 0 < a < b$ $r = a \pm b \sin \theta, \quad 0 < a < b$
Typical graph	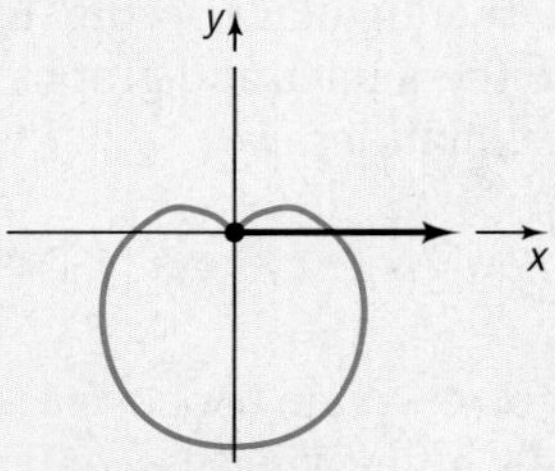	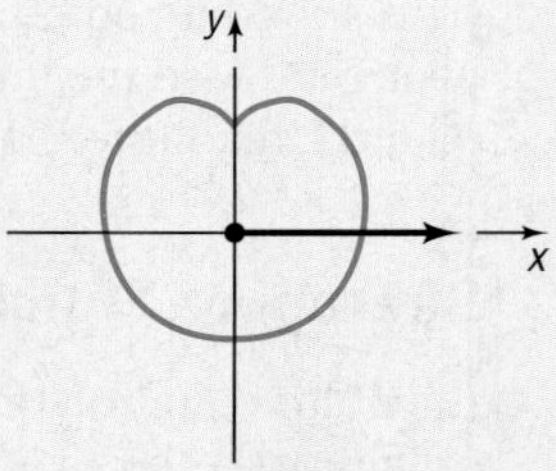	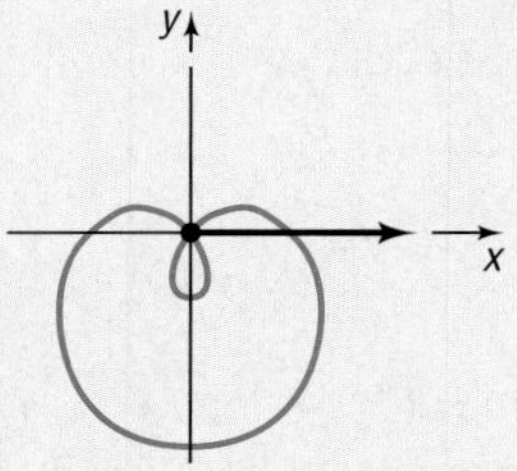
Name	Lemniscate	Rose with three petals	Rose with four petals
Polar equations	$r^2 = a^2 \cos(2\theta), \quad a > 0$ $r^2 = a^2 \sin(2\theta), \quad a > 0$	$r = a \sin(3\theta), \quad a > 0$ $r = a \cos(3\theta), \quad a > 0$	$r = a \sin(2\theta), \quad a > 0$ $r = a \cos(2\theta), \quad a > 0$
Typical graph	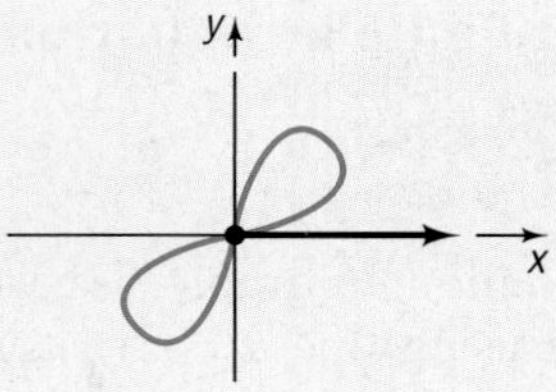	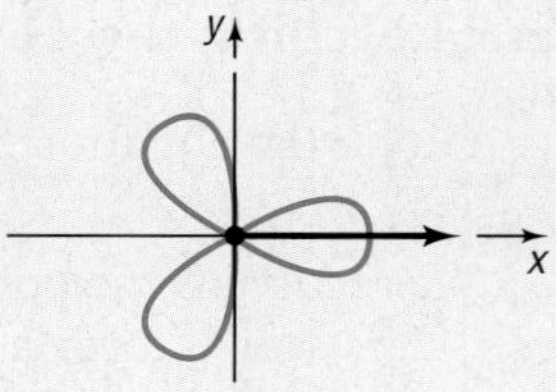	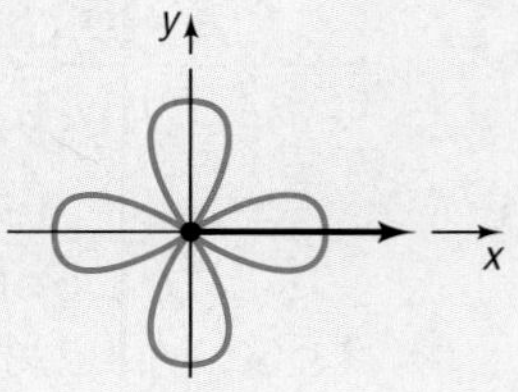

Table 8

θ	$r = 2 + 2\sin\theta$
0	$2 + 2(0) = 2$
$\frac{\pi}{2}$	$2 + 2(1) = 4$
π	$2 + 2(0) = 2$
$\frac{3\pi}{2}$	$2 + 2(-1) = 0$
2π	$2 + 2(0) = 2$

Figure 32

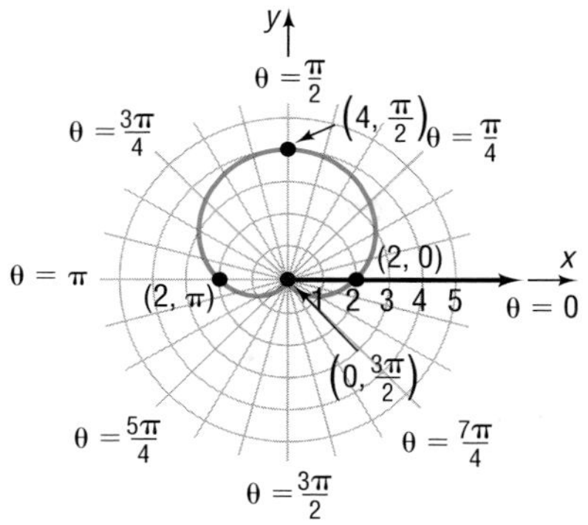

Calculus Comment

For those of you who are planning to study calculus, a comment about one important role of polar equations is in order.

In rectangular coordinates, the equation $x^2 + y^2 = 1$, whose graph is the unit circle, is not the graph of a function. In fact, it requires two functions to obtain the graph of the unit circle:

$$y_1 = \sqrt{1 - x^2} \quad \text{Upper semicircle} \qquad y_2 = -\sqrt{1 - x^2} \quad \text{Lower semicircle}$$

In polar coordinates, the equation $r = 1$, whose graph is also the unit circle, does define a function. That is, for each choice of θ there is only one corresponding value of r, namely, $r = 1$. Since many problems in calculus require the use of functions, the opportunity to express nonfunctions in rectangular coordinates as functions in polar coordinates becomes extremely useful.

Note also that the vertical-line test for functions is valid only for equations in rectangular coordinates.

Historical Feature

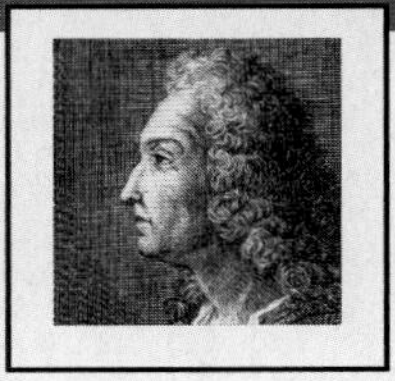

Jakob Bernoulli (1654–1705)

Polar coordinates seem to have been invented by Jakob Bernoulli (1654–1705) in about 1691, although, as with most such ideas, earlier traces of the notion exist. Early users of calculus remained committed to rectangular coordinates, and polar coordinates did not become widely used until the early 1800s. Even then, it was mostly geometers who used them for describing odd curves. Finally, about the mid-1800s, applied mathematicians realized the tremendous simplification that polar coordinates make possible in the description of objects with circular or cylindrical symmetry. From then on their use became widespread.

8.2 Assess Your Understanding

'Are You Prepared?' *Answers are given at the end of these exercises. If you get a wrong answer, read the pages listed in red.*

1. If the rectangular coordinates of a point are $(4, -6)$, the point symmetric to it with respect to the origin is __________. (pp. 9–16)
2. The difference formula for cosine is $\cos(\alpha - \beta) =$ __________. (pp. 438 and 441)
3. The standard equation of a circle with center at $(-2, 5)$ and radius 3 is __________. (pp. 16–19)
4. Is the sine function even, odd, or neither? (p. 366)
5. $\sin \frac{5\pi}{4} =$ __________. (pp. 342–351)
6. $\cos \frac{2\pi}{3} =$ __________. (pp. 342–351)

Concepts and Vocabulary

7. An equation whose variables are polar coordinates is called a __________ __________.

8. Using polar coordinates (r, θ), the circle $x^2 + y^2 = 2x$ takes the form __________.

9. A polar equation is symmetric with respect to the pole if an equivalent equation results when r is replaced by __________.

10. *True or False:* The tests for symmetry in polar coordinates are necessary, but not sufficient.

11. *True or False:* The graph of a cardioid never passes through the pole.

12. *True or False:* All polar equations have a symmetric feature.

Exercises

In Problems 13–28, transform each polar equation to an equation in rectangular coordinates. Then identify and graph the equation.

13. $r = 4$
14. $r = 2$
15. $\theta = \frac{\pi}{3}$
16. $\theta = -\frac{\pi}{4}$
17. $r \sin \theta = 4$
18. $r \cos \theta = 4$
19. $r \cos \theta = -2$
20. $r \sin \theta = -2$
21. $r = 2 \cos \theta$
22. $r = 2 \sin \theta$
23. $r = -4 \sin \theta$
24. $r = -4 \cos \theta$
25. $r \sec \theta = 4$
26. $r \csc \theta = 8$
27. $r \csc \theta = -2$
28. $r \sec \theta = -4$

In Problems 29–36, match each of the graphs (A) through (H) to one of the following polar equations.

29. $r = 2$
30. $\theta = \frac{\pi}{4}$
31. $r = 2 \cos \theta$
32. $r \cos \theta = 2$
33. $r = 1 + \cos \theta$
34. $r = 2 \sin \theta$
35. $\theta = \frac{3\pi}{4}$
36. $r \sin \theta = 2$

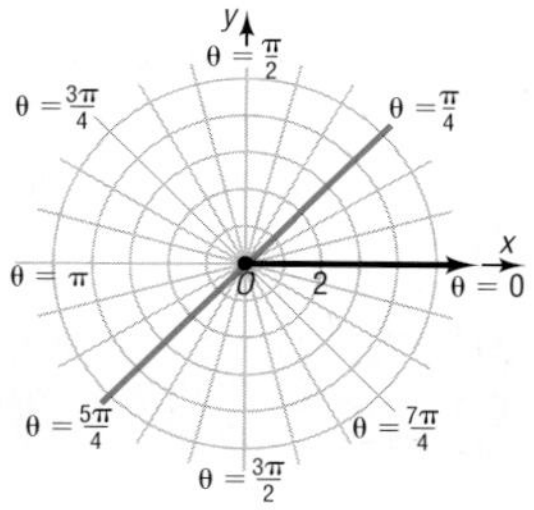

(A)

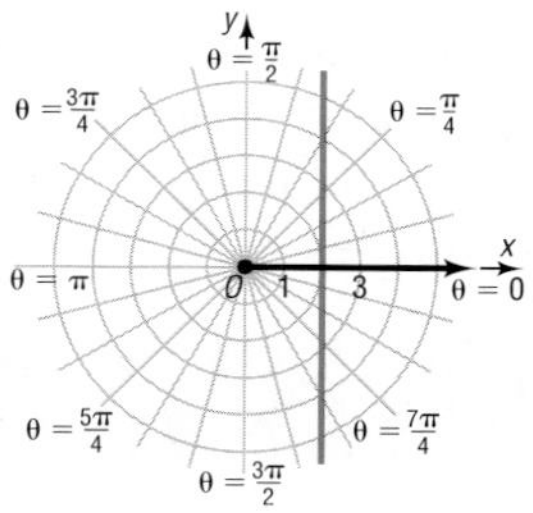

(B)

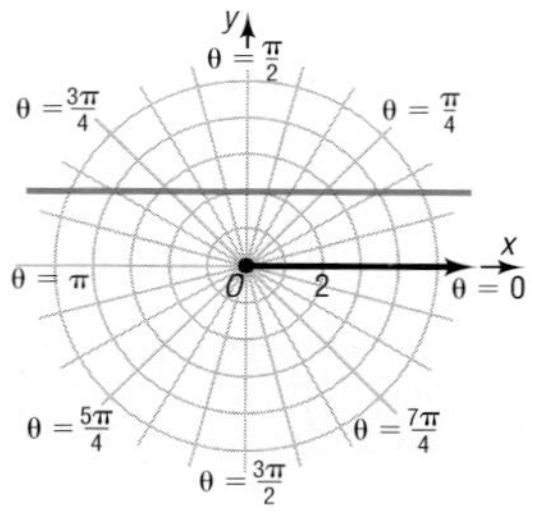

(C)

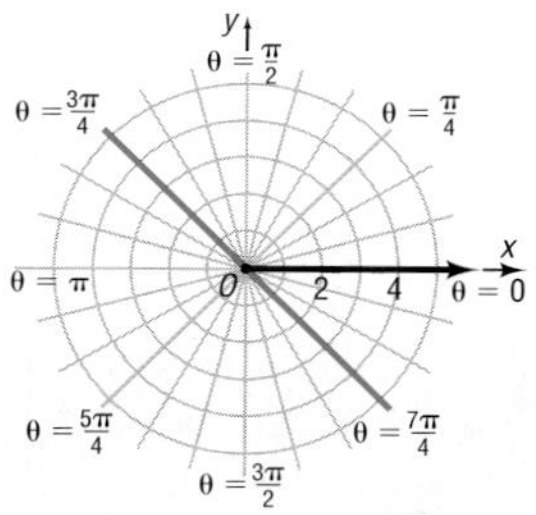

(D)

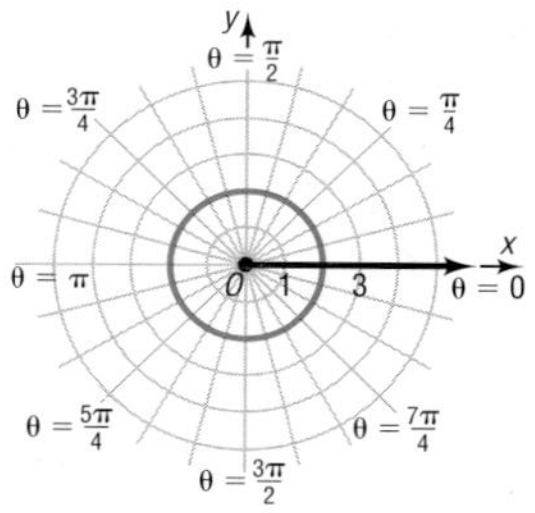

(E)

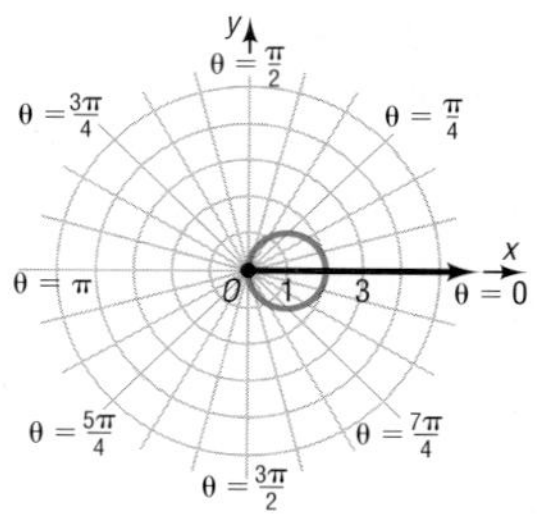

(F)

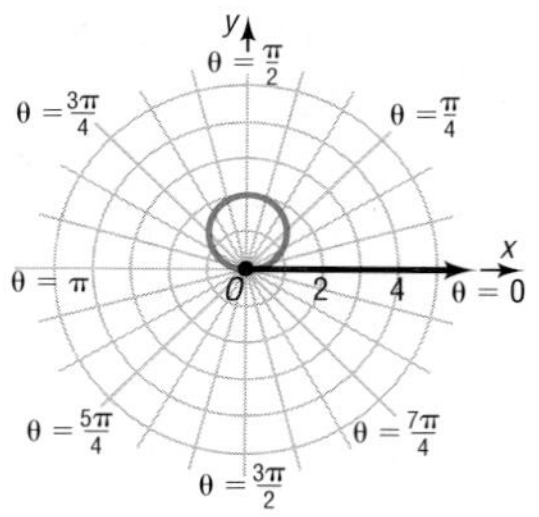

(G)

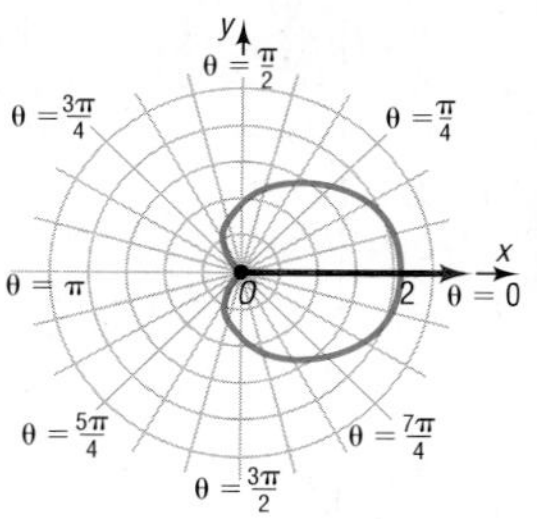

(H)

In Problems 37–60, identify and graph each polar equation.

37. $r = 2 + 2 \cos \theta$
38. $r = 1 + \sin \theta$
39. $r = 3 - 3 \sin \theta$
40. $r = 2 - 2 \cos \theta$
41. $r = 2 + \sin \theta$
42. $r = 2 - \cos \theta$
43. $r = 4 - 2 \cos \theta$
44. $r = 4 + 2 \sin \theta$
45. $r = 1 + 2 \sin \theta$
46. $r = 1 - 2 \sin \theta$
47. $r = 2 - 3 \cos \theta$
48. $r = 2 + 4 \cos \theta$
49. $r = 3 \cos(2\theta)$
50. $r = 2 \sin(3\theta)$
51. $r = 4 \sin(5\theta)$
52. $r = 3 \cos(4\theta)$
53. $r^2 = 9 \cos(2\theta)$
54. $r^2 = \sin(2\theta)$
55. $r = 2^\theta$
56. $r = 3^\theta$
57. $r = 1 - \cos \theta$
58. $r = 3 + \cos \theta$
59. $r = 1 - 3 \cos \theta$
60. $r = 4 \cos(3\theta)$

In Problems 61–64, the polar equation for each graph is either $r = a + b\cos\theta$ or $r = a + b\sin\theta$, $a > 0$, $b > 0$. Select the correct equation and find the values of a and b.

61.

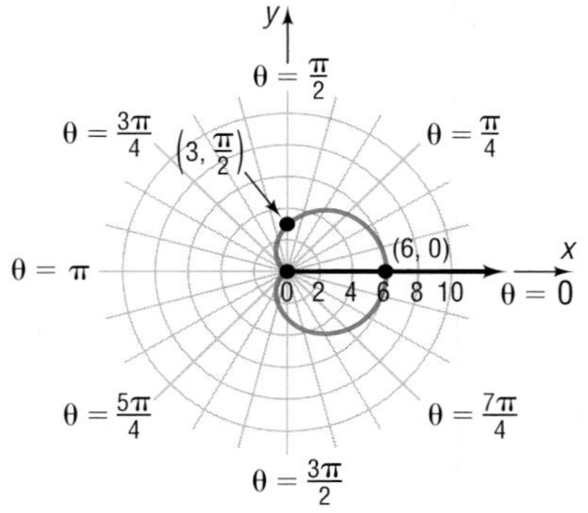

62.

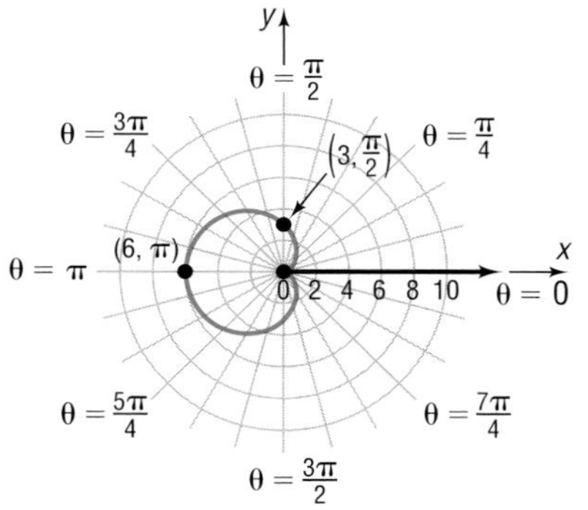

63.

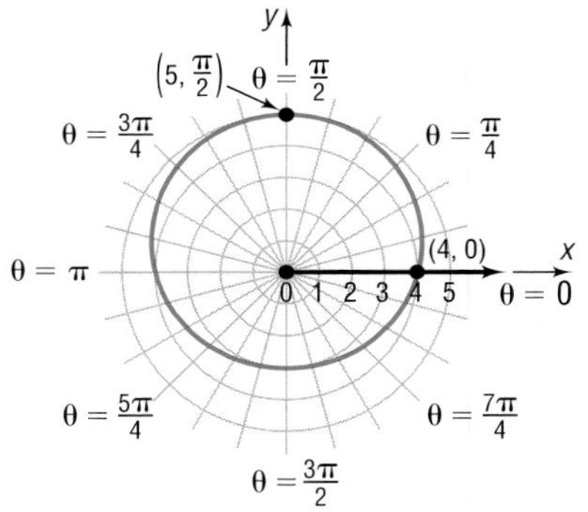

64.

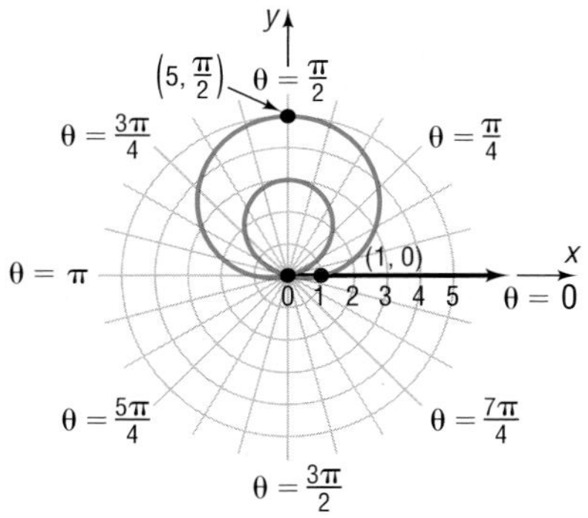

In Problems 65–74, graph each polar equation.

65. $r = \dfrac{2}{1 - \cos\theta}$ (parabola)

66. $r = \dfrac{2}{1 - 2\cos\theta}$ (hyperbola)

67. $r = \dfrac{1}{3 - 2\cos\theta}$ (ellipse)

68. $r = \dfrac{1}{1 - \cos\theta}$ (parabola)

69. $r = \theta, \quad \theta \ge 0$ (spiral of Archimedes)

70. $r = \dfrac{3}{\theta}$ (reciprocal spiral)

71. $r = \csc\theta - 2, \quad 0 < \theta < \pi$ (conchoid)

72. $r = \sin\theta\tan\theta$ (cissoid)

73. $r = \tan\theta, \quad -\dfrac{\pi}{2} < \theta < \dfrac{\pi}{2}$ (kappa curve)

74. $r = \cos\dfrac{\theta}{2}$

---

75. Show that the graph of the equation $r\sin\theta = a$ is a horizontal line a units above the pole if $a > 0$ and $|a|$ units below the pole if $a < 0$.

76. Show that the graph of the equation $r\cos\theta = a$ is a vertical line a units to the right of the pole if $a > 0$ and $|a|$ units to the left of the pole if $a < 0$.

77. Show that the graph of the equation $r = 2a\sin\theta$, $a > 0$, is a circle of radius a with center at $(0, a)$ in rectangular coordinates.

78. Show that the graph of the equation $r = -2a\sin\theta$, $a > 0$, is a circle of radius a with center at $(0, -a)$ in rectangular coordinates.

79. Show that the graph of the equation $r = 2a\cos\theta$, $a > 0$, is a circle of radius a with center at $(a, 0)$ in rectangular coordinates.

80. Show that the graph of the equation $r = -2a\cos\theta$, $a > 0$, is a circle of radius a with center at $(-a, 0)$ in rectangular coordinates.

81. Explain why the following test for symmetry is valid: Replace r by $-r$ and θ by $-\theta$ in a polar equation. If an equivalent equation results, the graph is symmetric with respect to the line $\theta = \frac{\pi}{2}$ (y-axis).
   (a) Show that the test on page 551 fails for $r^2 = \cos\theta$, but this new test works.
   (b) Show that the test on page 551 works for $r^2 = \sin\theta$, yet this new test fails.

82. Develop a new test for symmetry with respect to the pole.
   (a) Find a polar equation for which this new test fails, yet the test on page 551 works.
   (b) Find a polar equation for which the test on page 551 fails, yet the new test works.

83. Write down two different tests for symmetry with respect to the polar axis. Find examples in which one test works and the other fails. Which test do you prefer to use? Justify your answer.

'Are You Prepared?' Answers

1. $(-4, 6)$ **2.** $\cos\alpha\cos\beta + \sin\alpha\sin\beta$ **3.** $(x + 2)^2 + (y - 5)^2 = 9$ **4.** odd **5.** $\frac{\sqrt{2}}{2}$ **6.** $-\frac{1}{2}$

8.3 The Complex Plane; De Moivre's Theorem

PREPARING FOR THIS SECTION *Before getting started, review the following:*

- Complex Numbers (Appendix A, Section A.6, pp. 947–952)
- Value of the Sine and Cosine Functions at Certain Angles (Section 5.2, pp. 342–351)
- Sum and Difference Formulas for Sine and Cosine (Section 6.4, pp. 438 and 441)

Now work the 'Are You Prepared?' problems on page 568.

OBJECTIVES
1. Convert a Complex Number from Rectangular Form to Polar Form
2. Plot Points in the Complex Plane
3. Find Products and Quotients of Complex Numbers in Polar Form
4. Use De Moivre's Theorem
5. Find Complex Roots

When we first introduced complex numbers, we were not prepared to give a geometric interpretation of a complex number. Now we are ready. Although we could give several interpretations, the one that follows is the easiest to understand.

Figure 33
Complex plane

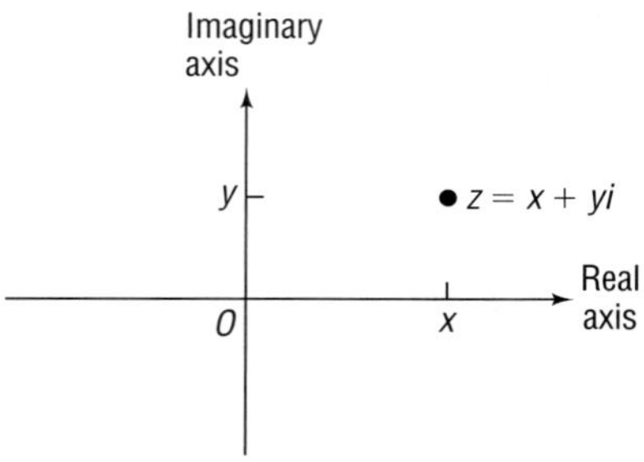

A complex number $z = x + yi$ can be interpreted geometrically as the point (x, y) in the xy-plane. Each point in the plane corresponds to a complex number and, conversely, each complex number corresponds to a point in the plane. We shall refer to the collection of such points as the **complex plane.** The x-axis will be referred to as the **real axis,** because any point that lies on the real axis is of the form $z = x + 0i = x$, a real number. The y-axis is called the **imaginary axis,** because any point that lies on it is of the form $z = 0 + yi = yi$, a pure imaginary number. See Figure 33.

Let $z = x + yi$ be a complex number. The **magnitude** or **modulus** of z, denoted by $|z|$, is defined as the distance from the origin to the point (x, y). That is,

$$|z| = \sqrt{x^2 + y^2} \quad \textbf{(1)}$$

See Figure 34 for an illustration.

Figure 34

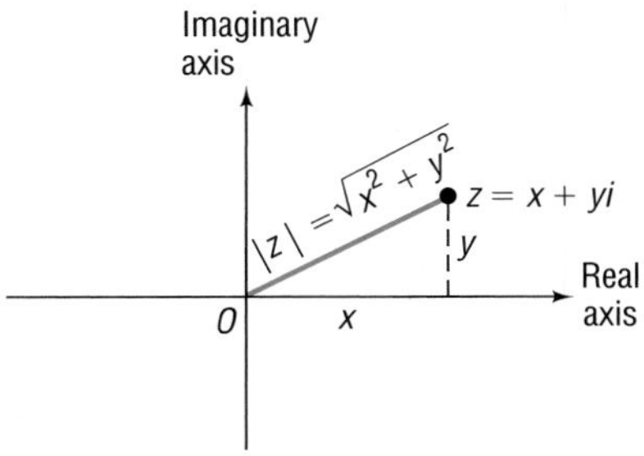

This definition for $|z|$ is consistent with the definition for the absolute value of a real number: If $z = x + yi$ is real, then $z = x + 0i$ and

$$|z| = \sqrt{x^2 + 0^2} = \sqrt{x^2} = |x|$$

For this reason, the magnitude of z is sometimes called the absolute value of z.

Recall that if $z = x + yi$ then its **conjugate,** denoted by $\bar{z}$, is $\bar{z} = x - yi$. Because $z\bar{z} = x^2 + y^2$, it follows from equation (1) that the magnitude of z can be written as

$$|z| = \sqrt{z\bar{z}} \qquad \textbf{(2)}$$

Polar Form of a Complex Number

1 When a complex number is written in the standard form $z = x + yi$, we say that it is in **rectangular,** or **Cartesian, form** because (x, y) are the rectangular coordinates of the corresponding point in the complex plane. Suppose that (r, θ) are the polar coordinates of this point. Then

$$x = r\cos\theta \qquad y = r\sin\theta \qquad \textbf{(3)}$$

If $r \geq 0$ and $0 \leq \theta < 2\pi$, the complex number $z = x + yi$ may be written in **polar form** as

$$z = x + yi = (r\cos\theta) + (r\sin\theta)i = r(\cos\theta + i\sin\theta) \qquad \textbf{(4)}$$

Figure 35

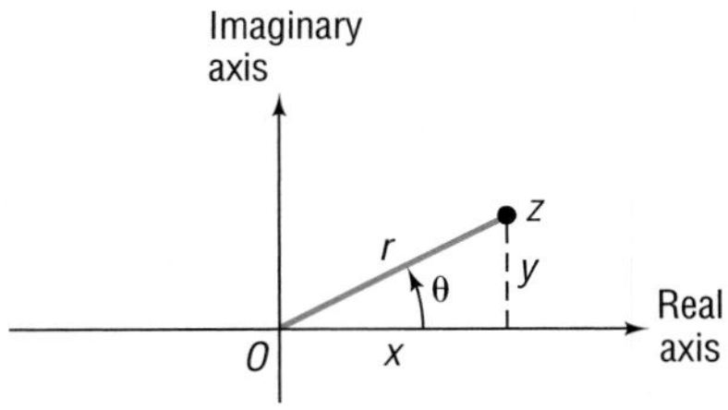

$z = x + yi = r(\cos\theta + i\sin\theta)$,
$r \geq 0, 0 \leq \theta < 2\pi$

See Figure 35.

If $z = r(\cos\theta + i\sin\theta)$ is the polar form of a complex number, the angle θ, $0 \leq \theta < 2\pi$, is called the **argument of z.**

Also, because $r \geq 0$, we have $r = \sqrt{x^2 + y^2}$. From equation (1) it follows that the magnitude of $z = r(\cos\theta + i\sin\theta)$ is

$$|z| = r$$

2 **EXAMPLE 1** **Plotting a Point in the Complex Plane and Writing a Complex Number in Polar Form**

Plot the point corresponding to $z = \sqrt{3} - i$ in the complex plane, and write an expression for z in polar form.

Solution The point corresponding to $z = \sqrt{3} - i$ has the rectangular coordinates $(\sqrt{3}, -1)$. The point, located in quadrant IV, is plotted in Figure 36. Because $x = \sqrt{3}$ and $y = -1$, it follows that

Figure 36

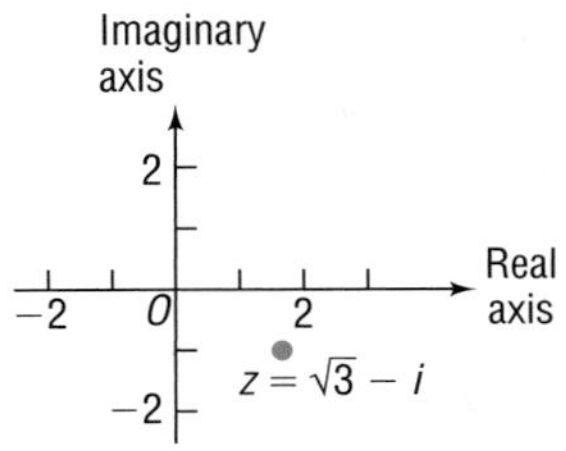

$$r = \sqrt{x^2 + y^2} = \sqrt{\left(\sqrt{3}\right)^2 + (-1)^2} = \sqrt{4} = 2$$

and

$$\sin\theta = \frac{y}{r} = \frac{-1}{2}, \qquad \cos\theta = \frac{x}{r} = \frac{\sqrt{3}}{2}, \qquad 0 \leq \theta < 2\pi$$

Then $\theta = \dfrac{11\pi}{6}$ and $r = 2$, so the polar form of $z = \sqrt{3} - i$ is

$$z = r(\cos\theta + i\sin\theta) = 2\left(\cos\frac{11\pi}{6} + i\sin\frac{11\pi}{6}\right)$$

◄

NOW WORK PROBLEM 11.

EXAMPLE 2 Plotting a Point in the Complex Plane and Converting from Polar to Rectangular Form

Plot the point corresponding to $z = 2(\cos 30° + i \sin 30°)$ in the complex plane, and write an expression for z in rectangular form.

Solution To plot the complex number $z = 2(\cos 30° + i \sin 30°)$, we plot the point whose polar coordinates are $(r, \theta) = (2, 30°)$, as shown in Figure 37. In rectangular form,

Figure 37

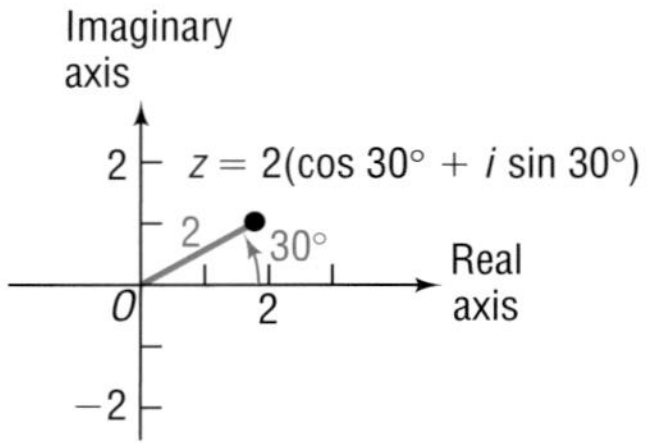

$$z = 2(\cos 30° + i \sin 30°) = 2\left(\frac{\sqrt{3}}{2} + \frac{1}{2}i\right) = \sqrt{3} + i$$ ◀

NOW WORK PROBLEM 23.

3 The polar form of a complex number provides an alternative method for finding products and quotients of complex numbers.

Theorem Let $z_1 = r_1(\cos \theta_1 + i \sin \theta_1)$ and $z_2 = r_2(\cos \theta_2 + i \sin \theta_2)$ be two complex numbers. Then

$$z_1 z_2 = r_1 r_2[\cos(\theta_1 + \theta_2) + i \sin(\theta_1 + \theta_2)] \quad \textbf{(5)}$$

If $z_2 \neq 0$, then

$$\frac{z_1}{z_2} = \frac{r_1}{r_2}[\cos(\theta_1 - \theta_2) + i \sin(\theta_1 - \theta_2)] \quad \textbf{(6)}$$

In Words
The magnitude of a complex number z is r and its argument is θ, so when $z = r(\cos \theta + i \sin \theta)$, the magnitude of the product (quotient) of two complex numbers equals the product (quotient) of their magnitudes; the argument of the product (quotient) of two complex numbers is determined by the sum (difference) of their arguments.

Proof We will prove formula (5). The proof of formula (6) is left as an exercise (see Problem 66).

$$\begin{aligned} z_1 z_2 &= [r_1(\cos \theta_1 + i \sin \theta_1)][r_2(\cos \theta_2 + i \sin \theta_2)] \\ &= r_1 r_2[(\cos \theta_1 + i \sin \theta_1)(\cos \theta_2 + i \sin \theta_2)] \\ &= r_1 r_2[(\cos \theta_1 \cos \theta_2 - \sin \theta_1 \sin \theta_2) + i(\sin \theta_1 \cos \theta_2 + \cos \theta_1 \sin \theta_2)] \\ &= r_1 r_2[\cos(\theta_1 + \theta_2) + i \sin(\theta_1 + \theta_2)] \end{aligned}$$ ■

Let's look at an example of how this theorem can be used.

EXAMPLE 3 Finding Products and Quotients of Complex Numbers in Polar Form

If $z = 3(\cos 20° + i \sin 20°)$ and $w = 5(\cos 100° + i \sin 100°)$, find the following (leave your answers in polar form):

(a) zw (b) $\dfrac{z}{w}$

Solution (a)
$$\begin{aligned} zw &= [3(\cos 20° + i \sin 20°)][5(\cos 100° + i \sin 100°)] \\ &= (3 \cdot 5)[\cos(20° + 100°) + i \sin(20° + 100°)] \\ &= 15(\cos 120° + i \sin 120°) \end{aligned}$$

(b) $$\frac{z}{w} = \frac{3(\cos 20^\circ + i \sin 20^\circ)}{5(\cos 100^\circ + i \sin 100^\circ)}$$
$$= \frac{3}{5}[\cos(20^\circ - 100^\circ) + i \sin(20^\circ - 100^\circ)]$$
$$= \frac{3}{5}[\cos(-80^\circ) + i \sin(-80^\circ)]$$
$$= \frac{3}{5}(\cos 280^\circ + i \sin 280^\circ)$$ Argument must lie between 0° and 360°. ◄

NOW WORK PROBLEM 33.

De Moivre's Theorem

4 De Moivre's Theorem, stated by Abraham De Moivre (1667–1754) in 1730, but already known to many people by 1710, is important for the following reason: The fundamental processes of algebra are the four operations of addition, subtraction, multiplication, and division, together with powers and the extraction of roots. De Moivre's Theorem allows these latter fundamental algebraic operations to be applied to complex numbers.

De Moivre's Theorem, in its most basic form, is a formula for raising a complex number z to the power n, where $n \geq 1$ is a positive integer. Let's see if we can guess the form of the result.

Let $z = r(\cos \theta + i \sin \theta)$ be a complex number. Then, based on equation (5), we have

$n = 2$: $z^2 = r^2[\cos(2\theta) + i \sin(2\theta)]$ Equation (5)

$n = 3$: $z^3 = z^2 \cdot z$
$$= \{r^2[\cos(2\theta) + i \sin(2\theta)]\}[r(\cos \theta + i \sin \theta)]$$
$$= r^3[\cos(3\theta) + i \sin(3\theta)]$$ Equation (5)

$n = 4$: $z^4 = z^3 \cdot z$
$$= \{r^3[\cos(3\theta) + i \sin(3\theta)]\}[r(\cos \theta + i \sin \theta)]$$
$$= r^4[\cos(4\theta) + i \sin(4\theta)]$$ Equation (5)

The pattern should now be clear.

Theorem

De Moivre's Theorem

If $z = r(\cos \theta + i \sin \theta)$ is a complex number, then

$$z^n = r^n[\cos(n\theta) + i \sin(n\theta)] \quad \textbf{(7)}$$

where $n \geq 1$ is a positive integer.

We will not prove De Moivre's Theorem because the proof requires mathematical induction (which is not discussed until Section 11.4).

Let's look at some examples.

EXAMPLE 4 **Using De Moivre's Theorem**

Write $[2(\cos 20^\circ + i \sin 20^\circ)]^3$ in the standard form $a + bi$.

Solution

$$\begin{aligned}[2(\cos 20^\circ + i \sin 20^\circ)]^3 &= 2^3[\cos(3 \cdot 20^\circ) + i \sin(3 \cdot 20^\circ)] \\ &= 8(\cos 60^\circ + i \sin 60^\circ) \\ &= 8\left(\frac{1}{2} + \frac{\sqrt{3}}{2} i\right) = 4 + 4\sqrt{3} i\end{aligned}$$

◀

NOW WORK PROBLEM 41.

EXAMPLE 5 **Using De Moivre's Theorem**

Write $(1 + i)^5$ in the standard form $a + bi$.

Solution To apply De Moivre's Theorem, we must first write the complex number in polar form. Since the magnitude of $1 + i$ is $\sqrt{1^2 + 1^2} = \sqrt{2}$, we begin by writing

$$1 + i = \sqrt{2}\left(\frac{1}{\sqrt{2}} + \frac{1}{\sqrt{2}} i\right) = \sqrt{2}\left(\cos \frac{\pi}{4} + i \sin \frac{\pi}{4}\right)$$

Now

$$\begin{aligned}(1 + i)^5 &= \left[\sqrt{2}\left(\cos \frac{\pi}{4} + i \sin \frac{\pi}{4}\right)\right]^5 \\ &= (\sqrt{2})^5\left[\cos\left(5 \cdot \frac{\pi}{4}\right) + i \sin\left(5 \cdot \frac{\pi}{4}\right)\right] \\ &= 4\sqrt{2}\left(\cos \frac{5\pi}{4} + i \sin \frac{5\pi}{4}\right) \\ &= 4\sqrt{2}\left[-\frac{1}{\sqrt{2}} + \left(-\frac{1}{\sqrt{2}}\right) i\right] = -4 - 4i\end{aligned}$$

◀

Complex Roots

5 Let w be a given complex number, and let $n \geq 2$ denote a positive integer. Any complex number z that satisfies the equation

$$z^n = w$$

is called a **complex nth root** of w. In keeping with previous usage, if $n = 2$, the solutions of the equation $z^2 = w$ are called **complex square roots** of w, and if $n = 3$, the solutions of the equation $z^3 = w$ are called **complex cube roots** of w.

Theorem

Finding Complex Roots

Let $w = r(\cos \theta_0 + i \sin \theta_0)$ be a complex number and let $n \geq 2$ be an integer. If $w \neq 0$, there are n distinct complex roots of w, given by the formula

$$z_k = \sqrt[n]{r}\left[\cos\left(\frac{\theta_0}{n} + \frac{2k\pi}{n}\right) + i \sin\left(\frac{\theta_0}{n} + \frac{2k\pi}{n}\right)\right] \quad \textbf{(8)}$$

where $k = 0, 1, 2, \ldots, n - 1$.

Proof (Outline) We will not prove this result in its entirety. Instead, we shall show only that each z_k in equation (8) satisfies the equation $z_k^n = w$, proving that each z_k is a complex nth root of w.

$$z_k^n = \left\{\sqrt[n]{r}\left[\cos\left(\frac{\theta_0}{n} + \frac{2k\pi}{n}\right) + i\sin\left(\frac{\theta_0}{n} + \frac{2k\pi}{n}\right)\right]\right\}^n$$

$$= (\sqrt[n]{r})^n\left\{\cos\left[n\left(\frac{\theta_0}{n} + \frac{2k\pi}{n}\right)\right] + i\sin\left[n\left(\frac{\theta_0}{n} + \frac{2k\pi}{n}\right)\right]\right\}$$ De Moivre's Theorem

$$= r[\cos(\theta_0 + 2k\pi) + i\sin(\theta_0 + 2k\pi)]$$ Simplify

$$= r(\cos\theta_0 + i\sin\theta_0) = w$$ Periodic Property

So, each z_k, $k = 0, 1, \ldots, n - 1$, is a complex nth root of w. To complete the proof, we would need to show that each z_k, $k = 0, 1, \ldots, n - 1$, is, in fact, distinct and that there are no complex nth roots of w other than those given by equation (8). ■

EXAMPLE 6 Finding Complex Cube Roots

Find the complex cube roots of $-1 + \sqrt{3}i$. Leave your answers in polar form, with the argument in degrees.

Solution First, we express $-1 + \sqrt{3}i$ in polar form using degrees.

$$-1 + \sqrt{3}i = 2\left(-\frac{1}{2} + \frac{\sqrt{3}}{2}i\right) = 2(\cos 120° + i\sin 120°)$$

So $r = 2$ and $\theta_0 = 120°$. The three complex cube roots of $-1 + \sqrt{3}i = 2(\cos 120° + i\sin 120°)$ are

$$z_k = \sqrt[3]{2}\left[\cos\left(\frac{120°}{3} + \frac{360°k}{3}\right) + \mathrm{i}\sin\left(\frac{120°}{3} + \frac{360°k}{3}\right)\right], \quad k = 0, 1, 2$$

$$= \sqrt[3]{2}[\cos(40° + 120°k) + i\sin(40° + 120°k)], \quad k = 0, 1, 2$$

So

$$z_0 = \sqrt[3]{2}[\cos(40° + 120°\cdot 0) + i\sin(40° + 120°\cdot 0)] = \sqrt[3]{2}(\cos 40° + i\sin 40°)$$

$$z_1 = \sqrt[3]{2}[\cos(40° + 120°\cdot 1) + i\sin(40° + 120°\cdot 1)] = \sqrt[3]{2}(\cos 160° + i\sin 160°)$$

$$z_2 = \sqrt[3]{2}[\cos(40° + 120°\cdot 2) + i\sin(40° + 120°\cdot 2)] = \sqrt[3]{2}(\cos 280° + i\sin 280°)$$ ◀

Notice that each of the three complex roots of $-1 + \sqrt{3}i$ has the same magnitude, $\sqrt[3]{2}$. This means that the points corresponding to each cube root lie the same distance from the origin; that is, the three points lie on a circle with center at the origin and radius $\sqrt[3]{2}$. Furthermore, the arguments of these cube roots are 40°, 160°, and 280°, the difference of consecutive pairs being $120° = \frac{360°}{3}$. This means that the three points are equally spaced on the circle, as shown in Figure 38. These results are not coincidental. In fact, you are asked to show that these results hold for complex nth roots in Problems 63 through 65.

Figure 38

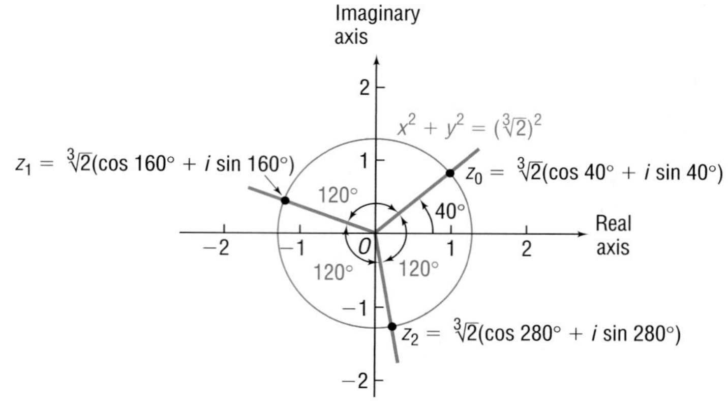

NOW WORK PROBLEM 53.

HISTORICAL FEATURE

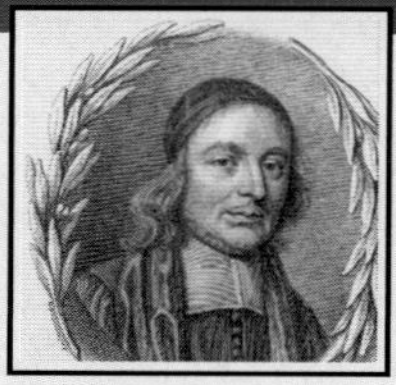

John Wallis

The Babylonians, Greeks, and Arabs considered square roots of negative quantities to be impossible and equations with complex solutions to be unsolvable. The first hint that there was some connection between real solutions of equations and complex numbers came when Girolamo Cardano (1501–1576) and Tartaglia (1499–1557) found *real* roots of cubic equations by taking cube roots of *complex* quantities. For centuries thereafter, mathematicians worked with complex numbers without much belief in their actual existence. In 1673, John Wallis appears to have been the first to suggest the graphical representation of complex numbers, a truly significant idea that was not pursued further until about 1800. Several people, including Karl Friedrich Gauss (1777–1855), then rediscovered the idea, and graphical representation helped to establish complex numbers as equal members of the number family. In practical applications, complex numbers have found their greatest uses in the area of alternating current, where they are a commonplace tool, and in the area of subatomic physics.

Historical Problems

1. The quadratic formula will work perfectly well if the coefficients are complex numbers. Solve the following using De Moivre's Theorem where necessary.

   [**Hint:** The answers are "nice."]

   (a) $z^2 - (2 + 5i)z - 3 + 5i = 0$ (b) $z^2 - (1 + i)z - 2 - i = 0$

8.3 Assess your Understanding

'Are You Prepared?' *Answers are given at the end of these exercises. If you get a wrong answer, read the pages listed in red.*

1. The conjugate of $-4 - 3i$ is ________. (pp. 947–952)
2. The sum formula for sine is $\sin(\alpha + \beta) =$ ________. (pp. 438 and 441)
3. The sum formula for cosine is $\cos(\alpha + \beta) =$ ________. (pp. 438 and 441)
4. $\sin(120°) =$ ________; $\cos(240°) =$ ________. (pp. 342–351)

Concepts and Vocabulary

5. When a complex number z is written in the polar form $z = r(\cos\theta + i\sin\theta)$, the nonnegative number r is the ________ or ________ of z, and the angle θ, $0 \le \theta < 2\pi$, is the ________ of z.

6. ________ Theorem can be used to raise a complex number to a power.

7. A complex number will, in general, have ________ cube roots.

8. *True of False:* De Moivre's Theorem is useful for raising a complex number to a positive integer power.

9. *True of False:* Using De Moivre's Theorem, the square of a complex number will have two answers.

10. *True of False:* The polar form of a complex number is unique.

Exercises

In Problems 11–22, plot each complex number in the complex plane and write it in polar form. Express the argument in degrees.

11. $1 + i$ **12.** $-1 + i$ **13.** $\sqrt{3} - i$ **14.** $1 - \sqrt{3}i$

15. $-3i$ **16.** -2 **17.** $4 - 4i$ **18.** $9\sqrt{3} + 9i$

19. $3 - 4i$ **20.** $2 + \sqrt{3}i$ **21.** $-2 + 3i$ **22.** $\sqrt{5} - i$

In Problems 23–32, write each complex number in rectangular form.

23. $2(\cos 120° + i\sin 120°)$ **24.** $3(\cos 210° + i\sin 210°)$ **25.** $4\left(\cos\frac{7\pi}{4} + i\sin\frac{7\pi}{4}\right)$

26. $2\left(\cos\frac{5\pi}{6} + i\sin\frac{5\pi}{6}\right)$ **27.** $3\left(\cos\frac{3\pi}{2} + i\sin\frac{3\pi}{2}\right)$ **28.** $4\left(\cos\frac{\pi}{2} + i\sin\frac{\pi}{2}\right)$

29. $0.2(\cos 100° + i\sin 100°)$ **30.** $0.4(\cos 200° + i\sin 200°)$ **31.** $2\left(\cos\frac{\pi}{18} + i\sin\frac{\pi}{18}\right)$

32. $3\left(\cos\frac{\pi}{10} + i\sin\frac{\pi}{10}\right)$

In Problems 33–40, find zw and $\frac{z}{w}$. Leave your answers in polar form.

33. $z = 2(\cos 40° + i\sin 40°)$, $w = 4(\cos 20° + i\sin 20°)$

34. $z = \cos 120° + i\sin 120°$, $w = \cos 100° + i\sin 100°$

35. $z = 3(\cos 130° + i\sin 130°)$, $w = 4(\cos 270° + i\sin 270°)$

36. $z = 2(\cos 80° + i\sin 80°)$, $w = 6(\cos 200° + i\sin 200°)$

37. $z = 2\left(\cos\frac{\pi}{8} + i\sin\frac{\pi}{8}\right)$, $w = 2\left(\cos\frac{\pi}{10} + i\sin\frac{\pi}{10}\right)$

38. $z = 4\left(\cos\frac{3\pi}{8} + i\sin\frac{3\pi}{8}\right)$, $w = 2\left(\cos\frac{9\pi}{16} + i\sin\frac{9\pi}{16}\right)$

39. $z = 2 + 2i$, $w = \sqrt{3} - i$

40. $z = 1 - i$, $w = 1 - \sqrt{3}i$

In Problems 41–52, write each expression in the standard form $a + bi$.

41. $[4(\cos 40° + i\sin 40°)]^3$ **42.** $[3(\cos 80° + i\sin 80°)]^3$ **43.** $\left[2\left(\cos\frac{\pi}{10} + i\sin\frac{\pi}{10}\right)\right]^5$

44. $\left[\sqrt{2}\left(\cos\frac{5\pi}{16} + i\sin\frac{5\pi}{16}\right)\right]^4$ **45.** $\left[\sqrt{3}(\cos 10° + i\sin 10°)\right]^6$ **46.** $\left[\frac{1}{2}(\cos 72° + i\sin 72°)\right]^5$

47. $\left[\sqrt{5}\left(\cos\frac{3\pi}{16} + i\sin\frac{3\pi}{16}\right)\right]^4$ **48.** $\left[\sqrt{3}\left(\cos\frac{5\pi}{18} + i\sin\frac{5\pi}{18}\right)\right]^6$ **49.** $(1 - i)^5$

50. $\left(\sqrt{3} - i\right)^6$ **51.** $\left(\sqrt{2} - i\right)^6$ **52.** $\left(1 - \sqrt{5}i\right)^8$

In Problems 53–60, find all the complex roots. Leave your answers in polar form with the argument in degrees.

53. The complex cube roots of $1 + i$

54. The complex fourth roots of $\sqrt{3} - i$

55. The complex fourth roots of $4 - 4\sqrt{3}i$

56. The complex cube roots of $-8 - 8i$

57. The complex fourth roots of $-16i$

58. The complex cube roots of -8

59. The complex fifth roots of i

60. The complex fifth roots of $-i$

61. Find the four complex fourth roots of unity (1) and plot them.

62. Find the six complex sixth roots of unity (1) and plot them.

63. Show that each complex nth root of a nonzero complex number w has the same magnitude.

64. Use the result of Problem 63 to draw the conclusion that each complex nth root lies on a circle with center at the origin. What is the radius of this circle?

65. Refer to Problem 64. Show that the complex nth roots of a nonzero complex number w are equally spaced on the circle.

66. Prove formula (6).

'Are You Prepared?' Answers

1. $-4 + 3i$

2. $\sin\alpha\cos\beta + \cos\alpha\sin\beta$

3. $\cos\alpha\cos\beta - \sin\alpha\sin\beta$

4. $\dfrac{\sqrt{3}}{2}; -\dfrac{1}{2}$

8.4 Vectors

OBJECTIVES
1 Graph Vectors
2 Find a Position Vector
3 Add and Subtract Vectors
4 Find a Scalar Product and the Magnitude of a Vector
5 Find a Unit Vector
6 Find a Vector from Its Direction and Magnitude
7 Work with Objects in Static Equilibrium

In simple terms, a **vector** (derived from the Latin *vehere*, meaning "to carry") is a quantity that has both magnitude and direction. It is customary to represent a vector by using an arrow. The length of the arrow represents the **magnitude** of the vector, and the arrowhead indicates the **direction** of the vector.

Figure 39

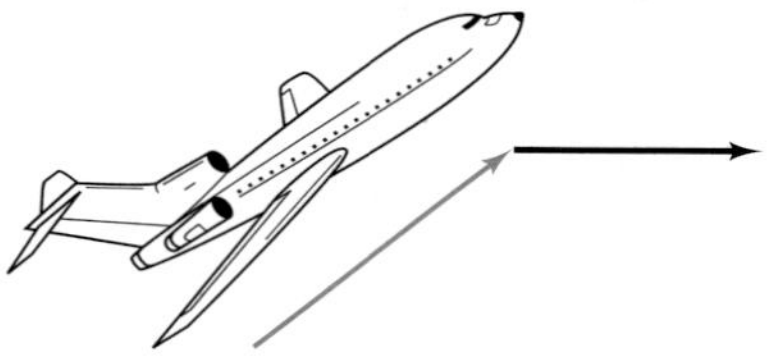

Many quantities in physics can be represented by vectors. For example, the velocity of an aircraft can be represented by an arrow that points in the direction of movement; the length of the arrow represents speed. If the aircraft speeds up, we lengthen the arrow; if the aircraft changes direction, we introduce an arrow in the new direction. See Figure 39. Based on this representation, it is not surprising that vectors and directed line segments are somehow related.

Geometric Vectors

If P and Q are two distinct points in the xy-plane, there is exactly one line containing both P and Q [Figure 40(a)]. The points on that part of the line that joins P to Q, including P and Q, form what is called the **line segment** $\overline{PQ}$ [Figure 40(b)]. If we order the points so that they proceed from P to Q, we have a **directed line segment** from P to Q, or a **geometric vector,** which we denote by $\overrightarrow{PQ}$. In a directed line segment $\overrightarrow{PQ}$, we call P the **initial point** and Q the **terminal point,** as indicated in Figure 40(c).

Figure 40

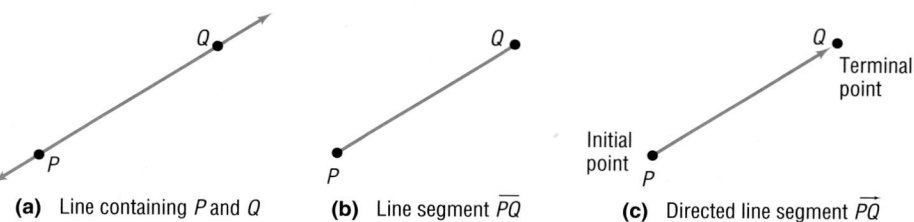

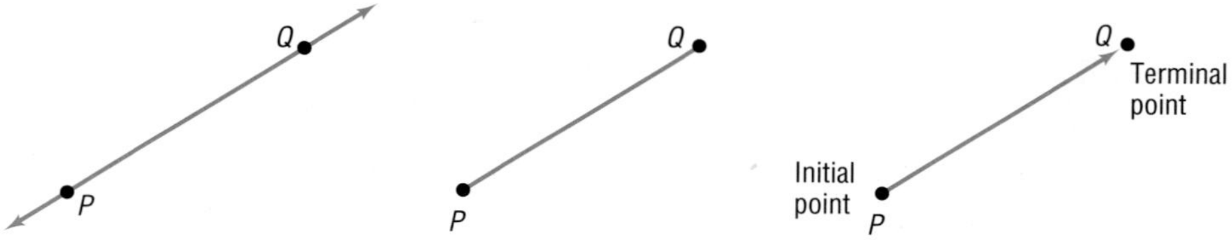

(a) Line containing P and Q (b) Line segment $\overline{PQ}$ (c) Directed line segment $\overrightarrow{PQ}$

The magnitude of the directed line segment $\overrightarrow{PQ}$ is the distance from the point P to the point Q; that is, it is the length of the line segment. The direction of $\overrightarrow{PQ}$ is from P to Q. If a vector $\mathbf{v}$* has the same magnitude and the same direction as the directed line segment $\overrightarrow{PQ}$, we write

$$\mathbf{v} = \overrightarrow{PQ}$$

The vector $\mathbf{v}$ whose magnitude is 0 is called the **zero vector, 0.** The zero vector is assigned no direction.

Two vectors $\mathbf{v}$ and $\mathbf{w}$ are **equal,** written

$$\mathbf{v} = \mathbf{w}$$

if they have the same magnitude and the same direction.

For example, the vectors shown in Figure 41 have the same magnitude and the same direction, so they are equal, even though they have different initial points and different terminal points. As a result, we find it useful to think of a vector simply as an arrow, keeping in mind that two arrows (vectors) are equal if they have the same direction and the same magnitude (length).

Figure 41

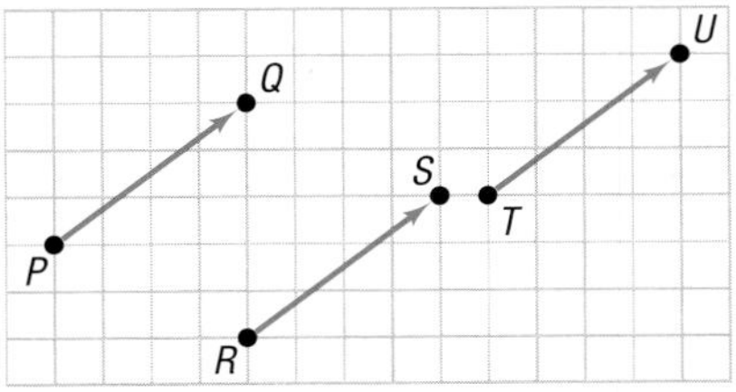

Adding Vectors

The **sum** $\mathbf{v} + \mathbf{w}$ of two vectors is defined as follows: We position the vectors $\mathbf{v}$ and $\mathbf{w}$ so that the terminal point of $\mathbf{v}$ coincides with the initial point of $\mathbf{w}$, as shown in Figure 42. The vector $\mathbf{v} + \mathbf{w}$ is then the unique vector whose initial point coincides with the initial point of $\mathbf{v}$ and whose terminal point coincides with the terminal point of $\mathbf{w}$.

Figure 42

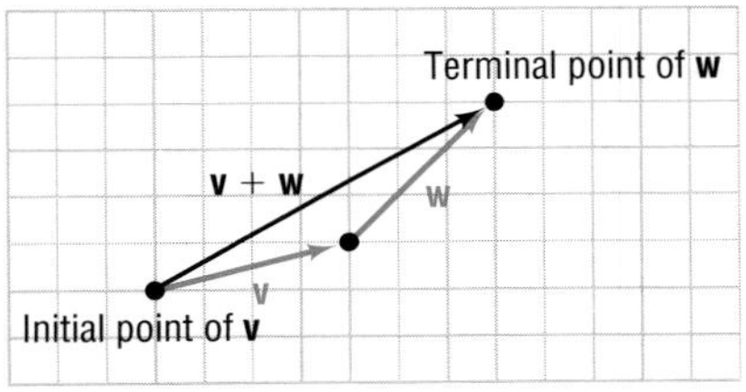

Vector addition is **commutative.** That is, if $\mathbf{v}$ and $\mathbf{w}$ are any two vectors, then

$$\mathbf{v} + \mathbf{w} = \mathbf{w} + \mathbf{v}$$

Figure 43 illustrates this fact. (Observe that the commutative property is another way of saying that opposite sides of a parallelogram are equal and parallel.)

Figure 43

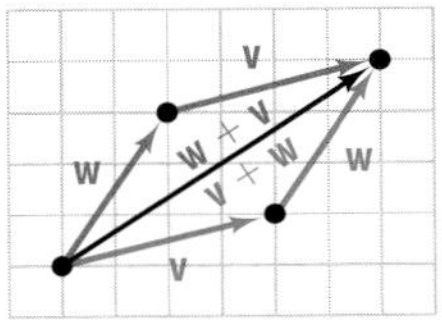

Vector addition is also **associative.** That is, if $\mathbf{u}$, $\mathbf{v}$, and $\mathbf{w}$ are vectors, then

$$\mathbf{u} + (\mathbf{v} + \mathbf{w}) = (\mathbf{u} + \mathbf{v}) + \mathbf{w}$$

Figure 44 illustrates the associative property for vectors.

Figure 44
$(\mathbf{u} + \mathbf{v}) + \mathbf{w} = \mathbf{u} + (\mathbf{v} + \mathbf{w})$

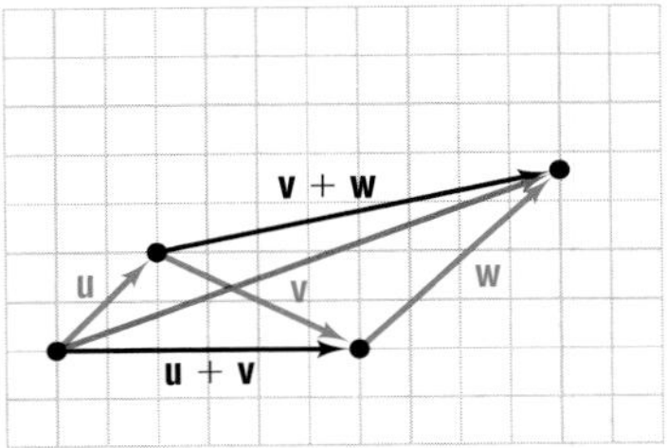

The zero vector has the property that

$$\mathbf{v} + \mathbf{0} = \mathbf{0} + \mathbf{v} = \mathbf{v}$$

for any vector $\mathbf{v}$.

If $\mathbf{v}$ is a vector, then $-\mathbf{v}$ is the vector having the same magnitude as $\mathbf{v}$, but whose direction is opposite to $\mathbf{v}$, as shown in Figure 45.

Figure 45

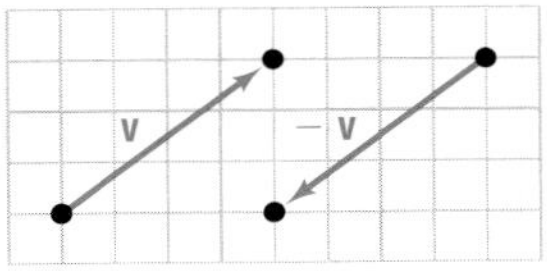

Furthermore,

$$\mathbf{v} + (-\mathbf{v}) = \mathbf{0}$$

If $\mathbf{v}$ and $\mathbf{w}$ are two vectors, we define the **difference** $\mathbf{v} - \mathbf{w}$ as

$$\mathbf{v} - \mathbf{w} = \mathbf{v} + (-\mathbf{w})$$

*Boldface letters will be used to denote vectors, in order to distinguish them from numbers. For handwritten work, an arrow is placed over the letter to signify a vector.

Figure 46

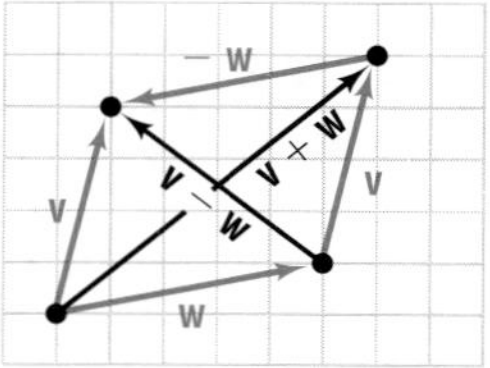

Figure 46 illustrates the relationships among **v**, **w**, **v** + **w**, and **v** − **w**.

Multiplying Vectors by Numbers

When dealing with vectors, we refer to real numbers as **scalars.** Scalars are quantities that have only magnitude. Examples from physics of scalar quantities are temperature, speed, and time. We now define how to multiply a vector by a scalar.

> If α is a scalar and **v** is a vector, the **scalar product** $\alpha\mathbf{v}$ is defined as follows:
>
> 1. If $\alpha > 0$, the product $\alpha\mathbf{v}$ is the vector whose magnitude is α times the magnitude of **v** and whose direction is the same as **v.**
> 2. If $\alpha < 0$, the product $\alpha\mathbf{v}$ is the vector whose magnitude is $|\alpha|$ times the magnitude of **v** and whose direction is opposite that of **v.**
> 3. If $\alpha = 0$ or if $\mathbf{v} = \mathbf{0}$, then $\alpha\mathbf{v} = \mathbf{0}$.

Figure 47

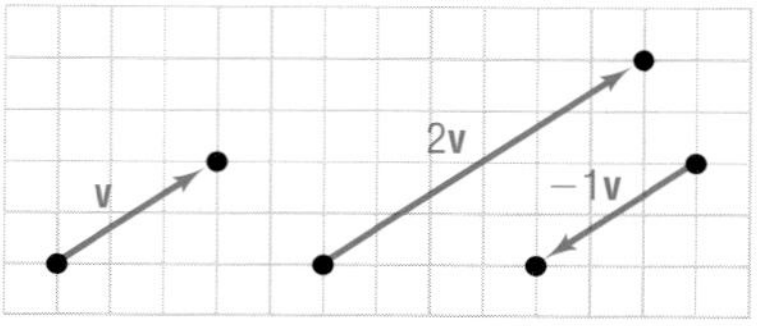

See Figure 47 for some illustrations.

For example, if **a** is the acceleration of an object of mass m due to a force **F** being exerted on it, then, by Newton's second law of motion, $\mathbf{F} = m\mathbf{a}$. Here, $m\mathbf{a}$ is the product of the scalar m and the vector **a**.

Scalar products have the following properties:

$$0\mathbf{v} = \mathbf{0} \qquad 1\mathbf{v} = \mathbf{v} \qquad -1\mathbf{v} = -\mathbf{v}$$
$$(\alpha + \beta)\mathbf{v} = \alpha\mathbf{v} + \beta\mathbf{v} \qquad \alpha(\mathbf{v} + \mathbf{w}) = \alpha\mathbf{v} + \alpha\mathbf{w}$$
$$\alpha(\beta\mathbf{v}) = (\alpha\beta)\mathbf{v}$$

1 **EXAMPLE 1** **Graphing Vectors**

Figure 48

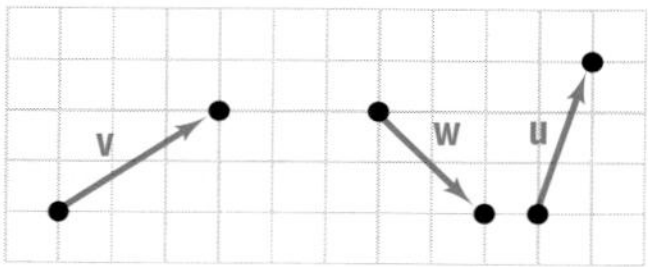

Use the vectors illustrated in Figure 48 to graph each of the following vectors:

(a) **v** − **w** (b) 2**v** + 3**w** (c) 2**v** − **w** + **u**

Solution Figure 49 illustrates each graph.

Figure 49

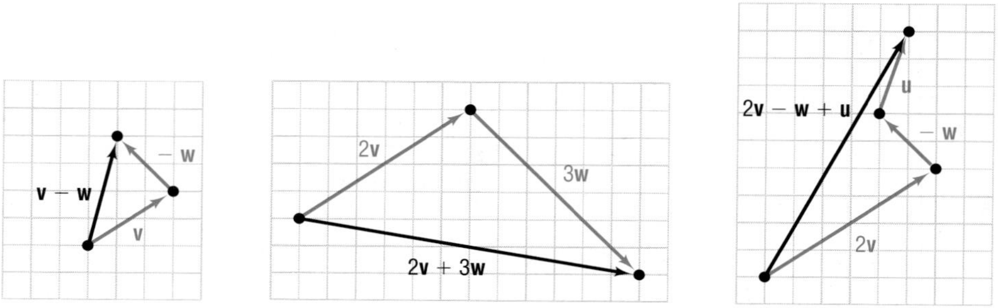

(a) **v** − **w** (b) 2**v** + 3**w** (c) 2**v** − **w** + **u**

NOW WORK PROBLEMS 7 AND 9.

Magnitudes of Vectors

If **v** is a vector, we use the symbol $\|\mathbf{v}\|$ to represent the **magnitude** of **v.** Since $\|\mathbf{v}\|$ equals the length of a directed line segment, it follows that $\|\mathbf{v}\|$ has the following properties:

Theorem

Properties of $\|\mathbf{v}\|$

If **v** is a vector and if α is a scalar, then

(a) $\|\mathbf{v}\| \geq 0$ (b) $\|\mathbf{v}\| = 0$ if and only if $\mathbf{v} = \mathbf{0}$

(c) $\|-\mathbf{v}\| = \|\mathbf{v}\|$ (d) $\|\alpha\mathbf{v}\| = |\alpha|\|\mathbf{v}\|$

Property (a) is a consequence of the fact that distance is a nonnegative number. Property (b) follows, because the length of the directed line segment $\overrightarrow{PQ}$ is positive unless P and Q are the same point, in which case the length is 0. Property (c) follows because the length of the line segment $\overline{PQ}$ equals the length of the line segment $\overline{QP}$. Property (d) is a direct consequence of the definition of a scalar product.

A vector **u** for which $\|\mathbf{u}\| = 1$ is called a **unit vector.**

To compute the magnitude and direction of a vector, we need an algebraic way of representing vectors.

Algebraic Vectors

2 An **algebraic vector v** is represented as

$$\mathbf{v} = \langle a, b \rangle$$

where a and b are real numbers (scalars) called the **components** of the vector **v.**

We use a rectangular coordinate system to represent algebraic vectors in the plane. If $\mathbf{v} = \langle a, b \rangle$ is an algebraic vector whose initial point is at the origin, then **v** is called a **position vector.** See Figure 50. Notice that the terminal point of the position vector $\mathbf{v} = \langle a, b \rangle$ is $P = (a, b)$.

Figure 50

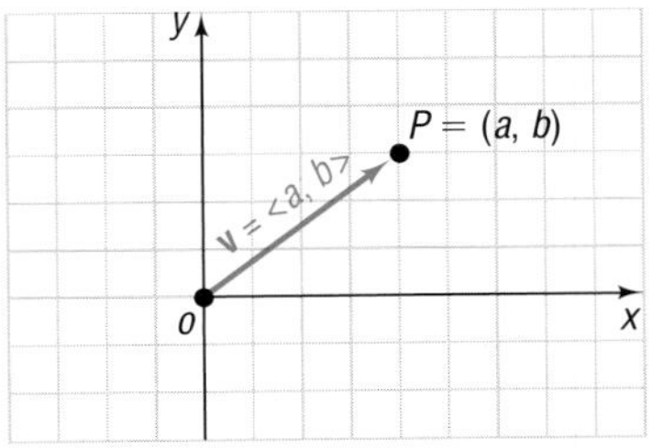

The next result states that any vector whose initial point is not at the origin is equal to a unique position vector.

Theorem

Suppose that **v** is a vector with initial point $P_1 = (x_1, y_1)$, not necessarily the origin, and terminal point $P_2 = (x_2, y_2)$. If $\mathbf{v} = \overrightarrow{P_1P_2}$, then **v** is equal to the position vector

$$\mathbf{v} = \langle x_2 - x_1, y_2 - y_1 \rangle \quad \textbf{(1)}$$

To see why this is true, look at Figure 51 on page 574. Triangle OPA and triangle P_1P_2Q are congruent. (Do you see why? The line segments have the same magnitude, so $d(O, P) = d(P_1, P_2)$; and they have the same direction, so $\angle POA = \angle P_2P_1Q$. Since the triangles are right triangles, we have

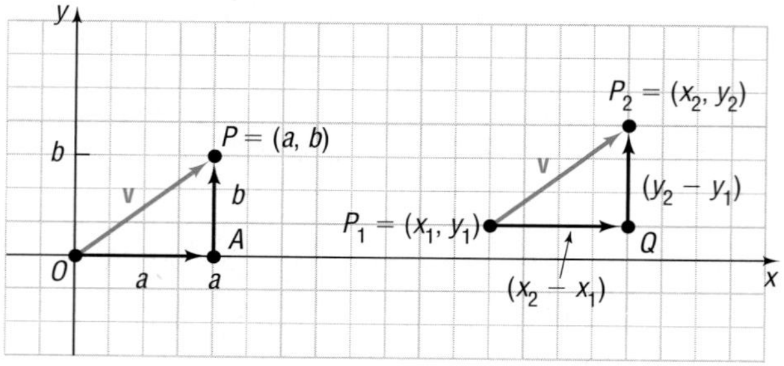

Figure 51
$\langle a, b \rangle = \langle x_2 - x_1, y_2 - y_1 \rangle$

angle–side–angle.) It follows that corresponding sides are equal. As a result, $x_2 - x_1 = a$ and $y_2 - y_1 = b$, so **v** may be written as

$$\mathbf{v} = \langle a, b \rangle = \langle x_2 - x_1, y_2 - y_1 \rangle$$

Because of this result, we can replace any algebraic vector by a unique position vector, and vice versa. This flexibility is one of the main reasons for the wide use of vectors.

EXAMPLE 2 Finding a Position Vector

Find the position vector of the vector $\mathbf{v} = \overrightarrow{P_1P_2}$ if $P_1 = (-1, 2)$ and $P_2 = (4, 6)$.

Solution By equation (1), the position vector equal to **v** is

$$\mathbf{v} = \langle 4 - (-1), 6 - 2 \rangle = \langle 5, 4 \rangle$$

See Figure 52.

Figure 52

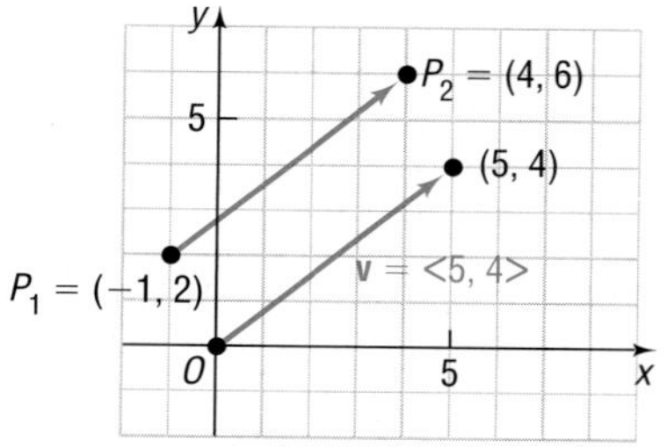

◀

Two position vectors **v** and **w** are equal if and only if the terminal point of **v** is the same as the terminal point of **w**. This leads to the following result:

Theorem

Equality of Vectors

Two vectors **v** and **w** are equal if and only if their corresponding components are equal. That is,

$$\text{If } \mathbf{v} = \langle a_1, b_1 \rangle \quad \text{and} \quad \mathbf{w} = \langle a_2, b_2 \rangle$$
$$\text{then} \quad \mathbf{v} = \mathbf{w} \quad \text{if and only if} \quad a_1 = a_2 \quad \text{and} \quad b_1 = b_2.$$

Figure 53

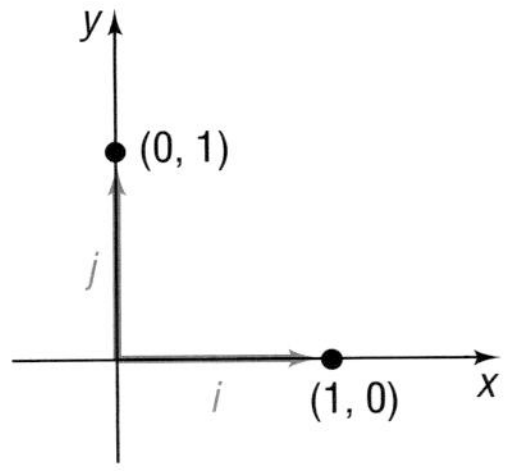

We now present an alternative representation of a vector in the plane that is common in the physical sciences. Let **i** denote the unit vector whose direction is along the positive x-axis; let **j** denote the unit vector whose direction is along the positive y-axis. Then $\mathbf{i} = \langle 1, 0 \rangle$ and $\mathbf{j} = \langle 0, 1 \rangle$, as shown in Figure 53. Any vector $\mathbf{v} = \langle a, b \rangle$ can be written using the unit vectors **i** and **j** as follows:

$$\mathbf{v} = \langle a, b \rangle = a\langle 1, 0 \rangle + b\langle 0, 1 \rangle = a\mathbf{i} + b\mathbf{j}$$

We call a and b the **horizontal** and **vertical components** of **v**, respectively.

NOW WORK PROBLEM 27.

We define addition, subtraction, scalar product, and magnitude in terms of the components of a vector.

Let $\mathbf{v} = a_1\mathbf{i} + b_1\mathbf{j} = \langle a_1, b_1 \rangle$ and $\mathbf{w} = a_2\mathbf{i} + b_2\mathbf{j} = \langle a_2, b_2 \rangle$ be two vectors, and let α be a scalar. Then

$$\mathbf{v} + \mathbf{w} = (a_1 + a_2)\mathbf{i} + (b_1 + b_2)\mathbf{j} = \langle a_1 + a_2, b_1 + b_2 \rangle \quad (2)$$

$$\mathbf{v} - \mathbf{w} = (a_1 - a_2)\mathbf{i} + (b_1 - b_2)\mathbf{j} = \langle a_1 - a_2, b_1 - b_2 \rangle \quad (3)$$

$$\alpha\mathbf{v} = (\alpha a_1)\mathbf{i} + (\alpha b_1)\mathbf{j} = \langle \alpha a_1, \alpha b_1 \rangle \quad (4)$$

$$\|\mathbf{v}\| = \sqrt{a_1^2 + b_1^2} \quad (5)$$

These definitions are compatible with the geometric definitions given earlier in this section. See Figure 54.

Figure 54

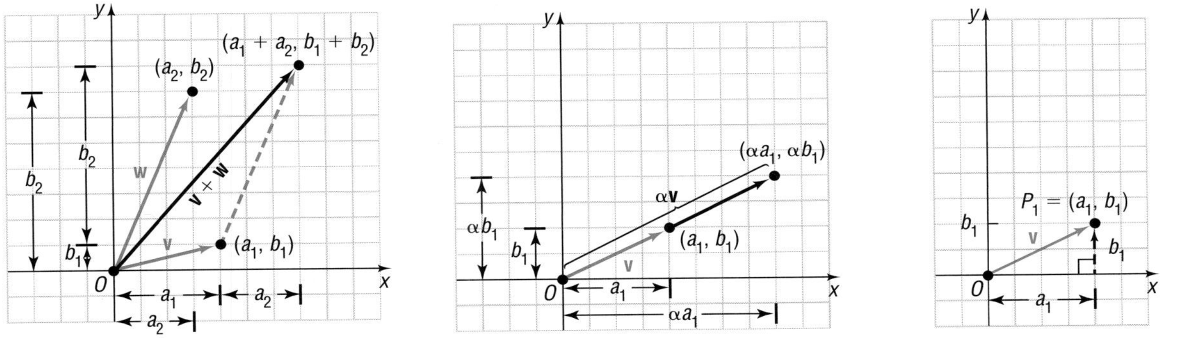

(a) Illustration of property (2)

(b) Illustration of property (4), $\alpha > 0$

(c) Illustration of property (5):
$\|\mathbf{v}\|$ = Distance from O to P_1
$\|\mathbf{v}\| = \sqrt{a_1^2 + b_1^2}$

To add two vectors, add corresponding components. To subtract two vectors, subtract corresponding components.

3 **EXAMPLE 3** **Adding and Subtracting Vectors**

If $\mathbf{v} = 2\mathbf{i} + 3\mathbf{j} = \langle 2, 3 \rangle$ and $\mathbf{w} = 3\mathbf{i} - 4\mathbf{j} = \langle 3, -4 \rangle$, find:

(a) $\mathbf{v} + \mathbf{w}$ (b) $\mathbf{v} - \mathbf{w}$

Solution (a) $\mathbf{v} + \mathbf{w} = (2\mathbf{i} + 3\mathbf{j}) + (3\mathbf{i} - 4\mathbf{j}) = (2 + 3)\mathbf{i} + (3 - 4)\mathbf{j} = 5\mathbf{i} - \mathbf{j}$

or

$\mathbf{v} + \mathbf{w} = \langle 2, 3 \rangle + \langle 3, -4 \rangle = \langle 2 + 3, 3 + (-4) \rangle = \langle 5, -1 \rangle$

(b) $\mathbf{v} - \mathbf{w} = (2\mathbf{i} + 3\mathbf{j}) - (3\mathbf{i} - 4\mathbf{j}) = (2 - 3)\mathbf{i} + [3 - (-4)]\mathbf{j} = -\mathbf{i} + 7\mathbf{j}$

or

$\mathbf{v} - \mathbf{w} = \langle 2, 3 \rangle - \langle 3, -4 \rangle = \langle 2 - 3, 3 - (-4) \rangle = \langle -1, 7 \rangle$ ◀

4 **EXAMPLE 4** **Finding Scalar Products and Magnitudes**

If $\mathbf{v} = 2\mathbf{i} + 3\mathbf{j} = \langle 2, 3 \rangle$ and $\mathbf{w} = 3\mathbf{i} - 4\mathbf{j} = \langle 3, -4 \rangle$, find:

(a) $3\mathbf{v}$ (b) $2\mathbf{v} - 3\mathbf{w}$ (c) $\|\mathbf{v}\|$

Solution (a) $3\mathbf{v} = 3(2\mathbf{i} + 3\mathbf{j}) = 6\mathbf{i} + 9\mathbf{j}$

or

$3\mathbf{v} = 3\langle 2, 3 \rangle = \langle 6, 9 \rangle$

(b) $2\mathbf{v} - 3\mathbf{w} = 2(2\mathbf{i} + 3\mathbf{j}) - 3(3\mathbf{i} - 4\mathbf{j}) = 4\mathbf{i} + 6\mathbf{j} - 9\mathbf{i} + 12\mathbf{j}$
$= -5\mathbf{i} + 18\mathbf{j}$

or

$2\mathbf{v} - 3\mathbf{w} = 2\langle 2, 3 \rangle - 3\langle 3, -4 \rangle = \langle 4, 6 \rangle - \langle 9, -12 \rangle$
$= \langle 4 - 9, 6 - (-12) \rangle = \langle -5, 18 \rangle$

(c) $\|\mathbf{v}\| = \|2\mathbf{i} + 3\mathbf{j}\| = \sqrt{2^2 + 3^2} = \sqrt{13}$ ◀

NOW WORK PROBLEMS 33 AND 39.

For the remainder of the section, we will express a vector $\mathbf{v}$ in the form $a\mathbf{i} + b\mathbf{j}$.

5 Recall that a unit vector $\mathbf{u}$ is a vector for which $\|\mathbf{u}\| = 1$. In many applications, it is useful to be able to find a unit vector $\mathbf{u}$ that has the same direction as a given vector $\mathbf{v}$.

Theorem

Unit Vector in the Direction of v

For any nonzero vector $\mathbf{v}$, the vector

$$\mathbf{u} = \frac{\mathbf{v}}{\|\mathbf{v}\|}$$

is a unit vector that has the same direction as $\mathbf{v}$.

Proof Let $\mathbf{v} = a\mathbf{i} + b\mathbf{j}$. Then $\|\mathbf{v}\| = \sqrt{a^2 + b^2}$ and

$$\mathbf{u} = \frac{\mathbf{v}}{\|\mathbf{v}\|} = \frac{a\mathbf{i} + b\mathbf{j}}{\sqrt{a^2 + b^2}} = \frac{a}{\sqrt{a^2 + b^2}}\mathbf{i} + \frac{b}{\sqrt{a^2 + b^2}}\mathbf{j}$$

The vector $\mathbf{u}$ is in the same direction as $\mathbf{v}$, since $\|\mathbf{v}\| > 0$. Furthermore,

$$\|\mathbf{u}\| = \sqrt{\frac{a^2}{a^2 + b^2} + \frac{b^2}{a^2 + b^2}} = \sqrt{\frac{a^2 + b^2}{a^2 + b^2}} = 1$$

Thus, $\mathbf{u}$ is a unit vector in the direction of $\mathbf{v}$. ■

As a consequence of this theorem, if **u** is a unit vector in the same direction as a vector **v**, then **v** may be expressed as

$$\mathbf{v} = \|\mathbf{v}\|\mathbf{u} \qquad (6)$$

This way of expressing a vector is useful in many applications.

EXAMPLE 5 **Finding a Unit Vector**

Find a unit vector in the same direction as $\mathbf{v} = 4\mathbf{i} - 3\mathbf{j}$.

Solution We find $\|\mathbf{v}\|$ first.

$$\|\mathbf{v}\| = \|4\mathbf{i} - 3\mathbf{j}\| = \sqrt{16 + 9} = 5$$

Now we multiply **v** by the scalar $\frac{1}{\|\mathbf{v}\|} = \frac{1}{5}$. A unit vector in the same direction as **v** is

$$\frac{\mathbf{v}}{\|\mathbf{v}\|} = \frac{4\mathbf{i} - 3\mathbf{j}}{5} = \frac{4}{5}\mathbf{i} - \frac{3}{5}\mathbf{j}$$

CHECK: This vector is, in fact, a unit vector because

$$\left(\frac{4}{5}\right)^2 + \left(-\frac{3}{5}\right)^2 = \frac{16}{25} + \frac{9}{25} = \frac{25}{25} = 1$$

■

NOW WORK PROBLEM 49.

Writing a Vector in Terms of Its Magnitude and Direction

6 If a vector represents the speed and direction of an object, it is called a **velocity vector.** If a vector represents the direction and amount of a force acting on an object, it is called a **force vector.** In many applications, a vector is described in terms of its magnitude and direction, rather than in terms of its components. For example, a ball thrown with an initial speed of 25 miles per hour at an angle 30° to the horizontal is a velocity vector.

Suppose that we are given the magnitude $\|\mathbf{v}\|$ of a nonzero vector **v** and the angle α, $0° \le \alpha < 360°$, between **v** and **i.** To express **v** in terms of $\|\mathbf{v}\|$ and α, we first find the unit vector **u** having the same direction as **v.**

$$\mathbf{u} = \frac{\mathbf{v}}{\|\mathbf{v}\|} \quad \text{or} \quad \mathbf{v} = \|\mathbf{v}\|\mathbf{u} \qquad (7)$$

Figure 55

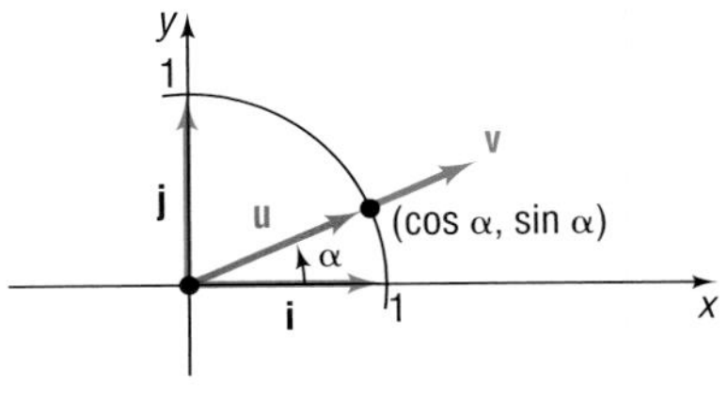

Look at Figure 55. The coordinates of the terminal point of **u** are $(\cos\alpha, \sin\alpha)$. Then $\mathbf{u} = \cos\alpha\mathbf{i} + \sin\alpha\mathbf{j}$ and, from (7),

$$\mathbf{v} = \|\mathbf{v}\|(\cos\alpha\mathbf{i} + \sin\alpha\mathbf{j}) \qquad (8)$$

where α is the angle between **v** and **i.**

EXAMPLE 6 Writing a Vector When Its Magnitude and Direction Are Given

A ball is thrown with an initial speed of 25 miles per hour in a direction that makes an angle of 30° with the positive x-axis. Express the velocity vector $\mathbf{v}$ in terms of $\mathbf{i}$ and $\mathbf{j}$. What is the initial speed in the horizontal direction? What is the initial speed in the vertical direction?

Solution The magnitude of $\mathbf{v}$ is $\|\mathbf{v}\| = 25$ miles per hour, and the angle between the direction of $\mathbf{v}$ and $\mathbf{i}$, the positive x-axis, is $\alpha = 30°$. By equation (8),

$$\mathbf{v} = \|\mathbf{v}\|(\cos\alpha\mathbf{i} + \sin\alpha\mathbf{j}) = 25(\cos 30°\mathbf{i} + \sin 30°\mathbf{j}) = 25\left(\frac{\sqrt{3}}{2}\mathbf{i} + \frac{1}{2}\mathbf{j}\right) = \frac{25\sqrt{3}}{2}\mathbf{i} + \frac{25}{2}\mathbf{j}$$

The initial speed of the ball in the horizontal direction is the horizontal component of $\mathbf{v}$, $\frac{25\sqrt{3}}{2} \approx 21.65$ miles per hour. The initial speed in the vertical direction is the vertical component of $\mathbf{v}$, $\frac{25}{2} = 12.5$ miles per hour. ◀

NOW WORK PROBLEM 61.

Application: Static Equilibrium

Figure 56

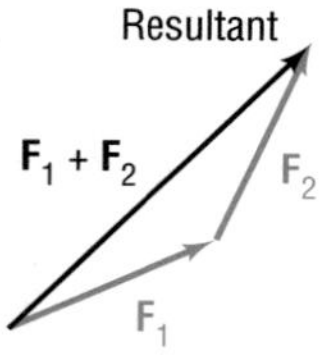

7 Because forces can be represented by vectors, two forces "combine" the way that vectors "add." If $\mathbf{F}_1$ and $\mathbf{F}_2$ are two forces simultaneously acting on an object, the vector sum $\mathbf{F}_1 + \mathbf{F}_2$ is the **resultant force.** The resultant force produces the same effect on the object as that obtained when the two forces $\mathbf{F}_1$ and $\mathbf{F}_2$ act on the object. See Figure 56. An application of this concept is *static equilibrium.* An object is said to be in **static equilibrium** if (1) the object is at rest and (2) the sum of all forces acting on the object is zero, that is, if the resultant force is 0.

EXAMPLE 7 An Object in Static Equilibrium

Figure 57

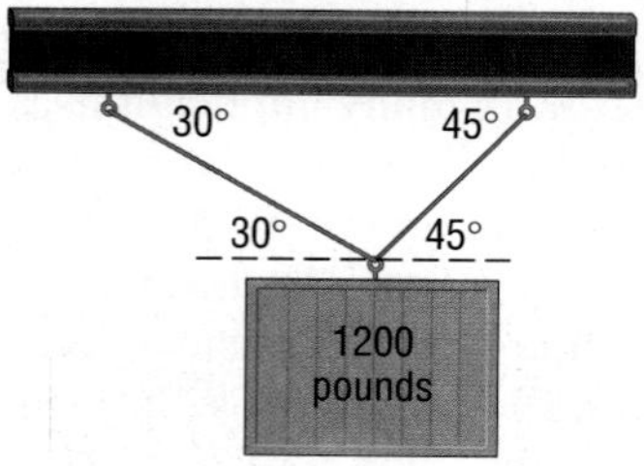

A box of supplies that weighs 1200 pounds is suspended by two cables attached to the ceiling, as shown in Figure 57. What is the tension in the two cables?

Solution We draw a force diagram using vectors shown in Figure 58. The tensions in the cables are the magnitudes $\|\mathbf{F}_1\|$ and $\|\mathbf{F}_2\|$ of the force vectors $\mathbf{F}_1$ and $\mathbf{F}_2$. The magnitude of the force vector $\mathbf{F}_3$ equals 1200 pounds, the weight of the box. Now write each force vector in terms of the unit vectors $\mathbf{i}$ and $\mathbf{j}$. For $\mathbf{F}_1$ and $\mathbf{F}_2$, we use equation (8). Remember that α is the angle between the vector and the positive x-axis.

Figure 58

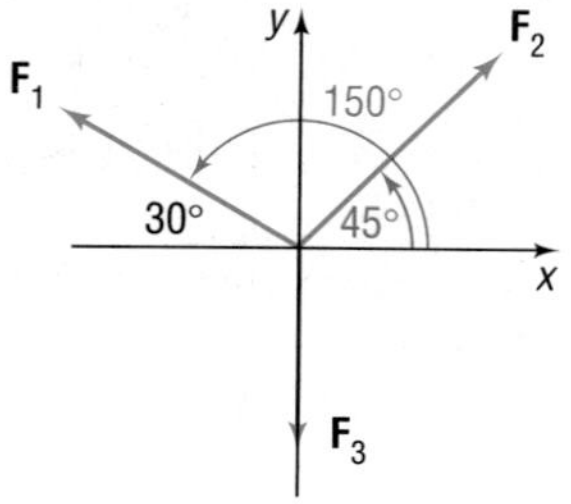

$$\mathbf{F}_1 = \|\mathbf{F}_1\|(\cos 150°\mathbf{i} + \sin 150°\mathbf{j}) = \|\mathbf{F}_1\|\left(-\frac{\sqrt{3}}{2}\mathbf{i} + \frac{1}{2}\mathbf{j}\right) = -\frac{\sqrt{3}}{2}\|\mathbf{F}_1\|\mathbf{i} + \frac{1}{2}\|\mathbf{F}_1\|\mathbf{j}$$

$$\mathbf{F}_2 = \|\mathbf{F}_2\|(\cos 45°\mathbf{i} + \sin 45°\mathbf{j}) = \|\mathbf{F}_2\|\left(\frac{\sqrt{2}}{2}\mathbf{i} + \frac{\sqrt{2}}{2}\mathbf{j}\right) = \frac{\sqrt{2}}{2}\|\mathbf{F}_2\|\mathbf{i} + \frac{\sqrt{2}}{2}\|\mathbf{F}_2\|\mathbf{j}$$

$$\mathbf{F}_3 = -1200\mathbf{j}$$

For static equilibrium, the sum of the force vectors must equal zero.

$$\mathbf{F}_1 + \mathbf{F}_2 + \mathbf{F}_3 = -\frac{\sqrt{3}}{2}\|\mathbf{F}_1\|\mathbf{i} + \frac{1}{2}\|\mathbf{F}_1\|\mathbf{j} + \frac{\sqrt{2}}{2}\|\mathbf{F}_2\|\mathbf{i} + \frac{\sqrt{2}}{2}\|\mathbf{F}_2\|\mathbf{j} - 1200\mathbf{j} = \mathbf{0}$$

The **i** component and **j** component will each equal zero. This results in the two equations

$$-\frac{\sqrt{3}}{2}\|\mathbf{F}_1\| + \frac{\sqrt{2}}{2}\|\mathbf{F}_2\| = 0 \qquad \textbf{(9)}$$

$$\frac{1}{2}\|\mathbf{F}_1\| + \frac{\sqrt{2}}{2}\|\mathbf{F}_2\| - 1200 = 0 \qquad \textbf{(10)}$$

We solve equation (9) for $\|\mathbf{F}_2\|$ and obtain

$$\|\mathbf{F}_2\| = \frac{\sqrt{3}}{\sqrt{2}}\|\mathbf{F}_1\| \qquad \textbf{(11)}$$

Substituting into equation (10) and solving for $\|\mathbf{F}_1\|$, we obtain

$$\frac{1}{2}\|\mathbf{F}_1\| + \frac{\sqrt{2}}{2}\left(\frac{\sqrt{3}}{\sqrt{2}}\|\mathbf{F}_1\|\right) - 1200 = 0$$

$$\frac{1}{2}\|\mathbf{F}_1\| + \frac{\sqrt{3}}{2}\|\mathbf{F}_1\| - 1200 = 0$$

$$\frac{1 + \sqrt{3}}{2}\|\mathbf{F}_1\| = 1200$$

$$\|\mathbf{F}_1\| = \frac{2400}{1 + \sqrt{3}} \approx 878.5 \text{ pounds}$$

Substituting this value into equation (11) yields $\|\mathbf{F}_2\|$.

$$\|\mathbf{F}_2\| = \frac{\sqrt{3}}{\sqrt{2}}\|\mathbf{F}_1\| = \frac{\sqrt{3}}{\sqrt{2}} \cdot \frac{2400}{1 + \sqrt{3}} \approx 1075.9 \text{ pounds}$$

The left cable has tension of approximately 878.5 pounds and the right cable has tension of approximately 1075.9 pounds. ◄

HISTORICAL FEATURE

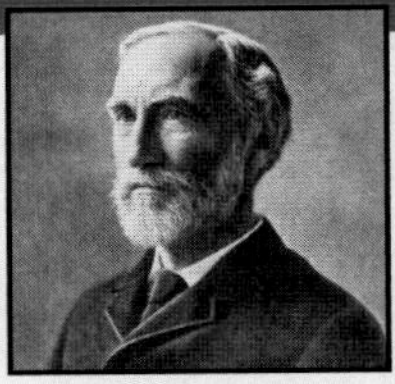

Josiah Gibbs (1839–1903)

The history of vectors is surprisingly complicated for such a natural concept. In the *xy*-plane, complex numbers do a good job of imitating vectors. About 1840, mathematicians became interested in finding a system that would do for three dimensions what the complex numbers do for two dimensions. Hermann Grassmann (1809–1877), in Germany, and William Rowan Hamilton (1805–1865), in Ireland, both attempted to find solutions.

Hamilton's system was the *quaternions*, which are best thought of as a real number plus a vector, and do for four dimensions what complex numbers do for two dimensions. In this system the order of multiplication matters; that is, **ab** ≠ **ba**. Also, two products of vectors emerged, the scalar (or dot) product and the vector (or cross) product.

Grassmann's abstract style, although easily read today, was almost impenetrable during the previous century, and only a few of his ideas were appreciated. Among those few were the same scalar and vector products that Hamilton had found.

About 1880, the American physicist Josiah Willard Gibbs (1839–1903) worked out an algebra involving only the simplest concepts: the vectors and the two products. He then added some calculus, and the resulting system was simple, flexible, and well adapted to expressing a large number of physical laws. This system remains in use essentially unchanged. Hamilton's and Grassmann's more extensive systems each gave birth to much interesting mathematics, but little of this mathematics is seen at elementary levels.

8.4 Assess Your Understanding

Concepts and Vocabulary

1. A vector whose magnitude is 1 is called a(n) _________ vector.
2. The product of a vector by a number is called a(n) _________ product.
3. If $\mathbf{v} = a\mathbf{i} + b\mathbf{j}$, then a is called the _________ component of $\mathbf{v}$ and b is the _________ component of $\mathbf{v}$.
4. *True or False:* Vectors are quantities that have magnitude and direction.
5. *True or False:* Force is a physical example of a vector.
6. *True or False:* Mass is a physical example of a vector.

Exercises

In Problems 7–14, use the vectors in the figure at the right to graph each of the following vectors.

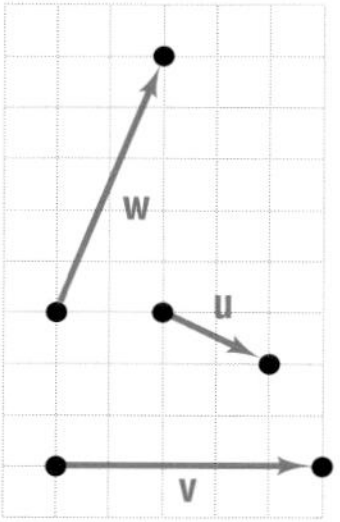

7. $\mathbf{v} + \mathbf{w}$
8. $\mathbf{u} + \mathbf{v}$
9. $3\mathbf{v}$
10. $4\mathbf{w}$
11. $\mathbf{v} - \mathbf{w}$
12. $\mathbf{u} - \mathbf{v}$
13. $3\mathbf{v} + \mathbf{u} - 2\mathbf{w}$
14. $2\mathbf{u} - 3\mathbf{v} + \mathbf{w}$

In Problems 15–22, use the figure at the right. Determine whether the given statement is true or false.

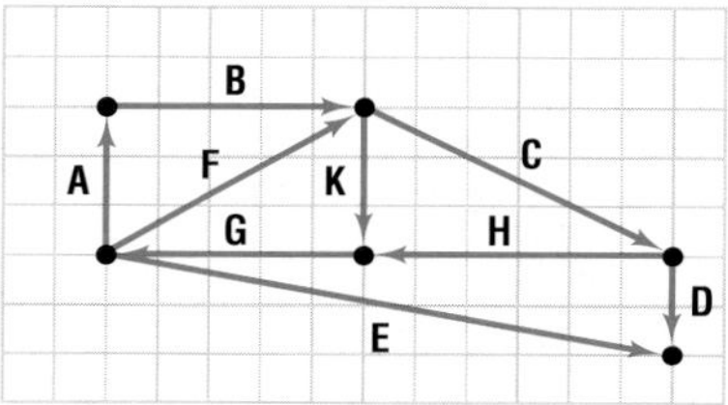

15. $\mathbf{A} + \mathbf{B} = \mathbf{F}$
16. $\mathbf{K} + \mathbf{G} = \mathbf{F}$
17. $\mathbf{C} = \mathbf{D} - \mathbf{E} + \mathbf{F}$
18. $\mathbf{G} + \mathbf{H} + \mathbf{E} = \mathbf{D}$
19. $\mathbf{E} + \mathbf{D} = \mathbf{G} + \mathbf{H}$
20. $\mathbf{H} - \mathbf{C} = \mathbf{G} - \mathbf{F}$
21. $\mathbf{A} + \mathbf{B} + \mathbf{K} + \mathbf{G} = \mathbf{0}$
22. $\mathbf{A} + \mathbf{B} + \mathbf{C} + \mathbf{H} + \mathbf{G} = \mathbf{0}$

23. If $\|\mathbf{v}\| = 4$, what is $\|3\mathbf{v}\|$?
24. If $\|\mathbf{v}\| = 2$, what is $\|-4\mathbf{v}\|$?

In Problems 25–32, the vector **v** *has initial point P and terminal point Q. Write* **v** *in the form* $a\mathbf{i} + b\mathbf{j}$; *that is, find its position vector.*

25. $P = (0, 0)$; $Q = (3, 4)$
26. $P = (0, 0)$; $Q = (-3, -5)$
27. $P = (3, 2)$; $Q = (5, 6)$
28. $P = (-3, 2)$; $Q = (6, 5)$
29. $P = (-2, -1)$; $Q = (6, -2)$
30. $P = (-1, 4)$; $Q = (6, 2)$
31. $P = (1, 0)$; $Q = (0, 1)$
32. $P = (1, 1)$; $Q = (2, 2)$

In Problems 33–38, find $\|\mathbf{v}\|$.

33. $\mathbf{v} = 3\mathbf{i} - 4\mathbf{j}$
34. $\mathbf{v} = -5\mathbf{i} + 12\mathbf{j}$
35. $\mathbf{v} = \mathbf{i} - \mathbf{j}$
36. $\mathbf{v} = -\mathbf{i} - \mathbf{j}$
37. $\mathbf{v} = -2\mathbf{i} + 3\mathbf{j}$
38. $\mathbf{v} = 6\mathbf{i} + 2\mathbf{j}$

In Problems 39–44, find each quantity if $\mathbf{v} = 3\mathbf{i} - 5\mathbf{j}$ *and* $\mathbf{w} = -2\mathbf{i} + 3\mathbf{j}$.

39. $2\mathbf{v} + 3\mathbf{w}$
40. $3\mathbf{v} - 2\mathbf{w}$
41. $\|\mathbf{v} - \mathbf{w}\|$
42. $\|\mathbf{v} + \mathbf{w}\|$
43. $\|\mathbf{v}\| - \|\mathbf{w}\|$
44. $\|\mathbf{v}\| + \|\mathbf{w}\|$

In Problems 45–50, find the unit vector having the same direction as **v**.

45. $\mathbf{v} = 5\mathbf{i}$
46. $\mathbf{v} = -3\mathbf{j}$
47. $\mathbf{v} = 3\mathbf{i} - 4\mathbf{j}$
48. $\mathbf{v} = -5\mathbf{i} + 12\mathbf{j}$
49. $\mathbf{v} = \mathbf{i} - \mathbf{j}$
50. $\mathbf{v} = 2\mathbf{i} - \mathbf{j}$

51. Find a vector $\mathbf{v}$ whose magnitude is 4 and whose component in the $\mathbf{i}$ direction is twice the component in the $\mathbf{j}$ direction.

52. Find a vector $\mathbf{v}$ whose magnitude is 3 and whose component in the $\mathbf{i}$ direction is equal to the component in the $\mathbf{j}$ direction.

53. If $\mathbf{v} = 2\mathbf{i} - \mathbf{j}$ and $\mathbf{w} = x\mathbf{i} + 3\mathbf{j}$, find all numbers x for which $\|\mathbf{v} + \mathbf{w}\| = 5$.

54. If $P = (-3, 1)$ and $Q = (x, 4)$, find all numbers x such that the vector represented by $\overrightarrow{PQ}$ has length 5.

In Problems 55–60, write the vector $\mathbf{v}$ *in the form* $a\mathbf{i} + b\mathbf{j}$*, given its magnitude* $\|\mathbf{v}\|$ *and the angle* α *it makes with the positive x-axis.*

55. $\|\mathbf{v}\| = 5, \quad \alpha = 60^\circ$

56. $\|\mathbf{v}\| = 8, \quad \alpha = 45^\circ$

57. $\|\mathbf{v}\| = 14, \quad \alpha = 120^\circ$

58. $\|\mathbf{v}\| = 3, \quad \alpha = 240^\circ$

59. $\|\mathbf{v}\| = 25, \quad \alpha = 330^\circ$

60. $\|\mathbf{v}\| = 15, \quad \alpha = 315^\circ$

61. A child pulls a wagon with a force of 40 pounds. The handle of the wagon makes an angle of 30° with the ground. Express the force vector $\mathbf{F}$ in terms of $\mathbf{i}$ and $\mathbf{j}$.

62. A man pushes a wheelbarrow up an incline of 20° with a force of 100 pounds. Express the force vector $\mathbf{F}$ in terms of $\mathbf{i}$ and $\mathbf{j}$.

63. Resultant Force Two forces of magnitude 40 newtons (N) and 60 newtons act on an object at angles of 30° and −45° with the positive x-axis as shown in the figure. Find the direction and magnitude of the resultant force; that is, find $\mathbf{F}_1 + \mathbf{F}_2$.

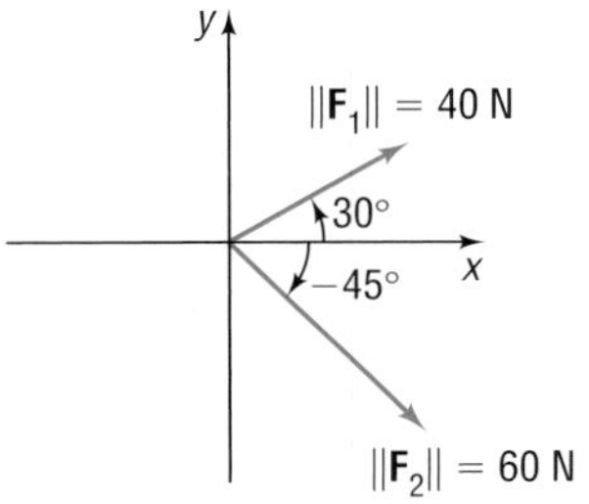

64. Resultant Force Two forces of magnitude 30 newtons (N) and 70 newtons act on an object at angles of 45° and 120° with the positive x-axis as shown in the figure. Find the direction and magnitude of the resultant force; that is, find $\mathbf{F}_1 + \mathbf{F}_2$.

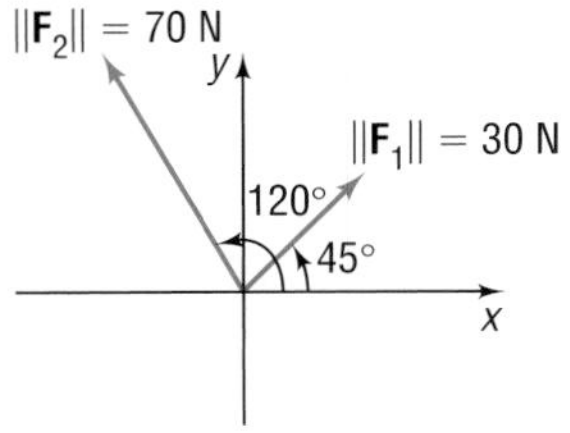

65. Static Equilibrium A weight of 1000 pounds is suspended from two cables as shown in the figure. What is the tension in the two cables?

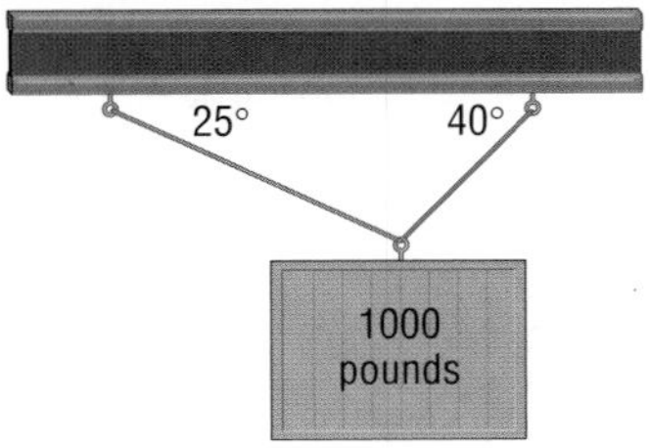

66. Static Equilibrium A weight of 800 pounds is suspended from two cables as shown in the figure. What is the tension in the two cables?

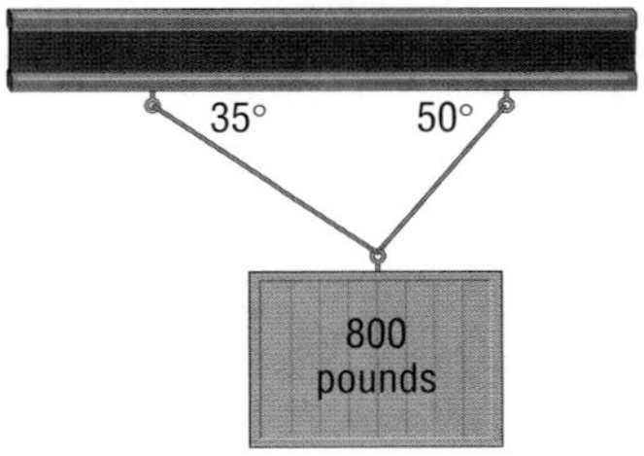

67. Static Equilibrium A tightrope walker located at a certain point deflects the rope as indicated in the figure. If the weight of the tightrope walker is 150 pounds, how much tension is in each part of the rope?

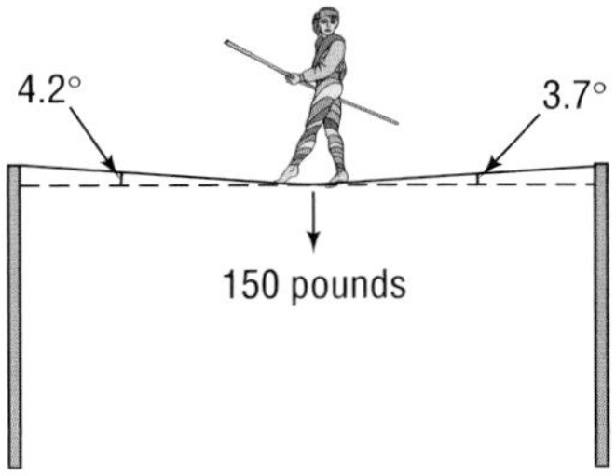

68. Static Equilibrium Repeat Problem 67 if the left angle is 3.8°, the right angle is 2.6°, and the weight of the tightrope walker is 135 pounds.

69. Show on the following graph the force needed for the object at P to be in static equilibrium.

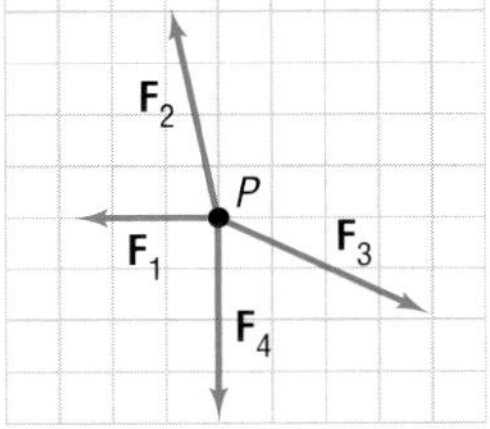

70. Explain in your own words what a vector is. Give an example of a vector.

71. Write a brief paragraph comparing the algebra of complex numbers and the algebra of vectors.

8.5 The Dot Product

PREPARING FOR THIS SECTION *Before getting started, review the following:*

- Law of Cosines (Section 7.3, p. 507)

Now work the 'Are You Prepared?' problem on page 588.

OBJECTIVES
1. Find the Dot Product of Two Vectors
2. Find the Angle between Two Vectors
3. Determine Whether Two Vectors Are Parallel
4. Determine Whether Two Vectors Are Orthogonal
5. Decompose a Vector into Two Orthogonal Vectors
6. Compute Work

1 The definition for a product of two vectors is somewhat unexpected. However, such a product has meaning in many geometric and physical applications.

If $\mathbf{v} = a_1\mathbf{i} + b_1\mathbf{j}$ and $\mathbf{w} = a_2\mathbf{i} + b_2\mathbf{j}$ are two vectors, the **dot product** $\mathbf{v} \cdot \mathbf{w}$ is defined as

$$\mathbf{v} \cdot \mathbf{w} = a_1a_2 + b_1b_2 \tag{1}$$

EXAMPLE 1 **Finding Dot Products**

If $\mathbf{v} = 2\mathbf{i} - 3\mathbf{j}$ and $\mathbf{w} = 5\mathbf{i} + 3\mathbf{j}$, find:

(a) $\mathbf{v} \cdot \mathbf{w}$ (b) $\mathbf{w} \cdot \mathbf{v}$ (c) $\mathbf{v} \cdot \mathbf{v}$

(d) $\mathbf{w} \cdot \mathbf{w}$ (e) $\|\mathbf{v}\|$ (f) $\|\mathbf{w}\|$

Solution

(a) $\mathbf{v} \cdot \mathbf{w} = 2(5) + (-3)3 = 1$ (b) $\mathbf{w} \cdot \mathbf{v} = 5(2) + 3(-3) = 1$

(c) $\mathbf{v} \cdot \mathbf{v} = 2(2) + (-3)(-3) = 13$ (d) $\mathbf{w} \cdot \mathbf{w} = 5(5) + 3(3) = 34$

(e) $\|\mathbf{v}\| = \sqrt{2^2 + (-3)^2} = \sqrt{13}$ (f) $\|\mathbf{w}\| = \sqrt{5^2 + 3^2} = \sqrt{34}$ ◀

Since the dot product $\mathbf{v} \cdot \mathbf{w}$ of two vectors $\mathbf{v}$ and $\mathbf{w}$ is a real number (scalar), we sometimes refer to it as the **scalar product.**

Properties

The results obtained in Example 1 suggest some general properties.

Theorem

Properties of the Dot Product

If **u, v,** and **w** are vectors, then

Commutative Property

$$\mathbf{u} \cdot \mathbf{v} = \mathbf{v} \cdot \mathbf{u} \tag{2}$$

Distributive Property

$$\mathbf{u} \cdot (\mathbf{v} + \mathbf{w}) = \mathbf{u} \cdot \mathbf{v} + \mathbf{u} \cdot \mathbf{w} \tag{3}$$

$$\mathbf{v} \cdot \mathbf{v} = \|\mathbf{v}\|^2 \tag{4}$$

$$\mathbf{0} \cdot \mathbf{v} = 0 \tag{5}$$

Proof We will prove properties (2) and (4) here and leave properties (3) and (5) as exercises (see Problems 39 and 40).

To prove property (2), we let $\mathbf{u} = a_1\mathbf{i} + b_1\mathbf{j}$ and $\mathbf{v} = a_2\mathbf{i} + b_2\mathbf{j}$. Then

$$\mathbf{u} \cdot \mathbf{v} = a_1a_2 + b_1b_2 = a_2a_1 + b_2b_1 = \mathbf{v} \cdot \mathbf{u}$$

To prove property (4), we let $\mathbf{v} = a\mathbf{i} + b\mathbf{j}$. Then

$$\mathbf{v} \cdot \mathbf{v} = a^2 + b^2 = \|\mathbf{v}\|^2$$ ■

One use of the dot product is to calculate the angle between two vectors.

Angle between Vectors

2 Let **u** and **v** be two vectors with the same initial point A. Then the vectors **u**, **v**, and $\mathbf{u} - \mathbf{v}$ form a triangle. The angle θ at vertex A of the triangle is the angle between the vectors **u** and **v**. See Figure 59. We wish to find a formula for calculating the angle θ.

Figure 59

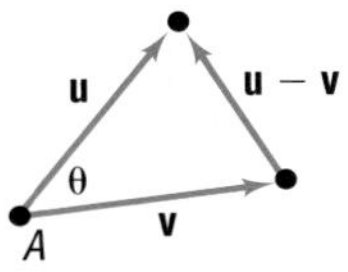

The sides of the triangle have lengths $\|\mathbf{v}\|$, $\|\mathbf{u}\|$, and $\|\mathbf{u} - \mathbf{v}\|$, and θ is the included angle between the sides of length $\|\mathbf{v}\|$ and $\|\mathbf{u}\|$. The Law of Cosines (Section 7.3) can be used to find the cosine of the included angle.

$$\|\mathbf{u} - \mathbf{v}\|^2 = \|\mathbf{u}\|^2 + \|\mathbf{v}\|^2 - 2\|\mathbf{u}\|\|\mathbf{v}\| \cos\theta$$

Now we use property (4) to rewrite this equation in terms of dot products.

$$(\mathbf{u} - \mathbf{v}) \cdot (\mathbf{u} - \mathbf{v}) = \mathbf{u} \cdot \mathbf{u} + \mathbf{v} \cdot \mathbf{v} - 2\|\mathbf{u}\|\|\mathbf{v}\| \cos\theta \tag{6}$$

Then we apply the distributive property (3) twice on the left side of (6) to obtain

$$\begin{aligned}(\mathbf{u} - \mathbf{v}) \cdot (\mathbf{u} - \mathbf{v}) &= \mathbf{u} \cdot (\mathbf{u} - \mathbf{v}) - \mathbf{v} \cdot (\mathbf{u} - \mathbf{v}) \\ &= \mathbf{u} \cdot \mathbf{u} - \mathbf{u} \cdot \mathbf{v} - \mathbf{v} \cdot \mathbf{u} + \mathbf{v} \cdot \mathbf{v} \\ &= \mathbf{u} \cdot \mathbf{u} + \mathbf{v} \cdot \mathbf{v} - 2\,\mathbf{u} \cdot \mathbf{v}\end{aligned} \tag{7}$$

↑ Property (2)

Combining equations (6) and (7), we have

$$\mathbf{u} \cdot \mathbf{u} + \mathbf{v} \cdot \mathbf{v} - 2\mathbf{u} \cdot \mathbf{v} = \mathbf{u} \cdot \mathbf{u} + \mathbf{v} \cdot \mathbf{v} - 2\|\mathbf{u}\|\|\mathbf{v}\| \cos\theta$$

$$\mathbf{u} \cdot \mathbf{v} = \|\mathbf{u}\|\|\mathbf{v}\| \cos\theta$$

We have proved the following result:

Theorem

Angle between Vectors

If $\mathbf{u}$ and $\mathbf{v}$ are two nonzero vectors, the angle θ, $0 \le \theta \le \pi$, between $\mathbf{u}$ and $\mathbf{v}$ is determined by the formula

$$\cos\theta = \frac{\mathbf{u}\cdot\mathbf{v}}{\|\mathbf{u}\|\|\mathbf{v}\|} \qquad \textbf{(8)}$$

EXAMPLE 2 Finding the Angle θ between Two Vectors

Find the angle θ between $\mathbf{u} = 4\mathbf{i} - 3\mathbf{j}$ and $\mathbf{v} = 2\mathbf{i} + 5\mathbf{j}$.

Solution We compute the quantities $\mathbf{u}\cdot\mathbf{v}$, $\|\mathbf{u}\|$, and $\|\mathbf{v}\|$.

Figure 60

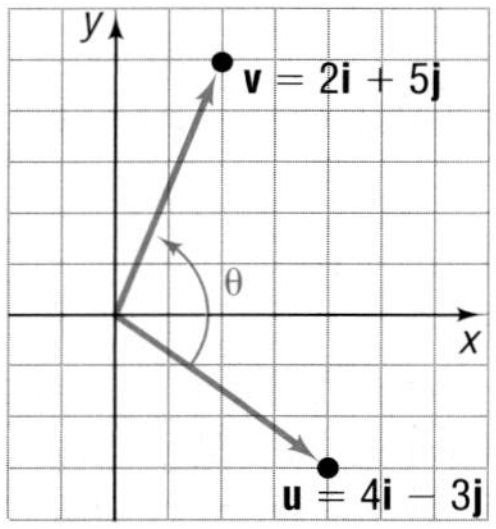

$$\mathbf{u}\cdot\mathbf{v} = 4(2) + (-3)(5) = -7$$
$$\|\mathbf{u}\| = \sqrt{4^2 + (-3)^2} = 5$$
$$\|\mathbf{v}\| = \sqrt{2^2 + 5^2} = \sqrt{29}$$

By formula (8), if θ is the angle between $\mathbf{u}$ and $\mathbf{v}$, then

$$\cos\theta = \frac{\mathbf{u}\cdot\mathbf{v}}{\|\mathbf{u}\|\|\mathbf{v}\|} = \frac{-7}{5\sqrt{29}} \approx -0.26$$

We find that $\theta \approx 105°$. See Figure 60. ◄

NOW WORK PROBLEM 7(a) AND (b).

EXAMPLE 3 Finding the Actual Speed and Direction of an Aircraft

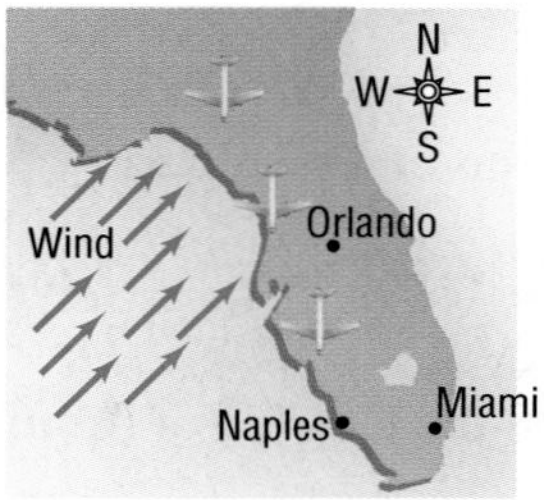

A Boeing 737 aircraft maintains a constant airspeed of 500 miles per hour in the direction due south. The velocity of the jet stream is 80 miles per hour in a northeasterly direction. Find the actual speed and direction of the aircraft relative to the ground.

Solution We set up a coordinate system in which north (N) is along the positive y-axis. See Figure 61. Let

Figure 61

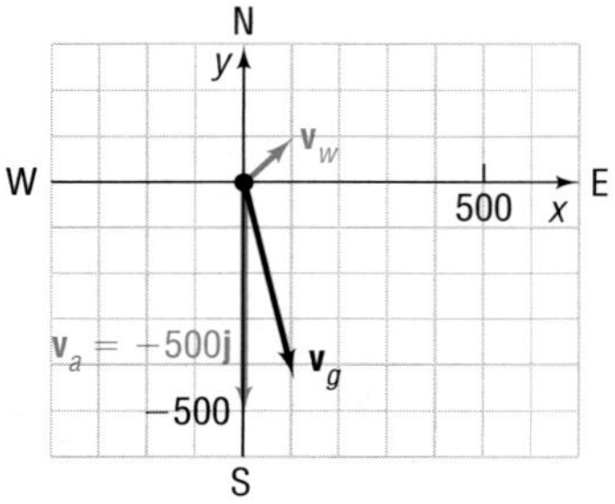

$$\mathbf{v}_a = \text{velocity of aircraft relative to the air} = -500\mathbf{j}$$
$$\mathbf{v}_w = \text{velocity of jet stream}$$
$$\mathbf{v}_g = \text{velocity of aircraft relative to ground}$$

The velocity of the jet stream $\mathbf{v}_w$ has magnitude 80 and direction NE (northeast), so $\alpha = 45°$. We express $\mathbf{v}_w$ in terms of $\mathbf{i}$ and $\mathbf{j}$ as

$$\mathbf{v}_w = 80(\cos 45°\mathbf{i} + \sin 45°\mathbf{j}) = 80\left(\frac{\sqrt{2}}{2}\mathbf{i} + \frac{\sqrt{2}}{2}\mathbf{j}\right) = 40\sqrt{2}(\mathbf{i} + \mathbf{j})$$

The velocity of the aircraft relative to the ground is

$$\mathbf{v}_g = \mathbf{v}_a + \mathbf{v}_w = -500\mathbf{j} + 40\sqrt{2}(\mathbf{i} + \mathbf{j}) = 40\sqrt{2}\mathbf{i} + \left(40\sqrt{2} - 500\right)\mathbf{j}$$

The actual speed of the aircraft is

$$\|\mathbf{v}_g\| = \sqrt{(40\sqrt{2})^2 + (40\sqrt{2} - 500)^2} \approx 447 \text{ miles per hour}$$

The angle θ between $\mathbf{v}_g$ and the vector $\mathbf{v}_a = -500\mathbf{j}$ (the velocity of the aircraft relative to the air) is determined by the equation

$$\cos\theta = \frac{\mathbf{v}_g \cdot \mathbf{v}_a}{\|\mathbf{v}_g\|\|\mathbf{v}_a\|} = \frac{(40\sqrt{2} - 500)(-500)}{(447)(500)} \approx 0.9920$$

$$\theta \approx 7.3°$$

The direction of the aircraft relative to the ground is approximately S7.3°E (about 7.3° east of south). ◀

NOW WORK PROBLEM 25.

Parallel and Orthogonal Vectors

3 Two vectors **v** and **w** are said to be **parallel** if there is a nonzero scalar α so that $\mathbf{v} = \alpha\mathbf{w}$. In this case, the angle θ between **v** and **w** is 0 or π.

EXAMPLE 4 Determining Whether Vectors Are Parallel

The vectors $\mathbf{v} = 3\mathbf{i} - \mathbf{j}$ and $\mathbf{w} = 6\mathbf{i} - 2\mathbf{j}$ are parallel, since $\mathbf{v} = \frac{1}{2}\mathbf{w}$. Furthermore, since

$$\cos\theta = \frac{\mathbf{v} \cdot \mathbf{w}}{\|\mathbf{v}\|\|\mathbf{w}\|} = \frac{18 + 2}{\sqrt{10}\sqrt{40}} = \frac{20}{\sqrt{400}} = 1$$

the angle θ between **v** and **w** is 0. ◀

4 If the angle θ between two nonzero vectors **v** and **w** is $\frac{\pi}{2}$, the vectors **v** and **w** are called **orthogonal.*** See Figure 62.

Figure 62
v is orthogonal to **w**

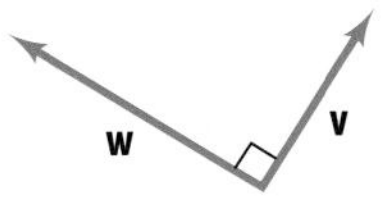

It follows from formula (8) that if **v** and **w** are orthogonal then $\mathbf{v} \cdot \mathbf{w} = 0$, since $\cos\frac{\pi}{2} = 0$.

On the other hand, if $\mathbf{v} \cdot \mathbf{w} = 0$, then either $\mathbf{v} = \mathbf{0}$ or $\mathbf{w} = \mathbf{0}$ or $\cos\theta = 0$. In the latter case, $\theta = \frac{\pi}{2}$, and **v** and **w** are orthogonal. If **v** or **w** is the zero vector, then, since the zero vector has no specific direction, we adopt the convention that the zero vector is orthogonal to every vector.

Theorem

Two vectors **v** and **w** are orthogonal if and only if

$$\mathbf{v} \cdot \mathbf{w} = 0$$

**Orthogonal, perpendicular*, and *normal* are all terms that mean "meet at a right angle." It is customary to refer to two vectors as being *orthogonal*, two lines as being *perpendicular*, and a line and a plane or a vector and a plane as being *normal*.

EXAMPLE 5 Determining Whether Two Vectors Are Orthogonal

Figure 63

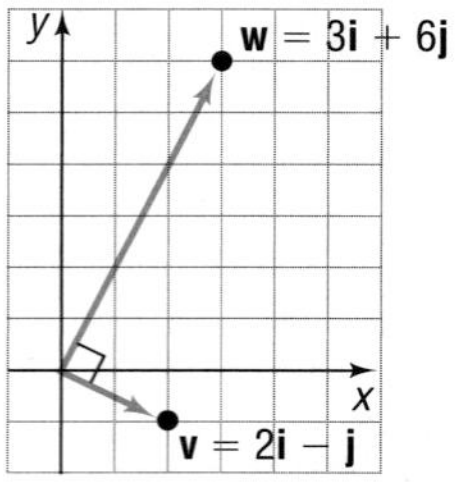

The vectors

$$\mathbf{v} = 2\mathbf{i} - \mathbf{j} \quad \text{and} \quad \mathbf{w} = 3\mathbf{i} + 6\mathbf{j}$$

are orthogonal, since

$$\mathbf{v} \cdot \mathbf{w} = 6 - 6 = 0$$

See Figure 63. ◀

NOW WORK PROBLEM 7(c).

Projection of a Vector onto Another Vector

Figure 64

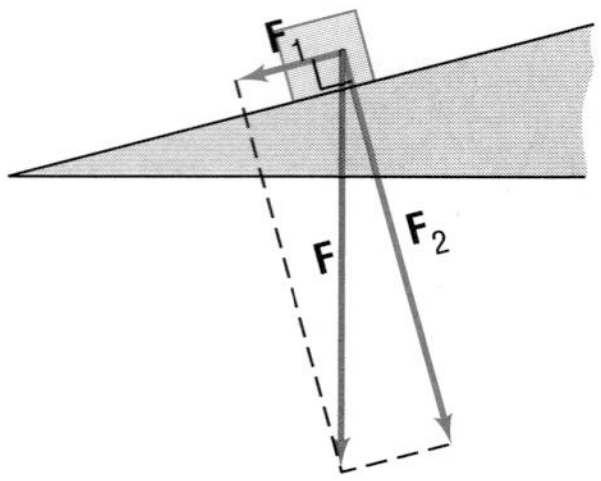

5 In many physical applications, it is necessary to find "how much" of a vector is applied in a given direction. Look at Figure 64. The force **F** due to gravity is pulling straight down (toward the center of Earth) on the block. To study the effect of gravity on the block, it is necessary to determine how much of **F** is actually pushing the block down the incline ($\mathbf{F}_1$) and how much is pressing the block against the incline ($\mathbf{F}_2$), at a right angle to the incline. Knowing the **decomposition** of **F** often will allow us to determine when friction is overcome and the block will slide down the incline.

Suppose that **v** and **w** are two nonzero vectors with the same initial point P. We seek to decompose **v** into two vectors: $\mathbf{v}_1$, which is parallel to **w**, and $\mathbf{v}_2$, which is orthogonal to **w**. See Figure 65(a) and (b). The vector $\mathbf{v}_1$ is called the **vector projection of v onto w.**

Figure 65

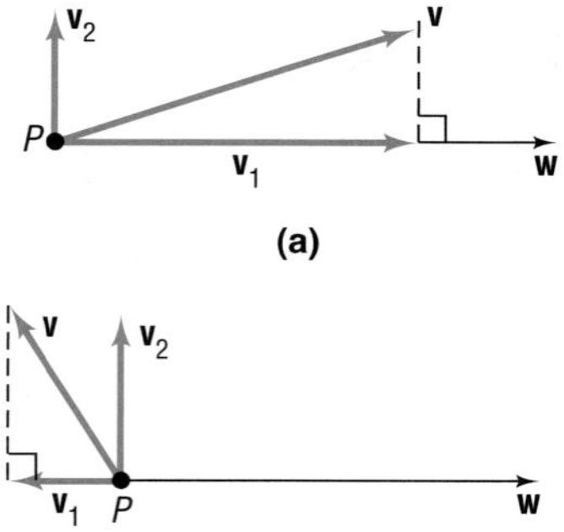

The vector $\mathbf{v}_1$ is obtained as follows: From the terminal point of **v**, drop a perpendicular to the line containing **w**. The vector $\mathbf{v}_1$ is the vector from P to the foot of this perpendicular. The vector $\mathbf{v}_2$ is given by $\mathbf{v}_2 = \mathbf{v} - \mathbf{v}_1$. Note that $\mathbf{v} = \mathbf{v}_1 + \mathbf{v}_2$, $\mathbf{v}_1$ is parallel to **w**, and $\mathbf{v}_2$ is orthogonal to **w**. This is the decomposition of **v** that we wanted.

Now we seek a formula for $\mathbf{v}_1$ that is based on a knowledge of the vectors **v** and **w**. Since $\mathbf{v} = \mathbf{v}_1 + \mathbf{v}_2$, we have

$$\mathbf{v} \cdot \mathbf{w} = (\mathbf{v}_1 + \mathbf{v}_2) \cdot \mathbf{w} = \mathbf{v}_1 \cdot \mathbf{w} + \mathbf{v}_2 \cdot \mathbf{w} \tag{9}$$

Since $\mathbf{v}_2$ is orthogonal to **w**, we have $\mathbf{v}_2 \cdot \mathbf{w} = 0$. Since $\mathbf{v}_1$ is parallel to **w**, we have $\mathbf{v}_1 = \alpha\mathbf{w}$ for some scalar α. Equation (9) can be written as

$$\mathbf{v} \cdot \mathbf{w} = \alpha\mathbf{w} \cdot \mathbf{w} = \alpha\|\mathbf{w}\|^2 \qquad \mathbf{v}_1 = \alpha\mathbf{w};\ \mathbf{v}_2 \cdot \mathbf{w} = 0$$

$$\alpha = \frac{\mathbf{v} \cdot \mathbf{w}}{\|\mathbf{w}\|^2}$$

Then

$$\mathbf{v}_1 = \alpha\mathbf{w} = \frac{\mathbf{v} \cdot \mathbf{w}}{\|\mathbf{w}\|^2}\mathbf{w}$$

Theorem

If **v** and **w** are two nonzero vectors, the vector projection of **v** onto **w** is

$$\mathbf{v}_1 = \frac{\mathbf{v} \cdot \mathbf{w}}{\|\mathbf{w}\|^2}\mathbf{w} \tag{10}$$

The decomposition of $\mathbf{v}$ into $\mathbf{v}_1$ and $\mathbf{v}_2$, where $\mathbf{v}_1$ is parallel to $\mathbf{w}$ and $\mathbf{v}_2$ is perpendicular to $\mathbf{w}$, is

$$\mathbf{v}_1 = \frac{\mathbf{v}\cdot\mathbf{w}}{\|\mathbf{w}\|^2}\mathbf{w} \qquad \mathbf{v}_2 = \mathbf{v} - \mathbf{v}_1 \tag{11}$$

EXAMPLE 6 **Decomposing a Vector into Two Orthogonal Vectors**

Find the vector projection of $\mathbf{v} = \mathbf{i} + 3\mathbf{j}$ onto $\mathbf{w} = \mathbf{i} + \mathbf{j}$. Decompose $\mathbf{v}$ into two vectors $\mathbf{v}_1$ and $\mathbf{v}_2$, where $\mathbf{v}_1$ is parallel to $\mathbf{w}$ and $\mathbf{v}_2$ is orthogonal to $\mathbf{w}$.

Solution We use formulas (10) and (11).

$$\mathbf{v}_1 = \frac{\mathbf{v}\cdot\mathbf{w}}{\|\mathbf{w}\|^2}\mathbf{w} = \frac{1+3}{(\sqrt{2})^2}\mathbf{w} = 2\mathbf{w} = 2(\mathbf{i}+\mathbf{j})$$

$$\mathbf{v}_2 = \mathbf{v} - \mathbf{v}_1 = (\mathbf{i}+3\mathbf{j}) - 2(\mathbf{i}+\mathbf{j}) = -\mathbf{i}+\mathbf{j}$$

See Figure 66. ◄

Figure 66

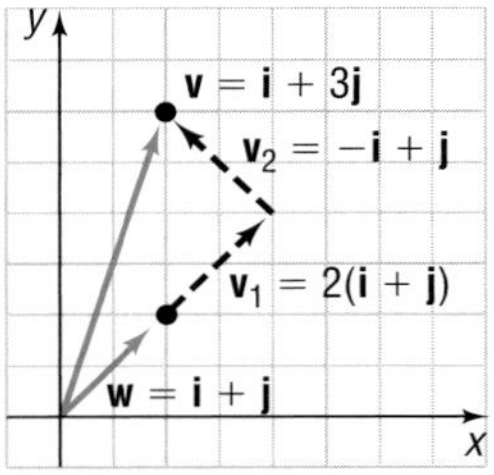

NOW WORK PROBLEM 19.

Work Done by a Constant Force

6 In elementary physics, the **work** W done by a constant force $\mathbf{F}$ in moving an object from a point A to a point B is defined as

$$W = (\text{magnitude of force})(\text{distance}) = \|\mathbf{F}\|\|\overrightarrow{AB}\|$$

Work is commonly measured in foot-pounds or in newton-meters (joules).

In this definition, it is assumed that the force $\mathbf{F}$ is applied along the line of motion. If the constant force $\mathbf{F}$ is not along the line of motion, but, instead, is at an angle θ to the direction of motion, as illustrated in Figure 67, then the **work** W **done by** $\mathbf{F}$ in moving an object from A to B is defined as

Figure 67

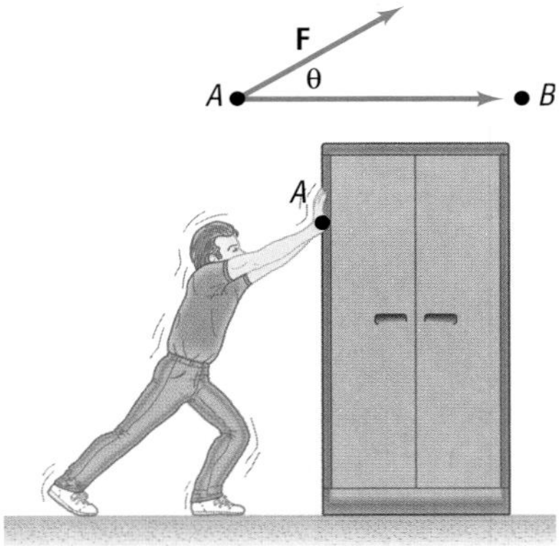

$$W = \mathbf{F}\cdot\overrightarrow{AB} \tag{12}$$

This definition is compatible with the force times distance definition given above, since

$$W = (\text{amount of force in the direction of } \overrightarrow{AB})(\text{distance})$$

$$= \|\text{projection of } \mathbf{F} \text{ on } AB\|\,\|\overrightarrow{AB}\| = \frac{\mathbf{F}\cdot\overrightarrow{AB}}{\|\overrightarrow{AB}\|^2}\|\overrightarrow{AB}\|\|\overrightarrow{AB}\| = \mathbf{F}\cdot\overrightarrow{AB}$$

EXAMPLE 7 **Computing Work**

Figure 68(a) shows a girl pulling a wagon with a force of 50 pounds. How much work is done in moving the wagon 100 feet if the handle makes an angle of 30° with the ground?

Figure 68

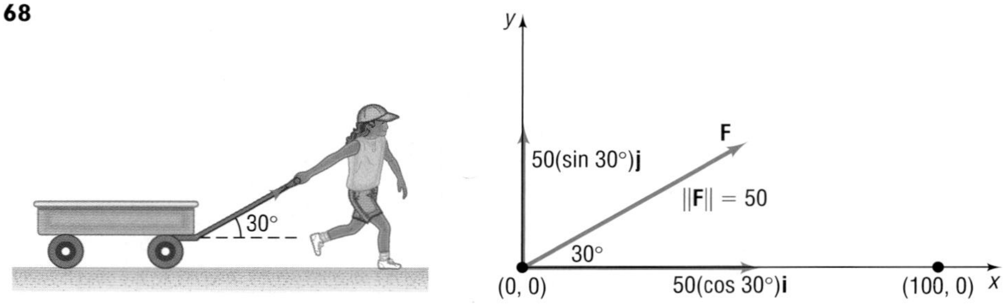

(a) (b)

Solution We position the vectors in a coordinate system in such a way that the wagon is moved from $(0, 0)$ to $(100, 0)$. The motion is from $A = (0, 0)$ to $B = (100, 0)$, so $\overrightarrow{AB} = 100\mathbf{i}$. The force vector $\mathbf{F}$, as shown in Figure 68(b), is

$$\mathbf{F} = 50(\cos 30°\mathbf{i} + \sin 30°\mathbf{j}) = 50\left(\frac{\sqrt{3}}{2}\mathbf{i} + \frac{1}{2}\mathbf{j}\right) = 25\left(\sqrt{3}\mathbf{i} + \mathbf{j}\right)$$

By formula (12), the work done is

$$W = \mathbf{F} \cdot \overrightarrow{AB} = 25\left(\sqrt{3}\mathbf{i} + \mathbf{j}\right) \cdot 100\mathbf{i} = 2500\sqrt{3} \text{ foot-pounds}$$ ◀

NOW WORK PROBLEM 35.

Historical Feature

1. We stated in an earlier Historical Feature that complex numbers were used as vectors in the plane before the general notion of a vector was clarified. Suppose that we make the correspondence

Vector ↔ Complex number

$a\mathbf{i} + b\mathbf{j} \leftrightarrow a + bi$

$c\mathbf{i} + d\mathbf{j} \leftrightarrow c + di$

Show that

$$(a\mathbf{i} + b\mathbf{j}) \cdot (c\mathbf{i} + d\mathbf{j}) = \text{real part}\left[\overline{(a + bi)}(c + di)\right]$$

This is how the dot product was found originally. The imaginary part is also interesting. It is a determinant (see Section 10.3) and represents the area of the parallelogram whose edges are the vectors. This is close to some of Hermann Grassmann's ideas and is also connected with the scalar triple product of three-dimensional vectors.

8.5 Assess Your Understanding

'Are You Prepared?' *Answer is given at the end of these exercises. If you get a wrong answer, read the page listed in red.*

1. In a triangle with sides a, b, c and angles α, β, γ, the Law of Cosines states that __________. (p. 588)

Concepts and Vocabulary

2. If $\mathbf{v} \cdot \mathbf{w} = 0$, then the two vectors $\mathbf{v}$ and $\mathbf{w}$ are __________.

3. If $\mathbf{v} = 3\mathbf{w}$, then the two vectors $\mathbf{v}$ and $\mathbf{w}$ are __________.

4. *True or False:* If $\mathbf{v}$ and $\mathbf{w}$ are parallel vectors, then $\mathbf{v} \cdot \mathbf{w} = 0$.

5. *True or False:* Given two nonzero vectors $\mathbf{v}$ and $\mathbf{w}$, it is always possible to decompose $\mathbf{v}$ into two vectors, one parallel to $\mathbf{w}$ and the other perpendicular to $\mathbf{w}$.

6. *True or False:* Work is a physical example of a vector.

Exercises

In Problems 7–16, (a) find the dot product $\mathbf{v} \cdot \mathbf{w}$; (b) find the angle between $\mathbf{v}$ and $\mathbf{w}$; (c) state whether the vectors are parallel, orthogonal, or neither.

7. $\mathbf{v} = \mathbf{i} - \mathbf{j}, \quad \mathbf{w} = \mathbf{i} + \mathbf{j}$

8. $\mathbf{v} = \mathbf{i} + \mathbf{j}, \quad \mathbf{w} = -\mathbf{i} + \mathbf{j}$

9. $\mathbf{v} = 2\mathbf{i} + \mathbf{j}, \quad \mathbf{w} = \mathbf{i} + 2\mathbf{j}$

10. $\mathbf{v} = 2\mathbf{i} + 2\mathbf{j}, \quad \mathbf{w} = \mathbf{i} + 2\mathbf{j}$

11. $\mathbf{v} = \sqrt{3}\mathbf{i} - \mathbf{j}, \quad \mathbf{w} = \mathbf{i} + \mathbf{j}$

12. $\mathbf{v} = \mathbf{i} + \sqrt{3}\mathbf{j}, \quad \mathbf{w} = \mathbf{i} - \mathbf{j}$

13. $\mathbf{v} = 3\mathbf{i} + 4\mathbf{j}, \quad \mathbf{w} = 4\mathbf{i} + 3\mathbf{j}$

14. $\mathbf{v} = 3\mathbf{i} - 4\mathbf{j}, \quad \mathbf{w} = 4\mathbf{i} - 3\mathbf{j}$

15. $\mathbf{v} = 4\mathbf{i}, \quad \mathbf{w} = \mathbf{j}$

16. $\mathbf{v} = \mathbf{i}, \quad \mathbf{w} = -3\mathbf{j}$

17. Find a so that the vectors $\mathbf{v} = \mathbf{i} - a\mathbf{j}$ and $\mathbf{w} = 2\mathbf{i} + 3\mathbf{j}$ are orthogonal.

18. Find b so that the vectors $\mathbf{v} = \mathbf{i} + \mathbf{j}$ and $\mathbf{w} = i + b\mathbf{j}$ are orthogonal.

In Problems 19–24, decompose $\mathbf{v}$ *into two vectors* $\mathbf{v}_1$ *and* $\mathbf{v}_2$, *where* $\mathbf{v}_1$ *is parallel to* $\mathbf{w}$ *and* $\mathbf{v}_2$ *is orthogonal to* $\mathbf{w}$.

19. $\mathbf{v} = 2\mathbf{i} - 3\mathbf{j}, \quad \mathbf{w} = \mathbf{i} - \mathbf{j}$

20. $\mathbf{v} = -3\mathbf{i} + 2\mathbf{j}, \quad \mathbf{w} = 2\mathbf{i} + \mathbf{j}$

21. $\mathbf{v} = \mathbf{i} - \mathbf{j}, \quad \mathbf{w} = \mathbf{i} + 2\mathbf{j}$

22. $\mathbf{v} = 2\mathbf{i} - \mathbf{j}, \quad \mathbf{w} = \mathbf{i} - 2\mathbf{j}$

23. $\mathbf{v} = 3\mathbf{i} + \mathbf{j}, \quad \mathbf{w} = -2\mathbf{i} - \mathbf{j}$

24. $\mathbf{v} = \mathbf{i} - 3\mathbf{j}, \quad \mathbf{w} = 4\mathbf{i} - \mathbf{j}$

---

25. Finding the Actual Speed and Direction of an Aircraft A Boeing 747 jumbo jet maintains an airspeed of 550 miles per hour in a southwesterly direction. The velocity of the jet stream is a constant 80 miles per hour from the west. Find the actual speed and direction of the aircraft.

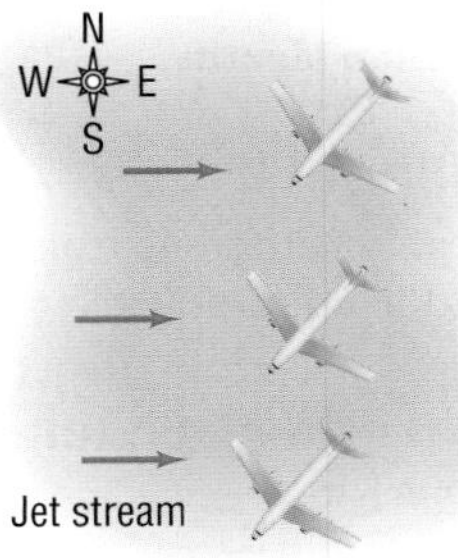

26. Finding the Correct Compass Heading The pilot of an aircraft wishes to head directly east, but is faced with a wind speed of 40 miles per hour from the northwest. If the pilot maintains an airspeed of 250 miles per hour, what compass heading should be maintained? What is the actual speed of the aircraft?

27. Correct Direction for Crossing a River A river has a constant current of 3 kilometers per hour. At what angle to a boat dock should a motorboat, capable of maintaining a constant speed of 20 kilometers per hour, be headed in order to reach a point directly opposite the dock? If the river is $\frac{1}{2}$ kilometer wide, how long will it take to cross?

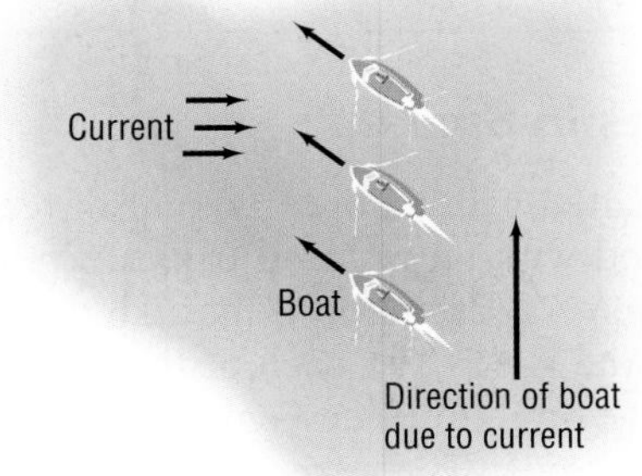

28. Correct Direction for Crossing a River Repeat Problem 27 if the current is 5 kilometers per hour.

29. Braking Load A Toyota Sienna with a gross weight of 5300 pounds is parked on a street with a slope of 8°. See the figure. Find the force required to keep the Sienna from rolling down the hill. What is the force perpendicular to the hill?

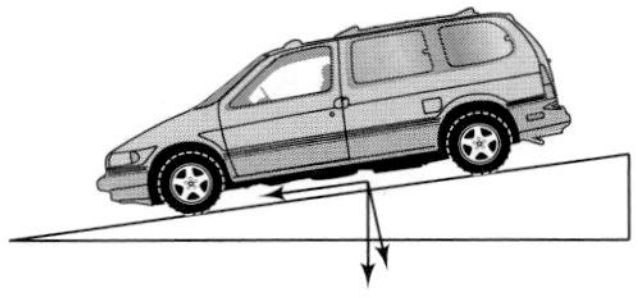

30. Braking Load A Pontiac Bonneville with a gross weight of 4500 pounds is parked on a street with a slope of 10°. Find the force required to keep the Bonneville from rolling down the hill. What is the force perpendicular to the hill?

31. Ground Speed and Direction of an Airplane An airplane has an airspeed of 500 kilometers per hour bearing N45°E. The wind velocity is 60 kilometers per hour in the direction N30°W. Find the resultant vector representing the path of the plane relative to the ground. What is the ground speed of the plane? What is its direction?

32. Ground Speed and Direction of an Airplane An airplane has an airspeed of 600 kilometers per hour bearing S30°E. The wind velocity is 40 kilometers per hour in the direction S45°E. Find the resultant vector representing the path of the plane relative to the ground. What is the ground speed of the plane? What is its direction?

33. Crossing a River A small motorboat in still water maintains a speed of 20 miles per hour. In heading directly across a river (that is, perpendicular to the current) whose current is 3 miles per hour, find a vector representing the speed and direction of the motorboat. What is the true speed of the motorboat? What is its direction?

34. Crossing a River A small motorboat in still water maintains a speed of 10 miles per hour. In heading directly across a river (that is, perpendicular to the current) whose current is 4 miles per hour, find a vector representing the speed and direction of the motorboat. What is the true speed of the motorboat? What is its direction?

35. Computing Work Find the work done by a force of 3 pounds acting in the direction 60° to the horizontal in moving an object 2 feet from $(0, 0)$ to $(2, 0)$.

36. Computing Work Find the work done by a force of 1 pound acting in the direction 45° to the horizontal in moving an object 5 feet from $(0, 0)$ to $(5, 0)$.

37. **Computing Work** A wagon is pulled horizontally by exerting a force of 20 pounds on the handle at an angle of 30° with the horizontal. How much work is done in moving the wagon 100 feet?

38. Find the acute angle that a constant unit force vector makes with the positive x-axis if the work done by the force in moving a particle from $(0, 0)$ to $(4, 0)$ equals 2.

39. Prove the distributive property:

$$\mathbf{u}\cdot(\mathbf{v} + \mathbf{w}) = \mathbf{u}\cdot\mathbf{v} + \mathbf{u}\cdot\mathbf{w}$$

40. Prove property (5), $\mathbf{0}\cdot\mathbf{v} = 0$.

41. If $\mathbf{v}$ is a unit vector and the angle between $\mathbf{v}$ and $\mathbf{i}$ is α, show that $\mathbf{v} = \cos\alpha\mathbf{i} + \sin\alpha\mathbf{j}$.

42. Suppose that $\mathbf{v}$ and $\mathbf{w}$ are unit vectors. If the angle between $\mathbf{v}$ and $\mathbf{i}$ is α and that between $\mathbf{w}$ and $\mathbf{i}$ is β, use the idea of the dot product $\mathbf{v}\cdot\mathbf{w}$ to prove that

$$\cos(\alpha - \beta) = \cos\alpha\cos\beta + \sin\alpha\sin\beta$$

43. Show that the projection of $\mathbf{v}$ onto $\mathbf{i}$ is $(\mathbf{v}\cdot\mathbf{i})\mathbf{i}$. In fact, show that we can always write a vector $\mathbf{v}$ as

$$\mathbf{v} = (\mathbf{v}\cdot\mathbf{i})\mathbf{i} + (\mathbf{v}\cdot\mathbf{j})\mathbf{j}$$

44. (a) If $\mathbf{u}$ and $\mathbf{v}$ have the same magnitude, show that $\mathbf{u} + \mathbf{v}$ and $\mathbf{u} - \mathbf{v}$ are orthogonal.

(b) Use this to prove that an angle inscribed in a semicircle is a right angle (see the figure).

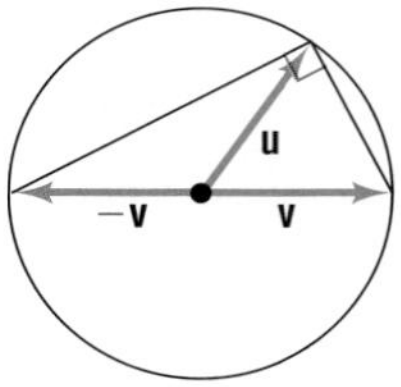

45. Let $\mathbf{v}$ and $\mathbf{w}$ denote two nonzero vectors. Show that the vector $\mathbf{v} - \alpha\mathbf{w}$ is orthogonal to $\mathbf{w}$ if $\alpha = (\mathbf{v}\cdot\mathbf{w})/\|\mathbf{w}\|^2$.

46. Let $\mathbf{v}$ and $\mathbf{w}$ denote two nonzero vectors. Show that the vectors $\|\mathbf{w}\|\mathbf{v} + \|\mathbf{v}\|\mathbf{w}$ and $\|\mathbf{w}\|\mathbf{v} - \|\mathbf{v}\|\mathbf{w}$ are orthogonal.

47. In the definition of work given in this section, what is the work done if $\mathbf{F}$ is orthogonal to $\overrightarrow{AB}$?

48. Prove the **polarization identity,**

$$\|\mathbf{u} + \mathbf{v}\|^2 - \|\mathbf{u} - \mathbf{v}\|^2 = 4(\mathbf{u}\cdot\mathbf{v}).$$

49. Create an application different from any found in the text that requires the dot product.

'Are You Prepared?' Answer

1. $c^2 = a^2 + b^2 - 2ab\cos\gamma$

8.6 Vectors in Space

PREPARING FOR THIS SECTION *Before getting started, review the following:*

- Distance Formula (Section 1.1, p. 3)

Now work the 'Are You Prepared?' problem on page 599.

OBJECTIVES
1 Find the Distance between Two Points
2 Find Position Vectors
3 Perform Operations on Vectors
4 Find the Dot Product
5 Find the Angle between Two Vectors
6 Find the Direction Angles of a Vector

Figure 69

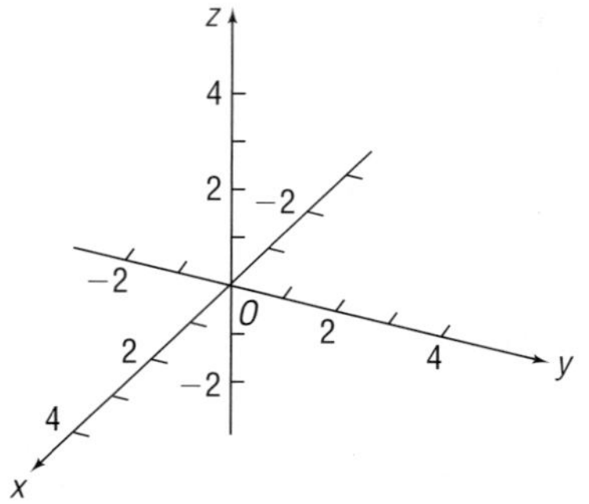

Rectangular Coordinates in Space

In the plane, each point is associated with an ordered pair of real numbers. In space, each point is associated with an ordered triple of real numbers. Through a fixed point, the **origin,** O, draw three mutually perpendicular lines, the x-axis, the y-axis, and the z-axis. On each of these axes, select an appropriate scale and the positive direction. See Figure 69.

The direction chosen for the positive z-axis in Figure 69 makes the system *right-handed*. This conforms to the *right-hand rule*, which states that if the index finger of the right hand points in the direction of the positive x-axis and the middle finger points in the direction of the positive y-axis then the thumb will point in the direction of the positive z-axis. See Figure 70.

Figure 70

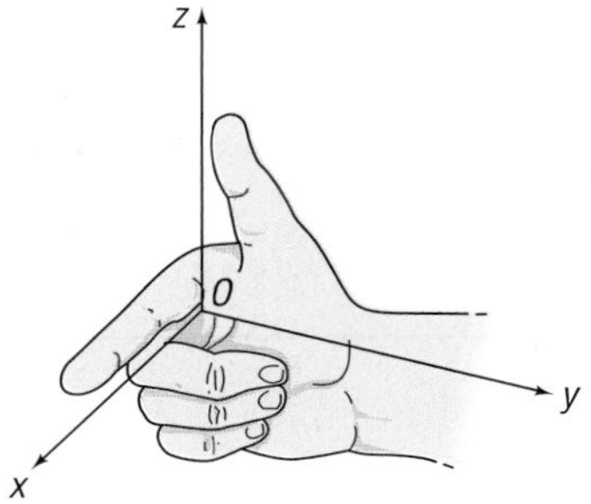

Figure 71

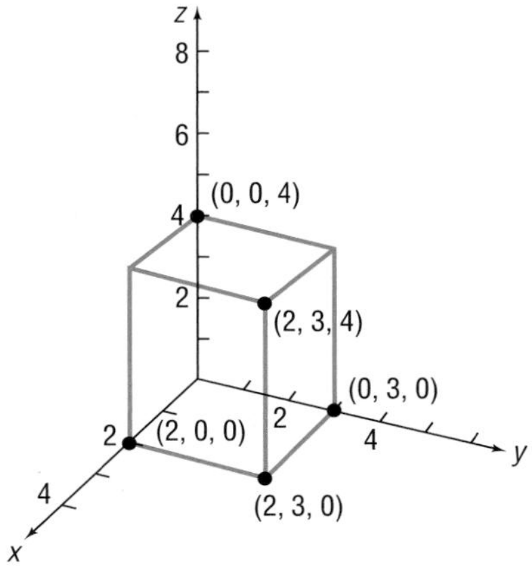

We associate with each point P an ordered triple (x, y, z) of real numbers, the **coordinates of *P*.** For example, the point $(2, 3, 4)$ is located by starting at the origin and moving 2 units along the positive x-axis, 3 units in the direction of the positive y-axis, and 4 units in the direction of the positive z-axis. See Figure 71.

Figure 71 also shows the location of the points $(2, 0, 0)$, $(0, 3, 0)$, $(0, 0, 4)$, and $(2, 3, 0)$. Points of the form $(x, 0, 0)$ lie on the x-axis, while points of the form $(0, y, 0)$ and $(0, 0, z)$ lie on the y-axis and z-axis, respectively. Points of the form $(x, y, 0)$ lie in a plane, called the ***xy*-plane.** Its equation is $z = 0$. Similarly, points of the form $(x, 0, z)$ lie in the ***xz*-plane** (equation $y = 0$) and points of the form $(0, y, z)$ lie in the ***yz*-plane** (equation $x = 0$). See Figure 72(a). By extension of these ideas, all points obeying the equation $z = 3$ will lie in a plane parallel to and 3 units above the xy-plane. The equation $y = 4$ represents a plane parallel to the xz-plane and 4 units to the right of the plane $y = 0$. See Figure 72(b).

Figure 72

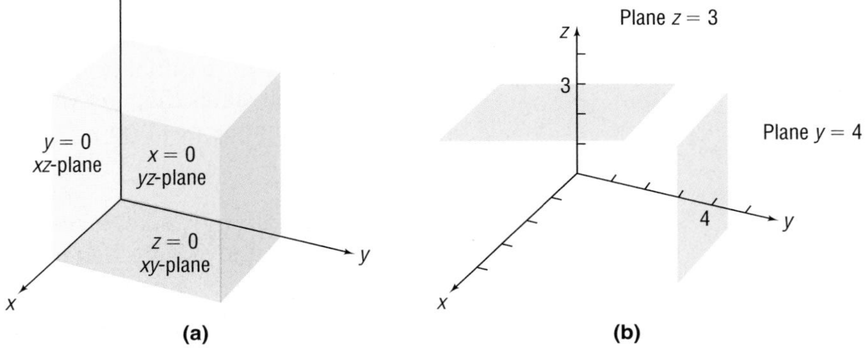

NOW WORK PROBLEM 9.

1 The formula for the distance between two points in space is an extension of the Distance Formula for points in the plane given in Chapter 1.

Theorem

Distance Formula in Space

If $P_1 = (x_1, y_1, z_1)$ and $P_2 = (x_2, y_2, z_2)$ are two points in space, the distance d from P_1 to P_2 is

$$d = \sqrt{(x_2 - x_1)^2 + (y_2 - y_1)^2 + (z_2 - z_1)^2} \quad \textbf{(1)}$$

The proof, which we omit, utilizes a double application of the Pythagorean Theorem.

EXAMPLE 1 **Using the Distance Formula**

Find the distance from $P_1 = (-1, 3, 2)$ to $P_2 = (4, -2, 5)$.

Solution $d = \sqrt{[4 - (-1)]^2 + [-2 - 3]^2 + [5 - 2]^2} = \sqrt{25 + 25 + 9} = \sqrt{59}$ ◀

NOW WORK PROBLEM 15.

Representing Vectors in Space

2 To represent vectors in space, we introduce the unit vectors **i**, **j**, and **k** whose directions are along the positive x-axis, positive y-axis, and positive z-axis, respectively. If **v** is a vector with initial point at the origin O and terminal point at $P = (a, b, c)$, then we can represent **v** in terms of the vectors **i**, **j**, and **k** as

Figure 73

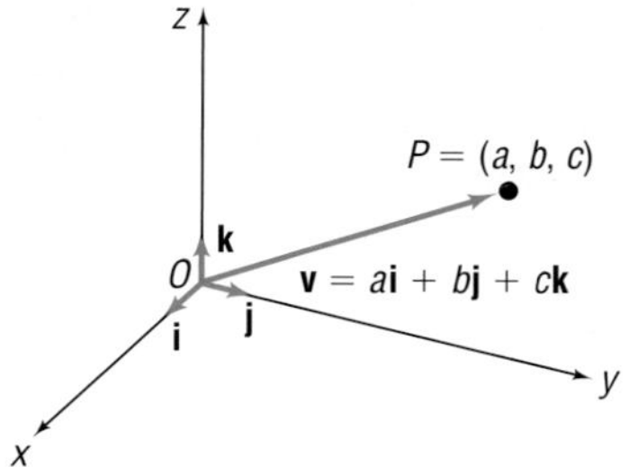

$$\mathbf{v} = a\mathbf{i} + b\mathbf{j} + c\mathbf{k}$$

See Figure 73.

The scalars a, b, and c are called the **components** of the vector $\mathbf{v} = a\mathbf{i} + b\mathbf{j} + c\mathbf{k}$, with a being the component in the direction **i**, b the component in the direction **j**, and c the component in the direction **k**.

A vector whose initial point is at the origin is called a **position vector.** The next result states that any vector whose initial point is not at the origin is equal to a unique position vector.

Theorem

Suppose that **v** is a vector with initial point $P_1 = (x_1, y_1, z_1)$, not necessarily the origin, and terminal point $P_2 = (x_2, y_2, z_2)$. If $\mathbf{v} = \overrightarrow{P_1P_2}$, then **v** is equal to the position vector

$$\mathbf{v} = (x_2 - x_1)\mathbf{i} + (y_2 - y_1)\mathbf{j} + (z_2 - z_1)\mathbf{k} \quad \textbf{(2)}$$

Figure 74 illustrates this result.

Figure 74

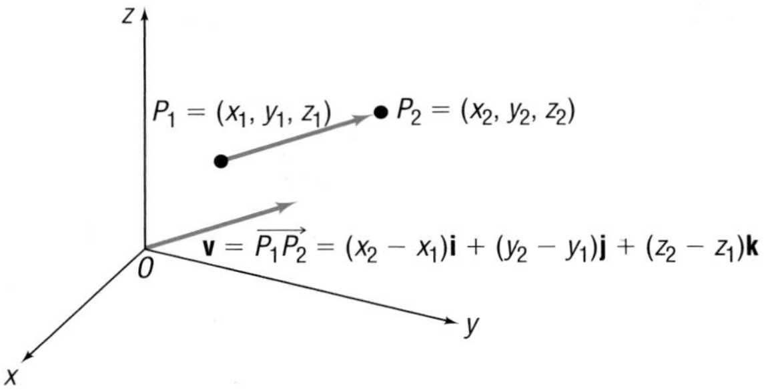

EXAMPLE 2 Finding a Position Vector

Find the position vector of the vector $\mathbf{v} = \overrightarrow{P_1P_2}$ if $P_1 = (-1, 2, 3)$ and $P_2 = (4, 6, 2)$.

Solution By equation (2), the position vector equal to $\mathbf{v}$ is

$$\mathbf{v} = [4 - (-1)]\mathbf{i} + (6 - 2)\mathbf{j} + (2 - 3)\mathbf{k} = 5\mathbf{i} + 4\mathbf{j} - \mathbf{k}$$ ◀

NOW WORK PROBLEM 29.

3 Next, we define equality, addition, subtraction, scalar product, and magnitude in terms of the components of a vector.

Let $\mathbf{v} = a_1\mathbf{i} + b_1\mathbf{j} + c_1\mathbf{k}$ and $\mathbf{w} = a_2\mathbf{i} + b_2\mathbf{j} + c_2\mathbf{k}$ be two vectors, and let α be a scalar. Then

$$\mathbf{v} = \mathbf{w} \quad \text{if and only if } a_1 = a_2, b_1 = b_2, \text{ and } c_1 = c_2$$
$$\mathbf{v} + \mathbf{w} = (a_1 + a_2)\mathbf{i} + (b_1 + b_2)\mathbf{j} + (c_1 + c_2)\mathbf{k}$$
$$\mathbf{v} - \mathbf{w} = (a_1 - a_2)\mathbf{i} + (b_1 - b_2)\mathbf{j} + (c_1 - c_2)\mathbf{k}$$
$$\alpha\mathbf{v} = (\alpha a_1)\mathbf{i} + (\alpha b_1)\mathbf{j} + (\alpha c_1)\mathbf{k}$$
$$\|\mathbf{v}\| = \sqrt{a_1^2 + b_1^2 + c_1^2}$$

These definitions are compatible with the geometric ones given earlier in Section 8.4.

EXAMPLE 3 Adding and Subtracting Vectors

If $\mathbf{v} = 2\mathbf{i} + 3\mathbf{j} - 2\mathbf{k}$ and $\mathbf{w} = 3\mathbf{i} - 4\mathbf{j} + 5\mathbf{k}$, find:

(a) $\mathbf{v} + \mathbf{w}$ (b) $\mathbf{v} - \mathbf{w}$

Solution (a)
$$\begin{aligned}\mathbf{v} + \mathbf{w} &= (2\mathbf{i} + 3\mathbf{j} - 2\mathbf{k}) + (3\mathbf{i} - 4\mathbf{j} + 5\mathbf{k})\\ &= (2 + 3)\mathbf{i} + (3 - 4)\mathbf{j} + (-2 + 5)\mathbf{k}\\ &= 5\mathbf{i} - \mathbf{j} + 3\mathbf{k}\end{aligned}$$

(b)
$$\begin{aligned}\mathbf{v} - \mathbf{w} &= (2\mathbf{i} + 3\mathbf{j} - 2\mathbf{k}) - (3\mathbf{i} - 4\mathbf{j} + 5\mathbf{k})\\ &= (2 - 3)\mathbf{i} + [3 - (-4)]\mathbf{j} + [-2 - 5]\mathbf{k}\\ &= -\mathbf{i} + 7\mathbf{j} - 7\mathbf{k}\end{aligned}$$ ◀

EXAMPLE 4 **Finding Scalar Products and Magnitudes**

If $\mathbf{v} = 2\mathbf{i} + 3\mathbf{j} - 2\mathbf{k}$ and $\mathbf{w} = 3\mathbf{i} - 4\mathbf{j} + 5\mathbf{k}$, find:

(a) $3\mathbf{v}$ (b) $2\mathbf{v} - 3\mathbf{w}$ (c) $\|\mathbf{v}\|$

Solution (a) $3\mathbf{v} = 3(2\mathbf{i} + 3\mathbf{j} - 2\mathbf{k}) = 6\mathbf{i} + 9\mathbf{j} - 6\mathbf{k}$

(b) $$\begin{aligned}2\mathbf{v} - 3\mathbf{w} &= 2(2\mathbf{i} + 3\mathbf{j} - 2\mathbf{k}) - 3(3\mathbf{i} - 4\mathbf{j} + 5\mathbf{k})\\ &= 4\mathbf{i} + 6\mathbf{j} - 4\mathbf{k} - 9\mathbf{i} + 12\mathbf{j} - 15\mathbf{k} = -5\mathbf{i} + 18\mathbf{j} - 19\mathbf{k}\end{aligned}$$

(c) $\|\mathbf{v}\| = \|2\mathbf{i} + 3\mathbf{j} - 2\mathbf{k}\| = \sqrt{2^2 + 3^2 + (-2)^2} = \sqrt{17}$ ◀

NOW WORK PROBLEMS 33 AND 39.

Recall that a unit vector $\mathbf{u}$ is one for which $\|\mathbf{u}\| = 1$. In many applications, it is useful to be able to find a unit vector $\mathbf{u}$ that has the same direction as a given vector $\mathbf{v}$.

Theorem **Unit Vector in the Direction of v**

For any nonzero vector $\mathbf{v}$, the vector

$$\mathbf{u} = \frac{\mathbf{v}}{\|\mathbf{v}\|}$$

is a unit vector that has the same direction as $\mathbf{v}$.

As a consequence of this theorem, if $\mathbf{u}$ is a unit vector in the same direction as a vector $\mathbf{v}$, then $\mathbf{v}$ may be expressed as

$$\mathbf{v} = \|\mathbf{v}\|\mathbf{u}$$

This way of expressing a vector is useful in many applications.

EXAMPLE 5 **Finding a Unit Vector**

Find a unit vector in the same direction as $\mathbf{v} = 2\mathbf{i} - 3\mathbf{j} - 6\mathbf{k}$.

Solution We find $\|\mathbf{v}\|$ first.

$$\|\mathbf{v}\| = \|2\mathbf{i} - 3\mathbf{j} - 6\mathbf{k}\| = \sqrt{4 + 9 + 36} = \sqrt{49} = 7$$

Now we multiply $\mathbf{v}$ by the scalar $\dfrac{1}{\|\mathbf{v}\|} = \dfrac{1}{7}$. The result is the unit vector

$$\mathbf{u} = \frac{\mathbf{v}}{\|\mathbf{v}\|} = \frac{2\mathbf{i} - 3\mathbf{j} - 6\mathbf{k}}{7} = \frac{2}{7}\mathbf{i} - \frac{3}{7}\mathbf{j} - \frac{6}{7}\mathbf{k}$$ ◀

NOW WORK PROBLEM 47.

Dot Product

4 The definition of *dot product* is an extension of the definition given for vectors in the plane.

If $\mathbf{v} = a_1\mathbf{i} + b_1\mathbf{j} + c_1\mathbf{k}$ and $\mathbf{w} = a_2\mathbf{i} + b_2\mathbf{j} + c_2\mathbf{k}$ are two vectors, the **dot product** $\mathbf{v} \cdot \mathbf{w}$ is defined as

$$\mathbf{v} \cdot \mathbf{w} = a_1a_2 + b_1b_2 + c_1c_2 \qquad \textbf{(3)}$$

EXAMPLE 6 **Finding Dot Products**

If $\mathbf{v} = 2\mathbf{i} - 3\mathbf{j} + 6\mathbf{k}$ and $\mathbf{w} = 5\mathbf{i} + 3\mathbf{j} - \mathbf{k}$, find:

(a) $\mathbf{v} \cdot \mathbf{w}$ (b) $\mathbf{w} \cdot \mathbf{v}$ (c) $\mathbf{v} \cdot \mathbf{v}$
(d) $\mathbf{w} \cdot \mathbf{w}$ (e) $\|\mathbf{v}\|$ (f) $\|\mathbf{w}\|$

Solution (a) $\mathbf{v} \cdot \mathbf{w} = 2(5) + (-3)3 + 6(-1) = -5$

(b) $\mathbf{w} \cdot \mathbf{v} = 5(2) + 3(-3) + (-1)(6) = -5$

(c) $\mathbf{v} \cdot \mathbf{v} = 2(2) + (-3)(-3) + 6(6) = 49$

(d) $\mathbf{w} \cdot \mathbf{w} = 5(5) + 3(3) + (-1)(-1) = 35$

(e) $\|\mathbf{v}\| = \sqrt{2^2 + (-3)^2 + 6^2} = \sqrt{49} = 7$

(f) $\|\mathbf{w}\| = \sqrt{5^2 + 3^2 + (-1)^2} = \sqrt{35}$ ◀

The dot product in space has the same properties as the dot product in the plane.

Theorem **Properties of the Dot Product**

If $\mathbf{u}$, $\mathbf{v}$, and $\mathbf{w}$ are vectors, then

Commutative Property

$$\mathbf{u} \cdot \mathbf{v} = \mathbf{v} \cdot \mathbf{u}$$

Distributive Property

$$\mathbf{u} \cdot (\mathbf{v} + \mathbf{w}) = \mathbf{u} \cdot \mathbf{v} + \mathbf{u} \cdot \mathbf{w}$$

$$\mathbf{v} \cdot \mathbf{v} = \|\mathbf{v}\|^2$$
$$\mathbf{0} \cdot \mathbf{v} = 0$$

5 The angle θ between two vectors in space follows the same formula as for two vectors in the plane.

Theorem **Angle between Vectors**

If $\mathbf{u}$ and $\mathbf{v}$ are two nonzero vectors, the angle θ, $0 \le \theta \le \pi$, between $\mathbf{u}$ and $\mathbf{v}$ is determined by the formula

$$\cos\theta = \frac{\mathbf{u} \cdot \mathbf{v}}{\|\mathbf{u}\|\,\|\mathbf{v}\|} \qquad \textbf{(4)}$$

EXAMPLE 7 **Finding the Angle θ between Two Vectors**

Find the angle θ between $\mathbf{u} = 2\mathbf{i} - 3\mathbf{j} + 6\mathbf{k}$ and $\mathbf{v} = 2\mathbf{i} + 5\mathbf{j} - \mathbf{k}$.

Solution We compute the quantities $\mathbf{u} \cdot \mathbf{v}$, $\|\mathbf{u}\|$, and $\|\mathbf{v}\|$.

$$\mathbf{u} \cdot \mathbf{v} = 2(2) + (-3)(5) + 6(-1) = -17$$
$$\|\mathbf{u}\| = \sqrt{2^2 + (-3)^2 + 6^2} = \sqrt{49} = 7$$
$$\|\mathbf{v}\| = \sqrt{2^2 + 5^2 + (-1)^2} = \sqrt{30}$$

By formula (4), if θ is the angle between **u** and **v**, then

$$\cos\theta = \frac{\mathbf{u}\cdot\mathbf{v}}{\|\mathbf{u}\|\,\|\mathbf{v}\|} = \frac{-17}{7\sqrt{30}} \approx -0.443$$

We find that $\theta \approx 116.3°$. ◀

NOW WORK PROBLEM 51.

Direction Angles of Vectors in Space

6 A nonzero vector **v** in space can be described by specifying its magnitude and its three **direction angles** α, β, and γ. These direction angles are defined as

α = angle between **v** and **i**, the positive x-axis, $0 \le \alpha \le \pi$

β = angle between **v** and **j**, the positive y-axis, $0 \le \beta \le \pi$

γ = angle between **v** and **k**, the positive z-axis, $0 \le \gamma \le \pi$

See Figure 75.

Figure 75

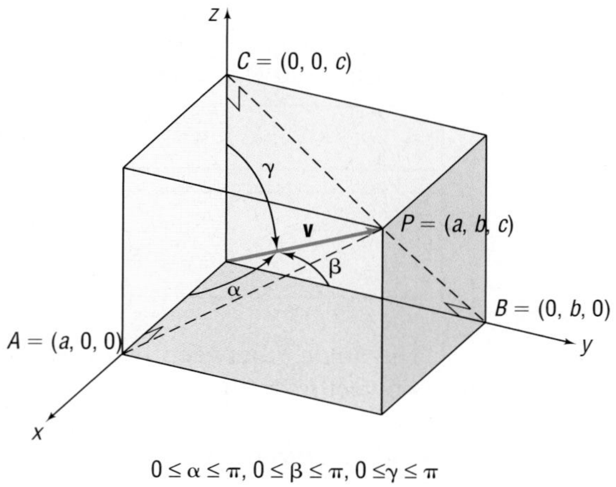

Our first goal is to find an expression for α, β, and γ in terms of the components of a vector. Let $\mathbf{v} = a\mathbf{i} + b\mathbf{j} + c\mathbf{k}$ denote a nonzero vector. The angle α between **v** and **i**, the positive x-axis, obeys

$$\cos\alpha = \frac{\mathbf{v}\cdot\mathbf{i}}{\|\mathbf{v}\|\,\|\mathbf{i}\|} = \frac{a}{\|\mathbf{v}\|}$$

Similarly,

$$\cos\beta = \frac{b}{\|\mathbf{v}\|} \qquad \cos\gamma = \frac{c}{\|\mathbf{v}\|}$$

Since $\|\mathbf{v}\| = \sqrt{a^2 + b^2 + c^2}$, we have the following result:

Theorem

Direction Angles

If $\mathbf{v} = a\mathbf{i} + b\mathbf{j} + c\mathbf{k}$ is a nonzero vector in space, the direction angles α, β, and γ obey

$$\cos\alpha = \frac{a}{\sqrt{a^2 + b^2 + c^2}} = \frac{a}{\|\mathbf{v}\|} \qquad \cos\beta = \frac{b}{\sqrt{a^2 + b^2 + c^2}} = \frac{b}{\|\mathbf{v}\|}$$
$$\cos\gamma = \frac{c}{\sqrt{a^2 + b^2 + c^2}} = \frac{c}{\|\mathbf{v}\|} \qquad \textbf{(5)}$$

The numbers $\cos\alpha$, $\cos\beta$, and $\cos\gamma$ are called the **direction cosines** of the vector $\mathbf{v}$. They play the same role in space as slope does in the plane.

EXAMPLE 8 Finding the Direction Angles of a Vector

Find the direction angles of $\mathbf{v} = -3\mathbf{i} + 2\mathbf{j} - 6\mathbf{k}$.

Solution

$$\|\mathbf{v}\| = \sqrt{(-3)^2 + 2^2 + (-6)^2} = \sqrt{49} = 7$$

Using the theorem on direction angles, we get

$$\cos\alpha = \frac{-3}{7} \qquad \cos\beta = \frac{2}{7} \qquad \cos\gamma = \frac{-6}{7}$$
$$\alpha \approx 115.4° \qquad \beta \approx 73.4° \qquad \gamma \approx 149.0°$$

◀

Theorem

Property of the Direction Cosines

If α, β, and γ are the direction angles of a nonzero vector $\mathbf{v}$ in space, then

$$\cos^2\alpha + \cos^2\beta + \cos^2\gamma = 1 \qquad \textbf{(6)}$$

The proof is a direct consequence of the equations in (5).

Based on equation (6), when two direction cosines are known, the third is determined up to its sign. Knowing two direction cosines is not sufficient to uniquely determine the direction of a vector in space.

EXAMPLE 9 Finding the Direction Angle of a Vector

The vector $\mathbf{v}$ makes an angle of $\alpha = \frac{\pi}{3}$ with the positive x-axis, an angle of $\beta = \frac{\pi}{3}$ with the positive y-axis, and an acute angle γ with the positive z-axis. Find γ.

Solution By equation (6), we have

$$\cos^2\left(\frac{\pi}{3}\right) + \cos^2\left(\frac{\pi}{3}\right) + \cos^2 \gamma = 1 \qquad 0 < \gamma < \frac{\pi}{2}$$

$$\left(\frac{1}{2}\right)^2 + \left(\frac{1}{2}\right)^2 + \cos^2 \gamma = 1$$

$$\cos^2 \gamma = \frac{1}{2}$$

$$\cos \gamma = \frac{\sqrt{2}}{2} \quad \text{or} \quad \cos \gamma = -\frac{\sqrt{2}}{2}$$

$$\gamma = \frac{\pi}{4} \quad \text{or} \quad \gamma = \frac{3\pi}{4}$$

Since we are requiring that γ be acute, the answer is $\gamma = \frac{\pi}{4}$. ◀

The direction cosines of a vector give information about only the direction of the vector; they provide no information about its magnitude. For example, *any* vector parallel to the xy-plane and making an angle of $\frac{\pi}{4}$ radian with the positive x- and y-axes has direction cosines

$$\cos \alpha = \frac{\sqrt{2}}{2} \qquad \cos \beta = \frac{\sqrt{2}}{2} \qquad \cos \gamma = 0$$

However, if the direction angles *and* the magnitude of a vector are known, then the vector is uniquely determined.

EXAMPLE 10 Writing a Vector in Terms of Its Magnitude and Direction Cosines

Show that any nonzero vector $\mathbf{v}$ in space can be written in terms of its magnitude and direction cosines as

$$\mathbf{v} = \|\mathbf{v}\|[(\cos \alpha)\mathbf{i} + (\cos \beta)\mathbf{j} + (\cos \gamma)\mathbf{k}] \tag{7}$$

Solution Let $\mathbf{v} = a\mathbf{i} + b\mathbf{j} + c\mathbf{k}$. From the equations in (5), we see that

$$a = \|\mathbf{v}\| \cos \alpha \qquad b = \|\mathbf{v}\| \cos \beta \qquad c = \|\mathbf{v}\| \cos \gamma$$

Substituting, we find that

$$\mathbf{v} = a\mathbf{i} + b\mathbf{j} + c\mathbf{k} = \|\mathbf{v}\|(\cos \alpha)\mathbf{i} + \|\mathbf{v}\|(\cos \beta)\mathbf{j} + \|\mathbf{v}\|(\cos \gamma)\mathbf{k}$$
$$= \|\mathbf{v}\|[(\cos \alpha)\mathbf{i} + (\cos \beta)\mathbf{j} + (\cos \gamma)\mathbf{k}]$$ ◀

NOW WORK PROBLEM 59.

Example 10 shows that the direction cosines of a vector $\mathbf{v}$ are also the components of the unit vector in the direction of $\mathbf{v}$.

8.6 Assess Your Understanding

'Are You Prepared?' *Answer is given at the end of these exercises. If you get a wrong answer, read the pages in red.*

1. The distance d from $P_1 = (x_1, y_1)$ to $P_2 = (x_2, y_2)$ is $d =$ ________ (p. 3)

Concepts and Vocabulary

2. In space, points of the form $(x, y, 0)$ lie in a plane called the ________.

3. If $\mathbf{v} = a\mathbf{i} + b\mathbf{j} + c\mathbf{k}$ is a vector in space, the scalars a, b, c are called the ________ of $\mathbf{v}$.

4. The sum of the squares of the direction cosines of a vector in space add up to ________.

5. *True or False:* In space, the dot product of two vectors is a positive number.

6. *True or False:* A vector in space may be described by specifying its magnitude and its direction angles.

Exercises

In Problems 7–14, describe the set of points (x, y, z) defined by the equation.

7. $y = 0$ **8.** $x = 0$ **9.** $z = 2$ **10.** $y = 3$

11. $x = -4$ **12.** $z = -3$ **13.** $x = 1$ and $y = 2$ **14.** $x = 3$ and $z = 1$

In Problems 15–20, find the distance from P_1 to P_2.

15. $P_1 = (0, 0, 0)$ and $P_2 = (4, 1, 2)$ **16.** $P_1 = (0, 0, 0)$ and $P_2 = (1, -2, 3)$

17. $P_1 = (-1, 2, -3)$ and $P_2 = (0, -2, 1)$ **18.** $P_1 = (-2, 2, 3)$ and $P_2 = (4, 0, -3)$

19. $P_1 = (4, -2, -2)$ and $P_2 = (3, 2, 1)$ **20.** $P_1 = (2, -3, -3)$ and $P_2 = (4, 1, -1)$

In Problems 21–26, opposite vertices of a rectangular box whose edges are parallel to the coordinate axes are given. List the coordinates of the other six vertices of the box.

21. $(0, 0, 0)$; $(2, 1, 3)$ **22.** $(0, 0, 0)$; $(4, 2, 2)$ **23.** $(1, 2, 3)$; $(3, 4, 5)$

24. $(5, 6, 1)$; $(3, 8, 2)$ **25.** $(-1, 0, 2)$; $(4, 2, 5)$ **26.** $(-2, -3, 0)$; $(-6, 7, 1)$

In Problems 27–32, the vector $\mathbf{v}$ *has initial point P and terminal point Q. Write* $\mathbf{v}$ *in the form $a\mathbf{i} + b\mathbf{j} + c\mathbf{k}$; that is, find its position vector.*

27. $P = (0, 0, 0)$; $Q = (3, 4, -1)$ **28.** $P = (0, 0, 0)$; $Q = (-3, -5, 4)$

29. $P = (3, 2, -1)$; $Q = (5, 6, 0)$ **30.** $P = (-3, 2, 0)$; $Q = (6, 5, -1)$

31. $P = (-2, -1, 4)$; $Q = (6, -2, 4)$ **32.** $P = (-1, 4, -2)$; $Q = (6, 2, 2)$

In Problems 33–38, find $\|\mathbf{v}\|$.

33. $\mathbf{v} = 3\mathbf{i} - 6\mathbf{j} - 2\mathbf{k}$ **34.** $\mathbf{v} = -6\mathbf{i} + 12\mathbf{j} + 4\mathbf{k}$ **35.** $\mathbf{v} = \mathbf{i} - \mathbf{j} + \mathbf{k}$

36. $\mathbf{v} = -\mathbf{i} - \mathbf{j} + \mathbf{k}$ **37.** $\mathbf{v} = -2\mathbf{i} + 3\mathbf{j} - 3\mathbf{k}$ **38.** $\mathbf{v} = 6\mathbf{i} + 2\mathbf{j} - 2\mathbf{k}$

In Problems 39–44, find each quantity if $\mathbf{v} = 3\mathbf{i} - 5\mathbf{j} + 2\mathbf{k}$ and $\mathbf{w} = -2\mathbf{i} + 3\mathbf{j} - 2\mathbf{k}$.

39. $2\mathbf{v} + 3\mathbf{w}$ **40.** $3\mathbf{v} - 2\mathbf{w}$ **41.** $\|\mathbf{v} - \mathbf{w}\|$

42. $\|\mathbf{v} + \mathbf{w}\|$ **43.** $\|\mathbf{v}\| - \|\mathbf{w}\|$ **44.** $\|\mathbf{v}\| + \|\mathbf{w}\|$

In Problems 45–50, find the unit vector having the same direction as $\mathbf{v}$.

45. $\mathbf{v} = 5\mathbf{i}$ **46.** $\mathbf{v} = -3\mathbf{j}$ **47.** $\mathbf{v} = 3\mathbf{i} - 6\mathbf{j} - 2\mathbf{k}$

48. $\mathbf{v} = -6\mathbf{i} + 12\mathbf{j} + 4\mathbf{k}$ **49.** $\mathbf{v} = \mathbf{i} + \mathbf{j} + \mathbf{k}$ **50.** $\mathbf{v} = 2\mathbf{i} - \mathbf{j} + \mathbf{k}$

In Problems 51–58, find the dot product $\mathbf{v} \cdot \mathbf{w}$ and the angle between $\mathbf{v}$ *and* $\mathbf{w}$.

51. $\mathbf{v} = \mathbf{i} - \mathbf{j}$, $\mathbf{w} = \mathbf{i} + \mathbf{j} + \mathbf{k}$ **52.** $\mathbf{v} = \mathbf{i} + \mathbf{j}$, $\mathbf{w} = -\mathbf{i} + \mathbf{j} - \mathbf{k}$

53. $\mathbf{v} = 2\mathbf{i} + \mathbf{j} - 3\mathbf{k}$, $\mathbf{w} = \mathbf{i} + 2\mathbf{j} + 2\mathbf{k}$ **54.** $\mathbf{v} = 2\mathbf{i} + 2\mathbf{j} - \mathbf{k}$, $\mathbf{w} = \mathbf{i} + 2\mathbf{j} + 3\mathbf{k}$

55. $\mathbf{v} = 3\mathbf{i} - \mathbf{j} + 2\mathbf{k}$, $\mathbf{w} = \mathbf{i} + \mathbf{j} - \mathbf{k}$ **56.** $\mathbf{v} = \mathbf{i} + 3\mathbf{j} + 2\mathbf{k}$, $\mathbf{w} = \mathbf{i} - \mathbf{j} + \mathbf{k}$

57. $\mathbf{v} = 3\mathbf{i} + 4\mathbf{j} + \mathbf{k}$, $\mathbf{w} = 6\mathbf{i} + 8\mathbf{j} + 2\mathbf{k}$ **58.** $\mathbf{v} = 3\mathbf{i} - 4\mathbf{j} + \mathbf{k}$, $\mathbf{w} = 6\mathbf{i} - 8\mathbf{j} + 2\mathbf{k}$

In Problems 59–66, find the direction angles of each vector. Write each vector in the form of equation (7).

59. $\mathbf{v} = 3\mathbf{i} - 6\mathbf{j} - 2\mathbf{k}$
60. $\mathbf{v} = -6\mathbf{i} + 12\mathbf{j} + 4\mathbf{k}$
61. $\mathbf{v} = \mathbf{i} + \mathbf{j} + \mathbf{k}$
62. $\mathbf{v} = \mathbf{i} - \mathbf{j} - \mathbf{k}$
63. $\mathbf{v} = \mathbf{i} + \mathbf{j}$
64. $\mathbf{v} = \mathbf{j} + \mathbf{k}$
65. $\mathbf{v} = 3\mathbf{i} - 5\mathbf{j} + 2\mathbf{k}$
66. $\mathbf{v} = 2\mathbf{i} + 3\mathbf{j} - 4\mathbf{k}$

67. **The Sphere** In space, the collection of all points that are the same distance from some fixed point is called a **sphere.** See the illustration. The constant distance is called the **radius,** and the fixed point is the **center** of the sphere. Show that the equation of a sphere with center at (x_0, y_0, z_0) and radius r is

$$(x - x_0)^2 + (y - y_0)^2 + (z - z_0)^2 = r^2$$

[**Hint:** Use the Distance Formula (1).]

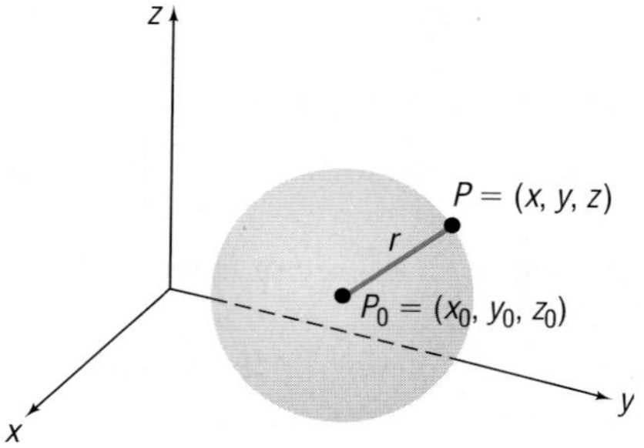

In Problems 68–70, find the equation of a sphere with radius r and center P_0.

68. $r = 1$; $P_0 = (3, 1, 1)$
69. $r = 2$; $P_0 = (1, 2, 2)$
70. $r = 3$; $P_0 = (-1, 1, 2)$

In Problems 71–76, find the radius and center of each sphere.
[**Hint**: *Complete the square in each variable.*]

71. $x^2 + y^2 + z^2 + 2x - 2y = 2$
72. $x^2 + y^2 + z^2 + 2x - 2z = -1$
73. $x^2 + y^2 + z^2 - 4x + 4y + 2z = 0$
74. $x^2 + y^2 + z^2 - 4x = 0$
75. $2x^2 + 2y^2 + 2z^2 - 8x + 4z = -1$
76. $3x^2 + 3y^2 + 3z^2 + 6x - 6y = 3$

The **work** *W done by a constant force* **F** *in moving an object from a point A in space to a point B in space is defined as* $W = \mathbf{F} \cdot \overrightarrow{AB}$. *Use this definition in Problems 77–79.*

77. **Work** Find the work done by a force of 3 newtons acting in the direction $2\mathbf{i} + \mathbf{j} + 2\mathbf{k}$ in moving an object 2 meters from $(0, 0, 0)$ to $(0, 2, 0)$.

78. **Work** Find the work done by a force of 1 newton acting in the direction $2\mathbf{i} + 2\mathbf{j} + \mathbf{k}$ in moving an object 3 meters from $(0, 0, 0)$ to $(1, 2, 2)$.

79. **Work** Find the work done in moving an object along a vector $\mathbf{u} = 3\mathbf{i} + 2\mathbf{j} - 5\mathbf{k}$ if the applied force is $\mathbf{F} = 2\mathbf{i} - \mathbf{j} - \mathbf{k}$.

'Are You Prepared?' Answer

1. $d = \sqrt{(x_2 - x_1)^2 + (y_2 - y_1)^2}$

8.7 The Cross Product

OBJECTIVES
1. Find the Cross Product of Two Vectors
2. Know Algebraic Properties of the Cross Product
3. Know Geometric Properties of the Cross Product
4. Find a Vector Orthogonal to Two Given Vectors
5. Find the Area of a Parallelogram

1 For vectors in space, and only for vectors in space, a second product of two vectors is defined, called the *cross product*. The cross product of two vectors in space is, in fact, also a vector that has applications in both geometry and physics.

If $\mathbf{v} = a_1\mathbf{i} + b_1\mathbf{j} + c_1\mathbf{k}$ and $\mathbf{w} = a_2\mathbf{i} + b_2\mathbf{j} + c_2\mathbf{k}$ are two vectors in space, the **cross product** $\mathbf{v} \times \mathbf{w}$ is defined as the vector

$$\mathbf{v} \times \mathbf{w} = (b_1c_2 - b_2c_1)\mathbf{i} - (a_1c_2 - a_2c_1)\mathbf{j} + (a_1b_2 - a_2b_1)\mathbf{k} \quad \textbf{(1)}$$

Notice that the cross product $\mathbf{v} \times \mathbf{w}$ of two vectors is a vector. Because of this, it is sometimes referred to as the **vector product.**

EXAMPLE 1 **Finding Cross Products Using Equation (1)**

If $\mathbf{v} = 2\mathbf{i} + 3\mathbf{j} + 5\mathbf{k}$ and $\mathbf{w} = \mathbf{i} + 2\mathbf{j} + 3\mathbf{k}$, then an application of equation (1) gives

$$\begin{aligned} \mathbf{v} \times \mathbf{w} &= (3 \cdot 3 - 2 \cdot 5)\mathbf{i} - (2 \cdot 3 - 1 \cdot 5)\mathbf{j} + (2 \cdot 2 - 1 \cdot 3)\mathbf{k} \\ &= (9 - 10)\mathbf{i} - (6 - 5)\mathbf{j} + (4 - 3)\mathbf{k} \\ &= -\mathbf{i} - \mathbf{j} + \mathbf{k} \end{aligned}$$

◄

Determinants* may be used as an aid in computing cross products. A **2 by 2 determinant,** symbolized by

$$\begin{vmatrix} a_1 & b_1 \\ a_2 & b_2 \end{vmatrix}$$

has the value $a_1b_2 - a_2b_1$; that is,

$$\begin{vmatrix} a_1 & b_1 \\ a_2 & b_2 \end{vmatrix} = a_1b_2 - a_2b_1$$

A **3 by 3 determinant** has the value

$$\begin{vmatrix} A & B & C \\ a_1 & b_1 & c_1 \\ a_2 & b_2 & c_2 \end{vmatrix} = \begin{vmatrix} b_1 & c_1 \\ b_2 & c_2 \end{vmatrix} A - \begin{vmatrix} a_1 & c_1 \\ a_2 & c_2 \end{vmatrix} B + \begin{vmatrix} a_1 & b_1 \\ a_2 & b_2 \end{vmatrix} C$$

EXAMPLE 2 **Evaluating Determinants**

(a) $\begin{vmatrix} 2 & 3 \\ 1 & 2 \end{vmatrix} = 2 \cdot 2 - 1 \cdot 3 = 4 - 3 = 1$

(b) $$\begin{aligned} \begin{vmatrix} A & B & C \\ 2 & 3 & 5 \\ 1 & 2 & 3 \end{vmatrix} &= \begin{vmatrix} 3 & 5 \\ 2 & 3 \end{vmatrix} A - \begin{vmatrix} 2 & 5 \\ 1 & 3 \end{vmatrix} B + \begin{vmatrix} 2 & 3 \\ 1 & 2 \end{vmatrix} C \\ &= (9 - 10)A - (6 - 5)B + (4 - 3)C \\ &= -A - B + C \end{aligned}$$

◄

NOW WORK PROBLEM 7.

The cross product of the vectors $\mathbf{v} = a_1\mathbf{i} + b_1\mathbf{j} + c_1\mathbf{k}$ and $\mathbf{w} = a_2\mathbf{i} + b_2\mathbf{j} + c_2\mathbf{k}$, that is,

$$\mathbf{v} \times \mathbf{w} = (b_1c_2 - b_2c_1)\mathbf{i} - (a_1c_2 - a_2c_1)\mathbf{j} + (a_1b_2 - a_2b_1)\mathbf{k}$$

*Determinants are discussed in detail in Section 10.3.

may be written symbolically using determinants as

$$\mathbf{v} \times \mathbf{w} = \begin{vmatrix} \mathbf{i} & \mathbf{j} & \mathbf{k} \\ a_1 & b_1 & c_1 \\ a_2 & b_2 & c_2 \end{vmatrix} = \begin{vmatrix} b_1 & c_1 \\ b_2 & c_2 \end{vmatrix}\mathbf{i} - \begin{vmatrix} a_1 & c_1 \\ a_2 & c_2 \end{vmatrix}\mathbf{j} + \begin{vmatrix} a_1 & b_1 \\ a_2 & b_2 \end{vmatrix}\mathbf{k}$$

EXAMPLE 3 Using Determinants to Find Cross Products

If $\mathbf{v} = 2\mathbf{i} + 3\mathbf{j} + 5\mathbf{k}$ and $\mathbf{w} = \mathbf{i} + 2\mathbf{j} + 3\mathbf{k}$, find:

(a) $\mathbf{v} \times \mathbf{w}$ (b) $\mathbf{w} \times \mathbf{v}$ (c) $\mathbf{v} \times \mathbf{v}$ (d) $\mathbf{w} \times \mathbf{w}$

Solution

(a) $$\mathbf{v} \times \mathbf{w} = \begin{vmatrix} \mathbf{i} & \mathbf{j} & \mathbf{k} \\ 2 & 3 & 5 \\ 1 & 2 & 3 \end{vmatrix} = \begin{vmatrix} 3 & 5 \\ 2 & 3 \end{vmatrix}\mathbf{i} - \begin{vmatrix} 2 & 5 \\ 1 & 3 \end{vmatrix}\mathbf{j} + \begin{vmatrix} 2 & 3 \\ 1 & 2 \end{vmatrix}\mathbf{k} = -\mathbf{i} - \mathbf{j} + \mathbf{k}$$

(b) $$\mathbf{w} \times \mathbf{v} = \begin{vmatrix} \mathbf{i} & \mathbf{j} & \mathbf{k} \\ 1 & 2 & 3 \\ 2 & 3 & 5 \end{vmatrix} = \begin{vmatrix} 2 & 3 \\ 3 & 5 \end{vmatrix}\mathbf{i} - \begin{vmatrix} 1 & 3 \\ 2 & 5 \end{vmatrix}\mathbf{j} + \begin{vmatrix} 1 & 2 \\ 2 & 3 \end{vmatrix}\mathbf{k} = \mathbf{i} + \mathbf{j} - \mathbf{k}$$

(c) $$\mathbf{v} \times \mathbf{v} = \begin{vmatrix} \mathbf{i} & \mathbf{j} & \mathbf{k} \\ 2 & 3 & 5 \\ 2 & 3 & 5 \end{vmatrix}$$

$$= \begin{vmatrix} 3 & 5 \\ 3 & 5 \end{vmatrix}\mathbf{i} - \begin{vmatrix} 2 & 5 \\ 2 & 5 \end{vmatrix}\mathbf{j} + \begin{vmatrix} 2 & 3 \\ 2 & 3 \end{vmatrix}\mathbf{k} = 0\mathbf{i} - 0\mathbf{j} + 0\mathbf{k} = \mathbf{0}$$

(d) $$\mathbf{w} \times \mathbf{w} = \begin{vmatrix} \mathbf{i} & \mathbf{j} & \mathbf{k} \\ 1 & 2 & 3 \\ 1 & 2 & 3 \end{vmatrix}$$

$$= \begin{vmatrix} 2 & 3 \\ 2 & 3 \end{vmatrix}\mathbf{i} - \begin{vmatrix} 1 & 3 \\ 1 & 3 \end{vmatrix}\mathbf{j} + \begin{vmatrix} 1 & 2 \\ 1 & 2 \end{vmatrix}\mathbf{k} = 0\mathbf{i} - 0\mathbf{j} + 0\mathbf{k} = \mathbf{0}$$

◀

NOW WORK PROBLEM 15.

Algebraic Properties of the Cross Product

2 Notice in Examples 3(a) and 3(b) that $\mathbf{v} \times \mathbf{w}$ and $\mathbf{w} \times \mathbf{v}$ are negatives of one another. From Examples 3(c) and 3(d), we might conjecture that the cross product of a vector with itself is the zero vector. These and other algebraic properties of the cross product are given next.

Theorem

Algebraic Properties of the Cross Product

If $\mathbf{u}$, $\mathbf{v}$, and $\mathbf{w}$ are vectors in space and if α is a scalar, then

$$\mathbf{u} \times \mathbf{u} = \mathbf{0} \quad (2)$$

$$\mathbf{u} \times \mathbf{v} = -(\mathbf{v} \times \mathbf{u}) \quad (3)$$

$$\alpha(\mathbf{u} \times \mathbf{v}) = (\alpha\mathbf{u}) \times \mathbf{v} = \mathbf{u} \times (\alpha\mathbf{v}) \quad (4)$$

$$\mathbf{u} \times (\mathbf{v} + \mathbf{w}) = (\mathbf{u} \times \mathbf{v}) + (\mathbf{u} \times \mathbf{w}) \quad (5)$$

Proof We will prove properties (2) and (4) here and leave properties (3) and (5) as exercises (see Problems 55 and 56).

To prove property (2), we let $\mathbf{u} = a_1\mathbf{i} + b_1\mathbf{j} + c_1\mathbf{k}$. Then

$$\mathbf{u} \times \mathbf{u} = \begin{vmatrix} \mathbf{i} & \mathbf{j} & \mathbf{k} \\ a_1 & b_1 & c_1 \\ a_1 & b_1 & c_1 \end{vmatrix} = \begin{vmatrix} b_1 & c_1 \\ b_1 & c_1 \end{vmatrix}\mathbf{i} - \begin{vmatrix} a_1 & c_1 \\ a_1 & c_1 \end{vmatrix}\mathbf{j} + \begin{vmatrix} a_1 & b_1 \\ a_1 & b_1 \end{vmatrix}\mathbf{k}$$
$$= 0\mathbf{i} - 0\mathbf{j} + 0\mathbf{k} = \mathbf{0}$$

To prove property (4), we let $\mathbf{u} = a_1\mathbf{i} + b_1\mathbf{j} + c_1\mathbf{k}$ and $\mathbf{v} = a_2\mathbf{i} + b_2\mathbf{j} + c_2\mathbf{k}$. Then

$$\alpha(\mathbf{u} \times \mathbf{v}) = \alpha[(b_1c_2 - b_2c_1)\mathbf{i} - (a_1c_2 - a_2c_1)\mathbf{j} + (a_1b_2 - a_2b_1)\mathbf{k}]$$

↑ Apply (1).

$$= \alpha(b_1c_2 - b_2c_1)\mathbf{i} - \alpha(a_1c_2 - a_2c_1)\mathbf{j} + \alpha(a_1b_2 - a_2b_1)\mathbf{k} \quad (6)$$

Since $\alpha\mathbf{u} = \alpha a_1\mathbf{i} + \alpha b_1\mathbf{j} + \alpha c_1\mathbf{k}$, we have

$$(\alpha\mathbf{u}) \times \mathbf{v} = (\alpha b_1c_2 - b_2\alpha c_1)\mathbf{i} - (\alpha a_1c_2 - a_2\alpha c_1)\mathbf{j} + (\alpha a_1b_2 - a_2\alpha b_1)\mathbf{k}$$
$$= \alpha(b_1c_2 - b_2c_1)\mathbf{i} - \alpha(a_1c_2 - a_2c_1)\mathbf{j} + \alpha(a_1b_2 - a_2b_1)\mathbf{k} \quad (7)$$

Based on equations (6) and (7), the first part of property (4) follows. The second part can be proved in like fashion. ■

NOW WORK PROBLEM 17.

3 Geometric Properties of the Cross Product

The cross product has several interesting geometric properties.

Theorem

Geometric Properties of the Cross Product

Let **u** and **v** be vectors in space.

$\mathbf{u} \times \mathbf{v}$ is orthogonal to both **u** and **v**. **(8)**

$\|\mathbf{u} \times \mathbf{v}\| = \|\mathbf{u}\|\,\|\mathbf{v}\| \sin\theta$, where θ is the angle between **u** and **v**. **(9)**

$\|\mathbf{u} \times \mathbf{v}\|$ is the area of the parallelogram having $\mathbf{u} \neq \mathbf{0}$ and $\mathbf{v} \neq \mathbf{0}$ as adjacent sides. **(10)**

$\mathbf{u} \times \mathbf{v} = \mathbf{0}$ if and only if **u** and **v** are parallel. **(11)**

Proof of Property (8) Let $\mathbf{u} = a_1\mathbf{i} + b_1\mathbf{j} + c_1\mathbf{k}$ and $\mathbf{v} = a_2\mathbf{i} + b_2\mathbf{j} + c_2\mathbf{k}$. Then

$$\mathbf{u} \times \mathbf{v} = (b_1c_2 - b_2c_1)\mathbf{i} - (a_1c_2 - a_2c_1)\mathbf{j} + (a_1b_2 - a_2b_1)\mathbf{k}$$

Now we compute the dot product $\mathbf{u} \cdot (\mathbf{u} \times \mathbf{v})$.

$$\mathbf{u} \cdot (\mathbf{u} \times \mathbf{v}) = (a_1\mathbf{i} + b_1\mathbf{j} + c_1\mathbf{k}) \cdot [(b_1c_2 - b_2c_1)\mathbf{i} - (a_1c_2 - a_2c_1)\mathbf{j} + (a_1b_2 - a_2b_1)\mathbf{k}]$$
$$= a_1(b_1c_2 - b_2c_1) - b_1(a_1c_2 - a_2c_1) + c_1(a_1b_2 - a_2b_1) = 0$$

Since two vectors are orthogonal if their dot product is zero, it follows that **u** and $\mathbf{u} \times \mathbf{v}$ are orthogonal. Similarly, $\mathbf{v} \cdot (\mathbf{u} \times \mathbf{v}) = 0$, so **v** and $\mathbf{u} \times \mathbf{v}$ are orthogonal. ■

4 As long as the vectors **u** and **v** are not parallel, they will form a plane in space. See Figure 76. Based on property (8), the vector $\mathbf{u} \times \mathbf{v}$ is normal to this plane. As Figure 76 illustrates, there are two vectors normal to the plane containing **u** and **v**. It can be shown that the vector $\mathbf{u} \times \mathbf{v}$ is the one determined by the thumb of the right hand when the other fingers of the right hand are cupped so that they point in a direction from **u** to **v**. See Figure 77.*

Figure 76

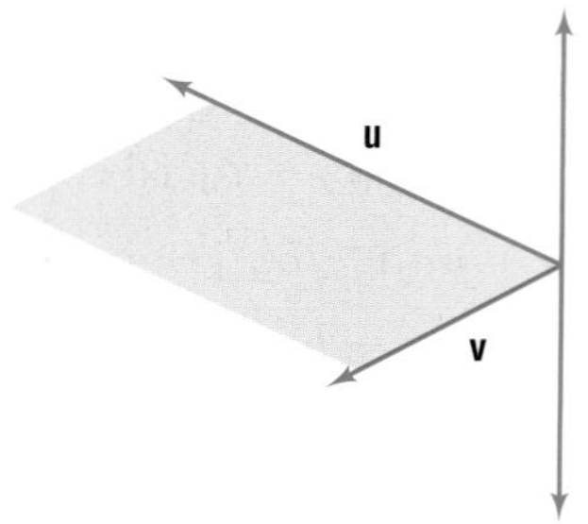

Figure 77

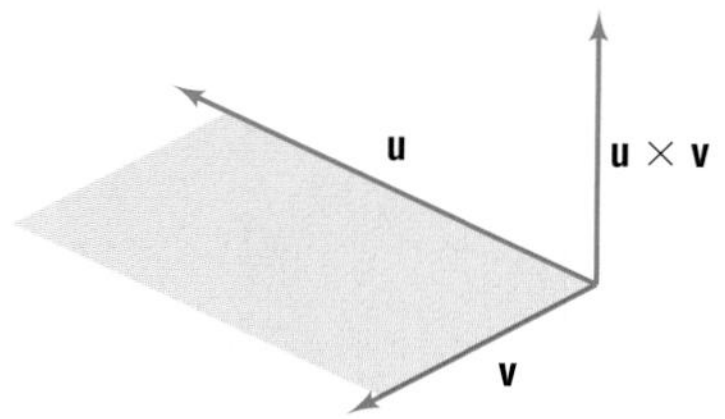

EXAMPLE 4 Finding a Vector Orthogonal to Two Given Vectors

Find a vector that is orthogonal to $\mathbf{u} = 3\mathbf{i} - 2\mathbf{j} + \mathbf{k}$ and $\mathbf{v} = -\mathbf{i} + 3\mathbf{j} - \mathbf{k}$.

Solution Based on property (8), such a vector is $\mathbf{u} \times \mathbf{v}$.

$$\mathbf{u} \times \mathbf{v} = \begin{vmatrix} \mathbf{i} & \mathbf{j} & \mathbf{k} \\ 3 & -2 & 1 \\ -1 & 3 & -1 \end{vmatrix} = (2 - 3)\mathbf{i} - [-3 - (-1)]\mathbf{j} + (9 - 2)\mathbf{k} = -\mathbf{i} + 2\mathbf{j} + 7\mathbf{k}$$

The vector $-\mathbf{i} + 2\mathbf{j} + 7\mathbf{k}$ is orthogonal to both **u** and **v**. ◀

CHECK: Two vectors are orthogonal if their dot product is zero.

$$\mathbf{u} \cdot (-\mathbf{i} + 2\mathbf{j} + 7\mathbf{k}) = (3\mathbf{i} - 2\mathbf{j} + \mathbf{k}) \cdot (-\mathbf{i} + 2\mathbf{j} + 7\mathbf{k}) = -3 - 4 + 7 = 0$$

$$\mathbf{v} \cdot (-\mathbf{i} + 2\mathbf{j} + 7\mathbf{k}) = (-\mathbf{i} + 3\mathbf{j} - \mathbf{k}) \cdot (-\mathbf{i} + 2\mathbf{j} + 7\mathbf{k}) = 1 + 6 - 7 = 0$$

NOW WORK PROBLEM 41.

The proof of property (9) is left as an exercise. See Problem 58.

Figure 78

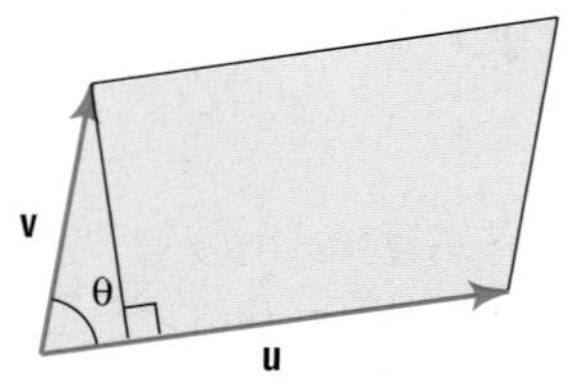

Proof of Property (10) Suppose that **u** and **v** are adjacent sides of a parallelogram. See Figure 78. Then the lengths of these sides are $\|\mathbf{u}\|$ and $\|\mathbf{v}\|$. If θ is the angle between **u** and **v**, then the height of the parallelogram is $\|\mathbf{v}\| \sin\theta$ and its area is

$$\text{Area of parallelogram} = \text{Base} \times \text{Height} = \|\mathbf{u}\|[\|\mathbf{v}\| \sin\theta] \underset{\uparrow \text{ Property (9)}}{=} \|\mathbf{u} \times \mathbf{v}\|$$

■

*This is a consequence of using a right-handed coordinate system.

5 **EXAMPLE 5** **Finding the Area of a Parallelogram**

Find the area of the parallelogram whose vertices are $P_1 = (0, 0, 0)$, $P_2 = (3, -2, 1)$, $P_3 = (-1, 3, -1)$, and $P_4 = (2, 1, 0)$.

Solution Two adjacent sides* of this parallelogram are

$$\mathbf{u} = \overrightarrow{P_1P_2} = 3\mathbf{i} - 2\mathbf{j} + \mathbf{k} \quad \text{and} \quad \mathbf{v} = \overrightarrow{P_1P_3} = -\mathbf{i} + 3\mathbf{j} - \mathbf{k}$$

Since $\mathbf{u} \times \mathbf{v} = -\mathbf{i} + 2\mathbf{j} + 7\mathbf{k}$ (Example 4), the area of the parallelogram is

$$\text{Area of parallelogram} = \|\mathbf{u} \times \mathbf{v}\| = \sqrt{1 + 4 + 49} = \sqrt{54} = 3\sqrt{6}$$ ◀

NOW WORK PROBLEM 49.

Proof of Property (11) The proof requires two parts. If **u** and **v** are parallel, then there is a scalar α such that $\mathbf{u} = \alpha\mathbf{v}$. Then

$$\mathbf{u} \times \mathbf{v} = (\alpha\mathbf{v}) \times \mathbf{v} \underset{\text{Property (4)}}{=} \alpha(\mathbf{v} \times \mathbf{v}) \underset{\text{Property (2)}}{=} \mathbf{0}$$

If $\mathbf{u} \times \mathbf{v} = \mathbf{0}$, then, by property (9), we have

$$\|\mathbf{u} \times \mathbf{v}\| = \|\mathbf{u}\|\,\|\mathbf{v}\| \sin\theta = 0$$

Since $\mathbf{u} \neq \mathbf{0}$ and $\mathbf{v} \neq \mathbf{0}$, then we must have $\sin\theta = 0$, so $\theta = 0$ or $\theta = \pi$. In either case, since θ is the angle between **u** and **v**, then **u** and **v** are parallel. ■

*Be careful! Not all pairs of vertices give rise to a side. For example, $\overrightarrow{P_1P_4}$ is a diagonal of the parallelogram since $\overrightarrow{P_1P_3} + \overrightarrow{P_3P_4} = \overrightarrow{P_1P_4}$. Also, $\overrightarrow{P_1P_3}$ and $\overrightarrow{P_2P_4}$ are not adjacent sides; they are parallel sides.

8.7 Assess Your Understanding

Concepts and Vocabulary

1. *True or False:* If **u** and **v** are parallel vectors, then $\mathbf{u} \times \mathbf{v} = \mathbf{0}$.
2. *True or False:* For any vector **v**, $\mathbf{v} \times \mathbf{v} = \mathbf{0}$.
3. *True or False:* If **u** and **v** are vectors, then $\mathbf{u} \times \mathbf{v} + \mathbf{v} \times \mathbf{u} = \mathbf{0}$.
4. *True or False:* $\mathbf{u} \times \mathbf{v}$ is a vector that is parallel to both **u** and **v**.
5. *True or False:* $\|\mathbf{u} \times \mathbf{v}\| = \|\mathbf{u}\|\,\|\mathbf{v}\| \cos\theta$, where θ is the angle between **u** and **v**.
6. *True or False:* The area of the parallelogram having **u** and **v** as adjacent sides is the magnitude of the cross product of **u** and **v**.

Exercises

In Problems 7–14, find the value of each determinant.

7. $\begin{vmatrix} 3 & 4 \\ 1 & 2 \end{vmatrix}$
8. $\begin{vmatrix} -2 & 5 \\ 2 & -3 \end{vmatrix}$
9. $\begin{vmatrix} 6 & 5 \\ -2 & -1 \end{vmatrix}$
10. $\begin{vmatrix} -4 & 0 \\ 5 & 3 \end{vmatrix}$
11. $\begin{vmatrix} A & B & C \\ 2 & 1 & 4 \\ 1 & 3 & 1 \end{vmatrix}$
12. $\begin{vmatrix} A & B & C \\ 0 & 2 & 4 \\ 3 & 1 & 3 \end{vmatrix}$
13. $\begin{vmatrix} A & B & C \\ -1 & 3 & 5 \\ 5 & 0 & -2 \end{vmatrix}$
14. $\begin{vmatrix} A & B & C \\ 1 & -2 & -3 \\ 0 & 2 & -2 \end{vmatrix}$

In Problems 15–22, find (a) $\mathbf{v} \times \mathbf{w}$, *(b)* $\mathbf{w} \times \mathbf{v}$, *(c)* $\mathbf{w} \times \mathbf{w}$, *and (d)* $\mathbf{v} \times \mathbf{v}$.

15. $\mathbf{v} = 2\mathbf{i} - 3\mathbf{j} + \mathbf{k}$, $\mathbf{w} = 3\mathbf{i} - 2\mathbf{j} - \mathbf{k}$

16. $\mathbf{v} = -\mathbf{i} + 3\mathbf{j} + 2\mathbf{k}$, $\mathbf{w} = 3\mathbf{i} - 2\mathbf{j} - \mathbf{k}$

17. $\mathbf{v} = \mathbf{i} + \mathbf{j}$, $\mathbf{w} = 2\mathbf{i} + \mathbf{j} + \mathbf{k}$

18. $\mathbf{v} = \mathbf{i} - 4\mathbf{j} + 2\mathbf{k}$, $\mathbf{w} = 3\mathbf{i} + 2\mathbf{j} + \mathbf{k}$

19. $\mathbf{v} = 2\mathbf{i} - \mathbf{j} + 2\mathbf{k}$, $\mathbf{w} = \mathbf{j} - \mathbf{k}$

20. $\mathbf{v} = 3\mathbf{i} + \mathbf{j} + 3\mathbf{k}$, $\mathbf{w} = \mathbf{i} - \mathbf{k}$

21. $\mathbf{v} = \mathbf{i} - \mathbf{j} - \mathbf{k}$, $\mathbf{w} = 4\mathbf{i} - 3\mathbf{k}$

22. $\mathbf{v} = 2\mathbf{i} - 3\mathbf{j}$, $\mathbf{w} = 3\mathbf{j} - 2\mathbf{k}$

In Problems 23–44, use the vectors **u**, **v**, *and* **w** *given below to find each expression.*

$$\mathbf{u} = 2\mathbf{i} - 3\mathbf{j} + \mathbf{k} \qquad \mathbf{v} = -3\mathbf{i} + 3\mathbf{j} + 2\mathbf{k} \qquad \mathbf{w} = \mathbf{i} + \mathbf{j} + 3\mathbf{k}$$

23. $\mathbf{u} \times \mathbf{v}$

24. $\mathbf{v} \times \mathbf{w}$

25. $\mathbf{v} \times \mathbf{u}$

26. $\mathbf{w} \times \mathbf{v}$

27. $\mathbf{v} \times \mathbf{v}$

28. $\mathbf{w} \times \mathbf{w}$

29. $(3\mathbf{u}) \times \mathbf{v}$

30. $\mathbf{v} \times (4\mathbf{w})$

31. $\mathbf{u} \times (2\mathbf{v})$

32. $(-3\mathbf{v}) \times \mathbf{w}$

33. $\mathbf{u} \cdot (\mathbf{u} \times \mathbf{v})$

34. $\mathbf{v} \cdot (\mathbf{v} \times \mathbf{w})$

35. $\mathbf{u} \cdot (\mathbf{v} \times \mathbf{w})$

36. $(\mathbf{u} \times \mathbf{v}) \cdot \mathbf{w}$

37. $\mathbf{v} \cdot (\mathbf{u} \times \mathbf{w})$

38. $(\mathbf{v} \times \mathbf{u}) \cdot \mathbf{w}$

39. $\mathbf{u} \times (\mathbf{v} \times \mathbf{v})$

40. $(\mathbf{w} \times \mathbf{w}) \times \mathbf{v}$

41. Find a vector orthogonal to both **u** and **v**.

42. Find a vector orthogonal to both **u** and **w**.

43. Find a vector orthogonal to both **u** and $\mathbf{i} + \mathbf{j}$.

44. Find a vector orthogonal to both **u** and $\mathbf{j} + \mathbf{k}$.

In Problems 45–48, find the area of the parallelogram with one corner at P_1 *and adjacent sides* $\overrightarrow{P_1P_2}$ *and* $\overrightarrow{P_1P_3}$.

45. $P_1 = (0, 0, 0)$, $P_2 = (1, 2, 3)$, $P_3 = (-2, 3, 0)$

46. $P_1 = (0, 0, 0)$, $P_2 = (2, 3, 1)$, $P_3 = (-2, 4, 1)$

47. $P_1 = (1, 2, 0)$, $P_2 = (-2, 3, 4)$, $P_3 = (0, -2, 3)$

48. $P_1 = (-2, 0, 2)$, $P_2 = (2, 1, -1)$, $P_3 = (2, -1, 2)$

In Problems 49–52, find the area of the parallelogram with vertices P_1, P_2, P_3, *and* P_4.

49. $P_1 = (1, 1, 2)$, $P_2 = (1, 2, 3)$, $P_3 = (-2, 3, 0)$, $P_4 = (-2, 4, 1)$

50. $P_1 = (2, 1, 1)$, $P_2 = (2, 3, 1)$, $P_3 = (-2, 4, 1)$, $P_4 = (-2, 6, 1)$

51. $P_1 = (1, 2, -1)$, $P_2 = (4, 2, -3)$, $P_3 = (6, -5, 2)$, $P_4 = (9, -5, 0)$

52. $P_1 = (-1, 1, 1)$, $P_2 = (-1, 2, 2)$, $P_3 = (-3, 4, -5)$, $P_4 = (-3, 5, -4)$

53. Find a unit vector normal to the plane containing $\mathbf{v} = \mathbf{i} + 3\mathbf{j} - 2\mathbf{k}$ and $\mathbf{w} = -2\mathbf{i} + \mathbf{j} + 3\mathbf{k}$.

54. Find a unit vector normal to the plane containing $\mathbf{v} = 2\mathbf{i} + 3\mathbf{j} - \mathbf{k}$ and $\mathbf{w} = -2\mathbf{i} - 4\mathbf{j} - 3\mathbf{k}$.

55. Prove property (3).

56. Prove property (5).

57. Prove for vectors **u** and **v** that

$$\|\mathbf{u} \times \mathbf{v}\|^2 = \|\mathbf{u}\|^2\|\mathbf{v}\|^2 - (\mathbf{u} \cdot \mathbf{v})^2.$$

[**Hint:** Proceed as in the proof of property (4), computing first the left side and then the right side.]

58. Prove property (9).

[**Hint:** Use the result of Problem 57 and the fact that if θ is the angle between **u** and **v** then $\mathbf{u} \cdot \mathbf{v} = \|\mathbf{u}\|\,\|\mathbf{v}\| \cos \theta$.]

59. Show that if **u** and **v** are orthogonal then

$$\|\mathbf{u} \times \mathbf{v}\| = \|\mathbf{u}\|\,\|\mathbf{v}\|.$$

60. Show that if **u** and **v** are orthogonal unit vectors then so is $\mathbf{u} \times \mathbf{v}$ a unit vector.

61. If $\mathbf{u} \cdot \mathbf{v} = 0$ and $\mathbf{u} \times \mathbf{v} = \mathbf{0}$, what can you conclude about **u** and **v**?

Chapter Review

Things to Know

Relationship between polar coordinates (r, θ) and rectangular coordinates (x, y) (pp. 539 and 542)	$x = r\cos\theta,\ y = r\sin\theta$ $r^2 = x^2 + y^2,\ \tan\theta = \dfrac{y}{x},\quad x \neq 0$
Polar form of a complex number (p. 563)	If $z = x + yi$, then $z = r(\cos\theta + i\sin\theta)$, where $r = \lvert z\rvert = \sqrt{x^2 + y^2}$, $\sin\theta = \dfrac{y}{r}$, $\cos\theta = \dfrac{x}{r}$, $0 \le \theta < 2\pi$.
De Moivre's Theorem (p. 565)	If $z = r(\cos\theta + i\sin\theta)$, then $z^n = r^n[\cos(n\theta) + i\sin(n\theta)]$, where $n \ge 1$ is a positive integer.
nth root of a complex number $z = r(\cos\theta_0 + i\sin\theta_0)$ (p. 566)	$\sqrt[n]{z} = \sqrt[n]{r}\left[\cos\left(\dfrac{\theta_0}{n} + \dfrac{2k\pi}{n}\right) + i\sin\left(\dfrac{\theta_0}{n} + \dfrac{2k\pi}{n}\right)\right],\quad k = 0, \ldots, n-1,$ where $n \ge 2$ is an integer.
Vector (p. 570)	Quantity having magnitude and direction; equivalent to a directed line segment $\overrightarrow{PQ}$
Position vector (p. 573)	Vector whose initial point is at the origin
Unit vector (pp. 573 and 594)	Vector whose magnitude is 1
Dot product (pp. 582 and 594)	If $\mathbf{v} = a_1\mathbf{i} + b_1\mathbf{j}$ and $\mathbf{w} = a_2\mathbf{i} + b_2\mathbf{j}$, then $\mathbf{v}\cdot\mathbf{w} = a_1a_2 + b_1b_2$. If $\mathbf{v} = a_1\mathbf{i} + b_1\mathbf{j} + c_1\mathbf{k}$ and $\mathbf{w} = a_2\mathbf{i} + b_2\mathbf{j} + c_2\mathbf{k}_2$, then $\mathbf{v}\cdot\mathbf{w} = a_1a_2 + b_1b_2 + c_1c_2$.
Angle θ between two nonzero vectors $\mathbf{u}$ and $\mathbf{v}$ (pp. 584 and 595)	$\cos\theta = \dfrac{\mathbf{u}\cdot\mathbf{v}}{\lVert\mathbf{u}\rVert\,\lVert\mathbf{v}\rVert}$
Direction angles of a vector in space (p. 597)	If $\mathbf{v} = a\mathbf{i} + b\mathbf{j} + c\mathbf{k}$, then $\mathbf{v} = \lVert\mathbf{v}\rVert[(\cos\alpha)\mathbf{i} + (\cos\beta)\mathbf{j} + (\cos\gamma)\mathbf{k}]$, where $\cos\alpha = \dfrac{a}{\lVert\mathbf{v}\rVert}$, $\cos\beta = \dfrac{b}{\lVert\mathbf{v}\rVert}$, $\cos\gamma = \dfrac{c}{\lVert\mathbf{v}\rVert}$.
Cross product (p. 600)	If $\mathbf{v} = a_1\mathbf{i} + b_1\mathbf{j} + c_1\mathbf{k}$ and $\mathbf{w} = a_2\mathbf{i} + b_2\mathbf{j} + c_2\mathbf{k}$, then $\mathbf{v}\times\mathbf{w} = [b_1c_2 - b_2c_1]\mathbf{i} - [a_1c_2 - a_2c_1]\mathbf{j} + [a_1b_2 - a_2b_1]\mathbf{k}$.
Area of a parallelogram (p. 603)	$\lVert\mathbf{u}\times\mathbf{v}\rVert = \lVert\mathbf{u}\rVert\,\lVert\mathbf{v}\rVert\sin\theta$, where θ is the angle between $\mathbf{u}$ and $\mathbf{v}$.

Objectives

Section		You should be able to . . .	Review Exercises
8.1	1	Plot points using polar coordinates (p. 536)	1–6
	2	Convert from polar coordinates to rectangular coordinates (p. 539)	1–6
	3	Convert from rectangular coordinates to polar coordinates (p. 540)	7–12
8.2	1	Graph and identify polar equations by converting to rectangular equations (p. 546)	13–18
	2	Test polar equations for symmetry (p. 550)	19–24
	3	Graph polar equations by plotting points (p. 551)	19–24
8.3	1	Convert a complex number from rectangular form to polar form (p. 563)	25–28
	2	Plot points in the complex plane (p. 563)	29–34
	3	Find products and quotients of complex numbers in polar form (p. 564)	35–40

	4	Use De Moivre's Theorem (p. 565)	41–48
	5	Find complex roots (p. 566)	49–50
8.4	1	Graph vectors (p. 572)	51–54
	2	Find a position vector (p. 573)	55–58
	3	Add and subtract vectors (p. 575)	59, 60
	4	Find a scalar product and the magnitude of a vector (p. 576)	61–66
	5	Find a unit vector (p. 576)	67, 68
	6	Find a vector from its direction and magnitude (p. 577)	69, 70
	7	Work with objects in static equilibrium (p. 578)	111
8.5	1	Find the dot product of two vectors (p. 582)	85–88
	2	Find the angle between two vectors (p. 583)	85–88, 109, 110, 112
	3	Determine whether two vectors are parallel (p. 585)	93–98
	4	Determine whether two vectors are orthogonal (p. 585)	93–98
	5	Decompose a vector into two orthogonal vectors (p. 586)	99, 100
	6	Compute work (p. 587)	113
8.6	1	Find the distance between two points (p. 592)	71, 72
	2	Find position vectors (p. 592)	73, 74
	3	Perform operations on vectors (p. 593)	75–80
	4	Find the dot product (p. 594)	89–92
	5	Find the angle between two vectors (p. 595)	89–92
	6	Find the direction angles of a vector (p. 596)	103, 104
8.7	1	Find the cross product of two vectors (p. 600)	81, 82
	2	Know algebraic properties of the cross product (p. 602)	107, 108
	3	Know geometric properties of the cross product (p. 603)	105, 106
	4	Find a vector orthogonal to two given vectors (p. 604)	84
	5	Find the area of a parallelogram (p. 605)	105, 106

Review Exercises *(Blue problem numbers indicate the author's suggestions for use in a Practice Test.)*

In Problems 1–6, plot each point given in polar coordinates, and find its rectangular coordinates.

1. $\left(3, \frac{\pi}{6}\right)$ **2.** $\left(4, \frac{2\pi}{3}\right)$ **3.** $\left(-2, \frac{4\pi}{3}\right)$ **4.** $\left(-1, \frac{5\pi}{4}\right)$ **5.** $\left(-3, -\frac{\pi}{2}\right)$ **6.** $\left(-4, -\frac{\pi}{4}\right)$

In Problems 7–12, the rectangular coordinates of a point are given. Find two pairs of polar coordinates (r, θ) for each point, one with $r > 0$ and the other with $r < 0$. Express θ in radians.

7. $(-3, 3)$ **8.** $(1, -1)$ **9.** $(0, -2)$ **10.** $(2, 0)$ **11.** $(3, 4)$ **12.** $(-5, 12)$

In Problems 13–18, the letters r and θ represent polar coordinates. Write each polar equation as an equation in rectangular coordinates (x, y). Identify the equation and graph it.

13. $r = 2 \sin \theta$ **14.** $3r = \sin \theta$ **15.** $r = 5$

16. $\theta = \frac{\pi}{4}$ **17.** $r \cos \theta + 3r \sin \theta = 6$ **18.** $r^2 + 4r \sin \theta - 8r \cos \theta = 5$

In Problems 19–24, sketch the graph of each polar equation. Be sure to test for symmetry.

19. $r = 4\cos\theta$

20. $r = 3\sin\theta$

21. $r = 3 - 3\sin\theta$

22. $r = 2 + \cos\theta$

23. $r = 4 - \cos\theta$

24. $r = 1 - 2\sin\theta$

In Problems 25–28, write each complex number in polar form. Express each argument in degrees.

25. $-1 - i$

26. $-\sqrt{3} + i$

27. $4 - 3i$

28. $3 - 2i$

In Problems 29–34, write each complex number in the standard form $a + bi$ and plot each in the complex plane.

29. $2(\cos 150° + i\sin 150°)$

30. $3(\cos 60° + i\sin 60°)$

31. $3\left(\cos\frac{2\pi}{3} + i\sin\frac{2\pi}{3}\right)$

32. $4\left(\cos\frac{3\pi}{4} + i\sin\frac{3\pi}{4}\right)$

33. $0.1(\cos 350° + i\sin 350°)$

34. $0.5(\cos 160° + i\sin 160°)$

In Problems 35–40, find zw and $\frac{z}{w}$. Leave your answers in polar form.

35. $z = \cos 80° + i\sin 80°$
$w = \cos 50° + i\sin 50°$

36. $z = \cos 205° + i\sin 205°$
$w = \cos 85° + i\sin 85°$

37. $z = 3\left(\cos\frac{9\pi}{5} + i\sin\frac{9\pi}{5}\right)$
$w = 2\left(\cos\frac{\pi}{5} + i\sin\frac{\pi}{5}\right)$

38. $z = 2\left(\cos\frac{5\pi}{3} + i\sin\frac{5\pi}{3}\right)$
$w = 3\left(\cos\frac{\pi}{3} + i\sin\frac{\pi}{3}\right)$

39. $z = 5(\cos 10° + i\sin 10°)$
$w = \cos 355° + i\sin 355°$

40. $z = 4(\cos 50° + i\sin 50°)$
$w = \cos 340° + i\sin 340°$

In Problems 41–48, write each expression in the standard form $a + bi$.

41. $[3(\cos 20° + i\sin 20°)]^3$

42. $[2(\cos 50° + i\sin 50°)]^3$

43. $\left[\sqrt{2}\left(\cos\frac{5\pi}{8} + i\sin\frac{5\pi}{8}\right)\right]^4$

44. $\left[2\left(\cos\frac{5\pi}{16} + i\sin\frac{5\pi}{16}\right)\right]^4$

45. $\left(1 - \sqrt{3}i\right)^6$

46. $(2 - 2i)^8$

47. $(3 + 4i)^4$

48. $(1 - 2i)^4$

49. Find all the complex cube roots of 27.

50. Find all the complex fourth roots of -16.

In Problems 51–54, use the figure to graph each of the following:

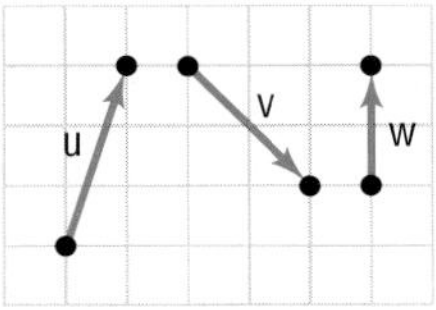

51. $\mathbf{u} + \mathbf{v}$

52. $\mathbf{v} + \mathbf{w}$

53. $2\mathbf{u} + 3\mathbf{v}$

54. $5\mathbf{v} - 2\mathbf{w}$

In Problems 55–58, the vector $\mathbf{v}$ is represented by the directed line segment $\overrightarrow{PQ}$. Write $\mathbf{v}$ in the form $a\mathbf{i} + b\mathbf{j}$ and find $\|\mathbf{v}\|$.

55. $P = (1, -2)$; $Q = (3, -6)$

56. $P = (-3, 1)$; $Q = (4, -2)$

57. $P = (0, -2)$; $Q = (-1, 1)$

58. $P = (3, -4)$; $Q = (-2, 0)$

In Problems 59–68, use the vectors $\mathbf{v} = -2\mathbf{i} + \mathbf{j}$ and $\mathbf{w} = 4\mathbf{i} - 3\mathbf{j}$ to find:

59. $\mathbf{v} + \mathbf{w}$

60. $\mathbf{v} - \mathbf{w}$

61. $4\mathbf{v} - 3\mathbf{w}$

62. $-\mathbf{v} + 2\mathbf{w}$

63. $\|\mathbf{v}\|$

64. $\|\mathbf{v} + \mathbf{w}\|$

65. $\|\mathbf{v}\| + \|\mathbf{w}\|$

66. $\|2\mathbf{v}\| - 3\|\mathbf{w}\|$

67. Find a unit vector in the same direction as $\mathbf{v}$.

68. Find a unit vector in the opposite direction of $\mathbf{w}$.

69. Find the vector **v** with magnitude 3 if the angle between **v** and **i** is 60°.

70. Find the vector **v** with magnitude 5 if the angle between **v** and **i** is 150°.

71. Find the distance from $P_1 = (1, 3, -2)$ to $P_2 = (4, -2, 1)$.

72. Find the distance from $P_1 = (0, -4, 3)$ to $P_2 = (6, -5, -1)$.

73. A vector **v** has initial point $P = (1, 3, -2)$ and terminal point $Q = (4, -2, 1)$. Write **v** in the form $\mathbf{v} = a\mathbf{i} + b\mathbf{j} + c\mathbf{k}$.

74. A vector **v** has initial point $P = (0, -4, 3)$ and terminal point $Q = (6, -5, -1)$. Write **v** in the form $\mathbf{v} = a\mathbf{i} + b\mathbf{j} + c\mathbf{k}$.

In Problems 75–84, use the vectors $\mathbf{v} = 3\mathbf{i} + \mathbf{j} - 2\mathbf{k}$ *and* $\mathbf{w} = -3\mathbf{i} + 2\mathbf{j} - \mathbf{k}$ *to find each expression.*

75. $4\mathbf{v} - 3\mathbf{w}$

76. $-\mathbf{v} + 2\mathbf{w}$

77. $\|\mathbf{v} - \mathbf{w}\|$

78. $\|\mathbf{v} + \mathbf{w}\|$

79. $\|\mathbf{v}\| - \|\mathbf{w}\|$

80. $\|\mathbf{v}\| + \|\mathbf{w}\|$

81. $\mathbf{v} \times \mathbf{w}$

82. $\mathbf{v} \cdot (\mathbf{v} \times \mathbf{w})$

83. Find a unit vector in the same direction as **v** and then in the opposite direction of **v**.

84. Find a unit vector orthogonal to both **v** and **w**.

In Problems 85–92, find the dot product $\mathbf{v} \cdot \mathbf{w}$ *and the angle between* **v** *and* **w**.

85. $\mathbf{v} = -2\mathbf{i} + \mathbf{j}, \quad \mathbf{w} = 4\mathbf{i} - 3\mathbf{j}$

86. $\mathbf{v} = 3\mathbf{i} - \mathbf{j}, \quad \mathbf{w} = \mathbf{i} + \mathbf{j}$

87. $\mathbf{v} = \mathbf{i} - 3\mathbf{j}, \quad \mathbf{w} = -\mathbf{i} + \mathbf{j}$

88. $\mathbf{v} = \mathbf{i} + 4\mathbf{j}, \quad \mathbf{w} = 3\mathbf{i} - 2\mathbf{j}$

89. $\mathbf{v} = \mathbf{i} + \mathbf{j} + \mathbf{k}, \quad \mathbf{w} = \mathbf{i} - \mathbf{j} + \mathbf{k}$

90. $\mathbf{v} = \mathbf{i} - \mathbf{j} + \mathbf{k}, \quad \mathbf{w} = 2\mathbf{i} + \mathbf{j} + \mathbf{k}$

91. $\mathbf{v} = 4\mathbf{i} - \mathbf{j} + 2\mathbf{k}, \quad \mathbf{w} = \mathbf{i} - 2\mathbf{j} - 3\mathbf{k}$

92. $\mathbf{v} = -\mathbf{i} - 2\mathbf{j} + 3\mathbf{k}, \quad \mathbf{w} = 5\mathbf{i} + \mathbf{j} + \mathbf{k}$

In Problems 93–98, determine whether **v** *and* **w** *are parallel, orthogonal, or neither.*

93. $\mathbf{v} = 2\mathbf{i} + 3\mathbf{j}; \quad \mathbf{w} = -4\mathbf{i} - 6\mathbf{j}$

94. $\mathbf{v} = -2\mathbf{i} - \mathbf{j}; \quad \mathbf{w} = 2\mathbf{i} + \mathbf{j}$

95. $\mathbf{v} = 3\mathbf{i} - 4\mathbf{j}; \quad \mathbf{w} = -3\mathbf{i} + 4\mathbf{j}$

96. $\mathbf{v} = -2\mathbf{i} + 2\mathbf{j}; \quad \mathbf{w} = -3\mathbf{i} + 2\mathbf{j}$

97. $\mathbf{v} = 3\mathbf{i} - 2\mathbf{j}; \quad \mathbf{w} = 4\mathbf{i} + 6\mathbf{j}$

98. $\mathbf{v} = -4\mathbf{i} + 2\mathbf{j}; \quad \mathbf{w} = 2\mathbf{i} + 4\mathbf{j}$

In Problems 99 and 100, decompose **v** *into two vectors, one parallel to* **w** *and the other orthogonal to* **w**.

99. $\mathbf{v} = 2\mathbf{i} + \mathbf{j}; \quad \mathbf{w} = -4\mathbf{i} + 3\mathbf{j}$

100. $\mathbf{v} = -3\mathbf{i} + 2\mathbf{j}; \quad \mathbf{w} = -2\mathbf{i} + \mathbf{j}$

101. Find the vector projection of $\mathbf{v} = 2\mathbf{i} + 3\mathbf{j}$ onto $\mathbf{w} = 3\mathbf{i} + \mathbf{j}$.

102. Find the vector projection of $\mathbf{v} = -\mathbf{i} + 2\mathbf{j}$ onto $\mathbf{w} = 3\mathbf{i} - \mathbf{j}$.

103. Find the direction angles of the vector $\mathbf{v} = 3\mathbf{i} - 4\mathbf{j} + 2\mathbf{k}$.

104. Find the direction angles of the vector $\mathbf{v} = \mathbf{i} - \mathbf{j} + 2\mathbf{k}$.

105. Find the area of the parallelogram with vertices $P_1 = (1, 1, 1)$, $P_2 = (2, 3, 4)$, $P_3 = (6, 5, 2)$, and $P_4 = (7, 7, 5)$.

106. Find the area of the parallelogram with vertices $P_1 = (2, -1, 1)$, $P_2 = (5, 1, 4)$, $P_3 = (0, 1, 1)$, and $P_4 = (3, 3, 4)$.

107. If $\mathbf{u} \times \mathbf{v} = 2\mathbf{i} - 3\mathbf{j} + \mathbf{k}$, what is $\mathbf{v} \times \mathbf{u}$?

108. Suppose that $\mathbf{u} = 3\mathbf{v}$. What is $\mathbf{u} \times \mathbf{v}$?

109. Actual Speed and Direction of a Swimmer A swimmer can maintain a constant speed of 5 miles per hour. If the swimmer heads directly across a river that has a current moving at the rate of 2 miles per hour, what is the actual speed of the swimmer? (See the figure.) If the river is 1 mile wide, how far downstream will the swimmer end up from the point directly across the river from the starting point?

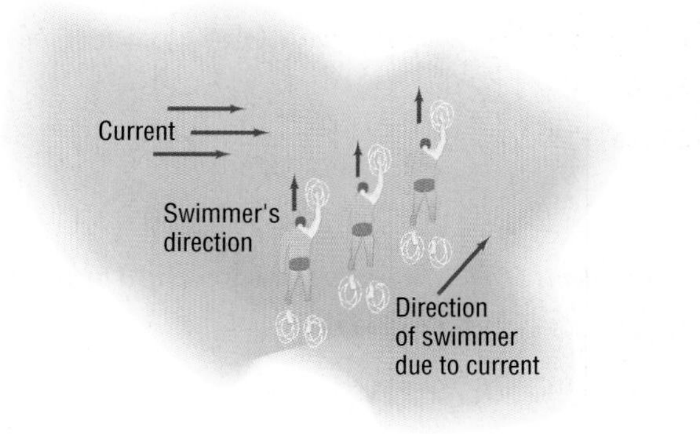

110. Actual Speed and Direction of an Airplane An airplane has an airspeed of 500 kilometers per hour in a northerly direction. The wind velocity is 60 kilometers per hour in a southeasterly direction. Find the actual speed and direction of the plane relative to the ground.

111. **Static Equilibrium** A weight of 2000 pounds is suspended from two cables as shown in the figure. What are the tensions in each cable?

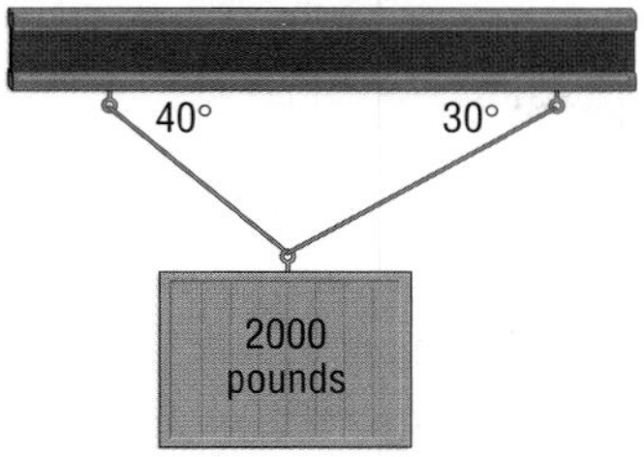

112. **Actual Speed and Distance of a Motorboat** A small motorboat is moving at a true speed of 11 miles per hour in a southerly direction. The current is known to be from the northeast at 3 miles per hour. What is the speed of the motorboat relative to the water? In what direction does the compass indicate that the boat is headed?

113. **Computing Work** Find the work done by a force of 5 pounds acting in the direction 60° to the horizontal in moving an object 20 feet from $(0, 0)$ to $(20, 0)$.

Chapter Projects

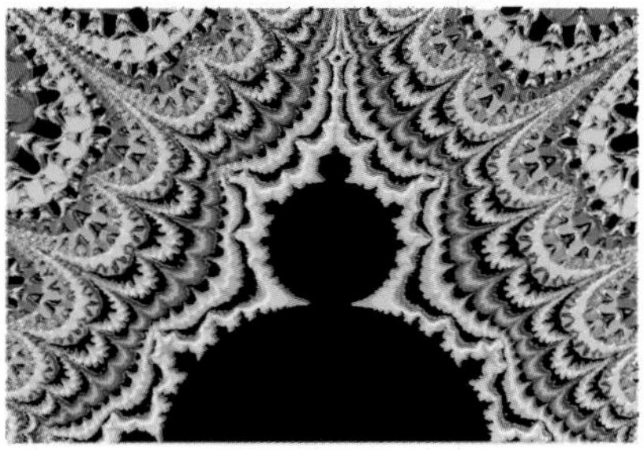

1. Mandelbrot Sets

(a) Let $z = x + yi$ be a complex number. We can plot complex numbers using a coordinate system called the complex plane. The x-axis will be referred to as the real axis, because any point that lies on the real axis is of the form $z = x + 0i = x$, a real number. The y-axis is called the imaginary axis, because any point that lies on it is of the form $z = 0 + yi = yi$, a pure imaginary number. To plot the complex number $z = x + yi$, plot the ordered pair (x, y) where x is the signed distance from the imaginary axis and y is the signed distance from the real axis. Draw a complex plane and plot the points $z_1 = 3 + 4i$, $z_2 = -2 + i$, $z_3 = 0 - 2i$, and $z_4 = -2$.

(b) Consider the expression $a_n = (a_{n-1})^2 + z$, where z is some complex number (called the **seed**) and $a_0 = z$. Compute $a_1 \, (= a_0^2 + z)$, $a_2 \, (= a_1^2 + z)$, $a_3(= a_2^2 + z)$, a_4, a_5, and a_6 for the following seeds: $z_1 = 0.1 - 0.4i$, $z_2 = 0.5 + 0.8i$, $z_3 = -0.9 + 0.7i$, $z_4 = -1.1 + 0.1i$, $z_5 = 0 - 1.3i$, and $z_6 = 1 + 1i$.

(c) The dark portion of the graph below, represents the set of all values $z = x + yi$ that are in the Mandelbrot set. Determine which complex numbers in part (b) are in this set by plotting them on the graph. Do the complex numbers that are not in the Mandelbrot set have any common characteristics regarding the values of a_6 found in part (b)?

(d) Compute $|z| = \sqrt{x^2 + y^2}$ for each of the complex numbers in part (b). Now compute $|a_6|$ for each of the complex numbers in part (b). For which complex numbers is $|a_6| \geq |z|$ and $|z| > 2$? Conclude that the criterion for a complex number to be excluded from the Mandelbrot set is that $|a_n| \geq |z|$ and $|z| > 2$.

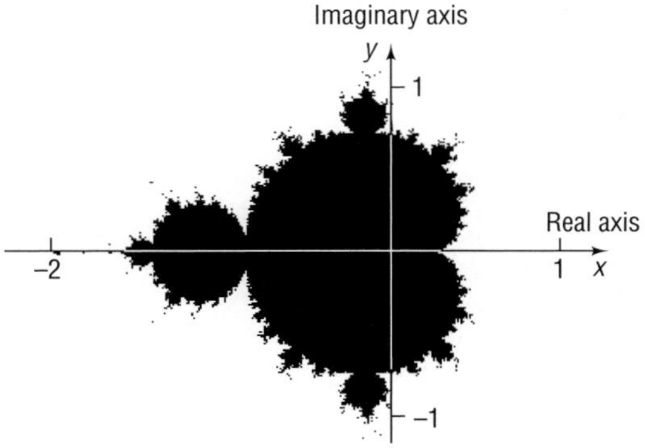

The following projects may be found at www.prenhall.com/sullivan7e.

2. **Project at Motorola** ***Signal Fades Due to Interference***
3. **Compound Interest**
4. **Complex Equations**

Cumulative Review

1. Find the real solutions, if any, of the equation $e^{x^2-9} = 1$.
2. Find an equation for the line containing the origin that makes an angle of 30° with the positive x-axis.
3. Find an equation for the circle with center at the point $(0, 1)$ and radius 3. Graph this circle.
4. What is the domain of the function $f(x) = \ln(1 - 2x)$?
5. Test the equation $x^2 + y^3 = 2x^4$ for symmetry with respect to the x-axis, the y-axis, and the origin.
6. Graph the function $y = |\ln x|$.
7. Graph the function $y = |\sin x|$.
8. Graph the function $y = \sin|x|$.
9. Find the exact value of $\sin^{-1}\left(-\frac{1}{2}\right)$.
10. Graph the equations $x = 3$ and $y = 4$ using the same set of rectangular coordinates.
11. Graph the equations $r = 2$ and $\theta = \frac{\pi}{3}$ using the same set of polar coordinates.

9 Analytic Geometry

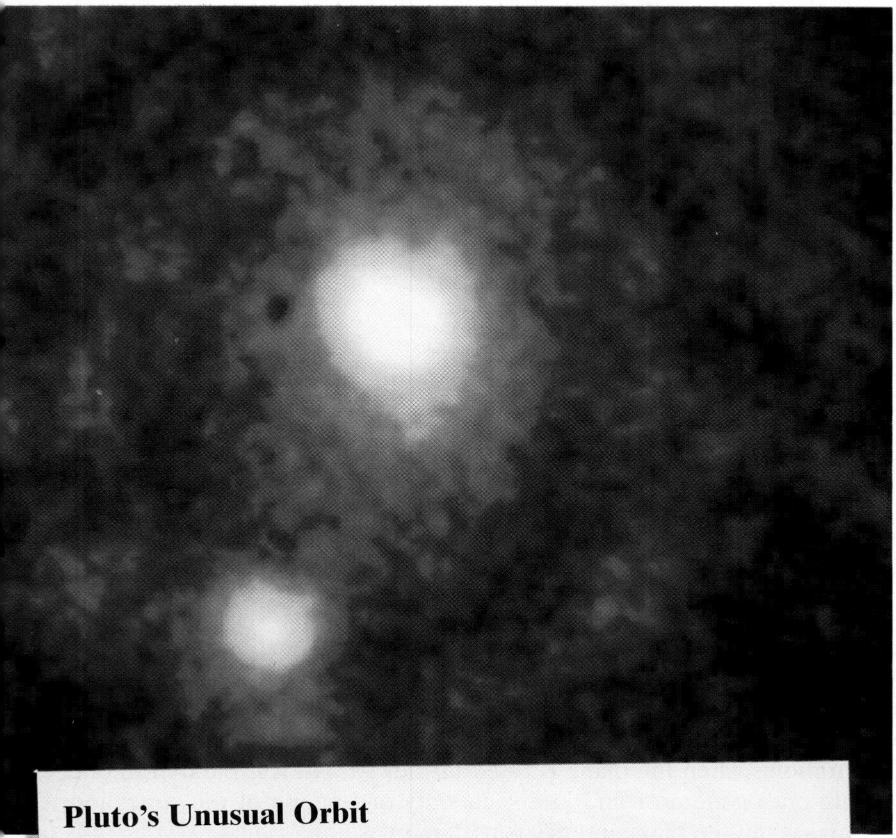

OUTLINE

Pluto's Unusual Orbit

Pluto is about 39 times as far from the sun as Earth is. Its average distance from the sun is about 3,647,240,000 miles (5,869,660,000 kilometers). Pluto travels around the sun in an elliptical (oval-shaped) orbit. At some point in its orbit, it comes closer to the sun than Neptune, usually the second farthest planet. It stays inside Neptune's orbit for about 20 Earth-years. This event occurs every 248 Earth-years, which is about the same number of Earth-years it takes Pluto to travel once around the sun. Pluto entered Neptune's orbit on Jan. 23, 1979, and remained there until Feb. 11, 1999. Pluto will remain the outermost planet until the year 2227.

SOURCE: Reprinted with permission of The Associated Press. http://www2.worldbook.com/features/features.asp?feature=outerplanets&page=html/pluto_orbit.html&direct=yes

—SEE CHAPTER PROJECT 1.

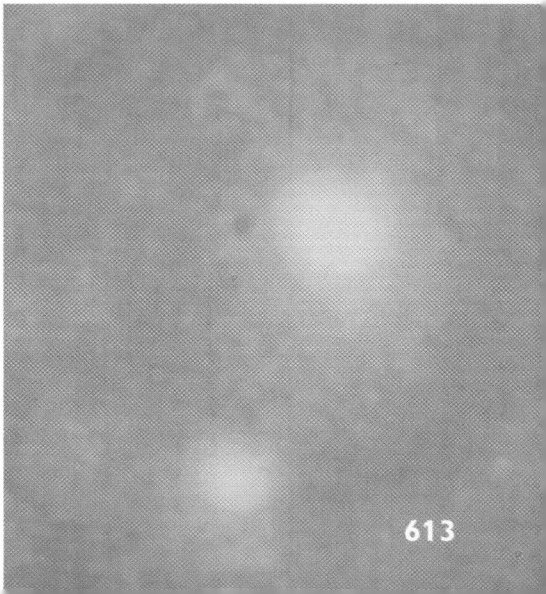

9.1 Conics

OBJECTIVE 1 Know the Names of the Conics

1 The word *conic* derives from the word *cone*, which is a geometric figure that can be constructed in the following way: Let a and g be two distinct lines that intersect at a point V. Keep the line a fixed. Now rotate the line g about a while maintaining the same angle between a and g. The collection of points swept out (generated) by the line g is called a **(right circular) cone**. See Figure 1. The fixed line a is called the **axis** of the cone; the point V is its **vertex;** the lines that pass through V and make the same angle with a as g are **generators** of the cone. Each generator is a line that lies entirely on the cone. The cone consists of two parts, called **nappes,** that intersect at the vertex.

Figure 1

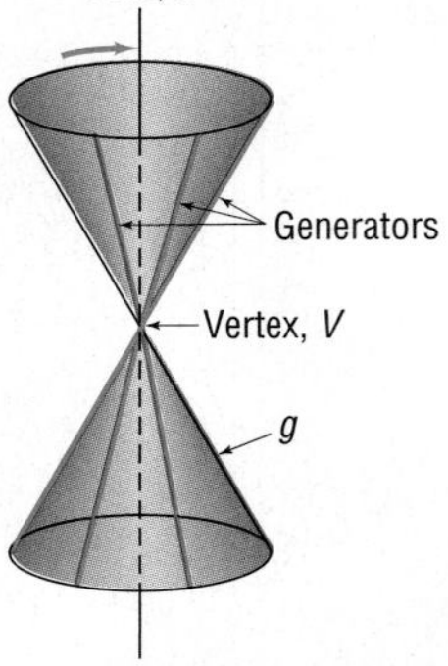

Conics, an abbreviation for **conic sections,** are curves that result from the intersection of a (right circular) cone and a plane. The conics we shall study arise when the plane does not contain the vertex, as shown in Figure 2. These conics are **circles** when the plane is perpendicular to the axis of the cone and intersects each generator; **ellipses** when the plane is tilted slightly so that it intersects each generator, but intersects only one nappe of the cone; **parabolas** when the plane is tilted farther so that it is parallel to one (and only one) generator and intersects only one nappe of the cone; and **hyperbolas** when the plane intersects both nappes.

If the plane does contain the vertex, the intersection of the plane and the cone is a point, a line, or a pair of intersecting lines. These are usually called **degenerate conics.**

Figure 2

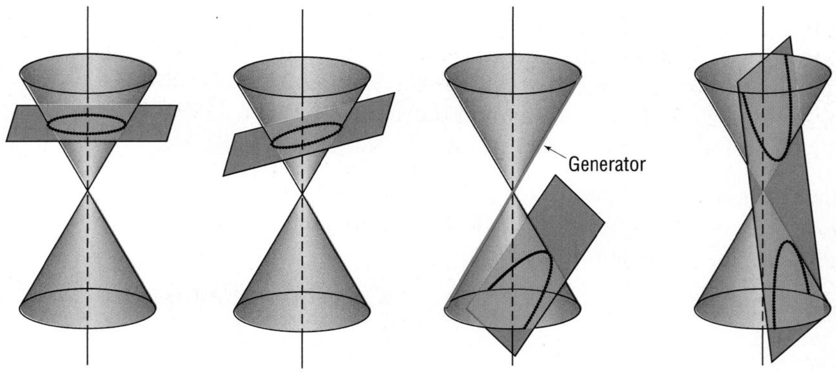

(a) Circle **(b)** Ellipse **(c)** Parabola **(d)** Hyperbola

9.2 The Parabola

PREPARING FOR THIS SECTION *Before getting started, review the following:*

- Distance Formula (Section 1.1, p. 3)
- Symmetry (Section 1.2, pp. 12–13)
- Square Root Method (Appendix A, Section A.5, pp. 940–941)
- Completing the Square (Appendix A, Section A.5, p. 941)
- Graphing Techniques: Transformations (Section 2.5, pp. 94–103)

Now work the 'Are You Prepared?' problems on page 622.

OBJECTIVES
1 Find the Equation of a Parabola
2 Graph Parabolas
3 Discuss the Equation of a Parabola
4 Work with Parabolas with Vertex at (h, k)
5 Solve Applied Problems Involving Parabolas

We stated earlier (Section 3.1) that the graph of a quadratic function is a parabola. In this section, we begin with a geometric definition of a parabola and use it to obtain an equation.

A **parabola** is the collection of all points P in the plane that are the same distance from a fixed point F as they are from a fixed line D. The point F is called the **focus** of the parabola, and the line D is its **directrix.** As a result, a parabola is the set of points P for which

$$d(F, P) = d(P, D) \tag{1}$$

1 Figure 3 shows a parabola. The line through the focus F and perpendicular to the directrix D is called the **axis of symmetry** of the parabola. The point of intersection of the parabola with its axis of symmetry is called the **vertex** V.

Figure 3

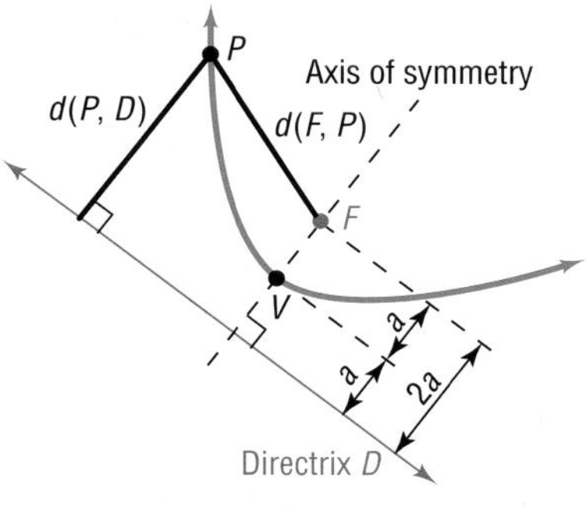

Because the vertex V lies on the parabola, it must satisfy equation (1): $d(F, V) = d(V, D)$. The vertex is midway between the focus and the directrix. We shall let a equal the distance $d(F, V)$ from F to V. Now we are ready to derive an equation for a parabola. To do this, we use a rectangular system of coordinates positioned so that the vertex V, focus F, and directrix D of the parabola are conveniently located. If we choose to locate the vertex V at the origin $(0, 0)$, then we can conveniently position the focus F on either the x-axis or the y-axis.

Figure 4
$y^2 = 4ax$

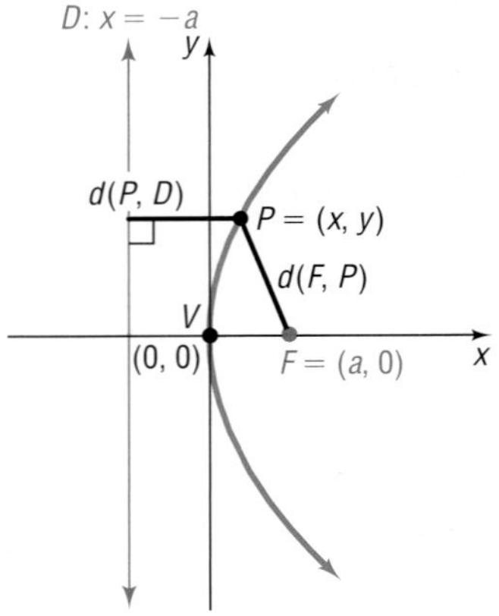

First, we consider the case where the focus F is on the positive x-axis, as shown in Figure 4. Because the distance from F to V is a, the coordinates of F will be $(a, 0)$ with $a > 0$. Similarly, because the distance from V to the directrix D is also a and, because D must be perpendicular to the x-axis (since the x-axis is the axis of symmetry), the equation of the directrix D must be $x = -a$.

Now, if $P = (x, y)$ is any point on the parabola, then P must obey equation (1):

$$d(F, P) = d(P, D)$$

So we have

$$\sqrt{(x-a)^2 + y^2} = |x + a| \quad \text{Use the distance formula.}$$
$$(x-a)^2 + y^2 = (x+a)^2 \quad \text{Square both sides.}$$
$$x^2 - 2ax + a^2 + y^2 = x^2 + 2ax + a^2 \quad \text{Remove parentheses.}$$
$$y^2 = 4ax \quad \text{Simplify.}$$

Theorem

Equation of a Parabola; Vertex at (0, 0), Focus at (a, 0), a > 0

The equation of a parabola with vertex at $(0, 0)$, focus at $(a, 0)$, and directrix $x = -a$, $a > 0$, is

$$y^2 = 4ax \qquad (2)$$

2 **EXAMPLE 1** **Finding the Equation of a Parabola and Graphing It**

Find an equation of the parabola with vertex at $(0, 0)$ and focus at $(3, 0)$. Graph the equation.

Solution The distance from the vertex $(0, 0)$ to the focus $(3, 0)$ is $a = 3$. Based on equation (2), the equation of this parabola is

$$y^2 = 4ax$$
$$y^2 = 12x \quad a = 3$$

To graph this parabola, it is helpful to plot the two points on the graph above and below the focus. To locate them, we let $x = 3$. Then

$$y^2 = 12x = 12(3) = 36$$
$$y = \pm 6 \quad \text{Solve for } y.$$

The points on the parabola above and below the focus are $(3, 6)$ and $(3, -6)$. These points help in graphing the parabola because they determine the "opening." See Figure 5. ◀

Figure 5

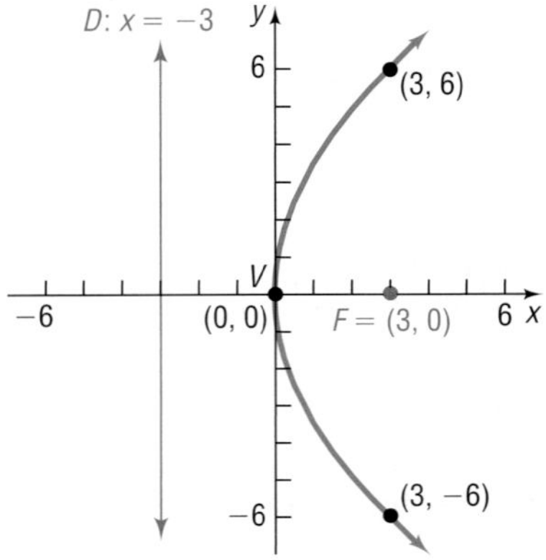

In general, the points on a parabola $y^2 = 4ax$ that lie above and below the focus $(a, 0)$ are each at a distance $2a$ from the focus. This follows from the fact that if $x = a$ then $y^2 = 4ax = 4a^2$, so $y = \pm 2a$. The line segment joining these two points is called the **latus rectum;** its length is $4a$.

COMMENT: To graph the parabola $y^2 = 12x$ discussed in Example 1, we need to graph the two functions $Y_1 = \sqrt{12x}$ and $Y_2 = -\sqrt{12x}$. Do this and compare what you see with Figure 5. ■

NOW WORK PROBLEM 19.

By reversing the steps we used to obtain equation (2), it follows that the graph of an equation of the form of equation (2), $y^2 = 4ax$, is a parabola; its vertex is at $(0, 0)$, its focus is at $(a, 0)$, its directrix is the line $x = -a$, and its axis of symmetry is the x-axis.

3 For the remainder of this section, the direction "Discuss the equation" will mean to find the vertex, focus, and directrix of the parabola and graph it.

EXAMPLE 2 Discussing the Equation of a Parabola

Discuss the equation: $y^2 = 8x$

Solution The equation $y^2 = 8x$ is of the form $y^2 = 4ax$, where $4a = 8$, so $a = 2$. Consequently, the graph of the equation is a parabola with vertex at $(0, 0)$ and focus on the positive x-axis at $(2, 0)$. The directrix is the vertical line $x = -2$. The two points defining the latus rectum are obtained by letting $x = 2$. Then $y^2 = 16$, so $y = \pm 4$. See Figure 6. ◀

Figure 6

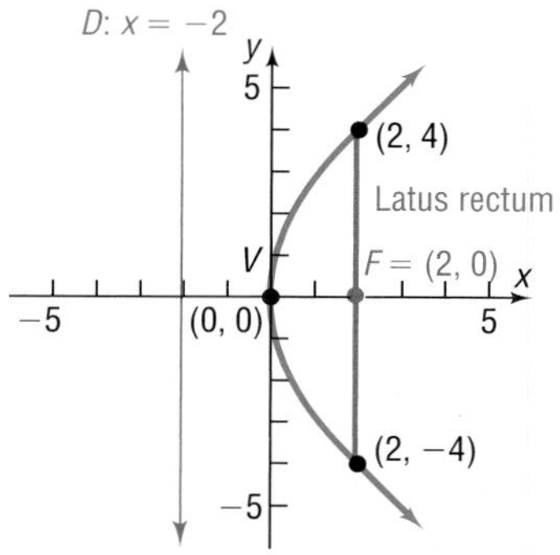

Recall that we arrived at equation (2) after placing the focus on the positive x-axis. If the focus is placed on the negative x-axis, positive y-axis, or negative y-axis, a different form of the equation for the parabola results. The four forms of the equation of a parabola with vertex at $(0, 0)$ and focus on a coordinate axis a distance a from $(0, 0)$ are given in Table 1, and their graphs are given in Figure 7. Notice that each graph is symmetric with respect to its axis of symmetry.

Table 1 **Equations of a Parabola: Vertex at (0, 0); Focus on an Axis; $a > 0$**

Vertex	Focus	Directrix	Equation	Description
(0, 0)	$(a, 0)$	$x = -a$	$y^2 = 4ax$	Parabola, axis of symmetry is the x-axis, opens to right
(0, 0)	$(-a, 0)$	$x = a$	$y^2 = -4ax$	Parabola, axis of symmetry is the x-axis, opens to left
(0, 0)	$(0, a)$	$y = -a$	$x^2 = 4ay$	Parabola, axis of symmetry is the y-axis, opens up
(0, 0)	$(0, -a)$	$y = a$	$x^2 = -4ay$	Parabola, axis of symmetry is the y-axis, opens down

Figure 7

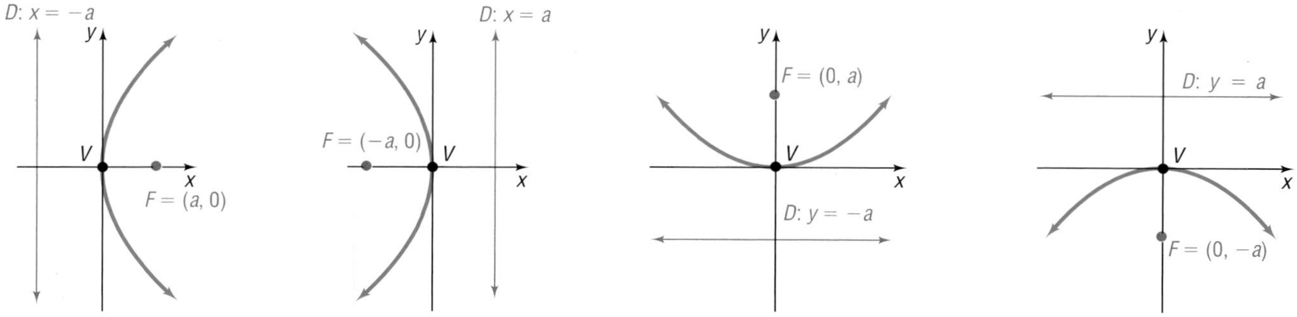

(a) $y^2 = 4ax$ **(b)** $y^2 = -4ax$ **(c)** $x^2 = 4ay$ **(d)** $x^2 = -4ay$

EXAMPLE 3 Discussing the Equation of a Parabola

Discuss the equation: $x^2 = -12y$

Figure 8

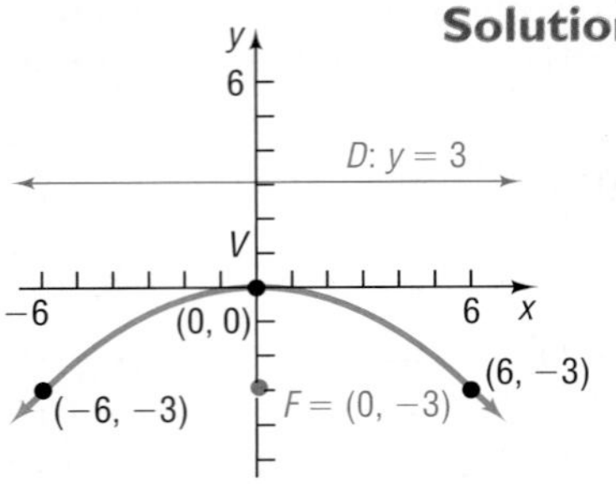

Solution The equation $x^2 = -12y$ is of the form $x^2 = -4ay$, with $a = 3$. Consequently, the graph of the equation is a parabola with vertex at $(0, 0)$, focus at $(0, -3)$ and directrix the line $y = 3$. The parabola opens down, and its axis of symmetry is the y-axis. To obtain the points defining the latus rectum, let $y = -3$. Then $x^2 = 36$, so $x = \pm 6$. See Figure 8. ◀

NOW WORK PROBLEM 39.

EXAMPLE 4 Finding the Equation of a Parabola

Find the equation of the parabola with focus at $(0, 4)$ and directrix the line $y = -4$. Graph the equation.

Figure 9

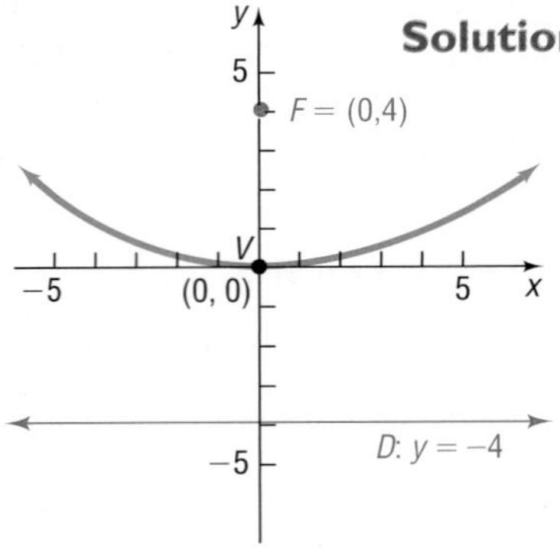

Solution A parabola whose focus is at $(0, 4)$ and whose directrix is the horizontal line $y = -4$ will have its vertex at $(0, 0)$. (Do you see why? The vertex is midway between the focus and the directrix.) Since the focus is on the positive y-axis at $(0, 4)$, the equation of this parabola is of the form $x^2 = 4ay$, with $a = 4$; that is,

$$x^2 = 4ay = \underset{\substack{\uparrow \\ a = 4}}{4(4)y} = 16y$$

Figure 9 shows the graph of $x^2 = 16y$. ◀

EXAMPLE 5 Finding the Equation of a Parabola

Find the equation of a parabola with vertex at $(0, 0)$ if its axis of symmetry is the x-axis and its graph contains the point $\left(-\frac{1}{2}, 2\right)$. Find its focus and directrix, and graph the equation.

Solution The vertex is at the origin, the axis of symmetry is the x-axis, and the graph contains a point in the second quadrant, so the parabola opens to the left. We see from Table 1 that the form of the equation is

$$y^2 = -4ax$$

Because the point $\left(-\frac{1}{2}, 2\right)$ is on the parabola, the coordinates $x = -\frac{1}{2}$, $y = 2$ must satisfy the equation. Substituting $x = -\frac{1}{2}$ and $y = 2$ into the equation, we find that

$$4 = -4a\left(-\frac{1}{2}\right) \qquad y^2 = -4ax;\ x = -\frac{1}{2},\ y = 2$$

$$a = 2$$

The equation of the parabola is

$$y^2 = -4(2)x = -8x$$

Figure 10

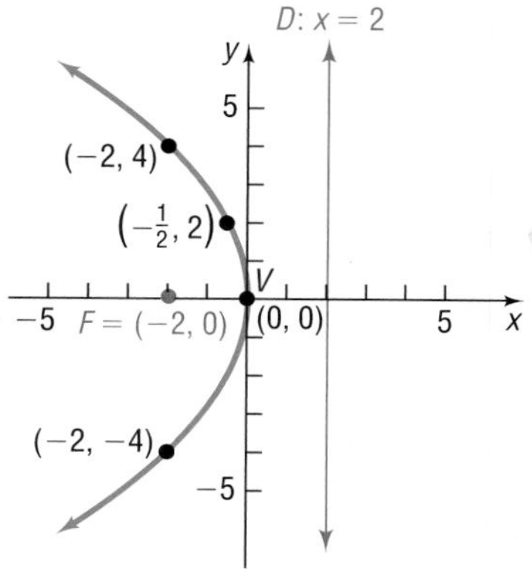

The focus is at $(-2, 0)$ and the directrix is the line $x = 2$. Letting $x = -2$, we find $y^2 = 16$, so $y = \pm 4$. The points $(-2, 4)$ and $(-2, -4)$ define the latus rectum. See Figure 10. ◀

NOW WORK PROBLEM 27.

4 Vertex at (h, k)

If a parabola with vertex at the origin and axis of symmetry along a coordinate axis is shifted horizontally h units and then vertically k units, the result is a parabola with vertex at (h, k) and axis of symmetry parallel to a coordinate axis. The equations of such parabolas have the same forms as those in Table 1, but with x replaced by $x - h$ (the horizontal shift) and y replaced by $y - k$ (the vertical shift). Table 2 gives the forms of the equations of such parabolas. Figures 11(a)–(d) illustrate the graphs for $h > 0, k > 0$.

Table 2 **Parabolas with Vertex at (h, k); Axis of Symmetry Parallel to a Coordinate Axis, $a > 0$**

Vertex	Focus	Directrix	Equation	Description
(h, k)	$(h + a, k)$	$x = h - a$	$(y - k)^2 = 4a(x - h)$	Parabola, axis of symmetry parallel to x-axis, opens to right
(h, k)	$(h - a, k)$	$x = h + a$	$(y - k)^2 = -4a(x - h)$	Parabola, axis of symmetry parallel to x-axis, opens to left
(h, k)	$(h, k + a)$	$y = k - a$	$(x - h)^2 = 4a(y - k)$	Parabola, axis of symmetry parallel to y-axis, opens up
(h, k)	$(h, k - a)$	$y = k + a$	$(x - h)^2 = -4a(y - k)$	Parabola, axis of symmetry parallel to y-axis, opens down

Figure 11

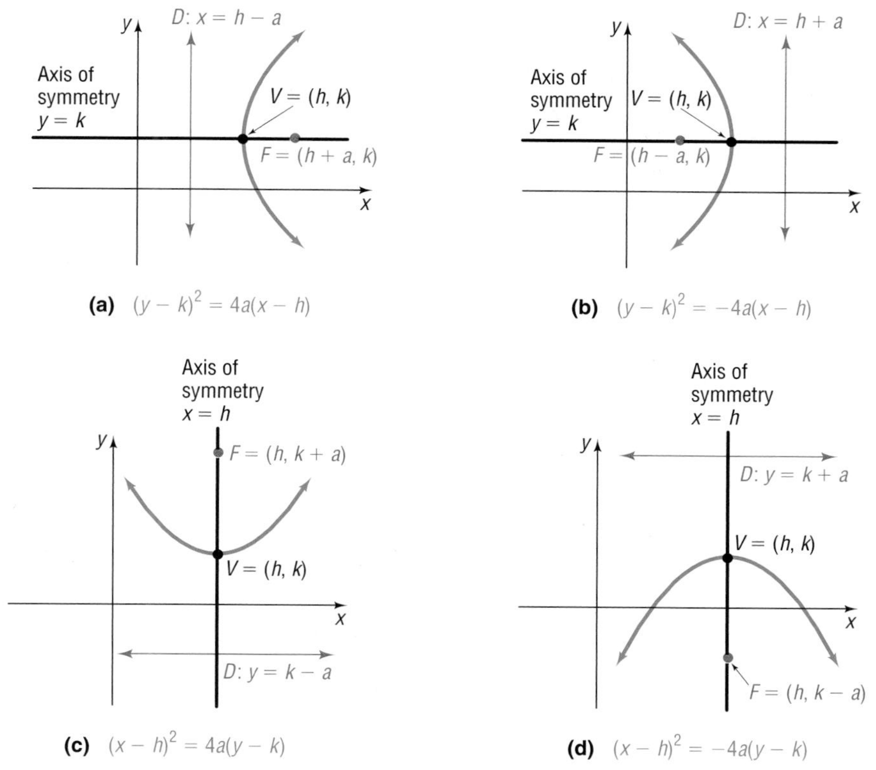

EXAMPLE 6 Finding the Equation of a Parabola, Vertex Not at Origin

Find an equation of the parabola with vertex at $(-2, 3)$ and focus at $(0, 3)$. Graph the equation.

Figure 12

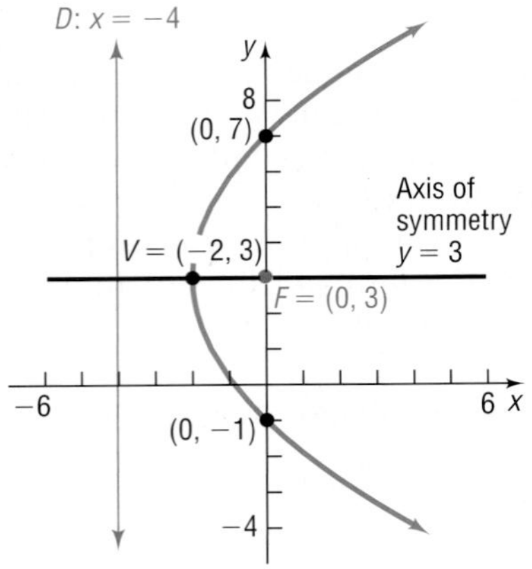

Solution The vertex $(-2, 3)$ and focus $(0, 3)$ both lie on the horizontal line $y = 3$ (the axis of symmetry). The distance a from the vertex $(-2, 3)$ to the focus $(0, 3)$ is $a = 2$. Also, because the focus lies to the right of the vertex, we know that the parabola opens to the right. Consequently, the form of the equation is

$$(y - k)^2 = 4a(x - h)$$

where $(h, k) = (-2, 3)$ and $a = 2$. Therefore, the equation is

$$(y - 3)^2 = 4 \cdot 2[x - (-2)]$$
$$(y - 3)^2 = 8(x + 2)$$

If $x = 0$, then $(y - 3)^2 = 16$. Then $y - 3 = \pm 4$, so $y = -1$ or $y = 7$. The points $(0, -1)$ and $(0, 7)$ define the latus rectum; the line $x = -4$ is the directrix. See Figure 12. ◀

NOW WORK PROBLEM 29.

Polynomial equations define parabolas whenever they involve two variables that are quadratic in one variable and linear in the other. To discuss this type of equation, we first complete the square of the variable that is quadratic.

EXAMPLE 7 Discussing the Equation of a Parabola

Discuss the equation: $x^2 + 4x - 4y = 0$

Figure 13

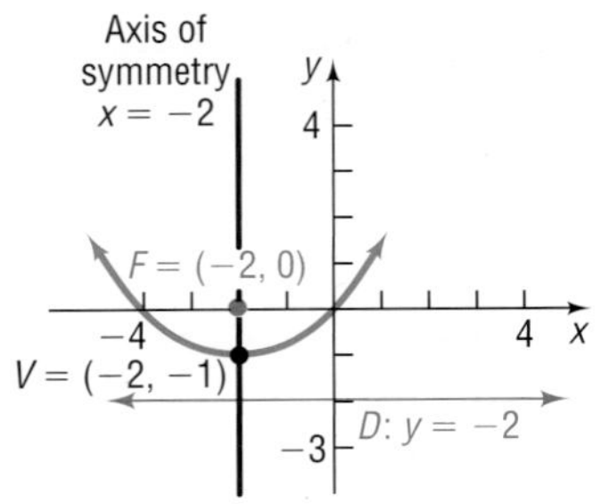

Solution To discuss the equation $x^2 + 4x - 4y = 0$, we complete the square involving the variable x.

$$x^2 + 4x - 4y = 0$$
$$x^2 + 4x = 4y \quad \text{Isolate the terms involving } x \text{ on the left side.}$$
$$x^2 + 4x + 4 = 4y + 4 \quad \text{Complete the square on the left side.}$$
$$(x + 2)^2 = 4(y + 1) \quad \text{Factor.}$$

This equation is of the form $(x - h)^2 = 4a(y - k)$, with $h = -2$, $k = -1$, and $a = 1$. The graph is a parabola with vertex at $(h, k) = (-2, -1)$ that opens up. The focus is at $(-2, 0)$, and the directrix is the line $y = -2$. See Figure 13. ◀

NOW WORK PROBLEM 47.

5 Parabolas find their way into many applications. For example, as we discussed in Section 3.1, suspension bridges have cables in the shape of a parabola. Another property of parabolas that is used in applications is their reflecting property.

Reflecting Property

Suppose that a mirror is shaped like a **paraboloid of revolution,** a surface formed by rotating a parabola about its axis of symmetry. If a light (or any other emitting source) is placed at the focus of the parabola, all the rays emanating from the light will reflect off the mirror in lines parallel to the axis of symmetry. This principle is used in the design of searchlights, flashlights, certain automobile headlights, and other such devices. See Figure 14.

Conversely, suppose that rays of light (or other signals) emanate from a distant source so that they are essentially parallel. When these rays strike the surface of a parabolic mirror whose axis of symmetry is parallel to these rays, they are reflected to a single point at the focus. This principle is used in the design of some solar energy devices, satellite dishes, and the mirrors used in some types of telescopes. See Figure 15.

Figure 14
Searchlight

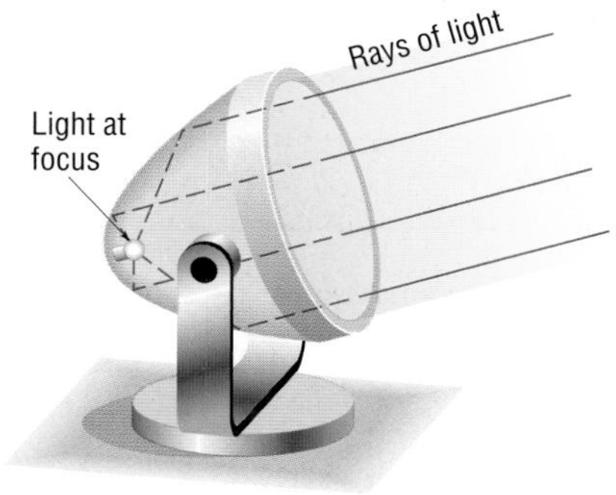

Figure 15
Telescope

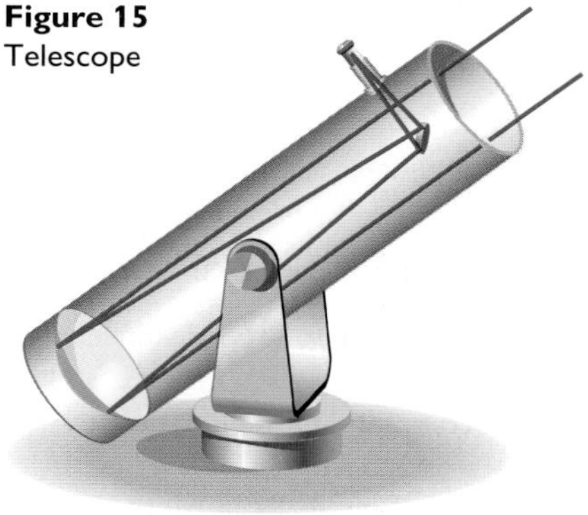

EXAMPLE 8 Satellite Dish

A satellite dish is shaped like a paraboloid of revolution. The signals that emanate from a satellite strike the surface of the dish and are reflected to a single point, where the receiver is located. If the dish is 8 feet across at its opening and 3 feet deep at its center, at what position should the receiver be placed?

Solution Figure 16(a) shows the satellite dish. We draw the parabola used to form the dish on a rectangular coordinate system so that the vertex of the parabola is at the origin and its focus is on the positive y-axis. See Figure 16(b).

Figure 16

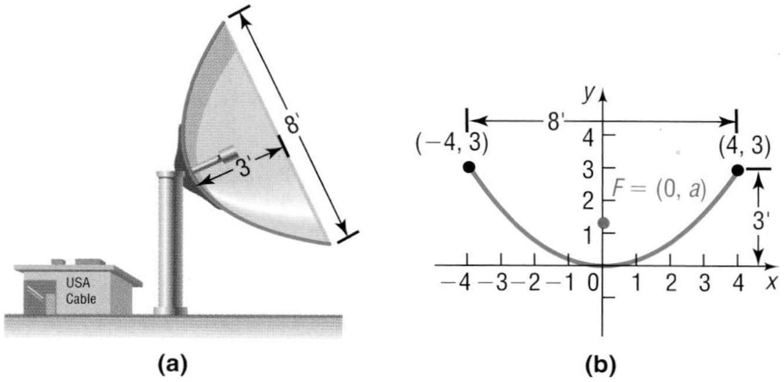

The form of the equation of the parabola is

$$x^2 = 4ay$$

and its focus is at $(0, a)$. Since $(4, 3)$ is a point on the graph, we have

$$4^2 = 4a(3)$$

$$a = \frac{4}{3}$$

The receiver should be located $1\frac{1}{3}$ feet from the base of the dish, along its axis of symmetry.

NOW WORK PROBLEM 63.

9.2 Assess Your Understanding

'Are You Prepared?' *Answers are given at the end of these exercises. If you get a wrong answer, read the pages listed in red.*

1. The formula for the distance d from $P_1 = (x_1, y_1)$ to $P_2 = (x_2, y_2)$ is $d =$ __________. (p. 3)

2. To complete the square of $x^2 - 4x$, add __________. (p. 941)

3. Use the Square Root Method to find the real solutions of $(x + 4)^2 = 9$. (pp. 940–941)

4. The point that is symmetric with respect to the x-axis to the point $(-2, 5)$ is __________. (pp. 12–13)

5. To graph $y = (x - 3)^2 + 1$, shift the graph of $y = x^2$ to the right __________ units and then __________ 1 unit. (pp. 94–103)

Concepts and Vocabulary

6. A(n) __________ is the collection of all points in the plane such that the distance from each point to a fixed point equals its distance to a fixed line.

7. The surface formed by rotating a parabola about its axis of symmetry is called a __________ __________ __________.

8. *True or False:* The vertex of a parabola is a point on the parabola that also is on its axis of symmetry.

9. *True or False:* If a light is placed at the focus of a parabola, all the rays reflected off the parabola will be parallel to the axis of symmetry.

10. *True or False:* The graph of a quadratic function is a parabola.

Exercises

In Problems 11–18, the graph of a parabola is given. Match each graph to its equation.

A. $y^2 = 4x$
B. $x^2 = 4y$
C. $y^2 = -4x$
D. $x^2 = -4y$
E. $(y - 1)^2 = 4(x - 1)$
F. $(x + 1)^2 = 4(y + 1)$
G. $(y - 1)^2 = -4(x - 1)$
H. $(x + 1)^2 = -4(y + 1)$

11.
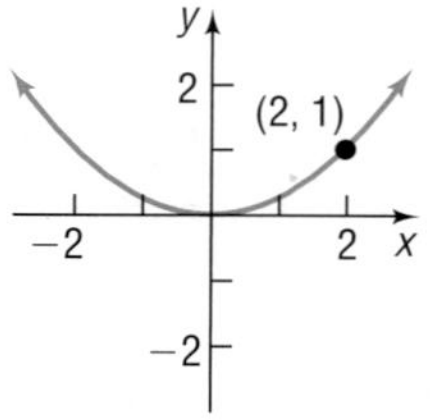

12.
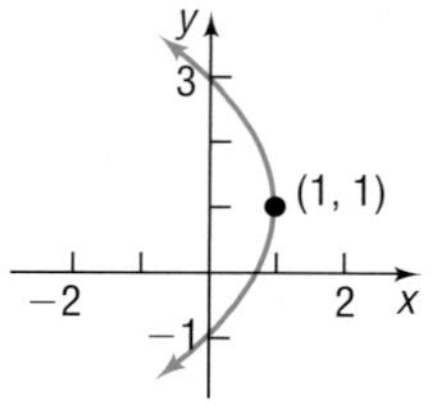

13.
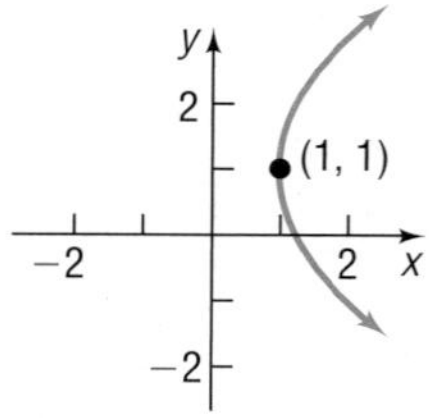

14.
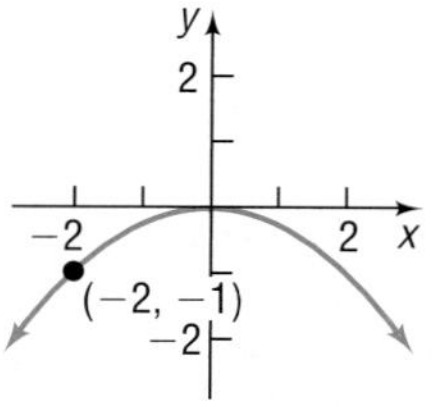

15.
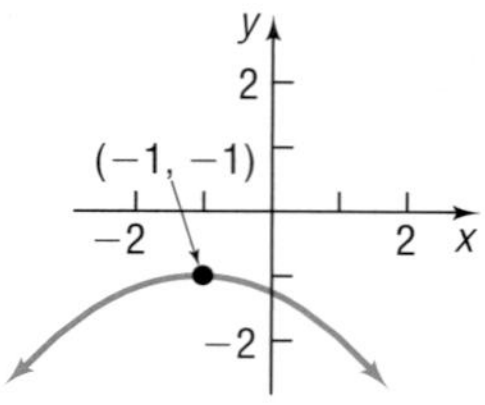

16.
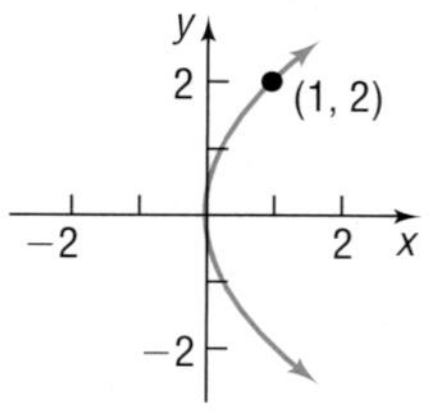

17.
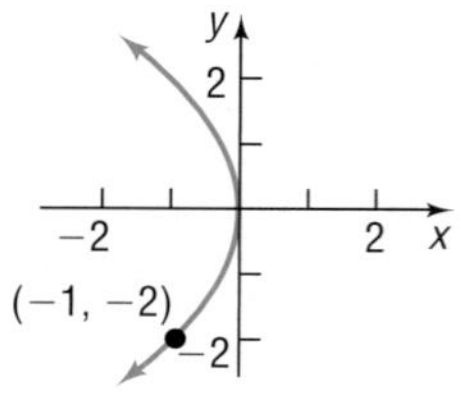

18.
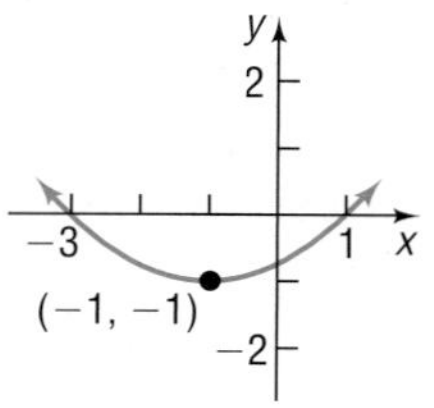

In Problems 19–36, find the equation of the parabola described. Find the two points that define the latus rectum, and graph the equation.

19. Focus at $(4, 0)$; vertex at $(0, 0)$

20. Focus at $(0, 2)$; vertex at $(0, 0)$

21. Focus at $(0, -3)$; vertex at $(0, 0)$

22. Focus at $(-4, 0)$; vertex at $(0, 0)$

23. Focus at $(-2, 0)$; directrix the line $x = 2$

24. Focus at $(0, -1)$; directrix the line $y = 1$

25. Directrix the line $y = -\frac{1}{2}$; vertex at $(0, 0)$

26. Directrix the line $x = -\frac{1}{2}$; vertex at $(0, 0)$

27. Vertex at $(0, 0)$; axis of symmetry the y-axis; containing the point $(2, 3)$

28. Vertex at $(0, 0)$; axis of symmetry the x-axis; containing the point $(2, 3)$

29. Vertex at $(2, -3)$; focus at $(2, -5)$

30. Vertex at $(4, -2)$; focus at $(6, -2)$

31. Vertex at $(-1, -2)$; focus at $(0, -2)$

32. Vertex at $(3, 0)$; focus at $(3, -2)$

33. Focus at $(-3, 4)$; directrix the line $y = 2$

34. Focus at $(2, 4)$; directrix the line $x = -4$

35. Focus at $(-3, -2)$; directrix the line $x = 1$

36. Focus at $(-4, 4)$; directrix the line $y = -2$

In Problems 37–54, find the vertex, focus, and directrix of each parabola. Graph the equation.

37. $x^2 = 4y$

38. $y^2 = 8x$

39. $y^2 = -16x$

40. $x^2 = -4y$

41. $(y - 2)^2 = 8(x + 1)$

42. $(x + 4)^2 = 16(y + 2)$

43. $(x - 3)^2 = -(y + 1)$

44. $(y + 1)^2 = -4(x - 2)$

45. $(y + 3)^2 = 8(x - 2)$

46. $(x - 2)^2 = 4(y - 3)$

47. $y^2 - 4y + 4x + 4 = 0$

48. $x^2 + 6x - 4y + 1 = 0$

49. $x^2 + 8x = 4y - 8$

50. $y^2 - 2y = 8x - 1$

51. $y^2 + 2y - x = 0$

52. $x^2 - 4x = 2y$

53. $x^2 - 4x = y + 4$

54. $y^2 + 12y = -x + 1$

In Problems 55–62, write an equation for each parabola.

55.

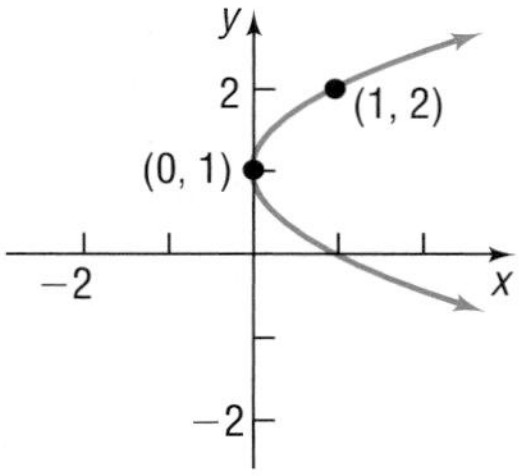

56.

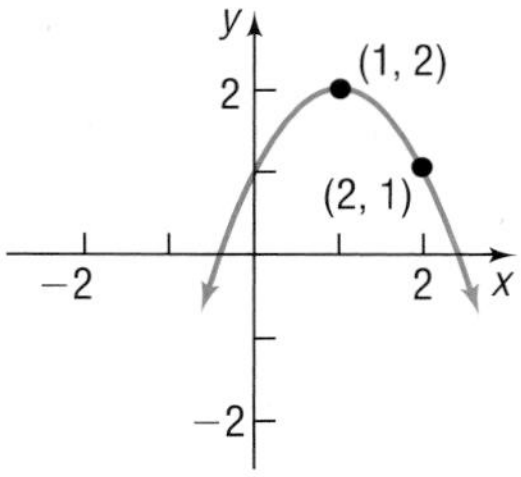

57.

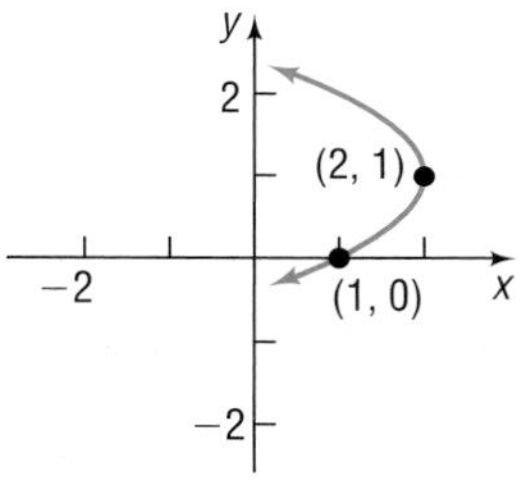

58.

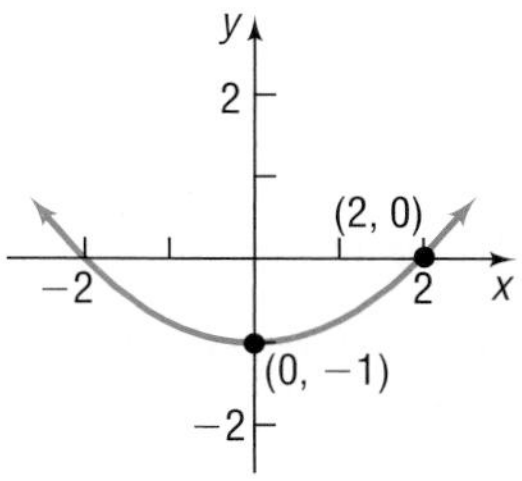

59.

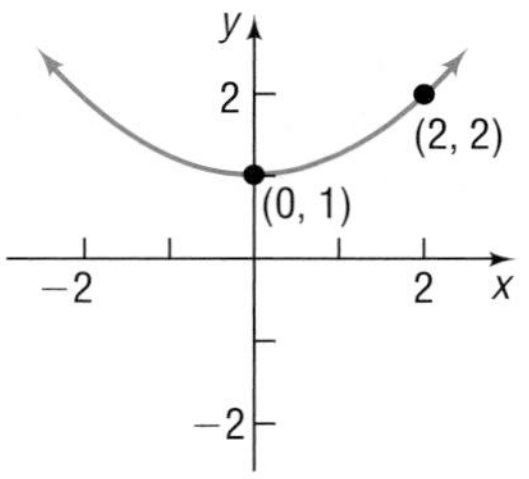

60.

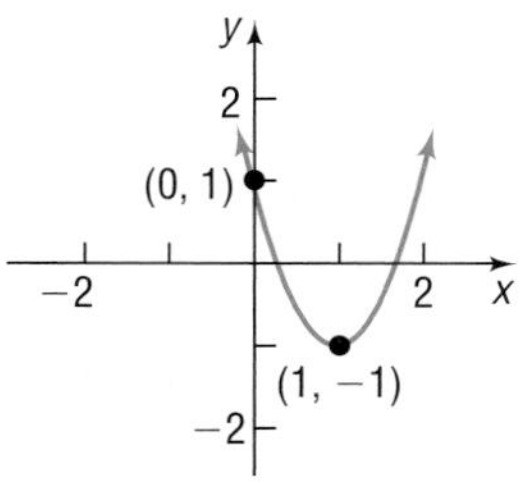

61.

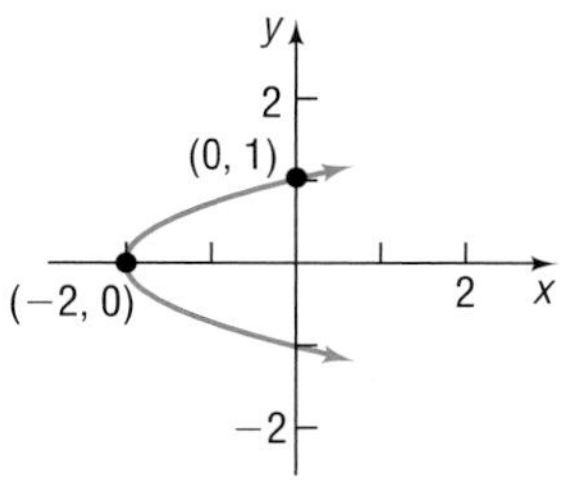

62.

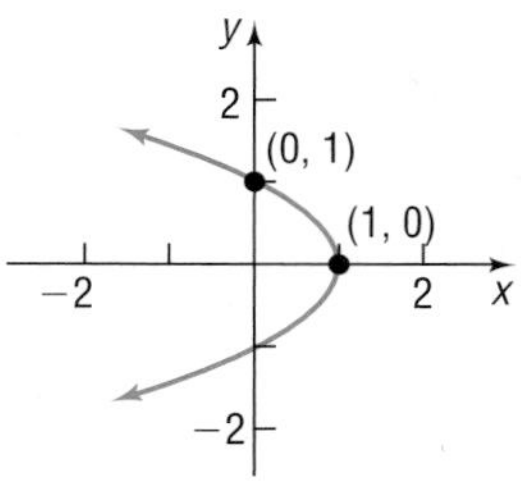

63. Satellite Dish A satellite dish is shaped like a paraboloid of revolution. The signals that emanate from a satellite strike the surface of the dish and are reflected to a single point, where the receiver is located. If the dish is 10 feet across at its opening and 4 feet deep at its center, at what position should the receiver be placed?

64. Constructing a TV Dish A cable TV receiving dish is in the shape of a paraboloid of revolution. Find the location of the receiver, which is placed at the focus, if the dish is 6 feet across at its opening and 2 feet deep.

65. Constructing a Flashlight The reflector of a flashlight is in the shape of a paraboloid of revolution. Its diameter is 4 inches and its depth is 1 inch. How far from the vertex should the light bulb be placed so that the rays will be reflected parallel to the axis?

66. Constructing a Headlight A sealed-beam headlight is in the shape of a paraboloid of revolution. The bulb, which is placed at the focus, is 1 inch from the vertex. If the depth is to be 2 inches, what is the diameter of the headlight at its opening?

67. Suspension Bridge The cables of a suspension bridge are in the shape of a parabola, as shown in the figure. The towers supporting the cable are 600 feet apart and 80 feet high. If the cables touch the road surface midway between the towers, what is the height of the cable at a point 150 feet from the center of the bridge?

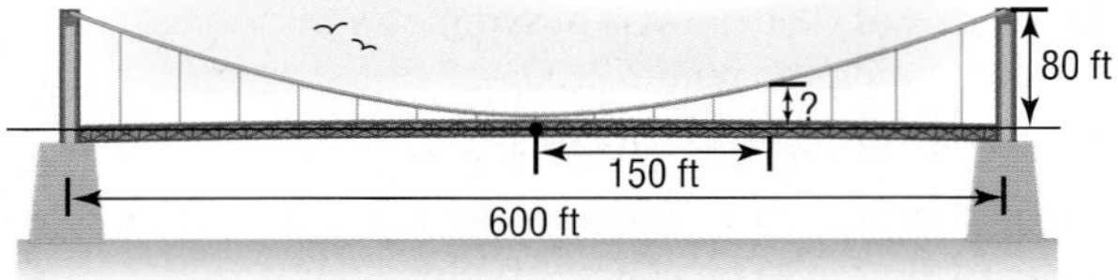

68. Suspension Bridge The cables of a suspension bridge are in the shape of a parabola. The towers supporting the cable are 400 feet apart and 100 feet high. If the cables are at a height of 10 feet midway between the towers, what is the height of the cable at a point 50 feet from the center of the bridge?

69. Searchlight A searchlight is shaped like a paraboloid of revolution. If the light source is located 2 feet from the base along the axis of symmetry and the opening is 5 feet across, how deep should the searchlight be?

70. Searchlight A searchlight is shaped like a paraboloid of revolution. If the light source is located 2 feet from the base along the axis of symmetry and the depth of the searchlight is 4 feet, what should the width of the opening be?

71. Solar Heat A mirror is shaped like a paraboloid of revolution and will be used to concentrate the rays of the sun at its focus, creating a heat source. (See the figure.) If the mirror is 20 feet across at its opening and is 6 feet deep, where will the heat source be concentrated?

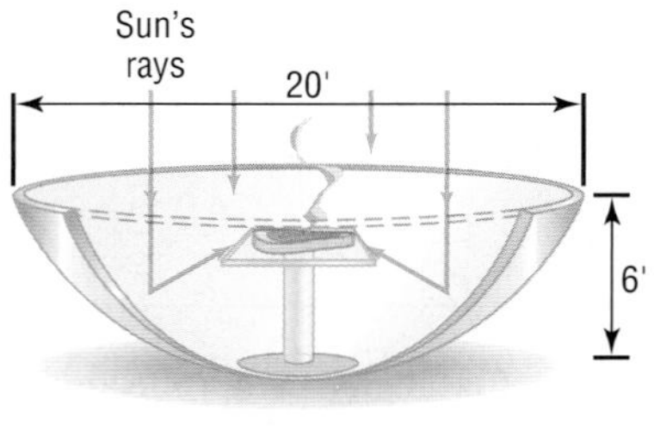

72. Reflecting Telescope A reflecting telescope contains a mirror shaped like a paraboloid of revolution. If the mirror is 4 inches across at its opening and is 3 feet deep, where will the collected light be concentrated?

73. Parabolic Arch Bridge A bridge is built in the shape of a parabolic arch. The bridge has a span of 120 feet and a maximum height of 25 feet. See the illustration. Choose a suitable rectangular coordinate system and find the height of the arch at distances of 10, 30, and 50 feet from the center.

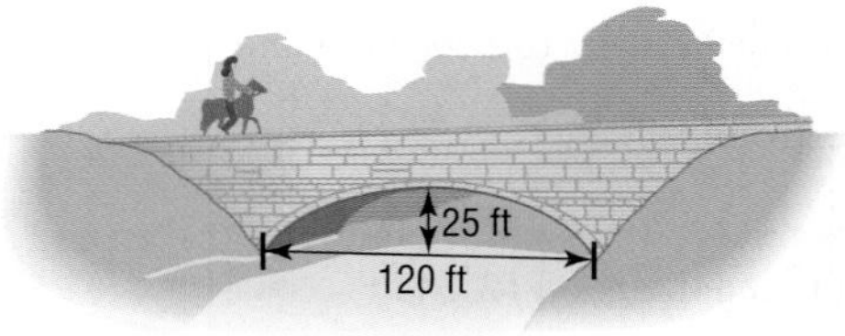

74. Parabolic Arch Bridge A bridge is to be built in the shape of a parabolic arch and is to have a span of 100 feet. The height of the arch a distance of 40 feet from the center is to be 10 feet. Find the height of the arch at its center.

75. Show that an equation of the form

$$Ax^2 + Ey = 0, \qquad A \neq 0, E \neq 0$$

is the equation of a parabola with vertex at $(0, 0)$ and axis of symmetry the y-axis. Find its focus and directrix.

76. Show that an equation of the form

$$Cy^2 + Dx = 0, \qquad C \neq 0, D \neq 0$$

is the equation of a parabola with vertex at $(0, 0)$ and axis of symmetry the x-axis. Find its focus and directrix.

77. Show that the graph of an equation of the form

$$Ax^2 + Dx + Ey + F = 0, \qquad A \neq 0$$

(a) Is a parabola if $E \neq 0$.
(b) Is a vertical line if $E = 0$ and $D^2 - 4AF = 0$.
(c) Is two vertical lines if $E = 0$ and $D^2 - 4AF > 0$.
(d) Contains no points if $E = 0$ and $D^2 - 4AF < 0$.

78. Show that the graph of an equation of the form

$$Cy^2 + Dx + Ey + F = 0, \qquad C \neq 0$$

(a) Is a parabola if $D \neq 0$.
(b) Is a horizontal line if $D = 0$ and $E^2 - 4CF = 0$.
(c) Is two horizontal lines if $D = 0$ and $E^2 - 4CF > 0$.
(d) Contains no points if $D = 0$ and $E^2 - 4CF < 0$.

'Are You Prepared?' Answers

1. $d = \sqrt{(x_2 - x_1)^2 + (y_2 - y_1)^2}$

2. 4

3. $x + 4 = \pm 3$; $\{-7, -1\}$

4. $(-2, -5)$

5. 3; up

9.3 The Ellipse

PREPARING FOR THIS SECTION *Before getting started, review the following:*

- Distance Formula (Section 1.1, p. 3)
- Completing the Square (Appendix A, Section A.5, p. 941)
- Intercepts (Section 1.2, pp. 11–12)
- Symmetry (Section 1.2, pp. 12–13)
- Circles (Section 1.2, pp. 16–19)
- Graphing Techniques: Transformations (Section 2.5, pp. 94–103)

Now work the 'Are You Prepared?' problems on page 632.

OBJECTIVES
1. Find the Equation of an Ellipse
2. Graph Ellipses
3. Discuss the Equation of an Ellipse
4. Work with Ellipses with Center at (h, k)
5. Solve Applied Problems Involving Ellipses

An **ellipse** is the collection of all points in the plane the sum of whose distances from two fixed points, called the **foci,** is a constant.

Figure 17

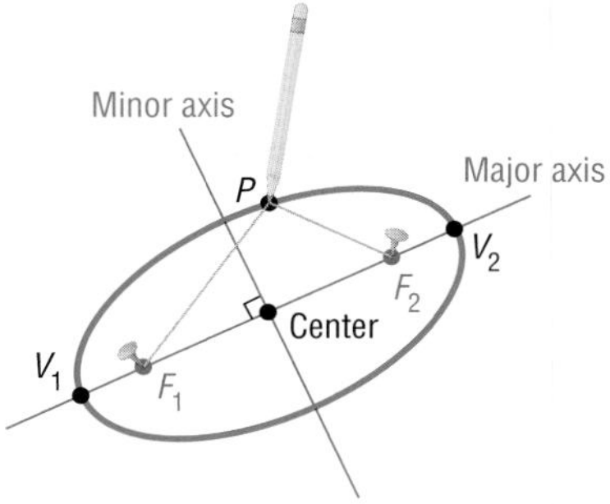

The definition actually contains within it a physical means for drawing an ellipse. Find a piece of string (the length of this string is the constant referred to in the definition). Then take two thumbtacks (the foci) and stick them on a piece of cardboard so that the distance between them is less than the length of the string. Now attach the ends of the string to the thumbtacks and, using the point of a pencil, pull the string taut. See Figure 17. Keeping the string taut, rotate the pencil around the two thumbtacks. The pencil traces out an ellipse, as shown in Figure 17.

In Figure 17, the foci are labeled F_1 and F_2. The line containing the foci is called the **major axis.** The midpoint of the line segment joining the foci is the **center** of the ellipse. The line through the center and perpendicular to the major axis is the **minor axis.**

Figure 18
$d(F_1, P) + d(F_2, P) = 2a$

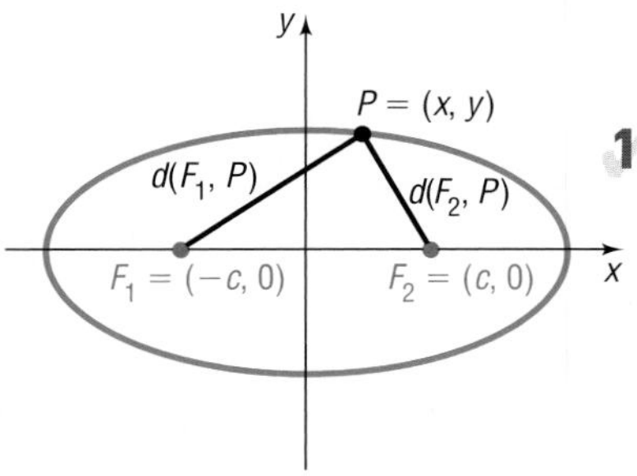

The two points of intersection of the ellipse and the major axis are the **vertices,** V_1 and V_2, of the ellipse. The distance from one vertex to the other is the **length of the major axis.** The ellipse is symmetric with respect to its major axis, with respect to its minor axis, and with respect to its center.

1 With these ideas in mind, we are now ready to find the equation of an ellipse in a rectangular coordinate system. First, we place the center of the ellipse at the origin. Second, we position the ellipse so that its major axis coincides with a coordinate axis. Suppose that the major axis coincides with the x-axis, as shown in Figure 18. If c is the distance from the center to a focus, then one focus will be at $F_1 = (-c, 0)$ and the other at $F_2 = (c, 0)$. As we shall see, it is convenient to let $2a$ denote the constant distance referred to in the definition. Then, if $P = (x, y)$ is any point on the ellipse, we have

$$d(F_1, P) + d(F_2, P) = 2a \qquad \text{Sum of the distances from } P \text{ to the foci equals a constant, } 2a.$$

$$\sqrt{(x + c)^2 + y^2} + \sqrt{(x - c)^2 + y^2} = 2a \qquad \text{Use the Distance Formula.}$$

$$\sqrt{(x + c)^2 + y^2} = 2a - \sqrt{(x - c)^2 + y^2} \qquad \text{Isolate one radical.}$$

$$(x + c)^2 + y^2 = 4a^2 - 4a\sqrt{(x - c)^2 + y^2} + (x - c)^2 + y^2 \qquad \text{Square both sides.}$$

$$x^2 + 2cx + c^2 + y^2 = 4a^2 - 4a\sqrt{(x - c)^2 + y^2} + x^2 - 2cx + c^2 + y^2 \qquad \text{Remove parentheses.}$$

$$4cx - 4a^2 = -4a\sqrt{(x - c)^2 + y^2} \qquad \text{Simplify; isolate the radical.}$$

$$cx - a^2 = -a\sqrt{(x - c)^2 + y^2} \qquad \text{Divide each side by 4.}$$

$$(cx - a^2)^2 = a^2[(x - c)^2 + y^2] \qquad \text{Square both sides again.}$$

$$c^2x^2 - 2a^2cx + a^4 = a^2(x^2 - 2cx + c^2 + y^2) \qquad \text{Remove parentheses.}$$

$$(c^2 - a^2)x^2 - a^2y^2 = a^2c^2 - a^4 \qquad \text{Rearrange the terms.}$$

$$(a^2 - c^2)x^2 + a^2y^2 = a^2(a^2 - c^2) \qquad \text{Multiply each side by } -1\text{; factor } a^2 \text{ on the right side.} \quad \textbf{(1)}$$

To obtain points on the ellipse off the x-axis, it must be that $a > c$. To see why, look again at Figure 18.

$$d(F_1, P) + d(F_2, P) > d(F_1, F_2) \qquad \text{The sum of the lengths of two sides of a triangle is greater than the length of the third side.}$$

$$2a > 2c \qquad d(F_1, P) + d(F_2, P) = 2a;\ d(F_1, F_2) = 2c$$

$$a > c$$

Since $a > c$, we also have $a^2 > c^2$, so $a^2 - c^2 > 0$. Let $b^2 = a^2 - c^2$, $b > 0$. Then $a > b$ and equation (1) can be written as

$$b^2x^2 + a^2y^2 = a^2b^2$$

$$\frac{x^2}{a^2} + \frac{y^2}{b^2} = 1 \qquad \text{Divide each side by } a^2b^2.$$

Theorem

Equation of an Ellipse; Center at (0, 0); Foci at (±c, 0); Major Axis along the x-Axis

An equation of the ellipse with center at $(0, 0)$ and foci at $(-c, 0)$ and $(c, 0)$ is

$$\frac{x^2}{a^2} + \frac{y^2}{b^2} = 1, \qquad \text{where } a > b > 0 \text{ and } b^2 = a^2 - c^2 \quad \textbf{(2)}$$

The major axis is the x-axis. The vertices are at $(-a, 0)$ and $(a, 0)$.

Figure 19

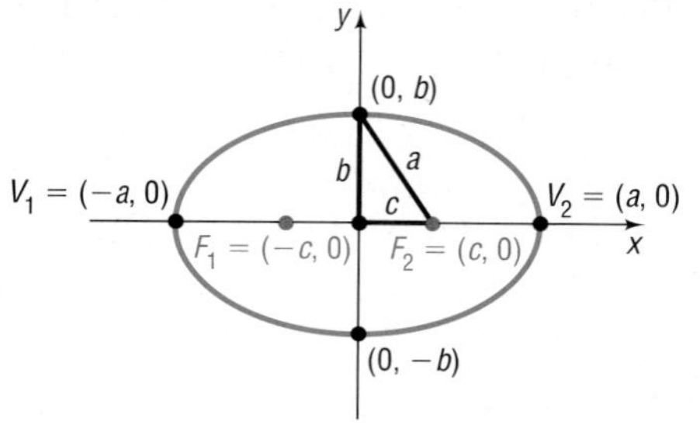

2 As you can verify, the ellipse defined by equation (2) is symmetric with respect to the x-axis, y-axis, and origin.

Because the major axis is the x-axis, we find the vertices of the ellipse defined by equation (2) by letting $y = 0$. The vertices satisfy the equation $\frac{x^2}{a^2} = 1$, the solutions of which are $x = \pm a$. Consequently, the vertices of the ellipse given by equation (2) are $V_1 = (-a, 0)$ and $V_2 = (a, 0)$. The y-intercepts of the ellipse, found by letting $x = 0$, have coordinates $(0, -b)$ and $(0, b)$. These four intercepts, $(a, 0)$, $(-a, 0)$, $(0, b)$, and $(0, -b)$, are used to graph the ellipse. See Figure 19.

Notice in Figure 19 the right triangle formed with the points $(0, 0)$, $(c, 0)$, and $(0, b)$. Because $b^2 = a^2 - c^2$ (or $b^2 + c^2 = a^2$), the distance from the focus at $(c, 0)$ to the point $(0, b)$ is a.

EXAMPLE 1 Finding an Equation of an Ellipse

Find an equation of the ellipse with center at the origin, one focus at $(3, 0)$, and a vertex at $(-4, 0)$. Graph the equation.

Figure 20

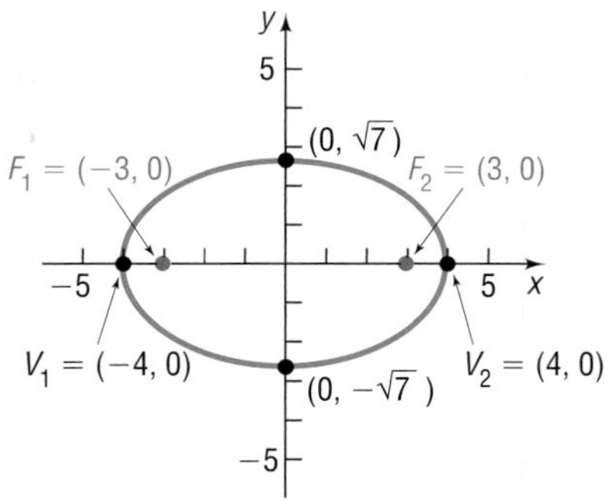

Solution The ellipse has its center at the origin and, since the given focus and vertex lie on the x-axis, the major axis is the x-axis. The distance from the center, $(0, 0)$, to one of the foci, $(3, 0)$, is $c = 3$. The distance from the center, $(0, 0)$, to one of the vertices, $(-4, 0)$, is $a = 4$. From equation (2), it follows that

$$b^2 = a^2 - c^2 = 16 - 9 = 7$$

so an equation of the ellipse is

$$\frac{x^2}{16} + \frac{y^2}{7} = 1$$

Figure 20 shows the graph. ◄

Notice in Figure 20 how we used the intercepts of the equation to graph the ellipse. Following this practice will make it easier for you to obtain an accurate graph of an ellipse.

COMMENT: The intercepts of the ellipse also provide information about how to set the viewing rectangle. To graph the ellipse

$$\frac{x^2}{16} + \frac{y^2}{7} = 1$$

discussed in Example 1, we would set the viewing rectangle using a square screen that includes the intercepts, perhaps $-4.5 \le x \le 4.5$, $-3 \le y \le 3$. Then we would proceed to solve the equation for y:

$$\frac{x^2}{16} + \frac{y^2}{7} = 1$$

$$\frac{y^2}{7} = 1 - \frac{x^2}{16} \quad \text{Subtract } \frac{x^2}{16} \text{ from each side.}$$

$$y^2 = 7\left(1 - \frac{x^2}{16}\right) \quad \text{Multiply both sides by 7.}$$

$$y = \pm\sqrt{7\left(1 - \frac{x^2}{16}\right)} \quad \text{Take the square root of each side.}$$

Figure 21

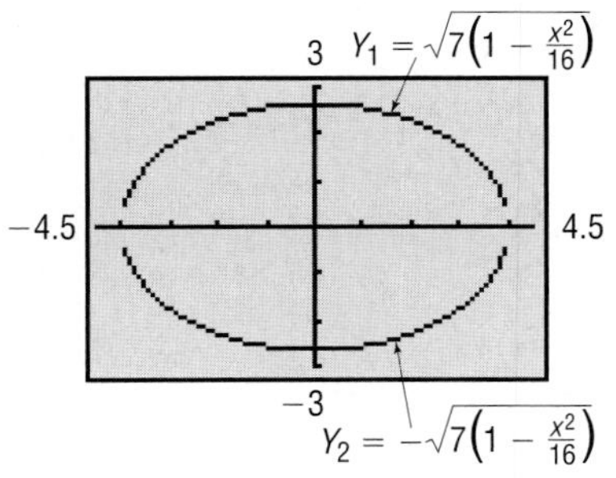

Now graph the two functions

$$Y_1 = \sqrt{7\left(1 - \frac{x^2}{16}\right)} \quad \text{and} \quad Y_2 = -\sqrt{7\left(1 - \frac{x^2}{16}\right)}$$

Figure 21 shows the result. ■

NOW WORK PROBLEM 27.

An equation of the form of equation (2), with $a > b$, is the equation of an ellipse with center at the origin, foci on the x-axis at $(-c, 0)$ and $(c, 0)$, where $c^2 = a^2 - b^2$, and major axis along the x-axis.

3 For the remainder of this section, the direction "Discuss the equation" will mean to find the center, major axis, foci, and vertices of the ellipse and graph it.

EXAMPLE 2 Discussing the Equation of an Ellipse

Discuss the equation: $\frac{x^2}{25} + \frac{y^2}{9} = 1$

Solution The given equation is of the form of equation (2), with $a^2 = 25$ and $b^2 = 9$. The equation is that of an ellipse with center $(0, 0)$ and major axis along the x-axis. The vertices are at $(\pm a, 0) = (\pm 5, 0)$. Because $b^2 = a^2 - c^2$, we find that

$$c^2 = a^2 - b^2 = 25 - 9 = 16$$

The foci are at $(\pm c, 0) = (\pm 4, 0)$. Figure 22 shows the graph.

Figure 22

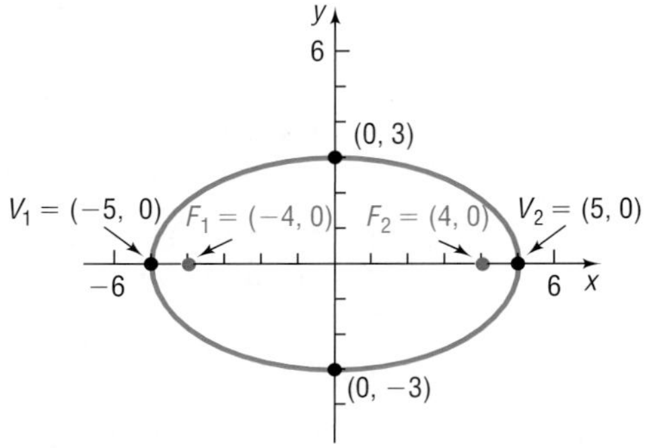

◀

NOW WORK PROBLEM 17.

If the major axis of an ellipse with center at $(0, 0)$ lies on the y-axis, then the foci are at $(0, -c)$ and $(0, c)$. Using the same steps as before, the definition of an ellipse leads to the following result:

Theorem

Equation of an Ellipse; Center at (0, 0); Foci at (0, ±c); Major Axis along the y-Axis

An equation of the ellipse with center at $(0, 0)$ and foci at $(0, -c)$ and $(0, c)$ is

$$\frac{x^2}{b^2} + \frac{y^2}{a^2} = 1, \qquad \text{where } a > b > 0 \text{ and } b^2 = a^2 - c^2 \qquad \textbf{(3)}$$

The major axis is the y-axis; the vertices are at $(0, -a)$ and $(0, a)$.

Figure 23

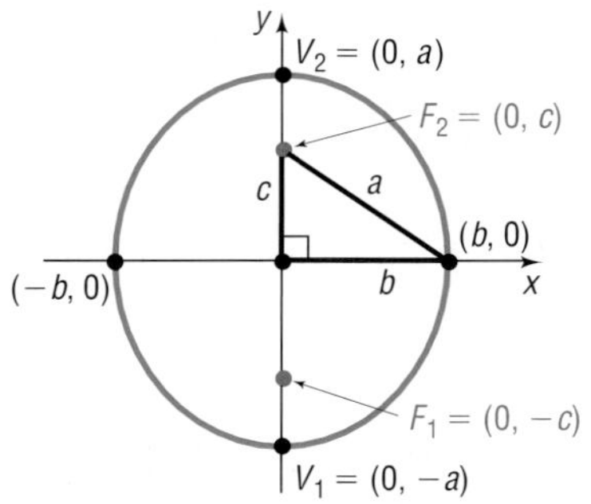

Figure 23 illustrates the graph of such an ellipse. Again, notice the right triangle with the points at $(0, 0)$, $(b, 0)$, and $(0, c)$.

Look closely at equations (2) and (3). Although they may look alike, there is a difference! In equation (2), the larger number, a^2, is in the denominator of the x^2-term, so the major axis of the ellipse is along the x-axis. In equation (3), the larger number, a^2, is in the denominator of the y^2-term, so the major axis is along the y-axis.

EXAMPLE 3 Discussing the Equation of an Ellipse

Discuss the equation: $9x^2 + y^2 = 9$

Figure 24

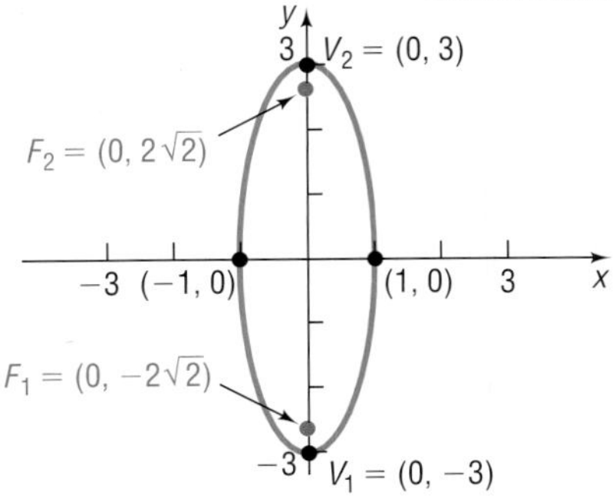

Solution To put the equation in proper form, we divide each side by 9.

$$x^2 + \frac{y^2}{9} = 1$$

The larger number, 9, is in the denominator of the y^2-term so, based on equation (3), this is the equation of an ellipse with center at the origin and major axis along the y-axis. Also, we conclude that $a^2 = 9$, $b^2 = 1$, and $c^2 = a^2 - b^2 = 9 - 1 = 8$. The vertices are at $(0, \pm a) = (0, \pm 3)$, and the foci are at $(0, \pm c) = (0, \pm 2\sqrt{2})$. The graph is given in Figure 24. ◀

NOW WORK PROBLEM 21.

EXAMPLE 4 Finding an Equation of an Ellipse

Find an equation of the ellipse having one focus at $(0, 2)$ and vertices at $(0, -3)$ and $(0, 3)$. Graph the equation.

Figure 25

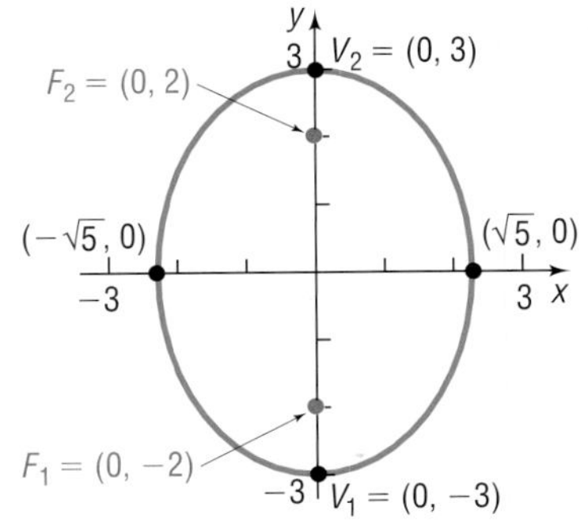

Solution Because the vertices are at $(0, -3)$ and $(0, 3)$, the center of this ellipse is at their midpoint, the origin. Also, its major axis lies on the y-axis. The distance from the center, $(0, 0)$, to one of the foci, $(0, 2)$, is $c = 2$. The distance from the center, $(0, 0)$, to one of the vertices, $(0, 3)$, is $a = 3$. So $b^2 = a^2 - c^2 = 9 - 4 = 5$. The form of the equation of this ellipse is given by equation (3).

$$\frac{x^2}{b^2} + \frac{y^2}{a^2} = 1$$

$$\frac{x^2}{5} + \frac{y^2}{9} = 1$$

Figure 25 shows the graph. ◀

NOW WORK PROBLEM 29.

The circle may be considered a special kind of ellipse. To see why, let $a = b$ in equation (2) or (3). Then

$$\frac{x^2}{a^2} + \frac{y^2}{a^2} = 1$$

$$x^2 + y^2 = a^2$$

This is the equation of a circle with center at the origin and radius a. The value of c is

$$c^2 = a^2 - b^2 = 0$$

We conclude that the closer the two foci of an ellipse are to the center, the more the ellipse will look like a circle.

Center at (h, k)

4 If an ellipse with center at the origin and major axis coinciding with a coordinate axis is shifted horizontally h units and then vertically k units, the result is an ellipse with center at (h, k) and major axis parallel to a coordinate axis. The equations of such ellipses have the same forms as those given in equations (2) and (3), except that x is replaced by $x - h$ (the horizontal shift) and y is replaced by $y - k$ (the vertical shift). Table 3 gives the forms of the equations of such ellipses, and Figure 26 shows their graphs.

Table 3 **Ellipses with Center at (h, k) and Major Axis Parallel to a Coordinate Axis**

Center	Major Axis	Foci	Vertices	Equation
(h, k)	Parallel to x-axis	$(h + c, k)$	$(h + a, k)$	$\frac{(x-h)^2}{a^2} + \frac{(y-k)^2}{b^2} = 1,$
		$(h - c, k)$	$(h - a, k)$	$a > b$ and $b^2 = a^2 - c^2$
(h, k)	Parallel to y-axis	$(h, k + c)$	$(h, k + a)$	$\frac{(x-h)^2}{b^2} + \frac{(y-k)^2}{a^2} = 1,$
		$(h, k - c)$	$(h, k - a)$	$a > b$ and $b^2 = a^2 - c^2$

Figure 26

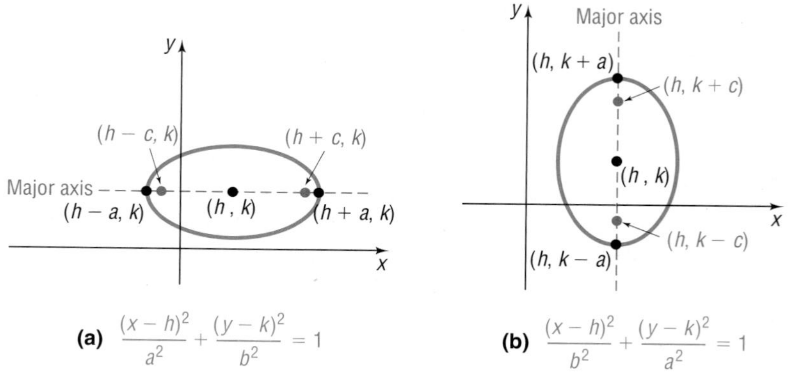

EXAMPLE 5 **Finding an Equation of an Ellipse, Center Not at the Origin**

Find an equation for the ellipse with center at $(2, -3)$, one focus at $(3, -3)$, and one vertex at $(5, -3)$. Graph the equation.

Solution The center is at $(h, k) = (2, -3)$, so $h = 2$ and $k = -3$. Since the center, focus, and vertex all lie on the line $y = -3$, the major axis is parallel to the x-axis. The distance from the center $(2, -3)$ to a focus $(3, -3)$ is $c = 1$; the distance from the center $(2, -3)$ to a vertex $(5, -3)$ is $a = 3$. Then $b^2 = a^2 - c^2 = 9 - 1 = 8$. The form of the equation is

$$\frac{(x-h)^2}{a^2} + \frac{(y-k)^2}{b^2} = 1$$

$$\frac{(x-2)^2}{9} + \frac{(y+3)^2}{8} = 1 \qquad h = 2, k = -3, a = 3, b = 2\sqrt{2}$$

Figure 27

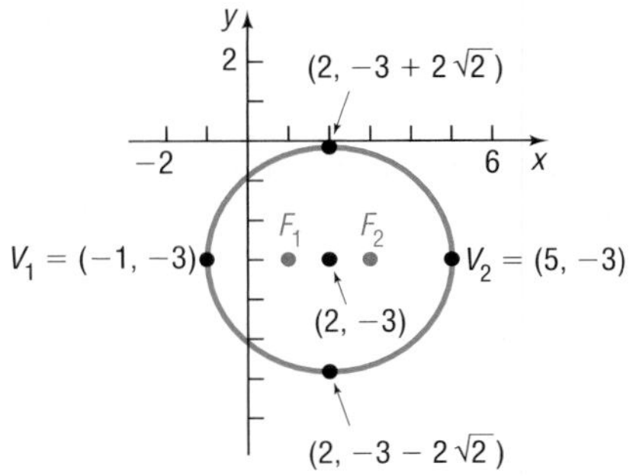

To graph the equation, we use the center $(h, k) = (2, -3)$ to locate the vertices. The major axis is parallel to the x-axis, so the vertices are $a = 3$ units left and right of the center $(2, -3)$. Therefore, the vertices are

$$V_1 = (2 - 3, -3) = (-1, -3) \quad \text{and} \quad V_2 = (2 + 3, -3) = (5, -3)$$

Since $c = 1$ and the major axis is parallel to the x-axis, the foci are 1 unit left and right of the center. Therefore, the foci are

$$F_1 = (2 - 1, -3) = (1, -3) \quad \text{and} \quad F_2 = (2 + 1, -3) = (3, -3)$$

Finally, we use the value of $b = 2\sqrt{2}$ to find the two points above and below the center.

$$\left(2, -3 - 2\sqrt{2}\right) \quad \text{and} \quad \left(2, -3 + 2\sqrt{2}\right)$$

Figure 27 shows the graph. ◀

NOW WORK PROBLEM 55.

EXAMPLE 6 Discussing the Equation of an Ellipse

Discuss the equation: $4x^2 + y^2 - 8x + 4y + 4 = 0$

Solution We proceed to complete the squares in x and in y.

$$4x^2 + y^2 - 8x + 4y + 4 = 0$$

$$4x^2 - 8x + y^2 + 4y = -4 \qquad \text{Group like variables; place the constant on the right side.}$$

$$4(x^2 - 2x) + (y^2 + 4y) = -4 \qquad \text{Factor out 4 from the first two terms.}$$

$$4(x^2 - 2x + 1) + (y^2 + 4y + 4) = -4 + 4 + 4 \qquad \text{Complete each square.}$$

$$4(x - 1)^2 + (y + 2)^2 = 4 \qquad \text{Factor.}$$

$$(x - 1)^2 + \frac{(y + 2)^2}{4} = 1 \qquad \text{Divide each side by 4.}$$

Figure 28

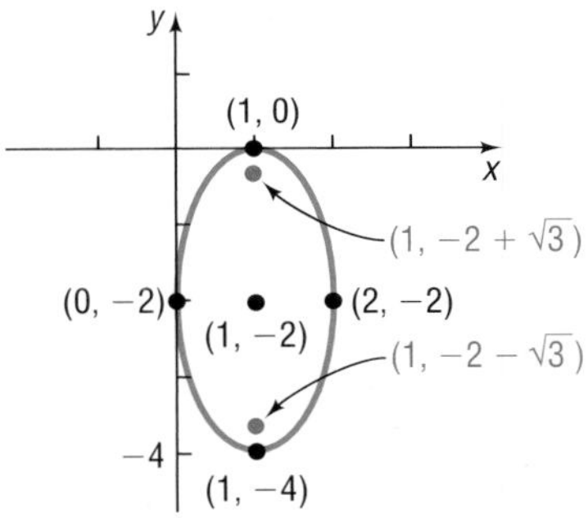

This is the equation of an ellipse with center at $(1, -2)$ and major axis parallel to the y-axis. Since $a^2 = 4$ and $b^2 = 1$, we have $c^2 = a^2 - b^2 = 4 - 1 = 3$. The vertices are at $(h, k \pm a) = (1, -2 \pm 2)$ or $(1, 0)$ and $(1, -4)$. The foci are at $(h, k \pm c) = (1, -2 \pm \sqrt{3})$ or $(1, -2 - \sqrt{3})$ and $(1, -2 + \sqrt{3})$. Figure 28 shows the graph. ◀

NOW WORK PROBLEM 47.

Applications

5 Ellipses are found in many applications in science and engineering. For example, the orbits of the planets around the Sun are elliptical, with the Sun's position at a focus. See Figure 29.

Figure 29

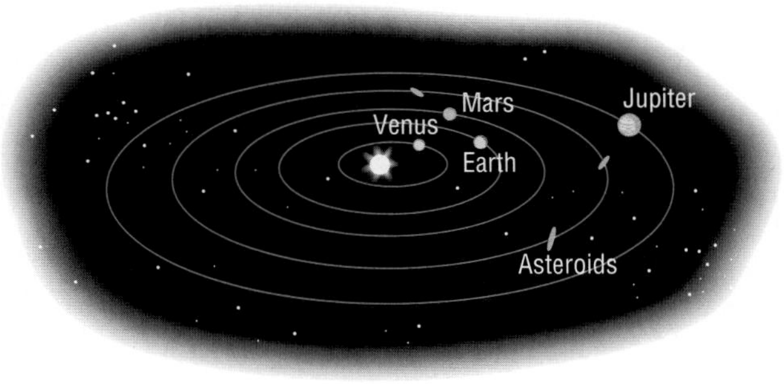

Stone and concrete bridges are often shaped as semielliptical arches. Elliptical gears are used in machinery when a variable rate of motion is required.

Ellipses also have an interesting reflection property. If a source of light (or sound) is placed at one focus, the waves transmitted by the source will reflect off the ellipse and concentrate at the other focus. This is the principle behind *whispering galleries*, which are rooms designed with elliptical ceilings. A person standing at one focus of the ellipse can whisper and be heard by a person standing at the other focus, because all the sound waves that reach the ceiling are reflected to the other person.

EXAMPLE 7 Whispering Galleries

Figure 30

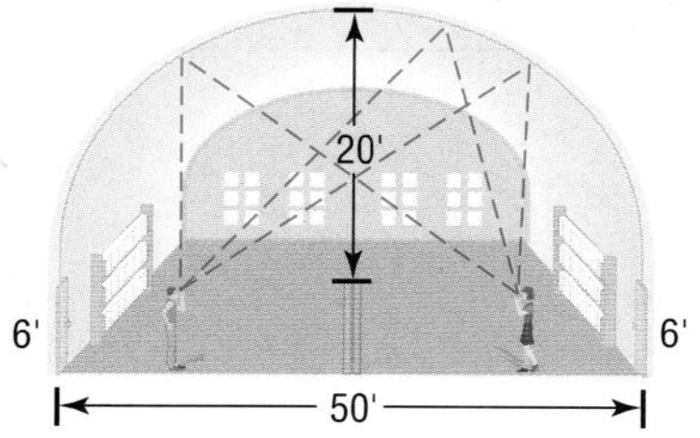

Figure 31

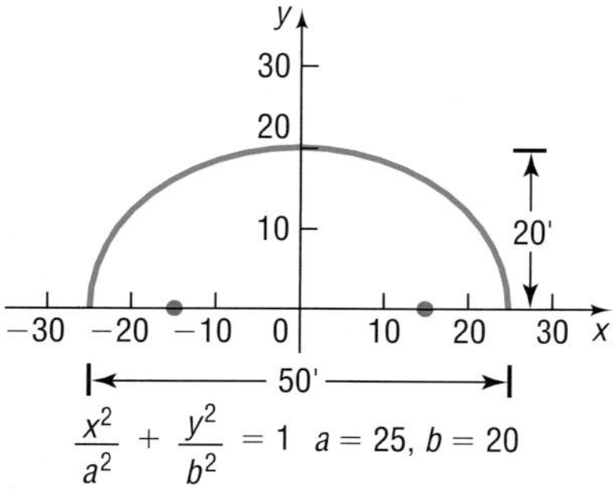

Figure 30 shows the specifications for an elliptical ceiling in a hall designed to be a whispering gallery. In a whispering gallery, a person standing at one focus of the ellipse can whisper and be heard by another person standing at the other focus, because all the sound waves that reach the ceiling from one focus are reflected to the other focus. Where are the foci located in the hall?

Solution We set up a rectangular coordinate system so that the center of the ellipse is at the origin and the major axis is along the x-axis. See Figure 31. The equation of the ellipse is

$$\frac{x^2}{a^2} + \frac{y^2}{b^2} = 1$$

where $a = 25$ and $b = 20$.

Then, since

$$c^2 = a^2 - b^2 = 25^2 - 20^2 = 625 - 400 = 225$$

we have $c = 15$. The foci are located 15 feet from the center of the ellipse along the major axis. ◀

9.3 Assess Your Understanding

'Are You Prepared?' *Answers are given at the end of these exercises. If you get a wrong answer, read the pages listed in red.*

1. The distance d from $P_1 = (2, -5)$ to $P_2 = (4, -2)$ is $d =$ ________. (p. 3)
2. To complete the square of $x^2 - 3x$, add ________. (p. 941)
3. Find the intercepts of the equation $y^2 = 16 - 4x^2$. (pp. 11–12)
4. The point that is symmetric with respect to the y-axis to the point $(-2, 5)$ is ________. (pp. 12–13)
5. To graph $y = (x + 1)^2 - 4$, shift the graph of $y = x^2$ to the left/right ________ unit(s) and then up/down ________ unit(s). (pp. 94–103)
6. The standard equation of a circle with center at $(2, -3)$ and radius 1 is ________. (pp. 16–19)

Concepts and Vocabulary

7. A(n) __________ is the collection of all points in the plane the sum of whose distances from two fixed points is a constant.

8. For an ellipse, the foci lie on a line called the __________ axis.

9. For the ellipse $\frac{x^2}{4} + \frac{y^2}{25} = 1$, the vertices are the points __________ and __________.

10. *True or False:* The foci, vertices, and center of an ellipse lie on a line called the axis of symmetry.

11. *True or False:* If the center of an ellipse is at the origin and the foci lie on the y-axis, the ellipse is symmetric with respect to the x-axis, the y-axis, and the origin.

12. *True or False:* A circle is a certain type of ellipse.

Exercises

In Problems 13–16, the graph of an ellipse is given. Match each graph to its equation.

A. $\frac{x^2}{4} + y^2 = 1$ *B.* $x^2 + \frac{y^2}{4} = 1$ *C.* $\frac{x^2}{16} + \frac{y^2}{4} = 1$ *D.* $\frac{x^2}{4} + \frac{y^2}{16} = 1$

13.

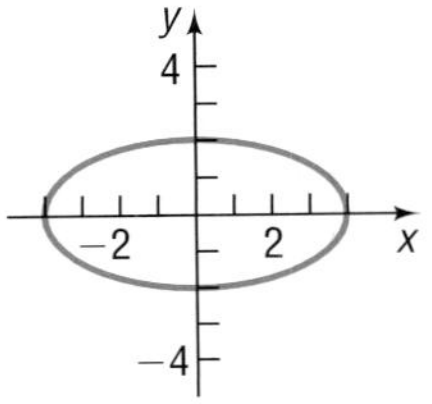

14.

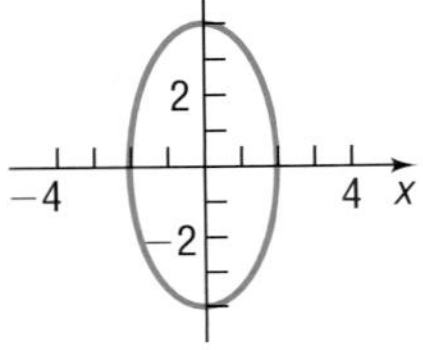

15.

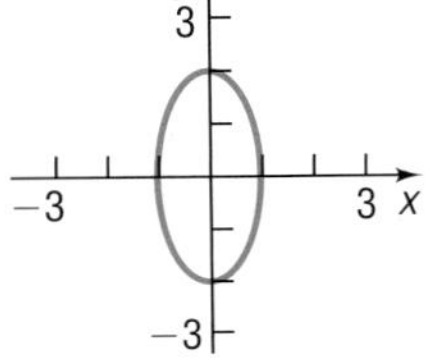

16.

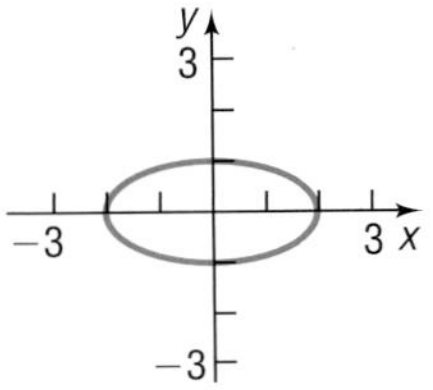

In Problems 17–26, find the vertices and foci of each ellipse. Graph each equation.

17. $\frac{x^2}{25} + \frac{y^2}{4} = 1$

18. $\frac{x^2}{9} + \frac{y^2}{4} = 1$

19. $\frac{x^2}{9} + \frac{y^2}{25} = 1$

20. $x^2 + \frac{y^2}{16} = 1$

21. $4x^2 + y^2 = 16$

22. $x^2 + 9y^2 = 18$

23. $4y^2 + x^2 = 8$

24. $4y^2 + 9x^2 = 36$

25. $x^2 + y^2 = 16$

26. $x^2 + y^2 = 4$

In Problems 27–38, find an equation for each ellipse. Graph the equation.

27. Center at $(0, 0)$; focus at $(3, 0)$; vertex at $(5, 0)$

28. Center at $(0, 0)$; focus at $(-1, 0)$; vertex at $(3, 0)$

29. Center at $(0, 0)$; focus at $(0, -4)$; vertex at $(0, 5)$

30. Center at $(0, 0)$; focus at $(0, 1)$; vertex at $(0, -2)$

31. Foci at $(\pm 2, 0)$; length of the major axis is 6

32. Foci at $(0, \pm 2)$; length of the major axis is 8

33. Focus at $(-4, 0)$; vertices at $(\pm 5, 0)$

34. Focus at $(0, -4)$; vertices at $(0, \pm 8)$

35. Foci at $(0, \pm 3)$; x-intercepts are ± 2

36. Vertices at $(\pm 4, 0)$; y-intercepts are ± 1

37. Center at $(0, 0)$; vertex at $(0, 4)$; $b = 1$

38. Vertices at $(\pm 5, 0)$; $c = 2$

In Problems 39–42, write an equation for each ellipse.

39.

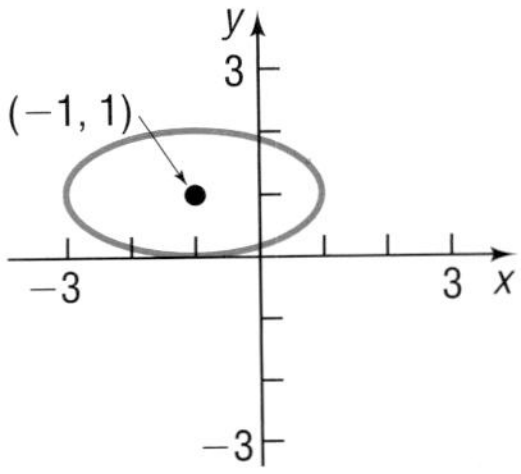

40.

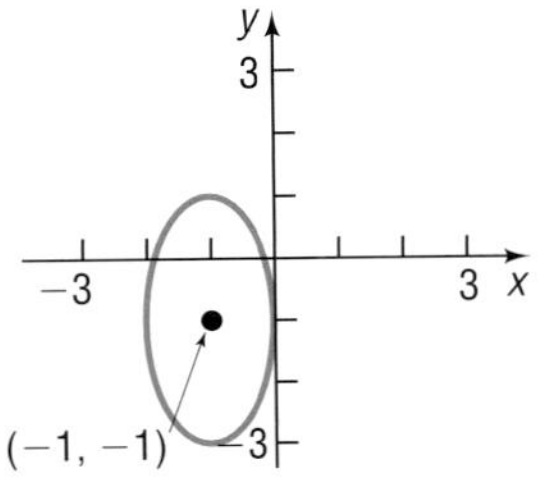

41.

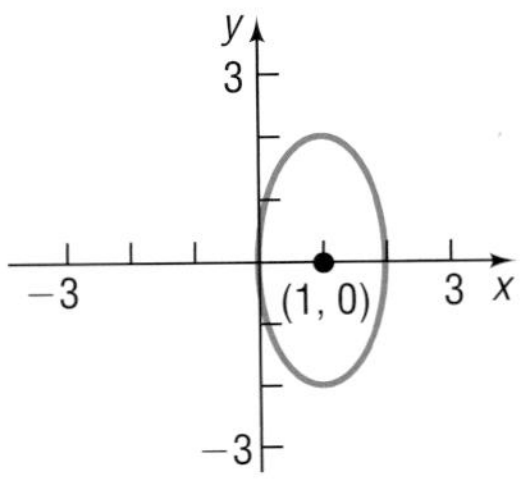

42.

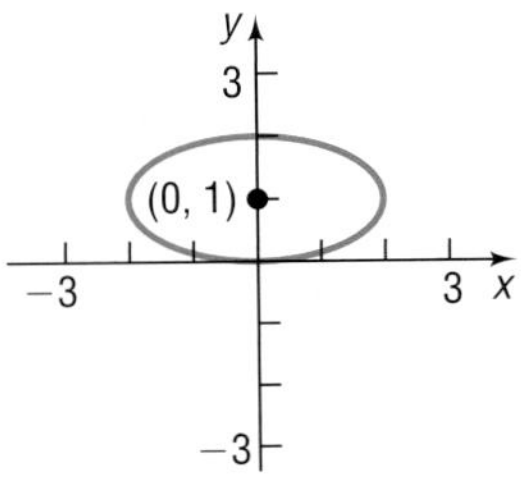

In Problems 43–54, discuss each equation; that is, find the center, foci, and vertices of each ellipse. Graph each equation.

43. $\frac{(x-3)^2}{4} + \frac{(y+1)^2}{9} = 1$

44. $\frac{(x+4)^2}{9} + \frac{(y+2)^2}{4} = 1$

45. $(x+5)^2 + 4(y-4)^2 = 16$

46. $9(x-3)^2 + (y+2)^2 = 18$

47. $x^2 + 4x + 4y^2 - 8y + 4 = 0$

48. $x^2 + 3y^2 - 12y + 9 = 0$

49. $2x^2 + 3y^2 - 8x + 6y + 5 = 0$

50. $4x^2 + 3y^2 + 8x - 6y = 5$

51. $9x^2 + 4y^2 - 18x + 16y - 11 = 0$

52. $x^2 + 9y^2 + 6x - 18y + 9 = 0$

53. $4x^2 + y^2 + 4y = 0$

54. $9x^2 + y^2 - 18x = 0$

In Problems 55–64, find an equation for each ellipse. Graph the equation.

55. Center at $(2, -2)$; vertex at $(7, -2)$; focus at $(4, -2)$

56. Center at $(-3, 1)$; vertex at $(-3, 3)$; focus at $(-3, 0)$

57. Vertices at $(4, 3)$ and $(4, 9)$; focus at $(4, 8)$

58. Foci at $(1, 2)$ and $(-3, 2)$; vertex at $(-4, 2)$

59. Foci at $(5, 1)$ and $(-1, 1)$; length of the major axis is 8

60. Vertices at $(2, 5)$ and $(2, -1)$; $c = 2$

61. Center at $(1, 2)$; focus at $(4, 2)$; contains the point $(1, 3)$

62. Center at $(1, 2)$; focus at $(1, 4)$; contains the point $(2, 2)$

63. Center at $(1, 2)$; vertex at $(4, 2)$; contains the point $(1, 3)$

64. Center at $(1, 2)$; vertex at $(1, 4)$; contains the point $(2, 2)$

In Problems 65–68, graph each function.
[**Hint:** Notice that each function is half an ellipse.]

65. $f(x) = \sqrt{16 - 4x^2}$

66. $f(x) = \sqrt{9 - 9x^2}$

67. $f(x) = -\sqrt{64 - 16x^2}$

68. $f(x) = -\sqrt{4 - 4x^2}$

69. Semielliptical Arch Bridge An arch in the shape of the upper half of an ellipse is used to support a bridge that is to span a river 20 meters wide. The center of the arch is 6 meters above the center of the river (see the figure). Write an equation for the ellipse in which the x-axis coincides with the water level and the y-axis passes through the center of the arch.

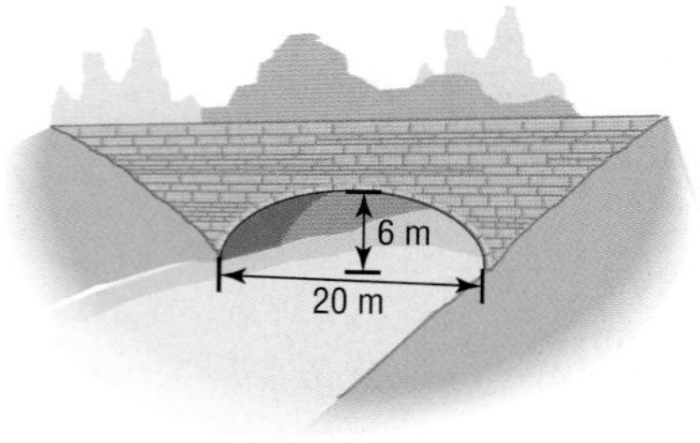

70. Semielliptical Arch Bridge The arch of a bridge is a semiellipse with a horizontal major axis. The span is 30 feet, and the top of the arch is 10 feet above the major axis. The roadway is horizontal and is 2 feet above the top of the arch. Find the vertical distance from the roadway to the arch at 5-foot intervals along the roadway.

71. Whispering Gallery A hall 100 feet in length is to be designed as a whispering gallery. If the foci are located 25 feet from the center, how high will the ceiling be at the center?

72. Whispering Gallery Jim, standing at one focus of a whispering gallery, is 6 feet from the nearest wall. His friend is standing at the other focus, 100 feet away. What is the length of this whispering gallery? How high is its elliptical ceiling at the center?

73. Semielliptical Arch Bridge A bridge is built in the shape of a semielliptical arch. The bridge has a span of 120 feet and a maximum height of 25 feet. Choose a suitable rectangular coordinate system and find the height of the arch at distances of 10, 30, and 50 feet from the center.

74. Semielliptical Arch Bridge A bridge is to be built in the shape of a semielliptical arch and is to have a span of 100 feet. The height of the arch, at a distance of 40 feet from the center, is to be 10 feet. Find the height of the arch at its center.

75. Semielliptical Arch An arch in the form of half an ellipse is 40 feet wide and 15 feet high at the center. Find the height of the arch at intervals of 10 feet along its width.

76. Semielliptical Arch Bridge An arch for a bridge over a highway is in the form of half an ellipse. The top of the arch is 20 feet above the ground level (the major axis). The highway has four lanes, each 12 feet wide; a center safety strip 8 feet wide; and two side strips, each 4 feet wide. What should the span of the bridge be (the length of its major axis) if the height 28 feet from the center is to be 13 feet?

In Problems 77–80, use the fact that the orbit of a planet about the Sun is an ellipse, with the Sun at one focus. The **aphelion** *of a planet is its greatest distance from the Sun, and the* **perihelion** *is its shortest distance. The* **mean distance** *of a planet from the Sun is the length of the semimajor axis of the elliptical orbit. See the illustration.*

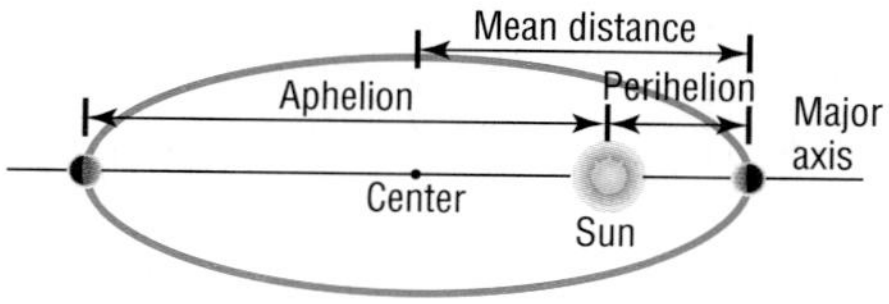

77. Earth The mean distance of Earth from the Sun is 93 million miles. If the aphelion of Earth is 94.5 million miles, what is the perihelion? Write an equation for the orbit of Earth around the Sun.

78. Mars The mean distance of Mars from the Sun is 142 million miles. If the perihelion of Mars is 128.5 million miles, what is the aphelion? Write an equation for the orbit of Mars about the Sun.

79. Jupiter The aphelion of Jupiter is 507 million miles. If the distance from the Sun to the center of its elliptical orbit is 23.2 million miles, what is the perihelion? What is the mean distance? Write an equation for the orbit of Jupiter around the Sun.

80. Pluto The perihelion of Pluto is 4551 million miles, and the distance of the Sun from the center of its elliptical orbit is 897.5 million miles. Find the aphelion of Pluto. What is the mean distance of Pluto from the Sun? Write an equation for the orbit of Pluto about the Sun.

81. Racetrack Design Consult the figure. A racetrack is in the shape of an ellipse, 100 feet long and 50 feet wide. What is the width 10 feet from a vertex?

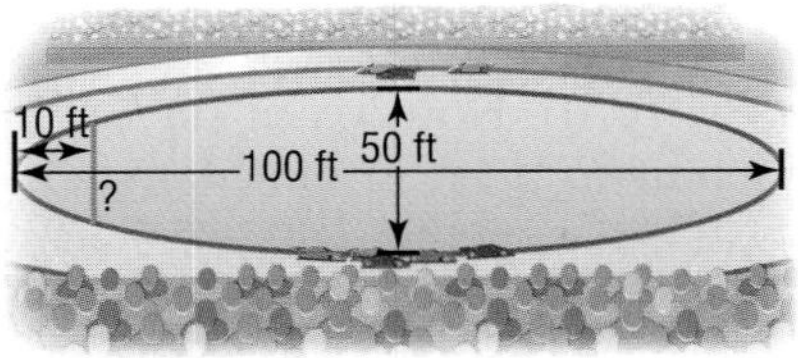

82. Racetrack Design A racetrack is in the shape of an ellipse 80 feet long and 40 feet wide. What is the width 10 feet from a vertex?

83. Show that an equation of the form

$$Ax^2 + Cy^2 + F = 0, \qquad A \neq 0, C \neq 0, F \neq 0$$

where A and C are of the same sign and F is of opposite sign,
(a) Is the equation of an ellipse with center at $(0, 0)$ if $A \neq C$.
(b) Is the equation of a circle with center $(0, 0)$ if $A = C$.

84. Show that the graph of an equation of the form

$$Ax^2 + Cy^2 + Dx + Ey + F = 0, \qquad A \neq 0, C \neq 0$$

where A and C are of the same sign,
(a) Is an ellipse if $\frac{D^2}{4A} + \frac{E^2}{4C} - F$ is the same sign as A.
(b) Is a point if $\frac{D^2}{4A} + \frac{E^2}{4C} - F = 0$.
(c) Contains no points if $\frac{D^2}{4A} + \frac{E^2}{4C} - F$ is of opposite sign to A.

85. The **eccentricity** e of an ellipse is defined as the number $\frac{c}{a}$, where a and c are the numbers given in equation (2). Because $a > c$, it follows that $e < 1$. Write a brief paragraph about the general shape of each of the following ellipses. Be sure to justify your conclusions.
(a) Eccentricity close to 0
(b) Eccentricity $= 0.5$
(c) Eccentricity close to 1

'Are You Prepared?' Answers

1. $\sqrt{13}$ **2.** $\frac{9}{4}$
3. $(-2, 0), (2, 0), (0, -4), (0, 4)$
4. $(2, 5)$ **5.** left; 1; down; 4
6. $(x - 2)^2 + (y + 3)^2 = 1$

9.4 The Hyperbola

PREPARING FOR THIS SECTION *Before getting started, review the following:*

- Distance Formula (Section 1.1, p. 3)
- Completing the Square (Appendix A, Section A.5, p. 941)
- Intercepts and Symmetry (Section 1.2, pp. 11–13)
- Asymptotes (Section 3.3, pp. 165–166)
- Graphing Techniques: Transformations (Section 2.5, pp. 94–103)
- Square Root Method (Appendix A, Section A.5, pp. 940–941)

Now work the 'Are You Prepared?' problems on page 647.

OBJECTIVES

1 Find the Equation of a Hyperbola
2 Graph Hyperbolas
3 Discuss the Equation of a Hyperbola
4 Find the Asymptotes of a Hyperbola
5 Work with Hyperbolas with Center at (h, k)
6 Solve Applied Problems Involving Hyperbolas

A **hyperbola** is the collection of all points in the plane the difference of whose distances from two fixed points, called the **foci,** is a constant.

Figure 32

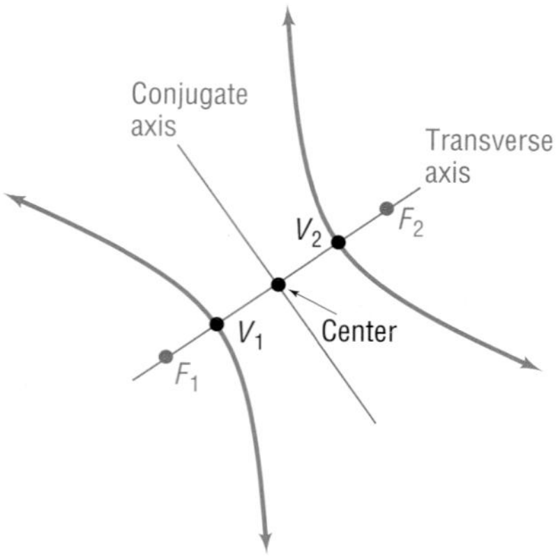

Figure 33
$d(F_1, P) - d(F_2, P) = \pm 2a$

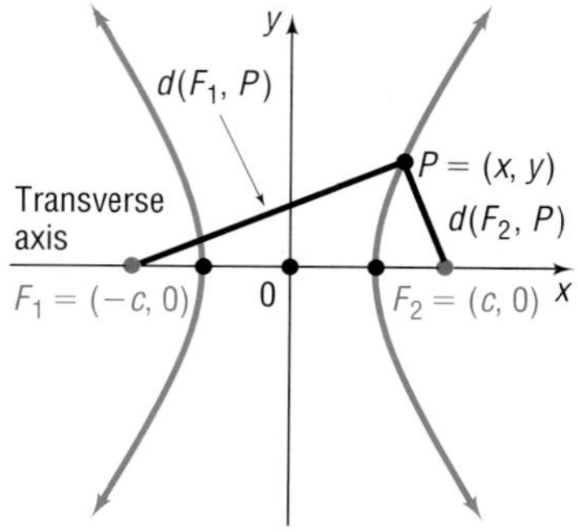

1 Figure 32 illustrates a hyperbola with foci F_1 and F_2. The line containing the foci is called the **transverse axis.** The midpoint of the line segment joining the foci is the **center** of the hyperbola. The line through the center and perpendicular to the transverse axis is the **conjugate axis.** The hyperbola consists of two separate curves, called **branches,** that are symmetric with respect to the transverse axis, conjugate axis, and center. The two points of intersection of the hyperbola and the transverse axis are the **vertices,** V_1 and V_2, of the hyperbola.

With these ideas in mind, we are now ready to find the equation of a hyperbola in a rectangular coordinate system. First, we place the center at the origin. Next, we position the hyperbola so that its transverse axis coincides with a coordinate axis. Suppose that the transverse axis coincides with the x-axis, as shown in Figure 33.

If c is the distance from the center to a focus, then one focus will be at $F_1 = (-c, 0)$ and the other at $F_2 = (c, 0)$. Now we let the constant difference of the distances from any point $P = (x, y)$ on the hyperbola to the foci F_1 and F_2 be denoted by $\pm 2a$. (If P is on the right branch, the $+$ sign is used; if P is on the left branch, the $-$ sign is used.) The coordinates of P must satisfy the equation

$$d(F_1, P) - d(F_2, P) = \pm 2a$$ Difference of the distances from P to the foci equals $\pm 2a$.

$$\sqrt{(x + c)^2 + y^2} - \sqrt{(x - c)^2 + y^2} = \pm 2a$$ Use the distance formula.

$$\sqrt{(x + c)^2 + y^2} = \pm 2a + \sqrt{(x - c)^2 + y^2}$$ Isolate one radical.

$$(x + c)^2 + y^2 = 4a^2 \pm 4a\sqrt{(x - c)^2 + y^2} + (x - c)^2 + y^2$$ Square both sides.

Next we remove the parentheses.

$$x^2 + 2cx + c^2 + y^2 = 4a^2 \pm 4a\sqrt{(x - c)^2 + y^2} + x^2 - 2cx + c^2 + y^2$$

$$4cx - 4a^2 = \pm 4a\sqrt{(x - c)^2 + y^2}$$ Simplify; isolate the radical.

$$cx - a^2 = \pm a\sqrt{(x - c)^2 + y^2}$$ Divide each side by 4.

$$(cx - a^2)^2 = a^2[(x - c)^2 + y^2]$$ Square both sides.

$$c^2x^2 - 2ca^2x + a^4 = a^2(x^2 - 2cx + c^2 + y^2)$$ Simplify.

$$c^2x^2 + a^4 = a^2x^2 + a^2c^2 + a^2y^2$$ Remove parentheses and simplify.

$$(c^2 - a^2)x^2 - a^2y^2 = a^2c^2 - a^4$$ Rearrange terms.

$$(c^2 - a^2)x^2 - a^2y^2 = a^2(c^2 - a^2) \quad \textbf{(1)}$$ Factor a^2 on the right side.

To obtain points on the hyperbola off the x-axis, it must be that $a < c$. To see why, look again at Figure 33.

$$d(F_1, P) < d(F_2, P) + d(F_1, F_2)$$ Use triangle F_1PF_2.

$$d(F_1, P) - d(F_2, P) < d(F_1, F_2)$$ P is on the right branch, so $d(F_1, P) - d(F_2, P) = 2a$.

$$2a < 2c$$
$$a < c$$

Since $a < c$, we also have $a^2 < c^2$, so $c^2 - a^2 > 0$. Let $b^2 = c^2 - a^2, b > 0$. Then equation (1) can be written as

$$b^2x^2 - a^2y^2 = a^2b^2$$

$$\frac{x^2}{a^2} - \frac{y^2}{b^2} = 1 \qquad \text{Divide each side by } a^2b^2.$$

To find the vertices of the hyperbola defined by this equation, let $y = 0$. The vertices satisfy the equation $\frac{x^2}{a^2} = 1$, the solutions of which are $x = \pm a$. Consequently, the vertices of the hyperbola are $V_1 = (-a, 0)$ and $V_2 = (a, 0)$. Notice that the distance from the center $(0, 0)$ to either vertex is a.

Theorem

Equation of a Hyperbola; Center at (0, 0); Foci at ($\pm c$, 0); Vertices at ($\pm a$, 0); Transverse Axis along the x-Axis

An equation of the hyperbola with center at $(0, 0)$, foci at $(-c, 0)$ and $(c, 0)$, and vertices at $(-a, 0)$ and $(a, 0)$ is

$$\frac{x^2}{a^2} - \frac{y^2}{b^2} = 1, \quad \text{where } b^2 = c^2 - a^2 \qquad \textbf{(2)}$$

The transverse axis is the x-axis.

2 See Figure 34. As you can verify, the hyperbola defined by equation (2) is symmetric with respect to the x-axis, y-axis, and origin. To find the y-intercepts, if any, let $x = 0$ in equation (2). This results in the equation $\frac{y^2}{b^2} = -1$, which has no real solution. We conclude that the hyperbola defined by equation (2) has no y-intercepts. In fact, since $\frac{x^2}{a^2} - 1 = \frac{y^2}{b^2} \geq 0$, it follows that $\frac{x^2}{a^2} \geq 1$. There are no points on the graph for $-a < x < a$.

Figure 34
$\frac{x^2}{a^2} - \frac{y^2}{b^2} = 1, \quad b^2 = c^2 - a^2$

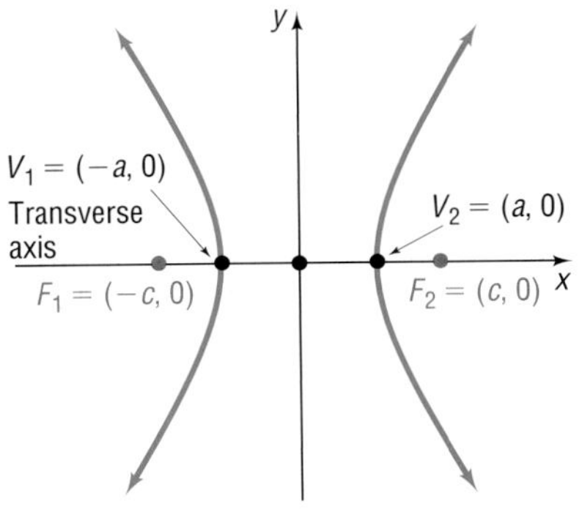

EXAMPLE 1 **Finding and Graphing an Equation of a Hyperbola**

Find an equation of the hyperbola with center at the origin, one focus at $(3, 0)$, and one vertex at $(-2, 0)$. Graph the equation.

Solution The hyperbola has its center at the origin, and the transverse axis coincides with the x-axis. One focus is at $(c, 0) = (3, 0)$, so $c = 3$. One vertex is at $(-a, 0) = (-2, 0)$, so $a = 2$. From equation (2), it follows that $b^2 = c^2 - a^2 = 9 - 4 = 5$, so an equation of the hyperbola is

$$\frac{x^2}{4} - \frac{y^2}{5} = 1$$

To graph a hyperbola, it is helpful to locate and plot other points on the graph. For example, to find the points above and below the foci, we let $x = \pm 3$. Then

$$\frac{x^2}{4} - \frac{y^2}{5} = 1$$

$$\frac{(\pm 3)^2}{4} - \frac{y^2}{5} = 1 \qquad x = \pm 3$$

$$\frac{9}{4} - \frac{y^2}{5} = 1$$

$$\frac{y^2}{5} = \frac{5}{4}$$

$$y^2 = \frac{25}{4}$$

$$y = \pm\frac{5}{2}$$

Figure 35

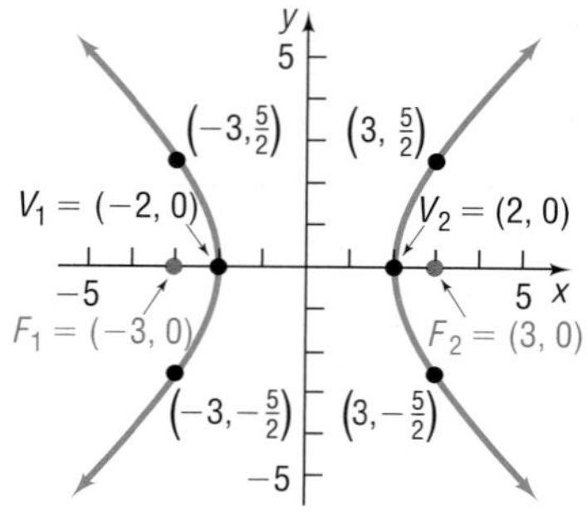

The points above and below the foci are $\left(\pm 3, \frac{5}{2}\right)$ and $\left(\pm 3, -\frac{5}{2}\right)$. These points help because they determine the "opening" of the hyperbola. See Figure 35. ◀

COMMENT: To graph the hyperbola $\frac{x^2}{4} - \frac{y^2}{5} = 1$ discussed in Example 1, we need to graph the two functions $Y_1 = \sqrt{5}\sqrt{\frac{x^2}{4} - 1}$ and $Y_2 = -\sqrt{5}\sqrt{\frac{x^2}{4} - 1}$. Do this and compare what you see with Figure 35. ■

NOW WORK PROBLEM 17.

An equation of the form of equation (2) is the equation of a hyperbola with center at the origin; foci on the x-axis at $(-c, 0)$ and $(c, 0)$, where $c^2 = a^2 + b^2$; and transverse axis along the x-axis.

3 For now, the direction "Discuss the equation" will mean to find the center, transverse axis, vertices, and foci of the hyperbola and graph it.

EXAMPLE 2 Discussing the Equation of a Hyperbola

Discuss the equation: $\frac{x^2}{16} - \frac{y^2}{4} = 1$

Solution The given equation is of the form of equation (2), with $a^2 = 16$ and $b^2 = 4$. The graph of the equation is a hyperbola with center at $(0, 0)$ and transverse axis along the x-axis. Also, we know that $c^2 = a^2 + b^2 = 16 + 4 = 20$. The vertices are at $(\pm a, 0) = (\pm 4, 0)$, and the foci are at $(\pm c, 0) = (\pm 2\sqrt{5}, 0)$.

To locate the points on the graph above and below the foci, we let $x = \pm 2\sqrt{5}$. Then

$$\frac{x^2}{16} - \frac{y^2}{4} = 1$$
$$\frac{(\pm 2\sqrt{5})^2}{16} - \frac{y^2}{4} = 1 \qquad x = \pm 2\sqrt{5}$$
$$\frac{20}{16} - \frac{y^2}{4} = 1$$
$$\frac{5}{4} - \frac{y^2}{4} = 1$$
$$\frac{y^2}{4} = \frac{1}{4}$$
$$y = \pm 1$$

The points above and below the foci are $(\pm 2\sqrt{5}, 1)$ and $(\pm 2\sqrt{5}, -1)$. See Figure 36.

Figure 36

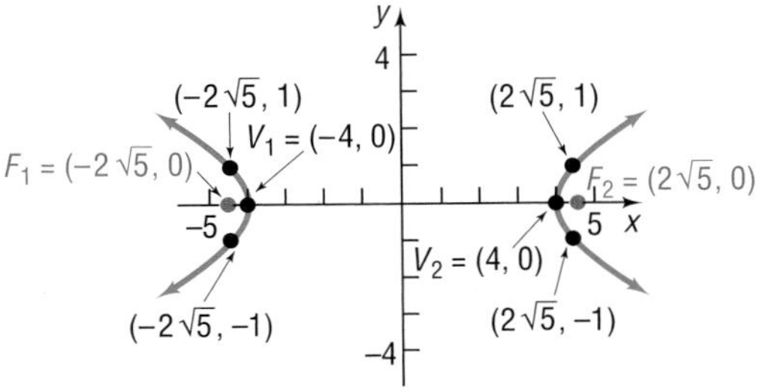

◄

NOW WORK PROBLEM 27.

The next result gives the form of the equation of a hyperbola with center at the origin and transverse axis along the y-axis.

Theorem

Equation of a Hyperbola; Center at (0, 0); Foci at (0, $\pm c$); Vertices at (0, $\pm a$); Transverse Axis along the y-Axis

An equation of the hyperbola with center at $(0, 0)$, foci at $(0, -c)$ and $(0, c)$, and vertices at $(0, -a)$ and $(0, a)$ is

$$\frac{y^2}{a^2} - \frac{x^2}{b^2} = 1, \quad \text{where } b^2 = c^2 - a^2 \tag{3}$$

The transverse axis is the y-axis.

Figure 37 $\frac{y^2}{a^2} - \frac{x^2}{b^2} = 1, \quad b^2 = c^2 - a^2$

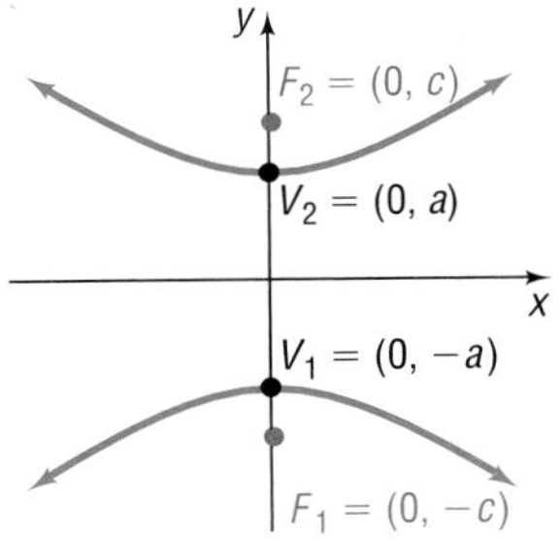

Figure 37 shows the graph of a typical hyperbola defined by equation (3).

An equation of the form of equation (2), $\frac{x^2}{a^2} - \frac{y^2}{b^2} = 1$, is the equation of a hyperbola with center at the origin; foci on the x-axis at $(-c, 0)$ and $(c, 0)$, where $c^2 = a^2 + b^2$; and transverse axis along the x-axis.

An equation of the form of equation (3), $\frac{y^2}{a^2} - \frac{x^2}{b^2} = 1$, is the equation of a hyperbola with center at the origin; foci on the y-axis at $(0, -c)$ and $(0, c)$, where $c^2 = a^2 + b^2$; and transverse axis along the y-axis.

Notice the difference in the forms of equations (2) and (3). When the y^2-term is subtracted from the x^2-term, the transverse axis is the x-axis. When the x^2-term is subtracted from the y^2-term, the transverse axis is the y-axis.

EXAMPLE 3 Discussing the Equation of a Hyperbola

Discuss the equation: $y^2 - 4x^2 = 4$

Solution To put the equation in proper form, we divide each side by 4:

$$\frac{y^2}{4} - x^2 = 1$$

Since the x^2-term is subtracted from the y^2-term, the equation is that of a hyperbola with center at the origin and transverse axis along the y-axis. Also, comparing the above equation to equation (3), we find $a^2 = 4$, $b^2 = 1$, and $c^2 = a^2 + b^2 = 5$. The vertices are at $(0, \pm a) = (0, \pm 2)$, and the foci are at $(0, \pm c) = (0, \pm\sqrt{5})$.

To locate other points on the graph, we let $x = \pm 2$. Then

$$\begin{aligned} y^2 - 4x^2 &= 4 \\ y^2 - 4(\pm 2)^2 &= 4 \qquad x = \pm 2 \\ y^2 - 16 &= 4 \\ y^2 &= 20 \\ y &= \pm 2\sqrt{5} \end{aligned}$$

Four other points on the graph are $(\pm 2, 2\sqrt{5})$ and $(\pm 2, -2\sqrt{5})$. See Figure 38.

Figure 38

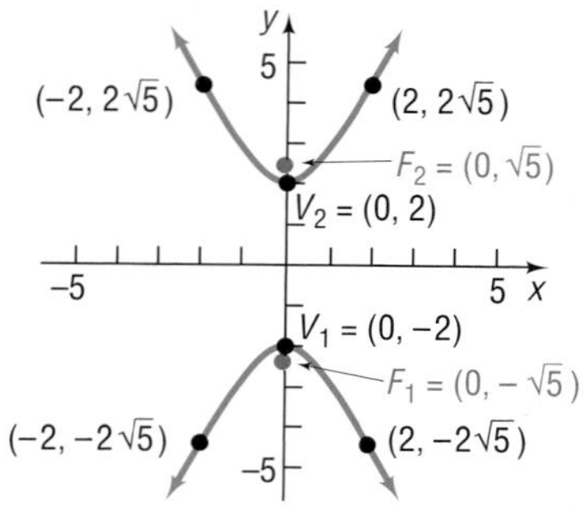

EXAMPLE 4 Finding an Equation of a Hyperbola

Find an equation of the hyperbola having one vertex at $(0, 2)$ and foci at $(0, -3)$ and $(0, 3)$.

Solution Since the foci are at $(0, -3)$ and $(0, 3)$, the center of the hyperbola is at their midpoint, the origin. Also, the transverse axis is along the y-axis. The given information also reveals that $c = 3$, $a = 2$, and $b^2 = c^2 - a^2 = 9 - 4 = 5$. The form of the equation of the hyperbola is given by equation (3):

$$\begin{aligned} \frac{y^2}{a^2} - \frac{x^2}{b^2} &= 1 \\ \frac{y^2}{4} - \frac{x^2}{5} &= 1 \end{aligned}$$

Let $y = \pm 3$ to obtain points on the graph across from the foci. See Figure 39.

Figure 39

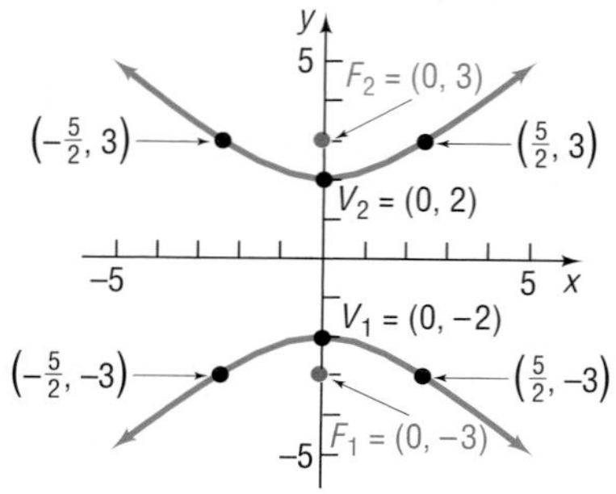

NOW WORK PROBLEM 21.

Look at the equations of the hyperbolas in Examples 2 and 4. For the hyperbola in Example 2, $a^2 = 16$ and $b^2 = 4$, so $a > b$; for the hyperbola in Example 4, $a^2 = 4$ and $b^2 = 5$, so $a < b$. We conclude that, for hyperbolas, there are no requirements involving the relative sizes of a and b. Contrast this situation to the case of an ellipse, in which the relative sizes of a and b dictate which axis is the major axis. Hyperbolas have another feature to distinguish them from ellipses and parabolas: Hyperbolas have asymptotes.

Asymptotes

4 Recall from Section 3.3 that a horizontal or oblique asymptote of a graph is a line with the property that the distance from the line to points on the graph approaches 0 as $x \to -\infty$ or as $x \to \infty$. The asymptotes provide information about the end behavior of the graph of a hyperbola.

Theorem

Asymptotes of a Hyperbola

The hyperbola $\dfrac{x^2}{a^2} - \dfrac{y^2}{b^2} = 1$ has the two oblique asymptotes

$$y = \frac{b}{a}x \quad \text{and} \quad y = -\frac{b}{a}x \tag{4}$$

Proof We begin by solving for y in the equation of the hyperbola.

$$\frac{x^2}{a^2} - \frac{y^2}{b^2} = 1$$

$$\frac{y^2}{b^2} = \frac{x^2}{a^2} - 1$$

$$y^2 = b^2\left(\frac{x^2}{a^2} - 1\right)$$

Since $x \neq 0$, we can rearrange the right side in the form

$$y^2 = \frac{b^2x^2}{a^2}\left(1 - \frac{a^2}{x^2}\right)$$

$$y = \pm\frac{bx}{a}\sqrt{1 - \frac{a^2}{x^2}}$$

Now, as $x \to -\infty$ or as $x \to \infty$, the term $\dfrac{a^2}{x^2}$ approaches 0, so the expression under the radical approaches 1. Thus, as $x \to -\infty$ or as $x \to \infty$, the value of y approaches $\pm\dfrac{bx}{a}$; that is, the graph of the hyperbola approaches the lines

$$y = -\frac{b}{a}x \quad \text{and} \quad y = \frac{b}{a}x$$

These lines are oblique asymptotes of the hyperbola. ■

The asymptotes of a hyperbola are not part of the hyperbola, but they do serve as a guide for graphing a hyperbola. For example, suppose that we want to graph the equation

$$\frac{x^2}{a^2} - \frac{y^2}{b^2} = 1$$

Figure 40
$\frac{x^2}{a^2} - \frac{y^2}{b^2} = 1$

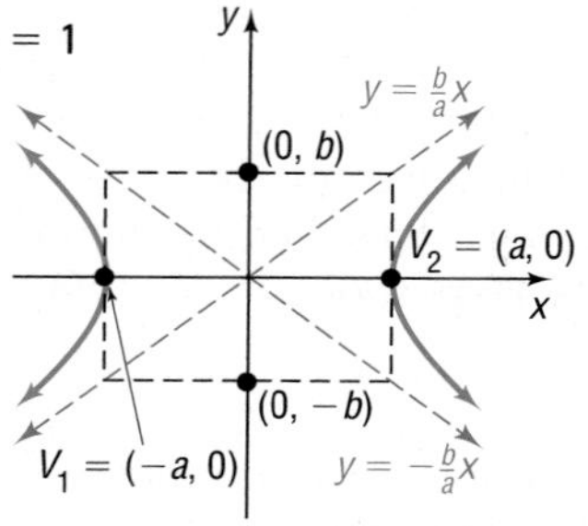

We begin by plotting the vertices $(-a, 0)$ and $(a, 0)$. Then we plot the points $(0, -b)$ and $(0, b)$ and use these four points to construct a rectangle, as shown in Figure 40. The diagonals of this rectangle have slopes $\frac{b}{a}$ and $\frac{-b}{a}$, and their extensions are the asymptotes $y = \frac{b}{a}x$ and $y = -\frac{b}{a}x$ of the hyperbola. If we graph the asymptotes, we can use them to establish the "opening" of the hyperbola and avoid plotting other points.

Theorem

Asymptotes of a Hyperbola

The hyperbola $\frac{y^2}{a^2} - \frac{x^2}{b^2} = 1$ has the two oblique asymptotes

$$y = \frac{a}{b}x \quad \text{and} \quad y = -\frac{a}{b}x \tag{5}$$

You are asked to prove this result in Problem 72.

For the remainder of this section, the direction "Discuss the equation" will mean to find the center, transverse axis, vertices, foci, and asymptotes of the hyperbola and graph it.

EXAMPLE 5 Discussing the Equation of a Hyperbola

Discuss the equation: $\frac{y^2}{4} - x^2 = 1$

Figure 41

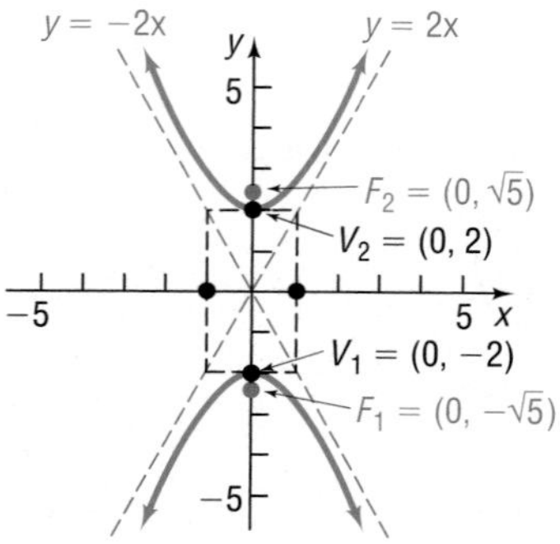

Solution Since the x^2-term is subtracted from the y^2-term, the equation is of the form of equation (3) and is a hyperbola with center at the origin and transverse axis along the y-axis. Also, comparing this equation to equation (3), we find that $a^2 = 4$, $b^2 = 1$, and $c^2 = a^2 + b^2 = 5$. The vertices are at $(0, \pm a) = (0, \pm 2)$, and the foci are at $(0, \pm c) = (0, \pm\sqrt{5})$. Using equation (5), the asymptotes are the lines $y = \frac{a}{b}x = 2x$ and $y = -\frac{a}{b}x = -2x$. Form the rectangle containing the points $(0, \pm a) = (0, \pm 2)$ and $(\pm b, 0) = (\pm 1, 0)$. The extensions of the diagonals of this rectangle are the asymptotes. Now graph the rectangle, the asymptotes, and the hyperbola. See Figure 41. ◀

EXAMPLE 6 Discussing the Equation of a Hyperbola

Discuss the equation: $9x^2 - 4y^2 = 36$

Solution Divide each side of the equation by 36 to put the equation in proper form.

$$\frac{x^2}{4} - \frac{y^2}{9} = 1$$

We now proceed to analyze the equation. The center of the hyperbola is the origin. Since the x^2-term is first in the equation, we know that the transverse axis is along the x-axis and the vertices and foci will lie on the x-axis. Using

Figure 42

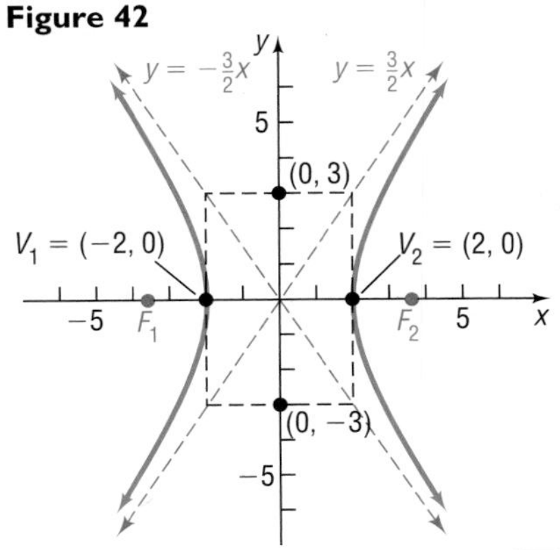

equation (2), we find $a^2 = 4$, $b^2 = 9$, and $c^2 = a^2 + b^2 = 13$. The vertices are $a = 2$ units left and right of the center at $(\pm a, 0) = (\pm 2, 0)$; the foci are $c = \sqrt{13}$ units left and right of the center at $(\pm c, 0) = (\pm\sqrt{13}, 0)$; and the asymptotes have the equations

$$y = \frac{b}{a}x = \frac{3}{2}x \quad \text{and} \quad y = -\frac{b}{a}x = -\frac{3}{2}x$$

To graph the hyperbola, form the rectangle containing the points $(\pm a, 0)$ and $(0, \pm b)$, that is, $(-2, 0)$, $(2, 0)$, $(0, -3)$, and $(0, 3)$. The extensions of the diagonals of this rectangle are the asymptotes. See Figure 42 for the graph. ◀

NOW WORK PROBLEM 29.

Exploration

Graph the upper portion of the hyperbola $9x^2 - 4y^2 = 36$ discussed in Example 6 and its asymptotes $y = \frac{3}{2}x$ and $y = -\frac{3}{2}x$. Now use ZOOM and TRACE to see what happens as x becomes unbounded in the positive direction. What happens as x becomes unbounded in the negative direction?

Center at (*h*, *k*)

5 If a hyperbola with center at the origin and transverse axis coinciding with a coordinate axis is shifted horizontally h units and then vertically k units, the result is a hyperbola with center at (h, k) and transverse axis parallel to a coordinate axis. The equations of such hyperbolas have the same forms as those given in equations (2) and (3), except that x is replaced by $x - h$ (the horizontal shift) and y is replaced by $y - k$ (the vertical shift). Table 4 gives the forms of the equations of such hyperbolas. See Figure 43 for the graphs.

Table 4 **Hyperbolas with Center at *(h, k)* and Transverse Axis Parallel to a Coordinate Axis**

Center	Transverse Axis	Foci	Vertices	Equation	Asymptotes
(h, k)	Parallel to x-axis	$(h \pm c, k)$	$(h \pm a, k)$	$\frac{(x-h)^2}{a^2} - \frac{(y-k)^2}{b^2} = 1, \quad b^2 = c^2 - a^2$	$y - k = \pm\frac{b}{a}(x - h)$
(h, k)	Parallel to y-axis	$(h, k \pm c)$	$(h, k \pm a)$	$\frac{(y-k)^2}{a^2} - \frac{(x-h)^2}{b^2} = 1, \quad b^2 = c^2 - a^2$	$y - k = \pm\frac{a}{b}(x - h)$

Figure 43

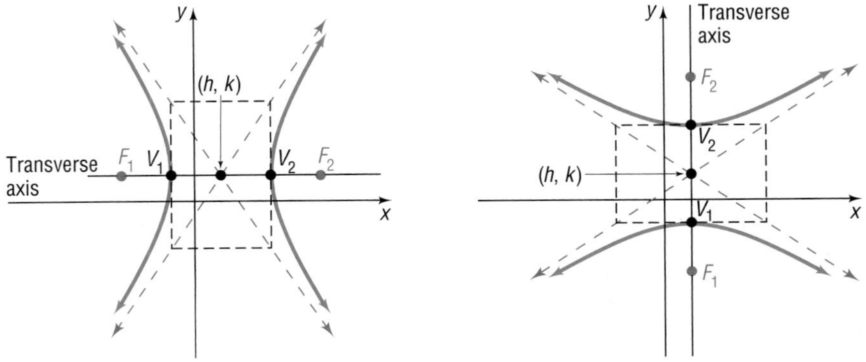

(a) $\frac{(x-h)^2}{a^2} - \frac{(y-k)^2}{b^2} = 1$ **(b)** $\frac{(y-k)^2}{a^2} - \frac{(x-h)^2}{b^2} = 1$

EXAMPLE 7 Finding an Equation of a Hyperbola, Center Not at the Origin

Find an equation for the hyperbola with center at $(1, -2)$, one focus at $(4, -2)$, and one vertex at $(3, -2)$. Graph the equation.

Solution The center is at $(h, k) = (1, -2)$, so $h = 1$ and $k = -2$. Since the center, focus, and vertex all lie on the line $y = -2$, the transverse axis is parallel to the x-axis. The distance from the center $(1, -2)$ to the focus $(4, -2)$ is $c = 3$; the distance from the center $(1, -2)$ to the vertex $(3, -2)$ is $a = 2$. Then, $b^2 = c^2 - a^2 = 9 - 4 = 5$. The equation is

$$\frac{(x-h)^2}{a^2} - \frac{(y-k)^2}{b^2} = 1$$

$$\frac{(x-1)^2}{4} - \frac{(y+2)^2}{5} = 1$$

See Figure 44.

Figure 44

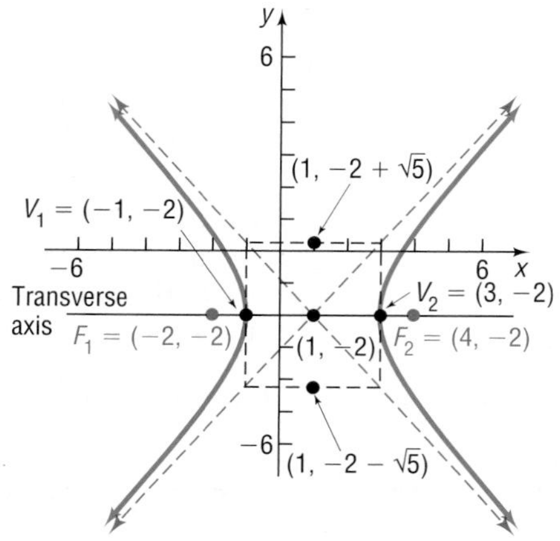

NOW WORK PROBLEM 39.

EXAMPLE 8 Discussing the Equation of a Hyperbola

Discuss the equation: $-x^2 + 4y^2 - 2x - 16y + 11 = 0$

Solution We complete the squares in x and in y.

$$-x^2 + 4y^2 - 2x - 16y + 11 = 0$$

$$-(x^2 + 2x) + 4(y^2 - 4y) = -11 \quad \text{Group terms.}$$

$$-(x^2 + 2x + 1) + 4(y^2 - 4y + 4) = -11 - 1 + 16 \quad \text{Complete each square.}$$

$$-(x+1)^2 + 4(y-2)^2 = 4$$

$$(y-2)^2 - \frac{(x+1)^2}{4} = 1 \quad \text{Divide each side by 4.}$$

This is the equation of a hyperbola with center at $(-1, 2)$ and transverse axis parallel to the y-axis. Also, $a^2 = 1$ and $b^2 = 4$, so $c^2 = a^2 + b^2 = 5$.

Figure 45

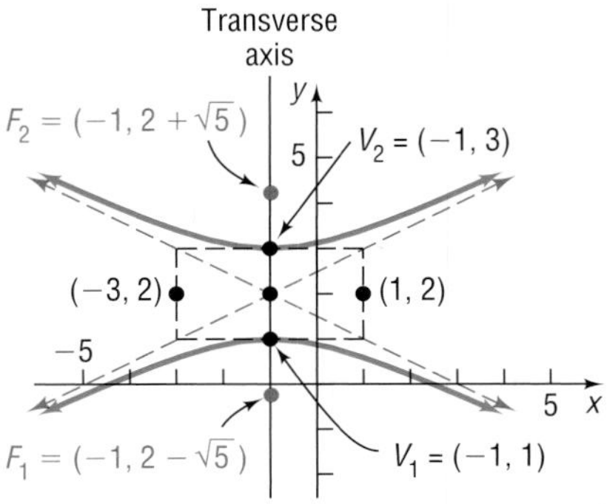

Since the transverse axis is parallel to the y-axis, the vertices and foci are located a and c units above and below the center, respectively. The vertices are at $(h, k \pm a) = (-1, 2 \pm 1)$, or $(-1, 1)$ and $(-1, 3)$. The foci are at $(h, k \pm c) = (-1, 2 \pm \sqrt{5})$. The asymptotes are $y - 2 = \frac{1}{2}(x + 1)$ and $y - 2 = -\frac{1}{2}(x + 1)$. Figure 45 shows the graph. ◀

NOW WORK PROBLEM 53.

Applications

Figure 46

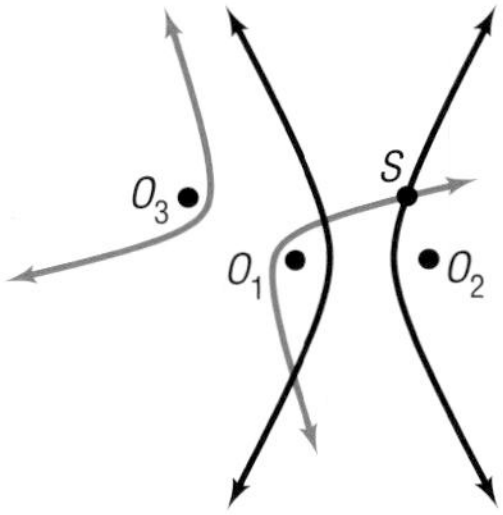

6 See Figure 46. Suppose that a gun is fired from an unknown source S. An observer at O_1 hears the report (sound of gun shot) 1 second after another observer at O_2. Because sound travels at about 1100 feet per second, it follows that the point S must be 1100 feet closer to O_2 than to O_1. S lies on one branch of a hyperbola with foci at O_1 and O_2. (Do you see why? The difference of the distances from S to O_1 and from S to O_2 is the constant 1100.) If a third observer at O_3 hears the same report 2 seconds after O_1 hears it, then S will lie on a branch of a second hyperbola with foci at O_1 and O_3. The intersection of the two hyperbolas will pinpoint the location of S.

LORAN

Figure 47

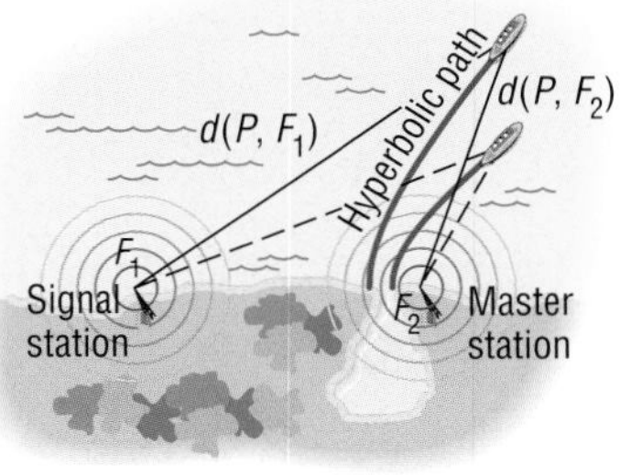

$d(P, F_1) - d(P, F_2) = \text{constant}$

In the LOng RAnge Navigation system (LORAN), a master radio sending station and a secondary sending station emit signals that can be received by a ship at sea. See Figure 47. Because a ship monitoring the two signals will usually be nearer to one of the two stations, there will be a difference in the distance that the two signals travel, which will register as a slight time difference between the signals. As long as the time difference remains constant, the difference of the two distances will also be constant. If the ship follows a path corresponding to the fixed time difference, it will follow the path of a hyperbola whose foci are located at the positions of the two sending stations. So for each time difference a different hyperbolic path results, each bringing the ship to a different shore location. Navigation charts show the various hyperbolic paths corresponding to different time differences.

EXAMPLE 9 **LORAN**

Two LORAN stations are positioned 250 miles apart along a straight shore.

(a) A ship records a time difference of 0.00054 second between the LORAN signals. Set up an appropriate rectangular coordinate system to determine where the ship would reach shore if it were to follow the hyperbola corresponding to this time difference.

(b) If the ship wants to enter a harbor located between the two stations 25 miles from the master station, what time difference should it be looking for?

(c) If the ship is 80 miles offshore when the desired time difference is obtained, what is the approximate location of the ship?

[*Note:* The speed of each radio signal is about 186,000 miles per second.]

Solution (a) We set up a rectangular coordinate system so that the two stations lie on the x-axis and the origin is midway between them. See Figure 48. The ship lies on a hyperbola whose foci are the locations of the two stations. The reason for this is that the constant time difference of the signals from each station results in a constant difference in the distance of the ship from each station. Since the time difference is 0.00054 second and the speed of the signal is 186,000 miles per second, the difference of the distances from the ship to each station (foci) is

Figure 48

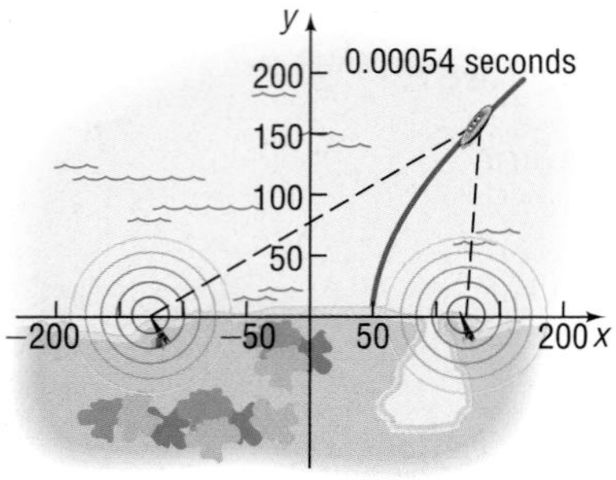

$$\text{Distance} = \text{Speed} \times \text{Time} = 186{,}000 \times 0.00054 \approx 100 \text{ miles}$$

The difference of the distances from the ship to each station, 100, equals $2a$, so $a = 50$ and the vertex of the corresponding hyperbola is at $(50, 0)$. Since the focus is at $(125, 0)$, following this hyperbola the ship would reach shore 75 miles from the master station.

(b) To reach shore 25 miles from the master station, the ship would follow a hyperbola with vertex at $(100, 0)$. For this hyperbola, $a = 100$, so the constant difference of the distances from the ship to each station is $2a = 200$. The time difference that the ship should look for is

$$\text{Time} = \frac{\text{Distance}}{\text{Speed}} = \frac{200}{186{,}000} \approx 0.001075 \text{ second}$$

(c) To find the approximate location of the ship, we need to find the equation of the hyperbola with vertex at $(100, 0)$ and a focus at $(125, 0)$. The form of the equation of this hyperbola is

$$\frac{x^2}{a^2} - \frac{y^2}{b^2} = 1$$

where $a = 100$. Since $c = 125$, we have

$$b^2 = c^2 - a^2 = 125^2 - 100^2 = 5625$$

The equation of the hyperbola is

$$\frac{x^2}{100^2} - \frac{y^2}{5625} = 1$$

Since the ship is 80 miles from shore, we use $y = 80$ in the equation and solve for x.

Figure 49

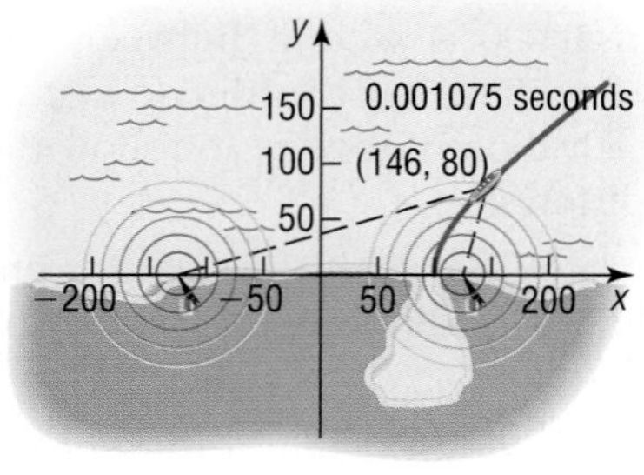

$$\frac{x^2}{100^2} - \frac{80^2}{5625} = 1$$

$$\frac{x^2}{100^2} = 1 + \frac{80^2}{5625} \approx 2.14$$

$$x^2 \approx 100^2(2.14)$$

$$x \approx 146$$

The ship is at the position $(146, 80)$. See Figure 49. ◄

NOW WORK PROBLEM 65.

9.4 Assess Your Understanding

'Are You Prepared?'

Answers are given at the end of these exercises. If you get a wrong answer, read the pages listed in red.

1. The distance d from $P_1 = (3, -4)$ to $P_2 = (-2, 1)$ is $d =$ ________. (p. 3)
2. To complete the square of $x^2 + 5x$, add ________. (p. 941)
3. Find the intercepts of the equation $y^2 = 9 + 4x^2$. (pp. 11–13)
4. *True or False:* The equation $y^2 = 9 + x^2$ is symmetric with respect to the x-axis, the y-axis, and the origin. (pp. 11–13)
5. To graph $y = (x - 5)^3 - 4$, shift the graph of $y = x^3$ to the left/right ________ unit(s) and then up/down ________ unit(s). (pp. 94–103)
6. Find the vertical asymptotes, if any, and the horizontal or oblique asymptotes, if any, of $y = \dfrac{x^2 - 9}{x^2 - 4}$. (pp. 165–166)

Concepts and Vocabulary

7. A(n) ________ is the collection of points in the plane the difference of whose distances from two fixed points is a constant.
8. For a hyperbola, the foci lie on a line called the ________ ________.
9. The asymptotes of the hyperbola $\dfrac{x^2}{4} - \dfrac{y^2}{9} = 1$ are ________ and ________.
10. *True or False:* The foci of a hyperbola lie on a line called the axis of symmetry.
11. *True or False:* Hyperbolas always have asymptotes.
12. *True or False:* A hyperbola will never intersect its transverse axis.

Exercises

In Problems 13–16, the graph of a hyperbola is given. Match each graph to its equation.

A. $\dfrac{x^2}{4} - y^2 = 1$ B. $x^2 - \dfrac{y^2}{4} = 1$ C. $\dfrac{y^2}{4} - x^2 = 1$ D. $y^2 - \dfrac{x^2}{4} = 1$

13.

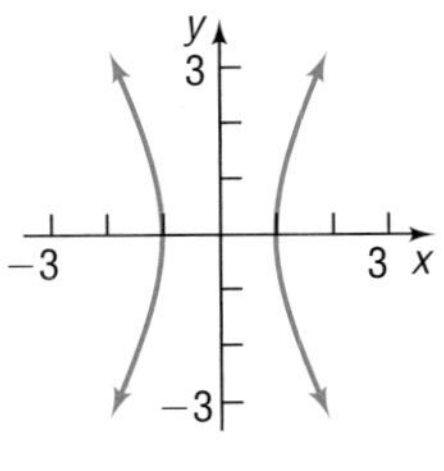

14.

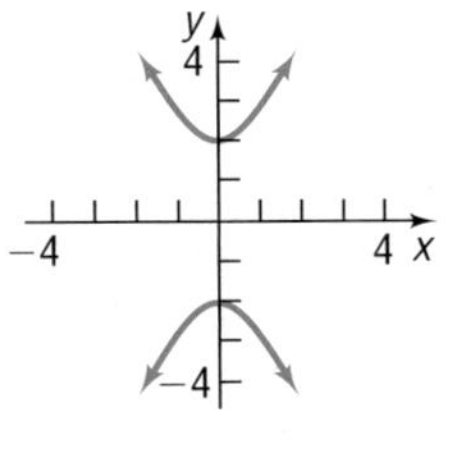

15.

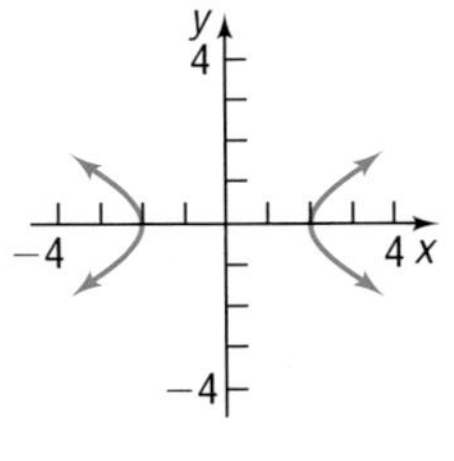

16.

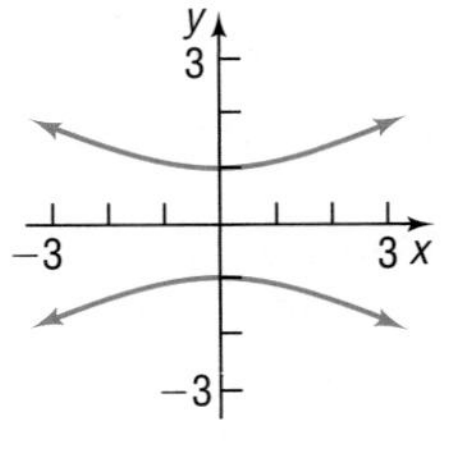

In Problems 17–26, find an equation for the hyperbola described. Graph the equation.

17. Center at $(0, 0)$; focus at $(3, 0)$; vertex at $(1, 0)$
18. Center at $(0, 0)$; focus at $(0, 5)$; vertex at $(0, 3)$
19. Center at $(0, 0)$; focus at $(0, -6)$; vertex at $(0, 4)$
20. Center at $(0, 0)$; focus at $(-3, 0)$; vertex at $(2, 0)$
21. Foci at $(-5, 0)$ and $(5, 0)$; vertex at $(3, 0)$
22. Focus at $(0, 6)$; vertices at $(0, -2)$ and $(0, 2)$
23. Vertices at $(0, -6)$ and $(0, 6)$; asymptote the line $y = 2x$
24. Vertices at $(-4, 0)$ and $(4, 0)$; asymptote the line $y = 2x$
25. Foci at $(-4, 0)$ and $(4, 0)$; asymptote the line $y = -x$
26. Foci at $(0, -2)$ and $(0, 2)$; asymptote the line $y = -x$

In Problems 27–34, find the center, transverse axis, vertices, foci, and asymptotes. Graph each equation.

27. $\dfrac{x^2}{25} - \dfrac{y^2}{9} = 1$
28. $\dfrac{y^2}{16} - \dfrac{x^2}{4} = 1$
29. $4x^2 - y^2 = 16$
30. $y^2 - 4x^2 = 16$
31. $y^2 - 9x^2 = 9$
32. $x^2 - y^2 = 4$
33. $y^2 - x^2 = 25$
34. $2x^2 - y^2 = 4$

In Problems 35–38, write an equation for each hyperbola.

35.

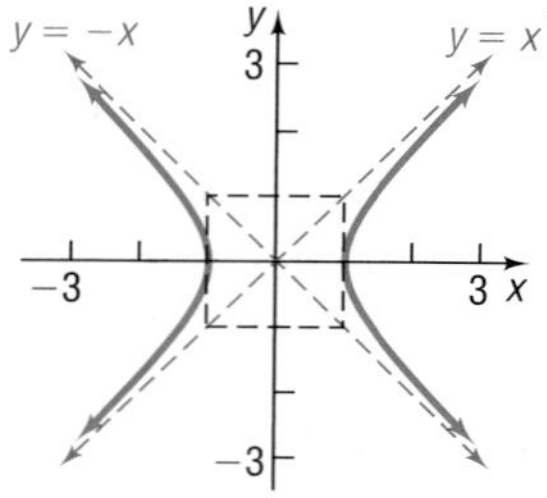

36.

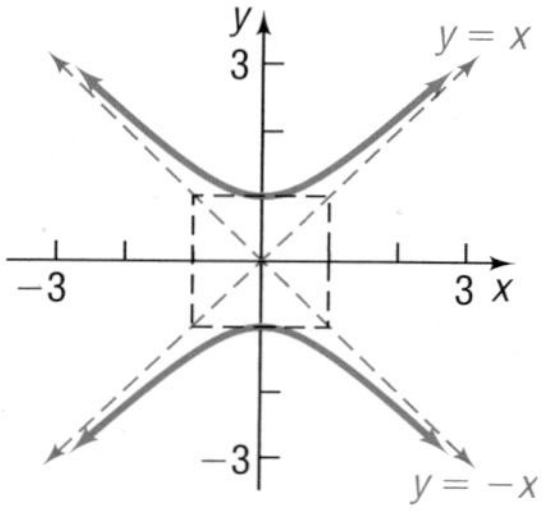

37.

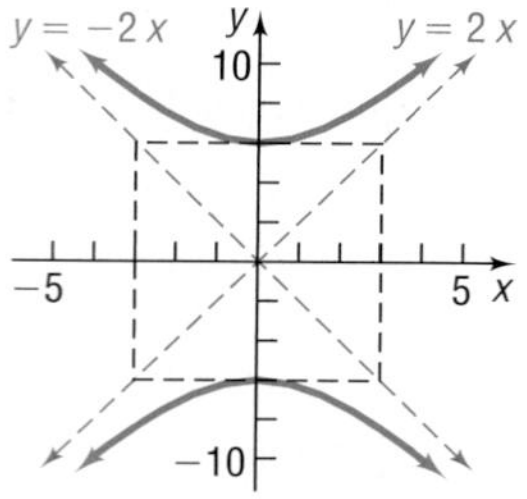

38.

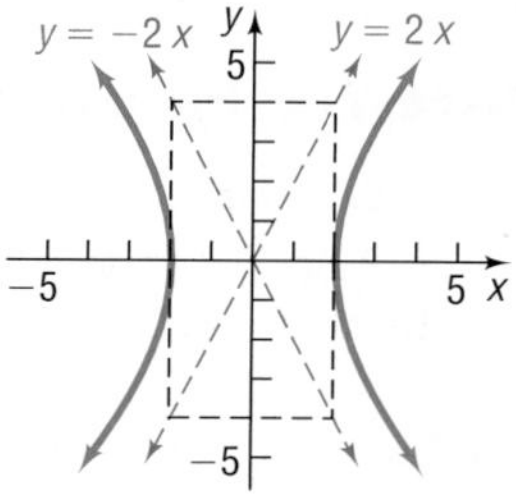

In Problems 39–46, find an equation for the hyperbola described. Graph the equation.

39. Center at $(4, -1)$; focus at $(7, -1)$; vertex at $(6, -1)$

40. Center at $(-3, 1)$; focus at $(-3, 6)$; vertex at $(-3, 4)$

41. Center at $(-3, -4)$; focus at $(-3, -8)$; vertex at $(-3, -2)$

42. Center at $(1, 4)$; focus at $(-2, 4)$; vertex at $(0, 4)$

43. Foci at $(3, 7)$ and $(7, 7)$; vertex at $(6, 7)$

44. Focus at $(-4, 0)$; vertices at $(-4, 4)$ and $(-4, 2)$

45. Vertices at $(-1, -1)$ and $(3, -1)$; asymptote the line $y + 1 = \frac{3}{2}(x - 1)$

46. Vertices at $(1, -3)$ and $(1, 1)$; asymptote the line $y + 1 = \frac{3}{2}(x - 1)$

In Problems 47–60, find the center, transverse axis, vertices, foci, and asymptotes. Graph each equation.

47. $\frac{(x-2)^2}{4} - \frac{(y+3)^2}{9} = 1$

48. $\frac{(y+3)^2}{4} - \frac{(x-2)^2}{9} = 1$

49. $(y-2)^2 - 4(x+2)^2 = 4$

50. $(x+4)^2 - 9(y-3)^2 = 9$

51. $(x+1)^2 - (y+2)^2 = 4$

52. $(y-3)^2 - (x+2)^2 = 4$

53. $x^2 - y^2 - 2x - 2y - 1 = 0$

54. $y^2 - x^2 - 4y + 4x - 1 = 0$

55. $y^2 - 4x^2 - 4y - 8x - 4 = 0$

56. $2x^2 - y^2 + 4x + 4y - 4 = 0$

57. $4x^2 - y^2 - 24x - 4y + 16 = 0$

58. $2y^2 - x^2 + 2x + 8y + 3 = 0$

59. $y^2 - 4x^2 - 16x - 2y - 19 = 0$

60. $x^2 - 3y^2 + 8x - 6y + 4 = 0$

In Problems 61–64, graph each function.

[**Hint:** Notice that each function is half a hyperbola.]

61. $f(x) = \sqrt{16 + 4x^2}$

62. $f(x) = -\sqrt{9 + 9x^2}$

63. $f(x) = -\sqrt{-25 + x^2}$

64. $f(x) = \sqrt{-1 + x^2}$

65. LORAN Two LORAN stations are positioned 200 miles apart along a straight shore.

(a) A ship records a time difference of 0.00038 second between the LORAN signals. Set up an appropriate rectangular coordinate system to determine where the ship would reach shore if it were to follow the hyperbola corresponding to this time difference.

(b) If the ship wants to enter a harbor located between the two stations 20 miles from the master station, what time difference should it be looking for?

(c) If the ship is 50 miles offshore when the desired time difference is obtained, what is the approximate location of the ship?

[*Note:* The speed of each radio signal is about 186,000 miles per second.]

66. LORAN Two LORAN stations are positioned 100 miles apart along a straight shore.

(a) A ship records a time difference of 0.00032 second between the LORAN signals. Set up an appropriate rectangular coordinate system to determine where the ship would reach shore if it were to follow the hyperbola corresponding to this time difference.

(b) If the ship wants to enter a harbor located between the two stations 10 miles from the master station, what time difference should it be looking for?

(c) If the ship is 20 miles offshore when the desired time difference is obtained, what is the approximate location of the ship?

[*Note:* The speed of each radio signal is about 186,000 miles per second.]

67. Calibrating Instruments In a test of their recording devices, a team of seismologists positioned two of the devices 2000 feet apart, with the device at point A to the west of the device at point B. At a point between the devices and 200 feet from point B, a small amount of explosive was detonated and a note made of the time at which the sound reached each device. A second explosion is to be carried out at a point directly north of point B.

(a) How far north should the site of the second explosion be chosen so that the measured time difference recorded by the devices for the second detonation is the same as that recorded for the first detonation?

(b) Explain why this experiment can be used to calibrate the instruments.

68. Explain in your own words the LORAN system of navigation.

69. The **eccentricity** e of a hyperbola is defined as the number $\frac{c}{a}$, where a and c are the numbers given in equation (2). Because $c > a$, it follows that $e > 1$. Describe the general shape of a hyperbola whose eccentricity is close to 1. What is the shape if e is very large?

70. A hyperbola for which $a = b$ is called an **equilateral hyperbola.** Find the eccentricity e of an equilateral hyperbola.

[*Note:* The eccentricity of a hyperbola is defined in Problem 69.]

71. Two hyperbolas that have the same set of asymptotes are called **conjugate.** Show that the hyperbolas

$$\frac{x^2}{4} - y^2 = 1 \quad \text{and} \quad y^2 - \frac{x^2}{4} = 1$$

are conjugate. Graph each hyperbola on the same set of coordinate axes.

72. Prove that the hyperbola

$$\frac{y^2}{a^2} - \frac{x^2}{b^2} = 1$$

has the two oblique asymptotes

$$y = \frac{a}{b}x \quad \text{and} \quad y = -\frac{a}{b}x$$

73. Show that the graph of an equation of the form

$$Ax^2 + Cy^2 + F = 0, \qquad A \neq 0, C \neq 0, F \neq 0$$

where A and C are of opposite sign, is a hyperbola with center at $(0, 0)$.

74. Show that the graph of an equation of the form

$$Ax^2 + Cy^2 + Dx + Ey + F = 0, \qquad A \neq 0, C \neq 0$$

where A and C are of opposite sign,

(a) Is a hyperbola if $\frac{D^2}{4A} + \frac{E^2}{4C} - F \neq 0$.

(b) Is two intersecting lines if

$$\frac{D^2}{4A} + \frac{E^2}{4C} - F = 0$$

'Are You Prepared?' Answers

1. $5\sqrt{2}$ **2.** $\frac{25}{4}$ **3.** $(0, -3), (0, 3)$ **4.** True **5.** right; 5; down; 4 **6.** Vertical: $x = -2, x = 2$; Horizontal: $y = 1$

9.5 Rotation of Axes; General Form of a Conic

PREPARING FOR THIS SECTION *Before getting started, review the following:*

- Sum Formulas for Sine and Cosine (Section 6.4, pp. 438 and 441)
- Half-angle Formulas for Sine and Cosine (Section 6.5, p. 452)
- Double-angle Formulas for Sine and Cosine (Section 6.5, p. 448)

Now work the 'Are You Prepared?' problems on page 656.

OBJECTIVES
1 Identify a Conic
2 Use a Rotation of Axes to Transform Equations
3 Discuss an Equation Using a Rotation of Axes
4 Identify Conics without a Rotation of Axes

In this section, we show that the graph of a general second-degree polynomial containing two variables x and y, that is, an equation of the form

$$Ax^2 + Bxy + Cy^2 + Dx + Ey + F = 0 \tag{1}$$

where A, B, and C are not simultaneously 0, is a conic. We shall not concern ourselves here with the degenerate cases of equation (1), such as $x^2 + y^2 = 0$, whose graph is a single point $(0, 0)$; or $x^2 + 3y^2 + 3 = 0$, whose graph contains no points; or $x^2 - 4y^2 = 0$, whose graph is two lines, $x - 2y = 0$ and $x + 2y = 0$.

We begin with the case where $B = 0$. In this case, the term containing xy is not present, so equation (1) has the form

$$Ax^2 + Cy^2 + Dx + Ey + F = 0$$

where either $A \neq 0$ or $C \neq 0$.

1 We have already discussed the procedure for identifying the graph of this kind of equation; we complete the squares of the quadratic expressions in x or y, or both. Once this has been done, the conic can be identified by comparing it to one of the forms studied in Sections 9.2 through 9.4.

In fact, though, we can identify the conic directly from the equation without completing the squares.

Theorem

Identifying Conics without Completing the Squares

Excluding degenerate cases, the equation

$$Ax^2 + Cy^2 + Dx + Ey + F = 0 \qquad \textbf{(2)}$$

where A and C cannot both equal zero:

(a) Defines a parabola if $AC = 0$.

(b) Defines an ellipse (or a circle) if $AC > 0$.

(c) Defines a hyperbola if $AC < 0$.

Proof

(a) If $AC = 0$, then either $A = 0$ or $C = 0$, but not both, so the form of equation (2) is either

$$Ax^2 + Dx + Ey + F = 0, \qquad A \neq 0$$

or

$$Cy^2 + Dx + Ey + F = 0, \qquad C \neq 0$$

Using the results of Problems 77 and 78 in Exercise 9.2, it follows that, except for the degenerate cases, the equation is a parabola.

(b) If $AC > 0$, then A and C are of the same sign. Using the results of Problem 84 in Exercise 9.3, except for the degenerate cases, the equation is an ellipse if $A \neq C$ or a circle if $A = C$.

(c) If $AC < 0$, then A and C are of opposite sign. Using the results of Problem 74 in Exercise 9.4, except for the degenerate cases, the equation is a hyperbola. ■

We will not be concerned with the degenerate cases of equation (2). However, in practice, you should be alert to the possibility of degeneracy.

EXAMPLE 1 Identifying a Conic without Completing the Squares

Identify each equation without completing the squares.

(a) $3x^2 + 6y^2 + 6x - 12y = 0$ (b) $2x^2 - 3y^2 + 6y + 4 = 0$

(c) $y^2 - 2x + 4 = 0$

Solution (a) We compare the given equation to equation (2) and conclude that $A = 3$ and $C = 6$. Since $AC = 18 > 0$, the equation is an ellipse.

(b) Here $A = 2$ and $C = -3$, so $AC = -6 < 0$. The equation is a hyperbola.

(c) Here $A = 0$ and $C = 1$, so $AC = 0$. The equation is a parabola. ◀

NOW WORK PROBLEM 11.

Although we can now identify the type of conic represented by any equation of the form of equation (2) without completing the squares, we will still need to complete the squares if we desire additional information about a conic.

Now we turn our attention to equations of the form of equation (1), where $B \neq 0$. To discuss this case, we first need to investigate a new procedure: *rotation of axes*.

Rotation of Axes

2 In a **rotation of axes,** the origin remains fixed while the x-axis and y-axis are rotated through an angle θ to a new position; the new positions of the x- and y-axes are denoted by x' and y', respectively, as shown in Figure 50(a).

Figure 50

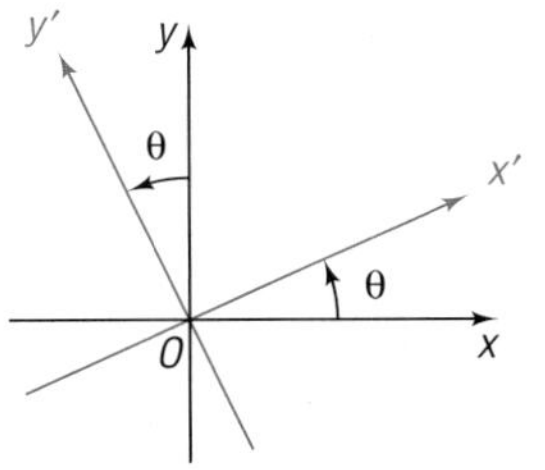

(a)

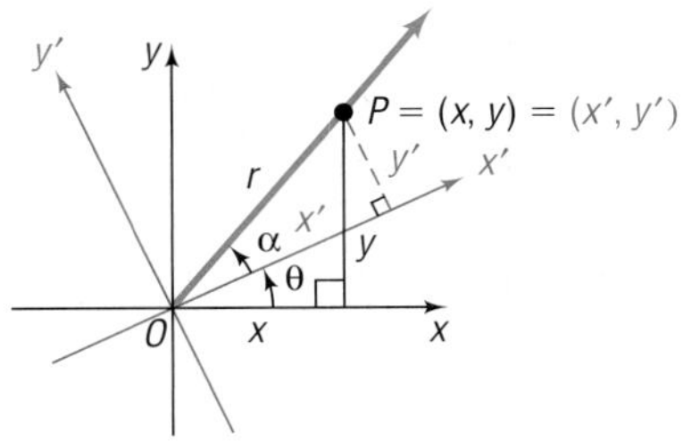

(b)

Now look at Figure 50(b). There the point P has the coordinates (x, y) relative to the xy-plane, while the same point P has coordinates (x', y') relative to the $x'y'$-plane. We seek relationships that will enable us to express x and y in terms of x', y', and θ.

As Figure 50(b) shows, r denotes the distance from the origin O to the point P, and α denotes the angle between the positive x'-axis and the ray from O through P. Then, using the definitions of sine and cosine, we have

$$x' = r\cos\alpha \qquad y' = r\sin\alpha \tag{3}$$

$$x = r\cos(\theta + \alpha) \qquad y = r\sin(\theta + \alpha) \tag{4}$$

Now

$$\begin{aligned} x &= r\cos(\theta + \alpha) \\ &= r(\cos\theta\cos\alpha - \sin\theta\sin\alpha) && \text{Sum Formula for cosine} \\ &= (r\cos\alpha)(\cos\theta) - (r\sin\alpha)(\sin\theta) \\ &= x'\cos\theta - y'\sin\theta && \text{By equation (3)} \end{aligned}$$

Similarly,

$$\begin{aligned} y &= r\sin(\theta + \alpha) \\ &= r(\sin\theta\cos\alpha + \cos\theta\sin\alpha) \\ &= x'\sin\theta + y'\cos\theta \end{aligned}$$

Theorem **Rotation Formulas**

If the x- and y-axes are rotated through an angle θ, the coordinates (x, y) of a point P relative to the xy-plane and the coordinates (x', y') of the same point relative to the new x'- and y'-axes are related by the formulas

$$x = x'\cos\theta - y'\sin\theta \qquad y = x'\sin\theta + y'\cos\theta \tag{5}$$

EXAMPLE 2 Rotating Axes

Express the equation $xy = 1$ in terms of new $x'y'$-coordinates by rotating the axes through a 45° angle. Discuss the new equation.

Solution Let $\theta = 45°$ in equation (5). Then

$$x = x' \cos 45° - y' \sin 45° = x' \frac{\sqrt{2}}{2} - y' \frac{\sqrt{2}}{2} = \frac{\sqrt{2}}{2}(x' - y')$$

$$y = x' \sin 45° + y' \cos 45° = x' \frac{\sqrt{2}}{2} + y' \frac{\sqrt{2}}{2} = \frac{\sqrt{2}}{2}(x' + y')$$

Substituting these expressions for x and y in $xy = 1$ gives

$$\left[\frac{\sqrt{2}}{2}(x' - y')\right]\left[\frac{\sqrt{2}}{2}(x' + y')\right] = 1$$

$$\frac{1}{2}(x'^2 - y'^2) = 1$$

$$\frac{x'^2}{2} - \frac{y'^2}{2} = 1$$

Figure 51

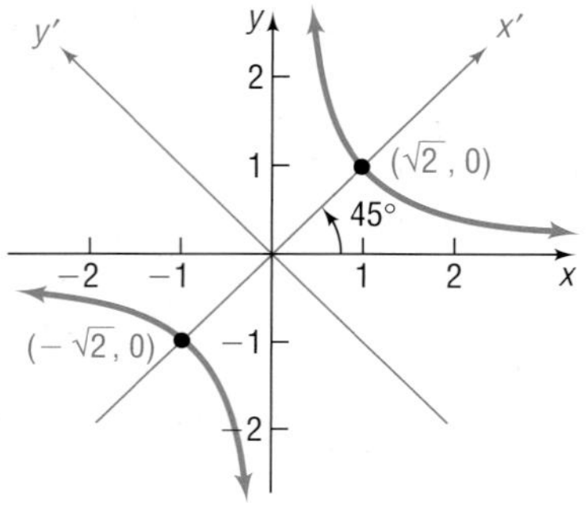

This is the equation of a hyperbola with center at (0, 0) and transverse axis along the x'-axis. The vertices are at $(\pm\sqrt{2}, 0)$ on the x'-axis; the asymptotes are $y' = x'$ and $y' = -x'$ (which correspond to the original x- and y-axes). See Figure 51 for the graph. ◀

As Example 2 illustrates, a rotation of axes through an appropriate angle can transform a second-degree equation in x and y containing an xy-term into one in x' and y' in which no $x'y'$-term appears. In fact, we will show that a rotation of axes through an appropriate angle will transform any equation of the form of equation (1) into an equation in x' and y' without an $x'y'$-term.

To find the formula for choosing an appropriate angle θ through which to rotate the axes, we begin with equation (1),

$$Ax^2 + Bxy + Cy^2 + Dx + Ey + F = 0, \qquad B \neq 0$$

Next we rotate through an angle θ using rotation formulas (5).

$$\begin{aligned} A(x' \cos\theta - y' \sin\theta)^2 &+ B(x' \cos\theta - y' \sin\theta)(x' \sin\theta + y' \cos\theta) \\ &+ C(x' \sin\theta + y' \cos\theta)^2 + D(x' \cos\theta - y' \sin\theta) \\ &+ E(x' \sin\theta + y' \cos\theta) + F = 0 \end{aligned}$$

By expanding and collecting like terms, we obtain

$$\begin{aligned} (A\cos^2\theta + B\sin\theta\cos\theta + C\sin^2\theta)x'^2 &+ [B(\cos^2\theta - \sin^2\theta) + 2(C - A)(\sin\theta\cos\theta)]x'y' \\ &+ (A\sin^2\theta - B\sin\theta\cos\theta + C\cos^2\theta)y'^2 \\ &+ (D\cos\theta + E\sin\theta)x' \\ &+ (-D\sin\theta + E\cos\theta)y' + F = 0 \end{aligned} \tag{6}$$

In equation (6), the coefficient of $x'y'$ is

$$B(\cos^2\theta - \sin^2\theta) + 2(C - A)(\sin\theta\cos\theta)$$

Since we want to eliminate the $x'y'$-term, we select an angle θ so that

$$\begin{aligned} B(\cos^2\theta - \sin^2\theta) + 2(C - A)(\sin\theta\cos\theta) &= 0 \\ B\cos(2\theta) + (C - A)\sin(2\theta) &= 0 \quad \text{Double-angle Formulas} \\ B\cos(2\theta) &= (A - C)\sin(2\theta) \\ \cot(2\theta) &= \frac{A - C}{B}, \qquad B \neq 0 \end{aligned}$$

Theorem

To transform the equation

$$Ax^2 + Bxy + Cy^2 + Dx + Ey + F = 0, \qquad B \neq 0$$

into an equation in x' and y' without an $x'y'$-term, rotate the axes through an angle θ that satisfies the equation

$$\cot(2\theta) = \frac{A - C}{B} \tag{7}$$

Equation (7) has an infinite number of solutions for θ. We shall adopt the convention of choosing the acute angle θ that satisfies (7). Then we have the following two possibilities:

If $\cot(2\theta) \geq 0$, then $0° < 2\theta \leq 90°$, so $0° < \theta \leq 45°$.

If $\cot(2\theta) < 0$, then $90° < 2\theta < 180°$, so $45° < \theta < 90°$.

Each of these results in a counterclockwise rotation of the axes through an acute angle θ.*

WARNING: Be careful if you use a calculator to solve equation (7).

1. If $\cot(2\theta) = 0$, then $2\theta = 90°$ and $\theta = 45°$.
2. If $\cot(2\theta) \neq 0$, first find $\cos(2\theta)$. Then use the inverse cosine function key(s) to obtain 2θ, $0° < 2\theta < 180°$. Finally, divide by 2 to obtain the correct acute angle θ. ■

3 **EXAMPLE 3** **Discussing an Equation Using a Rotation of Axes**

Discuss the equation: $x^2 + \sqrt{3}xy + 2y^2 - 10 = 0$

Solution Since an xy-term is present, we must rotate the axes. Using $A = 1$, $B = \sqrt{3}$, and $C = 2$ in equation (7), the appropriate acute angle θ through which to rotate the axes satisfies the equation

$$\cot(2\theta) = \frac{A - C}{B} = \frac{-1}{\sqrt{3}} = -\frac{\sqrt{3}}{3}, \qquad 0° < 2\theta < 180°$$

*Any rotation (clockwise or counterclockwise) through an angle θ that satisfies $\cot(2\theta) = \frac{A - C}{B}$ will eliminate the $x'y'$-term. However, the final form of the transformed equation may be different (but equivalent), depending on the angle chosen.

Since $\cot(2\theta) = -\frac{\sqrt{3}}{3}$, we find $2\theta = 120°$, so $\theta = 60°$. Using $\theta = 60°$ in rotation formulas (5), we find

$$x = \frac{1}{2}x' - \frac{\sqrt{3}}{2}y' = \frac{1}{2}\left(x' - \sqrt{3}y'\right)$$
$$y = \frac{\sqrt{3}}{2}x' + \frac{1}{2}y' = \frac{1}{2}\left(\sqrt{3}x' + y'\right)$$

Substituting these values into the original equation and simplifying, we have

$$x^2 + \sqrt{3}xy + 2y^2 - 10 = 0$$
$$\frac{1}{4}\left(x' - \sqrt{3}y'\right)^2 + \sqrt{3}\left[\frac{1}{2}\left(x' - \sqrt{3}y'\right)\right]\left[\frac{1}{2}\left(\sqrt{3}x' + y'\right)\right] + 2\left[\frac{1}{4}\left(\sqrt{3}x' + y'\right)^2\right] = 10$$

Multiply both sides by 4 and expand to obtain

$$x'^2 - 2\sqrt{3}x'y' + 3y'^2 + \sqrt{3}\left(\sqrt{3}x'^2 - 2x'y' - \sqrt{3}y'^2\right) + 2\left(3x'^2 + 2\sqrt{3}x'y' + y'^2\right) = 40$$
$$10x'^2 + 2y'^2 = 40$$
$$\frac{x'^2}{4} + \frac{y'^2}{20} = 1$$

Figure 52

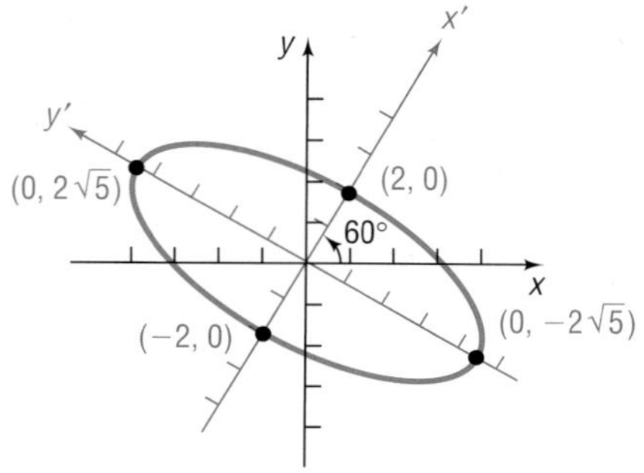

This is the equation of an ellipse with center at $(0, 0)$ and major axis along the y'-axis. The vertices are at $(0, \pm 2\sqrt{5})$ on the y'-axis. See Figure 52 for the graph. ◀

NOW WORK PROBLEM 31.

In Example 3, the acute angle θ through which to rotate the axes was easy to find because of the numbers that we used in the given equation. In general, the equation $\cot(2\theta) = \frac{A - C}{B}$ will not have such a "nice" solution. As the next example shows, we can still find the appropriate rotation formulas without using a calculator approximation by applying Half-angle Formulas.

EXAMPLE 4 Discussing an Equation Using a Rotation of Axes

Discuss the equation: $4x^2 - 4xy + y^2 + 5\sqrt{5}x + 5 = 0$

Solution Letting $A = 4$, $B = -4$, and $C = 1$ in equation (7), the appropriate angle θ through which to rotate the axes satisfies

$$\cot(2\theta) = \frac{A - C}{B} = \frac{3}{-4} = -\frac{3}{4}$$

To use rotation formulas (5), we need to know the values of $\sin\theta$ and $\cos\theta$. Since we seek an acute angle θ, we know that $\sin\theta > 0$ and $\cos\theta > 0$. We use the Half-angle Formulas in the form

$$\sin\theta = \sqrt{\frac{1 - \cos(2\theta)}{2}} \qquad \cos\theta = \sqrt{\frac{1 + \cos(2\theta)}{2}}$$

Now we need to find the value of $\cos(2\theta)$. Since $\cot(2\theta) = -\frac{3}{4}$, then $90^\circ < 2\theta < 180^\circ$ (Do you know why?), so $\cos(2\theta) = -\frac{3}{5}$. Then

$$\sin\theta = \sqrt{\frac{1-\cos(2\theta)}{2}} = \sqrt{\frac{1-\left(-\frac{3}{5}\right)}{2}} = \sqrt{\frac{4}{5}} = \frac{2}{\sqrt{5}} = \frac{2\sqrt{5}}{5}$$

$$\cos\theta = \sqrt{\frac{1+\cos(2\theta)}{2}} = \sqrt{\frac{1+\left(-\frac{3}{5}\right)}{2}} = \sqrt{\frac{1}{5}} = \frac{1}{\sqrt{5}} = \frac{\sqrt{5}}{5}$$

With these values, the rotation formulas (5) are

$$x = \frac{\sqrt{5}}{5}x' - \frac{2\sqrt{5}}{5}y' = \frac{\sqrt{5}}{5}(x' - 2y')$$
$$y = \frac{2\sqrt{5}}{5}x' + \frac{\sqrt{5}}{5}y' = \frac{\sqrt{5}}{5}(2x' + y')$$

Substituting these values in the original equation and simplifying, we obtain

$$4x^2 - 4xy + y^2 + 5\sqrt{5}x + 5 = 0$$
$$4\left[\frac{\sqrt{5}}{5}(x' - 2y')\right]^2 - 4\left[\frac{\sqrt{5}}{5}(x' - 2y')\right]\left[\frac{\sqrt{5}}{5}(2x' + y')\right] + \left[\frac{\sqrt{5}}{5}(2x' + y')\right]^2 + 5\sqrt{5}\left[\frac{\sqrt{5}}{5}(x' - 2y')\right] = -5$$

Multiply both sides by 5 and expand to obtain

$$4(x'^2 - 4x'y' + 4y'^2) - 4(2x'^2 - 3x'y' - 2y'^2) + 4x'^2 + 4x'y' + y'^2 + 25(x' - 2y') = -25$$
$$25y'^2 - 50y' + 25x' = -25 \quad \text{Combine like terms.}$$
$$y'^2 - 2y' + x' = -1 \quad \text{Divide by 25.}$$
$$y'^2 - 2y' + 1 = -x' \quad \text{Complete the square in } y'.$$
$$(y' - 1)^2 = -x'$$

Figure 53

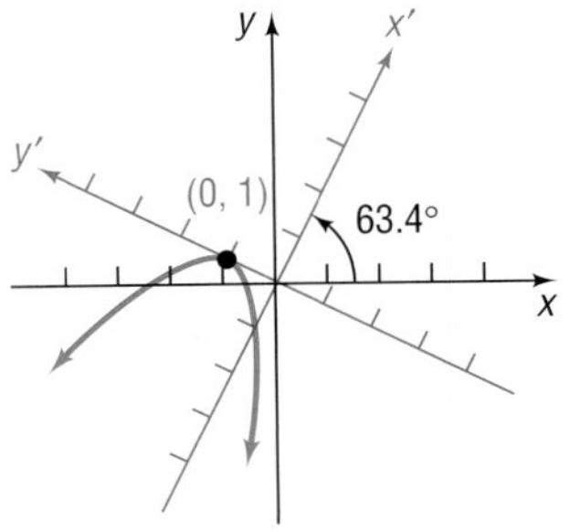

This is the equation of a parabola with vertex at $(0, 1)$ in the $x'y'$-plane. The axis of symmetry is parallel to the x'-axis. Using a calculator to solve $\sin\theta = \frac{2\sqrt{5}}{5}$, we find that $\theta \approx 63.4^\circ$. See Figure 53 for the graph. ◀

NOW WORK PROBLEM 37.

Identifying Conics without a Rotation of Axes

4 Suppose that we are required only to identify (rather than discuss) an equation of the form

$$Ax^2 + Bxy + Cy^2 + Dx + Ey + F = 0, \qquad B \neq 0 \tag{8}$$

If we apply rotation formulas (5) to this equation, we obtain an equation of the form

$$A'x'^2 + B'x'y' + C'y'^2 + D'x' + E'y' + F' = 0 \tag{9}$$

where A', B', C', D', E', and F' can be expressed in terms of A, B, C, D, E, F and the angle θ of rotation (see Problem 53). It can be shown that the value of $B^2 - 4AC$ in equation (8) and the value of $B'^2 - 4A'C'$ in equation (9) are equal no matter what angle θ of rotation is chosen (see Problem 55). In particular, if the angle θ of rotation satisfies equation (7), then $B' = 0$ in equation (9), and $B^2 - 4AC = -4A'C'$. Since equation (9) then has the form of equation (2),

$$A'x'^2 + C'y'^2 + D'x' + E'y' + F' = 0$$

we can identify it without completing the squares, as we did in the beginning of this section. In fact, now we can identify the conic described by any equation of the form of equation (8) without a rotation of axes.

Theorem

Identifying Conics without a Rotation of Axes

Except for degenerate cases, the equation

$$Ax^2 + Bxy + Cy^2 + Dx + Ey + F = 0$$

(a) Defines a parabola if $B^2 - 4AC = 0$.
(b) Defines an ellipse (or a circle) if $B^2 - 4AC < 0$.
(c) Defines a hyperbola if $B^2 - 4AC > 0$.

You are asked to prove this theorem in Problem 56.

EXAMPLE 5 **Identifying a Conic without a Rotation of Axes**

Identify the equation: $8x^2 - 12xy + 17y^2 - 4\sqrt{5}x - 2\sqrt{5}y - 15 = 0$

Solution Here $A = 8$, $B = -12$, and $C = 17$, so $B^2 - 4AC = -400$. Since $B^2 - 4AC < 0$, the equation defines an ellipse. ◀

NOW WORK PROBLEM 43.

9.5 Assess Your Understanding

'Are You Prepared?' *Answers are given at the end of these exercises. If you get a wrong answer, read the pages listed in red.*

1. The sum formula for the sine function is $\sin(\alpha + \beta) =$ ________. (pp. 438 and 441)
2. The Double-angle Formula for the sine function is $\sin(2\theta) =$ ________. (p. 448)
3. If θ is acute, the Half-angle Formula for the sine function is $\sin\left(\frac{\theta}{2}\right) =$ ________. (p. 452)
4. If θ is acute, the Half-angle Formula for the cosine function is $\cos\left(\frac{\theta}{2}\right) =$ ________. (p. 452)

Concepts and Vocabulary

5. To transform the equation

$$Ax^2 + Bxy + Cy^2 + Dx + Ey + F = 0, \qquad B \neq 0$$

into one in x' and y' without an $x'y'$-term, rotate the axes through an acute angle θ that satisfies the equation __________.

6. Identify the conic: $x^2 - 2y^2 - x - y - 18 = 0$. __________.

7. Identify the conic: $x^2 + 2xy + 3y^2 - 2x + 4y + 10 = 0$ __________.

8. *True or False:* The equation $ax^2 + 6y^2 - 12y = 0$ defines an ellipse if $a > 0$.

9. *True or False:* The equation $3x^2 + bxy + 12y^2 = 10$ defines a parabola if $b = -12$.

10. *True or False:* To eliminate the xy-term from the equation $x^2 - 2xy + y^2 - 2x + 3y + 5 = 0$, rotate the axes through an angle θ, where $\cot\theta = B^2 - 4AC$.

Exercises

In Problems 11–20, identify each equation without completing the squares.

11. $x^2 + 4x + y + 3 = 0$
12. $2y^2 - 3y + 3x = 0$
13. $6x^2 + 3y^2 - 12x + 6y = 0$
14. $2x^2 + y^2 - 8x + 4y + 2 = 0$
15. $3x^2 - 2y^2 + 6x + 4 = 0$
16. $4x^2 - 3y^2 - 8x + 6y + 1 = 0$
17. $2y^2 - x^2 - y + x = 0$
18. $y^2 - 8x^2 - 2x - y = 0$
19. $x^2 + y^2 - 8x + 4y = 0$
20. $2x^2 + 2y^2 - 8x + 8y = 0$

In Problems 21–30, determine the appropriate rotation formulas to use so that the new equation contains no xy-term.

21. $x^2 + 4xy + y^2 - 3 = 0$
22. $x^2 - 4xy + y^2 - 3 = 0$
23. $5x^2 + 6xy + 5y^2 - 8 = 0$
24. $3x^2 - 10xy + 3y^2 - 32 = 0$
25. $13x^2 - 6\sqrt{3}xy + 7y^2 - 16 = 0$
26. $11x^2 + 10\sqrt{3}xy + y^2 - 4 = 0$
27. $4x^2 - 4xy + y^2 - 8\sqrt{5}x - 16\sqrt{5}y = 0$
28. $x^2 + 4xy + 4y^2 + 5\sqrt{5}y + 5 = 0$
29. $25x^2 - 36xy + 40y^2 - 12\sqrt{13}x - 8\sqrt{13}y = 0$
30. $34x^2 - 24xy + 41y^2 - 25 = 0$

In Problems 31–42, rotate the axes so that the new equation contains no xy-term. Discuss and graph the new equation. Refer to Problems 21–30 for Problems 31–40.

31. $x^2 + 4xy + y^2 - 3 = 0$
32. $x^2 - 4xy + y^2 - 3 = 0$
33. $5x^2 + 6xy + 5y^2 - 8 = 0$
34. $3x^2 - 10xy + 3y^2 - 32 = 0$
35. $13x^2 - 6\sqrt{3}xy + 7y^2 - 16 = 0$
36. $11x^2 + 10\sqrt{3}xy + y^2 - 4 = 0$
37. $4x^2 - 4xy + y^2 - 8\sqrt{5}x - 16\sqrt{5}y = 0$
38. $x^2 + 4xy + 4y^2 + 5\sqrt{5}y + 5 = 0$
39. $25x^2 - 36xy + 40y^2 - 12\sqrt{13}x - 8\sqrt{13}y = 0$
40. $34x^2 - 24xy + 41y^2 - 25 = 0$
41. $16x^2 + 24xy + 9y^2 - 130x + 90y = 0$
42. $16x^2 + 24xy + 9y^2 - 60x + 80y = 0$

In Problems 43–52, identify each equation without applying a rotation of axes.

43. $x^2 + 3xy - 2y^2 + 3x + 2y + 5 = 0$
44. $2x^2 - 3xy + 4y^2 + 2x + 3y - 5 = 0$
45. $x^2 - 7xy + 3y^2 - y - 10 = 0$
46. $2x^2 - 3xy + 2y^2 - 4x - 2 = 0$
47. $9x^2 + 12xy + 4y^2 - x - y = 0$
48. $10x^2 + 12xy + 4y^2 - x - y + 10 = 0$
49. $10x^2 - 12xy + 4y^2 - x - y - 10 = 0$
50. $4x^2 + 12xy + 9y^2 - x - y = 0$
51. $3x^2 - 2xy + y^2 + 4x + 2y - 1 = 0$
52. $3x^2 + 2xy + y^2 + 4x - 2y + 10 = 0$

In Problems 53–56, apply rotation formulas (5) to

$$Ax^2 + Bxy + Cy^2 + Dx + Ey + F = 0$$

to obtain the equation

$$A'x'^2 + B'x'y' + C'y'^2 + D'x' + E'y' + F' = 0$$

53. Express A', B', C', D', E', and F' in terms of A, B, C, D, E, F, and the angle θ of rotation.

[***Hint:*** Refer to equation (6).]

54. Show that $A + C = A' + C'$, and thus show that $A + C$ is **invariant;** that is, its value does not change under a rotation of axes.

55. Refer to Problem 54. Show that $B^2 - 4AC$ is invariant.

56. Prove that, except for degenerate cases, the equation

$$Ax^2 + Bxy + Cy^2 + Dx + Ey + F = 0$$

(a) Defines a parabola if $B^2 - 4AC = 0$.
(b) Defines an ellipse (or a circle) if $B^2 - 4AC < 0$.
(c) Defines a hyperbola if $B^2 - 4AC > 0$.

57. Use rotation formulas (5) to show that distance is invariant under a rotation of axes. That is, show that the distance from $P_1 = (x_1, y_1)$ to $P_2 = (x_2, y_2)$ in the xy-plane equals the distance from $P_1 = (x_1', y_1')$ to $P_2 = (x_2', y_2')$ in the $x'y'$-plane.

58. Show that the graph of the equation $x^{1/2} + y^{1/2} = a^{1/2}$ is part of the graph of a parabola.

59. Formulate a strategy for discussing and graphing an equation of the form

$$Ax^2 + Cy^2 + Dx + Ey + F = 0$$

How does your strategy change if the equation is of the following form?

$$Ax^2 + Bxy + Cy^2 + Dx + Ey + F = 0$$

'Are You Prepared?' Answers

1. $\sin \alpha \cos \beta + \cos \alpha \sin \beta$

2. $2 \sin \theta \cos \theta$

3. $\sqrt{\dfrac{1 - \cos \theta}{2}}$ **4.** $\sqrt{\dfrac{1 + \cos \theta}{2}}$

9.6 Polar Equations of Conics

PREPARING FOR THIS SECTION *Before getting started, review the following:*

- Polar Coordinates (Section 8.1, pp. 536–543)

Now work the 'Are You Prepared?' problems on page 663.

OBJECTIVES 1 Discuss and Graph Polar Equations of Conics
2 Convert the Polar Equation of a Conic to a Rectangular Equation

1 In Sections 9.2 through 9.4, we gave separate definitions for the parabola, ellipse, and hyperbola based on geometric properties and the distance formula. In this section, we present an alternative definition that simultaneously defines all these conics. As we shall see, this approach is well suited to polar coordinate representation. (Refer to Section 8.1.)

Let D denote a fixed line called the **directrix;** let F denote a fixed point called the **focus,** which is not on D; and let e be a fixed positive number called the **eccentricity.** A **conic** is the set of points P in the plane such that the ratio of the distance from F to P to the distance from D to P equals e. That is, a conic is the collection of points P for which

$$\frac{d(F, P)}{d(D, P)} = e \qquad \textbf{(1)}$$

If $e = 1$, the conic is a **parabola.**
If $e < 1$, the conic is an **ellipse.**
If $e > 1$, the conic is a **hyperbola.**

Observe that if $e = 1$ the definition of a parabola in equation (1) is exactly the same as the definition used earlier in Section 9.2.

In the case of an ellipse, the **major axis** is a line through the focus perpendicular to the directrix. In the case of a hyperbola, the **transverse axis** is a line through the focus perpendicular to the directrix. For both an ellipse and a hyperbola, the eccentricity e satisfies

$$e = \frac{c}{a} \qquad \textbf{(2)}$$

where c is the distance from the center to the focus and a is the distance from the center to a vertex.

Figure 54

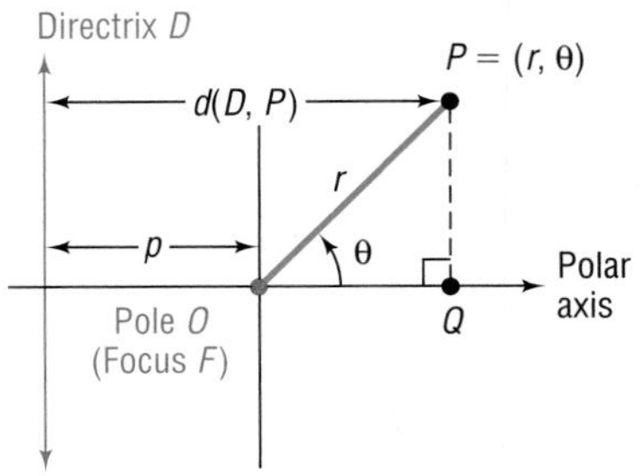

Just as we did earlier using rectangular coordinates, we derive equations for the conics in polar coordinates by choosing a convenient position for the focus F and the directrix D. The focus F is positioned at the pole, and the directrix D is either parallel or perpendicular to the polar axis.

Suppose that we start with the directrix D perpendicular to the polar axis at a distance p units to the left of the pole (the focus F). See Figure 54.

If $P = (r, \theta)$ is any point on the conic, then, by equation (1),

$$\frac{d(F, P)}{d(D, P)} = e \quad \text{or} \quad d(F, P) = e \cdot d(D, P) \tag{3}$$

Now we use the point Q obtained by dropping the perpendicular from P to the polar axis to calculate $d(D, P)$.

$$d(D, P) = p + d(O, Q) = p + r\cos\theta$$

Using this expression and the fact that $d(F, P) = d(O, P) = r$ in equation (3), we get

$$\begin{aligned} d(F, P) &= e \cdot d(D, P) \\ r &= e(p + r\cos\theta) \\ r &= ep + er\cos\theta \\ r - er\cos\theta &= ep \\ r(1 - e\cos\theta) &= ep \\ r &= \frac{ep}{1 - e\cos\theta} \end{aligned}$$

Theorem

Polar Equation of a Conic; Focus at Pole; Directrix Perpendicular to Polar Axis a Distance p to the Left of the Pole

The polar equation of a conic with focus at the pole and directrix perpendicular to the polar axis at a distance p to the left of the pole is

$$r = \frac{ep}{1 - e\cos\theta} \tag{4}$$

where e is the eccentricity of the conic.

EXAMPLE 1 Discussing and Graphing the Polar Equation of a Conic

Discuss and graph the equation: $r = \dfrac{4}{2 - \cos\theta}$

Solution The given equation is not quite in the form of equation (4), since the first term in the denominator is 2 instead of 1. We divide the numerator and denominator by 2 to obtain

$$r = \frac{2}{1 - \frac{1}{2}\cos\theta} \qquad r = \frac{ep}{1 - e\cos\theta}$$

This equation is in the form of equation (4), with

$$e = \frac{1}{2} \quad \text{and} \quad ep = 2, \quad \frac{1}{2}p = 2, \quad \text{so} \quad p = 4$$

We conclude that the conic is an ellipse, since $e = \frac{1}{2} < 1$. One focus is at the pole, and the directrix is perpendicular to the polar axis, a distance of $p = 4$ units to the left of the pole. It follows that the major axis is along the polar axis. To find the vertices, we let $\theta = 0$ and $\theta = \pi$. The vertices of the ellipse are $(4, 0)$ and $\left(\frac{4}{3}, \pi\right)$. The midpoint of the vertices, $\left(\frac{4}{3}, 0\right)$ in polar coordinates, is the center of the ellipse. [Do you see why? The vertices $(4, 0)$ and $\left(\frac{4}{3}, \pi\right)$ in polar coordinates are $(4, 0)$ and $\left(-\frac{4}{3}, 0\right)$ in rectangular coordinates. The midpoint in rectangular coordinates is $\left(\frac{4}{3}, 0\right)$, which is also $\left(\frac{4}{3}, 0\right)$ in polar coordinates.] Then a = distance from the center to a vertex $= \frac{8}{3}$. Using $a = \frac{8}{3}$ and $e = \frac{1}{2}$ in equation (2), $e = \frac{c}{a}$, we find $c = \frac{4}{3}$. Finally, using $a = \frac{8}{3}$ and $c = \frac{4}{3}$ in $b^2 = a^2 - c^2$, we have

$$b^2 = a^2 - c^2 = \frac{64}{9} - \frac{16}{9} = \frac{48}{9}$$

$$b = \frac{4\sqrt{3}}{3}$$

Figure 55 shows the graph. ◀

Figure 55

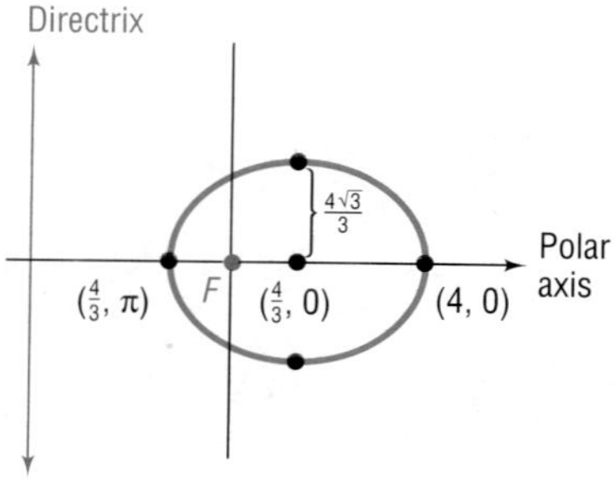

CHECK: In POLar mode with θ min $= 0$, θ max $= 2\pi$, and θ step $= \frac{\pi}{24}$, graph $r_1 = \frac{4}{2 - \cos\theta}$ and compare the result with Figure 55.

Exploration

Graph $r_1 = \frac{4}{2 + \cos\theta}$ and compare the result with Figure 55. What do you conclude? Clear the screen and graph $r_1 = \frac{4}{2 - \sin\theta}$ and then $r_1 = \frac{4}{2 + \sin\theta}$. Compare each of these graphs with Figure 55. What do you conclude?

NOW WORK PROBLEM 11.

Equation (4) was obtained under the assumption that the directrix was perpendicular to the polar axis at a distance p units to the left of the pole. A similar derivation (see Problem 43), in which the directrix is perpendicular to the polar axis at a distance p units to the right of the pole, results in the equation

$$r = \frac{ep}{1 + e\cos\theta}$$

In Problems 44 and 45, you are asked to derive the polar equations of conics with focus at the pole and directrix parallel to the polar axis. Table 5 summarizes the polar equations of conics.

Table 5 **Polar Equations of Conics (Focus at the Pole, Eccentricity e)**

Equation	Description
(a) $r = \dfrac{ep}{1 - e\cos\theta}$	Directrix is perpendicular to the polar axis at a distance p units to the left of the pole.
(b) $r = \dfrac{ep}{1 + e\cos\theta}$	Directrix is perpendicular to the polar axis at a distance p units to the right of the pole.
(c) $r = \dfrac{ep}{1 + e\sin\theta}$	Directrix is parallel to the polar axis at a distance p units above the pole.
(d) $r = \dfrac{ep}{1 - e\sin\theta}$	Directrix is parallel to the polar axis at a distance p units below the pole.

Eccentricity

If $e = 1$, the conic is a parabola; the axis of symmetry is perpendicular to the directrix.
If $e < 1$, the conic is an ellipse; the major axis is perpendicular to the directrix.
If $e > 1$, the conic is a hyperbola; the transverse axis is perpendicular to the directrix.

EXAMPLE 2 Discussing and Graphing the Polar Equation of a Conic

Discuss and graph the equation: $r = \dfrac{6}{3 + 3\sin\theta}$

Solution To place the equation in proper form, we divide the numerator and denominator by 3 to get

$$r = \frac{2}{1 + \sin\theta}$$

Referring to Table 5, we conclude that this equation is in the form of equation (c) with

$$e = 1 \quad \text{and} \quad ep = 2$$
$$p = 2 \quad e = 1$$

Figure 56

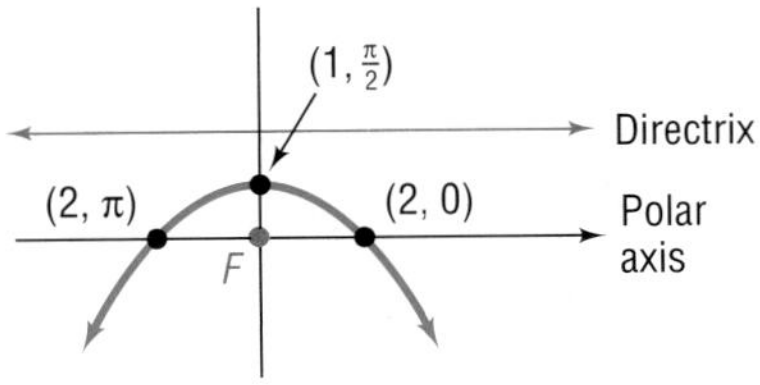

The conic is a parabola with focus at the pole. The directrix is parallel to the polar axis at a distance 2 units above the pole; the axis of symmetry is perpendicular to the polar axis. The vertex of the parabola is at $\left(1, \dfrac{\pi}{2}\right)$. (Do you see why?) See Figure 56 for the graph. Notice that we plotted two additional points, $(2, 0)$ and $(2, \pi)$, to assist in graphing. ◄

NOW WORK PROBLEM 13.

EXAMPLE 3 Discussing and Graphing the Polar Equation of a Conic

Discuss and graph the equation: $r = \dfrac{3}{1 + 3\cos\theta}$

Solution This equation is in the form of equation (b) in Table 5. We conclude that

$$e = 3 \quad \text{and} \quad ep = 3$$
$$p = 1 \quad e = 3$$

This is the equation of a hyperbola with a focus at the pole. The directrix is perpendicular to the polar axis, 1 unit to the right of the pole. The transverse axis is along the polar axis. To find the vertices, we let $\theta = 0$ and $\theta = \pi$. The vertices are $\left(\frac{3}{4}, 0\right)$ and $\left(-\frac{3}{2}, \pi\right)$. The center, which is at the midpoint of $\left(\frac{3}{4}, 0\right)$ and $\left(-\frac{3}{2}, \pi\right)$, is $\left(\frac{9}{8}, 0\right)$. Then c = distance from the center to a focus $= \frac{9}{8}$. Since $e = 3$, it follows from equation (2), $e = \frac{c}{a}$, that $a = \frac{3}{8}$. Finally, using $a = \frac{3}{8}$ and $c = \frac{9}{8}$ in $b^2 = c^2 - a^2$, we find

$$b^2 = c^2 - a^2 = \frac{81}{64} - \frac{9}{64} = \frac{72}{64} = \frac{9}{8}$$

$$b = \frac{3}{2\sqrt{2}} = \frac{3\sqrt{2}}{4}$$

Figure 57

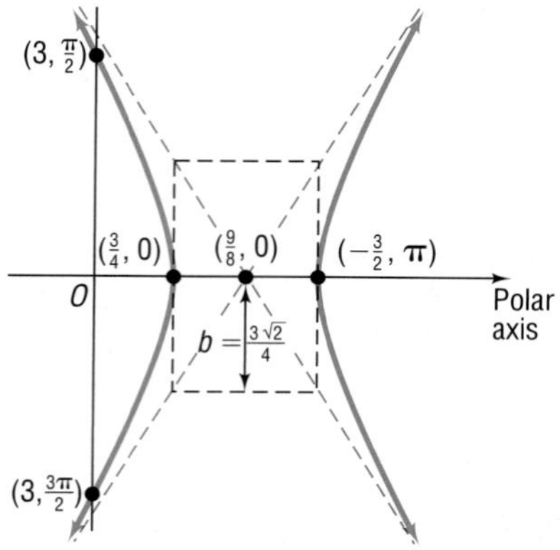

Figure 57 shows the graph. Notice that we plotted two additional points, $\left(3, \frac{\pi}{2}\right)$ and $\left(3, \frac{3\pi}{2}\right)$, on the left branch and used symmetry to obtain the right branch. The asymptotes of this hyperbola were found in the usual way by constructing the rectangle shown. ◀

CHECK: Graph $r_1 = \frac{3}{1 + 3\cos\theta}$ and compare the result with Figure 57.

NOW WORK PROBLEM 17.

2 **EXAMPLE 4** **Converting a Polar Equation to a Rectangular Equation**

Convert the polar equation

$$r = \frac{1}{3 - 3\cos\theta}$$

to a rectangular equation.

Solution The strategy here is first to rearrange the equation and square each side before using the transformation equations.

$$\begin{aligned} r &= \frac{1}{3 - 3\cos\theta} \\ 3r - 3r\cos\theta &= 1 \\ 3r &= 1 + 3r\cos\theta && \text{Rearrange the equation.} \\ 9r^2 &= (1 + 3r\cos\theta)^2 && \text{Square each side.} \\ 9(x^2 + y^2) &= (1 + 3x)^2 && x^2 + y^2 = r^2;\ x = r\cos\theta \\ 9x^2 + 9y^2 &= 9x^2 + 6x + 1 \\ 9y^2 &= 6x + 1 \end{aligned}$$

This is the equation of a parabola in rectangular coordinates. ◀

NOW WORK PROBLEM 25.

9.6 Assess Your Understanding

'Are You Prepared?' *Answers are given at the end of these exercises. If you get a wrong answer, read the pages listed in red.*

1. If (x, y) are the rectangular coordinates of a point P and (r, θ) are its polar coordinates, then $x =$ ________ and $y =$ ________. (pp. 536–543)

2. The point $(0, 0)$ is called the ________ in polar coordinates. (pp. 536–543)

Concepts and Vocabulary

3. The polar equation $r = \dfrac{8}{4 - 2\sin\theta}$ is a conic whose eccentricity is ________. It is a(n) ________ whose directrix is ________ to the polar axis at a distance ________ units ________ the pole.

4. The eccentricity e of a parabola is ________, of an ellipse it is ________, and of a hyperbola it is ________.

5. *True or False:* If (r, θ) are polar coordinates, the equation $r = \dfrac{2}{2 + 3\sin\theta}$ defines a hyperbola.

6. *True or False:* The eccentricity of any parabola is 1.

Exercises

In Problems 7–12, identify the conic that each polar equation represents. Also, give the position of the directrix.

7. $r = \dfrac{1}{1 + \cos\theta}$

8. $r = \dfrac{3}{1 - \sin\theta}$

9. $r = \dfrac{4}{2 - 3\sin\theta}$

10. $r = \dfrac{2}{1 + 2\cos\theta}$

11. $r = \dfrac{3}{4 - 2\cos\theta}$

12. $r = \dfrac{6}{8 + 2\sin\theta}$

In Problems 13–24, discuss each equation and graph it.

13. $r = \dfrac{1}{1 + \cos\theta}$

14. $r = \dfrac{3}{1 - \sin\theta}$

15. $r = \dfrac{8}{4 + 3\sin\theta}$

16. $r = \dfrac{10}{5 + 4\cos\theta}$

17. $r = \dfrac{9}{3 - 6\cos\theta}$

18. $r = \dfrac{12}{4 + 8\sin\theta}$

19. $r = \dfrac{8}{2 - \sin\theta}$

20. $r = \dfrac{8}{2 + 4\cos\theta}$

21. $r(3 - 2\sin\theta) = 6$

22. $r(2 - \cos\theta) = 2$

23. $r = \dfrac{6\sec\theta}{2\sec\theta - 1}$

24. $r = \dfrac{3\csc\theta}{\csc\theta - 1}$

In Problems 25–36, convert each polar equation to a rectangular equation.

25. $r = \dfrac{1}{1 + \cos\theta}$

26. $r = \dfrac{3}{1 - \sin\theta}$

27. $r = \dfrac{8}{4 + 3\sin\theta}$

28. $r = \dfrac{10}{5 + 4\cos\theta}$

29. $r = \dfrac{9}{3 - 6\cos\theta}$

30. $r = \dfrac{12}{4 + 8\sin\theta}$

31. $r = \dfrac{8}{2 - \sin\theta}$

32. $r = \dfrac{8}{2 + 4\cos\theta}$

33. $r(3 - 2\sin\theta) = 6$

34. $r(2 - \cos\theta) = 2$

35. $r = \dfrac{6\sec\theta}{2\sec\theta - 1}$

36. $r = \dfrac{3\csc\theta}{\csc\theta - 1}$

In Problems 37–42, find a polar equation for each conic. For each, a focus is at the pole.

37. $e = 1$; directrix is parallel to the polar axis 1 unit above the pole

38. $e = 1$; directrix is parallel to the polar axis 2 units below the pole

39. $e = \dfrac{4}{5}$; directrix is perpendicular to the polar axis 3 units to the left of the pole

40. $e = \dfrac{2}{3}$; directrix is parallel to the polar axis 3 units above the pole

41. $e = 6$; directrix is parallel to the polar axis 2 units below the pole

42. $e = 5$; directrix is perpendicular to the polar axis 5 units to the right of the pole

43. Derive equation (b) in Table 5:

$$r = \frac{ep}{1 + e\cos\theta}$$

44. Derive equation (c) in Table 5:

$$r = \frac{ep}{1 + e\sin\theta}$$

45. Derive equation (d) in Table 5:

$$r = \frac{ep}{1 - e\sin\theta}$$

46. Orbit of Mercury The planet Mercury travels around the Sun in an elliptical orbit given approximately by

$$r = \frac{(3.442)10^7}{1 - 0.206\cos\theta}$$

where r is measured in miles and the Sun is at the pole. Find the distance from Mercury to the Sun at *aphelion* (greatest distance from the Sun) and at *perihelion* (shortest distance from the Sun). See the figure. Use the aphelion and perihelion to graph the orbit of Mercury using a graphing utility.

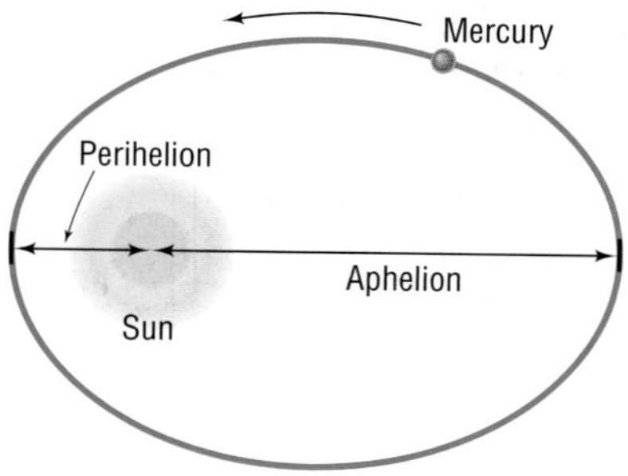

'Are You Prepared?' Answers

1. $r\cos\theta$; $r\sin\theta$

2. pole

9.7 Plane Curves and Parametric Equations

PREPARING FOR THIS SECTION *Before getting started, review the following:*

- Amplitude and Period of Sinusoidal Graphs (Section 5.4, p. 376)

Now work the 'Are You Prepared?' problems on page 674.

OBJECTIVES
1. Graph Parametric Equations
2. Find a Rectangular Equation for a Curve Defined Parametrically
3. Use Time as a Parameter in Parametric Equations
4. Find Parametric Equations for Curves Defined by Rectangular Equations

Equations of the form $y = f(x)$, where f is a function, have graphs that are intersected no more than once by any vertical line. The graphs of many of the conics and certain other, more complicated, graphs do not have this characteristic. Yet each graph, like the graph of a function, is a collection of points (x, y) in the xy-plane; that is, each is a *plane curve*. In this section, we discuss another way of representing such graphs.

Let $x = f(t)$ and $y = g(t)$, where f and g are two functions whose common domain is some interval I. The collection of points defined by

$$(x, y) = (f(t), g(t))$$

is called a **plane curve.** The equations

$$x = f(t) \qquad y = g(t)$$

where t is in I, are called **parametric equations** of the curve. The variable t is called a **parameter.**

1 Parametric equations are particularly useful in describing movement along a curve. Suppose that a curve is defined by the parametric equations

$$x = f(t), \qquad y = g(t), \qquad a \le t \le b$$

Figure 58

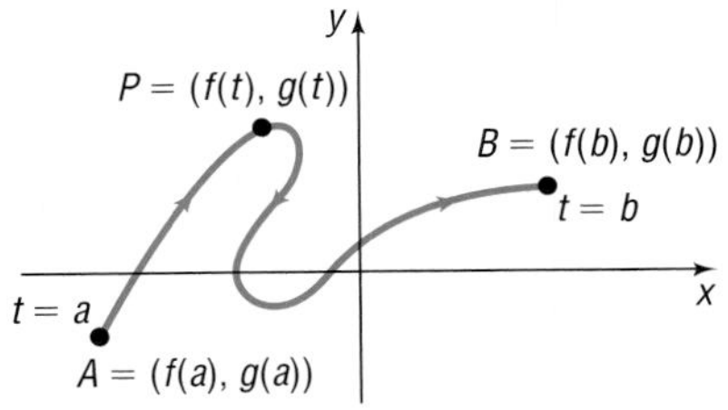

where f and g are each defined over the interval $a \le t \le b$. For a given value of t, we can find the value of $x = f(t)$ and $y = g(t)$, obtaining a point (x, y) on the curve. In fact, as t varies over the interval from $t = a$ to $t = b$, successive values of t give rise to a directed movement along the curve; that is, the curve is traced out in a certain direction by the corresponding succession of points (x, y). See Figure 58. The arrows show the direction, or **orientation,** along the curve as t varies from a to b.

EXAMPLE 1 Discussing a Curve Defined by Parametric Equations

Discuss the curve defined by the parametric equations

$$x = 3t^2, \qquad y = 2t, \qquad -2 \le t \le 2 \tag{1}$$

Solution For each number t, $-2 \le t \le 2$, there corresponds a number x and a number y. For example, when $t = -2$, then $x = 12$ and $y = -4$. When $t = 0$, then $x = 0$ and $y = 0$. Indeed, we can set up a table listing various choices of the parameter t and the corresponding values for x and y, as shown in Table 6. Plotting these points and connecting them with a smooth curve leads to Figure 59. The arrows in Figure 59 are used to indicate the orientation.

Table 6

t	x	y	(x, y)
−2	12	−4	(12, −4)
−1	3	−2	(3, −2)
0	0	0	(0, 0)
1	3	2	(3, 2)
2	12	4	(12, 4)

Figure 59

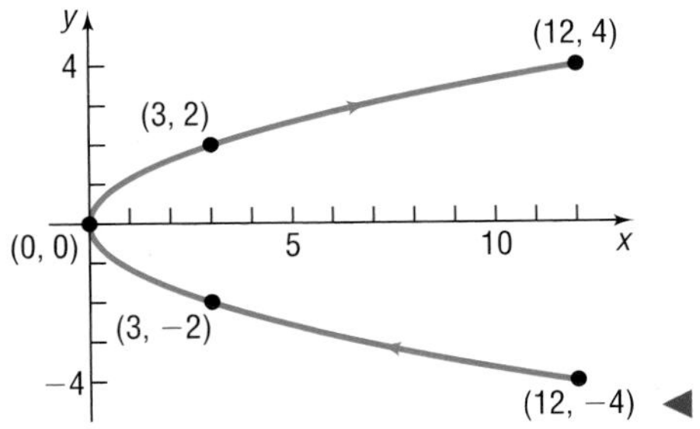

COMMENT: Most graphing utilities have the capability of graphing parametric equations. See Section B.9 in Appendix B. ■

2 The curve given in Example 1 should be familiar. To identify it accurately, we find the corresponding rectangular equation by eliminating the parameter t from the parametric equations (1) given in Example 1:

$$x = 3t^2, \qquad y = 2t, \qquad -2 \le t \le 2$$

Noting that we can readily solve for t in $y = 2t$, obtaining $t = \frac{y}{2}$, we substitute this expression in the other equation.

$$x = 3t^2 = 3\left(\frac{y}{2}\right)^2 = \frac{3y^2}{4}, \qquad -4 \le y \le 4$$

$$\uparrow$$
$$t = \frac{y}{2}$$

This equation, $x = \frac{3y^2}{4}$, is the equation of a parabola with vertex at $(0, 0)$ and axis of symmetry along the x-axis.

Note that the parameterized curve defined by equation (1) and shown in Figure 59 is only a part of the parabola $x = \frac{3y^2}{4}$. The graph of the rectangular equation obtained by eliminating the parameter will, in general, contain more points than the original parameterized curve. Care must therefore be taken when a parameterized curve is sketched by hand after eliminating the parameter. Even so, the process of eliminating the parameter t of a parameterized curve in order to identify it accurately is sometimes a better approach than merely plotting points. However, the elimination process sometimes requires a little ingenuity.

EXAMPLE 2 Finding the Rectangular Equation of a Curve Defined Parametrically

Find the rectangular equation of the curve whose parametric equations are

$$x = a \cos t \qquad y = a \sin t$$

where $a > 0$ is a constant. Graph this curve, indicating its orientation.

Solution The presence of sines and cosines in the parametric equations suggests that we use a Pythagorean identity. In fact, since

$$\cos t = \frac{x}{a} \qquad \sin t = \frac{y}{a}$$

we find that

$$\cos^2 t + \sin^2 t = 1$$
$$\left(\frac{x}{a}\right)^2 + \left(\frac{y}{a}\right)^2 = 1$$
$$x^2 + y^2 = a^2$$

Figure 60

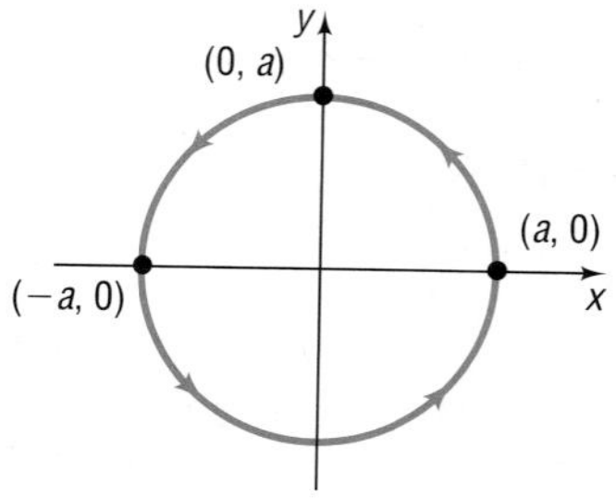

The curve is a circle with center at $(0, 0)$ and radius a. As the parameter t increases, say from $t = 0$ [the point $(a, 0)$] to $t = \frac{\pi}{2}$ [the point $(0, a)$] to $t = \pi$ [the point $(-a, 0)$], we see that the corresponding points are traced in a counterclockwise direction around the circle. The orientation is as indicated in Figure 60. ◀

NOW WORK PROBLEMS 7 AND 19.

Let's discuss the curve in Example 2 further. The domain of each parametric equation is $-\infty < t < \infty$. Thus, the graph in Figure 60 is actually being repeated each time that t increases by 2π.

If we wanted the curve to consist of exactly 1 revolution in the counterclockwise direction, we could write

$$x = a \cos t, \qquad y = a \sin t, \qquad 0 \leq t \leq 2\pi$$

This curve starts at $t = 0$ [the point $(a, 0)$] and, proceeding counterclockwise around the circle, ends at $t = 2\pi$ [also the point $(a, 0)$].

If we wanted the curve to consist of exactly three revolutions in the counterclockwise direction, we could write

$$x = a\cos t, \quad y = a\sin t, \quad -2\pi \le t \le 4\pi$$

or

$$x = a\cos t, \quad y = a\sin t, \quad 0 \le t \le 6\pi$$

or

$$x = a\cos t, \quad y = a\sin t, \quad 2\pi \le t \le 8\pi$$

EXAMPLE 3 Describing Parametric Equations

Find rectangular equations for and graph the curves defined by the following parametric equations.

(a) $x = a\cos t, \quad y = a\sin t, \quad 0 \le t \le \pi, \quad a > 0$

(b) $x = -a\sin t, \quad y = -a\cos t, \quad 0 \le t \le \pi, \quad a > 0$

Solution (a) We eliminate the parameter t using a Pythagorean identity.

$$\left(\frac{x}{a}\right)^2 + \left(\frac{y}{a}\right)^2 = \cos^2 t + \sin^2 t = 1$$

$$x^2 + y^2 = a^2$$

The curve defined by these parametric equations is a circle, with radius a and center at $(0, 0)$. The circle begins at the point $(a, 0)$, $t = 0$; passes through the point $(0, a)$, $t = \frac{\pi}{2}$; and ends at the point $(-a, 0)$, $t = \pi$. The parametric equations define an upper semicircle of radius a with a counterclockwise orientation. See Figure 61. The rectangular equation is

$$y = a\sqrt{1 - \left(\frac{x}{a}\right)^2}, \quad -a \le x \le a$$

Figure 61

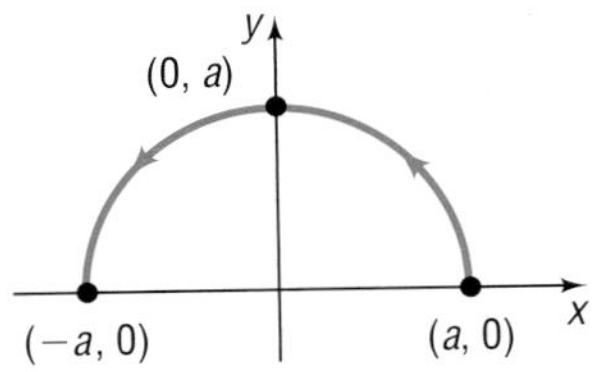

(b) We eliminate the parameter t using a Pythagorean identity.

$$\left(\frac{x}{-a}\right)^2 + \left(\frac{y}{-a}\right)^2 = \sin^2 t + \cos^2 t = 1$$

$$x^2 + y^2 = a^2$$

The curve defined by these parametric equations is a circle, with radius a and center at $(0, 0)$. The circle begins at the point $(0, -a)$, $t = 0$; passes through the point $(-a, 0)$, $t = \frac{\pi}{2}$; and ends at the point $(0, a)$, $t = \pi$. The parametric equations define a left semicircle of radius a with a clockwise orientation. See Figure 62. The rectangular equation is

$$x = -a\sqrt{1 - \left(\frac{y}{a}\right)^2}, \quad -a \le y \le a$$

Figure 62

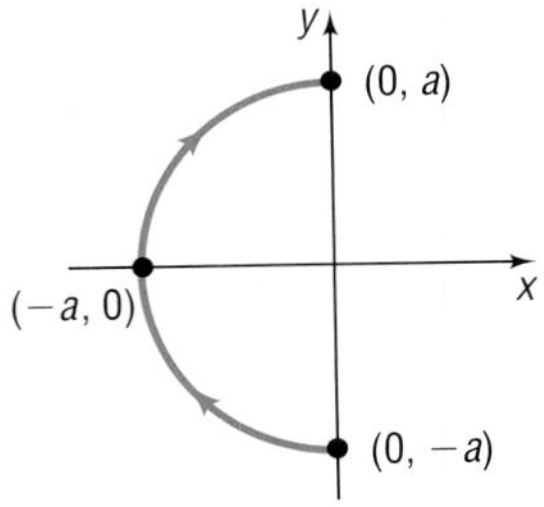

Example 3 illustrates the versatility of parametric equations for replacing complicated rectangular equations, while providing additional information about orientation. These characteristics make parametric equations very useful in applications, such as projectile motion.

Seeing the Concept

Graph $x = \cos t, y = \sin t$ for $0 \le t \le 2\pi$. Compare to Figure 60. Graph $x = \cos t, y = \sin t$ for $0 \le t \le \pi$. Compare to Figure 61. Graph $x = -\sin t, y = -\cos t$ for $0 \le t \le \pi$. Compare to Figure 62.

Time as a Parameter: Projectile Motion; Simulated Motion

3 If we think of the parameter t as time, then the parametric equations $x = f(t)$ and $y = g(t)$ of a curve C specify how the x- and y-coordinates of a moving point vary with time.

For example, we can use parametric equations to describe the motion of an object, sometimes referred to as **curvilinear motion.** Using parametric equations, we can specify not only where the object travels, that is, its location (x, y), but also when it gets there, that is, the time t.

When an object is propelled upward at an inclination θ to the horizontal with initial speed v_0, the resulting motion is called **projectile motion.** See Figure 63(a).

Figure 63

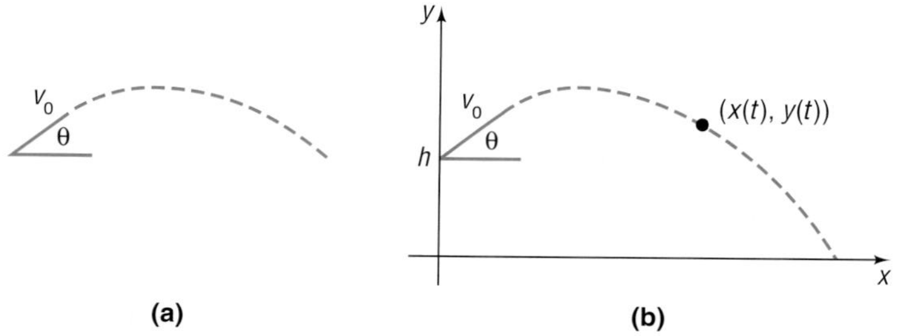

In calculus it is shown that the parametric equations of the path of a projectile fired at an inclination θ to the horizontal, with an initial speed v_0, from a height h above the horizontal are

$$x = (v_0 \cos \theta)t \qquad y = -\frac{1}{2}gt^2 + (v_0 \sin \theta)t + h \qquad \textbf{(2)}$$

where t is the time and g is the constant acceleration due to gravity (approximately 32 ft/sec/sec or 9.8 m/sec/sec). See Figure 63(b).

EXAMPLE 4 **Projectile Motion**

Suppose that Jim hit a golf ball with an initial velocity of 150 feet per second at an angle of 30° to the horizontal. See Figure 64.

Figure 64

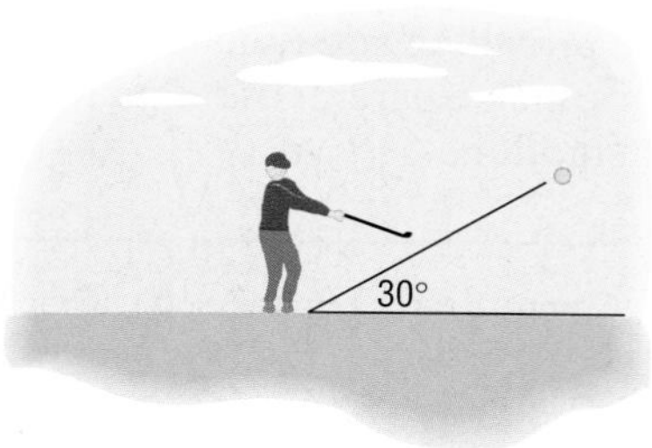

(a) Find parametric equations that describe the position of the ball as a function of time.

(b) How long is the golf ball in the air?

(c) When is the ball at its maximum height? Determine the maximum height of the ball.

(d) Determine the distance that the ball traveled in the air.

(e) Using a graphing utility, simulate the motion of the golf ball by simultaneously graphing the equations found in part (a).

Solution (a) We have $v_0 = 150$, $\theta = 30°$, $h = 0$ (the ball is on the ground), and $g = 32$ (since units are in feet and seconds). Substituting these values into equations (2), we find that

$$x = (v_0 \cos \theta)t = (150 \cos 30°)t = 75\sqrt{3}t$$

$$y = -\frac{1}{2}gt^2 + (v_0 \sin \theta)t + h = -\frac{1}{2}(32)t^2 + (150 \sin 30°)t + 0$$

$$= -16t^2 + 75t$$

(b) To determine the length of time that the ball is in the air, we solve the equation $y = 0$.

$$-16t^2 + 75t = 0$$

$$t(-16t + 75) = 0$$

$$t = 0 \text{ sec} \quad \text{or} \quad t = \frac{75}{16} = 4.6875 \text{ sec}$$

The ball will strike the ground after 4.6875 seconds.

(c) Notice that the height y of the ball is a quadratic function of t, so the maximum height of the ball can be found by determining the vertex of $y = -16t^2 + 75t$. The value of t at the vertex is

$$t = \frac{-b}{2a} = \frac{-75}{-32} = 2.34375 \text{ sec}$$

The ball is at its maximum height after 2.34375 seconds. The maximum height of the ball is found by evaluating the function y at $t = 2.34375$ seconds.

$$\text{Maximum height} = -16(2.34375)^2 + (75)2.34375 \approx 87.89 \text{ feet}$$

(d) Since the ball is in the air for 4.6875 seconds, the horizontal distance that the ball travels is

$$x = (75\sqrt{3})4.6875 \approx 608.92 \text{ feet}$$

(e) We enter the equations from part (a) into a graphing utility with $T\text{min} = 0$, $T\text{max} = 4.7$, and $T\text{step} = 0.1$. We use ZOOM-SQUARE to avoid any distortion to the angle of elevation. See Figure 65. ◀

Figure 65

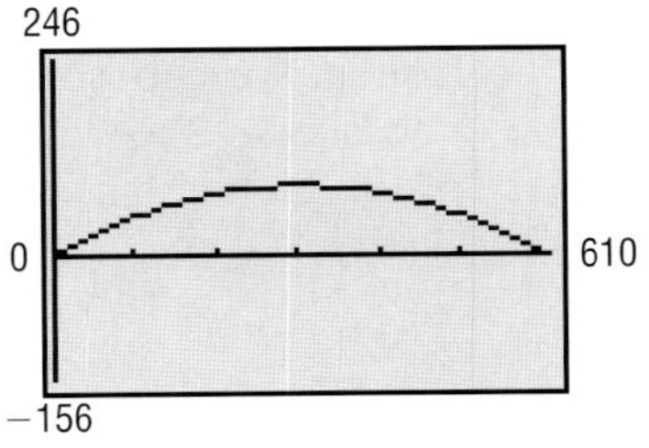

Exploration

Simulate the motion of a ball thrown straight up with an initial speed of 100 feet per second from a height of 5 feet above the ground. Use PARametric mode with $T\text{min} = 0$, $T\text{max} = 6.5$, $T\text{step} = 0.1$, $X\text{min} = 0$, $X\text{max} = 5$, $Y\text{min} = 0$, and $Y\text{max} = 180$. What happens to the speed with which the graph is drawn as the ball goes up and then comes back down? How do you interpret this physically? Repeat the experiment using other values for Tstep. How does this affect the experiment?

[**Hint:** In the projectile motion equations, let $\theta = 90°$, $v_0 = 100$, $h = 5$, and $g = 32$. We use $x = 3$ instead of $x = 0$ to see the vertical motion better.]

RESULT See Figure 66. In Figure 66(a) the ball is going up. In Figure 66(b) the ball is near its highest point. Finally, in Figure 66(c) the ball is coming back down.

Figure 66

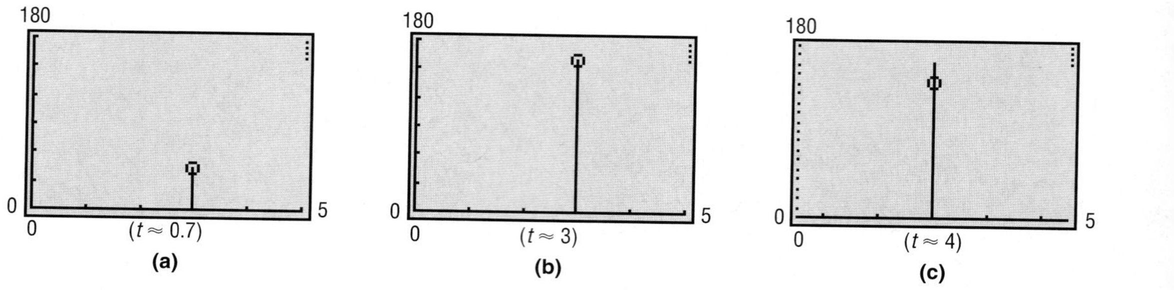

Notice that, as the ball goes up, its speed decreases, until at the highest point it is zero. Then the speed increases as the ball comes back down.

NOW WORK PROBLEM 33.

A graphing utility can be used to simulate other kinds of motion as well. Compare the solution below to that of Example 4 from Section A.7 of Appendix A (p. 960).

EXAMPLE 5 Simulating Motion

Tanya, who is a long distance runner, runs at an average velocity of 8 miles per hour. Two hours after Tanya leaves your house, you leave in your Honda and follow the same route. If your average velocity is 40 miles per hour, how long will it be before you catch up to Tanya? See Figure 67. Use a simulation of the two motions to verify the answer.

Figure 67

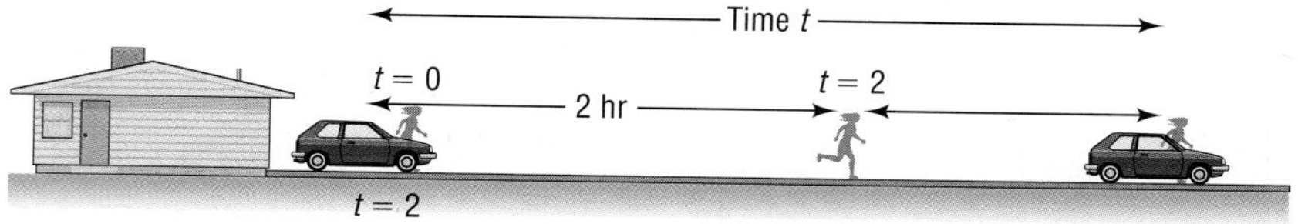

Solution We begin with two sets of parametric equations: one to describe Tanya's motion, the other to describe the motion of the Honda. We choose time $t = 0$ to be when Tanya leaves the house. If we choose $y_1 = 2$ as Tanya's path, then we can use $y_2 = 4$ as the parallel path of the Honda. The horizontal distances traversed in time t (Distance = Velocity × Time) are

$$\text{Tanya:} \quad x_1 = 8t \qquad \text{Honda:} \quad x_2 = 40(t - 2)$$

The Honda catches up to Tanya when $x_1 = x_2$.

$$\begin{aligned} 8t &= 40(t - 2) \\ 8t &= 40t - 80 \\ -32t &= -80 \\ t &= \frac{-80}{-32} = 2.5 \end{aligned}$$

The Honda catches up to Tanya 2.5 hours after Tanya leaves the house.

In PARametric mode with Tstep = 0.01, we simultaneously graph

$$\text{Tanya:} \quad \begin{aligned} x_1 &= 8t \\ y_1 &= 2 \end{aligned} \qquad \text{Honda:} \quad \begin{aligned} x_2 &= 40(t - 2) \\ y_2 &= 4 \end{aligned}$$

for $0 \le t \le 3$.

Figure 68 shows the relative position of Tanya and the Honda for $t = 0, t = 2, t = 2.25, t = 2.5$, and $t = 2.75$.

Figure 68

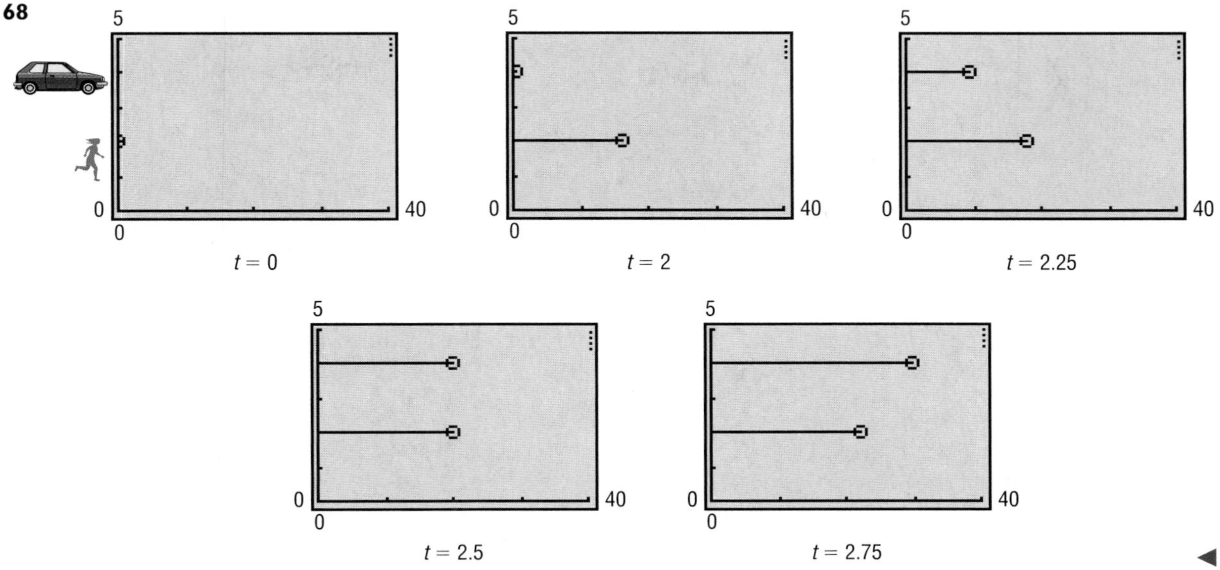

◀

Finding Parametric Equations

We now take up the question of how to find parametric equations of a given curve.

4 If a curve is defined by the equation $y = f(x)$, where f is a function, one way of finding parametric equations is to let $x = t$. Then $y = f(t)$ and

$$x = t, \quad y = f(t), \qquad t \text{ in the domain of } f$$

are parametric equations of the curve.

EXAMPLE 6 **Finding Parametric Equations for a Curve Defined by a Rectangular Equation**

Find parametric equations for the equation $y = x^2 - 4$.

Solution Let $x = t$. Then the parametric equations are

$$x = t, \quad y = t^2 - 4, \qquad -\infty < t < \infty$$

◀

Another less obvious approach to Example 6 is to let $x = t^3$. Then the parametric equations become

$$x = t^3, \quad y = t^6 - 4, \qquad -\infty < t < \infty$$

Care must be taken when using this approach, since the substitution for x must be a function that allows x to take on all the values stipulated by the domain of f. For example, letting $x = t^2$ so that $y = t^4 - 4$ does not result in equivalent parametric equations for $y = x^2 - 4$, since only points for which $x \geq 0$ are obtained.

EXAMPLE 7 Finding Parametric Equations for an Object in Motion

Find parametric equations for the ellipse

$$x^2 + \frac{y^2}{9} = 1$$

where the parameter t is time (in seconds) and

(a) The motion around the ellipse is clockwise, begins at the point $(0, 3)$, and requires 1 second for a complete revolution.

(b) The motion around the ellipse is counterclockwise, begins at the point $(1, 0)$, and requires 2 seconds for a complete revolution.

Figure 69

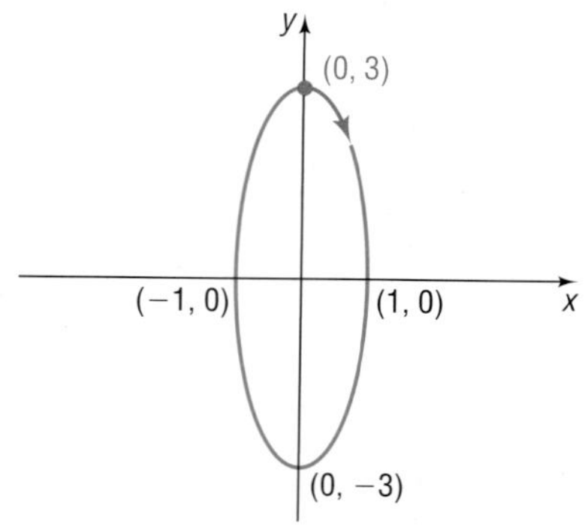

Solution (a) See Figure 69. Since the motion begins at the point $(0, 3)$, we want $x = 0$ and $y = 3$ when $t = 0$. Furthermore, since the given equation is an ellipse, we begin by letting

$$x = \sin(\omega t) \qquad \frac{y}{3} = \cos(\omega t)$$

for some constant ω. These parametric equations satisfy the equation of the ellipse. Furthermore, with this choice, when $t = 0$, we have $x = 0$ and $y = 3$.

For the motion to be clockwise, the motion will have to begin with the value of x increasing and y decreasing as t increases. This requires that $\omega > 0$. [Do you know why? If $\omega > 0$, then $x = \sin(\omega t)$ is increasing when $t > 0$ is near zero and $y = 3\cos(\omega t)$ is decreasing when $t > 0$ is near zero]. See the red part of the graph in Figure 69.

Finally, since 1 revolution requires 1 second, the period $\frac{2\pi}{\omega} = 1$, so $\omega = 2\pi$. Parametric equations that satisfy the conditions stipulated are

$$x = \sin(2\pi t), \quad y = 3\cos(2\pi t), \qquad 0 \le t \le 1 \qquad \textbf{(3)}$$

Figure 70

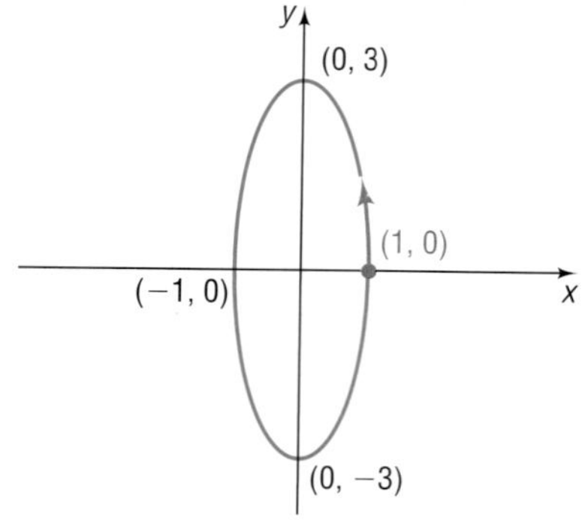

(b) See Figure 70. Since the motion begins at the point $(1, 0)$, we want $x = 1$ and $y = 0$ when $t = 0$. Furthermore, since the given equation is an ellipse, we begin by letting

$$x = \cos(\omega t) \qquad \frac{y}{3} = \sin(\omega t)$$

for some constant ω. These parametric equations satisfy the equation of the ellipse. Furthermore, with this choice, when $t = 0$, we have $x = 1$ and $y = 0$.

For the motion to be counterclockwise, the motion will have to begin with the value of x decreasing and y increasing as t increases. This requires that $\omega > 0$. [Do you know why?] Finally, since 1 revolution requires 2 seconds, the period is $\frac{2\pi}{\omega} = 2$, so $\omega = \pi$. The parametric equations that satisfy the conditions stipulated are

$$x = \cos(\pi t), \quad y = 3\sin(\pi t), \qquad 0 \le t \le 2 \qquad \textbf{(4)}$$

◀

Either of equations (3) or (4) can serve as parametric equations for the ellipse $x^2 + \frac{y^2}{9} = 1$ given in Example 7. The direction of the motion, the beginning point, and the time for 1 revolution merely serve to help us arrive at a particular parametric representation.

NOW WORK PROBLEM 49.

The Cycloid

Suppose that a circle of radius a rolls along a horizontal line without slipping. As the circle rolls along the line, a point P on the circle will trace out a curve called a **cycloid** (see Figure 71). We now seek parametric equations* for a cycloid.

We begin with a circle of radius a and take the fixed line on which the circle rolls as the x-axis. Let the origin be one of the points at which the point P comes in contact with the x-axis. Figure 71 illustrates the position of this point P after the circle has rolled somewhat. The angle t (in radians) measures the angle through which the circle has rolled.

Figure 71

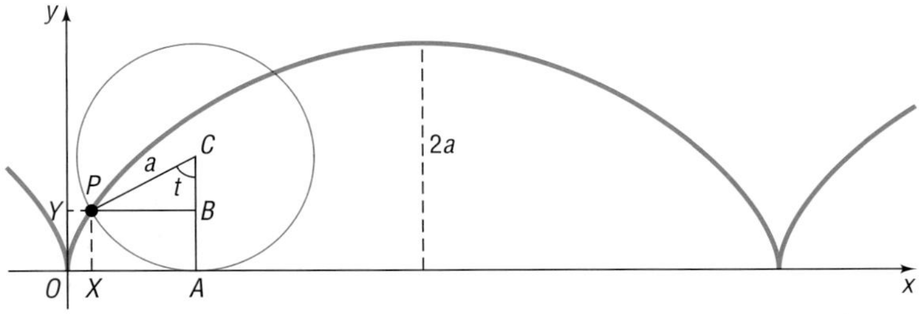

Since we require no slippage, it follows that

$$\text{Arc } AP = d(O, A)$$

The length of the arc AP is given by $s = r\theta$, where $r = a$ and $\theta = t$ radians. Then

$$at = d(O, A) \qquad s = r\theta, \text{ where } r = a \text{ and } \theta = t$$

The x-coordinate of the point P is

$$d(O, X) = d(O, A) - d(X, A) = at - a\sin t = a(t - \sin t)$$

The y-coordinate of the point P is equal to

$$d(O, Y) = d(A, C) - d(B, C) = a - a\cos t = a(1 - \cos t)$$

The parametric equations of the cycloid are

$$x = a(t - \sin t) \qquad y = a(1 - \cos t) \qquad \textbf{(5)}$$

Exploration

Graph $x = t - \sin t$, $y = 1 - \cos t$, $0 \le t \le 3\pi$, using your graphing utility with Tstep $= \frac{\pi}{36}$ and a square screen. Compare your results with Figure 71.

*Any attempt to derive the rectangular equation of a cycloid would soon demonstrate how complicated the task is.

Applications to Mechanics

If a is negative in equation (5), we obtain an inverted cycloid, as shown in Figure 72(a). The inverted cycloid occurs as a result of some remarkable applications in the field of mechanics. We shall mention two of them: the *brachistochrone* and the *tautochrone*.*

Figure 72

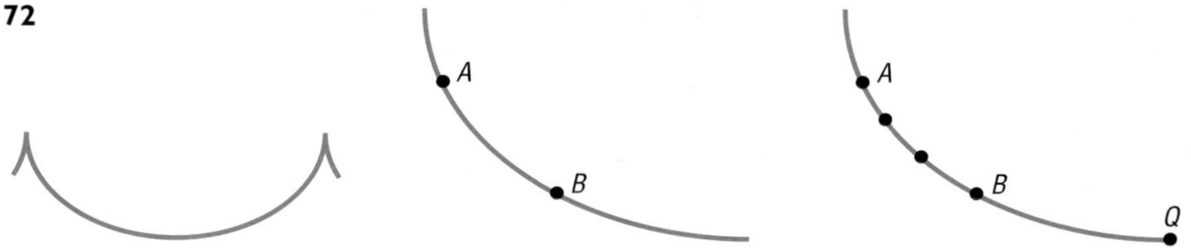

(a) Inverted cycloid (b) Curve of quickest descent (c) All reach Q at the same time

The **brachistochrone** is the curve of quickest descent. If a particle is constrained to follow some path from one point A to a lower point B (not on the same vertical line) and is acted on only by gravity, the time needed to make the descent is least if the path is an inverted cycloid. See Figure 72(b). This remarkable discovery, which is attributed to many famous mathematicians (including Johann Bernoulli and Blaise Pascal), was a significant step in creating the branch of mathematics known as the *calculus of variations*.

To define the **tautochrone,** let Q be the lowest point on an inverted cycloid. If several particles placed at various positions on an inverted cycloid simultaneously begin to slide down the cycloid, they will reach the point Q at the same time, as indicated in Figure 72(c). The tautochrone property of the cycloid was used by Christiaan Huygens (1629–1695), the Dutch mathematician, physicist, and astronomer, to construct a pendulum clock with a bob that swings along a cycloid (see Figure 73). In Huygen's clock, the bob was made to swing along a cycloid by suspending the bob on a thin wire constrained by two plates shaped like cycloids. In a clock of this design, the period of the pendulum is independent of its amplitude.

Figure 73

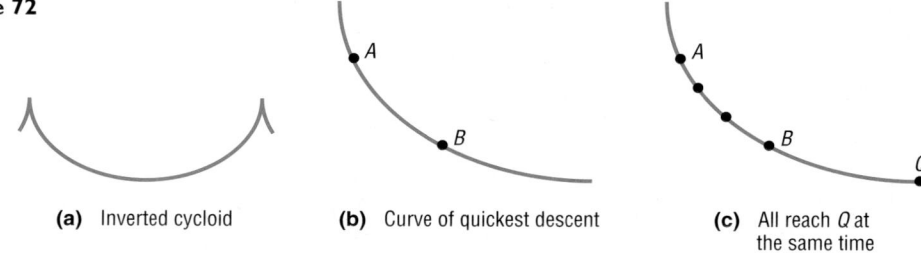

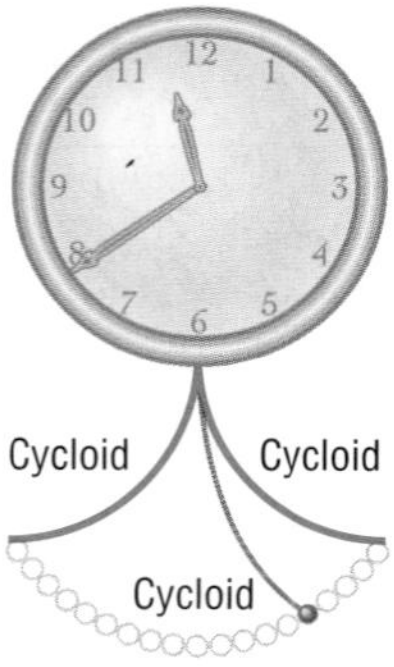

*In Greek, *brachistochrone* means "the shortest time" and *tautochrone* means "equal time."

9.7 Assess Your Understanding

'Are You Prepared?' *Answers are given at the end of these exercises. If you get a wrong answer, read the page listed in red.*

1. The function $f(x) = 3\sin(4x)$ has amplitude ________ and period ________. (p. 376)

Concepts and Vocabulary

2. Let $x = f(t)$ and $y = g(t)$, where f and g are two functions whose common domain is some interval I. The collection of points defined by $(x, y) = (f(t), g(t))$ is called a(n) ________ ________. The variable t is called a(n) ________.

3. The parametric equations $x = 2\sin t$, $y = 3\cos t$ define a(n) ________.

4. If a circle rolls along a horizontal line without slippage, a point P on the circle will trace out a curve called a(n) ________.

5. *True or False:* Parametric equations defining a curve are unique.

6. *True or False:* Curves defined using parametric equations have an orientation.

Exercises

In Problems 7–26, graph the curve whose parametric equations are given and show its orientation. Find the rectangular equation of each curve.

7. $x = 3t + 2, \quad y = t + 1; \quad 0 \leq t \leq 4$
8. $x = t - 3, \quad y = 2t + 4; \quad 0 \leq t \leq 2$
9. $x = t + 2, \quad y = \sqrt{t}; \quad t \geq 0$
10. $x = \sqrt{2t}, \quad y = 4t; \quad t \geq 0$
11. $x = t^2 + 4, \quad y = t^2 - 4; \quad -\infty < t < \infty$
12. $x = \sqrt{t} + 4, \quad y = \sqrt{t} - 4; \quad t \geq 0$
13. $x = 3t^2, \quad y = t + 1; \quad -\infty < t < \infty$
14. $x = 2t - 4, \quad y = 4t^2; \quad -\infty < t < \infty$
15. $x = 2e^t, \quad y = 1 + e^t; \quad t \geq 0$
16. $x = e^t, \quad y = e^{-t}; \quad t \geq 0$
17. $x = \sqrt{t}, \quad y = t^{3/2}; \quad t \geq 0$
18. $x = t^{3/2} + 1, \quad y = \sqrt{t}; \quad t \geq 0$
19. $x = 2\cos t, \quad y = 3\sin t; \quad 0 \leq t \leq 2\pi$
20. $x = 2\cos t, \quad y = 3\sin t; \quad 0 \leq t \leq \pi$
21. $x = 2\cos t, \quad y = 3\sin t; \quad -\pi \leq t \leq 0$
22. $x = 2\cos t, \quad y = \sin t; \quad 0 \leq t \leq \frac{\pi}{2}$
23. $x = \sec t, \quad y = \tan t; \quad 0 \leq t \leq \frac{\pi}{4}$
24. $x = \csc t, \quad y = \cot t; \quad \frac{\pi}{4} \leq t \leq \frac{\pi}{2}$
25. $x = \sin^2 t, \quad y = \cos^2 t; \quad 0 \leq t \leq 2\pi$
26. $x = t^2, \quad y = \ln t; \quad t > 0$

27. **Projectile Motion** Bob throws a ball straight up with an initial speed of 50 feet per second from a height of 6 feet.
    (a) Find parametric equations that describe the motion of the ball as a function of time.
    (b) How long is the ball in the air?
    (c) When is the ball at its maximum height? Determine the maximum height of the ball.
    (d) Simulate the motion of the ball by graphing the equations found in part (a).

28. **Projectile Motion** Alice throws a ball straight up with an initial speed of 40 feet per second from a height of 5 feet.
    (a) Find parametric equations that describe the motion of the ball as a function of time.
    (b) How long is the ball in the air?
    (c) When is the ball at its maximum height? Determine the maximum height of the ball.
    (d) Simulate the motion of the ball by graphing the equations found in part (a).

29. **Catching a Train** Bill's train leaves at 8:06 AM and accelerates at the rate of 2 meters per second per second. Bill, who can run 5 meters per second, arrives at the train station 5 seconds after the train has left.
    (a) Find parametric equations that describe the motion of the train and Bill as a function of time.
    [***Hint:*** The position s at time t of an object having acceleration a is $s = \frac{1}{2}at^2$.]
    (b) Determine algebraically whether Bill will catch the train. If so, when?
    (c) Simulate the motion of the train and Bill by simultaneously graphing the equations found in part (a).

30. **Catching a Bus** Jodi's bus leaves at 5:30 PM and accelerates at the rate of 3 meters per second per second. Jodi, who can run 5 meters per second, arrives at the bus station 2 seconds after the bus has left.
    (a) Find parametric equations that describe the motion of the bus and Jodi as a function of time.
    [***Hint:*** The position s at time t of an object having acceleration a is $s = \frac{1}{2}at^2$.]
    (b) Determine algebraically whether Jodi will catch the bus. If so, when?
    (c) Simulate the motion of the bus and Jodi by simultaneously graphing the equations found in part (a).

31. **Projectile Motion** Nolan Ryan throws a baseball with an initial speed of 145 feet per second at an angle of 20° to the horizontal. The ball leaves Nolan Ryan's hand at a height of 5 feet.
    (a) Find parametric equations that describe the position of the ball as a function of time.
    (b) How long is the ball in the air?
    (c) When is the ball at its maximum height? Determine the maximum height of the ball.
    (d) Determine the distance that the ball traveled.
    (e) Using a graphing utility, simultaneously graph the equations found in part (a).

32. **Projectile Motion** Mark McGwire hit a baseball with an initial speed of 180 feet per second at an angle of 40° to the horizontal. The ball was hit at a height of 3 feet off the ground.
    (a) Find parametric equations that describe the position of the ball as a function of time.
    (b) How long is the ball in the air?
    (c) When is the ball at its maximum height? Determine the maximum height of the ball.
    (d) Determine the distance that the ball traveled.
    (e) Using a graphing utility, simultaneously graph the equations found in part (a).

33. **Projectile Motion** Suppose that Adam throws a tennis ball off a cliff 300 meters high with an initial speed of 40 meters per second at an angle of 45° to the horizontal.
    (a) Find parametric equations that describe the position of the ball as a function of time.
    (b) How long is the ball in the air?
    (c) When is the ball at its maximum height? Determine the maximum height of the ball.
    (d) Determine the distance that the ball traveled.
    (e) Using a graphing utility, simultaneously graph the equations found in part (a).

34. Projectile Motion Suppose that Adam throws a tennis ball off a cliff 300 meters high with an initial speed of 40 meters per second at an angle of 45° to the horizontal on the Moon (gravity on the Moon is one-sixth of that on Earth).

(a) Find parametric equations that describe the position of the ball as a function of time.
(b) How long is the ball in the air?
(c) When is the ball at its maximum height? Determine the maximum height of the ball.
(d) Determine the distance that the ball traveled.
(e) Using a graphing utility, simultaneously graph the equations found in part (a).

35. Uniform Motion A Toyota Paseo (traveling east at 40 mph) and a Pontiac Bonneville (traveling north at 30 mph) are heading toward the same intersection. The Paseo is 5 miles from the intersection when the Bonneville is 4 miles from the intersection. See the figure.

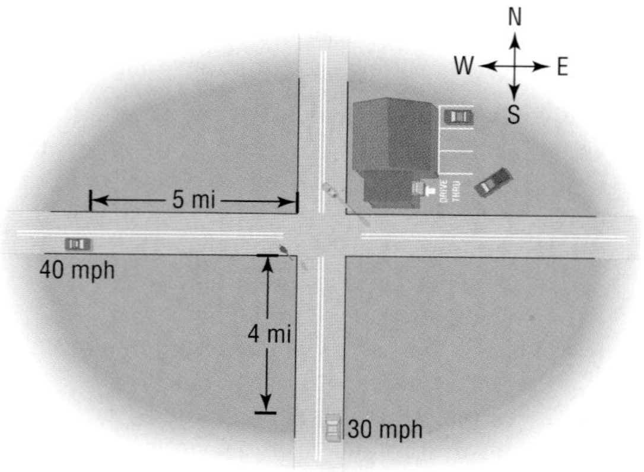

(a) Find parametric equations that describe the motion of the Paseo and Bonneville.
(b) Find a formula for the distance between the cars as a function of time.
(c) Graph the function in part (b) using a graphing utility.
(d) What is the minimum distance between the cars? When are the cars closest?
(e) Simulate the motion of the cars by simultaneously graphing the equations found in part (a).

36. Uniform Motion A Cessna (heading south at 120 mph) and a Boeing 747 (heading west at 600 mph) are flying toward the same point at the same altitude. The Cessna is 100 miles from the point where the flight patterns intersect, and the 747 is 550 miles from this intersection point. See the figure.

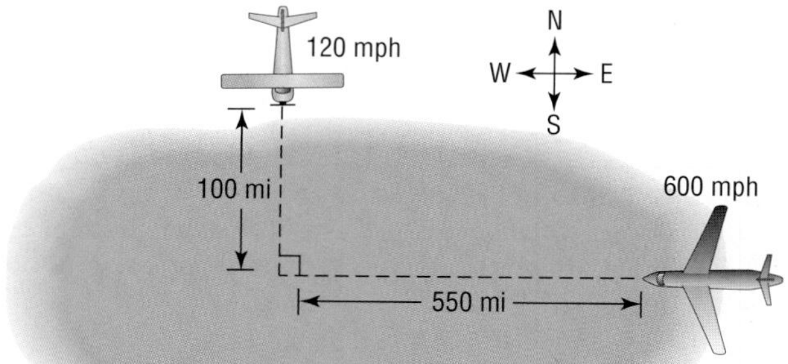

(a) Find parametric equations that describe the motion of the Cessna and 747.
(b) Find a formula for the distance between the planes as a function of time.
(c) Graph the function in part (b) using a graphing utility.
(d) What is the minimum distance between the planes? When are the planes closest?
(e) Simulate the motion of the planes by simultaneously graphing the equations found in part (a).

In Problems 37–44, find two different parametric equations for each rectangular equation.

37. $y = 4x - 1$ **38.** $y = -8x + 3$ **39.** $y = x^2 + 1$ **40.** $y = -2x^2 + 1$

41. $y = x^3$ **42.** $y = x^4 + 1$ **43.** $x = y^{3/2}$ **44.** $x = \sqrt{y}$

In Problems 45–48, find parametric equations that define the curve shown.

45.

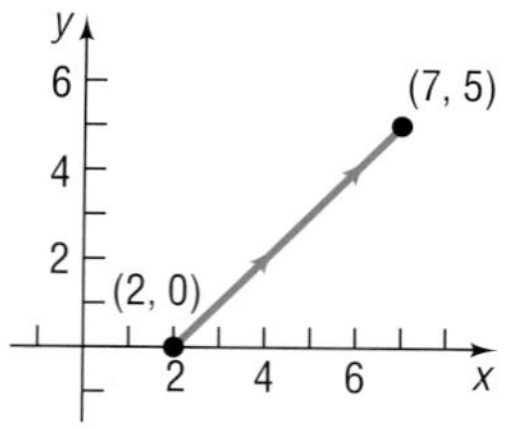

46.

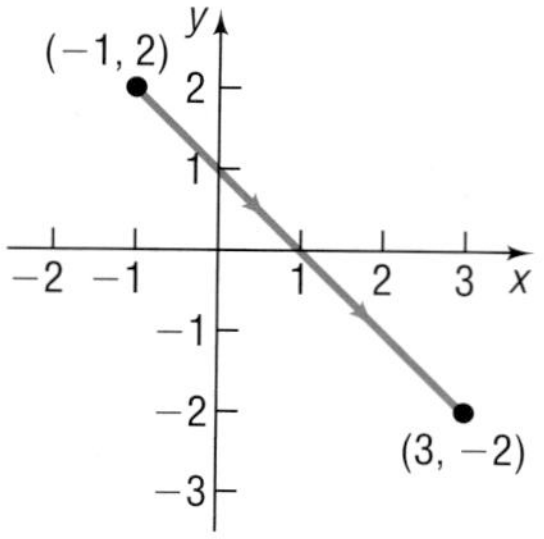

47.

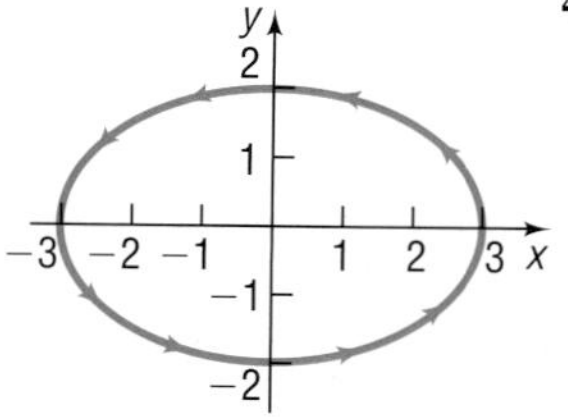

48.

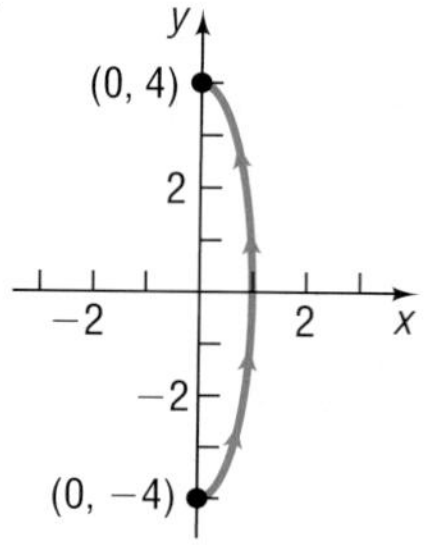

In Problems 49–52, find parametric equations for an object that moves along the ellipse $\frac{x^2}{4} + \frac{y^2}{9} = 1$ *with the motion described.*

49. The motion begins at $(2, 0)$, is clockwise, and requires 2 seconds for a complete revolution.

50. The motion begins at $(0, 3)$, is counterclockwise, and requires 1 second for a complete revolution.

51. The motion begins at $(0, 3)$, is clockwise, and requires 1 second for a complete revolution.

52. The motion begins at $(2, 0)$, is counterclockwise, and requires 3 seconds for a complete revolution.

In Problems 53 and 54, the parametric equations of four curves are given. Graph each of them, indicating the orientation.

53. C_1: $x = t, \quad y = t^2; \quad -4 \le t \le 4$
C_2: $x = \cos t, \quad y = 1 - \sin^2 t; \quad 0 \le t \le \pi$
C_3: $x = e^t, \quad y = e^{2t}; \quad 0 \le t \le \ln 4$
C_4: $x = \sqrt{t}, \quad y = t; \quad 0 \le t \le 16$

54. C_1: $x = t, \quad y = \sqrt{1 - t^2}; \quad -1 \le t \le 1$
C_2: $x = \sin t, \quad y = \cos t; \quad 0 \le t \le 2\pi$
C_3: $x = \cos t, \quad y = \sin t; \quad 0 \le t \le 2\pi$
C_4: $x = \sqrt{1 - t^2}, \quad y = t; \quad -1 \le t \le 1$

55. Show that the parametric equations for a line passing through the points (x_1, y_1) and (x_2, y_2) are

$$x = (x_2 - x_1)t + x_1$$
$$y = (y_2 - y_1)t + y_1, \quad -\infty < t < \infty$$

What is the orientation of this line?

56. Projectile Motion The position of a projectile fired with an initial velocity v_0 feet per second and at an angle θ to the horizontal at the end of t seconds is given by the parametric equations

$$x = (v_0 \cos \theta)t \qquad y = (v_0 \sin \theta)t - 16t^2$$

See the following illustration.

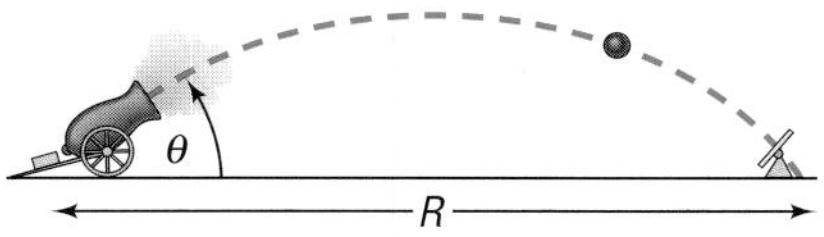

(a) Obtain the rectangular equation of the trajectory and identify the curve.

(b) Show that the projectile hits the ground $(y = 0)$ when $t = \frac{1}{16}v_0 \sin \theta$.

(c) How far has the projectile traveled (horizontally) when it strikes the ground? In other words, find the range R.

(d) Find the time t when $x = y$. Then find the horizontal distance x and the vertical distance y traveled by the projectile in this time. Then compute $\sqrt{x^2 + y^2}$. This is the distance R, the range, that the projectile travels up a plane inclined at 45° to the horizontal $(x = y)$. See the following illustration. (Also see Problem 83 in Exercise 6.5.)

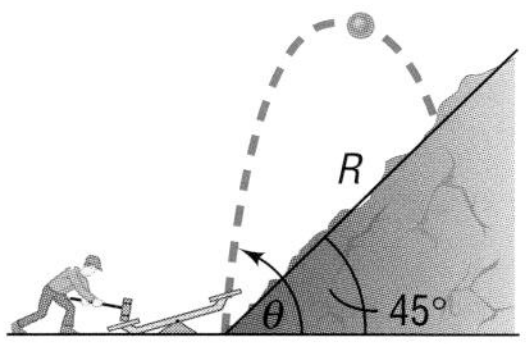

In Problems 57–60, use a graphing utility to graph the curve defined by the given parametric equations.

57. $x = t \sin t, \quad y = t \cos t$

58. $x = \sin t + \cos t, \quad y = \sin t - \cos t$

59. $x = 4 \sin t - 2 \sin(2t)$
$y = 4 \cos t - 2 \cos(2t)$

60. $x = 4 \sin t + 2 \sin(2t)$
$y = 4 \cos t + 2 \cos(2t)$

61. Hypocycloid The hypocycloid is a curve defined by the parametric equations

$$x(t) = \cos^3 t, \quad y(t) = \sin^3 t, \quad 0 \le t \le 2\pi$$

(a) Graph the hypocycloid using a graphing utility.
(b) Find rectangular equations of the hypocycloid.

62. In Problem 61, we graphed the hypocycloid. Now graph the rectangular equations of the hypocycloid. Did you obtain a complete graph? If not, experiment until you do.

63. Look up the curves called *hypocycloid* and *epicycloid*. Write a report on what you find. Be sure to draw comparisons with the cycloid.

'Are You Prepared?' Answers

1. $3; \frac{\pi}{2}$

Chapter Review

Things to Know

Equations

Parabola	See Tables 1 and 2 (pp. 617 and 619).	
Ellipse	See Table 3 (p. 630).	
Hyperbola	See Table 4 (p. 643).	
General equation of a conic (p. 656)	$Ax^2 + Bxy + Cy^2 + Dx + Ey + F = 0$	Parabola if $B^2 - 4AC = 0$ Ellipse (or circle) if $B^2 - 4AC < 0$ Hyperbola if $B^2 - 4AC > 0$
Polar equations of a conic with focus at the pole	See Table 5 (p. 661).	
Parametric equations of a curve (p. 664)	$x = f(t), y = g(t)$, t is the parameter	

Definitions

Parabola (p. 615)	Set of points P in the plane for which $d(F, P) = d(P, D)$, where F is the focus and D is the directrix	
Ellipse (p. 625)	Set of points P in the plane, the sum of whose distances from two fixed points (the foci) is a constant	
Hyperbola (p. 635)	Set of points P in the plane, the difference of whose distances from two fixed points (the foci) is a constant	
Conic in polar coordinates (p. 658)	$\dfrac{d(F, P)}{d(P, D)} = e$	Parabola if $e = 1$ Ellipse if $e < 1$ Hyperbola if $e > 1$

Formulas

Rotation formulas (p. 651)	$x = x' \cos\theta - y' \sin\theta$ $y = x' \sin\theta + y' \cos\theta$
Angle θ of rotation that eliminates the $x'y'$-term (p. 653)	$\cot(2\theta) = \dfrac{A - C}{B}, \quad 0° < \theta < 90°$

Objectives

Section		You should be able to . . .	Review Exercises
9.1	1	Know the names of the conics (p. 614)	1–20
9.2	1	Find the equation of a parabola (p. 615)	21, 24
	2	Graph parabolas (p. 616)	21, 24
	3	Discuss the equation of a parabola (p. 617)	1, 2
	4	Work with parabolas with vertex at (h, k) (p. 619)	7, 11, 12, 17, 18, 27, 30
	5	Solve applied problems involving parabolas (p. 620)	77, 78
9.3	1	Find the equation of an ellipse (p. 625)	22, 25
	2	Graph ellipses (p. 626)	22, 25
	3	Discuss the equation of an ellipse (p. 628)	5, 6, 10
	4	Work with ellipses with center at (h, k) (p. 630)	14–16, 19, 28, 31
	5	Solve applied problems involving ellipses (p. 631)	79, 80

9.4	1	Find the equation of a hyperbola (p. 636)	23, 26
	2	Graph hyperbolas (p. 637)	23, 26
	3	Discuss the equation of a hyperbola (p. 638)	3, 4, 8, 9
	4	Find the asymptotes of a hyperbola (p. 641)	3, 4, 8, 9
	5	Work with hyperbolas with center at (h, k) (p. 643)	13, 20, 29, 32–36
	6	Solve applied problems involving hyperbolas (p. 645)	81
9.5	1	Identify a conic (p. 650)	37–40
	2	Use a rotation of axes to transform equations (p. 651)	47–52
	3	Discuss an equation using a rotation of axes (p. 653)	47–52
	4	Identify conics without a rotation of axes (p. 655)	41–46
9.6	1	Discuss and graph polar equations of conics (p. 658)	53–58
	2	Convert the polar equation of a conic to a rectangular equation (p. 662)	59–62
9.7	1	Graph parametric equations (p. 665)	63–68
	2	Find a rectangular equation for a curve defined parametrically (p. 665)	63–68
	3	Use time as a parameter in parametric equations (p. 668)	82–83
	4	Find parametric equations for curves defined by rectangular equations (p. 671)	69–72

Review Exercises *(Blue problem numbers indicate the author's suggestions for use in a Practice Test.)*

In Problems 1–20, identify each equation. If it is a parabola, give its vertex, focus, and directrix; if it is an ellipse, give its center, vertices, and foci; if it is a hyperbola, give its center, vertices, foci, and asymptotes.

1. $y^2 = -16x$
2. $16x^2 = y$
3. $\frac{x^2}{25} - y^2 = 1$
4. $\frac{y^2}{25} - x^2 = 1$
5. $\frac{y^2}{25} + \frac{x^2}{16} = 1$
6. $\frac{x^2}{9} + \frac{y^2}{16} = 1$
7. $x^2 + 4y = 4$
8. $3y^2 - x^2 = 9$
9. $4x^2 - y^2 = 8$
10. $9x^2 + 4y^2 = 36$
11. $x^2 - 4x = 2y$
12. $2y^2 - 4y = x - 2$
13. $y^2 - 4y - 4x^2 + 8x = 4$
14. $4x^2 + y^2 + 8x - 4y + 4 = 0$
15. $4x^2 + 9y^2 - 16x - 18y = 11$
16. $4x^2 + 9y^2 - 16x + 18y = 11$
17. $4x^2 - 16x + 16y + 32 = 0$
18. $4y^2 + 3x - 16y + 19 = 0$
19. $9x^2 + 4y^2 - 18x + 8y = 23$
20. $x^2 - y^2 - 2x - 2y = 1$

In Problems 21–36, obtain an equation of the conic described. Graph the equation.

21. Parabola; focus at $(-2, 0)$; directrix the line $x = 2$
22. Ellipse; center at $(0, 0)$; focus at $(0, 3)$; vertex at $(0, 5)$
23. Hyperbola; center at $(0, 0)$; focus at $(0, 4)$; vertex at $(0, -2)$
24. Parabola; vertex at $(0, 0)$; directrix the line $y = -3$
25. Ellipse; foci at $(-3, 0)$ and $(3, 0)$; vertex at $(4, 0)$
26. Hyperbola; vertices at $(-2, 0)$ and $(2, 0)$; focus at $(4, 0)$
27. Parabola; vertex at $(2, -3)$; focus at $(2, -4)$
28. Ellipse; center at $(-1, 2)$; focus at $(0, 2)$; vertex at $(2, 2)$
29. Hyperbola; center at $(-2, -3)$; focus at $(-4, -3)$; vertex at $(-3, -3)$
30. Parabola; focus at $(3, 6)$; directrix the line $y = 8$
31. Ellipse; foci at $(-4, 2)$ and $(-4, 8)$; vertex at $(-4, 10)$
32. Hyperbola; vertices at $(-3, 3)$ and $(5, 3)$; focus at $(7, 3)$
33. Center at $(-1, 2)$; $a = 3$; $c = 4$; transverse axis parallel to the x-axis
34. Center at $(4, -2)$; $a = 1$; $c = 4$; transverse axis parallel to the y-axis
35. Vertices at $(0, 1)$ and $(6, 1)$; asymptote the line $3y + 2x = 9$
36. Vertices at $(4, 0)$ and $(4, 4)$; asymptote the line $y + 2x = 10$

In Problems 37–46, identify each conic without completing the squares and without applying a rotation of axes.

37. $y^2 + 4x + 3y - 8 = 0$

38. $2x^2 - y + 8x = 0$

39. $x^2 + 2y^2 + 4x - 8y + 2 = 0$

40. $x^2 - 8y^2 - x - 2y = 0$

41. $9x^2 - 12xy + 4y^2 + 8x + 12y = 0$

42. $4x^2 + 4xy + y^2 - 8\sqrt{5}x + 16\sqrt{5}y = 0$

43. $4x^2 + 10xy + 4y^2 - 9 = 0$

44. $4x^2 - 10xy + 4y^2 - 9 = 0$

45. $x^2 - 2xy + 3y^2 + 2x + 4y - 1 = 0$

46. $4x^2 + 12xy - 10y^2 + x + y - 10 = 0$

In Problems 47–52, rotate the axes so that the new equation contains no xy-term. Discuss and graph the new equation.

47. $2x^2 + 5xy + 2y^2 - \frac{9}{2} = 0$

48. $2x^2 - 5xy + 2y^2 - \frac{9}{2} = 0$

49. $6x^2 + 4xy + 9y^2 - 20 = 0$

50. $x^2 + 4xy + 4y^2 + 16\sqrt{5}x - 8\sqrt{5}y = 0$

51. $4x^2 - 12xy + 9y^2 + 12x + 8y = 0$

52. $9x^2 - 24xy + 16y^2 + 80x + 60y = 0$

In Problems 53–58, identify the conic that each polar equation represents and graph it.

53. $r = \frac{4}{1 - \cos\theta}$

54. $r = \frac{6}{1 + \sin\theta}$

55. $r = \frac{6}{2 - \sin\theta}$

56. $r = \frac{2}{3 + 2\cos\theta}$

57. $r = \frac{8}{4 + 8\cos\theta}$

58. $r = \frac{10}{5 + 20\sin\theta}$

In Problems 59–62, convert each polar equation to a rectangular equation.

59. $r = \frac{4}{1 - \cos\theta}$

60. $r = \frac{6}{2 - \sin\theta}$

61. $r = \frac{8}{4 + 8\cos\theta}$

62. $r = \frac{2}{3 + 2\cos\theta}$

In Problems 63–68, graph the curve whose parametric equations are given and show its orientation. Find the rectangular equation of each curve.

63. $x = 4t - 2, \quad y = 1 - t; \quad -\infty < t < \infty$

64. $x = 2t^2 + 6, \quad y = 5 - t; \quad -\infty < t < \infty$

65. $x = 3\sin t, \quad y = 4\cos t + 2; \quad 0 \le t \le 2\pi$

66. $x = \ln t, \quad y = t^3; \quad t > 0$

67. $x = \sec^2 t, \quad y = \tan^2 t; \quad 0 \le t \le \frac{\pi}{4}$

68. $x = t^{3/2}, \quad y = 2t + 4; \quad t \ge 0$

In Problems 69 and 70, find two different parametric equations for each rectangular equation.

69. $y = -2x + 4$

70. $y = 2x^2 - 8$

In Problems 71 and 72, find parametric equations for an object that moves along the ellipse $\frac{x^2}{16} + \frac{y^2}{9} = 1$ *with the motion described.*

71. The motion begins at $(4, 0)$, is counterclockwise, and requires 4 seconds for a complete revolution.

72. The motion begins at $(0, 3)$, is clockwise, and requires 5 seconds for a complete revolution.

---

73. Find an equation of the hyperbola whose foci are the vertices of the ellipse $4x^2 + 9y^2 = 36$ and whose vertices are the foci of this ellipse.

74. Find an equation of the ellipse whose foci are the vertices of the hyperbola $x^2 - 4y^2 = 16$ and whose vertices are the foci of this hyperbola.

75. Describe the collection of points in a plane so that the distance from each point to the point $(3, 0)$ is three-fourths of its distance from the line $x = \frac{16}{3}$.

76. Describe the collection of points in a plane so that the distance from each point to the point $(5, 0)$ is five-fourths of its distance from the line $x = \frac{16}{5}$.

77. Mirrors A mirror is shaped like a paraboloid of revolution. If a light source is located 1 foot from the base along the axis of symmetry and the opening is 2 feet across, how deep should the mirror be?

78. Parabolic Arch Bridge A bridge is built in the shape of a parabolic arch. The bridge has a span of 60 feet and a maximum height of 20 feet. Find the height of the arch at distances of 5, 10, and 20 feet from the center.

79. Semielliptical Arch Bridge A bridge is built in the shape of a semielliptical arch. The bridge has a span of 60 feet and a maximum height of 20 feet. Find the height of the arch at distances of 5, 10, and 20 feet from the center.

80. Whispering Galleries The figure shows the specifications for an elliptical ceiling in a hall designed to be a whispering gallery. Where are the foci located in the hall?

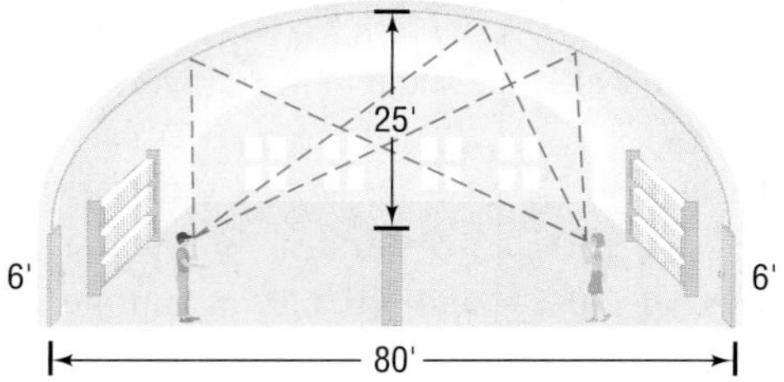

81. LORAN Two LORAN stations are positioned 150 miles apart along a straight shore.

(a) A ship records a time difference of 0.00032 second between the LORAN signals. Set up an appropriate rectangular coordinate system to determine where the ship would reach shore if it were to follow the hyperbola corresponding to this time difference.

(b) If the ship wants to enter a harbor located between the two stations 15 miles from the master station, what time difference should it be looking for?

(c) If the ship is 20 miles offshore when the desired time difference is obtained, what is the approximate location of the ship?

[*Note:* The speed of each radio signal is about 186,000 miles per second.]

82. Uniform Motion Mary's train leaves at 7:15 AM and accelerates at the rate of 3 meters per second per second. Mary, who can run 6 meters per second, arrives at the train station 2 seconds after the train has left.

(a) Find parametric equations that describe the motion of the train and Mary as a function of time.

[***Hint:*** The position s at time t of an object having acceleration a is $s = \frac{1}{2}at^2$.]

(b) Determine algebraically whether Mary will catch the train. If so, when?

(c) Simulate the motion of the train and Mary by simultaneously graphing the equations found in part (a).

83. Projectile Motion Drew Bledsoe throws a football with an initial speed of 100 feet per second at an angle of 35° to the horizontal. The ball leaves Drew Bledsoe's hand at a height of 6 feet.

(a) Find parametric equations that describe the position of the ball as a function of time.

(b) How long is the ball in the air?

(c) When is the ball at its maximum height? Determine the maximum height of the ball.

(d) Determine the distance that the ball travels.

(e) Using a graphing utility, simultaneously graph the equations found in part (a).

84. Formulate a strategy for discussing and graphing an equation of the form

$$Ax^2 + Bxy + Cy^2 + Dx + Ey + F = 0$$

Chapter Projects

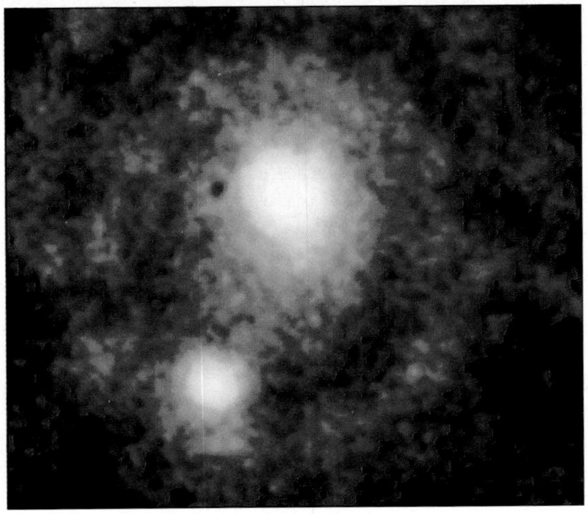

1. The Orbits of Neptune and Pluto The orbit of a planet about the Sun is an ellipse, with the Sun at one focus. The **aphelion** of a planet is its greatest distance from the Sun and the **perihelion** is its shortest distance. The **mean distance** of a planet from the Sun is the length of the semimajor axis of the elliptical orbit. See the illustration.

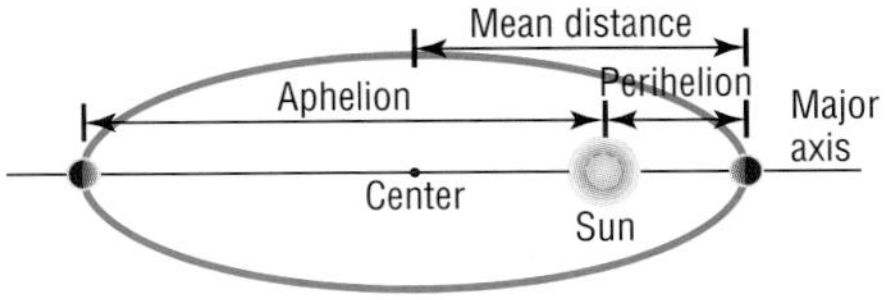

(a) The aphelion of Neptune is 4532.2×10^6 km and its perihelion is 4458.0×10^6 km. Find the equation for the orbit of Neptune around the Sun.

(b) The aphelion of Pluto is 7381.2×10^6 km and its perihelion is 4445.8×10^6 km. Find the equation for the orbit of Pluto around the Sun.

(c) Graph the orbits of Pluto and Neptune on a graphing utility. The orbits of the planets do not intersect! But the orbits *do* intersect. What is the explanation?

(d) The graphs of the orbits have the same center, so their foci lie in different locations. To see an accurate representation, the location of the Sun (a focus) needs to be the same for both graphs. This can be accomplished by shifting Pluto's orbit to the left. The shift amount is equal to Pluto's distance from the center [in the graph in part (c)] to the Sun minus Neptune's distance from the center to the Sun. Find the new equation representing the orbit of Pluto.

(e) Graph the equation for the orbit of Pluto found in part (d) along with the equation of the orbit of Neptune. Do you see that Pluto's orbit is sometimes inside Neptune's?

(f) Find the point(s) of intersection of the two orbits.

(g) Do you think two planets ever collide?

The following Chapter Projects may be found at www.prenhall.com/sullivan7e.

2. **Project at Motorola** ***Distorted Deployable Space Reflector Antennas***
3. **Constructing a Bridge over the East River**
4. **Systems of Parametric Equations**

Cumulative Review

1. Find all the solutions of the equation $\sin(2\theta) = 0.5$.

2. Find a polar equation for the line containing the origin that makes an angle of 30° with the positive x-axis.

3. Find a polar equation for the circle with center at the point $(0, 4)$ and radius 4. Graph this circle.

4. What is the domain of the function $f(x) = \dfrac{3}{\sin x + \cos x}$?

5. For $f(x) = -3x^2 + 5x - 2$, find
$$\frac{f(x+h) - f(x)}{h}, \; h \neq 0.$$

6. (a) Find the domain and range of $y = 3^x + 2$.
   (b) Find the inverse of $y = 3^x + 2$ and state its domain and range.

7. Solve the equation $9x^4 + 33x^3 - 71x^2 - 57x - 10 = 0$.

8. For what numbers x is $6 - x \geq x^2$?

9. Solve the equation $\cot(2\theta) = 1$, where $0° < \theta < 90°$.

10. Find an equation for each of the following graphs:

(a) Line:

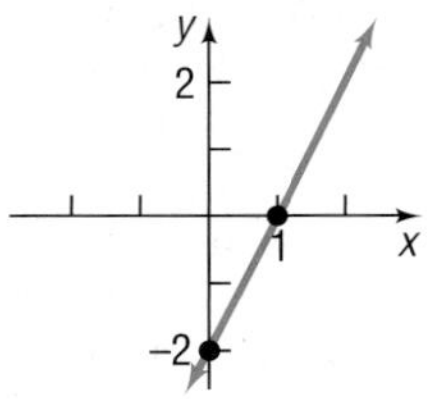

(b) Circle:

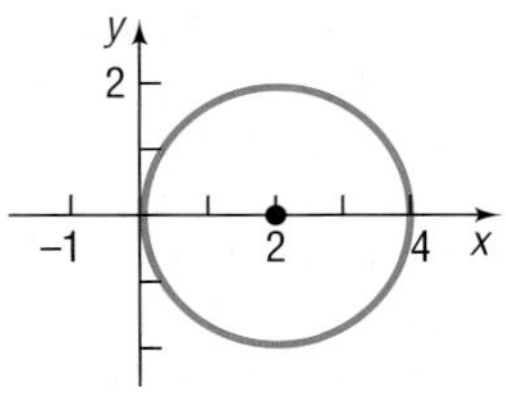

(c) Ellipse:

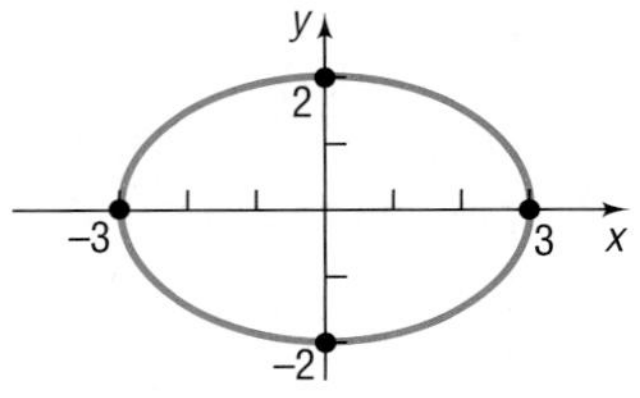

(d) Parabola:

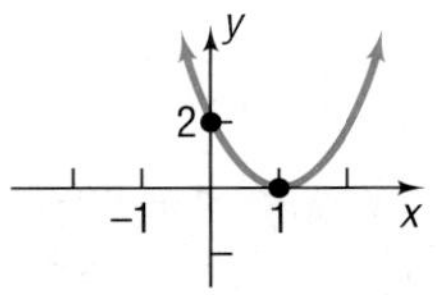

(e) Hyperbola:

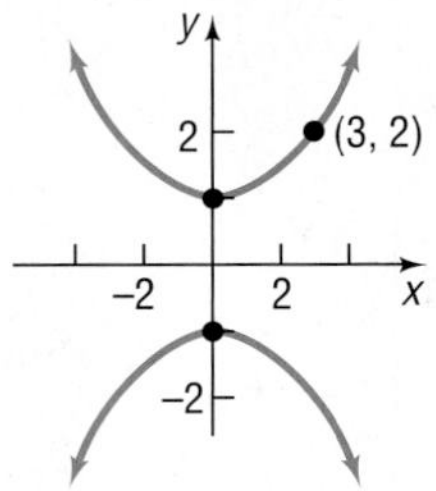

(f) Exponential:

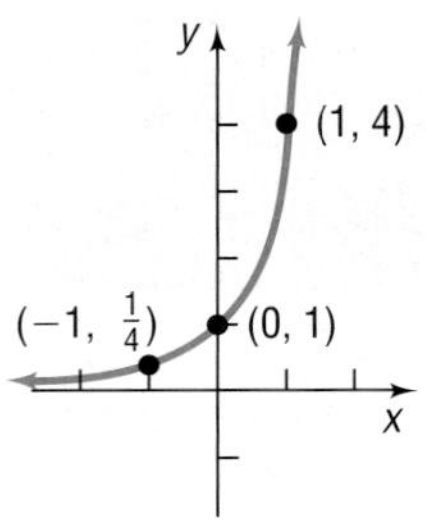

11. If $f(x) = \log_4(x - 2)$:
   (a) Solve $f(x) = 2$.
   (b) Solve $f(x) \leq 2$.

10 Systems of Equations and Inequalities

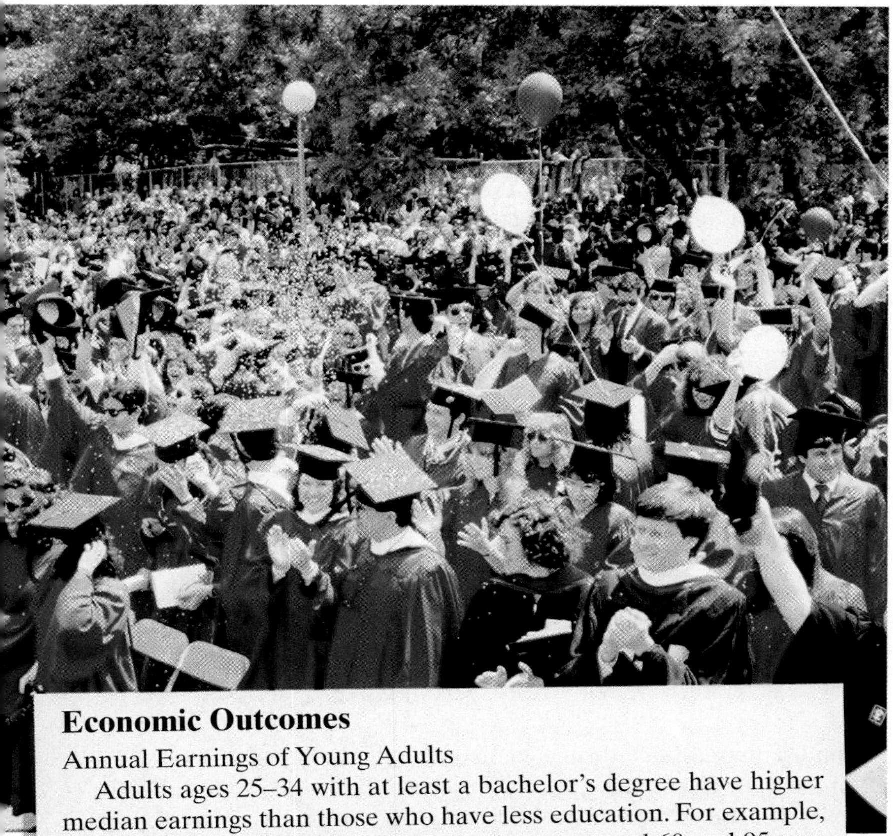

OUTLINE

Economic Outcomes

Annual Earnings of Young Adults

Adults ages 25–34 with at least a bachelor's degree have higher median earnings than those who have less education. For example, in 2000, male and female college graduates earned 60 and 95 percent more, respectively, than those who completed only high school or a General Education Development Certificate (GED). In contrast, males and females ages 25–34 who dropped out of high school earned 27 and 30 percent less, respectively, than their peers who had a high school diploma or GED.

Between 1980 and 2000, the median earnings of young adults who have at least a bachelor's degree increased relative to their counterparts with no more than a high school diploma or GED. This increase occurred for men and women, rising from a difference of 19 percent in 1980 to 60 percent in 2000 for men, and from 52 percent in 1980 to 95 percent in 2000 for women.

SOURCE: U.S. Department of Education, National Center for Education Statistics, *The Condition of Education*, 2002.

—SEE CHAPTER PROJECT 1.

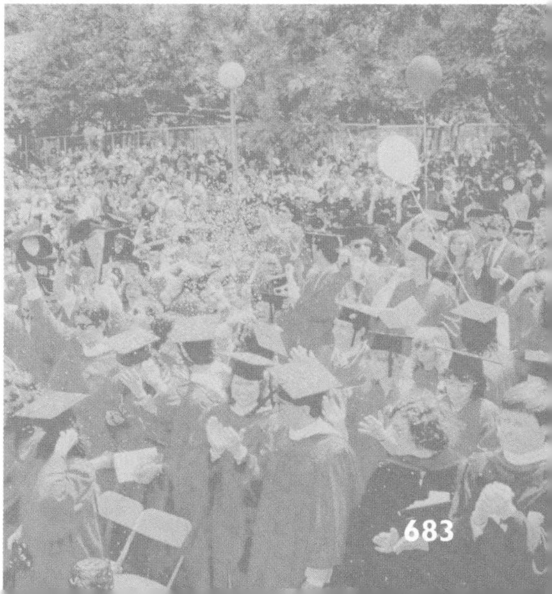

10.1 Systems of Linear Equations: Substitution and Elimination

PREPARING FOR THIS SECTION *Before getting started, review the following:*

- Solving Equations (Appendix A, Section A.5, pp. 936–938)
- Parallel Lines (Section 1.3, pp. 30–32)
- Lines (Section 1.3, pp 22–30)

Now work the 'Are You Prepared?' problems on page 696.

OBJECTIVES
1. Solve Systems of Equations by Substitution
2. Solve Systems of Equations by Elimination
3. Identify Inconsistent Systems of Equations Containing Two Variables
4. Express the Solution of a System of Dependent Equations Containing Two Variables
5. Solve Systems of Three Equations Containing Three Variables
6. Identify Inconsistent Systems of Equations Containing Three Variables
7. Express the Solution of a System of Dependent Equations Containing Three Variables

We begin with an example.

EXAMPLE 1 Movie Theater Ticket Sales

A movie theater sells tickets for \$8.00 each, with seniors receiving a discount of \$2.00. One evening the theater took in \$3580 in revenue. If x represents the number of tickets sold at \$8.00 and y the number of tickets sold at the discounted price of \$6.00, write an equation that relates these variables.

Solution Each nondiscounted ticket brings in \$8.00, so x tickets will bring in $8x$ dollars. Similarly, y discounted tickets bring in $6y$ dollars. Since the total brought in is \$3580, we must have

$$8x + 6y = 3580$$

◀

In Example 1, suppose that we also know that 525 tickets were sold that evening. Then we have another equation relating the variables x and y,

$$x + y = 525$$

The two equations

$$\begin{aligned} 8x + 6y &= 3580 \\ x + y &= 525 \end{aligned}$$

form a *system* of equations.

In general, a **system of equations** is a collection of two or more equations, each containing one or more variables. Example 2 gives some illustrations of systems of equations.

EXAMPLE 2 Examples of Systems of Equations

(a) $\begin{cases} 2x + y = 5 & (1) \\ -4x + 6y = -2 & (2) \end{cases}$ Two equations containing two variables, x and y

(b) $\begin{cases} x + y^2 = 5 & (1) \\ 2x + y = 4 & (2) \end{cases}$ Two equations containing two variables, x and y

(c) $\begin{cases} x + y + z = 6 & (1) \\ 3x - 2y + 4z = 9 & (2) \\ x - y - z = 0 & (3) \end{cases}$ Three equations containing three variables, x, y, and z

(d) $\begin{cases} x + y + z = 5 & (1) \\ x - y = 2 & (2) \end{cases}$ Two equations containing three variables, x, y, and z

(e) $\begin{cases} x + y + z = 6 & (1) \\ 2x + 2z = 4 & (2) \\ y + z = 2 & (3) \\ x = 4 & (4) \end{cases}$ Four equations containing three variables, x, y, and z

◄

We use a brace, as shown, to remind us that we are dealing with a system of equations. We also will find it convenient to number each equation in the system.

A **solution** of a system of equations consists of values for the variables that are solutions of each equation of the system. To **solve** a system of equations means to find all solutions of the system.

For example, $x = 2, y = 1$ is a solution of the system in Example 2(a), because

$$\begin{cases} 2x + y = 5 & (1) \\ -4x + 6y = -2 & (2) \end{cases} \qquad \begin{cases} 2(2) + 1 = 4 + 1 = 5 \\ -4(2) + 6(1) = -8 + 6 = -2 \end{cases}$$

A solution of the system in Example 2(b) is $x = 1, y = 2$, because

$$\begin{cases} x + y^2 = 5 & (1) \\ 2x + y = 4 & (2) \end{cases} \qquad \begin{cases} 1 + 2^2 = 1 + 4 = 5 \\ 2(1) + 2 = 2 + 2 = 4 \end{cases}$$

Another solution of the system in Example 2(b) is $x = \frac{11}{4}, y = -\frac{3}{2}$, which you can check for yourself.

A solution of the system in Example 2(c) is $x = 3, y = 2, z = 1$, because

$$\begin{cases} x + y + z = 6 & (1) \\ 3x - 2y + 4z = 9 & (2) \\ x - y - z = 0 & (3) \end{cases} \qquad \begin{cases} 3 + 2 + 1 = 6 & (1) \\ 3(3) - 2(2) + 4(1) = 9 - 4 + 4 = 9 & (2) \\ 3 - 2 - 1 = 0 & (3) \end{cases}$$

Note that $x = 3, y = 3, z = 0$ is not a solution of the system in Example 2(c).

$$\begin{cases} x + y + z = 6 & (1) \\ 3x - 2y + 4z = 9 & (2) \\ x - y - z = 0 & (3) \end{cases} \qquad \begin{cases} 3 + 3 + 0 = 6 & (1) \\ 3(3) - 2(3) + 4(0) = 3 \neq 9 & (2) \\ 3 - 3 - 0 = 0 & (3) \end{cases}$$

Although these values satisfy equations (1) and (3), they do not satisfy equation (2). Any solution of the system must satisfy *each* equation of the system.

NOW WORK PROBLEM 9.

When a system of equations has at least one solution, it is said to be **consistent;** otherwise, it is called **inconsistent.**

An equation in n variables is said to be **linear** if it is equivalent to an equation of the form

$$a_1x_1 + a_2x_2 + \cdots + a_nx_n = b$$

where $x_1, x_2, \dots, x_n$ are n distinct variables, $a_1, a_2, \dots, a_n, b$ are constants, and at least one of the a's is not 0.

Some examples of linear equations are

$$2x + 3y = 2 \qquad 5x - 2y + 3z = 10 \qquad 8x + 8y - 2z + 5w = 0$$

If each equation in a system of equations is linear, then we have a **system of linear equations.** The systems in Examples 2(a), (c), (d), and (e) are linear, whereas the system in Example 2(b) is nonlinear. In this chapter we shall solve linear systems in Sections 10.1–10.4. We discuss nonlinear systems in Section 10.6.

Two Linear Equations Containing Two Variables

We can view the problem of solving a system of two linear equations containing two variables as a geometry problem. The graph of each equation in such a system is a line. So, a system of two equations containing two variables represents a pair of lines. The lines either (1) intersect or (2) are parallel or (3) are **coincident** (that is, identical).

1. If the lines intersect, then the system of equations has one solution, given by the point of intersection. The system is **consistent** and the equations are **independent.** See Figure 1(a).
2. If the lines are parallel, then the system of equations has no solution, because the lines never intersect. The system is **inconsistent.** See Figure 1(b).
3. If the lines are coincident, then the system of equations has infinitely many solutions, represented by the totality of points on the line. The system is **consistent** and the equations are **dependent.** See Figure 1(c).

Figure 1

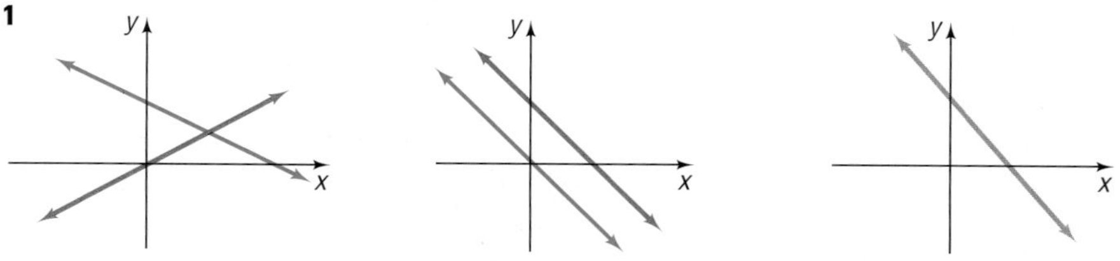

(a) Intersecting lines; system has one solution **(b)** Parallel lines; system has no solution **(c)** Coincident lines; system has infinitely many solutions

EXAMPLE 3 Graphing a System of Linear Equations

Graph the system: $\begin{cases} 2x + y = 5 & (1) \\ -4x + 6y = 12 & (2) \end{cases}$

Figure 2

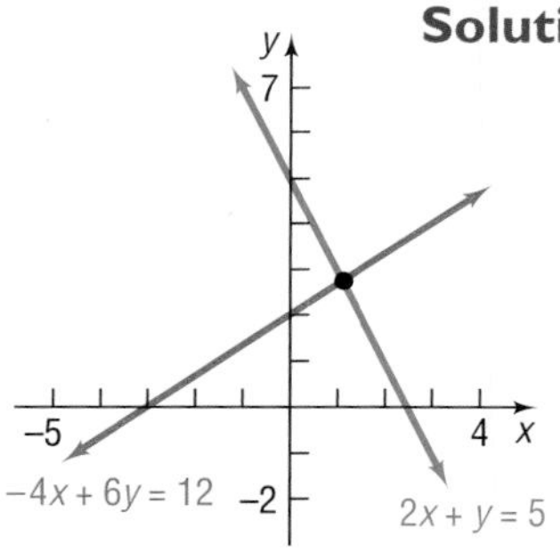

Solution Equation (1) in slope–intercept form is $y = -2x + 5$, which has slope -2 and y-intercept 5. Equation (2) in slope–intercept form is $y = \frac{2}{3}x + 2$, which has slope $\frac{2}{3}$ and y-intercept 2. Figure 2 shows their graphs. ◀

From the graph in Figure 2, we see that the lines intersect, so the system given in Example 3 is consistent. We can also use the graph as a means of approximating the solution. For this system, the solution appears to be close to the point $(1, 3)$. The actual solution, which you should verify, is $\left(\frac{9}{8}, \frac{11}{4}\right)$.

Seeing the Concept

Graph the lines $2x + y = 5$ ($Y_1 = -2x + 5$) and $-4x + 6y = 12$ $\left(Y_2 = \frac{2}{3}x + 2\right)$ and compare what you see with Figure 2. Use INTERSECT to verify that the point of intersection is (1.125, 2.75).

1 To obtain exact solutions, we use algebraic methods. The first algebraic method we take up is the *method of substitution.*

A number of methods are available to us for solving systems of linear equations algebraically. In this section, we introduce two methods: *substitution* and *elimination*. We illustrate the **method of substitution** by solving the system given in Example 3.

EXAMPLE 4 Solving a System of Linear Equations by Substitution

Solve: $\begin{cases} 2x + y = 5 & (1) \\ -4x + 6y = 12 & (2) \end{cases}$

Solution We solve the first equation for y, obtaining

$$2x + y = 5 \quad (1)$$
$$y = -2x + 5 \quad \text{Subtract } 2x \text{ from each side of (1).}$$

We substitute this result for y in the second equation. The result is an equation containing just the variable x, which we can then solve.

$$-4x + 6y = 12 \quad (2)$$
$$-4x + 6(-2x + 5) = 12 \quad \text{Substitute } y = -2x + 5 \text{ in (2).}$$
$$-4x - 12x + 30 = 12 \quad \text{Remove parentheses.}$$
$$-16x = -18 \quad \text{Combine like terms and subtract 30 from both sides.}$$
$$x = \frac{-18}{-16} = \frac{9}{8} \quad \text{Divide each side by } -16.$$

Once we know that $x = \frac{9}{8}$, we can easily find the value of y by **back-substitution,** that is, by substituting $\frac{9}{8}$ for x in one of the original equations. We will use the first equation.

$$2x + y = 5 \qquad (1)$$

$$y = -2x + 5 \qquad \text{Subtract } 2x \text{ from each side.}$$

$$y = -2\left(\frac{9}{8}\right) + 5 \qquad \text{Substitute } x = \frac{9}{8} \text{ in (1).}$$

$$= \frac{-9}{4} + \frac{20}{4} = \frac{11}{4}$$

The solution of the system is $x = \frac{9}{8} = 1.125$, $y = \frac{11}{4} = 2.75$.

✔ **CHECK:**

$$\begin{cases} 2x + y = 5: & 2\left(\frac{9}{8}\right) + \frac{11}{4} = \frac{9}{4} + \frac{11}{4} = \frac{20}{4} = 5 \\ -4x + 6y = 12: & -4\left(\frac{9}{8}\right) + 6\left(\frac{11}{4}\right) = -\frac{9}{2} + \frac{33}{2} = \frac{24}{2} = 12 \end{cases}$$ ◀

The method used to solve the system in Example 4 is called **substitution.** The steps to be used are outlined next.

Steps for Solving by Substitution

STEP 1: Pick one of the equations and solve for one of the variables in terms of the remaining variables.

STEP 2: Substitute the result in the remaining equations.

STEP 3: If one equation in one variable results, solve this equation. Otherwise, repeat Steps 1 and 2 until a single equation with one variable remains.

STEP 4: Find the values of the remaining variables by back-substitution.

STEP 5: Check the solution found.

NOW USE SUBSTITUTION TO WORK PROBLEM 19.

2 A second method for solving a system of linear equations is the *method of elimination*. This method is usually preferred over substitution if substitution leads to fractions or if the system contains more than two variables. Elimination also provides the necessary motivation for solving systems using matrices (the subject of Section 10.2).

The idea behind the method of elimination is to replace the original system of equations by an equivalent system so that adding two of the equations eliminates a variable. The rules for obtaining equivalent equations are the same as those studied earlier. However, we may also interchange any two equations of the system and/or replace any equation in the system by the sum (or difference) of that equation and a nonzero multiple of any other equation in the system.

Rules for Obtaining an Equivalent System of Equations

1. Interchange any two equations of the system.
2. Multiply (or divide) each side of an equation by the same nonzero constant.
3. Replace any equation in the system by the sum (or difference) of that equation and a nonzero constant multiple of any other equation in the system.

An example will give you the idea. As you work through the example, pay particular attention to the pattern being followed.

EXAMPLE 5 Solving a System of Linear Equations by Elimination

Solve: $\begin{cases} 2x + 3y = 1 & (1) \\ -x + y = -3 & (2) \end{cases}$

Solution We multiply each side of equation (2) by 2 so that the coefficients of x in the two equations are negatives of one another. The result is the equivalent system

$$\begin{cases} 2x + 3y = 1 & (1) \\ -2x + 2y = -6 & (2) \end{cases}$$

If we now replace equation (2) of this system by the sum of the two equations, we obtain an equation containing just the variable y, which we can solve.

$$\begin{cases} 2x + 3y = 1 & (1) \\ -2x + 2y = -6 & (2) \end{cases}$$
$$5y = -5 \quad \text{Add (1) and (2).}$$
$$y = -1 \quad \text{Solve for } y.$$

We back-substitute this value for y in equation (1) and simplify to get

$$2x + 3y = 1 \quad (1)$$
$$2x + 3(-1) = 1 \quad \text{Substitute } y = -1 \text{ in (1).}$$
$$2x = 4 \quad \text{Simplify.}$$
$$x = 2 \quad \text{Solve for } x.$$

The solution of the original system is $x = 2$, $y = -1$. We leave it to you to check the solution. ◀

The procedure used in Example 5 is called the **method of elimination.** Notice the pattern of the solution. First, we eliminated the variable x from the second equation. Then we back-substituted; that is, we substituted the value found for y back into the first equation to find x.

NOW USE ELIMINATION TO WORK PROBLEM 19.

Let's return to the movie theater example in Example 1.

EXAMPLE 6 Movie Theater Ticket Sales

A movie theater sells tickets for \$8.00 each, with seniors receiving a discount of \$2.00. One evening the theater sold 525 tickets and took in \$3580 in revenue. How many of each type of ticket were sold?

Solution If x represents the number of tickets sold at \$8.00 and y the number of tickets sold at the discounted price of \$6.00, then the given information results in the system of equations

$$\begin{cases} 8x + 6y = 3580 & (1) \\ x + y = 525 & (2) \end{cases}$$

We use elimination and multiply the second equation by -6 and then add the equations.

$$\begin{cases} 8x + 6y = 3580 \\ -6x - 6y = -3150 \end{cases}$$

$$2x = 430 \qquad \text{Add the equations.}$$

$$x = 215$$

Since $x + y = 525$, then $y = 525 - x = 525 - 215 = 310$. Thus, 215 nondiscounted tickets and 310 senior discount tickets were sold. ◀

3 The previous examples dealt with consistent systems of equations that had unique solutions. The next two examples deal with the other possibilities that may occur, the first being a system that has no solution.

EXAMPLE 7 An Inconsistent System of Linear Equations

Solve: $\begin{cases} 2x + y = 5 & (1) \\ 4x + 2y = 8 & (2) \end{cases}$

Solution We choose to use the method of substitution and solve equation (1) for y.

$$2x + y = 5 \qquad (1)$$

$$y = -2x + 5 \qquad \text{Subtract } 2x \text{ from each side.}$$

Now substitute $y = -2x + 5$ for y in equation (2) and solve for x.

$$4x + 2y = 8 \qquad (2)$$

$$4x + 2(-2x + 5) = 8 \qquad \text{Substitute } y = -2x + 5 \text{ in (2).}$$

$$4x - 4x + 10 = 8 \qquad \text{Remove parentheses.}$$

$$0 \cdot x = -2 \qquad \text{Subtract 10 from both sides.}$$

This equation has no solution. We conclude that the system itself has no solution and is therefore inconsistent. ◀

Figure 3

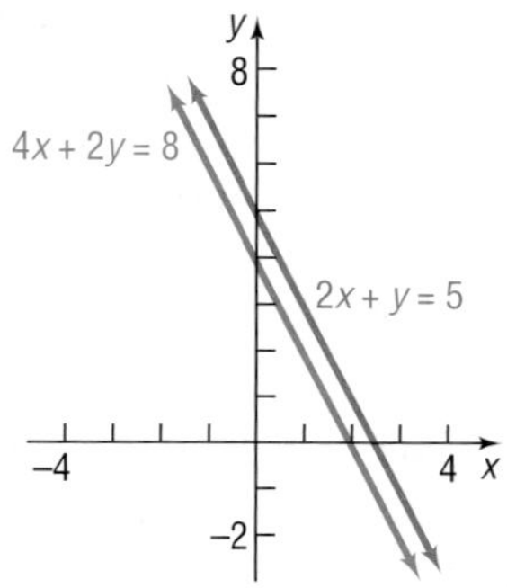

Figure 3 illustrates the pair of lines whose equations form the system in Example 7. Notice that the graphs of the two equations are lines, each with slope -2; one has a y-intercept of 5, the other a y-intercept of 4. The lines are parallel and have no point of intersection. This geometric statement is equivalent to the algebraic statement that the system has no solution.

Seeing the Concept

Graph the lines $2x + y = 5$ ($Y_1 = -2x + 5$) and $4x + 2y = 8$ ($Y_2 = -2x + 4$) and compare what you see with Figure 3. How can you be sure that the lines are parallel?

4 The next example is an illustration of a system with infinitely many solutions.

EXAMPLE 8 Solving a System of Dependent Equations

Solve: $\begin{cases} 2x + y = 4 & (1) \\ -6x - 3y = -12 & (2) \end{cases}$

Solution We choose to use the method of elimination.

$$\begin{cases} 2x + y = 4 & (1) \\ -6x - 3y = -12 & (2) \end{cases}$$

$$\begin{cases} 6x + 3y = 12 & (1) \\ -6x - 3y = -12 & (2) \end{cases}$$ Multiply each side of equation (1) by 3.

$$\begin{cases} 6x + 3y = 12 & (1) \\ 0 = 0 & (2) \end{cases}$$ Replace equation (2) by the sum of equations (1) and (2).

The original system is equivalent to a system containing one equation, so the equations are dependent. This means that any values of x and y for which $6x + 3y = 12$ or, equivalently, $2x + y = 4$ are solutions. For example, $x = 2, y = 0$; $x = 0, y = 4$; $x = -2, y = 8$; $x = 4, y = -4$; and so on, are solutions. There are, in fact, infinitely many values of x and y for which $2x + y = 4$, so the original system has infinitely many solutions. We will write the solutions of the original system either as

$$y = -2x + 4$$

where x can be any real number, or as

$$x = -\frac{1}{2}y + 2$$

where y can be any real number. ◀

Figure 4

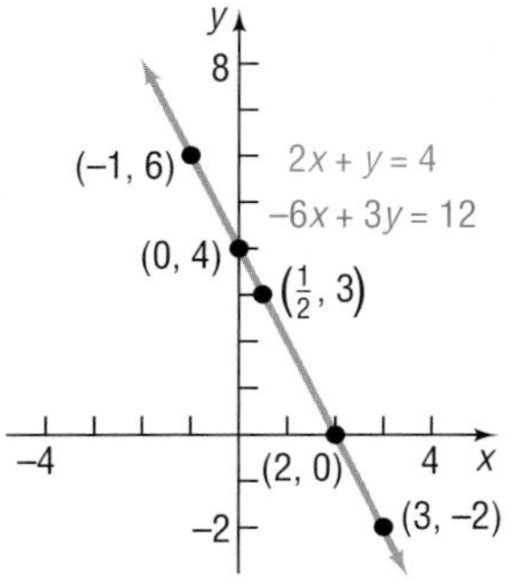

Figure 4 illustrates the situation presented in Example 8. Notice that the graphs of the two equations are lines, each with slope -2 and each with y-intercept 4. The lines are coincident. Notice also that equation (2) in the original system is just -3 times equation (1), indicating that the two equations are dependent.

For the system in Example 8, we can write down some of the infinite number of solutions by assigning values to x and then finding $y = -2x + 4$.

If $x = -2$, then $y = 8$.
If $x = 0$, then $y = 4$.
If $x = 2$, then $y = 0$.

The pairs (x, y) are points on the line in Figure 4.

Seeing the Concept

Graph the lines $2x + y = 4$ ($Y_1 = -2x + 4$) and $-6x - 3y = -12$ ($Y_2 = -2x + 4$) and compare what you see with Figure 4. How can you be sure that the lines are coincident?

NOW WORK PROBLEMS 25 AND 29.

Three Equations Containing Three Variables

Just as with a system of two linear equations containing two variables, a system of three linear equations containing three variables also has either (1) exactly one solution (a consistent system with independent equations), or (2) no solution (an inconsistent system), or (3) infinitely many solutions (a consistent system with dependent equations).

We can view the problem of solving a system of three linear equations containing three variables as a geometry problem. The graph of each equation in such a system is a plane in space. A system of three linear equations containing three variables represents three planes in space. Figure 5 illustrates some of the possibilities.

Figure 5

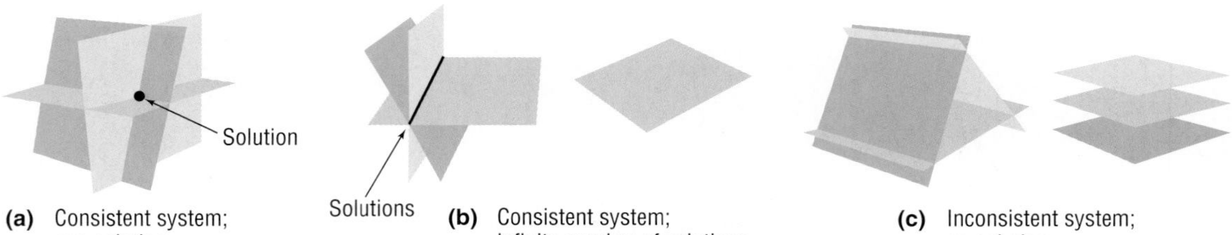

(a) Consistent system; one solution

(b) Consistent system; infinite number of solutions

(c) Inconsistent system; no solution

Recall that a **solution** to a system of equations consists of values for the variables that are solutions of each equation of the system. For example, $x = 3$, $y = -1$, $z = -5$ is a solution to the system of equations

$$\begin{cases} x + y + z = -3 & (1) \quad 3 + (-1) + (-5) = -3 \\ 2x - 3y + 6z = -21 & (2) \quad 2(3) - 3(-1) + 6(-5) = 6 + 3 - 30 = -21 \\ -3x + 5y = -14 & (3) \quad -3(3) + 5(-1) = -9 - 5 = -14 \end{cases}$$

because these values of the variables are solutions of each equation.

Typically, when solving a system of three linear equations containing three variables, we use the method of elimination. Recall that the idea behind the method of elimination is to form equivalent equations so that adding two of the equations eliminates a variable.

5 Let's see how elimination works on a system of three equations containing three variables.

EXAMPLE 9 Solving a System of Three Linear Equations With Three Variables

Use the method of elimination to solve the system of equations.

$$\begin{cases} x + y - z = -1 & (1) \\ 4x - 3y + 2z = 16 & (2) \\ 2x - 2y - 3z = 5 & (3) \end{cases}$$

Solution For a system of three equations, we attempt to eliminate one variable at a time, using pairs of equations, until an equation with a single variable remains. Our plan of attack on this system will be to use equation (1) to eliminate the variable x from equations (2) and (3).

We begin by multiplying each side of equation (1) by -4 and adding the result to equation (2). (Do you see why? The coefficients of x are now negatives of one another.) We also multiply equation (1) by -2 and add the result to equation (3). Notice that these two procedures result in the removal of the variable x from equations (2) and (3).

$$\begin{array}{ll} x + y - z = -1 \ (1) & \text{Multiply by } -4 \\ 4x - 3y + 2z = 16 \ (2) & \end{array} \qquad \begin{array}{r} -4x - 4y + 4z = 4 \ (1) \\ 4x - 3y + 2z = 16 \ (2) \\ \hline -7y + 6z = 20 \end{array} \ \text{Add.}$$

$$\begin{array}{ll} x + y - z = -1 \ (1) & \text{Multiply by } -2 \\ 2x - 2y - 3z = 5 \ (3) & \end{array} \qquad \begin{array}{r} -2x - 2y + 2z = 2 \ (1) \\ 2x - 2y - 3z = 5 \ (3) \\ \hline -4y - z = 7 \end{array} \ \text{Add.}$$

$$\begin{cases} x + y - z = -1 & (1) \\ -7y + 6z = 20 & (2) \\ -4y - z = 7 & (3) \end{cases}$$

We now concentrate on equations (2) and (3), treating them as a system of two equations containing two variables. It is easiest to eliminate z. We multiply each side of equation (3) by 6 and add equations (2) and (3). The result is the new equation (3).

$$\begin{array}{ll} -7y + 6z = 20 \ (2) & \\ -4y - z = 7 \ (3) & \text{Multiply by 6} \end{array} \qquad \begin{array}{r} -7y + 6z = 20 \ (2) \\ -24y - 6z = 42 \ (3) \\ \hline -31y = 62 \end{array} \ \text{Add.} \longrightarrow \begin{cases} x + y - z = -1 & (1) \\ -7y + 6z = 20 & (2) \\ -31y = 62 & (3) \end{cases}$$

We now solve equation (3) for y by dividing both sides of the equation by -31.

$$\begin{cases} x + y - z = -1 & (1) \\ -7y + 6z = 20 & (2) \\ y = -2 & (3) \end{cases}$$

Back-substitute $y = -2$ in equation (2) and solve for z.

$$\begin{aligned} -7y + 6z &= 20 && (2) \\ -7(-2) + 6z &= 20 && \text{Substitute } y = -2 \text{ in (2).} \\ 6z &= 6 && \text{Subtract 14 from both sides of the equation.} \\ z &= 1 && \text{Divide both sides of the equation by 6.} \end{aligned}$$

Finally, we back-substitute $y = -2$ and $z = 1$ in equation (1) and solve for x.

$$\begin{aligned} x + y - z &= -1 && (1) \\ x + (-2) - 1 &= -1 && \text{Substitute } y = -2 \text{ and } z = 1 \text{ in (1).} \\ x - 3 &= -1 && \text{Simplify.} \\ x &= 2 && \text{Add 3 to both sides.} \end{aligned}$$

The solution of the original system is $x = 2$, $y = -2$, $z = 1$. You should verify this solution. ◀

Look back over the solution given in Example 9. Note the pattern of removing one of the variables from two of the equations, followed by solving this system of two equations and two unknowns. Although which variables to remove is your choice, the methodology remains the same for all systems.

NOW WORK PROBLEM 43.

6 The previous example was a consistent system that had a unique solution. The next two examples deal with the other possibilities that may occur.

EXAMPLE 10 An Inconsistent System of Linear Equations

$$\text{Solve: } \begin{cases} 2x + y - z = -2 & (1) \\ x + 2y - z = -9 & (2) \\ x - 4y + z = 1 & (3) \end{cases}$$

Solution Our plan of attack is the same as in Example 9. However, in this system, it seems easiest to eliminate the variable z first. Do you see why?

Multiply each side of equation (1) by -1 and add the result to equation (2). Add equations (2) and (3).

$$\begin{array}{rl} -2x - y + z = 2 & (1) \text{ Multiply by } -1. \\ x + 2y - z = -9 & (2) \\ \hline -x + y = -7 & \text{Add.} \end{array}$$

$$\begin{array}{rl} x + 2y - z = -9 & (2) \\ x - 4y + z = 1 & (3) \\ \hline 2x - 2y = -8 & \text{Add.} \end{array}$$

$$\begin{cases} 2x + y - z = -2 & (1) \\ -x + y = -7 & (2) \\ 2x - 2y = -8 & (3) \end{cases}$$

We now concentrate on equations (2) and (3), treating them as a system of two equations containing two variables. Multiply each side of equation (2) by 2 and add the result to equation (3).

$$\begin{array}{rl} -x + y = -7 & (2) \text{ Multiply by 2.} \\ 2x - 2y = -8 & (3) \end{array} \quad \begin{array}{rl} -2x + 2y = -14 & (2) \\ 2x - 2y = -8 & (3) \\ \hline 0 = -22 & \text{Add.} \end{array} \quad \begin{cases} 2x + y - z = -2 & (1) \\ -x + y = -7 & (2) \\ 0 = -22 & (3) \end{cases}$$

Equation (3) has no solution and the system is inconsistent. ◀

7 Now let's look at a system of dependent equations.

EXAMPLE 11 Solving a System of Dependent Equations

$$\text{Solve: } \begin{cases} x - 2y - z = 8 & (1) \\ 2x - 3y + z = 23 & (2) \\ 4x - 5y + 5z = 53 & (3) \end{cases}$$

Solution Multiply each side of equation (1) by -2 and add the result to equation (2). Also, multiply each side of equation (1) by -4 and add the result to equation (3).

$$\begin{array}{rl} x - 2y - z = 8 & (1) \text{ Multiply by } -2. \\ 2x - 3y + z = 23 & (2) \end{array} \quad \begin{array}{rl} -2x + 4y + 2z = -16 & (1) \\ 2x - 3y + z = 23 & (2) \\ \hline y + 3z = 7 & \text{Add.} \end{array}$$

$$\begin{array}{rl} x - 2y - z = 8 & (1) \text{ Multiply by } -4. \\ 4x - 5y + 5z = 53 & (3) \end{array} \quad \begin{array}{rl} -4x + 8y + 4z = -32 & (1) \\ 4x - 5y + 5z = 53 & (2) \\ \hline 3y + 9z = 21 & \text{Add.} \end{array}$$

$$\begin{cases} x - 2y - z = 8 & (1) \\ y + 3z = 7 & (2) \\ 3y + 9z = 21 & (3) \end{cases}$$

Treat equations (2) and (3) as a system of two equations containing two variables, and eliminate the variable y by multiplying both sides of equation (2) by -3 and adding the result to equation (3).

$$\begin{aligned} y + 3z &= 7 \\ 3y + 9z &= 21 \end{aligned} \quad \text{Multiply by } -3. \quad \begin{aligned} -3y - 9z &= -21 \\ 3y + 9z &= 21 \quad \text{Add.} \\ \hline 0 &= 0 \end{aligned} \longrightarrow \begin{cases} x - 2y - z = 8 & (1) \\ y + 3z = 7 & (2) \\ 0 = 0 & (3) \end{cases}$$

The original system is equivalent to a system containing two equations, so the equations are dependent and the system has infinitely many solutions. If we solve equation (2) for y, we can express y in terms of z as $y = -3z + 7$. Substitute this expression into equation (1) to determine x in terms of z.

$$\begin{aligned} x - 2y - z &= 8 && (1) \\ x - 2(-3z + 7) - z &= 8 && \text{Substitute } y = -3z + 7 \text{ in (1).} \\ x + 6z - 14 - z &= 8 && \text{Remove parentheses.} \\ x + 5z &= 22 && \text{Combine like terms.} \\ x &= -5z + 22 && \text{Solve for } x. \end{aligned}$$

We will write the solution to the system as

$$\begin{cases} x = -5z + 22 \\ y = -3z + 7 \end{cases}$$

where z can be any real number.

This way of writing the solution makes it easier to find specific solutions of the system. To find specific solutions, choose any value of z and use the equations $x = -5z + 22$ and $y = -3z + 7$ to determine x and y. For example, if $z = 0$, then $x = 22$ and $y = 7$, and if $z = 1$, then $x = 17$ and $y = 4$. ◀

NOW WORK PROBLEM 45.

Two points in the Cartesian plane determine a unique line. Given three noncollinear points, we can find the (unique) quadratic function whose graph contains these three points.

EXAMPLE 12 Curve Fitting

Find real numbers a, b, and c so that the graph of the quadratic function $y = ax^2 + bx + c$ contains the points $(-1, -4)$, $(1, 6)$, and $(3, 0)$.

Solution We require that the three points satisfy the equation $y = ax^2 + bx + c$.

For the point $(-1, -4)$ we have:	$-4 = a(-1)^2 + b(-1) + c$	$-4 = a - b + c$
For the point $(1, 6)$ we have:	$6 = a(1)^2 + b(1) + c$	$6 = a + b + c$
For the point $(3, 0)$ we have:	$0 = a(3)^2 + b(3) + c$	$0 = 9a + 3b + c$

We wish to determine a, b, and c so that each equation is satisfied. That is, we want to solve the following system of three equations containing three variables:

$$\begin{cases} a - b + c = -4 & (1) \\ a + b + c = 6 & (2) \\ 9a + 3b + c = 0 & (3) \end{cases}$$

Figure 6

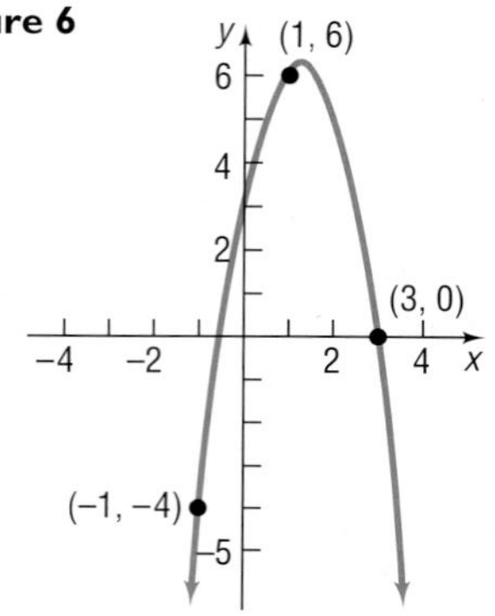

Solving this system of equations, we obtain $a = -2$, $b = 5$, and $c = 3$. So the quadratic function whose graph contains the points $(-1, -4)$, $(1, 6)$, and $(3, 0)$ is

$$y = -2x^2 + 5x + 3 \qquad y = ax^2 + bx + c, \quad a = -2, b = 5, c = 3$$

Figure 6 shows the graph of the function along with the three points. ◀

10.1 Assess Your Understanding

'Are You Prepared?' *Answers are given at the end of these exercises. If you get a wrong answer, read the pages listed in red.*

1. Solve the equation: $3x + 4 = 8 - x$. (pp. 936–938)

2. (a) Graph the line: $3x + 4y = 12$. (pp. 22–30)
(b) What is the slope of a line parallel to this line? (pp. 30–32)

Concepts and Vocabulary

3. If a system of equations has no solution, it is said to be ________.

4. If a system of equations has one or more solutions, the system is said to be ________.

5. *True or False:* A system of two linear equations containing two variables always has at least one solution.

6. *True or False:* A solution of a system of equations consists of values for the variables that are solutions of each equation of the system.

Exercises

In Problems 7–16, verify that the values of the variables listed are solutions of the system of equations.

7. $\begin{cases} 2x - y = 5 \\ 5x + 2y = 8 \end{cases}$ $x = 2, y = -1$

8. $\begin{cases} 3x + 2y = 2 \\ x - 7y = -30 \end{cases}$ $x = -2, y = 4$

9. $\begin{cases} 3x - 4y = 4 \\ \frac{1}{2}x - 3y = -\frac{1}{2} \end{cases}$ $x = 2, y = \frac{1}{2}$

10. $\begin{cases} 2x + \frac{1}{2}y = 0 \\ 3x - 4y = -\frac{19}{2} \end{cases}$ $x = -\frac{1}{2}, y = 2$

11. $\begin{cases} x - y = 3 \\ \frac{1}{2}x + y = 3 \end{cases}$ $x = 4, y = 1$

12. $\begin{cases} x - y = 3 \\ -3x + y = 1 \end{cases}$ $x = -2, y = -5$

13. $\begin{cases} 3x + 3y + 2z = 4 \\ x - y - z = 0 \\ 2y - 3z = -8 \end{cases}$ $x = 1, y = -1, z = 2$

14. $\begin{cases} 4x - z = 7 \\ 8x + 5y - z = 0 \\ -x - y + 5z = 6 \end{cases}$ $x = 2, y = -3, z = 1$

15. $\begin{cases} 3x + 3y + 2z = 4 \\ x - 3y + z = 10 \\ 5x - 2y - 3z = 8 \end{cases}$ $x = 2, y = -2, z = 2$

16. $\begin{cases} 4x - 5z = 6 \\ 5y - z = -17 \\ -x - 6y + 5z = 24 \end{cases}$ $x = 4, y = -3, z = 2$

In Problems 17–54, solve each system of equations. If the system has no solution, say that it is inconsistent.

17. $\begin{cases} x + y = 8 \\ x - y = 4 \end{cases}$

18. $\begin{cases} x + 2y = 5 \\ x + y = 3 \end{cases}$

19. $\begin{cases} 5x - y = 13 \\ 2x + 3y = 12 \end{cases}$

20. $\begin{cases} x + 3y = 5 \\ 2x - 3y = -8 \end{cases}$

21. $\begin{cases} 3x = 24 \\ x + 2y = 0 \end{cases}$

22. $\begin{cases} 4x + 5y = -3 \\ -2y = -4 \end{cases}$

23. $\begin{cases} 3x - 6y = 2 \\ 5x + 4y = 1 \end{cases}$

24. $\begin{cases} 2x + 4y = \frac{2}{3} \\ 3x - 5y = -10 \end{cases}$

25. $\begin{cases} 2x + y = 1 \\ 4x + 2y = 3 \end{cases}$

26. $\begin{cases} x - y = 5 \\ -3x + 3y = 2 \end{cases}$

27. $\begin{cases} 2x - y = 0 \\ 3x + 2y = 7 \end{cases}$

28. $\begin{cases} 3x + 3y = -1 \\ 4x + y = \frac{8}{3} \end{cases}$

29. $\begin{cases} x + 2y = 4 \\ 2x + 4y = 8 \end{cases}$

30. $\begin{cases} 3x - y = 7 \\ 9x - 3y = 21 \end{cases}$

31. $\begin{cases} 2x - 3y = -1 \\ 10x + y = 11 \end{cases}$

32. $\begin{cases} 3x - 2y = 0 \\ 5x + 10y = 4 \end{cases}$

33. $\begin{cases} 2x + 3y = 6 \\ x - y = \frac{1}{2} \end{cases}$

34. $\begin{cases} \frac{1}{2}x + y = -2 \\ x - 2y = 8 \end{cases}$

35. $\begin{cases} \frac{1}{2}x + \frac{1}{3}y = 3 \\ \frac{1}{4}x - \frac{2}{3}y = -1 \end{cases}$

36. $\begin{cases} \frac{1}{3}x - \frac{3}{2}y = -5 \\ \frac{3}{4}x + \frac{1}{3}y = 11 \end{cases}$

37. $\begin{cases} 3x - 5y = 3 \\ 15x + 5y = 21 \end{cases}$

38. $\begin{cases} 2x - y = -1 \\ x + \frac{1}{2}y = \frac{3}{2} \end{cases}$

39. $\begin{cases} \frac{1}{x} + \frac{1}{y} = 8 \\ \frac{3}{x} - \frac{5}{y} = 0 \end{cases}$

40. $\begin{cases} \frac{4}{x} - \frac{3}{y} = 0 \\ \frac{6}{x} + \frac{3}{2y} = 2 \end{cases}$

[**Hint:** Let $u = \frac{1}{x}$ and $v = \frac{1}{y}$, and solve for u and v. Then $x = \frac{1}{u}$ and $y = \frac{1}{v}$.]

41. $\begin{cases} x - y = 6 \\ 2x - 3z = 16 \\ 2y + z = 4 \end{cases}$

42. $\begin{cases} 2x + y = -4 \\ -2y + 4z = 0 \\ 3x - 2z = -11 \end{cases}$

43. $\begin{cases} x - 2y + 3z = 7 \\ 2x + y + z = 4 \\ -3x + 2y - 2z = -10 \end{cases}$

44. $\begin{cases} 2x + y - 3z = 0 \\ -2x + 2y + z = -7 \\ 3x - 4y - 3z = 7 \end{cases}$

45. $\begin{cases} x - y - z = 1 \\ 2x + 3y + z = 2 \\ 3x + 2y = 0 \end{cases}$

46. $\begin{cases} 2x - 3y - z = 0 \\ -x + 2y + z = 5 \\ 3x - 4y - z = 1 \end{cases}$

47. $\begin{cases} x - y - z = 1 \\ -x + 2y - 3z = -4 \\ 3x - 2y - 7z = 0 \end{cases}$

48. $\begin{cases} 2x - 3y - z = 0 \\ 3x + 2y + 2z = 2 \\ x + 5y + 3z = 2 \end{cases}$

49. $\begin{cases} 2x - 2y + 3z = 6 \\ 4x - 3y + 2z = 0 \\ -2x + 3y - 7z = 1 \end{cases}$

50. $\begin{cases} 3x - 2y + 2z = 6 \\ 7x - 3y + 2z = -1 \\ 2x - 3y + 4z = 0 \end{cases}$

51. $\begin{cases} x + y - z = 6 \\ 3x - 2y + z = -5 \\ x + 3y - 2z = 14 \end{cases}$

52. $\begin{cases} x - y + z = -4 \\ 2x - 3y + 4z = -15 \\ 5x + y - 2z = 12 \end{cases}$

53. $\begin{cases} x + 2y - z = -3 \\ 2x - 4y + z = -7 \\ -2x + 2y - 3z = 4 \end{cases}$

54. $\begin{cases} x + 4y - 3z = -8 \\ 3x - y + 3z = 12 \\ x + y + 6z = 1 \end{cases}$

55. The perimeter of a rectangular floor is 90 feet. Find the dimensions of the floor if the length is twice the width.

56. The length of fence required to enclose a rectangular field is 3000 meters. What are the dimensions of the field if it is known that the difference between its length and width is 50 meters?

57. **Cost of Fast Food** Four large cheeseburgers and two chocolate shakes cost a total of \$7.90. Two shakes cost 15¢ more than one cheeseburger. What is the cost of a cheeseburger? A shake?

58. **Movie Theater Tickets** A movie theater charges \$9.00 for adults and \$7.00 for senior citizens. On a day when 325 people paid an admission, the total receipts were \$2495. How many who paid were adults? How many were seniors?

59. **Mixing Nuts** A store sells cashews for \$5.00 per pound and peanuts for \$1.50 per pound. The manager decides to mix 30 pounds of peanuts with some cashews and sell the mixture for \$3.00 per pound. How many pounds of cashews should be mixed with the peanuts so that the mixture will produce the same revenue as would selling the nuts separately?

60. **Financial Planning** A recently retired couple need \$12,000 per year to supplement their Social Security. They have \$150,000 to invest to obtain this income. They have decided on two investment options: AA bonds yielding 10% per annum and a Bank Certificate yielding 5%.
(a) How much should be invested in each to realize exactly \$12,000?
(b) If, after two years, the couple requires \$14,000 per year in income, how should they reallocate their investment to achieve the new amount?

61. Computing Wind Speed With a tail wind, a small Piper aircraft can fly 600 miles in 3 hours. Against this same wind, the Piper can fly the same distance in 4 hours. Find the average wind speed and the average airspeed of the Piper.

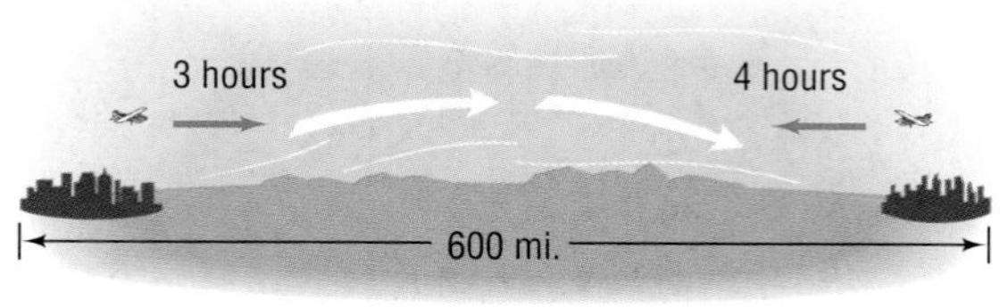

62. Computing Wind Speed The average airspeed of a single-engine aircraft is 150 miles per hour. If the aircraft flew the same distance in 2 hours with the wind as it flew in 3 hours against the wind, what was the wind speed?

63. Restaurant Management A restaurant manager wants to purchase 200 sets of dishes. One design costs \$25 per set, while another costs \$45 per set. If she only has \$7400 to spend, how many of each design should be ordered?

64. Cost of Fast Food One group of people purchased 10 hot dogs and 5 soft drinks at a cost of \$12.50. A second bought 7 hot dogs and 4 soft drinks at a cost of \$9.00. What is the cost of a single hot dog? A single soft drink?

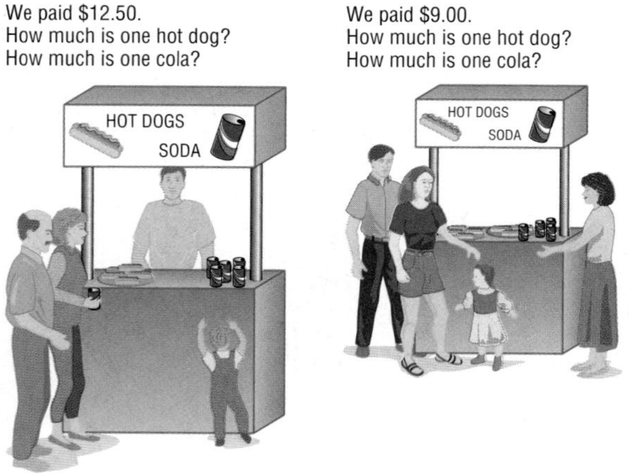

We paid $12.50.
How much is one hot dog?
How much is one cola?
HOT DOGS
SODA
We paid $9.00.
How much is one hot dog?
How much is one cola?
HOT DOGS
SODA

65. Computing a Refund The grocery store we use does not mark prices on its goods. My wife went to this store, bought three 1-pound packages of bacon and two cartons of eggs, and paid a total of \$7.45. Not knowing that she went to the store, I also went to the same store, purchased two 1-pound packages of bacon and three cartons of eggs, and paid a total of \$6.45. Now we want to return two 1-pound packages of bacon and two cartons of eggs. How much will be refunded?

66. Finding the Current of a Stream Pamela requires 3 hours to swim 15 miles downstream on the Illinois River. The return trip upstream takes 5 hours. Find Pamela's average speed in still water. How fast is the current? (Assume that Pamela's speed is the same in each direction.)

67. Pharmacy A doctor's prescription calls for a daily intake containing 40 mg of vitamin C and 30 mg of vitamin D. Your pharmacy stocks two compounds that can be used: one contains 20% vitamin C and 30% vitamin D, the other 40% vitamin C and 20% vitamin D. How many milligrams of each compound should be mixed to fill the prescription?

68. Pharmacy A doctor's prescription calls for the creation of pills that contain 12 units of vitamin B_{12} and 12 units of vitamin E. Your pharmacy stocks two powders that can be used to make these pills: one contains 20% vitamin B_{12} and 30% vitamin E, the other 40% vitamin B_{12} and 20% vitamin E. How many units of each powder should be mixed in each pill?

69. Curve Fitting Find real numbers a, b, and c so that the graph of the function $y = ax^2 + bx + c$ contains the points $(-1, 4)$, $(2, 3)$, and $(0, 1)$.

70. Curve Fitting Find real numbers a, b, and c so that the graph of the function $y = ax^2 + bx + c$ contains the points $(-1, -2)$, $(1, -4)$, and $(2, 4)$.

71. Electricity: Kirchhoff's Rules An application of Kirchhoff's Rules to the circuit shown results in the following system of equations:

$$\begin{cases} I_2 = I_1 + I_3 \\ 5 - 3I_1 - 5I_2 = 0 \\ 10 - 5I_2 - 7I_3 = 0 \end{cases}$$

Find the currents I_1, I_2, and I_3.*

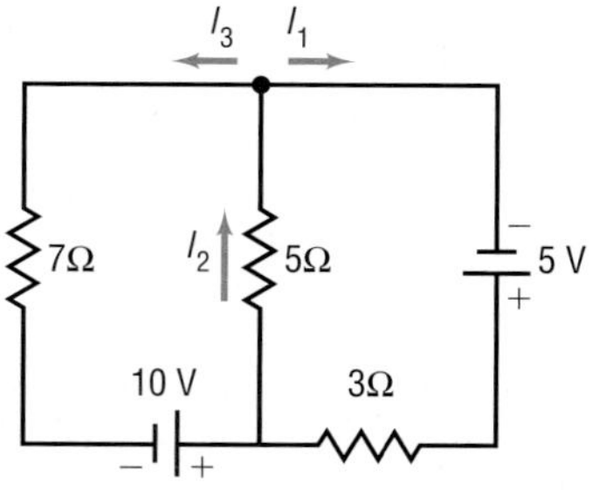

72. Electricity: Kirchhoff's Rules An application of Kirchhoff's Rules to the circuit shown results in the following system of equations:

$$\begin{cases} I_3 = I_1 + I_2 \\ 8 = 4I_3 + 6I_2 \\ 8I_1 = 4 + 6I_2 \end{cases}$$

*Source: Based on Raymond Serway, *Physics*, 3rd ed. (Philadelphia: Saunders, 1990), Prob. 26, p. 790.

Find the currents I_1, I_2, and I_3.*

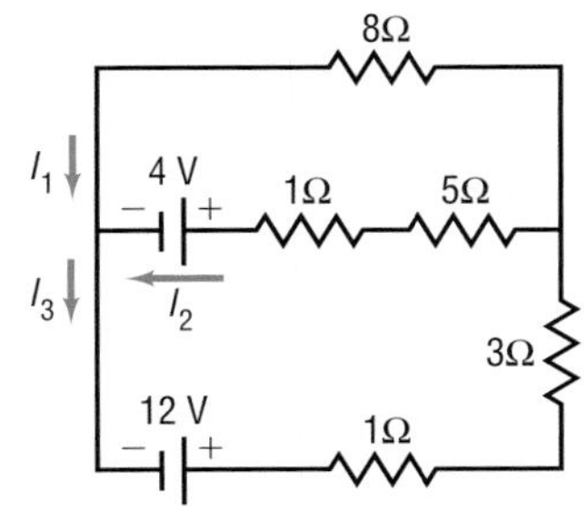

73. Theater Revenues A Broadway theater has 500 seats, divided into orchestra, main, and balcony seating. Orchestra seats sell for \$50, main seats for \$35, and balcony seats for \$25. If all the seats are sold, the gross revenue to the theater is \$17,100. If all the main and balcony seats are sold, but only half the orchestra seats are sold, the gross revenue is \$14,600. How many are there of each kind of seat?

74. Theater Revenues A movie theater charges \$8.00 for adults, \$4.50 for children, and \$6.00 for senior citizens. One day the theater sold 405 tickets and collected \$2320 in receipts. There were twice as many children's tickets sold as adult tickets. How many adults, children, and senior citizens went to the theater that day?

75. Nutrition A dietitian wishes a patient to have a meal that has 66 grams of protein, 94.5 grams of carbohydrates, and 910 milligrams of calcium. The hospital food service tells the dietitian that the dinner for today is chicken, corn, and 2% milk. Each serving of chicken has 30 grams of protein, 35 grams of carbohydrates, and 200 milligrams of calcium. Each serving of corn has 3 grams of protein, 16 grams of carbohydrates, and 10 milligrams of calcium. Each glass of 2% milk has 9 grams of protein, 13 grams of carbohydrates, and 300 milligrams of calcium. How many servings of each food should the dietitian provide for the patient?

76. Investments Kelly has \$20,000 to invest. As her financial planner, you recommend that she diversify into three investments: Treasury bills that yield 5% simple interest, Treasury bonds that yield 7% simple interest, and corporate bonds that yield 10% simple interest. Kelly wishes to earn \$1390 per year in income. Also, Kelly wants her investment in Treasury bills to be \$3000 more than her investment in corporate bonds. How much money should Kelly place in each investment?

77. Prices of Fast Food One group of customers bought 8 deluxe hamburgers, 6 orders of large fries, and 6 large colas for \$26.10. A second group ordered 10 deluxe hamburgers, 6 large fries, and 8 large colas and paid \$31.60. Is there sufficient information to determine the price of each food item? If not, construct a table showing the various possibilities. Assume that the hamburgers cost between \$1.75 and \$2.25, the fries between \$0.75 and \$1.00, and the colas between \$0.60 and \$0.90.

78. Prices of Fast Food Use the information given in Problem 77 and suppose that a third group purchased 3 deluxe hamburgers, 2 large fries, and 4 large colas for \$10.95. Now is there sufficient information to determine the price of each food item? If so, determine each price.

79. Painting a House Three painters, Beth, Bill, and Edie, working together, can paint the exterior of a home in 10 hours. Bill and Edie together have painted a similar house in 15 hours. One day, all three worked on this same kind of house for 4 hours, after which Edie left. Beth and Bill required 8 more hours to finish. Assuming no gain or loss in efficiency, how long should it take each person to complete such a job alone?

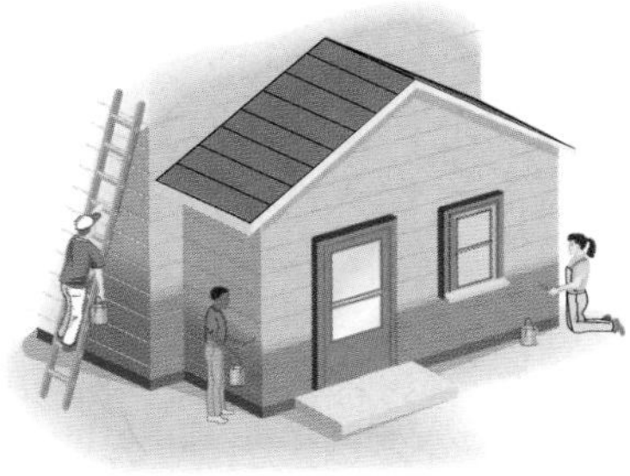

80. Create a system of two linear equations containing two variables that has
(a) No solution
(b) Exactly one solution
(c) Infinitely many solutions
Give the three systems to a friend to solve and critique.

81. Write a brief paragraph outlining your strategy for solving a system of two linear equations containing two variables.

82. Do you prefer the method of substitution or the method of elimination for solving a system of two linear equations containing two variables? Give reasons.

'Are You Prepared?' Answers

1. $\{1\}$

2. (a)

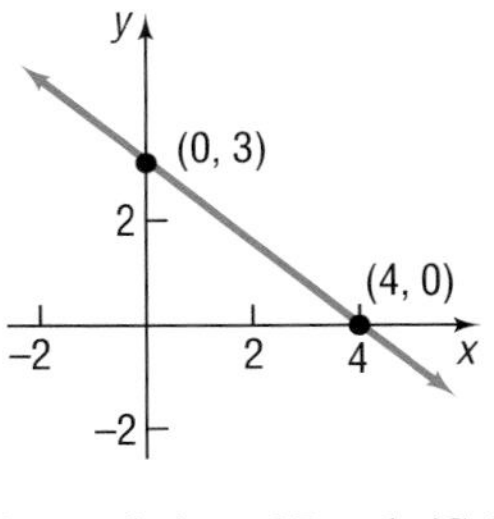

(b) $-\frac{3}{4}$

*SOURCE: *Physics for Scientists & Engineers* 3/E by Serway. ©1990. Reprinted with permission of Brooks/Cole, a division of Thomson Learning: www.thomsonrights.com. Fax 800-730-2215.

10.2 Systems of Linear Equations: Matrices

OBJECTIVES
1 Write the Augmented Matrix of a System of Linear Equations
2 Write the System from the Augmented Matrix
3 Perform Row Operations on a Matrix
4 Solve Systems of Linear Equations Using Matrices

The systematic approach of the method of elimination for solving a system of linear equations provides another method of solution that involves a simplified notation.

Consider the following system of linear equations:

$$\begin{cases} x + 4y = 14 \\ 3x - 2y = 0 \end{cases}$$

If we choose not to write the symbols used for the variables, we can represent this system as

$$\left[\begin{array}{cc|c} 1 & 4 & 14 \\ 3 & -2 & 0 \end{array}\right]$$

where it is understood that the first column represents the coefficients of the variable x, the second column the coefficients of y, and the third column the constants on the right side of the equal signs. The vertical line serves as a reminder of the equal signs. The large square brackets are the traditional symbols used to denote a *matrix* in algebra.

A **matrix** is defined as a rectangular array of numbers,

$$\begin{array}{l} \\ \text{Row } 1 \\ \text{Row } 2 \\ \vdots \\ \text{Row } i \\ \vdots \\ \text{Row } m \end{array} \begin{array}{c} \begin{array}{ccccccc} \text{Column 1} & \text{Column 2} & & \text{Column } j & & \text{Column } n \end{array} \\ \begin{bmatrix} a_{11} & a_{12} & \cdots & a_{1j} & \cdots & a_{1n} \\ a_{21} & a_{22} & \cdots & a_{2j} & \cdots & a_{2n} \\ \vdots & \vdots & & \vdots & & \vdots \\ a_{i1} & a_{i2} & \cdots & a_{ij} & \cdots & a_{in} \\ \vdots & \vdots & & \vdots & & \vdots \\ a_{m1} & a_{m2} & \cdots & a_{mj} & \cdots & a_{mn} \end{bmatrix} \end{array} \quad (1)$$

Each number a_{ij} of the matrix has two indexes: the **row index** i and the **column index** j. The matrix shown in display (1) has m rows and n columns. The numbers a_{ij} are usually referred to as the **entries** of the matrix. For example, a_{23} refers to the entry in the second row, third column.

1 Now we will use matrix notation to represent a system of linear equations. The matrices used to represent systems of linear equations are called **augmented matrices.** In writing the augmented matrix of a system, the variables of each equation must be on the left side of the equal sign and the constants on the right side. A variable that does not appear in an equation has a coefficient of 0.

EXAMPLE 1 Writing the Augmented Matrix of a System of Linear Equations

Write the augmented matrix of each system of equations.

(a) $\begin{cases} 3x - 4y = -6 & (1) \\ 2x - 3y = -5 & (2) \end{cases}$ (b) $\begin{cases} 2x - y + z = 0 & (1) \\ x + z - 1 = 0 & (2) \\ x + 2y - 8 = 0 & (3) \end{cases}$

Solution (a) The augmented matrix is

$$\left[\begin{array}{cc|c} 3 & -4 & -6 \\ 2 & -3 & -5 \end{array}\right]$$

(b) Care must be taken that the system be written so that the coefficients of all variables are present (if any variable is missing, its coefficient is 0). Also, all constants must be to the right of the equal sign. We need to rearrange the given system as follows:

$$\begin{cases} 2x - y + z = 0 & (1) \\ x + z - 1 = 0 & (2) \\ x + 2y - 8 = 0 & (3) \end{cases}$$

$$\begin{cases} 2x - y + z = 0 & (1) \\ x + 0 \cdot y + z = 1 & (2) \\ x + 2y + 0 \cdot z = 8 & (3) \end{cases}$$

The augmented matrix is

$$\left[\begin{array}{ccc|c} 2 & -1 & 1 & 0 \\ 1 & 0 & 1 & 1 \\ 1 & 2 & 0 & 8 \end{array}\right]$$

◀

If we do not include the constants to the right of the equal sign, that is, to the right of the vertical bar in the augmented matrix of a system of equations, the resulting matrix is called the **coefficient matrix** of the system. For the systems discussed in Example 1, the coefficient matrices are

$$\begin{bmatrix} 3 & -4 \\ 2 & -3 \end{bmatrix} \text{ and } \begin{bmatrix} 2 & -1 & 1 \\ 1 & 0 & 1 \\ 1 & 2 & 0 \end{bmatrix}$$

NOW WORK PROBLEM 7.

2

EXAMPLE 2 Writing the System of Linear Equations from the Augmented Matrix

Write the system of linear equations corresponding to each augmented matrix.

(a) $\left[\begin{array}{cc|c} 5 & 2 & 13 \\ -3 & 1 & -10 \end{array}\right]$ (b) $\left[\begin{array}{ccc|c} 3 & -1 & -1 & 7 \\ 2 & 0 & 2 & 8 \\ 0 & 1 & 1 & 0 \end{array}\right]$

Solution (a) The matrix has two rows and so represents a system of two equations. The two columns to the left of the vertical bar indicate that the system has two variables. If x and y are used to denote these variables, the system of equations is

$$\begin{cases} 5x + 2y = 13 & (1) \\ -3x + y = -10 & (2) \end{cases}$$

(b) Since the augmented matrix has three rows, it represents a system of three equations. Since there are three columns to the left of the vertical bar, the system contains three variables. If x, y, and z are the three variables, the system of equations is

$$\begin{cases} 3x - y - z = 7 & (1) \\ 2x + 2z = 8 & (2) \\ y + z = 0 & (3) \end{cases}$$

◀

Row Operations on a Matrix

3 **Row operations** on a matrix are used to solve systems of equations when the system is written as an augmented matrix. There are three basic row operations.

> **Row Operations**
>
> 1. Interchange any two rows.
> 2. Replace a row by a nonzero constant multiple of that row.
> 3. Replace a row by the sum of that row and a nonzero constant multiple of some other row.

These three row operations correspond to the three rules given earlier for obtaining an equivalent system of equations. When a row operation is performed on a matrix, the resulting matrix represents a system of equations equivalent to the system represented by the original matrix.

For example, consider the augmented matrix

$$\left[\begin{array}{cc|c} 1 & 2 & 3 \\ 4 & -1 & 2 \end{array}\right]$$

Suppose that we want to apply a row operation to this matrix that results in a matrix whose entry in row 2, column 1 is a 0. The row operation to use is

Multiply each entry in row 1 by -4 and add the result to the corresponding entries in row 2. **(2)**

If we use R_2 to represent the new entries in row 2 and we use r_1 and r_2 to represent the original entries in rows 1 and 2, respectively, then we can represent the row operation in statement (2) by

$$R_2 = -4r_1 + r_2$$

Then

$$\left[\begin{array}{cc|c} 1 & 2 & 3 \\ 4 & -1 & 2 \end{array}\right] \underset{R_2 = -4r_1 + r_2}{\rightarrow} \left[\begin{array}{cc|c} 1 & 2 & 3 \\ -4(1) + 4 & -4(2) + (-1) & -4(3) + 2 \end{array}\right] = \left[\begin{array}{cc|c} 1 & 2 & 3 \\ 0 & -9 & -10 \end{array}\right]$$

As desired, we now have the entry 0 in row 2, column 1.

EXAMPLE 3 Applying a Row Operation to an Augmented Matrix

Apply the row operation $R_2 = -3r_1 + r_2$ to the augmented matrix

$$\left[\begin{array}{cc|c} 1 & -2 & 2 \\ 3 & -5 & 9 \end{array}\right]$$

Solution The row operation $R_2 = -3r_1 + r_2$ tells us that the entries in row 2 are to be replaced by the entries obtained after multiplying each entry in row 1 by -3 and adding the result to the corresponding entries in row 2.

$$\left[\begin{array}{cc|c} 1 & -2 & 2 \\ 3 & -5 & 9 \end{array}\right] \underset{R_2 = -3r_1 + r_2}{\rightarrow} \left[\begin{array}{cc|c} 1 & -2 & 2 \\ -3(1) + 3 & (-3)(-2) + (-5) & -3(2) + 9 \end{array}\right] = \left[\begin{array}{cc|c} 1 & -2 & 2 \\ 0 & 1 & 3 \end{array}\right]$$

◀

NOW WORK PROBLEM 17.

EXAMPLE 4 Finding a Particular Row Operation

Using the matrix

$$\left[\begin{array}{cc|c} 1 & -2 & 2 \\ 0 & 1 & 3 \end{array}\right]$$

find a row operation that will result in this matrix having a 0 in row 1, column 2.

Solution We want a 0 in row 1, column 2. This result can be accomplished by multiplying row 2 by 2 and adding the result to row 1. That is, we apply the row operation $R_1 = 2r_2 + r_1$.

$$\left[\begin{array}{cc|c} 1 & -2 & 2 \\ 0 & 1 & 3 \end{array}\right] \underset{R_1 = 2r_2 + r_1}{\rightarrow} \left[\begin{array}{cc|c} 2(0) + 1 & 2(1) + (-2) & 2(3) + 2 \\ 0 & 1 & 3 \end{array}\right] = \left[\begin{array}{cc|c} 1 & 0 & 8 \\ 0 & 1 & 3 \end{array}\right]$$

◀

A word about the notation that we have introduced. A row operation such as $R_1 = 2r_2 + r_1$ changes the entries in row 1. Note also that for this type of row operation we change the entries in a given row by multiplying the entries in some other row by an appropriate nonzero number and adding the results to the original entries of the row to be changed.

Solving a System of Linear Equations Using Matrices

4 To solve a system of linear equations using matrices, we use row operations on the augmented matrix of the system to obtain a matrix that is in *row echelon form*.

A matrix is in **row echelon form** when

1. The entry in row 1, column 1 is a 1, and 0's appear below it.
2. The first nonzero entry in each row after the first row is a 1, 0's appear below it, and it appears to the right of the first nonzero entry in any row above.
3. Any rows that contain all 0's to the left of the vertical bar appear at the bottom.

For example, for a system of three equations containing three variables with a unique solution, the augmented matrix is in row echelon form if it is of the form

$$\left[\begin{array}{ccc|c} 1 & a & b & d \\ 0 & 1 & c & e \\ 0 & 0 & 1 & f \end{array}\right]$$

where a, b, c, d, e, and f are real numbers. The last row of the augmented matrix states that $z = f$. We can then determine the value of y using back-substitution with $z = f$, since row 2 represents the equation $y + cz = e$. Finally, x is determined using back-substitution again.

Two advantages of solving a system of equations by writing the augmented matrix in row echelon form are the following:

1. The process is algorithmic; that is, it consists of repetitive steps that can be programmed on a computer.
2. The process works on any system of linear equations, no matter how many equations or variables are present.

The next example shows how to write a matrix in row echelon form.

EXAMPLE 5 **Solving a System of Linear Equations Using Matrices (Row Echelon Form)**

Solve: $\begin{cases} 2x + 2y = 6 & (1) \\ x + y + z = 1 & (2) \\ 3x + 4y - z = 13 & (3) \end{cases}$

Solution First, we write the augmented matrix that represents this system.

$$\left[\begin{array}{ccc|c} 2 & 2 & 0 & 6 \\ 1 & 1 & 1 & 1 \\ 3 & 4 & -1 & 13 \end{array}\right]$$

The first step requires getting the entry 1 in row 1, column 1. An interchange of rows 1 and 2 is the easiest way to do this. [Note that this is equivalent to interchanging equations (1) and (2) of the system.]

$$\left[\begin{array}{ccc|c} 1 & 1 & 1 & 1 \\ 2 & 2 & 0 & 6 \\ 3 & 4 & -1 & 13 \end{array}\right]$$

Next, we want a 0 in row 2, column 1 and a 0 in row 3, column 1. We use the row operations $R_2 = -2r_1 + r_2$ and $R_3 = -3r_1 + r_3$ to accomplish this. Notice that row 1 is unchanged using these row operations. Also, do you see that performing these row operations simultaneously is the same as doing one followed by the other?

$$\left[\begin{array}{ccc|c} 1 & 1 & 1 & 1 \\ 2 & 2 & 0 & 6 \\ 3 & 4 & -1 & 13 \end{array}\right] \underset{\substack{R_2 = -2r_1 + r_2 \\ R_3 = -3r_1 + r_3}}{\rightarrow} \left[\begin{array}{ccc|c} 1 & 1 & 1 & 1 \\ 0 & 0 & -2 & 4 \\ 0 & 1 & -4 & 10 \end{array}\right]$$

Now we want the entry 1 in row 2, column 2. Interchanging rows 2 and 3 will accomplish this.

$$\left[\begin{array}{ccc|c} 1 & 1 & 1 & 1 \\ 0 & 0 & -2 & 4 \\ 0 & 1 & -4 & 10 \end{array}\right] \rightarrow \left[\begin{array}{ccc|c} 1 & 1 & 1 & 1 \\ 0 & 1 & -4 & 10 \\ 0 & 0 & -2 & 4 \end{array}\right]$$

Finally, we want a 1 in row 3, column 3. To obtain it, we use the row operation $R_3 = -\frac{1}{2}r_3$. The result is

$$\left[\begin{array}{ccc|c} 1 & 1 & 1 & 1 \\ 0 & 1 & -4 & 10 \\ 0 & 0 & -2 & 4 \end{array}\right] \underset{R_3 = -\frac{1}{2}r_3}{\rightarrow} \left[\begin{array}{ccc|c} 1 & 1 & 1 & 1 \\ 0 & 1 & -4 & 10 \\ 0 & 0 & 1 & -2 \end{array}\right]$$

This matrix is the row echelon form of the augmented matrix. The third row of this matrix represents the equation $z = -2$. Using $z = -2$, we back-substitute into the equation $y - 4z = 10$ (from the second row) and obtain

$$\begin{aligned} y - 4z &= 10 \\ y - 4(-2) &= 10 \qquad z = -2 \\ y &= 2 \qquad \text{Solve for } y. \end{aligned}$$

Finally, we back-substitute $y = 2$ and $z = -2$ into the equation $x + y + z = 1$ (from the first row) and obtain

$$\begin{aligned} x + y + z &= 1 \\ x + 2 + (-2) &= 1 \qquad y = 2, z = -2 \\ x &= 1 \qquad \text{Solve for } x. \end{aligned}$$

The solution of the system is $x = 1$, $y = 2$, $z = -2$. ◀

The steps that we used to solve the system of linear equations in Example 5 can be summarized as follows:

Matrix Method for Solving a System of Linear Equations (Row Echelon Form)

STEP 1: Write the augmented matrix that represents the system.

STEP 2: Perform row operations that place the entry 1 in row 1, column 1.

STEP 3: Perform row operations that leave the entry 1 in row 1, column 1 unchanged, while causing 0's to appear below it in column 1.

STEP 4: Perform row operations that place the entry 1 in row 2, column 2, but leave the entries in columns to the left unchanged. If it is impossible to place a 1 in row 2, column 2, then proceed to place a 1 in row 2, column 3. Once a 1 is in place, perform row operations to place 0's under it.
[If any rows are obtained that contain only 0's on the left side of the vertical bar, place such rows at the bottom of the matrix.]

STEP 5: Now repeat Step 4, placing a 1 in the next row, but one column to the right. Continue until the bottom row or the vertical bar is reached.

STEP 6: The matrix that results is the row echelon form of the augmented matrix. Analyze the system of equations corresponding to it to solve the original system.

EXAMPLE 6 Solving a System of Linear Equations Using Matrices (Row Echelon Form)

Solve: $\begin{cases} x - y + z = 8 & (1) \\ 2x + 3y - z = -2 & (2) \\ 3x - 2y - 9z = 9 & (3) \end{cases}$

Solution **STEP 1:** The augmented matrix of the system is

$$\left[\begin{array}{ccc|c} 1 & -1 & 1 & 8 \\ 2 & 3 & -1 & -2 \\ 3 & -2 & -9 & 9 \end{array}\right]$$

STEP 2: Because the entry 1 is already present in row 1, column 1, we can go to step 3.

STEP 3: Perform the row operations $R_2 = -2r_1 + r_2$ and $R_3 = -3r_1 + r_3$. Each of these leaves the entry 1 in row 1, column 1 unchanged, while causing 0's to appear under it.

$$\left[\begin{array}{ccc|c} 1 & -1 & 1 & 8 \\ 2 & 3 & -1 & -2 \\ 3 & -2 & -9 & 9 \end{array}\right] \rightarrow \left[\begin{array}{ccc|c} 1 & -1 & 1 & 8 \\ 0 & 5 & -3 & -18 \\ 0 & 1 & -12 & -15 \end{array}\right]$$

$R_2 = -2r_1 + r_2$
$R_3 = -3r_1 + r_3$

STEP 4: The easiest way to obtain the entry 1 in row 2, column 2 without altering column 1 is to interchange rows 2 and 3 (another way would be to multiply row 2 by $\frac{1}{5}$, but this introduces fractions).

$$\left[\begin{array}{rrr|r} 1 & -1 & 1 & 8 \\ 0 & 1 & -12 & -15 \\ 0 & 5 & -3 & -18 \end{array}\right]$$

To get a 0 under the 1 in row 2, column 2, perform the row operation $R_3 = -5r_2 + r_3$.

$$\left[\begin{array}{rrr|r} 1 & -1 & 1 & 8 \\ 0 & 1 & -12 & -15 \\ 0 & 5 & -3 & -18 \end{array}\right] \underset{R_3 = -5r_2 + r_3}{\longrightarrow} \left[\begin{array}{rrr|r} 1 & -1 & 1 & 8 \\ 0 & 1 & -12 & -15 \\ 0 & 0 & 57 & 57 \end{array}\right]$$

STEP 5: Continuing, we obtain a 1 in row 3, column 3 by using $R_3 = \frac{1}{57}r_3$.

$$\left[\begin{array}{rrr|r} 1 & -1 & 1 & 8 \\ 0 & 1 & -12 & -15 \\ 0 & 0 & 57 & 57 \end{array}\right] \underset{R_3 = \frac{1}{57}r_3}{\longrightarrow} \left[\begin{array}{rrr|r} 1 & -1 & 1 & 8 \\ 0 & 1 & -12 & -15 \\ 0 & 0 & 1 & 1 \end{array}\right]$$

STEP 6: The matrix on the right is the row echelon form of the augmented matrix. The system of equations represented by the matrix in row echelon form is

$$\begin{cases} x - y + \quad z = \ \ 8 & (1) \\ \quad\ \ y - 12z = -15 & (2) \\ \qquad\qquad z = \ \ 1 & (3) \end{cases}$$

Using $z = 1$, we back-substitute to get

$$\begin{cases} x - y + 1 \qquad = \ \ 8 & (1) \\ \quad\ \ y - 12(1) = -15 & (2) \end{cases} \underset{\text{Simplify.}}{\longrightarrow} \begin{cases} x - y = \ \ 7 & (1) \\ \quad\ \ y = -3 & (2) \end{cases}$$

We get $y = -3$, and back-substituting into $x - y = 7$, we find that $x = 4$. The solution of the system is $x = 4$, $y = -3$, $z = 1$. ◀

Sometimes it is advantageous to write a matrix in **reduced row echelon form.** In this form, row operations are used to obtain entries that are 0 above (as well as below) the leading 1 in a row. For example, the row echelon form obtained in Example 6 is

$$\left[\begin{array}{rrr|r} 1 & -1 & 1 & 8 \\ 0 & 1 & -12 & -15 \\ 0 & 0 & 1 & 1 \end{array}\right]$$

To write this matrix in reduced row echelon form, we proceed as follows:

$$\left[\begin{array}{rrr|r} 1 & -1 & 1 & 8 \\ 0 & 1 & -12 & -15 \\ 0 & 0 & 1 & 1 \end{array}\right] \underset{R_1 = r_2 + r_1}{\rightarrow} \left[\begin{array}{rrr|r} 1 & 0 & -11 & -7 \\ 0 & 1 & -12 & -15 \\ 0 & 0 & 1 & 1 \end{array}\right] \underset{\substack{R_1 = 11r_3 + r_1 \\ R_2 = 12r_3 + r_2}}{\rightarrow} \left[\begin{array}{rrr|r} 1 & 0 & 0 & 4 \\ 0 & 1 & 0 & -3 \\ 0 & 0 & 1 & 1 \end{array}\right]$$

The matrix is now written in reduced row echelon form. The advantage of writing the matrix in this form is that the solution to the system, $x = 4$, $y = -3$, $z = 1$, is readily found, without the need to back-substitute. Another advantage will be seen in Section 10.4, where the inverse of a matrix is discussed.

NOW WORK PROBLEMS 37 AND 47.

The matrix method for solving a system of linear equations also identifies systems that have infinitely many solutions and systems that are inconsistent. Let's see how.

EXAMPLE 7 Solving a System of Linear Equations Using Matrices

Solve: $\begin{cases} 6x - y - z = 4 & (1) \\ -12x + 2y + 2z = -8 & (2) \\ 5x + y - z = 3 & (3) \end{cases}$

Solution We start with the augmented matrix of the system.

$$\left[\begin{array}{rrr|r} 6 & -1 & -1 & 4 \\ -12 & 2 & 2 & -8 \\ 5 & 1 & -1 & 3 \end{array}\right] \underset{R_1 = -1r_3 + r_1}{\rightarrow} \left[\begin{array}{rrr|r} 1 & -2 & 0 & 1 \\ -12 & 2 & 2 & -8 \\ 5 & 1 & -1 & 3 \end{array}\right] \underset{\substack{R_2 = 12r_1 + r_2 \\ R_3 = -5r_1 + r_3}}{\rightarrow} \left[\begin{array}{rrr|r} 1 & -2 & 0 & 1 \\ 0 & -22 & 2 & 4 \\ 0 & 11 & -1 & -2 \end{array}\right]$$

Obtaining a 1 in row 2, column 2 without altering column 1 can be accomplished only by $R_2 = -\frac{1}{22}r_2$ or by $R_3 = -\frac{1}{11}r_3$ and interchanging rows or by $R_2 = \frac{23}{11}r_3 + r_2$. We shall use the first of these.

$$\left[\begin{array}{rrr|r} 1 & -2 & 0 & 1 \\ 0 & -22 & 2 & 4 \\ 0 & 11 & -1 & -2 \end{array}\right] \underset{R_2 = -\frac{1}{22}r_2}{\rightarrow} \left[\begin{array}{rrr|r} 1 & -2 & 0 & 1 \\ 0 & 1 & -\frac{1}{11} & -\frac{2}{11} \\ 0 & 11 & -1 & -2 \end{array}\right] \underset{R_3 = -11r_2 + r_3}{\rightarrow} \left[\begin{array}{rrr|r} 1 & -2 & 0 & 1 \\ 0 & 1 & -\frac{1}{11} & -\frac{2}{11} \\ 0 & 0 & 0 & 0 \end{array}\right]$$

This matrix is in row echelon form. Because the bottom row consists entirely of 0's, the system actually consists of only two equations.

$$\begin{cases} x - 2y = 1 & (1) \\ y - \frac{1}{11}z = -\frac{2}{11} & (2) \end{cases}$$

To make it easier to write down some of the solutions, we express both x and y in terms of z. From the second equation, $y = \frac{1}{11}z - \frac{2}{11}$. Now back-substitute this solution for y into the first equation to get

$$x = 2y + 1 = 2\left(\frac{1}{11}z - \frac{2}{11}\right) + 1 = \frac{2}{11}z + \frac{7}{11}$$

The original system is equivalent to the system

$$\begin{cases} x = \frac{2}{11}z + \frac{7}{11} & (1) \\ y = \frac{1}{11}z - \frac{2}{11} & (2) \end{cases}$$

where z can be any real number.

Let's look at the situation. The original system of three equations is equivalent to a system containing two equations. This means that any values of x, y, z that satisfy both

$$x = \frac{2}{11}z + \frac{7}{11} \quad \text{and} \quad y = \frac{1}{11}z - \frac{2}{11}$$

will be solutions. For example, $z = 0, x = \frac{7}{11}, y = -\frac{2}{11}$; $z = 1, x = \frac{9}{11}, y = -\frac{1}{11}$; and $z = -1, x = \frac{5}{11}, y = -\frac{3}{11}$ are some of the solutions of the original system. There are, in fact, infinitely many values of x, y, and z for which the two equations are satisfied. That is, the original system has infinitely many solutions. We will write the solution of the original system as

$$\begin{cases} x = \frac{2}{11}z + \frac{7}{11} \\ y = \frac{1}{11}z - \frac{2}{11} \end{cases}$$

where z can be any real number. ◀

We can also find the solution by writing the augmented matrix in reduced row echelon form. Starting with the row echelon form, we have

$$\left[\begin{array}{ccc|c} 1 & -2 & 0 & 1 \\ 0 & 1 & -\frac{1}{11} & -\frac{2}{11} \\ 0 & 0 & 0 & 0 \end{array}\right] \underset{R_1 = 2r_2 + r_1}{\longrightarrow} \left[\begin{array}{ccc|c} 1 & 0 & -\frac{2}{11} & \frac{7}{11} \\ 0 & 1 & -\frac{1}{11} & -\frac{2}{11} \\ 0 & 0 & 0 & 0 \end{array}\right]$$

The matrix on the right is in reduced row echelon form. The corresponding system of equations is

$$\begin{cases} x - \frac{2}{11}z = \frac{7}{11} & (1) \\ y - \frac{1}{11}z = -\frac{2}{11} & (2) \end{cases}$$

or, equivalently,

$$\begin{cases} x = \dfrac{2}{11}z + \dfrac{7}{11} & (1) \\ y = \dfrac{1}{11}z - \dfrac{2}{11} & (2) \end{cases}$$

where z can be any real number.

NOW WORK PROBLEM 53.

EXAMPLE 8 Solving a System of Linear Equations Using Matrices

Solve: $\begin{cases} x + y + z = 6 \\ 2x - y - z = 3 \\ x + 2y + 2z = 0 \end{cases}$

Solution We proceed as follows, beginning with the augmented matrix.

$$\left[\begin{array}{ccc|c} 1 & 1 & 1 & 6 \\ 2 & -1 & -1 & 3 \\ 1 & 2 & 2 & 0 \end{array}\right] \rightarrow \left[\begin{array}{ccc|c} 1 & 1 & 1 & 6 \\ 0 & -3 & -3 & -9 \\ 0 & 1 & 1 & -6 \end{array}\right] \rightarrow \left[\begin{array}{ccc|c} 1 & 1 & 1 & 6 \\ 0 & 1 & 1 & -6 \\ 0 & -3 & -3 & -9 \end{array}\right] \rightarrow \left[\begin{array}{ccc|c} 1 & 1 & 1 & 6 \\ 0 & 1 & 1 & -6 \\ 0 & 0 & 0 & -27 \end{array}\right]$$

$R_2 = -2r_1 + r_2$, $R_3 = -1r_1 + r_3$; Interchange rows 2 and 3.; $R_3 = 3r_2 + r_3$

This matrix is in row echelon form. The bottom row is equivalent to the equation

$$0x + 0y + 0z = -27$$

which has no solution. Hence, the original system is inconsistent. ◀

NOW WORK PROBLEM 27.

The matrix method is especially effective for systems of equations for which the number of equations and the number of variables are unequal. Here, too, such a system is either inconsistent or consistent. If it is consistent, it will have either exactly one solution or infinitely many solutions.

Let's look at a system of four equations containing three variables.

EXAMPLE 9 Solving a System of Linear Equations Using Matrices

Solve: $\begin{cases} x - 2y + z = 0 & (1) \\ 2x + 2y - 3z = -3 & (2) \\ y - z = -1 & (3) \\ -x + 4y + 2z = 13 & (4) \end{cases}$

Solution We proceed as follows, beginning with the augmented matrix.

$$\left[\begin{array}{rrr|r} 1 & -2 & 1 & 0 \\ 2 & 2 & -3 & -3 \\ 0 & 1 & -1 & -1 \\ -1 & 4 & 2 & 13 \end{array}\right] \rightarrow \left[\begin{array}{rrr|r} 1 & -2 & 1 & 0 \\ 0 & 6 & -5 & -3 \\ 0 & 1 & -1 & -1 \\ 0 & 2 & 3 & 13 \end{array}\right] \rightarrow \left[\begin{array}{rrr|r} 1 & -2 & 1 & 0 \\ 0 & 1 & -1 & -1 \\ 0 & 6 & -5 & -3 \\ 0 & 2 & 3 & 13 \end{array}\right]$$

$R_2 = -2r_1 + r_2$, $R_4 = r_1 + r_4$; Interchange rows 2 and 3.

$$\rightarrow \left[\begin{array}{rrr|r} 1 & -2 & 1 & 0 \\ 0 & 1 & -1 & -1 \\ 0 & 0 & 1 & 3 \\ 0 & 0 & 5 & 15 \end{array}\right] \rightarrow \left[\begin{array}{rrr|r} 1 & -2 & 1 & 0 \\ 0 & 1 & -1 & -1 \\ 0 & 0 & 1 & 3 \\ 0 & 0 & 0 & 0 \end{array}\right]$$

$R_3 = -6r_2 + r_3$, $R_4 = -2r_2 + r_4$; $R_4 = -5r_3 + r_4$

We could stop here, since the matrix is in row echelon form, and back-substitute $z = 3$ to find x and y. Or we can continue to obtain the reduced row echelon form.

$$\rightarrow \left[\begin{array}{rrr|r} 1 & 0 & -1 & -2 \\ 0 & 1 & -1 & -1 \\ 0 & 0 & 1 & 3 \\ 0 & 0 & 0 & 0 \end{array}\right] \rightarrow \left[\begin{array}{rrr|r} 1 & 0 & 0 & 1 \\ 0 & 1 & 0 & 2 \\ 0 & 0 & 1 & 3 \\ 0 & 0 & 0 & 0 \end{array}\right]$$

$R_1 = 2r_2 + r_1$; $R_1 = r_3 + r_1$, $R_2 = r_3 + r_2$

The matrix is now in reduced row echelon form, and we can see that the solution is $x = 1$, $y = 2$, $z = 3$. ◀

NOW WORK PROBLEM 69.

EXAMPLE 10 Nutrition

A dietitian at Cook County Hospital wants a patient to have a meal that has 65 grams of protein, 95 grams of carbohydrates, and 905 milligrams of calcium. The hospital food service tells the dietitian that the dinner for today is chicken *a la king*, baked potatoes, and 2% milk. Each serving of chicken *a la king* has 30 grams of protein, 35 grams of carbohydrates, and 200 milligrams of calcium. Each serving of baked potatoes contains 4 grams of protein, 33 grams of carbohydrates, and 10 milligrams of calcium. Each glass of 2% milk contains 9 grams of protein, 13 grams of carbohydrates, and 300 milligrams of calcium. How many servings of each food should the dietitian provide for the patient?

Solution Let c, p, and m represent the number of servings of chicken *a la king*, baked potatoes, and milk, respectively. The dietitian wants the patient to have 65 grams of protein. Each serving of chicken *a la king* has 30 grams of protein, so c servings will have $30c$ grams of protein. Each serving of baked potatoes contains 4 grams of protein, so p potatoes will have $4p$ grams of protein. Finally, each glass of milk has 9 grams of protein, so m glasses of milk will have $9m$ grams of protein. The same logic will result in equations for carbohydrates and calcium, and we have the following system of equations:

$$\begin{cases} 30c + 4p + 9m = 65 & \text{protein equation} \\ 35c + 33p + 13m = 95 & \text{carbohydrate equation} \\ 200c + 10p + 300m = 905 & \text{calcium equation} \end{cases}$$

We begin with the augmented matrix and proceed as follows:

$$\left[\begin{array}{ccc|c} 30 & 4 & 9 & 65 \\ 35 & 33 & 13 & 95 \\ 200 & 10 & 300 & 905 \end{array}\right] \underset{R_1 = \left(\frac{1}{30}\right) r_1}{\longrightarrow} \left[\begin{array}{ccc|c} 1 & \frac{2}{15} & \frac{3}{10} & \frac{13}{6} \\ 35 & 33 & 13 & 95 \\ 200 & 10 & 300 & 905 \end{array}\right] \underset{\substack{R_2 = -35r_1 + r_2 \\ R_3 = -200r_1 + r_3}}{\longrightarrow} \left[\begin{array}{ccc|c} 1 & \frac{2}{15} & \frac{3}{10} & \frac{13}{6} \\ 0 & \frac{85}{3} & \frac{5}{2} & \frac{115}{6} \\ 0 & -\frac{50}{3} & 240 & \frac{1415}{3} \end{array}\right]$$

$$\underset{R_2 = \left(\frac{3}{85}\right) r_2}{\longrightarrow} \left[\begin{array}{ccc|c} 1 & \frac{2}{15} & \frac{3}{10} & \frac{13}{6} \\ 0 & 1 & \frac{3}{34} & \frac{23}{34} \\ 0 & -\frac{50}{3} & 240 & \frac{1415}{3} \end{array}\right] \underset{R_3 = \left(\frac{50}{3}\right) r_2 + r_3}{\longrightarrow} \left[\begin{array}{ccc|c} 1 & \frac{2}{15} & \frac{3}{10} & \frac{13}{6} \\ 0 & 1 & \frac{3}{34} & \frac{23}{34} \\ 0 & 0 & \frac{4105}{17} & \frac{8210}{17} \end{array}\right] \underset{R_3 = \left(\frac{17}{4105}\right) r_3}{\longrightarrow} \left[\begin{array}{ccc|c} 1 & \frac{2}{15} & \frac{3}{10} & \frac{13}{6} \\ 0 & 1 & \frac{3}{34} & \frac{23}{34} \\ 0 & 0 & 1 & 2 \end{array}\right]$$

The matrix is now in echelon form. The final matrix represents the system

$$\begin{cases} c + \frac{2}{15}p + \frac{3}{10}m = \frac{13}{6} & (1) \\ p + \frac{3}{34}m = \frac{23}{34} & (2) \\ m = 2 & (3) \end{cases}$$

From (3), we determine that 2 glasses of milk should be served. Back-substitute $m = 2$ into equation (2) to find that $p = \frac{1}{2}$, so $\frac{1}{2}$ of a baked potato should be served. Back-substitute these values into equation (1) and find that $c = 1.5$, so 1.5 servings of chicken *a la king* should be given to the patient to meet the dietary requirements. ◀

COMMENT: Most graphing utilities have the capability to put an augmented matrix into row echelon form (ref) and also reduced row echelon form (rref). See Appendix B, Section B.7, for a discussion. ■

10.2 Assess Your Understanding

Concepts and Vocabulary

1. An m by n rectangular array of numbers is called a(n) __________.

2. The matrix used to represent a system of linear equations is called a(n) __________ matrix.

3. *True or False:* The augmented matrix of a system of two equations containing three variables has two rows and four columns.

4. *True or False:* The matrix $\left[\begin{array}{cc|c} 1 & 3 & -2 \\ 0 & 1 & 5 \\ 0 & 0 & 0 \end{array}\right]$ is in row echelon form.

Exercises

In Problems 5–16, write the augmented matrix of the given system of equations.

5. $\begin{cases} x - 5y = 5 \\ 4x + 3y = 6 \end{cases}$

6. $\begin{cases} 3x + 4y = 7 \\ 4x - 2y = 5 \end{cases}$

7. $\begin{cases} 2x + 3y - 6 = 0 \\ 4x - 6y + 2 = 0 \end{cases}$

8. $\begin{cases} 9x - y = 0 \\ 3x - y - 4 = 0 \end{cases}$

9. $\begin{cases} 0.01x - 0.03y = 0.06 \\ 0.13x + 0.10y = 0.20 \end{cases}$

10. $\begin{cases} \frac{4}{3}x - \frac{3}{2}y = \frac{3}{4} \\ -\frac{1}{4}x + \frac{1}{3}y = \frac{2}{3} \end{cases}$

11. $\begin{cases} x - y + z = 10 \\ 3x + 3y = 5 \\ x + y + 2z = 2 \end{cases}$

12. $\begin{cases} 5x - y - z = 0 \\ x + y = 5 \\ 2x - 3z = 2 \end{cases}$

13. $\begin{cases} x + y - z = 2 \\ 3x - 2y = 2 \\ 5x + 3y - z = 1 \end{cases}$

14. $\begin{cases} 2x + 3y - 4z = 0 \\ x - 5z + 2 = 0 \\ x + 2y - 3z = -2 \end{cases}$

15. $\begin{cases} x - y - z = 10 \\ 2x + y + 2z = -1 \\ -3x + 4y = 5 \\ 4x - 5y + z = 0 \end{cases}$

16. $\begin{cases} x - y + 2z - w = 5 \\ x + 3y - 4z + 2w = 2 \\ 3x - y - 5z - w = -1 \end{cases}$

In Problems 17–24, perform each row operation on the given augmented matrix.

17. $\left[\begin{array}{cc|c} 1 & -3 & -2 \\ 2 & -5 & 5 \end{array}\right]$ $R_2 = -2r_1 + r_2$

18. $\left[\begin{array}{cc|c} 1 & -3 & -3 \\ 2 & -5 & -4 \end{array}\right]$ $R_2 = -2r_1 + r_2$

19. $\left[\begin{array}{ccc|c} 1 & -3 & 4 & 3 \\ 2 & -5 & 6 & 6 \\ -3 & 3 & 4 & 6 \end{array}\right]$ (a) $R_2 = -2r_1 + r_2$ (b) $R_3 = 3r_1 + r_3$

20. $\left[\begin{array}{ccc|c} 1 & -3 & 3 & -5 \\ 2 & -5 & -3 & -5 \\ -3 & -2 & 4 & 6 \end{array}\right]$ (a) $R_2 = -2r_1 + r_2$ (b) $R_3 = 3r_1 + r_3$

21. $\left[\begin{array}{ccc|c} 1 & -3 & 2 & -6 \\ 2 & -5 & 3 & -4 \\ -3 & -6 & 4 & 6 \end{array}\right]$ (a) $R_2 = -2r_1 + r_2$ (b) $R_3 = 3r_1 + r_3$

22. $\left[\begin{array}{ccc|c} 1 & -3 & -4 & -6 \\ 2 & -5 & 6 & -6 \\ -3 & 1 & 4 & 6 \end{array}\right]$ (a) $R_2 = -2r_1 + r_2$ (b) $R_3 = 3r_1 + r_3$

23. $\left[\begin{array}{ccc|c} 1 & -3 & 1 & -2 \\ 2 & -5 & 6 & -2 \\ -3 & 1 & 4 & 6 \end{array}\right]$ (a) $R_2 = -2r_1 + r_2$ (b) $R_3 = 3r_1 + r_3$

24. $\left[\begin{array}{ccc|c} 1 & -3 & -1 & 2 \\ 2 & -5 & 2 & 6 \\ -3 & -6 & 4 & 6 \end{array}\right]$ (a) $R_2 = -2r_1 + r_2$ (b) $R_3 = 3r_1 + r_3$

In Problems 25–36, the reduced row echelon form of a system of linear equations is given. Write the system of equations corresponding to the given matrix. Use x, y; or x, y, z; or x_1, x_2, x_3, x_4 as variables. Determine whether the system is consistent or inconsistent. If it is consistent, give the solution.

25. $\left[\begin{array}{cc|c} 1 & 0 & 5 \\ 0 & 1 & -1 \end{array}\right]$

26. $\left[\begin{array}{cc|c} 1 & 0 & -4 \\ 0 & 1 & 0 \end{array}\right]$

27. $\left[\begin{array}{ccc|c} 1 & 0 & 0 & 1 \\ 0 & 1 & 0 & 2 \\ 0 & 0 & 0 & 3 \end{array}\right]$

28. $\left[\begin{array}{ccc|c} 1 & 0 & 0 & 0 \\ 0 & 1 & 0 & 0 \\ 0 & 0 & 0 & 2 \end{array}\right]$

29. $\left[\begin{array}{ccc|c} 1 & 0 & 2 & -1 \\ 0 & 1 & -4 & -2 \\ 0 & 0 & 0 & 0 \end{array}\right]$

30. $\left[\begin{array}{ccc|c} 1 & 0 & 4 & 4 \\ 0 & 1 & 3 & 2 \\ 0 & 0 & 0 & 0 \end{array}\right]$

31. $\left[\begin{array}{cccc|c} 1 & 0 & 0 & 0 & 1 \\ 0 & 1 & 0 & 1 & 2 \\ 0 & 0 & 1 & 2 & 3 \end{array}\right]$

32. $\left[\begin{array}{cccc|c} 1 & 0 & 0 & 0 & 1 \\ 0 & 1 & 0 & 2 & 2 \\ 0 & 0 & 1 & 3 & 0 \end{array}\right]$

33. $\left[\begin{array}{cccc|c} 1 & 0 & 0 & 4 & 2 \\ 0 & 1 & 1 & 3 & 3 \\ 0 & 0 & 0 & 0 & 0 \end{array}\right]$

34. $\left[\begin{array}{cccc|c} 1 & 0 & 0 & 0 & 1 \\ 0 & 1 & 0 & 0 & 2 \\ 0 & 0 & 1 & 2 & 3 \end{array}\right]$

35. $\left[\begin{array}{cccc|c} 1 & 0 & 0 & 1 & -2 \\ 0 & 1 & 0 & 2 & 2 \\ 0 & 0 & 1 & -1 & 0 \\ 0 & 0 & 0 & 0 & 0 \end{array}\right]$

36. $\left[\begin{array}{cccc|c} 1 & 0 & 0 & 0 & 1 \\ 0 & 1 & 0 & 0 & 2 \\ 0 & 0 & 1 & 0 & 3 \\ 0 & 0 & 0 & 1 & 0 \end{array}\right]$

In Problems 37–72, solve each system of equations using matrices (row operations). If the system has no solution, say that it is inconsistent.

37. $\begin{cases} x + y = 8 \\ x - y = 4 \end{cases}$

38. $\begin{cases} x + 2y = 5 \\ x + y = 3 \end{cases}$

39. $\begin{cases} 2x - 4y = -2 \\ 3x + 2y = 3 \end{cases}$

40. $\begin{cases} 3x + 3y = 3 \\ 4x + 2y = \dfrac{8}{3} \end{cases}$

41. $\begin{cases} x + 2y = 4 \\ 2x + 4y = 8 \end{cases}$

42. $\begin{cases} 3x - y = 7 \\ 9x - 3y = 21 \end{cases}$

43. $\begin{cases} 2x + 3y = 6 \\ x - y = \dfrac{1}{2} \end{cases}$

44. $\begin{cases} \dfrac{1}{2}x + y = -2 \\ x - 2y = 8 \end{cases}$

45. $\begin{cases} 3x - 5y = 3 \\ 15x + 5y = 21 \end{cases}$

46. $\begin{cases} 2x - y = -1 \\ x + \dfrac{1}{2}y = \dfrac{3}{2} \end{cases}$

47. $\begin{cases} x - y = 6 \\ 2x - 3z = 16 \\ 2y + z = 4 \end{cases}$

48. $\begin{cases} 2x + y = -4 \\ -2y + 4z = 0 \\ 3x - 2z = -11 \end{cases}$

49. $\begin{cases} x - 2y + 3z = 7 \\ 2x + y + z = 4 \\ -3x + 2y - 2z = -10 \end{cases}$

50. $\begin{cases} 2x + y - 3z = 0 \\ -2x + 2y + z = -7 \\ 3x - 4y - 3z = 7 \end{cases}$

51. $\begin{cases} 2x - 2y - 2z = 2 \\ 2x + 3y + z = 2 \\ 3x + 2y = 0 \end{cases}$

52. $\begin{cases} 2x - 3y - z = 0 \\ -x + 2y + z = 5 \\ 3x - 4y - z = 1 \end{cases}$

53. $\begin{cases} -x + y + z = -1 \\ -x + 2y - 3z = -4 \\ 3x - 2y - 7z = 0 \end{cases}$

54. $\begin{cases} 2x - 3y - z = 0 \\ 3x + 2y + 2z = 2 \\ x + 5y + 3z = 2 \end{cases}$

55. $\begin{cases} 2x - 2y + 3z = 6 \\ 4x - 3y + 2z = 0 \\ -2x + 3y - 7z = 1 \end{cases}$

56. $\begin{cases} 3x - 2y + 2z = 6 \\ 7x - 3y + 2z = -1 \\ 2x - 3y + 4z = 0 \end{cases}$

57. $\begin{cases} x + y - z = 6 \\ 3x - 2y + z = -5 \\ x + 3y - 2z = 14 \end{cases}$

58. $\begin{cases} x - y + z = -4 \\ 2x - 3y + 4z = -15 \\ 5x + y - 2z = 12 \end{cases}$

59. $\begin{cases} x + 2y - z = -3 \\ 2x - 4y + z = -7 \\ -2x + 2y - 3z = 4 \end{cases}$

60. $\begin{cases} x + 4y - 3z = -8 \\ 3x - y + 3z = 12 \\ x + y + 6z = 1 \end{cases}$

61. $\begin{cases} 3x + y - z = \dfrac{2}{3} \\ 2x - y + z = 1 \\ 4x + 2y = \dfrac{8}{3} \end{cases}$

62. $\begin{cases} x + y = 1 \\ 2x - y + z = 1 \\ x + 2y + z = \dfrac{8}{3} \end{cases}$

63. $\begin{cases} x + y + z + w = 4 \\ 2x - y + z = 0 \\ 3x + 2y + z - w = 6 \\ x - 2y - 2z + 2w = -1 \end{cases}$

64. $\begin{cases} x + y + z + w = 4 \\ -x + 2y + z = 0 \\ 2x + 3y + z - w = 6 \\ -2x + y - 2z + 2w = -1 \end{cases}$

65. $\begin{cases} x + 2y + z = 1 \\ 2x - y + 2z = 2 \\ 3x + y + 3z = 3 \end{cases}$

66. $\begin{cases} x + 2y - z = 3 \\ 2x - y + 2z = 6 \\ x - 3y + 3z = 4 \end{cases}$

67. $\begin{cases} x - y + z = 5 \\ 3x + 2y - 2z = 0 \end{cases}$

68. $\begin{cases} 2x + y - z = 4 \\ -x + y + 3z = 1 \end{cases}$

69. $\begin{cases} 2x + 3y - z = 3 \\ x - y - z = 0 \\ -x + y + z = 0 \\ x + y + 3z = 5 \end{cases}$

70. $\begin{cases} x - 3y + z = 1 \\ 2x - y - 4z = 0 \\ x - 3y + 2z = 1 \\ x - 2y = 5 \end{cases}$

71. $\begin{cases} 4x + y + z - w = 4 \\ x - y + 2z + 3w = 3 \end{cases}$

72. $\begin{cases} -4x + y = 5 \\ 2x - y + z - w = 5 \\ z + w = 4 \end{cases}$

73. **Curve Fitting** Find the function $y = ax^2 + bx + c$ whose graph contains the points $(1, 2)$, $(-2, -7)$, and $(2, -3)$.

74. **Curve Fitting** Find the function $y = ax^2 + bx + c$ whose graph contains the points $(1, -1)$, $(3, -1)$, and $(-2, 14)$.

75. **Curve Fitting** Find the function $f(x) = ax^3 + bx^2 + cx + d$ for which $f(-3) = -112$, $f(-1) = -2$, $f(1) = 4$, and $f(2) = 13$.

76. **Curve Fitting** Find the function $f(x) = ax^3 + bx^2 + cx + d$ for which $f(-2) = -10$, $f(-1) = 3$, $f(1) = 5$, and $f(3) = 15$.

77. **Nutrition** A dietitian at Palos Community Hospital wants a patient to have a meal that has 78 grams of protein, 59 grams of carbohydrates, and 75 milligrams of vitamin A. The hospital food service tells the dietitian that the dinner for today is salmon steak, baked eggs, and acorn squash. Each serving of salmon steak has 30 grams of protein, 20 grams of carbohydrates, and 2 milligrams of vitamin A. Each serving of baked eggs contains 15 grams of protein, 2 grams of carbohydrates, and 20 milligrams of vitamin A. Each serving of acorn squash contains 3 grams of protein, 25 grams of carbohydrates, and 32 milligrams of vitamin A. How many servings of each food should the dietitian provide for the patient?

78. **Nutrition** A dietitian at General Hospital wants a patient to have a meal that has 47 grams of protein, 58 grams of carbohydrates, and 630 milligrams of calcium. The hospital food service tells the dietitian that the dinner for today is pork chops, corn on the cob, and 2% milk. Each serving of pork chops has 23 grams of protein, 0 grams of carbohydrates, and 10 milligrams of calcium. Each serving of corn on the cob contains 3 grams of protein, 16 grams of carbohydrates, and 10 milligrams of calcium. Each glass of 2% milk contains 9 grams of protein, 13 grams of carbohydrates, and 300 milligrams of calcium. How many servings of each food should the dietitian provide for the patient?

79. **Financial Planning** Carletta has \$10,000 to invest. As her financial consultant, you recommend that she invest in Treasury bills that yield 6%, Treasury bonds that yield 7%, and corporate bonds that yield 8%. Carletta wants to have an annual income of \$680, and the amount invested in corporate bonds must be half that invested in Treasury bills. Find the amount in each investment.

80. **Financial Planning** John has \$20,000 to invest. As his financial consultant, you recommend that he invest in Treasury bills that yield 5%, Treasury bonds that yield 7%, and corporate bonds that yield 9%. John wants to have an annual income of \$1280, and the amount invested in Treasury bills must be two times the amount invested in corporate bonds. Find the amount in each investment.

81. **Production** To manufacture an automobile requires painting, drying, and polishing. Epsilon Motor Company produces three types of cars: the Delta, the Beta, and the Sigma. Each Delta requires 10 hours for painting, 3 hours for drying, and 2 hours for polishing. A Beta requires 16 hours of painting, 5 hours of drying, and 3 hours of polishing, while a Sigma requires 8 hours for painting, 2 hours for drying, and 1 hour for polishing. If the company has 240 hours for painting, 69 hours for drying, and 41 hours for polishing per month, how many of each type of car are produced?

82. **Production** A Florida juice company completes the preparation of its products by sterilizing, filling, and labeling bottles. Each case of orange juice requires 9 minutes for sterilizing, 6 minutes for filling, and 1 minute for labeling. Each case of grapefruit juice requires 10 minutes for sterilizing, 4 minutes for filling, and 2 minutes for labeling. Each case of tomato juice requires 12 minutes for sterilizing, 4 minutes for filling, and 1 minute for labeling. If the company runs the sterilizing machine for 398 minutes, the filling machine for 164 minutes, and the labeling machine for 58 minutes, how many cases of each type of juice are prepared?

83. **Electricity: Kirchhoff's Rules** An application of Kirchhoff's Rules to the circuit shown results in the following system of equations:

$$\begin{cases} -4 + 8 - 2I_2 = 0 \\ 8 = 5I_4 + I_1 \\ 4 = 3I_3 + I_1 \\ I_3 + I_4 = I_1 \end{cases}$$

Find the currents I_1, I_2, I_3, and I_4.*

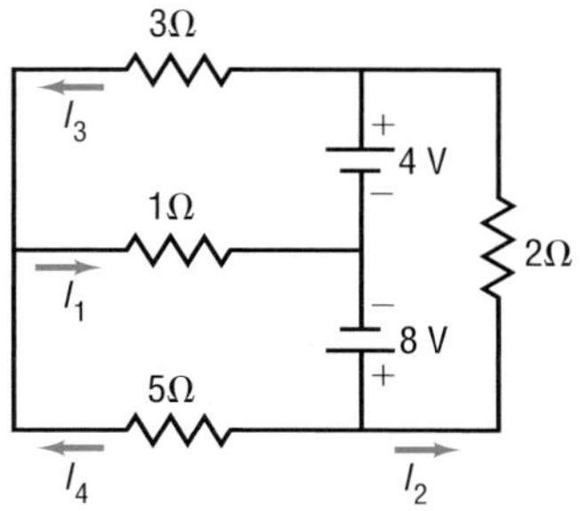

84. **Electricity: Kirchhoff's Rules** An application of Kirchhoff's Rules to the circuit shown results in the following system of equations:

$$\begin{cases} I_1 = I_3 + I_2 \\ 24 - 6I_1 - 3I_3 = 0 \\ 12 + 24 - 6I_1 - 6I_2 = 0 \end{cases}$$

*Source: Based on Raymond Serway, *Physics* 3rd ed. (Philadelphia: Saunders, 1990), Prob. 34, p. 790.

Find the currents I_1, I_2, and I_3.*

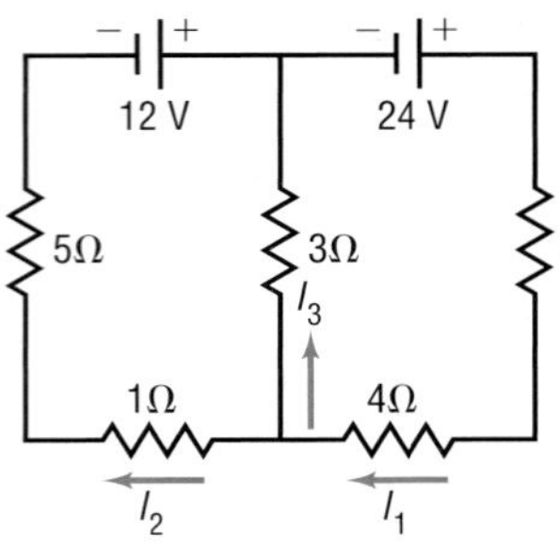

85. Financial Planning Three retired couples each require an additional annual income of $2000 per year. As their financial consultant, you recommend that they invest some money in Treasury bills that yield 7%, some money in corporate bonds that yield 9%, and some money in junk bonds that yield 11%. Prepare a table for each couple showing the various ways that their goals can be achieved:

(a) If the first couple has $20,000 to invest.

(b) If the second couple has $25,000 to invest.

(c) If the third couple has $30,000 to invest.

(d) What advice would you give each couple regarding the amount to invest and the choices available?

[**Hint:** Higher yields generally carry more risk.]

86. Financial Planning A young couple has $25,000 to invest. As their financial consultant, you recommend that they invest some money in Treasury bills that yield 7%, some money in corporate bonds that yield 9%, and some money in junk bonds that yield 11%. Prepare a table showing the various ways that this couple can achieve the following goals:

(a) The couple wants $1500 per year in income.

(b) The couple wants $2000 per year in income.

(c) The couple wants $2500 per year in income.

(d) What advice would you give this couple regarding the income that they require and the choices available?

[**Hint:** Higher yields generally carry more risk.]

87. Pharmacy A doctor's prescription calls for a daily intake containing 40 mg of vitamin C and 30 mg of vitamin D. Your pharmacy stocks three compounds that can be used: one contains 20% vitamin C and 30% vitamin D; a second, 40% vitamin C and 20% vitamin D; and a third, 30% vitamin C and 50% vitamin D. Create a table showing the possible combinations that could be used to fill the prescription.

88. Pharmacy A doctor's prescription calls for the creation of pills that contain 12 units of vitamin B_{12} and 12 units of vitamin E. Your pharmacy stocks three powders that can be used to make these pills: one contains 20% vitamin B_{12} and 30% vitamin E; a second, 40% vitamin B_{12} and 20% vitamin E; and a third, 30% vitamin B_{12} and 40% vitamin E. Create a table showing the possible combinations of each powder that could be mixed in each pill.

89. Write a brief paragraph or two that outlines your strategy for solving a system of linear equations using matrices.

90. When solving a system of linear equations using matrices, do you prefer to place the augmented matrix in row echelon form or in reduced row echelon form? Give reasons for your choice.

91. Create a system of three linear equations containing three variables that has:

(a) No solution

(b) Exactly one solution

(c) Infinitely many solutions

Give the three systems to a friend to solve and critique.

*Source: Ibid., Prob. 38, p. 791.

10.3 Systems of Linear Equations: Determinants

OBJECTIVES
1 Evaluate 2 by 2 Determinants
2 Use Cramer's Rule to Solve a System of Two Equations, Two Variables
3 Evaluate 3 by 3 Determinants
4 Use Cramer's Rule to Solve a System of Three Equations, Three Variables
5 Know Properties of Determinants

1 In the preceding section, we described a method of using matrices to solve a system of linear equations. This section deals with yet another method for solving systems of linear equations; however, it can be used only when the number of equations equals the number of variables. Although the method will work for any system (provided that the number of equations equals the

number of variables), it is most often used for systems of two equations containing two variables or three equations containing three variables. This method, called *Cramer's Rule*, is based on the concept of a *determinant*.

2 by 2 Determinants

If a, b, c, and d are four real numbers, the symbol

$$D = \begin{vmatrix} a & b \\ c & d \end{vmatrix}$$

is called a **2 by 2 determinant.** Its value is the number $ad - bc$; that is,

$$D = \begin{vmatrix} a & b \\ c & d \end{vmatrix} = ad - bc \qquad \textbf{(1)}$$

A device that may be helpful for remembering the value of a 2 by 2 determinant is the following:

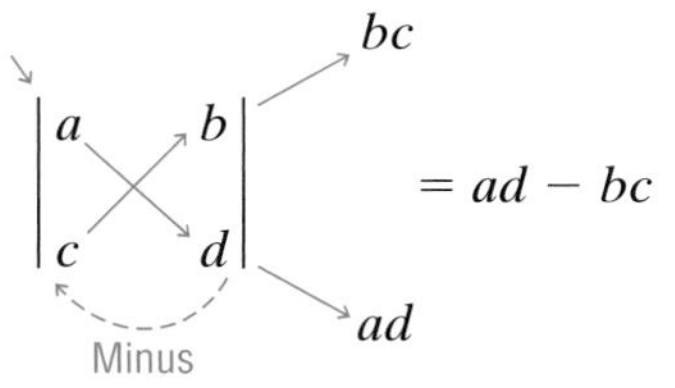

EXAMPLE 1 **Evaluating a 2 × 2 Determinant**

$$\begin{vmatrix} 3 & -2 \\ 6 & 1 \end{vmatrix} = (3)(1) - (6)(-2) = 3 - (-12) = 15$$

◄

NOW WORK PROBLEM 7.

Cramer's Rule

2 Let's now see the role that a 2 by 2 determinant plays in the solution of a system of two equations containing two variables. Consider the system

$$\begin{cases} ax + by = s & (1) \\ cx + dy = t & (2) \end{cases} \qquad \textbf{(2)}$$

We shall use the method of elimination to solve this system.

Provided $d \neq 0$ and $b \neq 0$, this system is equivalent to the system

$$\begin{cases} adx + bdy = sd & (1) \text{ Multiply by } d. \\ bcx + bdy = tb & (2) \text{ Multiply by } b. \end{cases}$$

On subtracting the second equation from the first equation, we get

$$\begin{cases} (ad - bc)x + 0 \cdot y = sd - tb & (1) \\ \qquad bcx + bdy = tb & (2) \end{cases}$$

Now, the first equation can be rewritten using determinant notation.

$$\begin{vmatrix} a & b \\ c & d \end{vmatrix} x = \begin{vmatrix} s & b \\ t & d \end{vmatrix}$$

If $D = \begin{vmatrix} a & b \\ c & d \end{vmatrix} = ad - bc \neq 0$, we can solve for x to get

$$x = \frac{\begin{vmatrix} s & b \\ t & d \end{vmatrix}}{\begin{vmatrix} a & b \\ c & d \end{vmatrix}} = \frac{\begin{vmatrix} s & b \\ t & d \end{vmatrix}}{D} \tag{3}$$

Return now to the original system (2). Provided that $a \neq 0$ and $c \neq 0$, the system is equivalent to

$$\begin{cases} acx + bcy = cs & (1) \quad \text{Multiply by } c. \\ acx + ady = at & (2) \quad \text{Multiply by } a. \end{cases}$$

On subtracting the first equation from the second equation, we get

$$\begin{cases} acx + \qquad bcy = \quad cs & (1) \\ 0 \cdot x + (ad - bc)y = at - cs & (2) \end{cases}$$

The second equation can now be rewritten using determinant notation.

$$\begin{vmatrix} a & b \\ c & d \end{vmatrix} y = \begin{vmatrix} a & s \\ c & t \end{vmatrix}$$

If $D = \begin{vmatrix} a & b \\ c & d \end{vmatrix} = ad - bc \neq 0$, we can solve for y to get

$$y = \frac{\begin{vmatrix} a & s \\ c & t \end{vmatrix}}{\begin{vmatrix} a & b \\ c & d \end{vmatrix}} = \frac{\begin{vmatrix} a & s \\ c & t \end{vmatrix}}{D} \tag{4}$$

Equations (3) and (4) lead us to the following result, called **Cramer's Rule.**

Theorem

Cramer's Rule for Two Equations Containing Two Variables

The solution to the system of equations

$$\begin{cases} ax + by = s & (1) \\ cx + dy = t & (2) \end{cases} \tag{5}$$

is given by

$$x = \frac{\begin{vmatrix} s & b \\ t & d \end{vmatrix}}{\begin{vmatrix} a & b \\ c & d \end{vmatrix}}, \qquad y = \frac{\begin{vmatrix} a & s \\ c & t \end{vmatrix}}{\begin{vmatrix} a & b \\ c & d \end{vmatrix}} \tag{6}$$

provided that

$$D = \begin{vmatrix} a & b \\ c & d \end{vmatrix} = ad - bc \neq 0$$

In the derivation given for Cramer's Rule above, we assumed that none of the numbers a, b, c, and d was 0. In Problem 60 you will be asked to complete the proof under the less stringent conditions that $D = ad - bc \neq 0$.

Now look carefully at the pattern in Cramer's Rule. The denominator in the solution (6) is the determinant of the coefficients of the variables.

$$\begin{cases} ax + by = s \\ cx + dy = t \end{cases} \qquad D = \begin{vmatrix} a & b \\ c & d \end{vmatrix}$$

In the solution for x, the numerator is the determinant, denoted by D_x, formed by replacing the entries in the first column (the coefficients of x) of D by the constants on the right side of the equal sign.

$$D_x = \begin{vmatrix} s & b \\ t & d \end{vmatrix}$$

In the solution for y, the numerator is the determinant, denoted by D_y, formed by replacing the entries in the second column (the coefficients of y) of D by the constants on the right side of the equal sign.

$$D_y = \begin{vmatrix} a & s \\ c & t \end{vmatrix}$$

Cramer's Rule then states that, if $D \neq 0$,

$$x = \frac{D_x}{D}, \qquad y = \frac{D_y}{D} \tag{7}$$

EXAMPLE 2 Solving a System of Linear Equations Using Determinants

Use Cramer's Rule, if applicable, to solve the system

$$\begin{cases} 3x - 2y = 4 & (1) \\ 6x + y = 13 & (2) \end{cases}$$

Solution The determinant D of the coefficients of the variables is

$$D = \begin{vmatrix} 3 & -2 \\ 6 & 1 \end{vmatrix} = (3)(1) - (6)(-2) = 15$$

Because $D \neq 0$, Cramer's Rule (7) can be used.

$$x = \frac{D_x}{D} = \frac{\begin{vmatrix} 4 & -2 \\ 13 & 1 \end{vmatrix}}{15} = \frac{30}{15} = 2 \qquad y = \frac{D_y}{D} = \frac{\begin{vmatrix} 3 & 4 \\ 6 & 13 \end{vmatrix}}{15} = \frac{15}{15} = 1$$

The solution is $x = 2$, $y = 1$. ◀

In attempting to use Cramer's Rule, if the determinant D of the coefficients of the variables is found to equal 0 (so that Cramer's Rule is not applicable), then the system is either inconsistent or has infinitely many solutions.

NOW WORK PROBLEM 15.

3 by 3 Determinants

3 To use Cramer's Rule to solve a system of three equations containing three variables, we need to define a 3 by 3 determinant.

A **3 by 3 determinant** is symbolized by

$$\begin{vmatrix} a_{11} & a_{12} & a_{13} \\ a_{21} & a_{22} & a_{23} \\ a_{31} & a_{32} & a_{33} \end{vmatrix} \tag{8}$$

in which $a_{11}, a_{12}, \ldots,$ are real numbers.

As with matrices, we use a double subscript to identify an entry by indicating its row and column numbers. For example, the entry a_{23} is in row 2, column 3.

The value of a 3 by 3 determinant may be defined in terms of 2 by 2 determinants by the following formula:

Minus

$$\begin{vmatrix} a_{11} & a_{12} & a_{13} \\ a_{21} & a_{22} & a_{23} \\ a_{31} & a_{32} & a_{33} \end{vmatrix} = a_{11}\begin{vmatrix} a_{22} & a_{23} \\ a_{32} & a_{33} \end{vmatrix} - a_{12}\begin{vmatrix} a_{21} & a_{23} \\ a_{31} & a_{33} \end{vmatrix} + a_{13}\begin{vmatrix} a_{21} & a_{22} \\ a_{31} & a_{32} \end{vmatrix} \tag{9}$$

2 by 2 determinant left after removing row and column containing a_{11}

2 by 2 determinant left after removing row and column containing a_{12}

2 by 2 determinant left after removing row and column containing a_{13}

The 2 by 2 determinants shown in formula (9) are called **minors** of the 3 by 3 determinant. For an n by n determinant, the **minor** M_{ij} of element a_{ij} is the determinant resulting from removing the ith row and jth column.

EXAMPLE 3 Finding Minors of a 3 by 3 Determinant

For the determinant $A = \begin{vmatrix} 2 & -1 & 3 \\ -2 & 5 & 1 \\ 0 & 6 & -9 \end{vmatrix}$, find: (a) M_{12} (b) M_{23}

Solution (a) M_{12} is the determinant that results from removing the first row and second column from A.

$$A = \begin{vmatrix} 2 & -1 & 3 \\ -2 & 5 & 1 \\ 0 & 6 & -9 \end{vmatrix} \qquad M_{12} = \begin{vmatrix} -2 & 1 \\ 0 & -9 \end{vmatrix} = (-2)(-9) - (0)(1) = 18$$

(b) M_{23} is the determinant that results from removing the second row and third column from A.

$$A = \begin{vmatrix} 2 & -1 & 3 \\ -2 & 5 & 1 \\ 0 & 6 & -9 \end{vmatrix} \qquad M_{23} = \begin{vmatrix} 2 & -1 \\ 0 & 6 \end{vmatrix} = (2)(6) - (0)(-1) = 12$$

◀

Referring back to formula (9), we see that each element a_{ij} is multiplied by its minor, but sometimes this term is added and other times subtracted. To determine whether to add or subtract a term, we must consider the *cofactor*.

For an n by n determinant A, the **cofactor** of element a_{ij}, denoted by A_{ij}, is given by

$$A_{ij} = (-1)^{i+j} M_{ij}$$

where M_{ij} is the minor of element a_{ij}.

The exponent of $(-1)^{i+j}$ is the sum of the row and column of the element a_{ij}; so if $i + j$ is even, $(-1)^{i+j}$ will equal 1, and if $i + j$ is odd, $(-1)^{i+j}$ will equal -1.

To find the value of a determinant, multiply each element in any row or column by its cofactor and sum the results. This process is referred to as *expanding across a row or down a column*. For example, the value of the 3 by 3 determinant in formula (9) was found by expanding across row 1.

If we choose to expand down column 2, we obtain

$$\begin{vmatrix} a_{11} & a_{12} & a_{13} \\ a_{21} & a_{22} & a_{23} \\ a_{31} & a_{32} & a_{33} \end{vmatrix} = (-1)^{1+2} a_{12} \begin{vmatrix} a_{21} & a_{23} \\ a_{31} & a_{33} \end{vmatrix} + (-1)^{2+2} a_{22} \begin{vmatrix} a_{11} & a_{13} \\ a_{31} & a_{33} \end{vmatrix} + (-1)^{3+2} a_{32} \begin{vmatrix} a_{11} & a_{13} \\ a_{21} & a_{23} \end{vmatrix}$$

Expand down column 2.

If we choose to expand across row 3, we obtain

$$\begin{vmatrix} a_{11} & a_{12} & a_{13} \\ a_{21} & a_{22} & a_{23} \\ a_{31} & a_{32} & a_{33} \end{vmatrix} = (-1)^{3+1} a_{31} \begin{vmatrix} a_{12} & a_{13} \\ a_{22} & a_{23} \end{vmatrix} + (-1)^{3+2} a_{32} \begin{vmatrix} a_{11} & a_{13} \\ a_{21} & a_{23} \end{vmatrix} + (-1)^{3+3} a_{33} \begin{vmatrix} a_{11} & a_{12} \\ a_{21} & a_{22} \end{vmatrix}$$

Expand across row 3.

It can be shown that the value of a determinant does not depend on the choice of the row or column used in the expansion. However, expanding across a row or column that has an element equal to 0 reduces the amount of work needed to compute the value of the determinant.

EXAMPLE 4 Evaluating a 3 × 3 Determinant

Find the value of the 3 by 3 determinant: $\begin{vmatrix} 3 & 4 & -1 \\ 4 & 6 & 2 \\ 8 & -2 & 3 \end{vmatrix}$

Solution We choose to expand across row 1.

$$\begin{aligned}\begin{vmatrix} 3 & 4 & -1 \\ 4 & 6 & 2 \\ 8 & -2 & 3 \end{vmatrix} &= (-1)^{1+1} 3 \begin{vmatrix} 6 & 2 \\ -2 & 3 \end{vmatrix} + (-1)^{1+2} 4 \begin{vmatrix} 4 & 2 \\ 8 & 3 \end{vmatrix} + (-1)^{1+3} (-1) \begin{vmatrix} 4 & 6 \\ 8 & -2 \end{vmatrix} \\ &= 3(18 + 4) - 4(12 - 16) + (-1)(-8 - 48) \\ &= 3(22) - 4(-4) + (-1)(-56) \\ &= 66 + 16 + 56 = 138\end{aligned}$$

◀

We could also find the value of the 3 by 3 determinant in Example 4 by expanding down column 3.

$$\begin{vmatrix} 3 & 4 & -1 \\ 4 & 6 & 2 \\ 8 & -2 & 3 \end{vmatrix} = (-1)^{1+3}(-1)\begin{vmatrix} 4 & 6 \\ 8 & -2 \end{vmatrix} + (-1)^{2+3}2\begin{vmatrix} 3 & 4 \\ 8 & -2 \end{vmatrix} + (-1)^{3+3}3\begin{vmatrix} 3 & 4 \\ 4 & 6 \end{vmatrix}$$

$$= -1(-8 - 48) - 2(-6 - 32) + 3(18 - 16)$$

$$= 56 + 76 + 6 = 138$$

COMMENT: A graphing utility can be used to evaluate determinants. Check your manual to see how. Then check the answer obtained in Example 4. ■

NOW WORK PROBLEM 11.

Systems of Three Equations Containing Three Variables

4 Consider the following system of three equations containing three variables.

$$\begin{cases} a_{11}x + a_{12}y + a_{13}z = c_1 \\ a_{21}x + a_{22}y + a_{23}z = c_2 \\ a_{31}x + a_{32}y + a_{33}z = c_3 \end{cases} \tag{10}$$

It can be shown that if the determinant D of the coefficients of the variables is not 0, that is, if

$$D = \begin{vmatrix} a_{11} & a_{12} & a_{13} \\ a_{21} & a_{22} & a_{23} \\ a_{31} & a_{32} & a_{33} \end{vmatrix} \neq 0$$

then the unique solution of system (10) is given by

Cramer's Rule for Three Equations Containing Three Variables

$$x = \frac{D_x}{D} \qquad y = \frac{D_y}{D} \qquad z = \frac{D_z}{D}, \qquad D \neq 0$$

where

$$D_x = \begin{vmatrix} c_1 & a_{12} & a_{13} \\ c_2 & a_{22} & a_{23} \\ c_3 & a_{32} & a_{33} \end{vmatrix} \qquad D_y = \begin{vmatrix} a_{11} & c_1 & a_{13} \\ a_{21} & c_2 & a_{23} \\ a_{31} & c_3 & a_{33} \end{vmatrix} \qquad D_z = \begin{vmatrix} a_{11} & a_{12} & c_1 \\ a_{21} & a_{22} & c_2 \\ a_{31} & a_{32} & c_3 \end{vmatrix}$$

The similarity of this pattern and the pattern observed earlier for a system of two equations containing two variables should be apparent.

EXAMPLE 5 Using Cramer's Rule

Use Cramer's Rule, if applicable, to solve the following system:

$$\begin{cases} 2x + y - z = 3 & (1) \\ -x + 2y + 4z = -3 & (2) \\ x - 2y - 3z = 4 & (3) \end{cases}$$

Solution The value of the determinant D of the coefficients of the variables is

$$D = \begin{vmatrix} 2 & 1 & -1 \\ -1 & 2 & 4 \\ 1 & -2 & -3 \end{vmatrix} = (-1)^{1+1}2\begin{vmatrix} 2 & 4 \\ -2 & -3 \end{vmatrix} + (-1)^{1+2}1\begin{vmatrix} -1 & 4 \\ 1 & -3 \end{vmatrix} + (-1)^{1+3}(-1)\begin{vmatrix} -1 & 2 \\ 1 & -2 \end{vmatrix}$$
$$= 2(2) - 1(-1) + (-1)(0)$$
$$= 4 + 1 = 5$$

Because $D \neq 0$, we proceed to find the values of D_x, D_y, D_z.

$$D_x = \begin{vmatrix} 3 & 1 & -1 \\ -3 & 2 & 4 \\ 4 & -2 & -3 \end{vmatrix} = (-1)^{1+1}3\begin{vmatrix} 2 & 4 \\ -2 & -3 \end{vmatrix} + (-1)^{1+2}1\begin{vmatrix} -3 & 4 \\ 4 & -3 \end{vmatrix} + (-1)^{1+3}(-1)\begin{vmatrix} -3 & 2 \\ 4 & -2 \end{vmatrix}$$
$$= 3(2) - 1(-7) + (-1)(-2) = 15$$

$$D_y = \begin{vmatrix} 2 & 3 & -1 \\ -1 & -3 & 4 \\ 1 & 4 & -3 \end{vmatrix} = (-1)^{1+1}2\begin{vmatrix} -3 & 4 \\ 4 & -3 \end{vmatrix} + (-1)^{1+2}3\begin{vmatrix} -1 & 4 \\ 1 & -3 \end{vmatrix} + (-1)^{1+3}(-1)\begin{vmatrix} -1 & -3 \\ 1 & 4 \end{vmatrix}$$
$$= 2(-7) - 3(-1) + (-1)(-1)$$
$$= -14 + 3 + 1 = -10$$

$$D_z = \begin{vmatrix} 2 & 1 & 3 \\ -1 & 2 & -3 \\ 1 & -2 & 4 \end{vmatrix} = (-1)^{1+1}2\begin{vmatrix} 2 & -3 \\ -2 & 4 \end{vmatrix} + (-1)^{1+2}1\begin{vmatrix} -1 & -3 \\ 1 & 4 \end{vmatrix} + (-1)^{1+3}3\begin{vmatrix} -1 & 2 \\ 1 & -2 \end{vmatrix}$$
$$= 2(2) - 1(-1) + 3(0) = 5$$

As a result,

$$x = \frac{D_x}{D} = \frac{15}{5} = 3 \qquad y = \frac{D_y}{D} = \frac{-10}{5} = -2 \qquad z = \frac{D_z}{D} = \frac{5}{5} = 1$$

The solution is $x = 3$, $y = -2$, $z = 1$. ◀

If the determinant of the coefficients of the variables of a system of three linear equations containing three variables is 0, then Cramer's Rule is not applicable. In such a case, the system is either inconsistent or has infinitely many solutions.

NOW WORK PROBLEM 33.

Properties of Determinants

5 Determinants have several properties that are sometimes helpful for obtaining their value. We list some of them here.

Theorem

The value of a determinant changes sign if any two rows (or any two columns) are interchanged. **(11)**

Proof for 2 by 2 Determinants

$$\begin{vmatrix} a & b \\ c & d \end{vmatrix} = ad - bc \quad \text{and} \quad \begin{vmatrix} c & d \\ a & b \end{vmatrix} = bc - ad = -(ad - bc)$$ ■

EXAMPLE 6 Demonstrating Theorem (11)

$$\begin{vmatrix} 3 & 4 \\ 1 & 2 \end{vmatrix} = 6 - 4 = 2 \qquad \begin{vmatrix} 1 & 2 \\ 3 & 4 \end{vmatrix} = 4 - 6 = -2$$

◀

Theorem If all the entries in any row (or any column) equal 0, the value of the determinant is 0. **(12)**

Proof Merely expand across the row (or down the column) containing the 0's. ■

Theorem If any two rows (or any two columns) of a determinant have corresponding entries that are equal, the value of the determinant is 0. **(13)**

You are asked to prove this result for a 3 by 3 determinant in which the entries in column 1 equal the entries in column 3 in Problem 63.

EXAMPLE 7 Demonstrating Theorem (13)

$$\begin{vmatrix} 1 & 2 & 3 \\ 1 & 2 & 3 \\ 4 & 5 & 6 \end{vmatrix} = (-1)^{1+1}1\begin{vmatrix} 2 & 3 \\ 5 & 6 \end{vmatrix} + (-1)^{1+2}2\begin{vmatrix} 1 & 3 \\ 4 & 6 \end{vmatrix} + (-1)^{1+3}3\begin{vmatrix} 1 & 2 \\ 4 & 5 \end{vmatrix}$$

$$= 1(-3) - 2(-6) + 3(-3)$$

$$= -3 + 12 - 9 = 0$$

◀

Theorem If any row (or any column) of a determinant is multiplied by a nonzero number k, the value of the determinant is also changed by a factor of k. **(14)**

You are asked to prove this result for a 3 by 3 determinant using row 2 in Problem 62.

EXAMPLE 8 Demonstrating Theorem (14)

$$\begin{vmatrix} 1 & 2 \\ 4 & 6 \end{vmatrix} = 6 - 8 = -2$$

$$\begin{vmatrix} k & 2k \\ 4 & 6 \end{vmatrix} = 6k - 8k = -2k = k(-2) = k\begin{vmatrix} 1 & 2 \\ 4 & 6 \end{vmatrix}$$

◀

Theorem If the entries of any row (or any column) of a determinant are multiplied by a nonzero number k and the result is added to the corresponding entries of another row (or column), the value of the determinant remains unchanged. **(15)**

In Problem 64, you are asked to prove this result for a 3 by 3 determinant using rows 1 and 2.

EXAMPLE 9 **Demonstrating Theorem (15)**

$$\begin{vmatrix} 3 & 4 \\ 5 & 2 \end{vmatrix} = -14 \qquad \begin{vmatrix} 3 & 4 \\ 5 & 2 \end{vmatrix} \rightarrow \begin{vmatrix} -7 & 0 \\ 5 & 2 \end{vmatrix} = -14$$

Multiply row 2 by -2 and add to row 1. ◀

10.3 Assess Your Understanding

Concepts and Vocabulary

1. Cramer's Rule uses __________ to solve a system of linear equations.
2. $D = \begin{vmatrix} a & b \\ c & d \end{vmatrix} =$ __________.
3. *True or False:* A 3 by 3 determinant can never equal 0.
4. *True or False:* The value of a determinant remains unchanged if any two rows or any two columns are interchanged.

Exercises

In Problems 5–14, find the value of each determinant.

5. $\begin{vmatrix} 3 & 1 \\ 4 & 2 \end{vmatrix}$
6. $\begin{vmatrix} 6 & 1 \\ 5 & 2 \end{vmatrix}$
7. $\begin{vmatrix} 6 & 4 \\ -1 & 3 \end{vmatrix}$
8. $\begin{vmatrix} 8 & -3 \\ 4 & 2 \end{vmatrix}$
9. $\begin{vmatrix} -3 & -1 \\ 4 & 2 \end{vmatrix}$
10. $\begin{vmatrix} -4 & 2 \\ -5 & 3 \end{vmatrix}$
11. $\begin{vmatrix} 3 & 4 & 2 \\ 1 & -1 & 5 \\ 1 & 2 & -2 \end{vmatrix}$
12. $\begin{vmatrix} 1 & 3 & -2 \\ 6 & 1 & -5 \\ 8 & 2 & 3 \end{vmatrix}$
13. $\begin{vmatrix} 4 & -1 & 2 \\ 6 & -1 & 0 \\ 1 & -3 & 4 \end{vmatrix}$
14. $\begin{vmatrix} 3 & -9 & 4 \\ 1 & 4 & 0 \\ 8 & -3 & 1 \end{vmatrix}$

In Problems 15–42, solve each system of equations using Cramer's Rule if it is applicable. If Cramer's Rule is not applicable, say so.

15. $\begin{cases} x + y = 8 \\ x - y = 4 \end{cases}$
16. $\begin{cases} x + 2y = 5 \\ x - y = 3 \end{cases}$
17. $\begin{cases} 5x - y = 13 \\ 2x + 3y = 12 \end{cases}$
18. $\begin{cases} x + 3y = 5 \\ 2x - 3y = -8 \end{cases}$
19. $\begin{cases} 3x = 24 \\ x + 2y = 0 \end{cases}$
20. $\begin{cases} 4x + 5y = -3 \\ -2y = -4 \end{cases}$
21. $\begin{cases} 3x - 6y = 24 \\ 5x + 4y = 12 \end{cases}$
22. $\begin{cases} 2x + 4y = 16 \\ 3x - 5y = -9 \end{cases}$
23. $\begin{cases} 3x - 2y = 4 \\ 6x - 4y = 0 \end{cases}$
24. $\begin{cases} -x + 2y = 5 \\ 4x - 8y = 6 \end{cases}$
25. $\begin{cases} 2x - 4y = -2 \\ 3x + 2y = 3 \end{cases}$
26. $\begin{cases} 3x + 3y = 3 \\ 4x + 2y = \dfrac{8}{3} \end{cases}$
27. $\begin{cases} 2x - 3y = -1 \\ 10x + 10y = 5 \end{cases}$
28. $\begin{cases} 3x - 2y = 0 \\ 5x + 10y = 4 \end{cases}$
29. $\begin{cases} 2x + 3y = 6 \\ x - y = \dfrac{1}{2} \end{cases}$
30. $\begin{cases} \dfrac{1}{2}x + y = -2 \\ x - 2y = 8 \end{cases}$
31. $\begin{cases} 3x - 5y = 3 \\ 15x + 5y = 21 \end{cases}$
32. $\begin{cases} 2x - y = -1 \\ x + \dfrac{1}{2}y = \dfrac{3}{2} \end{cases}$
33. $\begin{cases} x + y - z = 6 \\ 3x - 2y + z = -5 \\ x + 3y - 2z = 14 \end{cases}$
34. $\begin{cases} x - y + z = -4 \\ 2x - 3y + 4z = -15 \\ 5x + y - 2z = 12 \end{cases}$
35. $\begin{cases} x + 2y - z = -3 \\ 2x - 4y + z = -7 \\ -2x + 2y - 3z = 4 \end{cases}$
36. $\begin{cases} x + 4y - 3z = -8 \\ 3x - y + 3z = 12 \\ x + y + 6z = 1 \end{cases}$
37. $\begin{cases} x - 2y + 3z = 1 \\ 3x + y - 2z = 0 \\ 2x - 4y + 6z = 2 \end{cases}$
38. $\begin{cases} x - y + 2z = 5 \\ 3x + 2y = 4 \\ -2x + 2y - 4z = -10 \end{cases}$
39. $\begin{cases} x + 2y - z = 0 \\ 2x - 4y + z = 0 \\ -2x + 2y - 3z = 0 \end{cases}$
40. $\begin{cases} x + 4y - 3z = 0 \\ 3x - y + 3z = 0 \\ x + y + 6z = 0 \end{cases}$
41. $\begin{cases} x - 2y + 3z = 0 \\ 3x + y - 2z = 0 \\ 2x - 4y + 6z = 0 \end{cases}$
42. $\begin{cases} x - y + 2z = 0 \\ 3x + 2y = 0 \\ -2x + 2y - 4z = 0 \end{cases}$

In Problems 43–48, solve for x.

43. $\begin{vmatrix} x & x \\ 4 & 3 \end{vmatrix} = 5$

44. $\begin{vmatrix} x & 1 \\ 3 & x \end{vmatrix} = -2$

45. $\begin{vmatrix} x & 1 & 1 \\ 4 & 3 & 2 \\ -1 & 2 & 5 \end{vmatrix} = 2$

46. $\begin{vmatrix} 3 & 2 & 4 \\ 1 & x & 5 \\ 0 & 1 & -2 \end{vmatrix} = 0$

47. $\begin{vmatrix} x & 2 & 3 \\ 1 & x & 0 \\ 6 & 1 & -2 \end{vmatrix} = 7$

48. $\begin{vmatrix} x & 1 & 2 \\ 1 & x & 3 \\ 0 & 1 & 2 \end{vmatrix} = -4x$

In Problems 49–56, use properties of determinants to find the value of each determinant if it is known that

$$\begin{vmatrix} x & y & z \\ u & v & w \\ 1 & 2 & 3 \end{vmatrix} = 4$$

49. $\begin{vmatrix} 1 & 2 & 3 \\ u & v & w \\ x & y & z \end{vmatrix}$

50. $\begin{vmatrix} x & y & z \\ u & v & w \\ 2 & 4 & 6 \end{vmatrix}$

51. $\begin{vmatrix} x & y & z \\ -3 & -6 & -9 \\ u & v & w \end{vmatrix}$

52. $\begin{vmatrix} 1 & 2 & 3 \\ x-u & y-v & z-w \\ u & v & w \end{vmatrix}$

53. $\begin{vmatrix} 1 & 2 & 3 \\ x-3 & y-6 & z-9 \\ 2u & 2v & 2w \end{vmatrix}$

54. $\begin{vmatrix} x & y & z-x \\ u & v & w-u \\ 1 & 2 & 2 \end{vmatrix}$

55. $\begin{vmatrix} 1 & 2 & 3 \\ 2x & 2y & 2z \\ u-1 & v-2 & w-3 \end{vmatrix}$

56. $\begin{vmatrix} x+3 & y+6 & z+9 \\ 3u-1 & 3v-2 & 3w-3 \\ 1 & 2 & 3 \end{vmatrix}$

57. Geometry: Equation of a Line An equation of the line containing the two points (x_1, y_1) and (x_2, y_2) may be expressed as the determinant

$$\begin{vmatrix} x & y & 1 \\ x_1 & y_1 & 1 \\ x_2 & y_2 & 1 \end{vmatrix} = 0$$

Prove this result by expanding the determinant and comparing the result to the 2-point form of the equation of a line.

58. Geometry: Collinear Points Using the result obtained in Problem 57, show that three distinct points (x_1, y_1), (x_2, y_2), and (x_3, y_3) are collinear (lie on the same line) if and only if

$$\begin{vmatrix} x_1 & y_1 & 1 \\ x_2 & y_2 & 1 \\ x_3 & y_3 & 1 \end{vmatrix} = 0$$

59. Show that $\begin{vmatrix} x^2 & x & 1 \\ y^2 & y & 1 \\ z^2 & z & 1 \end{vmatrix} = (y - z)(x - y)(x - z)$.

60. Complete the proof of Cramer's Rule for two equations containing two variables.
[**Hint:** In system (5), page 718, if $a = 0$, then $b \neq 0$ and $c \neq 0$, since $D = -bc \neq 0$. Now show that equation (6) provides a solution of the system when $a = 0$. There are then three remaining cases: $b = 0$, $c = 0$, and $d = 0$.]

61. Interchange columns 1 and 3 of a 3 by 3 determinant. Show that the value of the new determinant is -1 times the value of the original determinant.

62. Multiply each entry in row 2 of a 3 by 3 determinant by the number k, $k \neq 0$. Show that the value of the new determinant is k times the value of the original determinant.

63. Prove that a 3 by 3 determinant in which the entries in column 1 equal those in column 3 has the value 0.

64. Prove that, if row 2 of a 3 by 3 determinant is multiplied by k, $k \neq 0$, and the result is added to the entries in row 1, there is no change in the value of the determinant.

10.4 Matrix Algebra

OBJECTIVES
1 Find the Sum and Difference of Two Matrices
2 Find Scalar Multiples of a Matrix
3 Find the Product of Two Matrices
4 Find the Inverse of a Matrix
5 Solve a System of Equations Using Inverse Matrices

In Section 10.2, we defined a matrix as a rectangular array of real numbers and used an augmented matrix to represent a system of linear equations. There is, however, a branch of mathematics, called **linear algebra,** that deals with matrices in such a way that an algebra of matrices is permitted. In this section, we provide a survey of how this **matrix algebra** is developed.

Before getting started, we restate the definition of a matrix.

A **matrix** is defined as a rectangular array of numbers:

$$\begin{array}{c} \\ \text{Row } 1 \\ \text{Row } 2 \\ \vdots \\ \text{Row } i \\ \vdots \\ \text{Row } m \end{array} \begin{array}{c} \begin{array}{cccccc} \text{Column } 1 & \text{Column } 2 & & \text{Column } j & & \text{Column } n \end{array} \\ \begin{bmatrix} a_{11} & a_{12} & \cdots & a_{1j} & \cdots & a_{1n} \\ a_{21} & a_{22} & \cdots & a_{2j} & \cdots & a_{2n} \\ \vdots & \vdots & & \vdots & & \vdots \\ a_{i1} & a_{i2} & \cdots & a_{ij} & \cdots & a_{in} \\ \vdots & \vdots & & \vdots & & \vdots \\ a_{m1} & a_{m2} & \cdots & a_{mj} & \cdots & a_{mn} \end{bmatrix} \end{array}$$

Each number a_{ij} of the matrix has two indexes: the **row index** i and the **column index** j. The matrix shown above has m rows and n columns. The numbers a_{ij} are usually referred to as the **entries** of the matrix. For example, a_{23} refers to the entry in the second row, third column.

Let's begin with an example that illustrates how matrices can be used to conveniently represent an array of information.

EXAMPLE 1 Arranging Data in a Matrix

In a survey of 900 people, the following information was obtained:

200 males	Thought federal defense spending was too high
150 males	Thought federal defense spending was too low
45 males	Had no opinion
315 females	Thought federal defense spending was too high
125 females	Thought federal defense spending was too low
65 females	Had no opinion

We can arrange these data in a rectangular array as follows:

	Too High	**Too Low**	**No Opinion**
Male	200	150	45
Female	315	125	65

or as the matrix

$$\begin{bmatrix} 200 & 150 & 45 \\ 315 & 125 & 65 \end{bmatrix}$$

This matrix has two rows (representing males and females) and three columns (representing "too high," "too low," and "no opinion"). ◀

The matrix we developed in Example 1 has 2 rows and 3 columns. In general, a matrix with m rows and n columns is called an ***m* by *n* matrix.** The matrix we developed in Example 1 is a 2 by 3 matrix and contains $2 \cdot 3 = 6$ entries. An m by n matrix will contain $m \cdot n$ entries.

If an m by n matrix has the same number of rows as columns, that is, if $m = n$, then the matrix is referred to as a **square matrix.**

EXAMPLE 2 Examples of Matrices

(a) $\begin{bmatrix} 5 & 0 \\ -6 & 1 \end{bmatrix}$ A 2 by 2 square matrix (b) $\begin{bmatrix} 1 & 0 & 3 \end{bmatrix}$ A 1 by 3 matrix

(c) $\begin{bmatrix} 6 & -2 & 4 \\ 4 & 3 & 5 \\ 8 & 0 & 1 \end{bmatrix}$ A 3 by 3 square matrix

◀

The Sum and Difference of Two Matrices

1 We begin our discussion of matrix algebra by first defining what is meant by two matrices being equal and then defining the operations of addition and subtraction. It is important to note that these definitions require each matrix to have the same number of rows *and* the same number of columns.

We usually represent matrices by capital letters, such as A, B, C, and so on.

Two m by n matrices A and B are said to be **equal,** written as

$$A = B$$

provided that each entry a_{ij} in A is equal to the corresponding entry b_{ij} in B.

For example,

$$\begin{bmatrix} 2 & 1 \\ 0.5 & -1 \end{bmatrix} = \begin{bmatrix} \sqrt{4} & 1 \\ \frac{1}{2} & -1 \end{bmatrix} \quad \text{and} \quad \begin{bmatrix} 3 & 2 & 1 \\ 0 & 1 & -2 \end{bmatrix} = \begin{bmatrix} \sqrt{9} & \sqrt{4} & 1 \\ 0 & 1 & \sqrt[3]{-8} \end{bmatrix}$$

$$\begin{bmatrix} 4 & 1 \\ 6 & 1 \end{bmatrix} \neq \begin{bmatrix} 4 & 0 \\ 6 & 1 \end{bmatrix}$$ Because the entries in row 1, column 2 are not equal

$$\begin{bmatrix} 4 & 1 & 2 \\ 6 & 1 & 2 \end{bmatrix} \neq \begin{bmatrix} 4 & 1 & 2 & 3 \\ 6 & 1 & 2 & 4 \end{bmatrix}$$ Because the matrix on the left is 2 by 3 and the matrix on the right is 2 by 4

Suppose that A and B represent two m by n matrices. We define their **sum $A + B$** to be the m by n matrix formed by adding the corresponding entries a_{ij} of A and b_{ij} of B. The **difference $A - B$** is defined as the m by n matrix formed by subtracting the entries b_{ij} in B from the corresponding entries a_{ij} in A. Addition and subtraction of matrices are allowed only for matrices having the same number m of rows and the same number n of columns. For example, a 2 by 3 matrix and a 2 by 4 matrix cannot be added or subtracted.

EXAMPLE 3 Adding and Subtracting Matrices

Suppose that

$$A = \begin{bmatrix} 2 & 4 & 8 & -3 \\ 0 & 1 & 2 & 3 \end{bmatrix} \quad \text{and} \quad B = \begin{bmatrix} -3 & 4 & 0 & 1 \\ 6 & 8 & 2 & 0 \end{bmatrix}$$

Find: (a) $A + B$ (b) $A - B$

Solution First, we observe that both A and B have 2 rows and 4 columns, so it is possible to find their sum and their difference.

$$\text{(a) } A + B = \begin{bmatrix} 2 & 4 & 8 & -3 \\ 0 & 1 & 2 & 3 \end{bmatrix} + \begin{bmatrix} -3 & 4 & 0 & 1 \\ 6 & 8 & 2 & 0 \end{bmatrix}$$

$$= \begin{bmatrix} 2 + (-3) & 4 + 4 & 8 + 0 & -3 + 1 \\ 0 + 6 & 1 + 8 & 2 + 2 & 3 + 0 \end{bmatrix} \quad \text{Add corresponding entries.}$$

$$= \begin{bmatrix} -1 & 8 & 8 & -2 \\ 6 & 9 & 4 & 3 \end{bmatrix}$$

$$\text{(b) } A - B = \begin{bmatrix} 2 & 4 & 8 & -3 \\ 0 & 1 & 2 & 3 \end{bmatrix} - \begin{bmatrix} -3 & 4 & 0 & 1 \\ 6 & 8 & 2 & 0 \end{bmatrix}$$

$$= \begin{bmatrix} 2 - (-3) & 4 - 4 & 8 - 0 & -3 - 1 \\ 0 - 6 & 1 - 8 & 2 - 2 & 3 - 0 \end{bmatrix} \quad \text{Subtract corresponding entries.}$$

$$= \begin{bmatrix} 5 & 0 & 8 & -4 \\ -6 & -7 & 0 & 3 \end{bmatrix}$$

◄

Figure 7

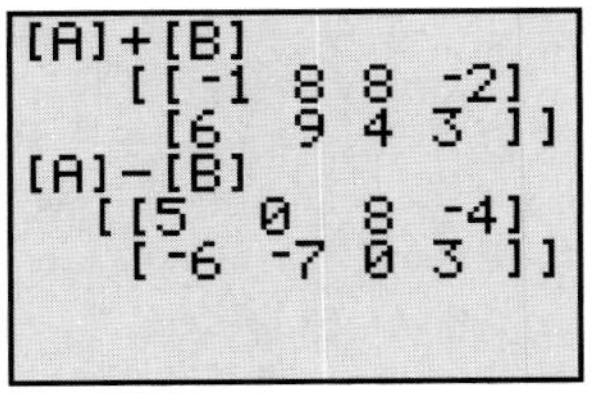

Seeing the Concept

Graphing utilities can make the sometimes tedious process of matrix algebra easy. In fact, most graphing calculators can handle matrices as large as 9 by 9, some even larger ones. Enter the matrices into a graphing utility. Name them [*A*] and [*B*]. Figure 7 shows the results of adding and subtracting [*A*] and [*B*].

NOW WORK PROBLEM 7.

Many of the algebraic properties of sums of real numbers are also true for sums of matrices. Suppose that A, B, and C are m by n matrices. Then matrix addition is **commutative.** That is,

Commutative Property

$$A + B = B + A$$

Matrix addition is also **associative.** That is,

Associative Property

$$(A + B) + C = A + (B + C)$$

Although we shall not prove these results, the proofs, as the following example illustrates, are based on the commutative and associative properties for real numbers.

EXAMPLE 4 Demonstrating the Commutative Property

$$\begin{bmatrix} 2 & 3 & -1 \\ 4 & 0 & 7 \end{bmatrix} + \begin{bmatrix} -1 & 2 & 1 \\ 5 & -3 & 4 \end{bmatrix} = \begin{bmatrix} 2 + (-1) & 3 + 2 & -1 + 1 \\ 4 + 5 & 0 + (-3) & 7 + 4 \end{bmatrix}$$

$$= \begin{bmatrix} -1 + 2 & 2 + 3 & 1 + (-1) \\ 5 + 4 & -3 + 0 & 4 + 7 \end{bmatrix}$$

$$= \begin{bmatrix} -1 & 2 & 1 \\ 5 & -3 & 4 \end{bmatrix} + \begin{bmatrix} 2 & 3 & -1 \\ 4 & 0 & 7 \end{bmatrix}$$

◄

A matrix whose entries are all equal to 0 is called a **zero matrix.** Each of the following matrices is a zero matrix.

$$\begin{bmatrix} 0 & 0 \\ 0 & 0 \end{bmatrix} \quad \text{2 by 2 square zero matrix} \qquad \begin{bmatrix} 0 & 0 & 0 \\ 0 & 0 & 0 \end{bmatrix} \quad \text{2 by 3 zero matrix} \qquad \begin{bmatrix} 0 & 0 & 0 \end{bmatrix} \quad \text{1 by 3 zero matrix}$$

Zero matrices have properties similar to the real number 0. If A is an m by n matrix and 0 is an m by n zero matrix, then

$$A + 0 = A$$

In other words, the zero matrix is the additive identity in matrix algebra.

Scalar Multiples of a Matrix

2 We can multiply a matrix by a real number. If k is a real number and A is an m by n matrix, the matrix kA is the m by n matrix formed by multiplying each entry a_{ij} in A by k. The number k is sometimes referred to as a **scalar,** and the matrix kA is called a **scalar multiple** of A.

EXAMPLE 5 Operations Using Matrices

Suppose that

$$A = \begin{bmatrix} 3 & 1 & 5 \\ -2 & 0 & 6 \end{bmatrix} \qquad B = \begin{bmatrix} 4 & 1 & 0 \\ 8 & 1 & -3 \end{bmatrix} \qquad C = \begin{bmatrix} 9 & 0 \\ -3 & 6 \end{bmatrix}$$

Find: (a) $4A$ (b) $\frac{1}{3}C$ (c) $3A - 2B$

Solution

(a) $$4A = 4\begin{bmatrix} 3 & 1 & 5 \\ -2 & 0 & 6 \end{bmatrix} = \begin{bmatrix} 4\cdot 3 & 4\cdot 1 & 4\cdot 5 \\ 4(-2) & 4\cdot 0 & 4\cdot 6 \end{bmatrix} = \begin{bmatrix} 12 & 4 & 20 \\ -8 & 0 & 24 \end{bmatrix}$$

(b) $$\frac{1}{3}C = \frac{1}{3}\begin{bmatrix} 9 & 0 \\ -3 & 6 \end{bmatrix} = \begin{bmatrix} \frac{1}{3}\cdot 9 & \frac{1}{3}\cdot 0 \\ \frac{1}{3}(-3) & \frac{1}{3}\cdot 6 \end{bmatrix} = \begin{bmatrix} 3 & 0 \\ -1 & 2 \end{bmatrix}$$

(c) $$\begin{aligned} 3A - 2B &= 3\begin{bmatrix} 3 & 1 & 5 \\ -2 & 0 & 6 \end{bmatrix} - 2\begin{bmatrix} 4 & 1 & 0 \\ 8 & 1 & -3 \end{bmatrix} \\ &= \begin{bmatrix} 3\cdot 3 & 3\cdot 1 & 3\cdot 5 \\ 3(-2) & 3\cdot 0 & 3\cdot 6 \end{bmatrix} - \begin{bmatrix} 2\cdot 4 & 2\cdot 1 & 2\cdot 0 \\ 2\cdot 8 & 2\cdot 1 & 2(-3) \end{bmatrix} \\ &= \begin{bmatrix} 9 & 3 & 15 \\ -6 & 0 & 18 \end{bmatrix} - \begin{bmatrix} 8 & 2 & 0 \\ 16 & 2 & -6 \end{bmatrix} \\ &= \begin{bmatrix} 9-8 & 3-2 & 15-0 \\ -6-16 & 0-2 & 18-(-6) \end{bmatrix} \\ &= \begin{bmatrix} 1 & 1 & 15 \\ -22 & -2 & 24 \end{bmatrix} \end{aligned}$$

◀

CHECK: Enter the matrices $[A]$, $[B]$, and $[C]$ into a graphing utility. Then find $4A$, $\frac{1}{3}C$, and $3A - 2B$.

NOW WORK PROBLEM 11.

We list next some of the algebraic properties of scalar multiplication. Let h and k be real numbers, and let A and B be m by n matrices. Then

Properties of Scalar Multiplication

$$k(hA) = (kh)A$$
$$(k + h)A = kA + hA$$
$$k(A + B) = kA + kB$$

The Product of Two Matrices

3 Unlike the straightforward definition for adding two matrices, the definition for multiplying two matrices is not what we might expect. In preparation for this definition, we need the following definitions:

A **row vector** R is a 1 by n matrix

$$R = [r_1 \quad r_2 \quad \cdots \quad r_n]$$

A **column vector** C is an n by 1 matrix

$$C = \begin{bmatrix} c_1 \\ c_2 \\ \vdots \\ c_n \end{bmatrix}$$

The **product** $\boldsymbol{RC}$ of R times C is defined as the number

$$RC = [r_1 \quad r_2 \quad \cdots \quad r_n]\begin{bmatrix} c_1 \\ c_2 \\ \vdots \\ c_n \end{bmatrix} = r_1c_1 + r_2c_2 + \cdots + r_nc_n$$

Notice that a row vector and a column vector can be multiplied only if they contain the same number of entries.

EXAMPLE 6 **The Product of a Row Vector by a Column Vector**

If $R = [3 \quad -5 \quad 2]$ and $C = \begin{bmatrix} 3 \\ 4 \\ -5 \end{bmatrix}$, then

$$RC = [3 \quad -5 \quad 2]\begin{bmatrix} 3 \\ 4 \\ -5 \end{bmatrix} = 3 \cdot 3 + (-5)4 + 2(-5)$$
$$= 9 - 20 - 10 = -21$$

Let's look at an application of the product of a row vector by a column vector.

EXAMPLE 7 **Using Matrices to Compute Revenue**

A clothing store sells men's shirts for \$25, silk ties for \$8, and wool suits for \$300. Last month, the store had sales consisting of 100 shirts, 200 ties, and 50 suits. What was the total revenue due to these sales?

Solution We set up a row vector R to represent the prices of each item and a column vector C to represent the corresponding number of items sold.
Then

$$
\begin{array}{cc}
\underset{\text{Shirts Ties Suits}}{\underset{\text{Prices}}{}} & \underset{\text{sold}}{\underset{\text{Number}}{}}
\end{array}
$$

$$
R = [25 \quad 8 \quad 300] \qquad C = \begin{bmatrix} 100 \\ 200 \\ 50 \end{bmatrix} \begin{array}{l} \text{Shirts} \\ \text{Ties} \\ \text{Suits} \end{array}
$$

The total revenue obtained is the product RC. That is,

$$
RC = [25 \ 8 \ 300] \begin{bmatrix} 100 \\ 200 \\ 50 \end{bmatrix}
$$

$$
= \underbrace{25 \cdot 100}_{\text{Shirt revenue}} + \underbrace{8 \cdot 200}_{\text{Tie revenue}} + \underbrace{300 \cdot 50}_{\text{Suit revenue}} = \underbrace{\$19,100}_{\text{Total revenue}}
$$

◀

The definition for multiplying two matrices is based on the definition of a row vector times a column vector.

> Let A denote an m by r matrix, and let B denote an r by n matrix. The **product** AB is defined as the m by n matrix whose entry in row i, column j is the product of the ith row of A and the jth column of B.

The definition of the product AB of two matrices A and B, in this order, requires that the number of columns of A equal the number of rows of B; otherwise, no product is defined.

$$
\underset{m \text{ by } r}{\overset{A}{}} \qquad\qquad \underset{r \text{ by } n}{\overset{B}{}}
$$

Must be same
for AB to be defined
AB is m by n.

An example will help to clarify the definition.

EXAMPLE 8 **Multiplying Two Matrices**

Find the product AB if

$$
A = \begin{bmatrix} 2 & 4 & -1 \\ 5 & 8 & 0 \end{bmatrix} \quad \text{and} \quad B = \begin{bmatrix} 2 & 5 & 1 & 4 \\ 4 & 8 & 0 & 6 \\ -3 & 1 & -2 & -1 \end{bmatrix}
$$

Solution First, we note that A is 2 by 3 and B is 3 by 4, so the product AB is defined and will be a 2 by 4 matrix.
Suppose that we want the entry in row 2, column 3 of AB. To find it, we find the product of the row vector from row 2 of A and the column vector from column 3 of B.

$$
\underset{\text{Row 2 of } A}{} \overset{\text{Column 3 of } B}{} \\
[5 \ 8 \ 0] \begin{bmatrix} 1 \\ 0 \\ -2 \end{bmatrix} = 5 \cdot 1 + 8 \cdot 0 + 0(-2) = 5
$$

So far we have

$$AB = \begin{bmatrix} \_\_ & \_\_ & \_\_ & \_\_ \\ \_\_ & \_\_ & 5 & \_\_ \end{bmatrix} \quad \begin{matrix} \text{Column 3 } \downarrow \\ \leftarrow \text{Row 2} \end{matrix}$$

Now, to find the entry in row 1, column 4 of AB, we find the product of row 1 of A and column 4 of B.

$$\underset{\text{Row 1 of } A}{\begin{bmatrix} 2 & 4 & -1 \end{bmatrix}} \underset{\text{Column 4 of } B}{\begin{bmatrix} 4 \\ 6 \\ -1 \end{bmatrix}} = 2 \cdot 4 + 4 \cdot 6 + (-1)(-1) = 33$$

Continuing in this fashion, we find AB.

$$AB = \begin{bmatrix} 2 & 4 & -1 \\ 5 & 8 & 0 \end{bmatrix} \begin{bmatrix} 2 & 5 & 1 & 4 \\ 4 & 8 & 0 & 6 \\ -3 & 1 & -2 & -1 \end{bmatrix}$$

$$= \begin{bmatrix} \text{Row 1 of } A \text{ times column 1 of } B & \text{Row 1 of } A \text{ times column 2 of } B & \text{Row 1 of } A \text{ times column 3 of } B & \text{Row 1 of } A \text{ times column 4 of } B \\ \text{Row 2 of } A \text{ times column 1 of } B & \text{Row 2 of } A \text{ times column 2 of } B & \text{Row 2 of } A \text{ times column 3 of } B & \text{Row 2 of } A \text{ times column 4 of } B \end{bmatrix}$$

$$= \begin{bmatrix} 2 \cdot 2 + 4 \cdot 4 + (-1)(-3) & 2 \cdot 5 + 4 \cdot 8 + (-1)1 & 2 \cdot 1 + 4 \cdot 0 + (-1)(-2) & 33 \text{ (from earlier)} \\ 5 \cdot 2 + 8 \cdot 4 + 0(-3) & 5 \cdot 5 + 8 \cdot 8 + 0 \cdot 1 & 5 \text{ (from earlier)} & 5 \cdot 4 + 8 \cdot 6 + 0(-1) \end{bmatrix}$$

$$= \begin{bmatrix} 23 & 41 & 4 & 33 \\ 42 & 89 & 5 & 68 \end{bmatrix}$$

◀

CHECK: Enter the matrices A and B. Then find AB. (See what happens if you try to find BA.)

NOW WORK PROBLEM 23.

Notice that for the matrices given in Example 8 the product BA is not defined, because B is 3 by 4 and A is 2 by 3.

Another result that can occur when multiplying two matrices is illustrated in the next example.

EXAMPLE 9 Multiplying Two Matrices

If

$$A = \begin{bmatrix} 2 & 1 & 3 \\ 1 & -1 & 0 \end{bmatrix} \quad \text{and} \quad B = \begin{bmatrix} 1 & 0 \\ 2 & 1 \\ 3 & 2 \end{bmatrix}$$

find: (a) AB (b) BA

Solution

(a) $AB = \begin{bmatrix} 2 & 1 & 3 \\ 1 & -1 & 0 \end{bmatrix} \begin{bmatrix} 1 & 0 \\ 2 & 1 \\ 3 & 2 \end{bmatrix} = \begin{bmatrix} 13 & 7 \\ -1 & -1 \end{bmatrix}$

2 by 3 3 by 2 2 by 2

(b) $BA = \begin{bmatrix} 1 & 0 \\ 2 & 1 \\ 3 & 2 \end{bmatrix} \begin{bmatrix} 2 & 1 & 3 \\ 1 & -1 & 0 \end{bmatrix} = \begin{bmatrix} 2 & 1 & 3 \\ 5 & 1 & 6 \\ 8 & 1 & 9 \end{bmatrix}$

3 by 2 2 by 3 3 by 3

◄

Notice in Example 9 that AB is 2 by 2 and BA is 3 by 3. It is possible for both AB and BA to be defined, yet be unequal. In fact, even if A and B are both n by n matrices, so that AB and BA are each defined and n by n, AB and BA will usually be unequal.

EXAMPLE 10 Multiplying Two Square Matrices

If

$$A = \begin{bmatrix} 2 & 1 \\ 0 & 4 \end{bmatrix} \quad \text{and} \quad B = \begin{bmatrix} -3 & 1 \\ 1 & 2 \end{bmatrix}$$

find: (a) AB (b) BA

Solution

(a) $AB = \begin{bmatrix} 2 & 1 \\ 0 & 4 \end{bmatrix} \begin{bmatrix} -3 & 1 \\ 1 & 2 \end{bmatrix} = \begin{bmatrix} -5 & 4 \\ 4 & 8 \end{bmatrix}$

(b) $BA = \begin{bmatrix} -3 & 1 \\ 1 & 2 \end{bmatrix} \begin{bmatrix} 2 & 1 \\ 0 & 4 \end{bmatrix} = \begin{bmatrix} -6 & 1 \\ 2 & 9 \end{bmatrix}$

◄

The preceding examples demonstrate that an important property of real numbers, the commutative property of multiplication, is not shared by matrices. In general:

Theorem Matrix multiplication is not commutative.

NOW WORK PROBLEMS 13 AND 15.

Next we give two of the properties of real numbers that are shared by matrices. Assuming that each product and sum is defined, we have the following:

Associative Property

$$A(BC) = (AB)C$$

Distributive Property

$$A(B + C) = AB + AC$$

The Identity Matrix

For an n by n square matrix, the entries located in row i, column i, $1 \le i \le n$, are called the **diagonal entries.** An n by n square matrix whose diagonal entries are 1's, while all other entries are 0's, is called the **identity matrix I_n.** For example,

$$I_2 = \begin{bmatrix} 1 & 0 \\ 0 & 1 \end{bmatrix} \qquad I_3 = \begin{bmatrix} 1 & 0 & 0 \\ 0 & 1 & 0 \\ 0 & 0 & 1 \end{bmatrix}$$

and so on.

EXAMPLE 11 **Multiplication with an Identity Matrix**

Let

$$A = \begin{bmatrix} -1 & 2 & 0 \\ 0 & 1 & 3 \end{bmatrix} \quad \text{and} \quad B = \begin{bmatrix} 3 & 2 \\ 4 & 6 \\ 5 & 2 \end{bmatrix}$$

Find: (a) AI_3 (b) I_2A (c) BI_2

Solution (a) $AI_3 = \begin{bmatrix} -1 & 2 & 0 \\ 0 & 1 & 3 \end{bmatrix}\begin{bmatrix} 1 & 0 & 0 \\ 0 & 1 & 0 \\ 0 & 0 & 1 \end{bmatrix} = \begin{bmatrix} -1 & 2 & 0 \\ 0 & 1 & 3 \end{bmatrix} = A$

(b) $I_2A = \begin{bmatrix} 1 & 0 \\ 0 & 1 \end{bmatrix}\begin{bmatrix} -1 & 2 & 0 \\ 0 & 1 & 3 \end{bmatrix} = \begin{bmatrix} -1 & 2 & 0 \\ 0 & 1 & 3 \end{bmatrix} = A$

(c) $BI_2 = \begin{bmatrix} 3 & 2 \\ 4 & 6 \\ 5 & 2 \end{bmatrix}\begin{bmatrix} 1 & 0 \\ 0 & 1 \end{bmatrix} = \begin{bmatrix} 3 & 2 \\ 4 & 6 \\ 5 & 2 \end{bmatrix} = B$ ◀

Example 11 demonstrates the following property:

Identity Property

If A is an m by n matrix, then

$$I_mA = A \quad \text{and} \quad AI_n = A$$

If A is an n by n square matrix, then $AI_n = I_nA = A$.

An identity matrix has properties analogous to those of the real number 1. In other words, the identity matrix is a multiplicative identity in matrix algebra.

The Inverse of a Matrix

4 Let A be a square n by n matrix. If there exists an n by n matrix A^{-1}, read "A inverse," for which

$$AA^{-1} = A^{-1}A = I_n$$

then A^{-1} is called the **inverse** of the matrix A.

As we shall soon see, not every square matrix has an inverse. When a matrix A does have an inverse A^{-1}, then A is said to be **nonsingular.** If a matrix A has no inverse, it is called **singular.***

EXAMPLE 12 Multiplying a Matrix by Its Inverse

Show that the inverse of

$$A = \begin{bmatrix} 3 & 1 \\ 2 & 1 \end{bmatrix} \quad \text{is} \quad A^{-1} = \begin{bmatrix} 1 & -1 \\ -2 & 3 \end{bmatrix}$$

Solution We need to show that $AA^{-1} = A^{-1}A = I_2$.

$$AA^{-1} = \begin{bmatrix} 3 & 1 \\ 2 & 1 \end{bmatrix}\begin{bmatrix} 1 & -1 \\ -2 & 3 \end{bmatrix} = \begin{bmatrix} 1 & 0 \\ 0 & 1 \end{bmatrix} = I_2$$

$$A^{-1}A = \begin{bmatrix} 1 & -1 \\ -2 & 3 \end{bmatrix}\begin{bmatrix} 3 & 1 \\ 2 & 1 \end{bmatrix} = \begin{bmatrix} 1 & 0 \\ 0 & 1 \end{bmatrix} = I_2$$

◀

We now show one way to find the inverse of

$$A = \begin{bmatrix} 3 & 1 \\ 2 & 1 \end{bmatrix}$$

Suppose that A^{-1} is given by

$$A^{-1} = \begin{bmatrix} x & y \\ z & w \end{bmatrix} \qquad \textbf{(1)}$$

where x, y, z, and w are four variables. Based on the definition of an inverse, if, indeed, A has an inverse, we have

$$AA^{-1} = I_2$$

$$\begin{bmatrix} 3 & 1 \\ 2 & 1 \end{bmatrix}\begin{bmatrix} x & y \\ z & w \end{bmatrix} = \begin{bmatrix} 1 & 0 \\ 0 & 1 \end{bmatrix}$$

$$\begin{bmatrix} 3x + z & 3y + w \\ 2x + z & 2y + w \end{bmatrix} = \begin{bmatrix} 1 & 0 \\ 0 & 1 \end{bmatrix}$$

Because corresponding entries must be equal, it follows that this matrix equation is equivalent to four ordinary equations.

$$\begin{cases} 3x + z = 1 \\ 2x + z = 0 \end{cases} \qquad \begin{cases} 3y + w = 0 \\ 2y + w = 1 \end{cases}$$

The augmented matrix of each system is

$$\left[\begin{array}{cc|c} 3 & 1 & 1 \\ 2 & 1 & 0 \end{array}\right] \qquad \left[\begin{array}{cc|c} 3 & 1 & 0 \\ 2 & 1 & 1 \end{array}\right] \qquad \textbf{(2)}$$

The usual procedure would be to transform each augmented matrix into reduced row echelon form. Notice, though, that the left sides of the augmented matrices are equal, so the same row operations (see Section 10.2) can be

*If the determinant of A is zero, then A is singular. (Refer to Section 10.3.)

used to reduce each side. We find it more efficient to combine the two augmented matrices (2) into a single matrix, as shown next, and then transform it into reduced row echelon form.

$$\left[\begin{array}{cc|cc} 3 & 1 & 1 & 0 \\ 2 & 1 & 0 & 1 \end{array}\right]$$

Now we attempt to transform the left side into an identity matrix.

$$\left[\begin{array}{cc|cc} 3 & 1 & 1 & 0 \\ 2 & 1 & 0 & 1 \end{array}\right] \underset{R_1 = -1r_2 + r_1}{\longrightarrow} \left[\begin{array}{cc|cc} 1 & 0 & 1 & -1 \\ 2 & 1 & 0 & 1 \end{array}\right]$$

$$\underset{R_2 = -2r_1 + r_2}{\longrightarrow} \left[\begin{array}{cc|cc} 1 & 0 & 1 & -1 \\ 0 & 1 & -2 & 3 \end{array}\right] \qquad \textbf{(3)}$$

Matrix (3) is in reduced row echelon form. Now we reverse the earlier step of combining the two augmented matrices in (2) and write the single matrix (3) as two augmented matrices.

$$\left[\begin{array}{cc|c} 1 & 0 & 1 \\ 0 & 1 & -2 \end{array}\right] \text{ and } \left[\begin{array}{cc|c} 1 & 0 & -1 \\ 0 & 1 & 3 \end{array}\right]$$

We conclude from these matrices that $x = 1$, $z = -2$, and $y = -1$, $w = 3$. Substituting these values into matrix (1), we find that

$$A^{-1} = \begin{bmatrix} 1 & -1 \\ -2 & 3 \end{bmatrix}$$

Notice in display (3) that the 2 by 2 matrix to the right of the vertical bar is, in fact, the inverse of A. Also notice that the identity matrix I_2 is the matrix that appears to the left of the vertical bar. These observations and the procedures followed above will work in general.

Procedure for Finding the Inverse of a Nonsingular Matrix

To find the inverse of an n by n nonsingular matrix A, proceed as follows:

Step 1: Form the matrix $[A|I_n]$.

Step 2: Transform the matrix $[A|I_n]$ into reduced row echelon form.

Step 3: The reduced row echelon form of $[A|I_n]$ will contain the identity matrix I_n on the left of the vertical bar; the n by n matrix on the right of the vertical bar is the inverse of A.

In other words, if A is nonsingular, we begin with the matrix $[A|I_n]$ and, after transforming it into reduced row echelon form, we end up with the matrix $[I_n|A^{-1}]$.

Let's look at another example.

EXAMPLE 13 **Finding the Inverse of a Matrix**

The matrix

$$A = \begin{bmatrix} 1 & 1 & 0 \\ -1 & 3 & 4 \\ 0 & 4 & 3 \end{bmatrix}$$

is nonsingular. Find its inverse.

Solution First, we form the matrix

$$[A|I_3] = \left[\begin{array}{ccc|ccc} 1 & 1 & 0 & 1 & 0 & 0 \\ -1 & 3 & 4 & 0 & 1 & 0 \\ 0 & 4 & 3 & 0 & 0 & 1 \end{array}\right]$$

Next, we use row operations to transform $[A|I_3]$ into reduced row echelon form.

$$\left[\begin{array}{ccc|ccc} 1 & 1 & 0 & 1 & 0 & 0 \\ -1 & 3 & 4 & 0 & 1 & 0 \\ 0 & 4 & 3 & 0 & 0 & 1 \end{array}\right] \underset{R_2 = r_1 + r_2}{\rightarrow} \left[\begin{array}{ccc|ccc} 1 & 1 & 0 & 1 & 0 & 0 \\ 0 & 4 & 4 & 1 & 1 & 0 \\ 0 & 4 & 3 & 0 & 0 & 1 \end{array}\right] \underset{R_2 = \frac{1}{4}r_2}{\rightarrow} \left[\begin{array}{ccc|ccc} 1 & 1 & 0 & 1 & 0 & 0 \\ 0 & 1 & 1 & \frac{1}{4} & \frac{1}{4} & 0 \\ 0 & 4 & 3 & 0 & 0 & 1 \end{array}\right]$$

$$\underset{\substack{R_1 = -1r_2 + r_1 \\ R_3 = -4r_2 + r_3}}{\rightarrow} \left[\begin{array}{ccc|ccc} 1 & 0 & -1 & \frac{3}{4} & -\frac{1}{4} & 0 \\ 0 & 1 & 1 & \frac{1}{4} & \frac{1}{4} & 0 \\ 0 & 0 & -1 & -1 & -1 & 1 \end{array}\right] \underset{R_3 = -1r_3}{\rightarrow} \left[\begin{array}{ccc|ccc} 1 & 0 & -1 & \frac{3}{4} & -\frac{1}{4} & 0 \\ 0 & 1 & 1 & \frac{1}{4} & \frac{1}{4} & 0 \\ 0 & 0 & 1 & 1 & 1 & -1 \end{array}\right]$$

$$\underset{\substack{R_1 = r_3 + r_1 \\ R_2 = -1r_3 + r_2}}{\rightarrow} \left[\begin{array}{ccc|ccc} 1 & 0 & 0 & \frac{7}{4} & \frac{3}{4} & -1 \\ 0 & 1 & 0 & -\frac{3}{4} & -\frac{3}{4} & 1 \\ 0 & 0 & 1 & 1 & 1 & -1 \end{array}\right]$$

The matrix $[A|I_3]$ is now in reduced row echelon form, and the identity matrix I_3 is on the left of the vertical bar. Hence, the inverse of A is

$$A^{-1} = \begin{bmatrix} \frac{7}{4} & \frac{3}{4} & -1 \\ -\frac{3}{4} & -\frac{3}{4} & 1 \\ 1 & 1 & -1 \end{bmatrix}$$

◀

You can (and should) verify that this is the correct inverse by showing that $AA^{-1} = A^{-1}A = I_3$.

Figure 8

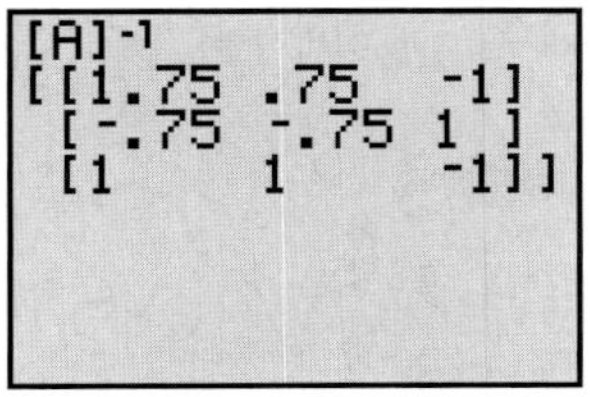

CHECK: Enter the matrix A into a graphing utility. Figure 8 shows A^{-1}.

NOW WORK PROBLEM 31.

If transforming the matrix $[A|I_n]$ into reduced row echelon form does not result in the identity matrix I_n to the left of the vertical bar, then A is singular and has no inverse. The next example demonstrates such a matrix.

EXAMPLE 14 Showing That a Matrix Has No Inverse

Show that the following matrix has no inverse.

$$A = \begin{bmatrix} 4 & 6 \\ 2 & 3 \end{bmatrix}$$

Solution Proceeding as in Example 13, we form the matrix

$$[A|I_2] = \left[\begin{array}{cc|cc} 4 & 6 & 1 & 0 \\ 2 & 3 & 0 & 1 \end{array}\right]$$

Then we use row operations to transform $[A|I_2]$ into reduced row echelon form.

$$[A|I_2] = \left[\begin{array}{cc|cc} 4 & 6 & 1 & 0 \\ 2 & 3 & 0 & 1 \end{array}\right] \underset{R_1 = \frac{1}{4}r_1}{\rightarrow} \left[\begin{array}{cc|cc} 1 & \frac{3}{2} & \frac{1}{4} & 0 \\ 2 & 3 & 0 & 1 \end{array}\right] \underset{R_2 = -2r_1 + r_2}{\rightarrow} \left[\begin{array}{cc|cc} 1 & \frac{3}{2} & \frac{1}{4} & 0 \\ 0 & 0 & -\frac{1}{2} & 1 \end{array}\right]$$

The matrix $[A|I_2]$ is sufficiently reduced for us to see that the identity matrix cannot appear to the left of the vertical bar. We conclude that A is singular and so has no inverse. ◀

CHECK: Enter the matrix A. Try to find its inverse. What happens?

NOW WORK PROBLEM 59.

Solving Systems of Linear Equations

5 Inverse matrices can be used to solve systems of equations in which the number of equations is the same as the number of variables.

EXAMPLE 15 Using the Inverse Matrix to Solve a System of Linear Equations

Solve the system of equations: $\begin{cases} x + y = 3 \\ -x + 3y + 4z = -3 \\ 4y + 3z = 2 \end{cases}$

Solution If we let

$$A = \begin{bmatrix} 1 & 1 & 0 \\ -1 & 3 & 4 \\ 0 & 4 & 3 \end{bmatrix} \quad X = \begin{bmatrix} x \\ y \\ z \end{bmatrix} \quad B = \begin{bmatrix} 3 \\ -3 \\ 2 \end{bmatrix}$$

then the original system of equations can be written compactly as the matrix equation

$$AX = B \tag{4}$$

We know from Example 13 that the matrix A has the inverse A^{-1}, so we multiply each side of equation (4) by A^{-1}.

$$\begin{aligned} AX &= B & & \\ A^{-1}(AX) &= A^{-1}B & & \text{Multiply both sides by } A^{-1}. \\ (A^{-1}A)X &= A^{-1}B & & \text{Associative property of multiplication} \\ I_3X &= A^{-1}B & & \text{Definition of inverse matrix} \\ X &= A^{-1}B & & \text{Property of identity matrix} \end{aligned} \tag{5}$$

Now we use (5) to find $X = \begin{bmatrix} x \\ y \\ z \end{bmatrix}$.

$$X = \begin{bmatrix} x \\ y \\ z \end{bmatrix} = A^{-1}B = \underbrace{\begin{bmatrix} \frac{7}{4} & \frac{3}{4} & -1 \\ -\frac{3}{4} & -\frac{3}{4} & 1 \\ 1 & 1 & -1 \end{bmatrix}}_{\text{Example 13}} \begin{bmatrix} 3 \\ -3 \\ 2 \end{bmatrix} = \begin{bmatrix} 1 \\ 2 \\ -2 \end{bmatrix}$$

Thus, $x = 1$, $y = 2$, $z = -2$. ◀

The method used in Example 15 to solve a system of equations is particularly useful when it is necessary to solve several systems of equations in which the constants appearing to the right of the equal signs change, while the coefficients of the variables on the left side remain the same. See Problems 39–58 for some illustrations. Be careful; this method can only be used if the inverse exists. If it does not exist, row reduction must be used, since the system is either inconsistent or dependent.

HISTORICAL FEATURE

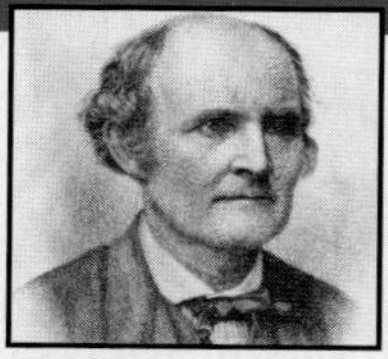

Arthur Cayley (1821–1895)

Matrices were invented in 1857 by Arthur Cayley (1821–1895) as a way of efficiently computing the result of substituting one linear system into another (see Historical Problem 2). The resulting system had incredible richness, in the sense that a very wide variety of mathematical systems could be mimicked by the matrices. Cayley and his friend James J. Sylvester (1814–1897) spent much of the rest of their lives elaborating the theory. The torch was then passed to Georg Frobenius (1849–1917), whose deep investigations established a central place for matrices in modern mathematics. In 1925, rather to the surprise of physicists, it was found that matrices (with complex numbers in them) were exactly the right tool for describing the behavior of atomic systems. Today, matrices are used in a wide variety of applications.

Historical Problems

1. **Matrices and Complex Numbers** Frobenius emphasized in his research how matrices could be used to mimic other mathematical systems. Here, we mimic the behavior of complex numbers using matrices. Mathematicians call such a relationship an isomorphism.

Complex number ⟷ Matrix

$$a + bi \longleftrightarrow \begin{bmatrix} a & b \\ -b & a \end{bmatrix}$$

Note that the complex number can be read off the top line of the matrix.

$$2 + 3i \longleftrightarrow \begin{bmatrix} 2 & 3 \\ -3 & 2 \end{bmatrix} \quad \text{and} \quad \begin{bmatrix} 4 & -2 \\ 2 & 4 \end{bmatrix} \longleftrightarrow 4 - 2i$$

(a) Find the matrices corresponding to $2 - 5i$ and $1 + 3i$.

(b) Multiply the two matrices.

(continued on page 741)

(c) Find the corresponding complex number for the matrix found in part (b).

(d) Multiply $2 - 5i$ by $1 + 3i$. The result should be the same as that found in part (c).

The process also works for addition and subtraction. Try it for yourself.

2. Cayley's Definition of Matrix Multiplication Cayley invented matrix multiplication to simplify the following problem:

$$\begin{cases} u = ar + bs \\ v = cr + ds \end{cases} \qquad \begin{cases} x = ku + lv \\ y = mu + nv \end{cases}$$

(a) Find x and y in terms of r and s by substituting u and v from the first system of equations into the second system of equations.

(b) Use the result of part (a) to find the 2 by 2 matrix A in

$$\begin{bmatrix} x \\ y \end{bmatrix} = A\begin{bmatrix} r \\ s \end{bmatrix}$$

(c) Now look at the following way to do it. Write the equations in matrix form.

$$\begin{bmatrix} u \\ v \end{bmatrix} = \begin{bmatrix} a & b \\ c & d \end{bmatrix}\begin{bmatrix} r \\ s \end{bmatrix} \qquad \begin{bmatrix} x \\ y \end{bmatrix} = \begin{bmatrix} k & l \\ m & n \end{bmatrix}\begin{bmatrix} u \\ v \end{bmatrix}$$

So

$$\begin{bmatrix} x \\ y \end{bmatrix} = \begin{bmatrix} k & l \\ m & n \end{bmatrix}\begin{bmatrix} a & b \\ c & d \end{bmatrix}\begin{bmatrix} r \\ s \end{bmatrix}$$

Do you see how Cayley defined matrix multiplication?

10.4 Assess Your Understanding

Concepts and Vocabulary

1. A matrix B, for which $AB = I_n$, the identity matrix, is called the ________ of A.

2. A matrix that has the same number of rows as columns is called a(n) ________ matrix.

3. In the algebra of matrices, the matrix that has properties similar to the number 1 is called the ________ matrix.

4. *True or False:* Every square matrix has an inverse.

5. *True or False:* Matrix multiplication is commutative.

6. *True or False:* Any pair of matrices can be multiplied.

Exercises

In Problems 7–22, use the following matrices to compute the given expression.

$$A = \begin{bmatrix} 0 & 3 & -5 \\ 1 & 2 & 6 \end{bmatrix} \qquad B = \begin{bmatrix} 4 & 1 & 0 \\ -2 & 3 & -2 \end{bmatrix} \qquad C = \begin{bmatrix} 4 & 1 \\ 6 & 2 \\ -2 & 3 \end{bmatrix}$$

7. $A + B$ **8.** $A - B$ **9.** $4A$ **10.** $-3B$

11. $3A - 2B$ **12.** $2A + 4B$ **13.** AC **14.** BC

15. CA **16.** CB **17.** $C(A + B)$ **18.** $(A + B)C$

19. $AC - 3I_2$ **20.** $CA + 5I_3$ **21.** $CA - CB$ **22.** $AC + BC$

In Problems 23–28, compute each product.

23. $\begin{bmatrix} 2 & -2 \\ 1 & 0 \end{bmatrix}\begin{bmatrix} 2 & 1 & 4 & 6 \\ 3 & -1 & 3 & 2 \end{bmatrix}$

24. $\begin{bmatrix} 4 & 1 \\ 2 & 1 \end{bmatrix}\begin{bmatrix} -6 & 6 & 1 & 0 \\ 2 & 5 & 4 & -1 \end{bmatrix}$

25. $\begin{bmatrix} 1 & 2 & 3 \\ 0 & -1 & 4 \end{bmatrix}\begin{bmatrix} 1 & 2 \\ -1 & 0 \\ 2 & 4 \end{bmatrix}$

26. $\begin{bmatrix} 1 & -1 \\ -3 & 2 \\ 0 & 5 \end{bmatrix}\begin{bmatrix} 2 & 8 & -1 \\ 3 & 6 & 0 \end{bmatrix}$

27. $\begin{bmatrix} 1 & 0 & 1 \\ 2 & 4 & 1 \\ 3 & 6 & 1 \end{bmatrix}\begin{bmatrix} 1 & 3 \\ 6 & 2 \\ 8 & -1 \end{bmatrix}$

28. $\begin{bmatrix} 4 & -2 & 3 \\ 0 & 1 & 2 \\ -1 & 0 & 1 \end{bmatrix}\begin{bmatrix} 2 & 6 \\ 1 & -1 \\ 0 & 2 \end{bmatrix}$

In Problems 29–38, each matrix is nonsingular. Find the inverse of each matrix. Be sure to check your answer.

29. $\begin{bmatrix} 2 & 1 \\ 1 & 1 \end{bmatrix}$

30. $\begin{bmatrix} 3 & -1 \\ -2 & 1 \end{bmatrix}$

31. $\begin{bmatrix} 6 & 5 \\ 2 & 2 \end{bmatrix}$

32. $\begin{bmatrix} -4 & 1 \\ 6 & -2 \end{bmatrix}$

33. $\begin{bmatrix} 2 & 1 \\ a & a \end{bmatrix}, \quad a \neq 0$

34. $\begin{bmatrix} b & 3 \\ b & 2 \end{bmatrix}, \quad b \neq 0$

35. $\begin{bmatrix} 1 & -1 & 1 \\ 0 & -2 & 1 \\ -2 & -3 & 0 \end{bmatrix}$

36. $\begin{bmatrix} 1 & 0 & 2 \\ -1 & 2 & 3 \\ 1 & -1 & 0 \end{bmatrix}$

37. $\begin{bmatrix} 1 & 1 & 1 \\ 3 & 2 & -1 \\ 3 & 1 & 2 \end{bmatrix}$

38. $\begin{bmatrix} 3 & 3 & 1 \\ 1 & 2 & 1 \\ 2 & -1 & 1 \end{bmatrix}$

In Problems 39–58, use the inverses found in Problems 29–38 to solve each system of equations.

39. $\begin{cases} 2x + y = 8 \\ x + y = 5 \end{cases}$

40. $\begin{cases} 3x - y = 8 \\ -2x + y = 4 \end{cases}$

41. $\begin{cases} 2x + y = 0 \\ x + y = 5 \end{cases}$

42. $\begin{cases} 3x - y = 4 \\ -2x + y = 5 \end{cases}$

43. $\begin{cases} 6x + 5y = 7 \\ 2x + 2y = 2 \end{cases}$

44. $\begin{cases} -4x + y = 0 \\ 6x - 2y = 14 \end{cases}$

45. $\begin{cases} 6x + 5y = 13 \\ 2x + 2y = 5 \end{cases}$

46. $\begin{cases} -4x + y = 5 \\ 6x - 2y = -9 \end{cases}$

47. $\begin{cases} 2x + y = -3 \\ ax + ay = -a \end{cases} \quad a \neq 0$

48. $\begin{cases} bx + 3y = 2b + 3 \\ bx + 2y = 2b + 2 \end{cases} \quad b \neq 0$

49. $\begin{cases} 2x + y = \dfrac{7}{a} \\ ax + ay = 5 \end{cases} \quad a \neq 0$

50. $\begin{cases} bx + 3y = 14 \\ bx + 2y = 10 \end{cases} \quad b \neq 0$

51. $\begin{cases} x - y + z = 0 \\ -2y + z = -1 \\ -2x - 3y = -5 \end{cases}$

52. $\begin{cases} x + 2z = 6 \\ -x + 2y + 3z = -5 \\ x - y = 6 \end{cases}$

53. $\begin{cases} x - y + z = 2 \\ -2y + z = 2 \\ -2x - 3y = \dfrac{1}{2} \end{cases}$

54. $\begin{cases} x + 2z = 2 \\ -x + 2y + 3z = -\dfrac{3}{2} \\ x - y = 2 \end{cases}$

55. $\begin{cases} x + y + z = 9 \\ 3x + 2y - z = 8 \\ 3x + y + 2z = 1 \end{cases}$

56. $\begin{cases} 3x + 3y + z = 8 \\ x + 2y + z = 5 \\ 2x - y + z = 4 \end{cases}$

57. $\begin{cases} x + y + z = 2 \\ 3x + 2y - z = \dfrac{7}{3} \\ 3x + y + 2z = \dfrac{10}{3} \end{cases}$

58. $\begin{cases} 3x + 3y + z = 1 \\ x + 2y + z = 0 \\ 2x - y + z = 4 \end{cases}$

In Problems 59–64, show that each matrix has no inverse.

59. $\begin{bmatrix} 4 & 2 \\ 2 & 1 \end{bmatrix}$

60. $\begin{bmatrix} -3 & \frac{1}{2} \\ 6 & -1 \end{bmatrix}$

61. $\begin{bmatrix} 15 & 3 \\ 10 & 2 \end{bmatrix}$

62. $\begin{bmatrix} -3 & 0 \\ 4 & 0 \end{bmatrix}$

63. $\begin{bmatrix} -3 & 1 & -1 \\ 1 & -4 & -7 \\ 1 & 2 & 5 \end{bmatrix}$

64. $\begin{bmatrix} 1 & 1 & -3 \\ 2 & -4 & 1 \\ -5 & 7 & 1 \end{bmatrix}$

In Problems 65–68, use a graphing utility to find the inverse, if it exists, of each matrix. Round answers to two decimal places.

65. $\begin{bmatrix} 25 & 61 & -12 \\ 18 & -2 & 4 \\ 8 & 35 & 21 \end{bmatrix}$

66. $\begin{bmatrix} 18 & -3 & 4 \\ 6 & -20 & 14 \\ 10 & 25 & -15 \end{bmatrix}$

67. $\begin{bmatrix} 44 & 21 & 18 & 6 \\ -2 & 10 & 15 & 5 \\ 21 & 12 & -12 & 4 \\ -8 & -16 & 4 & 9 \end{bmatrix}$

68. $\begin{bmatrix} 16 & 22 & -3 & 5 \\ 21 & -17 & 4 & 8 \\ 2 & 8 & 27 & 20 \\ 5 & 15 & -3 & -10 \end{bmatrix}$

In Problems 69–72, use the idea behind Example 15 with a graphing utility to solve the following systems of equations. Round answers to two decimal places.

69. $\begin{cases} 25x + 61y - 12z = 10 \\ 18x - 12y + 7y = -9 \\ 3x + 4y - z = 12 \end{cases}$

70. $\begin{cases} 25x + 61y - 12z = 15 \\ 18x - 12y + 7z = -3 \\ 3x + 4y - z = 12 \end{cases}$

71. $\begin{cases} 25x + 61y - 12z = 21 \\ 18x - 12y + 7z = 7 \\ 3x + 4y - z = -2 \end{cases}$

72. $\begin{cases} 25x + 61y - 12z = 25 \\ 18x - 12y + 7z = 10 \\ 3x + 4y - z = -4 \end{cases}$

---

73. Computing the Cost of Production The Acme Steel Company is a producer of stainless steel and aluminum containers. On a certain day, the following stainless steel containers were manufactured: 500 with 10-gallon capacity, 350 with 5-gallon capacity, and 400 with 1-gallon capacity. On the same day, the following aluminum containers were manufactured: 700 with 10-gallon capacity, 500 with 5-gallon capacity, and 850 with 1-gallon capacity.

(a) Find a 2 by 3 matrix representing the above data. Find a 3 by 2 matrix to represent the same data.

(b) If the amount of material used in the 10-gallon containers is 15 pounds, the amount used in the 5-gallon containers is 8 pounds, and the amount used in the 1-gallon containers is 3 pounds, find a 3 by 1 matrix representing the amount of material.

(c) Multiply the 2 by 3 matrix found in part (a) and the 3 by 1 matrix found in part (b) to get a 2 by 1 matrix showing the day's usage of material.

(d) If stainless steel costs Acme \$0.10 per pound and aluminum costs \$0.05 per pound, find a 1 by 2 matrix representing cost.

(e) Multiply the matrices found in parts (c) and (d) to determine the total cost of the day's production.

74. Computing Profit Rizza Ford has two locations, one in the city and the other in the suburbs. In January, the city location sold 400 subcompacts, 250 intermediate-size cars, and 50 SUV's; in February, it sold 350 subcompacts, 100 intermediates, and 30 SUV's. At the suburban location in January, 450 subcompacts, 200 intermediates, and 140 SUV's were sold. In February, the suburban location sold 350 subcompacts, 300 intermediates, and 100 SUV's.

(a) Find 2 by 3 matrices that summarize the sales data for each location for January and February (one matrix for each month).

(b) Use matrix addition to obtain total sales for the two-month period.

(c) The profit on each kind of car is \$100 per subcompact, \$150 per intermediate, and \$200 per SUV. Find a 3 by 1 matrix representing this profit.

(d) Multiply the matrices found in parts (b) and (c) to get a 2 by 1 matrix showing the profit at each location.

75. Consider the 2 by 2 square matrix

$$A = \begin{bmatrix} a & b \\ c & d \end{bmatrix}$$

If $D = ad - bc \neq 0$, show that A is nonsingular and that

$$A^{-1} = \frac{1}{D}\begin{bmatrix} d & -b \\ -c & a \end{bmatrix}$$

76. Create a situation different from any found in the text that can be represented by a matrix.

10.5 Partial Fraction Decomposition

PREPARING FOR THIS SECTION *Before getting started, review the following:*

- Identity (Appendix A, Section A.5, p. 936)
- Proper and Improper Rational Functions (Section 3.3, pp. 167–168)
- Factoring Polynomials (Appendix A, Section A.3, pp. 920–923)
- Fundamental Theorem of Algebra (Section 3.7, p. 209)

Now work the 'Are You Prepared?' problems on page 750.

OBJECTIVES
1 Decompose $\frac{P}{Q}$, Where Q Has Only Nonrepeated Linear Factors
2 Decompose $\frac{P}{Q}$, Where Q Has Repeated Linear Factors
3 Decompose $\frac{P}{Q}$, Where Q Has a Nonrepeated Irreducible Quadratic Factor
4 Decompose $\frac{P}{Q}$, Where Q Has Repeated Irreducible Quadratic Factors

Consider the problem of adding the two fractions:

$$\frac{3}{x+4} \quad \text{and} \quad \frac{2}{x-3}$$

The result is

$$\frac{3}{x+4} + \frac{2}{x-3} = \frac{3(x-3) + 2(x+4)}{(x+4)(x-3)} = \frac{5x-1}{x^2+x-12}$$

The reverse procedure, of starting with the rational expression $\frac{5x-1}{x^2+x-12}$ and writing it as the sum (or difference) of the two simpler fractions $\dfrac{3}{x+4}$ and $\dfrac{2}{x-3}$, is referred to as **partial fraction decomposition,** and the two simpler fractions are called **partial fractions.** Decomposing a rational expression into a sum of partial fractions is important in solving certain types of calculus problems. This section presents a systematic way to decompose rational expressions.

We begin by recalling that a rational expression is the ratio of two polynomials, say, P and $Q \neq 0$. We assume that P and Q have no common factors. Recall also that a rational expression $\dfrac{P}{Q}$ is called **proper** if the degree of the polynomial in the numerator is less than the degree of the polynomial in the denominator. Otherwise, the rational expression is termed **improper.**

Because any improper rational expression can be reduced by long division to a mixed form consisting of the sum of a polynomial and a proper rational expression, we shall restrict the discussion that follows to proper rational expressions.

The partial fraction decomposition of the rational expression $\dfrac{P}{Q}$ depends on the factors of the denominator Q. Recall (from Section 3.7) that any polynomial whose coefficients are real numbers can be factored (over the real numbers) into products of linear and/or irreducible quadratic factors. Thus, the denominator Q of the rational expression $\dfrac{P}{Q}$ will contain only factors of one or both of the following types:

1. *Linear factors* of the form $x - a$, where a is a real number.
2. *Irreducible quadratic factors* of the form $ax^2 + bx + c$, where a, b, and c are real numbers, $a \neq 0$, and $b^2 - 4ac < 0$ (which guarantees that $ax^2 + bx + c$ cannot be written as the product of two linear factors with real coefficients).

1 As it turns out, there are four cases to be examined. We begin with the case for which Q has only nonrepeated linear factors.

Case 1: Q has only nonrepeated linear factors.

Under the assumption that Q has only nonrepeated linear factors, the polynomial Q has the form

$$Q(x) = (x - a_1)(x - a_2) \cdot \cdots \cdot (x - a_n)$$

where none of the numbers $a_1, a_2, \ldots, a_n$ are equal. In this case, the partial fraction decomposition of $\dfrac{P}{Q}$ is of the form

$$\frac{P(x)}{Q(x)} = \frac{A_1}{x - a_1} + \frac{A_2}{x - a_2} + \cdots + \frac{A_n}{x - a_n} \qquad (1)$$

where the numbers $A_1, A_2, \ldots, A_n$ are to be determined.

We show how to find these numbers in the example that follows.

EXAMPLE 1 Nonrepeated Linear Factors

Write the partial fraction decomposition of $\dfrac{x}{x^2 - 5x + 6}$

Solution First, we factor the denominator,

$$x^2 - 5x + 6 = (x - 2)(x - 3)$$

and conclude that the denominator contains only nonrepeated linear factors. Then we decompose the rational expression according to equation (1):

$$\frac{x}{x^2 - 5x + 6} = \frac{A}{x - 2} + \frac{B}{x - 3} \quad \textbf{(2)}$$

where A and B are to be determined. To find A and B, we clear the fractions by multiplying each side by $(x - 2)(x - 3) = x^2 - 5x + 6$. The result is

$$x = A(x - 3) + B(x - 2) \quad \textbf{(3)}$$

or

$$x = (A + B)x + (-3A - 2B)$$

This equation is an identity in x. We equate the coefficients of like powers of x to get

$$\begin{cases} 1 = A + B & \text{Equate coefficients of } x\text{: } 1x = (A + B)x. \\ 0 = -3A - 2B & \text{Equate coefficients of } x^0\text{, the constants: } 0x^0 = (-3A - 2B)x^0. \end{cases}$$

This system of two equations containing two variables, A and B, can be solved using whatever method you wish. Solving it, we get

$$A = -2 \qquad B = 3$$

From equation (2), the partial fraction decomposition is

$$\frac{x}{x^2 - 5x + 6} = \frac{-2}{x - 2} + \frac{3}{x - 3}$$

◀

CHECK: The decomposition can be checked by adding the fractions.

$$\frac{-2}{x - 2} + \frac{3}{x - 3} = \frac{-2(x - 3) + 3(x - 2)}{(x - 2)(x - 3)} = \frac{x}{(x - 2)(x - 3)}$$

$$= \frac{x}{x^2 - 5x + 6}$$

NOW WORK PROBLEM 13.

The numbers to be found in the partial fraction decomposition can sometimes be found more readily by using suitable choices for x (which may include complex numbers) in the identity obtained after fractions have been cleared. In Example 1, the identity after clearing fractions, equation (3), is

$$x = A(x - 3) + B(x - 2)$$

If we let $x = 2$ in this expression, the term containing B drops out, leaving $2 = A(-1)$, or $A = -2$. Similarly, if we let $x = 3$, the term containing A drops out, leaving $3 = B$. As before, $A = -2$ and $B = 3$.

We use this method in the next example.

2

Case 2: Q has repeated linear factors.

If the polynomial Q has a repeated factor, say $(x - a)^n$, $n \geq 2$ an integer, then, in the partial fraction decomposition of $\frac{P}{Q}$, we allow for the terms

$$\frac{A_1}{x - a} + \frac{A_2}{(x - a)^2} + \cdots + \frac{A_n}{(x - a)^n}$$

where the numbers $A_1, A_2, \ldots, A_n$ are to be determined.

EXAMPLE 2 **Repeated Linear Factors**

Write the partial fraction decomposition of $\frac{x + 2}{x^3 - 2x^2 + x}$

Solution First, we factor the denominator,

$$x^3 - 2x^2 + x = x(x^2 - 2x + 1) = x(x - 1)^2$$

and find that the denominator has the nonrepeated linear factor x and the repeated linear factor $(x - 1)^2$. By Case 1, we must allow for the term $\frac{A}{x}$ in the decomposition; and, by Case 2, we must allow for the terms $\frac{B}{x - 1} + \frac{C}{(x - 1)^2}$ in the decomposition.

We write

$$\frac{x + 2}{x^3 - 2x^2 + x} = \frac{A}{x} + \frac{B}{x - 1} + \frac{C}{(x - 1)^2} \qquad \textbf{(4)}$$

Again, we clear fractions by multiplying each side by $x^3 - 2x^2 + x = x(x - 1)^2$. The result is the identity

$$x + 2 = A(x - 1)^2 + Bx(x - 1) + Cx \qquad \textbf{(5)}$$

If we let $x = 0$ in this expression, the terms containing B and C drop out, leaving $2 = A(-1)^2$, or $A = 2$. Similarly, if we let $x = 1$, the terms containing A and B drop out, leaving $3 = C$. Then, equation (5) becomes

$$x + 2 = 2(x - 1)^2 + Bx(x - 1) + 3x$$

Now let $x = 2$ (any choice other than 0 or 1 will work as well). The result is

$$\begin{aligned} 4 &= 2(1)^2 + B(2)(1) + 3(2) \\ 4 &= 2 + 2B + 6 \\ 2B &= -4 \\ B &= -2 \end{aligned}$$

We have $A = 2$, $B = -2$, and $C = 3$.

From equation (4), the partial fraction decomposition is

$$\frac{x + 2}{x^3 - 2x^2 + x} = \frac{2}{x} + \frac{-2}{x - 1} + \frac{3}{(x - 1)^2}$$

◀

EXAMPLE 3 **Repeated Linear Factors**

Write the partial fraction decomposition of $\dfrac{x^3 - 8}{x^2(x - 1)^3}$

Solution The denominator contains the repeated linear factor x^2 and the repeated linear factor $(x - 1)^3$. The partial fraction decomposition takes the form

$$\frac{x^3 - 8}{x^2(x - 1)^3} = \frac{A}{x} + \frac{B}{x^2} + \frac{C}{x - 1} + \frac{D}{(x - 1)^2} + \frac{E}{(x - 1)^3} \qquad \textbf{(6)}$$

As before, we clear fractions and obtain the identity

$$x^3 - 8 = Ax(x - 1)^3 + B(x - 1)^3 + Cx^2(x - 1)^2 + Dx^2(x - 1) + Ex^2 \qquad \textbf{(7)}$$

Let $x = 0$. (Do you see why this choice was made?) Then

$$-8 = B(-1)$$
$$B = 8$$

Now let $x = 1$ in equation (7). Then

$$-7 = E$$

Use $B = 8$ and $E = -7$ in equation (7) and collect like terms.

$$\begin{aligned}
x^3 - 8 &= Ax(x - 1)^3 + 8(x - 1)^3 \\
&\quad + Cx^2(x - 1)^2 + Dx^2(x - 1) - 7x^2 \\
x^3 - 8 - 8(x^3 - 3x^2 + 3x - 1) + 7x^2 &= Ax(x - 1)^3 + Cx^2(x - 1)^2 + Dx^2(x - 1) \\
-7x^3 + 31x^2 - 24x &= x(x - 1)[A(x - 1)^2 + Cx(x - 1) + Dx] \\
x(x - 1)(-7x + 24) &= x(x - 1)[A(x - 1)^2 + Cx(x - 1) + Dx] \\
-7x + 24 &= A(x - 1)^2 + Cx(x - 1) + Dx \qquad \textbf{(8)}
\end{aligned}$$

We now work with equation (8). Let $x = 0$. Then

$$24 = A$$

Now let $x = 1$ in equation (8). Then

$$17 = D$$

Use $A = 24$ and $D = 17$ in equation (8) and collect like terms.

$$-7x + 24 = 24(x - 1)^2 + Cx(x - 1) + 17x$$

Now let $x = 2$. Then

$$\begin{aligned}
-14 + 24 &= 24 + C(2) + 34 \\
-48 &= 2C \\
-24 &= C
\end{aligned}$$

We now know all the numbers A, B, C, D, and E, so, from equation (6), we have the decomposition

$$\frac{x^3 - 8}{x^2(x - 1)^3} = \frac{24}{x} + \frac{8}{x^2} + \frac{-24}{x - 1} + \frac{17}{(x - 1)^2} + \frac{-7}{(x - 1)^3}$$ ◀

The method employed in Example 3, although somewhat tedious, is still preferable to solving the system of five equations containing five variables that the expansion of equation (6) leads to.

NOW WORK PROBLEM 19.

3 The final two cases involve irreducible quadratic factors. A quadratic factor is irreducible if it cannot be factored into linear factors with real coefficients. A quadratic expression $ax^2 + bx + c$ is irreducible whenever $b^2 - 4ac < 0$. For example, $x^2 + x + 1$ and $x^2 + 4$ are irreducible.

Case 3: Q contains a nonrepeated irreducible quadratic factor.

If Q contains a nonrepeated irreducible quadratic factor of the form $ax^2 + bx + c$, then, in the partial fraction decomposition of $\frac{P}{Q}$, allow for the term

$$\frac{Ax + B}{ax^2 + bx + c}$$

where the numbers A and B are to be determined.

EXAMPLE 4 Nonrepeated Irreducible Quadratic Factor

Write the partial fraction decomposition of $\frac{3x - 5}{x^3 - 1}$

Solution We factor the denominator,

$$x^3 - 1 = (x - 1)(x^2 + x + 1)$$

and find that it has a nonrepeated linear factor $x - 1$ and a nonrepeated irreducible quadratic factor $x^2 + x + 1$. We allow for the term $\frac{A}{x - 1}$ by Case 1, and we allow for the term $\frac{Bx + C}{x^2 + x + 1}$ by Case 3. We write

$$\frac{3x - 5}{x^3 - 1} = \frac{A}{x - 1} + \frac{Bx + C}{x^2 + x + 1} \tag{9}$$

We clear fractions by multiplying each side of equation (9) by $x^3 - 1 = (x - 1)(x^2 + x + 1)$ to obtain the identity

$$3x - 5 = A(x^2 + x + 1) + (Bx + C)(x - 1) \tag{10}$$

Now let $x = 1$. Then equation (10) gives $-2 = A(3)$, or $A = -\frac{2}{3}$. We use this value of A in equation (10) and simplify.

$$3x - 5 = -\frac{2}{3}(x^2 + x + 1) + (Bx + C)(x - 1)$$

$$3(3x - 5) = -2(x^2 + x + 1) + 3(Bx + C)(x - 1) \quad \text{Multiply each side by 3.}$$

$$9x - 15 = -2x^2 - 2x - 2 + 3(Bx + C)(x - 1)$$

$$2x^2 + 11x - 13 = 3(Bx + C)(x - 1) \quad \text{Collect terms.}$$

$$(2x + 13)(x - 1) = 3(Bx + C)(x - 1) \quad \text{Factor the left side.}$$
$$2x + 13 = 3Bx + 3C \quad \text{Cancel } x - 1 \text{ on each side.}$$
$$2 = 3B \quad \text{and} \quad 13 = 3C \quad \text{Equate coefficients.}$$
$$B = \frac{2}{3} \qquad C = \frac{13}{3}$$

From equation (9), we see that

$$\frac{3x - 5}{x^3 - 1} = \frac{-\frac{2}{3}}{x - 1} + \frac{\frac{2}{3}x + \frac{13}{3}}{x^2 + x + 1}$$

NOW WORK PROBLEM 21.

4

Case 4: Q contains repeated irreducible quadratic factors.

If the polynomial Q contains a repeated irreducible quadratic factor $(ax^2 + bx + c)^n$, $n \geq 2$, n an integer, then, in the partial fraction decomposition of $\frac{P}{Q}$, allow for the terms

$$\frac{A_1x + B_1}{ax^2 + bx + c} + \frac{A_2x + B_2}{(ax^2 + bx + c)^2} + \cdots + \frac{A_nx + B_n}{(ax^2 + bx + c)^n}$$

where the numbers $A_1, B_1, A_2, B_2, \ldots, A_n, B_n$ are to be determined.

EXAMPLE 5 Repeated Irreducible Quadratic Factor

Write the partial fraction decomposition of $\dfrac{x^3 + x^2}{(x^2 + 4)^2}$

Solution The denominator contains the repeated irreducible quadratic factor $(x^2 + 4)^2$, so we write

$$\frac{x^3 + x^2}{(x^2 + 4)^2} = \frac{Ax + B}{x^2 + 4} + \frac{Cx + D}{(x^2 + 4)^2} \qquad \textbf{(11)}$$

We clear fractions to obtain

$$x^3 + x^2 = (Ax + B)(x^2 + 4) + Cx + D$$

Collecting like terms yields

$$x^3 + x^2 = Ax^3 + Bx^2 + (4A + C)x + D + 4B$$

Equating coefficients, we arrive at the system

$$\begin{cases} A = 1 \\ B = 1 \\ 4A + C = 0 \\ D + 4B = 0 \end{cases}$$

The solution is $A = 1, B = 1, C = -4, D = -4$. From equation (11),

$$\frac{x^3 + x^2}{(x^2 + 4)^2} = \frac{x + 1}{x^2 + 4} + \frac{-4x - 4}{(x^2 + 4)^2}$$

NOW WORK PROBLEM 35.

10.5 Assess Your Understanding

'Are You Prepared?' *Answers are given at the end of these exercises. If you get a wrong answer, read the pages listed in red.*

1. *True or False:* The equation $(x - 1)^2 - 1 = x(x - 2)$ is an example of an identity. (p. 936)
2. *True or False:* The rational expression $\frac{5x^2 - 1}{x^3 + 1}$ is proper. (pp. 167–168)
3. Factor completely: $3x^4 + 6x^3 + 3x^2$. (pp. 920–923)
4. *True or False:* Every polynomial with real numbers as coefficients can be factored into products of linear and/or irreducible quadratic factors. (p. 209)

Exercises

In Problems 5–12, tell whether the given rational expression is proper or improper. If improper, rewrite it as the sum of a polynomial and a proper rational expression.

5. $\frac{x}{x^2 - 1}$
6. $\frac{5x + 2}{x^3 - 1}$
7. $\frac{x^2 + 5}{x^2 - 4}$
8. $\frac{3x^2 - 2}{x^2 - 1}$
9. $\frac{5x^3 + 2x - 1}{x^2 - 4}$
10. $\frac{3x^4 + x^2 - 2}{x^3 + 8}$
11. $\frac{x(x - 1)}{(x + 4)(x - 3)}$
12. $\frac{2x(x^2 + 4)}{x^2 + 1}$

In Problems 13–46, write the partial fraction decomposition of each rational expression.

13. $\frac{4}{x(x - 1)}$
14. $\frac{3x}{(x + 2)(x - 1)}$
15. $\frac{1}{x(x^2 + 1)}$
16. $\frac{1}{(x + 1)(x^2 + 4)}$
17. $\frac{x}{(x - 1)(x - 2)}$
18. $\frac{3x}{(x + 2)(x - 4)}$
19. $\frac{x^2}{(x - 1)^2(x + 1)}$
20. $\frac{x + 1}{x^2(x - 2)}$
21. $\frac{1}{x^3 - 8}$
22. $\frac{2x + 4}{x^3 - 1}$
23. $\frac{x^2}{(x - 1)^2(x + 1)^2}$
24. $\frac{x + 1}{x^2(x - 2)^2}$
25. $\frac{x - 3}{(x + 2)(x + 1)^2}$
26. $\frac{x^2 + x}{(x + 2)(x - 1)^2}$
27. $\frac{x + 4}{x^2(x^2 + 4)}$
28. $\frac{10x^2 + 2x}{(x - 1)^2(x^2 + 2)}$
29. $\frac{x^2 + 2x + 3}{(x + 1)(x^2 + 2x + 4)}$
30. $\frac{x^2 - 11x - 18}{x(x^2 + 3x + 3)}$
31. $\frac{x}{(3x - 2)(2x + 1)}$
32. $\frac{1}{(2x + 3)(4x - 1)}$
33. $\frac{x}{x^2 + 2x - 3}$
34. $\frac{x^2 - x - 8}{(x + 1)(x^2 + 5x + 6)}$
35. $\frac{x^2 + 2x + 3}{(x^2 + 4)^2}$
36. $\frac{x^3 + 1}{(x^2 + 16)^2}$
37. $\frac{7x + 3}{x^3 - 2x^2 - 3x}$
38. $\frac{x^5 + 1}{x^6 - x^4}$
39. $\frac{x^2}{x^3 - 4x^2 + 5x - 2}$
40. $\frac{x^2 + 1}{x^3 + x^2 - 5x + 3}$
41. $\frac{x^3}{(x^2 + 16)^3}$
42. $\frac{x^2}{(x^2 + 4)^3}$
43. $\frac{4}{2x^2 - 5x - 3}$
44. $\frac{4x}{2x^2 + 3x - 2}$
45. $\frac{2x + 3}{x^4 - 9x^2}$
46. $\frac{x^2 + 9}{x^4 - 2x^2 - 8}$

'Are You Prepared?' Answers

1. True **2.** True **3.** $3x^2(x + 1)^2$ **4.** True

10.6 Systems of Nonlinear Equations

PREPARING FOR THIS SECTION *Before getting started, review the following:*

- Lines (Section 1.3, pp. 22–30)
- Parabolas (Section 9.2, pp. 615–622)
- Hyperbolas (Section 9.4, pp. 635–646)
- Circles (Section 1.2, pp. 16–19)
- Ellipses (Section 9.3, pp. 625–632)

Now work the 'Are You Prepared?' problems on page 756.

OBJECTIVES 1 Solve a System of Nonlinear Equations Using Substitution
2 Solve a System of Nonlinear Equations Using Elimination

1 In Section 10.1 we observed that the solution to a system of linear equations could be found geometrically by determining the point(s) of intersection (if any) of the equations in the system. Similarly, when solving systems of nonlinear equations, the solution(s) also represent the point(s) of intersection (if any) of the graphs of the equations.

There is no general methodology for solving a system of nonlinear equations. There are times when substitution is best; other times, elimination is best; and there are times when neither of these methods works. Experience and a certain degree of imagination are your allies here.

Before we begin, two comments are in order.

1. If the system contains two variables and if the equations in the system are easy to graph, then graph them. By graphing each equation in the system, you can get an idea of how many solutions a system has and approximately where they are located.
2. Extraneous solutions can creep in when solving nonlinear systems, so it is imperative that all apparent solutions be checked.

EXAMPLE 1 **Solving a System of Nonlinear Equations**

Solve the following system of equations:

$$\begin{cases} 3x - y = -2 & (1) \text{ A line} \\ 2x^2 - y = 0 & (2) \text{ A parabola} \end{cases}$$

Solution Using Substitution

Figure 9

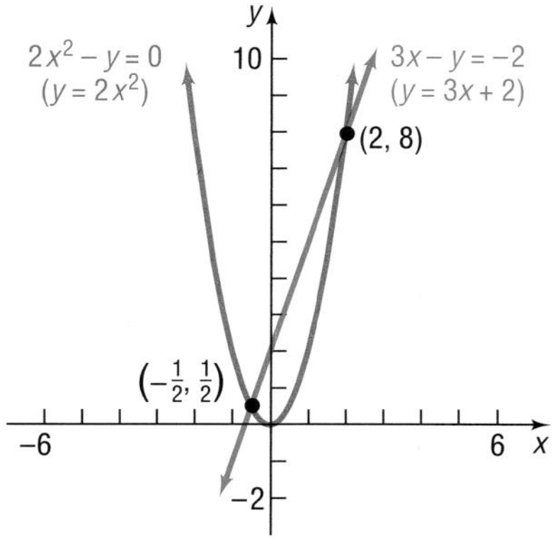

First, we notice that the system contains two variables and that we know how to graph each equation. In Figure 9, we see that the system apparently has two solutions.

We will use substitution to solve the system. Equation (1) is easily solved for y.

$$3x - y = -2 \quad (1)$$
$$y = 3x + 2$$

We substitute this expression for y in equation (2). The result is an equation containing just the variable x, which we can then solve.

$$2x^2 - y = 0 \quad (2)$$
$$2x^2 - (3x + 2) = 0 \quad \text{Substitute } y = 3x + 2.$$
$$2x^2 - 3x - 2 = 0 \quad \text{Remove parentheses.}$$
$$(2x + 1)(x - 2) = 0 \quad \text{Factor.}$$
$$2x + 1 = 0 \quad \text{or} \quad x - 2 = 0 \quad \text{Apply the Zero-Product Property.}$$
$$x = -\frac{1}{2} \quad \text{or} \quad x = 2$$

Using these values for x in $y = 3x + 2$, we find

$$y = 3\left(-\frac{1}{2}\right) + 2 = \frac{1}{2} \quad \text{or} \quad y = 3(2) + 2 = 8$$

The apparent solutions are $x = -\frac{1}{2}, y = \frac{1}{2}$ and $x = 2, y = 8$.

CHECK: For $x = -\frac{1}{2}, y = \frac{1}{2}$:

$$\begin{cases} 3\left(-\frac{1}{2}\right) - \frac{1}{2} = -\frac{3}{2} - \frac{1}{2} = -2 & (1) \\ 2\left(-\frac{1}{2}\right)^2 - \frac{1}{2} = 2\left(\frac{1}{4}\right) - \frac{1}{2} = 0 & (2) \end{cases}$$

For $x = 2, y = 8$:

$$\begin{cases} 3(2) - 8 = 6 - 8 = -2 & (1) \\ 2(2)^2 - 8 = 2(4) - 8 = 0 & (2) \end{cases}$$

Each solution checks. Now we know that the graphs in Figure 9 intersect at $\left(-\frac{1}{2}, \frac{1}{2}\right)$ and at $(2, 8)$. ◀

CHECK: Graph $3x - y = -2$ ($Y_1 = 3x + 2$) and $2x^2 - y = 0$ ($Y_2 = 2x^2$) and compare what you see with Figure 9. Use INTERSECT (twice) to find the two points of intersection.

NOW WORK PROBLEM 15 USING SUBSTITUTION.

2 Our next example illustrates how the method of elimination works for nonlinear systems.

EXAMPLE 2 Solving a System of Nonlinear Equations

Solve: $\begin{cases} x^2 + y^2 = 13 & (1) \text{ A circle} \\ x^2 - y = 7 & (2) \text{ A parabola} \end{cases}$

Solution Using Elimination

Figure 10

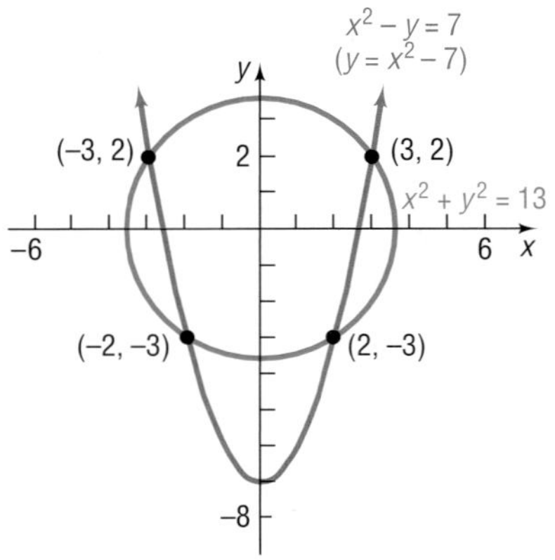

First, we graph each equation, as shown in Figure 10. Based on the graph, we expect four solutions. By subtracting equation (2) from equation (1), the variable x can be eliminated.

$$\begin{cases} x^2 + y^2 = 13 \\ x^2 - y = 7 \end{cases}$$
$$y^2 + y = 6 \quad \text{Subtract.}$$

This quadratic equation in y is easily solved by factoring.

$$y^2 + y - 6 = 0$$
$$(y + 3)(y - 2) = 0$$
$$y = -3 \quad \text{or} \quad y = 2$$

We use these values for y in equation (2) to find x.

If $y = 2$, then $x^2 = y + 7 = 9$, so $x = 3$ or -3.

If $y = -3$, then $x^2 = y + 7 = 4$, so $x = 2$ or -2.

We have four solutions: $x = 3, y = 2$; $x = -3, y = 2$; $x = 2, y = -3$; and $x = -2, y = -3$.

You should verify that, in fact, these four solutions also satisfy equation (1), so all four are solutions of the system. The four points, $(3, 2)$, $(-3, 2)$, $(2, -3)$, and $(-2, -3)$, are the points of intersection of the graphs. Look again at Figure 10. ◀

CHECK: Graph $x^2 + y^2 = 13$ and $x^2 - y = 7$. (Remember that to graph $x^2 + y^2 = 13$ requires two functions: $Y_1 = \sqrt{13 - x^2}$ and $Y_2 = -\sqrt{13 - x^2}$.) Compare what you see with Figure 10. Use INTERSECT to find the four points of intersection.

NOW WORK PROBLEM 13 USING ELIMINATION.

EXAMPLE 3 Solving a System of Nonlinear Equations

Solve: $$\begin{cases} x^2 + x + y^2 - 3y + 2 = 0 & (1) \\ x + 1 + \dfrac{y^2 - y}{x} = 0 & (2) \end{cases}$$

Solution Using Elimination First, we multiply equation (2) by x to eliminate the fraction. The result is an equivalent system because x cannot be 0. [Look at equation (2) to see why].

$$\begin{cases} x^2 + x + y^2 - 3y + 2 = 0 & (1) \\ x^2 + x + y^2 - y = 0 & (2) \end{cases}$$

Now subtract equation (2) from equation (1) to eliminate x. The result is

$$-2y + 2 = 0$$
$$y = 1 \quad \text{Solve for } y.$$

To find x, we back-substitute $y = 1$ in equation (1).

$$x^2 + x + 1 - 3 + 2 = 0$$
$$x^2 + x = 0$$
$$x(x + 1) = 0$$
$$x = 0 \quad \text{or} \quad x = -1$$

Because x cannot be 0, the value $x = 0$ is extraneous, and we discard it. We proceed to check the solution $x = -1, y = 1$.

CHECK: $$\begin{cases} (-1)^2 + (-1) + 1^2 - 3(1) + 2 = 1 - 1 + 1 - 3 + 2 = 0 & (1) \\ -1 + 1 + \dfrac{1^2 - 1}{-1} = 0 + \dfrac{0}{-1} = 0 & (2) \end{cases}$$

The only solution to the system is $x = -1, y = 1$. ◀

NOW WORK PROBLEMS 29 AND 49.

EXAMPLE 4 **Solving a System of Nonlinear Equations**

Solve: $\begin{cases} x^2 - y^2 = 4 & (1)\ \text{A hyperbola} \\ y = x^2 & (2)\ \text{A parabola} \end{cases}$

Figure 11

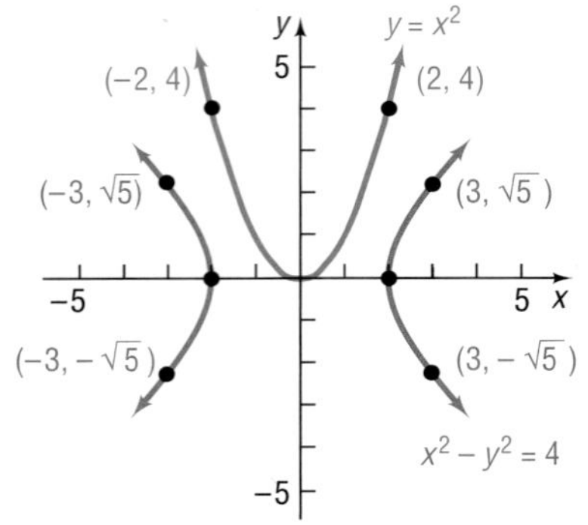

Solution Either substitution or elimination can be used here. We use substitution and replace x^2 by y in equation (1). The result is

$$y - y^2 = 4$$
$$y^2 - y + 4 = 0$$

This is a quadratic equation whose discriminant is $(-1)^2 - 4 \cdot 1 \cdot 4 = 1 - 4 \cdot 4 = -15 < 0$. The equation has no real solutions, so the system is inconsistent. The graphs of these two equations do not intersect. See Figure 11. ◀

EXAMPLE 5 **Solving a System of Nonlinear Equations**

Solve: $\begin{cases} 3xy - 2y^2 = -2 & (1) \\ 9x^2 + 4y^2 = 10 & (2) \end{cases}$

Solution We multiply equation (1) by 2 and add the result to equation (2) to eliminate the y^2 terms.

$$\begin{cases} 6xy - 4y^2 = -4 & (1) \\ 9x^2 + 4y^2 = 10 & (2)\ \text{Add.} \end{cases}$$
$$9x^2 + 6xy = 6$$
$$3x^2 + 2xy = 2 \qquad \text{Divide each side by 3.}$$

Since $x \neq 0$ (do you see why?), we can solve for y in this equation to get

$$y = \frac{2 - 3x^2}{2x}, \qquad x \neq 0 \qquad \textbf{(1)}$$

Now substitute for y in equation (2) of the system.

$$9x^2 + 4y^2 = 10 \qquad (2)$$
$$9x^2 + 4\left(\frac{2 - 3x^2}{2x}\right)^2 = 10 \qquad \text{Substitute } y = \frac{2 - 3x^2}{2x}.$$
$$9x^2 + \frac{4 - 12x^2 + 9x^4}{x^2} = 10$$
$$9x^4 + 4 - 12x^2 + 9x^4 = 10x^2 \qquad \text{Multiply both sides by } x^2.$$
$$18x^4 - 22x^2 + 4 = 0 \qquad \text{Subtract } 10x^2 \text{ from both sides.}$$
$$9x^4 - 11x^2 + 2 = 0 \qquad \text{Divide both sides by 2.}$$

This quadratic equation (in x^2) can be factored.

$$(9x^2 - 2)(x^2 - 1) = 0$$

$$9x^2 - 2 = 0 \quad \text{or} \quad x^2 - 1 = 0$$

$$x^2 = \frac{2}{9} \quad \text{or} \quad x^2 = 1$$

$$x = \pm\sqrt{\frac{2}{9}} = \pm\frac{\sqrt{2}}{3} \quad \text{or} \quad x = \pm 1$$

To find y, we use equation (1):

$$\text{If } x = \frac{\sqrt{2}}{3}: \quad y = \frac{2 - 3x^2}{2x} = \frac{2 - \frac{2}{3}}{2\left(\frac{\sqrt{2}}{3}\right)} = \frac{4}{2\sqrt{2}} = \sqrt{2}$$

$$\text{If } x = -\frac{\sqrt{2}}{3}: \quad y = \frac{2 - 3x^2}{2x} = \frac{2 - \frac{2}{3}}{2\left(-\frac{\sqrt{2}}{3}\right)} = \frac{4}{-2\sqrt{2}} = -\sqrt{2}$$

$$\text{If } x = 1: \quad y = \frac{2 - 3x^2}{2x} = \frac{2 - 3}{2} = -\frac{1}{2}$$

$$\text{If } x = -1: \quad y = \frac{2 - 3x^2}{2x} = \frac{2 - 3}{-2} = \frac{1}{2}$$

The system has four solutions. Check them for yourself. ◀

NOW WORK PROBLEM 47.

EXAMPLE 6 Running Long Distance Races

In a 50-mile race, the winner crosses the finish line 1 mile ahead of the second-place runner and 4 miles ahead of the third-place runner. Assuming that each runner maintains a constant speed throughout the race, by how many miles does the second-place runner beat the third-place runner?

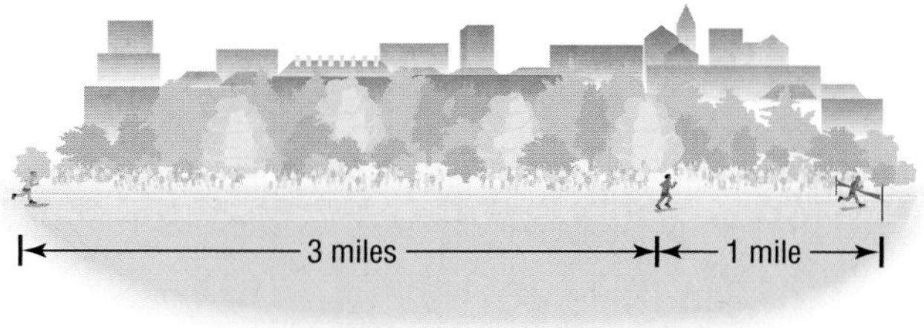

Solution Let v_1, v_2, and v_3 denote the speeds of the first-, second-, and third-place runners, respectively. Let t_1 and t_2 denote the times (in hours) required for the first-place runner and second-place runner to finish the race. Then we have the system of equations

$$\begin{cases} 50 = v_1 t_1 & \text{(1) First-place runner goes 50 miles in } t_1 \text{ hours.} \\ 49 = v_2 t_1 & \text{(2) Second-place runner goes 49 miles in } t_1 \text{ hours.} \\ 46 = v_3 t_1 & \text{(3) Third-place runner goes 46 miles in } t_1 \text{ hours.} \\ 50 = v_2 t_2 & \text{(4) Second-place runner goes 50 miles in } t_2 \text{ hours.} \end{cases}$$

We seek the distance of the third-place runner from the finish at time t_2. At time t_2, the third-place runner has gone a distance of $v_3 t_2$ miles, so the distance remaining is $50 - v_3 t_2$. Now

$$
\begin{aligned}
50 - v_3 t_2 &= 50 - v_3\left(t_1 \cdot \frac{t_2}{t_1}\right) \\
&= 50 - (v_3 t_1) \cdot \frac{t_2}{t_1} \\
&= 50 - 46 \cdot \frac{\frac{50}{v_2}}{\frac{50}{v_1}} && \left\{\begin{array}{l}\text{From (3), } v_3 t_1 = 46; \\ \text{from (4), } t_2 = \frac{50}{v_2}; \\ \text{from (1), } t_1 = \frac{50}{v_1}.\end{array}\right. \\
&= 50 - 46 \cdot \frac{v_1}{v_2} \\
&= 50 - 46 \cdot \frac{50}{49} && \text{Form the quotient of (1) and (2).} \\
&\approx 3.06 \text{ miles}
\end{aligned}
$$

When the second-place runner crosses the finish line, the third-place runner is about 3.06 miles behind. ◀

Historical Feature

In the beginning of this section, we said that imagination and experience are important in solving simultaneous nonlinear equations. Indeed, these kinds of problems lead into some of the deepest and most difficult parts of modern mathematics. Look again at the graphs in Examples 1 and 2 of this section (Figures 9 and 10). We see that Example 1 has two solutions, and Example 2 has four solutions. We might conjecture that the number of solutions is equal to the product of the degrees of the equations involved. This conjecture was indeed made by Etienne Bezout (1730–1783), but working out the details took about 150 years. It turns out that, to arrive at the correct number of intersections, we must count not only the complex number intersections, but also those intersections that, in a certain sense, lie at infinity. For example, a parabola and a line lying on the axis of the parabola intersect at the vertex and at infinity. This topic is part of the study of algebraic geometry.

Historical Problem

A papyrus dating back to 1950 BC contains the following problem: "A given surface area of 100 units of area shall be represented as the sum of two squares whose sides are to each other as $1:\frac{3}{4}$." Solve for the sides by solving the system of equations

$$\begin{cases} x^2 + y^2 = 100 \\ x = \frac{3}{4}y \end{cases}$$

10.6 Assess Your Understanding

'Are You Prepared?' *Answers are given at the end of these exercises. If you get a wrong answer, read the pages listed in red.*

1. Graph the equation: $y = 3x + 2$. (pp. 22–30)
2. Graph the equation: $y = x^2 - 4$. (pp. 615–622)
3. Graph the equation: $y^2 = x^2 - 1$. (pp. 635–646)
4. Graph the equation: $x^2 + 4y^2 = 4$. (pp. 625–632)

Exercises

In Problems 5–24, graph each equation of the system. Then solve the system to find the points of intersection.

5. $\begin{cases} y = x^2 + 1 \\ y = x + 1 \end{cases}$ **6.** $\begin{cases} y = x^2 + 1 \\ y = 4x + 1 \end{cases}$ **7.** $\begin{cases} y = \sqrt{36 - x^2} \\ y = 8 - x \end{cases}$ **8.** $\begin{cases} y = \sqrt{4 - x^2} \\ y = 2x + 4 \end{cases}$

9. $\begin{cases} y = \sqrt{x} \\ y = 2 - x \end{cases}$ **10.** $\begin{cases} y = \sqrt{x} \\ y = 6 - x \end{cases}$ **11.** $\begin{cases} x = 2y \\ x = y^2 - 2y \end{cases}$ **12.** $\begin{cases} y = x - 1 \\ y = x^2 - 6x + 9 \end{cases}$

13. $\begin{cases} x^2 + y^2 = 4 \\ x^2 + 2x + y^2 = 0 \end{cases}$ **14.** $\begin{cases} x^2 + y^2 = 8 \\ x^2 + y^2 + 4y = 0 \end{cases}$ **15.** $\begin{cases} y = 3x - 5 \\ x^2 + y^2 = 5 \end{cases}$ **16.** $\begin{cases} x^2 + y^2 = 10 \\ y = x + 2 \end{cases}$

17. $\begin{cases} x^2 + y^2 = 4 \\ y^2 - x = 4 \end{cases}$ **18.** $\begin{cases} x^2 + y^2 = 16 \\ x^2 - 2y = 8 \end{cases}$ **19.** $\begin{cases} xy = 4 \\ x^2 + y^2 = 8 \end{cases}$ **20.** $\begin{cases} x^2 = y \\ xy = 1 \end{cases}$

21. $\begin{cases} x^2 + y^2 = 4 \\ y = x^2 - 9 \end{cases}$ **22.** $\begin{cases} xy = 1 \\ y = 2x + 1 \end{cases}$ **23.** $\begin{cases} y = x^2 - 4 \\ y = 6x - 13 \end{cases}$ **24.** $\begin{cases} x^2 + y^2 = 10 \\ xy = 3 \end{cases}$

In Problems 25–54, solve each system. Use any method you wish.

25. $\begin{cases} 2x^2 + y^2 = 18 \\ xy = 4 \end{cases}$ **26.** $\begin{cases} x^2 - y^2 = 21 \\ x + y = 7 \end{cases}$ **27.** $\begin{cases} y = 2x + 1 \\ 2x^2 + y^2 = 1 \end{cases}$

28. $\begin{cases} x^2 - 4y^2 = 16 \\ 2y - x = 2 \end{cases}$ **29.** $\begin{cases} x + y + 1 = 0 \\ x^2 + y^2 + 6y - x = -5 \end{cases}$ **30.** $\begin{cases} 2x^2 - xy + y^2 = 8 \\ xy = 4 \end{cases}$

31. $\begin{cases} 4x^2 - 3xy + 9y^2 = 15 \\ 2x + 3y = 5 \end{cases}$ **32.** $\begin{cases} 2y^2 - 3xy + 6y + 2x + 4 = 0 \\ 2x - 3y + 4 = 0 \end{cases}$ **33.** $\begin{cases} x^2 - 4y^2 + 7 = 0 \\ 3x^2 + y^2 = 31 \end{cases}$

34. $\begin{cases} 3x^2 - 2y^2 + 5 = 0 \\ 2x^2 - y^2 + 2 = 0 \end{cases}$ **35.** $\begin{cases} 7x^2 - 3y^2 + 5 = 0 \\ 3x^2 + 5y^2 = 12 \end{cases}$ **36.** $\begin{cases} x^2 - 3y^2 + 1 = 0 \\ 2x^2 - 7y^2 + 5 = 0 \end{cases}$

37. $\begin{cases} x^2 + 2xy = 10 \\ 3x^2 - xy = 2 \end{cases}$ **38.** $\begin{cases} 5xy + 13y^2 + 36 = 0 \\ xy + 7y^2 = 6 \end{cases}$ **39.** $\begin{cases} 2x^2 + y^2 = 2 \\ x^2 - 2y^2 + 8 = 0 \end{cases}$

40. $\begin{cases} y^2 - x^2 + 4 = 0 \\ 2x^2 + 3y^2 = 6 \end{cases}$ **41.** $\begin{cases} x^2 + 2y^2 = 16 \\ 4x^2 - y^2 = 24 \end{cases}$ **42.** $\begin{cases} 4x^2 + 3y^2 = 4 \\ 2x^2 - 6y^2 = -3 \end{cases}$

43. $\begin{cases} \dfrac{5}{x^2} - \dfrac{2}{y^2} + 3 = 0 \\ \dfrac{3}{x^2} + \dfrac{1}{y^2} = 7 \end{cases}$ **44.** $\begin{cases} \dfrac{2}{x^2} - \dfrac{3}{y^2} + 1 = 0 \\ \dfrac{6}{x^2} - \dfrac{7}{y^2} + 2 = 0 \end{cases}$ **45.** $\begin{cases} \dfrac{1}{x^4} + \dfrac{6}{y^4} = 6 \\ \dfrac{2}{x^4} - \dfrac{2}{y^4} = 19 \end{cases}$

46. $\begin{cases} \dfrac{1}{x^4} - \dfrac{1}{y^4} = 1 \\ \dfrac{1}{x^4} + \dfrac{1}{y^4} = 4 \end{cases}$ **47.** $\begin{cases} x^2 - 3xy + 2y^2 = 0 \\ x^2 + xy = 6 \end{cases}$ **48.** $\begin{cases} x^2 - xy - 2y^2 = 0 \\ xy + x + 6 = 0 \end{cases}$

49. $\begin{cases} y^2 + y + x^2 - x - 2 = 0 \\ y + 1 + \dfrac{x - 2}{y} = 0 \end{cases}$ **50.** $\begin{cases} x^3 - 2x^2 + y^2 + 3y - 4 = 0 \\ x - 2 + \dfrac{y^2 - y}{x^2} = 0 \end{cases}$ **51.** $\begin{cases} \log_x y = 3 \\ \log_x (4y) = 5 \end{cases}$

52. $\begin{cases} \log_x (2y) = 3 \\ \log_x (4y) = 2 \end{cases}$ **53.** $\begin{cases} \ln x = 4 \ln y \\ \log_3 x = 2 + 2 \log_3 y \end{cases}$ **54.** $\begin{cases} \ln x = 5 \ln y \\ \log_2 x = 3 + 2 \log_2 y \end{cases}$

In Problems 55–60, graph each equation and find the point(s) of intersection, if any.

55. The line $x + 2y = 0$ and the circle $(x - 1)^2 + (y - 1)^2 = 5$

56. The line $x + 2y + 6 = 0$ and the circle $(x + 1)^2 + (y + 1)^2 = 5$

57. The circle $(x - 1)^2 + (y + 2)^2 = 4$ and the parabola $y^2 + 4y - x + 1 = 0$

58. The circle $(x + 2)^2 + (y - 1)^2 = 4$ and the parabola $y^2 - 2y - x - 5 = 0$

59. The graph of $y = \dfrac{4}{x - 3}$ and the circle $x^2 - 6x + y^2 + 1 = 0$

60. The graph of $y = \dfrac{4}{x + 2}$ and the circle $x^2 + 4x + y^2 - 4 = 0$

In Problems 61–68, use a graphing utility to solve each system of equations. Express the solution(s) rounded to two decimal places.

61. $\begin{cases} y = x^{2/3} \\ y = e^{-x} \end{cases}$

62. $\begin{cases} y = x^{3/2} \\ y = e^{-x} \end{cases}$

63. $\begin{cases} x^2 + y^3 = 2 \\ x^3 y = 4 \end{cases}$

64. $\begin{cases} x^3 + y^2 = 2 \\ x^2 y = 4 \end{cases}$

65. $\begin{cases} x^4 + y^4 = 12 \\ xy^2 = 2 \end{cases}$

66. $\begin{cases} x^4 + y^4 = 6 \\ xy = 1 \end{cases}$

67. $\begin{cases} xy = 2 \\ y = \ln x \end{cases}$

68. $\begin{cases} x^2 + y^2 = 4 \\ y = \ln x \end{cases}$

69. The difference of two numbers is 2 and the sum of their squares is 10. Find the numbers.

70. The sum of two numbers is 7 and the difference of their squares is 21. Find the numbers.

71. The product of two numbers is 4 and the sum of their squares is 8. Find the numbers.

72. The product of two numbers is 10 and the difference of their squares is 21. Find the numbers.

73. The difference of two numbers is the same as their product, and the sum of their reciprocals is 5. Find the numbers.

74. The sum of two numbers is the same as their product, and the difference of their reciprocals is 3. Find the numbers.

75. The ratio of a to b is $\frac{2}{3}$. The sum of a and b is 10. What is the ratio of $a + b$ to $b - a$?

76. The ratio of a to b is 4:3. The sum of a and b is 14. What is the ratio of $a - b$ to $a + b$?

77. Geometry The perimeter of a rectangle is 16 inches and its area is 15 square inches. What are its dimensions?

78. Geometry An area of 52 square feet is to be enclosed by two squares whose sides are in the ratio of 2 : 3. Find the sides of the squares.

79. Geometry Two circles have circumferences that add up to 12π centimeters and areas that add up to 20π square centimeters. Find the radius of each circle.

80. Geometry The altitude of an isosceles triangle drawn to its base is 3 centimeters, and its perimeter is 18 centimeters. Find the length of its base.

81. The Tortoise and the Hare In a 21-meter race between a tortoise and a hare, the tortoise leaves 9 minutes before the hare. The hare, by running at an average speed of 0.5 meter per hour faster than the tortoise, crosses the finish line 3 minutes before the tortoise. What are the average speeds of the tortoise and the hare?

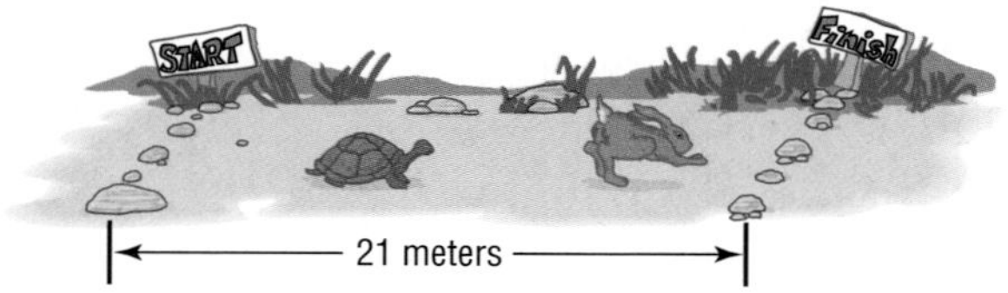

82. Running a Race In a 1-mile race, the winner crosses the finish line 10 feet ahead of the second-place runner and 20 feet ahead of the third-place runner. Assuming that each runner maintains a constant speed throughout the race, by how many feet does the second-place runner beat the third-place runner?

83. Constructing a Box A rectangular piece of cardboard, whose area is 216 square centimeters, is made into an open box by cutting a 2-centimeter square from each corner and turning up the sides. See the figure. If the box is to have a volume of 224 cubic centimeters, what size cardboard should you start with?

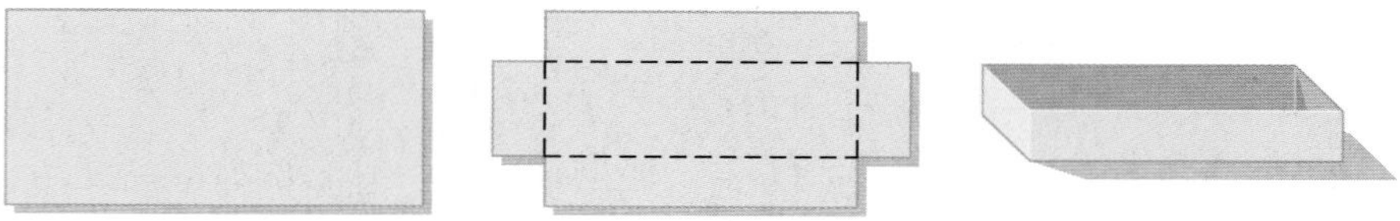

84. Constructing a Cylindrical Tube A rectangular piece of cardboard, whose area is 216 square centimeters, is made into a cylindrical tube by joining together two sides of the rectangle. See the figure. If the tube is to have a volume of 224 cubic centimeters, what size cardboard should you start with?

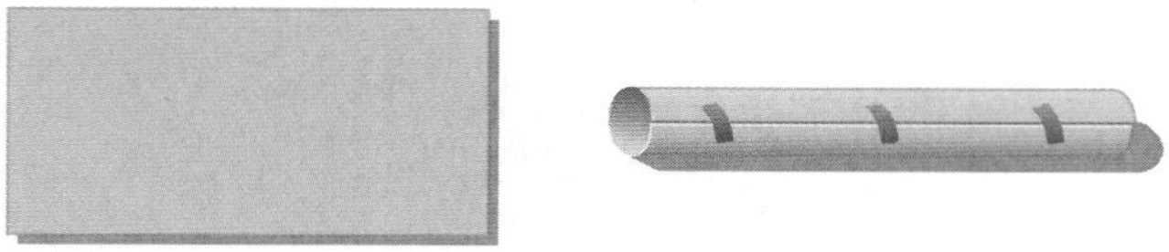

85. Fencing A farmer has 300 feet of fence available to enclose a 4500 square foot region in the shape of adjoining squares, with sides of length x and y. See the figure. Find x and y.

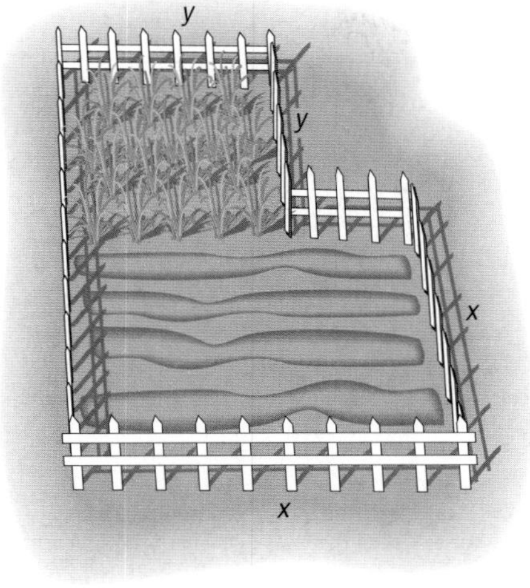

86. Bending Wire A wire 60 feet long is cut into two pieces. Is it possible to bend one piece into the shape of a square and the other into the shape of a circle so that the total area enclosed by the two pieces is 100 square feet? If this is possible, find the length of the side of the square and the radius of the circle.

87. Geometry Find formulas for the length l and width w of a rectangle in terms of its area A and perimeter P.

88. Geometry Find formulas for the base b and one of the equal sides l of an isosceles triangle in terms of its altitude h and perimeter P.

89. Descartes's Method of Equal Roots Descartes's method for finding tangents depends on the idea that, for many graphs, the tangent line at a given point is the *unique* line that intersects the graph at that point only. We will apply his method to find an equation of the tangent line to the parabola $y = x^2$ at the point $(2, 4)$. See the figure. First, we know that the equation of the tangent line must be in the form $y = mx + b$. Using the fact that the point $(2, 4)$ is on the line, we can solve for b in terms of m and get the equation $y = mx + (4 - 2m)$. Now we want $(2, 4)$ to be the *unique* solution to the system

$$\begin{cases} y = x^2 \\ y = mx + 4 - 2m \end{cases}$$

From this system, we get $x^2 = mx + 4 - 2m$ or $x^2 - mx + (2m - 4) = 0$. By using the quadratic formula, we get

$$x = \frac{m \pm \sqrt{m^2 - 4(2m - 4)}}{2}$$

To obtain a unique solution for x, the two roots must be equal; in other words, the discriminant $m^2 - 4(2m - 4)$ must be 0. Complete the work to get m, and write an equation of the tangent line.

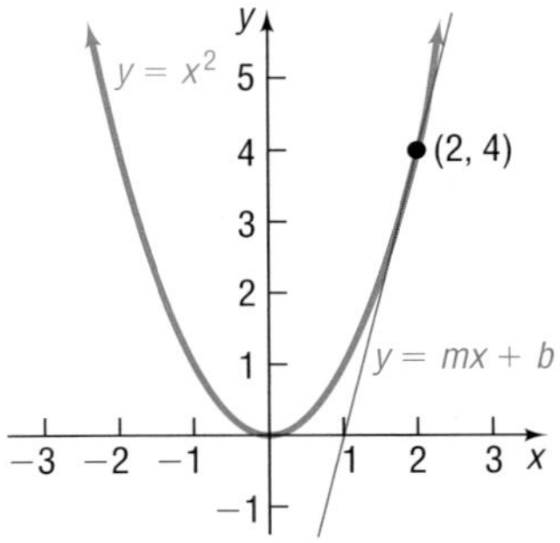

In Problems 90–96, use Descartes's method from Problem 89 to find the equation of the line tangent to each graph at the given point.

90. $x^2 + y^2 = 10$; at $(1, 3)$

91. $y = x^2 + 2$; at $(1, 3)$

92. $x^2 + y = 5$; at $(-2, 1)$

93. $2x^2 + 3y^2 = 14$; at $(1, 2)$

94. $3x^2 + y^2 = 7$; at $(-1, 2)$

95. $x^2 - y^2 = 3$; at $(2, 1)$

96. $2y^2 - x^2 = 14$; at $(2, 3)$

97. If r_1 and r_2 are two solutions of a quadratic equation $ax^2 + bx + c = 0$, then it can be shown that

$$r_1 + r_2 = -\frac{b}{a} \quad \text{and} \quad r_1 r_2 = \frac{c}{a}$$

Solve this system of equations for r_1 and r_2.

98. A circle and a line intersect at most twice. A circle and a parabola intersect at most four times. Deduce that a circle and the graph of a polynomial of degree 3 intersect at most six times. What do you conjecture about a polynomial of degree 4? What about a polynomial of degree n? Can you explain your conclusions using an algebraic argument?

99. Suppose that you are the manager of a sheet metal shop. A customer asks you to manufacture 10,000 boxes, each box being open on top. The boxes are required to have a square base and a 9-cubic-foot capacity. You construct the boxes by cutting out a square from each corner of a square piece of sheet metal and folding along the edges.
(a) What are the dimensions of the square to be cut if the area of the square piece of sheet metal is 100 square feet?
(b) Could you make the box using a smaller piece of sheet metal? Make a list of the dimensions of the box for various pieces of sheet metal.

'Are You Prepared?' Answers

1.

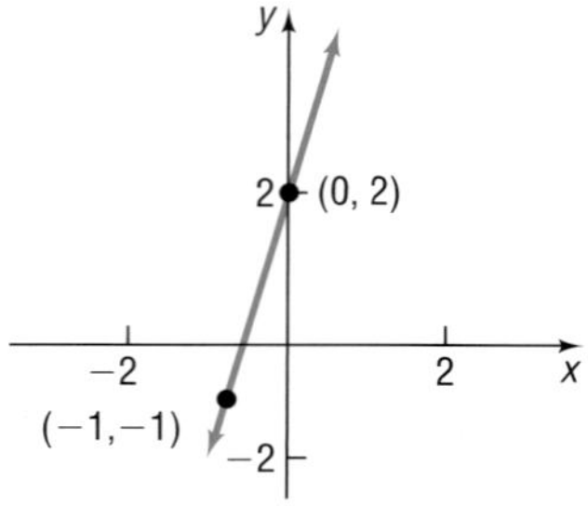

2.

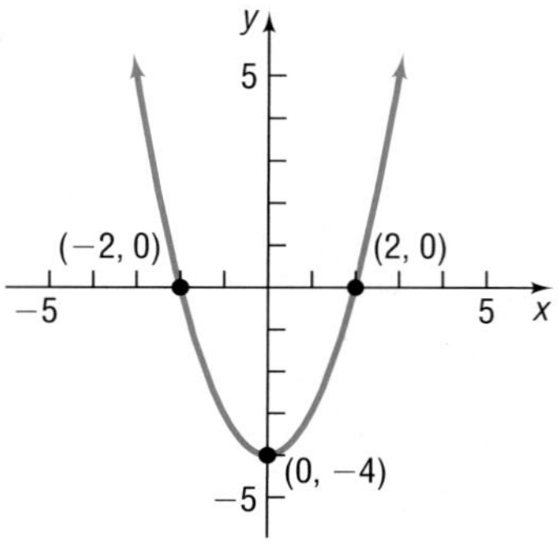

3.

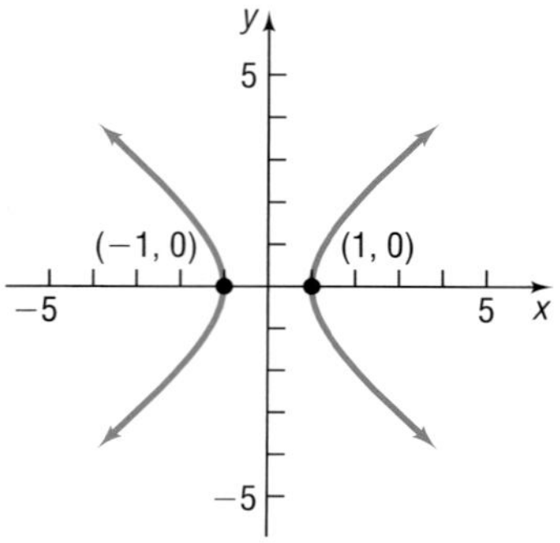

4.

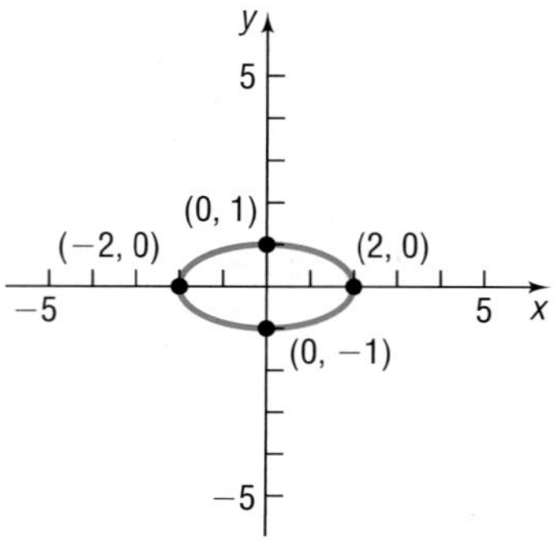

10.7 Systems of Inequalities

PREPARING FOR THIS SECTION *Before getting started, review the following:*

- Solving Inequalities (Appendix A, Section A.8, pp. 971–974)
- Lines (Section 1.3, pp. 22–32)
- Circles (Section 1.2, pp. 16–19)
- Graphing Techniques: Transformations (Section 2.5, pp. 94–103)

Now work the 'Are You Prepared?' problems on page 766.

OBJECTIVES 1 Graph an Inequality
2 Graph a System of Inequalities

In Appendix A, Section A.8, we discussed inequalities in one variable. In this section, we discuss inequalities in two variables.

EXAMPLE 1 **Examples of Inequalities in Two Variables**

(a) $3x + y \leq 6$ (b) $x^2 + y^2 < 4$ (c) $y^2 \leq x$ ◀

1 An inequality in two variables x and y is **satisfied** by an ordered pair (a, b) if, when x is replaced by a and y by b, a true statement results. The **graph of an inequality in two variables** x and y consists of all points (x, y) whose coordinates satisfy the inequality.

Let's look at an example.

EXAMPLE 2 Graphing a Linear Inequality

Graph the linear inequality: $3x + y \leq 6$

Solution We begin with the associated problem of the graph of the linear equation

$$3x + y = 6$$

formed by replacing (for now) the $\leq$ symbol with an $=$ sign. The graph of the linear equation is a line. See Figure 12(a). This line is part of the graph of the inequality that we seek because the inequality is nonstrict. (Do you see why? We are seeking points for which $3x + y$ is less than *or equal to* 6.)

Now let's test a few randomly selected points to see whether they belong to the graph of the inequality.

	$3x + y \leq 6$	**Conclusion**
$(4, -1)$	$3(4) + (-1) = 11 > 6$	Does not belong to graph
$(5, 5)$	$3(5) + 5 = 20 > 6$	Does not belong to graph
$(-1, 2)$	$3(-1) + 2 = -1 \leq 6$	Belongs to graph
$(-2, -2)$	$3(-2) + (-2) = -8 \leq 6$	Belongs to graph

Look again at Figure 12(a). Notice that the two points that belong to the graph both lie on the same side of the line, and the two points that do not belong to the graph lie on the opposite side. As it turns out, this is always the case. The graph we seek consists of all points that lie on the same side of the line as $(-1, 2)$ and $(-2, -2)$. The graph we seek is the shaded region in Figure 12(b).

Figure 12

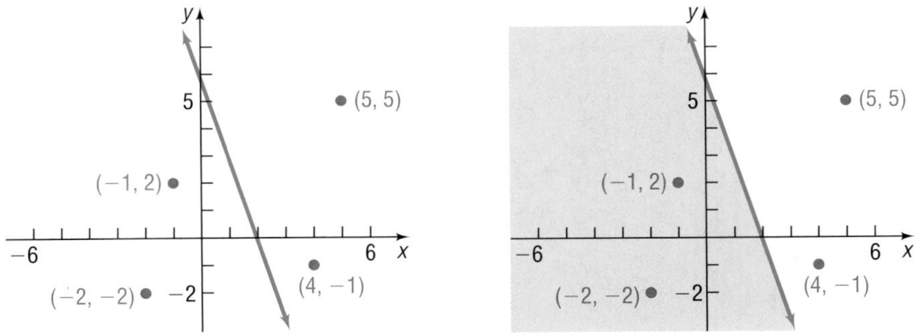

(a) $3x + y = 6$ **(b)** Graph of $3x + y \leq 6$

The graph of any inequality in two variables may be obtained in a like way. First, the equation corresponding to the inequality is graphed, using dashes if the inequality is strict and solid marks if it is nonstrict. This graph, in almost every case, will separate the xy-plane into two or more regions. In each region, either all points satisfy the inequality or no points satisfy the inequality. The use of a single test point in each region is all that is required to determine whether the points of that region are part of the graph. The steps to follow are given next.

Steps for Graphing an Inequality

STEP 1: Replace the inequality symbol by an equal sign and graph the resulting equation. If the inequality is strict, use dashes; if it is nonstrict, use a solid mark. This graph separates the xy-plane into two or more regions.

STEP 2: In each of the regions, select a test point P.

(a) If the coordinates of P satisfy the inequality, then so do all the points in that region. Indicate this by shading the region.

(b) If the coordinates of P do not satisfy the inequality, then none of the points in that region do.

EXAMPLE 3 Graphing an Inequality

Graph: $x^2 + y^2 \le 4$

Figure 13

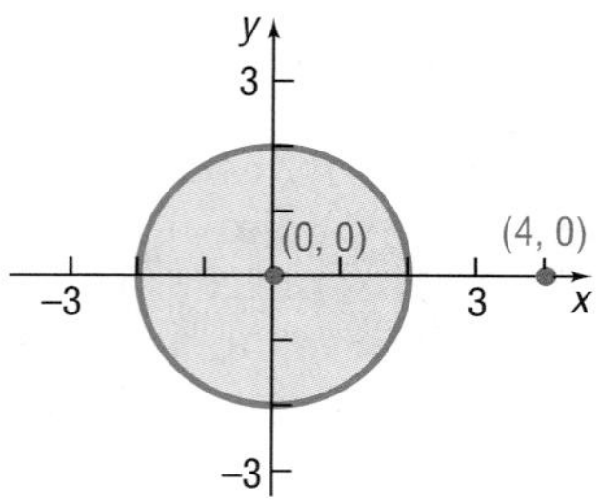

Solution First, we graph the equation $x^2 + y^2 = 4$, a circle of radius 2, center at the origin. A solid circle will be used because the inequality is not strict. We use two test points, one inside the circle, the other outside.

Inside	$(0, 0)$:	$x^2 + y^2 = 0^2 + 0^2 = 0 \le 4$	Belongs to the graph
Outside	$(4, 0)$:	$x^2 + y^2 = 4^2 + 0^2 = 16 > 4$	Does not belong to graph

All the points inside and on the circle satisfy the inequality. See Figure 13. ◀

NOW WORK PROBLEM 17.

Figure 14

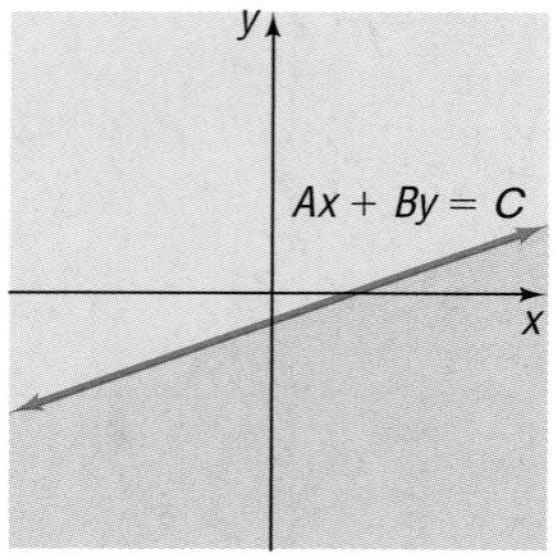

Linear inequalities are inequalities in one of the forms

$$Ax + By < C \qquad Ax + By > C \qquad Ax + By \le C \qquad Ax + By \ge C$$

where A and B are not both zero.

The graph of the corresponding equation of a linear inequality is a line, which separates the xy-plane into two regions, called **half-planes.** See Figure 14.

As shown, $Ax + By = C$ is the equation of the boundary line and it divides the plane into two half-planes: one for which $Ax + By < C$ and the other for which $Ax + By > C$. Because of this, for linear inequalities, only one test point is required.

EXAMPLE 4 Graphing Linear Inequalities

Graph: (a) $y < 2$ (b) $y \ge 2x$

Solution (a) The graph of the equation $y = 2$ is a horizontal line and is not part of the graph of the inequality. Since $(0, 0)$ satisfies the inequality, the graph consists of the half-plane below the line $y = 2$. See Figure 15.

(b) The graph of the equation $y = 2x$ is a line and is part of the graph of the inequality. Using $(3, 0)$ as a test point, we find it does not satisfy the inequality $[0 < 2 \cdot 3]$. Points in the half-plane on the opposite side of $y = 2x$ satisfy the inequality. See Figure 16.

Figure 15

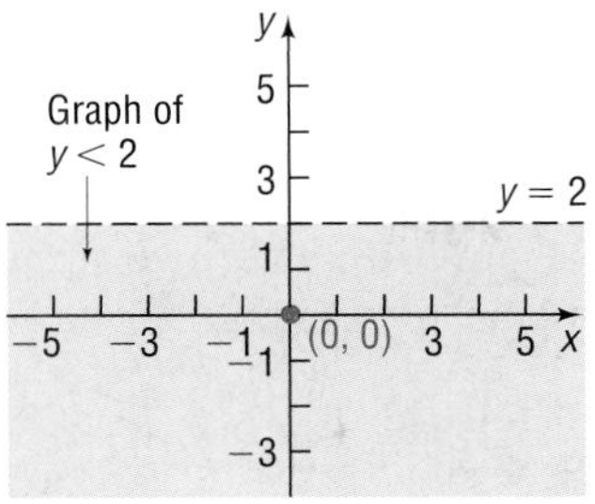

Figure 16

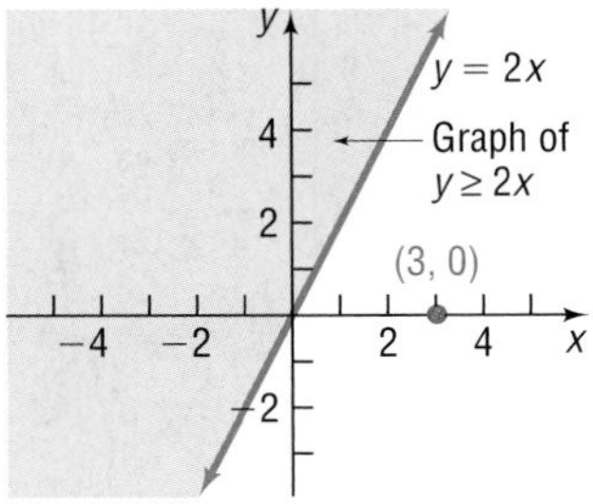

COMMENT: A graphing utility can be used to graph inequalities. To see how, read Section B.6 in Appendix B.

NOW WORK PROBLEM 13.

Systems of Inequalities in Two Variables

2 The **graph of a system of inequalities** in two variables x and y is the set of all points (x, y) that simultaneously satisfy *each* of the inequalities in the system. The graph of a system of inequalities can be obtained by graphing each inequality individually and then determining where, if at all, they intersect.

EXAMPLE 5 Graphing a System of Linear Inequalities

Graph the system: $\begin{cases} x + y \geq 2 \\ 2x - y \leq 4 \end{cases}$

Solution First, we graph the inequality $x + y \geq 2$ as the shaded region in Figure 17(a). Next, we graph the inequality $2x - y \leq 4$ as the shaded region in Figure 17(b). Now, superimpose the two graphs, as shown in Figure 17(c). The points that are in both shaded regions [the overlapping, darker region in Figure 17(c)] are the solutions we seek to the system, because they simultaneously satisfy each linear inequality.

Figure 17

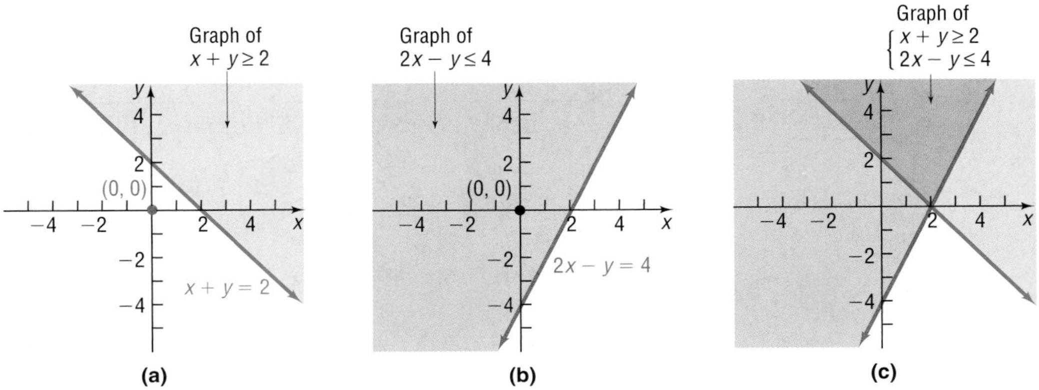

EXAMPLE 6 Graphing a System of Linear Inequalities

Graph the system: $\begin{cases} x + y \le 2 \\ x + y \ge 0 \end{cases}$

Solution See Figure 18. The overlapping, darker shaded region between the two boundary lines is the graph of the system.

Figure 18

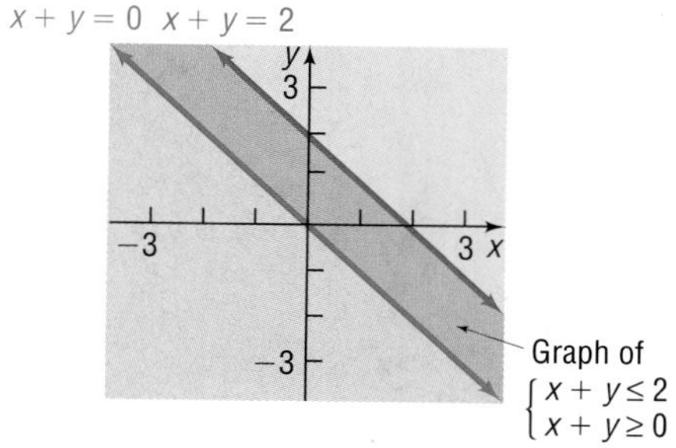

NOW WORK PROBLEM 35.

EXAMPLE 7 Graphing a System of Linear Inequalities

Graph the system: $\begin{cases} 2x - y \ge 2 \\ 2x - y \ge 0 \end{cases}$

Solution See Figure 19. The overlapping, darker shaded region is the graph of the system. Note that the graph of the system is identical to the graph of the single inequality $2x - y \ge 2$.

Figure 19

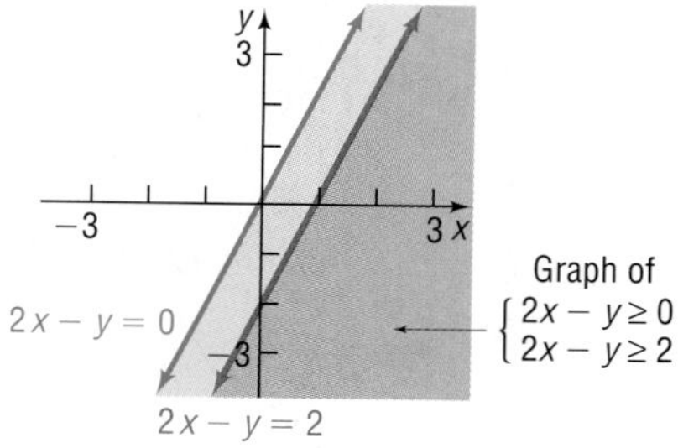

EXAMPLE 8 Graphing a System of Linear Inequalities

Graph the system: $\begin{cases} x + 2y \le 2 \\ x + 2y \ge 6 \end{cases}$

Solution See Figure 20. Because no overlapping region results, there are no points in the xy-plane that simultaneously satisfy each inequality. Hence, the system has no solution.

Figure 20

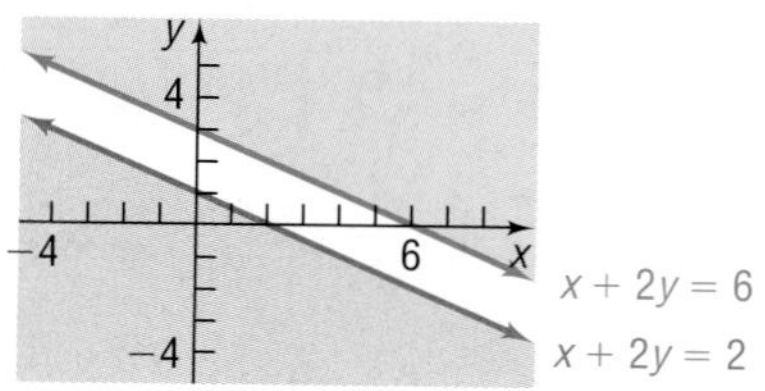

The following example is important for doing certain kinds of calculus problems.

EXAMPLE 9 Graphing a System of Inequalities

Graph the region below the graph of $x + y = 2$ and above the graph of $y = x^2 - 4$ by graphing the system:

$$\begin{cases} y \geq x^2 - 4 \\ x + y \leq 2 \end{cases}$$

Label all points of intersection.

Figure 21

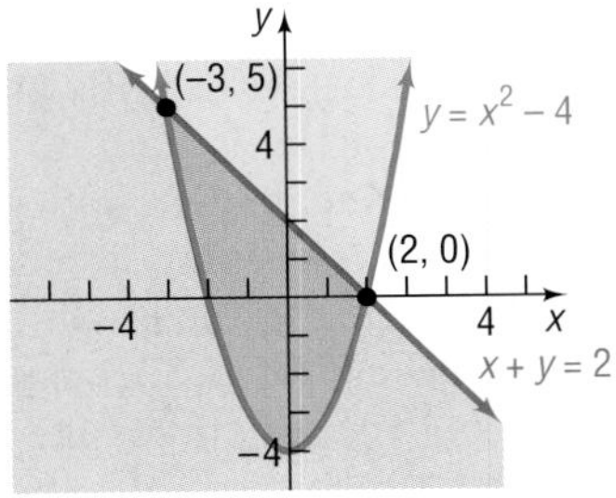

Solution Figure 21 shows the graph of the region above the graph of the parabola $y = x^2 - 4$ and below the graph of the line $x + y = 2$. The points of intersection are found by solving the system of equations

$$\begin{cases} y = x^2 - 4 \\ x + y = 2 \end{cases}$$

Using substitution, we find

$$\begin{aligned} x + (x^2 - 4) &= 2 \\ x^2 + x - 6 &= 0 \\ (x + 3)(x - 2) &= 0 \\ x = -3 \qquad x &= 2 \end{aligned}$$

The two points of intersection are $(-3, 5)$ and $(2, 0)$. ◀

NOW WORK PROBLEM 55.

EXAMPLE 10 Graphing a System of Four Linear Inequalities

Figure 22

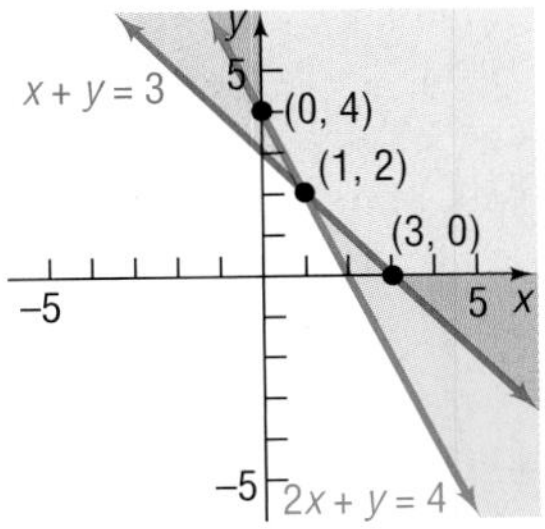

Graph the system: $\begin{cases} x + y \geq 3 \\ 2x + y \geq 4 \\ x \geq 0 \\ y \geq 0 \end{cases}$

Solution The two inequalities $x \geq 0$ and $y \geq 0$ require the graph of the system to be in quadrant I. We concentrate on the remaining two inequalities. The intersection of the graphs of these two inequalities and quadrant I, shown in gray in Figure 22, is the graph of the system. ◀

EXAMPLE 11 Financial Planning

A retired couple has up to \$25,000 to invest. As their financial adviser, you recommend that they place at least \$15,000 in Treasury bills yielding 6% and at most \$5000 in corporate bonds yielding 9%.

(a) Using x to denote the amount of money invested in Treasury bills and y the amount invested in corporate bonds, write a system of linear inequalities that describes the possible amounts of each investment. We shall assume that x and y are in thousands of dollars.

(b) Graph the system.

Solution (a) The system of linear inequalities is

$$\begin{cases} x \geq 0 \\ y \geq 0 \\ x + y \leq 25 \\ x \geq 15 \\ y \leq 5 \end{cases}$$

$x \geq 0$, $y \geq 0$: x and y are nonnegative variables since they represent money invested in thousands of dollars.

$x + y \leq 25$: The total of the two investments, x + y, cannot exceed $25,000.

$x \geq 15$: At least $15,000 in Treasury bills

$y \leq 5$: At most $5000 in corporate bonds

(b) See the shaded region in Figure 23. Note that the inequalities $x \geq 0$ and $y \geq 0$ again require that the graph of the system be in quadrant I.

Figure 23

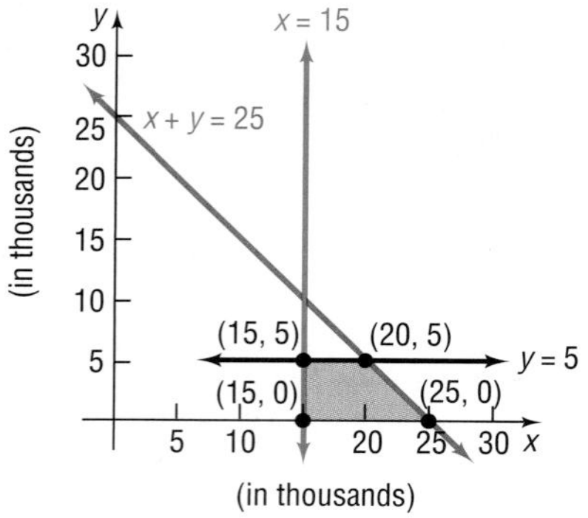

The graph of the system of linear inequalities in Figure 23 is said to be **bounded,** because it can be contained within some circle of sufficiently large radius. A graph that cannot be contained in any circle is said to be **unbounded.** For example, the graph of the system of linear inequalities in Figure 22 is unbounded, since it extends indefinitely in a particular direction.

Notice in Figures 22 and 23 that those points belonging to the graph that are also points of intersection of boundary lines have been plotted. Such points are referred to as **vertices** or **corner points** of the graph. The system graphed in Figure 22 has three corner points: (0, 4), (1, 2), and (3, 0). The system graphed in Figure 23 has four corner points: (15, 0), (25, 0), (20, 5), and (15, 5).

These ideas will be used in the next section in developing a method for solving linear programming problems, an important application of linear inequalities.

NOW WORK PROBLEM 43.

10.7 Assess Your Understanding

'Are You Prepared?' *Answers are given at the end of these exercises. If you get a wrong answer, read the pages listed in red.*

1. Solve the inequality: $3x + 4 < 8 - x$. (pp. 971–974)
2. Graph the equation: $3x - 2y = 6$. (pp. 22–30)
3. Graph the equation: $x^2 + y^2 = 9$. (pp. 16–19)
4. Graph the equation: $y = x^2 + 4$. (pp. 94–103)
5. *True or False:* The lines $2x + y = 4$ and $4x + 2y = 0$ are parallel. (pp. 30–32)
6. The graph of $y = (x - 2)^2$ may be obtained by shifting the graph of ________ to the left/right a distance of ______ units. (pp. 94–103)

Concepts and Vocabulary

7. An inequality in two variables x and y is __________ by an ordered pair (a, b) if, when x is replaced by a and y by b, a true statement results.
8. The graph of a linear inequality is called a(n) ________.
9. *True or False:* The graph of a linear inequality is a line.
10. *True or False:* The graph of a system of linear inequalities is sometimes unbounded.

Exercises

In Problems 11–22, graph each inequality.

11. $x \geq 0$
12. $y \geq 0$
13. $x \geq 4$
14. $y \leq 2$
15. $2x + y \geq 6$
16. $3x + 2y \leq 6$
17. $x^2 + y^2 > 1$
18. $x^2 + y^2 \leq 9$
19. $y \leq x^2 - 1$
20. $y > x^2 + 2$
21. $xy \geq 4$
22. $xy \leq 1$

In Problems 23–40, graph each system of inequalities.

23. $\begin{cases} x + y \leq 2 \\ 2x + y \geq 4 \end{cases}$
24. $\begin{cases} 3x - y \geq 6 \\ x + 2y \leq 2 \end{cases}$
25. $\begin{cases} 2x - y \leq 4 \\ 3x + 2y \geq -6 \end{cases}$
26. $\begin{cases} 4x - 5y \leq 0 \\ 2x - y \geq 2 \end{cases}$

27. $\begin{cases} 2x - 3y \le 0 \\ 3x + 2y \le 6 \end{cases}$

28. $\begin{cases} 4x - y \ge 2 \\ x + 2y \ge 2 \end{cases}$

29. $\begin{cases} x^2 + y^2 \le 9 \\ x + y \ge 3 \end{cases}$

30. $\begin{cases} x^2 + y^2 \ge 9 \\ x + y \le 3 \end{cases}$

31. $\begin{cases} y \ge x^2 - 4 \\ y \le x - 2 \end{cases}$

32. $\begin{cases} y^2 \le x \\ y \ge x \end{cases}$

33. $\begin{cases} xy \ge 4 \\ y \ge x^2 + 1 \end{cases}$

34. $\begin{cases} y + x^2 \le 1 \\ y \ge x^2 - 1 \end{cases}$

35. $\begin{cases} x - 2y \le 6 \\ 2x - 4y \ge 0 \end{cases}$

36. $\begin{cases} x + 4y \le 8 \\ x + 4y \ge 4 \end{cases}$

37. $\begin{cases} 2x + y \ge -2 \\ 2x + y \ge 2 \end{cases}$

38. $\begin{cases} x - 4y \le 4 \\ x - 4y \ge 0 \end{cases}$

39. $\begin{cases} 2x + 3y \ge 6 \\ 2x + 3y \le 0 \end{cases}$

40. $\begin{cases} 2x + y \ge 0 \\ 2x + y \ge 2 \end{cases}$

In Problems 41–50, graph each system of linear inequalities. Tell whether the graph is bounded or unbounded, and label the corner points.

41. $\begin{cases} x \ge 0 \\ y \ge 0 \\ 2x + y \le 6 \\ x + 2y \le 6 \end{cases}$

42. $\begin{cases} x \ge 0 \\ y \ge 0 \\ x + y \ge 4 \\ 2x + 3y \ge 6 \end{cases}$

43. $\begin{cases} x \ge 0 \\ y \ge 0 \\ x + y \ge 2 \\ 2x + y \ge 4 \end{cases}$

44. $\begin{cases} x \ge 0 \\ y \ge 0 \\ 3x + y \le 6 \\ 2x + y \le 2 \end{cases}$

45. $\begin{cases} x \ge 0 \\ y \ge 0 \\ x + y \ge 2 \\ 2x + 3y \le 12 \\ 3x + y \le 12 \end{cases}$

46. $\begin{cases} x \ge 0 \\ y \ge 0 \\ x + y \ge 2 \\ x + y \le 10 \\ 2x + y \le 3 \end{cases}$

47. $\begin{cases} x \ge 0 \\ y \ge 0 \\ x + y \ge 2 \\ x + y \le 8 \\ 2x + y \le 10 \end{cases}$

48. $\begin{cases} x \ge 0 \\ y \ge 0 \\ x + y \ge 2 \\ x + y \le 8 \\ x + 2y \ge 1 \end{cases}$

49. $\begin{cases} x \ge 0 \\ y \ge 0 \\ x + 2y \ge 1 \\ x + 2y \le 10 \end{cases}$

50. $\begin{cases} x \ge 0 \\ y \ge 0 \\ x + 2y \ge 1 \\ x + 2y \le 10 \\ x + y \ge 2 \\ x + y \le 8 \end{cases}$

In Problems 51–54, write a system of linear inequalities that has the given graph.

51.

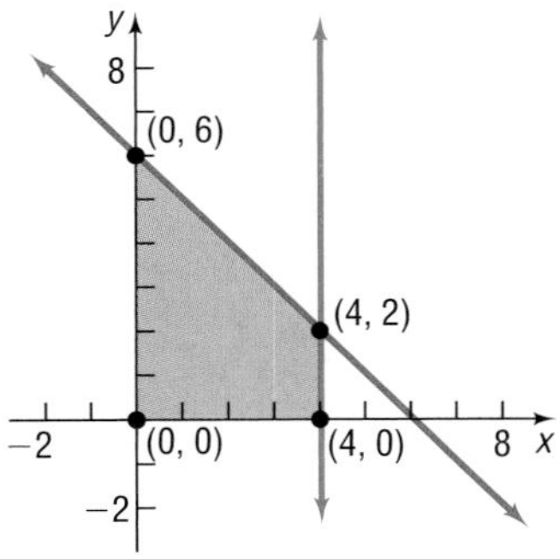

52.

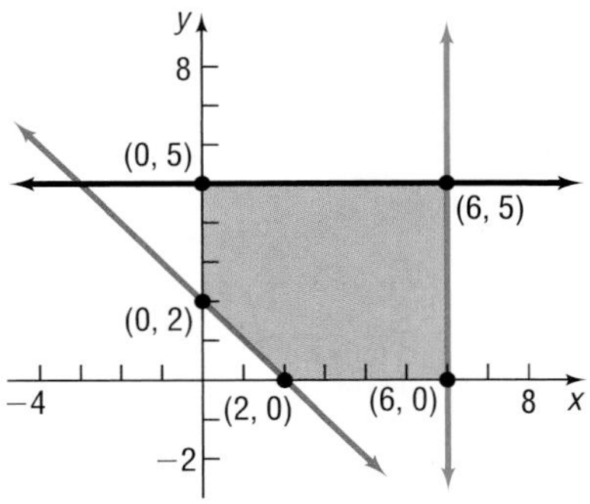

53.

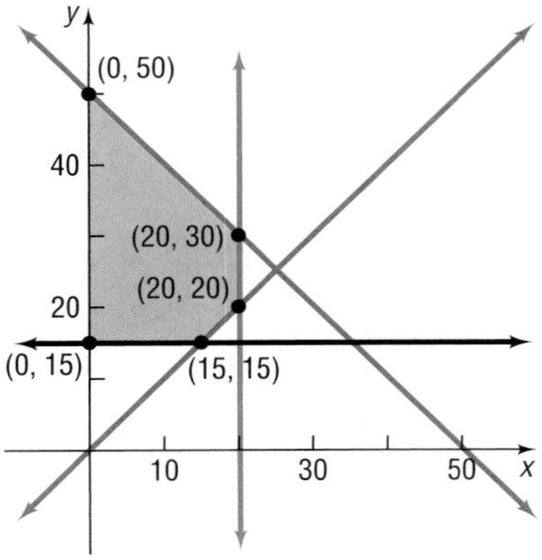

54.

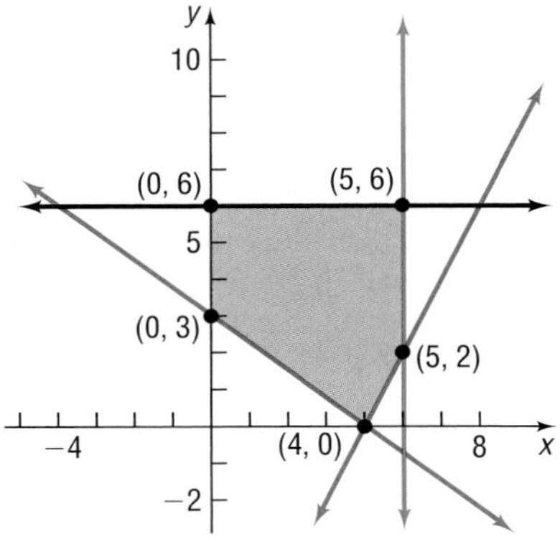

In Problems 55–60, graph the region indicated by graphing the system of inequalities. Label all points of intersection.

55. Above $y = x + 2$ Below $y = -x^2 + 4$ $\begin{cases} y \geq x + 2 \\ y \leq -x^2 + 4 \end{cases}$

56. Above $y = x^2 + 1$ Below $y = 2$ $\begin{cases} y \geq x^2 + 1 \\ y \leq 2 \end{cases}$

57. Above $y = x^2 - 3$ Below $y = -2$ $\begin{cases} y \geq x^2 - 3 \\ y \leq -2 \end{cases}$

58. Above $y = x^2 - 1$ Below $y = x + 1$ $\begin{cases} y \geq x^2 - 1 \\ y \leq x + 1 \end{cases}$

59. Above $y = (x - 2)^2$ Below $y = -x^2 + 4$ $\begin{cases} y \geq (x - 2)^2 \\ y \leq -x^2 + 4 \end{cases}$

60. Above $y = x^2 - 2$ Below $y = -x^2$ $\begin{cases} y \geq x^2 - 2 \\ y \leq -x^2 \end{cases}$

61. Financial Planning A retired couple has up to \$50,000 to invest. As their financial adviser, you recommend that they place at least \$35,000 in Treasury bills yielding 7% and at most \$10,000 in corporate bonds yielding 10%.
(a) Using x to denote the amount of money invested in Treasury bills and y the amount invested in corporate bonds, write a system of linear inequalities that describes the possible amounts of each investment.
(b) Graph the system and label the corner points.

62. Manufacturing Trucks Mike's Toy Truck Company manufacturers two models of toy trucks, a standard model and a deluxe model. Each standard model requires 2 hours for painting and 3 hours for detail work; each deluxe model requires 3 hours for painting and 4 hours for detail work. Two painters and three detail workers are employed by the company, and each works 40 hours per week.
(a) Using x to denote the number of standard model trucks and y to denote the number of deluxe model trucks, write a system of linear inequalities that describes the possible number of each model of truck that can be manufactured in a week.
(b) Graph the system and label the corner points.

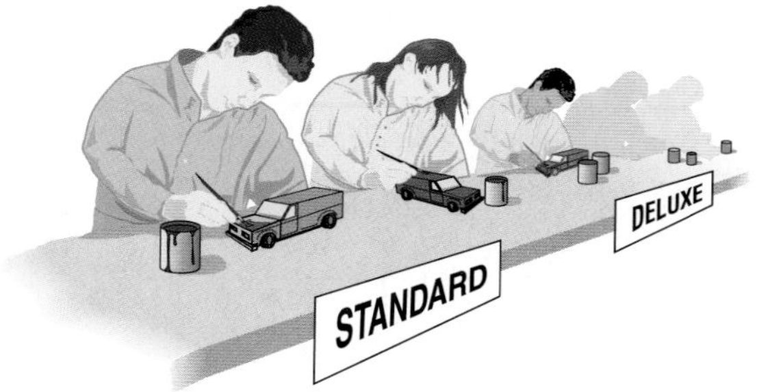

63. Blending Coffee Bill's Coffee House, a store that specializes in coffee, has available 75 pounds of A grade coffee and 120 pounds of B grade coffee. These will be blended into 1 pound packages as follows: An economy blend that contains 4 ounces of A grade coffee and 12 ounces of B grade coffee and a superior blend that contains 8 ounces of A grade coffee and 8 ounces of B grade coffee.
(a) Using x to denote the number of packages of the economy blend and y to denote the number of packages of the superior blend, write a system of linear inequalities that describes the possible number of packages of each kind of blend.
(b) Graph the system and label the corner points.

64. Mixed Nuts Nola's Nuts, a store that specializes in selling nuts, has available 90 pounds of cashews and 120 pounds of peanuts. These are to be mixed in 12 ounce packages as follows: a lower priced package containing 8 ounces of peanuts and 4 ounces of cashews and a quality package containing 6 ounces of peanuts and 6 ounces of cashews.
(a) Use x to denote the number of lower priced packages and use y to denote the number of quality packages. Write a system of linear inequalities that describes the possible number of each kind of package.
(b) Graph the system and label the corner points.

65. Transporting Goods A small truck can carry no more than 1600 pounds of cargo or 150 cubic feet of cargo. A printer weighs 20 pounds and occupies 3 cubic feet of space. A microwave oven weighs 30 pounds and occupies 2 cubic feet of space.
(a) Using x to represent the number of microwave ovens and y to represent the number of printers, write a system of linear inequalities that describes the number of ovens and printers that can be hauled by the truck.
(b) Graph the system and label the corner points.

'Are You Prepared?' Answers

1. $\{x|x < 1\}$

2.

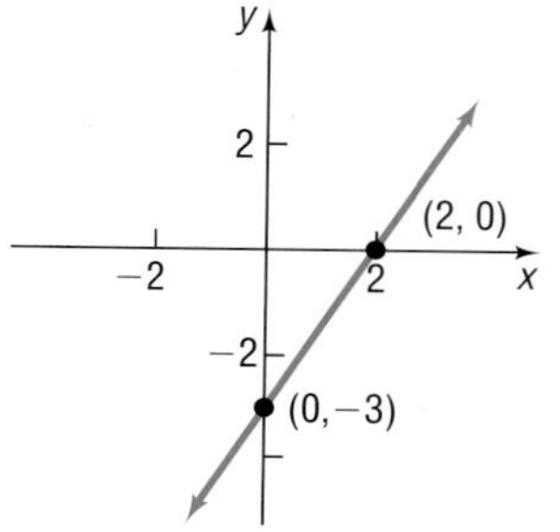

3.

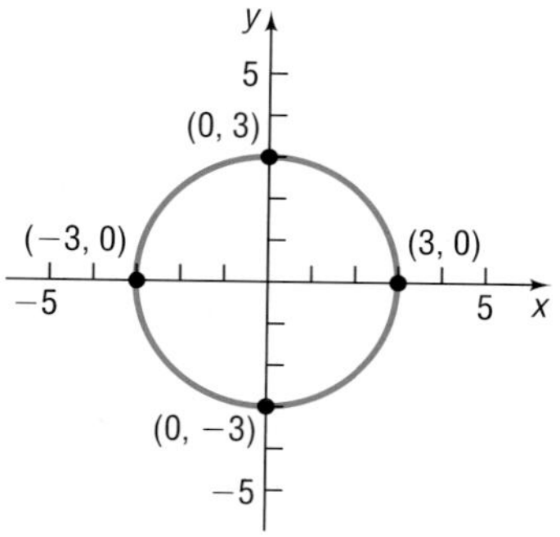

4.

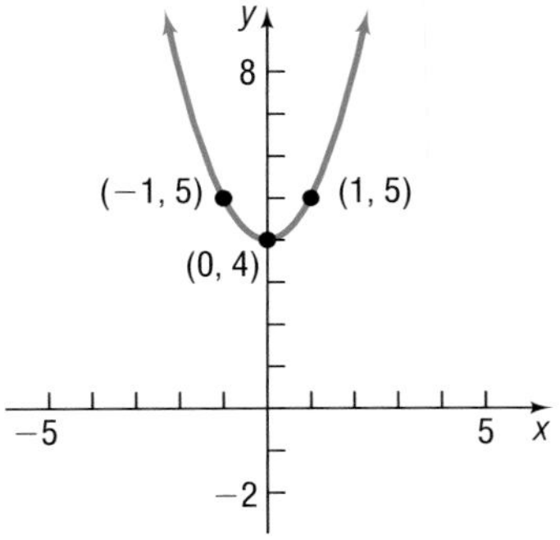

5. True **6.** $y = x^2$; right; 2

10.8 Linear Programming

OBJECTIVES 1 Set up a Linear Programming Problem
2 Solve a Linear Programming Problem

1 Historically, linear programming evolved as a technique for solving problems involving resource allocation of goods and materials for the U.S. Air Force during World War II. Today, linear programming techniques are used to solve a wide variety of problems, such as optimizing airline scheduling and establishing telephone lines. Although most practical linear programming problems involve systems of several hundred linear inequalities containing several hundred variables, we will limit our discussion to problems containing only two variables, because we can solve such problems using graphing techniques.*

We begin by returning to Example 11 of the previous section.

EXAMPLE 1 Financial Planning

A retired couple has up to \$25,000 to invest. As their financial adviser, you recommend that they place at least \$15,000 in Treasury bills yielding 6% and at most \$5000 in corporate bonds yielding 9%. How much money should be placed in each investment so that income is maximized? ◀

The problem given in Example 1 is typical of a *linear programming problem.* The problem requires that a certain linear expression, the income, be maximized. If I represents income, x the amount invested in Treasury bills at 6%, and y the amount invested in corporate bonds at 9%, then

$$I = 0.06x + 0.09y$$

We shall assume, as before, that I, x, and y are in thousands of dollars.

The linear expression $I = 0.06x + 0.09y$ is called the **objective function.** Furthermore, the problem requires that the maximum income be

*The **simplex method** is a way to solve linear programming problems involving many inequalities and variables. This method was developed by George Dantzig in 1946 and is particularly well suited for computerization. In 1984, Narendra Karmarkar of Bell Laboratories discovered a way of solving large linear programming problems that improves on the simplex method.

achieved under certain conditions or **constraints,** each of which is a linear inequality involving the variables. (See Example 11 in Section 10.7.) The linear programming problem given in Example 1 may be restated as

$$\text{Maximize} \qquad I = 0.06x + 0.09y$$

subject to the conditions that

$$\begin{aligned} x \geq 0, \qquad y &\geq 0 \\ x + y &\leq 25 \\ x &\geq 15 \\ y &\leq 5 \end{aligned}$$

In general, every linear programming problem has two components:

1. A linear objective function that is to be maximized or minimized.
2. A collection of linear inequalities that must be satisfied simultaneously.

A **linear programming problem** in two variables x and y consists of maximizing (or minimizing) a linear objective function

$$z = Ax + By, \quad A \text{ and } B \text{ are real numbers, not both } 0$$

subject to certain conditions, or constraints, expressible as linear inequalities in x and y.

To maximize (or minimize) the quantity $z = Ax + By$, we need to identify points (x, y) that make the expression for z the largest (or smallest) possible. But not all points (x, y) are eligible; only those that also satisfy each linear inequality (constraint) can be used. We refer to each point (x, y) that satisfies the system of linear inequalities (the constraints) as a **feasible point.** In a linear programming problem, we seek the feasible point(s) that maximizes (or minimizes) the objective function.

Let's look again at the linear programming problem in Example 1.

EXAMPLE 2 Analyzing a Linear Programming Problem

Consider the linear programming problem

$$\text{Maximize} \qquad I = 0.06x + 0.09y$$

subject to the conditions that

$$\begin{aligned} x \geq 0, \qquad y &\geq 0 \\ x + y &\leq 25 \\ x &\geq 15 \\ y &\leq 5 \end{aligned}$$

Graph the constraints. Then graph the objective function for $I = 0, 0.9, 1.35, 1.65$, and 1.8.

Solution Figure 24 shows the graph of the constraints. We superimpose on this graph the graph of the objective function for the given values of I.

For $I = 0$, the objective function is the line $0 = 0.06x + 0.09y$.

For $I = 0.9$, the objective function is the line $0.9 = 0.06x + 0.09y$.

Figure 24

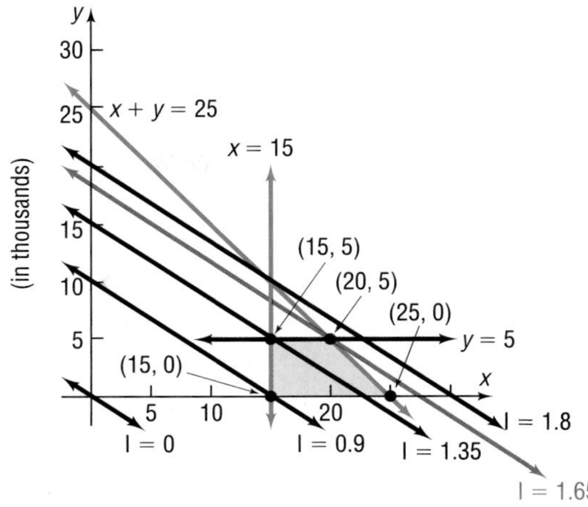

For $I = 1.35$, the objective function is the line $1.35 = 0.06x + 0.09y$.

For $I = 1.65$, the objective function is the line $1.65 = 0.06x + 0.09y$.

For $I = 1.8$, the objective function is the line $1.8 = 0.06x + 0.09y$. ◀

A **solution** to a linear programming problem consists of a feasible point that maximizes (or minimizes) the objective function, together with the corresponding value of the objective function.

One condition for a linear programming problem in two variables to have a solution is that the graph of the feasible points be bounded. (Refer to page 766.)

If none of the feasible points maximizes (or minimizes) the objective function or if there are no feasible points, then the linear programming problem has no solution.

Consider the linear programming problem stated in Example 2, and look again at Figure 24. The feasible points are the points that lie in the shaded region. For example, (20, 3) is a feasible point, as are (15, 5), (20, 5), (18, 4), and so on. To find the solution of the problem requires that we find a feasible point (x, y) that makes $I = 0.06x + 0.09y$ as large as possible. Notice that as I increases in value from $I = 0$ to $I = 0.9$ to $I = 1.35$ to $I = 1.65$ to $I = 1.8$, we obtain a collection of parallel lines. Furthermore, notice that the largest value of I that can be obtained using feasible points is $I = 1.65$, which corresponds to the line $1.65 = 0.06x + 0.09y$. Any larger value of I results in a line that does not pass through any feasible points. Finally, notice that the feasible point that yields $I = 1.65$ is the point (20, 5), a corner point. These observations form the basis of the following result, which we state without proof.

Theorem

Location of the Solution of a Linear Programming Problem

If a linear programming problem has a solution, it is located at a corner point of the graph of the feasible points.

If a linear programming problem has multiple solutions, at least one of them is located at a corner point of the graph of the feasible points.

In either case, the corresponding value of the objective function is unique.

We shall not consider here linear programming problems that have no solution. As a result, we can outline the procedure for solving a linear programming problem as follows:

> **Procedure for Solving a Linear Programming Problem**
>
> **STEP 1:** Write an expression for the quantity to be maximized (or minimized). This expression is the objective function.
>
> **STEP 2:** Write all the constraints as a system of linear inequalities and graph the system.
>
> **STEP 3:** List the corner points of the graph of the feasible points.
>
> **STEP 4:** List the corresponding values of the objective function at each corner point. The largest (or smallest) of these is the solution.

2 **EXAMPLE 3** **Solving a Minimum Linear Programming Problem**

Minimize the expression

$$z = 2x + 3y$$

subject to the constraints

$$y \leq 5, \qquad x \leq 6 \qquad x + y \geq 2, \qquad x \geq 0, \qquad y \geq 0$$

Solution The objective function is $z = 2x + 3y$. We seek the smallest value of z that can occur if x and y are solutions of the system of linear inequalities

$$\begin{cases} y \leq 5 \\ x \leq 6 \\ x + y \geq 2 \\ x \geq 0 \\ y \geq 0 \end{cases}$$

The graph of this system (the set of feasible points) is shown as the shaded region in Figure 25. We have also plotted the corner points. Table 1 lists the corner points and the corresponding values of the objective function. From the table, we can see that the minimum value of z is 4, and it occurs at the point (2, 0).

Figure 25

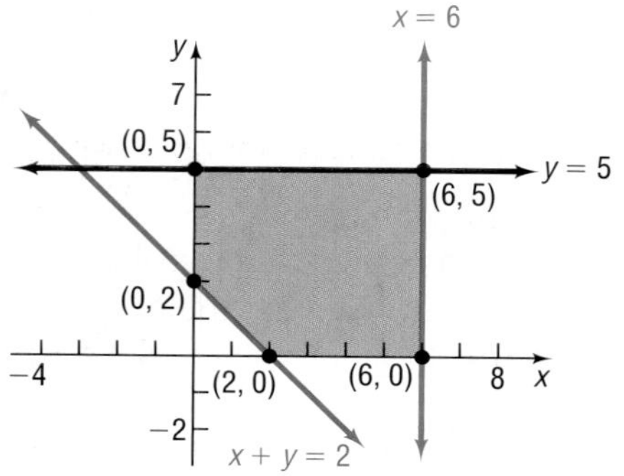

Table 1

Corner Point (x, y)	Value of the Objective Function $z = 2x + 3y$
(0, 2)	$z = 2(0) + 3(2) = 6$
(0, 5)	$z = 2(0) + 3(5) = 15$
(6, 5)	$z = 2(6) + 3(5) = 27$
(6, 0)	$z = 2(6) + 3(0) = 12$
(2, 0)	$z = 2(2) + 3(0) = 4$

◀

NOW WORK PROBLEMS 5 AND 11.

EXAMPLE 4 **Maximizing Profit**

At the end of every month, after filling orders for its regular customers, a coffee company has some pure Colombian coffee and some special-blend coffee remaining. The practice of the company has been to package a mixture of the two coffees into 1-pound packages as follows: a low-grade mixture containing 4 ounces of Colombian coffee and 12 ounces of special-blend coffee and a high-grade mixture containing 8 ounces of Colombian and 8 ounces of special-blend coffee. A profit of \$0.30 per package is made on the low-grade mixture, whereas a profit of \$0.40 per package is made on the high-grade mixture. This month, 120 pounds of special-blend coffee and 100 pounds of pure Colombian coffee remain. How many packages of each mixture should be prepared to achieve a maximum profit? Assume that all packages prepared can be sold.

Solution We begin by assigning symbols for the two variables.

x = Number of packages of the low-grade mixture

y = Number of packages of the high-grade mixture

If P denotes the profit, then

$$P = \$0.30x + \$0.40y$$

This expression is the objective function. We seek to maximize P subject to certain constraints on x and y. Because x and y represent numbers of packages, the only meaningful values for x and y are nonnegative integers. Thus, we have the two constraints

$$x \geq 0, \qquad y \geq 0 \qquad \text{Nonnegative constraints}$$

We also have only so much of each type of coffee available. For example, the total amount of Colombian coffee used in the two mixtures cannot exceed 100 pounds, or 1600 ounces. Because we use 4 ounces in each low-grade package and 8 ounces in each high-grade package, we are led to the constraint

$$4x + 8y \leq 1600 \qquad \text{Colombian coffee constraint}$$

Similarly, the supply of 120 pounds, or 1920 ounces, of special-blend coffee leads to the constraint

$$12x + 8y \leq 1920 \qquad \text{Special-blend coffee constraint}$$

The linear programming problem may be stated as

$$\text{Maximize} \qquad P = 0.3x + 0.4y$$

subject to the constraints

$$x \geq 0, \qquad y \geq 0, \qquad 4x + 8y \leq 1600, \qquad 12x + 8y \leq 1920$$

Figure 26

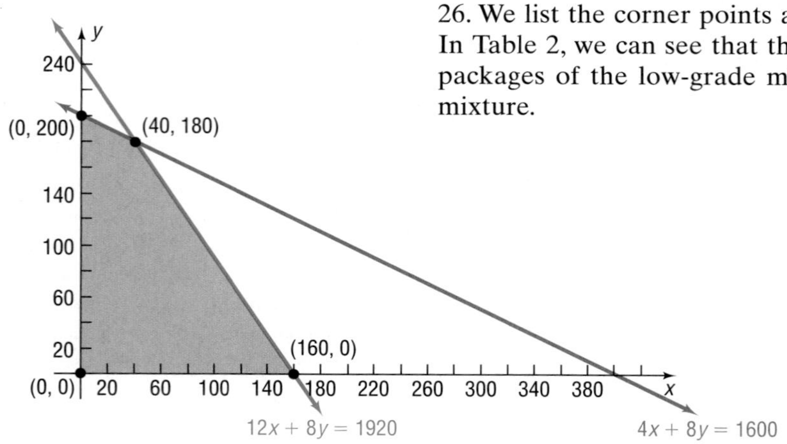

The graph of the constraints (the feasible points) is illustrated in Figure 26. We list the corner points and evaluate the objective function at each. In Table 2, we can see that the maximum profit, \$84, is achieved with 40 packages of the low-grade mixture and 180 packages of the high-grade mixture.

Table 2

Corner Point (x, y)	Value of Profit $P = 0.3x + 0.4y$
(0, 0)	$P = 0$
(0, 200)	$P = 0.3(0) + 0.4(200) = \$80$
(40, 180)	$P = 0.3(40) + 0.4(180) = \$84$
(160, 0)	$P = 0.3(160) + 0.4(0) = \$48$

NOW WORK PROBLEM 19.

10.8 Assess Your Understanding

Concepts and Vocabulary

1. A linear programming problem requires that a linear expression, called the __________ __________, be maximized or minimized.

2. *True or False:* If a linear programming problem has a solution, it is located at a corner point of the graph of the feasible points.

Exercises

In Problems 3–8, find the maximum and minimum value of the given objective function of a linear programming problem. The figure illustrates the graph of the feasible points.

3. $z = x + y$
4. $z = 2x + 3y$
5. $z = x + 10y$
6. $z = 10x + y$
7. $z = 5x + 7y$
8. $z = 7x + 5y$

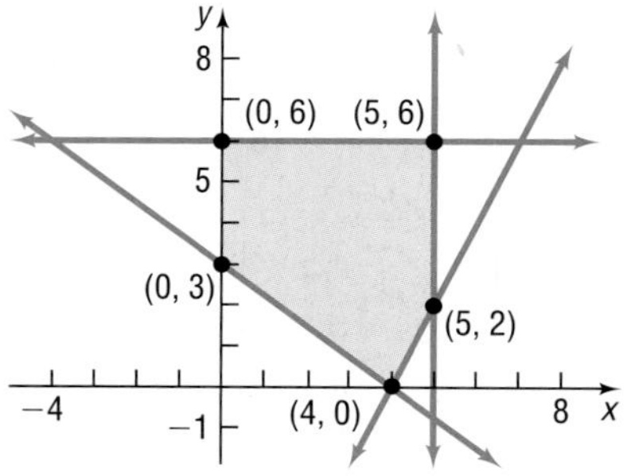

In Problems 9–18, solve each linear programming problem.

9. Maximize $z = 2x + y$ subject to $x \geq 0$, $y \geq 0$, $x + y \leq 6$, $x + y \geq 1$
10. Maximize $z = x + 3y$ subject to $x \geq 0$, $y \geq 0$, $x + y \geq 3$, $x \leq 5$, $y \leq 7$
11. Minimize $z = 2x + 5y$ subject to $x \geq 0$, $y \geq 0$, $x + y \geq 2$, $x \leq 5$, $y \leq 3$
12. Minimize $z = 3x + 4y$ subject to $x \geq 0$, $y \geq 0$, $2x + 3y \geq 6$, $x + y \leq 8$
13. Maximize $z = 3x + 5y$ subject to $x \geq 0$, $y \geq 0$, $x + y \geq 2$, $2x + 3y \leq 12$, $3x + 2y \leq 12$
14. Maximize $z = 5x + 3y$ subject to $x \geq 0$, $y \geq 0$, $x + y \geq 2$, $x + y \leq 8$, $2x + y \leq 10$
15. Minimize $z = 5x + 4y$ subject to $x \geq 0$, $y \geq 0$, $x + y \geq 2$, $2x + 3y \leq 12$, $3x + y \leq 12$
16. Minimize $z = 2x + 3y$ subject to $x \geq 0$, $y \geq 0$, $x + y \geq 3$, $x + y \leq 9$, $x + 3y \geq 6$
17. Maximize $z = 5x + 2y$ subject to $x \geq 0$, $y \geq 0$, $x + y \leq 10$, $2x + y \geq 10$, $x + 2y \geq 10$
18. Maximize $z = 2x + 4y$ subject to $x \geq 0$, $y \geq 0$, $2x + y \geq 4$, $x + y \leq 9$

19. **Maximizing Profit** A manufacturer of skis produces two types: downhill and cross-country. Use the following table to determine how many of each kind of ski should be produced to achieve a maximum profit. What is the maximum profit? What would the maximum profit be if the maximum time available for manufacturing is increased to 48 hours?

	Downhill	Cross-country	Maximum Time Available
Manufacturing time per ski	2 hours	1 hour	40 hours
Finishing time per ski	1 hour	1 hour	32 hours
Profit per ski	$70	$50	

20. **Farm Management** A farmer has 70 acres of land available for planting either soybeans or wheat. The cost of preparing the soil, the workdays required, and the expected profit per acre planted for each type of crop are given in the following table:

	Soybeans	Wheat
Preparation cost per acre	$60	$30
Workdays required per acre	3	4
Profit per acre	$180	$100

The farmer cannot spend more than $1800 in preparation costs nor use more than a total of 120 workdays. How many acres of each crop should be planted in order to maximize the profit? What is the maximum profit? What is the maximum profit if the farmer is willing to spend no more than $2400 on preparation?

21. **Farm Management** A small farm in Illinois has 100 acres of land available on which to grow corn and soybeans. The following table shows the cultivation cost per acre, the labor cost per acre, and the expected profit per acre. The column on the right shows the amount of money available for each of these expenses. Find the number of acres of each crop that should be planted in order to maximize profit.

	Soybeans	Corn	Money Available
Cultivation cost per acre	$40	$60	$1800
Labor cost per acre	$60	$60	$2400
Profit per acre	$200	$250	

22. **Dietary Requirements** A certain diet requires at least 60 units of carbohydrates, 45 units of protein, and 30 units of fat each day. Each ounce of Supplement A provides 5 units of carbohydrates, 3 units of protein, and 4 units of fat. Each ounce of Supplement B provides 2 units of carbohydrates, 2 units of protein, and 1 unit of fat. If Supplement A costs $1.50 per ounce and Supplement B costs $1.00 per ounce, how many ounces of each supplement should be taken daily to minimize the cost of the diet?

23. **Production Scheduling** In a factory, machine 1 produces 8-inch pliers at the rate of 60 units per hour and 6-inch pliers at the rate of 70 units per hour. Machine 2 produces 8-inch pliers at the rate of 40 units per hour and 6-inch pliers at the rate of 20 units per hour. It costs $50 per hour to operate machine 1, and machine 2 costs $30 per hour to operate. The production schedule requires that at least 240 units of 8-inch pliers and at least 140 units of 6-inch pliers be produced during each 10-hour day. Which combination of machines will cost the least money to operate?

24. **Farm Management** An owner of a fruit orchard hires a crew of workers to prune at least 25 of his 50 fruit trees. Each newer tree requires one hour to prune, while each older tree needs one-and-a-half hours. The crew contracts to work for at least 30 hours and charge $15 for each newer tree and $20 for each older tree. To minimize his cost, how many of each kind of tree will the orchard owner have pruned? What will be the cost?

25. **Managing a Meat Market** A meat market combines ground beef and ground pork in a single package for meat loaf. The ground beef is 75% lean (75% beef, 25% fat) and costs the market $0.75 per pound. The ground pork is 60% lean and costs the market $0.45 per pound. The meat loaf must be at least 70% lean. If the market wants to use at least 50 lb of its available pork, but no more than 200 lb of its available ground beef, how much ground beef should be mixed with ground pork so that the cost is minimized?

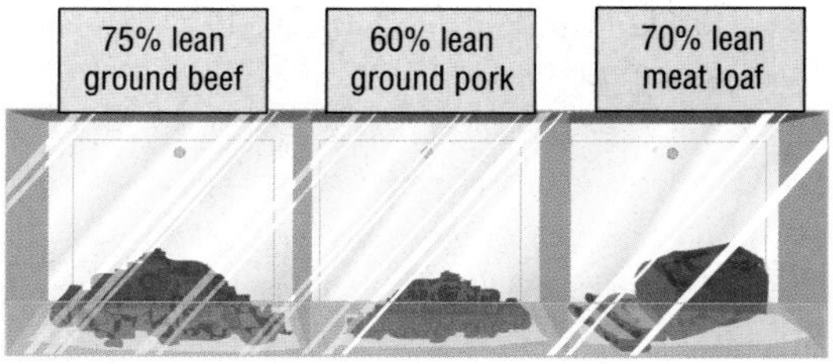

26. **Return on Investment** An investment broker is instructed by her client to invest up to $20,000, some in a junk bond yielding 9% per annum and some in Treasury bills yielding 7% per annum. The client wants to invest at least $8000 in T-bills and no more than $12,000 in the junk bond.

(a) How much should the broker recommend that the client place in each investment to maximize income if the client insists that the amount invested in T-bills must equal or exceed the amount placed in junk bonds?

(b) How much should the broker recommend that the client place in each investment to maximize income if the client insists that the amount invested in T-bills must not exceed the amount placed in junk bonds?

27. Maximizing Profit on Ice Skates A factory manufactures two kinds of ice skates: racing skates and figure skates. The racing skates require 6 work-hours in the fabrication department, whereas the figure skates require 4 work-hours there. The racing skates require 1 work-hour in the finishing department, whereas the figure skates require 2 work-hours there. The fabricating department has available at most 120 work-hours per day, and the finishing department has no more than 40 work-hours per day available. If the profit on each racing skate is \$10 and the profit on each figure skate is \$12, how many of each should be manufactured each day to maximize profit? (Assume that all skates made are sold.)

28. Financial Planning A retired couple has up to \$50,000 to place in fixed-income securities. Their financial adviser suggests two securities to them: one is an AAA bond that yields 8% per annum; the other is a Certificate of Deposit (CD) that yields 4%. After careful consideration of the alternatives, the couple decides to place at most \$20,000 in the AAA bond and at least \$15,000 in the CD. They also instruct the financial adviser to place at least as much in the CD as in the AAA bond. How should the financial adviser proceed to maximize the return on their investment?

29. Product Design An entrepreneur is having a design group produce at least six samples of a new kind of fastener that he wants to market. It costs \$9.00 to produce each metal fastener and \$4.00 to produce each plastic fastener. He wants to have at least two of each version of the fastener and needs to have all the samples 24 hours from now. It takes 4 hours to produce each metal sample and 2 hours to produce each plastic sample. To minimize the cost of the samples, how many of each kind should the entrepreneur order? What will be the cost of the samples?

30. Animal Nutrition Kevin's dog Amadeus likes two kinds of canned dog food. "Gourmet Dog" costs 40 cents a can and has 20 units of a vitamin complex; the calorie content is 75 calories. "Chow Hound" costs 32 cents a can and has 35 units of vitamins and 50 calories. Kevin likes Amadeus to have at least 1175 units of vitamins a month and at least 2375 calories during the same time period. Kevin has space to store only 60 cans of dog food at a time. How much of each kind of dog food should Kevin buy each month in order to minimize his cost?

31. Airline Revenue An airline has two classes of service: first class and coach. Management's experience has been that each aircraft should have at least 8 but no more than 16 first-class seats and at least 80 but not more than 120 coach seats.

(a) If management decides that the ratio of first class to coach seats should never exceed 1:12, with how many of each type of seat should an aircraft be configured to maximize revenue?

(b) If management decides that the ratio of first class to coach seats should never exceed 1:8, with how many of each type of seat should an aircraft be configured to maximize revenue?

(c) If you were management, what would you do?

[**Hint:** Assume that the airline charges \$$C$ for a coach seat and \$$F$ for a first-class seat; $C > 0$, $F > C$.]

32. Minimizing Cost A farm that specializes in raising frying chickens supplements the regular chicken feed with four vitamins. The owner wants the supplemental food to contain at least 50 units of vitamin I, 90 units of vitamin II, 60 units of vitamin III, and 100 units of vitamin IV per 100 ounces of feed. Two supplements are available: supplement A, which contains 5 units of vitamin I, 25 units of vitamin II, 10 units of vitamin III, and 35 units of vitamin IV per ounce, and supplement B, which contains 25 units of vitamin I, 10 units of vitamin II, 10 units of vitamin III, and 20 units of vitamin IV per ounce. If supplement A costs \$0.06 per ounce and supplement B costs \$0.08 per ounce, how much of each supplement should the manager of the farm buy to add to each 100 ounces of feed in order to keep the total cost at a minimum, while still meeting the owner's vitamin specifications?

33. Explain what a linear programming problem is and how it can be solved.

Chapter Review

Things to Know

Systems of equations (p. 686)

Systems with no solutions are inconsistent. Systems with a solution are consistent.

Consistent systems of linear equations have either a unique solution or an infinite number of solutions.

Determinants and Cramer's Rule (pp. 718 and 722)

Matrix (pp. 700 and 727)	Rectangular array of numbers, called entries
m by n matrix (p. 727)	Matrix with m rows and n columns
Identity matrix I (p. 735)	Square matrix whose diagonal entries are 1's, while all other entries are 0's
Inverse of a matrix (p. 735)	A^{-1} is the inverse of A if $AA^{-1} = A^{-1}A = I$
Nonsingular matrix (p. 736)	A square matrix that has an inverse

Linear programming (p. 770)

Maximize (or minimize) a linear objective function, $z = Ax + By$, subject to certain conditions, or constraints, expressible as linear inequalities in x and y. A feasible point (x, y) is a point that satisfies the constraints of a linear programming problem.

Location of solution (p. 771)

If a linear programming problem has a solution, it is located at a corner point of the graph of the feasible points.
If a linear programming problem has multiple solutions, at least one of them is located at a corner point of the graph of the feasible points.
In either case, the corresponding value of the objective function is unique.

Objectives

Section		You should be able to . . .	Review Exercises
10.1	1	Solve systems of equations by substitution (p. 687)	1–14, 99, 100, 103–105
	2	Solve systems of equations by elimination (p. 688)	1–14, 99, 100, 103–105
	3	Identify inconsistent systems of equations containing two variables (p. 690)	9, 10, 13, 96
	4	Express the solution of a system of dependent equations containing two variables (p. 691)	14, 95
	5	Solve systems of three equations containing three variables (p. 692)	15–18, 97, 98, 101
	6	Identify inconsistent systems of equations containing three variables (p. 693)	18
	7	Express the solution of a system of dependent equations containing three variables (p. 694)	17
10.2	1	Write the augmented matrix of a system of linear equations (p. 700)	35–44
	2	Write the system from the augmented matrix (p. 701)	19, 20
	3	Perform row operations on a matrix (p. 702)	35–44
	4	Solve systems of linear equations using matrices (p. 704)	35–44
10.3	1	Evaluate 2 by 2 determinants (p. 716)	45, 46
	2	Use Cramer's Rule to solve a system of two equations, two variables (p. 717)	51–54
	3	Evaluate 3 by 3 determinants (p. 720)	47–50
	4	Use Cramer's Rule to solve a system of three equations, three variables (p. 722)	55, 56
	5	Know properties of determinants (p. 723)	57, 58
10.4	1	Find the sum and difference of two matrices (p. 728)	21, 22
	2	Find scalar multiples of a matrix (p. 730)	23, 24
	3	Find the product of two matrices (p. 731)	25–28
	4	Find the inverse of a matrix (p. 735)	29–34
	5	Solve a system of equations using inverse matrices (p. 739)	35–37, 39, 40, 43, 44

10.5	1	Decompose $\frac{P}{Q}$, where Q has only nonrepeated linear factors (p. 744)	59, 60
	2	Decompose $\frac{P}{Q}$, where Q has repeated linear factors (p. 746)	61, 62
	3	Decompose $\frac{P}{Q}$, where Q has a nonrepeated irreducible quadratic factor (p. 748)	63, 64, 67, 68
	4	Decompose $\frac{P}{Q}$, where Q has repeated irreducible quadratic factors (p. 749)	65, 66
10.6	1	Solve a system of nonlinear equations using substitution (p. 751)	69–78
	2	Solve a system of nonlinear equations using elimination (p. 752)	69–78
10.7	1	Graph an inequality (p. 760)	79, 80
	2	Graph a system of inequalities (p. 763)	81–90, 102
10.8	1	Set up a linear programming problem (p. 769)	106, 107
	2	Solve a linear programming problem (p. 772)	91–94, 106, 107

Review Exercises

(Blue problem numbers indicate the author's suggestions for use in a Practice Test.)

In Problems 1–18, solve each system of equations using the method of substitution or the method of elimination. If the system has no solution, say that it is inconsistent.

1. $\begin{cases} 2x - y = 5 \\ 5x + 2y = 8 \end{cases}$

2. $\begin{cases} 2x + 3y = 2 \\ 7x - y = 3 \end{cases}$

3. $\begin{cases} 3x - 4y = 4 \\ x - 3y = \frac{1}{2} \end{cases}$

4. $\begin{cases} 2x + y = 0 \\ 5x - 4y = -\frac{13}{2} \end{cases}$

5. $\begin{cases} x - 2y - 4 = 0 \\ 3x + 2y - 4 = 0 \end{cases}$

6. $\begin{cases} x - 3y + 5 = 0 \\ 2x + 3y - 5 = 0 \end{cases}$

7. $\begin{cases} y = 2x - 5 \\ x = 3y + 4 \end{cases}$

8. $\begin{cases} x = 5y + 2 \\ y = 5x + 2 \end{cases}$

9. $\begin{cases} x - 3y + 4 = 0 \\ \frac{1}{2}x - \frac{3}{2}y + \frac{4}{3} = 0 \end{cases}$

10. $\begin{cases} x + \frac{1}{4}y = 2 \\ y + 4x + 2 = 0 \end{cases}$

11. $\begin{cases} 2x + 3y - 13 = 0 \\ 3x - 2y = 0 \end{cases}$

12. $\begin{cases} 4x + 5y = 21 \\ 5x + 6y = 42 \end{cases}$

13. $\begin{cases} 3x - 2y = 8 \\ x - \frac{2}{3}y = 12 \end{cases}$

14. $\begin{cases} 2x + 5y = 10 \\ 4x + 10y = 20 \end{cases}$

15. $\begin{cases} x + 2y - z = 6 \\ 2x - y + 3z = -13 \\ 3x - 2y + 3z = -16 \end{cases}$

16. $\begin{cases} x + 5y - z = 2 \\ 2x + y + z = 7 \\ x - y + 2z = 11 \end{cases}$

17. $\begin{cases} 2x - 4y + z = -15 \\ x + 2y - 4z = 27 \\ 5x - 6y - 2z = -3 \end{cases}$

18. $\begin{cases} x - 4y + 3z = 15 \\ -3x + y - 5z = -5 \\ -7x - 5y - 9z = 10 \end{cases}$

In Problems 19 and 20, write the system of equations corresponding to the given augmented matrix.

19. $\left[\begin{array}{cc|c} 3 & 2 & 8 \\ 1 & 4 & -1 \end{array}\right]$

20. $\left[\begin{array}{ccc|c} 1 & 2 & 5 & -2 \\ 5 & 0 & -3 & 8 \\ 2 & -1 & 0 & 0 \end{array}\right]$

In Problems 21–28, use the following matrices to compute each expression.

$$A = \begin{bmatrix} 1 & 0 \\ 2 & 4 \\ -1 & 2 \end{bmatrix}, \quad B = \begin{bmatrix} 4 & -3 & 0 \\ 1 & 1 & -2 \end{bmatrix}, \quad C = \begin{bmatrix} 3 & -4 \\ 1 & 5 \\ 5 & 2 \end{bmatrix}$$

21. $A + C$ **22.** $A - C$ **23.** $6A$ **24.** $-4B$

25. AB **26.** BA **27.** CB **28.** BC

In Problems 29–34, find the inverse of each matrix, if there is one. If there is not an inverse, say that the matrix is singular.

29. $\begin{bmatrix} 4 & 6 \\ 1 & 3 \end{bmatrix}$

30. $\begin{bmatrix} -3 & 2 \\ 1 & -2 \end{bmatrix}$

31. $\begin{bmatrix} 1 & 3 & 3 \\ 1 & 2 & 1 \\ 1 & -1 & 2 \end{bmatrix}$

32. $\begin{bmatrix} 3 & 1 & 2 \\ 3 & 2 & -1 \\ 1 & 1 & 1 \end{bmatrix}$ **33.** $\begin{bmatrix} 4 & -8 \\ -1 & 2 \end{bmatrix}$ **34.** $\begin{bmatrix} -3 & 1 \\ -6 & 2 \end{bmatrix}$

In Problems 35–44, solve each system of equations using matrices. If the system has no solution, say that it is inconsistent.

35. $\begin{cases} 3x - 2y = 1 \\ 10x + 10y = 5 \end{cases}$ **36.** $\begin{cases} 3x + 2y = 6 \\ x - y = -\frac{1}{2} \end{cases}$ **37.** $\begin{cases} 5x + 6y - 3z = 6 \\ 4x - 7y - 2z = -3 \\ 3x + y - 7z = 1 \end{cases}$

38. $\begin{cases} 2x + y + z = 5 \\ 4x - y - 3z = 1 \\ 8x + y - z = 5 \end{cases}$ **39.** $\begin{cases} x - 2z = 1 \\ 2x + 3y = -3 \\ 4x - 3y - 4z = 3 \end{cases}$ **40.** $\begin{cases} x + 2y - z = 2 \\ 2x - 2y + z = -1 \\ 6x + 4y + 3z = 5 \end{cases}$

41. $\begin{cases} x - y + z = 0 \\ x - y - 5z - 6 = 0 \\ 2x - 2y + z - 1 = 0 \end{cases}$ **42.** $\begin{cases} 4x - 3y + 5z = 0 \\ 2x + 4y - 3z = 0 \\ 6x + 2y + z = 0 \end{cases}$ **43.** $\begin{cases} x - y - z - t = 1 \\ 2x + y - z + 2t = 3 \\ x - 2y - 2z - 3t = 0 \\ 3x - 4y + z + 5t = -3 \end{cases}$

44. $\begin{cases} x - 3y + 3z - t = 4 \\ x + 2y - z = -3 \\ x + 3z + 2t = 3 \\ x + y + 5z = 6 \end{cases}$

In Problems 45–50, find the value of each determinant.

45. $\begin{vmatrix} 3 & 4 \\ 1 & 3 \end{vmatrix}$ **46.** $\begin{vmatrix} -4 & 0 \\ 1 & 3 \end{vmatrix}$ **47.** $\begin{vmatrix} 1 & 4 & 0 \\ -1 & 2 & 6 \\ 4 & 1 & 3 \end{vmatrix}$

48. $\begin{vmatrix} 2 & 3 & 10 \\ 0 & 1 & 5 \\ -1 & 2 & 3 \end{vmatrix}$ **49.** $\begin{vmatrix} 2 & 1 & -3 \\ 5 & 0 & 1 \\ 2 & 6 & 0 \end{vmatrix}$ **50.** $\begin{vmatrix} -2 & 1 & 0 \\ 1 & 2 & 3 \\ -1 & 4 & 2 \end{vmatrix}$

In Problems 51–56, use Cramer's Rule, if applicable, to solve each system.

51. $\begin{cases} x - 2y = 4 \\ 3x + 2y = 4 \end{cases}$ **52.** $\begin{cases} x - 3y = -5 \\ 2x + 3y = 5 \end{cases}$ **53.** $\begin{cases} 2x + 3y - 13 = 0 \\ 3x - 2y = 0 \end{cases}$

54. $\begin{cases} 3x - 4y - 12 = 0 \\ 5x + 2y + 6 = 0 \end{cases}$ **55.** $\begin{cases} x + 2y - z = 6 \\ 2x - y + 3z = -13 \\ 3x - 2y + 3z = -16 \end{cases}$ **56.** $\begin{cases} x - y + z = 8 \\ 2x + 3y - z = -2 \\ 3x - y - 9z = 9 \end{cases}$

In Problems 57 and 58, use properties of determinants to find the value of each determinant if it is known that

$$\begin{vmatrix} x & y \\ a & b \end{vmatrix} = 8$$

57. $\begin{vmatrix} 2x & y \\ 2a & b \end{vmatrix}$ **58.** $\begin{vmatrix} y & x \\ b & a \end{vmatrix}$

In Problems 59–68, write the partial fraction decomposition of each rational expression.

59. $\dfrac{6}{x(x-4)}$ **60.** $\dfrac{x}{(x+2)(x-3)}$ **61.** $\dfrac{x-4}{x^2(x-1)}$ **62.** $\dfrac{2x-6}{(x-2)^2(x-1)}$ **63.** $\dfrac{x}{(x^2+9)(x+1)}$

64. $\dfrac{3x}{(x-2)(x^2+1)}$ **65.** $\dfrac{x^3}{(x^2+4)^2}$ **66.** $\dfrac{x^3+1}{(x^2+16)^2}$ **67.** $\dfrac{x^2}{(x^2+1)(x^2-1)}$ **68.** $\dfrac{4}{(x^2+4)(x^2-1)}$

In Problems 69–78, solve each system of nonlinear equations.

69. $\begin{cases} 2x + y + 3 = 0 \\ x^2 + y^2 = 5 \end{cases}$ **70.** $\begin{cases} x^2 + y^2 = 16 \\ 2x - y^2 = -8 \end{cases}$ **71.** $\begin{cases} 2xy + y^2 = 10 \\ 3y^2 - xy = 2 \end{cases}$ **72.** $\begin{cases} 3x^2 - y^2 = 1 \\ 7x^2 - 2y^2 - 5 = 0 \end{cases}$

73. $\begin{cases} x^2 + y^2 = 6y \\ x^2 = 3y \end{cases}$ **74.** $\begin{cases} 2x^2 + y^2 = 9 \\ x^2 + y^2 = 9 \end{cases}$ **75.** $\begin{cases} 3x^2 + 4xy + 5y^2 = 8 \\ x^2 + 3xy + 2y^2 = 0 \end{cases}$ **76.** $\begin{cases} 3x^2 + 2xy - 2y^2 = 6 \\ xy - 2y^2 + 4 = 0 \end{cases}$

77. $\begin{cases} x^2 - 3x + y^2 + y = -2 \\ \dfrac{x^2 - x}{y} + y + 1 = 0 \end{cases}$ **78.** $\begin{cases} x^2 + x + y^2 = y + 2 \\ x + 1 = \dfrac{2 - y}{x} \end{cases}$

In Problems 79 and 80, graph each inequality.

79. $3x + 4y \le 12$ **80.** $2x - 3y \ge 6$

In Problems 81–86, graph each system of inequalities. Tell whether the graph is bounded or unbounded, and label the corner points.

81. $\begin{cases} -2x + y \le 2 \\ x + y \ge 2 \end{cases}$ **82.** $\begin{cases} x - 2y \le 6 \\ 2x + y \ge 2 \end{cases}$ 83. $\begin{cases} x \ge 0 \\ y \ge 0 \\ x + y \le 4 \\ 2x + 3y \le 6 \end{cases}$

84. $\begin{cases} x \ge 0 \\ y \ge 0 \\ 3x + y \ge 6 \\ 2x + y \ge 2 \end{cases}$ **85.** $\begin{cases} x \ge 0 \\ y \ge 0 \\ 2x + y \le 8 \\ x + 2y \ge 2 \end{cases}$ **86.** $\begin{cases} x \ge 0 \\ y \ge 0 \\ 3x + y \le 9 \\ 2x + 3y \ge 6 \end{cases}$

In Problems 87–90, graph each system of inequalities.

87. $\begin{cases} x^2 + y^2 \le 16 \\ x + y \ge 2 \end{cases}$ **88.** $\begin{cases} y^2 \le x - 1 \\ x - y \le 3 \end{cases}$ 89. $\begin{cases} y \le x^2 \\ xy \le 4 \end{cases}$ **90.** $\begin{cases} x^2 + y^2 \ge 1 \\ x^2 + y^2 \le 4 \end{cases}$

In Problems 91–94, solve each linear programming problem.

91. Maximize $z = 3x + 4y$ subject to $x \ge 0, y \ge 0, 3x + 2y \ge 6, x + y \le 8$

92. Maximize $z = 2x + 4y$ subject to $x \ge 0, y \ge 0, x + y \le 6, x \ge 2$

93. Minimize $z = 3x + 5y$ subject to $x \ge 0, y \ge 0, x + y \ge 1, 3x + 2y \le 12, x + 3y \le 12$

94. Minimize $z = 3x + y$ subject to $x \ge 0, y \ge 0, x \le 8, y \le 6, 2x + y \ge 4$

---

95. Find A such that the system of equations has infinitely many solutions.

$$\begin{cases} 2x + 5y = 5 \\ 4x + 10y = A \end{cases}$$

96. Find A such that the system in Problem 95 is inconsistent.

97. Curve Fitting Find the quadratic function $y = ax^2 + bx + c$ that passes through the three points $(0, 1)$, $(1, 0)$, and $(-2, 1)$.

98. Curve Fitting Find the general equation of the circle that passes through the three points $(0, 1)$, $(1, 0)$, and $(-2, 1)$.

[**Hint:** The general equation of a circle is $x^2 + y^2 + Dx + Ey + F = 0$.]

99. Blending Coffee A coffee distributor is blending a new coffee that will cost \$3.90 per pound. It will consist of a blend of \$3.00 per pound coffee and \$6.00 per pound coffee. What amounts of each type of coffee should be mixed to achieve the desired blend?

[**Hint:** Assume that the weight of the blended coffee is 100 pounds].

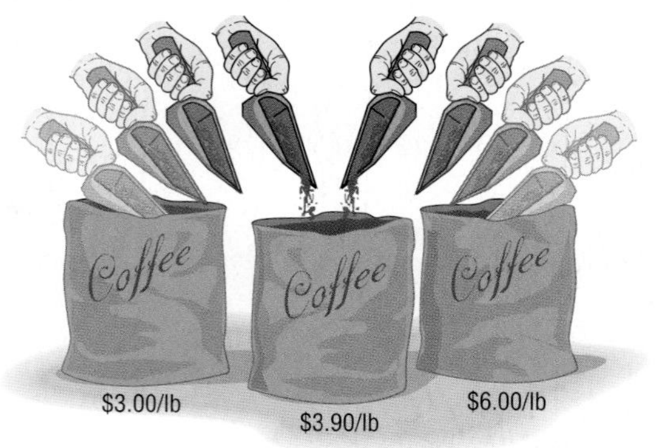

100. Farming A 1000-acre farm in Illinois is used to grow corn and soy beans. The cost per acre for raising corn is \$65, and the cost per acre for soy beans is \$45. If \$54,325 has been budgeted for costs and all the acreage is to be used, how many acres should be allocated for each crop?

101. Cookie Orders A cookie company makes three kinds of cookies, oatmeal raisin, chocolate chip, and shortbread, packaged in small, medium, and large boxes. The small box contains 1 dozen oatmeal raisin and 1 dozen chocolate chip; the medium box has 2 dozen oatmeal raisin, 1 dozen chocolate chip, and 1 dozen shortbread;

the large box contains 2 dozen oatmeal raisin, 2 dozen chocolate chip, and 3 dozen shortbread. If you require exactly 15 dozen oatmeal raisin, 10 dozen chocolate chip, and 11 dozen shortbread, how many of each size box should you buy?

102. Mixed Nuts A store that specializes in selling nuts has available 72 pounds of cashews and 120 pounds of peanuts. These are to be mixed in 12-ounce packages as follows: a lower-priced package containing 8 ounces of peanuts and 4 ounces of cashews and a quality package containing 6 ounces of peanuts and 6 ounces of cashews.

(a) Use x to denote the number of lower-priced packages and use y to denote the number of quality packages. Write a system of linear inequalities that describes the possible number of each kind of package.

(b) Graph the system and label the corner points.

103. Determining the Speed of the Current of the Aguarico River On a recent trip to the Cuyabeno Wildlife Reserve in the Amazon region of Ecuador, a 100-kilometer trip by speedboat was taken down the Aguarico River from Chiritza to the Flotel Orellana. As I watched the Amazon unfold, I wondered how fast the speedboat was going and how fast the current of the white-water Aguarico River was. I timed the trip downstream at 2.5 hours and the return trip at 3 hours. What were the two speeds?

104. Finding the Speed of the Jet Stream On a flight between Midway Airport in Chicago and Ft. Lauderdale, Florida, a Boeing 737 jet maintains an airspeed of 475 miles per hour. If the trip from Chicago to Ft. Lauderdale takes 2 hours, 30 minutes and the return flight takes 2 hours, 50 minutes, what is the speed of the jet stream? (Assume that the speed of the jet stream remains constant at the various altitudes of the plane and that the plane flies with the jet stream one way and against it the other way.)

105. Constant Rate Jobs If Bruce and Bryce work together for 1 hour and 20 minutes, they will finish a certain job. If Bryce and Marty work together for 1 hour and 36 minutes, the same job can be finished. If Marty and Bruce work together, they can complete this job in 2 hours and 40 minutes. How long will it take each of them working alone to finish the job?

106. Maximizing Profit on Figurines A factory manufactures two kinds of ceramic figurines: a dancing girl and a mermaid. Each requires three processes: molding, painting, and glazing. The daily labor available for molding is no more than 90 work-hours, labor available for painting does not exceed 120 work-hours, and labor available for glazing is no more than 60 work-hours. The dancing girl requires 3 work-hours for molding, 6 work-hours for painting, and 2 work-hours for glazing. The mermaid requires 3 work-hours for molding, 4 work-hours for painting, and 3 work-hours for glazing. If the profit on each figurine is $25 for dancing girls and $30 for mermaids, how many of each should be produced each day to maximize profit? If management decides to produce the number of each figurine that maximizes profit, determine which of these processes has excess work-hours assigned to it.

107. Minimizing Production Cost A factory produces gasoline engines and diesel engines. Each week the factory is obligated to deliver at least 20 gasoline engines and at least 15 diesel engines. Due to physical limitations, however, the factory cannot make more than 60 gasoline engines nor more than 40 diesel engines. Finally, to prevent layoffs, a total of at least 50 engines must be produced. If gasoline engines cost $450 each to produce and diesel engines cost $550 each to produce, how many of each should be produced per week to minimize the cost? What is the excess capacity of the factory; that is, how many of each kind of engine is being produced in excess of the number that the factory is obligated to deliver?

108. Describe four ways of solving a system of three linear equations containing three variables. Which method do you prefer? Why?

Chapter Projects

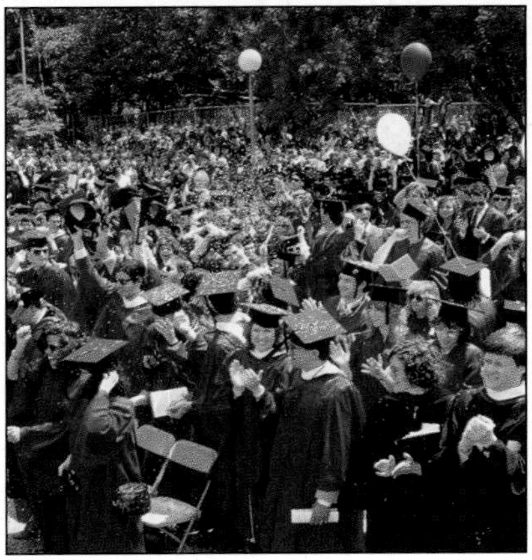

1. Markov Chains A **Markov chain** (or process) is one in which future outcomes are determined by a current state. Future outcomes are based on probabilities. The probability of moving to a certain state depends only on the state previously occupied and does not vary with time. An example of a Markov chain would be the maximum education achieved by children based on the highest education attained by their parents, where the states are (1) earned college degree, (2) high-school diploma only, (3) elementary school only. If p_{ij} is the probability of moving from state i to state j, then the **transition matrix** is the $m \times m$ matrix

$$P = \begin{bmatrix} p_{11} & p_{12} & \cdots & p_{1m} \\ \vdots & \vdots & & \vdots \\ p_{m1} & p_{m2} & \cdots & p_{mm} \end{bmatrix}$$

The table below represents the probabilities of the highest educational level of children based on the highest educational level of their parents. For example, the table shows that the probability p_{21} is 40% that parents with a high-school education (row 2) will have children with a college education (column 1).

(a) Convert the percentages to decimals.
(b) What is the transition matrix?
(c) Sum across the rows. What do you notice? Why do you think that you obtained this result?
(d) If P is the transition matrix of a Markov chain, then the (i, j)th entry of P^n (nth power of P) gives the probability of passing from state i to state j in n stages. What is the probability that a grandchild of a college graduate is a college graduate?
(e) What is the probability that the grandchild of a high school graduate finishes college?
(f) The row vector $v^{(0)} = [0.267 \quad 0.574 \quad 0.159]$ represents the proportion of the U.S. population that has college, high school, and elementary school, respectively, as the highest educational level in 2002.* In a Markov chain the probability distribution $v^{(k)}$ after k stages is $v^{(k)} = v^{(0)}P^k$, where P^k is the kth power of the transition matrix. What will be the distribution of highest educational attainment of the grandchildren of the current population?
(g) Calculate $P^3, P^4, P^5, \ldots$. Continue until the matrix does not change. This is called the long-run distribution. What is the long-run distribution of highest educational attainment of the population?

*SOURCE: U.S. Census Bureau.

Highest Educational Level of Parents	Maximum Education That Children Achieve		
	College	High School	Elementary
College	80%	18%	2%
High school	40%	50%	10%
Elementary	20%	60%	20%

The following Chapter Projects may be found at www.prenhall.com/sullivan7e.

2. **Project at Motorola** ***Error Control Codings***
3. **Using Matrices to Find the Line of Best Fit**
4. **CBL Experiment**

Cumulative Review

In Problems 1–6, solve each equation.

1. $2x^2 - x = 0$
2. $\sqrt{3x + 1} = 4$
3. $2x^3 - 3x^2 - 8x - 3 = 0$
4. $3^x = 9^{x+1}$
5. $\log_3(x - 1) + \log_3(2x + 1) = 2$
6. $3^x = e$

7. Determine whether the function $g(x) = \dfrac{2x^3}{x^4 + 1}$ is even, odd, or neither. Is the graph of g symmetric with respect to the x-axis, y-axis, or origin?

8. Find the center and radius of the circle

$$x^2 + y^2 - 2x + 4y - 11 = 0.$$

Graph the circle.

9. Graph $f(x) = 3^{x-2} + 1$ using transformations. What is the domain, range, and horizontal asymptote of f?

10. The function $f(x) = \dfrac{5}{x + 2}$ is one-to-one. Find f^{-1}. Find the domain and the range of f and the domain and the range of f^{-1}.

11. Graph each equation:
(a) $y = 3x + 6$
(b) $x^2 + y^2 = 4$
(c) $x^2 - y^2 = 4$
(d) $9x^2 + y^2 = 9$
(e) $y^2 = 3x + 6$
(f) $y = x^3$
(g) $y = \dfrac{1}{x}$
(h) $y = \sqrt{x}$
(i) $y = e^x$
(j) $y = \ln x$
(k) $y = \sin x$
(l) $y = \cos x$

12. Solve each equation:
(a) $\sin x = \dfrac{1}{2}, 0 \le x < 2\pi$
(b) $\cos(3x) = \dfrac{1}{2}, 0 \le x < 2\pi$

11 Sequences; Induction; the Binomial Theorem

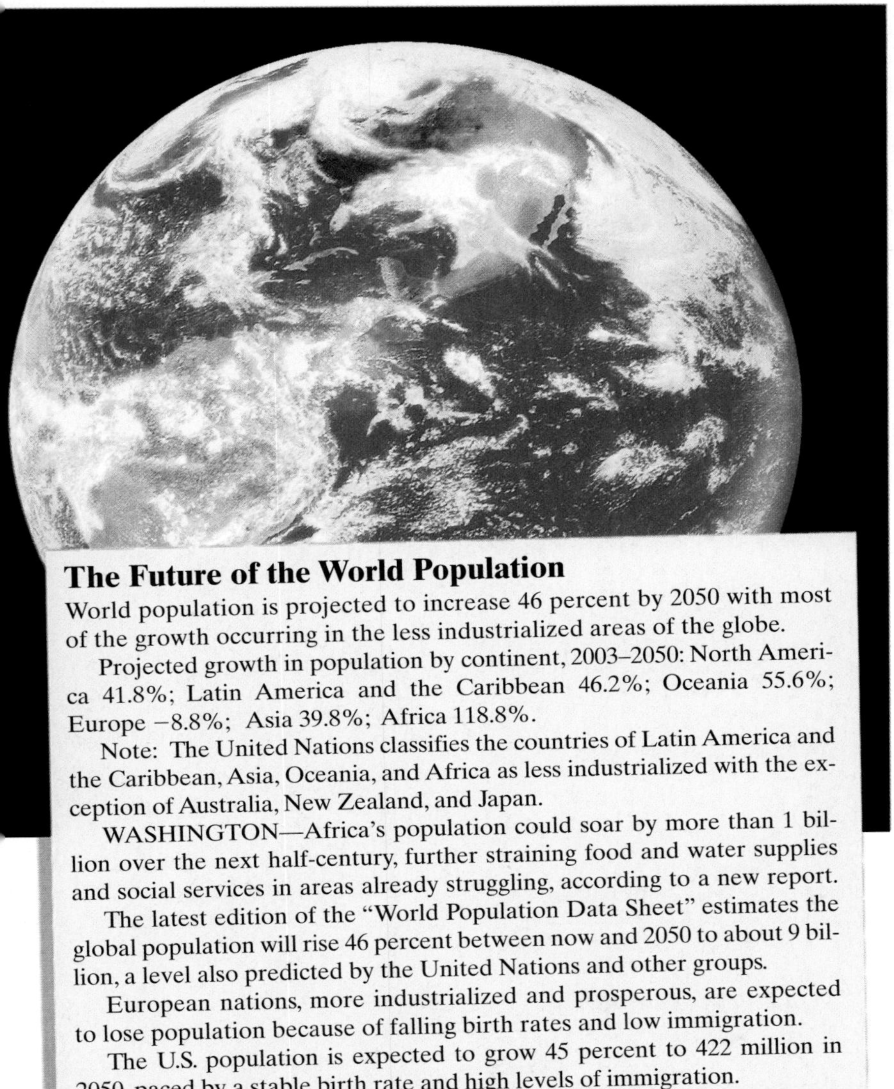

O U T L I N E

The Future of the World Population

World population is projected to increase 46 percent by 2050 with most of the growth occurring in the less industrialized areas of the globe.

Projected growth in population by continent, 2003–2050: North America 41.8%; Latin America and the Caribbean 46.2%; Oceania 55.6%; Europe −8.8%; Asia 39.8%; Africa 118.8%.

Note: The United Nations classifies the countries of Latin America and the Caribbean, Asia, Oceania, and Africa as less industrialized with the exception of Australia, New Zealand, and Japan.

WASHINGTON—Africa's population could soar by more than 1 billion over the next half-century, further straining food and water supplies and social services in areas already struggling, according to a new report.

The latest edition of the "World Population Data Sheet" estimates the global population will rise 46 percent between now and 2050 to about 9 billion, a level also predicted by the United Nations and other groups.

European nations, more industrialized and prosperous, are expected to lose population because of falling birth rates and low immigration.

The U.S. population is expected to grow 45 percent to 422 million in 2050, paced by a stable birth rate and high levels of immigration.

But most of the world's growth will be in developing nations. India's population is estimated to grow 52 percent to 1.6 billion by 2050, when it will surpass China as the world's largest country.

Africa is predicted to more than double in population to 1.9 billion by midcentury.

SOURCE: *The Houston Chronicle* (Houston, TX), July 23, 2003, p. 12.

—SEE CHAPTER PROJECT 1.

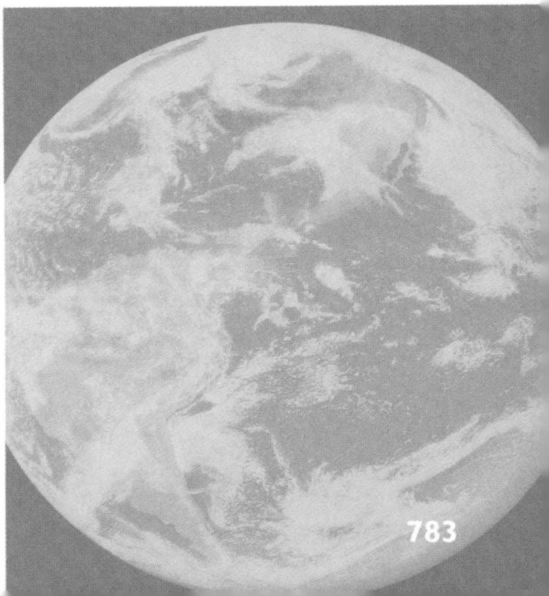

11.1 Sequences

PREPARING FOR THIS SECTION *Before getting started, review the following concept:*

- Functions (Section 2.1, pp. 50–60)

Now work the 'Are You Prepared?' problems on page 791.

OBJECTIVES
1. Write the First Several Terms of a Sequence
2. Write the Terms of a Sequence Defined by a Recursive Formula
3. Use Summation Notation
4. Find the Sum of a Sequence

A **sequence** is a function whose domain is the set of positive integers.

Because a sequence is a function, it will have a graph. In Figure 1(a), you will recognize the graph of the function $f(x) = \frac{1}{x}$, $x > 0$. If all the points on this graph were removed except those whose x-coordinates are positive integers, that is, if all points were removed except $(1, 1)$, $\left(2, \frac{1}{2}\right)$, $\left(3, \frac{1}{3}\right)$, and so on, the remaining points would be the graph of the sequence $f(n) = \frac{1}{n}$, as shown in Figure 1(b).

Figure 1

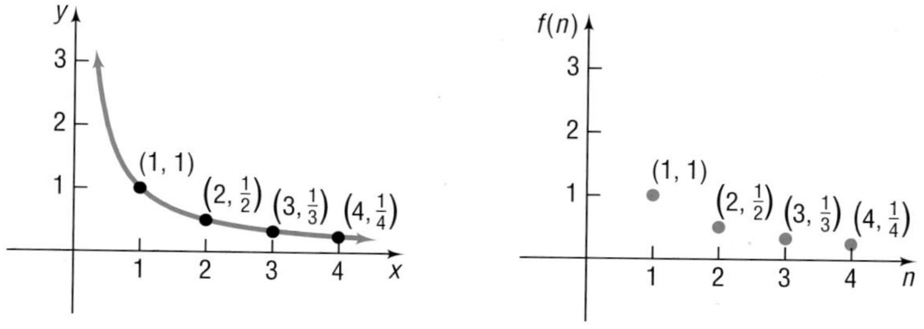

(a) $f(x) = \frac{1}{x}, x > 0$ **(b)** $f(n) = \frac{1}{n}$, n a positive integer

A sequence is usually represented by listing its values in order. For example, the sequence whose graph is given in Figure 1(b) might be represented as

$$f(1), f(2), f(3), f(4), \ldots \quad \text{or} \quad 1, \frac{1}{2}, \frac{1}{3}, \frac{1}{4}, \ldots$$

The list never ends, as the ellipsis indicates. The numbers in this ordered list are called the **terms** of the sequence.

1 In dealing with sequences, we usually use subscripted letters, such as a_1, to represent the first term, a_2 for the second term, a_3 for the third term, and so on. For the sequence $f(n) = \frac{1}{n}$, we write

$$a_1 = f(1) = 1, \quad a_2 = f(2) = \frac{1}{2}, \quad a_3 = f(3) = \frac{1}{3}, \quad a_4 = f(4) = \frac{1}{4}, \ldots, \quad a_n = f(n) = \frac{1}{n}, \ldots$$

In other words, we usually do not use the traditional function notation $f(n)$ for sequences. For this particular sequence, we have a rule for the nth term, which is $a_n = \frac{1}{n}$, so it is easy to find any term of the sequence.

When a formula for the nth term (sometimes called the **general term**) of a sequence is known, rather than write out the terms of the sequence, we usually represent the entire sequence by placing braces around the formula for the nth term. For example, the sequence whose nth term is $b_n = \left(\frac{1}{2}\right)^n$ may be represented as

$$\{b_n\} = \left\{\left(\frac{1}{2}\right)^n\right\}$$

or by

$$b_1 = \frac{1}{2}, \quad b_2 = \frac{1}{4}, \quad b_3 = \frac{1}{8}, \ldots, \quad b_n = \left(\frac{1}{2}\right)^n, \ldots$$

EXAMPLE 1 Writing the First Several Terms of a Sequence

Write down the first six terms of the following sequence and graph it.

$$\{a_n\} = \left\{\frac{n-1}{n}\right\}$$

Solution The first six terms of the sequence are

$$a_1 = 0, \quad a_2 = \frac{1}{2}, \quad a_3 = \frac{2}{3}, \quad a_4 = \frac{3}{4}, \quad a_5 = \frac{4}{5}, \quad a_6 = \frac{5}{6}$$

See Figure 2 for the graph. ◀

Figure 2

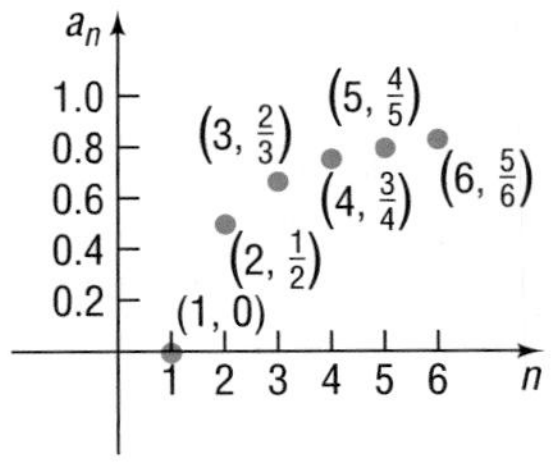

COMMENT: Graphing utilities can be used to write the terms of a sequence and graph them. Figure 3 shows the sequence given in Example 1 generated on a TI-83 graphing calculator. We can see the first few terms of the sequence on the viewing window. You need to press the right arrow key to scroll right in order to see the remaining terms of the sequence. Figure 4 shows a graph of the sequence. Notice that the first term of the sequence is not visible since it lies on the x-axis. TRACEing the graph will allow you to see the terms of the sequence. The TABLE feature can also be used to generate the terms of the sequence. See Table 1.

Figure 3

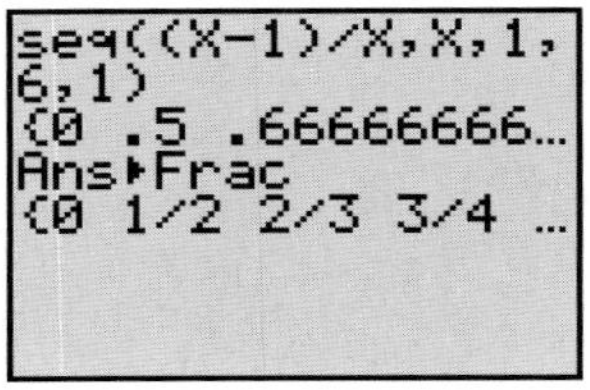

Figure 4

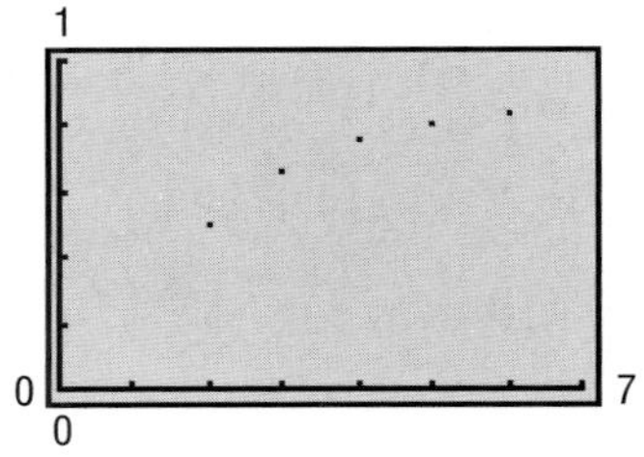

Table 1

n	u(n)
1	0
2	.5
3	.66667
4	.75
5	.8
6	.83333
7	.85714

u(n)■(n−1)/n

■

EXAMPLE 2 Writing the First Several Terms of a Sequence

Figure 5

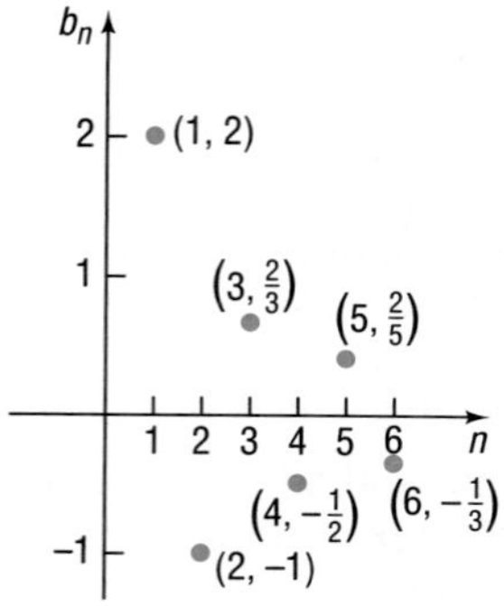

Write down the first six terms of the following sequence and graph it.

$$\{b_n\} = \left\{(-1)^{n+1}\left(\frac{2}{n}\right)\right\}$$

Solution The first six terms of the sequence are

$$b_1 = 2, \quad b_2 = -1, \quad b_3 = \frac{2}{3}, \quad b_4 = -\frac{1}{2}, \quad b_5 = \frac{2}{5}, \quad b_6 = -\frac{1}{3}$$

See Figure 5 for the graph. ◄

Notice in the sequence $\{b_n\}$ in Example 2 that the signs of the terms **alternate.** When this occurs, we use factors such as $(-1)^{n+1}$, which equals 1 if n is odd and -1 if n is even, or $(-1)^n$, which equals -1 if n is odd and 1 if n is even.

EXAMPLE 3 Writing the First Several Terms of a Sequence

Figure 6

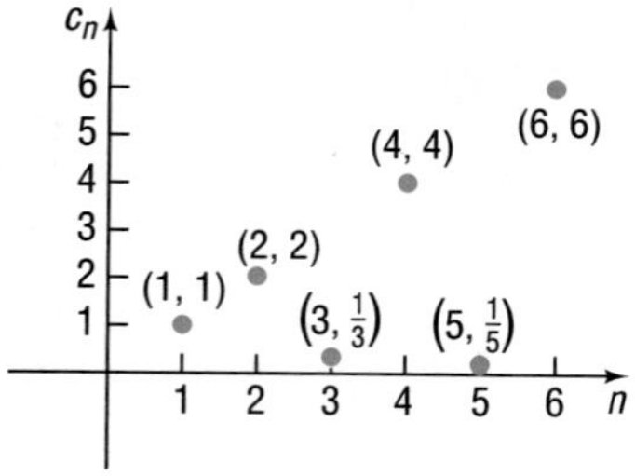

Write down the first six terms of the following sequence and graph it.

$$\{c_n\} = \left\{\begin{matrix} n & \text{if } n \text{ is even} \\ \frac{1}{n} & \text{if } n \text{ is odd} \end{matrix}\right\}$$

Solution The first six terms of the sequence are

$$c_1 = 1, \quad c_2 = 2, \quad c_3 = \frac{1}{3}, \quad c_4 = 4, \quad c_5 = \frac{1}{5}, \quad c_6 = 6$$

See Figure 6 for the graph. ◄

NOW WORK PROBLEMS 11 AND 13.

Sometimes a sequence is indicated by an observed pattern in the first few terms that makes it possible to infer the makeup of the nth term. In the example that follows, a sufficient number of terms of the sequence is given so that a natural choice for the nth term is suggested.

EXAMPLE 4 Determining a Sequence from a Pattern

(a) $e, \frac{e^2}{2}, \frac{e^3}{3}, \frac{e^4}{4}, \ldots$ $\quad a_n = \frac{e^n}{n}$

(b) $1, \frac{1}{3}, \frac{1}{9}, \frac{1}{27}, \ldots$ $\quad b_n = \frac{1}{3^{n-1}}$

(c) $1, 3, 5, 7, \ldots$ $\quad c_n = 2n - 1$

(d) $1, 4, 9, 16, 25, \ldots$ $\quad d_n = n^2$

(e) $1, -\frac{1}{2}, \frac{1}{3}, -\frac{1}{4}, \frac{1}{5}, \ldots$ $\quad e_n = (-1)^{n+1}\left(\frac{1}{n}\right)$ ◄

NOW WORK PROBLEM 21.

The Factorial Symbol

If $n \geq 0$ is an integer, the **factorial symbol $n!$** is defined as follows:

$$0! = 1 \qquad 1! = 1$$
$$n! = n(n-1)\cdot\ldots\cdot 3\cdot 2\cdot 1 \qquad \text{if } n \geq 2$$

For example, $2! = 2\cdot 1 = 2$, $3! = 3\cdot 2\cdot 1 = 6$, $4! = 4\cdot 3\cdot 2\cdot 1 = 24$, and so on. Table 2 lists the values of $n!$ for $0 \leq n \leq 6$.

Table 2

n	0	1	2	3	4	5	6
$n!$	1	1	2	6	24	120	720

Because

$$n! = n\underbrace{(n-1)(n-2)\cdot\ldots\cdot 3\cdot 2\cdot 1}_{(n-1)!}$$

we can use the formula

$$n! = n(n-1)!$$

to find successive factorials. For example, because $6! = 720$, we have

$$7! = 7\cdot 6! = 7(720) = 5040$$

and

$$8! = 8\cdot 7! = 8(5040) = 40{,}320$$

COMMENT: Your calculator has a factorial key. Use it to see how fast factorials increase in value. Find the value of 69!. What happens when you try to find 70!? In fact, 70! is larger than 10^{100} (a **googol**), the largest number most calculators can display. ■

Recursive Formulas

2 A second way of defining a sequence is to assign a value to the first (or the first few) term(s) and specify the nth term by a formula or equation that involves one or more of the terms preceding it. Sequences defined this way are said to be defined **recursively,** and the rule or formula is called a **recursive formula.**

EXAMPLE 5 Writing the Terms of a Recursively Defined Sequence

Write down the first five terms of the following recursively defined sequence.

$$s_1 = 1, \qquad s_n = 4s_{n-1}$$

Solution The first term is given as $s_1 = 1$. To get the second term, we use $n = 2$ in the formula to get $s_2 = 4s_1 = 4\cdot 1 = 4$. To get the third term, we use $n = 3$ in the formula to get $s_3 = 4s_2 = 4\cdot 4 = 16$. To get a new term requires that we know the value of the preceding term. The first five terms are

$$s_1 = 1$$
$$s_2 = 4\cdot 1 = 4$$
$$s_3 = 4\cdot 4 = 16$$
$$s_4 = 4\cdot 16 = 64$$
$$s_5 = 4\cdot 64 = 256$$

◀

EXAMPLE 6 **Writing the Terms of a Recursively Defined Sequence**

Write down the first five terms of the following recursively defined sequence.

$$f_1 = 1, \qquad f_n = nf_{n-1}$$

Solution Here

$$\begin{aligned} f_1 &= 1 \\ f_2 &= 2f_1 = 2 \cdot 1 = 2 \\ f_3 &= 3f_2 = 3 \cdot 2 = 6 \\ f_4 &= 4f_3 = 4 \cdot 6 = 24 \\ f_5 &= 5f_4 = 5 \cdot 24 = 120 \end{aligned}$$

◀

You should recognize the nth term of the sequence in Example 6 as $n!$.

EXAMPLE 7 **Writing the Terms of a Recursively Defined Sequence**

Write down the first five terms of the following recursively defined sequence.

$$u_1 = 1, \qquad u_2 = 1, \qquad u_{n+2} = u_n + u_{n+1}$$

Solution We are given the first two terms. To get the third term requires that we know each of the previous two terms. Thus,

$$\begin{aligned} u_1 &= 1 \\ u_2 &= 1 \\ u_3 &= u_1 + u_2 = 1 + 1 = 2 \\ u_4 &= u_2 + u_3 = 1 + 2 = 3 \\ u_5 &= u_3 + u_4 = 2 + 3 = 5 \end{aligned}$$

◀

The sequence defined in Example 7 is called a **Fibonacci sequence,** and the terms of this sequence are called **Fibonacci numbers.** These numbers appear in a wide variety of applications (see Problems 79–82).

NOW WORK PROBLEMS 29 AND 37.

Summation Notation

3 It is often important to be able to find the sum of the first n terms of a sequence $\{a_n\}$, that is,

$$a_1 + a_2 + a_3 + \cdots + a_n$$

Rather than write down all these terms, we introduce a more concise way to express the sum, called **summation notation.** Using summation notation, we would write the sum as

$$a_1 + a_2 + a_3 + \cdots + a_n = \sum_{k=1}^{n} a_k$$

The symbol Σ (a stylized version of the Greek letter sigma, which is an S in our alphabet) is simply an instruction to sum, or add up, the terms. The integer k is called the **index** of the sum; it tells you where to start the sum and where to end it. The expression

$$\sum_{k=1}^{n} a_k$$

is an instruction to add the terms a_k of the sequence $\{a_n\}$ from $k = 1$ through $k = n$. We read the expression as "the sum of a_k from $k = 1$ to $k = n$."

EXAMPLE 8 Expanding Summation Notation

Write out each sum.

(a) $\sum_{k=1}^{n} \frac{1}{k}$ (b) $\sum_{k=1}^{n} k!$

Solution (a) $\sum_{k=1}^{n} \frac{1}{k} = 1 + \frac{1}{2} + \frac{1}{3} + \cdots + \frac{1}{n}$ (b) $\sum_{k=1}^{n} k! = 1! + 2! + \cdots + n!$ ◄

EXAMPLE 9 Writing a Sum in Summation Notation

Express each sum using summation notation.

(a) $1^2 + 2^2 + 3^2 + \cdots + 9^2$ (b) $1 + \frac{1}{2} + \frac{1}{4} + \frac{1}{8} + \cdots + \frac{1}{2^{n-1}}$

Solution (a) The sum $1^2 + 2^2 + 3^2 + \cdots + 9^2$ has 9 terms, each of the form k^2, and starts at $k = 1$ and ends at $k = 9$:

$$1^2 + 2^2 + 3^2 + \cdots + 9^2 = \sum_{k=1}^{9} k^2$$

(b) The sum

$$1 + \frac{1}{2} + \frac{1}{4} + \frac{1}{8} + \cdots + \frac{1}{2^{n-1}}$$

has n terms, each of the form $\frac{1}{2^{k-1}}$, and starts at $k = 1$ and ends at $k = n$:

$$1 + \frac{1}{2} + \frac{1}{4} + \frac{1}{8} + \cdots + \frac{1}{2^{n-1}} = \sum_{k=1}^{n} \frac{1}{2^{k-1}}$$ ◄

The index of summation need not always begin at 1 or end at n; for example, we could have expressed the sum in Example 9(b) as

$$\sum_{k=0}^{n-1} \frac{1}{2^k} = 1 + \frac{1}{2} + \frac{1}{4} + \cdots + \frac{1}{2^{n-1}}$$

Letters other than k may be used as the index. For example,

$$\sum_{j=1}^{n} j! \quad \text{and} \quad \sum_{i=1}^{n} i!$$

each represent the same sum as the one given in Example 8(b).

NOW WORK PROBLEMS 57 AND 67.

Adding the First n Terms of a Sequence

4 Next, we list some properties of sequences using summation notation. These properties are useful for adding the terms of a sequence.

Theorem

Properties of Sequences

If $\{a_n\}$ and $\{b_n\}$ are two sequences and c is a real number, then:

$$\sum_{k=1}^{n} c = \underbrace{c + c + c + \cdots + c}_{n \text{ terms}} = cn \tag{1}$$

$$\sum_{k=1}^{n} (ca_k) = ca_1 + ca_2 + \cdots + ca_n = c(a_1 + a_2 + \cdots + a_n) = c\sum_{k=1}^{n} a_k \tag{2}$$

$$\sum_{k=1}^{n} (a_k + b_k) = \sum_{k=1}^{n} a_k + \sum_{k=1}^{n} b_k \tag{3}$$

$$\sum_{k=1}^{n} (a_k - b_k) = \sum_{k=1}^{n} a_k - \sum_{k=1}^{n} b_k \tag{4}$$

$$\sum_{k=1}^{n} a_k = \sum_{k=1}^{j} a_k + \sum_{k=j+1}^{n} a_k, \quad \text{where } 0 < j < n \tag{5}$$

$$\sum_{k=1}^{n} k = 1 + 2 + 3 + \cdots + n = \frac{n(n+1)}{2} \tag{6}$$

$$\sum_{k=1}^{n} k^2 = 1^2 + 2^2 + 3^2 + \cdots + n^2 = \frac{n(n+1)(2n+1)}{6} \tag{7}$$

$$\sum_{k=1}^{n} k^3 = 1^3 + 2^3 + 3^3 + \cdots + n^3 = \left(\frac{n(n+1)}{2}\right)^2 \tag{8}$$

We shall not prove these properties. The proofs of (1) through (5) are based on properties of real numbers; the proofs of (7) and (8) require mathematical induction, which is discussed in Section 11.4. See Problem 83 for a derivation of (6).

EXAMPLE 10 **Finding the Sum of a Sequence**

Find the sum of each sequence.

(a) $\sum_{k=1}^{5} (3k)$ (b) $\sum_{k=1}^{3} (k^3 + 1)$ (c) $\sum_{k=1}^{4} (k^2 - 7k + 2)$

Solution (a) $$\sum_{k=1}^{5} (3k) \underset{\text{Property (2)}}{=} 3\sum_{k=1}^{5} k \underset{\text{Property (6)}}{=} 3\left(\frac{5(5+1)}{2}\right) = 3(15) = 45$$

(b)
$$\begin{aligned}\sum_{k=1}^{3} (k^3 + 1) &= \sum_{k=1}^{3} k^3 + \sum_{k=1}^{3} 1 && \text{Property (3)}\\ &= \left(\frac{3(3+1)}{2}\right)^2 + 1(3) && \text{Properties (1) and (8)}\\ &= 36 + 3\\ &= 39\end{aligned}$$

(c) $\sum_{k=1}^{4}(k^2 - 7k + 2) = \sum_{k=1}^{4} k^2 - \sum_{k=1}^{4}(7k) + \sum_{k=1}^{4} 2$ Properties (3) and (4)

$= \sum_{k=1}^{4} k^2 - 7\sum_{k=1}^{4} k + \sum_{k=1}^{4} 2$ Property (2)

$= \frac{4(4+1)(2\cdot 4+1)}{6} - 7\left(\frac{4(4+1)}{2}\right) + 2(4)$ Properties (1), (6), and (7)

$= 30 - 70 + 8$

$= -32$ ◄

NOW WORK PROBLEM 47.

11.1 Assess Your Understanding

'Are You Prepared?' *Answers are given at the end of these exercises. If you get a wrong answer, read the pages listed in red.*

1. For the function $f(x) = \frac{x-1}{x}$, find $f(2)$ and $f(3)$. (pp. 50–60)

2. *True or False:* A function is a relation between two sets D and R so that each element x in the first set D is related to exactly one element y in the second set R. (pp. 50–60)

Concepts and Vocabulary

3. A(n) __________ is a function whose domain is the set of positive integers.

4. For the sequence $\{s_n\} = \{4n - 1\}$, the first term is $s_1 =$ __________ and the fourth term is $s_4 =$ __________.

5. $\sum_{k=1}^{4}(2k) =$ __________.

6. *True or False:* Sequences are sometimes defined recursively.

7. *True or False:* A sequence is a function.

8. *True or False:* $\sum_{k=1}^{2} k = 3$

Exercises

In Problems 9–20, write down the first five terms of each sequence.

9. $\{n\}$ **10.** $\{n^2 + 1\}$ **11.** $\left\{\frac{n}{n+2}\right\}$ **12.** $\left\{\frac{2n+1}{2n}\right\}$

13. $\{(-1)^{n+1}n^2\}$ **14.** $\left\{(-1)^{n-1}\left(\frac{n}{2n-1}\right)\right\}$ **15.** $\left\{\frac{2^n}{3^n+1}\right\}$ **16.** $\left\{\left(\frac{4}{3}\right)^n\right\}$

17. $\left\{\frac{(-1)^n}{(n+1)(n+2)}\right\}$ **18.** $\left\{\frac{3^n}{n}\right\}$ **19.** $\left\{\frac{n}{e^n}\right\}$ **20.** $\left\{\frac{n^2}{2^n}\right\}$

In Problems 21–28, the given pattern continues. Write down the nth term of each sequence suggested by the pattern.

21. $\frac{1}{2}, \frac{2}{3}, \frac{3}{4}, \frac{4}{5}, \dots$ **22.** $\frac{1}{1\cdot 2}, \frac{1}{2\cdot 3}, \frac{1}{3\cdot 4}, \frac{1}{4\cdot 5}, \dots$ **23.** $1, \frac{1}{2}, \frac{1}{4}, \frac{1}{8}, \dots$ **24.** $\frac{2}{3}, \frac{4}{9}, \frac{8}{27}, \frac{16}{81}, \dots$

25. $1, -1, 1, -1, 1, -1, \dots$ **26.** $1, \frac{1}{2}, 3, \frac{1}{4}, 5, \frac{1}{6}, 7, \frac{1}{8}, \dots$ **27.** $1, -2, 3, -4, 5, -6, \dots$ **28.** $2, -4, 6, -8, 10, \dots$

In Problems 29–42, a sequence is defined recursively. Write the first five terms.

29. $a_1 = 2; \quad a_n = 3 + a_{n-1}$

30. $a_1 = 3; \quad a_n = 4 - a_{n-1}$

31. $a_1 = -2; \quad a_n = n + a_{n-1}$

32. $a_1 = 1; \quad a_n = n - a_{n-1}$

33. $a_1 = 5; \quad a_n = 2a_{n-1}$

34. $a_1 = 2; \quad a_n = -a_{n-1}$

35. $a_1 = 3; \quad a_n = \dfrac{a_{n-1}}{n}$

36. $a_1 = -2; \quad a_n = n + 3a_{n-1}$

37. $a_1 = 1; \quad a_2 = 2; \quad a_n = a_{n-1} \cdot a_{n-2}$

38. $a_1 = -1; \quad a_2 = 1; \quad a_n = a_{n-2} + na_{n-1}$

39. $a_1 = A; \quad a_n = a_{n-1} + d$

40. $a_1 = A; \quad a_n = ra_{n-1}, \quad r \neq 0$

41. $a_1 = \sqrt{2}; \quad a_n = \sqrt{2 + a_{n-1}}$

42. $a_1 = \sqrt{2}; \quad a_n = \sqrt{\dfrac{a_{n-1}}{2}}$

In Problems 43–54, find the sum of each sequence.

43. $\sum_{k=1}^{10} 5$

44. $\sum_{k=1}^{20} 8$

45. $\sum_{k=1}^{6} k$

46. $\sum_{k=1}^{4} (-k)$

47. $\sum_{k=1}^{5} (5k + 3)$

48. $\sum_{k=1}^{6} (3k - 7)$

49. $\sum_{k=1}^{3} (k^2 + 4)$

50. $\sum_{k=0}^{4} (k^2 - 4)$

51. $\sum_{k=1}^{6} (-1)^k 2^k$

52. $\sum_{k=1}^{4} (-1)^k 3^k$

53. $\sum_{k=1}^{4} (k^3 - 1)$

54. $\sum_{k=0}^{3} (k^3 + 2)$

In Problems 55–64, write out each sum.

55. $\sum_{k=1}^{n} (k + 2)$

56. $\sum_{k=1}^{n} (2k + 1)$

57. $\sum_{k=1}^{n} \dfrac{k^2}{2}$

58. $\sum_{k=1}^{n} (k + 1)^2$

59. $\sum_{k=0}^{n} \dfrac{1}{3^k}$

60. $\sum_{k=0}^{n} \left(\dfrac{3}{2}\right)^k$

61. $\sum_{k=0}^{n-1} \dfrac{1}{3^{k+1}}$

62. $\sum_{k=0}^{n-1} (2k + 1)$

63. $\sum_{k=2}^{n} (-1)^k \ln k$

64. $\sum_{k=3}^{n} (-1)^{k+1} 2^k$

In Problems 65–74, express each sum using summation notation.

65. $1 + 2 + 3 + \cdots + 20$

66. $1^3 + 2^3 + 3^3 + \cdots + 8^3$

67. $\dfrac{1}{2} + \dfrac{2}{3} + \dfrac{3}{4} + \cdots + \dfrac{13}{13 + 1}$

68. $1 + 3 + 5 + 7 + \cdots + [2(12) - 1]$

69. $1 - \dfrac{1}{3} + \dfrac{1}{9} - \dfrac{1}{27} + \cdots + (-1)^6\left(\dfrac{1}{3^6}\right)$

70. $\dfrac{2}{3} - \dfrac{4}{9} + \dfrac{8}{27} - \cdots + (-1)^{11+1}\left(\dfrac{2}{3}\right)^{11}$

71. $3 + \dfrac{3^2}{2} + \dfrac{3^3}{3} + \cdots + \dfrac{3^n}{n}$

72. $\dfrac{1}{e} + \dfrac{2}{e^2} + \dfrac{3}{e^3} + \cdots + \dfrac{n}{e^n}$

73. $a + (a + d) + (a + 2d) + \cdots + (a + nd)$

74. $a + ar + ar^2 + \cdots + ar^{n-1}$

75. Credit Card Debt John has a balance of \$3000 on his Discover card that charges 1% interest per month on any unpaid balance. John can afford to pay \$100 toward the balance each month. His balance each month after making a \$100 payment is given by the recursively defined sequence

$$B_0 = \$3000, \qquad B_n = 1.01B_{n-1} - 100$$

Determine John's balance after making the first payment. That is, determine B_1.

76. Car Loans Phil bought a car by taking out a loan for \$18,500 at 0.5% interest per month. Phil's normal monthly payment is \$434.47 per month, but he decides that he can afford to pay \$100 extra toward the balance each month. His balance each month is given by the recursively defined sequence

$$B_0 = \$18{,}500, \qquad B_n = 1.005B_{n-1} - 534.47$$

Determine Phil's balance after making the first payment. That is, determine B_1.

77. Trout Population A pond currently has 2000 trout in it. A fish hatchery decides to add an additional 20 trout each month. In addition, it is known that the trout population is growing 3% per month. The size of the population after n months is given by the recursively defined sequence

$$p_0 = 2000, \qquad p_n = 1.03p_{n-1} + 20$$

How many trout are in the pond after two months? That is, what is p_2?

78. Environmental Control The Environmental Protection Agency (EPA) determines that Maple Lake has 250 tons of pollutant as a result of industrial waste and that 10% of the pollutant present is neutralized by solar oxidation every year. The EPA imposes new pollution control laws that result in 15 tons of new pollutant entering the lake each year. The amount of pollutant in the lake after n years is given by the recursively defined sequence

$$p_0 = 250, \qquad p_n = 0.9p_{n-1} + 15$$

Determine the amount of pollutant in the lake after two years. That is, determine p_2.

79. Growth of a Rabbit Colony A colony of rabbits begins with one pair of mature rabbits, which will produce a pair of offspring (one male, one female) each month. Assume that all rabbits mature in 1 month and produce a pair of offspring (one male, one female) after 2 months. If no rabbits ever die, how many pairs of mature rabbits are there after 7 months?

[**Hint:** A Fibonacci sequence models this colony. Do you see why?]

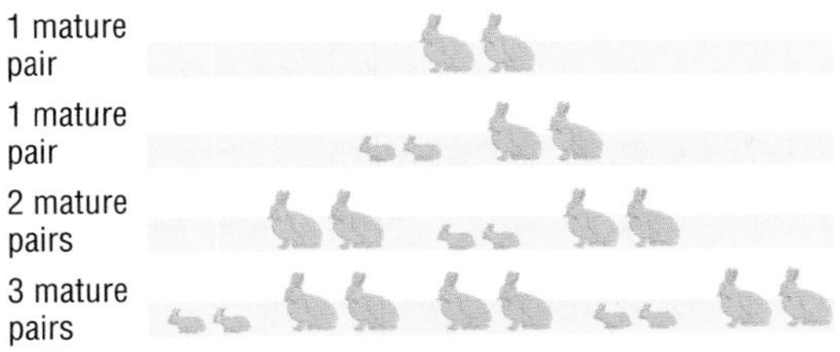

80. Fibonacci Sequence Let

$$u_n = \frac{\left(1 + \sqrt{5}\right)^n - \left(1 - \sqrt{5}\right)^n}{2^n\sqrt{5}}$$

define the nth term of a sequence.

(a) Show that $u_1 = 1$ and $u_2 = 1$.

(b) Show that $u_{n+2} = u_{n+1} + u_n$.

(c) Draw the conclusion that $\{u_n\}$ is a Fibonacci sequence.

81. Pascal's Triangle Divide the triangular array shown (called Pascal's triangle) using diagonal lines as indicated. Find the sum of the numbers in each of these diagonal rows. Do you recognize this sequence?

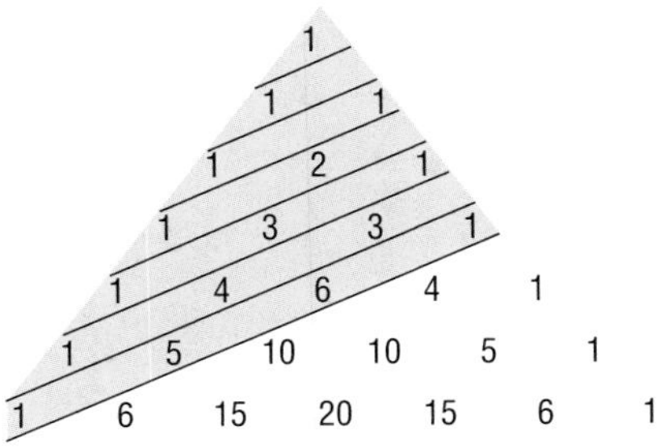

82. Fibonacci Sequence Use the result of Problem 80 to do the following problems:

(a) Write the first 10 terms of the Fibonacci sequence.

(b) Compute the ratio $\dfrac{u_{n+1}}{u_n}$ for the first 10 terms.

(c) As n gets large, what number does the ratio approach? This number is referred to as the **golden ratio.** Rectangles whose sides are in this ratio were considered pleasing to the eye by the Greeks. For example, the facade of the Parthenon was constructed using the golden ratio.

(d) Compute the ratio $\dfrac{u_n}{u_{n+1}}$ for the first 10 terms.

(e) As n gets large, what number does the ratio approach? This number is also referred to as the **golden ratio.** This ratio is believed to have been used in the construction of the Great Pyramid in Egypt. The ratio equals the sum of the areas of the four face triangles divided by the total surface area of the Great Pyramid.

83. Show that

$$1 + 2 + \cdots + (n - 1) + n = \frac{n(n + 1)}{2}$$

[**Hint:** Let

$$S = 1 + 2 + \cdots + (n - 1) + n$$
$$S = n + (n - 1) + (n - 2) + \cdots + 1$$

Add these equations. Then

$$2S = \underbrace{[1 + n] + [2 + (n - 1)] + \cdots + [n + 1]}_{n \text{ terms in brackets}}$$

Now complete the derivation.]

84. Investigate various applications that lead to a Fibonacci sequence, such as art, architecture, or financial markets. Write an essay on these applications.

'Are You Prepared?' Answers

1. $f(2) = \frac{1}{2}$; $f(3) = \frac{2}{3}$

2. True

11.2 Arithmetic Sequences

OBJECTIVES
1 Determine If a Sequence Is Arithmetic
2 Find a Formula for an Arithmetic Sequence
3 Find the Sum of an Arithmetic Sequence

1 When the difference between successive terms of a sequence is always the same number, the sequence is called **arithmetic.** An **arithmetic sequence***

*Sometimes called an **arithmetic progression.**

may be defined recursively as $a_1 = a$, $a_n - a_{n-1} = d$, or as

$$a_1 = a, \qquad a_n = a_{n-1} + d \tag{1}$$

where $a = a_1$ and d are real numbers. The number a is the first term, and the number d is called the **common difference.**

The terms of an arithmetic sequence with first term a and common difference d follow the pattern

$$a, \quad a + d, \quad a + 2d, \quad a + 3d, \dots$$

EXAMPLE 1 Determining If a Sequence Is Arithmetic

The sequence

$$4, \quad 7, \quad 10, \quad 13, \dots$$

is arithmetic since the difference of successive terms is 3. The first term is 4, and the common difference is 3. ◀

EXAMPLE 2 Determining If a Sequence Is Arithmetic

Show that the following sequence is arithmetic. Find the first term and the common difference.

$$\{s_n\} = \{3n + 5\}$$

Solution The first term is $s_1 = 3 \cdot 1 + 5 = 8$. The nth and $(n - 1)$st terms of the sequence $\{s_n\}$ are

$$s_n = 3n + 5 \quad \text{and} \quad s_{n-1} = 3(n - 1) + 5 = 3n + 2$$

Their difference is

$$s_n - s_{n-1} = (3n + 5) - (3n + 2) = 5 - 2 = 3$$

Since the difference of two successive terms is constant, the sequence is arithmetic and the common difference is 3. ◀

EXAMPLE 3 Determining If a Sequence Is Arithmetic

Show that the sequence $\{t_n\} = \{4 - n\}$ is arithmetic. Find the first term and the common difference.

Solution The first term is $t_1 = 4 - 1 = 3$. The nth and $(n - 1)$st terms are

$$t_n = 4 - n \quad \text{and} \quad t_{n-1} = 4 - (n - 1) = 5 - n$$

Their difference is

$$t_n - t_{n-1} = (4 - n) - (5 - n) = 4 - 5 = -1$$

Since the difference of two successive terms is constant, $\{t_n\}$ is an arithmetic sequence whose common difference is -1. ◀

NOW WORK PROBLEM 5.

2 Suppose that a is the first term of an arithmetic sequence whose common difference is d. We seek a formula for the nth term, a_n. To see the pattern, we write down the first few terms.

$$\begin{aligned}
a_1 &= a \\
a_2 &= a_1 + d = a + 1\cdot d \\
a_3 &= a_2 + d = (a + d) + d = a + 2\cdot d \\
a_4 &= a_3 + d = (a + 2\cdot d) + d = a + 3\cdot d \\
a_5 &= a_4 + d = (a + 3\cdot d) + d = a + 4\cdot d \\
&\vdots \\
a_n &= a_{n-1} + d = [a + (n-2)d] + d = a + (n-1)d
\end{aligned}$$

We are led to the following result:

Theorem

nth Term of an Arithmetic Sequence

For an arithmetic sequence $\{a_n\}$ whose first term is a and whose common difference is d, the nth term is determined by the formula

$$a_n = a + (n-1)d \tag{2}$$

EXAMPLE 4 Finding a Particular Term of an Arithmetic Sequence

Find the thirteenth term of the arithmetic sequence: 2, 6, 10, 14, 18, . . .

Solution The first term of this arithmetic sequence is $a = 2$, and the common difference is 4. By formula (2), the nth term is

$$a_n = 2 + (n-1)4$$

Hence, the thirteenth term is

$$a_{13} = 2 + 12\cdot 4 = 50$$ ◀

Exploration

Use a graphing utility to find the thirteenth term of the sequence given in Example 4. Use it to find the twentieth and fiftieth terms.

EXAMPLE 5 Finding a Recursive Formula for an Arithmetic Sequence

The eighth term of an arithmetic sequence is 75, and the twentieth term is 39. Find the first term and the common difference. Give a recursive formula for the sequence. What is the nth term of the sequence?

Solution By formula (2), we know that $a_n = a + (n-1)d$. As a result,

$$\begin{cases} a_8 = a + 7d = 75 \\ a_{20} = a + 19d = 39 \end{cases}$$

This is a system of two linear equations containing two variables, a and d, which we can solve by elimination. Subtracting the second equation from the first equation, we get

$$\begin{aligned} -12d &= 36 \\ d &= -3 \end{aligned}$$

With $d = -3$, we find that $a = 75 - 7d = 75 - 7(-3) = 96$. The first term is $a = 96$, and the common difference is $d = -3$. A recursive formula for this sequence is found using formula (1).

$$a_1 = 96, \qquad a_n = a_{n-1} - 3$$

Based on formula (2), a formula for the nth term of the sequence $\{a_n\}$ is

$$a_n = a + (n - 1)d = 96 + (n - 1)(-3) = 99 - 3n$$ ◀

NOW WORK PROBLEMS 21 AND 27.

Exploration

Graph the recursive formula from Example 5, $a_1 = 96$, $a_n = a_{n-1} - 3$, using a graphing utility. Conclude that the graph of the recursive formula behaves like the graph of a linear function. How is d, the common difference, related to m, the slope of a line?

Adding the First *n* Terms of an Arithmetic Sequence

3 The next result gives a formula for finding the sum of the first n terms of an arithmetic sequence.

Theorem

Sum of *n* Terms of an Arithmetic Sequence

Let $\{a_n\}$ be an arithmetic sequence with first term a and common difference d. The sum S_n of the first n terms of $\{a_n\}$ is

$$S_n = \frac{n}{2}[2a + (n - 1)d] = \frac{n}{2}(a + a_n) \tag{3}$$

Proof

$$\begin{aligned} S_n &= a_1 + a_2 + a_3 + \cdots + a_n && \text{Sum of first } n \text{ terms} \\ &= a + (a + d) + (a + 2d) + \cdots + [a + (n - 1)d] && \text{Formula (2)} \\ &= \underbrace{(a + a + \cdots + a)}_{n \text{ terms}} + [d + 2d + \cdots + (n - 1)d] && \text{Rearrange terms} \\ &= na + d[1 + 2 + \cdots + (n - 1)] \\ &= na + d\left[\frac{(n - 1)n}{2}\right] && \text{Property 6, Section 11.1} \\ &= na + \frac{n}{2}(n - 1)d \\ &= \frac{n}{2}[2a + (n - 1)d] && \text{Factor out } \frac{n}{2} \qquad (4) \\ &= \frac{n}{2}[a + a + (n - 1)d] \\ &= \frac{n}{2}(a + a_n) && \text{Formula (2)} \qquad (5) \end{aligned}$$

■

Formula (3) provides two ways to find the sum of the first n terms of an arithmetic sequence. Notice that formula (4) involves the first term and common difference, whereas formula (5) involves the first term and the nth term. Use whichever form is easier.

EXAMPLE 6 Finding the Sum of *n* Terms of an Arithmetic Sequence

Find the sum S_n of the first n terms of the sequence $\{3n + 5\}$; that is, find

$$8 + 11 + 14 + \cdots + (3n + 5)$$

Solution The sequence $\{3n + 5\}$ is an arithmetic sequence with first term $a = 8$ and the nth term $(3n + 5)$. To find the sum S_n, we use formula (5).

$$S_n = \frac{n}{2}(a + a_n) = \frac{n}{2}[8 + (3n + 5)] = \frac{n}{2}(3n + 13)$$ ◀

NOW WORK PROBLEM 35.

EXAMPLE 7 Using a Graphing Utility to Find the Sum of 20 Terms of an Arithmetic Sequence

Figure 7

Use a graphing utility to find the sum of the first 20 terms of the sequence $\{9.5n + 2.6\}$.

Solution Figure 7 shows the results obtained using a TI-83 graphing calculator.

The sum of the first 20 terms of the sequence $\{9.5n + 2.6\}$ is 2047. ◀

WORK EXAMPLE 7 USING FORMULA (3).

NOW WORK PROBLEM 43.

EXAMPLE 8 Creating a Floor Design

A ceramic tile floor is designed in the shape of a trapezoid 20 feet wide at the base and 10 feet wide at the top. See Figure 8. The tiles, 12 inches by 12 inches, are to be placed so that each successive row contains one less tile than the preceding row. How many tiles will be required?

Figure 8

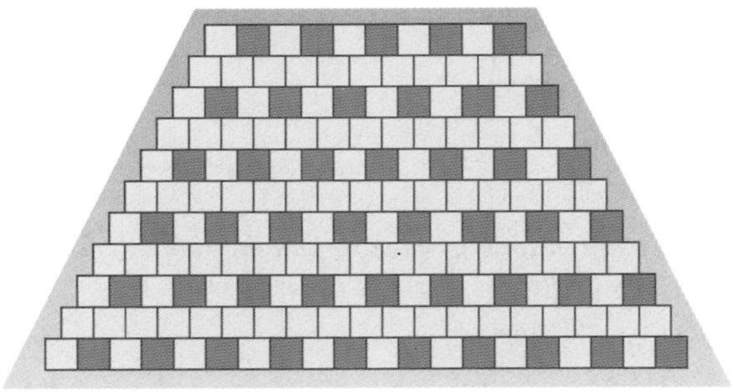

Solution The bottom row requires 20 tiles and the top row, 10 tiles. Since each successive row requires one less tile, the total number of tiles required is

$$S = 20 + 19 + 18 + \cdots + 11 + 10$$

This is the sum of an arithmetic sequence; the common difference is -1. The number of terms to be added is $n = 11$, with the first term $a = 20$ and the last term $a_{11} = 10$. The sum S is

$$S = \frac{n}{2}(a + a_{11}) = \frac{11}{2}(20 + 10) = 165$$

In all, 165 tiles will be required. ◀

11.2 Assess Your Understanding

Concepts and Vocabulary

1. In a(n) __________ sequence, the difference between successive terms is a constant.

2. *True or False:* In an arithmetic sequence the sum of the first and last terms equals twice the sum of all the terms.

Exercises

In Problems 3–12, an arithmetic sequence is given. Find the common difference and write out the first four terms.

3. $\{n + 4\}$ **4.** $\{n - 5\}$ **5.** $\{2n - 5\}$ **6.** $\{3n + 1\}$ **7.** $\{6 - 2n\}$

8. $\{4 - 2n\}$ **9.** $\left\{\frac{1}{2} - \frac{1}{3}n\right\}$ **10.** $\left\{\frac{2}{3} + \frac{n}{4}\right\}$ **11.** $\{\ln 3^n\}$ **12.** $\{e^{\ln n}\}$

In Problems 13–20, find the nth term of the arithmetic sequence whose initial term a and common difference d are given. What is the fifth term?

13. $a = 2;\ d = 3$ **14.** $a = -2;\ d = 4$ **15.** $a = 5;\ d = -3$ **16.** $a = 6;\ d = -2$

17. $a = 0;\ d = \frac{1}{2}$ **18.** $a = 1;\ d = -\frac{1}{3}$ **19.** $a = \sqrt{2};\ d = \sqrt{2}$ **20.** $a = 0;\ d = \pi$

In Problems 21–26, find the indicated term in each arithmetic sequence.

21. 12th term of $2, 4, 6, \ldots$

22. 8th term of $-1, 1, 3, \ldots$

23. 10th term of $1, -2, -5, \ldots$

24. 9th term of $5, 0, -5, \ldots$

25. 8th term of $a, a + b, a + 2b, \ldots$

26. 7th term of $2\sqrt{5}, 4\sqrt{5}, 6\sqrt{5}, \ldots$

In Problems 27–34, find the first term and the common difference of the arithmetic sequence described. Give a recursive formula for the sequence.

27. 8th term is 8; 20th term is 44

28. 4th term is 3; 20th term is 35

29. 9th term is -5; 15th term is 31

30. 8th term is 4; 18th term is -96

31. 15th term is 0; 40th term is -50

32. 5th term is -2; 13th term is 30

33. 14th term is -1; 18th term is -9

34. 12th term is 4; 18th term is 28

In Problems 35–42, find each sum.

35. $1 + 3 + 5 + \cdots + (2n - 1)$

36. $2 + 4 + 6 + \cdots + 2n$

37. $7 + 12 + 17 + \cdots + (2 + 5n)$

38. $-1 + 3 + 7 + \cdots + (4n - 5)$

39. $2 + 4 + 6 + \cdots + 70$

40. $1 + 3 + 5 + \cdots + 59$

41. $5 + 9 + 13 + \cdots + 49$

42. $2 + 5 + 8 + \cdots + 41$

For Problems 43–48, use a graphing utility to find the sum of each sequence.

43. $\{3.45n + 4.12\},\ n = 20$

44. $\{2.67n - 1.23\},\ n = 25$

45. $2.8 + 5.2 + 7.6 + \cdots + 36.4$

46. $5.4 + 7.3 + 9.2 + \cdots + 32$

47. $4.9 + 7.48 + 10.06 + \cdots + 66.82$

48. $3.71 + 6.9 + 10.09 + \cdots + 80.27$

49. Find x so that $x + 3, 2x + 1$, and $5x + 2$ are consecutive terms of an arithmetic sequence.

50. Find x so that $2x$, $3x + 2$, and $5x + 3$ are consecutive terms of an arithmetic sequence.

51. Drury Lane Theater The Drury Lane Theater has 25 seats in the first row and 30 rows in all. Each successive row contains one additional seat. How many seats are in the theater?

52. Football Stadium The corner section of a football stadium has 15 seats in the first row and 40 rows in all. Each successive row contains two additional seats. How many seats are in this section?

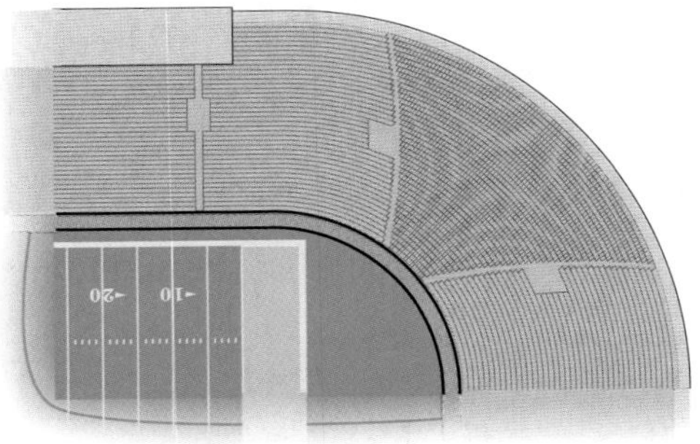

53. Creating a Mosaic A mosaic is designed in the shape of an equilateral triangle, 20 feet on each side. Each tile in the mosaic is in the shape of an equilateral triangle, 12 inches to a side. The tiles are to alternate in color as shown in the illustration. How many tiles of each color will be required?

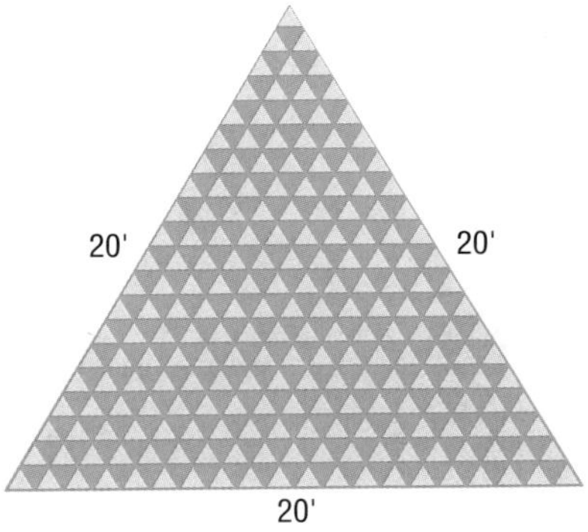

54. Constructing a Brick Staircase A brick staircase has a total of 30 steps. The bottom step requires 100 bricks. Each successive step requires two less bricks than the prior step.
(a) How many bricks are required for the top step?
(b) How many bricks are required to build the staircase?

55. Stadium Construction How many rows are in the corner section of a stadium containing 2040 seats if the first row has 10 seats and each successive row has 4 additional seats?

56. Salary Suppose that you just received a job offer with a starting salary of \$35,000 per year and a guaranteed raise of \$1400 per year. How many years will it take before your aggregate salary is \$280,000?

[**Hint:** Your aggregate salary after two years is \$35,000 + (\$35,000 + \$1400).]

57. Make up an arithmetic sequence. Give it to a friend and ask for its twentieth term.

11.3 Geometric Sequences; Geometric Series

PREPARING FOR THIS SECTION *Before getting started, review the following:*

- Compound Interest (Section 4.7, pp. 287–294)

Now work the 'Are You Prepared?' problems on page 808.

OBJECTIVES
1 Determine If a Sequence Is Geometric
2 Find a Formula for a Geometric Sequence
3 Find the Sum of a Geometric Sequence
4 Find the Sum of a Geometric Series
5 Solve Annuity Problems

1 When the ratio of successive terms of a sequence is always the same nonzero number, the sequence is called **geometric.** A **geometric sequence*** may be defined recursively as $a_1 = a$, $\dfrac{a_n}{a_{n-1}} = r$, or as

$$a_1 = a, \qquad a_n = ra_{n-1} \tag{1}$$

*Sometimes called a **geometric progression.**

where $a_1 = a$ and $r \neq 0$ are real numbers. The number a is the first term, and the nonzero number r is called the **common ratio.**

The terms of a geometric sequence with first term a and common ratio r follow the pattern

$$a, \quad ar, \quad ar^2, \quad ar^3, \ldots$$

EXAMPLE 1 Determining If a Sequence Is Geometric

The sequence

$$2, 6, 18, 54, 162, \ldots$$

is geometric since the ratio of successive terms is 3 $\left(\frac{6}{2} = \frac{18}{6} = \cdots = 3\right)$. The first term is 2, and the common ratio is 3. ◀

EXAMPLE 2 Determining If a Sequence Is Geometric

Show that the following sequence is geometric. Find the first term and the common ratio.

$$\{s_n\} = 2^{-n}$$

Solution The first term is $s_1 = 2^{-1} = \frac{1}{2}$. The nth and $(n - 1)$st terms of the sequence $\{s_n\}$ are

$$s_n = 2^{-n} \quad \text{and} \quad s_{n-1} = 2^{-(n-1)}$$

Their ratio is

$$\frac{s_n}{s_{n-1}} = \frac{2^{-n}}{2^{-(n-1)}} = 2^{-n+(n-1)} = 2^{-1} = \frac{1}{2}$$

Because the ratio of successive terms is a nonzero constant, the sequence $\{s_n\}$ is geometric with common ratio $\frac{1}{2}$. ◀

EXAMPLE 3 Determining If a Sequence Is Geometric

Show that the following sequence is geometric. Find the first term and the common ratio.

$$\{t_n\} = \{4^n\}$$

Solution The first term is $t_1 = 4^1 = 4$. The nth and $(n - 1)$st terms are

$$t_n = 4^n \quad \text{and} \quad t_{n-1} = 4^{n-1}$$

Their ratio is

$$\frac{t_n}{t_{n-1}} = \frac{4^n}{4^{n-1}} = 4^{n-(n-1)} = 4$$

The sequence, $\{t_n\}$ is a geometric sequence with common ratio 4. ◀

NOW WORK PROBLEM 11.

2 Suppose that a is the first term of a geometric sequence with common ratio $r \neq 0$. We seek a formula for the nth term a_n. To see the pattern, we write down the first few terms:

$$\begin{aligned}
a_1 &= 1a = ar^0\\
a_2 &= ra_1 = ar^1\\
a_3 &= ra_2 = r(ar) = ar^2\\
a_4 &= ra_3 = r(ar^2) = ar^3\\
a_5 &= ra_4 = r(ar^3) = ar^4\\
&\vdots\\
a_n &= ra_{n-1} = r(ar^{n-2}) = ar^{n-1}
\end{aligned}$$

We are led to the following result:

Theorem

_n_th Term of a Geometric Sequence

For a geometric sequence $\{a_n\}$ whose first term is a and whose common ratio is r, the nth term is determined by the formula

$$a_n = ar^{n-1}, \qquad r \neq 0 \tag{2}$$

EXAMPLE 4 Finding a Particular Term of a Geometric Sequence

(a) Find the ninth term of the geometric sequence: $10, 9, \frac{81}{10}, \frac{729}{100} \ldots$

(b) Find a recursive formula for this sequence.

Solution (a) The first term of this geometric sequence is $a = 10$ and the common ratio is $\frac{9}{10}$. (Use $\frac{9}{10}$, or $\frac{\frac{81}{10}}{9} = \frac{9}{10}$, or any two successive terms.) By formula (2), the nth term is

$$a_n = 10\left(\frac{9}{10}\right)^{n-1}$$

The ninth term is

$$a_9 = 10\left(\frac{9}{10}\right)^{9-1} = 10\left(\frac{9}{10}\right)^8 = 4.3046721$$

(b) The first term in the sequence is 10 and the common ratio is $r = \frac{9}{10}$.

Using formula (1), the recursive formula is $a_1 = 10$, $a_n = \frac{9}{10}a_{n-1}$. ◀

Exploration

Use a graphing utility to find the ninth term of the sequence given in Example 4. Use it to find the twentieth and fiftieth terms. Now use a graphing utility to graph the recursive formula found in Example 4(b). Conclude that the graph of the recursive formula behaves like the graph of an exponential function. How is r, the common ratio, related to a, the base of the exponential function $y = a^x$?

NOW WORK PROBLEMS 33 AND 41.

Adding the First n Terms of a Geometric Sequence

3 The next result gives us a formula for finding the sum of the first n terms of a geometric sequence.

Theorem

Sum of n Terms of a Geometric Sequence

Let $\{a_n\}$ be a geometric sequence with first term a and common ratio r where $r \neq 0, r \neq 1$. The sum S_n of the first n terms of $\{a_n\}$ is

$$S_n = a\frac{1 - r^n}{1 - r}, \qquad r \neq 0, 1 \tag{3}$$

Proof The sum S_n of the first n terms of $\{a_n\} = \{ar^{n-1}\}$ is

$$S_n = a + ar + \cdots + ar^{n-1} \tag{4}$$

Multiply each side by r to obtain

$$rS_n = ar + ar^2 + \cdots + ar^n \tag{5}$$

Now, subtract (5) from (4). The result is

$$S_n - rS_n = a - ar^n$$
$$(1 - r)S_n = a(1 - r^n)$$

Since $r \neq 1$, we can solve for S_n.

$$S_n = a\frac{1 - r^n}{1 - r}$$

■

EXAMPLE 5 **Finding the Sum of n Terms of a Geometric Sequence**

Find the sum S_n of the first n terms of the sequence $\left\{\left(\frac{1}{2}\right)^n\right\}$; that is, find

$$\frac{1}{2} + \frac{1}{4} + \frac{1}{8} + \cdots + \left(\frac{1}{2}\right)^n$$

Solution The sequence $\left\{\left(\frac{1}{2}\right)^n\right\}$ is a geometric sequence with $a = \frac{1}{2}$ and $r = \frac{1}{2}$. The sum S_n that we seek is the sum of the first n terms of the sequence, so we use formula (3) to get

$$S_n = \sum_{k=1}^{n}\left(\frac{1}{2}\right)^k = \frac{1}{2} + \frac{1}{4} + \frac{1}{8} + \cdots + \left(\frac{1}{2}\right)^n$$

$$= \frac{1}{2}\left[\frac{1-\left(\frac{1}{2}\right)^n}{1-\frac{1}{2}}\right] \quad \text{Formula (3)}$$

$$= \frac{1}{2}\left[\frac{1-\left(\frac{1}{2}\right)^n}{\frac{1}{2}}\right]$$

$$= 1 - \left(\frac{1}{2}\right)^n$$

◀

NOW WORK PROBLEM 47.

EXAMPLE 6 Using a Graphing Utility to Find the Sum of a Geometric Sequence

Use a graphing utility to find the sum of the first 15 terms of the sequence $\left\{\left(\frac{1}{3}\right)^n\right\}$; that is, find

$$\frac{1}{3} + \frac{1}{9} + \frac{1}{27} + \cdots + \left(\frac{1}{3}\right)^{15}$$

Figure 9

Solution Figure 9 shows the result obtained using a TI-83 graphing calculator. The sum of the first 15 terms of the sequence $\left\{\left(\frac{1}{3}\right)^n\right\}$ is 0.4999999652. ◀

NOW WORK PROBLEM 53.

Geometric Series

An infinite sum of the form

$$a + ar + ar^2 + \cdots + ar^{n-1} + \cdots$$

with first term a and common ratio r, is called an **infinite geometric series** and is denoted by

$$\sum_{k=1}^{\infty} ar^{k-1}$$

4 Based on formula (3), the sum S_n of the first n terms of a geometric series is

$$S_n = a\frac{1-r^n}{1-r} = \frac{a}{1-r} - \frac{ar^n}{1-r} \qquad \textbf{(6)}$$

If this finite sum S_n approaches a number L as $n \to \infty$, then we call L the **sum of the infinite geometric series,** and we write

$$L = \sum_{k=1}^{\infty} ar^{k-1}$$

Theorem

Sum of an Infinite Geometric Series

If $|r| < 1$, the sum of the infinite geometric series $\sum_{k=1}^{\infty} ar^{k-1}$ is

$$\sum_{k=1}^{\infty} ar^{k-1} = \frac{a}{1-r} \qquad (7)$$

Intuitive Proof Since $|r| < 1$, it follows that $|r^n|$ approaches 0 as $n \to \infty$. Then, based on formula (6), the sum S_n approaches $\frac{a}{1-r}$ as $n \to \infty$. ■

EXAMPLE 7 Finding the Sum of a Geometric Series

Find the sum of the geometric series: $2 + \frac{4}{3} + \frac{8}{9} + \cdots$

Solution The first term is $a = 2$, and the common ratio is

$$r = \frac{\frac{4}{3}}{2} = \frac{4}{6} = \frac{2}{3}$$

Since $|r| < 1$, we use formula (7) to find that

$$2 + \frac{4}{3} + \frac{8}{9} + \cdots = \frac{2}{1 - \frac{2}{3}} = 6$$

◀

NOW WORK PROBLEM 59.

Exploration

Use a graphing utility to graph $U_n = 2\left(\frac{2}{3}\right)^{n-1}$ in sequence mode. TRACE the graph for large values of n. What happens to the value of U_n as n increases without bound? What can you conclude about $\sum_{n=1}^{\infty} 2\left(\frac{2}{3}\right)^{(n-1)}$?

EXAMPLE 8 Repeating Decimals

Show that the repeating decimal 0.999 ... equals 1.

Solution

$$0.999\ldots = \frac{9}{10} + \frac{9}{100} + \frac{9}{1000} + \cdots$$

The decimal 0.999... is a geometric series with first term $\frac{9}{10}$ and common ratio $\frac{1}{10}$. Using formula (7), we find

$$0.999\ldots = \frac{\frac{9}{10}}{1 - \frac{1}{10}} = \frac{\frac{9}{10}}{\frac{9}{10}} = 1$$

◀

EXAMPLE 9 Pendulum Swings

Figure 10

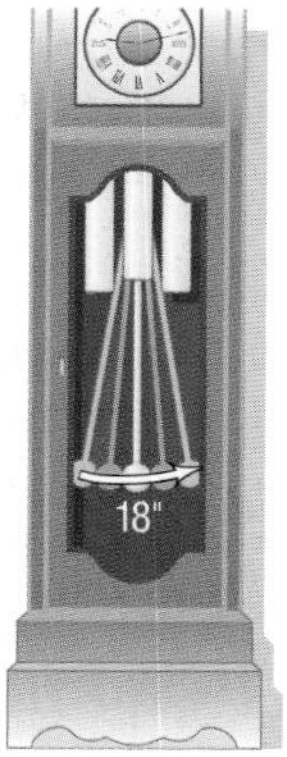

Initially, a pendulum swings through an arc of 18 inches. See Figure 10. On each successive swing, the length of the arc is 0.98 of the previous length.

(a) What is the length of the arc of the 10th swing?
(b) On which swing is the length of the arc first less than 12 inches?
(c) After 15 swings, what total distance will the pendulum have swung?
(d) When it stops, what total distance will the pendulum have swung?

Solution (a) The length of the first swing is 18 inches.
The length of the second swing is 0.98(18) inches.
The length of the third swing is $0.98(0.98)(18) = 0.98^2(18)$ inches.
The length of the arc of the 10th swing is

$$(0.98)^9(18) = 15.007 \text{ inches}$$

(b) The length of the arc of the nth swing is $(0.98)^{n-1}(18)$. For this to be exactly 12 inches requires that

$$(0.98)^{n-1}(18) = 12$$

$$(0.98)^{n-1} = \frac{12}{18} = \frac{2}{3} \qquad \text{Divide both sides by 18.}$$

$$n - 1 = \log_{0.98}\left(\frac{2}{3}\right) \qquad \text{Express as a logarithm.}$$

$$n = 1 + \frac{\ln\left(\frac{2}{3}\right)}{\ln 0.98} \approx 1 + 20.07 \approx 21.07 \qquad \text{Solve for } n\text{; use the Change of Base Formula}$$

The length of the arc of the pendulum exceeds 12 inches on the 21st swing and is first less than 12 inches on the 22nd swing.

(c) After 15 swings, the pendulum will have swung the following total distance L:

$$L = \underset{\text{1st}}{18} + \underset{\text{2nd}}{0.98(18)} + \underset{\text{3rd}}{(0.98)^2(18)} + \underset{\text{4th}}{(0.98)^3(18)} + \cdots + \underset{\text{15th}}{(0.98)^{14}(18)}$$

This is the sum of a geometric sequence. The common ratio is 0.98; the first term is 18. The sum has 15 terms, so

$$L = 18\frac{1 - 0.98^{15}}{1 - 0.98} \approx 18(13.07) \approx 235.3 \text{ inches}$$

The pendulum will have swung through 235.3 inches after 15 swings.

(d) When the pendulum stops, it will have swung the following total distance T:

$$T = 18 + 0.98(18) + (0.98)^2(18) + (0.98)^3(18) + \cdots$$

This is the sum of a geometric series. The common ratio is $r = 0.98$; the first term is $a = 18$. The sum is

$$T = \frac{a}{1 - r} = \frac{18}{1 - 0.98} = 900$$

The pendulum will have swung a total of 900 inches when it finally stops. ◀

NOW WORK PROBLEM 73.

Annuities

5 In Section 4.7 we developed the compound interest formula that gives the future value when a fixed amount of money is deposited in an account that pays interest compounded periodically. Often, though, money is invested in small amounts at periodic intervals. An **annuity** is a sequence of equal periodic deposits. The periodic deposits may be made annually, quarterly, monthly, or daily.

When deposits are made at the same time that the interest is credited, the annuity is called **ordinary.** We will only deal with ordinary annuities here. The **amount of an annuity** is the sum of all deposits made plus all interest paid.

Suppose that the interest that an account earns is i percent per payment period (expressed as a decimal). For example, if an account pays 12% compounded monthly (12 times a year), then $i = \frac{0.12}{12} = 0.01$. If an account pays 8% compounded quarterly (4 times a year) then $i = \frac{0.08}{4} = 0.02$. To develop a formula for the amount of an annuity, suppose that $\$P$ is deposited each payment period for n payment periods in an account that earns i percent per payment period. When the last deposit is made at the nth payment period, the first deposit of $\$P$ has earned interest compounded for $n - 1$ payment periods, the second deposit of $\$P$ has earned interest compounded for $n - 2$ payment periods, and so on. Table 3 shows the value of each deposit after n deposits have been made.

Table 3

Deposit	1	2	3	...	$n-1$	n
Amount	$P(1+i)^{n-1}$	$P(1+i)^{n-2}$	$P(1+i)^{n-3}$	...	$P(1+i)$	P

The amount A of the annuity is the sum of the amounts shown in Table 3; that is,

$$\begin{aligned} A &= P(1+i)^{n-1} + P(1+i)^{n-2} + \cdots + P(1+i) + P \\ &= P[1 + (1+i) + \cdots + (1+i)^{n-1}] \end{aligned}$$

The expression in brackets is the sum of a geometric sequence with n terms and a common ratio of $(1 + i)$. As a result,

$$\begin{aligned} A &= P[1 + (1+i) + \cdots + (1+i)^{n-2} + (1+i)^{n-1}] \\ &= P\frac{1-(1+i)^n}{1-(1+i)} = P\frac{1-(1+i)^n}{-i} = P\frac{(1+i)^n - 1}{i} \end{aligned}$$

We have established the following result:

Theorem

Amount of an Annuity

Suppose P is the deposit in dollars made at each payment period for an annuity paying i percent interest per payment period. The amount A of the annuity after n deposits is

$$A = P\frac{(1+i)^n - 1}{i} \tag{8}$$

Note: In using formula (8), remember that when the nth deposit is made, the first deposit has earned interest for $n - 1$ compounding periods.

EXAMPLE 10 Determining the Amount of an Annuity

To save for retirement, Brett decides to place \$2000 into an Individual Retirement Account (IRA) each year for the next 30 years. What will the value of the IRA be when Brett makes his 30th deposit? Assume that the rate of return of the IRA is 10% per annum compounded annually.

Solution This is an ordinary annuity with $n = 30$ annual deposits of $P = \$2000$. The rate of interest per payment period is $i = \frac{0.10}{1} = 0.10$. The amount A of the annuity after 30 deposits is

$$A = 2000\left\{\frac{(1 + 0.10)^{30} - 1}{0.10}\right\} = \$2000(164.494023) = \$328{,}988.05$$

◀

EXAMPLE 11 Determining the Amount of an Annuity

To save for her daughter's college education, Ms. Miranda decides to put \$50 aside every month in a credit union account paying 10% interest compounded monthly. She begins this savings program when her daughter is 3 years old. How much will she have saved by the time she makes the 180th deposit? How old is her daughter at this time?

Solution This is an annuity with $P = \$50$, $n = 180$, and $i = \frac{0.10}{12}$. The amount A saved is

$$A = 50\left[\frac{\left(1 + \frac{0.10}{12}\right)^{180} - 1}{\frac{0.10}{12}}\right] = \$50(414.47035) = \$20{,}723.52$$

Since there are 12 deposits per year, when the 180th deposit is made $\frac{180}{12} = 15$ years have passed and Ms. Miranda's daughter is 18 years old. ◀

NOW WORK PROBLEM 77.

HISTORICAL FEATURE

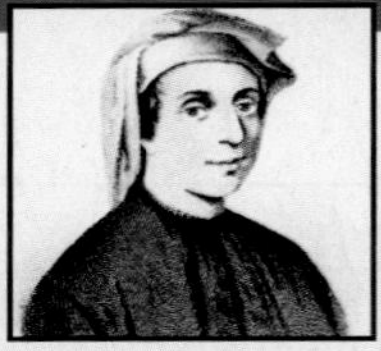
Fibonacci

Sequences are among the oldest objects of mathematical investigation, having been studied for over 3500 years. After the initial steps, however, little progress was made until about 1600.

Arithmetic and geometric sequences appear in the Rhind papyrus, a mathematical text containing 85 problems copied around 1650 BC by the Egyptian scribe Ahmes from an earlier work (see Historical Problem 1). Fibonacci (AD 1220) wrote about problems similar to those found in the Rhind papyrus, leading one to suspect that Fibonacci may have had material available that is now lost. This material would have been in the non-Euclidean Greek tradition of Heron (about AD 75) and Diophantus (about AD 250). One problem, again modified slightly, is still with us in the familiar puzzle rhyme "As I was going to St. Ives . . ." (see Historical Problem 2).

The Rhind papyrus indicates that the Egyptians knew how to add up the terms of an arithmetic or geometric sequence, as did the Babylonians. The rule for summing up a geometric sequence is found in Euclid's *Elements* (Book IX, 35, 36), where, like all Euclid's algebra, it is presented in a geometric form.

Investigations of other kinds of sequences began in the 1500s, when algebra became sufficiently developed to handle the more complicated problems. The development of calculus in the 1600s added a powerful new tool, especially for finding the sum of infinite series, and the subject continues to flourish today.

Historical Problems

1. *Arithmetic sequence problem from the Rhind papyrus (statement modified slightly for clarity)* One hundred loaves of bread are to be divided among five people so that the amounts that they receive form an arithmetic sequence. The first two together receive one-seventh of what the last three receive. How many loaves does each receive? [*Partial answer:* First person receives $1\frac{2}{3}$ loaves.]

2. The following old English children's rhyme resembles one of the Rhind papyrus problems.

   As I was going to St. Ives
   I met a man with seven wives
   Each wife had seven sacks
   Each sack had seven cats
   Each cat had seven kits [kittens]
   Kits, cats, sacks, wives
   How many were going to St. Ives?

   (a) Assuming that the speaker and the cat fanciers met by traveling in opposite directions, what is the answer?
   (b) How many kittens are being transported?
   (c) Kits, cats, sacks, wives; how many?

   [**Hint:** It is easier to include the man, find the sum with the formula, and then subtract 1 for the man.]

11.3 Assess Your Understanding

'Are You Prepared?' *Answers are given at the end of these exercises. If you get a wrong answer, read the pages listed in red.*

1. If \$1000 is invested at 4% per annum compounded semiannually, how much is in the account after two years? (pp. 287–294)

2. How much do you need to invest now at 5% per annum compounded monthly so that in one year you will have \$10,000? (pp. 287–294)

Concepts and Vocabulary

3. In a(n) __________ sequence the ratio of successive terms is a constant.

4. If $|r| < 1$, the sum of the geometric series $\sum_{k=1}^{\infty} ar^{k-1}$ is __________.

5. A sequence of equal periodic deposits is called a(n) __________.

6. *True or False:* A geometric sequence may be defined recursively.

7. *True or False:* In a geometric sequence the common ratio is always a positive number.

8. *True or False:* For a geometric sequence with first term a and common ratio r, where $r \neq 0, r \neq 1$, the sum of the first n terms is $S_n = a \cdot \frac{1 - r^n}{1 - r}$.

Exercises

In Problems 9–18, a geometric sequence is given. Find the common ratio and write out the first four terms.

9. $\{3^n\}$ **10.** $\{(-5)^n\}$ **11.** $\left\{-3\left(\frac{1}{2}\right)^n\right\}$ **12.** $\left\{\left(\frac{5}{2}\right)^n\right\}$ **13.** $\left\{\frac{2^{n-1}}{4}\right\}$

14. $\left\{\frac{3^n}{9}\right\}$ **15.** $\{2^{n/3}\}$ **16.** $\{3^{2n}\}$ **17.** $\left\{\frac{3^{n-1}}{2^n}\right\}$ **18.** $\left\{\frac{2^n}{3^{n-1}}\right\}$

In Problems 19–32, determine whether the given sequence is arithmetic, geometric, or neither. If the sequence is arithmetic, find the common difference; if it is geometric, find the common ratio.

19. $\{n + 2\}$ **20.** $\{2n - 5\}$ **21.** $\{4n^2\}$ **22.** $\{5n^2 + 1\}$ **23.** $\left\{3 - \frac{2}{3}n\right\}$

24. $\left\{8 - \frac{3}{4}n\right\}$ **25.** $1, 3, 6, 10, \ldots$ **26.** $2, 4, 6, 8, \ldots$ **27.** $\left\{\left(\frac{2}{3}\right)^n\right\}$ **28.** $\left\{\left(\frac{5}{4}\right)^n\right\}$

29. $-1, -2, -4, -8, \ldots$ **30.** $1, 1, 2, 3, 5, 8, \ldots$ **31.** $\{3^{n/2}\}$ **32.** $\{(-1)^n\}$

In Problems 33–40, find the fifth term and the nth term of the geometric sequence whose initial term a and common ratio r are given.

33. $a = 2;\quad r = 3$ **34.** $a = -2;\quad r = 4$ **35.** $a = 5;\quad r = -1$ **36.** $a = 6;\quad r = -2$

37. $a = 0;\quad r = \frac{1}{2}$ **38.** $a = 1;\quad r = -\frac{1}{3}$ **39.** $a = \sqrt{2};\quad r = \sqrt{2}$ **40.** $a = 0;\quad r = \frac{1}{\pi}$

In Problems 41–46, find the indicated term of each geometric sequence.

41. 7th term of $1, \frac{1}{2}, \frac{1}{4}, \ldots$ **42.** 8th term of $1, 3, 9, \ldots$ **43.** 9th term of $1, -1, 1, \ldots$

44. 10th term of $-1, 2, -4, \ldots$ **45.** 8th term of $0.4, 0.04, 0.004, \ldots$ **46.** 7th term of $0.1, 1.0, 10.0, \ldots$

In Problems 47–52, find each sum.

47. $\frac{1}{4} + \frac{2}{4} + \frac{2^2}{4} + \frac{2^3}{4} + \cdots + \frac{2^{n-1}}{4}$ **48.** $\frac{3}{9} + \frac{3^2}{9} + \frac{3^3}{9} + \cdots + \frac{3^n}{9}$ **49.** $\sum_{k=1}^{n} \left(\frac{2}{3}\right)^k$

50. $\sum_{k=1}^{n} 4 \cdot 3^{k-1}$ **51.** $-1 - 2 - 4 - 8 - \cdots - (2^{n-1})$ **52.** $2 + \frac{6}{5} + \frac{18}{25} + \cdots + 2\left(\frac{3}{5}\right)^{n-1}$

For Problems 53–58, use a graphing utility to find the sum of each geometric sequence.

53. $\frac{1}{4} + \frac{2}{4} + \frac{2^2}{4} + \frac{2^3}{4} + \cdots + \frac{2^{14}}{4}$ **54.** $\frac{3}{9} + \frac{3^2}{9} + \frac{3^3}{9} + \cdots + \frac{3^{15}}{9}$ **55.** $\sum_{n=1}^{15} \left(\frac{2}{3}\right)^n$

56. $\sum_{n=1}^{15} 4 \cdot 3^{n-1}$ **57.** $-1 - 2 - 4 - 8 - \cdots - 2^{14}$ **58.** $2 + \frac{6}{5} + \frac{18}{25} + \cdots + 2\left(\frac{3}{5}\right)^{15}$

In Problems 59–68, find the sum of each infinite geometric series.

59. $1 + \frac{1}{3} + \frac{1}{9} + \cdots$ **60.** $2 + \frac{4}{3} + \frac{8}{9} + \cdots$ **61.** $8 + 4 + 2 + \cdots$ **62.** $6 + 2 + \frac{2}{3} + \cdots$

63. $2 - \frac{1}{2} + \frac{1}{8} - \frac{1}{32} + \cdots$ **64.** $1 - \frac{3}{4} + \frac{9}{16} - \frac{27}{64} + \cdots$ **65.** $\sum_{k=1}^{\infty} 5\left(\frac{1}{4}\right)^{k-1}$ **66.** $\sum_{k=1}^{\infty} 8\left(\frac{1}{3}\right)^{k-1}$

67. $\sum_{k=1}^{\infty} 6\left(-\frac{2}{3}\right)^{k-1}$ **68.** $\sum_{k=1}^{\infty} 4\left(-\frac{1}{2}\right)^{k-1}$

69. Find x so that x, $x + 2$, and $x + 3$ are consecutive terms of a geometric sequence.

70. Find x so that $x - 1$, x, and $x + 2$ are consecutive terms of a geometric sequence.

71. Salary Increases Suppose that you have just been hired at an annual salary of \$18,000 and expect to receive annual increases of 5%. What will your salary be when you begin your fifth year?

72. Equipment Depreciation A new piece of equipment cost a company \$15,000. Each year, for tax purposes, the company depreciates the value by 15%. What value should the company give the equipment after 5 years?

73. **Pendulum Swings** Initially, a pendulum swings through an arc of 2 feet. On each successive swing, the length of the arc is 0.9 of the previous length.
(a) What is the length of the arc of the 10th swing?
(b) On which swing is the length of the arc first less than 1 foot?
(c) After 15 swings, what total length will the pendulum have swung?
(d) When it stops, what total length will the pendulum have swung?

74. Bouncing Balls A ball is dropped from a height of 30 feet. Each time it strikes the ground, it bounces up to 0.8 of the previous height.

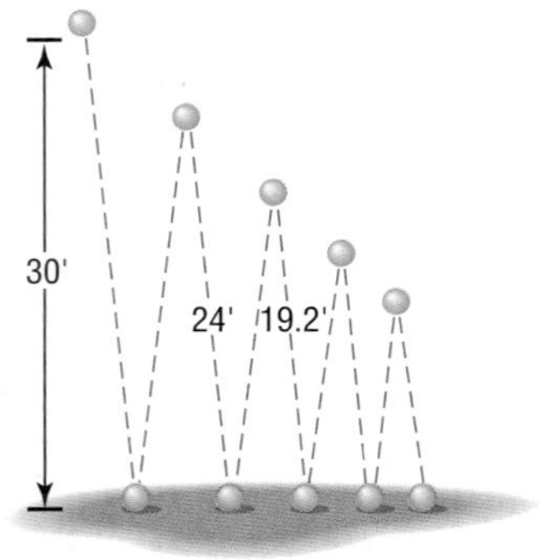

(a) What height will the ball bounce up to after it strikes the ground for the third time?
(b) How high will it bounce after it strikes the ground for the nth time?
(c) How many times does the ball need to strike the ground before its bounce is less than 6 inches?
(d) What total distance does the ball travel before it stops bouncing?

75. Retirement Christine contributes \$100 each month to her 401(k). What will be the value of Christine's 401(k) after the 360th deposit (30 years) if the per annum rate of return is assumed to be 12% compounded monthly?

76. Saving for a Home Jolene wants to purchase a new home. Suppose that she invests \$400 per month into a mutual fund. If the per annum rate of return of the mutual fund is assumed to be 10% compounded monthly, how much will Jolene have for a down payment after the 36th deposit (3 years)?

77. Tax Sheltered Annuity Don contributes \$500 at the end of each quarter to a Tax Sheltered Annuity (TSA). What will the value of the TSA be after the 80th deposit (20 years) if the per annum rate of return is assumed to be 8% compounded quarterly?

78. Retirement Ray contributes \$1000 to an Individual Retirement Account (IRA) semiannually. What will the value of the IRA be when Ray makes his 30th deposit (after 15 years) if the per annum rate of return is assumed to be 10% compounded semiannually?

79. Sinking Fund Scott and Alice want to purchase a vacation home in 10 years and need \$50,000 for a down payment. How much should they place in a savings account each month if the per annum rate of return is assumed to be 6% compounded monthly?

80. Sinking Fund For a child born in 1996, a 4-year college education at a public university is projected to be \$150,000. Assuming an 8% per annum rate of return compounded monthly, how much must be contributed to a college fund every month in order to have \$150,000 in 18 years when the child begins college?

81. Critical Thinking You are interviewing for a job and receive two offers:

A: \$20,000 to start, with guaranteed annual increases of 6% for the first 5 years
B: \$22,000 to start, with guaranteed annual increases of 3% for the first 5 years

Which offer is best if your goal is to be making as much as possible after 5 years? Which is best if your goal is to make as much money as possible over the contract (5 years)?

82. Critical Thinking Which of the following choices, A or B, results in more money?

A: To receive \$1000 on day 1, \$999 on day 2, \$998 on day 3, with the process to end after 1000 days
B: To receive \$1 on day 1, \$2 on day 2, \$4 on day 3, for 19 days

83. Critical Thinking You have just signed a 7-year professional football league contract with a beginning salary of \$2,000,000 per year. Management gives you the following options with regard to your salary over the 7 years.

1. A bonus of \$100,000 each year
2. An annual increase of 4.5% per year beginning after 1 year
3. An annual increase of \$95,000 per year beginning after 1 year

Which option provides the most money over the 7-year period? Which the least? Which would you choose? Why?

84. A Rich Man's Promise A rich man promises to give you \$1000 on September 1, 2001. Each day thereafter he will give you $\frac{9}{10}$ of what he gave you the previous day. What is the first date on which the amount you receive is less than 1¢? How much have you received when this happens?

85. Grains of Wheat on a Chess Board In an old fable, a commoner who had saved the king's life was told he could ask the king for any just reward. Being a shrewd man, the commoner said, "A simple wish, sire. Place one grain of wheat on the first square of a chessboard, two grains on the second square, four grains on the third square, continuing until you have filled the board. This is all I seek." Compute the total number of grains needed to do this to see why the request, seemingly simple, could not be granted. (A chessboard consists of $8 \times 8 = 64$ squares.)

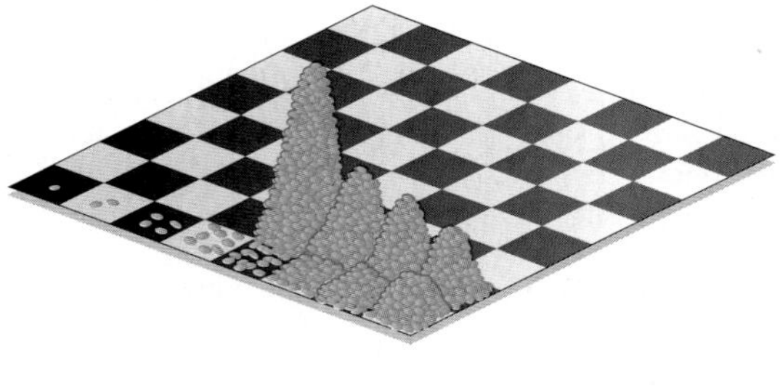

86. Look at the figure below. What fraction of the square is eventually shaded if the indicated shading process continues indefinitely?

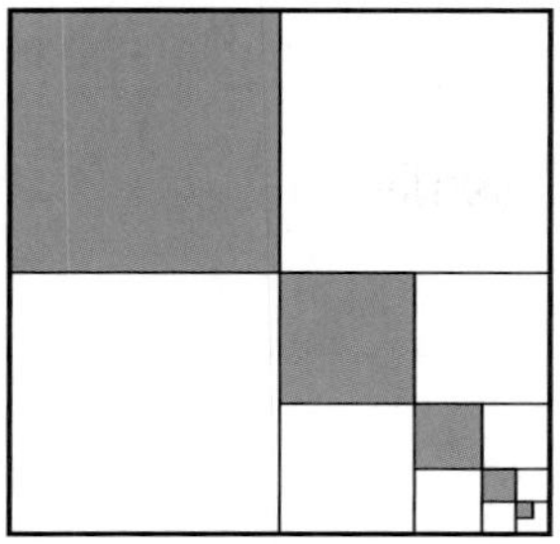

87. Multiplier Suppose that, throughout the U.S. economy, individuals spend 90% of every additional dollar that they earn. Economists would say that an individual's **marginal propensity to consume** is 0.90. For example, if Jane earns an additional dollar, she will spend $0.9(1) = \$0.90$ of it. The individual that earns \$0.90 (from Jane) will spend 90% of it or \$0.81. This process of spending continues and results in an infinite geometric series as follows:

$$1, 0.90, 0.90^2, 0.90^3, 0.90^4, \ldots$$

The sum of this infinite geometric series is called the **multiplier.** What is the multiplier if individuals spend 90% of every additional dollar that they earn?

88. Multiplier Refer to Problem 87. Suppose that the marginal propensity to consume throughout the U.S. economy is 0.95. What is the multiplier for the U.S. economy?

89. Stock Price One method of pricing a stock is to discount the stream of future dividends of the stock. Suppose that a stock pays $\$P$ per year in dividends and, historically, the dividend has been increased $i\%$ per year. If you desire an annual rate of return of $r\%$, this method of pricing a stock states that the price that you should pay is the present value of an infinite stream of payments:

$$\text{Price} = P + P\frac{1+i}{1+r} + P\left(\frac{1+i}{1+r}\right)^2 + P\left(\frac{1+i}{1+r}\right)^3 + \cdots$$

The price of the stock is the sum of an infinite geometric series. Suppose that a stock pays an annual dividend of \$4.00 and, historically, the dividend has been increased 3% per year. You desire an annual rate of return of 9%. What is the most you should pay for the stock?

90. Stock Price Refer to Problem 89. Suppose that a stock pays an annual dividend of \$2.50 and, historically, the dividend has increased 4% per year. You desire an annual rate of return of 11%. What is the most that you should pay for the stock?

91. Can a sequence be both arithmetic and geometric? Give reasons for your answer.

92. Make up a geometric sequence. Give it to a friend and ask for its 20th term.

93. Make up two infinite geometric series, one that has a sum and one that does not. Give them to a friend and ask for the sum of each series.

'Are You Prepared?' Answers

1. \$1082.43 **2.** \$9513.28

11.4 Mathematical Induction

OBJECTIVE 1 Prove Statements Using Mathematical Induction

1 *Mathematical induction* is a method for proving that statements involving natural numbers are true for all natural numbers.* For example, the statement "$2n$ is always an even integer" can be proved true for all natural numbers n by using mathematical induction. Also, the statement "the sum of the first n positive odd integers equals n^2," that is,

$$1 + 3 + 5 + \cdots + (2n - 1) = n^2 \qquad \textbf{(1)}$$

can be proved for all natural numbers n by using mathematical induction.

Before stating the method of mathematical induction, let's try to gain a sense of the power of the method. We shall use the statement in equation (1) for this purpose by restating it for various values of $n = 1, 2, 3, \ldots$.

*Recall that the natural numbers are the numbers $1, 2, 3, 4, \ldots$. In other words, the terms *natural numbers* and *positive integers* are synonymous.

$n = 1$ The sum of the first positive odd integer is 1^2; $1 = 1^2$.

$n = 2$ The sum of the first 2 positive odd integers is 2^2; $1 + 3 = 4 = 2^2$.

$n = 3$ The sum of the first 3 positive odd integers is 3^2; $1 + 3 + 5 = 9 = 3^2$.

$n = 4$ The sum of the first 4 positive odd integers is 4^2; $1 + 3 + 5 + 7 = 16 = 4^2$.

Although from this pattern we might conjecture that statement (1) is true for any choice of n, can we really be sure that it does not fail for some choice of n? The method of proof by mathematical induction will, in fact, prove that the statement is true for all n.

Theorem

The Principle of Mathematical Induction

Suppose that the following two conditions are satisfied with regard to a statement about natural numbers:

CONDITION I: The statement is true for the natural number 1.

CONDITION II: If the statement is true for some natural number k, it is also true for the next natural number $k + 1$.

Then the statement is true for all natural numbers.

We shall not prove this principle. However, we can provide a physical interpretation that will help us to see why the principle works. Think of a collection of natural numbers obeying a statement as a collection of infinitely many dominoes (see Figure 11).

Figure 11

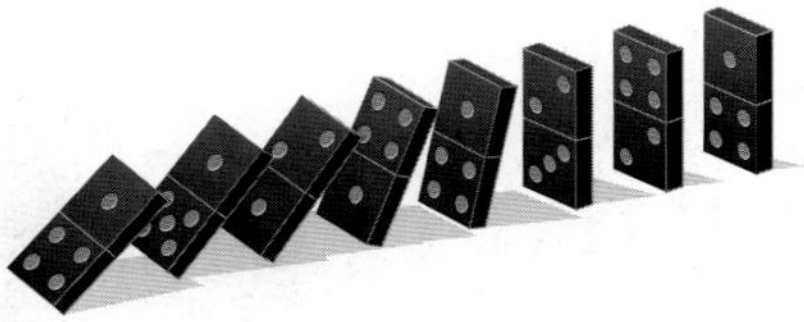

Now, suppose that we are told two facts:

1. The first domino is pushed over.
2. If one domino falls over, say the kth domino, then so will the next one, the $(k + 1)$st domino.

Is it safe to conclude that *all* the dominoes fall over? The answer is yes, because if the first one falls (Condition I), then the second one does also (by Condition II); and if the second one falls, then so does the third (by Condition II); and so on.

Now let's prove some statements about natural numbers using mathematical induction.

EXAMPLE 1 Using Mathematical Induction

Show that the following statement is true for all natural numbers n.

$$1 + 3 + 5 + \cdots + (2n - 1) = n^2 \qquad \textbf{(2)}$$

Solution We need to show first that statement (2) holds for $n = 1$. Because $1 = 1^2$, statement (2) is true for $n = 1$. Condition I holds.

Next, we need to show that Condition II holds. Suppose that we know for some k that

$$1 + 3 + \cdots + (2k - 1) = k^2 \qquad \textbf{(3)}$$

We wish to show that, based on equation (3), statement (2) holds for $k + 1$. We look at the sum of the first $k + 1$ positive odd integers to determine whether this sum equals $(k + 1)^2$.

$$\begin{aligned} 1 + 3 + \cdots + (2k - 1) + (2k + 1) &= \underbrace{[1 + 3 + \cdots + (2k - 1)]}_{= k^2 \text{ by equation (3)}} + (2k + 1) \\ &= k^2 + (2k + 1) \\ &= k^2 + 2k + 1 = (k + 1)^2 \end{aligned}$$

Conditions I and II are satisfied; by the Principle of Mathematical Induction, statement (2) is true for all natural numbers n. ◀

EXAMPLE 2 Using Mathematical Induction

Show that the following statement is true for all natural numbers n.

$$2^n > n$$

Solution First, we show that the statement $2^n > n$ holds when $n = 1$. Because $2^1 = 2 > 1$, the inequality is true for $n = 1$. Condition I holds.

Next, we assume, for some natural number k, that $2^k > k$. We wish to show that the formula holds for $k + 1$; that is, we wish to show that $2^{k+1} > k + 1$. Now

$$2^{k+1} = 2 \cdot 2^k \underset{\substack{\uparrow \\ \text{We know that} \\ 2^k > k.}}{>} 2 \cdot k = k + k \underset{\substack{\uparrow \\ k \geq 1}}{\geq} k + 1$$

If $2^k > k$, then $2^{k+1} > k + 1$, so Condition II of the Principle of Mathematical Induction is satisfied. The statement $2^n > n$ is true for all natural numbers n. ◀

EXAMPLE 3 Using Mathematical Induction

Show that the following formula is true for all natural numbers n.

$$1 + 2 + 3 + \cdots + n = \frac{n(n + 1)}{2} \qquad \textbf{(4)}$$

Solution First, we show that formula (4) is true when $n = 1$. Because

$$\frac{1(1 + 1)}{2} = \frac{1(2)}{2} = 1$$

Condition I of the Principle of Mathematical Induction holds.

Next, we assume that formula (4) holds for some k, and we determine whether the formula then holds for $k + 1$. We assume that

$$1 + 2 + 3 + \cdots + k = \frac{k(k + 1)}{2} \quad \text{for some } k \qquad \textbf{(5)}$$

Now we need to show that

$$1 + 2 + 3 + \cdots + k + (k + 1) = \frac{(k + 1)(k + 1 + 1)}{2} = \frac{(k + 1)(k + 2)}{2}$$

We do this as follows:

$$1 + 2 + 3 + \cdots + k + (k + 1) = \underbrace{[1 + 2 + 3 + \cdots + k]}_{= \frac{k(k+1)}{2} \text{ by equation (5)}} + (k + 1)$$
$$= \frac{k(k + 1)}{2} + (k + 1)$$
$$= \frac{k^2 + k + 2k + 2}{2}$$
$$= \frac{k^2 + 3k + 2}{2} = \frac{(k + 1)(k + 2)}{2}$$

Condition II also holds. As a result, formula (4) is true for all natural numbers n. ◀

NOW WORK PROBLEM 1.

EXAMPLE 4 Using Mathematical Induction

Show that $3^n - 1$ is divisible by 2 for all natural numbers n.

Solution First, we show that the statement is true when $n = 1$. Because $3^1 - 1 = 3 - 1 = 2$ is divisible by 2, the statement is true when $n = 1$. Condition I is satisfied.

Next, we assume that the statement holds for some k, and we determine whether the statement then holds for $k + 1$. We assume that $3^k - 1$ is divisible by 2 for some k. We need to show that $3^{k+1} - 1$ is divisible by 2. Now

$$3^{k+1} - 1 = 3^{k+1} - 3^k + 3^k - 1 \quad \text{Subtract and add } 3^k.$$
$$= 3^k(3 - 1) + (3^k - 1) = 3^k \cdot 2 + (3^k - 1)$$

Because $3^k \cdot 2$ is divisible by 2 and $3^k - 1$ is divisible by 2, it follows that $3^k \cdot 2 + (3^k - 1) = 3^{k+1} - 1$ is divisible by 2. Condition II is also satisfied. As a result, the statement "$3^n - 1$ is divisible by 2" is true for all natural numbers n. ◀

WARNING: The conclusion that a statement involving natural numbers is true for all natural numbers is made only after *both* Conditions I and II of the Principle of Mathematical Induction have been satisfied. Problem 27 demonstrates a statement for which only Condition I holds, but the statement is *not* true for all natural numbers. Problem 28 demonstrates a statement for which only Condition II holds, but the statement is *not* true for any natural number. ■

11.4 Assess Your Understanding

Exercises

In Problems 1–26, use the Principle of Mathematical Induction to show that the given statement is true for all natural numbers n.

1. $2 + 4 + 6 + \cdots + 2n = n(n + 1)$
2. $1 + 5 + 9 + \cdots + (4n - 3) = n(2n - 1)$
3. $3 + 4 + 5 + \cdots + (n + 2) = \frac{1}{2}n(n + 5)$
4. $3 + 5 + 7 + \cdots + (2n + 1) = n(n + 2)$
5. $2 + 5 + 8 + \cdots + (3n - 1) = \frac{1}{2}n(3n + 1)$
6. $1 + 4 + 7 + \cdots + (3n - 2) = \frac{1}{2}n(3n - 1)$
7. $1 + 2 + 2^2 + \cdots + 2^{n-1} = 2^n - 1$
8. $1 + 3 + 3^2 + \cdots + 3^{n-1} = \frac{1}{2}(3^n - 1)$

9. $1 + 4 + 4^2 + \cdots + 4^{n-1} = \frac{1}{3}(4^n - 1)$

10. $1 + 5 + 5^2 + \cdots + 5^{n-1} = \frac{1}{4}(5^n - 1)$

11. $\frac{1}{1\cdot 2} + \frac{1}{2\cdot 3} + \frac{1}{3\cdot 4} + \cdots + \frac{1}{n(n+1)} = \frac{n}{n+1}$

12. $\frac{1}{1\cdot 3} + \frac{1}{3\cdot 5} + \frac{1}{5\cdot 7} + \cdots + \frac{1}{(2n-1)(2n+1)} = \frac{n}{2n+1}$

13. $1^2 + 2^2 + 3^2 + \cdots + n^2 = \frac{1}{6}n(n+1)(2n+1)$

14. $1^3 + 2^3 + 3^3 + \cdots + n^3 = \frac{1}{4}n^2(n+1)^2$

15. $4 + 3 + 2 + \cdots + (5 - n) = \frac{1}{2}n(9 - n)$

16. $-2 - 3 - 4 - \cdots - (n+1) = -\frac{1}{2}n(n+3)$

17. $1\cdot 2 + 2\cdot 3 + 3\cdot 4 + \cdots + n(n+1) = \frac{1}{3}n(n+1)(n+2)$

18. $1\cdot 2 + 3\cdot 4 + 5\cdot 6 + \cdots + (2n-1)(2n) = \frac{1}{3}n(n+1)(4n-1)$

19. $n^2 + n$ is divisible by 2.

20. $n^3 + 2n$ is divisible by 3.

21. $n^2 - n + 2$ is divisible by 2.

22. $n(n+1)(n+2)$ is divisible by 6.

23. If $x > 1$, then $x^n > 1$.

24. If $0 < x < 1$, then $0 < x^n < 1$.

25. $a - b$ is a factor of $a^n - b^n$.
[**Hint:** $a^{k+1} - b^{k+1} = a(a^k - b^k) + b^k(a - b)$]

26. $a + b$ is a factor of $a^{2n+1} + b^{2n+1}$.

---

27. Show that the statement "$n^2 - n + 41$ is a prime number" is true for $n = 1$, but is not true for $n = 41$.

28. Show that the formula

$$2 + 4 + 6 + \cdots + 2n = n^2 + n + 2$$

obeys Condition II of the Principle of Mathematical Induction. That is, show that if the formula is true for some k it is also true for $k + 1$. Then show that the formula is false for $n = 1$ (or for any other choice of n).

29. Use mathematical induction to prove that if $r \neq 1$ then

$$a + ar + ar^2 + \cdots + ar^{n-1} = a\frac{1 - r^n}{1 - r}$$

30. Use mathematical induction to prove that

$$a + (a + d) + (a + 2d) + \cdots + [a + (n-1)d] = na + d\frac{n(n-1)}{2}$$

31. Extended Principle of Mathematical Induction The Extended Principle of Mathematical Induction states that if Conditions I and II hold, that is,

(I) A statement is true for a natural number j.

(II) If the statement is true for some natural number $k \geq j$, then it is also true for the next natural number $k + 1$.

then the statement is true for all natural numbers $\geq j$.

Use the Extended Principle of Mathematical Induction to show that the number of diagonals in a convex polygon of n sides is $\frac{1}{2}n(n - 3)$.

[**Hint:** Begin by showing that the result is true when $n = 4$ (Condition I).]

32. Geometry Use the Extended Principle of Mathematical Induction to show that the sum of the interior angles of a convex polygon of n sides equals $(n - 2)\cdot 180°$.

33. How would you explain the Principle of Mathematical Induction to a friend?

11.5 The Binomial Theorem

OBJECTIVES
1 Evaluate a Binomial Coefficient
2 Expand a Binomial

Formulas have been given for expanding $(x + a)^n$ for $n = 2$ and $n = 3$. The *Binomial Theorem** is a formula for the expansion of $(x + a)^n$ for any positive integer n. If $n = 1, 2, 3,$ and 4, the expansion of $(x + a)^n$ is straightforward.

*The name *binomial* is derived from the fact that $x + a$ is a binomial; that is, it contains two terms.

$(x + a)^1 = x + a$ Two terms, beginning with x^1 and ending with a^1

$(x + a)^2 = x^2 + 2ax + a^2$ Three terms, beginning with x^2 and ending with a^2

$(x + a)^3 = x^3 + 3ax^2 + 3a^2x + a^3$ Four terms, beginning with x^3 and ending with a^3

$(x + a)^4 = x^4 + 4ax^3 + 6a^2x^2 + 4a^3x + a^4$ Five terms, beginning with x^4 and ending with a^4

Notice that each expansion of $(x + a)^n$ begins with x^n and ends with a^n. As you read from left to right, the powers of x are decreasing, while the powers of a are increasing. Also, the number of terms that appears equals $n + 1$. Notice, too, that the degree of each monomial in the expansion equals n. For example, in the expansion of $(x + a)^3$, each monomial $(x^3, 3ax^2, 3a^2x, a^3)$ is of degree 3. As a result, we might conjecture that the expansion of $(x + a)^n$ would look like this:

$$(x + a)^n = x^n + \_\,ax^{n-1} + \_\,a^2x^{n-2} + \cdots + \_\,a^{n-1}x + a^n$$

where the blanks are numbers to be found. This is, in fact, the case, as we shall see shortly.

First, we need to introduce a symbol.

The Symbol $\binom{n}{j}$

1 We define the symbol $\binom{n}{j}$, read "n taken j at a time," as follows:

If j and n are integers with $0 \le j \le n$, the symbol $\binom{n}{j}$ is defined as

$$\binom{n}{j} = \frac{n!}{j!(n - j)!} \qquad (1)$$

COMMENT: On a graphing calculator, the symbol $\binom{n}{j}$ may be denoted by the key nCr. ■

EXAMPLE 1 **Evaluating $\binom{n}{j}$**

Find:

(a) $\binom{3}{1}$ (b) $\binom{4}{2}$ (c) $\binom{8}{7}$ (d) $\binom{65}{15}$

Solution (a) $\binom{3}{1} = \dfrac{3!}{1!(3 - 1)!} = \dfrac{3!}{1!\,2!} = \dfrac{3 \cdot 2 \cdot 1}{1(2 \cdot 1)} = \dfrac{6}{2} = 3$

(b) $\binom{4}{2} = \dfrac{4!}{2!(4 - 2)!} = \dfrac{4!}{2!\,2!} = \dfrac{4 \cdot 3 \cdot 2 \cdot 1}{(2 \cdot 1)(2 \cdot 1)} = \dfrac{24}{4} = 6$

(c) $\binom{8}{7} = \dfrac{8!}{7!(8 - 7)!} = \dfrac{8!}{7!\,1!} = \dfrac{8 \cdot \cancel{7!}}{\cancel{7!} \cdot 1!} = \dfrac{8}{1} = 8$

↑ $8! = 8 \cdot 7!$

Figure 12

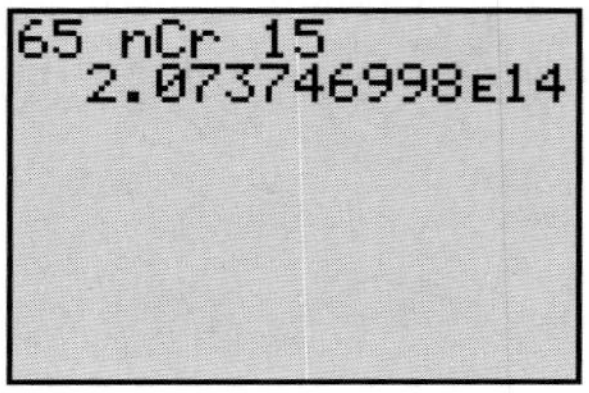

(d) Figure 12 shows the solution using a TI-83 graphing calculator: $\binom{65}{15} = 2.073746998 \times 10^{14}$.

NOW WORK PROBLEM 5.

Four useful formulas involving the symbol $\binom{n}{j}$ are

$$\binom{n}{0} = 1 \qquad \binom{n}{1} = n \qquad \binom{n}{n-1} = n \qquad \binom{n}{n} = 1$$

Proof $\binom{n}{0} = \dfrac{n!}{0!(n-0)!} = \dfrac{n!}{0!n!} = \dfrac{1}{1} = 1$

$$\binom{n}{1} = \frac{n!}{1!(n-1)!} = \frac{n!}{(n-1)!} = \frac{n\cancel{(n-1)!}}{\cancel{(n-1)!}} = n$$

You are asked to show the remaining two formulas in Problem 45. ■

Suppose that we arrange the various values of the symbol $\binom{n}{j}$ in a triangular display, as shown next and in Figure 13.

$$\binom{0}{0}$$

$$\binom{1}{0} \quad \binom{1}{1}$$

$$\binom{2}{0} \quad \binom{2}{1} \quad \binom{2}{2}$$

$$\binom{3}{0} \quad \binom{3}{1} \quad \binom{3}{2} \quad \binom{3}{3}$$

$$\binom{4}{0} \quad \binom{4}{1} \quad \binom{4}{2} \quad \binom{4}{3} \quad \binom{4}{4}$$

$$\binom{5}{0} \quad \binom{5}{1} \quad \binom{5}{2} \quad \binom{5}{3} \quad \binom{5}{4} \quad \binom{5}{5}$$

Figure 13
Pascal triangle

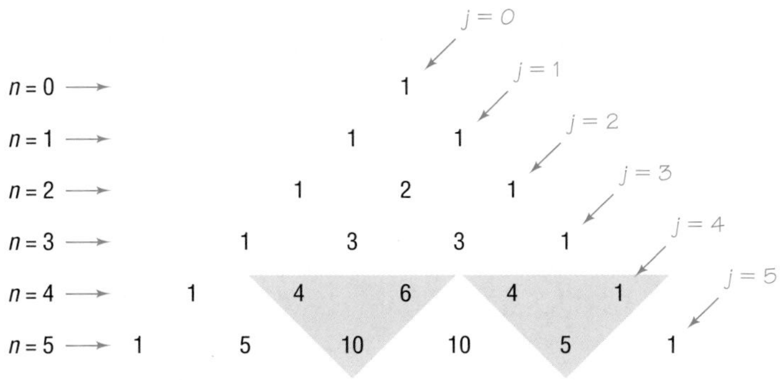

This display is called the **Pascal triangle,** named after Blaise Pascal (1623–1662), a French mathematician.

The Pascal triangle has 1's down the sides. To get any other entry, merely add the two nearest entries in the row above it. The shaded triangles in Figure 13 illustrate this feature of the Pascal triangle. Based on this feature, the row corresponding to $n = 6$ is found as follows:

$$\begin{array}{lccccccccccccc} n=5\rightarrow & & 1 & & 5 & & 10 & & 10 & & 5 & & 1 & \\ n=6\rightarrow & 1 & & 6 & & 15 & & 20 & & 15 & & 6 & & 1 \end{array}$$

Later we shall prove that this addition always works (see the theorem on page 820).

Although the Pascal triangle provides an interesting and organized display of the symbol $\binom{n}{j}$, in practice it is not all that helpful. For example, if you wanted to know the value of $\binom{12}{5}$, you would need to produce 13 rows of the triangle before seeing the answer. It is much faster to use the definition (1).

Binomial Theorem

2 Now we are ready to state the **Binomial Theorem.**

Theorem

Binomial Theorem

Let x and a be real numbers. For any positive integer n, we have

$$(x+a)^n = \binom{n}{0}x^n + \binom{n}{1}ax^{n-1} + \cdots + \binom{n}{j}a^j x^{n-j} + \cdots + \binom{n}{n}a^n$$
$$= \sum_{j=0}^{n} \binom{n}{j} x^{n-j} a^j \qquad \textbf{(2)}$$

Now you know why we needed to introduce the symbol $\binom{n}{j}$; these symbols are the numerical coefficients that appear in the expansion of $(x + a)^n$. Because of this, the symbol $\binom{n}{j}$ is called the **binomial coefficient.**

EXAMPLE 2 **Expanding a Binomial**

Use the Binomial Theorem to expand $(x + 2)^5$.

Solution In the Binomial Theorem, let $a = 2$ and $n = 5$. Then

$$(x+2)^5 = \binom{5}{0}x^5 + \binom{5}{1}2x^4 + \binom{5}{2}2^2x^3 + \binom{5}{3}2^3x^2 + \binom{5}{4}2^4x + \binom{5}{5}2^5$$

↑ Use equation (2).

$$= 1\cdot x^5 + 5\cdot 2x^4 + 10\cdot 4x^3 + 10\cdot 8x^2 + 5\cdot 16x + 1\cdot 32$$

↑ Use row $n = 5$ of the Pascal triangle or formula (1) for $\binom{n}{j}$.

$$= x^5 + 10x^4 + 40x^3 + 80x^2 + 80x + 32$$

◀

EXAMPLE 3 Expanding a Binomial

Expand $(2y - 3)^4$ using the Binomial Theorem.

Solution First, we rewrite the expression $(2y - 3)^4$ as $[2y + (-3)]^4$. Now we use the Binomial Theorem with $n = 4$, $x = 2y$, and $a = -3$.

$$[2y + (-3)]^4 = \binom{4}{0}(2y)^4 + \binom{4}{1}(-3)(2y)^3 + \binom{4}{2}(-3)^2(2y)^2 + \binom{4}{3}(-3)^3(2y) + \binom{4}{4}(-3)^4$$

$$= 1 \cdot 16y^4 + 4(-3)8y^3 + 6 \cdot 9 \cdot 4y^2 + 4(-27)2y + 1 \cdot 81$$

↑ Use row $n = 4$ of the Pascal triangle or formula (1) for $\binom{n}{j}$.

$$= 16y^4 - 96y^3 + 216y^2 - 216y + 81$$

In this expansion, note that the signs alternate due to the fact that $a = -3 < 0$. ◀

NOW WORK PROBLEM 21.

EXAMPLE 4 Finding a Particular Coefficient in a Binomial Expansion

Find the coefficient of y^8 in the expansion of $(2y + 3)^{10}$.

Solution We write out the expansion using the Binomial Theorem.

$$(2y + 3)^{10} = \binom{10}{0}(2y)^{10} + \binom{10}{1}(2y)^9(3)^1 + \binom{10}{2}(2y)^8(3)^2 + \binom{10}{3}(2y)^7(3)^3 + \binom{10}{4}(2y)^6(3)^4 + \cdots + \binom{10}{9}(2y)(3)^9 + \binom{10}{10}(3)^{10}$$

From the third term in the expansion, the coefficient of y^8 is

$$\binom{10}{2}(2)^8(3)^2 = \frac{10!}{2!8!} \cdot 2^8 \cdot 9 = \frac{10 \cdot 9 \cdot \cancel{8!}}{2 \cdot \cancel{8!}} \cdot 2^8 \cdot 9 = 103{,}680$$ ◀

As this solution demonstrates, we can use the Binomial Theorem to find a particular term in an expansion without writing the entire expansion.

Based on the expansion of $(x + a)^n$, the term containing x^j is

$$\binom{n}{n-j}a^{n-j}x^j \qquad (3)$$

For example, we can solve Example 4 by using formula (3) with $n = 10$, $a = 3$, $x = 2y$, and $j = 8$. Then the term containing y^8 is

$$\binom{10}{10-8}3^{10-8}(2y)^8 = \binom{10}{2} \cdot 3^2 \cdot 2^8 \cdot y^8 = \frac{10!}{2!8!} \cdot 9 \cdot 2^8 y^8$$

$$= \frac{10 \cdot 9 \cdot \cancel{8!}}{2 \cdot \cancel{8!}} \cdot 9 \cdot 2^8 y^8 = 103{,}680y^8$$

EXAMPLE 5 Finding a Particular Term in a Binomial Expansion

Find the sixth term in the expansion of $(x + 2)^9$.

Solution A We expand using the Binomial Theorem until the sixth term is reached.

$$(x + 2)^9 = \binom{9}{0}x^9 + \binom{9}{1}x^8 \cdot 2 + \binom{9}{2}x^7 \cdot 2^2 + \binom{9}{3}x^6 \cdot 2^3 + \binom{9}{4}x^5 \cdot 2^4 + \binom{9}{5}x^4 \cdot 2^5 + \cdots$$

The sixth term is

$$\binom{9}{5}x^4 \cdot 2^5 = \frac{9!}{5!4!} \cdot x^4 \cdot 32 = 4032x^4$$

Solution B The sixth term in the expansion of $(x + 2)^9$, which has 10 terms total, contains x^4. (Do you see why?) By formula (3), the sixth term is

$$\binom{9}{9-4}2^{9-4}x^4 = \binom{9}{5}2^5x^4 = \frac{9!}{5!4!} \cdot 32x^4 = 4032x^4$$ ◄

NOW WORK PROBLEMS 29 AND 35.

Next we show that the *triangular addition* feature of the Pascal triangle illustrated in Figure 13 always works.

Theorem If n and j are integers with $1 \le j \le n$, then

$$\binom{n}{j-1} + \binom{n}{j} = \binom{n+1}{j} \quad (4)$$

Proof

$$\binom{n}{j-1} + \binom{n}{j} = \frac{n!}{(j-1)![n-(j-1)]!} + \frac{n!}{j!(n-j)!}$$

$$= \frac{n!}{(j-1)!(n-j+1)!} + \frac{n!}{j!(n-j)!}$$

$$= \frac{jn!}{j(j-1)!(n-j+1)!} + \frac{(n-j+1)n!}{j!(n-j+1)(n-j)!}$$

Multiply the first term by $\frac{j}{j}$ and the second term by $\frac{n-j+1}{n-j+1}$.

$$= \frac{jn!}{j!(n-j+1)!} + \frac{(n-j+1)n!}{j!(n-j+1)!}$$

Now the denominators are equal.

$$= \frac{jn! + (n-j+1)n!}{j!(n-j+1)!}$$

$$= \frac{n!(j+n-j+1)}{j!(n-j+1)!}$$

$$= \frac{n!(n+1)}{j!(n-j+1)!} = \frac{(n+1)!}{j![(n+1)-j]!} = \binom{n+1}{j}$$ ■

Historical Feature

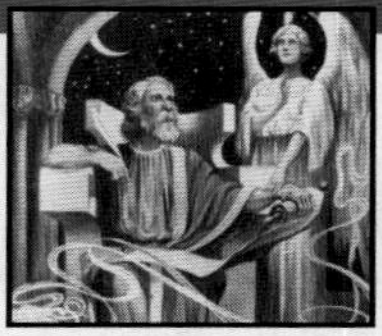

Omar Khayyám (1050–1123)

The case $n = 2$ of the Binomial Theorem, $(a + b)^2$, was known to Euclid in 300 BC, but the general law seems to have been discovered by the Persian mathematician and astronomer Omar Khayyám (1050–1123), who is also well known as the author of the *Rubáiyát*, a collection of four-line poems making observations on the human condition. Omar Khayyám did not state the Binomial Theorem explicitly, but he claimed to have a method for extracting third, fourth, fifth roots, and so on. A little study shows that one must know the Binomial Theorem to create such a method.

The heart of the Binomial Theorem is the formula for the numerical coefficients, and, as we saw, they can be written out in a symmetric triangular form. The Pascal triangle appears first in the books of Yang Hui (about 1270) and Chu Shih-chieh (1303). Pascal's name is attached to the triangle because of the many applications he made of it, especially to counting and probability. In establishing these results, he was one of the earliest users of mathematical induction.

Many people worked on the proof of the Binomial Theorem, which was finally completed for all n (including complex numbers) by Niels Abel (1802–1829).

11.5 Assess Your Understanding

Concepts and Vocabulary

1. The __________ __________ is a triangular display of the binomial coefficients.

2. $\binom{6}{2} =$ __________

3. *True of False:* $\binom{n}{j} = \dfrac{j!}{(n-j)!n!}$

4. The __________ __________ can be used to expand expressions like $(2x + 3)^6$.

Exercises

In Problems 5–16, evaluate each expression.

5. $\binom{5}{3}$	**6.** $\binom{7}{3}$	**7.** $\binom{7}{5}$	**8.** $\binom{9}{7}$
9. $\binom{50}{49}$	**10.** $\binom{100}{98}$	**11.** $\binom{1000}{1000}$	**12.** $\binom{1000}{0}$
13. $\binom{55}{23}$	**14.** $\binom{60}{20}$	**15.** $\binom{47}{25}$	**16.** $\binom{37}{19}$

In Problems 17–28, expand each expression using the Binomial Theorem.

17. $(x + 1)^5$	**18.** $(x - 1)^5$	**19.** $(x - 2)^6$	**20.** $(x + 3)^5$
21. $(3x + 1)^4$	**22.** $(2x + 3)^5$	**23.** $(x^2 + y^2)^5$	**24.** $(x^2 - y^2)^6$
25. $(\sqrt{x} + \sqrt{2})^6$	**26.** $(\sqrt{x} - \sqrt{3})^4$	**27.** $(ax + by)^5$	**28.** $(ax - by)^4$

In Problems 29–42, use the Binomial Theorem to find the indicated coefficient or term.

29. The coefficient of x^6 in the expansion of $(x + 3)^{10}$

30. The coefficient of x^3 in the expansion of $(x - 3)^{10}$

31. The coefficient of x^7 in the expansion of $(2x - 1)^{12}$

32. The coefficient of x^3 in the expansion of $(2x + 1)^{12}$

33. The coefficient of x^7 in the expansion of $(2x + 3)^9$

34. The coefficient of x^2 in the expansion of $(2x - 3)^9$

35. The fifth term in the expansion of $(x + 3)^7$

36. The third term in the expansion of $(x - 3)^7$

37. The third term in the expansion of $(3x - 2)^9$

38. The sixth term in the expansion of $(3x + 2)^8$

39. The coefficient of x^0 in the expansion of $\left(x^2 + \dfrac{1}{x}\right)^{12}$

40. The coefficient of x^0 in the expansion of $\left(x - \dfrac{1}{x^2}\right)^9$

41. The coefficient of x^4 in the expansion of $\left(x - \dfrac{2}{\sqrt{x}}\right)^{10}$

42. The coefficient of x^2 in the expansion of $\left(\sqrt{x} + \dfrac{3}{\sqrt{x}}\right)^8$

43. Use the Binomial Theorem to find the numerical value of $(1.001)^5$ correct to five decimal places.
[**Hint:** $(1.001)^5 = (1 + 10^{-3})^5$]

44. Use the Binomial Theorem to find the numerical value of $(0.998)^6$ correct to five decimal places.

45. Show that $\binom{n}{n-1} = n$ and $\binom{n}{n} = 1$.

46. Show that if n and j are integers with $0 \le j \le n$ then

$$\binom{n}{j} = \binom{n}{n-j}$$

Conclude that the Pascal triangle is symmetric with respect to a vertical line drawn from the topmost entry.

47. If n is a positive integer, show that

$$\binom{n}{0} + \binom{n}{1} + \cdots + \binom{n}{n} = 2^n$$

[**Hint:** $2^n = (1 + 1)^n$; now use the Binomial Theorem.]

48. If n is a positive integer, show that

$$\binom{n}{0} - \binom{n}{1} + \binom{n}{2} - \cdots + (-1)^n\binom{n}{n} = 0$$

49. $$\binom{5}{0}\left(\frac{1}{4}\right)^5 + \binom{5}{1}\left(\frac{1}{4}\right)^4\left(\frac{3}{4}\right) + \binom{5}{2}\left(\frac{1}{4}\right)^3\left(\frac{3}{4}\right)^2 + \binom{5}{3}\left(\frac{1}{4}\right)^2\left(\frac{3}{4}\right)^3 + \binom{5}{4}\left(\frac{1}{4}\right)\left(\frac{3}{4}\right)^4 + \binom{5}{5}\left(\frac{3}{4}\right)^5 = ?$$

50. Stirling's Formula An approximation for $n!$, when n is large, is given by

$$n! \approx \sqrt{2n\pi}\left(\frac{n}{e}\right)^n\left(1 + \frac{1}{12n - 1}\right)$$

Calculate $12!$, $20!$, and $25!$ on your calculator. Then use Stirling's formula to approximate $12!$, $20!$, and $25!$.

Chapter Review

Things to Know

Sequence (p. 784)	A function whose domain is the set of positive integers.
Factorials (p. 787)	$0! = 1, 1! = 1, n! = n(n-1)\cdot\ldots\cdot 3\cdot 2\cdot 1$ if $n \ge 2$
Arithmetic sequence (pp. 794 and 795)	$a_1 = a$, $a_n = a_{n-1} + d$, where a = first term, d = common difference, $a_n = a + (n-1)d$
Sum of the first n terms of an arithmetic sequence (p. 796)	$S_n = \frac{n}{2}[2a + (n-1)d] = \frac{n}{2}(a + a_n)$
Geometric sequence (pp. 799 and 801)	$a_1 = a$, $a_n = ra_{n-1}$, where a = first term, r = common ratio; $a_n = ar^{n-1}$, $r \ne 0$
Sum of the first n terms of a geometric sequence (p. 802)	$S_n = a\frac{1 - r^n}{1 - r}$, $r \ne 0, 1$
Infinite geometric series (p. 803)	$a + ar + \cdots + ar^{n-1} + \cdots = \sum_{k=1}^{\infty} ar^{k-1}$
Sum of an infinite geometric series (p. 804)	$\sum_{k=1}^{\infty} ar^{k-1} = \frac{a}{1-r}$, $\lvert r\rvert < 1$
Amount of an annuity (p. 806)	$A = P\frac{(1+i)^n - 1}{i}$
Principle of Mathematical Induction (p. 812)	Suppose the following two conditions are satisfied. Condition I: The statement is true for the natural number 1. Condition II: If the statement is true for some natural number k, it is also true for $k + 1$. Then the statement is true for all natural numbers n.
Binomial coefficient (p. 816)	$\binom{n}{j} = \frac{n!}{j!(n-j)!}$
Pascal triangle (p. 817)	See Figure 13.
Binomial Theorem (p. 818)	$(x + a)^n = \binom{n}{0}x^n + \binom{n}{1}ax^{n-1} + \cdots + \binom{n}{j}a^jx^{n-j} + \cdots + \binom{n}{n}a^n$

Objectives

Section		You should be able to . . .	Review Exercises
11.1	1	Write the first several terms of a sequence (p. 785)	1–4
	2	Write the terms of a sequence defined by a recursive formula (p. 787)	5–8
	3	Use summation notation (p. 788)	9–12
	4	Find the sum of a sequence (p. 790)	25–30
11.2	1	Determine if a sequence is arithmetic (p. 793)	13–24
	2	Find a formula for an arithmetic sequence (p. 795)	31, 32, 35, 37–40, 63, 64
	3	Find the sum of an arithmetic sequence (p. 796)	13, 14, 19, 20, 63, 64
11.3	1	Determine if a sequence is geometric (p. 799)	13–24
	2	Find a formula for geometric sequence (p. 801)	33, 34, 36, 65, 68
	3	Find the sum of a geometric sequence (p. 802)	17, 18, 21, 22
	4	Find the sum of a geometric series (p. 803)	41–46, 65(d)
	5	Solve annuity problems (p. 806)	66, 67
11.4	1	Prove statements using mathematical induction (p. 811)	47–52
11.5	1	Evaluate a binomial coefficient (p. 816)	53, 54
	2	Expand a binomial (p. 818)	55–62

Review Exercises

(Blue problem numbers indicate the author's suggestions for use in a Practice Test.)

In Problems 1–8, write down the first five terms of each sequence.

1. $\left\{(-1)^n\left(\frac{n+3}{n+2}\right)\right\}$ **2.** $\{(-1)^{n+1}(2n+3)\}$ 3. $\left\{\frac{2^n}{n^2}\right\}$ **4.** $\left\{\frac{e^n}{n}\right\}$

5. $a_1 = 3; \quad a_n = \frac{2}{3}a_{n-1}$ **6.** $a_1 = 4; \quad a_n = -\frac{1}{4}a_{n-1}$ 7. $a_1 = 2; \quad a_n = 2 - a_{n-1}$ **8.** $a_1 = -3; \quad a_n = 4 + a_{n-1}$

In Problems 9 and 10, write out each sum.

9. $\sum_{k=1}^{4}(4k+2)$ **10.** $\sum_{k=1}^{3}(3-k^2)$

In Problems 11 and 12, express each sum using summation notation.

11. $1 - \frac{1}{2} + \frac{1}{3} - \frac{1}{4} + \cdots + \frac{1}{13}$ **12.** $2 + \frac{2^2}{3} + \frac{2^3}{3^2} + \cdots + \frac{2^{n+1}}{3^n}$

In Problems 13–24, determine whether the given sequence is arithmetic, geometric, or neither. If the sequence is arithmetic, find the common difference and the sum of the first n terms. If the sequence is geometric, find the common ratio and the sum of the first n terms.

13. $\{n+5\}$ **14.** $\{4n+3\}$ **15.** $\{2n^3\}$ **16.** $\{2n^2-1\}$

17. $\{2^{3n}\}$ **18.** $\{3^{2n}\}$ 19. $0, 4, 8, 12, \ldots$ **20.** $1, -3, -7, -11, \ldots$

21. $3, \frac{3}{2}, \frac{3}{4}, \frac{3}{8}, \frac{3}{16}, \ldots$ **22.** $5, -\frac{5}{3}, \frac{5}{9}, -\frac{5}{27}, \frac{5}{81}, \ldots$ **23.** $\frac{2}{3}, \frac{3}{4}, \frac{4}{5}, \frac{5}{6}, \ldots$ **24.** $\frac{3}{2}, \frac{5}{4}, \frac{7}{6}, \frac{9}{8}, \frac{11}{10}, \ldots$

In Problems 25–30, evaluate each sum.

25. $\sum_{k=1}^{5}(k^2 + 12)$ **26.** $\sum_{k=1}^{3}(k + 2)^2$ 27. $\sum_{k=1}^{10}(3k - 9)$

28. $\sum_{k=1}^{9}(-2k + 8)$ 29. $\sum_{k=1}^{7}\left(\frac{1}{3}\right)^k$ **30.** $\sum_{k=1}^{10}(-2)^k$

In Problems 31–36, find the indicated term in each sequence.
[**Hint:** Find the general term first.]

31. 9th term of $3, 7, 11, 15, \ldots$ **32.** 8th term of $1, -1, -3, -5, \ldots$ **33.** 11th term of $1, \frac{1}{10}, \frac{1}{100}, \ldots$

34. 11th term of $1, 2, 4, 8, \ldots$ **35.** 9th term of $\sqrt{2}, 2\sqrt{2}, 3\sqrt{2}, \ldots$ **36.** 9th term of $\sqrt{2}, 2, 2^{3/2}, \ldots$

In Problems 37–40, find a general formula for each arithmetic sequence.

37. 7th term is 31; 20th term is 96 **38.** 8th term is -20; 17th term is -47

39. 10th term is 0; 18th term is 8 **40.** 12th term is 30; 22nd term is 50

In Problems 41–46, find the sum of each infinite geometric series.

41. $3 + 1 + \frac{1}{3} + \frac{1}{9} + \cdots$ **42.** $2 + 1 + \frac{1}{2} + \frac{1}{4} + \cdots$ **43.** $2 - 1 + \frac{1}{2} - \frac{1}{4} + \cdots$

44. $6 - 4 + \frac{8}{3} - \frac{16}{9} + \cdots$ 45. $\sum_{k=1}^{\infty} 4\left(\frac{1}{2}\right)^{k-1}$ **46.** $\sum_{k=1}^{\infty} 3\left(-\frac{3}{4}\right)^{k-1}$

In Problems 47–52, use the Principle of Mathematical Induction to show that the given statement is true for all natural numbers.

47. $3 + 6 + 9 + \cdots + 3n = \frac{3n}{2}(n + 1)$ **48.** $2 + 6 + 10 + \cdots + (4n - 2) = 2n^2$

49. $2 + 6 + 18 + \cdots + 2 \cdot 3^{n-1} = 3^n - 1$ **50.** $3 + 6 + 12 + \cdots + 3 \cdot 2^{n-1} = 3(2^n - 1)$

51. $1^2 + 4^2 + 7^2 + \cdots + (3n - 2)^2 = \frac{1}{2}n(6n^2 - 3n - 1)$ **52.** $1 \cdot 3 + 2 \cdot 4 + 3 \cdot 5 + \cdots + n(n + 2) = \frac{n}{6}(n + 1)(2n + 7)$

In Problems 53 and 54, evaluate each binomial coefficient.

53. $\binom{5}{2}$ **54.** $\binom{8}{6}$

In Problems 55–58, expand each expression using the Binomial Theorem.

55. $(x + 2)^5$ **56.** $(x - 3)^4$ 57. $(2x + 3)^5$ **58.** $(3x - 4)^4$

59. Find the coefficient of x^7 in the expansion of $(x + 2)^9$.

60. Find the coefficient of x^3 in the expansion of $(x - 3)^8$.

61. Find the coefficient of x^2 in the expansion of $(2x + 1)^7$.

62. Find the coefficient of x^6 in the expansion of $(2x + 1)^8$.

63. **Constructing a Brick Staircase** A brick staircase has a total of 25 steps. The bottom step requires 80 bricks. Each successive step requires three less bricks than the prior step.
(a) How many bricks are required for the top step?
(b) How many bricks are required to build the staircase?

64. Creating a Floor Design A mosaic tile floor is designed in the shape of a trapezoid 30 feet wide at the base and 15 feet wide at the top. The tiles, 12 inches by 12 inches, are to be placed so that each successive row contains one less tile than the row below. How many tiles will be required?

[**Hint:** Refer to Figure 8.]

65. Bouncing Balls A ball is dropped from a height of 20 feet. Each time it strikes the ground, it bounces up to three-quarters of the previous height.
(a) What height will the ball bounce up to after it strikes the ground for the third time?

(b) How high will it bounce after it strikes the ground for the nth time?
(c) How many times does the ball need to strike the ground before its bounce is less than 6 inches?
(d) What total distance does the ball travel before it stops bouncing?

66. Retirement Planning Chris gets paid once a month and contributes \$200 each pay period into his 401(k). If Chris plans on retiring in 20 years, what will be the value of his 401(k) if the per annum rate of return of the 401(k) is 10% compounded monthly?

67. Retirement Planning Jacky contributes \$500 every quarter to an IRA. If Jacky plans on retiring in 30 years, what will be the value of the IRA if the per annum rate of return of the IRA is 8% compounded quarterly?

68. Salary Increases Your friend has just been hired at an annual salary of \$20,000. If she expects to receive annual increases of 4%, what will be her salary as she begins her fifth year?

Chapter Projects

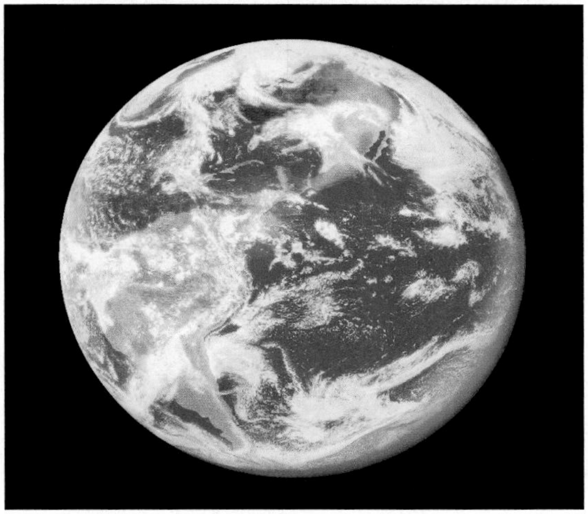

1. **Population Growth** The size of the population of the United States essentially depends on its current population, the birth and death rates of the population, and immigration. Suppose that b represents the birth rate of the U.S. population and d represents its death rate. Then $r = b - d$ represents the growth rate of the population, where r varies from year to year. The U.S. population after n years can be modeled using the recursive function

$$p_n = (1 + r)p_{n-1} + I$$

where I represents net immigration into the United States.

(a) Using data from the National Center for Health Statistics *http://www.fedstats.gov*, determine the birth and death rates for all races for the most recent year that data are available. Birth rates are given as the number of live births per 1000 population, while death rates are given as the number of deaths per 100,000 population. Each must be computed as the number of births (deaths) per individual. For example, in 1990, the birth rate was 16.7 per 1000 and the death rate was 863.8 per 100,000, so $b = \frac{16.7}{1000} = 0.0167$, while $d = \frac{863.8}{100{,}000} = 0.008638$.

Next, using data from the Immigration and Naturalization Service *http://www.fedstats.gov*, determine the immigration to the United States for the same year used to obtain b and d in part (a).

(b) Determine the value of r, the growth rate of the population.
(c) Find a recursive formula for the population of the United States.
(d) Use the recursive formula to predict the population of the United States in the following year. In other words, if data are available up to the year 2003, predict the U.S. population in 2004.
(e) Compare your prediction to actual data.
(f) Do you think the recursive formula found in part (c) will be useful in predicting future populations? Why or why not?

The following Chapter Projects may be found at www.prenhall.com/Sullivan7e.

2. **Project at Motorola** ***Digital Wireless Communication***
3. **Economics**
4. **Standardized Tests**

Cumulative Review

1. Find all the solutions, real and complex, of the equation $|x^2| = 9$.
2. (a) Graph the circle $x^2 + y^2 = 100$ and the parabola $y = 3x^2$.

   (b) Solve the system of equations: $\begin{cases} x^2 + y^2 = 100 \\ y = 3x^2 \end{cases}$

   (c) Where do the circle and the parabola intersect?
3. Solve the equation $2e^x = 5$.
4. Find an equation of the line with slope 5 and x-intercept 2.
5. Find the general equation of the circle whose center is the point $(-1, 2)$ if $(3, 5)$ is a point on the circle.
6. $f(x) = \dfrac{3x}{x - 2}$, $g(x) = 2x + 1$

   Find:

   (a) $(f \circ g)(2)$
   (b) $(g \circ f)(4)$
   (c) $(f \circ g)(x)$
   (d) The domain of $(f \circ g)(x)$
   (e) $(g \circ f)(x)$
   (f) The domain of $(g \circ f)(x)$
   (g) The function g^{-1} and its domain
   (h) The function f^{-1} and its domain
7. Find the equation of an ellipse with center at the origin, a focus at $(0, 3)$, and a vertex at $(0, 4)$.
8. Find the equation of a parabola with vertex at $(-1, 2)$ and focus at $(-1, 3)$.
9. Find the polar equation of a circle with center at $(0, 4)$ that passes through the pole. What is its rectangular equation?
10. Solve the equation $2\sin^2 x - \sin x - 3 = 0$, $0 < x < 2\pi$.
11. Find the exact value of $\cos^{-1}(-0.5)$.
12. If $\sin\theta = \dfrac{1}{4}$ and θ is in the second quadrant, find:

    (a) $\cos\theta$ (b) $\tan\theta$
    (c) $\sin(2\theta)$ (d) $\cos(2\theta)$
    (e) $\sin\left(\dfrac{1}{2}\theta\right)$

12 Counting and Probability

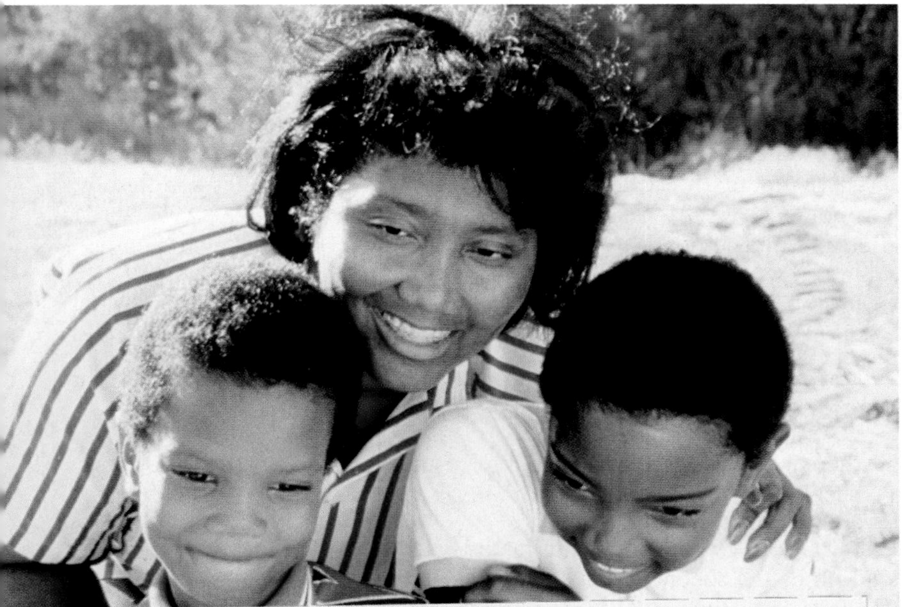

OUTLINE

The Two-children Problem

PROBLEM: A woman and a man (unrelated) each have two children. At least one of the woman's children is a boy and the man's older child is a boy. Do the chances that the woman has two boys equal the chances that the man has two boys?

The above problem was posed to Marilyn vos Savant in her column *Ask Marilyn*. Her original answer, based on theoretical probabilities, was that the chances that the woman has two boys are 1 in 3 and the chances that the man has two boys are 1 in 2. This is found by looking at the sample space of two-child families: BB, BG, GB, GG. In the case of the man, we know that his older child is a boy and the sample space reduces to BB and BG. Hence the probability that he has two boys is 1 out of 2. In the case of the woman, since we only know that she has at least one boy, the sample space reduces to BB, BG, and GB. Thus, her chances of having two boys is 1 out of 3.

The answer about the woman's chances created quite a bit of controversy resulting in many letters that challenged the correctness of her answer (*Parade*, July 27, 1997). Marilyn proposed that readers with exactly two children and at least one boy write in and tell the sex of both their children. (Reprinted with permission from PARADE and Marilyn vos Savant, copyright © 1997.)

—SEE CHAPTER PROJECT 1.

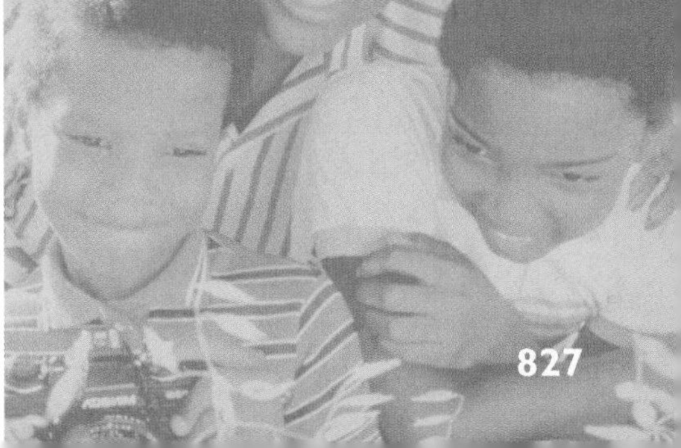

12.1 Sets and Counting

OBJECTIVES
1. Find All the Subsets of a Set
2. Find the Intersection and Union of Sets
3. Find the Complement of a Set
4. Count the Number of Elements in a Set

Sets

A **set** is a well-defined collection of distinct objects. The objects of a set are called its **elements.** By **well-defined,** we mean that there is a rule that enables us to determine whether a given object is an element of the set. If a set has no elements, it is called the **empty set,** or **null set,** and is denoted by the symbol $\emptyset$.

Because the elements of a set are distinct, we never repeat elements. For example, we would never write $\{1, 2, 3, 2\}$; the correct listing is $\{1, 2, 3\}$. Because a set is a collection, the order in which the elements are listed is immaterial. $\{1, 2, 3\}$, $\{1, 3, 2\}$, $\{2, 1, 3\}$, and so on, all represent the same set.

EXAMPLE 1 **Writing the Elements of a Set**

Write the set consisting of the possible results (outcomes) from tossing a coin twice. Use H for *heads* and T for *tails*.

Solution In tossing a coin twice, we can get heads each time, HH; or heads the first time and tails the second, HT; or tails the first time and heads the second, TH; or tails each time, TT. Because no other possibilities exist, the set of outcomes is

$$\{\text{HH}, \text{HT}, \text{TH}, \text{TT}\}$$ ◀

1 We now look at ways that two sets can be compared, beginning with set equality.

If two sets A and B have precisely the same elements, we say that A and B are **equal** and write $A = B$.

If each element of a set A is also an element of a set B, we say that A is a **subset** of B and write $A \subseteq B$.

If $A \subseteq B$ and $A \neq B$, then we say that A is a **proper subset** of B and write $A \subset B$.

If $A \subseteq B$, every element in set A is also in set B, but B may or may not have additional elements. If $A \subset B$, every element in A is also in B, and B has at least one element not found in A.

Finally, we agree that the empty set is a subset of every set; that is,

$$\emptyset \subseteq A, \qquad \text{for any set } A$$

EXAMPLE 2 **Finding All the Subsets of a Set**

Write down all the subsets of the set $\{a, b, c\}$.

Solution To organize our work, we write down all the subsets with no elements, then those with one element, then those with two elements, and finally those with three elements. These will give us all the subsets. Do you see why?

0 Elements	1 Element	2 Elements	3 Elements
$\emptyset$	$\{a\}, \{b\}, \{c\}$	$\{a, b\}, \{b, c\}, \{a, c\}$	$\{a, b, c\}$

NOW WORK PROBLEM 25.

2 If A and B are sets, the **intersection** of A with B, denoted $A \cap B$, is the set consisting of elements that belong to both A and B. The **union** of A with B, denoted $A \cup B$, is the set consisting of elements that belong to either A or B, or both.

EXAMPLE 3 Finding the Intersection and Union of Sets

Let $A = \{1, 3, 5, 8\}$, $B = \{3, 5, 7\}$, and $C = \{2, 4, 6, 8\}$. Find:

(a) $A \cap B$ (b) $A \cup B$ (c) $B \cap (A \cup C)$

Solution (a) $A \cap B = \{1, 3, 5, 8\} \cap \{3, 5, 7\} = \{3, 5\}$

(b) $A \cup B = \{1, 3, 5, 8\} \cup \{3, 5, 7\} = \{1, 3, 5, 7, 8\}$

(c) $B \cap (A \cup C) = \{3, 5, 7\} \cap [\{1, 3, 5, 8\} \cup \{2, 4, 6, 8\}]$
$= \{3, 5, 7\} \cap \{1, 2, 3, 4, 5, 6, 8\} = \{3, 5\}$

NOW WORK PROBLEM 9.

3 Usually, in working with sets, we designate a **universal set** U, the set consisting of all the elements that we wish to consider. Once a universal set has been designated, we can consider elements of the universal set not found in a given set.

If A is a set, the **complement** of A, denoted $\overline{A}$, is the set consisting of all the elements in the universal set that are not in A.

Note: Some books use the notation A' for the complement of A.

EXAMPLE 4 Finding the Complement of a Set

If the universal set is $U = \{1, 2, 3, 4, 5, 6, 7, 8, 9\}$, and if $A = \{1, 3, 5, 7, 9\}$, then $\overline{A} = \{2, 4, 6, 8\}$.

It follows that $A \cup \overline{A} = U$ and $A \cap \overline{A} = \emptyset$. Do you see why?

NOW WORK PROBLEM 17.

It is often helpful to draw pictures of sets. Such pictures, called **Venn diagrams,** represent sets as circles enclosed in a rectangle, which represents the universal set. Such diagrams often help us to visualize various relationships among sets. See Figure 1.

Figure 1

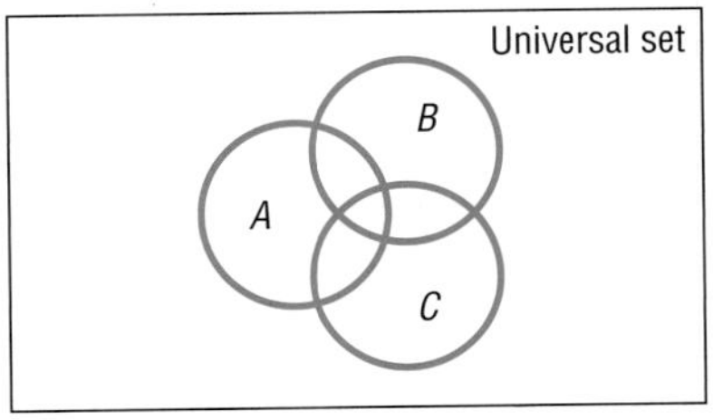

If we know that $A \subset B$, we might use the Venn diagram in Figure 2(a). If we know that A and B have no elements in common, that is, if $A \cap B = \varnothing$, we might use the Venn diagram in Figure 2(b). The sets A and B in Figure 2(b) are said to be **disjoint.**

Figure 2

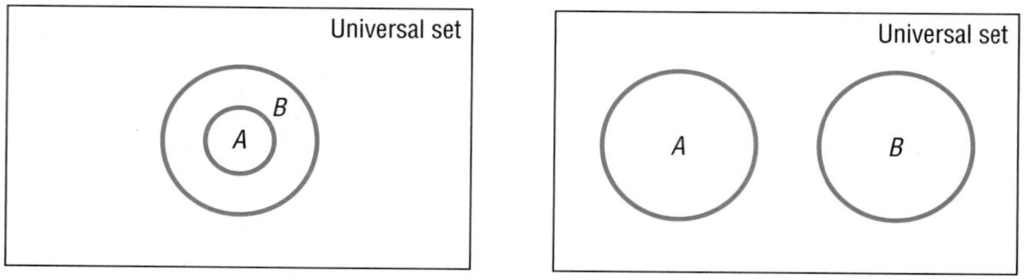

(a) $A \subset B$
proper subset

(b) $A \cap B = \varnothing$
disjoint sets

Figures 3(a), 3(b), and 3(c) use Venn diagrams to illustrate the definitions of intersection, union, and complement, respectively.

Figure 3

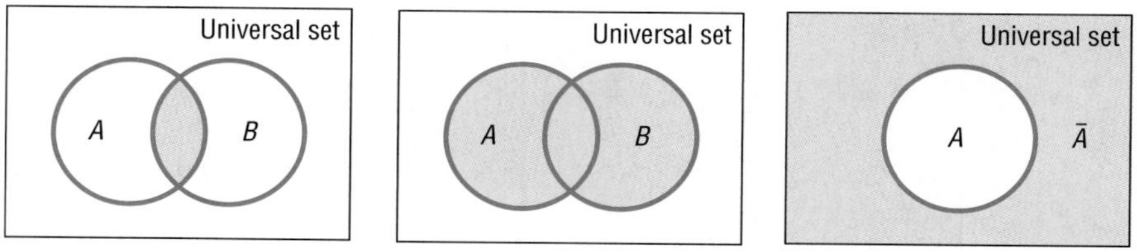

(a) $A \cap B$
intersection

(b) $A \cup B$
union

(c) $\bar{A}$
complement

Counting

4 As you count the number of students in a classroom or the number of pennies in your pocket, what you are really doing is matching, on a one-to-one basis, each object to be counted with the set of counting numbers $1, 2, 3, \ldots, n$, for some number n. If a set A matched up in this fashion with the set $\{1, 2, \ldots, 25\}$, you would conclude that there are 25 elements in the set A. We use the notation $n(A) = 25$ to indicate that there are 25 elements in the set A.

Because the empty set has no elements, we write

$$n(\varnothing) = 0$$

If the number of elements in a set is a nonnegative integer, we say that the set is **finite.** Otherwise, it is **infinite.** We shall concern ourselves only with finite sets.

Look again at Example 2. A set with 3 elements has $2^3 = 8$ subsets. This result can be generalized.

If A is a set with n elements, then A has 2^n subsets.

For example, the set $\{a, b, c, d, e\}$ has $2^5 = 32$ subsets.

EXAMPLE 5 Analyzing Survey Data

In a survey of 100 college students, 35 were registered in College Algebra, 52 were registered in Computer Science I, and 18 were registered in both courses.

(a) How many students were registered in College Algebra or Computer Science I?

(b) How many were registered in neither course?

Solution (a) First, let A = set of students in College Algebra
B = set of students in Computer Science I

Then the given information tells us that

$$n(A) = 35 \qquad n(B) = 52 \qquad n(A \cap B) = 18$$

Figure 4

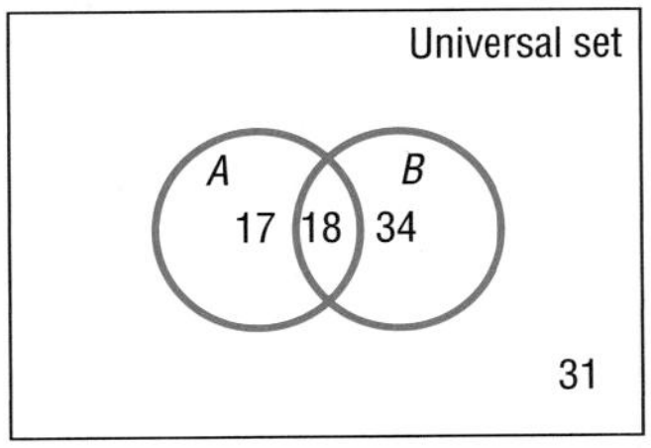

Refer to Figure 4. Since $n(A \cap B) = 18$, we know that the common part of the circles representing set A and set B has 18 elements. In addition, we know that the remaining portion of the circle representing set A will have $35 - 18 = 17$ elements. Similarly, we know that the remaining portion of the circle representing set B has $52 - 18 = 34$ elements. We conclude that $17 + 18 + 34 = 69$ students were registered in College Algebra or Computer Science I.

(b) Since 100 students were surveyed, it follows that $100 - 69 = 31$ were registered in neither course. ◀

NOW WORK PROBLEM 39

The solution to Example 5 contains the basis for a general counting formula. If we count the elements in each of two sets A and B, we necessarily count twice any elements that are in both A and B, that is, those elements in $A \cap B$. To count correctly the elements that are in A or B, that is, to find $n(A \cup B)$, we need to subtract those in $A \cap B$ from $n(A) + n(B)$.

Theorem

Counting Formula

If A and B are finite sets, then

$$n(A \cup B) = n(A) + n(B) - n(A \cap B) \qquad \textbf{(1)}$$

Refer back to Example 5. Using (1), we have

$$\begin{aligned} n(A \cup B) &= n(A) + n(B) - n(A \cap B) \\ &= 35 + 52 - 18 \\ &= 69 \end{aligned}$$

There are 69 students registered in College Algebra or Computer Science I.

A special case of the counting formula (1) occurs if A and B have no elements in common. In this case, $A \cap B = \varnothing$, so $n(A \cap B) = 0$.

Theorem

Addition Principle of Counting

If two sets A and B have no elements in common, that is,

$$\text{if } A \cap B = \varnothing, \text{ then } n(A \cup B) = n(A) + n(B) \qquad \textbf{(2)}$$

We can generalize formula (2).

Theorem

General Addition Principle of Counting

If, for n sets $A_1, A_2, \ldots, A_n$, no two have elements in common, then

$$n(A_1 \cup A_2 \cup \cdots \cup A_n) = n(A_1) + n(A_2) + \cdots + n(A_n) \qquad (3)$$

EXAMPLE 6 Counting

As of June, 2002, Federal agencies employed 93,445 full-time personnel authorized to make arrests and to carry firearms. Table 1 lists the type of law-enforcement officer and the corresponding number of full-time officers. No officer is classified as more than one type of officer.

Table 1

Type of Officer	Number of Full-time Federal Officers
Criminal (investigation/enforcement)	37,208
Police response and patrol	20,955
Corrections	16,915
Noncriminal (investigation/inspection)	12,801
Court operations	4,090
Security/protection	1,320
Other	156

SOURCE: Bureau of Justice Statistics

(a) How many full-time law-enforcement officers in the United States federal government were criminal officers or corrections officers?

(b) How many full-time law-enforcement officers in the United States federal government were criminal officers, corrections officers, or noncriminal officers?

Solution Let A represent the set of criminal officers, B represent the set of corrections officers, and C represent the set of noncriminal officers. No two of the sets A, B, and C have elements in common since a single officer cannot be classified as more than one type of officer. Then

$$n(A) = 37{,}208 \qquad n(B) = 16{,}915 \qquad n(C) = 12{,}801$$

(a) Using formula (2), we have

$$n(A \cup B) = n(A) + n(B) = 37{,}208 + 16{,}915 = 54{,}123$$

There were 54,123 officers that were criminal or corrections officers.

(b) Using formula (3), we have

$$n(A \cup B \cup C) = n(A) + n(B) + n(C) = 37{,}208 + 16{,}915 + 12{,}801 = 66{,}924$$

There were 66,924 officers that were criminal, corrections, or noncriminal officers. ◀

NOW WORK PROBLEM 43

12.1 Assess Your Understanding

Concepts and Vocabulary

1. The __________ of A and B consists of all elements in either A or B or both.
2. The __________ of A with B consists of all elements in both A and B.
3. *True or False:* The intersection of two sets is always a subset of their union.
4. *True or False:* If A is a set, the complement of A is the set of all the elements in the universal set that are not in A.

Exercises

In Problems 5–14, use $A = \{1, 3, 5, 7, 9\}$, $B = \{1, 5, 6, 7\}$, *and* $C = \{1, 2, 4, 6, 8, 9\}$ *to find each set.*

5. $A \cup B$
6. $A \cup C$
7. $A \cap B$
8. $A \cap C$
9. $(A \cup B) \cap C$
10. $(A \cap C) \cup (B \cap C)$
11. $(A \cap B) \cup C$
12. $(A \cup B) \cup C$
13. $(A \cup C) \cap (B \cup C)$
14. $(A \cap B) \cap C$

In Problems 15–24, use U = *universal set* $= \{0, 1, 2, 3, 4, 5, 6, 7, 8, 9\}$, $A = \{1, 3, 4, 5, 9\}$, $B = \{2, 4, 6, 7, 8\}$, *and* $C = \{1, 3, 4, 6\}$ *to find each set.*

15. $\overline{A}$
16. $\overline{C}$
17. $\overline{A \cap B}$
18. $\overline{B \cup C}$
19. $\overline{A} \cup \overline{B}$
20. $\overline{B} \cap \overline{C}$
21. $\overline{A \cap \overline{C}}$
22. $\overline{\overline{B} \cup C}$
23. $\overline{A \cup B \cup C}$
24. $\overline{A \cap B \cap C}$

25. Write down all the subsets of $\{a, b, c, d\}$.
26. Write down all the subsets of $\{a, b, c, d, e\}$.
27. If $n(A) = 15$, $n(B) = 20$, and $n(A \cap B) = 10$, find $n(A \cup B)$.
28. If $n(A) = 30$, $n(B) = 40$, and $n(A \cup B) = 45$, find $n(A \cap B)$.
29. If $n(A \cup B) = 50$, $n(A \cap B) = 10$, and $n(B) = 20$, find $n(A)$.
30. If $n(A \cup B) = 60$, $n(A \cap B) = 40$, and $n(A) = n(B)$, find $n(A)$.

In Problems 31–38, use the information given in the figure.

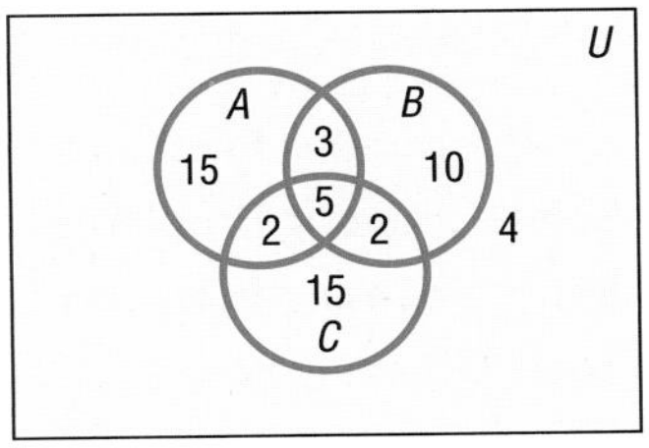

31. How many are in set A?
32. How many are in set B?
33. How many are in A or B?
34. How many are in A and B?
35. How many are in A but not C?
36. How many are not in A?
37. How many are in A and B and C?
38. How many are in A or B or C?

39. **Analyzing Survey Data** In a consumer survey of 500 people, 200 indicated that they would be buying a major appliance within the next month, 150 indicated that they would buy a car, and 25 said that they would purchase both a major appliance and a car. How many will purchase neither? How many will purchase only a car?

40. **Analyzing Survey Data** In a student survey, 200 indicated that they would attend Summer Session I and 150 indicated Summer Session II. If 75 students plan to attend both summer sessions and 275 indicated that they would attend neither session, how many students participated in the survey?

41. **Analyzing Survey Data** In a survey of 100 investors in the stock market,
    50 owned shares in IBM
    40 owned shares in AT&T
    45 owned shares in GE
    20 owned shares in both IBM and GE
    15 owned shares in both AT&T and GE
    20 owned shares in both IBM and AT&T
    5 owned shares in all three
    (a) How many of the investors surveyed did not have shares in any of the three companies?
    (b) How many owned just IBM shares?
    (c) How many owned just GE shares?
    (d) How many owned neither IBM nor GE?
    (e) How many owned either IBM or AT&T but no GE?

42. **Classifying Blood Types** Human blood is classified as either Rh+ or Rh−. Blood is also classified by type: A, if it contains an A antigen; B, if it contains a B antigen; AB, if it contains both A and B antigens; and O, if it contains neither antigen. Draw a Venn diagram illustrating the various blood types. Based on this classification, how many different kinds of blood are there?

43. The following data represent the marital status of males 18 years old and older in March 1997.

Marital Status	Number (in thousands)
Married, spouse present	54,654
Married, spouse absent	3,232
Widowed	2,686
Divorced	8,208
Never married	25,375

SOURCE: Current Population Survey

(a) Determine the number of males 18 years old and older who are married.
(b) Determine the number of males 18 years old and older who are widowed or divorced.
(c) Determine the number of males 18 years old and older who are married, spouse absent, widowed, or divorced.

44. The following data represent the marital status of females 18 years old and older in March 1997.

Marital Status	Number (in thousands)
Married, spouse present	54,626
Married, spouse absent	4,122
Widowed	11,056
Divorced	11,107
Never married	20,503

SOURCE: Current Population Survey

(a) Determine the number of females 18 years old and older who are married.
(b) Determine the number of females 18 years old and older who are widowed or divorced.
(c) Determine the number of females 18 years old and older who are married, spouse absent, widowed, or divorced.

45. Make up a problem different from any found in the text that requires the addition principle of counting to solve. Give it to a friend to solve and critique.

46. Investigate the notion of counting as it relates to infinite sets. Write an essay on your findings.

12.2 Permutations and Combinations

PREPARING FOR THIS SECTION *Before getting started, review the following:*

- Factorial (Section 11.1, p. 787)

Now work the 'Are You Prepared?' problems on page 842.

OBJECTIVES
1. Solve Counting Problems Using the Multiplication Principle
2. Solve Counting Problems Using Permutations
3. Solve Counting Problems Using Combinations
4. Solve Counting Problems Using Permutations Involving n Nondistinct Objects

1 Counting plays a major role in many diverse areas, such as probability, statistics, and computer science; counting techniques are a part of a branch of mathematics called **combinatorics.** In this section we shall look at special types of counting problems and develop general formulas for solving them.

We begin with an example that will demonstrate a general counting principle.

EXAMPLE 1 Counting the Number of Possible Meals

The fixed-price dinner at Mabenka Restaurant provides the following choices:

Appetizer:	soup or salad
Entree:	baked chicken, broiled beef patty, baby beef liver, or roast beef au jus
Dessert:	ice cream or cheese cake

How many different meals can be ordered?

Solution Ordering such a meal requires three separate decisions:

Choose an Appetizer	**Choose an Entree**	**Choose a Dessert**
2 choices	4 choices	2 choices

Look at the **tree diagram** in Figure 5. We see that, for each choice of appetizer, there are 4 choices of entrees. And for each of these $2 \cdot 4 = 8$ choices, there are 2 choices for dessert. A total of

$$2 \cdot 4 \cdot 2 = 16$$

different meals can be ordered.

Figure 5

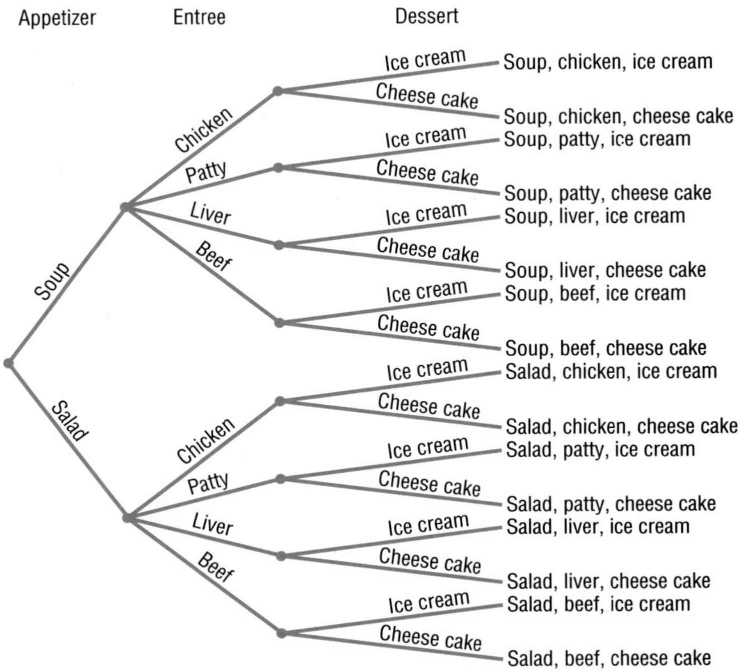

◄

Theorem

Multiplication Principle of Counting

If a task consists of a sequence of choices in which there are p selections for the first choice, q selections for the second choice, r selections for the third choice, and so on, then the task of making these selections can be done in

$$p \cdot q \cdot r \cdot \ldots$$

different ways.

EXAMPLE 2 Forming Codes

How many two-symbol codewords can be formed if the first symbol is a letter (uppercase) and the second symbol is a digit?

Solution It sometimes helps to begin by listing some of the possibilities. The code consists of a letter (uppercase) followed by a digit, so some possibilities are A1, A2, B3, X0, and so on. The task consists of making two selections: the first selection requires choosing an uppercase letter (26 choices) and the second task requires choosing a digit (10 choices). By the Multiplication Principle, there are

$$26 \cdot 10 = 260$$

different codewords of the type described. ◀

NOW WORK PROBLEM 31.

Permutations

2 We begin with a definition.

> A **permutation** is an ordered arrangement of r objects chosen from n objects.

We discuss three types of permutations:

1. The n objects are distinct (different), and repetition is allowed in the selection of r of them. [Distinct, with repetition]
2. The n objects are distinct (different), and repetition is not allowed in the selection of r of them, where $r \leq n$. [Distinct, without repetition]
3. The n objects are not distinct, and we use all of them in the arrangement. [Not distinct]

We take up the first two types here and deal with the third type at the end of this section.

The first type of permutation is handled using the Multiplication Principle.

EXAMPLE 3 Counting Airport Codes [Permutation: Distinct, with Repetition]

The International Airline Transportation Association (IATA) assigns three-letter codes to represent airport locations. For example, the airport code for Ft. Lauderdale, Florida, is FLL. Notice that repetition is allowed in forming this code. How many airport codes are possible?

Solution We are choosing 3 letters from 26 letters and arranging them in order. In the ordered arrangement a letter may be repeated. This is an example of a permutation with repetition in which 3 objects are chosen from 26 distinct objects.

The task of counting the number of such arrangements consists of making three selections. Each selection requires choosing a letter of the alphabet (26 choices). By the Multiplication Principle, there are

$$26 \cdot 26 \cdot 26 = 17{,}576$$

different airport codes. ◀

The solution given to Example 3 can be generalized.

Theorem

Permutations: Distinct Objects with Repetition

The number of ordered arrangements of r objects chosen from n objects, in which the n objects are distinct and repetition is allowed, is n^r.

NOW WORK PROBLEM 35.

We begin the discussion of permutations in which the objects are distinct and repetition is not allowed with an example.

EXAMPLE 4 Forming Codes [Permutation: Distinct, without Repetition]

Suppose that we wish to establish a three-letter code using any of the 26 uppercase letters of the alphabet, but we require that no letter be used more than once. How many different three-letter codes are there?

Solution Some of the possibilities are: ABC, ABD, ABZ, ACB, CBA, and so on. The task consists of making three selections. The first selection requires choosing from 26 letters. Because no letter can be used more than once, the second selection requires choosing from 25 letters. The third selection requires choosing from 24 letters. (Do you see why?) By the Multiplication Principle, there are

$$26 \cdot 25 \cdot 24 = 15{,}600$$

different three-letter codes with no letter repeated. ◀

For the second type of permutation, we introduce the following symbol.

The symbol $\boldsymbol{P(n, r)}$ represents the number of ordered arrangements of r objects chosen from n distinct objects, where $r \leq n$ and repetition is not allowed.

For example, the question posed in Example 4 asks for the number of ways that the 26 letters of the alphabet can be arranged in order using three nonrepeated letters. The answer is

$$P(26, 3) = 26 \cdot 25 \cdot 24 = 15{,}600$$

EXAMPLE 5 Lining Up People

In how many ways can 5 people be lined up?

Solution The 5 people are distinct. Once a person is in line, that person will not be repeated elsewhere in the line; and, in lining up people, order is important. We have a permutation of 5 objects taken 5 at a time. We can line up 5 people in

$$P(5, 5) = \underbrace{5 \cdot 4 \cdot 3 \cdot 2 \cdot 1}_{\text{5 factors}} = 120 \text{ ways}$$

◀

NOW WORK PROBLEM 37.

To arrive at a formula for $P(n, r)$, we note that the task of obtaining an ordered arrangement of n objects in which only $r \le n$ of them are used, without repeating any of them, requires making r selections. For the first selection, there are n choices; for the second selection, there are $n - 1$ choices; for the third selection, there are $n - 2$ choices; ...; for the rth selection, there are $n - (r - 1)$ choices. By the Multiplication Principle, we have

$$\begin{aligned} & \quad\ \ \text{1st} \quad \text{2nd} \qquad \text{3rd} \qquad\qquad\qquad \textit{r}\text{th} \\ P(n, r) &= n \cdot (n - 1) \cdot (n - 2) \cdot \ldots \cdot [n - (r - 1)] \\ &= n \cdot (n - 1) \cdot (n - 2) \cdot \ldots \cdot (n - r + 1) \end{aligned}$$

This formula for $P(n, r)$ can be compactly written using factorial notation.*

$$\begin{aligned} P(n, r) &= n \cdot (n - 1) \cdot (n - 2) \cdot \ldots \cdot (n - r + 1) \\ &= n \cdot (n - 1) \cdot (n - 2) \cdot \ldots \cdot (n - r + 1) \cdot \frac{(n - r) \cdot \ldots \cdot 3 \cdot 2 \cdot 1}{(n - r) \cdot \ldots \cdot 3 \cdot 2 \cdot 1} = \frac{n!}{(n - r)!} \end{aligned}$$

Theorem

Permutations of *r* Objects Chosen from *n* Distinct Objects without Repetition

The number of arrangements of n objects using $r \le n$ of them, in which

1. the n objects are distinct,
2. once an object is used it cannot be repeated, and
3. order is important,

is given by the formula

$$P(n, r) = \frac{n!}{(n - r)!} \qquad \textbf{(1)}$$

EXAMPLE 6 **Computing Permutations**

Evaluate: (a) $P(7, 3)$ (b) $P(6, 1)$ (c) $P(52, 5)$

Solution We shall work parts (a) and (b) in two ways.

(a) $P(7, 3) = \underbrace{7 \cdot 6 \cdot 5}_{\text{3 factors}} = 210$

or

$$P(7, 3) = \frac{7!}{(7 - 3)!} = \frac{7!}{4!} = \frac{7 \cdot 6 \cdot 5 \cdot \cancel{4!}}{\cancel{4!}} = 210$$

*Recall that $0! = 1, 1! = 1, 2! = 2 \cdot 1, \ldots, n! = n(n - 1) \cdot \ldots \cdot 3 \cdot 2 \cdot 1$.

Figure 6

(b) $P(6, 1) = \underbrace{6}_{\text{1 factor}} = 6$

or

$$P(6, 1) = \frac{6!}{(6-1)!} = \frac{6!}{5!} = \frac{6 \cdot \cancel{5!}}{\cancel{5!}} = 6$$

(c) Figure 6 shows the solution using a TI-83 graphing calculator: $P(52, 5) = 311{,}875{,}200$. ◀

NOW WORK PROBLEM 7.

EXAMPLE 7 The Birthday Problem

All we know about Shannon, Patrick, and Ryan is that they have different birthdays. If we listed all the possible ways this could occur, how many would there be? Assume that there are 365 days in a year.

Solution This is an example of a permutation in which 3 birthdays are selected from a possible 365 days, and no birthday may repeat itself. The number of ways that this can occur is

$$P(365, 3) = \frac{365!}{(365-3)!} = \frac{365 \cdot 364 \cdot 363 \cdot \cancel{362!}}{\cancel{362!}} = 365 \cdot 364 \cdot 363 = 48{,}228{,}180$$

There are 48,228,180 ways in a group of three people that each has a different birthday. ◀

NOW WORK PROBLEM 53.

Combinations

3 In a permutation, order is important. For example, the arrangements ABC, CAB, $BAC, \ldots$ are considered different arrangements of the letters A, B, and C. In many situations, though, order is unimportant. For example, in the card game of poker, the order in which the cards are received does not matter; it is the *combination* of the cards that matters.

A **combination** is an arrangement, without regard to order, of r objects selected from n distinct objects without repetition, where $r \le n$. The symbol $\boldsymbol{C(n, r)}$ represents the number of combinations of n distinct objects using r of them.

EXAMPLE 8 Listing Combinations

List all the combinations of the 4 objects a, b, c, d taken 2 at a time. What is $C(4, 2)$?

Solution One combination of a, b, c, d taken 2 at a time is

$$ab$$

We exclude ba from the list because order is not important in a combination. The list of all such combinations (convince yourself of this) is

$$ab, \quad ac, \quad ad, \quad bc, \quad bd, \quad cd$$

so

$$C(4, 2) = 6$$ ◀

We can find a formula for $C(n, r)$ by noting that the only difference between a permutation of type 2 (distinct, without repetition) and a combination is that we disregard order in combinations. To determine $C(n, r)$, we need only eliminate from the formula for $P(n, r)$ the number of permutations that were simply rearrangements of a given set of r objects. This can be determined from the formula for $P(n, r)$ by calculating $P(r, r) = r!$. So, if we divide $P(n, r)$ by $r!$, we will have the desired formula for $C(n, r)$:

$$C(n, r) = \frac{P(n, r)}{r!} \underset{\uparrow}{=} \frac{\dfrac{n!}{(n-r)!}}{r!} = \frac{n!}{(n-r)!\,r!}$$

Use formula (1).

We have proved the following result:

Theorem

Number of Combinations of *n* Distinct Objects Taken *r* at a Time

The number of arrangements of n objects using $r \le n$ of them, in which

1. the n objects are distinct,
2. once an object is used, it cannot be repeated, and
3. order is not important,

is given by the formula

$$C(n, r) = \frac{n!}{(n-r)!\,r!} \qquad \textbf{(2)}$$

Based on formula (2), we discover that the symbol $C(n, r)$ and the symbol $\binom{n}{r}$ for the binomial coefficients are, in fact, the same. The Pascal triangle (see Section 11.5) can be used to find the value of $C(n, r)$. However, because it is more practical and convenient, we will use formula (2) instead.

EXAMPLE 9 Using Formula (2)

Use formula (2) to find the value of each expression.

(a) $C(3, 1)$ (b) $C(6, 3)$ (c) $C(n, n)$ (d) $C(n, 0)$ (e) $C(52, 5)$

Solution

(a) $C(3, 1) = \dfrac{3!}{(3-1)!\,1!} = \dfrac{3!}{2!\,1!} = \dfrac{3 \cdot \cancel{2} \cdot 1}{\cancel{2} \cdot 1 \cdot 1} = 3$

(b) $C(6, 3) = \dfrac{6!}{(6-3)!\,3!} = \dfrac{6 \cdot 5 \cdot 4 \cdot \cancel{3!}}{3! \cdot \cancel{3!}} = \dfrac{\cancel{6} \cdot 5 \cdot 4}{\cancel{6}} = 20$

(c) $C(n, n) = \dfrac{n!}{(n-n)!\,n!} = \dfrac{\cancel{n!}}{0!\,\cancel{n!}} = \dfrac{1}{1} = 1$

(d) $C(n, 0) = \dfrac{n!}{(n-0)!\,0!} = \dfrac{\cancel{n!}}{\cancel{n!}\,0!} = \dfrac{1}{1} = 1$

(e) Figure 7 shows the solution using a TI-83 graphing calculator: $C(52, 5) = 2{,}598{,}960$.

Figure 7

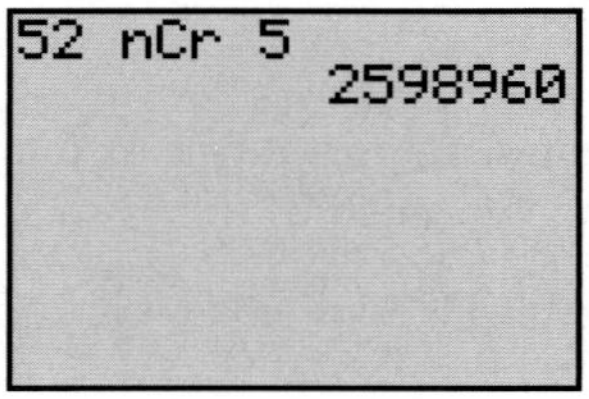

NOW WORK PROBLEM 15.

EXAMPLE 10 **Forming Committees**

How many different committees of 3 people can be formed from a pool of 7 people?

Solution The 7 people are distinct. More important, though, is the observation that the order of being selected for a committee is not significant. The problem asks for the number of combinations of 7 objects taken 3 at a time.

$$C(7, 3) = \frac{7!}{4!\,3!} = \frac{7 \cdot 6 \cdot 5 \cdot \cancel{4!}}{\cancel{4!}3!} = \frac{7 \cdot \cancel{6} \cdot 5}{\cancel{6}} = 35$$

◀

EXAMPLE 11 **Forming Committees**

In how many ways can a committee consisting of 2 faculty members and 3 students be formed if 6 faculty members and 10 students are eligible to serve on the committee?

Solution The problem can be separated into two parts: the number of ways that the faculty members can be chosen, $C(6, 2)$, and the number of ways that the student members can be chosen, $C(10, 3)$. By the Multiplication Principle, the committee can be formed in

$$C(6, 2) \cdot C(10, 3) = \frac{6!}{4!\,2!} \cdot \frac{10!}{7!\,3!} = \frac{6 \cdot 5 \cdot \cancel{4!}}{\cancel{4!}2!} \cdot \frac{10 \cdot 9 \cdot 8 \cdot \cancel{7!}}{\cancel{7!}3!}$$
$$= \frac{30}{2} \cdot \frac{720}{6} = 1800 \text{ ways}$$

◀

NOW WORK PROBLEM 55.

Permutations Involving *n* Objects That Are Not Distinct

4 We begin with an example.

EXAMPLE 12 **Forming Different Words**

How many different words (real or imaginary) can be formed using all the letters in the word REARRANGE?

Solution Each word formed will have 9 letters: 3 R's, 2 A's, 2 E's, 1 N, and 1 G. To construct each word, we need to fill in 9 positions with the 9 letters:

$$\overline{1} \quad \overline{2} \quad \overline{3} \quad \overline{4} \quad \overline{5} \quad \overline{6} \quad \overline{7} \quad \overline{8} \quad \overline{9}$$

The process of forming a word consists of five tasks:

Task 1: Choose the positions for the 3 R's.
Task 2: Choose the positions for the 2 A's.
Task 3: Choose the positions for the 2 E's.
Task 4: Choose the position for the 1 N.
Task 5: Choose the position for the 1 G.

Task 1 can be done in $C(9, 3)$ ways. There then remain 6 positions to be filled, so Task 2 can be done in $C(6, 2)$ ways. There remain 4 positions to be filled, so Task 3 can be done in $C(4, 2)$ ways. There remain 2 positions to be filled, so Task 4 can be done in $C(2, 1)$ ways. The last position can be filled in $C(1, 1)$

way. Using the Multiplication Principle, the number of possible words that can be formed is

$$C(9,3)\cdot C(6,2)\cdot C(4,2)\cdot C(2,1)\cdot C(1,1) = \frac{9!}{3!\cdot \cancel{6!}}\cdot\frac{\cancel{6!}}{2!\cdot\cancel{4!}}\cdot\frac{\cancel{4!}}{2!\cdot\cancel{2!}}\cdot\frac{\cancel{2!}}{1!\cdot\cancel{1!}}\cdot\frac{\cancel{1!}}{0!\cdot 1!}$$

$$= \frac{9!}{3!\cdot 2!\cdot 2!\cdot 1!\cdot 1!}$$ ◀

The form of the answer to Example 12 is suggestive of a general result. Had the letters in REARRANGE each been different, there would have been $P(9, 9) = 9!$ possible words formed. This is the numerator of the answer. The presence of 3 R's, 2 A's, and 2 E's reduces the number of different words, as the entries in the denominator illustrate. We are led to the following result:

Theorem

Permutations Involving *n* Objects That Are Not Distinct

The number of permutations of n objects of which n_1 are of one kind, n_2 are of a second kind, . . . , and n_k are of a kth kind is given by

$$\frac{n!}{n_1!\cdot n_2!\cdot \ldots \cdot n_k!} \qquad \textbf{(3)}$$

where $n = n_1 + n_2 + \cdots + n_k$.

EXAMPLE 13 **Arranging Flags**

How many different vertical arrangements are there of 8 flags if 4 are white, 3 are blue, and 1 is red?

Solution We seek the number of permutations of 8 objects, of which 4 are of one kind, 3 of a second kind, and 1 of a third kind. Using formula (3), we find that there are

$$\frac{8!}{4!\cdot 3!\cdot 1!} = \frac{8\cdot 7\cdot 6\cdot 5\cdot \cancel{4!}}{\cancel{4!}\cdot 3!\cdot 1!} = 280 \text{ different arrangements}$$ ◀

NOW WORK PROBLEM 57.

12.2 Assess Your Understanding

'Are You Prepared?' *Answers are given at the end of these exercises. If you get a wrong answer, read the pages listed in red.*

1. $0! =$ ________; $1! =$ ________. (p. 787)

2. *True or False:* $n! = \dfrac{(n+1)!}{n}$. (p. 787)

Concepts and Vocabulary

3. A(n) ________ is an ordered arrangement of r objects chosen from n objects.

4. A(n) ________ is an arrangement of r objects chosen from n distinct objects, without repetition and without regard to order.

5. *True or False:* In a combination problem, order is not important.

6. *True or False:* In some permutation problems, once an object is used, it cannot be repeated.

Exercises

In Problems 7–14, find the value of each permutation.

7. $P(6, 2)$
8. $P(7, 2)$
9. $P(4, 4)$
10. $P(8, 8)$
11. $P(7, 0)$
12. $P(9, 0)$
13. $P(8, 4)$
14. $P(8, 3)$

In Problems 15–22, use formula (2) to find the value of each combination.

15. $C(8, 2)$
16. $C(8, 6)$
17. $C(7, 4)$
18. $C(6, 2)$
19. $C(15, 15)$
20. $C(18, 1)$
21. $C(26, 13)$
22. $C(18, 9)$

23. List all the ordered arrangements of 5 objects $a, b, c, d,$ and e choosing 3 at a time without repetition. What is $P(5, 3)$?
24. List all the ordered arrangements of 5 objects $a, b, c, d,$ and e choosing 2 at a time without repetition. What is $P(5, 2)$?
25. List all the ordered arrangements of 4 objects 1, 2, 3, and 4 choosing 3 at a time without repetition. What is $P(4, 3)$?
26. List all the ordered arrangements of 6 objects 1, 2, 3, 4, 5, and 6 choosing 3 at a time without repetition. What is $P(6, 3)$?
27. List all the combinations of 5 objects a, b, c, d, and e taken 3 at a time. What is $C(5, 3)$?
28. List all the combinations of 5 objects a, b, c, d, and e taken 2 at a time. What is $C(5, 2)$?
29. List all the combinations of 4 objects 1, 2, 3, and 4 taken 3 at a time. What is $C(4, 3)$?
30. List all the combinations of 6 objects 1, 2, 3, 4, 5, and 6 taken 3 at a time. What is $C(6, 3)$?
31. **Shirts and Ties** A man has 5 shirts and 3 ties. How many different shirt and tie arrangements can he wear?
32. **Blouses and Skirts** A woman has 3 blouses and 5 skirts. How many different outfits can she wear?
33. **Forming Codes** How many two-letter codes can be formed using the letters A, B, C, and D? Repeated letters are allowed.
34. **Forming Codes** How many two-letter codes can be formed using the letters $A, B, C, D,$ and E? Repeated letters are allowed.
35. **Forming Numbers** How many three-digit numbers can be formed using the digits 0 and 1? Repeated digits are allowed.
36. **Forming Numbers** How many three-digit numbers can be formed using the digits 0, 1, 2, 3, 4, 5, 6, 7, 8, and 9? Repeated digits are allowed.
37. **Lining People Up** In how many ways can 4 people be lined up?
38. **Stacking Boxes** In how many ways can 5 different boxes be stacked?
39. **Forming Codes** How many different three-letter codes are there if only the letters $A, B, C, D,$ and E can be used and no letter can be used more than once?
40. **Forming Codes** How many different four-letter codes are there if only the letters A, B, C, D, E, and F can be used and no letter can be used more than once?
41. **Stocks on the NYSE** Companies whose stocks are listed on the New York Stock Exchange (NYSE) have their company name represented by either 1, 2, or 3 letters (repetition of letters is allowed). What is the maximum number of companies that can be listed on the NYSE?
42. **Stocks on the NASDAQ** Companies whose stocks are listed on the NASDAQ stock exchange have their company name represented by either 4 or 5 letters (repetition of letters is allowed). What is the maximum number of companies that can be listed on the NASDAQ?
43. **Establishing Committees** In how many ways can a committee of 4 students be formed from a pool of 7 students?
44. **Establishing Committees** In how many ways can a committee of 3 professors be formed from a department having 8 professors?
45. **Possible Answers on a True/False Test** How many arrangements of answers are possible for a true/false test with 10 questions?
46. **Possible Answers on a Multiple-choice Test** How many arrangements of answers are possible in a multiple-choice test with 5 questions, each of which has 4 possible answers?
47. **Four-digit Numbers** How many four-digit numbers can be formed using the digits 0, 1, 2, 3, 4, 5, 6, 7, 8, and 9 if the first digit cannot be 0? Repeated digits are allowed.
48. **Five-digit Numbers** How many five-digit numbers can be formed using the digits 0, 1, 2, 3, 4, 5, 6, 7, 8, and 9 if the first digit cannot be 0 or 1? Repeated digits are allowed.
49. **Arranging Books** Five different mathematics books are to be arranged on a student's desk. How many arrangements are possible?

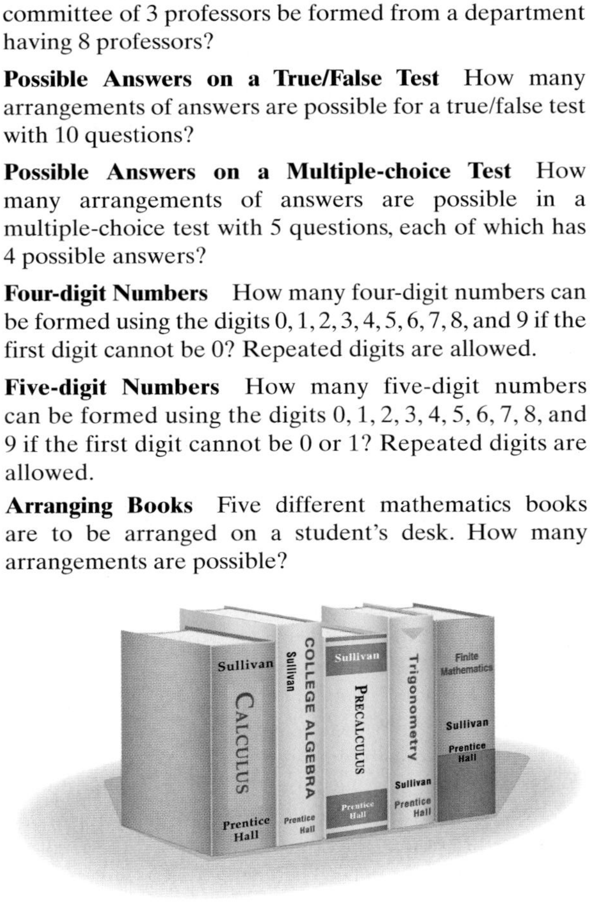

50. **Forming License Plate Numbers** How many different license plate numbers can be made using 2 letters followed by 4 digits selected from the digits 0 through 9, if
    (a) Letters and digits may be repeated?
    (b) Letters may be repeated, but digits may not be repeated?
    (c) Neither letters nor digits may be repeated?

51. **Stock Portfolios** As a financial planner, you are asked to select one stock each from the following groups: 8 DOW stocks, 15 NASDAQ stocks, and 4 global stocks. How many different portfolios are possible?

52. **Combination Locks** A combination lock displays 50 numbers. To open it, you turn to a number, then rotate clockwise to a second number, and then counterclockwise to the third number. How many different lock combinations are there?

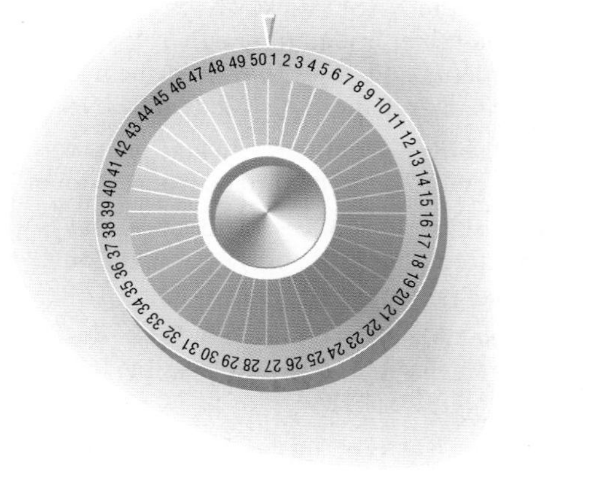

53. **Birthday Problem** In how many ways can 2 people each have different birthdays? Assume that there are 365 days in a year.

54. **Birthday Problem** In how many ways can 5 people each have different birthdays? Assume that there are 365 days in a year.

55. **Forming a Committee** A student dance committee is to be formed consisting of 2 boys and 3 girls. If the membership is to be chosen from 4 boys and 8 girls, how many different committees are possible?

56. **Forming a Committee** The student relations committee of a college consists of 2 administrators, 3 faculty members, and 5 students. Four administrators, 8 faculty members, and 20 students are eligible to serve. How many different committees are possible?

57. **Forming Words** How many different 9-letter words (real or imaginary) can be formed from the letters in the word ECONOMICS?

58. **Forming Words** How many different 11-letter words (real or imaginary) can be formed from the letters in the word MATHEMATICS?

59. **Selecting Objects** An urn contains 7 white balls and 3 red balls. Three balls are selected. In how many ways can the 3 balls be drawn from the total of 10 balls:
    (a) If 2 balls are white and 1 is red?
    (b) If all 3 balls are white?
    (c) If all 3 balls are red?

60. **Selecting Objects** An urn contains 15 red balls and 10 white balls. Five balls are selected. In how many ways can the 5 balls be drawn from the total of 25 balls:
    (a) If all 5 balls are red?
    (b) If 3 balls are red and 2 are white?
    (c) If at least 4 are red balls?

61. **Senate Committees** The U.S. Senate has 100 members. Suppose that it is desired to place each senator on exactly 1 of 7 possible committees. The first committee has 22 members, the second has 13, the third has 10, the fourth has 5, the fifth has 16, and the sixth and seventh have 17 apiece. In how many ways can these committees be formed?

62. **Football Teams** A defensive football squad consists of 25 players. Of these, 10 are linemen, 10 are linebackers, and 5 are safeties. How many different teams of 5 linemen, 3 linebackers, and 3 safeties can be formed?

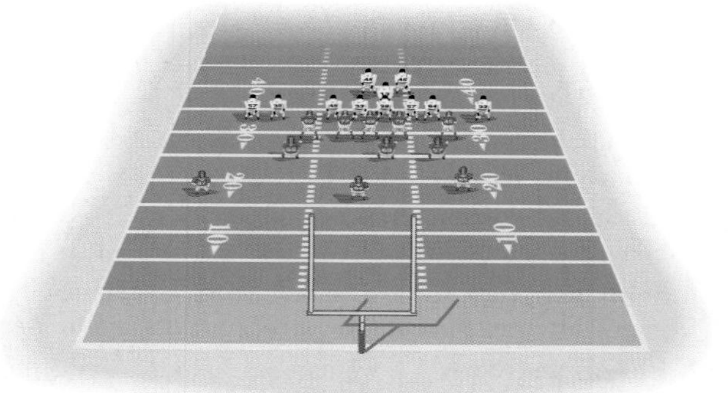

63. **Baseball** In the American Baseball League, a designated hitter may be used. How many batting orders is it possible for a manager to use? (There are 9 regular players on a team.)

64. **Baseball** In the National Baseball League, the pitcher usually bats ninth. If this is the case, how many batting orders is it possible for a manager to use?

65. **Baseball Teams** A baseball team has 15 members. Four of the players are pitchers, and the remaining 11 members can play any position. How many different teams of 9 players can be formed?

66. **World Series** In the World Series the American League team (A) and the National League team (N) play until one team wins four games. If the sequence of winners is designated by letters (for example, $NAAAA$ means that the National League team won the first game and the American League won the next four), how many different sequences are possible?

67. **Basketball Teams** A basketball team has 6 players who play guard (2 of 5 starting positions). How many different teams are possible, assuming that the remaining 3 positions are filled and it is not possible to distinguish a left guard from a right guard?

68. Basketball Teams On a basketball team of 12 players, 2 only play center, 3 only play guard, and the rest play forward (5 players on a team: 2 forwards, 2 guards, and 1 center). How many different teams are possible, assuming that it is not possible to distinguish left and right guards and left and right forwards?

69. Create a problem different from any found in the text that requires the Multiplication Principle to solve. Give it to a friend to solve and critique.

70. Create a problem different from any found in the text that requires a permutation to solve. Give it to a friend to solve and critique.

71. Create a problem different from any found in the text that requires a combination to solve. Give it to a friend to solve and critique.

72. Explain the difference between a permutation and a combination. Give an example to illustrate your explanation.

'Are You Prepared?' Answers

1. 1; 1

2. False

12.3 Probability

OBJECTIVES
1 Construct Probability Models
2 Compute Probabilities of Equally Likely Outcomes
3 Use the Addition Rule to Find Probabilities
4 Use the Complement Rule to Find Probabilities

Probability is an area of mathematics that deals with experiments that yield random results, yet admit a certain regularity. Such experiments do not always produce the same result or outcome, so the result of any one observation is not predictable. However, the results of the experiment over a long period do produce regular patterns that enable us to predict with remarkable accuracy.

EXAMPLE 1 **Tossing a Fair Coin**

In tossing a fair coin, we know that the outcome is either a head or a tail. On any particular throw, we cannot predict what will happen, but, if we toss the coin many times, we observe that the number of times that a head comes up is approximately equal to the number of times that we get a tail. It seems reasonable, therefore, to assign a probability of $\frac{1}{2}$ that a head comes up and a probability of $\frac{1}{2}$ that a tail comes up. ◀

Probability Models

1 The discussion in Example 1 constitutes the construction of a **probability model** for the experiment of tossing a fair coin once. A probability model has two components: a sample space and an assignment of probabilities. A **sample space** S is a set whose elements represent all the possibilities that can occur as a result of the experiment. Each element of S is called an **outcome.** To each outcome, we assign a number, called the **probability** of that outcome, which has two properties:

1. The probability assigned to each outcome is nonnegative.
2. The sum of all the probabilities equals 1.

If a probability model has the sample space

$$S = \{e_1, e_2, \dots, e_n\}$$

where $e_1, e_2, \dots, e_n$ are the possible outcomes, and if $P(e_1), P(e_2), \dots, P(e_n)$ denote the respective probabilities of these outcomes, then

$$P(e_1) \geq 0, P(e_2) \geq 0, \dots, P(e_n) \geq 0 \quad \textbf{(1)}$$

$$\sum_{i=1}^{n} P(e_i) = P(e_1) + P(e_2) + \cdots + P(e_n) = 1 \quad \textbf{(2)}$$

EXAMPLE 2 Determining Probability Models

In a bag of M&Ms, the candies are colored red, green, blue, brown, yellow, and orange. Suppose that a candy is drawn from the bag and the color is recorded. The sample space of this experiment is {red, green, blue, brown, yellow, orange}. Determine which of the following are probability models.

(a)

Outcome	Probability
{red}	0.3
{green}	0.15
{blue}	0
{brown}	0.15
{yellow}	0.2
{orange}	0.2

(b)

Outcome	Probability
{red}	0.1
{green}	0.1
{blue}	0.1
{brown}	0.4
{yellow}	0.2
{orange}	0.3

(c)

Outcome	Probability
{red}	0.3
{green}	−0.3
{blue}	0.2
{brown}	0.4
{yellow}	0.2
{orange}	0.2

(d)

Outcome	Probability
{red}	0
{green}	0
{blue}	0
{brown}	0
{yellow}	1
{orange}	0

Solution

(a) This model is a probability model since all the outcomes have probabilities that are nonnegative and the sum of the probabilities is 1.

(b) This model is not a probability model because the sum of the probabilities is not 1.

(c) This model is not a probability model because $P(\text{green})$ is less than 0. Recall, all probabilities must be nonnegative.

(d) This model is a probability model because all the outcomes have probabilities that are nonnegative, and the sum of the probabilities is 1. Notice that $P(\text{yellow}) = 1$, meaning that this outcome will occur with 100% certainty each time that the experiment is repeated. This means that the bag of M&Ms has only yellow candies. ◀

NOW WORK PROBLEM 7.

Let's look at an example of constructing a probability model.

EXAMPLE 3 **Constructing a Probability Model**

An experiment consists of rolling a fair die once.* Construct a probability model for this experiment.

Solution A sample space S consists of all the possibilities that can occur. Because rolling the die will result in one of six faces showing, the sample space S consists of

$$S = \{1, 2, 3, 4, 5, 6\}$$

Because the die is fair, one face is no more likely to occur than another. As a result, our assignment of probabilities is

$$P(1) = \frac{1}{6} \qquad P(2) = \frac{1}{6}$$
$$P(3) = \frac{1}{6} \qquad P(4) = \frac{1}{6}$$
$$P(5) = \frac{1}{6} \qquad P(6) = \frac{1}{6}$$

Now suppose that a die is loaded (weighted) so that the probability assignments are

$$P(1) = 0, \quad P(2) = 0, \quad P(3) = \frac{1}{3}, \quad P(4) = \frac{2}{3}, \quad P(5) = 0, \quad P(6) = 0$$

This assignment would be made if the die were loaded so that only a 3 or 4 could occur and the 4 is twice as likely as the 3 to occur. This assignment is consistent with the definition, since each assignment is nonnegative and the sum of all the probability assignments equals 1.

NOW WORK PROBLEM 23.

EXAMPLE 4 **Constructing a Probability Model**

An experiment consists of tossing a coin. The coin is weighted so that heads (H) is three times as likely to occur as tails (T). Construct a probability model for this experiment.

Solution The sample space S is $S = \{\text{H}, \text{T}\}$. If x denotes the probability that a tail occurs, then

$$P(\text{T}) = x \quad \text{and} \quad P(\text{H}) = 3x$$

Since the sum of the probabilities of the possible outcomes must equal 1, we have

$$P(\text{T}) + P(\text{H}) = x + 3x = 1$$
$$4x = 1$$
$$x = \frac{1}{4}$$

Figure 8

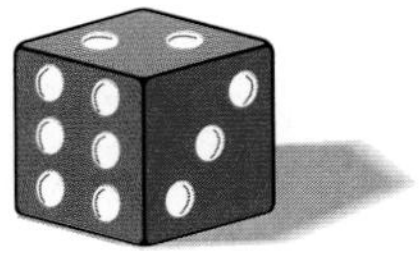

*A die is a cube with each face having either 1, 2, 3, 4, 5, or 6 dots on it. See Figure 8.

We assign the probabilities

$$P(\text{T}) = \frac{1}{4} \qquad P(\text{H}) = \frac{3}{4}$$

◄

NOW WORK PROBLEM 27.

In working with probability models, the term **event** is used to describe a set of possible outcomes of the experiment. An event E is some subset of the sample space S. The **probability of an event** E, $E \neq \emptyset$, denoted by $P(E)$, is defined as the sum of the probabilities of the outcomes in E. We can also think of the probability of an event E as the likelihood that the event E occurs. If $E = \emptyset$, then $P(E) = 0$; if $E = S$, then $P(E) = P(S) = 1$.

Equally Likely Outcomes

2 When the same probability is assigned to each outcome of the sample space, the experiment is said to have **equally likely outcomes.**

Theorem

Probability for Equally Likely Outcomes

If an experiment has n equally likely outcomes and if the number of ways that an event E can occur is m, then the probability of E is

$$P(E) = \frac{\text{Number of ways that } E \text{ can occur}}{\text{Number of all logical possibilities}} = \frac{m}{n} \qquad \textbf{(3)}$$

If S is the sample space of this experiment, then

$$P(E) = \frac{n(E)}{n(S)} \qquad \textbf{(4)}$$

EXAMPLE 5 **Calculating Probabilities of Events Involving Equally Likely Outcomes**

Calculate the probability that in a 3-child family there are 2 boys and 1 girl. Assume equally likely outcomes.

Solution We begin by constructing a tree diagram to help in listing the possible outcomes of the experiment. See Figure 9, where B stands for boy and G for girl. The sample space S of this experiment is

Figure 9

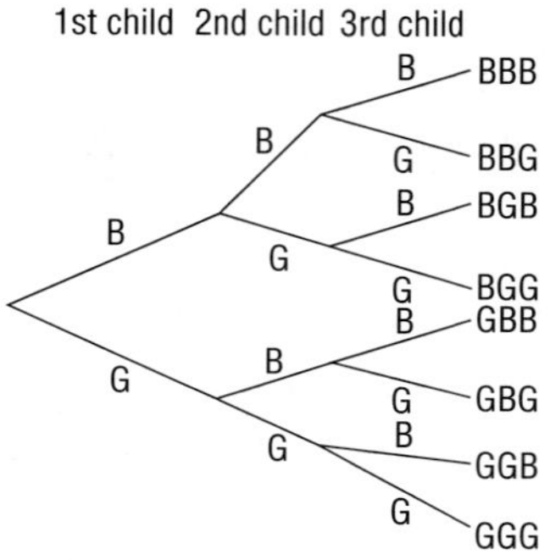

$$S = \{\text{BBB, BBG, BGB, BGG, GBB, GBG, GGB, GGG}\}$$

so $n(S) = 8$.

We wish to know the probability of the event E: "having two boys and one girl." From Figure 9, we conclude that $E = \{\text{BBG, BGB, GBB}\}$, so $n(E) = 3$. Since the outcomes are equally likely, the probability of E is

$$P(E) = \frac{n(E)}{n(S)} = \frac{3}{8}$$

◄

NOW WORK PROBLEM 37.

Compound Probabilities

So far, we have calculated probabilities of single events. We will now compute probabilities of multiple events, called **compound probabilities.**

EXAMPLE 6 Computing Compound Probabilities

Consider the experiment of rolling a single fair die. Let E represent the event "roll an odd number," and let F represent the event "roll a 1 or 2."

(a) Write the event E and F.
(b) Write the event E or F.
(c) Compute $P(E)$ and $P(F)$.
(d) Compute $P(E \cap F)$.
(e) Compute $P(E \cup F)$.

Solution The sample space S of the experiment is $\{1, 2, 3, 4, 5, 6\}$, so $n(S) = 6$. Since the die is fair, the outcomes are equally likely. The event E: "roll an odd number" is $\{1, 3, 5\}$, and the event F: "roll a 1 or 2" is $\{1, 2\}$, so $n(E) = 3$ and $n(F) = 2$.

(a) The word *and* in probability means the intersection of two events. The event E and F is

$$E \cap F = \{1, 3, 5\} \cap \{1, 2\} = \{1\} \qquad n(E \cap F) = 1$$

(b) The word *or* in probability means the union of the two events. The event E or F is

$$E \cup F = \{1, 3, 5\} \cup \{1, 2\} = \{1, 2, 3, 5\} \qquad n(E \cup F) = 4$$

(c) We use formula (4).

$$P(E) = \frac{n(E)}{n(S)} = \frac{3}{6} = \frac{1}{2} \qquad P(F) = \frac{n(F)}{n(S)} = \frac{2}{6} = \frac{1}{3}$$

(d) $P(E \cap F) = \dfrac{n(E \cap F)}{n(S)} = \dfrac{1}{6}$

(e) $P(E \cup F) = \dfrac{n(E \cup F)}{n(S)} = \dfrac{4}{6} = \dfrac{2}{3}$ ◀

3 The **Addition Rule** can be used to find the probability of the union of two events.

Theorem

Addition Rule

For any two events E and F,

$$P(E \cup F) = P(E) + P(F) - P(E \cap F) \qquad \textbf{(5)}$$

For example, we can use the Addition Rule to find $P(E \cup F)$ in Example 6(e). Then

$$P(E \cup F) = P(E) + P(F) - P(E \cap F) = \frac{1}{2} + \frac{1}{3} - \frac{1}{6} = \frac{3}{6} + \frac{2}{6} - \frac{1}{6} = \frac{4}{6} = \frac{2}{3}$$

as before.

EXAMPLE 7 Computing Probabilities of Compound Events Using the Addition Rule

If $P(E) = 0.2$, $P(F) = 0.3$, and $P(E \cap F) = 0.1$, find the probability of E or F; that is, find $P(E \cup F)$.

Solution We use the Addition Rule, formula (5).

$$\text{Probability of } E \text{ or } F = P(E \cup F) = P(E) + P(F) - P(E \cap F)$$
$$= 0.2 + 0.3 - 0.1 = 0.4$$

A Venn diagram can sometimes be used to obtain probabilities. To construct a Venn diagram representing the information in Example 7, we draw two sets E and F. We begin with the fact that $P(E \cap F) = 0.1$. See Figure 10(a). Then, since $P(E) = 0.2$ and $P(F) = 0.3$, we fill in E with $0.2 - 0.1 = 0.1$ and F with $0.3 - 0.1 = 0.2$. See Figure 10(b). Since $P(S) = 1$, we complete the diagram by inserting $1 - (0.1 + 0.1 + 0.2) = 0.6$. See Figure 10(c). Now it is easy to see, for example, that the probability of F, but not E, is 0.2. Also, the probability of neither E nor F is 0.6.

Figure 10

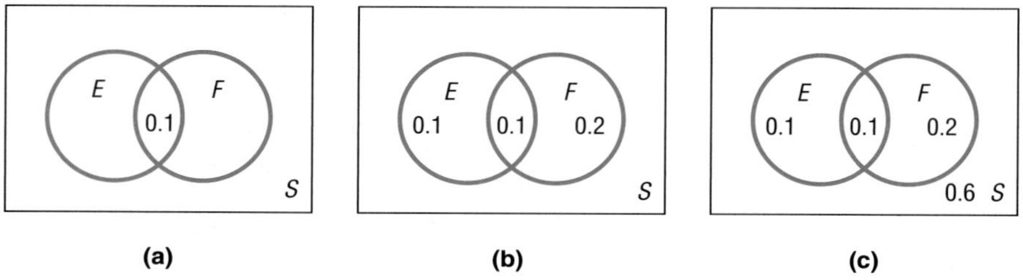

NOW WORK PROBLEM 45.

If events E and F are disjoint so that $E \cap F = \emptyset$, we say they are **mutually exclusive.** In this case, $P(E \cap F) = 0$, and the Addition Rule takes the following form:

Theorem

Mutually Exclusive Events

If E and F are **mutually exclusive events,** then

$$P(E \cup F) = P(E) + P(F) \qquad \textbf{(6)}$$

EXAMPLE 8 Computing Compound Probabilities of Mutually Exclusive Events

If $P(E) = 0.4$ and $P(F) = 0.25$, and E and F are mutually exclusive, find $P(E \cup F)$.

Solution Since E and F are mutually exclusive, we use formula (6).

$$P(E \cup F) = P(E) + P(F) = 0.4 + 0.25 = 0.65$$

NOW WORK PROBLEM 47.

Complements

4 Recall, if A is a set, the complement of A, denoted $\overline{A}$, is the set of all elements in the universal set U not in A. We similarly define the complement of an event.

Complement of an Event

Let S denote the sample space of an experiment, and let E denote an event. The **complement of E,** denoted $\overline{E}$, is the set of all outcomes in the sample space S that are not outcomes in the event E.

The complement of an event E, that is, $\overline{E}$, in a sample space S has the following two properties:

$$E \cap \overline{E} = \emptyset \qquad E \cup \overline{E} = S$$

Since E and $\overline{E}$ are mutually exclusive, it follows from (6) that

$$P(E \cup \overline{E}) = P(S) = 1 \qquad P(E) + P(\overline{E}) = 1 \qquad P(\overline{E}) = 1 - P(E)$$

We have the following result:

Theorem

Computing Probabilities of Complementary Events

If E represents any event and $\overline{E}$ represents the complement of E, then

$$P(\overline{E}) = 1 - P(E) \qquad \textbf{(7)}$$

EXAMPLE 9 **Computing Probabilities Using Complements**

On the local news the weather reporter stated that the probability of rain tomorrow is 40%. What is the probability that it will not rain?

Solution The complement of the event "rain" is "no rain."

$$P(\text{no rain}) = 1 - P(\text{rain}) = 1 - 0.4 = 0.6$$

There is a 60% chance of no rain tomorrow. ◀

NOW WORK PROBLEM 51.

EXAMPLE 10 **Birthday Problem**

What is the probability that in a group of 10 people at least 2 people have the same birthday? Assume that there are 365 days in a year.

Solution We assume that a person is as likely to be born on one day as another, so we have equally likely outcomes.

We first determine the number of outcomes in the sample space S. There are 365 possibilities for each person's birthday. Since there are 10 people in the group, there are 365^{10} possibilities for the birthdays. [For one person in the group, there are 365 days on which his or her birthday can fall; for two people, there are $(365)(365) = 365^2$ pairs of days; and, in general, using the Multiplication Principle, for n people there are 365^n possibilities.] So

$$n(S) = 365^{10}$$

We wish to find the probability of the event E: "at least two people have the same birthday." It is difficult to count the elements in this set; it is much easier to count the elements of the complementary event $\overline{E}$: "no two people have the same birthday."

We find $n(\overline{E})$ as follows: Choose one person at random. There are 365 possibilities for his or her birthday. Choose a second person. There are 364 possibilities for this birthday, if no two people are to have the same birthday. Choose a third person. There are 363 possibilities left for this birthday. Finally, we arrive at the tenth person. There are 356 possibilities left for this birthday. By the Multiplication Principle, the total number of possibilities is

$$n(\overline{E}) = 365 \cdot 364 \cdot 363 \cdot \ldots \cdot 356$$

Hence, the probability of event $\overline{E}$ is

$$P(\overline{E}) = \frac{n(\overline{E})}{n(S)} \approx \frac{365 \cdot 364 \cdot 363 \cdot \ldots \cdot 356}{365^{10}} \approx 0.883$$

The probability of two or more people in a group of 10 people having the same birthday is then

$$P(E) = 1 - P(\overline{E}) \approx 1 - 0.883 = 0.117$$ ◀

The birthday problem can be solved for any group size. The following table gives the probabilities for two or more people having the same birthday for various group sizes. Notice that the probability is greater than $\frac{1}{2}$ for any group of 23 or more people.

	Number of People															
	5	10	15	20	21	22	23	24	25	30	40	50	60	70	80	90
Probability That Two or More Have the Same Birthday	0.027	0.117	0.253	0.411	0.444	0.476	0.507	0.538	0.569	0.706	0.891	0.970	0.994	0.99916	0.99991	0.99999

NOW WORK PROBLEM 69.

Historical Feature

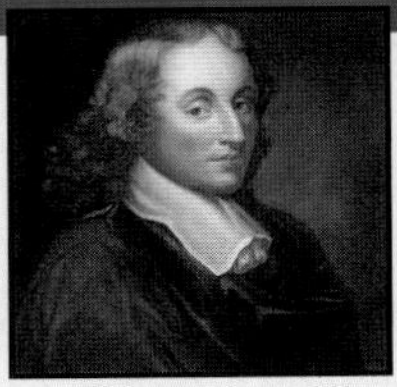

Blaise Pascal (1623–1662)

Set theory, counting, and probability first took form as a systematic theory in an exchange of letters (1654) between Pierre de Fermat (1601–1665) and Blaise Pascal (1623–1662). They discussed the problem of how to divide the stakes in a game that is interrupted before completion, knowing how many points each player needs to win. Fermat solved the problem by listing all possibilities and counting the favorable ones, whereas Pascal made use of the triangle that now bears his name. As mentioned in the text, the entries in Pascal's triangle are equivalent to $C(n, r)$. This recognition of the role of $C(n, r)$ in counting is the foundation of all further developments.

The first book on probability, the work of Christiaan Huygens (1629–1695), appeared in 1657. In it, the notion of mathematical expectation is explored. This allows the calculation of the profit or loss that a gambler might expect, knowing the probabilities involved in the game (see the Historical Problems that follow).

Although Girolamo Cardano (1501–1576) wrote a treatise on probability, it was not published until 1663 in Cardano's collected works, and this was too late to have any effect on the development of the theory.

In 1713, the posthumously published *Ars Conjectandi* of Jakob Bernoulli (1654–1705) gave the theory the form it would have until 1900. Recently, both combinatorics (counting) and probability have undergone rapid development due to the use of computers.

A final comment about notation. The notations $C(n, r)$ and $P(n, r)$ are variants of a form of notation developed in England after 1830. The notation $\binom{n}{r}$ for $C(n, r)$ goes back to Leonhard Euler (1707–1783), but is now losing ground because it has no clearly related symbolism of the same type for permutations. The set symbols $\cup$ and $\cap$ were introduced by Giuseppe Peano (1858–1932) in 1888 in a slightly different context. The inclusion symbol $\subset$ was introduced by E. Schroeder (1841–1902) about 1890. The treatment of set theory in the text is due to George Boole (1815–1864), who wrote $A + B$ for $A \cup B$ and AB for $A \cap B$ (statisticians still use AB for $A \cap B$).

Historical Problems

1. **The Problem Discussed by Fermat and Pascal** A game between two equally skilled players, A and B, is interrupted when A needs 2 points to win and B needs 3 points. In what proportion would the stakes be divided?

   (a) ***Fermat's solution*** List all possible outcomes that can occur as a result of four more plays. The probabilities for A to win and B to win then determine how the stakes should be divided.

   (b) ***Pascal's solution*** Use combinations to determine the number of ways that the 2 points needed for A to win could occur in four plays. Then use combinations to determine the number of ways that the 3 points needed for B to win could occur. This is trickier than it looks, since A can win with 2 points in either two plays, three plays, or four plays. Compute the probabilities and compare with the results in part (a).

2. **Huygen's Mathematical Expectation** In a game with n possible outcomes with probabilities $p_1, p_2, \ldots, p_n$, suppose that the *net* winnings are $w_1, w_2, \ldots, w_n$, respectively. Then the mathematical expectation is

$$E = p_1w_1 + p_2w_2 + \cdots + p_nw_n$$

   The number E represents the profit or loss per game in the long run. The following problems are a modification of those of Huygens.

   (a) A fair die is tossed. A gambler wins \$3 if he throws a 6 and \$6 if he throws a 5. What is his expectation? [**Hint:** $w_1 = w_2 = w_3 = w_4 = 0$]

   (b) A gambler plays the same game as in part (a), but now the gambler must pay \$1 to play. This means that $w_5 = \$5$, $w_6 = \$2$, and $w_1 = w_2 = w_3 = w_4 = -\$1$. What is the expectation?

12.3 Assess Your Understanding

Concepts and Vocabulary

1. When the same probability is assigned to each outcome of a sample space, the experiment is said to have __________ __________ outcomes.

2. The __________ of an event E is the set of all outcomes in the sample space S that are not outcomes in the event E.

3. *True or False:* The probability of an event can never equal 0.

4. *True or False:* In a probability model, the sum of all probabilities is 1.

Exercises

5. In a probability model, which of the following numbers could be the probability of an outcome:

$$0, \quad 0.01, \quad 0.35, \quad -0.4, \quad 1, \quad 1.4?$$

6. In a probability model, which of the following numbers could be the probability of an outcome:

$$1.5, \quad \frac{1}{2}, \quad \frac{3}{4}, \quad \frac{2}{3}, \quad 0, \quad -\frac{1}{4}?$$

7. Determine whether the following is a probability model.

Outcome	Probability
{1}	0.2
{2}	0.3
{3}	0.1
{4}	0.4

8. Determine whether the following is a probability model.

Outcome	Probability
{Jim}	0.4
{Bob}	0.3
{Faye}	0.1
{Patricia}	0.2

9. Determine whether the following is a probability model.

Outcome	Probability
{Linda}	0.3
{Jean}	0.2
{Grant}	0.1
{Ron}	0.3

10. Determine whether the following is a probability model.

Outcome	Probability
{Lanny}	0.3
{Joanne}	0.2
{Nelson}	0.1
{Rich}	0.5
{Judy}	−0.1

In Problems 11–16, construct a probability model for each experiment.

11. Tossing a fair coin twice

12. Tossing two fair coins once

13. Tossing two fair coins, then a fair die

14. Tossing a fair coin, a fair die, and then a fair coin

15. Tossing three fair coins once

16. Tossing one fair coin three times

In Problems 17–22, use the following spinners to construct a probability model for each experiment.

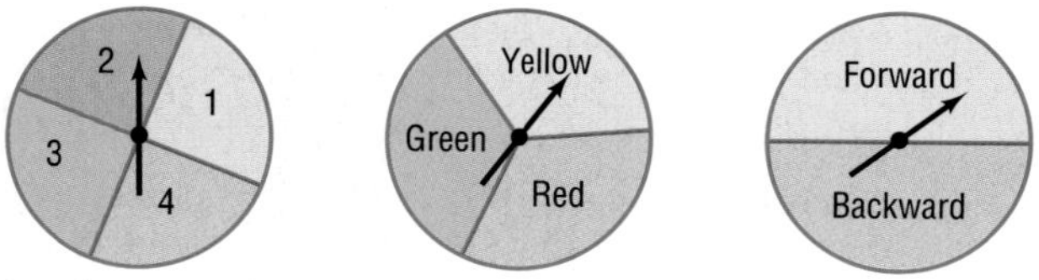

Spinner I Spinner II Spinner III

17. Spin spinner I, then spinner II. What is the probability of getting a 2 or a 4, followed by Red?

18. Spin spinner III, then spinner II. What is the probability of getting Forward, followed by Yellow or Green?

19. Spin spinner I, then II, then III. What is the probability of getting a 1, followed by Red or Green, followed by Backward?

20. Spin spinner II, then I, then III. What is the probability of getting Yellow, followed by a 2 or a 4, followed by Forward?

21. Spin spinner I twice, then spinner II. What is the probability of getting a 2, followed by a 2 or a 4, followed by Red or Green?

22. Spin spinner III, then spinner I twice. What is the probability of getting Forward, followed by a 1 or a 3, followed by a 2 or a 4?

In Problems 23–26, consider the experiment of tossing a coin twice. The table below lists six possible assignments of probabilities for this experiment. Using this table, answer the following questions.

Assignments	Sample Space			
	HH	HT	TH	TT
A	$\frac{1}{4}$	$\frac{1}{4}$	$\frac{1}{4}$	$\frac{1}{4}$
B	0	0	0	1
C	$\frac{3}{16}$	$\frac{5}{16}$	$\frac{5}{16}$	$\frac{3}{16}$
D	$\frac{1}{2}$	$\frac{1}{2}$	$-\frac{1}{2}$	$\frac{1}{2}$
E	$\frac{1}{4}$	$\frac{1}{4}$	$\frac{1}{4}$	$\frac{1}{8}$
F	$\frac{1}{9}$	$\frac{2}{9}$	$\frac{2}{9}$	$\frac{4}{9}$

23. Which of the assignments of probabilities are consistent with the definition of a probability model?

24. Which of the assignments of probabilities should be used if the coin is known to be fair?

25. Which of the assignments of probabilities should be used if the coin is known to always come up tails?

26. Which of the assignments of probabilities should be used if tails is twice as likely as heads to occur?

27. **Assigning Probabilities** A coin is weighted so that heads is four times as likely as tails to occur. What probability should we assign to heads? to tails?

28. **Assigning Probabilities** A coin is weighted so that tails is twice as likely as heads to occur. What probability should we assign to heads? to tails?

29. **Assigning Probabilities** A die is weighted so that an odd-numbered face is twice as likely to occur as an even-numbered face. What probability should we assign to each face?

30. **Assigning Probabilities** A die is weighted so that a six cannot appear. The other faces occur with the same probability. What probability should we assign to each face?

For Problems 31–34, let the sample space be $S = \{1, 2, 3, 4, 5, 6, 7, 8, 9, 10\}$. *Suppose that the outcomes are equally likely.*

31. Compute the probability of the event $E = \{1, 2, 3\}$.

32. Compute the probability of the event $F = \{3, 5, 9, 10\}$.

33. Compute the probability of the event E: "an even number."

34. Compute the probability of the event F: "an odd number."

For Problems 35 and 36, an urn contains 5 white marbles, 10 green marbles, 8 yellow marbles, and 7 black marbles.

35. If one marble is selected, determine the probability that it is white.

36. If one marble is selected, determine the probability that it is black.

In Problems 37–40, assume equally likely outcomes.

37. Determine the probability of having 3 boys in a 3-child family.

38. Determine the probability of having 3 girls in a 3-child family.

39. Determine the probability of having 1 girl and 3 boys in a 4-child family.

40. Determine the probability of having 2 girls and 2 boys in a 4-child family.

For Problems 41–44, two fair dice are rolled.

41. Determine the probability that the sum of the two dice is 7.

42. Determine the probability that the sum of the two dice is 11.

43. Determine the probability that the sum of the two dice is 3.

44. Determine the probability that the sum of the two dice is 12.

In Problems 45–48, find the probability of the indicated event if $P(A) = 0.25$ *and* $P(B) = 0.45$.

45. $P(A \cup B)$ if $P(A \cap B) = 0.15$

46. $P(A \cap B)$ if $P(A \cup B) = 0.6$

47. $P(A \cup B)$ if A, B are mutually exclusive

48. $P(A \cap B)$ if A, B are mutually exclusive

49. If $P(A) = 0.60$, $P(A \cup B) = 0.85$, and $P(A \cap B) = 0.05$, find $P(B)$.

50. If $P(B) = 0.30$, $P(A \cup B) = 0.65$, and $P(A \cap B) = 0.15$, find $P(A)$.

51. According to the Federal Bureau of Investigation, in 2002 there was a 26.5% probability of theft from a motor vehicle. If a victim of theft is randomly selected, what is the probability that he or she was not a victim of theft from a motor vehicle?

52. According to the Federal Bureau of Investigation, in 2002 there was a 3.9% probability of theft involving a bicycle. If a victim of theft is randomly selected, what is the probability that he or she was not a victim of bicycle theft?

53. In Chicago, there is a 30% probability that Memorial Day will have a high temperature in the 70s. What is the probability that next Memorial Day will not have a high temperature in the 70s in Chicago?

54. In Chicago, there is a 4% probability that Memorial Day will have a low temperature in the 30s. What is the probability that next Memorial Day will not have a low temperature in the 30s in Chicago?

For Problems 55–58, a golf ball is selected at random from a container. If the container has 9 white balls, 8 green balls, and 3 orange balls, find the probability of each event.

55. The golf ball is white or green.

56. The golf ball is white or orange.

57. The golf ball is not white.

58. The golf ball is not green.

59. On the "Price is Right" there is a game in which a bag is filled with 3 strike chips and 5 numbers. Let's say that the numbers in the bag are 0, 1, 3, 6, and 9. What is the probability of selecting a strike chip or the number 1?

60. Another game on the "Price is Right" requires the contestant to spin a wheel with numbers 5, 10, 15, 20, ..., 100. What is the probability that the contestant spins 100 or 30?

Problems 61–64, are based on a consumer survey of annual incomes in 100 households. The following table gives the data.

Income	\$0–9999	\$10,000–19,999	\$20,000–29,999	\$30,000–39,999	\$40,000 or more
Number of households	5	35	30	20	10

61. What is the probability that a household has an annual income of \$30,000 or more?

62. What is the probability that a household has an annual income between \$10,000 and \$29,999, inclusive?

63. What is the probability that a household has an annual income of less than \$20,000?

64. What is the probability that a household has an annual income of \$20,000 or more?

65. Surveys In a survey about the number of TV sets in a house, the following probability table was constructed:

Number of TV sets	0	1	2	3	4 or more
Probability	0.05	0.24	0.33	0.21	0.17

Find the probability of a house having:
(a) 1 or 2 TV sets
(b) 1 or more TV sets
(c) 3 or fewer TV sets
(d) 3 or more TV sets
(e) Less than 2 TV sets
(f) Less than 1 TV set
(g) 1, 2, or 3 TV sets
(h) 2 or more TV sets

66. Checkout Lines Through observation it has been determined that the probability for a given number of people waiting in line at the "5 items or less" checkout register of a supermarket is as follows:

Number waiting in line	0	1	2	3	4 or more
Probability	0.10	0.15	0.20	0.24	0.31

Find the probability of:
(a) At most 2 people in line
(b) At least 2 people in line
(c) At least 1 person in line

67. In a certain Precalculus class, there are 18 freshmen and 15 sophomores. Of the 18 freshmen, 10 are male, and of the 15 sophomores, 8 are male. Find the probability that a randomly selected student is:
(a) A freshman or female
(b) A sophomore or male

68. The faculty of the mathematics department at Joliet Junior College is composed of 4 females and 9 males. Of the 4 females, 2 are under the age of 40, and 3 of the males are under age 40. Find the probability that a randomly selected faculty member is:
(a) Female or under age 40
(b) Male or over age 40

69. Birthday Problem What is the probability that at least 2 people have the same birthday in a group of 12 people? Assume that there are 365 days in a year.

70. Birthday Problem What is the probability that at least 2 people have the same birthday in a group of 35 people? Assume that there are 365 days in a year.

71. Winning a Lottery In a certain lottery, there are ten balls, numbered 1, 2, 3, 4, 5, 6, 7, 8, 9, 10. Of these, five are drawn in order. If you pick five numbers that match those drawn in the correct order, you win \$1,000,000. What is the probability of winning such a lottery?

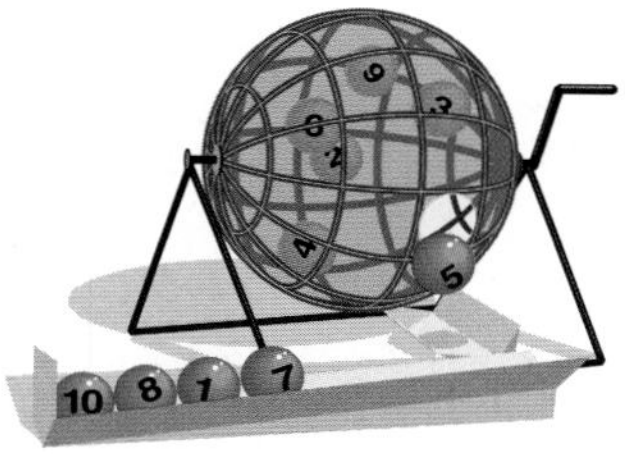

Chapter Review

Things to Know

Set (p. 828) — Well-defined collection of distinct objects, called elements

Null set (p. 828)	$\varnothing$	Set that has no elements
Equality (p. 828)	$A = B$	A and B have the same elements.
Subset (p. 828)	$A \subseteq B$	Each element of A is also an element of B.
Intersection (p. 829)	$A \cap B$	Set consisting of elements that belong to both A and B
Union (p. 829)	$A \cup B$	Set consisting of elements that belong to either A or B, or both
Universal set (p. 829)	U	Set consisting of all the elements that we wish to consider
Complement (p. 829)	$\overline{A}$	Set consisting of elements of the universal set that are not in A
Finite set (p. 830)		The number of elements in the set is a nonnegative integer.
Infinite set (p. 830)		A set that is not finite

Counting formula (p. 831)	$n(A \cup B) = n(A) + n(B) - n(A \cap B)$
Addition Principle (p. 831)	If $A \cap B = \emptyset$, then $n(A \cup B) = n(A) + n(B)$.
Multiplication Principle (p. 835)	If a task consists of a sequence of choices in which there are p selections for the first choice, q selections for the second choice, and so on, then the task of making these selections can be done in $p \cdot q \cdot \ldots$ different ways.
Permutation (p. 836)	An ordered arrangement of r objects chosen from n objects
Permutation: Distinct, with repetition (p. 837)	n^r The n objects are distinct (different) and repetition is allowed in the selection of r of them.
Permutation: Distinct, without repetition (p. 838)	$P(n, r) = n(n-1) \cdot \ldots \cdot [n-(r-1)] = \dfrac{n!}{(n-r)!}$ An ordered arrangement of n distinct objects without repetition
Combination (pp. 839 and 840)	$C(n, r) = \dfrac{P(n, r)}{r!} = \dfrac{n!}{(n-r)!r!}$ An arrangement, without regard to order, of n distinct objects without repetition
Permutation: Not distinct, (p. 842)	$\dfrac{n!}{n_1!n_2! \cdots n_k!}$ The number of permutations of n objects of which n_1 are of one kind, n_2 are of a second kind, ..., and n_k are of a kth kind, where $n = n_1 + n_2 + \cdots + n_k$
Sample space (p. 845)	Set whose elements represent all the logical possibilities that can occur as a result of an experiment
Probability (p. 845)	A nonnegative number assigned to each outcome of a sample space; the sum of all the probabilities of the outcomes equals 1.
Equally likely outcomes (p. 848)	$P(E) = \dfrac{n(E)}{n(S)}$ The same probability is assigned to each outcome.
Addition Rule (p. 849)	$P(E \cup F) = P(E) + P(F) - P(E \cap F)$
Complement of an event (p. 851)	$P(\overline{E}) = 1 - P(E)$

Objectives

Section		You should be able to . . .	Review Exercises
12.1	1	Find all the subsets of a set (p. 828)	1, 2
	2	Find the intersection and union of sets (p. 829)	3–6
	3	Find the complement of a set (p. 829)	7–10
	4	Count the number of elements in a set (p. 830)	11–18
12.2	1	Solve counting problems using the Multiplication Principle (p. 834)	23–26, 32–36
	2	Solve counting problems using permutations (p. 836)	19, 20, 27, 28, 41(a)
	3	Solve counting problems using combinations (p. 839)	21, 22, 29–31, 39–40
	4	Solve counting problems using permutations involving n nondistinct objects (p. 841)	37, 38
12.3	1	Construct probability models (p. 845)	41(b)
	2	Compute probabilities of equally likely outcomes (p. 848)	41(b), 42(a), 43(a), 44–47
	3	Use the Addition Rule to find probabilities (p. 849)	48
	4	Use the Complement Rule to find probabilities (p. 851)	41(c), 42(b), 43(b), 44

Review Exercises *(Blue problem numbers indicate the author's suggestions for use in a Practice Test.)*

1. Write down all the subsets of the set {Dave, Joanne, Erica}.
2. Write down all the subsets of the set {Green, Blue, Red}.

In Problems 3–10, use U = *universal set* = $\{1, 2, 3, 4, 5, 6, 7, 8, 9\}$, $A = \{1, 3, 5, 7\}$, $B = \{3, 5, 6, 7, 8\}$, *and* $C = \{2, 3, 7, 8, 9\}$ *to find each set.*

3. $A \cup B$
4. $B \cup C$
5. $A \cap C$
6. $A \cap B$
7. $\overline{A} \cup \overline{B}$
8. $\overline{B} \cap \overline{C}$
9. $\overline{B \cap C}$
10. $\overline{A \cup B}$
11. If $n(A) = 8$, $n(B) = 12$, and $n(A \cap B) = 3$, find $n(A \cup B)$.
12. If $n(A) = 12$, $n(A \cup B) = 30$, and $n(A \cap B) = 6$, find $n(B)$.

In Problems 13–18, use the information supplied in the figure:

13. How many are in A?
14. How many are in A or B?
15. How many are in A and C?
16. How many are not in B?
17. How many are in neither A nor C?
18. How many are in B but not in C?

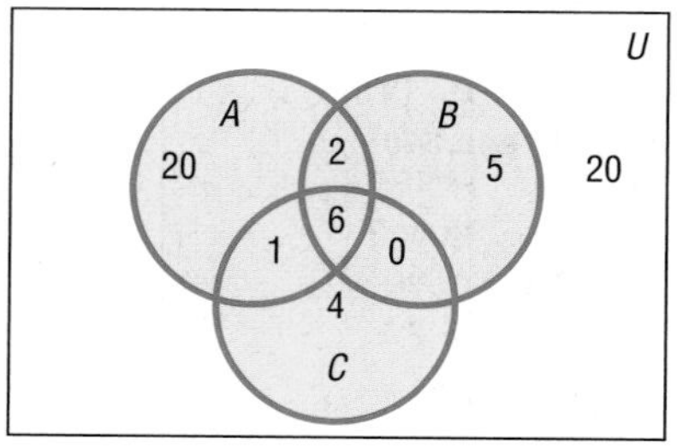

In Problems 19–22, compute the given expression.

19. $P(8, 3)$
20. $P(7, 3)$
21. $C(8, 3)$
22. $C(7, 3)$

23. A clothing store sells pure wool and polyester–wool suits. Each suit comes in 3 colors and 10 sizes. How many suits are required for a complete assortment?
24. In connecting a certain electrical device, 5 wires are to be connected to 5 different terminals. How many different wirings are possible if 1 wire is connected to each terminal?
25. **Baseball** On a given day, the American Baseball League schedules 7 games. How many different outcomes are possible, assuming that each game is played to completion?
26. **Baseball** On a given day, the National Baseball League schedules 6 games. How many different outcomes are possible, assuming that each game is played to completion?
27. If 4 people enter a bus having 9 vacant seats, in how many ways can they be seated?
28. How many different arrangements are there of the letters in the word ROSE?
29. In how many ways can a squad of 4 relay runners be chosen from a track team of 8 runners?
30. A professor has 10 similar problems to put on a test with 3 problems. How many different tests can she design?
31. **Baseball** In how many ways can 2 teams from 14 teams in the American League be chosen without regard to which team is at home?
32. **Arranging Books on a Shelf** There are 5 different French books and 5 different Spanish books. How many ways are there to arrange them on a shelf if:
    (a) Books of the same language must be grouped together, French on the left, Spanish on the right?
    (b) French and Spanish books must alternate in the grouping, beginning with a French book?
33. **Telephone Numbers** Using the digits 0, 1, 2, . . . , 9, how many 7-digit numbers can be formed if the first digit cannot be 0 or 9 and if the last digit is greater than or equal to 2 and less than or equal to 3? Repeated digits are allowed.
34. **Home Choices** A contractor constructs homes with 5 different choices of exterior finish, 3 different roof arrangements, and 4 different window designs. How many different types of homes can be built?
35. **License Plate Possibilities** A license plate consists of 1 letter, excluding O and I, followed by a 4-digit number that cannot have a 0 in the lead position. How many different plates are possible?
36. Using the digits 0 and 1, how many different numbers consisting of 8 digits can be formed?
37. **Forming Different Words** How many different words, real on imaginary, can be formed using all the letters in the word MISSING?
38. **Arranging Flags** How many different vertical arrangements are there of 10 flags if 4 are white, 3 are blue, 2 are green, and 1 is red?
39. **Forming Committees** A group of 9 people is going to be formed into committees of 4, 3, and 2 people. How many committees can be formed if:
    (a) A person can serve on any number of committees?
    (b) No person can serve on more than one committee?
40. **Forming Committees** A group consists of 5 men and 8 women. A committee of 4 is to be formed from this group, and policy dictates that at least 1 woman be on this committee.
    (a) How many committees can be formed that contain exactly 1 man?

(b) How many committees can be formed that contain exactly 2 women?
(c) How many committees can be formed that contain at least 1 man?

41. Birthday Problem For this problem, assume that a year has 365 days.
(a) How many ways can 18 people have different birthdays?
(b) What is the probability that nobody has the same birthday in a group of 18 people?
(c) What is the probability in a group of 18 people that at least 2 people have the same birthday?

42. Death Rates According to the U.S. National Center for Health Statistics, 29% of all deaths in 2001 were due to heart disease.
(a) What is the probability that a randomly selected death in 2001 was due to heart disease?
(b) What is the probability that a randomly selected death in 2001 was not due to heart disease?

43. Unemployment According to the U.S. Bureau of Labor Statistics, 5.8% of the U.S. labor force was unemployed in 2002.
(a) What is the probability that a randomly selected member of the U.S. labor force was unemployed in 2002?
(b) What is the probability that a randomly selected member of the U.S. labor force was not unemployed in 2002?

44. From a box containing three 40-watt bulbs, six 60-watt bulbs, and eleven 75-watt bulbs, a bulb is drawn at random. What is the probability that the bulb is 40 watts? What is the probability that it is not a 75-watt bulb?

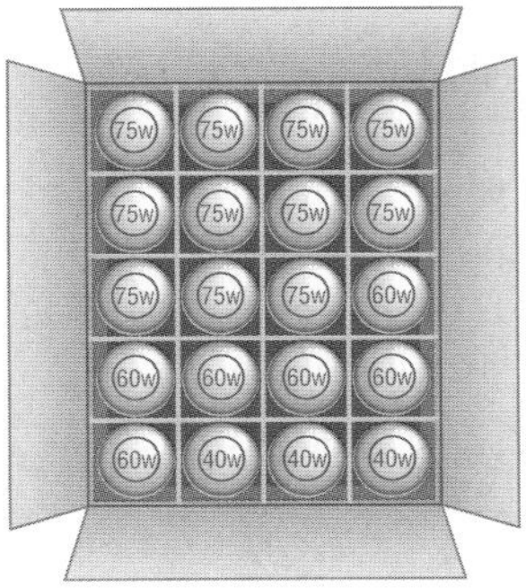

45. You have four $1 bills, three $5 bills, and two $10 bills in your wallet. If you pick a bill at random, what is the probability that it will be a $1 bill?

46. Each of the letters in the word ROSE is written on an index card and the cards are then shuffled. What is the probability that, when the cards are dealt out, they spell the word ROSE?

47. Each of the numbers, 1, 2, . . . , 100 is written on an index card and the cards are then shuffled. If a card is selected at random, what is the probability that the number on the card is divisible by 5? What is the probability that the card selected is either a 1 or names a prime number?

48. At the Milex tune-up and brake repair shop, the manager has found that a car will require a tune-up with a probability of 0.6, a brake job with a probability of 0.1, and both with a probability of 0.02.
(a) What is the probability that a car requires either a tune-up or a brake job?
(b) What is the probability that a car requires a tune-up but not a brake job?
(c) What is the probability that a car requires neither a tune-up nor a brake job?

Chapter Projects

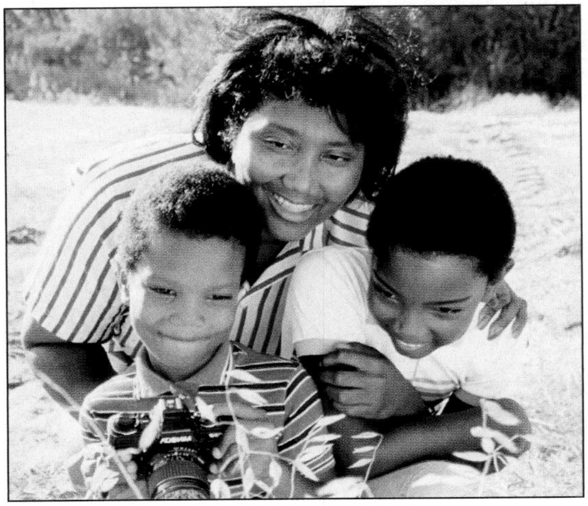

1. Simulation In the Winter 1998 edition of *Eightysomething!*, Mike Koehler uses simulation to calculate the following probabilities: "A woman and man (unrelated) each have two children. At least one of the woman's children is a boy, and the man's older child is a boy. Do the chances that the woman has two boys equal the chances that the man has two boys?" Perform a simulation to answer the question.

The following Chapter Projects are available at www.prenhall.com/Sullivan7e.

2. Project at Motorola ***Probability of Error in Digital Wireless Communications***

3. Surveys

4. Law of Large Numbers

Cumulative Review

1. Solve $3x^2 - 2x = -1$.

2. Graph $f(x) = x^2 + 4x - 5$ by determining whether the graph opens up or down and by finding the vertex, axis of symmetry, and intercepts.

3. Graph $f(x) = 2(x + 1)^2 - 4$ using transformations.

4. Solve $|x - 4| \leq 0.01$.

5. Find the complex zeros of $f(x) = 5x^4 - 9x^3 - 7x^2 - 31x - 6$.

6. Graph $g(x) = 3^{x-1} + 5$ using transformations. Determine the domain, range, and horizontal asymptote of g.

7. What is the exact value of $\log_3 9$?

8. Solve $\log_2(3x - 2) + \log_2 x = 4$.

9. Solve the system: $\begin{cases} x - 2y + z = 15 \\ 3x + y - 3z = -8 \\ -2x + 4y - z = -27 \end{cases}$

10. What is the 33rd term in the sequence $-3, 1, 5, 9, \ldots$? What is the sum of the first 20 terms?

11. Graph $y = 3\sin(2x + \pi)$.

12. Solve the following triangle and determine its area.

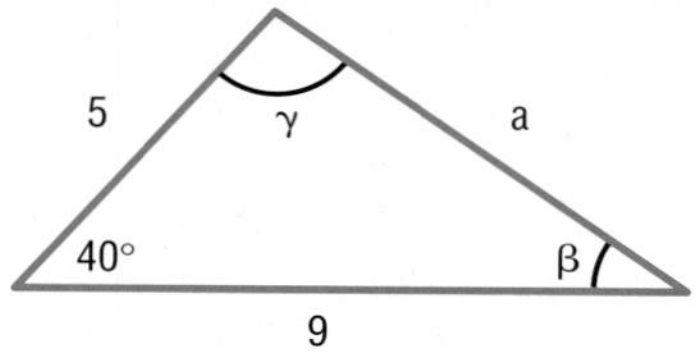

13 A Preview of Calculus: The Limit, Derivative, and Integral of a Function

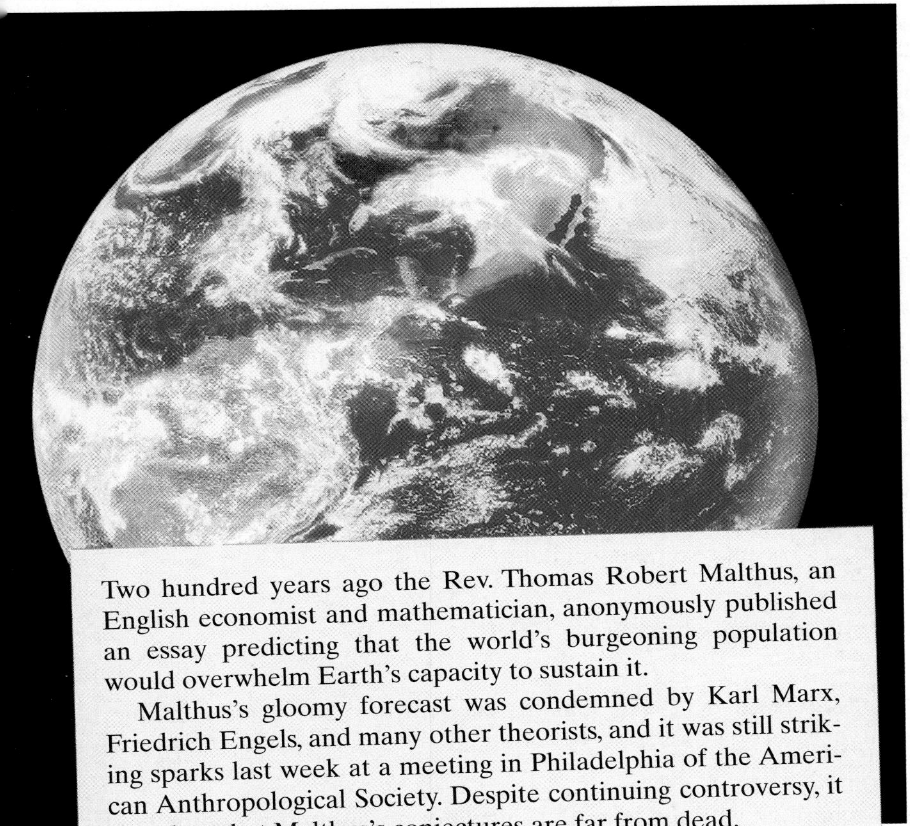

OUTLINE

Two hundred years ago the Rev. Thomas Robert Malthus, an English economist and mathematician, anonymously published an essay predicting that the world's burgeoning population would overwhelm Earth's capacity to sustain it.

Malthus's gloomy forecast was condemned by Karl Marx, Friedrich Engels, and many other theorists, and it was still striking sparks last week at a meeting in Philadelphia of the American Anthropological Society. Despite continuing controversy, it was clear that Malthus's conjectures are far from dead.

Among the scores of special conferences organized for the 5,000 participating anthropologists, many touched directly and indirectly on the Malthusian dilemma: Although global food supplies increase arithmetically, the population increases geometrically—a vastly faster rate.

("Will Humans Overwhelm the Earth? The Debate Continues," Malcolm W. Browne, *New York Times*, December 8, 1998.)

—SEE CHAPTER PROJECT 1.

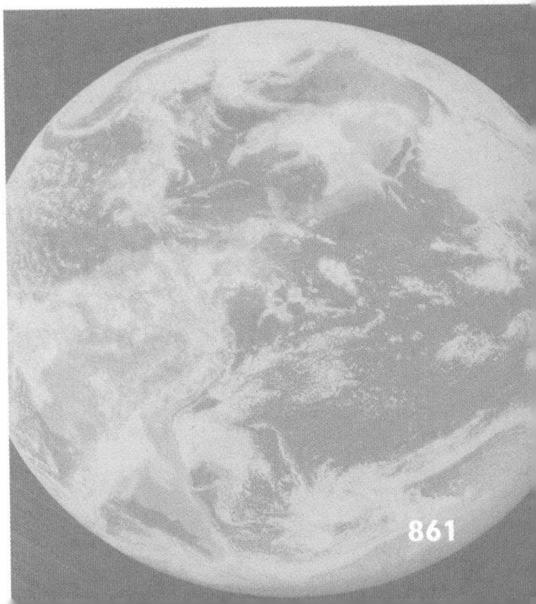

13.1 Finding Limits Using Tables and Graphs

PREPARING FOR THIS SECTION *Before getting started, review the following:*

- Piecewise–defined Functions (Section 2.4, pp. 88–90)

Now work the 'Are You Prepared?' problems on page 866.

OBJECTIVES 1 Find a Limit Using a Table
2 Find a Limit Using a Graph

1 The idea of the limit of a function is what connects algebra and geometry to the mathematics of calculus. In working with the limit of a function, we encounter notation of the form

$$\lim_{x \to c} f(x) = N$$

This is read as "the limit of $f(x)$ as x approaches c equals the number N." Here f is a function defined on some open interval containing the number c; f need not be defined at c, however.

We may describe the meaning of $\lim_{x \to c} f(x) = N$ as follows:

For all x approximately equal to c, with $x \neq c$, the corresponding value $f(x)$ is approximately equal to N.

Another description of $\lim_{x \to c} f(x) = N$ is

As x gets closer to c, but remains unequal to c, the corresponding value of $f(x)$ gets closer to N.

Tables generated with the help of a calculator are useful for finding limits.

EXAMPLE 1 Finding a Limit Using a Table

Find: $\lim_{x \to 3} (5x^2)$

Solution Here $f(x) = 5x^2$ and $c = 3$. We choose values of x close to 3, arbitrarily starting with 2.99. Then we select additional numbers that get closer to 3, but remain less than 3. Next we choose values of x greater than 3, starting with 3.01, that get closer to 3. Finally, we evaluate f at each choice to obtain Table 1.

Table 1

x	2.99	2.999	2.9999 →	← 3.0001	3.001	3.01
$f(x) = 5x^2$	44.701	44.97	44.997 →	← 45.003	45.03	45.301

From Table 1, we infer that as x gets closer to 3 the value of $f(x) = 5x^2$ gets closer to 45. That is,

$$\lim_{x \to 3}(5x^2) = 45$$

◀

When choosing the values of x in a table, the number to start with and the subsequent entries are arbitrary. However, the entries should be chosen so that the table makes it clear what the corresponding values of f are getting close to.

Table 2

X	Y1
2.99	44.701
2.999	44.97
2.9999	44.997
3.0001	45.003
3.001	45.03
3.01	45.301

Y1=5X²

COMMENT: A graphing utility with a TABLE feature can be used to generate the entries. Table 2 shows the result using a TI-83 Plus. ■

NOW WORK PROBLEM 7.

EXAMPLE 2 Finding a Limit Using a Table

Find: (a) $\lim_{x \to 2} \dfrac{x^2 - 4}{x - 2}$ (b) $\lim_{x \to 2}(x + 2)$

Solution (a) Here $f(x) = \dfrac{x^2 - 4}{x - 2}$ and $c = 2$. Notice that the domain of f is $\{x \mid x \neq 2\}$, so f is not defined at 2. We proceed to choose values of x close to 2 and evaluate f at each choice, as shown in Table 3.

Table 3

x	1.99	1.999	1.9999 →	← 2.0001	2.001	2.01
$f(x) = \dfrac{x^2 - 4}{x - 2}$	3.99	3.999	3.9999 →	← 4.0001	4.001	4.01

We infer that as x gets closer to 2 the value of $f(x) = \dfrac{x^2 - 4}{x - 2}$ gets closer to 4. That is,

$$\lim_{x \to 2} \frac{x^2 - 4}{x - 2} = 4$$

(b) Here $g(x) = x + 2$ and $c = 2$. The domain of g is all real numbers. See Table 4.

Table 4

x	1.99	1.999	1.9999 →	← 2.0001	2.001	2.01
$f(x) = x + 2$	3.99	3.999	3.9999 →	← 4.0001	4.001	4.01

We infer that as x gets closer to 2 the value of $g(x)$ gets closer to 4. That is,

$$\lim_{x \to 2}(x + 2) = 4$$

◀

Check: Use a graphing utility with a TABLE feature to verify the results obtained in Example 2.

The conclusion that $\lim_{x \to 2}(x + 2) = 4$ could have been obtained without the use of Table 4; as x gets closer to 2, it follows that $x + 2$ will get closer to $2 + 2 = 4$.

Also, for part (a), you are right if you make the observation that, since $x \neq 2$, then

$$f(x) = \frac{x^2 - 4}{x - 2} = \frac{\cancel{(x-2)}(x+2)}{\cancel{x-2}} = x + 2, \qquad x \neq 2$$

Now it is easy to conclude that

$$\lim_{x \to 2} \frac{x^2 - 4}{x - 2} = \lim_{x \to 2}(x + 2) = 4$$

Let's look at an example for which the factoring technique used above does not work.

EXAMPLE 3 Finding a Limit Using a Table

Find: $\lim_{x \to 0} \frac{\sin x}{x}$

Solution First, we observe that the domain of the function $f(x) = \frac{\sin x}{x}$ is $\{x \mid x \neq 0\}$. We create Table 5, where x is measured in radians.

Table 5

x (radians)	−0.03	−0.02	−0.01	→	← 0.01	0.02	0.03
$f(x) = \frac{\sin x}{x}$	0.99985	0.99993	0.99998	→	← 0.99998	0.99993	0.99985

We infer from Table 5 that $\lim_{x \to 0} \frac{\sin x}{x} = 1$. ◄

Check: Use a graphing utility with a TABLE feature to verify the result obtained in Example 3.

2 The graph of a function f can also be of help in finding limits. See Figure 1. In each graph, notice that, as x gets closer to c, the value of f gets closer to the number N. We conclude that

$$\lim_{x \to c} f(x) = N$$

This is the conclusion regardless of the value of f at c. In Figure 1(a), $f(c) = N$, and in Figure 1(b), $f(c) \neq N$. Figure 1(c) illustrates that $\lim_{x \to c} f(x) = N$, even if f is not defined at c.

Figure 1

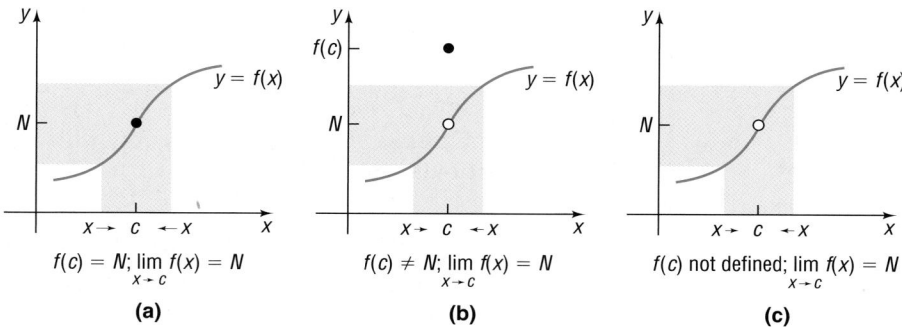

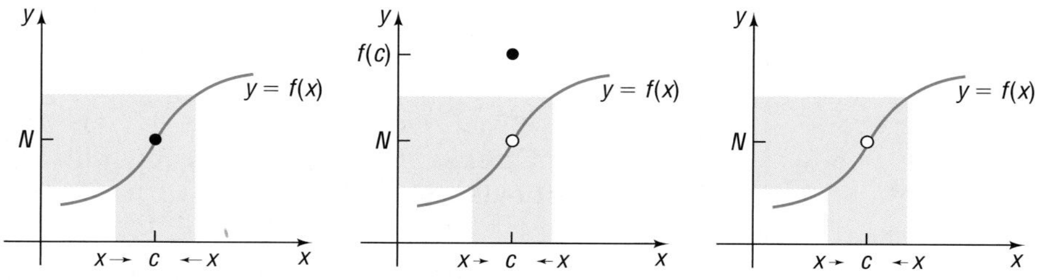

$f(c) = N$; $\lim_{x \to c} f(x) = N$ (a) $f(c) \neq N$; $\lim_{x \to c} f(x) = N$ (b) $f(c)$ not defined; $\lim_{x \to c} f(x) = N$ (c)

EXAMPLE 4 Finding a Limit by Graphing

Find: $\lim_{x\to 2} f(x)$ if $f(x) = \begin{cases} 3x - 2 & \text{if } x \neq 2 \\ 3 & \text{if } x = 2 \end{cases}$

Figure 2

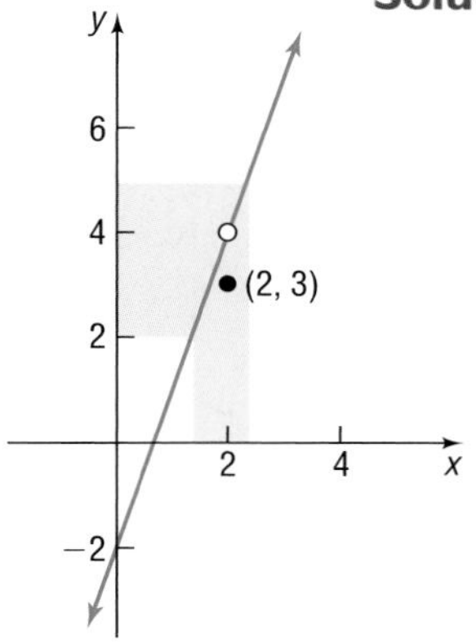

Solution The function f is a piecewise-defined function. Its graph is shown in Figure 2. We conclude from the graph that $\lim_{x\to 2} f(x) = 4$. ◀

Notice in Example 4 that the value of f at 2, that is, $f(2) = 3$, plays no role in the conclusion that $\lim_{x\to 2} f(x) = 4$. In fact, even if f were undefined at 2, it would still happen that $\lim_{x\to 2} f(x) = 4$.

NOW WORK PROBLEM 23.

Sometimes there is no *single* number that the values of f get closer to as x gets closer to c. In this case, we say that **f has no limit as x approaches c** or that $\lim_{x\to c} f(x)$ **does not exist.**

EXAMPLE 5 A Function That Has No Limit at 0

Find: $\lim_{x\to 0} f(x)$ if $f(x) = \begin{cases} x & \text{if } x \le 0 \\ 1 & \text{if } x > 0 \end{cases}$

Figure 3

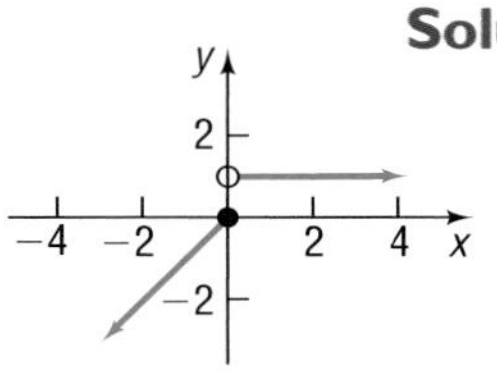

Solution See Figure 3. As x gets closer to 0, but remains negative, the value of f also gets closer to 0. As x gets closer to 0, but remains positive, the value of f always equals 1. Since there is no single number that the values of f are close to when x is close to 0, we conclude that $\lim_{x\to 0} f(x)$ does not exist. ◀

NOW WORK PROBLEMS 17 AND 37.

EXAMPLE 6 Using a Graphing Utility to Find a Limit

Find: $\lim_{x\to 2} \dfrac{x^3 - 2x^2 + 4x - 8}{x^4 - 2x^3 + x - 2}$

Solution Table 6 shows the solution, from which we conclude that

$$\lim_{x\to 2} \frac{x^3 - 2x^2 + 4x - 8}{x^4 - 2x^3 + x - 2} = 0.889$$

rounded to three decimal places. ◀

Table 6

X	Y1
1	2.5
1.5	1.4286
1.8	1.0597
1.9	.96832
1.99	.89635
1.999	.88963
1.9999	.88896

Y1■(X^3−2X²+4X−...

X	Y1
3	.46429
2.5	.61654
2.3	.70555
2.1	.81961
2.01	.88153
2.001	.88815
2.0001	.88881

Y1■(X^3−2X²+4X−...

NOW WORK PROBLEM 43.

In the next section, we will see how algebra can be used to obtain exact solutions to limits like the one in Example 6.

13.1 Assess Your Understanding

'Are You Prepared?' *Answers are given at the end of these exercises. If you get a wrong answer, read the pages listed in red.*

1. Graph $f(x) = \begin{cases} 3x - 2 & \text{if } x \neq 2 \\ 3 & \text{if } x = 2 \end{cases}$ (pp. 88–90)

2. If $f(x) = \begin{cases} x & \text{if } x \leq 0 \\ 1 & \text{if } x > 0 \end{cases}$ what is $f(0)$? (pp. 88–90)

Concepts and Vocabulary

3. The limit of a function $f(x)$ as x approaches c is denoted by the symbol __________.

4. If a function f has no limit as x approaches c, then we say that $\lim_{x \to c} f(x)$ __________ __________ __________.

5. *True or False:* $\lim_{x \to c} f(x) = N$ may be described by saying that the value of $f(x)$ gets closer to N as x gets closer to c, but remains unequal to c.

6. *True or False:* $\lim_{x \to c} f(x)$ exists and equals some number for any function f as long as c is in the domain of f.

Exercises

In Problems 7–16, use a table to find the indicated limit.

7. $\lim_{x \to 2} (4x^3)$

8. $\lim_{x \to 3} (2x^2 + 1)$

9. $\lim_{x \to 0} \dfrac{x + 1}{x^2 + 1}$

10. $\lim_{x \to 0} \dfrac{2 - x}{x^2 + 4}$

11. $\lim_{x \to 4} \dfrac{x^2 - 4x}{x - 4}$

12. $\lim_{x \to 3} \dfrac{x^2 - 9}{x^2 - 3x}$

13. $\lim_{x \to 0} (e^x + 1)$

14. $\lim_{x \to 0} \dfrac{e^x - e^{-x}}{2}$

15. $\lim_{x \to 0} \dfrac{\cos x - 1}{x}$, x in radians

16. $\lim_{x \to 0} \dfrac{\tan x}{x}$, x in radians

In Problems 17–22, use the graph shown to determine if the limit exists. If it does, find its value.

17.

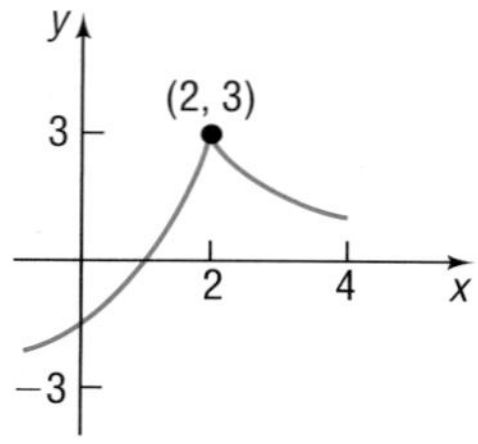

$\lim_{x \to 2} f(x)$

18.

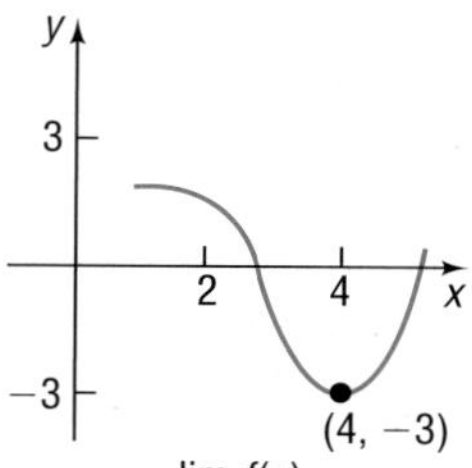

$\lim_{x \to 4} f(x)$

19.

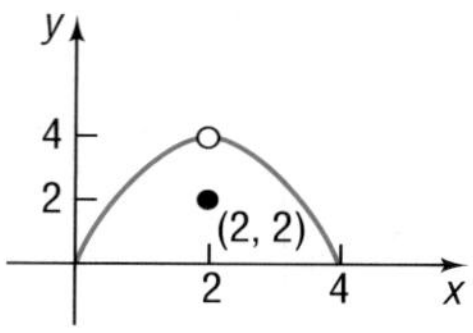

$\lim_{x \to 2} f(x)$

20.

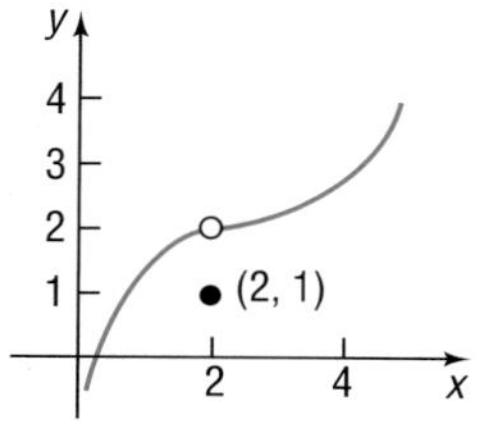

$\lim_{x \to 2} f(x)$

21.

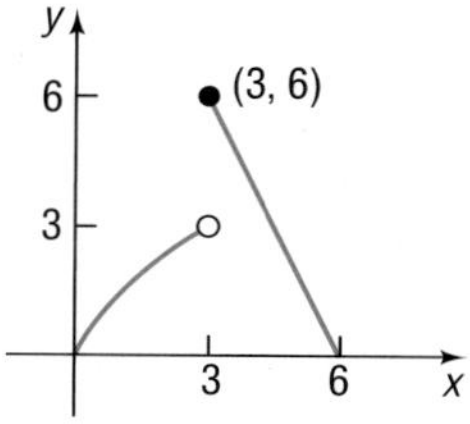

$\lim_{x \to 3} f(x)$

22.

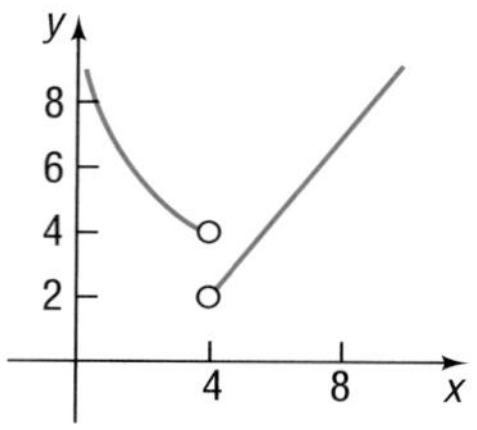

$\lim_{x \to 4} f(x)$

In Problems 23–42, graph each function. Use the graph to find the indicated limit, if it exists.

23. $\lim_{x \to 4} f(x)$, $f(x) = 3x + 1$

24. $\lim_{x \to -1} f(x)$, $f(x) = 2x - 1$

25. $\lim_{x \to 2} f(x)$, $f(x) = 1 - x^2$

26. $\lim_{x \to -1} f(x)$, $f(x) = x^3 - 1$

27. $\lim_{x \to -3} f(x)$, $f(x) = |2x|$

28. $\lim_{x \to 4} f(x)$, $f(x) = 3\sqrt{x}$

29. $\lim_{x \to \pi/2} f(x)$, $f(x) = \sin x$ **30.** $\lim_{x \to \pi} f(x)$, $f(x) = \cos x$ **31.** $\lim_{x \to 0} f(x)$, $f(x) = e^x$

32. $\lim_{x \to 1} f(x)$, $f(x) = \ln x$ **33.** $\lim_{x \to -1} f(x)$, $f(x) = \frac{1}{x}$ **34.** $\lim_{x \to 2} f(x)$, $f(x) = \frac{1}{x^2}$

35. $\lim_{x \to 0} f(x)$, $f(x) = \begin{cases} x^2 & x \geq 0 \\ 2x & x < 0 \end{cases}$ **36.** $\lim_{x \to 0} f(x)$, $f(x) = \begin{cases} x - 1 & x < 0 \\ 3x - 1 & x \geq 0 \end{cases}$

37. $\lim_{x \to 1} f(x)$, $f(x) = \begin{cases} 3x & x \leq 1 \\ x + 1 & x > 1 \end{cases}$ **38.** $\lim_{x \to 2} f(x)$, $f(x) = \begin{cases} x^2 & x \leq 2 \\ 2x - 1 & x > 2 \end{cases}$

39. $\lim_{x \to 0} f(x)$, $f(x) = \begin{cases} x & x < 0 \\ 1 & x = 0 \\ 3x & x > 0 \end{cases}$ **40.** $\lim_{x \to 0} f(x)$, $f(x) = \begin{cases} 1 & x < 0 \\ -1 & x > 0 \end{cases}$

41. $\lim_{x \to 0} f(x)$, $f(x) = \begin{cases} \sin x & x \leq 0 \\ x^2 & x > 0 \end{cases}$ **42.** $\lim_{x \to 0} f(x)$, $f(x) = \begin{cases} e^x & x > 0 \\ 1 - x & x \leq 0 \end{cases}$

In Problems 43–48, use a graphing utility to find the indicated limit rounded to two decimal places.

43. $\lim_{x \to 1} \frac{x^3 - x^2 + x - 1}{x^4 - x^3 + 2x - 2}$ **44.** $\lim_{x \to -1} \frac{x^3 + x^2 + 3x + 3}{x^4 + x^3 + 2x + 2}$ **45.** $\lim_{x \to 2} \frac{x^3 - 2x^2 + 4x - 8}{x^2 + x - 6}$

46. $\lim_{x \to 1} \frac{x^3 - x^2 + 3x - 3}{x^2 + 3x - 4}$ **47.** $\lim_{x \to -1} \frac{x^3 + 2x^2 + x}{x^4 + x^3 + 2x + 2}$ **48.** $\lim_{x \to 3} \frac{x^3 - 3x^2 + 4x - 12}{x^4 - 3x^3 + x - 3}$

'Are You Prepared?' Answers

1. See Figure 2 on page 865. **2.** $f(0) = 0$

13.2 Algebra Techniques for Finding Limits

OBJECTIVES
1. Find the Limit of a Sum, a Difference, a Product, and a Quotient
2. Find the Limit of a Polynomial
3. Find the Limit of a Power or a Root
4. Find the Limit of an Average Rate of Change

We mentioned in the previous section that algebra can sometimes be used to find the exact value of a limit. This is accomplished by developing two formulas involving limits and several properties of limits.

Theorem

Two Formulas: $\lim_{x \to c} b$ and $\lim_{x \to c} x$

Limit of a Constant

For the constant function $f(x) = b$,

$$\lim_{x \to c} f(x) = \lim_{x \to c} b = b \qquad (1)$$

where c is any number.

In Words
The limit of a constant is the constant.

Limit of x

For the identity function $f(x) = x$,

$$\lim_{x \to c} f(x) = \lim_{x \to c} x = c \qquad (2)$$

where c is any number.

In Words
The limit of x as x approaches c is c.

We use graphs to establish formulas (1) and (2). Since the graph of a constant function is a horizontal line, it follows that, no matter how close x is to c, the corresponding value of $f(x)$ equals b. That is, $\lim_{x \to c} b = b$. See Figure 4.

See Figure 5. For any choice of c, as x gets closer to c, the corresponding value of $f(x)$ is x, which is just as close to c. That is, $\lim_{x \to c} x = c$.

Figure 4

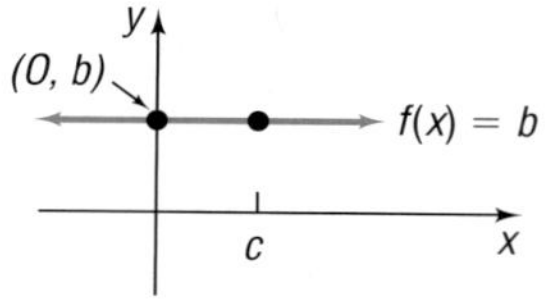

Figure 5

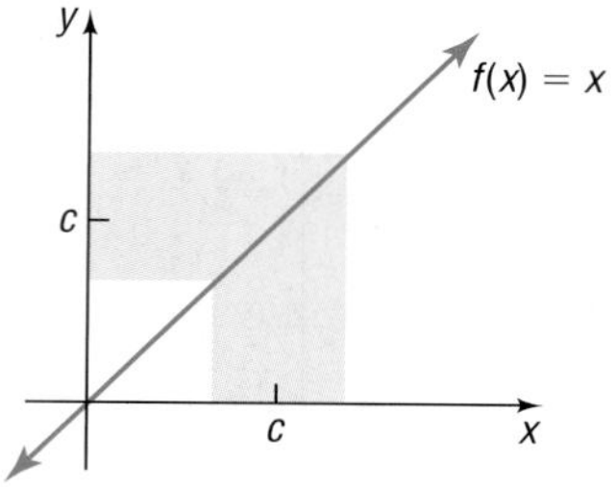

EXAMPLE 1 Using Formulas (1) and (2)

(a) $\lim_{x \to 3} 5 = 5$ (b) $\lim_{x \to 3} x = 3$ (c) $\lim_{x \to 0} -8 = -8$ (d) $\lim_{x \to -1/2} x = -\frac{1}{2}$ ◀

NOW WORK PROBLEM 7.

Formulas (1) and (2), when used with the properties that follow, enable us to evaluate limits of more complicated functions.

Properties of Limits

1 In the following properties, we assume that f and g are two functions for which both $\lim_{x \to c} f(x)$ and $\lim_{x \to c} g(x)$ exist.

Theorem

Limit of a Sum

$$\lim_{x \to c}[f(x) + g(x)] = \lim_{x \to c} f(x) + \lim_{x \to c} g(x) \quad (3)$$

In Words
The limit of the sum of two functions equals the sum of their limits.

EXAMPLE 2 Finding the Limit of a Sum

Find: $\lim_{x \to -3}(x + 4)$

Solution The limit we seek is the sum of two functions $f(x) = x$ and $g(x) = 4$. From formulas (1) and (2), we know that

$$\lim_{x \to -3} f(x) = \lim_{x \to -3} x = -3 \quad \text{and} \quad \lim_{x \to -3} g(x) = \lim_{x \to -3} 4 = 4$$

From formula (3), it follows that

$$\lim_{x \to -3}(x + 4) = \lim_{x \to -3} x + \lim_{x \to -3} 4 = -3 + 4 = 1$$ ◀

Theorem **Limit of a Difference**

In Words
The limit of the difference of two functions equals the difference of their limits.

$$\lim_{x \to c}[f(x) - g(x)] = \lim_{x \to c} f(x) - \lim_{x \to c} g(x) \qquad \textbf{(4)}$$

EXAMPLE 3 **Finding the Limit of a Difference**

Find: $\lim_{x \to 4}(6 - x)$

Solution The limit we seek is the difference of two functions $f(x) = 6$ and $g(x) = x$. From formulas (1) and (2), we know that

$$\lim_{x \to 4} f(x) = \lim_{x \to 4} 6 = 6 \quad \text{and} \quad \lim_{x \to 4} g(x) = \lim_{x \to 4} x = 4$$

From formula (4), it follows that

$$\lim_{x \to 4}(6 - x) = \lim_{x \to 4} 6 - \lim_{x \to 4} x = 6 - 4 = 2$$

◀

Theorem **Limit of a Product**

In Words
The limit of the product of two functions equals the product of their limits.

$$\lim_{x \to c}[f(x) \cdot g(x)] = [\lim_{x \to c} f(x)][\lim_{x \to c} g(x)] \qquad \textbf{(5)}$$

EXAMPLE 4 **Finding the Limit of a Product**

Find: $\lim_{x \to -5}(-4x)$

Solution The limit we seek is the product of two functions $f(x) = -4$ and $g(x) = x$. From formulas (1) and (2), we know that

$$\lim_{x \to -5} f(x) = \lim_{x \to -5} (-4) = -4 \quad \text{and} \quad \lim_{x \to -5} g(x) = \lim_{x \to -5} x = -5$$

From formula (5), it follows that

$$\lim_{x \to -5}(-4x) = [\lim_{x \to -5} -4][\lim_{x \to -5} x] = (-4)(-5) = 20$$

◀

EXAMPLE 5 **Finding Limits Using Algebraic Properties**

Find: (a) $\lim_{x \to -2}(3x - 5)$ (b) $\lim_{x \to 2}(5x^2)$

Solution (a) $\lim_{x \to -2}(3x - 5) = \lim_{x \to -2}(3x) - \lim_{x \to -2} 5 = [\lim_{x \to -2} 3][\lim_{x \to -2} x] - \lim_{x \to -2} 5$

$$= (3)(-2) - 5 = -6 - 5 = -11$$

(b) $\lim_{x \to 2}(5x^2) = [\lim_{x \to 2} 5][\lim_{x \to 2} x^2] = 5 \lim_{x \to 2}(x \cdot x) = 5[\lim_{x \to 2} x][\lim_{x \to 2} x]$

$= 5 \cdot 2 \cdot 2 = 20$ ◀

NOW WORK PROBLEM 11.

Notice in the solution to part (b) that $\lim_{x \to 2}(5x^2) = 5 \cdot 2^2$.

Theorem

Limit of a Monomial

If $n \geq 1$ is a positive integer and a is a constant, then

$$\lim_{x \to c}(ax^n) = ac^n \qquad (6)$$

for any number c.

Proof

$$\lim_{x \to c}(ax^n) = [\lim_{x \to c} a][\lim_{x \to c} x^n] = a[\lim_{x \to c} \underbrace{(x \cdot x \cdot x \cdot \ldots \cdot x)}_{n \text{ factors}}]$$

$$= a\underbrace{[\lim_{x \to c} x][\lim_{x \to c} x][\lim_{x \to c} x] \ldots [\lim_{x \to c} x]}_{n \text{ factors}}$$

$$= a \cdot \underbrace{c \cdot c \cdot c \cdot \ldots \cdot c}_{n \text{ factors}} = ac^n$$

■

EXAMPLE 6 **Finding the Limit of a Monomial**

Find: $\lim_{x \to 2}(-4x^3)$

Solution $\lim_{x \to 2}(-4x^3) = -4 \cdot 2^3 = -4 \cdot 8 = -32$ ◀

2 Since a polynomial is a sum of monomials, we can use formula (6) and repeated use of formula (3) to obtain the following result:

Theorem

Limit of a Polynomial

If P is a polynomial function, then

$$\lim_{x \to c} P(x) = P(c) \qquad (7)$$

for any number c.

Proof If P is a polynomial function, that is, if

$$P(x) = a_n x^n + a_{n-1}x^{n-1} + \cdots + a_1 x + a_0$$

then

$$\begin{aligned}\lim_{x\to c} P(x) &= \lim_{x\to c}[a_n x^n + a_{n-1}x^{n-1} + \cdots + a_1 x + a_0] \\ &= \lim_{x\to c}(a_n x^n) + \lim_{x\to c}(a_{n-1}x^{n-1}) + \cdots + \lim_{x\to c}(a_1 x) + \lim_{x\to c} a_0 \\ &= a_n c^n + a_{n-1}c^{n-1} + \cdots + a_1 c + a_0 \\ &= P(c)\end{aligned}$$

■

Formula (7) states that to find the limit of a polynomial as x approaches c all we need to do is to evaluate the polynomial at c.

EXAMPLE 7 Finding the Limit of a Polynomial

Find: $\lim_{x\to 2}[5x^4 - 6x^3 + 3x^2 + 4x - 2]$

Solution

$$\begin{aligned}\lim_{x\to 2}[5x^4 - 6x^3 + 3x^2 + 4x - 2] &= 5\cdot 2^4 - 6\cdot 2^3 + 3\cdot 2^2 + 4\cdot 2 - 2 \\ &= 5\cdot 16 - 6\cdot 8 + 3\cdot 4 + 8 - 2 \\ &= 80 - 48 + 12 + 6 = 50\end{aligned}$$

◀

NOW WORK PROBLEM 13.

Theorem

Limit of a Power or Root

If $\lim_{x\to c} f(x)$ exists and if $n \geq 2$ is a positive integer, then

$$\lim_{x\to c}[f(x)]^n = [\lim_{x\to c} f(x)]^n \quad \textbf{(8)}$$

and

$$\lim_{x\to c}\sqrt[n]{f(x)} = \sqrt[n]{\lim_{x\to c} f(x)} \quad \textbf{(9)}$$

In formula (9), we require that both $\sqrt[n]{f(x)}$ and $\sqrt[n]{\lim_{x\to c} f(x)}$ be defined.

3

EXAMPLE 8 Finding the Limit of a Power or a Root

Find:

(a) $\lim_{x\to 1}(3x - 5)^4$ (b) $\lim_{x\to 0}\sqrt{5x^2 + 8}$ (c) $\lim_{x\to -1}(5x^3 - x + 3)^{4/3}$

Solution

(a) $\lim_{x\to 1}(3x - 5)^4 = [\lim_{x\to 1}(3x - 5)]^4 = (-2)^4 = 16$

(b) $\lim_{x\to 0}\sqrt{5x^2 + 8} = \sqrt{\lim_{x\to 0}(5x^2 + 8)} = \sqrt{8} = 2\sqrt{2}$

(c) $$\begin{aligned}\lim_{x\to -1}(5x^3 - x + 3)^{4/3} &= \sqrt[3]{\lim_{x\to -1}(5x^3 - x + 3)^4} \\ &= \sqrt[3]{[\lim_{x\to -1}(5x^3 - x + 3)]^4} = \sqrt[3]{(-1)^4} = \sqrt[3]{1} = 1\end{aligned}$$

◀

NOW WORK PROBLEM 23.

Theorem

In Words
The limit of the quotient of two functions equals the quotient of their limits, provided that the limit of the denominator is not zero.

Limit of a Quotient

$$\lim_{x\to c}\left[\frac{f(x)}{g(x)}\right] = \frac{\lim_{x\to c} f(x)}{\lim_{x\to c} g(x)} \tag{10}$$

provided that $\lim_{x\to c} g(x) \neq 0$.

EXAMPLE 9 Finding the Limit of a Quotient

Find: $\lim_{x\to 1} \dfrac{5x^3 - x + 2}{3x + 4}$

Solution The limit we seek is the quotient of two functions: $f(x) = 5x^3 - x + 2$ and $g(x) = 3x + 4$. First, we find the limit of the denominator $g(x)$.

$$\lim_{x\to 1} g(x) = \lim_{x\to 1}(3x + 4) = 7$$

Since the limit of the denominator is not zero, we can proceed to use formula (10).

$$\lim_{x\to 1} \frac{5x^3 - x + 2}{3x + 4} = \frac{\lim_{x\to 1}(5x^3 - x + 2)}{\lim_{x\to 1}(3x + 4)} = \frac{6}{7}$$

◀

NOW WORK PROBLEM 21.

When the limit of the denominator is zero, formula (10) cannot be used. In such cases, other strategies need to be used. Let's look at two examples.

EXAMPLE 10 Finding the Limit of a Quotient

Find: (a) $\lim_{x\to 3} \dfrac{x^2 - x - 6}{x^2 - 9}$ (b) $\lim_{x\to 0} \dfrac{5x - \sin x}{x}$

Solution (a) The limit of the denominator equals zero, so formula (10) cannot be used. Instead, we notice that the expression can be factored as

$$\frac{x^2 - x - 6}{x^2 - 9} = \frac{(x - 3)(x + 2)}{(x - 3)(x + 3)}$$

When we compute a limit as x approaches 3, we are interested in the values of the function when x is close to 3, but unequal to 3. Since $x \neq 3$, we can cancel the $(x - 3)$'s. Formula (10) can then be used.

$$\lim_{x\to 3} \frac{x^2 - x - 6}{x^2 - 9} = \lim_{x\to 3} \frac{\cancel{(x - 3)}(x + 2)}{\cancel{(x - 3)}(x + 3)} = \frac{\lim_{x\to 3}(x + 2)}{\lim_{x\to 3}(x + 3)} = \frac{5}{6}$$

(b) Again, the limit of the denominator is zero. In this situation, we perform the indicated operation and divide by x.

$$\lim_{x \to 0} \frac{5x - \sin x}{x} = \lim_{x \to 0}\left[\frac{5x}{x} - \frac{\sin x}{x}\right] = \lim_{x \to 0} \frac{5\cancel{x}}{\cancel{x}} - \lim_{x \to 0} \frac{\sin x}{x} = 5 - 1 = 4$$

↑ Refer to Example 3, Section 13.1 ◀

Let's work Example 6 of Section 13.1 using algebra.

EXAMPLE 11 Finding Limits Using Algebraic Properties

Find: $\lim_{x \to 2} \frac{x^3 - 2x^2 + 4x - 8}{x^4 - 2x^3 + x - 2}$

Solution The limit of the denominator is zero, so formula (10) cannot be used. We factor the expression.

$$\frac{x^3 - 2x^2 + 4x - 8}{x^4 - 2x^3 + x - 2} = \frac{x^2(x - 2) + 4(x - 2)}{x^3(x - 2) + 1(x - 2)} = \frac{(x^2 + 4)(x - 2)}{(x^3 + 1)(x - 2)}$$

↑ Factor by grouping

Then

$$\lim_{x \to 2} \frac{x^3 - 2x^2 + 4x - 8}{x^4 - 2x^3 + x - 2} = \lim_{x \to 2} \frac{(x^2 + 4)\cancel{(x - 2)}}{(x^3 + 1)\cancel{(x - 2)}} = \frac{8}{9}$$

which is exact. ◀

Compare the exact solution above with the approximate solution found in Example 6 of Section 13.1.

4 EXAMPLE 12 Finding the Limit of an Average Rate of Change

Find the limit as x approaches 2 of the average rate of change of the function

$$f(x) = x^2 + 3x$$

from 2 to x.

Solution The average rate of change of f from 2 to x is

$$\frac{\Delta y}{\Delta x} = \frac{f(x) - f(2)}{x - 2} = \frac{(x^2 + 3x) - 10}{x - 2} = \frac{(x + 5)(x - 2)}{x - 2}$$

The limit of the average rate of change is

$$\lim_{x \to 2} \frac{f(x) - f(2)}{x - 2} = \lim_{x \to 2} \frac{(x^2 + 3x) - 10}{x - 2} = \lim_{x \to 2} \frac{(x + 5)\cancel{(x - 2)}}{\cancel{x - 2}} = 7$$ ◀

Summary

To find exact values for $\lim_{x \to c} f(x)$, try the following:

1. If f is a polynomial function, then $\lim_{x \to c} f(x) = f(c)$ (formula 7).
2. If f is a polynomial raised to a power or is the root of a polynomial, use formulas (8) and (9) with formula (7).
3. If f is a quotient and the limit of the denominator is not zero, use the fact that the limit of a quotient is the quotient of the limits.
4. If f is a quotient and the limit of the denominator is zero, use other techniques, such as factoring.

13.2 Assess Your Understanding

Concepts and Vocabulary

1. The limit of the product of two functions equals the ________ of their limits.

2. $\lim_{x \to 0} 5 =$ ________.

3. $\lim_{x \to 1} x =$ ________.

4. *True or False:* The limit of a polynomial function as x approaches 5 equals the value of the polynomial at 5.

5. *True or False:* The limit of a rational function at 5 equals the value of the rational function at 5.

6. *True or False:* The limit of a quotient equals the quotient of the limits.

Exercises

In Problems 7–38, find the limit algebraically.

7. $\lim_{x \to 1} 5$

8. $\lim_{x \to 1} (-3)$

9. $\lim_{x \to 4} x$

10. $\lim_{x \to -3} x$

11. $\lim_{x \to 2} (3x + 2)$

12. $\lim_{x \to 3} (2 - 5x)$

13. $\lim_{x \to -1} (3x^2 - 5x)$

14. $\lim_{x \to 2} (8x^2 - 4)$

15. $\lim_{x \to 1} (5x^4 - 3x^2 + 6x - 9)$

16. $\lim_{x \to -1} (8x^5 - 7x^3 + 8x^2 + x - 4)$

17. $\lim_{x \to 1} (x^2 + 1)^3$

18. $\lim_{x \to 2} (3x - 4)^2$

19. $\lim_{x \to 1} \sqrt{5x + 4}$

20. $\lim_{x \to 0} \sqrt{1 - 2x}$

21. $\lim_{x \to 0} \dfrac{x^2 - 4}{x^2 + 4}$

22. $\lim_{x \to 2} \dfrac{3x + 4}{x^2 + x}$

23. $\lim_{x \to 2} (3x - 2)^{5/2}$

24. $\lim_{x \to -1} (2x + 1)^{5/3}$

25. $\lim_{x \to 2} \dfrac{x^2 - 4}{x^2 - 2x}$

26. $\lim_{x \to -1} \dfrac{x^2 + x}{x^2 - 1}$

27. $\lim_{x \to -3} \dfrac{x^2 - x - 12}{x^2 - 9}$

28. $\lim_{x \to -3} \dfrac{x^2 + x - 6}{x^2 + 2x - 3}$

29. $\lim_{x \to 1} \dfrac{x^3 - 1}{x - 1}$

30. $\lim_{x \to 1} \dfrac{x^4 - 1}{x - 1}$

31. $\lim_{x \to -1} \dfrac{(x + 1)^2}{x^2 - 1}$

32. $\lim_{x \to 2} \dfrac{x^3 - 8}{x^2 - 4}$

33. $\lim_{x \to 1} \dfrac{x^3 - x^2 + x - 1}{x^4 - x^3 + 2x - 2}$

34. $\lim_{x \to -1} \dfrac{x^3 + x^2 + 3x + 3}{x^4 + x^3 + 2x + 2}$

35. $\lim_{x \to 2} \dfrac{x^3 - 2x^2 + 4x - 8}{x^2 + x - 6}$

36. $\lim_{x \to 1} \dfrac{x^3 - x^2 + 3x - 3}{x^2 + 3x - 4}$

37. $\lim_{x \to -1} \dfrac{x^3 + 2x^2 + x}{x^4 + x^3 + 2x + 2}$

38. $\lim_{x \to 3} \dfrac{x^3 - 3x^2 + 4x - 12}{x^4 - 3x^3 + x - 3}$

In Problems 39–48, find the limit as x approaches c of the average rate of change of each function from c to x.

39. $c = 2;\ f(x) = 5x - 3$

40. $c = -2;\ f(x) = 4 - 3x$

41. $c = 3;\ f(x) = x^2$

42. $c = 3;\quad f(x) = x^3$

43. $c = -1;\quad f(x) = x^2 + 2x$

44. $c = -1;\quad f(x) = 2x^2 - 3x$

45. $c = 0;\quad f(x) = 3x^3 - 2x^2 + 4$

46. $c = 0;\quad f(x) = 4x^3 - 5x + 8$

47. $c = 1;\quad f(x) = \dfrac{1}{x}$

48. $c = 1;\quad f(x) = \dfrac{1}{x^2}$

In Problems 49–52, use the properties of limits and the facts that

$$\lim_{x\to 0}\frac{\sin x}{x} = 1 \qquad \lim_{x\to 0}\frac{\cos x - 1}{x} = 0 \qquad \lim_{x\to 0}\sin x = 0 \qquad \lim_{x\to 0}\cos x = 1$$

where x is in radians, to find each limit.

49. $\displaystyle\lim_{x\to 0}\frac{\tan x}{x}$

50. $\displaystyle\lim_{x\to 0}\frac{\sin(2x)}{x}$

[**Hint:** Use a Double-angle Formula.]

51. $\displaystyle\lim_{x\to 0}\frac{3\sin x + \cos x - 1}{4x}$

52. $\displaystyle\lim_{x\to 0}\frac{\sin^2 x + \sin x(\cos x - 1)}{x^2}$

13.3 One-sided Limits; Continuous Functions

PREPARING FOR THIS SECTION *Before getting started, review the following:*

- Piecewise-defined Functions (Section 2.4, pp. 88–90)
- Library of Functions (Section 2.4, pp. 83–88)
- Properties of the Logarithmic Function (Section 4.4, p. 269)
- Polynomial and Rational Functions (Sections 3.2 and 3.3, pp. 146 and 181–183)
- Properties of the Exponential Function (Section 4.3, p. 255)
- Properties of the Trigonometric Functions (Section 5.3, pp. 355–366)

Now work the 'Are You Prepared?' problems on page 880.

OBJECTIVES 1 Find the One-sided Limits of a Function
2 Determine Whether a Function Is Continuous

1 Earlier we described $\lim_{x\to c} f(x) = N$ by saying that as x gets closer to c, but remains unequal to c, the corresponding values of $f(x)$ get closer to N. Whether we use a numerical argument or the graph of the function f, the variable x can get closer to c in only two ways: either by approaching c from the left, through numbers less than c, or by approaching c from the right, through numbers greater than c.

If we only approach c from one side, we have a **one-sided limit.** The notation

$$\lim_{x\to c^-} f(x) = L$$

sometimes called the **left limit,** read as "the limit of $f(x)$ as x approaches c from the left equals L," may be described by the following statement:

As x gets closer to c, but remains less than c, the corresponding value of $f(x)$ gets closer to L.

In Words
$x \to c^-$ means $x < c$.

The notation $x \to c^-$ is used to remind us that x is less than c.

The notation

$$\lim_{x \to c^+} f(x) = R$$

sometimes called the **right limit,** read as "the limit of $f(x)$ as x approaches c from the right equals R," may be described by the following statement:

As x gets closer to c, but remains greater than c, the corresponding value of $f(x)$ gets closer to R.

In Words
$x \to c^+$ means $x > c$.

The notation $x \to c^+$ is used to remind us that x is greater than c.

Figure 6 illustrates left and right limits.

Figure 6

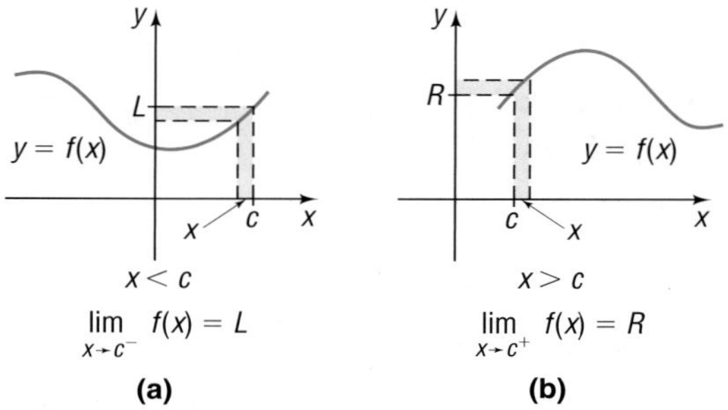

The left and right limits can be used to determine whether $\lim_{x \to c} f(x)$ exists. See Figure 7.

Figure 7

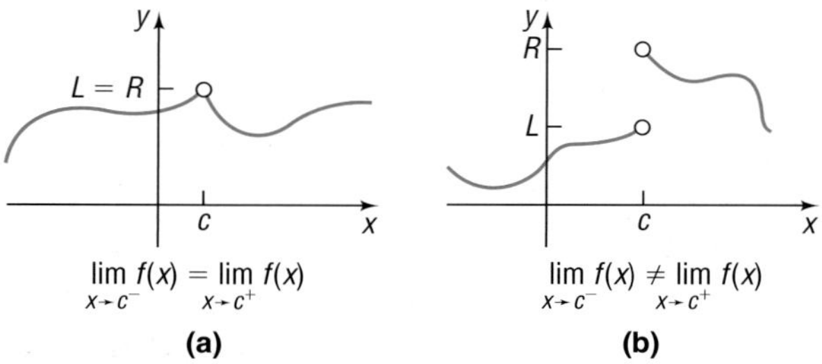

As Figure 7(a) illustrates, $\lim_{x \to c} f(x)$ exists and equals the common value of the left limit and the right limit ($L = R$). In Figure 7(b), we see that $\lim_{x \to c} f(x)$ does not exist because $L \neq R$. This leads us to the following result:

Theorem

Suppose that $\lim_{x \to c^-} f(x) = L$ and $\lim_{x \to c^+} f(x) = R$. Then $\lim_{x \to c} f(x)$ exists if and only if $L = R$. Furthermore, if $L = R$, then $\lim_{x \to c} f(x) = L\ (=R)$.

Collectively, the left and right limits of a function are called **one-sided limits** of the function.

| **EXAMPLE 1** | **Finding One-sided Limits of a Function** |

For the function

$$f(x) = \begin{cases} 2x - 1 & \text{if } x < 2 \\ 1 & \text{if } x = 2 \\ x - 2 & \text{if } x > 2 \end{cases}$$

find: (a) $\lim\limits_{x \to 2^-} f(x)$ (b) $\lim\limits_{x \to 2^+} f(x)$ (c) $\lim\limits_{x \to 2} f(x)$

Figure 8

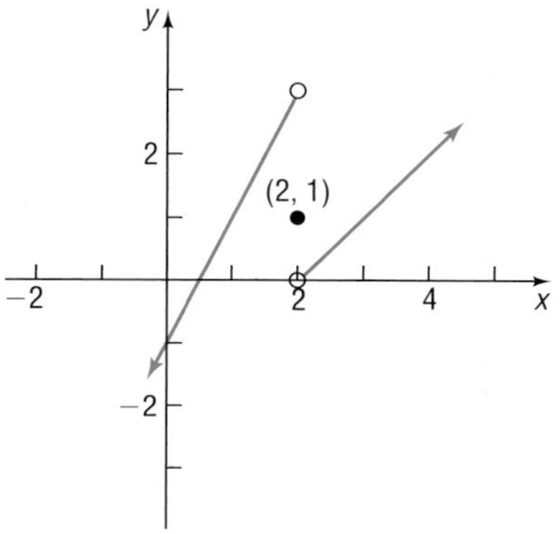

Solution Figure 8 shows the graph of f.

(a) To find $\lim\limits_{x \to 2^-} f(x)$, we look at the values of f when x is close to 2, but less than 2. Since $f(x) = 2x - 1$ for such numbers, we conclude that

$$\lim_{x \to 2^-} f(x) = \lim_{x \to 2^-} (2x - 1) = 3$$

(b) To find $\lim\limits_{x \to 2^+} f(x)$, we look at the values of f when x is close to 2, but greater than 2. Since $f(x) = x - 2$ for such numbers, we conclude that

$$\lim_{x \to 2^+} f(x) = \lim_{x \to 2^+} (x - 2) = 0$$

(c) Since the left and right limits are unequal, $\lim\limits_{x \to 2} f(x)$ does not exist. ◀

NOW WORK PROBLEMS 21 AND 35.

Continuous Functions

2 We have observed that the value of a function f at c, $f(c)$, plays no role in determining the one-sided limits of f at c. What is the role of the value of a function at c and its one-sided limits at c? Let's look at some of the possibilities. See Figure 9.

Figure 9

$\lim\limits_{x \to c^-} f(x) = \lim\limits_{x \to c^+} f(x)$, so $\lim\limits_{x \to c} f(x)$ exists;

$\lim\limits_{x \to c} f(x) = f(c)$

(a)

$\lim\limits_{x \to c^-} f(x) = \lim\limits_{x \to c^+} f(x)$, so $\lim\limits_{x \to c} f(x)$ exists;

$\lim\limits_{x \to c} f(x) \neq f(c)$

(b)

$\lim\limits_{x \to c^-} f(x) = \lim\limits_{x \to c^+} f(x)$, so $\lim\limits_{x \to c} f(x)$ exists;

$f(c)$ is not defined

(c)

$\lim\limits_{x \to c^-} f(x) \neq \lim\limits_{x \to c^+} f(x)$, so $\lim\limits_{x \to c} f(x)$ does not exist;

$f(c)$ is defined

(d)

$\lim\limits_{x \to c^-} f(x) \neq \lim\limits_{x \to c^+} f(x)$, so $\lim\limits_{x \to c} f(x)$ does not exist;

$f(c)$ is not defined

(e)

$\lim\limits_{x \to c^-} f(x) = f(c) \neq \lim\limits_{x \to c^+} f(x)$

so $\lim\limits_{x \to c} f(x)$ does not exist

$f(c)$ is defined **(f)**

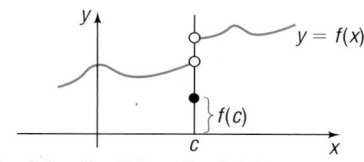

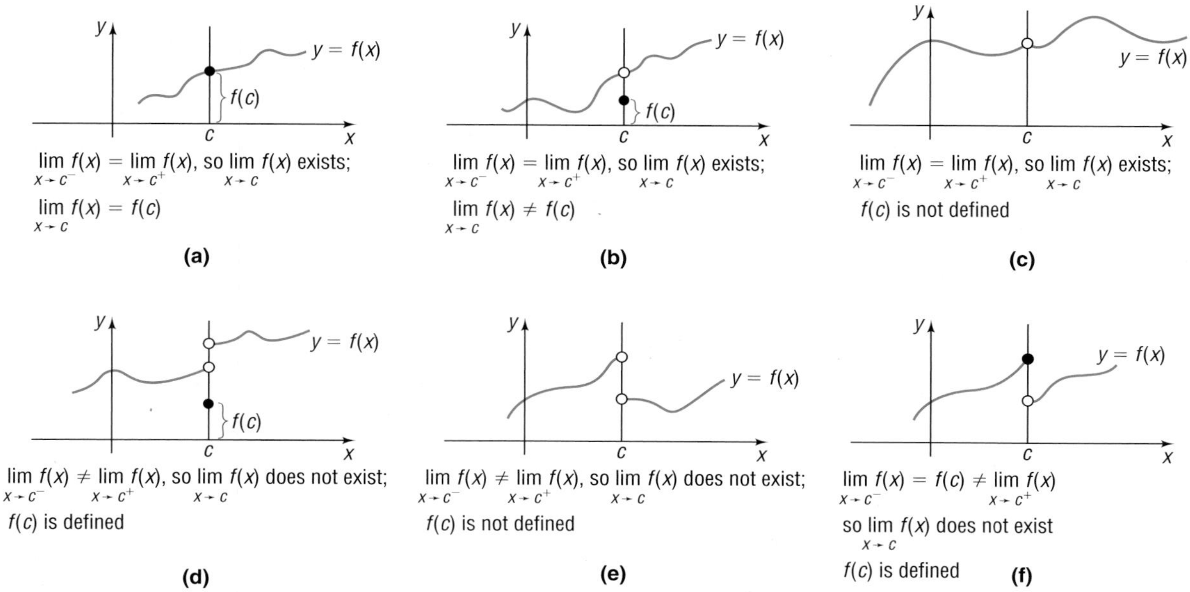

Much earlier in this book, we said that a function f was *continuous* if its graph could be drawn without lifting pencil from paper. In looking at Figure 9, the only graph that has this characteristic is the graph in Figure 9(a), for which the one-sided limits at c each exist and are equal to the value of f at c. This leads us to the following definition:

A function f is **continuous** at c if:

1. f is defined at c; that is, c is in the domain of f so that $f(c)$ equals a number.
2. $\lim_{x \to c^-} f(x) = f(c)$
3. $\lim_{x \to c^+} f(x) = f(c)$

In other words, a function f is continuous at c if

$$\lim_{x \to c} f(x) = f(c)$$

If f is not continuous at c, we say that f is **discontinuous at *c*.** Each of the functions whose graphs appear in Figures 9(b) to 9(f) is discontinuous at c.

Look again at formula (7) on page 870. Based on (7), we conclude that a polynomial function is continuous at every number.

Look at formula (10) on page 872. We conclude that a rational function is continuous at every number, except any at which it is not defined. At numbers where a rational function is not defined, either a hole appears in the graph or else an asymptote appears.

NOW WORK PROBLEM 27.

EXAMPLE 2 Determining the Numbers at Which a Rational Function Is Continuous

(a) Determine the numbers at which the rational function

$$R(x) = \frac{x - 2}{x^2 - 6x + 8}$$

is continuous.

(b) Use limits to analyze the graph of R near 2 and near 4.

(c) Graph R.

Solution (a) Since $R(x) = \dfrac{x - 2}{(x - 2)(x - 4)}$, the domain of R is $\{x | x \neq 2, x \neq 4\}$. We conclude that R is discontinuous at both 2 and 4. (Condition 1 of the definition is violated.) Based on formula (10) (page 872), R is continuous at every number except 2 and 4.

(b) To determine the behavior of the graph near 2 and near 4, we look at $\lim_{x \to 2} R(x)$ and $\lim_{x \to 4} R(x)$.

For $\lim_{x \to 2} R(x)$, we have

$$\lim_{x \to 2} R(x) = \lim_{x \to 2} \frac{\cancel{x - 2}}{\cancel{(x - 2)}(x - 4)} = \lim_{x \to 2} \frac{1}{x - 4} = -\frac{1}{2}$$

As x gets closer to 2, the graph of R gets closer to $-\frac{1}{2}$. Since R is not defined at 2, the graph will have a hole at $\left(2, -\frac{1}{2}\right)$.

For $\lim_{x \to 4} R(x)$, we have

$$\lim_{x \to 4} R(x) = \lim_{x \to 4} \frac{\cancel{x-2}}{\cancel{(x-2)}(x-4)} = \lim_{x \to 4} \frac{1}{x-4}$$

If $x < 4$ and x is getting closer to 4, the value of $\frac{1}{x-4} < 0$ and is becoming unbounded; that is, $\lim_{x \to 4^-} R(x) = -\infty$.

If $x > 4$ and x is getting closer to 4, the value of $\frac{1}{x-4} > 0$ and is becoming unbounded; that is, $\lim_{x \to 4^+} R(x) = \infty$.

Figure 10

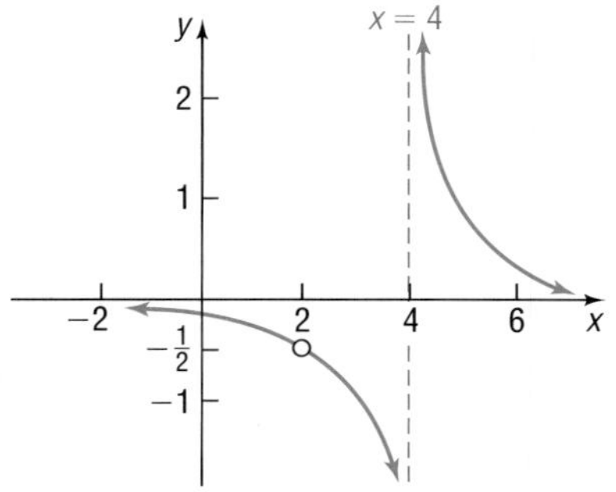

Since $|R(x)| \to \infty$ for x close to 4, the graph of R will have a vertical asymptote at $x = 4$.

(c) It is easiest to graph R by observing that

$$\text{if } x \neq 2, \qquad \text{then } R(x) = \frac{\cancel{x-2}}{\cancel{(x-2)}(x-4)} = \frac{1}{x-4}$$

So the graph of R is the graph of $y = \frac{1}{x}$ shifted to the right 4 units with a hole at $\left(2, -\frac{1}{2}\right)$. See Figure 10. ◀

NOW WORK PROBLEM 73.

The exponential, logarithmic, sine, and cosine functions are continuous at every number in their domain. The tangent, cotangent, secant, and cosecant functions are continuous except at numbers for which they are not defined, where asymptotes occur. The square root function and absolute value function are continuous at every number in their domain. The function $f(x) = \text{int}(x)$ is continuous except for $x =$ an integer, where a jump occurs in the graph.

Piecewise-defined functions require special attention.

EXAMPLE 3 Determining Where a Piecewise-defined Function Is Continuous

Determine the numbers at which the following function is continuous.

$$f(x) = \begin{cases} x^2 & \text{if } x \leq 0 \\ x + 1 & \text{if } 0 < x < 2 \\ 5 - x & \text{if } 2 \leq x \leq 5 \end{cases}$$

Solution The "pieces" of f, that is, $y = x^2$, $y = x + 1$, and $y = 5 - x$, are each continuous for every number since they are polynomials. In other words, when we graph the pieces, we will not lift our pencil. When we graph the function f, however, we have to be careful, because the pieces change at $x = 0$ and at $x = 2$. So the numbers we need to investigate for f are $x = 0$ and $x = 2$.

$$\text{At } x = 0: \qquad f(0) = 0^2 = 0$$

$$\lim_{x \to 0^-} f(x) = \lim_{x \to 0^-} x^2 = 0$$

$$\lim_{x \to 0^+} f(x) = \lim_{x \to 0^+} (x + 1) = 1$$

Figure 11

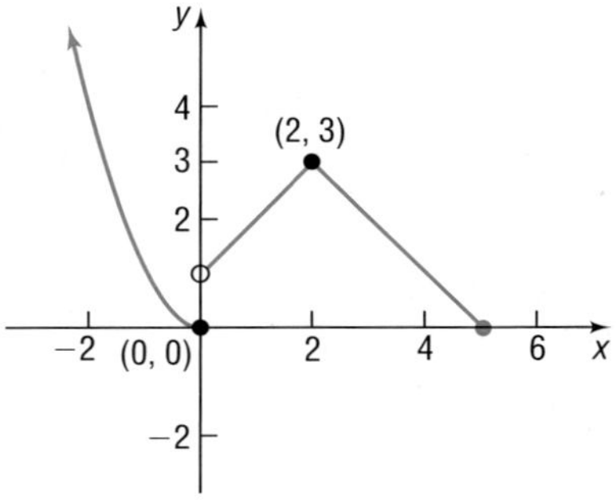

Since $\lim_{x \to 0^+} f(x) \neq f(0)$, we conclude that f is not continuous at $x = 0$.

$$\text{At } x = 2: \quad f(2) = 5 - 2 = 3$$

$$\lim_{x \to 2^-} f(x) = \lim_{x \to 2^-} (x + 1) = 3$$

$$\lim_{x \to 2^+} f(x) = \lim_{x \to 2^+} (5 - x) = 3$$

We conclude that f is continuous at $x = 2$.

The graph of f, given in Figure 11, demonstrates the conclusions drawn above. ◀

NOW WORK PROBLEMS 53 AND 61.

Summary

Library of Functions: Continuity Properties

Function	Domain	Property
Polynomial function	All real numbers	Continuous at every number in the domain
Rational function $R(x) = \dfrac{P(x)}{Q(x)}$, P, Q are polynomials	$\{x \mid Q(x) \neq 0\}$	Continuous at every number in the domain Hole or vertical asymptote where R is undefined
Exponential function	All real numbers	Continuous at every number in the domain
Logarithmic function	Positive real numbers	Continuous at every number in the domain
Sine and cosine functions	All real numbers	Continuous at every number in the domain
Tangent and secant functions	All real numbers, except odd multiples of $\dfrac{\pi}{2}$	Continuous at every number in the domain Vertical asymptotes at odd multiples of $\dfrac{\pi}{2}$
Cotangent and cosecant functions	All real numbers, except multiples of π	Continuous at every number in the domain Vertical asymptotes at multiples of π

13.3 Assess Your Understanding

'Are You Prepared?' *Answers are given at the end of these exercises. If you get a wrong answer, read the pages listed in red.*

1. For the function $f(x) = \begin{cases} x^2 & \text{if } x \leq 0 \\ x + 1 & \text{if } 0 < x < 2, \\ 5 - x & \text{if } 2 \leq x \leq 5 \end{cases}$ find $f(0)$ and $f(2)$. (pp. 88–90)

2. What is the domain and range of $f(x) = \ln x$? (p. 269)

3. *True or False:* The exponential function $f(x) = e^x$ is increasing on the interval $(-\infty, \infty)$. (p. 255)

4. Name the trigonometric functions that have asymptotes. (pp. 355–366)

5. *True or False:* Some rational functions have holes in their graph. (pp. 181–183)

6. *True or False:* Every polynomial function has a graph that can be traced without lifting pencil from paper. (p. 146)

Concepts and Vocabulary

7. If we only approach c from one side, then we have a(n) __________ limit.

8. The notation __________ is used to describe the fact that as x gets closer to c, but remains greater than c, the value of $f(x)$ gets closer to R.

9. If $\lim_{x \to c} f(x) = f(c)$, then f is __________ at __________.

10. *True or False:* For any function f, $\lim_{x \to c^-} f(x) = \lim_{x \to c^+} f(x)$.

11. *True or False:* If f is continuous at c, then $\lim_{x \to c^+} f(x) = f(c)$.

12. *True or False:* Every polynomial function is continuous at every real number.

Exercises

In Problems 13–32, use the accompanying graph of $y = f(x)$.

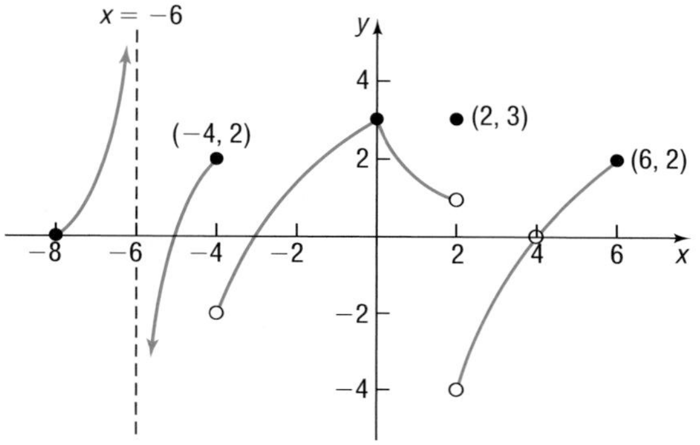

13. What is the domain of f?

14. What is the range of f?

15. Find the x-intercept(s), if any, of f.

16. Find the y-intercept(s), if any, of f.

17. Find $f(-8)$ and $f(-4)$.

18. Find $f(2)$ and $f(6)$.

19. Find $\lim_{x \to -6^-} f(x)$.

20. Find $\lim_{x \to -6^+} f(x)$.

21. Find $\lim_{x \to -4^-} f(x)$.

22. Find $\lim_{x \to -4^+} f(x)$.

23. Find $\lim_{x \to 2^-} f(x)$.

24. Find $\lim_{x \to 2^+} f(x)$.

25. Does $\lim_{x \to 4} f(x)$ exist? If it does, what is it?

26. Does $\lim_{x \to 0} f(x)$ exist? If it does, what is it?

27. Is f continuous at -6?

28. Is f continuous at -4?

29. Is f continuous at 0?

30. Is f continuous at 2?

31. Is f continuous at 4?

32. Is f continuous at 5?

In Problems 33–44, find the one-sided limit.

33. $\lim_{x \to 1^+} (2x + 3)$

34. $\lim_{x \to 2^-} (4 - 2x)$

35. $\lim_{x \to 1^-} (2x^3 + 5x)$

36. $\lim_{x \to -2^+} (3x^2 - 8)$

37. $\lim_{x \to \pi/2^+} \sin x$

38. $\lim_{x \to \pi^-} (3 \cos x)$

39. $\lim_{x \to 2^+} \dfrac{x^2 - 4}{x - 2}$

40. $\lim_{x \to 1^-} \dfrac{x^3 - x}{x - 1}$

41. $\lim_{x \to -1^-} \dfrac{x^2 - 1}{x^3 + 1}$

42. $\lim_{x \to 0^+} \dfrac{x^3 - x^2}{x^4 + x^2}$

43. $\lim_{x \to -2^+} \dfrac{x^2 + x - 2}{x^2 + 2x}$

44. $\lim_{x \to -4^-} \dfrac{x^2 + x - 12}{x^2 + 4x}$

In Problems 45–60, determine whether f is continuous at c.

45. $f(x) = x^3 - 3x^2 + 2x - 6, \quad c = 2$

46. $f(x) = 3x^2 - 6x + 5, \quad c = -3$

47. $f(x) = \dfrac{x^2 + 5}{x - 6}, \quad c = 3$

48. $f(x) = \dfrac{x^3 - 8}{x^2 + 4}, \quad c = 2$

49. $f(x) = \dfrac{x + 3}{x - 3}, \quad c = 3$

50. $f(x) = \dfrac{x - 6}{x + 6}, \quad c = -6$

51. $f(x) = \dfrac{x^3 + 3x}{x^2 - 3x}, \quad c = 0$

52. $f(x) = \dfrac{x^2 - 6x}{x^2 + 6x}, \quad c = 0$

53. $f(x) = \begin{cases} \dfrac{x^3 + 3x}{x^2 - 3x} & \text{if } x \neq 0 \\ 1 & \text{if } x = 0 \end{cases}, \quad c = 0$

54. $f(x) = \begin{cases} \dfrac{x^2 - 6x}{x^2 + 6x} & \text{if } x \neq 0 \\ -2 & \text{if } x = 0 \end{cases}, \quad c = 0$

55. $f(x) = \begin{cases} \dfrac{x^3 + 3x}{x^2 - 3x} & \text{if } x \neq 0 \\ -1 & \text{if } x = 0 \end{cases} \quad c = 0$

56. $f(x) = \begin{cases} \dfrac{x^2 - 6x}{x^2 + 6x} & \text{if } x \neq 0 \\ -1 & \text{if } x = 0 \end{cases} \quad c = 0$

57. $f(x) = \begin{cases} \dfrac{x^3 - 1}{x^2 - 1} & \text{if } x < 1 \\ 2 & \text{if } x = 1 \\ \dfrac{3}{x + 1} & \text{if } x > 1 \end{cases} \quad c = 1$

58. $f(x) = \begin{cases} \dfrac{x^2 - 2x}{x - 2} & \text{if } x < 2 \\ 2 & \text{if } x = 2 \\ \dfrac{x - 4}{x - 1} & \text{if } x > 2 \end{cases} \quad c = 2$

59. $f(x) = \begin{cases} 2e^x & \text{if } x < 0 \\ 2 & \text{if } x = 0 \\ \dfrac{x^3 + 2x^2}{x^2} & \text{if } x > 0 \end{cases} \quad c = 0$

60. $f(x) = \begin{cases} 3\cos x & \text{if } x < 0 \\ 3 & \text{if } x = 0 \\ \dfrac{x^3 + 3x^2}{x^2} & \text{if } x > 0 \end{cases} \quad c = 0$

In Problems 61–72, find the numbers at which f is continuous. At which numbers is f discontinuous?

61. $f(x) = 2x + 3$

62. $f(x) = 4 - 3x$

63. $f(x) = 3x^2 + x$

64. $f(x) = -3x^3 + 7$

65. $f(x) = 4 \sin x$

66. $f(x) = -2 \cos x$

67. $f(x) = 2 \tan x$

68. $f(x) = 4 \csc x$

69. $f(x) = \dfrac{2x + 5}{x^2 - 4}$

70. $f(x) = \dfrac{x^2 - 4}{x^2 - 9}$

71. $f(x) = \dfrac{x - 3}{\ln x}$

72. $f(x) = \dfrac{\ln x}{x - 3}$

In Problems 73–76, discuss whether R is continuous at c. Use limits to analyze the graph of R at c. Graph R.

73. $R(x) = \dfrac{x - 1}{x^2 - 1}, c = -1 \text{ and } c = 1$

74. $R(x) = \dfrac{3x + 6}{x^2 - 4}, c = -2 \text{ and } c = 2$

75. $R(x) = \dfrac{x^2 + x}{x^2 - 1}, c = -1 \text{ and } c = 1$

76. $R(x) = \dfrac{x^2 + 4x}{x^2 - 16}, c = -4 \text{ and } c = 4$

In Problems 77–82, determine where each rational function is undefined. Determine whether an asymptote or a hole appears at such numbers.

77. $R(x) = \dfrac{x^3 - x^2 + x - 1}{x^4 - x^3 + 2x - 2}$

78. $R(x) = \dfrac{x^3 + x^2 + 3x + 3}{x^4 + x^3 + 2x + 2}$

79. $R(x) = \dfrac{x^3 - 2x^2 + 4x - 8}{x^2 + x - 6}$

80. $R(x) = \dfrac{x^3 - x^2 + 3x - 3}{x^2 + 3x - 4}$

81. $R(x) = \dfrac{x^3 + 2x^2 + x}{x^4 + x^3 + 2x + 2}$

82. $R(x) = \dfrac{x^3 - 3x^2 + 4x - 12}{x^4 - 3x^3 + x - 3}$

For Problems 83–88, graph the functions R given in Problems 77–82 to verify the solutions found above.

89. Name three functions that are continuous at every real number.

90. Create a function that is not continuous at the number 5.

'Are You Prepared?' Answers

1. $f(0) = 0; f(2) = 3$

2. Domain: $\{x | x > 0\}$; range $\{y | -\infty < y < \infty\}$

3. True

4. Secant, cosecant, tangent, cotangent

5. True

6. True

13.4 The Tangent Problem; The Derivative

PREPARING FOR THIS SECTION *Before getting started, review the following:*

- Point–slope Form of a Line (Section 1.3, p. 27)
- Average Rate of Change (Section 2.3, pp. 78–79)

Now work the 'Are You Prepared?' problems on page 889.

OBJECTIVES
1 Find an Equation of the Tangent Line to the Graph of a Function
2 Find the Derivative of a Function
3 Find Instantaneous Rates of Change
4 Find the Instantaneous Speed of a Particle

Tangent Problem

One question that motivated the development of calculus was a geometry problem, the **tangent problem.** This problem asks, "What is the slope of the tangent line to the graph of a function $y = f(x)$ at a point P on its graph?" See Figure 12.

Figure 12

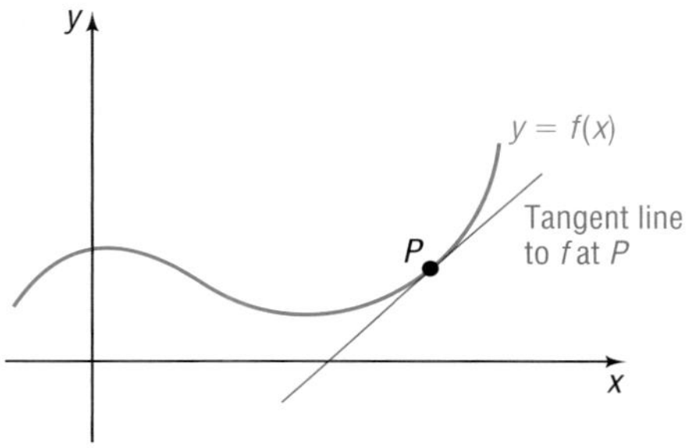

Figure 13

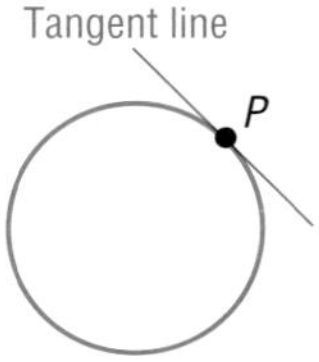

We first need to define what we mean by a *tangent* line. In high school geometry, the tangent line to a circle is defined as the line that intersects the graph in exactly one point. Look at Figure 13. Notice that the tangent line just touches the graph of the circle.

This definition, however, does not work in general. Look at Figure 14. The lines L_1 and L_2 only intersect the graph in one point P, but neither touches the graph at P. Additionally, the tangent line L_T shown in Figure 15 touches the graph of f at P, but also intersects the graph elsewhere. So how should we define the tangent line to the graph of f at a point P?

Figure 14

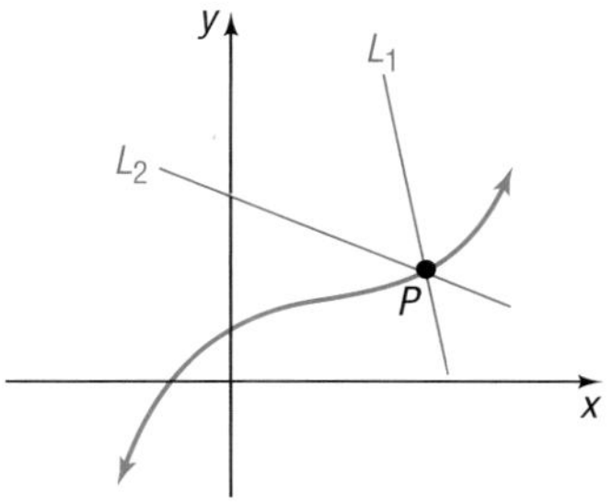

Figure 15

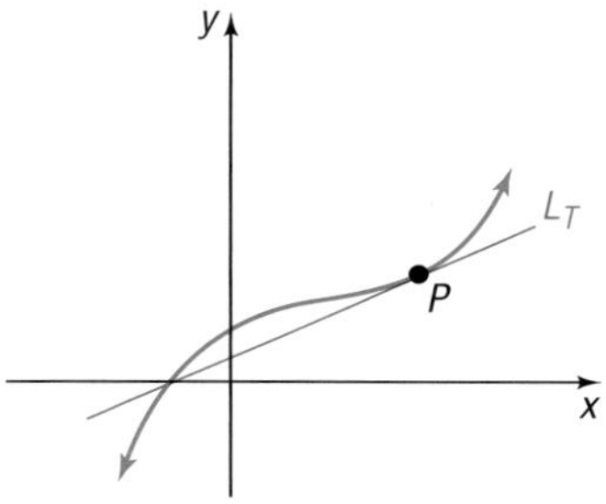

1 The tangent line L_T to the graph of a function $y = f(x)$ at a point P necessarily contains the point P. To find an equation for L_T using the point–slope form of the equation of a line, it remains to find the slope $m_{\tan}$ of the tangent line.

Suppose that the coordinates of the point P are $(c, f(c))$. Locate another point $Q = (x, f(x))$ on the graph of f. The line containing P and Q is a secant line. (Refer to Section 2.3.) The slope $m_{\sec}$ of the secant line is

$$m_{\sec} = \frac{f(x) - f(c)}{x - c}$$

Figure 16

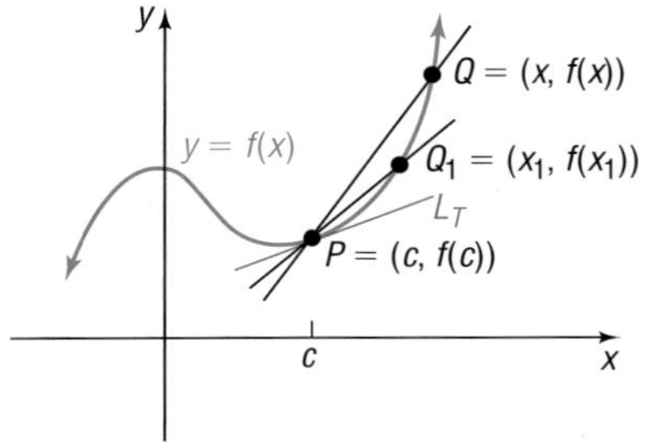

Now look at Figure 16.

As we move along the graph of f from Q toward P, we obtain a succession of secant lines. The closer we get to P, the closer the secant line is to the tangent line L_T. The limiting position of these secant lines is the tangent line L_T. Therefore, the limiting value of the slopes of these secant lines equals the slope of the tangent line. But, as we move from Q toward P, the values of x get closer to c. Therefore,

$$m_{\tan} = \lim_{x \to c} m_{\sec} = \lim_{x \to c} \frac{f(x) - f(c)}{x - c}$$

The **tangent line** to the graph of a function $y = f(x)$ at a point $P = (c, f(c))$ on its graph is defined as the line containing the point P whose slope is

$$m_{\tan} = \lim_{x \to c} \frac{f(x) - f(c)}{x - c} \tag{1}$$

provided that this limit exists. If $m_{\tan}$ exists, an equation of the tangent line is

$$y - f(c) = m_{\tan}(x - c) \tag{2}$$

EXAMPLE 1 **Finding an Equation of the Tangent Line**

Find an equation of the tangent line to the graph of $f(x) = \dfrac{x^2}{4}$ at the point $\left(1, \dfrac{1}{4}\right)$. Graph f and the tangent line.

Solution The tangent line contains the point $\left(1, \dfrac{1}{4}\right)$. The slope of the tangent line to the graph of $f(x) = \dfrac{x^2}{4}$ at $\left(1, \dfrac{1}{4}\right)$ is

$$m_{\tan} = \lim_{x \to 1} \frac{f(x) - f(1)}{x - 1} = \lim_{x \to 1} \frac{\frac{x^2}{4} - \frac{1}{4}}{x - 1} = \lim_{x \to 1} \frac{\cancel{(x - 1)}(x + 1)}{4\cancel{(x - 1)}}$$

$$= \lim_{x \to 1} \frac{x + 1}{4} = \frac{1}{2}$$

An equation of the tangent line is

$$y - \frac{1}{4} = \frac{1}{2}(x - 1) \qquad y - f(c) = m_{\tan}(x - c)$$

$$y = \frac{1}{2}x - \frac{1}{4}$$

Figure 17

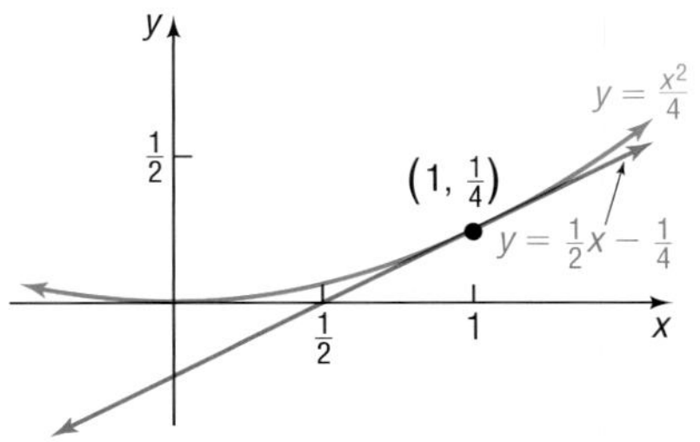

Figure 17 shows the graph of $y = \frac{x^2}{4}$ and the tangent line at $\left(1, \frac{1}{4}\right)$. ◀

NOW WORK PROBLEM 11.

2 The limit in formula (1) has an important generalization: it is called the *derivative of f at c.*

Let $y = f(x)$ denote a function f. If c is a number in the domain of f, the **derivative of f at c,** denoted by $f'(c)$, read "f prime of c," is defined as

$$f'(c) = \lim_{x \to c} \frac{f(x) - f(c)}{x - c} \qquad \textbf{(3)}$$

provided that this limit exists.

EXAMPLE 2 Finding the Derivative of a Function

Find the derivative of $f(x) = 2x^2 - 5x$ at 2. That is, find $f'(2)$.

Solution Since $f(2) = 2(4) - 5(2) = -2$, we have

$$\frac{f(x) - f(2)}{x - 2} = \frac{(2x^2 - 5x) - (-2)}{x - 2} = \frac{2x^2 - 5x + 2}{x - 2} = \frac{(2x - 1)(x - 2)}{x - 2}$$

The derivative of f at 2 is

$$f'(2) = \lim_{x \to 2} \frac{f(x) - f(2)}{x - 2} = \lim_{x \to 2} \frac{(2x - 1)\cancel{(x - 2)}}{\cancel{x - 2}} = 3$$ ◀

NOW WORK PROBLEM 21.

Example 2 provides a way of finding the derivative at 2 analytically. Graphing utilities have built-in procedures to approximate the derivative of a function at any number c. Consult your owner's manual for the appropriate keystrokes.

EXAMPLE 3 Finding the Derivative of a Function Using a Graphing Utility

Figure 18

Use a graphing utility to find the derivative of $f(x) = 2x^2 - 5x$ at 2. That is, find $f'(2)$.

Solution Figure 18 shows the solution using a TI-83 Plus graphing calculator. So $f'(2) = 3$. ◀

NOW WORK PROBLEM 33.

EXAMPLE 4 Finding the Derivative of a Function

Find the derivative of $f(x) = x^2$ at c. That is, find $f'(c)$.

Solution Since $f(c) = c^2$, we have

$$\frac{f(x) - f(c)}{x - c} = \frac{x^2 - c^2}{x - c} = \frac{(x + c)(x - c)}{x - c}$$

The derivative of f at c is

$$f'(c) = \lim_{x \to c} \frac{f(x) - f(c)}{x - c} = \lim_{x \to c} \frac{(x + c)\cancel{(x - c)}}{\cancel{x - c}} = 2c$$ ◀

As Example 4 illustrates, the derivative of $f(x) = x^2$ exists and equals $2c$ for any number c. In other words, the derivative is itself a function and, using x for the independent variable, we can write $f'(x) = 2x$. The function f' is called the **derivative function of f** or the **derivative of f.** We also say that f is **differentiable.** The instruction "differentiate f" means "find the derivative of f."

Interpretations of the Derivative

3 In Chapter 2 we defined the average rate of change of a function f from c to x as

$$\frac{\Delta y}{\Delta x} = \frac{f(x) - f(c)}{x - c}$$

The limit as x approaches c of the average rate of change of f, based on formula (3), is the derivative of f at c. As a result, we call the derivative of f at c the **instantaneous rate of change of f with respect to x at c.** That is,

$$\begin{pmatrix} \text{Instantaneous rate of} \\ \text{change of } f \text{ with respect to } x \text{ at } c \end{pmatrix} = f'(c) = \lim_{x \to c} \frac{f(x) - f(c)}{x - c} \qquad \textbf{(4)}$$

EXAMPLE 5 **Finding the Instantaneous Rate of Change**

The volume V of a right circular cone of height $h = 6$ feet and radius r feet is $V = V(r) = \frac{1}{3}\pi r^2 h = 2\pi r^2$. If r is changing, find the instantaneous rate of change of the volume V with respect to the radius r at $r = 3$.

Solution The instantaneous rate of change of V with respect to r at $r = 3$ is the derivative $V'(3)$.

$$\begin{aligned} V'(3) &= \lim_{r \to 3} \frac{V(r) - V(3)}{r - 3} = \lim_{r \to 3} \frac{2\pi r^2 - 18\pi}{r - 3} = \lim_{r \to 3} \frac{2\pi(r^2 - 9)}{r - 3} \\ &= \lim_{r \to 3}[2\pi(r + 3)] = 12\pi \end{aligned}$$

At the instant $r = 3$ feet, the volume of the cone is increasing with respect to r at a rate of $12\pi \approx 37.699$ cubic feet per 1-foot change in the radius. ◀

NOW WORK PROBLEM 43.

4 If $s = f(t)$ denotes the position of a particle at time t, then the average speed of the particle from c to t is

$$\frac{\text{Change in position}}{\text{Change in time}} = \frac{\Delta s}{\Delta t} = \frac{f(t) - f(c)}{t - c} \qquad \textbf{(5)}$$

The limit as t approaches c of the expression in formula (5) is the **instantaneous speed of the particle at c** or the **velocity of the particle at c.** That is,

$$\begin{pmatrix} \text{Instantaneous speed of} \\ \text{a particle at time } c \end{pmatrix} = f'(c) = \lim_{t \to c} \frac{f(t) - f(c)}{t - c} \qquad \textbf{(6)}$$

EXAMPLE 6 Finding the Instantaneous Speed of a Particle

In physics it is shown that the height s of a ball thrown straight up with an initial speed of 80 feet per second (ft/sec) from a rooftop 96 feet high is

$$s = s(t) = -16t^2 + 80t + 96$$

where t is the elapsed time that the ball is in the air. The ball misses the rooftop on its way down and eventually strikes the ground. See Figure 19.

Figure 19

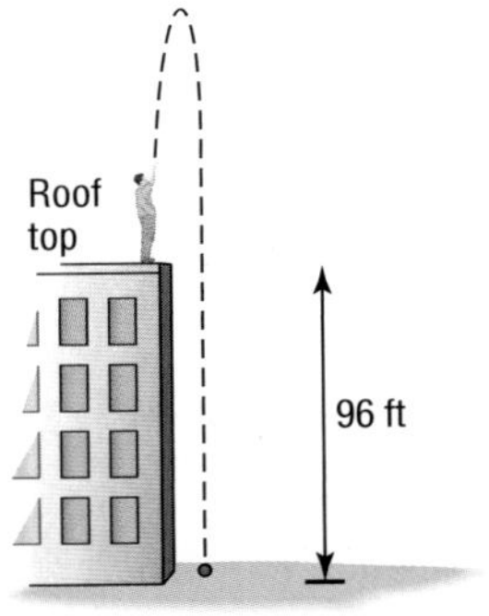

(a) When does the ball strike the ground? That is, how long is the ball in the air?

(b) At what time t will the ball pass the rooftop on its way down?

(c) What is the average speed of the ball from $t = 0$ to $t = 2$?

(d) What is the instantaneous speed of the ball at time t_0?

(e) What is the instantaneous speed of the ball at $t = 2$?

(f) When is the instantaneous speed of the ball equal to zero?

(g) What is the instantaneous speed of the ball as it passes the rooftop on the way down?

(h) What is the instantaneous speed of the ball when it strikes the ground?

Solution (a) The ball strikes the ground when $s = s(t) = 0$.

$$\begin{aligned} -16t^2 + 80t + 96 &= 0 \\ t^2 - 5t - 6 &= 0 \\ (t - 6)(t + 1) &= 0 \\ t = 6 \quad \text{or} \quad t &= -1 \end{aligned}$$

We discard the solution $t = -1$. The ball strikes the ground after 6 sec.

(b) The ball passes the rooftop when $s = s(t) = 96$.

$$\begin{aligned} -16t^2 + 80t + 96 &= 96 \\ t^2 - 5t &= 0 \\ t(t - 5) &= 0 \\ t = 0 \quad \text{or} \quad t &= 5 \end{aligned}$$

We discard the solution $t = 0$. The ball passes the rooftop on the way down after 5 sec.

(c) The average speed of the ball from $t = 0$ to $t = 2$ is

$$\frac{\Delta s}{\Delta t} = \frac{s(2) - s(0)}{2 - 0} = \frac{192 - 96}{2} = 48 \text{ ft/sec}$$

(d) The instantaneous speed of the ball at time t_0 is the derivative $s'(t_0)$; that is,

$$\begin{aligned} s'(t_0) &= \lim_{t \to t_0} \frac{s(t) - s(t_0)}{t - t_0} \\ &= \lim_{t \to t_0} \frac{(-16t^2 + 80t + 96) - (-16t_0^2 + 80t_0 + 96)}{t - t_0} \\ &= \lim_{t \to t_0} \frac{-16[t^2 - t_0^2 - 5t + 5t_0]}{t - t_0} \end{aligned}$$

$$= \lim_{t \to t_0} \frac{-16[(t + t_0)(t - t_0) - 5(t - t_0)]}{t - t_0}$$

$$= \lim_{t \to t_0} \frac{-16[(t + t_0 - 5)\cancel{(t - t_0)}]}{\cancel{t - t_0}} = \lim_{t \to t_0}[-16(t + t_0 - 5)]$$

$$= -16(2t_0 - 5) \quad \text{ft/sec}$$

The instantaneous speed of the ball at time t is

$$s'(t) = -16(2t - 5) \text{ ft/sec}$$

(e) At $t = 2$ sec, the instantaneous speed of the ball is

$$s'(2) = -16(-1) = 16 \text{ ft/sec}$$

(f) The instantaneous speed of the ball is zero when

$$s'(t) = 0$$

$$-16(2t - 5) = 0$$

$$t = \frac{5}{2} = 2.5 \text{ sec}$$

(g) The ball passes the rooftop on the way down when $t = 5$. The instantaneous speed at $t = 5$ is

$$s'(5) = -16(10 - 5) = -80 \text{ ft/sec}$$

At $t = 5$ sec, the ball is traveling -80 ft/sec. When the instantaneous rate of change is negative, it means that the direction of the object is downward. The ball is traveling 80 ft/sec in the downward direction when $t = 5$ sec.

(h) The ball strikes the ground when $t = 6$. The instantaneous speed when $t = 6$ is

$$s'(6) = -16(12 - 5) = -112 \text{ ft/sec}$$

The speed of the ball at $t = 6$ sec is -112 ft/sec. Again, the negative value implies that the ball is traveling downward. ◀

Exploration

Determine the vertex of the quadratic function given in Example 6. What do you conclude about instantaneous velocity when $s(t)$ is a maximum?

Summary

The derivative of a function $y = f(x)$ at c is defined as

$$f'(c) = \lim_{x \to c} \frac{f(x) - f(c)}{x - c}$$

In geometry, $f'(c)$ equals the slope of the tangent line to the graph of f at the point $(c, f(c))$.

In physics, $f'(c)$ equals the instantaneous speed (velocity) of a particle at time c, where $s = f(t)$ is the position of the particle at time t.

In applications, if two variables are related by the function $y = f(x)$, then $f'(c)$ equals the instantaneous rate of change of f with respect to x at c.

13.4 Assess Your Understanding

'Are You Prepared?' *Answers are given at the end of these exercises. If you get a wrong answer, read the pages listed in red.*

1. Find an equation of the line with slope 5 containing the point $(2, -4)$. (p. 27)

2. *True or False:* If c is in the domain of a function f, the average rate of change of f from c to x is

$$\frac{f(x) + f(c)}{x + c}$$ (pp. 78–79)

Concepts and Vocabulary

3. If $\lim_{x \to c} \frac{f(x) - f(c)}{x - c}$ exists, it equals the slope of the __________ __________ to the graph of f at the point $(c, f(c))$.

4. If $\lim_{x \to c} \frac{f(x) - f(c)}{x - c}$ exists, it is called the __________ of f at c.

5. If $s = f(t)$ denotes the position of a particle at time t, the derivative $f'(c)$ is the __________ of the particle at c.

6. *True or False:* The tangent line to a function is the limiting position of a secant line.

7. *True or False:* The slope of the tangent line to the graph of f at $(c, f(c))$ is the derivative of f at c.

8. *True or False:* The velocity of a particle whose position at time t is $s(t)$ is the derivative $s'(t)$.

Exercises

In Problems 9–20, find the slope of the tangent line to the graph of f at the given point. Graph f and the tangent line.

9. $f(x) = 3x + 5$ at $(1, 8)$

10. $f(x) = -2x + 1$ at $(-1, 3)$

11. $f(x) = x^2 + 2$ at $(-1, 3)$

12. $f(x) = 3 - x^2$ at $(1, 2)$

13. $f(x) = 3x^2$ at $(2, 12)$

14. $f(x) = -4x^2$ at $(-2, -16)$

15. $f(x) = 2x^2 + x$ at $(1, 3)$

16. $f(x) = 3x^2 - x$ at $(0, 0)$

17. $f(x) = x^2 - 2x + 3$ at $(-1, 6)$

18. $f(x) = -2x^2 + x - 3$ at $(1, -4)$

19. $f(x) = x^3 + x$ at $(2, 10)$

20. $f(x) = x^3 - x^2$ at $(1, 0)$

In Problems 21–32, find the derivative of each function at the given number.

21. $f(x) = -4x + 5$ at 3

22. $f(x) = -4 + 3x$ at 1

23. $f(x) = x^2 - 3$ at 0

24. $f(x) = 2x^2 + 1$ at -1

25. $f(x) = 2x^2 + 3x$ at 1

26. $f(x) = 3x^2 - 4x$ at 2

27. $f(x) = x^3 + 4x$ at -1

28. $f(x) = 2x^3 - x^2$ at 2

29. $f(x) = x^3 + x^2 - 2x$ at 1

30. $f(x) = x^3 - 2x^2 + x$ at -1

31. $f(x) = \sin x$ at 0

32. $f(x) = \cos x$ at 0

In Problems 33–42, find the derivative of each function at the given number using a graphing utility.

33. $f(x) = 3x^3 - 6x^2 + 2$ at -2

34. $f(x) = -5x^4 + 6x^2 - 10$ at 5

35. $f(x) = \frac{-x^3 + 1}{x^2 + 5x + 7}$ at 8

36. $f(x) = \frac{-5x^4 + 9x + 3}{x^3 + 5x^2 - 6}$ at -3

37. $f(x) = x \sin x$ at $\frac{\pi}{3}$

38. $f(x) = x \sin x$ at $\frac{\pi}{4}$

39. $f(x) = x^2 \sin x$ at $\frac{\pi}{3}$

40. $f(x) = x^2 \sin x$ at $\frac{\pi}{4}$

41. $f(x) = e^x \sin x$ at 2

42. $f(x) = e^{-x} \sin x$ at 2

43. **Instantaneous Rate of Change** The volume V of a right circular cylinder of height 3 feet and radius r feet is $V = V(r) = 3\pi r^2$. Find the instantaneous rate of change of the volume with respect to the radius r at $r = 3$.

44. **Instantaneous Rate of Change** The surface area S of a sphere of radius r feet is $S = S(r) = 4\pi r^2$. Find the instantaneous rate of change of the surface area with respect to the radius r at $r = 2$.

45. **Instantaneous Rate of Change** The volume V of a sphere of radius r feet is $V = V(r) = \frac{4}{3}\pi r^3$. Find the instantaneous rate of change of the volume with respect to the radius r at $r = 2$.

46. **Instantaneous Rate of Change** The volume V of a cube of side x meters is $V = V(x) = x^3$. Find the instantaneous rate of change of the volume with respect to the side x at $x = 3$.

47. **Instantaneous Speed of a Ball** In physics it is shown that the height s of a ball thrown straight up with an initial speed of 96 ft/sec from ground level is

$$s = s(t) = -16t^2 + 96t$$

where t is the elapsed time that the ball is in the air.
(a) When does the ball strike the ground? That is, how long is the ball in the air?
(b) What is the average speed of the ball from $t = 0$ to $t = 2$?
(c) What is the instantaneous speed of the ball at time t?
(d) What is the instantaneous speed of the ball at $t = 2$?
(e) When is the instantaneous speed of the ball equal to zero?
(f) How high is the ball when its instantaneous speed equals zero?
(g) What is the instantaneous speed of the ball when it strikes the ground?

48. **Instantaneous Speed of a Ball** In physics it is shown that the height s of a ball thrown straight down with an initial speed of 48 ft/sec from a rooftop 160 feet high is

$$s = s(t) = -16t^2 - 48t + 160$$

where t is the elapsed time that the ball is in the air.
(a) When does the ball strike the ground? That is, how long is the ball in the air?
(b) What is the average speed of the ball from $t = 0$ to $t = 1$?
(c) What is the instantaneous speed of the ball at time t?
(d) What is the instantaneous speed of the ball at $t = 1$?
(e) What is the instantaneous speed of the ball when it strikes the ground?

49. **Instantaneous Speed on the Moon** Neil Armstrong throws a ball down into a crater on the moon. The height s (in feet) of the ball from the bottom of the crater after t seconds is given in the following table:

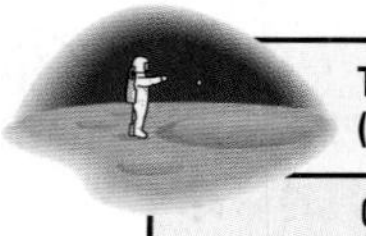

Time, t (in Seconds)	Distance, s (in Feet)
0	1000
1	987
2	969
3	945
4	917
5	883
6	843
7	800
8	749

(a) Find the average speed from $t = 1$ to $t = 4$ seconds.
(b) Find the average speed from $t = 1$ to $t = 3$ seconds.
(c) Find the average speed from $t = 1$ to $t = 2$ seconds.
(d) Using a graphing utility, find the quadratic function of best fit.
(e) Using the function found in part (d), determine the instantaneous speed at $t = 1$ second.

50. **Instantaneous Rate of Change** The following data represent the total revenue, R (in dollars), received from selling x bicycles at Tunney's Bicycle Shop.

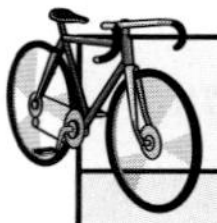

Number of Bicycles, x	Total Revenue, R (in Dollars)
0	0
25	28,000
60	45,000
102	53,400
150	59,160
190	62,360
223	64,835
249	66,525

(a) Find the average rate in change in revenue from $x = 25$ to $x = 150$ bicycles.
(b) Find the average rate of change in revenue from $x = 25$ to $x = 102$ bicycles.
(c) Find the average rate of change in revenue from $x = 25$ to $x = 60$ bicycles.
(d) Using a graphing utility, find the quadratic function of best fit.
(e) Using the function found in part (d), determine the instantaneous rate of change of revenue at $x = 25$ bicycles.

'Are You Prepared?' Answers

1. $y = 5x - 14$
2. False

13.5 The Area Problem; The Integral

PREPARING FOR THIS SECTION *Before getting started, review the following:*

- Geometry Formulas (Appendix A, Section A.2, pp. 915–916)
- Summation Notation (Section 11.1, pp. 788–789)

Now work the 'Are You Prepared?' problems on page 896.

OBJECTIVES 1 Approximate the Area under the Graph of a Function
2 Approximate Integrals Using a Graphing Utility

The development of the integral, like that of the derivative, was originally motivated to a large extent by a problem in geometry: the *area problem*.

Figure 20

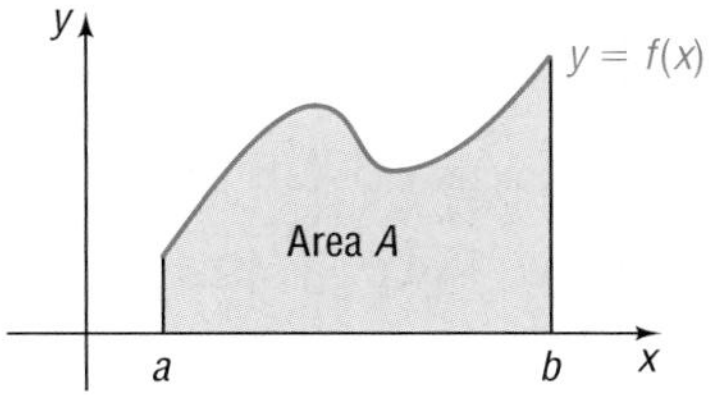

Area A = area under the graph of f from a to b

Area Problem

Suppose that $y = f(x)$ is a function whose domain is a closed interval $[a, b]$. We assume that $f(x) \geq 0$ for all x in $[a, b]$. Find the area enclosed by the graph of f, the x-axis, and the vertical lines $x = a$ and $x = b$.

Figure 20 illustrates the area problem. We refer to the area A shown in Figure 20 as the area under the graph of f from a to b.

For a constant function $f(x) = k$ and for a linear function $f(x) = mx + B$, we can solve the area problem using formulas from geometry. See Figures 21(a) and (b).

Figure 21

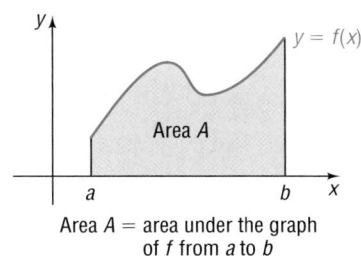

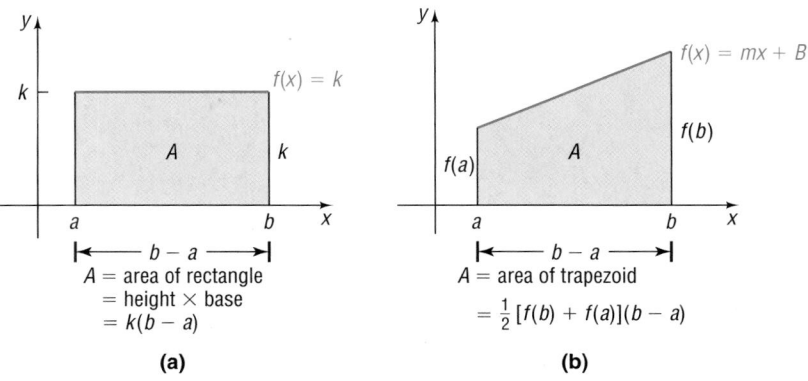

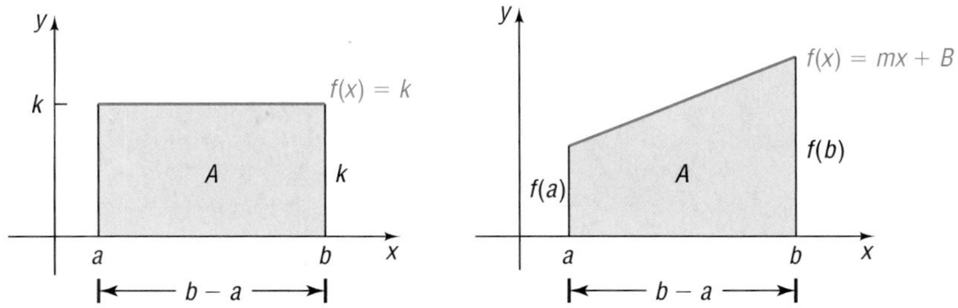

(a) A = area of rectangle = height × base = $k(b - a)$

(b) A = area of trapezoid = $\frac{1}{2}[f(b) + f(a)](b - a)$

For most other functions, no formulas from geometry are available.

We begin by discussing a way to approximate the area under the graph of a function f from a to b.

Approximating Area

1 We use rectangles to approximate the area under the graph of a function f. We do this by *partitioning* or dividing the interval $[a, b]$ into subintervals of equal length. On each subinterval, we form a rectangle whose base is the length of the subinterval and whose height is $f(u)$ for some number u in the subinterval. Look at Figure 22.

Figure 22

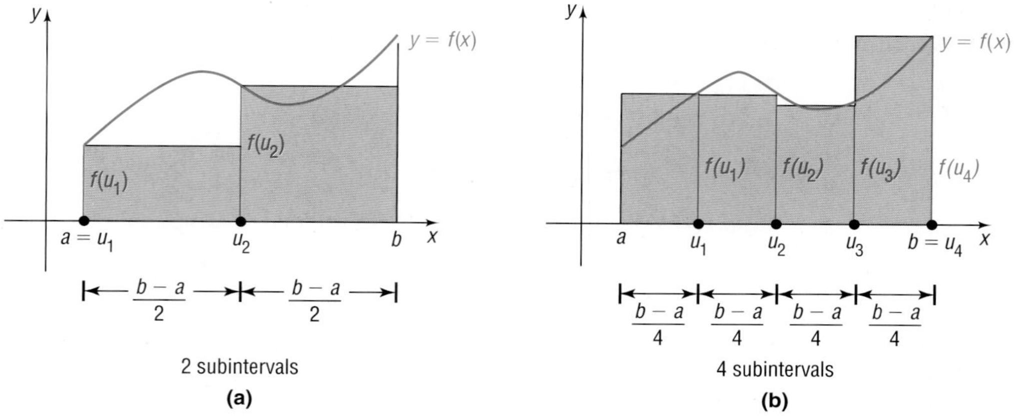

In Figure 22(a), the interval $[a, b]$ is partitioned into two subintervals, each of length $\frac{b-a}{2}$, and the number u is chosen as the left endpoint of each subinterval. In Figure 22(b), the interval $[a, b]$ is partitioned into four subintervals, each of length $\frac{b-a}{4}$, and the number u is chosen as the right endpoint of each subinterval. We approximate the area A under f from a to b by adding the areas of the rectangles formed by the partition.

Using Figure 22(a),

$$\begin{aligned}\text{Area } A &\approx \text{area of first rectangle} + \text{area of second rectangle}\\ &= f(u_1)\frac{b-a}{2} + f(u_2)\frac{b-a}{2}\end{aligned}$$

Using Figure 22(b),

$$\begin{aligned}\text{Area } A \approx\ &\text{area of first rectangle} + \text{area of second rectangle}\\ &+ \text{area of third rectangle} + \text{area of fourth rectangle}\\ =\ & f(u_1)\frac{b-a}{4} + f(u_2)\frac{b-a}{4} + f(u_3)\frac{b-a}{4} + f(u_4)\frac{b-a}{4}\end{aligned}$$

In approximating the area under the graph of a function f from a to b, the choice of the number u in each subinterval is arbitrary. For convenience, we shall always pick u as either the left endpoint of each subinterval or the right endpoint. The choice of how many subintervals to use is also arbitrary. In general, the more subintervals used, the better the approximation will be. Let's look at a specific example.

EXAMPLE 1 Approximating the Area under the Graph of $f(x) = 2x$ from 0 to 1

Approximate the area A under the graph of $f(x) = 2x$ from 0 to 1 as follows:

(a) By partitioning $[0, 1]$ into two subintervals of equal length and choosing u as the left endpoint.

(b) By partitioning $[0, 1]$ into two subintervals of equal length and choosing u as the right endpoint.

Figure 23

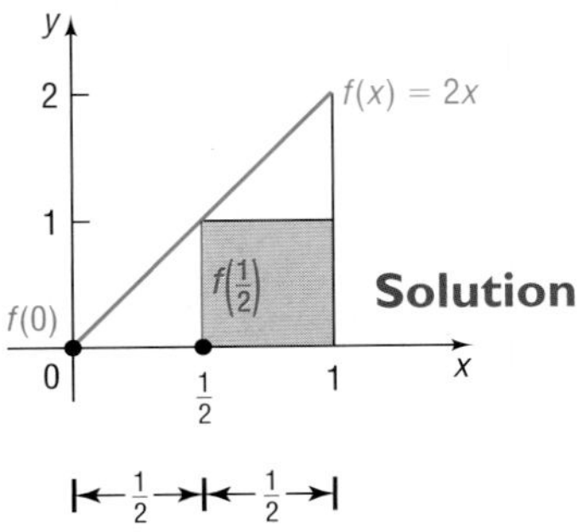

2 subintervals; *u*'s are left endpoints

(a)

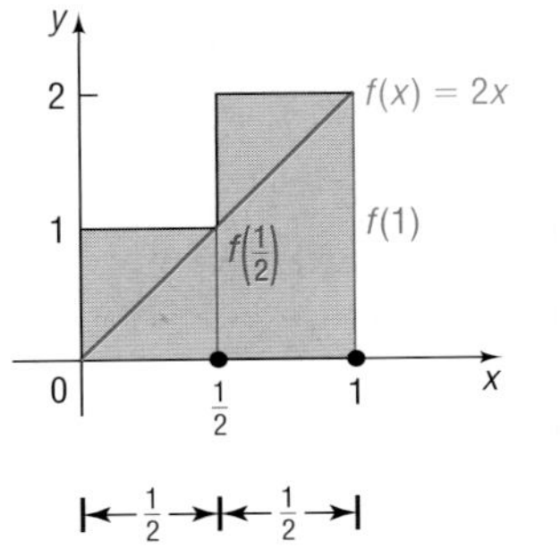

2 subintervals; *u*'s are right endpoints

(b)

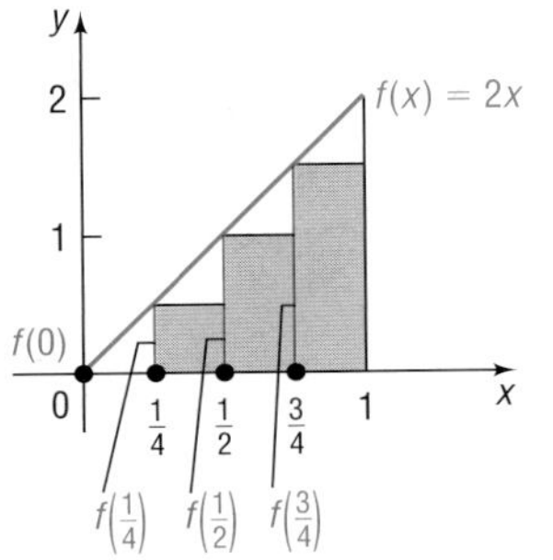

4 subintervals; *u*'s are left endpoints

(c)

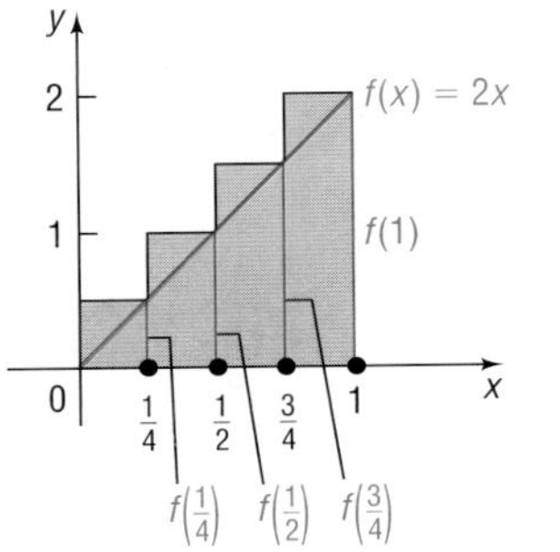

4 subintervals; *u*'s are right endpoints

(d)

(c) By partitioning $[0, 1]$ into four subintervals of equal length and choosing u as the left endpoint.

(d) By partitioning $[0, 1]$ into four subintervals of equal length and choosing u as the right endpoint.

(e) Compare the approximations found in parts (a)–(d) with the actual area.

Solution

(a) We partition $[0, 1]$ into two subintervals, each of length $\frac{1}{2}$, and choose u as the left endpoint. See Figure 23(a). The area A is approximated as

$$\begin{aligned} A &\approx f(0)\left(\frac{1}{2}\right) + f\left(\frac{1}{2}\right)\left(\frac{1}{2}\right) \\ &= (0)\left(\frac{1}{2}\right) + (1)\left(\frac{1}{2}\right) \\ &= \frac{1}{2} = 0.5 \end{aligned}$$

(b) We partition $[0, 1]$ into two subintervals, each of length $\frac{1}{2}$, and choose u as the right endpoint. See Figure 23(b). The area A is approximated as

$$\begin{aligned} A &\approx f\left(\frac{1}{2}\right)\left(\frac{1}{2}\right) + f(1)\left(\frac{1}{2}\right) \\ &= (1)\left(\frac{1}{2}\right) + (2)\left(\frac{1}{2}\right) \\ &= \frac{3}{2} = 1.5 \end{aligned}$$

(c) We partition $[0, 1]$ into four subintervals, each of length $\frac{1}{4}$, and choose u as the left endpoint. See Figure 23(c). The area A is approximated as

$$\begin{aligned} A &\approx f(0)\left(\frac{1}{4}\right) + f\left(\frac{1}{4}\right)\left(\frac{1}{4}\right) + f\left(\frac{1}{2}\right)\left(\frac{1}{4}\right) + f\left(\frac{3}{4}\right)\left(\frac{1}{4}\right) \\ &= (0)\left(\frac{1}{4}\right) + \left(\frac{1}{2}\right)\left(\frac{1}{4}\right) + (1)\left(\frac{1}{4}\right) + \left(\frac{3}{2}\right)\left(\frac{1}{4}\right) \\ &= \frac{3}{4} = 0.75 \end{aligned}$$

(d) We partition $[0, 1]$ into four subintervals, each of length $\frac{1}{4}$, and choose u as the right endpoint. See Figure 23(d). The area A is approximated as

$$\begin{aligned} A &\approx f\left(\frac{1}{4}\right)\left(\frac{1}{4}\right) + f\left(\frac{1}{2}\right)\left(\frac{1}{4}\right) + f\left(\frac{3}{4}\right)\left(\frac{1}{4}\right) + f(1)\left(\frac{1}{4}\right) \\ &= \left(\frac{1}{2}\right)\left(\frac{1}{4}\right) + (1)\left(\frac{1}{4}\right) + \left(\frac{3}{2}\right)\left(\frac{1}{4}\right) + (2)\left(\frac{1}{4}\right) \\ &= \frac{5}{4} = 1.25 \end{aligned}$$

(e) The actual area under the graph of $f(x) = 2x$ from 0 to 1 is the area of a right triangle whose base is of length 1 and whose height is 2. The actual area A is therefore

$$A = \frac{1}{2}\text{base} \times \text{height} = \left(\frac{1}{2}\right)(1)(2) = 1$$

◀

Now look at Table 7, which shows the approximations to the area under the graph of $f(x) = 2x$ from 0 to 1 for $n = 2, 4, 10,$ and 100 subintervals. Notice that the approximations to the actual area improve as the number of subintervals increases.

Table 7

Using left endpoints:	n	2	4	10	100
	Area	0.5	0.75	0.9	0.99
Using right endpoints:	n	2	4	10	100
	Area	1.5	1.25	1.1	1.01

You are asked to confirm the entries in Table 7 in Problem 31.

There is another useful observation about Example 1. Look again at Figures 23(a)–(d) and at Table 7. Since the graph of $f(x) = 2x$ is increasing on $[0, 1]$, the choice of u as left endpoint gives a lower estimate to the actual area, while choosing u as the right endpoint gives an upper estimate. Do you see why?

NOW WORK PROBLEM 9.

EXAMPLE 2 Approximating the Area Under the Graph of $f(x) = x^2$

Approximate the area under the graph of $f(x) = x^2$ from 1 to 5 as follows:

(a) Using four subintervals of equal length.

(b) Using eight subintervals of equal length.

In each case, choose the number u to be the left endpoint of each subinterval.

Figure 24

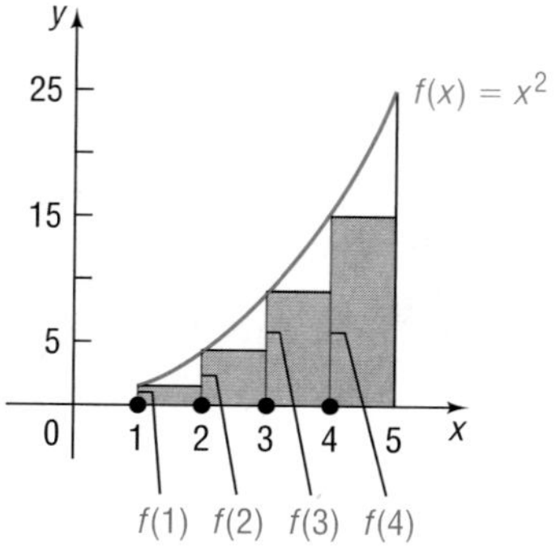

4 subintervals; each of length 1

(a)

Solution (a) See Figure 24(a). Using four subintervals of equal length, the interval $[1, 5]$ is partitioned into subintervals of length $\frac{5-1}{4} = 1$ as follows:

$$[1, 2] \quad [2, 3] \quad [3, 4] \quad [4, 5]$$

Each of these subintervals is of length 1. Choosing u as the left endpoint of each subinterval, the area A under the graph of $f(x) = x^2$ is approximated by

$$\begin{aligned}\text{Area } A &\approx f(1)(1) + f(2)(1) + f(3)(1) + f(4)(1)\\ &= 1 + 4 + 9 + 16 = 30\end{aligned}$$

(b) See Figure 24(b). Using eight subintervals of equal length, the interval $[1, 5]$ is partitioned into subintervals of length $\frac{5-1}{8} = 0.5$ as follows:

$$[1, 1.5] \quad [1.5, 2] \quad [2, 2.5] \quad [2.5, 3] \quad [3, 3.5] \quad [3.5, 4] \quad [4, 4.5] \quad [4.5, 5]$$

Figure 24

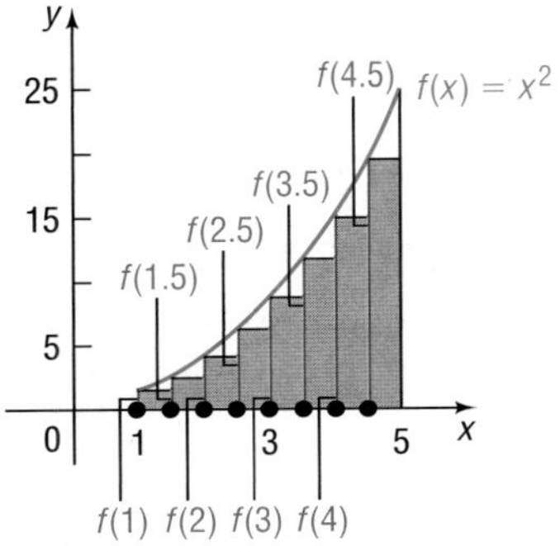

8 subintervals; each of length 1/2

(b)

Each of these subintervals is of length 0.5. Choosing u as the left endpoint of each subinterval, the area A under the graph of $f(x) = x^2$ is approximated by

$$\begin{aligned}\text{Area } A &\approx f(1)(0.5) + f(1.5)(0.5) + f(2)(0.5) + f(2.5)(0.5)\\ &\quad + f(3)(0.5) + f(3.5)(0.5) + f(4)(0.5) + f(4.5)(0.5)\\ &= [f(1) + f(1.5) + f(2) + f(2.5) + f(3) + f(3.5) + f(4) + f(4.5)](0.5)\\ &= [1 + 2.25 + 4 + 6.25 + 9 + 12.25 + 16 + 20.25](0.5)\\ &= 35.5\end{aligned}$$

◀

In general, we approximate the area under the graph of a function $y = f(x)$ from a to b as follows:

1. Partition the interval $[a, b]$ into n subintervals of equal length. The length Δx of each subinterval is then

$$\Delta x = \frac{b - a}{n}$$

2. In each of these subintervals, pick a number u and evaluate the function f at each u. This results in n numbers $u_1, u_2, \ldots, u_n$, and n functional values $f(u_1), f(u_2), \ldots, f(u_n)$.

3. Form n rectangles with base equal to Δx, the length of each subinterval, and with height equal to the functional value $f(u_i)$, $i = 1, 2, \ldots, n$. See Figure 25.

4. Add up the areas of the n rectangles.

$$\begin{aligned}A_1 + A_2 + \cdots + A_n &= f(u_1)\Delta x + f(u_2)\Delta x + \cdots + f(u_n)\,\Delta x\\ &= \sum_{i=1}^{n} f(u_i)\Delta x\end{aligned}$$

Figure 25

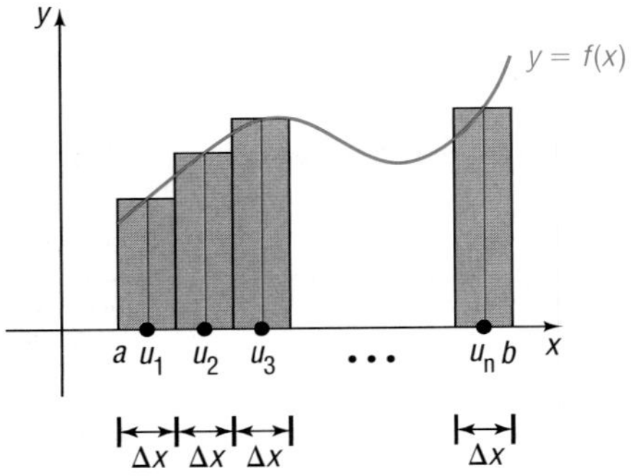

This number is the approximation to the area under the graph of f from a to b.

Definition of Area

We have observed that the larger the number n of subintervals used, the better the approximation to the area. If we let n become unbounded, we obtain the exact area under the graph of f from a to b.

Area under a Graph

Let f denote a function whose domain is a closed interval $[a, b]$. Partition $[a, b]$ into n subintervals, each of length $\Delta x = \dfrac{b - a}{n}$. In each subinterval, pick a number u_i, $i = 1, 2, \ldots, n$, and evaluate $f(u_i)$. Form the products $f(u_i)\Delta x$ and add them up obtaining the sum

$$\sum_{i=1}^{n} f(u_i)\Delta x$$

If the limit of this sum exists as $n \to \infty$, that is,

$$\text{if} \qquad \lim_{n\to\infty} \sum_{i=1}^{n} f(u_i)\Delta x \qquad \text{exists}$$

it is defined as the area under the graph of f from a to b. If this limit exists, it is denoted by the symbol

$$\int_a^b f(x)\,dx$$

read as "the integral from a to b of $f(x)$."

2 We can use a graphing utility to approximate integrals.

EXAMPLE 3 Using a Graphing Utility to Approximate an Integral

Use a graphing utility to approximate the area under the graph of $f(x) = x^2$ from 1 to 5. That is, evaluate the integral

$$\int_1^5 x^2\,dx$$

Figure 26

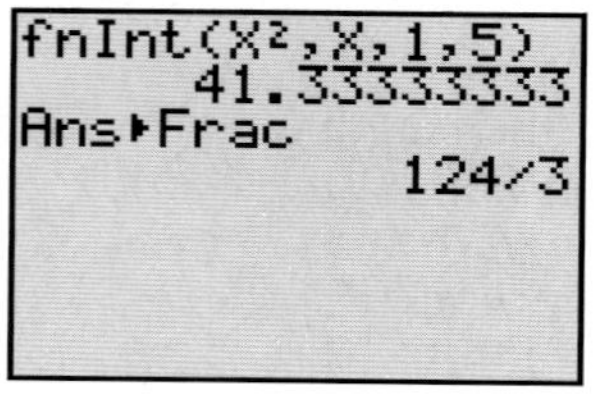

Solution Figure 26 shows the result using a TI-83 Plus calculator. Consult your owner's manual for the proper keystrokes. ◀

In calculus, techniques are given for evaluating integrals to obtain exact answers.

13.5 Assess Your Understanding

'Are You Prepared?' *Answers are given at the end of these exercises. If you get a wrong answer, read the pages listed in red.*

1. The formula for the area A of a rectangle of length l and width w is ________. (pp. 915–916)

2. $\sum_{k=1}^{4}(2k + 1) =$ ________. (pp. 788–789)

Concepts and Vocabulary

3. The integral from a to b of $f(x)$ is denoted by the symbol ________.

4. The area under the graph of f from a to b is denoted by the symbol ________.

Exercises

In Problems 5 and 6, refer to the illustration. The interval $[1, 3]$ *is partitioned into two subintervals* $[1, 2]$ *and* $[2, 3]$.

5. Approximate the area A choosing u as the left endpoint of each subinterval.

6. Approximate the area A choosing u as the right endpoint of each subinterval.

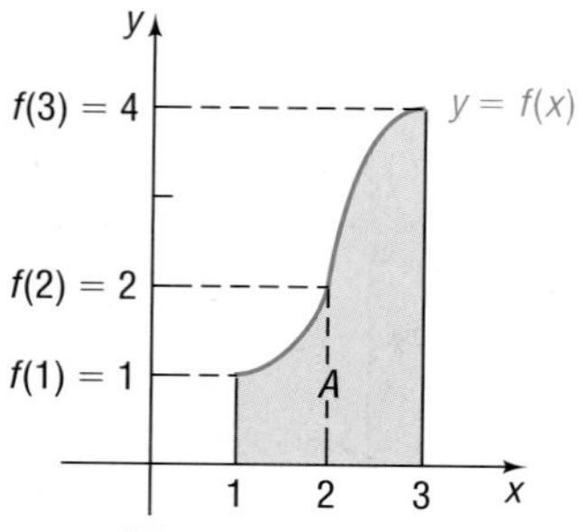

$f(1) = 1$, $f(2) = 2$, $f(3) = 4$

In Problems 7 and 8, refer to the illustration. The interval $[0, 8]$ is partitioned into four subintervals $[0, 2], [2, 4], [4, 6]$, and $[6, 8]$.

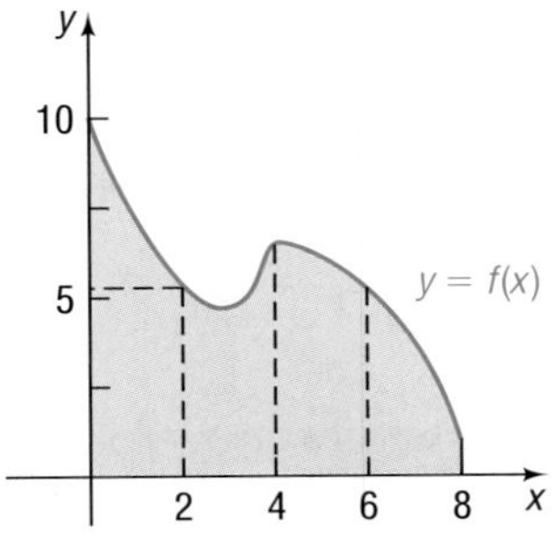

$f(0) = 10, f(2) = 6, f(4) = 7, f(6) = 5, f(8) = 1$

7. Approximate the area A choosing u as the left endpoint of each subinterval.

8. Approximate the area A choosing u as the right endpoint of each subinterval.

9. The function $f(x) = 3x$ is defined on the interval $[0, 6]$.
(a) Graph f.
In (b)–(e), approximate the area A under f from 0 to 6 as follows:
(b) By partitioning $[0, 6]$ into three subintervals of equal length and choosing u as the left endpoint of each subinterval.
(c) By partitioning $[0, 6]$ into three subintervals of equal length and choosing u as the right endpoint of each subinterval.
(d) By partitioning $[0, 6]$ into six subintervals of equal length and choosing u as the left endpoint of each subinterval.
(e) By partitioning $[0, 6]$ into six subintervals of equal length and choosing u as the right endpoint of each subinterval.
(f) What is the actual area A?

10. Repeat Problem 9 for $f(x) = 4x$.

11. The function $f(x) = -3x + 9$ is defined on the interval $[0, 3]$.
(a) Graph f.
In (b)–(e), approximate the area A under f from 0 to 3 as follows:
(b) By partitioning $[0, 3]$ into three subintervals of equal length and choosing u as the left endpoint of each subinterval.
(c) By partitioning $[0, 3]$ into three subintervals of equal length and choosing u as the right endpoint of each subinterval.
(d) By partitioning $[0, 3]$ into six subintervals of equal length and choosing u as the left endpoint of each subinterval.
(e) By partitioning $[0, 3]$ into six subintervals of equal length and choosing u as the right endpoint of each subinterval.
(f) What is the actual area A?

12. Repeat Problem 11 for $f(x) = -2x + 8$.

In Problems 13–22, a function f is defined over an interval $[a, b]$.
(a) *Graph f, indicating the area A under f from a to b.*
(b) *Approximate the area A by partitioning $[a, b]$ into four subintervals of equal length and choosing u as the left endpoint of each subinterval.*
(c) *Approximate the area A by partitioning $[a, b]$ into eight subintervals of equal length and choosing u as the left endpoint of each subinterval.*
(d) *Express the area A as an integral.*
(e) *Use a graphing utility to approximate the integral.*

13. $f(x) = x^2 + 2, \quad [0, 4]$ **14.** $f(x) = x^2 - 4, \quad [2, 6]$ **15.** $f(x) = x^3, \quad [0, 4]$ **16.** $f(x) = x^3, \quad [1, 5]$

17. $f(x) = \frac{1}{x}, \quad [1, 5]$ **18.** $f(x) = \sqrt{x}, \quad [0, 4]$ **19.** $f(x) = e^x, \quad [-1, 3]$ **20.** $f(x) = \ln x, \quad [3, 7]$

21. $f(x) = \sin x, \quad [0, \pi]$ **22.** $f(x) = \cos x, \quad \left[0, \frac{\pi}{2}\right]$

In Problems 23–30, an integral is given.
(a) *What area does the integral represent?*
(b) *Provide a graph that illustrates this area.*
(c) *Use a graphing utility to approximate this area.*

23. $\int_0^4 (3x + 1)\, dx$ **24.** $\int_1^3 (-2x + 7)\, dx$ **25.** $\int_2^5 (x^2 - 1)\, dx$ **26.** $\int_0^4 (16 - x^2)\, dx$

27. $\int_0^{\pi/2} \sin x\, dx$ **28.** $\int_{-\pi/4}^{\pi/4} \cos x\, dx$ **29.** $\int_0^2 e^x\, dx$ **30.** $\int_e^{2e} \ln x\, dx$

31. Confirm the entries in Table 7.

[**Hint:** Review the formula for the sum of an arithmetic sequence.]

32. Consider the function $f(x) = \sqrt{1 - x^2}$ whose domain is the interval $[-1, 1]$.

(a) Graph f.

(b) Approximate the area under the graph of f from -1 to 1 by dividing $[-1, 1]$ into five subintervals, each of equal length.

(c) Approximate the area under the graph of f from -1 to 1 by dividing $[-1, 1]$ into ten subintervals each of equal length.

(d) Express the area as an integral.

(e) Evaluate the integral using a graphing utility.

(f) What is the actual area?

'Are You Prepared?' Answers

1. $A = lw$

2. 24

Chapter Review

Things to Know

Limit (p. 862)

$\lim_{x \to c} f(x) = N$ — As x gets closer to c, $x \neq c$, the value of f gets closer to N.

Limit Formulas (p. 867)

$\lim_{x \to c} b = b$ — The limit of a constant is the constant.

$\lim_{x \to c} x = c$ — The limit of x as x approaches c is c.

Limit Properties

$\lim_{x \to c}[f(x) + g(x)] = \lim_{x \to c} f(x) + \lim_{x \to c} g(x)$ (p. 868) — The limit of a sum equals the sum of the limits.

$\lim_{x \to c}[f(x) - g(x)] = \lim_{x \to c} f(x) - \lim_{x \to c} g(x)$ (p. 869) — The limit of a difference equals the difference of the limits.

$\lim_{x \to c}[f(x) \cdot g(x)] = \lim_{x \to c} f(x) \cdot \lim_{x \to c} g(x)$ (p. 869) — The limit of a product equals the product of the limits.

$\lim_{x \to c}\left[\dfrac{f(x)}{g(x)}\right] = \dfrac{\lim_{x \to c} f(x)}{\lim_{x \to c} g(x)}$ (p. 872) provided that $\lim_{x \to c} g(x) \neq 0$ — The limit of a quotient equals the quotient of the limits, provided that the limit of the denominator is not zero.

Limit of a Polynomial (p. 870)

$\lim_{x \to c} P(x) = P(c)$, where P is a polynomial

Continuous Function (p. 878)

$\lim_{x \to c} f(x) = f(c)$

Derivative of a Function (p. 885)

$f'(c) = \lim_{x \to c} \dfrac{f(x) - f(c)}{x - c}$, provided that the limit exists

Area under a Graph (pp. 895–896)

$\int_a^b f(x)\, dx = \lim_{n \to \infty} \sum_{i=1}^{n} f(u_i)\, \Delta x$, provided that the limit exists

Objectives

Section		You should be able to . . .	Review Exercises
13.1	1	Find a limit using a table (p. 862)	1–22
	2	Find a limit using a graph (p. 864)	43, 44
13.2	1	Find the limit of a sum, a difference, a product, and a quotient (p. 868)	1, 2, 11–22
	2	Find the limit of a polynomial (p. 870)	1–4

	3	Find the limit of a power or a root (p. 871)	3–10
	4	Find the limit of an average rate of change (p. 873)	55–60
13.3	1	Find the one-sided limits of a function (p. 875)	37–42
	2	Determine whether a function is continuous (p. 877)	23–30, 45–50, 51–54
13.4	1	Find an equation of the tangent line to the graph of a function (p. 884)	55–60
	2	Find the derivative of a function (p. 885)	61–70
	3	Find instantaneous rates of change (p. 886)	72, 73
	4	Find the instantaneous speed of a particle (p. 886)	71, 74
13.5	1	Approximate the area under the graph of a function (p. 891)	75–80
	2	Approximate integrals using a graphing utility (p. 896)	77(e)–80(e); 81(c)–84(c)

Review Exercises *(Blue problem numbers indicate the author's suggestions for use in a Practice Test.)*

In Problems 1–22, find the limit.

1. $\lim_{x\to 2}(3x^2 - 2x + 1)$
2. $\lim_{x\to 1}(-2x^3 + x + 4)$
3. $\lim_{x\to -2}(x^2 + 1)^2$
4. $\lim_{x\to -2}(x^3 + 1)^2$
5. $\lim_{x\to 3}\sqrt{x^2 + 7}$
6. $\lim_{x\to -2}\sqrt[3]{x + 10}$
7. $\lim_{x\to 1^-}\sqrt{1 - x^2}$
8. $\lim_{x\to 2^+}\sqrt{3x - 2}$
9. $\lim_{x\to 2}(5x + 6)^{3/2}$
10. $\lim_{x\to -3}(15 - 3x)^{-3/2}$
11. $\lim_{x\to -1}\dfrac{x^2 + x + 2}{x^2 - 9}$
12. $\lim_{x\to 3}\dfrac{3x + 4}{x^2 + 1}$
13. $\lim_{x\to 1}\dfrac{x - 1}{x^3 - 1}$
14. $\lim_{x\to -1}\dfrac{x^2 - 1}{x^2 + x}$
15. $\lim_{x\to -3}\dfrac{x^2 - 9}{x^2 - x - 12}$
16. $\lim_{x\to -3}\dfrac{x^2 + 2x - 3}{x^2 - 9}$
17. $\lim_{x\to -1^-}\dfrac{x^2 - 1}{x^3 - 1}$
18. $\lim_{x\to 2^+}\dfrac{x^2 - 4}{x^3 - 8}$
19. $\lim_{x\to 2}\dfrac{x^3 - 8}{x^3 - 2x^2 + 4x - 8}$
20. $\lim_{x\to 1}\dfrac{x^3 - 1}{x^3 - x^2 + 3x - 3}$
21. $\lim_{x\to 3}\dfrac{x^4 - 3x^3 + x - 3}{x^3 - 3x^2 + 2x - 6}$
22. $\lim_{x\to -1}\dfrac{x^4 + x^3 + 2x + 2}{x^3 + x^2}$

In Problems 23–30, determine whether f is continuous at c.

23. $f(x) = 3x^4 - x^2 + 2, \quad c = 5$
24. $f(x) = \dfrac{x^2 - 9}{x + 10} \quad c = 2$
25. $f(x) = \dfrac{x^2 - 4}{x + 2} \quad c = -2$
26. $f(x) = \dfrac{x^2 + 6x}{x^2 - 6x} \quad c = 0$
27. $f(x) = \begin{cases} \dfrac{x^2 - 4}{x + 2} & \text{if } x \neq -2 \\ 4 & \text{if } x = -2 \end{cases} \quad c = -2$
28. $f(x) = \begin{cases} \dfrac{x^2 + 6x}{x^2 - 6x} & \text{if } x \neq 0 \\ 1 & \text{if } x = 0 \end{cases} \quad c = 0$
29. $f(x) = \begin{cases} \dfrac{x^2 - 4}{x + 2} & \text{if } x \neq -2 \\ -4 & \text{if } x = -2 \end{cases} \quad c = -2$
30. $f(x) = \begin{cases} \dfrac{x^2 + 6x}{x^2 - 6x} & \text{if } x \neq 0 \\ -1 & \text{if } x = 0 \end{cases} \quad c = 0$

In Problems 31–50, use the accompanying graph of $y = f(x)$.

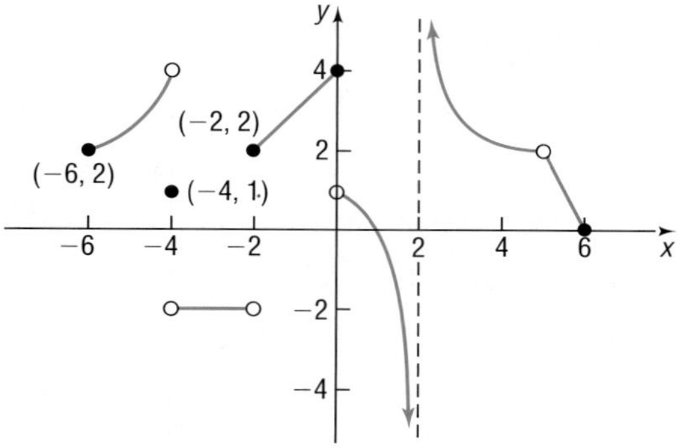

31. What is the domain of f?

32. What is the range of f?

33. Find the x-intercept(s), if any, of f.

34. Find the y-intercept(s), if any, of f.

35. Find $f(-6)$ and $f(-4)$.

36. Find $f(-2)$ and $f(6)$.

37. Find $\lim_{x \to -4^-} f(x)$.

38. Find $\lim_{x \to -4^+} f(x)$.

39. Find $\lim_{x \to -2^-} f(x)$.

40. Find $\lim_{x \to -2^+} f(x)$.

41. Find $\lim_{x \to 2^-} f(x)$.

42. Find $\lim_{x \to 2^+} f(x)$.

43. Does $\lim_{x \to 0} f(x)$ exist? If it does, what is it?

44. Does $\lim_{x \to 2} f(x)$ exist? If it does, what is it?

45. Is f continuous at -2?

46. Is f continuous at -4?

47. Is f continuous at 0?

48. Is f continuous at 2?

49. Is f continuous at 4?

50. Is f continuous at 5?

In Problems 51 and 52, discuss whether R is continuous at c. Use limits to analyze the graph of R at c.

51. $R(x) = \dfrac{x + 4}{x^2 - 16}$ at $c = -4$ and $c = 4$

52. $R(x) = \dfrac{3x^2 + 6x}{x^2 - 4}$ at $c = -2$ and $c = 2$

In Problems 53 and 54, determine where each rational function is undefined. Determine whether an asymptote or a hole appears at such numbers.

53. $R(x) = \dfrac{x^3 - 2x^2 + 4x - 8}{x^2 - 11x + 18}$

54. $R(x) = \dfrac{x^3 + 3x^2 - 2x - 6}{x^2 + x - 6}$

In Problems 55–60, find the slope of the tangent line to the graph of f at the given point. Graph f and the tangent line.

55. $f(x) = 2x^2 + 8x$ at $(1, 10)$

56. $f(x) = 3x^2 - 6x$ at $(0, 0)$

57. $f(x) = x^2 + 2x - 3$ at $(-1, -4)$

58. $f(x) = 2x^2 + 5x - 3$ at $(1, 4)$

59. $f(x) = x^3 + x^2$ at $(2, 12)$

60. $f(x) = x^3 - x^2$ at $(1, 0)$

In Problems 61–66, find the derivative of each function at the number indicated.

61. $f(x) = -4x^2 + 5$ at 3

62. $f(x) = -4 + 3x^2$ at 1

63. $f(x) = x^2 - 3x$ at 0

64. $f(x) = 2x^2 + 4x$ at -1

65. $f(x) = 2x^2 + 3x + 2$ at 1

66. $f(x) = 3x^2 - 4x + 1$ at 2

In Problems 67–70, find the derivative of each function at the number indicated using a graphing utility.

67. $f(x) = 4x^4 - 3x^3 + 6x - 9$ at -2

68. $f(x) = \dfrac{-6x^3 + 9x - 2}{8x^2 + 6x - 1}$ at 5

69. $f(x) = x^3 \tan x$ at $\dfrac{\pi}{6}$

70. $f(x) = x \sec x$ at $\dfrac{\pi}{6}$

---

71. Instantaneous Speed of a Ball In physics it is shown that the height s of a ball thrown straight up with an initial speed of 96 ft/sec from a rooftop 112 feet high is

$$s = s(t) = -16t^2 + 96t + 112$$

where t is the elapsed time that the ball is in the air. The ball misses the rooftop on its way down and eventually strikes the ground.

(a) When does the ball strike the ground? That is, how long is the ball in the air?

(b) At what time t will the ball pass the rooftop on its way down?

(c) What is the average speed of the ball from $t = 0$ to $t = 2$?

(d) What is the instantaneous speed of the ball at time t?

(e) What is the instantaneous speed of the ball at $t = 2$?

(f) When is the instantaneous speed of the ball equal to zero?

(g) What is the instantaneous speed of the ball as it passes the rooftop on the way down?

(h) What is the instantaneous speed of the ball when it strikes the ground?

72. Finding an Instantaneous Rate of Change The area A of a circle is πr^2. Find the instantaneous rate of change of area with respect to r at $r = 2$ feet. What is the average rate of change from $r = 2$ to $r = 3$? What is the average rate of change from $r = 2$ to $r = 2.5$? From $r = 2$ to $r = 2.1$?

73. Instantaneous Rate of Change The following data represent the revenue R (in dollars) received from selling x wristwatches at Wilk's Watch Shop.

(a) Find the average rate of change of revenue from $x = 25$ to $x = 130$ wristwatches.

(b) Find the average rate of change of revenue from $x = 25$ to $x = 90$ wristwatches.

(c) Find the average rate of change of revenue from $x = 25$ to $x = 50$ wristwatches.

(d) Using a graphing utility, find the quadratic function of best fit.

(e) Using the function found in part (d), determine the instantaneous rate of change of revenue at $x = 25$ wristwatches.

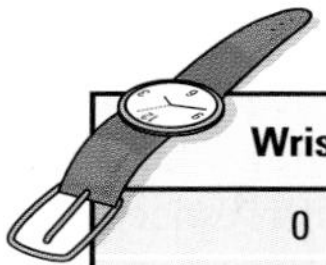

Wristwatches, x	Revenue, R
0	0
25	2340
40	3600
50	4375
90	6975
130	8775
160	9600
200	10,000
220	9900
250	9375

74. Instantaneous Speed of a Parachutist The following data represent the distance s (in feet) that a parachutist has fallen over time t (in seconds).

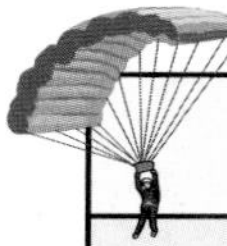

Time, t (in Seconds)	Distance, s (in Feet)
1	16
2	64
3	144
4	256
5	400

(a) Find the average speed from $t = 1$ to $t = 4$ seconds.

(b) Find the average speed from $t = 1$ to $t = 3$ seconds.

(c) Find the average speed from $t = 1$ to $t = 2$ seconds.

(d) Using a graphing utility, find the power function of best fit.

(e) Using the function found in part (d), determine the instantaneous speed at $t = 1$ seconds.

75. The function $f(x) = 2x + 3$ is defined on the interval $[0, 4]$.

(a) Graph f.

In (b)–(e), approximate the area A under f from $x = 0$ to $x = 4$ as follows:

(b) By partitioning $[0, 4]$ into four subintervals of equal length and choosing u as the left endpoint of each subinterval.

(c) By partitioning $[0, 4]$ into four subintervals of equal length and choosing u as the right endpoint of each subinterval.

(d) By partitioning $[0, 4]$ into eight subintervals of equal length and choosing u as the left endpoint of each subinterval.

(e) By partitioning $[0, 4]$ into eight subintervals of equal length and choosing u as the right endpoint of each subinterval.

(f) What is the actual area A?

76. Repeat Problem 75 for $f(x) = -2x + 8$.

In Problems 77–80, a function f is defined over an interval $[a, b]$.

(a) *Graph f, indicating the area A under f from a to b.*

(b) *Approximate the area A by partitioning $[a, b]$ into three subintervals of equal length and choosing u as the left endpoint of each subinterval.*

(c) *Approximate the area A by partitioning $[a, b]$ into six subintervals of equal length and choosing u as the left endpoint of each subinterval.*

(d) *Express the area A as an integral.*

(e) *Use a graphing utility to approximate the integral.*

77. $f(x) = 4 - x^2, \quad [-1, 2]$ **78.** $f(x) = x^2 + 3, \quad [0, 6]$ **79.** $f(x) = \dfrac{1}{x^2}, \quad [1, 4]$ **80.** $f(x) = e^x, \quad [0, 6]$

In Problems 81–84, an integral is given.

(a) *What area does the integral represent?*
(b) *Provide a graph that illustrates this area.*
(c) *Use a graphing utility to approximate this area.*

81. $\int_{-1}^{3} (9 - x^2)\, dx$ **82.** $\int_{1}^{4} \sqrt{x}\, dx$ **83.** $\int_{-1}^{1} e^x\, dx$ **84.** $\int_{\pi/3}^{2\pi/3} \sin x\, dx$

Chapter Projects

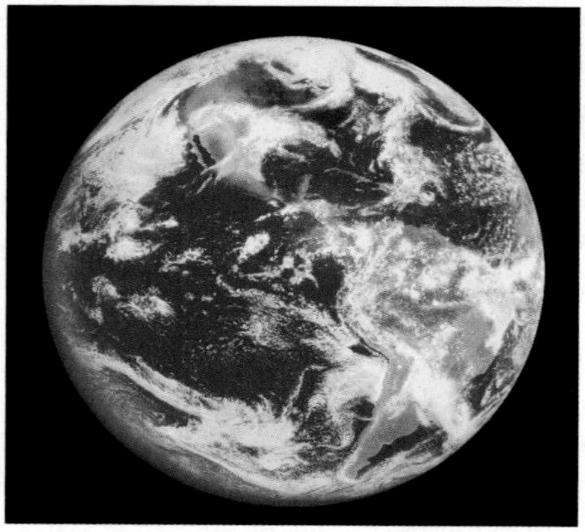

1. **World Population** Thomas Malthus believed that "population, when unchecked, increases in a geometrical progression of such nature as to double itself every twenty-five years." However, the growth of population is limited because the resources available to us are limited in supply. If Malthus's conjecture were true, then geometric growth of the world's population would imply that

$$\frac{P_t}{P_{t-1}} = r + 1, \text{ where } r \text{ is the growth rate}$$

(a) Using *world population data* and a graphing utility, find the logistic growth function of best fit, treating the year as the independent variable. Let $t = 0$ represent 1950, $t = 1$ represent 1951, and so on, until you have entered all the years and the corresponding population up to the current year.
(b) Graph $Y_1 = f(t)$, where $f(t)$ represents the logistic growth function of best fit found in part (a).
(c) Determine the instantaneous rate of growth of population in 1960 using the numerical derivative function on your graphing utility.
(d) Use the result from part (c) to predict the population in 1961. What was the actual population in 1961?
(e) Determine the instantaneous growth of population in 1970, 1980, and 1990. What is happening to the instantaneous growth rate as time passes? Is Malthus's contention of a geometric growth rate accurate?
(f) Using the numerical derivative function on your graphing utility, graph $Y_2 = f'(t)$, where $f'(t)$ represents the derivative of $f(t)$ with respect to time. Y_2 is the growth rate of the population at any time t.
(g) Using the MAXIMUM function on your graphing utility, determine the year in which the growth rate of the population is largest. What is happening to the growth rate in the years following the maximum? Find this point on the graph of $Y_1 = f(t)$.
(h) Evaluate $\lim_{t\to\infty} f(t)$. This limiting value is the carrying capacity of Earth. What is the carrying capacity of Earth?
(i) What do you think will happen if the population of Earth exceeds the carrying capacity? Do you think that agricultural output will continue to increase at the same rate as population growth? What effect will urban sprawl have on agricultural output?

The following Chapter Projects are available at www.prenhall.com/sullivan7e.

2. **Project at Motorola** ***Curing Scalar***
3. **Finding the Profit-maximizing Level of Output**

APPENDIX A

Review

Outline

A.1 Algebra Review

PREPARING FOR THIS BOOK *Before getting started, read "To the Student" at the beginning of this book.*

OBJECTIVES
1. Evaluate Algebraic Expressions
2. Determine the Domain of a Variable
3. Graph Inequalities
4. Find Distance on the Real Number Line
5. Use the Laws of Exponents
6. Evaluate Square Roots

Sets

When we want to treat a collection of similar but distinct objects as a whole, we use the idea of a **set.** For example, the set of *digits* consists of the collection of numbers 0, 1, 2, 3, 4, 5, 6, 7, 8, and 9. If we use the symbol D to denote the set of digits, then we can write

$$D = \{0, 1, 2, 3, 4, 5, 6, 7, 8, 9\}$$

In this notation, the braces $\{\ \}$ are used to enclose the objects, or **elements,** in the set. This method of denoting a set is called the **roster method.** A second way to denote a set is to use **set-builder notation,** where the set D of digits is written as

$$D = \{ \quad x \quad | \quad x \text{ is a digit}\}$$

Read as "D is the set of all x such that x is a digit."

EXAMPLE 1 **Using Set-builder Notation and the Roster Method**

(a) $E = \{x | x \text{ is an even digit}\} = \{0, 2, 4, 6, 8\}$

(b) $O = \{x | x \text{ is an odd digit}\} = \{1, 3, 5, 7, 9\}$ ◀

In listing the elements of a set, we do not list an element more than once because the elements of a set are distinct. Also, the order in which the elements are listed is not relevant. For example, $\{2, 3\}$ and $\{3, 2\}$ both represent the same set.

If every element of a set A is also an element of a set B, then we say that A is a **subset** of B. If two sets A and B have the same elements, then we say that A **equals** B. For example, $\{1, 2, 3\}$ is a subset of $\{1, 2, 3, 4, 5\}$; and $\{1, 2, 3\}$ equals $\{2, 3, 1\}$.

Finally, if a set has no elements, it is called the **empty set,** or the **null set,** and it is denoted by the symbol $\varnothing$.

Real Numbers

Real numbers are represented by symbols such as

$$25, \quad 0, \quad -3, \quad \frac{1}{2}, \quad -\frac{5}{4}, \quad 0.125, \quad \sqrt{2}, \quad \pi, \quad \sqrt[3]{-2}, \quad 0.666\ldots$$

The set of **counting numbers,** or **natural numbers,** contains the numbers in the set $\{1, 2, 3, 4, \ldots\}$. (The three dots, called an **ellipsis,** indicate that the pattern continues indefinitely.) The set of **integers** contains the numbers in the set $\{\ldots, -3, -2, -1, 0, 1, 2, 3, \ldots\}$. A **rational number** is a number that can be expressed as a *quotient* $\frac{a}{b}$ of two integers, where the integer b cannot be 0. Examples of rational numbers are $\frac{3}{4}, \frac{5}{2}, \frac{0}{4}$, and $-\frac{2}{3}$. Since $\frac{a}{1} = a$ for any integer a, every integer is also a rational number. Real numbers that are not rational are called **irrational.** Examples of irrational numbers are $\sqrt{2}$ and π (the Greek letter pi), which equals the constant ratio of the circumference to the diameter of a circle. See Figure 1.

Figure 1 $\pi = \frac{C}{d}$

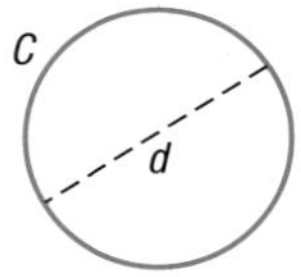

Real numbers can be represented as **decimals.** Rational real numbers have decimal representations that either **terminate** or are nonterminating with **repeating** blocks of digits. For example, $\frac{3}{4} = 0.75$, which terminates; and $\frac{2}{3} = 0.666\ldots$, in which the digit 6 repeats indefinitely. Irrational real numbers have decimal representations that neither repeat nor terminate. For example, $\sqrt{2} = 1.414213\ldots$ and $\pi = 3.14159\ldots$. In practice, the decimal representation of an irrational number is given as an approximation. We use the symbol $\approx$ (read as "approximately equal to") to write $\sqrt{2} \approx 1.4142$ and $\pi \approx 3.1416$.

Two properties of real numbers that we shall use often are given next. Suppose that a, b, and c are real numbers.

Distributive Property

$$a \cdot (b + c) = ab + ac$$

Zero-Product Property

If $ab = 0$, then either $a = 0$ or $b = 0$ or both equal 0.

The Distributive Property can be used to remove parentheses: $2(x + 3) = 2x + 2 \cdot 3 = 2x + 6$.

The Zero-Product Property will be used to solve equations (Section A.5). If $2x = 0$, then $2 = 0$ or $x = 0$. Since $2 \neq 0$, it follows that $x = 0$.

Constants and Variables

In algebra we use letters to represent numbers. If the letter used is to represent *any* number from a given set of numbers, it is called a **variable.** A **constant** is either a fixed number, such as 5 or $\sqrt{3}$, or a letter that represents a fixed (possibly unspecified) number.

Constants and variables are combined using the operations of addition, subtraction, multiplication. and division to form *algebraic expressions.* Examples of algebraic expressions include

$$x + 3 \qquad \frac{3}{1 - t} \qquad 7x - 2y$$

1 To evaluate an algebraic expression, substitute for each variable its numerical value.

EXAMPLE 2 Evaluating an Algebraic Expression

Evaluate each expression if $x = 3$ and $y = -1$.

(a) $x + 3y$ (b) $5xy$ (c) $\dfrac{3y}{2 - 2x}$

Solution (a) Substitute 3 for x and -1 for y in the expression $x + 3y$.

$$x + 3y = 3 + 3(-1) = 3 + (-3) = 0$$

↑ $x = 3, y = -1$

(b) If $x = 3$ and $y = -1$, then

$$5xy = 5(3)(-1) = -15$$

(c) If $x = 3$ and $y = -1$, then

$$\frac{3y}{2 - 2x} = \frac{3(-1)}{2 - 2(3)} = \frac{-3}{2 - 6} = \frac{-3}{-4} = \frac{3}{4}$$ ◀

NOW WORK PROBLEM 9.

2 In working with expressions or formulas involving variables, the variables may be allowed to take on values from only a certain set of numbers. For example, in the formula for the area A of a circle of radius r, $A = \pi r^2$, the variable r is necessarily restricted to the positive real numbers. In the expression $\frac{1}{x}$, the variable x cannot take on the value 0, since division by 0 is not defined.

The set of values that a variable in an expression may assume is called the **domain of the variable.**

EXAMPLE 3 **Finding the Domain of a Variable**

The domain of the variable x in the rational expression

$$\frac{5}{x-2}$$

is $\{x|x \neq 2\}$, since, if $x = 2$, the denominator becomes 0, which is not defined. ◀

EXAMPLE 4 **Circumference of a Circle**

In the formula for the circumference C of a circle of radius r,

$$C = 2\pi r$$

the domain of the variable r, representing the radius of the circle, is the set of positive real numbers. The domain of the variable C, representing the circumference of the circle, is also the set of positive real numbers. ◀

In describing the domain of a variable, we may use either set notation or words, whichever is more convenient.

NOW WORK PROBLEM 17.

The Real Number Line

The real numbers can be represented by points on a line called the **real number line.** There is a one-to-one correspondence between real numbers and points on a line. That is, every real number corresponds to a point on the line, and each point on the line has a unique real number associated with it.

Pick a point on the line somewhere in the center, and label it O. This point, called the **origin,** corresponds to the real number 0. See Figure 2. The point 1 unit to the right of O corresponds to the number 1. The distance between 0 and 1 determines the scale of the number line. For example, the point associated with the number 2 is twice as far from O as 1 is. Notice that an arrowhead on the right end of the line indicates the direction in which the numbers increase. Figure 2 also shows the points associated with the irrational numbers $\sqrt{2}$ and π. Points to the left of the origin correspond to the real numbers -1, -2, and so on.

Figure 2 Real number line

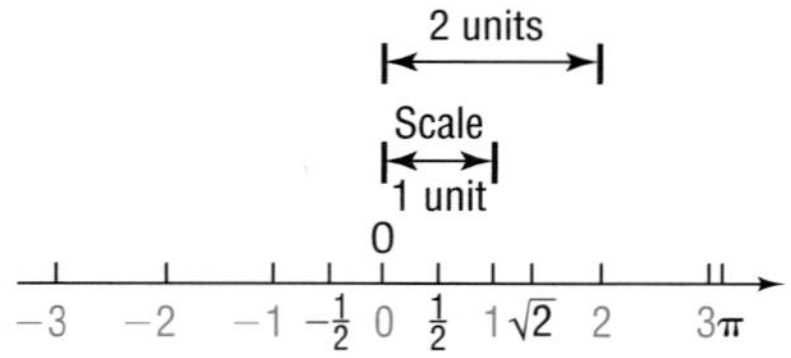

The real number associated with a point P is called the **coordinate** of P, and the line whose points have been assigned coordinates is called the **real number line.**

NOW WORK PROBLEM 29.

The real number line consists of three classes of real numbers, as shown in Figure 3.

Figure 3

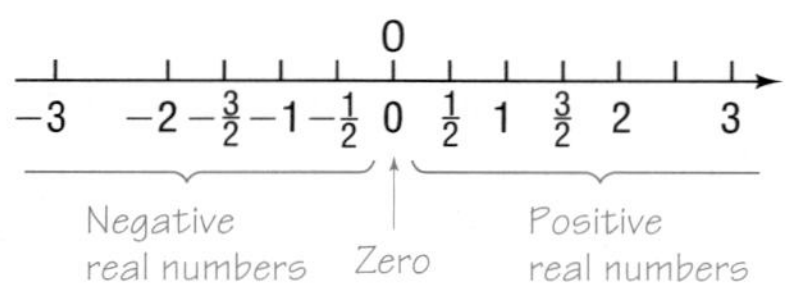

1. The **negative real numbers** are the coordinates of points to the left of the origin O.
2. The real number **zero** is the coordinate of the origin O.
3. The **positive real numbers** are the coordinates of points to the right of the origin O.

Inequalities

An important property of the real number line follows from the fact that, given two numbers (points) a and b, either a is to the left of b, a is at the same location as b, or a is to the right of b. See Figure 4.

Figure 4

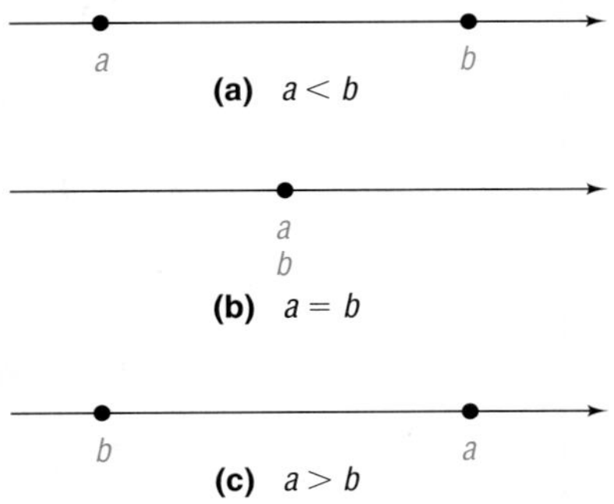

If a is to the left of b, we say that "a is less than b" and write $a < b$. If a is to the right of b, we say that "a is greater than b" and write $a > b$. If a is at the same location as b, then $a = b$. If a is either less than or equal to b, we write $a \leq b$. Similarly, $a \geq b$ means that a is either greater than or equal to b. Collectively, the symbols $<, >, \leq$, and $\geq$ are called **inequality symbols.**

Note that $a < b$ and $b > a$ mean the same thing. It does not matter whether we write $2 < 3$ or $3 > 2$.

Furthermore, if $a < b$ or if $b > a$, then the difference $b - a$ is positive. Do you see why?

An **inequality** is a statement in which two expressions are related by an inequality symbol. The expressions are referred to as the **sides** of the inequality. Statements of the form $a < b$ or $b > a$ are called **strict inequalities,** while statements of the form $a \leq b$ or $b \geq a$ are called **nonstrict inequalities.**

Based on the discussion thus far, we conclude that

$a > 0$	is equivalent to	a is positive
$a < 0$	is equivalent to	a is negative

We sometimes read $a > 0$ by saying that "a is positive." If $a \geq 0$, then either $a > 0$ or $a = 0$, and we may read this as "a is nonnegative."

NOW WORK PROBLEMS 33 AND 43.

3 We shall find it useful in later work to graph inequalities on the real number line.

EXAMPLE 5 **Graphing Inequalities**

(a) On the real number line, graph all numbers x for which $x > 4$.

(b) On the real number line, graph all numbers x for which $x \leq 5$.

Solution (a) See Figure 5. Notice that we use a left parenthesis to indicate that the number 4 is not part of the graph.

Figure 5

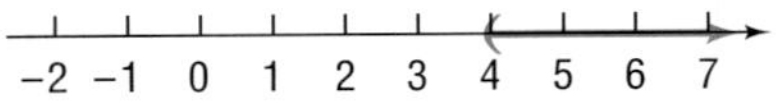

(b) See Figure 6. Notice that we use a right bracket to indicate that the number 5 is part of the graph. ◀

Figure 6

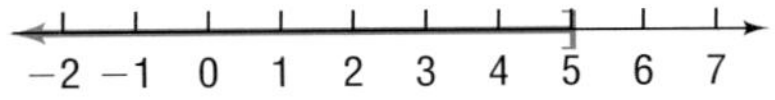

NOW WORK PROBLEM 49.

Absolute Value

The *absolute value* of a number a is the distance from 0 to a on the number line. For example, -4 is 4 units from 0; and 3 is 3 units from 0. See Figure 7. Thus, the absolute value of -4 is 4, and the absolute value of 3 is 3.

Figure 7

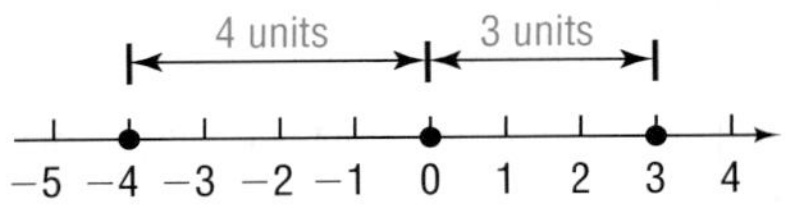

A more formal definition of absolute value is given next.

The **absolute value** of a real number a, denoted by the symbol $|a|$, is defined by the rules

$$|a| = a \quad \text{if } a \geq 0 \qquad \text{and} \qquad |a| = -a \quad \text{if } a < 0$$

For example, since $-4 < 0$, the second rule must be used to get $|-4| = -(-4) = 4$.

EXAMPLE 6 **Computing Absolute Value**

(a) $|8| = 8$ (b) $|0| = 0$ (c) $|-15| = -(-15) = 15$ ◀

NOW WORK PROBLEM 51.

4 Look again at Figure 7. The distance from -4 to 3 is 7 units. This distance is the difference $3 - (-4)$, obtained by subtracting the smaller coordinate from the larger. However, since $|3 - (-4)| = |7| = 7$ and $|-4 - 3| = |-7| = 7$, we can use absolute value to calculate the distance between two points without being concerned about which is smaller.

If P and Q are two points on a real number line with coordinates a and b, respectively, the **distance between P and Q,** denoted by $d(P, Q)$, is

$$d(P, Q) = |b - a|$$

Since $|b - a| = |a - b|$, it follows that $d(P, Q) = d(Q, P)$.

EXAMPLE 7 **Finding Distance on a Number Line**

Let P, Q, and R be points on a real number line with coordinates -5, 7, and -3, respectively. Find the distance

(a) between P and Q (b) between Q and R

Solution See Figure 8.

Figure 8

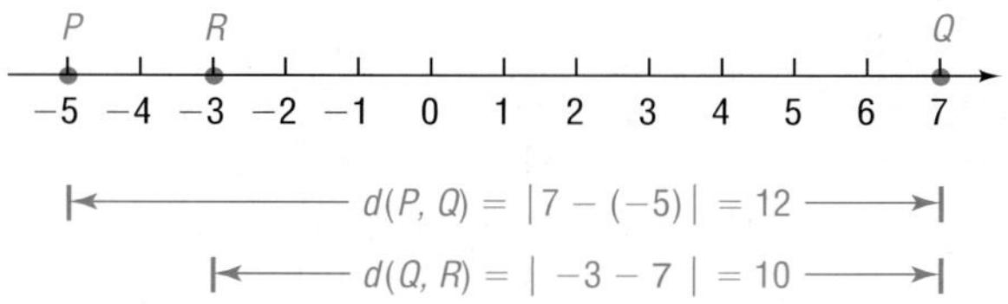

(a) $d(P, Q) = |7 - (-5)| = |12| = 12$

(b) $d(Q, R) = |-3 - 7| = |-10| = 10$ ◀

NOW WORK PROBLEM 65.

Exponents

5 Integer exponents provide a shorthand device for representing repeated multiplications of a real number.

If a is a real number and n is a positive integer, then the symbol a^n represents the product of n factors of a. That is,

$$a^n = \underbrace{a \cdot a \cdot \ldots \cdot a}_{n \text{ factors}}$$

where it is understood that $a^1 = a$. Then, $a^2 = a \cdot a$, $a^3 = a \cdot a \cdot a$, and so on. In the expression a^n, a is called the **base** and n is called the **exponent,** or **power.** We read a^n as "a raised to the power n" or as "a to the nth power." We usually read a^2 as "a squared" and a^3 as "a cubed."

In working with exponents, the operation of *raising to a power* is performed before any other operation. For example,

$$4 \cdot 3^2 + 5 = 4 \cdot 9 + 5 = 36 + 5 = 41 \qquad -2^4 = -16 \qquad 2^3 + 3^2 = 4 + 9 = 13$$

Parentheses are used to indicate operations to be performed first. For example,

$$(-2)^4 = (-2)(-2)(-2)(-2) = 16 \qquad (2 + 3)^2 = 5^2 = 25$$

If $a \neq 0$, we define

$$a^0 = 1 \qquad \text{if } a \neq 0$$

If $a \neq 0$ and if n is a positive integer, then we define

$$a^{-n} = \frac{1}{a^n} \qquad \text{if } a \neq 0$$

With these definitions, the symbol a^n is defined for any integer n.

The following properties, called the **laws of exponents,** can be proved using the preceding definitions. In the list, a and b are real numbers, and m and n are integers.

Laws of Exponents

$$a^m a^n = a^{m+n} \qquad (a^m)^n = a^{mn} \qquad (ab)^n = a^n b^n$$

$$\frac{a^m}{a^n} = a^{m-n} = \frac{1}{a^{n-m}}, \quad \text{if } a \neq 0 \qquad \left(\frac{a}{b}\right)^n = \frac{a^n}{b^n}, \quad \text{if } b \neq 0$$

EXAMPLE 8 Using the Laws of Exponents

Write each expression so that all exponents are positive.

(a) $\dfrac{x^5 y^{-2}}{x^3 y}$, $x \neq 0, y \neq 0$ (b) $\left(\dfrac{x^{-3}}{3y^{-1}}\right)^{-2}$, $x \neq 0, y \neq 0$

Solution

(a) $$\frac{x^5 y^{-2}}{x^3 y} = \frac{x^5}{x^3} \cdot \frac{y^{-2}}{y} = x^{5-3} \cdot y^{-2-1} = x^2 y^{-3} = x^2 \cdot \frac{1}{y^3} = \frac{x^2}{y^3}$$

(b) $$\left(\frac{x^{-3}}{3y^{-1}}\right)^{-2} = \frac{(x^{-3})^{-2}}{(3y^{-1})^{-2}} = \frac{x^6}{3^{-2}(y^{-1})^{-2}} = \frac{x^6}{\frac{1}{9}y^2} = \frac{9x^6}{y^2}$$

◀

NOW WORK PROBLEMS 71 AND 81.

Square Roots

6 A real number is squared when it is raised to the power 2. The inverse of squaring is finding a **square root.** For example, since $6^2 = 36$ and $(-6)^2 = 36$, the numbers 6 and -6 are square roots of 36.

The symbol $\sqrt{\ }$, called a **radical sign,** is used to denote the **principal,** or nonnegative, square root. Thus, $\sqrt{36} = 6$.

In general, if a is a nonnegative real number, the nonnegative number b such that $b^2 = a$ is **the principal square root** of a and is denoted by $b = \sqrt{a}$.

The following comments are noteworthy:

1. Negative numbers do not have square roots (in the real number system), because the square of any real number is *nonnegative*. For example, $\sqrt{-4}$ is not a real number, because there is no real number whose square is -4.
2. The principal square root of 0 is 0, since $0^2 = 0$. That is, $\sqrt{0} = 0$.
3. The principal square root of a positive number is positive.
4. If $c \geq 0$, then $(\sqrt{c})^2 = c$. For example, $\left(\sqrt{2}\right)^2 = 2$ and $\left(\sqrt{3}\right)^2 = 3$.

EXAMPLE 9 **Evaluating Square Roots**

(a) $\sqrt{64} = 8$ (b) $\sqrt{\dfrac{1}{16}} = \dfrac{1}{4}$ (c) $\left(\sqrt{1.4}\right)^2 = 1.4$ (d) $\sqrt{(-3)^2} = |-3| = 3$ ◀

Examples 9(a) and (b) are examples of square roots of perfect squares, since $64 = 8^2$ and $\dfrac{1}{16} = \left(\dfrac{1}{4}\right)^2$.

Notice the need for the absolute value in Example 9(d). Since $a^2 \geq 0$, the principal square root of a^2 is defined whether $a > 0$ or $a < 0$. However, since the principal square root is nonnegative, we need the absolute value to ensure the nonnegative result.

In general, we have

$$\sqrt{a^2} = |a| \qquad \textbf{(1)}$$

EXAMPLE 10 **Using Equation (1)**

(a) $\sqrt{(2.3)^2} = |2.3| = 2.3$ (b) $\sqrt{(-2.3)^2} = |-2.3| = 2.3$ (c) $\sqrt{x^2} = |x|$ ◀

NOW WORK PROBLEM 77.

Calculators

Calculators are finite machines. As a result, they are incapable of displaying decimals that contain a large number of digits. For example, some calculators are capable of displaying only eight digits. When a number requires more than eight digits, the calculator either truncates or rounds. To see how your calculator handles decimals, divide 2 by 3. How many digits do you see? Is the last digit a 6 or a 7? If it is a 6, your calculator truncates; if it is a 7, your calculator rounds.

There are different kinds of calculators. An **arithmetic** calculator can only add, subtract, multiply, and divide numbers; therefore, this type is not adequate for this course. **Scientific** calculators have all the capabilities of arithmetic calculators and also contain **function keys** labeled ln, log, sin, cos, tan, x^y, inv, and so on. As you proceed through this text, you will discover how to use many of the function keys. **Graphing** calculators have all the capabilities of scientific calculators and contain a screen on which graphs can be displayed.

For those who have access to a graphing calculator, we have included comments, examples, and exercises marked with a , indicating that a graphing calculator is required. We have also included Appendix B that explains some of the capabilities of a graphing calculator. The comments, examples, and exercises may be omitted without loss of continuity, if so desired.

A.1 Assess Your Understanding

Concepts and Vocabulary

1. A ________ is a letter used in algebra to represent any number from a given set of numbers.

2. On the real number line, the real number zero is the coordinate of the ________.

3. An inequality of the form $a > b$ is called a(n) ________ inequality.

4. In the expression 2^4, the number 2 is called the ________ and 4 is called the ________.

5. $\sqrt{(-5)^2} =$ ________.

6. *True or False:* The distance between two points on the real number line is always greater than zero.

7. *True or False:* The absolute value of a real number is always greater than zero.

8. *True or False:* To multiply two expressions having the same base, retain the base and multiply the exponents.

Exercises

In Problems 9–16, find the value of each expression if $x = -2$ and $y = 3$.

9. $x + 2y$ **10.** $3x + y$ **11.** $5xy + 2$ **12.** $-2x + xy$

13. $\dfrac{2x}{x - y}$ **14.** $\dfrac{x + y}{x - y}$ **15.** $\dfrac{3x + 2y}{2 + y}$ **16.** $\dfrac{2x - 3}{y}$

In Problems 17–24, determine which of the value(s) given below, if any, must be excluded from the domain of the variable in each expression.

(a) $x = 3$ (b) $x = 1$ (c) $x = 0$ (d) $x = -1$

17. $\dfrac{x^2 - 1}{x}$ **18.** $\dfrac{x^2 + 1}{x}$ **19.** $\dfrac{x}{x^2 - 9}$ **20.** $\dfrac{x}{x^2 + 9}$

21. $\dfrac{x^2}{x^2 + 1}$ **22.** $\dfrac{x^3}{x^2 - 1}$ **23.** $\dfrac{x^2 + 5x - 10}{x^3 - x}$ **24.** $\dfrac{-9x^2 - x + 1}{x^3 + x}$

In Problems 25–28, determine the domain of the variable x in each expression.

25. $\dfrac{4}{x - 5}$ **26.** $\dfrac{-6}{x + 4}$ **27.** $\dfrac{x}{x + 4}$ **28.** $\dfrac{x - 2}{x - 6}$

29. On the real number line, label the points with coordinates $0, 1, -1, \frac{5}{2}, -2.5, \frac{3}{4}$, and 0.25.

30. Repeat Problem 29 for the coordinates $0, -2, 2, -1.5, \frac{3}{2}, \frac{1}{3}$, and $\frac{2}{3}$.

In Problems 31–40, replace the question mark by <, >, or =, whichever is correct.

31. $\frac{1}{2}$? 0 **32.** 5 ? 6 **33.** -1 ? -2 **34.** -3 ? $-\frac{5}{2}$ **35.** π ? 3.14

36. $\sqrt{2}$? 1.41 **37.** $\frac{1}{2}$? 0.5 **38.** $\frac{1}{3}$? 0.33 **39.** $\frac{2}{3}$? 0.67 **40.** $\frac{1}{4}$? 0.25

In Problems 41–46, write each statement as an inequality.

41. x is positive. **42.** z is negative. **43.** x is less than 2.

44. y is greater than -5. **45.** x is less than or equal to 1. **46.** x is greater than or equal to 2.

In Problems 47–50, graph the numbers x on the real number line.

47. $x \geq -2$ **48.** $x < 4$ **49.** $x > -1$ **50.** $x \leq 7$

In Problems 51–60, find the value of each expression if $x = 3$ and $y = -2$.

51. $|x + y|$ **52.** $|x - y|$ **53.** $|x| + |y|$ **54.** $|x| - |y|$ **55.** $\frac{|x|}{x}$

56. $\frac{|y|}{y}$ **57.** $|4x - 5y|$ **58.** $|3x + 2y|$ **59.** $||4x| - |5y||$ **60.** $3|x| + 2|y|$

In Problems 61–66, use the real number line below to compute each distance.

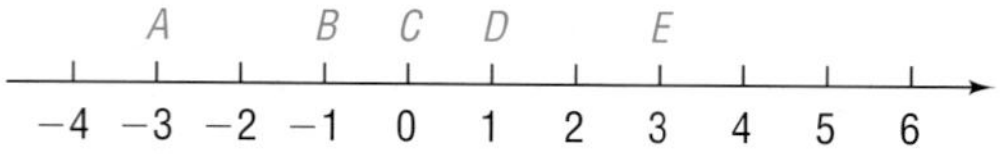

61. $d(C, D)$ **62.** $d(C, A)$ **63.** $d(D, E)$ **64.** $d(C, E)$ **65.** $d(A, E)$ **66.** $d(D, B)$

In Problems 67–78, simplify each expression.

67. $(-4)^2$ **68.** -4^2 **69.** 4^{-2} **70.** -4^{-2} **71.** $3^{-6} \cdot 3^4$ **72.** $4^{-2} \cdot 4^3$

73. $(3^{-2})^{-1}$ **74.** $(2^{-1})^{-3}$ **75.** $\sqrt{25}$ **76.** $\sqrt{36}$ **77.** $\sqrt{(-4)^2}$ **78.** $\sqrt{(-3)^2}$

In Problems 79–88, simplify each expression. Express the answer so that all exponents are positive. Whenever an exponent is 0 or negative, we assume that the base is not 0.

79. $(8x^3)^{-2}$ **80.** $(-4x^2)^{-1}$ **81.** $(x^2y^{-1})^2$ **82.** $(x^{-1}y)^3$ **83.** $\frac{x^{-2}y^3}{xy^4}$

84. $\frac{x^{-2}y}{xy^2}$ **85.** $\frac{(-2)^3x^4(yz)^2}{3^2xy^3z}$ **86.** $\frac{4x^{-2}(yz)^{-1}}{2^3x^4y}$ **87.** $\left(\frac{3x^{-1}}{4y^{-1}}\right)^{-2}$ **88.** $\left(\frac{5x^{-2}}{6y^{-2}}\right)^{-3}$

In Problems 89–100, express each statement as an equation involving the indicated variables.

89. Area of a Rectangle The area A of a rectangle is the product of its length l times its width w.

90. Perimeter of a Rectangle The perimeter P of a rectangle is twice the sum of its length l and its width w.

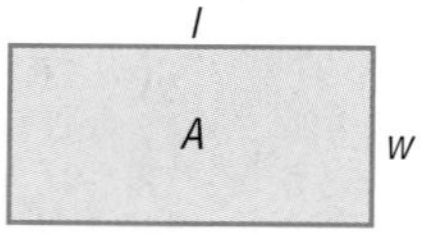

91. Circumference of a Circle The circumference C of a circle is the product of π times its diameter d.

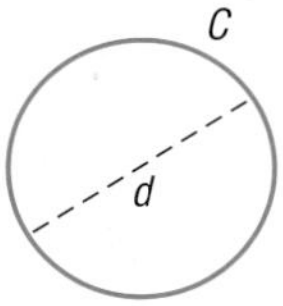

92. Area of a Triangle The area A of a triangle is one-half the product of its base b and its height h.

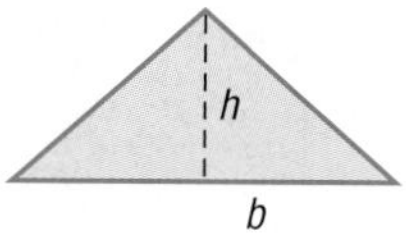

93. Area of an Equilateral Triangle The area A of an equilateral triangle is $\frac{\sqrt{3}}{4}$ times the square of the length x of one side.

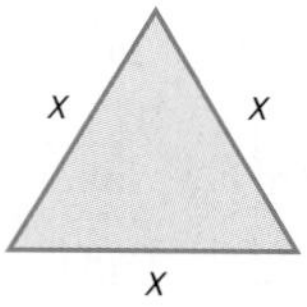

94. Perimeter of an Equilateral Triangle The perimeter P of an equilateral triangle is 3 times the length x of one side.

95. Volume of a Sphere The volume V of a sphere is $\frac{4}{3}$ times π times the cube of the radius r.

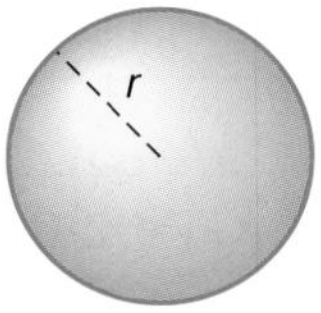

96. Surface Area of a Sphere The surface area S of a sphere is 4 times π times the square of the radius r.

97. Volume of a Cube The volume V of a cube is the cube of the length x of a side.

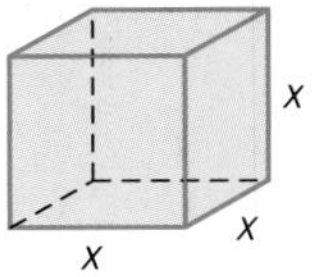

98. Surface Area of a Cube The surface area S of a cube is 6 times the square of the length x of a side.

99. U.S. Voltage In the United States, normal household voltage is 115 volts. It is acceptable for the actual voltage x to differ from normal by at most 5 volts. A formula that describes this is

$$|x - 115| \leq 5$$

(a) Show that a voltage of 113 volts is acceptable.
(b) Show that a voltage of 109 volts is not acceptable.

100. Foreign Voltage In countries other than the United States, normal household voltage is 220 volts. It is acceptable for the actual voltage x to differ from normal by at most 8 volts. A formula that describes this is

$$|x - 220| \leq 8$$

(a) Show that a voltage of 214 volts is acceptable.
(b) Show that a voltage of 209 volts is not acceptable.

101. Making Precision Ball Bearings The FireBall Company manufactures ball bearings for precision equipment. One of their products is a ball bearing with a stated radius of 3 centimeters (cm). Only ball bearings with a radius within 0.01 cm of this stated radius are acceptable. If x is the radius of a ball bearing, a formula describing this situation is

$$|x - 3| \leq 0.01$$

(a) Is a ball bearing of radius $x = 2.999$ acceptable?
(b) Is a ball bearing of radius $x = 2.89$ acceptable?

102. Body Temperature Normal human body temperature is 98.6°F. A temperature x that differs from normal by at least 1.5°F is considered unhealthy. A formula that describes this is

$$|x - 98.6| \geq 1.5$$

(a) Show that a temperature of 97°F is unhealthy.
(b) Show that a temperature of 100°F is not unhealthy.

103. Does $\frac{1}{3}$ equal 0.333? If not, which is larger? By how much?

104. Does $\frac{2}{3}$ equal 0.666? If not, which is larger? By how much?

105. Is there a positive real number "closest" to 0?

106. I'm thinking of a number! It lies between 1 and 10; its square is rational and lies between 1 and 10. The number is larger than π. Correct to two decimal places, name the number. Now think of your own number, describe it, and challenge a fellow student to name it.

107. Write a brief paragraph that illustrates the similarities and differences between "less than" ($<$) and "less than or equal" ($\leq$).

A.2 Geometry Review

OBJECTIVES 1 Use the Pythagorean Theorem and Its Converse
2 Know Geometry Formulas

In this section we review some topics studied in geometry that we shall need for our study of algebra.

Pythagorean Theorem

Figure 9

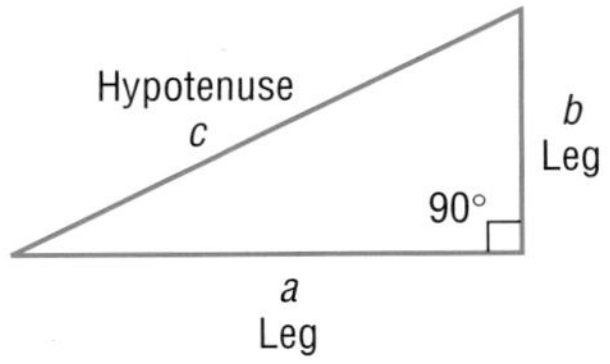

1 The *Pythagorean Theorem* is a statement about *right triangles*. A **right triangle** is one that contains a **right angle,** that is, an angle of 90°. The side of the triangle opposite the 90° angle is called the **hypotenuse;** the remaining two sides are called **legs.** In Figure 9 we have used c to represent the length of the hypotenuse and a and b to represent the lengths of the legs. Notice the use of the symbol ⌐ to show the 90° angle. We now state the Pythagorean Theorem.

Pythagorean Theorem

In a right triangle, the square of the length of the hypotenuse is equal to the sum of the squares of the lengths of the legs. That is, in the right triangle shown in Figure 9,

$$c^2 = a^2 + b^2 \quad (1)$$

EXAMPLE 1 Finding the Hypotenuse of a Right Triangle

In a right triangle, one leg is of length 4 and the other is of length 3. What is the length of the hypotenuse?

Solution Since the triangle is a right triangle, we use the Pythagorean Theorem with $a = 4$ and $b = 3$ to find the length c of the hypotenuse. From equation (1), we have

$$c^2 = a^2 + b^2$$
$$c^2 = 4^2 + 3^2 = 16 + 9 = 25$$
$$c = \sqrt{25} = 5$$

◀

NOW WORK PROBLEM 9.

The converse of the Pythagorean Theorem is also true.

Converse of the Pythagorean Theorem

In a triangle, if the square of the length of one side equals the sum of the squares of the lengths of the other two sides, then the triangle is a right triangle. The 90° angle is opposite the longest side.

EXAMPLE 2 Verifying That a Triangle Is a Right Triangle

Show that a triangle whose sides are of lengths 5, 12, and 13 is a right triangle. Identify the hypotenuse.

Solution We square the lengths of the sides.

$$5^2 = 25, \qquad 12^2 = 144, \qquad 13^2 = 169$$

Notice that the sum of the first two squares (25 and 144) equals the third square (169). Hence, the triangle is a right triangle. The longest side, 13, is the hypotenuse. See Figure 10. ◀

Figure 10

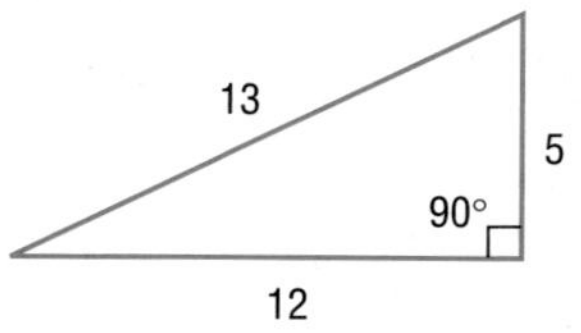

NOW WORK PROBLEM 17.

EXAMPLE 3 Applying the Pythagorean Theorem

The tallest inhabited building in the world is the Sears Tower in Chicago. If the observation tower is 1450 feet above ground level, how far can a person standing in the observation tower see (with the aid of a telescope)? Use 3960 miles for the radius of Earth. See Figure 11.

SOURCE: Council on Tall Buildings and Urban Habitat (1997); Sears Tower No. 1 for tallest roof (1450 ft) and tallest occupied floor (1431 ft).

Figure 11

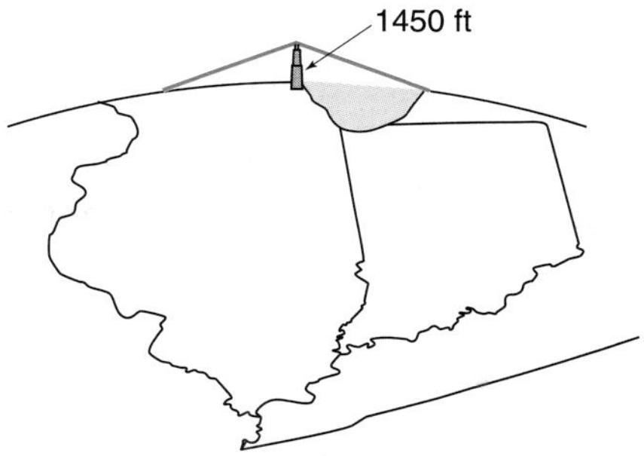

Solution From the center of Earth, draw two radii: one through the Sears Tower and the other to the farthest point a person can see from the tower. See Figure 12. Apply the Pythagorean Theorem to the right triangle.

Figure 12

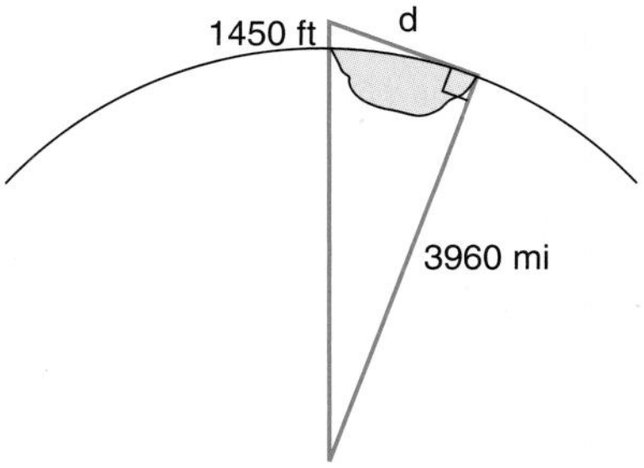

Since 1 mile $=$ 5280 feet, then 1450 feet $= \frac{1450}{5280}$ miles. So we have

$$d^2 + (3960)^2 = \left(3960 + \frac{1450}{5280}\right)^2$$

$$d^2 = \left(3960 + \frac{1450}{5280}\right)^2 - (3960)^2 \approx 2175.08$$

$$d \approx 46.64$$

A person can see about 47 miles from the observation tower. ◀

NOW WORK PROBLEM 43.

Geometry Formulas

2 Certain formulas from geometry are useful in solving algebra problems. We list some of these formulas next.

For a rectangle of length l and width w,

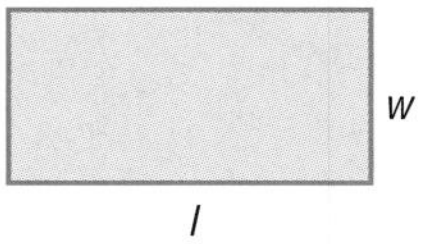

$$\text{Area} = lw \qquad \text{Perimeter} = 2l + 2w$$

For a triangle with base b and altitude h,

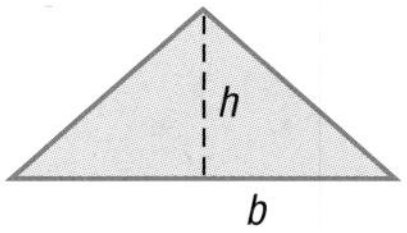

$$\text{Area} = \frac{1}{2}bh$$

For a circle of radius r (diameter $d = 2r$),

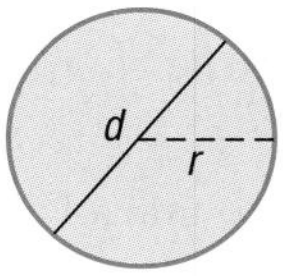

$$\text{Area} = \pi r^2 \qquad \text{Circumference} = 2\pi r = \pi d$$

For a closed rectangular box of length l, width w, and height h,

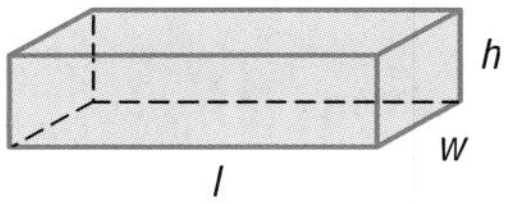

$$\text{Volume} = lwh \qquad \text{Surface area} = 2lh + 2wh + 2lh$$

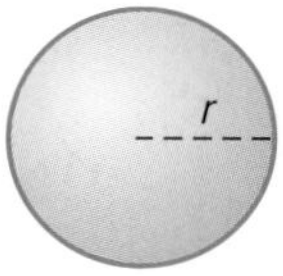

For a sphere of radius r,

$$\text{Volume} = \frac{4}{3}\pi r^3 \qquad \text{Surface area} = 4\pi r^2$$

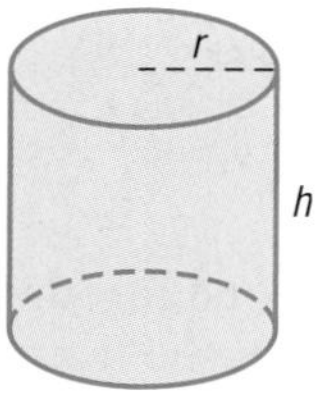

For a right circular cylinder of height h and radius r,

$$\text{Volume} = \pi r^2 h \qquad \text{Surface area} = 2\pi r^2 + 2\pi r h$$

NOW WORK PROBLEM 25.

EXAMPLE 4 Using Geometry Formulas

A Christmas tree ornament is in the shape of a semicircle on top of a triangle. How many square centimeters (cm) of copper is required to make the ornament if the height of the triangle is 6 cm and the base is 4 cm?

Solution See Figure 13. The amount of copper required equals the shaded area. This area is the sum of the area of the triangle and the semicircle. The triangle has height $h = 6$ and base $b = 4$. The semicircle has diameter $d = 4$, so its radius is $r = 2$.

Figure 13

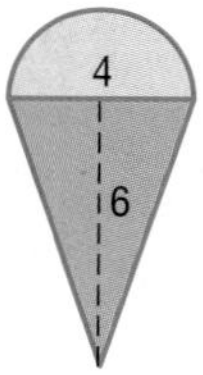

$$\text{Area} = \text{Area of triangle} + \text{Area of semicircle}$$

$$= \frac{1}{2}bh + \frac{1}{2}\pi r^2 = \frac{1}{2}(4)(6) + \frac{1}{2}\pi \cdot 2^2 \qquad b = 4; h = 6; r = 2$$

$$= 12 + 2\pi \approx 18.28 \text{ cm}^2$$

About 18.28 cm^2 of copper is required. ◀

NOW WORK PROBLEM 39.

A.2 Assess Your Understanding

Concepts and Vocabulary

1. A __________ triangle is one that contains an angle of 90 degrees. The longest side is called the __________.
2. For a triangle with base b and altitude h, a formula for the area A is __________.
3. The formula for the circumference C of a circle of radius r is __________.
4. *True or False:* In a right triangle, the square of the length of the longest side equals the sum of the squares of the lengths of the other two sides.
5. *True or False:* The triangle with sides of length 6, 8, and 10 is a right triangle.
6. *True or False:* The volume of a sphere of radius r is $\frac{4}{3}\pi r^2$.

Exercises

In Problems 7–12, the lengths of the legs of a right triangle are given. Find the hypotenuse.

7. $a = 5, \quad b = 12$
8. $a = 6, \quad b = 8$
9. $a = 10, \quad b = 24$
10. $a = 4, \quad b = 3$
11. $a = 7, \quad b = 24$
12. $a = 14, \quad b = 48$

In Problems 13–20, the lengths of the sides of a triangle are given. Determine which are right triangles. For those that are, identify the hypotenuse.

13. 3, 4, 5 **14.** 6, 8, 10 **15.** 4, 5, 6 **16.** 2, 2, 3

17. 7, 24, 25 **18.** 10, 24, 26 **19.** 6, 4, 3 **20.** 5, 4, 7

21. Find the area A of a rectangle with length 4 inches and width 2 inches.

22. Find the area A of a rectangle with length 9 centimeters and width 4 centimeters.

23. Find the area A of a triangle with height 4 inches and base 2 inches.

24. Find the area A of a triangle with height 9 centimeters and base 4 centimeters.

25. Find the area A and circumference C of a circle of radius 5 meters.

26. Find the area A and circumference C of a circle of radius 2 feet.

27. Find the volume V and surface area S of a rectangular box with length 8 feet, width 4 feet, and height 7 feet.

28. Find the volume V and surface area S of a rectangular box with length 9 inches, width 4 inches, and height 8 inches.

29. Find the volume V and surface area S of a sphere of radius 4 centimeters.

30. Find the volume V and surface area S of a sphere of radius 3 feet.

31. Find the volume V and surface area S of a right circular cylinder with radius 9 inches and height 8 inches.

32. Find the volume V and surface area S of a right circular cylinder with radius 8 inches and height 9 inches.

In Problems 33–36, find the area of the shaded region.

33.

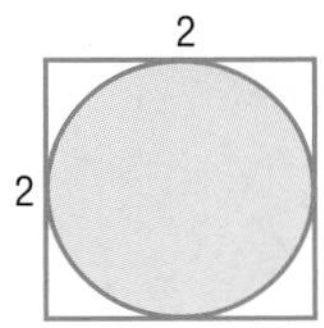

34.

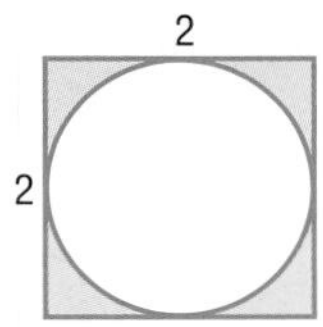

35.

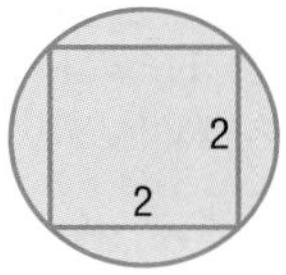

36.

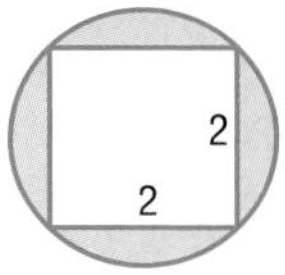

37. How many feet does a wheel with a diameter of 16 inches travel after four revolutions?

38. How many revolutions will a circular disk with a diameter of 4 feet have completed after it has rolled 20 feet?

39. In the figure shown, $ABCD$ is a square, with each side of length 6 feet. The width of the border (shaded portion) between the outer square $EFGH$ and $ABCD$ is 2 feet. Find the area of the border.

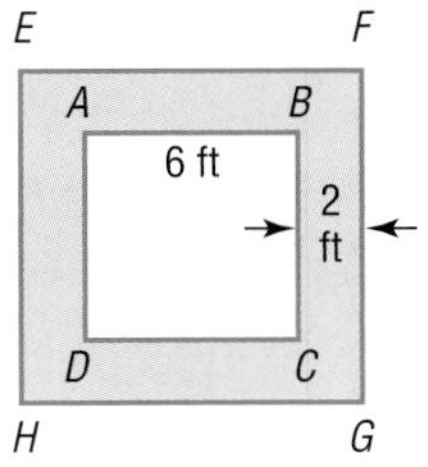

40. Refer to the figure. Square $ABCD$ has an area of 100 square feet; square $BEFG$ has an area of 16 square feet. What is the area of the triangle CGF?

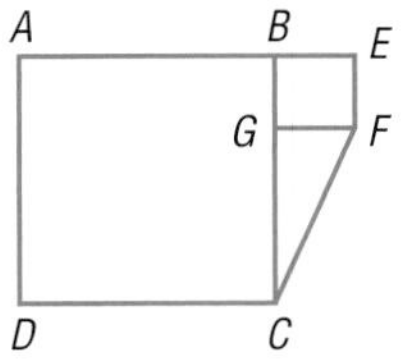

41. Architecture A Norman window consists of a rectangle surmounted by a semicircle. Find the area of the Norman window shown in the illustration. How much wood frame is needed to enclose the window?

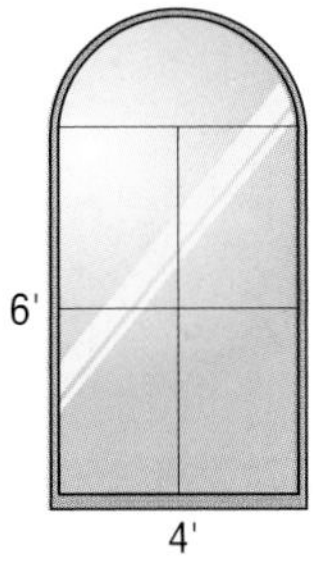

42. Construction A circular swimming pool, 20 feet in diameter, is enclosed by a wooden deck that is 3 feet wide. What is the area of the deck? How much fence is required to enclose the deck?

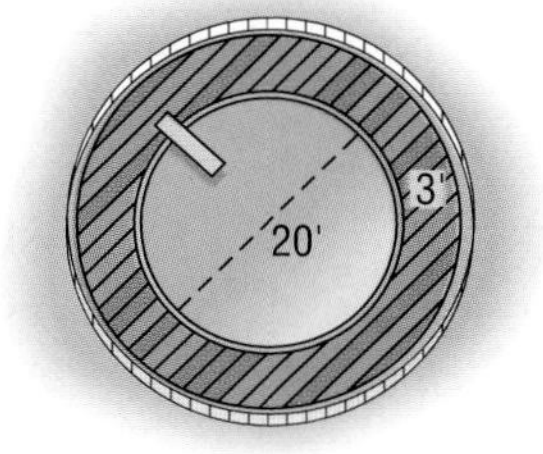

In Problems 43–45, use the facts that the radius of Earth is 3960 miles and 1 mile = 5280 feet.

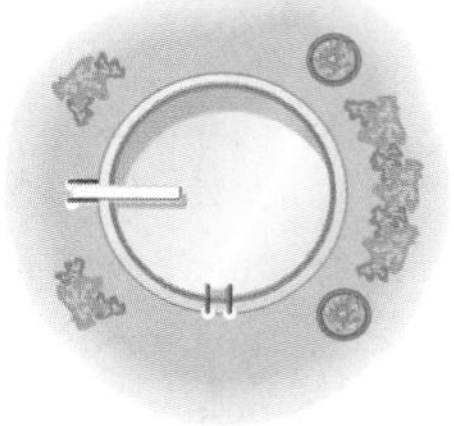

43. **How Far Can You See?** The conning tower of the U.S.S. *Silversides*, a World War II submarine now permanently stationed in Muskegon, Michigan, is approximately 20 feet above sea level. How far can you see from the conning tower?

44. **How Far Can You See?** A person who is 6 feet tall is standing on the beach in Fort Lauderdale, Florida, and looks out onto the Atlantic Ocean. Suddenly, a ship appears on the horizon. How far is the ship from shore?

45. **How Far Can You See?** The deck of a destroyer is 100 feet above sea level. How far can a person see from the deck? How far can a person see from the bridge, which is 150 feet above sea level?

46. Suppose that m and n are positive integers with $m > n$. If $a = m^2 - n^2$, $b = 2mn$, and $c = m^2 + n^2$, show that a, b, and c are the lengths of the sides of a right triangle. (This formula can be used to find the sides of a right triangle that are integers, such as 3, 4, 5; 5, 12, 13; and so on. Such triplets of integers are called **Pythagorean triples.**)

47. You have 1000 feet of flexible pool siding and wish to construct a swimming pool. Experiment with rectangular-shaped pools with perimeters of 1000 feet. How do their areas vary? What is the shape of the rectangle with the largest area? Now compute the area enclosed by a circular pool with a perimeter (circumference) of 1000 feet. What would be your choice of shape for the pool? If rectangular, what is your preference for dimensions? Justify your choice. If your only consideration is to have a pool that encloses the most area, what shape should you use?

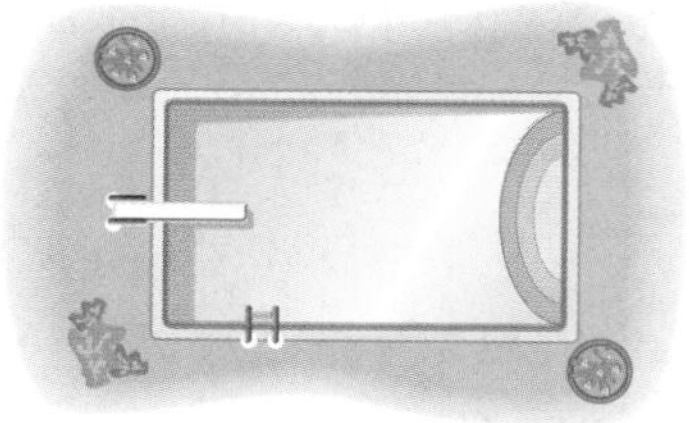

48. **The Gibb's Hill Lighthouse, Southampton, Bermuda,** in operation since 1846, stands 117 feet high on a hill 245 feet high, so its beam of light is 362 feet above sea level. A brochure states that the light itself can be seen on the horizon about 26 miles distant. Verify the correctness of this information. The brochure further states that ships 40 miles away can see the light and planes flying at 10,000 feet can see it 120 miles away. Verify the accuracy of these statements. What assumption did the brochure make about the height of the ship?

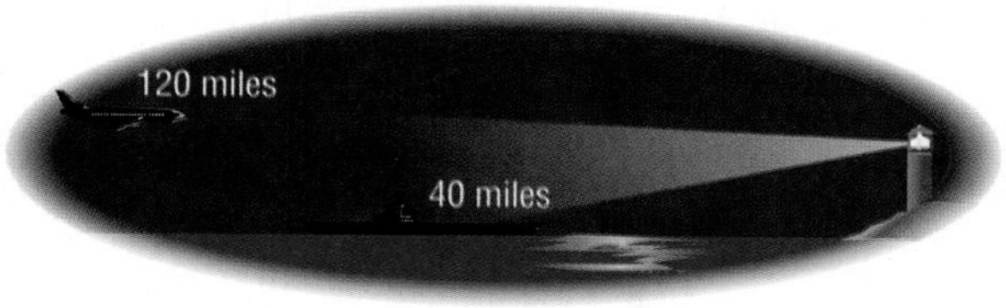

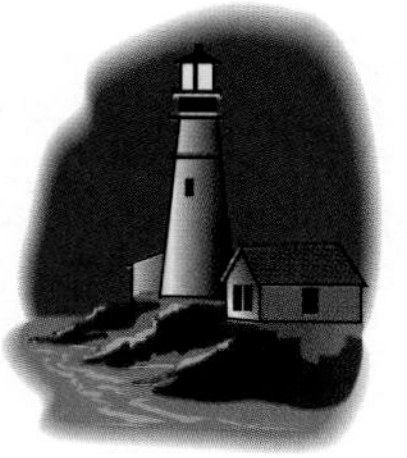

A.3 Polynomials and Rational Expressions

OBJECTIVES
1. Recognize Special Products
2. Factor Polynomials
3. Simplify Rational Expressions
4. Use the LCM to Add Rational Expressions

As we said earlier, in algebra we use letters to represent real numbers. We shall use the letters at the end of the alphabet, such as x, y, and z, to represent variables and the letters at the beginning of the alphabet, such as a, b, and c, to represent constants. In the expressions $3x + 5$ and $ax + b$, it is understood that x is a variable and that a and b are constants, even though the constants a and b are unspecified. As you will find out, the context usually makes the intended meaning clear.

Now we introduce some basic vocabulary.

A **monomial** in one variable is the product of a constant and a variable raised to a nonnegative integer power. Thus, a monomial is of the form

$$ax^k$$

where a is a constant, x is a variable, and $k \geq 0$ is an integer. The constant a is called the **coefficient** of the monomial. If $a \neq 0$, then k is called the **degree** of the monomial.

Examples of monomials follow:

Monomial	Coefficient	Degree	
$6x^2$	6	2	
$-\sqrt{2}x^3$	$-\sqrt{2}$	3	
3	3	0	Since $3 = 3 \cdot 1 = 3x^0$
$-5x$	-5	1	Since $-5x = -5x^1$
x^4	1	4	Since $x^4 = 1 \cdot x^4$

Two monomials ax^k and bx^k with the same degree and the same variable are called **like terms.** Such monomials when added or subtracted can be combined into a single monomial by using the distributive property. For example,

$$2x^2 + 5x^2 = (2 + 5)x^2 = 7x^2 \quad \text{and} \quad 8x^3 - 5x^3 = (8 - 5)x^3 = 3x^3$$

The sum or difference of two monomials having different degrees is called a **binomial.** The sum or difference of three monomials with three different degrees is called a **trinomial.** For example,

$x^2 - 2$ is a binomial.

$x^3 - 3x + 5$ is a trinomial.

$2x^2 + 5x^2 + 2 = 7x^2 + 2$ is a binomial.

In Words

A polynomial is a sum of monomials

A **polynomial** in one variable is an algebraic expression of the form

$$a_n x^n + a_{n-1}x^{n-1} + \cdots + a_1 x + a_0 \quad (1)$$

where $a_n, a_{n-1}, \ldots, a_1, a_0$ are constants* called the **coefficients** of the polynomial, $n \geq 0$ is an integer, and x is a variable. If $a_n \neq 0$, it is called the **leading coefficient,** and n is called the **degree** of the polynomial.

The monomials that make up a polynomial are called its **terms.** If all the coefficients are 0, the polynomial is called the **zero polynomial,** which has no degree.

*The notation a_n is read as "a sub n." The number n is called a **subscript** and should not be confused with an exponent. We use subscripts in order to distinguish one constant from another when a large or undetermined number of constants is required.

Polynomials are usually written in **standard form,** beginning with the nonzero term of highest degree and continuing with terms in descending order according to degree. If a power of x is missing, it is because its coefficient is zero. Examples of polynomials follow:

Polynomial	Coefficients	Degree
$3x^2 - 5 = 3x^2 + 0 \cdot x + (-5)$	3, 0, −5	2
$8 - 2x + x^2 = 1 \cdot x^2 - 2x + 8$	1, −2, 8	2
$5x + \sqrt{2} = 5x^1 + \sqrt{2}$	5, $\sqrt{2}$	1
$3 = 3 \cdot 1 = 3 \cdot x^0$	3	0
0	0	No degree

Although we have been using x to represent the variable, letters such as y or z are also commonly used.

$3x^4 - x^2 + 2$ is a polynomial (in x) of degree 4.

$9y^3 - 2y^2 + y - 3$ is a polynomial (in y) of degree 3.

$z^5 + \pi$ is a polynomial (in z) of degree 5.

Algebraic expressions such as

$$\frac{1}{x} \quad \text{and} \quad \frac{x^2 + 1}{x + 5}$$

are not polynomials. The first is not a polynomial because $\frac{1}{x} = x^{-1}$ has an exponent that is not a nonnegative integer. Although the second expression is the quotient of two polynomials, the polynomial in the denominator has degree greater than 0, so the expression cannot be a polynomial.

1 Certain products, which we call **special products,** occur frequently in algebra. In the list that follows, x, a, b, c, and d are real numbers.

Difference of Two Squares

$$(x - a)(x + a) = x^2 - a^2 \qquad \textbf{(2)}$$

Squares of Binomials, or Perfect Squares

$$(x + a)^2 = x^2 + 2ax + a^2 \qquad \textbf{(3a)}$$

$$(x - a)^2 = x^2 - 2ax + a^2 \qquad \textbf{(3b)}$$

Miscellaneous Trinomials

$$(x + a)(x + b) = x^2 + (a + b)x + ab \qquad \textbf{(4a)}$$

$$(ax + b)(cx + d) = acx^2 + (ad + bc)x + bd \qquad \textbf{(4b)}$$

Cubes of Binomials, or Perfect Cubes

$$(x + a)^3 = x^3 + 3ax^2 + 3a^2x + a^3 \tag{5a}$$

$$(x - a)^3 = x^3 - 3ax^2 + 3a^2x - a^3 \tag{5b}$$

Difference of Two Cubes

$$(x - a)(x^2 + ax + a^2) = x^3 - a^3 \tag{6}$$

Sum of Two Cubes

$$(x + a)(x^2 - ax + a^2) = x^3 + a^3 \tag{7}$$

The special product formulas in equations (2) through (7) are used often, and their patterns should be committed to memory. But if you forget one or are unsure of its form, you should be able to derive it as needed.

EXAMPLE 1 **Using Special Formulas**

(a) $(x - 4)(x + 4) = x^2 - 4^2 = x^2 - 16$

(b) $(2x + 5)(3x - 1) = 6x^2 - 2x + 15x - 5 = 6x^2 + 13x - 5$

(c) $(x - 2)^3 = x^3 - 3(2)x^2 + 3(2)^2x - (2)^3 = x^3 - 6x^2 + 12x - 8$ ◀

NOW WORK PROBLEM 17.

Factoring

2 Consider the following product:

$$(2x + 3)(x - 4) = 2x^2 - 5x - 12$$

The two polynomials on the left are called **factors** of the polynomial on the right. Expressing a given polynomial as a product of other polynomials, that is, finding the factors of a polynomial, is called **factoring.**

We shall restrict our discussion here to factoring polynomials in one variable into products of polynomials in one variable, where all coefficients are integers. We call this **factoring over the integers.**

Any polynomial can be written as the product of 1 times itself or as -1 times its additive inverse. If a polynomial cannot be written as the product of two other polynomials (excluding 1 and -1), then the polynomial is said to be **prime.** When a polynomial has been written as a product consisting only of prime factors, it is said to be **factored completely.** Examples of prime polynomials are

$$2, \quad 3, \quad 5, \quad x, \quad x + 1, \quad x - 1, \quad 3x + 4$$

The first factor to look for in a factoring problem is a common monomial factor present in each term of the polynomial. If one is present, use the distributive property to factor it out. For example,

Polynomial	Common Monomial Factor	Remaining Factor	Factored Form
$2x + 4$	2	$x + 2$	$2x + 4 = 2(x + 2)$
$3x - 6$	3	$x - 2$	$3x - 6 = 3(x - 2)$
$2x^2 - 4x + 8$	2	$x^2 - 2x + 4$	$2x^2 - 4x + 8 = 2(x^2 - 2x + 4)$
$8x - 12$	4	$2x - 3$	$8x - 12 = 4(2x - 3)$
$x^2 + x$	x	$x + 1$	$x^2 + x = x(x + 1)$
$x^3 - 3x^2$	x^2	$x - 3$	$x^3 - 3x^2 = x^2(x - 3)$
$6x^2 + 9x$	$3x$	$2x + 3$	$6x^2 + 9x = 3x(2x + 3)$

The list of special products (2) through (7) given earlier provides a list of factoring formulas when the equations are read from right to left. For example, equation (2) states that if the polynomial is the difference of two squares, $x^2 - a^2$, it can be factored into $(x - a)(x + a)$. The following example illustrates several factoring techniques.

EXAMPLE 2 Factoring Polynomials

Factor completely each polynomial.

(a) $x^4 - 16$ (b) $x^3 - 1$ (c) $9x^2 - 6x + 1$

(d) $x^2 + 4x - 12$ (e) $3x^2 + 10x - 8$ (f) $x^3 - 4x^2 + 2x - 8$

Solution (a) $x^4 - 16 = (x^2 - 4)(x^2 + 4) = (x - 2)(x + 2)(x^2 + 4)$

↑ Difference of squares ↑ Difference of squares

(b) $x^3 - 1 = (x - 1)(x^2 + x + 1)$

↑ Difference of cubes

(c) $9x^2 - 6x + 1 = (3x - 1)^2$

↑ Perfect square

(d) $x^2 + 4x - 12 = (x + 6)(x - 2)$

↑ The product of 6 and −2 is −12, and the sum of 6 and −2 is 4.

(e) $12x - 2x = 10x$

$3x^2 + 10x - 8 = (3x - 2)(x + 4)$

$3x^2$ -8

(f) $x^3 - 4x^2 + 2x - 8 = (x^3 - 4x^2) + (2x - 8)$

↑ Regroup

$= x^2(x - 4) + 2(x - 4) = (x^2 + 2)(x - 4)$

↑ Distributive property ↑ Distributive property ◄

The technique used in Example 2(f) is called **factoring by grouping.**

NOW WORK PROBLEMS 31, 47, AND 79.

Rational Expressions

If we form the quotient of two polynomials, the result is called a **rational expression.** Some examples of rational expressions are

$$\text{(a)}\ \frac{x^3+1}{x} \qquad \text{(b)}\ \frac{3x^3+x-2}{x^5+5} \qquad \text{(c)}\ \frac{x}{x^2-1} \qquad \text{(d)}\ \frac{xy^2}{(x-y)^2}$$

Expressions (a), (b), and (c) are rational expressions in one variable, x, whereas (d) is a rational expression in two variables, x and y.

Rational expressions are described in the same manner as rational numbers. Thus, in expression (a), the polynomial x^3+1 is called the **numerator,** and x is called the **denominator.** When the numerator and denominator of a rational expression contain no common factors (except 1 and -1), we say that the rational expression is **reduced to lowest terms,** or **simplified.**

3 A rational expression is reduced to lowest terms by completely factoring the numerator and the denominator and canceling any common factors by using the cancellation property.

$$\frac{ac}{bc} = \frac{a}{b}, \qquad b \neq 0, \quad c \neq 0$$

We shall follow the common practice of using a slash mark to indicate cancellation. For example,

$$\frac{x^2-1}{x^2-2x-3} = \frac{(x-1)\cancel{(x+1)}}{(x-3)\cancel{(x+1)}} = \frac{x-1}{x-3}$$

EXAMPLE 3 Simplifying Rational Expressions

Reduce each rational expression to lowest terms.

$$\text{(a)}\ \frac{x^2+4x+4}{x^2+3x+2} \qquad \text{(b)}\ \frac{x^3-8}{x^3-2x^2} \qquad \text{(c)}\ \frac{8-2x}{x^2-x-12}$$

Solution (a) $\dfrac{x^2+4x+4}{x^2+3x+2} = \dfrac{\cancel{(x+2)}(x+2)}{\cancel{(x+2)}(x+1)} = \dfrac{x+2}{x+1}, \qquad x \neq -2, -1$

(b) $\dfrac{x^3-8}{x^3-2x^2} = \dfrac{\cancel{(x-2)}(x^2+2x+4)}{x^2\cancel{(x-2)}} = \dfrac{x^2+2x+4}{x^2}, \qquad x \neq 0, 2$

(c) $\dfrac{8-2x}{x^2-x-12} = \dfrac{2(4-x)}{(x-4)(x+3)} = \dfrac{2(-1)\cancel{(x-4)}}{\cancel{(x-4)}(x+3)} = \dfrac{-2}{x+3}, \quad x \neq -3, 4$

◄

NOW WORK PROBLEM 89.

The rules for multiplying and dividing rational expressions are the same as the rules for multiplying and dividing rational numbers.

$$\frac{a}{b}\cdot\frac{c}{d}=\frac{ac}{bd}, \quad \text{if } b \neq 0, d \neq 0 \qquad \textbf{(8)}$$

$$\frac{\frac{a}{b}}{\frac{c}{d}}=\frac{a}{b}\cdot\frac{d}{c}=\frac{ad}{bc}, \quad \text{if } b \neq 0, c \neq 0, d \neq 0 \qquad \textbf{(9)}$$

In using equations (8) and (9) with rational expressions, be sure first to factor each polynomial completely so that common factors can be canceled. We shall follow the practice of leaving our answers in factored form.

EXAMPLE 4 Finding Products and Quotients of Rational Expressions

Perform the indicated operation and simplify the result. Leave your answer in factored form.

(a) $\dfrac{x^2-2x+1}{x^3+x}\cdot\dfrac{4x^2+4}{x^2+x-2}$ (b) $\dfrac{\frac{x+3}{x^2-4}}{\frac{x^2-x-12}{x^3-8}}$

Solution (a)

$$\frac{x^2-2x+1}{x^3+x}\cdot\frac{4x^2+4}{x^2+x-2}=\frac{(x-1)^2}{x(x^2+1)}\cdot\frac{4(x^2+1)}{(x+2)(x-1)}$$

$$=\frac{(x-1)^{\cancel{2}}(4)\cancel{(x^2+1)}}{x\cancel{(x^2+1)}(x+2)\cancel{(x-1)}}=\frac{4(x-1)}{x(x+2)}, \quad x \neq -2, 0, 1$$

(b)

$$\frac{\frac{x+3}{x^2-4}}{\frac{x^2-x-12}{x^3-8}}=\frac{x+3}{x^2-4}\cdot\frac{x^3-8}{x^2-x-12}$$

$$=\frac{x+3}{(x-2)(x+2)}\cdot\frac{(x-2)(x^2+2x+4)}{(x-4)(x+3)}$$

$$=\frac{\cancel{(x+3)}\cancel{(x-2)}(x^2+2x+4)}{\cancel{(x-2)}(x+2)(x-4)\cancel{(x+3)}}=\frac{x^2+2x+4}{(x+2)(x-4)}, \ x \neq -3, -2, 2, 4$$

◀

Note: Slanting the cancellation marks in different directions for different factors, as in Example 4, is a good practice to follow, since it will help in checking for errors.

NOW WORK PROBLEM 67.

In Words
Keep the common denominator and add (subtract) the numerators.

If the denominators of two rational expressions to be added (or subtracted) are equal, we add (or subtract) the numerators and keep the common denominator. That is, if $\frac{a}{b}$ and $\frac{c}{b}$ are two rational expressions, then

$$\frac{a}{b}+\frac{c}{b}=\frac{a+c}{b} \qquad \frac{a}{b}-\frac{c}{b}=\frac{a-c}{b}, \quad \text{if } b \neq 0 \qquad \textbf{(10)}$$

EXAMPLE 5 **Finding the Sum of Two Rational Expressions**

Perform the indicated operation and simplify the result. Leave your answer in factored form.

$$\frac{2x^2 - 4}{2x + 5} + \frac{x + 3}{2x + 5}, \qquad x \neq -\frac{5}{2}$$

Solution

$$\frac{2x^2 - 4}{2x + 5} + \frac{x + 3}{2x + 5} = \frac{(2x^2 - 4) + (x + 3)}{2x + 5}$$
$$= \frac{2x^2 + x - 1}{2x + 5} = \frac{(2x - 1)(x + 1)}{2x + 5}$$

◀

If the denominators of two rational expressions to be added or subtracted are not equal, we can use the general formulas for adding and subtracting quotients.

$$\frac{a}{b} + \frac{c}{d} = \frac{a \cdot d}{b \cdot d} + \frac{b \cdot c}{b \cdot d} = \frac{ad + bc}{bd}, \qquad \text{if } b \neq 0, d \neq 0$$
$$\frac{a}{b} - \frac{c}{d} = \frac{a \cdot d}{b \cdot d} - \frac{b \cdot c}{b \cdot d} = \frac{ad - bc}{bd}, \qquad \text{if } b \neq 0, d \neq 0 \qquad \textbf{(11)}$$

EXAMPLE 6 **Finding the Difference of Two Rational Expressions**

Perform the indicated operation and simplify the result. Leave your answer in factored form.

$$\frac{x^2}{x^2 - 4} - \frac{1}{x}, \qquad x \neq -2, 0, 2$$

Solution

$$\frac{x^2}{x^2 - 4} - \frac{1}{x} = \frac{x^2(x) - (x^2 - 4)(1)}{(x^2 - 4)(x)} = \frac{x^3 - x^2 + 4}{(x - 2)(x + 2)(x)}$$

◀

Least Common Multiple (LCM)

4 If the denominators of two rational expressions to be added (or subtracted) have common factors, we usually do not use the general rules given by equation (11), since, in doing so, we make the problem more complicated than it needs to be. Instead, just as with fractions, we apply the **least common multiple (LCM) method** by using the polynomial of least degree that contains each denominator polynomial as a factor. Then we rewrite each rational expression using the LCM as the common denominator and use equation (10) to do the addition (or subtraction).

To find the least common multiple of two or more polynomials, first factor completely each polynomial. The LCM is the product of the different prime factors of each polynomial, each factor appearing the greatest number of times it occurs in each polynomial. The next example will give you the idea.

EXAMPLE 7 **Finding the Least Common Multiple**

Find the least common multiple of the following pair of polynomials:

$$x(x-1)^2(x+1) \quad \text{and} \quad 4(x-1)(x+1)^3$$

Solution The polynomials are already factored completely as

$$x(x-1)^2(x+1) \quad \text{and} \quad 4(x-1)(x+1)^3$$

Start by writing the factors of the left-hand polynomial. (Alternatively, you could start with the one on the right.)

$$x(x-1)^2(x+1)$$

Now look at the right-hand polynomial. Its first factor, 4, does not appear in our list, so we insert it:

$$4x(x-1)^2(x+1)$$

The next factor, $x - 1$, is already in our list, so no change is necessary. The final factor is $(x+1)^3$. Since our list has $x + 1$ to the first power only, we replace $x + 1$ in the list by $(x+1)^3$. The LCM is

$$4x(x-1)^2(x+1)^3$$

Notice that the LCM is, in fact, the polynomial of least degree that contains $x(x-1)^2(x+1)$ and $4(x-1)(x+1)^3$ as factors. ◀

The next example illustrates how the LCM is used for adding and subtracting rational expressions.

EXAMPLE 8 **Using the LCM to Add Rational Expressions**

Perform the indicated operation and simplify the result. Leave your answer in factored form.

$$\frac{x}{x^2+3x+2} + \frac{2x-3}{x^2-1}, \qquad x \neq -2, -1, 1$$

Solution First, we find the LCM of the denominators.

$$x^2 + 3x + 2 = (x+2)(x+1)$$
$$x^2 - 1 = (x-1)(x+1)$$

The LCM is $(x+2)(x+1)(x-1)$. Next, we rewrite each rational expression using the LCM as the denominator.

$$\frac{x}{x^2+3x+2} = \frac{x}{(x+2)(x+1)} = \frac{x(x-1)}{(x+2)(x+1)(x-1)}$$

↑ Multiply numerator and denominator by $x - 1$ to get the LCM in the denominator.

$$\frac{2x-3}{x^2-1} = \frac{2x-3}{(x-1)(x+1)} = \frac{(2x-3)(x+2)}{(x-1)(x+1)(x+2)}$$

↑ Multiply numerator and denominator by $x + 2$ to get the LCM in the denominator.

Now we can add using equation (10).

$$\frac{x}{x^2 + 3x + 2} + \frac{2x - 3}{x^2 - 1} = \frac{x(x - 1)}{(x + 2)(x + 1)(x - 1)} + \frac{(2x - 3)(x + 2)}{(x + 2)(x + 1)(x - 1)}$$

$$= \frac{(x^2 - x) + (2x^2 + x - 6)}{(x + 2)(x + 1)(x - 1)}$$

$$= \frac{3x^2 - 6}{(x + 2)(x + 1)(x - 1)} = \frac{3(x^2 - 2)}{(x + 2)(x + 1)(x - 1)} \blacktriangleleft$$

If we had not used the LCM technique to add the quotients in Example 8, but decided instead to use the general rule of equation (11), we would have obtained a more complicated expression, as follows:

$$\frac{x}{x^2 + 3x + 2} + \frac{2x - 3}{x^2 - 1} = \frac{x(x^2 - 1) + (x^2 + 3x + 2)(2x - 3)}{(x^2 + 3x + 2)(x^2 - 1)}$$

$$= \frac{3x^3 + 3x^2 - 6x - 6}{(x^2 + 3x + 2)(x^2 - 1)} = \frac{3(x^3 + x^2 - 2x - 2)}{(x^2 + 3x + 2)(x^2 - 1)}$$

Now we are faced with a more complicated problem of expressing this quotient in lowest terms. It is always best to first look for common factors in the denominators of expressions to be added or subtracted and to use the LCM if any common factors are found.

NOW WORK PROBLEM 71.

A.3 Assess Your Understanding

Concepts and Vocabulary

1. The polynomial $3x^4 - 2x^3 + 13x^2 - 5$ is of degree ________. The leading coefficient is ________.
2. $(x^2 - 4)(x^2 + 4) =$ ________.
3. $(x - 2)(x^2 + 2x + 4) =$ ________.
4. *True or False:* $4x^{-2}$ is a monomial of degree -2.
5. *True or False:* The degree of the product of two nonzero polynomials equals the sum of their degrees.
6. *True or False:* $(x + a)(x^2 + ax + a) = x^3 + a^3$.
7. If factored completely, $3x^3 - 12x =$ ________.
8. If a polynomial cannot be written as the product of two other polynomials (excluding 1 and -1), then the polynomial is said to be ________.
9. *True or False:* The polynomial $x^2 + 4$ is prime.
10. *True or False:* $3x^3 - 2x^2 - 6x + 4 = (3x - 2)(x^3 + 2)$.
11. When the numerator and denominator of a rational expression contain no common factors (except 1 and -1), the rational expression is ________.
12. LCM is an abbreviation for ________ ________ ________.
13. *True or False:* The rational expression $\frac{2x^3 - 4x}{x - 2}$ is reduced to lowest terms.
14. *True or False:* The LCM of $2x^3 + 6x^2$ and $6x^4 + 4x^3$ is $4x^3(x + 1)$.

Exercises

In Problems 15–24, perform the indicated operations. Express each answer as a polynomial written in standard form.

15. $(10x^5 - 8x^2) + (3x^3 - 2x^2 + 6)$
16. $3(x^2 - 3x + 1) + 2(3x^2 + x - 4)$
17. $(x + a)^2 - x^2$
18. $(x - a)^2 - x^2$

19. $(x + 8)(2x + 1)$

20. $(2x - 1)(x + 2)$

21. $(x^2 + x - 1)(x^2 - x + 1)$

22. $(x^2 + 2x + 1)(x^2 - 3x + 4)$

23. $(x + 1)^3 - (x - 1)^3$

24. $(x + 1)^3 - (x + 2)^3$

In Problems 25–66, factor completely each polynomial. If the polynomial cannot be factored, say it is prime.

25. $x^2 - 36$

26. $x^2 - 9$

27. $1 - 4x^2$

28. $1 - 9x^2$

29. $x^2 + 7x + 10$

30. $x^2 + 5x + 4$

31. $x^2 - 2x + 8$

32. $x^2 - 4x + 5$

33. $x^2 + 4x + 16$

34. $x^2 + 12x + 36$

35. $15 + 2x - x^2$

36. $14 + 6x - x^2$

37. $3x^2 - 12x - 36$

38. $x^3 + 8x^2 - 20x$

39. $y^4 + 11y^3 + 30y^2$

40. $3y^3 - 18y^2 - 48y$

41. $4x^2 + 12x + 9$

42. $9x^2 - 12x + 4$

43. $3x^2 + 4x + 1$

44. $4x^2 + 3x - 1$

45. $x^4 - 81$

46. $x^4 - 1$

47. $x^6 - 2x^3 + 1$

48. $x^6 + 2x^3 + 1$

49. $x^7 - x^5$

50. $x^8 - x^5$

51. $5 + 16x - 16x^2$

52. $5 + 11x - 16x^2$

53. $4y^2 - 16y + 15$

54. $9y^2 + 9y - 4$

55. $1 - 8x^2 - 9x^4$

56. $4 - 14x^2 - 8x^4$

57. $x(x + 3) - 6(x + 3)$

58. $5(3x - 7) + x(3x - 7)$

59. $(x + 2)^2 - 5(x + 2)$

60. $(x - 1)^2 - 2(x - 1)$

61. $6x(2 - x)^4 - 9x^2(2 - x)^3$

62. $6x(1 - x^2)^4 - 24x^3(1 - x^2)^3$

63. $x^3 + 2x^2 - x - 2$

64. $x^3 - 3x^2 - x + 3$

65. $x^4 - x^3 + x - 1$

66. $x^4 + x^3 + x + 1$

In Problems 67–78, perform the indicated operation and simplify the result. Leave your answer in factored form.

67. $\dfrac{3x - 6}{5x} \cdot \dfrac{x^2 - x - 6}{x^2 - 4}$

68. $\dfrac{9x^2 - 25}{2x - 2} \cdot \dfrac{1 - x^2}{6x - 10}$

69. $\dfrac{4x^2 - 1}{x^2 - 16} \cdot \dfrac{x^2 - 4x}{2x + 1}$

70. $\dfrac{12}{x^2 - x} \cdot \dfrac{x^2 - 1}{4x - 2}$

71. $\dfrac{x}{x^2 - 7x + 6} - \dfrac{x}{x^2 - 2x - 24}$

72. $\dfrac{x}{x - 3} - \dfrac{x + 1}{x^2 + 5x - 24}$

73. $\dfrac{4}{x^2 - 4} - \dfrac{2}{x^2 + x - 6}$

74. $\dfrac{3}{x - 1} - \dfrac{x - 4}{x^2 - 2x + 1}$

75. $\dfrac{1}{x} - \dfrac{2}{x^2 + x} + \dfrac{3}{x^3 - x^2}$

76. $\dfrac{x}{(x - 1)^2} + \dfrac{2}{x} - \dfrac{x + 1}{x^3 - x^2}$

77. $\dfrac{1}{h}\left(\dfrac{1}{x + h} - \dfrac{1}{x}\right)$

78. $\dfrac{1}{h}\left[\dfrac{1}{(x + h)^2} - \dfrac{1}{x^2}\right]$

In Problems 79–88, expressions that occur in calculus are given. Factor completely each expression.

79. $2(3x + 4)^2 + (2x + 3) \cdot 2(3x + 4) \cdot 3$

80. $5(2x + 1)^2 + (5x - 6) \cdot 2(2x + 1) \cdot 2$

81. $2x(2x + 5) + x^2 \cdot 2$

82. $3x^2(8x - 3) + x^3 \cdot 8$

83. $2(x + 3)(x - 2)^3 + (x + 3)^2 \cdot 3(x - 2)^2$

84. $4(x + 5)^3(x - 1)^2 + (x + 5)^4 \cdot 2(x - 1)$

85. $(4x - 3)^2 + x \cdot 2(4x - 3) \cdot 4$

86. $3x^2(3x + 4)^2 + x^3 \cdot 2(3x + 4) \cdot 3$

87. $2(3x - 5) \cdot 3(2x + 1)^3 + (3x - 5)^2 \cdot 3(2x + 1)^2 \cdot 2$

88. $3(4x + 5)^2 \cdot 4(5x + 1)^2 + (4x + 5)^3 \cdot 2(5x + 1) \cdot 5$

In Problems 89–96, expressions that occur in calculus are given. Reduce each expression to lowest terms.

89. $\dfrac{(2x + 3) \cdot 3 - (3x - 5) \cdot 2}{(3x - 5)^2}$

90. $\dfrac{(4x + 1) \cdot 5 - (5x - 2) \cdot 4}{(5x - 2)^2}$

91. $\dfrac{x \cdot 2x - (x^2 + 1) \cdot 1}{(x^2 + 1)^2}$

92. $\dfrac{x \cdot 2x - (x^2 - 4) \cdot 1}{(x^2 - 4)^2}$

93. $\dfrac{(3x + 1) \cdot 2x - x^2 \cdot 3}{(3x + 1)^2}$

94. $\dfrac{(2x - 5) \cdot 3x^2 - x^3 \cdot 2}{(2x - 5)^2}$

95. $\dfrac{(x^2 + 1) \cdot 3 - (3x + 4) \cdot 2x}{(x^2 + 1)^2}$

96. $\dfrac{(x^2 + 9) \cdot 2 - (2x - 5) \cdot 2x}{(x^2 + 9)^2}$

A.4 Polynomial Division; Synthetic Division

OBJECTIVES 1 Divide Polynomials Using Long Division
2 Divide Polynomials Using Synthetic Division

Long Division

1 The procedure for dividing two polynomials is similar to the procedure for dividing two integers.

EXAMPLE 1 **Dividing Two Integers**

Divide 842 by 15.

Solution

$$\begin{array}{r l} 56 & \leftarrow \text{Quotient} \\ \text{Divisor} \rightarrow 15\overline{)842} & \leftarrow \text{Dividend} \\ \underline{75} & \leftarrow 5 \cdot 15 \text{ (Subtract)} \\ 92 & \\ \underline{90} & \leftarrow 6 \cdot 15 \text{ (Subtract)} \\ 2 & \leftarrow \text{Remainder} \end{array}$$

So, $\dfrac{842}{15} = 56 + \dfrac{2}{15}$. ◀

In the long division process detailed in Example 1, the number 15 is called the **divisor,** the number 842 is called the **dividend,** the number 56 is called the **quotient,** and the number 2 is called the **remainder.**

To check the answer obtained in a division problem, multiply the quotient by the divisor and add the remainder. The answer should be the dividend.

(Quotient)(Divisor) + Remainder = Dividend

For example, we can check the results obtained in Example 1 as follows:

$$(56)(15) + 2 = 840 + 2 = 842$$

To divide two polynomials, we first must write each polynomial in standard form. The process then follows a pattern similar to that of Example 1. The next example illustrates the procedure.

EXAMPLE 2 **Dividing Two Polynomials**

Find the quotient and the remainder when

$$3x^3 + 4x^2 + x + 7 \quad \text{is divided by} \quad x^2 + 1$$

Solution Each polynomial is in standard form. The dividend is $3x^3 + 4x^2 + x + 7$, and the divisor is $x^2 + 1$.

STEP 1: Divide the leading term of the dividend, $3x^3$, by the leading term of the divisor, x^2. Enter the result, $3x$, over the term $3x^3$, as follows:

$$\begin{array}{r} 3x \\ x^2 + 1 \overline{)3x^3 + 4x^2 + x + 7} \end{array}$$

STEP 2: Multiply $3x$ by $x^2 + 1$ and enter the result below the dividend.

$$\begin{array}{rl} 3x & \\ x^2 + 1 \overline{)3x^3 + 4x^2 + x + 7} & \\ \underline{3x^3 + 3x } & \leftarrow 3x \cdot (x^2 + 1) = 3x^3 + 3x \end{array}$$

↑ Notice that we align the $3x$ term under the x to make the next step easier.

STEP 3: Subtract and bring down the remaining terms.

$$\begin{array}{rl} 3x & \\ x^2 + 1 \overline{)3x^3 + 4x^2 + x + 7} & \\ \underline{3x^3 + 3x } & \leftarrow \text{Subtract (change the signs and add)} \\ 4x^2 - 2x + 7 & \leftarrow \text{Bring down the } 4x^2 \text{ and the 7.} \end{array}$$

STEP 4: Repeat Steps 1–3 using $4x^2 - 2x + 7$ as the dividend.

$$\begin{array}{rl} 3x + 4 & \\ x^2 + 1 \overline{)3x^3 + 4x^2 + x + 7} & \\ \underline{3x^3 + 3x } & \\ 4x^2 - 2x + 7 & \leftarrow \text{Divide } 4x^2 \text{ by } x^2 \text{ to get 4.} \\ \underline{4x^2 + 4} & \leftarrow \text{Multiply } x^2 + 1 \text{ by 4; subtract.} \\ -2x + 3 & \end{array}$$

Since x^2 does not divide $-2x$ evenly (that is, the result is not a monomial), the process ends. The quotient is $3x + 4$, and the remainder is $-2x + 3$.

✔ **CHECK:** (Quotient)(Divisor) + Remainder

$$= (3x + 4)(x^2 + 1) + (-2x + 3)$$
$$= 3x^3 + 4x^2 + 3x + 4 + (-2x + 3)$$
$$= 3x^3 + 4x^2 + x + 7 = \text{Dividend}$$

Then

$$\frac{3x^3 + 4x^2 + x + 7}{x^2 + 1} = 3x + 4 + \frac{-2x + 3}{x^2 + 1}$$ ◀

The next example combines the steps involved in long division.

EXAMPLE 3 Dividing Two Polynomials

Find the quotient and the remainder when

$$x^4 - 3x^3 + 2x - 5 \quad \text{is divided by} \quad x^2 - x + 1$$

Solution In setting up this division problem, it is necessary to leave a space for the missing x^2 term in the dividend.

$$
\begin{array}{lrl}
 & x^2 - 2x - 3 & \leftarrow \text{Quotient} \\
\text{Divisor} \rightarrow & x^2 - x + 1 \overline{)\, x^4 - 3x^3 \qquad + 2x - 5} & \leftarrow \text{Dividend} \\
\text{Subtract} \rightarrow & \underline{x^4 - x^3 + x^2} & \\
 & -2x^3 - x^2 + 2x - 5 & \\
\text{Subtract} \rightarrow & \underline{-2x^3 + 2x^2 - 2x} & \\
 & -3x^2 + 4x - 5 & \\
\text{Subtract} \rightarrow & \underline{-3x^2 + 3x - 3} & \\
 & x - 2 & \leftarrow \text{Remainder}
\end{array}
$$

✔ **CHECK:** (Quotient)(Divisor) + Remainder

$$= (x^2 - 2x - 3)(x^2 - x + 1) + x - 2$$
$$= x^4 - x^3 + x^2 - 2x^3 + 2x^2 - 2x - 3x^2 + 3x - 3 + x - 2$$
$$= x^4 - 3x^3 + 2x - 5 = \text{Dividend}$$

As a result,

$$\frac{x^4 - 3x^3 + 2x - 5}{x^2 - x + 1} = x^2 - 2x - 3 + \frac{x - 2}{x^2 - x + 1}$$ ◀

The process of dividing two polynomials leads to the following result:

Theorem Let Q be a polynomial of positive degree and let P be a polynomial whose degree is greater than the degree of Q. The remainder after dividing P by Q is either the zero polynomial or a polynomial whose degree is less than the degree of the divisor Q.

NOW WORK PROBLEM 9.

Synthetic Division

To find the quotient as well as the remainder when a polynomial of degree 1 or higher is divided by $x - c$, a shortened version of long division, called **synthetic division,** makes the task simpler.

2 To see how synthetic division works, we will use long division to divide the polynomial $2x^3 - x^2 + 3$ by $x - 3$.

$$
\begin{array}{rl}
2x^2 + 5x + 15 & \leftarrow \text{Quotient} \\
x - 3 \overline{)\, 2x^3 - x^2 \qquad + 3} & \\
\underline{2x^3 - 6x^2} & \\
5x^2 & \\
\underline{5x^2 - 15x} & \\
15x + 3 & \\
\underline{15x - 45} & \\
48 & \leftarrow \text{Remainder}
\end{array}
$$

✔ **CHECK:** (Divisor) · (Quotient) + Remainder

$$= (x - 3)(2x^2 + 5x + 15) + 48$$
$$= 2x^3 + 5x^2 + 15x - 6x^2 - 15x - 45 + 48$$
$$= 2x^3 - x^2 + 3$$

The process of synthetic division arises from rewriting the long division in a more compact form, using simpler notation. For example, in the long division on p. 931, the terms in blue are not really necessary because they are identical to the terms directly above them. With these terms removed, we have

$$
\begin{array}{r|llll}
 & 2x^2 & +\,5x & +\,15 & \\ \hline
x - 3 & 2x^3 & -\,x^2 & & +\,3 \\
 & & -\,6x^2 & & \\ \cline{2-3}
 & & 5x^2 & & \\
 & & & -\,15x & \\ \cline{3-5}
 & & & 15x & \\
 & & & & -\,45 \\ \cline{4-5}
 & & & & 48
\end{array}
$$

Most of the x's that appear in this process can also be removed, provided that we are careful about positioning each coefficient. In this regard, we will need to use 0 as the coefficient of x in the dividend, because that power of x is missing. Now we have

$$
\begin{array}{r|rrrr}
 & 2x^2 & +\,5x & +\,15 & \\ \hline
x - 3 & 2 & -\,1 & 0 & 3 \\
 & & -\,6 & & \\ \cline{2-3}
 & & 5 & & \\
 & & & -\,15 & \\ \cline{3-5}
 & & & 15 & \\
 & & & & -\,45 \\ \cline{4-5}
 & & & & 48
\end{array}
$$

We can make this display more compact by moving the lines up until the numbers in color align horizontally.

$$
\begin{array}{r|rrrrl}
 & 2x^2 & +\,5x & +\,15 & & \text{Row 1} \\ \hline
x - 3 & 2 & -\,1 & 0 & 3 & \text{Row 2} \\
 & & -\,6 & -\,15 & -\,45 & \text{Row 3} \\ \cline{2-5}
 & \bigcirc & 5 & 15 & 48 & \text{Row 4}
\end{array}
$$

Because the leading coefficient of the divisor is always 1, we know that the leading coefficient of the dividend will also be the leading coefficient of the quotient. So we place the leading coefficient of the quotient, 2, in the circled position. Now, the first three numbers in row 4 are precisely the coefficients of the quotient, and the last number in row 4 is the remainder. Thus, row 1 is not really needed, so we can compress the process to three rows, where the bottom row contains both the coefficients of the quotient and the remainder.

$$
\begin{array}{r|rrrrl}
x - 3 & 2 & -\,1 & 0 & 3 & \text{Row 1} \\
 & & -\,6 & -\,15 & -\,45 & \text{Row 2 (subtract)} \\ \cline{2-5}
 & 2 & 5 & 15 & 48 & \text{Row 3}
\end{array}
$$

Recall that the entries in row 3 are obtained by subtracting the entries in row 2 from those in row 1. Rather than subtracting the entries in row 2, we can change the sign of each entry and add. With this modification, our display will look like this:

$$\begin{array}{r|rrrrl} x-3) & 2 & -1 & 0 & 3 & \text{Row 1} \\ & & 6 & 15 & 45 & \text{Row 2 (add)} \\ \hline & 2 & 5 & 15 & 48 & \text{Row 3} \end{array}$$

Notice that the entries in row 2 are three times the prior entries in row 3. Our last modification to the display replaces the $x - 3$ by 3. The entries in row 3 give the quotient and the remainder, as shown next.

$$\begin{array}{r|rrrrl} 3) & 2 & -1 & 0 & 3 & \text{Row 1} \\ & & 6 & 15 & 45 & \text{Row 2 (add)} \\ \hline & 2 & 5 & 15 & 48 & \text{Row 3} \end{array}$$

Quotient: $2x^2 + 5x + 15$ Remainder: 48

Let's go through an example step by step.

EXAMPLE 4 Using Synthetic Division to Find the Quotient and Remainder

Use synthetic division to find the quotient and remainder when

$$x^3 - 4x^2 - 5 \quad \text{is divided by} \quad x - 3$$

Solution **Step 1:** Write the dividend in descending powers of x. Then copy the coefficients, remembering to insert a 0 for any missing powers of x.

$$\begin{array}{rrrrl} 1 & -4 & 0 & -5 & \text{Row 1} \end{array}$$

Step 2: Insert the usual division symbol. In synthetic division, the divisor is of the form $x - c$, and c is the number placed to the left of the division symbol. Here, since the divisor is $x - 3$, we insert 3 to the left of the division symbol.

$$\begin{array}{r|rrrrl} 3) & 1 & -4 & 0 & -5 & \text{Row 1} \end{array}$$

Step 3: Bring the 1 down two rows, and enter it in row 3.

$$\begin{array}{r|rrrrl} 3) & 1 & -4 & 0 & -5 & \text{Row 1} \\ & \downarrow & & & & \text{Row 2} \\ \hline & 1 & & & & \text{Row 3} \end{array}$$

Step 4: Multiply the latest entry in row 3 by 3, and place the result in row 2, one column over to the right.

$$\begin{array}{r|rrrrl} 3) & 1 & -4 & 0 & -5 & \text{Row 1} \\ & & 3 & & & \text{Row 2} \\ \hline & 1 & & & & \text{Row 3} \end{array}$$

Step 5: Add the entry in row 2 to the entry above it in row 1, and enter the sum in row 3.

$$\begin{array}{r|rrrrl} 3) & 1 & -4 & 0 & -5 & \text{Row 1} \\ & & 3 & & & \text{Row 2} \\ \hline & 1 & -1 & & & \text{Row 3} \end{array}$$

STEP 6: Repeat Steps 4 and 5 until no more entries are available in row 1.

3)	1	−4	0	−5	Row 1
		3	−3	−9	Row 2
	1	−1	−3	−14	Row 3

STEP 7: The final entry in row 3, the −14, is the remainder; the other entries in row 3, the 1, −1, and −3, are the coefficients (in descending order) of a polynomial whose degree is 1 less than that of the dividend. This is the quotient. Thus,

$$\text{Quotient} = x^2 - x - 3 \qquad \text{Remainder} = -14$$

✔ **CHECK:** (Divisor)(Quotient) + Remainder

$$= (x - 3)(x^2 - x - 3) + (-14)$$
$$= (x^3 - x^2 - 3x - 3x^2 + 3x + 9) + (-14)$$
$$= x^3 - 4x^2 - 5 = \text{Dividend}$$ ◀

Let's do an example in which all seven steps are combined.

EXAMPLE 5 **Using Synthetic Division to Verify a Factor**

Use synthetic division to show that $x + 3$ is a factor of

$$2x^5 + 5x^4 - 2x^3 + 2x^2 - 2x + 3$$

Solution The divisor is $x + 3 = x - (-3)$, so we place −3 to the left of the division symbol. Then the row 3 entries will be multiplied by −3, entered in row 2, and added to row 1.

−3)	2	5	−2	2	−2	3	Row 1
		−6	3	−3	3	−3	Row 2
	2	−1	1	−1	1	0	Row 3

Because the remainder is 0, we have

(Divisor)(Quotient) + Remainder

$$= (x + 3)(2x^4 - x^3 + x^2 - x + 1) = 2x^5 + 5x^4 - 2x^3 + 2x^2 - 2x + 3$$

As we see, $x + 3$ is a factor of $2x^5 + 5x^4 - 2x^3 + 2x^2 - 2x + 3$. ◀

As Example 5 illustrates, the remainder after division gives information about whether the divisor is, or is not, a factor.

NOW WORK PROBLEMS 23 AND 33.

A.4 Assess Your Understanding

Concepts and Vocabulary

1. To check division, the expression that is being divided, the dividend, should equal the product of the __________ and the __________ plus the __________.
2. To divide $2x^3 - 5x + 1$ by $x + 3$ using synthetic division, the first step is to write ______ $\overline{)\ \_\_\_\_\_\_}$.
3. *True or False:* In using synthetic division, the divisor is always a polynomial of degree 1, whose leading coefficient is 1.

4. *True or False:* $-2\overline{)5 \quad 3 \quad 2 \quad 1}$ $\begin{array}{rrrr} & -10 & 14 & -32 \\ \hline 5 & -7 & 16 & -31 \end{array}$ means $\dfrac{5x^3 + 3x^2 + 2x + 1}{x + 2} = 5x^2 - 7x + 16 + \dfrac{-31}{x + 2}$.

Exercises

In Problems 5–20, find the quotient and the remainder. Check your work by verifying that

$$(\textit{Quotient})(\textit{Divisor}) + \textit{Remainder} = \textit{Dividend}$$

5. $4x^3 - 3x^2 + x + 1$ divided by $x + 2$

6. $3x^3 - x^2 + x - 2$ divided by $x + 2$

7. $4x^3 - 3x^2 + x + 1$ divided by x^2

8. $3x^3 - x^2 + x - 2$ divided by x^2

9. $5x^4 - 3x^2 + x + 1$ divided by $x^2 + 2$

10. $5x^4 - x^2 + x - 2$ divided by $x^2 + 2$

11. $4x^5 - 3x^2 + x + 1$ divided by $2x^3 - 1$

12. $3x^5 - x^2 + x - 2$ divided by $3x^3 - 1$

13. $2x^4 - 3x^3 + x + 1$ divided by $2x^2 + x + 1$

14. $3x^4 - x^3 + x - 2$ divided by $3x^2 + x + 1$

15. $-4x^3 + x^2 - 4$ divided by $x - 1$

16. $-3x^4 - 2x - 1$ divided by $x - 1$

17. $1 - x^2 + x^4$ divided by $x^2 + x + 1$

18. $1 - x^2 + x^4$ divided by $x^2 - x + 1$

19. $x^3 - a^3$ divided by $x - a$

20. $x^5 - a^5$ divided by $x - a$

In Problems 21–32, use synthetic division to find the quotient and remainder when $f(x)$ is divided by $g(x)$.

21. $x^3 - x^2 + 2x + 4$ divided by $x - 2$

22. $x^3 + 2x^2 - 3x + 1$ divided by $x + 1$

23. $3x^3 + 2x^2 - x + 3$ divided by $x - 3$

24. $-4x^3 + 2x^2 - x + 1$ divided by $x + 2$

25. $x^5 - 4x^3 + x$ divided by $x + 3$

26. $x^4 + x^2 + 2$ divided by $x - 2$

27. $4x^6 - 3x^4 + x^2 + 5$ divided by $x - 1$

28. $x^5 + 5x^3 - 10$ divided by $x + 1$

29. $0.1x^3 + 0.2x$ divided by $x + 1.1$

30. $0.1x^2 - 0.2$ divided by $x + 2.1$

31. $x^5 - 1$ divided by $x - 1$

32. $x^5 + 1$ divided by $x + 1$

In Problems 33–42, use synthetic division to determine whether $x - c$ is a factor of the given polynomial.

33. $4x^3 - 3x^2 - 8x + 4;\ x - 2$

34. $-4x^3 + 5x^2 + 8;\ x + 3$

35. $3x^4 - 6x^3 - 5x + 10;\ x - 2$

36. $4x^4 - 15x^2 - 4;\ x - 2$

37. $3x^6 + 82x^3 + 27;\ x + 3$

38. $2x^6 - 18x^4 + x^2 - 9;\ x + 3$

39. $4x^6 - 64x^4 + x^2 - 15;\ x + 4$

40. $x^6 - 16x^4 + x^2 - 16;\ x + 4$

41. $2x^4 - x^3 + 2x - 1;\ x - \dfrac{1}{2}$

42. $3x^4 + x^3 - 3x + 1;\ x + \dfrac{1}{3}$

43. Find the sum of a, b, c, and d if

$$\frac{x^3 - 2x^2 + 3x + 5}{x + 2} = ax^2 + bx + c + \frac{d}{x + 2}$$

44. When dividing a polynomial by $x - c$, do you prefer to use long division or synthetic division? Does the value of c make a difference to you in choosing? Give reasons.

A.5 Solving Equations

PREPARING FOR THIS SECTION *Before getting started, review the following:*

- Factoring Polynomials (Appendix A, Section A.3, pp. 920–923)
- Zero-Product Property (Appendix A, Section A.1, p. 905)
- Square Roots (Appendix A, Section A.1, p. 910)
- Absolute Value (Appendix A, Section A.1, pp. 907–908)

Now work the 'Are You Prepared?' problems on page 945.

OBJECTIVES
1. Solve Equations by Factoring
2. Solve Quadratic Equations by Factoring
3. Know How to Complete the Square
4. Solve a Quadratic Equation by Completing the Square
5. Solve a Quadratic Equation Using the Quadratic Formula

An **equation in one variable** is a statement in which two expressions, at least one containing the variable, are equal. The expressions are called the **sides** of the equation. Since an equation is a statement, it may be true or false, depending on the value of the variable. Unless otherwise restricted, the admissible values of the variable are those in the domain of the variable. Those admissible values of the variable, if any, that result in a true statement are called **solutions,** or **roots,** of the equation. To **solve an equation** means to find all the solutions of the equation.

For example, the following are all equations in one variable, x:

$$x + 5 = 9 \qquad x^2 + 5x = 2x - 2 \qquad \frac{x^2 - 4}{x + 1} = 0 \qquad \sqrt{x^2 + 9} = 5$$

The first of these statements, $x + 5 = 9$, is true when $x = 4$ and false for any other choice of x. Thus, 4 is a solution of the equation $x + 5 = 9$. We also say that 4 **satisfies** the equation $x + 5 = 9$, because, when we substitute 4 for x, a true statement results.

Sometimes an equation will have more than one solution. For example, the equation

$$\frac{x^2 - 4}{x + 1} = 0$$

has $x = -2$ and $x = 2$ as solutions.

Usually, we will write the solution of an equation in set notation. This set is called the **solution set** of the equation. For example, the solution set of the equation $x^2 - 9 = 0$ is $\{-3, 3\}$.

Some equations have no real solution. For example, $x^2 + 9 = 5$ has no real solution, because there is no real number whose square when added to 9 equals 5.

An equation that is satisfied for every choice of the variable for which both sides are defined is called an **identity.** For example, the equation

$$3x + 5 = x + 3 + 2x + 2$$

is an identity, because this statement is true for any real number x.

Two or more equations that have precisely the same solution set are called **equivalent equations.**

For example, all the following equations are equivalent, because each has only the solution $x = 5$:

$$2x + 3 = 13$$
$$2x = 10$$
$$x = 5$$

These three equations illustrate one method for solving many types of equations: Replace the original equation by an equivalent equation, and continue until an equation with an obvious solution, such as $x = 5$, is reached. The question, though, is "How do I obtain an equivalent equation?" In general, there are five ways to do so.

Procedures That Result in Equivalent Equations

1. Interchange the two sides of the equation:

Replace $3 = x$ by $x = 3$

2. Simplify the sides of the equation by combining like terms, eliminating parentheses, and so on:

Replace $(x + 2) + 6 = 2x + (x + 1)$

by $x + 8 = 3x + 1$

3. Add or subtract the same expression on both sides of the equation:

Replace $3x - 5 = 4$

by $(3x - 5) + 5 = 4 + 5$

4. Multiply or divide both sides of the equation by the same nonzero expression:

Replace $\dfrac{3x}{x-1} = \dfrac{6}{x-1}, \quad x \neq 1$

by $\dfrac{3x}{x-1} \cdot (x - 1) = \dfrac{6}{x-1} \cdot (x - 1)$

5. If one side of the equation is 0 and the other side can be factored, then we may use the Zero-Product Property* and set each factor equal to 0:

Replace $x(x - 3) = 0$

by $x = 0$ or $x - 3 = 0$

WARNING: Squaring both sides of an equation does not necessarily lead to an equivalent equation. ■

*The Zero-Product Property says that if $ab = 0$ then $a = 0$ or $b = 0$ or both equal 0.

Whenever it is possible to solve an equation in your head, do so. For example:

The solution of $2x = 8$ is $x = 4$.

The solution of $3x - 15 = 0$ is $x = 5$.

Often, though, some rearrangement is necessary.

EXAMPLE 1 Solving an Equation

Solve the equation: $3x - 5 = 4$

Solution We replace the original equation by a succession of equivalent equations.

$$
\begin{aligned}
3x - 5 &= 4 \\
(3x - 5) + 5 &= 4 + 5 && \text{Add 5 to both sides.} \\
3x &= 9 && \text{Simplify.} \\
\frac{3x}{3} &= \frac{9}{3} && \text{Divide both sides by 3.} \\
x &= 3 && \text{Simplify.}
\end{aligned}
$$

The last equation, $x = 3$, has the single solution 3. All these equations are equivalent, so 3 is the only solution of the original equation, $3x - 5 = 4$.

✔ **CHECK:** It is a good practice to check the solution by substituting 3 for x in the original equation.

$$
\begin{aligned}
3x - 5 &= 4 \\
3(3) - 5 &\stackrel{?}{=} 4 \\
9 - 5 &\stackrel{?}{=} 4 \\
4 &= 4
\end{aligned}
$$

The solution checks. ◀

NOW WORK PROBLEMS 27 AND 33.

1 In the next examples, we use the Zero-Product Property.

EXAMPLE 2 Solving Equations by Factoring

Solve the equations: (a) $x^2 = 4x$ (b) $x^3 - x^2 - 4x + 4 = 0$

Solution (a) We begin by collecting all terms on one side. This results in 0 on one side and an expression to be factored on the other.

$$
\begin{aligned}
x^2 &= 4x \\
x^2 - 4x &= 0 \\
x(x - 4) &= 0 && \text{Factor.} \\
x = 0 \quad \text{or} \quad x - 4 &= 0 && \text{Apply the Zero-Product Property.} \\
x &= 4
\end{aligned}
$$

The solution set is $\{0, 4\}$.

✔ **CHECK:** $x = 0$: $0^2 = 4 \cdot 0$ So 0 is a solution.

$x = 4$: $4^2 = 4 \cdot 4$ So 4 is a solution.

(b) We group the terms of $x^3 - x^2 - 4x + 4 = 0$ as follows:

$$(x^3 - x^2) - (4x - 4) = 0$$

Factor out x^2 from the first grouping and 4 from the second.

$$x^2(x - 1) - 4(x - 1) = 0$$

This reveals the common factor $(x - 1)$, so we have

$$(x^2 - 4)(x - 1) = 0$$

$$(x - 2)(x + 2)(x - 1) = 0 \quad \text{Factor again.}$$

$$x - 2 = 0 \quad \text{or} \quad x + 2 = 0 \quad \text{or} \quad x - 1 = 0 \quad \text{Set each factor equal to 0.}$$

$$x = 2 \qquad x = -2 \qquad x = 1 \quad \text{Solve.}$$

The solution set is $\{-2, 1, 2\}$.

✔ **CHECK:** $x = -2$: $(-2)^3 - (-2)^2 - 4(-2) + 4 = -8 - 4 + 8 + 4 = 0$ −2 is a solution.

$x = 1$: $1^3 - 1^2 - 4(1) + 4 = 1 - 1 - 4 + 4 = 0$ 1 is a solution.

$x = 2$: $2^3 - 2^2 - 4(2) + 4 = 8 - 4 - 8 + 4 = 0$ 2 is a solution. ◀

NOW WORK PROBLEM 37.

There are two points whose distance from the origin is 5 units, −5 and 5, so the equation $|x| = 5$ will have the solution set $\{-5, 5\}$.

EXAMPLE 3 **Solving an Equation Involving Absolute Value**

Solve the equation: $|x + 4| = 13$

Solution There are two possibilities:

$$x + 4 = 13 \quad \text{or} \quad x + 4 = -13$$

$$x = 9 \qquad\qquad x = -17$$

The solution set is $\{-17, 9\}$. ◀

NOW WORK PROBLEM 49.

Quadratic Equations

A **quadratic equation** is an equation equivalent to one written in the **standard form** $ax^2 + bx + c = 0$, where a, b, and c are real numbers and $a \neq 0$.

2 When a quadratic equation is written in the standard form, $ax^2 + bx + c = 0$, it may be possible to factor the expression on the left side as the product of two first-degree polynomials.

EXAMPLE 4 **Solving a Quadratic Equation by Factoring**

Solve the equation: $2x^2 = x + 3$

Solution We put the equation in standard form by adding $-x - 3$ to both sides.

$$2x^2 = x + 3 \qquad \text{Add } -x - 3 \text{ to both sides.}$$
$$2x^2 - x - 3 = 0$$

The left side may now be factored as

$$(2x - 3)(x + 1) = 0$$

so that

$$2x - 3 = 0 \quad \text{or} \quad x + 1 = 0$$
$$x = \frac{3}{2} \qquad\qquad x = -1$$

The solution set is $\left\{-1, \frac{3}{2}\right\}$.

When the left side factors into two linear equations with the same solution, the quadratic equation is said to have a **repeated solution.** We also call this solution a **root of multiplicity 2,** or a **double root.**

EXAMPLE 5 **Solving a Quadratic Equation by Factoring**

Solve the equation: $9x^2 - 6x + 1 = 0$

Solution This equation is already in standard form, and the left side can be factored.

$$9x^2 - 6x + 1 = 0$$
$$(3x - 1)(3x - 1) = 0$$

so

$$x = \frac{1}{3} \quad \text{or} \quad x = \frac{1}{3}$$

This equation has only the repeated solution $\frac{1}{3}$.

NOW WORK PROBLEM 67.

The Square Root Method

Suppose that we wish to solve the quadratic equation

$$x^2 = p \tag{1}$$

where $p \geq 0$ is a nonnegative number. We proceed as in the earlier examples.

$$x^2 - p = 0 \qquad \text{Put in standard form.}$$
$$(x - \sqrt{p})(x + \sqrt{p}) = 0 \qquad \text{Factor (over the real numbers).}$$
$$x = \sqrt{p} \quad \text{or} \quad x = -\sqrt{p} \qquad \text{Solve.}$$

We have the following result:

$$\text{If } x^2 = p \text{ and } p \geq 0, \text{ then } x = \sqrt{p} \text{ or } x = -\sqrt{p}. \qquad \textbf{(2)}$$

When statement (2) is used, it is called the **Square Root Method.** In statement (2), note that if $p > 0$ the equation $x^2 = p$ has two solutions, $x = \sqrt{p}$ and $x = -\sqrt{p}$. We usually abbreviate these solutions as $x = \pm\sqrt{p}$, read as "x equals plus or minus the square root of p."

For example, the two solutions of the equation

$$x^2 = 4$$

are

$$x = \pm\sqrt{4} \quad \text{Use the Square Root Method.}$$

and, since $\sqrt{4} = 2$, we have

$$x = \pm 2$$

The solution set is $\{-2, 2\}$.

NOW WORK PROBLEM 81.

Completing the Square

3 We now introduce the method of **completing the square.** The idea behind this method is to *adjust* the left side of a quadratic equation, $ax^2 + bx + c = 0$, so that it becomes a perfect square, that is, the square of a first-degree polynomial. For example, $x^2 + 6x + 9$ and $x^2 - 4x + 4$ are perfect squares because

$$x^2 + 6x + 9 = (x + 3)^2 \quad \text{and} \quad x^2 - 4x + 4 = (x - 2)^2$$

How do we adjust the left side? We do it by adding the appropriate number to the left side to create a perfect square. For example, to make $x^2 + 6x$ a perfect square, we add 9.

Let's look at several examples of completing the square when the coefficient of x^2 is 1:

Start	Add	Result
$x^2 + 4x$	4	$x^2 + 4x + 4 = (x + 2)^2$
$x^2 + 12x$	36	$x^2 + 12x + 36 = (x + 6)^2$
$x^2 - 6x$	9	$x^2 - 6x + 9 = (x - 3)^2$
$x^2 + x$	$\frac{1}{4}$	$x^2 + x + \frac{1}{4} = \left(x + \frac{1}{2}\right)^2$

Do you see the pattern? Provided that the coefficient of x^2 is 1, we complete the square by adding the square of $\frac{1}{2}$ of the coefficient of x.

Procedure for Completing a Square

Start	Add	Result
$x^2 + mx$	$\left(\frac{m}{2}\right)^2$	$x^2 + mx + \left(\frac{m}{2}\right)^2 = \left(x + \frac{m}{2}\right)^2$

NOW WORK PROBLEM 85.

4 The next example illustrates how the procedure of completing the square can be used to solve a quadratic equation.

EXAMPLE 6 **Solving a Quadratic Equation by Completing the Square**

Solve by completing the square: $2x^2 - 8x - 5 = 0$

Solution First, we rewrite the equation.

$$2x^2 - 8x - 5 = 0$$
$$2x^2 - 8x = 5$$

Next, we divide both sides by 2 so that the coefficient of x^2 is 1. (This enables us to complete the square at the next step.)

$$x^2 - 4x = \frac{5}{2}$$

Finally, we complete the square by adding 4 to both sides.

$$x^2 - 4x + 4 = \frac{5}{2} + 4$$
$$(x - 2)^2 = \frac{13}{2}$$
$$x - 2 = \pm\sqrt{\frac{13}{2}} \qquad \text{Use the Square Root Method.}$$
$$x - 2 = \pm\frac{\sqrt{26}}{2} \qquad \sqrt{\frac{13}{2}} = \frac{\sqrt{13}}{\sqrt{2}} \cdot \frac{\sqrt{2}}{\sqrt{2}} = \frac{\sqrt{26}}{2}$$
$$x = 2 \pm \frac{\sqrt{26}}{2}$$

The solution set is $\left\{2 - \frac{\sqrt{26}}{2}, 2 + \frac{\sqrt{26}}{2}\right\}$ ◀

Note: If we wanted an approximation, say rounded to two decimal places, of these solutions, we would use a calculator to get $\{-0.55, 4.55\}$.

NOW WORK PROBLEM 91.

The Quadratic Formula

5 We can use the method of completing the square to obtain a general formula for solving the quadratic equation.

$$ax^2 + bx + c = 0, \qquad a \neq 0$$

Note: There is no loss in generality to assume that $a > 0$, since if $a < 0$ we can multiply by -1 to obtain an equivalent equation with a positive leading coefficient.

As in Example 6, we rearrange the terms as

$$ax^2 + bx = -c \qquad a > 0$$

Since $a > 0$, we can divide both sides by a to get

$$x^2 + \frac{b}{a}x = -\frac{c}{a}$$

Now the coefficient of x^2 is 1. To complete the square on the left side, add the square of $\frac{1}{2}$ of the coefficient of x; that is, add

$$\left(\frac{1}{2} \cdot \frac{b}{a}\right)^2 = \frac{b^2}{4a^2}$$

to both sides. Then

$$x^2 + \frac{b}{a}x + \frac{b^2}{4a^2} = \frac{b^2}{4a^2} - \frac{c}{a}$$

$$\left(x + \frac{b}{2a}\right)^2 = \frac{b^2 - 4ac}{4a^2} \qquad \frac{b^2}{4a^2} - \frac{c}{a} = \frac{b^2}{4a^2} - \frac{4ac}{4a^2} = \frac{b^2 - 4ac}{4a^2} \qquad \textbf{(3)}$$

Provided that $b^2 - 4ac \geq 0$, we now can use the Square Root Method to get

$$x + \frac{b}{2a} = \pm\sqrt{\frac{b^2 - 4ac}{4a^2}}$$

$$x + \frac{b}{2a} = \frac{\pm\sqrt{b^2 - 4ac}}{2a}$$ The square root of a quotient equals the quotient of the square roots. Also, $\sqrt{4a^2} = 2a$ since $a > 0$.

$$x = -\frac{b}{2a} \pm \frac{\sqrt{b^2 - 4ac}}{2a}$$ Add $-\frac{b}{2a}$ to both sides.

$$x = \frac{-b \pm \sqrt{b^2 - 4ac}}{2a}$$ Combine the quotients on the right.

What if $b^2 - 4ac$ is negative? Then equation (3) states that the left expression (a real number squared) equals the right expression (a negative number). Since this occurrence is impossible for real numbers, we conclude that if $b^2 - 4ac < 0$ the quadratic equation has no *real* solution. (We discuss quadratic equations for which the quantity $b^2 - 4ac < 0$ in detail in the next section.)

We now state the *quadratic formula*.

Theorem

Consider the quadratic equation

$$ax^2 + bx + c = 0, \qquad a \neq 0$$

If $b^2 - 4ac < 0$, this equation has no real solution.

If $b^2 - 4ac \geq 0$, the real solution(s) of this equation is (are) given by the **quadratic formula.**

Quadratic Formula

$$x = \frac{-b \pm \sqrt{b^2 - 4ac}}{2a} \qquad \textbf{(4)}$$

The quantity $\boldsymbol{b^2 - 4ac}$ is called the **discriminant** of the quadratic equation, because its value tells us whether the equation has real solutions. In fact, it also tells us how many solutions to expect.

Discriminant of a Quadratic Equation

For a quadratic equation $ax^2 + bx + c = 0$:

1. If $b^2 - 4ac > 0$, there are two unequal real solutions.
2. If $b^2 - 4ac = 0$, there is a repeated solution, a root of multiplicity 2.
3. If $b^2 - 4ac < 0$, there is no real solution.

When asked to find the real solutions, if any, of a quadratic equation, always evaluate the discriminant first to see how many real solutions there are.

EXAMPLE 7 Solving a Quadratic Equation Using the Quadratic Formula

Use the quadratic formula to find the real solutions, if any, of the equation

$$3x^2 - 5x + 1 = 0$$

Solution The equation is in standard form, so we compare it to $ax^2 + bx + c = 0$ to find a, b, and c.

$$3x^2 - 5x + 1 = 0$$
$$ax^2 + bx + c = 0 \qquad a = 3, b = -5, c = 1$$

With $a = 3$, $b = -5$, and $c = 1$, we evaluate the discriminant $b^2 - 4ac$.

$$b^2 - 4ac = (-5)^2 - 4(3)(1) = 25 - 12 = 13$$

Since $b^2 - 4ac > 0$, there are two real solutions, which can be found using the quadratic formula.

$$x = \frac{-b \pm \sqrt{b^2 - 4ac}}{2a} = \frac{-(-5) \pm \sqrt{13}}{2(3)} = \frac{5 \pm \sqrt{13}}{6}$$

The solution set is $\left\{\frac{5 - \sqrt{13}}{6}, \frac{5 + \sqrt{13}}{6}\right\}$.

EXAMPLE 8 Solving a Quadratic Equation Using the Quadratic Formula

Use the quadratic formula to find the real solutions, if any, of the equation

$$3x^2 + 2 = 4x$$

Solution The equation, as given, is not in standard form.

$$3x^2 + 2 = 4x$$
$$3x^2 - 4x + 2 = 0 \qquad \text{Put in standard form.}$$
$$ax^2 + bx + c = 0 \qquad \text{Compare to standard form.}$$

With $a = 3$, $b = -4$, and $c = 2$, we find

$$b^2 - 4ac = (-4)^2 - 4(3)(2) = 16 - 24 = -8$$

Since $b^2 - 4ac < 0$, the equation has no real solution.

NOW WORK PROBLEMS 97 AND 103.

Summary

Procedure for Solving a Quadratic Equation

To solve a quadratic equation, first put it in standard form:

$$ax^2 + bx + c = 0, \quad a \neq 0$$

Then:

Step 1: Identify a, b, and c.

Step 2: Evaluate the discriminant, $b^2 - 4ac$.

Step 3: (a) If the discriminant is negative, the equation has no real solution.

(b) If the discriminant is zero, the equation has one real solution, a repeated root.

(c) If the discriminant is positive, the equation has two distinct real solutions. If you can easily spot factors, use the factoring method to solve the equation. Otherwise, use the quadratic formula or the method of completing the square.

A.5 Assess Your Understanding

'Are You Prepared?' *Answers are given at the end of these exercises. If you get a wrong answer, read the pages listed in red.*

1. Factor $x^2 - 5x - 6$. (pp. 920–923)
2. Factor $2x^2 - x - 3$. (pp. 920–923)
3. The solution set of the equation $(x - 3)(3x + 5) = 0$ is __________. (p. 905)
4. *True or False:* $\sqrt{x^2} = |x|$. (pp. 907–908, 910)

Concepts and Vocabulary

5. Two equations that have the same solution set are called __________.
6. An equation that is satisfied for every choice of the variable for which both sides are defined is called a(n) __________.
7. *True or False:* The solution of the equation $3x - 8 = 0$ is $\frac{3}{8}$.
8. *True or False:* Some equations have no solution.
9. To complete the square of the expression $x^2 + 5x$, you would __________ the number __________.
10. The quantity $b^2 - 4ac$ is called the __________ of a quadratic equation. If it is __________, the equation has no real solution.
11. *True or False:* Quadratic equations always have two real solutions.
12. *True or False:* If the discriminant of a quadratic equation is positive, then the equation has two solutions that are negatives of one another.

Exercises

In Problems 13–78, solve each equation.

13. $3x = 21$ **14.** $3x = -24$ **15.** $5x + 15 = 0$ **16.** $3x + 18 = 0$

17. $2x - 3 = 5$ **18.** $3x + 4 = -8$ **19.** $\frac{1}{3}x = \frac{5}{12}$ **20.** $\frac{2}{3}x = \frac{9}{2}$

21. $6 - x = 2x + 9$ **22.** $3 - 2x = 2 - x$ **23.** $2(3 + 2x) = 3(x - 4)$ **24.** $3(2 - x) = 2x - 1$

25. $8x - (2x + 1) = 3x - 10$ **26.** $5 - (2x - 1) = 10$ 27. $\frac{1}{2}x - 4 = \frac{3}{4}x$ **28.** $1 - \frac{1}{2}x = 5$

29. $0.9t = 0.4 + 0.1t$ **30.** $0.9t = 1 + t$ **31.** $\frac{2}{y} + \frac{4}{y} = 3$ **32.** $\frac{4}{y} - 5 = \frac{5}{2y}$

33. $(x + 7)(x - 1) = (x + 1)^2$ **34.** $(x + 2)(x - 3) = (x - 3)^2$ **35.** $z(z^2 + 1) = 3 + z^3$

36. $w(4 - w^2) = 8 - w^3$ 37. $x^2 = 9x$ **38.** $x^3 = x^2$ **39.** $t^3 - 9t^2 = 0$

40. $4z^3 - 8z^2 = 0$ **41.** $\frac{3}{2x - 3} = \frac{2}{x + 5}$ **42.** $\frac{-2}{x + 4} = \frac{-3}{x + 1}$ **43.** $(x + 2)(3x) = (x + 2)(6)$

44. $(x - 5)(2x) = (x - 5)(4)$ **45.** $\frac{2}{x - 2} = \frac{3}{x + 5} + \frac{10}{(x + 5)(x - 2)}$ **46.** $\frac{1}{2x + 3} + \frac{1}{x - 1} = \frac{1}{(2x + 3)(x - 1)}$

47. $|2x| = 6$ **48.** $|3x| = 12$ **49.** $|2x + 3| = 5$ **50.** $|3x - 1| = 2$

51. $|1 - 4t| = 5$ **52.** $|1 - 2z| = 3$ **53.** $|-2x| = 8$ **54.** $|-x| = 1$

55. $|-2|x = 4$ **56.** $|3|x = 9$ **57.** $|x - 2| = -\frac{1}{2}$ **58.** $|2 - x| = -1$

59. $|x^2 - 4| = 0$ **60.** $|x^2 - 9| = 0$ **61.** $|x^2 - 2x| = 3$ **62.** $|x^2 + x| = 12$

63. $|x^2 + x - 1| = 1$ **64.** $|x^2 + 3x - 2| = 2$ **65.** $x^2 = 4x$ **66.** $x^2 = -8x$

67. $z^2 + 4z - 12 = 0$ **68.** $v^2 + 7v + 12 = 0$ **69.** $2x^2 - 5x - 3 = 0$ **70.** $3x^2 + 5x + 2 = 0$

71. $x(x - 7) + 12 = 0$ **72.** $x(x + 1) = 12$ **73.** $4x^2 + 9 = 12x$ **74.** $25x^2 + 16 = 40x$

75. $6x - 5 = \frac{6}{x}$ **76.** $x + \frac{12}{x} = 7$ **77.** $\frac{4(x - 2)}{x - 3} + \frac{3}{x} = \frac{-3}{x(x - 3)}$ **78.** $\frac{5}{x + 4} = 4 + \frac{3}{x - 2}$

In Problems 79–84, solve each equation by the Square Root Method.

79. $x^2 = 25$ **80.** $x^2 = 36$ **81.** $(x - 1)^2 = 4$

82. $(x + 2)^2 = 1$ **83.** $(2x + 3)^2 = 9$ **84.** $(3x - 2)^2 = 4$

In Problems 85–90, what number should be added to complete the square of each expression?

85. $x^2 + 8x$ **86.** $x^2 - 4x$ **87.** $x^2 + \frac{1}{2}x$

88. $x^2 - \frac{1}{3}x$ **89.** $x^2 - \frac{2}{3}x$ **90.** $x^2 - \frac{2}{5}x$

In Problems 91–96, solve each equation by completing the square.

91. $x^2 + 4x = 21$ **92.** $x^2 - 6x = 13$ **93.** $x^2 - \frac{1}{2}x - \frac{3}{16} = 0$

94. $x^2 + \frac{2}{3}x - \frac{1}{3} = 0$ **95.** $3x^2 + x - \frac{1}{2} = 0$ **96.** $2x^2 - 3x - 1 = 0$

In Problems 97–108, find the real solutions, if any, of each equation. Use the quadratic formula.

97. $x^2 - 4x + 2 = 0$ **98.** $x^2 + 4x + 2 = 0$ **99.** $x^2 - 5x - 1 = 0$

100. $x^2 + 5x + 3 = 0$ **101.** $2x^2 - 5x + 3 = 0$ **102.** $2x^2 + 5x + 3 = 0$

103. $4y^2 - y + 2 = 0$ **104.** $4t^2 + t + 1 = 0$ **105.** $4x^2 = 1 - 2x$

106. $2x^2 = 1 - 2x$ **107.** $x^2 + \sqrt{3}x - 3 = 0$ **108.** $x^2 + \sqrt{2}x - 2 = 0$

In Problems 109–114, use the discriminant to determine whether each quadratic equation has two unequal real solutions, a repeated real solution, or no real solution without solving the equation.

109. $x^2 - 5x + 7 = 0$ **110.** $x^2 + 5x + 7 = 0$ **111.** $9x^2 - 30x + 25 = 0$

112. $25x^2 - 20x + 4 = 0$ **113.** $3x^2 + 5x - 8 = 0$ **114.** $2x^2 - 3x - 4 = 0$

In Problems 115–120, solve each equation. The letters a, b, and c are constants.

115. $ax - b = c, \quad a \neq 0$ **116.** $1 - ax = b, \quad a \neq 0$ **117.** $\frac{x}{a} + \frac{x}{b} = c, \quad a \neq 0, b \neq 0, a \neq -b$

118. $\frac{a}{x} + \frac{b}{x} = c, \quad c \neq 0$ **119.** $\frac{1}{x - a} + \frac{1}{x + a} = \frac{2}{x - 1}$ **120.** $\frac{b + c}{x + a} = \frac{b - c}{x - a}, \quad c \neq 0, a \neq 0$

Problems 121–126 list some formulas that occur in applications. Solve each formula for the indicated variable.

121. Electricity $\frac{1}{R} = \frac{1}{R_1} + \frac{1}{R_2}$ for R **122. Finance** $A = P(1 + rt)$ for r

123. Mechanics $F = \frac{mv^2}{R}$ for R **124. Chemistry** $PV = nRT$ for T

125. Mathematics $S = \frac{a}{1 - r}$ for r **126. Mechanics** $v = -gt + v_0$ for t

127. Show that the sum of the roots of a quadratic equation is $-\frac{b}{a}$.

128. Show that the product of the roots of a quadratic equation is $\frac{c}{a}$.

129. Find k such that the equation $kx^2 + x + k = 0$ has a repeated real solution.

130. Find k such that the equation $x^2 - kx + 4 = 0$ has a repeated real solution.

131. Show that the real solutions of the equation $ax^2 + bx + c = 0$ are the negatives of the real solutions of the equation $ax^2 - bx + c = 0$. Assume that $b^2 - 4ac \geq 0$.

132. Show that the real solutions of the equation $ax^2 + bx + c = 0$ are the reciprocals of the real solutions of the equation $cx^2 + bx + a = 0$. Assume that $b^2 - 4ac \geq 0$.

133. Which of the following pairs of equations are equivalent? Explain.

(a) $x^2 = 9$; $x = 3$

(b) $x = \sqrt{9}$; $x = 3$

(c) $(x - 1)(x - 2) = (x - 1)^2$; $x - 2 = x - 1$

134. The equation

$$\frac{5}{x+3} + 3 = \frac{8+x}{x+3}$$

has no solution, yet when we go through the process of solving it we obtain $x = -3$. Write a brief paragraph to explain what causes this to happen.

135. Make up an equation that has no solution and give it to a fellow student to solve. Ask the fellow student to write a critique of your equation.

136. Describe three ways you might solve a quadratic equation. State your preferred method; explain why you chose it.

137. Explain the benefits of evaluating the discriminant of a quadratic equation before attempting to solve it.

138. Make up three quadratic equations: one having two distinct solutions, one having no real solution, and one having exactly one real solution.

139. The word *quadratic* seems to imply four (*quad*), yet a quadratic equation is an equation that involves a polynomial of degree 2. Investigate the origin of the term *quadratic* as it is used in the expression *quadratic equation*. Write a brief essay on your findings.

'Are You Prepared?' Answers

1. $(x - 6)(x + 1)$ **2.** $(2x - 3)(x + 1)$

3. $\left\{-\frac{5}{3}, 3\right\}$ **4.** True

A.6 Complex Numbers; Quadratic Equations in the Complex Number System

OBJECTIVES 1 Add, Subtract, Multiply, and Divide Complex Numbers
2 Solve Quadratic Equations in the Complex Number System

Complex Numbers

One property of a real number is that its square is nonnegative. For example, there is no real number x for which

$$x^2 = -1$$

To remedy this situation, we introduce a number called the **imaginary unit,** which we denote by $\boldsymbol{i}$ and whose square is -1:

$$i^2 = -1$$

This should not surprise you. If our universe were to consist only of integers, there would be no number x for which $2x = 1$. This unfortunate circumstance was remedied by introducing numbers such as $\frac{1}{2}$ and $\frac{2}{3}$, the *rational numbers*. If our universe were to consist only of rational numbers, there would be no x whose square equals 2. That is, there would be no number x for which $x^2 = 2$. To remedy this, we introduced numbers such as $\sqrt{2}$ and $\sqrt[3]{5}$, the *irrational numbers*. The *real numbers*, you will recall, consist of the rational numbers and the irrational numbers. Now, if our universe were to consist only of real numbers, then there would be no number x whose square is -1. To remedy this, we introduce a number i, whose square is -1.

In the progression outlined, each time that we encountered a situation that was unsuitable, we introduced a new number system to remedy this situation. And each new number system contained the earlier number system as a subset. The number system that results from introducing the number i is called the **complex number system.**

Complex numbers are numbers of the form $a + bi$, where a and b are real numbers. The real number a is called the **real part** of the number $a + bi$; the real number b is called the **imaginary part** of $a + bi$; and i is the imaginary unit, so $i^2 = -1$.

For example, the complex number $-5 + 6i$ has the real part -5 and the imaginary part 6.

When a complex number is written in the form $a + bi$, where a and b are real numbers, we say it is in **standard form.** However, if the imaginary part of a complex number is negative, such as in the complex number $3 + (-2)i$, we agree to write it instead in the form $3 - 2i$.

Also, the complex number $a + 0i$ is usually written merely as a. This serves to remind us that the real numbers are a subset of the complex numbers. The complex number $0 + bi$ is usually written as bi. Sometimes the complex number bi is called a **pure imaginary number.**

1 Equality, addition, subtraction, and multiplication of complex numbers are defined so as to preserve the familiar rules of algebra for real numbers. Two complex numbers are equal if and only if their real parts are equal and their imaginary parts are equal. That is,

Equality of Complex Numbers

$$a + bi = c + di \qquad \text{if and only if } a = c \text{ and } b = d \qquad \textbf{(1)}$$

Two complex numbers are added by forming the complex number whose real part is the sum of the real parts and whose imaginary part is the sum of the imaginary parts. That is,

Sum of Complex Numbers

$$(a + bi) + (c + di) = (a + c) + (b + d)i \qquad \textbf{(2)}$$

To subtract two complex numbers, we use this rule:

Difference of Complex Numbers

$$(a + bi) - (c + di) = (a - c) + (b - d)i \qquad \textbf{(3)}$$

EXAMPLE 1 Adding and Subtracting Complex Numbers

(a) $(3 + 5i) + (-2 + 3i) = [3 + (-2)] + (5 + 3)i = 1 + 8i$

(b) $(6 + 4i) - (3 + 6i) = (6 - 3) + (4 - 6)i = 3 + (-2)i = 3 - 2i$ ◀

NOW WORK PROBLEM 13.

Products of complex numbers are calculated as illustrated in Example 2.

EXAMPLE 2 Multiplying Complex Numbers

$$(5 + 3i)\cdot(2 + 7i) = \underset{\uparrow\ \text{Distributive Property}}{5\cdot(2 + 7i) + 3i(2 + 7i)} \underset{\uparrow\ \text{Distributive Property}}{=} 10 + 35i + 6i + 21i^2$$

$$\underset{\uparrow\ i^2 = -1}{=} 10 + 41i + 21(-1)$$

$$= -11 + 41i$$ ◀

Based on the procedure of Example 2, we define the **product** of two complex numbers by the following formula:

Product of Complex Numbers

$$(a + bi)\cdot(c + di) = (ac - bd) + (ad + bc)i \qquad \textbf{(4)}$$

Do not bother to memorize formula (4). Instead, whenever it is necessary to multiply two complex numbers, follow the usual rules for multiplying two binomials, as in Example 2, remembering that $i^2 = -1$. For example,

$$(2i)(2i) = 4i^2 = -4$$

$$(2 + i)(1 - i) = 2 - 2i + i - i^2 = 3 - i$$

NOW WORK PROBLEM 19.

Algebraic properties for addition and multiplication, such as the commutative, associative, and distributive properties, hold for complex numbers. The property that every nonzero complex number has a multiplicative inverse, or reciprocal, requires a closer look.

Conjugate

If $z = a + bi$ is a complex number, then its **conjugate,** denoted by $\bar{z}$, is defined as

$$\bar{z} = \overline{a + bi} = a - bi$$

For example, $\overline{2 + 3i} = 2 - 3i$ and $\overline{-6 - 2i} = -6 + 2i$.

EXAMPLE 3 Multiplying a Complex Number by Its Conjugate

Find the product of the complex number $z = 3 + 4i$ and its conjugate $\bar{z}$.

Solution Since $\bar{z} = 3 - 4i$, we have

$$z\bar{z} = (3 + 4i)(3 - 4i) = 9 - 12i + 12i - 16i^2 = 9 + 16 = 25$$ ◀

The result obtained in Example 3 has an important generalization.

Theorem The product of a complex number and its conjugate is a nonnegative real number. That is, if $z = a + bi$, then

$$z\bar{z} = a^2 + b^2 \tag{5}$$

Proof If $z = a + bi$, then

$$z\bar{z} = (a + bi)(a - bi) = a^2 - (bi)^2 = a^2 - b^2i^2 = a^2 + b^2 \quad \blacksquare$$

To express the reciprocal of a nonzero complex number z in standard form, multiply the numerator and denominator of $\frac{1}{z}$ by $\bar{z}$. That is, if $z = a + bi$ is a nonzero complex number, then

$$\frac{1}{a + bi} = \frac{1}{z} = \frac{1}{z} \cdot \frac{\bar{z}}{\bar{z}} = \frac{\bar{z}}{z\bar{z}} \underset{\text{Use (5).}}{=} \frac{a - bi}{a^2 + b^2}$$

$$= \frac{a}{a^2 + b^2} - \frac{b}{a^2 + b^2}i$$

EXAMPLE 4 Writing the Reciprocal of a Complex Number in Standard Form

Write $\frac{1}{3 + 4i}$ in standard form $a + bi$; that is, find the reciprocal of $3 + 4i$.

Solution The idea is to multiply the numerator and denominator by the conjugate of $3 + 4i$, that is, the complex number $3 - 4i$. The result is

$$\frac{1}{3 + 4i} = \frac{1}{3 + 4i} \cdot \frac{3 - 4i}{3 - 4i} = \frac{3 - 4i}{9 + 16} = \frac{3}{25} - \frac{4}{25}i \quad \blacktriangleleft$$

To express the quotient of two complex numbers in standard form, we multiply the numerator and denominator of the quotient by the conjugate of the denominator.

EXAMPLE 5 Writing the Quotient of Complex Numbers in Standard Form

Write each of the following in standard form.

(a) $\frac{1 + 4i}{5 - 12i}$ (b) $\frac{2 - 3i}{4 - 3i}$

Solution (a) $\frac{1 + 4i}{5 - 12i} = \frac{1 + 4i}{5 - 12i} \cdot \frac{5 + 12i}{5 + 12i} = \frac{5 + 12i + 20i + 48i^2}{25 + 144}$

$$= \frac{-43 + 32i}{169} = -\frac{43}{169} + \frac{32}{169}i$$

(b) $\dfrac{2-3i}{4-3i} = \dfrac{2-3i}{4-3i} \cdot \dfrac{4+3i}{4+3i} = \dfrac{8+6i-12i-9i^2}{16+9}$

$= \dfrac{17-6i}{25} = \dfrac{17}{25} - \dfrac{6}{25}i$ ◀

NOW WORK PROBLEM 27.

EXAMPLE 6 Writing Other Expressions in Standard Form

If $z = 2 - 3i$ and $w = 5 + 2i$, write each of the following expressions in standard form.

(a) $\dfrac{z}{w}$ (b) $\overline{z + w}$ (c) $z + \overline{z}$

Solution (a) $\dfrac{z}{w} = \dfrac{z \cdot \overline{w}}{w \cdot \overline{w}} = \dfrac{(2-3i)(5-2i)}{(5+2i)(5-2i)} = \dfrac{10-4i-15i+6i^2}{25+4}$

$= \dfrac{4-19i}{29} = \dfrac{4}{29} - \dfrac{19}{29}i$

(b) $\overline{z + w} = \overline{(2-3i)+(5+2i)} = \overline{7-i} = 7 + i$

(c) $z + \overline{z} = (2-3i) + (2+3i) = 4$ ◀

The conjugate of a complex number has certain general properties that we shall find useful later.

For a real number $a = a + 0i$, the conjugate is $\overline{a} = \overline{a + 0i} = a - 0i = a$. That is,

Theorem The conjugate of a real number is the real number itself.

Other properties of the conjugate that are direct consequences of the definition are given next. In each statement, z and w represent complex numbers.

Theorem The conjugate of the conjugate of a complex number is the complex number itself.

$$(\overline{\overline{z}}) = z \quad \textbf{(6)}$$

The conjugate of the sum of two complex numbers equals the sum of their conjugates.

$$\overline{z + w} = \overline{z} + \overline{w} \quad \textbf{(7)}$$

The conjugate of the product of two complex numbers equals the product of their conjugates.

$$\overline{z \cdot w} = \overline{z} \cdot \overline{w} \quad \textbf{(8)}$$

We leave the proofs of equations (6), (7), and (8) as exercises.

Powers of *i*

The **powers of *i*** follow a pattern that is useful to know.

$$i^1 = i \qquad i^5 = i^4 \cdot i = 1 \cdot i = i$$
$$i^2 = -1 \qquad i^6 = i^4 \cdot i^2 = -1$$
$$i^3 = i^2 \cdot i = -1 \cdot i = -i \qquad i^7 = i^4 \cdot i^3 = -i$$
$$i^4 = i^2 \cdot i^2 = (-1)(-1) = 1 \qquad i^8 = i^4 \cdot i^4 = 1$$

And so on. The powers of i repeat with every fourth power.

EXAMPLE 7 **Evaluating Powers of *i***

(a) $i^{27} = i^{24} \cdot i^3 = (i^4)^6 \cdot i^3 = 1^6 \cdot i^3 = -i$
(b) $i^{101} = i^{100} \cdot i^1 = (i^4)^{25} \cdot i = 1^{25} \cdot i = i$ ◀

EXAMPLE 8 **Writing the Power of a Complex Number in Standard Form**

Write $(2 + i)^3$ in standard form.

Solution We use the special product formula for $(x + a)^3$.

$$(x + a)^3 = x^3 + 3ax^2 + 3a^2x + a^3$$

Using this special product formula,

$$\begin{aligned}(2 + i)^3 &= 2^3 + 3 \cdot i \cdot 2^2 + 3 \cdot i^2 \cdot 2 + i^3 \\ &= 8 + 12i + 6(-1) + (-i) \\ &= 2 + 11i.\end{aligned}$$ ◀

NOW WORK PROBLEM 41.

Quadratic Equations with a Negative Discriminant

2 Quadratic equations with a negative discriminant have no real number solution. However, if we extend our number system to allow complex numbers, quadratic equations will always have a solution. Since the solution to a quadratic equation involves the square root of the discriminant, we begin with a discussion of square roots of negative numbers.

If N is a positive real number, we define the **principal square root of $-N$,** denoted by $\sqrt{-N}$, as

$$\sqrt{-N} = \sqrt{N}\,i$$

where i is the imaginary unit and $i^2 = -1$.

WARNING: In writing $\sqrt{-N} = \sqrt{N}\,i$, be sure to place i outside the $\sqrt{\ }$ symbol. ■

EXAMPLE 9 **Evaluating the Square Root of a Negative Number**

(a) $\sqrt{-1} = \sqrt{1}\,i = i$ (b) $\sqrt{-4} = \sqrt{4}\,i = 2i$
(c) $\sqrt{-8} = \sqrt{8}\,i = 2\sqrt{2}\,i$ ◀

EXAMPLE 10 Solving Equations

Solve each equation in the complex number system.

(a) $x^2 = 4$ (b) $x^2 = -9$

Solution (a) $x^2 = 4$

$x = \pm\sqrt{4} = \pm 2$

The equation has two solutions, -2 and 2.

(b) $x^2 = -9$

$x = \pm\sqrt{-9} = \pm\sqrt{9}\,i = \pm 3i$

The equation has two solutions, $-3i$ and $3i$. ◀

NOW WORK PROBLEMS 49 AND 53.

WARNING: When working with square roots of negative numbers, do not set the square root of a product equal to the product of the square roots (which can be done with positive numbers). To see why, look at this calculation: We know that $\sqrt{100} = 10$. However, it is also true that $100 = (-25)(-4)$, so

$$10 = \sqrt{100} = \sqrt{(-25)(-4)} \neq \sqrt{-25}\,\sqrt{-4} = \left(\sqrt{25}\,i\right)\left(\sqrt{4}\,i\right) = (5i)(2i) = 10i^2 = -10$$

↑ Here is the error. ■

Because we have defined the square root of a negative number, we can now restate the quadratic formula without restriction.

Theorem In the complex number system, the solutions of the quadratic equation $ax^2 + bx + c = 0$, where a, b, and c are real numbers and $a \neq 0$, are given by the formula

$$x = \frac{-b \pm \sqrt{b^2 - 4ac}}{2a} \qquad \textbf{(9)}$$

EXAMPLE 11 Solving Quadratic Equations in the Complex Number System

Solve the equation $x^2 - 4x + 8 = 0$ in the complex number system.

Solution Here $a = 1, b = -4, c = 8$, and $b^2 - 4ac = 16 - 4(1)(8) = -16$. Using equation (9), we find that

$$x = \frac{-(-4) \pm \sqrt{-16}}{2(1)} = \frac{4 \pm \sqrt{16}\,i}{2} = \frac{4 \pm 4i}{2} = 2 \pm 2i$$

The equation has the solution set $\{2 - 2i, 2 + 2i\}$. ◀

✔ **CHECK:**

$2 + 2i$: $(2 + 2i)^2 - 4(2 + 2i) + 8 = 4 + \cancel{8i} + 4i^2 - \cancel{8} - \cancel{8i} + \cancel{8}$
$= 4 - 4 = 0$

$2 - 2i$: $(2 - 2i)^2 - 4(2 - 2i) + 8 = 4 - \cancel{8i} + 4i^2 - \cancel{8} + \cancel{8i} + \cancel{8}$
$= 4 - 4 = 0$

NOW WORK PROBLEM 59.

The discriminant $b^2 - 4ac$ of a quadratic equation still serves as a way to determine the character of the solutions.

Character of the Solutions of a Quadratic Equation

In the complex number system, consider a quadratic equation $ax^2 + bx + c = 0$ with real coefficients.

1. If $b^2 - 4ac > 0$, the equation has two unequal real solutions.
2. If $b^2 - 4ac = 0$, the equation has a repeated real solution, a double root.
3. If $b^2 - 4ac < 0$, the equation has two complex solutions that are not real. The solutions are conjugates of each other.

The third conclusion in the display is a consequence of the fact that if $b^2 - 4ac = -N < 0$ then, by the quadratic formula, the solutions are

$$x = \frac{-b + \sqrt{b^2 - 4ac}}{2a} = \frac{-b + \sqrt{-N}}{2a} = \frac{-b + \sqrt{N}\,i}{2a} = \frac{-b}{2a} + \frac{\sqrt{N}}{2a}i$$

and

$$x = \frac{-b - \sqrt{b^2 - 4ac}}{2a} = \frac{-b - \sqrt{-N}}{2a} = \frac{-b - \sqrt{N}\,i}{2a} = \frac{-b}{2a} - \frac{\sqrt{N}}{2a}i$$

which are conjugates of each other.

EXAMPLE 12 **Determining the Character of the Solution of a Quadratic Equation**

Without solving, determine the character of the solution of each equation.

(a) $3x^2 + 4x + 5 = 0$ (b) $2x^2 + 4x + 1 = 0$

(c) $9x^2 - 6x + 1 = 0$

Solution

(a) Here $a = 3, b = 4$, and $c = 5$, so $b^2 - 4ac = 16 - 4(3)(5) = -44$. The solutions are two complex numbers that are not real and are conjugates of each other.

(b) Here $a = 2, b = 4$, and $c = 1$, so $b^2 - 4ac = 16 - 8 = 8$. The solutions are two unequal real numbers.

(c) Here $a = 9, b = -6$, and $c = 1$, so $b^2 - 4ac = 36 - 4(9)(1) = 0$. The solution is a repeated real number, that is, a double root. ◀

NOW WORK PROBLEM 73.

A.6 Assess Your Understanding

Concepts and Vocabulary

1. *True or False:* The square of a complex number is sometimes negative.

2. $(2 + i)(2 - i) =$ ________.

3. *True or False:* In the complex number system, a quadratic equation has four solutions.

4. In the complex number $5 + 2i$, the number 5 is called the ________ part; the number 2 is called the ________ part; the number i is called the ________ ________.

5. The equation $x^2 = -4$ has the solution set ________.

6. *True or False:* The conjugate of $2 + 5i$ is $-2 - 5i$.

7. *True or False:* All real numbers are complex numbers.

8. *True or False:* If $2 - 3i$ is a solution of a quadratic equation with real coefficients, then $-2 + 3i$ is also a solution.

Exercises

In Problems 9–46, write each expression in the standard form a + bi.

9. $(2 - 3i) + (6 + 8i)$
10. $(4 + 5i) + (-8 + 2i)$
11. $(-3 + 2i) - (4 - 4i)$
12. $(3 - 4i) - (-3 - 4i)$
13. $(2 - 5i) - (8 + 6i)$
14. $(-8 + 4i) - (2 - 2i)$
15. $3(2 - 6i)$
16. $-4(2 + 8i)$
17. $2i(2 - 3i)$
18. $3i(-3 + 4i)$
19. $(3 - 4i)(2 + i)$
20. $(5 + 3i)(2 - i)$
21. $(-6 + i)(-6 - i)$
22. $(-3 + i)(3 + i)$
23. $\dfrac{10}{3 - 4i}$
24. $\dfrac{13}{5 - 12i}$
25. $\dfrac{2 + i}{i}$
26. $\dfrac{2 - i}{-2i}$
27. $\dfrac{6 - i}{1 + i}$
28. $\dfrac{2 + 3i}{1 - i}$
29. $\left(\dfrac{1}{2} + \dfrac{\sqrt{3}}{2}i\right)^2$
30. $\left(\dfrac{\sqrt{3}}{2} - \dfrac{1}{2}i\right)^2$
31. $(1 + i)^2$
32. $(1 - i)^2$
33. i^{23}
34. i^{14}
35. i^{-15}
36. i^{-23}
37. $i^6 - 5$
38. $4 + i^3$
39. $6i^3 - 4i^5$
40. $4i^3 - 2i^2 + 1$
41. $(1 + i)^3$
42. $(3i)^4 + 1$
43. $i^7(1 + i^2)$
44. $2i^4(1 + i^2)$
45. $i^6 + i^4 + i^2 + 1$
46. $i^7 + i^5 + i^3 + i$

In Problems 47–52, perform the indicated operations and express your answer in the form a + bi.

47. $\sqrt{-4}$
48. $\sqrt{-9}$
49. $\sqrt{-25}$
50. $\sqrt{-64}$
51. $\sqrt{(3 + 4i)(4i - 3)}$
52. $\sqrt{(4 + 3i)(3i - 4)}$

In Problems 53–72, solve each equation in the complex number system.

53. $x^2 + 4 = 0$
54. $x^2 - 4 = 0$
55. $x^2 - 16 = 0$
56. $x^2 + 25 = 0$
57. $x^2 - 6x + 13 = 0$
58. $x^2 + 4x + 8 = 0$
59. $x^2 - 6x + 10 = 0$
60. $x^2 - 2x + 5 = 0$
61. $8x^2 - 4x + 1 = 0$
62. $10x^2 + 6x + 1 = 0$
63. $5x^2 + 1 = 2x$
64. $13x^2 + 1 = 6x$
65. $x^2 + x + 1 = 0$
66. $x^2 - x + 1 = 0$
67. $x^3 - 8 = 0$
68. $x^3 + 27 = 0$
69. $x^4 = 16$
70. $x^4 = 1$
71. $x^4 + 13x^2 + 36 = 0$
72. $x^4 + 3x^2 - 4 = 0$

In Problems 73–78, without solving, determine the character of the solutions of each equation in the complex number system.

73. $3x^2 - 3x + 4 = 0$
74. $2x^2 - 4x + 1 = 0$
75. $2x^2 + 3x = 4$
76. $x^2 + 6 = 2x$
77. $9x^2 - 12x + 4 = 0$
78. $4x^2 + 12x + 9 = 0$

79. $2 + 3i$ is a solution of a quadratic equation with real coefficients. Find the other solution.

80. $4 - i$ is a solution of a quadratic equation with real coefficients. Find the other solution.

In Problems 81–84, $z = 3 - 4i$ and $w = 8 + 3i$. Write each expression in the standard form a + bi.

81. $z + \bar{z}$
82. $w - \bar{w}$
83. $z\bar{z}$
84. $\overline{z - w}$

85. Use $z = a + bi$ to show that $z + \bar{z} = 2a$ and $z - \bar{z} = 2bi$.

86. Use $z = a + bi$ to show that $\bar{\bar{z}} = z$.

87. Use $z = a + bi$ and $w = c + di$ to show that $\overline{z + w} = \bar{z} + \bar{w}$.

88. Use $z = a + bi$ and $w = c + di$ to show that $\overline{z \cdot w} = \bar{z} \cdot \bar{w}$.

89. Explain to a friend how you would add two complex numbers and how you would multiply two complex numbers. Explain any differences in the two explanations.

90. Write a brief paragraph that compares the method used to rationalize the denominator of a rational expression and the method used to write the quotient of two complex numbers in standard form.

A.7 Setting Up Equations; Applications

OBJECTIVES
1. Translate Verbal Descriptions into Mathematical Expressions
2. Set up Applied Problems
3. Solve Interest Problems
4. Solve Mixture Problems
5. Solve Uniform Motion Problems
6. Solve Constant Rate Job Problems

Applied (word) problems do not come in the form "Solve the equation. . . ." Instead, they supply information using words, a verbal description of the real problem. So, to solve applied problems, we must be able to translate the verbal description into the language of mathematics. We do this by using variables to represent unknown quantities and then finding relationships (such as equations) that involve these variables. The process of doing all this is called **mathematical modeling.**

Any solution to the mathematical problem must be checked against the mathematical problem, the verbal description, and the real problem. See Figure 14 for an illustration of the **modeling process.**

Figure 14

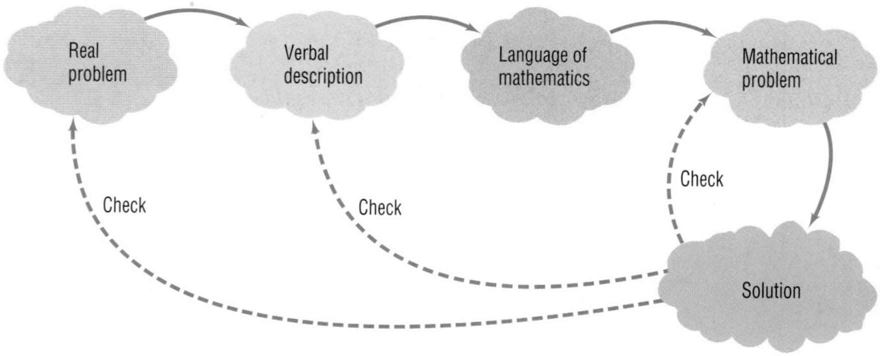

1 Let's look at a few examples that will help you to translate certain words into mathematical symbols.

EXAMPLE 1

Translating Verbal Descriptions into Mathematical Expressions

(a) The area of a rectangle is the product of its length and its width.

Translation: If A is used to represent the area, l the length, and w the width, then $A = lw$.

(b) For uniform motion, the velocity of an object equals the distance traveled divided by the time required.

Translation: If v is the velocity, s the distance, and t the time, then $v = \frac{s}{t}$.

(c) A total of \$5000 is invested, some in stocks and some in bonds. If the amount invested in stocks is x, express the amount invested in bonds in terms of x.

Translation: If x is the amount invested in stocks, then the amount invested in bonds is $5000 - x$, since their sum is $x + (5000 - x) = 5000$.

(d) Let x denote a number.

The number 5 times as large as x is $5x$.

The number 3 less than x is $x - 3$.

The number that exceeds x by 4 is $x + 4$.

The number that, when added to x, gives 5 is $5 - x$. ◀

NOW WORK PROBLEM 7.

2 Always check the units used to measure the variables of an applied problem. In Example 1(a), if l is measured in feet, then w also must be expressed in feet, and A will be expressed in square feet. In Example 1(b), if v is measured in miles per hour, then the distance s must be expressed in miles and the time t must be expressed in hours. It is a good practice to check units to be sure that they are consistent and make sense.

Although each situation has unique features, we can provide an outline of the steps to follow in setting up applied problems.

Steps for Setting Up Applied Problems

STEP 1: Read the problem carefully, perhaps two or three times. Pay particular attention to the question being asked in order to identify what you are looking for. If you can, determine realistic possibilities for the answer.

STEP 2: Assign a letter (variable) to represent what you are looking for, and, if necessary, express any remaining unknown quantities in terms of this variable.

STEP 3: Make a list of all the known facts, and translate them into mathematical expressions. These may take the form of an equation (or, later, an inequality) involving the variable. If possible, draw an appropriately labeled diagram to assist you. Sometimes a table or chart helps.

STEP 4: Solve the equation for the variable, and then answer the question, usually using a complete sentence.

STEP 5: Check the answer with the facts in the problem. If it agrees, congratulations! If it does not agree, try again.

Interest

3 **Interest** is money paid for the use of money. The total amount borrowed (whether by an individual from a bank in the form of a loan or by a bank from an individual in the form of a savings account) is called the **principal.** The **rate of interest,** expressed as a percent, is the amount charged for the use of the principal for a given period of time, usually on a yearly (that is, per annum) basis.

Simple Interest Formula

If a principal of P dollars is borrowed for a period of t years at a per annum interest rate r, expressed as a decimal, the interest I charged is

$$I = Prt \quad \textbf{(1)}$$

Interest charged according to formula (1) is called **simple interest.** In using formula (1), be sure to express r as a decimal.

EXAMPLE 2 Financial Planning

Candy has \$70,000 to invest and requires an overall rate of return of 9%. She can invest in a safe, government-insured certificate of deposit, but it only pays 8%. To obtain 9%, she agrees to invest some of her money in non-insured corporate bonds paying 12%. How much should be placed in each investment to achieve her goal?

Solution

STEP 1: The question is asking for two dollar amounts: the principal to invest in the corporate bonds and the principal to invest in the certificate of deposit.

STEP 2: We let x represent the amount (in dollars) to be invested in the bonds. Then $70{,}000 - x$ is the amount that will be invested in the certificate. (Do you see why?)

STEP 3: We set up a table:

	Principal (\$)	Rate	Time (yr)	Interest (\$)
Bonds	x	12% = 0.12	1	$0.12x$
Certificate	$70{,}000 - x$	8% = 0.08	1	$0.08(70{,}000 - x)$
Total	70,000	9% = 0.09	1	$0.09(70{,}000) = 6300$

Since the total interest from the investments is equal to $0.09(70{,}000) = 6300$, we must have the equation

$$0.12x + 0.08(70{,}000 - x) = 6300$$

(Note that the units are consistent: the unit is dollars on each side.)

STEP 4: $0.12x + 5600 - 0.08x = 6300$

$$0.04x = 700$$

$$x = 17{,}500$$

Candy should place \$17,500 in the bonds and \$70,000 − \$17,500 = \$52,500 in the certificate.

STEP 5: The interest on the bonds after 1 year is 0.12(\$17,500) = \$2100; the interest on the certificate after 1 year is 0.08(\$52,500) = \$4200. The total annual interest is \$6300, the required amount. ◀

NOW WORK PROBLEMS 17 AND 23.

Mixture Problems

4 Oil refineries sometimes produce gasoline that is a blend of two or more types of fuel; bakeries occasionally blend two or more types of flour for their bread. These problems are referred to as **mixture problems** because they combine two or more quantities to form a mixture.

EXAMPLE 3 Blending Coffees

The manager of a Starbucks store decides to experiment with a new blend of coffee. She will mix some B grade Colombian coffee that sells for \$5 per pound with some A grade Arabica coffee that sells for \$10 per pound to get 100 pounds of the new blend. The selling price of the new blend is to be \$7 per pound, and there is to be no difference in revenue from selling the new blend versus selling the other types. How many pounds of the B grade Colombian and A grade Arabica coffees are required?

Solution Let x represent the number of pounds of the B grade Colombian coffee. Then $100 - x$ equals the number of pounds of the A grade Arabica coffee. See Figure 15.

Figure 15

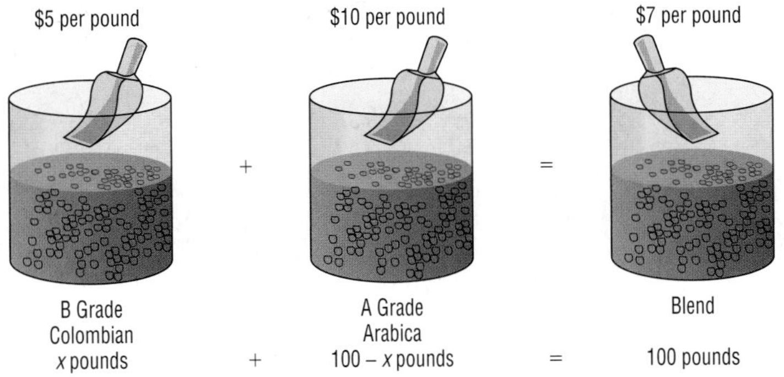

Since there is to be no difference in revenue between selling the A and B grades separately versus the blend, we have

$$\begin{Bmatrix}\text{Price per pound}\\ \text{of B grade}\end{Bmatrix}\begin{Bmatrix}\text{Pounds}\\ \text{of B grade}\end{Bmatrix} + \begin{Bmatrix}\text{Price per pound}\\ \text{of A grade}\end{Bmatrix}\begin{Bmatrix}\text{Pounds}\\ \text{of A grade}\end{Bmatrix} = \begin{Bmatrix}\text{Price per pound}\\ \text{of blend}\end{Bmatrix}\begin{Bmatrix}\text{Pounds}\\ \text{of blend}\end{Bmatrix}$$

$$\$5 \cdot x + \$10 \cdot (100 - x) = \$7 \cdot 100$$

We have the equation

$$\begin{aligned} 5x + 10(100 - x) &= 700 \\ 5x + 1000 - 10x &= 700 \\ -5x &= -300 \\ x &= 60 \end{aligned}$$

The manager should blend 60 pounds of B grade Colombian coffee with $100 - 60 = 40$ pounds of A grade Arabica coffee to get the desired blend.

✔ **CHECK:** The 60 pounds of B grade coffee would sell for $(\$5)(60) = \$300$, and the 40 pounds of A grade coffee would sell for $(\$10)(40) = \$400$; the total revenue, \$700, equals the revenue obtained from selling the blend, as desired. ◀

NOW WORK PROBLEM 27.

Uniform Motion

5 Objects that move at a constant velocity are said to be in **uniform motion.** When the average velocity of an object is known, it can be interpreted as its constant velocity. For example, a bicyclist traveling at an average velocity of 25 miles per hour is in uniform motion.

Uniform Motion Formula

If an object moves at an average velocity v, the distance s covered in time t is given by the formula

$$s = vt \qquad (2)$$

That is, Distance = Velocity · Time.

EXAMPLE 4 Physics: Uniform Motion

Tanya, who is a long-distance runner, runs at an average velocity of 8 miles per hour (mi/hr). Two hours after Tanya leaves your house, you leave in your Honda and follow the same route. If your average velocity is 40 mi/hr, how long will it be before you catch up to Tanya? How far will each of you be from your home?

Solution Refer to Figure 16. We use t to represent the time (in hours) that it takes the Honda to catch up to Tanya. When this occurs, the total time elapsed for Tanya is $t + 2$ hours.

Figure 16

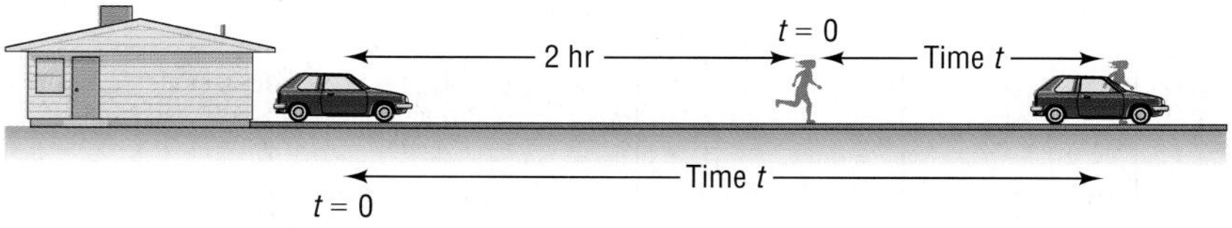

Set up the following table:

	Velocity mi/hr	Time hr	Distance mi
Tanya	8	$t + 2$	$8(t + 2)$
Honda	40	t	$40t$

Since the distance traveled is the same, we are led to the following equation:

$$\begin{aligned} 8(t + 2) &= 40t \\ 8t + 16 &= 40t \\ 32t &= 16 \\ t &= \frac{1}{2} \text{ hour} \end{aligned}$$

It will take the Honda $\frac{1}{2}$ hour to catch up to Tanya. Each of you will have gone 20 miles.

✔ **CHECK:** In 2.5 hours, Tanya travels a distance of $(2.5)(8) = 20$ miles. In $\frac{1}{2}$ hour, the Honda travels a distance of $\left(\frac{1}{2}\right)(40) = 20$ miles. ◀

EXAMPLE 5 Physics: Uniform Motion

A motorboat heads upstream a distance of 24 miles on the Illinois River, whose current is running at 3 miles per hour (mi/hr). The trip up and back takes 6 hours. Assuming that the motorboat maintained a constant speed relative to the water, what was its speed?

Solution See Figure 17. We use v to represent the constant speed of the motorboat relative to the water. Then the true speed going upstream is $v - 3$ mi/hr, and the true speed going downstream is $v + 3$ mi/hr. Since Distance = Velocity × Time, then Time $= \dfrac{\text{Distance}}{\text{Velocity}}$. We set up a table.

Figure 17

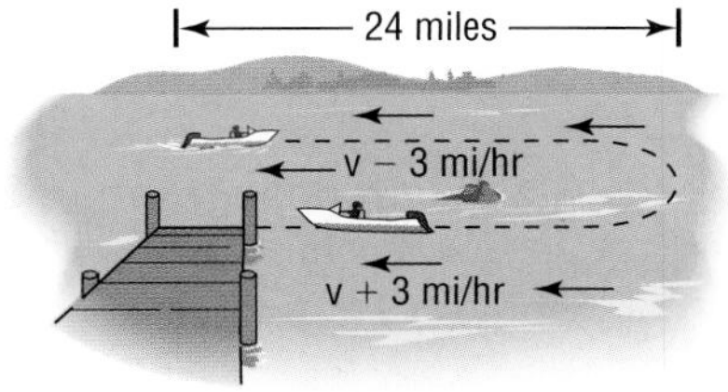

	Velocity (mi/hr)	Distance (mi)	Time = $\frac{\text{Distance}}{\text{Velocity}}$ (hr)
Upstream	$v - 3$	24	$\frac{24}{v-3}$
Downstream	$v + 3$	24	$\frac{24}{v+3}$

Since the total time up and back is 6 hours, we have

$$\frac{24}{v-3} + \frac{24}{v+3} = 6$$

$$\frac{24(v+3) + 24(v-3)}{(v-3)(v+3)} = 6 \qquad \text{Add the quotients on the left.}$$

$$\frac{48v}{v^2 - 9} = 6 \qquad \text{Simplify.}$$

$$48v = 6(v^2 - 9) \qquad \text{Clear fractions.}$$

$$8v = v^2 - 9 \qquad \text{Divide each side by 6.}$$

$$v^2 - 8v - 9 = 0 \qquad \text{Place in standard form.}$$

$$(v - 9)(v + 1) = 0 \qquad \text{Factor.}$$

$$v = 9 \quad \text{or} \quad v = -1 \qquad \text{Apply the Zero-Product Property and solve.}$$

We discard the solution $v = -1$ mi/hr, so the speed of the motorboat relative to the water is 9 mi/hr. ◀

NOW WORK PROBLEM 31.

Constant Rate Jobs

6 This section involves jobs that are performed at a **constant rate.** Our assumption is that, if a job can be done in t units of time, $\frac{1}{t}$ of the job is done in 1 unit of time. Let's look at an example.

EXAMPLE 6 Working Together to Do a Job

At 10 AM Danny is asked by his father to weed the garden. From past experience, Danny knows that this will take him 4 hours, working alone. His older brother, Mike, when it is his turn to do this job, requires 6 hours. Since Mike wants to go golfing with Danny and has a reservation for 1 PM, he agrees to help Danny. Assuming no gain or loss of efficiency, when will they finish if they work together? Can they make the golf date?

Solution We set up Table 1. In 1 hour, Danny does $\frac{1}{4}$ of the job, and in 1 hour, Mike does $\frac{1}{6}$ of the job. Let t be the time (in hours) that it takes them to do the job together. In 1 hour, then, $\frac{1}{t}$ of the job is completed. We reason as follows:

Table 1

	Hours to Do Job	Part of Job Done in 1 Hour
Danny	4	$\frac{1}{4}$
Mike	6	$\frac{1}{6}$
Together	t	$\frac{1}{t}$

$$\begin{pmatrix}\text{Part done by Danny}\\ \text{in 1 hour}\end{pmatrix} + \begin{pmatrix}\text{Part done by Mike}\\ \text{in 1 hour}\end{pmatrix} = \begin{pmatrix}\text{Part done together}\\ \text{in 1 hour}\end{pmatrix}$$

From Table 1,

$$\begin{aligned} \frac{1}{4} + \frac{1}{6} &= \frac{1}{t} \\ \frac{3+2}{12} &= \frac{1}{t} \\ \frac{5}{12} &= \frac{1}{t} \\ 5t &= 12 \\ t &= \frac{12}{5} \end{aligned}$$

Working together, the job can be done in $\frac{12}{5}$ hours, or 2 hours, 24 minutes. They should make the golf date, since they will finish at 12:24 PM. ◀

NOW WORK PROBLEM 35.

Here is an applied problem that you will probably see again in a slightly different form if you study calculus.

EXAMPLE 7 Constructing a Box

From each corner of a square piece of sheet metal, remove a square of side 9 centimeters. Turn up the edges to form an open box. If the box is to hold 144 cubic centimeters (cm^3), what should be the dimensions of the piece of sheet metal?

Solution We use Figure 18 as a guide. We have labeled by x the length of a side of the square piece of sheet metal. The box will be of height 9 centimeters, and its square base will have $x - 18$ as the length of a side. The volume (Length × Width × Height) of the box is therefore

$$9(x - 18)(x - 18) = 9(x - 18)^2$$

Figure 18

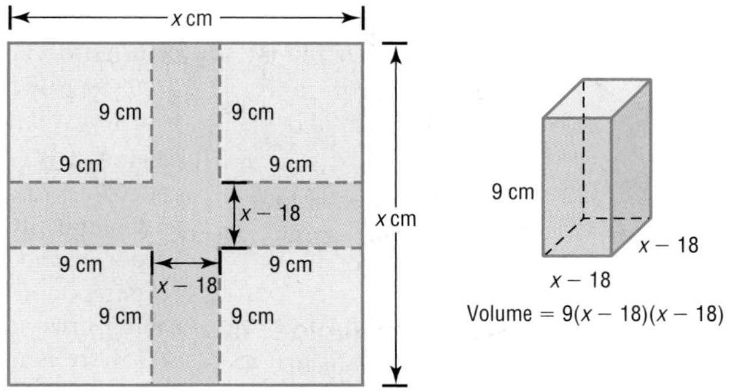

Since the volume of the box is to be 144 cm^3, we have

$$\begin{aligned} 9(x - 18)^2 &= 144 \\ (x - 18)^2 &= 16 && \text{Divide each side by 9.} \\ x - 18 &= \pm 4 && \text{Use the Square Root Method.} \\ x &= 18 \pm 4 \\ x &= 22 \quad \text{or} \quad x = 14 \end{aligned}$$

We discard the solution $x = 14$ (do you see why?) and conclude that the sheet metal should be 22 centimeters by 22 centimeters. ◀

✔ **CHECK:** If we begin with a piece of sheet metal 22 centimeters by 22 centimeters, cut out a 9 centimeter square from each corner, and fold up the edges, we get a box whose dimensions are 9 by 4 by 4, with volume $9 \times 4 \times 4 = 144$ cm^3, as required.

NOW WORK PROBLEM 37.

A.7 Assess Your Understanding

Concepts and Vocabulary

1. In applied problems, we translate a verbal description into the language of mathematics. Then we use variables to represent unknown quantities and find relationships that involve these variables. This process is referred to as __________ __________.

2. The money paid for the use of money is __________.

3. Objects that move at a constant velocity are said to be in __________ __________.

4. *True or False:* The amount charged for the use of principal for a given period of time is called the rate of interest.

5. *True or False:* If an object moves at an average velocity v, the distance s covered in time t is given by the formula $s = vt$.

6. Suppose that you want to mix two coffees in order to obtain 100 pounds of the blend. If x represents the number of pounds of coffee A, write an algebraic expression that represents the number of pounds of coffee B.

Exercises

In Problems 7–16, translate each sentence into a mathematical equation. Be sure to identify the meaning of all symbols.

7. **Geometry** The area of a circle is the product of the number π and the square of the radius.
8. **Geometry** The circumference of a circle is the product of the number π and twice the radius.
9. **Geometry** The area of a square is the square of the length of a side.
10. **Geometry** The perimeter of a square is four times the length of a side.
11. **Physics** Force equals the product of mass and acceleration.
12. **Physics** Pressure is force per unit area.
13. **Physics** Work equals force times distance.
14. **Physics** Kinetic energy is one-half the product of the mass and the square of the velocity.
15. **Business** The total variable cost of manufacturing x dishwashers is $150 per dishwasher times the number of dishwashers manufactured.
16. **Business** The total revenue derived from selling x dishwashers is $250 per dishwasher times the number of dishwashers sold.
17. **Finance** A total of $20,000 is to be invested, some in bonds and some in certificates of deposit (CDs). If the amount invested in bonds is to exceed that in CDs by $3000, how much will be invested in each type of investment?
18. **Finance** A total of $10,000 is to be divided between Sean and George, with George to receive $3000 less than Sean. How much will each receive?
19. **Computing Grades** Going into the final exam, which will count as two tests, Brooke has test scores of 80, 83, 71, 61, and 95. What score does Brooke need on the final in order to have an average score of 80?
20. **Computing Grades** Going into the final exam, which will count as two-thirds of the final grade, Mike has test scores of 86, 80, 84, and 90. What score does Mike need on the final in order to earn a B, which requires an average score of 80? What does he need to earn an A, which requires an average of 90?
21. **Geometry** The perimeter of a rectangle is 60 feet. Find its length and width if the length is 8 feet longer than the width.
22. **Geometry** The perimeter of a rectangle is 42 meters. Find its length and width if the length is twice the width.
23. **Financial Planning** Betsy, a recent retiree, requires $6000 per year in extra income. She has $50,000 to invest and can invest in B-rated bonds paying 15% per year or in a certificate of deposit (CD) paying 7% per year. How much money should be invested in each to realize exactly $6000 in interest per year?
24. **Financial Planning** After 2 years, Betsy (see Problem 23) finds that she will now require $7000 per year. Assuming that the remaining information is the same, how should the money be reinvested?
25. **Banking** A bank loaned out $12,000, part of it at the rate of 8% per year and the rest at the rate of 18% per year. If the interest received totaled $1000, how much was loaned at 8%?
26. **Banking** Wendy, a loan officer at a bank, has $1,000,000 to lend and is required to obtain an average return of 18% per year. If she can lend at the rate of 19% or at the rate of 16%, how much can she lend at the 16% rate and still meet her requirement?
27. **Business: Blending Teas** The manager of a store that specializes in selling tea decides to experiment with a new blend. She will mix some Earl Grey tea that sells for $5 per pound with some Orange Pekoe tea that sells for $3 per pound to get 100 pounds of the new blend. The selling price of the new blend is to be $4.50 per pound, and there is to be no difference in revenue from selling the new blend versus selling the other types. How many pounds of the Earl Grey tea and Orange Pekoe tea are required?
28. **Business: Blending Coffee** A coffee manufacturer wants to market a new blend of coffee that sells for $3.90 per pound by mixing two coffees that sell for $2.75 and $5 per pound, respectively. What amounts of each coffee should be blended to obtain the desired mixture?

    [**Hint:** Assume that the total weight of the desired blend is 100 pounds.]
29. **Business: Mixing Nuts** A nut store normally sells cashews for $4.00 per pound and peanuts for $1.50 per pound. But at the end of the month the peanuts had not sold well, so, in order to sell 60 pounds of peanuts, the manager decided to mix the 60 pounds of peanuts with some cashews and sell the mixture for $2.50 per pound. How many pounds of cashews should be mixed with the peanuts to ensure no change in the profit?
30. **Business: Mixing Candy** A candy store sells boxes of candy containing caramels and cremes. Each box sells for $12.50 and holds 30 pieces of candy (all pieces are the same size). If the caramels cost $0.25 to produce and the cremes cost $0.45 to produce, how many of each should be in a box to make a profit of $3?
31. **Physics: Uniform Motion** A motorboat can maintain a constant speed of 16 miles per hour relative to the water. The boat makes a trip upstream to a certain point in 20 minutes; the return trip takes 15 minutes. What is the speed of the current? (See the figure.)

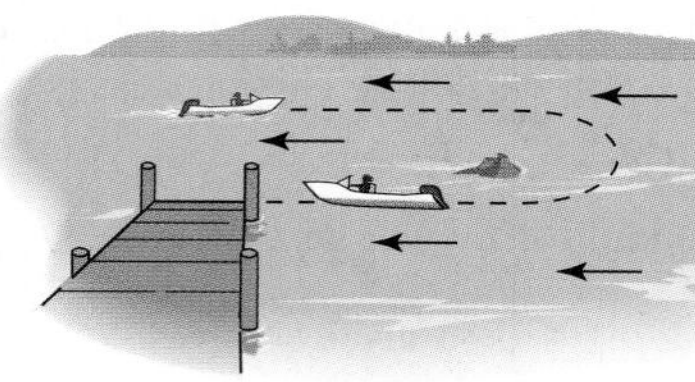

32. Physics: Uniform Motion A motorboat heads upstream on a river that has a current of 3 miles per hour. The trip upstream takes 5 hours, and the return trip takes 2.5 hours. What is the speed of the motorboat? (Assume that the motorboat maintains a constant speed relative to the water.)

33. Physics: Uniform Motion A Metra commuter train leaves Union Station in Chicago at 12 noon. Two hours later, an Amtrak train leaves on the same track, traveling at an average speed that is 50 miles per hour faster than the Metra train. At 3 PM the Amtrak train is 10 miles behind the commuter train. How fast is each going?

34. Physics: Uniform Motion Two cars enter the Florida Turnpike at Commercial Boulevard at 8:00 AM, each heading for Wildwood. One car's average speed is 10 miles per hour more than the other's. The faster car arrives at Wildwood at 11:00 AM, $\frac{1}{2}$ hour before the other car. What is the average speed of each car? How far did each travel?

35. Working Together on a Job Trent can deliver his newspapers in 30 minutes. It takes Lois 20 minutes to do the same route. How long would it take them to deliver the newspapers if they work together?

36. Working Together on a Job Patrice, by himself, can paint four rooms in 10 hours. If he hires April to help, they can do the same job together in 6 hours. If he lets April work alone, how long will it take her to paint four rooms?

37. Constructing a Box An open box is to be constructed from a square piece of sheet metal by removing a square of side 1 foot from each corner and turning up the edges. If the box is to hold 4 cubic feet, what should be the dimension of the sheet metal?

38. Constructing a Box Rework Problem 37 if the piece of sheet metal is a rectangle whose length is twice its width.

39. Physics: Uniform Motion A motorboat maintained a constant speed of 15 miles per hour relative to the water in going 10 miles upstream and then returning. The total time for the trip was 1.5 hours. Use this information to find the speed of the current.

40. Dimension of a Patio A contractor orders 8 cubic yards of premixed cement, all of which is to be used to pour a rectangular patio that will be 4 inches thick. If the length of the patio is specified to be twice the width, what will be the patio dimension? (1 cubic yard = 27 cubic feet)

41. Dimensions of a Window The area of the opening of a rectangular window is to be 143 square feet. If the length is to be 2 feet more than the width, what are the dimensions?

42. Dimensions of a Window The area of a rectangular window is to be 306 square centimeters. If the length exceeds the width by 1 centimeter, what are the dimensions?

43. Geometry Find the dimensions of a rectangle whose perimeter is 26 meters and whose area is 40 square meters.

44. Watering a Field An adjustable water sprinkler that sprays water in a circular pattern is placed at the center of a square field whose area is 1250 square feet (see the figure). What is the shortest radius setting that can be used if the field is to be completely enclosed within the circle?

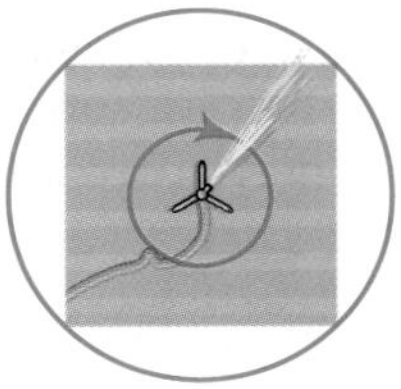

45. Physics A ball is thrown vertically upward from the top of a building 96 feet tall with an initial velocity of 80 feet per second. The distance s (in feet) of the ball from the ground after t seconds is $s = 96 + 80t - 16t^2$.
(a) After how many seconds does the ball strike the ground?
(b) After how many seconds will the ball pass the top of the building on its way down?

46. Physics An object is propelled vertically upward with an initial velocity of 20 meters per second. The distance s (in meters) of the object from the ground after t seconds is $s = -4.9t^2 + 20t$.
(a) When will the object be 15 meters above the ground?
(b) When will it strike the ground?
(c) Will the object reach a height of 100 meters?
(d) What is the maximum height?

47. Enclosing a Garden A gardener has 46 feet of fencing to be used to enclose a rectangular garden that has a border 2 feet wide surrounding it (see the figure).
(a) If the length of the garden is to be twice its width, what will be the dimensions of the garden?
(b) What is the area of the garden?
(c) If the length and width of the garden are to be the same, what would be the dimensions of the garden?
(d) What would be the area of the square garden?

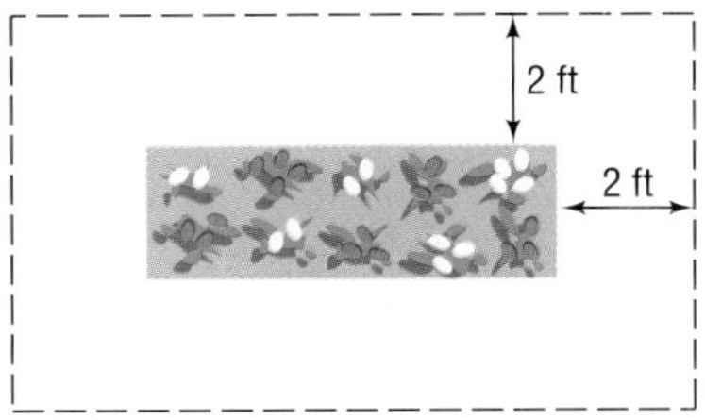

48. Construction A pond is enclosed by a wooden deck that is 3 feet wide. The fence surrounding the deck is 100 feet long.
(a) If the pond is square, what are its dimensions?
(b) If the pond is rectangular and the length of the pond is to be three times its width, what are its dimensions?
(c) If the pond is circular, what is its diameter?
(d) Which pond has the most area?

49. Constructing a Border around a Pool A pool in the shape of a circle measures 10 feet across. One cubic yard of concrete is to be used to create a circular border of uniform width around the pool. If the border is to have a depth of 3 inches, how wide will the border be? (1 cubic yard = 27 cubic feet)

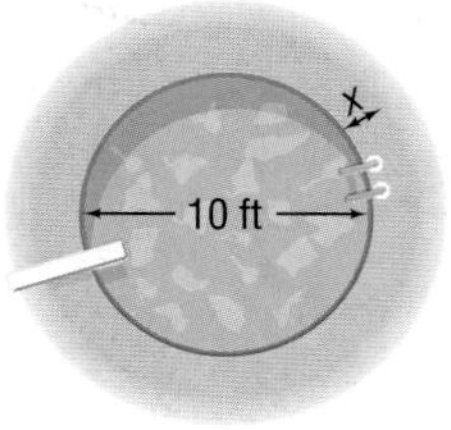

50. Constructing a Border around a Pool Rework Problem 49 if the depth of the border is 4 inches.

51. Constructing a Border around a Garden A landscaper, who just completed a rectangular flower garden measuring 6 feet by 10 feet, orders 1 cubic yard of premixed cement, all of which is to be used to create a border of uniform width around the garden. If the border is to have a depth of 3 inches, how wide will the border be? (1 cubic yard = 27 cubic feet)

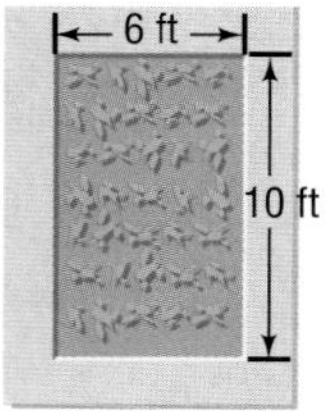

52. Constructing a Coffee Can A 39 ounce can of Hill Bros.® coffee requires 188.5 square inches of aluminum. If its height is 7 inches, what is its radius? (The surface area S of a right cylinder is $S = 2\pi r^2 + 2\pi r h$, where r is the radius and h is the height.)

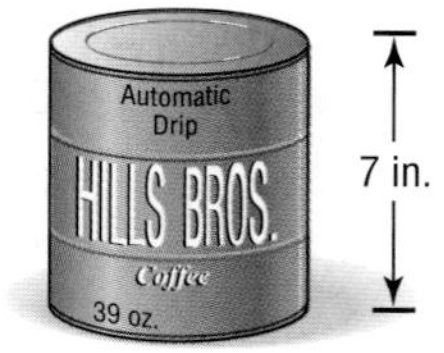

53. Mixing Water and Antifreeze How much water should be added to 1 gallon of pure antifreeze to obtain a solution that is 60% antifreeze?

54. Mixing Water and Antifreeze The cooling system of a certain foreign-made car has a capacity of 15 liters. If the system is filled with a mixture that is 40% antifreeze, how much of this mixture should be drained and replaced by pure antifreeze so that the system is filled with a solution that is 60% antifreeze?

55. Chemistry: Salt Solutions How much water must be evaporated from 32 ounces of a 4% salt solution to make a 6% salt solution?

56. Chemistry: Salt Solutions How much water must be evaporated from 240 gallons of a 3% salt solution to produce a 5% salt solution?

57. Purity of Gold The purity of gold is measured in karats, with pure gold being 24 karats. Other purities of gold are expressed as proportional parts of pure gold. Thus, 18 karat gold is $\frac{18}{24}$, or 75% pure gold; 12 karat gold is $\frac{12}{24}$, or 50% pure gold; and so on. How much 12 karat gold should be mixed with pure gold to obtain 60 grams of 16 karat gold?

58. Chemistry: Sugar Molecules A sugar molecule has twice as many atoms of hydrogen as it does oxygen and one more atom of carbon than oxygen. If a sugar molecule has a total of 45 atoms, how many are oxygen? How many are hydrogen?

59. Running a Race Mike can run the mile in 6 minutes, and Dan can run the mile in 9 minutes. If Mike gives Dan a head start of 1 minute, how far from the start will Mike pass Dan? (See the figure.) How long does it take?

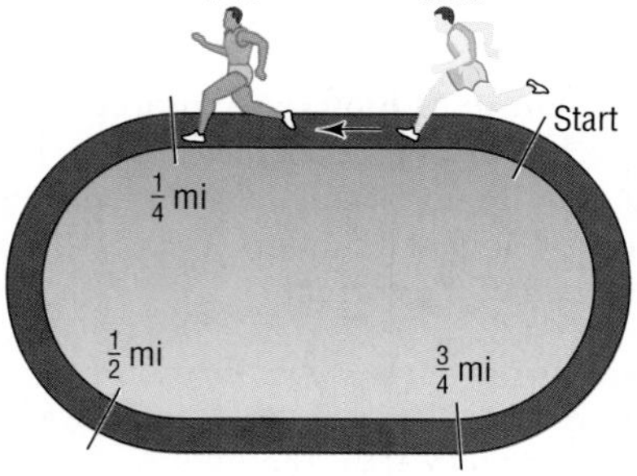

60. Range of an Airplane An air rescue plane averages 300 miles per hour in still air. It carries enough fuel for 5 hours of flying time. If, upon takeoff, it encounters a head wind of 30 mi/hr, how far can it fly and return safely? (Assume that the wind remains constant.)

61. Emptying Oil Tankers An oil tanker can be emptied by the main pump in 4 hours. An auxiliary pump can empty the tanker in 9 hours. If the main pump is started at 9 AM, when should the auxiliary pump be started so that the tanker is emptied by noon?

62. Cement Mix A 20-pound bag of Economy brand cement mix contains 25% cement and 75% sand. How much pure cement must be added to produce a cement mix that is 40% cement?

63. Filling a Tub A bathroom tub will fill in 15 minutes with both faucets open and the stopper in place. With both faucets closed and the stopper removed, the tub will empty in 20 minutes. How long will it take for the tub to fill if both faucets are open and the stopper is removed?

64. Using Two Pumps A 5 horsepower (hp) pump can empty a pool in 5 hours. A smaller, 2 hp pump empties the same pool in 8 hours. The pumps are used together to begin emptying this pool. After two hours, the 2 hp pump breaks down. How long will it take the larger pump to empty the pool?

65. Comparing Olympic Heroes In the 1984 Olympics, Carl Lewis of the United States won the gold medal in the 100 meter race with a time of 9.99 seconds. In the 1896 Olympics, Thomas Burke, also of the United States, won the gold medal in the 100 meter race in 12.0 seconds. If they ran in the same race repeating their respective times, by how many meters would Lewis beat Burke?

66. Football A tight end can run the 100 yard dash in 12 seconds. A defensive back can do it in 10 seconds. The tight end catches a pass at his own 20 yard line with the defensive back at the 15 yard line. (See the figure.) If no other players are nearby, at what yard line will the defensive back catch up to the tight end?

[**Hint:** At time $t = 0$, the defensive back is 5 yards behind the tight end.]

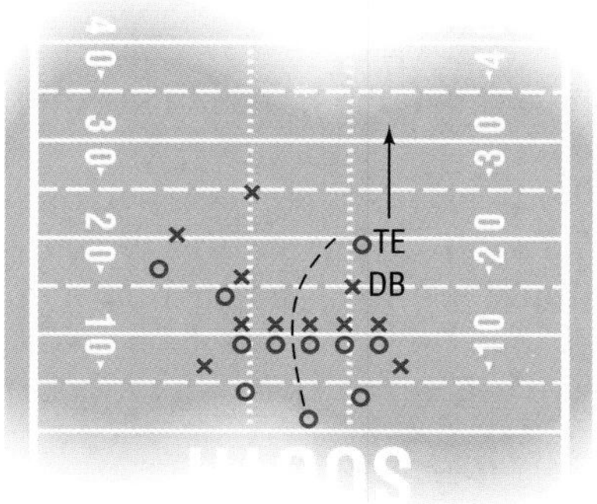

67. Computing Business Expense Therese, an outside sales-person, uses her car for both business and pleasure. Last year, she traveled 30,000 miles, using 900 gallons of gasoline. Her car gets 40 miles per gallon on the highway and 25 in the city. She can deduct all highway travel, but no city travel, on her taxes. How many miles should Therese be allowed as a business expense?

68. Summing Consecutive Integers The sum of the consecutive integers $1, 2, 3, \dots, n$ is given by the formula $\frac{1}{2}n(n + 1)$. How many consecutive integers, starting with 1, must be added to get a sum of 666?

69. Geometry If a polygon of n sides has $\frac{1}{2}n(n - 3)$ diagonals, how many sides will a polygon with 65 diagonals have? Is there a polygon with 80 diagonals?

70. Computing Average Speed In going from Chicago to Atlanta, a car averages 45 miles per hour, and in going from Atlanta to Miami, it averages 55 miles per hour. If Atlanta is halfway between Chicago and Miami, what is the average speed from Chicago to Miami? Discuss an intuitive solution. Write a paragraph defending your intuitive solution. Then solve the problem algebraically. Is your intuitive solution the same as the algebraic one? If not, find the flaw.

71. Speed of a Plane On a recent flight from Phoenix to Kansas City, a distance of 919 nautical miles, the plane arrived 20 minutes early. On leaving the aircraft, I asked the captain, "What was our tail wind?" He replied, "I don't know, but our ground speed was 550 knots." How can you determine if enough information is provided to find the tail wind? If possible, find the tail wind. (1 knot = 1 nautical mile per hour)

72. Critical Thinking You are the manager of a clothing store and have just purchased 100 dress shirts for \$20.00 each. After 1 month of selling the shirts at the regular price, you plan to have a sale giving 40% off the original selling price. However, you still want to make a profit of \$4 on each shirt at the sale price. What should you price the shirts at initially to ensure this? If, instead of 40% off at the sale, you give 50% off, by how much is your profit reduced?

73. Critical Thinking Make up a word problem that requires solving a linear equation as part of its solution. Exchange problems with a friend. Write a critique of your friend's problem.

74. Critical Thinking Without solving, explain what is wrong with the following mixture problem: How many liters of 25% ethanol should be added to 20 liters of 48% ethanol to obtain a solution of 58% ethanol? Now go through an algebraic solution. What happens?

A.8 Interval Notation; Solving Inequalities

PREPARING FOR THIS SECTION *Before getting started, review the following:*

- Real Number Line, Inequalities, Absolute Value (Appendix A, Section A.1, pp. 906–908)

Now work the 'Are You Prepared?' problems on page 974.

OBJECTIVES
1. Use Interval Notation
2. Use Properties of Inequalities
3. Solve Inequalities
4. Solve Combined Inequalities
5. Solve Inequalities Involving Absolute Value

Suppose that a and b are two real numbers and $a < b$. We shall use the notation $a < x < b$ to mean that x is a number *between* a and b. The expression $a < x < b$ is equivalent to the two inequalities $a < x$ and $x < b$. Similarly, the expression $a \le x \le b$ is equivalent to the two inequalities $a \le x$ and $x \le b$. The remaining two possibilities, $a \le x < b$ and $a < x \le b$, are defined similarly.

Although it is acceptable to write $3 \ge x \ge 2$, it is preferable to reverse the inequality symbols and write instead $2 \le x \le 3$ so that, as you read from left to right, the values go from smaller to larger.

A statement such as $2 \le x \le 1$ is false because there is no number x for which $2 \le x$ and $x \le 1$. Finally, we never mix inequality symbols, as in $2 \le x \ge 3$.

Intervals

1 Let a and b represent two real numbers with $a < b$.

> A **closed interval,** denoted by $[a, b]$, consists of all real numbers x for which $a \le x \le b$.
>
> An **open interval,** denoted by (a, b), consists of all real numbers x for which $a < x < b$.
>
> The **half-open,** or **half-closed, intervals** are $(a, b]$, consisting of all real numbers x for which $a < x \le b$, and $[a, b)$, consisting of all real numbers x for which $a \le x < b$.

In each of these definitions, a is called the **left endpoint** and b the **right endpoint** of the interval.

The symbol ∞ (read as "infinity") is not a real number, but a notational device used to indicate unboundedness in the positive direction. The symbol $-\infty$ (read as "negative infinity") also is not a real number, but a notational device used to indicate unboundedness in the negative direction. Using the symbols ∞ and $-\infty$, we can define five other kinds of intervals:

$[a, \infty)$	Consists of all real numbers x for which $x \ge a$ $(a \le x < \infty)$
(a, ∞)	Consists of all real numbers x for which $x > a$ $(a < x < \infty)$
$(-\infty, a]$	Consists of all real numbers x for which $x \le a$ $(-\infty < x \le a)$
$(-\infty, a)$	Consists of all real numbers x for which $x < a$ $(-\infty < x < a)$
$(-\infty, \infty)$	Consists of all real numbers x $(-\infty < x < \infty)$

Note that ∞ and $-\infty$ are never included as endpoints, since neither is a real number.

Table 2 summarizes interval notation, corresponding inequality notation, and their graphs.

Table 2

Interval	Inequality	Graph
The open interval (a, b)	$a < x < b$	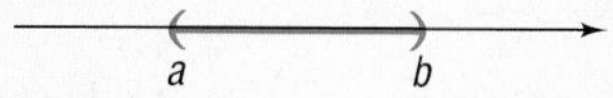
The closed interval $[a, b]$	$a \leq x \leq b$	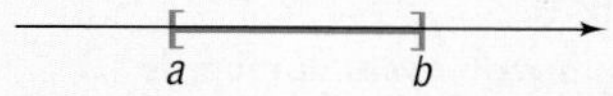
The half-open interval $[a, b)$	$a \leq x < b$	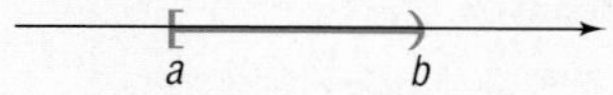
The half-open interval $(a, b]$	$a < x \leq b$	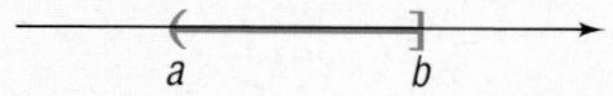
The interval $[a, \infty)$	$x \geq a$	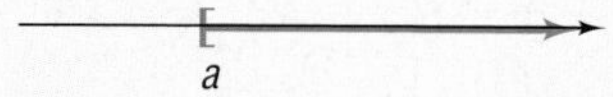
The interval (a, ∞)	$x > a$	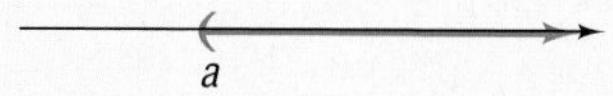
The interval $(-\infty, a]$	$x \leq a$	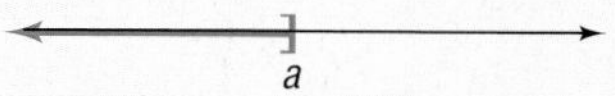
The interval $(-\infty, a)$	$x < a$	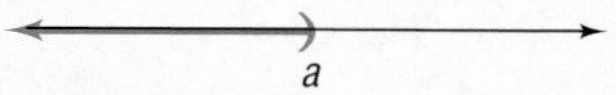
The interval $(-\infty, \infty)$	All real numbers	

EXAMPLE 1 Writing Inequalities Using Interval Notation

Write each inequality using interval notation.

(a) $1 \leq x \leq 3$ (b) $-4 < x < 0$ (c) $x > 5$ (d) $x \leq 1$

Solution

(a) $1 \leq x \leq 3$ describes all numbers x between 1 and 3, inclusive. In interval notation, we write $[1, 3]$.

(b) In interval notation, $-4 < x < 0$ is written $(-4, 0)$.

(c) $x > 5$ consists of all numbers x greater than 5. In interval notation, we write $(5, \infty)$.

(d) In interval notation, $x \leq 1$ is written $(-\infty, 1]$. ◀

EXAMPLE 2 Writing Intervals Using Inequality Notation

Write each interval as an inequality involving x.

(a) $[1, 4)$ (b) $(2, \infty)$ (c) $[2, 3]$ (d) $(-\infty, -3]$

Solution

(a) $[1, 4)$ consists of all numbers x for which $1 \leq x < 4$.

(b) $(2, \infty)$ consists of all numbers x for which $x > 2$ $(2 < x < \infty)$.

(c) $[2, 3]$ consists of all numbers x for which $2 \leq x \leq 3$.

(d) $(-\infty, -3]$ consists of all numbers x for which $x \leq -3$ $(-\infty < x \leq -3)$. ◀

NOW WORK PROBLEMS 11, 17, AND 25.

Properties of Inequalities

2 The product of two positive real numbers is positive, the product of two negative real numbers is positive, and the product of 0 and 0 is 0. For any real number a, the value of a^2 is 0 or positive; that is, a^2 is nonnegative. This is called the **nonnegative property.**

For any real number a, we have the following:

In words
The square of a real number is never negative.

Nonnegative Property

$$a^2 \geq 0 \tag{1}$$

If we add the same number to both sides of an inequality, we obtain an equivalent inequality. For example, since $3 < 5$, then $3 + 4 < 5 + 4$ or $7 < 9$. This is called the **addition property** of inequalities.

Addition Property of Inequalities

If $a < b$, then $a + c < b + c$. **(2a)**
If $a > b$, then $a + c > b + c$. **(2b)**

The addition property states that the sense, or direction, of an inequality remains unchanged if the same number is added to each side. Now let's see what happens if we multiply each side of an inequality by a nonzero number.

Begin with $3 < 7$ and multiply each side by 2. The numbers 6 and 14 that result yield the inequality $6 < 14$.

Begin with $9 > 2$ and multiply each side by -4. The numbers -36 and -8 that result yield the inequality $-36 < -8$.

Note that the effect of multiplying both sides of $9 > 2$ by the negative number -4 is that the direction of the inequality symbol is reversed. We are led to the following general **multiplication properties** for inequalities:

In Words
Multiplying by a negative number reverses the inequality.

Multiplication Properties for Inequalities

If $a < b$ and if $c > 0$, then $ac < bc$.
If $a < b$ and if $c < 0$, then $ac > bc$. **(3a)**
If $a > b$ and if $c > 0$, then $ac > bc$.
If $a > b$ and if $c < 0$, then $ac < bc$. **(3b)**

The multiplication properties state that the sense, or direction, of an inequality *remains the same* if each side is multiplied by a *positive* real number, whereas the direction is *reversed* if each side is multiplied by a *negative* real number.

NOW WORK PROBLEMS 39 AND 45.

The **reciprocal property** states that the reciprocal of a positive real number is positive and that the reciprocal of a negative real number is negative.

Reciprocal Property for Inequalities

If $a > 0$, then $\frac{1}{a} > 0$. **(4a)**

If $a < 0$, then $\frac{1}{a} < 0$. **(4b)**

Solving Inequalities

3 An **inequality in one variable** is a statement involving two expressions, at least one containing the variable, separated by one of the inequality symbols $<, \leq, >$, or $\geq$. To **solve an inequality** means to find all values of the variable for which the statement is true. These values are called **solutions** of the inequality.

For example, the following are all inequalities involving one variable, x:

$$x + 5 < 8 \qquad 2x - 3 \geq 4 \qquad x^2 - 1 \leq 3 \qquad \frac{x + 1}{x - 2} > 0$$

Two inequalities having exactly the same solution set are called **equivalent inequalities.** As with equations, one method for solving an inequality is to replace it by a series of equivalent inequalities until an inequality with an obvious solution, such as $x < 3$, is obtained. We obtain equivalent inequalities by applying some of the same properties as those used to find equivalent equations. The addition property and the multiplication properties form the basis for the following procedures.

Procedures That Leave the Inequality Symbol Unchanged

1. Simplify both sides of the inequality by combining like terms and eliminating parentheses:
   Replace $(x + 2) + 6 > 2x + 5(x + 1)$
   by $x + 8 > 7x + 5$
2. Add or subtract the same expression on both sides of the inequality:
   Replace $3x - 5 < 4$
   by $(3x - 5) + 5 < 4 + 5$
3. Multiply or divide both sides of the inequality by the same positive expression:
   Replace $4x > 16$ by $\frac{4x}{4} > \frac{16}{4}$

Procedures That Reverse the Sense or Direction of the Inequality Symbol

1. Interchange the two sides of the inequality:
   Replace $3 < x$ by $x > 3$
2. Multiply or divide both sides of the inequality by the same *negative* expression.
   Replace $-2x > 6$ by $\frac{-2x}{-2} < \frac{6}{-2}$

As the examples that follow illustrate, we solve inequalities using many of the same steps that we would use to solve equations. In writing the solution of an inequality, we may use either set notation or interval notation, whichever is more convenient.

EXAMPLE 3 Solving an Inequality

Solve the inequality: $4x + 7 \geq 2x - 3$
Graph the solution set.

Solution

$$\begin{aligned} 4x + 7 &\geq 2x - 3 && \\ 4x + 7 - 7 &\geq 2x - 3 - 7 && \text{Subtract 7 from both sides.} \\ 4x &\geq 2x - 10 && \text{Simplify.} \\ 4x - 2x &\geq 2x - 10 - 2x && \text{Subtract } 2x \text{ from both sides.} \\ 2x &\geq -10 && \text{Simplify.} \\ \frac{2x}{2} &\geq \frac{-10}{2} && \text{Divide both sides by 2. (The sense of the inequality symbol is unchanged.)} \\ x &\geq -5 && \text{Simplify.} \end{aligned}$$

Figure 19

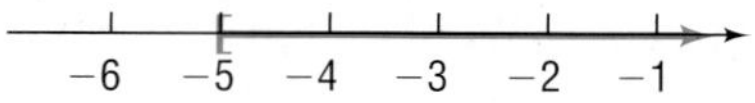

The solution set is $\{x | x \geq -5\}$ or, using interval notation, all numbers in the interval $[-5, \infty)$. See Figure 19 for the graph. ◀

NOW WORK PROBLEM 53.

4

EXAMPLE 4 Solving Combined Inequalities

Solve the inequality: $-5 < 3x - 2 < 1$
Graph the solution set.

Solution Recall that the inequality

$$-5 < 3x - 2 < 1$$

is equivalent to the two inequalities

$$-5 < 3x - 2 \quad \text{and} \quad 3x - 2 < 1$$

We will solve each of these inequalities separately.

$$\begin{aligned} -5 &< 3x - 2 && && 3x - 2 &< 1 \\ -5 + 2 &< 3x - 2 + 2 && \text{Add 2 to both sides.} && 3x - 2 + 2 &< 1 + 2 \\ -3 &< 3x && \text{Simplify.} && 3x &< 3 \\ \frac{-3}{3} &< \frac{3x}{3} && \text{Divide both sides by 3.} && \frac{3x}{3} &< \frac{3}{3} \\ -1 &< x && \text{Simplify.} && x &< 1 \end{aligned}$$

The solution set of the original pair of inequalities consists of all x for which

$$-1 < x \quad \text{and} \quad x < 1$$

Figure 20

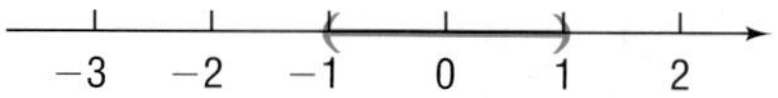

This may be written more compactly as $\{x | -1 < x < 1\}$. In interval notation, the solution is $(-1, 1)$. See Figure 20 for the graph. ◀

NOW WORK PROBLEM 73.

EXAMPLE 5 Using the Reciprocal Property to Solve an Inequality

Solve the inequality: $(4x - 1)^{-1} > 0$
Graph the solution set.

Solution Since $(4x - 1)^{-1} = \dfrac{1}{4x - 1}$ and since the Reciprocal Property states that when $\dfrac{1}{a} > 0$ then $a > 0$, we have

$$\begin{aligned} (4x - 1)^{-1} &> 0 \\ \frac{1}{4x - 1} &> 0 \\ 4x - 1 &> 0 \quad \text{Reciprocal Property} \\ 4x &> 1 \\ x &> \frac{1}{4} \end{aligned}$$

Figure 21

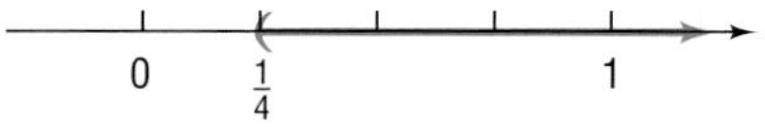

The solution set is $\left\{x \middle| x > \frac{1}{4}\right\}$, that is, all x in the interval $\left(\frac{1}{4}, \infty\right)$. Figure 21 illustrates the graph. ◀

NOW WORK PROBLEM 77.

5

EXAMPLE 6 Solving an Inequality Involving Absolute Value

Solve the inequality $|x| < 4$, and graph the solution set.

Figure 22

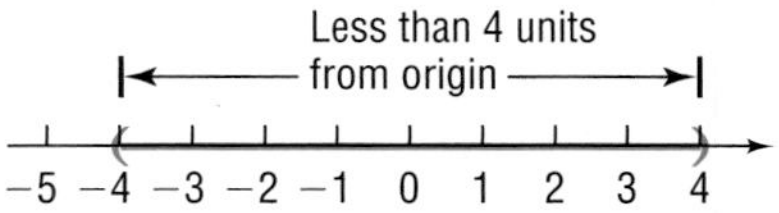

Solution We are looking for all points whose coordinate x is a distance less than 4 units from the origin. See Figure 22 for an illustration. Because any x between -4 and 4 satisfies the condition $|x| < 4$, the solution set consists of all numbers x for which $-4 < x < 4$, that is, all x in $(-4, 4)$. ◀

EXAMPLE 7 Solving an Inequality Involving Absolute Value

Solve the inequality $|x| > 3$, and graph the solution set.

Figure 23

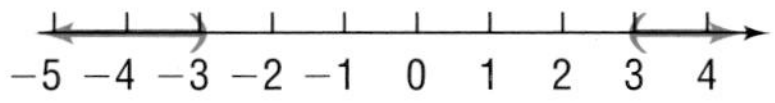

Solution We are looking for all points whose coordinate x is a distance greater than 3 units from the origin. Figure 23 illustrates the situation. We conclude that any x less than -3 or greater than 3 satisfies the condition $|x| > 3$. Consequently, the solution set consists of all numbers x for which $x < -3$ or $x > 3$, that is, all x in the interval $(-\infty, -3)$ or in the interval $(3, \infty)$. ◀

We are led to the following results:

Theorem If a is any positive number, then

$\lvert u\rvert < a$	is equivalent to	$-a < u < a$	**(5)**
$\lvert u\rvert \le a$	is equivalent to	$-a \le u \le a$	**(6)**
$\lvert u\rvert > a$	is equivalent to	$u < -a$ or $u > a$	**(7)**
$\lvert u\rvert \ge a$	is equivalent to	$u \le -a$ or $u \ge a$	**(8)**

EXAMPLE 8 Solving an Inequality Involving Absolute Value

Solve the inequality $|2x + 4| \leq 3$, and graph the solution set.

Solution

$$|2x+4| \leq 3$$
This follows the form of equation (6); the expression $u = 2x + 4$ is inside the absolute value bars.

$$-3 \leq 2x + 4 \leq 3$$
Apply equation (6).

$$-3 - 4 \leq 2x + 4 - 4 \leq 3 - 4$$
Subtract 4 from each part.

$$-7 \leq 2x \leq -1$$
Simplify.

$$\frac{-7}{2} \leq \frac{2x}{2} \leq \frac{-1}{2}$$
Divide each part by 2.

$$-\frac{7}{2} \leq x \leq -\frac{1}{2}$$
Simplify.

The solution set is $\left\{x \,\middle|\, -\frac{7}{2} \leq x \leq -\frac{1}{2}\right\}$, that is, all x in the interval $\left[-\frac{7}{2}, -\frac{1}{2}\right]$. See Figure 24. ◀

Figure 24

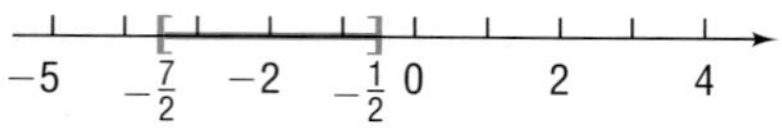

NOW WORK PROBLEM 81.

EXAMPLE 9 Solving an Inequality Involving Absolute Value

Solve the inequality $|2x - 5| > 3$, and graph the solution set.

Solution

$$|2x - 5| > 3$$
This follows the form of equation (7); the expression $u = 2x - 5$ is inside the absolute value bars.

$$2x - 5 < -3 \quad \text{or} \quad 2x - 5 > 3$$
Apply equation (7).

$$2x - 5 + 5 < -3 + 5 \quad \text{or} \quad 2x - 5 + 5 > 3 + 5$$
Add 5 to each part.

$$2x < 2 \quad \text{or} \quad 2x > 8$$
Simplify.

$$\frac{2x}{2} < \frac{2}{2} \quad \text{or} \quad \frac{2x}{2} > \frac{8}{2}$$
Divide each part by 2.

$$x < 1 \quad \text{or} \quad x > 4$$
Simplify.

Figure 25

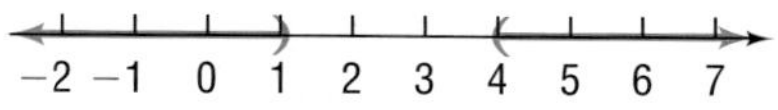

The solution set is $\{x | x < 1 \text{ or } x > 4\}$, that is, all x in the interval $(-\infty, 1)$ or in the interval $(4, \infty)$. See Figure 25. ◀

WARNING A common error to be avoided is to attempt to write the solution $x < 1$ or $x > 4$ as the combined inequality $1 > x > 4$, which is incorrect, since there are no numbers x for which $x < 1$ *and* $x > 4$. Another common error is to "mix" the symbols and write $1 < x > 4$, which makes no sense. ■

NOW WORK PROBLEM 87.

A.8 Assess Your Understanding

'Are You Prepared?' *Answers are given at the end of these exercises. If you get a wrong answer, read the pages listed in red.*

1. Graph the inequality $x \geq -2$. (pp. 906–908)

2. *True or False:* $-5 > -3$. (pp. 906–908)

3. $|-2| =$ ________. (pp. 906–908)

4. *True or False:* $|x| \geq 0$ for any real number x. (pp. 906–908)

Concepts and Vocabulary

5. If each side of an inequality is multiplied by a(n) __________ number, then the sense of the inequality symbol is reversed.

6. A(n) __________ __________, denoted $[a, b]$, consists of all real number x for which $a \le x \le b$.

7. The solution set of the equation $|x| = 5$ is {__________}.

8. The solution set of the inequality $|x| < 5$ is $\{x|$__________$\}$

9. *True or False:* The equation $|x| = -2$ has no solution.

10. *True or False:* The inequality $|x| \ge -2$ has the set of real numbers as solution set.

Exercises

In Problems 11–16, express the graph shown in color using interval notation. Also express each as an inequality involving x.

11. 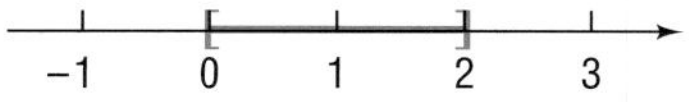

12. 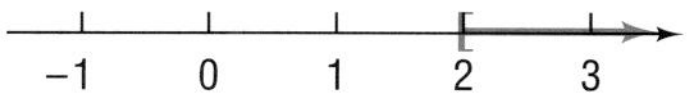

13. 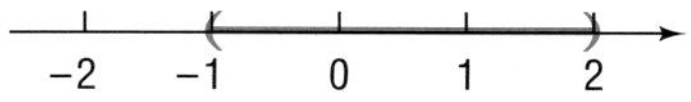

14. 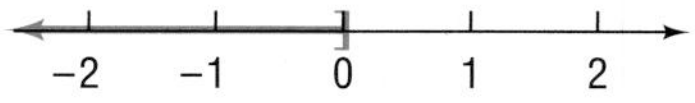

15. 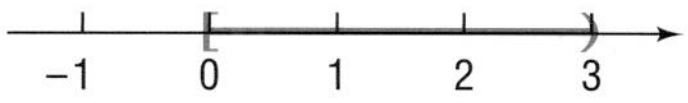

16. 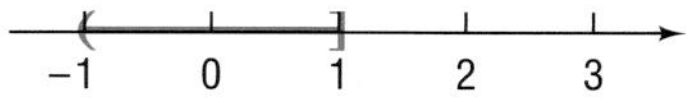

In Problems 17–24, write each inequality using interval notation, and illustrate each inequality using the real number line.

17. $0 \le x \le 4$ **18.** $-1 < x < 5$ **19.** $4 \le x < 6$ **20.** $-2 < x < 0$

21. $x \ge 4$ **22.** $x \le 5$ **23.** $x < -4$ **24.** $x > 1$

In Problems 25–32, write each interval as an inequality involving x, and illustrate each inequality using the real number line.

25. $[2, 5]$ **26.** $(1, 2)$ **27.** $(-3, -2)$ **28.** $[0, 1)$

29. $[4, \infty)$ **30.** $(-\infty, 2]$ **31.** $(-\infty, -3)$ **32.** $(-8, \infty)$

In Problems 33–38, an inequality is given. Write the inequality obtained by:
(a) *Adding 3 to each side of the given inequality.*
(b) *Subtracting 5 from each side of the given inequality.*
(c) *Multiplying each side of the given inequality by 3.*
(d) *Multiplying each side of the given inequality by −2.*

33. $3 < 5$ **34.** $2 > 1$ **35.** $4 > -3$

36. $-3 > -5$ **37.** $2x + 1 < 2$ **38.** $1 - 2x > 5$

In Problems 39–52, fill in the blank with the correct inequality symbol.

39. If $x < 5$, then $x - 5$ __________ 0.

40. If $x < -4$, then $x + 4$ __________ 0.

41. If $x > -4$, then $x + 4$ __________ 0.

42. If $x > 6$, then $x - 6$ __________ 0.

43. If $x \ge -4$, then $3x$ __________ -12.

44. If $x \le 3$, then $2x$ __________ 6.

45. If $x > 6$, then $-2x$ __________ -12.

46. If $x > -2$, then $-4x$ __________ 8.

47. If $x \ge 5$, then $-4x$ __________ -20.

48. If $x \le -4$, then $-3x$ __________ 12.

49. If $2x > 6$, then x __________ 3.

50. If $3x \le 12$, then x __________ 4.

51. If $-\frac{1}{2}x \le 3$, then x __________ -6.

52. If $-\frac{1}{4}x > 1$, then x __________ -4.

In Problems 53–92, solve each inequality. Express your answer using set notation or interval notation. Graph the solution set.

53. $x + 1 < 5$ **54.** $x - 6 < 1$ **55.** $1 - 2x \le 3$ **56.** $2 - 3x \le 5$

57. $3x - 7 > 2$ **58.** $2x + 5 > 1$ **59.** $3x - 1 \ge 3 + x$ **60.** $2x - 2 \ge 3 + x$

61. $-2(x + 3) < 8$ **62.** $-3(1 - x) < 12$ **63.** $4 - 3(1 - x) \le 3$ **64.** $8 - 4(2 - x) \le -2x$

65. $\frac{1}{2}(x - 4) > x + 8$ **66.** $3x + 4 > \frac{1}{3}(x - 2)$ **67.** $\frac{x}{2} \ge 1 - \frac{x}{4}$ **68.** $\frac{x}{3} \ge 2 + \frac{x}{6}$

69. $0 \le 2x - 6 \le 4$ **70.** $4 \le 2x + 2 \le 10$ **71.** $-5 \le 4 - 3x \le 2$ **72.** $-3 \le 3 - 2x \le 9$

73. $-3 < \frac{2x - 1}{4} < 0$ **74.** $0 < \frac{3x + 2}{2} < 4$ **75.** $1 < 1 - \frac{1}{2}x < 4$ **76.** $0 < 1 - \frac{1}{3}x < 1$

77. $(4x + 2)^{-1} < 0$
78. $(2x - 1)^{-1} > 0$
79. $0 < \frac{2}{x} < \frac{3}{5}$
80. $0 < \frac{4}{x} < \frac{2}{3}$
81. $|2x| < 8$
82. $|3x| < 12$
83. $|3x| > 12$
84. $|2x| > 6$
85. $|2x - 1| \le 1$
86. $|2x + 5| \le 7$
87. $|1 - 2x| > 3$
88. $|2 - 3x| > 1$
89. $|-4x| + |-5| \le 9$
90. $|-x| - |4| \le 2$
91. $|-2x| \ge |-4|$
92. $|-x - 2| \ge 1$

93. Express the fact that x differs from 2 by less than $\frac{1}{2}$ as an inequality involving an absolute value. Solve for x.
94. Express the fact that x differs from -1 by less than 1 as an inequality involving an absolute value. Solve for x.
95. Express the fact that x differs from -3 by more than 2 as an inequality involving an absolute value. Solve for x.
96. Express the fact that x differs from 2 by more than 3 as an inequality involving an absolute value. Solve for x.
97. A young adult may be defined as someone older than 21, but less than 30 years of age. Express this statement using inequalities.
98. Middle-aged may be defined as being 40 or more and less than 60. Express the statement using inequalities.
99. **Body Temperature** "Normal" human body temperature is 98.6°F. If a temperature x that differs from normal by at least 1.5° is considered unhealthy, write the condition for an unhealthy temperature x as an inequality involving an absolute value, and solve for x.
100. **Household Voltage** In the United States, normal household voltage is 115 volts. However, it is not uncommon for actual voltage to differ from normal voltage by at most 5 volts. Express this situation as an inequality involving an absolute value. Use x as the actual voltage and solve for x.
101. **Life Expectancy** Metropolitan Life Insurance Co. reported that an average 25-year-old male in 1996 could expect to live at least 48.4 more years and an average 25-year-old female in 1996 could expect to live at least 54.7 more years.

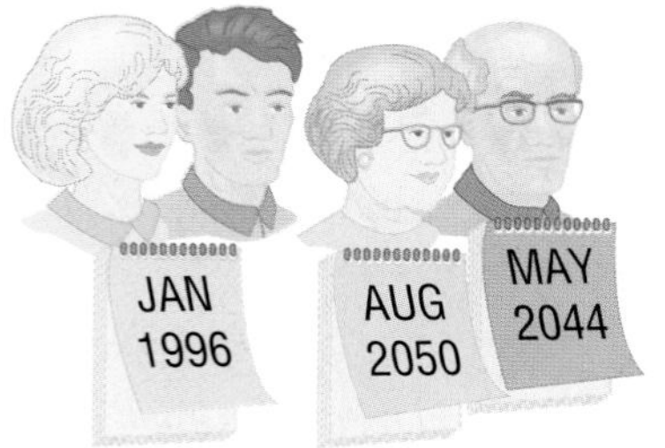

(a) To what age can an average 25-year-old male expect to live? Express your answer as an inequality.
(b) To what age can an average 25-year-old female expect to live? Express your answer as an inequality.
(c) Who can expect to live longer, a male or a female? By how many years?

102. **General Chemistry** For a certain ideal gas, the volume V (in cubic centimeters) equals 20 times the temperature T (in degrees Celsius). If the temperature varies from 80° to 120° C inclusive, what is the corresponding range of the volume of the gas?
103. **Real Estate** A real estate agent agrees to sell a large apartment complex according to the following commission schedule: \$45,000 plus 25% of the selling price in excess of \$900,000. Assuming that the complex will sell at some price between \$900,000 and \$1,100,000 inclusive, over what range does the agent's commission vary? How does the commission vary as a percent of selling price?
104. **Sales Commission** A used car salesperson is paid a commission of \$25 plus 40% of the selling price in excess of owner's cost. The owner claims that used cars typically sell for at least owner's cost plus \$70 and at most owner's cost plus \$300. For each sale made, over what range can the salesperson expect the commission to vary?
105. **Federal Tax Withholding** The percentage method of withholding for federal income tax (2004) states that a single person whose weekly wages, after subtracting withholding allowances, are over \$592, but not over \$1317, shall have \$74.35 plus 25% of the excess over \$592 withheld. Over what range does the amount withheld vary if the weekly wages vary from \$600 to \$800 inclusive? SOURCE: *Employer's Tax Guide*. Department of the Treasury, Internal Revenue Service, 2004.
106. **Federal Tax Withholding** Rework Problem 105 if the weekly wages vary from \$800 to \$1000 inclusive.
107. **Electricity Rates** Commonwealth Edison Company's charge for electricity in May 2003 is 8.275¢ per kilowatt-hour. In addition, each monthly bill contains a customer charge of \$7.58. If last summer's bills ranged from a low of \$63.47 to a high of \$214.53, over what range did usage vary (in kilowatt-hours)? SOURCE: Commonwealth Edison Co., Chicago, Illinois, 2003.
108. **Water Bills** The Village of Oak Lawn charges homeowners \$27.18 per quarter-year plus \$1.79 per 1000 gallons for water usage in excess of 12,000 gallons. In 2003, one homeowner's quarterly bill ranged from a high of \$76.52 to a low of \$34.78. Over what range did water usage vary? SOURCE: Village of Oak Lawn, Illinois, 2003.
109. **Markup of a New Car** The markup over dealer's cost of a new car ranges from 12% to 18%. If the sticker price is \$8800, over what range will the dealer's cost vary?
110. **IQ Tests** A standard intelligence test has an average score of 100. According to statistical theory, of the people who take the test, the 2.5% with the highest scores

will have scores of more than 1.96σ above the average, where σ (sigma, a number called the **standard deviation**) depends on the nature of the test. If $\sigma = 12$ for this test and there is (in principle) no upper limit to the score possible on the test, write the interval of possible test scores of the people in the top 2.5%.

111. Computing Grades In your Economics 101 class, you have scores of 68, 82, 87, and 89 on the first four of five tests. To get a grade of B, the average of the first five test scores must be greater than or equal to 80 and less than 90. Solve an inequality to find the range of the score that you need on the last test to get a B.

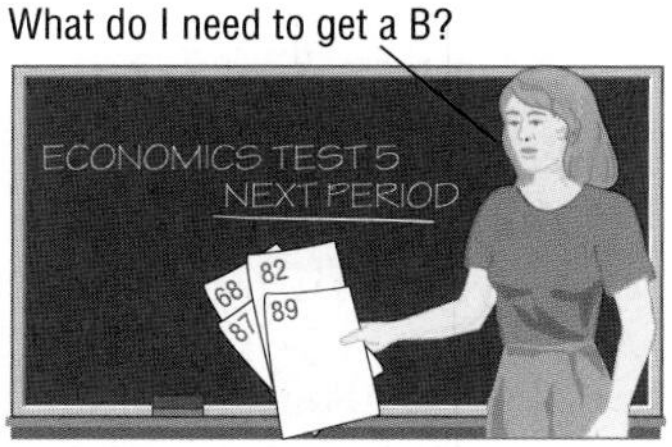

112. Computing Grades Repeat Problem 111 if the fifth test counts double.

113. Arithmetic Mean If $a < b$, show that $a < \frac{a+b}{2} < b$. The number $\frac{a+b}{2}$ is called the **arithmetic mean** of a and b.

114. Refer to Problem 113. Show that the arithmetic mean of a and b is equidistant from a and b.

115. Geometric Mean If $0 < a < b$, show that $a < \sqrt{ab} < b$. The number $\sqrt{ab}$ is called the **geometric mean** of a and b.

116. Refer to Problems 113 and 115. Show that the geometric mean of a and b is less than the arithmetic mean of a and b.

117. Harmonic Mean For $0 < a < b$, let h be defined by

$$\frac{1}{h} = \frac{1}{2}\left(\frac{1}{a} + \frac{1}{b}\right)$$

Show that $a < h < b$. The number h is called the **harmonic mean** of a and b.

118. Refer to Problems 113, 115, and 117. Show that the harmonic mean of a and b equals the geometric mean squared divided by the arithmetic mean.

119. Make up an inequality that has no solution. Make up one that has exactly one solution.

120. The inequality $x^2 + 1 < -5$ has no solution. Explain why.

121. Do you prefer to use inequality notation or interval notation to express the solution to an inequality? Give your reasons. Are there particular circumstances when you prefer one to the other? Cite examples.

122. How would you explain to a fellow student the underlying reason for the multiplication properties for inequalities (page 970); that is, the sense or direction of an inequality remains the same if each side is multiplied by a positive real number, whereas the direction is reversed if each side is multiplied by a negative real number.

'Are You Prepared?' Answers

1. 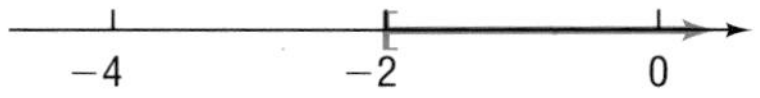

2. False

3. 2

4. True

A.9 *n*th Roots; Rational Exponents; Radical Equations

PREPARING FOR THIS SECTION *Before getting started, review the following:*

- Exponents, Square Roots (Appendix A, Section A.1, pp. 908–910)

Now work the 'Are You Prepared?' problems on page 982.

OBJECTIVES
1 Work with *n*th Roots
2 Simplify Radicals
3 Rationalize Denominators
4 Solve Radical Equations
5 Simplify Expressions with Rational Exponents

*n*th Roots

The **principal *n*th root of a number *a***, symbolized by $\sqrt[n]{a}$, where $n \geq 2$ is an integer, is defined as follows:

$$\sqrt[n]{a} = b \quad \text{means} \quad a = b^n$$

where $a \geq 0$ and $b \geq 0$ if $n \geq 2$ is even, and a, b are any real numbers if $n \geq 3$ is odd.

Notice that if a is negative and n is even then $\sqrt[n]{a}$ is not defined. When it is defined, the principal nth root of a number is unique.

1 The symbol $\sqrt[n]{a}$ for the principal nth root of a is sometimes called a **radical;** the integer n is called the **index,** and a is called the **radicand.** If the index of a radical is 2, we call $\sqrt[n]{a}$ the **square root** of a and omit the index 2 by simply writing $\sqrt{a}$. If the index is 3, we call $\sqrt[3]{a}$ the **cube root** of a.

EXAMPLE 1

Evaluating Principal *n*th Roots

(a) $\sqrt[3]{8} = \sqrt[3]{2^3} = 2$ (b) $\sqrt[3]{-64} = \sqrt[3]{(-4)^3} = -4$

(c) $\sqrt[4]{\frac{1}{16}} = \sqrt[4]{\left(\frac{1}{2}\right)^4} = \frac{1}{2}$ (d) $\sqrt[6]{(-2)^6} = |-2| = 2$ ◀

These are examples of **perfect roots,** since each simplifies to a rational number. Notice the absolute value in Example 1(d). If n is even, the principal nth root must be nonnegative.

In general, if $n \geq 2$ is a positive integer and a is a real number, we have

$$\sqrt[n]{a^n} = a, \quad \text{if } n \geq 3 \text{ is odd} \qquad \textbf{(1a)}$$

$$\sqrt[n]{a^n} = |a|, \quad \text{if } n \geq 2 \text{ is even} \qquad \textbf{(1b)}$$

NOW WORK PROBLEM 7.

Properties of Radicals

Let $n \geq 2$ and $m \geq 2$ denote positive integers, and let a and b represent real numbers. Assuming that all radicals are defined, we have the following properties:

$$\sqrt[n]{ab} = \sqrt[n]{a}\,\sqrt[n]{b} \qquad \textbf{(2a)}$$

$$\sqrt[n]{\frac{a}{b}} = \frac{\sqrt[n]{a}}{\sqrt[n]{b}} \qquad \textbf{(2b)}$$

$$\sqrt[n]{a^m} = (\sqrt[n]{a})^m \qquad \textbf{(2c)}$$

2 When used in reference to radicals, the direction to "simplify" will mean to remove from the radicals any perfect roots that occur as factors. Let's look at some examples of how the preceding rules are applied to simplify radicals.

EXAMPLE 2

Simplifying Radicals

(a) $\sqrt{32} = \sqrt{16 \cdot 2} = \sqrt{16} \cdot \sqrt{2} = 4\sqrt{2}$

↑ 16 is a perfect square.

(b) $\sqrt[3]{16} = \sqrt[3]{8 \cdot 2} = \sqrt[3]{8} \cdot \sqrt[3]{2} = 2\sqrt[3]{2}$

↑ 8 is a perfect cube. ↑ (2a)

(c) $\sqrt[3]{-16x^4} = \sqrt[3]{-8 \cdot 2 \cdot x^3 \cdot x} = \sqrt[3]{(-8x^3)(2x)}$

↑ Factor perfect cubes inside radical. ↑ Combine perfect cubes.

$= \sqrt[3]{(-2x)^3 \cdot 2x} = \sqrt[3]{(-2x)^3} \cdot \sqrt[3]{2x}$

↑ (2a)

$= -2x\sqrt[3]{2x}$ ◀

NOW WORK PROBLEM 13.

EXAMPLE 3 **Combining Like Radicals**

(a) $-8\sqrt{12} + \sqrt{3} = -8\sqrt{4 \cdot 3} + \sqrt{3} = -8 \cdot \sqrt{4}\sqrt{3} + \sqrt{3}$
$= -16\sqrt{3} + \sqrt{3} = -15\sqrt{3}$

(b) $\sqrt[3]{8x^4} + \sqrt[3]{-x} + 4\sqrt[3]{27x} = \sqrt[3]{2^3x^3x} + \sqrt[3]{-1 \cdot x} + 4\sqrt[3]{3^3x}$
$= \sqrt[3]{(2x)^3} \cdot \sqrt[3]{x} + \sqrt[3]{-1} \cdot \sqrt[3]{x} + 4\sqrt[3]{3^3} \cdot \sqrt[3]{x}$
$= 2x\sqrt[3]{x} - 1 \cdot \sqrt[3]{x} + 12\sqrt[3]{x}$
$= (2x + 11)\sqrt[3]{x}$ ◀

NOW WORK PROBLEM 31.

Rationalizing

3 When radicals occur in quotients, it is customary to rewrite the quotient so that the denominator contains no square roots. This process is referred to as **rationalizing the denominator.**

The idea is to multiply by an appropriate expression so that the new denominator contains no radicals. For example:

If Denominator Contains the Factor	Multiply by	To Obtain Denominator Free of Radicals
$\sqrt{3}$	$\sqrt{3}$	$(\sqrt{3})^2 = 3$
$\sqrt{3} + 1$	$\sqrt{3} - 1$	$(\sqrt{3})^2 - 1^2 = 3 - 1 = 2$
$\sqrt{2} - 3$	$\sqrt{2} + 3$	$(\sqrt{2})^2 - 3^2 = 2 - 9 = -7$
$\sqrt{5} - \sqrt{3}$	$\sqrt{5} + \sqrt{3}$	$(\sqrt{5})^2 - (\sqrt{3})^2 = 5 - 3 = 2$
$\sqrt[3]{4}$	$\sqrt[3]{2}$	$\sqrt[3]{4} \cdot \sqrt[3]{2} = \sqrt[3]{8} = 2$

In rationalizing the denominator of a quotient, be sure to multiply both the numerator and the denominator by the expression.

EXAMPLE 4 **Rationalizing Denominators**

Rationalize the denominator of each expression.

(a) $\dfrac{4}{\sqrt{2}}$ (b) $\dfrac{\sqrt{3}}{\sqrt[3]{2}}$ (c) $\dfrac{\sqrt{x} - 2}{\sqrt{x} + 2}, \quad x \geq 0$

Solution

(a) $$\frac{4}{\sqrt{2}} = \frac{4}{\sqrt{2}} \cdot \frac{\sqrt{2}}{\sqrt{2}} = \frac{4\sqrt{2}}{(\sqrt{2})^2} = \frac{4\sqrt{2}}{2} = 2\sqrt{2}$$

Multiply by $\frac{\sqrt{2}}{\sqrt{2}}$.

(b) $$\frac{\sqrt{3}}{\sqrt[3]{2}} = \frac{\sqrt{3}}{\sqrt[3]{2}} \cdot \frac{\sqrt[3]{4}}{\sqrt[3]{4}} = \frac{\sqrt{3}\sqrt[3]{4}}{\sqrt[3]{8}} = \frac{\sqrt{3}\sqrt[3]{4}}{2}$$

Multiply by $\frac{\sqrt[3]{4}}{\sqrt[3]{4}}$.

(c) $$\frac{\sqrt{x}-2}{\sqrt{x}+2} = \frac{\sqrt{x}-2}{\sqrt{x}+2} \cdot \frac{\sqrt{x}-2}{\sqrt{x}-2} = \frac{(\sqrt{x}-2)^2}{(\sqrt{x})^2 - 2^2}$$

$$= \frac{(\sqrt{x})^2 - 4\sqrt{x} + 4}{x-4} = \frac{x - 4\sqrt{x} + 4}{x-4}$$ ◀

NOW WORK PROBLEM 39.

Equations Containing Radicals

4 When the variable in an equation occurs in a square root, cube root, and so on, that is, when it occurs under a radical, the equation is called a **radical equation.** Sometimes a suitable operation will change a radical equation to one that is linear or quadratic. The most commonly used procedure is to isolate the most complicated radical on one side of the equation and then eliminate it by raising each side to a power equal to the index of the radical. Care must be taken, because extraneous solutions may result. Thus, when working with radical equations, we always check apparent solutions. Let's look at an example.

EXAMPLE 5 Solving Radical Equations

Solve the equation: $\sqrt[3]{2x-4} - 2 = 0$

Solution The equation contains a radical whose index is 3. We isolate it on the left side.

$$\sqrt[3]{2x-4} - 2 = 0$$
$$\sqrt[3]{2x-4} = 2$$

Now raise each side to the third power (since the index of the radical is 3) and solve.

$$\left(\sqrt[3]{2x-4}\right)^3 = 2^3$$ Raise each side to the 3rd power.
$$2x - 4 = 8$$ Simplify.
$$2x = 12$$ Solve for x.
$$x = 6$$

✔ **CHECK:** $\sqrt[3]{2(6)-4} - 2 = \sqrt[3]{12-4} - 2 = \sqrt[3]{8} - 2 = 2 - 2 = 0.$

The solution is $x = 6$. ◀

NOW WORK PROBLEM 47.

Rational Exponents

Radicals are used to define rational exponents.

If a is a real number and $n \geq 2$ is an integer, then

$$a^{1/n} = \sqrt[n]{a} \tag{3}$$

provided that $\sqrt[n]{a}$ exists.

Note that if n is even and $a < 0$, then $\sqrt[n]{a}$ and $a^{1/n}$ do not exist.

EXAMPLE 6 Using Equation (3)

(a) $4^{1/2} = \sqrt{4} = 2$ (b) $(-27)^{1/3} = \sqrt[3]{-27} = -3$

(c) $8^{1/2} = \sqrt{8} = 2\sqrt{2}$ (d) $16^{1/3} = \sqrt[3]{16} = 2\sqrt[3]{2}$ ◀

If a is a real number and m and n are integers containing no common factors with $n \geq 2$, then

$$a^{m/n} = \sqrt[n]{a^m} = (\sqrt[n]{a})^m \tag{4}$$

provided that $\sqrt[n]{a}$ exists.

We have two comments about equation (4):

1. The exponent $\dfrac{m}{n}$ must be in lowest terms and n must be positive.
2. In simplifying $a^{m/n}$, either $\sqrt[n]{a^m}$ or $(\sqrt[n]{a})^m$ may be used. Generally, taking the root first, as in $(\sqrt[n]{a})^m$, is easier.

EXAMPLE 7 Using Equation (4)

(a) $4^{3/2} = \left(\sqrt{4}\right)^3 = 2^3 = 8$ (b) $(-8)^{4/3} = \left(\sqrt[3]{-8}\right)^4 = (-2)^4 = 16$

(c) $(32)^{-2/5} = \left(\sqrt[5]{32}\right)^{-2} = 2^{-2} = \dfrac{1}{4}$ ◀

NOW WORK PROBLEMS 51 AND 57.

It can be shown that the laws of exponents hold for rational exponents.

5
EXAMPLE 8 Simplifying Expressions with Rational Exponents

Simplify each expression. Express your answer so that only positive exponents occur. Assume that the variables are positive.

(a) $\left(\dfrac{2x^{1/3}}{y^{2/3}}\right)^{-3}$ (b) $(x^{2/3}y)(x^{-2}y)^{1/2}$

Solution (a) $\left(\dfrac{2x^{1/3}}{y^{2/3}}\right)^{-3} = \left(\dfrac{y^{2/3}}{2x^{1/3}}\right)^3 = \dfrac{(y^{2/3})^3}{(2x^{1/3})^3} = \dfrac{y^2}{2^3(x^{1/3})^3} = \dfrac{y^2}{8x}$

(b) $(x^{2/3}y)(x^{-2}y)^{1/2} = (x^{2/3}y)[(x^{-2})^{1/2}y^{1/2}]$

$$= x^{2/3}yx^{-1}y^{1/2} = (x^{2/3}x^{-1})(y \cdot y^{1/2})$$

$$= x^{-1/3}y^{3/2} = \frac{y^{3/2}}{x^{1/3}}$$ ◀

NOW WORK PROBLEM 67.

The next two examples illustrate some algebra that you will need to know for certain calculus problems.

EXAMPLE 9 Writing an Expression as a Single Quotient

Write the following expression as a single quotient in which only positive exponents appear.

$$(x^2 + 1)^{1/2} + x \cdot \frac{1}{2}(x^2 + 1)^{-1/2} \cdot 2x$$

Solution

$$(x^2 + 1)^{1/2} + x \cdot \frac{1}{2}(x^2 + 1)^{-1/2} \cdot 2x = (x^2 + 1)^{1/2} + \frac{x^2}{(x^2 + 1)^{1/2}}$$

$$= \frac{(x^2 + 1)^{1/2}(x^2 + 1)^{1/2} + x^2}{(x^2 + 1)^{1/2}}$$

$$= \frac{(x^2 + 1) + x^2}{(x^2 + 1)^{1/2}}$$

$$= \frac{2x^2 + 1}{(x^2 + 1)^{1/2}}$$ ◀

NOW WORK PROBLEM 73.

EXAMPLE 10 Factoring an Expression Containing Rational Exponents

Factor: $4x^{1/3}(2x + 1) + 2x^{4/3}$

Solution We begin by looking for factors that are common to the two terms. Notice that 2 and $x^{1/3}$ are common factors. Then

$$4x^{1/3}(2x + 1) + 2x^{4/3} = 2x^{1/3}[2(2x + 1) + x]$$

$$= 2x^{1/3}(5x + 2)$$ ◀

A.9 Assess Your Understanding

'Are You Prepared?' *Answers are given at the end of these exercises. If you get a wrong answer, read the pages listed in red.*

1. *True or False:* One of the Laws of Exponents states that $a^n + b^n = (a + b)^n$ (pp. 908–910)

2. $\sqrt{(-4)^2} =$ ________. (pp. 908–910)

Concepts and Vocabulary

3. In the symbol $\sqrt[n]{a}$, the integer n is called the _________.

4. We call $\sqrt[3]{a}$ the _________ _________ of a.

5. *True or False:* $\sqrt[5]{-32} = -2$

6. *True or False:* $\sqrt[4]{(-3)^4} = -3$

Exercises

In Problems 7–34, simplify each expression. Assume that all variables are positive when they appear.

7. $\sqrt[3]{27}$ 8. $\sqrt[4]{16}$ 9. $\sqrt[3]{-8}$ 10. $\sqrt[3]{-1}$ 11. $\sqrt{8}$ 12. $\sqrt[3]{54}$

13. $\sqrt[3]{-8x^4}$ 14. $\sqrt[4]{48x^5}$ 15. $\sqrt[4]{x^{12}y^8}$ 16. $\sqrt[5]{x^{10}y^5}$ 17. $\sqrt[4]{\dfrac{x^9y^7}{xy^3}}$ 18. $\sqrt[3]{\dfrac{3xy^2}{81x^4y^2}}$

19. $\sqrt{36x}$ 20. $\sqrt{9x^5}$ 21. $\sqrt{3x^2}\sqrt{12x}$ 22. $\sqrt{5x}\sqrt{20x^3}$

23. $(\sqrt{5}\sqrt[3]{9})^2$ 24. $(\sqrt[3]{3}\sqrt{10})^4$ 25. $(3\sqrt{6})(2\sqrt{2})$ 26. $(5\sqrt{8})(-3\sqrt{3})$

27. $(\sqrt{3}+3)(\sqrt{3}-1)$ 28. $(\sqrt{5}-2)(\sqrt{5}+3)$ 29. $(\sqrt{x}-1)^2$ 30. $(\sqrt{x}+\sqrt{5})^2$

31. $3\sqrt{2}-4\sqrt{8}$ 32. $\sqrt[3]{-x^4}+\sqrt[3]{8x}$ 33. $\sqrt[3]{16x^4}-\sqrt[3]{2x}$ 34. $\sqrt[4]{32x}+\sqrt[4]{2x^5}$

In Problems 35–46, rationalize the denominator of each expression. Assume that all variables are positive when they appear.

35. $\dfrac{1}{\sqrt{2}}$ 36. $\dfrac{6}{\sqrt[3]{4}}$ 37. $\dfrac{-\sqrt{3}}{\sqrt{5}}$ 38. $\dfrac{-\sqrt[3]{3}}{\sqrt{8}}$

39. $\dfrac{\sqrt{3}}{5-\sqrt{2}}$ 40. $\dfrac{\sqrt{2}}{\sqrt{7}+2}$ 41. $\dfrac{2-\sqrt{5}}{2+3\sqrt{5}}$ 42. $\dfrac{\sqrt{3}-1}{2\sqrt{3}+3}$

43. $\dfrac{5}{\sqrt[3]{2}}$ 44. $\dfrac{-2}{\sqrt[3]{9}}$ 45. $\dfrac{\sqrt{x+h}-\sqrt{x}}{\sqrt{x+h}+\sqrt{x}}$ 46. $\dfrac{\sqrt{x+h}+\sqrt{x-h}}{\sqrt{x+h}-\sqrt{x-h}}$

In Problems 47–50, solve each equation.

47. $\sqrt[3]{2t-1}=2$ 48. $\sqrt[3]{3t+1}=-2$ 49. $\sqrt{15-2x}=x$ 50. $\sqrt{12-x}=x$

In Problems 51–62, simplify each expression.

51. $8^{2/3}$ 52. $4^{3/2}$ 53. $(-27)^{1/3}$ 54. $16^{3/4}$ 55. $16^{3/2}$ 56. $64^{3/2}$

57. $9^{-3/2}$ 58. $25^{-5/2}$ 59. $\left(\dfrac{9}{8}\right)^{3/2}$ 60. $\left(\dfrac{27}{8}\right)^{2/3}$ 61. $\left(\dfrac{8}{9}\right)^{-3/2}$ 62. $\left(\dfrac{8}{27}\right)^{-2/3}$

In Problems 63–70, simplify each expression. Express your answer so that only positive exponents occur. Assume that the variables are positive.

63. $x^{3/4}x^{1/3}x^{-1/2}$ 64. $x^{2/3}x^{1/2}x^{-1/4}$ 65. $(x^3y^6)^{1/3}$ 66. $(x^4y^8)^{3/4}$

67. $(x^2y)^{1/3}(xy^2)^{2/3}$ 68. $(xy)^{1/4}(x^2y^2)^{1/2}$ 69. $(16x^2y^{-1/3})^{3/4}$ 70. $(4x^{-1}y^{1/3})^{3/2}$

In Problems 71–84, expressions that occur in calculus are given. Write each expression as a single quotient in which only positive exponents and/or radicals appear.

71. $\dfrac{x}{(1+x)^{1/2}}+2(1+x)^{1/2},\quad x>-1$

72. $\dfrac{1+x}{2x^{1/2}}+x^{1/2},\quad x>0$

73. $2x(x^2+1)^{1/2}+x^2\cdot\dfrac{1}{2}(x^2+1)^{-1/2}\cdot 2x$

74. $(x+1)^{1/3}+x\cdot\dfrac{1}{3}(x+1)^{-2/3},\quad x\neq -1$

75. $\sqrt{4x+3}\cdot\dfrac{1}{2\sqrt{x-5}}+\sqrt{x-5}\cdot\dfrac{1}{5\sqrt{4x+3}},\ x>5$

76. $\dfrac{\sqrt[3]{8x+1}}{3\sqrt[3]{(x-2)^2}}+\dfrac{\sqrt[3]{x-2}}{24\sqrt[3]{(8x+1)^2}},\ x\neq 2,\ x\neq -\dfrac{1}{8}$

77. $\dfrac{\sqrt{1+x}-x\cdot\dfrac{1}{2\sqrt{1+x}}}{1+x},\quad x>-1$

78. $\dfrac{\sqrt{x^2+1}-x\cdot\dfrac{2x}{2\sqrt{x^2+1}}}{x^2+1}$

79. $\dfrac{(x+4)^{1/2} - 2x(x+4)^{-1/2}}{x+4}, \quad x > -4$

80. $\dfrac{(9-x^2)^{1/2} + x^2(9-x^2)^{-1/2}}{9-x^2}, \quad -3 < x < 3$

81. $\dfrac{\dfrac{x^2}{(x^2-1)^{1/2}} - (x^2-1)^{1/2}}{x^2}, \quad x < -1 \text{ or } x > 1$

82. $\dfrac{(x^2+4)^{1/2} - x^2(x^2+4)^{-1/2}}{x^2+4}$

83. $\dfrac{\dfrac{1+x^2}{2\sqrt{x}} - 2x\sqrt{x}}{(1+x^2)^2}, \quad x > 0$

84. $\dfrac{2x(1-x^2)^{1/3} + \dfrac{2}{3}x^3(1-x^2)^{-2/3}}{(1-x^2)^{2/3}}, \quad x \neq -1, x \neq 1$

In Problems 85–96, expressions that occur in calculus are given. Factor each expression, express your answer so that only positive exponents occur.

85. $(x+1)^{3/2} + x \cdot \dfrac{3}{2}(x+1)^{1/2}, \quad x \geq -1$

86. $(x^2+4)^{4/3} + x \cdot \dfrac{4}{3}(x^2+4)^{1/3} \cdot 2x$

87. $6x^{1/2}(x^2+x) - 8x^{3/2} - 8x^{1/2}, \quad x \geq 0$

88. $6x^{1/2}(2x+3) + x^{3/2} \cdot 8, \quad x \geq 0$

89. $3(x^2+4)^{4/3} + x \cdot 4(x^2+4)^{1/3} \cdot 2x$

90. $2x(3x+4)^{4/3} + x^2 \cdot 4(3x+4)^{1/3}$

91. $4(3x+5)^{1/3}(2x+3)^{3/2} + 3(3x+5)^{4/3}(2x+3)^{1/2}, \quad x \geq -\dfrac{3}{2}$

92. $6(6x+1)^{1/3}(4x-3)^{3/2} + 6(6x+1)^{4/3}(4x-3)^{1/2}, \quad x \geq \dfrac{3}{4}$

93. $3x^{-1/2} + \dfrac{3}{2}x^{1/2}, \quad x > 0$

94. $8x^{1/3} - 4x^{-2/3}, \quad x \neq 0$

95. $x\left(\dfrac{1}{2}\right)(8-x^2)^{-1/2}(-2x) + (8-x^2)^{1/2}$

96. $2x(1-x^2)^{3/2} + x^2\left(\dfrac{3}{2}\right)(1-x^2)^{1/2}(-2x)$

'Are You Prepared?' Answers

1. False **2.** 4

APPENDIX **B**

Graphing Utilities

Outline

B.1 The Viewing Rectangle

Figure 1
$y = 2x$

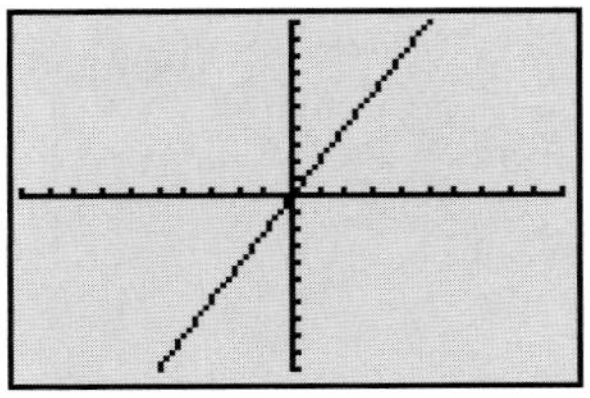

Figure 2

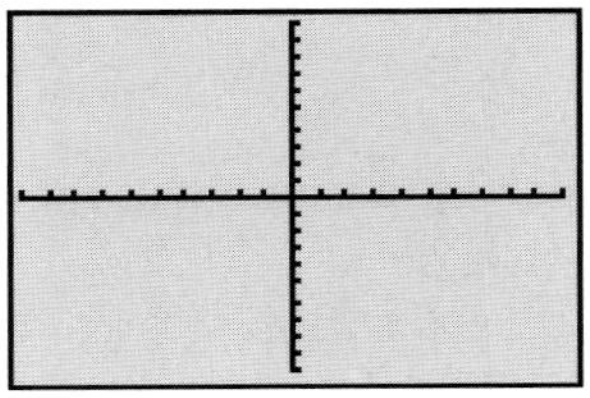

All graphing utilities, that is, all graphing calculators and all computer software graphing packages, graph equations by plotting points on a screen. The screen itself actually consists of small rectangles, called **pixels.** The more pixels the screen has, the better the resolution. Most graphing calculators have 2048 pixels per square inch; most computer screens have 4096 to 8192 pixels per square inch. When a point to be plotted lies inside a pixel, the pixel is turned on (lights up). The graph of an equation is a collection of pixels. Figure 1 shows how the graph of $y = 2x$ looks on a TI-83 graphing calculator.

The screen of a graphing utility will display the coordinate axes of a rectangular coordinate system. However, you must set the scale on each axis. You must also include the smallest and largest values of x and y that you want included in the graph. This is called **setting the viewing rectangle** or **viewing window.** Figure 2 illustrates a typical viewing window.

To select the viewing window, we must give values to the following expressions:

Xmin:	the smallest value of x
Xmax:	the largest value of x
Xscl:	the number of units per tick mark on the x-axis
Ymin:	the smallest value of y
Ymax:	the largest value of y
Yscl:	the number of units per tick mark on the y-axis

Figure 3 illustrates these settings and their relation to the Cartesian coordinate system.

Figure 3

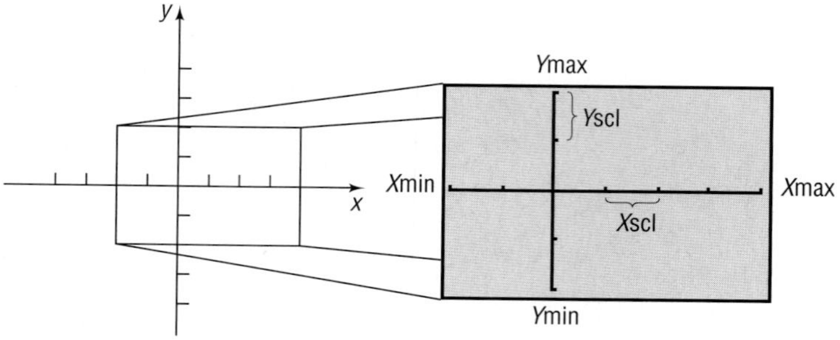

If the scale used on each axis is known, we can determine the minimum and maximum values of x and y shown on the screen by counting the tick marks. Look again at Figure 2. For a scale of 1 on each axis, the minimum and maximum values of x are -10 and 10, respectively; the minimum and maximum values of y are also -10 and 10. If the scale is 2 on each axis, then the minimum and maximum values of x are -20 and 20, respectively; and the minimum and maximum values of y are -20 and 20, respectively.

Conversely, if we know the minimum and maximum values of x and y, we can determine the scales being used by counting the tick marks displayed. We shall follow the practice of showing the minimum and maximum values of x and y in our illustrations so that you will know how the viewing window was set. See Figure 4.

Figure 4

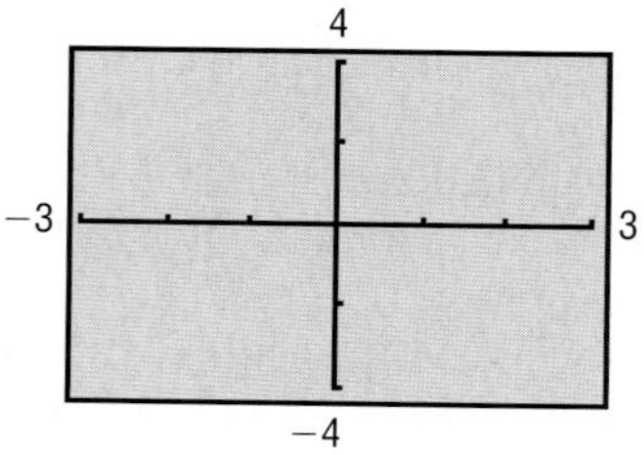

means $X\text{min} = -3 \quad Y\text{min} = -4$
$X\text{max} = 3 \quad Y\text{max} = 4$
$X\text{scl} = 1 \quad Y\text{scl} = 2$

EXAMPLE 1 Finding the Coordinates of a Point Shown on a Graphing Utility Screen

Find the coordinates of the point shown in Figure 5. Assume that the coordinates are integers.

Figure 5

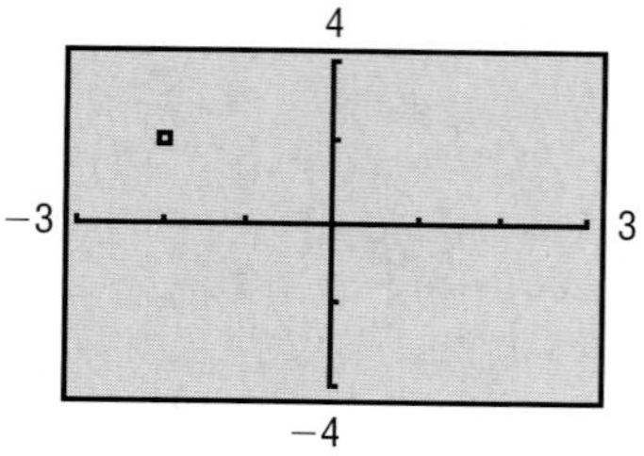

Solution First we note that the viewing window used in Figure 5 is

$$X\text{min} = -3 \quad Y\text{min} = -4$$
$$X\text{max} = 3 \quad Y\text{max} = 4$$
$$X\text{scl} = 1 \quad Y\text{scl} = 2$$

The point shown is 2 tick units to the left on the horizontal axis (scale $= 1$) and 1 tick up on the vertical axis (scale $= 2$). The coordinates of the point shown are $(-2, 2)$. ◀

B.1 Exercises

In Problems 1–4, determine the coordinates of the points shown. Tell in which quadrant each point lies. Assume that the coordinates are integers.

1.

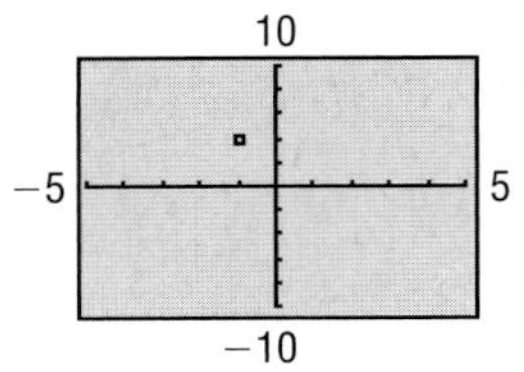

2.

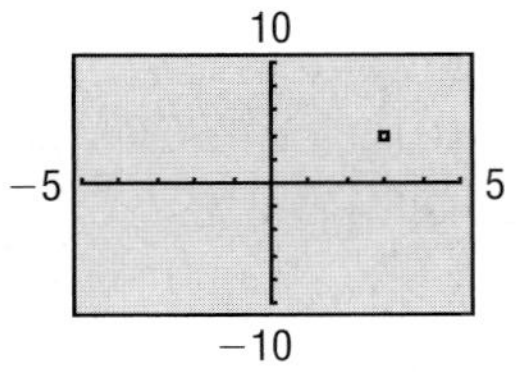

3.

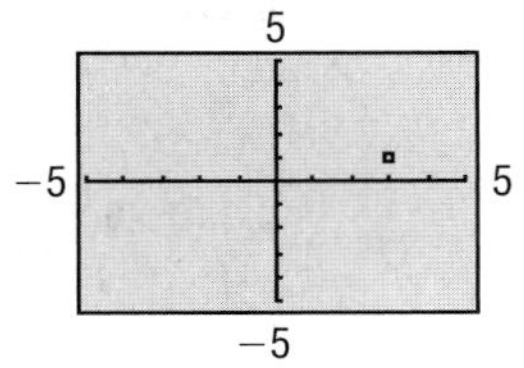

4.

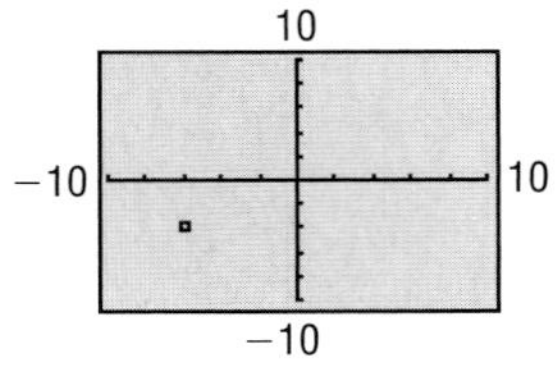

In Problems 5–10, determine the viewing window used.

5.

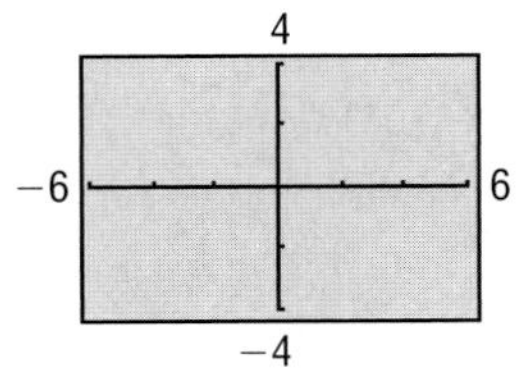

6.

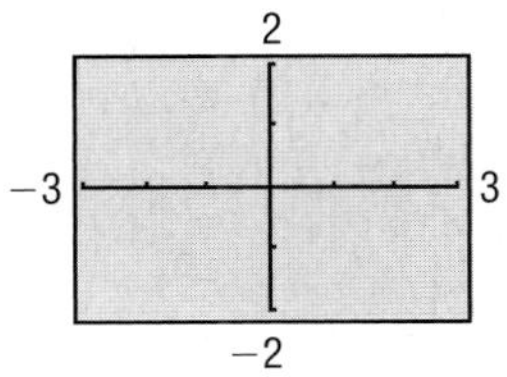

7.

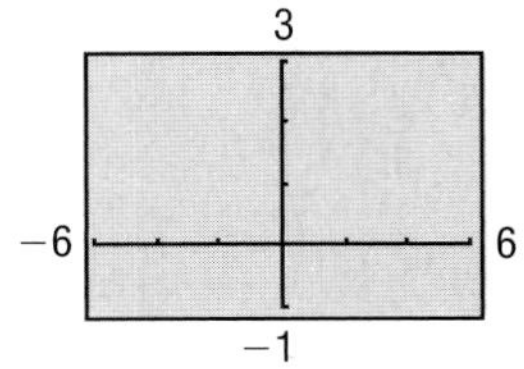

8.

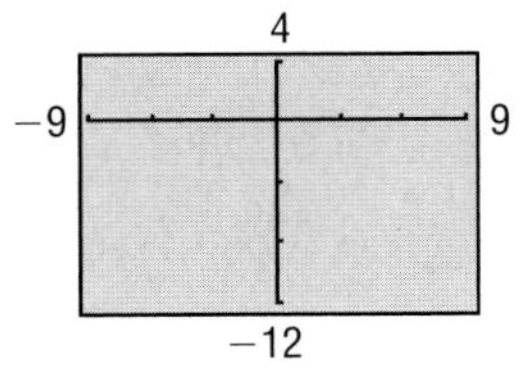

9.

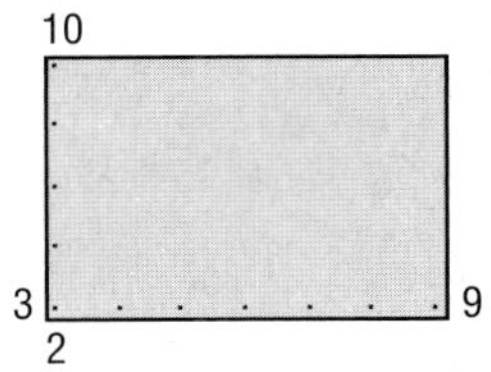

10.

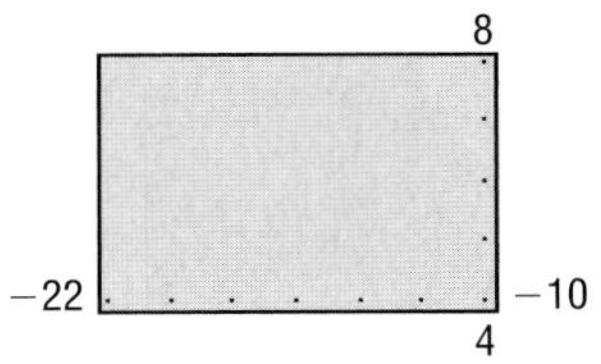

In Problems 11–16, select a setting so that each of the given points will lie within the viewing rectangle.

11. $(-10, 5), (3, -2), (4, -1)$

12. $(5, 0), (6, 8), (-2, -3)$

13. $(40, 20), (-20, -80), (10, 40)$

14. $(-80, 60), (20, -30), (-20, -40)$

15. $(0, 0), (100, 5), (5, 150)$

16. $(0, -1), (100, 50), (-10, 30)$

In Problems 17–20, find the length of the line segment. Assume that the endpoints of each line segment have integer coordinates.

17.

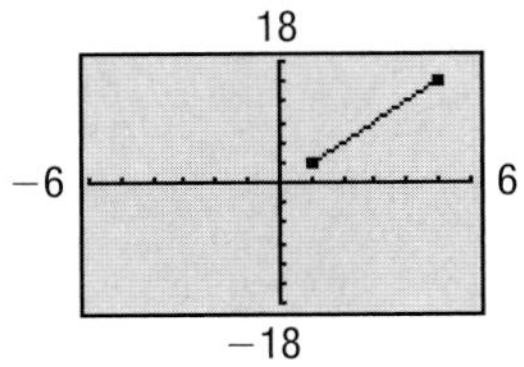

18.

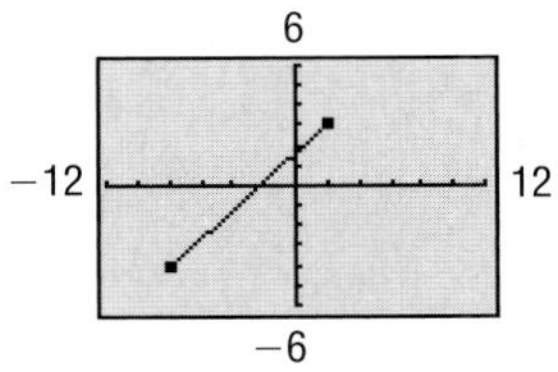

19.

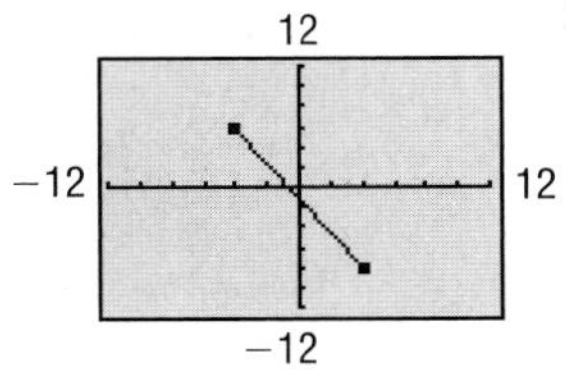

20.

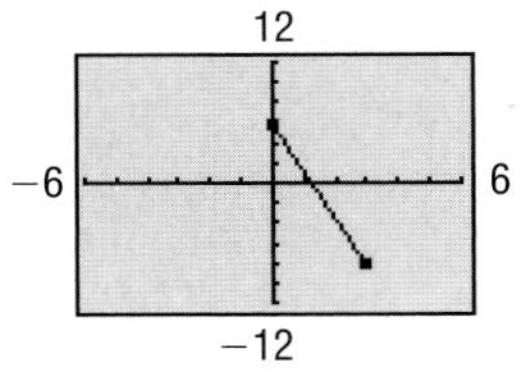

B.2 Using a Graphing Utility to Graph Equations

From Examples 2 and 3 of Section 1.2, we see that a graph can be obtained by plotting points in a rectangular coordinate system and connecting them. Graphing utilities perform these same steps when graphing an equation. For example, the TI-83 determines 95 evenly spaced input values,* starting at Xmin and ending at Xmax, uses the equation to determine the output values, plots these points on the screen, and finally (if in the connected mode) draws a line between consecutive points.

*These input values depend on the values of Xmin and Xmax. For example, if $X\text{min} = -10$ and $X\text{max} = 10$, then the first input value will be -10 and the next input value will be $-10 + \frac{10 - (-10)}{94} = -9.7872$, and so on.

To graph an equation in two variables x and y using a graphing utility requires that the equation be written in the form $y = \{\text{expression in } x\}$. If the original equation is not in this form, replace it by equivalent equations until the form $y = \{\text{expression in } x\}$ is obtained. In general, there are four ways to obtain equivalent equations.

Procedures That Result in Equivalent Equations

1. Interchange the two sides of the equation:

   Replace $3x + 5 = y$ by $y = 3x + 5$

2. Simplify the sides of the equation by combining like terms, eliminating parentheses, and so on:

   Replace $(2y + 2) + 6 = 2x + 5(x + 1)$

   by $2y + 8 = 7x + 5$

3. Add or subtract the same expression on both sides of the equation:

   Replace $y + 3x - 5 = 4$

   by $y + 3x - 5 + 5 = 4 + 5$

4. Multiply or divide both sides of the equation by the same nonzero expression:

   Replace $3y = 6 - 2x$

   by $\frac{1}{3} \cdot 3y = \frac{1}{3}(6 - 2x)$

EXAMPLE 1 **Expressing an Equation in the Form $y = \{\text{expression in } x\}$**

Solve for y: $2y + 3x - 5 = 4$

Solution We replace the original equation by a succession of equivalent equations.

$$2y + 3x - 5 = 4$$

$$2y + 3x - 5 + 5 = 4 + 5 \quad \text{Add 5 to both sides.}$$

$$2y + 3x = 9 \quad \text{Simplify.}$$

$$2y + 3x - 3x = 9 - 3x \quad \text{Subtract } 3x \text{ from both sides.}$$

$$2y = 9 - 3x \quad \text{Simplify.}$$

$$\frac{2y}{2} = \frac{9 - 3x}{2} \quad \text{Divide both sides by 2.}$$

$$y = \frac{9 - 3x}{2} \quad \text{Simplify.}$$

◀

Now we are ready to graph equations using a graphing utility. Most graphing utilities require the following steps:

Steps for Graphing an Equation Using a Graphing Utility

STEP 1: Solve the equation for y in terms of x.

STEP 2: Get into the graphing mode of your graphing utility. The screen will usually display $Y =$, prompting you to enter the expression involving x that you found in Step 1. (Consult your manual for the correct way to enter the expression; for example, $y = x^2$ might be entered as x^2 or as x*x or as $x\ x^Y\ 2$).

STEP 3: Select the viewing window. Without prior knowledge about the behavior of the graph of the equation, it is common to select the **standard viewing window*** initially. The viewing window is then adjusted based on the graph that appears. In this text the standard viewing window is

$$X\text{min} = -10 \qquad Y\text{min} = -10$$
$$X\text{max} = 10 \qquad Y\text{max} = 10$$
$$X\text{scl} = 1 \qquad Y\text{scl} = 1$$

STEP 4: Graph.

STEP 5: Adjust the viewing window until a complete graph is obtained.

EXAMPLE 2 Graphing an Equation on a Graphing Utility

Graph the equation: $6x^2 + 3y = 36$

Solution **STEP 1:** We solve for y in terms of x.

$$6x^2 + 3y = 36$$
$$3y = -6x^2 + 36 \qquad \text{Subtract } 6x^2 \text{ from both sides of the equation.}$$
$$y = -2x^2 + 12 \qquad \text{Divide both sides of the equation by 3 and simplify.}$$

STEP 2: From the $Y =$ screen, enter the expression $-2x^2 + 12$ after the prompt $Y_1 =$.

STEP 3: Set the viewing window to the standard viewing window.

STEP 4: Graph. The screen should look like Figure 6.

STEP 5: The graph of $y = -2x^2 + 12$ is not complete. The value of Ymax must be increased so that the top portion of the graph is visible. After increasing the value of Ymax to 12, we obtain the graph in Figure 7. The graph is now complete.

Figure 6

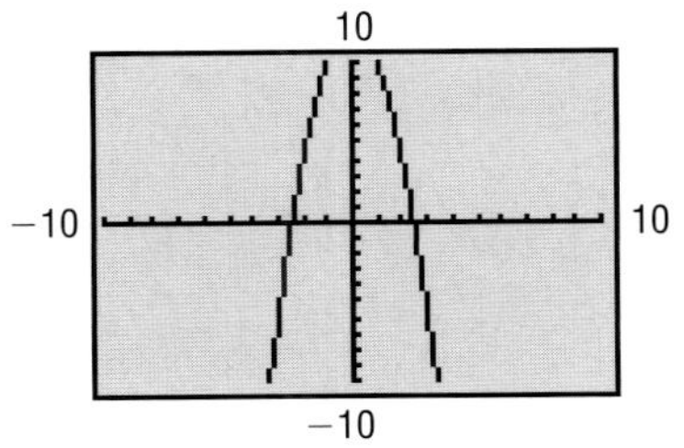

Figure 7

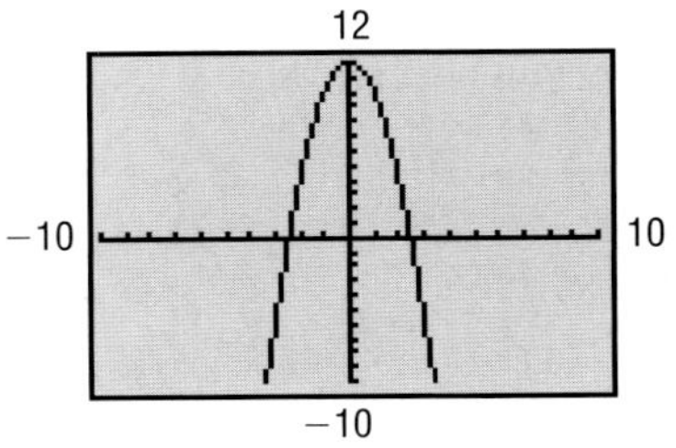

*Some graphing utilities have a ZOOM-STANDARD feature that automatically sets the viewing window to the standard viewing window and graphs the equation.

Look again at Figure 7. Although a complete graph is shown, the graph might be improved by adjusting the values of Xmin and Xmax. Figure 8 shows the graph of $y = -2x^2 + 12$ using $X\text{min} = -4$ and $X\text{max} = 4$. Do you think this is a better choice for the viewing window?

Figure 8

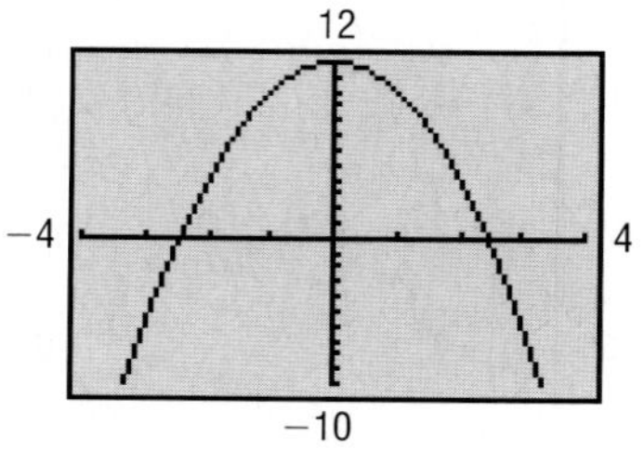

EXAMPLE 3 Creating a Table and Graphing an Equation

Create a table and graph the equation: $y = x^3$

Solution Most graphing utilities have the capability of creating a table of values for an equation. (Check your manual to see if your graphing utility has this capability.) Table 1 illustrates a table of values for $y = x^3$ on a TI-83. See Figure 9 for the graph.

Table 1

X	Y1
-3	-27
-2	-8
-1	-1
0	0
1	1
2	8
3	27

Y1=X^3

Figure 9

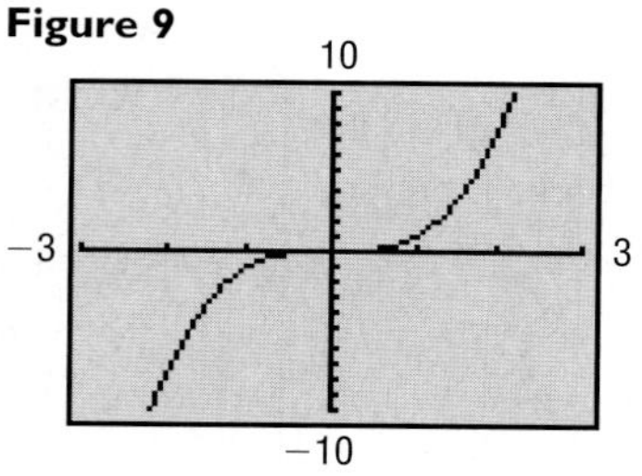

B.2 Exercises

In Problems 1–16, graph each equation using the following viewing windows:

(a) $X\text{min} = -5$, $X\text{max} = 5$, $X\text{scl} = 1$, $Y\text{min} = -4$, $Y\text{max} = 4$, $Y\text{scl} = 1$

(b) $X\text{min} = -10$, $X\text{max} = 10$, $X\text{scl} = 1$, $Y\text{min} = -8$, $Y\text{max} = 8$, $Y\text{scl} = 1$

(c) $X\text{min} = -10$, $X\text{max} = 10$, $X\text{scl} = 2$, $Y\text{min} = -8$, $Y\text{max} = 8$, $Y\text{scl} = 2$

(d) $X\text{min} = -5$, $X\text{max} = 5$, $X\text{scl} = 1$, $Y\text{min} = -20$, $Y\text{max} = 20$, $Y\text{scl} = 5$

1. $y = x + 2$ **2.** $y = x - 2$ **3.** $y = -x + 2$ **4.** $y = -x - 2$

5. $y = 2x + 2$ **6.** $y = 2x - 2$ **7.** $y = -2x + 2$ **8.** $y = -2x - 2$

9. $y = x^2 + 2$ **10.** $y = x^2 - 2$ **11.** $y = -x^2 + 2$ **12.** $y = -x^2 - 2$

13. $3x + 2y = 6$ **14.** $3x - 2y = 6$ **15.** $-3x + 2y = 6$ **16.** $-3x - 2y = 6$

17–32. *For each of the above equations, create a table, $-3 \leq x \leq 3$, and list points on the graph.*

B.3 Using a Graphing Utility to Locate Intercepts and Check for Symmetry

Value and Zero (or Root)

Most graphing utilities have an eVALUEate feature that, given a value of x, determines the value of y for an equation. We can use this feature to evaluate an equation at $x = 0$ to determine the y-intercept. Most graphing utilities also have a ZERO (or ROOT) feature that can be used to determine the x-intercept(s) of an equation.

EXAMPLE 1 **Finding Intercepts Using a Graphing Utility**

Use a graphing utility to find the intercepts of the equation $y = x^3 - 8$.

Solution Figure 10(a) shows the graph of $y = x^3 - 8$.

Figure 10

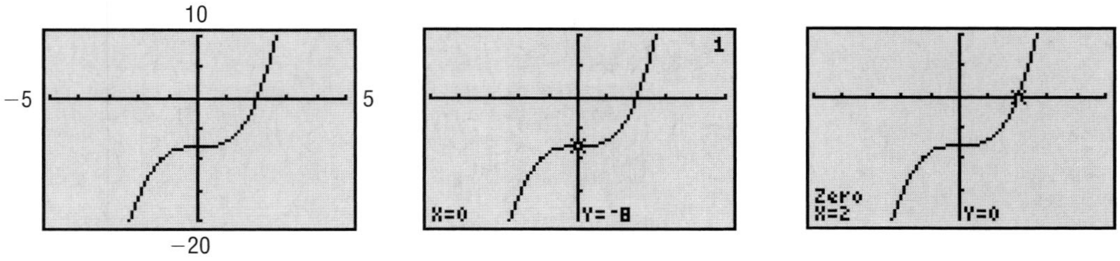

(a) (b) (c)

The eVALUEate feature of a TI-83 graphing calculator accepts as input a value of x and determines the value of y. If we let $x = 0$, we find that the y-intercept is -8. See Figure 10(b).

The ZERO feature of a TI-83 is used to find the x-intercept(s). See Figure 10(c). The x-intercept is 2. ◀

Trace

Most graphing utilities allow you to move from point to point along the graph, displaying on the screen the coordinates of each point. This feature is called TRACE.

EXAMPLE 2 **Using TRACE to Locate Intercepts**

Graph the equation $y = x^3 - 8$. Use TRACE to locate the intercepts.

Solution Figure 11 shows the graph of $y = x^3 - 8$.

Figure 11

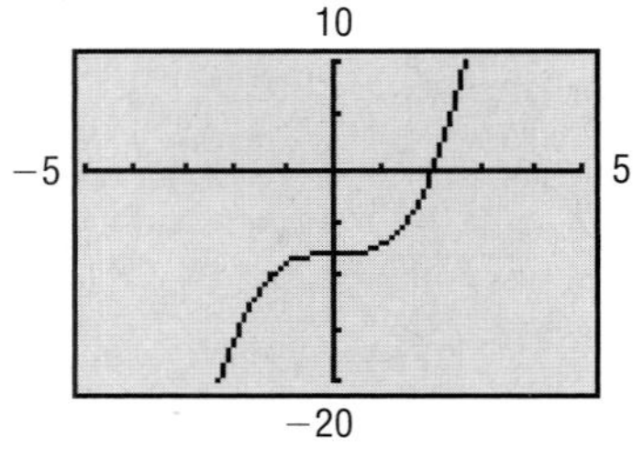

Activate the TRACE feature. As you move the cursor along the graph, you will see the coordinates of each point displayed. When the cursor is on the y-axis, we find that the y-intercept is -8. See Figure 12.

Figure 12

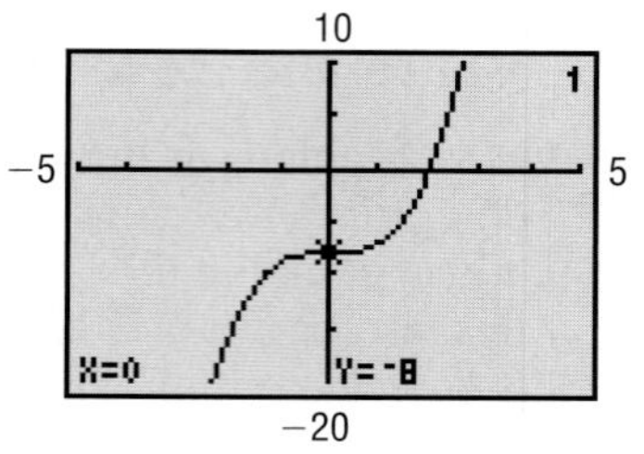

Continue moving the cursor along the graph. Just before you get to the x-axis, the display will look like the one in Figure 13(a). (Due to differences in graphing utilities, your display may be slightly different from the one shown here.)

Figure 13

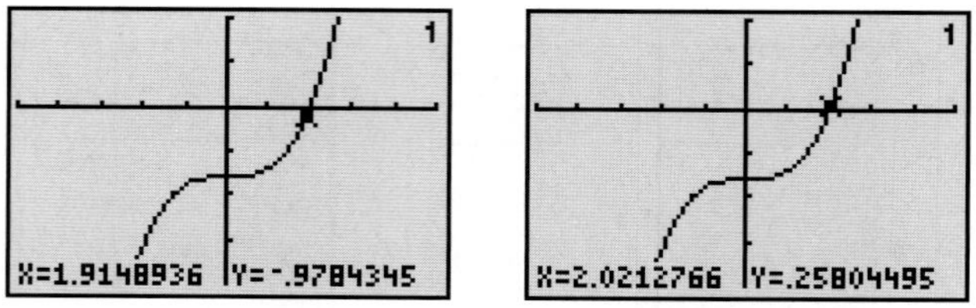

(a) (b)

In Figure 13(a), the negative value of the y-coordinate indicates that we are still below the x-axis. The next position of the cursor is shown in Figure 13(b). The positive value of the y-coordinate indicates that we are now above the x-axis. This means that between these two points the x-axis was crossed. The x-intercept lies between 1.9148936 and 2.0212766. ◀

EXAMPLE 3 Graphing the Equation $y = \frac{1}{x}$

Figure 14

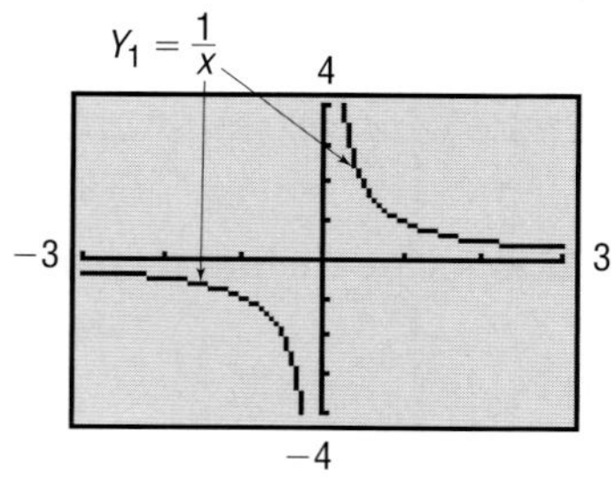

Graph the equation: $y = \frac{1}{x}$

With the viewing window set as

$$X\text{min} = -3 \qquad Y\text{min} = -4$$
$$X\text{max} = 3 \qquad Y\text{max} = 4$$
$$X\text{scl} = 1 \qquad Y\text{scl} = 1$$

use TRACE to infer information about the intercepts and symmetry.

Solution Figure 14 illustrates the graph. We infer from the graph that there are no intercepts; we may also infer that symmetry with respect to the origin is a possibility. The TRACE feature on a graphing utility can provide further evidence of symmetry with respect to the origin. Using TRACE, we observe that for any ordered pair (x, y) the ordered pair $(-x, -y)$ is also a point on the graph. For example, the points $(0.95744681, 1.0444444)$ and $(-0.95744681, -1.0444444)$ both lie on the graph. ◀

B.3 Exercises

In Problems 1–6, use ZERO (or ROOT) to approximate the smaller of the two x-intercepts of each equation. Express the answer rounded to two decimal places.

1. $y = x^2 + 4x + 2$

2. $y = x^2 + 4x - 3$

3. $y = 2x^2 + 4x + 1$

4. $y = 3x^2 + 5x + 1$

5. $y = 2x^2 - 3x - 1$

6. $y = 2x^2 - 4x - 1$

In Problems 7–14, use ZERO (or ROOT) to approximate the ***positive*** *x-intercepts of each equation. Express each answer rounded to two decimal places.*

7. $y = x^3 + 3.2x^2 - 16.83x - 5.31$

8. $y = x^3 + 3.2x^2 - 7.25x - 6.3$

9. $y = x^4 - 1.4x^3 - 33.71x^2 + 23.94x + 292.41$

10. $y = x^4 + 1.2x^3 - 7.46x^2 - 4.692x + 15.2881$

11. $y = \pi x^3 - (8.88\pi + 1)x^2 - (42.066\pi - 8.88)x + 42.066$

12. $y = \pi x^3 - (5.63\pi + 2)x^2 - (108.392\pi - 11.26)x + 216.784$

13. $y = x^3 + 19.5x^2 - 1021x + 1000.5$

14. $y = x^3 + 14.2x^2 - 4.8x - 12.4$

B.4 Using a Graphing Utility to Solve Equations

For many equations, there are no algebraic techniques that lead to a solution. For such equations, a graphing utility can often be used to investigate possible solutions. When a graphing utility is used to solve an equation, usually *approximate* solutions are obtained. Unless otherwise stated, we shall follow the practice of giving approximate solutions *rounded to two decimal places.*

The ZERO (or ROOT) feature of a graphing utility can be used to find the solutions of an equation when one side of the equation is 0. In using this feature to solve equations, we make use of the fact that the x-intercepts (or zeros) of the graph of an equation are found by letting $y = 0$ and solving the equation for x. Solving an equation for x when one side of the equation is 0 is equivalent to finding where the graph of the corresponding equation crosses or touches the x-axis.

EXAMPLE 1 Using ZERO (or ROOT) to Approximate Solutions of an Equation

Find the solution(s) of the equation $x^2 - 6x + 7 = 0$. Round answers to two decimal places.

Solution The solutions of the equation $x^2 - 6x + 7 = 0$ are the same as the x-intercepts of the graph of $Y_1 = x^2 - 6x + 7$. We begin by graphing the equation. See Figure 15(a).

Figure 15

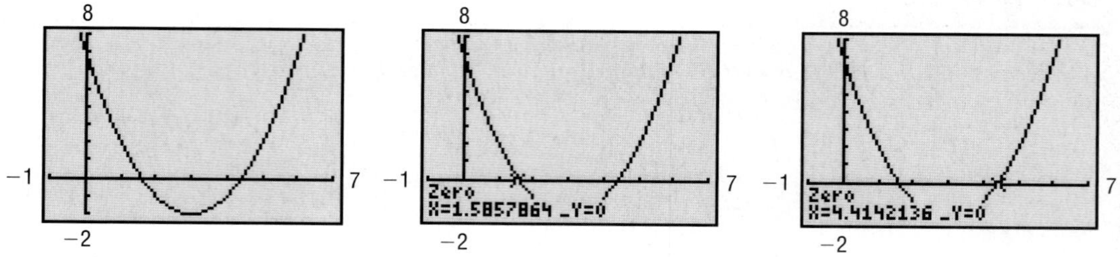

(a) (b) (c)

From the graph there appear to be two x-intercepts (solutions to the equation): one between 1 and 2, the other between 4 and 5.

Using the ZERO (or ROOT) feature of our graphing utility, we determine that the x-intercepts, and so the solutions to the equation, are $x = 1.59$ and $x = 4.41$, rounded to two decimal places. See Figures 15(b) and (c). ◀

A second method for solving equations using a graphing utility involves the INTERSECT feature of the graphing utility. This feature is used most effectively when one side of the equation is not 0.

EXAMPLE 2 Using INTERSECT to Approximate Solutions of an Equation

Find the solution(s) to the equation $3(x - 2) = 5(x - 1)$. Round answers to two decimal places.

Solution We begin by graphing each side of the equation as follows: graph $Y_1 = 3(x - 2)$ and $Y_2 = 5(x - 1)$. See Figure 16(a).

Figure 16

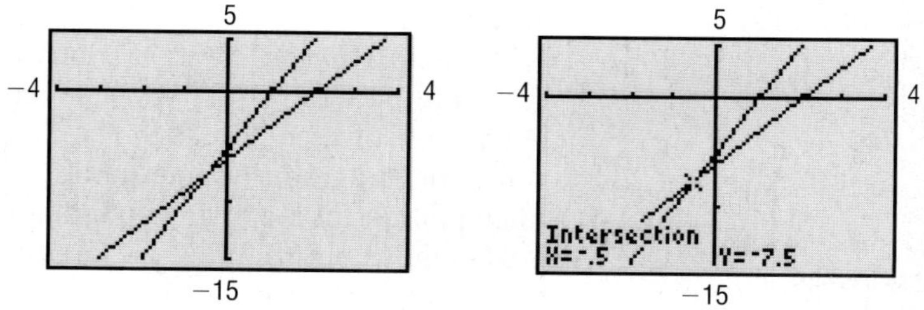

(a) (b)

At the point of intersection of the graphs, the value of the y-coordinate is the same. We conclude that the x-coordinate of the point of intersection represents the solution to the equation. Do you see why? The INTERSECT feature on a graphing utility determines the point of intersection of the graphs. Using this feature, we find that the graphs intersect at $(-0.5, -7.5)$. See Figure 16(b). The solution of the equation is therefore $x = -0.5$.

Figure 17

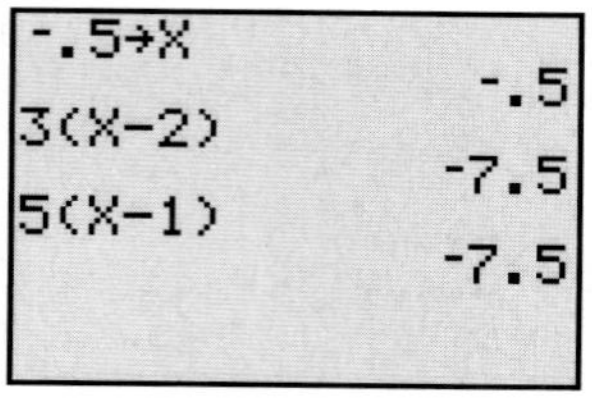

CHECK: We can verify our solution by evaluating each side of the equation with -0.5 STOred in x. See Figure 17. Since the left side of the equation equals the right side of the equation, the solution checks. ◀

Summary

The steps to follow for approximating solutions of equations are given next.

Steps for Approximating Solutions of Equations Using ZERO (or ROOT)

STEP 1: Write the equation in the form {expression in x} = 0.

STEP 2: Graph Y_1 = {expression in x}.

Be sure that the graph is complete. That is, be sure that all the intercepts are shown on the screen.

STEP 3: Use ZERO (or ROOT) to determine each x-intercept of the graph.

Steps for Approximating Solutions of Equations Using INTERSECT

STEP 1: Graph Y_1 = {expression in x on left side of equation}.

Graph Y_2 = {expression in x on right side of equation}.

STEP 2: Use INTERSECT to determine each x-coordinate of the point(s) of intersection, if any.

Be sure that the graphs are complete. That is, be sure that all the points of intersection are shown on the screen.

EXAMPLE 3 Solving a Radical Equation

Find the real solutions of the equation $\sqrt[3]{2x-4} - 2 = 0$.

Figure 18

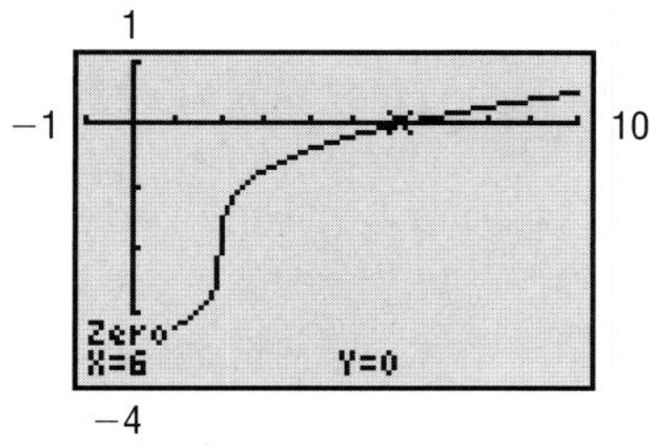

Solution Figure 18 shows the graph of the equation $Y_1 = \sqrt[3]{2x-4} - 2$. From the graph, we see one x-intercept near 6. Using ZERO (or ROOT), we find that the x-intercept is 6. The only solution is $x = 6$.

◀

B.5 Square Screens

Figure 19

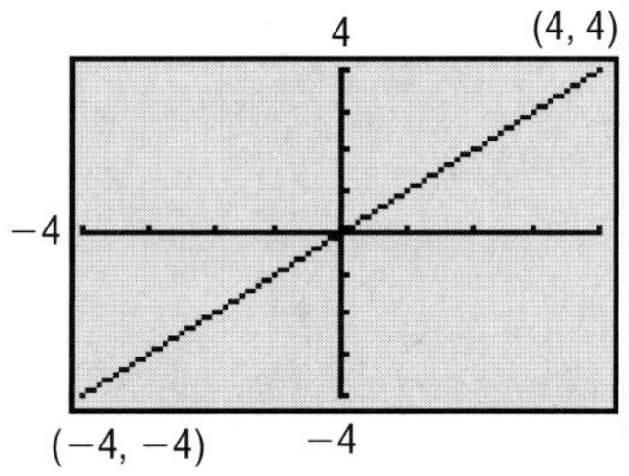

Most graphing utilities have a rectangular screen. Because of this, using the same settings for both x and y will result in a distorted view. For example, Figure 19 shows the graph of the line $y = x$ connecting the points $(-4, -4)$ and $(4, 4)$.

We expect the line to bisect the first and third quadrants, but it doesn't. We need to adjust the selections for Xmin, Xmax, Ymin, and Ymax so that a **square screen** results. On most graphing utilities, this is accomplished by setting the ratio of x to y at 3:2.* For example, if

$$X\text{min} = -6 \qquad Y\text{min} = -4$$
$$X\text{max} = 6 \qquad Y\text{max} = 4$$

*Some graphing utilities have a built-in function that automatically squares the screen. For example, the TI-85 has a ZSQR function that does this. Some graphing utilities require a ratio other than 3:2 to square the screen. For example, the HP 48G requires the ratio of x to y to be 2:1 for a square screen. Consult your manual.

then the ratio of x to y is

$$\frac{X\text{max} - X\text{min}}{Y\text{max} - Y\text{min}} = \frac{6 - (-6)}{4 - (-4)} = \frac{12}{8} = \frac{3}{2}$$

for a ratio of 3 : 2, resulting in a square screen.

EXAMPLE 1 **Examples of Viewing Rectangles That Result in Square Screens**

(a)	(b)	(c)
$X\text{min} = -3$	$X\text{min} = -6$	$X\text{min} = -6$
$X\text{max} = 3$	$X\text{max} = 6$	$X\text{max} = 6$
$X\text{scl} = 1$	$X\text{scl} = 1$	$X\text{scl} = 1$
$Y\text{min} = -2$	$Y\text{min} = -4$	$Y\text{min} = -4$
$Y\text{max} = 2$	$Y\text{max} = 4$	$Y\text{max} = 4$
$Y\text{scl} = 1$	$Y\text{scl} = 1$	$Y\text{scl} = 2$ ◀

Figure 20

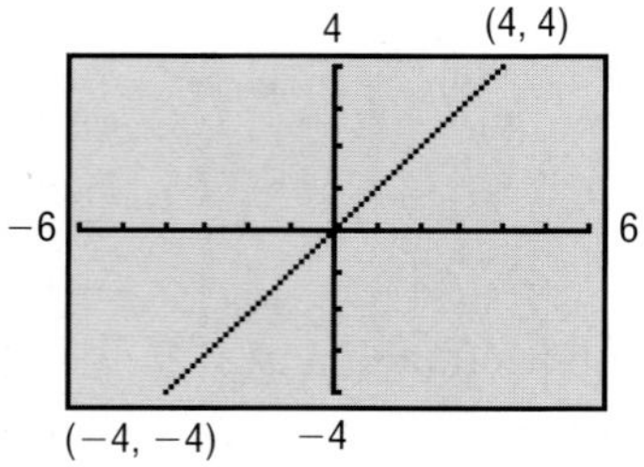

Figure 20 shows the graph of the line $y = x$ on a square screen using the viewing rectangle given in example (b). Notice that the line now bisects the first and third quadrants. Compare this illustration to Figure 19.

B.5 Exercises

In Problems 1–8, determine which of the given viewing rectangles result in a square screen.

1.	2.	3.	4.
$X\text{min} = -3$	$X\text{min} = -5$	$X\text{min} = 0$	$X\text{min} = -6$
$X\text{max} = 3$	$X\text{max} = 5$	$X\text{max} = 9$	$X\text{max} = 6$
$X\text{scl} = 2$	$X\text{scl} = 1$	$X\text{scl} = 3$	$X\text{scl} = 1$
$Y\text{min} = -2$	$Y\text{min} = -4$	$Y\text{min} = -2$	$Y\text{min} = -4$
$Y\text{max} = 2$	$Y\text{max} = 4$	$Y\text{max} = 4$	$Y\text{max} = 4$
$Y\text{scl} = 2$	$Y\text{scl} = 1$	$Y\text{scl} = 2$	$Y\text{scl} = 2$

5.	6.	7.	8.
$X\text{min} = -6$	$X\text{min} = -6$	$X\text{min} = 0$	$X\text{min} = -6$
$X\text{max} = 6$	$X\text{max} = 6$	$X\text{max} = 9$	$X\text{max} = 6$
$X\text{scl} = 1$	$X\text{scl} = 2$	$X\text{scl} = 1$	$X\text{scl} = 2$
$Y\text{min} = -2$	$Y\text{min} = -4$	$Y\text{min} = -2$	$Y\text{min} = -4$
$Y\text{max} = 2$	$Y\text{max} = 4$	$Y\text{max} = 4$	$Y\text{max} = 4$
$Y\text{scl} = 0.5$	$Y\text{scl} = 1$	$Y\text{scl} = 1$	$Y\text{scl} = 2$

9. If $X\text{min} = -4$, $X\text{max} = 8$, and $X\text{scl} = 1$, how should $Y\text{min}$, $Y\text{max}$, and $Y\text{scl}$ be selected so that the viewing rectangle contains the point $(4, 8)$ and the screen is square?

10. If $X\text{min} = -6$, $X\text{max} = 12$, and $X\text{scl} = 2$, how should $Y\text{min}$, $Y\text{max}$, and $Y\text{scl}$ be selected so that the viewing rectangle contains the point $(4, 8)$ and the screen is square?

B.6 Using a Graphing Utility to Graph Inequalities

It is easiest to begin with an example.

EXAMPLE 1 **Graphing an Inequality Using a Graphing Utility**

Use a graphing utility to graph: $3x + y - 6 \le 0$

Solution We begin by graphing the equation $3x + y - 6 = 0$ ($Y_1 = -3x + 6$). See Figure 21.

Figure 21

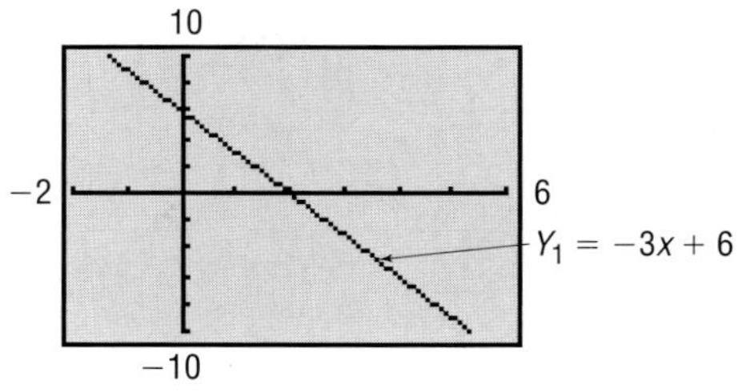

As with graphing by hand, we need to test points selected from each region and determine whether they satisfy the inequality. To test the point $(-1, 2)$, for example, enter $3(-1) + 2 - 6 \leq 0$. See Figure 22(a). The 1 that appears indicates that the statement entered (the inequality) is true. When the point $(5, 5)$ is tested, a 0 appears, indicating that the statement entered is false. Thus, $(-1, 2)$ is a part of the graph of the inequality and $(5, 5)$ is not. Figure 22(b) shows the graph of the inequality on a TI-83.*

Figure 22

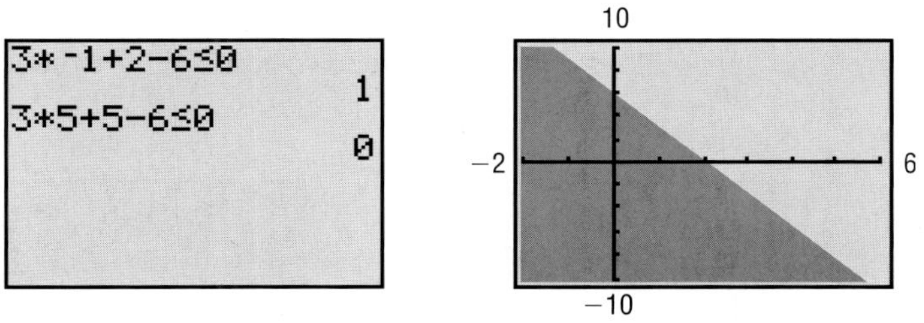

(a) (b) ◀

The steps to follow to graph an inequality using a graphing utility are given next.

> **Steps for Graphing an Inequality Using a Graphing Utility**
>
> **STEP 1:** Replace the inequality symbol by an equal sign, solve the equation for y, and graph the equation.
>
> **STEP 2:** In each of the regions, select a test point P and determine if the coordinates of P satisfy the inequality.
>
> (a) If the test point satisfies the inequality, then so do all the points in the region. Indicate this by using the graphing utility to shade the region.
>
> (b) If the coordinates of P do not satisfy the inequality, then none of the points in that region does.

*Consult your owner's manual for shading techniques.

B.7 Using a Graphing Utility to Solve Systems of Linear Equations

Most graphing utilities have the capability to put the augmented matrix of a system of linear equations in row echelon form. The next example, Example 6 from Section 10.2, demonstrates this feature using a TI-83 graphing calculator.

EXAMPLE 1 Solving a System of Linear Equations Using a Graphing Utility

$$\text{Solve:}\quad \begin{cases} x - y + z = 8 & (1) \\ 2x + 3y - z = -2 & (2) \\ 3x - 2y - 9z = 9 & (3) \end{cases}$$

Solution The augmented matrix of the system is

$$\left[\begin{array}{ccc|c} 1 & -1 & 1 & 8 \\ 2 & 3 & -1 & -2 \\ 3 & -2 & -9 & 9 \end{array}\right]$$

We enter this matrix into our graphing utility and name it A. See Figure 23(a). Using the REF (row echelon form) command on matrix A, we obtain the results shown in Figure 23(b). Since the entire matrix does not fit on the screen, we need to scroll right to see the rest of it. See Figure 23(c).

Figure 23

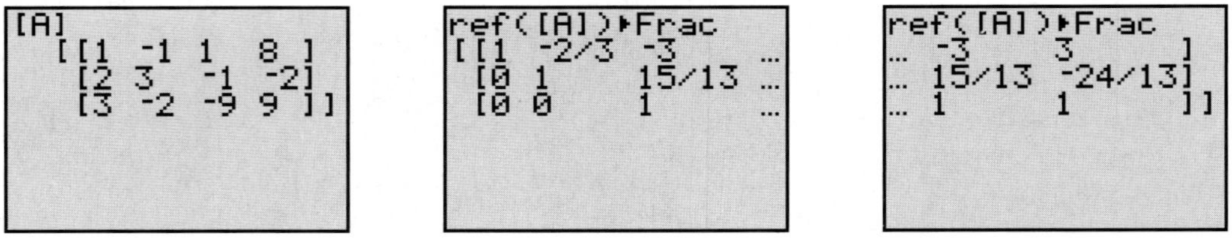

(a) (b) (c)

The system of equations represented by the matrix in row echelon form is

$$\left[\begin{array}{ccc|c} 1 & -\frac{2}{3} & -3 & 3 \\ 0 & 1 & \frac{15}{13} & -\frac{24}{13} \\ 0 & 0 & 1 & 1 \end{array}\right] \quad \begin{cases} x - \frac{2}{3}y - 3z = 3 & (1) \\ y + \frac{15}{13}z = -\frac{24}{13} & (2) \\ z = 1 & (3) \end{cases}$$

Using $z = 1$, we back-substitute to get

$$\begin{cases} x - \frac{2}{3}y - 3(1) = 3 & (1) \\ y + \frac{15}{13}(1) = -\frac{24}{13} & (2) \end{cases} \xrightarrow{\text{Simplify.}} \begin{cases} x - \frac{2}{3}y = 6 & (1) \\ y = \frac{-39}{13} = -3 & (2) \end{cases}$$

Solving the second equation for y, we find that $y = -3$. Back-substituting $y = -3$ into $x - \frac{2}{3}y = 6$, we find that $x = 4$. The solution of the system is $x = 4$, $y = -3$, $z = 1$. ◄

Figure 24

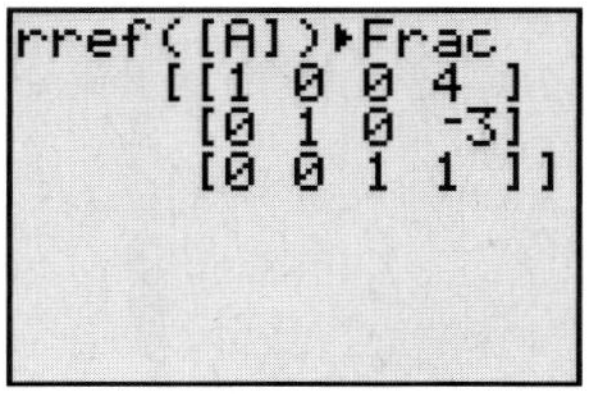

Notice that the row echelon form of the augmented matrix using the graphing utility differs from the row echelon form in our solution (p. 707), yet both matrices provide the same solution! This is because the two solutions used different row operations to obtain the row echelon form. In all likelihood, the two solutions parted ways in Step 4 of the algebraic solution, where we avoided introducing fractions by interchanging rows 2 and 3.

Most graphing utilities also have the ability to put a matrix in reduced row echelon form. Figure 24 shows the reduced row echelon form of the augmented matrix from Example 1 using the RREF command on a TI-83 graphing calculator. Using this command, we see that the solution of the system is $x = 4$, $y = -3$, $z = 1$.

B.8 Using a Graphing Utility to Graph a Polar Equation

Most graphing utilities require the following steps in order to obtain the graph of a polar equation. Be sure to be in POLar mode.

Graphing a Polar Equation Using a Graphing Utility

STEP 1: Set the mode to POLar. Solve the equation for r in terms of θ.

STEP 2: Select the viewing rectangle in polar mode. Besides setting Xmin, Xmax, Xscl, and so forth, the viewing rectangle in polar mode requires setting the minimum and maximum values for θ and an increment setting for θ (θstep). In addition, a square screen and radian measure should be used.

STEP 3: Enter the expression involving θ that you found in Step 1. (Consult your manual for the correct way to enter the expression.)

STEP 4: Graph.

EXAMPLE 1 **Graphing a Polar Equation Using a Graphing Utility**

Use a graphing utility to graph the polar equation $r \sin\theta = 2$.

Solution **STEP 1:** We solve the equation for r in terms of θ.

$$r \sin\theta = 2$$
$$r = \frac{2}{\sin\theta}$$

STEP 2: From the POLar mode, select the viewing rectangle. We will use the one given next.

$$\theta\text{min} = 0 \qquad X\text{min} = -9 \qquad Y\text{min} = -6$$
$$\theta\text{max} = 2\pi \qquad X\text{max} = 9 \qquad Y\text{max} = 6$$
$$\theta\text{step} = \frac{\pi}{24} \qquad X\text{scl} = 1 \qquad Y\text{scl} = 1$$

θstep determines the number of points that the graphing utility will plot. For example, if θstep is $\frac{\pi}{24}$, then the graphing utility will evaluate r at $\theta = 0(\theta\text{min}), \frac{\pi}{24}, \frac{2\pi}{24}, \frac{3\pi}{24}$, and so forth, up to $2\pi(\theta\text{max})$. The smaller θstep is, the more points that the graphing utility will plot. The student is encouraged to experiment with different values for θmin, θmax, and θstep to see how the graph is affected.

Figure 25

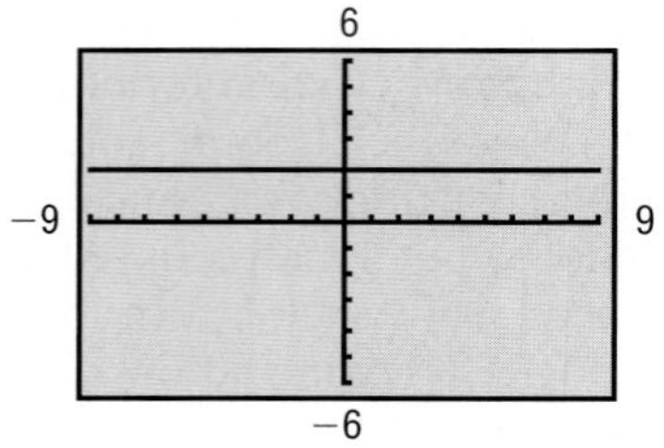

Step 3: Enter the expression $\frac{2}{\sin\theta}$ after the prompt $r_1 =$.

Step 4: Graph.

The graph is shown in Figure 25. ◀

B.9 Using a Graphing Utility to Graph Parametric Equations

Most graphing utilities have the capability of graphing parametric equations. The following steps are usually required in order to obtain the graph of parametric equations. Check your owner's manual to see how yours works.

Graphing Parametric Equations Using a Graphing Utility

Step 1: Set the mode to PARametric. Enter $x(t)$ and $y(t)$.

Step 2: Select the viewing window. In addition to setting Xmin, Xmax, Xscl, and so on, the viewing window in parametric mode requires setting minimum and maximum values for the parameter t and an increment setting for t (Tstep).

Step 3: Graph.

EXAMPLE 1 **Graphing a Curve Defined by Parametric Equations Using a Graphing Utility**

Graph the curve defined by the parametric equations

$$x = 3t^2, \qquad y = 2t, \qquad -2 \le t \le 2$$

Solution **Step 1:** Enter the equations $x(t) = 3t^2$, $y(t) = 2t$ with the graphing utility in PARametric mode.

Step 2: Select the viewing window. The interval is $-2 \le t \le 2$, so we select the following square viewing window:

$$T\text{min} = -2 \qquad X\text{min} = 0 \qquad Y\text{min} = -5$$
$$T\text{max} = 2 \qquad X\text{max} = 15 \qquad Y\text{max} = 5$$
$$T\text{step} = 0.1 \qquad X\text{scl} = 1 \qquad Y\text{scl} = 1$$

Figure 26

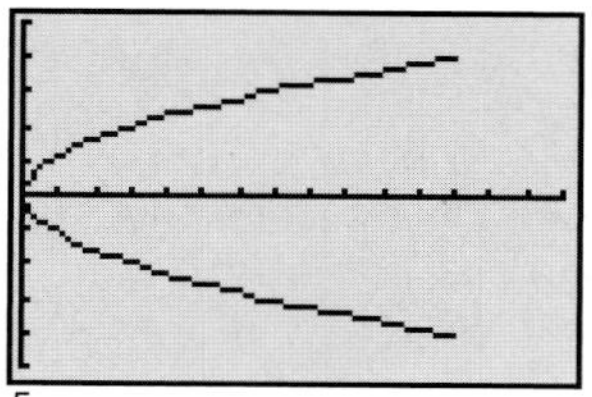

We choose Tmin = −2 and Tmax = 2 because $-2 \le t \le 2$. Finally, the choice for Tstep will determine the number of points that the graphing utility will plot. For example, with Tstep at 0.1, the graphing utility will evaluate x and y at $t = -2, -1.9, -1.8$, and so on. The smaller the Tstep, the more points the graphing utility will plot. The reader is encouraged to experiment with different values of Tstep to see how the graph is affected.

STEP 3: Graph. Notice the direction in which the graph is drawn. This direction shows the orientation of the curve.

The graph shown in Figure 26 is complete. ◀

Exploration

Graph the following parametric equations using a graphing utility with Xmin = 0, Xmax = 15, Ymin = −5, Ymax = 5, and Tstep = 0.1.

1. $x = \dfrac{3t^2}{4}, \quad y = t, \quad -4 \le t \le 4$

2. $x = 3t^2 + 12t + 12, \quad y = 2t + 4, \quad -4 \le t \le 0$

3. $x = 3t^{2/3}, \quad y = 2\sqrt[3]{t}, \quad -8 \le t \le 8$

Compare these graphs to the graph in Figure 26. Conclude that parametric equations defining a curve are not unique; that is, different parametric equations can represent the same graph.

Exploration

In FUNCtion mode, graph $x = \dfrac{3y^2}{4}\left(Y_1 = \sqrt{\dfrac{4x}{3}} \text{ and } Y_2 = -\sqrt{\dfrac{4x}{3}}\right)$ with Xmin = 0, Xmax = 15, Ymin = −5, Ymax = 5. Compare this graph with Figure 26. Why do the graphs differ?

ANSWERS

CHAPTER 1 Graphs

1.1 Concepts and Vocabulary *(page 6)*

4. x-coordinate or abscissa; y-coordinate or ordinate **5.** quadrants **6.** midpoint **7.** x **8.** False **9.** False **10.** True

1.1 Exercises *(page 6)*

11. (a) Quadrant II **(b)** Positive x-axis **(c)** Quadrant III **(d)** Quadrant I **(e)** Negative y-axis **(f)** Quadrant IV

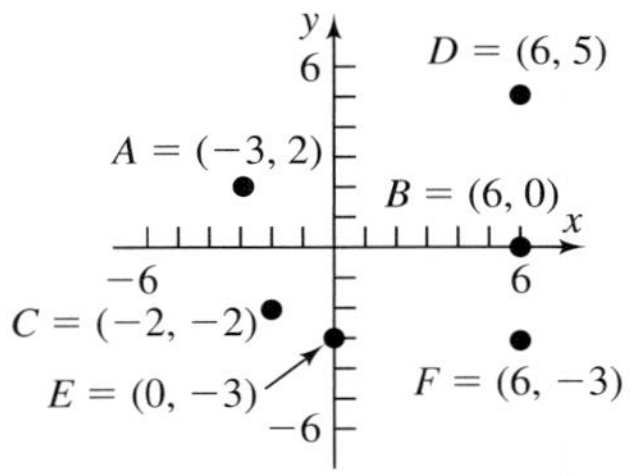

12. (a) Quadrant I **(b)** Quadrant III **(c)** Quadrant II **(d)** Quadrant I **(e)** positive y-axis **(f)** negative x-axis

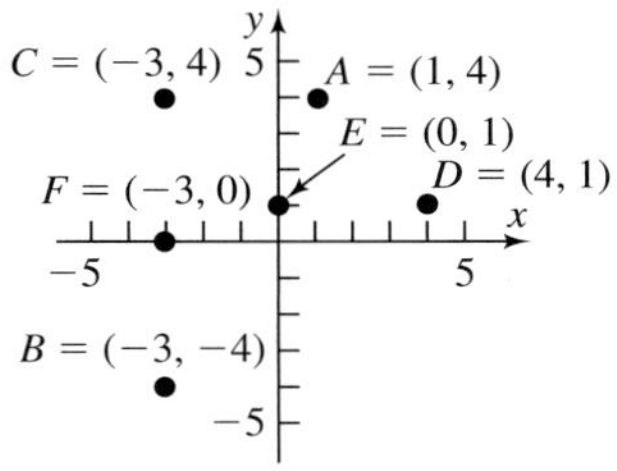

13. The points will be on a vertical line that is 2 units to the right of the y-axis.

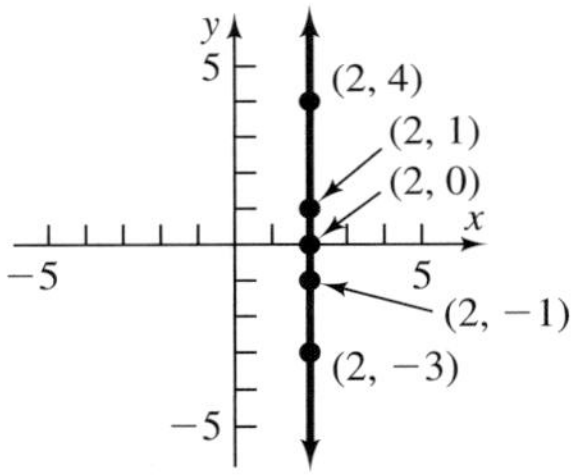

14. The points will be on a horizontal line that is 3 units above the x-axis.

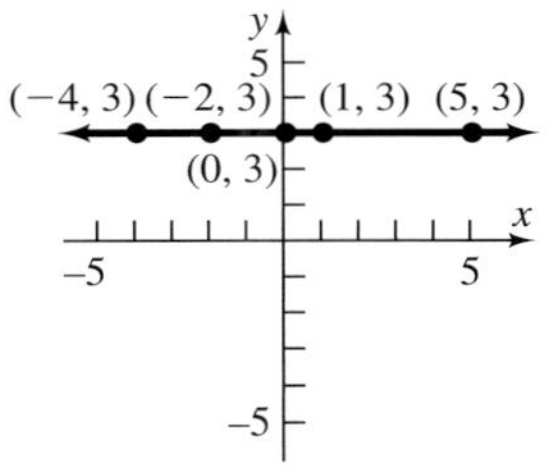

15. $\sqrt{5}$ **16.** $\sqrt{5}$ **17.** $\sqrt{10}$ **18.** $\sqrt{10}$ **19.** $2\sqrt{17}$ **20.** 5 **21.** $\sqrt{85}$ **22.** $\sqrt{29}$ **23.** $\sqrt{53}$ **24.** $5\sqrt{5}$ **25.** $\sqrt{6.89} \approx 2.625$ **26.** $\sqrt{3.69} \approx 1.921$ **27.** $\sqrt{a^2 + b^2}$ **28.** $\sqrt{2}|a|$

29. $d(A, B) = \sqrt{13}$
$d(B, C) = \sqrt{13}$
$d(A, C) = \sqrt{26}$
$(\sqrt{13})^2 + (\sqrt{13})^2 = (\sqrt{26})^2$
Area $= \frac{13}{2}$ square units

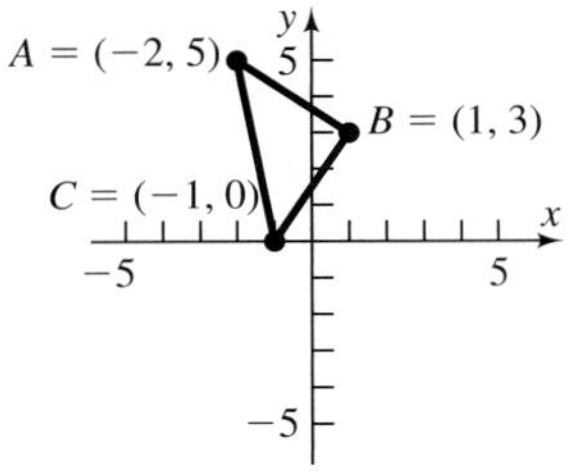

30. $d(A, B) = 10\sqrt{2}$
$d(B, C) = 10\sqrt{2}$
$d(A, C) = 20$
$20^2 = (10\sqrt{2})^2 + (10\sqrt{2})^2$
Area = 100 square units

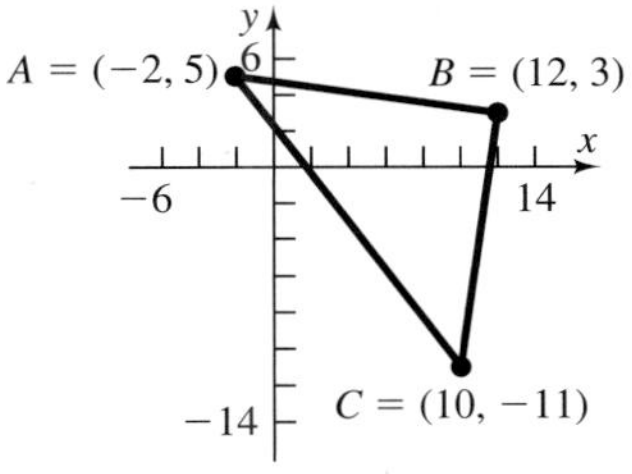

31. $d(A, B) = \sqrt{130}$
$d(B, C) = \sqrt{26}$
$d(A, C) = 2\sqrt{26}$
$(\sqrt{26})^2 + (2\sqrt{26})^2 = (\sqrt{130})^2$
Area = 26 square units

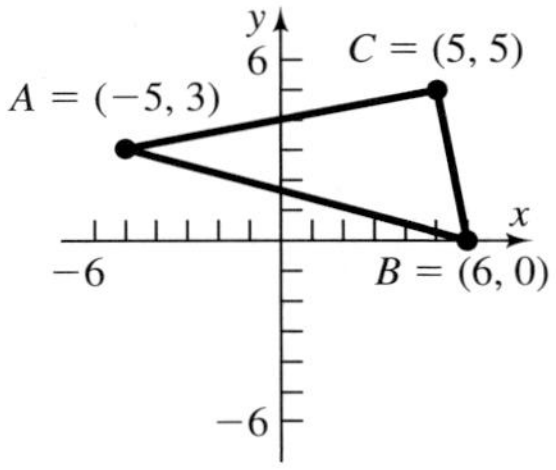

32. $d(A, B) = \sqrt{145}$
$d(B, C) = 2\sqrt{29}$
$d(A, C) = \sqrt{29}$
$(\sqrt{145})^2 = (2\sqrt{29})^2 = (\sqrt{29})^2$
Area = 29 square units

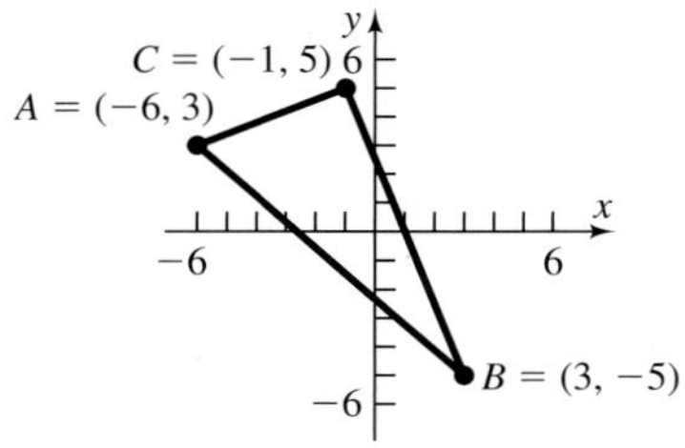

33. $d(A, B) = 4$
$d(B, C) = \sqrt{41}$
$d(A, C) = 5$
$4^2 + 5^2 = (\sqrt{41})^2$
Area = 10 square units

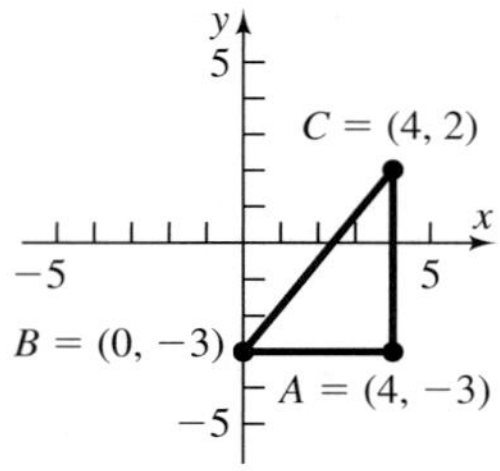

34. $d(A, B) = 4$
$d(B, C) = 2$
$d(A, C) = 2\sqrt{5}$
$(2\sqrt{5})^2 = 4^2 + 2^2$
Area = 4 square units

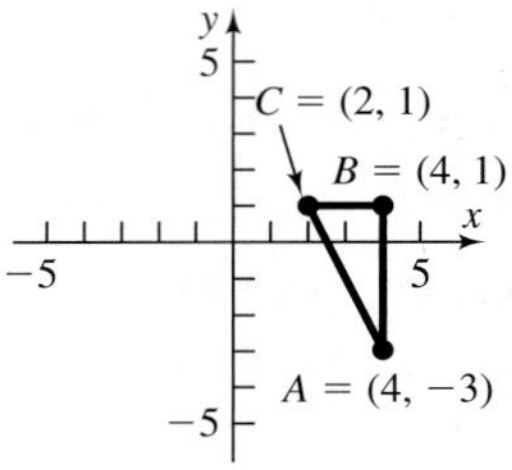

35. (2, 2); (2, −4) **36.** (13, −3); (−11, −3) **37.** (0, 0); (8, 0) **38.** (0, 1); (0, 7) **39.** (4, −1) **40.** $\left(\frac{1}{2}, 2\right)$ **41.** $\left(\frac{3}{2}, 1\right)$ **42.** $\left(3, -\frac{1}{2}\right)$ **43.** (5, −1) **44.** $\left(-1, -\frac{1}{2}\right)$ **45.** (1.05, 0.7) **46.** (0.45, 1.7) **47.** $\left(\frac{a}{2}, \frac{b}{2}\right)$ **48.** $\left(\frac{a}{2}, \frac{a}{2}\right)$ **49.** $\sqrt{17}; 2\sqrt{5}; \sqrt{29}$
50. Two triangles are possible. The third vertex is $(2\sqrt{3}, 2)$ or $(-2\sqrt{3}, 2)$. **51.** $d(P_1, P_2) = 6; d(P_2, P_3) = 4; d(P_1, P_3) = 2\sqrt{13}$; right triangle
52. $d(P_1, P_2) = \sqrt{53}; d(P_2, P_3) = \sqrt{53}; d(P_1, P_3) = \sqrt{106}$; isosceles right triangle
53. $d(P_1, P_2) = 2\sqrt{17}; d(P_2, P_3) = \sqrt{34}; d(P_1, P_3) = \sqrt{34}$; isosceles right triangle
54. $d(P_1, P_2) = 5\sqrt{5}; d(P_2, P_3) = 10; d(P_1, P_3) = 5$; right triangle **55.** $4\sqrt{10}$ **56.** $\sqrt{149}$ **57.** $2\sqrt{65}$ **58.** $\sqrt{205}$
59. $\left(\frac{s}{2}, \frac{s}{2}\right)$ **60.** Let $P_1 = (0, 0)$, $P_2 = (a, 0)$, $P_3 = \left(\frac{a}{2}, \frac{\sqrt{3}a}{2}\right)$; $d(P_1, P_2) = a$, $d(P_2, P_3) = a$, $d(P_1, P_3) = a$, an equilateral triangle.

Midpoints $P_4 = \left(\frac{a}{2}, 0\right), P_5 = \left(\frac{3a}{4}, \frac{\sqrt{3}a}{4}\right), P_6 = \left(\frac{a}{4}, \frac{\sqrt{3}a}{4}\right)$

$d(P_4, P_5) = \frac{a}{2}$, $d(P_5, P_6) = \frac{a}{2}$, $d(P_4, P_6) = \frac{a}{2}$, an equilateral triangle

61. $90\sqrt{2} \approx 127.28$ ft **62.** $60\sqrt{2} \approx 84.85$ ft **63.** **(a)** (90, 0), (90, 90), (0, 90) **(b)** $5\sqrt{2161}$ ft or ≈ 232.4 ft **(c)** $30\sqrt{149}$ ft or ≈ 366.2 ft
64. **(a)** First base (60, 0); second base (60, 60); third base (0, 60) **(b)** $40\sqrt{10}$ ft or ≈ 126.5 ft **(c)** $20\sqrt{185}$ ft or ≈ 272.0 ft
65. $d = 50t$ **66.** $d = 2\sqrt{2500 + 121t^2}$

1.2 Concepts and Vocabulary *(page 19)*

3. intercepts **4.** True **5.** (3, −4) **6.** radius **7.** True **8.** (2, −5); 6

1.2 Exercises *(page 20)*

9.

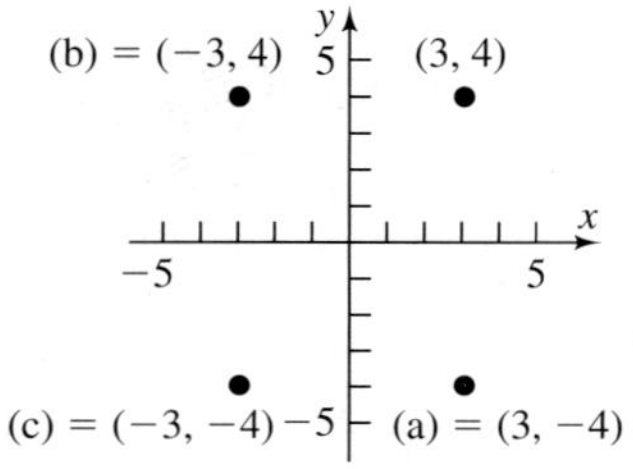

10.

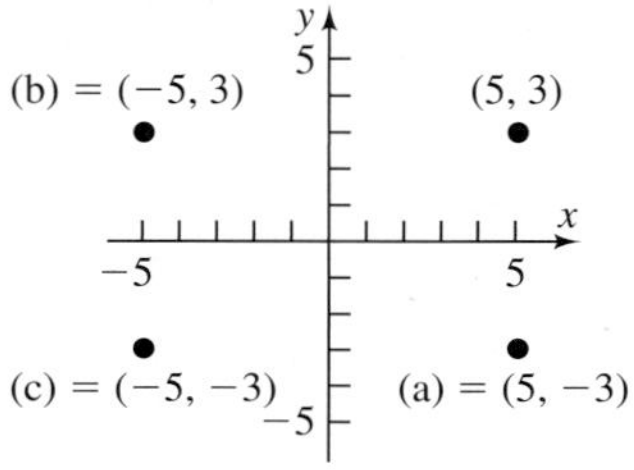

11.

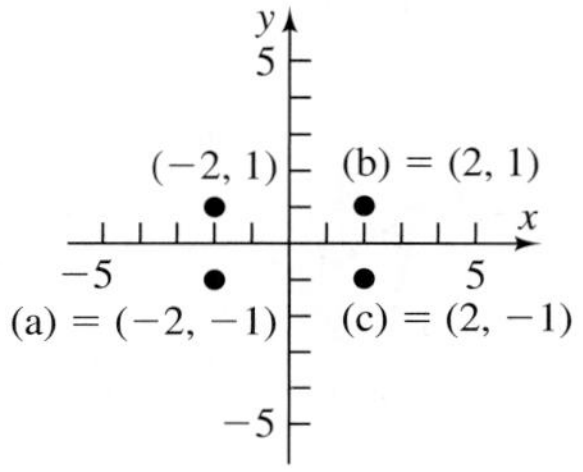

12.

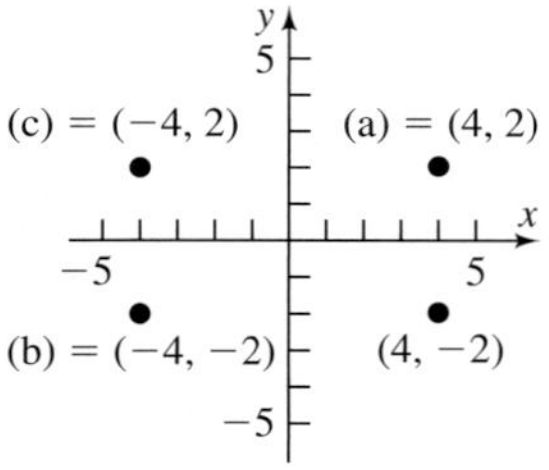

13.

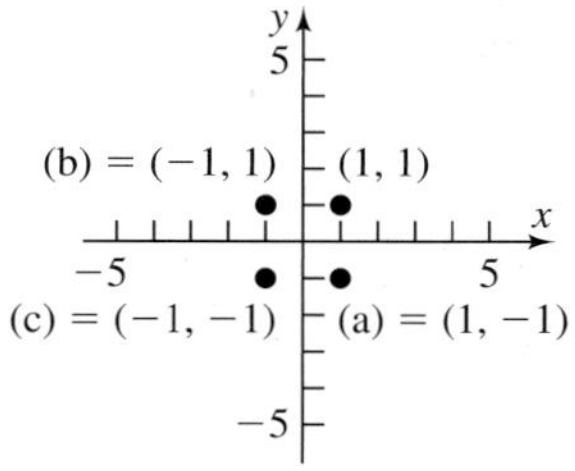

14.

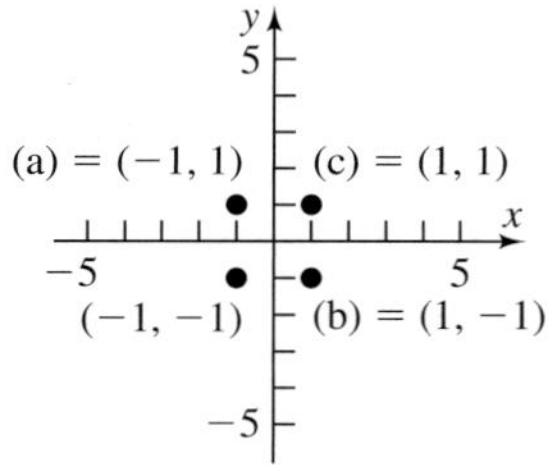

15.
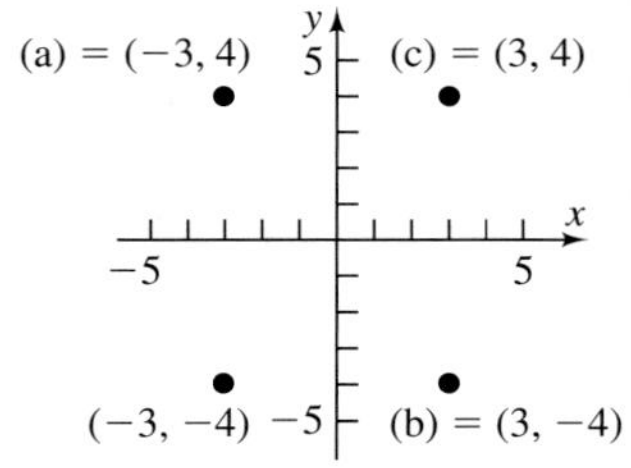

16.
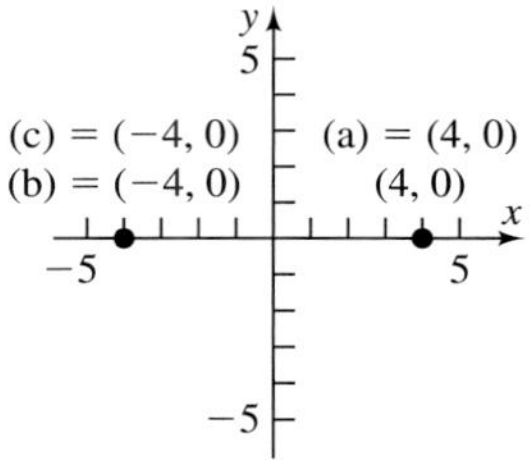

17.
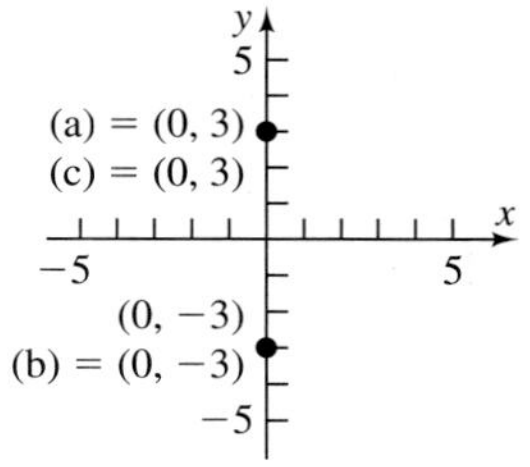

18.
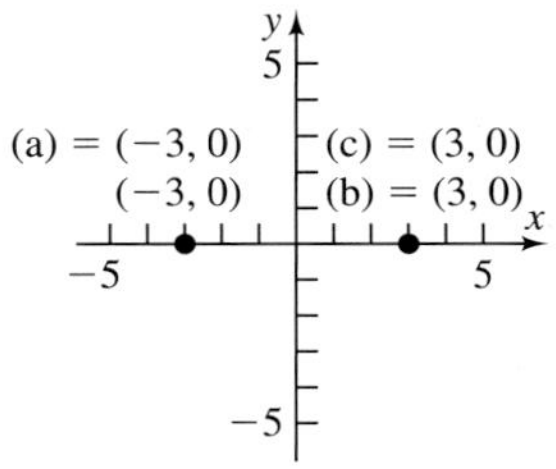

19. **(a)** $(-1, 0), (1, 0)$ **(b)** x-axis, y-axis, origin **20.** **(a)** $(0, 1)$ **(b)** none

21. **(a)** $\left(-\frac{\pi}{2}, 0\right), (0, 1), \left(\frac{\pi}{2}, 0\right)$ **(b)** y-axis **22.** **(a)** $(-2, 0), (0, -3), (2, 0)$ **(b)** y-axis **23.** **(a)** $(0, 0)$ **(b)** x-axis
24. **(a)** $(-2, 0), (2, 0), (0, -2), (0, 2)$ **(b)** x-axis, y-axis, origin **25.** **(a)** $(1, 0)$ **(b)** none **26.** **(a)** $(0, 0)$ **(b)** none
27. **(a)** $(-1.5, 0), (0, -2), (1.5, 0)$ **(b)** y-axis **28.** **(a)** $(0, 0)$ **(b)** origin **29.** **(a)** none **(b)** origin **30.** **(a)** none **(b)** x-axis
31. $(0, 0)$ is on the graph. **32.** $(0, 0)$ and $(1, -1)$ are on the graph. **33.** $(0, 3)$ is on the graph. **34.** $(0, 1)$ and $(-1, 0)$ are on the graph.
35. $(0, 2)$ and $(\sqrt{2}, \sqrt{2})$ are on the graph. **36.** $(0, 1)$ and $(2, 0)$ are on the graph. **37.** $(0, 0)$; symmetric with respect to the y-axis
38. $(0, 0)$; symmetric with respect to the x-axis **39.** $(0, 0)$; symmetric with respect to the origin **40.** $(0, 0)$; symmetric with respect to the origin
41. $(0, 9), (3, 0), (-3, 0)$; symmetric with respect to the y-axis **42.** $(-4, 0), (0, -2), (0, 2)$; symmetric with respect to the x-axis
43. $(-2, 0), (2, 0), (0, -3), (0, 3)$; symmetric with respect to the x-axis, y-axis, and origin **44.** $(-1, 0), (1, 0), (0, -2), (0, 2)$; symmetric with respect to the x-axis, y-axis, and origin **45.** $(0, -27), (3, 0)$; no symmetry **46.** $(-1, 0), (1, 0), (0, -1)$; symmetric with respect to the y-axis
47. $(0, -4), (4, 0), (-1, 0)$; no symmetry **48.** $(0, 4)$; symmetric with respect to the y-axis **49.** $(0, 0)$; symmetric with respect to the origin
50. $(2, 0), (-2, 0)$; symmetric with respect to the origin **51.** $(0, 0)$; symmetric with respect to the origin **52.** No intercepts; symmetric with respect to the origin

53.
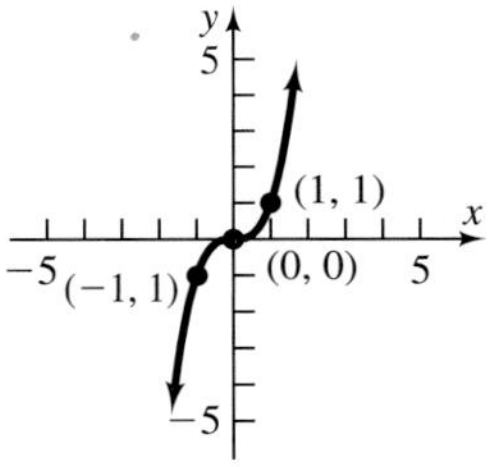

54.
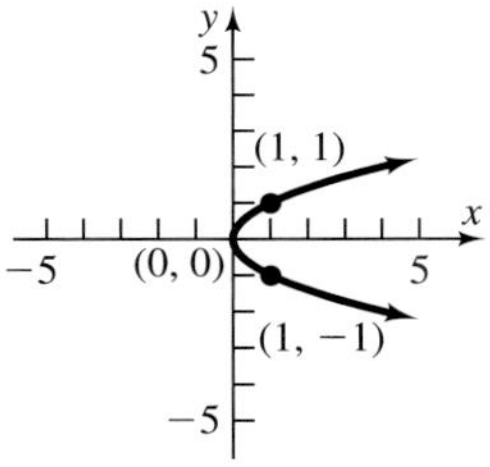

55.
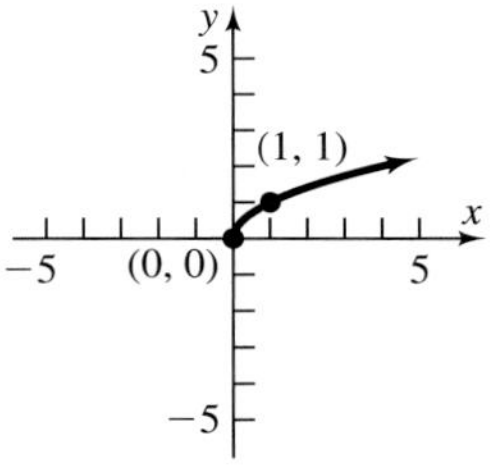

56.
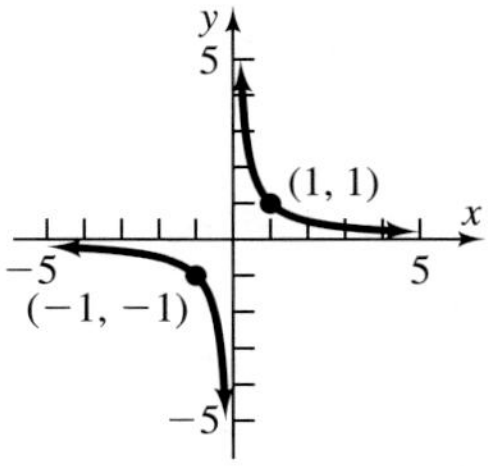

57. $a = -1$ **58.** $b = 12$ **59.** $2a + 3b = 6$ **60.** $m = -\frac{5}{2}, b = 5$ **61.** Center $(2, 1)$; Radius 2; $(x - 2)^2 + (y - 1)^2 = 4$

62. Center $(1, 2)$; Radius $= 2$; $(x - 1)^2 + (y - 2)^2 = 4$ **63.** Center $\left(\frac{5}{2}, 2\right)$; Radius $\frac{3}{2}$; $\left(x - \frac{5}{2}\right)^2 + (y - 2)^2 = \frac{9}{4}$

64. Center $(1, 2)$; Radius $= \sqrt{2}$; $(x - 1)^2 + (y - 2)^2 = 2$

65. $x^2 + y^2 = 4$; $x^2 + y^2 - 4 = 0$
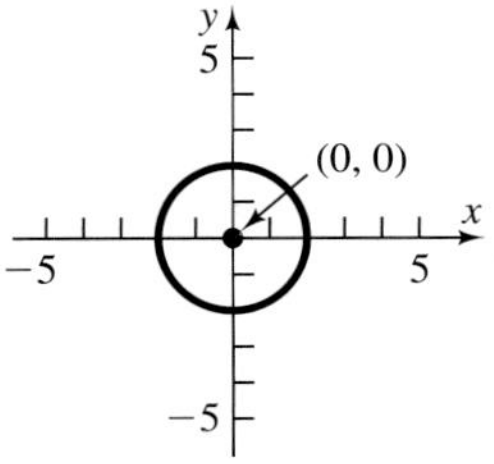

66. $x^2 + y^2 = 9$; $x^2 + y^2 - 9 = 0$
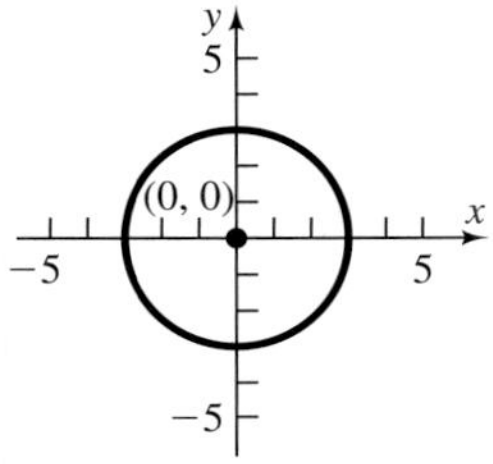

67. $(x - 1)^2 + (y + 1)^2 = 1$; $x^2 + y^2 - 2x + 2y + 1 = 0$
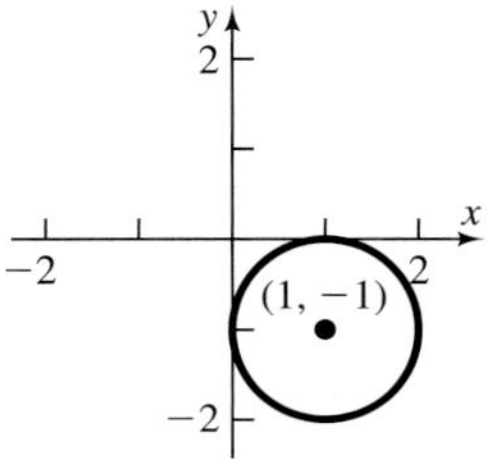

68. $(x + 2)^2 + (y - 1)^2 = 4$; $x^2 + y^2 + 4x - 2y + 1 = 0$
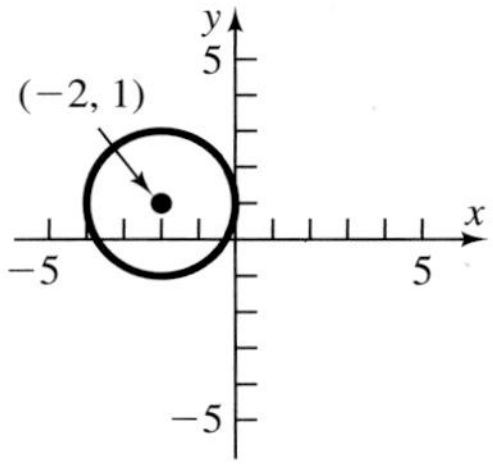

69. $x^2 + (y - 2)^2 = 4$; $x^2 + y^2 - 4y = 0$

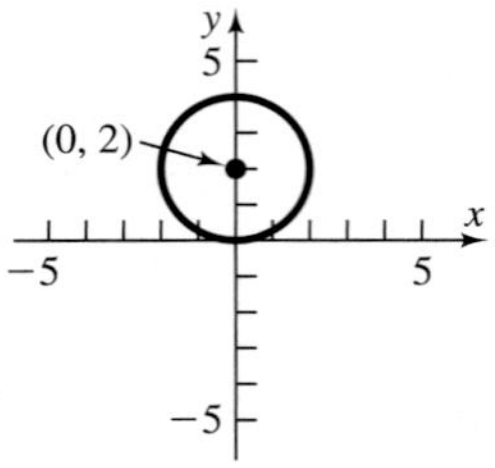

70. $(x - 1)^2 + y^2 = 9$; $x^2 + y^2 - 2x - 8 = 0$

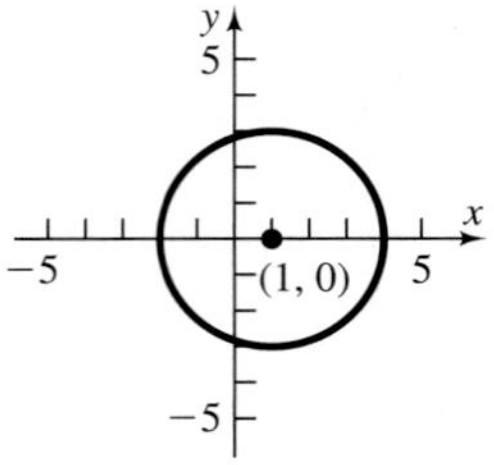

71. $(x - 4)^2 + (y + 3)^2 = 25$; $x^2 + y^2 - 8x + 6y = 0$

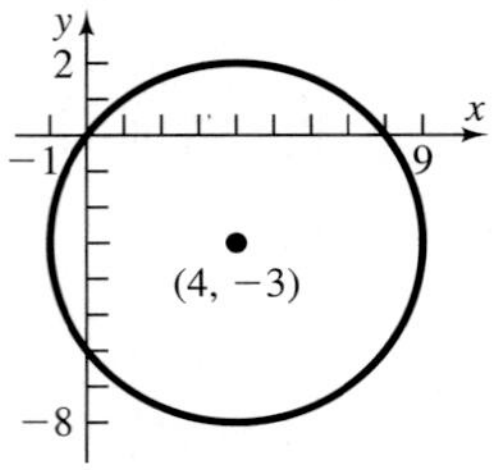

72. $(x - 2)^2 + (y + 3)^2 = 16$; $x^2 + y^2 - 4x + 6y - 3 = 0$

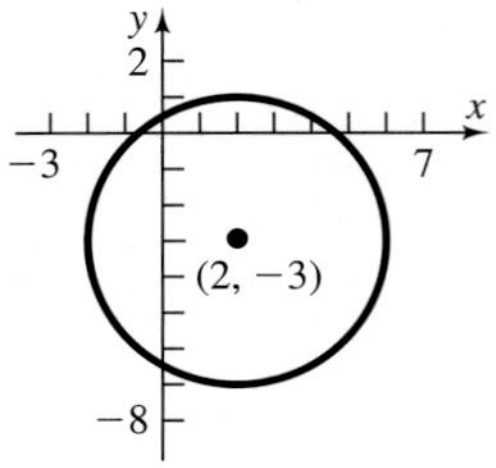

73. $(x + 3)^2 + (y + 6)^2 = 36$
$x^2 + y^2 + 6x + 12y + 9 = 0$

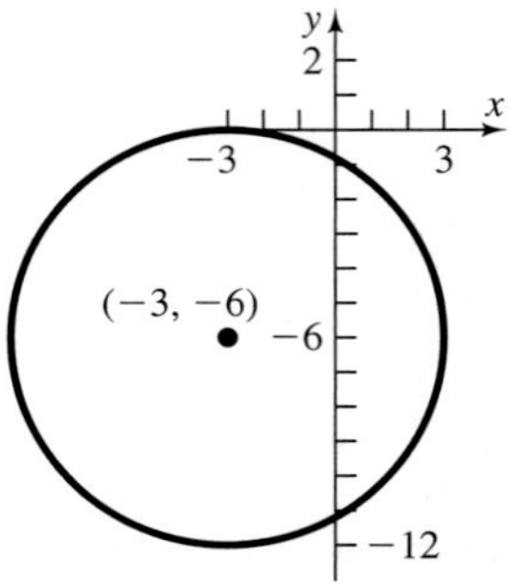

74. $(x + 5)^2 + (y - 2)^2 = 25$
$x^2 + y^2 + 10x - 4y + 4 = 0$

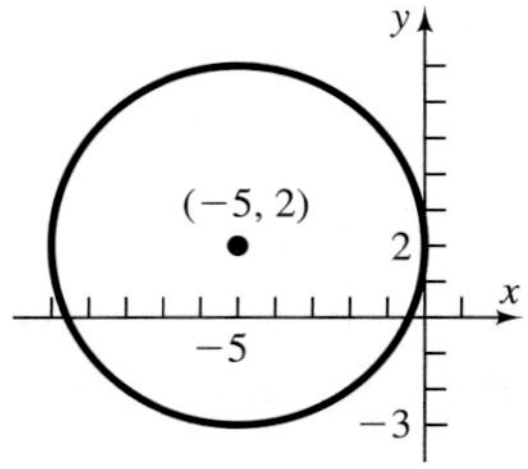

75. $x^2 + (y + 3)^2 = 9$
$x^2 + y^2 + 6y = 0$

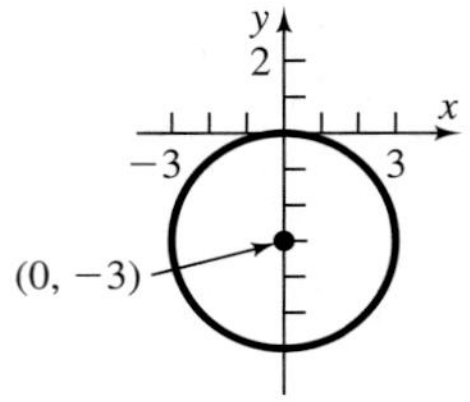

76. $(x^2 + 2)^2 + y^2 = 4$
$x^2 + y^2 + 4x = 0$

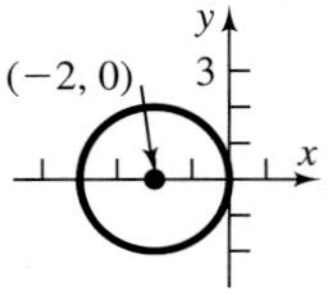

77. **(a)** $(h, k) = (0, 0); r = 2$
(b)

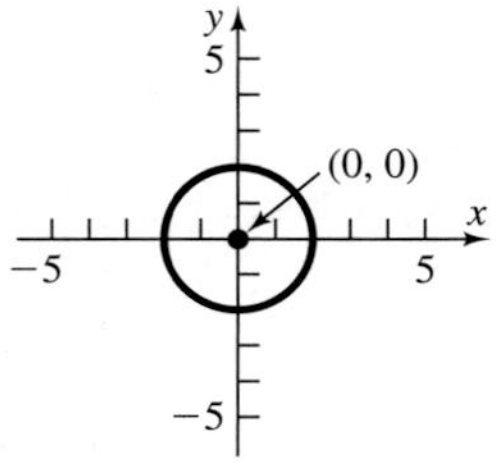

(c) $(\pm 2, 0); (0, \pm 2)$

78. **(a)** $(h, k) = (0, 1); r = 1$
(b)

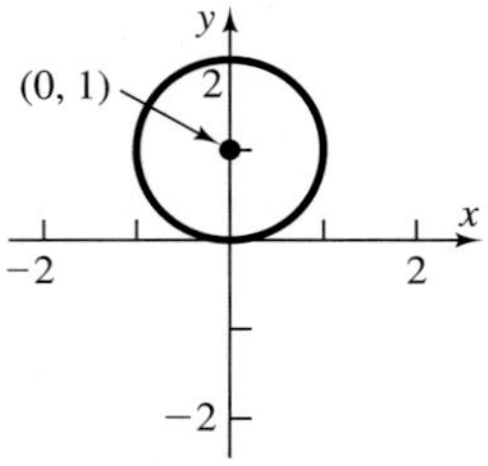

(c) $(0, 0); (0, 2)$

79. **(a)** $(h, k) = (3, 0); r = 2$
(b)

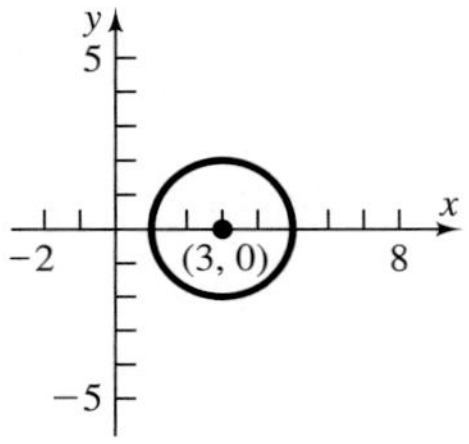

(c) $(1, 0); (5, 0)$

80. **(a)** $(h, k) = (-1, 1); r = \sqrt{2}$
(b)

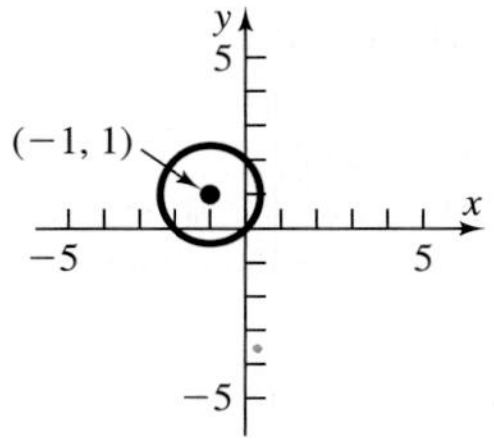

(c) $(0, 0); (0, 2); (-2, 0)$

81. **(a)** $(h, k) = (-2, 2); r = 3$
(b)

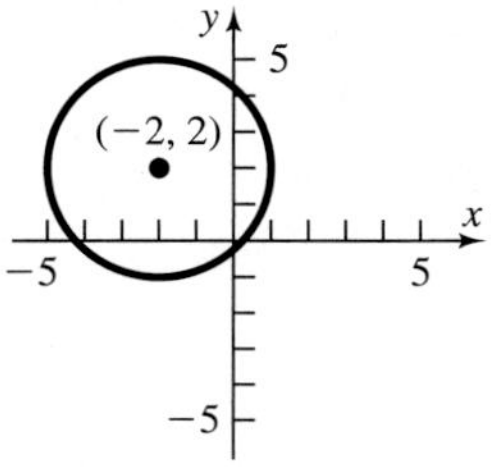

(c) $(-2 \pm \sqrt{5}, 0), (0, 2 \pm \sqrt{5})$

82. **(a)** $(h, k) = (3, -1); r = 1$
(b)

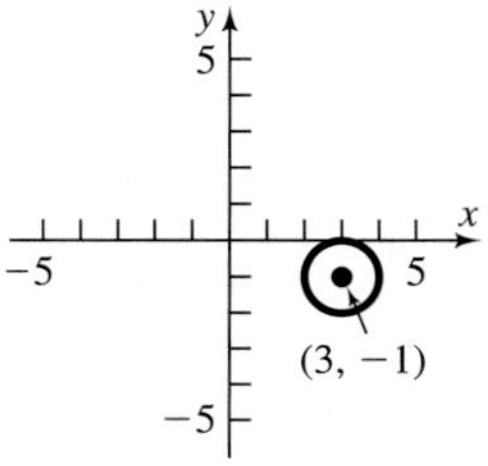

(c) $(3, 0)$

83. **(a)** $(h, k) = (1, -2); r = 3$
(b)

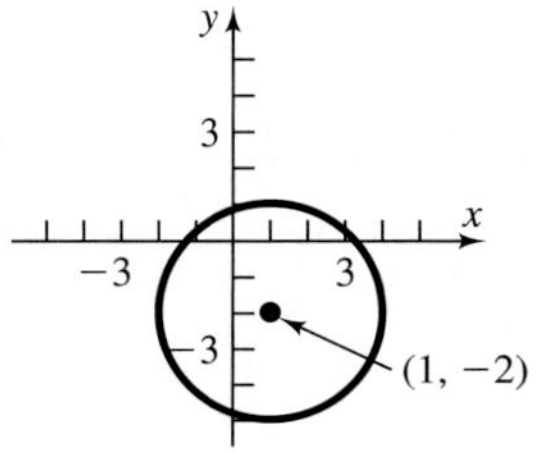

(c) $(1 \pm \sqrt{5}, 0); (0, -2 \pm 2\sqrt{2})$

84. (a) $(h, k) = (-2, -1); r = 5$

(b)

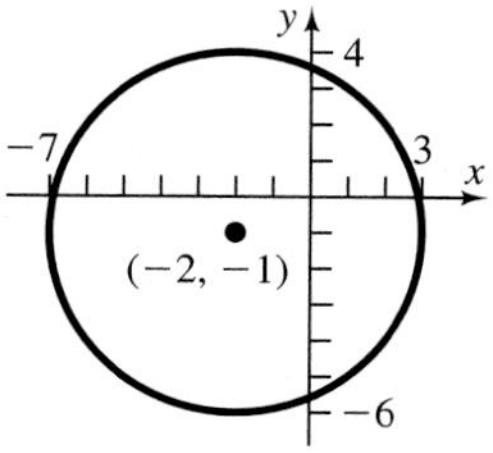

(c) $(-2 \pm 2\sqrt{6}, 0); (0, -1 \pm \sqrt{21})$

85. (a) $(h, k) = \left(\frac{1}{2}, -1\right); r = \frac{1}{2}$

(b)

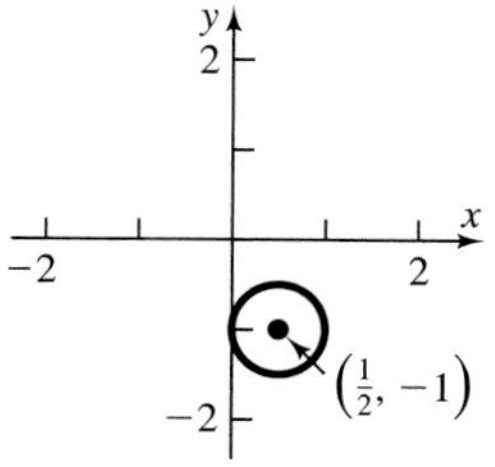

(c) $(0, -1)$

86. (a) $(h, k) = \left(-\frac{1}{2}, -\frac{1}{2}\right); r = 1$

(b)

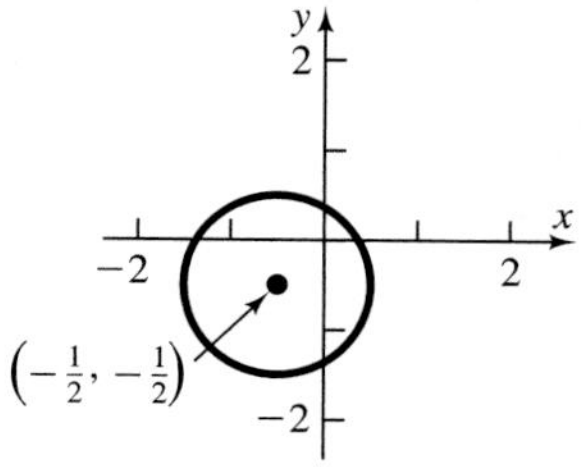

(c) $\left(\frac{-1 \pm \sqrt{3}}{2}, 0\right), \left(0, \frac{-1 \pm \sqrt{3}}{2}\right)$

87. (a) $(h, k) = (3, -2); r = 5$

(b)

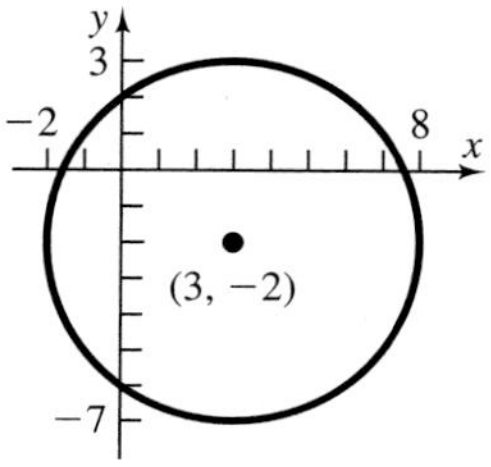

(c) $(3 \pm \sqrt{21}, 0); (0, -6), (0, 2)$

88. (a) $(h, k) = (-2, 0); r = \frac{\sqrt{2}}{2}$

(b)

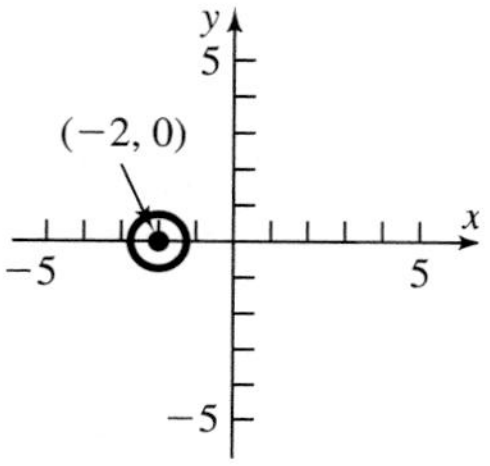

(c) $\left(-2 \pm \frac{\sqrt{2}}{2}, 0\right)$

89. $x^2 + y^2 - 13 = 0$ **90.** $x^2 + y^2 - 2x - 17 = 0$

91. $x^2 + y^2 - 4x - 6y + 4 = 0$
92. $x^2 + y^2 + 6x - 2y + 1 = 0$
93. $x^2 + y^2 + 2x - 6y + 5 = 0$
94. $x^2 + y^2 - 4x - 4y + 3 = 0$
95. c **96.** d **97.** b **98.** a **99.** b, c, e, g **100.** b, e, g
101. $x^2 + y^2 + 2x + 4y - 4168.16 = 0$

102. (a)

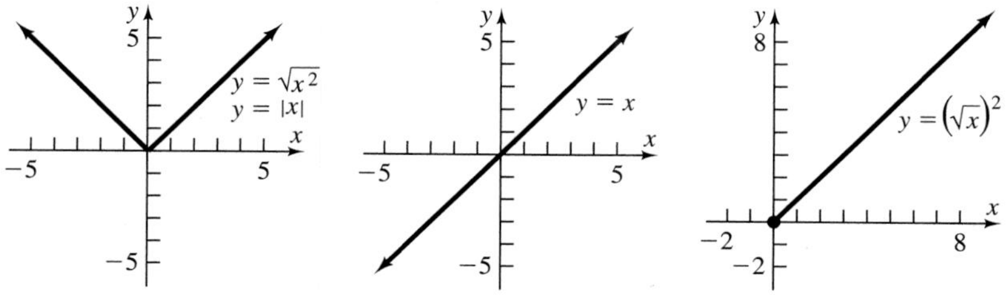

(b) Since $\sqrt{x^2} = |x|$, for all x, the graphs of $y = \sqrt{x^2}$ and $y = |x|$ are the same.

(c) For $y = (\sqrt{x})^2$, the domain of the variable x is $x \geq 0$; for $y = x$, the domain of the variable x is all real numbers. Thus, $(\sqrt{x})^2 = x$ only for $x \geq 0$.

(d) For $y = \sqrt{x^2}$, the range of the variable y is $y \geq 0$; for $y = x$, the range of the variable y is all real numbers. Also, $\sqrt{x^2} = |x|$, which equals x only if $x \geq 0$.

1.3 Concepts and Vocabulary *(page 34)*

1. undefined; 0 **2.** 3; 2 **3.** $y = b$; y-intercept **4.** True **5.** False **6.** True **7.** $m_1 = m_2$; y-intercepts; $m_1 m_2 = -1$ **8.** 2 **9.** $-\frac{1}{2}$
10. False

1.3 Exercises *(page 34)*

11. (a) $\frac{1}{2}$ **(b)** For every 2-unit increase in x, y will increase by 1 unit. **12. (a)** $-\frac{1}{2}$ **(b)** For every 2-unit increase in x, y will decrease by 1 unit.

13. (a) $-\frac{1}{3}$ **(b)** For every 3-unit increase in x, y will decrease by 1 unit. **14. (a)** $\frac{1}{3}$ **(b)** For every 3-unit increase in x, y will increase by 1 unit.

15. Slope $= -\frac{3}{2}$

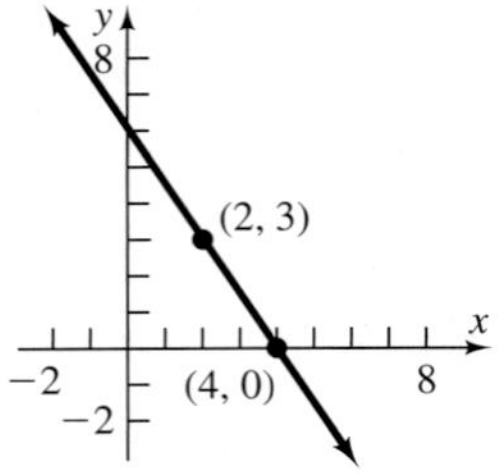

16. Slope $= -2$

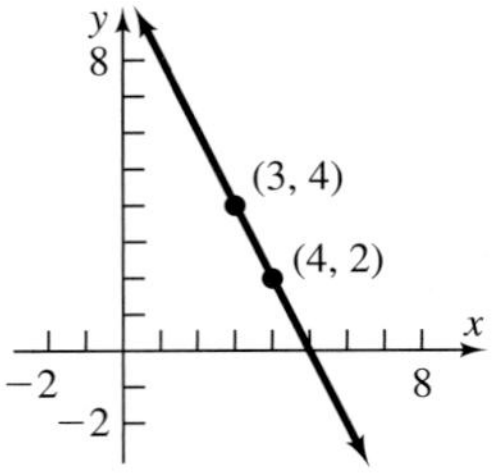

17. Slope $= -\frac{1}{2}$

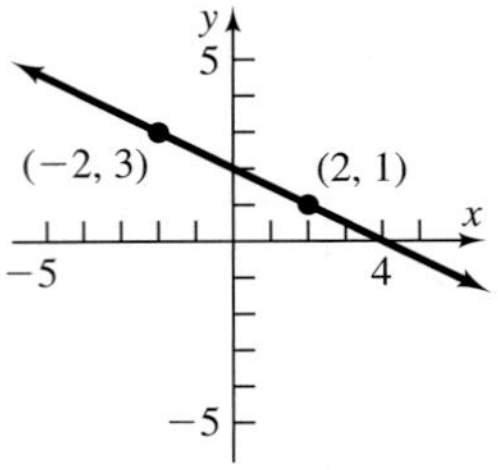

18. Slope $= \frac{2}{3}$

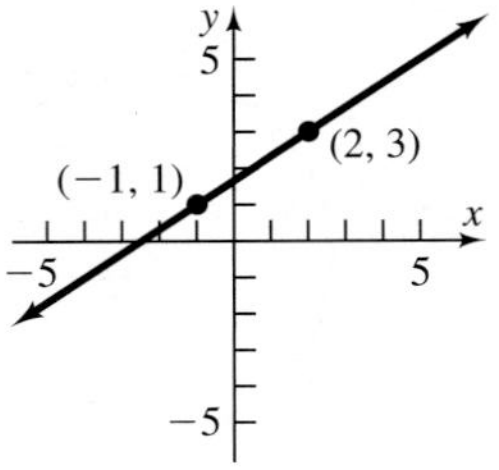

19. Slope $= 0$

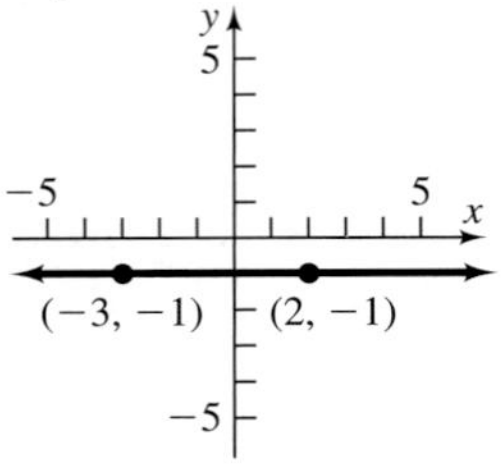

20. Slope $= 0$

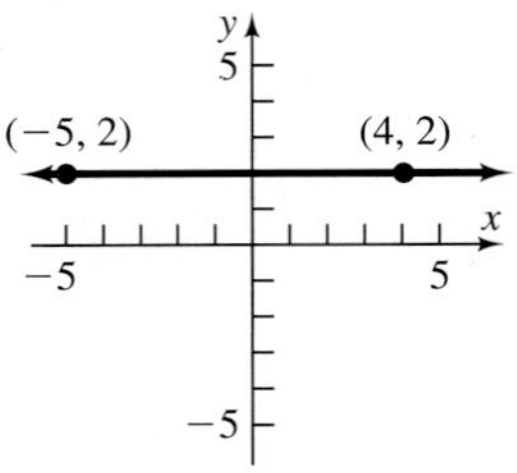

21. Slope undefined

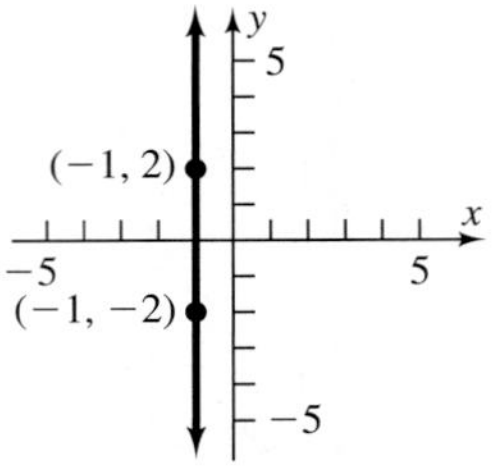

22. Slope undefined

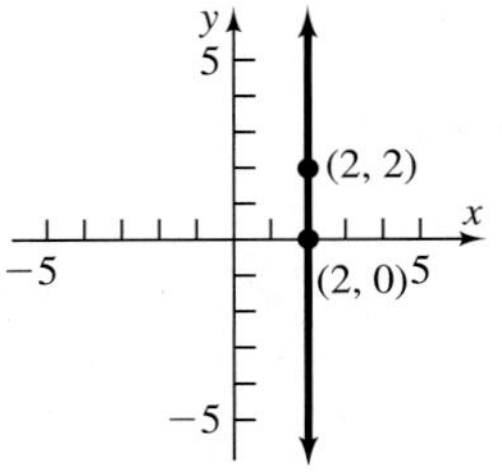

23.

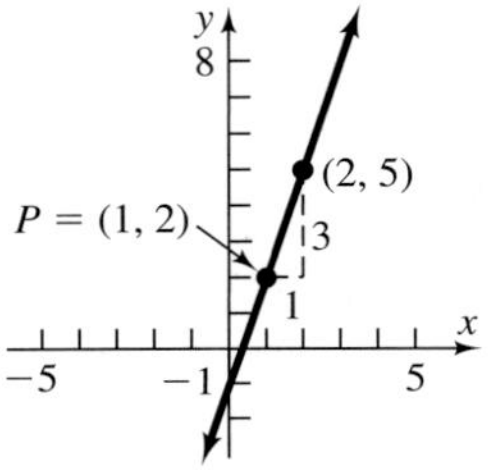

24.

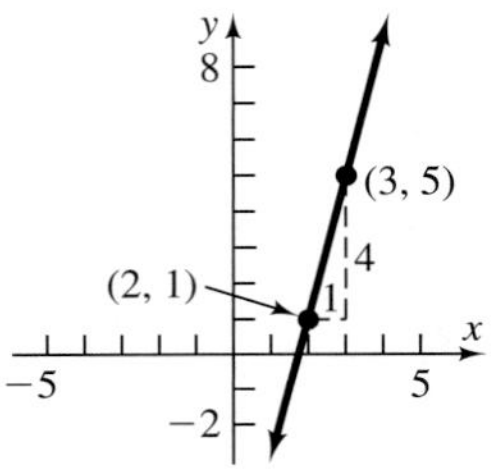

25.

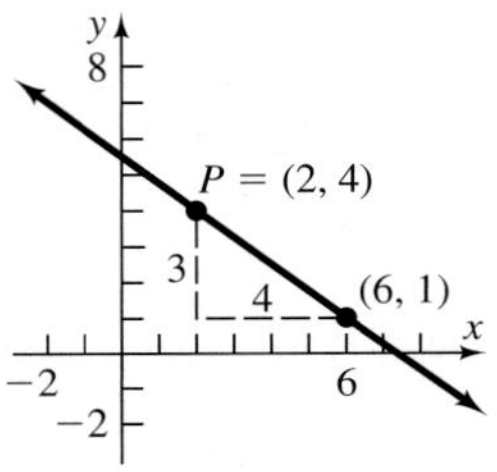

26.

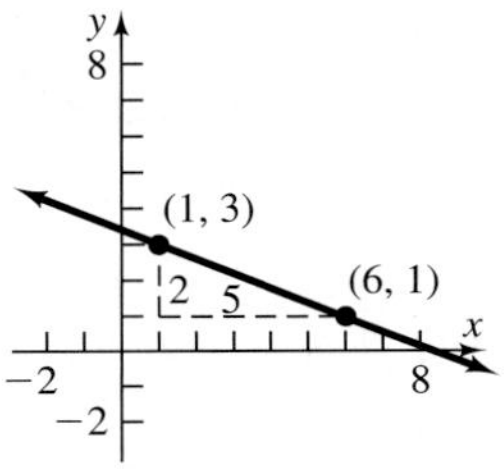

27.

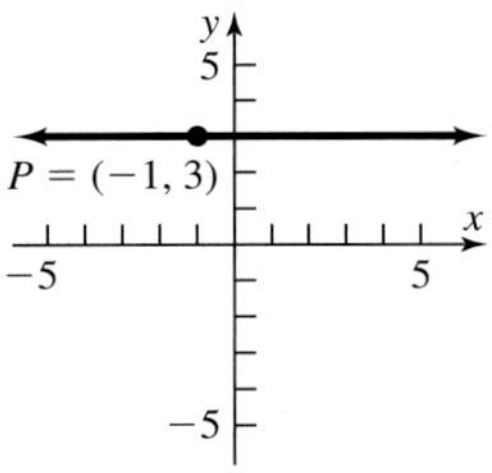

28.

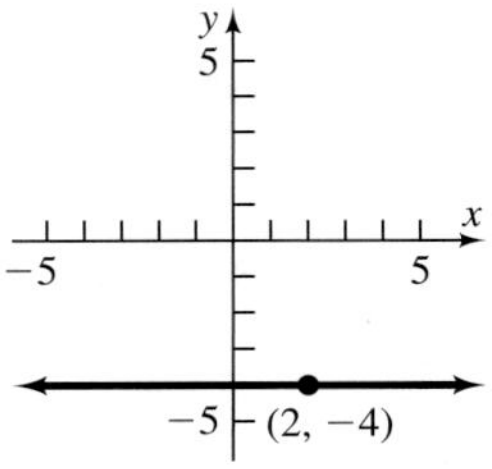

29.

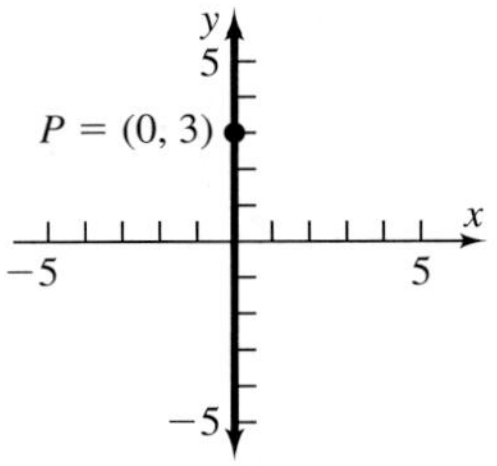

30.

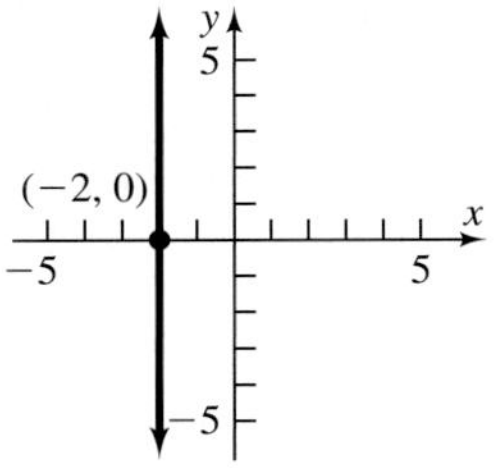

31. $x - 2y = 0$ or $y = \frac{1}{2}x$ **32.** $x + 2y = 0$ or $y = -\frac{1}{2}x$ **33.** $x + 3y = 4$ or $y = -\frac{1}{3}x + \frac{4}{3}$ **34.** $x - 3y = -4$ or $y = \frac{1}{3}x + \frac{4}{3}$

35. $2x - y = 3$ or $y = 2x - 3$ **36.** $x + y = 3$ or $y = -x + 3$ **37.** $x + 2y = 5$ or $y = -\frac{1}{2}x + \frac{5}{2}$ **38.** $x - y = -2$ or $y = x + 2$

39. $3x - y = -9$ or $y = 3x + 9$ **40.** $2x - y = 11$ or $y = 2x - 11$ **41.** $2x + 3y = -1$ or $y = -\frac{2}{3}x - \frac{1}{3}$

42. $x - 2y = 1$ or $y = \frac{1}{2}x - \frac{1}{2}$ **43.** $x - 2y = -5$ or $y = \frac{1}{2}x + \frac{5}{2}$ **44.** $x - 5y = -23$ or $y = \frac{1}{5}x + \frac{23}{5}$

45. $3x + y = 3$ or $y = -3x + 3$ **46.** $2x + y = -2$ or $y = -2x - 2$ **47.** $x - 2y = 2$ or $y = \frac{1}{2}x - 1$ **48.** $x - y = -4$ or $y = x + 4$

49. $x = 2$; no slope–intercept form **50.** $x = 3$; no slope–intercept form **51.** $y - 2 = 0$ or $y = 2$ **52.** $x = 4$; no slope–intercept form

53. $2x - y = -4$ or $y = 2x + 4$ **54.** $3x + y = -1$ or $y = -3x - 1$ **55.** $2x - y = 0$ or $y = 2x$ **56.** $x - 2y = 0$ or $y = \frac{1}{2}x$

57. $x = 4$; no slope–intercept form **58.** $y = 2$ or $y = 2$ **59.** $2x + y = 0$ or $y = -2x$ **60.** $x + 2y = -3$ or $y = -\frac{1}{2}x - \frac{3}{2}$

61. $x - 2y = -3$ or $y = \frac{1}{2}x + \frac{3}{2}$ **62.** $2x + y = 4$ or $y = -2x + 4$ **63.** $y = 4$ or $y = 4$ **64.** $x = 3$; no slope–intercept form

65. Slope = 2; y-intercept = 3

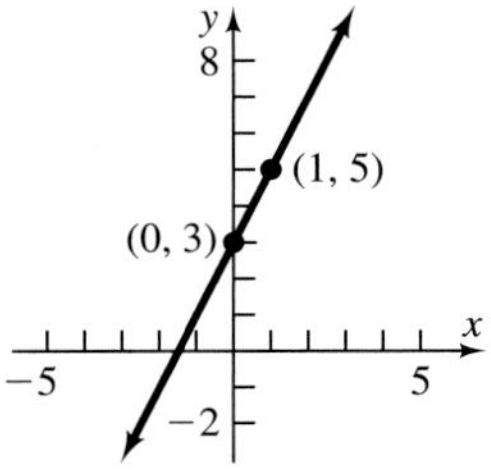

66. Slope = −3; y-intercept = 4

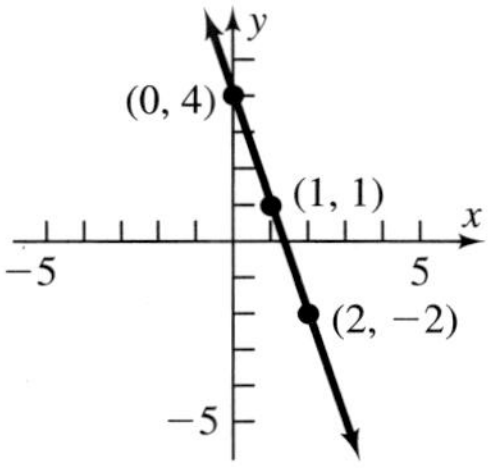

67. $y = 2x - 2$; Slope = 2; y-intercept = −2

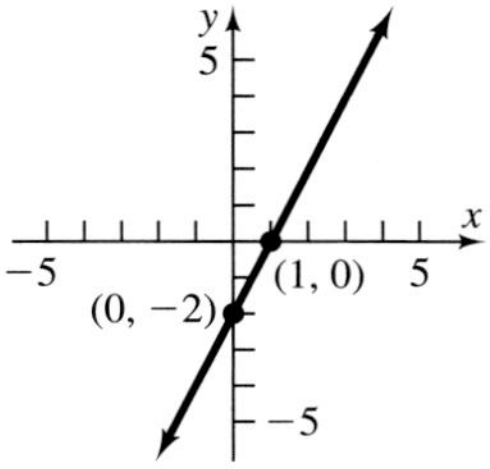

68. $y = -\frac{1}{3}x + 2$; Slope = $-\frac{1}{3}$; y-intercept = 2

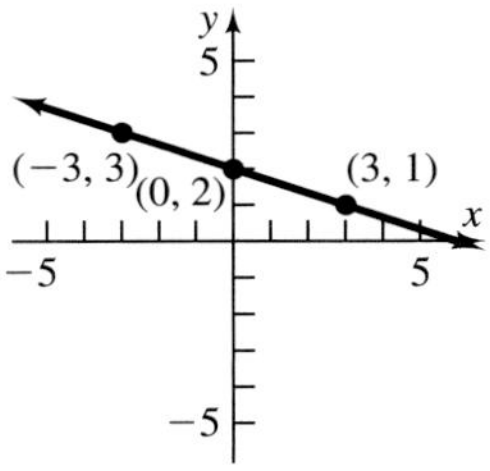

69. Slope = $\frac{1}{2}$; y-intercept = 2

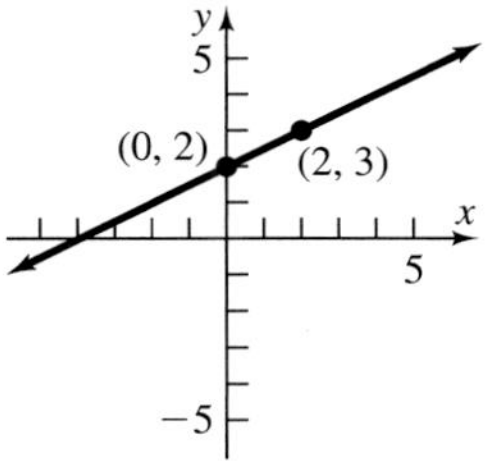

70. Slope = 2; y-intercept = $\frac{1}{2}$

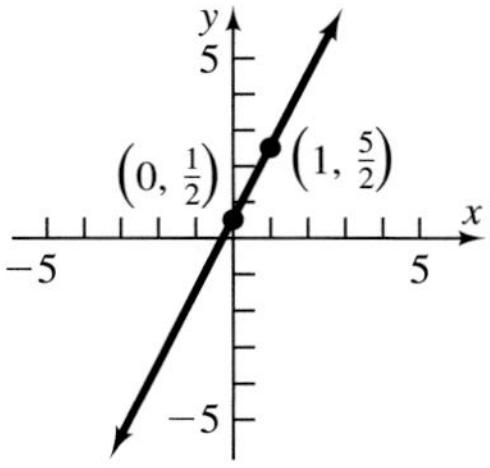

71. $y = -\frac{1}{2}x + 2$; Slope = $-\frac{1}{2}$; y-intercept = 2

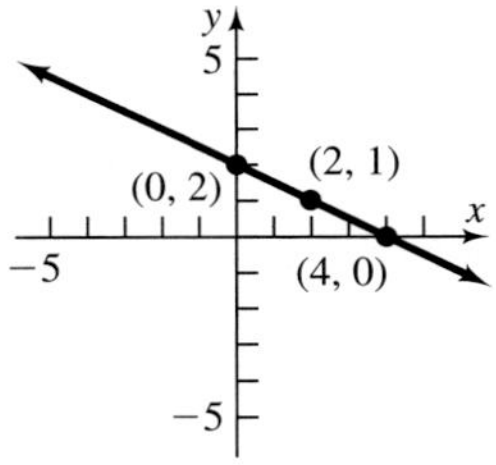

72. $y = \frac{1}{3}x + 2$; Slope = $\frac{1}{3}$; y-intercept = 2

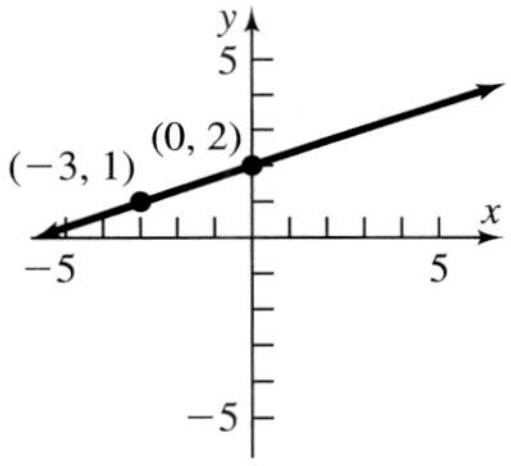

73. $y = \frac{2}{3}x - 2$; Slope = $\frac{2}{3}$; y-intercept = −2

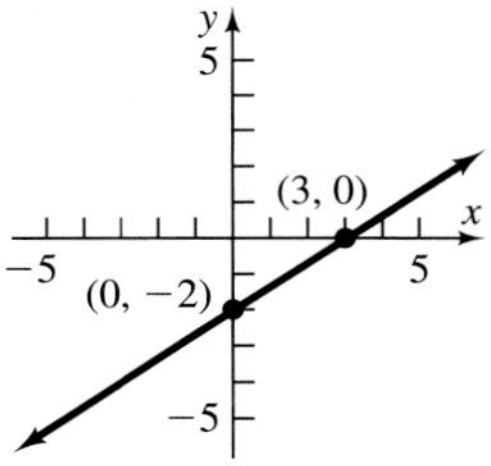

74. $y = -\frac{3}{2}x + 3$; Slope = $-\frac{3}{2}$; y-intercept = 3

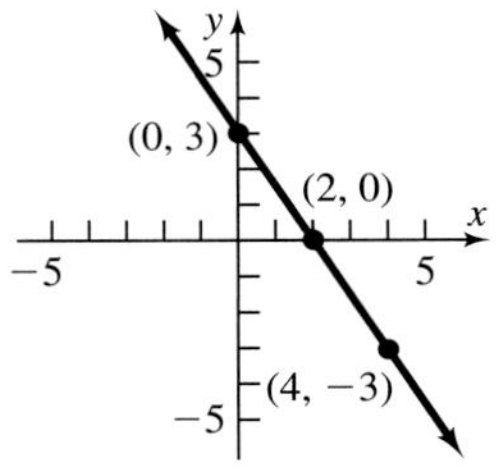

75. $y = -x + 1$; Slope = −1; y-intercept = 1

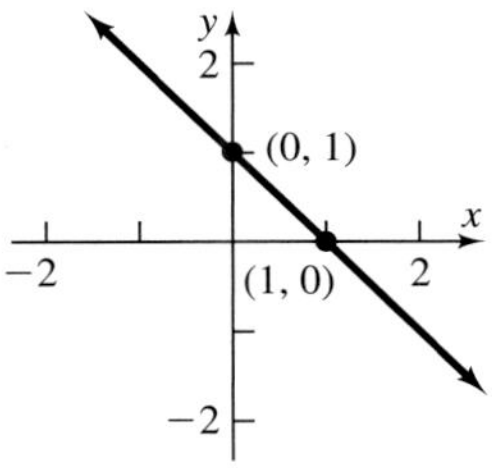

76. $y = x - 2$; Slope = 1; y-intercept = −2

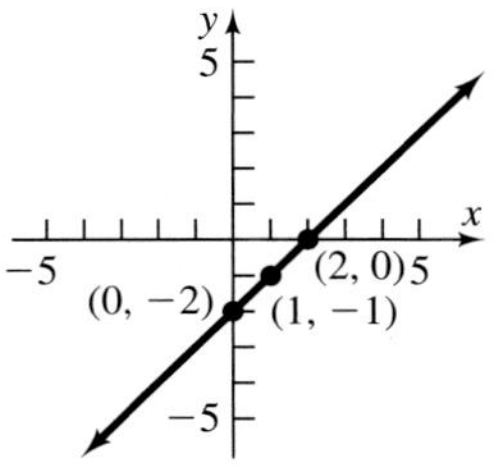

77. Slope undefined; no y-intercept

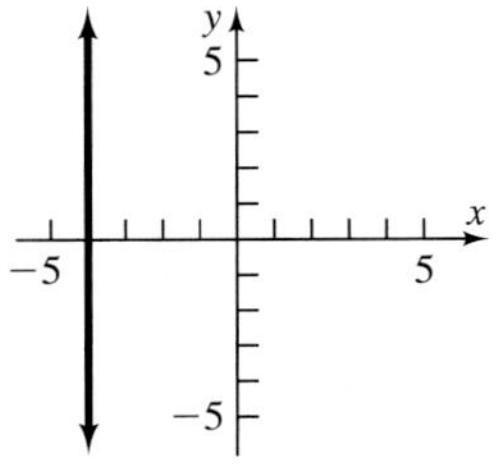

78. Slope = 0; y-intercept = −1

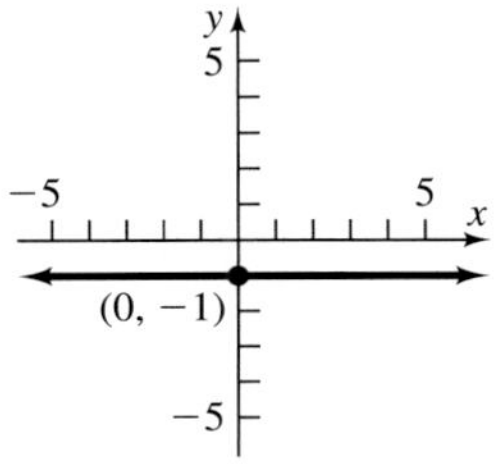

79. Slope = 0; y-intercept = 5

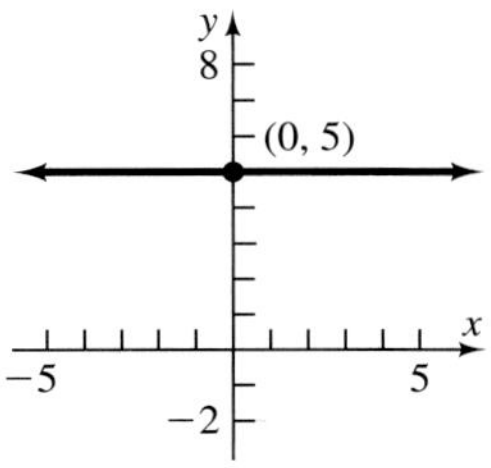

80. Slope undefined; no y-intercept

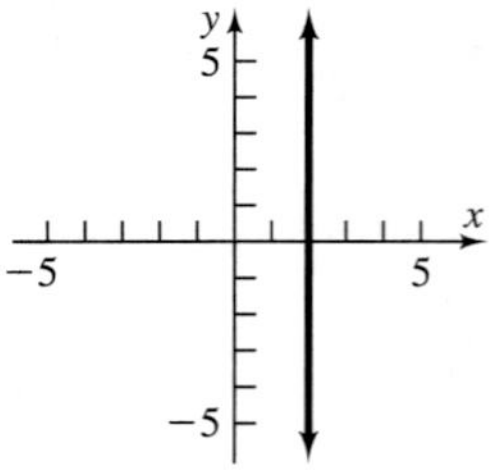

81. $y = x$; Slope $= 1$; y-intercept $= 0$

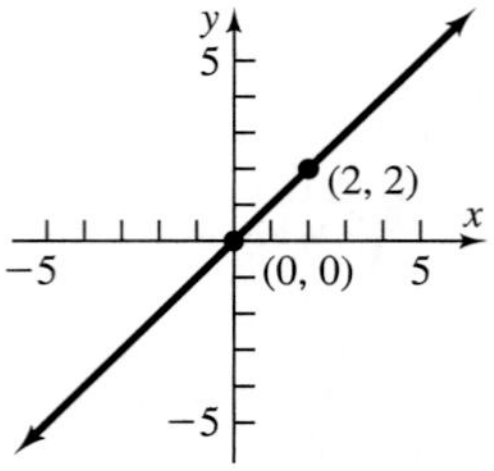

82. $y = -x$; Slope $= -1$; y-intercept $= 0$

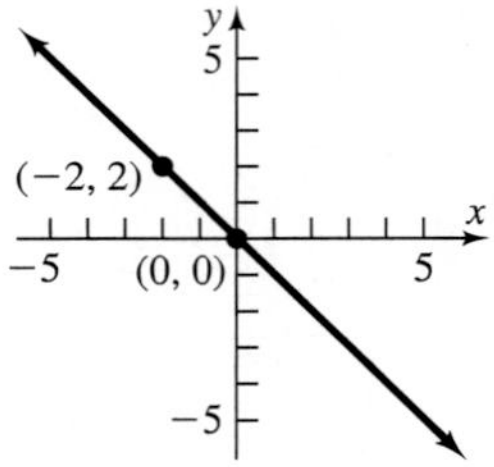

83. $y = \frac{3}{2}x$; Slope $= \frac{3}{2}$; y-intercept $= 0$

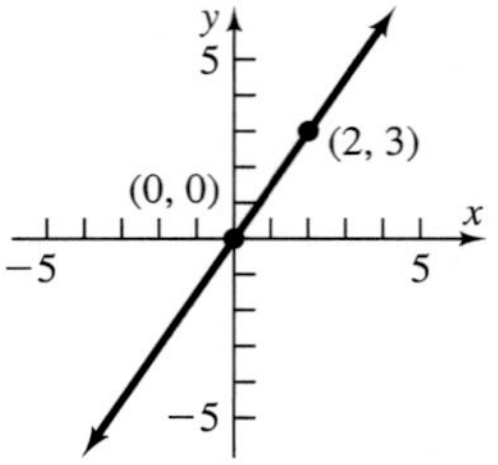

84. $y = -\frac{3}{2}x$; Slope $= -\frac{3}{2}$; y-intercept $= 0$

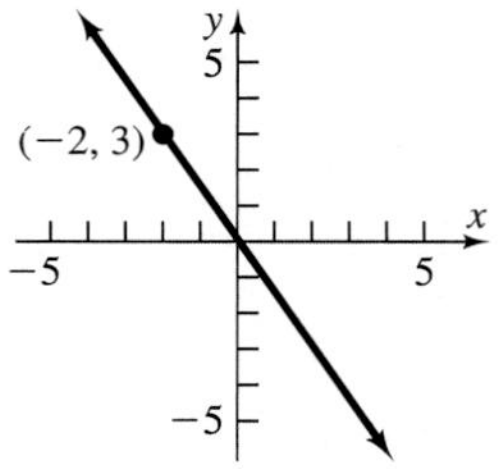

85. parallel

86. perpendicular

87. neither

88. neither

89. $P_1 = (-2, 5), P_2 = (1, 3), m_1 = -\frac{2}{3}; P_2 = (1, 3), P_3 = (-1, 0), m_2 = \frac{3}{2}$; because $m_1 m_2 = -1$, the lines are perpendicular; thus, the points P_1, P_2, and P_3 are the vertices of a right triangle.

90. $P_1 = (1, -1), P_3 = (2, 2), m = 3; P_2 = (4, 1), P_4 = (5, 4), m = 3; P_3 = (2, 2), P_4 = (5, 4), m = \frac{2}{3}$; $P_1 = (1, -1), P_2 = (4, 1), m = \frac{2}{3}$; opposite sides are parallel; the points are the vertices of a parallelogram.

91. $P_1 = (-1, 0), P_2 = (2, 3), m = 1; P_3 = (1, -2), P_4 = (4, 1), m = 1; P_1 = (-1, 0), P_3 = (1, -2), m = -1$; $P_2 = (2, 3), P_4 = (4, 1), m = -1$; opposite sides are parallel, and adjacent sides are perpendicular; the points are the vertices of a rectangle.

92. $P_1 = (0, 0), P_2 = (1, 3), m = 3; P_3 = (4, 2), P_4 = (3, -1), m = 3; P_2 = (1, 3), P_3 = (4, 2), m = -\frac{1}{3}; P_1 = (0, 0), P_4 = (3, -1), m = -\frac{1}{3}$; opposite sides are parallel; adjacent sides are perpendicular;
$d(P_1, P_2) = \sqrt{1^2 + 3^2} = \sqrt{10}; d(P_2, P_3) = \sqrt{(4-1)^2 + (2-3)^2} = \sqrt{3^2 + (-1)^2} = \sqrt{10}$;
$d(P_3, P_4) = \sqrt{(3-4)^2 + (-1-2)^2} = \sqrt{(-1)^2 + (-3)^2} = \sqrt{10}; d(P_1, P_4) = \sqrt{3^2 + (-1)^2} = \sqrt{10}$
all sides have equal lengths; the quadrilateral is a square.

93. $x - y = -2$ or $y = x + 2$ **94.** $x + y = 1$ or $y = -x + 1$

95. $x + 3y = 3$ or $y = -\frac{1}{3}x + 1$ **96.** $x + 2y = -2$ or $y = -\frac{1}{2}x - 1$ **97.** $y = 0$ **98.** $x = 0$ **99.** $C = 0.07x + 29$; \$36.70; \$45.10

100. $C = 8x + 500$; \$3700; \$6420 **101.** $C = 0.53x + 1{,}070{,}000$ **102.** $S = 0.05x + 375$

103. **(a)** $C = 0.08275x + 7.58, 0 \le x \le 400$ **(b)**

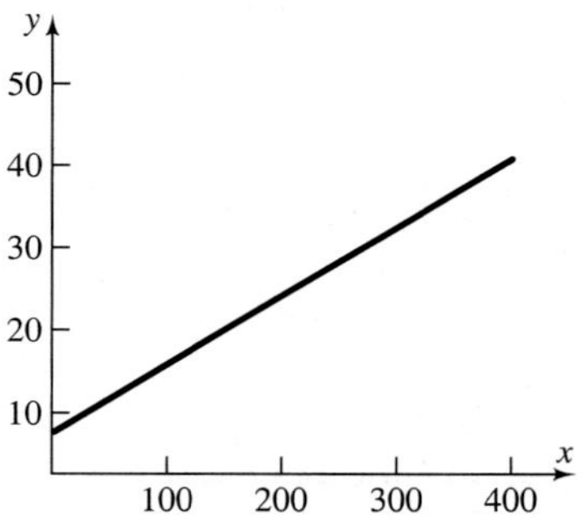

(c) \$15.86 **(d)** \$32.41

(e) Each kw-hr more that is used will add \$0.08275 to the bill.

104. **(a)** $C = 0.06787x + 5.25, 0 \le x \le 750$ **(b)**

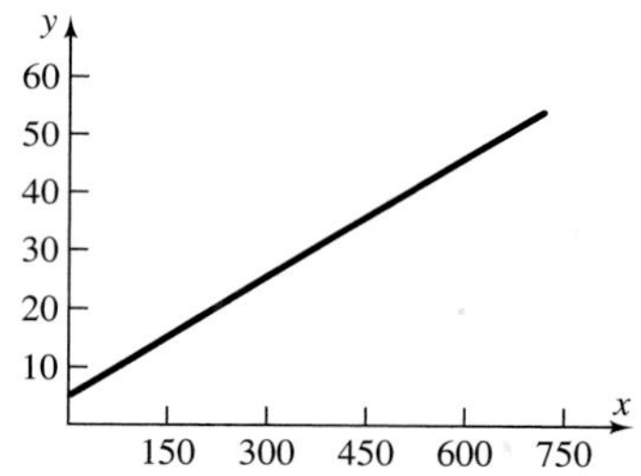

(c) \$18.82 **(d)** \$39.19 **(e)** Each kw-hr more that is used will add \$0.06787 to the bill.

105. $°C = \frac{5}{9}(°F - 32)$; approximately 21°C **106.** **(a)** $K = °C + 273$ **(b)** $K = \frac{5}{9}°F + \frac{2297}{9}$ **107.** **(a)** $A = \frac{1}{5}x + 20{,}000$ **(b)** \$80,000 **(c)** Each additional \$1 spent on advertising generates 5 additional boxes sold.

108. $y - 0 = \frac{-b}{a}(x - a)$ **109.** b, c, e, g **110.** a, c, f, g **111.** c **112.** d

$$y = \frac{-bx}{a} + b$$
$$\frac{bx}{a} + y = b$$
$$\frac{x}{a} + \frac{y}{b} = 1$$

113. **(a)** $x^2 + (mx + b)^2 = r^2$

$(1 + m^2)x^2 + 2mbx + b^2 - r^2 = 0$

One solution if and only if discriminant $= 0$

$(2mb)^2 - 4(1 + m^2)(b^2 - r^2) = 0$

$-4b^2 + 4r^2 + 4m^2r^2 = 0$

$r^2(1 + m^2) = b^2$

(b) $x = \frac{-2mb}{2(1 + m^2)} = \frac{-2mb}{2b^2/r^2} = -\frac{r^2m}{b}$

$y = m\left(-\frac{r^2m}{b}\right) + b = -\frac{r^2m^2}{b} + b = \frac{-r^2m^2 + b^2}{b} = \frac{r^2}{b}$

(c) Slope of tangent line $= m$

Slope of line joining center to point of tangency $= \frac{r^2/b}{-r^2m/b} = -\frac{1}{m}$

114. $\sqrt{2}x + 4y = 9\sqrt{2}$ **115.** $\sqrt{2}x + 4y = 11\sqrt{2} - 12$ **116.** $(1, 0)$ **117.** $x + 5y = -13$

118. Slope from (a, b) to (b, a) is $\frac{a - b}{b - a} = -1$. Slope of the line $y = x$ is 1. Since $-1 \cdot 1 = -1$, the line containing the points (a, b) and (b, a) is perpendicular to the line $y = x$. The midpoint of (a, b) and $(b, a) = \left(\frac{a + b}{2}, \frac{b + a}{2}\right)$. Since $\frac{b + a}{2} = \frac{a + b}{2}$, the midpoint lies on the line $y = x$.

119. All have the same slope, 2; the lines are parallel.

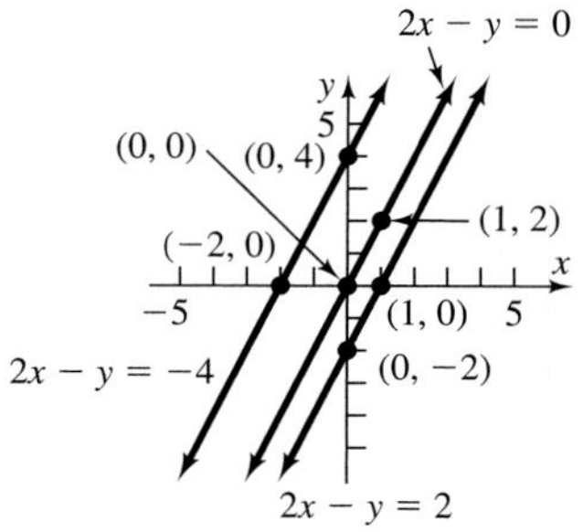

120. The family of lines $Cx + y = -4$ intersect at the point $(0, -4)$.

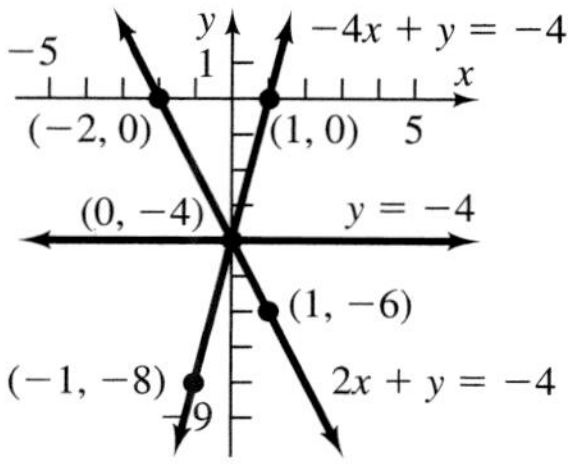

121. $y = 2$

122. In Figure 48, $[d(O, A)]^2 = 1 + m_2^2$, $[d(O, B)]^2 = 1 + m_1^2$, $[d(A, B)]^2 = m_2^2 - 2m_1m_2 + m_1^2$ (See top of p. 33.) If $m_1m_2 = -1$, then $[d(A, B)]^2 = m_1^2 + m_2^2 + 2$ and $[d(O, A)]^2 + [d(O, B)]^2 = [d(A, B)]^2$. By the converse of the Pythagorean Theorem, $\triangle ABO$ is a right triangle and $\angle O$ is a right angle. **123.** No, not if the line is vertical **124.** No, if the intercepts are $(0, 0)$ or the line is horizontal or vertical. **125.** They are the same line. **126.** They are the same line. **127.** No **128.** Yes, if the y-intercept is 0

1.4 Concepts and Vocabulary *(page 42)*

1. Scatter diagram **2.** True

1.4 Exercises *(page 42)*

3. Linear **4.** Nonlinear **5.** Linear **6.** No relation **7.** Nonlinear **8.** Nonlinear

9. (a), (c)

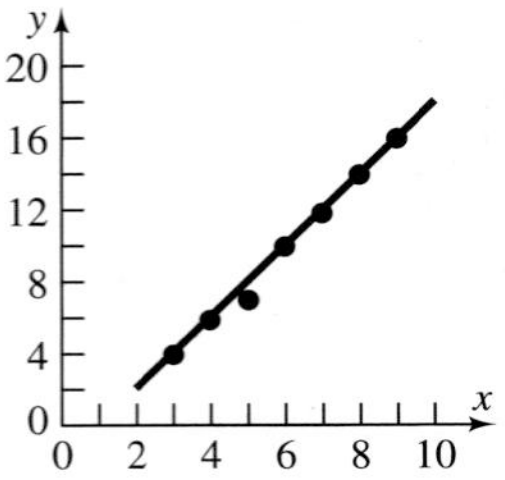

(b) Using (3, 4) and (9, 16), $y = 2x - 2$.

(d) $y = 2.0357x - 2.3571$ **(e)**

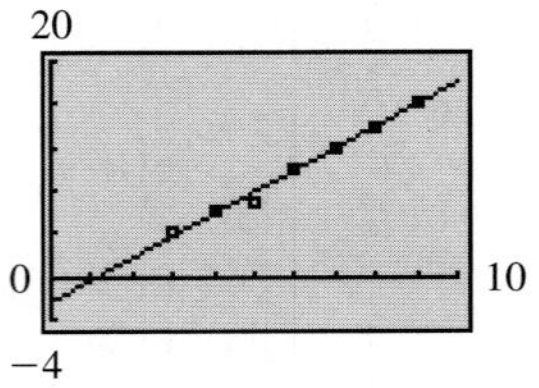

10. (a), (c)

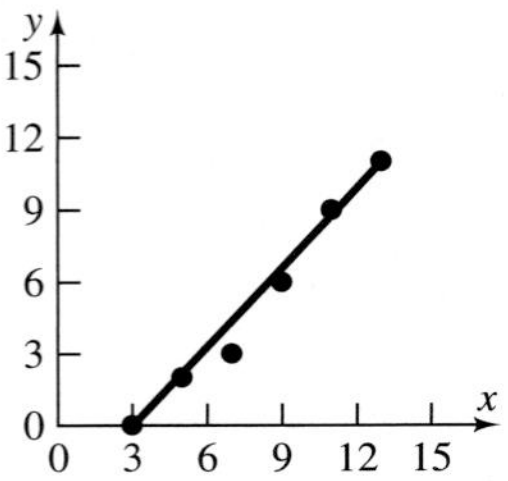

(b) Using (3, 0) and (13, 11), $y = \frac{11}{10}x - \frac{33}{10}$.

(d) $y = 1.129x - 3.862$ **(e)**

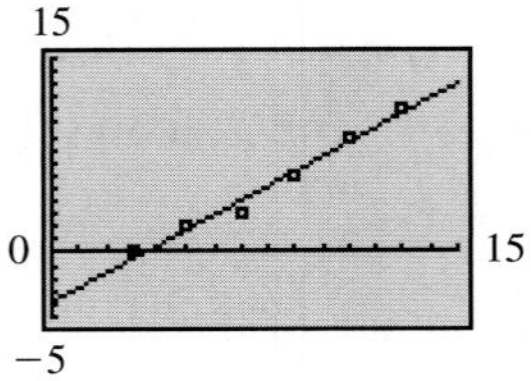

11. (a), (c)

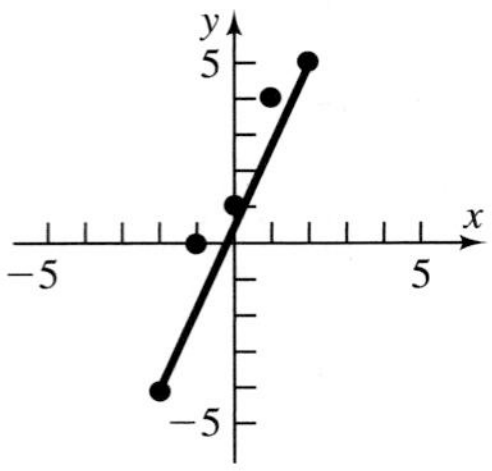

(b) Using (−2, −4) and (2, 5), $y = \frac{9}{4}x + \frac{1}{2}$.

(d) $y = 2.2x + 1.2$ **(e)**

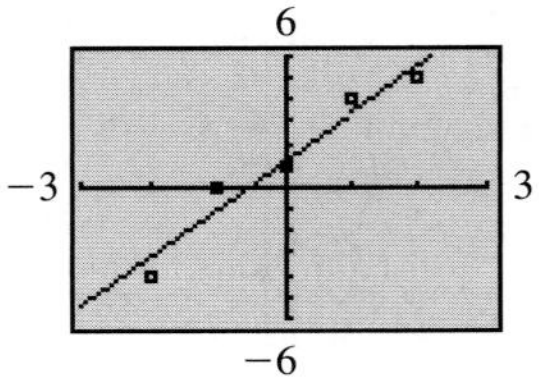

12. (a), (c)

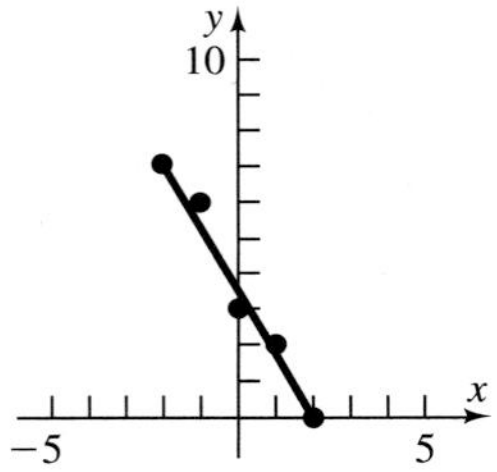

(b) Using (−2, 7) and (2, 0), $y = -\frac{7}{4}x + \frac{7}{2}$.

(d) $y = -1.8x + 3.6$ **(e)**

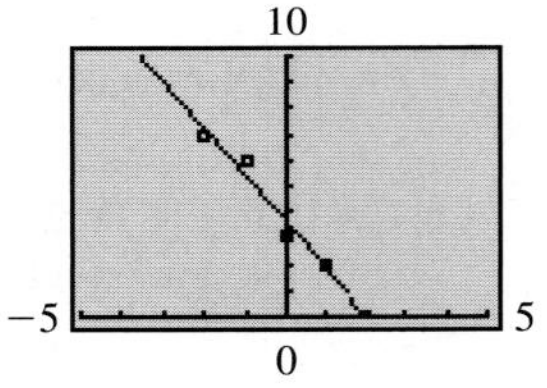

13. (a), (c)

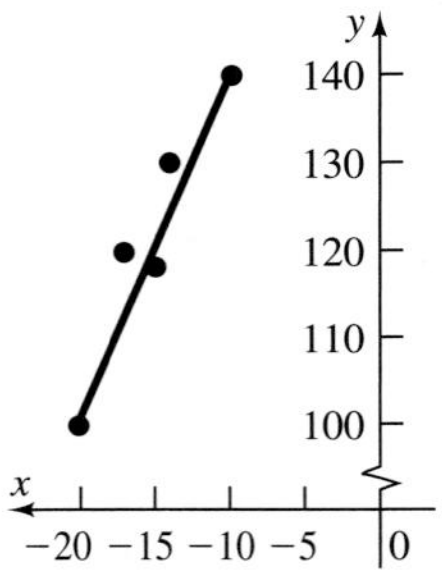

(b) Using (−20, 100) and (−10, 140), $y = 4x + 180$.

(d) $y = 3.8613x + 180.292$ **(e)**

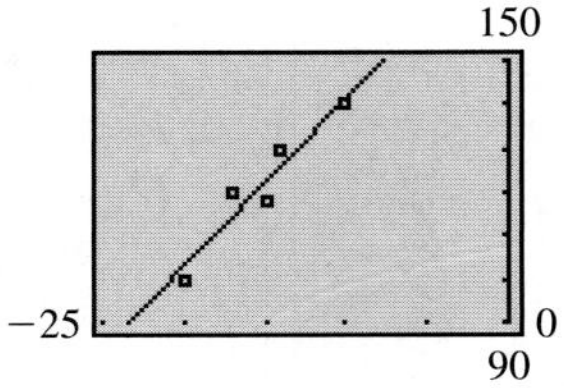

14. (a)

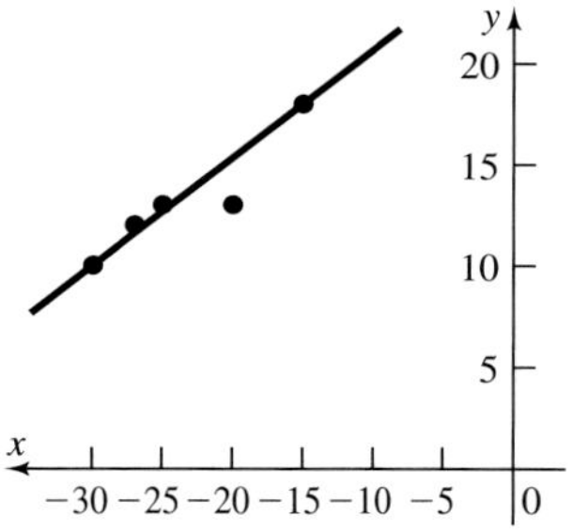

(b) Using $(-30, 10)$ and $(-14, 18)$, $y = \frac{1}{2}x + 25$.

(d) $y = 0.442x + 23.456$ **(e)**

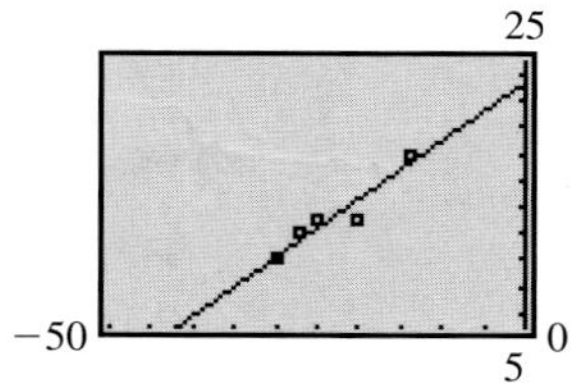

15. (a)

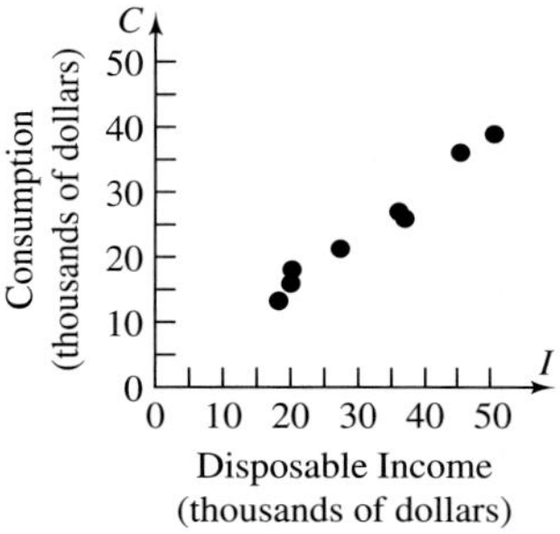

(b) Using $(20, 16)$ and $(50, 39)$, $C = \frac{23}{30}I + \frac{2}{3}$.

(c) As disposable income increases by \$1, consumption increases by about \$0.77.

(d) \$32,867

(e) $C = 0.755I + 0.6266$

16. (a)

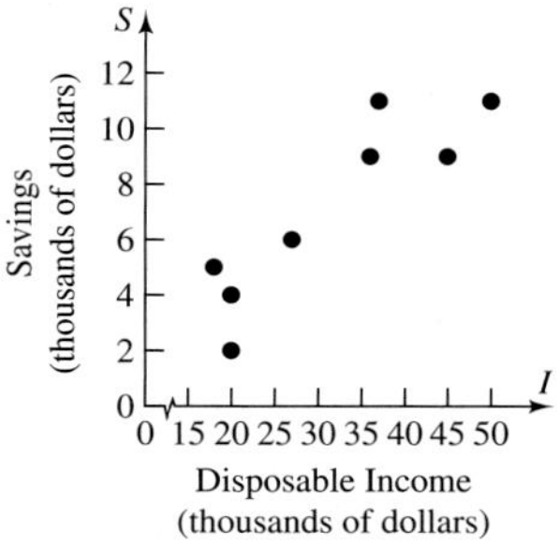

(b) Using $(20, 4)$ and $(50, 11)$, $S = \frac{7}{30}I - \frac{2}{3}$.

(c) Savings increases by \$233 for every \$1000 extra in disposable income.

(d) \$9133

(e) $S = 0.245I - 0.627$

17. (a)

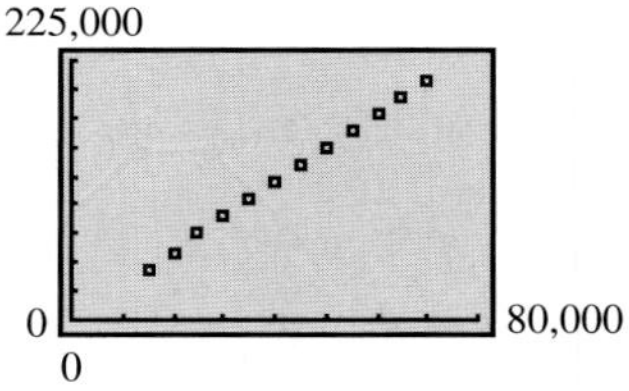

(b) $L = 2.98I - 76.11$

(c)

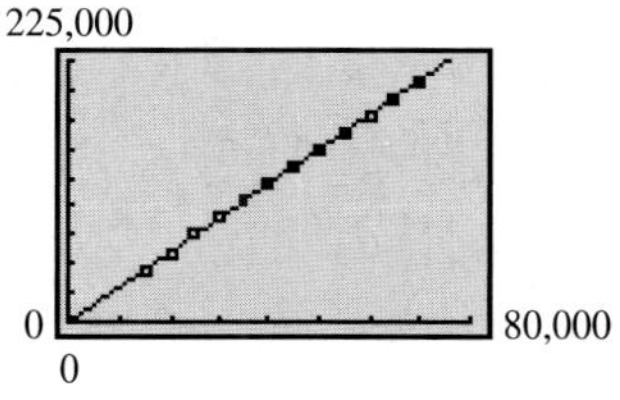

(d) For each additional dollar of income, the amount that the institution will loan you increases by \$2.98.

(e) \$125,143

18. (a)

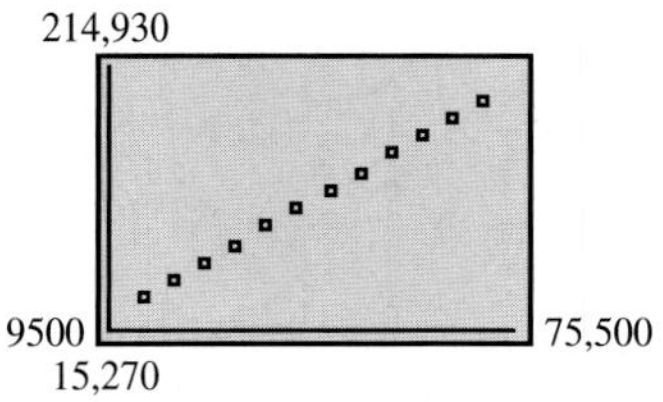

(b) $L = 2.71I - 66.08$

(c)

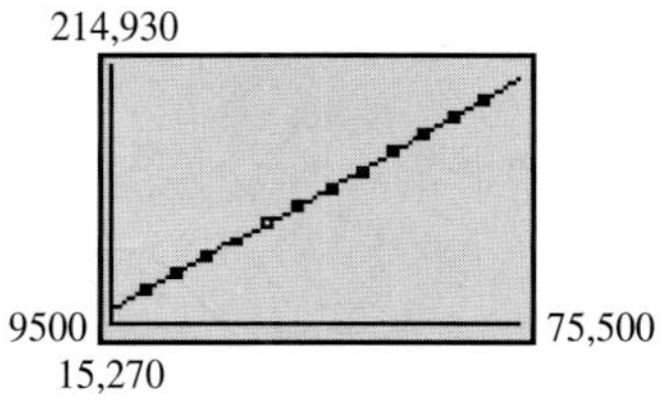

(d) For each additional dollar of income, the amount the institution will loan you increases by \$2.71.

(e) \$113,753.92

19. (a)

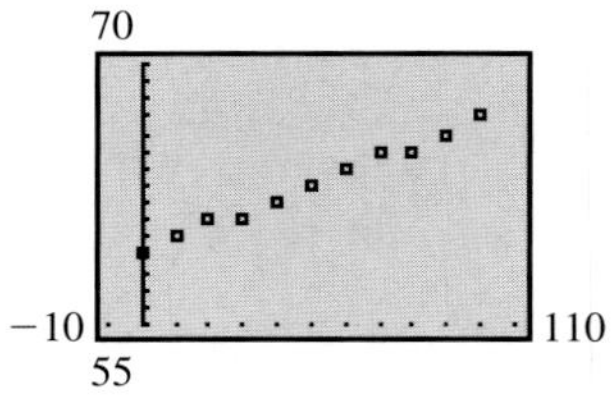

(b) $T = 0.0782h + 59.091$

(c)

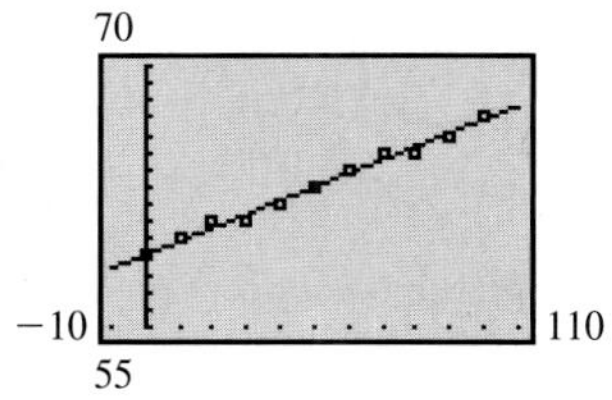

(d) If relative humidity increases by 1%, the apparent temperature increases by 0.0782 degrees Fahrenheit.

(e) About 65°F

20. (a)

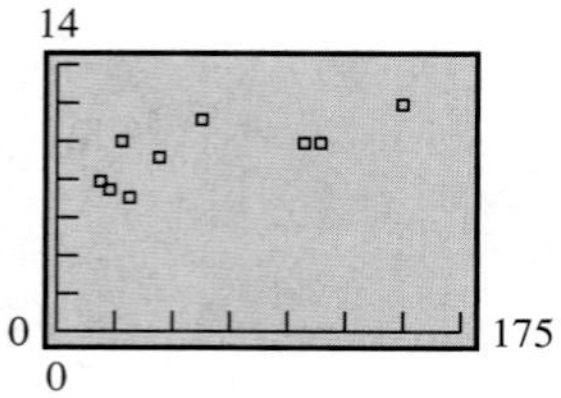

(b) $L(G) = 0.0261G + 7.8738$
(c) If gestation period increases by 1 day, then life expectancy increases by about 0.0261 year.
(d) About 10.2 years

Review Exercises

1. (a) $2\sqrt{5}$ **(b)** $(2, 1)$ **(c)** $\frac{1}{2}$ **(d)** For each run of 2, there is a rise of 1.

2. (a) $2\sqrt{13}$ **(b)** $(-2, 3)$ **(c)** $-\frac{3}{2}$ **(d)** For each run of 2, there is a rise of -3.

3. (a) 5 **(b)** $\left(-\frac{1}{2}, 1\right)$ **(c)** $-\frac{4}{3}$ **(d)** For each run of 3, there is a rise of -4.

4. (a) $\sqrt{13}$ **(b)** $\left(-\frac{1}{2}, 3\right)$ **(c)** $\frac{2}{3}$ **(d)** For each run of 3, there is a rise of 2.

5. (a) 12 **(b)** $(4, 2)$ **(c)** Undefined **(d)** No change in x **6. (a)** 5 **(b)** $\left(-\frac{1}{2}, 4\right)$ **(c)** 0 **(d)** No change in y

7.

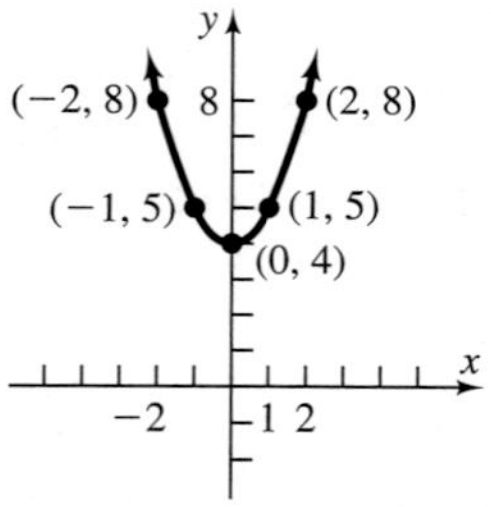

8 $(-4, 0), (0, 2), (0, 0), (0, -2), (2, 0)$ **9.** $(0, 0)$; symmetric with respect to the x-axis
10. $(0, 0)$; symmetric with respect to the origin
11. $(\pm 4, 0), (0, \pm 2)$; symmetric with respect to the x-axis, y-axis, and origin
12. $(\pm 1, 0)$; symmetric with respect to the x-axis, y-axis, and origin
13. $(0, 1)$; symmetric with respect to the y-axis.
14. $(0, 0), (\pm 1, 0)$; symmetric with respect to the origin
15. $(90, 0), (-1, 0), (0, -2)$; no symmetry **16.** $(0, 0), (-4, 0), (0, 2)$; no symmetry
17. $(x + 2)^2 + (y - 3)^2 = 16$ **18.** $(x - 3)^2 + (y - 4)^2 = 16$
19. $(x + 1)^2 + (y + 2)^2 = 1$ **20.** $(x - 2)^2 + (y + 4)^2 = 9$

21. Center $(0, 1)$; Radius $= 2$

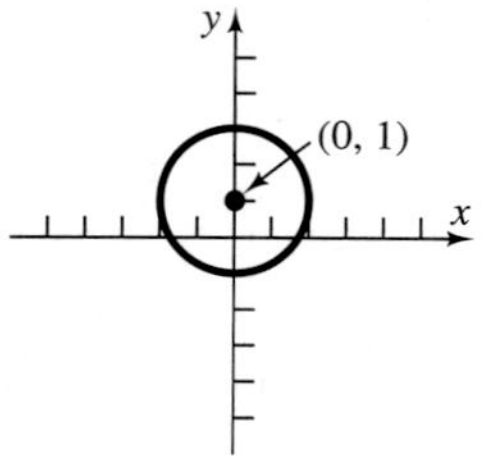

22. Center $(-2, 0)$; Radius $= 3$

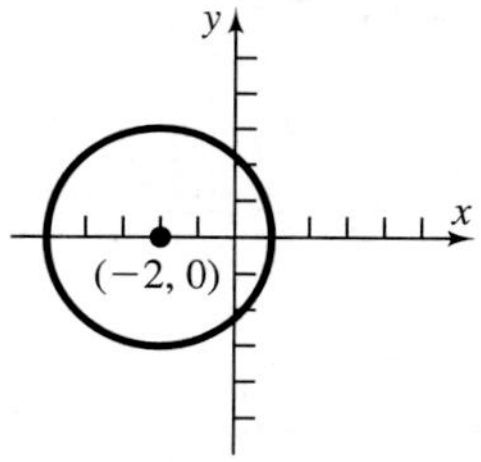

23. Center $(1, -2)$; Radius $= 3$

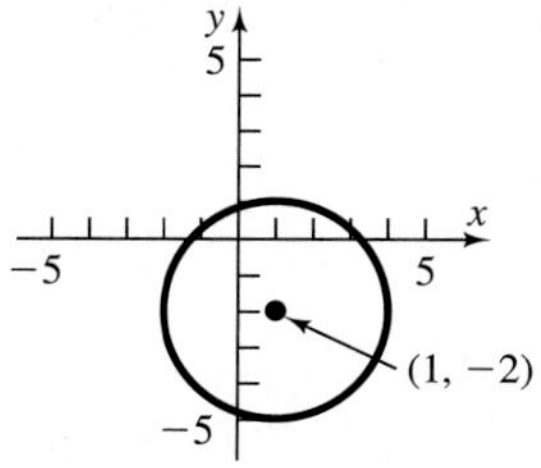

24. Center $(-2, 2)$, Radius $= 3$

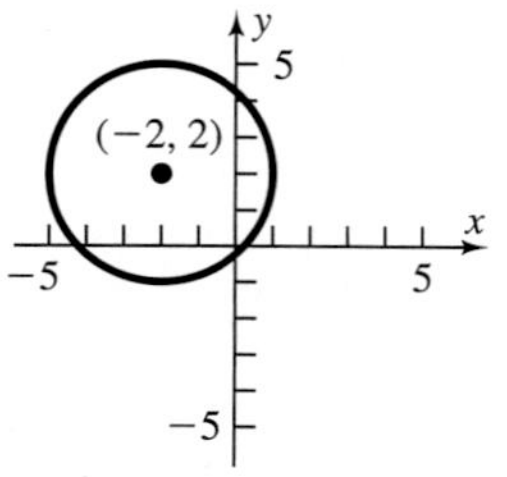

25. Center $(1, -2)$; Radius $= \sqrt{5}$

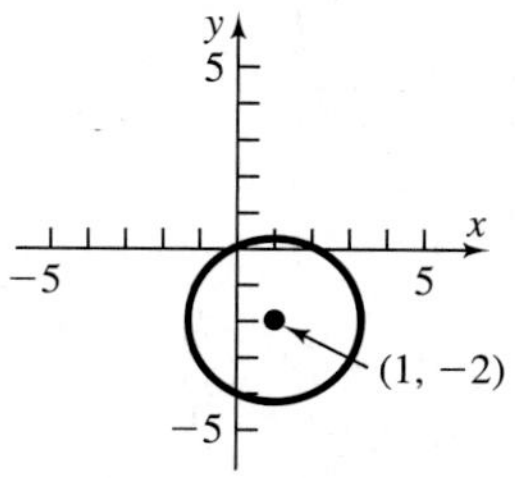

26. Center $(1, 0)$, Radius $= 1$

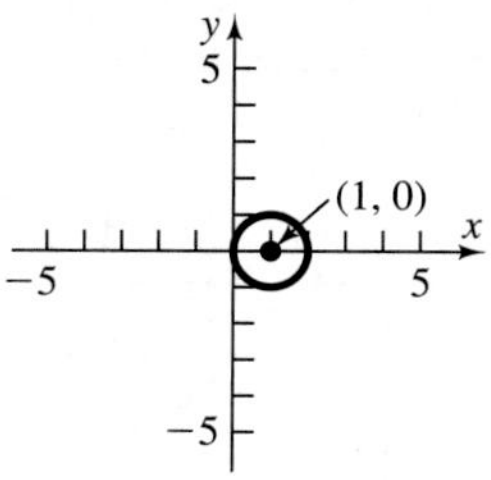

27. $2x + y = 5$ or $y = -2x + 5$ **28.** $y = 4$ **29.** $x = -3$; no slope–intercept form **30.** $5x + 2y = 10$ or $y = -\frac{5}{2}x + 5$

31. $x + 5y = -10$ or $y = -\frac{1}{5}x - 2$ **32.** $5x + y = 11$ or $y = -5x + 11$ **33.** $2x - 3y = -19$ or $y = \frac{2}{3}x + \frac{19}{3}$

34. $x + y = -2$ or $y = -x - 2$ **35.** $x - y = 7$ or $y = x - 7$ **36.** $x + 3y = 10$ or $y = -\frac{1}{3}x + \frac{10}{3}$

37. $m = \frac{4}{5}, b = 4$ **38.** $m = -\frac{3}{4}, b = 3$ **39.** $m = \frac{3}{2}, b = \frac{1}{2}$

40. $m = \frac{3}{2}, b = 0$ **41.**

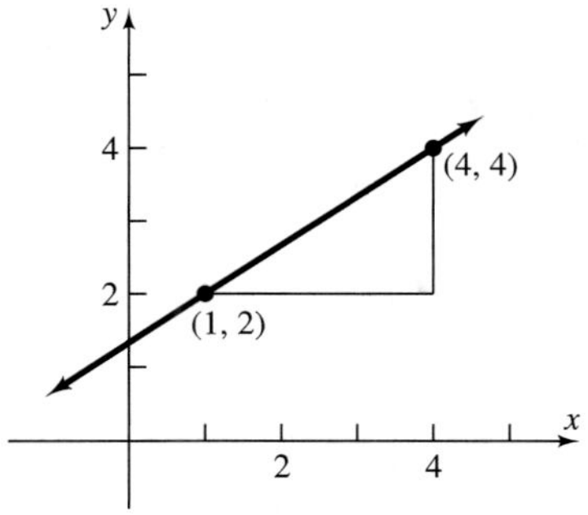

42. $d(A, B) = \sqrt{(1 - 3)^2 + (1 - 4)^2} = \sqrt{13}$ and $d(B, C) = \sqrt{(-2 - 1)^2 + (3 - 1)^2} = \sqrt{13}$

43. **(a)** $d(A, B) = 2\sqrt{5}; d(B, C) = \sqrt{145}; d(A, C) = 5\sqrt{5}; [d(B, C)]^2 = [d(A, B)]^2 + [d(A, C)]^2$

(b) Slope from A to B is -2; slope from A to C is $\frac{1}{2}$.

44. Center $= (1, -2)$; Radius $= 4\sqrt{2}; x^2 + y^2 - 2x + 4y - 27 = 0$ **45.** Slope from A to B is -1; slope from A to C is -1.

46. **(a)**

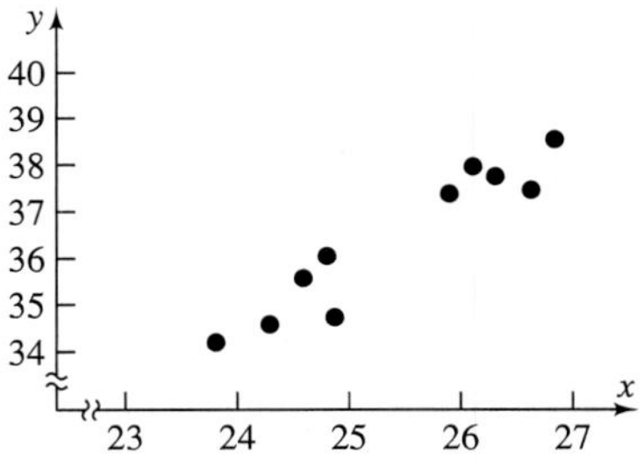

(b) Yes; the two variables appear to be linearly related.

(c) $y = 1.3902x + 1.114$ **(d)** 38.0 mm

Chapter 2 Functions and Their Graphs

2.1 Concepts and Vocabulary *(page 61)*

5. independent; dependent **6.** range **7.** $[0, 5]$ **8.** $\neq; f; g$ **9.** $(g - f)(x)$ **10.** False **11.** True **12.** True **13.** False **14.** False

2.1 Exercises *(page 61)*

15. Function; Domain: {Dad, Colleen, Kaleigh, Marissa}, Range: {January 8, March 15, September 17}
16. Function; Domain: {Bob, Dave, John, Chuck}; Range: {Beth, Diane, Linda, Marcia} **17.** Not a function
18. Function; Domain: {Bob, Dave, John, Chuck}; Range: {Diane, Linda, Marcia} **19.** Not a function
20. Function; Domain: $\{-2, -1, 3, 4\}$, Range: $\{3, 5, 7, 12\}$ **21.** Function; Domain: $\{1, 2, 3, 4\}$; Range: $\{3\}$
22. Function; Domain: $\{0, 1, 2, 3\}$, Range: $\{-2, 3, 7\}$ **23.** Not a function **24.** Not a function
25. Function; Domain: $\{-2, -1, 0, 1\}$, Range: $\{0, 1, 4\}$ **26.** Function; Domain: $\{-2, -1, 0, 1\}$, Range: $\{3, 4, 16\}$
27. **(a)** -4 **(b)** 1 **(c)** -3 **(d)** $3x^2 - 2x - 4$ **(e)** $-3x^2 - 2x + 4$ **(f)** $3x^2 + 8x + 1$ **(g)** $12x^2 + 4x - 4$ **(h)** $3x^2 + 6xh + 3h^2 + 2x + 2h - 4$ **28.** **(a)** -1 **(b)** -2 **(c)** -4 **(d)** $-2x^2 - x - 1$ **(e)** $2x^2 - x + 1$ **(f)** $-2x^2 - 3x - 2$ **(g)** $-8x^2 + 2x - 1$ **(h)** $-2x^2 - 4xh - 2h^2 + x + h - 1$
29. **(a)** 0 **(b)** $\frac{1}{2}$ **(c)** $-\frac{1}{2}$ **(d)** $\frac{-x}{x^2 + 1}$ **(e)** $\frac{-x}{x^2 + 1}$ **(f)** $\frac{x + 1}{x^2 + 2x + 2}$ **(g)** $\frac{2x}{4x^2 + 1}$ **(h)** $\frac{x + h}{x^2 + 2xh + h^2 + 1}$
30. **(a)** $-\frac{1}{4}$ **(b)** 0 **(c)** 0 **(d)** $\frac{x^2 - 1}{-x + 4}$ **(e)** $\frac{-x^2 + 1}{x + 4}$ **(f)** $\frac{x^2 + 2x}{x + 5}$ **(g)** $\frac{4x^2 - 1}{2x + 4}$ **(h)** $\frac{x^2 + 2xh + h^2 - 1}{x + h + 4}$
31. **(a)** 4 **(b)** 5 **(c)** 5 **(d)** $|x| + 4$ **(e)** $-|x| - 4$ **(f)** $|x + 1| + 4$ **(g)** $2|x| + 4$ **(h)** $|x + h| + 4$
32. **(a)** 0 **(b)** $\sqrt{2}$ **(c)** 0 **(d)** $\sqrt{x^2 - x}$ **(e)** $-\sqrt{x^2 + x}$ **(f)** $\sqrt{x^2 + 3x + 2}$ **(g)** $\sqrt{4x^2 + 2x}$ **(h)** $\sqrt{x^2 + 2xh + h^2 + x + h}$
33. **(a)** $-\frac{1}{5}$ **(b)** $-\frac{3}{2}$ **(c)** $\frac{1}{8}$ **(d)** $\frac{-2x + 1}{-3x - 5}$ **(e)** $\frac{-2x - 1}{3x - 5}$ **(f)** $\frac{2x + 3}{3x - 2}$ **(g)** $\frac{4x + 1}{6x - 5}$ **(h)** $\frac{2x + 2h + 1}{3x + 3h - 5}$

34. **(a)** $\frac{3}{4}$ **(b)** $\frac{8}{9}$ **(c)** 0 **(d)** $1 - \frac{1}{(-x + 2)^2}$ **(e)** $-1 + \frac{1}{(x + 2)^2}$ **(f)** $1 - \frac{1}{(x + 3)^2}$ **(g)** $1 - \frac{1}{(2x + 2)^2}$ **(h)** $1 - \frac{1}{(x + h + 2)^2}$

35. Function **36.** Function **37.** Function **38.** Function **39.** Not a function **40.** Not a function **41.** Not a function **42.** Not a function **43.** Function **44.** Function **45.** Not a function **46.** Not a function **47.** All real numbers **48.** All real numbers **49.** All real numbers **50.** All real numbers **51.** $\{x|x \neq -4, x \neq 4\}$ **52.** $\{x|x \neq -2, x \neq 2\}$ **53.** $\{x|x \neq 0\}$ **54.** $\{x|x \neq -2, x \neq 0, x \neq 2\}$ **55.** $\{x|x \geq 4\}$ **56.** $\{x|x \leq 1\}$ **57.** $\{x|x > 9\}$ **58.** $\{x|x > 4\}$ **59.** $\{x|x > 1\}$ **60.** $\{x|x \leq -2\}$

61. **(a)** $(f + g)(x) = 5x + 1$; All real numbers **(b)** $(f - g)(x) = x + 7$; All real numbers **(c)** $(f \cdot g)(x) = 6x^2 - x - 12$; All real numbers **(d)** $\left(\frac{f}{g}\right)(x) = \frac{3x + 4}{2x - 3}$; $\left\{x \middle| x \neq \frac{3}{2}\right\}$

62. **(a)** $(f + g)(x) = 5x - 1$; All real numbers **(b)** $(f - g)(x) = -x + 3$; All real numbers **(c)** $(f \cdot g)(x) = 6x^2 - x - 2$; All real numbers **(d)** $\left(\frac{f}{g}\right)(x) = \frac{2x + 1}{3x - 2}$; $\left\{x|x \neq \frac{2}{3}\right\}$

63. **(a)** $(f + g)(x) = 2x^2 + x - 1$; All real numbers **(b)** $(f - g)(x) = -2x^2 + x - 1$; All real numbers **(c)** $(f \cdot g)(x) = 2x^3 - 2x^2$; All real numbers **(d)** $\left(\frac{f}{g}\right)(x) = \frac{x - 1}{2x^2}$; $\{x|x \neq 0\}$

64. **(a)** $(f + g)(x) = 4x^3 + 2x^2 + 4$; All real numbers **(b)** $(f - g)(x) = -4x^3 + 2x^2 + 2$; All real numbers **(c)** $(f \cdot g)(x) = 8x^5 + 12x^3 + 2x^2 + 3$; All real numbers **(d)** $\left(\frac{f}{g}\right)(x) = \frac{2x^2 + 3}{4x^3 + 1}$; $\left\{x \middle| x \neq -\frac{\sqrt[3]{16}}{4}\right\}$

65. **(a)** $(f + g)(x) = \sqrt{x} + 3x - 5$; $\{x|x \geq 0\}$ **(b)** $(f - g)(x) = \sqrt{x} - 3x + 5$; $\{x|x \geq 0\}$ **(c)** $(f \cdot g)(x) = 3x\sqrt{x} - 5\sqrt{x}$; $\{x|x \geq 0\}$ **(d)** $\left(\frac{f}{g}\right)(x) = \frac{\sqrt{x}}{3x - 5}$; $\left\{x|x \geq 0, x \neq \frac{5}{3}\right\}$

66. **(a)** $(f + g)(x) = |x| + x$; All real numbers **(b)** $(f - g)(x) = |x| - x$; All real numbers **(c)** $(f \cdot g)(x) = x|x|$; All real numbers **(d)** $\left(\frac{f}{g}\right)(x) = \frac{|x|}{x}$; $\{x|x \neq 0\}$

67. **(a)** $(f + g)(x) = 1 + \frac{2}{x}$; $\{x|x \neq 0\}$ **(b)** $(f - g)(x) = 1$; $\{x|x \neq 0\}$ **(c)** $(f \cdot g)(x) = \frac{1}{x} + \frac{1}{x^2}$; $\{x|x \neq 0\}$ **(d)** $\left(\frac{f}{g}\right)(x) = x + 1$; $\{x|x \neq 0\}$ **68.** **(a)** $(f + g)(x) = \sqrt{x - 2} + \sqrt{4 - x}$; $\{x|2 \leq x \leq 4\}$ **(b)** $(f - g)(x) = \sqrt{x - 2} - \sqrt{4 - x}$; $\{x|2 \leq x \leq 4\}$ **(c)** $(f \cdot g)(x) = \sqrt{-x^2 + 6x - 8}$; $\{x|2 \leq x \leq 4\}$ **(d)** $\left(\frac{f}{g}\right)(x) = \sqrt{\frac{x - 2}{4 - x}}$; $\{x|2 \leq x < 4\}$ **69.** **(a)** $(f + g)(x) = \frac{6x + 3}{3x - 2}$; $\left\{x \middle| x \neq \frac{2}{3}\right\}$ **(b)** $(f - g)(x) = \frac{-2x + 3}{3x - 2}$; $\left\{x \middle| x \neq \frac{2}{3}\right\}$ **(c)** $(f \cdot g)(x) = \frac{8x^2 + 12x}{(3x - 2)^2}$; $\left\{x \middle| x \neq \frac{2}{3}\right\}$ **(d)** $\left(\frac{f}{g}\right)(x) = \frac{2x + 3}{4x}$; $\left\{x \middle| x \neq 0, x \neq \frac{2}{3}\right\}$ **70.** **(a)** $(f + g)(x) = \sqrt{x + 1} + \frac{2}{x}$; $\{x|x \geq -1, x \neq 0\}$ **(b)** $(f - g)(x) = \sqrt{x + 1} - \frac{2}{x}$; $\{x|x \geq -1, x \neq 0\}$ **(c)** $(f \cdot g)(x) = \frac{2}{x}\sqrt{x + 1}$; $\{x|x \geq -1, x \neq 0\}$ **(d)** $\left(\frac{f}{g}\right)(x) = \frac{x\sqrt{x + 1}}{2}$; $\{x|x \geq -1, x \neq 0\}$ **71.** $g(x) = 5 - \frac{7}{2}x$ **72.** $g(x) = \frac{x - 1}{x + 1}$ **73.** 4 **74.** -3 **75.** $2x + h - 1$ **76.** $2x + h + 5$ **77.** $3x^2 + 3xh + h^2$ **78.** $\frac{-1}{(x + h + 3)(x + 3)}$ **79.** $A = -\frac{7}{2}$ **80.** $B = 5$ **81.** $A = -4$ **82.** $B = -1$ **83.** $A = 8$; undefined at $x = 3$ **84.** $A = 1, B = 2$ **85.** $A(x) = \frac{1}{2}x^2$ **86.** $A(x) = \frac{1}{2}x^2$ **87.** $G(x) = 10x$ **88.** $G(x) = 100 + 10x$

89. **(a)** 15.1 m, 14.07 m, 12.94 m, 11.72 m **(b)** 1.01 sec, 1.43 sec, 1.75 sec **(c)** 2.02 sec **90.** **(a)** 7 m, 4.27 m, 1.28 m **(b)** 0.62 sec, 0.88 sec, 1.07 sec **(c)** 1.24 sec **91.** **(a)** \$222 **(b)** \$225 **(c)** \$220 **(d)** \$230 **92.** **(a)** 1.26 ft^2 **(b)** 1.73 ft^2 **(c)** 1.99 ft^2

93. $R(x) = \frac{L(x)}{P(x)}$ **94.** $T(x) = V(x) + P(x)$ **95.** $H(x) = P(x) \cdot I(x)$ **96.** $N(x) = I(x) - T(x)$ **97.** Only $h(x) = 2x$

98. No; f has a domain of all real numbers, while g has a domain of $\{x|x \neq -1\}$.

2.2 Concepts and Vocabulary *(page 68)*

3. vertical **4.** 5; -3 **5.** $a = -2$ **6.** False **7.** False **8.** True

2.2 Exercises *(page 68)*

9. **(a)** $f(0) = 3; f(-6) = -3$ **(b)** $f(6) = 0; f(11) = 1$ **(c)** Positive **(d)** Negative **(e)** $-3, 6$, and 10 **(f)** $-3 < x < 6; 10 < x \leq 11$ **(g)** $\{x|-6 \leq x \leq 11\}$ **(h)** $\{y|-3 \leq y \leq 4\}$ **(i)** $-3, 6, 10$ **(j)** 3 **(k)** 3 times **(l)** once **(m)** 0, 4 **(n)** $-5, 8$

10. **(a)** $f(0) = 0; f(6) = 0$ **(b)** $f(2) = -2; f(-2) = 1$ **(c)** Negative **(d)** Positive **(e)** 0, 4, and 6 **(f)** $0 < x < 4$ **(g)** $\{x \mid -4 \le x \le 6\}$ **(h)** $\{y \mid -2 \le y \le 3\}$ **(i)** 0, 4, 6 **(j)** 0 **(k)** twice **(l)** once **(m)** 5 **(n)** 2 **11.** Not a function
12. Function **(a)** Domain: all real numbers; Range: $\{y \mid 0 < y < \infty\}$ **(b)** $(0, 1)$ **(c)** None
13. Function **(a)** Domain: $\{x \mid -\pi \le x \le \pi\}$; Range: $\{y \mid -1 \le y \le 1\}$ **(b)** $\left(-\frac{\pi}{2}, 0\right), \left(\frac{\pi}{2}, 0\right), (0, 1)$ **(c)** y-axis
14. Function **(a)** Domain: $\{x \mid -\pi \le x \le \pi\}$; Range: $\{y \mid -1 \le y \le 1\}$ **(b)** $(-\pi, 0), (\pi, 0), (0, 0)$ **(c)** origin **15.** Not a function
16. Not a function **17.** Function **(a)** Domain: $\{x \mid x > 0\}$; Range: all real numbers **(b)** $(1, 0)$ **(c)** None
18. Function **(a)** Domain: $\{x \mid 0 \le x \le 4\}$; Range: $\{y \mid 0 \le y \le 3\}$ **(b)** $(0, 0)$ **(c)** None
19. Function **(a)** Domain: all real numbers; Range: $\{y \mid y \le 2\}$ **(b)** $(-3, 0), (3, 0), (0, 2)$ **(c)** y-axis
20. Function **(a)** Domain: $\{x \mid x \ge -3\}$; Range: $\{y \mid y \ge 0\}$ **(b)** $(-3, 0), (0, 2), (2, 0)$ **(c)** None
21. Function **(a)** Domain: all real numbers; Range: $\{y \mid y \ge -3\}$ **(b)** $(1, 0), (3, 0), (0, 9)$ **(c)** None
22. Function **(a)** Domain: all real numbers; Range: $\{y \mid y \le 5\}$ **(b)** $(-1, 0), (2, 0), (0, 4)$ **(c)** None
23. **(a)** Yes **(b)** $f(-2) = 9; (-2, 9)$ **(c)** $0, \frac{1}{2}; (0, -1), \left(\frac{1}{2}, -1\right)$ **(d)** All real numbers **(e)** $-\frac{1}{2}, 1$ **(f)** -1
24. **(a)** No **(b)** $f(-2) = -22; (-2, -22)$ **(c)** $-\frac{1}{3}, 2; \left(-\frac{1}{3}, -2\right), (2, -2)$ **(d)** All real numbers **(e)** $0, \frac{5}{3}$ **(f)** 0
25. **(a)** No **(b)** $f(4) = -3; (4, -3)$ **(c)** $14; (14, 2)$ **(d)** $\{x \mid x \ne 6\}$ **(e)** -2 **(f)** $-\frac{1}{3}$ **26.** **(a)** Yes **(b)** $f(0) = \frac{1}{2}; \left(0, \frac{1}{2}\right)$ **(c)** $0, \frac{1}{2}; \left(0, \frac{1}{2}\right), \left(\frac{1}{2}, \frac{1}{2}\right)$ **(d)** $\{x \mid x \ne -4\}$ **(e)** none **(f)** $\frac{1}{2}$ **27.** **(a)** Yes **(b)** $f(2) = \frac{8}{17}; \left(2, \frac{8}{17}\right)$ **(c)** $-1, 1; (-1, 1), (1, 1)$ **(d)** All real numbers **(e)** 0 **(f)** 0 **28.** **(a)** Yes **(b)** $f(4) = 4; (4, 4)$ **(c)** $-2; (-2, 1)$ **(d)** $\{x \mid x \ne 2\}$ **(e)** 0 **(f)** 0

29. **(a)** About 81.07 ft **(b)** About 129.59 ft **(c)** About 26.63 ft **(d)** About 528.13 ft
(e)

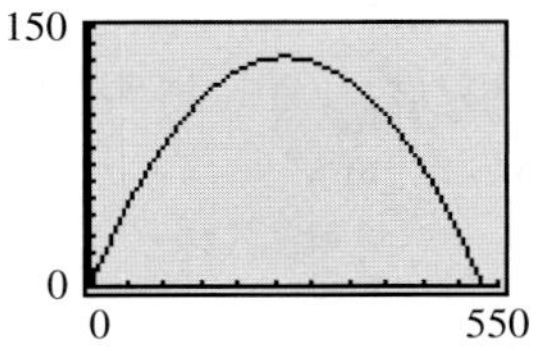

(f) 115.07 ft and 413.05 ft
(g) 275 ft; maximum height shown in the table is 131.8 ft
(h) 264 ft

30. **(a)** About 119.84 pounds
(b)

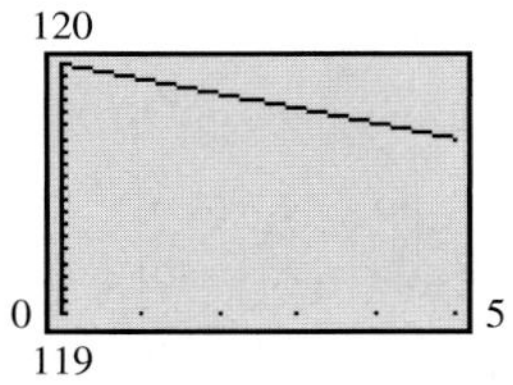

(c) The weight W varies from 120 pounds to about 119.7 pounds.
(d) About 0.83 miles

31. **(a)** III **(b)** IV **(c)** I **(d)** V **(e)** II **32.** **(a)** II **(b)** V **(c)** IV **(d)** III **(e)** I

33.

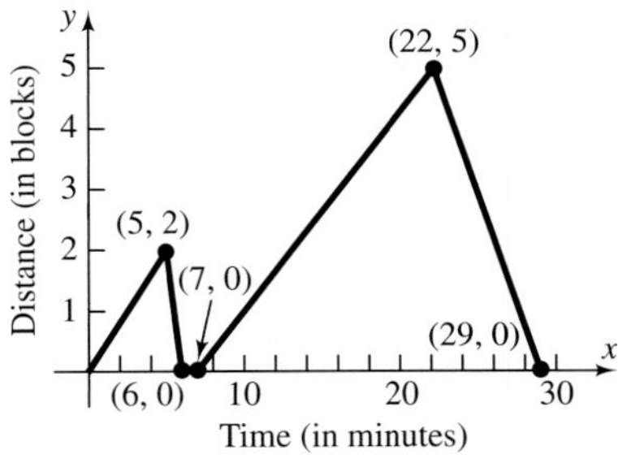

34.

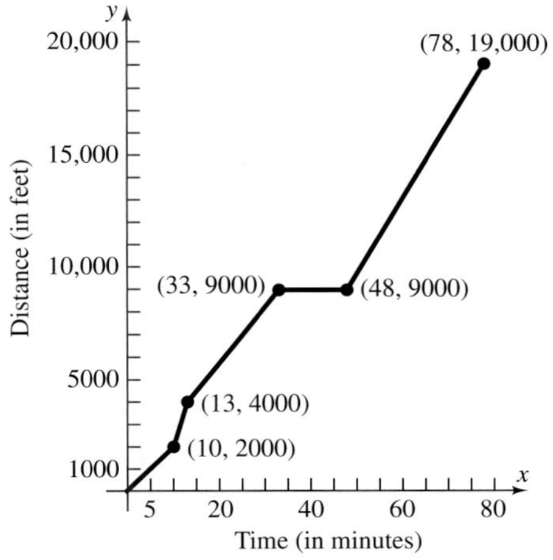

35. **(a)** 2 hr elapsed during which Kevin was between 0 and 3 mi from home. **(b)** 0.5 hr elapsed during which Kevin was 3 mi from home. **(c)** 0.3 hr elapsed during which Kevin was between 0 and 3 mi from home. **(d)** 0.2 hr elapsed during which Kevin was 0 mi from home. **(e)** 0.9 hr elapsed during which Kevin was between 0 and 2.8 mi from home. **(f)** 0.3 hr elapsed during which Kevin was 2.8 mi from home. **(g)** 1.1 hr elapsed during which Kevin was between 0 and 2.8 mi from home. **(h)** 3 mi **(i)** 2 times

36. (a) $(7, 7.4)$ **(b)** $(4.2, 6)$ **(c)** Increasing from 0 to 30 mi/hr **(d)** 0 mi/hr **(e)** 50 mi/hr **(f)** $(2, 4)$, $(4.2, 6)$, $(7, 7.4)$, $(7.6, 8)$ **37.** All points $(5, y)$ and $(x, 0)$ are excluded. **39.** The x-intercepts can number anywhere from 0 to infinitely many. There is at most one y-intercept. **40.** yes **41.** yes; $f(x) = 0$

2.3 Concepts and Vocabulary *(page 80)*

6. increasing **7.** even; odd **8.** True **9.** True **10.** False

2.3 Exercises *(page 80)*

11. Yes **12.** No **13.** No **14.** Yes **15.** $(-8, -2)$; $(0, 2)$; $(5, \infty)$ **16.** $(-\infty, -8)$, $(-2, 0)$, $(2, 5)$ **17.** Yes; 10 **18.** No **19.** $-2, 2; 6, 10$ **20.** $-8, 0, 5; -4, 0, 0$ **21. (a)** $(-2, 0)$, $(0, 3)$, $(2, 0)$ **(b)** Domain: $\{x \mid -4 \le x \le 4\}$ or $[-4, 4]$; Range: $\{y \mid 0 \le y \le 3\}$ or $[0, 3]$ **(c)** Increasing on $(-2, 0)$ and $(2, 4)$; Decreasing on $(-4, -2)$ and $(0, 2)$ **(d)** Even **22. (a)** $(-1, 0)$, $(0, 2)$, $(1, 0)$ **(b)** Domain: $\{x \mid -3 \le x \le 3\}$ or $[-3, 3]$; Range: $\{y \mid 0 \le y \le 3\}$ or $[0, 3]$ **(c)** Increasing on $(-1, 0)$ and $(1, 3)$; Decreasing on $(-3, -1)$ and $(0, 1)$ **(d)** Even **23. (a)** $(0, 1)$ **(b)** Domain: all real numbers; Range: $\{y \mid y > 0\}$ or $(0, \infty)$ **(c)** Increasing on $(-\infty, \infty)$ **(d)** Neither **24. (a)** $(1, 0)$ **(b)** Domain: $\{x \mid x > 0\}$ or $(0, \infty)$; Range: all real numbers **(c)** Increasing on $(0, \infty)$ **(d)** Neither **25. (a)** $(-\pi, 0)$, $(0, 0)$, $(\pi, 0)$ **(b)** Domain: $\{x \mid -\pi \le x \le \pi\}$ or $[-\pi, \pi]$; Range: $\{y \mid -1 \le y \le 1\}$ or $[-1, 1]$ **(c)** Increasing on $\left(-\frac{\pi}{2}, \frac{\pi}{2}\right)$; Decreasing on $\left(-\pi, -\frac{\pi}{2}\right)$ and $\left(\frac{\pi}{2}, \pi\right)$ **(d)** Odd **26. (a)** $\left(-\frac{\pi}{2}, 0\right)$, $(0, 1)$, $\left(\frac{\pi}{2}, 0\right)$ **(b)** Domain: $\{x \mid -\pi \le x \le \pi\}$ or $[-\pi, \pi]$; Range: $\{y \mid -1 \le y \le 1\}$ or $[-1, 1]$ **(c)** Increasing on $(-\pi, 0)$; Decreasing on $(0, \pi)$ **(d)** Even **27. (a)** $\left(0, \frac{1}{2}\right)$, $\left(\frac{1}{2}, 0\right)$, $\left(\frac{5}{2}, 0\right)$ **(b)** Domain: $\{x \mid -3 \le x \le 3\}$ or $[-3, 3]$; Range: $\{y \mid -1 \le y \le 2\}$ or $[-1, 2]$ **(c)** Increasing on $(2, 3)$; Decreasing on $(-1, 1)$; Constant on $(-3, -1)$ and $(1, 2)$ **(d)** Neither **28. (a)** $(-2.3, 0)$, $(0, 1)$, $(3, 0)$ **(b)** Domain: $\{x \mid -3 \le x \le 3\}$ or $[-3, 3]$; Range: $\{y \mid -2 \le y \le 2\}$ or $[-2, 2]$ **(c)** Increasing on $(-3, -2)$ and $(0, 2)$; Decreasing on $(2, 3)$; Constant on $(-2, 0)$ **(d)** Neither

29. (a) $0; 3$ **(b)** $-2, 2; 0, 0$ **30. (a)** $0; 2$ **(b)** $-1, 1; 0, 0$ **31. (a)** $\frac{\pi}{2}; 1$ **(b)** $-\frac{\pi}{2}; -1$ **32. (a)** $0; 1$ **(b)** $-\pi, \pi; -1, -1$ **33. (a)** -4 **(b)** -8 **(c)** -10 **34. (a)** -4 **(b)** -13 **(c)** -1 **35. (a)** 5 **(b)** 5 **(c)** $y = 5x$ **36. (a)** -4 **(b)** -4 **(c)** $y = -4x$ **37. (a)** -3 **(b)** -3 **(c)** $y = -3x + 1$ **38. (a)** $x + 1$ **(b)** 3 **(c)** $y = 3x - 1$ **39. (a)** $x - 1$ **(b)** 1 **(c)** $y = x - 2$ **40. (a)** $-2x - 1$ **(b)** -5 **(c)** $y = -5x + 4$ **41. (a)** $x(x + 1)$ **(b)** 6 **(c)** $y = 6x - 6$ **42. (a)** $x^2 + x + 2$ **(b)** 8 **(c)** $y = 8x - 6$ **43. (a)** $\frac{-1}{x + 1}$ **(b)** $-\frac{1}{3}$ **(c)** $y = -\frac{1}{3}x + \frac{4}{3}$ **44. (a)** $-\frac{x + 1}{x^2}$ **(b)** $-\frac{3}{4}$ **(c)** $y = -\frac{3}{4}x + \frac{7}{4}$ **45. (a)** $\frac{\sqrt{x} - 1}{x - 1}$ **(b)** $\sqrt{2} - 1$ **(c)** $(\sqrt{2} - 1)x - \sqrt{2} + 2$ **46. (a)** $\frac{\sqrt{x + 3} - 2}{x - 1}$ **(b)** $\sqrt{5} - 2$ **(c)** $y = (\sqrt{5} - 2)x - \sqrt{5} + 4$ **47.** Odd **48.** Even **49.** Even **50.** Neither **51.** Odd **52.** Neither **53.** Neither **54.** Even **55.** Even **56.** Odd **57.** Odd **58.** Odd

59.

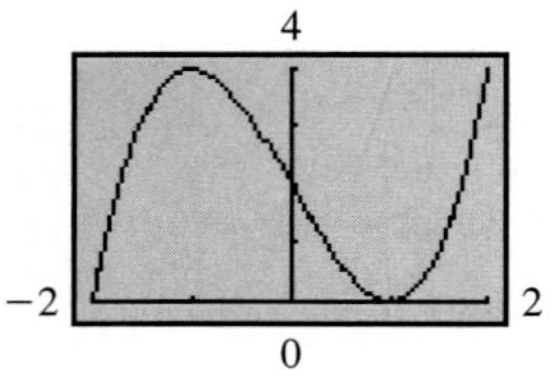

Increasing: $(-2, -1)$, $(1, 2)$
Decreasing: $(-1, 1)$
Local maximum: $(-1, 4)$
Local minimum: $(1, 0)$

60.

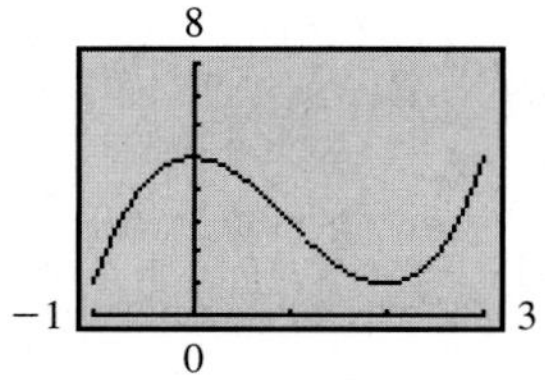

Increasing: $(-1, 0)$, $(2, 3)$
Decreasing: $(0, 2)$
Local maximum: $(0, 5)$
Local minimum: $(2, 1)$

61.

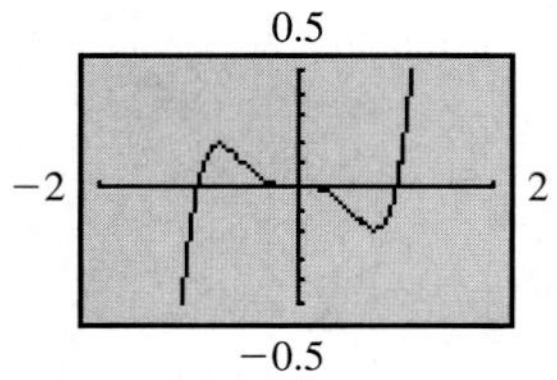

Increasing: $(-2, -0.77)$, $(0.77, 2)$
Decreasing: $(-0.77, 0.77)$
Local maximum: $(-0.77, 0.19)$
Local minimum: $(0.77, -0.19)$

62.

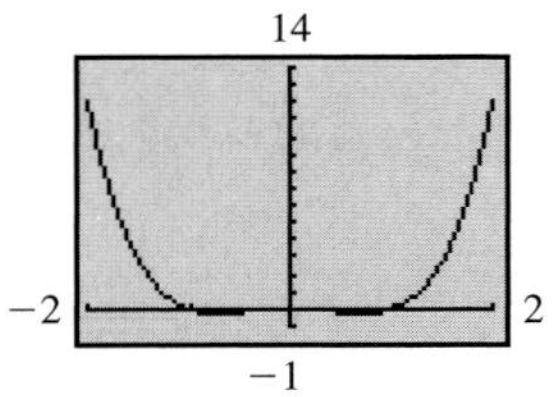

Increasing: (−0.71, 0), (0.71, 2)
Decreasing: (−2, −0.71), (0, 0.71)
Local maximum: (0, 0)
Local minima: (−0.71, −0.25), (0.71, −0.25)

63.

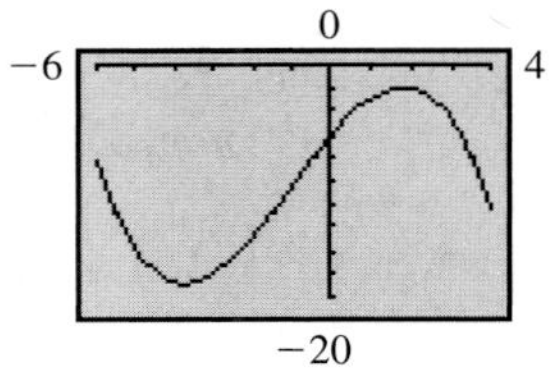

Increasing: (−3.77, 1.77)
Decreasing: (−6, −3.77), (1.77, 4)
Local maximum: (1.77, −1.91)
Local minimum: (−3.77, −18.89)

64.

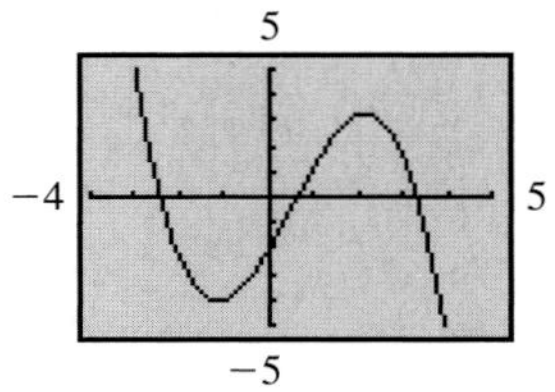

Increasing: (−1.16, 2.16)
Decreasing: (−4, −1.16), (2.16, 5)
Local maximum: (2.16, 3.25)
Local minimum: (−1.16, −4.05)

65.

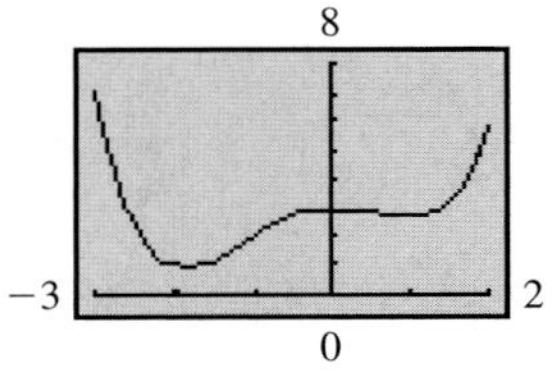

Increasing: (−1.87, 0), (0.97, 2)
Decreasing: (−3, −1.87), (0, 0.97)
Local maximum: (0, 3)
Local minima: (−1.87, 0.95), (0.97, 2.65)

66.

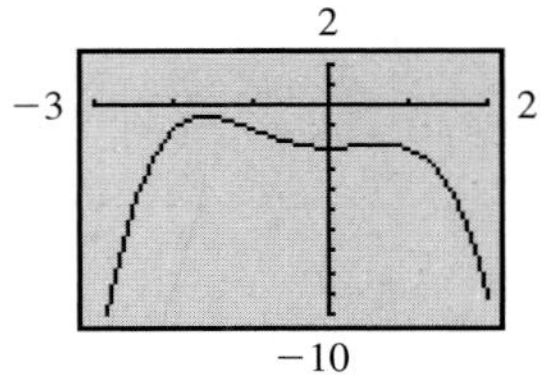

Increasing: (−3, −1.57), (0, 0.64)
Decreasing: (−1.57, 0), (0.64, 2)
Local maxima: (−1.57, −0.52), (0.64, −1.87)
Local minimum: (0, −2)

67. **(a)** $V(x) = x(24 - 2x)^2$
(b) 972 in.^{3} **(c)** 160 in.^{3}
(d) V is largest when $x = 4$.

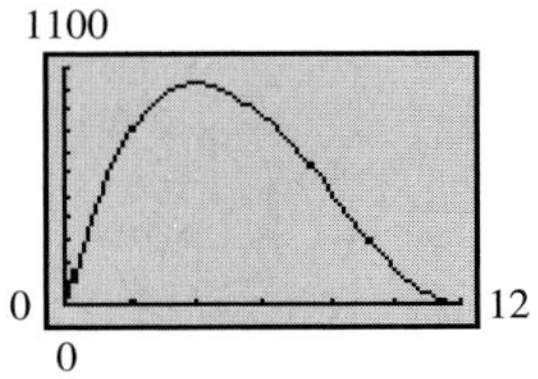

68. **(a)** $A(x) = x^2 + \dfrac{40}{x}$
(b) 41 ft^2 **(c)** 24 ft^2
(d) A is smallest when $x \approx 2.71$ ft.

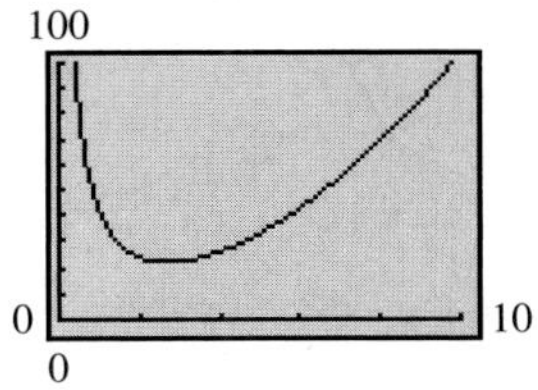

69. **(a)**

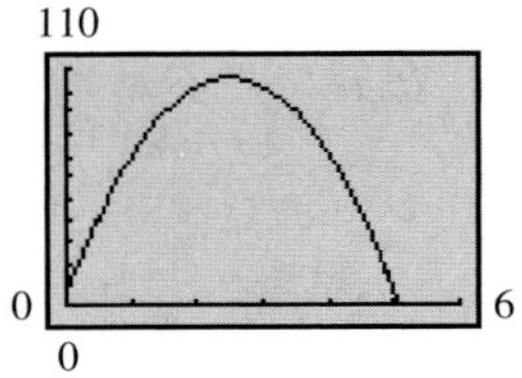

(b) 2.5 sec
(c) 106 ft

70. **(a)**

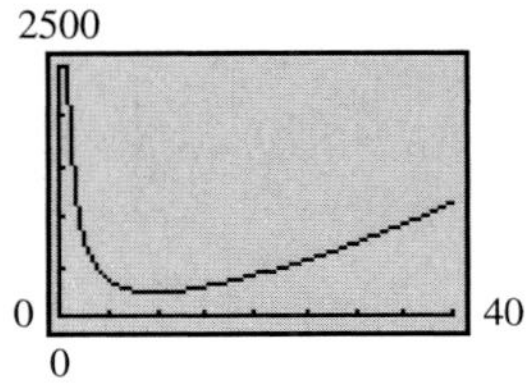

(b) 9.66 riding lawn mowers/hr
(c) \$238.65/mower

71. **(a)** 1 **(b)** 0.5 **(c)** 0.1 **(d)** 0.01 **(e)** 0.001 **(f)**

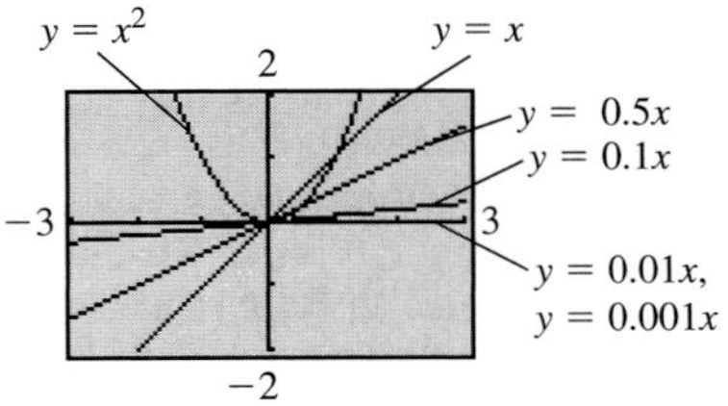

(g) They are getting closer to the tangent line at (0, 0).
(h) They are getting closer to 0.

72. **(a)** 3 **(b)** 2.5 **(c)** 2.1 **(d)** 2.01 **(e)** 2.001 **(f)**

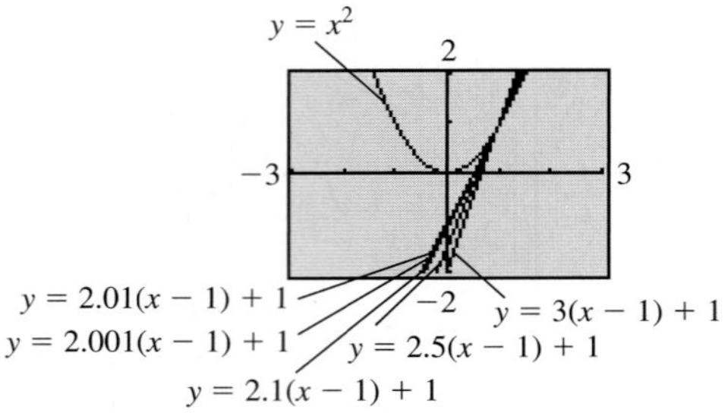

(g) They are getting closer to the tangent line at $(1, 1)$.
(h) They are getting closer to 2.

75. At most one **77.** Yes; the function $f(x) = 0$ is both even and odd.

2.4 Concepts and Vocabulary *(page 90)*

4. less **5.** piecewise defined **6.** True **7.** False **8.** False

2.4 Exercises *(page 90)*

9. C **10.** A **11.** E **12.** G **13.** B **14.** D **15.** F **16.** H

17.

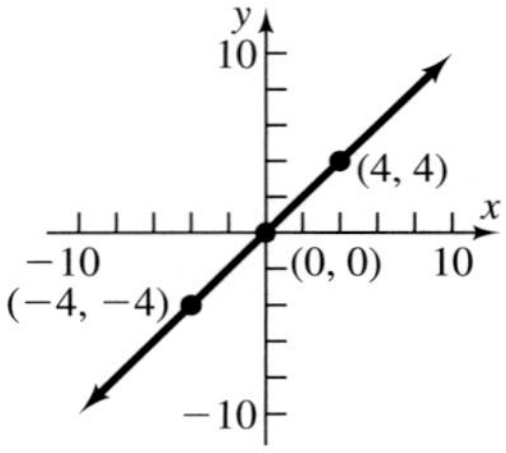

18.

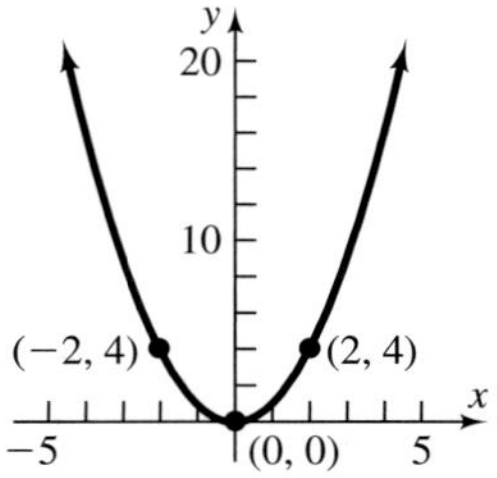

19.

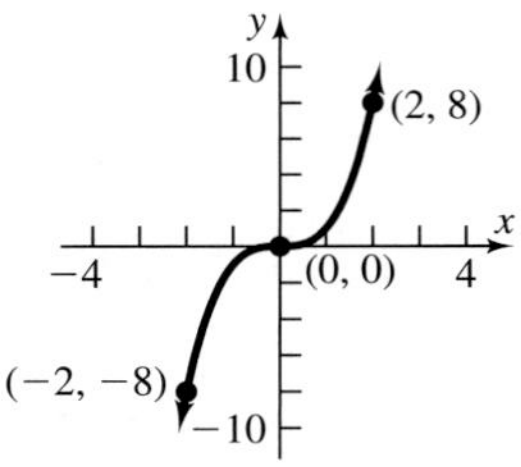

20.

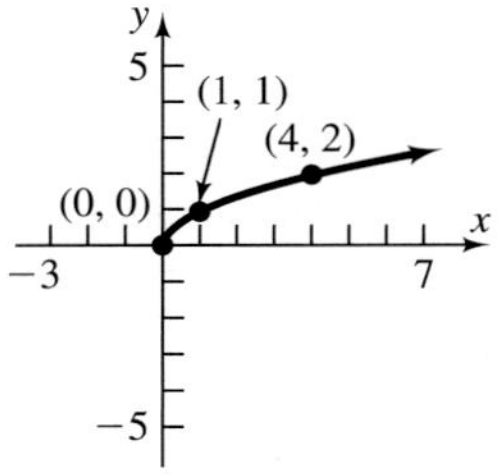

21.

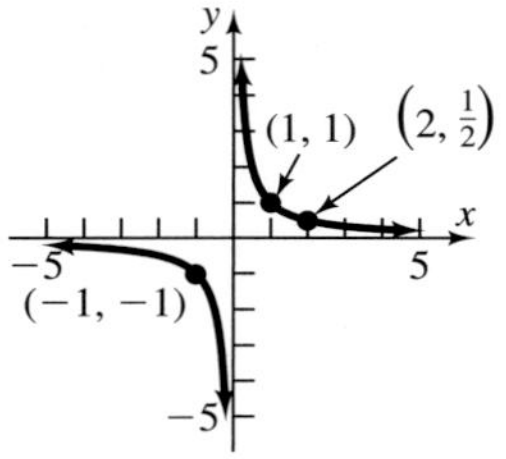

22.

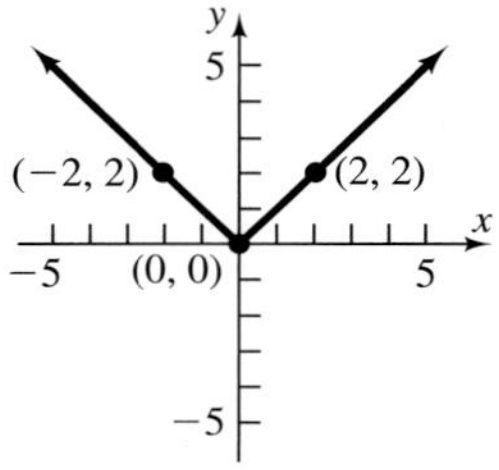

23.

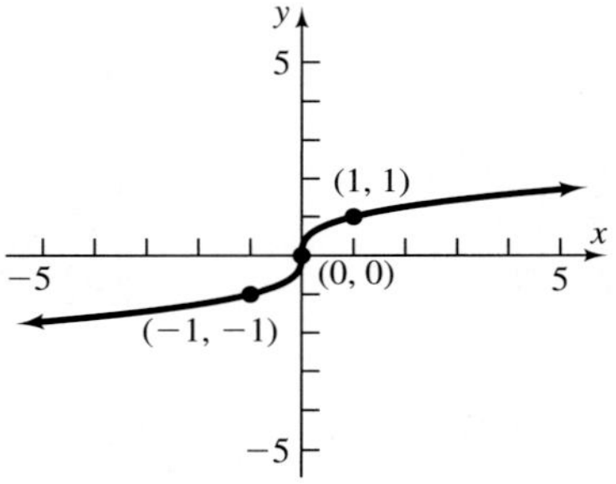

24.

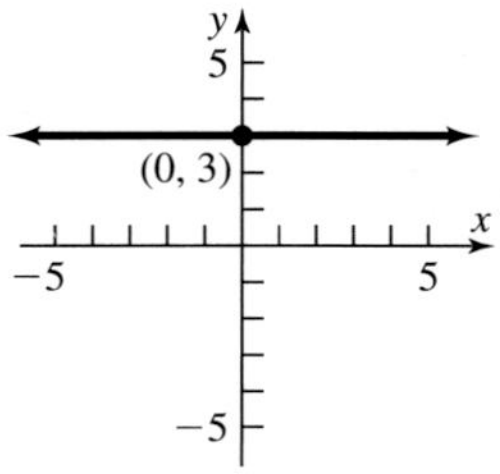

25. **(a)** 4 **(b)** 2 **(c)** 5
26. **(a)** −1 **(b)** 2 **(c)** 5
27. **(a)** 2 **(b)** 3 **(c)** −4
28. **(a)** 0 **(b)** 0 **(c)** −1

29. **(a)** All real numbers
(b) $(0, 1)$
(c)

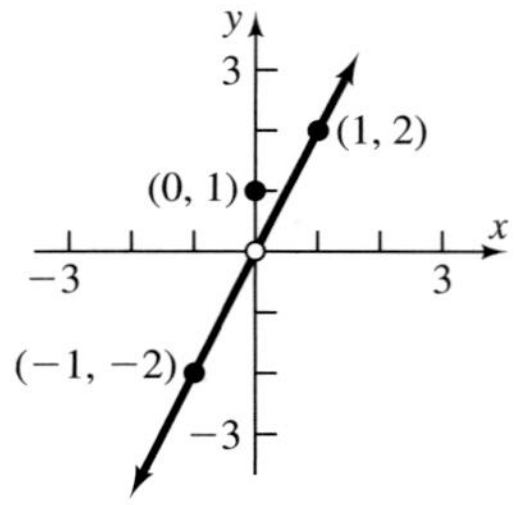

(d) $\{y | y \neq 0\}$; $(-\infty, 0)$ or $(0, \infty)$

30. **(a)** All real numbers
(b) $(0, 4)$
(c)

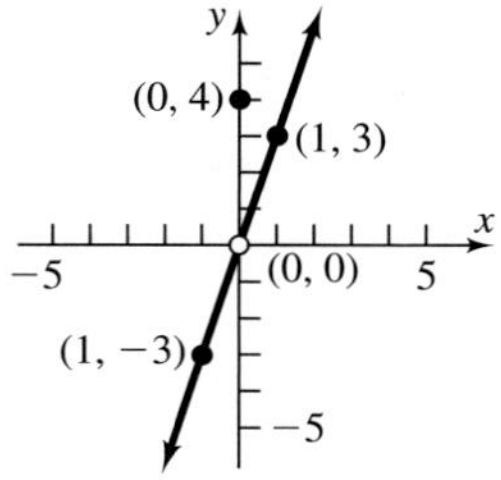

(d) $\{y | y \neq 0\}$; $(-\infty, 0)$ or $(0, \infty)$

31. **(a)** All real numbers
(b) $(0, 3)$
(c)

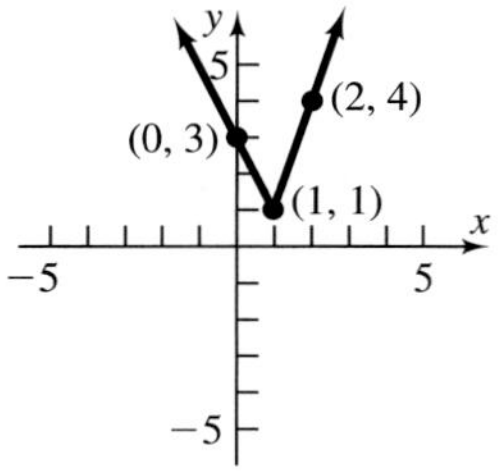

(d) $\{y | y \geq 1\}$; $[1, \infty)$

32. **(a)** All real numbers
(b) $(-3, 0)$, $\left(-\frac{3}{2}, 0\right)$, $(0, -3)$
(c)
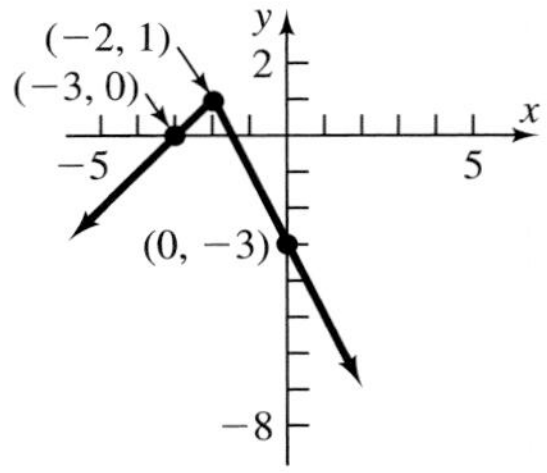

(d) $\{y|y \leq 1\}$; $(-\infty, 1]$

33. **(a)** $\{x|x \geq -2\}$; $[-2, \infty)$
(b) $(0, 3)$, $(2, 0)$
(c)
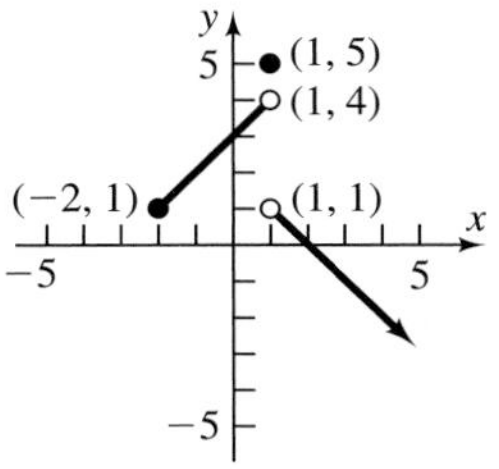

(d) $\{y|y < 4, y = 5\}$; $(-\infty, 4)$ or $\{5\}$

34. **(a)** $\{x|x \geq -3\}$; $[-3, \infty)$
(b) $\left(-\frac{5}{2}, 0\right)$, $(0, -3)$
(c)
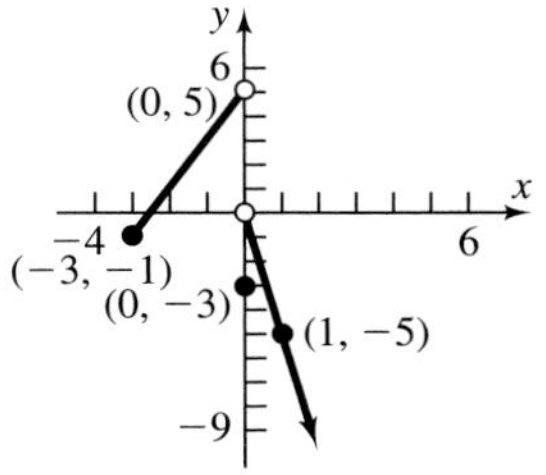

(d) $\{y|y < 5\}$; $(-\infty, 5)$

35. **(a)** All real numbers
(b) $(-1, 0)$, $(0, 0)$
(c)
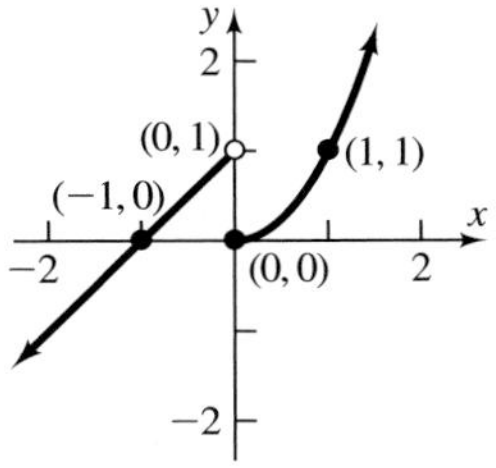

(d) All real numbers

36. **(a)** All real numbers
(b) $(0, 0)$
(c)
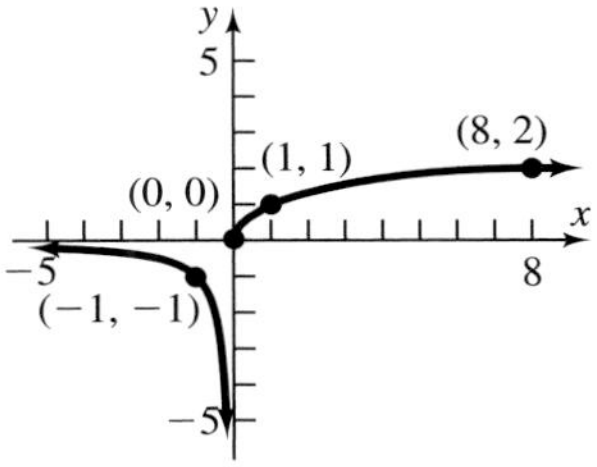

(d) All real numbers

37. **(a)** $\{x|x \geq -2\}$; $[-2, \infty)$
(b) $(0, 1)$
(c)
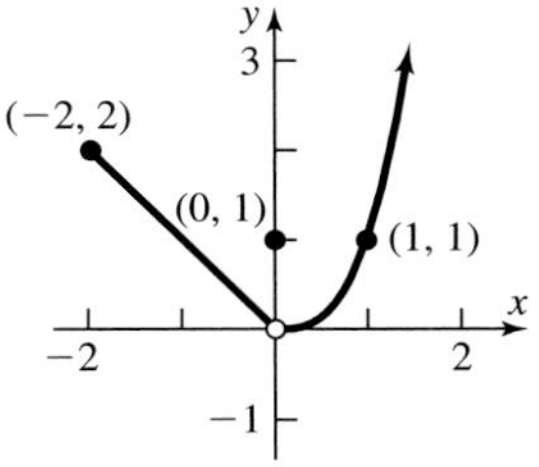

(d) $\{y|y > 0\}$; $(0, \infty)$

38. **(a)** $\{x|x \geq -3\}$; $[-3, \infty)$
(b) $(-3, 0)$, $(0, 3)$
(c)
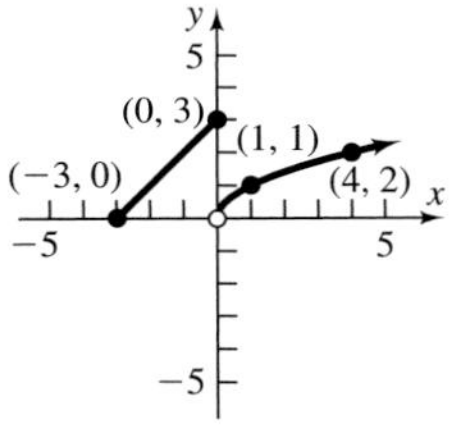

(d) $\{y|y \geq 0\}$; $[0, \infty)$

39. **(a)** All real numbers
(b) $(x, 0)$ for $0 \leq x < 1$
(c)
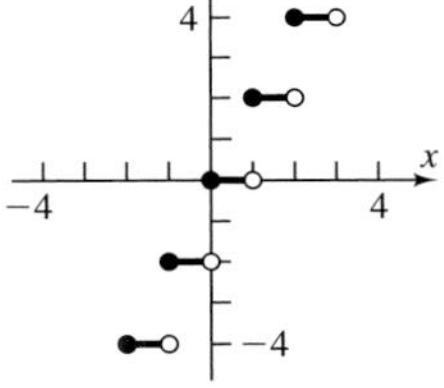

(d) Set of even integers

40. **(a)** All real numbers
(b) $(x, 0)$ for $0 \leq x < \frac{1}{2}$
(c)
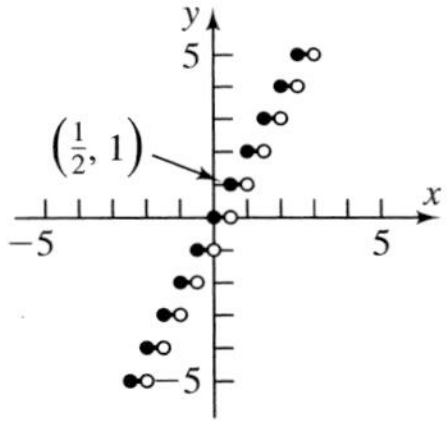

(d) Set of integers

41. $f(x) = \begin{cases} -x & \text{if } -1 \leq x \leq 0 \\ \frac{1}{2}x & \text{if } 0 < x \leq 2 \end{cases}$ (Other answers are possible.)

42. $f(x) = \begin{cases} x & \text{if } -1 \leq x \leq 0 \\ 1 & \text{if } 0 < x \leq 2 \end{cases}$ (Other answers are possible.)

43. $f(x) = \begin{cases} -x & \text{if } x \leq 0 \\ -x + 2 & \text{if } 0 < x \leq 2 \end{cases}$ (Other answers are possible.)

44. $f(x) = \begin{cases} 2x + 2 & \text{if } -1 \leq x \leq 0 \\ x & \text{if } x > 0 \end{cases}$ (Other answers are possible.)

45. **(a)** \$39.99 **(b)** \$43.74 **(c)** \$40.24 **46.** **(a)** \$1.29 **(b)** \$0.37 **(c)** \$3.13

47. **(a)** \$59.33 **(b)** \$396.04 **(c)** $C = \begin{cases} 0.99755x + 9.45 & \text{if } 0 \leq x \leq 50 \\ 0.74825x + 21.915 & \text{if } x > 50 \end{cases}$ **(d)**
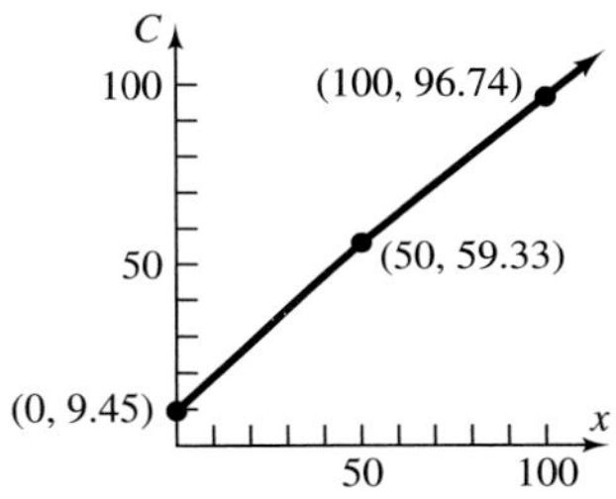

48. **(a)** \$41.78 **(b)** \$166.32 **(c)** $C = \begin{cases} 0.928x + 6.45 & \text{if } 0 \le x \le 20 \\ 0.8385x + 8.24 & \text{if } 20 < x \le 50 \\ 0.7642x + 11.955 & \text{if } x > 50 \end{cases}$ **(d)**

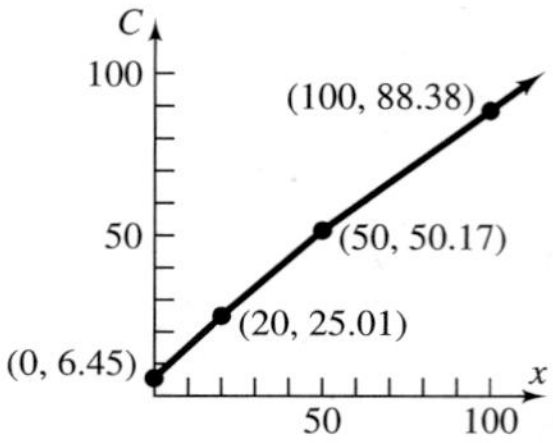

49. For schedule X: $f(x) = \begin{cases} 0.10x & \text{if } 0 < x \le 7000 \\ 700.00 + 0.15(x - 7000) & \text{if } 7000 < x \le 28{,}400 \\ 3910 + 0.25(x - 28{,}400) & \text{if } 28{,}400 < x \le 68{,}800 \\ 14{,}010 + 0.28(x - 68{,}800) & \text{if } 68{,}800 < x \le 143{,}500 \\ 34{,}926 + 0.33(x - 143{,}500) & \text{if } 143{,}500 < x \le 311{,}950 \\ 90{,}514.50 + 0.35(x - 311{,}950) & \text{if } x > 311{,}950 \end{cases}$

50. For schedule Y-1: $f(x) = \begin{cases} 0.10x & \text{if } 0 < x \le 14{,}000 \\ 1400 + 0.15(x - 14{,}000) & \text{if } 14{,}000 < x \le 56{,}800 \\ 7820 + 0.25(x - 56{,}800) & \text{if } 56{,}800 < x \le 114{,}650 \\ 22{,}282.50 + 0.28(x - 114{,}650) & \text{if } 114{,}650 < x \le 174{,}700 \\ 39{,}096.50 + 0.33(x - 174{,}700) & \text{if } 174{,}700 < x \le 311{,}950 \\ 84{,}389 + 0.35(x - 311{,}950) & \text{if } x > 311{,}950 \end{cases}$

51. **(a)**

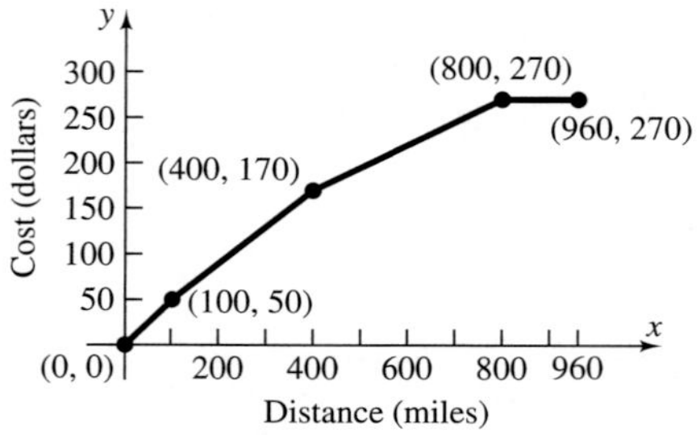

(b) $C = 50 + 0.4(x - 100)$
(c) $C = 170 + 0.25(x - 400)$

52. $C = \begin{cases} 95 & \text{if } x = 7 \\ 119 & \text{if } 7 < x \le 8 \\ 143 & \text{if } 8 < x \le 9 \\ 167 & \text{if } 9 < x \le 10 \\ 190 & \text{if } 10 < x \le 14 \end{cases}$

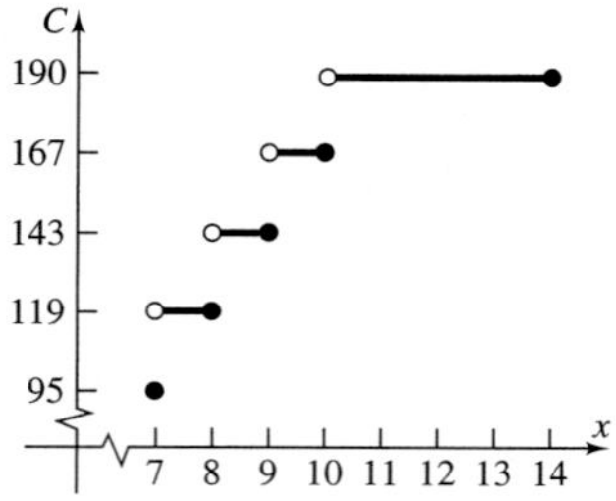

53. $f(x) = \begin{cases} x & \text{if } 0 \le x < 10 \\ 10 & \text{if } 10 \le x < 500 \\ 30 & \text{if } 500 \le x < 1000 \\ 50 & \text{if } 1000 \le x < 1500 \\ 70 & \text{if } 1500 \le x \end{cases}$

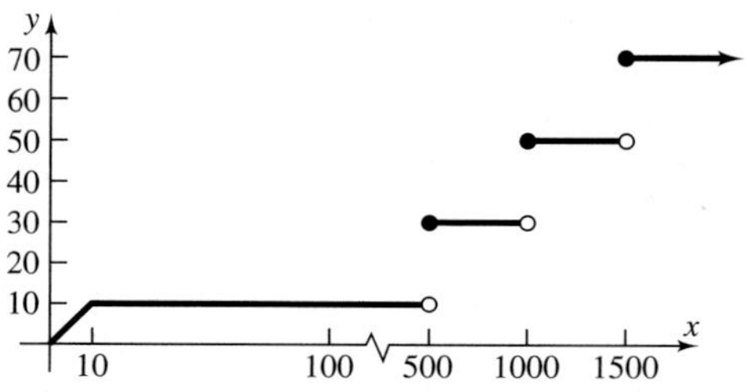

54. $g(x) = \begin{cases} 0.015x & \text{if } 0 \le x \le 1000 \\ 5 + 0.01x & \text{if } x > 1000 \end{cases}$

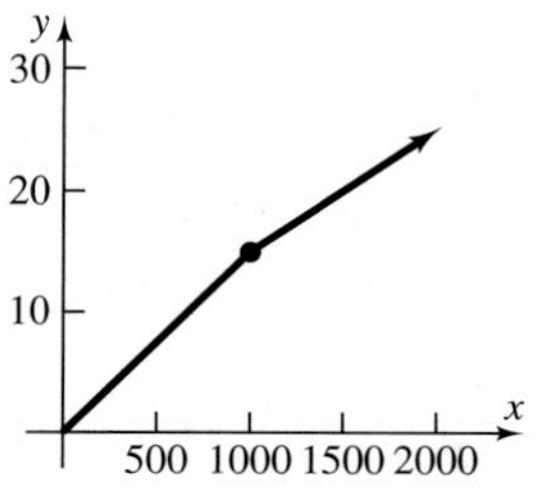

55. **(a)** 10°C **(b)** 4°C **(c)** −3°C **(d)** −4°C **(e)** The wind chill is equal to the air temperature. **(f)** At wind speed greater than 20 m/sec, the wind chill factor depends only on the air temperature. **56.** **(a)** −10°C **(b)** −21°C **(c)** −34°C **(d)** −36°C
57. Each graph is that of $y = x^2$, but shifted vertically. If $y = x^2 + k$, $k > 0$, the shift is up k units; if $y = x^2 - k$, $k > 0$, the shift is down k units. **58.** Each graph is that of $y = x^2$, but shifted horizontally. If $y = (x - k)^2$, $k > 0$, the shift is right k units; if $y = (x + k)^2$, $k > 0$, the shift is left k units. **59.** Each graph is that of $y = |x|$, but either compressed or stretched. If $y = k|x|$ and $k > 1$, the graph is stretched vertically; if $y = k|x|$, $0 < k < 1$, the graph is compressed vertically. **60.** The graph of $y = -f(x)$ is the reflection about the x-axis of the graph of $y = f(x)$.

61. The graph of $y = f(-x)$ is the reflection about the y-axis of the graph of $y = f(x)$. **62.** Yes. The graph of $y = (x - 1)^3 + 2$ is the graph of $y = x^3$ shifted right 1 unit and up 2 units. **63.** They are all ∪-shaped and open upward. All three go through the points $(-1, 1)$, $(0, 0)$ and $(1, 1)$. As the exponent increases, the steepness of the curve increases (except near $x = 0$). **64.** They all have the same general shape. All three go through the points $(-1, -1)$, $(0, 0)$, and $(1, 1)$. As the exponent increases, the steepness of the curve increases (except near $x = 0$).

2.5 Concepts and Vocabulary *(page 103)*

1. horizontal; right **2.** y **3.** –5, –2, 2 **4.** True **5.** False **6.** True

2.5 Exercises *(page 104)*

7. B **8.** E **9.** H **10.** D **11.** I **12.** A **13.** L **14.** C **15.** F **16.** J **17.** G **18.** K **19.** $y = (x - 4)^3$ **20.** $y = (x + 4)^3$

21. $y = x^3 + 4$ **22.** $y = x^3 - 4$ **23.** $y = -x^3$ **24.** $y = -x^3$ **25.** $y = 4x^3$ **26.** $y = \left(\frac{1}{4}x\right)^3 = \frac{1}{64}x^3$

27. (1) $y = \sqrt{x} + 2$; (2) $y = -(\sqrt{x} + 2)$; (3) $y = -(\sqrt{-x} + 2)$ **28.** (1) $y = -\sqrt{x}$; (2) $y = -\sqrt{x - 3}$; (3) $y = -\sqrt{x - 3} - 2$

29. (1) $y = -\sqrt{x}$; (2) $y = -\sqrt{x} + 2$; (3) $y = -\sqrt{x + 3} + 2$ **30.** (1) $y = \sqrt{x} + 2$; (2) $y = \sqrt{-x} + 2$; (3) $y = \sqrt{-x - 3} + 2$

31. (c) **32.** (d) **33.** (c) **34.** (a)

35.
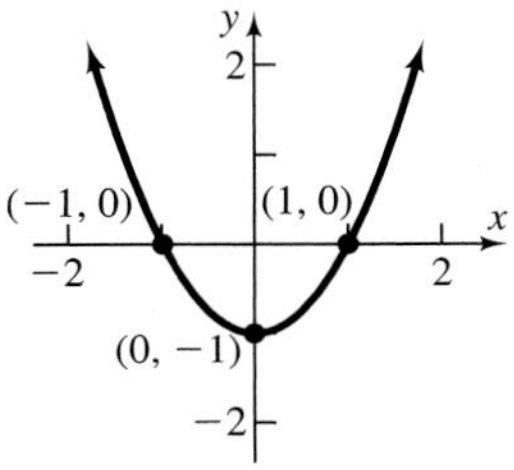

36.
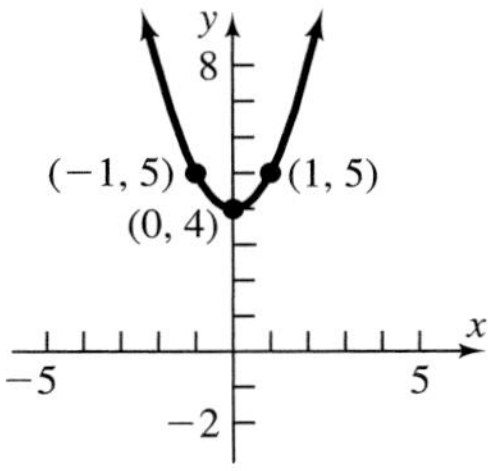

37.
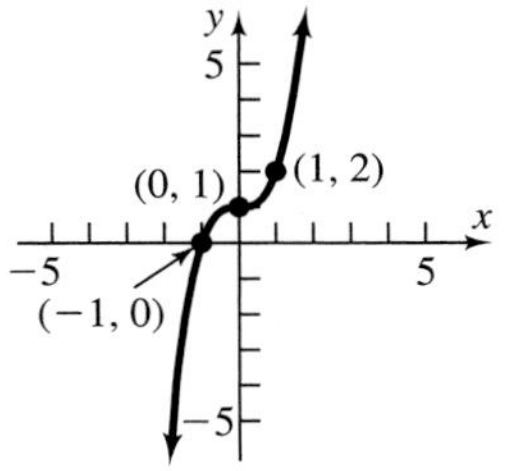

38.
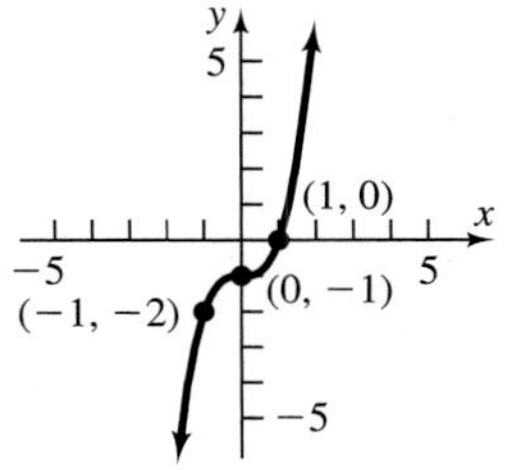

39.
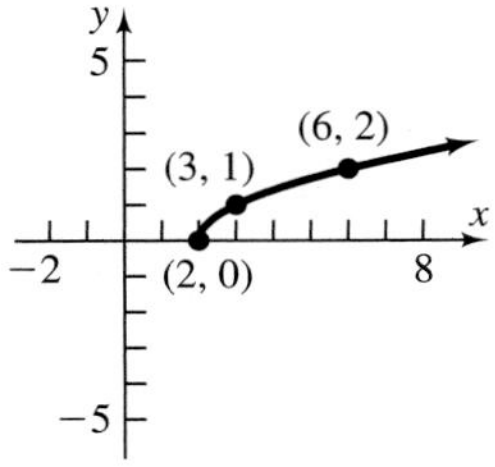

40.
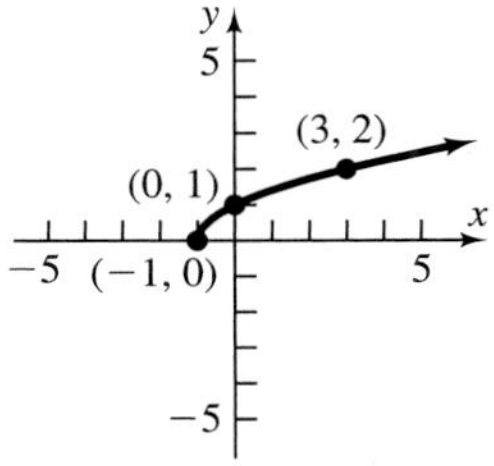

41.
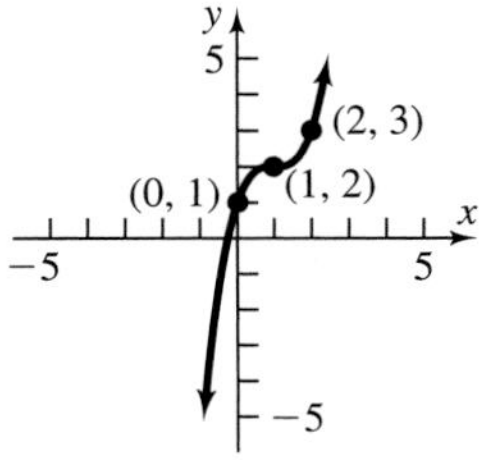

42.
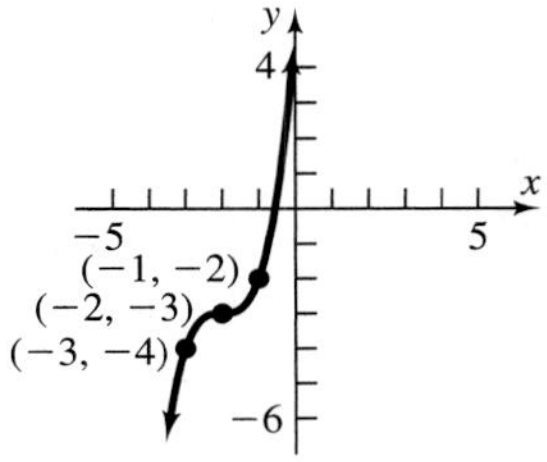

43.
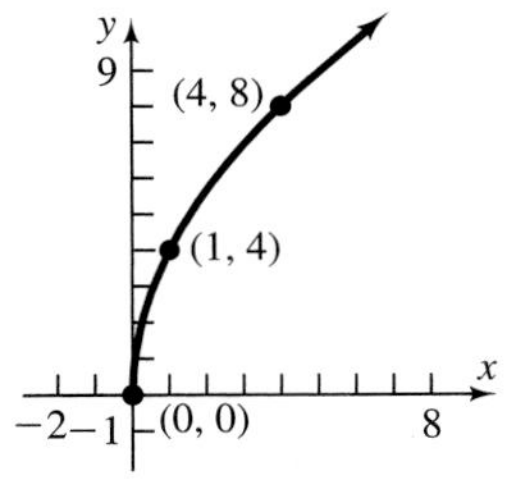

44.
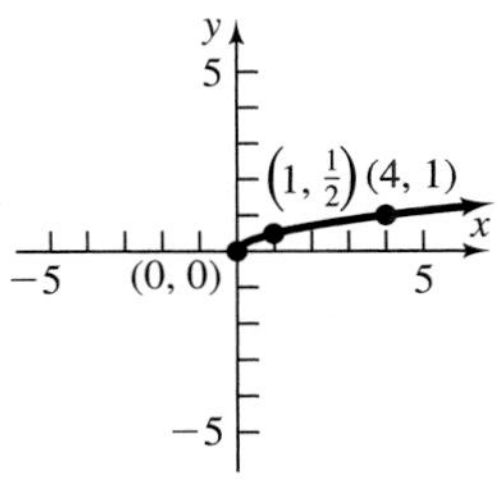

45.
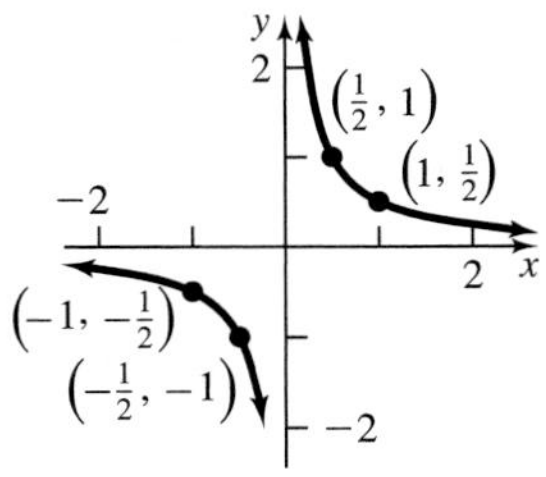

46.
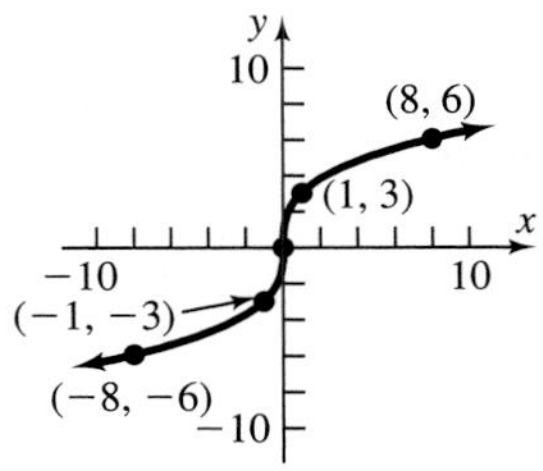

47.
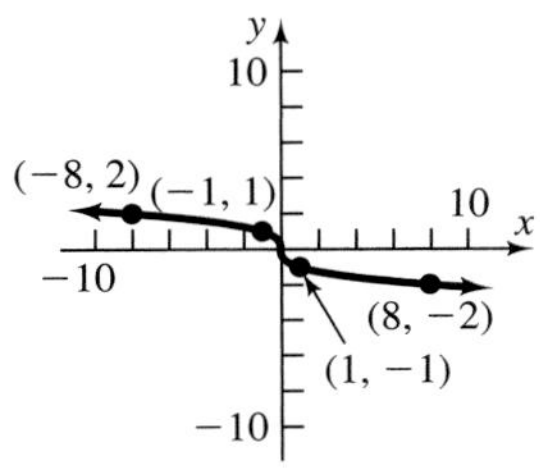

48.
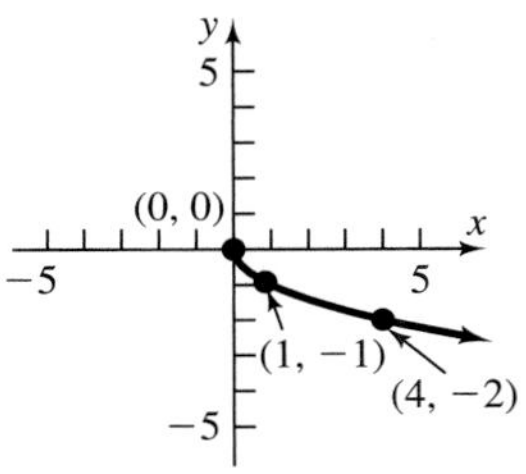

49.
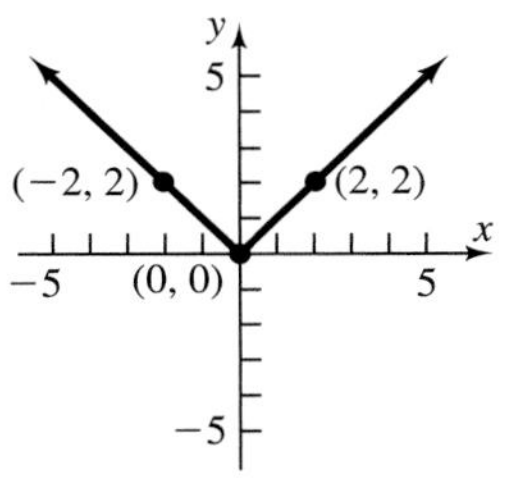

50.
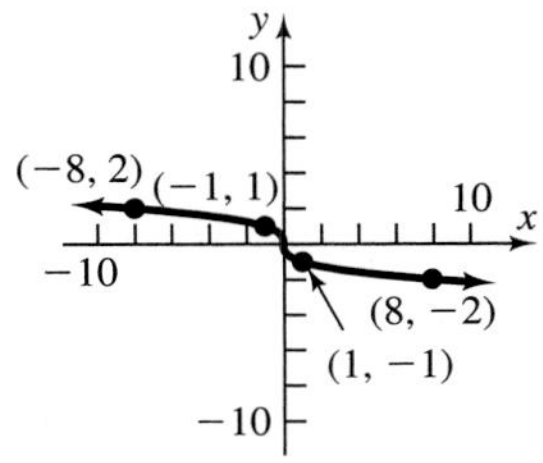

51.

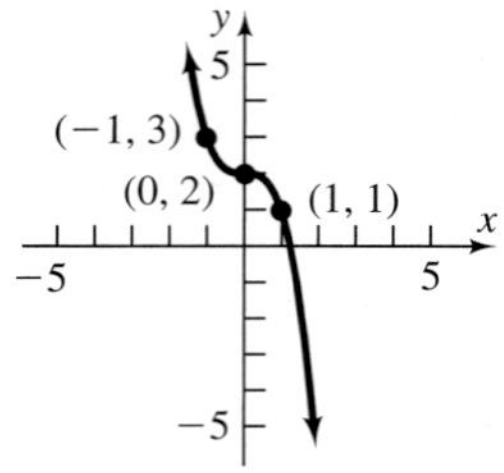

52.

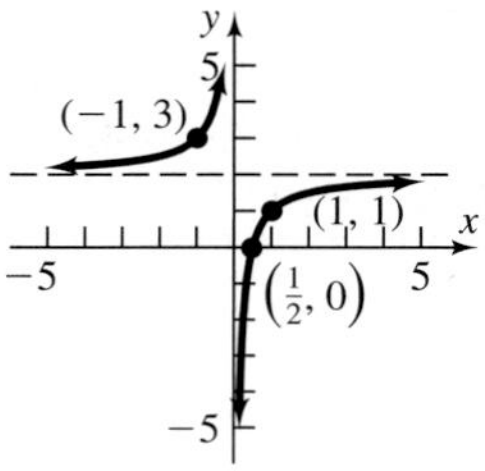

53.

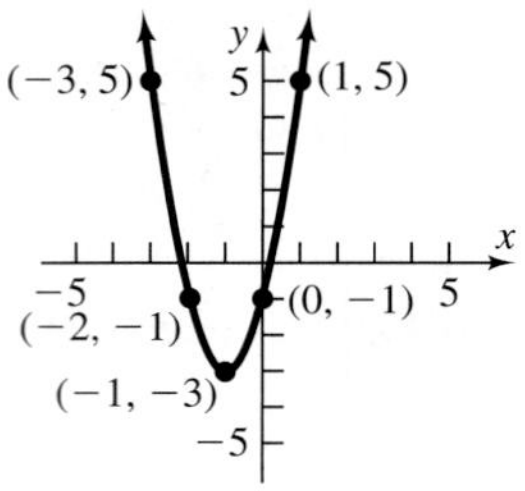

54.

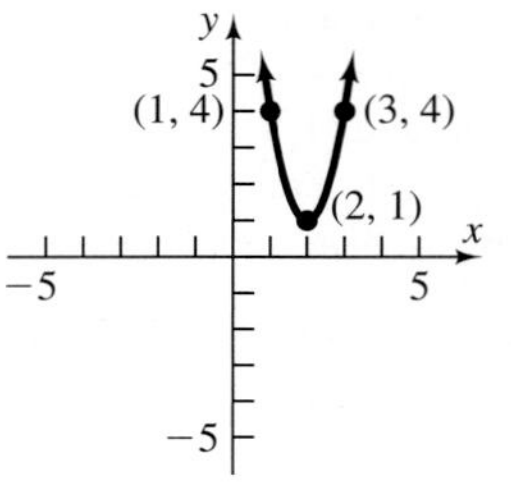

55.

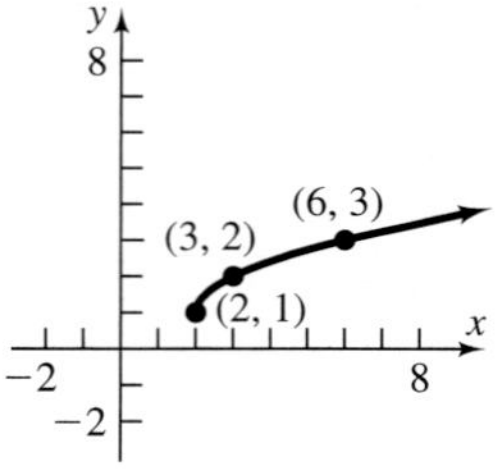

56.

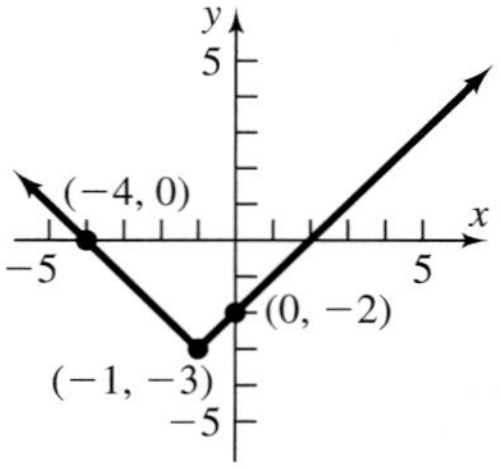

57.

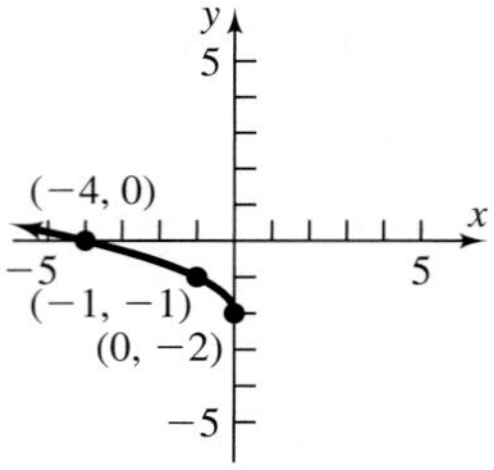

58.

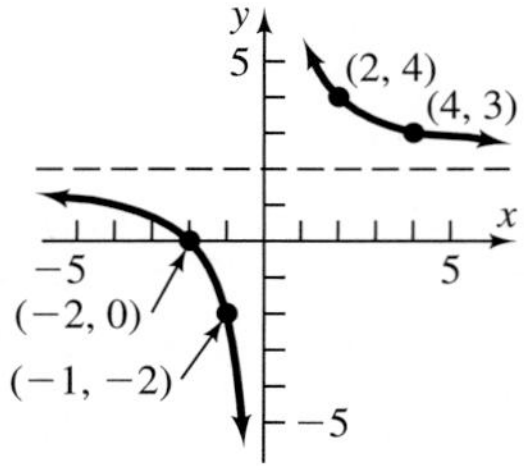

59.

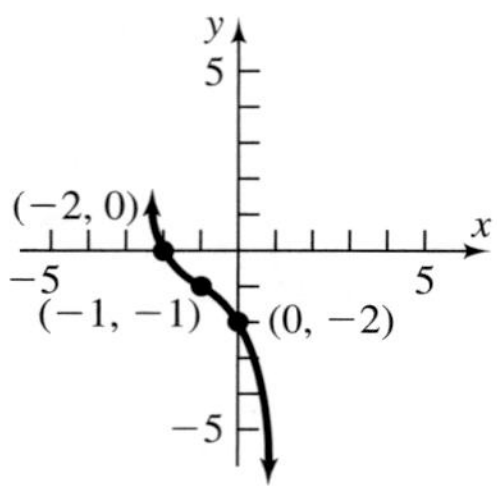

60.

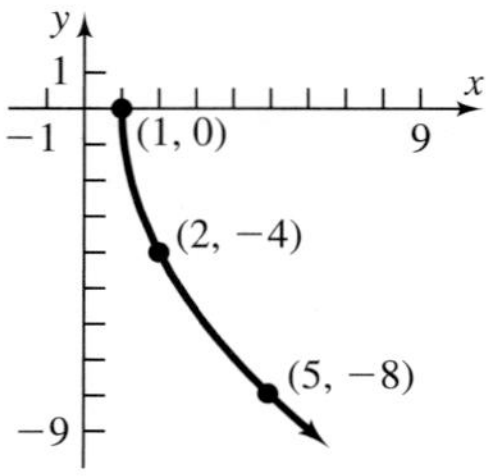

61.

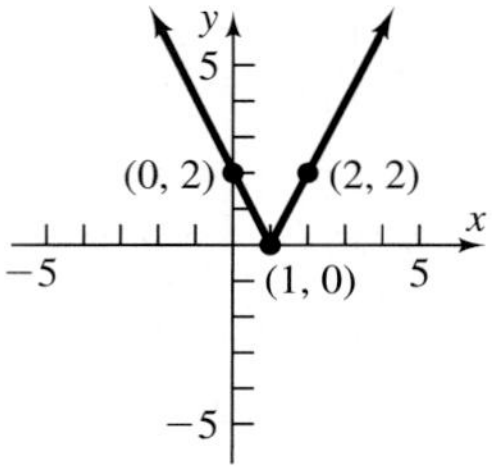

62.

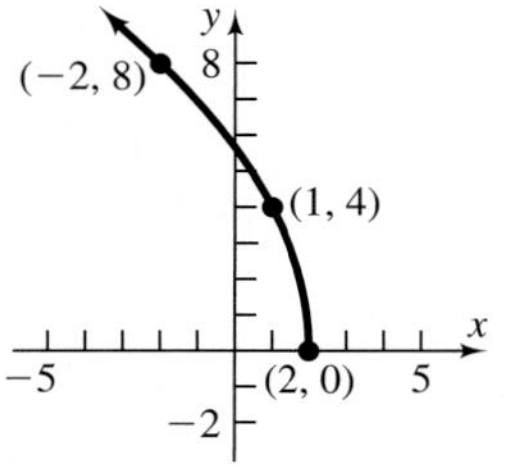

63.

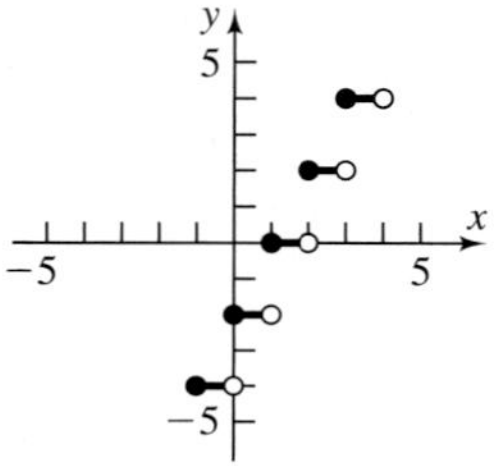

64.

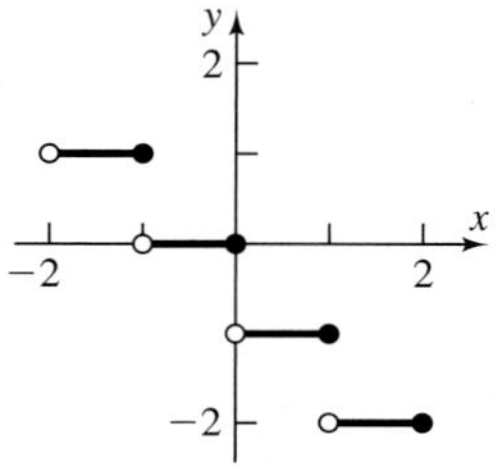

65. **(a)** $F(x) = f(x) + 3$

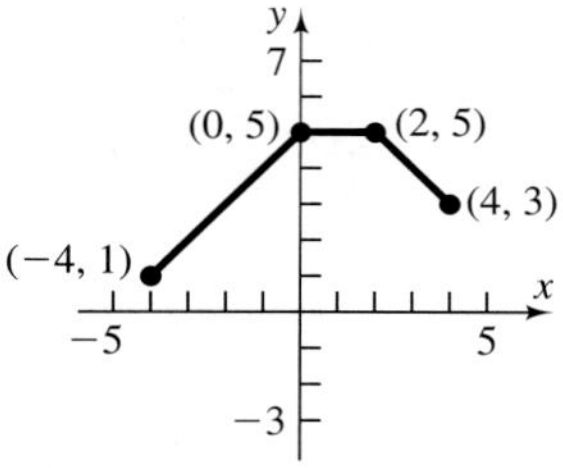

(b) $G(x) = f(x + 2)$

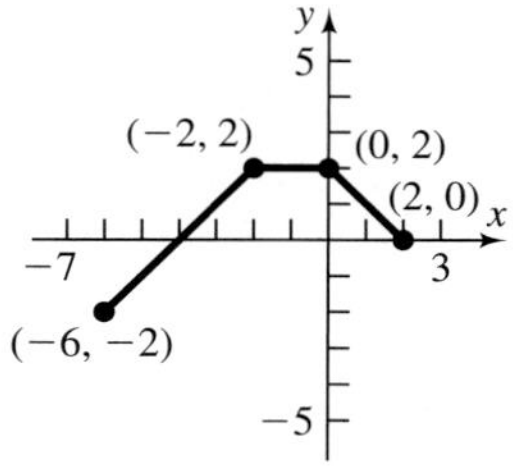

(c) $P(x) = -f(x)$

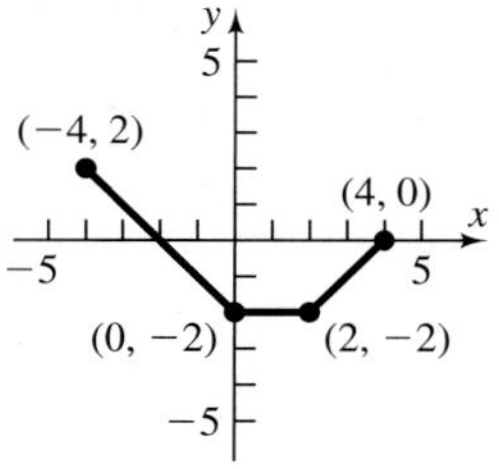

(d) $H(x) = f(x + 1) - 2$

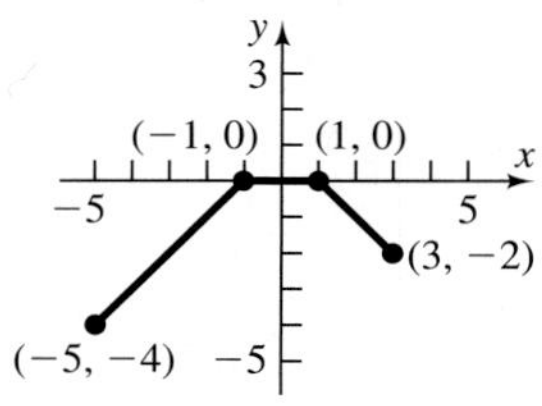

(e) $Q(x) = \frac{1}{2}f(x)$

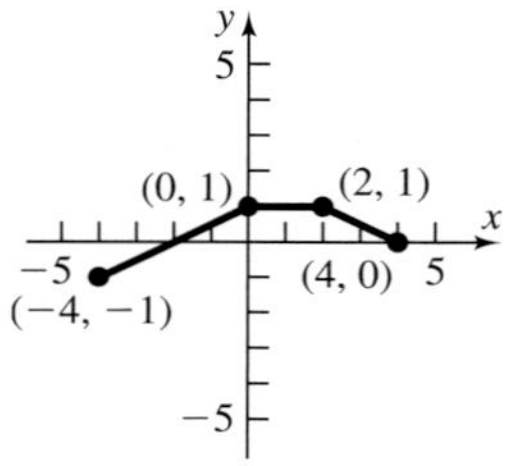

(f) $g(x) = f(-x)$

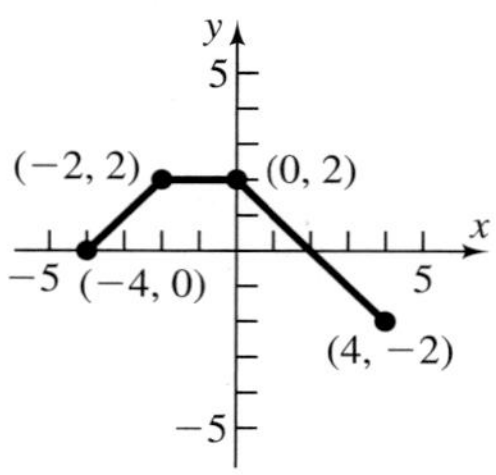

(g) $h(x) = f(2x)$

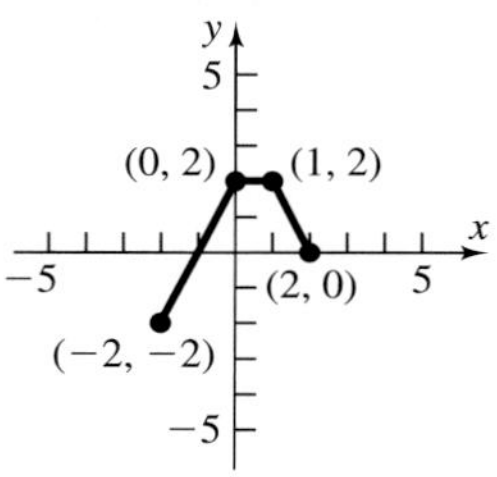

66. **(a)** $F(x) = f(x) + 3$

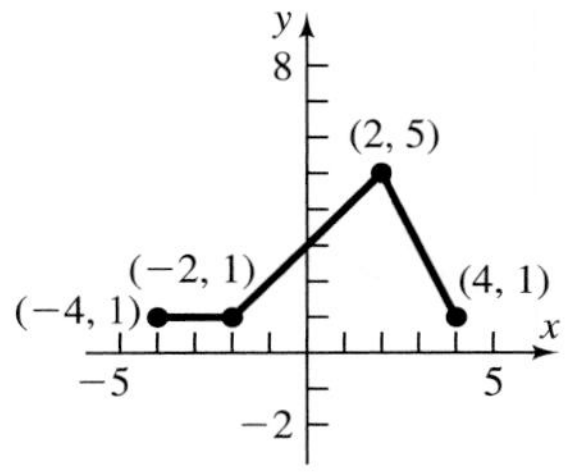

(b) $G(x) = f(x + 2)$

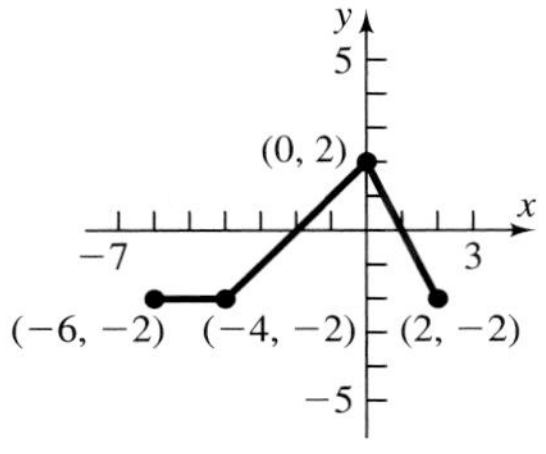

(c) $P(x) = -f(x)$

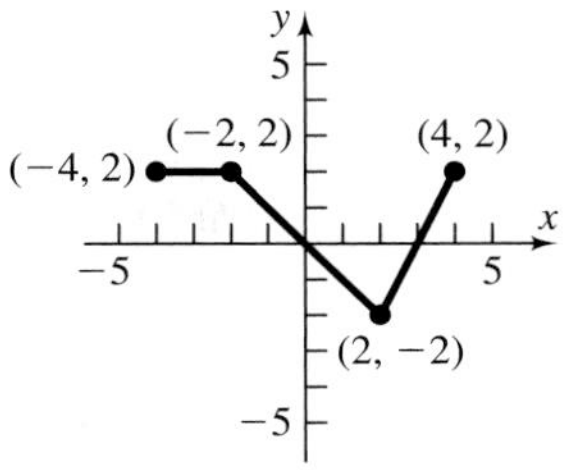

(d) $H(x) = f(x + 1) - 2$

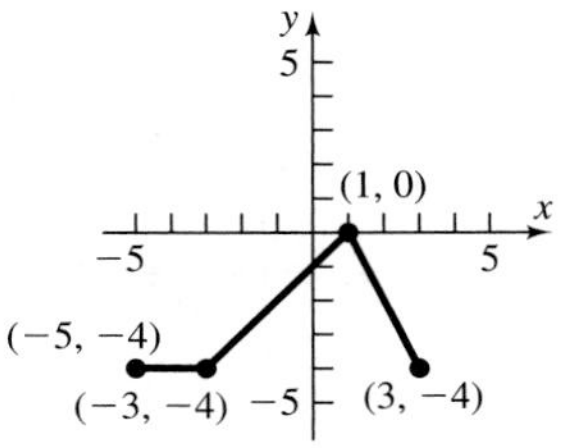

(e) $Q(x) = \frac{1}{2}f(x)$

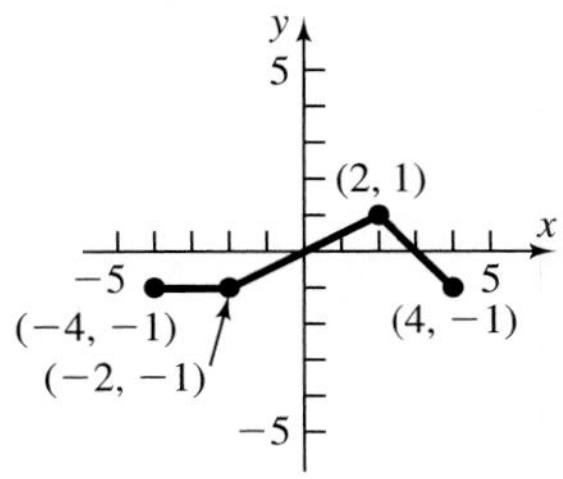

(f) $g(x) = f(-x)$

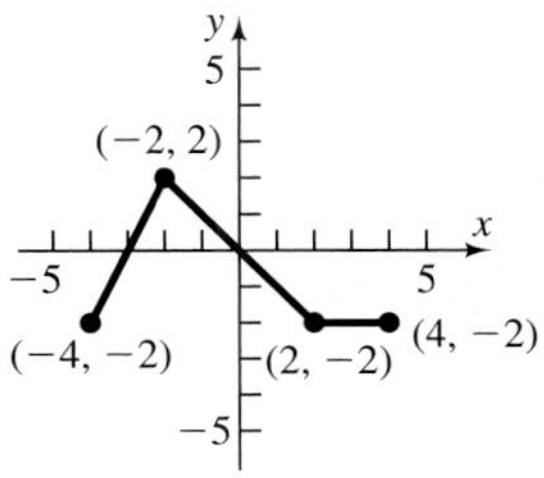

(g) $h(x) = f(2x)$

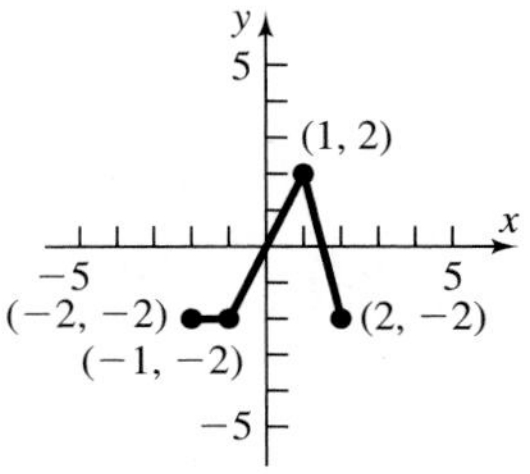

67. **(a)** $F(x) = f(x) + 3$

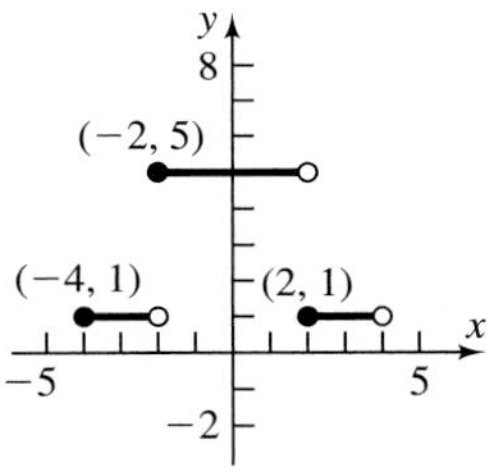

(b) $G(x) = f(x + 2)$

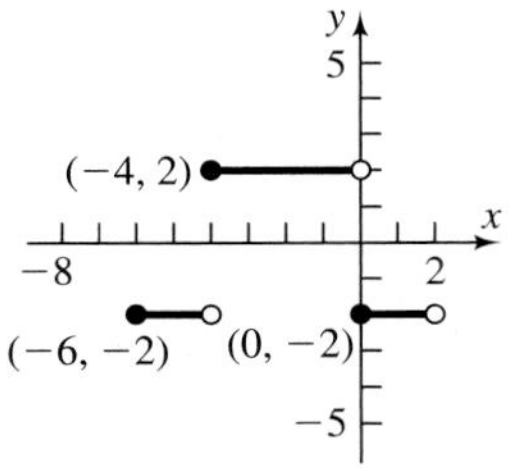

(c) $P(x) = -f(x)$

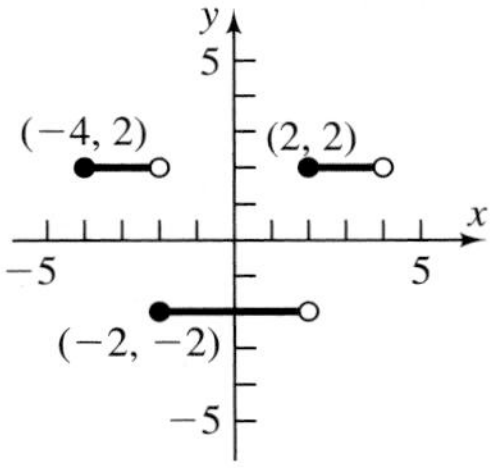

(d) $H(x) = f(x + 1) - 2$

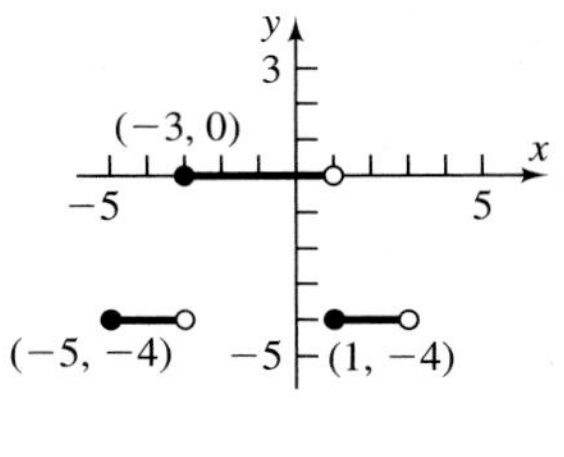

(e) $Q(x) = \frac{1}{2}f(x)$

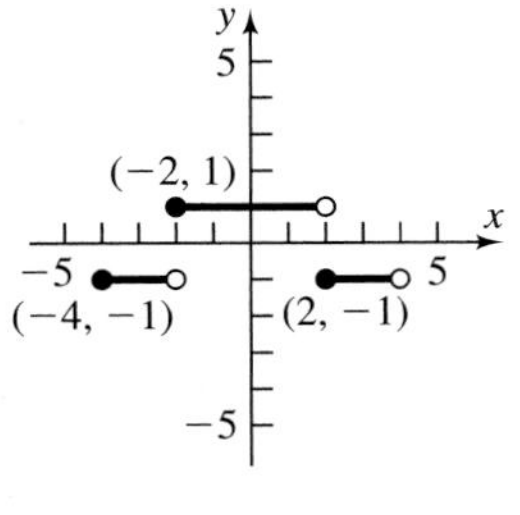

(f) $g(x) = f(-x)$

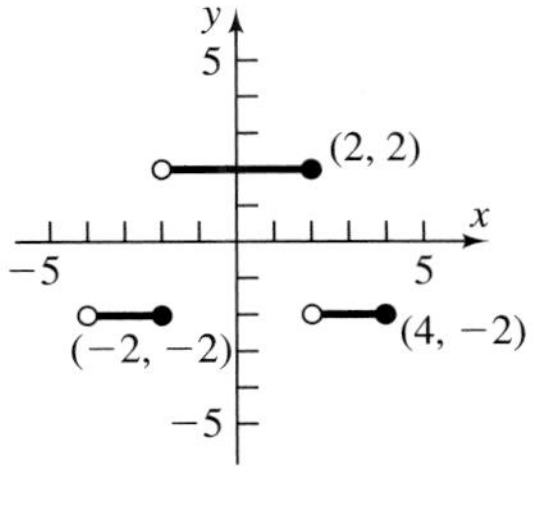

(g) $h(x) = f(2x)$

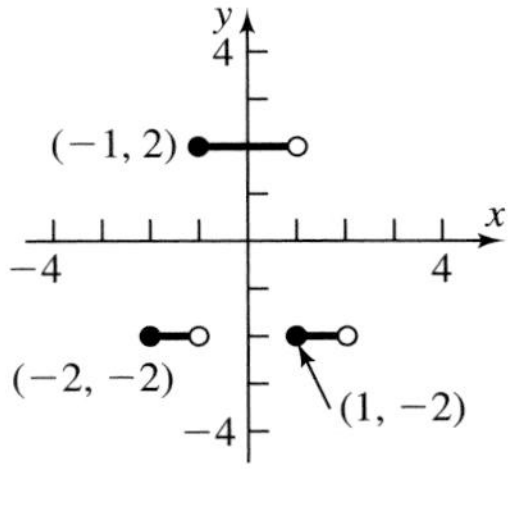

68. **(a)** $F(x) = f(x) + 3$

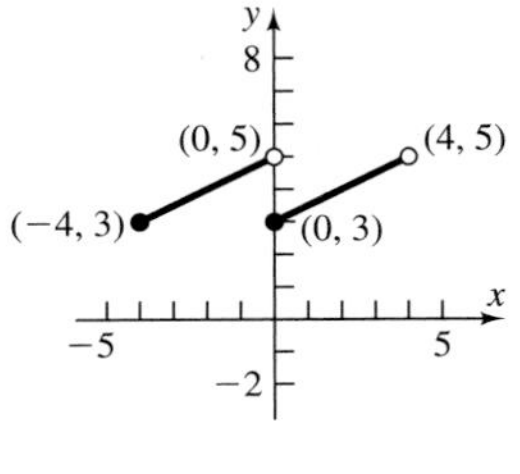

(b) $G(x) = f(x + 2)$

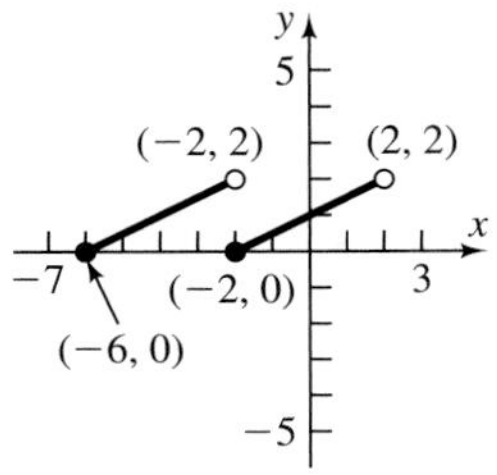

(c) $P(x) = -f(x)$

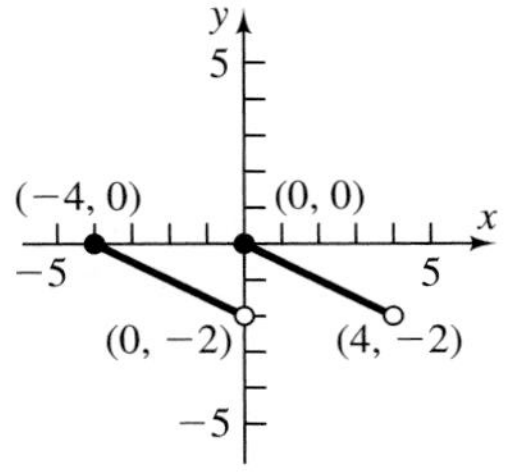

(d) $H(x) = f(x + 1) - 2$

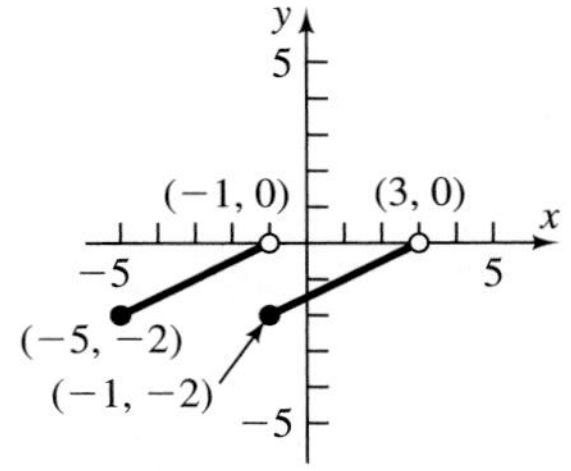

(e) $Q(x) = \frac{1}{2}f(x)$

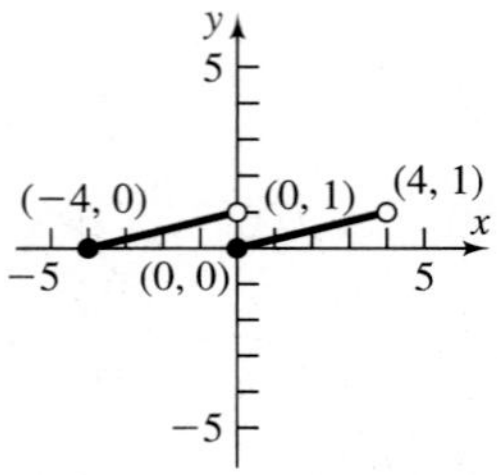

(f) $g(x) = f(-x)$

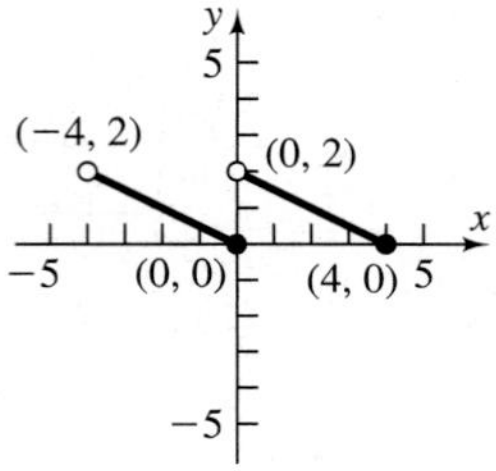

(g) $h(x) = f(2x)$

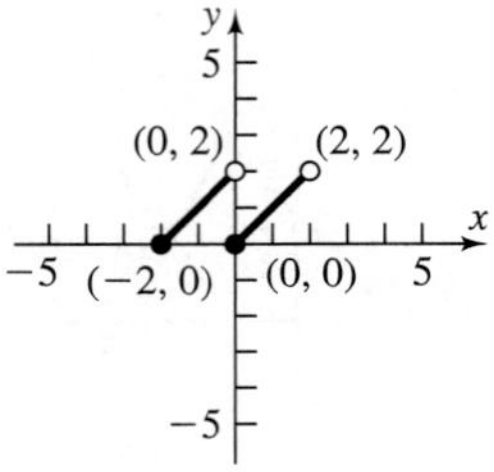

69. (a) $F(x) = f(x) + 3$

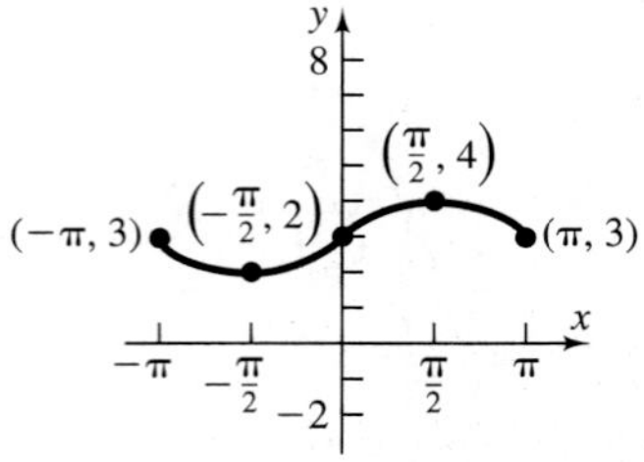

(b) $G(x) = f(x + 2)$

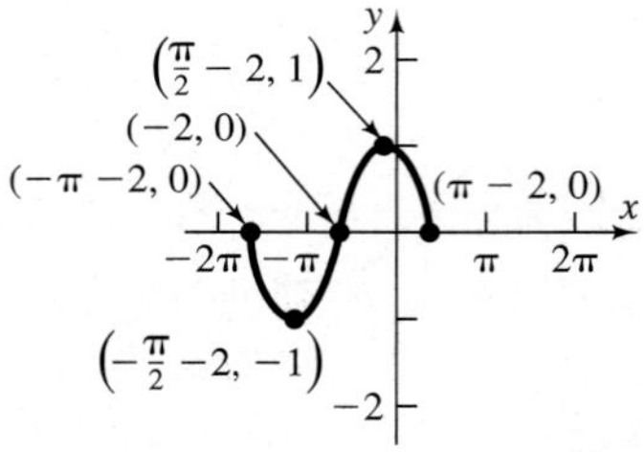

(c) $P(x) = -f(x)$

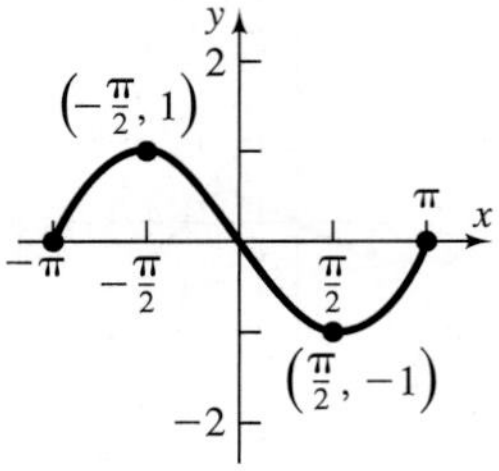

(d) $H(x) = f(x + 1) - 2$

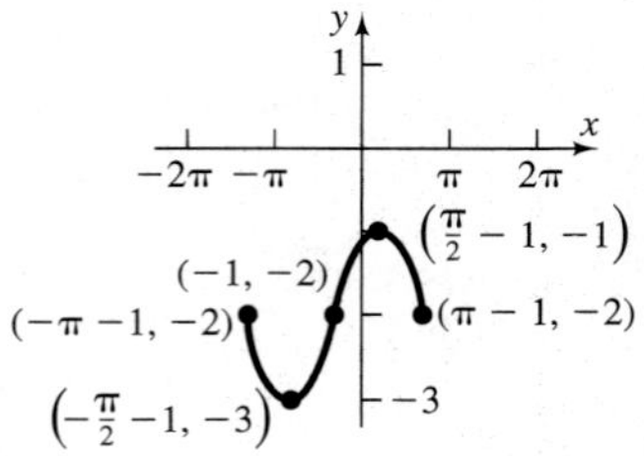

(e) $Q(x) = \frac{1}{2}f(x)$

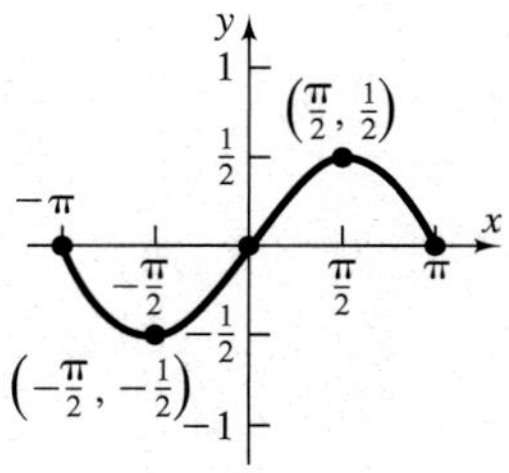

(f) $g(x) = f(-x)$

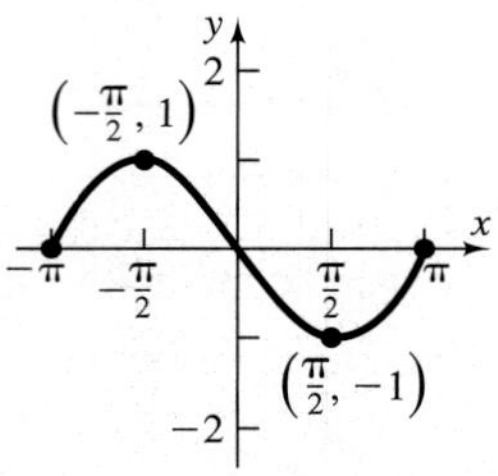

(g) $h(x) = f(2x)$

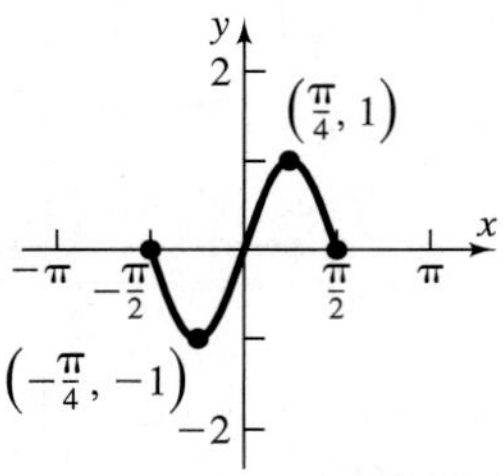

70. (a) $F(x) = f(x) + 3$

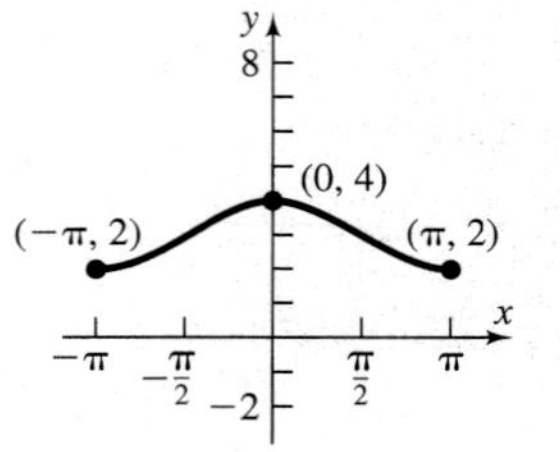

(b) $G(x) = f(x + 2)$

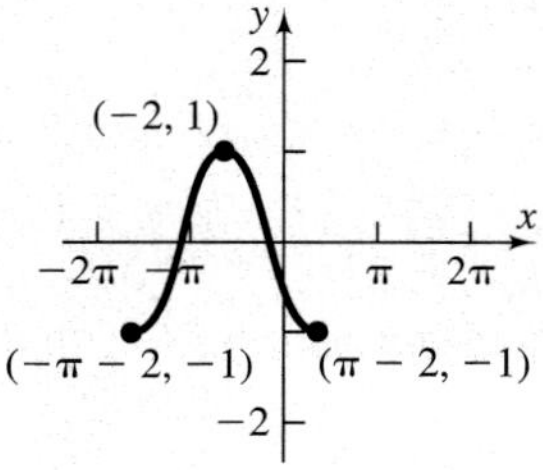

(c) $P(x) = -f(x)$

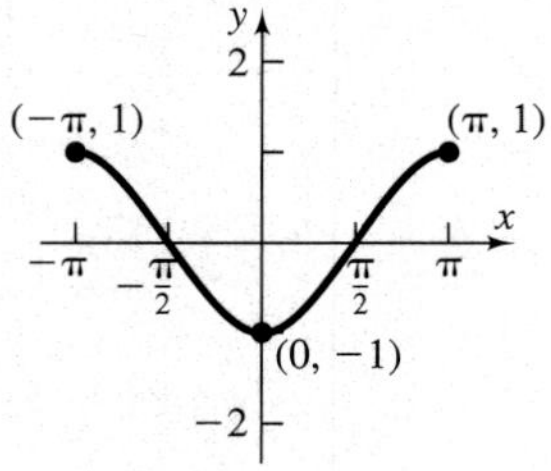

(d) $H(x) = f(x + 1) - 2$

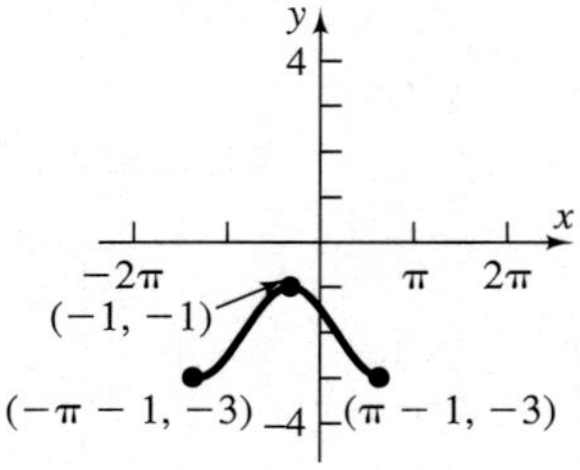

(e) $Q(x) = \frac{1}{2}f(x)$

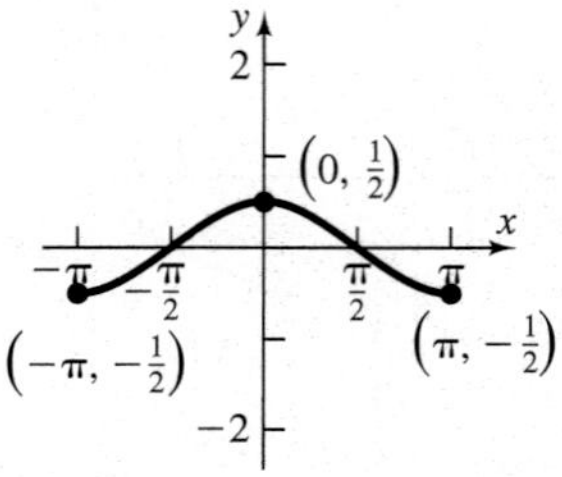

(f) $g(x) = f(-x)$

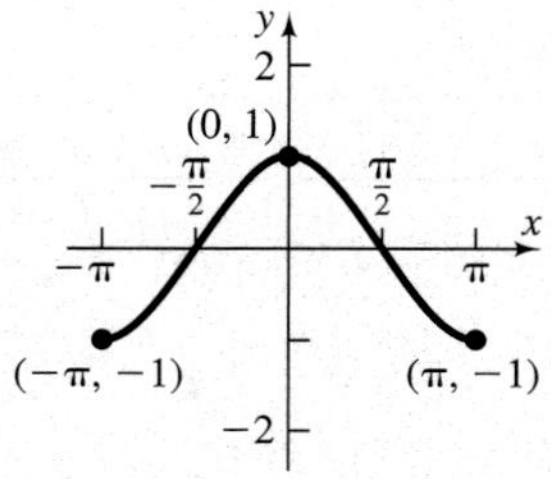

(g) $h(x) = f(2x)$

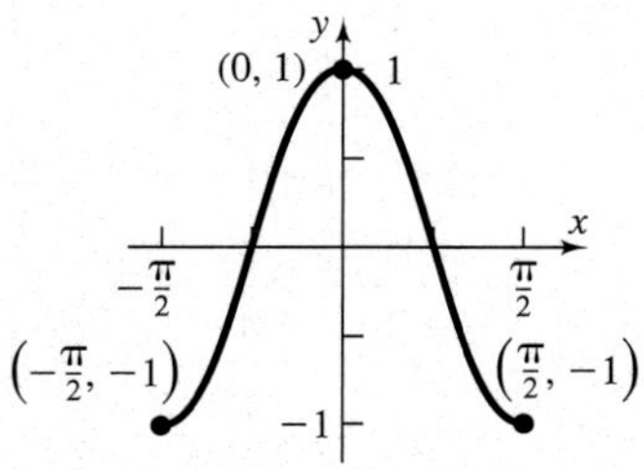

71. (a)

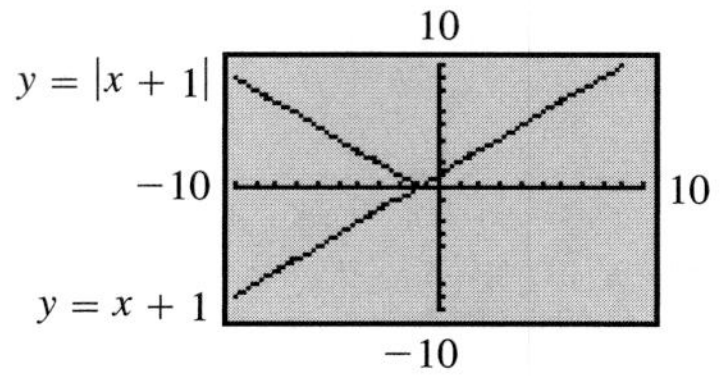

(b)

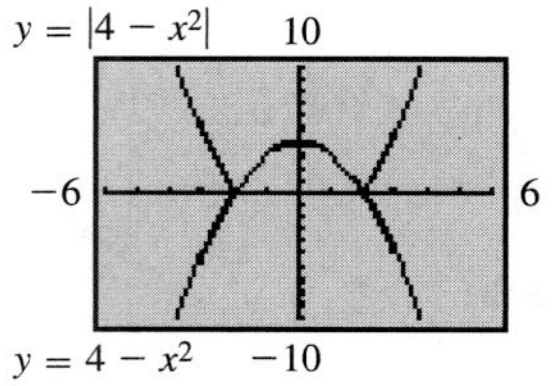

(c)

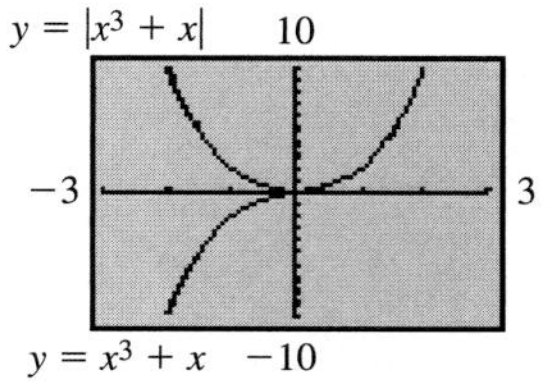

(d) Any part of the graph of $y = f(x)$ that lies below the x-axis is reflected about the x-axis to obtain the graph of $y = |f(x)|$.

72. (a)

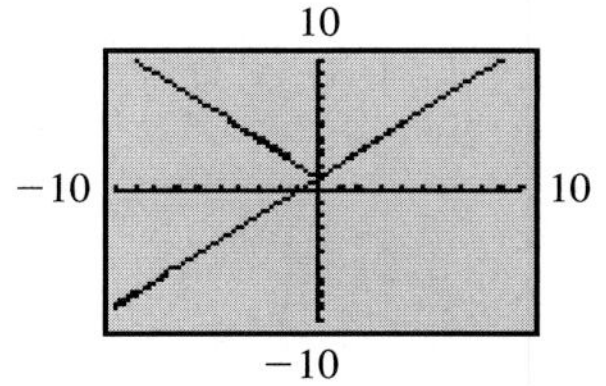

(b)

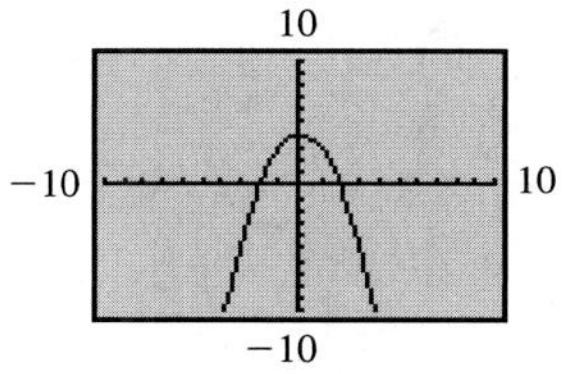

(c)

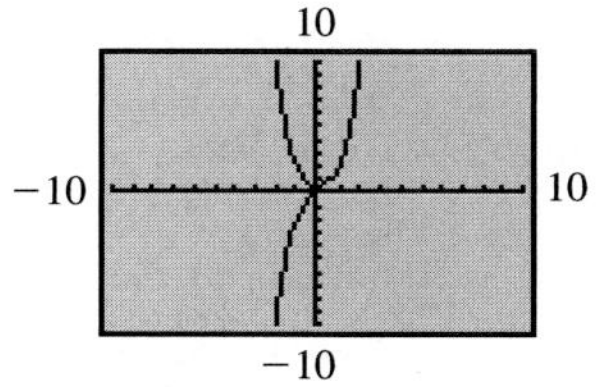

(d) The graph of $y = f(|x|)$ is the same as the graph of $y = f(x)$ for $x \geq 0$. For $x < 0$, the graph of $y = f(|x|)$ is obtained by reflecting across the y-axis the part of the graph for $x > 0$.

73. (a)

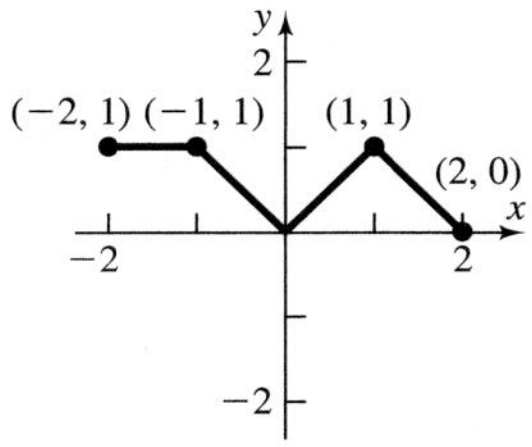

(b)

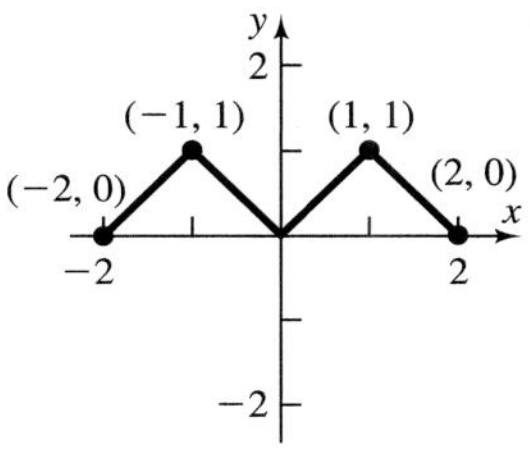

74. (a)

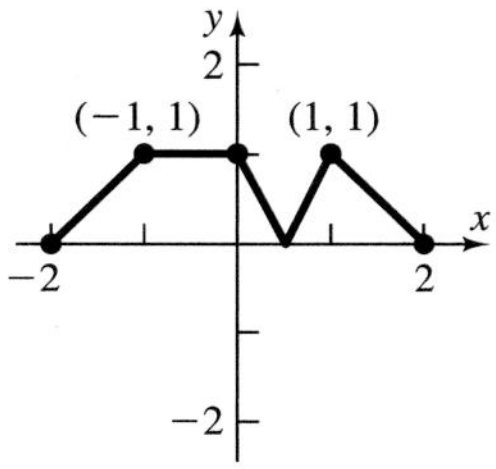

(b)

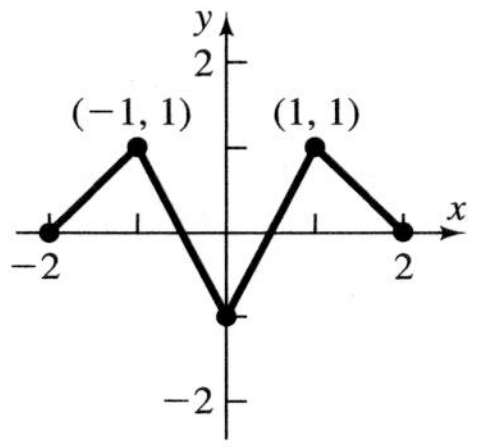

75. $f(x) = (x + 1)^2 - 1$

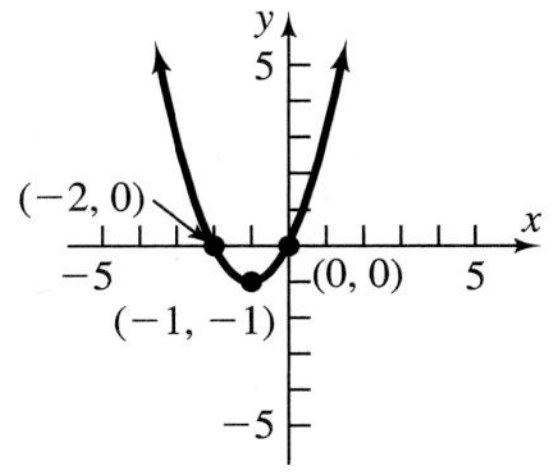

76. $f(x) = (x - 3)^2 - 9$

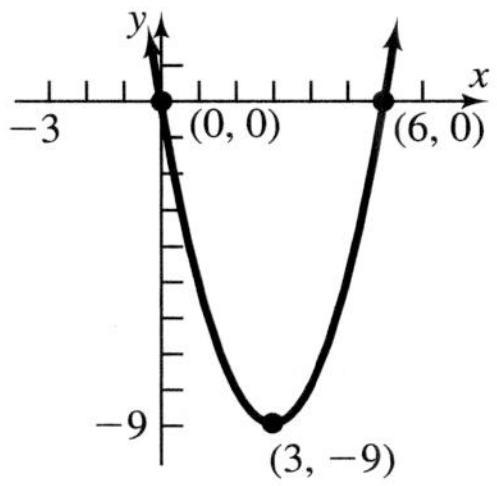

77. $f(x) = (x - 4)^2 - 15$

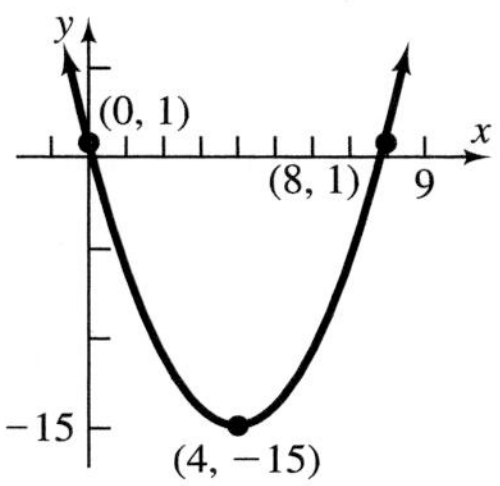

78. $f(x) = (x + 2)^2 - 2$

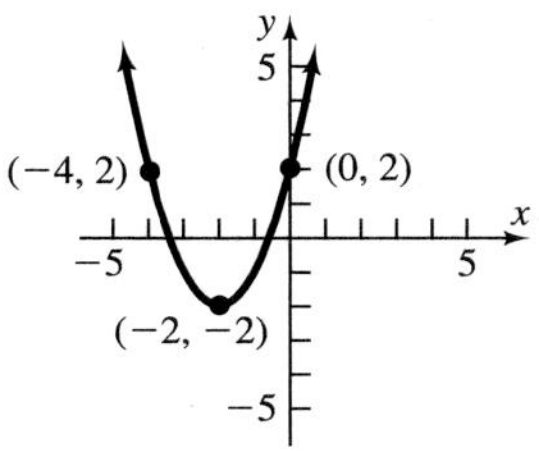

79. $f(x) = \left(x + \frac{1}{2}\right)^2 + \frac{3}{4}$

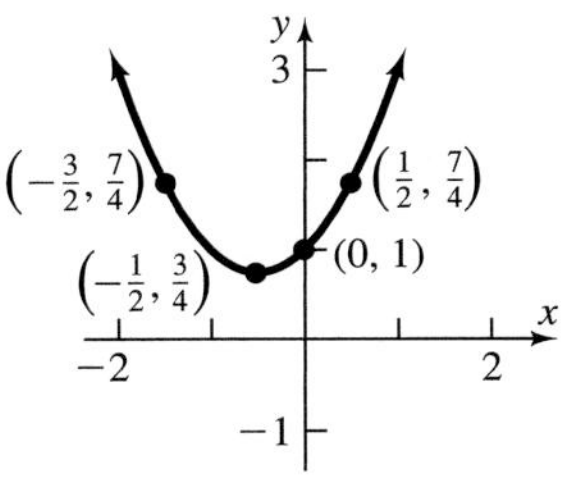

80. $f(x) = \left(x - \frac{1}{2}\right)^2 + \frac{3}{4}$

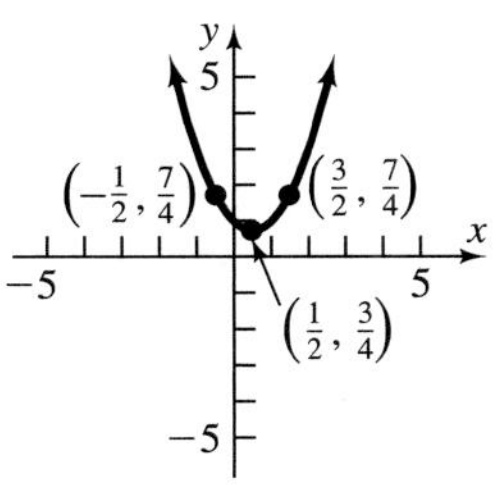

81.

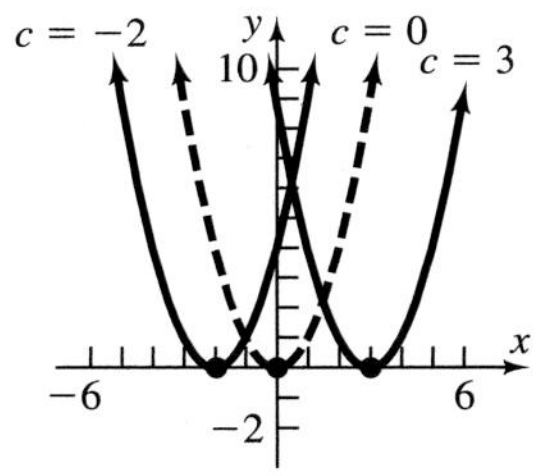

82.

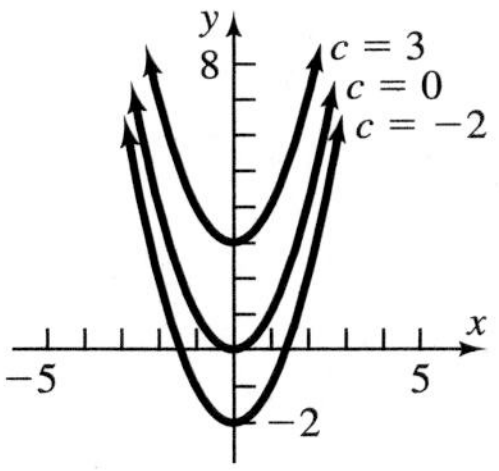

83.

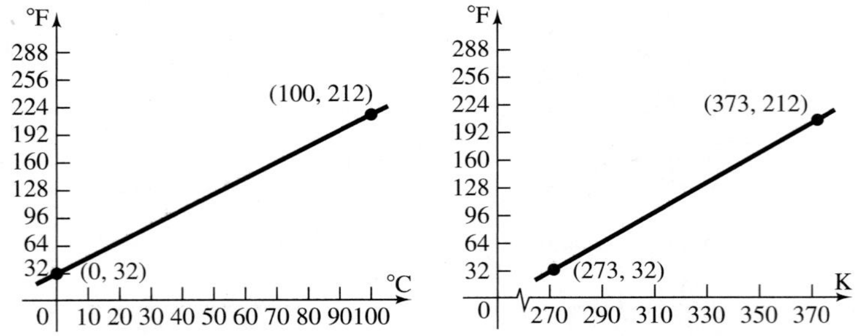

84. (a)

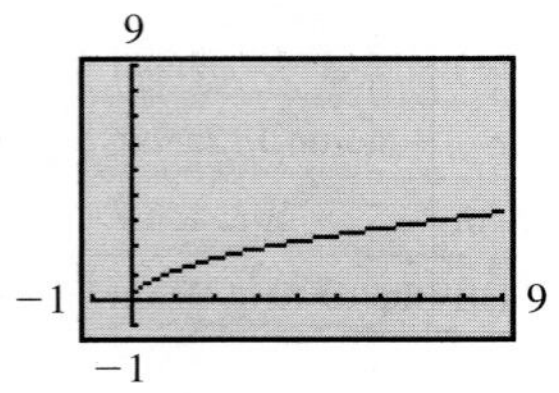

(b)

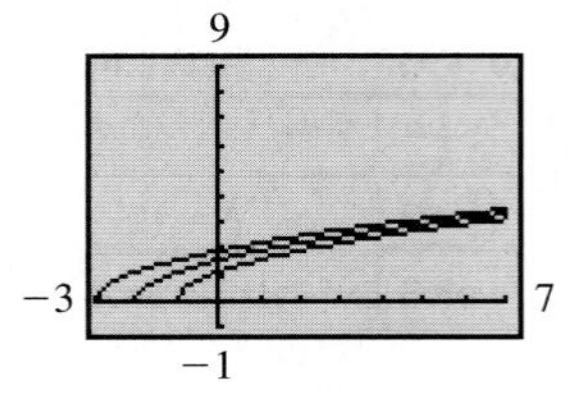

(d)

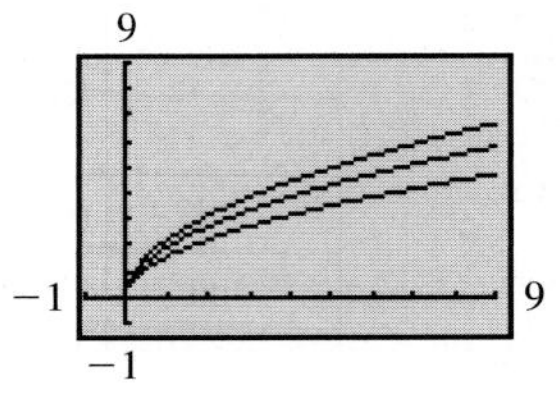

85. (a)

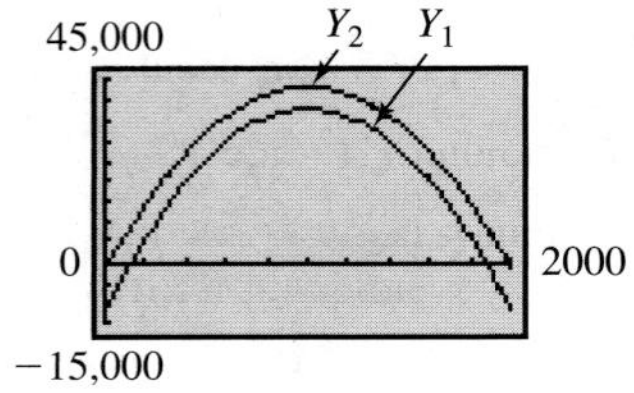

(b) 10% tax
(c) Y_1 is the graph of $p(x)$ shifted down vertically 10,000 units.
Y_2 is the graph of $p(x)$ vertically compressed by a factor of 0.9.
(d) 10% tax

2.6 Exercises *(page 112)*

1. $V(r) = 2\pi r^3$ **2.** $V(r) = \frac{2}{3}\pi r^3$

3. (a) $R(x) = -\frac{1}{6}x^2 + 100x$
(b) \$13,333.33
(c)

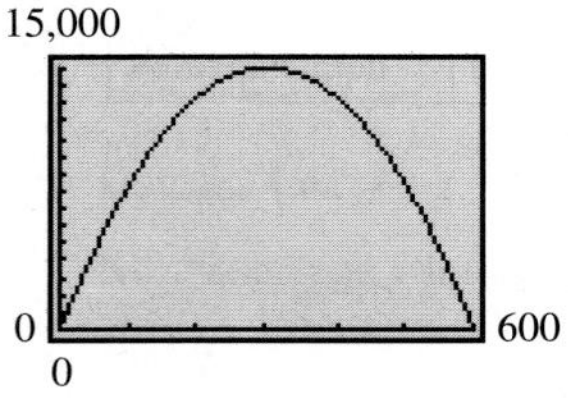

(d) 300; \$15,000 **(e)** \$50

4. (a) $R(x) = -\frac{1}{3}x^2 + 100x$
(b) \$6666.67
(c)

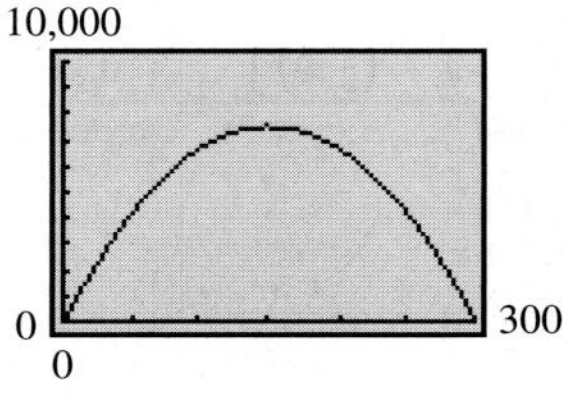

(d) 150; \$7500 **(e)** \$50

5. (a) $R(x) = -\frac{1}{5}x^2 + 20x$
(b) \$255
(c)

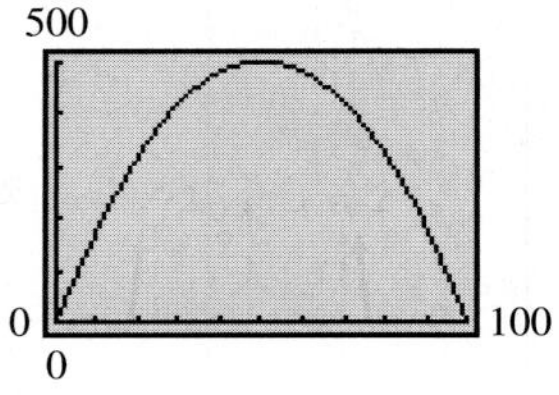

(d) 50; \$500 **(e)** \$10

6. (a) $R(x) = -\frac{1}{20}x^2 + 25x$
(b) \$480
(c)

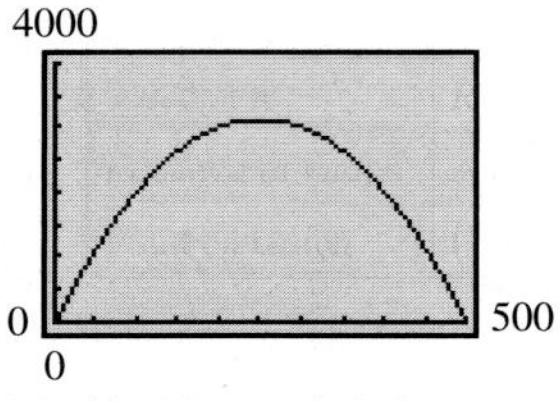

(d) 250; \$3125 **(e)** \$12.50

7. (a) $A(x) = -x^2 + 200x$
(b) $0 < x < 200$
(c) A is largest when $x = 100$ yd.

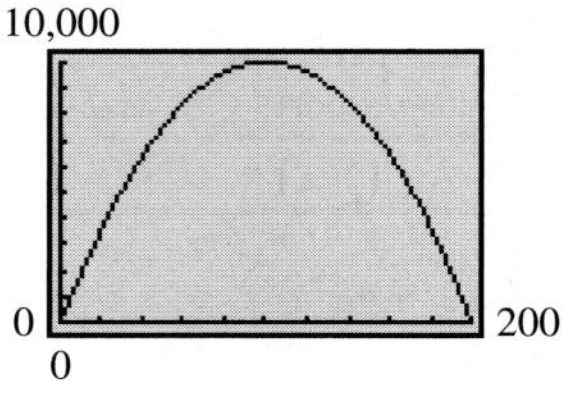

8. (a) $A(x) = -\frac{1}{2}x^2 + 1500x$
(b) A is largest when $x = 1500$.

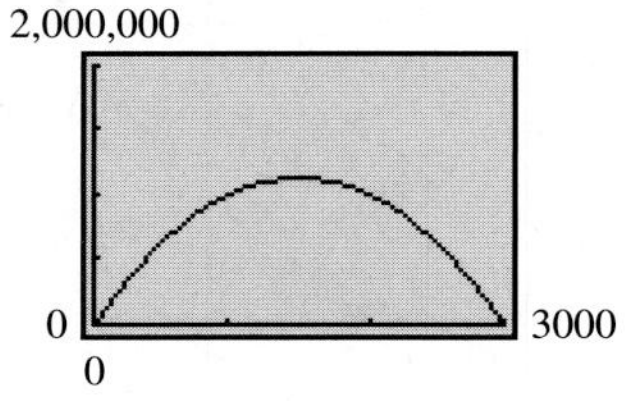

9. **(a)** $d(x) = \sqrt{x^4 - 15x^2 + 64}$
(b) $d(0) = 8$
(c) $d(1) = \sqrt{50} \approx 7.07$
(d)

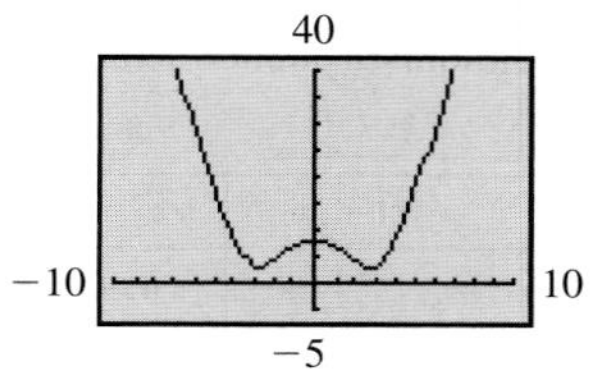

(e) d is smallest when $x \approx -2.74$ or $x \approx 2.74$.

10. **(a)** $d(x) = \sqrt{x^4 - 13x^2 + 49}$
(b) $d(0) = 7$
(c) $d(1) = \sqrt{37} \approx 6.08$
(d)

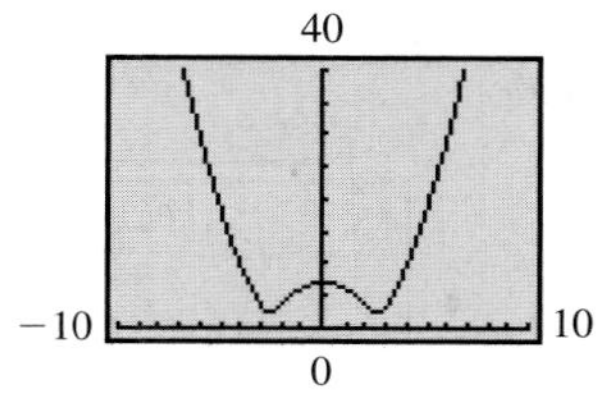

(e) d is smallest when $x \approx -2.55$ or $x \approx 2.55$.

11. **(a)** $d(x) = \sqrt{x^2 - x + 1}$
(b)

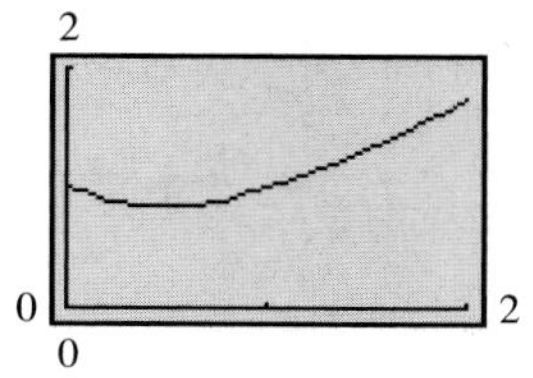

(c) $x = \frac{1}{2}$

12. **(a)** $d(x) = \sqrt{x^2 + \frac{1}{x^2}}$
(b)

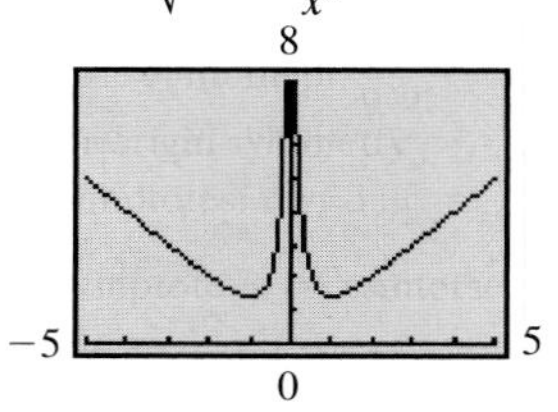

(c) d is smallest when $x = -1$ or $x = 1$.

13. $A(x) = \frac{1}{2}x^4$ **14.** $A(x) = \frac{9}{2}x - \frac{1}{2}x^3$

15. **(a)** $A(x) = x(16 - x^2)$
(b) Domain: $\{x \mid 0 < x < 4\}$
(c) The area is largest when $x \approx 2.31$.

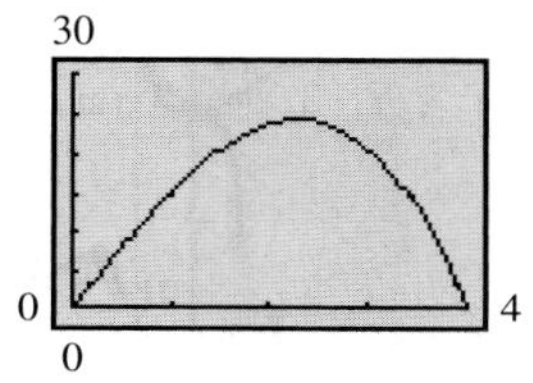

16. **(a)** $A(x) = 2x\sqrt{4 - x^2}$
(b) $p(x) = 4x + 2\sqrt{4 - x^2}$
(c) A is largest when $x \approx 1.41$.

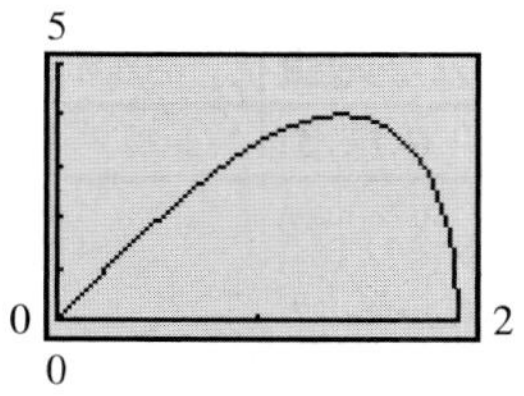

(d) p is largest when $x \approx 1.79$.

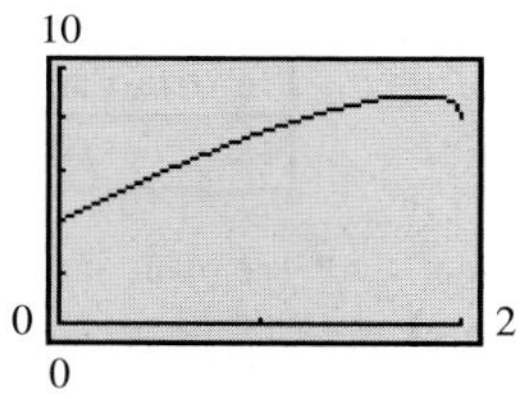

17. **(a)** $A(x) = 4x\sqrt{4 - x^2}$
(b) $p(x) = 4x + 4\sqrt{4 - x^2}$
(c) A is largest when $x \approx 1.41$.

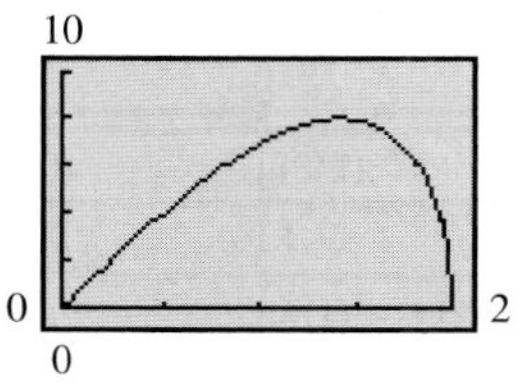

(d) p is largest when $x \approx 1.41$.

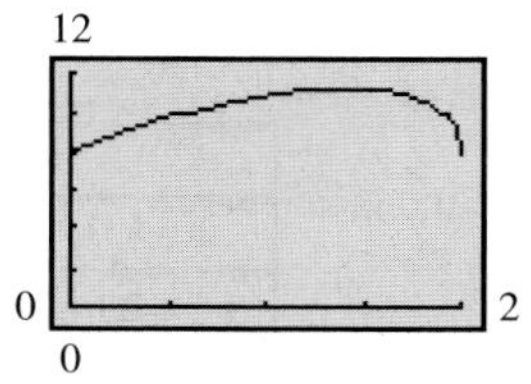

18. **(a)** $A(r) = 4r^2$ **(b)** $p(r) = 8r$

19. **(a)** $A(x) = x^2 + \frac{25 - 20x + 4x^2}{\pi}$
(b) Domain: $\{x \mid 0 < x < 2.5\}$
(c) A is smallest when $x \approx 1.40$ m.

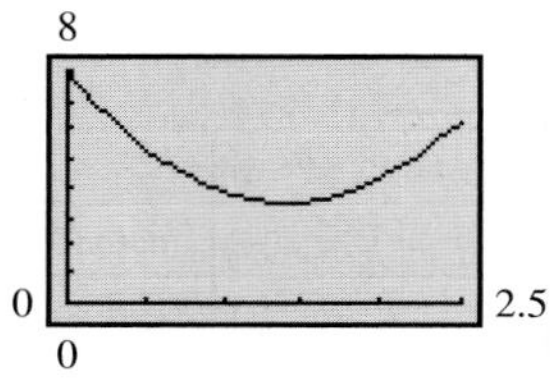

20. **(a)** $A(x) = \frac{\sqrt{3}}{4}x^2 + \frac{100 - 60x + 9x^2}{4\pi}$
(b) $0 < x < \frac{10}{3}$
(c) A is smallest when $x \approx 2.08$.

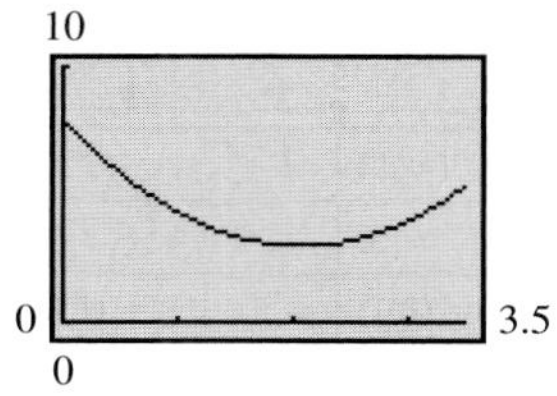

21. **(a)** $C(x) = x$ **(b)** $A(x) = \frac{x^2}{4\pi}$

22. **(a)** $p(x) = x$ **(b)** $A(x) = \frac{1}{16}x^2$

23. **(a)** $A(r) = 2r^2$ **(b)** $p(r) = 6r$

24. $C(x) = \frac{2\sqrt{3}}{3}\pi x$

25. $A(x) = \left(\frac{\pi}{3} - \frac{\sqrt{3}}{4}\right)x^2$

26. $d(t) = 50t$

27. (a) $d(t) = \sqrt{2500t^2 - 360t + 13}$
(b) d is smallest when $t \approx 0.07$.

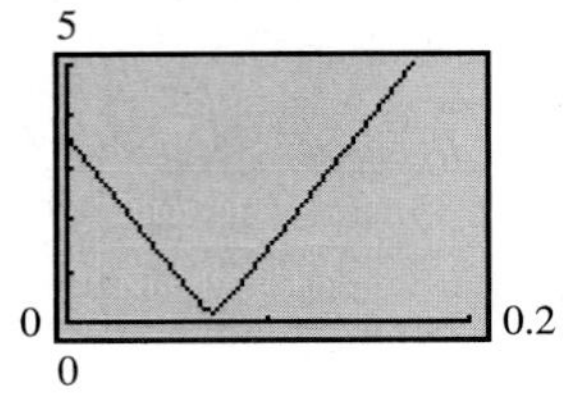

28. $V(h) = \pi h\left(R^2 - \frac{h^2}{4}\right)$ **29.** $V(r) = \frac{\pi H(R - r)r^2}{R}$

30. (a) $C(x) = 10x + 14\sqrt{x^2 - 10x + 29}, 0 \le x \le 5$
(b) $C(1) = \$72.61$ **(c)** $C(3) = \$69.60$
(d)

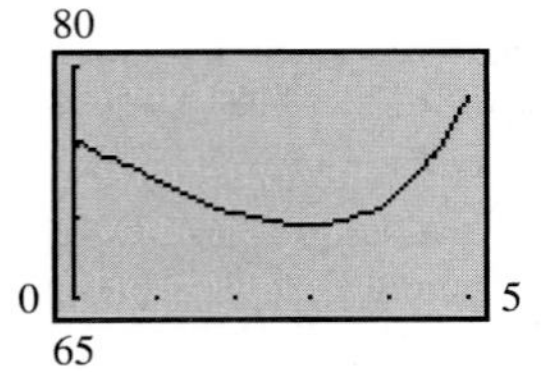

(e) The least cost occurs when $x \approx 2.96$ mi.

31. (a) $T(x) = \frac{12 - x}{5} + \frac{\sqrt{x^2 + 4}}{3}$
(b) $\{x|0 \le x \le 12\}$
(c) 3.09 hr
(d) 3.55 hr

32. $V(h) = \frac{\pi}{48}h^3$

Review Exercises *(page 118)*

1. Function; domain $\{-1, 2, 4\}$, range $\{0, 3\}$

2. Not a function **3. (a)** 2 **(b)** -2 **(c)** $-\frac{3x}{x^2 - 1}$ **(d)** $-\frac{3x}{x^2 - 1}$ **(e)** $\frac{3(x - 2)}{x^2 - 4x + 3}$ **(f)** $\frac{6x}{4x^2 - 1}$

4. (a) $\frac{4}{3}$ **(b)** -4 **(c)** $\frac{x^2}{-x + 1}$ **(d)** $-\frac{x^2}{x + 1}$ **(e)** $\frac{(x - 2)^2}{x - 1}$ **(f)** $\frac{4x^2}{2x + 1}$

5. (a) 0 **(b)** 0 **(c)** $\sqrt{x^2 - 4}$ **(d)** $-\sqrt{x^2 - 4}$ **(e)** $\sqrt{x^2 - 4x}$ **(f)** $2\sqrt{x^2 - 1}$

6. (a) 0 **(b)** 0 **(c)** $|x^2 - 4|$ **(d)** $-|x^2 - 4|$ **(e)** $|x^2 - 4x|$ **(f)** $2|x^2 - 1|$ **7. (a)** 0 **(b)** 0 **(c)** $\frac{x^2 - 4}{x^2}$ **(d)** $-\frac{x^2 - 4}{x^2}$
(e) $\frac{x(x - 4)}{(x - 2)^2}$ **(f)** $\frac{x^2 - 1}{x^2}$ **8. (a)** $-\frac{8}{5}$ **(b)** $\frac{8}{5}$ **(c)** $-\frac{x^3}{x^2 - 9}$ **(d)** $-\frac{x^3}{x^2 - 9}$ **(e)** $\frac{(x - 2)^3}{x^2 - 4x - 5}$ **(f)** $\frac{8x^3}{4x^2 - 9}$ **9.** $\{x|x \ne -3, x \ne 3\}$

10. $\{x|x \ne 2\}$ **11.** $\{x|x \le 2\}$ **12.** $\{x|x \ge -2\}$ **13.** $\{x|x > 0\}$ **14.** $\{x|x \ne 0\}$ **15.** $\{x|x \ne -3, x \ne 1\}$ **16.** $\{x|x \ne -1, x \ne 4\}$

17. $(f + g)(x) = 2x + 3$; Domain: all real numbers
$(f - g)(x) = -4x + 1$; Domain: all real numbers
$(f \cdot g)(x) = -3x^2 + 5x + 2$; Domain: all real numbers
$\left(\frac{f}{g}\right)(x) = \frac{2 - x}{3x + 1}$; Domain: $\left\{x \middle| x \ne -\frac{1}{3}\right\}$

18. $(f + g)(x) = 4x$; Domain: all real numbers
$(f - g)(x) = -2$; Domain: all real numbers
$(f \cdot g)(x) = 4x^2 - 1$; Domain: all real numbers
$\left(\frac{f}{g}\right)(x) = \frac{2x - 1}{2x + 1}$; Domain: $\left\{x \middle| x \ne -\frac{1}{2}\right\}$

19. $(f + g)(x) = 3x^2 + 4x + 1$; Domain: all real numbers
$(f - g)(x) = 3x^2 - 2x + 1$; Domain: all real numbers
$(f \cdot g)(x) = 9x^3 + 3x^2 + 3x$; Domain: all real numbers
$\left(\frac{f}{g}\right)(x) = \frac{3x^2 + x + 1}{3x}$; Domain: $\{x|x \ne 0\}$

20. $(f + g)(x) = 1 + 4x + x^2$; Domain: all real numbers
$(f - g)(x) = -1 + 2x - x^2$; Domain: all real numbers
$(f \cdot g)(x) = 3x + 3x^2 + 3x^3$; Domain: all real numbers
$\left(\frac{f}{g}\right)(x) = \frac{3x}{1 + x + x^2}$; Domain: all real numbers

21. $(f + g)(x) = \frac{x^2 + 2x - 1}{x(x - 1)}$; Domain: $\{x|x \ne 0, 1\}$
$(f - g)(x) = \frac{x^2 + 1}{x(x - 1)}$; Domain: $\{x|x \ne 0, 1\}$
$(f \cdot g)(x) = \frac{x + 1}{x(x - 1)}$; Domain: $\{x|x \ne 0, 1\}$
$\left(\frac{f}{g}\right)(x) = \frac{x(x + 1)}{x - 1}$; Domain: $\{x|x \ne 0, 1\}$

22. $(f + g)(x) = \frac{4x - 9}{x(x - 3)}$; Domain: $\{x|x \ne 0, 3\}$
$(f - g)(x) = \frac{-2x + 9}{x(x - 3)}$; Domain: $\{x|x \ne 0, 3\}$
$(f \cdot g)(x) = \frac{3}{x(x - 3)}$; Domain: $\{x|x \ne 0, 3\}$
$\left(\frac{f}{g}\right)(x) = \frac{x}{3(x - 3)}$; Domain: $\{x|x \ne 0, 3\}$

23. $-4x + 1 - 2h$ **24.** $6x - 2 + 3h$ **25.** **(a)** Domain: $\{x|-4 \le x \le 3\}$; Range: $\{y|-3 \le y \le 3\}$ **(b)** $(0, 0)$ **(c)** -1 **(d)** -4

(e) $\{x|0 < x \le 3\}$ **(f)**

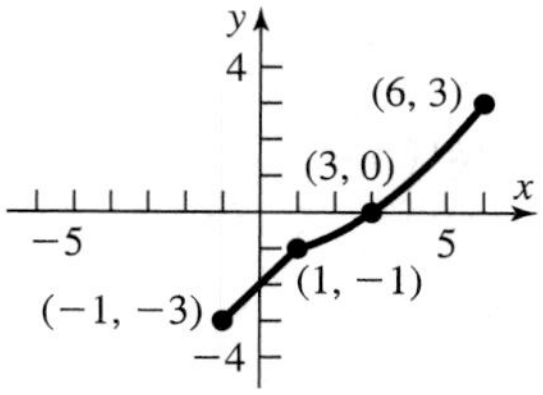

(g)

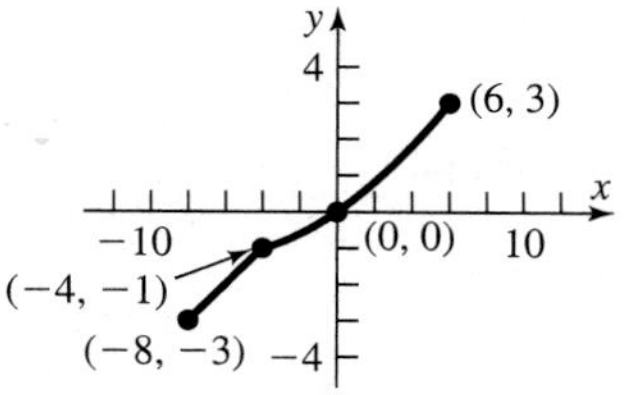

(h)

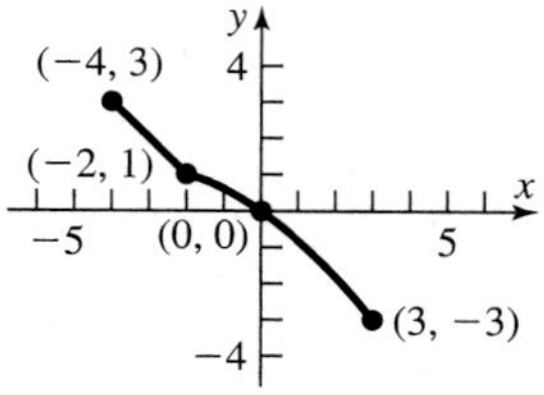

26. **(a)** Domain: $\{x|-5 \le x \le 4\}$; Range: $\{y|-3 \le y \le 1\}$ **(b)** 1 **(c)** $(0, 0), (4, 0)$ **(d)** 3

(e) $\{x|-5 \le x < 0\}$ **(f)**

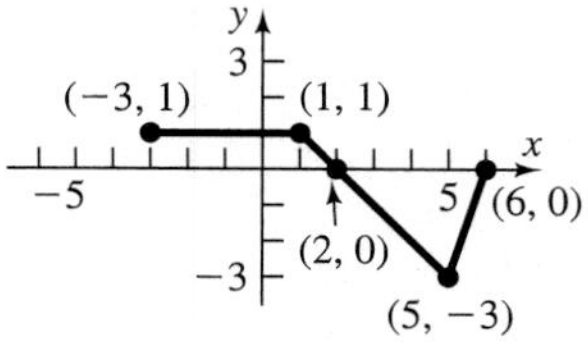

(g)

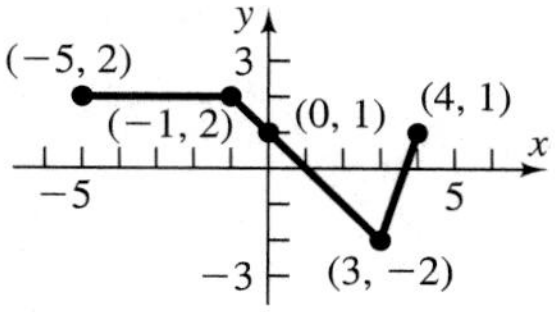

(h)

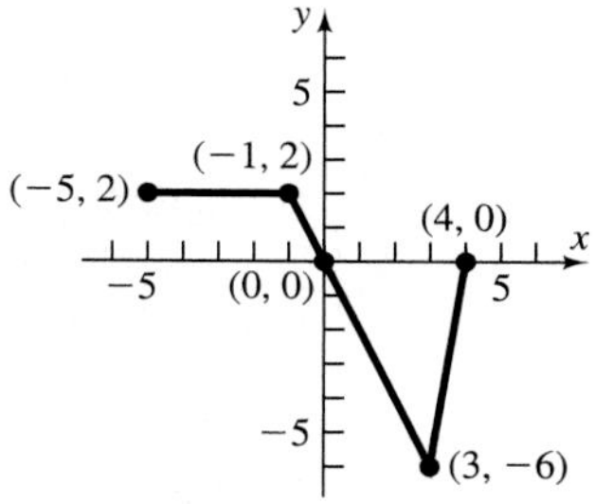

27. **(a)** Domain: $\{x|-4 \le x \le 4\}$ or $[-4, 4]$
Range: $\{y|-3 \le y \le 1\}$ or $[-3, 1]$
(b) Increasing on $(-4, -1)$ and $(3, 4)$;
Decreasing on $(-1, 3)$
(c) Local maximum is 1 and occurs at $x = -1$;
Local minimum is -3 and occurs at $x = 3$
(d) No symmetry
(e) Neither
(f) x-intercepts: $-2, 0, 4$; y-intercept: 0

28. **(a)** Domain: all real numbers or $(-\infty, \infty)$
Range: all real numbers or $(-\infty, \infty)$
(b) Increasing on $(-\infty, -2)$ and $(2, \infty)$;
Decreasing on $(-2, 2)$
(c) Local maximum is 1 and occurs at $x = -2$;
Local minimum is -1 and occurs at $x = 2$
(d) Origin
(e) Odd
(f) x-intercepts: $-3, 0, 3$; y-intercept: 0

29. Odd **30.** Even **31.** Even **32.** Neither **33.** Neither **34.** Neither **35.** Odd **36.** Odd

37.

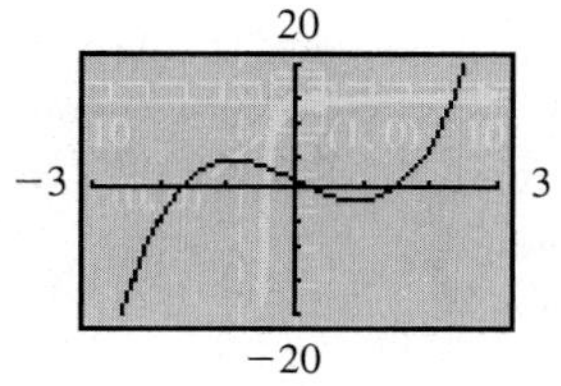

Local maximum: $(-0.91, 4.04)$
Local minimum: $(0.91, -2.04)$
Increasing: $(-3, -0.91)$; $(0.91, 3)$
Decreasing: $(-0.91, 0.91)$

38.

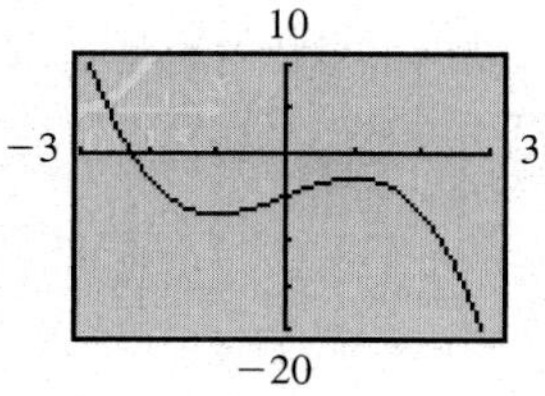

Local maximum: $(1, -3)$
Local minimum: $(-1, -7)$
Increasing: $(-1, 1)$
Decreasing: $(-3, -1)$; $(1, 3)$

39.

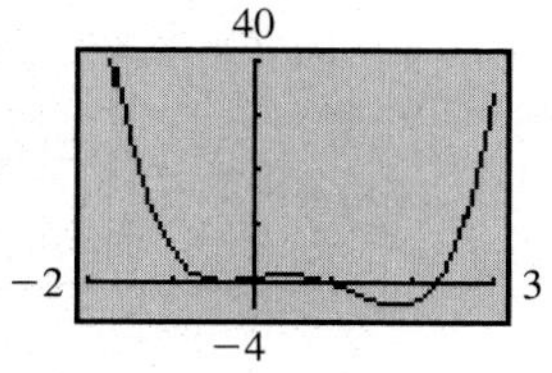

Local maximum: $(0.41, 1.53)$
Local minimum: $(-0.34, 0.54)$; $(1.80, -3.56)$
Increasing: $(-0.34, 0.41)$; $(1.80, 3)$
Decreasing: $(-2, -0.34)$; $(0.41, 1.80)$

40.

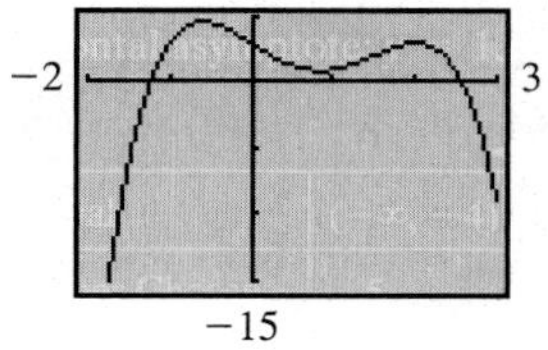

Local maxima: $(-0.59, 4.62)$; $(2, 3)$
Local minimum: $(0.84, 0.92)$
Increasing: $(-2, -0.59)$; $(0.84, 2)$
Decreasing: $(-0.59, 0.84)$; $(2, 3)$

41. **(a)** 23 **(b)** 7 **(c)** 47 **42.** **(a)** 15 **(b)** 3 **(c)** 57 **43.** -5 **44.** $2(x + 2)$ **45.** $-4x - 5$
46. $x - 1$ **47.** (b), (c), (d)

48.

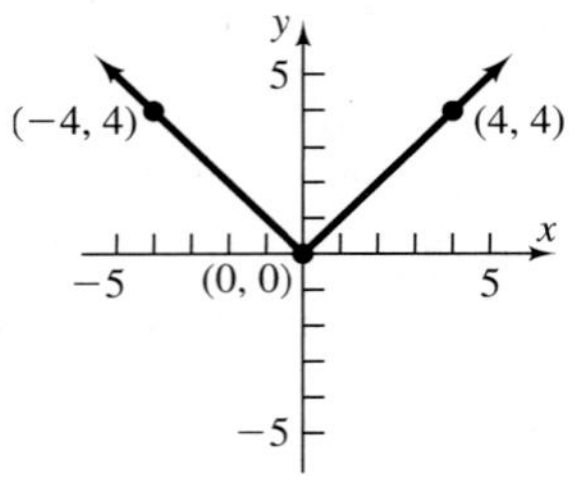

49.

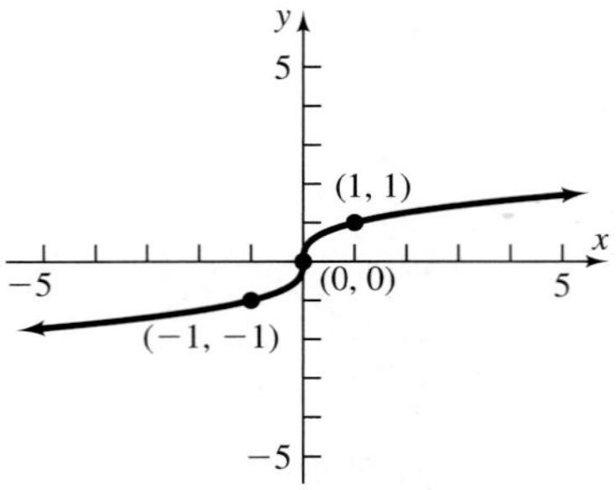

50.

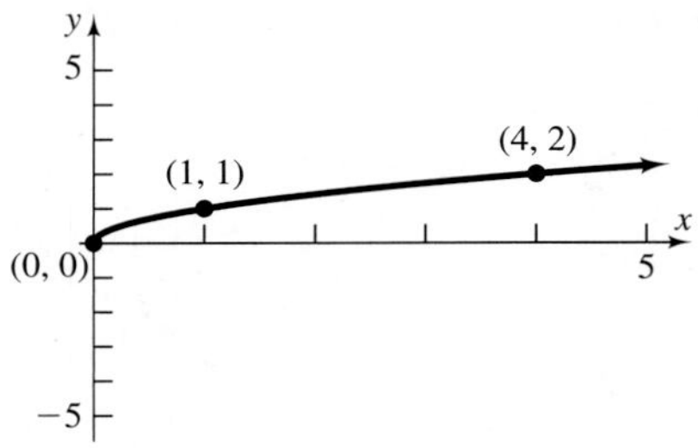

51.

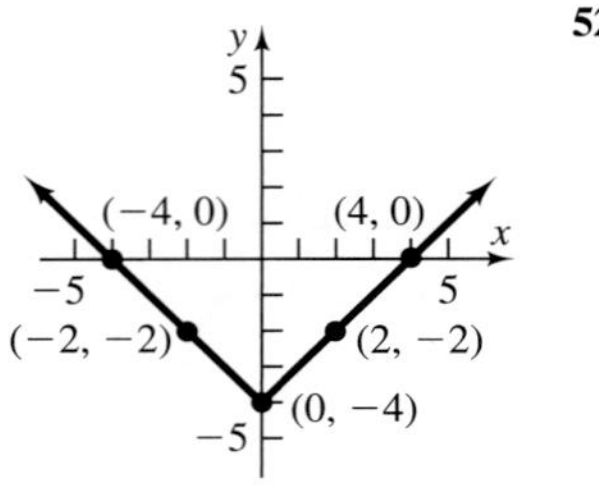

Intercepts: $(-4, 0), (4, 0), (0, -4)$
Domain: all real numbers
Range: $\{y|y \geq -4\}$ or $[-4, \infty)$

52.

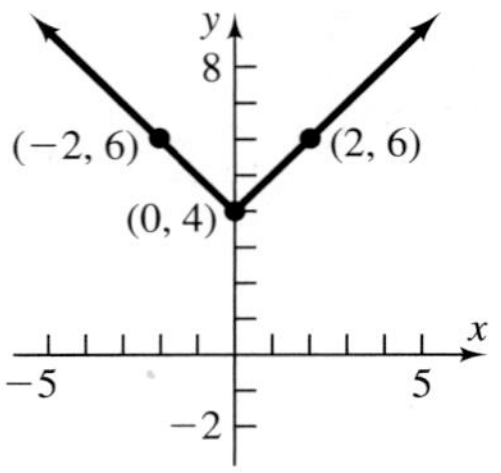

Intercept: $(0, 4)$
Domain: all real numbers
Range: $\{y|y \geq 4\}$ or $[4, \infty)$

53.

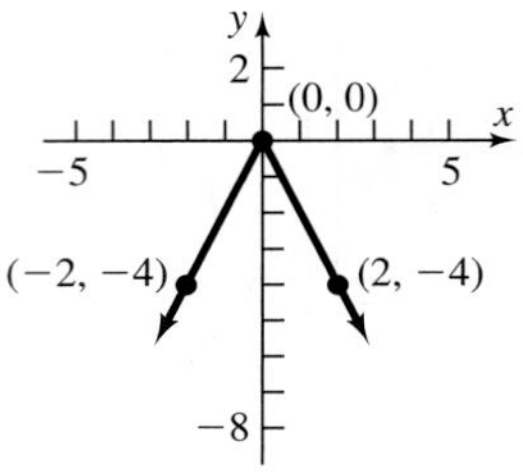

Intercept: $(0, 0)$
Domain: all real numbers
Range: $\{y|y \leq 0\}$ or $(-\infty, 0]$

54.

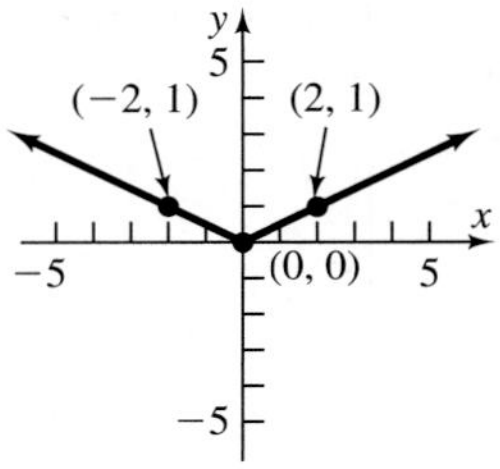

Intercept: $(0, 0)$
Domain: all real numbers
Range: $\{y|y \geq 0\}$ or $[0, \infty)$

55.

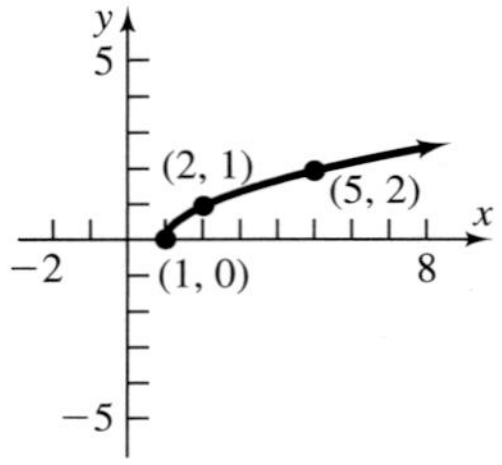

Intercept: $(1, 0)$
Domain: $\{x|x \geq 1\}$ or $[1, \infty)$
Range: $\{y|y \geq 0\}$ or $[0, \infty)$

56.

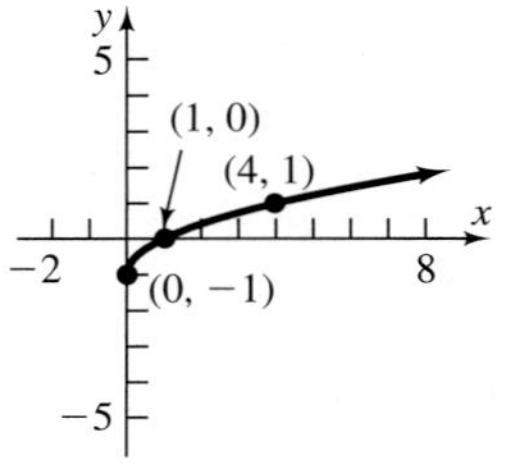

Intercepts: $(0, -1), (1, 0)$
Domain: $\{x|x \geq 0\}$ or $[0, \infty)$
Range: $\{y|y \geq -1\}$ or $[-1, \infty)$

57.

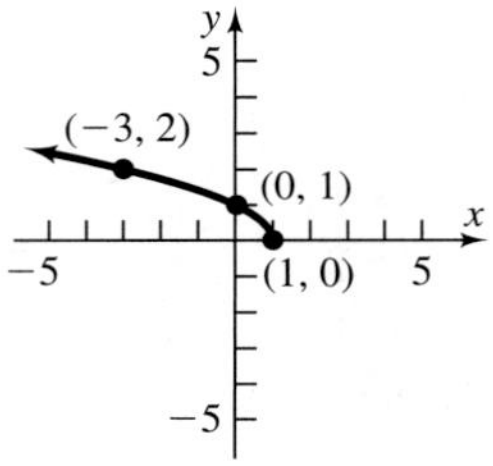

Intercepts: $(0, 1), (1, 0)$
Domain: $\{x|\ x \leq 1\}$ or $(-\infty, 1]$
Range: $\{y|y \geq 0\}$ or $[0, \infty)$

58.

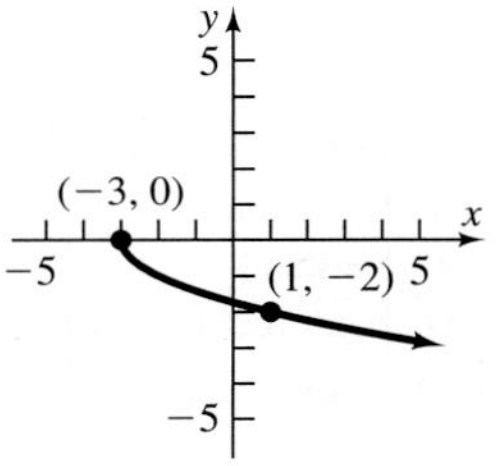

Intercepts: $(-3, 0), (0, -\sqrt{3})$
Domain: $\{x|x \geq -3\}$ or $[-3, \infty)$
Range: $\{y|y \leq 0\}$ or $(-\infty, 0]$

59.

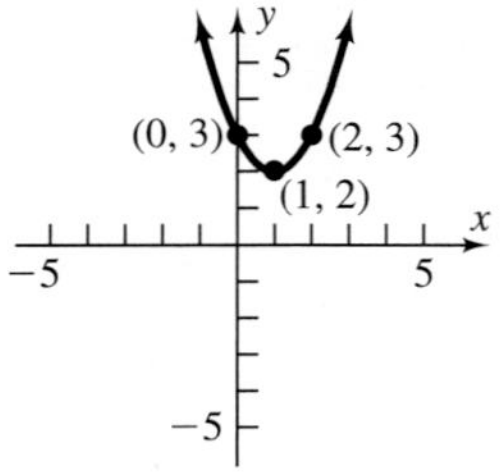

Intercept: $(0, 3)$

Domain: all real numbers

Range: $\{y|y \geq 2\}$ or $[2, \infty)$

60.

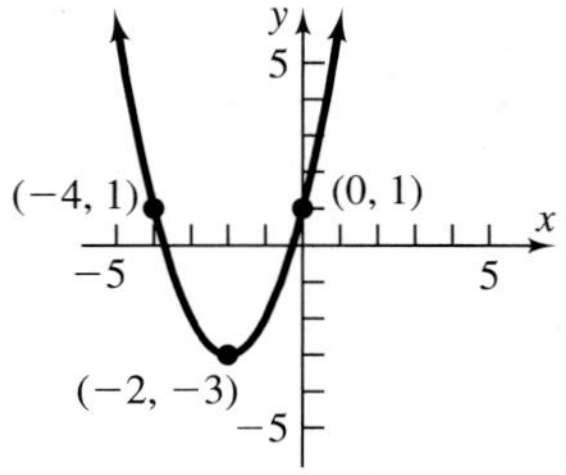

Intercepts: $(-2 - \sqrt{3}, 0)$,

$(-2 + \sqrt{3}, 0), (0, 1)$

Domain: all real numbers
Range: $\{y|y \geq -3\}$ or $[-3, \infty)$

61.

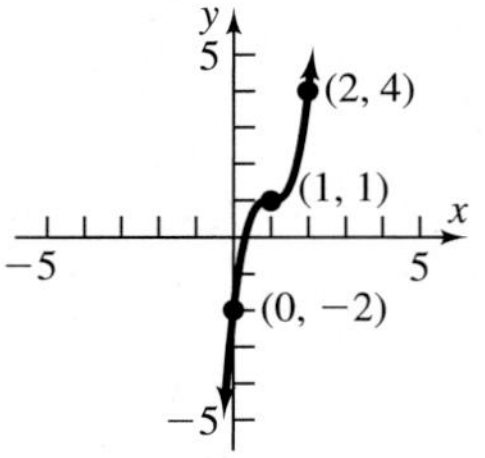

Intercepts: $(0, -2)$,

$\left(1 + \dfrac{\sqrt[3]{-9}}{3}, 0\right)$ or about $(0.3, 0)$

Domain: all real numbers
Range: all real numbers

62.

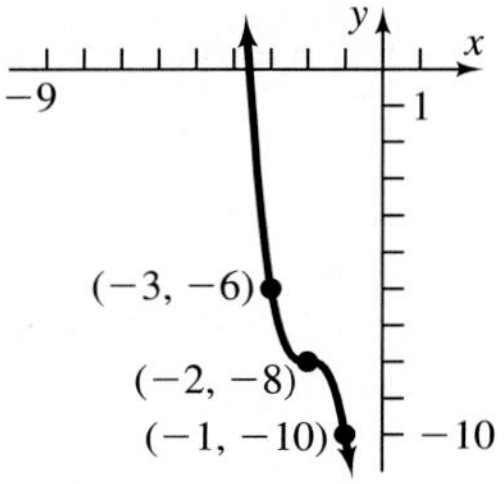

Intercepts: $(0, -24)$,

$(-2 + \sqrt[3]{-4}, 0)$ or about

$(-3.6, 0)$
Domain: all real numbers
Range: all real numbers

63. **(a)** $\{x|x > -2\}; (-2, \infty)$
(b) $(0, 0)$
(c)

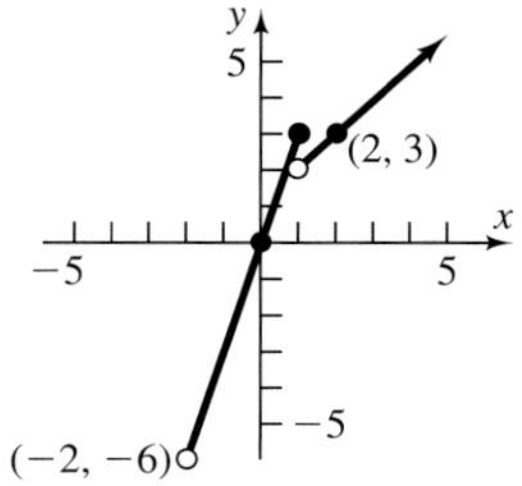

(d) $\{y|y > -6\}; (-6, \infty)$

64. **(a)** $\{x|x > -3\}; (-3, \infty)$
(b) $(0, -1), \left(\frac{1}{3}, 0\right)$
(c)

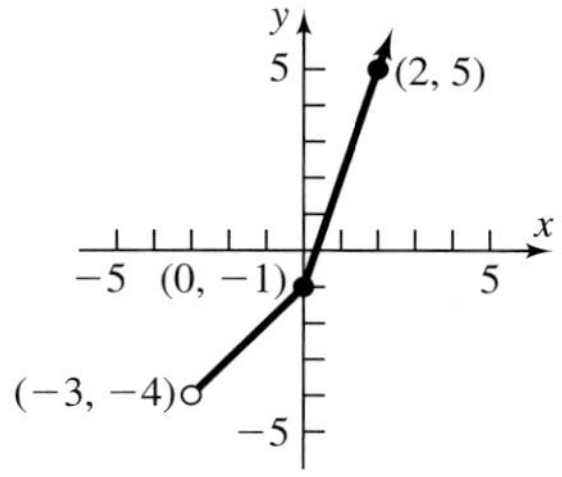

(d) $\{y|y > -4\}; (-4, \infty)$

65. **(a)** $\{x|x \geq -4\}; [-4, \infty)$
(b) $(0, 1)$
(c)

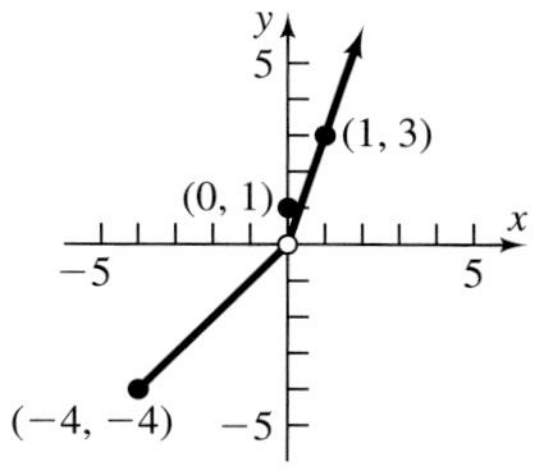

(d) $\{y|y \geq -4\}; [-4, \infty)$

66. **(a)** $\{x|x \geq -2\}; [-2, \infty)$ **(b)** $(0, 0)$
(c)

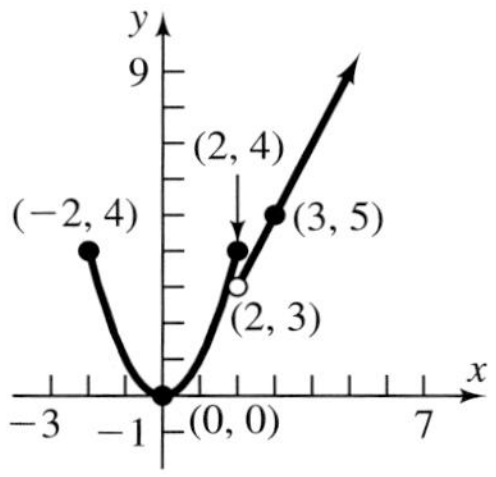

(d) $\{y|y \geq 0\}; [0, \infty)$

67. $f(x) = -2x + 3$
68. $g(x) = -4x - 6$
69. $A = 11$
70. $A = 8$

71. $T(h) = -0.0025h + 30, 0 \leq x \leq 10{,}000$ **72.** 130 ft/sec; $v(t) = 5t - 20$

73. $V(S) = \frac{S}{6}\sqrt{\frac{S}{\pi}}$; If the surface area doubles, the volume increases by a factor of $2\sqrt{2}$.

74. **(a)** $A(x) = (8.5 - 2x)(11 - 2x)$ **(b)** $0 \leq x \leq 4.25, 0 \leq A \leq 93.5$ **(c)** $A(1) = 58.5$ in^2, $A(1.2) = 52.46$ in^2, $A(1.5) = 44$ in^2
(d)

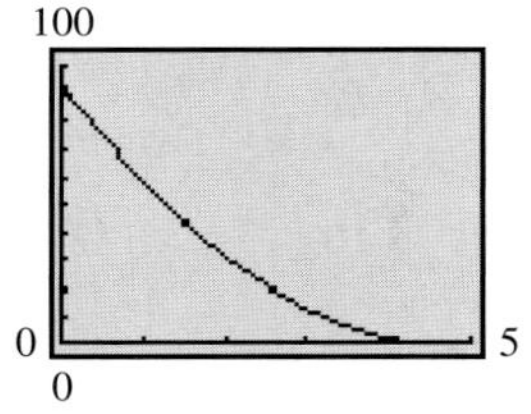

(e) $A(x) \approx 70$ when $x \approx 0.65$ in.;
$A(x) \approx 50$ when $x \approx 1.28$ in.

75. $S(x) = kx(36 - x^2)^{3/2}$; Domain: $\{x|\ 0 < x < 6\}$

76. **(a)** $A = 2\pi r^2 + \frac{200}{r}$ **(b)** 123.22 sq ft
(c) 150.53 sq ft **(d)** 197.08 sq ft
(e) A is smallest when $r \approx 2.52$ ft.

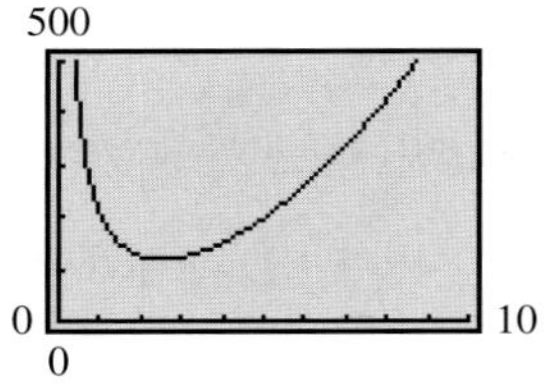

77. **(a)** $C(r) = 0.12\pi r^2 + \frac{40}{r}$ **(b)** \$16.03
(c) \$29.13
(d) The cost is least for $r \approx 3.76$ cm.

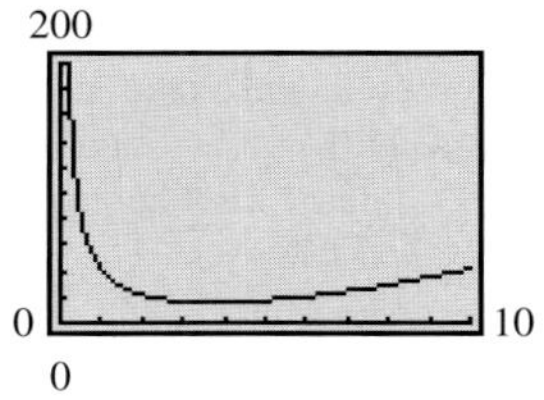

78. **(a)** $A(x) = 2x^2 + \frac{40}{x}$
(b) 42 ft^2 **(c)** 28 ft^2
(d) A is smallest when $x \approx 2.15$ ft.

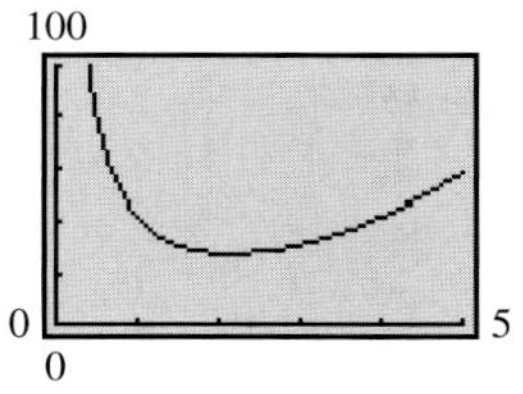

Cumulative Review *(page 122)*

1. $\left\{\frac{4}{5}\right\}$ **2.** $\{3, 4\}$ **3.** $\left\{-\frac{1}{3}, 2\right\}$ **4.** $\left\{-\frac{1}{2}\right\}$ **5.** no real solution **6.** $\{-7\}$ **7.** $\{-31\}$ **8.** $\left\{\frac{1}{3}, 1\right\}$ **9.** $\left\{\frac{1 - \sqrt{15}i}{4}, \frac{1 + \sqrt{15}i}{4}\right\}$

10. $\{x|1 < x < 4\}$

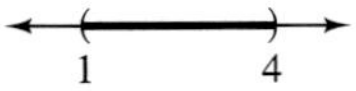

11.

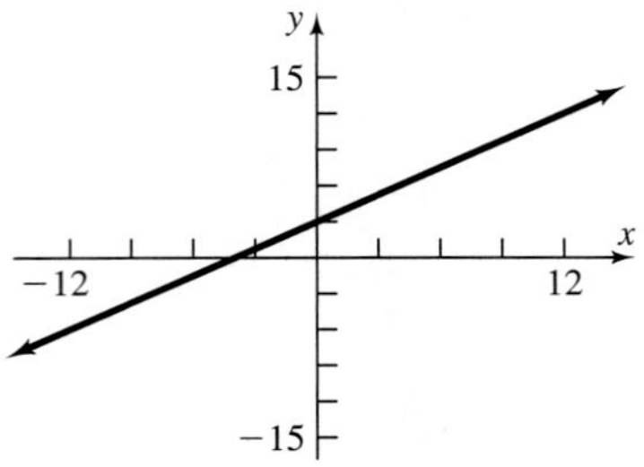

12.

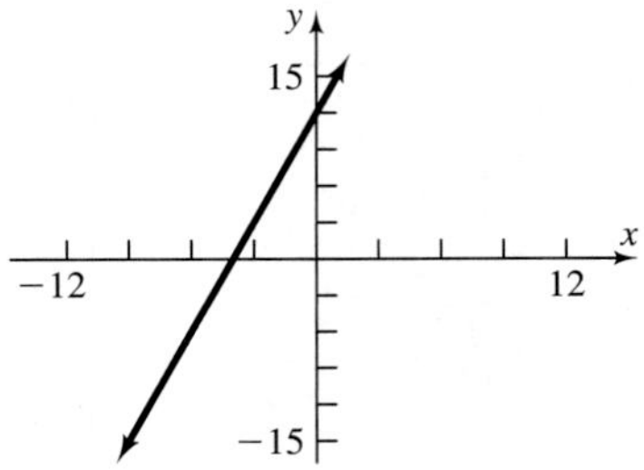

13.

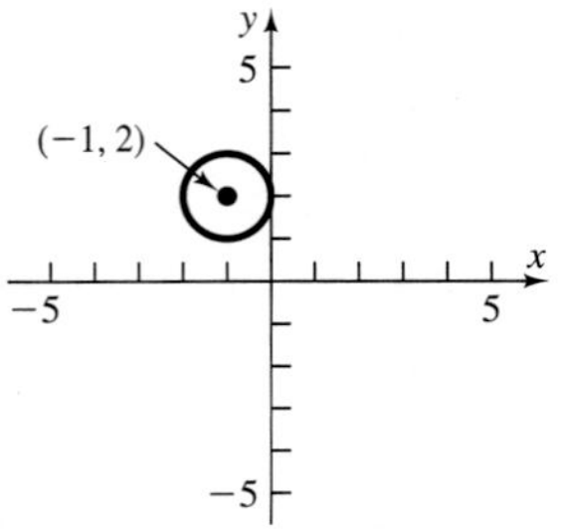

14.

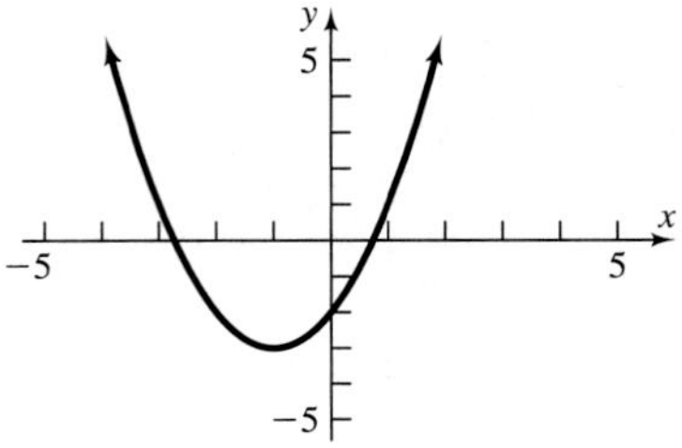

15. **(a)** Domain: $\{x|-4 \le x \le 4\}$; Range: $\{y|0 \le y \le 3\}$ **(b)** $(-1, 0), (0, -1), (1, 0)$ **(c)** y-axis **(d)** 1 **(e)** -4 and 4

(f) $\{x|-1 < x < 1\}$ **(g)**

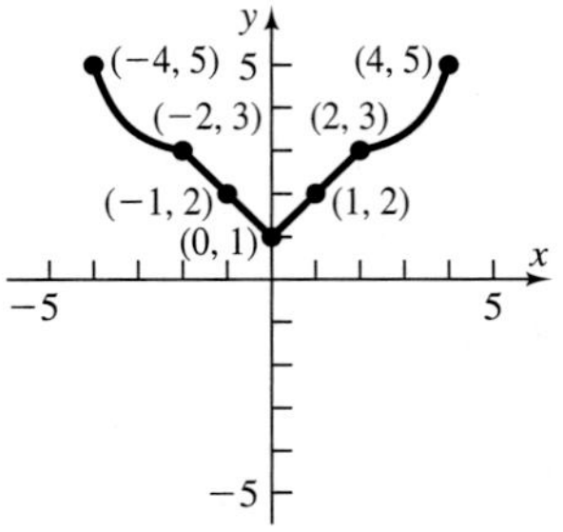

(h)

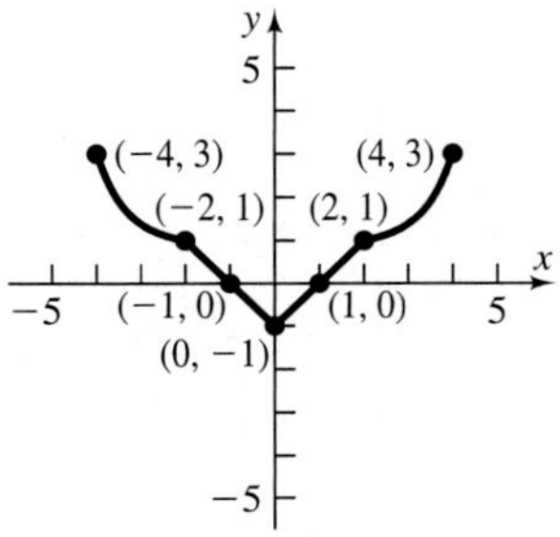

(i)

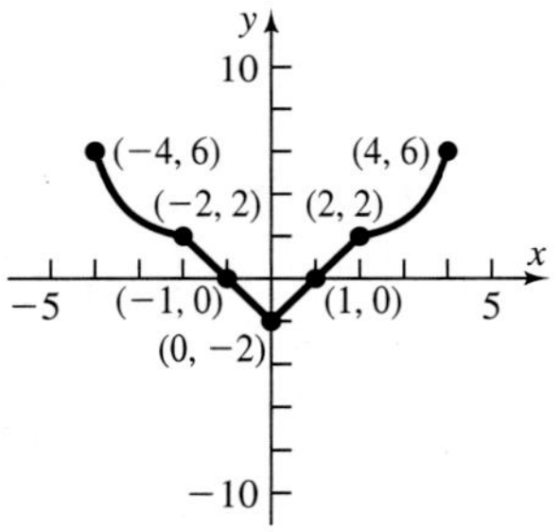

(j) even **(k)** $(0, 4)$ **(l)** $(-4, 0)$ **(m)** f has a local minimum of -1 at $x = 0$ **(n)** 1

16. **(a), (b), (e)**

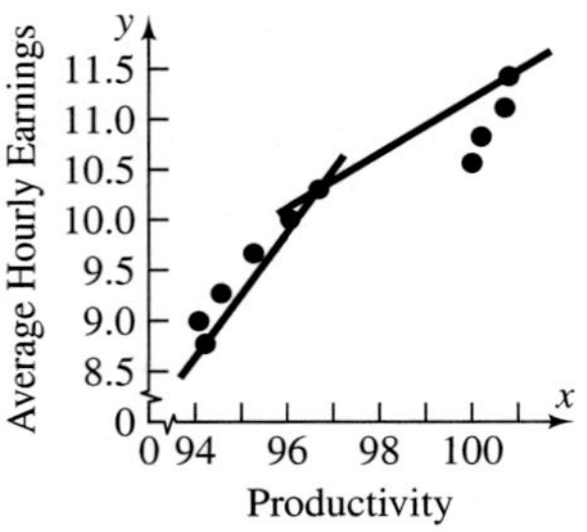

(c) 0.624 dollars/productivity unit
(d) For each 1-unit increase in productivity, earnings increased by an average of \$0.62.
(f) 0.273 dollars/productivity unit
(g) For each 1-unit increase in productivity, earnings increased by an average of \$0.27.
(h) It is decreasing.

CHAPTER 3 Polynomial and Rational Functions

3.1 Concepts and Vocabulary *(page 138)*

5. parabola **6.** axis or axis of symmetry **7.** $-\dfrac{b}{2a}$ **8.** True **9.** True **10.** True

3.1 Exercises *(page 138)*

11. *C* **12.** *E* **13.** *F* **14.** *A* **15.** *G* **16.** *B* **17.** *H* **18.** *D*

19.

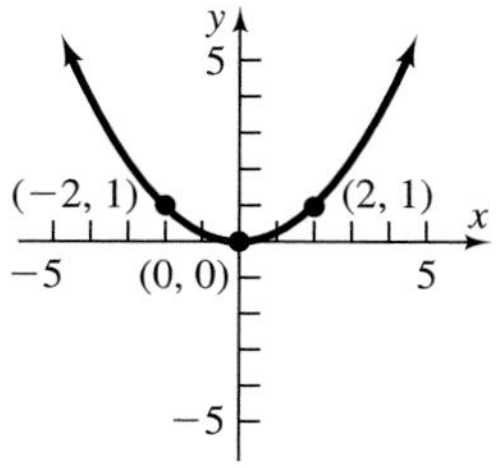

20.

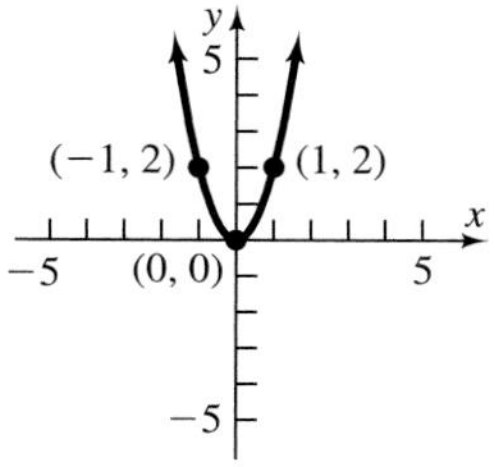

21.

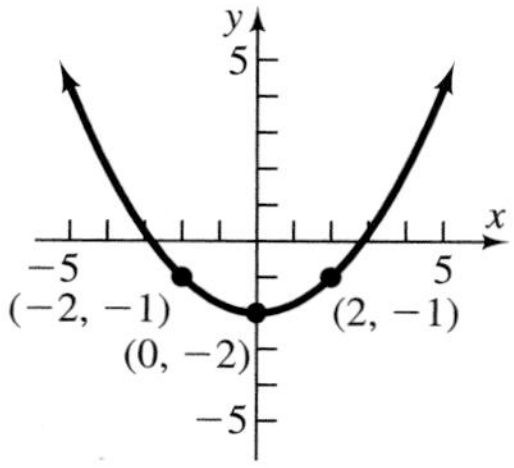

22.

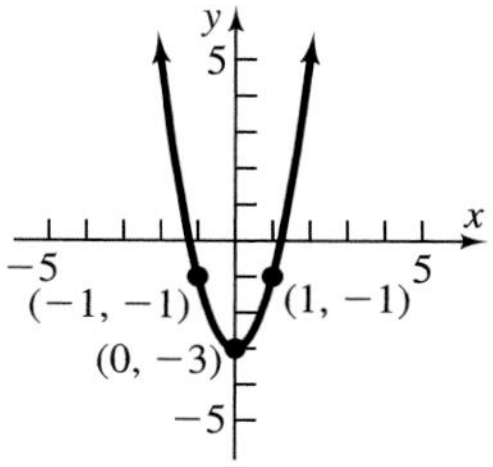

23.

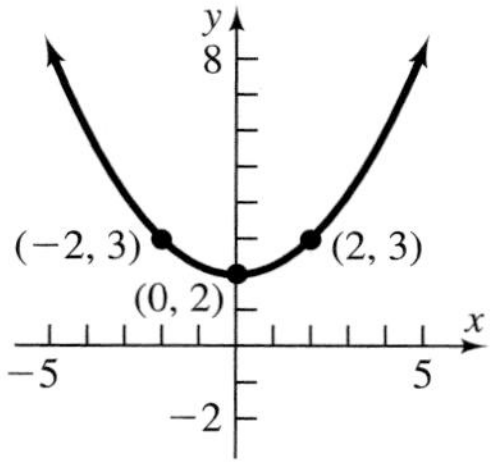

24.

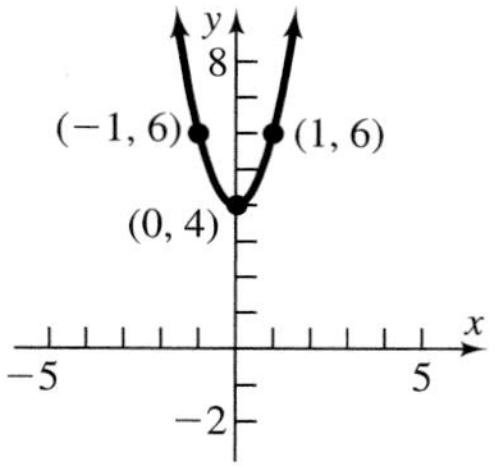

25.

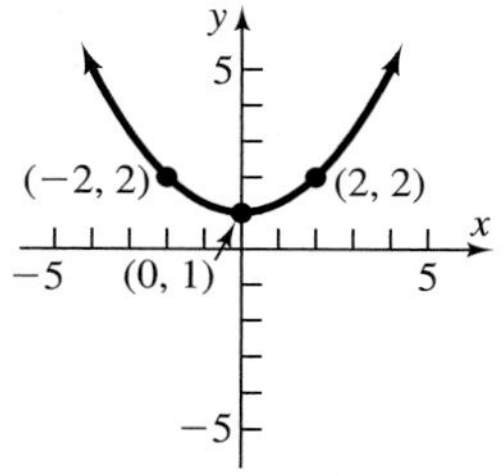

26.

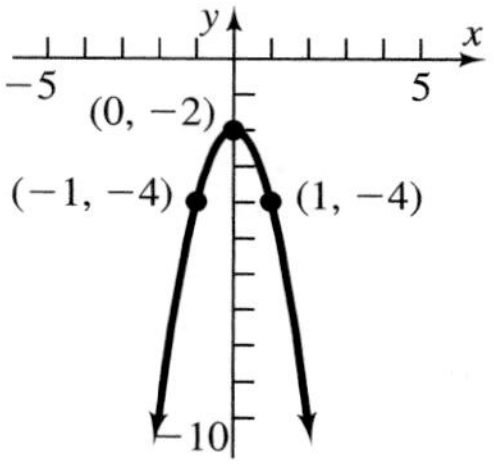

27. $f(x) = (x + 2)^2 - 2$

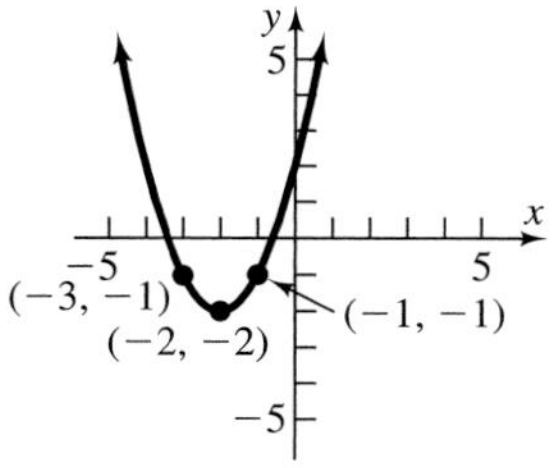

28. $f(x) = (x - 3)^2 - 10$

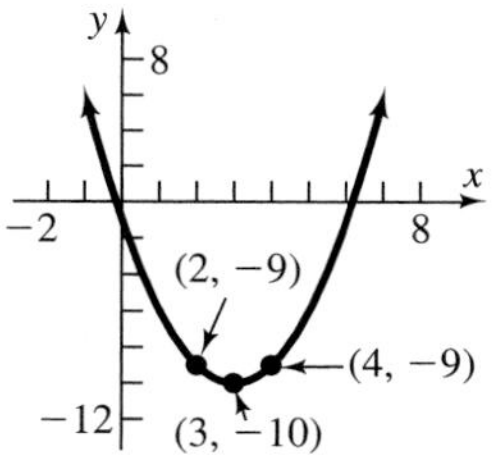

29. $f(x) = 2(x - 1)^2 - 1$

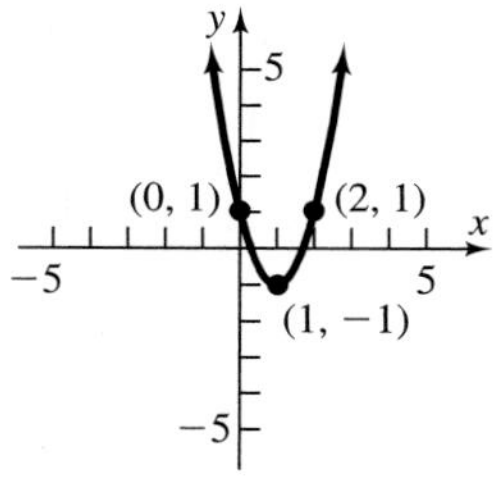

30. $f(x) = 3(x + 1)^2 - 3$

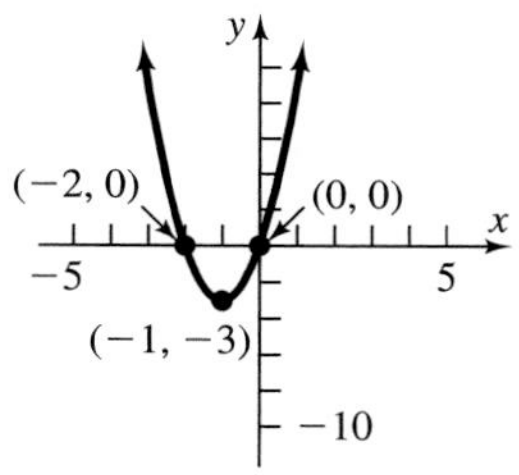

31. $f(x) = -(x + 1)^2 + 1$

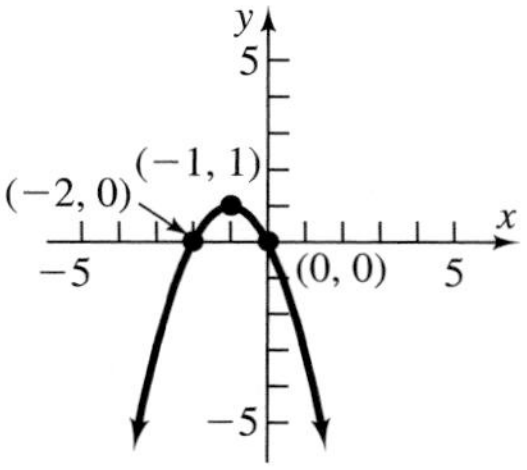

32. $f(x) = -2\left(x - \dfrac{3}{2}\right)^2 + 6\dfrac{1}{2}$

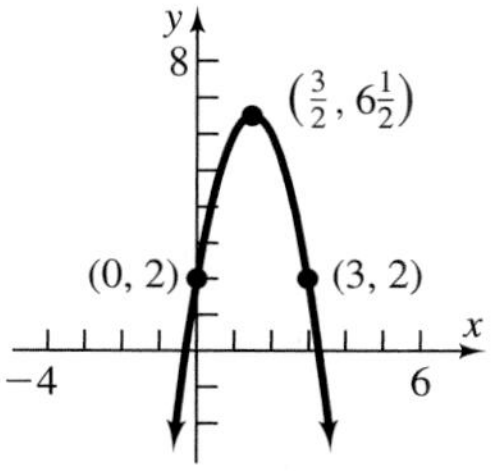

33. $f(x) = \dfrac{1}{2}(x + 1)^2 - \dfrac{3}{2}$

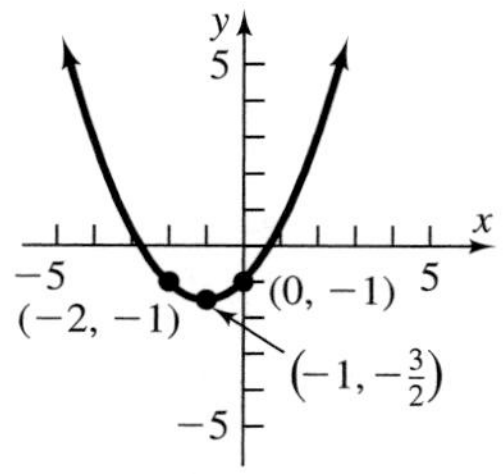

34. $f(x) = \dfrac{2}{3}(x + 1)^2 - \dfrac{5}{3}$

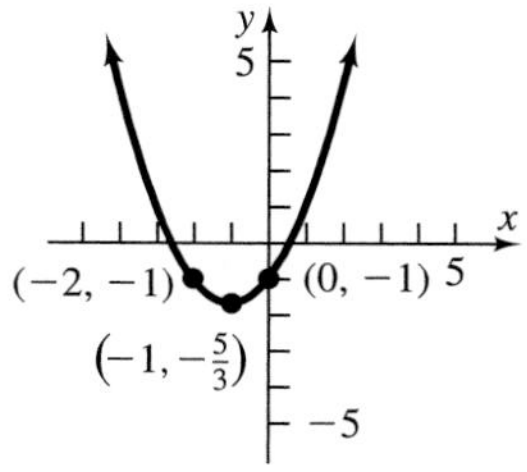

35.

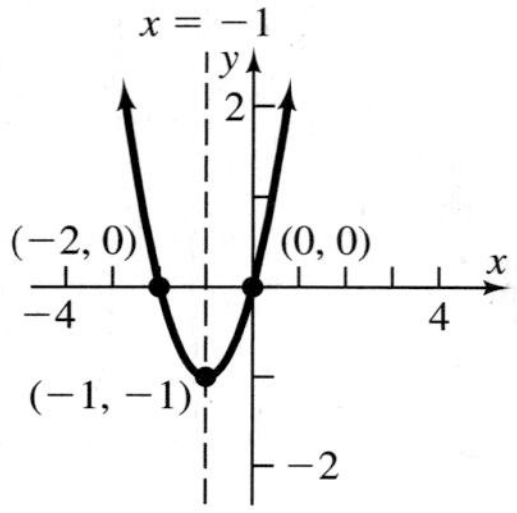

Domain: $(-\infty, \infty)$
Range: $[-1, \infty)$
Dec: $(-\infty, -1)$; Inc: $(-1, \infty)$

36.

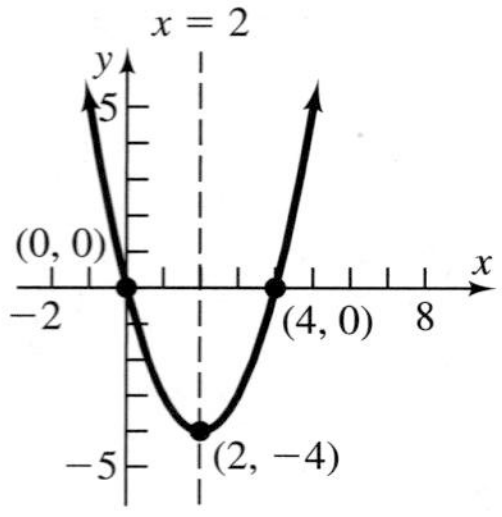

Domain: $(-\infty, \infty)$
Range: $[-4, \infty)$
Dec: $(-\infty, 2)$; Inc: $(2, \infty)$

37.

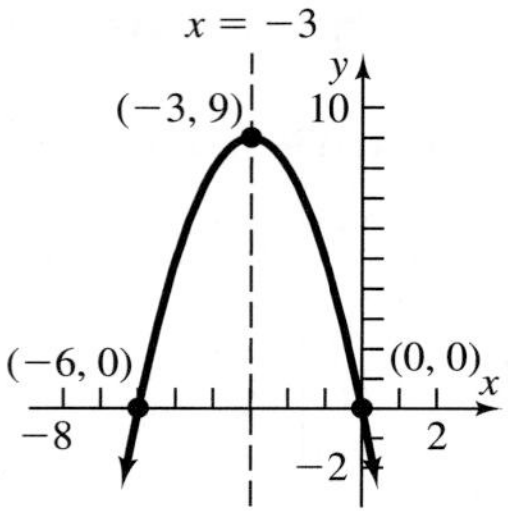

Domain: $(-\infty, \infty)$
Range: $(-\infty, 9]$
Inc: $(-\infty, -3)$; Dec: $(-3, \infty)$

38.

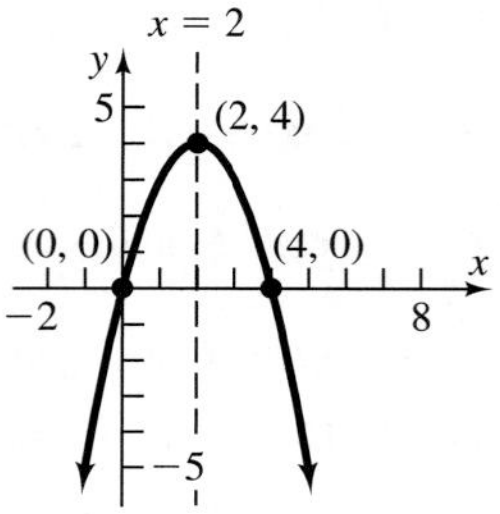

Domain: $(-\infty, \infty)$
Range: $(-\infty, 4]$
Inc: $(-\infty, 2)$; Dec: $(2, \infty)$

39.

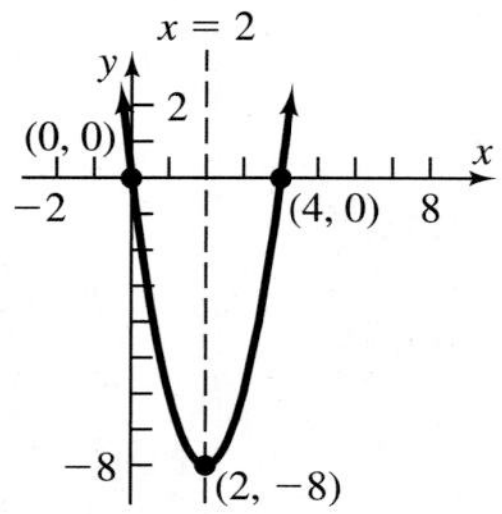

Domain: $(-\infty, \infty)$
Range: $[-8, \infty)$
Dec: $(-\infty, 2)$; Inc: $(2, \infty)$

40.

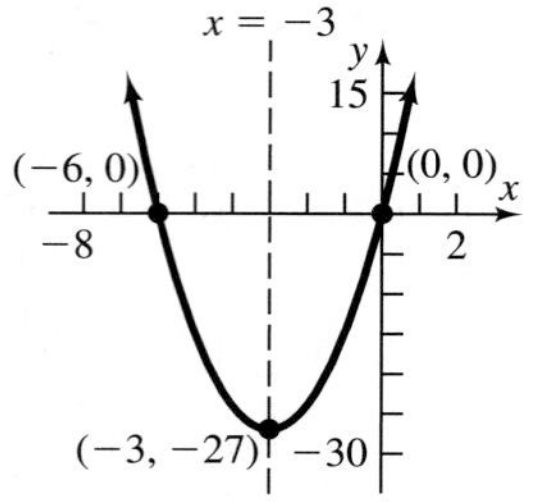

Domain: $(-\infty, \infty)$
Range: $[-27, \infty)$
Dec: $(-\infty, -3)$; Inc: $(-3, \infty)$

41.

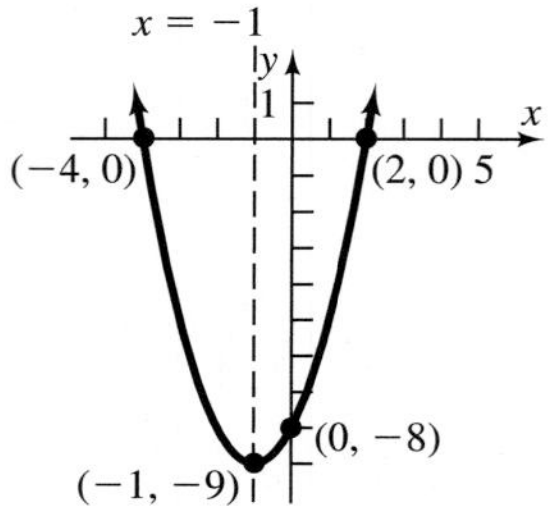

Domain: $(-\infty, \infty)$
Range: $[-9, \infty)$
Dec: $(-\infty, -1)$; Inc: $(-1, \infty)$

42.

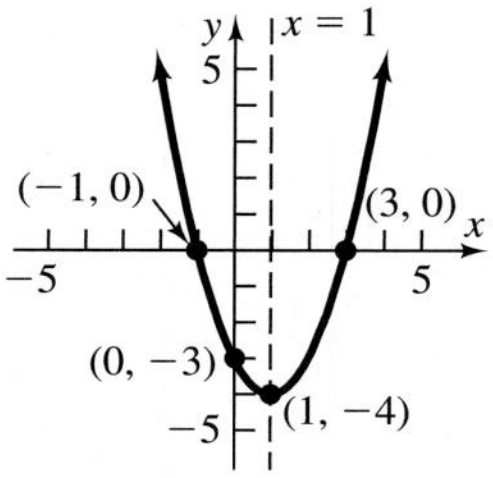

Domain: $(-\infty, \infty)$
Range: $[-4, \infty)$
Dec: $(-\infty, 1)$; Inc: $(1, \infty)$

43.

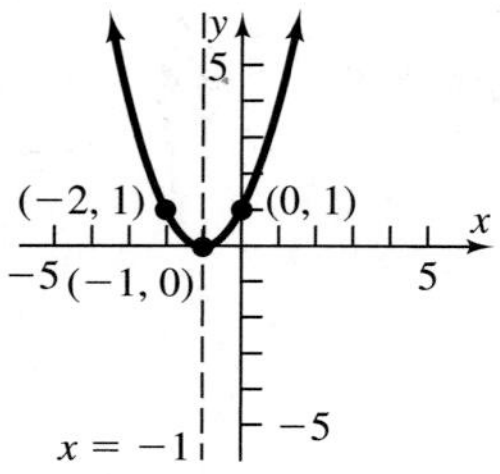

Domain: $(-\infty, \infty)$

Range: $[0, \infty)$

Dec: $(-\infty, -1)$; Inc: $(-1, \infty)$

44.

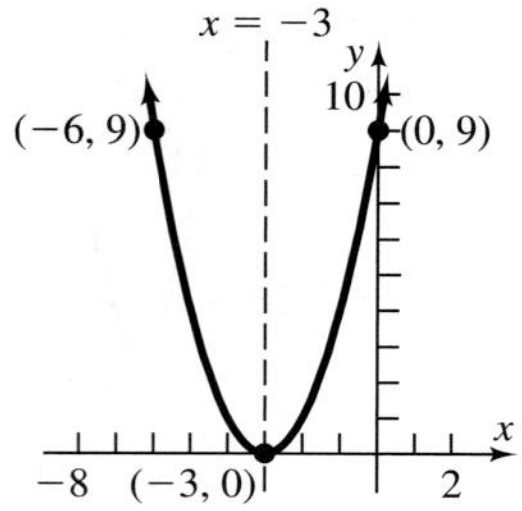

Domain: $(-\infty, \infty)$

Range: $[0, \infty)$

Dec: $(-\infty, -3)$; Inc: $(-3, \infty)$

45.

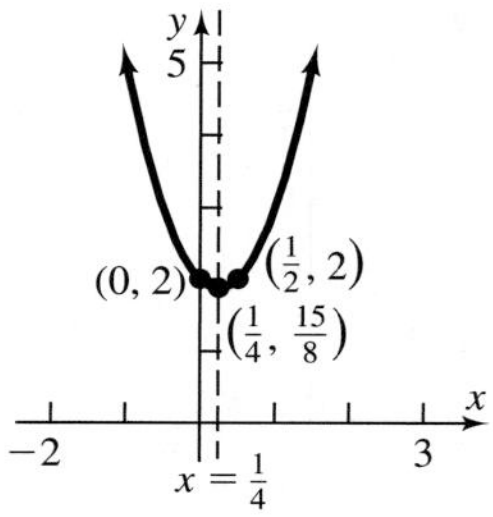

Domain: $(-\infty, \infty)$

Range: $\left[\frac{15}{8}, \infty\right)$

Dec: $\left(-\infty, \frac{1}{4}\right)$; Inc: $\left(\frac{1}{4}, \infty\right)$

46.

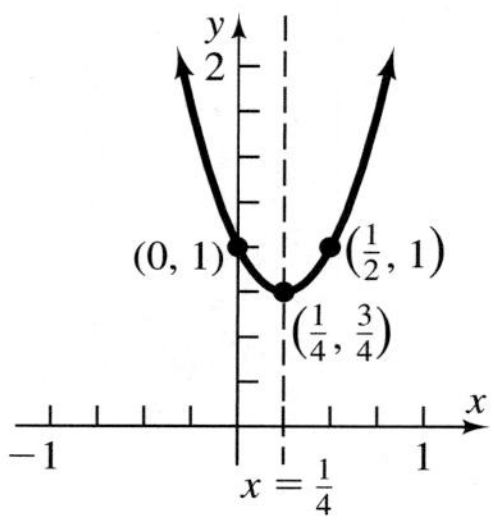

Domain: $(-\infty, \infty)$

Range: $\left[\frac{3}{4}, \infty\right)$

Dec: $\left(-\infty, \frac{1}{4}\right)$; Inc: $\left(\frac{1}{4}, \infty\right)$

47.

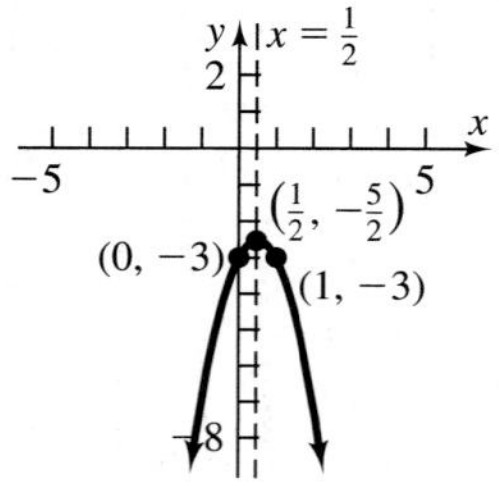

Domain: $(-\infty, \infty)$
Range: $\left(-\infty, -\frac{5}{2}\right]$
Dec: $\left(\frac{1}{2}, \infty\right)$; Inc: $\left(-\infty, \frac{1}{2}\right)$

48.

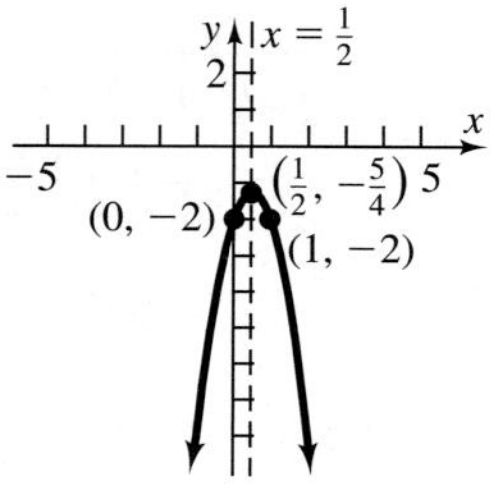

Domain: $(-\infty, \infty)$
Range: $\left(-\infty, -\frac{5}{4}\right]$
Dec: $\left(\frac{1}{2}, \infty\right)$; Inc: $\left(-\infty, \frac{1}{2}\right)$

49.

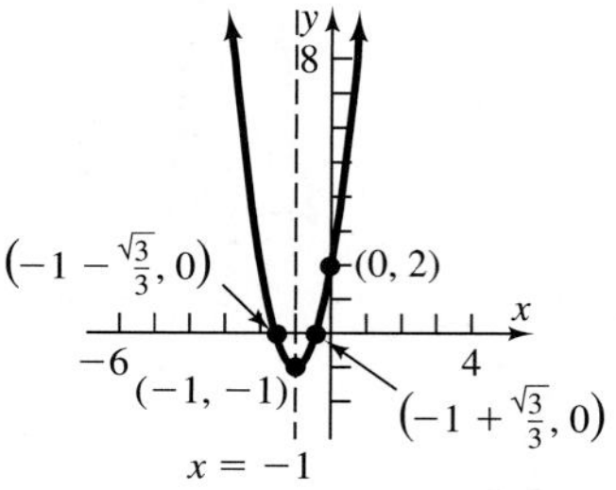

Domain: $(-\infty, \infty)$
Range: $[-1, \infty)$
Dec: $(-\infty, -1)$; Inc: $(-1, \infty)$

50.

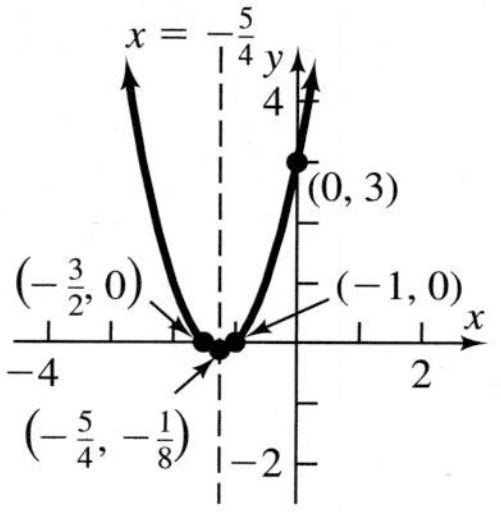

Domain: $(-\infty, \infty)$
Range: $\left[-\frac{1}{8}, \infty\right)$
Dec: $\left(-\infty, -\frac{5}{4}\right)$; Inc: $\left(-\frac{5}{4}, \infty\right)$

51.

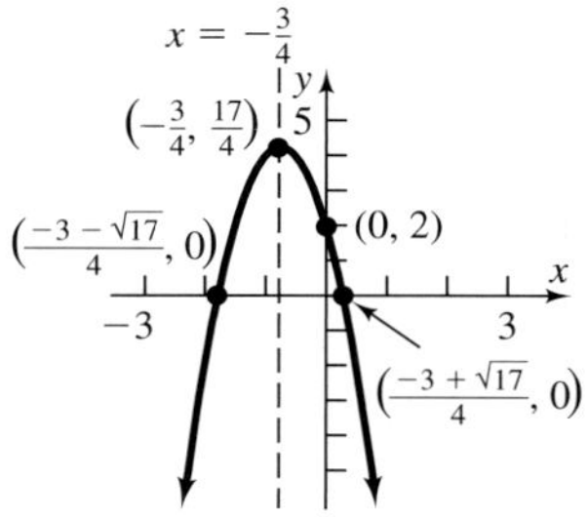

Domain: $(-\infty, \infty)$

Range: $\left(-\infty, \frac{17}{4}\right]$

Dec: $\left(-\frac{3}{4}, \infty\right)$; Inc: $\left(-\infty, -\frac{3}{4}\right)$

52.

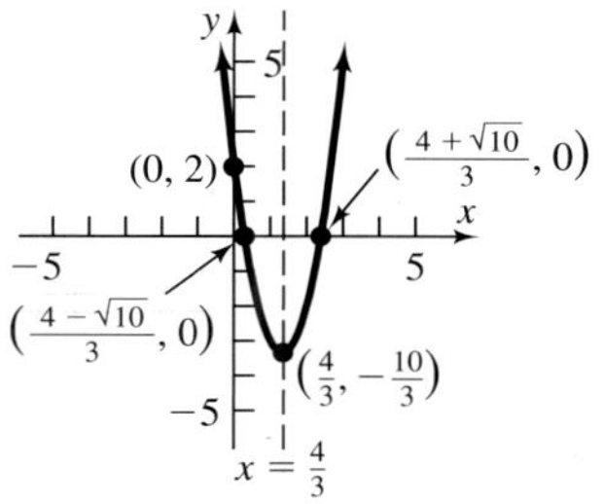

Domain: $(-\infty, \infty)$

Range: $\left[-\frac{10}{3}, \infty\right)$

Dec: $\left(-\infty, \frac{4}{3}\right)$; Inc: $\left(\frac{4}{3}, \infty\right)$

53. $f(x) = (x + 1)^2 - 2 = x^2 + 2x - 1$
54. $f(x) = (x - 2)^2 + 1 = x^2 - 4x + 5$
55. $f(x) = -(x + 3)^2 + 5 = -x^2 - 6x - 4$
56. $f(x) = -(x - 2)^2 + 3 = -x^2 + 4x - 1$
57. $f(x) = 2(x - 1)^2 - 3 = 2x^2 - 4x - 1$
58. $f(x) = -2(x + 2)^2 + 6 = -2x^2 - 8x - 2$

59. Minimum value; −18 **60.** Maximum value; 18 **61.** Minimum value; −21 **62.** Minimum value; −1 **63.** Maximum value; 21 **64.** Maximum value; 11 **65.** Maximum value; 13 **66.** Minimum value; −1

67. **(a)** $a = 1$: $f(x) = (x + 3)(x - 1) = x^2 + 2x - 3$
$a = 2$: $f(x) = 2(x + 3)(x - 1) = 2x^2 + 4x - 6$
$a = -2$: $f(x) = -2(x + 3)(x - 1) = -2x^2 - 4x + 6$
$a = 5$: $f(x) = 5(x + 3)(x - 1) = 5x^2 + 10x - 15$
(b) The value of a does not affect the x-intercepts but it changes the y-intercept by a factor of a.
(c) The value of a does not affect the axis of symmetry. It is $x = -1$ for all values of a.
(d) The value of a does not affect the x-coordinate of the vertex. However, the y-coordinate of the vertex is multiplied by a.
(e) The midpoint of the x-intercepts is the x-coordinate of the vertex.

68. **(a)** $a = 1$; $f(x) = (x + 5)(x - 3) = x^2 + 2x - 15$
$a = 2$; $f(x) = 2(x + 5)(x - 3) = 2x^2 + 4x - 30$
$a = -2$; $f(x) = -2(x + 5)(x - 3) = -2x^2 - 4x + 30$
$a = 5$; $f(x) = 5(x + 5)(x - 3) = 5x^2 + 10x - 75$
(b) The value of a does not affect the x-intercepts but it changes the y-intercept by a factor of a.
(c) The value of a does not affect the axis of symmetry. It is $x = -1$ for all values of a.
(d) The value of a does not affect the x-coordinate of the vertex. However, the y-coordinate of the vertex is multiplied by a.
(e) The midpoint of the x-intercepts is the x-coordinate of the vertex.

69. \$500; \$1,000,000 **70.** \$1900; \$1,805,000 **71.** **(a)** $R(x) = -\frac{1}{6}x^2 + 100x$ **(b)** \$13,333.33 **(c)** 300; \$15,000 **(d)** \$50

72. **(a)** $R(x) = -\frac{1}{3}x^2 + 100x$ **(b)** \$6666.67 **(c)** 150; \$7500 **(d)** \$50

73. **(a)** $R(x) = -\frac{1}{5}x^2 + 20x$ **(b)** \$255 **(c)** 50; \$500 **(d)** \$10

74. **(a)** $R(x) = -\frac{1}{20}x^2 + 25x$ **(b)** \$480 **(c)** 250; \$3125 **(d)** \$12.50

75. **(a)** $A(w) = -w^2 + 200w$ **(b)** A is largest when $w = 100$ yd. **(c)** 10,000 sq yd
76. **(a)** $A(x) = -x^2 + 1500x$ **(b)** A is largest when $x = 750$ ft. **(c)** 562,500 ft^2 **77.** 2,000,000 m^2 **78.** 500,000 m^2

79. **(a)** $\frac{625}{16} \approx 39$ ft **(b)** 219.5 ft **(c)** 170 ft

(d)

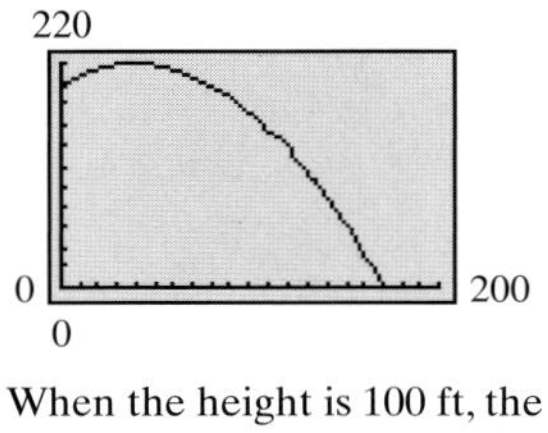

(e) When the height is 100 ft, the projectile is 135.7 ft from the cliff.

80. **(a)** 156.25 ft **(b)** 78.125 ft **(c)** 312.5 ft

(d)

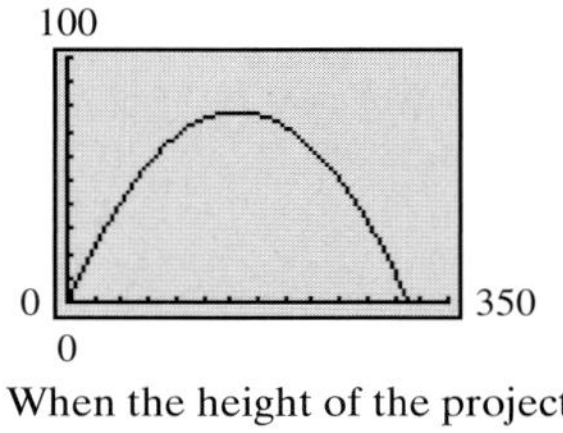

(e) When the height of the projectile is 50 ft, the projectile has traveled either 62.5 ft or 250 ft horizontally.

81. 18.75 m

82. Let the point (0, 25) represent the vertex of the arch. Then the equation of the parabola is
$$h(x) = \frac{-25}{(60)^2}x^2 + 25;\ h(10) \approx 24.3 \text{ ft};\ h(20) \approx 22.2 \text{ ft};\ h(40) \approx 13.9 \text{ ft}$$

83. 3 in. **84.** The width should be approximately 5.6 ft and the total height of the window should be approximately 5.6 ft.

85. $\frac{750}{\pi}$ by 375 m **86.** $x = \frac{-16}{-6 + \sqrt{3}} \approx 3.75$ ft, total height $= 8 + \frac{24}{-6 + \sqrt{3}} - \frac{8\sqrt{3}}{-6 + \sqrt{3}} \approx 5.62$ ft

87. **(a)** \$56,600; 3685 hunters
(b)

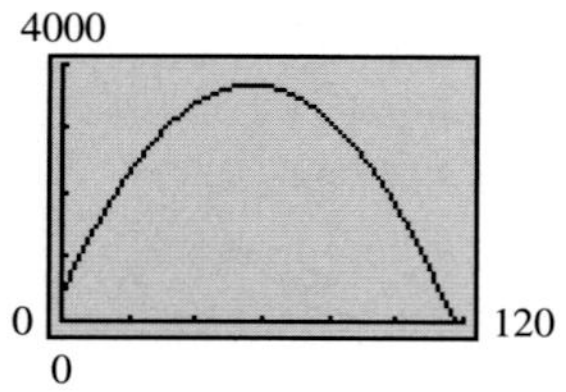

Increasing

88. **(a)** 54 years; 11.7%
(b)

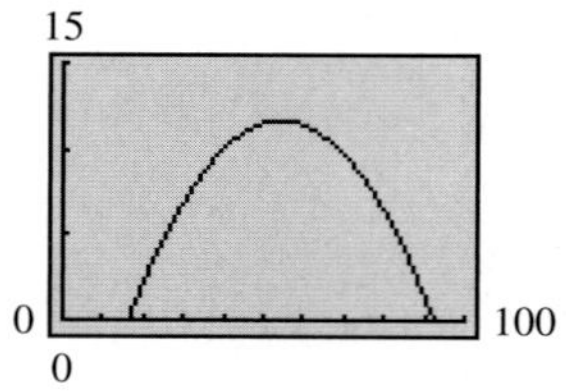

Increasing

89. **(a)** 1795 **(b)** 28 years old
(c)

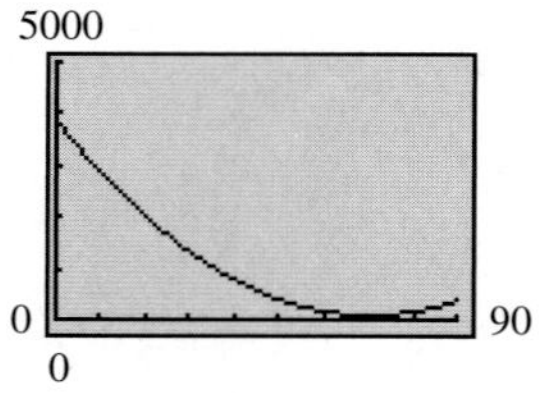

(d) The number of victims initially decreases, then begins to increase.

90. **(a)** 4.6% **(b)** 67 years
(c)

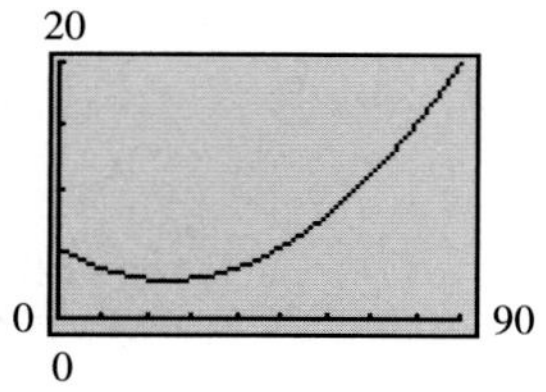

(d) Initially, it declines, but then it increases.

91. **(a)** Quadratic, $a < 0$

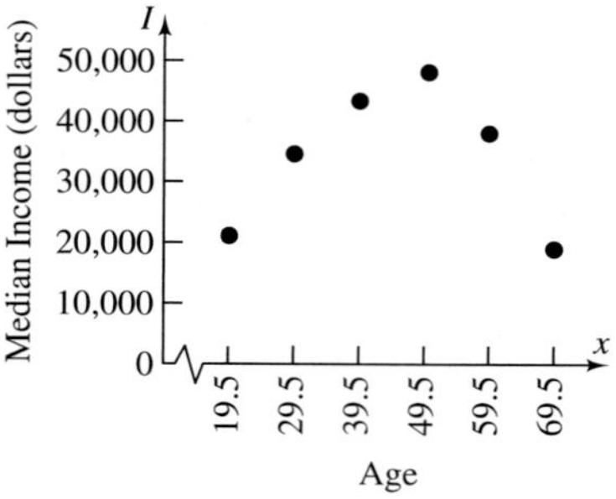

(b) 44.7 yr old
(c) \$46,484
(e)

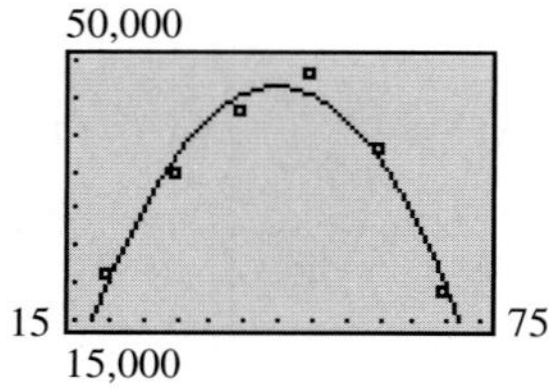

92. **(a)** Quadratic, $a < 0$

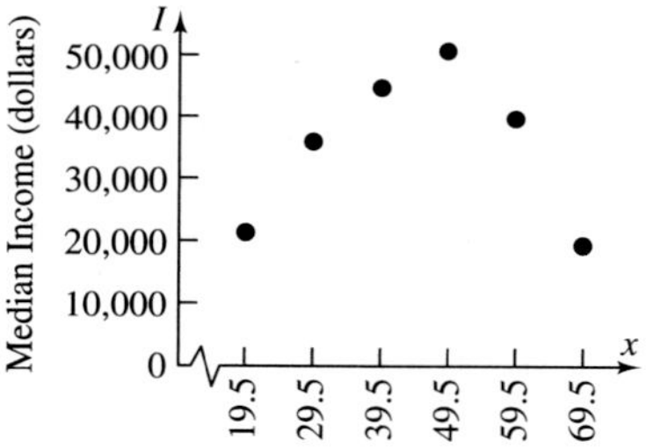

(b) 44.7 yr old
(c) \$48,296
(e)

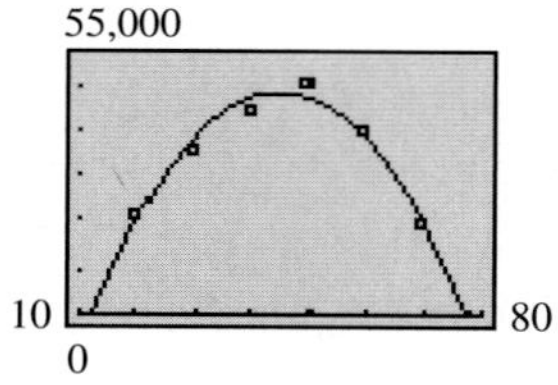

93. **(a)** Quadratic, $a < 0$

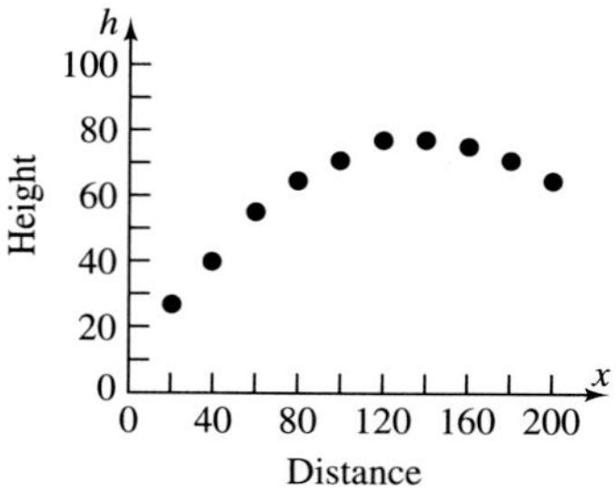

(b) 139.2 ft
(c) 77.4 ft
(e)

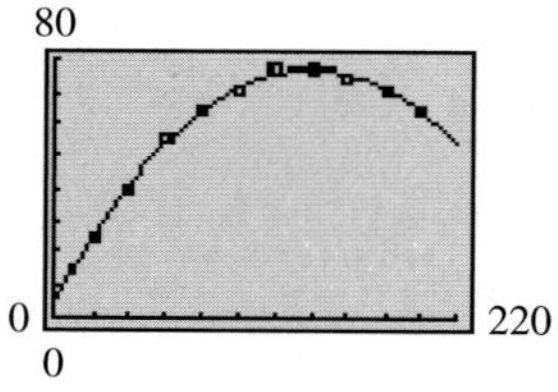

94. **(a)** Quadratic, $a < 0$

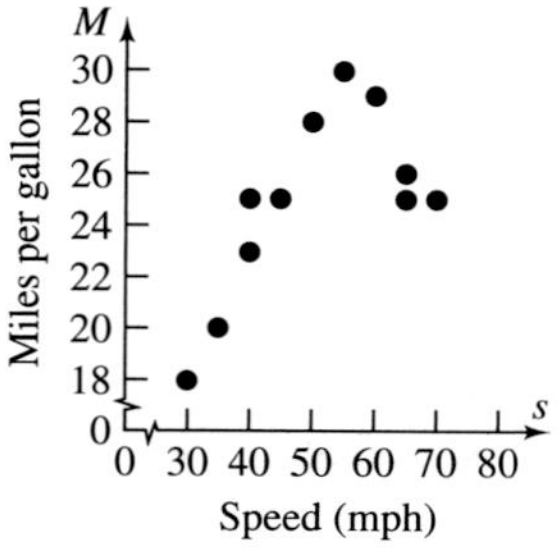

(b) Approximately 53.61 mi/hr
(c) Approximately 24.81 mi per gal
(e)

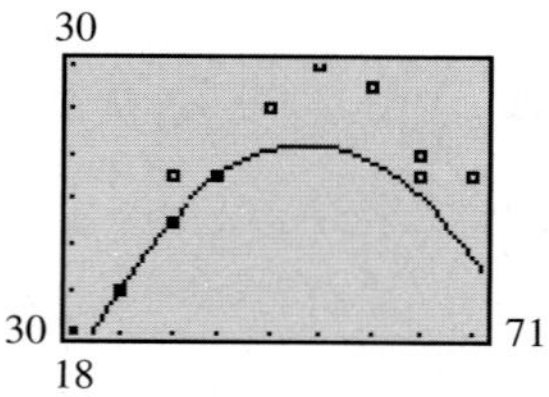

95. $x = \frac{a}{2}$ **96.** $\left.\begin{aligned} ah^2 - bh + c &= y_0 \\ c &= y_1 \\ ah^2 + bh + c &= y_2 \end{aligned}\right\} \begin{aligned} y_0 + y_2 &= 2ah^2 + 2c \\ 4y_1 &= 4c \end{aligned}\Bigg\}$ Area $= \frac{h}{3}(2ah^2 + 6c) = \frac{h}{3}(y_0 + 4y_1 + y_2)$ **97.** $\frac{38}{3}$ **98.** $\frac{128}{3}$ **99.** $\frac{248}{3}$

100. $\frac{22}{3}$ **101.** 25 square units **102.** If x is even, then ax^2 and bx are even and $ax^2 + bx$ is even, which means that $ax^2 + bx + c$ is odd. If x is odd, then ax^2 and bx are odd and $ax^2 + bx$ is even, which means that $ax^2 + bx + c$ is odd. In either case, $f(x)$ is odd.

3.2 Concepts and Vocabulary *(page 158)*

5. smooth; continuous **6.** zero or root **7.** touches **8.** True **9.** False **10.** False

3.2 Exercises *(page 159)*

11. Yes; degree 3 **12.** Yes; degree 4 **13.** Yes; degree 2 **14.** Yes; degree 1 **15.** No; x is raised to the -1 power. **16.** Yes; degree 2 **17.** No; x is raised to the $\frac{3}{2}$ power. **18.** No; x is raised to the $\frac{1}{2}$ power. **19.** Yes; degree 4 **20.** No; it is the ratio of two polynomials and the polynomial in the denominator is of positive degree. **21.** Yes; degree 4 **22.** Yes; degree 5

23.

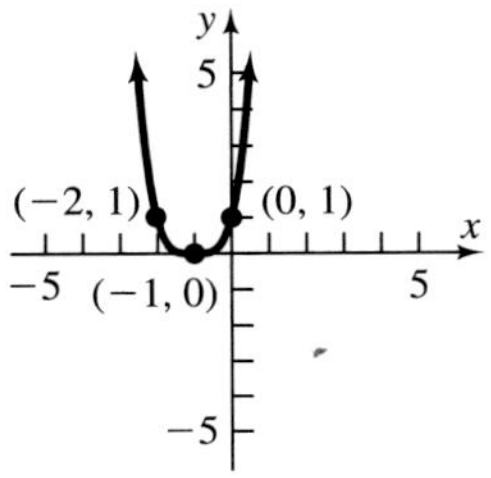

24.

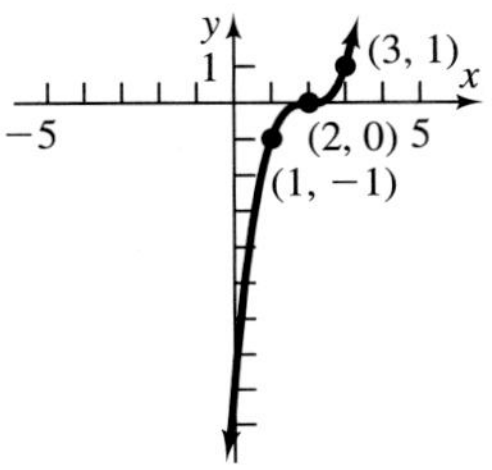

25.

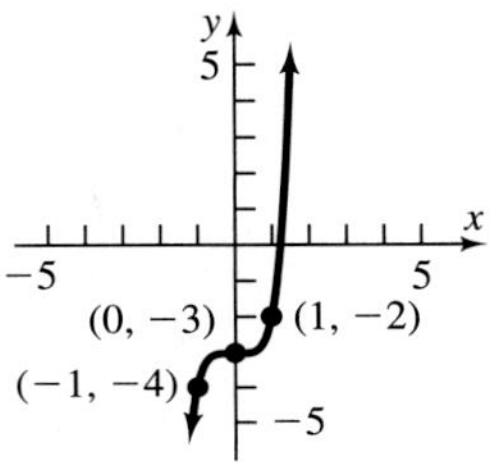

26.

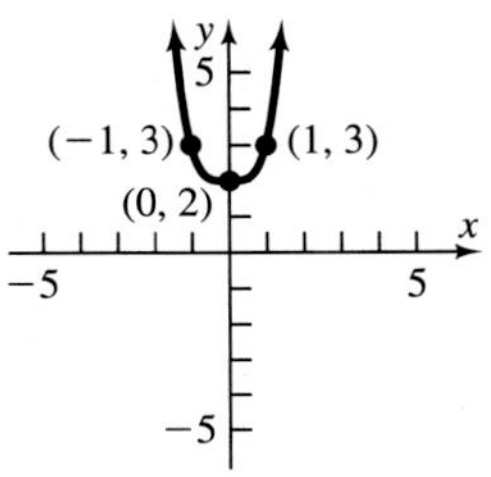

27.

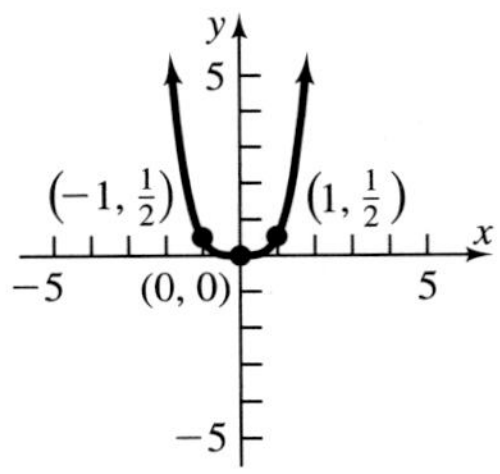

28.

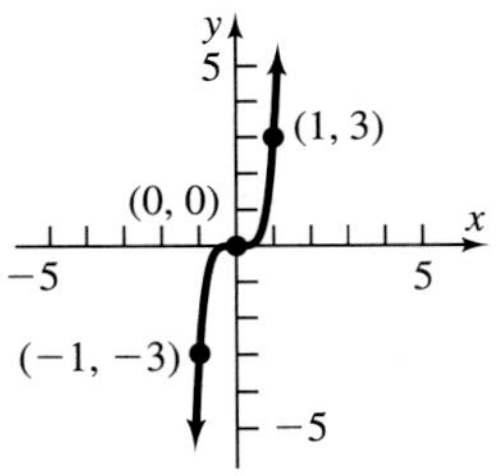

29.

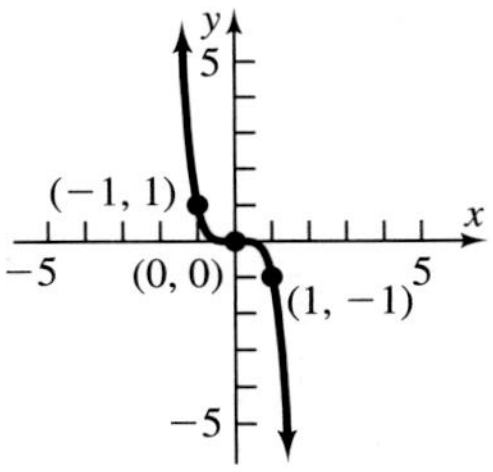

30.

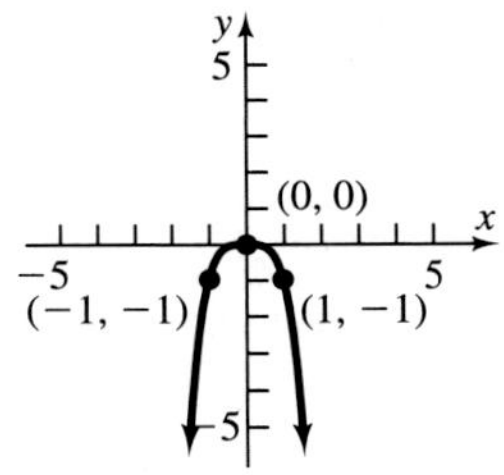

31.

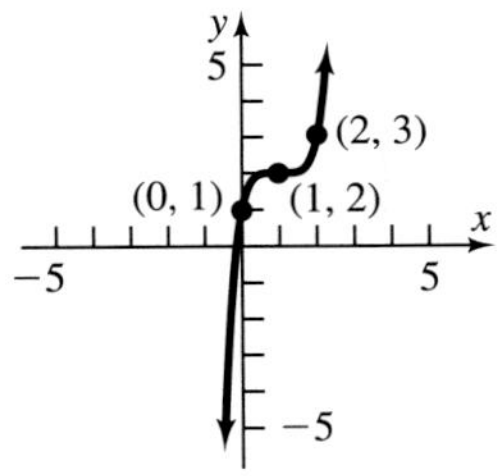

32.

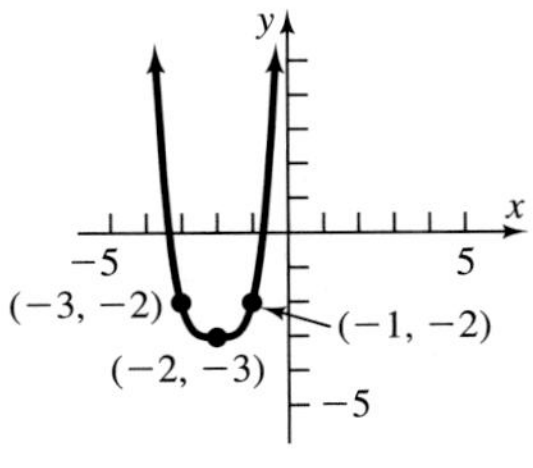

33.

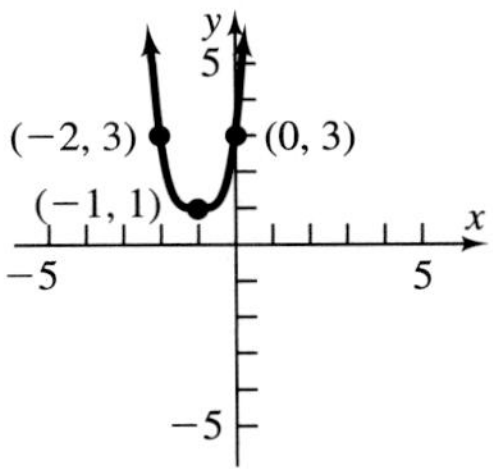

34.

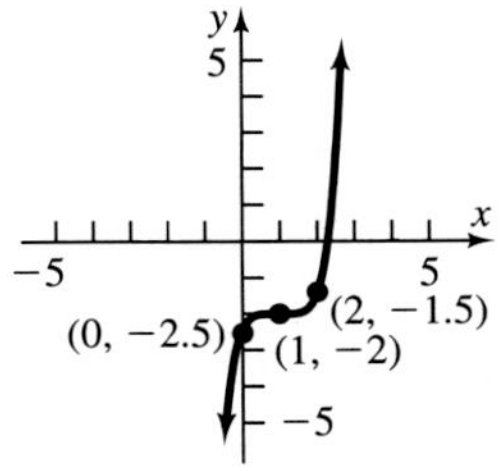

35.

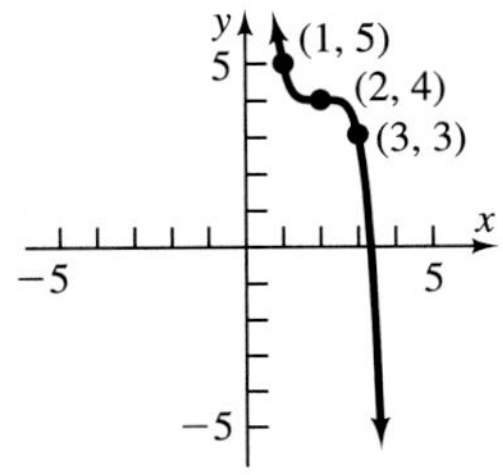

36.

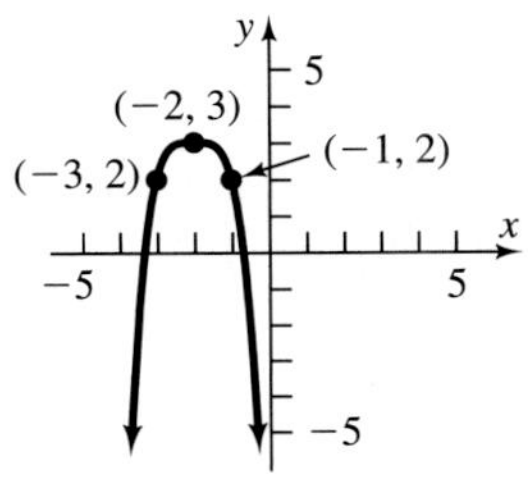

37. $f(x) = x^3 - 3x^2 - x + 3$ for $a = 1$
38. $f(x) = x^3 - 3x^2 - 4x + 12$ for $a = 1$
39. $f(x) = x^3 - x^2 - 12x$ for $a = 1$
40. $f(x) = x^3 + 2x^2 - 8x$ for $a = 1$
41. $f(x) = x^4 - 15x^2 + 10x + 24$ for $a = 1$
42. $f(x) = x^4 - 3x^3 - 15x^2 + 19x + 30$ for $a = 1$
43. $f(x) = x^3 - 5x^2 + 3x + 9$ for $a = 1$
44. $f(x) = x^3 - 12x - 16$ for $a = 1$

45. (a) 7, multiplicity 1; -3, multiplicity 2 **(b)** graph touches the x-axis at -3 and crosses it at 7 **(c)** $y = 3x^3$
46. (a) -4, multiplicity 1; -3, multiplicity 3 **(b)** graph crosses the x-axis at -4 and -3 **(c)** $y = 4x^4$ **47. (a)** 2, multiplicity 3 **(b)** graph crosses the x-axis at 2 **(c)** $y = 4x^5$ **48. (a)** 3, multiplicity 1; -4, multiplicity 3 **(b)** graph crosses the x-axis at -4 and 3 **(c)** $y = 2x^4$ **49. (a)** $-\frac{1}{2}$, multiplicity 2 **(b)** graph touches the x-axis at $-\frac{1}{2}$ **(c)** $y = -2x^6$

50. **(a)** $\frac{1}{3}$, multiplicity 2; 1, multiplicity 3 **(b)** graph touches the x-axis at $\frac{1}{3}$ and crosses it at 1 **(c)** $y = x^5$

51. **(a)** 5, multiplicity 3; -4, multiplicity 2 **(b)** graph touches the x-axis at -4 and crosses it at 5 **(c)** $y = x^5$

52. **(a)** $-\sqrt{3}$, multiplicity 2; 2, multiplicity 4 **(b)** graph touches the x-axis at $-\sqrt{3}$ and 2 **(c)** $y = x^6$

53. **(a)** no real zeros **(b)** graph neither crosses nor touches the x-axis **(c)** $y = 3x^6$

54. **(a)** no real zeros **(b)** graph neither crosses nor touches the x-axis **(c)** $y = -2x^6$

55. **(a)** 0, multiplicity 2; $-\sqrt{2}$, $\sqrt{2}$, multiplicity 1 **(b)** graph touches the x-axis at 0 and crosses at $-\sqrt{2}$ and $\sqrt{2}$ **(c)** $y = -2x^4$

56. **(a)** 0, multiplicity 1; $-\sqrt{3}$, $\sqrt{3}$, multiplicity 1 **(b)** graph crosses the x-axis at $-\sqrt{3}$, 0, and $\sqrt{3}$ **(c)** $y = 4x^3$

57. **(a)** x-intercept: 1; y-intercept: 1
(b) touches at 1
(c) $y = x^2$
(d) 1
(e)

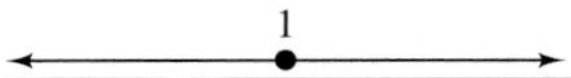

Interval	$(-\infty, 1)$	$(1, \infty)$
Number Chosen	-1	2
Value of f	$f(-1) = 4$	$f(2) = 1$
Location of Graph	Above x-axis	Above x-axis
Point on Graph	$(-1, 4)$	$(2, 1)$

(f)

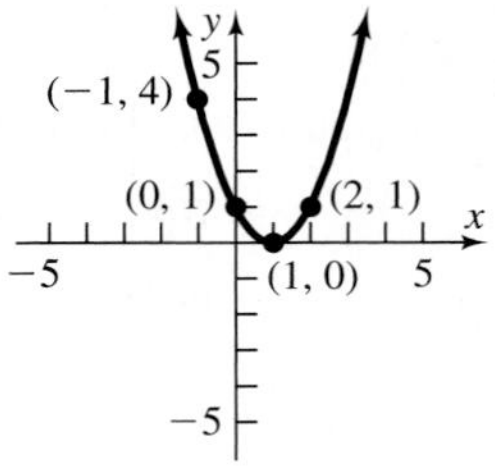

58. **(a)** x-intercept: 2; y-intercept: -8
(b) Crosses at 2
(c) $y = x^3$
(d) 2
(e)

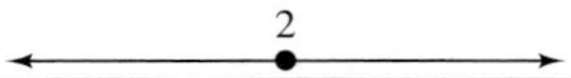

Interval	$(-\infty, 2)$	$(2, \infty)$
Number Chosen	1	3
Value of f	$f(1) = -1$	$f(3) = 1$
Location of Graph	Below x-axis	Above x-axis
Point on Graph	$(1, -1)$	$(3, 1)$

(f)

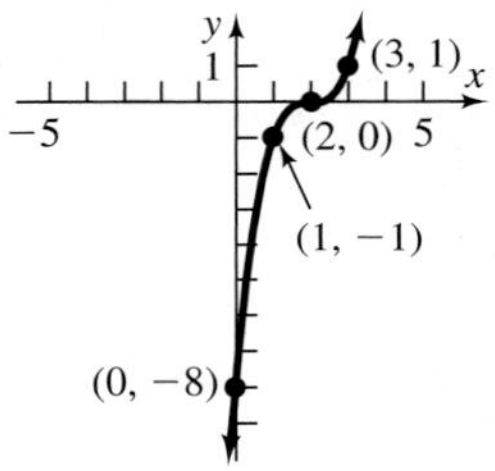

59. **(a)** x-intercepts: 0, 3; y-intercept: 0
(b) Touches at 0; crosses at 3
(c) $y = x^3$
(d) 2
(e)

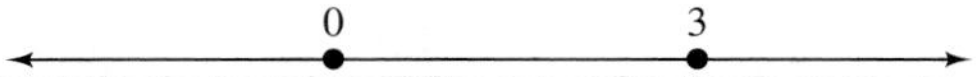

Interval	$(-\infty, 0)$	$(0, 3)$	$(3, \infty)$
Number Chosen	-1	2	4
Value of f	$f(-1) = -4$	$f(2) = -4$	$f(4) = 16$
Location of Graph	Below x-axis	Below x-axis	Above x-axis
Point on Graph	$(-1, -4)$	$(2, -4)$	$(4, 16)$

(f)

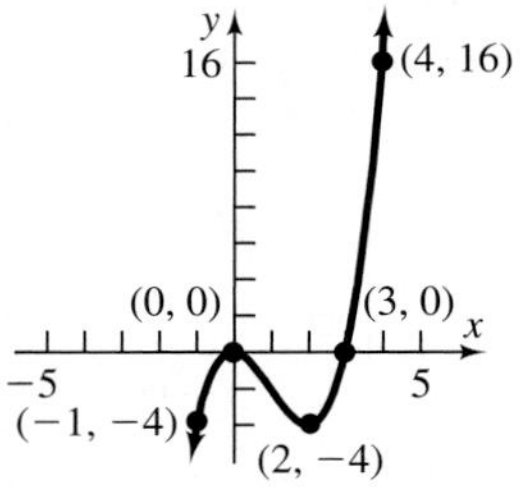

60. **(a)** x-intercepts: -2, 0; y-intercept: 0
(b) Touches at -2; crosses at 0
(c) $y = x^3$
(d) 2
(e)

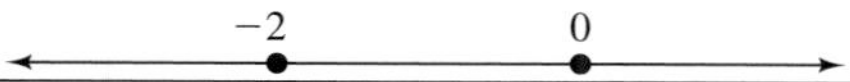

Interval	$(-\infty, -2)$	$(-2, 0)$	$(0, \infty)$
Number Chosen	-3	-1	1
Value of f	$f(-3) = -3$	$f(-1) = -1$	$f(1) = 9$
Location of Graph	Below x-axis	Below x-axis	Above x-axis
Point on Graph	$(-3, -3)$	$(-1, -1)$	$(1, 9)$

(f)

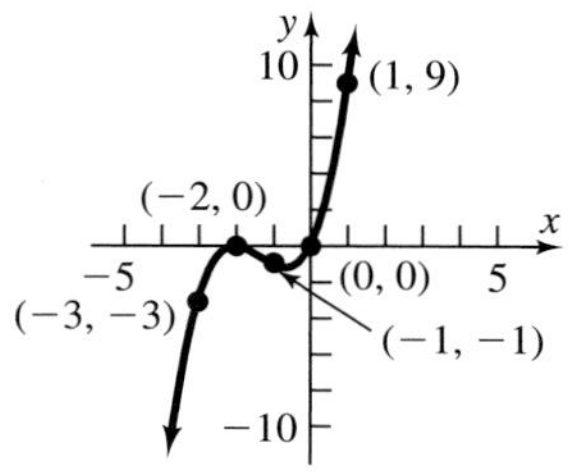

61. **(a)** x-intercepts: -4, 0; y-intercept: 0
(b) Crosses at -4, 0
(c) $y = 6x^4$
(d) 3
(e)

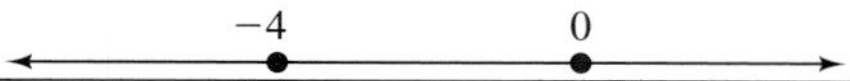

Interval	$(-\infty, -4)$	$(-4, 0)$	$(0, \infty)$
Number Chosen	-5	-2	1
Value of f	$f(-5) = 750$	$f(-2) = -96$	$f(1) = 30$
Location of Graph	Above x-axis	Below x-axis	Above x-axis
Point on Graph	$(-5, 750)$	$(-2, -96)$	$(1, 30)$

(f)

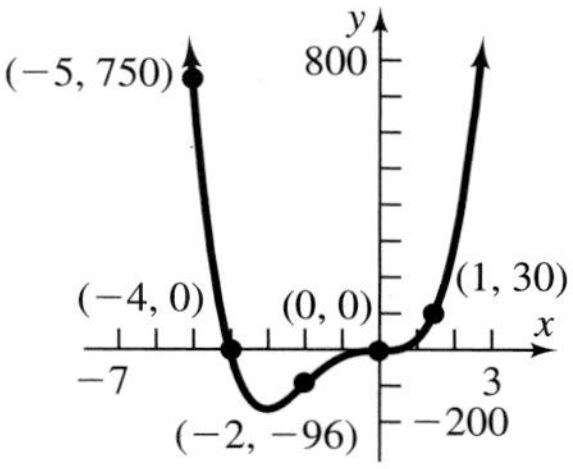

62. **(a)** x-intercepts: 0, 1; y-intercept: 0
(b) Crosses at 0 and 1
(c) $y = 5x^4$
(d) 3
(e)

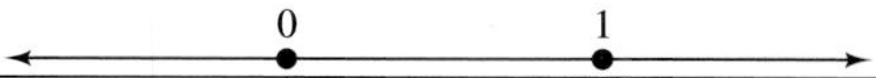

Interval	$(-\infty, 0)$	$(0, 1)$	$(1, \infty)$
Number Chosen	-1	$\frac{1}{2}$	2
Value of f	$f(-1) = 40$	$f\left(\frac{1}{2}\right) = -\frac{5}{16}$	$f(2) = 10$
Location of Graph	Above x-axis	Below x-axis	Above x-axis
Point on Graph	$(-1, 40)$	$\left(\frac{1}{2}, -\frac{5}{16}\right)$	$(2, 10)$

(f)

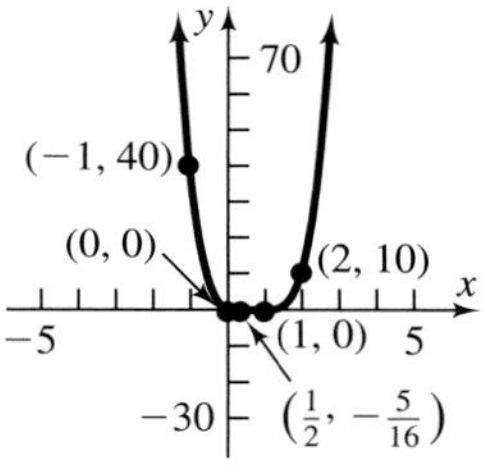

63. **(a)** x-intercepts: -2, 0; y-intercept: 0
(b) Crosses at -2; touches at 0
(c) $y = -4x^3$
(d) 2
(e)

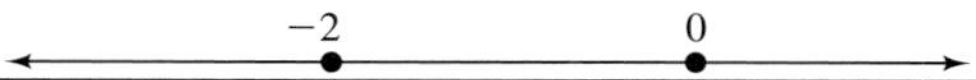

Interval	$(-\infty, -2)$	$(-2, 0)$	$(0, \infty)$
Number Chosen	-3	-1	1
Value of f	$f(-3) = 36$	$f(-1) = -4$	$f(1) = -12$
Location of Graph	Above x-axis	Below x-axis	Below x-axis
Point on Graph	$(-3, 36)$	$(-1, -4)$	$(1, -12)$

(f)

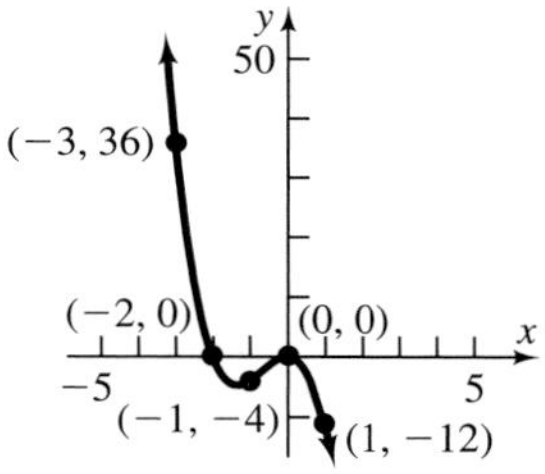

64. (a) x-intercepts: $-4, 0$; y-intercept: 0
(b) Crosses at -4 and 0
(c) $y = -\frac{1}{2}x^4$
(d) 3
(e)

-4 | 0 (number line)

Interval	$(-\infty, -4)$	$(-4, 0)$	$(0, \infty)$
Number Chosen	-5	-2	1
Value of f	$f(-5) = -\frac{125}{2}$	$f(-2) = 8$	$f(1) = -\frac{5}{2}$
Location of Graph	Below x-axis	Above x-axis	Below x-axis
Point on Graph	$\left(-5, -\frac{125}{2}\right)$	$(-2, 8)$	$\left(1, -\frac{5}{2}\right)$

(f)

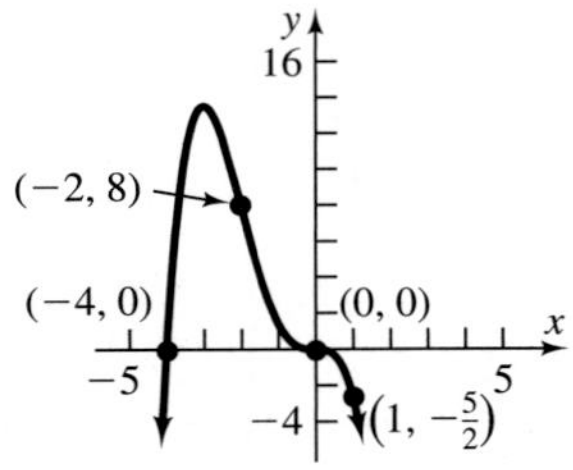

65. (a) x-intercepts: $-4, 1, 2$; y-intercept: 8
(b) Crosses at $-4, 1, 2$
(c) $y = x^3$
(d) 2
(e)

-4 | 1 | 2 (number line)

Interval	$(-\infty, -4)$	$(-4, 1)$	$(1, 2)$	$(2, \infty)$
Number Chosen	-5	-2	$\frac{3}{2}$	3
Value of f	$f(-5) = -42$	$f(-2) = 24$	$f\left(\frac{3}{2}\right) = -\frac{11}{8}$	$f(3) = 14$
Location of Graph	Below x-axis	Above x-axis	Below x-axis	Above x-axis
Point on Graph	$(-5, -42)$	$(-2, 24)$	$\left(\frac{3}{2}, -\frac{11}{8}\right)$	$(3, 14)$

(f)

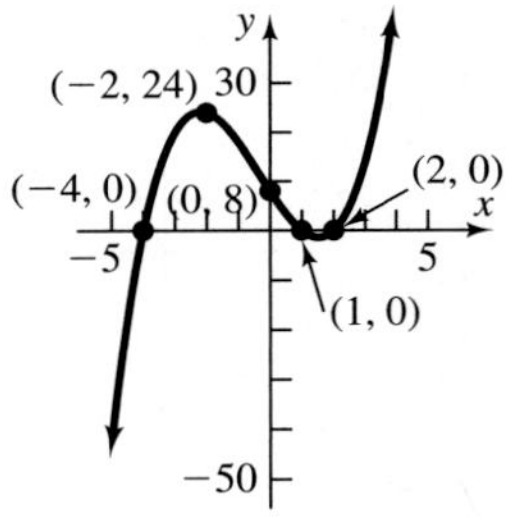

66. (a) x-intercepts: $-4, -1, 3$; y-intercept: -12
(b) Crosses at $-4, -1$, and 3
(c) $y = x^3$
(d) 2
(e)

-4 | -1 | 3 (number line)

Interval	$(-\infty, -4)$	$(-4, -1)$	$(-1, 3)$	$(3, \infty)$
Number Chosen	-5	-2	1	4
Value of f	$f(-5) = -32$	$f(-2) = 10$	$f(1) = -20$	$f(4) = 40$
Location of Graph	Below x-axis	Above x-axis	Below x-axis	Above x-axis
Point on Graph	$(-5, -32)$	$(-2, 10)$	$(1, -20)$	$(4, 40)$

(f)

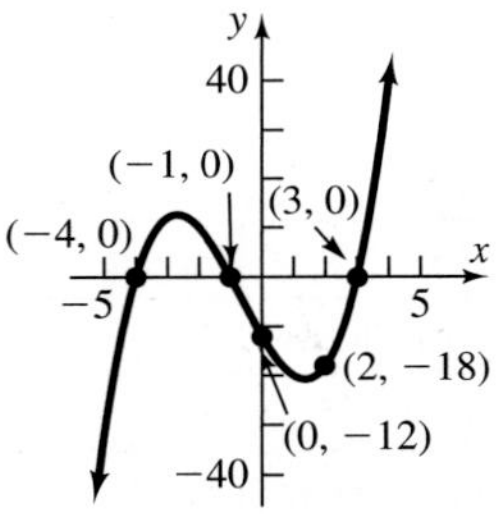

67. $f(x) = 4x - x^3 = -x(x^2 - 4)$
$= -x(x + 2)(x - 2)$
(a) x-intercepts: $-2, 0, 2$; y-intercept: 0
(b) Crosses at $-2, 0$, and 2
(c) $y = -x^3$
(d) 2
(e)

-2 | 0 | 2 (number line)

Interval	$(-\infty, -2)$	$(-2, 0)$	$(0, 2)$	$(2, \infty)$
Number Chosen	-3	-1	1	3
Value of f	$f(-3) = 15$	$f(-1) = -3$	$f(1) = 3$	$f(3) = -15$
Location of Graph	Above x-axis	Below x-axis	Above x-axis	Below x-axis
Point on Graph	$(-3, 15)$	$(-1, -3)$	$(1, 3)$	$(3, -15)$

(f)

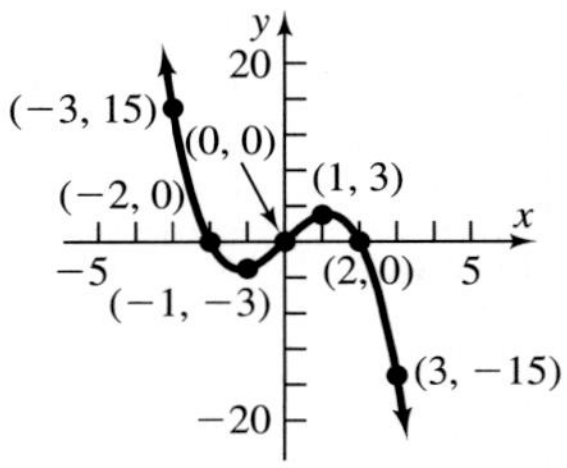

68. $f(x) = x - x^3 = -x(x^2 - 1)$
$= -x(x - 1)(x + 1)$

(a) x-intercepts: $-1, 0, 1$; y-intercept: 0
(b) Crosses at $-1, 0$, and 1
(c) $y = -x^3$
(d) 2
(e)

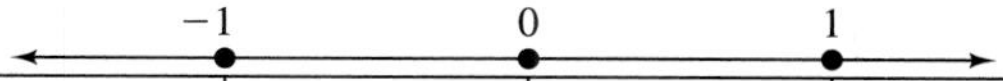

Interval	$(-\infty, -1)$	$(-1, 0)$	$(0, 1)$	$(1, \infty)$
Number Chosen	-2	$-\frac{1}{2}$	$\frac{1}{2}$	2
Value of f	$f(-2) = 6$	$f\left(-\frac{1}{2}\right) = -\frac{3}{8}$	$f\left(\frac{1}{2}\right) = \frac{3}{8}$	$f(2) = -6$
Location of Graph	Above x-axis	Below x-axis	Above x-axis	Below x-axis
Point on Graph	$(-2, 6)$	$\left(-\frac{1}{2}, -\frac{3}{8}\right)$	$\left(\frac{1}{2}, \frac{3}{8}\right)$	$(2, -6)$

(f)

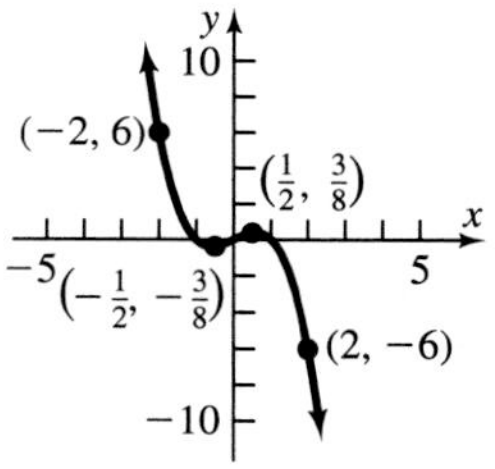

69. **(a)** x-intercepts: $-2, 0, 2$; y-intercept: 0
(b) Crosses at $-2, 2$; touches at 0
(c) $y = x^4$
(d) 3
(e)

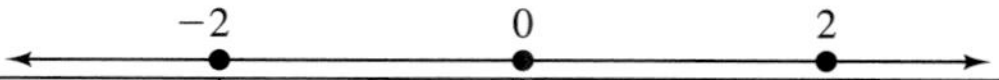

Interval	$(-\infty, -2)$	$(-2, 0)$	$(0, 2)$	$(2, \infty)$
Number Chosen	-3	-1	1	3
Value of f	$f(-3) = 45$	$f(-1) = -3$	$f(1) = -3$	$f(3) = 45$
Location of Graph	Above x-axis	Below x-axis	Below x-axis	Above x-axis
Point on Graph	$(-3, 45)$	$(-1, -3)$	$(1, -3)$	$(3, 45)$

(f)

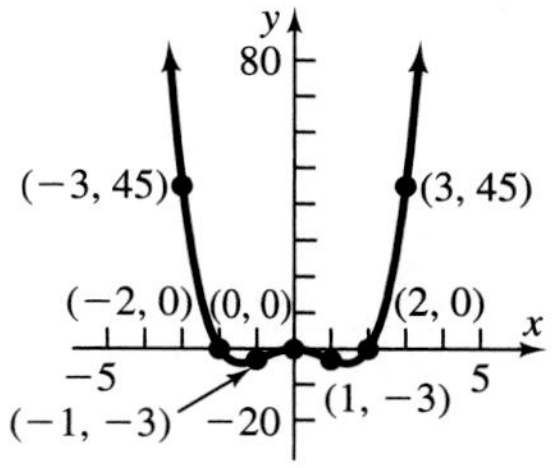

70. **(a)** x-intercepts: $-4, 0, 3$; y-intercept: 0
(b) Crosses at -4 and 3; touches at 0
(c) $y = x^4$
(d) 3
(e)

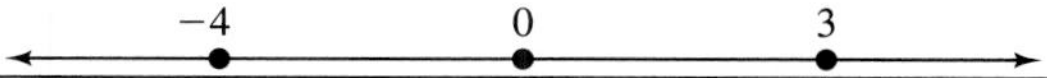

Interval	$(-\infty, -4)$	$(-4, 0)$	$(0, 3)$	$(3, \infty)$
Number Chosen	-5	-2	1	4
Value of f	$f(-5) = 200$	$f(-2) = -40$	$f(1) = -10$	$f(4) = 128$
Location of Graph	Above x-axis	Below x-axis	Below x-axis	Above x-axis
Point on Graph	$(-5, 200)$	$(-2, -40)$	$(1, -10)$	$(4, 128)$

(f)

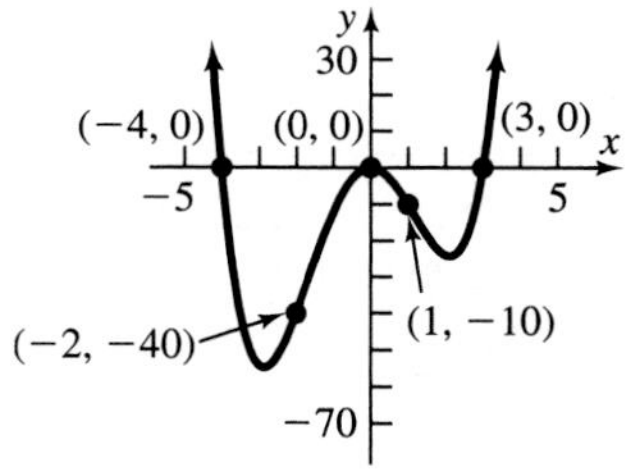

71. **(a)** x-intercepts: $-2, 2$; y-intercept: 16
(b) Touches at $-2, 2$
(c) $y = x^4$
(d) 3
(e)

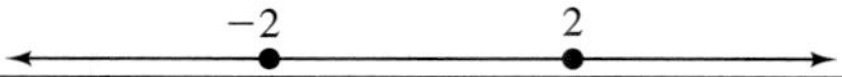

Interval	$(-\infty, -2)$	$(-2, 2)$	$(2, \infty)$
Number Chosen	-3	0	3
Value of f	$f(-3) = 25$	$f(0) = 16$	$f(3) = 25$
Location of Graph	Above x-axis	Above x-axis	Above x-axis
Point on Graph	$(-3, 25)$	$(0, 16)$	$(3, 25)$

(f)

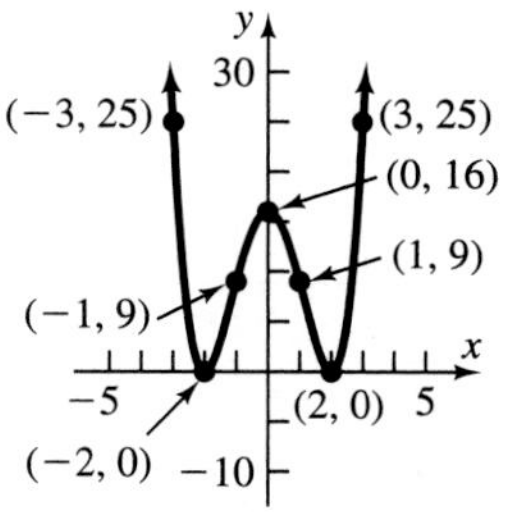

72. **(a)** x-intercepts: $-1, 3$; y-intercept: -3
(b) Crosses at -1 and 3
(c) $y = x^4$
(d) 3
(e)

Interval	$(-\infty, -1)$	$(-1, 3)$	$(3, \infty)$
Number Chosen	-2	1	4
Value of f	$f(-2) = 5$	$f(1) = -16$	$f(4) = 125$
Location of Graph	Above x-axis	Below x-axis	Above x-axis
Point on Graph	$(-2, 5)$	$(1, -16)$	$(4, 125)$

(f)

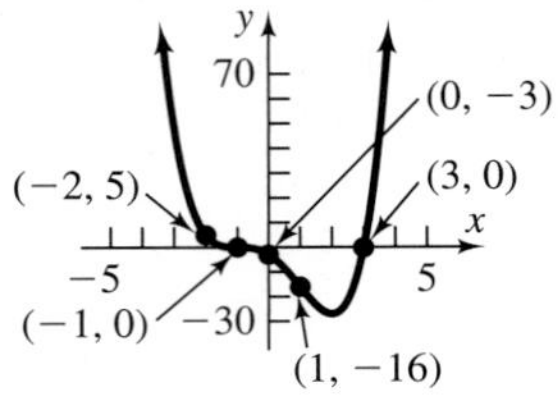

73. **(a)** x-intercepts: $-1, 1, 3$; y-intercept: -3
(b) Crosses at $-1, 3$; touches at 1
(c) $y = x^4$
(d) 3
(e)

Interval	$(-\infty, -1)$	$(-1, 1)$	$(1, 3)$	$(3, \infty)$
Number Chosen	-2	0	2	4
Value of f	$f(-2) = 45$	$f(0) = -3$	$f(2) = -3$	$f(4) = 45$
Location of Graph	Above x-axis	Below x-axis	Below x-axis	Above x-axis
Point on Graph	$(-2, 45)$	$(0, -3)$	$(2, -3)$	$(4, 45)$

(f)

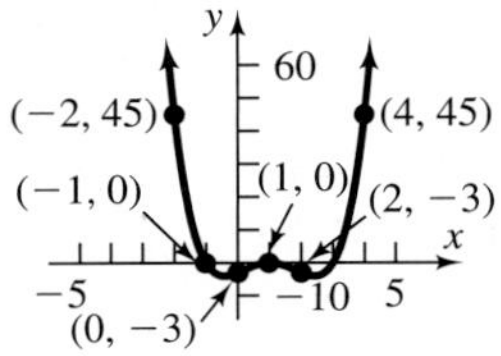

74. **(a)** x-intercepts: $-1, 1$, and 3; y-intercept: 3
(b) Touches at -1; crosses at 1 and 3
(c) $y = x^4$
(d) 3
(e)

Interval	$(-\infty, -1)$	$(-1, 1)$	$(1, 3)$	$(3, \infty)$
Number Chosen	-2	0	2	4
Value of f	$f(-2) = 15$	$f(0) = 3$	$f(2) = -9$	$f(4) = 75$
Location of Graph	Above x-axis	Above x-axis	Below x-axis	Above x-axis
Point on Graph	$(-2, 15)$	$(0, 3)$	$(2, -9)$	$(4, 75)$

(f)

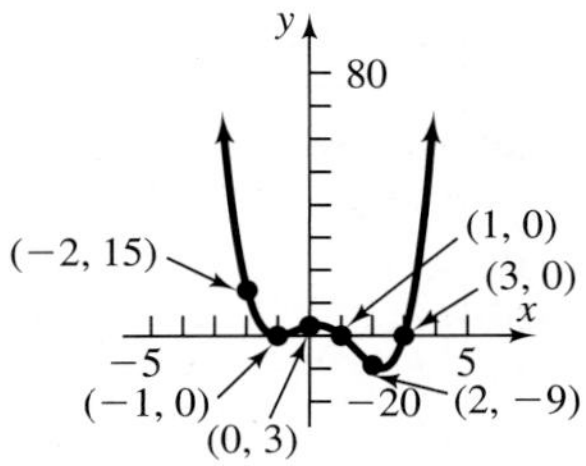

75. **(a)** x-intercepts: $-2, 4$; y-intercept: 64
(b) Touches at -2 and 4
(c) $y = x^4$
(d) 3
(e)

Interval	$(-\infty, -2)$	$(-2, 4)$	$(4, \infty)$
Number Chosen	-3	0	5
Value of f	$f(-3) = 49$	$f(0) = 64$	$f(5) = 49$
Location of Graph	Above x-axis	Above x-axis	Above x-axis
Point on Graph	$(-3, 49)$	$(0, 64)$	$(5, 49)$

(f)

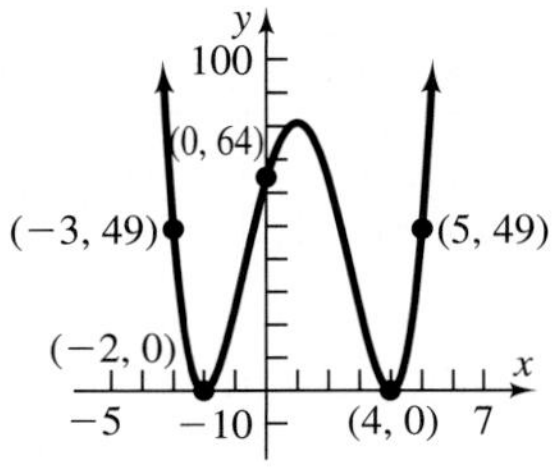

76. (a) x-intercepts: $-4, -2, 2$; y-intercept: 32
(b) Crosses at $-4, -2$; touches at 2
(c) $y = x^4$
(d) 3
(e)

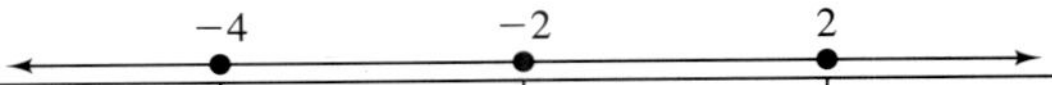

Interval	$(-\infty, -4)$	$(-4, -2)$	$(-2, 2)$	$(2, \infty)$
Number Chosen	-5	-3	0	3
Value of f	$f(-5) = 147$	$f(-3) = -25$	$f(0) = 32$	$f(3) = 35$
Location of Graph	Above x-axis	Below x-axis	Above x-axis	Above x-axis
Point of Graph	$(-5, 147)$	$(-3, -25)$	$(0, 32)$	$(3, 35)$

(f)

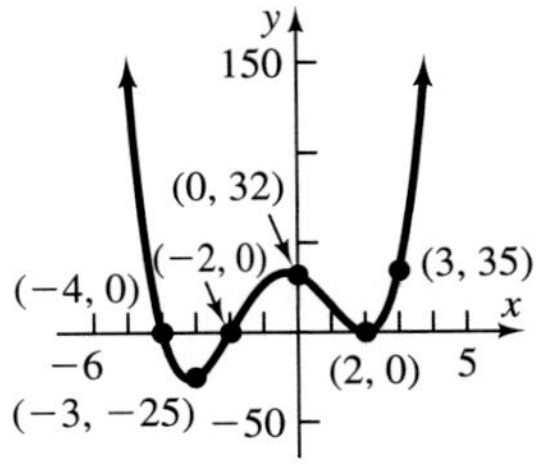

77. (a) x-intercepts: 0, 2; y-intercept: 0
(b) Touches at 0; crosses at 2
(c) $y = x^5$
(d) 4
(e)

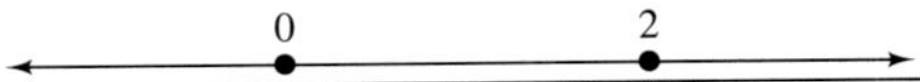

Interval	$(-\infty, 0)$	$(0, 2)$	$(2, \infty)$
Number Chosen	-1	1	3
Value of f	$f(-1) = -12$	$f(1) = -4$	$f(3) = 108$
Location of Graph	Below x-axis	Below x-axis	Above x-axis
Point on Graph	$(-1, -12)$	$(1, -4)$	$(3, 108)$

(f)

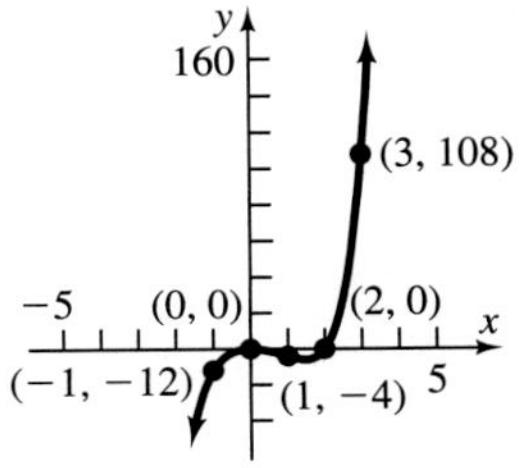

78. (a) x-intercepts: $-4, 0$; y-intercept: 0
(b) Crosses at -4; touches at 0
(c) $y = x^5$
(d) 4
(e)

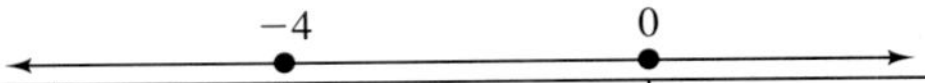

Interval	$(-\infty, -4)$	$(-4, 0)$	$(0, \infty)$
Number Chosen	-5	-2	1
Value of f	$f(-5) = -650$	$f(-2) = 40$	$f(1) = 10$
Location of Graph	Below x-axis	Above x-axis	Above x-axis
Point on Graph	$(-5, -650)$	$(-2, 40)$	$(1, 10)$

(f)

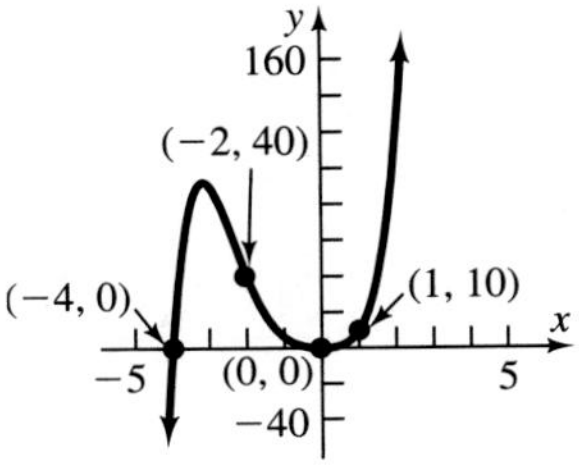

79. (a) x-intercepts: $-1, 0, 1$; y-intercept: 0
(b) Crosses at 1; touches at -1 and 0
(c) $y = -x^5$
(d) 4
(e)

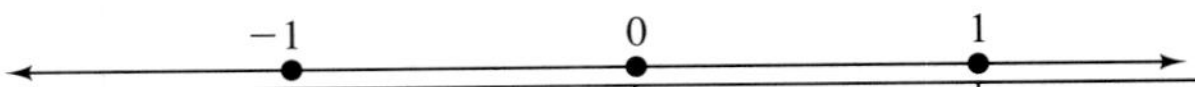

Interval	$(-\infty, -1)$	$(-1, 0)$	$(0, 1)$	$(1, \infty)$
Number Chosen	-2	$-\frac{1}{2}$	$\frac{1}{2}$	2
Value of f	$f(-2) = 12$	$f(-\frac{1}{2}) = \frac{3}{32}$	$f(\frac{1}{2}) = \frac{9}{32}$	$f(2) = -36$
Location of Graph	Above x-axis	Above x-axis	Above x-axis	Below x-axis
Point on Graph	$(-2, 12)$	$(-\frac{1}{2}, \frac{3}{32})$	$(\frac{1}{2}, \frac{9}{32})$	$(2, -36)$

(f)

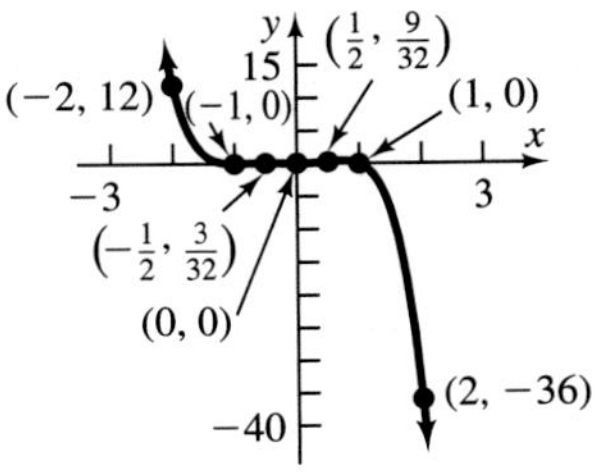

80. **(a)** x-intercepts: $-2, 0, 2, 5$; y-intercept: 0
(b) Touches at 0; crosses at -2, 2, and 5
(c) $y = -x^5$
(d) 4
(e)

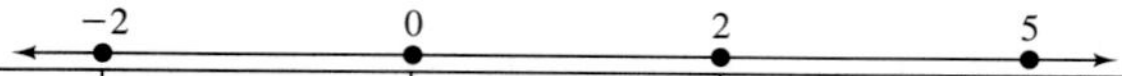

Interval	$(-\infty, -2)$	$(-2, 0)$	$(0, 2)$	$(2, 5)$	$(5, \infty)$
Number Chosen	-3	-1	1	3	6
Value of f	$f(-3) = 360$	$f(-1) = -18$	$f(1) = -12$	$f(3) = 90$	$f(6) = -1152$
Location of Graph	Above x-axis	Below x-axis	Below x-axis	Above x-axis	Below x-axis
Point on Graph	$(-3, 360)$	$(-1, -18)$	$(1, -12)$	$(3, 90)$	$(6, -1152)$

(f)

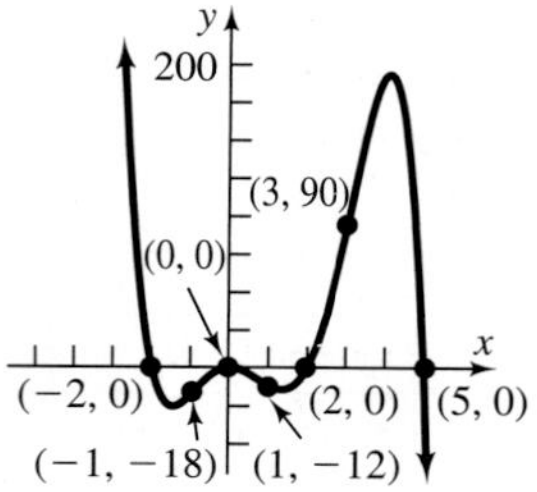

81. c, e, f **82.** c, e, f **83.** c, e **84.** d, f

85. **(a)** Degree 3; $y = x^3$
(b)

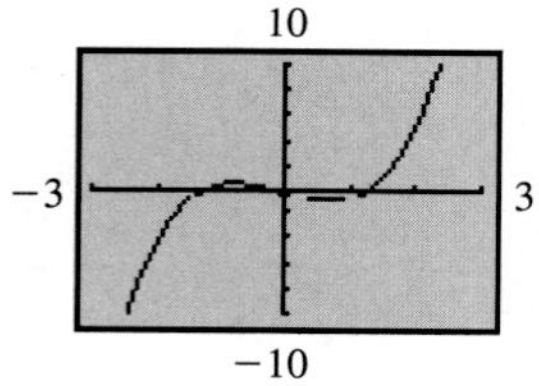

(c) x-intercepts: $-1.26, -0.20, 1.26$; y-intercept: -0.31752
(d) Above on $(-1.26, -0.20)$ and $(1.26, \infty)$; Below on $(-\infty, -1.26)$ and $(-0.20, 1.26)$

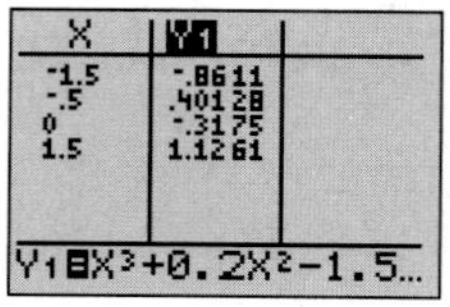

(e) Local maximum at $(-0.80, 0.57)$; local minimum at $(0.66, -0.99)$
(f)

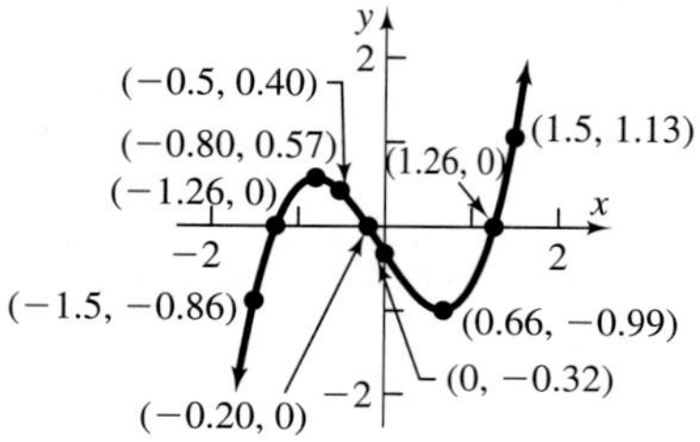

(g) Increasing on $(-\infty, -0.80)$ and $(0.66, \infty)$; decreasing on $(-0.80, 0.66)$

86. **(a)** Degree 3; $y = x^3$
(b)

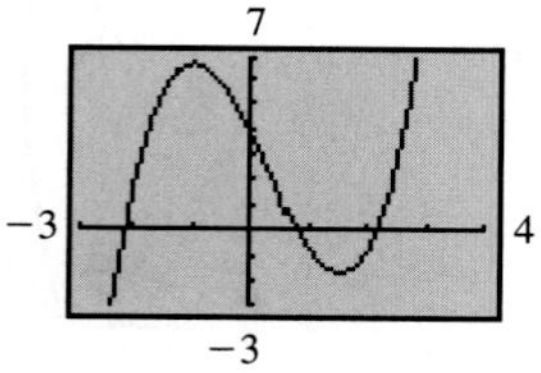

(c) x-intercepts: $-2.16, 0.8, 2.16$
y-intercept: 3.73248
(d)

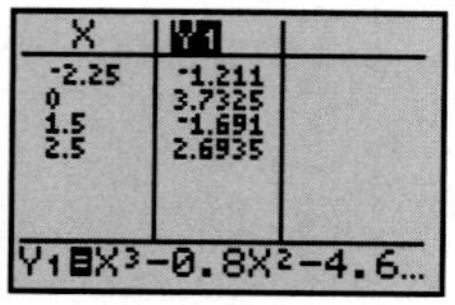

Above on $(-2.16, 0.8)$ and $(2.16, \infty)$
Below on $(-\infty, -2.16)$ and $(0.8, 2.16)$
(e) Local maximum at $(-1.01, 6.60)$
Local minimum at $(1.54, -1.70)$
(f)

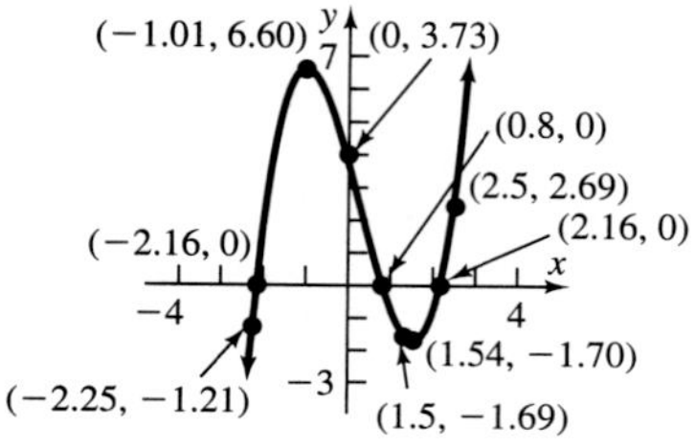

(g) Increasing on $(-\infty, -1.01)$ and $(1.54, \infty)$; decreasing on $(-1.01, 1.54)$

87. **(a)** Degree 3; $y = x^3$
(b)

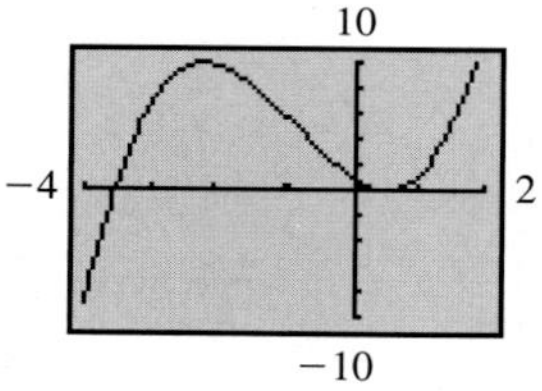

(c) x-intercepts: $-3.56, 0.50$
y-intercept: 0.89
(d)

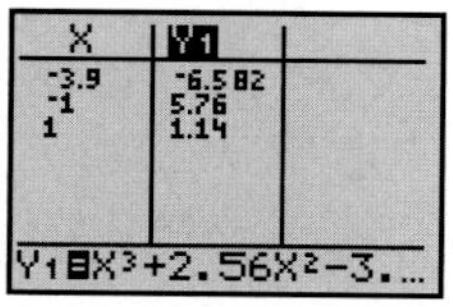

Above on $(-3.56, 0.5)$ and $(0.5, \infty)$
Below on $(-\infty, -3.56)$
(e) Local maximum at $(-2.21, 9.91)$
Local minimum at $(0.50, 0)$
(f)

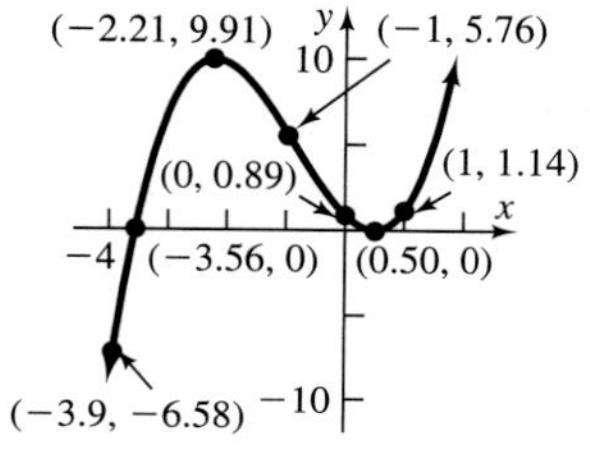

(g) Increasing on $(-\infty, -2.21)$ and $(0.50, \infty)$; decreasing on $(-2.21, 0.50)$

88. **(a)** Degree 3; $y = x^3$
(b)

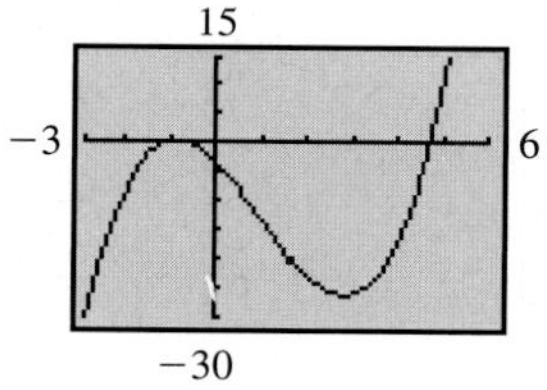

(c) x-intercepts: -0.9; 4.71
y-intercept: -3.8151
(d)

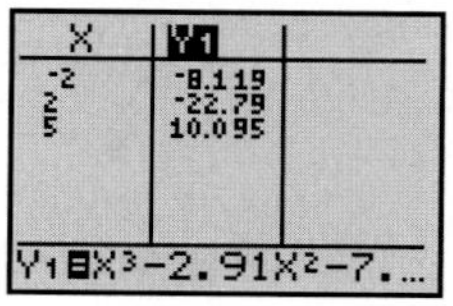

Above on $(4.71, \infty)$; below on $(-\infty, -0.9)$ and $(-0.9, 4.71)$
(e) Local maximum at $(-0.9, 0)$
Local minimum at $(2.84, -26.16)$
(f)

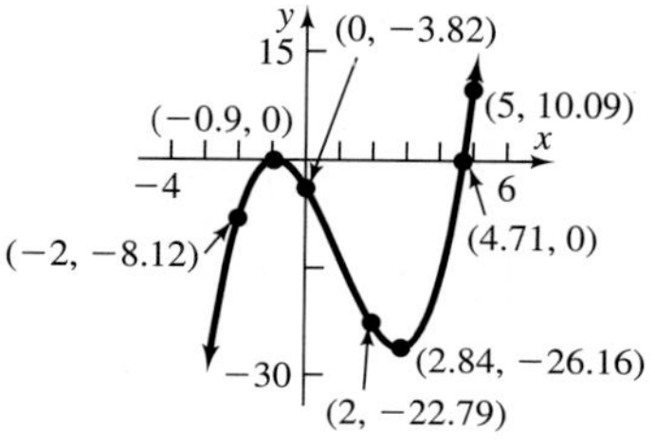

(g) Increasing on $(-\infty, -0.9)$ and $(2.84, \infty)$; decreasing on $(-0.9, 2.84)$

89. **(a)** Degree 4; $y = x^4$

(b)

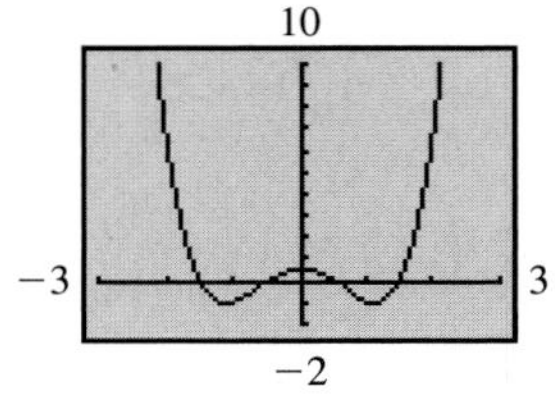

(c) x-intercepts: −1.5, −0.5, 0.5, 1.5
y-intercept: 0.5625

(d)

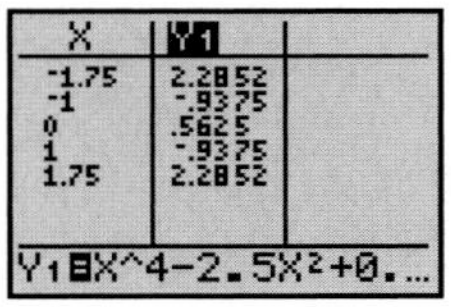

Above on $(-\infty, -1.5)$, $(-0.5, 0.5)$, and $(1.5, \infty)$; below on $(-1.5, -0.5)$ and $(0.5, 1.5)$

(e) Local minima at $(-1.12, -1)$, $(1.12, -1)$; local maximum at $(0, 0.56)$

(f)

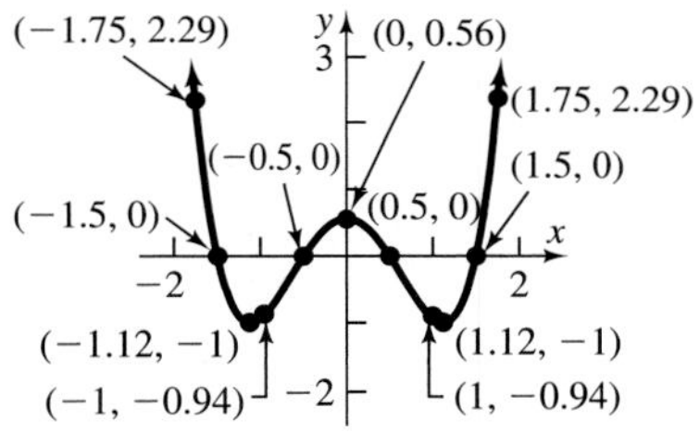

(g) Increasing on $(-1.12, 0)$ and $(1.12, \infty)$
Decreasing on $(-\infty, -1.12)$ and $(0, 1.12)$

90. **(a)** Degree 4; $y = x^4$

(b)

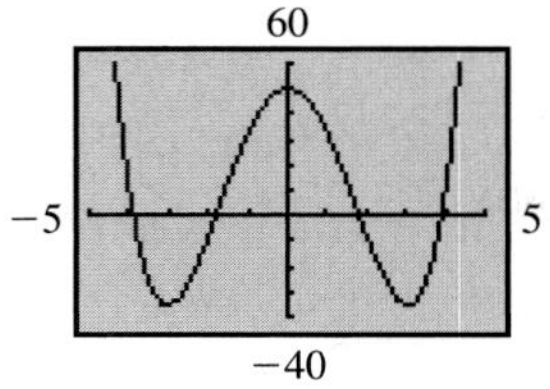

(c) x-intercepts: −3.90, −1.82, 1.82, 3.90
y-intercept: 50.2619

(d)

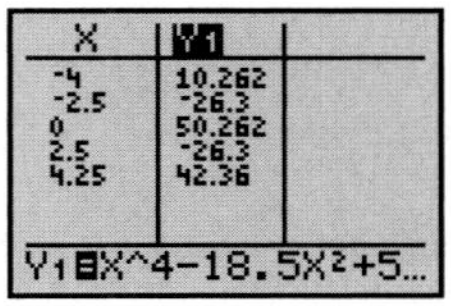

Above on $(-\infty, -3.90)$, $(-1.82, 1.82)$, and $(3.90, \infty)$; below on $(-3.90, -1.82)$ and $(1.82, 3.90)$

(e) Local minima at $(-3.04, -35.30)$, $(3.04, -35.30)$; local maximum at $(0, 50.26)$

(f)

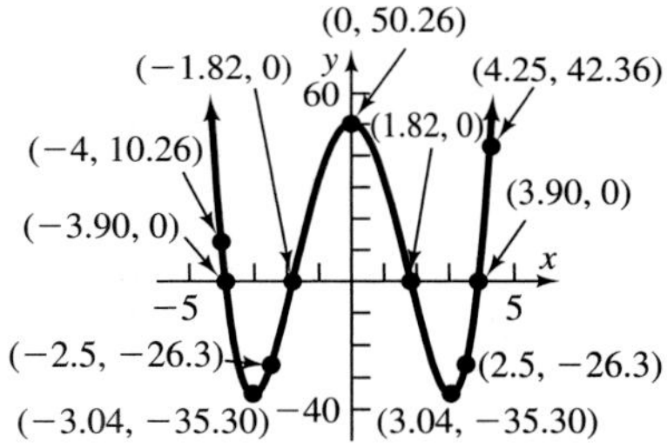

(g) Increasing on $(-3.04, 0)$ and $(3.04, \infty)$
Decreasing on $(-\infty, -3.04)$ and $(0, 3.04)$

91. **(a)** Degree 4; $y = 2x^4$

(b)

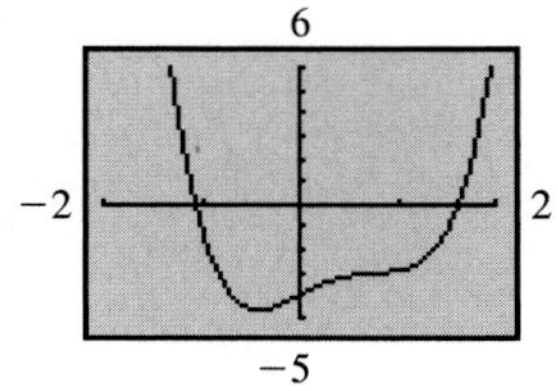

(c) x-intercepts: −1.07, 1.62
y-intercept: −4

(d)

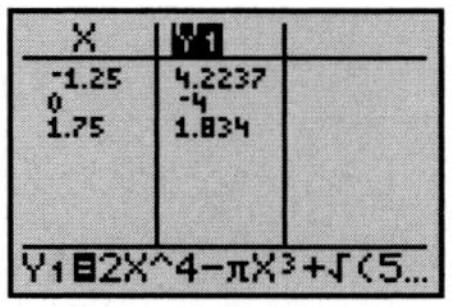

Above on $(-\infty, -1.07)$ and $(1.62, \infty)$; below on $(-1.07, 1.62)$

(e) Local minimum at $(-0.42, -4.64)$

(f)

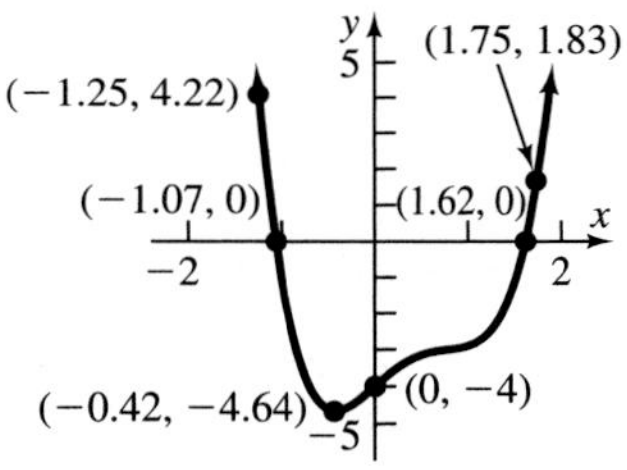

(g) Increasing on $(-0.42, \infty)$
Decreasing on $(-\infty, -0.42)$

92. **(a)** Degree 4; $y = -1.2x^4$ **(b)**

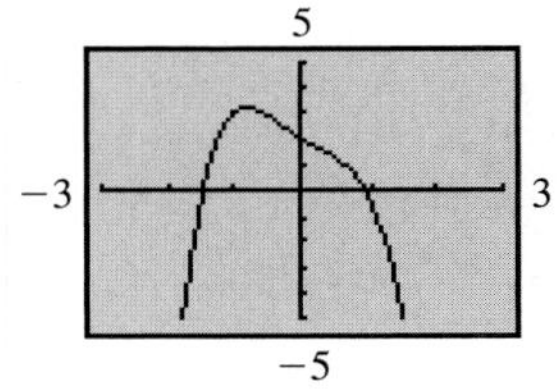

(c) x-intercepts: −1.47, 0.91; y-intercept: 2

(d)

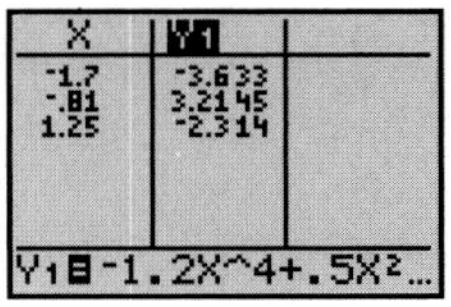

Above on $(-1.47, 0.91)$
Below on $(-\infty, -1.47)$ and $(0.91, \infty)$

(e) Local maximum at $(-0.81, 3.21)$

(f)

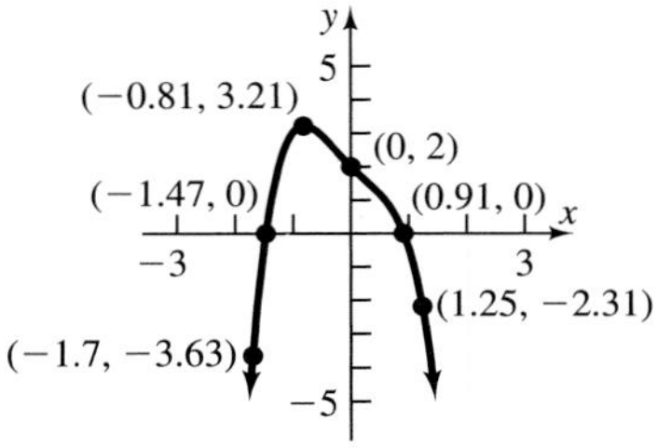

(g) Increasing on $(-\infty, -0.81)$; decreasing on $(-0.81, \infty)$

93. (a) Degree 5; $y = -2x^5$

(b)

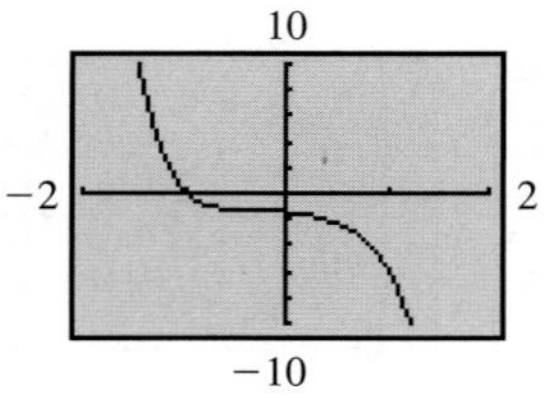

(c) x-intercept: -0.98; y-intercept: $-\sqrt{2}$

(d)

(e) None **(f)**

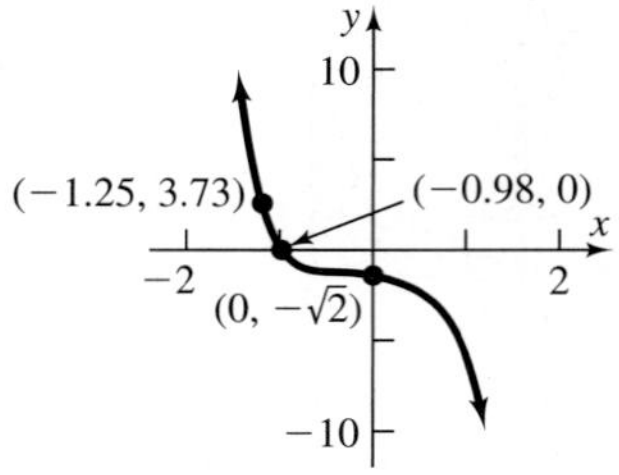

(g) Decreasing on $(-\infty, \infty)$

94. (a) Degree 5; $y = \pi x^5$

(b)

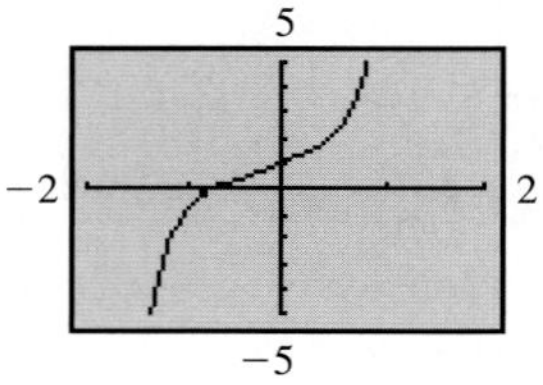

(c) x-intercept: -0.71; y-intercept: 1

(d)

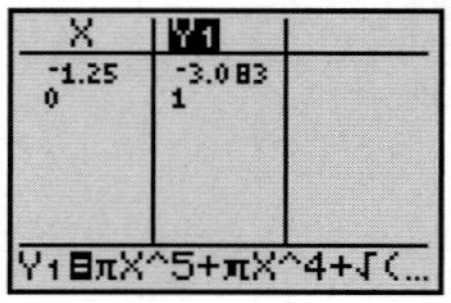

(e) None **(f)**

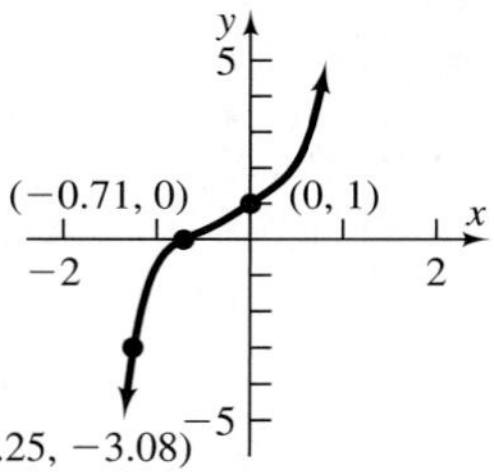

(g) Increasing on $(-\infty, \infty)$

Above on $(-0.71, \infty)$
Below on $(-\infty, -0.71)$
Above on $(-\infty, -0.98)$; below on $(0.98, \infty)$

95. (a) Cubic, $a > 0$

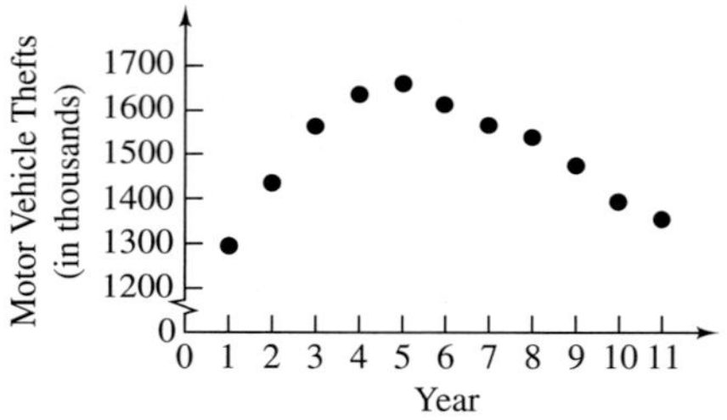

(b) $\approx$ 1,524,220 motor vehicle thefts

(d)

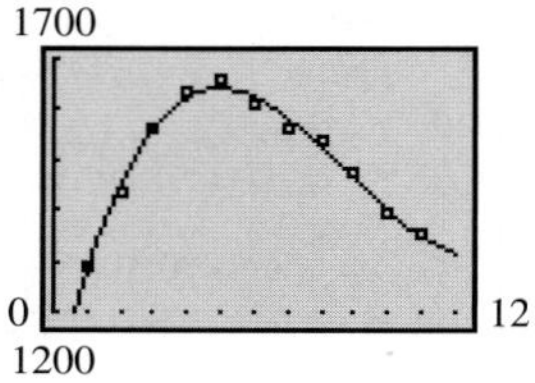

96. (a) Cubic, $a > 0$

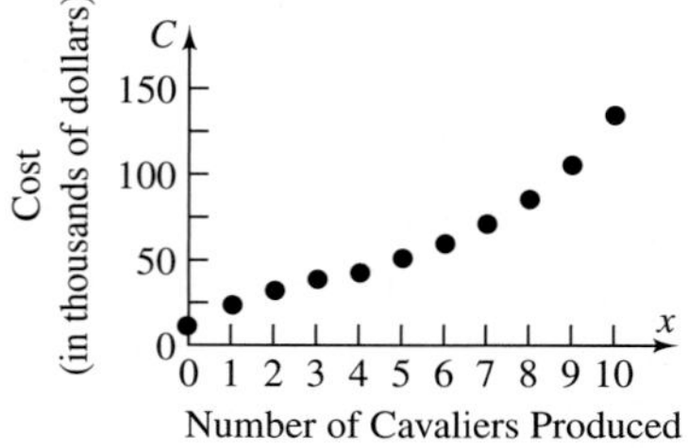

(b) 7 thousand dollars
(c) 20 thousand dollars
(d) $\approx$ \$155,000
(f)

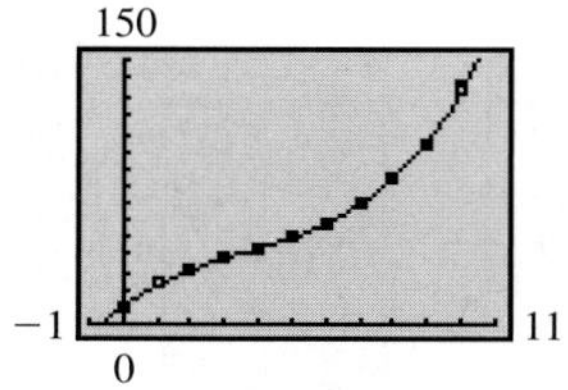

(g) Fixed costs of \$10,200

97. (a) Cubic, $a > 0$

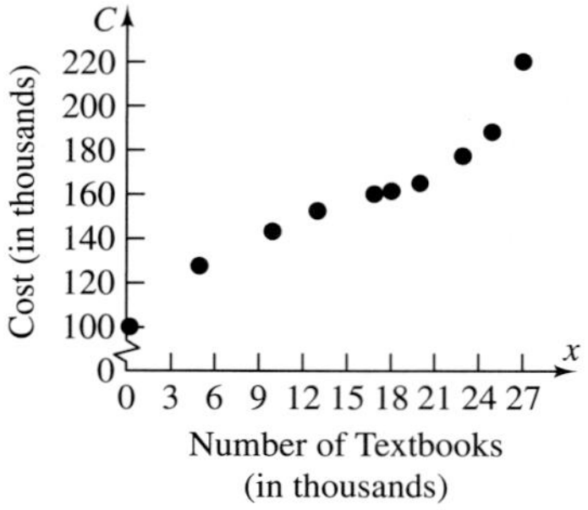

(b) Approximately 3.17 dollars/textbook
(c) 1.85 dollars/textbook
(d) \$171,470
(f)

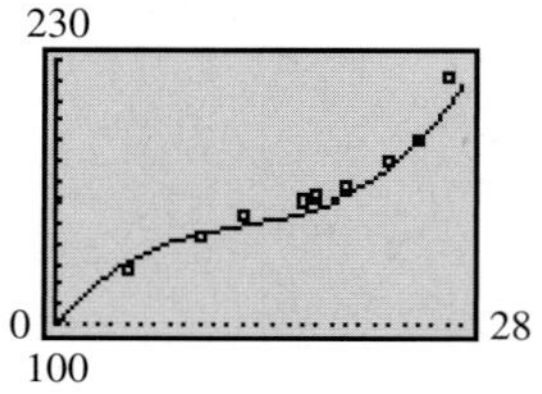

(g) Fixed costs of \$98,430

98. **(a)** Cubic, $a < 0$ **(b)** $S(x) = -74.4520x^3 + 1048.7554x^2 - 3565.7688x + 49{,}124.0714$ **99.** No; yes

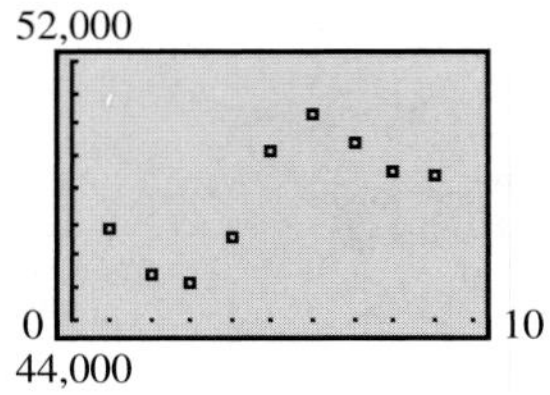

(c)

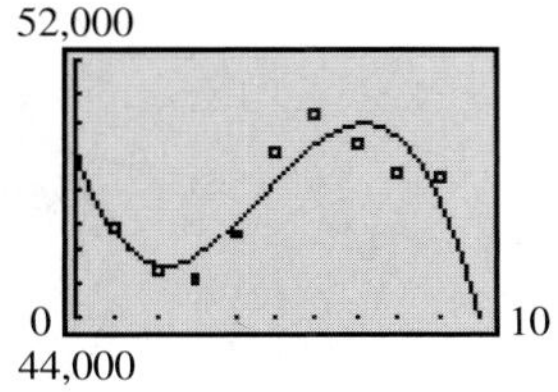

(d) About 43,890,000 sales

104. a, b, c, d

3.3 Concepts and Vocabulary *(page 171)*

5. $y = 1$ **6.** $x = -1$ **7.** proper **8.** False **9.** True **10.** True

3.3 Exercises *(page 171)*

11. All real numbers except 3; $\{x|x \neq 3\}$ **12.** All real numbers except -3; $\{x|x \neq -3\}$
13. All real numbers except 2 and -4; $\{x|x \neq 2, x \neq -4\}$ **14.** All real numbers except -3 and 4; $\{x|x \neq -3, x \neq 4\}$
15. All real numbers except $-\frac{1}{2}$ and 3; $\left\{x \middle| x \neq -\frac{1}{2}, x \neq 3\right\}$ **16.** All real numbers except $\frac{1}{3}$ and -2; $\left\{x \middle| x \neq \frac{1}{3}, x \neq -2\right\}$
17. All real numbers except 2; $\{x|x \neq 2\}$ **18.** All real numbers except -1 and 1; $\{x|x \neq -1, x \neq 1\}$ **19.** All real numbers
20. All real numbers **21.** All real numbers except -3 and 3; $\{x|x \neq -3, x \neq 3\}$ **22.** All real numbers except -2; $\{x|x \neq -2\}$
23. **(a)** Domain: $\{x|x \neq 2\}$; Range: $\{y|y \neq 1\}$ **(b)** $(0, 0)$ **(c)** $y = 1$ **(d)** $x = 2$ **(e)** None
24. **(a)** Domain: $\{x|x \neq -1\}$; Range: $\{y|y > 0\}$ **(b)** $(0, 2)$ **(c)** $y = 0$ **(d)** $x = -1$ **(e)** None
25. **(a)** Domain: $\{x|x \neq 0\}$; Range: all real numbers **(b)** $(-1, 0), (1, 0)$ **(c)** None **(d)** $x = 0$ **(e)** $y = 2x$
26. **(a)** Domain: $\{x|x \neq 0\}$; Range: $\{y|y \leq -2, y \geq 2\}$ **(b)** None **(c)** None **(d)** $x = 0$ **(e)** $y = -x$
27. **(a)** Domain: $\{x|x \neq -2, x \neq 2\}$; Range: $\{y|y \leq 0, y > 1\}$ **(b)** $(0, 0)$ **(c)** $y = 1$ **(d)** $x = -2, x = 2$ **(e)** None
28. **(a)** Domain: $\{x|x \neq -1, x \neq 1\}$; Range: All real numbers **(b)** $(0, 0)$ **(c)** $y = 0$ **(d)** $x = -1, x = 1$ **(e)** None

29.

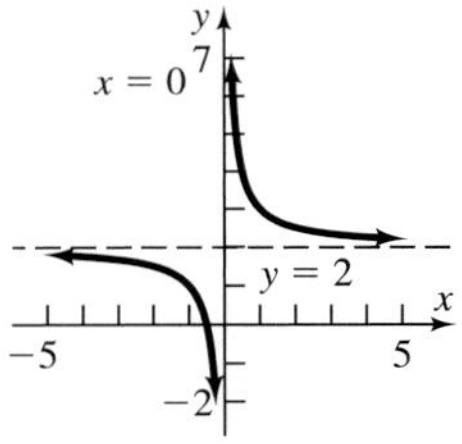

30.

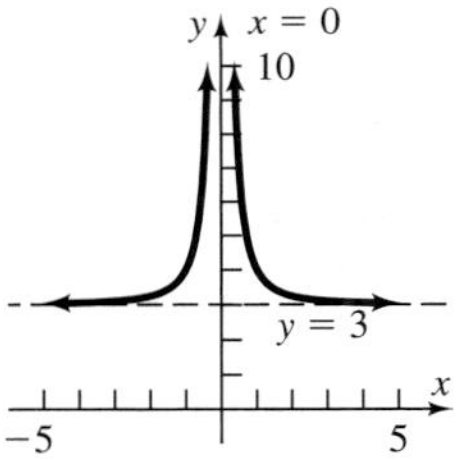

31.

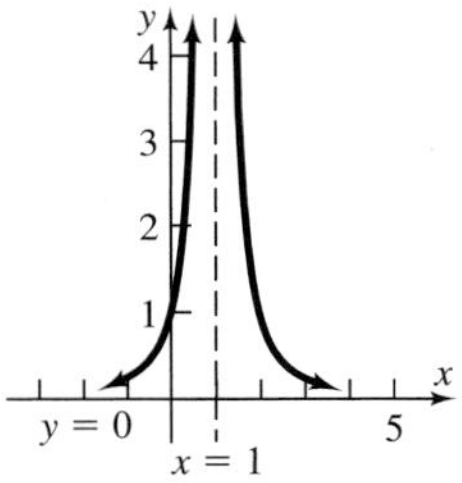

32.

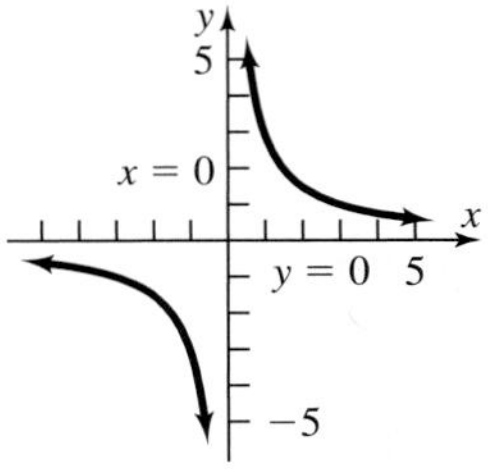

33.

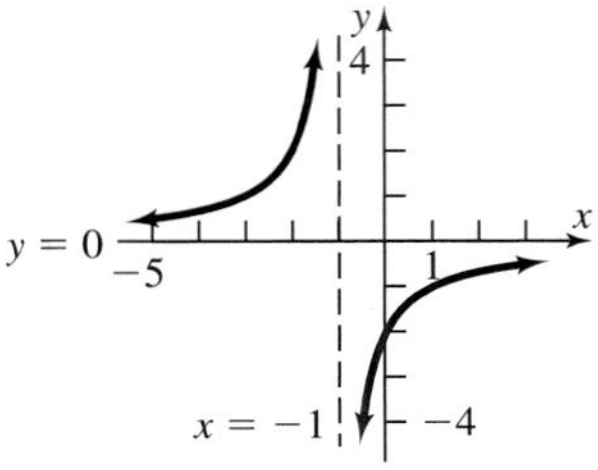

34.

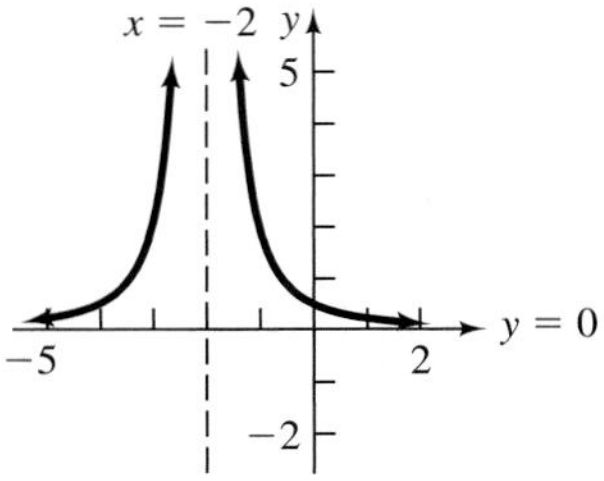

35.

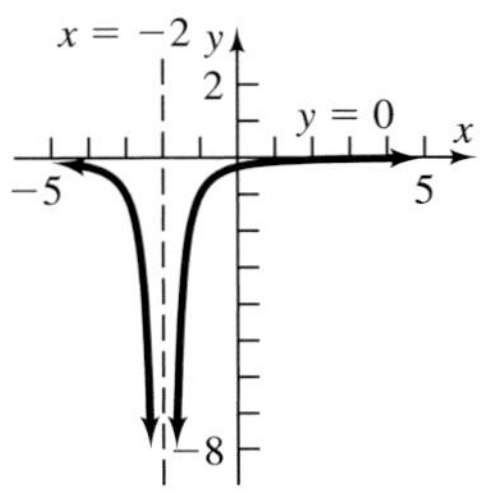

36.

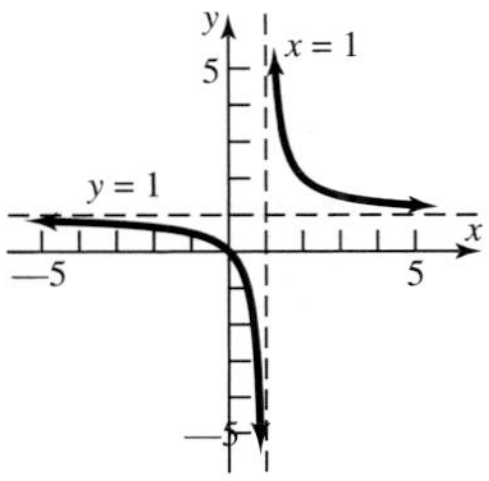

37.

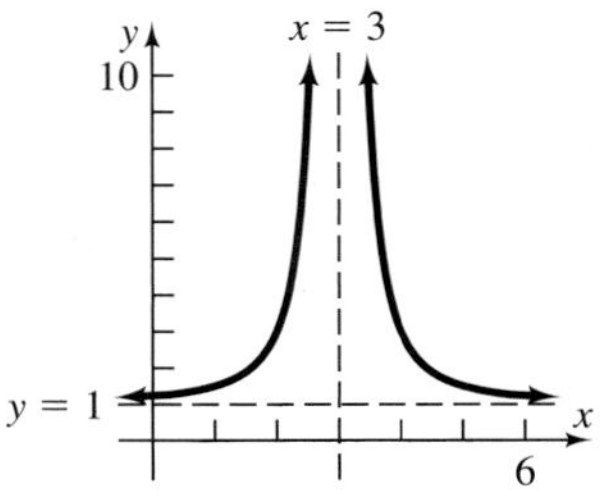

38.

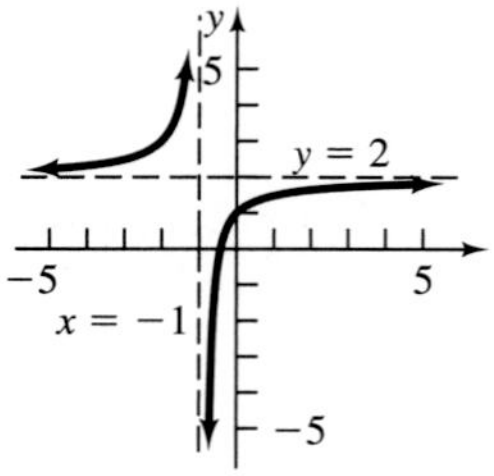

39.

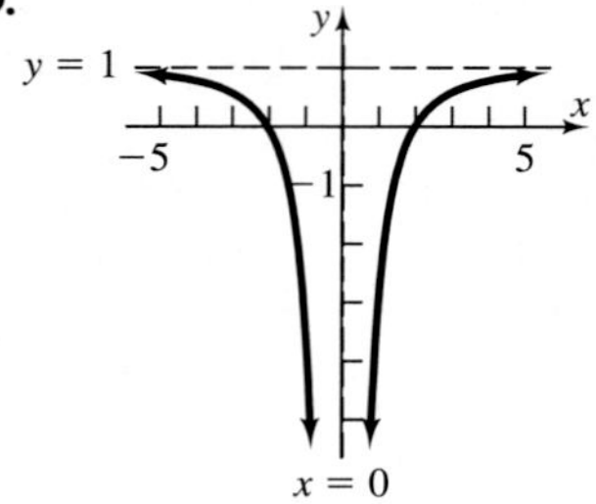

40.

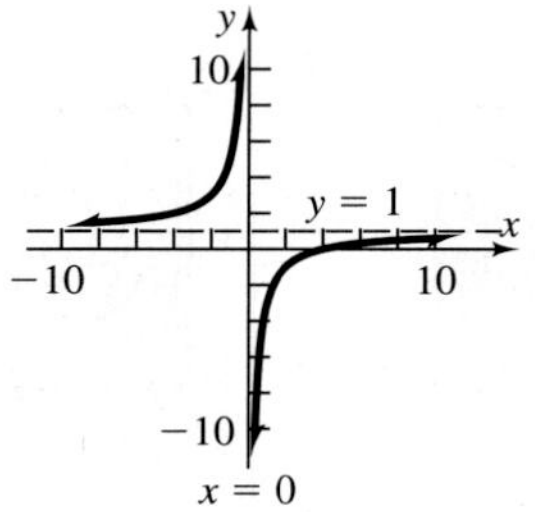

41. Horizontal asymptote: $y = 3$; vertical asymptote: $x = -4$ **42.** Horizontal asymptote: $y = 3$; vertical asymptote: $x = 6$
43. No asymptotes **44.** Vertical asymptote: $x = -5$; oblique asymptote: $y = -x + 5$
45. Horizontal asymptote: $y = 0$; vertical asymptotes: $x = 1$, $x = -1$ **46.** Vertical asymptote: $x = 1$
47. Horizontal asymptote: $y = 0$; vertical asymptote: $x = 0$ **48.** Horizontal asymptote: $y = 0$; vertical asymptotes: $x = 0$, $x = -2$
49. Oblique asymptote: $y = 3x$; vertical asymptote: $x = 0$ **50.** Horizontal asymptote: $y = 2$; vertical asymptotes: $x = -\frac{1}{3}$, $x = 2$
51. Oblique asymptote: $y = -(x + 1)$; vertical asymptote: $x = 0$ **52.** Horizontal asymptote: $y = 0$; vertical asymptotes $x = -1$, $x = 0$
53. **(a)** 9.8209 m/sec^2 **(b)** 9.8195 m/sec^2 **(c)** 9.7936 m/sec^2 **(d)** h-axis
54. **(a)** 25 **(b)** Approximately 596 **(c)** 2500

3.4 Concepts and Vocabulary *(page 185)*

3. in lowest terms **4.** False **5.** False **6.** True

3.4 Exercises *(page 185)*

7. 1. Domain: $\{x \mid x \neq 0, x \neq -4\}$
2. x-intercept: -1; no y-intercept
3. No y-axis or origin symmetry
4. Vertical asymptotes: $x = 0$, $x = -4$
5. Horizontal asymptote: $y = 0$, intersected at $(-1, 0)$
6.

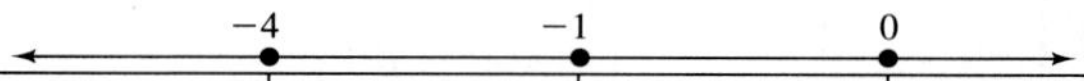

Interval	$(-\infty, -4)$	$(-4, -1)$	$(-1, 0)$	$(0, \infty)$
Number Chosen	-5	-2	$-\frac{1}{2}$	1
Value of R	$R(-5) = -\frac{4}{5}$	$R(-2) = \frac{1}{4}$	$R(-\frac{1}{2}) = -\frac{2}{7}$	$R(1) = \frac{2}{5}$
Location of Graph	Below x-axis	Above x-axis	Below x-axis	Above x-axis
Point on Graph	$(-5, -\frac{4}{5})$	$(-2, \frac{1}{4})$	$(-\frac{1}{2}, -\frac{2}{7})$	$(1, \frac{2}{5})$

7.

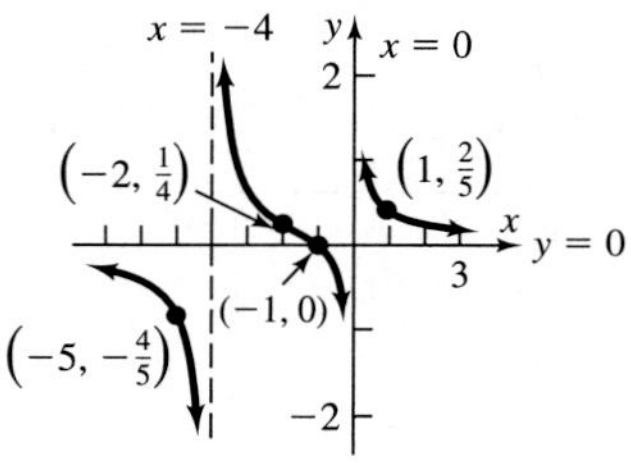

8. 1. Domain $\{x \mid x \neq -2, x \neq 1\}$
2. x-intercept: 0; y-intercept: 0
3. No y-axis or origin symmetry
4. Vertical asymptotes: $x = 1$, $x = -2$
5. Horizontal asymptotes: $y = 0$, intersected at $(0, 0)$
6.

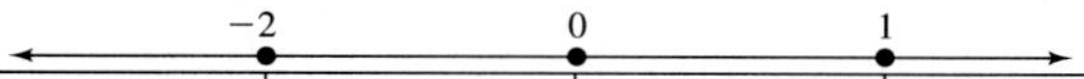

Interval	$(-\infty, -2)$	$(-2, 0)$	$(0, 1)$	$(1, \infty)$
Number Chosen	-3	-1	$\frac{1}{2}$	2
Value of R	$R(-3) = -\frac{3}{4}$	$R(-1) = \frac{1}{2}$	$R(\frac{1}{2}) = -\frac{2}{5}$	$R(2) = \frac{1}{2}$
Location of Graph	Below x-axis	Above x-axis	Below x-axis	Above x-axis
Point on Graph	$(-3, -\frac{3}{4})$	$(-1, \frac{1}{2})$	$(\frac{1}{2}, -\frac{2}{5})$	$(2, \frac{1}{2})$

7.

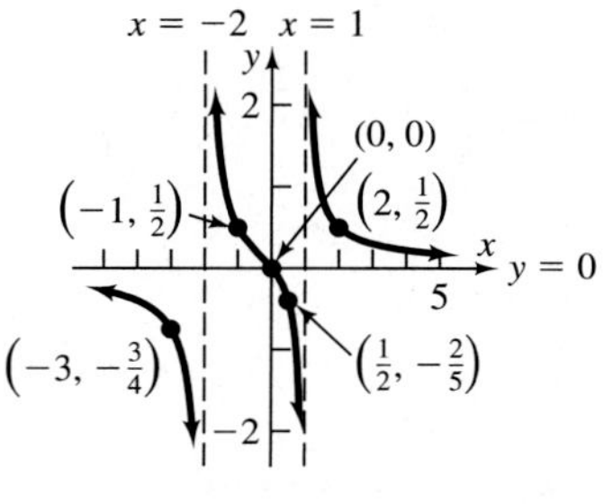

9. 1. Domain: $\{x|x \neq -2\}$

2. x-intercept: -1; y-intercept: $\frac{3}{4}$
3. No y-axis or origin symmetry
4. Vertical asymptote: $x = -2$
5. Horizontal asymptote: $y = \frac{3}{2}$, not intersected
6.

Interval	$(-\infty, -2)$	$(-2, -1)$	$(-1, \infty)$
Number Chosen	-3	$-\frac{3}{2}$	0
Value of R	$R(-3) = 3$	$R\left(-\frac{3}{2}\right) = -\frac{3}{2}$	$R(0) = \frac{3}{4}$
Location of Graph	Above x-axis	Below x-axis	Above x-axis
Point on Graph	$(-3, 3)$	$\left(-\frac{3}{2}, -\frac{3}{2}\right)$	$\left(0, \frac{3}{4}\right)$

7. 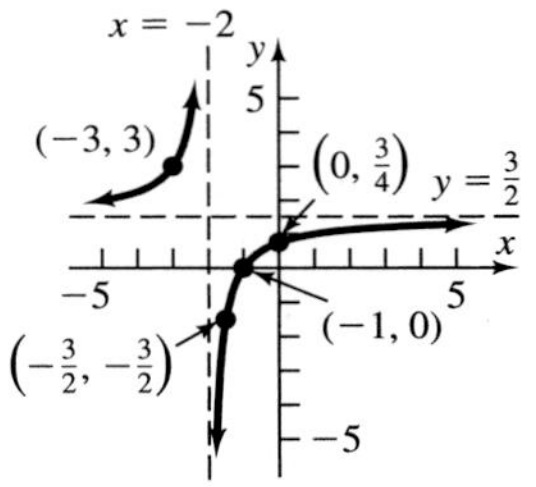

10. 1. Domain $\{x|x \neq 1\}$

2. x-intercept: -2; y-intercept: -4
3. No y-axis or origin symmetry
4. Vertical asymptotes: $x = 1$
5. Horizontal asymptotes: $y = 2$, not intersected
6.

Interval	$(-\infty, -2)$	$(-2, 1)$	$(1, \infty)$
Number Chosen	-3	0	2
Value of R	$R(-3) = \frac{1}{2}$	$R(0) = -4$	$R(2) = 8$
Location of Graph	Above x-axis	Below x-axis	Above x-axis
Point on Graph	$\left(-3, \frac{1}{2}\right)$	$(0, -4)$	$(2, 8)$

7. 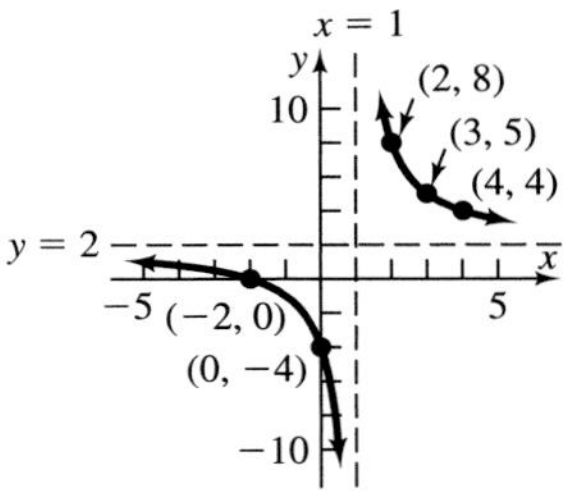

11. 1. Domain: $\{x|x \neq -2, x \neq 2\}$

2. No x-intercept; y-intercept: $-\frac{3}{4}$
3. Symmetric with respect to y-axis
4. Vertical asymptotes: $x = 2$, $x = -2$
5. Horizontal asymptote: $y = 0$, not intersected
6.

Interval	$(-\infty, -2)$	$(-2, 2)$	$(2, \infty)$
Number Chosen	-3	0	3
Value of R	$R(-3) = \frac{3}{5}$	$R(0) = -\frac{3}{4}$	$R(3) = \frac{3}{5}$
Location of Graph	Above x-axis	Below x-axis	Above x-axis
Point on Graph	$\left(-3, \frac{3}{5}\right)$	$\left(0, -\frac{3}{4}\right)$	$\left(3, \frac{3}{5}\right)$

7. 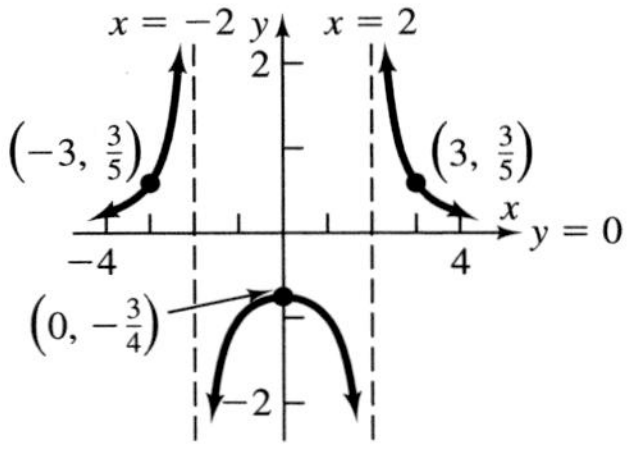

12. 1. Domain $\{x|x \neq -2, x \neq 3\}$

2. No x-intercept; y-intercept: -1
3. No y-axis or origin symmetry
4. Vertical asymptotes: $x = -2$, $x = 3$
5. Horizontal asymptotes: $y = 0$, not intersected
6.

Interval	$(-\infty, -2)$	$(-2, 3)$	$(3, \infty)$
Number Chosen	-3	0	4
Value of R	$R(-3) = 1$	$R(0) = -1$	$R(4) = 1$
Location of Graph	Above x-axis	Below x-axis	Above x-axis
Point on Graph	$(-3, 1)$	$(0, -1)$	$(4, 1)$

7. 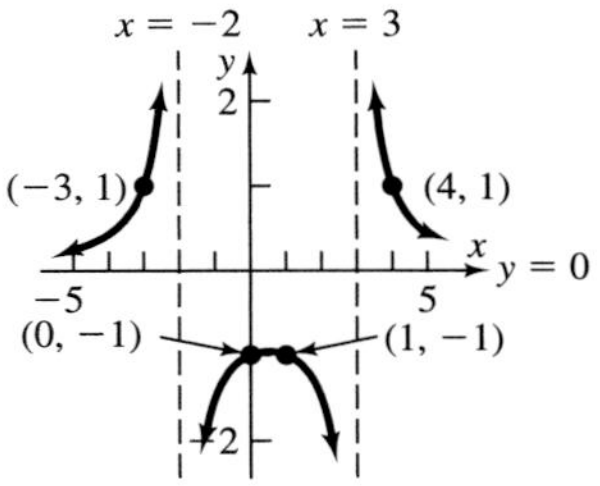

13. 1. Domain: $\{x|x \neq -1, x \neq 1\}$
2. No x-intercept; y-intercept: -1
3. Symmetric with respect to y-axis
4. Vertical asymptotes: $x = -1$, $x = 1$
5. No horizontal or oblique asymptotes
6. Number line: -1, 1

Interval	$(-\infty, -1)$	$(-1, 1)$	$(1, \infty)$
Number Chosen	-2	0	2
Value of P	$P(-2) = 7$	$P(0) = -1$	$P(2) = 7$
Location of Graph	Above x-axis	Below x-axis	Above x-axis
Point on Graph	$(-2, 7)$	$(0, -1)$	$(2, 7)$

7.
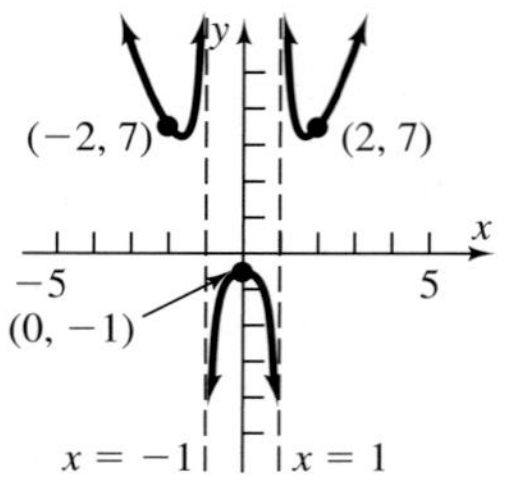

14. 1. Domain $\{x|x \neq -2, x \neq 2\}$
2. x-intercepts: -1, 1; y-intercept: $\frac{1}{4}$
3. Symmetric with respect to y-axis
4. Vertical asymptotes: $x = -2$, $x = 2$
5. No horizontal or oblique asymptotes
6. Number line: -2, -1, 1, 2

Interval	$(-\infty, -2)$	$(-2, -1)$	$(-1, 1)$	$(1, 2)$	$(2, \infty)$
Number Chosen	-3	$-\frac{3}{2}$	0	$\frac{3}{2}$	3
Value of Q	$Q(-3) = 16$	$Q\left(-\frac{3}{2}\right) \approx -2.3$	$Q(0) = \frac{1}{4}$	$Q\left(\frac{3}{2}\right) \approx -2.3$	$Q(3) = 16$
Location of Graph	Above x-axis	Below x-axis	Above x-axis	Below x-axis	Above x-axis
Point on Graph	$(-3, 16)$	$\left(-\frac{3}{2}, -2.3\right)$	$\left(0, \frac{1}{4}\right)$	$\left(\frac{3}{2}, -2.3\right)$	$(3, 16)$

7.
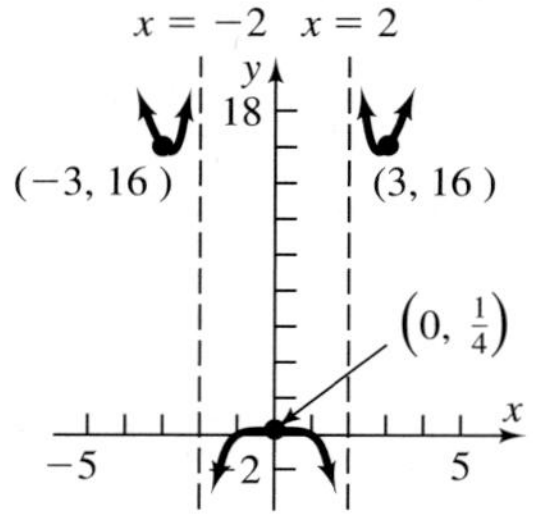

15. 1. Domain: $\{x|x \neq -3, x \neq 3\}$
2. x-intercept: 1; y-intercept: $\frac{1}{9}$
3. No y-axis or origin symmetry
4. Vertical asymptotes: $x = 3$, $x = -3$
5. Oblique asymptote: $y = x$, intersected at $\left(\frac{1}{9}, \frac{1}{9}\right)$
6. Number line: -3, 1, 3

Interval	$(-\infty, -3)$	$(-3, 1)$	$(1, 3)$	$(3, \infty)$
Number Chosen	-4	0	2	4
Value of H	$H(-4) \approx -9.3$	$H(0) = \frac{1}{9}$	$H(2) = -1.4$	$H(4) = 9$
Location of Graph	Below x-axis	Above x-axis	Below x-axis	Above x-axis
Point on Graph	$(-4, -9.3)$	$\left(0, \frac{1}{9}\right)$	$(2, -1.4)$	$(4, 9)$

7.
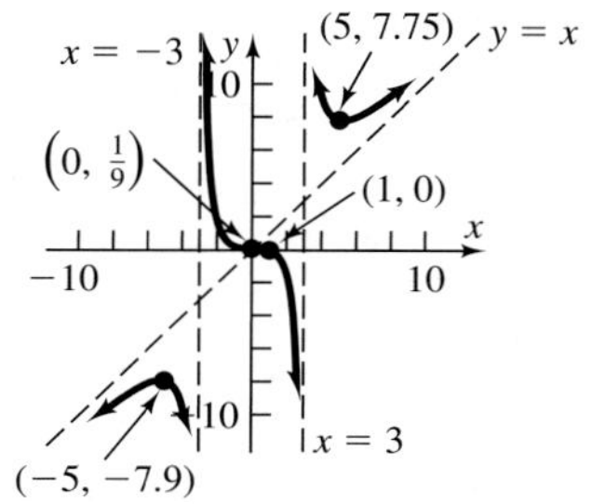

16. 1. Domain $\{x|x \neq -2, x \neq 0\}$
2. x-intercept: -1; no y-intercept
3. No y-axis or origin symmetry
4. Vertical asymptotes: $x = 0$, $x = -2$
5. Oblique asymptote: $y = x - 2$, intersected at $(-0.25, -2.25)$
6. Number line: -2, -1, 0

Interval	$(-\infty, -2)$	$(-2, -1)$	$(-1, 0)$	$(0, \infty)$
Number Chosen	-4	-1.5	-0.25	1
Value of G	$G(-4) = -7.875$	$G(-1.5) \approx 3.2$	$G(-0.25) = -2.25$	$G(1) = \frac{2}{3}$
Location of Graph	Below x-axis	Above x-axis	Below x-axis	Above x-axis
Point on Graph	$(-4, -7.875)$	$(-1.5, 3.2)$	$(-0.25, -2.25)$	$\left(1, \frac{2}{3}\right)$

7.
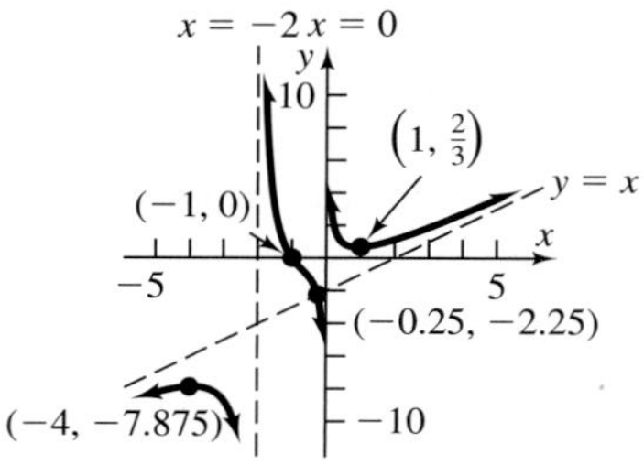

17. 1. Domain: $\{x \neq -3, x \neq 2\}$
2. x-intercept: 0; y-intercept: 0
3. No y-axis or origin symmetry
4. Vertical asymptotes: $x = 2$, $x = -3$
5. Horizontal asymptote: $y = 1$, intersected at $(6, 1)$
6. (Number line: −3, 0, 2)

Interval	$(-\infty, -3)$	$(-3, 0)$	$(0, 2)$	$(2, \infty)$
Number Chosen	-6	-1	1	3
Value of R	$R(-6) = 1.5$	$R(-1) = -\frac{1}{6}$	$R(1) = -0.25$	$R(3) = 1.5$
Location of Graph	Above x-axis	Below x-axis	Below x-axis	Above x-axis
Point on Graph	$(-6, 1.5)$	$\left(-1, -\frac{1}{6}\right)$	$(1, -0.25)$	$(3, 1.5)$

7.
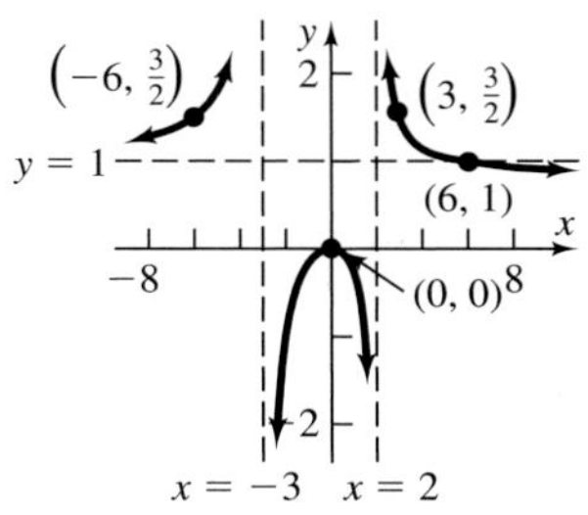

18. 1. Domain $\{x \mid x \neq -2, x \neq 2\}$
2. x-intercepts: 3, −4; y-intercept: 3
3. No y-axis or origin symmetry
4. Vertical asymptotes: $x = -2$, $x = 2$
5. Horizontal asymptote: $y = 1$, intersected at $(8, 1)$
6. (Number line: −4, −2, 2, 3)

Interval	$(-\infty, -4)$	$(-4, -2)$	$(-2, 2)$	$(2, 3)$	$(3, \infty)$
Number Chosen	-7	-3	0	2.5	8
Value of R	$R(-7) = \frac{2}{3}$	$R(-3) = -1.2$	$R(0) = 3$	$R(2.5) \approx -1.44$	$R(8) = 1$
Location of Graph	Above x-axis	Below x-axis	Above x-axis	Below x-axis	Above x-axis
Point on Graph	$\left(-7, \frac{2}{3}\right)$	$(-3, -1.2)$	$(0, 3)$	$(2.5, -1.44)$	$(8, 1)$

7.
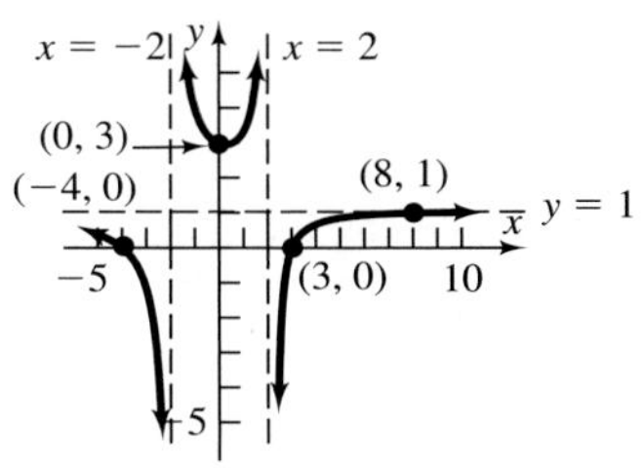

19. 1. Domain: $\{x \mid x \neq -2, x \neq 2\}$
2. x-intercept: 0; y-intercept: 0
3. Symmetry with respect to origin
4. Vertical asymptotes: $x = -2$, $x = 2$
5. Horizontal asymptote: $y = 0$, intersected at $(0, 0)$
6. (Number line: −2, 0, 2)

Interval	$(-\infty, -2)$	$(-2, 0)$	$(0, 2)$	$(2, \infty)$
Number Chosen	-3	-1	1	3
Value of G	$G(-3) = -\frac{3}{5}$	$G(-1) = \frac{1}{3}$	$G(1) = -\frac{1}{3}$	$G(3) = \frac{3}{5}$
Location of Graph	Below x-axis	Above x-axis	Below x-axis	Above x-axis
Point on Graph	$\left(-3, -\frac{3}{5}\right)$	$\left(-1, \frac{1}{3}\right)$	$\left(1, -\frac{1}{3}\right)$	$\left(3, \frac{3}{5}\right)$

7.
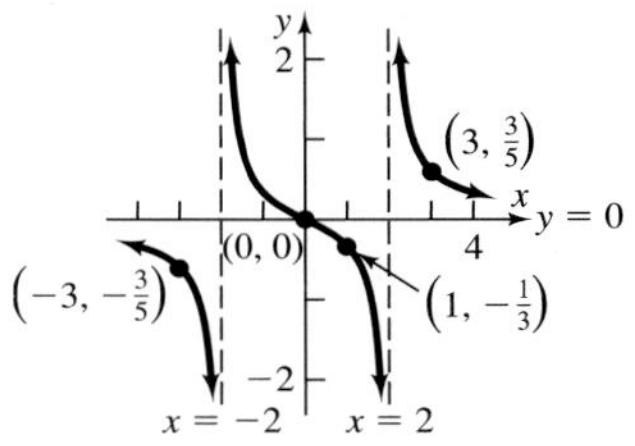

20. 1. Domain $\{x \mid x \neq -1, x \neq 1\}$
2. x-intercept: 0; y-intercept: 0
3. Symmetric with respect to origin
4. Vertical asymptotes: $x = -1$, $x = 1$
5. Horizontal asymptote: $y = 0$, intersected at $(0, 0)$
6. (Number line: −1, 0, 1)

Interval	$(-\infty, -1)$	$(-1, 0)$	$(0, 1)$	$(1, \infty)$
Number Chosen	-2	$-\frac{1}{2}$	$\frac{1}{2}$	2
Value of G	$G(-2) = -2$	$G\left(-\frac{1}{2}\right) = 2$	$G\left(\frac{1}{2}\right) = -2$	$G(2) = 2$
Location of Graph	Below x-axis	Above x-axis	Below x-axis	Above x-axis
Point on Graph	$(-2, -2)$	$\left(-\frac{1}{2}, 2\right)$	$\left(\frac{1}{2}, -2\right)$	$(2, 2)$

7.
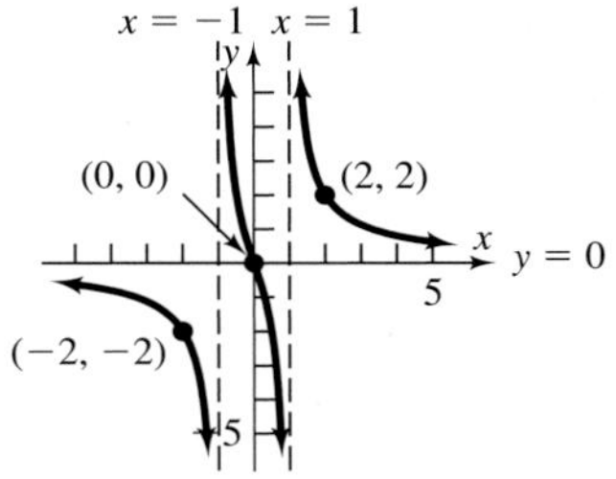

21. 1. Domain: $\{x|x \neq 1, x \neq -2, x \neq 2\}$

2. No x-intercept; y-intercept: $\frac{3}{4}$
3. No y-axis or origin symmetry
4. Vertical asymptotes: $x = -2$, $x = 1$, $x = 2$
5. Horizontal asymptote: $y = 0$, not intersected
6.

Interval	$(-\infty, -2)$	$(-2, 1)$	$(1, 2)$	$(2, \infty)$
Number Chosen	-3	0	1.5	3
Value of R	$R(-3) = -\frac{3}{20}$	$R(0) = \frac{3}{4}$	$R(1.5) = -\frac{24}{7}$	$R(3) = \frac{3}{10}$
Location of Graph	Below x-axis	Above x-axis	Below x-axis	Above x-axis
Point on Graph	$\left(-3, -\frac{3}{20}\right)$	$\left(0, \frac{3}{4}\right)$	$\left(1.5, -\frac{24}{7}\right)$	$\left(3, \frac{3}{10}\right)$

7.

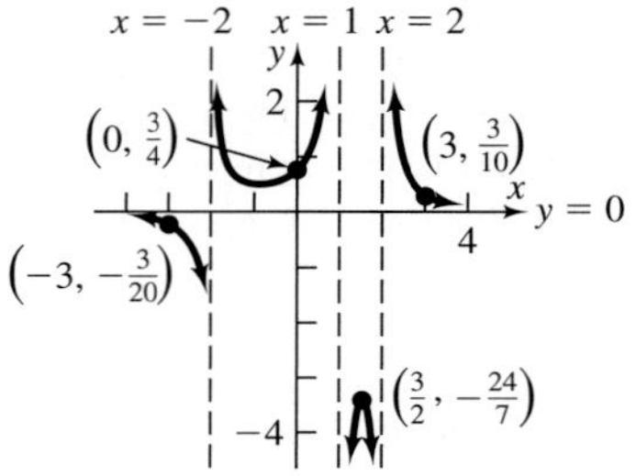

22. 1. Domain $\{x|x \neq -1, x \neq -3, x \neq 3\}$

2. No x-intercept; y-intercept: $\frac{4}{9}$
3. No y-axis or origin symmetry
4. Vertical asymptotes: $x = -3$, $x = -1$, $x = 3$
5. Horizontal asymptote: $y = 0$, not intersected
6.

Interval	$(-\infty, -3)$	$(-3, -1)$	$(-1, 3)$	$(3, \infty)$
Number Chosen	-4	-2	0	4
Value of R	$R(-4) \approx 0.19$	$R(-2) = -0.8$	$R(0) = \frac{4}{9}$	$R(4) \approx -0.11$
Location of Graph	Above x-axis	Below x-axis	Above x-axis	Below x-axis
Point on Graph	$(-4, 0.19)$	$(-2, -0.8)$	$\left(0, \frac{4}{9}\right)$	$(4, -0.11)$

7.

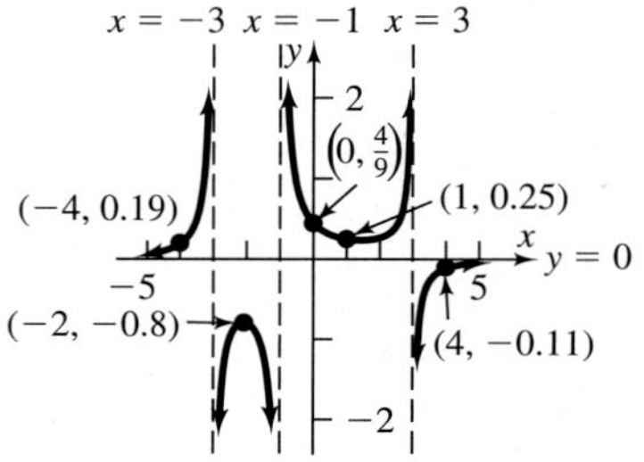

23. 1. Domain: $\{x|x \neq -2, x \neq 2\}$

2. x-intercepts: -1, 1; y-intercept: $\frac{1}{4}$
3. Symmetry with respect to y-axis
4. Vertical asymptotes: $x = -2$, $x = 2$
5. Horizontal asymptote: $y = 0$, intersected at $(-1, 0)$ and $(1, 0)$
6.

Interval	$(-\infty, -2)$	$(-2, -1)$	$(-1, 1)$	$(1, 2)$	$(2, \infty)$
Number Chosen	-3	-1.5	0	1.5	3
Value of H	$H(-3) \approx 0.49$	$H(-1.5) \approx -0.46$	$H(0) = \frac{1}{4}$	$H(1.5) \approx -0.46$	$H(3) \approx 0.49$
Location of Graph	Above x-axis	Below x-axis	Above x-axis	Below x-axis	Above x-axis
Point on Graph	$(-3, 0.49)$	$(-1.5, -0.46)$	$\left(0, \frac{1}{4}\right)$	$(1.5, -0.46)$	$(3, 0.49)$

7.

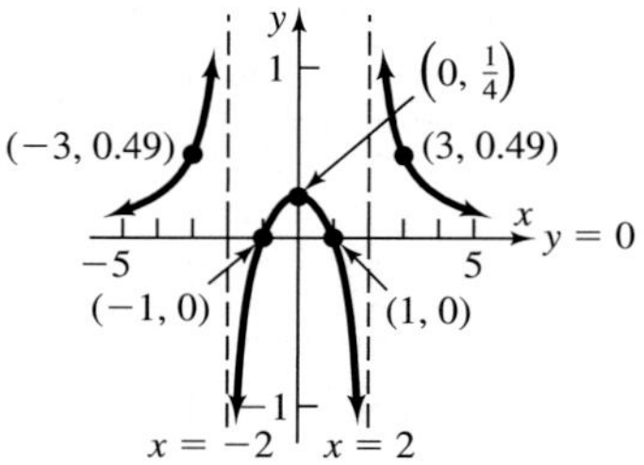

24. 1. Domain $\{x|x \neq -1, x \neq 1\}$

2. No x-intercept; y-intercept: -4
3. Symmetric with respect to y-axis
4. Vertical asymptotes: $x = -1$, $x = 1$
5. Horizontal asymptote: $y = 0$, not intersected
6.

Interval	$(-\infty, -1)$	$(-1, 1)$	$(1, \infty)$
Number Chosen	-2	0	2
Value of H	$H(-2) = \frac{8}{15}$	$H(0) = -4$	$H(2) = \frac{8}{15}$
Location of Graph	Above x-axis	Below x-axis	Above x-axis
Point on Graph	$\left(-2, \frac{8}{15}\right)$	$(0, -4)$	$\left(2, \frac{8}{15}\right)$

7.

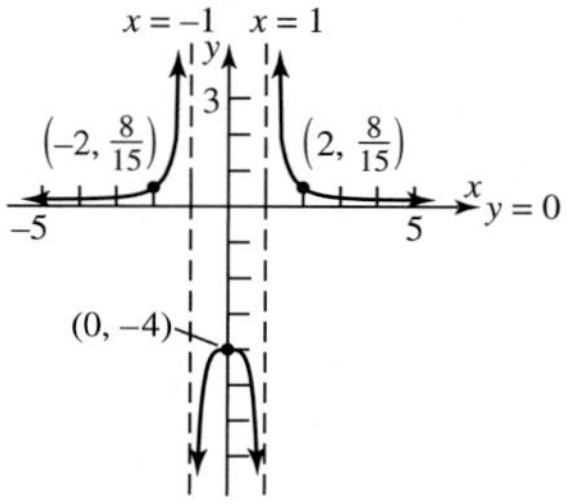

25. 1. Domain: $\{x|x \neq -2\}$
2. x-intercepts: -1, 4; y-intercept: -2
3. No y-axis or origin symmetry
4. Vertical asymptote: $x = -2$
5. Oblique asymptote: $y = x - 5$, not intersected
6. Number line points: -2, -1, 4

Interval	$(-\infty, -2)$	$(-2, -1)$	$(-1, 4)$	$(4, \infty)$
Number Chosen	-3	-1.5	0	5
Value of F	$F(-3) = -14$	$F(-1.5) = 5.5$	$F(0) = -2$	$F(5) \approx 0.86$
Location of Graph	Below x-axis	Above x-axis	Below x-axis	Above x-axis
Point on Graph	$(-3, -14)$	$(-1.5, 5.5)$	$(0, -2)$	$(5, 0.86)$

7.
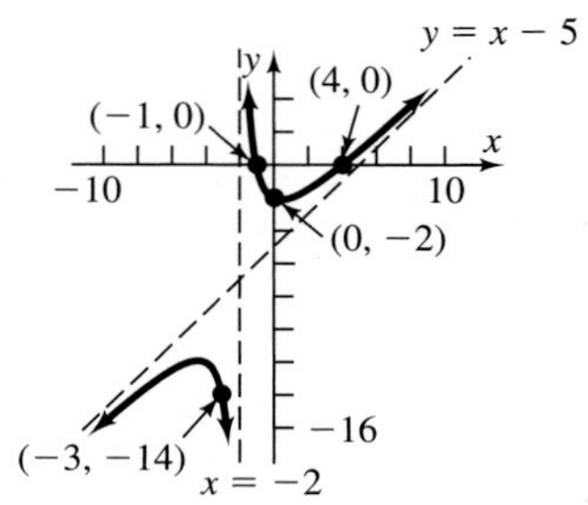

26. 1. Domain $\{x|x \neq 1\}$
2. x-intercepts: $-1, -2$; y-intercept: -2
3. No y-axis or origin symmetry
4. Vertical asymptote: $x = 1$
5. Oblique asymptote: $y = x + 4$, not intersected
6. Number line points: -2, -1, 1

Interval	$(-\infty, -2)$	$(-2, -1)$	$(-1, 1)$	$(1, \infty)$
Number Chosen	-3	-1.5	0	2
Value of F	$F(-3) = -0.5$	$F(-1.5) = 0.1$	$F(0) = -2$	$F(2) = 12$
Location of Graph	Below x-axis	Above x-axis	Below x-axis	Above x-axis
Point on Graph	$(-3, -0.5)$	$(-1.5, 0.1)$	$(0, -2)$	$(2, 12)$

7.
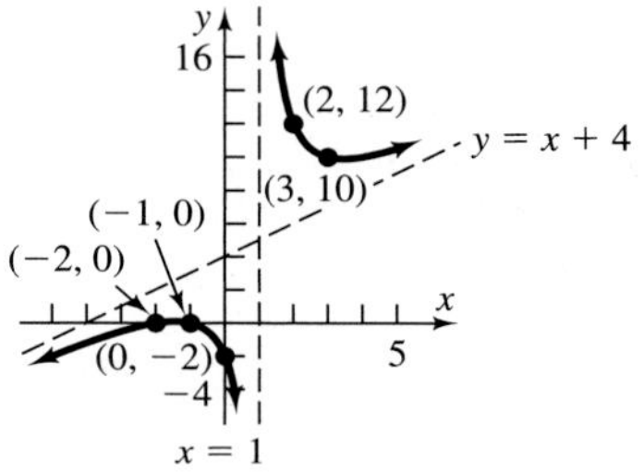

27. 1. Domain: $\{x|x \neq 4\}$
2. x-intercepts: -4, 3; y-intercept: 3
3. No y-axis or origin symmetry
4. Vertical asymptote: $x = 4$
5. Oblique asymptote: $y = x + 5$, not intersected
6. Number line points: -4, 3, 4

Interval	$(-\infty, -4)$	$(-4, 3)$	$(3, 4)$	$(4, \infty)$
Number Chosen	-5	0	3.5	5
Value of R	$R(-5) = -\frac{8}{9}$	$R(0) = 3$	$R(3.5) = -7.5$	$R(5) = 18$
Location of Graph	Below x-axis	Above x-axis	Below x-axis	Above x-axis
Point on Graph	$\left(-5, -\frac{8}{9}\right)$	$(0, 3)$	$(3.5, -7.5)$	$(5, 18)$

7.
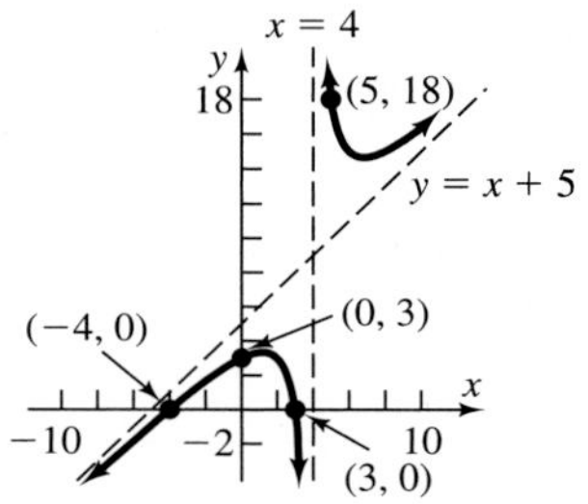

28. 1. Domain $\{x|x \neq -5\}$
2. x-intercepts: -3, 4; y-intercept: $-\frac{12}{5}$
3. No y-axis or origin symmetry
4. Vertical asymptote: $x = -5$
5. Oblique asymptote: $y = x - 6$, not intersected
6. Number line points: -5, -3, 4

Interval	$(-\infty, -5)$	$(-5, -3)$	$(-3, 4)$	$(4, \infty)$
Number Chosen	-7	-4	0	5
Value of R	$R(-7) = -22$	$R(-4) = 8$	$R(0) = -2.4$	$R(5) = 0.8$
Location of Graph	Below x-axis	Above x-axis	Below x-axis	Above x-axis
Point on Graph	$(-7, -22)$	$(-4, 8)$	$(0, -2.4)$	$(5, 0.8)$

7.
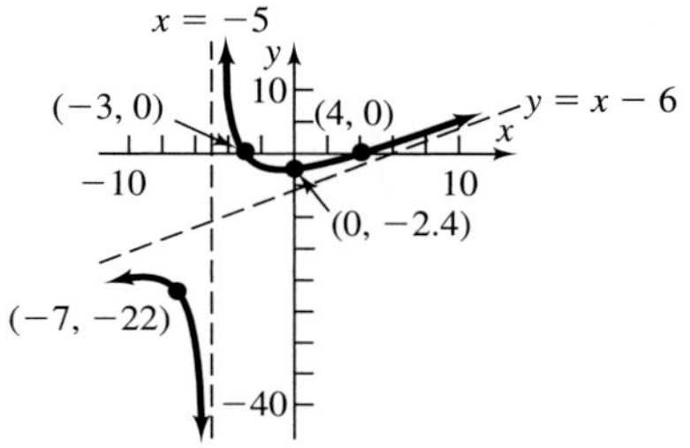

29. 1. Domain: $\{x \mid x \neq -2\}$
2. x-intercepts: -4, 3; y-intercept: -6
3. No y-axis or origin symmetry
4. Vertical asymptote: $x = -2$
5. Oblique asymptote: $y = x - 1$, not intersected
6. Number line points: -4, -2, 3

Interval	$(-\infty, -4)$	$(-4, -2)$	$(-2, 3)$	$(3, \infty)$
Number Chosen	-5	-3	0	4
Value of F	$F(-5) = -\frac{8}{3}$	$F(-3) = 6$	$F(0) = -6$	$F(4) = \frac{4}{3}$
Location of Graph	Below x-axis	Above x-axis	Below x-axis	Above x-axis
Point on Graph	$\left(-5, -\frac{8}{3}\right)$	$(-3, 6)$	$(0, -6)$	$\left(4, \frac{4}{3}\right)$

7.

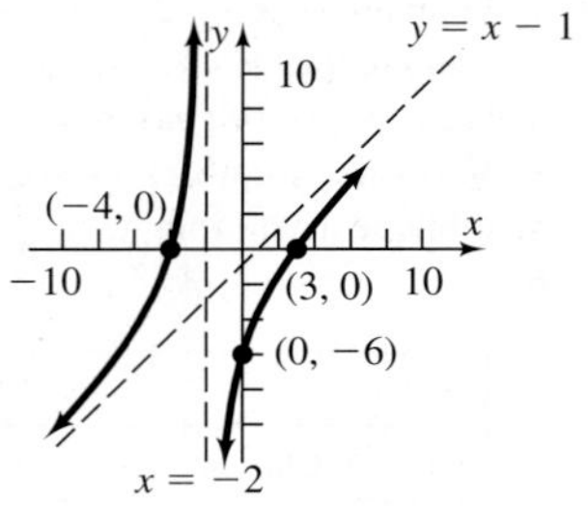

30. 1. Domain $\{x \mid x \neq -1\}$
2. x-intercepts: -3, 4; y-intercept: -12
3. No y-axis or origin symmetry
4. Vertical asymptote: $x = -1$
5. Oblique asymptote: $y = x - 2$, not intersected
6. Number line points: -3, -1, 4

Interval	$(-\infty, -3)$	$(-3, -1)$	$(-1, 4)$	$(4, \infty)$
Number Chosen	-4	-2	0	5
Value of G	$G(-4) = -\frac{8}{3}$	$G(-2) = 6$	$G(0) = -12$	$G(5) = \frac{4}{3}$
Location of Graph	Below x-axis	Above x-axis	Below x-axis	Above x-axis
Point on Graph	$\left(-4, -\frac{8}{3}\right)$	$(-2, 6)$	$(0, -12)$	$\left(5, \frac{4}{3}\right)$

7.

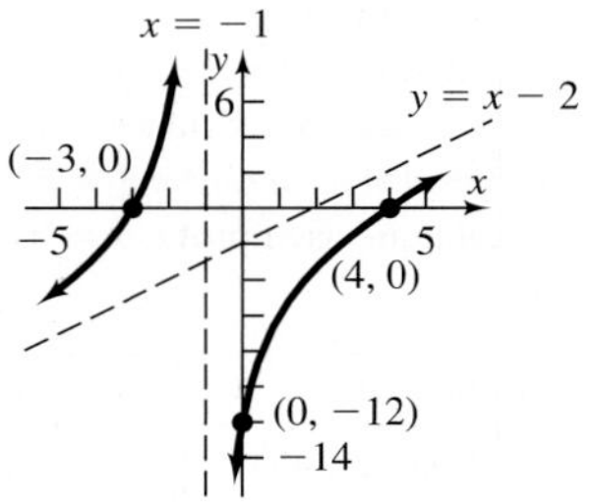

31. 1. Domain: $\{x \mid x \neq -3\}$
2. x-intercepts: 0, 1; y-intercept: 0
3. No y-axis or origin symmetry
4. Vertical asymptote: $x = -3$
5. Horizontal asymptote: $y = 1$, not intersected
6. Number line points: -3, 0, 1

Interval	$(-\infty, -3)$	$(-3, 0)$	$(0, 1)$	$(1, \infty)$
Number Chosen	-4	-1	$\frac{1}{2}$	2
Value of R	$R(-4) = 100$	$R(-1) = -0.5$	$R\left(\frac{1}{2}\right) \approx 0.003$	$R(2) = 0.016$
Location of Graph	Above x-axis	Below x-axis	Above x-axis	Above x-axis
Point on Graph	$(-4, 100)$	$(-1, -0.5)$	$\left(\frac{1}{2}, 0.003\right)$	$(2, 0.016)$

7.

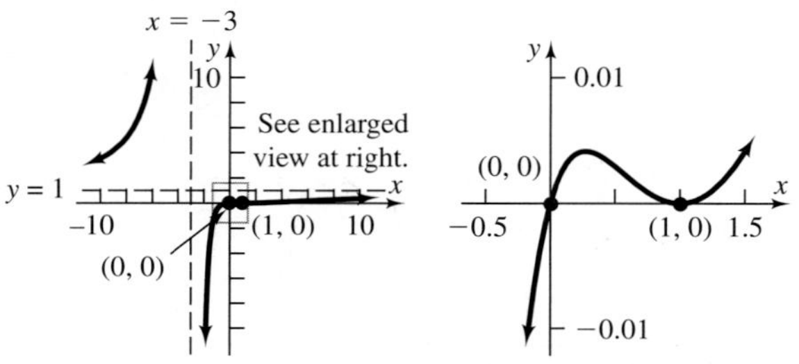

Enlarged view

32. 1. Domain $\{x | x \neq 0, x \neq 4\}$
2. x-intercepts: 1, −2, 3; no y-intercept
3. No y-axis or origin symmetry
4. Vertical asymptotes: $x = 0$, $x = 4$
5. Horizontal asymptote: $y = 1$, intersected at (3.186, 1), (0.314, 1)
6.

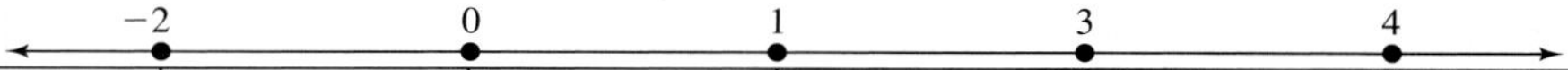

Interval	$(-\infty, -2)$	$(-2, 0)$	$(0, 1)$	$(1, 3)$	$(3, 4)$	$(4, \infty)$
Number Chosen	−3	−1	$\frac{1}{2}$	2	3.186	6
Value of R	$R(-3) \approx 0.16$	$R(-1) = -0.32$	$R\left(\frac{1}{2}\right) \approx 0.51$	$R(2) = -0.5$	$R(3.186) \approx 1$	$R(6) = 5$
Location of Graph	Above x-axis	Below x-axis	Above x-axis	Below x-axis	Above x-axis	Above x-axis
Point on Graph	$(-3, 0.16)$	$(-1, -0.32)$	$\left(\frac{1}{2}, 0.51\right)$	$(2, -0.5)$	$(3.186, 1)$	$(6, 5)$

7.

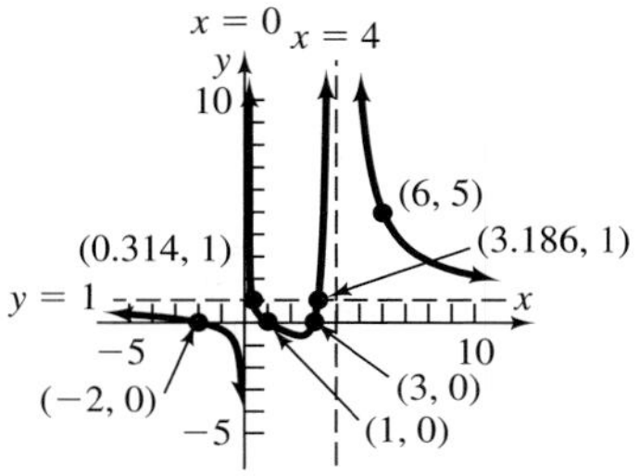

33. 1. Domain: $\{x | x \neq -2, x \neq 3\}$
2. x-intercept: −4; y-intercept: 2
3. No y-axis or origin symmetry
4. Vertical asymptote: $x = -2$; hole at $\left(3, \frac{7}{5}\right)$
5. Horizontal asymptote: $y = 1$, not intersected
6.

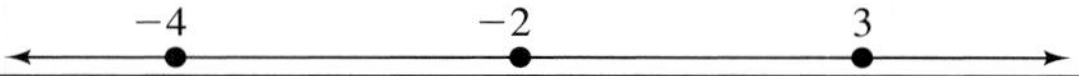

Interval	$(-\infty, -4)$	$(-4, -2)$	$(-2, 3)$	$(3, \infty)$
Number Chosen	−5	−3	0	4
Value of R	$R(-5) = \frac{1}{3}$	$R(-3) = -1$	$R(0) = 2$	$R(4) = \frac{4}{3}$
Location of Graph	Above x-axis	Below x-axis	Above x-axis	Above x-axis
Point on Graph	$\left(-5, \frac{1}{3}\right)$	$(-3, -1)$	$(0, 2)$	$\left(4, \frac{4}{3}\right)$

7.

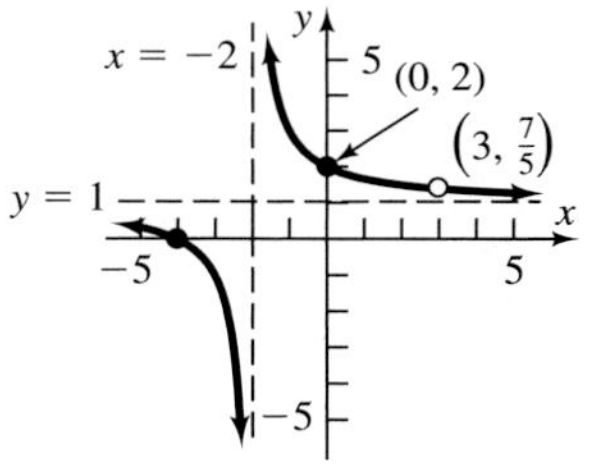

34. 1. Domain $\{x | x \neq -3, x \neq -5\}$
2. x-intercept: 2; y-intercept: $-\frac{2}{3}$
3. No y-axis or origin symmetry
4. Vertical asymptote: $x = -3$; hole at (−5, 3.5)
5. Horizontal asymptote: $y = 1$, not intersected
6.

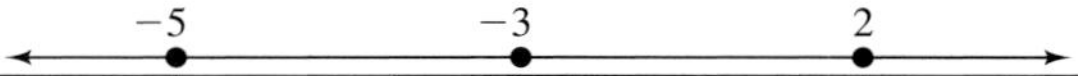

Interval	$(-\infty, -5)$	$(-5, -3)$	$(-3, 2)$	$(2, \infty)$
Number Chosen	−6	−4	0	3
Value of R	$R(-6) = \frac{8}{3}$	$R(-4) = 6$	$R(0) = -\frac{2}{3}$	$R(3) = \frac{1}{6}$
Location of Graph	Above x-axis	Above x-axis	Below x-axis	Above x-axis
Point on Graph	$\left(-6, \frac{8}{3}\right)$	$(-4, 6)$	$\left(0, -\frac{2}{3}\right)$	$\left(3, \frac{1}{6}\right)$

7.

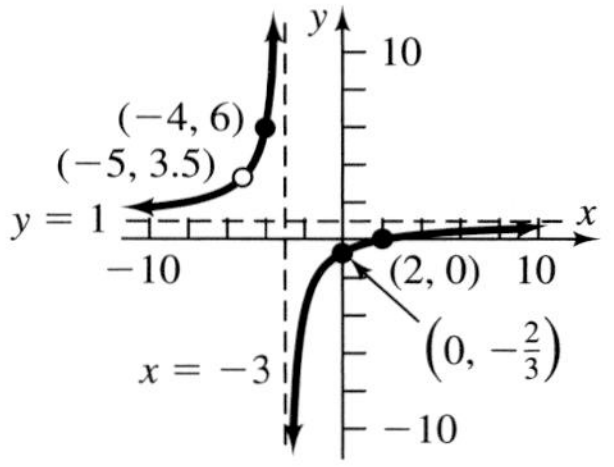

35. 1. Domain: $\left\{x \middle| x \neq \dfrac{3}{2}, x \neq 2\right\}$

2. x-intercept: $-\dfrac{1}{3}$; y-intercept: $-\dfrac{1}{2}$
3. No y-axis or origin symmetry
4. Vertical asymptote: $x = 2$; hole at $\left(\dfrac{3}{2}, -11\right)$
5. Horizontal asymptote: $y = 3$, not intersected
6.

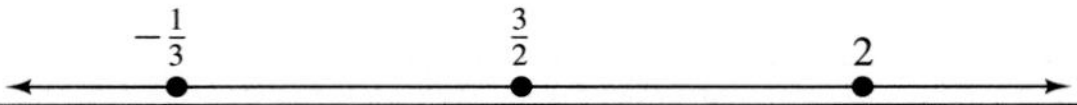

Interval	$\left(-\infty, -\frac{1}{3}\right)$	$\left(-\frac{1}{3}, \frac{3}{2}\right)$	$\left(\frac{3}{2}, 2\right)$	$(2, \infty)$
Number Chosen	-1	0	1.7	6
Value of R	$R(-1) = \frac{2}{3}$	$R(0) = -\frac{1}{2}$	$R(1.7) \approx -20.3$	$R(6) = 4.75$
Location of Graph	Above x-axis	Below x-axis	Below x-axis	Above x-axis
Point on Graph	$\left(-1, \frac{2}{3}\right)$	$\left(0, -\frac{1}{2}\right)$	$(1.7, -20.3)$	$(6, 4.75)$

7.

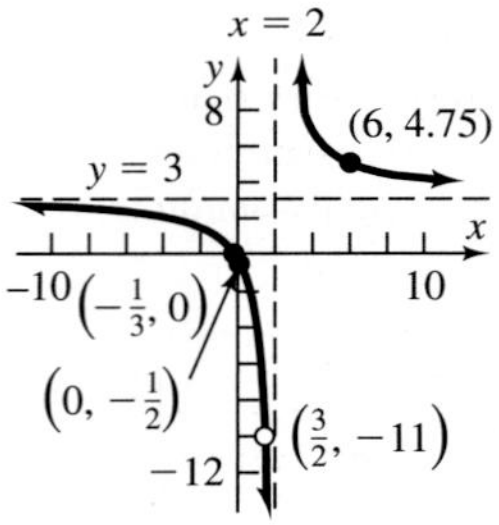

36. 1. Domain $\left\{x \middle| x \neq -\dfrac{5}{2}, x \neq 3\right\}$

2. x-intercept: $-\dfrac{3}{4}$; y-intercept: -1
3. No y-axis or origin symmetry
4. Vertical asymptote: $x = 3$, hole at $\left(-\dfrac{5}{2}, \dfrac{14}{11}\right)$
5. Horizontal asymptote: $y = 4$
6.

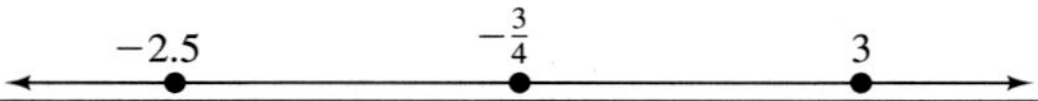

Interval	$(-\infty, -2.5)$	$\left(-2.5, -\frac{3}{4}\right)$	$\left(-\frac{3}{4}, 3\right)$	$(3, \infty)$
Number Chosen	-3	-1	0	6
Value of R	$R(-3) = \frac{3}{2}$	$R(-1) = \frac{1}{4}$	$R(0) = -1$	$R(6) = 9$
Location of Graph	Above x-axis	Above x-axis	Below x-axis	Above x-axis
Point on Graph	$\left(-3, \frac{3}{2}\right)$	$\left(-1, \frac{1}{4}\right)$	$(0, -1)$	$(6, 9)$

7.

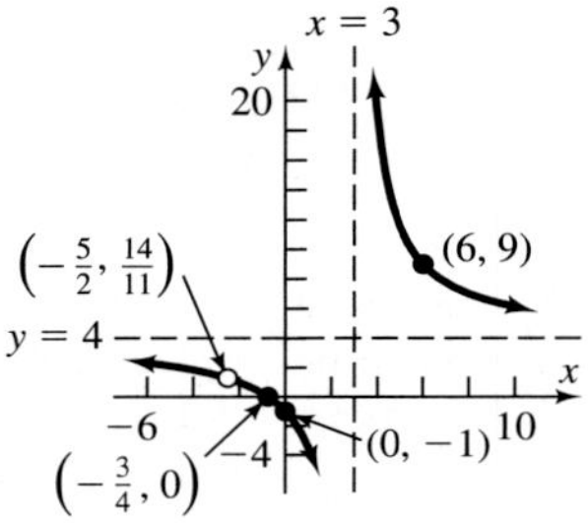

37. 1. Domain: $\{x | x \neq -3\}$
2. x-intercept: -2; y-intercept: 2
3. No y-axis or origin symmetry
4. Vertical asymptote: None; hole at $(-3, -1)$
5. Oblique asymptote: $y = x + 2$ intersected at all points except $x = -3$
6.

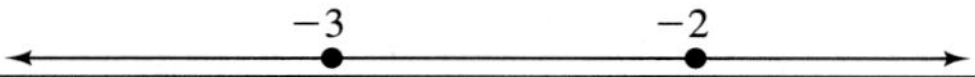

Interval	$(-\infty, -3)$	$(-3, -2)$	$(-2, \infty)$
Number Chosen	-4	-2.5	0
Value of R	$R(-4) = -2$	$R(-2.5) = -\frac{1}{2}$	$R(0) = 2$
Location of Graph	Below x-axis	Below x-axis	Above x-axis
Point on Graph	$(-4, -2)$	$\left(-2.5, -\frac{1}{2}\right)$	$(0, 2)$

7.

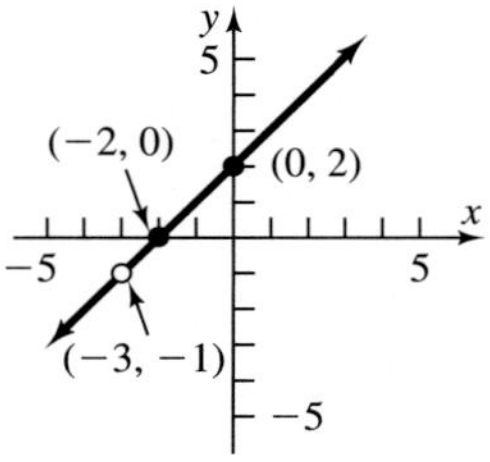

38. 1. Domain $\{x \mid x \neq -6\}$
2. x-intercept: 5; y-intercept: -5
3. No y-axis or origin symmetry
4. Vertical asymptote: None; hole at $(-6, -11)$
5. Oblique asymptote: $y = x - 5$ intersected at all points except $x = -6$
6.

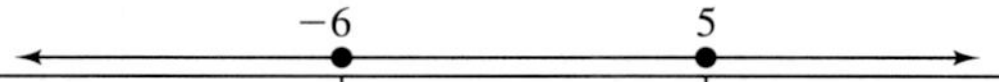

Interval	$(-\infty, -6)$	$(-6, 5)$	$(5, \infty)$
Number Chosen	-7	0	6
Value of R	$R(-7) = -12$	$R(0) = -5$	$R(6) = 1$
Location of Graph	Below x-axis	Below x-axis	Above x-axis
Point on Graph	$(-7, -12)$	$(0, -5)$	$(6, 1)$

7.

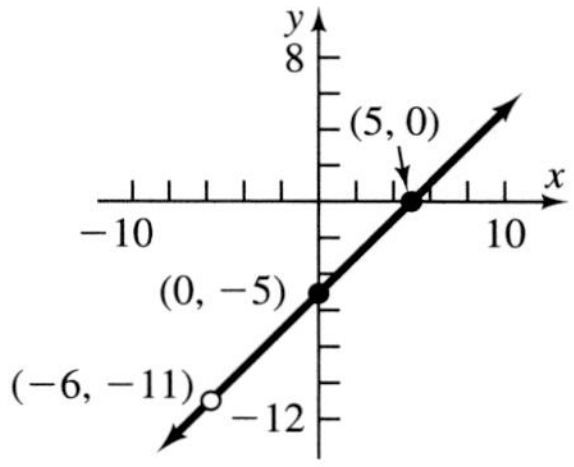

39. 1. Domain: $\{x \mid x \neq 0\}$
2. No x-intercepts; no y-intercepts
3. Symmetric about the origin
4. Vertical asymptote: $x = 0$
5. Oblique asymptotes: $y = x$, not intersected
6.

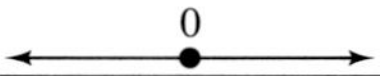

Interval	$(-\infty, 0)$	$(0, \infty)$
Number Chosen	-1	1
Value of f	$f(-1) = -2$	$f(1) = 2$
Location of Graph	Below x-axis	Above x-axis
Point on Graph	$(-1, -2)$	$(1, 2)$

7.

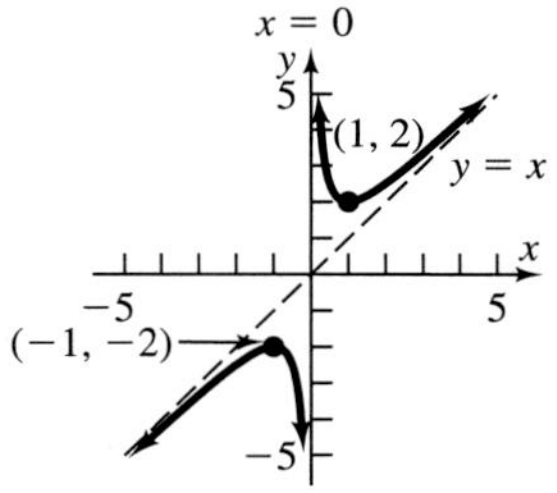

40. 1. Domain $\{x \mid x \neq 0\}$
2. No x-intercepts; no y-intercept
3. Symmetric about the origin
4. Vertical asymptote: $x = 0$
5. Oblique asymptote: $y = 2x$, not intersected
6.

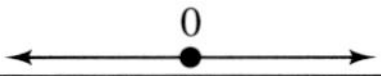

Interval	$(-\infty, 0)$	$(0, \infty)$
Number Chosen	-1	1
Value of f	$f(-1) = -11$	$f(1) = 11$
Location of Graph	Below x-axis	Above x-axis
Point on Graph	$(-1, -11)$	$(1, 11)$

7.

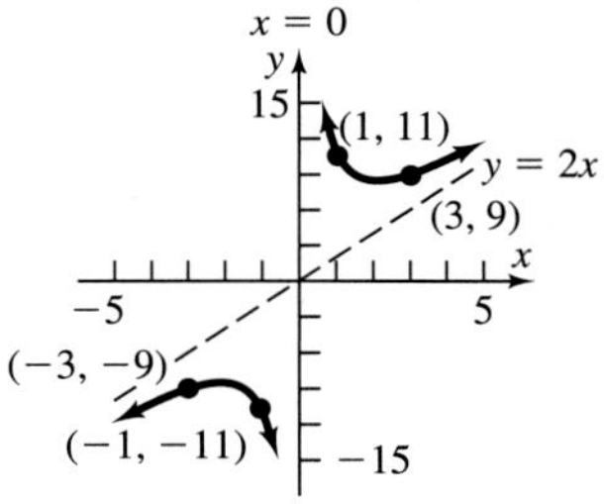

41. 1. Domain: $\{x \mid x \neq 0\}$
2. x-intercept: -1; no y-intercepts
3. No y-axis or origin symmetry
4. Vertical asymptote: $x = 0$
5. No horizontal or oblique asymptotes
6.

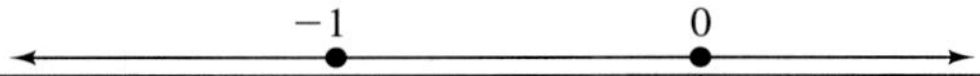

Interval	$(-\infty, -1)$	$(-1, 0)$	$(0, \infty)$
Number Chosen	-2	$-\frac{1}{2}$	1
Value of f	$f(-2) = 3.5$	$f\left(-\frac{1}{2}\right) = -1.75$	$f(1) = 2$
Location of Graph	Above x-axis	Below x-axis	Above x-axis
Point on Graph	$(-2, 3.5)$	$\left(-\frac{1}{2}, -1.75\right)$	$(1, 2)$

7.

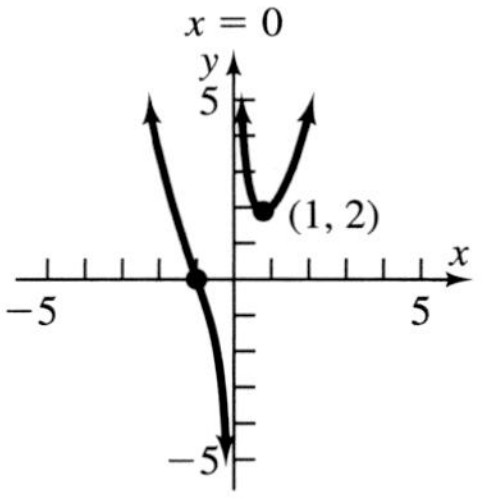

42. 1. Domain $\{x|x \neq 0\}$
2. x-intercept: -1.65; no y-intercept
3. No y-axis or origin symmetry
4. Vertical asymptote: $x = 0$
5. No oblique or horizontal asymptotes
6.

-1.65 0

Interval	$(-\infty, -1.65)$	$(-1.65, 0)$	$(0, \infty)$
Number Chosen	-2	-1	1
Value of f	$f(-2) = 3.5$	$f(-1) = -7$	$f(1) = 11$
Location of Graph	Above x-axis	Below x-axis	Above x-axis
Point on Graph	$(-2, 3.5)$	$(-1, -7)$	$(1, 11)$

7.

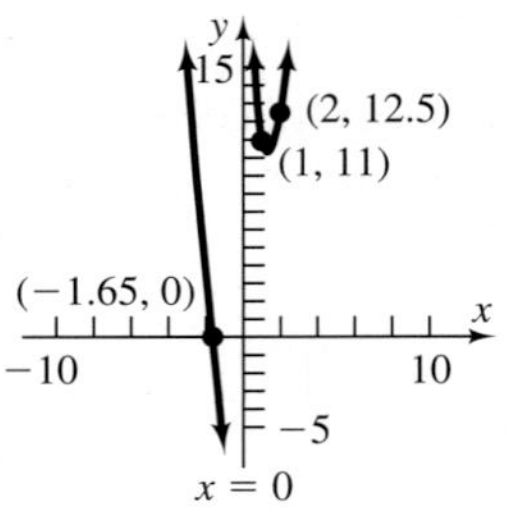

43. 1. Domain: $\{x|x \neq 0\}$
2. No x-intercepts; no y-intercepts
3. Symmetric about the origin
4. Vertical asymptote: $x = 0$
5. Oblique asymptote: $y = x$, not intersected
6.

0

Interval	$(-\infty, 0)$	$(0, \infty)$
Number Chosen	-1	1
Value of f	$f(-1) = -2$	$f(1) = 2$
Location of Graph	Below x-axis	Above x-axis
Point on Graph	$(-1, -2)$	$(1, 2)$

7.

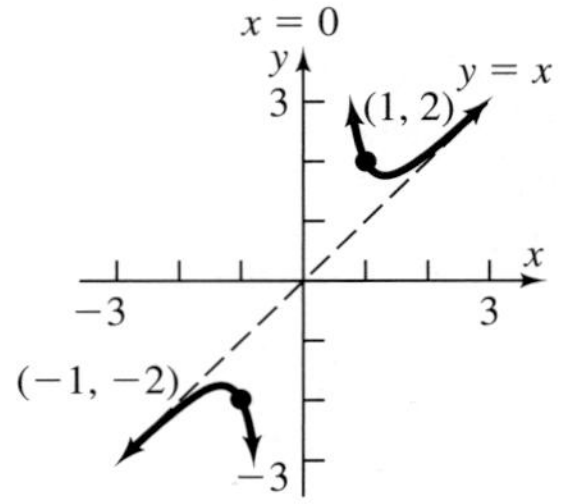

44. 1. Domain $\{x|x \neq 0\}$
2. No x-intercepts; no y-intercept
3. Symmetric about the origin
4. Vertical asymptote: $x = 0$
5. Oblique asymptote: $y = 2x$, not intersected
6.

0

Interval	$(-\infty, 0)$	$(0, \infty)$
Number Chosen	-2	2
Value of f	$f(-2) = -5.125$	$f(2) = 5.125$
Location of Graph	Below x-axis	Above x-axis
Point on Graph	$(-2, -5.125)$	$(2, 5.125)$

7.

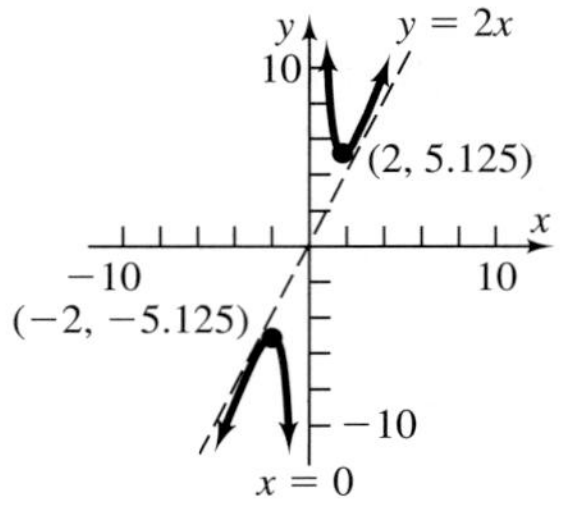

45. One possibility: $R(x) = \dfrac{x^2}{x^2 - 4}$ **46.** One possibility: $R(x) = -\dfrac{x}{x^2 - 1}$

47. One possibility: $R(x) = \dfrac{(x-1)(x-3)(x^2 + \frac{4}{3})}{(x+1)^2(x-2)^2}$ **48.** One possibility: $R(x) = \dfrac{3(x+2)(x-1)^2}{(x+3)(x-4)^2}$

49. **(a)** t-axis; $C(t) \to 0$

(b)

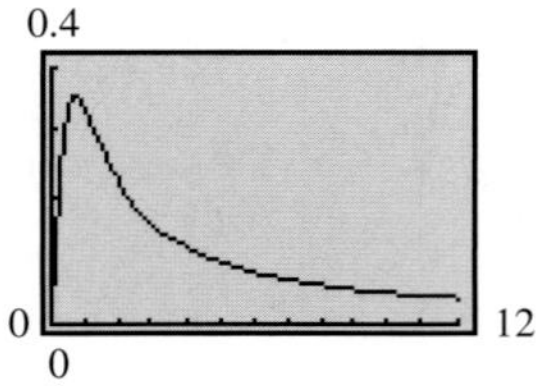

(c) 0.71 hr after injection

50. **(a)** t-axis; $C(t) \to 0$

(b)

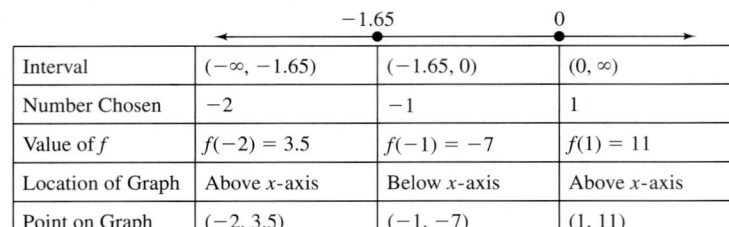

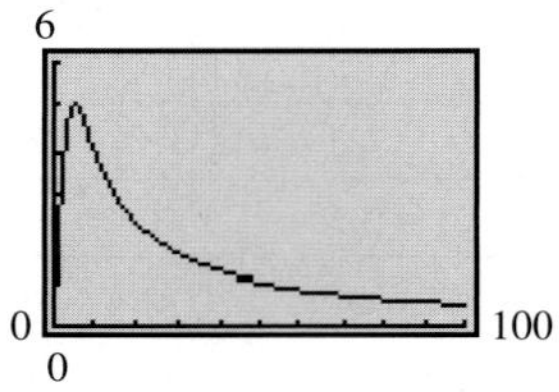

(c) 5 min after injection

51. **(a)** $\overline{C}(x) = \dfrac{0.2x^3 - 2.3x^2 + 14.3x + 10.2}{x}$
(b) \$9400
(c) $\approx$ \$10,933
(d)
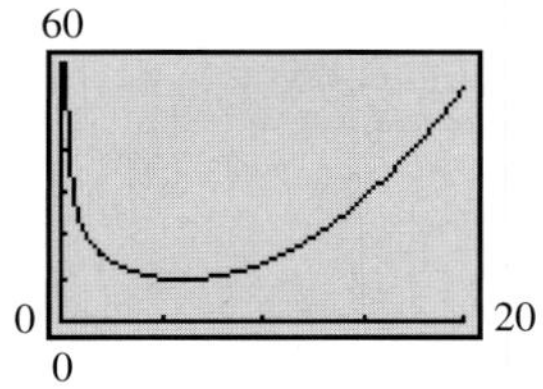

(e) 6
(f) \$9400

52. **(a)** $\overline{C}(x) = \dfrac{0.015x^3 - 0.595x^2 + 9.15x + 98.43}{x}$
(b) $\approx$ \$11.52
(c) $\approx$ \$7.59
(d)
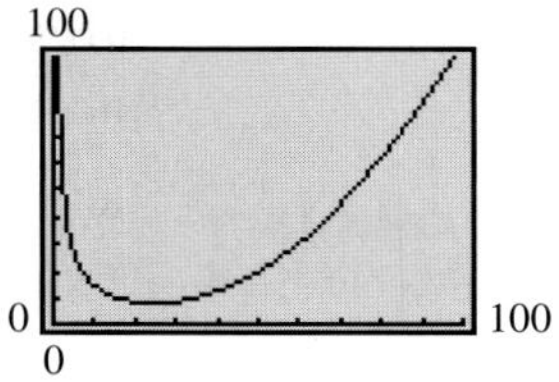

(e) $\approx$ 25,058 books
(f) $\approx$ \$7.59

53. **(a)** $S(x) = 2x^2 + \dfrac{40{,}000}{x}$
(b)
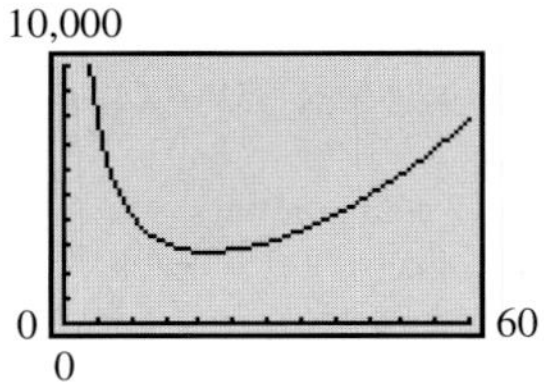

(c) 2784.95 sq in.
(d) 21.54 in. $\times$ 21.54 in. $\times$ 21.54 in.
(e) To minimize the cost of material needed for construction

54. **(a)** $S(x) = 2x^2 + \dfrac{20{,}000}{x}$
(b)
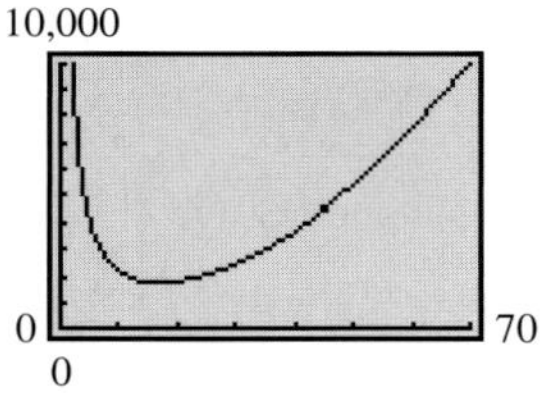

(c) 1754.41 sq in.
(d) 17.10 in. $\times$ 17.10 in. $\times$ 17.10 in.
(e) To minimize the cost of material needed for construction

55. **(a)** $C(r) = 12\pi r^2 + \dfrac{4000}{r}$
(b)
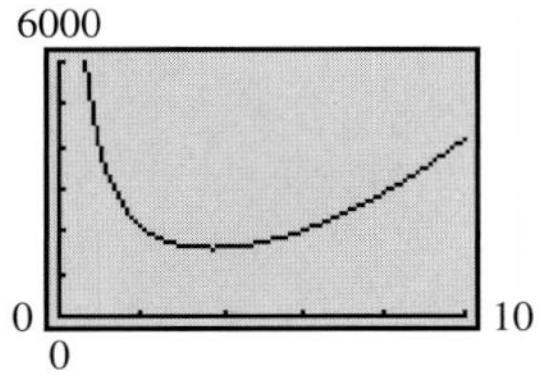

The cost is smallest when $r \approx 3.76$ cm.

56. **(a)** $A(r) = 2\pi r^2 + \dfrac{200}{r}$ **(b)** Approximately 123.22 sq ft
(c) Approximately 150.53 sq ft **(d)** Approximately 197.08 sq ft
(e)
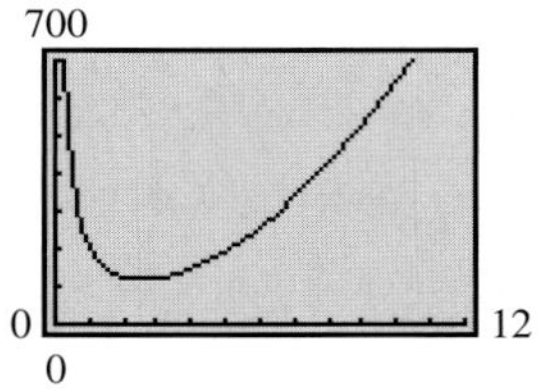

A is smallest when $r \approx 2.52$ ft.

57. No. Each function is a quotient of polynomials, but it is not written in lowest terms. Each function is undefined for $x = 1$; each graph has a hole at $x = 1$.

58. All four graphs have a vertical asymptote at $x = 1$; $y = \dfrac{x^2}{x - 1}$ has an oblique asymptote at $y = x$.

59. Minimum value: 2.00 at $x = 1.00$
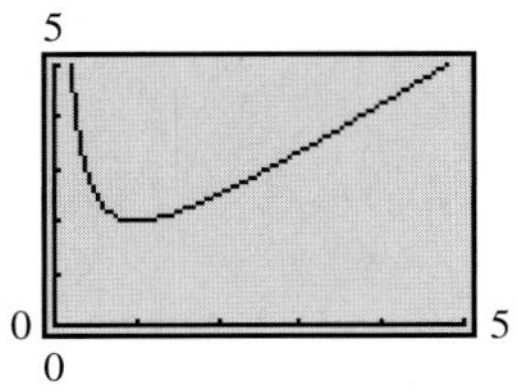

60. Minimum value: 8.49 at $x = 2.12$
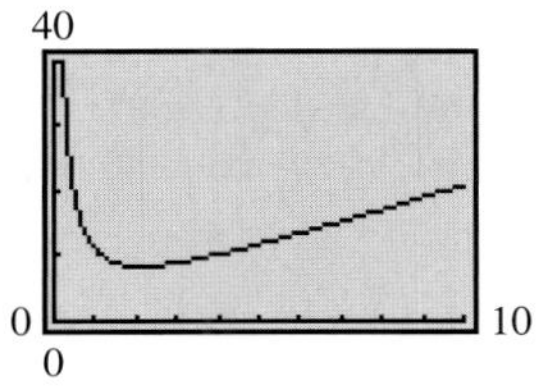

61. Minimum value: 1.89 at $x = 0.79$
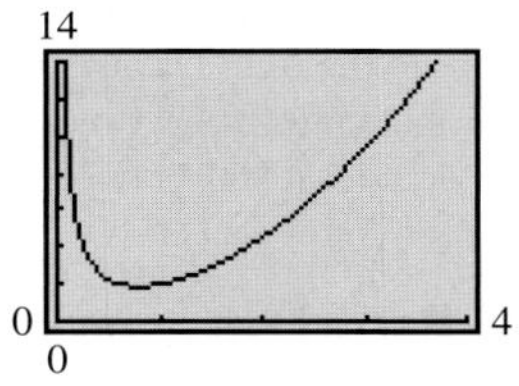

62. Minimum value: 10.30 at $x = 1.31$

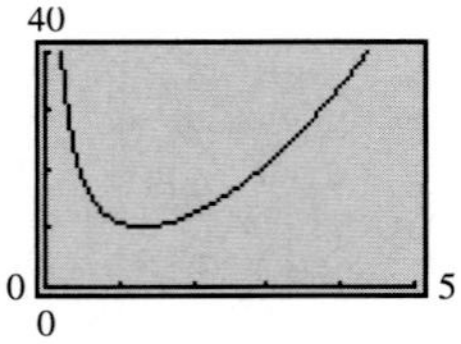

63. Minimum value: 1.75 at $x = 1.32$

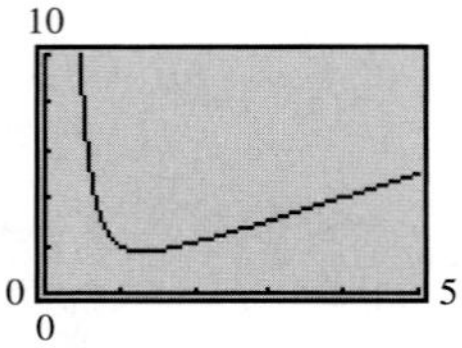

64. Minimum value: 5.11 at $x = 1.92$

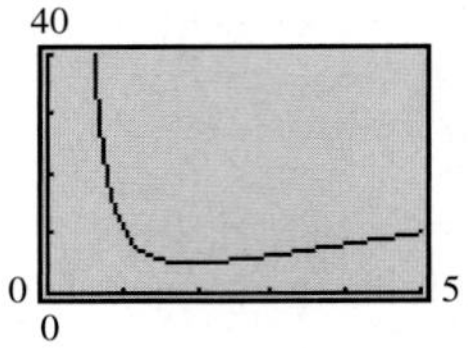

3.5 Concepts and Vocabulary *(page 192)*

2. True

3.5 Exercises *(page 193)*

3. $\{x|-2 < x < 5\}; (-2, 5)$ **4.** $\{x|x < -2 \text{ or } x > 5\}; (-\infty, -2) \text{ or } (5, \infty)$ **5.** $\{x|x \leq 0 \text{ or } x \geq 4\}; (-\infty, 0] \text{ or } [4, \infty)$
6. $\{x|x \leq -8 \text{ or } x \geq 0\}; (-\infty, -8] \text{ or } [0, \infty)$ **7.** $\{x|-3 < x < 3\}; (-3, 3)$ **8.** $\{x|-1 < x < 1\}; (-1, 1)$
9. $\{x|x \leq -2 \text{ or } x \geq 1\}; (-\infty, -2] \text{ or } [1, \infty)$ **10.** $\{x|-4 \leq x \leq -3\}; [-4, -3]$ **11.** $\left\{x\middle|-\frac{1}{2} \leq x \leq 3\right\}; \left[-\frac{1}{2}, 3\right]$
12. $\left\{x\middle|-\frac{2}{3} \leq x \leq \frac{3}{2}\right\}; \left[-\frac{2}{3}, \frac{3}{2}\right]$ **13.** $\{x|x < -1 \text{ or } x > 8\}; (-\infty, -1) \text{ or } (8, \infty)$ **14.** $\{x|x < -5 \text{ or } x > 4\}; (-\infty, -5) \text{ or } (4, \infty)$
15. No real solution **16.** No real solution **17.** $\left\{x\middle|x < -\frac{2}{3} \text{ or } x > \frac{3}{2}\right\}; \left(-\infty, -\frac{2}{3}\right) \text{ or } \left(\frac{3}{2}, \infty\right)$ **18.** All real numbers
19. $\{x|x \geq 1\}; [1, \infty)$ **20.** $\{x|x \geq -2\}; [-2, \infty)$ **21.** $\{x|x \leq 1 \text{ or } 2 \leq x \leq 3\}; (-\infty, 1] \text{ or } [2, 3]$
22. $\{x|x \leq -3 \text{ or } -2 \leq x \leq -1\}; (-\infty, -3] \text{ or } [-2, -1]$ **23.** $\{x|-1 < x < 0 \text{ or } x > 3\}; (-1, 0) \text{ or } (3, \infty)$
24. $\{x|-3 < x < 0 \text{ or } x > 1\}; (-3, 0) \text{ or } (1, \infty)$ **25.** $\{x|x < -1 \text{ or } x > 1\}; (-\infty, -1) \text{ or } (1, \infty)$
26. $\{x|-2 < x < 0 \text{ or } 0 < x < 2\}; (-2, 0) \text{ or } (0, 2)$ **27.** $\{x|x = 0 \text{ or } x \geq 4\}; 0 \text{ or } [4, \infty)$ **28.** $\{x|x \leq 9\}; (-\infty, 9]$
29. $\{x|x < -1 \text{ or } x > 1\}; (-\infty, -1) \text{ or } (1, \infty)$ **30.** $\{x|x > 1\}; (1, \infty)$ **31.** $\{x|x < -1 \text{ or } x > 1\}; (-\infty, -1) \text{ or } (1, \infty)$
32. $\{x|x < -1 \text{ or } x > 3\}; (-\infty, -1) \text{ or } (3, \infty)$ **33.** $\{x|x \leq -1 \text{ or } 0 < x \leq 1\}; (-\infty, -1] \text{ or } (0, 1]$
34. $\{x|x \leq -2 \text{ or } 1 < x \leq 3\}; (-\infty, -2] \text{ or } (1, 3]$ **35.** $\{x|x < -1 \text{ or } x > 1\}; (-\infty, -1) \text{ or } (1, \infty)$
36. $\{x|x < -2 \text{ or } x > 2\}; (-\infty, -2) \text{ or } (2, \infty)$ **37.** $\left\{x\middle|x < -\frac{2}{3} \text{ or } 0 < x < \frac{3}{2}\right\}; \left(-\infty, -\frac{2}{3}\right) \text{ or } \left(0, \frac{3}{2}\right)$
38. $\{x|x < 0 \text{ or } 3 < x < 4\}; (-\infty, 0) \text{ or } (3, 4)$ **39.** $\{x|x < 2\}; (-\infty, 2)$ **40.** $\{x|x > 4\}; (4, \infty)$ **41.** $\{x|-2 < x \leq 9\}; (-2, 9]$
42. $\{x|-8 \leq x < -2\}; [-8, -2)$ **43.** $\{x|x < 2 \text{ or } 3 < x < 5\}; (-\infty, 2) \text{ or } (3, 5)$ **44.** $\{x|-7 < x < -1 \text{ or } x > 3\}; (-7, -1) \text{ or } (3, \infty)$
45. $\{x|x < -3 \text{ or } -1 < x < 1 \text{ or } x > 2\}; (-\infty, -3) \text{ or } (-1, 1) \text{ or } (2, \infty)$ **46.** $\left\{x\middle|x < -\frac{5}{2} \text{ or } -2 < x < -1\right\}; \left(-\infty, -\frac{5}{2}\right) \text{ or } (-2, -1)$
47. $\{x|x < -5 \text{ or } -4 \leq x \leq -3 \text{ or } x = 0 \text{ or } x > 1\}; (-\infty, -5) \text{ or } [-4, -3] \text{ or } 0 \text{ or } (1, \infty)$
48. $\{x|x < -1 \text{ or } 0 \leq x < 1 \text{ or } x \geq 2\}; (-\infty, -1) \text{ or } [0, 1) \text{ or } [2, \infty)$
49. $\left\{x\middle|-\frac{1}{2} < x < 1 \text{ or } x > 3\right\}$ **50.** $\left\{x\middle|-1 < x < \frac{2}{3} \text{ or } x > 2\right\}$
51. $\{x|x > 4\}; (4, \infty)$ **52.** $\{x|x > 2\}; (2, \infty)$
53. $\{x|x \leq -4 \text{ or } x \geq 4\}; (-\infty, -4] \text{ or } [4, \infty)$ **54.** $\{x|x = 0 \text{ or } x \geq 3\}; 0 \text{ or } [3, \infty)$
55. $\{x|x < -4 \text{ or } x \geq 2\}; (-\infty, -4) \text{ or } [2, \infty)$ **56.** $\{x|x < -4 \text{ or } x \geq 1\}; (-\infty, -4) \text{ or } [1, \infty)$
57. The ball is more than 96 ft above the ground for time t between 2 and 3 sec, $2 < t < 3$.
58. The ball is more than 112 ft above the ground for time t between $3 - \sqrt{2} \approx 1.59$ sec and $3 + \sqrt{2} \approx 4.41$ sec.
59. For a profit of at least \$50, between 8 and 32 watches must be sold, $8 \leq x \leq 32$.
60. For a profit of at least \$60, between 30 and 40 boxes must be sold, $30 \leq x \leq 40$ **61.** $-2 < k < 2$ **62.** $k < 1$

Historical Problems *(page 206)*

1.
$$\left(x - \frac{b}{3}\right)^3 + b\left(x - \frac{b}{3}\right)^2 + c\left(x - \frac{b}{3}\right) + d = 0$$
$$x^3 - bx^2 + \frac{b^2x}{3} - \frac{b^3}{27} + bx^2 - \frac{2b^2x}{3} + \frac{b^3}{9} + cx - \frac{bc}{3} + d = 0$$
$$x^3 + \left(c - \frac{b^2}{3}\right)x + \left(\frac{2b^3}{27} - \frac{bc}{3} + d\right) = 0$$
Let $p = c - \frac{b^2}{3}$ and $q = \frac{2b^3}{27} - \frac{bc}{3} + d$. Then $x^3 + px + q = 0$.

2.
$$(H + K)^3 + p(H + K) + q = 0$$
$$H^3 + 3H^2K + 3HK^2 + K^3 + pH + pK + q = 0$$
Let $3HK = -p$.
$$H^3 - pH - pK + K^3 + pH + pK + q = 0$$
$$H^3 + K^3 = -q$$

3.
$$3HK = -p$$
$$K = -\frac{p}{3H}$$
$$H^3 + \left(-\frac{p}{3H}\right)^3 = -q$$
$$H^3 - \frac{p^3}{27H^3} = -q$$
$$27H^6 - p^3 = -27qH^3$$
$$27H^6 + 27qH^3 - p^3 = 0$$
$$H^3 = \frac{-27q \pm \sqrt{(27q)^2 - 4(27)(-p^3)}}{2 \cdot 27}$$
$$H^3 = \frac{-q}{2} \pm \sqrt{\frac{27^2q^2}{2^2(27^2)} + \frac{4(27)p^3}{2^2(27^2)}}$$
$$H^3 = \frac{-q}{2} \pm \sqrt{\frac{q^2}{4} + \frac{p^3}{27}}$$
Choose the positive root for now.
$$H = \sqrt[3]{\frac{-q}{2} + \sqrt{\frac{q^2}{4} + \frac{p^3}{27}}}$$

4. $H^3 + K^3 = -q$
$$K^3 = -q - H^3$$
$$K^3 = -q - \left[\frac{-q}{2} + \sqrt{\frac{q^2}{4} + \frac{p^3}{27}}\right]$$
$$K^3 = \frac{-q}{2} - \sqrt{\frac{q^2}{4} + \frac{p^3}{27}}$$
$$K = \sqrt[3]{\frac{-q}{2} - \sqrt{\frac{q^2}{4} + \frac{p^3}{27}}}$$

5. $x = H + K$
$$x = \sqrt[3]{\frac{-q}{2} + \sqrt{\frac{q^2}{4} + \frac{p^3}{27}}} + \sqrt[3]{\frac{-q}{2} - \sqrt{\frac{q^2}{4} + \frac{p^3}{27}}}$$
(Note that if we had used the negative root in 3 the result would be the same.)

6. $x = 3$ **7.** $x = 2$ **8.** $x = 2$

3.6 Concepts and Vocabulary *(page 207)*

5. Remainder; Dividend **6.** $f(c)$ **7.** -4 **8.** False **9.** False **10.** True

3.6 Exercises *(page 207)*

11. No; $f(2) = 8$ **12.** No; $f(-3) = 161$ **13.** Yes; $f(2) = 0$ **14.** Yes; $f(2) = 0$ **15.** Yes; $f(-3) = 0$ **16.** Yes; $f(-3) = 0$ **17.** No; $f(-4) = 1$ **18.** Yes; $f(-4) = 0$ **19.** Yes; $f\left(\frac{1}{2}\right) = 0$ **20.** No; $f\left(-\frac{1}{3}\right) = 2$ **21.** 7; 3 or 1 positive; 2 or 0 negative **22.** 4; 1 positive; 1 negative **23.** 6; 2 or 0 positive; 2 or 0 negative **24.** 5; 1 positive; 0 negative **25.** 3; 2 or 0 positive; 1 negative **26.** 3; 1 positive; 2 or 0 negative **27.** 4; 2 or 0 positive; 2 or 0 negative **28.** 4; 1 positive; 1 negative **29.** 5; 0 positive; 3 or 1 negative **30.** 5; 5, 3 or 1 positive; 0 negative **31.** 6; 1 positive; 1 negative **32.** 6; no positive; no negative **33.** $\pm 1, \pm\frac{1}{3}$ **34.** $\pm 1, \pm 3$ **35.** $\pm 1, \pm 3$ **36.** $\pm 1, \pm\frac{1}{2}$ **37.** $\pm 1, \pm 2, \pm\frac{1}{4}, \pm\frac{1}{2}$ **38.** $\pm 1, \pm 2, \pm\frac{1}{2}, \pm\frac{1}{3}, \pm\frac{1}{6}, \pm\frac{2}{3}$ **39.** $\pm 1, \pm 3, \pm 9, \pm\frac{1}{2}, \pm\frac{1}{3}, \pm\frac{1}{6}, \pm\frac{3}{2}, \pm\frac{9}{2}$ **40.** $\pm 1, \pm 2, \pm 3, \pm 6, \pm\frac{1}{4}, \pm\frac{1}{2}, \pm\frac{3}{4}, \pm\frac{3}{2}$ **41.** $\pm 1, \pm 2, \pm 3, \pm 4, \pm 6, \pm 12, \pm\frac{1}{2}, \pm\frac{3}{2}$ **42.** $\pm 1, \pm 2, \pm 3, \pm 6, \pm 9, \pm 18, \pm\frac{1}{3}, \pm\frac{2}{3}$ **43.** $\pm 1, \pm 2, \pm 4, \pm 5, \pm 10, \pm 20, \pm\frac{1}{2}, \pm\frac{5}{2}, \pm\frac{1}{3}, \pm\frac{2}{3}, \pm\frac{4}{3}, \pm\frac{5}{3}, \pm\frac{10}{3}, \pm\frac{20}{3}, \pm\frac{1}{6}, \pm\frac{5}{6}$ **44.** $\pm 1, \pm 2, \pm 5, \pm 10, \pm\frac{1}{2}, \pm\frac{1}{3}, \pm\frac{1}{6}, \pm\frac{2}{3}, \pm\frac{5}{3}, \pm\frac{5}{2}, \pm\frac{5}{6}, \pm\frac{10}{3}$ **45.** $-3, -1, 2$; $f(x) = (x + 3)(x + 1)(x - 2)$ **46.** $-4, -5, 1$; $f(x) = (x - 1)(x + 5)(x + 4)$ **47.** $\frac{1}{2}$; $f(x) = 2\left(x - \frac{1}{2}\right)(x^2 + 1)$

48. $-\frac{1}{2}$; $f(x) = 2\left(x + \frac{1}{2}\right)(x^2 + 1)$ **49.** $-1, 1$; $f(x) = (x + 1)(x - 1)(x^2 + 2)$ **50.** $2, -2$; $f(x) = (x - 2)(x + 2)(x^2 + 1)$

51. $-\frac{1}{2}, \frac{1}{2}$; $f(x) = 4\left(x + \frac{1}{2}\right)\left(x - \frac{1}{2}\right)(x^2 + 2)$ **52.** $\frac{1}{2}, -\frac{1}{2}$; $f(x) = 4\left(x - \frac{1}{2}\right)\left(x + \frac{1}{2}\right)(x^2 + 4)$

53. 1, multiplicity 2; $-2, -1$; $f(x) = (x + 2)(x + 1)(x - 1)^2$ **54.** 2, multiplicity 2; $-1, -2$; $f(x) = (x + 1)(x + 2)(x - 2)^2$

55. $-\frac{\sqrt{2}}{2}, \frac{\sqrt{2}}{2}, 2$; $f(x) = 4\left(x + \frac{\sqrt{2}}{2}\right)\left(x - \frac{\sqrt{2}}{2}\right)(x - 2)\left(x^2 + \frac{1}{2}\right)$

56. $-3, -\frac{\sqrt{2}}{2}, \frac{\sqrt{2}}{2}$; $f(x) = 4(x + 3)\left(x + \frac{\sqrt{2}}{2}\right)\left(x - \frac{\sqrt{2}}{2}\right)\left(x^2 + \frac{1}{2}\right)$ **57.** $\{-1, 2\}$ **58.** $\left\{-\frac{3}{2}\right\}$ **59.** $\left\{\frac{2}{3}, -1 + \sqrt{2}, -1 - \sqrt{2}\right\}$

60. $\left\{\frac{5}{2}\right\}$ **61.** $\left\{\frac{1}{3}, \sqrt{5}, -\sqrt{5}\right\}$ **62.** $\left\{-\frac{1}{2}, 2, 4\right\}$ **63.** $\{-3, -2\}$ **64.** $\{1\}$ **65.** $\left\{-\frac{1}{3}\right\}$ **66.** $\left\{\frac{1}{2}\right\}$ **67.** $\left\{\frac{1}{2}, 2, 5\right\}$ **68.** $\left\{-4, -\frac{1}{2}, 2\right\}$

69. y-intercept: -6; x-intercepts: $-3, -1, 2$
$(-\infty, -3)$, $f(-4) = -18$, below x-axis
$(-3, -1)$, $f(-2) = 4$, above x-axis
$(-1, 2)$, $f(0) = -6$, below x-axis
$(2, \infty)$, $f(3) = 24$, above x-axis

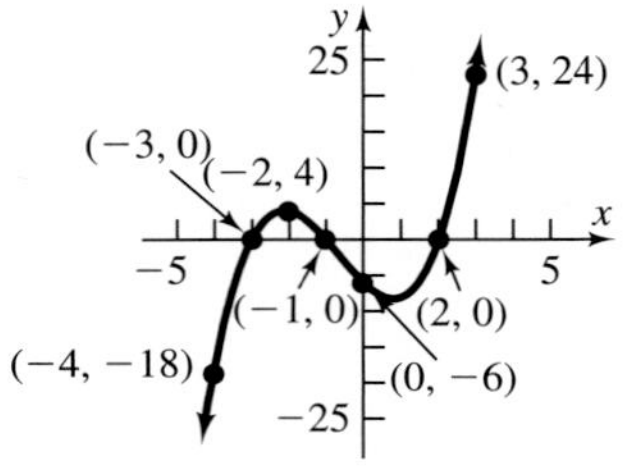

70. y-intercept: -20; x-intercepts: $-5, -4, 1$
$(-\infty, -5)$, $f(-6) = -14$, below x-axis
$(-5, -4)$, $f(-4.5) = 1.375$, above x-axis
$(-4, 1)$, $f(0) = -20$, below x-axis
$(1, \infty)$, $f(2) = 42$, above x-axis

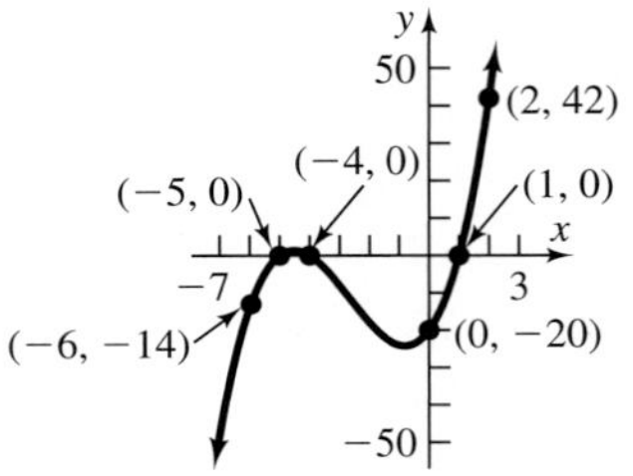

71. y-intercept: -1; x-intercept: $\frac{1}{2}$
$\left(-\infty, \frac{1}{2}\right)$, $f(0) = -1$, below x-axis
$\left(\frac{1}{2}, \infty\right)$, $f(1) = 2$, above x-axis

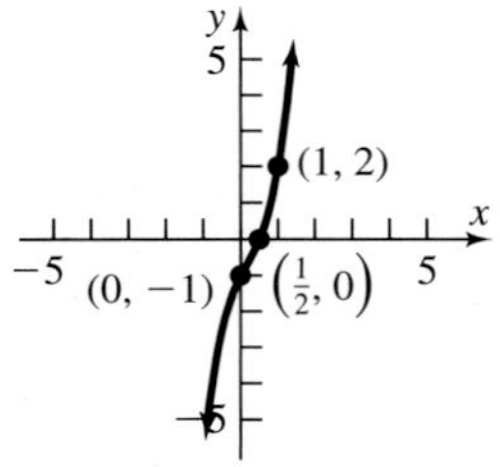

72. y-intercept: 1; x-intercept: $-\frac{1}{2}$
$\left(-\infty, -\frac{1}{2}\right)$, $f(-1) = -2$, below x-axis
$\left(-\frac{1}{2}, \infty\right)$, $f(0) = 1$, above x-axis

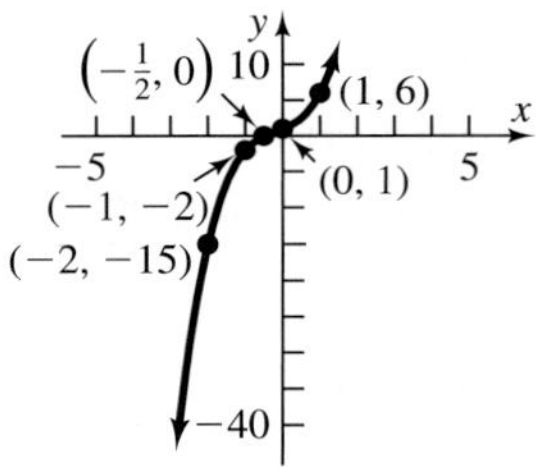

73. y-intercept: -2; x-intercepts: $-1, 1$
$(-\infty, -1)$, $f(-2) = 18$, above x-axis
$(-1, 1)$, $f(0) = -2$, below x-axis
$(1, \infty)$, $f(2) = 18$, above x-axis
(symmetric with respect to the y-axis)

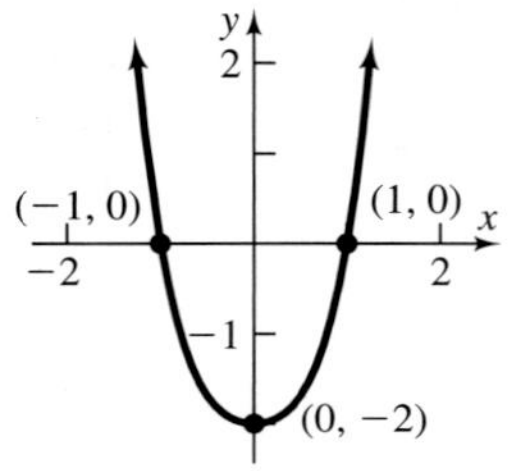

74. y-intercept: -4; x-intercepts: $-2, 2$
$(-\infty, -2)$, $f(-3) = 50$, above x-axis
$(-2, 2)$, $f(0) = -4$, below x-axis
$(2, \infty)$, $f(3) = 50$, above x-axis
(symmetric with respect to the y-axis)

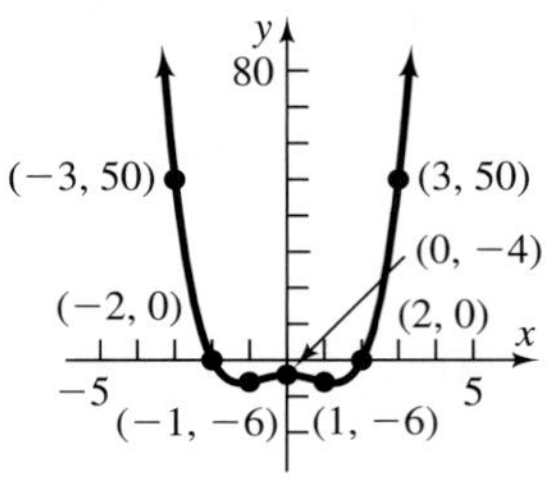

75. y-intercept: -2; x-intercepts: $-\frac{1}{2}, \frac{1}{2}$

$\left(-\infty, -\frac{1}{2}\right), f(-1) = 9$, above x-axis

$\left(-\frac{1}{2}, \frac{1}{2}\right), f(0) = -2$, below x-axis

$\left(\frac{1}{2}, \infty\right), f(1) = 9$, above x-axis

(symmetric with respect to the y-axis)

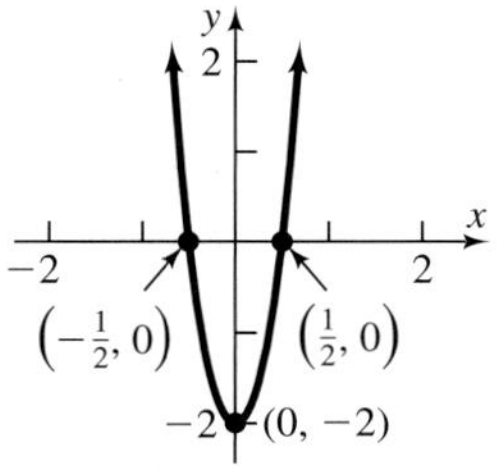

76. y-intercept: -4; x-intercepts: $-\frac{1}{2}, \frac{1}{2}$

$\left(-\infty, -\frac{1}{2}\right), f(-1) = 15$, above x-axis

$\left(-\frac{1}{2}, \frac{1}{2}\right), f(0) = -4$, below x-axis

$\left(\frac{1}{2}, \infty\right), f(1) = 15$, above x-axis

(symmetric with respect to the y-axis)

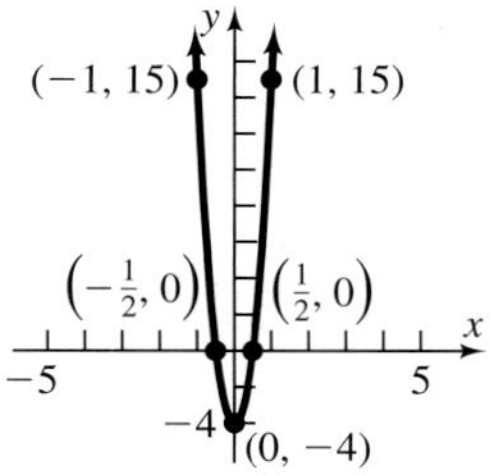

77. y-intercept: 2; x-intercepts: $-2, -1, 1$

$(-\infty, -2), f(-3) = 32$, above x-axis

$(-2, -1), f(-1.5) \approx -1.6$, below x-axis

$(-1, 1), f(0) = 2$, above x-axis

$(1, \infty), f(2) = 12$, above x-axis

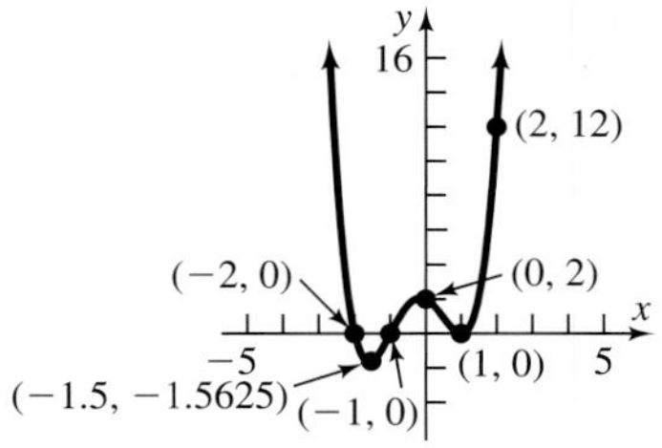

78. y-intercept: 8; x-intercepts: $-2, -1, 2$

$(-\infty, -2), f(-3) = 50$, above x-axis

$(-2, -1), f(-1.5) \approx -3$, below x-axis

$(-1, 2), f(0) = 8$, above x-axis

$(2, \infty), f(3) = 20$, above x-axis

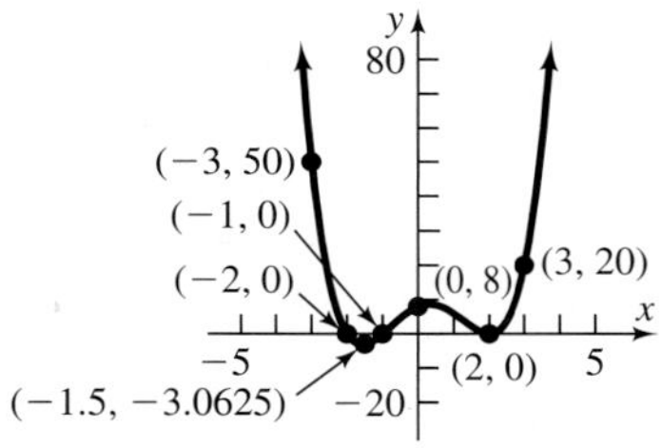

79. y-intercept: 2; x-intercepts: $-\frac{\sqrt{2}}{2}, \frac{\sqrt{2}}{2}, 2$

$\left(-\infty, -\frac{\sqrt{2}}{2}\right), f(-1) = -9$, below x-axis

$\left(-\frac{\sqrt{2}}{2}, \frac{\sqrt{2}}{2}\right), f(0) = 2$, above x-axis

$\left(\frac{\sqrt{2}}{2}, 2\right), f(1) = -3$, below x-axis

$(2, \infty), f(3) = 323$, above x-axis

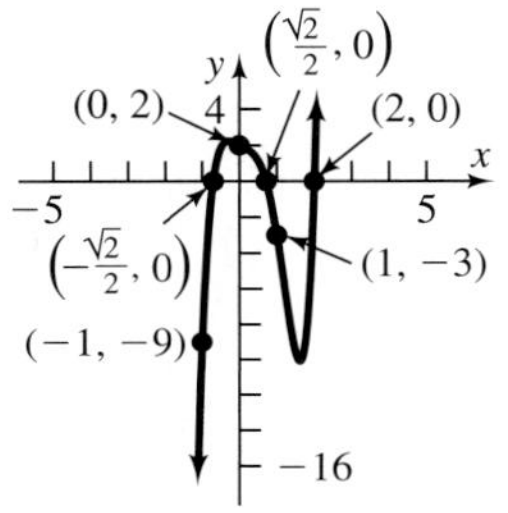

80. y-intercept: -3; x-intercepts: $-3, -\frac{\sqrt{2}}{2}, \frac{\sqrt{2}}{2}$

$(-\infty, -3), f(-3.5) \approx -300$, below x-axis

$\left(-3, -\frac{\sqrt{2}}{2}\right), f(-2) = 63$, above x-axis

$\left(-\frac{\sqrt{2}}{2}, \frac{\sqrt{2}}{2}\right), f(0) = -3$, below x-axis

$\left(\frac{\sqrt{2}}{2}, \infty\right), f(1) = 12$, above x-axis

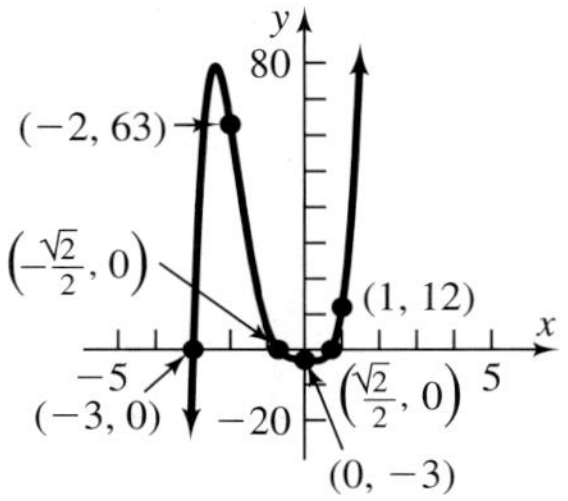

81. 5 **82.** 37 **83.** 2 **84.** 2 **85.** 5 **86.** 13 **87.** $\frac{3}{2}$ **88.** $\frac{3}{2}$ **89.** $f(0) = -1; f(1) = 10$ **90.** $f(-1) = -6; f(0) = 2$

91. $f(-5) = -58; f(-4) = 2$ **92.** $f(-3) = -42; f(-2) = 5$ **93.** $f(1.4) = -0.17536; f(1.5) = 1.40625$

94. $f(1.7) = 0.35627; f(1.8) = -1.02112$ **95.** 0.21 **96.** -0.60 **97.** -4.04 **98.** -2.17 **99.** 1.15 **100.** 0.70 **101.** 2.53 **102.** 2.13

103. $k = 5$ **104.** $k = -\frac{17}{12}$ **105.** -7 **106.** 1 **107.** If $f(x) = x^n - c^n$, then $f(c) = c^n - c^n = 0$, so $x - c$ is a factor of f.
108. If $n \geq 1$ is odd, then $(-c)^n = -c^n$, so $f(-c) = (-c)^n + c^n = -c^n + c^n = 0$. Thus, $x + c$ is a factor of f when n is odd.
109. 5 **110.** -3 **111.** No, by the Rational Zeros Theorem, $\frac{1}{3}$ is not a potential rational zero. **112.** No, by the Rational Zeros Theorem, $\frac{1}{3}$ is not a potential rational zero. **113.** No, by the Rational Zeros Theorem, $\frac{3}{5}$ is not a potential rational zero.
114. No, by the Rational Zeros Theorem, $\frac{2}{3}$ is not a potential rational zero. **115.** 7 in. **116.** 6 cm or 12 cm
117. All the potential rational zeros are integers. Hence, r is either an integer or is not a rational zero (and is therefore irrational).
118. Let $\frac{p}{q}$, where p and q have no common factors except 1 and -1, be a solution of the polynomial

$$f(x) = a_n x^n + a_{n-1}x^{n-1} + \cdots + a_1 x + a_0$$

whose coefficients are all integers. Then

$$f\left(\frac{p}{q}\right) = a_n\left(\frac{p}{q}\right)^n + a_{n-1}\left(\frac{p}{q}\right)^{n-1} + \cdots + a_1\left(\frac{p}{q}\right) + a_0 = 0$$

$$a_n p^n + a_{n-1}p^{n-1}q + \cdots + a_1 p q^{n-1} + a_0 q^n = 0$$

Because p is a factor of the first n terms of this equation, p must also be a factor of $a_0 q^n$.
Since p is not a factor of q (p and q have no common factors except 1 and -1),
p must be a factor of a_0. Similarly, q must be a factor of a_n.

3.7 Concepts and Vocabulary *(page 214)*

3. one **4.** $3 - 4i$ **5.** True **6.** False

3.7 Exercises *(page 214)*

7. $4 + i$ **8.** $3 - i$ **9.** $-i, 1 - i$ **10.** $2 - i$ **11.** $-i, -2i$ **12.** $-i$ **13.** $-i$ **14.** $2 + i, i$ **15.** $2 - i, -3 + i$ **16.** $-i, 3 + 2i, -2 - i$
17. $f(x) = x^4 - 14x^3 + 77x^2 - 200x + 208; a = 1$ **18.** $f(x) = x^4 - 2x^3 + 6x^2 - 2x + 5; a = 1$
19. $f(x) = x^5 - 4x^4 + 7x^3 - 8x^2 + 6x - 4; a = 1$ **20.** $f(x) = x^6 - 12x^5 + 55x^4 - 120x^3 + 139x^2 - 108x + 85; a = 1$
21. $f(x) = x^4 - 6x^3 + 10x^2 - 6x + 9; a = 1$ **22.** $f(x) = x^5 - 5x^4 + 11x^3 - 13x^2 + 8x - 2; a = 1$ **23.** $-2i, 4$ **24.** $5i, -3$
25. $2i, -3, \frac{1}{2}$ **26.** $-3i, -2, \frac{1}{3}$ **27.** $3 + 2i, -2, 5$ **28.** $1 - 3i, -1, 6$ **29.** $4i, -\sqrt{11}, \sqrt{11}, -\frac{2}{3}$ **30.** $-3i, -3, 4, \frac{1}{2}$
31. $1, -\frac{1}{2} - \frac{\sqrt{3}}{2}i, -\frac{1}{2} + \frac{\sqrt{3}}{2}i; f(x) = (x - 1)\left(x + \frac{1}{2} + \frac{\sqrt{3}}{2}i\right)\left(x + \frac{1}{2} - \frac{\sqrt{3}}{2}i\right)$ **32.** $-i, i, -1, 1; f(x) = (x + i)(x - i)(x + 1)(x - 1)$
33. $2, 3 - 2i, 3 + 2i; f(x) = (x - 2)(x - 3 + 2i)(x - 3 - 2i)$ **34.** $-5, -4 + i, -4 - i; f(x) = (x + 5)(x + 4 - i)(x + 4 + i)$
35. $-i, i, -2i, 2i; f(x) = (x + i)(x - i)(x + 2i)(x - 2i)$ **36.** $-2i, 2i, -3i, 3i; f(x) = (x + 2i)(x - 2i)(x + 3i)(x - 3i)$
37. $-5i, 5i, -3, 1; f(x) = (x + 5i)(x - 5i)(x + 3)(x - 1)$ **38.** $-3i, 3i, -7, 4; f(x) = (x + 3i)(x - 3i)(x + 7)(x - 4)$
39. $-4, \frac{1}{3}, 2 - 3i, 2 + 3i; f(x) = 3(x + 4)\left(x - \frac{1}{3}\right)(x - 2 + 3i)(x - 2 - 3i)$
40. $-3 - 2i, -3 + 2i, \frac{1}{2}, 5; f(x) = 2(x + 3 + 2i)(x + 3 - 2i)\left(x - \frac{1}{2}\right)(x - 5)$ **41.** Zeros that are complex numbers must occur in conjugate pairs; or a polynomial with real coefficients of odd degree must have at least one real zero.
42. Zeros that are complex numbers must occur in conjugate pairs. **43.** If the remaining zero were a complex number, then its conjugate would also be a zero, creating a polynomial of degree 5. **44.** A missing zero is $4 + i$. If the remaining zero were a complex number, then its conjugate would also be a zero, creating a polynomial of degree 5.

Review Exercises *(page 216)*

1.

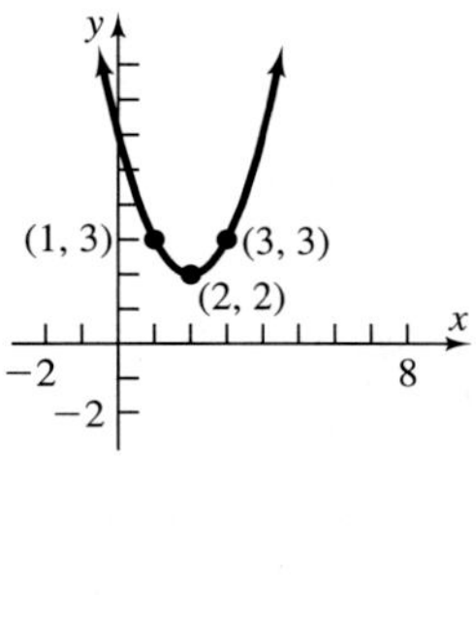

2.

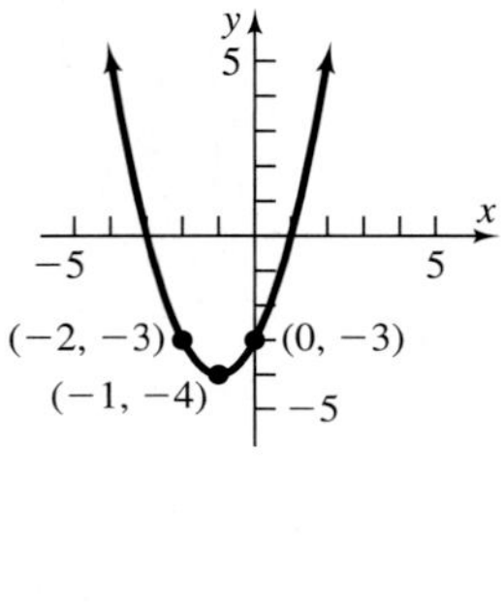

3.

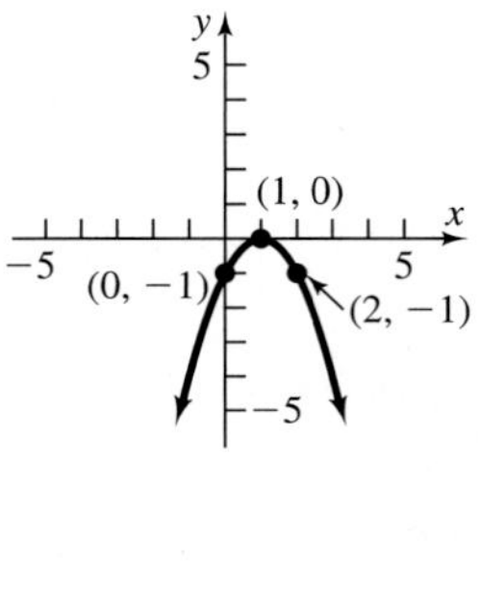

4.

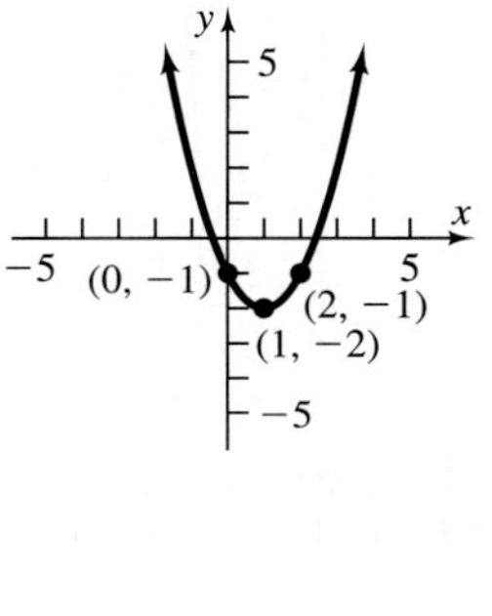

5.
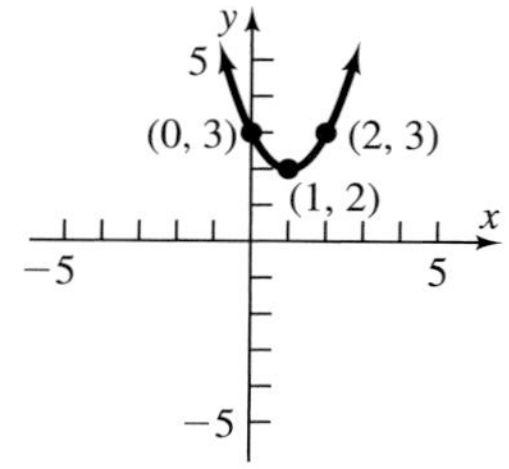

6.
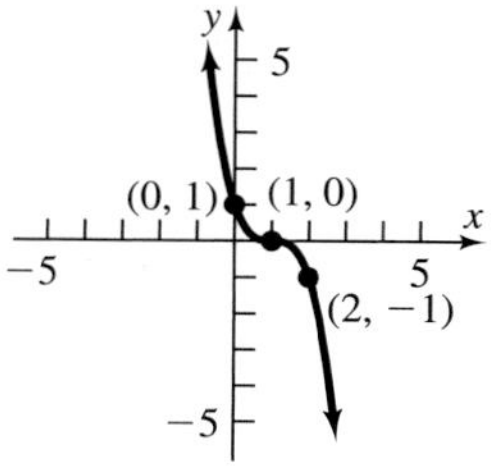

7.
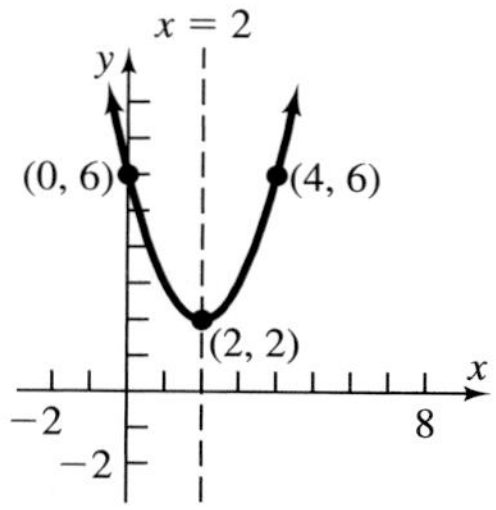

8.
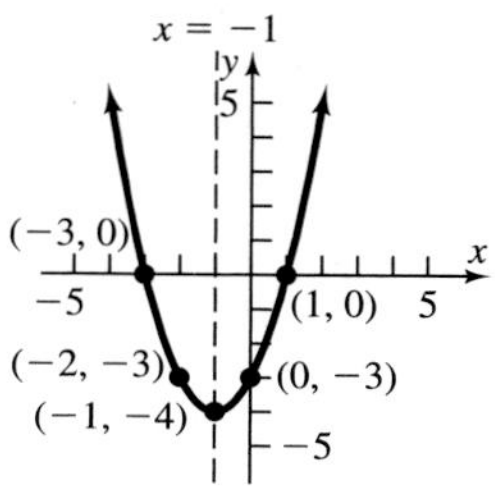

9.
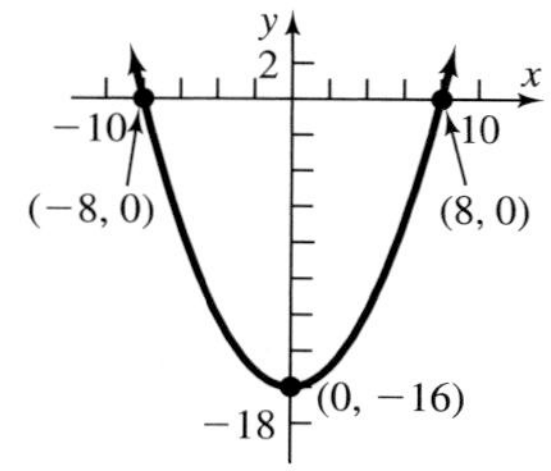

10.
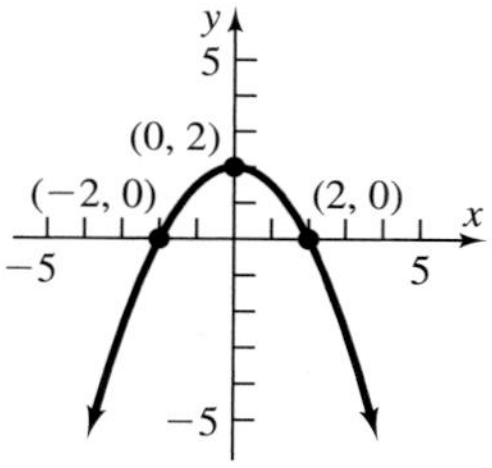

11.
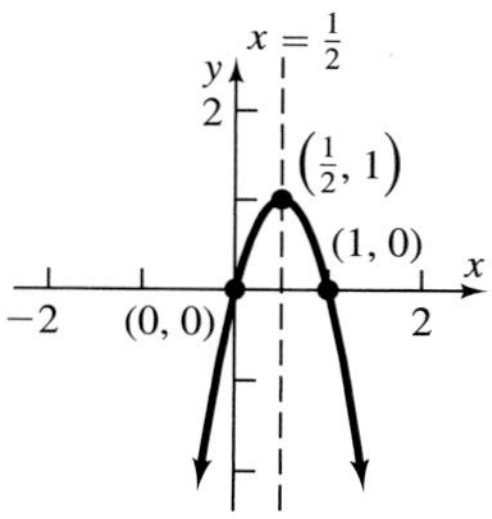

12.
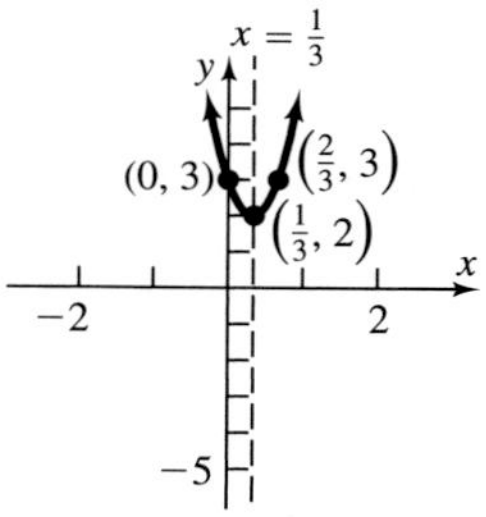

13.
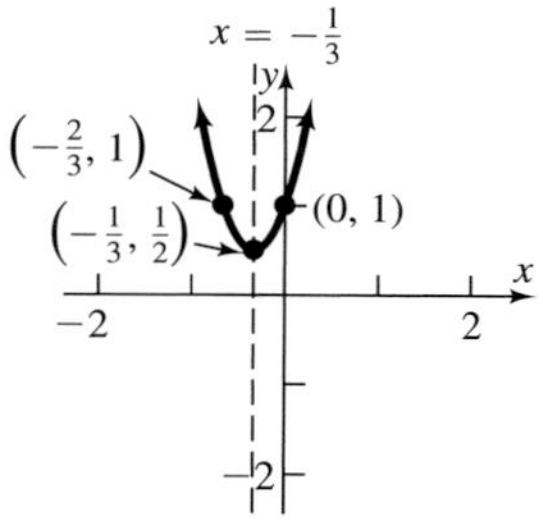

14.
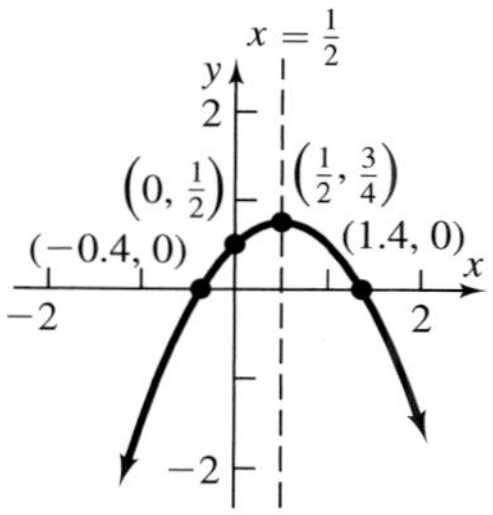

15.
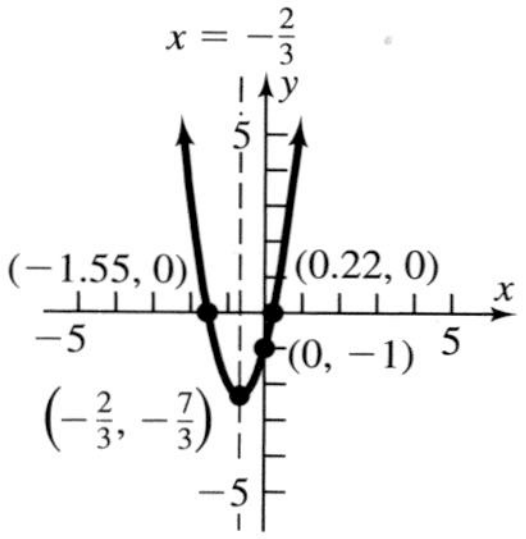

16.
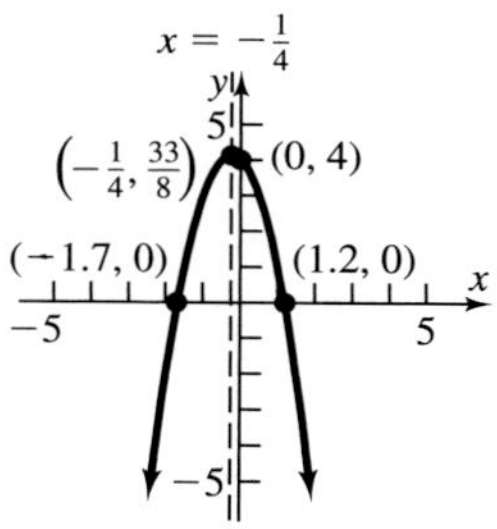

17. Minimum value; 1 **18.** Minimum value; −3 **19.** Maximum value; 12 **20.** Maximum value; 22 **21.** Maximum value; 16 **22.** Maximum value; 4 **23.** Polynomial of degree 5 **24.** Not a polynomial **25.** Not a polynomial **26.** Polynomial of degree 0

27.
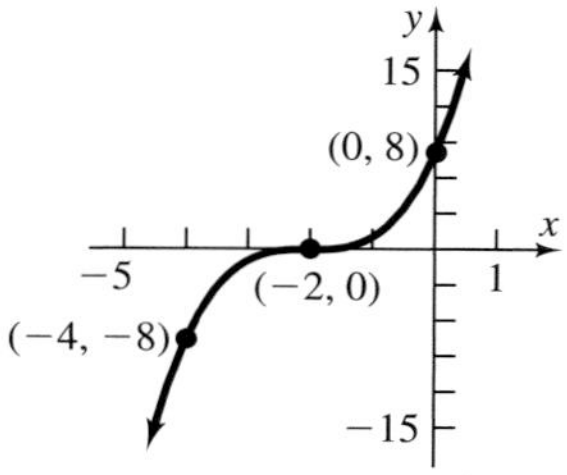

28.
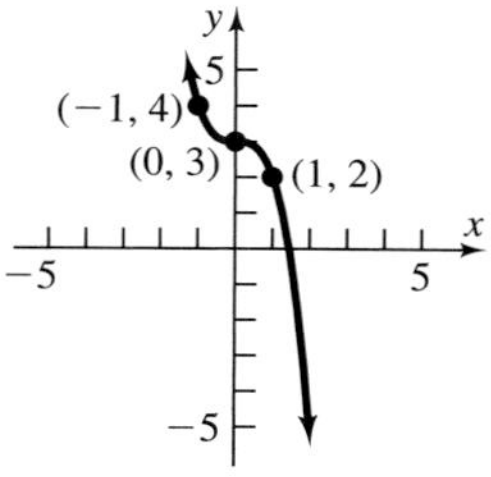

29.
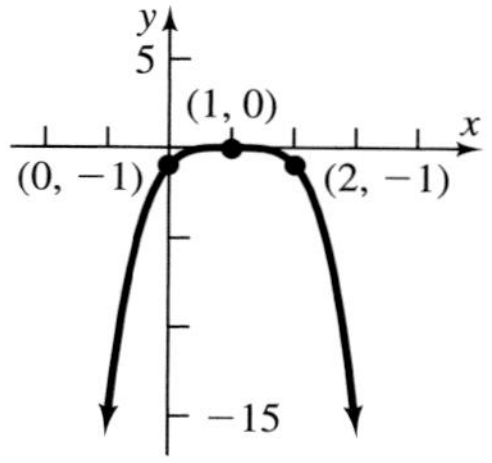

30.
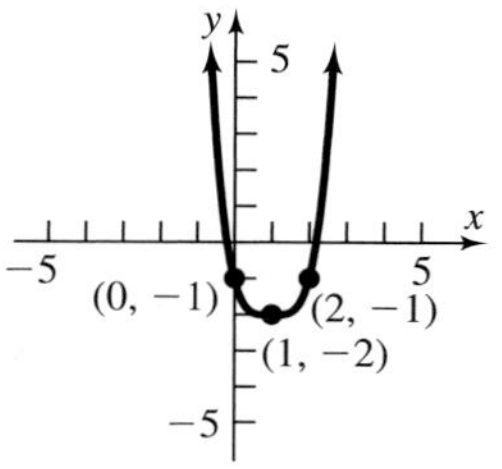

31.
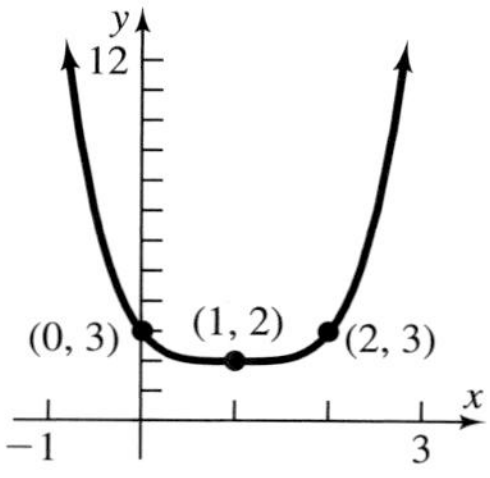

32.
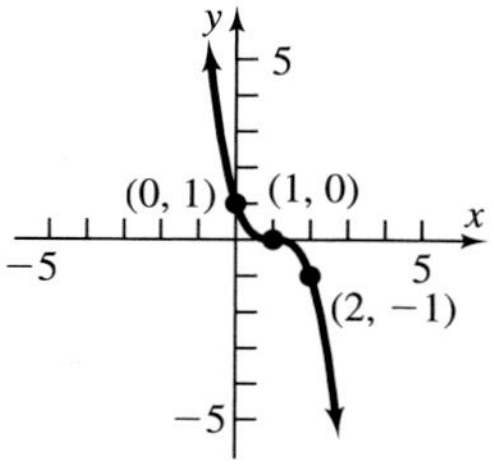

33. **(a)** x-intercepts: $-4, -2, 0$; y-intercept: 0
(b) Crosses at -4, -2, 0
(c) $y = x^3$
(d) 2
(e)

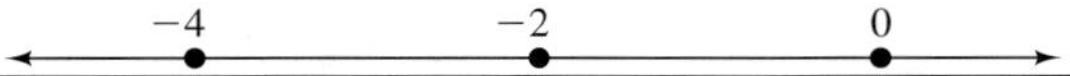

Interval	$(-\infty, -4)$	$(-4, -2)$	$(-2, 0)$	$(0, \infty)$
Number Chosen	-5	-3	-1	1
Value of f	$f(-5) = -15$	$f(-3) = 3$	$f(-1) = -3$	$f(1) = 15$
Location of Graph	Below x-axis	Above x-axis	Below x-axis	Above x-axis
Point on Graph	$(-5, -15)$	$(-3, 3)$	$(-1, -3)$	$(1, 15)$

(f)

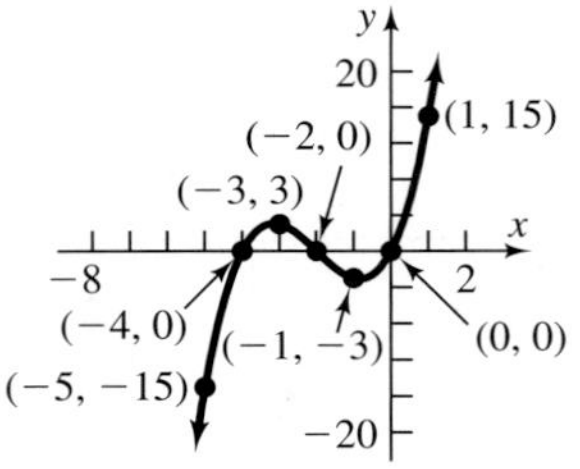

34. **(a)** x-intercepts: 0, 2, 4; y-intercept: 0
(b) Crosses at 0, 2, and 4
(c) $y = x^3$
(d) 2
(e)

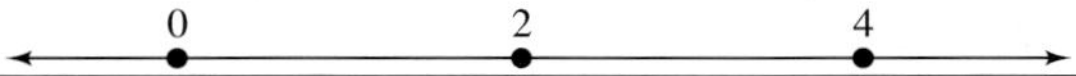

Interval	$(-\infty, 0)$	$(0, 2)$	$(2, 4)$	$(4, \infty)$
Number Chosen	-1	1	3	5
Value of f	$f(-1) = -15$	$f(1) = 3$	$f(3) = -3$	$f(5) = 15$
Location of Graph	Below x-axis	Above x-axis	Below x-axis	Above x-axis
Point on Graph	$(-1, -15)$	$(1, 3)$	$(3, -3)$	$(5, 15)$

(f)

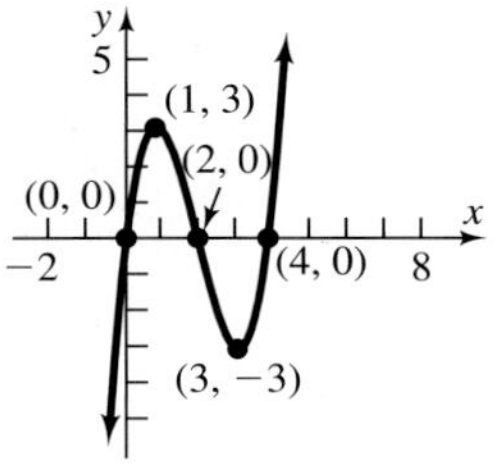

35. **(a)** x-intercepts: $-4, 2$; y-intercept: 16
(b) Crosses at -4; touches at 2
(c) $y = x^3$
(d) 2
(e)

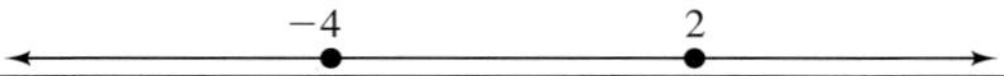

Interval	$(-\infty, -4)$	$(-4, 2)$	$(2, \infty)$
Number Chosen	-5	-2	3
Value of f	$f(-5) = -49$	$f(-2) = 32$	$f(3) = 7$
Location of Graph	Below x-axis	Above x-axis	Above x-axis
Point on Graph	$(-5, -49)$	$(-2, 32)$	$(3, 7)$

(f)

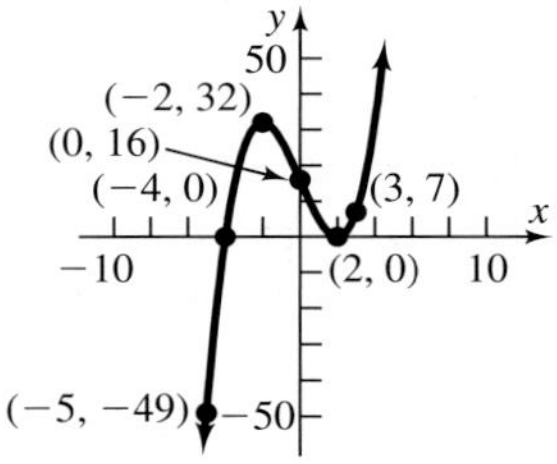

36. **(a)** x-intercepts: $-4, 2$; y-intercept: -32
(b) Touches at -4; crosses at 2
(c) $y = x^3$
(d) 2
(e)

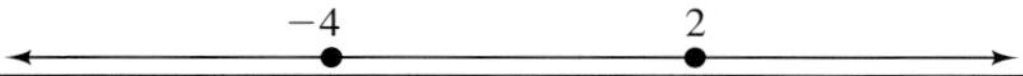

Interval	$(-\infty, -4)$	$(-4, 2)$	$(2, \infty)$
Number Chosen	-5	0	3
Value of f	$f(-5) = -7$	$f(0) = -32$	$f(3) = 49$
Location of Graph	Below x-axis	Below x-axis	Above x-axis
Point on Graph	$(-5, -7)$	$(0, -32)$	$(3, 49)$

(f)

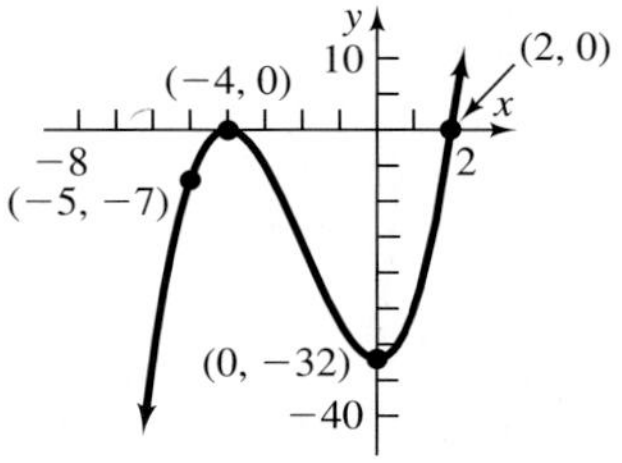

37. (a) x-intercepts: 0, 2; y-intercept: 0
(b) Touches at 0; crosses at 2
(c) $y = -2x^3$
(d) 2
(e)

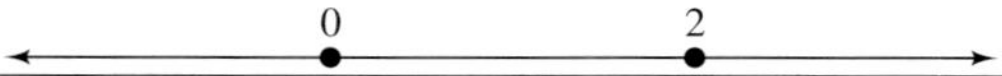

Interval	$(-\infty, 0)$	$(0, 2)$	$(2, \infty)$
Number Chosen	-1	1	3
Value of f	$f(-1) = 6$	$f(1) = 2$	$f(3) = -18$
Location of Graph	Above x-axis	Above x-axis	Below x-axis
Point on Graph	$(-1, 6)$	$(1, 2)$	$(3, -18)$

(f)

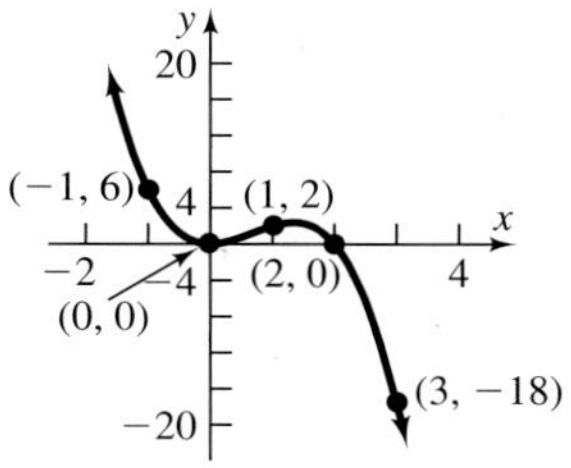

38. (a) x-intercepts: $-1, 0, 1$; y-intercept: 0
(b) Crosses at $-1, 0, 1$
(c) $y = -4x^3$
(d) 2
(e)

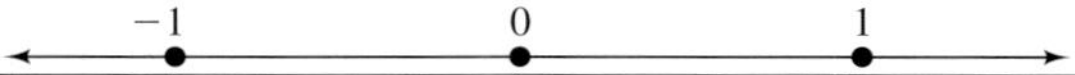

Interval	$(-\infty, -1)$	$(-1, 0)$	$(0, 1)$	$(1, \infty)$
Number Chosen	-2	$-\frac{1}{2}$	$\frac{1}{2}$	2
Value of f	$f(-2) = 24$	$f\left(-\frac{1}{2}\right) = -\frac{3}{2}$	$f\left(\frac{1}{2}\right) = \frac{3}{2}$	$f(2) = -24$
Location of Graph	Above x-axis	Below x-axis	Above x-axis	Below x-axis
Point on Graph	$(-2, 24)$	$\left(-\frac{1}{2}, -\frac{3}{2}\right)$	$\left(\frac{1}{2}, \frac{3}{2}\right)$	$(2, -24)$

(f)

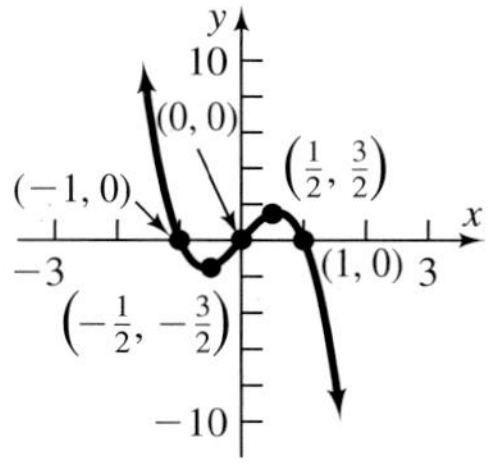

39. (a) x-intercepts: $-3, -1, 1$; y-intercept: 3
(b) Crosses at $-3, -1$; touches at 1
(c) $y = x^4$
(d) 3
(e)

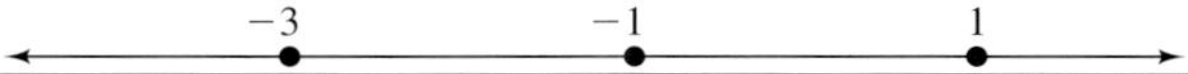

Interval	$(-\infty, -3)$	$(-3, -1)$	$(-1, 1)$	$(1, \infty)$
Number Chosen	-4	-2	0	2
Value of f	$f(-4) = 75$	$f(-2) = -9$	$f(0) = 3$	$f(2) = 15$
Location of Graph	Above x-axis	Below x-axis	Above x-axis	Above x-axis
Point on Graph	$(-4, 75)$	$(-2, -9)$	$(0, 3)$	$(2, 15)$

(f)

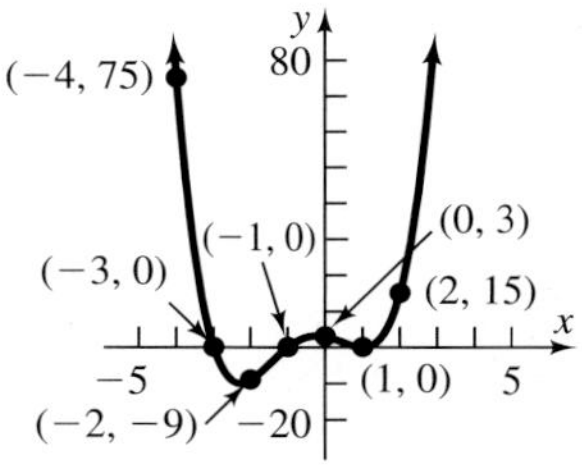

40. (a) x-intercepts: $-2, 2, 4$; y-intercept: 32
(b) Crosses at 2 and 4; touches at -2
(c) $y = x^4$
(d) 3
(e)

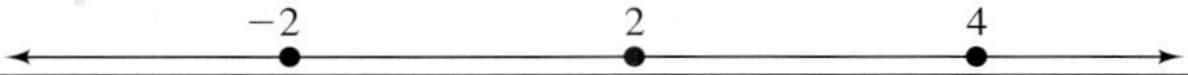

Interval	$(-\infty, -2)$	$(-2, 2)$	$(2, 4)$	$(4, \infty)$
Number Chosen	-3	0	3	5
Value of f	$f(-3) = 35$	$f(0) = 32$	$f(3) = -25$	$f(5) = 147$
Location of Graph	Above x-axis	Above x-axis	Below x-axis	Above x-axis
Point on Graph	$(-3, 35)$	$(0, 32)$	$(3, -25)$	$(5, 147)$

(f)

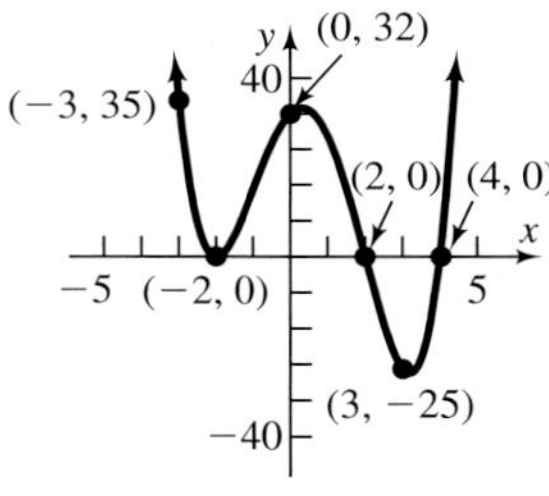

41. Domain: $\{x|x \neq -3, x \neq 3\}$; Horizontal Asymptote: $y = 0$; Vertical Asymptotes: $x = -3$, $x = 3$
42. Domain: $\{x|x \neq 2\}$; Oblique Asymptote: $y = x + 2$; Vertical Asymptote: $x = 2$
43. Domain: $\{x|x \neq -2\}$; Horizontal Asymptote: $y = 1$; Vertical Asymptote: $x = -2$
44. Domain: $\{x|x \neq 1\}$; Horizontal Asymptote: $y = 1$; Vertical Asymptote: $x = 1$
45. 1. Domain: $\{x|x \neq 0\}$
2. x-intercept: 3; no y-intercept
3. No y-axis or origin symmetry
4. Vertical asymptote: $x = 0$
5. Horizontal asymptote: $y = 2$; not intersected
6.

Interval	$(-\infty, 0)$	$(0, 3)$	$(3, \infty)$
Number Chosen	-2	1	4
Value of R	$R(-2) = 5$	$R(1) = -4$	$R(4) = \frac{1}{2}$
Location of Graph	Above x-axis	Below x-axis	Above x-axis
Point on Graph	$(-2, 5)$	$(1, -4)$	$\left(4, \frac{1}{2}\right)$

7.

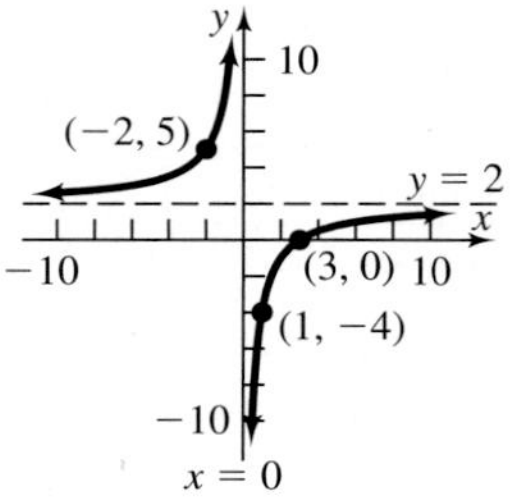

46. 1. Domain: $\{x|x \neq 0\}$
2. x-intercept: 4; no y-intercept
3. No y-axis or origin symmetry
4. Vertical asymptote: $x = 0$
5. Horizontal asymptote: $y = -1$; not intersected
6.

Interval	$(-\infty, 0)$	$(0, 4)$	$(4, \infty)$
Number Chosen	-1	1	5
Value of R	$R(-1) = -5$	$R(1) = 3$	$R(5) = -0.2$
Location of Graph	Below x-axis	Above x-axis	Below x-axis
Point on Graph	$(-1, -5)$	$(1, 3)$	$(5, -0.2)$

7.

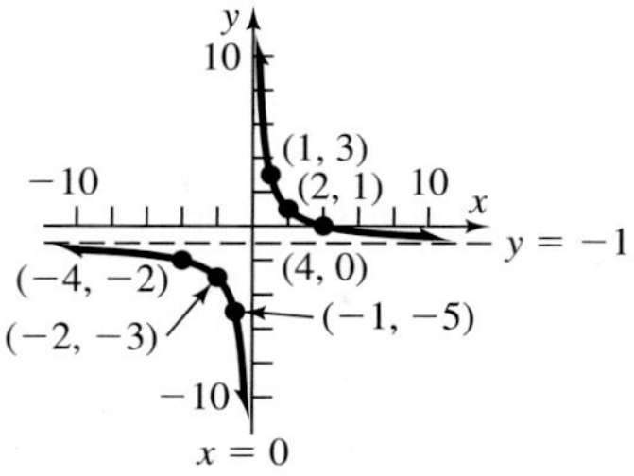

47. 1. Domain: $\{x|x \neq 0, x \neq 2\}$
2. x-intercept: -2; no y-intercept
3. No y-axis or origin symmetry
4. Vertical asymptotes: $x = 0$, $x = 2$
5. Horizontal asymptote: $y = 0$; intersected at $(-2, 0)$
6.

Interval	$(-\infty, -2)$	$(-2, 0)$	$(0, 2)$	$(2, \infty)$
Number Chosen	-3	-1	1	3
Value of H	$H(-3) = -\frac{1}{15}$	$H(-1) = \frac{1}{3}$	$H(1) = -3$	$H(3) = \frac{5}{3}$
Location of Graph	Below x-axis	Above x-axis	Below x-axis	Above x-axis
Point on Graph	$\left(-3, -\frac{1}{15}\right)$	$\left(-1, \frac{1}{3}\right)$	$(1, -3)$	$\left(3, \frac{5}{3}\right)$

7.

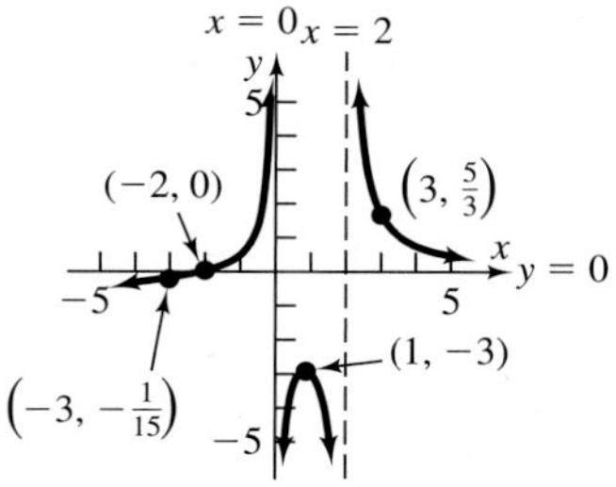

48. 1. Domain: $\{x|x \neq -1, x \neq 1\}$
2. x-intercept: 0; y-intercept: 0
3. Symmetric with respect to the origin
4. Vertical asymptotes: $x = -1$ and $x = 1$
5. Horizontal asymptote: $y = 0$; intersected at $(0, 0)$
6. Number line: -1, 0, 1

Interval	$(-\infty, -1)$	$(-1, 0)$	$(0, 1)$	$(1, \infty)$
Number Chosen	-2	$-\frac{1}{2}$	$\frac{1}{2}$	2
Value of H	$H(-2) = -\frac{2}{3}$	$H\left(-\frac{1}{2}\right) = \frac{2}{3}$	$H\left(\frac{1}{2}\right) = -\frac{2}{3}$	$H(2) = \frac{2}{3}$
Location of Graph	Below x-axis	Above x-axis	Below x-axis	Above x-axis
Point on Graph	$\left(-2, -\frac{2}{3}\right)$	$\left(-\frac{1}{2}, \frac{2}{3}\right)$	$\left(\frac{1}{2}, -\frac{2}{3}\right)$	$\left(2, \frac{2}{3}\right)$

7.
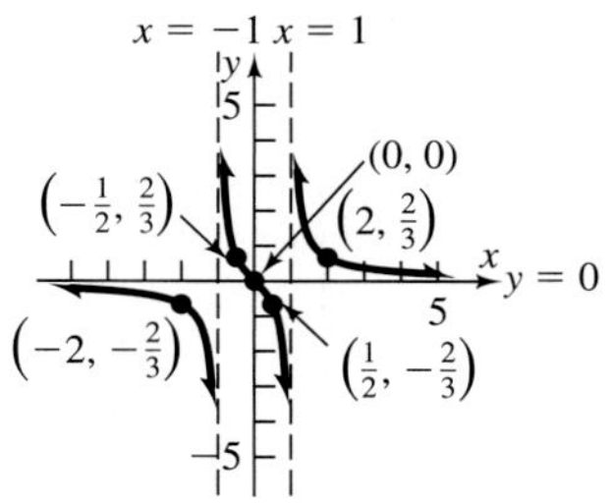

49. 1. Domain: $\{x|x \neq -2, x \neq 3\}$
2. x-intercepts: -3, 2; y-intercept: 1
3. No y-axis or origin symmetry
4. Vertical asymptote: $x = -2$, $x = 3$
5. Horizontal asymptote: $y = 1$; intersected at $(0, 1)$
6. Number line: -3, -2, 2, 3

Interval	$(-\infty, -3)$	$(-3, -2)$	$(-2, 2)$	$(2, 3)$	$(3, \infty)$
Number Chosen	-4	-2.5	0	2.5	4
Value of R	$R(-4) \approx 0.43$	$R(-2.5) \approx -0.82$	$R(0) = 1$	$R(2.5) \approx -1.22$	$R(4) = \frac{7}{3}$
Location of Graph	Above x-axis	Below x-axis	Above x-axis	Below x-axis	Above x-axis
Point on Graph	$(-4, 0.43)$	$(-2.5, -0.82)$	$(0, 1)$	$(2.5, -1.22)$	$\left(4, \frac{7}{3}\right)$

7.
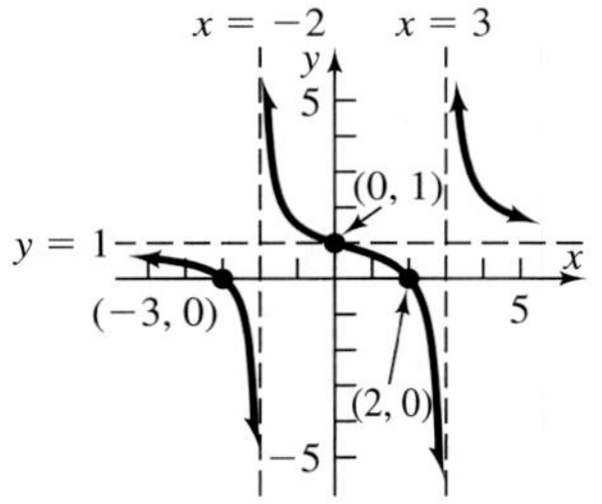

50. 1. Domain: $\{x|x \neq 0\}$
2. x-intercept: 3; no y-intercept
3. No y-axis or origin symmetry
4. Vertical asymptote: $x = 0$
5. Horizontal asymptote: $y = 1$; intersected at $\left(\frac{3}{2}, 1\right)$
6. Number line: 0, 3

Interval	$(-\infty, 0)$	$(0, 3)$	$(3, \infty)$
Number Chosen	-3	1	4
Value of R	$R(-3) = 4$	$R(1) = 4$	$R(4) = \frac{1}{16}$
Location of Graph	Above x-axis	Above x-axis	Above x-axis
Point on Graph	$(-3, 4)$	$(1, 4)$	$\left(4, \frac{1}{16}\right)$

7.
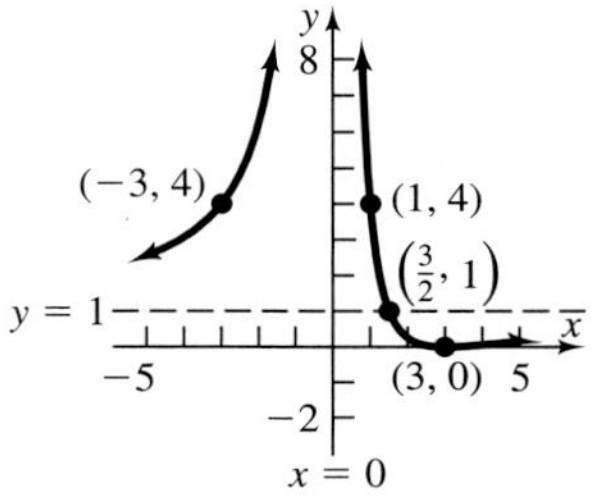

51. 1. Domain: $\{x|x \neq -2, x \neq 2\}$
2. x-intercept: 0; y-intercept: 0
3. Symmetric with respect to the origin
4. Vertical asymptotes: $x = -2$, $x = 2$
5. Oblique asymptote: $y = x$; intersected at $(0, 0)$
6.

$-2 \quad 0 \quad 2$

Interval	$(-\infty, -2)$	$(-2, 0)$	$(0, 2)$	$(2, \infty)$
Number Chosen	-3	-1	1	3
Value of F	$F(-3) = -\frac{27}{5}$	$F(-1) = \frac{1}{3}$	$F(1) = -\frac{1}{3}$	$F(3) = \frac{27}{5}$
Location of Graph	Below x-axis	Above x-axis	Below x-axis	Above x-axis
Point on Graph	$\left(-3, -\frac{27}{5}\right)$	$\left(-1, \frac{1}{3}\right)$	$\left(1, -\frac{1}{3}\right)$	$\left(3, \frac{27}{5}\right)$

7.

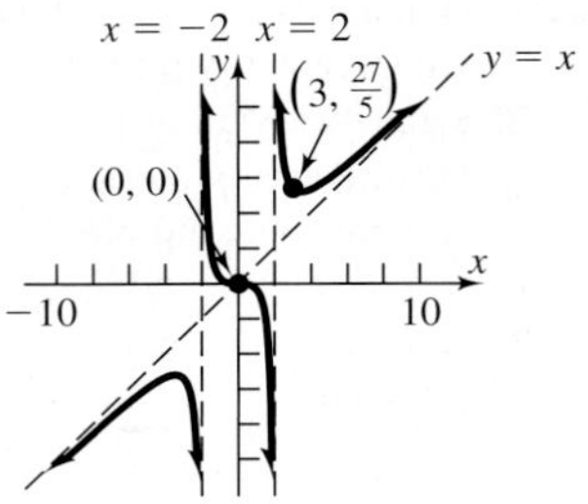

52. 1. Domain: $\{x|x \neq 1\}$
2. x-intercept: 0; y-intercept: 0
3. No y-axis or origin symmetry
4. Vertical asymptote: $x = 1$
5. Oblique asymptote: $y = 3x + 6$; intersected at $\left(\frac{2}{3}, 8\right)$
6.

$0 \quad 1$

Interval	$(-\infty, 0)$	$(0, 1)$	$(1, \infty)$
Number Chosen	-1	$\frac{1}{2}$	2
Value of F	$F(-1) = -\frac{3}{4}$	$F\left(\frac{1}{2}\right) = 1.5$	$F(2) = 24$
Location of Graph	Below x-axis	Above x-axis	Above x-axis
Point on Graph	$\left(-1, -\frac{3}{4}\right)$	$\left(\frac{1}{2}, 1.5\right)$	$(2, 24)$

7.

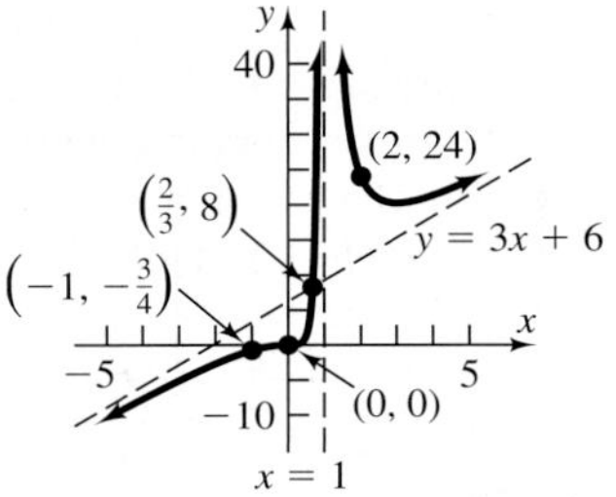

53. 1. Domain: $\{x|x \neq 1\}$
2. x-intercept: 0; y-intercept: 0
3. No y-axis or origin symmetry
4. Vertical asymptote: $x = 1$
5. No oblique or horizontal asymptote
6.

$0 \quad 1$

Interval	$(-\infty, 0)$	$(0, 1)$	$(1, \infty)$
Number Chosen	-2	$\frac{1}{2}$	2
Value of R	$R(-2) \approx \frac{32}{9}$	$R\left(\frac{1}{2}\right) = \frac{1}{2}$	$R(2) = 32$
Location of Graph	Above x-axis	Above x-axis	Above x-axis
Point on Graph	$\left(-2, \frac{32}{9}\right)$	$\left(\frac{1}{2}, \frac{1}{2}\right)$	$(2, 32)$

7.

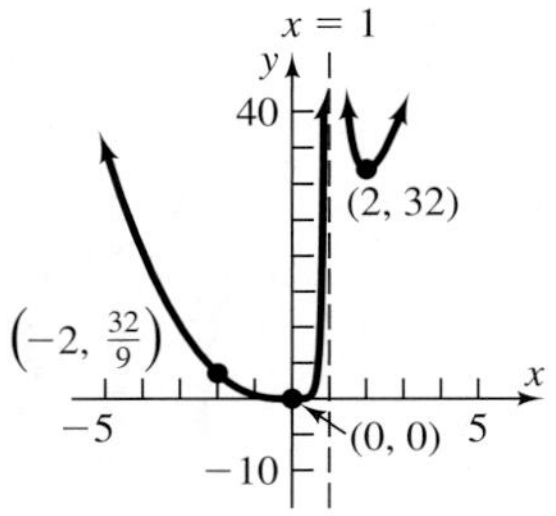

54. 1. Domain: $\{x|x \neq -3, x \neq 3\}$
2. x-intercept: 0; y-intercept: 0
3. Symmetric with respect to the y-axis
4. Vertical asymptotes: $x = 3$ and $x = -3$
5. No oblique or horizontal asymptote
6.

$-3 \quad 0 \quad 3$

Interval	$(-\infty, -3)$	$(-3, 0)$	$(0, 3)$	$(3, \infty)$
Number Chosen	-4	-1	1	4
Value of R	$R(-4) \approx 37$	$R(-1) = -\frac{1}{8}$	$R(1) = -\frac{1}{8}$	$R(4) \approx 37$
Location of Graph	Above x-axis	Below x-axis	Below x-axis	Above x-axis
Point on Graph	$(-4, 37)$	$\left(-1, -\frac{1}{8}\right)$	$\left(1, -\frac{1}{8}\right)$	$(4, 37)$

7.

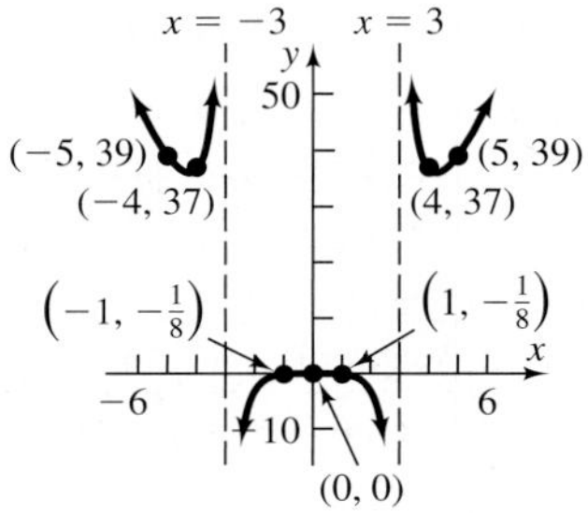

55. 1. Domain: $\{x | x \neq -1, x \neq 2\}$
2. x-intercept: -2; y-intercept: 2
3. No y-axis or origin symmetry
4. Vertical asymptote: $x = -1$; hole at $\left(2, \frac{4}{3}\right)$
5. Horizontal asymptote: $y = 1$, not intersected
6. -2, -1, 2

Interval	$(-\infty, -2)$	$(-2, -1)$	$(-1, 2)$	$(2, \infty)$
Number Chosen	-3	-1.5	0	3
Value of G	$G(-3) = \frac{1}{2}$	$G(-1.5) = -1$	$G(0) = 2$	$G(3) = 1.25$
Location of Graph	Above x-axis	Below x-axis	Above x-axis	Above x-axis
Point on Graph	$\left(-3, \frac{1}{2}\right)$	$(-1.5, -1)$	$(0, 2)$	$(3, 1.25)$

7.

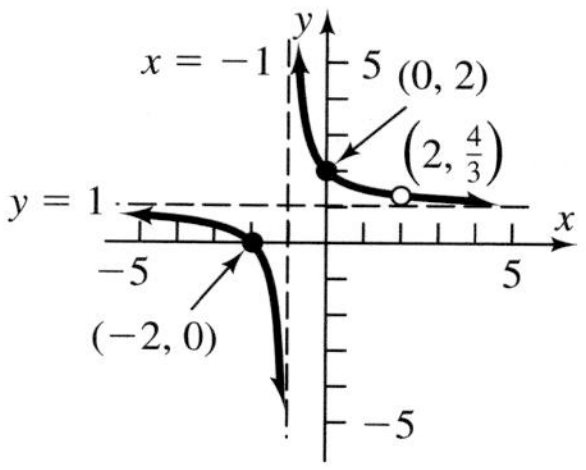

56. 1. Domain: $\{x | x \neq -1, x \neq 1\}$
2. No x-intercepts; y-intercept: -1
3. No y-axis or origin symmetry
4. Vertical asymptote: $x = -1$; hole at $(1, 0)$
5. Horizontal asymptote: $y = 1$, not intersected
6. -1, 1

Interval	$(-\infty, -1$	$(-1, 1)$	$(1, \infty)$
Number Chosen	-2	0	3
Value of F	$F(-2) = 3$	$F(0) = -1$	$F(3) = \frac{1}{2}$
Location of Graph	Above x-axis	Below x-axis	Above x-axis
Point on Graph	$(-2, 3)$	$(0, -1)$	$\left(3, \frac{1}{2}\right)$

7.

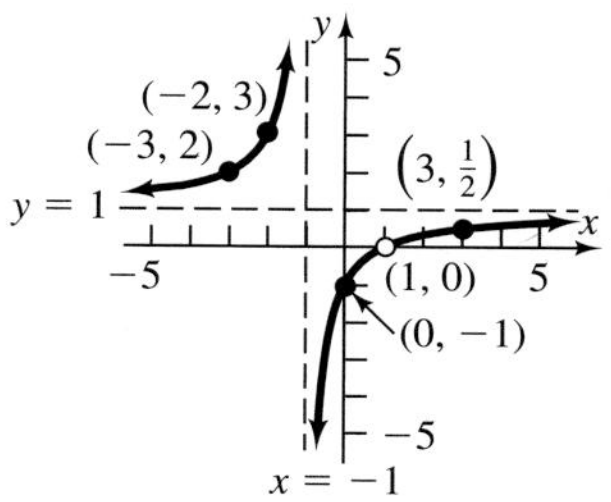

57. $\left\{x \middle| -4 < x < \frac{3}{2}\right\}; \left(-4, \frac{3}{2}\right)$ **58.** $\left\{x \middle| x \leq -\frac{1}{3} \text{ or } x \geq 1\right\}; \left(-\infty, -\frac{1}{3}\right]$ or $[1, \infty)$ **59.** $\{x | -3 < x \leq 3\}; (-3, 3]$

60. $\left\{x \middle| x < \frac{1}{3} \text{ or } x > 1\right\}; \left(-\infty, \frac{1}{3}\right)$ or $(1, \infty)$ **61.** $\{x | x < 1 \text{ or } x > 2\}; (-\infty, 1)$ or $(2, \infty)$ **62.** $\left\{x \middle| -\frac{5}{2} < x \leq -\frac{7}{6}\right\}; \left(-\frac{5}{2}, -\frac{7}{6}\right]$

63. $\{x | 1 \leq x \leq 2 \text{ or } x > 3\}; [1, 2]$ or $(3, \infty)$ **64.** $\{x | x \leq -1 \text{ or } 0 < x < 5\}; (-\infty, -1]$ or $(0, 5)$

65. $\{x | x < -4 \text{ or } 2 < x < 4 \text{ or } x > 6\}; (-\infty, -4)$ or $(2, 4)$ or $(6, \infty)$

66. $\{x | x < -5 \text{ or } -4 < x \leq -2 \text{ or } 0 \leq x \leq 1\}; (-\infty, -5)$ or $(-4, -2]$ or $[0, 1]$

67. $q(x) = 8x^2 + 5x + 6; R = 10$; g is not a factor of f. **68.** $q(x) = 2x^2 + 12x + 19; R = 43$; g is not a factor of f.

69. $q(x) = x^3 - 4x^2 + 8x - 1; R = 0$; g is a factor of f. **70.** $q(x) = x^3 - x^2 + 2; R = 0$; g is a factor of f.

71. $f(4) = 47{,}105$ **72.** $f(-2) = 204$ **73.** 4, 2, or 0 positive; 2 or 0 negative **74.** 1 positive; 2 or 0 negative

75. $\pm 1, \pm 3, \pm \frac{1}{2}, \pm \frac{3}{2}, \pm \frac{1}{3}, \pm \frac{1}{4}, \pm \frac{3}{4}, \pm \frac{1}{6}, \pm \frac{1}{12}$ **76.** $\pm 1, \pm \frac{1}{2}, \pm \frac{1}{3}, \pm \frac{1}{6}$ **77.** $-2, 1, 4; f(x) = (x + 2)(x - 1)(x - 4)$

78. $-1, 4, -2; f(x) = (x + 1)(x - 4)(x + 2)$ **79.** $\frac{1}{2}$, multiplicity 2; -2; $f(x) = 4\left(x - \frac{1}{2}\right)^2(x + 2)$

80. $-\frac{1}{2}$, multiplicity 2; 2; $f(x) = 4(x - 2)\left(x + \frac{1}{2}\right)^2$ **81.** 2, multiplicity 2; $f(x) = (x - 2)^2(x^2 + 5)$

82. -3, multiplicity 2; $f(x) = (x + 3)^2(x^2 + 2)$ **83.** $\{-3, 2\}$ **84.** $\{-3, 2\}$ **85.** $\left\{-3, -1, -\frac{1}{2}, 1\right\}$ **86.** $\left\{-3, -2, -\frac{1}{2}, 2\right\}$

87. $-2, 1, 4; f(x) = (x + 2)(x - 1)(x - 4)$ **88.** $-2, -1, 4; f(x) = (x + 2)(x + 1)(x - 4)$

89. $-2; \frac{1}{2}$ (multiplicity 2); $f(x) = 4(x + 2)\left(x - \frac{1}{2}\right)^2$ **90.** $-\frac{1}{2}$ (multiplicity 2), 2; $f(x) = 4\left(x + \frac{1}{2}\right)^2(x - 2)$

91. 2 (multiplicity 2), $-\sqrt{5}i, \sqrt{5}i; f(x) = (x + \sqrt{5}i)(x - \sqrt{5}i)(x - 2)^2$

92. -3 (multiplicity 2), $-\sqrt{2}i, \sqrt{2}i; f(x) = (x + 3)^2(x + \sqrt{2}i)(x - \sqrt{2}i)$

93. $-3, 2, -\frac{\sqrt{2}}{2}i, \frac{\sqrt{2}}{2}i; f(x) = 2(x + 3)(x - 2)\left(x + \frac{\sqrt{2}}{2}i\right)\left(x - \frac{\sqrt{2}}{2}i\right)$

94. $-3, 2, -\frac{\sqrt{3}}{3}i, \frac{\sqrt{3}}{3}i; f(x) = 3(x + 3)(x - 2)\left(x + \frac{\sqrt{3}}{3}i\right)\left(x - \frac{\sqrt{3}}{3}i\right)$ **95.** 5 **96.** 11 **97.** $\frac{37}{2}$ **98.** $\frac{17}{3}$ **99.** $f(0) = -1; f(1) = 1$

100. $f(1) = -2, f(2) = 9$ **101.** $f(0) = -1; f(1) = 1$ **102.** $f(1) = -3, f(2) = 62$ **103.** 1.52 **104.** 1.33 **105.** 0.93 **106.** 1.14 **107.** $4 - i$ **108.** $3 - 4i$ **109.** $-i, 1 - i$ **110.** $1 - i$ **111.** $(2, 2)$ **112.** 50 ft by 50 ft **113.** 4,166,666.7 m^2 **114.** 8 square units

115. The side with the semicircles should be $\frac{50}{\pi}$ ft; the other side should be 25 ft. **116.** 3.6 ft **117.** **(a)** 63 clubs **(b)** $151.90

118. **(a)** 1992 **(b)** 1901 violent crimes **(c)**

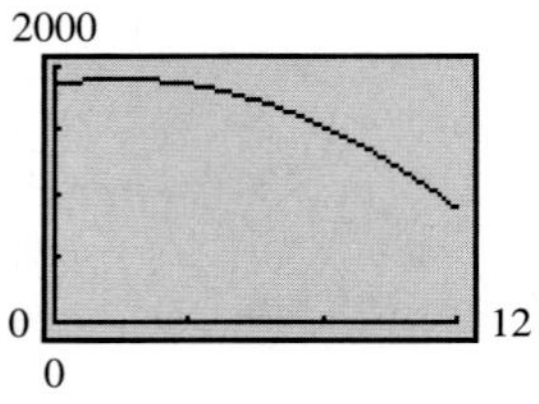

Decreasing

119. **(a)**

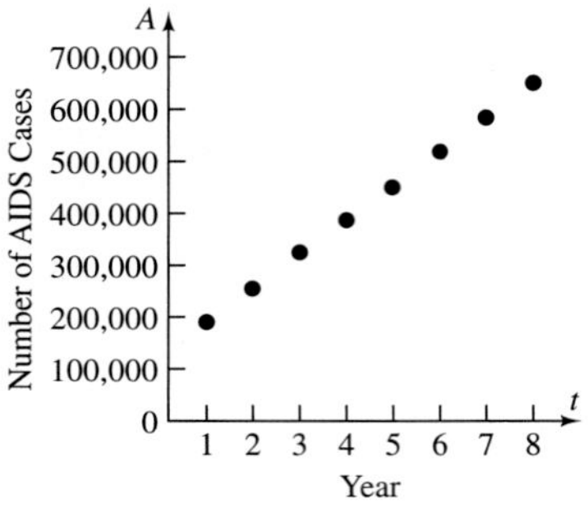

(b) 796,999 **(d)**

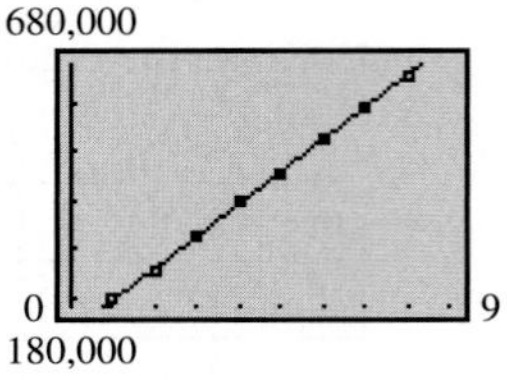

120. **(a)** $A(r) = 2\pi r^2 + \frac{500}{r}$ **(b)** 223.22 cm^2 **(c)** 257.08 cm^2 **(d)**

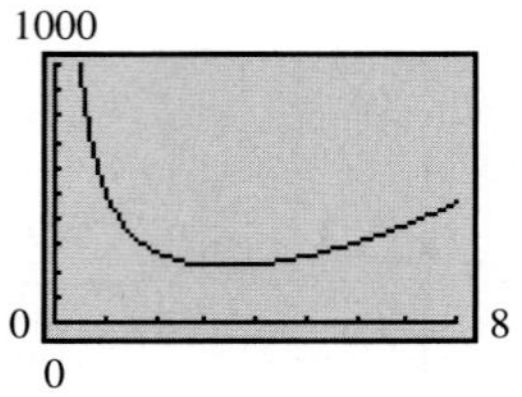

; A is smallest when $r \approx 3.41$ cm.

123. **(a)** even **(b)** positive **(c)** even **(d)** The graph touches the x-axis at $x = 0$, but does not cross it there. **(e)** 8

Cumulative Review *(page 221)*

1. $\sqrt{26}$

2. $\{x | x \le 0 \text{ or } x \ge 1\}$ or $(-\infty, 0] \cup [1, \infty)$

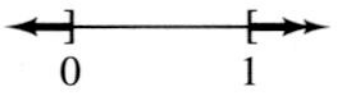

3. $\{x | -1 < x < 4\}$ or $(-1, 4)$

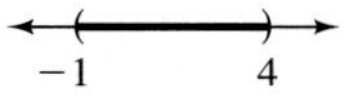

4. $f(x) = -3x + 1$

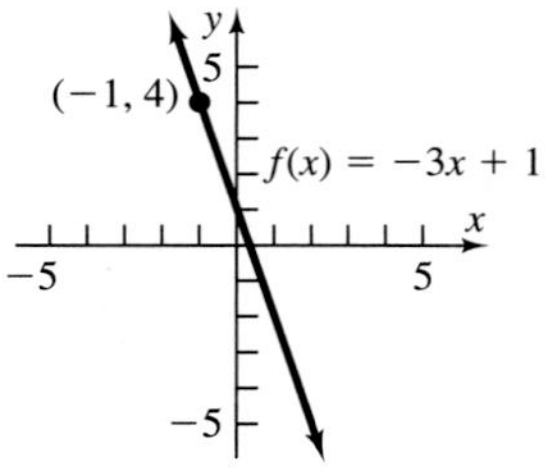

5. $y = 2x - 1$

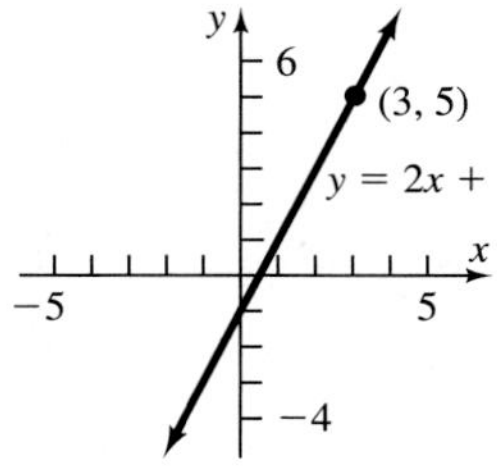

6.

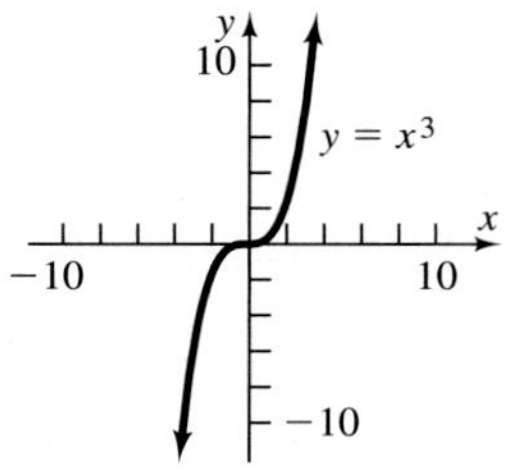

7. Not a function; 3 has two images **8.** $\{0, 2, 4\}$ **9.** $\left\{x \middle| x \ge \frac{3}{2}\right\}; \left[\frac{3}{2}, \infty\right)$

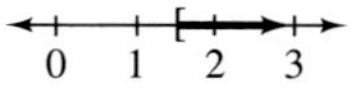

10. Center: $(-2, 1)$; Radius: 3

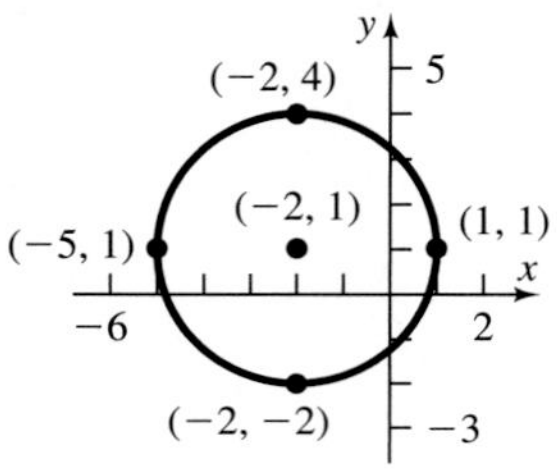

11. x-intercepts: $-3, 0, 3$
y-intercept: 0
Symmetric with respect to the origin

12. $y = -\frac{2}{3}x + \frac{17}{3}$

13. Not a function; it fails the Vertical Line Test.

14. **(a)** 22 **(b)** $x^2 - 5x - 2$ **(c)** $-x^2 - 5x + 2$ **(d)** $9x^2 + 15x - 2$ **(e)** $2x + h + 5$

15. **(a)** $\{x|x \neq 1\}$ **(b)** No, (2, 7) is on the graph. **(c)** 4; (3, 4) is on the graph. **(d)** $\frac{7}{4}$; $\left(\frac{7}{4}, 9\right)$ is on the graph.

16.

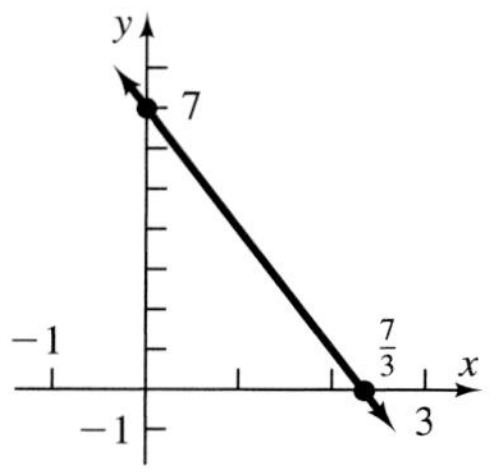

17.

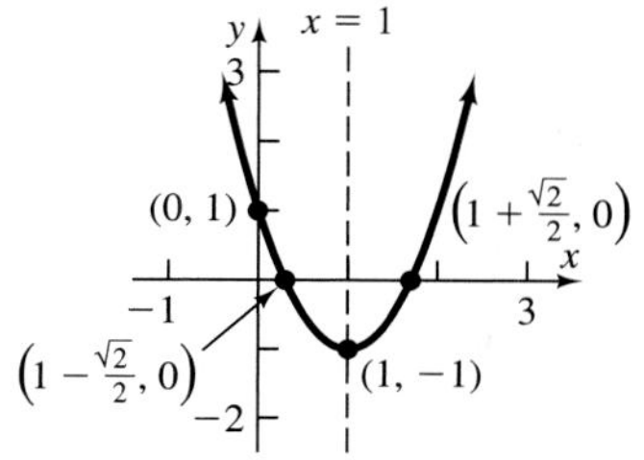

18. $x + 4$; $m_{sec} = 6$

19. **(a)** x-intercepts: $-5, -1, 5$; y-intercept: -3
(b) No symmetry
(c) Neither
(d) Increasing: $(-\infty, -3)$ and $(2, \infty)$
Decreasing: $(-3, 2)$
(e) Local maximum is 5 and occurs at $x = -3$.
(f) Local minimum is -6 and occurs at $x = 2$.

20. Odd **21.** **(a)** Domain: $\{x|-3 < x\}$ or $(-3, \infty)$
(b) x-intercept: $-\frac{1}{2}$; y-intercept: 1
(c)

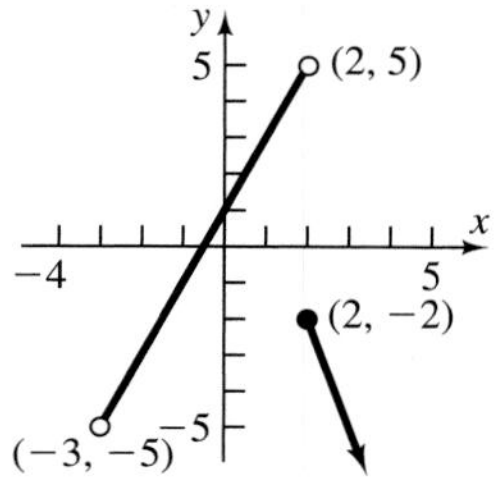

(d) Range: $\{y|y < 5\}$ or $(-\infty, 5)$

22.

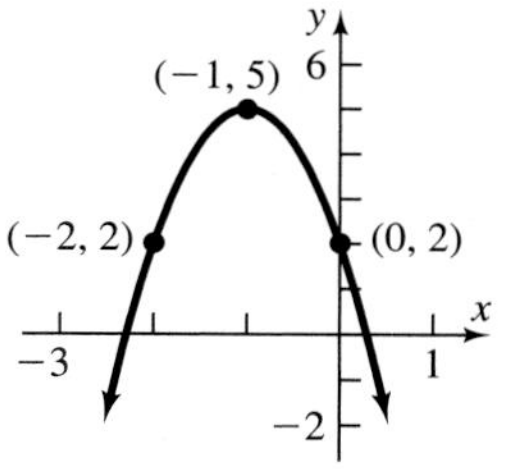

23. **(a)** $(f + g)(x) = x^2 - 9x - 6$; domain: all real numbers **(b)** $\left(\frac{f}{g}\right)(x) = \frac{x^2 - 5x + 1}{-4x - 7}$; domain: $\left\{x \middle| x \neq -\frac{7}{4}\right\}$

24. **(a)** $R(x) = -\frac{1}{10}x^2 + 150x$ **(b)** \$14,000 **(c)** 750; \$56,250 **(d)** \$75

CHAPTER 4 Exponential and Logarithmic Functions

4.1 Concepts and Vocabulary *(page 229)*

4. $(g \circ f)(x)$ **5.** False **6.** False

4.1 Exercises *(page 229)*

7. **(a)** -1 **(b)** -1 **(c)** 8 **(d)** 0 **(e)** 8 **(f)** -7
8. **(a)** 5 **(b)** 11 **(c)** 0 **(d)** 0 **(e)** 1 **(f)** 7
9. **(a)** 98 **(b)** 49 **(c)** 4 **(d)** 4 **10.** **(a)** 95 **(b)** 127 **(c)** 17 **(d)** 1 **11.** **(a)** 97 **(b)** $-\frac{163}{2}$ **(c)** 1 **(d)** $-\frac{3}{2}$
12. **(a)** 4418 **(b)** -191 **(c)** 8 **(d)** -2 **13.** **(a)** $2\sqrt{2}$ **(b)** $2\sqrt{2}$ **(c)** 1 **(d)** 0 **14.** **(a)** $\sqrt{13}$ **(b)** $3\sqrt{3}$ **(c)** $\sqrt{\sqrt{2} + 1}$ **(d)** 0
15. **(a)** $\frac{1}{17}$ **(b)** $\frac{1}{5}$ **(c)** 1 **(d)** $\frac{1}{2}$ **16.** **(a)** $\frac{11}{6}$ **(b)** $\frac{3}{2}$ **(c)** 1 **(d)** $\frac{12}{17}$ **17.** **(a)** $\frac{3}{\sqrt[3]{4} + 1}$ **(b)** 1 **(c)** $\frac{6}{5}$ **(d)** 0 **18.** **(a)** $\frac{2\sqrt{10}}{25}$
(b) $\frac{2}{2\sqrt{2} + 1}$ **(c)** 1 **(d)** $\frac{2}{3}$ **19.** $\{x|x \neq 0, x \neq 2\}$ **20.** $\left\{x \middle| x \neq 0, x \neq \frac{2}{3}\right\}$ **21.** $\{x|x \neq -4, x \neq 0\}$ **22.** $\left\{x \middle| x \neq -\frac{2}{3}, x \neq 0\right\}$
23. $\left\{x \middle| x \geq -\frac{3}{2}\right\}$ **24.** $\{x|x \leq 1\}$ **25.** $\{x|x \geq 1\}$ **26.** $\{x|x \geq 2\}$ **27.** **(a)** $(f \circ g)(x) = 6x + 3$; All real numbers
(b) $(g \circ f)(x) = 6x + 9$; All real numbers **(c)** $(f \circ f)(x) = 4x + 9$; All real numbers **(d)** $(g \circ g)(x) = 9x$; All real numbers
28. **(a)** $(f \circ g)(x) = -2x + 4$; All real numbers **(b)** $(g \circ f)(x) = -2x - 4$; All real numbers **(c)** $(f \circ f)(x) = x$; All real numbers
(d) $(g \circ g)(x) = 4x - 12$; All real numbers **29.** **(a)** $(f \circ g)(x) = 3x^2 + 1$; All real numbers
(b) $(g \circ f)(x) = 9x^2 + 6x + 1$; All real numbers **(c)** $(f \circ f)(x) = 9x + 4$; All real numbers **(d)** $(g \circ g)(x) = x^4$; All real numbers
30. **(a)** $(f \circ g)(x) = x^2 + 5$; All real numbers **(b)** $(g \circ f)(x) = x^2 + 2x + 5$; All real numbers
(c) $(f \circ f)(x) = x + 2$; All real numbers **(d)** $(g \circ g)(x) = x^4 + 8x^2 + 20$; All real numbers
31. **(a)** $(f \circ g)(x) = x^4 + 8x^2 + 16$; All real numbers **(b)** $(g \circ f)(x) = x^4 + 4$; All real numbers **(c)** $(f \circ f)(x) = x^4$; All real numbers
(d) $(g \circ g)(x) = x^4 + 8x^2 + 20$; All real numbers **32.** **(a)** $(f \circ g)(x) = 4x^4 + 12x^2 + 10$; All real numbers
(b) $(g \circ f)(x) = 2x^4 + 4x^2 + 5$; All real numbers **(c)** $(f \circ f)(x) = x^4 + 2x^2 + 2$; All real numbers
(d) $(g \circ g)(x) = 8x^4 + 24x^2 + 21$; All real numbers

33. **(a)** $(f \circ g)(x) = \dfrac{3x}{2-x}$; $\{x|x \neq 0, x \neq 2\}$ **(b)** $(g \circ f)(x) = \dfrac{2(x-1)}{3}$; $\{x|x \neq 1\}$
(c) $(f \circ f)(x) = \dfrac{3(x-1)}{4-x}$; $\{x|x \neq 1, x \neq 4\}$ **(d)** $(g \circ g)(x) = x$; $\{x|x \neq 0\}$
34. **(a)** $(f \circ g)(x) = \dfrac{x}{-2+3x}$; $\left\{x \middle| x \neq 0, x \neq \dfrac{2}{3}\right\}$ **(b)** $(g \circ f)(x) = -2(x+3)$; $\{x|x \neq -3\}$
(c) $(f \circ f)(x) = \dfrac{x+3}{3x+10}$; $\left\{x \middle| x \neq -\dfrac{10}{3}, x \neq -3\right\}$ **(d)** $(g \circ g)(x) = x$; $\{x|x \neq 0\}$ **35.** **(a)** $(f \circ g)(x) = \dfrac{4}{4+x}$; $\{x|x \neq -4, x \neq 0\}$
(b) $(g \circ f)(x) = \dfrac{-4(x-1)}{x}$; $\{x|x \neq 0, x \neq 1\}$ **(c)** $(f \circ f)(x) = x$; $\{x|x \neq 1\}$ **(d)** $(g \circ g)(x) = x$; $\{x|x \neq 0\}$
36. **(a)** $(f \circ g)(x) = \dfrac{2}{2+3x}$; $\left\{x \middle| x \neq 0, x \neq -\dfrac{2}{3}\right\}$ **(b)** $(g \circ f)(x) = \dfrac{2(x+3)}{x}$; $\{x|x \neq -3, x \neq 0\}$
(c) $(f \circ f)(x) = \dfrac{x}{4x+9}$; $\left\{x \middle| x \neq -3, x \neq -\dfrac{9}{4}\right\}$ **(d)** $(g \circ g)(x) = x$; $\{x|x \neq 0\}$ **37.** **(a)** $(f \circ g)(x) = \sqrt{2x+3}$; $\left\{x \middle| x \geq -\dfrac{3}{2}\right\}$
(b) $(g \circ f)(x) = 2\sqrt{x} + 3$; $\{x|x \geq 0\}$ **(c)** $(f \circ f)(x) = \sqrt[4]{x}$; $\{x|x \geq 0\}$ **(d)** $(g \circ g)(x) = 4x + 9$; All real numbers
38. **(a)** $(f \circ g)(x) = \sqrt{-2x-1}$; $\left\{x \middle| x \leq -\dfrac{1}{2}\right\}$ **(b)** $(g \circ f)(x) = 1 - 2\sqrt{x-2}$; $\{x|x \geq 2\}$ **(c)** $(f \circ f)(x) = \sqrt{\sqrt{x-2}-2}$; $\{x|x \geq 6\}$
(d) $(g \circ g)(x) = -1 + 4x$; All real numbers **39.** **(a)** $(f \circ g)(x) = x$; $\{x|x \geq 1\}$ **(b)** $(g \circ f)(x) = |x|$; All real numbers
(c) $(f \circ f)(x) = x^4 + 2x^2 + 2$; All real numbers **(d)** $(g \circ g)(x) = \sqrt{\sqrt{x-1}-1}$; $\{x|x \geq 2\}$ **40.** **(a)** $(f \circ g)(x) = x + 2$; $\{x|x \geq 2\}$
(b) $(g \circ f)(x) = \sqrt{x^2+2}$; All real numbers **(c)** $(f \circ f)(x) = x^4 + 8x^2 + 20$; All real numbers **(d)** $(g \circ g)(x) = \sqrt{\sqrt{x-2}-2}$; $\{x|x \geq 6\}$ **41.** **(a)** $(f \circ g)(x) = acx + ad + b$; All real numbers **(b)** $(g \circ f)(x) = acx + bc + d$; All real numbers
(c) $(f \circ f)(x) = a^2x + ab + b$; All real numbers **(d)** $(g \circ g)(x) = c^2x + cd + d$; All real numbers
42. **(a)** $(f \circ g)(x) = \dfrac{amx+b}{cmx+d}$; $\left\{x \middle| x \neq -\dfrac{d}{cm}\right\}$ **(b)** $(g \circ f)(x) = \dfrac{m(ax+b)}{cx+d}$; $\left\{x \middle| x \neq -\dfrac{d}{c}\right\}$
(c) $(f \circ f)(x) = \dfrac{(a^2+bc)x + ab + bd}{(ac+cd)x + bc + d^2}$; $\left\{x \middle| x \neq -\dfrac{d}{c}, x \neq -\dfrac{bc+d^2}{ac+cd}\right\}$ **(d)** $(g \circ g)(x) = m^2x$; All real numbers

43. $(f \circ g)(x) = f(g(x)) = f\left(\dfrac{1}{2}x\right) = 2\left(\dfrac{1}{2}x\right) = x$; $(g \circ f)(x) = g(f(x)) = g(2x) = \dfrac{1}{2}(2x) = x$
44. $(f \circ g)(x) = f(g(x)) = f\left(\dfrac{1}{4}x\right) = 4\left(\dfrac{1}{4}x\right) = x$; $(g \circ f)(x) = g(f(x)) = g(4x) = \dfrac{1}{4}(4x) = x$
45. $(f \circ g)(x) = f(g(x)) = f(\sqrt[3]{x}) = (\sqrt[3]{x})^3 = x$; $(g \circ f)(x) = g(f(x)) = g(x^3) = \sqrt[3]{x^3} = x$
46. $(f \circ g)(x) = f(g(x)) = f(x-5) = x - 5 + 5 = x$; $(g \circ f)(x) = g(f(x)) = g(x+5) = x + 5 - 5 = x$
47. $(f \circ g)(x) = f(g(x)) = f\left(\dfrac{1}{2}(x+6)\right) = 2\left[\dfrac{1}{2}(x+6)\right] - 6 = x + 6 - 6 = x$;

$(g \circ f)(x) = g(f(x)) = g(2x-6) = \dfrac{1}{2}(2x - 6 + 6) = x$

48. $(f \circ g)(x) = f(g(x)) = f\left(\dfrac{1}{3}(4-x)\right) = 4 - 3\left[\dfrac{1}{3}(4-x)\right] = 4 - (4-x) = x$;

$(g \circ f)(x) = g(f(x)) = g(4-3x) = \dfrac{1}{3}[4 - (4-3x)] = \dfrac{1}{3}(3x) = x$

49. $(f \circ g)(x) = f(g(x)) = f\left(\dfrac{1}{a}(x-b)\right) = a\left[\dfrac{1}{a}(x-b)\right] + b = x$; $(g \circ f)(x) = g(f(x)) = g(ax+b) = \dfrac{1}{a}(ax+b-b) = x$
50. $(f \circ g)(x) = f(g(x)) = f\left(\dfrac{1}{x}\right) = \dfrac{1}{\frac{1}{x}} = x$; $(g \circ f)(x) = g(f(x)) = g\left(\dfrac{1}{x}\right) = \dfrac{1}{\frac{1}{x}} = x$

51. $f(x) = x^4$; $g(x) = 2x + 3$ (Other answers are possible.) **52.** $f(x) = x^3$; $g(x) = 1 + x^2$ (Other answers are possible.)
53. $f(x) = \sqrt{x}$; $g(x) = x^2 + 1$ (Other answers are possible.) **54.** $f(x) = \sqrt{x}$; $g(x) = 1 - x^2$ (Other answers are possible.)
55. $f(x) = |x|$; $g(x) = 2x + 1$ (Other answers are possible.) **56.** $f(x) = |x|$; $g(x) = 2x^2 + 3$ (Other answers are possible.)
57. $(f \circ g)(x) = 11$; $(g \circ f)(x) = 2$ **58.** $(f \circ f)(x) = x$ **59.** $-3, 3$ **60.** $-5, 5$ **61.** $S(t) = \dfrac{16}{9}\pi t^6$ **62.** $V(t) = \dfrac{32}{81}\pi t^9$
63. $C(t) = 15{,}000 + 800{,}000t - 40{,}000t^2$ **64.** $A(t) = 40{,}000\pi t$ **65.** $C(p) = \dfrac{2\sqrt{100-p}}{25} + 600, 0 \leq p \leq 100$
66. $C(p) = \dfrac{\sqrt{1000-5p}}{10} + 400, 0 \leq p \leq 200$ **67.** $V(r) = 2\pi r^3$ **68.** $V(r) = \dfrac{2}{3}\pi r^3$
69. **(a)** $f(x) = 0.857118x$ **(b)** $g(x) = 128.6054x$ **(c)** $g(f(x)) = g(0.857118x) = 110.23x$ **(d)** 110,230.00 yen

70. f and g are odd functions, so $f(-x) = -f(x)$ and $g(-x) = -g(x)$.
Then $(f \circ g)(-x) = f(g(-x)) = f(-g(x)) = -f(g(x)) = -(f \circ g)(x)$. So $f \circ g$ is also odd.

71. f is an odd function, so $f(-x) = -f(x)$. g is an even function, so $g(-x) = g(x)$.
Then $(f \circ g)(-x) = f(g(-x)) = f(g(x)) = (f \circ g)(x)$. So $f \circ g$ is even.
Also, $(g \circ f)(-x) = g(f(-x)) = g(-f(x)) = g(f(x)) = (g \circ f)(x)$, so $g \circ f$ is even.

4.2 Concepts and Vocabulary *(page 241)*

4. One-to-one **5.** $y = x$ **6.** $[4, \infty)$ **7.** False **8.** True

4.2 Exercises *(page 241)*

9. (a)

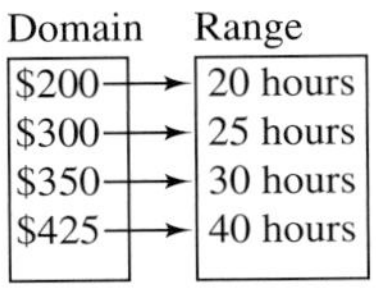

(b) Inverse is a function.

10. (a)

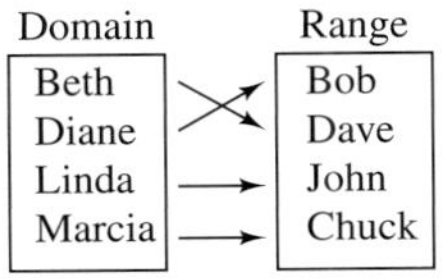

(b) Inverse is a function.

11. (a)

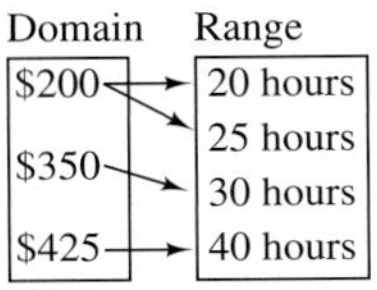

(b) Inverse is not a function.

12. (a)

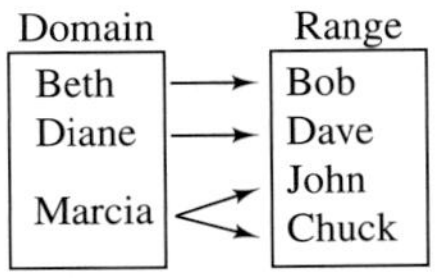

(b) Inverse is not a function.

13. (a) $\{(6, 2), (6, -3), (9, 4), (10, 1)\}$ **(b)** Inverse is not a function. **14. (a)** $\{(5, -2), (3, -1), (7, 3), (12, 4)\}$ **(b)** Inverse is a function. **15. (a)** $\{(0, 0), (1, 1), (16, 2), (81, 3)\}$ **(b)** Inverse is a function. **16. (a)** $\{(2, 1), (8, 2), (18, 3), (32, 4)\}$ **(b)** Inverse is a function. **17.** One-to-one **18.** One-to-one **19.** Not one-to-one **20.** Not one-to-one **21.** One-to-one **22.** Not one-to-one

23.

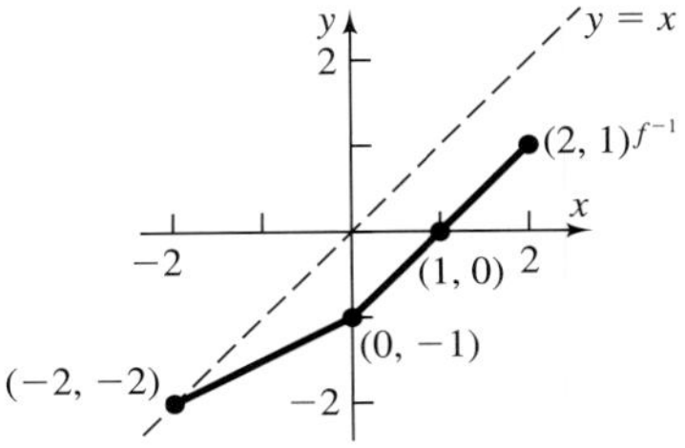

24.

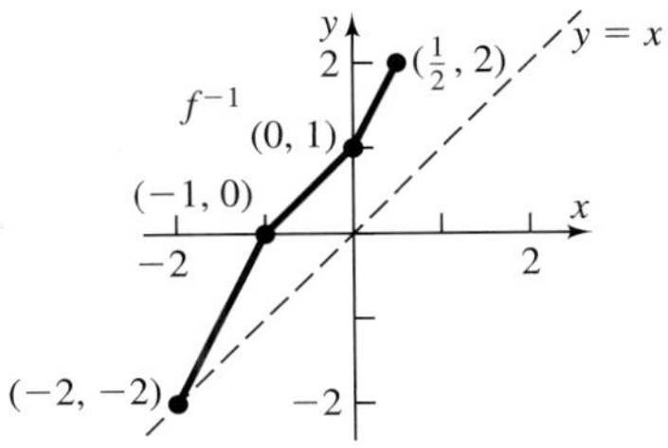

25.

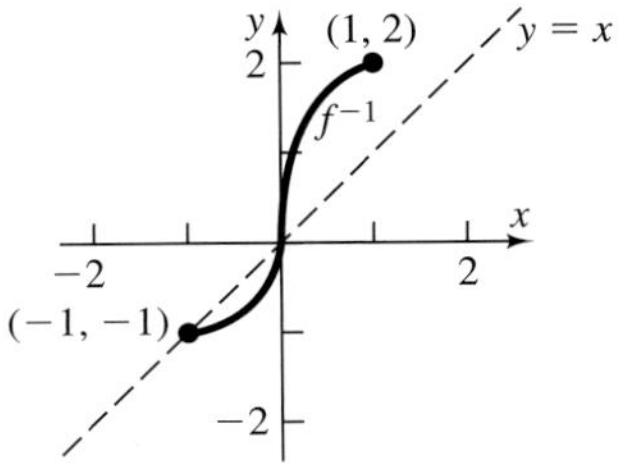

26.

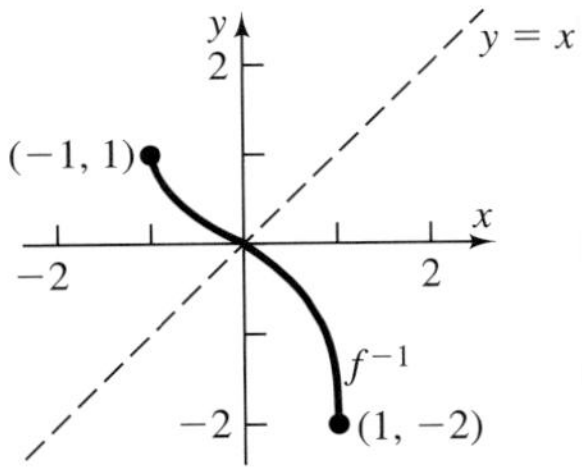

27.

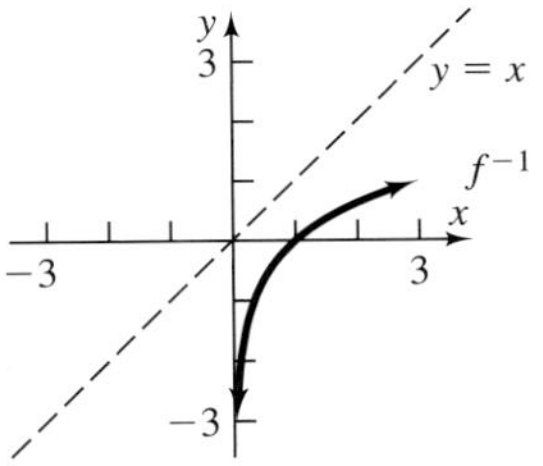

28.

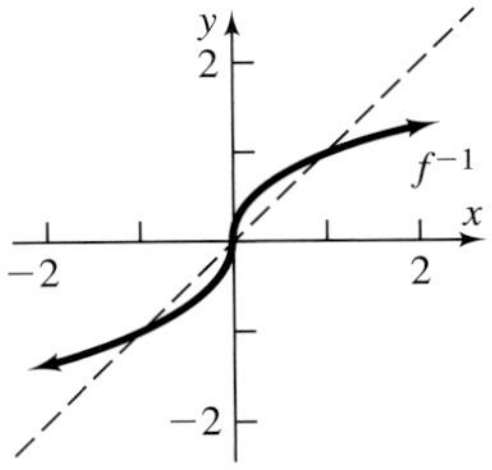

29. $f(g(x)) = f\left(\frac{1}{3}(x - 4)\right) = 3\left[\frac{1}{3}(x - 4)\right] + 4$
$= (x - 4) + 4 = x;$
$g(f(x)) = g(3x + 4) = \frac{1}{3}[(3x + 4) - 4] = \frac{1}{3}3x = x$

30. $f(g(x)) = f\left(-\frac{1}{2}(x - 3)\right) = 3 - 2\left(\frac{-1}{2}(x - 3)\right)$
$= 3 + (x - 3) = x$
$g(f(x)) = g(3 - 2x) = -\frac{1}{2}[(3 - 2x) - 3]$
$= -\frac{1}{2}(-2x) = x$

31. $f(g(x)) = 4\left[\frac{x}{4} + 2\right] - 8 = (x + 8) - 8 = x;$
$g(f(x)) = \frac{4x - 8}{4} + 2 = (x - 2) + 2 = x$

32. $f(g(x)) = f\left(\frac{1}{2}x - 3\right) = 2\left(\frac{1}{2}x - 3\right) + 6$
$= (x - 6) + 6 = x$
$g(f(x)) = g(2x + 6) = \frac{1}{2}(2x + 6) - 3$
$= (x + 3) - 3 = x$

33. $f(g(x)) = (\sqrt[3]{x + 8})^3 - 8 = (x + 8) - 8 = x;$
$g(f(x)) = \sqrt[3]{(x^3 - 8) + 8} = \sqrt[3]{x^3} = x$

34. $f(g(x)) = f(\sqrt{x} + 2) = (\sqrt{x} + 2 - 2)^2 = (\sqrt{x})^2 = x$
$g(f(x)) = g((x - 2)^2) = \sqrt{(x - 2)^2} + 2 = x - 2 + 2 = x$

35. $f(g(x)) = \dfrac{1}{\left(\dfrac{1}{x}\right)} = x; g(f(x)) = \dfrac{1}{\left(\dfrac{1}{x}\right)} = x$

36. $f(g(x)) = f(x) = x$
$g(f(x)) = g(x) = x$

37. $f(g(x)) = \dfrac{2\left(\dfrac{4x-3}{2-x}\right)+3}{\dfrac{4x-3}{2-x}+4} = \dfrac{2(4x-3)+3(2-x)}{4x-3+4(2-x)}$
$= \dfrac{5x}{5} = x;$
$g(f(x)) = \dfrac{4\left(\dfrac{2x+3}{x+4}\right)-3}{2-\dfrac{2x+3}{x+4}} = \dfrac{4(2x+3)-3(x+4)}{2(x+4)-(2x+3)}$
$= \dfrac{5x}{5} = x$

38. $f(g(x)) = f\left(\dfrac{3x+5}{1-2x}\right) = \dfrac{\dfrac{3x+5}{1-2x}-5}{2\left(\dfrac{3x+5}{1-2x}\right)+3}$
$= \dfrac{3x+5-5(1-2x)}{2(3x+5)+3(1-2x)} = \dfrac{13x}{13} = x$
$g(f(x)) = g\left(\dfrac{x-5}{2x+3}\right) = \dfrac{3\left(\dfrac{x-5}{2x+3}\right)+5}{1-2\left(\dfrac{x-5}{2x+3}\right)}$
$= \dfrac{3(x-5)+5(2x+3)}{(2x+3)-2(x-5)} = \dfrac{13x}{13} = x$

39. $f^{-1}(x) = \frac{1}{3}x$
$f(f^{-1}(x)) = 3\left(\frac{1}{3}x\right) = x$
$f^{-1}(f(x)) = \frac{1}{3}(3x) = x$
Domain f = Range f^{-1} = All real numbers
Range f = Domain f^{-1} = All real numbers

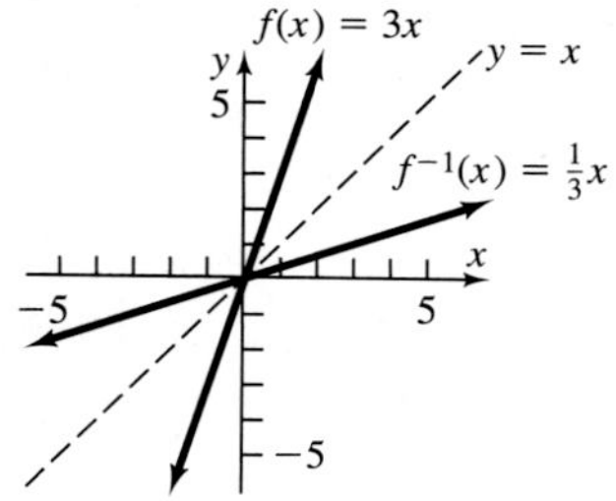

40. $f^{-1}(x) = -\frac{1}{4}x$
$f(f^{-1}(x)) = f\left(-\frac{1}{4}x\right) = -4\left(-\frac{1}{4}x\right) = x$
$f^{-1}(f(x)) = f^{-1}(-4x) = -\frac{1}{4}(-4x) = x$
Domain f = Range f^{-1} = All real numbers
Range f = Domain f^{-1} = All real numbers

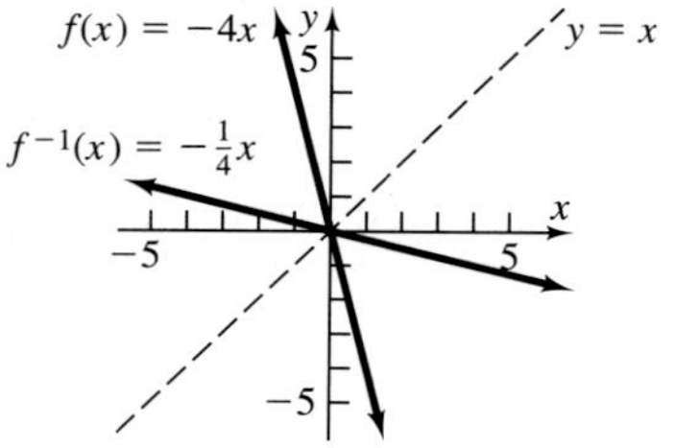

41. $f^{-1}(x) = \frac{x}{4} - \frac{1}{2}$
$f(f^{-1}(x)) = 4\left(\frac{x}{4} - \frac{1}{2}\right) + 2 = (x-2) + 2 = x$
$f^{-1}(f(x)) = \frac{4x+2}{4} - \frac{1}{2} = \left(x + \frac{1}{2}\right) - \frac{1}{2} = x$
Domain f = Range f^{-1} = All real numbers
Range f = Domain f^{-1} = All real numbers

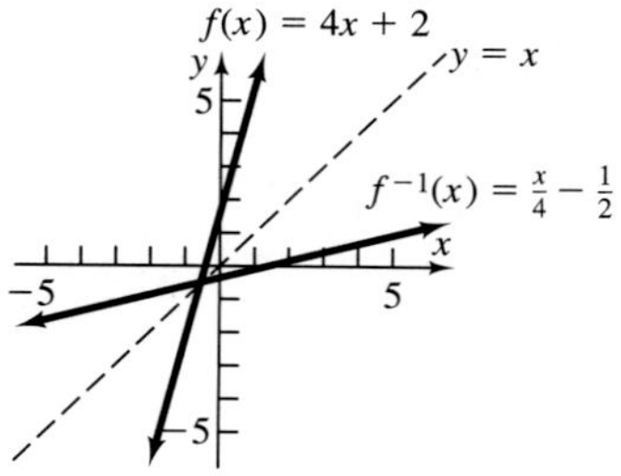

42. $f^{-1}(x) = \dfrac{1-x}{3}$
$f(f^{-1}(x)) = f\left(\dfrac{1-x}{3}\right) = 1 - 3\left(\dfrac{1-x}{3}\right) = 1 - (1-x) = x$
$f^{-1}(f(x)) = f^{-1}(1-3x) = \dfrac{1-(1-3x)}{3} = \dfrac{3x}{3} = x$
Domain f = Range f^{-1} = All real numbers
Range f = Domain f^{-1} = All real numbers

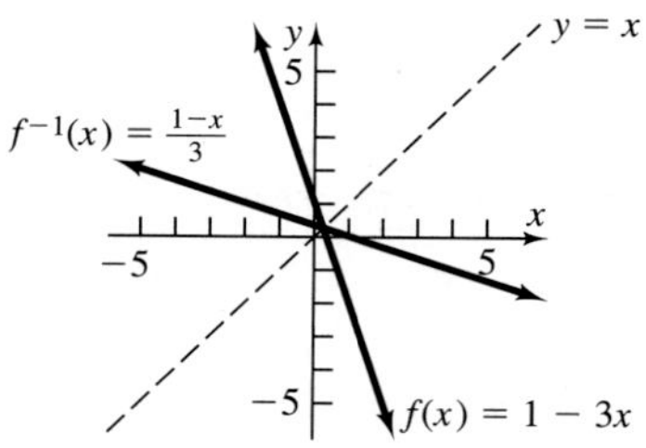

43. $f^{-1}(x) = \sqrt[3]{x+1}$

$f(f^{-1}(x)) = (\sqrt[3]{x+1})^3 - 1 = x$

$f^{-1}(f(x)) = \sqrt[3]{(x^3 - 1) + 1} = x$

Domain f = Range f^{-1} = All real numbers

Range f = Domain f^{-1} = All real numbers

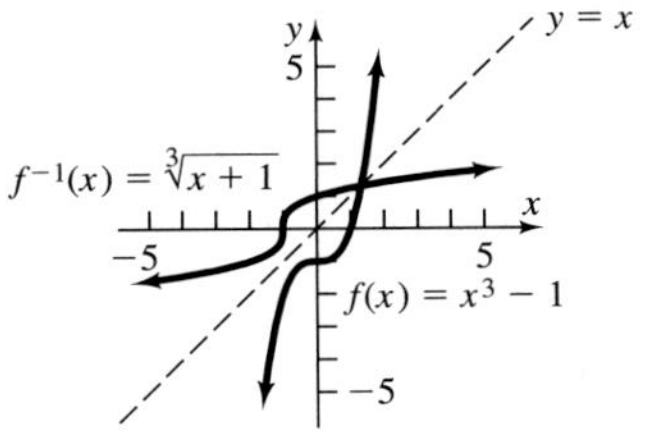

44. $f^{-1}(x) = (x - 1)^{1/3}$

$f(f^{-1}(x)) = f((x-1)^{1/3}) = ((x-1)^{1/3})^3 + 1$

$= x - 1 + 1 = x$

$f^{-1}(f(x)) = f^{-1}(x^3 + 1) = (x^3 + 1 - 1)^{1/3} = (x^3)^{1/3} = x$

Domain f = Range f^{-1} = All real numbers

Range f = Domain f^{-1} = All real numbers

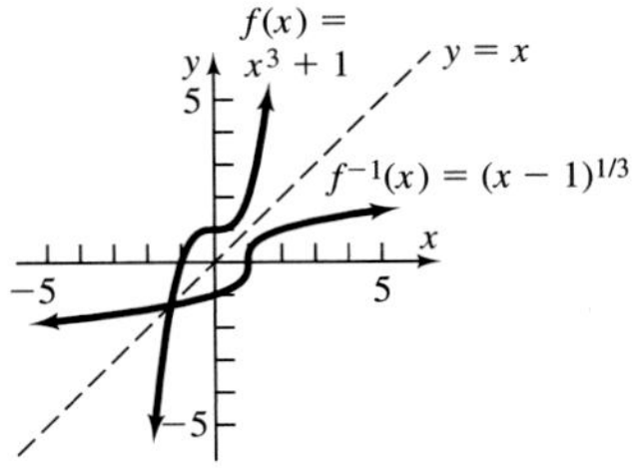

45. $f^{-1}(x) = \sqrt{x - 4},\ x \geq 4$

$f(f^{-1}(x)) = (\sqrt{x-4})^2 + 4 = x$

$f^{-1}(f(x)) = \sqrt{(x^2 + 4) - 4} = \sqrt{x^2} = x,\ x \geq 0$

Domain f = Range f^{-1} = $\{x|x \geq 0\}$ or $[0, \infty)$

Range f = Domain f^{-1} = $\{x|x \geq 4\}$ or $[4, \infty)$

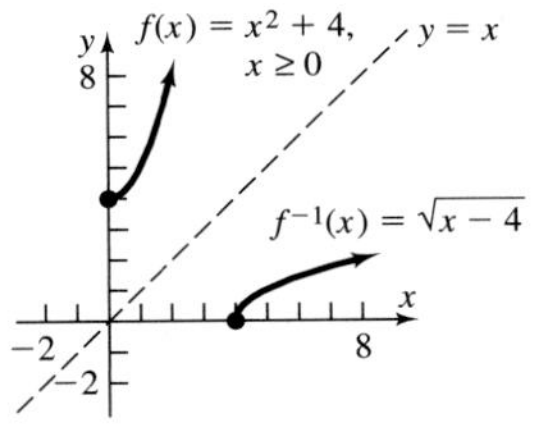

46. $f^{-1}(x) = \sqrt{x - 9},\ x \geq 9$

$f(f^{-1}(x)) = f(\sqrt{x-9}) = (\sqrt{x-9})^2 + 9 = x - 9 + 9 = x$

$f^{-1}(f(x)) = f^{-1}(x^2 + 9) = \sqrt{x^2 + 9 - 9} = \sqrt{x^2} = x,\ x \geq 0$

Domain f = Range f^{-1} = $\{x|x \geq 0\}$ or $[0, \infty)$

Range f = Domain f^{-1} = $\{x|x \geq 9\}$ or $[9, \infty)$

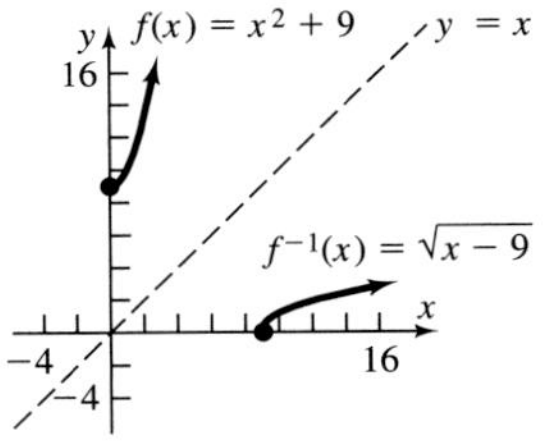

47. $f^{-1}(x) = \dfrac{4}{x}$

$$f(f^{-1}(x)) = \frac{4}{\left(\frac{4}{x}\right)} = x$$

$$f^{-1}(f(x)) = \frac{4}{\left(\frac{4}{x}\right)} = x$$

Domain f = Range f^{-1} = All real numbers except 0

Range f = Domain f^{-1} = All real numbers except 0

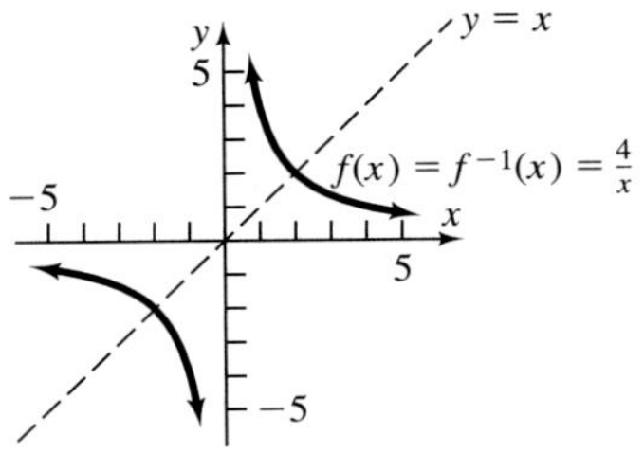

48. $f^{-1}(x) = -\dfrac{3}{x}$

$$f(f^{-1}(x)) = f\left(-\frac{3}{x}\right) = -\frac{3}{\left(-\frac{3}{x}\right)} = x$$

$$f^{-1}(f(x)) = f^{-1}\left(-\frac{3}{x}\right) = -\frac{3}{\left(-\frac{3}{x}\right)} = x$$

Domain f = Range f^{-1} = All real numbers except 0

Range f = Domain f^{-1} = All real numbers except 0

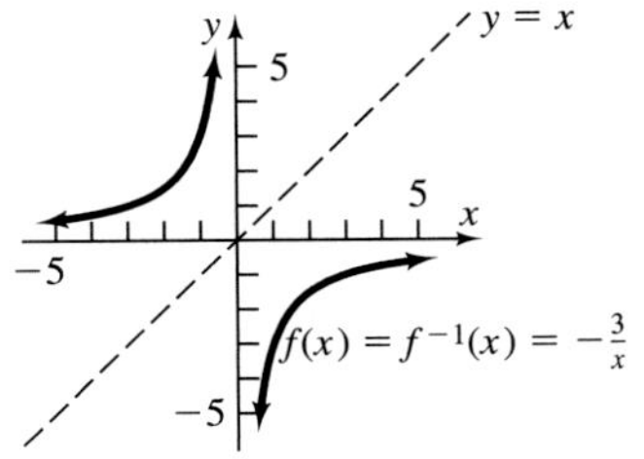

49. $f^{-1}(x) = \dfrac{2x+1}{x}$

$$f(f^{-1}(x)) = \frac{1}{\dfrac{2x+1}{x} - 2} = \frac{x}{(2x+1) - 2x} = x$$

$$f^{-1}(f(x)) = \frac{2\left(\dfrac{1}{x-2}\right) + 1}{\dfrac{1}{x-2}} = \frac{2 + (x-2)}{1} = x$$

Domain f = Range f^{-1} = All real numbers except 2

Range f = Domain f^{-1} = All real numbers except 0

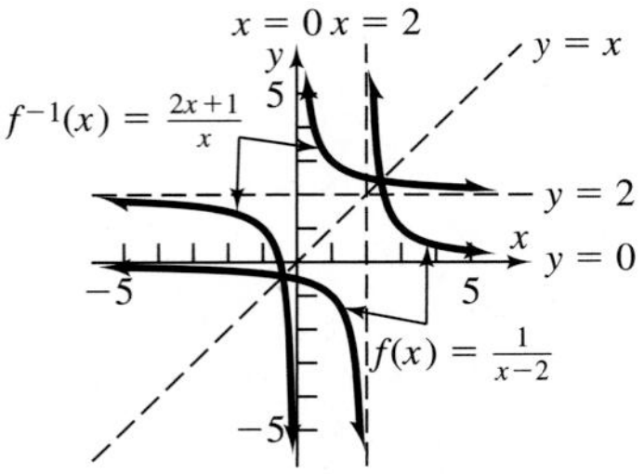

50. $f^{-1}(x) = \dfrac{4}{x} - 2$

$$f(f^{-1}(x)) = f\left(\frac{4}{x} - 2\right) = \frac{4}{\left(\dfrac{4}{x} - 2\right) + 2} = \frac{4}{\left(\dfrac{4}{x}\right)} = x$$

$$f^{-1}(f(x)) = f^{-1}\left(\frac{4}{x+2}\right) = \frac{4}{\left(\dfrac{4}{x+2}\right)} - 2 = x + 2 - 2 = x$$

Domain f = Range f^{-1} = All real numbers except -2

Range f = Domain f^{-1} = All real numbers except 0

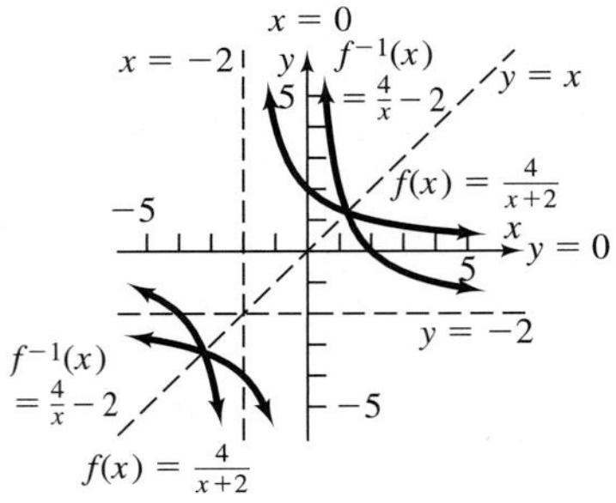

51. $f^{-1}(x) = \dfrac{2-3x}{x}$

$$f(f^{-1}(x)) = \frac{2}{3 + \dfrac{2-3x}{x}} = \frac{2x}{3x + 2 - 3x} = \frac{2x}{2} = x$$

$$f^{-1}(f(x)) = \frac{2 - 3\left(\dfrac{2}{3+x}\right)}{\dfrac{2}{3+x}} = \frac{2(3+x) - 3 \cdot 2}{2} = \frac{2x}{2} = x$$

Domain f = All real numbers except -3

Range f = Domain f^{-1} = All real numbers except 0

52. $f^{-1}(x) = 2 - \dfrac{4}{x}$

$$f(f^{-1}(x)) = f\left(2 - \frac{4}{x}\right) = \frac{4}{2 - \left(2 - \dfrac{4}{x}\right)} = \frac{4}{\left(\dfrac{4}{x}\right)} = x$$

$$f^{-1}(f(x)) = f^{-1}\left(\frac{4}{2-x}\right) = 2 - \frac{4}{\left(\dfrac{4}{2-x}\right)}$$

$$= 2 - (2 - x) = x$$

Domain f = All real numbers except 2

Range f = Domain f^{-1} = All real numbers except 0

53. $f^{-1}(x) = \dfrac{-2x}{x-3}$

$$f(f^{-1}(x)) = \frac{3\left(\dfrac{-2x}{x-3}\right)}{\dfrac{-2x}{x-3} + 2} = \frac{3(-2x)}{-2x + 2(x-3)} = \frac{-6x}{-6} = x$$

$$f^{-1}(f(x)) = \frac{-2\left(\dfrac{3x}{x+2}\right)}{\dfrac{3x}{x+2} - 3} = \frac{-2(3x)}{3x - 3(x+2)} = \frac{-6x}{-6} = x$$

Domain f = All real numbers except -2

Range f = Domain f^{-1} = All real numbers except 3

54. $f^{-1}(x) = \dfrac{x}{x+2}$

$$f(f^{-1}(x)) = \frac{-2\left(\dfrac{x}{x+2}\right)}{\dfrac{x}{x+2} - 1} = \frac{-2x}{x - (x+2)} = \frac{-2x}{-2} = x$$

$$f^{-1}(f(x)) = \frac{\dfrac{-2x}{x-1}}{\dfrac{-2x}{x-1} + 2} = \frac{-2x}{-2x + 2(x-1)} = \frac{-2x}{-2} = x$$

Domain f = All real numbers except 1

Range f = Domain f^{-1} = All real numbers except -2

55. $f^{-1}(x) = \dfrac{x}{3x - 2}$

$$f(f^{-1}(x)) = \frac{2\left(\dfrac{x}{3x-2}\right)}{3\left(\dfrac{x}{3x-2}\right) - 1} = \frac{2x}{3x - (3x - 2)} = \frac{2x}{2} = x$$

$$f^{-1}(f(x)) = \frac{\dfrac{2x}{3x-1}}{3\left(\dfrac{2x}{3x-1}\right) - 2} = \frac{2x}{6x - 2(3x - 1)} = \frac{2x}{2} = x$$

Domain f = All real numbers except $\frac{1}{3}$

Range f = Domain f^{-1} = All real numbers except $\frac{2}{3}$

56. $f^{-1}(x) = \dfrac{-1}{x + 3}$

$$f(f^{-1}(x)) = f\left(\frac{-1}{x+3}\right) = \frac{3\left(\dfrac{-1}{x+3}\right) + 1}{-\left(\dfrac{-1}{x+3}\right)} = \frac{-3 + (x + 3)}{1} = x$$

$$f^{-1}(f(x)) = f^{-1}\left(\frac{3x+1}{-x}\right) = \frac{-1}{\dfrac{3x+1}{-x} + 3} = \frac{x}{3x + 1 - 3x} = x$$

Domain f = All real numbers except 0

Range f = Domain f^{-1} = All real numbers except -3

57. $f^{-1}(x) = \dfrac{3x + 4}{2x - 3}$

$$f(f^{-1}(x)) = \frac{3\left(\dfrac{3x+4}{2x-3}\right) + 4}{2\left(\dfrac{3x+4}{2x-3}\right) - 3} = \frac{3(3x + 4) + 4(2x - 3)}{2(3x + 4) - 3(2x - 3)} = \frac{17x}{17} = x$$

$$f^{-1}(f(x)) = \frac{3\left(\dfrac{3x+4}{2x-3}\right) + 4}{2\left(\dfrac{3x+4}{2x-3}\right) - 3} = \frac{3(3x + 4) + 4(2x - 3)}{2(3x + 4) - 3(2x - 3)} = \frac{17x}{17} = x$$

Domain f = All real numbers except $\frac{3}{2}$

Range f = Domain f^{-1} = All real numbers except $\frac{3}{2}$

58. $f^{-1}(x) = \dfrac{3 + 4x}{2 - x}$

$$f(f^{-1}(x)) = f\left(\frac{3+4x}{2-x}\right) = \frac{2\left(\dfrac{3+4x}{2-x}\right) - 3}{\left(\dfrac{3+4x}{2-x}\right) + 4} = \frac{2(3 + 4x) - 3(2 - x)}{(3 + 4x) + 4(2 - x)} = \frac{11x}{11} = x$$

$$f^{-1}(f(x)) = f^{-1}\left(\frac{2x-3}{x+4}\right) = \frac{3 + 4\left(\dfrac{2x-3}{x+4}\right)}{2 - \left(\dfrac{2x-3}{x+4}\right)} = \frac{3(x + 4) + 4(2x - 3)}{2(x + 4) - (2x - 3)} = \frac{11x}{11} = x$$

Domain f = All real numbers except -4

Range f = Domain f^{-1} = All real numbers except 2

59. $f^{-1}(x) = \dfrac{-2x + 3}{x - 2}$

$$f(f^{-1}(x)) = \frac{2\left(\dfrac{-2x+3}{x-2}\right) + 3}{\dfrac{-2x+3}{x-2} + 2} = \frac{2(-2x + 3) + 3(x - 2)}{-2x + 3 + 2(x - 2)} = \frac{-x}{-1} = x$$

$$f^{-1}(f(x)) = \frac{-2\left(\dfrac{2x+3}{x+2}\right) + 3}{\dfrac{2x+3}{x+2} - 2} = \frac{-2(2x + 3) + 3(x + 2)}{2x + 3 - 2(x + 2)} = \frac{-x}{-1} = x$$

Domain f = All real numbers except -2

Range f = Domain f^{-1} = All real numbers except 2

60. $f^{-1}(x) = \dfrac{2x - 4}{x + 3}$

$$f(f^{-1}(x)) = f\left(\frac{2x-4}{x+3}\right) = \frac{-3\left(\dfrac{2x-4}{x+3}\right) - 4}{\left(\dfrac{2x-4}{x+3}\right) - 2} = \frac{-3(2x - 4) - 4(x + 3)}{(2x - 4) - 2(x + 3)} = \frac{-10x}{-10} = x$$

$$f^{-1}(f(x)) = f^{-1}\left(\frac{-3x-4}{x-2}\right) = \frac{2\left(\dfrac{-3x-4}{x-2}\right) - 4}{\left(\dfrac{-3x-4}{x-2}\right) + 3} = \frac{2(-3x - 4) - 4(x - 2)}{(-3x - 4) + 3(x - 2)} = \frac{-10x}{-10} = x$$

Domain f = All real numbers except 2

Range f = Domain f^{-1} = All real numbers except -3

61. $f^{-1}(x) = \dfrac{2}{\sqrt{1-2x}}$

$$f(f^{-1}(x)) = \frac{\dfrac{4}{1-2x} - 4}{2 \cdot \dfrac{4}{1-2x}} = \frac{4 - 4(1-2x)}{2 \cdot 4} = \frac{8x}{8} = x$$

$$f^{-1}(f(x)) = \frac{2}{\sqrt{1 - 2\left(\dfrac{x^2-4}{2x^2}\right)}} = \frac{2}{\sqrt{\dfrac{4}{x^2}}} = \sqrt{x^2} = x, \text{ since } x > 0.$$

Domain $f = \{x|x > 0\}$ or $(0, \infty)$

Range f = Domain $f^{-1} = \left\{x \middle| x < \frac{1}{2}\right\}$ or $\left(-\infty, \frac{1}{2}\right)$

62. $f^{-1}(x) = \sqrt{\dfrac{3}{3x-1}}$

$$f(f^{-1}(x)) = \frac{\dfrac{3}{3x-1} + 3}{3\left(\dfrac{3}{3x-1}\right)} = \frac{3 + 3(3x-1)}{9} = \frac{9x}{9} = x$$

$$f^{-1}(f(x)) = \sqrt{\frac{3}{3\left(\dfrac{x^2+3}{3x^2}\right) - 1}} = \sqrt{\frac{3x^2}{(x^2+3) - x^2}} = \sqrt{\frac{3x^2}{3}} = \sqrt{x^2} = x, \text{ since } x > 0$$

Domain $f = \{x|x > 0\}$

Range f = Domain $f^{-1} = \left\{x \middle| x > \frac{1}{3}\right\}$

63. **(a)** 0 **(b)** 2 **(c)** 0 **(d)** 1 **64.** **(a)** $\frac{1}{2}$ **(b)** 0 **(c)** 1 **(d)** 0 **65.** $f^{-1}(x) = \frac{1}{m}(x - b), m \neq 0$

66. $f^{-1}(x) = \sqrt{r^2 - x^2}, 0 \le x \le r$ **67.** Quadrant I **68.** Quadrant IV

69. Possible answer: $f(x) = |x|, x \ge 0$, is one-to-one; $f^{-1}(x) = x, x \ge 0$

70. Possible answer: $f(x) = x^4, x \ge 0$, is one-to-one; $f^{-1}(x) = x^{1/4}, x \ge 0$ **71.** **(a)** $C(H) = \dfrac{H + 10.53}{2.15}$ **(b)** 16.99 in.

72. $f(g(x)) = \frac{9}{5}\left[\frac{5}{9}(x - 32)\right] + 32 = x$; $g(f(x)) = \frac{5}{9}\left[\left(\frac{9}{5}x + 32\right) - 32\right] = x$ **73.** $x(p) = \dfrac{300 - p}{50}, p \le 300$ **74.** $l(T) = \dfrac{gT^2}{4\pi^2}, T > 0$

75. $f^{-1}(x) = \dfrac{-dx + b}{cx - a}$; $f = f^{-1}$ if $a = -d$ **76.** Yes, the graph must be symmetric about the line $y = x$; $y = \frac{1}{x}$ is an example.

78. Possible answer: $y = \begin{cases} \frac{1}{x}, & x < 0 \\ x, & x \ge 0 \end{cases}$ is neither an increasing nor a decreasing function on its domain, but is one-to-one.

79. Yes, if the domain is $\{x|x = 0\}$. **80.** No **81.** On the line $y = x$. No. No.

4.3 Concepts and Vocabulary *(page 255)*

6. $\left(-1, \frac{1}{a}\right), (0, 1), (1, a)$ **7.** 1 **8.** 4 **9.** False **10.** False

4.3 Exercises *(page 255)*

11. **(a)** 11.212 **(b)** 11.587 **(c)** 11.664 **(d)** 11.665 **12.** **(a)** 15.426 **(b)** 16.189 **(c)** 16.241 **(d)** 16.242
13. **(a)** 8.815 **(b)** 8.821 **(c)** 8.824 **(d)** 8.825 **14.** **(a)** 6.498 **(b)** 6.543 **(c)** 6.580 **(d)** 6.581
15. **(a)** 21.217 **(b)** 22.217 **(c)** 22.440 **(d)** 22.459 **16.** **(a)** 21.738 **(b)** 22.884 **(c)** 23.119 **(d)** 23.141
17. 3.320 **18.** 0.273 **19.** 0.427 **20.** 8.166 **21.** Not exponential **22.** Not exponential **23.** Exponential; $a = 4$
24. Exponential; $a = \frac{3}{2}$ **25.** Exponential; $a = 2$ **26.** Not exponential **27.** Not exponential **28.** Exponential; $a = \frac{1}{2}$

29. *B* **30.** *F* **31.** *D* **32.** *H* **33.** *A* **34.** *C* **35.** *E* **36.** *G*

37.

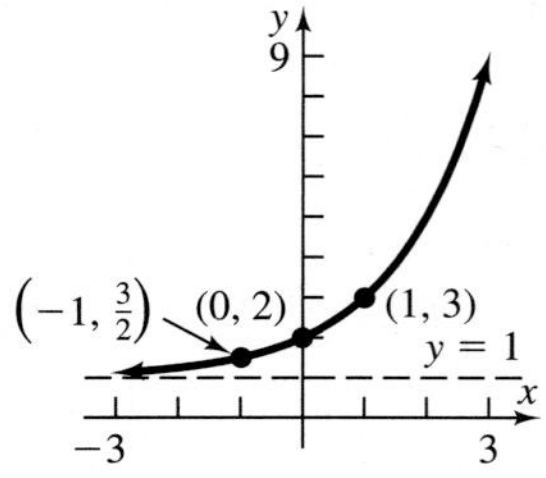

Domain: All real numbers
Range: $\{y|y > 1\}$ or $(1, \infty)$
Horizontal asymptote: $y = 1$

38.

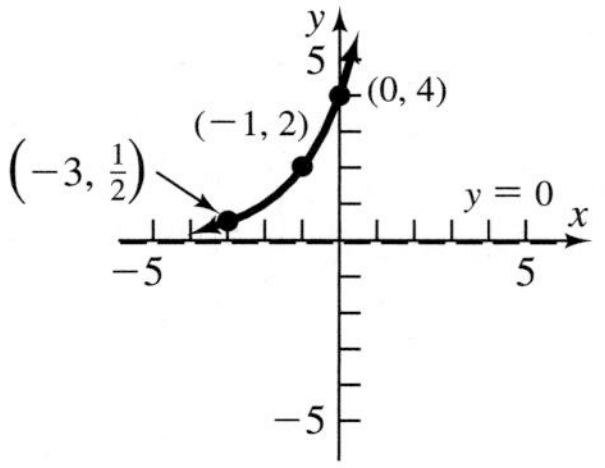

Domain: All real numbers
Range: $\{y|y > 0\}$ or $(0, \infty)$
Horizontal asymptote: $y = 0$

39.

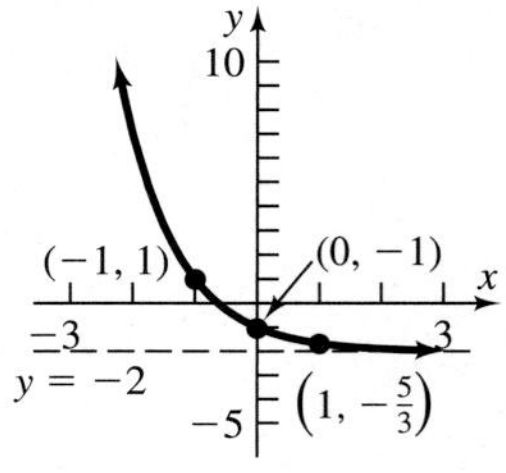

Domain: All real numbers
Range: $\{y|y > -2\}$ or $(-2, \infty)$
Horizontal asymptote: $y = -2$

40.

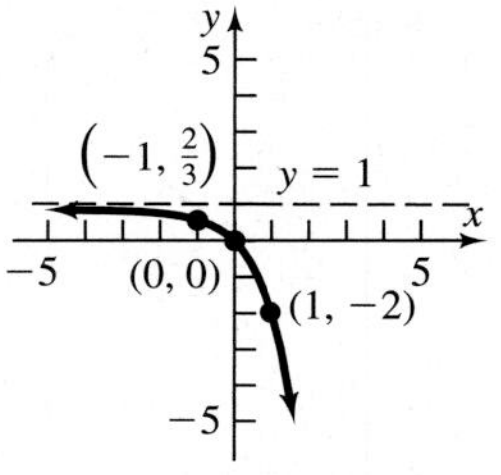

Domain: All real numbers
Range: $\{y|y < 1\}$ or $(-\infty, 1)$
Horizontal asymptote: $y = 1$

41.

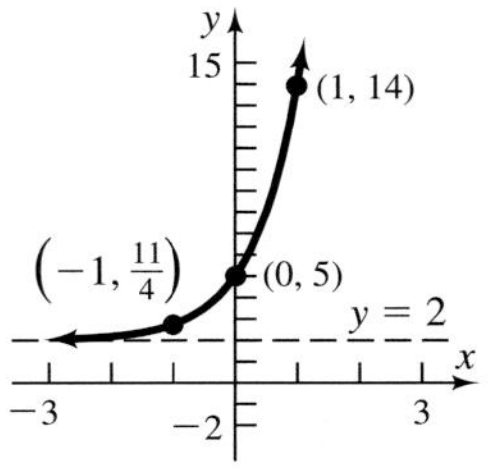

Domain: All real numbers
Range: $\{y|y > 2\}$ or $(2, \infty)$
Horizontal asymptote: $y = 2$

42.

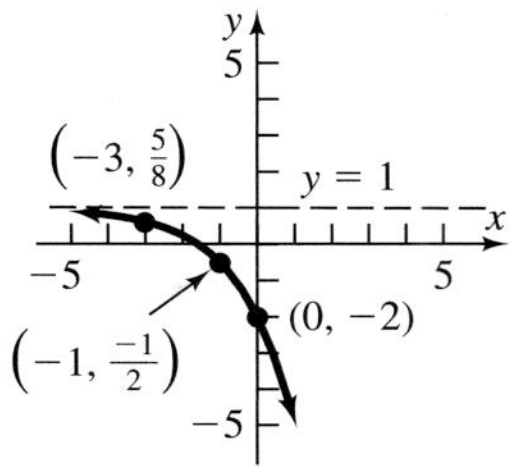

Domain: All real numbers
Range: $\{y|y < 1\}$ or $(-\infty, 1)$
Horizontal asymptote: $y = 1$

43.

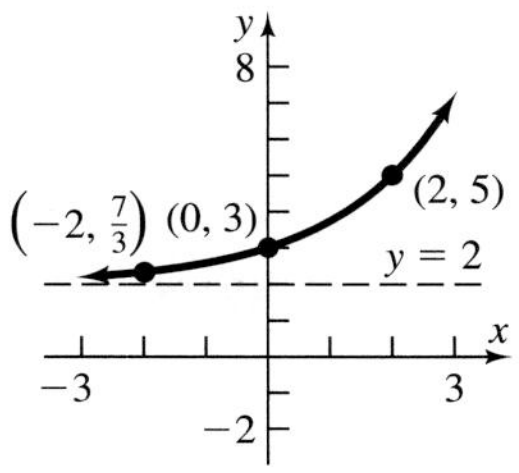

Domain: All real numbers
Range: $\{y|y > 2\}$ or $(2, \infty)$
Horizontal asymptote: $y = 2$

44.

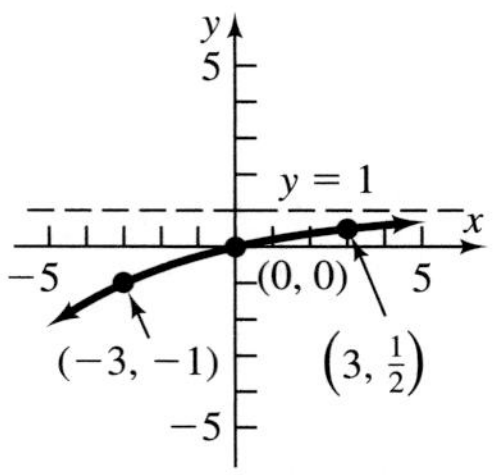

Domain: All real numbers
Range: $\{y|y < 1\}$ or $(-\infty, 1)$
Horizontal asymptote: $y = 1$

45.

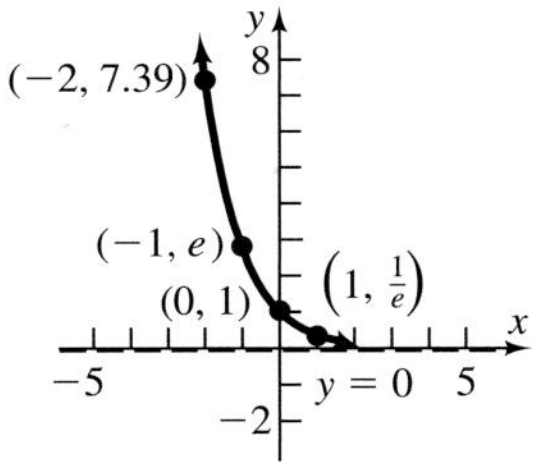

Domain: All real numbers
Range: $\{y|y > 0\}$ or $(0, \infty)$
Horizontal asymptote: $y = 0$

46.

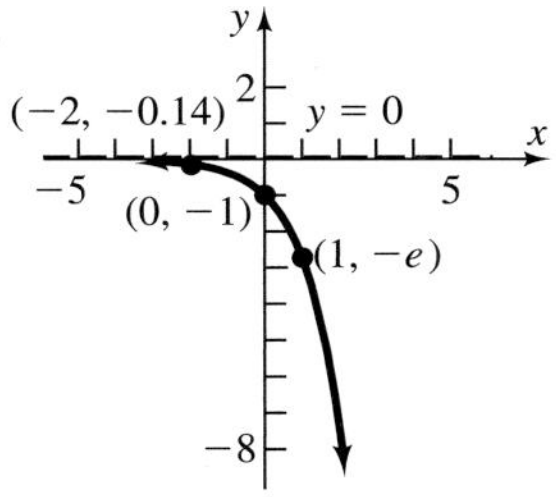

Domain: All real numbers
Range: $\{y|y < 0\}$ or $(-\infty, 0)$
Horizontal asymptote: $y = 0$

47.

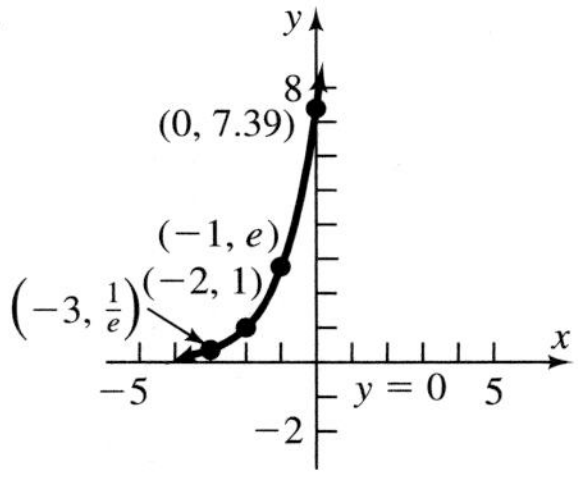

Domain: All real numbers
Range: $\{y|y > 0\}$ or $(0, \infty)$
Horizontal asymptote: $y = 0$

48.

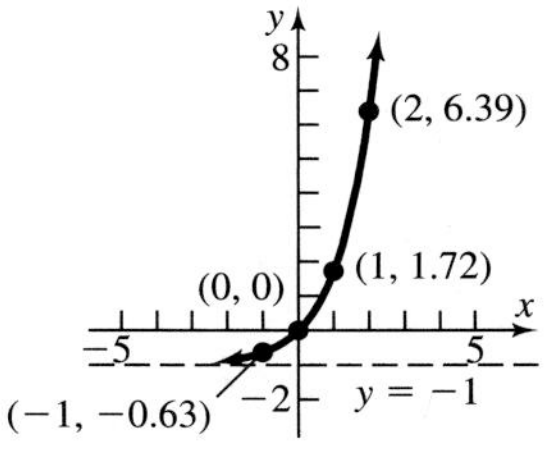

Domain: All real numbers
Range: $\{y|y > -1\}$ or $(-1, \infty)$
Horizontal asymptote: $y = -1$

49.

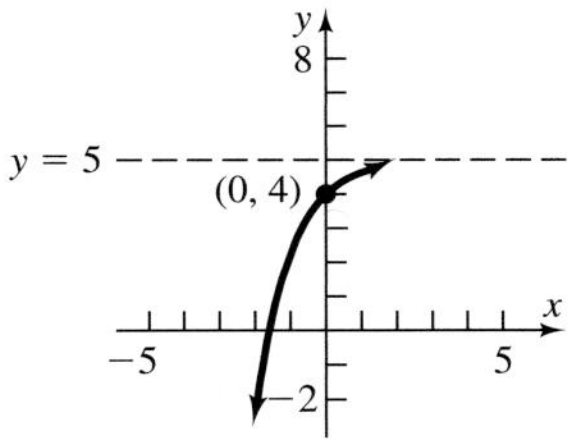

Domain: All real numbers
Range: $\{y|y < 5\}$ or $(-\infty, 5)$
Horizontal asymptote: $y = 5$

50.

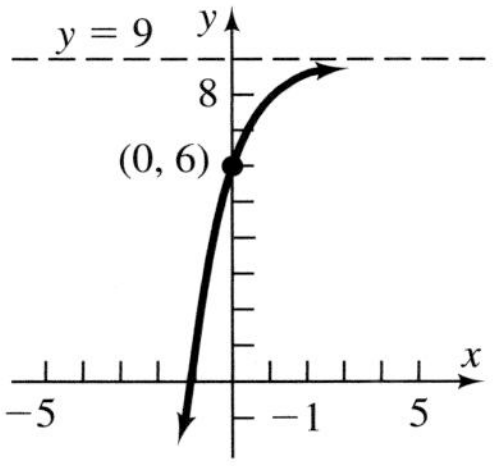

Domain: All real numbers
Range: $\{y|y < 9\}$ or $(-\infty, 9)$
Horizontal asymptote: $y = 9$

51.

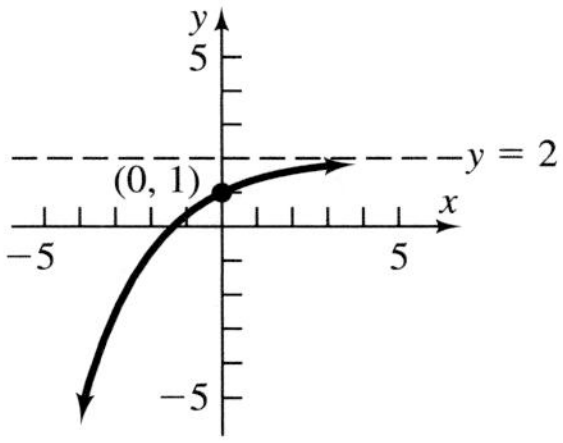

Domain: All real numbers
Range: $\{y|y < 2\}$ or $(-\infty, 2)$
Horizontal asymptote: $y = 2$

52.

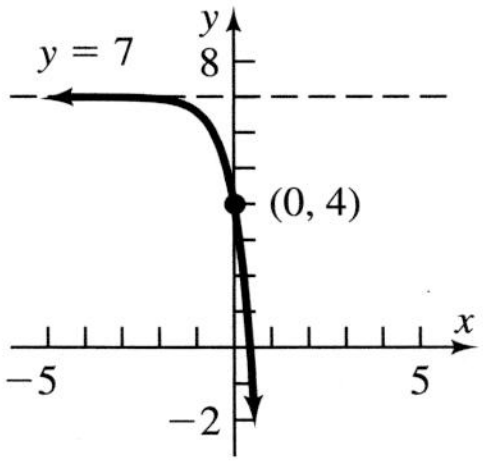

Domain: All real numbers
Range: $\{y|y < 7\}$ or $(-\infty, 7)$
Horizontal asymptote: $y = 7$

53. $\frac{1}{2}$ **54.** 1 **55.** $\{-\sqrt{2}, 0, \sqrt{2}\}$ **56.** $\left\{0, \frac{1}{2}\right\}$ **57.** $\left\{1 - \frac{\sqrt{6}}{3}, 1 + \frac{\sqrt{6}}{3}\right\}$ **58.** $\frac{1}{2}$ **59.** 0 **60.** 3 **61.** 4 **62.** 0 **63.** $\frac{3}{2}$ **64.** $\frac{3}{4}$

65. $\{1, 2\}$ **66.** $\{-6, 2\}$ **67.** $\frac{1}{49}$ **68.** $\frac{1}{9}$ **69.** $\frac{1}{4}$ **70.** $\frac{1}{27}$

71. $f(x) = 3^x$ **72.** $f(x) = 5^x$ **73.** $f(x) = -6^x$ **74.** $f(x) = -e^x$

75. **(a)** 74% **(b)** 47% **76.** **(a)** 568.68 mm Hg **(b)** 178.27 mm Hg **77.** **(a)** 44.3 watts **(b)** 11.6 watts

78. **(a)** 34.99 sq mm **(b)** 3.02 sq mm **79.** 3.35 mg; 0.45 mg **80.** 362 students

81. **(a)** 0.63 **(b)** 0.98 **(c)** 1
(d)

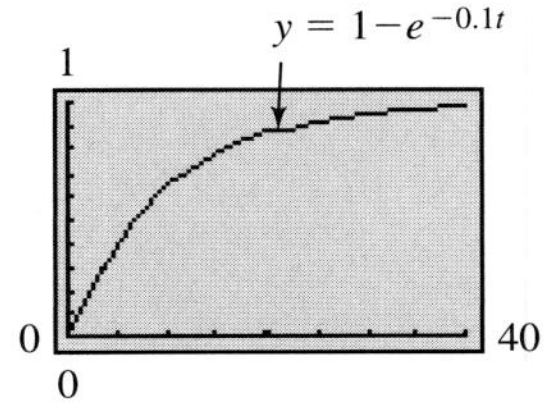

(e) About 7 min

82. **(a)** 0.895 **(b)** 0.989 **(c)** 1
(d)

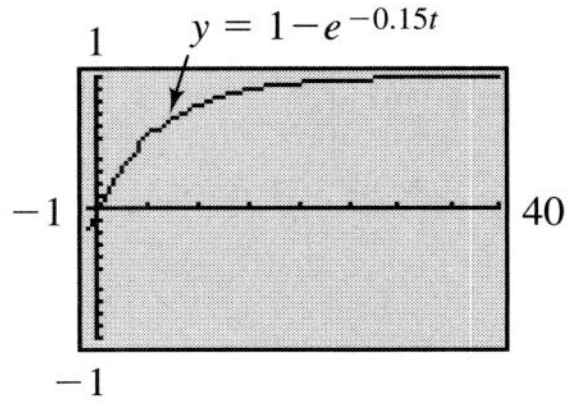

(e) About 6 min

83. **(a)** 5.16% **(b)** 8.88%
84. **(a)** 0.156 **(b)** 0.030
85. **(a)** $12, 123 **(b)** $6443
86. **(a)** 83 words **(b)** 153 words

87. **(a)** 5.414 amp, 7.585 amp, 10.376 amp **(b)** 12 amp **(d)** 3.343 amp, 5.309 amp, 9.443 amp **(e)** 24 amp

(c), (f)

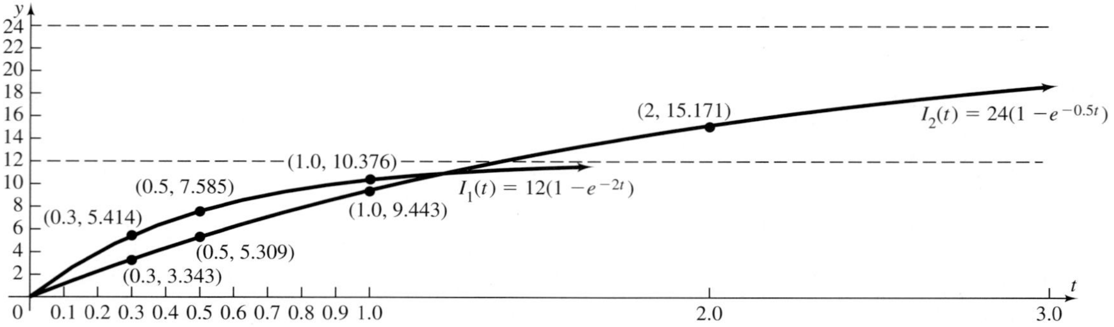

88. **(a)** 0.06 amp; 0.0364 amp; 0.0134 amp **(b)** 0.06 amp

(c)

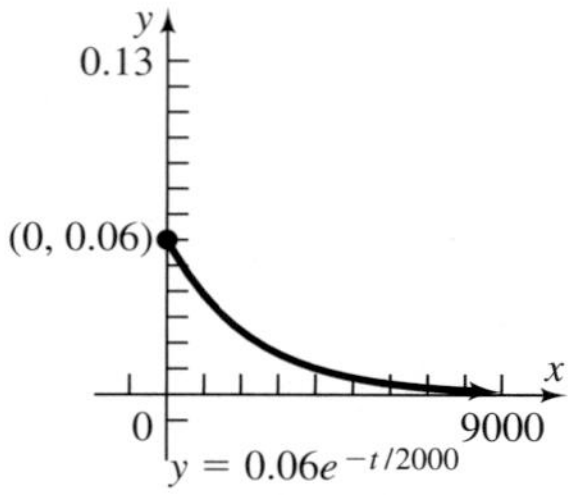

(d) 0.12 amp; 0.0728 amp; 0.0268 amp

(e) 0.12 amp

(f)

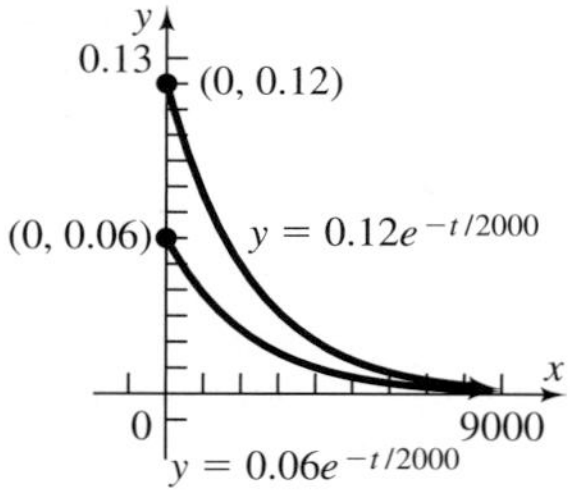

89. $n = 4: 2.7083; n = 6: 2.7181;$
$n = 8: 2.7182788; n = 10: 2.7182818$

90. 2.71826332
$e \approx 2.718281828$

91. $\dfrac{f(x+h) - f(x)}{h} = \dfrac{a^{x+h} - a^x}{h} = \dfrac{a^x a^h - a^x}{h} = \dfrac{a^x(a^h - 1)}{h}$ **92.** $f(A + B) = a^{A+B} = a^A \cdot a^B = f(A) \cdot f(B)$

93. $f(-x) = a^{-x} = \dfrac{1}{a^x} = \dfrac{1}{f(x)}$ **94.** $f(\alpha x) = a^{\alpha x} = (a^x)^\alpha = [f(x)]^\alpha$

95. **(a)** 70.95% **(b)** 72.62% **(c)** 100%

96. $f(1) = 5, f(2) = 17, f(3) = 257, f(4) = 65{,}537,$
$f(5) = 4{,}294{,}967{,}297 = 641 \times 6{,}700{,}417$

97. **(a)** $f(-x) = \dfrac{1}{2}(e^{-x} - e^{-(-x)}) = \dfrac{1}{2}(e^{-x} - e^x)$
$= -\dfrac{1}{2}(e^x - e^{-x}) = -f(x)$

(b)

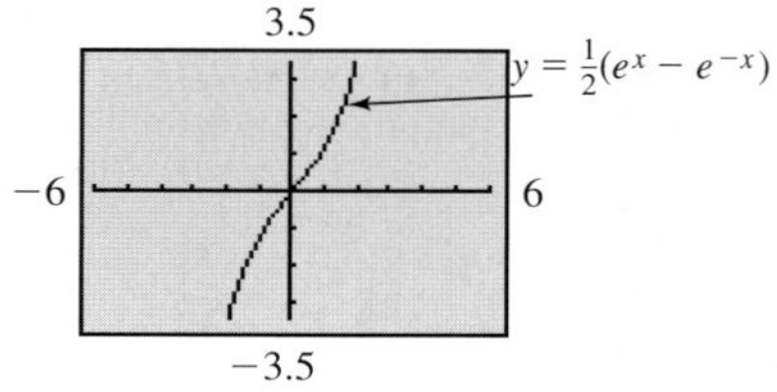

98. **(a)** $f(-x) = \dfrac{1}{2}(e^{-x} + e^{-(-x)}) = \dfrac{1}{2}(e^{-x} + e^x)$
$= \dfrac{1}{2}(e^x + e^{-x}) = f(x)$

(b)

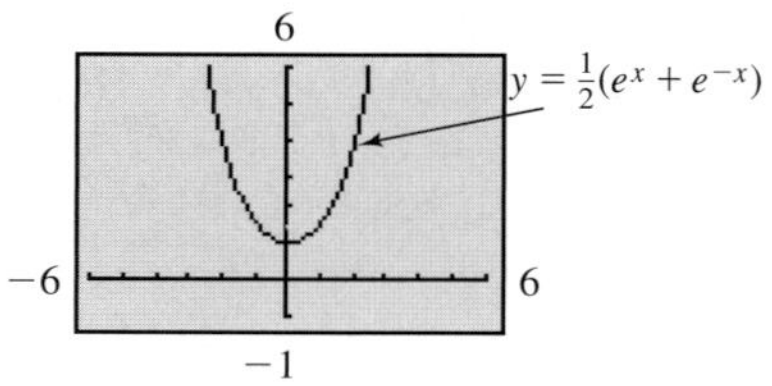

(c) $(\cosh x)^2 - (\sinh x)^2 = \left[\dfrac{1}{2}(e^x + e^{-x})\right]^2 - \left[\dfrac{1}{2}(e^x - e^{-x})\right]^2$

$$= \frac{1}{4}[e^{2x} + 2 + e^{-2x} - e^{2x} + 2 - e^{-2x}]$$

$$= \frac{1}{4}[4] = 1$$

99. 59 minutes

4.4 Concepts and Vocabulary *(page 269)*

4. $\{x|x > 0\}$ or $(0, \infty)$ **5.** $\left(\frac{1}{a}, -1\right)$, $(1, 0)$, $(a, 1)$ **6.** 1 **7.** False **8.** True

4.4 Exercises *(page 269)*

9. $2 = \log_3 9$ **10.** $\log_4 16 = 2$ **11.** $2 = \log_a 1.6$ **12.** $\log_a 2.1 = 3$ **13.** $2 = \log_{1.1} M$ **14.** $\log_{2.2} N = 3$ **15.** $x = \log_2 7.2$ **16.** $\log_3 4.6 = x$ **17.** $\sqrt{2} = \log_x \pi$ **18.** $\log_x e = \pi$ **19.** $x = \ln 8$ **20.** $\ln M = 2.2$ **21.** $2^3 = 8$ **22.** $3^{-2} = \frac{1}{9}$ **23.** $a^6 = 3$ **24.** $b^2 = 4$ **25.** $3^x = 2$ **26.** $2^x = 6$ **27.** $2^{1.3} = M$ **28.** $3^{2.1} = N$ **29.** $(\sqrt{2})^x = \pi$ **30.** $\pi^{1/2} = x$ **31.** $e^x = 4$ **32.** $e^4 = x$ **33.** 0 **34.** 1 **35.** 2 **36.** -2 **37.** -4 **38.** -2 **39.** $\frac{1}{2}$ **40.** $\frac{2}{3}$ **41.** 4 **42.** 4 **43.** $\frac{1}{2}$ **44.** 3 **45.** $\{x|x > 3\}; (3, \infty)$ **46.** $\{x|x > 1\}; (1, \infty)$ **47.** All real numbers except 0; $\{x|x \neq 0\}$ **48.** $\{x|x > 0\}; (0, \infty)$ **49.** $\{x|x > 0\}; (0, \infty)$ **50.** $\{x|x > 0\}; (0, \infty)$ **51.** $\{x|x > -1\}; (-1, \infty)$ **52.** $\{x|x > 5\}; (5, \infty)$ **53.** $\{x|x < -1 \text{ or } x > 0\}; (-\infty, -1)$ or $(0, \infty)$ **54.** $\{x|x < 0 \text{ or } x > 1\}; (-\infty, 0)$ or $(1, \infty)$ **55.** $\{x|x \geq 1\}; [1, \infty)$ **56.** $\{x|x > 0, x \neq 1\}$ **57.** 0.511 **58.** 0.536 **59.** 30.099 **60.** 4.055 **61.** $\sqrt{2}$ **62.** $a = 2^{1/4} \approx 1.189$

63.

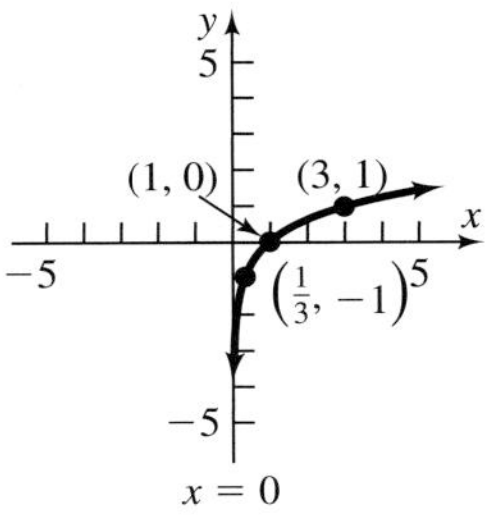

64.

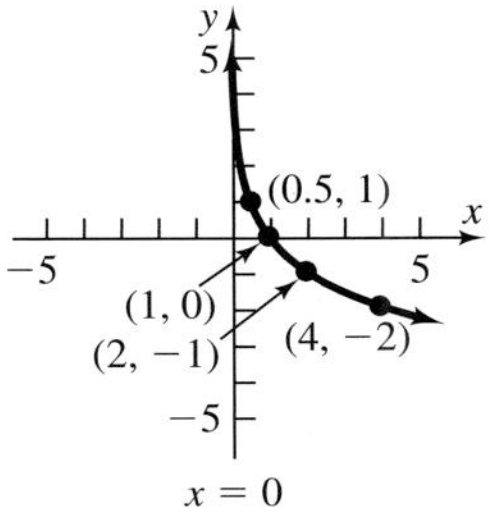

65.

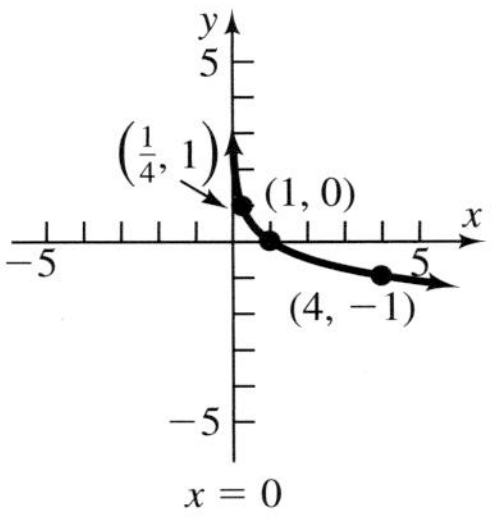

66.

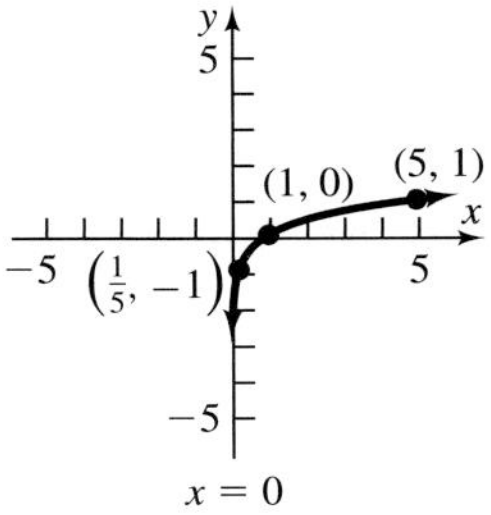

67. B **68.** F **69.** D **70.** H **71.** A **72.** C **73.** E **74.** G

75.

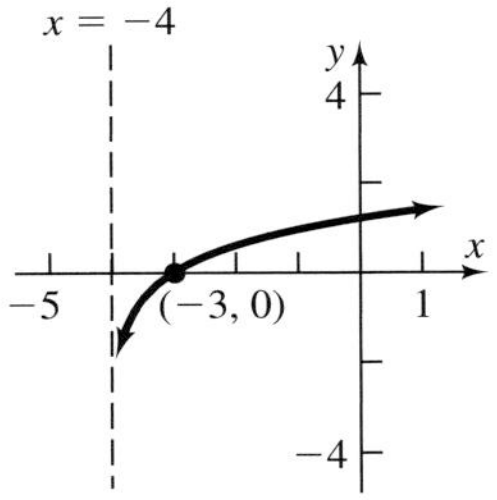

Domain: $(-4, \infty)$
Range: $(-\infty, \infty)$
Vertical asymptote: $x = -4$

76.

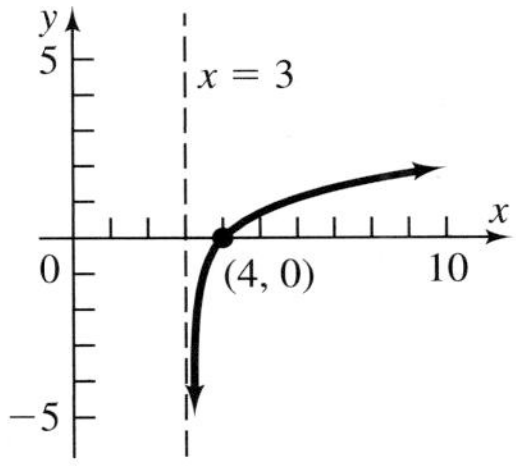

Domain: $(3, \infty)$
Range: $(-\infty, \infty)$
Vertical asymptote: $x = 3$

77.

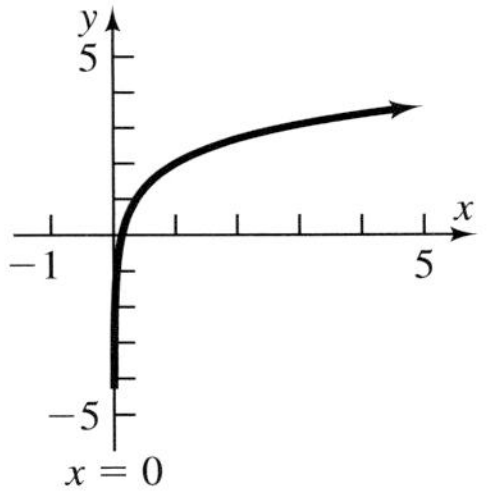

Domain: $(0, \infty)$
Range: $(-\infty, \infty)$
Vertical asymptote: $x = 0$

78.

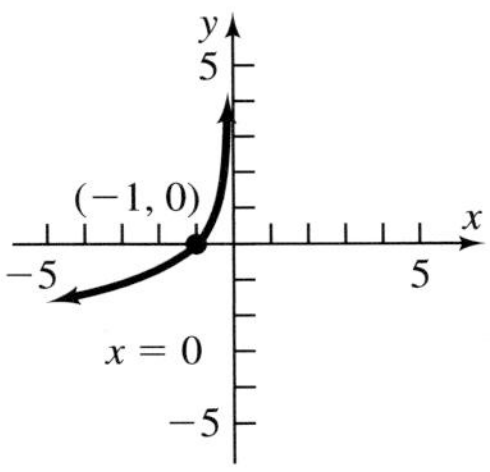

Domain: $(-\infty, 0)$
Range: $(-\infty, \infty)$
Vertical asymptote: $x = 0$

79.

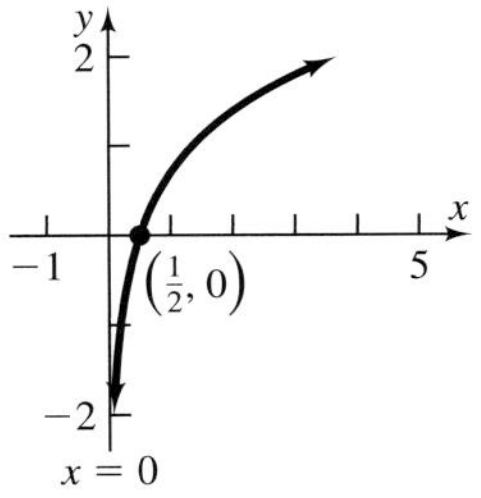

Domain: $(0, \infty)$
Range: $(-\infty, \infty)$
Vertical asymptote: $x = 0$

80.

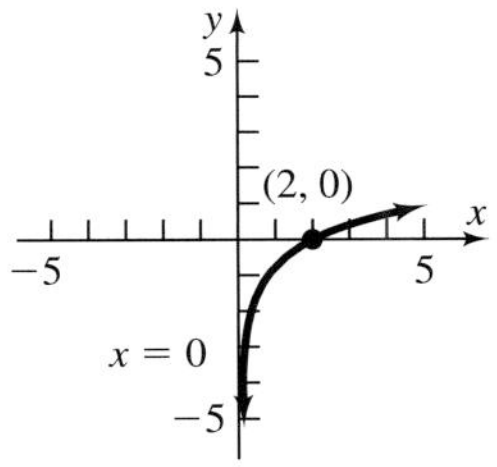

Domain: $(0, \infty)$
Range: $(-\infty, \infty)$
Vertical asymptote: $x = 0$

81.

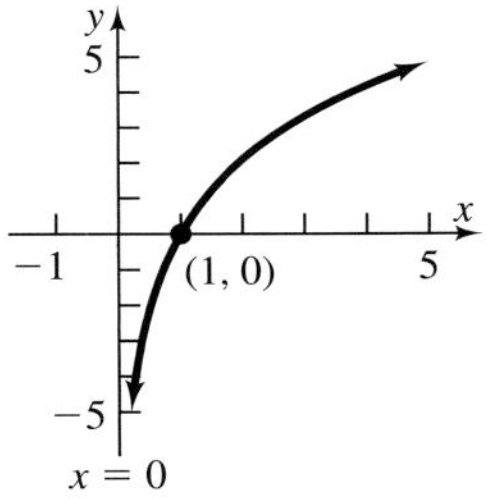

Domain: $(0, \infty)$
Range: $(-\infty, \infty)$
Vertical asymptote: $x = 0$

82.

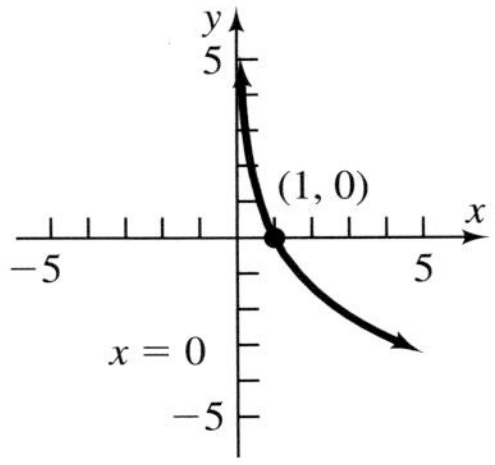

Domain: $(0, \infty)$
Range: $(-\infty, \infty)$
Vertical asymptote: $x = 0$

83.

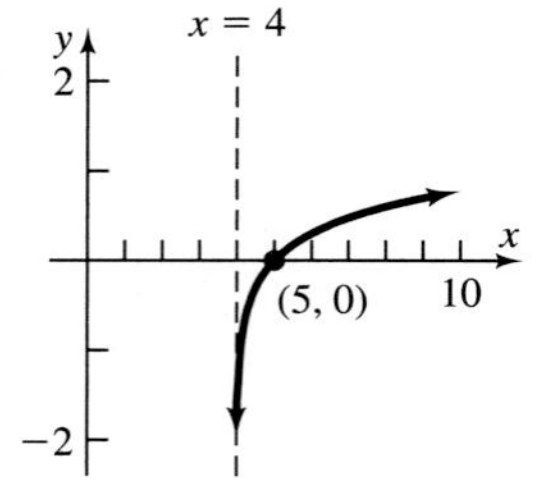

Domain: $(4, \infty)$
Range: $(-\infty, \infty)$
Vertical asymptote: $x = 4$

84.

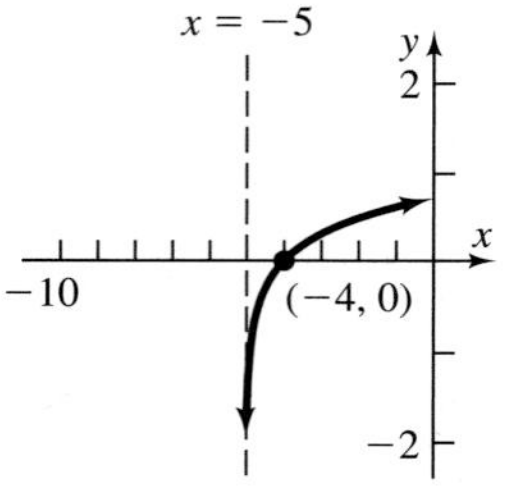

Domain: $(-5, \infty)$
Range: $(-\infty, \infty)$
Vertical asymptote: $x = -5$

85.

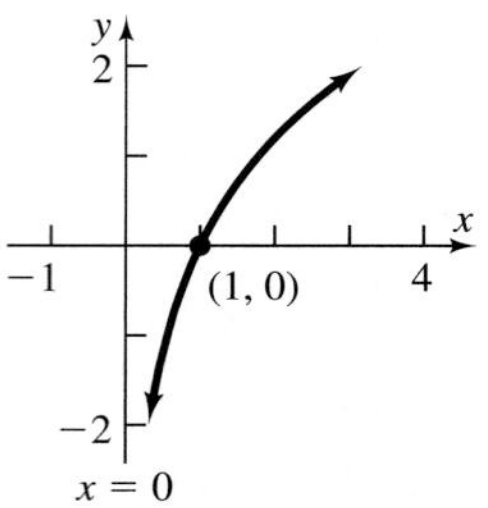

Domain: $(0, \infty)$
Range: $(-\infty, \infty)$
Vertical asymptote: $x = 0$

86.

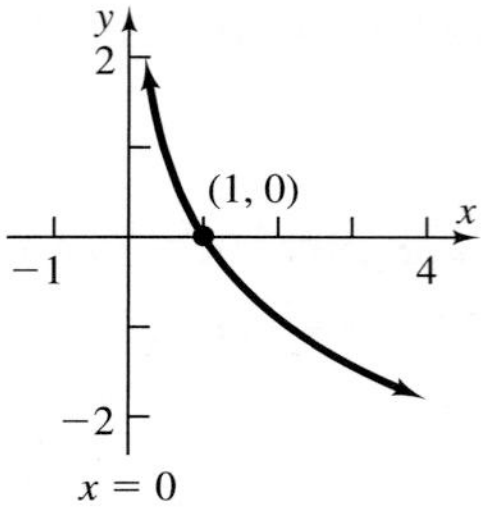

Domain: $(0, \infty)$
Range: $(-\infty, \infty)$
Vertical asymptote: $x = 0$

87.

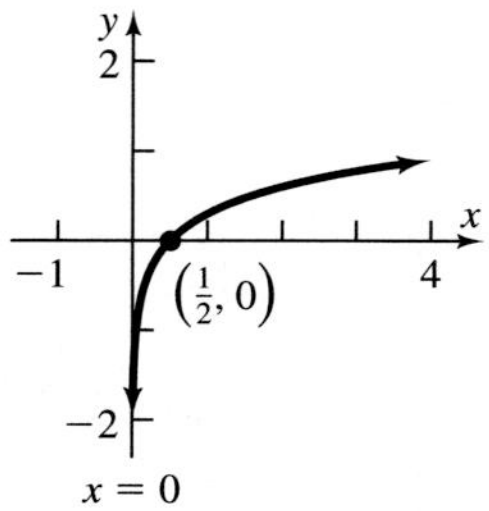

Domain: $(0, \infty)$
Range: $(-\infty, \infty)$
Vertical asymptote: $x = 0$

88.

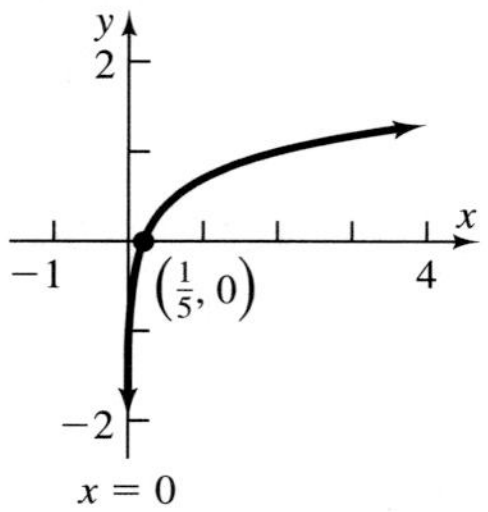

Domain: $(0, \infty)$
Range: $(-\infty, \infty)$
Vertical asymptote: $x = 0$

89.

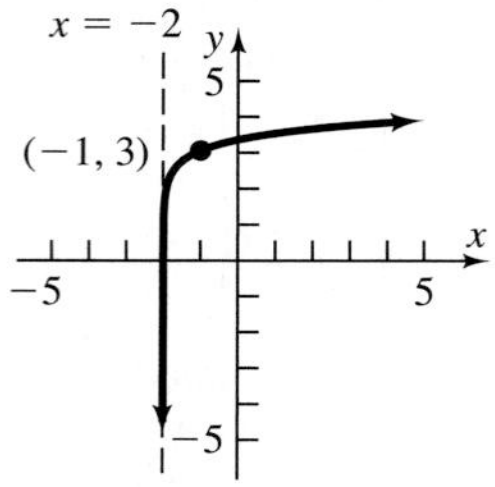

Domain: $(-2, \infty)$
Range: $(-\infty, \infty)$
Vertical asymptote: $x = -2$

90.

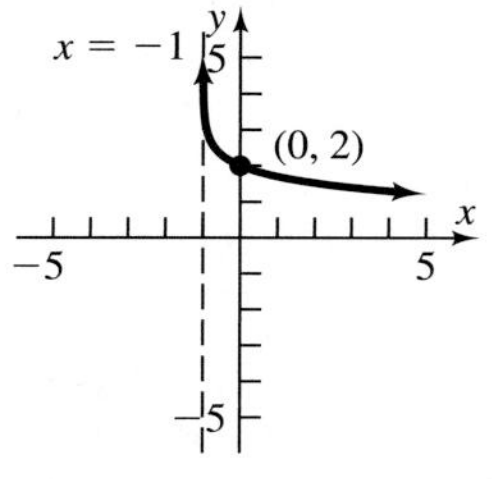

Domain: $(-1, \infty)$
Range: $(-\infty, \infty)$
Vertical asymptote: $x = -1$

91. $\{9\}$ **92.** $\{125\}$ **93.** $\left\{\frac{7}{2}\right\}$ **94.** $\left\{\frac{11}{3}\right\}$ **95.** $\{2\}$ **96.** $\left\{\frac{1}{2}\right\}$ **97.** $\{5\}$ **98.** $\{-4\}$ **99.** $\{3\}$ **100.** $\{4\}$ **101.** $\{2\}$ **102.** $\left\{-\frac{1}{5}\right\}$
103. $\left\{\frac{\ln 10}{3}\right\}$ **104.** $\left\{\frac{\ln 3}{2}\right\}$ **105.** $\left\{\frac{\ln 8 - 5}{2}\right\}$ **106.** $\left\{\frac{1 - \ln 13}{2}\right\}$ **107.** $\{-2\sqrt{2}, 2\sqrt{2}\}$ **108.** $\left\{\frac{-1 + \sqrt{85}}{2}, \frac{-1 - \sqrt{85}}{2}\right\}$
109. $\{-1\}$ **110.** $\{-1\}$ **111.** **(a)** 1 **(b)** 2 **(c)** 3 **(d)** It increases. **(e)** ≈ 0.000316 **(f)** $\approx 3.981 \times 10^{-8}$
112. **(a)** 0.4283 **(b)** 0.7782 **(c)** 0.5504 **113.** **(a)** 5.97 km **(b)** 0.90 km
114. **(a)** 1.98 days or slightly less than 2 days **(b)** 6.58 days or about 6 days and 14 hr
115. **(a)** 6.93 min **(b)** 16.09 min **(c)** No, since $F(t)$ can never equal 1
116. **(a)** 4.62 min or about 4 min and 37 sec **(b)** 10.73 min or about 10 min and 44 sec
117. $h \approx 2.29$, so the time between injections is about 2 hr, 17 min **118.** 3.99 days or about 4 days
119. 0.2695 sec
0.8959 sec

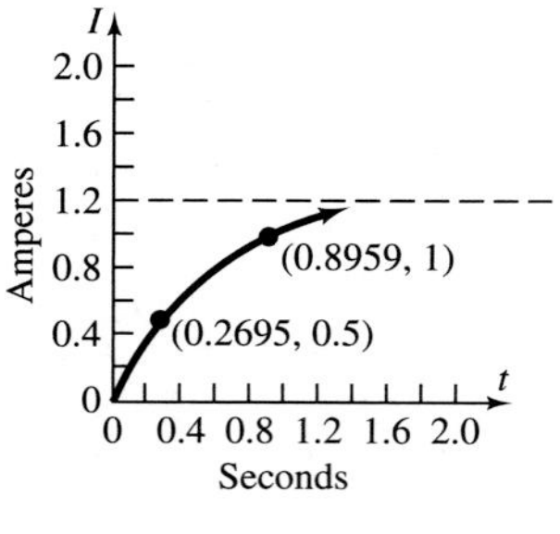

120. **(a)** 0.0211 **(b)** 38 words **(c)** 54 words **(d)** 109.27 min
121. 50 decibels (dB)
122. 90 dB
123. 110 dB
124. 80 dB
125. 8.1
126. 6.9
127. **(a)** $k = 20.07$ **(b)** 91% **(c)** 0.175 **(d)** 0.08
129. Because $y = \log_1 x$ means $1^y = 1 = x$, which cannot be true for $x \neq 1$

4.5 Concepts and Vocabulary *(page 280)*

1. Sum **2.** 7 **3.** $r \log_a M$ **4.** False **5.** False **6.** True

4.5 Exercises *(page 281)*

7. 71 **8.** −13 **9.** −4 **10.** $\sqrt{2}$ **11.** 7 **12.** 8 **13.** 1 **14.** 2 **15.** 1 **16.** 1 **17.** 2 **18.** 2 **19.** $\frac{5}{4}$ **20.** 42 **21.** 4 **22.** 3 **23.** $a + b$
24. $a - b$ **25.** $b - a$ **26.** $-a$ **27.** $3a$ **28.** $3b$ **29.** $\frac{1}{5}(a + b)$ **30.** $\frac{1}{4}(a - b)$ **31.** $2 + \log_5 x$ **32.** $\log_3 x - 2$ **33.** $3 \log_2 z$
34. $5 \log_7 x$ **35.** $1 + \ln x$ **36.** $1 - \ln x$ **37.** $\ln x + x$ **38.** $\ln x - x$ **39.** $2 \log_a u + 3 \log_a v$ **40.** $\log_2 a - 2 \log_2 b$

41. $2\ln x + \frac{1}{2}\ln(1-x)$ **42.** $\ln x + \frac{1}{2}\ln(1+x^2)$ **43.** $3\log_2 x - \log_2(x-3)$ **44.** $\frac{1}{3}\log_5(x^2+1) - \log_5(x+1) - \log_5(x-1)$

45. $\log x + \log(x+2) - 2\log(x+3)$ **46.** $3\log x + \frac{1}{2}\log(x+1) - 2\log(x-2)$ **47.** $\frac{1}{3}\ln(x-2) + \frac{1}{3}\ln(x+1) - \frac{2}{3}\ln(x+4)$

48. $\frac{4}{3}\ln(x-4) - \frac{2}{3}\ln(x+1) - \frac{2}{3}\ln(x-1)$ **49.** $\ln 5 + \ln x + \frac{1}{2}\ln(1+3x) - 3\ln(x-4)$

50. $\ln 5 + 2\ln x + \frac{1}{3}\ln(1-x) - \ln 4 - 2\ln(x+1)$ **51.** $\log_5 u^3v^4$ **52.** $\log_3\left(\frac{u^2}{v}\right)$ **53.** $-\frac{5}{2}\log_3 x$ **54.** $\log_2\left(\frac{1}{x^3}\right)$

55. $\log_4 \frac{x-1}{(x+1)^4}$ **56.** $\log \frac{x+2}{x+1}$ **57.** $-2\ln(x-1)$ **58.** $\log\left[\frac{x^2+2x-3}{(x-2)(x^2+7x+6)}\right]$ **59.** $\log_2[x(3x-2)^4]$ **60.** $\log_3 x^9$

61. $\log_a\left(\frac{25x^6}{\sqrt{2x+3}}\right)$ **62.** $\log(\sqrt[3]{x^3+1}\sqrt{x^2+1})$ **63.** $\log_2 \frac{(x+1)^2}{(x+3)(x-1)}$ **64.** $\log_5 \frac{(3x+1)^3}{x(2x-1)^2}$

65. 2.771 **66.** 1.796 **67.** −3.880 **68.** −3.907 **69.** 5.615 **70.** 2.584 **71.** 0.874 **72.** 0.303

73. $y = \frac{\log x}{\log 4}$

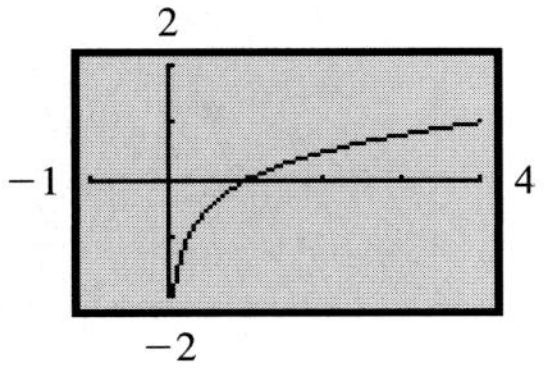

74. $y = \frac{\log x}{\log 5}$

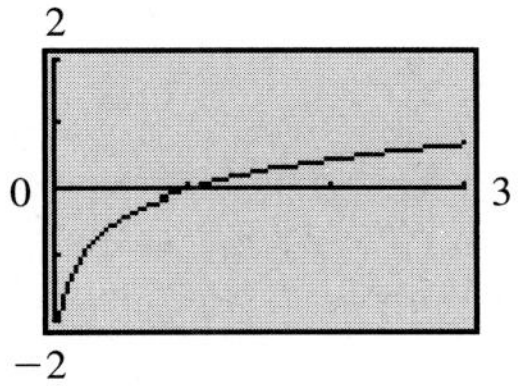

75. $y = \frac{\log(x+2)}{\log 2}$

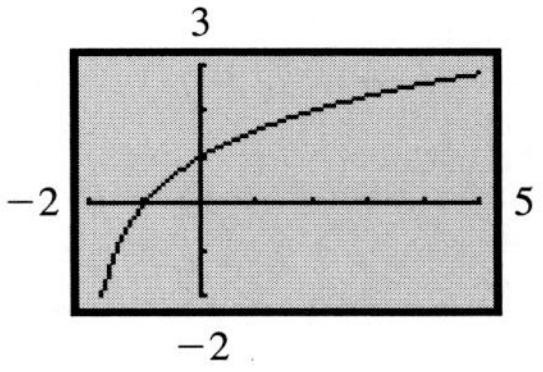

76. $y = \frac{\log(x-3)}{\log 4}$

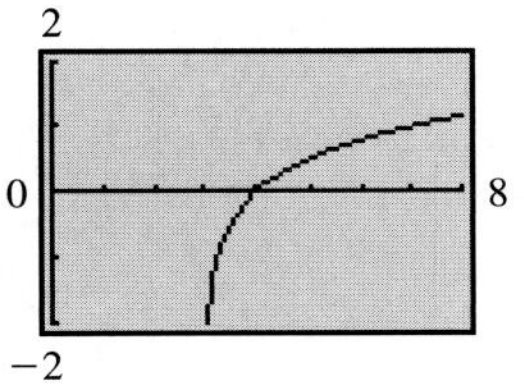

77. $y = \frac{\log(x+1)}{\log(x-1)}$

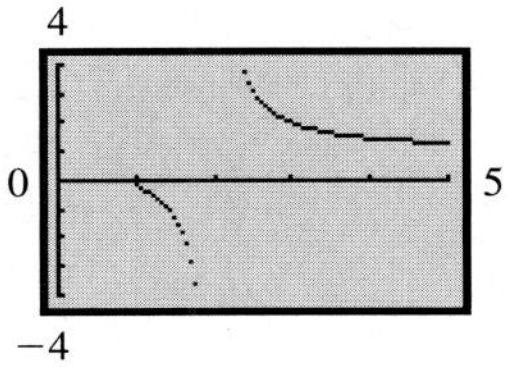

78. $y = \frac{\log(x-2)}{\log(x+2)}$

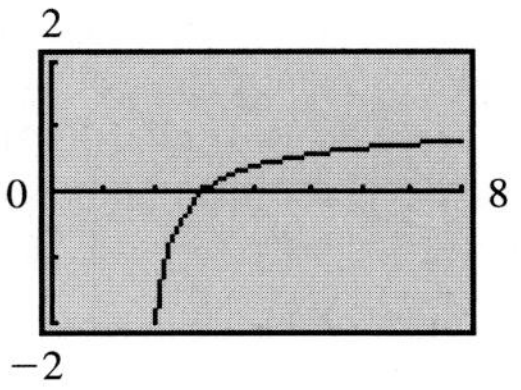

79. $y = Cx$ **80.** $y = x + C$ **81.** $y = Cx(x+1)$ **82.** $y = \frac{Cx^2}{x+1}$ **83.** $y = Ce^{3x}$ **84.** $y = Ce^{-2x}$ **85.** $y = Ce^{-4x} + 3$

86. $y = Ce^{5x} - 4$ **87.** $y = \frac{\sqrt[3]{C}(2x+1)^{1/6}}{(x+4)^{1/9}}$ **88.** $y = \frac{\sqrt{C}(x^2+1)^{1/6}}{x^{1/4}}$ **89.** 3 **90.** 3 **91.** 1 **92.** $n!$

93. $\log_a(x + \sqrt{x^2-1}) + \log_a(x - \sqrt{x^2-1}) = \log_a[(x + \sqrt{x^2-1})(x - \sqrt{x^2-1})]$
$= \log_a[x^2 - (x^2-1)] = \log_a 1 = 0$

94. $\log_a(\sqrt{x} + \sqrt{x-1}) + \log_a(\sqrt{x} - \sqrt{x-1})$
$= \log_a[(\sqrt{x} + \sqrt{x-1})(\sqrt{x} - \sqrt{x-1})]$
$= \log_a[x - (x-1)]$
$= \log_a 1$
$= 0$

95. $\ln(1+e^{2x}) = \ln[e^{2x}(e^{-2x}+1)] = \ln e^{2x} + \ln(e^{-2x}+1) = 2x + \ln(1+e^{-2x})$

96. $\frac{f(x+h) - f(x)}{h} = \frac{\log_a(x+h) - \log_a(x)}{h}$
$= \frac{1}{h}\log_a\left(\frac{x+h}{x}\right) = \log_a\left(1 + \frac{h}{x}\right)^{1/h}, h \neq 0$

97. $y = f(x) = \log_a x$; $a^y = x$ implies $a^y = \left(\frac{1}{a}\right)^{-y} = x$, so $-y = \log_{1/a} x = -f(x)$.

98. $f(AB) = \log_a(AB) = \log_a(A) + \log_a(B) = f(A) + f(B)$ **99.** $f(x) = \log_a x$; $f\left(\frac{1}{x}\right) = \log_a \frac{1}{x} = \log_a 1 - \log_a x = -f(x)$

100. $f(x^\alpha) = \log_a(x^\alpha) = \alpha\log_a(x) = \alpha f(x)$ **101.** $\log_a \frac{M}{N} = \log_a(M \cdot N^{-1}) = \log_a M + \log_a N^{-1} = \log_a M - \log_a N$,
since $a^{\log_a N^{-1}} = N^{-1}$ implies $a^{-\log_a N^{-1}} = N$; i.e., $\log_a N = -\log_a N^{-1}$.

102. $\log_a \frac{1}{N} = \log_a 1 - \log_a N = -\log_a N$

4.6 Exercises *(page 286)*

1. 6 **2.** 0 **3.** 16 **4.** $\frac{1}{3}$ **5.** 8 **6.** $\frac{1}{3}$ **7.** 3 **8.** 5 **9.** 5 **10.** 4 **11.** $-1 + \sqrt{1 + e^4} \approx 6.456$

12. $\frac{1}{e^2 - 1} \approx 0.157$ **13.** $\frac{\ln 3}{\ln 2} \approx 1.585$ **14.** 0 **15.** 0 **16.** 1 **17.** $\frac{\ln 10}{\ln 2} \approx 3.322$ **18.** $\log_3 14 = \frac{\ln 14}{\ln 3} \approx 2.402$ **19.** $-\frac{\ln 1.2}{\ln 8} \approx -0.088$

20. $\log_2\left(\frac{2}{3}\right) = \frac{\ln\left(\frac{2}{3}\right)}{\ln 2} \approx -0.585$ **21.** $\frac{\ln 3}{2 \ln 3 + \ln 4} \approx 0.307$ **22.** $\frac{\ln(2.5)}{\ln(50)} \approx 0.234$ **23.** $\frac{\ln 7}{\ln 0.6 + \ln 7} \approx 1.356$ **24.** $\frac{\ln\left(\frac{4}{3}\right)}{\ln\left(\frac{20}{3}\right)} \approx 0.152$

25. 0 **26.** $\frac{\ln(0.51)}{2 \ln 1.7 - \ln 0.3} \approx -0.297$ **27.** $\frac{\ln \pi}{1 + \ln \pi} \approx 0.534$ **28.** $\frac{3}{\ln(\pi) - 1} \approx 20.728$ **29.** $\frac{\ln 1.6}{3 \ln 2} \approx 0.226$ **30.** $\frac{\ln\left(\frac{2}{3}\right)}{0.2 \ln 4} \approx -1.462$

31. $\frac{9}{2}$ **32.** 4 **33.** 2 **34.** 67 **35.** 1 **36.** $\left\{\frac{26 + 8\sqrt{10}}{9}, \frac{26 - 8\sqrt{10}}{9}\right\} \approx \{5.7, 0.078\}$ **37.** 16 **38.** 81 **39.** $\left\{-1, \frac{2}{3}\right\}$ **40.** $\left\{4, \frac{1}{4}\right\}$ **41.** 0

42. $\{\ln(3 + 2\sqrt{2}), \ln(3 - 2\sqrt{2}) \approx \{1.763, -1.763\}$ **43.** $\ln(2 + \sqrt{5}) \approx 1.444$ **44.** $\ln(-2 + \sqrt{5}) \approx -1.444$ **45.** 1.92 **46.** 4.48 **47.** 2.79 **48.** 12.15 **49.** -0.57 **50.** $\{0.45, -1.98\}$ **51.** -0.70 **52.** $\{4.54, 1.86\}$ **53.** 0.57 **54.** 1.16 **55.** $\{0.39, 1.00\}$ **56.** 0.65 **57.** 1.32 **58.** $\{1.48, 0.05\}$ **59.** 1.31 **60.** 0.57

4.7 Exercises *(page 294)*

3. \$108.29 **4.** \$59.83 **5.** \$609.50 **6.** \$358.84 **7.** \$697.09 **8.** \$789.24 **9.** \$12.46 **10.** \$49.35 **11.** \$125.23 **12.** \$156.83 **13.** \$88.72 **14.** \$59.14 **15.** \$860.72 **16.** \$626.61 **17.** \$554.09 **18.** \$266.08 **19.** \$59.71 **20.** \$654.98 **21.** \$361.93 **22.** \$886.92 **23.** 5.35%

24. 6.82% **25.** 26% **26.** 7.18% **27.** $6\frac{1}{4}$% compounded annually **28.** 9% compounded quarterly **29.** 9% compounded monthly

30. 7.9% compounded daily **31.** 104.32 mo (about 8.7 yr); 103.97 mo (about 8.66 yr) **32.** 6.96 yr or 83.52 mo; 6.93 yr or 83.18 mo **33.** 61.02 mo; 60.82 mo **34.** 5.62 yr or 67.43 mo; 5.60 yr or 67.15 mo **35.** 15.27 yr or 15 yr, 4 mo **36.** 16.62 yr or 16 yr, 8 mo **37.** \$104,335 **38.** \$215.48 **39.** \$12,910.62 **40.** \$2955.39 **41.** About \$30.17 per share or \$3017 **42.** 15.47% **43.** 9.35% **44.** \$2315.97 **45.** Not quite. Jim will have \$1057.60. The second bank gives a better deal, since Jim will have \$1060.62 after 1 yr. **46.** \$1021.60 **47.** Will has \$11,632.73; Henry has \$10,947.89. **48.** It is better to give \$1000 now for investment to earn \$1349.86 in 3 yr. **49. (a)** Interest is \$30,000 **(b)** Interest is \$38,613.59 **(c)** Interest is \$37,752.73. Simple interest at 12% is best. **50. (a)** 4.34% **(b)** 4.34% **51. (a)** \$1364.62 **(b)** \$1353.35 **52.** \$10,810.76 **53.** \$4631.93 **54.** 9.07%

55. (a) 6.1 yr **(b)** 18.45 yr **(c)**

$$mP = P\left(1 + \frac{r}{n}\right)^{nt}$$
$$m = \left(1 + \frac{r}{n}\right)^{nt}$$
$$\ln m = \ln\left(1 + \frac{r}{n}\right)^{nt} = nt \ln\left(1 + \frac{r}{n}\right)$$
$$t = \frac{\ln m}{n \ln\left(1 + \frac{r}{n}\right)}$$

56. (a) 20.79 yr **(b)** 7.74% **(c)**

$$A = Pe^{rt}$$
$$\ln A = \ln P + rt$$
$$rt = \ln A - \ln P$$
$$t = \frac{\ln A - \ln P}{r}$$

4.8 Exercises *(page 304)*

1. (a) 500 insects **(b)** 0.02 = 2% **(c)** ≈ 611 insects **(d)** After about 23.5 days **(e)** After about 34.7 days
2. (a) 1000 **(b)** 0.01 = 1% **(c)** ≈ 1041 **(d)** After about 53.1 hr **(e)** After about 69.3 hr
3. (a) $-0.0244 = -2.44\%$ **(b)** About 391.7 g **(c)** After about 9.1 yr **(d)** 28.4 yr
4. (a) $-0.087 = -8.7\%$ **(b)** About 45.7 g **(c)** After about 4.1 days **(d)** About 7.97 days **5.** 5832; 3.9 days
6. About 5243; About 7.85 hr later. **7.** 25,198 **8.** About 711,111 **9.** 9.784 g **10.** 9.999999467 g; 9.999994668 g **11.** 9727 yr ago
12. 2882 yr old **13. (a)** 5:18 PM **(b)** After 14.3 min, the pizza will be 160°F. **(c)** As time passes, the temperature of the pizza gets closer to 70°F.
14. (a) 45.41°F **(b)** About 16.20 min **(c)** About 7.26 min **(d)** The temperature approaches 38°F. **15.** 18.63°C; 25.1°C
16. 45.7°F; 28.45 min **17.** 7.34 kg; 76.6 hr **18.** 1.25 V **19.** 26.6 days **20.** At about 9:42 PM **21. (a)** 0.1286 **(b)** 0.9 **(c)** 1996
22. (a) 0.20 **(b)** 0.90 **(c)** About 8.44 mo later **23. (a)** 1000 **(b)** 30 **(c)** 11.076 hr
24. (a) 500 **(b)** About 117 **(c)** About 29.8 yr later

4.9 Exercises *(page 311)*

1. (a)

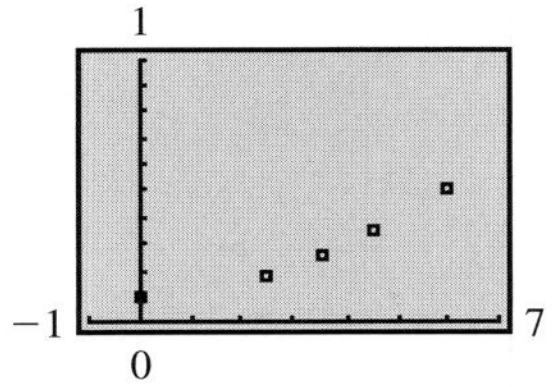

(b) $y = 0.0903(1.3384)^x$
(c) $N(t) = 0.0903e^{0.2915t}$
(d)

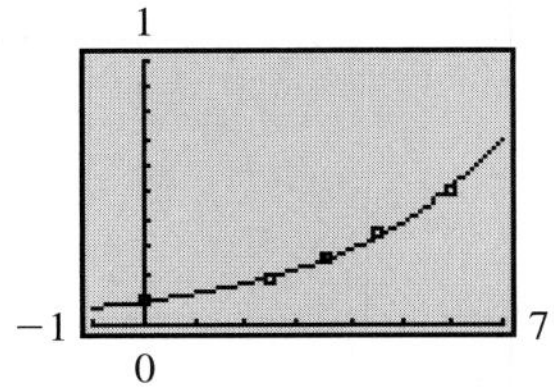

(e) 0.69 **(f)** After about 7.26 h

2. (a)

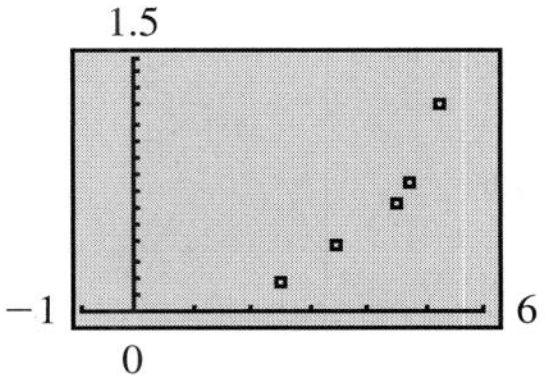

(b) $y = 0.0339(1.9474)^x$
(c) $N(t) = 0.0339e^{0.6665t}$
(d)

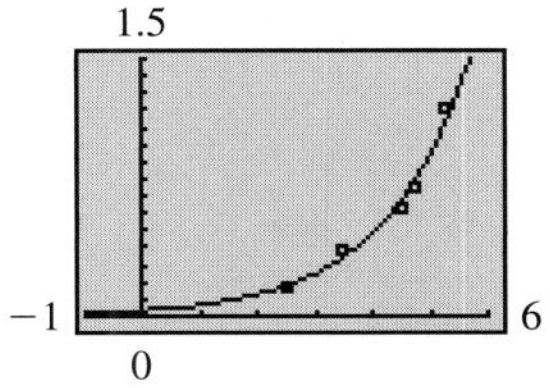

(e) 1.85 **(f)** 6.19 h

3. (a)

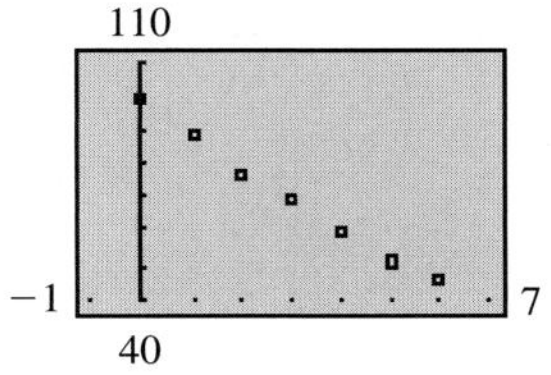

(b) $y = 100.326(0.8769)^x$
(c) $A = 100.326e^{-0.1314t}$
(d)

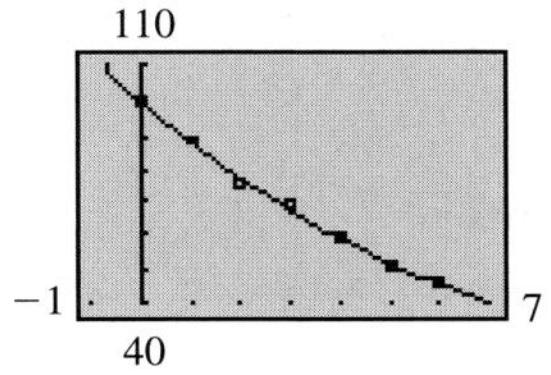

(e) 5.3 weeks **(f)** 0.14 g
(g) After about 12.3 weeks

4. (a)

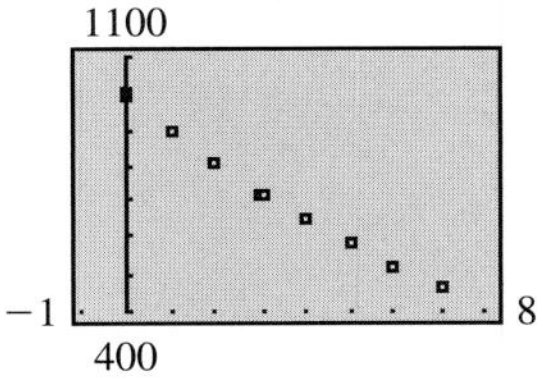

(b) $y = 998.91(0.8976)^t$
(c) $y = 998.91e^{-0.1081t}$
(d)

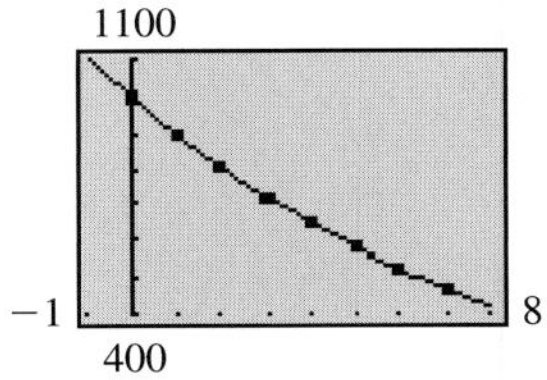

(e) 6.41 days **(f)** 115 g **(g)** After about 14.9 days

5. (a)

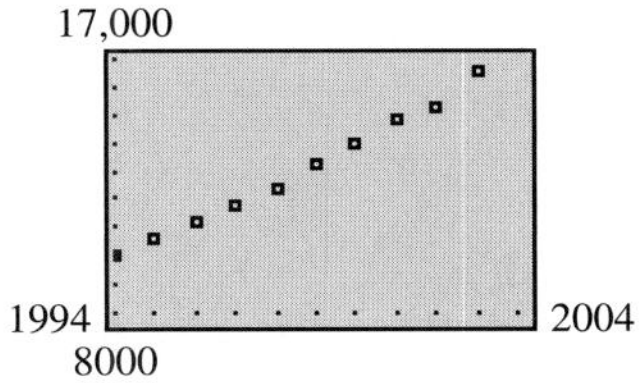

(b) $y = 2.2908 \times 10^{-44}(1.056554737)^x$
(c) 5.66%
(d) $44,230.54
(e) In 2023

6. (a)

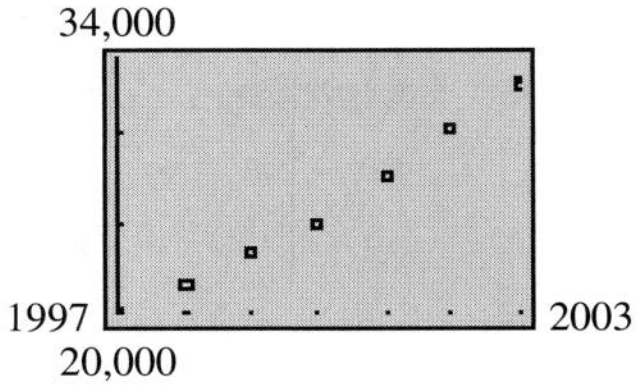

(b) $y = 1.4014 \times 10^{-67}(1.0855)^x$
(c) 8.55%
(d) $131,466.24
(e) In 2014

7. (a)

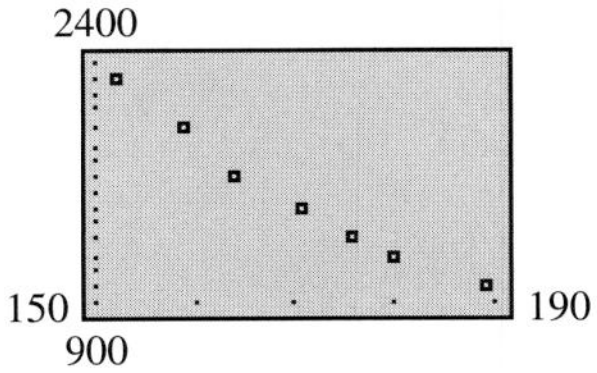

(b) $y = 32{,}741.02 - 6070.96 \ln x$
(c)

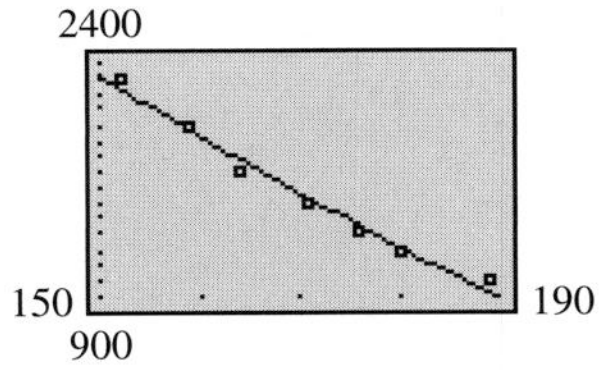

(d) Approximately 168 computers

8. (a)

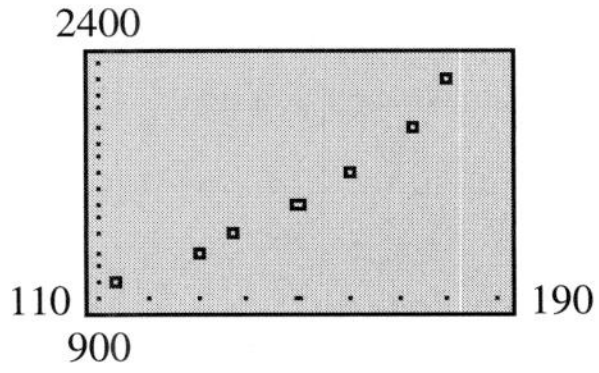

(b) $y = -11{,}850.72 + 2688.5 \ln x$
(c)

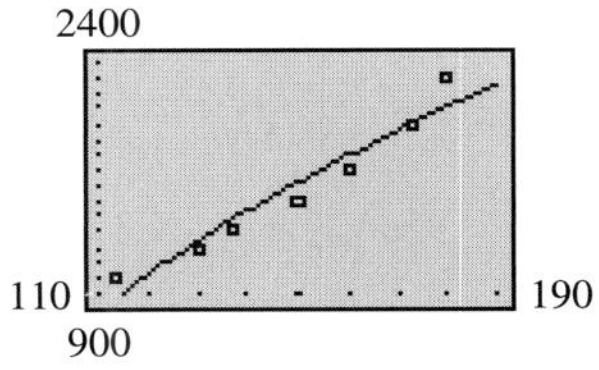

(d) Approximately 152 computers

9. (a)

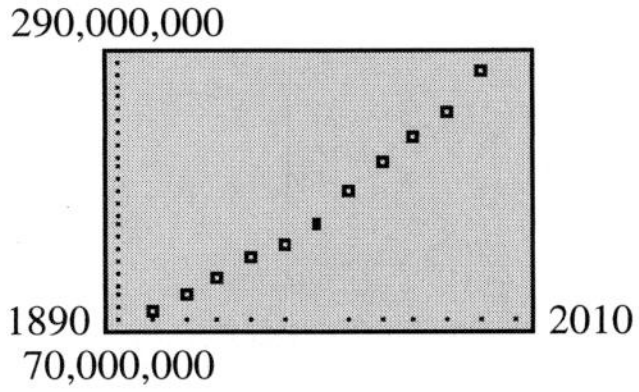

(b) $y = \dfrac{799{,}475{,}916.5}{1 + 1.56344 \times 10^{14}e^{-0.0160x}}$
(c)

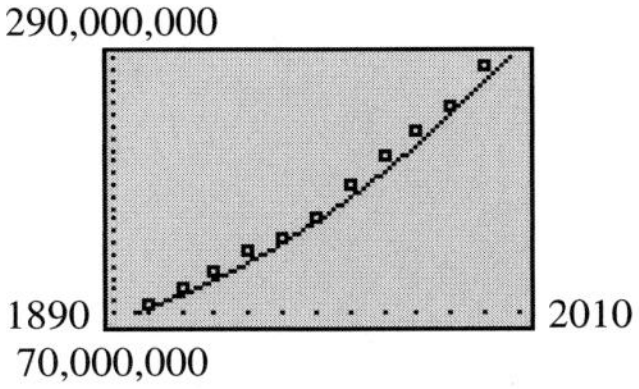

(d) 799,475,917
(e) Approximately 279,809,184 **(f)** 2011

10. **(a)** $x = 1$ corresponds to 1993.

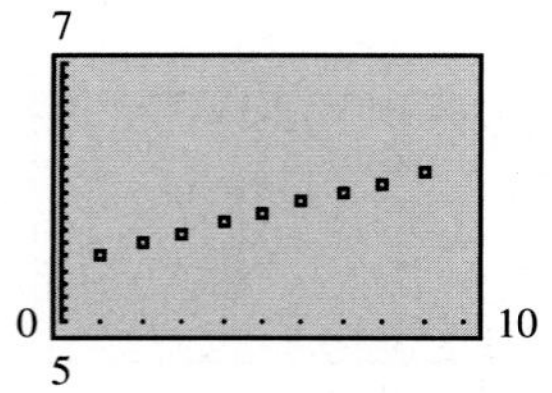

(b) $y = \dfrac{10.05267}{1 + 0.8439e^{-0.0320x}}$

(c)

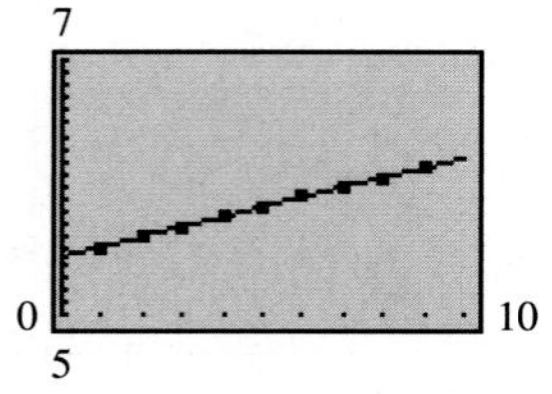

(d) 10.053 billion people
(e) Approximately 6.38 billion people
(f) 2013

11. **(a)**

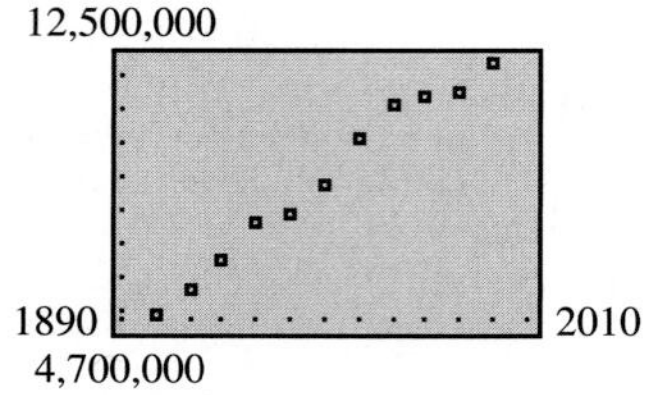

(b) $y = \dfrac{14{,}471{,}245.24}{1 + 3.860 \times 10^{20}\, e^{-0.0246x}}$

(c)

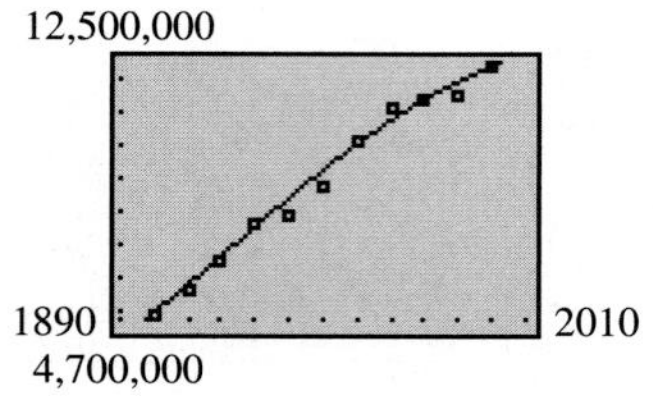

(d) 14,471,245
(e) Approximately 12,811,429

12. **(a)**

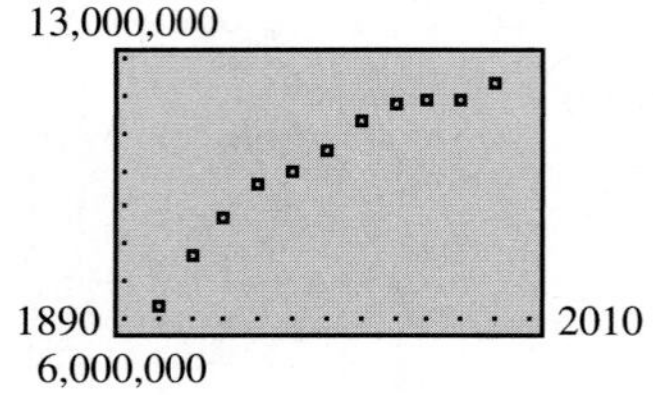

(b) $y = \dfrac{12{,}529{,}186.49}{1 + 6.0614 \times 10^{28}\, e^{-0.0349x}}$

(c)

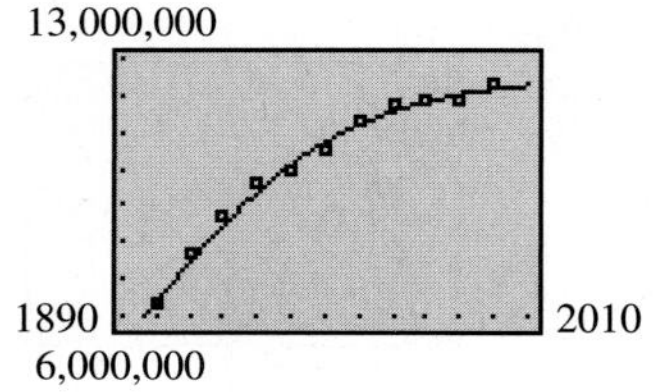

(d) 12,529,186
(e) Approximately 12,274,358

Review Exercises *(page 316)*

1. **(a)** -26 **(b)** -241 **(c)** 16 **(d)** -1 **2.** **(a)** -1 **(b)** 37 **(c)** 4 **(d)** 5

3. **(a)** $\sqrt{11}$ **(b)** 1 **(c)** $\sqrt{\sqrt{6} + 2}$ **(d)** 19 **4.** **(a)** -5 **(b)** $\sqrt{15}$ **(c)** -6626 **(d)** $\sqrt{4 - \sqrt{5}}$

5. **(a)** e^4 **(b)** $3e^{-2} - 2$ **(c)** e^{e^4} **(d)** -17 **6.** **(a)** $\frac{2}{73}$ **(b)** $\frac{2}{3}$ **(c)** $\frac{2178}{1097}$ **(d)** -9

7. $(f \circ g)(x) = 1 - 3x$, all real numbers; $(g \circ f)(x) = 7 - 3x$, all real numbers; $(f \circ f)(x) = x$, all real numbers; $(g \circ g)(x) = 9x + 4$, all real numbers

8. $(f \circ g)(x) = 4x + 1$, all real numbers; $(g \circ f)(x) = 4x - 1$, all real numbers; $(f \circ f)(x) = 4x - 3$, all real numbers; $(g \circ g)(x) = 4x + 3$, all real numbers

9. $(f \circ g)(x) = 27x^2 + 3|x| + 1$, all real numbers; $(g \circ f)(x) = 3|3x^2 + x + 1|$, all real numbers; $(f \circ f)(x) = 3(3x^2 + x + 1)^2 + 3x^2 + x + 2$, all real numbers; $(g \circ g)(x) = 9|x|$, all real numbers

10. $(f \circ g)(x) = \sqrt{3(1 + x + x^2)}$, all real numbers; $(g \circ f)(x) = 1 + \sqrt{3x} + 3x$, $\{x | x \geq 0\}$; $(f \circ f)(x) = \sqrt{3\sqrt{3x}}$, $\{x | x \geq 0\}$; $(g \circ g)(x) = 3 + 3x + 4x^2 + 2x^3 + x^4$, all real numbers

11. $(f \circ g)(x) = \dfrac{1 + x}{1 - x}$, $\{x | x \neq 0, x \neq 1\}$; $(g \circ f)(x) = \dfrac{x - 1}{x + 1}$, $\{x | x \neq -1, x \neq 1\}$; $(f \circ f)(x) = x$, $\{x | x \neq 1\}$; $(g \circ g)(x) = x$, $\{x | x \neq 0\}$

12. $(f \circ g)(x) = \sqrt{\dfrac{3 - 3x}{x}}$, $\{x | 0 < x \leq 1\}$; $(g \circ f)(x) = \dfrac{3}{\sqrt{x - 3}}$, $\{x | x > 3\}$; $(f \circ f)(x) = \sqrt{\sqrt{x - 3} - 3}$, $\{x | x \geq 12\}$; $(g \circ g)(x) = x$, $\{x | x \neq 0\}$

13. **(a)** $\{(2, 1), (5, 3), (8, 5), (10, 6)\}$ **(b)** Inverse is a function. **14.** **(a)** $\{(4, -1), (2, 0), (4, 1), (7, 3)\}$ **(b)** Inverse is not a function.

15.

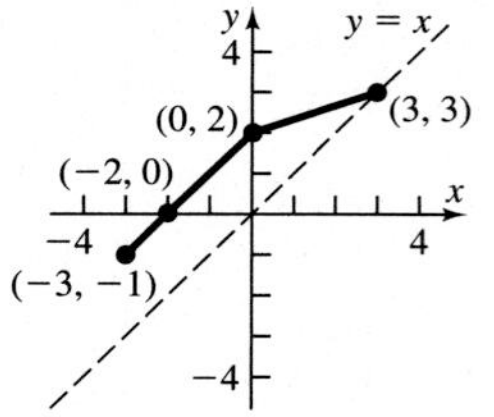

16.

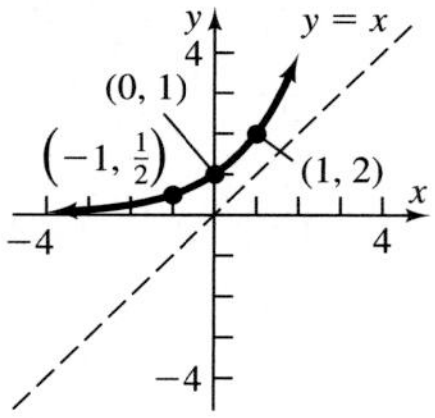

17. $f^{-1}(x) = \dfrac{2x+3}{5x-2}$

$f(f^{-1}(x)) = \dfrac{2\left(\dfrac{2x+3}{5x-2}\right)+3}{5\left(\dfrac{2x+3}{5x-2}\right)-2} = x$

$f^{-1}(f(x)) = \dfrac{2\left(\dfrac{2x+3}{5x-2}\right)+3}{5\left(\dfrac{2x+3}{5x-2}\right)-2} = x$

Domain of f = Range of f^{-1} = all real numbers except $\frac{2}{5}$

Range of f = Domain of f^{-1} = all real numbers except $\frac{2}{5}$

18. $f^{-1}(x) = \dfrac{2-3x}{1+x}$

$f(f^{-1}(x)) = \dfrac{2-\dfrac{2-3x}{1+x}}{3+\dfrac{2-3x}{1+x}} = x$

$f^{-1}(f(x)) = \dfrac{2-3\left(\dfrac{2-x}{3+x}\right)}{1+\left(\dfrac{2-x}{3+x}\right)} = x$

Domain of f = Range of f^{-1} = all real numbers except -3
Range of f = Domain of f^{-1} = all real numbers except -1

19. $f^{-1}(x) = \dfrac{x+1}{x}$

$f(f^{-1}(x)) = \dfrac{1}{\dfrac{x+1}{x}-1} = x$

$f^{-1}(f(x)) = \dfrac{\dfrac{1}{x-1}+1}{\dfrac{1}{x-1}} = x$

Domain of f = Range of f^{-1} = all real numbers except 1
Range of f = Domain of f^{-1} = all real numbers except 0

20. $f^{-1}(x) = x^2+2, x > 0$
$f(f^{-1}(x)) = \sqrt{x^2+2-2} = |x| = x, x \geq 0$
$f^{-1}(f(x)) = (\sqrt{x-2})^2 + 2 = x$
Domain of f = Range of f^{-1} = $[2, \infty)$
Range of f = Domain of f^{-1} = $[0, \infty)$

21. $f^{-1}(x) = \dfrac{27}{x^3}$

$f(f^{-1}(x)) = \dfrac{3}{\left(\dfrac{27}{x^3}\right)^{1/3}} = x$

$f^{-1}(f(x)) = \dfrac{27}{\left(\dfrac{3}{x^{1/3}}\right)^3} = x$

Domain of f = Range of f^{-1} = all real numbers except 0
Range of f = Domain of f^{-1} = all real numbers except 0

22. $f^{-1}(x) = (x-1)^3$
$f(f^{-1}(x)) = ((x-1)^3)^{1/3} + 1 = x$
$f^{-1}(f(x)) = (x^{1/3}+1-1)^3 = x$
Domain of f = Range of f^{-1} = $(-\infty, \infty)$
Range of f = Domain of f^{-1} = $(-\infty, \infty)$

23. (a) 81 **(b)** 2 **(c)** $\frac{1}{9}$ **(d)** -3 **24. (a)** 3 **(b)** 4 **(c)** $\frac{1}{81}$ **(d)** -5 **25.** $\log_5 z = 2$ **26.** $\log_a m = 5$ **27.** $5^{13} = u$ **28.** $a^3 = 4$

29. $\left\{x \middle| x > \frac{2}{3}\right\}; \left(\frac{2}{3}, \infty\right)$ **30.** $\left\{x \middle| x > -\frac{1}{2}\right\}; \left(-\frac{1}{2}, \infty\right)$ **31.** $\left\{x \middle| x < 1 \text{ or } x > 2\right\}; (-\infty, 1) \text{ or } (2, \infty)$

32. $\{x | x < -3 \text{ or } x > 3\}; (-\infty, -3) \text{ or } (3, \infty)$ **33.** -3 **34.** 4 **35.** $\sqrt{2}$ **36.** 0.1 **37.** 0.4 **38.** $\sqrt{3}$ **39.** $\log_3 u + 2\log_3 v - \log_3 w$

40. $8\log_2 a + 2\log_2 b$ **41.** $2\log x + \frac{1}{2}\log(x^3+1)$ **42.** $2\log_5(x+1) - 2\log_5 x$ **43.** $\ln x + \frac{1}{3}\ln(x^2+1) - \ln(x-3)$

44. $2\ln(2x+3) - 2\ln(x-1) - 2\ln(x-2)$ **45.** $\frac{25}{4}\log_4 x$ **46.** $\frac{13}{6}\log_3 x$ **47.** $-2\ln(x+1)$ **48.** $\log\left(\frac{x-3}{x+4}\right)$

49. $\log\left(\dfrac{4x^3}{[(x+3)(x-2)]^{1/2}}\right)$ **50.** $\ln\left[\dfrac{16\sqrt{x^2+1}}{\sqrt{x}(x-4)}\right]$ **51.** 2.124 **52.** 4.392

53.

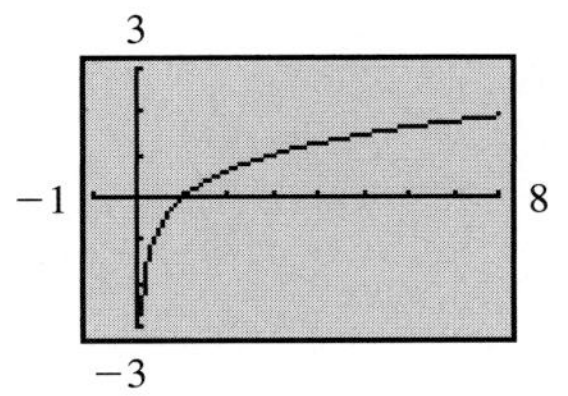

Domain: $(-\infty, \infty)$

54.

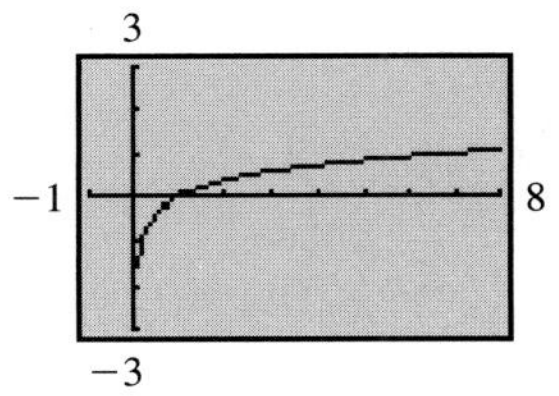

Domain: $(-\infty, \infty)$

55.

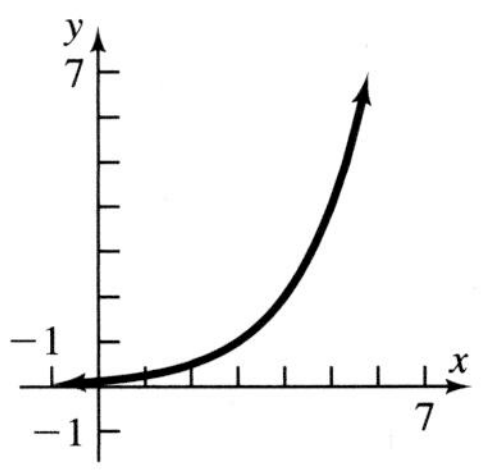

Range: $(0, \infty)$
Horizontal asymptote: $y = 0$

56.

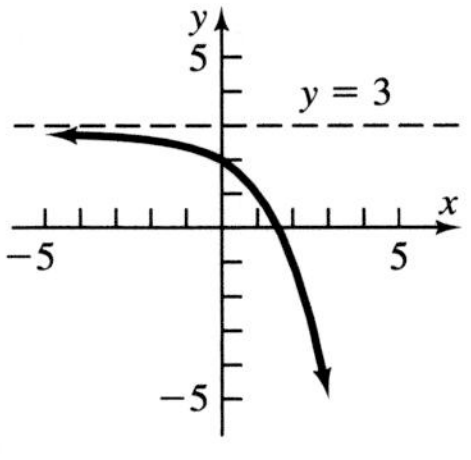

Range: $(-\infty, 3)$
Horizontal asymptote: $y = 3$

57.

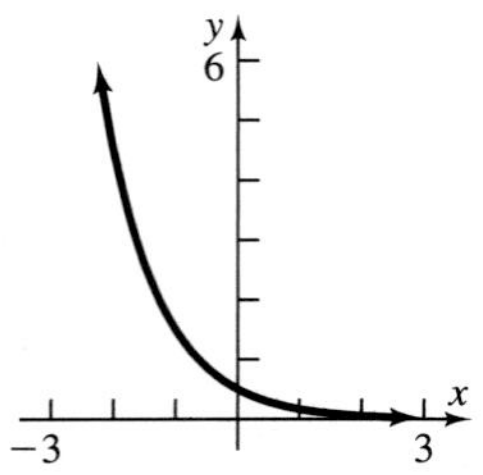

Domain: $(-\infty, \infty)$
Range: $(0, \infty)$
Horizontal asymptote: $y = 0$

58.

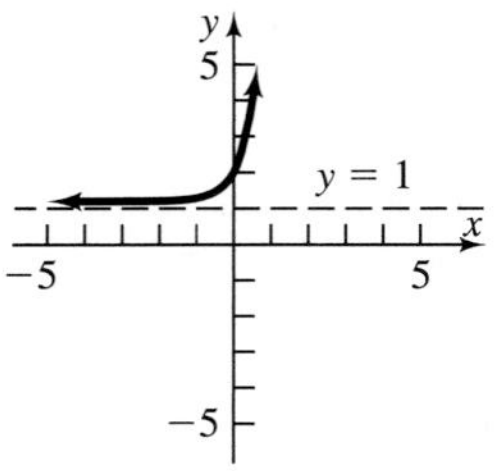

Domain: $(-\infty, \infty)$
Range: $(1, \infty)$
Horizontal asymptote: $y = 1$

59.

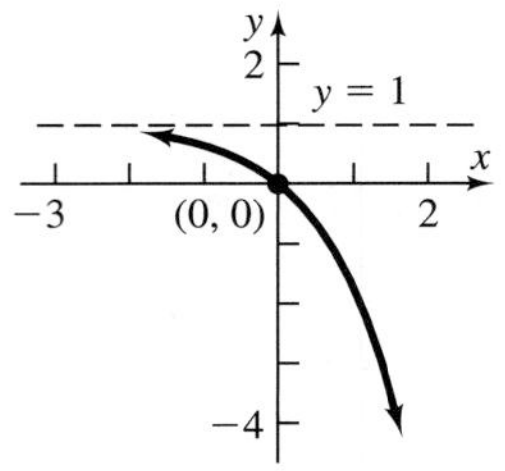

Domain: $(-\infty, \infty)$
Range: $(-\infty, 1)$
Horizontal asymptote: $y = 1$

60.

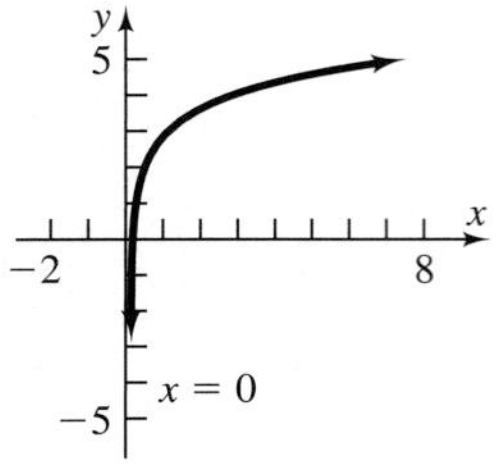

Domain: $(0, \infty)$
Range: $(-\infty, \infty)$
Vertical asymptote: $x = 0$

61.

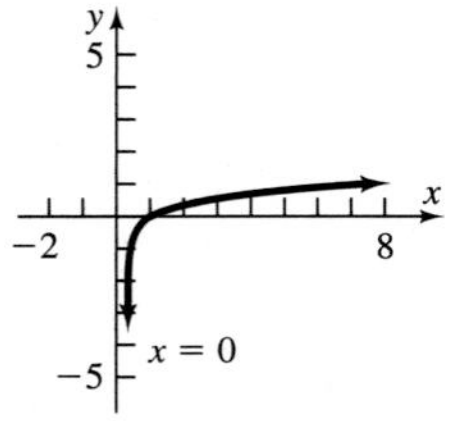

Domain: $(0, \infty)$
Range: $(-\infty, \infty)$
Vertical asymptote: $x = 0$

62.

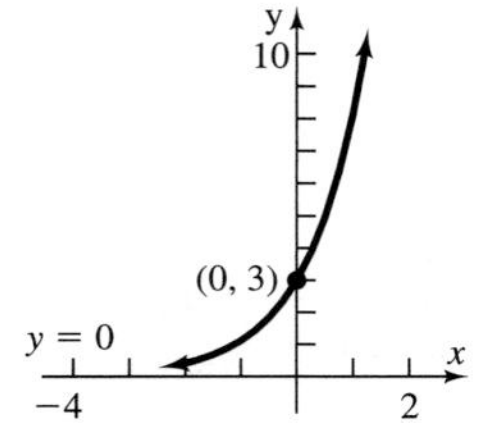

Domain: $(-\infty, \infty)$
Range: $(0, \infty)$
Horizontal asymptote: $y = 0$

63.

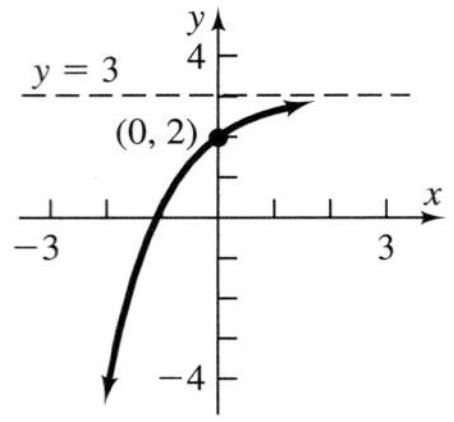

Domain: $(-\infty, \infty)$
Range: $(-\infty, 3)$
Horizontal asymptote: $y = 3$

64.

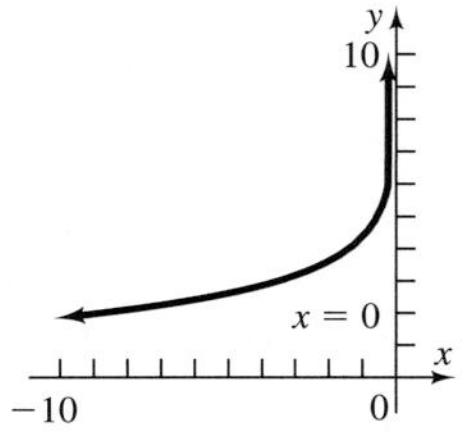

Domain: $(-\infty, 0)$
Range: $(-\infty, \infty)$
Vertical asymptote: $x = 0$

65. $\left\{\frac{1}{4}\right\}$ **66.** $\left\{-\frac{16}{9}\right\}$ **67.** $\left\{\frac{-1-\sqrt{3}}{2}, \frac{-1+\sqrt{3}}{2}\right\}$ **68.** $\left\{\frac{1-\sqrt{3}}{2}, \frac{1+\sqrt{3}}{2}\right\}$ **69.** $\left\{\frac{1}{4}\right\}$ **70.** $\left\{\frac{1}{8}\right\}$ **71.** $\left\{\frac{2\ln 3}{\ln 5 - \ln 3} \approx 4.301\right\}$

72. $\left\{\frac{2(\ln 7 + \ln 5)}{\ln 7 - \ln 5} \approx 21.133\right\}$ **73.** $\left\{\frac{12}{5}\right\}$ **74.** $\{-2, 6\}$ **75.** $\{83\}$ **76.** $\left\{-\frac{1}{2}\right\}$ **77.** $\left\{\frac{1}{2}, -3\right\}$ **78.** $\{1\}$ **79.** $\{-1\}$ **80.** $\{3, 4\}$

81. $\{1 - \ln 5 \approx -0.609\}$ **82.** $\left\{\frac{1-\ln 4}{2} \approx -0.193\right\}$ **83.** $\left\{\frac{\ln 3}{3\ln 2 - 2\ln 3} \approx -9.327\right\}$ **84.** $\left\{0, \frac{\ln 3}{\ln 2} \approx 1.585\right\}$ **85.** 3229.5 m

86. 1482 m **87.** **(a)** 37.3 W **(b)** 6.9 dB **88.** **(a)** 11.77 **(b)** 9.56 in. **89.** **(a)** 9.85 yr **(b)** 4.27 yr **90.** $20,398.87; 4.04; 17.5 yr

91. $41,668.97 **92.** **(a)** 10.436% **(b)** $32,249.24 **93.** 24,203 yr ago **94.** 55.22 min or 55 min and 13 sec **95.** 6,835,600,129

96. 7.204 g; 0.519 g

97. **(a)** 0.3

(b) 0.8

(c)

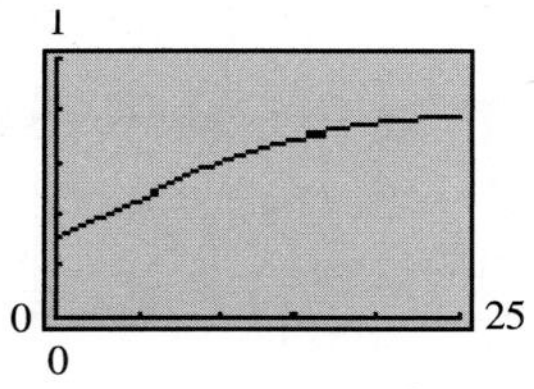

(d) In 2023

98. **(a)**

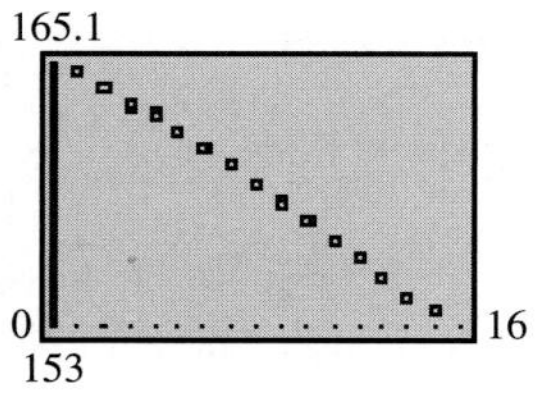

(b) $y = 165.73(0.9951)^x$

(c)

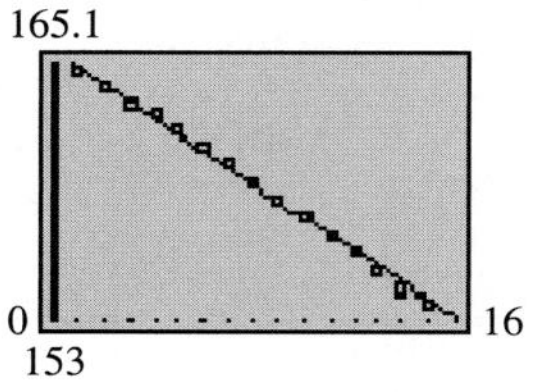

(d) Approximately 83 sec

99. **(a)**

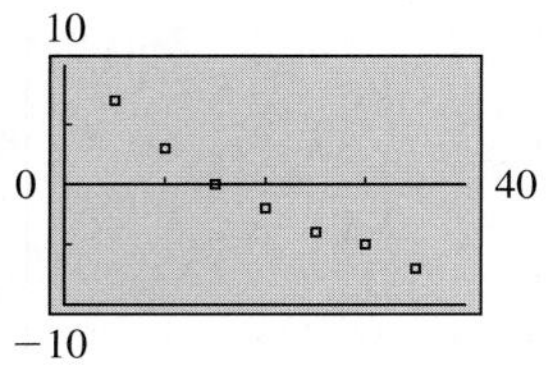

(b) Wind chill $= 18.921 - 7.096 \ln(\text{wind speed})$

(c)

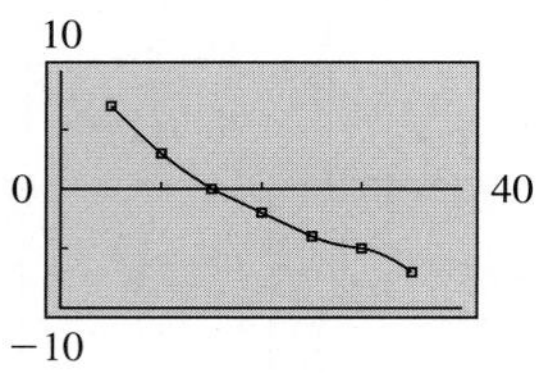

(d) Approximately -3°F

100. (a)

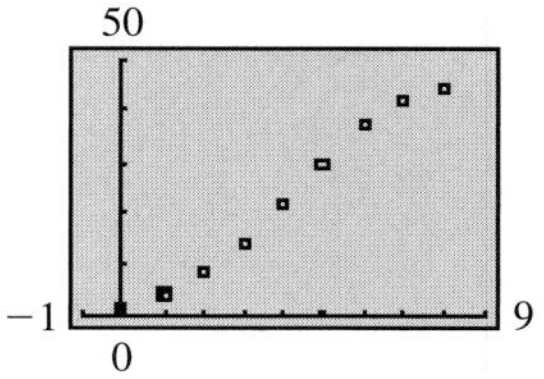

(b) $C = \dfrac{46.93}{1 + 21.273e^{-0.7306t}}$ **(c)**

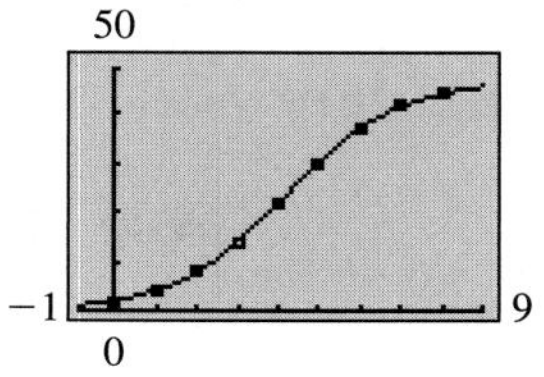

(d) About 47 people; 50 people **(e)** 2.4 days; during the 10th hour of the 3rd day **(f)** 9.5 days

Cumulative Review *(page 321)*

1. Yes; no **2. (a)** 10 **(b)** $2x^2 + 3x + 1$ **(c)** $2x^2 + 4xh + 2h^2 - 3x - 3h + 1$ **3.** $\left(\frac{1}{2}, \frac{\sqrt{3}}{2}\right)$ is on the graph **4.** -26

5.

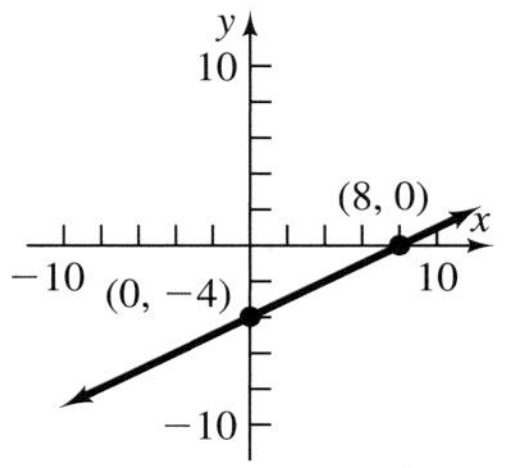

6. (a)

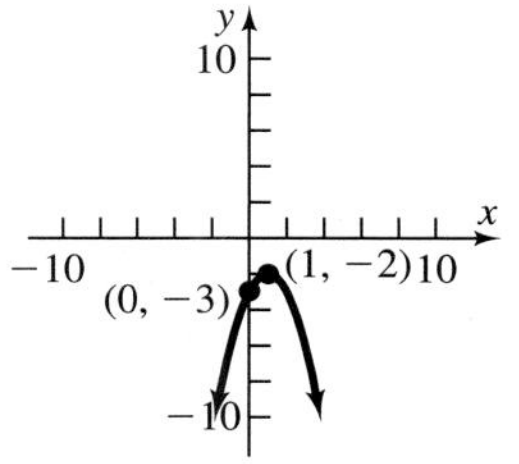

(b) $\{x \mid -\infty < x < \infty\}$

7. $f(x) = 2(x - 4)^2 - 8 = 2x^2 - 16x + 24$

8.

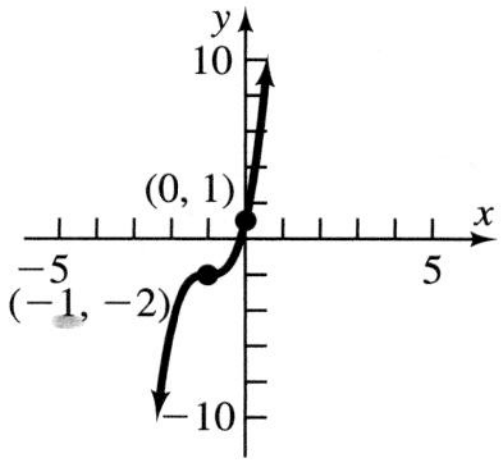

9. $f(g(x)) = \dfrac{4}{(x - 3)^2} + 2$; domain: $\{x \mid x \neq 3\}$; 3

10. (a) Zeros: $-4, -\frac{1}{4}, 2$

(b) x-intercepts: $-4, -\frac{1}{4}, 2$; y-intercept: -8

(c) Local maximum of 60.75 occurs at $x = -2.5$;
Local minimum of -25 occurs at $x = 1$

(d)

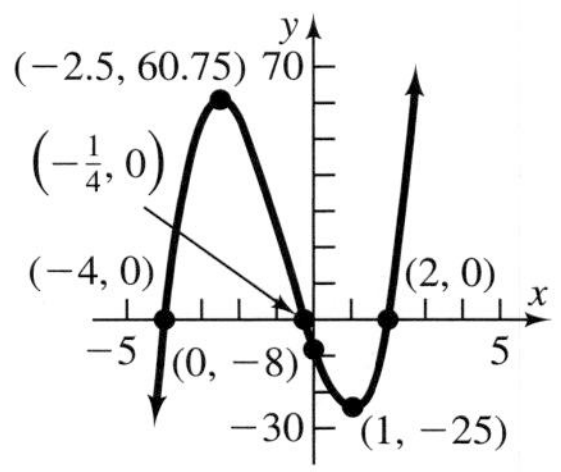

11. (a), (c)

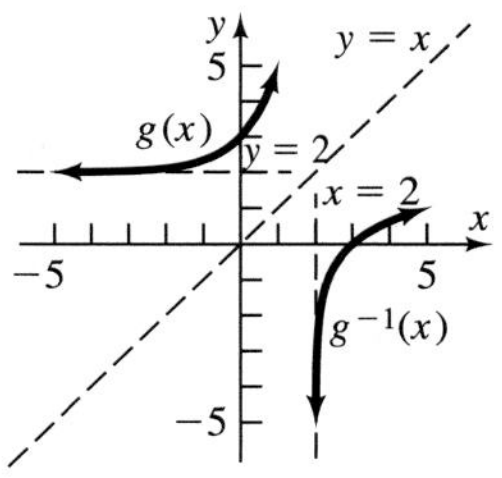

Domain g = Range g^{-1} = $(-\infty, \infty)$
Range g = Domain g^{-1} = $(2, \infty)$

(b) $g^{-1}(x) = \log_3(x - 2)$

12. $-\frac{3}{2}$ **13.** 2 **14. (a)** -1 **(b)** $\{x \mid x > -1\}$ or $(-1, \infty)$ **(c)** 25 **15. (a)**

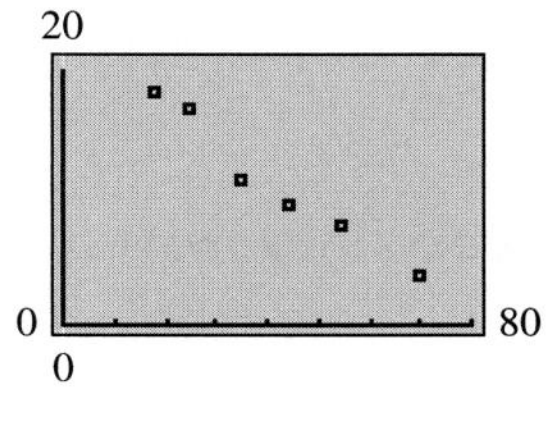

(b) Answers will vary.

CHAPTER 5 Trigonometric Functions

5.1 Concepts and Vocabulary *(page 334)*

3. standard position **4.** $r\theta$; $\frac{1}{2}r^2\theta$ **5.** $\frac{s}{t}$; $\frac{\theta}{t}$ **6.** False **7.** True **8.** True **9.** True **10.** False

5.1 Exercises *(page 335)*

11.
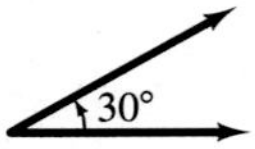

12.
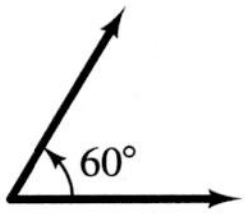

13.
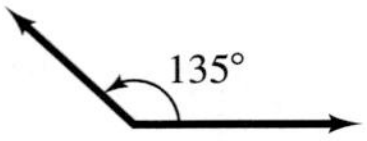

14.
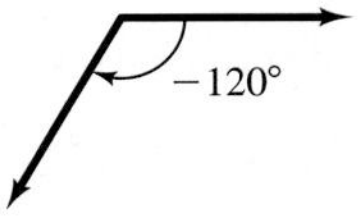

15.
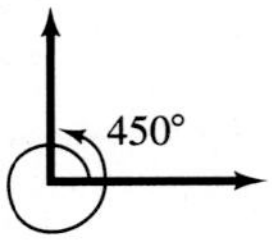

16.
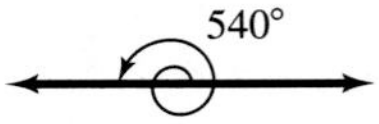

17.
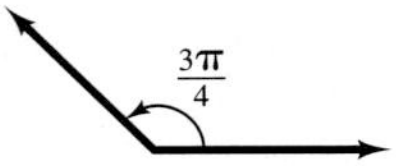

18.
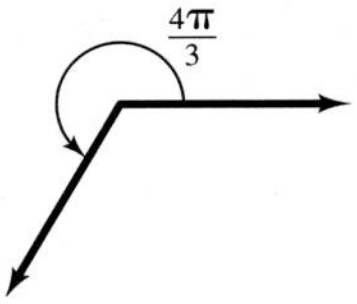

19.
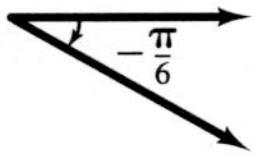

20.
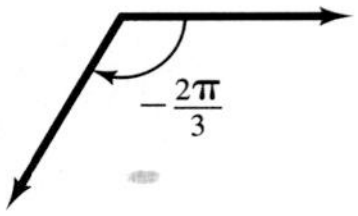

21.
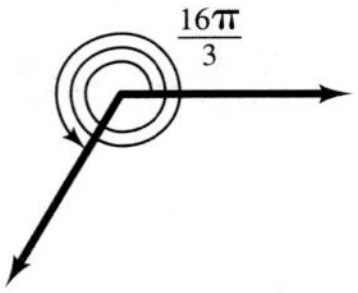

22.
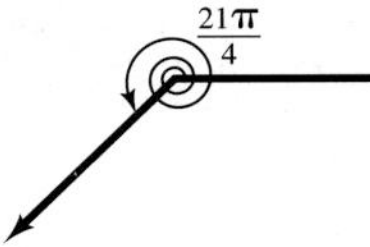

23. 40.17° **24.** 61.71° **25.** 1.03° **26.** 73.68° **27.** 9.15° **28.** 98.38° **29.** 40°19′12″ **30.** 61°14′24″ **31.** 18°15′18″ **32.** 29°24′40″ **33.** 19°59′24″ **34.** 44°0′36″ **35.** $\frac{\pi}{6}$ **36.** $\frac{2\pi}{3}$ **37.** $\frac{4\pi}{3}$ **38.** $\frac{11\pi}{6}$ **39.** $-\frac{\pi}{3}$ **40.** $-\frac{\pi}{6}$ **41.** π **42.** $\frac{3\pi}{2}$ **43.** $-\frac{3\pi}{4}$ **44.** $-\frac{5\pi}{4}$ **45.** $-\frac{\pi}{2}$ **46.** $-\pi$ **47.** 60° **48.** 150° **49.** −225° **50.** −120° **51.** 90° **52.** 720° **53.** 15° **54.** 75° **55.** −90° **56.** −180° **57.** −30° **58.** −135° **59.** 0.30 **60.** 1.27 **61.** −0.70 **62.** −0.89 **63.** 2.18 **64.** 6.11 **65.** 179.91° **66.** 42.97° **67.** 114.59° **68.** 171.89° **69.** 362.11° **70.** 81.03° **71.** 5 m **72.** 12 ft **73.** 6 ft **74.** 24 cm **75.** 0.6 radian **76.** $\frac{4}{3} \approx 1.333$ radians **77.** $\frac{\pi}{3} \approx 1.047$ in. **78.** $2\pi \approx 6.283$ m **79.** 25 m^2 **80.** 36 ft^2 **81.** $2\sqrt{3} \approx 3.464$ ft **82.** $4\sqrt{3} \approx 6.928$ cm **83.** 0.24 radian **84.** $\frac{4}{9} \approx 0.444$ radian **85.** $\frac{\pi}{3} \approx 1.047$ in^2 **86.** $3\pi \approx 9.425$ m^2 **87.** $s = 2.094$ ft; $A = 2.094$ ft^2 **88.** $s = 2.094$ m; $A = 4.189$ m^2 **89.** $s = 14.661$ yd; $A = 87.965$ yd^2 **90.** $s = 7.854$ cm; $A = 35.343$ cm^2 **91.** $3\pi \approx 9.4248$ in.; $5\pi \approx 15.7080$ in. **92.** $\frac{40\pi}{9} \approx 13.9626$ in. **93.** $2\pi \approx 6.28$ m^2 **94.** $\frac{3\pi}{2} \approx 4.71$ cm^2 **95.** $\frac{675\pi}{2} \approx 1060.29$ ft^2 **96.** $\frac{72}{5\pi} \approx 4.58°$ **97.** $\omega = \frac{1}{60}$ radian/sec; $v = \frac{1}{12}$ cm/sec **98.** $\omega = \frac{1}{8}$ radian/sec; $v = \frac{1}{4}$ m/sec
99. Approximately 452.5 rpm **100.** Approximately 282.7 in/sec; approximately 16.1 mi/hr **101.** Approximately 359 mi
102. Approximately 554 mi **103.** Approximately 898 mi/hr **104.** Approximately 794 mi/hr **105.** Approximately 2292 mi/hr
106. Approximately 66,633 mi/hr **107.** $\frac{3}{4}$ rpm **108.** Approximately 2.69 ft/sec; approximately 0.09 radians/sec
109. Approximately 2.86 mi/hr **110.** Approximately 37.13 mi/hr; approximately 1034.26 rpm **111.** Approximately 31.47 rpm
112. Approximately 5.8 min **113.** Approximately 1037 mi/hr **114.** Approximately 1.15 mi **115.** $v_1 = r_1\omega_1$, $v_2 = r_2\omega_2$, and $v_1 = v_2$, so $r_1\omega_1 = r_2\omega_2 \Rightarrow \frac{r_1}{r_2} = \frac{\omega_2}{\omega_1}$.

5.2 Concepts and Vocabulary *(page 352)*

7. $\frac{3}{2}$ **8.** 0.91 **9.** True **10.** False

5.2 Exercises *(page 352)*

11. $\sin t = \frac{1}{2}$; $\cos t = \frac{\sqrt{3}}{2}$; $\tan t = \frac{\sqrt{3}}{3}$; $\csc t = 2$; $\sec t = \frac{2\sqrt{3}}{3}$; $\cot t = \sqrt{3}$ **12.** $\sin t = -\frac{\sqrt{3}}{2}$; $\cos t = \frac{1}{2}$; $\tan t = -\sqrt{3}$; $\csc t = -\frac{2\sqrt{3}}{3}$; $\sec t = 2$; $\cot t = -\frac{\sqrt{3}}{3}$ **13.** $\sin t = \frac{\sqrt{21}}{5}$; $\cos t = -\frac{2}{5}$; $\tan t = -\frac{\sqrt{21}}{2}$; $\csc t = \frac{5\sqrt{21}}{21}$; $\sec t = -\frac{5}{2}$; $\cot t = -\frac{2\sqrt{21}}{21}$

14. $\sin t = \frac{2\sqrt{6}}{5}$; $\cos t = -\frac{1}{5}$; $\tan t = -2\sqrt{6}$; $\csc t = \frac{5\sqrt{6}}{12}$; $\sec t = -5$; $\cot t = -\frac{\sqrt{6}}{12}$ **15.** $\sin t = \frac{\sqrt{2}}{2}$; $\cos t = -\frac{\sqrt{2}}{2}$; $\tan t = -1$; $\csc t = \sqrt{2}$; $\sec t = -\sqrt{2}$; $\cot t = -1$ **16.** $\sin t = \frac{\sqrt{2}}{2}$; $\cos t = \frac{\sqrt{2}}{2}$; $\tan t = 1$; $\csc t = \sqrt{2}$; $\sec t = \sqrt{2}$; $\cot t = 1$
17. $\sin t = -\frac{1}{3}$; $\cos t = \frac{2\sqrt{2}}{3}$; $\tan t = -\frac{\sqrt{2}}{4}$; $\csc t = -3$; $\sec t = \frac{3\sqrt{2}}{4}$; $\cot t = -2\sqrt{2}$ **18.** $\sin t = -\frac{2}{3}$; $\cos t = -\frac{\sqrt{5}}{3}$; $\tan t = \frac{2\sqrt{5}}{5}$; $\csc t = -\frac{3}{2}$; $\sec t = -\frac{3\sqrt{5}}{5}$; $\cot t = \frac{\sqrt{5}}{2}$ **19.** -1 **20.** -1 **21.** 0 **22.** 0 **23.** -1 **24.** 1 **25.** 0 **26.** 0 **27.** -1 **28.** 0 **29.** $\frac{1}{2}(\sqrt{2}+1)$
30. $\frac{1}{2}(1-\sqrt{2})$ **31.** 2 **32.** -1 **33.** $\frac{1}{2}$ **34.** $\frac{\sqrt{3}}{2}$ **35.** $\sqrt{6}$ **36.** $\frac{2\sqrt{3}}{3}$ **37.** 4 **38.** 8 **39.** 0 **40.** $\sqrt{2}+3$ **41.** 0 **42.** $\sqrt{3}+\frac{1}{2}$
43. $2\sqrt{2}+\frac{4\sqrt{3}}{3}$ **44.** $2\sqrt{3}+1$ **45.** -1 **46.** -1 **47.** 1 **48.** -2 **49.** $\sin\frac{2\pi}{3} = \frac{\sqrt{3}}{2}$; $\cos\frac{2\pi}{3} = -\frac{1}{2}$; $\tan\frac{2\pi}{3} = -\sqrt{3}$; $\csc\frac{2\pi}{3} = \frac{2\sqrt{3}}{3}$; $\sec\frac{2\pi}{3} = -2$; $\cot\frac{2\pi}{3} = -\frac{\sqrt{3}}{3}$ **50.** $\sin\frac{5\pi}{6} = \frac{1}{2}$; $\cos\frac{5\pi}{6} = -\frac{\sqrt{3}}{2}$; $\tan\frac{5\pi}{6} = -\frac{\sqrt{3}}{3}$; $\csc\frac{5\pi}{6} = 2$; $\sec\frac{5\pi}{6} = -\frac{2\sqrt{3}}{3}$; $\cot\frac{5\pi}{6} = -\sqrt{3}$
51. $\sin 210° = -\frac{1}{2}$; $\cos 210° = -\frac{\sqrt{3}}{2}$; $\tan 210° = \frac{\sqrt{3}}{3}$; $\csc 210° = -2$; $\sec 210° = -\frac{2\sqrt{3}}{3}$; $\cot 210° = \sqrt{3}$
52. $\sin 240° = -\frac{\sqrt{3}}{2}$; $\cos 240° = -\frac{1}{2}$; $\tan 240° = \sqrt{3}$; $\csc 240° = -\frac{2\sqrt{3}}{3}$; $\sec 240° = -2$; $\cot 240° = \frac{\sqrt{3}}{3}$
53. $\sin\frac{3\pi}{4} = \frac{\sqrt{2}}{2}$; $\cos\frac{3\pi}{4} = -\frac{\sqrt{2}}{2}$; $\tan\frac{3\pi}{4} = -1$; $\csc\frac{3\pi}{4} = \sqrt{2}$; $\sec\frac{3\pi}{4} = -\sqrt{2}$; $\cot\frac{3\pi}{4} = -1$ **54.** $\sin\frac{11\pi}{4} = \frac{\sqrt{2}}{2}$; $\cos\frac{11\pi}{4} = -\frac{\sqrt{2}}{2}$; $\tan\frac{11\pi}{4} = -1$; $\csc\frac{11\pi}{4} = \sqrt{2}$; $\sec\frac{11\pi}{4} = -\sqrt{2}$; $\cot\frac{11\pi}{4} = -1$ **55.** $\sin\frac{8\pi}{3} = \frac{\sqrt{3}}{2}$; $\cos\frac{8\pi}{3} = -\frac{1}{2}$; $\tan\frac{8\pi}{3} = -\sqrt{3}$; $\csc\frac{8\pi}{3} = \frac{2\sqrt{3}}{3}$; $\sec\frac{8\pi}{3} = -2$; $\cot\frac{8\pi}{3} = -\frac{\sqrt{3}}{3}$ **56.** $\sin\frac{13\pi}{6} = \frac{1}{2}$; $\cos\frac{13\pi}{6} = \frac{\sqrt{3}}{2}$; $\tan\frac{13\pi}{6} = \frac{\sqrt{3}}{3}$; $\csc\frac{13\pi}{6} = 2$; $\sec\frac{13\pi}{6} = \frac{2\sqrt{3}}{3}$; $\cot\frac{13\pi}{6} = \sqrt{3}$
57. $\sin 405° = \frac{\sqrt{2}}{2}$; $\cos 405° = \frac{\sqrt{2}}{2}$; $\tan 405° = 1$; $\csc 405° = \sqrt{2}$; $\sec 405° = \sqrt{2}$; $\cot 405° = 1$
58. $\sin 390° = \frac{1}{2}$; $\cos 390° = \frac{\sqrt{3}}{2}$; $\tan 390° = \frac{\sqrt{3}}{3}$; $\csc 390° = 2$; $\sec 390° = \frac{2\sqrt{3}}{3}$; $\cot 390° = \sqrt{3}$
59. $\sin\left(-\frac{\pi}{6}\right) = -\frac{1}{2}$; $\cos\left(-\frac{\pi}{6}\right) = \frac{\sqrt{3}}{2}$; $\tan\left(-\frac{\pi}{6}\right) = -\frac{\sqrt{3}}{3}$; $\csc\left(-\frac{\pi}{6}\right) = -2$; $\sec\left(-\frac{\pi}{6}\right) = \frac{2\sqrt{3}}{3}$; $\cot\left(-\frac{\pi}{6}\right) = -\sqrt{3}$
60. $\sin\left(-\frac{\pi}{3}\right) = -\frac{\sqrt{3}}{2}$; $\cos\left(-\frac{\pi}{3}\right) = \frac{1}{2}$; $\tan\left(-\frac{\pi}{3}\right) = -\sqrt{3}$; $\csc\left(-\frac{\pi}{3}\right) = -\frac{2\sqrt{3}}{3}$; $\sec\left(-\frac{\pi}{3}\right) = 2$; $\cot\left(-\frac{\pi}{3}\right) = -\frac{\sqrt{3}}{3}$
61. $\sin(-45°) = -\frac{\sqrt{2}}{2}$; $\cos(-45°) = \frac{\sqrt{2}}{2}$; $\tan(-45°) = -1$; $\csc(-45°) = -\sqrt{2}$; $\sec(-45°) = \sqrt{2}$; $\cot(-45°) = -1$
62. $\sin(-60°) = -\frac{\sqrt{3}}{2}$; $\cos(-60°) = \frac{1}{2}$; $\tan(-60°) = -\sqrt{3}$; $\csc(-60°) = -\frac{2\sqrt{3}}{3}$; $\sec(-60°) = 2$; $\cot(-60°) = -\frac{\sqrt{3}}{3}$
63. $\sin\frac{5\pi}{2} = 1$; $\cos\frac{5\pi}{2} = 0$; $\tan\frac{5\pi}{2}$ is undefined; $\csc\frac{5\pi}{2} = 1$; $\sec\frac{5\pi}{2}$ is undefined; $\cot\frac{5\pi}{2} = 0$
64. $\sin(5\pi) = 0$; $\cos(5\pi) = -1$; $\tan(5\pi) = 0$; $\csc(5\pi)$ is undefined; $\sec(5\pi) = -1$; $\cot(5\pi)$ is undefined **65.** $\sin 720° = 0$; $\cos 720° = 1$; $\tan 720° = 0$; $\csc 720°$ is undefined; $\sec 720° = 1$; $\cot 720°$ is undefined **66.** $\sin 630° = -1$; $\cos 630° = 0$; $\tan 630°$ is undefined; $\csc 630° = -1$; $\sec 630°$ is undefined; $\cot 630° = 0$ **67.** 0.47 **68.** 0.97 **69.** 0.38 **70.** 0.36 **71.** 1.33 **72.** 1.22 **73.** 0.31 **74.** 0.92
75. 3.73 **76.** 5.67 **77.** 1.04 **78.** 1.07 **79.** 0.84 **80.** 1.56 **81.** 0.02 **82.** 0.02 **83.** $\sin\theta = \frac{4}{5}$; $\cos\theta = -\frac{3}{5}$; $\tan\theta = -\frac{4}{3}$; $\csc\theta = \frac{5}{4}$; $\sec\theta = -\frac{5}{3}$; $\cot\theta = -\frac{3}{4}$ **84.** $\sin\theta = -\frac{12}{13}$; $\cos\theta = \frac{5}{13}$; $\tan\theta = -\frac{12}{5}$; $\csc\theta = -\frac{13}{12}$; $\sec\theta = \frac{13}{5}$; $\cot\theta = -\frac{5}{12}$
85. $\sin\theta = -\frac{3\sqrt{13}}{13}$; $\cos\theta = \frac{2\sqrt{13}}{13}$; $\tan\theta = -\frac{3}{2}$; $\csc\theta = -\frac{\sqrt{13}}{3}$; $\sec\theta = \frac{\sqrt{13}}{2}$; $\cot\theta = -\frac{2}{3}$ **86.** $\sin\theta = -\frac{2\sqrt{5}}{5}$; $\cos\theta = -\frac{\sqrt{5}}{5}$; $\tan\theta = 2$; $\csc\theta = -\frac{\sqrt{5}}{2}$; $\sec\theta = -\sqrt{5}$; $\cot\theta = \frac{1}{2}$ **87.** $\sin\theta = -\frac{\sqrt{2}}{2}$; $\cos\theta = -\frac{\sqrt{2}}{2}$; $\tan\theta = 1$; $\csc\theta = -\sqrt{2}$; $\sec\theta = -\sqrt{2}$; $\cot\theta = 1$
88. $\sin\theta = -\frac{\sqrt{2}}{2}$; $\cos\theta = \frac{\sqrt{2}}{2}$; $\tan\theta = -1$; $\csc\theta = -\sqrt{2}$; $\sec\theta = \sqrt{2}$; $\cot\theta = -1$ **89.** $\sin\theta = -\frac{2\sqrt{13}}{13}$; $\cos\theta = -\frac{3\sqrt{13}}{13}$; $\tan\theta = \frac{2}{3}$; $\csc\theta = -\frac{\sqrt{13}}{2}$; $\sec\theta = -\frac{\sqrt{13}}{3}$; $\cot\theta = \frac{3}{2}$ **90.** $\sin\theta = \frac{\sqrt{2}}{2}$; $\cos\theta = \frac{\sqrt{2}}{2}$; $\tan\theta = 1$; $\csc\theta = \sqrt{2}$; $\sec\theta = \sqrt{2}$; $\cot\theta = 1$
91. $\sin\theta = -\frac{3}{5}$; $\cos\theta = \frac{4}{5}$; $\tan\theta = -\frac{3}{4}$; $\csc\theta = -\frac{5}{3}$; $\sec\theta = \frac{5}{4}$; $\cot\theta = -\frac{4}{3}$ **92.** $\sin\theta = -\frac{4}{5}$; $\cos\theta = -\frac{3}{5}$; $\tan\theta = \frac{4}{3}$; $\csc\theta = -\frac{5}{4}$; $\sec\theta = -\frac{5}{3}$; $\cot\theta = \frac{3}{4}$ **93.** 0 **94.** $\frac{2\sqrt{3}}{3}$ **95.** -0.1 **96.** -0.3 **97.** 3 **98.** -2 **99.** 5 **100.** $\frac{3}{2}$ **101.** $\frac{\sqrt{3}}{2}$ **102.** $\frac{1}{2}$ **103.** $\frac{1}{2}$ **104.** $\frac{\sqrt{3}}{2}$

105. $\frac{3}{4}$ **106.** $\frac{1}{4}$ **107.** $\frac{\sqrt{3}}{2}$ **108.** $-\frac{1}{2}$ **109.** $\sqrt{3}$ **110.** 1 **111.** $-\frac{\sqrt{3}}{2}$ **112.** $\frac{1}{2}$

113.

θ	0.5	0.4	0.2	0.1	0.01	0.001	0.0001	0.00001
$\sin\theta$	0.4794	0.3894	0.1987	0.0998	0.0100	0.0010	0.0001	0.00001
$\frac{\sin\theta}{\theta}$	0.9589	0.9735	0.9933	0.9983	1.0000	1.0000	1.0000	1.0000

$\frac{\sin\theta}{\theta}$ approaches 1 as θ approaches 0.

114.

θ	0.5	0.4	0.2	0.1	0.01	0.001	0.0001	0.00001
$\cos\theta - 1$	−0.1224	−0.0789	−0.0199	−0.0050	−0.00005	0.0000	0.0000	0.0000
$\frac{\cos\theta - 1}{\theta}$	−0.2448	−0.1973	−0.0997	−0.0500	−0.0050	−0.0005	−0.00005	−0.000005

$\frac{\cos\theta - 1}{\theta}$ approaches 0 as θ approaches 0.

115. $R \approx 310.56$ ft; $H \approx 77.64$ ft **116.** $R \approx 1988.32$ m; $H \approx 286.99$ m **117.** $R \approx 19{,}542$ m; $H \approx 2278$ m
118. $R \approx 1223.36$ ft; $H \approx 364.49$ ft **119. (a)** 1.20 s **(b)** 1.12 s **(c)** 1.20 s **120.** 4.90 cm; 4.71 cm
121. (a) 1.9 hr; 0.57 hr **(b)** 1.69 hr; 0.75 hr **(c)** 1.63 hr; 0.86 hr **(d)** 1.67 hr; tan 90° is undefined **122.** 251.42 cm^3; 117.88 cm^3; 75.4 cm^3
123. (a) 16.56 ft **(b)**

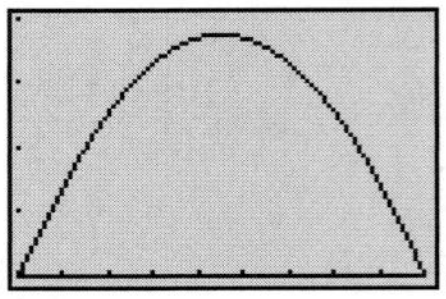

(c) 67.5° **124.** Since L and L^* are parallel, $m_L = m_{L^*} = \frac{\sin\theta - 0}{\cos\theta - 0} = \tan\theta$.

5.3 Concepts and Vocabulary *(page 367)*

5. 2π; π **6.** all real numbers except odd multiples of $\frac{\pi}{2}$ **7.** $[-1, 1]$ **8.** True **9.** False **10.** False

5.3 Exercises *(page 367)*

11. $\frac{\sqrt{2}}{2}$ **12.** $\frac{1}{2}$ **13.** 1 **14.** $\frac{1}{2}$ **15.** 1 **16.** −1 **17.** $\sqrt{3}$ **18.** 2 **19.** $\frac{\sqrt{2}}{2}$ **20.** $\frac{\sqrt{2}}{2}$ **21.** 0 **22.** 1 **23.** $\sqrt{2}$ **24.** 1 **25.** $\frac{\sqrt{3}}{3}$ **26.** $\frac{2\sqrt{3}}{3}$
27. II **28.** IV **29.** IV **30.** I **31.** IV **32.** III **33.** II **34.** II **35.** $\tan\theta = -\frac{3}{4}$; $\cot\theta = -\frac{4}{3}$; $\sec\theta = \frac{5}{4}$; $\csc\theta = -\frac{5}{3}$
36. $\tan\theta = -\frac{4}{3}$; $\cot\theta = -\frac{3}{4}$; $\sec\theta = -\frac{5}{3}$; $\csc\theta = \frac{5}{4}$ **37.** $\tan\theta = 2$; $\cot\theta = \frac{1}{2}$; $\sec\theta = \sqrt{5}$; $\csc\theta = \frac{\sqrt{5}}{2}$ **38.** $\tan\theta = \frac{1}{2}$; $\cot\theta = 2$; $\sec\theta = -\frac{\sqrt{5}}{2}$; $\csc\theta = -\sqrt{5}$ **39.** $\tan\theta = \frac{\sqrt{3}}{3}$; $\cot\theta = \sqrt{3}$; $\sec\theta = \frac{2\sqrt{3}}{3}$; $\csc\theta = 2$ **40.** $\tan\theta = \sqrt{3}$; $\cot\theta = \frac{\sqrt{3}}{3}$; $\sec\theta = 2$; $\csc\theta = \frac{2\sqrt{3}}{3}$ **41.** $\tan\theta = -\frac{\sqrt{2}}{4}$; $\cot\theta = -2\sqrt{2}$; $\sec\theta = \frac{3\sqrt{2}}{4}$; $\csc\theta = -3$ **42.** $\tan\theta = -2\sqrt{2}$; $\cot\theta = -\frac{\sqrt{2}}{4}$; $\sec\theta = -3$; $\csc\theta = \frac{3\sqrt{2}}{4}$ **43.** $\cos\theta = -\frac{5}{13}$; $\tan\theta = -\frac{12}{5}$; $\csc\theta = \frac{13}{12}$; $\sec\theta = -\frac{13}{5}$; $\cot\theta = -\frac{5}{12}$ **44.** $\sin\theta = -\frac{4}{5}$; $\tan\theta = -\frac{4}{3}$; $\csc\theta = -\frac{5}{4}$; $\sec\theta = \frac{5}{3}$; $\cot\theta = -\frac{3}{4}$ **45.** $\sin\theta = -\frac{3}{5}$; $\tan\theta = \frac{3}{4}$; $\csc\theta = -\frac{5}{3}$; $\sec\theta = -\frac{5}{4}$; $\cot\theta = \frac{4}{3}$ **46.** $\cos\theta = -\frac{12}{13}$; $\tan\theta = \frac{5}{12}$; $\csc\theta = -\frac{13}{5}$; $\sec\theta = -\frac{13}{12}$; $\cot\theta = \frac{12}{5}$ **47.** $\cos\theta = -\frac{12}{13}$; $\tan\theta = -\frac{5}{12}$; $\csc\theta = \frac{13}{5}$; $\sec\theta = -\frac{13}{12}$; $\cot\theta = -\frac{12}{5}$ **48.** $\sin\theta = -\frac{3}{5}$; $\tan\theta = -\frac{3}{4}$; $\csc\theta = -\frac{5}{3}$; $\sec\theta = \frac{5}{4}$; $\cot\theta = -\frac{4}{3}$ **49.** $\sin\theta = \frac{2\sqrt{2}}{3}$; $\tan\theta = -2\sqrt{2}$; $\csc\theta = \frac{3\sqrt{2}}{4}$; $\sec\theta = -3$; $\cot\theta = -\frac{\sqrt{2}}{4}$ **50.** $\cos\theta = -\frac{\sqrt{5}}{3}$; $\tan\theta = \frac{2\sqrt{5}}{5}$; $\csc\theta = -\frac{3}{2}$; $\sec\theta = -\frac{3\sqrt{5}}{5}$; $\cot\theta = \frac{\sqrt{5}}{2}$ **51.** $\cos\theta = -\frac{\sqrt{5}}{3}$; $\tan\theta = -\frac{2\sqrt{5}}{5}$; $\csc\theta = \frac{3}{2}$; $\sec\theta = -\frac{3\sqrt{5}}{5}$; $\cot\theta = -\frac{\sqrt{5}}{2}$
52. $\sin\theta = -\frac{\sqrt{15}}{4}$; $\tan\theta = \sqrt{15}$; $\csc\theta = -\frac{4\sqrt{15}}{15}$; $\sec\theta = -4$; $\cot\theta = \frac{\sqrt{15}}{15}$ **53.** $\sin\theta = -\frac{\sqrt{3}}{2}$; $\cos\theta = \frac{1}{2}$; $\tan\theta = -\sqrt{3}$; $\csc\theta = -\frac{2\sqrt{3}}{3}$; $\cot\theta = -\frac{\sqrt{3}}{3}$ **54.** $\sin\theta = \frac{1}{3}$; $\cos\theta = -\frac{2\sqrt{2}}{3}$; $\tan\theta = -\frac{\sqrt{2}}{4}$; $\sec\theta = -\frac{3\sqrt{2}}{4}$; $\cot\theta = -2\sqrt{2}$

55. $\sin\theta = -\frac{3}{5}$; $\cos\theta = -\frac{4}{5}$; $\csc\theta = -\frac{5}{3}$; $\sec\theta = -\frac{5}{4}$; $\cot\theta = \frac{4}{3}$ **56.** $\sin\theta = -\frac{3}{5}$; $\cos\theta = -\frac{4}{5}$; $\tan\theta = \frac{3}{4}$; $\csc\theta = -\frac{5}{3}$; $\sec\theta = -\frac{5}{4}$

57. $\sin\theta = \frac{\sqrt{10}}{10}$; $\cos\theta = -\frac{3\sqrt{10}}{10}$; $\csc\theta = \sqrt{10}$; $\sec\theta = -\frac{\sqrt{10}}{3}$; $\cot\theta = -3$ **58.** $\sin\theta = -\frac{\sqrt{3}}{2}$; $\cos\theta = -\frac{1}{2}$; $\tan\theta = \sqrt{3}$; $\csc\theta = -\frac{2\sqrt{3}}{3}$; $\cot\theta = \frac{\sqrt{3}}{3}$ **59.** $-\frac{\sqrt{3}}{2}$ **60.** $\frac{\sqrt{3}}{2}$ **61.** $-\frac{\sqrt{3}}{3}$ **62.** $-\frac{\sqrt{2}}{2}$ **63.** 2 **64.** -2 **65.** -1 **66.** 0 **67.** -1 **68.** 0 **69.** $\frac{\sqrt{2}}{2}$

70. $-\frac{\sqrt{3}}{2}$ **71.** 0 **72.** 1 **73.** $-\sqrt{2}$ **74.** -1 **75.** $\frac{2\sqrt{3}}{3}$ **76.** $-\frac{2\sqrt{3}}{3}$ **77.** 1 **78.** 1 **79.** 1 **80.** 1 **81.** 0 **82.** 0 **83.** 1 **84.** 1

85. -1 **86.** 1 **87.** 0 **88.** 0 **89.** 0.9 **90.** 0.6 **91.** 9 **92.** -6 **93.** 0 **94.** -1 **95.** All real numbers **96.** All real numbers

97. Odd multiples of $\frac{\pi}{2}$ **98.** Multiples of π **99.** Odd multiples of $\frac{\pi}{2}$ **100.** Multiples of π **101.** $-1 \le y \le 1$ **102.** $-1 \le y \le 1$

103. All real numbers **104.** All real numbers **105.** $|y| \ge 1$ **106.** $|y| \ge 1$ **107.** Odd; yes; origin **108.** Even; yes; y-axis **109.** Odd; yes; origin **110.** Odd; yes; origin **111.** Even; yes; y-axis **112.** Odd; yes; origin **113.** **(a)** $-\frac{1}{3}$ **(b)** 1 **114.** **(a)** $\frac{1}{4}$ **(b)** $\frac{3}{4}$ **115.** **(a)** -2 **(b)** 6

116. **(a)** 3 **(b)** -9 **117.** **(a)** -4 **(b)** -12 **118.** **(a)** -2 **(b)** 6 **119.** About 15.81 min **120.** About 2.75 hr

121. Let a be a real number and $P = (x, y)$ be the point on the unit circle that corresponds to t. Consider the equation $\tan t = \frac{y}{x} = a$. Then $y = ax$. But $x^2 + y^2 = 1$ so $x^2 + a^2x^2 = 1$. Thus, $x = \pm\frac{1}{\sqrt{1 + a^2}}$ and $y = \pm\frac{a}{\sqrt{1 + a^2}}$; that is, for any real number a, there is a point $P = (x, y)$ on the unit circle for which $\tan t = a$. In other words, the range of the tangent function is the set of all real numbers.

122. Let a be a real number and $P = (x, y)$ be the point on the unit circle that corresponds to t. Consider the equation $\cot t = \frac{x}{y} = a$. Then $x = ay$. But $x^2 + y^2 = 1$ so $a^2y^2 + y^2 = 1$. Thus, $y = \pm\frac{1}{\sqrt{1 + a^2}}$ and $x = \pm\frac{a}{\sqrt{1 + a^2}}$; that is, for any real number a, there is a point $P = (x, y)$ on the unit circle for which $\cot t = a$. In other words, the range of the cotangent function is the set of all real numbers.

123. Suppose that there is a number p, $0 < p < 2\pi$, for which $\sin(\theta + p) = \sin\theta$ for all θ. If $\theta = 0$, then $\sin(0 + p) = \sin p = \sin 0 = 0$, so $p = \pi$. If $\theta = \frac{\pi}{2}$, then $\sin\left(\frac{\pi}{2} + p\right) = \sin\left(\frac{\pi}{2}\right)$. But $p = \pi$. Thus, $\sin\left(\frac{3\pi}{2}\right) = -1 = \sin\left(\frac{\pi}{2}\right) = 1$. This is impossible. Therefore, the smallest positive number p for which $\sin(\theta + p) = \sin\theta$ for all θ is 2π.

124. Suppose that there is a number p, $0 < p < 2\pi$, for which $\cos(\theta + p) = \cos\theta$ for all θ. If $\theta = 0$, then $\cos(0 + p) = \cos p = \cos 0 = 1$. But on the interval $(0, 2\pi)$, $\cos p$ never equals 1. Therefore, the smallest positive number p for which $\cos(\theta + p) = \cos\theta$ for all θ is 2π.

125. $\sec\theta = \frac{1}{\cos\theta}$; since $\cos\theta$ has period 2π, so does $\sec\theta$. **126.** $\csc\theta = \frac{1}{\sin\theta}$; since $\sin\theta$ has period 2π, so does $\csc\theta$.

127. If $P = (a, b)$ is the point on the unit circle corresponding to θ, then $Q = (-a, -b)$ is the point on the unit circle corresponding to $\theta + \pi$. Thus, $\tan(\theta + \pi) = \frac{-b}{-a} = \frac{b}{a} = \tan\theta$. Suppose that there exists a number p, $0 < p < \pi$, for which $\tan(\theta + p) = \tan\theta$ for all θ. Then, if $\theta = 0$, then $\tan p = \tan 0 = 0$. But this means that p is a multiple of π. Since no multiple of π exists in the interval $(0, \pi)$, this is a contradiction. Therefore, the period of $f(\theta) = \tan\theta$ is π.

128. $\cot\theta = \frac{1}{\tan\theta}$; since $\tan\theta$ has period π, so does $\cot\theta$. **129.** Let $P = (a, b)$ be the point on the unit circle corresponding to θ. Then $\csc\theta = \frac{1}{b} = \frac{1}{\sin\theta}$; $\sec\theta = \frac{1}{a} = \frac{1}{\cos\theta}$; $\cot\theta = \frac{a}{b} = \frac{1}{b/a} = \frac{1}{\tan\theta}$.

130. If $P = (a, b)$ is the point on the unit circle corresponding to θ, then $\tan\theta = \frac{b}{a} = \frac{\sin\theta}{\cos\theta}$, and $\cot\theta = \frac{a}{b} = \frac{\cos\theta}{\sin\theta}$.

131.
$$\begin{aligned}(\sin\theta\cos\phi)^2 + (\sin\theta\sin\phi)^2 + \cos^2\theta &= \sin^2\theta\cos^2\phi + \sin^2\theta\sin^2\phi + \cos^2\theta \\ &= \sin^2\theta(\cos^2\phi + \sin^2\phi) + \cos^2\theta = \sin^2\theta + \cos^2\theta = 1\end{aligned}$$

5.4 Concepts and Vocabulary *(page 381)*

3. 1; $\frac{\pi}{2} + 2\pi k$, k any integer **4.** 3; π **5.** 3; $\frac{\pi}{3}$ **6.** True **7.** False **8.** True

5.4 Exercises *(page 382)*

9. 0 **10.** 1 **11.** $-\frac{\pi}{2} \le x \le \frac{\pi}{2}$ **12.** $0 \le x \le \pi$ **13.** 1 **14.** -1 **15.** $0, \pi, 2\pi$ **16.** $\frac{\pi}{2}, \frac{3\pi}{2}$ **17.** $\sin x = 1$ for $x = -\frac{3\pi}{2}, \frac{\pi}{2}$; $\sin x = -1$ for $x = -\frac{\pi}{2}, \frac{3\pi}{2}$ **18.** $\cos x = 1$ for $-2\pi, 0, 2\pi$; $\cos x = -1$ for $x = -\pi, \pi$ **19.** B, C, F **20.** A, D, E

21.
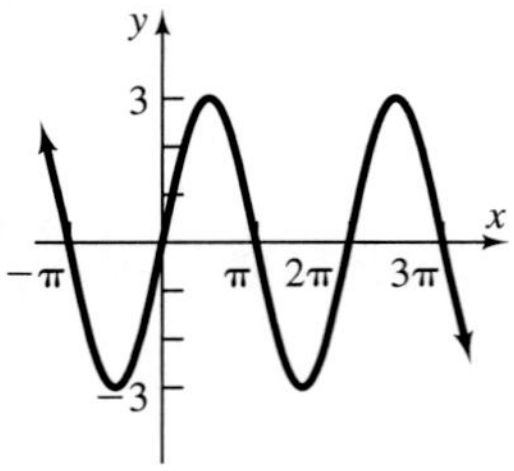

22.
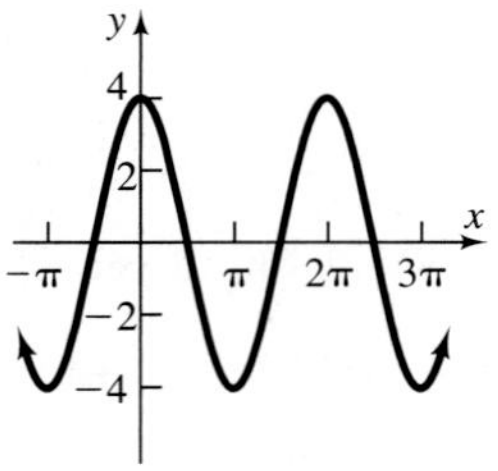

23.
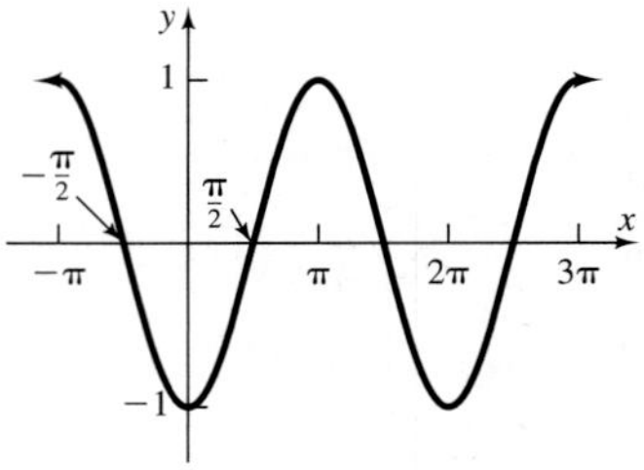

24.
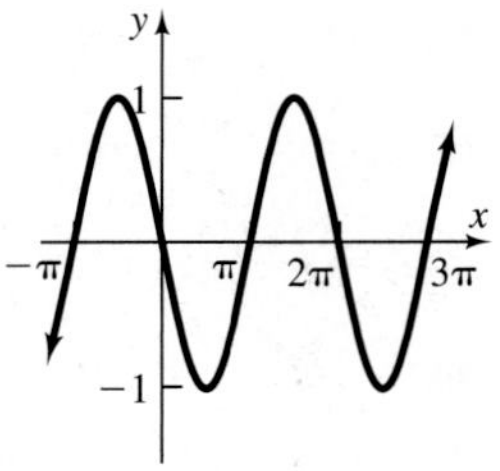

25.
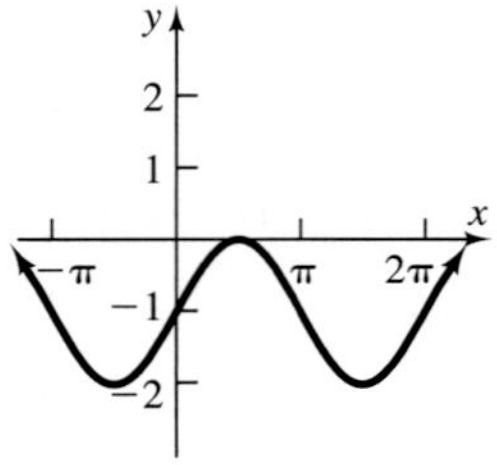

26.
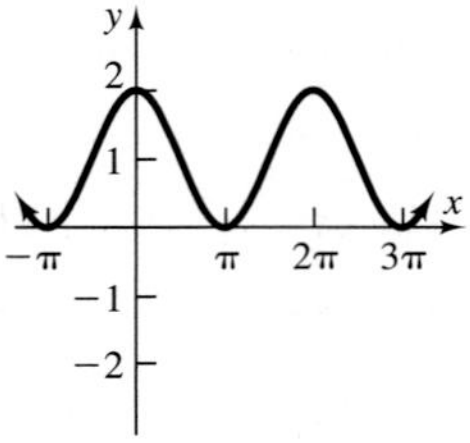

27.
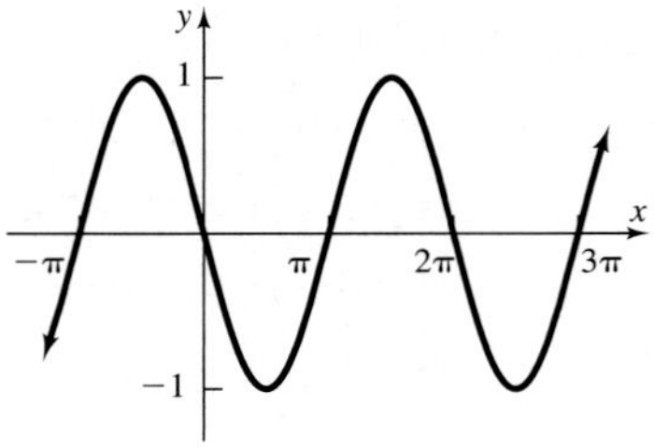

28.
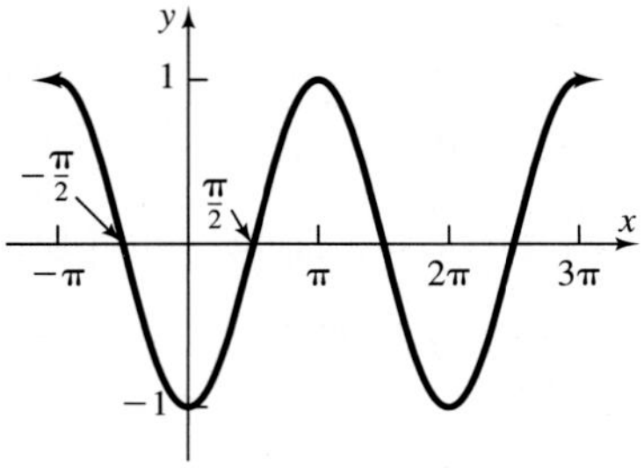

29.
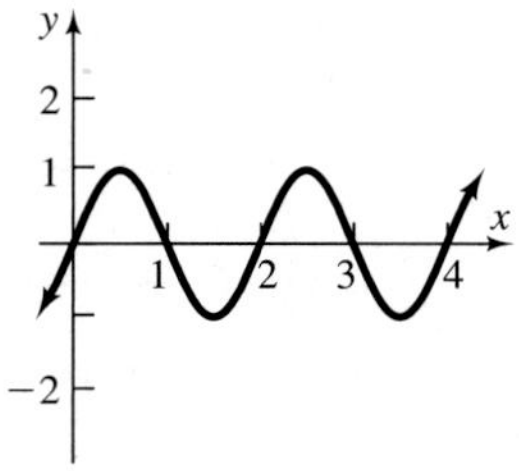

30.
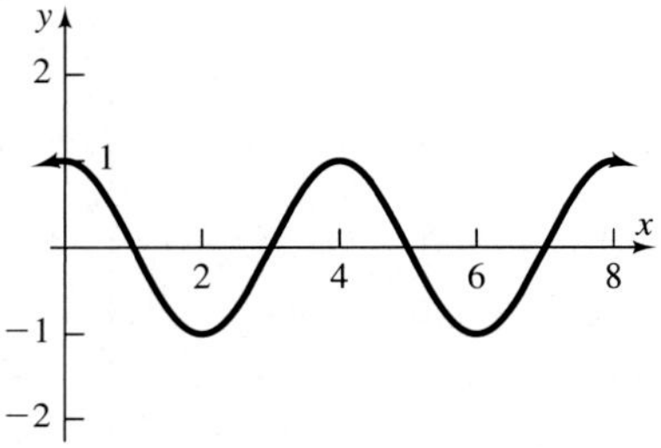

31.
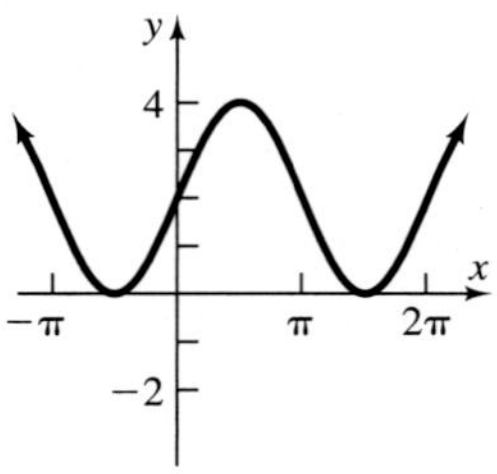

32.
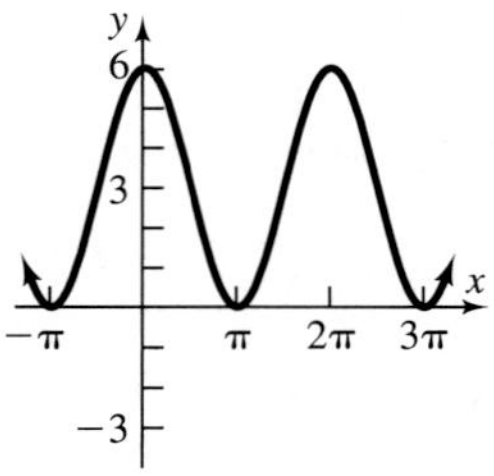

33.
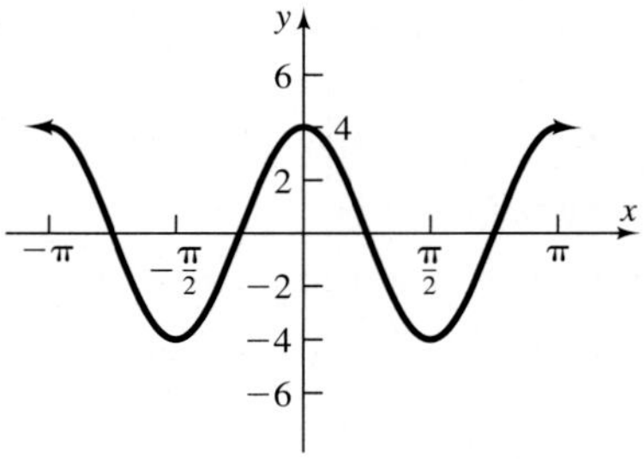

34.
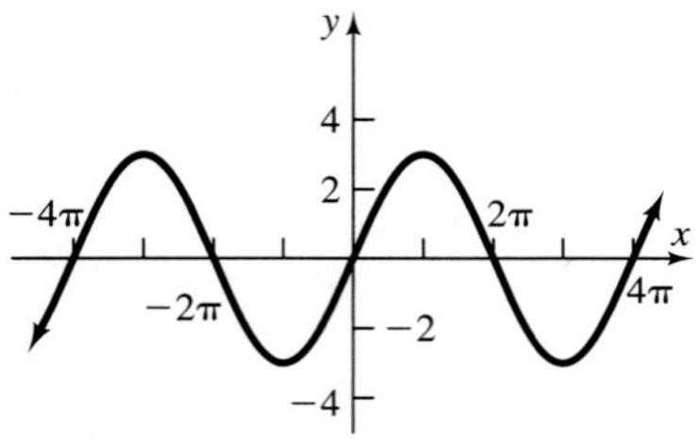

35.
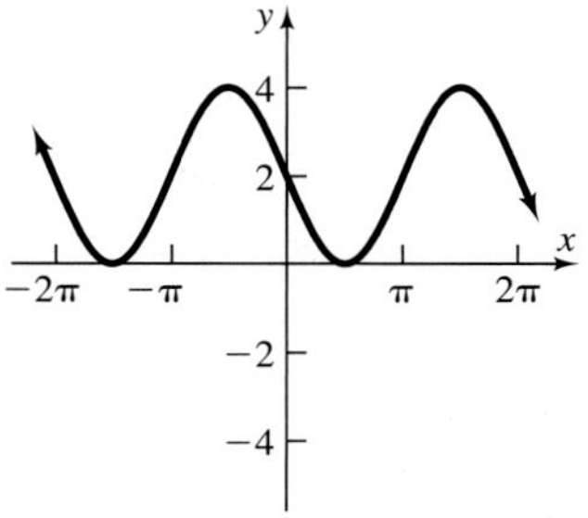

36.
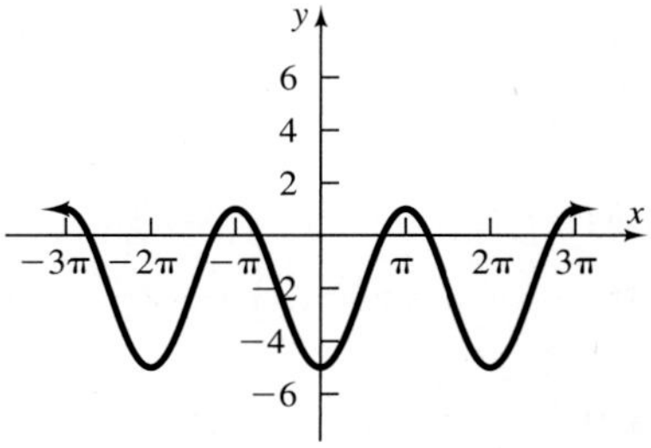

37. Amplitude = 2; Period = 2π **38.** Amplitude = 3; Period = 2π **39.** Amplitude = 4; Period = π

40. Amplitude = 1; Period = 4π **41.** Amplitude = 6; Period = 2 **42.** Amplitude = 3; Period = $\frac{2\pi}{3}$ **43.** Amplitude = $\frac{1}{2}$; Period = $\frac{4\pi}{3}$

44. Amplitude = $\frac{4}{3}$; Period = 3π **45.** Amplitude = $\frac{5}{3}$; Period = 3 **46.** Amplitude = $\frac{9}{5}$; Period = $\frac{4}{3}$

47. F **48.** E **49.** A **50.** I **51.** H **52.** B **53.** C **54.** G **55.** J **56.** D **57.** A **58.** C **59.** B **60.** D

61.
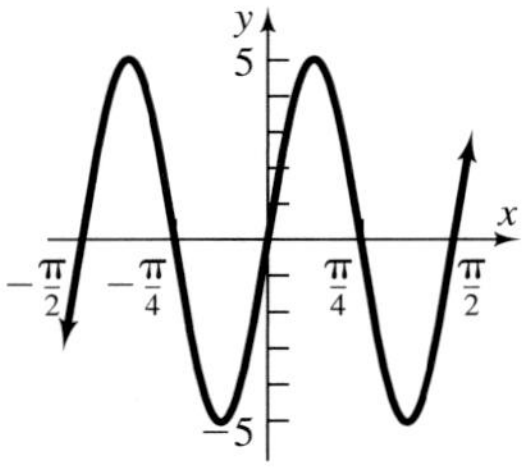

62.
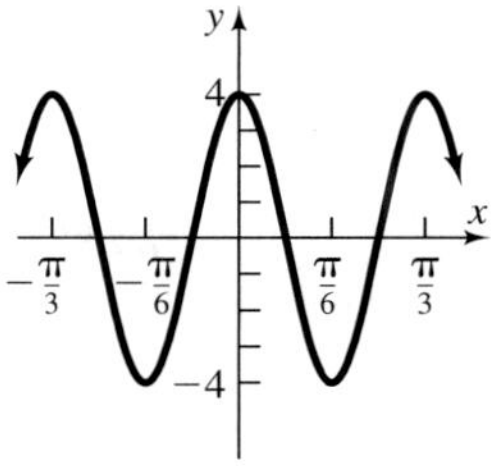

63.
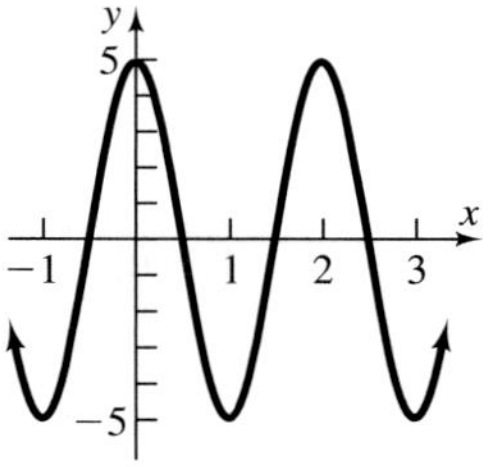

64.
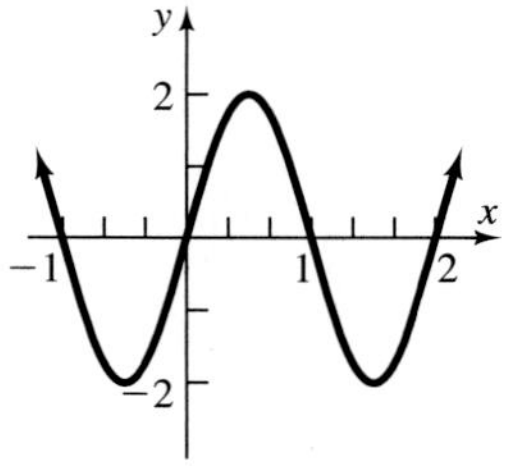

65.
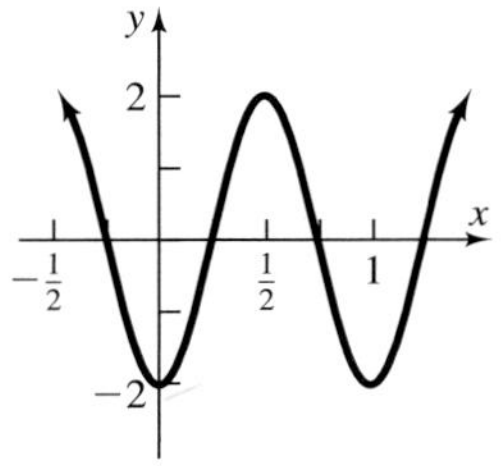

66.
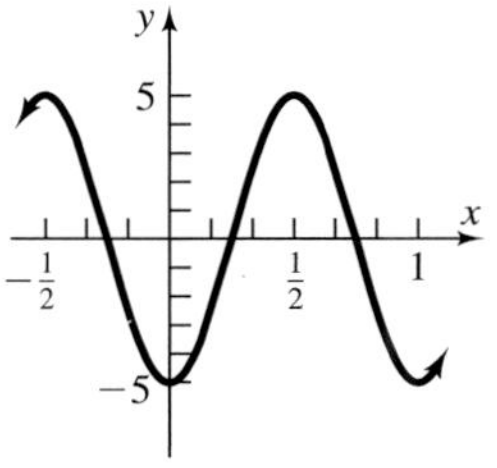

67.
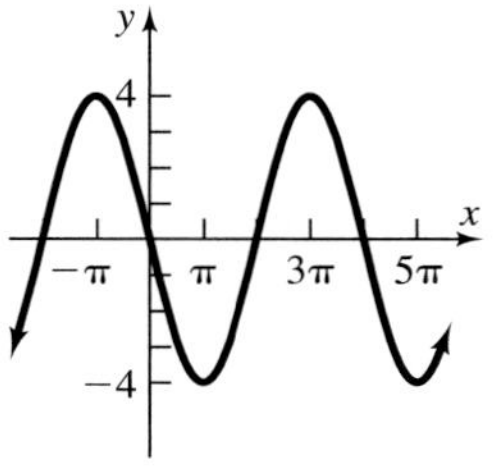

68.
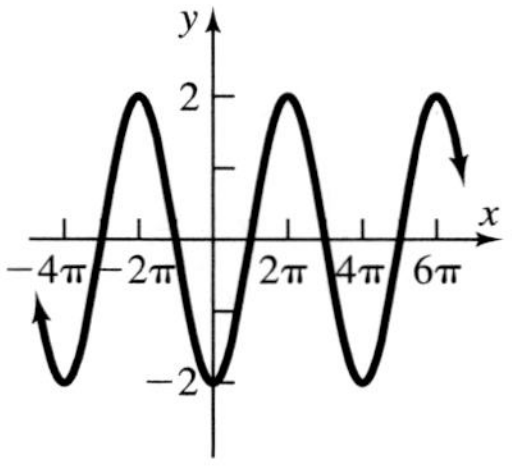

69.
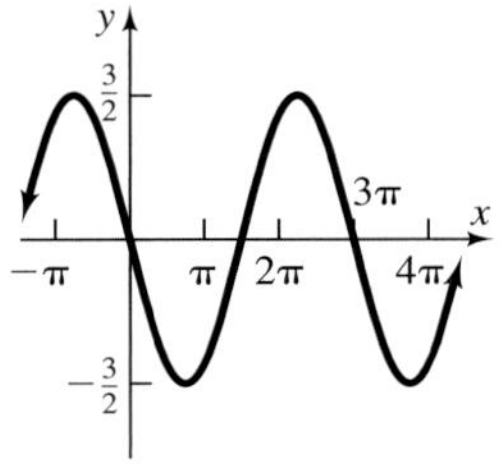

70.
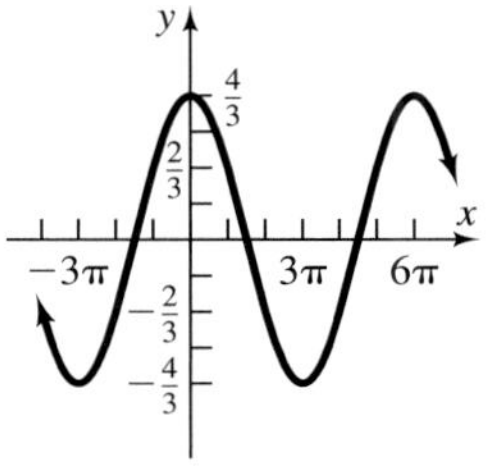

71. $y = \pm 3\sin(2x)$ **72.** $y = \pm 2\sin\left(\frac{1}{2}x\right)$ **73.** $y = \pm 3\sin(\pi x)$ **74.** $y = \pm 4\sin(2\pi x)$ **75.** $y = 5\cos\left(\frac{\pi}{4}x\right)$ **76.** $y = 4\sin\left(\frac{1}{4}x\right)$ **77.** $y = -3\cos\left(\frac{1}{2}x\right)$ **78.** $y = -2\sin\left(\frac{\pi}{2}x\right)$ **79.** $y = \frac{3}{4}\sin(2\pi x)$ **80.** $y = -\frac{5}{2}\cos(\pi x)$ **81.** $y = -\sin\left(\frac{3}{2}x\right)$ **82.** $y = -\pi\cos x$

83. $y = -\cos\left(\frac{4\pi}{3}x\right) + 1$ **84.** $y = -\frac{1}{2}\sin\left(\frac{3}{2}x\right) - 1$ **85.** $y = 3\sin\left(\frac{\pi}{2}x\right)$ **86.** $y = -2\cos(\pi x)$ **87.** $y = -4\cos(3x)$ **88.** $y = 4\sin(2x)$

89. Period $= \frac{1}{30}$; Amplitude $= 220$

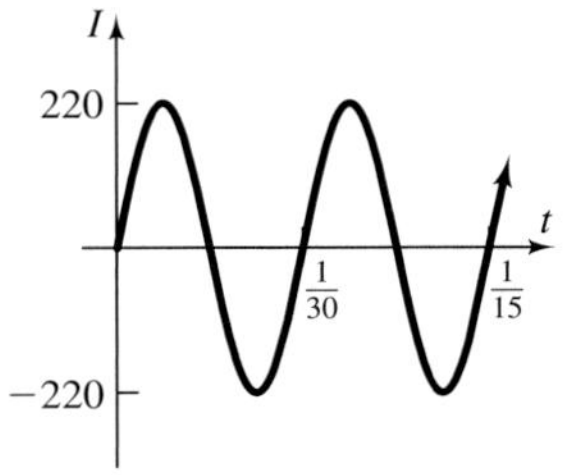

90. Period $= \frac{1}{15}$; Amplitude $= 120$

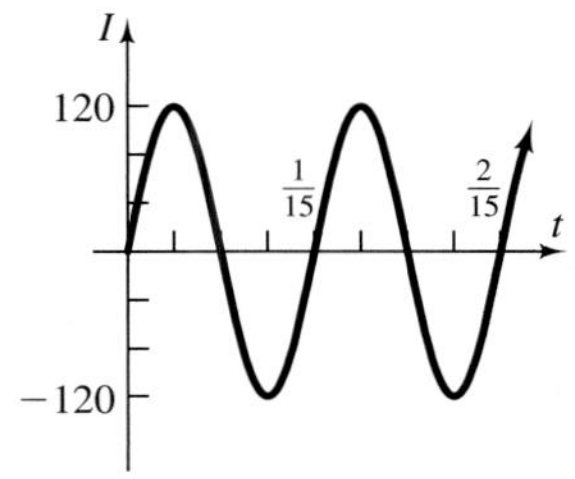

91. **(a)** Amplitude $= 220$; Period $= \frac{1}{60}$

(b), (e)

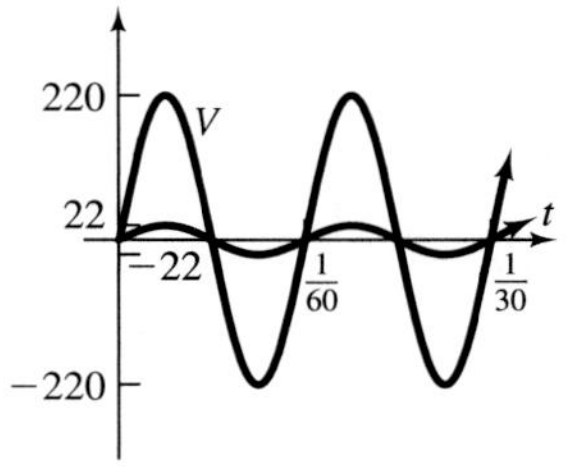

(c) $I = 22\sin(120\pi t)$

(d) Amplitude $= 22$; Period $= \frac{1}{60}$

92. **(a)** Amplitude $= 120$; Period $= \frac{1}{60}$

(b), (e)

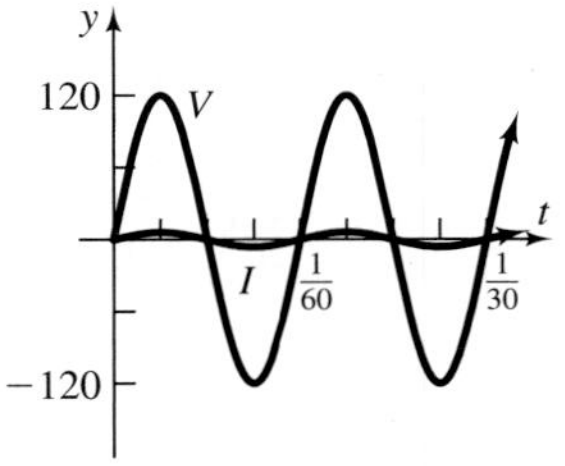

(c) $I = 6\sin(120\pi t)$

(d) Amplitude $= 6$; Period $= \frac{1}{60}$

93. **(a)** $P = \frac{[V_0\sin(2\pi ft)]^2}{R} = \frac{V_0^2}{R}\sin^2[2\pi ft]$

(b) Since the graph of P has amplitude $\frac{V_0^2}{2R}$ and period $\frac{1}{2f}$ and is of the form $y = A\cos(\omega t) + B$, then $A = -\frac{V_0^2}{2R}$ and $B = \frac{V_0^2}{2R}$. Since $\frac{1}{2f} = \frac{2\pi}{\omega}$, then $\omega = 4\pi f$. Therefore,
$P = -\frac{V_0^2}{2R}\cos(4\pi ft) + \frac{V_0^2}{2R} = \frac{V_0^2}{2R}[1 - \cos(4\pi ft)]$.

94. **(a)** Physical potential: $\omega = \frac{2\pi}{23}$; Emotional potential: $\omega = \frac{\pi}{14}$; Intellectual potential: $\omega = \frac{2\pi}{33}$

(b)

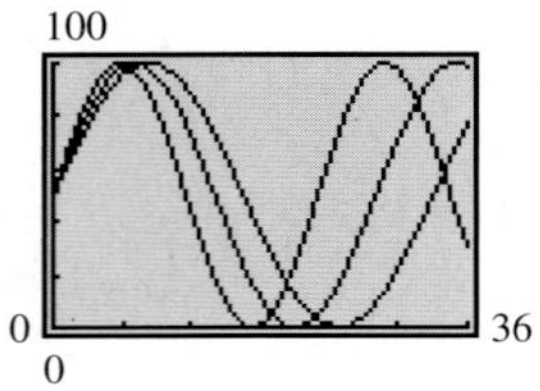

(c) No

(d) Physical potential peaks at 15 days after 20th birthday. Emotional potential is 50% at 17 days, with a maximum at 10 days and a minimum at 24 days. Intellectual potential starts fairly high, drops to a minimum at 13 days, and rises to a maximum at 29 days.

95.

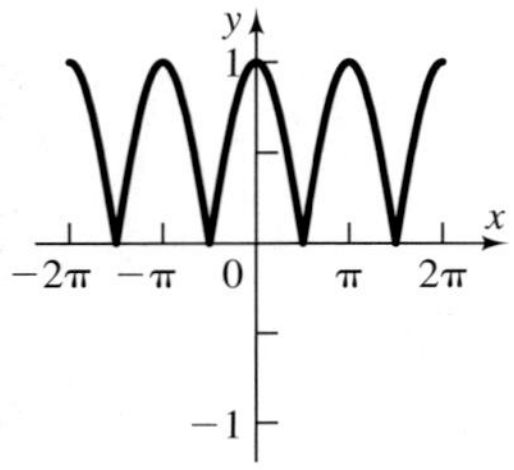

96.

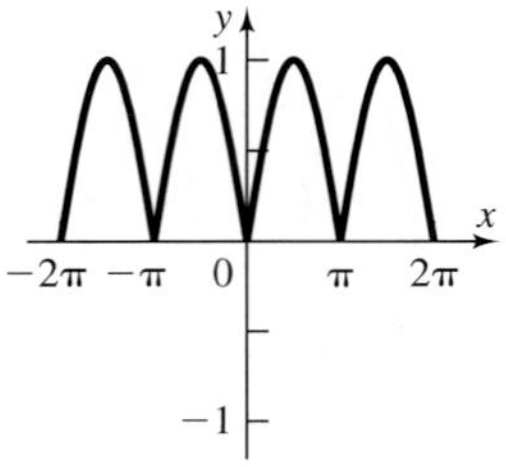

5.5 Concepts and Vocabulary *(page 392)*

3. origin; odd multiples of $\frac{\pi}{2}$ **4.** y-axis; odd multiples of $\frac{\pi}{2}$ **5.** $y = \cos x$ **6.** True

5.5 Exercises *(page 392)*

7. 0 **8.** No y-intercept **9.** 1 **10.** No y-intercept **11.** $\sec x = 1$ for $x = -2\pi, 0, 2\pi$; $\sec x = -1$ for $x = -\pi, \pi$

12. $\csc x = 1$ for $x = -\frac{3\pi}{2}, \frac{\pi}{2}$; $\csc x = -1$ for $x = -\frac{\pi}{2}, \frac{3\pi}{2}$ **13.** $-\frac{3\pi}{2}, -\frac{\pi}{2}, \frac{\pi}{2}, \frac{3\pi}{2}$ **14.** $-2\pi, -\pi, 0, \pi, 2\pi$ **15.** $-\frac{3\pi}{2}, -\frac{\pi}{2}, \frac{\pi}{2}, \frac{3\pi}{2}$

16. $-2\pi, -\pi, 0, \pi, 2\pi$ **17.** D **18.** C **19.** B **20.** A

21.

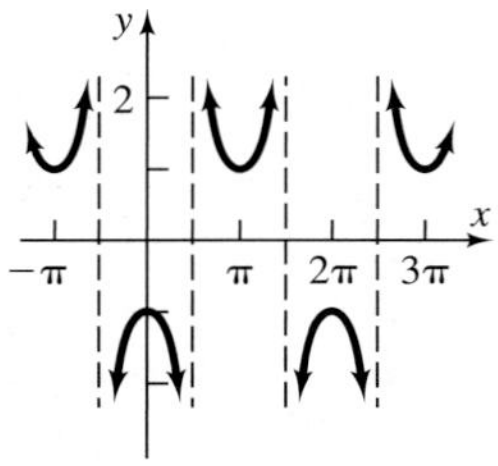

22.

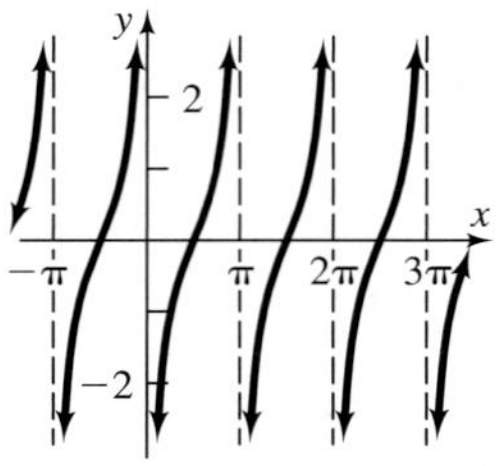

23.

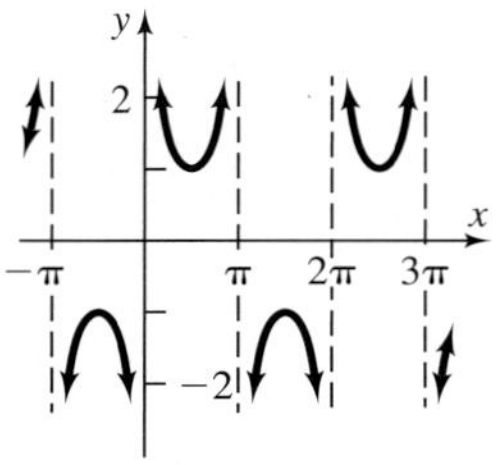

24.

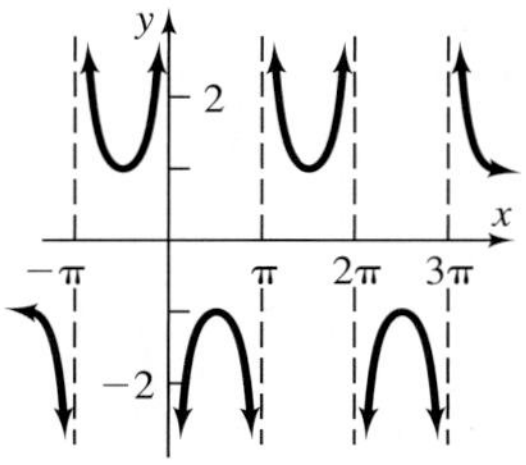

25.

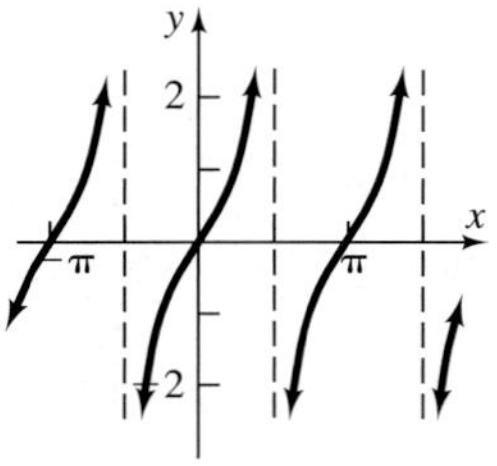

26.

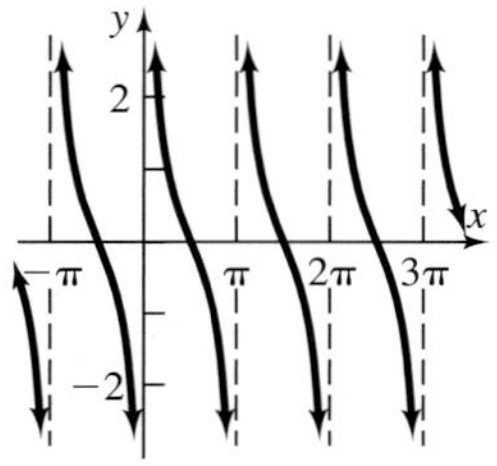

27.

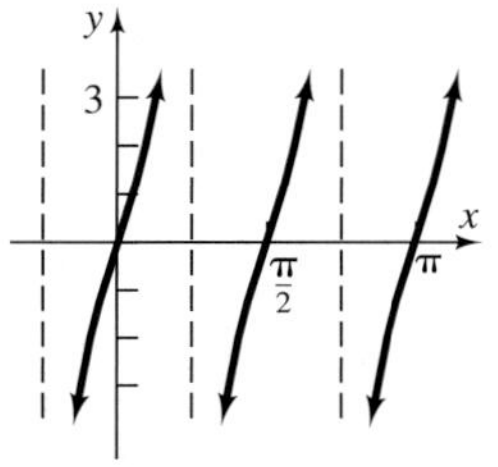

28.

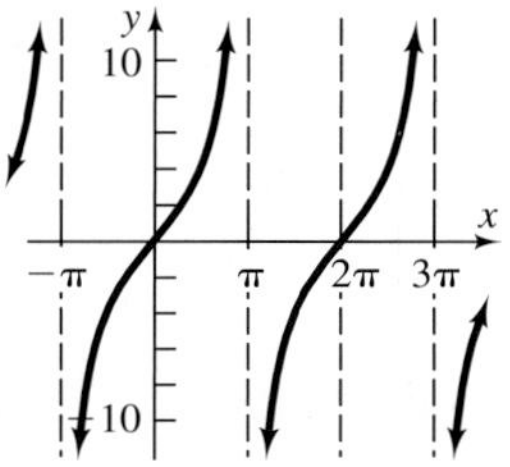

29.

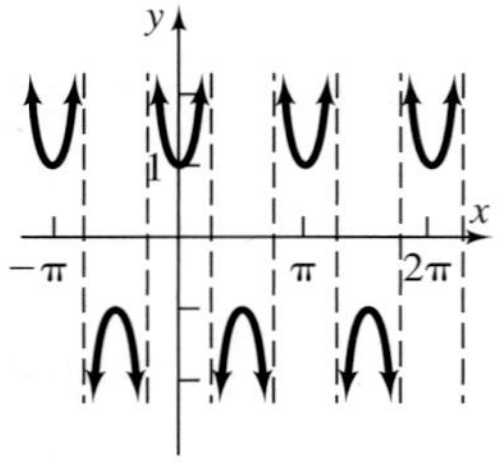

30.

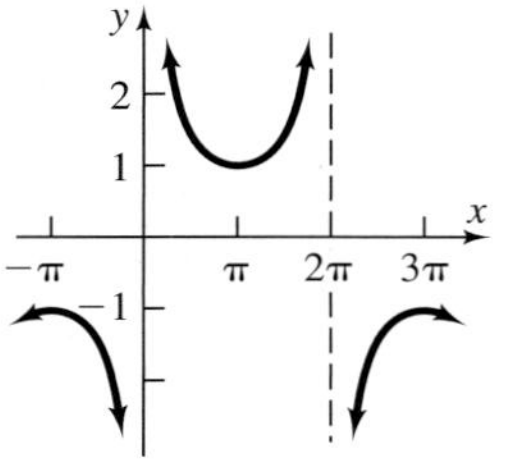

31.

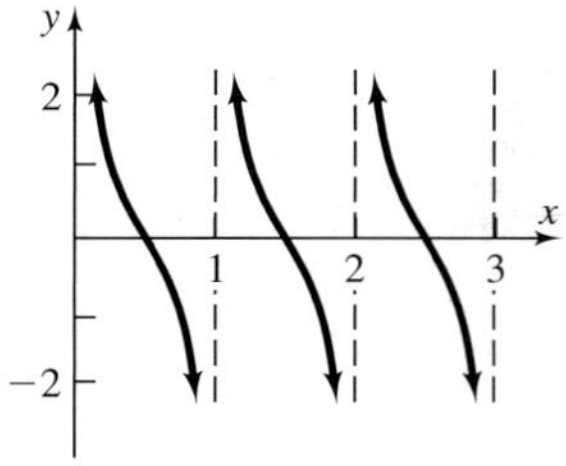

32.

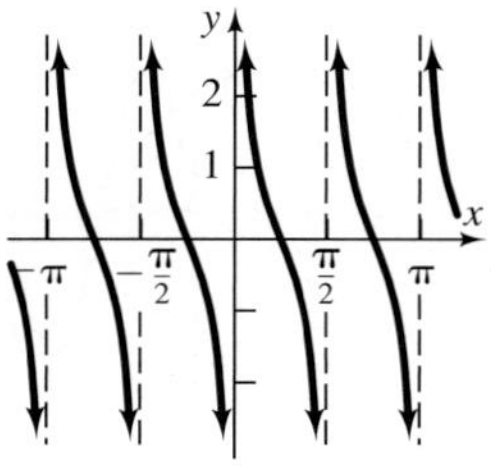

33.

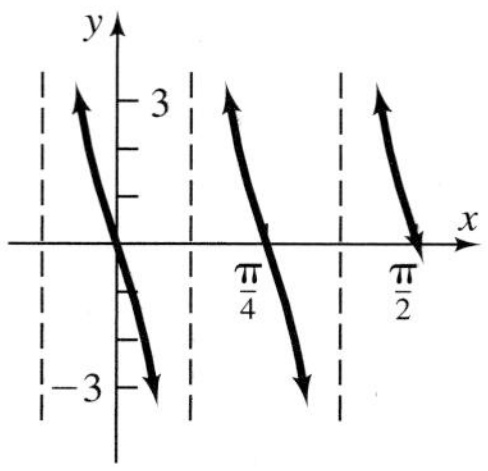

34.

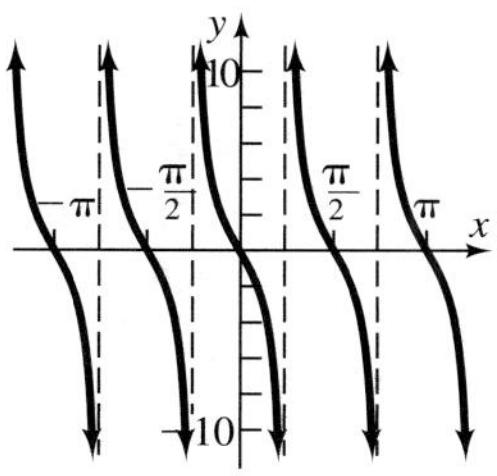

35.

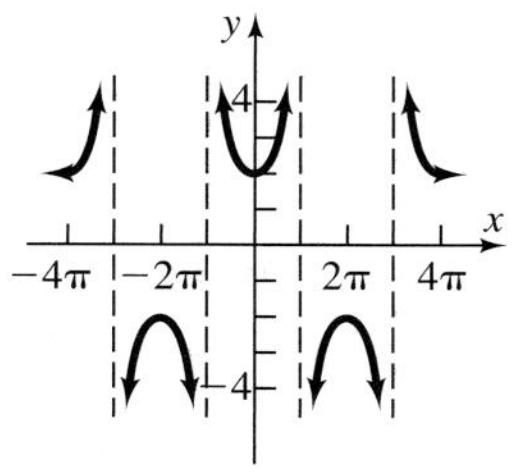

36.

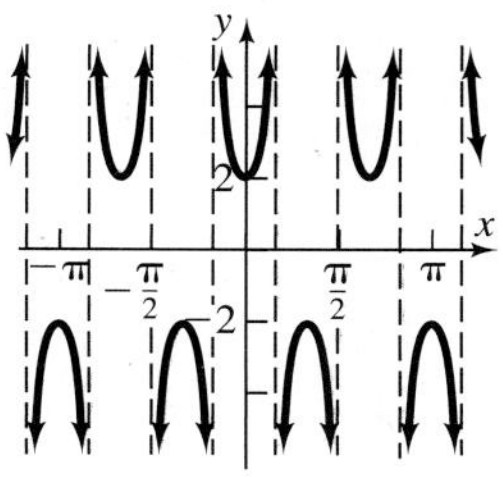

37.

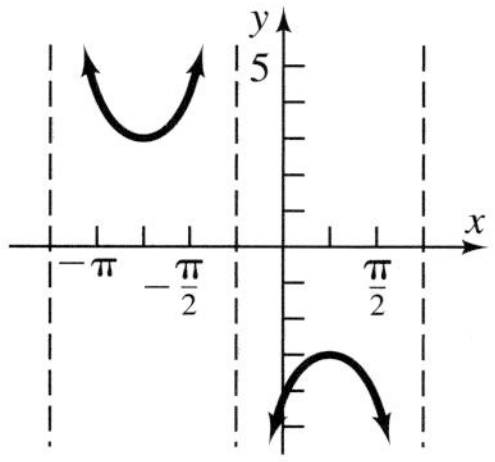

38.

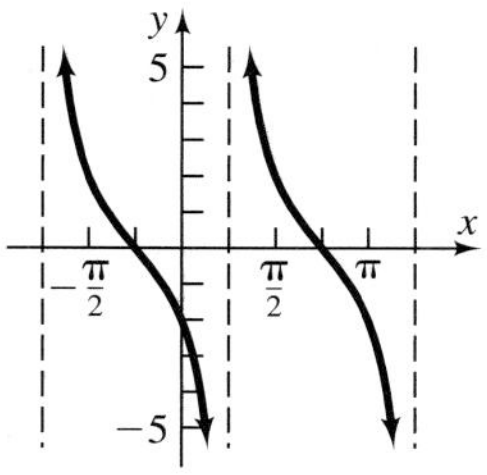

39.

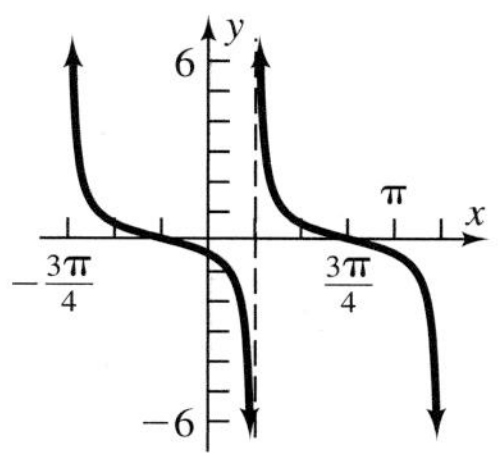

40.

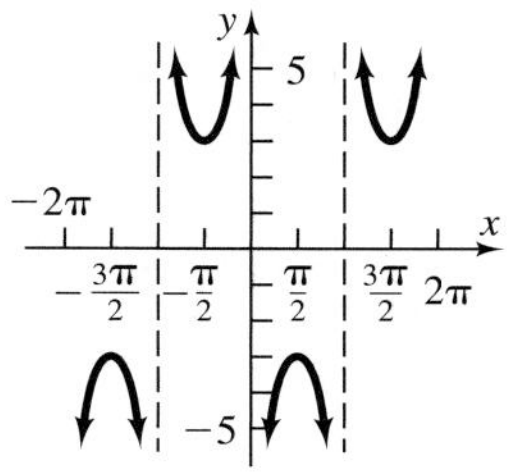

41. **(a)** $L(\theta) = \dfrac{3}{\cos\theta} + \dfrac{4}{\sin\theta} = 3\sec\theta + 4\csc\theta$

(b)

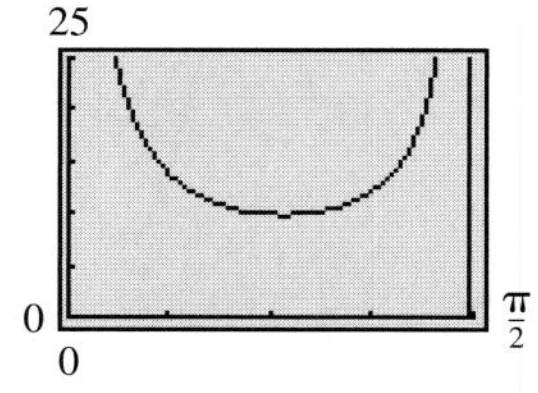

(c) 0.83
(d) 9.86 ft

42.

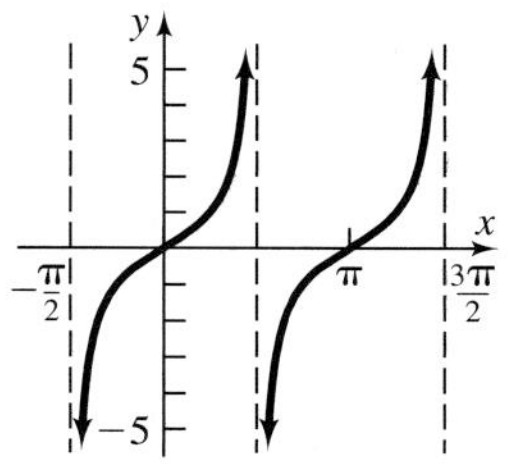

$y = \tan x$

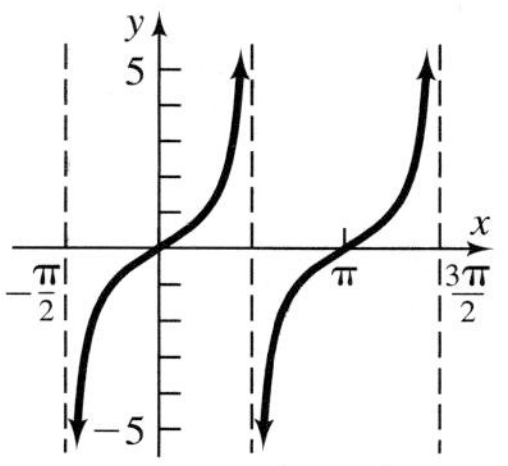

$y = -\cot\left(x + \frac{\pi}{2}\right)$

5.6 Concepts and Vocabulary *(page 402)*

1. phase shift **2.** False

5.6 Exercises *(page 402)*

3. Amplitude $= 4$

Period $= \pi$

Phase shift $= \dfrac{\pi}{2}$

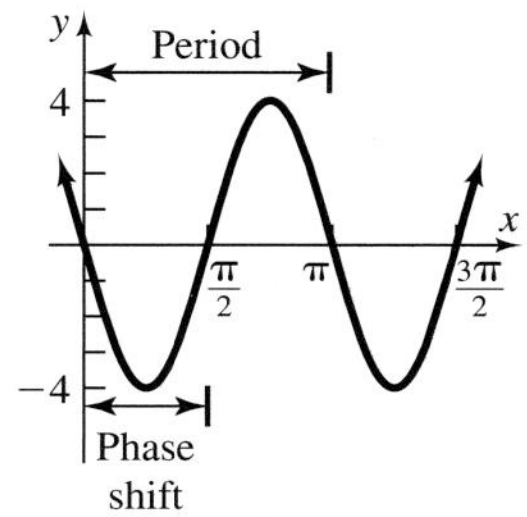

4. Amplitude $= 3$

Period $= \dfrac{2\pi}{3}$

Phase shift $= \dfrac{\pi}{3}$

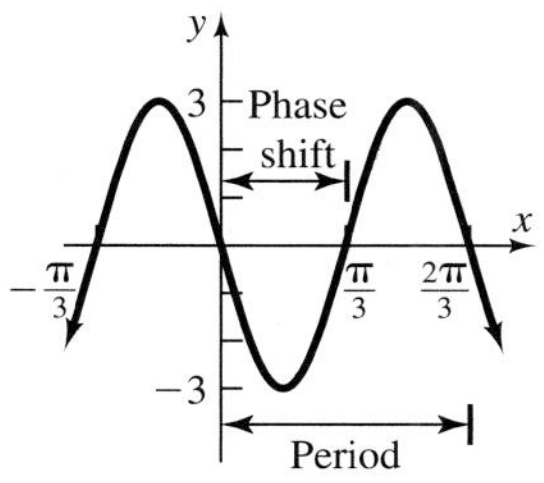

5. Amplitude $= 2$

Period $= \dfrac{2\pi}{3}$

Phase shift $= -\dfrac{\pi}{6}$

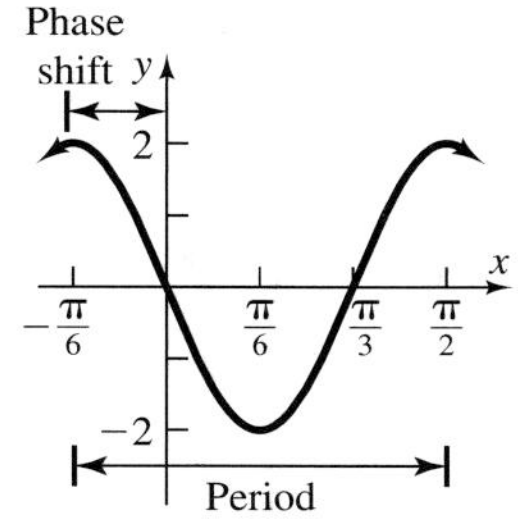

6. Amplitude $= 3$

Period $= \pi$

Phase shift $= -\dfrac{\pi}{2}$

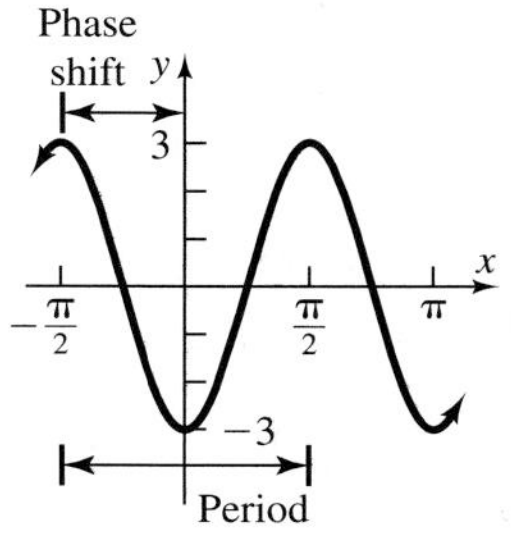

7. Amplitude = 3
Period = π
Phase shift = $-\dfrac{\pi}{4}$

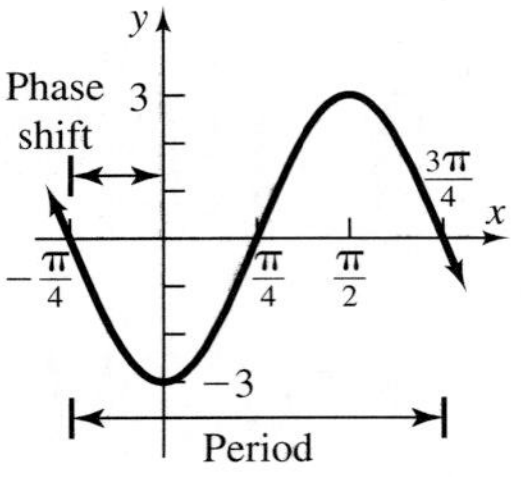

8. Amplitude = 2
Period = π
Phase shift = $\dfrac{\pi}{4}$

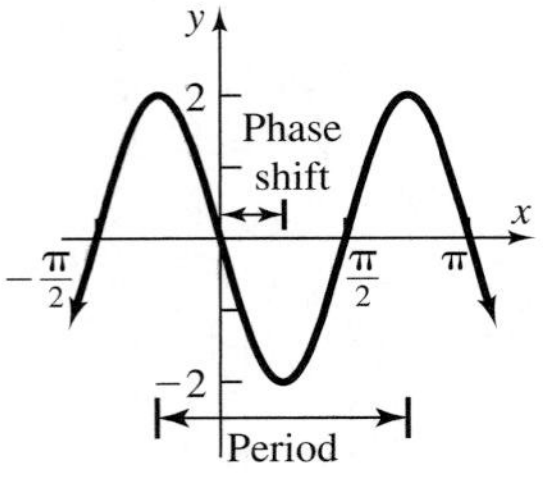

9. Amplitude = 4
Period = 2
Phase shift = $-\dfrac{2}{\pi}$

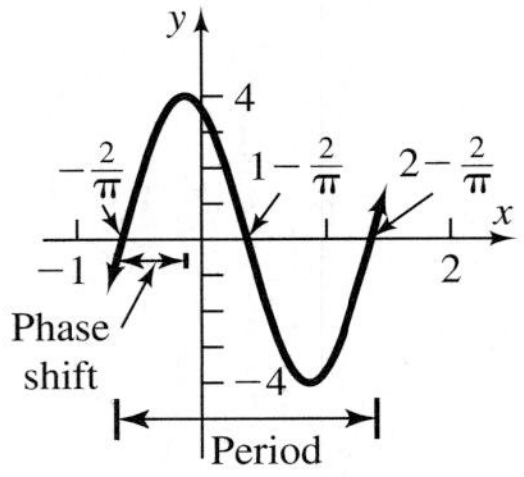

10. Amplitude = 2
Period = 1
Phase shift = $-\dfrac{2}{\pi}$

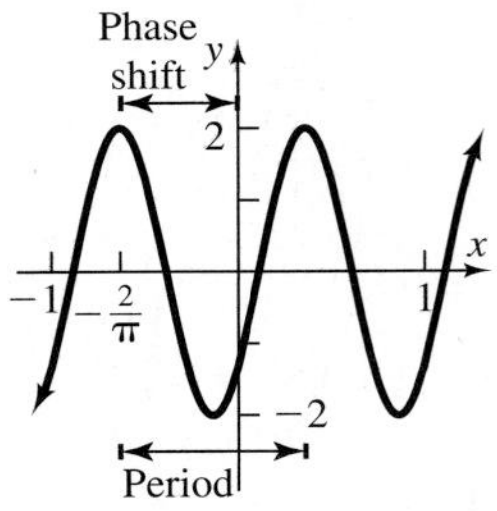

11. Amplitude = 3
Period = 2
Phase shift = $\dfrac{2}{\pi}$

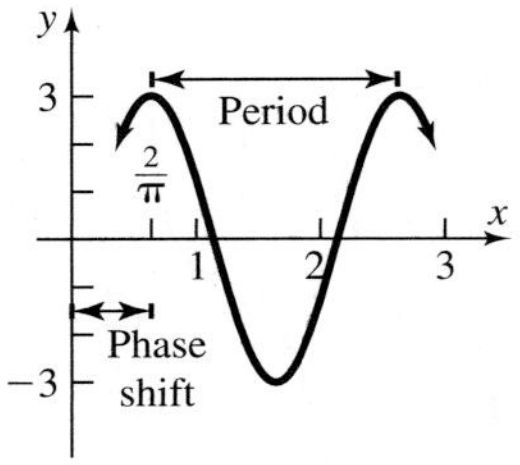

12. Amplitude = 2
Period = 1
Phase shift = $\dfrac{2}{\pi}$

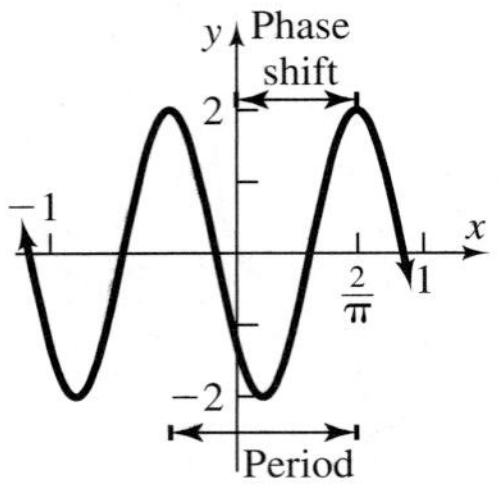

13. Amplitude = 3
Period = π
Phase shift = $\dfrac{\pi}{4}$

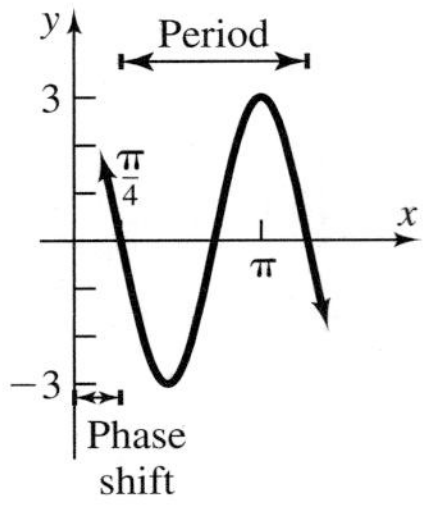

14. Amplitude = 3
Period = π
Phase shift = $\dfrac{\pi}{4}$

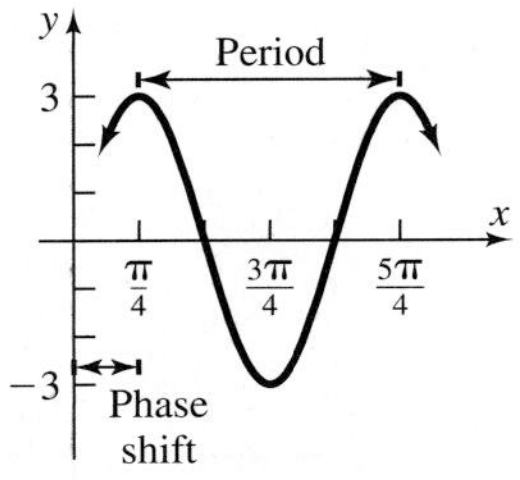

15. $y = 2\sin\left[2\left(x - \frac{1}{2}\right)\right]$ or $y = 2\sin(2x - 1)$ **16.** $y = 3\sin\left[4(x - 2)\right]$ or $y = 3\sin(4x - 8)$

17. $y = 3\sin\left[\frac{2}{3}\left(x + \frac{1}{3}\right)\right]$ or $y = 3\sin\left(\frac{2}{3}x + \frac{2}{9}\right)$ **18.** $y = 2\sin[2(x + 2)]$ or $y = 2\sin(2x + 4)$

19. Period = $\dfrac{1}{15}$; Amplitude = 120; Phase shift = $\dfrac{1}{90}$

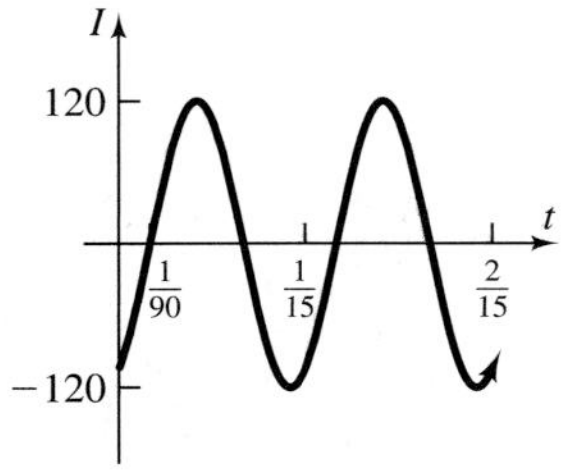

20. Period = $\dfrac{1}{30}$; Amplitude = 220; Phase shift = $\dfrac{1}{360}$

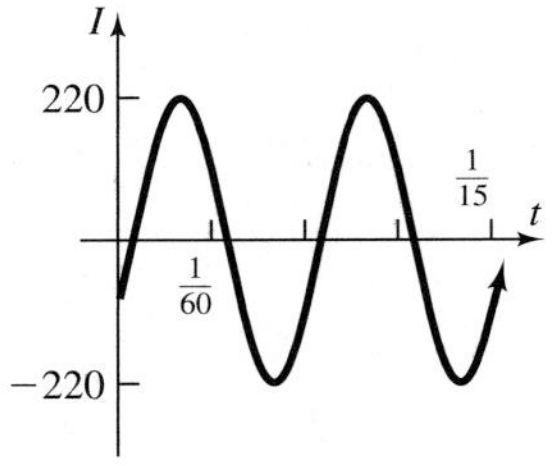

21. (a)

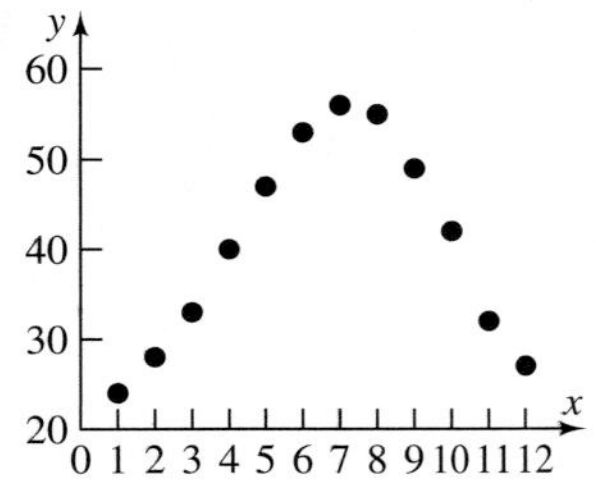

(b) $y = 15.9\sin\left(\frac{\pi}{6}x - \frac{2\pi}{3}\right) + 40.1$

(c)

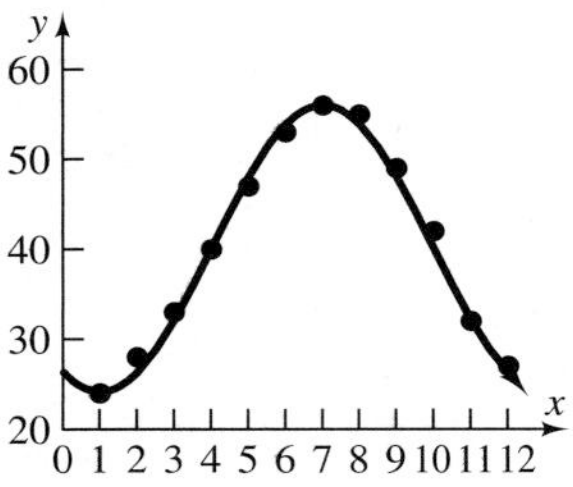

(d) $y = 15.62\sin(0.517x - 2.096) + 40.377$

(e)

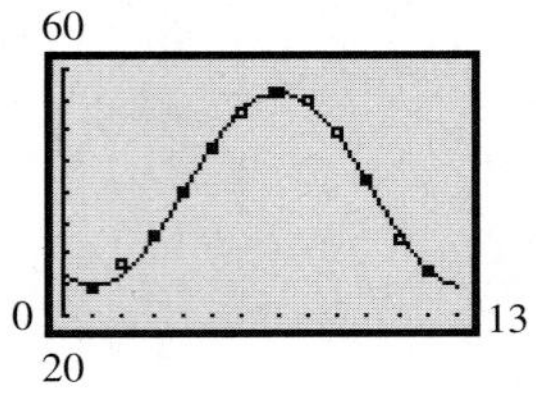

22. (a)

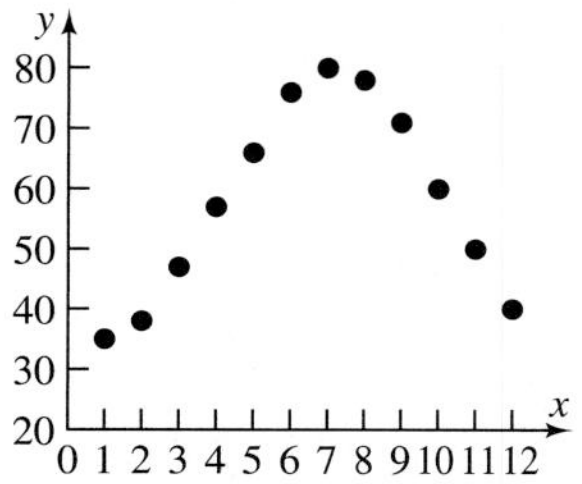

(b) $y = 22.7\sin\left(\frac{\pi}{6}x - \frac{2\pi}{3}\right) + 57.3$

(c)

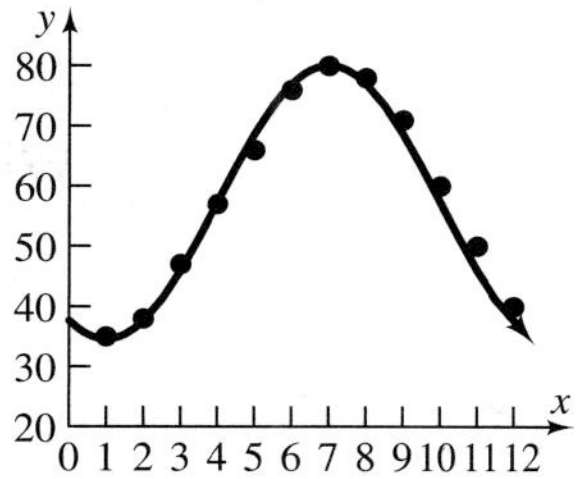

(d) $y = 22.61\sin(0.503x - 2.038) + 57.17$

(e)

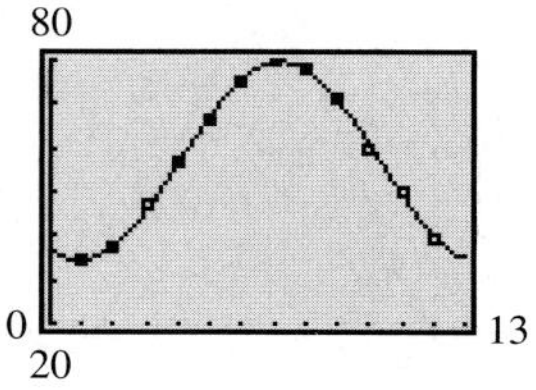

23. (a)

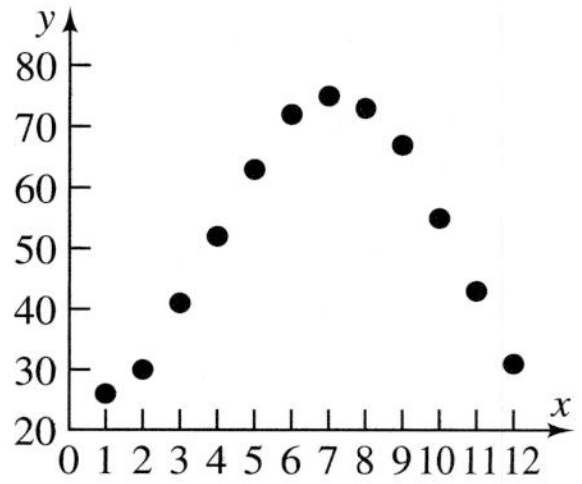

(b) $y = 24.95\sin\left(\frac{\pi}{6}x - \frac{2\pi}{3}\right) + 50.45$

(c)

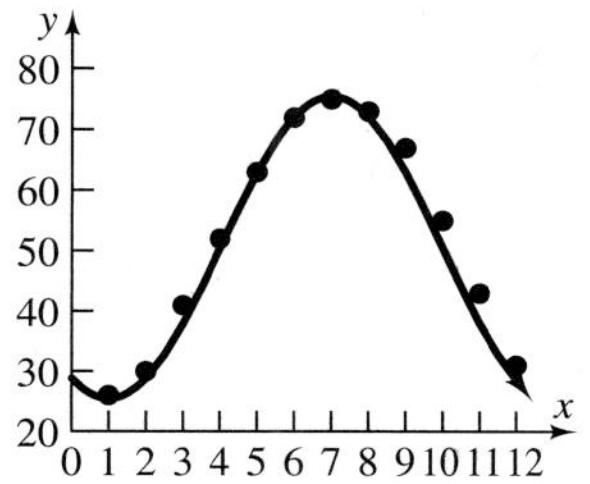

(d) $y = 25.693\sin(0.476x - 1.814) + 49.854$

(e)

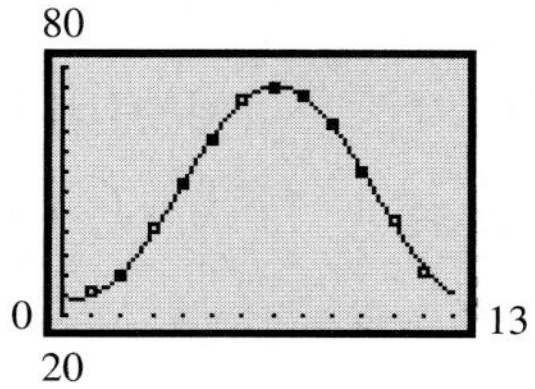

24. (a)

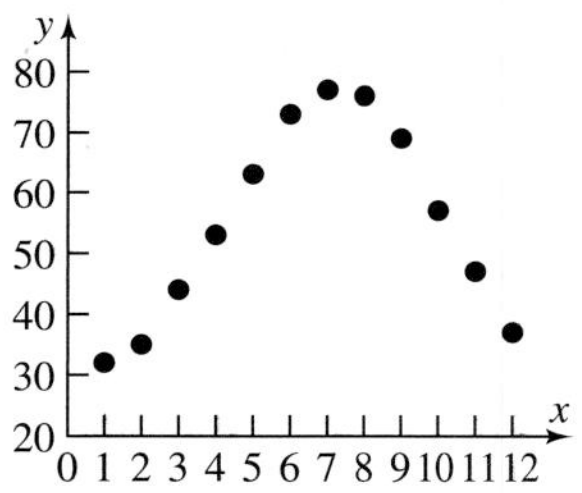

(b) $y = 22.6\sin\left(\frac{\pi}{6}x - \frac{2\pi}{3}\right) + 54.4$

(c)

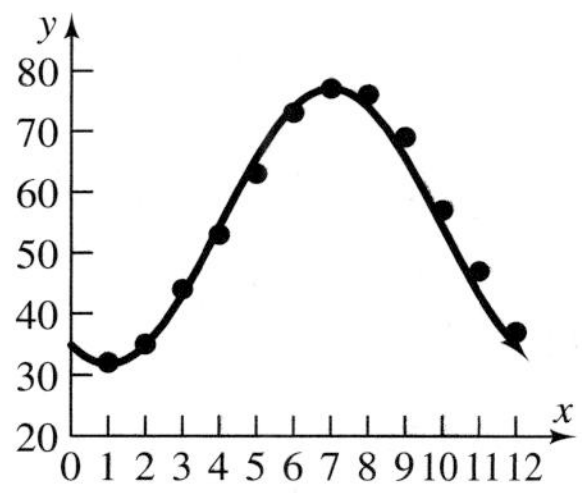

(d) $y = 22.46\sin(0.506x - 2.060) + 54.35$

(e)

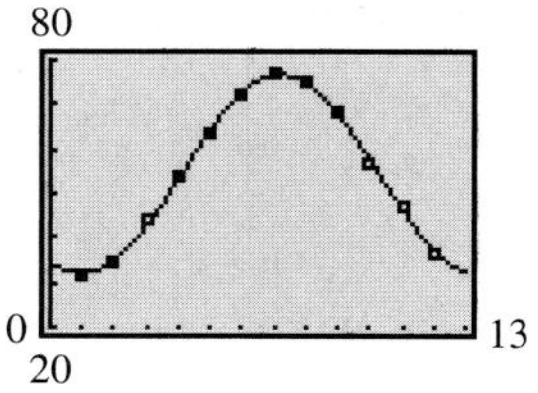

25. (a) 4:08 PM

(b) $y = 4.4\sin\left(\frac{4\pi}{25}x - 6.6643\right) + 3.8$

(c)

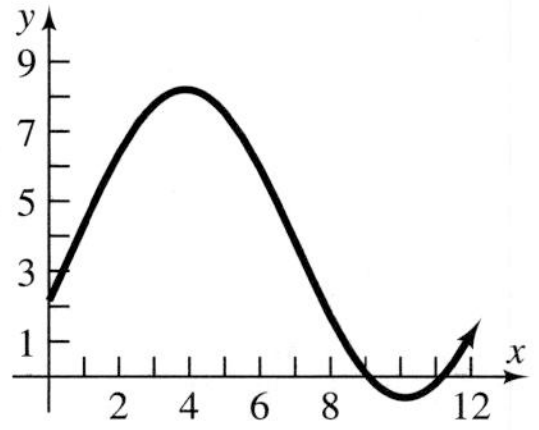

(d) 8.2 ft

26. (a) 8:41 PM

(b) $y = 5.5\sin\left(\frac{4\pi}{25}x - 8.7252\right) + 7.7$

(c)

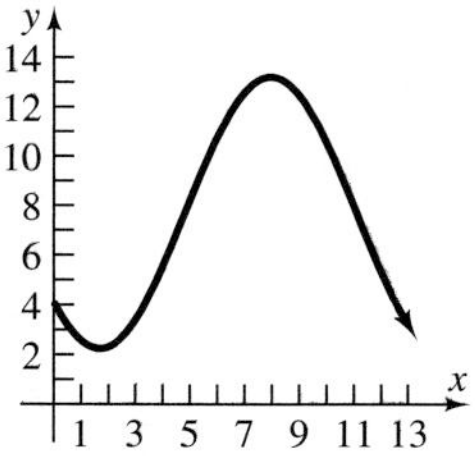

(d) 13.2 ft

27. (a) $y = 1.0835\sin\left(\frac{2\pi}{365}x - 2.45\pi\right) + 11.6665$

(b) 11.83 hr **(c)**

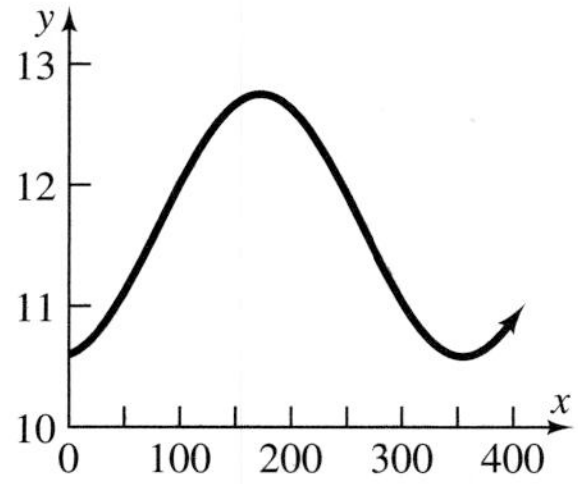

28. (a) $y = 2.2915\sin\left(\frac{2\pi}{365}x - 2.45\pi\right) + 11.3585$

(b) 11.71 hr **(c)**

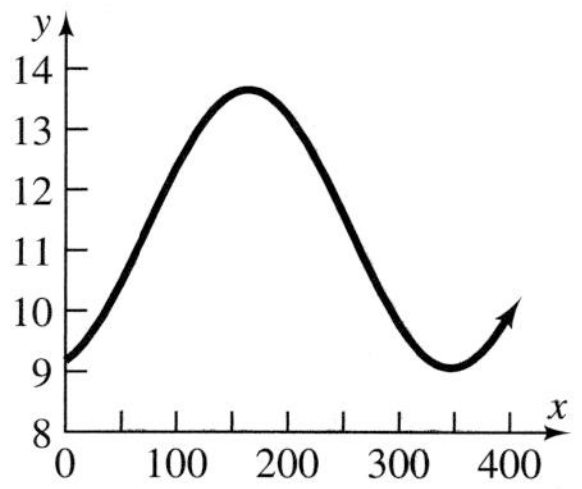

29. **(a)** $y = 5.3915 \sin\left(\frac{2\pi}{365}x - 2.45\pi\right) + 10.8415$
(b) 11.66 hr **(c)**

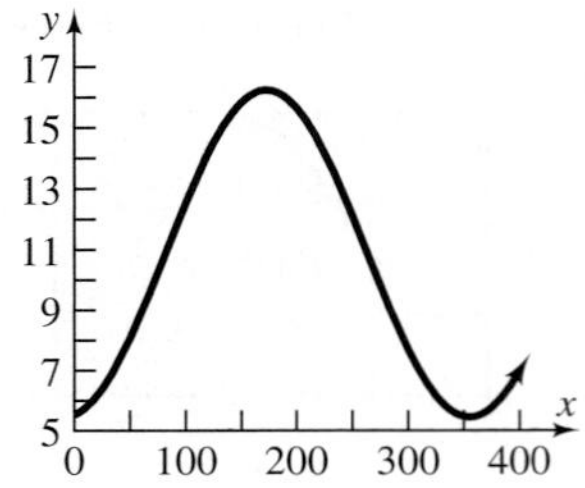

30. **(a)** $y = 0.992 \sin\left(\frac{2\pi}{365}x - 2.45\pi\right) + 11.775$
(b) 11.93 hr **(c)**

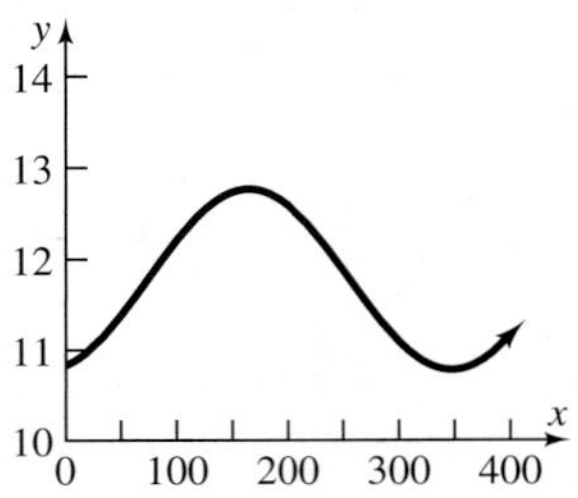

Review Exercises *(page 408)*

1. $\frac{3\pi}{4}$ **2.** $\frac{7\pi}{6}$ **3.** $\frac{\pi}{10}$ **4.** $\frac{\pi}{12}$ **5.** $135°$ **6.** $120°$ **7.** $-450°$ **8.** $-270°$ **9.** $\frac{1}{2}$ **10.** $\frac{1}{2} + \frac{\sqrt{2}}{2}$ **11.** $\frac{3\sqrt{2}}{2} - \frac{4\sqrt{3}}{3}$ **12.** $2 + 3\sqrt{3}$ **13.** $-3\sqrt{2} - 2\sqrt{3}$ **14.** $\frac{3\sqrt{3}}{2}$ **15.** 3 **16.** $1 + 4\sqrt{2}$ **17.** 0 **18.** 1 **19.** 0 **20.** -2 **21.** 1 **22.** 1 **23.** 1 **24.** 1 **25.** 1 **26.** 1 **27.** -1 **28.** -1 **29.** 1 **30.** -1 **31.** $\cos\theta = \frac{3}{5}$; $\tan\theta = \frac{4}{3}$; $\csc\theta = \frac{5}{4}$; $\sec\theta = \frac{5}{3}$; $\cot\theta = \frac{3}{4}$
32. $\sin\theta = \frac{4}{5}$; $\tan\theta = \frac{4}{3}$; $\csc\theta = \frac{5}{4}$; $\sec\theta = \frac{5}{3}$; $\cot\theta = \frac{3}{4}$ **33.** $\sin\theta = -\frac{12}{13}$; $\cos\theta = -\frac{5}{13}$; $\csc\theta = -\frac{13}{12}$; $\sec\theta = -\frac{13}{5}$; $\cot\theta = \frac{5}{12}$
34. $\sin\theta = -\frac{5}{13}$; $\cos\theta = -\frac{12}{13}$; $\tan\theta = \frac{5}{12}$; $\csc\theta = -\frac{13}{5}$; $\sec\theta = -\frac{13}{12}$ **35.** $\sin\theta = \frac{3}{5}$; $\cos\theta = -\frac{4}{5}$; $\tan\theta = -\frac{3}{4}$; $\csc\theta = \frac{5}{3}$; $\cot\theta = -\frac{4}{3}$
36. $\sin\theta = -\frac{3}{5}$; $\cos\theta = \frac{4}{5}$; $\tan\theta = -\frac{3}{4}$; $\sec\theta = \frac{5}{4}$; $\cot\theta = -\frac{4}{3}$ **37.** $\cos\theta = -\frac{5}{13}$; $\tan\theta = -\frac{12}{5}$; $\csc\theta = \frac{13}{12}$; $\sec\theta = -\frac{13}{5}$; $\cot\theta = -\frac{5}{12}$
38. $\sin\theta = -\frac{4}{5}$; $\tan\theta = \frac{4}{3}$; $\csc\theta = -\frac{5}{4}$; $\sec\theta = -\frac{5}{3}$; $\cot\theta = \frac{3}{4}$ **39.** $\cos\theta = \frac{12}{13}$; $\tan\theta = -\frac{5}{12}$; $\csc\theta = -\frac{13}{5}$; $\sec\theta = \frac{13}{12}$; $\cot\theta = -\frac{12}{5}$
40. $\sin\theta = -\frac{5}{13}$; $\tan\theta = -\frac{5}{12}$; $\csc\theta = -\frac{13}{5}$; $\sec\theta = \frac{13}{12}$; $\cot\theta = -\frac{12}{5}$ **41.** $\sin\theta = -\frac{\sqrt{10}}{10}$; $\cos\theta = -\frac{3\sqrt{10}}{10}$; $\csc\theta = -\sqrt{10}$; $\sec\theta = -\frac{\sqrt{10}}{3}$; $\cot\theta = 3$ **42.** $\sin\theta = \frac{2\sqrt{13}}{13}$; $\cos\theta = -\frac{3\sqrt{13}}{13}$; $\csc\theta = \frac{\sqrt{13}}{2}$; $\sec\theta = -\frac{\sqrt{13}}{3}$; $\cot\theta = -\frac{3}{2}$
43. $\sin\theta = -\frac{2\sqrt{2}}{3}$; $\cos\theta = \frac{1}{3}$; $\tan\theta = -2\sqrt{2}$; $\csc\theta = -\frac{3\sqrt{2}}{4}$; $\cot\theta = -\frac{\sqrt{2}}{4}$
44. $\sin\theta = -\frac{1}{4}$; $\cos\theta = -\frac{\sqrt{15}}{4}$; $\tan\theta = \frac{\sqrt{15}}{15}$; $\sec\theta = -\frac{4\sqrt{15}}{15}$; $\cot\theta = \sqrt{15}$
45. $\sin\theta = \frac{\sqrt{5}}{5}$; $\cos\theta = -\frac{2\sqrt{5}}{5}$; $\tan\theta = -\frac{1}{2}$; $\csc\theta = \sqrt{5}$; $\sec\theta = -\frac{\sqrt{5}}{2}$
46. $\sin\theta = -\frac{2\sqrt{5}}{5}$; $\cos\theta = \frac{\sqrt{5}}{5}$; $\csc\theta = -\frac{\sqrt{5}}{2}$; $\sec\theta = \sqrt{5}$; $\cot\theta = -\frac{1}{2}$

47.

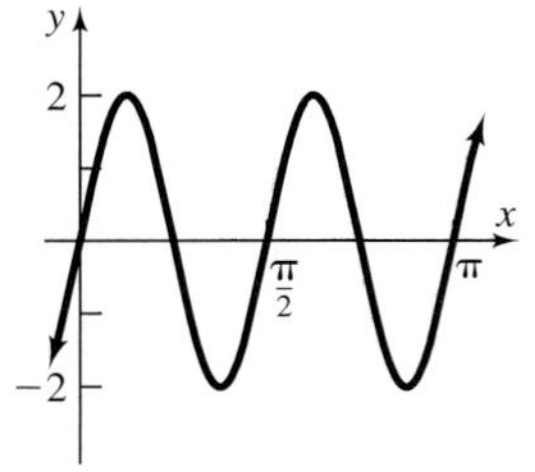

48.

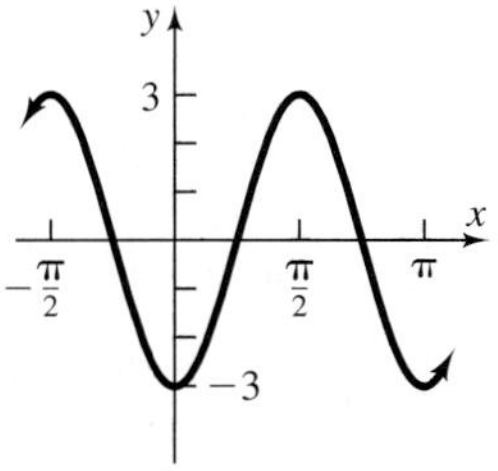

49.

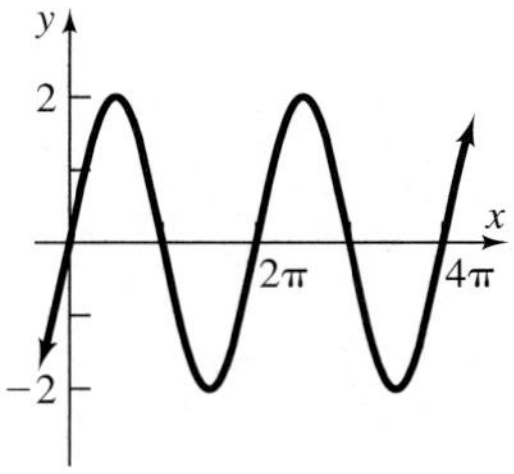

50.

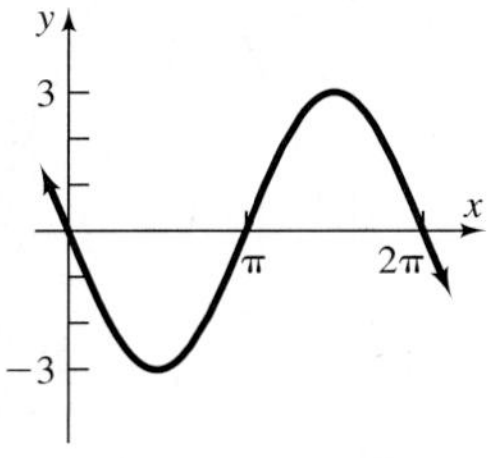

51.
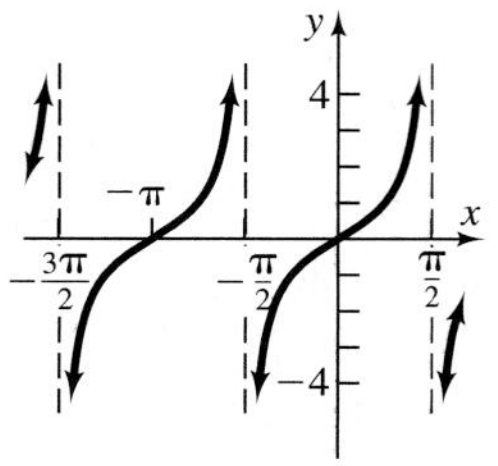

52.
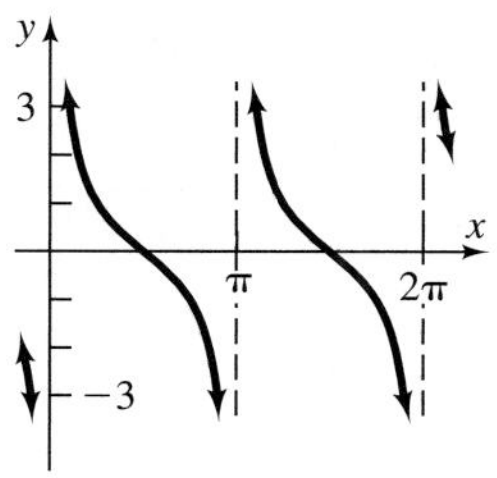

53.
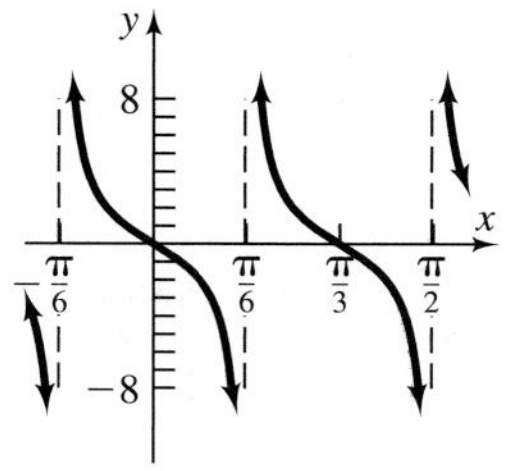

54.
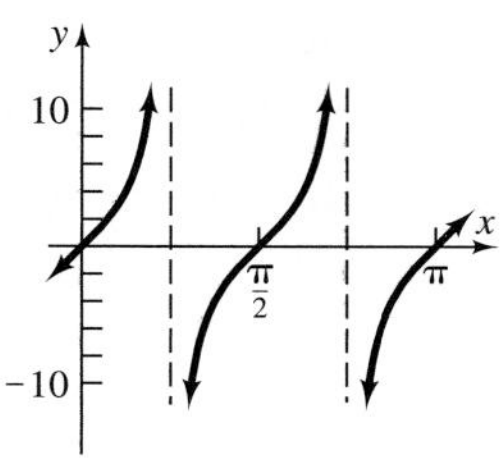

55.
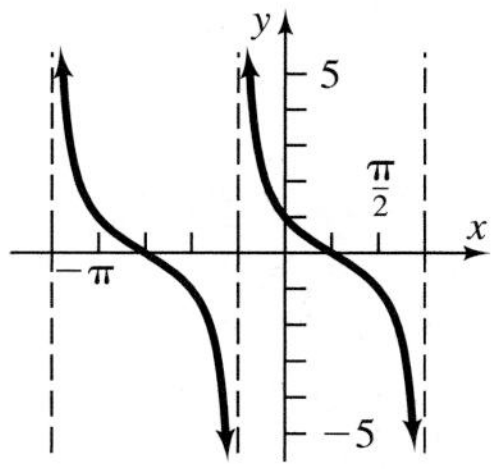

56.
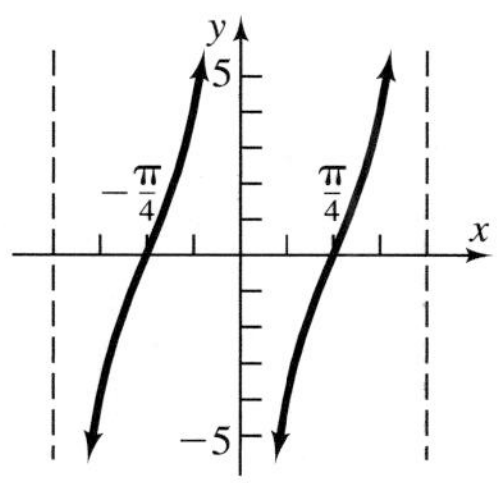

57.
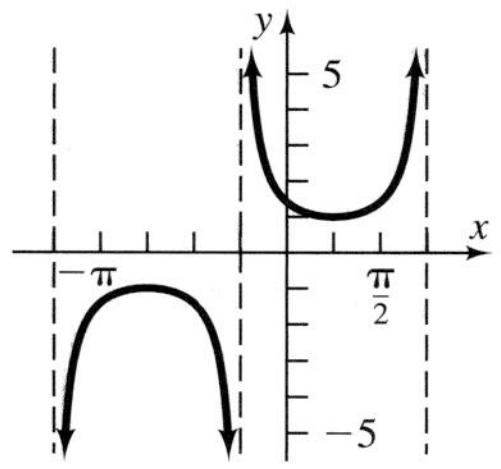

58.
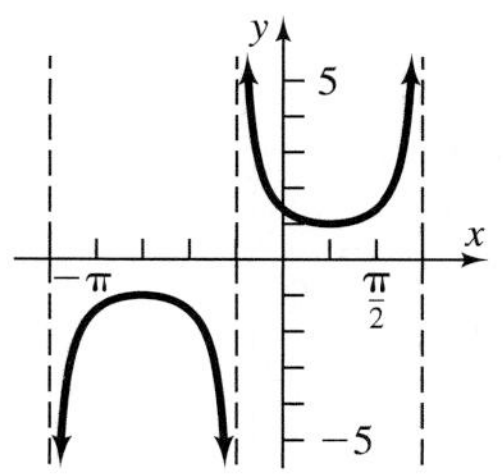

59. Amplitude = 4; period = 2π **60.** Amplitude = 1; period = π **61.** Amplitude = 8; period = 4 **62.** Amplitude = 2; period = $\frac{2}{3}$

63. Amplitude = 4
Period = $\frac{2\pi}{3}$
Phase shift = 0

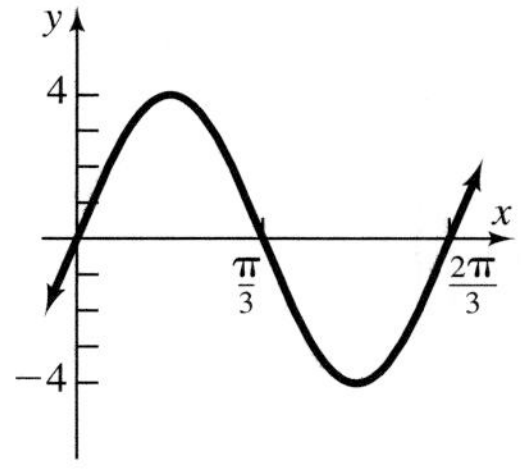

64. Amplitude = 2
Period = 6π
Phase shift = 0

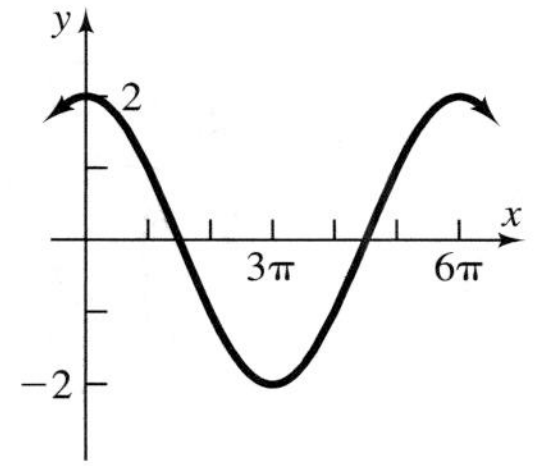

65. Amplitude = 2
Period = π
Phase Shift = $\frac{\pi}{2}$

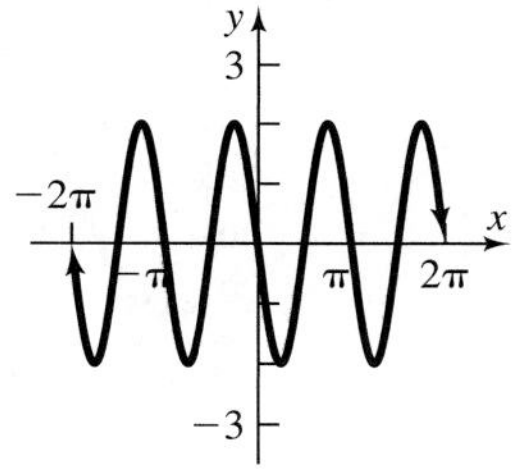

66. Amplitude = 1
Period = 4π
Phase Shift = $-\pi$

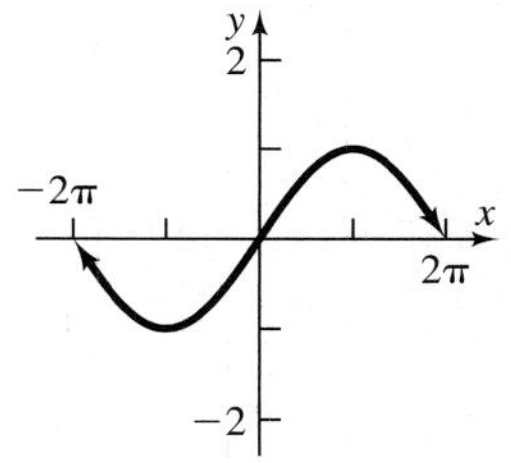

67. Amplitude = $\frac{1}{2}$
Period = $\frac{4\pi}{3}$
Phase shift = $\frac{2\pi}{3}$

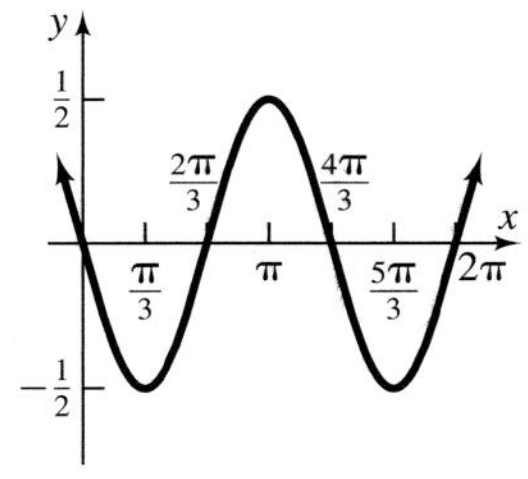

68. Amplitude = $\frac{3}{2}$
Period = $\frac{\pi}{3}$
Phase shift = $-\frac{\pi}{2}$

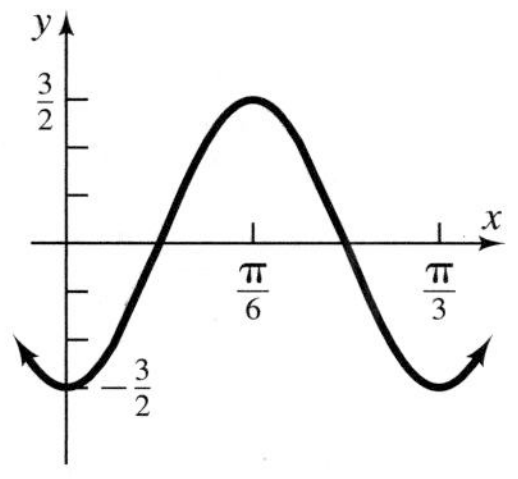

69. Amplitude = $\frac{2}{3}$
Period = 2
Phase shift = $\frac{6}{\pi}$

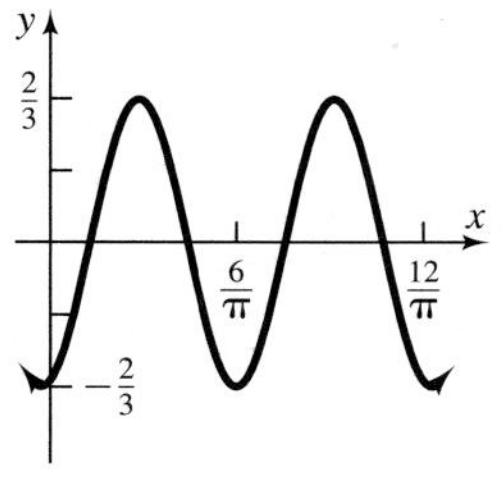

70. Amplitude = 7
Period = 6
Phase shift = $-\frac{4}{\pi}$

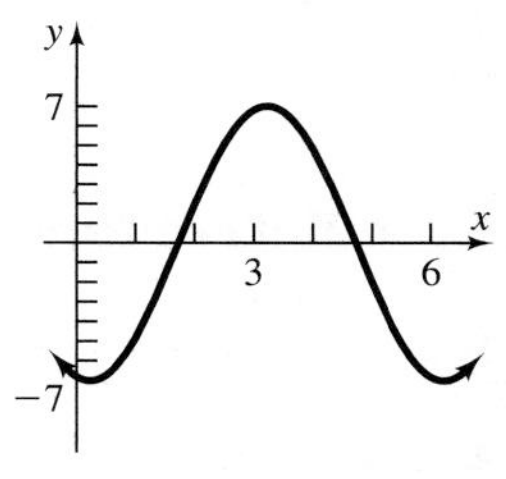

71. $y = 5\cos\frac{x}{4}$ **72.** $y = 4\sin\frac{x}{4}$ **73.** $y = -6\cos\left(\frac{\pi}{4}x\right)$ **74.** $y = -7\sin\left(\frac{\pi}{4}x\right)$ **75.** 0.38 **76.** 1.02

77. Sine, Cosine, Cosecant and Secant: Negative; Tangent and Cotangent: Positive **78.** Quadrant IV

79. $\sin\theta = \frac{2\sqrt{2}}{3}$; $\cos\theta = -\frac{1}{3}$; $\tan\theta = -2\sqrt{2}$; $\csc\theta = \frac{3\sqrt{2}}{4}$; $\sec\theta = -3$; $\cot\theta = -\frac{\sqrt{2}}{4}$

80. $\sin t = \frac{4}{5}$, $\cos t = -\frac{3}{5}$, $\tan t = -\frac{4}{3}$ **81.** Domain: $\left\{x \,\middle|\, x \neq \text{odd multiple of } \frac{\pi}{2}\right\}$; range: $\{y \mid |y| \geq 1\}$; period $= 2(\pi)$

82. **(a)** 32.34° **(b)** 63°10′48″ **83.** $\frac{\pi}{3} \approx 1.05$ ft; $\frac{\pi}{3} \approx 1.05\ \text{ft}^2$ **84.** $8\pi \approx 25.13$ in.; $\frac{16\pi}{3} \approx 16.76$ in.

85. Approximately 114.59 revolutions/hr **86.** Approximately 5.24 ft/s **87.** 0.1 revolution/s $= \dfrac{\pi}{5}$ radian/s

88. Approximately 945.38 rpm; yes; approximately 1080.43 rpm

89. **(a)** 120 **(b)** $\dfrac{1}{60}$ **(c)**

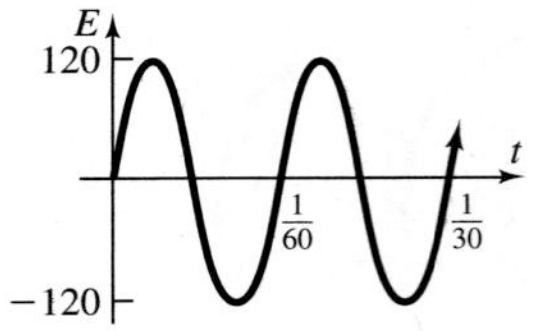

90. **(a)** $\dfrac{1}{15}$ **(b)** 220 **(c)** $-\dfrac{1}{180}$ **(d)**

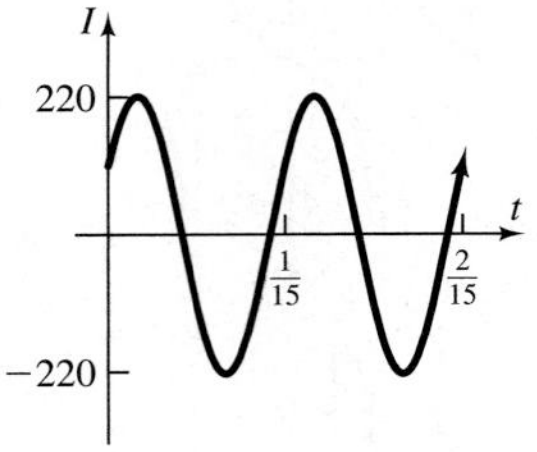

91. **(a)**

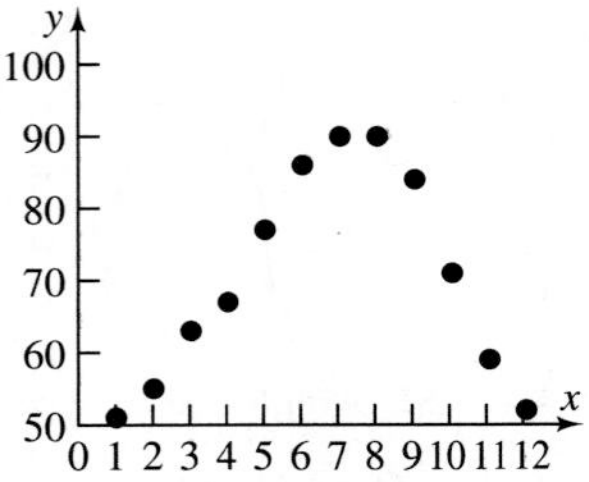

(b) $y = 19.5 \sin\left(\dfrac{\pi}{6}x - \dfrac{2\pi}{3}\right) + 70.5$

(c)

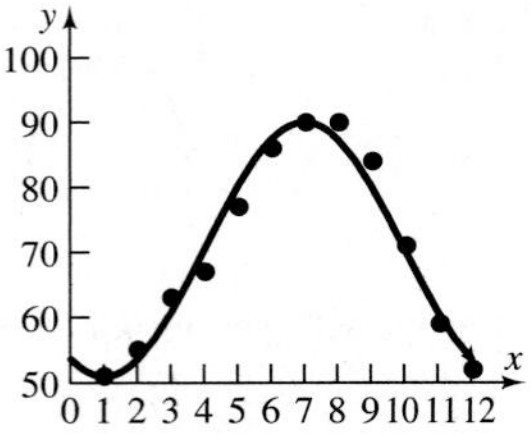

(d) $y = 19.518 \sin(0.541x - 2.283) + 71.014$

(e)

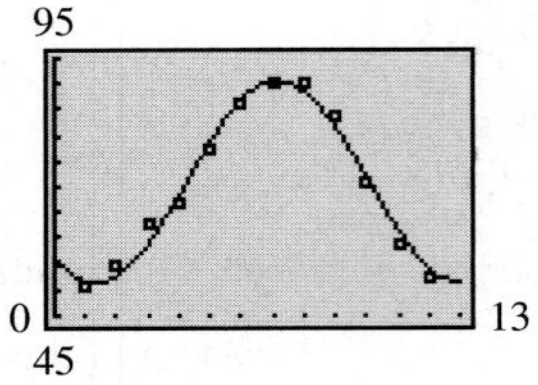

92. **(a)**

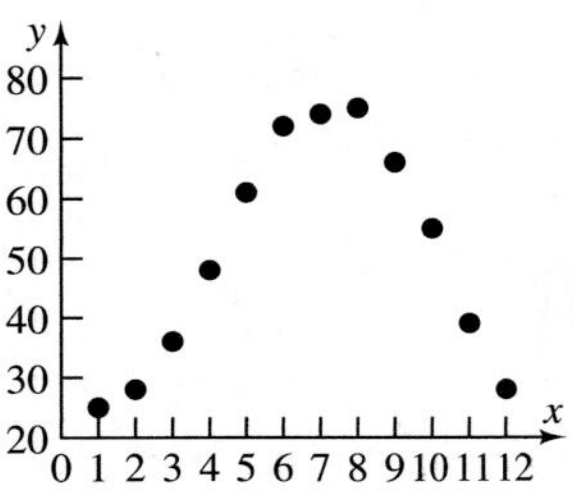

(b) $y = 25 \sin\left(\dfrac{\pi}{6}x - \dfrac{2\pi}{3}\right) + 50$

(c)

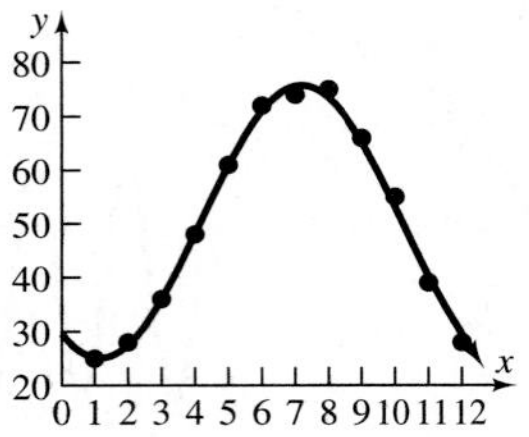

(d) $y = 25.815 \sin(0.521x - 2.175) + 50.46$

(e)

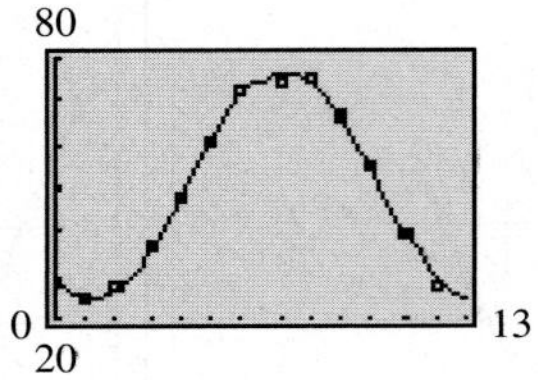

93. **(a)** $y = 1.85 \sin\left(\dfrac{2\pi}{365}x - \dfrac{357}{146}\pi\right) + 11.517$

(b)

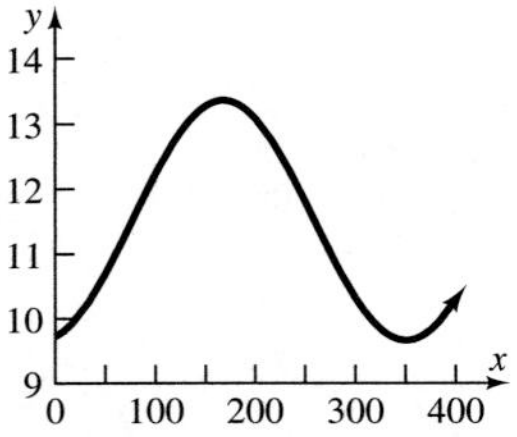

(c) 11.83 hr

94. **(a)** $y = 2.775 \sin\left(\dfrac{2\pi}{365}x - \dfrac{357}{146}\pi\right) + 11.192$

(b)

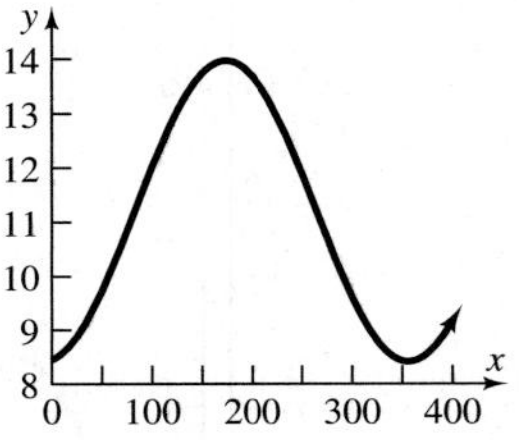

(c) 11.66 hr

Cumulative Review *(page 412)*

1. $\left\{-1, \dfrac{1}{2}\right\}$ **2.** $y - 5 = -3(x + 2)$ or $y = -3x - 1$ **3.** $x^2 + (y + 2)^2 = 16$

4. A line. Slope $\dfrac{2}{3}$; intercepts $(6, 0)$ and $(0, -4)$

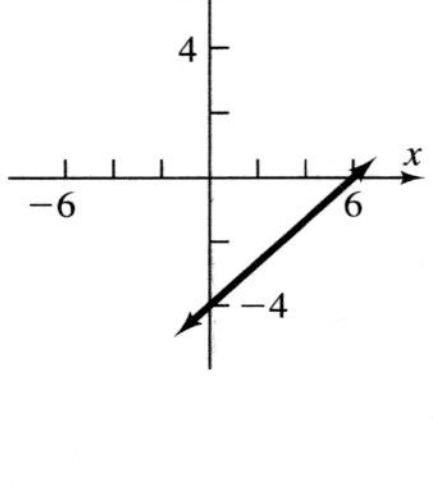

5. A circle. Center $(1, -2)$; Radius 3

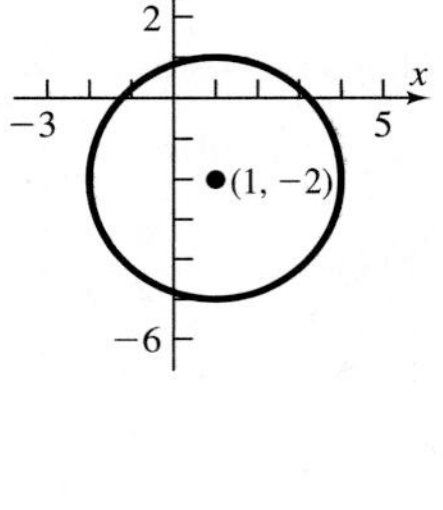

6.

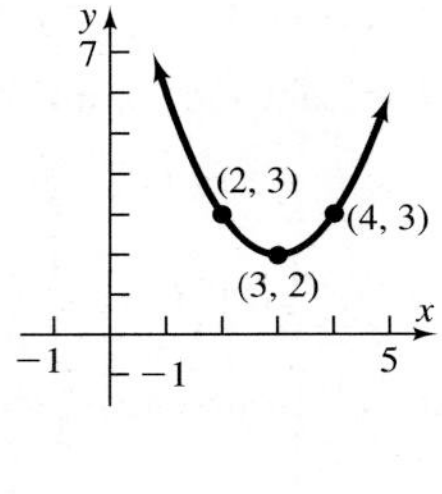

7. (a)

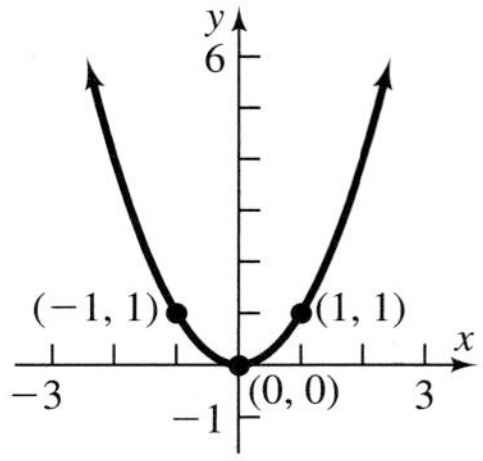

(b)

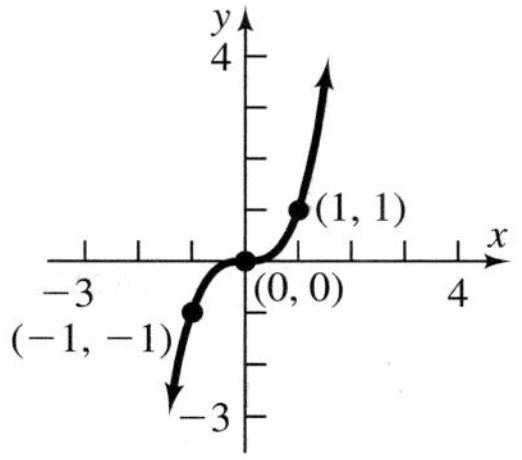

(c)

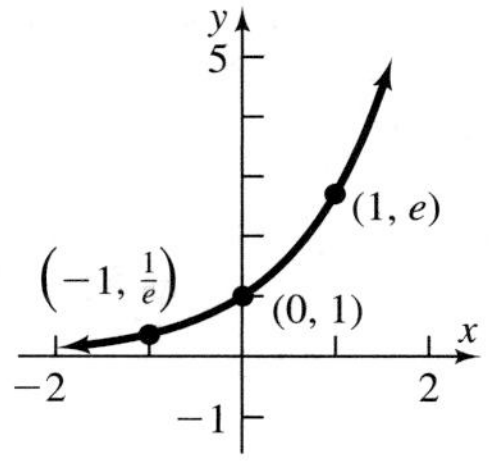

(d)

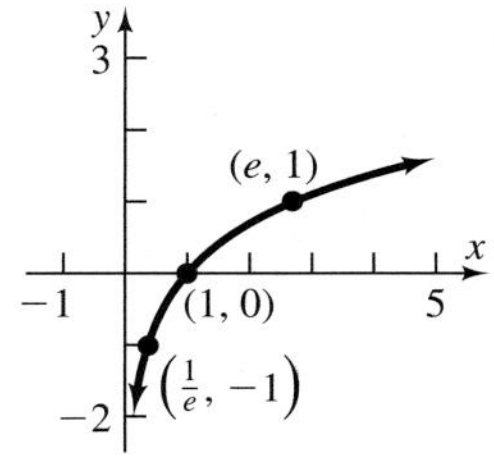

(e)

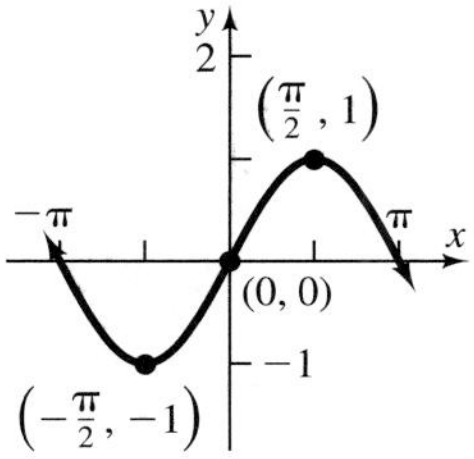

(f)

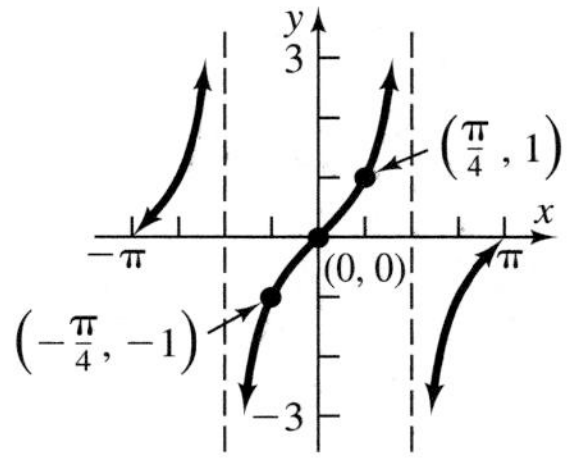

8. $f^{-1}(x) = \frac{1}{3}(x + 2)$ **9.** -2 **10.**

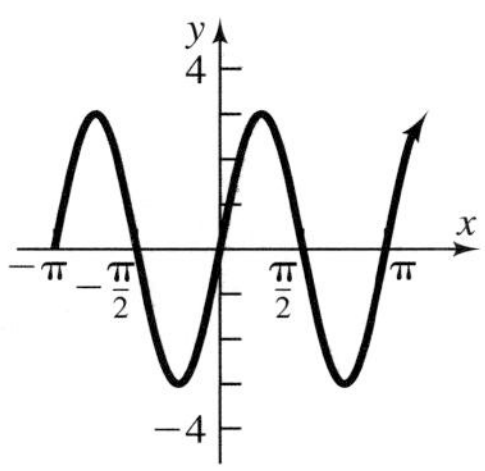

11. $3 - \frac{3\sqrt{3}}{2}$ **12.** $y = 2(3^x)$ **13.** $y = 3\cos\left(\frac{\pi}{6}x\right)$

14. (a) $f(x) = -3x - 3; m = -3; (-1, 0), (0, -3)$ **(b)** $f(x) = (x - 1)^2 - 6; (0, -5), (-\sqrt{6} + 1, 0); (\sqrt{6} + 1, 0)$

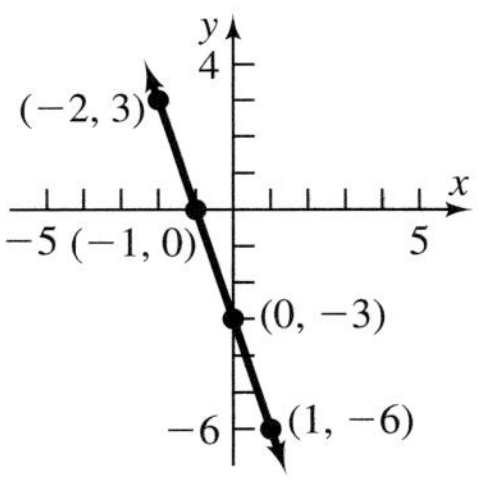

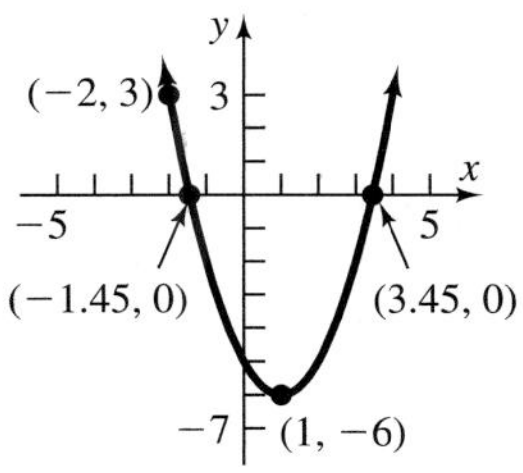

(c) We have that $y = 3$ when $x = -2$ and $y = -6$ when $x = 1$. Both points satisfy $y = ae^x$. Therefore, for $(-2, 3)$ we have $3 = ae^{-2}$ which implies that $a = 3e^2$. But for $(1, -6)$ we have $-6 = ae^1$, which implies that $a = -6e^{-1}$. Therefore, there is no exponential function $y = ae^x$ that contains $(-2, 3)$ and $(1, -6)$.

15. (a) $f(x) = \frac{1}{6}(x + 2)(x - 3)(x - 5)$ **(b)** $R(x) = -\frac{(x + 2)(x - 3)(x - 5)}{3(x - 2)}$

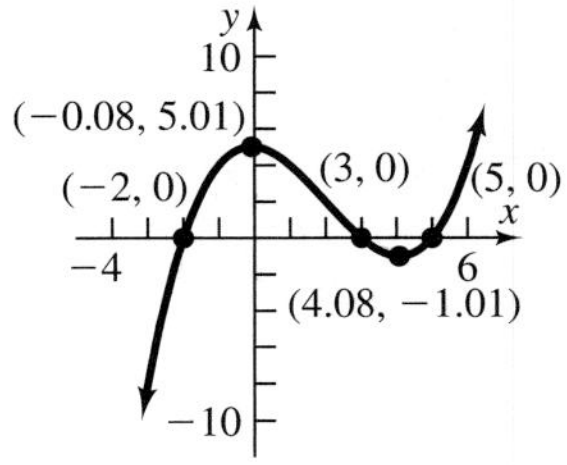

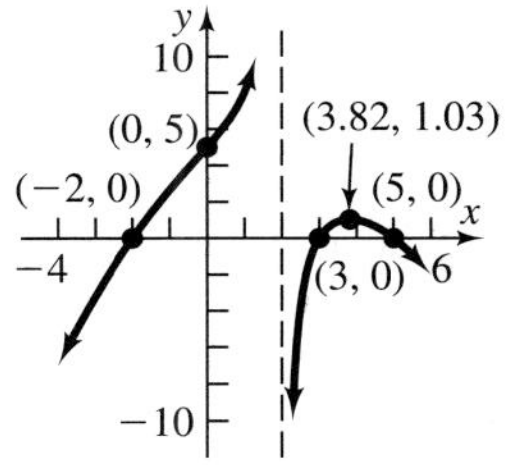

CHAPTER 6 Analytic Trigonometry

6.1 Concepts and Vocabulary *(page 423)*

7. $x = \sin y$ **8.** 0 **9.** $\frac{\pi}{5}$ **10.** False **11.** True **12.** True

6.1 Exercises *(page 423)*

13. 0 **14.** 0 **15.** $-\frac{\pi}{2}$ **16.** π **17.** 0 **18.** $-\frac{\pi}{4}$ **19.** $\frac{\pi}{4}$ **20.** $\frac{\pi}{6}$ **21.** $\frac{\pi}{3}$ **22.** $-\frac{\pi}{3}$ **23.** $\frac{5\pi}{6}$ **24.** $-\frac{\pi}{4}$ **25.** 0.10 **26.** 0.93 **27.** 1.37 **28.** 0.20 **29.** 0.51 **30.** 0.13 **31.** −0.38 **32.** −1.25 **33.** −0.12 **34.** 2.03 **35.** 1.08 **36.** 0.35 **37.** 0.54 **38.** 7.4 **39.** $\frac{4\pi}{5}$ **40.** $-\frac{\pi}{10}$ **41.** −3.5 **42.** −0.05 **43.** $-\frac{3\pi}{7}$ **44.** $\frac{2\pi}{5}$ **45.** Yes; $-\frac{\pi}{6}$ lies in the interval $\left[-\frac{\pi}{2}, \frac{\pi}{2}\right]$. **46.** No; $\frac{2\pi}{3}$ does not lie in the interval $\left[-\frac{\pi}{2}, \frac{\pi}{2}\right]$. **47.** No; 2 is not in the domain of $\sin^{-1} x$. **48.** Yes; $-\frac{1}{2}$ is in the domain of $\sin^{-1} x$. **49.** No; $-\frac{\pi}{6}$ does not lie in the interval $[0, \pi]$. **50.** Yes; $\frac{2\pi}{3}$ lies in the interval $[0, \pi]$. **51.** Yes, $-\frac{1}{2}$ is in the domain of $\cos^{-1} x$. **52.** No; 2 is not in the domain of $\cos^{-1} x$. **53.** Yes; $-\frac{\pi}{3}$ lies in the interval $\left(-\frac{\pi}{2}, \frac{\pi}{2}\right)$. **54.** No; $\frac{2\pi}{3}$ does not lie in the interval $\left(-\frac{\pi}{2}, \frac{\pi}{2}\right)$. **55.** Yes; 2 is in the domain of $\tan^{-1} x$. **56.** Yes; $-\frac{1}{2}$ is in the domain of $\tan^{-1} x$. **57.** **(a)** 13.92 hr or 13 hr, 55 min **(b)** 12 hr **(c)** 13.85 hr or 13 hr, 51 min **58.** **(a)** 14.93 hr or 14 hr, 56 min **(b)** 12 hr **(c)** 14.83 hr or 14 hr, 50 min **59.** **(a)** 13.3 hr or 13 hr, 18 min **(b)** 12 hr **(c)** 13.26 hr or 13 hr, 15 min **60.** **(a)** 18.96 hr or 18 hr, 57 min **(b)** 12 hr **(c)** 18.64 hr or 18 hr, 38 min **61.** **(a)** 12 hr **(b)** 12 hr **(c)** 12 hr **(d)** It's 12 hr. **62.** **(a)** 24 hr **(b)** 12 hr **(c)** 22.02 hr or 22 hr, 1 min **(d)** 0 hr; ranges from around-the-clock daylight to no daylight **63.** 3.35 min

6.2 Concepts and Vocabulary *(page 429)*

4. $x = \sec y$; ≥ 1; 0; π **5.** $\frac{\sqrt{2}}{2}$ **6.** False **7.** True **8.** True

6.2 Exercises *(page 429)*

9. $\frac{\sqrt{2}}{2}$ **10.** $\frac{\sqrt{3}}{2}$ **11.** $-\frac{\sqrt{3}}{3}$ **12.** $-\frac{\sqrt{3}}{3}$ **13.** 2 **14.** $-\sqrt{3}$ **15.** $\sqrt{2}$ **16.** 2 **17.** $-\frac{\sqrt{2}}{2}$ **18.** $\frac{1}{2}$ **19.** $\frac{2\sqrt{3}}{3}$ **20.** 2 **21.** $\frac{3\pi}{4}$ **22.** $-\frac{\pi}{3}$ **23.** $\frac{\pi}{6}$ **24.** $\frac{\pi}{3}$ **25.** $\frac{\sqrt{2}}{4}$ **26.** $2\sqrt{2}$ **27.** $\frac{\sqrt{5}}{2}$ **28.** $\frac{\sqrt{7}}{3}$ **29.** $-\frac{\sqrt{14}}{2}$ **30.** $-\frac{\sqrt{5}}{2}$ **31.** $-\frac{3\sqrt{10}}{10}$ **32.** $-\frac{\sqrt{2}}{2}$ **33.** $\sqrt{5}$ **34.** $\sqrt{5}$ **35.** $-\frac{\pi}{4}$ **36.** $\frac{2\pi}{3}$ **37.** $\frac{\pi}{6}$ **38.** $\frac{\pi}{4}$ **39.** $-\frac{\pi}{2}$ **40.** $\frac{\pi}{4}$ **41.** $\frac{\pi}{6}$ **42.** $\frac{2\pi}{3}$ **43.** $\frac{2\pi}{3}$ **44.** $-\frac{\pi}{3}$ **45.** 1.32 **46.** 0.20 **47.** 0.46 **48.** 1.91 **49.** −0.34 **50.** 2.03 **51.** 2.72 **52.** 3.02 **53.** −0.73 **54.** 2.42 **55.** 2.55 **56.** 2.84

57.

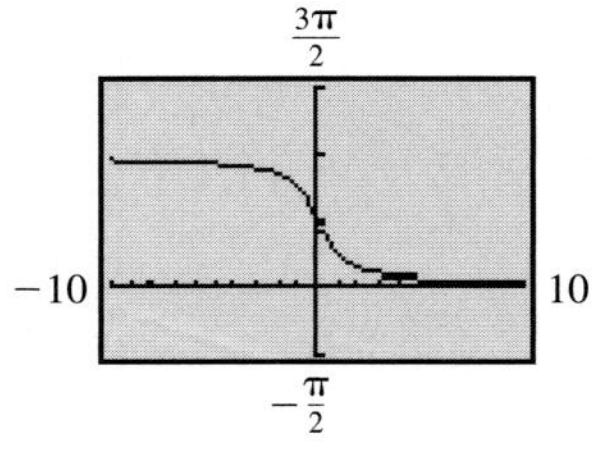

58.

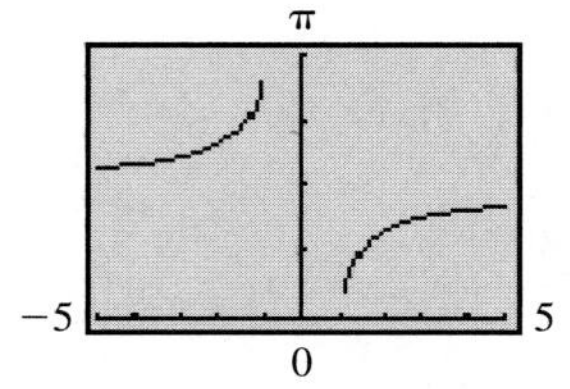

59.

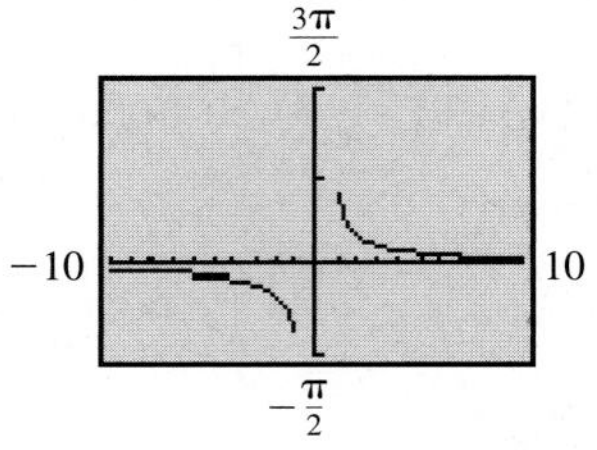

6.3 Concepts and Vocabulary *(page 435)*

3. identity; conditional **4.** −1 **5.** 0 **6.** True **7.** True **8.** True

6.3 Exercises *(page 435)*

9. $\frac{1}{\cos\theta}$ **10.** $\frac{1}{\sin\theta}$ **11.** $\frac{1 + \sin\theta}{\cos\theta}$ **12.** $\frac{1 - \cos\theta}{\sin\theta}$ **13.** $\frac{1}{\sin\theta\cos\theta}$ **14.** $\frac{2}{\sin^2\theta}$ **15.** 2 **16.** 2 **17.** $\frac{3\sin\theta + 1}{\sin\theta + 1}$ **18.** $\frac{\cos\theta + 1}{\cos\theta}$ **19.** $\csc\theta \cdot \cos\theta = \frac{1}{\sin\theta} \cdot \cos\theta = \frac{\cos\theta}{\sin\theta} = \cot\theta$ **20.** $\sec\theta \cdot \sin\theta = \frac{1}{\cos\theta} \cdot \sin\theta = \frac{\sin\theta}{\cos\theta} = \tan\theta$

21. $1 + \tan^2(-\theta) = 1 + (-\tan\theta)^2 = 1 + \tan^2\theta = \sec^2\theta$ **22.** $1 + \cot^2(-\theta) = 1 + (-\cot\theta)^2 = 1 + \cot^2\theta = \csc^2\theta$

23. $\cos\theta(\tan\theta + \cot\theta) = \cos\theta\left(\frac{\sin\theta}{\cos\theta} + \frac{\cos\theta}{\sin\theta}\right) = \cos\theta\left(\frac{\sin^2\theta + \cos^2\theta}{\cos\theta\sin\theta}\right) = \cos\theta\left(\frac{1}{\cos\theta\sin\theta}\right) = \frac{1}{\sin\theta} = \csc\theta$

24. $\sin\theta(\cot\theta + \tan\theta) = \sin\theta\left(\frac{\cos\theta}{\sin\theta} + \frac{\sin\theta}{\cos\theta}\right) = \sin\theta\left(\frac{\cos^2\theta + \sin^2\theta}{\sin\theta\cos\theta}\right) = \sin\theta\left(\frac{1}{\sin\theta\cos\theta}\right) = \frac{1}{\cos\theta} = \sec\theta$

25. $\tan\theta\cot\theta - \cos^2\theta = \frac{\sin\theta}{\cos\theta}\cdot\frac{\cos\theta}{\sin\theta} - \cos^2\theta = 1 - \cos^2\theta = \sin^2\theta$ **26.** $\sin\theta\csc\theta - \cos^2\theta = \sin\theta\cdot\frac{1}{\sin\theta} - \cos^2\theta = 1 - \cos^2\theta = \sin^2\theta$

27. $(\sec\theta - 1)(\sec\theta + 1) = \sec^2\theta - 1 = \tan^2\theta$ **28.** $(\csc\theta - 1)(\csc\theta + 1) = \csc^2\theta - 1 = \cot^2\theta$

29. $(\sec\theta + \tan\theta)(\sec\theta - \tan\theta) = \sec^2\theta - \tan^2\theta = 1$ **30.** $(\csc\theta + \cot\theta)(\csc\theta - \cot\theta) = \csc^2\theta - \cot^2\theta = 1$

31. $\cos^2\theta(1 + \tan^2\theta) = \cos^2\theta + \cos^2\theta\tan^2\theta = \cos^2\theta + \cos^2\theta\cdot\frac{\sin^2\theta}{\cos^2\theta} = \cos^2\theta + \sin^2\theta = 1$

32. $(1 - \cos^2\theta)(1 + \cot^2\theta) = \sin^2\theta\cdot\csc^2\theta = \sin^2\theta\cdot\frac{1}{\sin^2\theta} = 1$

33. $(\sin\theta + \cos\theta)^2 + (\sin\theta - \cos\theta)^2 = \sin^2\theta + 2\sin\theta\cos\theta + \cos^2\theta + \sin^2\theta - 2\sin\theta\cos\theta + \cos^2\theta = \sin^2\theta + \cos^2\theta + \sin^2\theta + \cos^2\theta$
$= 1 + 1 = 2$

34. $\tan^2\theta\cos^2\theta + \cot^2\theta\sin^2\theta = \frac{\sin^2\theta}{\cos^2\theta}\cdot\cos^2\theta + \frac{\cos^2\theta}{\sin^2\theta}\cdot\sin^2\theta = \sin^2\theta + \cos^2\theta = 1$

35. $\sec^4\theta - \sec^2\theta = \sec^2\theta(\sec^2\theta - 1) = (1 + \tan^2\theta)\tan^2\theta = \tan^4\theta + \tan^2\theta$

36. $\csc^4\theta - \csc^2\theta = \csc^2\theta(\csc^2\theta - 1) = (1 + \cot^2\theta)\cot^2\theta = \cot^4\theta + \cot^2\theta$

37. $\sec\theta - \tan\theta = \frac{1}{\cos\theta} - \frac{\sin\theta}{\cos\theta} = \frac{1 - \sin\theta}{\cos\theta}\cdot\frac{1 + \sin\theta}{1 + \sin\theta} = \frac{1 - \sin^2\theta}{\cos\theta(1 + \sin\theta)} = \frac{\cos^2\theta}{\cos\theta(1 + \sin\theta)} = \frac{\cos\theta}{1 + \sin\theta}$

38. $\csc\theta - \cot\theta = \frac{1}{\sin\theta} - \frac{\cos\theta}{\sin\theta} = \frac{1 - \cos\theta}{\sin\theta}\cdot\frac{1 + \cos\theta}{1 + \cos\theta} = \frac{1 - \cos^2\theta}{\sin\theta(1 + \cos\theta)} = \frac{\sin^2\theta}{\sin\theta(1 + \cos\theta)} = \frac{\sin\theta}{1 + \cos\theta}$

39. $3\sin^2\theta + 4\cos^2\theta = 3\sin^2\theta + 3\cos^2\theta + \cos^2\theta = 3(\sin^2\theta + \cos^2\theta) + \cos^2\theta = 3 + \cos^2\theta$

40. $9\sec^2\theta - 5\tan^2\theta = 4\sec^2\theta + 5\sec^2\theta - 5\tan^2\theta = 4\sec^2\theta + 5(\sec^2\theta - \tan^2\theta) = 5 + 4\sec^2\theta$

41. $1 - \frac{\cos^2\theta}{1 + \sin\theta} = 1 - \frac{1 - \sin^2\theta}{1 + \sin\theta} = 1 - \frac{(1 + \sin\theta)(1 - \sin\theta)}{1 + \sin\theta} = 1 - (1 - \sin\theta) = \sin\theta$

42. $1 - \frac{\sin^2\theta}{1 - \cos\theta} = 1 - \frac{1 - \cos^2\theta}{1 - \cos\theta} = 1 - \frac{(1 + \cos\theta)(1 - \cos\theta)}{1 - \cos\theta} = 1 - (1 + \cos\theta) = -\cos\theta$

43. $\frac{1 + \tan\theta}{1 - \tan\theta} = \frac{1 + \frac{1}{\cot\theta}}{1 - \frac{1}{\cot\theta}} = \frac{\frac{\cot\theta + 1}{\cot\theta}}{\frac{\cot\theta - 1}{\cot\theta}} = \frac{\cot\theta + 1}{\cot\theta - 1}$ **44.** $\frac{\csc\theta - 1}{\csc\theta + 1} = \frac{\frac{1}{\sin\theta} - 1}{\frac{1}{\sin\theta} + 1} = \frac{\frac{1 - \sin\theta}{\sin\theta}}{\frac{1 + \sin\theta}{\sin\theta}} = \frac{1 - \sin\theta}{1 + \sin\theta}$

45. $\frac{\sec\theta}{\csc\theta} + \frac{\sin\theta}{\cos\theta} = \frac{\frac{1}{\cos\theta}}{\frac{1}{\sin\theta}} + \tan\theta = \frac{\sin\theta}{\cos\theta} + \tan\theta = \tan\theta + \tan\theta = 2\tan\theta$

46. $\frac{\csc\theta - 1}{\cot\theta} = \frac{\csc\theta - 1}{\cot\theta}\cdot\frac{\csc\theta + 1}{\csc\theta + 1} = \frac{\csc^2\theta - 1}{\cot\theta(\csc\theta + 1)} = \frac{\cot^2\theta}{\cot\theta(\csc\theta + 1)} = \frac{\cot\theta}{\csc\theta + 1}$

47. $\frac{1 + \sin\theta}{1 - \sin\theta} = \frac{1 + \frac{1}{\csc\theta}}{1 - \frac{1}{\csc\theta}} = \frac{\frac{\csc\theta + 1}{\csc\theta}}{\frac{\csc\theta - 1}{\csc\theta}} = \frac{\csc\theta + 1}{\csc\theta - 1}$

48. $\frac{\cos\theta + 1}{\cos\theta - 1} = \frac{\frac{1}{\sec\theta} + 1}{\frac{1}{\sec\theta} - 1} = \frac{\frac{1 + \sec\theta}{\sec\theta}}{\frac{1 - \sec\theta}{\sec\theta}} = \frac{1 + \sec\theta}{1 - \sec\theta}$

49. $\frac{1 - \sin\theta}{\cos\theta} + \frac{\cos\theta}{1 - \sin\theta} = \frac{(1 - \sin\theta)^2 + \cos^2\theta}{\cos\theta(1 - \sin\theta)} = \frac{1 - 2\sin\theta + \sin^2\theta + \cos^2\theta}{\cos\theta(1 - \sin\theta)} = \frac{2 - 2\sin\theta}{\cos\theta(1 - \sin\theta)} = \frac{2(1 - \sin\theta)}{\cos\theta(1 - \sin\theta)}$
$= \frac{2}{\cos\theta} = 2\sec\theta$

50. $\frac{\cos\theta}{1 + \sin\theta} + \frac{1 + \sin\theta}{\cos\theta} = \frac{\cos^2\theta + (1 + \sin\theta)^2}{\cos\theta(1 + \sin\theta)} = \frac{\cos^2\theta + 1 + 2\sin\theta + \sin^2\theta}{\cos\theta(1 + \sin\theta)} = \frac{2 + 2\sin\theta}{\cos\theta(1 + \sin\theta)} = \frac{2(1 + \sin\theta)}{\cos\theta(1 + \sin\theta)}$
$= \frac{2}{\cos\theta} = 2\sec\theta$

51. $\dfrac{\sin\theta}{\sin\theta-\cos\theta}=\dfrac{1}{\dfrac{\sin\theta-\cos\theta}{\sin\theta}}=\dfrac{1}{1-\dfrac{\cos\theta}{\sin\theta}}=\dfrac{1}{1-\cot\theta}$

52. $1-\dfrac{\sin^2\theta}{1+\cos\theta}=1-\dfrac{1-\cos^2\theta}{1+\cos\theta}=1-\dfrac{(1-\cos\theta)(1+\cos\theta)}{1+\cos\theta}=1-(1-\cos\theta)=\cos\theta$

53. $(\sec\theta-\tan\theta)^2=\sec^2\theta-2\sec\theta\tan\theta+\tan^2\theta=\dfrac{1}{\cos^2\theta}-\dfrac{2\sin\theta}{\cos^2\theta}+\dfrac{\sin^2\theta}{\cos^2\theta}=\dfrac{1-2\sin\theta+\sin^2\theta}{\cos^2\theta}=\dfrac{(1-\sin\theta)^2}{1-\sin^2\theta}$
$=\dfrac{(1-\sin\theta)^2}{(1-\sin\theta)(1+\sin\theta)}=\dfrac{1-\sin\theta}{1+\sin\theta}$

54. $(\csc\theta-\cot\theta)^2=\csc^2\theta-2\csc\theta\cot\theta+\cot^2\theta=\dfrac{1}{\sin^2\theta}-\dfrac{2\cos\theta}{\sin^2\theta}+\dfrac{\cos^2\theta}{\sin^2\theta}=\dfrac{1-2\cos\theta+\cos^2\theta}{\sin^2\theta}=\dfrac{(1-\cos\theta)^2}{1-\cos^2\theta}$
$=\dfrac{(1-\cos\theta)^2}{(1-\cos\theta)(1+\cos\theta)}=\dfrac{1-\cos\theta}{1+\cos\theta}$

55. $\dfrac{\cos\theta}{1-\tan\theta}+\dfrac{\sin\theta}{1-\cot\theta}=\dfrac{\cos\theta}{1-\dfrac{\sin\theta}{\cos\theta}}+\dfrac{\sin\theta}{1-\dfrac{\cos\theta}{\sin\theta}}=\dfrac{\cos\theta}{\dfrac{\cos\theta-\sin\theta}{\cos\theta}}+\dfrac{\sin\theta}{\dfrac{\sin\theta-\cos\theta}{\sin\theta}}=\dfrac{\cos^2\theta}{\cos\theta-\sin\theta}+\dfrac{\sin^2\theta}{\sin\theta-\cos\theta}$
$=\dfrac{\cos^2\theta-\sin^2\theta}{\cos\theta-\sin\theta}=\dfrac{(\cos\theta-\sin\theta)(\cos\theta+\sin\theta)}{\cos\theta-\sin\theta}=\sin\theta+\cos\theta$

56. $\dfrac{\cot\theta}{1-\tan\theta}+\dfrac{\tan\theta}{1-\cot\theta}=\dfrac{\cot\theta}{1-\dfrac{\sin\theta}{\cos\theta}}+\dfrac{\tan\theta}{1-\dfrac{\cos\theta}{\sin\theta}}=\dfrac{\cot\theta}{\dfrac{\cos\theta-\sin\theta}{\cos\theta}}+\dfrac{\tan\theta}{\dfrac{\sin\theta-\cos\theta}{\sin\theta}}=\dfrac{\dfrac{\cos\theta}{\sin\theta}\cdot\cos\theta}{\cos\theta-\sin\theta}-\dfrac{\dfrac{\sin\theta}{\cos\theta}\cdot\sin\theta}{\cos\theta-\sin\theta}$
$=\dfrac{\dfrac{\cos^2\theta}{\sin\theta}-\dfrac{\sin^2\theta}{\cos\theta}}{\cos\theta-\sin\theta}=\dfrac{\dfrac{\cos^3\theta-\sin^3\theta}{\sin\theta\cos\theta}}{\cos\theta-\sin\theta}=\dfrac{\dfrac{(\cos\theta-\sin\theta)(\cos^2\theta+\sin\theta\cos\theta+\sin^2\theta)}{\sin\theta\cos\theta}}{\cos\theta-\sin\theta}=\dfrac{\cos^2\theta+\sin\theta\cos\theta+\sin^2\theta}{\sin\theta\cos\theta}$
$=\dfrac{\cos^2\theta}{\sin\theta\cos\theta}+\dfrac{\sin\theta\cos\theta}{\sin\theta\cos\theta}+\dfrac{\sin^2\theta}{\sin\theta\cos\theta}=\cot\theta+1+\tan\theta$

57. $\tan\theta+\dfrac{\cos\theta}{1+\sin\theta}=\dfrac{\sin\theta}{\cos\theta}+\dfrac{\cos\theta}{1+\sin\theta}=\dfrac{\sin\theta(1+\sin\theta)+\cos^2\theta}{\cos\theta(1+\sin\theta)}=\dfrac{\sin\theta+\sin^2\theta+\cos^2\theta}{\cos\theta(1+\sin\theta)}=\dfrac{\sin\theta+1}{\cos\theta(1+\sin\theta)}=\dfrac{1}{\cos\theta}=\sec\theta$

58. $\dfrac{\sin\theta\cos\theta}{\cos^2\theta-\sin^2\theta}=\dfrac{\dfrac{\sin\theta\cos\theta}{\cos^2\theta}}{\dfrac{\cos^2\theta-\sin^2\theta}{\cos^2\theta}}=\dfrac{\dfrac{\sin\theta}{\cos\theta}}{1-\dfrac{\sin^2\theta}{\cos^2\theta}}=\dfrac{\tan\theta}{1-\tan^2\theta}$

59. $\dfrac{\tan\theta+\sec\theta-1}{\tan\theta-\sec\theta+1}=\dfrac{\tan\theta+(\sec\theta-1)}{\tan\theta-(\sec\theta-1)}\cdot\dfrac{\tan\theta+(\sec\theta-1)}{\tan\theta+(\sec\theta-1)}=\dfrac{\tan^2\theta+2\tan\theta(\sec\theta-1)+\sec^2\theta-2\sec\theta+1}{\tan^2\theta-(\sec^2\theta-2\sec\theta+1)}$
$=\dfrac{\sec^2\theta-1+2\tan\theta(\sec\theta-1)+\sec^2\theta-2\sec\theta+1}{\sec^2\theta-1-\sec^2\theta+2\sec\theta-1}=\dfrac{2\sec^2\theta-2\sec\theta+2\tan\theta(\sec\theta-1)}{-2+2\sec\theta}$
$=\dfrac{2\sec\theta(\sec\theta-1)+2\tan\theta(\sec\theta-1)}{2(\sec\theta-1)}=\dfrac{2(\sec\theta-1)(\sec\theta+\tan\theta)}{2(\sec\theta-1)}=\tan\theta+\sec\theta$

60. $\dfrac{\sin\theta-\cos\theta+1}{\sin\theta+\cos\theta-1}=\dfrac{\sin\theta-(\cos\theta-1)}{\sin\theta+(\cos\theta-1)}\cdot\dfrac{\sin\theta-(\cos\theta-1)}{\sin\theta-(\cos\theta-1)}=\dfrac{\sin^2\theta-2\sin\theta(\cos\theta-1)+\cos^2\theta-2\cos\theta+1}{\sin^2\theta-(\cos^2\theta-2\cos\theta+1)}$
$=\dfrac{1-\cos^2\theta-2\sin\theta(\cos\theta-1)+\cos^2\theta-2\cos\theta+1}{\sin^2\theta-\cos^2\theta+2\cos\theta-1}=\dfrac{2-2\sin\theta(\cos\theta-1)-2\cos\theta}{1-\cos^2\theta-\cos^2\theta+2\cos\theta-1}$
$=\dfrac{2(1-\cos\theta)+2\sin\theta(1-\cos\theta)}{2\cos\theta-2\cos^2\theta}=\dfrac{2(1+\sin\theta)(1-\cos\theta)}{2\cos\theta(1-\cos\theta)}=\dfrac{\sin\theta+1}{\cos\theta}$

61. $\dfrac{\tan\theta-\cot\theta}{\tan\theta+\cot\theta}=\dfrac{\dfrac{\sin\theta}{\cos\theta}-\dfrac{\cos\theta}{\sin\theta}}{\dfrac{\sin\theta}{\cos\theta}+\dfrac{\cos\theta}{\sin\theta}}=\dfrac{\dfrac{\sin^2\theta-\cos^2\theta}{\cos\theta\sin\theta}}{\dfrac{\sin^2\theta+\cos^2\theta}{\cos\theta\sin\theta}}=\dfrac{\sin^2\theta-\cos^2\theta}{1}=\sin^2\theta-\cos^2\theta$

62. $\dfrac{\sec\theta-\cos\theta}{\sec\theta+\cos\theta}=\dfrac{\dfrac{1}{\cos\theta}-\cos\theta}{\dfrac{1}{\cos\theta}+\cos\theta}=\dfrac{\dfrac{1-\cos^2\theta}{\cos\theta}}{\dfrac{1+\cos^2\theta}{\cos\theta}}=\dfrac{1-\cos^2\theta}{1+\cos^2\theta}=\dfrac{\sin^2\theta}{1+\cos^2\theta}$

63. $\dfrac{\tan\theta-\cot\theta}{\tan\theta+\cot\theta}+1=\dfrac{\dfrac{\sin\theta}{\cos\theta}-\dfrac{\cos\theta}{\sin\theta}}{\dfrac{\sin\theta}{\cos\theta}+\dfrac{\cos\theta}{\sin\theta}}+1=\dfrac{\dfrac{\sin^2\theta-\cos^2\theta}{\cos\theta\sin\theta}}{\dfrac{\sin^2\theta+\cos^2\theta}{\cos\theta\sin\theta}}+1=\sin^2\theta-\cos^2\theta+1=\sin^2\theta+(1-\cos^2\theta)=2\sin^2\theta$

64. $\dfrac{\tan\theta-\cot\theta}{\tan\theta+\cot\theta}+2\cos^2\theta=\dfrac{\dfrac{\sin\theta}{\cos\theta}-\dfrac{\cos\theta}{\sin\theta}}{\dfrac{\sin\theta}{\cos\theta}+\dfrac{\cos\theta}{\sin\theta}}+2\cos^2\theta=\dfrac{\dfrac{\sin^2\theta-\cos^2\theta}{\sin\theta\cos\theta}}{\dfrac{\sin^2\theta+\cos^2\theta}{\sin\theta\cos\theta}}+2\cos^2\theta=\sin^2\theta-\cos^2\theta+2\cos^2\theta$

$=\sin^2\theta+\cos^2\theta=1$

65. $\dfrac{\sec\theta+\tan\theta}{\cot\theta+\cos\theta}=\dfrac{\dfrac{1}{\cos\theta}+\dfrac{\sin\theta}{\cos\theta}}{\dfrac{\cos\theta}{\sin\theta}+\cos\theta}=\dfrac{\dfrac{1+\sin\theta}{\cos\theta}}{\dfrac{\cos\theta+\cos\theta\sin\theta}{\sin\theta}}=\dfrac{1+\sin\theta}{\cos\theta}\cdot\dfrac{\sin\theta}{\cos\theta(1+\sin\theta)}=\dfrac{\sin\theta}{\cos\theta}\cdot\dfrac{1}{\cos\theta}=\tan\theta\sec\theta$

66. $\dfrac{\sec\theta}{1+\sec\theta}=\dfrac{\dfrac{1}{\cos\theta}}{1+\dfrac{1}{\cos\theta}}=\dfrac{\dfrac{1}{\cos\theta}}{\dfrac{\cos\theta+1}{\cos\theta}}=\dfrac{1}{\cos\theta+1}\cdot\dfrac{\cos\theta-1}{\cos\theta-1}==\dfrac{\cos\theta-1}{\cos^2\theta-1}=\dfrac{1-\cos\theta}{1-\cos^2\theta}=\dfrac{1-\cos\theta}{\sin^2\theta}$

67. $\dfrac{1-\tan^2\theta}{1+\tan^2\theta}+1=\dfrac{1-\tan^2\theta}{\sec^2\theta}+1=\dfrac{1}{\sec^2\theta}-\dfrac{\tan^2\theta}{\sec^2\theta}+1=\cos^2\theta-\dfrac{\dfrac{\sin^2\theta}{\cos^2\theta}}{\dfrac{1}{\cos^2\theta}}+1=\cos^2\theta-\sin^2\theta+1$

$=\cos^2\theta+(1-\sin^2\theta)=2\cos^2\theta$

68. $\dfrac{1-\cot^2\theta}{1+\cot^2\theta}+2\cos^2\theta=\dfrac{1-\cot^2\theta}{\csc^2\theta}+2\cos^2\theta=\dfrac{1}{\csc^2\theta}-\dfrac{\cot^2\theta}{\csc^2\theta}+2\cos^2\theta=\sin^2\theta-\dfrac{\dfrac{\cos^2\theta}{\sin^2\theta}}{\dfrac{1}{\sin^2\theta}}+2\cos^2\theta$

$=\sin^2\theta-\cos^2\theta+2\cos^2\theta=\sin^2\theta+\cos^2\theta=1$

69. $\dfrac{\sec\theta-\csc\theta}{\sec\theta\csc\theta}=\dfrac{\sec\theta}{\sec\theta\csc\theta}-\dfrac{\csc\theta}{\sec\theta\csc\theta}=\dfrac{1}{\csc\theta}-\dfrac{1}{\sec\theta}=\sin\theta-\cos\theta$

70. $\dfrac{\sin^2\theta-\tan\theta}{\cos^2\theta-\cot\theta}=\dfrac{\sin^2\theta-\dfrac{\sin\theta}{\cos\theta}}{\cos^2\theta-\dfrac{\cos\theta}{\sin\theta}}=\dfrac{\dfrac{\sin^2\theta\cos\theta-\sin\theta}{\cos\theta}}{\dfrac{\sin\theta\cos^2\theta-\cos\theta}{\sin\theta}}=\dfrac{\sin\theta(\sin\theta\cos\theta-1)}{\cos\theta}\cdot\dfrac{\sin\theta}{\cos\theta(\sin\theta\cos\theta-1)}=\dfrac{\sin^2\theta}{\cos^2\theta}=\tan^2\theta$

71. $\sec\theta-\cos\theta=\dfrac{1}{\cos\theta}-\cos\theta=\dfrac{1-\cos^2\theta}{\cos\theta}=\dfrac{\sin^2\theta}{\cos\theta}=\sin\theta\cdot\dfrac{\sin\theta}{\cos\theta}=\sin\theta\tan\theta$

72. $\tan\theta+\cot\theta=\dfrac{\sin\theta}{\cos\theta}+\dfrac{\cos\theta}{\sin\theta}=\dfrac{\sin^2\theta+\cos^2\theta}{\sin\theta\cos\theta}=\dfrac{1}{\sin\theta\cos\theta}=\dfrac{1}{\cos\theta}\cdot\dfrac{1}{\sin\theta}=\sec\theta\csc\theta$

73. $\dfrac{1}{1-\sin\theta}+\dfrac{1}{1+\sin\theta}=\dfrac{1+\sin\theta+1-\sin\theta}{(1+\sin\theta)(1-\sin\theta)}=\dfrac{2}{1-\sin^2\theta}=\dfrac{2}{\cos^2\theta}=2\sec^2\theta$

74. $\dfrac{1+\sin\theta}{1-\sin\theta}-\dfrac{1-\sin\theta}{1+\sin\theta}=\dfrac{(1+\sin\theta)^2-(1-\sin\theta)^2}{1-\sin^2\theta}=\dfrac{1+2\sin\theta+\sin^2\theta-(1-2\sin\theta+\sin^2\theta)}{\cos^2\theta}$

$=\dfrac{4\sin\theta}{\cos^2\theta}=4\cdot\dfrac{\sin\theta}{\cos\theta}\cdot\dfrac{1}{\cos\theta}=4\tan\theta\sec\theta$

75. $\dfrac{\sec\theta}{1-\sin\theta}=\dfrac{\sec\theta}{1-\sin\theta}\cdot\dfrac{1+\sin\theta}{1+\sin\theta}=\dfrac{\sec\theta(1+\sin\theta)}{1-\sin^2\theta}=\dfrac{\sec\theta(1+\sin\theta)}{\cos^2\theta}=\dfrac{1+\sin\theta}{\cos^3\theta}$

76. $\dfrac{1-\sin\theta}{1+\sin\theta}=\dfrac{1-\sin\theta}{1+\sin\theta}\cdot\dfrac{1-\sin\theta}{1-\sin\theta}=\dfrac{(1-\sin\theta)^2}{1-\sin^2\theta}=\dfrac{(1-\sin\theta)^2}{\cos^2\theta}=\left(\dfrac{1-\sin\theta}{\cos\theta}\right)^2=\left(\dfrac{1}{\cos\theta}-\dfrac{\sin\theta}{\cos\theta}\right)^2=(\sec\theta-\tan\theta)^2$

77. $\dfrac{(\sec\theta-\tan\theta)^2+1}{\csc\theta(\sec\theta-\tan\theta)}=\dfrac{\sec^2\theta-2\sec\theta\tan\theta+\tan^2\theta+1}{\dfrac{1}{\sin\theta}\left(\dfrac{1}{\cos\theta}-\dfrac{\sin\theta}{\cos\theta}\right)}=\dfrac{2\sec^2\theta-2\sec\theta\tan\theta}{\dfrac{1}{\sin\theta}\left(\dfrac{1-\sin\theta}{\cos\theta}\right)}=\dfrac{\dfrac{2}{\cos^2\theta}-\dfrac{2\sin\theta}{\cos^2\theta}}{\dfrac{1-\sin\theta}{\sin\theta\cos\theta}}=\dfrac{2-2\sin\theta}{\cos^2\theta}\cdot\dfrac{\sin\theta\cos\theta}{1-\sin\theta}$

$=\dfrac{2(1-\sin\theta)}{\cos\theta}\cdot\dfrac{\sin\theta}{1-\sin\theta}=\dfrac{2\sin\theta}{\cos\theta}=2\tan\theta$

78. $\dfrac{\sec^2\theta-\tan^2\theta+\tan\theta}{\sec\theta}=\dfrac{\sec^2\theta}{\sec\theta}-\dfrac{\tan^2\theta}{\sec\theta}+\dfrac{\tan\theta}{\sec\theta}=\sec\theta-\dfrac{\dfrac{\sin^2\theta}{\cos^2\theta}}{\dfrac{1}{\cos\theta}}+\dfrac{\dfrac{\sin\theta}{\cos\theta}}{\dfrac{1}{\cos\theta}}=\dfrac{1}{\cos\theta}-\dfrac{\sin^2\theta}{\cos\theta}+\sin\theta$

$=\dfrac{1}{\cos\theta}-\dfrac{1-\cos^2\theta}{\cos\theta}+\sin\theta=\dfrac{1}{\cos\theta}-\dfrac{1}{\cos\theta}+\dfrac{\cos^2\theta}{\cos\theta}+\sin\theta=\sin\theta+\cos\theta$

79. $\dfrac{\sin\theta+\cos\theta}{\cos\theta}-\dfrac{\sin\theta-\cos\theta}{\sin\theta}=\dfrac{\sin\theta}{\cos\theta}+1-1+\dfrac{\cos\theta}{\sin\theta}=\dfrac{\sin^2\theta+\cos^2\theta}{\cos\theta\sin\theta}=\dfrac{1}{\cos\theta\sin\theta}=\sec\theta\csc\theta$

80. $\dfrac{\sin\theta+\cos\theta}{\sin\theta}-\dfrac{\cos\theta-\sin\theta}{\cos\theta}=1+\dfrac{\cos\theta}{\sin\theta}-1+\dfrac{\sin\theta}{\cos\theta}=\dfrac{\cos^2\theta+\sin^2\theta}{\sin\theta\cos\theta}=\dfrac{1}{\sin\theta\cos\theta}=\sec\theta\csc\theta$

81. $\dfrac{\sin^3\theta+\cos^3\theta}{\sin\theta+\cos\theta}=\dfrac{(\sin\theta+\cos\theta)(\sin^2\theta-\sin\theta\cos\theta+\cos^2\theta)}{\sin\theta+\cos\theta}=\sin^2\theta+\cos^2\theta-\sin\theta\cos\theta=1-\sin\theta\cos\theta$

82. $\dfrac{\sin^3\theta+\cos^3\theta}{1-2\cos^2\theta}=\dfrac{(\sin\theta+\cos\theta)(\sin^2\theta-\sin\theta\cos\theta+\cos^2\theta)}{1-\cos^2\theta-\cos^2\theta}=\dfrac{(\sin\theta+\cos\theta)(1-\sin\theta\cos\theta)}{\sin^2\theta-\cos^2\theta}$

$=\dfrac{(\sin\theta+\cos\theta)(1-\sin\theta\cos\theta)}{(\sin\theta+\cos\theta)(\sin\theta-\cos\theta)}=\dfrac{1-\sin\theta\cos\theta}{\sin\theta-\cos\theta}\cdot\dfrac{\dfrac{1}{\cos\theta}}{\dfrac{1}{\cos\theta}}=\dfrac{\dfrac{1}{\cos\theta}-\sin\theta}{\dfrac{\sin\theta}{\cos\theta}-1}=\dfrac{\sec\theta-\sin\theta}{\tan\theta-1}$

83. $\dfrac{\cos^2\theta-\sin^2\theta}{1-\tan^2\theta}=\dfrac{\cos^2\theta-\sin^2\theta}{1-\dfrac{\sin^2\theta}{\cos^2\theta}}=\dfrac{\cos^2\theta-\sin^2\theta}{\dfrac{\cos^2\theta-\sin^2\theta}{\cos^2\theta}}=\cos^2\theta$

84. $\dfrac{\cos\theta+\sin\theta-\sin^3\theta}{\sin\theta}=\dfrac{\cos\theta}{\sin\theta}+\dfrac{\sin\theta\,(1-\sin^2\theta)}{\sin\theta}=\cot\theta+(1-\sin^2\theta)=\cot\theta+\cos^2\theta$

85. $\dfrac{(2\cos^2\theta-1)^2}{\cos^4\theta-\sin^4\theta}=\dfrac{[2\cos^2\theta-(\sin^2\theta+\cos^2\theta)]^2}{(\cos^2\theta-\sin^2\theta)(\cos^2\theta+\sin^2\theta)}=\dfrac{(\cos^2\theta-\sin^2\theta)^2}{\cos^2\theta-\sin^2\theta}=\cos^2\theta-\sin^2\theta=(1-\sin^2\theta)-\sin^2\theta=1-2\sin^2\theta$

86. $\dfrac{1-2\cos^2\theta}{\sin\theta\cos\theta}=\dfrac{1-\cos^2\theta-\cos^2\theta}{\sin\theta\cos\theta}=\dfrac{\sin^2\theta-\cos^2\theta}{\sin\theta\cos\theta}=\dfrac{\sin^2\theta}{\sin\theta\cos\theta}-\dfrac{\cos^2\theta}{\sin\theta\cos\theta}=\dfrac{\sin\theta}{\cos\theta}-\dfrac{\cos\theta}{\sin\theta}=\tan\theta-\cot\theta$

87. $\dfrac{1+\sin\theta+\cos\theta}{1+\sin\theta-\cos\theta}=\dfrac{(1+\sin\theta)+\cos\theta}{(1+\sin\theta)-\cos\theta}\cdot\dfrac{(1+\sin\theta)+\cos\theta}{(1+\sin\theta)+\cos\theta}=\dfrac{1+2\sin\theta+\sin^2\theta+2(1+\sin\theta)\cos\theta+\cos^2\theta}{1+2\sin\theta+\sin^2\theta-\cos^2\theta}$

$=\dfrac{1+2\sin\theta+\sin^2\theta+2(1+\sin\theta)(\cos\theta)+(1-\sin^2\theta)}{1+2\sin\theta+\sin^2\theta-(1-\sin^2\theta)}=\dfrac{2+2\sin\theta+2(1+\sin\theta)(\cos\theta)}{2\sin\theta+2\sin^2\theta}$

$=\dfrac{2(1+\sin\theta)+2(1+\sin\theta)(\cos\theta)}{2\sin\theta(1+\sin\theta)}=\dfrac{2(1+\sin\theta)(1+\cos\theta)}{2\sin\theta(1+\sin\theta)}=\dfrac{1+\cos\theta}{\sin\theta}$

88. $\dfrac{1+\cos\theta+\sin\theta}{1+\cos\theta-\sin\theta}=\dfrac{(1+\cos\theta)+\sin\theta}{(1+\cos\theta)-\sin\theta}\cdot\dfrac{(1+\cos\theta)+\sin\theta}{(1+\cos\theta)+\sin\theta}=\dfrac{1+2\cos\theta+\cos^2\theta+2(1+\cos\theta)\sin\theta+\sin^2\theta}{1+2\cos\theta+\cos^2\theta-\sin^2\theta}$

$=\dfrac{1+2\cos\theta+\cos^2\theta+2(1+\cos\theta)\sin\theta+(1-\cos^2\theta)}{1+2\cos\theta+\cos^2\theta-(1-\cos^2\theta)}=\dfrac{2+2\cos\theta+2(1+\cos\theta)\sin\theta}{2\cos\theta+2\cos^2\theta}$

$=\dfrac{2(1+\cos\theta)+2(1+\cos\theta)\sin\theta}{2\cos\theta(1+\cos\theta)}=\dfrac{2(1+\sin\theta)(1+\cos\theta)}{2\cos\theta\,(1+\cos\theta)}=\dfrac{1+\sin\theta}{\cos\theta}=\dfrac{1}{\cos\theta}+\dfrac{\sin\theta}{\cos\theta}=\sec\theta+\tan\theta$

89. $(a\sin\theta+b\cos\theta)^2+(a\cos\theta-b\sin\theta)^2=a^2\sin^2\theta+2ab\sin\theta\cos\theta+b^2\cos^2\theta+a^2\cos^2\theta-2ab\sin\theta\cos\theta+b^2\sin^2\theta$
$=a^2(\sin^2\theta+\cos^2\theta)+b^2(\cos^2\theta+\sin^2\theta)=a^2+b^2$

90. $(2a\sin\theta\cos\theta)^2+a^2(\cos^2\theta-\sin^2\theta)^2=4a^2\sin^2\theta\cos^2\theta+a^2(\cos^4\theta-2\sin^2\theta\cos^2\theta+\sin^4\theta)$
$=a^2(\cos^4\theta+2\sin^2\theta\cos^2\theta+\sin^4\theta)=a^2(\cos^2\theta+\sin^2\theta)^2=a^2$

91. $\dfrac{\tan\alpha+\tan\beta}{\cot\alpha+\cot\beta}=\dfrac{\tan\alpha+\tan\beta}{\dfrac{1}{\tan\alpha}+\dfrac{1}{\tan\beta}}=\dfrac{\tan\alpha+\tan\beta}{\dfrac{\tan\beta+\tan\alpha}{\tan\alpha\tan\beta}}=(\tan\alpha+\tan\beta)\cdot\dfrac{\tan\alpha\tan\beta}{\tan\alpha+\tan\beta}=\tan\alpha\tan\beta$

92. $(\tan\alpha+\tan\beta)(1-\cot\alpha\cot\beta)+(\cot\alpha+\cot\beta)(1-\tan\alpha\tan\beta)$
$=\tan\alpha-\tan\alpha\cot\alpha\cot\beta+\tan\beta-\cot\alpha\cot\beta\tan\beta+\cot\alpha-\cot\alpha\tan\alpha\tan\beta+\cot\beta-\tan\alpha\tan\beta\cot\beta$
$=\tan\alpha-\cot\beta+\tan\beta-\cot\alpha+\cot\alpha-\tan\beta+\cot\beta-\tan\alpha$
$=(\tan\alpha-\tan\alpha)+(\cot\beta-\cot\beta)+(\tan\beta-\tan\beta)+(\cot\alpha-\cot\alpha)=0$

93. $(\sin\alpha+\cos\beta)^2+(\cos\beta+\sin\alpha)(\cos\beta-\sin\alpha)=(\sin^2\alpha+2\sin\alpha\cos\beta+\cos^2\beta)\,1+(\cos^2\beta-\sin^2\alpha)$
$=2\cos^2\beta+2\sin\alpha\cos\beta=2\cos\beta(\cos\beta+\sin\alpha)$

94. $(\sin\alpha-\cos\beta)^2+(\cos\beta+\sin\alpha)(\cos\beta-\sin\alpha)=\sin^2\alpha-2\sin\alpha\cos\beta+\cos^2\beta+\cos^2\beta-\sin^2\alpha=-2\sin\alpha\cos\beta+2\cos^2\beta$
$=-2\cos\beta(\sin\alpha-\cos\beta)$

95. $\ln|\sec\theta|=\ln|\cos\theta|^{-1}=-\ln|\cos\theta|$ **96.** $\ln|\tan\theta|=\ln\left|\dfrac{\sin\theta}{\cos\theta}\right|=\ln|\sin\theta|-\ln|\cos\theta|$

97. $\ln|1+\cos\theta|+\ln|1-\cos\theta|=\ln(|1+\cos\theta||1-\cos\theta|)=\ln|1-\cos^2\theta|=\ln|\sin^2\theta|=2\ln|\sin\theta|$

98. $\ln|\sec\theta+\tan\theta|+\ln|\sec\theta-\tan\theta|=\ln(|\sec\theta+\tan\theta||\sec\theta-\tan\theta|)=\ln|\sec^2\theta-\tan^2\theta|=\ln|\tan^2\theta+1-\tan^2\theta|=\ln|1|=0$

99. Let $\theta=\tan^{-1}v$. Then $\tan\theta=v,-\dfrac{\pi}{2}<\theta<\dfrac{\pi}{2}$. Now, $\sec\theta>0$ and $\tan^2\theta+1=\sec^2\theta$. Thus $\sec(\tan^{-1}v)=\sec\theta=\sqrt{1+v^2}$.

100. Let $\theta=\sin^{-1}v$. Then $\sin\theta=v,-\dfrac{\pi}{2}\le\theta\le\dfrac{\pi}{2}$, and $\tan(\sin^{-1}v)=\tan\theta=\dfrac{\sin\theta}{\cos\theta}=\dfrac{\sin\theta}{\sqrt{1-\sin^2\theta}}=\dfrac{v}{\sqrt{1-v^2}}$

101. Let $\theta=\cos^{-1}v$. Then $\cos\theta=v, 0\le\theta\le\pi$, and $\tan(\cos^{-1}v)=\tan\theta=\dfrac{\sin\theta}{\cos\theta}=\dfrac{\sqrt{1-\cos^2\theta}}{\cos\theta}=\dfrac{\sqrt{1-v^2}}{v}$.

102. Let $\theta=\cos^{-1}v$. Then $\cos\theta=v, 0\le\theta\le\pi$, and $\sin(\cos^{-1}v)=\sin\theta=\sqrt{1-\cos^2\theta}=\sqrt{1-v^2}$.

103. Let $\theta=\sin^{-1}v$. Then $\sin\theta=v,-\dfrac{\pi}{2}\le\theta\le\dfrac{\pi}{2}$, and $\cos(\sin^{-1}v)=\cos\theta=\sqrt{1-\sin^2\theta}=\sqrt{1-v^2}$.

104. Let $\theta=\tan^{-1}v$. Then $\tan\theta=v,-\dfrac{\pi}{2}<\theta<\dfrac{\pi}{2}$, and $\cos(\tan^{-1}v)=\cos\theta=\dfrac{1}{\sec\theta}=\dfrac{1}{\sqrt{\sec^2\theta}}=\dfrac{1}{\sqrt{1+\tan^2\theta}}=\dfrac{1}{\sqrt{1+v^2}}$

6.4 Concepts and Vocabulary *(page 445)*

4. – **5.** – **6.** False **7.** False **8.** False

6.4 Exercises *(page 445)*

9. $\frac{1}{4}(\sqrt{6}+\sqrt{2})$ **10.** $\frac{1}{4}(\sqrt{6}-\sqrt{2})$ **11.** $\frac{1}{4}(\sqrt{2}-\sqrt{6})$ **12.** $-2-\sqrt{3}$ **13.** $-\frac{1}{4}(\sqrt{2}+\sqrt{6})$ **14.** $\frac{1}{4}(\sqrt{6}+\sqrt{2})$ **15.** $2-\sqrt{3}$
16. $2-\sqrt{3}$ **17.** $-\frac{1}{4}(\sqrt{6}+\sqrt{2})$ **18.** $-2-\sqrt{3}$ **19.** $\sqrt{6}-\sqrt{2}$ **20.** $-2+\sqrt{3}$ **21.** $\frac{1}{2}$ **22.** $-\frac{\sqrt{3}}{2}$ **23.** 0 **24.** $\frac{\sqrt{3}}{2}$ **25.** 1 **26.** $\frac{\sqrt{3}}{3}$
27. -1 **28.** -1 **29.** $\frac{1}{2}$ **30.** $\frac{\sqrt{3}}{2}$ **31.** **(a)** $\frac{2\sqrt{5}}{25}$ **(b)** $\frac{11\sqrt{5}}{25}$ **(c)** $\frac{2\sqrt{5}}{5}$ **(d)** 2 **32.** **(a)** $\frac{2\sqrt{5}}{25}$ **(b)** $\frac{11\sqrt{5}}{25}$ **(c)** $\frac{2\sqrt{5}}{5}$ **(d)** -2
33. **(a)** $\frac{4-3\sqrt{3}}{10}$ **(b)** $\frac{-3-4\sqrt{3}}{10}$ **(c)** $\frac{4+3\sqrt{3}}{10}$ **(d)** $\frac{25\sqrt{3}+48}{39}$ **34.** **(a)** $\frac{12+5\sqrt{3}}{26}$ **(b)** $\frac{-5+12\sqrt{3}}{26}$ **(c)** $\frac{-12+5\sqrt{3}}{26}$
(d) $\frac{240-169\sqrt{3}}{407}$ **35.** **(a)** $-\frac{5+12\sqrt{3}}{26}$ **(b)** $\frac{12-5\sqrt{3}}{26}$ **(c)** $\frac{-5+12\sqrt{3}}{26}$ **(d)** $\frac{-240+169\sqrt{3}}{69}$ **36.** **(a)** $\frac{1-2\sqrt{6}}{6}$ **(b)** $\frac{\sqrt{3}+2\sqrt{2}}{6}$
(c) $\frac{-1-2\sqrt{6}}{6}$ **(d)** $\frac{-8\sqrt{2}-9\sqrt{3}}{5}$ **37.** **(a)** $-\frac{2\sqrt{2}}{3}$ **(b)** $\frac{-2\sqrt{2}+\sqrt{3}}{6}$ **(c)** $\frac{-2\sqrt{2}+\sqrt{3}}{6}$ **(d)** $\frac{9-4\sqrt{2}}{7}$ **38.** **(a)** $-\frac{\sqrt{15}}{4}$
(b) $\frac{-1-3\sqrt{5}}{8}$ **(c)** $\frac{1+3\sqrt{5}}{8}$ **(d)** $\frac{8+\sqrt{15}}{7}$ **39.** $\sin\left(\frac{\pi}{2}+\theta\right)=\sin\frac{\pi}{2}\cos\theta+\cos\frac{\pi}{2}\sin\theta=1\cdot\cos\theta+0\cdot\sin\theta=\cos\theta$

40. $\cos\left(\frac{\pi}{2}+\theta\right)=\cos\frac{\pi}{2}\cos\theta-\sin\frac{\pi}{2}\sin\theta=0\cdot\cos\theta-1\cdot\sin\theta=-\sin\theta$

41. $\sin(\pi-\theta)=\sin\pi\cos\theta-\cos\pi\sin\theta=0\cdot\cos\theta-(-1)\sin\theta=\sin\theta$
42. $\cos(\pi-\theta)=\cos\pi\cos\theta+\sin\pi\sin\theta=-1\cdot\cos\theta+0\cdot\sin\theta=-\cos\theta$
43. $\sin(\pi+\theta)=\sin\pi\cos\theta+\cos\pi\sin\theta=0\cdot\cos\theta+(-1)\sin\theta=-\sin\theta$
44. $\cos(\pi+\theta)=\cos\pi\cos\theta-\sin\pi\sin\theta=-1\cdot\cos\theta-0\cdot\sin\theta=-\cos\theta$

45. $\tan(\pi-\theta)=\dfrac{\tan\pi-\tan\theta}{1+\tan\pi\tan\theta}=\dfrac{0-\tan\theta}{1+0\cdot\tan\theta}=-\tan\theta$ **46.** $\tan(2\pi-\theta)=\dfrac{\tan 2\pi-\tan\theta}{1+\tan 2\pi\tan\theta}=\dfrac{0-\tan\theta}{1+0\cdot\tan\theta}=-\tan\theta$

47. $\sin\left(\frac{3\pi}{2}+\theta\right)=\sin\frac{3\pi}{2}\cos\theta+\cos\frac{3\pi}{2}\sin\theta=(-1)\cos\theta+0\cdot\sin\theta=-\cos\theta$

48. $\cos\left(\frac{3\pi}{2} + \theta\right) = \cos\frac{3\pi}{2}\cos\theta - \sin\frac{3\pi}{2}\sin\theta = 0 \cdot \cos\theta - (-1)\sin\theta = \sin\theta$

49. $\sin(\alpha + \beta) + \sin(\alpha - \beta) = \sin\alpha\cos\beta + \cos\alpha\sin\beta + \sin\alpha\cos\beta - \cos\alpha\sin\beta = 2\sin\alpha\cos\beta$

50. $\cos(\alpha + \beta) + \cos(\alpha - \beta) = \cos\alpha\cos\beta - \sin\alpha\sin\beta + \cos\alpha\cos\beta + \sin\alpha\sin\beta = 2\cos\alpha\cos\beta$

51. $\frac{\sin(\alpha + \beta)}{\sin\alpha\cos\beta} = \frac{\sin\alpha\cos\beta + \cos\alpha\sin\beta}{\sin\alpha\cos\beta} = \frac{\sin\alpha\cos\beta}{\sin\alpha\cos\beta} + \frac{\cos\alpha\sin\beta}{\sin\alpha\cos\beta} = 1 + \cot\alpha\tan\beta$

52. $\frac{\sin(\alpha + \beta)}{\cos\alpha\cos\beta} = \frac{\sin\alpha\cos\beta + \cos\alpha\sin\beta}{\cos\alpha\cos\beta} = \frac{\sin\alpha\cos\beta}{\cos\alpha\cos\beta} + \frac{\cos\alpha\sin\beta}{\cos\alpha\cos\beta} = \tan\alpha + \tan\beta$

53. $\frac{\cos(\alpha + \beta)}{\cos\alpha\cos\beta} = \frac{\cos\alpha\cos\beta - \sin\alpha\sin\beta}{\cos\alpha\cos\beta} = \frac{\cos\alpha\cos\beta}{\cos\alpha\cos\beta} - \frac{\sin\alpha\sin\beta}{\cos\alpha\cos\beta} = 1 - \tan\alpha\tan\beta$

54. $\frac{\cos(\alpha - \beta)}{\sin\alpha\cos\beta} = \frac{\cos\alpha\cos\beta + \sin\alpha\sin\beta}{\sin\alpha\cos\beta} = \frac{\cos\alpha\cos\beta}{\sin\alpha\cos\beta} + \frac{\sin\alpha\sin\beta}{\sin\alpha\cos\beta} = \cot\alpha + \tan\beta$

55. $\frac{\sin(\alpha + \beta)}{\sin(\alpha - \beta)} = \frac{\sin\alpha\cos\beta + \cos\alpha\sin\beta}{\sin\alpha\cos\beta - \cos\alpha\sin\beta} = \frac{\frac{\sin\alpha\cos\beta + \cos\alpha\sin\beta}{\cos\alpha\cos\beta}}{\frac{\sin\alpha\cos\beta - \cos\alpha\sin\beta}{\cos\alpha\cos\beta}} = \frac{\frac{\sin\alpha\cos\beta}{\cos\alpha\cos\beta} + \frac{\cos\alpha\sin\beta}{\cos\alpha\cos\beta}}{\frac{\sin\alpha\cos\beta}{\cos\alpha\cos\beta} - \frac{\cos\alpha\sin\beta}{\cos\alpha\cos\beta}} = \frac{\tan\alpha + \tan\beta}{\tan\alpha - \tan\beta}$

56. $\frac{\cos(\alpha + \beta)}{\cos(\alpha - \beta)} = \frac{\cos\alpha\cos\beta - \sin\alpha\sin\beta}{\cos\alpha\cos\beta + \sin\alpha\sin\beta} = \frac{\frac{\cos\alpha\cos\beta - \sin\alpha\sin\beta}{\cos\alpha\cos\beta}}{\frac{\cos\alpha\cos\beta + \sin\alpha\sin\beta}{\cos\alpha\cos\beta}} = \frac{\frac{\cos\alpha\cos\beta}{\cos\alpha\cos\beta} - \frac{\sin\alpha\sin\beta}{\cos\alpha\cos\beta}}{\frac{\cos\alpha\cos\beta}{\cos\alpha\cos\beta} + \frac{\sin\alpha\sin\beta}{\cos\alpha\cos\beta}} = \frac{1 - \tan\alpha\tan\beta}{1 + \tan\alpha\tan\beta}$

57. $\cot(\alpha + \beta) = \frac{\cos(\alpha + \beta)}{\sin(\alpha + \beta)} = \frac{\cos\alpha\cos\beta - \sin\alpha\sin\beta}{\sin\alpha\cos\beta + \cos\alpha\sin\beta} = \frac{\frac{\cos\alpha\cos\beta - \sin\alpha\sin\beta}{\sin\alpha\sin\beta}}{\frac{\sin\alpha\cos\beta + \cos\alpha\sin\beta}{\sin\alpha\sin\beta}} = \frac{\frac{\cos\alpha\cos\beta}{\sin\alpha\sin\beta} - \frac{\sin\alpha\sin\beta}{\sin\alpha\sin\beta}}{\frac{\sin\alpha\cos\beta}{\sin\alpha\sin\beta} + \frac{\cos\alpha\sin\beta}{\sin\alpha\sin\beta}} = \frac{\cot\alpha\cot\beta - 1}{\cot\beta + \cot\alpha}$

58. $\cot(\alpha - \beta) = \frac{\cos(\alpha - \beta)}{\sin(\alpha - \beta)} = \frac{\cos\alpha\cos\beta + \sin\alpha\sin\beta}{\sin\alpha\cos\beta - \cos\alpha\sin\beta} = \frac{\frac{\cos\alpha\cos\beta + \sin\alpha\sin\beta}{\sin\alpha\sin\beta}}{\frac{\sin\alpha\cos\beta - \cos\alpha\sin\beta}{\sin\alpha\sin\beta}} = \frac{\frac{\cos\alpha\cos\beta}{\sin\alpha\sin\beta} + \frac{\sin\alpha\sin\beta}{\sin\alpha\sin\beta}}{\frac{\sin\alpha\cos\beta}{\sin\alpha\sin\beta} - \frac{\cos\alpha\sin\beta}{\sin\alpha\sin\beta}} = \frac{\cot\alpha\cot\beta + 1}{\cot\beta - \cot\alpha}$

59. $\sec(\alpha + \beta) = \frac{1}{\cos(\alpha + \beta)} = \frac{1}{\cos\alpha\cos\beta - \sin\alpha\sin\beta} = \frac{\frac{1}{\sin\alpha\sin\beta}}{\frac{\cos\alpha\cos\beta - \sin\alpha\sin\beta}{\sin\alpha\sin\beta}} = \frac{\frac{1}{\sin\alpha}\cdot\frac{1}{\sin\beta}}{\frac{\cos\alpha\cos\beta}{\sin\alpha\sin\beta} - \frac{\sin\alpha\sin\beta}{\sin\alpha\sin\beta}} = \frac{\csc\alpha\csc\beta}{\cot\alpha\cot\beta - 1}$

60. $\sec(\alpha - \beta) = \frac{1}{\cos(\alpha - \beta)} = \frac{1}{\cos\alpha\cos\beta + \sin\alpha\sin\beta} = \frac{\frac{1}{\cos\alpha\cos\beta}}{\frac{\cos\alpha\cos\beta + \sin\alpha\sin\beta}{\cos\alpha\cos\beta}} = \frac{\frac{1}{\cos\alpha}\cdot\frac{1}{\cos\beta}}{\frac{\cos\alpha\cos\beta}{\cos\alpha\cos\beta} + \frac{\sin\alpha\sin\beta}{\cos\alpha\cos\beta}} = \frac{\sec\alpha\sec\beta}{1 + \tan\alpha\tan\beta}$

61. $\sin(\alpha - \beta)\sin(\alpha + \beta) = (\sin\alpha\cos\beta - \cos\alpha\sin\beta)(\sin\alpha\cos\beta + \cos\alpha\sin\beta) = \sin^2\alpha\cos^2\beta - \cos^2\alpha\sin^2\beta$
$= (\sin^2\alpha)(1 - \sin^2\beta) - (1 - \sin^2\alpha)(\sin^2\beta) = \sin^2\alpha - \sin^2\beta$

62. $\cos(\alpha - \beta)\cos(\alpha + \beta) = (\cos\alpha\cos\beta + \sin\alpha\sin\beta)(\cos\alpha\cos\beta - \sin\alpha\sin\beta) = \cos^2\alpha\cos^2\beta - \sin^2\alpha\sin^2\beta$
$= \cos^2\alpha(1 - \sin^2\beta) - \sin^2\alpha\sin^2\beta = \cos^2\alpha - \cos^2\alpha\sin^2\beta - \sin^2\alpha\sin^2\beta = \cos^2\alpha - \sin^2\beta(\cos^2\alpha + \sin^2\alpha) = \cos^2\alpha - \sin^2\beta$

63. $\sin(\theta + k\pi) = \sin\theta\cos k\pi + \cos\theta\sin k\pi = (\sin\theta)(-1)^k + (\cos\theta)(0) = (-1)^k\sin\theta$, k any integer

64. $\cos(\theta + k\pi) = \cos\theta\cos k\pi - \sin\theta\sin k\pi = (\cos\theta)(-1)^k - (\sin\theta)(0) = (-1)^k\cos\theta$, k any integer

65. $\frac{\sqrt{3}}{2}$ **66.** $\frac{\sqrt{3}}{2}$ **67.** $-\frac{24}{25}$ **68.** -1 **69.** $-\frac{33}{65}$ **70.** $\frac{33}{65}$ **71.** $\frac{63}{65}$ **72.** $\frac{16}{65}$ **73.** $\frac{48 + 25\sqrt{3}}{39}$ **74.** $-\frac{1}{7}$ **75.** $\frac{4}{3}$ **76.** $-\frac{4}{3}$

77. $u\sqrt{1 - v^2} - v\sqrt{1 - u^2}$ **78.** $uv - \sqrt{1 - u^2}\sqrt{1 - v^2}$ **79.** $\frac{u\sqrt{1 - v^2} - v}{\sqrt{1 + u^2}}$ **80.** $\frac{1 - uv}{\sqrt{1 + u^2}\sqrt{1 + v^2}}$

81. $\frac{uv - \sqrt{1 - u^2}\sqrt{1 - v^2}}{v\sqrt{1 - u^2} + u\sqrt{1 - v^2}}$ **82.** $\frac{\sqrt{1 + u^2}}{v - u\sqrt{1 - v^2}}$

83. Let $\alpha = \sin^{-1} v$ and $\beta = \cos^{-1} v$. Then $\sin\alpha = \cos\beta = v$, and since $\sin\alpha = \cos\left(\frac{\pi}{2} - \alpha\right)$, $\cos\left(\frac{\pi}{2} - \alpha\right) = \cos\beta$.

If $v \geq 0$, then $0 \leq \alpha \leq \frac{\pi}{2}$, so $\left(\frac{\pi}{2} - \alpha\right)$ and β both lie on $\left[0, \frac{\pi}{2}\right]$. If $v < 0$, then $-\frac{\pi}{2} \leq \alpha < 0$, so $\left(\frac{\pi}{2} - \alpha\right)$ and β both lie on $\left(\frac{\pi}{2}, \pi\right]$. Either way, $\cos\left(\frac{\pi}{2} - \alpha\right) = \cos\beta$ implies $\frac{\pi}{2} - \alpha = \beta$, or $\alpha + \beta = \frac{\pi}{2}$.

84. Let $\alpha = \tan^{-1} v$ and $\beta = \cot^{-1} v$. Then $\tan\alpha = \cot\beta = v$, and since $\tan\alpha = \cot\left(\frac{\pi}{2} - \alpha\right)$, $\cot\left(\frac{\pi}{2} - \alpha\right) = \cot\beta$.
If $v \geq 0$, then $0 \leq \alpha < \frac{\pi}{2}$, so $\left(\frac{\pi}{2} - \alpha\right)$ and β both lie on $\left(0, \frac{\pi}{2}\right]$. If $v < 0$, then $-\frac{\pi}{2} < \alpha < 0$, so $\left(\frac{\pi}{2} - \alpha\right)$ and β both lie on $\left(\frac{\pi}{2}, \pi\right)$. Either way, $\cot\left(\frac{\pi}{2} - \alpha\right) = \cot\beta$ implies $\frac{\pi}{2} - \alpha = \beta$, or $\alpha + \beta = \frac{\pi}{2}$.

85. Let $\alpha = \tan^{-1}\frac{1}{v}$, and $\beta = \tan^{-1} v$. Because $v \neq 0$, $\alpha, \beta \neq 0$. Then $\tan\alpha = \frac{1}{v} = \frac{1}{\tan\beta} = \cot\beta$, and since $\tan\alpha = \cot\left(\frac{\pi}{2} - \alpha\right)$, $\cot\left(\frac{\pi}{2} - \alpha\right) = \cot\beta$. Because $v > 0$, $0 < \alpha < \frac{\pi}{2}$, and so $\left(\frac{\pi}{2} - \alpha\right)$ and β both lie on $\left(0, \frac{\pi}{2}\right)$.
Then $\cot\left(\frac{\pi}{2} - \alpha\right) = \cot\beta$ implies $\frac{\pi}{2} - \alpha = \beta$, or $\alpha = \frac{\pi}{2} - \beta$.

86. Let $\theta = \tan^{-1} e^{-v}$. Then $\tan\theta = e^{-v}$, so $\cot\theta = \frac{1}{e^{-v}} = e^{v}$. Since $0 < \theta < \frac{\pi}{2}$ (because $e^{-v} > 0$), $\cot^{-1} e^{v} = \cot^{-1}(\cot\theta) = \theta = \tan^{-1} e^{-v}$.

87. $\sin(\sin^{-1} v + \cos^{-1} v) = \sin(\sin^{-1} v)\cos(\cos^{-1} v) + \cos(\sin^{-1} v)\sin(\cos^{-1} v) = (v)(v) + \sqrt{1 - v^2}\sqrt{1 - v^2} = v^2 + 1 - v^2 = 1$

88. $\cos(\sin^{-1} v + \cos^{-1} v) = \cos(\sin^{-1} v)\cos(\cos^{-1} v) - \sin(\sin^{-1} v)\sin(\cos^{-1} v) = (\sqrt{1 - v^2})v - v\sqrt{1 - v^2} = 0$

89. $\frac{\sin(x + h) - \sin x}{h} = \frac{\sin x\cos h + \cos x\sin h - \sin x}{h} = \frac{\cos x\sin h - \sin x(1 - \cos h)}{h} = \cos x \cdot \frac{\sin h}{h} - \sin x \cdot \frac{1 - \cos h}{h}$

90. $\frac{\cos(x + h) - \cos x}{h} = \frac{\cos x\cos h - \sin x\sin h - \cos x}{h} = \frac{-\sin x\sin h - \cos x(1 - \cos h)}{h} = -\sin x \cdot \frac{\sin h}{h} - \cos x \cdot \frac{1 - \cos h}{h}$

91. $\tan\frac{\pi}{2}$ is not defined; $\tan\left(\frac{\pi}{2} - \theta\right) = \frac{\sin\left(\frac{\pi}{2} - \theta\right)}{\cos\left(\frac{\pi}{2} - \theta\right)} = \frac{\cos\theta}{\sin\theta} = \cot\theta$

92. $2\cot(\alpha - \beta) = \frac{2}{\tan(\alpha - \beta)} = 2\left(\frac{1 + \tan\alpha\tan\beta}{\tan\alpha - \tan\beta}\right) = 2\left(\frac{1 + (x + 1)(x - 1)}{(x + 1) - (x - 1)}\right) = 2\left(\frac{1 + x^2 - 1}{x + 1 - x + 1}\right) = \frac{2x^2}{2} = x^2$

93. $\tan\theta = \tan(\theta_2 - \theta_1) = \frac{\tan\theta_2 - \tan\theta_1}{1 + \tan\theta_1\tan\theta_2} = \frac{m_2 - m_1}{1 + m_1 m_2}$

94. $\sin(\alpha - \theta) = \sin\alpha\cos\theta - \cos\alpha\sin\theta = \sin\theta[\sin\alpha\cot\theta - \cos\alpha] = \sin\theta\sin\alpha(\cot\theta - \cot\alpha) = \sin\theta\sin\alpha(\cot\beta + \cot\gamma)$
$= \sin\theta\sin\alpha\left(\frac{\cos\beta}{\sin\beta} + \frac{\cos\gamma}{\sin\gamma}\right) = \sin\theta\sin\alpha\left(\frac{\sin\gamma\cos\beta + \sin\beta\cos\gamma}{\sin\beta\sin\gamma}\right) = \sin\theta\sin\alpha\frac{\sin(\beta + \gamma)}{\sin\beta\sin\gamma} = \frac{\sin\theta\sin^2\alpha}{\sin\beta\sin\gamma}$
Similarly, $\sin(\beta - \theta) = \frac{\sin\theta\sin^2\beta}{\sin\alpha\sin\gamma}$ and $\sin(\gamma - \theta) = \frac{\sin\theta\sin^2\gamma}{\sin\alpha\sin\beta}$. Thus, $\sin(\alpha - \theta)\sin(\beta - \theta)\sin(\gamma - \theta)$
$= \frac{\sin\theta\sin^2\alpha}{\sin\beta\sin\gamma} \cdot \frac{\sin\theta\sin^2\beta}{\sin\alpha\sin\gamma} \cdot \frac{\sin\theta\sin^2\gamma}{\sin\alpha\sin\beta} = \sin^3\theta$

95. No; $\tan\frac{\pi}{2}$ is undefined.

6.5 Concepts and Vocabulary *(page 455)*

1. $\sin^2\theta$; $2\cos^2\theta$; $2\sin^2\theta$; **2.** $1 - \cos\theta$ **3.** $\sin\theta$ **4.** True **5.** False **6.** False

6.5 Exercises *(page 455)*

7. (a) $\frac{24}{25}$ **(b)** $\frac{7}{25}$ **(c)** $\frac{\sqrt{10}}{10}$ **(d)** $\frac{3\sqrt{10}}{10}$ **8. (a)** $\frac{24}{25}$ **(b)** $-\frac{7}{25}$ **(c)** $\frac{\sqrt{5}}{5}$ **(d)** $\frac{2\sqrt{5}}{5}$ **9. (a)** $\frac{24}{25}$ **(b)** $-\frac{7}{25}$ **(c)** $\frac{2\sqrt{5}}{5}$ **(d)** $-\frac{\sqrt{5}}{5}$
10. (a) $\frac{4}{5}$ **(b)** $\frac{3}{5}$ **(c)** $\sqrt{\frac{5 + 2\sqrt{5}}{10}}$ **(d)** $-\sqrt{\frac{5 - 2\sqrt{5}}{10}}$ **11. (a)** $-\frac{2\sqrt{2}}{3}$ **(b)** $\frac{1}{3}$ **(c)** $\sqrt{\frac{3 + \sqrt{6}}{6}}$ **(d)** $\sqrt{\frac{3 - \sqrt{6}}{6}}$ **12. (a)** $-\frac{2\sqrt{2}}{3}$ **(b)** $\frac{1}{3}$
(c) $\sqrt{\frac{3 - \sqrt{6}}{6}}$ **(d)** $-\sqrt{\frac{3 + \sqrt{6}}{6}}$ **13. (a)** $\frac{4\sqrt{2}}{9}$ **(b)** $-\frac{7}{9}$ **(c)** $\frac{\sqrt{3}}{3}$ **(d)** $\frac{\sqrt{6}}{3}$ **14. (a)** $\frac{4}{5}$ **(b)** $\frac{3}{5}$ **(c)** $\sqrt{\frac{5 + 2\sqrt{5}}{10}}$ **(d)** $-\sqrt{\frac{5 - 2\sqrt{5}}{10}}$
15. (a) $-\frac{4}{5}$ **(b)** $\frac{3}{5}$ **(c)** $\sqrt{\frac{5 + 2\sqrt{5}}{10}}$ **(d)** $\sqrt{\frac{5 - 2\sqrt{5}}{10}}$ **16. (a)** $-\frac{\sqrt{3}}{2}$ **(b)** $-\frac{1}{2}$ **(c)** $\frac{1}{2}$ **(d)** $-\frac{\sqrt{3}}{2}$ **17. (a)** $-\frac{3}{5}$ **(b)** $-\frac{4}{5}$
(c) $\frac{1}{2}\sqrt{\frac{10 - \sqrt{10}}{5}}$ **(d)** $-\frac{1}{2}\sqrt{\frac{10 + \sqrt{10}}{5}}$ **18. (a)** $\frac{3}{5}$ **(b)** $\frac{4}{5}$ **(c)** $\frac{1}{2}\sqrt{\frac{10 + 3\sqrt{10}}{5}}$ **(d)** $-\frac{1}{2}\sqrt{\frac{10 - 3\sqrt{10}}{5}}$ **19.** $\frac{\sqrt{2 - \sqrt{2}}}{2}$
20. $\frac{\sqrt{2 + \sqrt{2}}}{2}$ **21.** $1 - \sqrt{2}$ **22.** $-1 + \sqrt{2}$ **23.** $-\frac{\sqrt{2 + \sqrt{3}}}{2}$ **24.** $-\frac{\sqrt{2 - \sqrt{3}}}{2}$ **25.** $\frac{2}{\sqrt{2 + \sqrt{2}}} = (2 - \sqrt{2})\sqrt{2 + \sqrt{2}}$

26. $\dfrac{2}{\sqrt{2-\sqrt{2}}} = (2+\sqrt{2})\sqrt{2-\sqrt{2}}$ **27.** $-\dfrac{\sqrt{2-\sqrt{2}}}{2}$ **28.** $\dfrac{\sqrt{2-\sqrt{2}}}{2}$

29. $\sin^4\theta = (\sin^2\theta)^2 = \left(\dfrac{1-\cos(2\theta)}{2}\right)^2 = \dfrac{1}{4}[1 - 2\cos(2\theta) + \cos^2(2\theta)] = \dfrac{1}{4} - \dfrac{1}{2}\cos(2\theta) + \dfrac{1}{4}\cos^2(2\theta)$
$= \dfrac{1}{4} - \dfrac{1}{2}\cos(2\theta) + \dfrac{1}{4}\left(\dfrac{1+\cos(4\theta)}{2}\right) = \dfrac{1}{4} - \dfrac{1}{2}\cos(2\theta) + \dfrac{1}{8} + \dfrac{1}{8}\cos(4\theta) = \dfrac{3}{8} - \dfrac{1}{2}\cos(2\theta) + \dfrac{1}{8}\cos(4\theta)$

30. $\cos(3\theta) = 4\cos^3\theta - 3\cos\theta$

31. $\sin(4\theta) = \sin[2(2\theta)] = 2\sin(2\theta)\cos(2\theta) = (4\sin\theta\cos\theta)(1 - 2\sin^2\theta) = 4\sin\theta\cos\theta - 8\sin^3\theta\cos\theta = (\cos\theta)(4\sin\theta - 8\sin^3\theta)$

32. $\cos(4\theta) = 8\cos^4\theta - 8\cos^2\theta + 1$ **33.** $\sin(5\theta) = 16\sin^5\theta - 20\sin^3\theta + 5\sin\theta$ **34.** $\cos(5\theta) = 16\cos^5\theta - 20\cos^3\theta + 5\cos\theta$

35. $\cos^4\theta - \sin^4\theta = (\cos^2\theta + \sin^2\theta)(\cos^2\theta - \sin^2\theta) = \cos(2\theta)$

36. $\dfrac{\cot\theta - \tan\theta}{\cot\theta + \tan\theta} = \dfrac{\dfrac{\cos\theta}{\sin\theta} - \dfrac{\sin\theta}{\cos\theta}}{\dfrac{\cos\theta}{\sin\theta} + \dfrac{\sin\theta}{\cos\theta}} = \dfrac{\dfrac{\cos^2\theta - \sin^2\theta}{\sin\theta\cos\theta}}{\dfrac{\cos^2\theta + \sin^2\theta}{\sin\theta\cos\theta}} = \dfrac{\cos^2\theta - \sin^2\theta}{\cos^2\theta + \sin^2\theta} = \cos^2\theta - \sin^2\theta = \cos(2\theta)$

37. $\cot(2\theta) = \dfrac{1}{\tan(2\theta)} = \dfrac{1-\tan^2\theta}{2\tan\theta} = \dfrac{1 - \dfrac{1}{\cot^2\theta}}{2\left(\dfrac{1}{\cot\theta}\right)} = \dfrac{\dfrac{\cot^2\theta - 1}{\cot^2\theta}}{\dfrac{2}{\cot\theta}} = \dfrac{\cot^2\theta - 1}{\cot^2\theta}\cdot\dfrac{\cot\theta}{2} = \dfrac{\cot^2\theta - 1}{2\cot\theta}$

38. $\cot(2\theta) = \dfrac{1}{\tan(2\theta)} = \dfrac{1-\tan^2\theta}{2\tan\theta} = \dfrac{\dfrac{1}{\tan\theta} - \dfrac{\tan^2\theta}{\tan\theta}}{\dfrac{2\tan\theta}{\tan\theta}} = \dfrac{1}{2}(\cot\theta - \tan\theta)$

39. $\sec(2\theta) = \dfrac{1}{\cos(2\theta)} = \dfrac{1}{2\cos^2\theta - 1} = \dfrac{1}{\dfrac{2}{\sec^2\theta} - 1} = \dfrac{1}{\dfrac{2-\sec^2\theta}{\sec^2\theta}} = \dfrac{\sec^2\theta}{2-\sec^2\theta}$

40. $\csc(2\theta) = \dfrac{1}{\sin(2\theta)} = \dfrac{1}{2\sin\theta\cos\theta} = \dfrac{1}{2}\cdot\dfrac{1}{\sin\theta}\cdot\dfrac{1}{\cos\theta} = \dfrac{1}{2}\csc\theta\sec\theta$ **41.** $\cos^2(2\theta) - \sin^2(2\theta) = \cos[2(2\theta)] = \cos(4\theta)$

42. $\sin(4\theta) = \sin(2\cdot 2\theta) = 2\sin(2\theta)\cos(2\theta) = 2(2\sin\theta\cos\theta)(1 - 2\sin^2\theta) = (4\sin\theta\cos\theta)(1 - 2\sin^2\theta)$

43. $\dfrac{\cos(2\theta)}{1+\sin(2\theta)} = \dfrac{\cos^2\theta - \sin^2\theta}{1 + 2\sin\theta\cos\theta} = \dfrac{(\cos\theta - \sin\theta)(\cos\theta + \sin\theta)}{\sin^2\theta + \cos^2\theta + 2\sin\theta\cos\theta} = \dfrac{(\cos\theta - \sin\theta)(\cos\theta + \sin\theta)}{(\sin\theta + \cos\theta)(\sin\theta + \cos\theta)} = \dfrac{\cos\theta - \sin\theta}{\cos\theta + \sin\theta}$
$= \dfrac{\dfrac{\cos\theta - \sin\theta}{\sin\theta}}{\dfrac{\cos\theta + \sin\theta}{\sin\theta}} = \dfrac{\dfrac{\cos\theta}{\sin\theta} - \dfrac{\sin\theta}{\sin\theta}}{\dfrac{\cos\theta}{\sin\theta} + \dfrac{\sin\theta}{\sin\theta}} = \dfrac{\cot\theta - 1}{\cot\theta + 1}$

44. $\dfrac{1}{8}[1 - \cos(4\theta)] = \dfrac{1}{8}[1 - \cos(2\cdot 2\theta)] = \dfrac{1}{8}[1 - (1 - 2\sin^2(2\theta))] = \dfrac{1}{4}\sin^2(2\theta) = \dfrac{1}{4}(2\sin\theta\cos\theta)^2$
$= \dfrac{1}{4}(4\sin^2\theta\cos^2\theta) = \sin^2\theta\cos^2\theta$

45. $\sec^2\dfrac{\theta}{2} = \dfrac{1}{\cos^2\left(\dfrac{\theta}{2}\right)} = \dfrac{1}{\dfrac{1+\cos\theta}{2}} = \dfrac{2}{1+\cos\theta}$ **46.** $\csc^2\dfrac{\theta}{2} = \dfrac{1}{\sin^2\left(\dfrac{\theta}{2}\right)} = \dfrac{1}{\dfrac{1-\cos\theta}{2}} = \dfrac{2}{1-\cos\theta}$

47. $\cot^2\dfrac{\theta}{2} = \dfrac{1}{\tan^2\left(\dfrac{\theta}{2}\right)} = \dfrac{1}{\dfrac{1-\cos\theta}{1+\cos\theta}} = \dfrac{1+\cos\theta}{1-\cos\theta} = \dfrac{1 + \dfrac{1}{\sec\theta}}{1 - \dfrac{1}{\sec\theta}} = \dfrac{\dfrac{\sec\theta + 1}{\sec\theta}}{\dfrac{\sec\theta - 1}{\sec\theta}} = \dfrac{\sec\theta + 1}{\sec\theta}\cdot\dfrac{\sec\theta}{\sec\theta - 1} = \dfrac{\sec\theta + 1}{\sec\theta - 1}$

48. $\tan\dfrac{\theta}{2} = \dfrac{1-\cos\theta}{\sin\theta} = \dfrac{1}{\sin\theta} - \dfrac{\cos\theta}{\sin\theta} = \csc\theta - \cot\theta$

49. $\dfrac{1 - \tan^2\left(\dfrac{\theta}{2}\right)}{1 + \tan^2\left(\dfrac{\theta}{2}\right)} = \dfrac{1 - \dfrac{1-\cos\theta}{1+\cos\theta}}{1 + \dfrac{1-\cos\theta}{1+\cos\theta}} = \dfrac{\dfrac{1 + \cos\theta - (1 - \cos\theta)}{1+\cos\theta}}{\dfrac{1 + \cos\theta + 1 - \cos\theta}{1+\cos\theta}} = \dfrac{2\cos\theta}{1+\cos\theta}\cdot\dfrac{1+\cos\theta}{2} = \cos\theta$

50. $\dfrac{\sin^3\theta + \cos^3\theta}{\sin\theta + \cos\theta} = \sin^2\theta - \sin\theta\cos\theta + \cos^2\theta = 1 - \sin\theta\cos\theta = 1 - \dfrac{1}{2}(2\sin\theta\cos\theta) = 1 - \dfrac{1}{2}\sin(2\theta)$

51. $\dfrac{\sin(3\theta)}{\sin\theta} - \dfrac{\cos(3\theta)}{\cos\theta} = \dfrac{\sin(3\theta)\cos\theta - \cos(3\theta)\sin\theta}{\sin\theta\cos\theta} = \dfrac{\sin(3\theta - \theta)}{\frac{1}{2}(2\sin\theta\cos\theta)} = \dfrac{2\sin(2\theta)}{\sin(2\theta)} = 2$

52. $\dfrac{\cos\theta + \sin\theta}{\cos\theta - \sin\theta} - \dfrac{\cos\theta - \sin\theta}{\cos\theta + \sin\theta} = \dfrac{(\cos\theta + \sin\theta)^2 - (\cos\theta - \sin\theta)^2}{(\cos\theta - \sin\theta)(\cos\theta + \sin\theta)}$

$= \dfrac{\cos^2 + 2\sin\theta\cos\theta + \sin^2\theta - \cos^2\theta + 2\sin\theta\cos\theta - \sin^2\theta}{\cos^2\theta - \sin^2\theta} = \dfrac{4\sin\theta\cos\theta}{\cos^2\theta - \sin^2\theta} = 2\left(\dfrac{2\sin\theta\cos\theta}{\cos^2\theta - \sin^2\theta}\right)$

$= 2\left(\dfrac{\sin(2\theta)}{\cos(2\theta)}\right) = 2\tan(2\theta)$

53. $\tan(3\theta) = \tan(\theta + 2\theta) = \dfrac{\tan\theta + \tan(2\theta)}{1 - \tan\theta\tan(2\theta)} = \dfrac{\tan\theta + \dfrac{2\tan\theta}{1 - \tan^2\theta}}{1 - \dfrac{\tan\theta(2\tan\theta)}{1 - \tan^2\theta}} = \dfrac{\tan\theta - \tan^3\theta + 2\tan\theta}{1 - \tan^2\theta - 2\tan^2\theta} = \dfrac{3\tan\theta - \tan^3\theta}{1 - 3\tan^2\theta}$

54. $\tan\theta + \tan(\theta + 120^\circ) + \tan(\theta + 240^\circ) = \tan\theta + \dfrac{\tan\theta + \tan 120^\circ}{1 - \tan\theta\tan 120^\circ} + \dfrac{\tan\theta + \tan 240^\circ}{1 - \tan\theta\tan 240^\circ}$

$= \tan\theta + \dfrac{\tan\theta - \sqrt{3}}{1 + \sqrt{3}\tan\theta} + \dfrac{\tan\theta + \sqrt{3}}{1 - \sqrt{3}\tan\theta} = \tan\theta + \dfrac{(\tan\theta - \sqrt{3})(1 - \sqrt{3}\tan\theta) + (\tan\theta + \sqrt{3})(1 + \sqrt{3}\tan\theta)}{1 - 3\tan^2\theta}$

$= \dfrac{\tan\theta - 3\tan^3\theta + \tan\theta - \sqrt{3}\tan^2\theta - \sqrt{3} + 3\tan\theta + \tan\theta + \sqrt{3}\tan^2\theta + \sqrt{3} + 3\tan\theta}{1 - 3\tan^2\theta}$

$= 3\left(\dfrac{3\tan\theta - \tan^3\theta}{1 - 3\tan^2\theta}\right) = 3\tan(3\theta)$ (from Problem 53)

55. $\frac{1}{2}(\ln|1 - \cos(2\theta)| - \ln 2) = \ln\left(\dfrac{|1 - \cos(2\theta)|}{2}\right)^{1/2} = \ln|\sin^2\theta|^{1/2} = \ln|\sin\theta|$

56. $\frac{1}{2}(\ln|1 + \cos(2\theta)| - \ln 2) = \ln\left(\dfrac{|1 + \cos(2\theta)|}{2}\right)^{1/2} = \ln|\cos^2\theta|^{1/2} = \ln|\cos\theta|$ **57.** $\dfrac{\sqrt{3}}{2}$ **58.** $\dfrac{\sqrt{3}}{2}$ **59.** $\dfrac{7}{25}$ **60.** $\dfrac{7}{25}$ **61.** $\dfrac{24}{7}$ **62.** $\dfrac{24}{7}$

63. $\dfrac{24}{25}$ **64.** $-\dfrac{7}{25}$ **65.** $\dfrac{1}{5}$ **66.** $\dfrac{9}{10}$ **67.** $\dfrac{25}{7}$ **68.** $-\dfrac{25}{24}$ **69.** $\sin(2\theta) = \dfrac{4x}{4 + x^2}$ **70.** $\cos(2\theta) = \dfrac{4 - x^2}{4 + x^2}$ **71.** $-\dfrac{1}{4}$ **72.** $-\dfrac{1}{4}$

73. $\dfrac{2z}{1 + z^2} = \dfrac{2\tan\left(\frac{\alpha}{2}\right)}{1 + \tan^2\left(\frac{\alpha}{2}\right)} = \dfrac{2\tan\left(\frac{\alpha}{2}\right)}{\sec^2\left(\frac{\alpha}{2}\right)} = \dfrac{\dfrac{2\sin\left(\frac{\alpha}{2}\right)}{\cos\left(\frac{\alpha}{2}\right)}}{\dfrac{1}{\cos^2\left(\frac{\alpha}{2}\right)}} = 2\sin\left(\frac{\alpha}{2}\right)\cos\left(\frac{\alpha}{2}\right) = \sin\left(2\cdot\frac{\alpha}{2}\right) = \sin\alpha$

74. $\dfrac{1 - z^2}{1 + z^2} = \dfrac{1 - \tan^2\left(\frac{\alpha}{2}\right)}{1 + \tan^2\left(\frac{\alpha}{2}\right)} = \dfrac{1 - \dfrac{1 - \cos\alpha}{1 + \cos\alpha}}{1 + \dfrac{1 - \cos\alpha}{1 + \cos\alpha}} = \dfrac{\dfrac{1 + \cos\alpha - 1 + \cos\alpha}{1 + \cos\alpha}}{\dfrac{1 + \cos\alpha + 1 - \cos\alpha}{1 + \cos\alpha}} = \dfrac{2\cos\alpha}{2} = \cos\alpha$

75. $A = \frac{1}{2}h(\text{base}) = h\left(\frac{1}{2}\text{base}\right) = s\cos\frac{\theta}{2}\cdot s\sin\frac{\theta}{2} = \frac{1}{2}s^2\sin\theta$

76. **(a)** $A = 2\sin\theta\cos\theta$ **(b)** $2\sin\theta\cos\theta = \sin(2\theta)$ **(c)** $\theta = 45^\circ$ **(d)** $\sqrt{2}$ by $\dfrac{\sqrt{2}}{2}$

77. 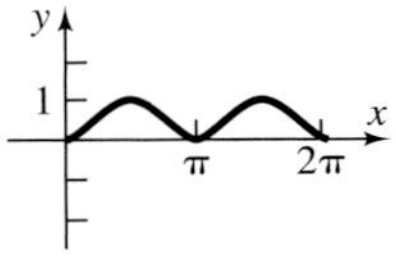

78. 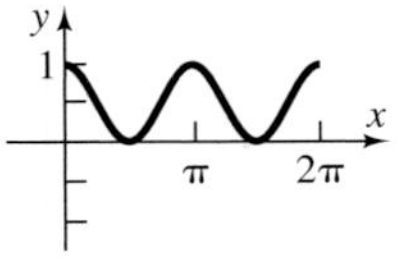

79. $\sin\dfrac{\pi}{24} = \dfrac{\sqrt{2}}{4}\sqrt{4 - \sqrt{6} - \sqrt{2}}$; $\cos\dfrac{\pi}{24} = \dfrac{\sqrt{2}}{4}\sqrt{4 + \sqrt{6} + \sqrt{2}}$

80. $\cos\dfrac{\pi}{8} = \cos\left(\dfrac{\frac{\pi}{4}}{2}\right) = \sqrt{\dfrac{1 + \cos\frac{\pi}{4}}{2}} = \sqrt{\dfrac{1 + \frac{\sqrt{2}}{2}}{2}} = \sqrt{\dfrac{2 + \sqrt{2}}{4}} = \dfrac{\sqrt{2 + \sqrt{2}}}{2}$; $\sin\left(\dfrac{\pi}{16}\right) = \dfrac{\sqrt{2 - \sqrt{2 + \sqrt{2}}}}{2}$;

$\cos\left(\dfrac{\pi}{16}\right) = \dfrac{\sqrt{2 + \sqrt{2 + \sqrt{2}}}}{2}$

81. $\sin^3\theta + \sin^3(\theta + 120°) + \sin^3(\theta + 240°) = \sin^3\theta + (\sin\theta\cos 120° + \cos\theta\sin 120°)^3 + (\sin\theta\cos 240° + \cos\theta\sin 240°)^3$

$= \sin^3\theta + \left(-\frac{1}{2}\sin\theta + \frac{\sqrt{3}}{2}\cos\theta\right)^3 + \left(-\frac{1}{2}\sin\theta - \frac{\sqrt{3}}{2}\cos\theta\right)^3$

$= \sin^3\theta + \frac{1}{8}(3\sqrt{3}\cos^3\theta - 9\cos^2\theta\sin\theta + 3\sqrt{3}\cos\theta\sin^2\theta - \sin^3\theta) - \frac{1}{8}(\sin^3\theta + 3\sqrt{3}\sin^2\theta\cos\theta + 9\sin\theta\cos^2\theta + 3\sqrt{3}\cos^3\theta)$

$= \frac{3}{4}\sin^3\theta - \frac{9}{4}\cos^2\theta\sin\theta = \frac{3}{4}[\sin^3\theta - 3\sin\theta(1 - \sin^2\theta)] = \frac{3}{4}(4\sin^3\theta - 3\sin\theta) = -\frac{3}{4}\sin(3\theta)$ (from Example 2)

82. $\tan\theta = \tan\left(3\cdot\frac{\theta}{3}\right) = \dfrac{3\tan\frac{\theta}{3} - \tan^3\frac{\theta}{3}}{1 - 3\tan^2\frac{\theta}{3}}$ (from Problem 53)

$a\tan\frac{\theta}{3} = \dfrac{\tan\frac{\theta}{3}\left(3 - \tan^2\frac{\theta}{3}\right)}{1 - 3\tan^2\frac{\theta}{3}} \Rightarrow a - 3a\tan^2\frac{\theta}{3} = 3 - \tan^2\frac{\theta}{3} \Rightarrow a - 3 = 3a\tan^2\frac{\theta}{3} - \tan^2\frac{\theta}{3} \Rightarrow a - 3 = \tan^2\frac{\theta}{3}(3a - 1)$

$\Rightarrow \tan^2\frac{\theta}{3} = \frac{a-3}{3a-1} \Rightarrow \tan\frac{\theta}{3} = \pm\sqrt{\frac{a-3}{3a-1}}$

83. **(a)** $R = \frac{v_0^2\sqrt{2}}{16}(\sin\theta\cos\theta - \cos^2\theta)$

$= \frac{v_0^2\sqrt{2}}{16}\left[\frac{1}{2}\sin(2\theta) - \frac{1 + \cos(2\theta)}{2}\right]$

$= \frac{v_0^2\sqrt{2}}{32}[\sin(2\theta) - \cos(2\theta) - 1]$

(b)

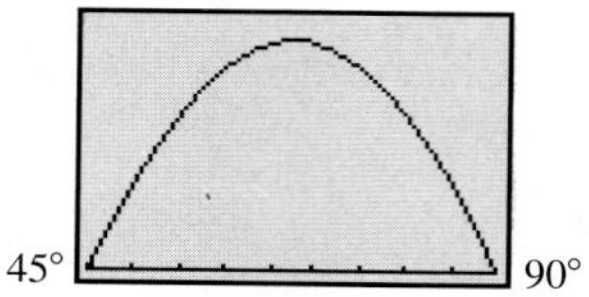

(c) $\theta = 67.5°$ makes R largest.

84. $y = \frac{1}{2}\sin(2\pi x) + \frac{1}{4}\sin(4\pi x) = \frac{1}{2}\sin(2\pi x) + \frac{1}{4}\sin(2\cdot 2\pi x) = \frac{1}{2}\sin(2\pi x) + \frac{1}{4}[2\sin(2\pi x)\cos(2\pi x)]$

$= \frac{1}{2}\sin(2\pi x) + \frac{1}{2}[\sin(2\pi x)\cdot\cos(2\cdot\pi x)] = \frac{1}{2}\sin(2\pi x) + \frac{1}{2}[\sin(2\pi x)\cdot(2\cos^2(\pi x) - 1)]$

$= \frac{1}{2}\sin(2\pi x) - \frac{1}{2}\sin(2\pi x) + \sin(2\pi x)\cos^2(\pi x) = \sin(2\pi x)\cos^2(\pi x)$

6.6 Exercises *(page 459)*

1. $\frac{1}{2}[\cos(2\theta) - \cos(6\theta)]$ **2.** $\frac{1}{2}[\cos(2\theta) + \cos(6\theta)]$ **3.** $\frac{1}{2}[\sin(6\theta) + \sin(2\theta)]$ **4.** $\frac{1}{2}[\cos(2\theta) - \cos(8\theta)]$ **5.** $\frac{1}{2}[\cos(2\theta) + \cos(8\theta)]$

6. $\frac{1}{2}[\sin(10\theta) - \sin(2\theta)]$ **7.** $\frac{1}{2}[\cos\theta - \cos(3\theta)]$ **8.** $\frac{1}{2}[\cos\theta + \cos(7\theta)]$ **9.** $\frac{1}{2}[\sin(2\theta) + \sin\theta]$ **10.** $\frac{1}{2}[\sin(3\theta) - \sin(2\theta)]$

11. $2\sin\theta\cos(3\theta)$ **12.** $2\sin(3\theta)\cos\theta$ **13.** $2\cos(3\theta)\cos\theta$ **14.** $-2\sin(4\theta)\sin\theta$ **15.** $2\sin(2\theta)\cos\theta$ **16.** $2\cos(2\theta)\cos\theta$

17. $2\sin\theta\sin\frac{\theta}{2}$ **18.** $-2\sin\frac{\theta}{2}\cos\theta$ **19.** $\dfrac{\sin\theta + \sin(3\theta)}{2\sin(2\theta)} = \dfrac{2\sin(2\theta)\cos\theta}{2\sin(2\theta)} = \cos\theta$

20. $\dfrac{\cos\theta + \cos(3\theta)}{2\cos(2\theta)} = \dfrac{2\cos(2\theta)\cos\theta}{2\cos(2\theta)} = \cos\theta$ **21.** $\dfrac{\sin(4\theta) + \sin(2\theta)}{\cos(4\theta) + \cos(2\theta)} = \dfrac{2\sin(3\theta)\cos\theta}{2\cos(3\theta)\cos\theta} = \dfrac{\sin(3\theta)}{\cos(3\theta)} = \tan(3\theta)$

22. $\dfrac{\cos\theta - \cos(3\theta)}{\sin(3\theta) - \sin\theta} = \dfrac{2\sin(2\theta)\sin\theta}{2\sin\theta\cos(2\theta)} = \dfrac{\sin(2\theta)}{\cos(2\theta)} = \tan(2\theta)$ **23.** $\dfrac{\cos\theta - \cos(3\theta)}{\sin\theta + \sin(3\theta)} = \dfrac{2\sin(2\theta)\sin\theta}{2\sin(2\theta)\cos\theta} = \dfrac{\sin\theta}{\cos\theta} = \tan\theta$

24. $\dfrac{\cos\theta - \cos(5\theta)}{\sin\theta + \sin(5\theta)} = \dfrac{2\sin(3\theta)\sin(2\theta)}{2\sin(3\theta)\cos(2\theta)} = \dfrac{\sin(2\theta)}{\cos(2\theta)} = \tan(2\theta)$

25. $\sin\theta[\sin\theta + \sin(3\theta)] = \sin\theta[2\sin(2\theta)\cos\theta] = \cos\theta[2\sin(2\theta)\sin\theta] = \cos\theta\left[2\cdot\frac{1}{2}[\cos\theta - \cos(3\theta)]\right] = \cos\theta[\cos\theta - \cos(3\theta)]$

26. $\sin\theta[\sin(3\theta) + \sin(5\theta)] = \sin\theta[2\sin(4\theta)\cos\theta] = \cos\theta[2\sin(4\theta)\sin\theta] = \cos\theta\left[2\cdot\frac{1}{2}[\cos(3\theta) - \cos(5\theta)]\right] = \cos\theta[\cos(3\theta) - \cos(5\theta)]$

27. $\dfrac{\sin(4\theta) + \sin(8\theta)}{\cos(4\theta) + \cos(8\theta)} = \dfrac{2\sin(6\theta)\cos(2\theta)}{2\cos(6\theta)\cos(2\theta)} = \dfrac{\sin(6\theta)}{\cos(6\theta)} = \tan(6\theta)$

28. $\dfrac{\sin(4\theta) - \sin(8\theta)}{\cos(4\theta) - \cos(8\theta)} = \dfrac{2\sin(-2\theta)\cos(6\theta)}{-2\sin(6\theta)\sin(-2\theta)} = -\dfrac{\cos(6\theta)}{\sin(6\theta)} = -\cot(6\theta)$

29. $\dfrac{\sin(4\theta) + \sin(8\theta)}{\sin(4\theta) - \sin(8\theta)} = \dfrac{2\sin(6\theta)\cos(-2\theta)}{2\sin(-2\theta)\cos(6\theta)} = \dfrac{\sin(6\theta)}{\cos(6\theta)}\cdot\dfrac{\cos(2\theta)}{-\sin(2\theta)} = \tan(6\theta)[-\cot(2\theta)] = -\dfrac{\tan(6\theta)}{\tan(2\theta)}$

30. $\dfrac{\cos(4\theta)-\cos(8\theta)}{\cos(4\theta)+\cos(8\theta)} = \dfrac{-2\sin(6\theta)\sin(-2\theta)}{2\cos(6\theta)\cos(-2\theta)} = \dfrac{\sin(2\theta)}{\cos(2\theta)}\cdot\dfrac{\sin(6\theta)}{\cos(6\theta)} = \tan(2\theta)\tan(6\theta)$

31. $\dfrac{\sin\alpha+\sin\beta}{\sin\alpha-\sin\beta} = \dfrac{2\sin\frac{\alpha+\beta}{2}\cos\frac{\alpha-\beta}{2}}{2\sin\frac{\alpha-\beta}{2}\cos\frac{\alpha+\beta}{2}} = \dfrac{\sin\frac{\alpha+\beta}{2}}{\cos\frac{\alpha+\beta}{2}}\cdot\dfrac{\cos\frac{\alpha-\beta}{2}}{\sin\frac{\alpha-\beta}{2}} = \tan\dfrac{\alpha+\beta}{2}\cot\dfrac{\alpha-\beta}{2}$

32. $\dfrac{\cos\alpha+\cos\beta}{\cos\alpha-\cos\beta} = \dfrac{2\cos\frac{\alpha+\beta}{2}\cos\frac{\alpha-\beta}{2}}{-2\sin\frac{\alpha+\beta}{2}\sin\frac{\alpha-\beta}{2}} = -\dfrac{\cos\frac{\alpha+\beta}{2}}{\sin\frac{\alpha+\beta}{2}}\cdot\dfrac{\cos\frac{\alpha-\beta}{2}}{\sin\frac{\alpha-\beta}{2}} = -\cot\dfrac{\alpha+\beta}{2}\cot\dfrac{\alpha-\beta}{2}$

33. $\dfrac{\sin\alpha+\sin\beta}{\cos\alpha+\cos\beta} = \dfrac{2\sin\frac{\alpha+\beta}{2}\cos\frac{\alpha-\beta}{2}}{2\cos\frac{\alpha+\beta}{2}\cos\frac{\alpha-\beta}{2}} = \dfrac{\sin\frac{\alpha+\beta}{2}}{\cos\frac{\alpha+\beta}{2}} = \tan\dfrac{\alpha+\beta}{2}$

34. $\dfrac{\sin\alpha-\sin\beta}{\cos\alpha-\cos\beta} = \dfrac{2\sin\frac{\alpha-\beta}{2}\cos\frac{\alpha+\beta}{2}}{-2\sin\frac{\alpha+\beta}{2}\sin\frac{\alpha-\beta}{2}} = -\dfrac{\cos\frac{\alpha+\beta}{2}}{\sin\frac{\alpha+\beta}{2}} = -\cot\dfrac{\alpha+\beta}{2}$

35. $1+\cos(2\theta)+\cos(4\theta)+\cos(6\theta) = [1+\cos(6\theta)]+[\cos(2\theta)+\cos(4\theta)] = 2\cos^2(3\theta)+2\cos(3\theta)\cos(-\theta)$
$= 2\cos(3\theta)[\cos(3\theta)+\cos\theta] = 2\cos(3\theta)[2\cos(2\theta)\cos\theta] = 4\cos\theta\cos(2\theta)\cos(3\theta)$

36. $1-\cos(2\theta)+\cos(4\theta)-\cos(6\theta) = [1-\cos(6\theta)]+[\cos(4\theta)-\cos(2\theta)] = 2\sin^2(3\theta)+[-2\sin(3\theta)\sin\theta]$
$= 2\sin(3\theta)[\sin(3\theta)-\sin\theta] = 2\sin(3\theta)[2\sin\theta\cos(2\theta)] = 4\sin\theta\cos(2\theta)\sin(3\theta)$

37. **(a)** $y = 2\sin(2061\pi t)\cos(357\pi t)$ **(b)** $y_{\max} = 2$
(c)

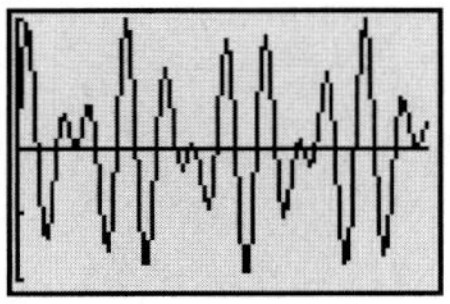

38. **(a)** $y = 2\sin(2418\pi t)\cos(536\pi t)$ **(b)** $y_{\max} = 2$
(c)

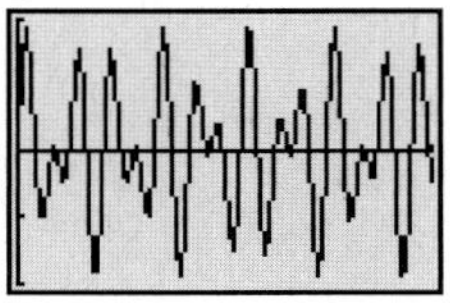

39. $\sin(2\alpha)+\sin(2\beta)+\sin(2\gamma) = 2\sin(\alpha+\beta)\cos(\alpha-\beta)+\sin(2\gamma) = 2\sin(\alpha+\beta)\cos(\alpha-\beta)+2\sin\gamma\cos\gamma$
$= 2\sin(\pi-\gamma)\cos(\alpha-\beta)+2\sin\gamma\cos\gamma = 2\sin\gamma\cos(\alpha-\beta)+2\sin\gamma\cos\gamma = 2\sin\gamma[\cos(\alpha-\beta)+\cos\gamma]$
$= 2\sin\gamma\left(2\cos\dfrac{\alpha-\beta+\gamma}{2}\cos\dfrac{\alpha-\beta-\gamma}{2}\right) = 4\sin\gamma\cos\dfrac{\pi-2\beta}{2}\cos\dfrac{2\alpha-\pi}{2} = 4\sin\gamma\cos\left(\dfrac{\pi}{2}-\beta\right)\cos\left(\alpha-\dfrac{\pi}{2}\right)$
$= 4\sin\gamma\sin\beta\sin\alpha$

40. $\tan\alpha+\tan\beta+\tan\gamma = \dfrac{\sin\alpha}{\cos\alpha}+\dfrac{\sin\beta}{\cos\beta}+\dfrac{\sin\gamma}{\cos\gamma} = \dfrac{\sin\alpha\cos\beta\cos\gamma+\sin\beta\cos\alpha\cos\gamma+\sin\gamma\cos\alpha\cos\beta}{\cos\alpha\cos\beta\cos\gamma}$
$= \dfrac{\cos\gamma\sin(\alpha+\beta)+\sin\gamma\cos\alpha\cos\beta}{\cos\alpha\cos\beta\cos\gamma} = \dfrac{\sin\gamma(\cos\gamma+\cos\alpha\cos\beta)}{\cos\alpha\cos\beta\cos\gamma} = \dfrac{\sin\gamma[-\cos(\alpha+\beta)+\cos\alpha\cos\beta]}{\cos\alpha\cos\beta\cos\gamma} = \dfrac{\sin\gamma(\sin\alpha\sin\beta)}{\cos\alpha\cos\beta\cos\gamma}$
$= \tan\alpha\tan\beta\tan\gamma$ [Note that $\alpha+\beta = \pi-\gamma$, $\sin(\alpha+\beta) = \sin\gamma$, $\cos\gamma = \cos[\pi-(\alpha+\beta)] = -\cos(\alpha+\beta)$.]

41.
$$\begin{aligned} \sin(\alpha-\beta) &= \sin\alpha\cos\beta-\cos\alpha\sin\beta \\ \sin(\alpha+\beta) &= \sin\alpha\cos\beta+\cos\alpha\sin\beta \\ \sin(\alpha-\beta)+\sin(\alpha+\beta) &= 2\sin\alpha\cos\beta \\ \sin\alpha\cos\beta &= \frac{1}{2}[\sin(\alpha+\beta)+\sin(\alpha-\beta)] \end{aligned}$$

42. $2\sin\dfrac{\alpha-\beta}{2}\cos\dfrac{\alpha+\beta}{2} = 2\cdot\dfrac{1}{2}\left[\sin\left(\dfrac{\alpha-\beta}{2}+\dfrac{\alpha+\beta}{2}\right)+\sin\left(\dfrac{\alpha-\beta}{2}-\dfrac{\alpha+\beta}{2}\right)\right] = \sin\dfrac{2\alpha}{2}+\sin\left(-\dfrac{2\beta}{2}\right) = \sin\alpha-\sin\beta$

43. $2\cos\dfrac{\alpha+\beta}{2}\cos\dfrac{\alpha-\beta}{2} = 2\cdot\dfrac{1}{2}\left[\cos\left(\dfrac{\alpha+\beta}{2}+\dfrac{\alpha-\beta}{2}\right)+\cos\left(\dfrac{\alpha+\beta}{2}-\dfrac{\alpha-\beta}{2}\right)\right] = \cos\dfrac{2\alpha}{2}+\cos\dfrac{2\beta}{2} = \cos\alpha+\cos\beta$

44. $-2\sin\dfrac{\alpha+\beta}{2}\sin\dfrac{\alpha-\beta}{2} = -2\cdot\dfrac{1}{2}\left[\cos\left(\dfrac{\alpha+\beta}{2}-\dfrac{\alpha-\beta}{2}\right)-\cos\left(\dfrac{\alpha+\beta}{2}+\dfrac{\alpha-\beta}{2}\right)\right] = -\cos\dfrac{2\beta}{2}+\cos\dfrac{2\alpha}{2} = \cos\alpha-\cos\beta$

6.7 Concepts and Vocabulary *(page 465)*

3. $\left\{\dfrac{\pi}{6},\dfrac{5\pi}{6}\right\}$ **4.** $\left\{\theta \,\middle|\, \theta = \dfrac{\pi}{6}+2\pi k, \theta = \dfrac{5\pi}{6}+2\pi k, k \text{ any integer}\right\}$ **5.** False **6.** False

6.7 Exercises *(page 465)*

7. $\left\{\frac{7\pi}{6}, \frac{11\pi}{6}\right\}$ **8.** $\left\{\frac{\pi}{3}, \frac{5\pi}{3}\right\}$ **9.** $\left\{\frac{\pi}{3}, \frac{2\pi}{3}, \frac{4\pi}{3}, \frac{5\pi}{3}\right\}$ **10.** $\left\{\frac{\pi}{6}, \frac{5\pi}{6}, \frac{7\pi}{6}, \frac{11\pi}{6}\right\}$ **11.** $\left\{\frac{\pi}{4}, \frac{3\pi}{4}, \frac{5\pi}{4}, \frac{7\pi}{4}\right\}$ **12.** $\left\{\frac{\pi}{6}, \frac{5\pi}{6}, \frac{7\pi}{6}, \frac{11\pi}{6}\right\}$

13. $\left\{\frac{\pi}{2}, \frac{7\pi}{6}, \frac{11\pi}{6}\right\}$ **14.** $\left\{\frac{2\pi}{3}\right\}$ **15.** $\left\{\frac{\pi}{3}, \frac{2\pi}{3}, \frac{4\pi}{3}, \frac{5\pi}{3}\right\}$ **16.** $\left\{\frac{3\pi}{8}, \frac{7\pi}{8}, \frac{11\pi}{8}, \frac{15\pi}{8}\right\}$ **17.** $\left\{\frac{4\pi}{9}, \frac{8\pi}{9}, \frac{16\pi}{9}\right\}$ **18.** $\left\{\frac{5\pi}{4}\right\}$

19. $\left\{\frac{7\pi}{6}, \frac{11\pi}{6}\right\}$ **20.** $\{\pi\}$ **21.** $\left\{\frac{3\pi}{4}, \frac{7\pi}{4}\right\}$ **22.** $\left\{\frac{2\pi}{3}, \frac{5\pi}{3}\right\}$ **23.** $\left\{\frac{2\pi}{3}, \frac{4\pi}{3}\right\}$ **24.** $\left\{\frac{\pi}{2}\right\}$ **25.** $\left\{\frac{3\pi}{4}, \frac{5\pi}{4}\right\}$ **26.** $\left\{\frac{4\pi}{3}, \frac{5\pi}{3}\right\}$

27. $\left\{\frac{3\pi}{4}, \frac{7\pi}{4}\right\}$ **28.** $\left\{\frac{4\pi}{27}, \frac{22\pi}{27}, \frac{40\pi}{27}\right\}$ **29.** $\left\{\frac{11\pi}{6}\right\}$ **30.** $\left\{\frac{7\pi}{4}\right\}$

31. $\left\{\theta \middle| \theta = \frac{\pi}{6} + 2k\pi, \theta = \frac{5\pi}{6} + 2k\pi\right\}; \frac{\pi}{6}, \frac{5\pi}{6}, \frac{13\pi}{6}, \frac{17\pi}{6}, \frac{25\pi}{6}, \frac{29\pi}{6}$ **32.** $\left\{\theta \middle| \theta = \frac{\pi}{4} + k\pi\right\}; \frac{\pi}{4}, \frac{5\pi}{4}, \frac{9\pi}{4}, \frac{13\pi}{4}, \frac{17\pi}{4}, \frac{21\pi}{4}$

33. $\left\{\theta \middle| \theta = \frac{5\pi}{6} + k\pi\right\}; \frac{5\pi}{6}, \frac{11\pi}{6}, \frac{17\pi}{6}, \frac{23\pi}{6}, \frac{29\pi}{6}, \frac{35\pi}{6}$ **34.** $\left\{\theta \middle| \theta = \frac{5\pi}{6} + 2k\pi, \theta = \frac{7\pi}{6} + 2k\pi\right\}; \frac{5\pi}{6}, \frac{7\pi}{6}, \frac{17\pi}{6}, \frac{19\pi}{6}, \frac{29\pi}{6}, \frac{31\pi}{6}$

35. $\left\{\theta \middle| \theta = \frac{\pi}{2} + 2k\pi, \theta = \frac{3\pi}{2} + 2k\pi\right\}; \frac{\pi}{2}, \frac{3\pi}{2}, \frac{5\pi}{2}, \frac{7\pi}{2}, \frac{9\pi}{2}, \frac{11\pi}{2}$ **36.** $\left\{\theta \middle| \theta = \frac{\pi}{4} + 2k\pi, \theta = \frac{3\pi}{4} + 2k\pi\right\}; \frac{\pi}{4}, \frac{3\pi}{4}, \frac{9\pi}{4}, \frac{11\pi}{4}, \frac{17\pi}{4}, \frac{19\pi}{4}$

37. $\left\{\theta \middle| \theta = \frac{\pi}{3} + k\pi, \theta = \frac{2\pi}{3} + k\pi\right\}; \frac{\pi}{3}, \frac{2\pi}{3}, \frac{4\pi}{3}, \frac{5\pi}{3}, \frac{7\pi}{3}, \frac{8\pi}{3}$ **38.** $\left\{\theta \middle| \theta = \frac{3\pi}{4} + k\pi\right\}; \frac{3\pi}{4}, \frac{7\pi}{4}, \frac{11\pi}{4}, \frac{15\pi}{4}, \frac{19\pi}{4}, \frac{23\pi}{4}$

39. $\left\{\theta \middle| \theta = \frac{8\pi}{3} + 4k\pi, \theta = \frac{10\pi}{3} + 4k\pi\right\}; \frac{8\pi}{3}, \frac{10\pi}{3}, \frac{20\pi}{3}, \frac{22\pi}{3}, \frac{32\pi}{3}, \frac{34\pi}{3}$ **40.** $\left\{\theta \middle| \theta = \frac{3\pi}{2} + 2k\pi\right\}; \frac{3\pi}{2}, \frac{7\pi}{2}, \frac{11\pi}{2}, \frac{15\pi}{2}, \frac{19\pi}{2}, \frac{23\pi}{2}$

41. $\{0.41, 2.73\}$ **42.** $\{0.93, 5.36\}$ **43.** $\{1.37, 4.51\}$ **44.** $\{0.46, 3.61\}$ **45.** $\{2.69, 3.59\}$ **46.** $\{3.34, 6.08\}$

47. $\{1.82, 4.46\}$ **48.** $\{3.48, 5.94\}$ **49.** $\{2.08, 5.22\}$ **50.** $\{2.47, 5.61\}$ **51.** $\{0.73, 2.41\}$ **52.** $\{2.42, 3.86\}$

53. (a) $\left\{x \middle| x = \frac{\pi}{6} + 2\pi k, x = \frac{5\pi}{6} + 2\pi k\right\}$ **(b)** $\frac{\pi}{6} < x < \frac{5\pi}{6}$ or $\left(\frac{\pi}{6}, \frac{5\pi}{6}\right)$

54. (a) $\left\{x \middle| x = \frac{5\pi}{6} + 2\pi k, x = \frac{7\pi}{6} + 2\pi k\right\}$ **(b)** $\frac{5\pi}{6} < x < \frac{7\pi}{6}$ or $\left(\frac{5\pi}{6}, \frac{7\pi}{6}\right)$ **55. (a)** $\left\{x \middle| x = -\frac{\pi}{4} + k\pi\right\}$

(b) $-\frac{\pi}{2} < x < -\frac{\pi}{4}$ or $\left(-\frac{\pi}{2}, -\frac{\pi}{4}\right)$ **56. (a)** $\left\{x \middle| x = \frac{5\pi}{6} + k\pi\right\}$ **(b)** $0 < x < \frac{5\pi}{6}$ or $\left(0, \frac{5\pi}{6}\right)$

57. (a) 10 sec; 30 sec **(b)** 20 sec; 60 sec **(c)** $10 < x < 30$ or $(10, 30)$ **58. (a)** 0.125, 0.375, 0.625, 0.875 sec

(b) 0.083, 0.417, 0.583, 0.917 sec **(c)** $0.125 < t < 0.375; 0.625 < t < 0.875$ **59. (a)** 150 mi **(b)** 6.06, 8.44, 15.72, 18.11 min **(c)** Before 6.06 min, between 8.44 and 15.72 min, and after 18.11 min **(d)** No **60. (a)** 21.02° or 68.98° **(b)** 26.74° or 63.26° **(c)** Between 22.79° and 67.21° **(d)** No **61.** 28.90° **62.** 27.48° **63.** Yes; it varies from 1.28 to 1.34. **64.** 1.56 **65.** 1.47 **66.** 19.20°

67. If θ is the original angle of incidence and ϕ is the angle of refraction, then $\frac{\sin\theta}{\sin\phi} = n_2$. The angle of incidence of the emerging beam is also ϕ, and the index of refraction is $\frac{1}{n_2}$. Thus, θ is the angle of refraction of the emerging beam.

6.8 Exercises *(page 473)*

3. $\frac{\pi}{2}, \frac{2\pi}{3}, \frac{4\pi}{3}, \frac{3\pi}{2}$ **4.** $\frac{\pi}{2}, \frac{3\pi}{2}$ **5.** $\frac{\pi}{2}, \frac{7\pi}{6}, \frac{11\pi}{6}$ **6.** $\frac{\pi}{3}, \pi, \frac{5\pi}{3}$ **7.** $0, \frac{\pi}{4}, \frac{5\pi}{4}$ **8.** $\frac{3\pi}{4}, \frac{7\pi}{4}$ **9.** $\frac{\pi}{2}, \frac{2\pi}{3}, \frac{4\pi}{3}, \frac{3\pi}{2}$ **10.** $\frac{\pi}{2}, \frac{7\pi}{6}, \frac{11\pi}{6}$ **11.** π

12. $0, \frac{\pi}{3}, \frac{5\pi}{3}$ **13.** $\frac{\pi}{3}, \frac{2\pi}{3}, \frac{4\pi}{3}, \frac{5\pi}{3}$ **14.** No solution **15.** $\frac{\pi}{4}, \frac{5\pi}{4}$ **16.** $\frac{3\pi}{4}, \frac{7\pi}{4}$ **17.** $0, \frac{\pi}{3}, \pi, \frac{5\pi}{3}$ **18.** $\frac{\pi}{6}, \frac{\pi}{2}, \frac{5\pi}{6}, \frac{3\pi}{2}$ **19.** $\frac{\pi}{2}, \frac{3\pi}{2}$

20. $\frac{\pi}{4}, \frac{3\pi}{4}, \frac{5\pi}{4}, \frac{7\pi}{4}$ **21.** $0, \frac{2\pi}{3}, \frac{4\pi}{3}$ **22.** $\frac{\pi}{4}, \frac{\pi}{2}, \frac{3\pi}{4}, \frac{5\pi}{4}, \frac{3\pi}{2}, \frac{7\pi}{4}$ **23.** $0, \frac{\pi}{3}, \frac{\pi}{2}, \frac{2\pi}{3}, \pi, \frac{4\pi}{3}, \frac{3\pi}{2}, \frac{5\pi}{3}$ **24.** $\frac{\pi}{6}, \frac{\pi}{2}, \frac{5\pi}{6}, \frac{7\pi}{6}, \frac{3\pi}{2}, \frac{11\pi}{6}$

25. $0, \frac{\pi}{5}, \frac{2\pi}{5}, \frac{3\pi}{5}, \frac{4\pi}{5}, \pi, \frac{6\pi}{5}, \frac{7\pi}{5}, \frac{8\pi}{5}, \frac{9\pi}{5}$ **26.** $0, \frac{\pi}{10}, \frac{3\pi}{10}, \frac{5\pi}{10}, \frac{7\pi}{10}, \frac{9\pi}{10}, \pi, \frac{11\pi}{10}, \frac{13\pi}{10}, \frac{15\pi}{10}, \frac{17\pi}{10}, \frac{19\pi}{10}$ **27.** $\frac{\pi}{6}, \frac{5\pi}{6}, \frac{3\pi}{2}$ **28.** π **29.** $\frac{\pi}{2}$

30. $\frac{2\pi}{3}, \frac{4\pi}{3}$ **31.** 0 **32.** $\frac{3\pi}{2}$ **33.** $\frac{\pi}{3}, \frac{5\pi}{3}$ **34.** $\frac{\pi}{4}, \frac{\pi}{2}, \frac{5\pi}{4}, \frac{3\pi}{2}$ **35.** No real solutions **36.** $\frac{2\pi}{3}, \frac{4\pi}{3}$ **37.** No real solutions

38. No real solutions **39.** $\frac{\pi}{2}, \frac{7\pi}{6}$ **40.** $0, \frac{2\pi}{3}$ **41.** $0, \frac{\pi}{3}, \pi, \frac{5\pi}{3}$ **42.** $\frac{\pi}{2}, \frac{7\pi}{6}, \frac{3\pi}{2}, \frac{11\pi}{6}$ **43.** $\frac{\pi}{4}$ **44.** $\frac{5\pi}{4}$

45. $-1.29, 0$

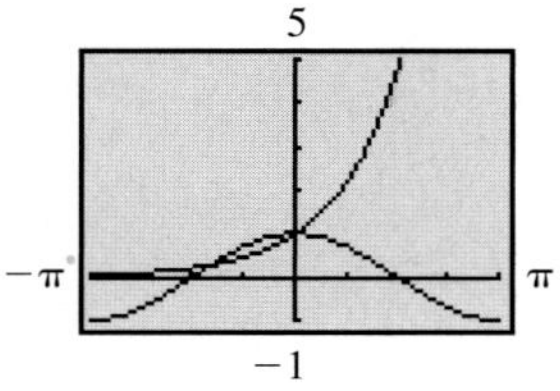

46. $-1.29, 0$

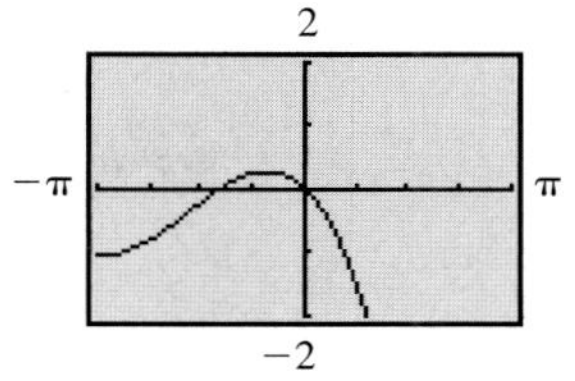

47. $-2.24, 0, 2.24$

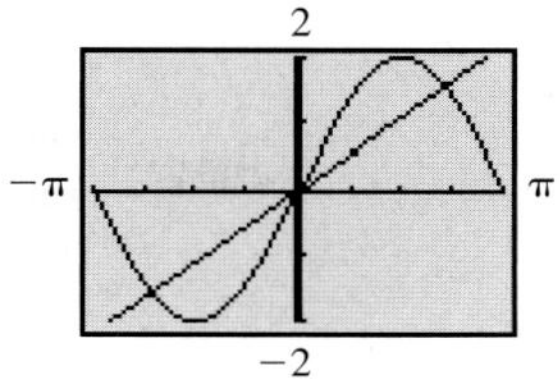

48. $-2.24, 0, 2.24$

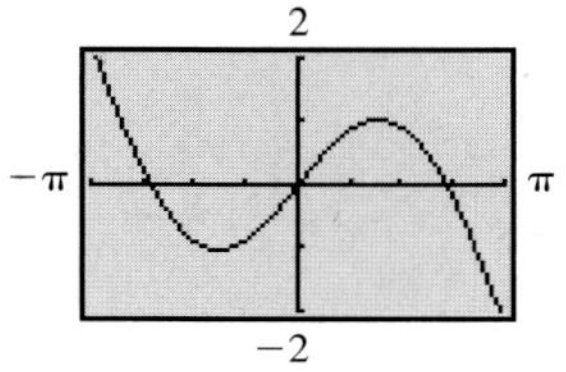

49. $-0.82, 0.82$

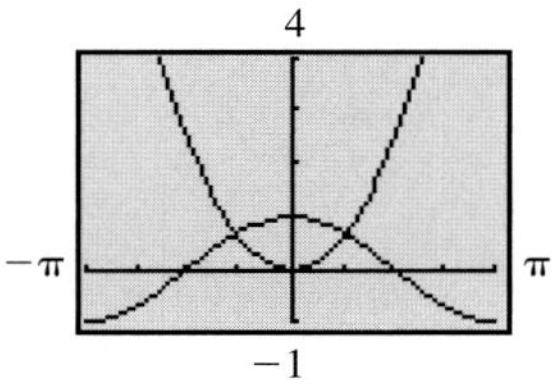

50. $-0.82, 0.82$

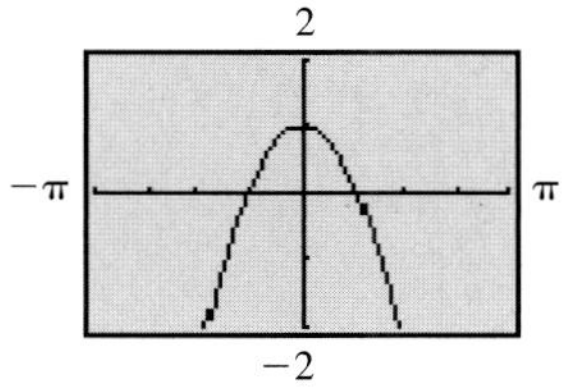

51. $-1.31, 1.98, 3.84$ **52.** $-2.47, 0, 2.47$
53. 0.52 **54.** -0.30
55. 1.26 **56.** -1.26
57. $-1.02, 1.02$ **58.** $-1.72, 0$
59. $0, 2.15$ **60.** $-0.62, 0.81$
61. $0.76, 1.35$ **62.** 0.31

63. **(a)** 60° **(b)** 60°
(c) $A(60°) = 12\sqrt{3}$ sq in.
(d)

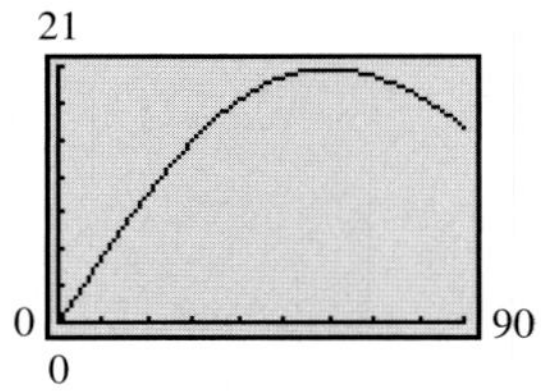

$\theta_{max} = 60°$
Maximum area $= 20.78$ sq in.

64. **(a)** 67.5° **(b)** 67.5°
(c) $R = 18.75$ ft
(d)

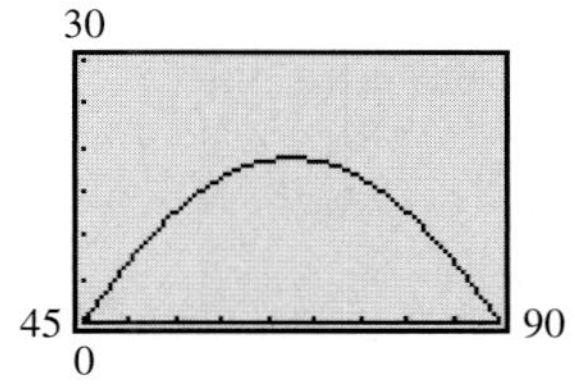

$\theta_{max} = 67.5°$
Maximum distance $= 18.75$ ft

65. $2.03, 4.91$

66. **(a)** $L = \dfrac{4}{\sin\theta} + \dfrac{3}{\cos\theta}$
(b) 47.74° **(c)** $L = 9.87$ ft
(d)

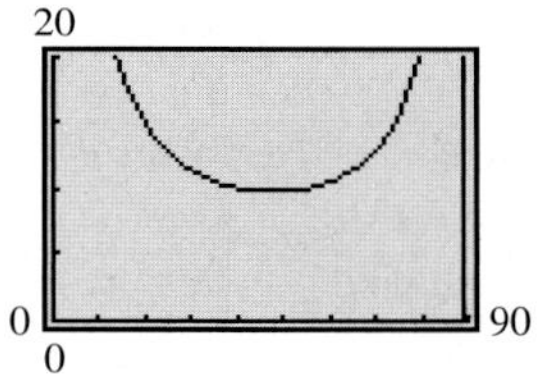

$\theta_{min} = 47.74°$

67. **(a)** 29.99° or 60.01° **(b)** 123.6 m
(c)

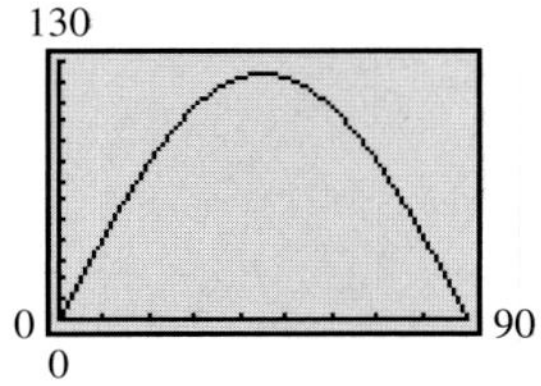

68. **(a)** 21.18° or 68.82° **(b)** 163.3 m
(c)

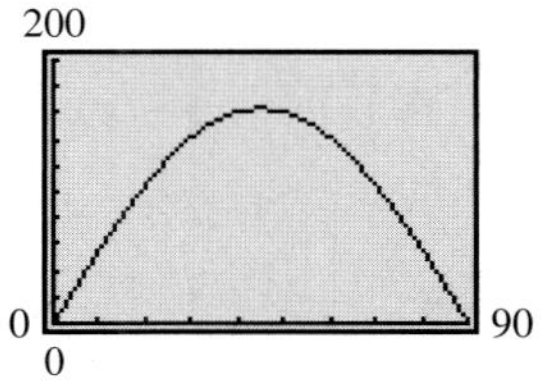

Review Exercises *(page 477)*

1. $\dfrac{\pi}{2}$ **2.** $\dfrac{\pi}{2}$ **3.** $\dfrac{\pi}{4}$ **4.** $-\dfrac{\pi}{6}$ **5.** $\dfrac{5\pi}{6}$ **6.** $-\dfrac{\pi}{3}$ **7.** $\dfrac{\pi}{4}$ **8.** $\dfrac{3\pi}{4}$ **9.** $-\sqrt{3}$ **10.** $-\sqrt{3}$ **11.** $\dfrac{2\sqrt{3}}{3}$ **12.** $\dfrac{2\sqrt{3}}{3}$ **13.** $\dfrac{3}{5}$ **14.** $\dfrac{4}{5}$ **15.** $-\dfrac{4}{3}$ **16.** $-\dfrac{4}{3}$
17. $-\dfrac{\pi}{6}$ **18.** π **19.** $-\dfrac{\pi}{4}$ **20.** $\dfrac{5\pi}{6}$ **21.** $\tan\theta\cot\theta - \sin^2\theta = 1 - \sin^2\theta = \cos^2\theta$ **22.** $\sin\theta\csc\theta - \sin^2\theta = 1 - \sin^2\theta = \cos^2\theta$
23. $\cos^2\theta(1 + \tan^2\theta) = \cos^2\theta\sec^2\theta = 1$ **24.** $(1 - \cos^2\theta)(1 + \cot^2\theta) = \sin^2\theta\csc^2\theta = 1$
25. $4\cos^2\theta + 3\sin^2\theta = \cos^2\theta + 3(\cos^2\theta + \sin^2\theta) = 3 + \cos^2\theta$
26. $4\sin^2\theta + 2\cos^2\theta = 4(1 - \cos^2\theta) + 2\cos^2\theta = 4 - 4\cos^2\theta + 2\cos^2\theta = 4 - 2\cos^2\theta$
27. $\dfrac{1 - \cos\theta}{\sin\theta} + \dfrac{\sin\theta}{1 - \cos\theta} = \dfrac{(1 - \cos\theta)^2 + \sin^2\theta}{\sin\theta(1 - \cos\theta)} = \dfrac{1 - 2\cos\theta + \cos^2\theta + \sin^2\theta}{\sin\theta(1 - \cos\theta)} = \dfrac{2(1 - \cos\theta)}{\sin\theta(1 - \cos\theta)} = 2\csc\theta$
28. $\dfrac{\sin\theta}{1 + \cos\theta} + \dfrac{1 + \cos\theta}{\sin\theta} = \dfrac{\sin^2\theta + (1 + \cos\theta)^2}{\sin\theta(1 + \cos\theta)} = \dfrac{\sin^2\theta + 1 + 2\cos\theta + \cos^2\theta}{\sin\theta(1 + \cos\theta)} = \dfrac{2(1 + \cos\theta)}{\sin\theta(1 + \cos\theta)} = 2\csc\theta$
29. $\dfrac{\cos\theta}{\cos\theta - \sin\theta} = \dfrac{\dfrac{\cos\theta}{\cos\theta}}{\dfrac{\cos\theta - \sin\theta}{\cos\theta}} = \dfrac{1}{1 - \dfrac{\sin\theta}{\cos\theta}} = \dfrac{1}{1 - \tan\theta}$

30. $1 - \dfrac{\cos^2\theta}{1+\sin\theta} = 1 - \dfrac{1-\sin^2\theta}{1+\sin\theta} = 1 - \dfrac{(1+\sin\theta)(1-\sin\theta)}{1+\sin\theta} = 1 - (1-\sin\theta) = \sin\theta$

31. $\dfrac{\csc\theta}{1+\csc\theta} = \dfrac{\frac{1}{\sin\theta}}{1+\frac{1}{\sin\theta}} = \dfrac{1}{1+\sin\theta} = \dfrac{1}{1+\sin\theta}\cdot\dfrac{1-\sin\theta}{1-\sin\theta} = \dfrac{1-\sin\theta}{1-\sin^2\theta} = \dfrac{1-\sin\theta}{\cos^2\theta}$

32. $\dfrac{1+\sec\theta}{\sec\theta} = \dfrac{1+\frac{1}{\cos\theta}}{\frac{1}{\cos\theta}} = (1+\cos\theta)\cdot\dfrac{1-\cos\theta}{1-\cos\theta} = \dfrac{1-\cos^2\theta}{1-\cos\theta} = \dfrac{\sin^2\theta}{1-\cos\theta}$

33. $\csc\theta - \sin\theta = \dfrac{1}{\sin\theta} - \sin\theta = \dfrac{1-\sin^2\theta}{\sin\theta} = \dfrac{\cos^2\theta}{\sin\theta} = \cos\theta\cdot\dfrac{\cos\theta}{\sin\theta} = \cos\theta\cot\theta$

34. $\dfrac{\csc\theta}{1-\cos\theta} = \dfrac{\frac{1}{\sin\theta}}{1-\cos\theta}\cdot\dfrac{1+\cos\theta}{1+\cos\theta} = \dfrac{1+\cos\theta}{\sin\theta(1-\cos^2\theta)} = \dfrac{1+\cos\theta}{\sin^3\theta}$

35. $\dfrac{1-\sin\theta}{\sec\theta} = \cos\theta(1-\sin\theta)\cdot\dfrac{1+\sin\theta}{1+\sin\theta} = \dfrac{\cos\theta(1-\sin^2\theta)}{1+\sin\theta} = \dfrac{\cos^3\theta}{1+\sin\theta}$

36. $\dfrac{1-\cos\theta}{1+\cos\theta}\cdot\dfrac{1-\cos\theta}{1-\cos\theta} = \dfrac{(1-\cos\theta)^2}{1-\cos^2\theta} = \dfrac{(1-\cos\theta)^2}{\sin^2\theta} = \left(\dfrac{1}{\sin\theta} - \dfrac{\cos\theta}{\sin\theta}\right)^2 = (\csc\theta - \cot\theta)^2$

37. $\cot\theta - \tan\theta = \dfrac{\cos\theta}{\sin\theta} - \dfrac{\sin\theta}{\cos\theta} = \dfrac{\cos^2\theta - \sin^2\theta}{\sin\theta\cos\theta} = \dfrac{1-2\sin^2\theta}{\sin\theta\cos\theta}$

38. $\dfrac{(2\sin^2\theta - 1)^2}{\sin^4\theta - \cos^4\theta} = \dfrac{(-\cos(2\theta))^2}{(\sin^2\theta + \cos^2\theta)(\sin^2\theta - \cos^2\theta)} = \dfrac{(\cos(2\theta))^2}{1(-\cos(2\theta))} = -\cos(2\theta) = 1 - 2\cos^2\theta$

39. $\dfrac{\cos(\alpha+\beta)}{\cos\alpha\sin\beta} = \dfrac{\cos\alpha\cos\beta - \sin\alpha\sin\beta}{\cos\alpha\sin\beta} = \dfrac{\cos\alpha\cos\beta}{\cos\alpha\sin\beta} - \dfrac{\sin\alpha\sin\beta}{\cos\alpha\sin\beta} = \cot\beta - \tan\alpha$

40. $\dfrac{\sin(\alpha-\beta)}{\sin\alpha\cos\beta} = \dfrac{\sin\alpha\cos\beta - \cos\alpha\sin\beta}{\sin\alpha\cos\beta} = \dfrac{\sin\alpha\cos\beta}{\sin\alpha\cos\beta} - \dfrac{\cos\alpha\sin\beta}{\sin\alpha\cos\beta} = 1 - \cot\alpha\tan\beta$

41. $\dfrac{\cos(\alpha-\beta)}{\cos\alpha\cos\beta} = \dfrac{\cos\alpha\cos\beta + \sin\alpha\sin\beta}{\cos\alpha\cos\beta} = \dfrac{\cos\alpha\cos\beta}{\cos\alpha\cos\beta} + \dfrac{\sin\alpha\sin\beta}{\cos\alpha\cos\beta} = 1 + \tan\alpha\tan\beta$

42. $\dfrac{\cos(\alpha+\beta)}{\sin\alpha\cos\beta} = \dfrac{\cos\alpha\cos\beta - \sin\alpha\sin\beta}{\sin\alpha\cos\beta} = \dfrac{\cos\alpha\cos\beta}{\sin\alpha\cos\beta} - \dfrac{\sin\alpha\sin\beta}{\sin\alpha\cos\beta} = \cot\alpha - \tan\beta$

43. $(1+\cos\theta)\left(\tan\dfrac{\theta}{2}\right) = \left(2\cos^2\dfrac{\theta}{2}\right)\dfrac{\sin\left(\frac{\theta}{2}\right)}{\cos\left(\frac{\theta}{2}\right)} = 2\sin\dfrac{\theta}{2}\cos\dfrac{\theta}{2} = \sin\theta$

44. $\sin\theta\tan\dfrac{\theta}{2} = 2\sin\dfrac{\theta}{2}\cos\dfrac{\theta}{2}\left(\dfrac{\sin\frac{\theta}{2}}{\cos\frac{\theta}{2}}\right) = 2\sin^2\dfrac{\theta}{2} = 2\left(\dfrac{1-\cos\theta}{2}\right) = 1 - \cos\theta$

45. $2\cot\theta\cot 2\theta = 2\left(\dfrac{\cos\theta}{\sin\theta}\right)\left(\dfrac{\cos 2\theta}{\sin 2\theta}\right) = \dfrac{2\cos\theta(\cos^2\theta - \sin^2\theta)}{2\sin^2\theta\cos\theta} = \dfrac{\cos^2\theta - \sin^2\theta}{\sin^2\theta} = \cot^2\theta - 1$

46. $2\sin(2\theta)(1 - 2\sin^2\theta) = 2\sin(2\theta)\cdot\cos(2\theta) = \sin(4\theta)$ **47.** $1 - 8\sin^2\theta\cos^2\theta = 1 - 2(2\sin\theta\cos\theta)^2 = 1 - 2\sin^2(2\theta) = \cos(4\theta)$

48. $\dfrac{\sin(3\theta)\cos\theta - \sin\theta\cos(3\theta)}{\sin(2\theta)} = \dfrac{\sin(2\theta)}{\sin(2\theta)} = 1$ **49.** $\dfrac{\sin(2\theta) + \sin(4\theta)}{\cos(2\theta) + \cos(4\theta)} = \dfrac{2\sin(3\theta)\cos(-\theta)}{2\cos(3\theta)\cos(-\theta)} = \tan(3\theta)$

50. $\dfrac{\sin(2\theta) + \sin(4\theta)}{\sin(2\theta) - \sin(4\theta)} + \dfrac{\tan(3\theta)}{\tan\theta} = \dfrac{2\sin(3\theta)\cos(-\theta)}{2\sin(-\theta)\cos(3\theta)} + \dfrac{\tan(3\theta)}{\tan\theta} = -\dfrac{\tan(3\theta)}{\tan\theta} + \dfrac{\tan(3\theta)}{\tan\theta} = 0$

51. $\dfrac{\cos(2\theta) - \cos(4\theta)}{\cos(2\theta) + \cos(4\theta)} - \tan\theta\tan(3\theta) = \dfrac{-2\sin(3\theta)\sin(-\theta)}{2\cos(3\theta)\cos(-\theta)} - \tan\theta\tan(3\theta) = \tan(3\theta)\tan\theta - \tan\theta\tan(3\theta) = 0$

52. $\tan(4\theta)[\sin(2\theta) + \sin(10\theta)] = \tan(4\theta)[2\sin(6\theta)\cos(-4\theta)] = 2\sin(4\theta)\sin(6\theta) = 2\cdot\dfrac{1}{2}[\cos(-2\theta) - \cos(10\theta)]$
$= \cos(2\theta) - \cos(10\theta)$

53. $\frac{1}{4}(\sqrt{6}-\sqrt{2})$ **54.** $-2-\sqrt{3}$ **55.** $\frac{1}{4}(\sqrt{6}-\sqrt{2})$ **56.** $\frac{1}{4}(\sqrt{2}-\sqrt{6})$ **57.** $\frac{1}{2}$ **58.** $\frac{1}{2}$ **59.** $\sqrt{2}-1$ **60.** $\frac{\sqrt{2+\sqrt{2}}}{2}$

61. (a) $-\frac{33}{65}$ **(b)** $-\frac{56}{65}$ **(c)** $-\frac{63}{65}$ **(d)** $\frac{33}{56}$ **(e)** $\frac{24}{25}$ **(f)** $\frac{119}{169}$ **(g)** $\frac{5\sqrt{26}}{26}$ **(h)** $\frac{2\sqrt{5}}{5}$ **62. (a)** $-\frac{33}{65}$ **(b)** $\frac{56}{65}$ **(c)** $\frac{63}{65}$ **(d)** $-\frac{33}{56}$ **(e)** $\frac{24}{25}$ **(f)** $-\frac{119}{169}$ **(g)** $-\frac{2\sqrt{13}}{13}$ **(h)** $\frac{3\sqrt{10}}{10}$ **63. (a)** $-\frac{16}{65}$ **(b)** $-\frac{63}{65}$ **(c)** $-\frac{56}{65}$ **(d)** $\frac{16}{63}$ **(e)** $\frac{24}{25}$ **(f)** $\frac{119}{169}$ **(g)** $\frac{\sqrt{26}}{26}$ **(h)** $-\frac{\sqrt{10}}{10}$

64. (a) $\frac{56}{65}$ **(b)** $\frac{33}{65}$ **(c)** $-\frac{16}{65}$ **(d)** $\frac{56}{33}$ **(e)** $-\frac{24}{25}$ **(f)** $-\frac{119}{169}$ **(g)** $\frac{3\sqrt{13}}{13}$ **(h)** $\frac{2\sqrt{5}}{5}$ **65. (a)** $-\frac{63}{65}$ **(b)** $\frac{16}{65}$ **(c)** $\frac{33}{65}$ **(d)** $-\frac{63}{16}$ **(e)** $\frac{24}{25}$ **(f)** $-\frac{119}{169}$ **(g)** $\frac{2\sqrt{13}}{13}$ **(h)** $-\frac{\sqrt{10}}{10}$ **66. (a)** $-\frac{33}{65}$ **(b)** $\frac{56}{65}$ **(c)** $-\frac{63}{65}$ **(d)** $-\frac{33}{56}$ **(e)** $-\frac{24}{25}$ **(f)** $\frac{119}{169}$ **(g)** $\frac{5\sqrt{26}}{26}$ **(h)** $\frac{\sqrt{5}}{5}$

67. (a) $\frac{-\sqrt{3}-2\sqrt{2}}{6}$ **(b)** $\frac{1-2\sqrt{6}}{6}$ **(c)** $\frac{-\sqrt{3}+2\sqrt{2}}{6}$ **(d)** $\frac{8\sqrt{2}+9\sqrt{3}}{23}$ **(e)** $-\frac{\sqrt{3}}{2}$ **(f)** $-\frac{7}{9}$ **(g)** $\frac{\sqrt{3}}{3}$ **(h)** $\frac{\sqrt{3}}{2}$

68. (a) $\frac{-1-2\sqrt{6}}{6}$ **(b)** $\frac{\sqrt{3}-2\sqrt{2}}{6}$ **(c)** $\frac{-1+2\sqrt{6}}{6}$ **(d)** $\frac{9\sqrt{3}+8\sqrt{2}}{5}$ **(e)** $-\frac{\sqrt{3}}{2}$ **(f)** $-\frac{7}{9}$ **(g)** $\frac{\sqrt{6}}{3}$ **(h)** $\frac{\sqrt{2-\sqrt{3}}}{2}$

69. (a) 1 **(b)** 0 **(c)** $-\frac{1}{9}$ **(d)** Not defined **(e)** $\frac{4\sqrt{5}}{9}$ **(f)** $-\frac{1}{9}$ **(g)** $\frac{\sqrt{30}}{6}$ **(h)** $-\frac{\sqrt{6}\sqrt{3-\sqrt{5}}}{6}$

70. (a) -1 **(b)** 0 **(c)** $-\frac{3}{5}$ **(d)** Not defined **(e)** $-\frac{4}{5}$ **(f)** $\frac{3}{5}$ **(g)** $\frac{\sqrt{10}\sqrt{5+2\sqrt{5}}}{10}$ **(h)** $\frac{\sqrt{10}\sqrt{5-\sqrt{5}}}{10}$ **71.** $\frac{4+3\sqrt{3}}{10}$ **72.** $\frac{33}{65}$

73. $-\frac{48+25\sqrt{3}}{39}$ **74.** $-\frac{\sqrt{2}}{10}$ **75.** $-\frac{24}{25}$ **76.** $-\frac{7}{25}$ **77.** $\left\{\frac{\pi}{3}, \frac{5\pi}{3}\right\}$ **78.** $\left\{\frac{4\pi}{3}, \frac{5\pi}{3}\right\}$ **79.** $\left\{\frac{3\pi}{4}, \frac{5\pi}{4}\right\}$ **80.** $\left\{\frac{2\pi}{3}, \frac{5\pi}{3}\right\}$ **81.** $\left\{\frac{3\pi}{4}, \frac{7\pi}{4}\right\}$

82. $\left\{\frac{\pi}{4}, \frac{3\pi}{4}, \frac{5\pi}{4}, \frac{7\pi}{4}\right\}$ **83.** $\left\{0, \frac{\pi}{2}, \pi, \frac{3\pi}{2}\right\}$ **84.** $\left\{\frac{\pi}{6}, \frac{5\pi}{6}, \frac{3\pi}{2}\right\}$ **85.** $\left\{\frac{\pi}{3}, \frac{2\pi}{3}, \frac{4\pi}{3}, \frac{5\pi}{3}\right\}$ **86.** $\left\{\frac{\pi}{2}, \frac{3\pi}{2}\right\}$ **87.** $\{0, \pi\}$ **88.** $\{0, \pi\}$

89. $\left\{0, \frac{2\pi}{3}, \pi, \frac{4\pi}{3}\right\}$ **90.** $\left\{\frac{\pi}{6}, \frac{5\pi}{6}, \frac{3\pi}{2}\right\}$ **91.** $\left\{0, \frac{\pi}{6}, \frac{5\pi}{6}\right\}$ **92.** $\left\{\frac{\pi}{3}, \frac{\pi}{2}, \frac{5\pi}{3}\right\}$ **93.** $\left\{\frac{\pi}{6}, \frac{\pi}{2}, \frac{5\pi}{6}\right\}$ **94.** $\left\{\frac{\pi}{3}, \pi, \frac{5\pi}{3}\right\}$ **95.** $\left\{\frac{\pi}{3}, \frac{5\pi}{3}\right\}$

96. $\left\{\frac{\pi}{4}, \frac{3\pi}{4}, \frac{5\pi}{4}, \frac{7\pi}{4}\right\}$ **97.** $\left\{\frac{\pi}{4}, \frac{\pi}{2}, \frac{3\pi}{4}, \frac{3\pi}{2}\right\}$ **98.** $\left\{\frac{\pi}{2}, \frac{5\pi}{6}, \frac{7\pi}{6}, \frac{3\pi}{2}\right\}$ **99.** $\left\{\frac{\pi}{2}, \pi\right\}$ **100.** $\left\{\frac{5\pi}{6}\right\}$ **101.** 0.78 **102.** 0.64 **103.** -1.11

104. 1.77 **105.** 1.23 **106.** 2.90 **107.** $\{1.11\}$ **108.** $\{0, 2.13\}$ **109.** $\{0.87\}$ **110.** $\{1.89, 3.07\}$ **111.** $\{2.22\}$ **112.** $\{0.59, 3.10\}$

113. $\sin\left(\frac{30°}{2}\right) = \frac{\sqrt{2-\sqrt{3}}}{2}$; $\sin(45° - 30°) = \frac{1}{4}(\sqrt{6}-\sqrt{2})$ **114.** $2\cos^2\theta - 1$

Cumulative Review *(page 480)*

1. $\left\{\frac{-1-\sqrt{13}}{6}, \frac{-1+\sqrt{13}}{6}\right\}$ **2.** $y + 1 = -1(x - 4)$ or $x + y = 3$; $6\sqrt{2}$; $(1, 2)$ **3.** x-axis symmetry; $(0, -3)$, $(0, 3)$, $(3, 0)$

4.

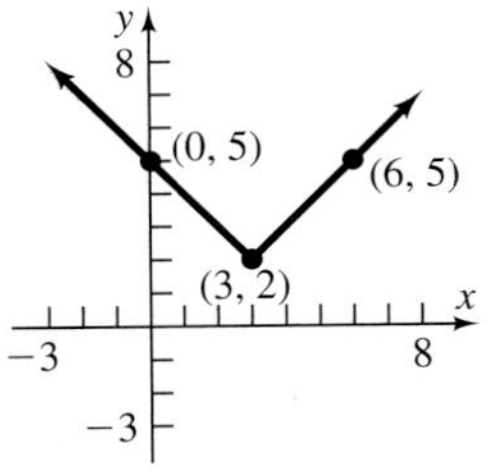

5.

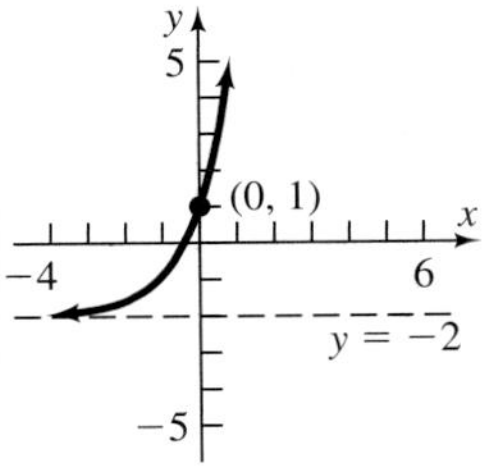

6.

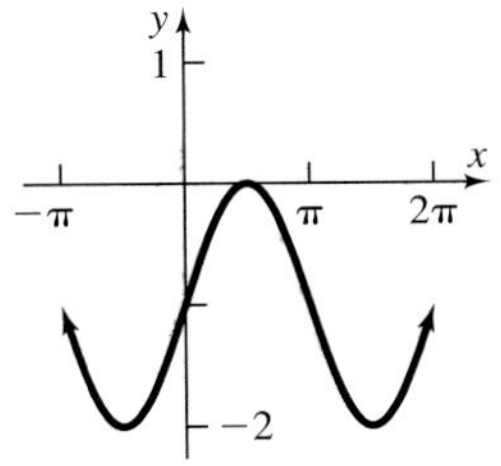

7. (a)

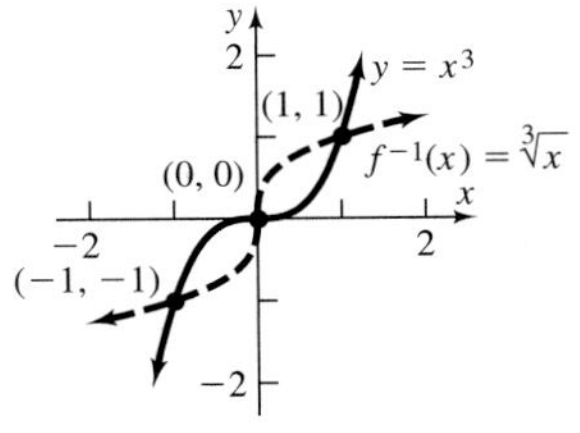

(b)

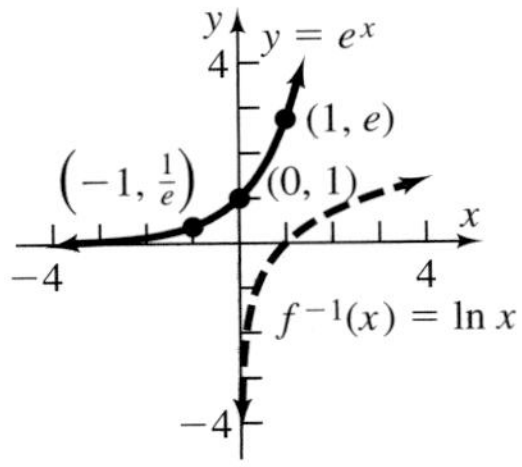

(c)

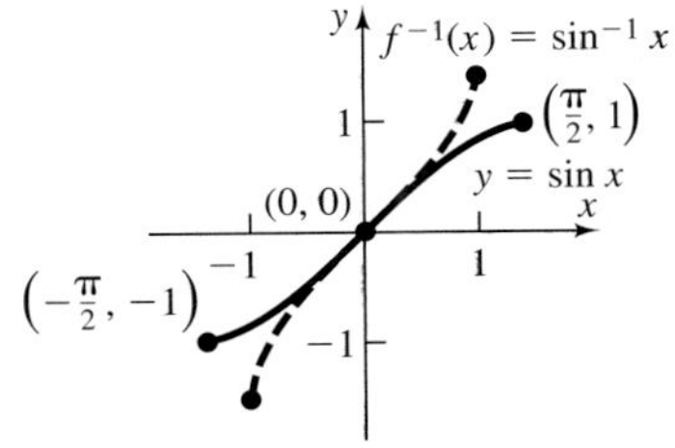

(d)

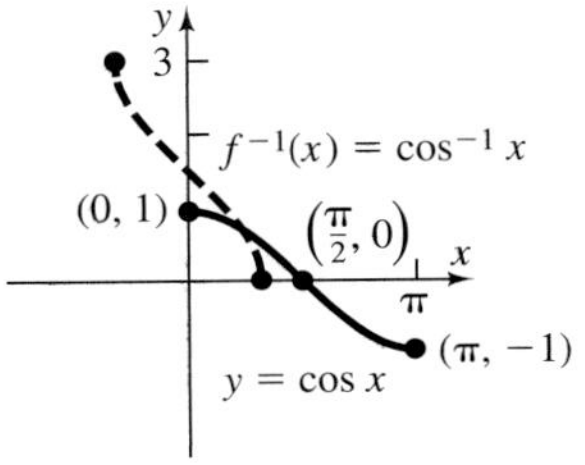

8. (a) $-\frac{2\sqrt{2}}{3}$ **(b)** $\frac{\sqrt{2}}{4}$ **(c)** $\frac{4\sqrt{2}}{9}$ **(d)** $\frac{7}{9}$ **(e)** $\sqrt{\frac{3+2\sqrt{2}}{6}}$ **(f)** $\sqrt{\frac{3-2\sqrt{2}}{6}}$ **9.** $\frac{\sqrt{5}}{5}$

10. (a) $-\frac{2\sqrt{2}}{3}$ **(b)** $-\frac{2\sqrt{2}}{3}$ **(c)** $\frac{7}{9}$ **(d)** $\frac{4\sqrt{2}}{9}$ **(e)** $\frac{\sqrt{6}}{3}$

11. (a) $f(x) = (2x-1)(x-1)^2(x+1)^2$; $\frac{1}{2}$ multiplicity 1; 1 and -1 multiplicity 2

(b) $(0,-1)$; $\left(\frac{1}{2}, 0\right)$; $(-1, 0)$; $(1, 0)$ **(c)** $y = 2x^5$

(d)

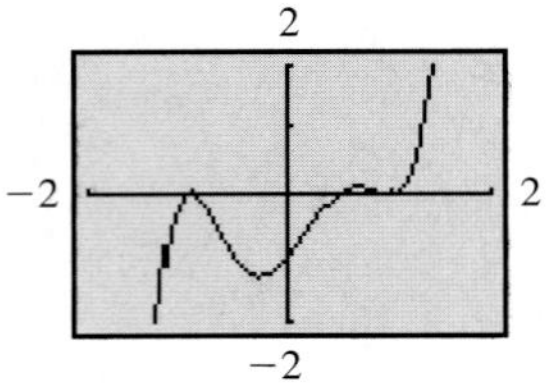

(e) Minima $(-0.29, -1.33)$, $(1, 0)$; maxima $(-1, 0)$, $(0.69, 0.10)$

(f)

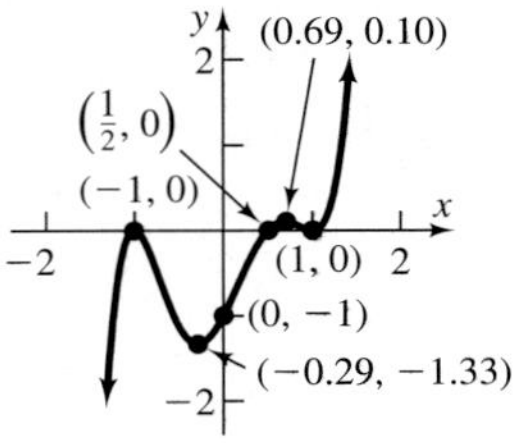

(g) Increasing: $(-\infty, -1)$, $(-0.29, 0.69)$, $(1, \infty)$; Decreasing: $(-1, -0.29)$, $(0.69, 1)$

12. (a) $\left\{-1, -\frac{1}{2}\right\}$ **(b)** $\{-1, 1\}$ **(c)** $(-\infty, -1)$ or $\left(-\frac{1}{2}, \infty\right)$ **(d)** $(-\infty, -1]$, $[1, \infty)$

CHAPTER 7 Applications of Trigonometric Functions

7.1 Concepts and Vocabulary *(page 490)*

3. False **4.** True **5.** angle of elevation **6.** angle of depression **7.** True **8.** False

7.1 Exercises *(page 490)*

9. $\sin\theta = \frac{5}{13}$; $\cos\theta = \frac{12}{13}$; $\tan\theta = \frac{5}{12}$; $\cot\theta = \frac{12}{5}$; $\sec\theta = \frac{13}{12}$; $\csc\theta = \frac{13}{5}$ **10.** $\sin\theta = \frac{3}{5}$; $\cos\theta = \frac{4}{5}$; $\tan\theta = \frac{3}{4}$; $\cot\theta = \frac{4}{3}$; $\sec\theta = \frac{5}{4}$; $\csc\theta = \frac{5}{3}$

11. $\sin\theta = \frac{2\sqrt{13}}{13}$; $\cos\theta = \frac{3\sqrt{13}}{13}$; $\tan\theta = \frac{2}{3}$; $\cot\theta = \frac{3}{2}$; $\sec\theta = \frac{\sqrt{13}}{3}$; $\csc\theta = \frac{\sqrt{13}}{2}$

12. $\sin\theta = \frac{\sqrt{2}}{2}$; $\cos\theta = \frac{\sqrt{2}}{2}$; $\tan\theta = 1$; $\cot\theta = 1$; $\sec\theta = \sqrt{2}$; $\csc\theta = \sqrt{2}$

13. $\sin\theta = \frac{\sqrt{3}}{2}$; $\cos\theta = \frac{1}{2}$; $\tan\theta = \sqrt{3}$; $\cot\theta = \frac{\sqrt{3}}{3}$; $\sec\theta = 2$; $\csc\theta = \frac{2\sqrt{3}}{3}$

14. $\sin\theta = \frac{3}{4}$; $\cos\theta = \frac{\sqrt{7}}{4}$; $\tan\theta = \frac{3\sqrt{7}}{7}$; $\cot\theta = \frac{\sqrt{7}}{3}$; $\sec\theta = \frac{4\sqrt{7}}{7}$; $\csc\theta = \frac{4}{3}$

15. $\sin\theta = \frac{\sqrt{6}}{3}$; $\cos\theta = \frac{\sqrt{3}}{3}$; $\tan\theta = \sqrt{2}$; $\cot\theta = \frac{\sqrt{2}}{2}$; $\sec\theta = \sqrt{3}$; $\csc\theta = \frac{\sqrt{6}}{2}$

16. $\sin\theta = \frac{2\sqrt{7}}{7}$; $\cos\theta = \frac{\sqrt{21}}{7}$; $\tan\theta = \frac{2\sqrt{3}}{3}$; $\cot\theta = \frac{\sqrt{3}}{2}$; $\sec\theta = \frac{\sqrt{21}}{3}$; $\csc\theta = \frac{\sqrt{7}}{2}$

17. $\sin\theta = \frac{\sqrt{5}}{5}$; $\cos\theta = \frac{2\sqrt{5}}{5}$; $\tan\theta = \frac{1}{2}$; $\cot\theta = 2$; $\sec\theta = \frac{\sqrt{5}}{2}$; $\csc\theta = \sqrt{5}$

18. $\sin\theta = \frac{\sqrt{5}}{5}$; $\cos\theta = \frac{2\sqrt{5}}{5}$; $\tan\theta = \frac{1}{2}$; $\cot\theta = 2$; $\sec\theta = \frac{\sqrt{5}}{2}$; $\csc\theta = \sqrt{5}$

19. 0 **20.** 0 **21.** 1 **22.** 1 **23.** 0 **24.** 0 **25.** 0 **26.** 0 **27.** 1 **28.** 1 **29.** $a \approx 13.74$, $c \approx 14.62$, $\alpha = 70°$ **30.** $a \approx 22.69$, $c \approx 23.04$, $\alpha = 80°$ **31.** $b \approx 5.03$, $c \approx 7.83$, $\alpha = 50°$ **32.** $b \approx 8.34$, $c \approx 10.89$, $\alpha = 40°$ **33.** $a \approx 0.71$, $c \approx 4.06$, $\beta = 80°$ **34.** $a \approx 2.18$, $c \approx 6.39$, $\beta = 70°$ **35.** $b \approx 10.72$, $c \approx 11.83$, $\beta = 65°$ **36.** $b \approx 7.15$, $c \approx 9.33$, $\beta = 50°$ **37.** $b \approx 3.08$, $a \approx 8.46$, $\alpha = 70°$ **38.** $a \approx 6.43$, $b \approx 7.66$, $\beta = 50°$ **39.** $c \approx 5.83$, $\alpha \approx 59.0°$, $\beta = 31.0°$ **40.** $c \approx 8.25$, $\alpha \approx 14.0°$, $\beta \approx 76.0°$ **41.** $b \approx 4.58$, $\alpha \approx 23.6°$, $\beta = 66.4°$ **42.** $a \approx 4.47$, $\alpha \approx 48.2°$, $\beta \approx 41.8°$ **43.** 4.59 in., 6.55 in. **44.** 6.43 cm, 7.66 cm **45.** 5.52 in. or 11.83 in. **46.** 3.25 m or 7.84 m **47.** 23.6° and 66.4° **48.** 19.5° and 70.5° **49.** 70.02 ft **50.** 83.91 ft **51.** 985.91 ft **52.** 214.45 ft **53.** 137.37 m **54.** 837.98 ft **55.** 20.67 ft **56.** 122.37 ft **57.** 1978.09 ft **58.** 80.5° **59.** 60.27 ft **60.** 15.9° **61.** 530.18 ft **62.** 33.81 ft **63.** 554.52 ft **64.** 7524.67 ft **65. (a)** 111.96 ft/sec or 76.3 mi/hr **(b)** 82.42 ft/sec or 56.2 mi/hr **(c)** Under 18.8° **66.** 14.0° **67.** S76.6°E **68.** S16.6°E **69.** 69.0° **70.** 14.9° **71.** 3.83 mi **72. (a)** $\cos\frac{\theta}{2} = \frac{3960}{3960 + h}$ **(b)** $d = 3960\,\theta$ **(c)** $\cos\frac{d}{7920} = \frac{3960}{3960 + h}$ **(d)** 206 mi **(e)** 2990 mi

73. No; move the tripod back about 1 ft. **74.** 19.32 ft

7.2 Concepts and Vocabulary *(page 502)*

4. oblique **5.** $\frac{\sin\alpha}{a} = \frac{\sin\beta}{b} = \frac{\sin\gamma}{c}$ **6.** False **7.** True **8.** False

7.2 Exercises *(page 502)*

9. $a \approx 3.23, b \approx 3.55, \alpha = 40°$ **10.** $a \approx 2.84, b \approx 2.58, \gamma = 95°$ **11.** $a \approx 3.25, c \approx 4.23, \beta = 45°$ **12.** $a \approx 8.45, c \approx 16.38, \alpha = 25°$ **13.** $\gamma = 95°, c \approx 9.86, a \approx 6.36$ **14.** $a \approx 3.35, b \approx 1.68, \gamma = 165°$ **15.** $\alpha = 40°, a = 2, c \approx 3.06$ **16.** $\alpha \approx 11.82, c \approx 9.19, \gamma = 50°$ **17.** $\gamma = 120°, b \approx 1.06, c \approx 2.69$ **18.** $b \approx 3.68, c \approx 1.34, \beta = 110°$ **19.** $\alpha = 100°, a \approx 5.24, c \approx 0.92$ **20.** $a \approx 4.91, b \approx 4.52, \gamma = 50°$ **21.** $\beta = 40°, a \approx 5.64, b \approx 3.86$ **22.** $a \approx 10.82, c \approx 11.34, \alpha = 70°$ **23.** $\gamma = 100°, a \approx 1.31, b \approx 1.31$ **24.** $b \approx 0.34, c \approx 0.94, \alpha = 90°$ **25.** One triangle; $\beta \approx 30.7°, \gamma \approx 99.3°, c \approx 3.86$ **26.** One triangle; $a \approx 5.80, \alpha \approx 111.2°, \gamma \approx 28.8°$ **27.** One triangle; $\gamma \approx 36.2°, \alpha \approx 43.8°, a \approx 3.51$ **28.** One triangle; $b \approx 1.30, \beta \approx 34.3°, \gamma \approx 25.7°$ **29.** No triangle **30.** Two triangles; $a_1 \approx 2.83, \alpha_1 \approx 65.4°, \gamma_1 \approx 74.6°$ or $a_2 \approx 1.77, \alpha_2 \approx 34.6°, \gamma_2 \approx 105.4°$ **31.** Two triangles; $\gamma_1 \approx 30.9°, \alpha_1 \approx 129.1°, a_1 \approx 9.07$ or $\gamma_2 \approx 149.1°, \alpha_2 \approx 10.9°, a_2 \approx 2.20$ **32.** No triangle **33.** No triangle **34.** No triangle **35.** Two triangles; $\alpha_1 \approx 57.7°, \beta_1 \approx 97.3°, b_1 \approx 2.35$ or $\alpha_2 \approx 122.3°, \beta_2 \approx 32.7°, b_2 \approx 1.28$ **36.** Two triangles; $a_1 = 6.21, \alpha_1 \approx 86.5°, \gamma_1 \approx 53.5°$ or $a_2 = 1.45, \alpha_2 \approx 13.5°, \gamma_2 \approx 126.5°$ **37. (a)** Station Able is about 143.33 mi from the ship; Station Baker is about 135.58 mi from the ship. **(b)** Approximately 41 min **38.** 76.60 ft **39.** 1490.48 ft **40.** 629.90 ft **41.** 381.69 ft **42.** 1053.15 ft **43. (a)** 169.18 mi **(b)** 161.3° **44.** 18.7 sec **45.** 84.7°; 183.72 ft **46.** 6.80 in. **47.** 2.64 mi **48. (a)** 3.21 mi **(b)** 3.78 mi **(c)** 3.10 mi **49.** 38.5 in. **50.** 110.01 ft **51.** 449.36 ft **52.** 1953.37 ft **53.** 187,600,000 km or 101,440,000 km **54.** 42,300,000 km or 252,300,000 km **55.** 39.39 ft **56.** 14.86 ft **57.** 29.97 ft

58. $$\frac{a+b}{c} = \frac{a}{c} + \frac{b}{c} = \frac{\sin\alpha}{\sin\gamma} + \frac{\sin\beta}{\sin\gamma} = \frac{\sin\alpha + \sin\beta}{\sin\gamma} = \frac{2\sin\left(\frac{\alpha+\beta}{2}\right)\cos\left(\frac{\alpha-\beta}{2}\right)}{2\sin\frac{\gamma}{2}\cos\frac{\gamma}{2}} = \frac{\sin\left(\frac{\pi}{2} - \frac{\gamma}{2}\right)\cos\left(\frac{\alpha-\beta}{2}\right)}{\sin\frac{\gamma}{2}\cos\frac{\gamma}{2}} = \frac{\cos\left(\frac{\alpha-\beta}{2}\right)}{\sin\frac{\gamma}{2}}$$

59. $$\frac{a-b}{c} = \frac{a}{c} - \frac{b}{c} = \frac{\sin\alpha}{\sin\gamma} - \frac{\sin\beta}{\sin\gamma} = \frac{\sin\alpha - \sin\beta}{\sin\gamma} = \frac{2\sin\left(\frac{\alpha-\beta}{2}\right)\cos\left(\frac{\alpha+\beta}{2}\right)}{2\sin\frac{\gamma}{2}\cos\frac{\gamma}{2}} = \frac{\sin\left(\frac{\alpha-\beta}{2}\right)\cos\left(\frac{\pi}{2} - \frac{\gamma}{2}\right)}{\sin\frac{\gamma}{2}\cos\frac{\gamma}{2}} = \frac{\sin\left(\frac{\alpha-\beta}{2}\right)}{\cos\frac{\gamma}{2}}$$

60. $$a = \frac{b\sin\alpha}{\sin\beta} = \frac{b\sin[180° - (\beta+\gamma)]}{\sin\beta} = \frac{b}{\sin\beta}(\sin\beta\cos\gamma + \cos\beta\sin\gamma) = b\cos\gamma + \frac{b\sin\gamma}{\sin\beta}\cos\beta = b\cos\gamma + c\cos\beta$$

61. $$\frac{a-b}{a+b} = \frac{\frac{a-b}{c}}{\frac{a+b}{c}} = \frac{\dfrac{\sin\left[\frac{1}{2}(\alpha-\beta)\right]}{\cos\frac{\gamma}{2}}}{\dfrac{\cos\left[\frac{1}{2}(\alpha-\beta)\right]}{\sin\frac{\gamma}{2}}} = \frac{\tan\left[\frac{1}{2}(\alpha-\beta)\right]}{\cot\frac{\gamma}{2}} = \frac{\tan\left[\frac{1}{2}(\alpha-\beta)\right]}{\tan\left(\frac{\pi}{2} - \frac{\gamma}{2}\right)} = \frac{\tan\left[\frac{1}{2}(\alpha-\beta)\right]}{\tan\left[\frac{1}{2}(\alpha+\beta)\right]}$$

62. $\sin\beta = \sin(\text{angle } ABC) = \sin(\text{angle } AB'C) = \frac{b}{2r}; \frac{\sin\beta}{b} = \frac{1}{2r}$; the result follows using the Law of Sines.

7.3 Concepts and Vocabulary *(page 510)*

3. Cosines **4.** Sines **5.** Cosines **6.** False **7.** False **8.** True

7.3 Exercises *(page 510)*

9. $b \approx 2.95, \alpha \approx 28.7°, \gamma \approx 106.3°$ **10.** $a \approx 2.05, \beta \approx 46.9°, \gamma \approx 103.1°$ **11.** $c \approx 3.75, \alpha \approx 32.1°, \beta \approx 52.9°$ **12.** $b \approx 3.19, \alpha \approx 12.4°, \gamma \approx 147.6°$ **13.** $\alpha \approx 48.5°, \beta \approx 38.6°, \gamma \approx 92.9°$ **14.** $\alpha \approx 125.1°, \beta \approx 30.8°, \gamma \approx 24.1°$ **15.** $\alpha \approx 127.2°, \beta \approx 32.1°, \gamma \approx 20.7°$ **16.** $\alpha \approx 68.0°, \beta \approx 44.0°, \gamma \approx 68.0°$ **17.** $c \approx 2.57, \alpha \approx 48.6°, \beta \approx 91.4°$ **18.** $b \approx 1.03, \alpha \approx 160.3°, \gamma \approx 9.7°$ **19.** $a \approx 2.99, \beta \approx 19.2°, \gamma \approx 80.8°$ **20.** $c \approx 5.29, \alpha \approx 79.1°, \beta \approx 40.9°$ **21.** $b \approx 4.14, \alpha \approx 43.0°, \gamma \approx 27.0°$ **22.** $a \approx 4.58, \beta \approx 49.1°, \gamma \approx 10.9°$ **23.** $c \approx 1.69, \alpha \approx 65.0°, \beta \approx 65.0°$ **24.** $b \approx 3.61, \alpha \approx 56.3°, \gamma \approx 33.7°$ **25.** $\alpha \approx 67.4°, \beta = 90°, \gamma \approx 22.6°$ **26.** $\alpha \approx 53.1°, \beta = 90°, \gamma \approx 36.9°$ **27.** $\alpha = 60°, \beta = 60°, \gamma = 60°$ **28.** $\alpha \approx 70.5°, \beta \approx 70.5°, \gamma \approx 38.9°$ **29.** $\alpha \approx 33.6°, \beta \approx 62.2°, \gamma \approx 84.3°$ **30.** $\alpha \approx 36.3°, \beta \approx 26.4°, \gamma \approx 117.3°$ **31.** $\alpha \approx 97.9°, \beta \approx 52.4°, \gamma \approx 29.7°$ **32.** $\alpha \approx 60.9°, \beta \approx 42.8°, \gamma \approx 76.2°$ **33.** 70.75 ft **34. (a)** 227.56 mi **(b)** 149.7° **35. (a)** 26.4° **(b)** 30.8 hr **36. (a)** 12.0° **(b)** 220.7 mph **37. (a)** 63.7 ft **(b)** 66.8 ft **(c)** 92.8° **38. (a)** 42.58 ft **(b)** 38.85 ft **(c)** 85.2° **39. (a)** 492.6 ft **(b)** 269.3 ft **40.** 501.28 ft, 518.38 ft **41.** 342.3 ft **42.** 241.33 ft

43. Using the Law of Cosines:
$L^2 = x^2 + r^2 - 2rx\cos\theta$
$x^2 - 2rx\cos\theta + r^2 - L^2 = 0$
Then, using the quadratic formula:
$x = r\cos\theta + \sqrt{r^2\cos^2\theta + L^2 - r^2}$

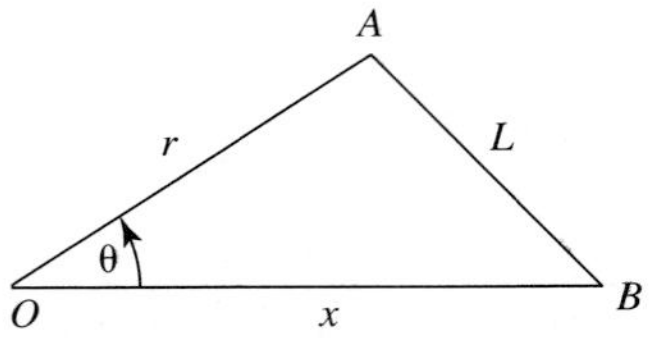

44. Using the Law of Cosines: $d^2 = r^2 + r^2 - 2r^2\cos\theta = 4r^2\left(\frac{1-\cos\theta}{2}\right) \Rightarrow d = 2r\sqrt{\frac{1-\cos\theta}{2}} = 2r\sin\frac{\theta}{2}$, by a half-angle identity and since $0 < \theta < 2\pi$ can be assumed. Since for any angle in $(0, 2\pi)$ d is strictly less than the length of the arc subtended by θ, that is, $d < r\theta$, then $2r\sin\frac{\theta}{2} < r\theta$, or $2\sin\frac{\theta}{2} < \theta$. Since $\cos\frac{\theta}{2} < 1$, then, for $0 < \theta < 2\pi$, $\sin\theta = 2\sin\frac{\theta}{2}\cos\frac{\theta}{2} < 2\sin\frac{\theta}{2} < \theta$.
This inequality $\sin\theta < \theta$ holds for values $\theta \geq 2\pi > 1 \geq \sin\theta$; thus $\sin\theta < \theta$ for all $\theta > 0$.

45. $\cos\frac{\gamma}{2} = \sqrt{\frac{1+\cos\gamma}{2}} = \sqrt{\frac{1+\frac{a^2+b^2-c^2}{2ab}}{2}} = \sqrt{\frac{2ab+a^2+b^2-c^2}{4ab}} = \sqrt{\frac{(a+b)^2-c^2}{4ab}} = \sqrt{\frac{(a+b+c)(a+b-c)}{4ab}}$
$= \sqrt{\frac{2s(2s-2c)}{4ab}} = \sqrt{\frac{s(s-c)}{ab}}$

46. $\sin\frac{\gamma}{2} = \sqrt{\frac{1-\cos\theta}{2}} = \sqrt{\frac{1-\frac{a^2+b^2-c^2}{2ab}}{2}} = \sqrt{\frac{2ab-a^2-b^2+c^2}{4ab}} = \sqrt{\frac{c^2-(a-b)^2}{4ab}} = \sqrt{\frac{(c+a-b)(c+b-a)}{4ab}}$
$= \sqrt{\frac{(2s-2b)(2s-2a)}{4ab}} = \sqrt{\frac{(s-a)(s-b)}{ab}}$

47. $\frac{\cos\alpha}{a} + \frac{\cos\beta}{b} + \frac{\cos\gamma}{c} = \frac{b^2+c^2-a^2}{2abc} + \frac{a^2+c^2-b^2}{2abc} + \frac{a^2+b^2-c^2}{2abc} = \frac{b^2+c^2-a^2+a^2+c^2-b^2+a^2+b^2-c^2}{2abc}$
$= \frac{a^2+b^2+c^2}{2abc}$

7.4 Concepts and Vocabulary *(page 516)*

2. Heron's **3.** False **4.** True

7.4 Exercises *(page 516)*

5. 2.83 **6.** 3 **7.** 2.99 **8.** 1.71 **9.** 14.98 **10.** 8.18 **11.** 9.56 **12.** 5.56 **13.** 3.86 **14.** 0.17 **15.** 1.48 **16.** 10.39 **17.** 2.82 **18.** 1.73 **19.** 30 **20.** 6 **21.** 1.73 **22.** 2.83 **23.** 19.90 **24.** 5.33 **25.** $A = \frac{1}{2}ab\sin\gamma = \frac{1}{2}a\sin\gamma\left(\frac{a\sin\beta}{\sin\alpha}\right) = \frac{a^2\sin\beta\sin\gamma}{2\sin\alpha}$

26. $A = \frac{1}{2}ab\sin\gamma = \frac{1}{2}b\sin\gamma\left(\frac{b\sin\alpha}{\sin\beta}\right) = \frac{b^2\sin\alpha\sin\gamma}{2\sin\beta}$; $A = \frac{1}{2}bc\sin\alpha = \frac{1}{2}c\sin\alpha\left(\frac{c\sin\beta}{\sin\gamma}\right) = \frac{c^2\sin\alpha\sin\beta}{2\sin\gamma}$ **27.** 0.92 **28.** 1.89 **29.** 2.27

30. 8.50 **31.** 5.44 **32.** 10.66 **33.** 9.03 sq ft **34.** 0.69 sq in. **35.** \$5446.38 **36.** 326.73 sq ft **37.** 9.26 sq cm **38.** 15.27 sq in.

39. $A = \frac{1}{2}r^2(\theta + \sin\theta)$ **40.** 1,645.14 sq ft **41.** **(a)** $A(\theta) = 2\sin\theta\cos\theta$ **(b)** From Double-angle Formula, since $2\sin\theta\cos\theta = \sin(2\theta)$

(c) $\theta = 45°$ **(d)** $\frac{\sqrt{2}}{2}$ by $\sqrt{2}$ **42.** Area of a triangle equals $\frac{1}{2}bh$. $\sin\frac{\theta}{2} = \frac{\frac{1}{2}b}{s}$ while $\cos\frac{\theta}{2} = \frac{h}{s}$. Solve for b and h and use the Double-angle Formula $\sin(2\alpha) = 2\cos\alpha\sin\alpha$, with $\alpha = \frac{\theta}{2}$.

43. **(a)** Area $\triangle OAC = \frac{1}{2}|OC||AC| = \frac{1}{2}\cdot\frac{|OC|}{1}\cdot\frac{|AC|}{1} = \frac{1}{2}\sin\alpha\cos\alpha$

(b) Area $\triangle OCB = \frac{1}{2}|BC||OC| = \frac{1}{2}|OB|^2\frac{|BC|}{|OB|}\cdot\frac{|OC|}{|OB|} = \frac{1}{2}|OB|^2\sin\beta\cos\beta$

(c) Area $\triangle OAB = \frac{1}{2}|BD||OA| = \frac{1}{2}|OB|\frac{|BD|}{|OB|} = \frac{1}{2}|OB|\sin(\alpha+\beta)$

(d) $\frac{\cos\alpha}{\cos\beta} = \frac{\frac{|OC|}{1}}{\frac{|OC|}{|OB|}} = |OB|$ **(e)** Use the hint and above results.

44. (a) The area of ΔOBC equals $\frac{\sin\theta}{2}$ since the length of the base is 1, and the height of the triangle is $\sin\theta$.

(b) The height of ΔOBD equals $\tan\theta$, so, since the base equals 1, the area of ΔOBD equals $\frac{\tan\theta}{2} = \frac{\sin\theta}{2\cos\theta}$.

(c) Since Area $\Delta OBC <$ Area $\widehat{OBC} <$ Area ΔOBD, then, from parts (a) and (b), $\frac{\sin\theta}{2} < \frac{\theta}{2} < \frac{\sin\theta}{2\cos\theta}$. Since θ is acute, then $\sin\theta > 0$, so multiplying by $\frac{2}{\sin\theta}$ yields $1 < \frac{\theta}{\sin\theta} < \frac{1}{\cos\theta}$. **45.** 31,145.15 sq ft **46.** 30,972.76 sq ft

47. $h_1 = 2\frac{K}{a}$, $h_2 = 2\frac{K}{b}$, $h_3 = 2\frac{K}{c}$. Then $\frac{1}{h_1} + \frac{1}{h_2} + \frac{1}{h_3} = \frac{a}{2K} + \frac{b}{2K} + \frac{c}{2K} = \frac{a+b+c}{2K} = \frac{2s}{2K} = \frac{s}{K}$.

48. $A = \frac{1}{2}ah = \frac{1}{2}ab\sin\gamma \Rightarrow h = b\sin\gamma = \frac{a\sin\beta\sin\gamma}{\sin\alpha}$

49. Angle AOB measures $180° - \left(\frac{\alpha}{2} + \frac{\beta}{2}\right) = 180° - \frac{1}{2}(180° - \gamma) = 90° + \frac{\gamma}{2}$, and $\sin\left(90° + \frac{\gamma}{2}\right) = \cos\left(-\frac{\gamma}{2}\right) = \cos\frac{\gamma}{2}$, since cosine is an even function. Therefore, $r = \frac{c\sin\frac{\alpha}{2}\sin\frac{\beta}{2}}{\sin\left(90° + \frac{\gamma}{2}\right)} = \frac{c\sin\frac{\alpha}{2}\sin\frac{\beta}{2}}{\cos\frac{\gamma}{2}}$.

50. $$\cot\frac{\gamma}{2} = \frac{\cos\frac{\gamma}{2}}{\sin\frac{\gamma}{2}} = \frac{\dfrac{c\sin\frac{\alpha}{2}\sin\frac{\beta}{2}}{r}}{\sqrt{\dfrac{(s-a)(s-b)}{ab}}} = \frac{c\sqrt{\dfrac{(s-b)(s-c)}{bc}}\sqrt{\dfrac{(s-a)(s-c)}{ac}}}{r\sqrt{\dfrac{(s-a)(s-b)}{ab}}} = \frac{c}{r}\sqrt{\frac{(s-c)^2}{c^2}} = \frac{s-c}{r}$$

51. $$\cot\frac{\alpha}{2} + \cot\frac{\beta}{2} + \cot\frac{\gamma}{2} = \frac{s-a}{r} + \frac{s-b}{r} + \frac{s-c}{r} = \frac{3s-(a+b+c)}{r} = \frac{3s-2s}{r} = \frac{s}{r}$$

52. K = area of triangle BOC + area of COA + area of AOB $= \frac{1}{2}ar + \frac{1}{2}br + \frac{1}{2}cr = r\frac{1}{2}(a+b+c) = rs$, so $r = \frac{K}{s} = \frac{\sqrt{s(s-a)(s-b)(s-c)}}{s} = \sqrt{\frac{(s-a)(s-b)(s-c)}{s}}$.

7.5 Concepts and Vocabulary *(page 526)*

2. harmonic motion; amplitude **3.** simple harmonic motion; damped motion **4.** True

7.5 Exercises *(page 526)*

5. $d = -5\cos(\pi t)$ **6.** $d = -10\cos\left(\frac{2\pi}{3}t\right)$ **7.** $d = -6\cos(2t)$ **8.** $d = -4\cos(4t)$ **9.** $d = -5\sin(\pi t)$ **10.** $d = -10\sin\left(\frac{2\pi}{3}t\right)$ **11.** $d = -6\sin(2t)$ **12.** $d = -4\sin(4t)$ **13. (a)** Simple harmonic **(b)** 5 m **(c)** $\frac{2\pi}{3}$ sec **(d)** $\frac{3}{2\pi}$ oscillation/sec

14. (a) Simple harmonic **(b)** 4 m **(c)** π sec **(d)** $\frac{1}{\pi}$ oscillation/sec **15. (a)** Simple harmonic **(b)** 6 m **(c)** 2 sec **(d)** $\frac{1}{2}$ oscillation/sec

16. (a) Simple harmonic **(b)** 5 m **(c)** 4 sec **(d)** $\frac{1}{4}$ oscillation/sec **17. (a)** Simple harmonic **(b)** 3 m **(c)** 4π sec **(d)** $\frac{1}{4\pi}$ oscillation/sec

18. (a) Simple harmonic **(b)** 2 m **(c)** π sec **(d)** $\frac{1}{\pi}$ oscillation/sec **19. (a)** Simple harmonic **(b)** 2 m **(c)** 1 sec **(d)** 1 oscillation/sec

20. (a) Simple harmonic **(b)** 3 m **(c)** 2 sec **(d)** $\frac{1}{2}$ oscillation/sec

21.

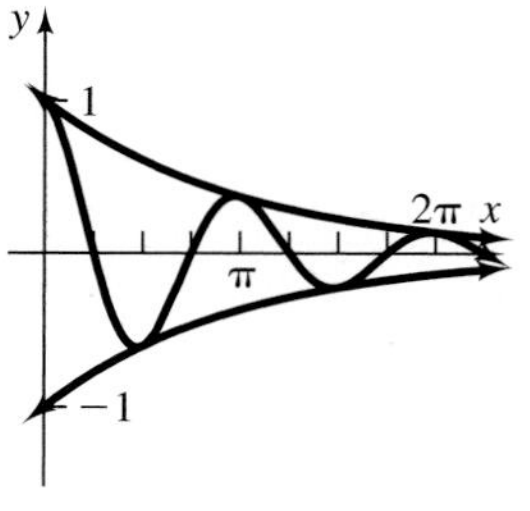

22.

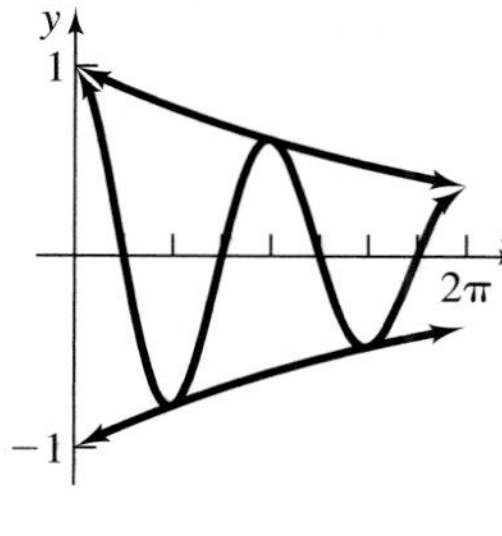

23.

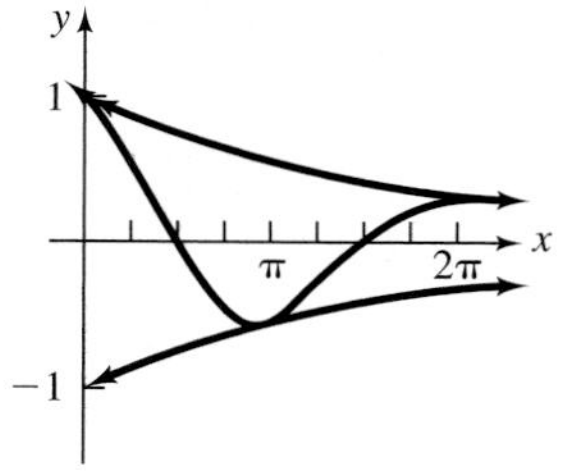

24.

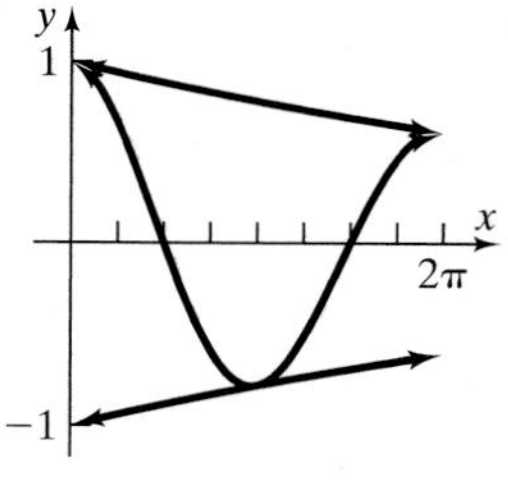

25.
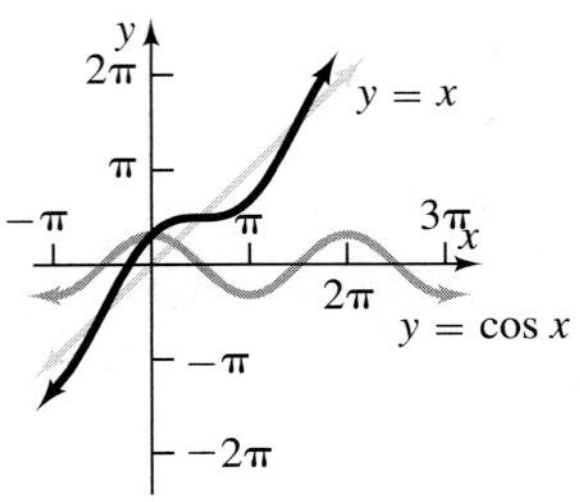

26.
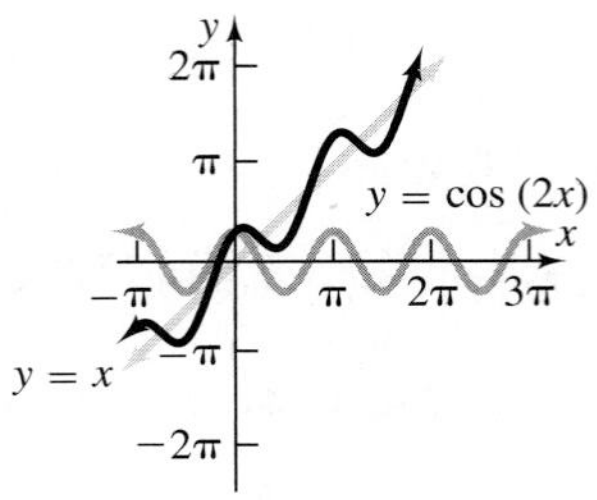

27.
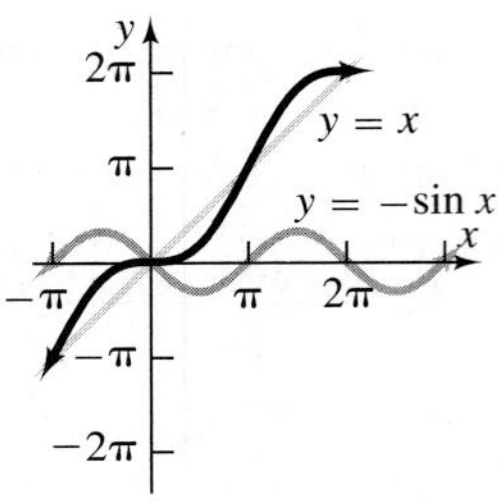

28.
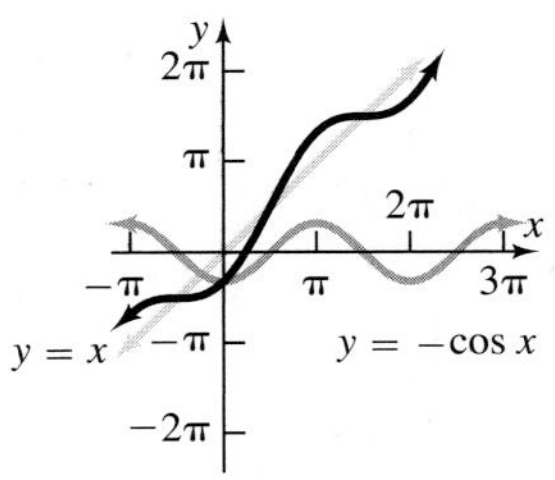

29.
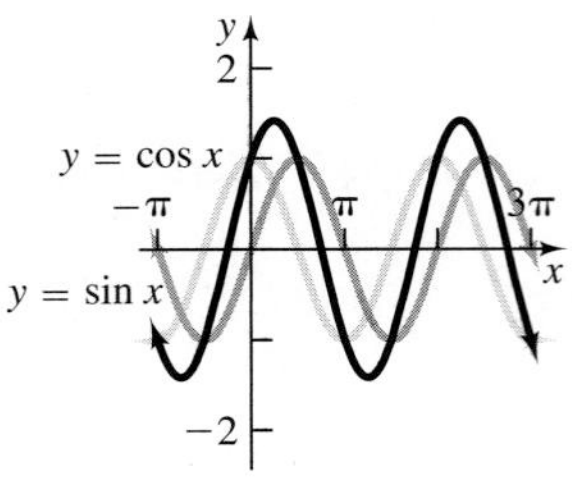

30.
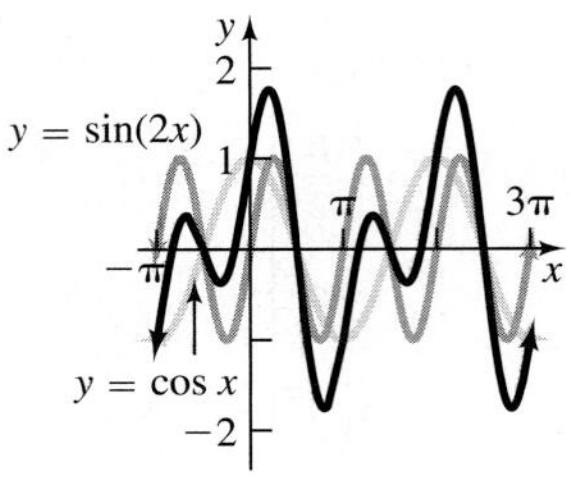

31.
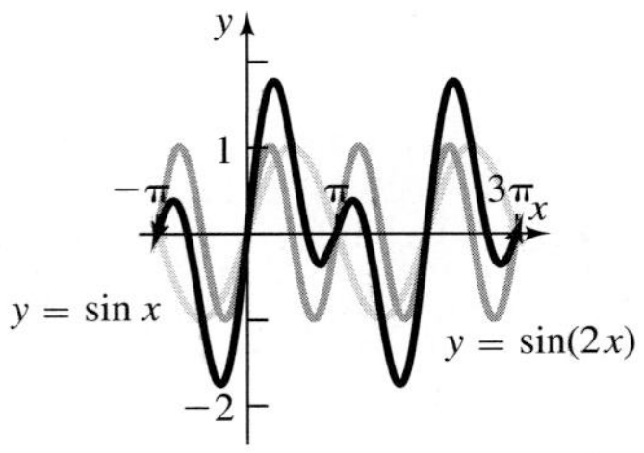

32.
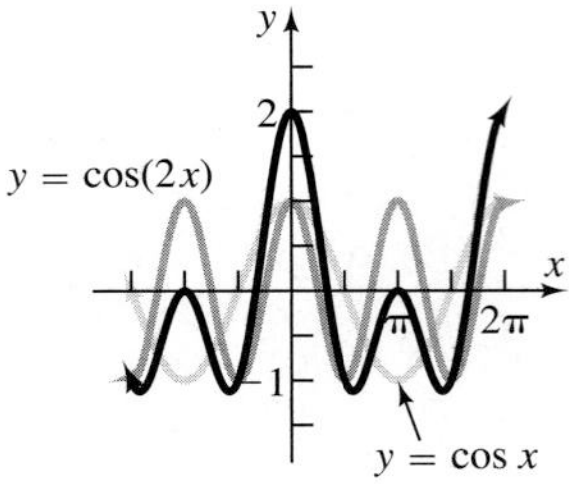

33.(a)
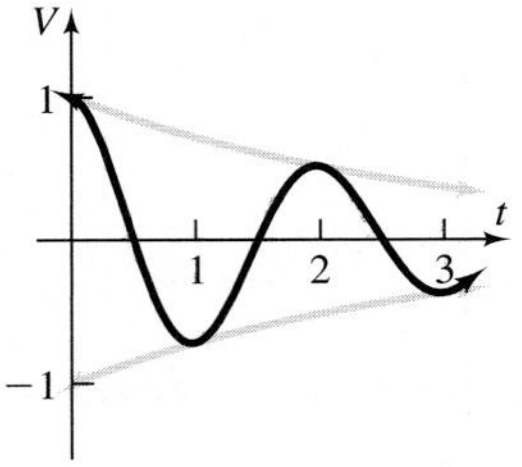

(b) At $t = 0, 2$; at $t = 1, t = 3$
(c) Between about 1.28 and 1.75 sec

34. (a)
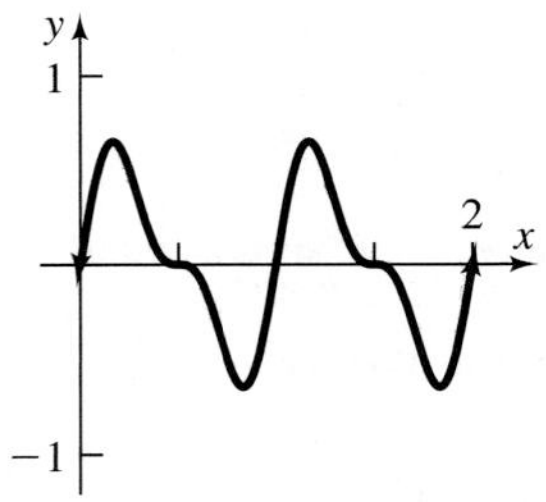

(b)
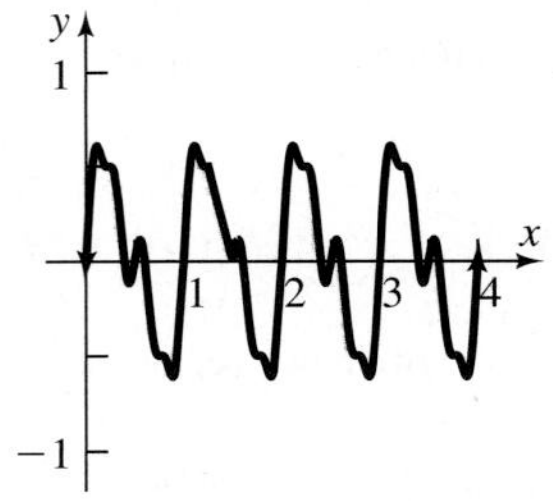

(c)
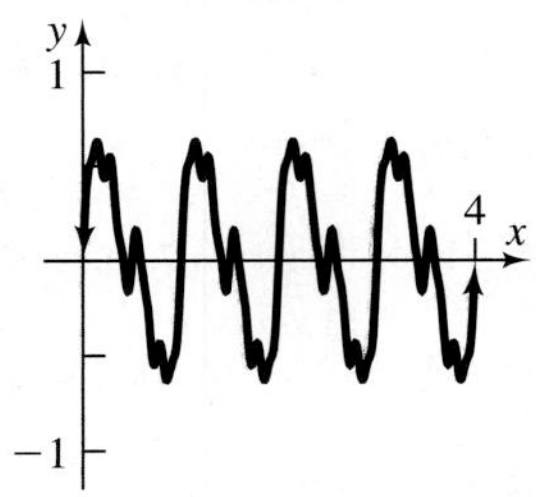

35.
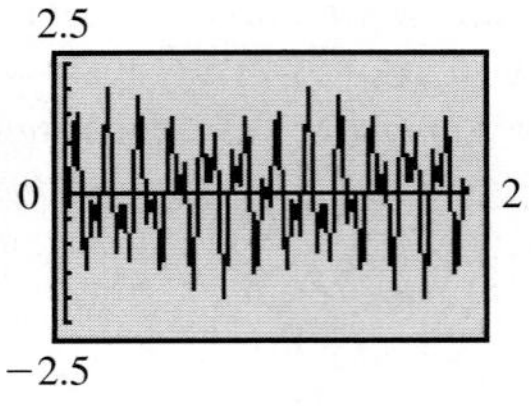

36.
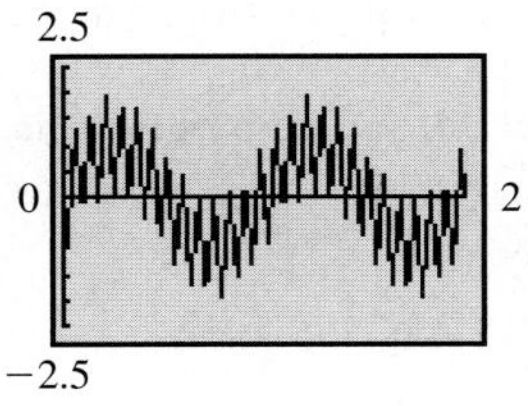

37.
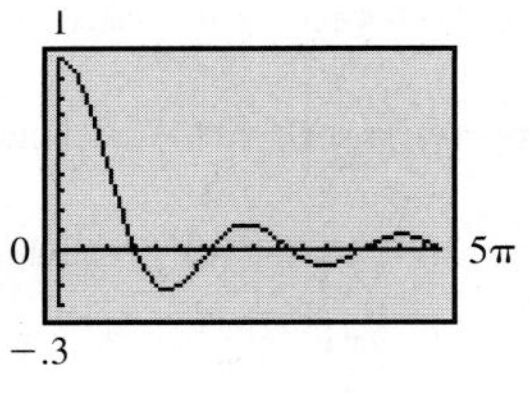

38. $y = x \sin x$

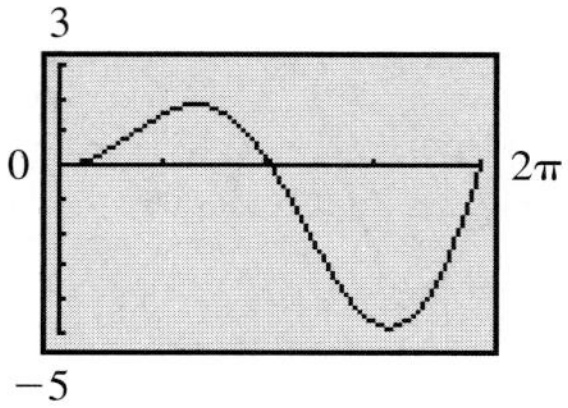

$y = x^2 \sin x$

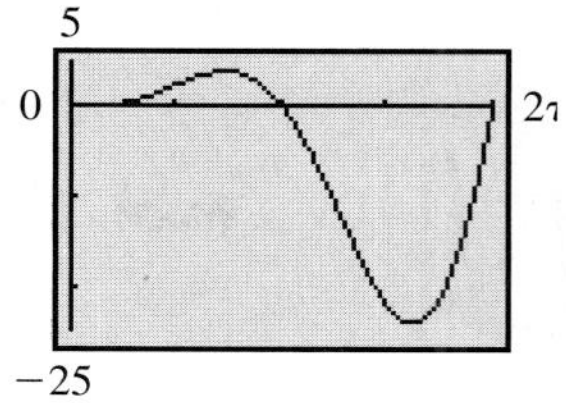

$y = x^3 \sin x$

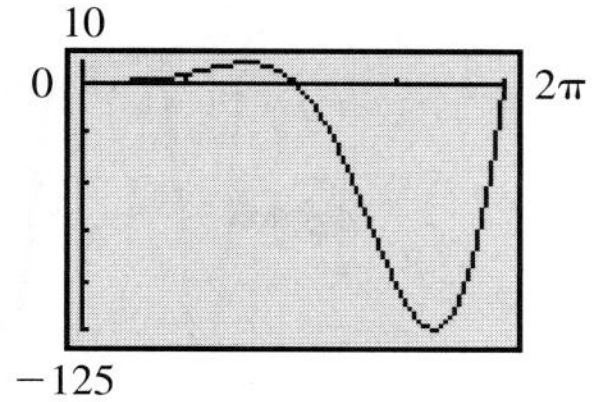

39. $y = \dfrac{1}{x} \sin x$

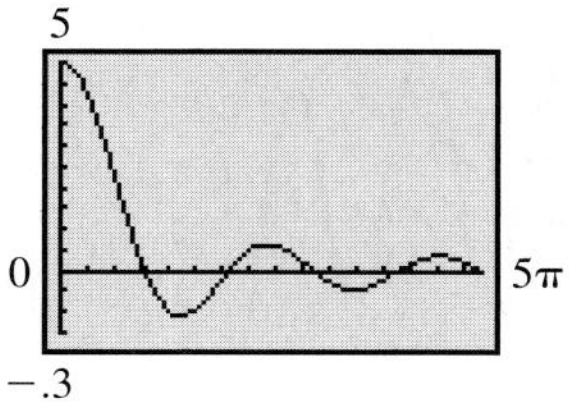

$y = \dfrac{1}{x^2} \sin x$

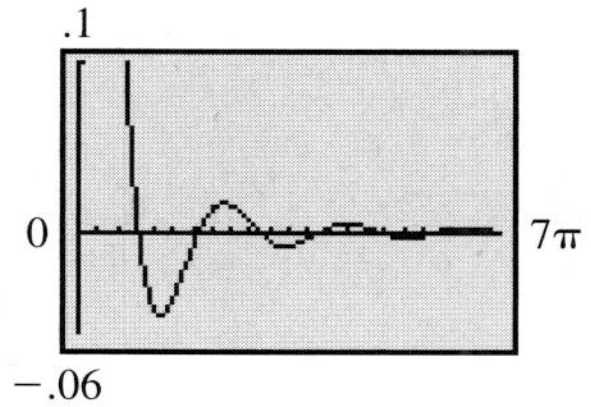

$y = \dfrac{1}{x^3} \sin x$

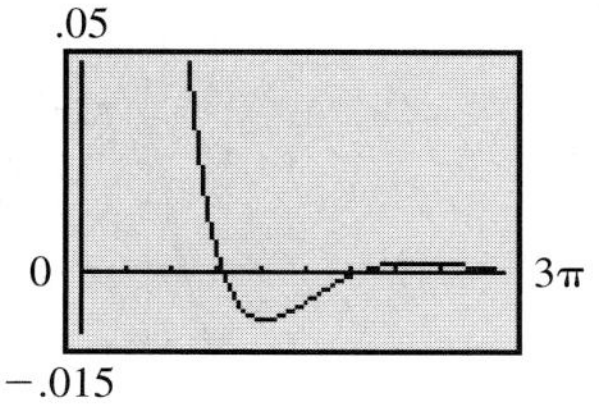

Review Exercises *(page 529)*

1. $\sin\theta = \frac{4}{5}$; $\cos\theta = \frac{3}{5}$; $\tan\theta = \frac{4}{3}$; $\cot\theta = \frac{3}{4}$; $\sec\theta = \frac{5}{3}$; $\csc\theta = \frac{5}{4}$
2. $\sin\theta = \frac{3\sqrt{34}}{34}$; $\cos\theta = \frac{5\sqrt{34}}{34}$; $\tan\theta = \frac{3}{5}$; $\cot\theta = \frac{5}{3}$; $\sec\theta = \frac{\sqrt{34}}{5}$; $\csc\theta = \frac{\sqrt{34}}{3}$
3. $\sin\theta = \frac{\sqrt{3}}{2}$; $\cos\theta = \frac{1}{2}$; $\tan\theta = \sqrt{3}$; $\cot\theta = \frac{\sqrt{3}}{3}$; $\sec\theta = 2$; $\csc\theta = \frac{2\sqrt{3}}{3}$
4. $\sin\theta = \frac{\sqrt{2}}{2}$; $\cos\theta = \frac{\sqrt{2}}{2}$; $\tan\theta = 1$; $\cot\theta = 1$; $\sec\theta = \sqrt{2}$; $\csc\theta = \sqrt{2}$ **5.** 0 **6.** 0 **7.** 1 **8.** 1 **9.** 1 **10.** −1 **11.** $\alpha = 70°, b \approx 3.42, a \approx 9.4$
12. $b \approx 7.14, c \approx 8.72, \beta = 55°$ **13.** $a \approx 4.58, \alpha = 66.4°, \beta \approx 23.6°$ **14.** $c \approx 3.16, \alpha \approx 71.6°, \beta \approx 18.4°$ **15.** $\gamma = 100°, b \approx 0.65, c \approx 1.29$
16. $a \approx 0.54, b \approx 2.38, \beta \approx 130°$ **17.** $\beta \approx 56.8°, \gamma \approx 23.2°, b \approx 4.25$ **18.** No triangle **19.** No triangle **20.** No triangle
21. $b \approx 3.32, \alpha \approx 62.8°, \gamma \approx 17.2°$ **22.** $\alpha \approx 36.2°, \gamma \approx 63.8°, c \approx 4.55$ **23.** No triangle **24.** $\alpha \approx 83.3°, \beta \approx 44.0°, \gamma \approx 52.6°$
25. $c \approx 2.32, \alpha \approx 16.1°, \beta \approx 123.9°$ **26.** $c \approx 4.29, \alpha \approx 66.7°, \beta \approx 13.3°$ **27.** $\beta = 36.2°, \gamma = 63.8°, c = 4.55$
28. Two triangles: $\beta_1 \approx 30.9°, \gamma_1 \approx 129.1°, c_1 \approx 4.54$ or $\beta_2 \approx 149.1°, \gamma_2 \approx 10.9°, c_2 \approx 1.10$ **29.** $\alpha = 39.6°, \beta = 18.6°, \gamma = 121.9°$
30. $\alpha \approx 97.2°, \beta \approx 41.4°, \gamma \approx 41.4°$ **31.** Two triangles: $\beta_1 \approx 13.4°, \gamma_1 \approx 156.6°, c_1 \approx 6.86$ or $\beta_2 \approx 166.6°, \gamma_2 \approx 3.4°, c_2 \approx 1.02$
32. $b \approx 11.52, c \approx 10.13, \gamma = 60°$ **33.** $a \approx 5.23, \beta = 46.0°, \gamma = 64.0°$ **34.** $c \approx 1.73, \alpha = 30°, \beta = 90°$ **35.** 1.93 **36.** 4.28 **37.** 18.79
38. 0.98 **39.** 6 **40.** 27.81 **41.** 3.80 **42.** 1.98 **43.** 0.32 **44.** 0.93 **45.** 839.10 ft **46.** 132.55 ft/min **47.** 23.32 ft **48.** 37.31 ft
49. 2.15 mi **50.** Approximately −12.7° **51.** 204.07 mi **52. (a)** 6.5° **(b)** 420.64 mi/hr **53. (a)** 2.59 mi **(b)** 2.92 mi **(c)** 2.53 mi
54. 6.22 mi **55. (a)** 131.8 mi **(b)** 23.1° **(c)** 0.21 hr **56.** 71.12 ft **57.** 8798.67 sq ft **58.** $222,983.51 **59.** 1.92 sq in. **60.** S4.0°E
61. 76.94 in. **62.** 79.69 in. **63.** $d = -3\cos\left[\frac{\pi}{2}t\right]$ **64.** $d = -5\cos\left[\frac{\pi}{3}t\right]$ **65. (a)** Simple harmonic **(b)** 6 ft **(c)** π sec
(d) $\frac{1}{\pi}$ oscillation/sec **66. (a)** Simple harmonic **(b)** 2 ft **(c)** $\frac{\pi}{2}$ sec **(d)** $\frac{2}{\pi}$ oscillation/sec **67. (a)** Simple harmonic **(b)** 2 ft **(c)** 2 sec
(d) $\frac{1}{2}$ oscillation/sec **68. (a)** Simple harmonic **(b)** 3 ft **(c)** 4 sec **(d)** $\frac{1}{4}$ oscillation/sec

69.

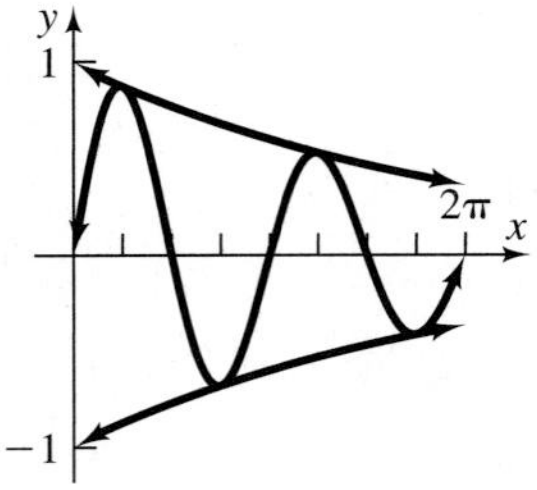

70.

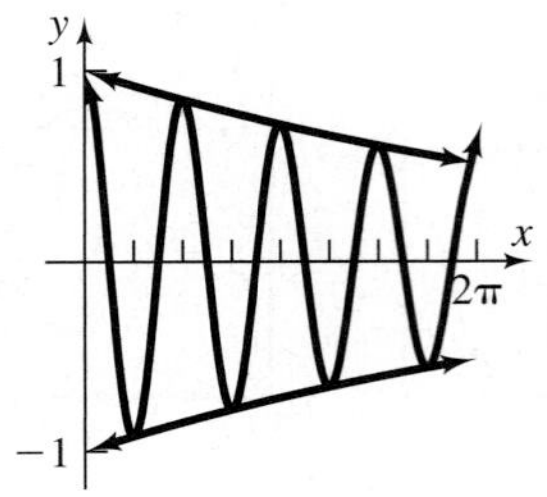

71.

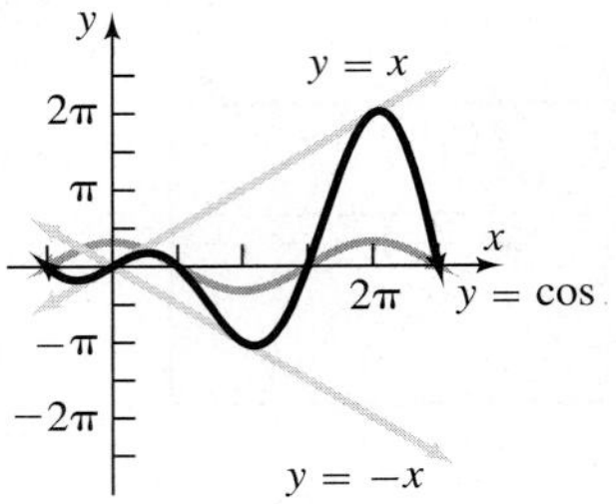

72.

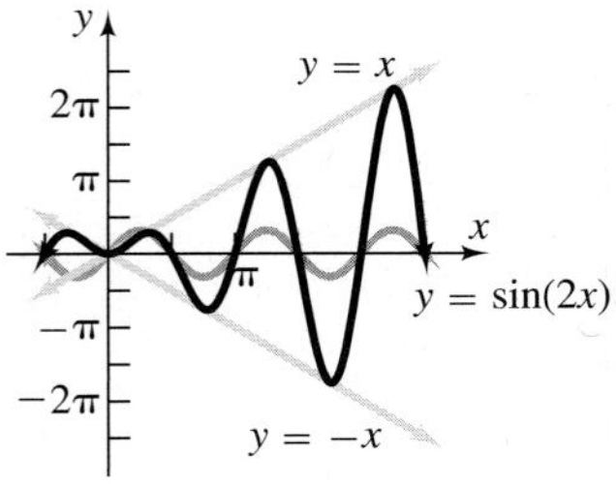

73.

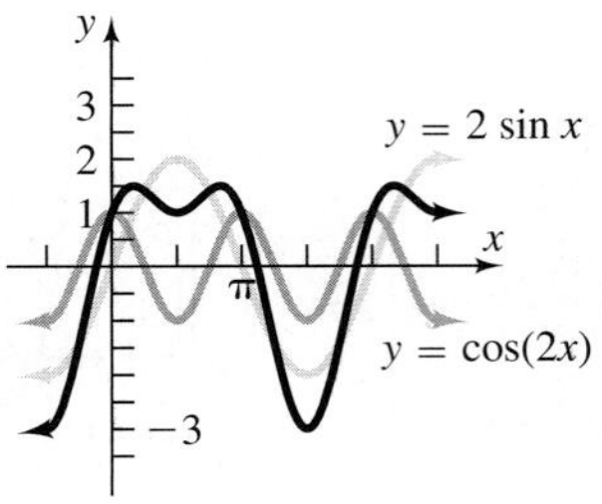

74.

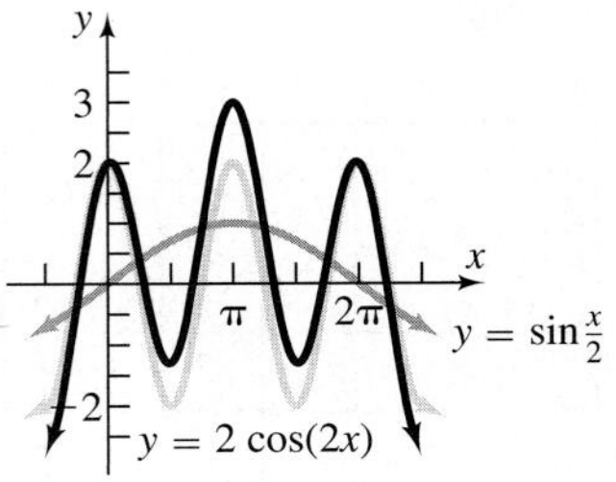

Cumulative Review *(page 534)*

1. $\left\{\frac{1}{3}, 1\right\}$ **2.** $(x + 5)^2 + (y - 1)^2 = 9$

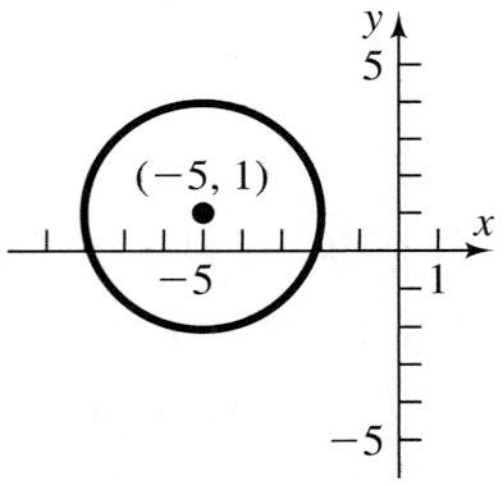

3. $\{x | x \leq -1 \text{ or } x \geq 4\}$ **4.**

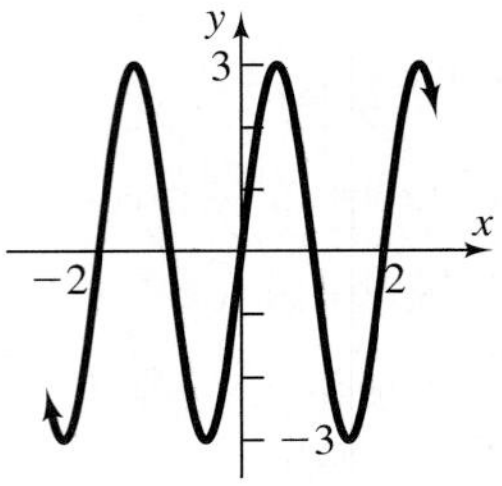

5.

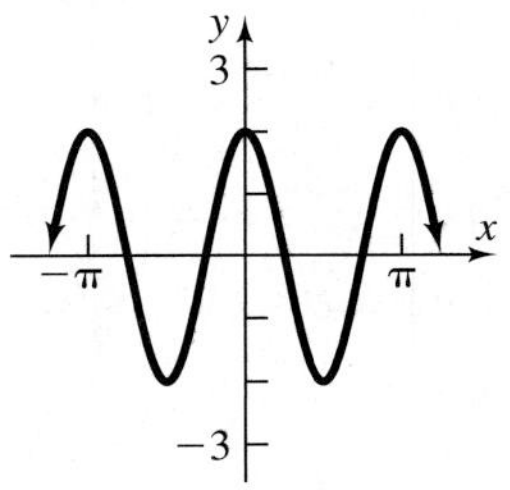

6. (a) $-\frac{2\sqrt{5}}{5}$ **(b)** $\frac{\sqrt{5}}{5}$ **(c)** $-\frac{4}{5}$ **(d)** $-\frac{3}{5}$ **(e)** $\sqrt{\frac{5 - \sqrt{5}}{10}}$ **(f)** $-\sqrt{\frac{5 + \sqrt{5}}{10}}$

7. (a)

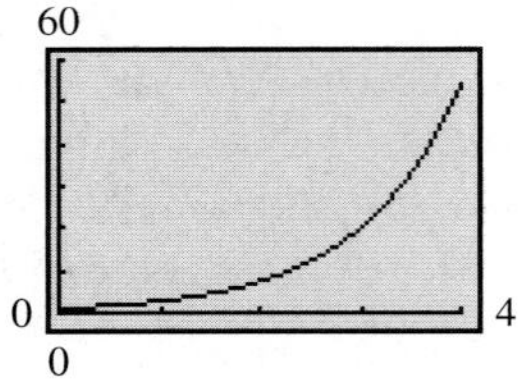

(b)

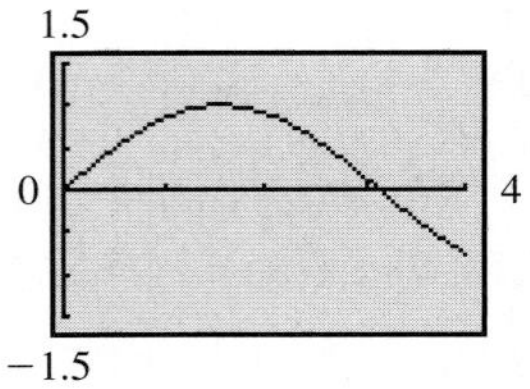

(c)

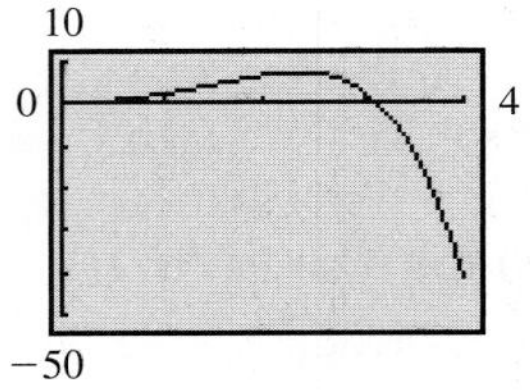

(d)

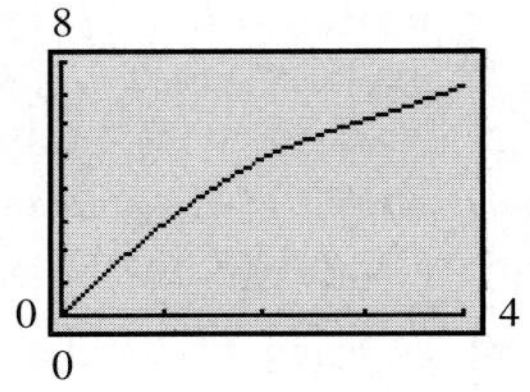

8. (a)
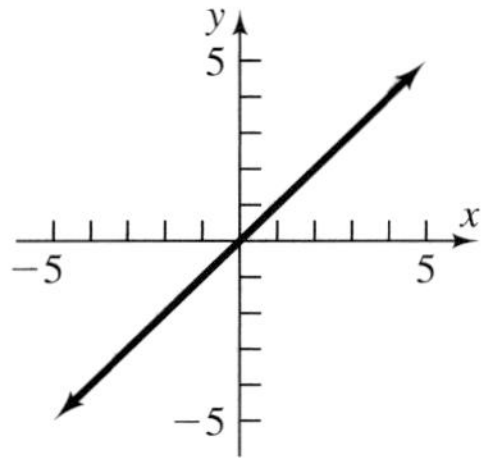

(b)
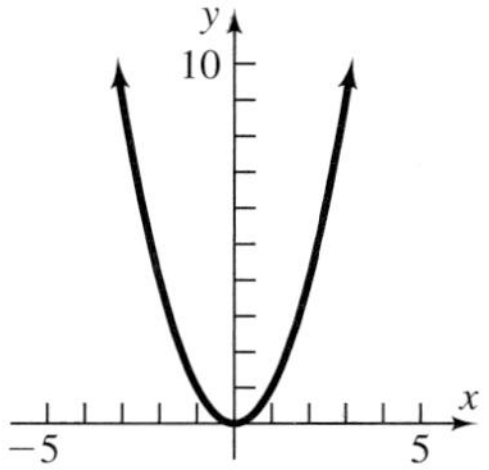

(c)
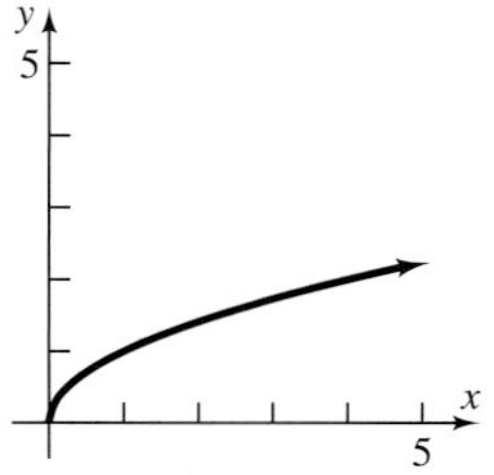

(d)
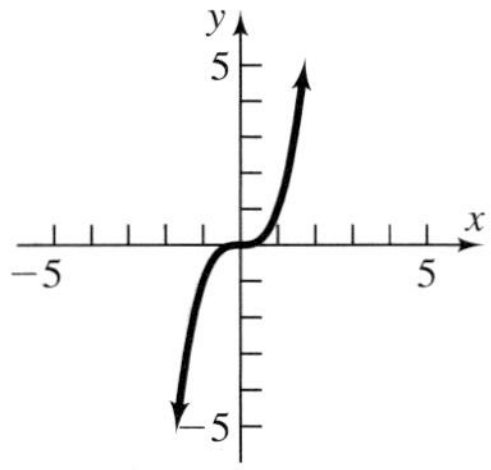

(e)
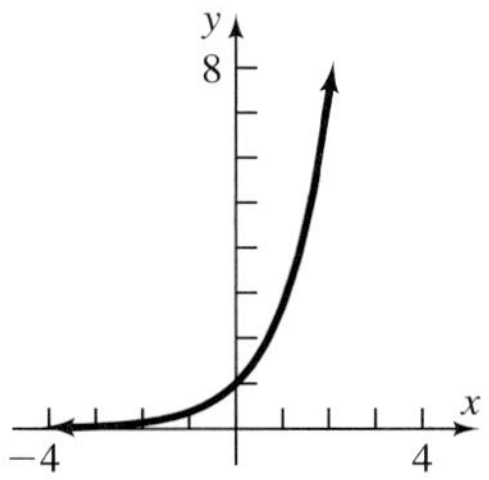

(f)
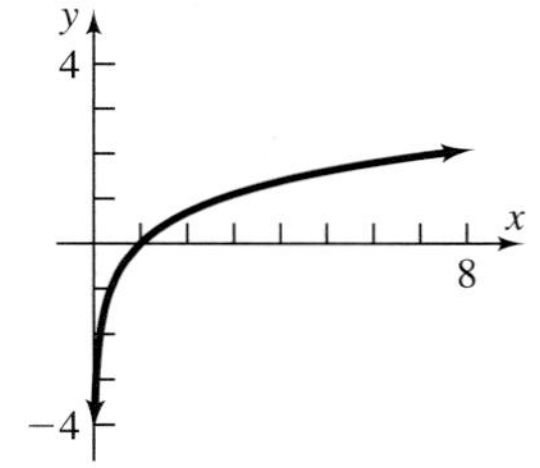

(g)
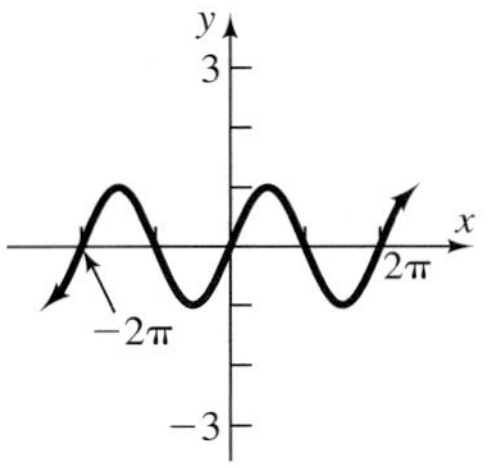

(h)
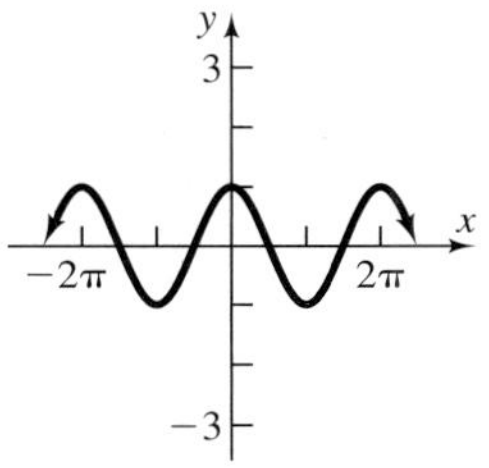

(i)
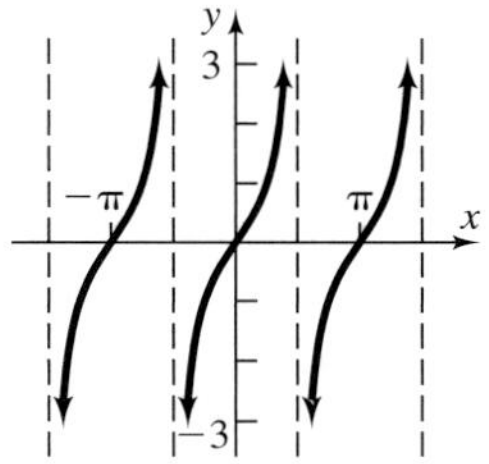

9. Two triangles: $\alpha_1 \approx 59.0°, \beta_1 \approx 81.0°, b_1 \approx 23.05$ or $\alpha_2 \approx 121.0°, \beta_2 \approx 19.0°, b_2 \approx 7.59$ **10.** $\left\{-2i, 2i, \frac{1}{3}, 1, 2\right\}$

11. $R(x) = \frac{(2x + 1)(x - 4)}{(x + 5)(x - 3)}$; domain: $\{x|x \neq -5, x \neq 3\}$

intercepts: $\left(-\frac{1}{2}, 0\right), (4, 0), \left(0, \frac{4}{15}\right)$

no symmetry

vertical asymptotes: $x = -5, x = 3$

horizontal asymptote: $y = 2$

intersects: $\left(\frac{26}{11}, 2\right)$

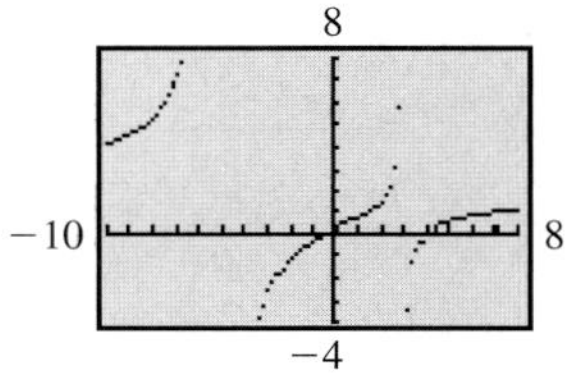

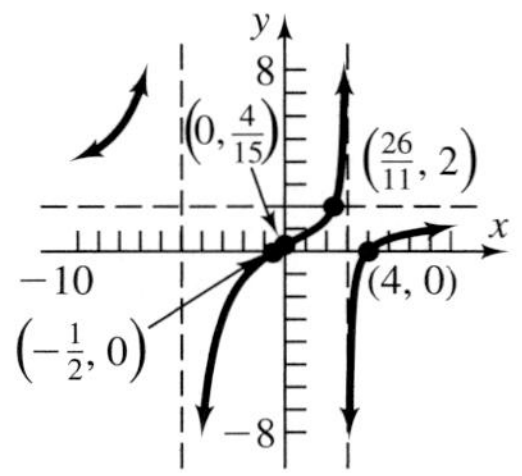

12. $\{2.26\}$ **13.** $\{1\}$ **14. (a)** $\left\{-\frac{5}{4}\right\}$ **(b)** $\{2\}$ **(c)** $\left\{\frac{-1 - \sqrt{117}}{2}, \frac{-1 + \sqrt{117}}{2}\right\}$ **(d)** $\left\{x \middle| x > -\frac{5}{4}\right\}$ or $\left(-\frac{5}{4}, \infty\right)$

(e) $\left\{x \middle| -8 \leq x \leq 3\right\}$ or $[-8, 3]$ **(f)**

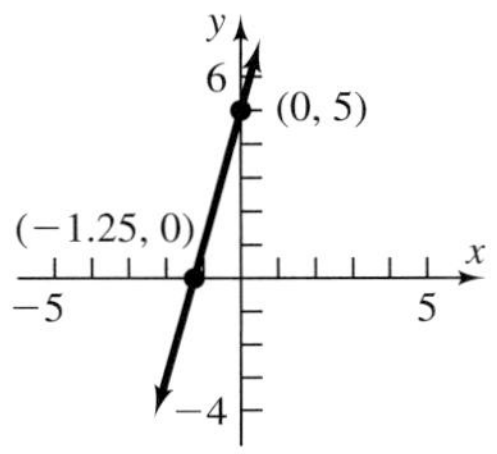

(g)
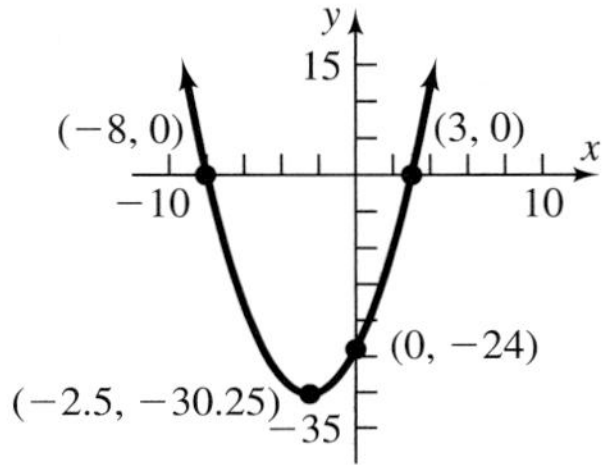

CHAPTER 8 Polar Coordinates; Vectors

8.1 Concepts and Vocabulary *(page 543)*

5. pole; polar axis **6.** -2 **7.** $(-\sqrt{3}, -1)$ **8.** False **9.** True **10.** True

8.1 Exercises *(page 543)*

11. A **12.** B **13.** C **14.** C **15.** B **16.** D **17.** A **18.** D

19.

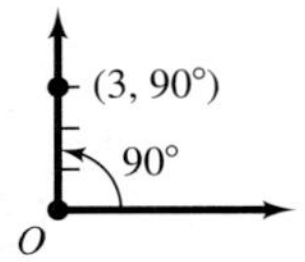

20.

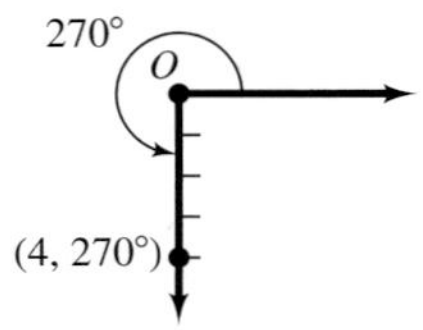

21.

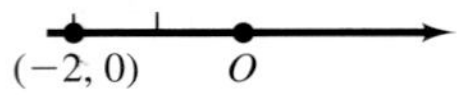

22.

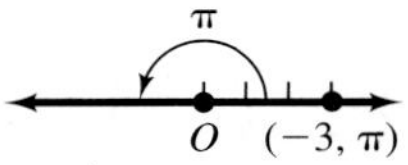

23.

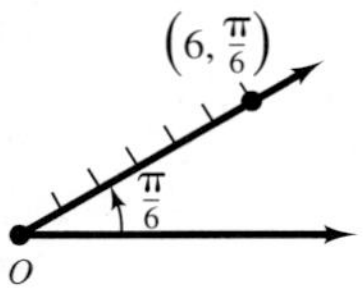

24.

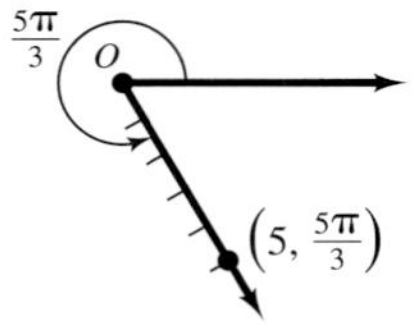

25.

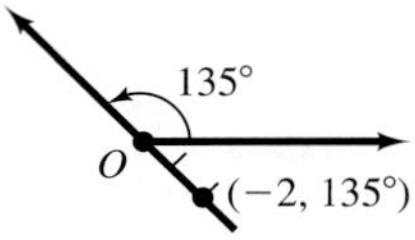

26.

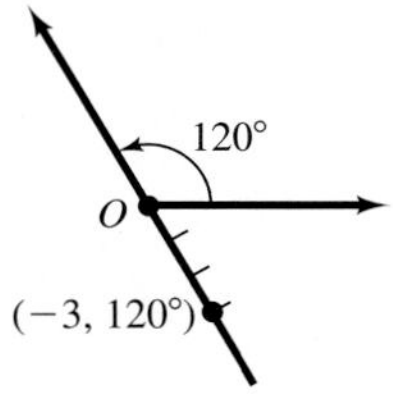

27.

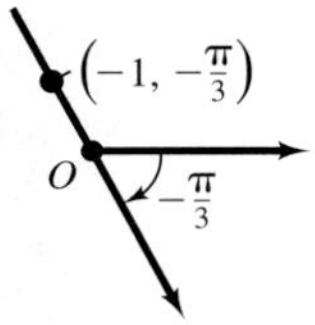

28.

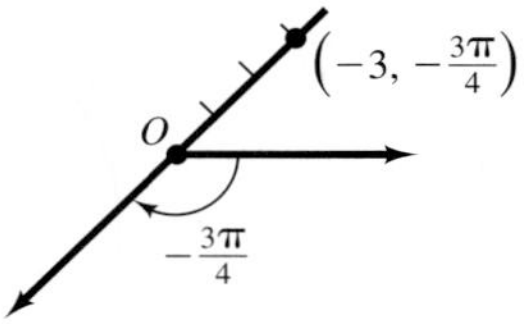

29.

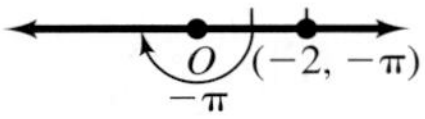

30.

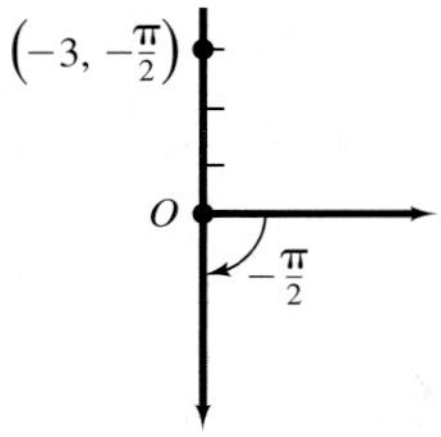

31.

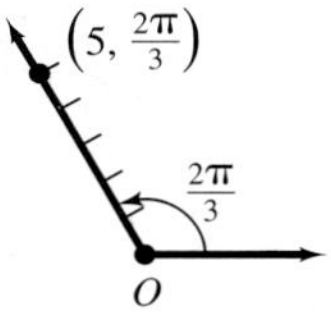

(a) $\left(5, -\frac{4\pi}{3}\right)$
(b) $\left(-5, \frac{5\pi}{3}\right)$
(c) $\left(5, \frac{8\pi}{3}\right)$

32.

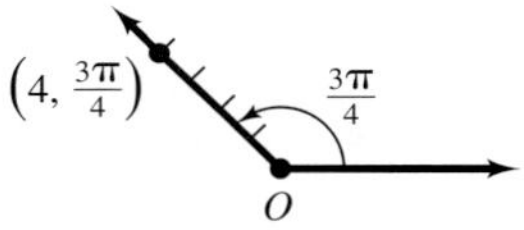

(a) $\left(4, -\frac{5\pi}{4}\right)$
(b) $\left(-4, \frac{7\pi}{4}\right)$
(c) $\left(4, \frac{11\pi}{4}\right)$

33.

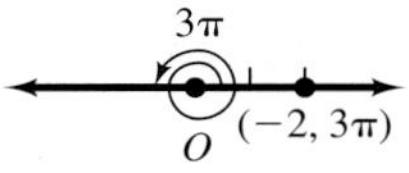

(a) $(2, -2\pi)$
(b) $(-2, \pi)$
(c) $(2, 2\pi)$

34.

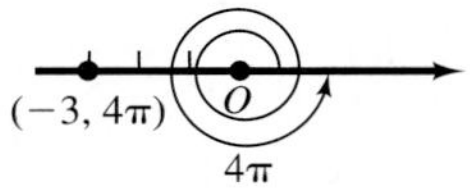

(a) $(3, -\pi)$
(b) $(-3, 0)$
(c) $(3, 3\pi)$

35.

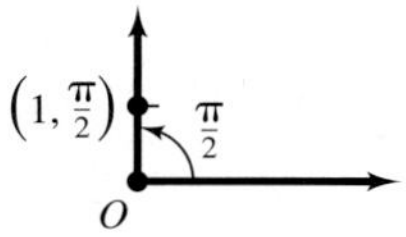

(a) $\left(1, -\frac{3\pi}{2}\right)$
(b) $\left(-1, \frac{3\pi}{2}\right)$
(c) $\left(1, \frac{5\pi}{2}\right)$

36.

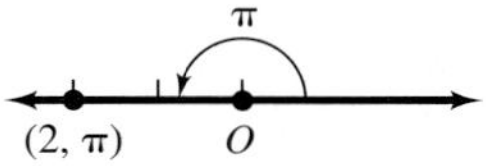

(a) $(2, -\pi)$
(b) $(-2, 0)$
(c) $(2, 3\pi)$

37.

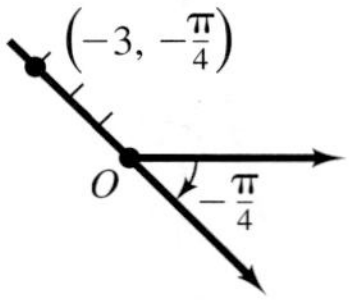

(a) $\left(3, -\frac{5\pi}{4}\right)$
(b) $\left(-3, \frac{7\pi}{4}\right)$
(c) $\left(3, \frac{11\pi}{4}\right)$

38.

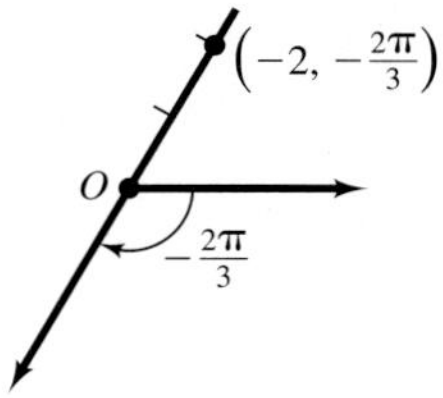

(a) $\left(2, -\frac{5\pi}{3}\right)$
(b) $\left(-2, \frac{4\pi}{3}\right)$
(c) $\left(2, \frac{7\pi}{3}\right)$

39. $(0, 3)$ **40.** $(0, -4)$ **41.** $(-2, 0)$ **42.** $(3, 0)$ **43.** $(-3\sqrt{3}, 3)$ **44.** $\left(\frac{5}{2}, -\frac{5\sqrt{3}}{2}\right)$ **45.** $(\sqrt{2}, -\sqrt{2})$ **46.** $(1, -\sqrt{3})$ **47.** $\left(-\frac{1}{2}, \frac{\sqrt{3}}{2}\right)$

48. $\left(\frac{3\sqrt{2}}{2}, \frac{3\sqrt{2}}{2}\right)$ **49.** $(2, 0)$ **50.** $(0, 3)$ **51.** $(-2.57, 7.05)$ **52.** $(3.10, 0.11)$ **53.** $(-4.98, -3.85)$ **54.** $(3.79, -7.16)$ **55.** $(3, 0)$ **56.** $\left(2, \frac{\pi}{2}\right)$ **57.** $(1, \pi)$ **58.** $\left(2, -\frac{\pi}{2}\right)$ **59.** $\left(\sqrt{2}, -\frac{\pi}{4}\right)$ **60.** $\left(3\sqrt{2}, \frac{3\pi}{4}\right)$ **61.** $\left(2, \frac{\pi}{6}\right)$ **62.** $\left(4, -\frac{2\pi}{3}\right)$ **63.** $(2.47, -1.02)$ **64.** $(2.25, -1.93)$ **65.** $(9.30, 0.47)$ **66.** $(2.31, 3.05)$ **67.** $r^2 = \frac{3}{2}$ **68.** $r = \cos\theta$ **69.** $r^2\cos^2\theta - 4r\sin\theta = 0$ **70.** $r^2\sin^2\theta - 2r\cos\theta = 0$ **71.** $r^2\sin 2\theta = 1$ **72.** $r^3(\sin\theta)(\cos^2\theta) = \frac{1}{4}$ **73.** $r\cos\theta = 4$ **74.** $r\sin\theta = -3$ **75.** $x^2 + y^2 - x = 0$ or $\left(x - \frac{1}{2}\right)^2 + y^2 = \frac{1}{4}$ **76.** $x^2 + y^2 = y + \sqrt{x^2 + y^2}$ **77.** $(x^2 + y^2)^{3/2} - x = 0$ **78.** $x^2 + y^2 = y - x$ **79.** $x^2 + y^2 = 4$ **80.** $x^2 + y^2 = 16$ **81.** $y^2 = 8(x + 2)$ **82.** $64\left(x - \frac{3}{8}\right)^2 + 72y^2 = 81$

83.
$$\begin{aligned} d &= \sqrt{(r_2\cos\theta_2 - r_1\cos\theta_1)^2 + (r_2\sin\theta_2 - r_1\sin\theta_1)^2} \\ &= \sqrt{(r_2^2\cos^2\theta_2 - 2r_2\cos\theta_2\, r_1\cos\theta_1 + r_1^2\cos^2\theta_1) + (r_2^2\sin^2\theta_2 - 2r_2\sin\theta_2\, r_1\sin\theta_1 + r_1^2\sin^2\theta_1)} \\ &= \sqrt{r_1^2 + r_2^2 - 2r_1r_2(\cos\theta_2\cos\theta_1 + \sin\theta_2\sin\theta_1)} \\ &= \sqrt{r_1^2 + r_2^2 - 2r_1r_2\cos(\theta_2 - \theta_1)} \end{aligned}$$

8.2 Concepts and Vocabulary *(page 560)*

7. polar equation **8.** $r = 2\cos\theta$ **9.** $-r$ **10.** False **11.** False **12.** False

8.2 Exercises *(page 560)*

13. $x^2 + y^2 = 16$;

Circle, radius 4, center at pole

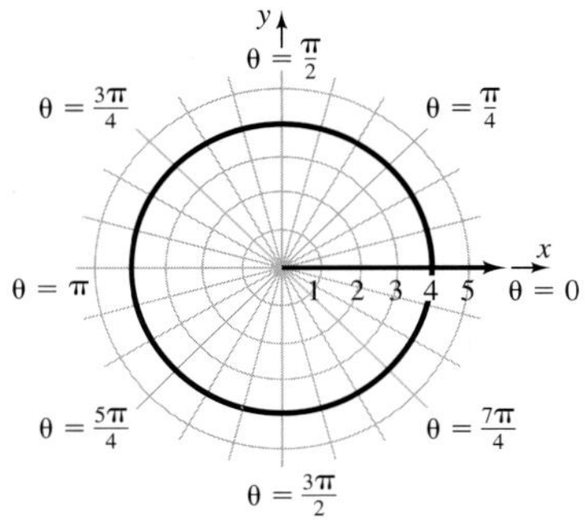

14. $x^2 + y^2 = 4$;

Circle, radius 2, center at pole

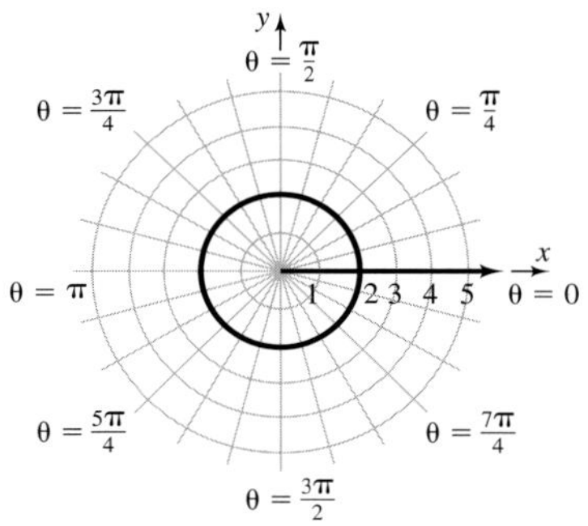

15. $y = \sqrt{3}x$; Line through pole, making an angle of $\frac{\pi}{3}$ with polar axis

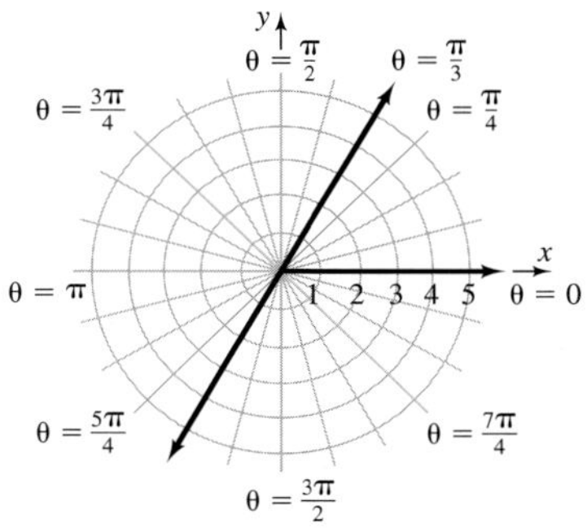

16. $y = -x$; Line through pole, making an angle of $-\frac{\pi}{4}$ or $\left(\frac{3\pi}{4}\right)$ with polar axis

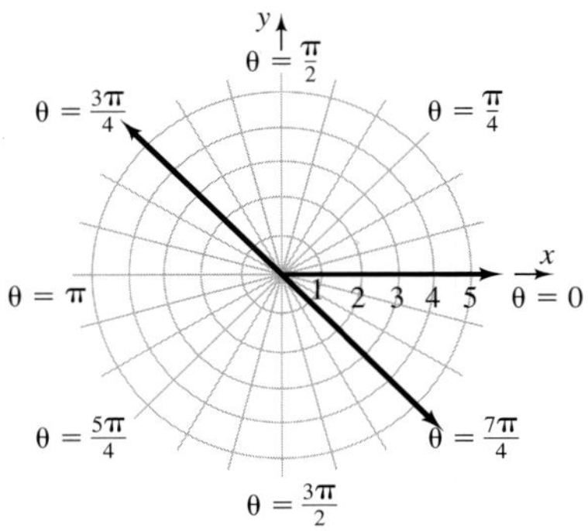

17. $y = 4$; Horizontal line 4 units above the pole

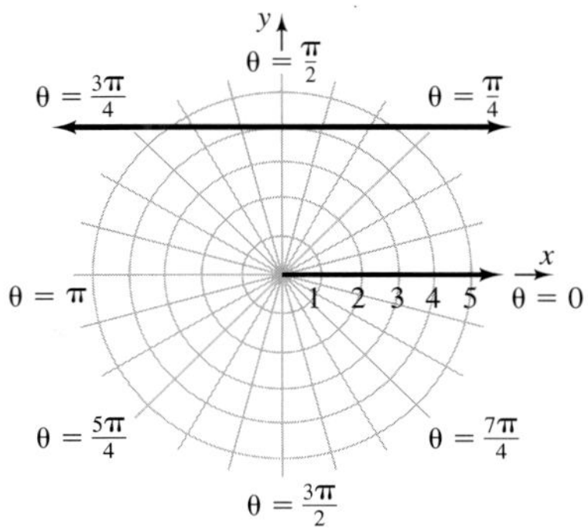

18. $x = 4$; Vertical line 4 units to the right of the pole

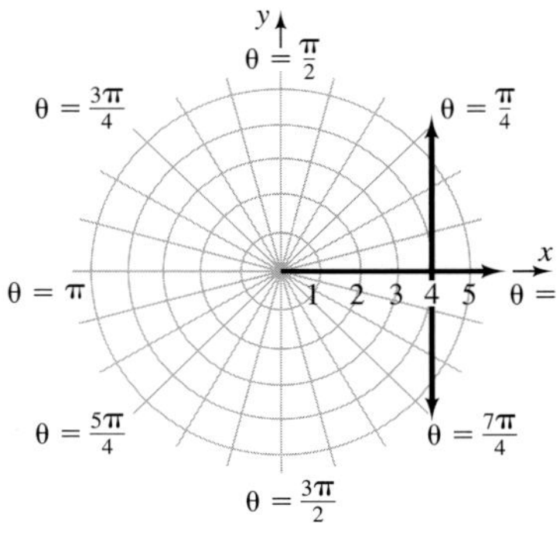

19. $x = -2$; Vertical line 2 units to the left of the pole

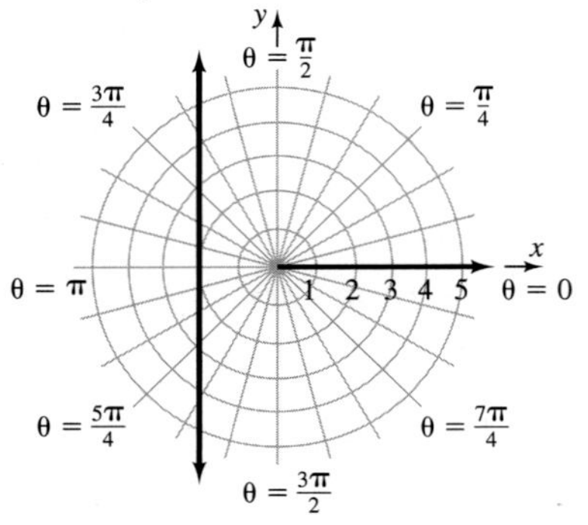

20. $y = -2$; Horizontal line 2 units below the pole

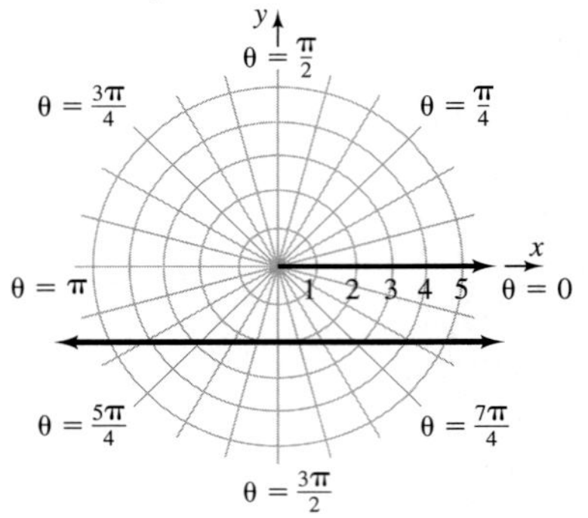

21. $(x - 1)^2 + y^2 = 1$; Circle, radius 1, center $(1, 0)$ in rectangular coordinates

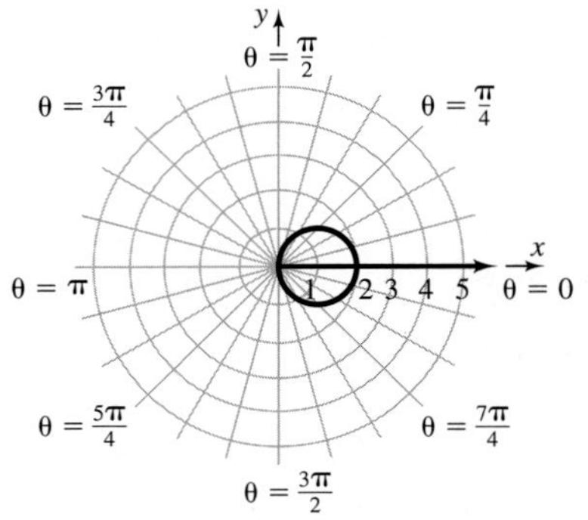

22. $x^2 + (y - 1)^2 = 1$; Circle, radius 1, center $(0, 1)$ in rectangular coordinates

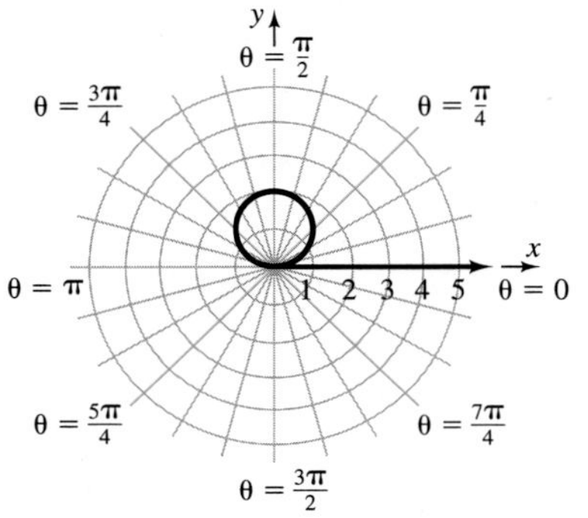

23. $x^2 + (y + 2)^2 = 4$; Circle, radius 2, center at $(0, -2)$ in rectangular coordinates

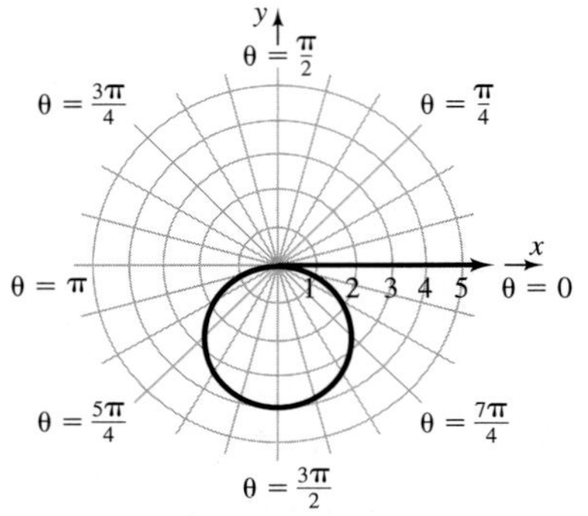

24. $(x + 2)^2 + y^2 = 4$; Circle, radius 2, center $(-2, 0)$ in rectangular coordinates

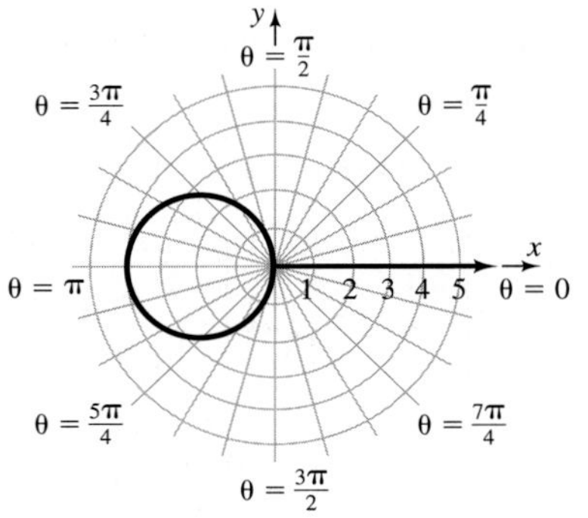

25. $(x - 2)^2 + y^2 = 4$; Circle, radius 2, center at $(2, 0)$ in rectangular coordinates

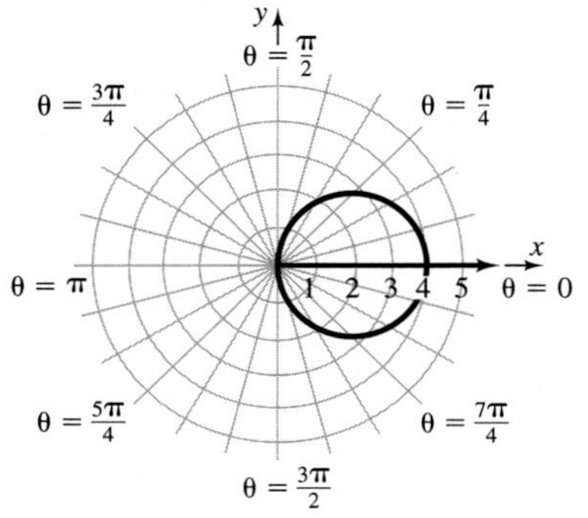

26. $x^2 + (y - 4)^2 = 16$; Circle, radius 4, center $(0, 4)$ in rectangular coordinates

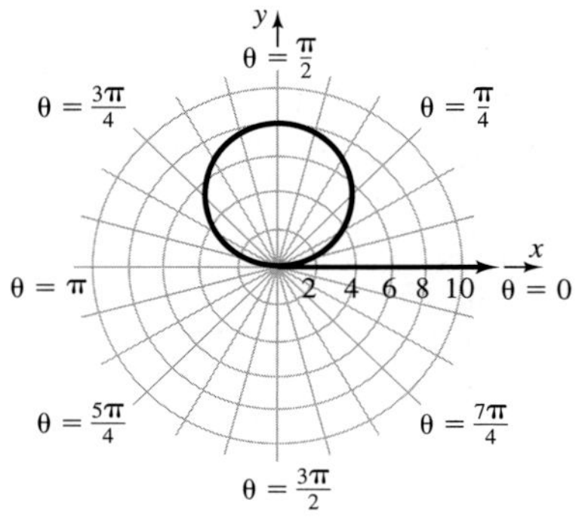

27. $x^2 + (y + 1)^2 = 1$; Circle, radius 1, center at $(0, -1)$ in rectangular coordinates

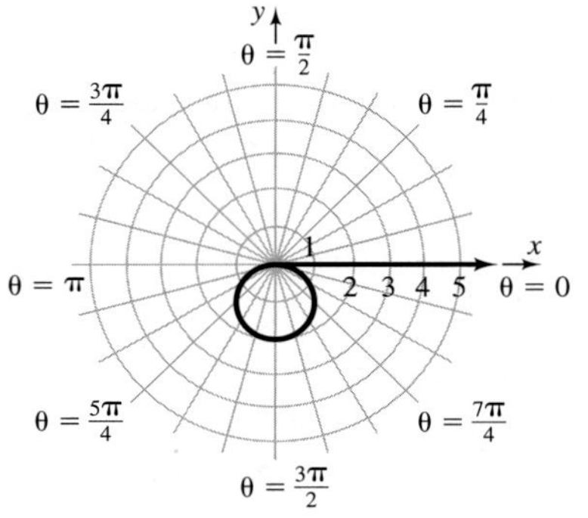

28. $(x + 2)^2 + y^2 = 4$;
Circle, radius 2, center $(-2, 0)$ in rectangular coordinates

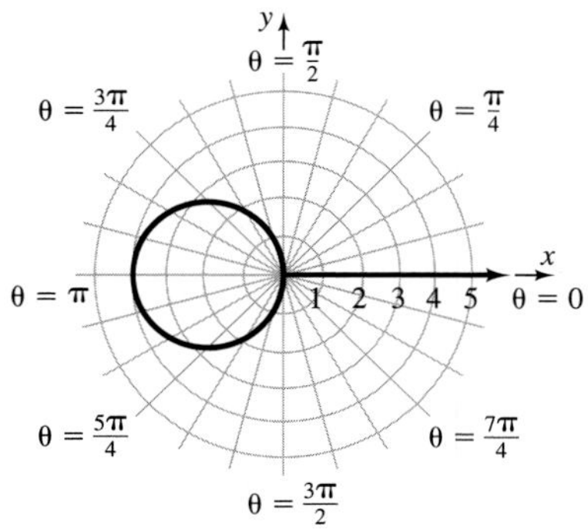

29. E **30.** A **31.** F **32.** B **33.** H **34.** G **35.** D **36.** C

37. Cardioid

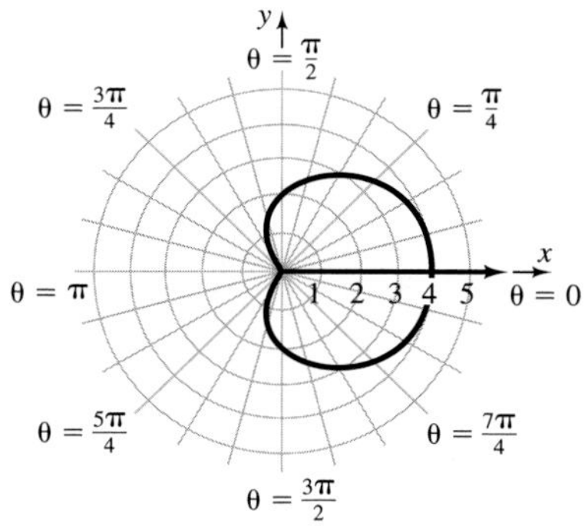

38. Cardioid

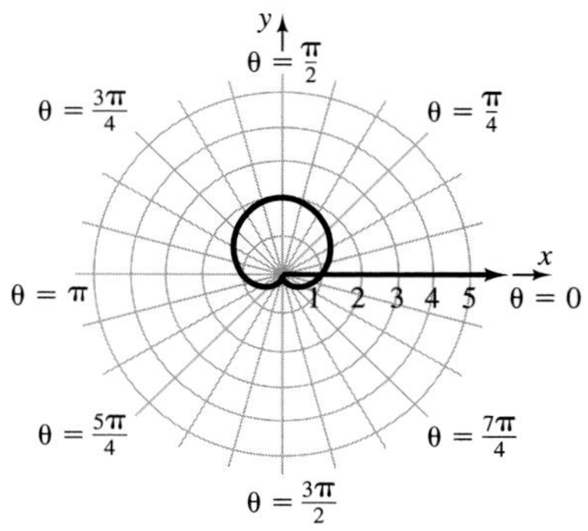

39. Cardioid

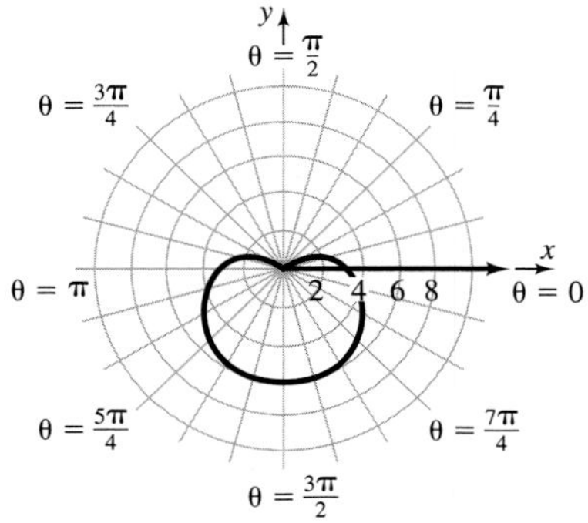

40. Cardioid

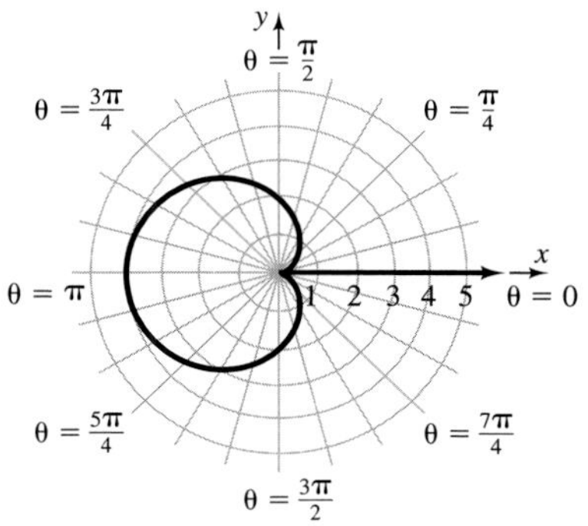

41. Limaçon without inner loop

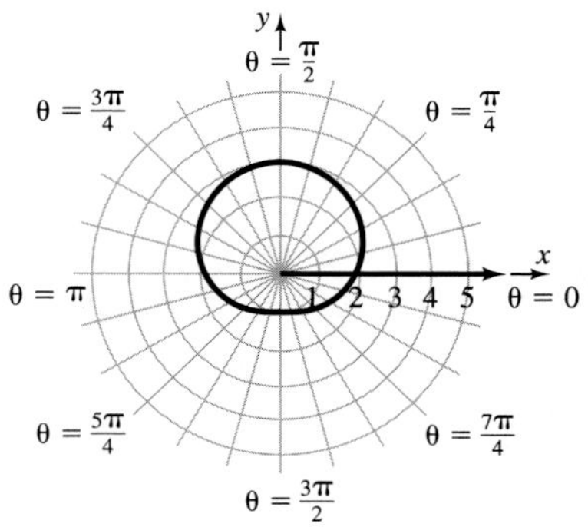

42. Limaçon without inner loop

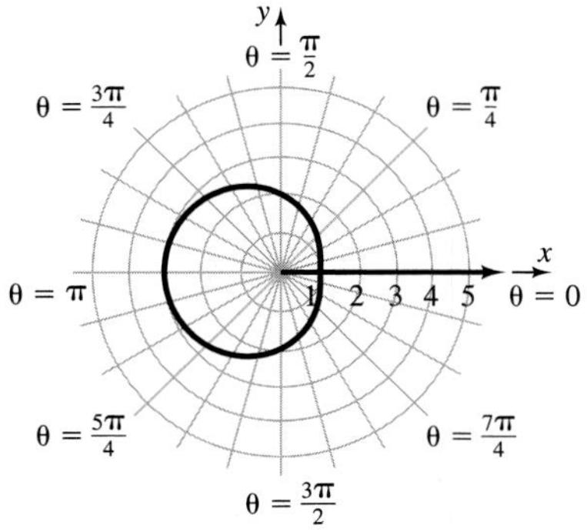

43. Limaçon without inner loop

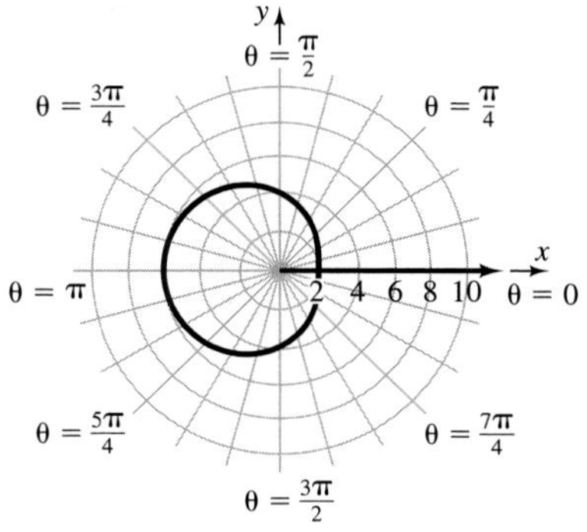

44. Limaçon without inner loop

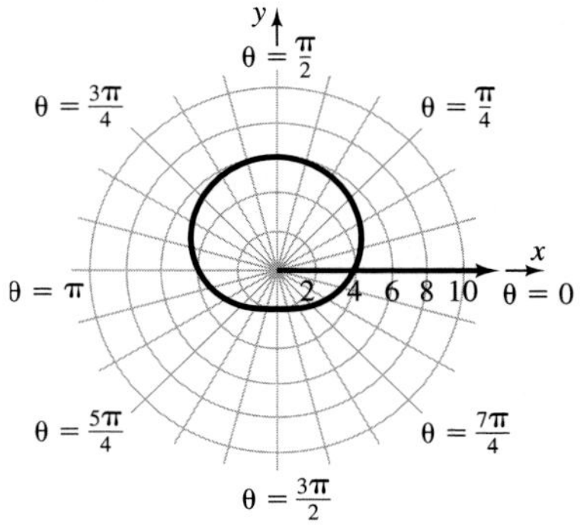

45. Limaçon with inner loop

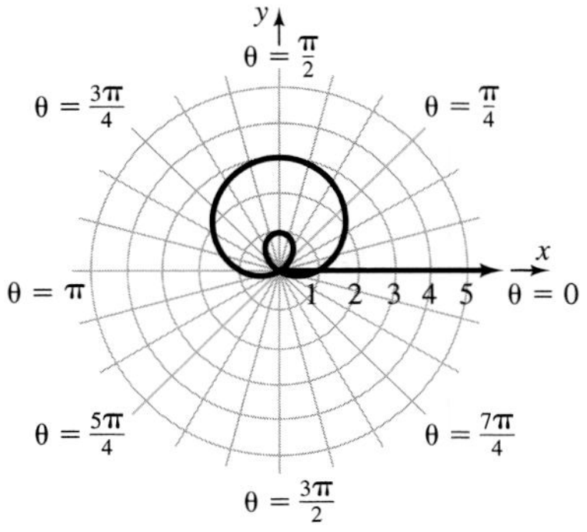

46. Limaçon with inner loop

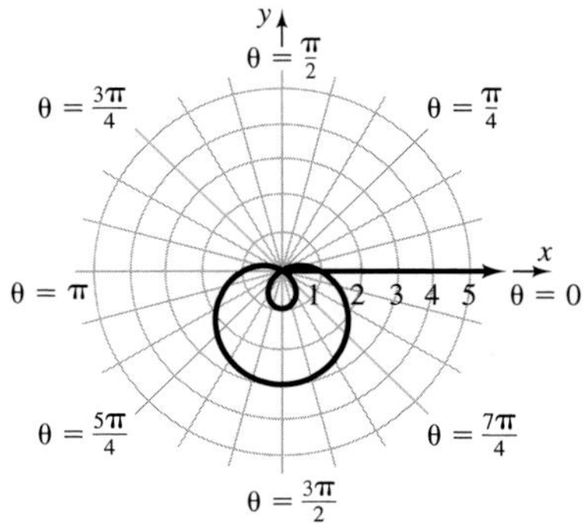

47. Limaçon with inner loop

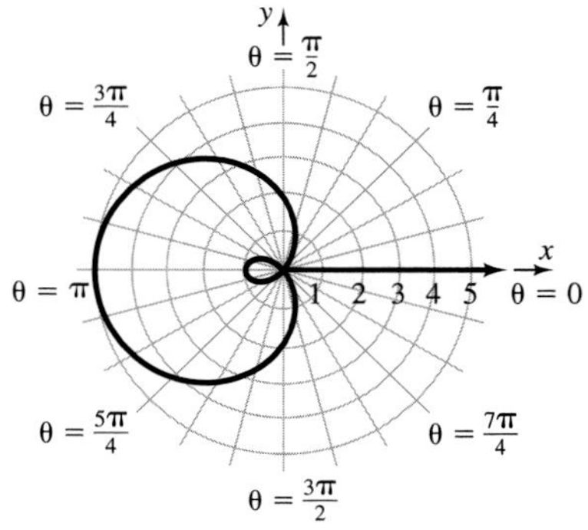

48. Limaçon with inner loop

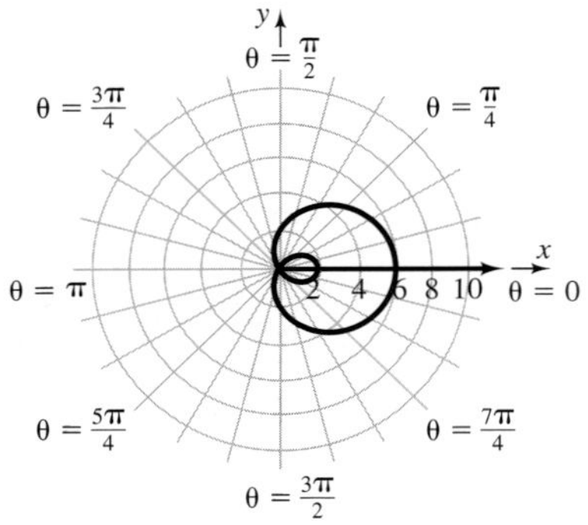

49. Rose

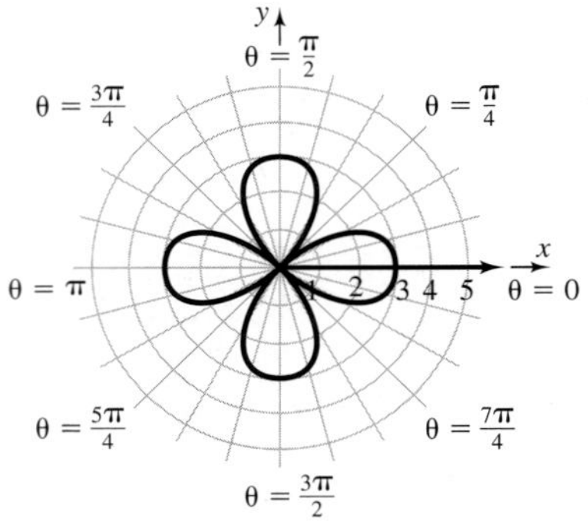

50. Rose

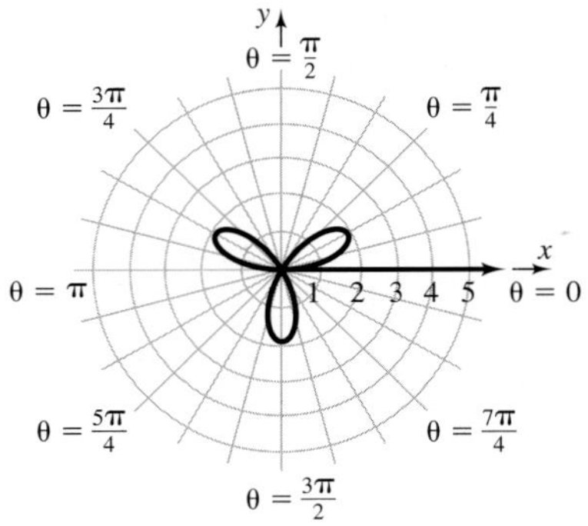

51. Rose

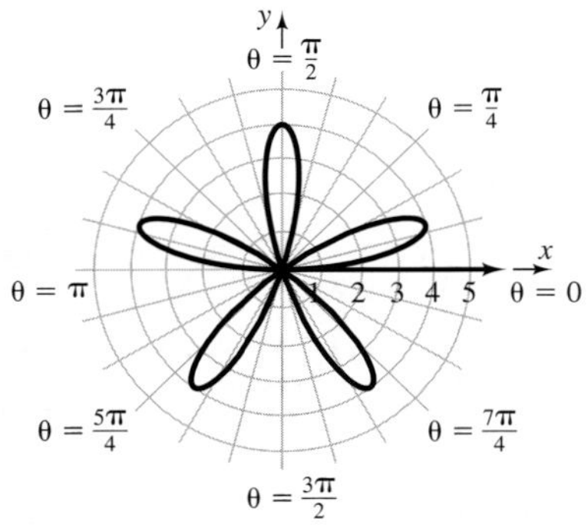

52. Rose

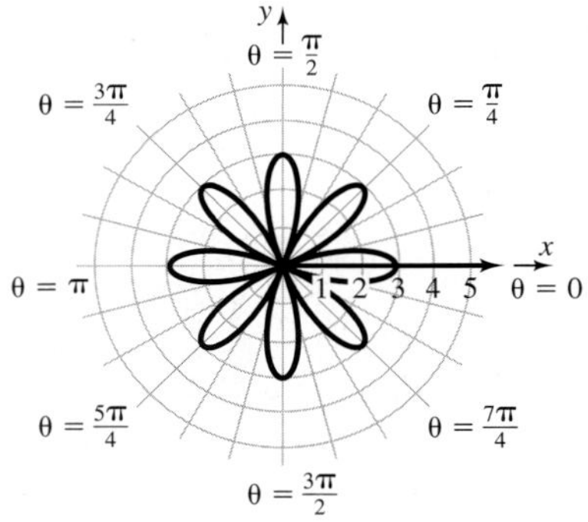

53. Lemniscate

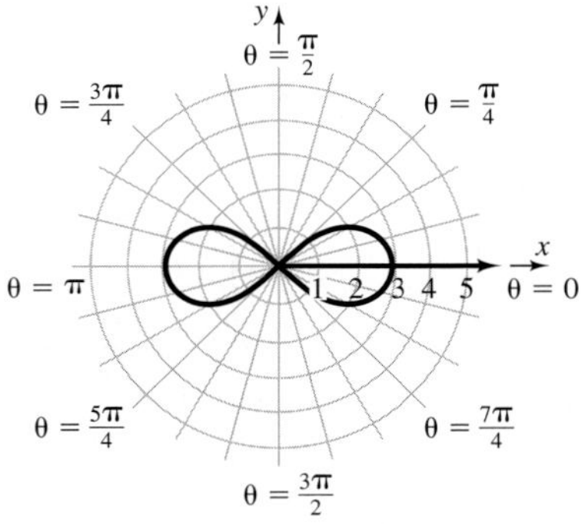

54. Lemniscate

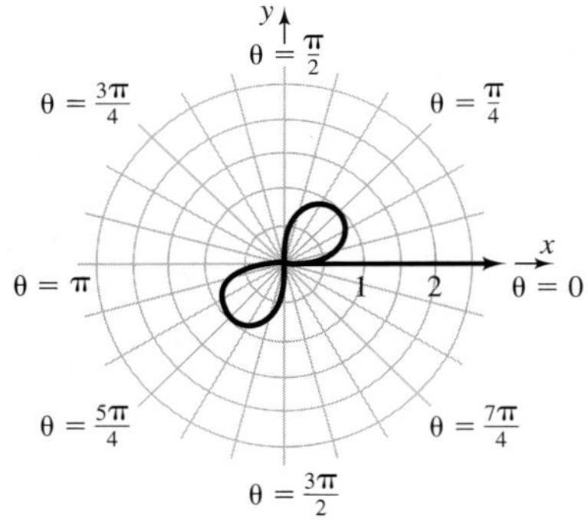

55. Spiral

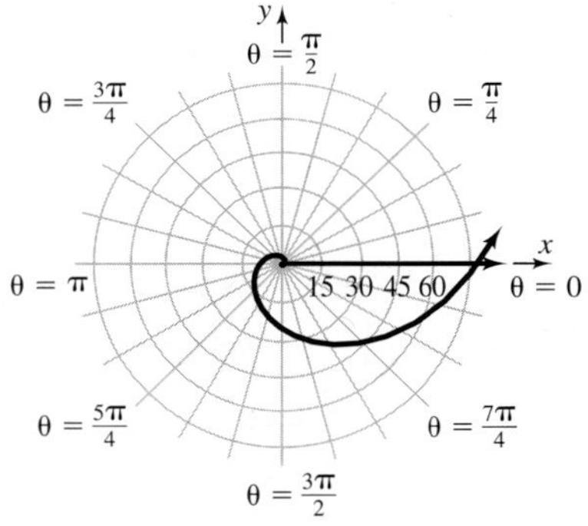

56. Spiral

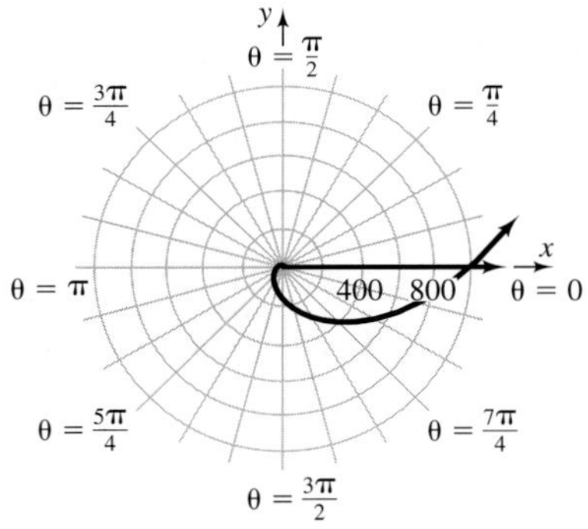

57. Cardioid

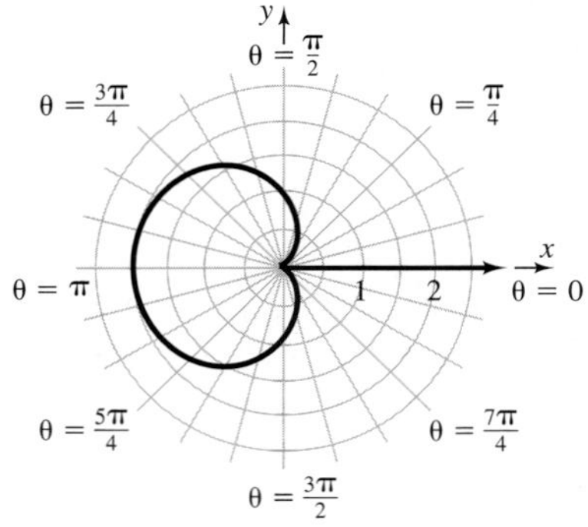

58. Limaçon without inner loop

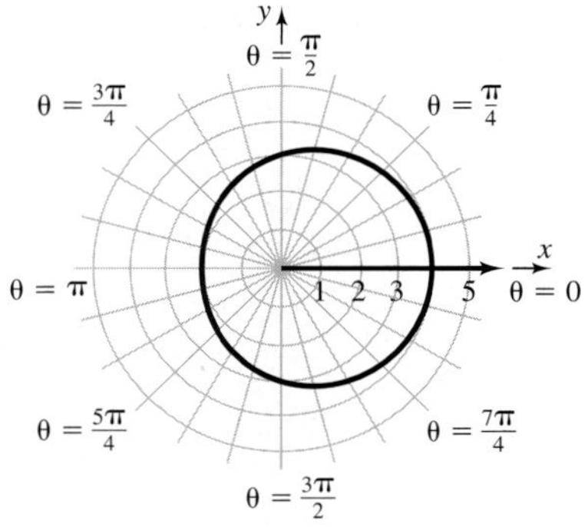

59. Limaçon with inner loop

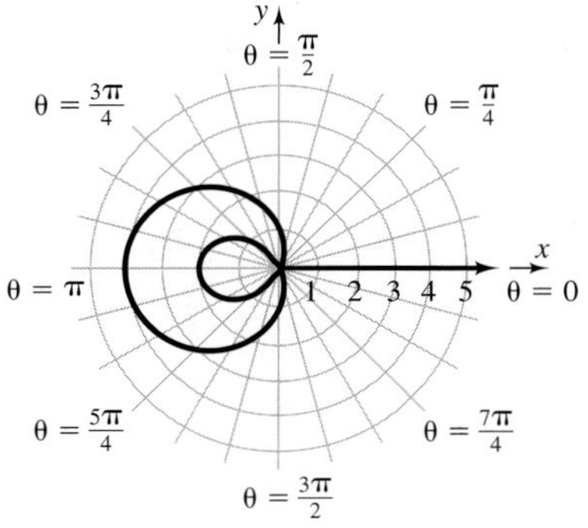

60. Rose

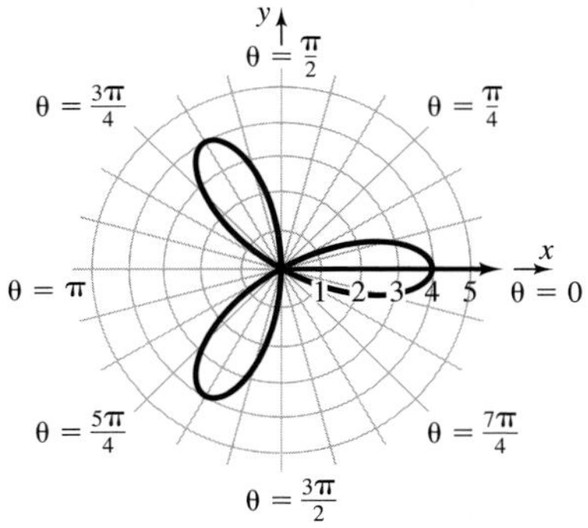

61. $r = 3 + 3 \cos \theta$

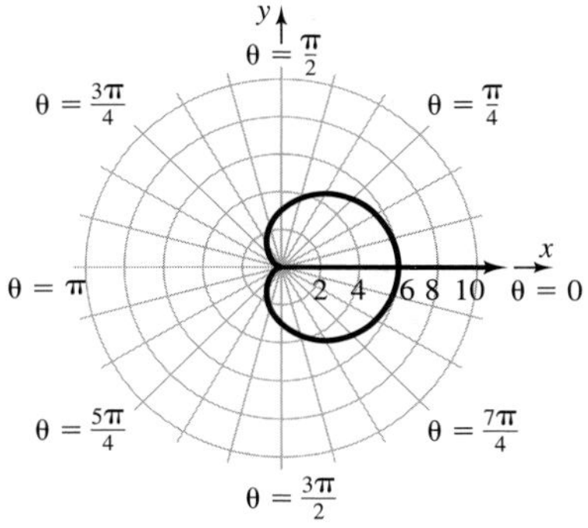

62. $r = 3 - 3 \cos \theta$

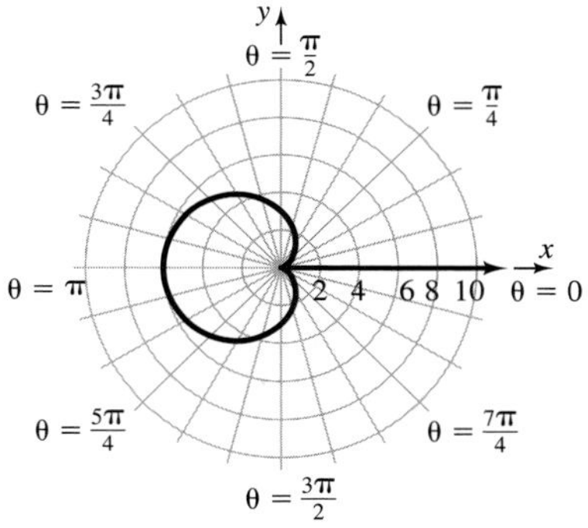

63. $r = 4 + \sin \theta$

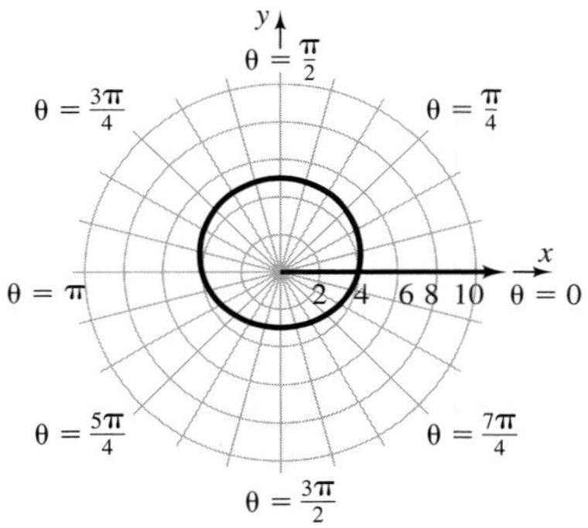

64. $r = 1 + 4 \sin \theta$

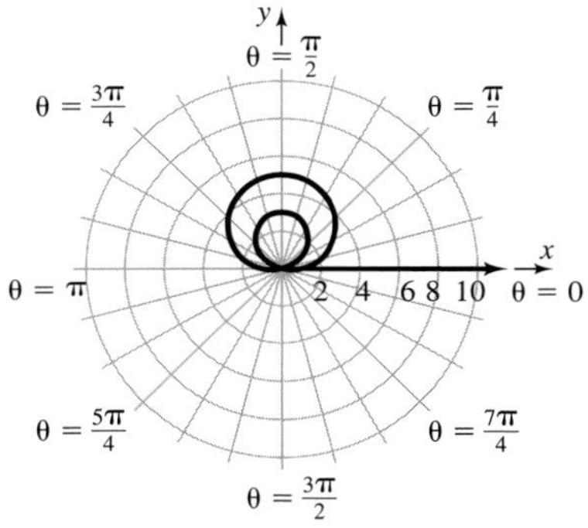

65.

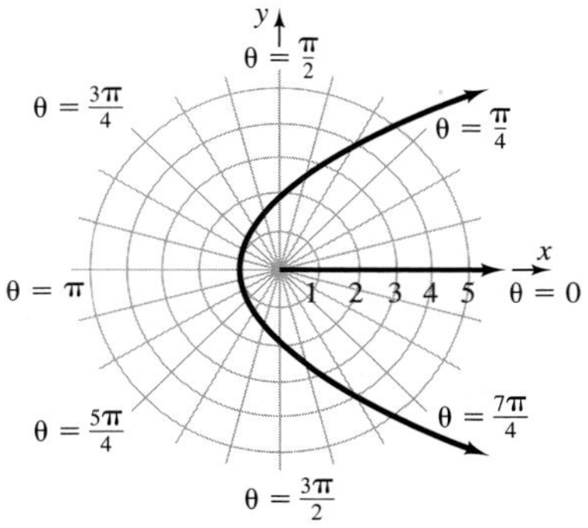

66.

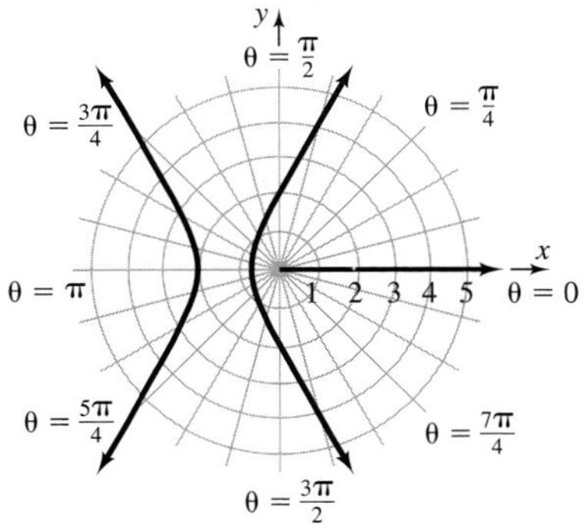

67.

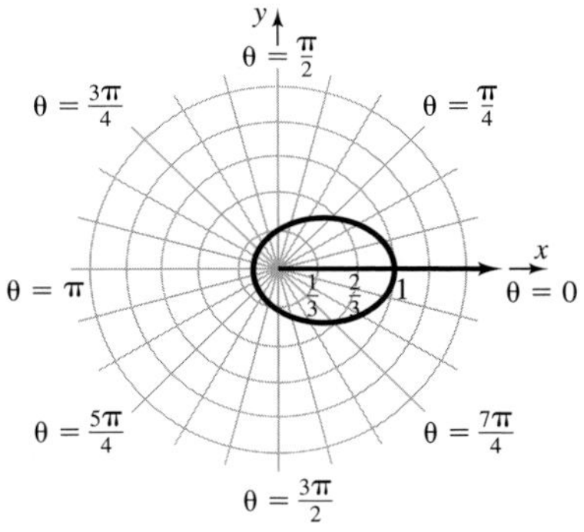

68.

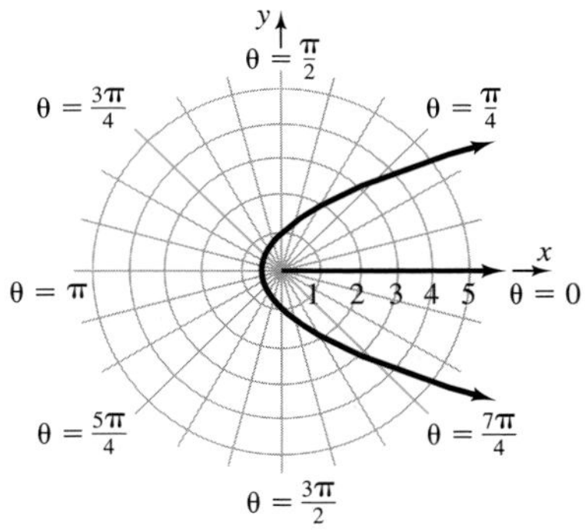

69.

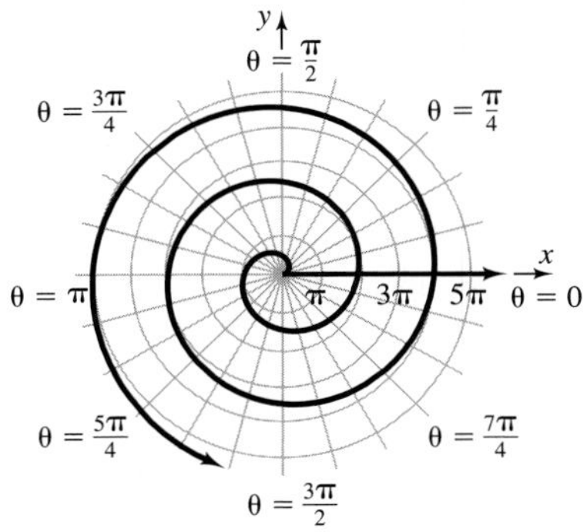

70.

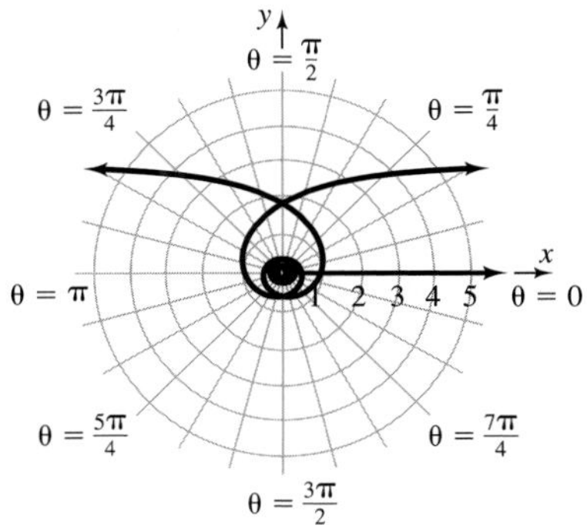

71.

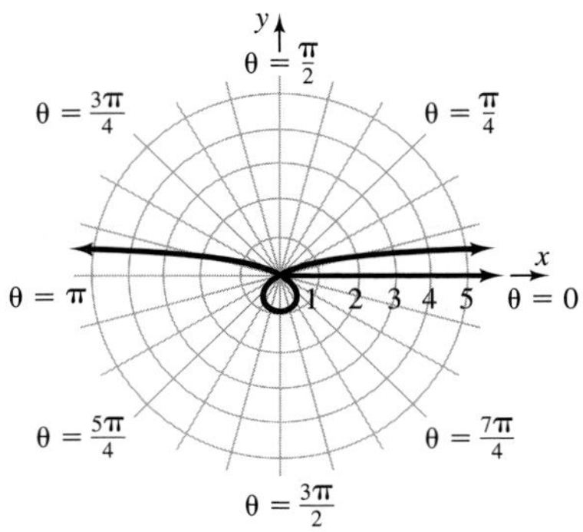

72.

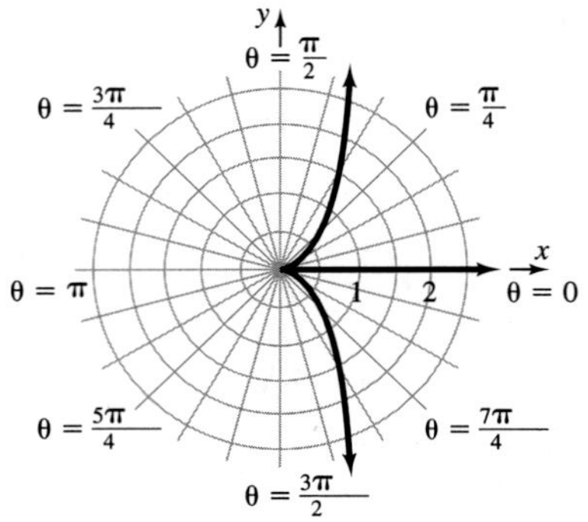

73.

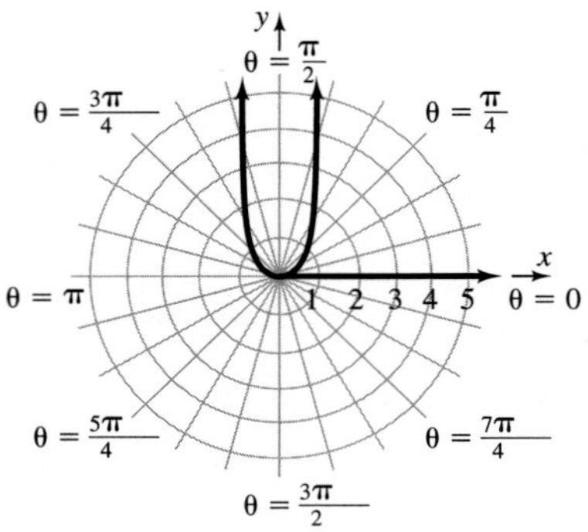

74.

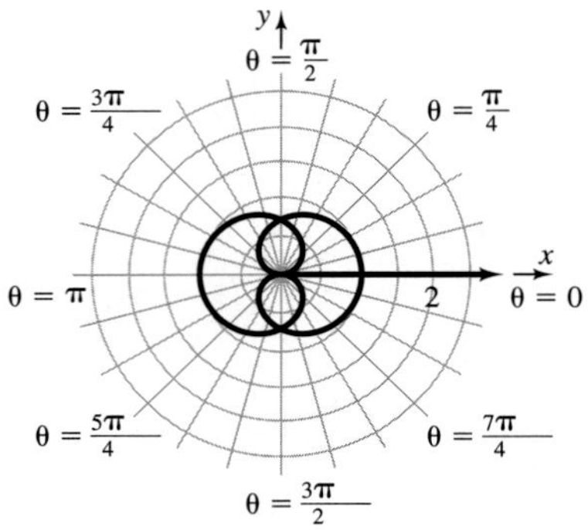

75. $r \sin\theta = a$
$y = a$

76. $r \cos\theta = a$
$x = a$

77. $r = 2a\sin\theta$
$r^2 = 2ar\sin\theta$
$x^2 + y^2 = 2ay$
$x^2 + y^2 - 2ay = 0$
$x^2 + (y - a)^2 = a^2$
Circle, radius a, center at $(0, a)$ in rectangular coordinates

78. $r = -2a\sin\theta$
$r^2 = -2ar\sin\theta$
$x^2 + y^2 = -2ay$
$x^2 + y^2 + 2ay = 0$
$x^2 + (y + a)^2 = a^2$
Circle, radius a, center at $(0, -a)$ in rectangular coordinates

79. $r = 2a\cos\theta$
$r^2 = 2ar\cos\theta$
$x^2 + y^2 = 2ax$
$x^2 - 2ax + y^2 = 0$
$(x - a)^2 + y^2 = a^2$
Circle, radius a, center at $(a, 0)$ in rectangular coordinates

80. $r = -2a\cos\theta$
$r^2 = -2ar\cos\theta$
$x^2 + y^2 = -2ax$
$x^2 + 2ax + y^2 = 0$
$(x + a)^2 + y^2 = a^2$
Circle, radius a, center at $(-a, 0)$ in rectangular coordinates

81. **(a)** $r^2 = \cos\theta: r^2 = \cos(\pi - \theta)$
$r^2 = -\cos\theta$
Not equivalent; test fails.
$(-r)^2 = \cos(-\theta)$
$r^2 = \cos\theta$
New test works.

(b) $r^2 = \sin\theta: r^2 = \sin(\pi - \theta)$
$r^2 = \sin\theta$
Test works.
$(-r)^2 = \sin(-\theta)$
$r^2 = -\sin\theta$
Not equivalent; new test fails.

82. Replace θ by $\theta + \pi$. If an equivalent equation results, the graph is symmetric with respect to the pole.
(a) $\theta = 0: \theta = 0$
(r has been "replaced" by $-r$.)
Test works.
$\theta + \pi = 0$
$\theta = -\pi$
Not equivalent; new test fails.

(b) $r = 1: -r = 1$
$r = -1$
Not equivalent; test fails.
$r = 1$
(θ has been "replaced" by $\theta + \pi$.)
New test works.

Historical Problems *(page 568)*

1. **(a)** $1 + 4i, 1 + i$ **(b)** $-1, 2 + i$

8.3 Concepts and Vocabulary *(page 568)*

5. magnitude or modulus; argument **6.** De Moivre's **7.** three **8.** True **9.** False **10.** True

8.3 Exercises *(page 569)*

11.

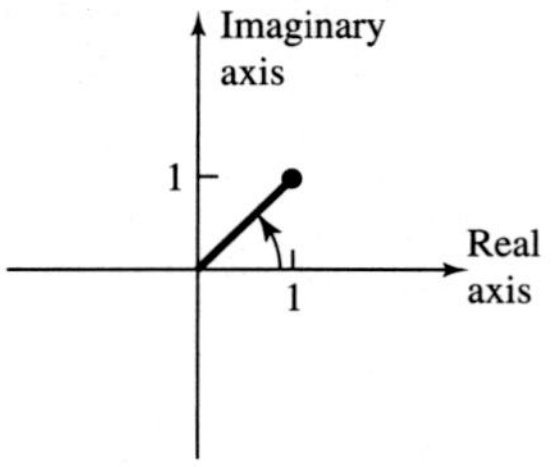

$\sqrt{2}(\cos 45° + i \sin 45°)$

12.

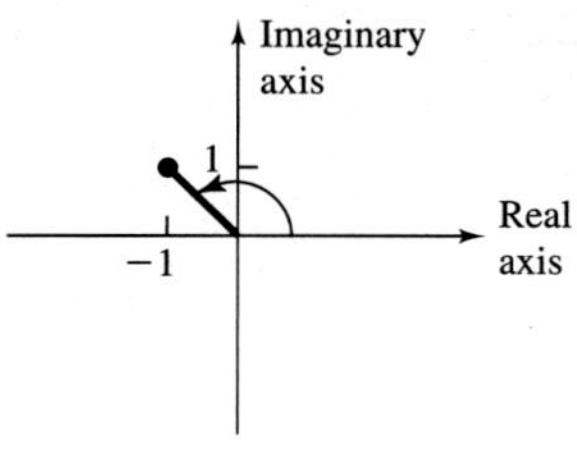

$\sqrt{2}(\cos 135° + i \sin 135°)$

13.

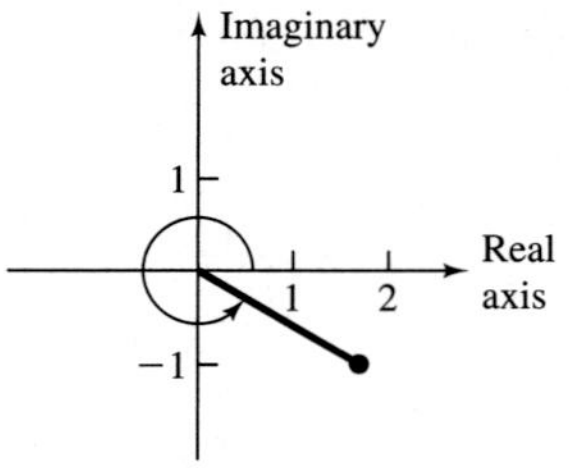

$2(\cos 330° + i \sin 330°)$

14.

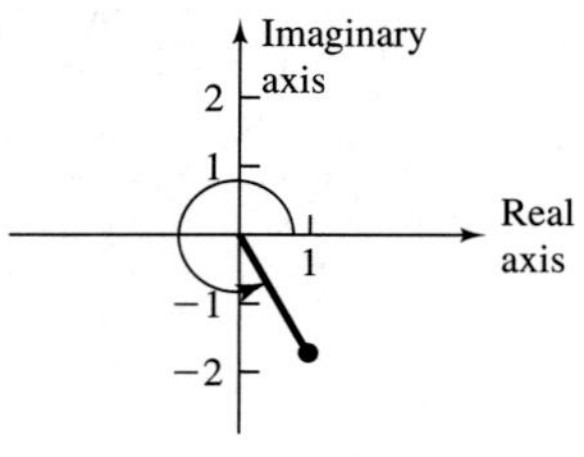

$2(\cos 300° + i \sin 300°)$

15.
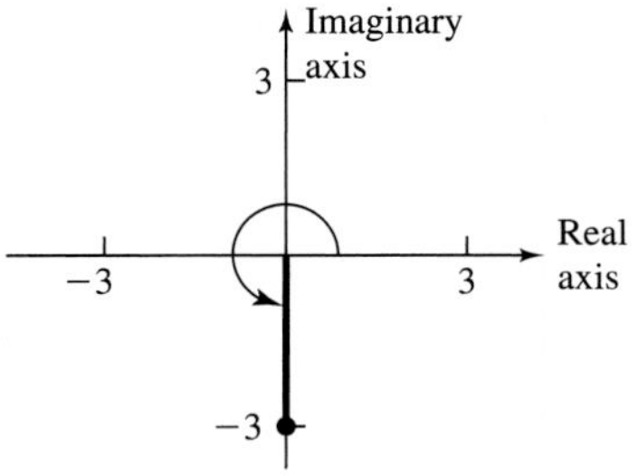

$3(\cos 270° + i \sin 270°)$

16.
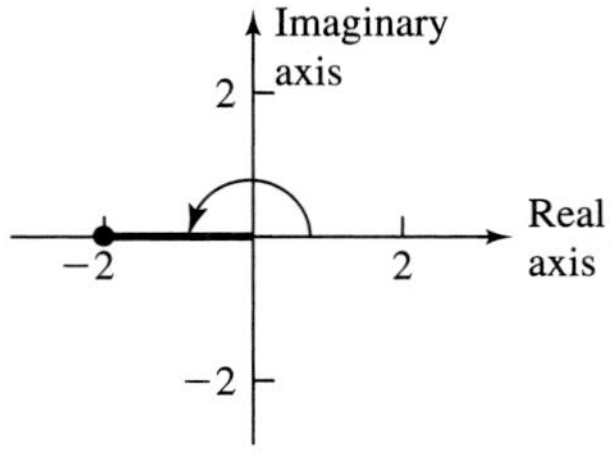

$2(\cos 180° + i \sin 180°)$

17.
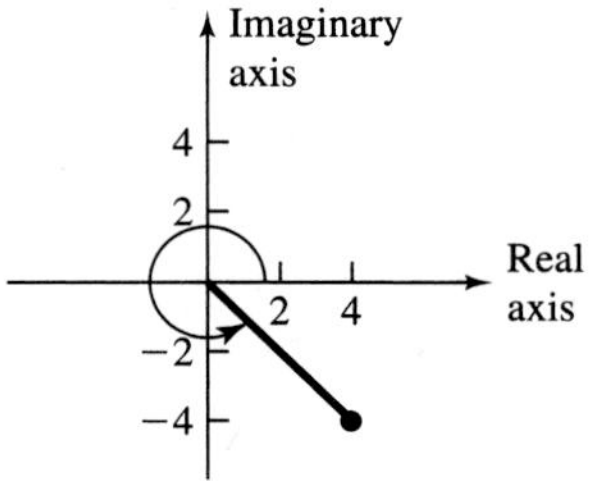

$4\sqrt{2}(\cos 315° + i \sin 315°)$

18.
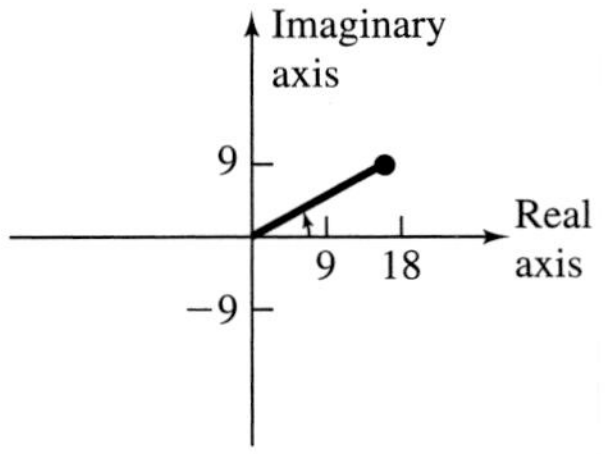

$18(\cos 30° + i \sin 30°)$

19.
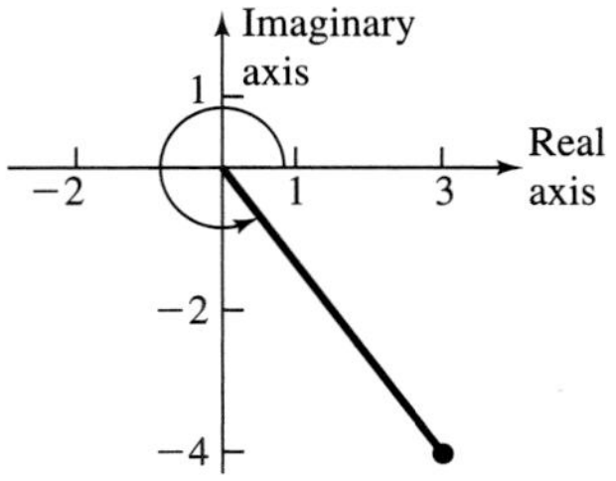

$5(\cos 306.9° + i \sin 306.9°)$

20.
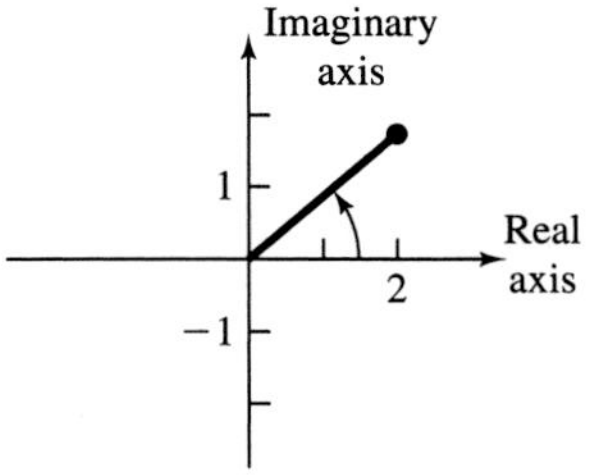

$\sqrt{7}(\cos 40.9° + i \sin 40.9°)$

21.
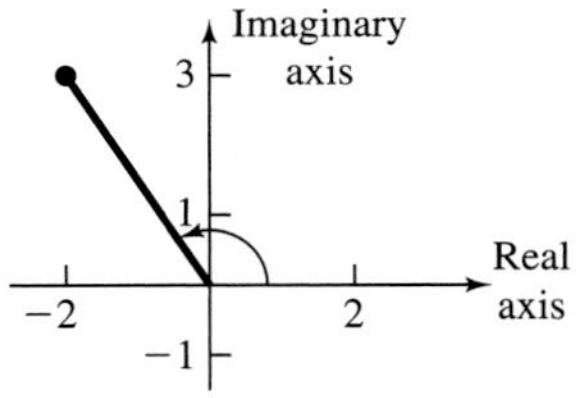

$\sqrt{13}(\cos 123.7° + i \sin 123.7°)$

22.
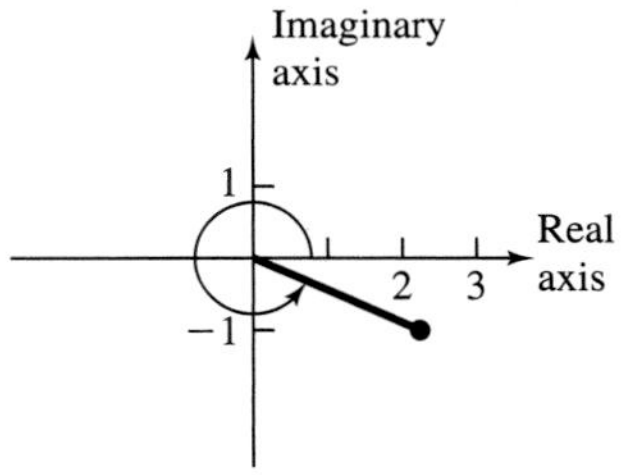

$\sqrt{6}(\cos 335.9° + i \sin 335.9°)$

23. $-1 + \sqrt{3}i$
24. $-\dfrac{3\sqrt{3}}{2} - \dfrac{3}{2}i$
25. $2\sqrt{2} - 2\sqrt{2}i$
26. $-\sqrt{3} + i$
27. $-3i$
28. $4i$
29. $-0.035 + 0.197i$
30. $-0.376 - 0.137i$

31. $1.970 + 0.347i$ **32.** $2.853 + 0.927i$ **33.** $zw = 8(\cos 60° + i \sin 60°); \dfrac{z}{w} = \dfrac{1}{2}(\cos 20° + i \sin 20°)$

34. $zw = \cos 220° + i \sin 220°; \dfrac{z}{w} = \cos 20° + i \sin 20°$ **35.** $zw = 12(\cos 40° + i \sin 40°); \dfrac{z}{w} = \dfrac{3}{4}(\cos 220° + i \sin 220°)$

36. $zw = 12(\cos 280° + i \sin 280°); \dfrac{z}{w} = \dfrac{1}{3}(\cos 240° + i \sin 240°)$ **37.** $zw = 4\left(\cos \dfrac{9\pi}{40} + i \sin \dfrac{9\pi}{40}\right); \dfrac{z}{w} = \cos \dfrac{\pi}{40} + i \sin \dfrac{\pi}{40}$

38. $zw = 8\left(\cos \dfrac{15\pi}{16} + i \sin \dfrac{15\pi}{16}\right); \dfrac{z}{w} = 2\left(\cos \dfrac{29\pi}{16} + i \sin \dfrac{29\pi}{16}\right)$ **39.** $zw = 4\sqrt{2}(\cos 15° + i \sin 15°); \dfrac{z}{w} = \sqrt{2}(\cos 75° + i \sin 75°)$

40. $zw = 2\sqrt{2}(\cos 255° + i \sin 255°); \dfrac{z}{w} = \dfrac{\sqrt{2}}{2}(\cos 15° + i \sin 15°)$ **41.** $-32 + 32\sqrt{3}i$ **42.** $-\dfrac{27}{2} - \dfrac{27\sqrt{3}}{2}i$ **43.** $32i$

44. $-2\sqrt{2} - 2\sqrt{2}i$ **45.** $\dfrac{27}{2} + \dfrac{27\sqrt{3}}{2}i$ **46.** $\dfrac{1}{32}$ **47.** $-\dfrac{25\sqrt{2}}{2} + \dfrac{25\sqrt{2}}{2}i$ **48.** $\dfrac{27}{2} - \dfrac{27\sqrt{3}}{2}i$ **49.** $-4 + 4i$ **50.** -64 **51.** $-23 + 14.142i$

52. $-1264 - 286.217i$ **53.** $\sqrt[6]{2}(\cos 15° + i \sin 15°), \sqrt[6]{2}(\cos 135° + i \sin 135°), \sqrt[6]{2}(\cos 255° + i \sin 255°)$

54. $\sqrt[4]{2}(\cos 82.5° + i \sin 82.5°), \sqrt[4]{2}(\cos 172.5° + i \sin 172.5°), \sqrt[4]{2}(\cos 262.5° + i \sin 262.5°), \sqrt[4]{2}(\cos 352.5° + i \sin 352.5°)$

55. $\sqrt[4]{8}(\cos 75° + i \sin 75°), \sqrt[4]{8}(\cos 165° + i \sin 165°), \sqrt[4]{8}(\cos 255° + i \sin 255°), \sqrt[4]{8}(\cos 345° + i \sin 345°)$

56. $2^{7/6}(\cos 75° + i \sin 75°), 2^{7/6}(\cos 195° + i \sin 195°), 2^{7/6}(\cos 315° + i \sin 315°)$

57. $2(\cos 67.5° + i \sin 67.5°), 2(\cos 157.5° + i \sin 157.5°), 2(\cos 247.5° + i \sin 247.5°), 2(\cos 337.5° + i \sin 337.5°)$

58. $2(\cos 60° + i \sin 60°), 2(\cos 180° + i \sin 180°), 2(\cos 300° + i \sin 300°)$

59. $\cos 18° + i \sin 18°, \cos 90° + i \sin 90°, \cos 162° + i \sin 162°, \cos 234° + i \sin 234°, \cos 306° + i \sin 306°$

60. $\cos 54° + i \sin 54°, \cos 126° + i \sin 126°, \cos 198° + i \sin 198°, \cos 270° + i \sin 270°, \cos 342° + i \sin 342°$

61. $1, i, -1, -i$

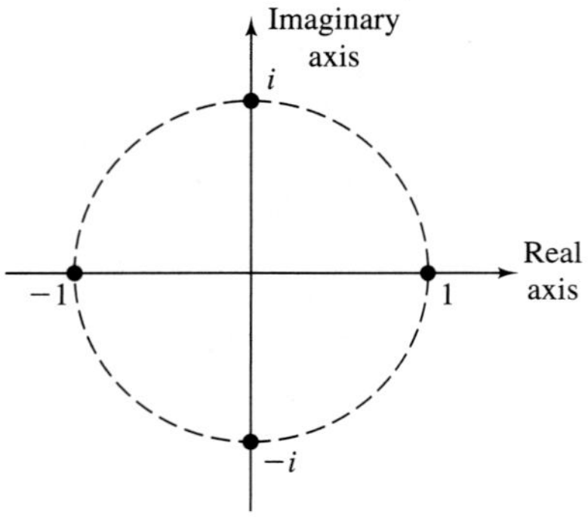

62. $1, \frac{1}{2} + \frac{\sqrt{3}}{2}i, -\frac{1}{2} + \frac{\sqrt{3}}{2}i, -1, -\frac{1}{2} - \frac{\sqrt{3}}{2}i, \frac{1}{2} - \frac{\sqrt{3}}{2}i$

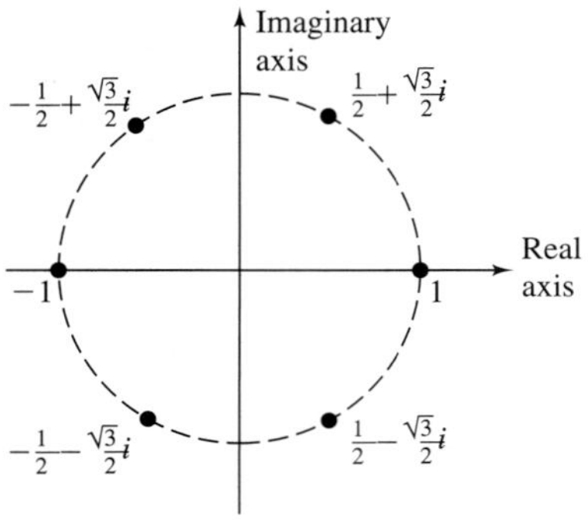

63. Look at formula (8); $|z_k| = \sqrt[n]{r}$ for all k.

64. The radius is $\sqrt[n]{|w|} = \sqrt[n]{r}$, where w is the original complex number.

65. Look at formula (8). The z_k are spaced apart by an angle of $\frac{2\pi}{n}$.

66.
$$\begin{aligned}\frac{z_1}{z_2} &= \frac{r_1(\cos\theta_1 + i\sin\theta_1)}{r_2(\cos\theta_2 + i\sin\theta_2)}\\ &= \left(\frac{r_1}{r_2}\right)\frac{(\cos\theta_1 + i\sin\theta_1)(\cos\theta_2 - i\sin\theta_2)}{(\cos\theta_2 + i\sin\theta_2)(\cos\theta_2 - i\sin\theta_2)}\\ &= \left(\frac{r_1}{r_2}\right)\frac{\cos\theta_1\cos\theta_2 + \sin\theta_1\sin\theta_2 + i(\sin\theta_1\cos\theta_2 - \cos\theta_1\sin\theta_2)}{\cos^2\theta_2 + \sin^2\theta_2}\\ &= \frac{r_1}{r_2}[\cos(\theta_1 - \theta_2) + i\sin(\theta_1 - \theta_2)]\end{aligned}$$

8.4 Concepts and Vocabulary *(page 580)*

1. unit **2.** scalar **3.** horizontal; vertical **4.** True **5.** True **6.** False

8.4 Exercises *(page 580)*

7.

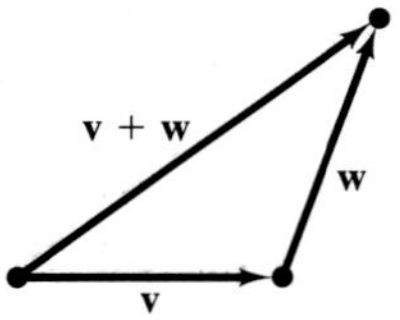

8.

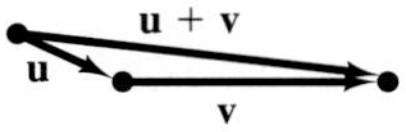

9.

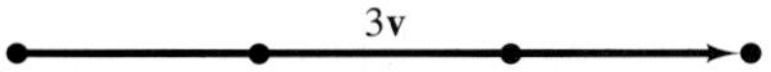

10.

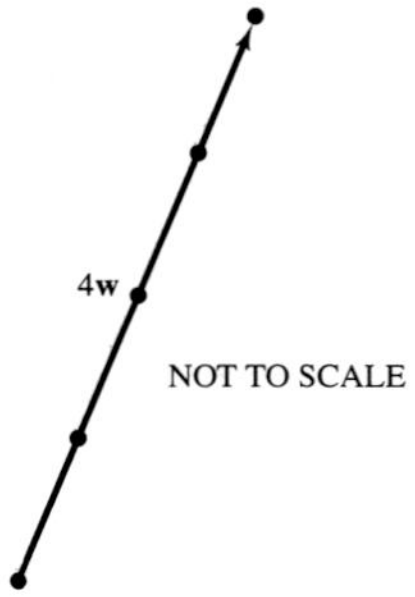

11.

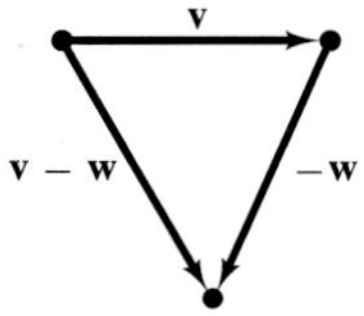

12.

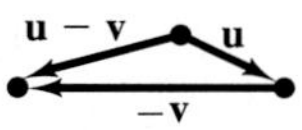

13.

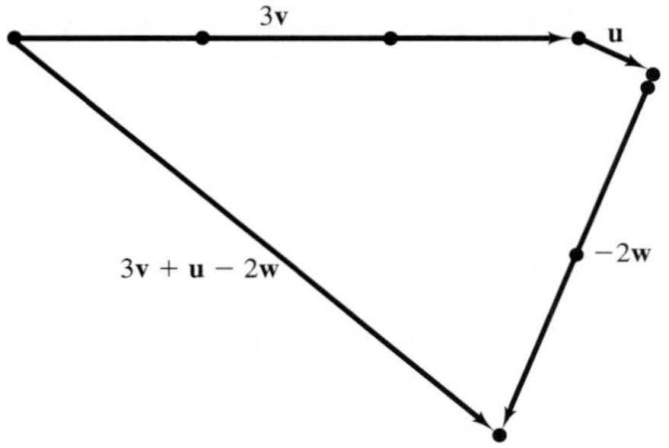

14.

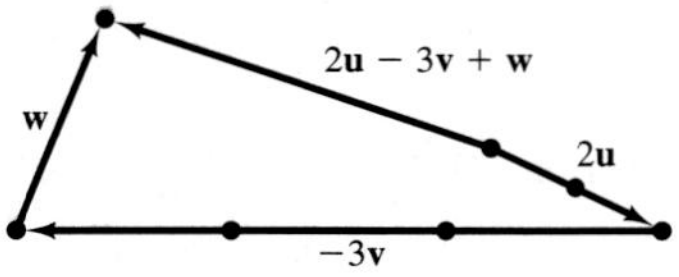

15. T **16.** F **17.** F **18.** T **19.** F **20.** F **21.** T **22.** T **23.** 12 **24.** 8 **25.** $\mathbf{v} = 3\mathbf{i} + 4\mathbf{j}$ **26.** $\mathbf{v} = -3\mathbf{i} - 5\mathbf{j}$ **27.** $\mathbf{v} = 2\mathbf{i} + 4\mathbf{j}$ **28.** $\mathbf{v} = 9\mathbf{i} + 3\mathbf{j}$ **29.** $\mathbf{v} = 8\mathbf{i} - \mathbf{j}$ **30.** $\mathbf{v} = 7\mathbf{i} - 2\mathbf{j}$ **31.** $\mathbf{v} = -\mathbf{i} + \mathbf{j}$ **32.** $\mathbf{v} = \mathbf{i} + \mathbf{j}$ **33.** 5 **34.** 13 **35.** $\sqrt{2}$ **36.** $\sqrt{2}$ **37.** $\sqrt{13}$ **38.** $2\sqrt{10}$ **39.** $-\mathbf{j}$ **40.** $13\mathbf{i} - 21\mathbf{j}$ **41.** $\sqrt{89}$ **42.** $\sqrt{5}$ **43.** $\sqrt{34} - \sqrt{13}$ **44.** $\sqrt{34} + \sqrt{13}$ **45.** $\mathbf{i}$ **46.** $-\mathbf{j}$

47. $\frac{3}{5}\mathbf{i} - \frac{4}{5}\mathbf{j}$ **48.** $-\frac{5}{13}\mathbf{i} + \frac{12}{13}\mathbf{j}$ **49.** $\frac{\sqrt{2}}{2}\mathbf{i} - \frac{\sqrt{2}}{2}\mathbf{j}$ **50.** $\frac{2\sqrt{5}}{5}\mathbf{i} - \frac{\sqrt{5}}{5}\mathbf{j}$ **51.** $\mathbf{v} = \frac{8\sqrt{5}}{5}\mathbf{i} + \frac{4\sqrt{5}}{5}\mathbf{j}$ or $\mathbf{v} = -\frac{8\sqrt{5}}{5}\mathbf{i} - \frac{4\sqrt{5}}{5}\mathbf{j}$

52. $\mathbf{v} = \frac{3\sqrt{2}}{2}\mathbf{i} + \frac{3\sqrt{2}}{2}\mathbf{j}$ or $\mathbf{v} = -\frac{3\sqrt{2}}{2}\mathbf{i} - \frac{3\sqrt{2}}{2}\mathbf{j}$ **53.** $\{-2 + \sqrt{21}, -2 - \sqrt{21}\}$ **54.** $\{-7, 1\}$ **55.** $\mathbf{v} = \frac{5}{2}\mathbf{i} + \frac{5\sqrt{3}}{2}\mathbf{j}$

56. $\mathbf{v} = 4\sqrt{2}\,\mathbf{i} + 4\sqrt{2}\,\mathbf{j}$ **57.** $\mathbf{v} = -7\mathbf{i} + 7\sqrt{3}\,\mathbf{j}$ **58.** $\mathbf{v} = -\frac{3}{2}\mathbf{i} + \frac{-3\sqrt{3}}{2}\mathbf{j}$ **59.** $\mathbf{v} = \frac{25\sqrt{3}}{2}\mathbf{i} - \frac{25}{2}\mathbf{j}$ **60.** $\mathbf{v} = \frac{15\sqrt{2}}{2}\mathbf{i} - \frac{15\sqrt{2}}{2}\mathbf{j}$

61. $\mathbf{F} = 20(\sqrt{3}\mathbf{i} + \mathbf{j})$ **62.** $\mathbf{F} = 100(\cos 20°\,\mathbf{i} + \sin 20°\,\mathbf{j})$ **63.** $\mathbf{F} = (20\sqrt{3} + 30\sqrt{2})\mathbf{i} + (20 - 30\sqrt{2})\mathbf{j}$

64. $\mathbf{F} = (15\sqrt{2} - 35)\mathbf{i} + (15\sqrt{2} + 35\sqrt{3})\mathbf{j}$

65. Tension in right cable: 1000 lb; tension in left cable: 845.2 lb **66.** Tension in right cable: 657.8 lb; tension in left cable: 516.2 lb

67. Tension in right part: 1088.4 lb; tension in left part: 1089.1 lb **68.** Tension in right part: 1208.4 lb; tension in left part: 1209.9 lb

69.

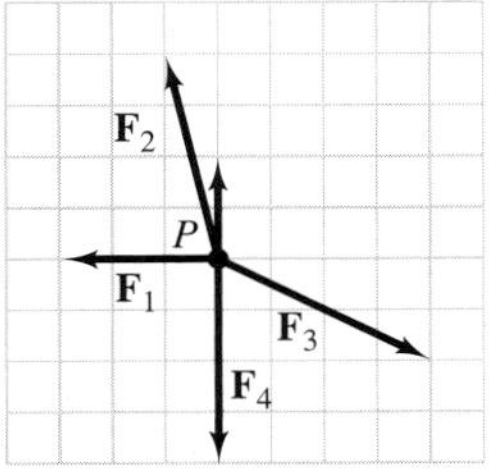

Historical Problems *(page 588)*

1. $(a\mathbf{i} + b\mathbf{j}) \cdot (c\mathbf{i} + d\mathbf{j}) = ac + bd$

Real part $[(\overline{a + bi})(c + di)]$ = real part$[(a - bi)(c + di)]$ = real part$[ac + adi - bci - bdi^2] = ac + bd$

8.5 Concepts and Vocabulary *(page 588)*

2. orthogonal **3.** parallel **4.** False **5.** True **6.** False

8.5 Exercises *(page 588)*

7. 0; 90°; orthogonal **8.** 0; 90°; orthogonal **9.** 4; 36.9°; neither **10.** 6; 18.4°; neither **11.** $\sqrt{3} - 1$; 75°; neither **12.** $1 - \sqrt{3}$; 105°; neither **13.** 24; 16.3°; neither **14.** 24; 16.3°; neither **15.** 0; 90°; orthogonal **16.** 0; 90°; orthogonal **17.** $\frac{2}{3}$ **18.** -1 **19.** $\mathbf{v}_1 = \frac{5}{2}\mathbf{i} - \frac{5}{2}\mathbf{j}$, $\mathbf{v}_2 = -\frac{1}{2}\mathbf{i} - \frac{1}{2}\mathbf{j}$ **20.** $\mathbf{v}_1 = -\frac{8}{5}\mathbf{i} - \frac{4}{5}\mathbf{j}$, $\mathbf{v}_2 = -\frac{7}{5}\mathbf{i} + \frac{14}{5}\mathbf{j}$ **21.** $\mathbf{v}_1 = -\frac{1}{5}\mathbf{i} - \frac{2}{5}\mathbf{j}$, $\mathbf{v}_2 = \frac{6}{5}\mathbf{i} - \frac{3}{5}\mathbf{j}$ **22.** $\mathbf{v}_1 = \frac{4}{5}\mathbf{i} - \frac{8}{5}\mathbf{j}$, $\mathbf{v}_2 = \frac{6}{5}\mathbf{i} + \frac{3}{5}\mathbf{j}$ **23.** $\mathbf{v}_1 = \frac{14}{5}\mathbf{i} + \frac{7}{5}\mathbf{j}$, $\mathbf{v}_2 = \frac{1}{5}\mathbf{i} - \frac{2}{5}\mathbf{j}$ **24.** $\mathbf{v}_1 = \frac{28}{17}\mathbf{i} - \frac{7}{17}\mathbf{j}$, $\mathbf{v}_2 = -\frac{11}{17}\mathbf{i} - \frac{44}{17}\mathbf{j}$ **25.** 496.7 mi/hr; 38.5° west of south **26.** 6.5° north of west; 276.7 mi/hr **27.** 8.6° off direct heading across the current, upstream; 1.52 min **28.** 14.5° off direct heading across the current; 1.55 min **29.** Force required to keep Sienna from rolling down the hill: 737.6 lb; force perpendicular to the hill: 5248.4 lb **30.** Force required to keep Bonneville from rolling down the hill: 781.4 lb; force perpendicular to the hill: 4431.6 lb **31.** $\mathbf{v} = (250\sqrt{2} - 30)\mathbf{i} + (250\sqrt{2} + 30\sqrt{3})\mathbf{j}$; 518.8 km/hr; N38.6°E **32.** $\mathbf{v} = (300 + 20\sqrt{2})\mathbf{i} - (300\sqrt{3} + 20\sqrt{2})\mathbf{j}$; 638.7 km/hr; S30.9°E **33.** $\mathbf{v} = 3\mathbf{i} + 20\mathbf{j}$; 20.2 mi/hr; N8.5°E (assuming boat traveling north and current traveling east) **34.** $\mathbf{v} = 4\mathbf{i} + 10\mathbf{j}$; 10.8 mi/hr; N21.8°E (assuming boat traveling north and current traveling east) **35.** 3 ft-lb **36.** $\frac{5}{2}\sqrt{2} \approx 3.54$ ft-lb **37.** $1000\sqrt{3}$ ft-lb ≈ 1732 ft-lb **38.** 60° **39.** Let $\mathbf{u} = a_1\mathbf{i} + b_1\mathbf{j}$, $\mathbf{v} = a_2\mathbf{i} + b_2\mathbf{j}$, $\mathbf{w} = a_3\mathbf{i} + b_3\mathbf{j}$. Compute $\mathbf{u} \cdot (\mathbf{v} + \mathbf{w})$ and $\mathbf{u} \cdot \mathbf{v} + \mathbf{u} \cdot \mathbf{w}$. **40.** Let $\mathbf{v} = a\mathbf{i} + b\mathbf{j}$. Then $\mathbf{0} \cdot \mathbf{v} = 0a + 0b = 0$.

41. $\cos\alpha = \frac{\mathbf{v} \cdot \mathbf{i}}{\|\mathbf{v}\|\,\|\mathbf{i}\|} = \mathbf{v} \cdot \mathbf{i}$; if $\mathbf{v} = x\mathbf{i} + y\mathbf{j}$, then $\mathbf{v} \cdot \mathbf{i} = x = \cos\alpha$ and $\mathbf{v} \cdot \mathbf{j} = y = \cos\left(\frac{\pi}{2} - \alpha\right) = \sin\alpha$.

42. By the definition of an angle between two vectors, the angle between $\mathbf{v}$ and $\mathbf{w}$ lies in the interval $[0, \pi]$ and thus equals either $|\alpha - \beta|$ or $2\pi - |\alpha - \beta|$. In either case, by the periodicity of cosine and because cosine is an even function, cosine of the angle between $\mathbf{v}$ and $\mathbf{w}$ equals $\cos(\alpha - \beta)$. Therefore, since $\mathbf{v}$ and $\mathbf{w}$ are each unit vectors, then $\cos(\alpha - \beta) = \mathbf{v} \cdot \mathbf{w}$. From Exercise 41, since $\mathbf{v} = \cos\alpha\mathbf{i} + \sin\alpha\mathbf{j}$ and $\mathbf{w} = \cos\beta\mathbf{i} + \sin\beta\mathbf{j}$, then $\cos(\alpha - \beta) = \mathbf{v} \cdot \mathbf{w} = \cos\alpha\cos\beta + \sin\alpha\sin\beta$.

43. $\mathbf{v} = a\mathbf{i} + b\mathbf{j}$; the vector projection of $\mathbf{v}$ onto $\mathbf{i}$ is $\frac{\mathbf{v} \cdot \mathbf{i}}{\|\mathbf{i}\|^2}\mathbf{i} = (\mathbf{v} \cdot \mathbf{i})\mathbf{i}$; $\mathbf{v} \cdot \mathbf{i} = a$, $\mathbf{v} \cdot \mathbf{j} = b$, so $\mathbf{v} = (\mathbf{v} \cdot \mathbf{i})\mathbf{i} + (\mathbf{v} \cdot \mathbf{j})\mathbf{j}$.

44. **(a)** If $\mathbf{u} = a_1\mathbf{i} + b_1\mathbf{j}$ and $\mathbf{v} = a_2\mathbf{i} + b_2\mathbf{j}$, then, since $\|\mathbf{u}\| = \|\mathbf{v}\|$, $a_1^2 + b_1^2 = \|\mathbf{u}\|^2 = \|\mathbf{v}\|^2 = a_2^2 + b_2^2$, and $(\mathbf{u} + \mathbf{v}) \cdot (\mathbf{u} - \mathbf{v}) = (a_1 + a_2)(a_1 - a_2) + (b_1 + b_2)(b_1 - b_2) = (a_1^2 + b_1^2) - (a_2^2 + b_2^2) = 0$.
(b) The legs of the angle can be made to correspond to vectors $\mathbf{u} + \mathbf{v}$ and $\mathbf{u} - \mathbf{v}$.

45. $(\mathbf{v} - \alpha\mathbf{w}) \cdot \mathbf{w} = \mathbf{v} \cdot \mathbf{w} - \alpha\mathbf{w} \cdot \mathbf{w} = \alpha\|\mathbf{w}\|^2 - \alpha\|\mathbf{w}\|^2 = 0$ since the dot product of any vector with itself equals to the square of its magnitude.

46. $(\|\mathbf{w}\|\mathbf{v} + \|\mathbf{v}\|\mathbf{w}) \cdot (\|\mathbf{w}\|\mathbf{v} - \|\mathbf{v}\|\mathbf{w}) = \|\mathbf{w}\|^2\mathbf{v} \cdot \mathbf{v} - \|\mathbf{w}\|\,\|\mathbf{v}\|\mathbf{v} \cdot \mathbf{w} + \|\mathbf{v}\|\,\|\mathbf{w}\|\mathbf{w} \cdot \mathbf{v} - \|\mathbf{v}\|^2\mathbf{w} \cdot \mathbf{w} = \|\mathbf{w}\|^2\mathbf{v} \cdot \mathbf{v} - \|\mathbf{v}\|^2\mathbf{w} \cdot \mathbf{w}$
$= \|\mathbf{w}\|^2\|\mathbf{v}\|^2 - \|\mathbf{v}\|^2\|\mathbf{w}\|^2 = 0$

47. $W = \mathbf{F} \cdot \overrightarrow{AB} = 0$ when $\mathbf{F}$ is orthogonal to $\overrightarrow{AB}$.

48. $\|\mathbf{u} + \mathbf{v}\|^2 - \|\mathbf{u} - \mathbf{v}\|^2 = (\mathbf{u} + \mathbf{v}) \cdot (\mathbf{u} + \mathbf{v}) - (\mathbf{u} - \mathbf{v}) \cdot (\mathbf{u} - \mathbf{v})$
$= (\mathbf{u} \cdot \mathbf{u} + \mathbf{u} \cdot \mathbf{v} + \mathbf{v} \cdot \mathbf{u} + \mathbf{v} \cdot \mathbf{v}) - (\mathbf{u} \cdot \mathbf{u} - \mathbf{u} \cdot \mathbf{v} - \mathbf{v} \cdot \mathbf{u} + \mathbf{v} \cdot \mathbf{v}) = 2(\mathbf{u} \cdot \mathbf{v}) + 2(\mathbf{v} \cdot \mathbf{u}) = 4(\mathbf{u} \cdot \mathbf{v})$

8.6 Concepts and Vocabulary *(page 599)*

2. x-y plane **3.** components **4.** 1 **5.** False **6.** True

8.6 Exercises *(page 599)*

7. All points of the form $(x, 0, z)$ **8.** All points of the form $(0, y, z)$ **9.** All points of the form $(x, y, 2)$ **10.** All points of the form $(x, 3, z)$ **11.** All points of the form $(-4, y, z)$ **12.** All points of the form $(x, y, -3)$ **13.** All points of the form $(1, 2, z)$ **14.** All points of the form $(3, y, 1)$ **15.** $\sqrt{21}$ **16.** $\sqrt{14}$ **17.** $\sqrt{33}$ **18.** $2\sqrt{19}$ **19.** $\sqrt{26}$ **20.** $2\sqrt{6}$ **21.** $(2,0,0); (2,1,0); (0,1,0); (2,0,3); (0,1,3); (0,0,3)$ **22.** $(4,0,0); (0,2,0); (0,0,2); (4,2,0); (0,2,2); (4,0,2)$ **23.** $(1,4,3); (3,2,3); (3,4,3); (3,2,5); (1,4,5); (1,2,5)$ **24.** $(5,6,2); (5,8,1); (3,6,1); (3,8,1); (3,6,2); (5,8,2)$ **25.** $(-1,2,2); (4,0,2); (4,2,2); (-1,2,5); (4,0,5); (-1,0,5)$ **26.** $(-2,-3,1); (-2,7,0); (-6,-3,0); (-2,7,1); (-6,-3,1); (-6,7,0)$ **27.** $\mathbf{v} = 3\mathbf{i} + 4\mathbf{j} - \mathbf{k}$ **28.** $\mathbf{v} = -3\mathbf{i} - 5\mathbf{j} + 4\mathbf{k}$ **29.** $\mathbf{v} = 2\mathbf{i} + 4\mathbf{j} + \mathbf{k}$ **30.** $\mathbf{v} = 9\mathbf{i} + 3\mathbf{j} - \mathbf{k}$ **31.** $\mathbf{v} = 8\mathbf{i} - \mathbf{j}$ **32.** $\mathbf{v} = 7\mathbf{i} - 2\mathbf{j} + 4\mathbf{k}$ **33.** 7 **34.** 14 **35.** $\sqrt{3}$ **36.** $\sqrt{3}$ **37.** $\sqrt{22}$ **38.** $2\sqrt{11}$ **39.** $-\mathbf{j} - 2\mathbf{k}$ **40.** $13\mathbf{i} - 21\mathbf{j} + 10\mathbf{k}$ **41.** $\sqrt{105}$ **42.** $\sqrt{5}$ **43.** $\sqrt{38} - \sqrt{17}$ **44.** $\sqrt{38} + \sqrt{17}$ **45.** $\frac{\mathbf{v}}{\|\mathbf{v}\|} = \mathbf{i}$ **46.** $\frac{\mathbf{v}}{\|\mathbf{v}\|} = -\mathbf{j}$ **47.** $\frac{\mathbf{v}}{\|\mathbf{v}\|} = \frac{3}{7}\mathbf{i} - \frac{6}{7}\mathbf{j} - \frac{2}{7}\mathbf{k}$ **48.** $\frac{\mathbf{v}}{\|\mathbf{v}\|} = -\frac{3}{7}\mathbf{i} + \frac{6}{7}\mathbf{j} + \frac{2}{7}\mathbf{k}$ **49.** $\frac{\mathbf{v}}{\|\mathbf{v}\|} = \frac{\sqrt{3}}{3}\mathbf{i} + \frac{\sqrt{3}}{3}\mathbf{j} + \frac{\sqrt{3}}{3}\mathbf{k}$ **50.** $\frac{\mathbf{v}}{\|\mathbf{v}\|} = \frac{\sqrt{6}}{3}\mathbf{i} - \frac{\sqrt{6}}{6}\mathbf{j} + \frac{\sqrt{6}}{6}\mathbf{k}$ **51.** $\mathbf{v} \cdot \mathbf{w} = 0;\ \theta = 90°$ **52.** $\mathbf{v} \cdot \mathbf{w} = 0;\ \theta = 90°$ **53.** $\mathbf{v} \cdot \mathbf{w} = -2,\ \theta \approx 100.3°$ **54.** $\mathbf{v} \cdot \mathbf{w} = 3;\ \theta \approx 74.5°$ **55.** $\mathbf{v} \cdot \mathbf{w} = 0;\ \theta = 90°$ **56.** $\mathbf{v} \cdot \mathbf{w} = 0;\ \theta = 90°$ **57.** $\mathbf{v} \cdot \mathbf{w} = 52;\ \theta = 0°$ **58.** $\mathbf{v} \cdot \mathbf{w} = 52;\ \theta = 0°$ **59.** $\alpha \approx 64.6°; \beta \approx 149.0°;\ \gamma \approx 106.6°;$ $\mathbf{v} = 7(\cos 64.6°\mathbf{i} + \cos 149.0°\mathbf{j} + \cos 106.6°\mathbf{k})$ **60.** $\alpha \approx 115.4°; \beta \approx 31.0°;\ \gamma \approx 73.4°; \mathbf{v} = 14(\cos 115.4°\mathbf{i} + \cos 31.0°\mathbf{j} + \cos 73.4°\mathbf{k})$

61. $\alpha = \beta = \gamma \approx 54.7°; \mathbf{v} = \sqrt{3}(\cos 54.7°\mathbf{i} + \cos 54.7°\mathbf{j} + \cos 54.7°\mathbf{k})$

62. $\alpha = 54.7°; \beta = \gamma \approx 125.3°;\ \mathbf{v} = \sqrt{3}(\cos 54.7°\mathbf{i} + \cos 125.3°\mathbf{j} + \cos 125.3°\mathbf{k})$

63. $\alpha = \beta = 45°; \gamma = 90°; \mathbf{v} = \sqrt{2}(\cos 45°\mathbf{i} + \cos 45°\mathbf{j} + \cos 90°\mathbf{k})$ **64.** $\alpha = 90°; \beta = \gamma = 45°;\ \mathbf{v} = \sqrt{2}(\cos 90°\mathbf{i} + \cos 45°\mathbf{j} + \cos 45°\mathbf{k})$ **65.** $\alpha \approx 60.9°; \beta \approx 144.2°;\ \gamma \approx 71.1°; \mathbf{v} = \sqrt{38}(\cos 60.9°\mathbf{i} + \cos 144.2°\mathbf{j} + \cos 71.1°\mathbf{k})$ **66.** $\alpha \approx 68.2°; \beta \approx 56.1°;\ \gamma \approx 138.0°;$ $\mathbf{v} = \sqrt{29}(\cos 68.2°\mathbf{i} + \cos 56.1°\mathbf{j} + \cos 138.0°\mathbf{k})$ **67.** If the point $P = (x, y, z)$ is on the sphere with center $C = (x_0, y_0, z_0)$ and radius r, then the distance between P and C is $r = \sqrt{(x - x_0)^2 + (y - y_0)^2 + (z - z_0)^2}$. Therefore, the equation for a sphere is $(x - x_0)^2 + (y - y_0)^2 + (z - z_0)^2 = r^2$. **68.** $(x - 3)^2 + (y - 1)^2 + (z - 1)^2 = 1$ **69.** $(x - 1)^2 + (y - 2)^2 + (z - 2)^2 = 4$ **70.** $(x + 1)^2 + (y - 1)^2 + (z - 2)^2 = 9$ **71.** radius $= 2$, center $(-1, 1, 0)$ **72.** radius $= 1$, center $(-1, 0, 1)$ **73.** radius $= 3$, center $(2, -2, -1)$ **74.** radius $= 2$, center $(2, 0, 0)$ **75.** radius $= \frac{3\sqrt{2}}{2}$, center $(2, 0, -1)$ **76.** radius $= \sqrt{3}$, center $(-1, 1, 0)$ **77.** 2 joules **78.** $\frac{8}{3}$ joules **79.** 9

8.7 Concepts and Vocabulary *(page 605)*

1. True **2.** True **3.** True **4.** False **5.** False **6.** True

8.7 Exercises *(page 605)*

7. 2 **8.** -4 **9.** 4 **10.** -12 **11.** $-11A + 2B + 5C$ **12.** $2A + 12B - 6C$ **13.** $-6A + 23B - 15C$ **14.** $10A + 2B + 2C$ **15.** **(a)** $5\mathbf{i} + 5\mathbf{j} + 5\mathbf{k}$ **(b)** $-5\mathbf{i} - 5\mathbf{j} - 5\mathbf{k}$ **(c)** $\mathbf{0}$ **(d)** $\mathbf{0}$ **16.** **(a)** $\mathbf{i} + 5\mathbf{j} - 7\mathbf{k}$ **(b)** $-\mathbf{i} - 5\mathbf{j} + 7\mathbf{k}$ **(c)** $\mathbf{0}$ **(d)** $\mathbf{0}$ **17.** **(a)** $\mathbf{i} - \mathbf{j} - \mathbf{k}$ **(b)** $-\mathbf{i} + \mathbf{j} + \mathbf{k}$ **(c)** $\mathbf{0}$ **(d)** $\mathbf{0}$ **18.** **(a)** $-8\mathbf{i} + 5\mathbf{j} + 14\mathbf{k}$ **(b)** $8\mathbf{i} - 5\mathbf{j} - 14\mathbf{k}$ **(c)** $\mathbf{0}$ **(d)** $\mathbf{0}$ **19.** **(a)** $-\mathbf{i} + 2\mathbf{j} + 2\mathbf{k}$ **(b)** $\mathbf{i} - 2\mathbf{j} - 2\mathbf{k}$ **(c)** $\mathbf{0}$ **(d)** $\mathbf{0}$ **20.** **(a)** $-\mathbf{i} + 6\mathbf{j} - \mathbf{k}$ **(b)** $\mathbf{i} - 6\mathbf{j} + \mathbf{k}$ **(c)** $\mathbf{0}$ **(d)** $\mathbf{0}$ **21.** **(a)** $3\mathbf{i} - \mathbf{j} + 4\mathbf{k}$ **(b)** $-3\mathbf{i} + \mathbf{j} - 4\mathbf{k}$ **(c)** $\mathbf{0}$ **(d)** $\mathbf{0}$ **22.** **(a)** $6\mathbf{i} + 4\mathbf{j} + 6\mathbf{k}$ **(b)** $-6\mathbf{i} - 4\mathbf{j} - 6\mathbf{k}$ **(c)** $\mathbf{0}$ **(d)** $\mathbf{0}$ **23.** $-9\mathbf{i} - 7\mathbf{j} - 3\mathbf{k}$ **24.** $7\mathbf{i} + 11\mathbf{j} - 6\mathbf{k}$ **25.** $9\mathbf{i} + 7\mathbf{j} + 3\mathbf{k}$ **26.** $-7\mathbf{i} - 11\mathbf{j} + 6\mathbf{k}$ **27.** $\mathbf{0}$ **28.** $\mathbf{0}$ **29.** $-27\mathbf{i} - 21\mathbf{j} - 9\mathbf{k}$ **30.** $28\mathbf{i} + 44\mathbf{j} - 24\mathbf{k}$ **31.** $-18\mathbf{i} - 14\mathbf{j} - 6\mathbf{k}$ **32.** $-21\mathbf{i} - 33\mathbf{j} + 18\mathbf{k}$ **33.** 0 **34.** $\mathbf{0}$ **35.** -25 **36.** -25 **37.** 25 **38.** 25 **39.** $\mathbf{0}$ **40.** $\mathbf{0}$ **41.** Any vector of the form $c(-9\mathbf{i} - 7\mathbf{j} - 3\mathbf{k})$, where c is a nonzero scalar **42.** Any vector of the form $c(-10\mathbf{i} - 5\mathbf{j} + 5\mathbf{k})$, where c is a nonzero scalar **43.** Any vector of the form $c(-\mathbf{i} + \mathbf{j} + 5\mathbf{k})$, where c is a nonzero scalar **44.** Any vector of the form $c(-4\mathbf{i} - 2\mathbf{j} + 2\mathbf{k})$, where c is a nonzero scalar **45.** $\sqrt{166}$ **46.** $\sqrt{213}$ **47.** $\sqrt{555}$ **48.** $\sqrt{217}$ **49.** $\sqrt{34}$ **50.** 8 **51.** $\sqrt{998}$ **52.** $\sqrt{89}$ **53.** $\frac{11\sqrt{19}}{57}\mathbf{i} + \frac{\sqrt{19}}{57}\mathbf{j} + \frac{7\sqrt{19}}{57}\mathbf{k}$ or $-\frac{11\sqrt{19}}{57}\mathbf{i} - \frac{\sqrt{19}}{57}\mathbf{j} - \frac{7\sqrt{19}}{57}\mathbf{k}$

54. $-\frac{13\sqrt{237}}{237}\mathbf{i} + \frac{8\sqrt{237}}{237}\mathbf{j} - \frac{2\sqrt{237}}{237}\mathbf{k}$ or $\frac{13\sqrt{237}}{237}\mathbf{i} - \frac{8\sqrt{237}}{237}\mathbf{j} + \frac{2\sqrt{237}}{237}\mathbf{k}$

55.
$$\mathbf{u} \times \mathbf{v} = \begin{vmatrix} \mathbf{i} & \mathbf{j} & \mathbf{k} \\ a_1 & b_1 & c_1 \\ a_2 & b_2 & c_2 \end{vmatrix} = (b_1c_2 - b_2c_1)\mathbf{i} - (a_1c_2 - a_2c_1)\mathbf{j} + (a_1b_2 - a_2b_1)\mathbf{k} = -[(b_2c_1 - b_1c_2)\mathbf{i} - (a_2c_1 - a_1c_2)\mathbf{j} + (a_2b_1 - a_1b_2)\mathbf{k}]$$
$$= -\begin{vmatrix} \mathbf{i} & \mathbf{j} & \mathbf{k} \\ a_2 & b_2 & c_2 \\ a_1 & b_1 & c_1 \end{vmatrix} = -(\mathbf{v} \times \mathbf{u})$$

56. Assume that $\mathbf{u} = a\mathbf{i} + b\mathbf{j} + c\mathbf{k}$, $\mathbf{v} = d\mathbf{i} + e\mathbf{j} + f\mathbf{k}$, and $\mathbf{w} = l\mathbf{i} + m\mathbf{j} + n\mathbf{k}$. Then $\mathbf{u} \times \mathbf{v} = (bf - ec)\mathbf{i} - (af - dc)\mathbf{j} + (ae - db)\mathbf{k}$, $\mathbf{u} \times \mathbf{w} = (bn - mc)\mathbf{i} - (an - lc)\mathbf{j} + (am - lb)\mathbf{k}$, and $\mathbf{v} + \mathbf{w} = (d + l)\mathbf{i} + (e + m)\mathbf{j} + (f + n)\mathbf{k}$. Therefore, $(\mathbf{u} \times \mathbf{v}) + (\mathbf{u} \times \mathbf{w}) = (bf - ec + bn - mc)\mathbf{i} - (af - dc + an - lc)\mathbf{j} + (ae - db + am - lb)\mathbf{k}$ and $\mathbf{u} \times (\mathbf{v} + \mathbf{w})$
$= [b(f + n) - (e + m)c]\mathbf{i} - [a(f + n) - (d + l)c]\mathbf{j} + [a(e + m) - (d + l)b]\mathbf{k}$
$= (bf - ec + bn - mc)\mathbf{i} - (af - dc + an - lc)\mathbf{j} + (ae - db + am - lb)\mathbf{k}$, which equals $(\mathbf{u} \times \mathbf{v}) + (\mathbf{u} \times \mathbf{w})$.

57. $\mathbf{u} \times \mathbf{v} = \begin{vmatrix} \mathbf{i} & \mathbf{j} & \mathbf{k} \\ a_1 & b_1 & c_1 \\ a_2 & b_2 & c_2 \end{vmatrix} = (b_1c_2 - b_2c_1)\mathbf{i} - (a_1c_2 - a_2c_1)\mathbf{j} + (a_1b_2 - a_2b_1)\mathbf{k}$

$\|\mathbf{u} \times \mathbf{v}\|^2 = (\sqrt{(b_1c_2 - b_2c_1)^2 + (a_1c_2 - a_2c_1)^2 + (a_1b_2 - a_2b_1)^2})^2$
$= b_1^2c_2^2 - 2b_1b_2c_1c_2 + b_2^2c_1^2 + a_1^2c_2^2 - 2a_1a_2c_1c_2 + a_2^2c_1^2 + a_1^2b_2^2 - 2a_1a_2b_1b_2 + a_2^2b_1^2$

$\|\mathbf{u}\|^2 = a_1^2 + b_1^2 + c_1^2$, $\|\mathbf{v}\|^2 = a_2^2 + b_2^2 + c_2^2$

$\|\mathbf{u}\|^2\|\mathbf{v}\|^2 = (a_1^2 + b_1^2 + c_1^2)(a_2^2 + b_2^2 + c_2^2) = a_1^2a_2^2 + a_1^2b_2^2 + a_1^2c_2^2 + b_1^2a_2^2 + b_1^2b_2^2 + b_1^2c_2^2 + a_2^2c_1^2 + b_2^2c_1^2 + c_1^2c_2^2$

$(\mathbf{u} \cdot \mathbf{v})^2 = (a_1a_2 + b_1b_2 + c_1c_2)^2 = (a_1a_2 + b_1b_2 + c_1c_2)(a_1a_2 + b_1b_2 + c_1c_2)$
$= a_1^2a_2^2 + a_1a_2b_1b_2 + a_1a_2c_1c_2 + b_1b_2c_1c_2 + b_1b_2a_1a_2 + b_1^2b_2^2 + b_1b_2c_1c_2 + a_1a_2c_1c_2 + c_1^2c_2^2$
$= a_1^2a_2^2 + b_1^2b_2^2 + c_1^2c_2^2 + 2a_1a_2b_1b_2 + 2b_1b_2c_1c_2 + 2a_1a_2c_1c_2$

$\|\mathbf{u}\|^2\|\mathbf{v}\|^2 - (\mathbf{u} \cdot \mathbf{v})^2 = a_1^2b_2^2 + a_1^2c_2^2 + b_1^2a_2^2 + a_2^2c_1^2 + b_2^2c_1^2 + c_1^2b_2^2 - 2a_1a_2b_1b_2 - 2b_1b_2c_1c_2 - 2a_1a_2c_1c_2$, which equals $\|\mathbf{u} \times \mathbf{v}\|^2$.

58. By Exercise 57, $\|\mathbf{u} \times \mathbf{v}\|^2 = \|\mathbf{u}\|^2\|\mathbf{v}\|^2 - (\mathbf{u} \cdot \mathbf{v})^2$. Since $\mathbf{u} \cdot \mathbf{v} = \|\mathbf{u}\|\,\|\mathbf{v}\| \cos\theta$, where θ is the angle between $\mathbf{u}$ and $\mathbf{v}$, then $\|\mathbf{u} \times \mathbf{v}\|^2 = \|\mathbf{u}\|^2\|\mathbf{v}\|^2 - (\|\mathbf{u}\|\,\|\mathbf{v}\| \cos\theta)^2 = \|\mathbf{u}\|^2\|\mathbf{v}\|^2 - \|\mathbf{u}\|^2\|\mathbf{v}\|^2\cos^2\theta = \|\mathbf{u}\|^2\|\mathbf{v}\|^2(1 - \cos^2\theta) = \|\mathbf{u}\|^2\|\mathbf{v}\|^2\sin^2\theta$. Since $0 \le \theta \le \pi$, then $\sin\theta \ge 0$. Therefore, $\|\mathbf{u} \times \mathbf{v}\| = \|\mathbf{u}\|\,\|\mathbf{v}\| \sin\theta$.

59. We know for any two vectors that $\|\mathbf{u} \times \mathbf{v}\| = \|\mathbf{u}\|\,\|\mathbf{v}\| \sin\theta$, where θ is the angle between **u** and **v**; so that if **u** and **v** are orthogonal, then $\theta = 90°$, and so the result follows.

60. By Exercise 59, since **u** and **v** are orthogonal, $\|\mathbf{u} \times \mathbf{v}\| = \|\mathbf{u}\|\,\|\mathbf{v}\|$. If in addition, **u** and **v** are unit vectors, then $\|\mathbf{u} \times \mathbf{v}\| = 1 \cdot 1 = 1$.

Review Exercises *(page 608)*

1. $\left(\frac{3\sqrt{3}}{2}, \frac{3}{2}\right)$

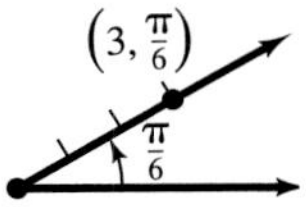

2. $(-2, 2\sqrt{3})$

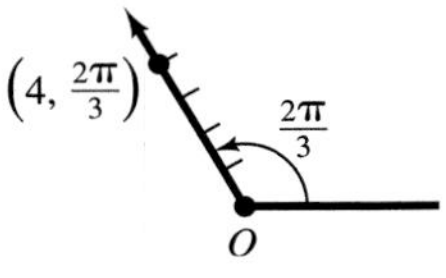

3. $(1, \sqrt{3})$

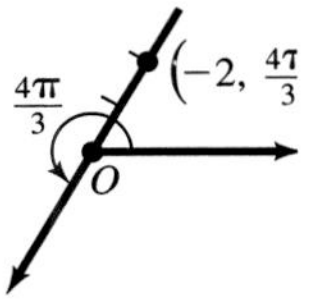

4. $\left(\frac{\sqrt{2}}{2}, \frac{\sqrt{2}}{2}\right)$

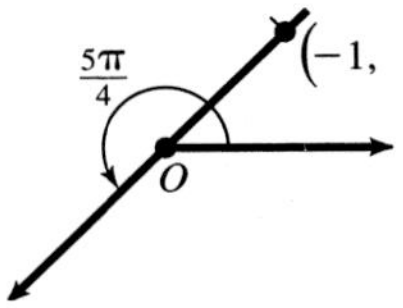

5. $(0, 3)$

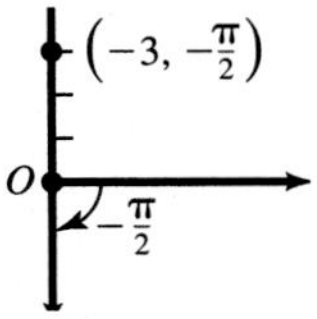

6. $(-2\sqrt{2}, 2\sqrt{2})$

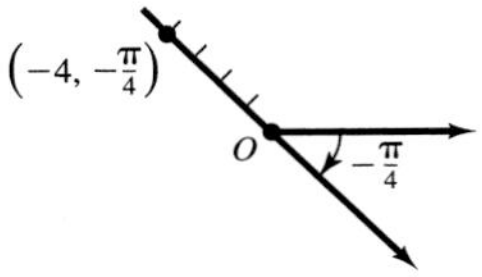

7. $\left(3\sqrt{2}, \frac{3\pi}{4}\right), \left(-3\sqrt{2}, -\frac{\pi}{4}\right)$ **8.** $\left(\sqrt{2}, -\frac{\pi}{4}\right), \left(-\sqrt{2}, \frac{3\pi}{4}\right)$

9. $\left(2, -\frac{\pi}{2}\right), \left(-2, \frac{\pi}{2}\right)$ **10.** $(2, 0), (-2, \pi)$

11. $(5, 0.93), (-5, 4.07)$ **12.** $(13, 1.97), (-13, 5.11)$

13. $x^2 + (y - 1)^2 = 1$;

Circle, radius 1, center $(0, 1)$ in rectangular coordinates

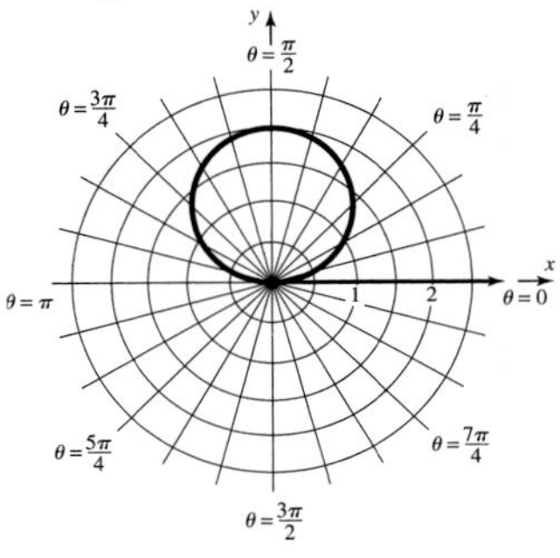

14. $x^2 + \left(y - \frac{1}{6}\right)^2 = \frac{1}{36}$;

Circle, radius $\frac{1}{6}$, center $\left(0, \frac{1}{6}\right)$ in rectangular coordinates

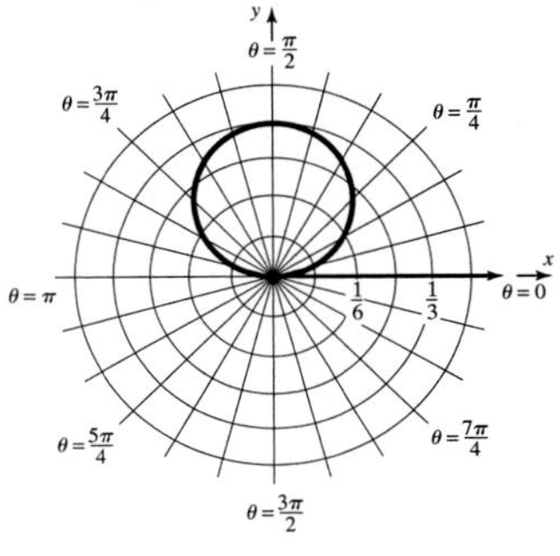

15. $x^2 + y^2 = 25$;

Circle, radius 5, center at pole

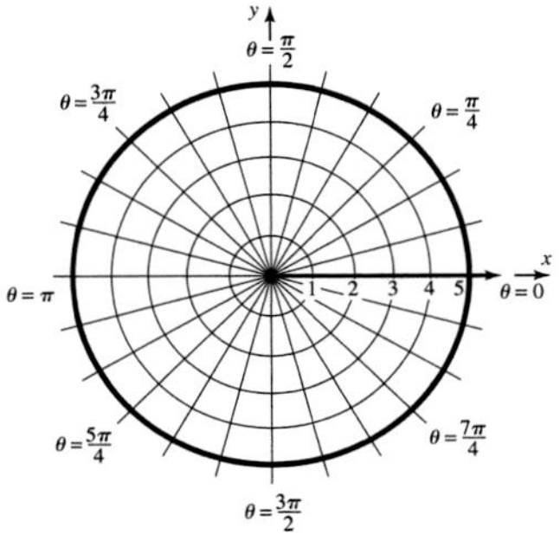

16. $x - y = 0$;
Line through pole, making an angle of $\frac{\pi}{4}$ with polar axis

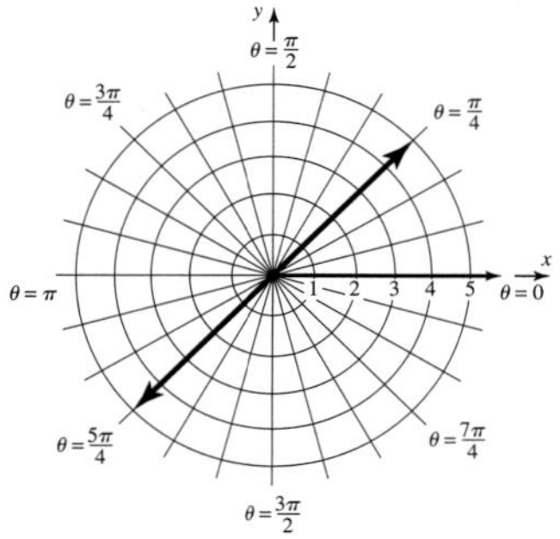

17. $x + 3y = 6$;
Line through $(6, 0)$ and $(0, 2)$ in rectangular coordinates

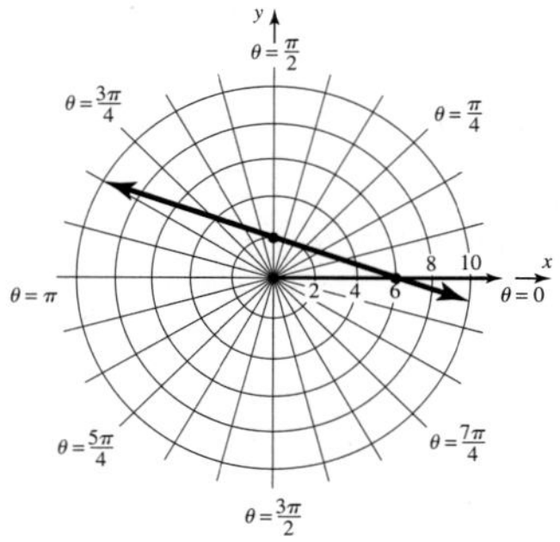

18. $(x - 4)^2 + (y + 2)^2 = 25$;
Circle, radius 5, center $(4, -2)$ in rectangular coordinates

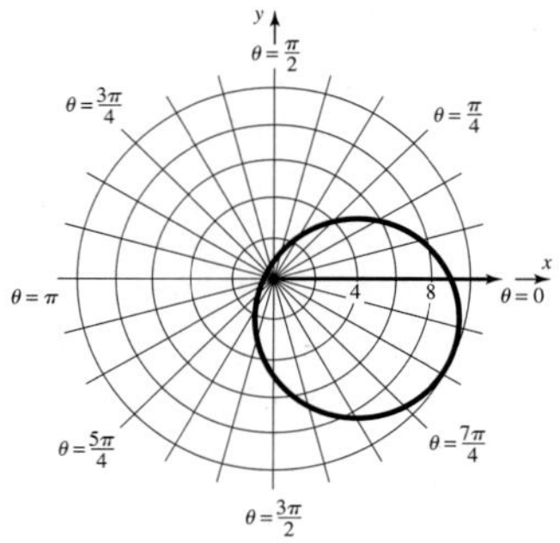

19. Circle; radius 2, center at $(2, 0)$ in rectangular coordinates; symmetric with respect to the polar axis

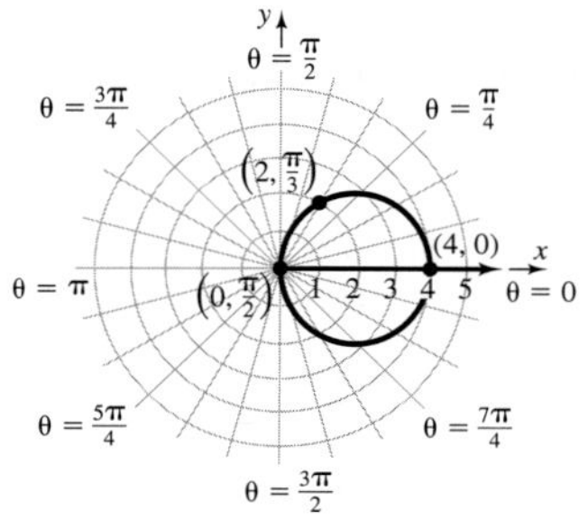

20. Circle; radius $\frac{3}{2}$, center at $\left(0, \frac{3}{2}\right)$ in rectangular coordinates; symmetric with respect to the line $\theta = \frac{\pi}{2}$

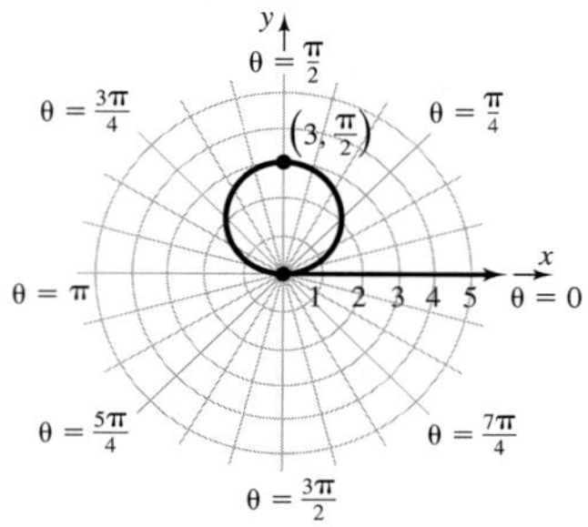

21. Cardioid; symmetric with respect to the line $\theta = \frac{\pi}{2}$

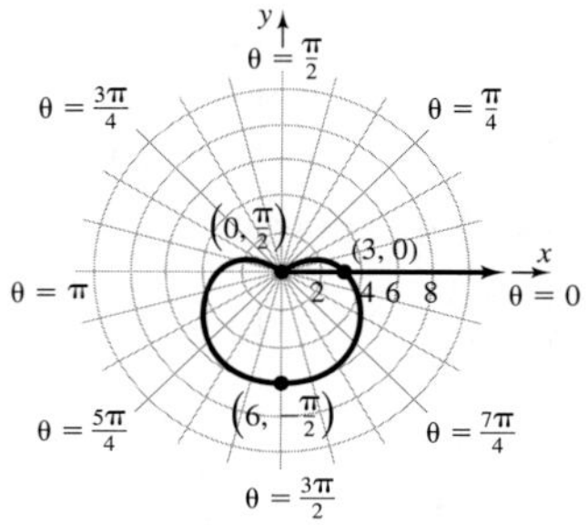

22. Limaçon without inner loop; symmetric with respect to the polar axis

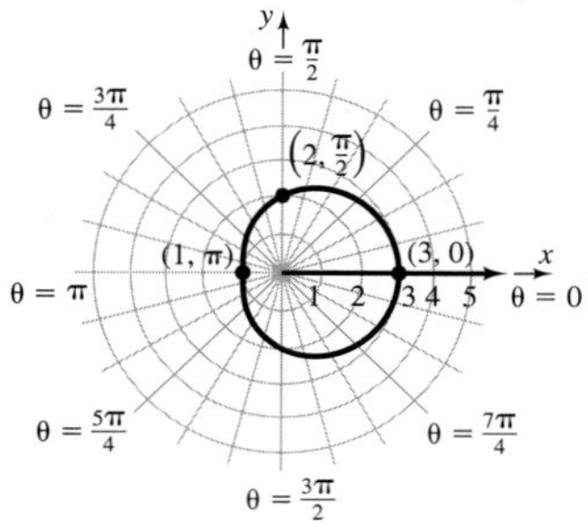

23. Limaçon without inner loop; symmetric with respect to the polar axis

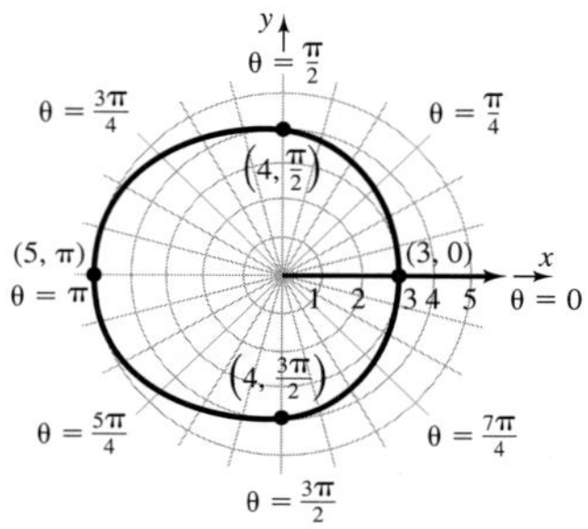

24. Limaçon with inner loop; symmetric with respect to the line $\theta = \frac{\pi}{2}$

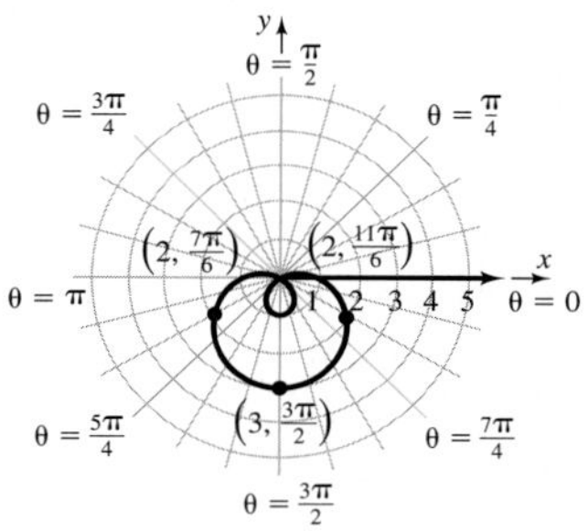

25. $\sqrt{2}(\cos 225^\circ + i \sin 225^\circ)$ **26.** $2(\cos 150^\circ + i \sin 150^\circ)$ **27.** $5(\cos 323.1^\circ + i \sin 323.1^\circ)$ **28.** $\sqrt{13}(\cos 326.3^\circ + i \sin 326.3^\circ)$

29. $-\sqrt{3} + i$

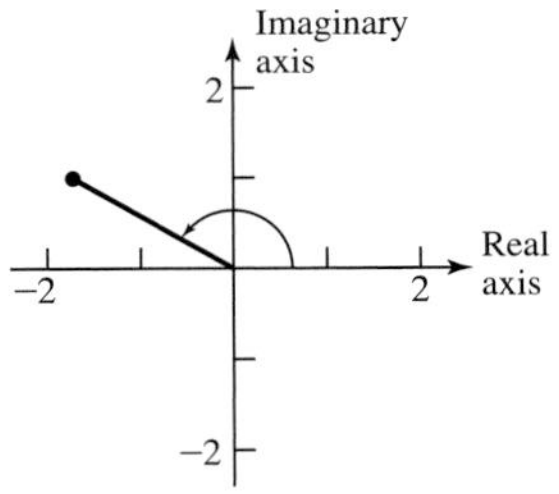

30. $\frac{3}{2} + \frac{3\sqrt{3}}{2}i$

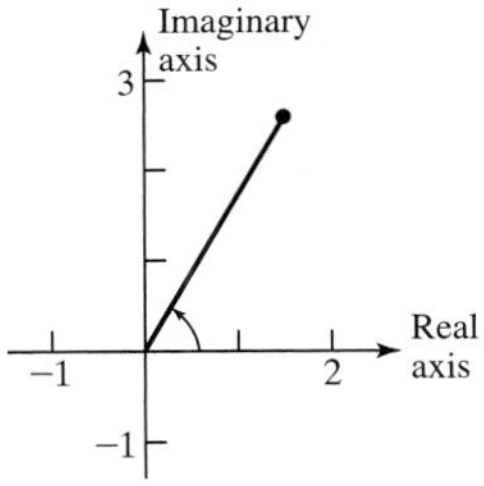

31. $-\frac{3}{2} + \left(\frac{3\sqrt{3}}{2}\right)i$

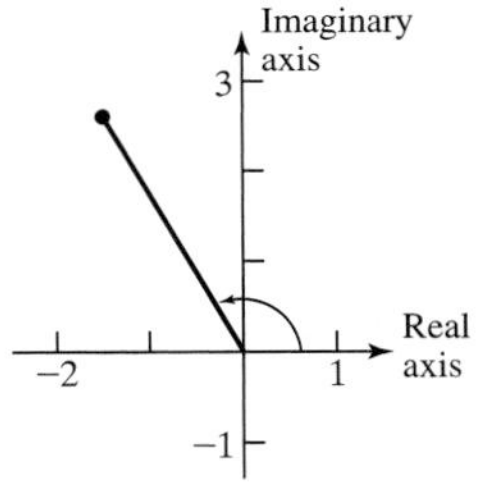

32. $-2\sqrt{2} + 2\sqrt{2}i$

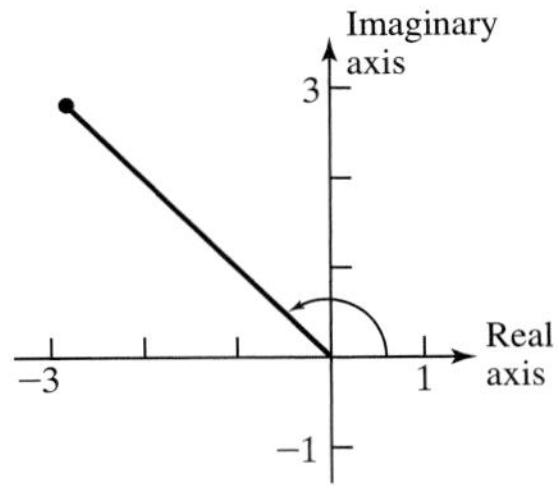

33. $0.10 - 0.02i$

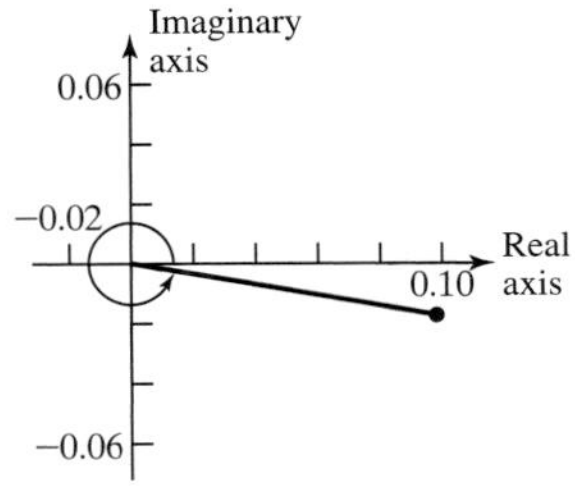

34. $-0.47 + 0.17i$

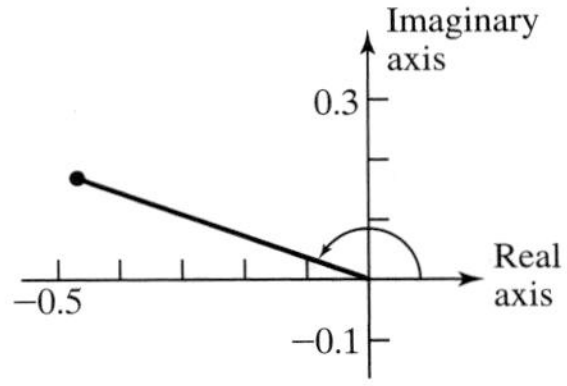

35. $zw = \cos 130° + i \sin 130°; \frac{z}{w} = \cos 30° + i \sin 30°$ **36.** $zw = \cos 290° + i \sin 290°; \frac{z}{w} = \cos 120° + i \sin 120°$
37. $zw = 6(\cos 0 + i \sin 0) = 6; \frac{z}{w} = \frac{3}{2}\left(\cos \frac{8\pi}{5} + i \sin \frac{8\pi}{5}\right)$ **38.** $zw = 6(\cos 0 + i \sin 0) = 6; \frac{z}{w} = \frac{2}{3}\left(\cos \frac{4\pi}{3} + i \sin \frac{4\pi}{3}\right)$
39. $zw = 5(\cos 5° + i \sin 5°); \frac{z}{w} = 5(\cos 15° + i \sin 15°)$ **40.** $zw = 4(\cos 30° + i \sin 30°); \frac{z}{w} = 4(\cos 70° + i \sin 70°)$ **41.** $\frac{27}{2} + \frac{27\sqrt{3}}{2}i$
42. $-4\sqrt{3} + 4i$ **43.** $4i$ **44.** $-8\sqrt{2} - 8\sqrt{2}i$ **45.** 64 **46.** 4096 **47.** $-527 - 336i$ **48.** $-7 + 24i$ **49.** $3, 3(\cos 120° + i \sin 120°), 3(\cos 240° + i \sin 240°)$ **50.** $2(\cos 45° + i \sin 45°), 2(\cos 135° + i \sin 135°), 2(\cos 225° + i \sin 225°), 2(\cos 315° + i \sin 315°)$

51.

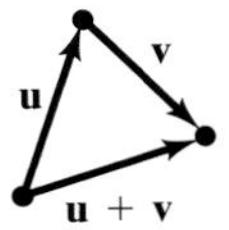

52.

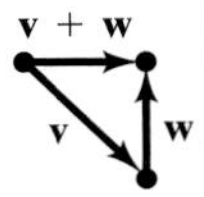

53.

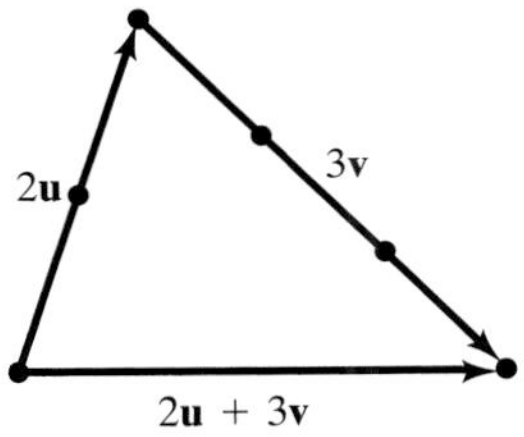

54.

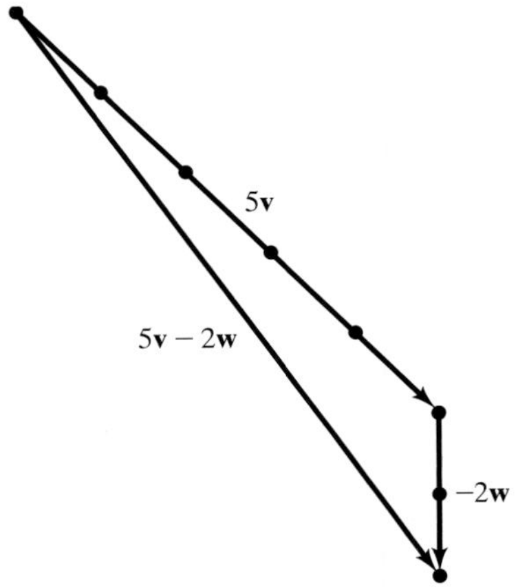

55. $\mathbf{v} = 2\mathbf{i} - 4\mathbf{j}; \|\mathbf{v}\| = 2\sqrt{5}$ **56.** $\mathbf{v} = 7\mathbf{i} - 3\mathbf{j}; \|\mathbf{v}\| = \sqrt{58}$
57. $\mathbf{v} = -\mathbf{i} + 3\mathbf{j}; \|\mathbf{v}\| = \sqrt{10}$ **58.** $\mathbf{v} = -5\mathbf{i} + 4\mathbf{j}; \|\mathbf{v}\| = \sqrt{41}$
59. $2\mathbf{i} - 2\mathbf{j}$ **60.** $-6\mathbf{i} + 4\mathbf{j}$ **61.** $-20\mathbf{i} + 13\mathbf{j}$ **62.** $10\mathbf{i} - 7\mathbf{j}$
63. $\sqrt{5}$ **64.** $2\sqrt{2}$ **65.** $\sqrt{5} + 5 \approx 7.24$ **66.** $2\sqrt{5} - 15 \approx -10.53$
67. $-\frac{2\sqrt{5}}{5}\mathbf{i} + \frac{\sqrt{5}}{5}\mathbf{j}$ **68.** $-\frac{4}{5}\mathbf{i} + \frac{3}{5}\mathbf{j}$ **69.** $\frac{3}{2} + \frac{3\sqrt{3}}{2}\mathbf{i}$
70. $-\frac{5\sqrt{3}}{2}\mathbf{i} + \frac{5}{2}\mathbf{j}$ **71.** $\sqrt{43} \approx 6.56$ **72.** $\sqrt{53} \approx 7.28$
73. $\mathbf{v} = 3\mathbf{i} - 5\mathbf{j} + 3\mathbf{k}$ **74.** $\mathbf{v} = 6\mathbf{i} - \mathbf{j} - 4\mathbf{k}$
75. $21\mathbf{i} - 2\mathbf{j} - 5\mathbf{k}$ **76.** $-9\mathbf{i} + 3\mathbf{j}$ **77.** $\sqrt{38}$ **78.** $3\sqrt{2}$
79. 0 **80.** $2\sqrt{14}$ **81.** $3\mathbf{i} + 9\mathbf{j} + 9\mathbf{k}$ **82.** 0
83. $\frac{3\sqrt{14}}{14}\mathbf{i} + \frac{\sqrt{14}}{14}\mathbf{j} - \frac{\sqrt{14}}{7}\mathbf{k}; -\frac{3\sqrt{14}}{14}\mathbf{i} - \frac{\sqrt{14}}{14}\mathbf{j} + \frac{\sqrt{14}}{7}\mathbf{k}$
84. $\frac{\sqrt{19}}{19}\mathbf{i} + \frac{3\sqrt{19}}{19}\mathbf{j} + \frac{3\sqrt{19}}{19}\mathbf{k}$ or $-\frac{\sqrt{19}}{19}\mathbf{i} - \frac{3\sqrt{19}}{19}\mathbf{j} - \frac{3\sqrt{19}}{19}\mathbf{k}$

85. $\mathbf{v} \cdot \mathbf{w} = -11; \theta \approx 169.7°$ **86.** $\mathbf{v} \cdot \mathbf{w} = 2; \theta \approx 63.4°$ **87.** $\mathbf{v} \cdot \mathbf{w} = -4; \theta \approx 153.4°$ **88.** $\mathbf{v} \cdot \mathbf{w} = -5; \theta \approx 109.7°$
89. $\mathbf{v} \cdot \mathbf{w} = 1; \theta \approx 70.5°$ **90.** $\mathbf{v} \cdot \mathbf{w} = 2; \theta \approx 61.9°$ **91.** $\mathbf{v} \cdot \mathbf{w} = 0; \theta = 90°$ **92.** $\mathbf{v} \cdot \mathbf{w} = -4; \theta \approx 101.9°$ **93.** Parallel **94.** Parallel

95. Parallel **96.** Neither **97.** Orthogonal **98.** Orthogonal **99.** $\mathbf{v}_1 = \frac{4}{5}\mathbf{i} - \frac{3}{5}\mathbf{j}$; $\mathbf{v}_2 = \frac{6}{5}\mathbf{i} + \frac{8}{5}\mathbf{j}$

100. $\mathbf{v}_1 = -\frac{16}{5}\mathbf{i} + \frac{8}{5}\mathbf{j}$; $\mathbf{v}_2 = \frac{1}{5}\mathbf{i} + \frac{2}{5}\mathbf{j}$ **101.** $\mathbf{v}_1 = \frac{9}{10}(3\mathbf{i} + \mathbf{j})$ **102.** $\mathbf{v}_1 = -\frac{1}{2}(3\mathbf{i} - \mathbf{j})$ **103.** $\alpha \approx 56.1°; \beta \approx 138°; \gamma \approx 68.2°$

104. $\alpha \approx 65.9°; \beta \approx 114.1°; \gamma \approx 35.3°$ **105.** $2\sqrt{83}$ **106.** $2\sqrt{43}$ **107.** $-2\mathbf{i} + 3\mathbf{j} - \mathbf{k}$ **108.** 0 **109.** $\sqrt{29} \approx 5.39$ mi/hr; 0.4 mi

110. 459.54 km/hr, at an angle 5.3° east of north **111.** Left cable: 1843.21 lb; right cable: 1630.41 lb **112.** 13.29 mi/hr; 9.18° west of south

113. 50 foot-pounds

Cumulative Review *(page 612)*

1. $\{-3, 3\}$ **2.** $y = \frac{\sqrt{3}}{3}x$

3. $x^2 + (y - 1)^2 = 9$

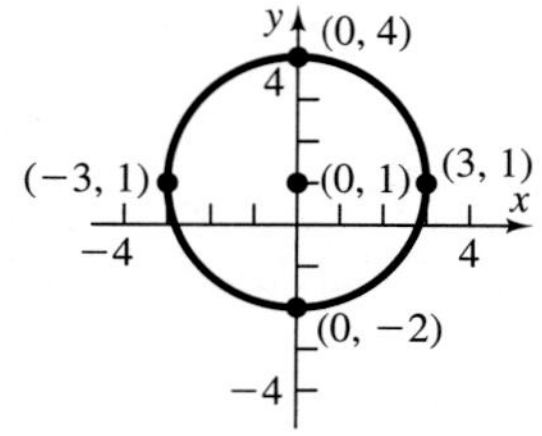

4. $\left\{x \middle| x < \frac{1}{2}\right\}$ or $\left(-\infty, \frac{1}{2}\right)$ **5.** Symmetry with respect to the y-axis

6.

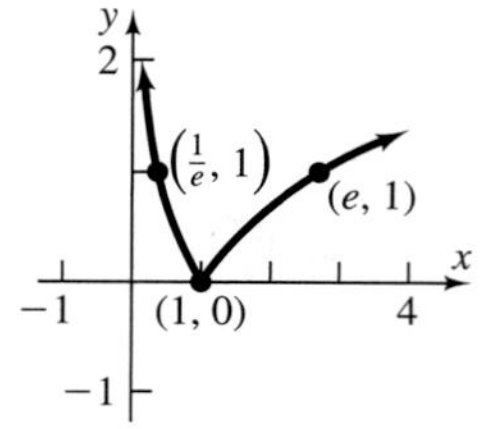

7.

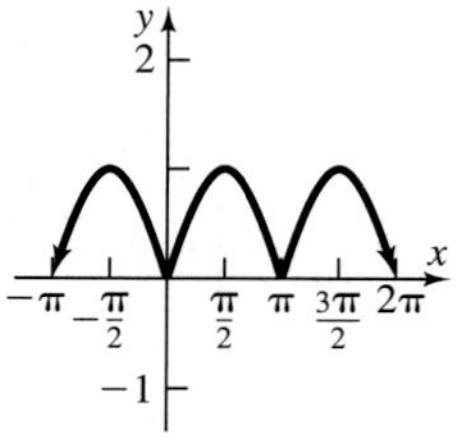

8.

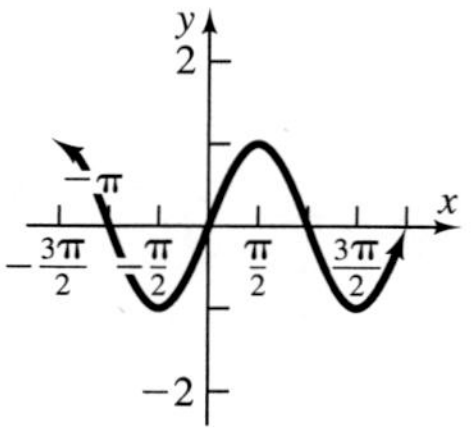

9. $-\frac{\pi}{6}$ **10.**

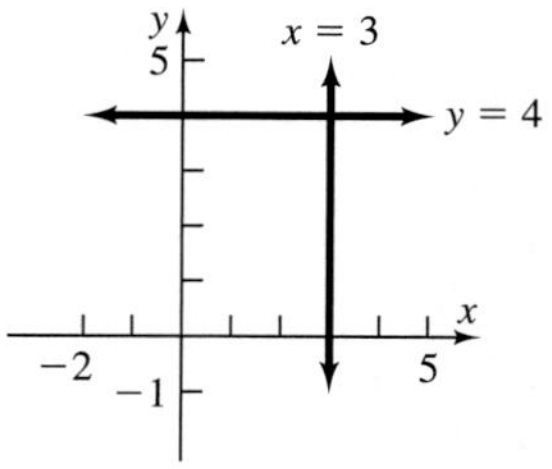

11.

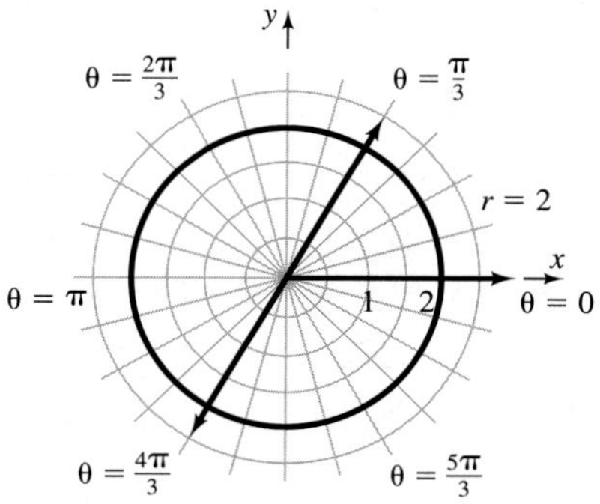

CHAPTER 9 Analytic Geometry

9.2 Concepts and Vocabulary *(page 622)*

6. parabola **7.** paraboloid of revolution **8.** True **9.** True **10.** True

9.2 Exercises *(page 622)*

11. *B* **12.** *G* **13.** *E* **14.** *D* **15.** *H* **16.** *A* **17.** *C* **18.** *F*

19. $y^2 = 16x$

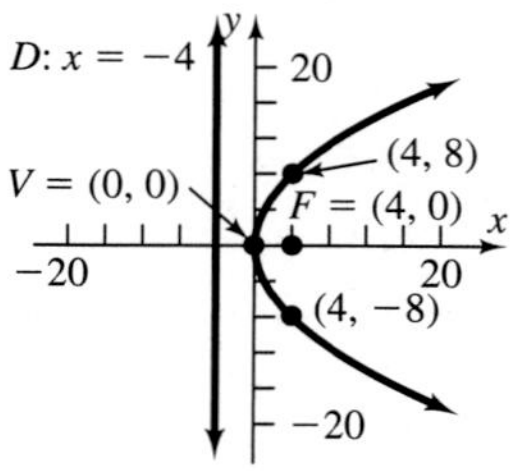

20. $x^2 = 8y$

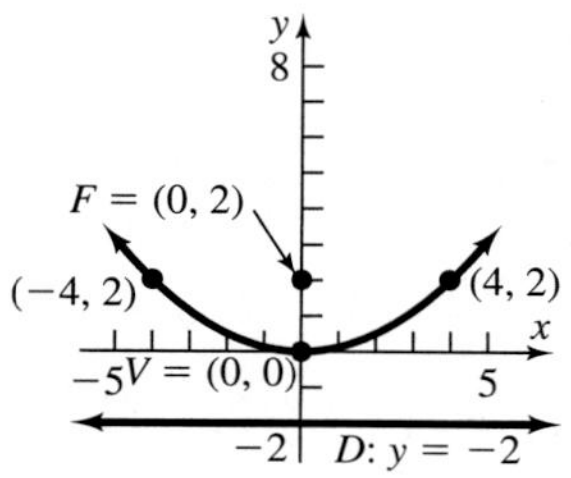

21. $x^2 = -12y$

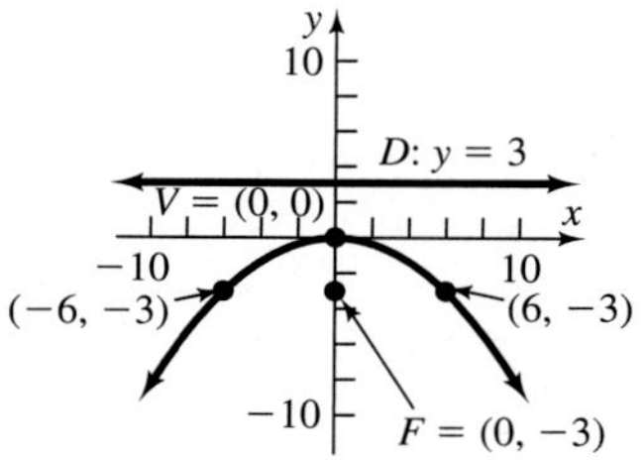

22. $y^2 = -16x$

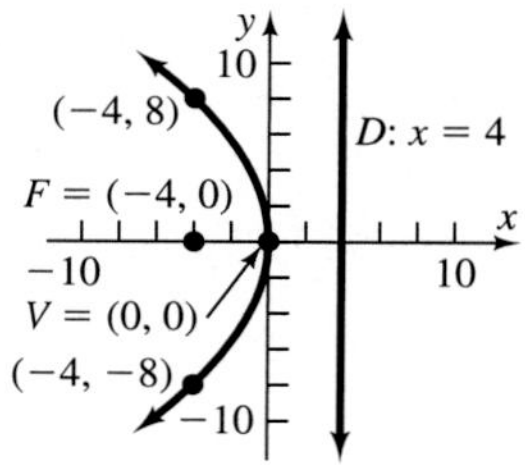

23. $y^2 = -8x$

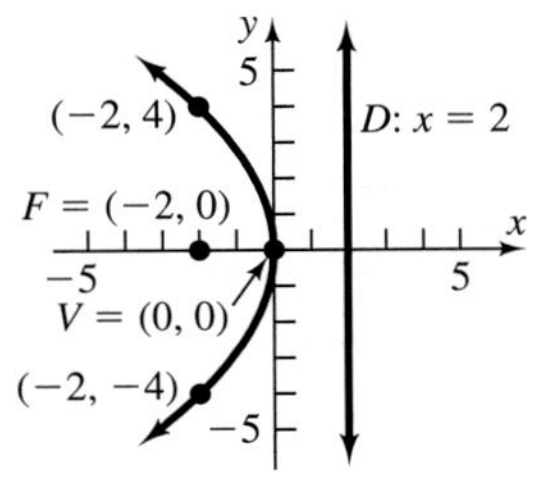

24. $x^2 = -4y$

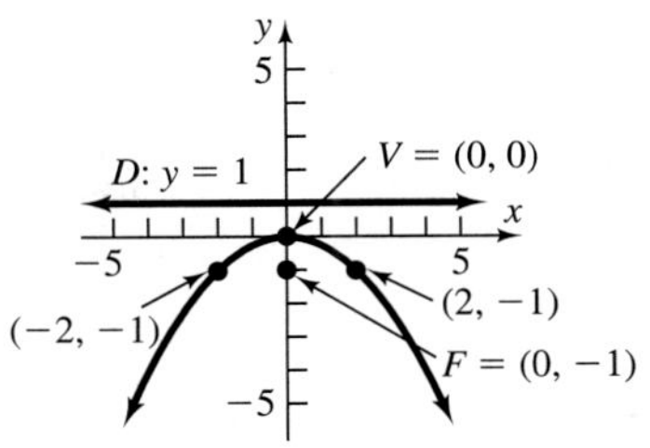

25. $x^2 = 2y$

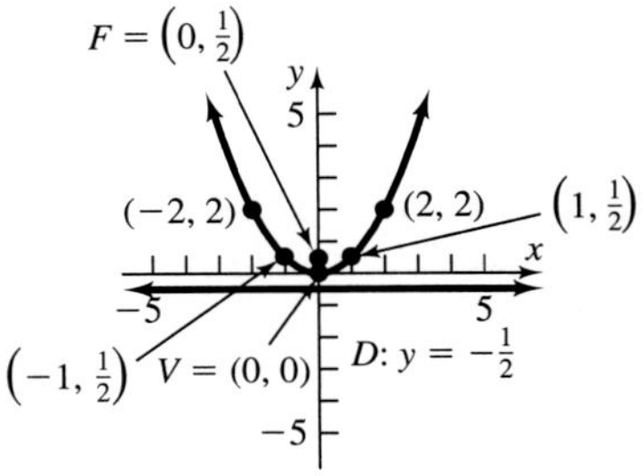

26. $y^2 = 2x$

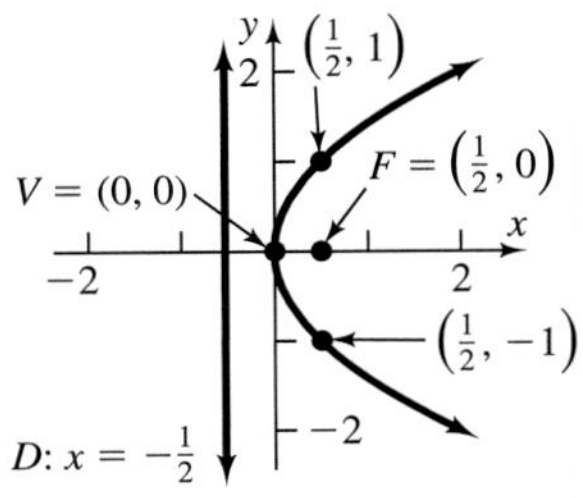

27. $x^2 = \frac{4}{3}y$

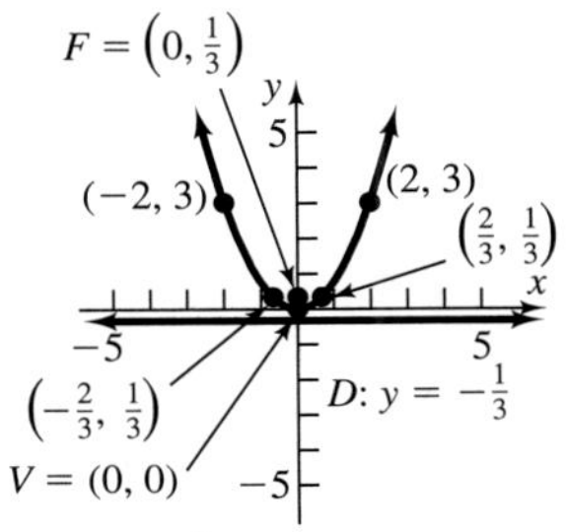

28. $y^2 = \frac{9}{2}x$

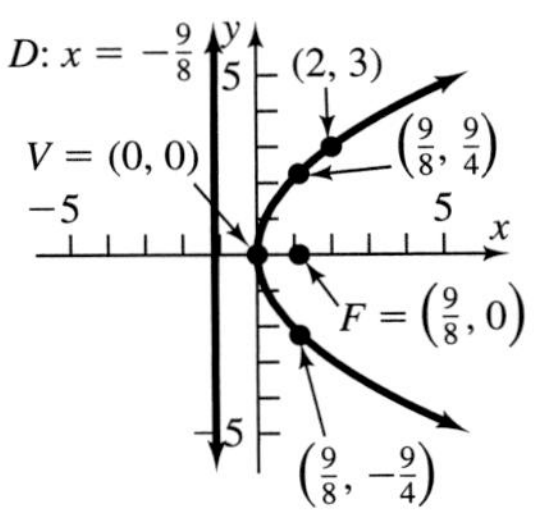

29. $(x - 2)^2 = -8(y + 3)$

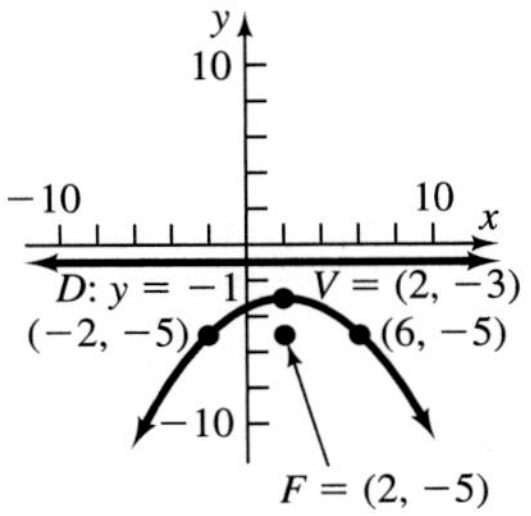

30. $(y + 2)^2 = 8(x - 4)$

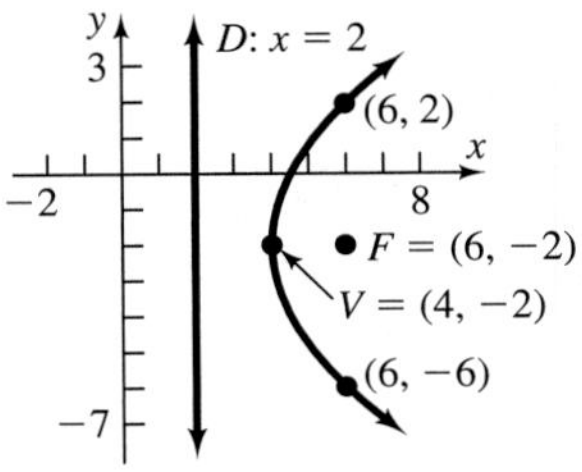

31. $(y + 2)^2 = 4(x + 1)$

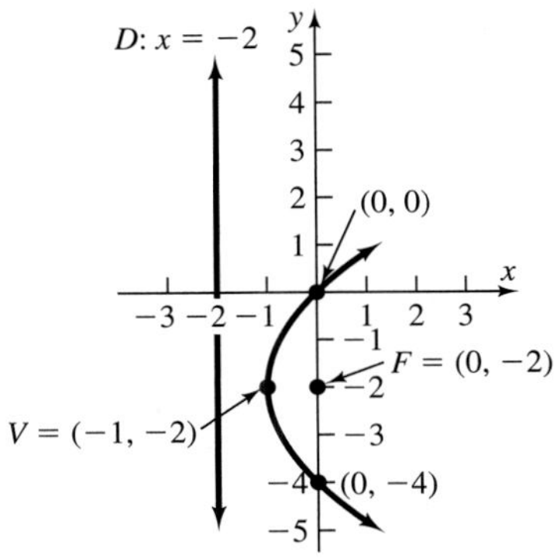

32. $(x - 3)^2 = -8y$

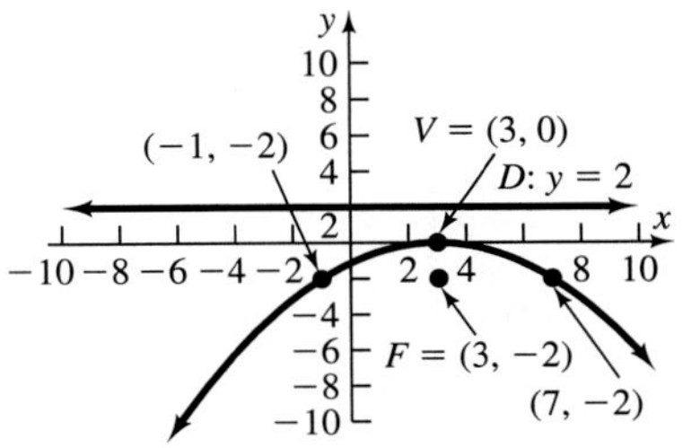

33. $(x + 3)^2 = 4(y - 3)$

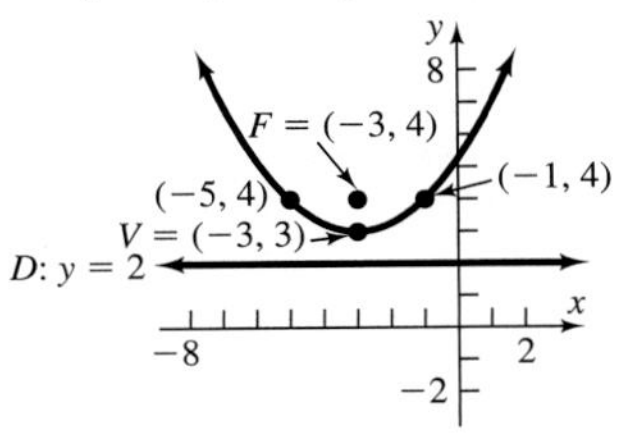

34. $(y - 4)^2 = 12(x + 1)$

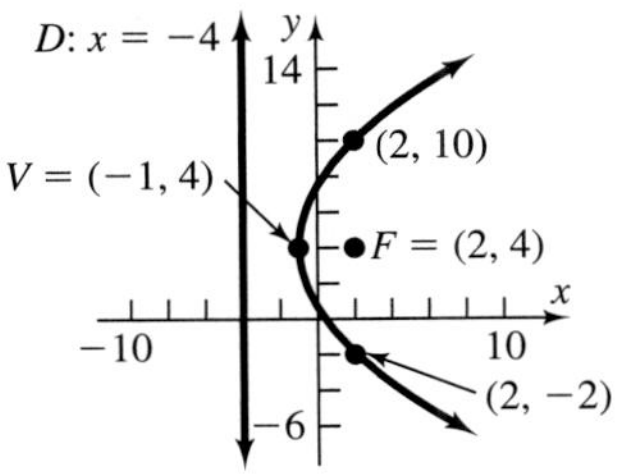

35. $(y + 2)^2 = -8(x + 1)$

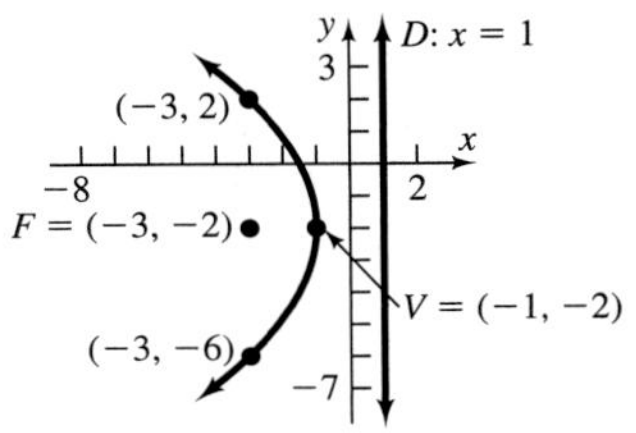

36. $(x + 4)^2 = 12(y - 1)$

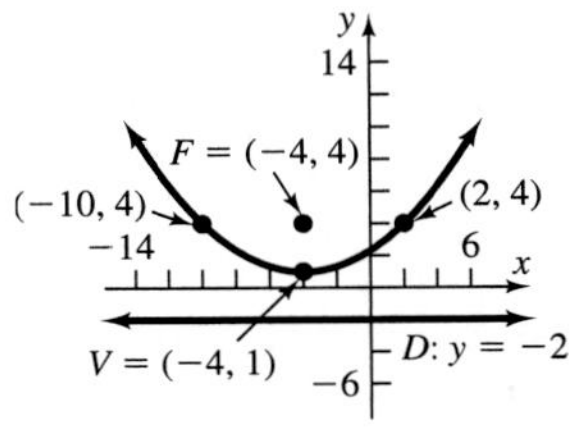

37. Vertex: (0, 0); Focus: (0, 1); Directrix: $y = -1$

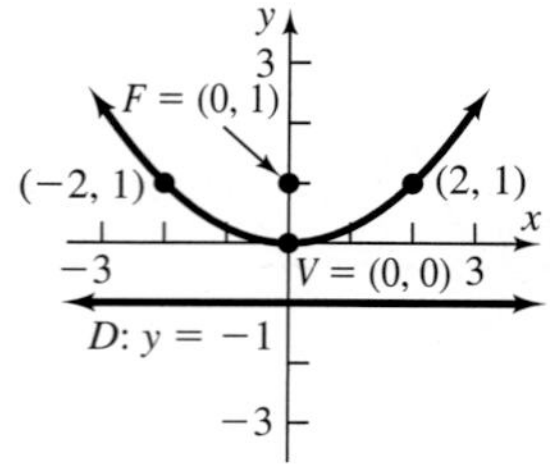

38. Vertex: (0, 0); Focus: (2, 0); Directrix: $x = -2$

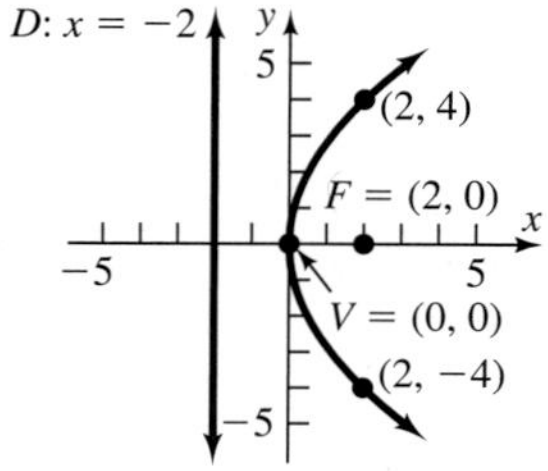

39. Vertex: (0, 0); Focus: (−4, 0); Directrix: $x = 4$

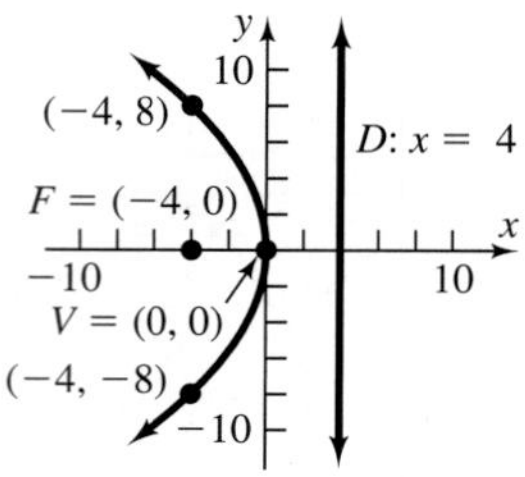

40. Vertex: (0, 0); Focus: (0, −1); Directrix: $y = 1$

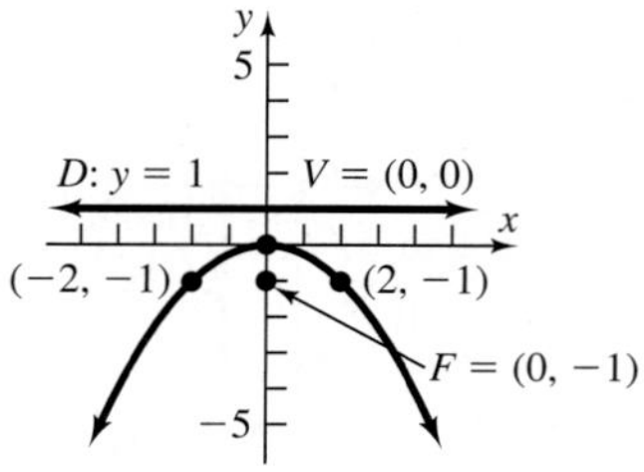

41. Vertex: (−1, 2); Focus: (1, 2);

Directrix: $x = -3$

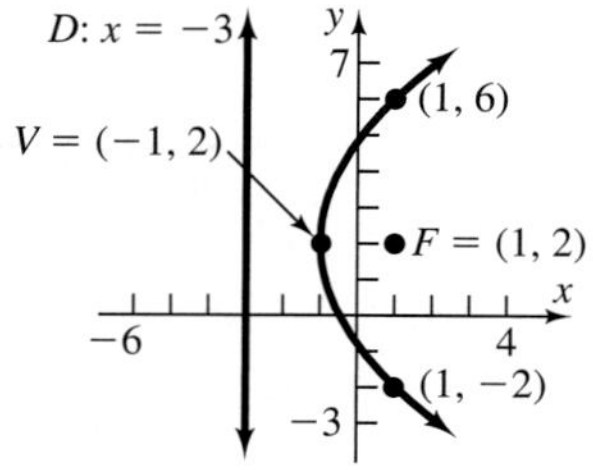

42. Vertex: (−4, −2); Focus: (−4, 2);

Directrix: $y = -6$

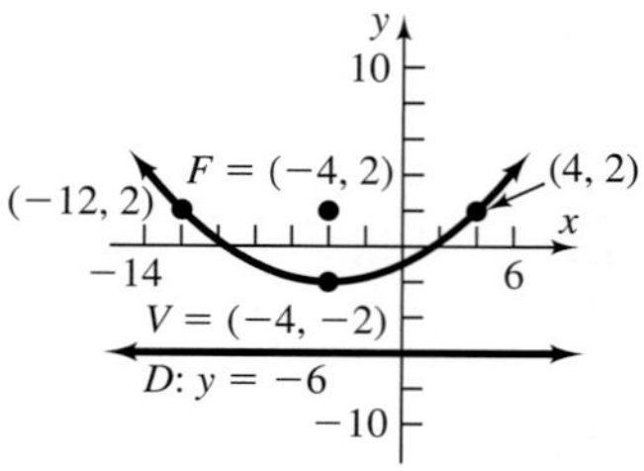

43. Vertex: (3, −1); Focus: $\left(3, -\frac{5}{4}\right)$;

Directrix: $y = -\frac{3}{4}$

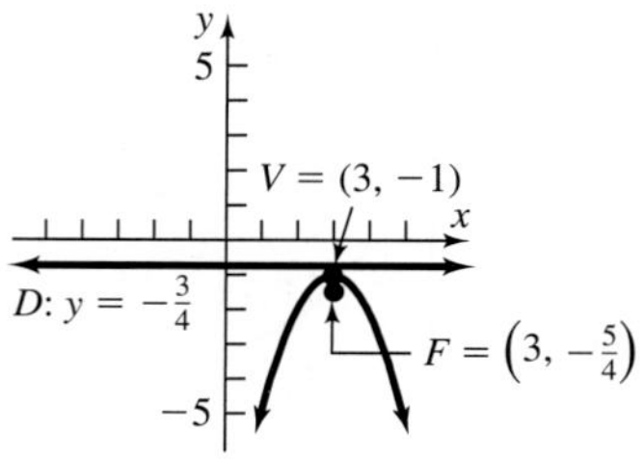

44. Vertex: (2, −1); Focus: (1, −1); Directrix: $x = 3$

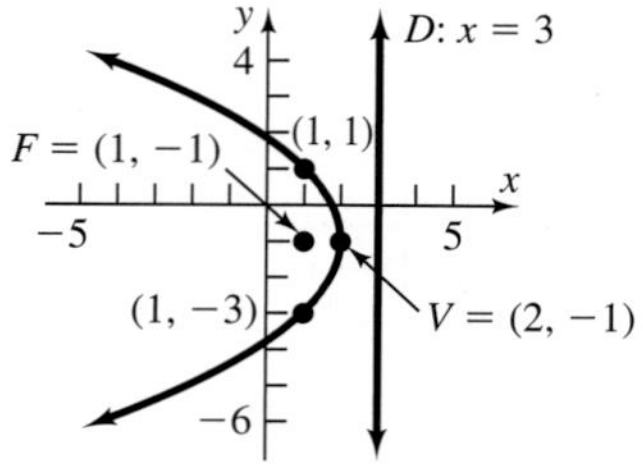

45. Vertex: (2, −3); Focus: (4, −3); Directrix: $x = 0$

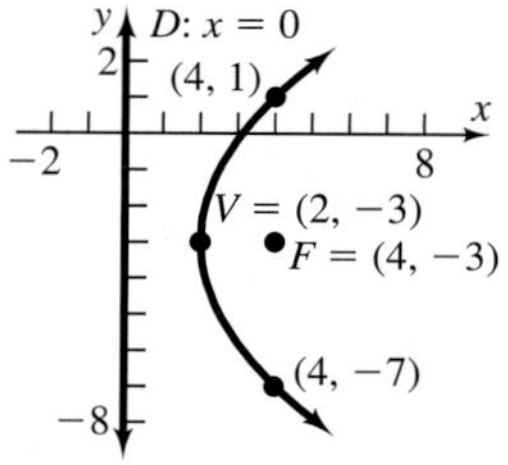

46. Vertex: (2, 3); Focus: (2, 4); Directrix: $y = 2$

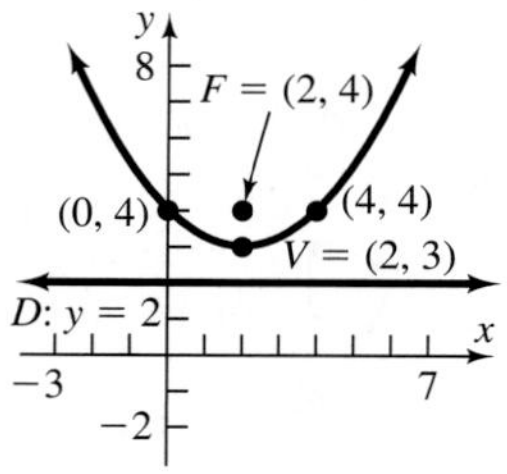

47. Vertex: (0, 2); Focus: (−1, 2); Directrix: $x = 1$

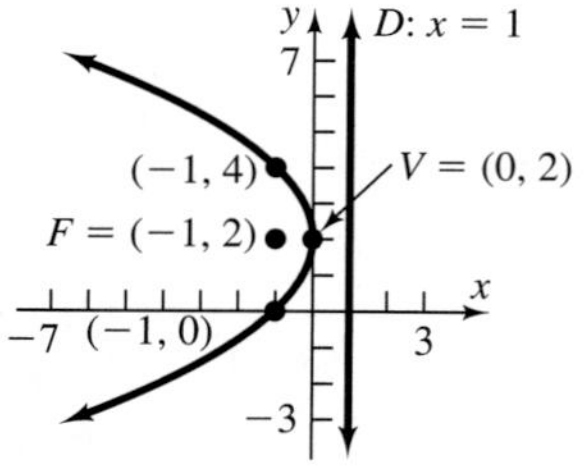

48. Vertex: (−3, −2); Focus: (−3, −1); Directrix: $y = -3$

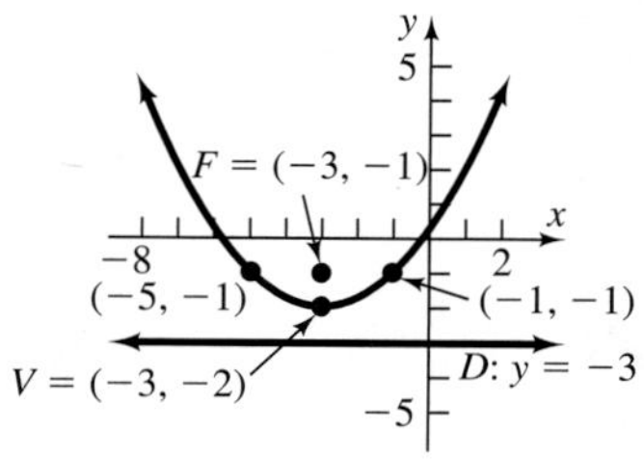

49. Vertex: (−4, −2); Focus: (−4, −1); Directrix: $y = -3$

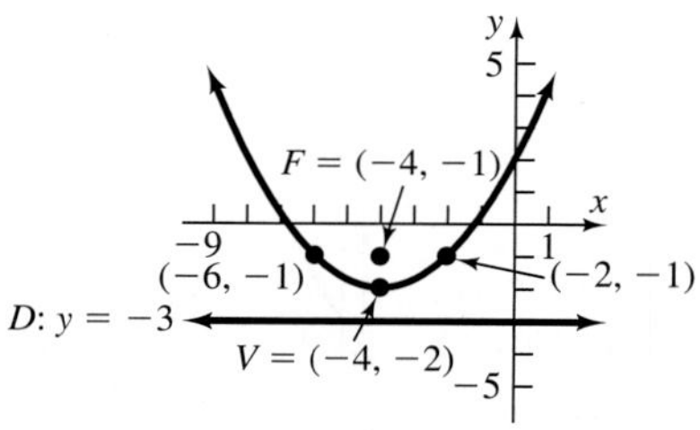

50. Vertex: (0, 1); Focus: (2, 1);

Directrix: $x = -2$

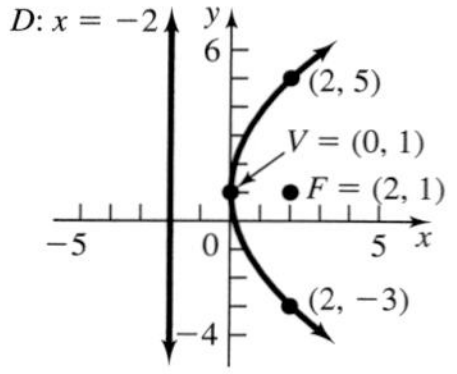

51. Vertex: (−1, −1); Focus: $\left(-\frac{3}{4}, -1\right)$;

Directrix: $x = -\frac{5}{4}$

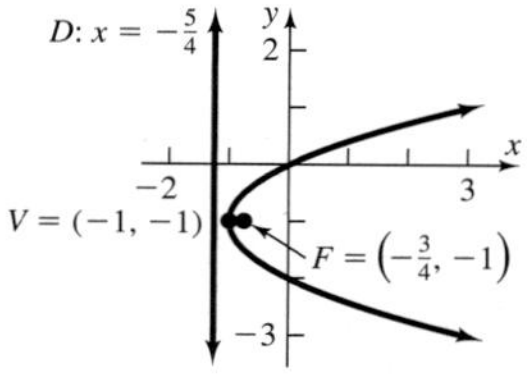

52. Vertex: (2, −2); Focus: $\left(2, -\frac{3}{2}\right)$;

Directrix: $y = -\frac{5}{2}$

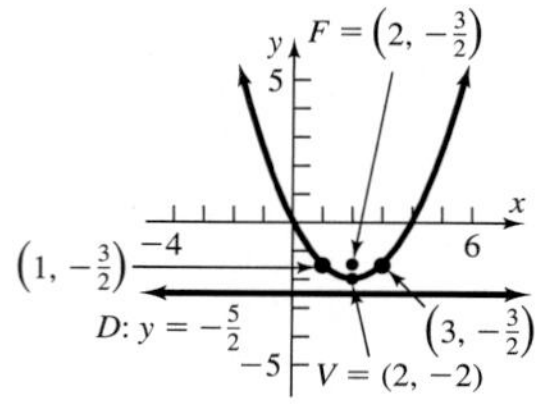

53. Vertex: $(2, -8)$; Focus: $\left(2, -\frac{31}{4}\right)$; Directrix: $y = -\frac{33}{4}$

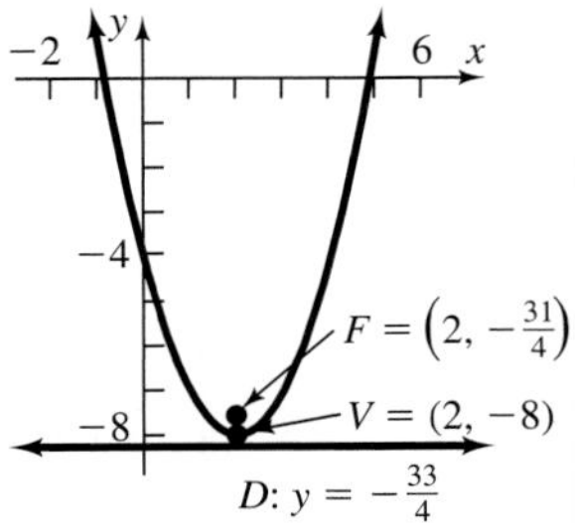

54. Vertex: $(37, -6)$; Focus: $\left(\frac{147}{4}, -6\right)$; Directrix: $x = \frac{149}{4}$

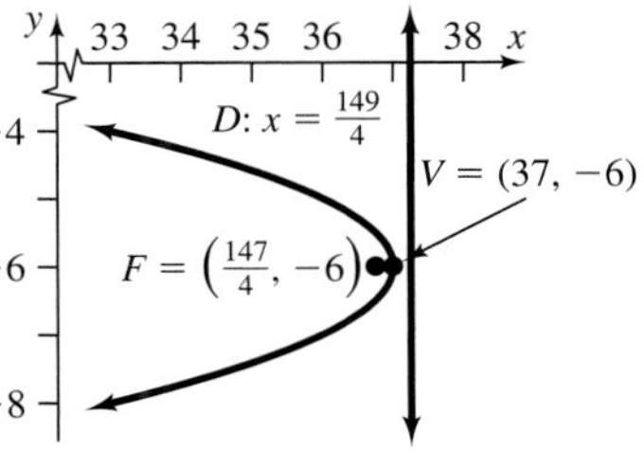

55. $(y - 1)^2 = x$

56. $(x - 1)^2 = -(y - 2)$

57. $(y - 1)^2 = -(x - 2)$

58. $x^2 = 4(y + 1)$

59. $x^2 = 4(y - 1)$

60. $(x - 1)^2 = \frac{1}{2}(y + 1)$

61. $y^2 = \frac{1}{2}(x + 2)$

62. $y^2 = -(x - 1)$

63. 1.5625 ft from the base of the dish, along the axis of symmetry **64.** 1.125 ft or 13.5 in. from the base of the dish **65.** 1 in. from the vertex **66.** $4\sqrt{2}$ in. **67.** 20 ft **68.** 15.625 ft **69.** 0.78125 ft **70.** $8\sqrt{2}$ ft **71.** 4.17 ft from the base along the axis of symmetry **72.** 0.0278 in. from the base of the mirror **73.** 24.31 ft, 18.75 ft, 7.64 ft **74.** 27.78 ft

75. $Ax^2 + Ey = 0,\ A \neq 0,\ E \neq 0$

$$Ax^2 = -Ey$$

$$x^2 = -\frac{E}{A}y$$

This is the equation of a parabola with vertex at $(0, 0)$ and axis of symmetry the y-axis. The focus is $\left(0, -\frac{E}{4A}\right)$; the directrix is the line $y = \frac{E}{4A}$. The parabola opens up if $-\frac{E}{A} > 0$ and down if $-\frac{E}{A} < 0$.

76. $Cy^2 + Dx = 0,\ C \neq 0,\ D \neq 0$

$$Cy^2 = -Dx$$

$$y^2 = -\frac{D}{C}x$$

This is the equation of a parabola with vertex at $(0, 0)$ and axis of symmetry the x-axis. The focus is $\left(-\frac{D}{4C}, 0\right)$; the directrix is the line $x = \frac{D}{4C}$. The parabola opens to the right if $-\frac{D}{C} > 0$ and to the left if $-\frac{D}{C} < 0$.

77. $Ax^2 + Dx + Ey + F = 0,\ A \neq 0$

$$Ax^2 + Dx = -Ey - F$$

$$x^2 + \frac{D}{A}x = -\frac{E}{A}y - \frac{F}{A}$$

$$\left(x + \frac{D}{2A}\right)^2 = -\frac{E}{A}y - \frac{F}{A} + \frac{D^2}{4A^2}$$

$$\left(x + \frac{D}{2A}\right)^2 = -\frac{E}{A}y + \frac{D^2 - 4AF}{4A^2}$$

(a) If $E \neq 0$, then the equation may be written as

$$\left(x + \frac{D}{2A}\right)^2 = -\frac{E}{A}\left(y - \frac{D^2 - 4AF}{4AE}\right)$$

This is the equation of a parabola with vertex at $\left(-\frac{D}{2A}, \frac{D^2 - 4AF}{4AE}\right)$ and axis of symmetry parallel to the y-axis.

(b)–(d) If $E = 0$, the graph of the equation contains no points if $D^2 - 4AF < 0$, is a single vertical line if $D^2 - 4AF = 0$, and is two vertical lines if $D^2 - 4AF > 0$.

78. $Cy^2 + Dx + Ey + F = 0,\ C \neq 0$

$$Cy^2 + Ey = -Dx - F$$

$$y^2 + \frac{E}{C}y = -\frac{D}{C}x - \frac{F}{C}$$

$$\left(y + \frac{E}{2C}\right)^2 = -\frac{D}{C}x - \frac{F}{C} + \frac{E^2}{4C^2}$$

$$\left(y + \frac{E}{2C}\right)^2 = -\frac{D}{C}x + \frac{E^2 - 4CF}{4C^2}$$

(a) If $D \neq 0$, then the equation may be written as

$$\left(y + \frac{E}{2C}\right)^2 = -\frac{D}{C}\left(x - \frac{E^2 - 4CF}{4CD}\right).$$

This is the equation of a parabola with vertex at $\left(\frac{E^2 - 4CF}{4CD}, -\frac{E}{2C}\right)$ and axis of symmetry parallel to the x-axis.

(b)–(d) If $D \neq 0$, the graph of the equation contains no points if $E^2 - 4CF < 0$, is a single vertical line if $E^2 - 4CF = 0$, and is two horizontal lines if $E^2 - 4CF > 0$.

9.3 Concepts and Vocabulary *(page 633)*

7. ellipse **8.** major **9.** $(0, -5)$; $(0, 5)$ **10.** False **11.** True **12.** True

9.3 Exercises *(page 633)*

13. *C* **14.** *D* **15.** *B* **16.** *A*

17. Vertices: $(-5, 0)$, $(5, 0)$
Foci: $(-\sqrt{21}, 0)$, $(\sqrt{21}, 0)$

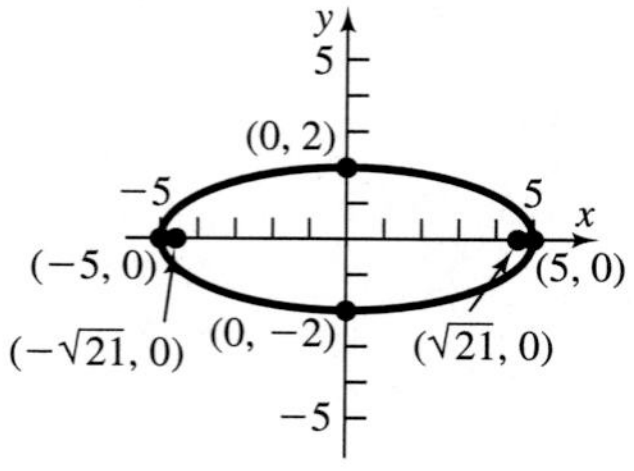

18. Vertices: $(-3, 0)$, $(3, 0)$
Foci: $(-\sqrt{5}, 0)$, $(\sqrt{5}, 0)$

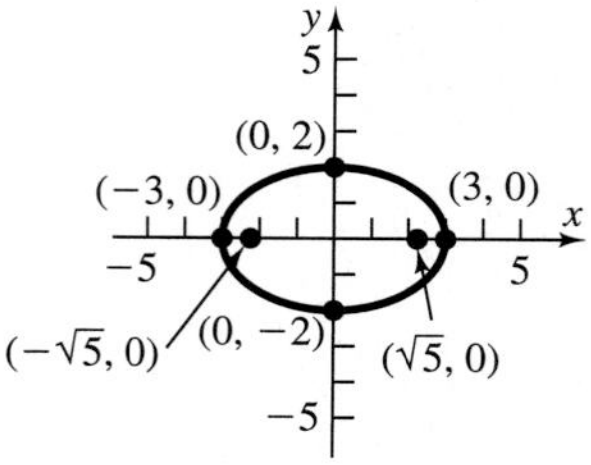

19. Vertices: $(0, -5)$, $(0, 5)$
Foci: $(0, -4)$, $(0, 4)$

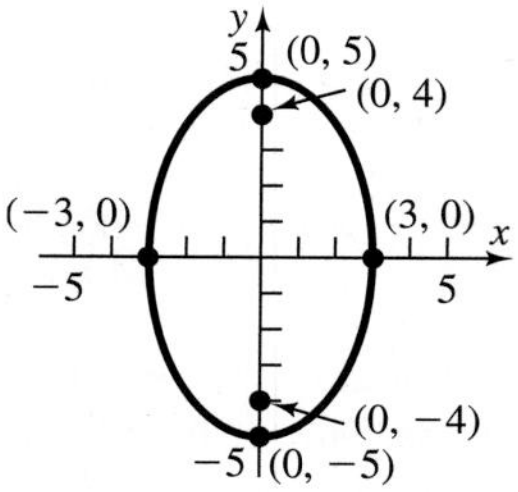

20. Vertices: $(0, -4)$, $(0, 4)$
Foci: $(0, -\sqrt{15})$, $(0, \sqrt{15})$

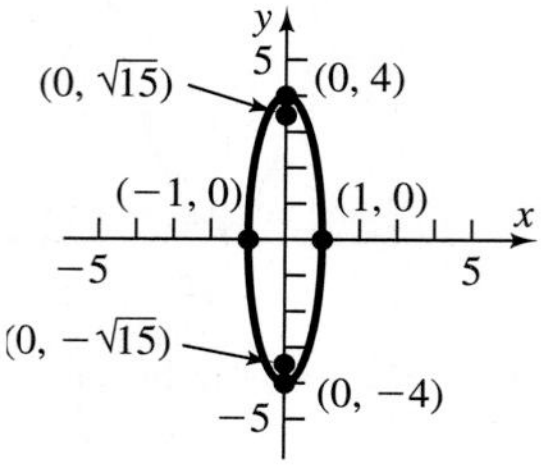

21. $\dfrac{x^2}{4} + \dfrac{y^2}{16} = 1$
Vertices: $(0, -4)$, $(0, 4)$
Foci: $(0, -2\sqrt{3})$, $(0, 2\sqrt{3})$

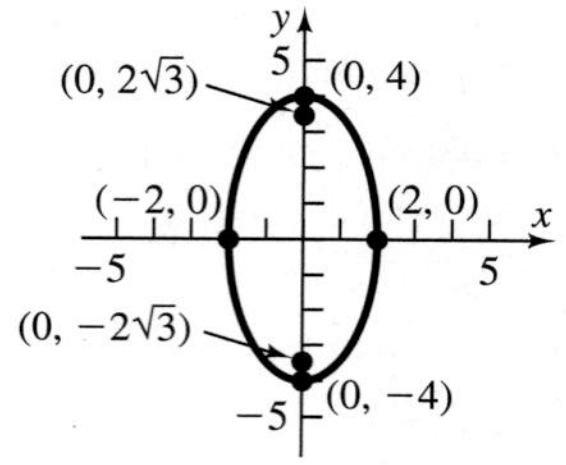

22. $\dfrac{x^2}{18} + \dfrac{y^2}{2} = 1$
Vertices: $(-3\sqrt{2}, 0)$, $(3\sqrt{2}, 0)$
Foci: $(-4, 0)$, $(4, 0)$

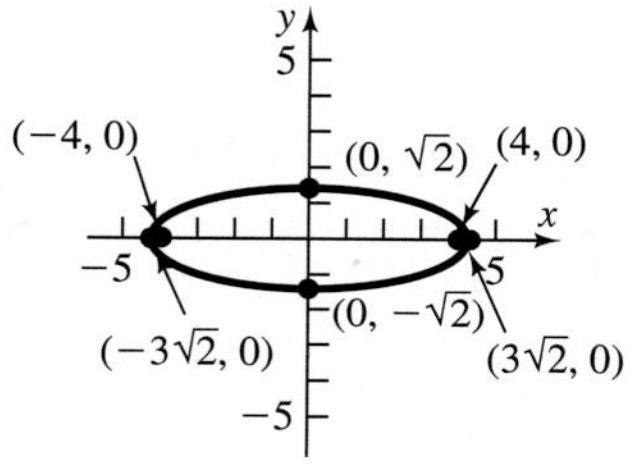

23. $\dfrac{x^2}{8} + \dfrac{y^2}{2} = 1$
Vertices: $(-2\sqrt{2}, 0)$, $(2\sqrt{2}, 0)$
Foci: $(-\sqrt{6}, 0)$, $(\sqrt{6}, 0)$

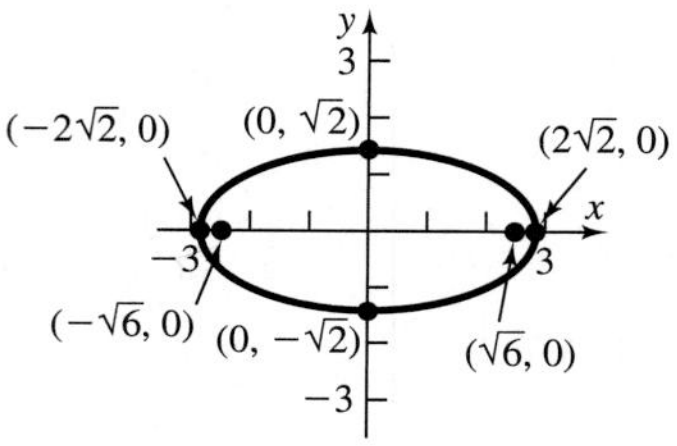

24. $\dfrac{x^2}{4} + \dfrac{y^2}{9} = 1$
Vertices: $(0, -3)$, $(0, 3)$
Foci: $(0, -\sqrt{5})$, $(0, \sqrt{5})$

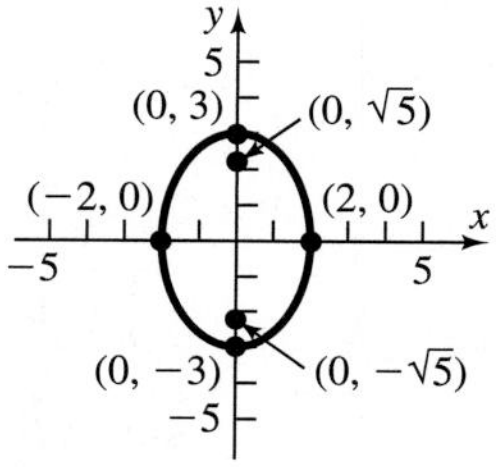

25. $\dfrac{x^2}{16} + \dfrac{y^2}{16} = 1$
Vertices: $(-4, 0)$, $(4, 0)$, $(0, -4)$, $(0, 4)$
Focus: $(0, 0)$

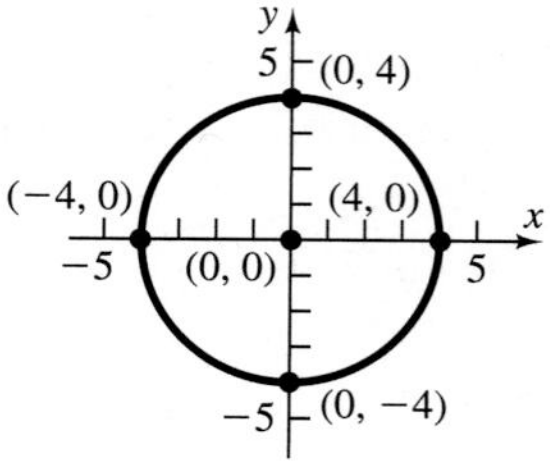

26. $\dfrac{x^2}{4} + \dfrac{y^2}{4} = 1$
Vertices: $(-2, 0)$, $(2, 0)$, $(0, -2)$, $(0, 2)$
Focus: $(0, 0)$

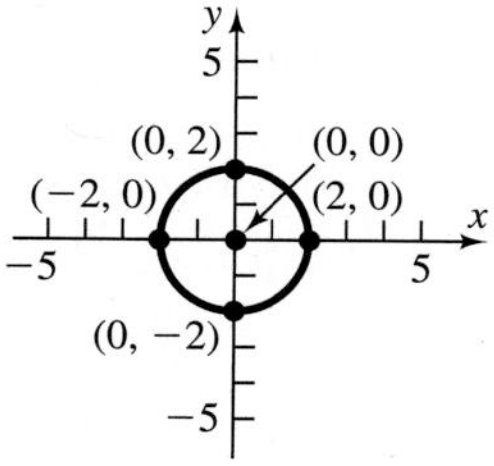

27. $\dfrac{x^2}{25} + \dfrac{y^2}{16} = 1$

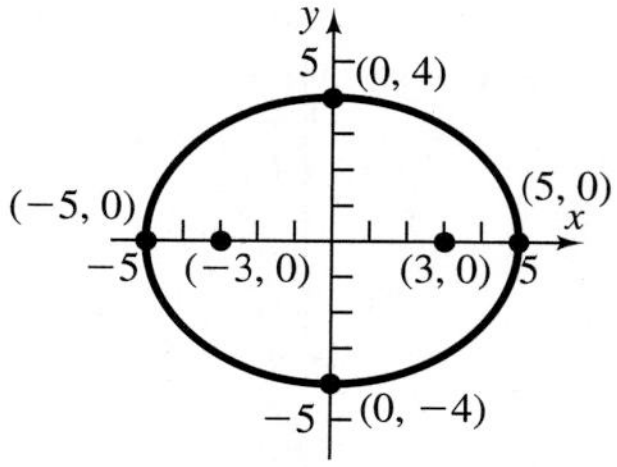

28. $\dfrac{x^2}{9} + \dfrac{y^2}{8} = 1$

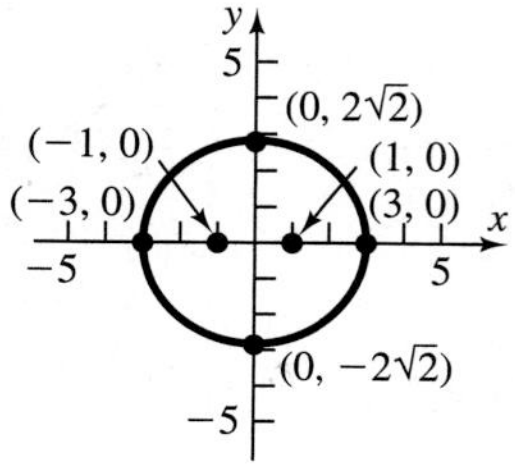

29. $\dfrac{x^2}{9} + \dfrac{y^2}{25} = 1$

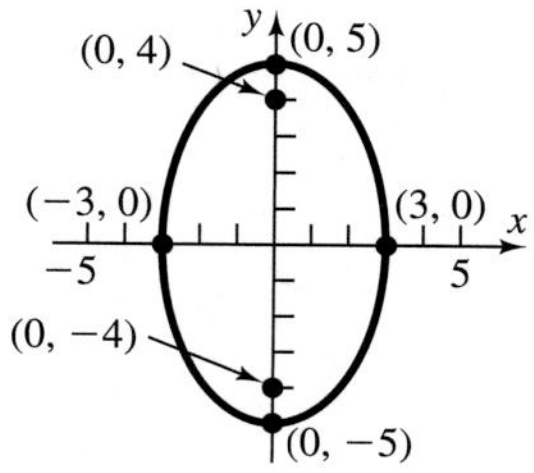

30. $\dfrac{x^2}{3} + \dfrac{y^2}{4} = 1$

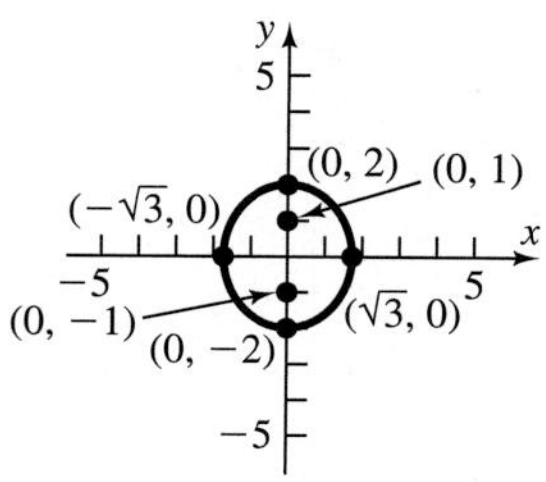

31. $\frac{x^2}{9} + \frac{y^2}{5} = 1$

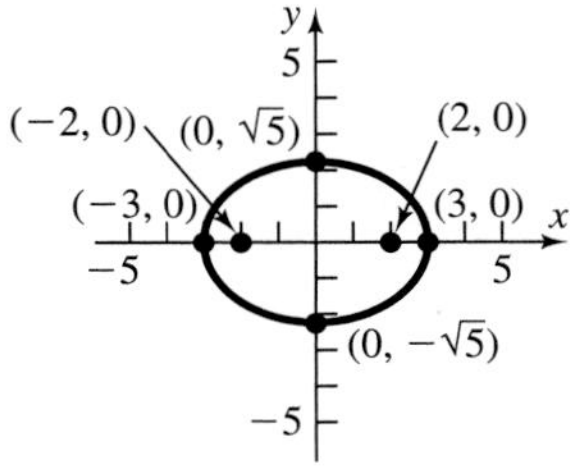

32. $\frac{x^2}{12} + \frac{y^2}{16} = 1$

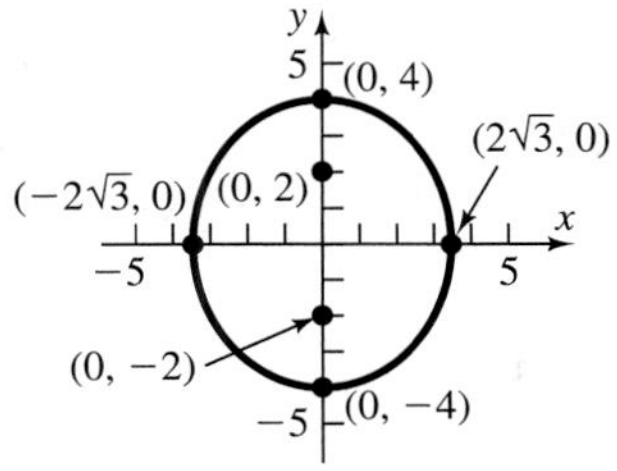

33. $\frac{x^2}{25} + \frac{y^2}{9} = 1$

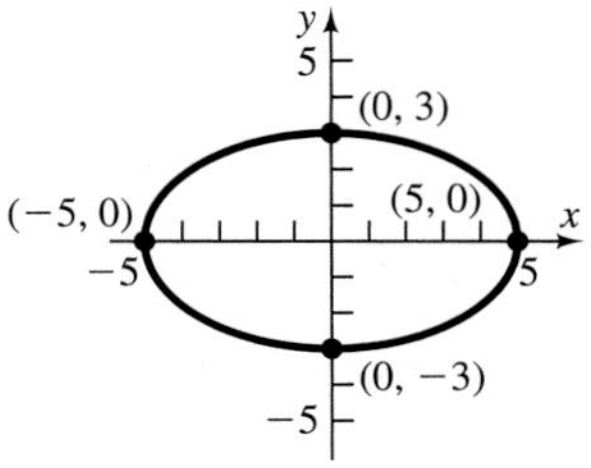

34. $\frac{x^2}{48} + \frac{y^2}{64} = 1$

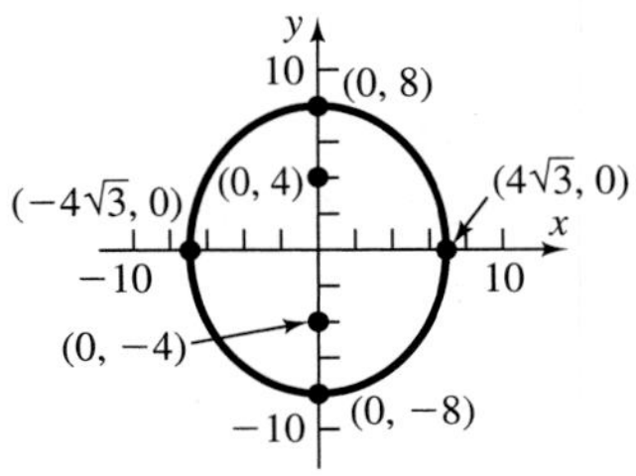

35. $\frac{x^2}{4} + \frac{y^2}{13} = 1$

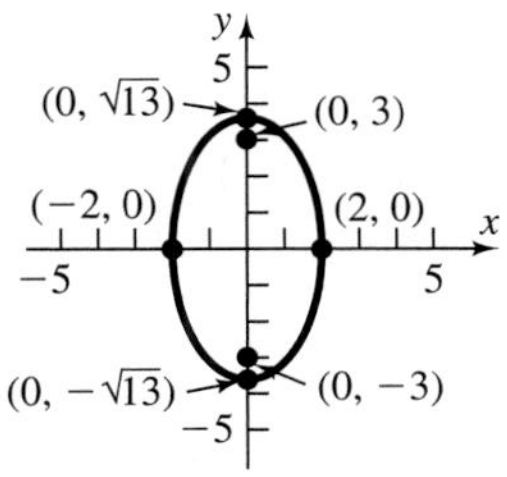

36. $\frac{x^2}{16} + y^2 = 1$

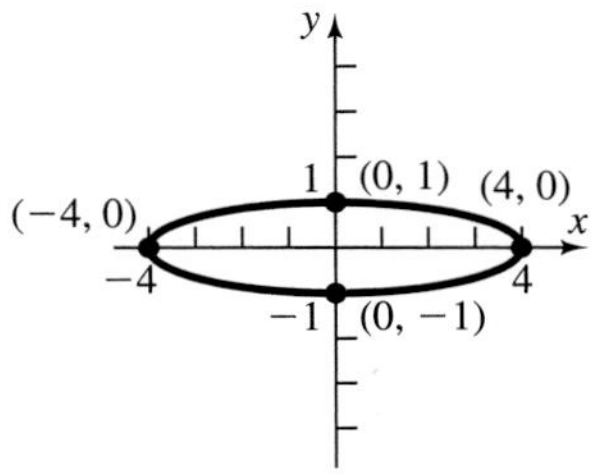

37. $x^2 + \frac{y^2}{16} = 1$

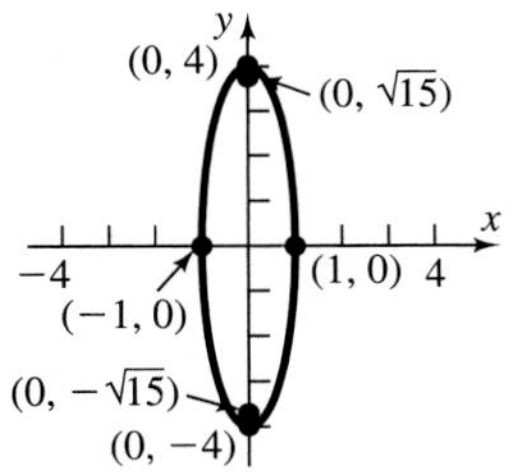

38. $\frac{x^2}{25} + \frac{y^2}{21} = 1$

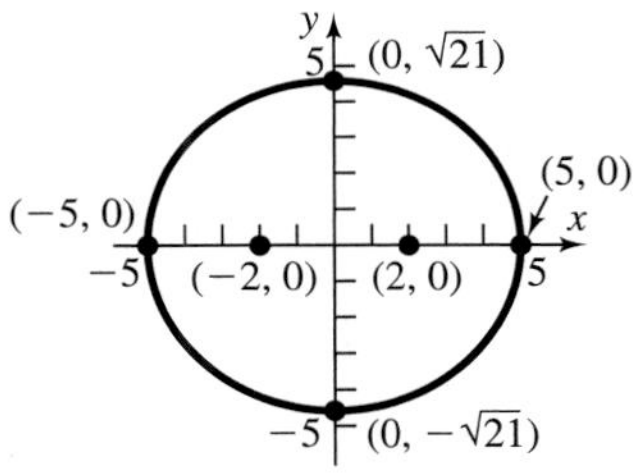

39. $\frac{(x + 1)^2}{4} + (y - 1)^2 = 1$

40. $(x + 1)^2 + \frac{(y + 1)^2}{4} = 1$

41. $(x - 1)^2 + \frac{y^2}{4} = 1$

42. $\frac{x^2}{4} + (y - 1)^2 = 1$

43. Center: $(3, -1)$; Vertices: $(3, -4)$, $(3, 2)$
Foci: $(3, -1 - \sqrt{5})$, $(3, -1 + \sqrt{5})$

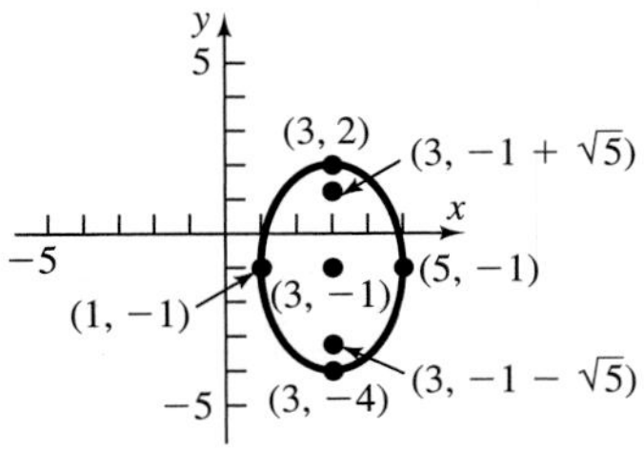

44. Center: $(-4, -2)$; Vertices: $(-7, -2)$, $(-1, -2)$
Foci: $(-4 - \sqrt{5}, -2)$, $(-4 + \sqrt{5}, -2)$

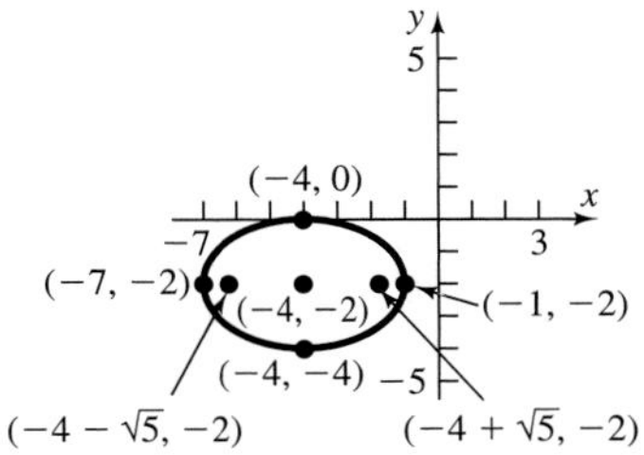

45. $\dfrac{(x+5)^2}{16}+\dfrac{(y-4)^2}{4}=1$

Center: $(-5, 4)$; Vertices: $(-9, 4)$, $(-1, 4)$

Foci: $(-5-2\sqrt{3}, 4)$, $(-5+2\sqrt{3}, 4)$

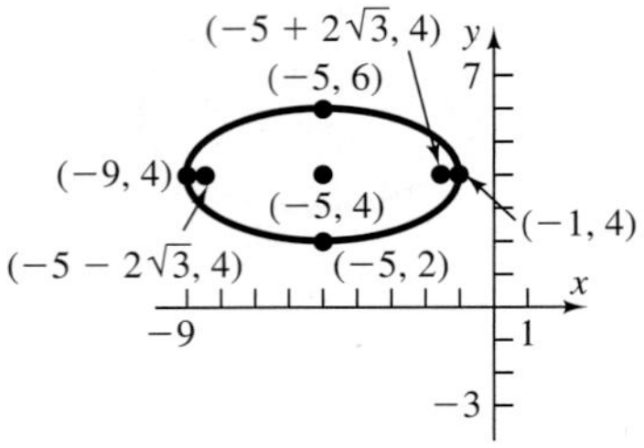

46. $\dfrac{(x-3)^2}{2}+\dfrac{(y+2)^2}{18}=1$

Center: $(3, -2)$; Vertices: $(3, -2-3\sqrt{2})$, $(3, -2+3\sqrt{2})$

Foci: $(3, -6)$, $(3, 2)$

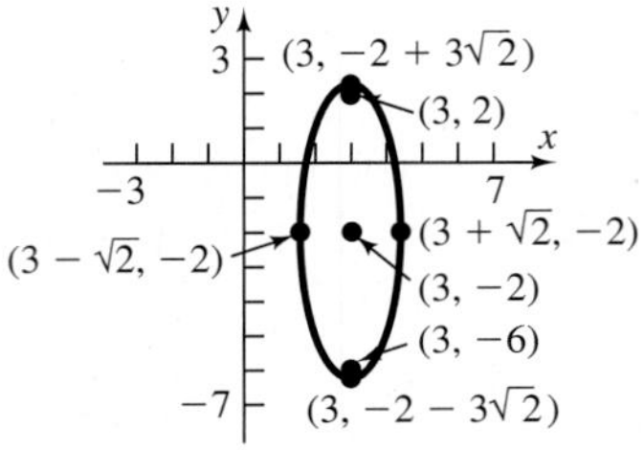

47. $\dfrac{(x+2)^2}{4}+(y-1)^2=1$

Center: $(-2, 1)$; Vertices: $(-4, 1)$, $(0, 1)$

Foci: $(-2-\sqrt{3}, 1)$, $(-2+\sqrt{3}, 1)$

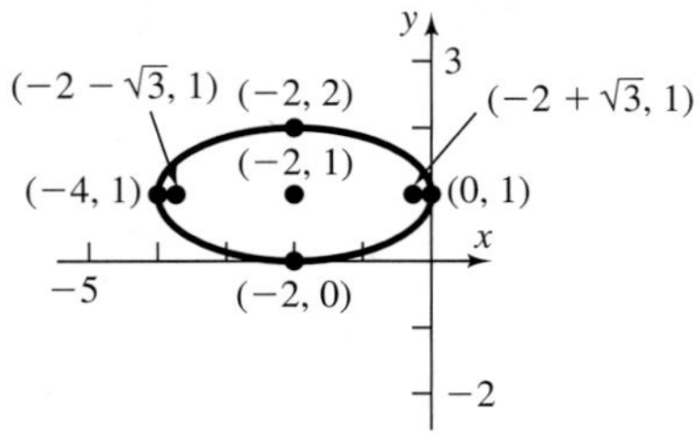

48. $\dfrac{x^2}{3}+(y-2)^2=1$

Center: $(0, 2)$; Vertices: $(-\sqrt{3}, 2)$, $(\sqrt{3}, 2)$

Foci: $(-\sqrt{2}, 2)$, $(\sqrt{2}, 2)$

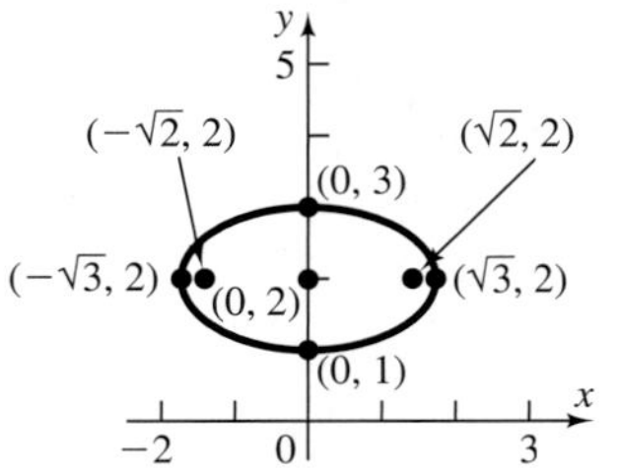

49. $\dfrac{(x-2)^2}{3}+\dfrac{(y+1)^2}{2}=1$

Center: $(2, -1)$; Vertices: $(2-\sqrt{3}, -1)$, $(2+\sqrt{3}, -1)$; Foci: $(1, -1)$, $(3, -1)$

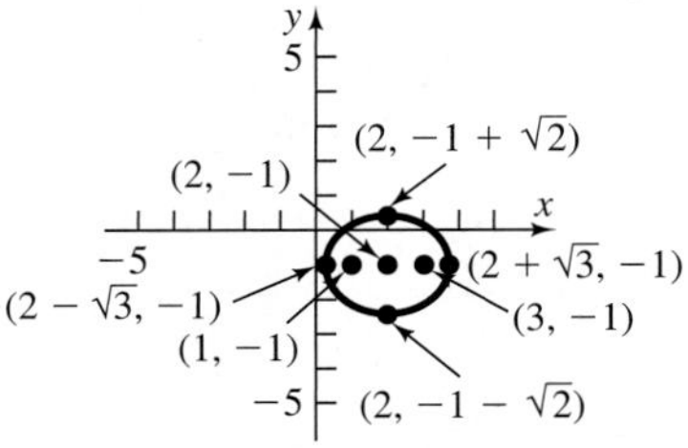

50. $\dfrac{(x+1)^2}{3}+\dfrac{(y-1)^2}{4}=1$

Center: $(-1, 1)$; Vertices: $(-1, -1)$, $(-1, 3)$

Foci: $(-1, 0)$, $(-1, 2)$

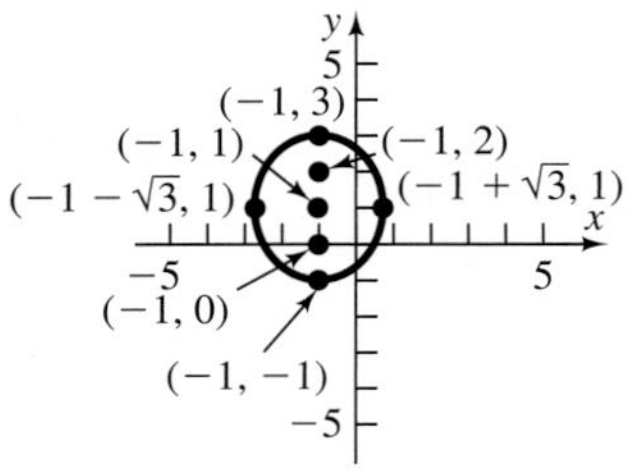

51. $\dfrac{(x-1)^2}{4}+\dfrac{(y+2)^2}{9}=1$

Center: $(1, -2)$; Vertices: $(1, -5)$, $(1, 1)$

Foci: $(1, -2-\sqrt{5})$, $(1, -2+\sqrt{5})$

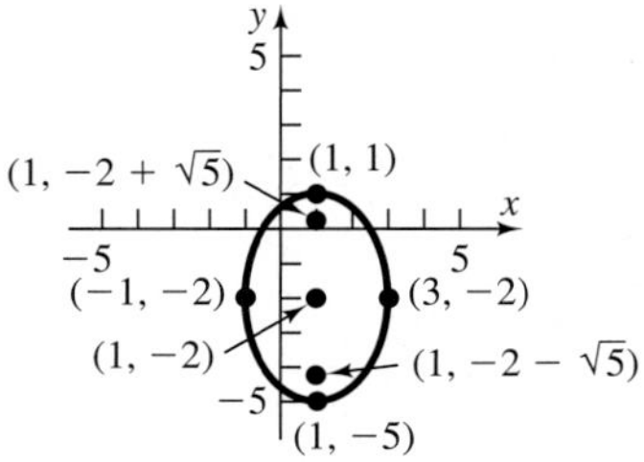

52. $\dfrac{(x+3)^2}{9}+(y-1)^2=1$

Center: $(-3, 1)$; Vertices: $(-6, 1)$, $(0, 1)$

Foci: $(-3-2\sqrt{2}, 1)$, $(-3+2\sqrt{2}, 1)$

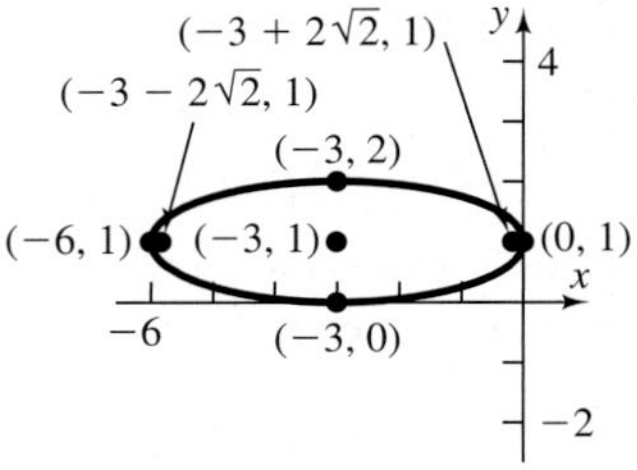

53. $x^2+\dfrac{(y+2)^2}{4}=1$

Center: $(0, -2)$; Vertices: $(0, -4)$, $(0, 0)$

Foci: $(0, -2-\sqrt{3})$, $(0, -2+\sqrt{3})$

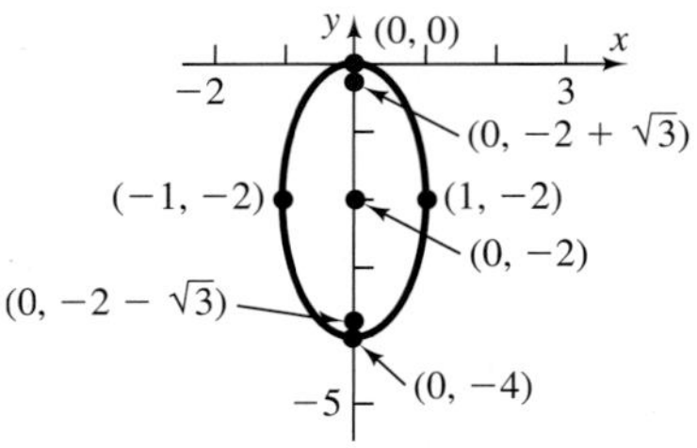

54. $(x-1)^2+\dfrac{y^2}{9}=1$

Center: $(1, 0)$; Vertices: $(1, -3)$, $(1, 3)$

Foci: $(1, -2\sqrt{2})$, $(1, 2\sqrt{2})$

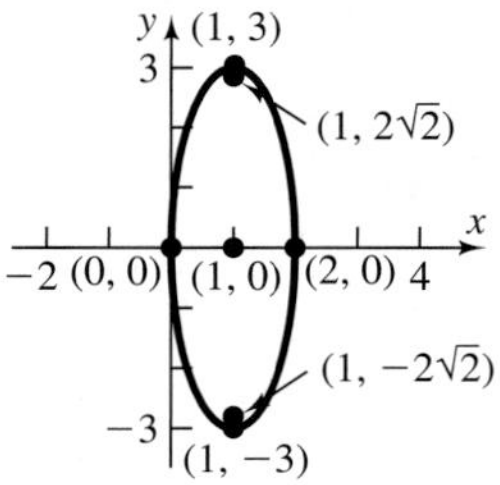

55. $\dfrac{(x-2)^2}{25}+\dfrac{(y+2)^2}{21}=1$

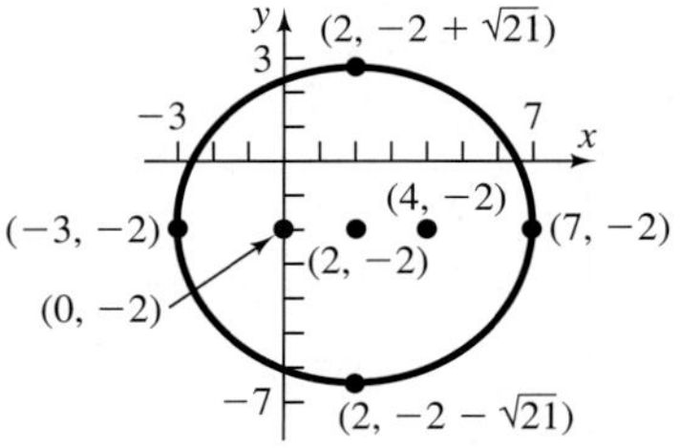

56. $\dfrac{(x+3)^2}{3} + \dfrac{(y-1)^2}{4} = 1$

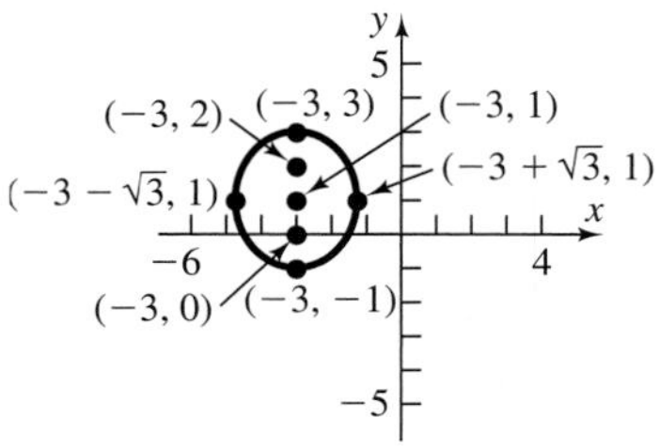

57. $\dfrac{(x-4)^2}{5} + \dfrac{(y-6)^2}{9} = 1$

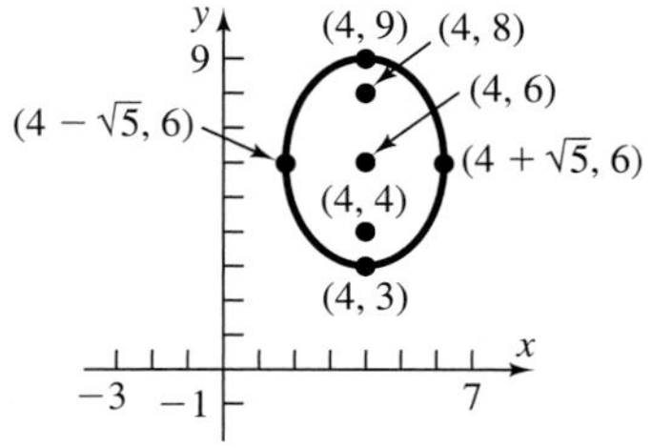

58. $\dfrac{(x+1)^2}{9} + \dfrac{(y-2)^2}{5} = 1$

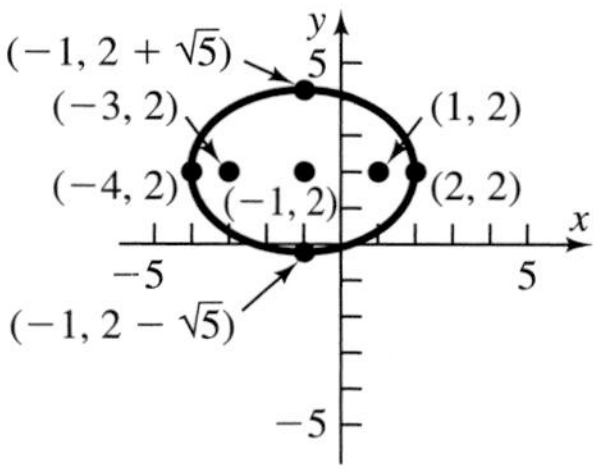

59. $\dfrac{(x-2)^2}{16} + \dfrac{(y-1)^2}{7} = 1$

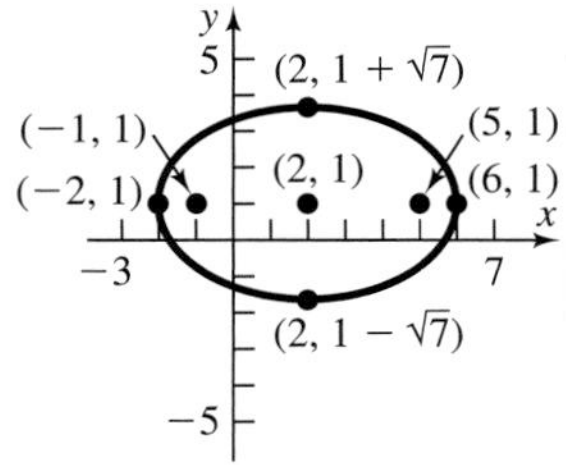

60. $\dfrac{(x-2)^2}{5} + \dfrac{(y-2)^2}{9} = 1$

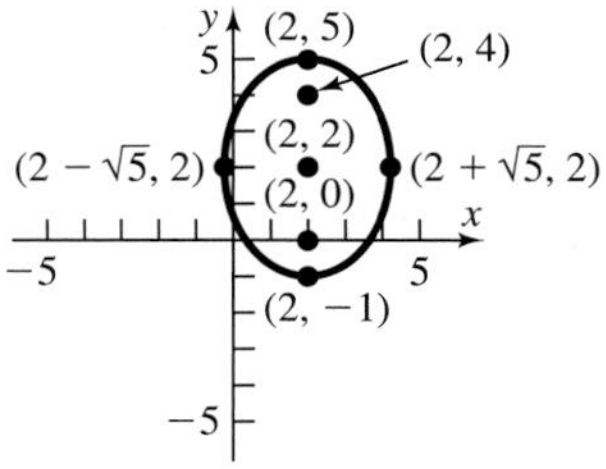

61. $\dfrac{(x-1)^2}{10} + (y-2)^2 = 1$

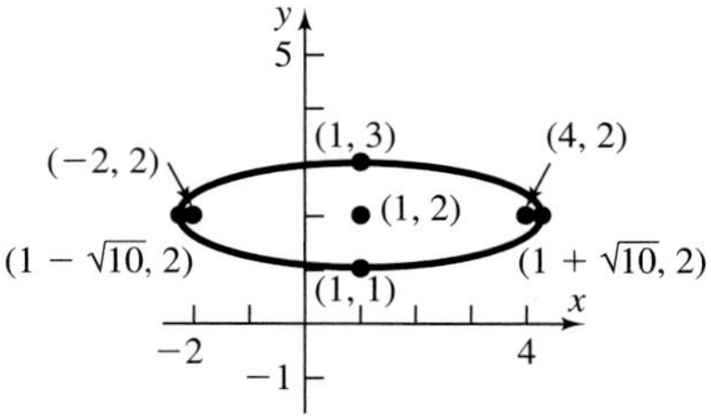

62. $(x-1)^2 + \dfrac{(y-2)^2}{5} = 1$

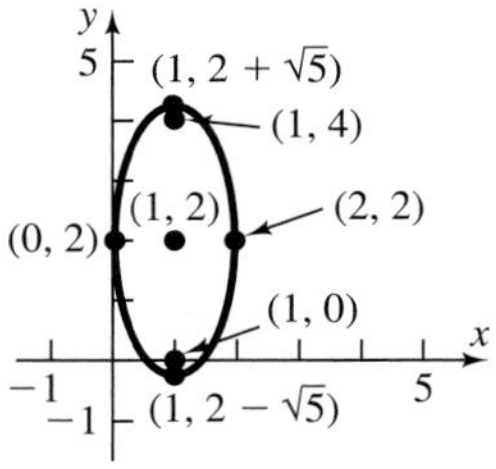

63. $\dfrac{(x-1)^2}{9} + (y-2)^2 = 1$

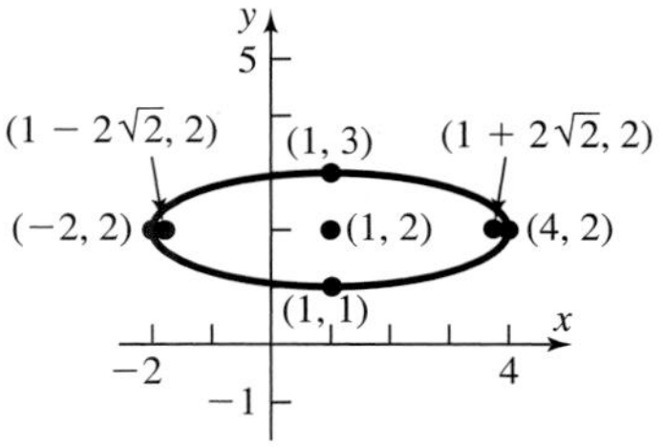

64. $(x-1)^2 + \dfrac{(y-2)^2}{4} = 1$

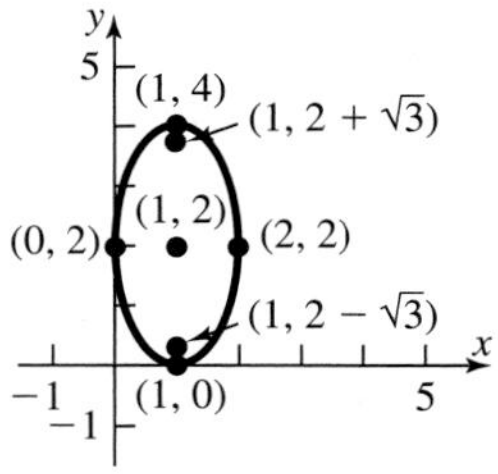

65.

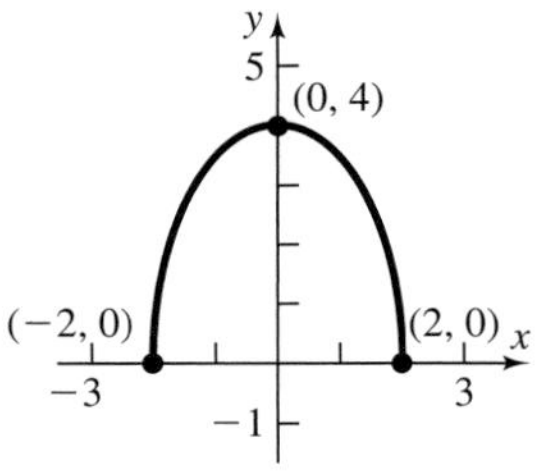

66.

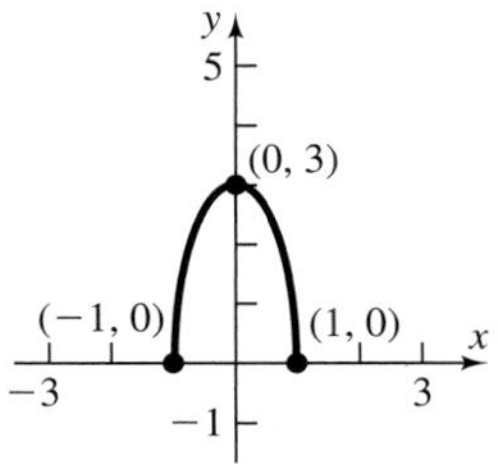

67.

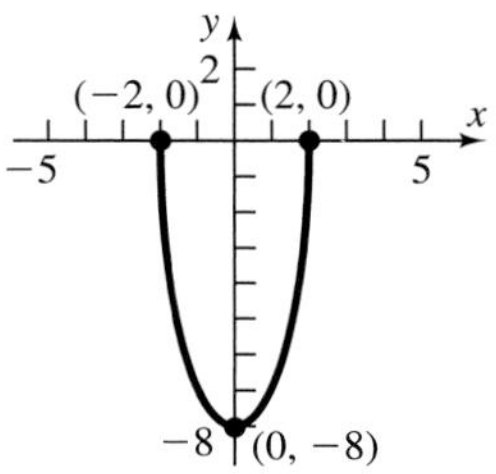

68.

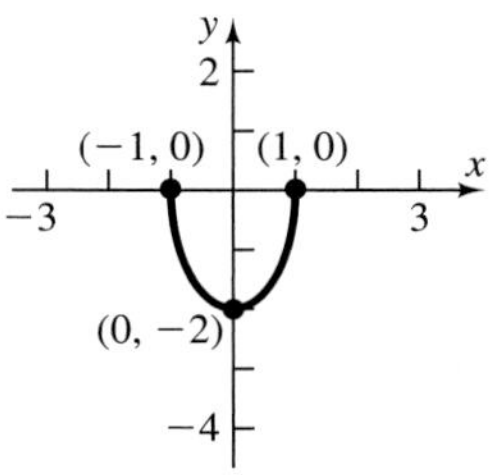

69. $\dfrac{x^2}{100} + \dfrac{y^2}{36} = 1$ **70.** 5 ft from center of bridge, distance is 2.57 ft; 10 ft from center, distance is 4.55 ft; 15 ft from center, distance is 12 ft
71. 43.3 ft **72.** 112 ft; 25.2 ft **73.** 24.65 ft, 21.65 ft, 13.82 ft **74.** 16.67 ft **75.** 0 ft, 12.99 ft, 15 ft, 12.99 ft, 0 ft **76.** 73.69 ft
77. 91.5 million mi; $\dfrac{x^2}{(93)^2} + \dfrac{y^2}{8646.75} = 1$ **78.** 155.5 million mi; $\dfrac{x^2}{(142)^2} + \dfrac{y^2}{19{,}981.75} = 1$

79. perihelion: 460.6 million mi; mean distance: 483.8 million mi; $\dfrac{x^2}{(483.8)^2} + \dfrac{y^2}{233{,}524.2} = 1$

80. aphelion: 6346 million mi; mean distance: 5448.5 million mi; $\dfrac{x^2}{(5448.5)^2} + \dfrac{y^2}{28{,}880{,}646} = 1$ **81.** 30 ft **82.** $10\sqrt{7}$ ft

83. (a) $Ax^2 + Cy^2 + F = 0$
$Ax^2 + Cy^2 = -F$

If A and C are of the same and F is of opposite sign, then the equation takes the form $\dfrac{x^2}{\left(-\dfrac{F}{A}\right)} + \dfrac{y^2}{\left(-\dfrac{F}{C}\right)} = 1$, where $-\dfrac{F}{A}$ and $-\dfrac{F}{C}$ are positive. This is the equation of an ellipse with center at (0, 0).

(b) If $A = C$, the equation may be written as $x^2 + y^2 = -\dfrac{F}{A}$.

This is the equation of a circle with center at (0, 0) and radius equal to $\sqrt{-\dfrac{F}{A}}$.

84. $Ax^2 + Cy^2 + Dx + Ey + F = 0$ $\quad A \neq 0, C \neq 0$

$Ax^2 + Dx + Cy^2 + Ey = -F$

$A\left(x^2 + \dfrac{D}{A}x\right) + C\left(y^2 + \dfrac{E}{C}y\right) = -F$

$A\left(x + \dfrac{D}{2A}\right)^2 + C\left(y + \dfrac{E}{2C}\right)^2 = -F + \dfrac{D^2}{4A} + \dfrac{E^2}{4C}$

(a) If $\dfrac{D^2}{4A} + \dfrac{E^2}{4C} - F$ is of the same sign as A (and C), this is the equation of an ellipse with center at $\left(\dfrac{-D}{2A}, \dfrac{-E}{2C}\right)$.

(b) If $\dfrac{D^2}{4A} + \dfrac{E^2}{4C} - F = 0$, the graph is the single point $\left(\dfrac{-D}{2A}, \dfrac{-E}{2C}\right)$.

(c) If $\dfrac{D^2}{4A} + \dfrac{E^2}{4C} - F$ is of the opposite sign to A (and C), the graph contains no points since, in this case, the left side has a sign opposite that of the right side.

9.4 Concepts and Vocabulary *(page 647)*

7. hyperbola **8.** transverse axis **9.** $3x = -2y, 3x = 2y$ **10.** False **11.** True **12.** False

9.4 Exercises *(page 647)*

13. B **14.** C **15.** A **16.** D

17. $x^2 - \dfrac{y^2}{8} = 1$

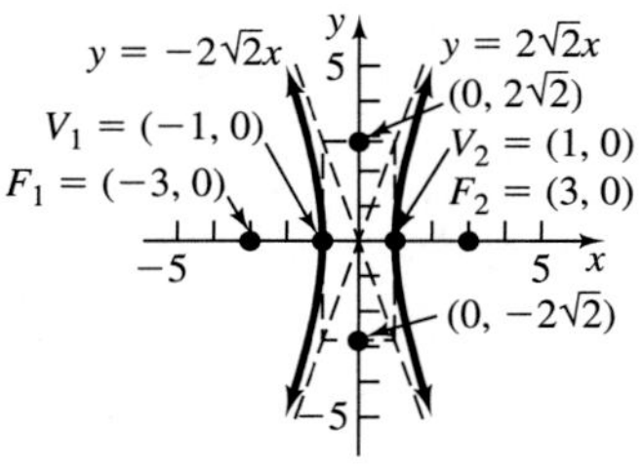

18. $\dfrac{y^2}{9} - \dfrac{x^2}{16} = 1$

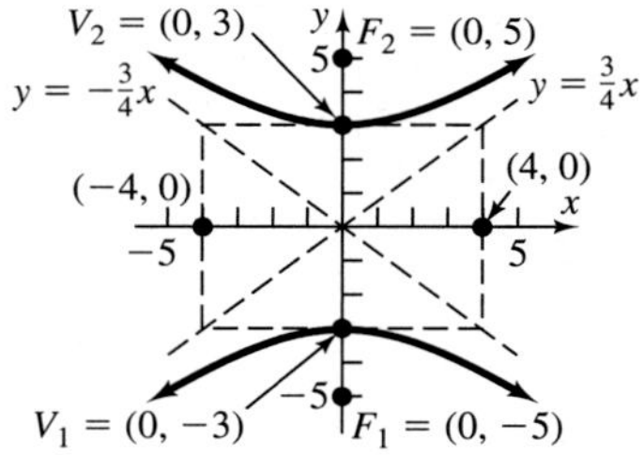

19. $\dfrac{y^2}{16} - \dfrac{x^2}{20} = 1$

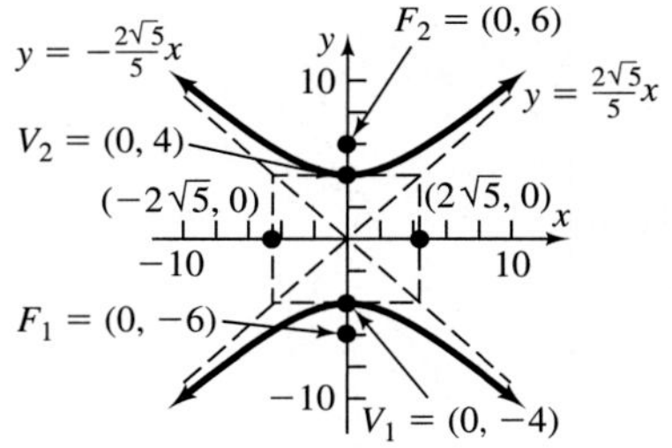

20. $\dfrac{x^2}{4} - \dfrac{y^2}{5} = 1$

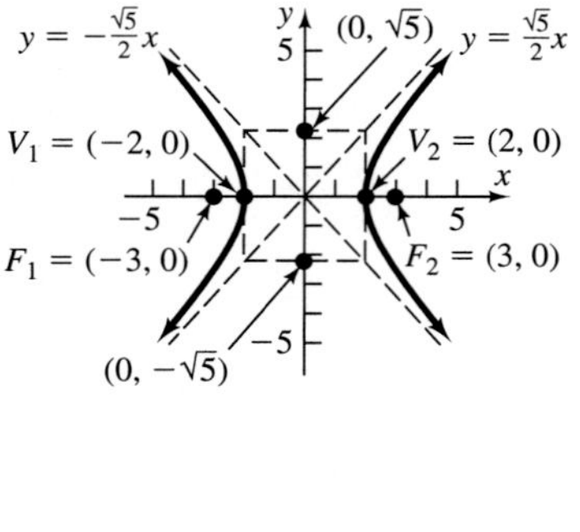

21. $\dfrac{x^2}{9} - \dfrac{y^2}{16} = 1$

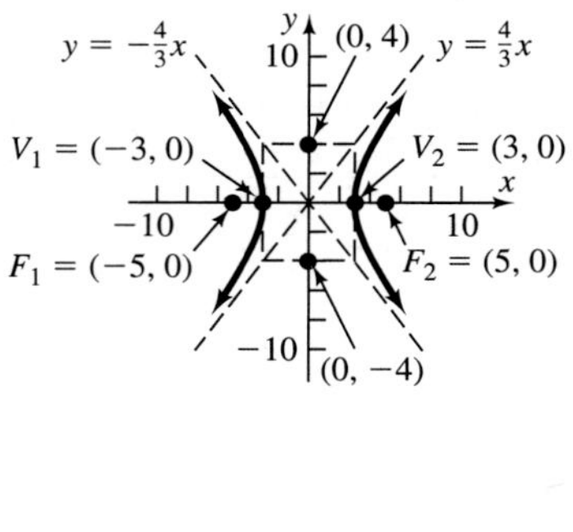

22. $\dfrac{y^2}{4} - \dfrac{x^2}{32} = 1$

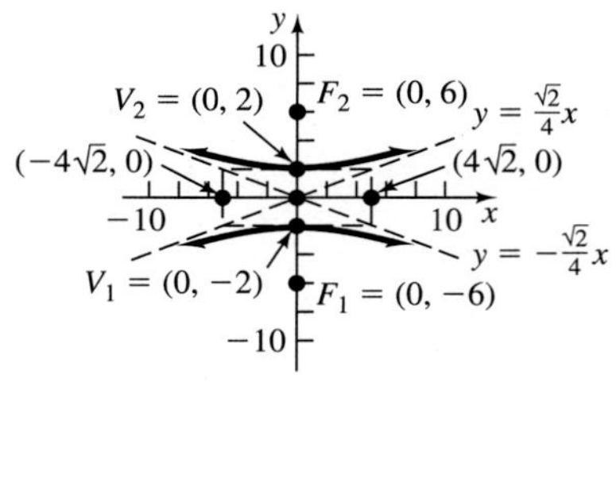

23. $\frac{y^2}{36} - \frac{x^2}{9} = 1$

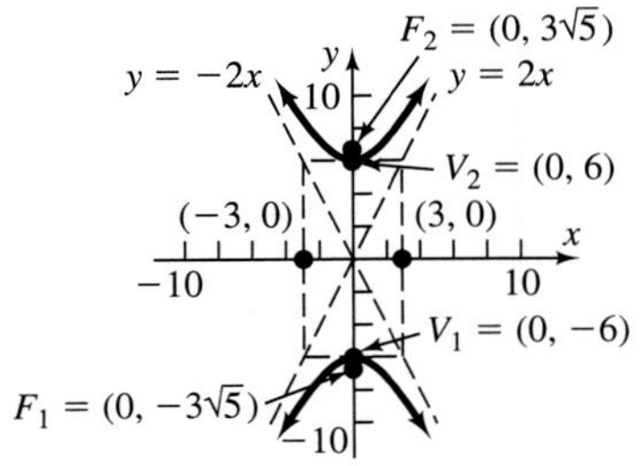

24. $\frac{x^2}{16} - \frac{y^2}{64} = 1$

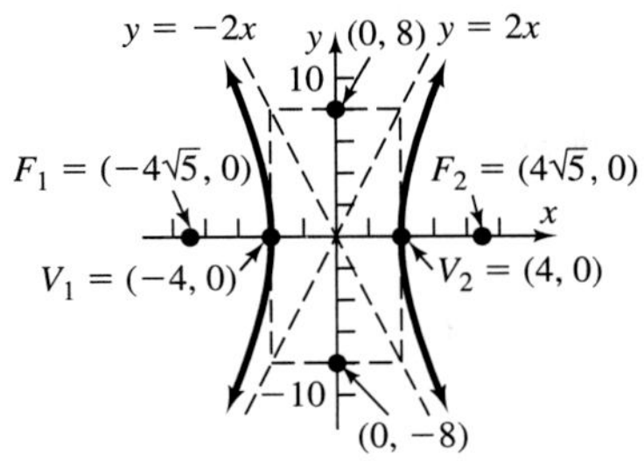

25. $\frac{x^2}{8} - \frac{y^2}{8} = 1$

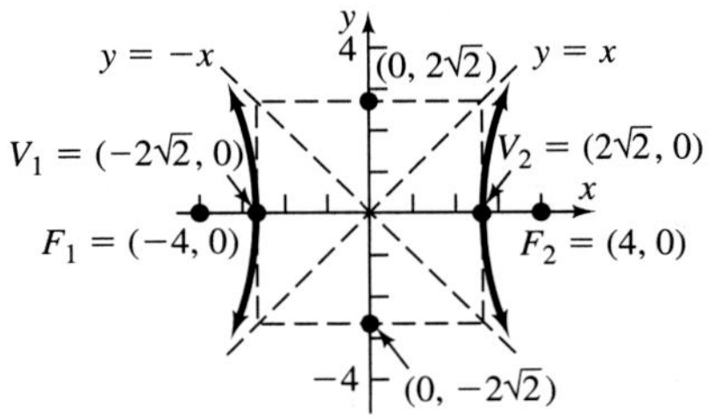

26. $\frac{y^2}{2} - \frac{x^2}{2} = 1$

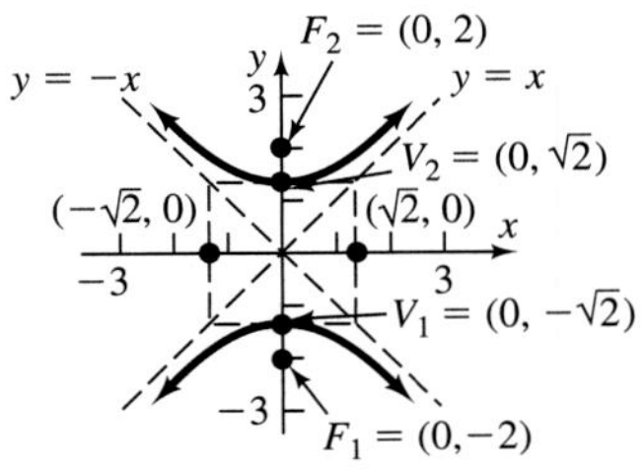

27. $\frac{x^2}{25} - \frac{y^2}{9} = 1$

Center: $(0, 0)$

Transverse axis: x-axis

Vertices: $(-5, 0)$, $(5, 0)$

Foci: $(-\sqrt{34}, 0)$, $(\sqrt{34}, 0)$

Asymptotes: $y = \pm\frac{3}{5}x$

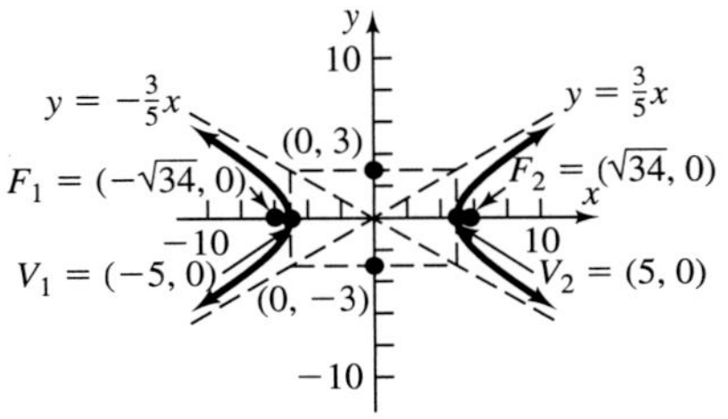

28. $\frac{y^2}{16} - \frac{x^2}{4} = 1$

Center: $(0, 0)$

Transverse axis: y-axis

Vertices: $(0, -4)$, $(0, 4)$

Foci: $(0, -2\sqrt{5})$, $(0, 2\sqrt{5})$

Asymptotes: $y = \pm 2x$

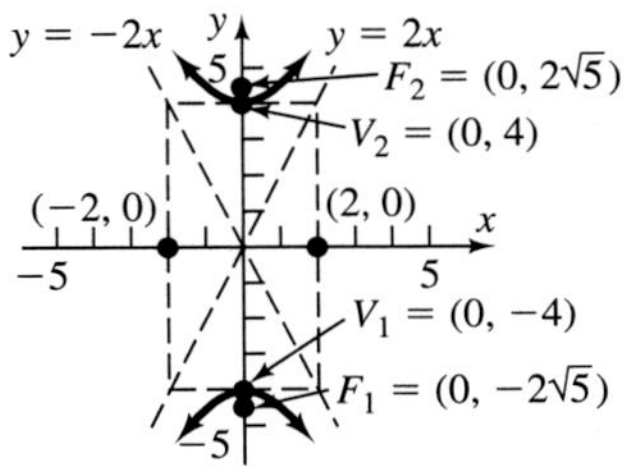

29. $\frac{x^2}{4} - \frac{y^2}{16} = 1$

Center: $(0, 0)$

Transverse axis: x-axis

Vertices: $(-2, 0)$, $(2, 0)$

Foci: $(-2\sqrt{5}, 0)$, $(2\sqrt{5}, 0)$

Asymptotes: $y = \pm 2x$

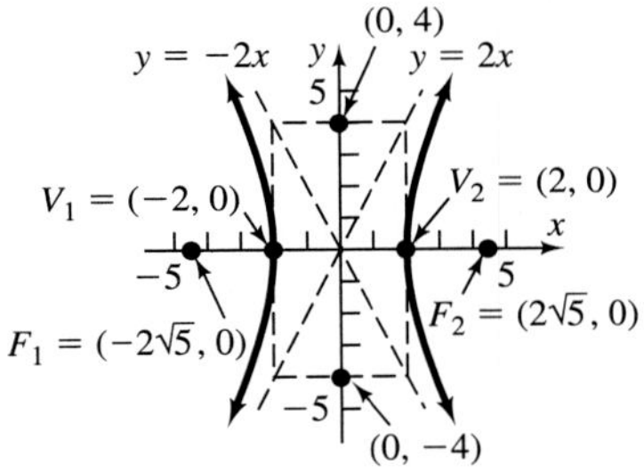

30. $\frac{y^2}{4} - \frac{x^2}{16} = 1$

Center: $(0, 0)$

Transverse axis: y-axis

Vertices: $(0, -2)$, $(0, 2)$

Foci: $(0, -2\sqrt{5})$, $(0, 2\sqrt{5})$

Asymptotes: $y = \pm\frac{1}{2}x$

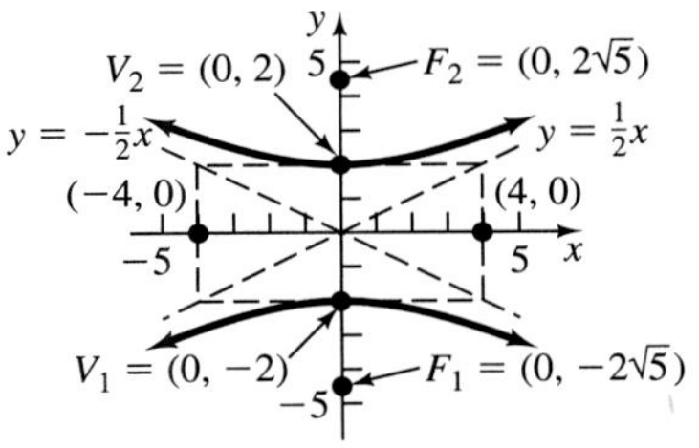

31. $\frac{y^2}{9} - x^2 = 1$

Center: $(0, 0)$

Transverse axis: y-axis

Vertices: $(0, -3)$, $(0, 3)$

Foci: $(0, -\sqrt{10})$, $(0, \sqrt{10})$

Asymptotes: $y = \pm 3x$

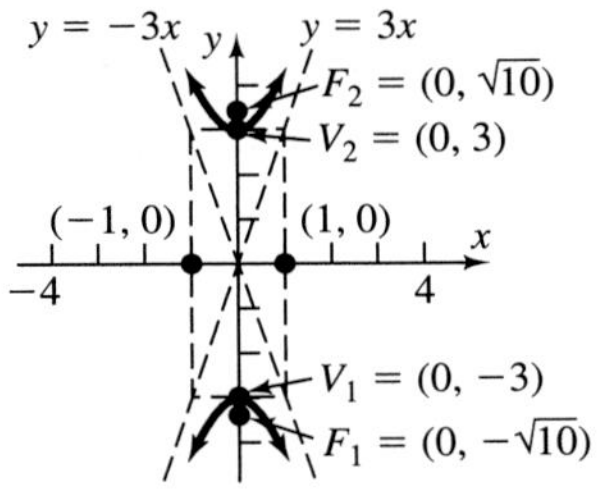

32. $\frac{x^2}{4} - \frac{y^2}{4} = 1$
Center: (0, 0)
Transverse axis: x-axis
Vertices: $(-2, 0)$, $(2, 0)$
Foci: $(-2\sqrt{2}, 0)$, $(2\sqrt{2}, 0)$
Asymptotes: $y = \pm x$

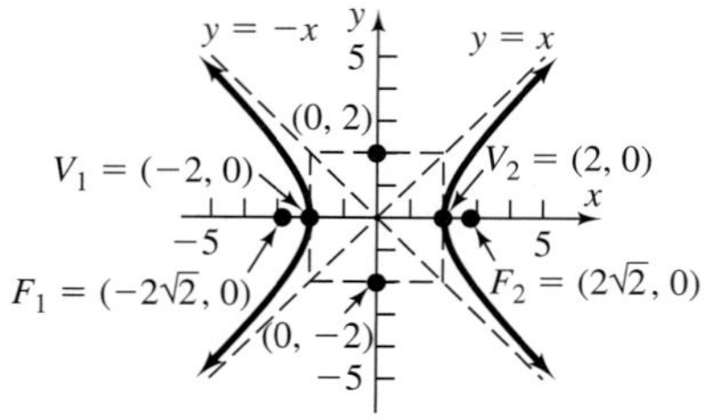

33. $\frac{y^2}{25} - \frac{x^2}{25} = 1$
Center: (0, 0)
Transverse axis: y-axis
Vertices: $(0, -5)$, $(0, 5)$
Foci: $(0, -5\sqrt{2})$, $(0, 5\sqrt{2})$
Asymptotes: $y = \pm x$

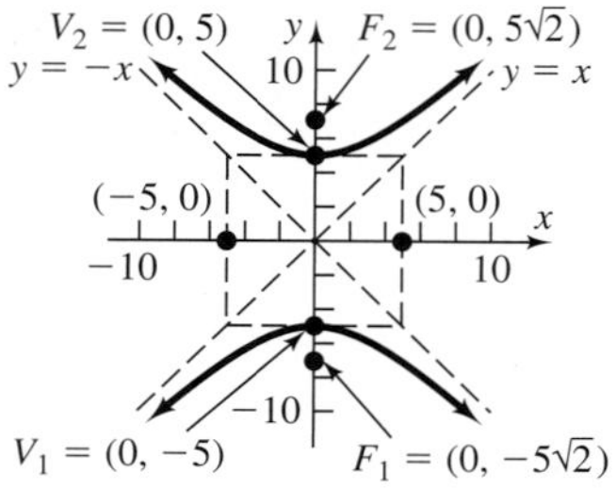

34. $\frac{x^2}{2} - \frac{y^2}{4} = 1$
Center: (0, 0)
Transverse axis: x-axis
Vertices: $(-\sqrt{2}, 0)$, $(\sqrt{2}, 0)$
Foci: $(-\sqrt{6}, 0)$, $(\sqrt{6}, 0)$
Asymptotes: $y = \pm\sqrt{2}x$

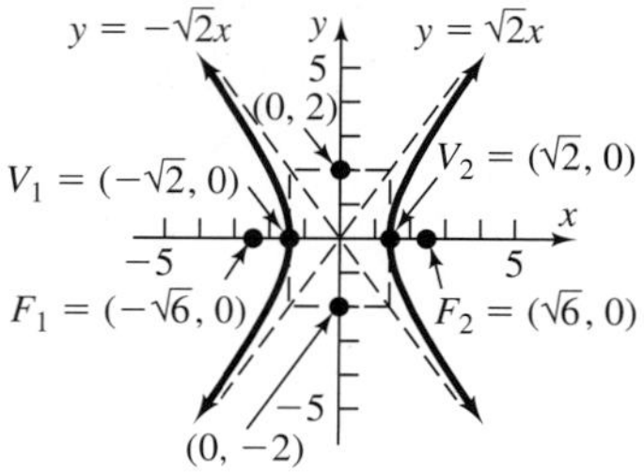

35. $x^2 - y^2 = 1$ **36.** $y^2 - x^2 = 1$ **37.** $\frac{y^2}{36} - \frac{x^2}{9} = 1$ **38.** $\frac{x^2}{4} - \frac{y^2}{16} = 1$

39. $\frac{(x-4)^2}{4} - \frac{(y+1)^2}{5} = 1$

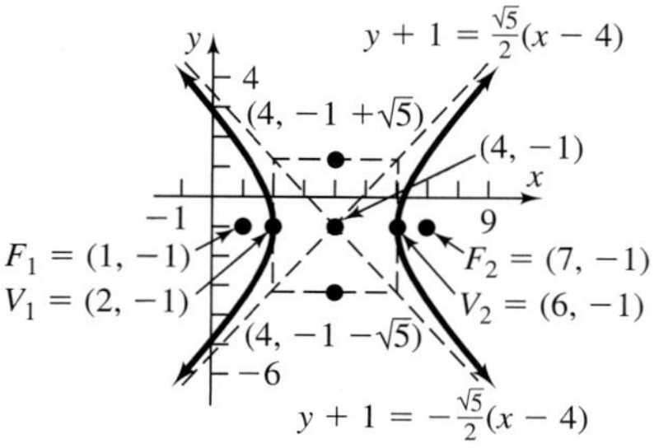

40. $\frac{(y-1)^2}{9} - \frac{(x+3)^2}{16} = 1$

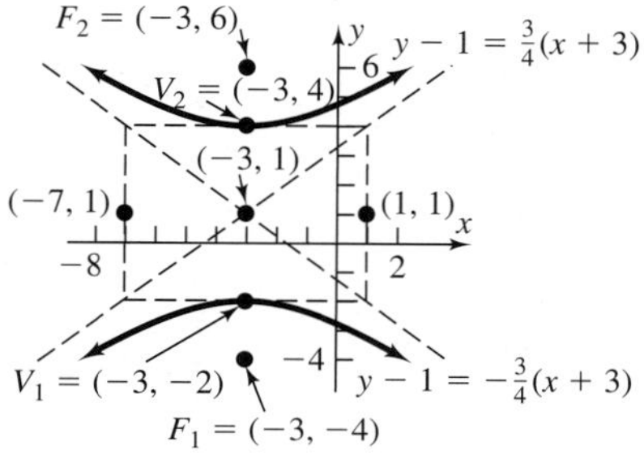

41. $\frac{(y+4)^2}{4} - \frac{(x+3)^2}{12} = 1$

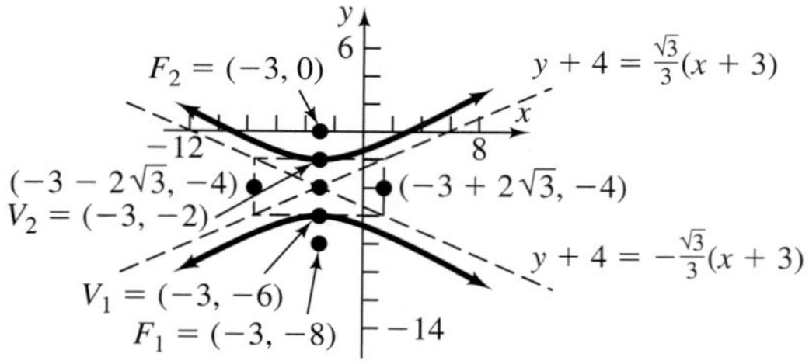

42. $(x-1)^2 - \frac{(y-4)^2}{8} = 1$

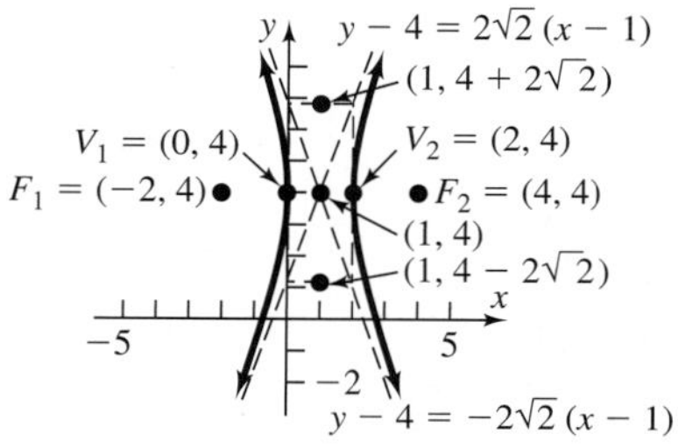

43. $(x-5)^2 - \frac{(y-7)^2}{3} = 1$

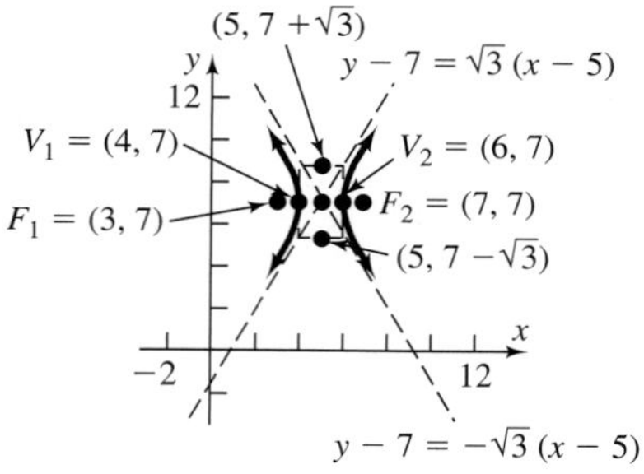

44. $(y-3)^2 - \frac{(x+4)^2}{8} = 1$

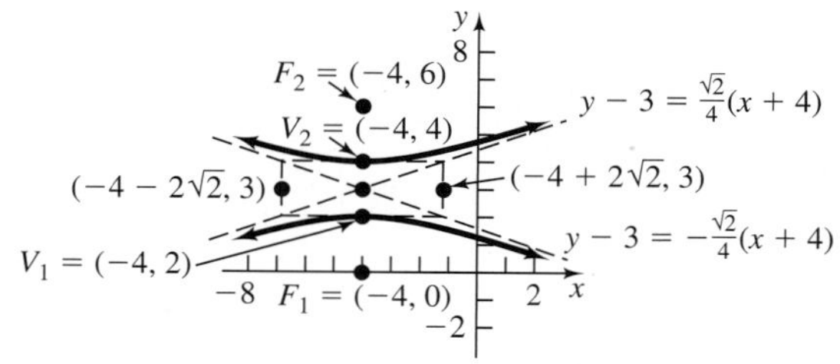

45. $\dfrac{(x-1)^2}{4} - \dfrac{(y+1)^2}{9} = 1$

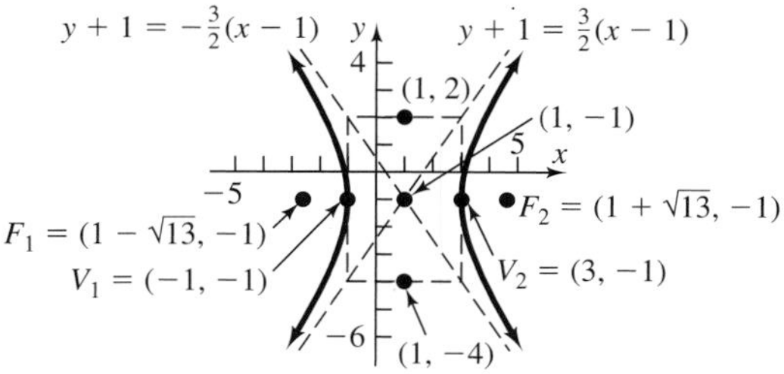

46. $\dfrac{(y+1)^2}{4} - \dfrac{9(x-1)^2}{16} = 1$

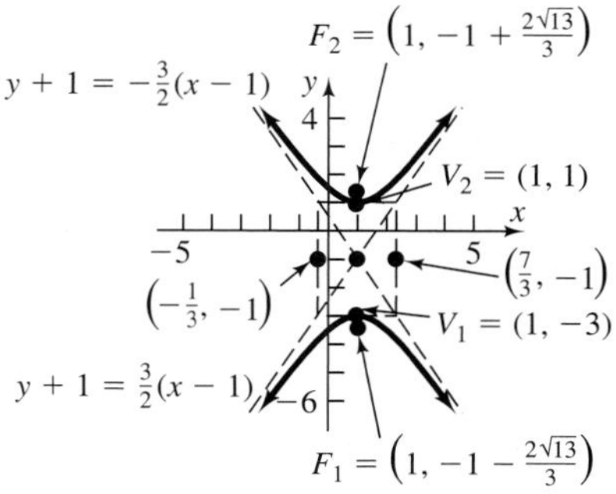

47. $\dfrac{(x-2)^2}{4} - \dfrac{(y+3)^2}{9} = 1$

Center: $(2, -3)$
Transverse axis: Parallel to x-axis
Vertices: $(0, -3)$, $(4, -3)$
Foci: $(2 - \sqrt{13}, -3)$, $(2 + \sqrt{13}, -3)$
Asymptotes: $y + 3 = \pm\dfrac{3}{2}(x - 2)$

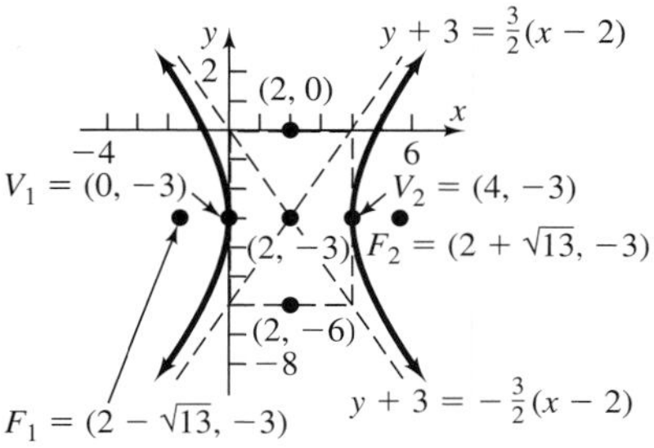

48. $\dfrac{(y+3)^2}{4} - \dfrac{(x-2)^2}{9} = 1$

Center: $(2, -3)$
Transverse axis: Parallel to y-axis
Vertices: $(2, -5)$, $(2, -1)$
Foci: $(2, -3 - \sqrt{13})$, $(2, -3 + \sqrt{13})$
Asymptotes: $y + 3 = \pm\dfrac{2}{3}(x - 2)$

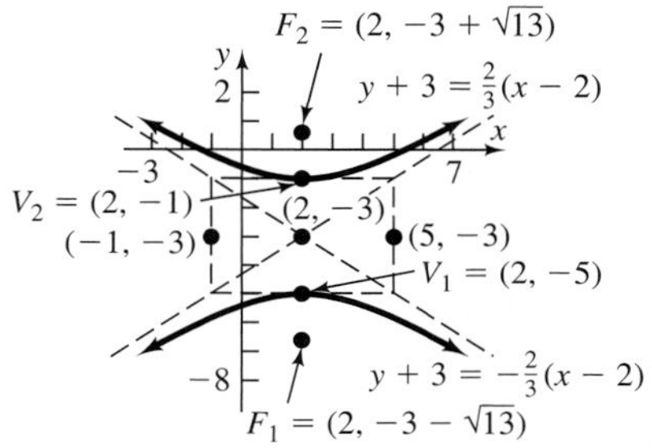

49. $\dfrac{(y-2)^2}{4} - (x+2)^2 = 1$

Center: $(-2, 2)$
Transverse axis: Parallel to y-axis
Vertices: $(-2, 0)$, $(-2, 4)$
Foci: $(-2, 2 - \sqrt{5})$, $(-2, 2 + \sqrt{5})$
Asymptotes: $y - 2 = \pm 2(x + 2)$

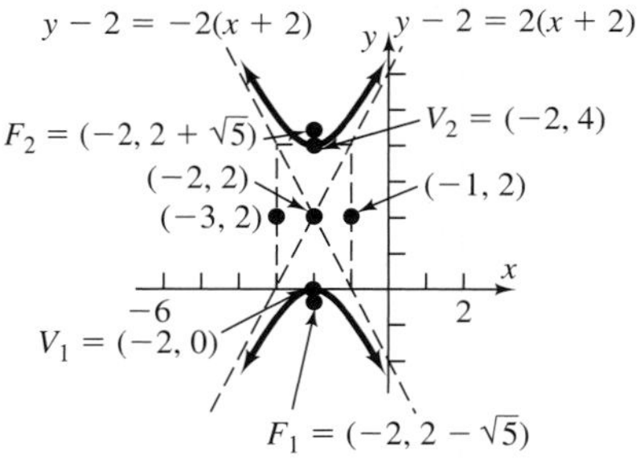

50. $\dfrac{(x+4)^2}{9} - (y-3)^2 = 1$

Center: $(-4, 3)$
Transverse axis: Parallel to x-axis
Vertices: $(-7, 3)$, $(-1, 3)$
Foci: $(-4 - \sqrt{10}, 3)$, $(-4 + \sqrt{10}, 3)$
Asymptotes: $y - 3 = \pm\dfrac{1}{3}(x + 4)$

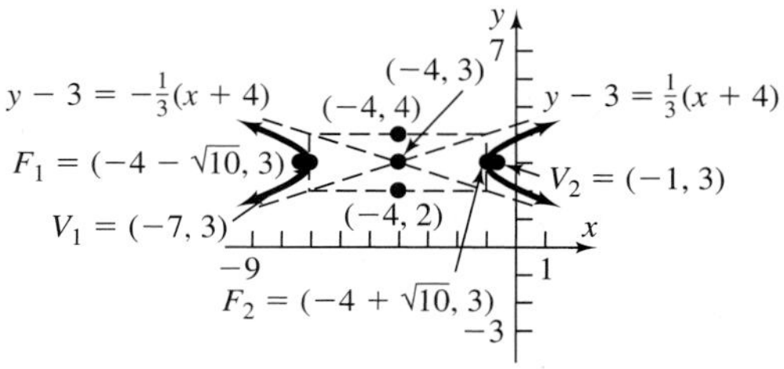

51. $\dfrac{(x+1)^2}{4} - \dfrac{(y+2)^2}{4} = 1$

Center: $(-1, -2)$

Transverse axis: Parallel to x-axis

Vertices: $(-3, -2)$, $(1, -2)$

Foci: $(-1 - 2\sqrt{2}, -2)$, $(-1 + 2\sqrt{2}, -2)$

Asymptotes: $y + 2 = \pm(x + 1)$

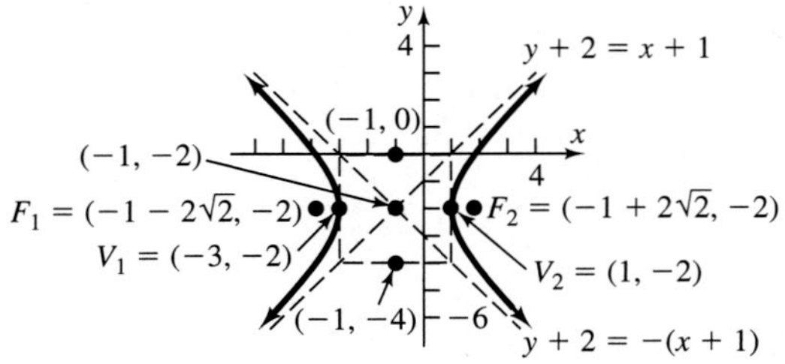

52. $\dfrac{(y-3)^2}{4} - \dfrac{(x+2)^2}{4} = 1$

Center: $(-2, 3)$

Transverse axis: Parallel to y-axis

Vertices: $(-2, 1)$, $(-2, 5)$

Foci: $(-2, 3 - 2\sqrt{2})$, $(-2, 3 + 2\sqrt{2})$

Asymptotes: $y - 3 = \pm(x + 2)$

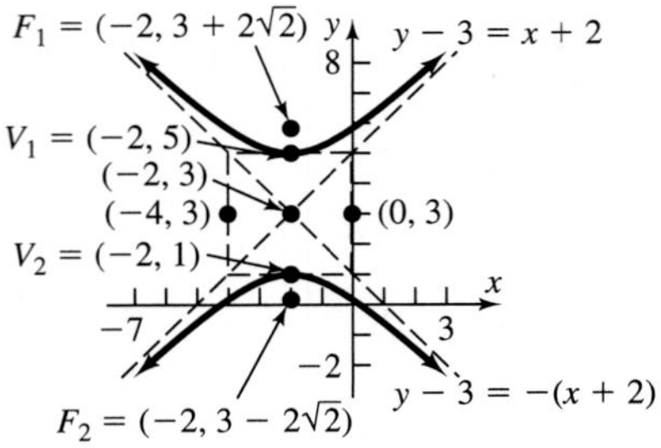

53. $(x-1)^2 - (y+1)^2 = 1$

Center: $(1, -1)$

Transverse axis: Parallel to x-axis

Vertices: $(0, -1)$, $(2, -1)$

Foci: $(1 - \sqrt{2}, -1)$, $(1 + \sqrt{2}, -1)$

Asymptotes: $y + 1 = \pm(x - 1)$

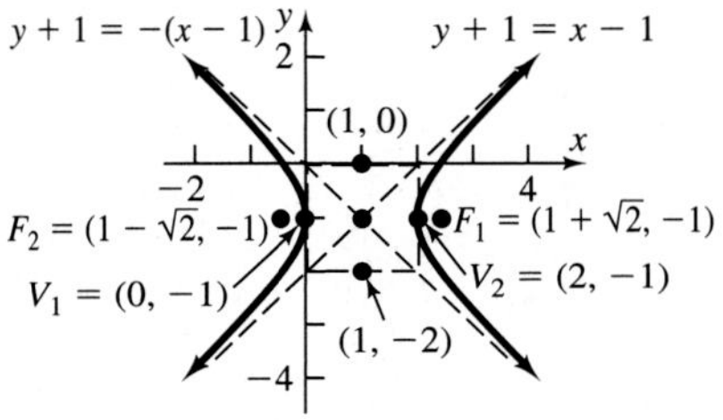

54. $(y-2)^2 - (x-2)^2 = 1$

Center: $(2, 2)$

Transverse axis: Parallel to y-axis

Vertices: $(2, 1)$, $(2, 3)$

Foci: $(2, 2 - \sqrt{2})$, $(2, 2 + \sqrt{2})$

Asymptotes: $y - 2 = \pm(x - 2)$

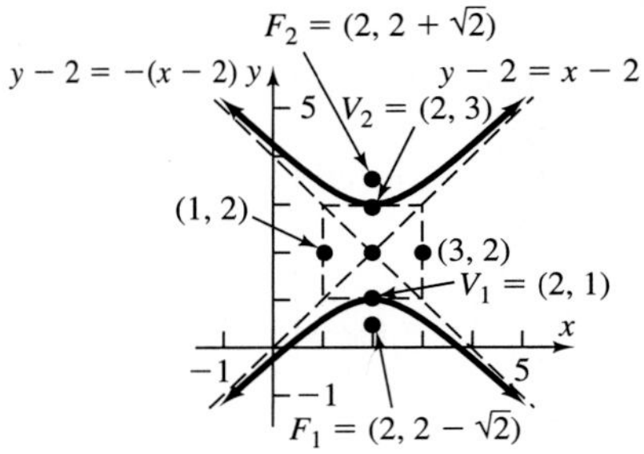

55. $\dfrac{(y-2)^2}{4} - (x+1)^2 = 1$

Center: $(-1, 2)$

Transverse axis: Parallel to y-axis

Vertices: $(-1, 0)$, $(-1, 4)$

Foci: $(-1, 2 - \sqrt{5})$, $(-1, 2 + \sqrt{5})$

Asymptotes: $y - 2 = \pm 2(x + 1)$

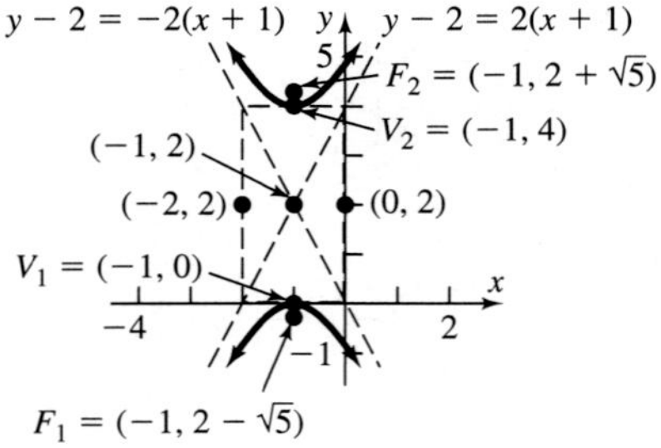

56. $(x+1)^2 - \dfrac{(y-2)^2}{2} = 1$

Center: $(-1, 2)$

Transverse axis: Parallel to x-axis

Vertices: $(-2, 2)$, $(0, 2)$

Foci: $(-1 - \sqrt{3}, 2)$, $(-1 + \sqrt{3}, 2)$

Asymptotes: $y - 2 = \pm\sqrt{2}(x + 1)$

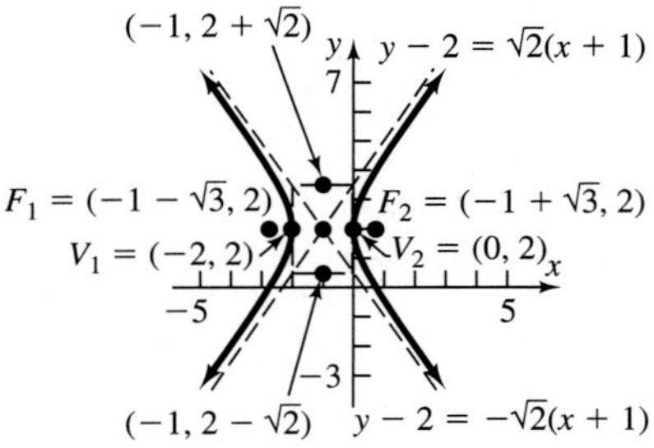

57. $\dfrac{(x-3)^2}{4} - \dfrac{(y+2)^2}{16} = 1$

Center: $(3, -2)$
Transverse axis: Parallel to x-axis
Vertices: $(1, -2)$, $(5, -2)$
Foci: $(3 - 2\sqrt{5}, -2)$, $(3 + 2\sqrt{5}, -2)$
Asymptotes: $y + 2 = \pm 2(x - 3)$

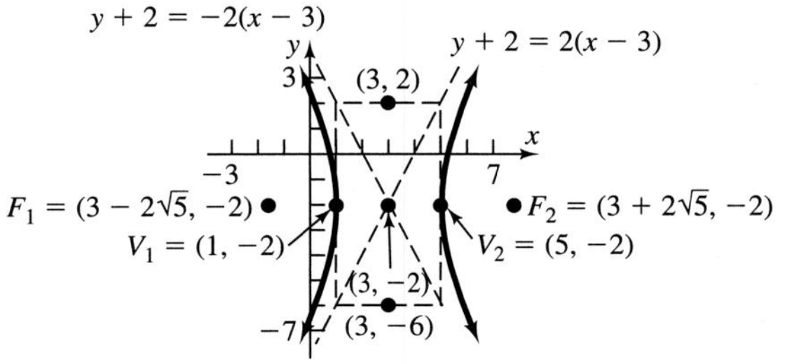

58. $\dfrac{(y+2)^2}{2} - \dfrac{(x-1)^2}{4} = 1$

Center: $(1, -2)$
Transverse axis: Parallel to y-axis
Vertices: $(1, -2 - \sqrt{2})$, $(1, -2 + \sqrt{2})$
Foci: $(1, -2 - \sqrt{6})$, $(1, -2 + \sqrt{6})$
Asymptotes: $y + 2 = \pm \dfrac{\sqrt{2}}{2}(x - 1)$

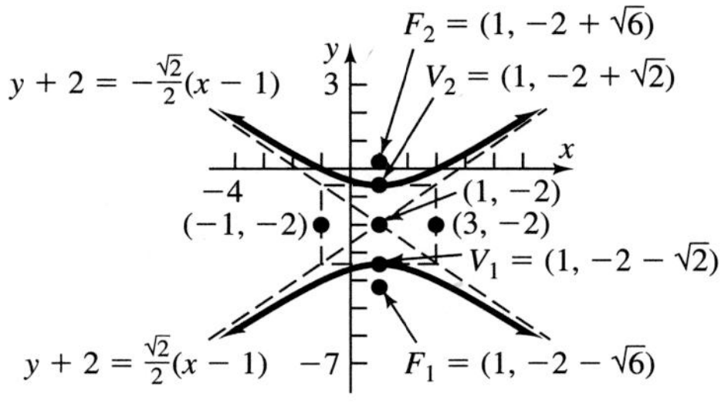

59. $\dfrac{(y-1)^2}{4} - (x+2)^2 = 1$

Center: $(-2, 1)$
Transverse axis: Parallel to y-axis
Vertices: $(-2, -1)$, $(-2, 3)$
Foci: $(-2, 1 - \sqrt{5})$, $(-2, 1 + \sqrt{5})$
Asymptotes: $y - 1 = \pm 2(x + 2)$

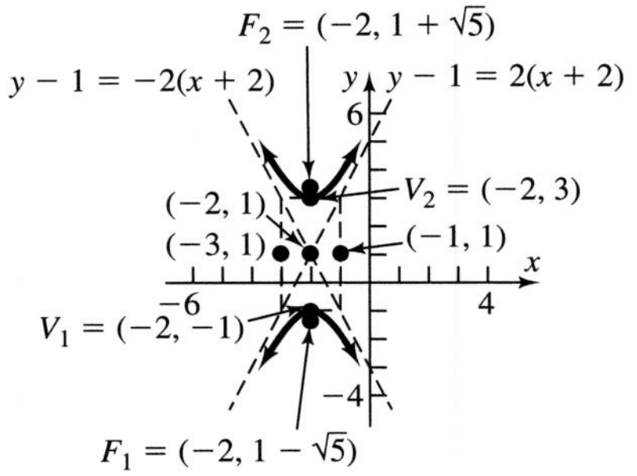

60. $\dfrac{(x+4)^2}{9} - \dfrac{(y+1)^2}{3} = 1$

Center: $(-4, -1)$
Transverse axis: Parallel to x-axis
Vertices: $(-7, -1)$, $(-1, -1)$
Foci: $(-4 - 2\sqrt{3}, -1)$, $(-4 + 2\sqrt{3}, -1)$
Asymptotes: $y + 1 = \pm \dfrac{\sqrt{3}}{3}(x + 4)$

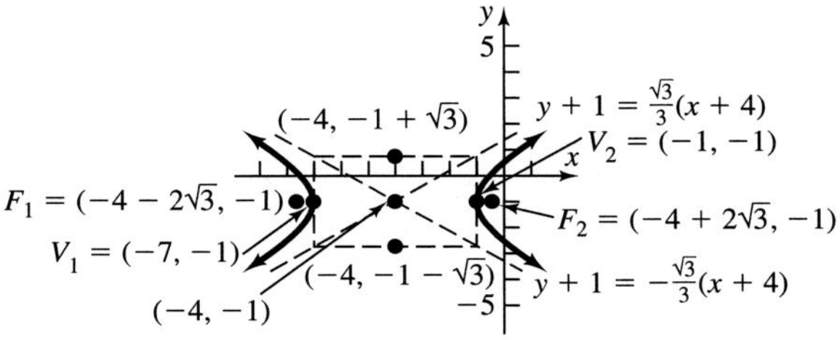

61.

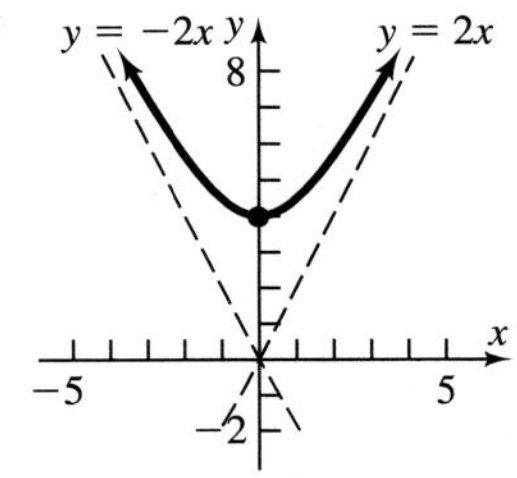

62.

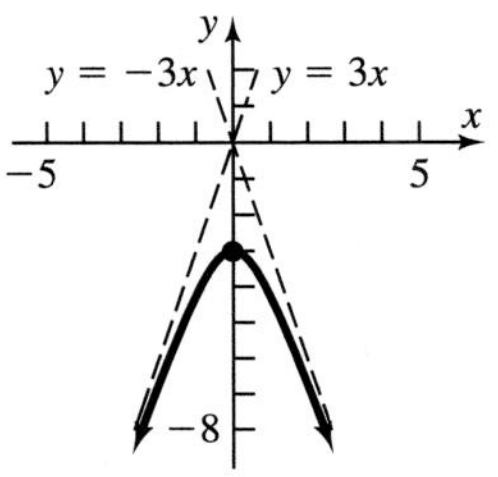

63.

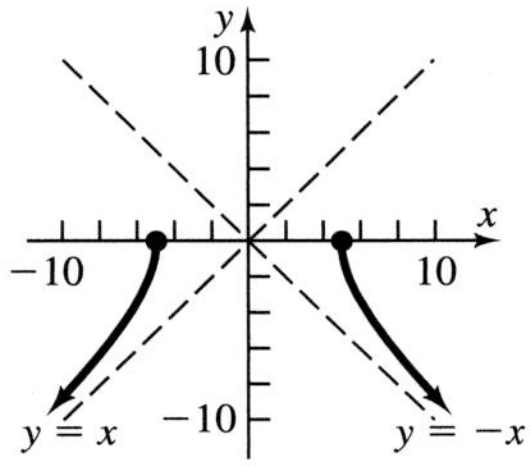

64.

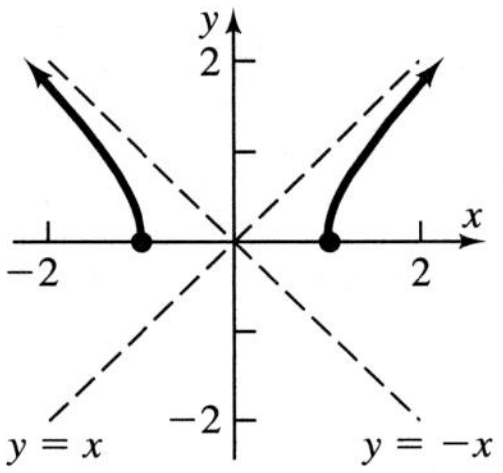

65. **(a)** The ship will reach shore at a point 64.66 mi from the master station. **(b)** 0.00086 sec **(c)** $(104, 50)$
66. **(a)** The ship would reach shore 20.24 mi from the master station. **(b)** 0.0004301 sec **(c)** $(48, 20)$
67. **(a)** 450 ft **69.** If e is close to 1, narrow hyperbola; if e is very large, wide hyperbola **70.** $\sqrt{2}$

71. $\frac{x^2}{4} - y^2 = 1$: asymptotes $y = \pm\frac{1}{2}x$; $y^2 - \frac{x^2}{4} = 1$: asymptotes $y = \pm\frac{1}{2}x$

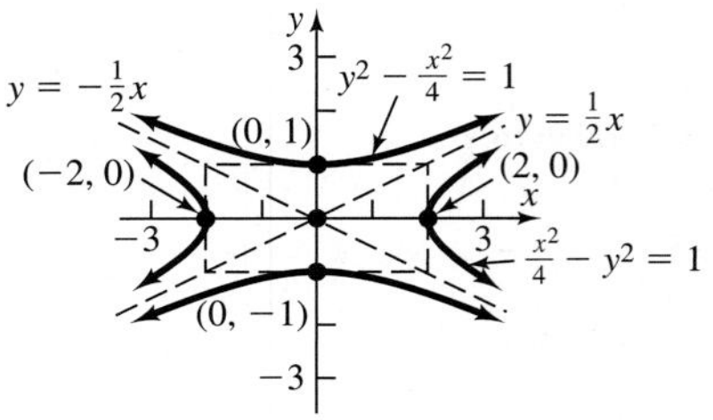

72. $\frac{y^2}{a^2} - \frac{x^2}{b^2} = 1$

$$\frac{y^2}{a^2} = 1 + \frac{x^2}{b^2}$$

$$y^2 = a^2\left(1 + \frac{x^2}{b^2}\right)$$

$$y^2 = \frac{a^2x^2}{b^2}\left(\frac{b^2}{x^2} + 1\right)$$

$$y = \pm\frac{ax}{b}\sqrt{\frac{b^2}{x^2} + 1}$$

As $x \to -\infty$ or as $x \to \infty$, the term $\frac{b^2}{x^2}$ gets close to 0, so the expression under the radical gets closer to 1.

Thus, the graph of the hyperbola gets closer to the lines

$y = -\frac{a}{b}x$ and $y = \frac{a}{b}x$

The lines are asymptotes of the hyperbola.

73. $Ax^2 + Cy^2 + F = 0$

$Ax^2 + Cy^2 = -F$

If A and C are of opposite sign and $F \neq 0$, this equation may be written as $\frac{x^2}{\left(-\frac{F}{A}\right)} + \frac{y^2}{\left(-\frac{F}{C}\right)} = 1$, where $-\frac{F}{A}$ and $-\frac{F}{C}$ are opposite in sign. This is the equation of a hyperbola with center (0, 0).

The transverse axis is the x-axis if $-\frac{F}{A} > 0$; the transverse axis is the y-axis if $-\frac{F}{A} < 0$.

74. $Ax^2 + Cy^2 + Dx + Ey + F = 0$

where A and C are of opposite sign

$$Ax^2 + Dx + Cy^2 + Ey = -F$$

$$A\left(x^2 + \frac{D}{A}x\right) + C\left(y^2 + \frac{E}{C}y\right) = -F$$

$$A\left(x + \frac{D}{2A}\right)^2 + C\left(y + \frac{E}{2C}\right)^2 = -F + \frac{D^2}{4A} + \frac{E^2}{4C}$$

(a) If $\frac{D^2}{4A} + \frac{E^2}{4C} - F \neq 0$, this is the equation of hyperbola with center at $\left(-\frac{D}{2A}, -\frac{E}{2C}\right)$.

(b) If $\frac{D^2}{4A} + \frac{E^2}{4C} - F = 0$, the graph is intersecting lines through the point $\left(-\frac{D}{2A}, -\frac{E}{2C}\right)$ with slopes $\pm\sqrt{\left|\frac{A}{C}\right|}$.

9.5 Concepts and Vocabulary *(page 657)*

5. $\cot(2\theta) = \frac{A - C}{B}$ **6.** hyperbola **7.** ellipse **8.** True **9.** True **10.** False

9.5 Exercises *(page 657)*

11. Parabola **12.** Parabola **13.** Ellipse **14.** Ellipse **15.** Hyperbola **16.** Hyperbola **17.** Hyperbola **18.** Hyperbola **19.** Circle **20.** Circle **21.** $x = \frac{\sqrt{2}}{2}(x' - y'), y = \frac{\sqrt{2}}{2}(x' + y')$ **22.** $x = \frac{\sqrt{2}}{2}(x' - y'), y = \frac{\sqrt{2}}{2}(x' + y')$ **23.** $x = \frac{\sqrt{2}}{2}(x' - y'), y = \frac{\sqrt{2}}{2}(x' + y')$ **24.** $x = \frac{\sqrt{2}}{2}(x' - y'), y = \frac{\sqrt{2}}{2}(x' + y')$ **25.** $x = \frac{1}{2}(x' - \sqrt{3}y'), y = \frac{1}{2}(\sqrt{3}x' + y')$ **26.** $x = \frac{1}{2}(\sqrt{3}x' - y'), y = \frac{1}{2}(x' + \sqrt{3}y')$ **27.** $x = \frac{\sqrt{5}}{5}(x' - 2y'), y = \frac{\sqrt{5}}{5}(2x' + y')$ **28.** $x = \frac{\sqrt{5}}{5}(x' - 2y'), y = \frac{\sqrt{5}}{5}(2x' + y')$ **29.** $x = \frac{\sqrt{13}}{13}(3x' - 2y'), y = \frac{\sqrt{13}}{13}(2x' + 3y')$ **30.** $x = \frac{1}{5}(4x' - 3y'), y = \frac{1}{5}(3x' + 4y')$

31. $\theta = 45°$ (see Problem 21)
Hyperbola
Center at origin
Transverse axis is the x'-axis.
Vertices at $(\pm 1, 0)$

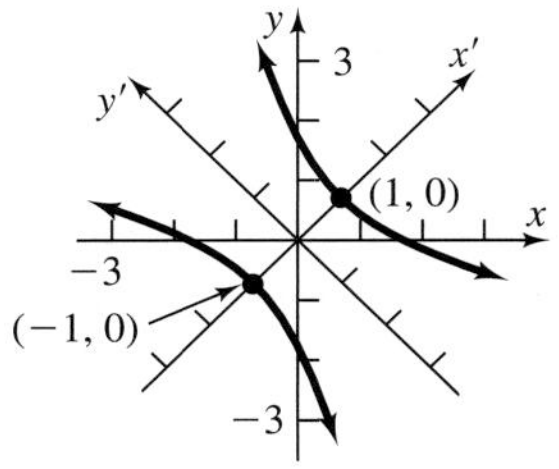

32. $\theta = 45°$ (see Problem 22)
Hyperbola
Center at origin
Transverse axis is the y'-axis.
Vertices at $(0, \pm 1)$

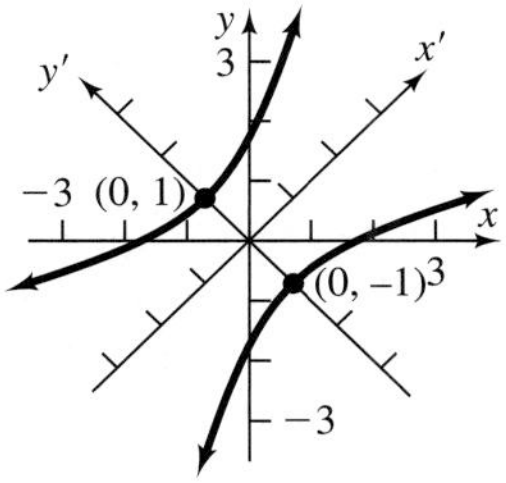

33. $\theta = 45°$ (see Problem 23)
Ellipse
Center at $(0, 0)$
Major axis is the y'-axis.
Vertices at $(0, \pm 2)$

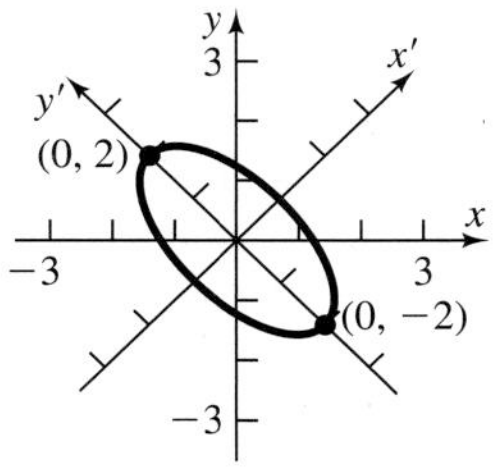

34. $\theta = 45°$ (see Problem 24)
Hyperbola
Center at $(0, 0)$
Transverse axis is the y'-axis.
Vertices at $(0, \pm 2)$

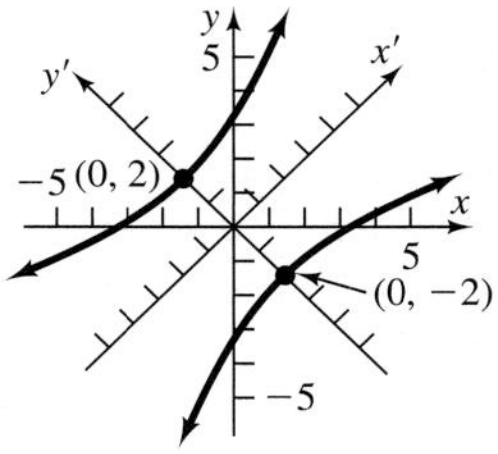

35. $\theta = 60°$ (see Problem 25)
$\frac{x'^2}{4} + y'^2 = 1$
Ellipse
Center at $(0, 0)$
Major axis is the x'-axis.
Vertices at $(\pm 2, 0)$

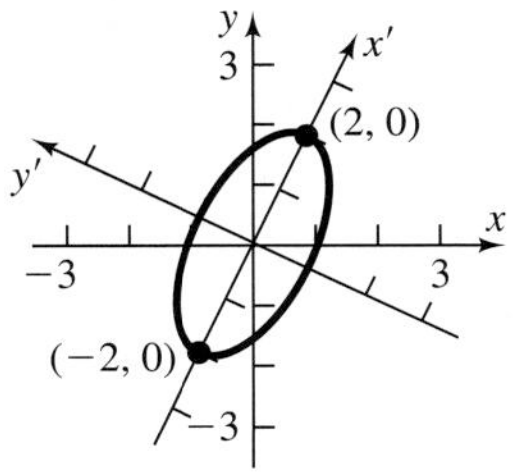

36. $\theta = 30°$ (see Problem 26)
$4x'^2 - y'^2 = 1$
Hyperbola
Center at $(0, 0)$
Transverse axis is the x'-axis.
Vertices at $\left(\pm \frac{1}{2}, 0\right)$

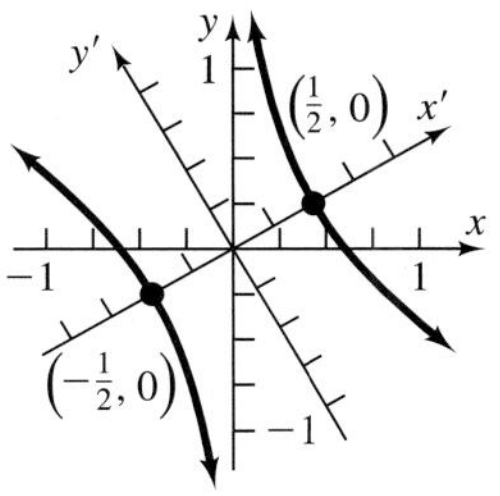

37. $\theta \approx 63°$ (see Problem 27)
$y'^2 = 8x'$
Parabola
Vertex at $(0, 0)$
Focus at $(2, 0)$

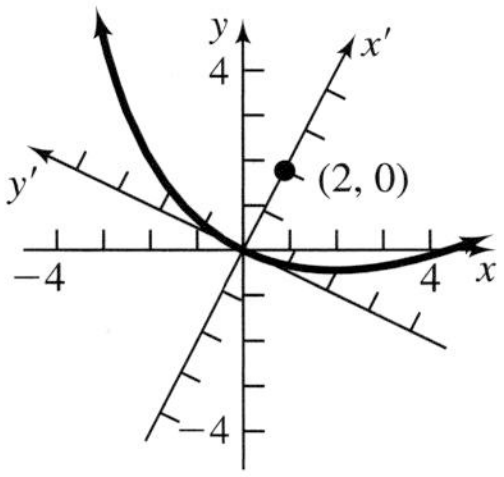

38. $\theta \approx 63°$ (see Problem 28)
$(x' + 1)^2 = -y'$
Parabola
Vertex at $(-1, 0)$
Focus at $\left(-1, -\frac{1}{4}\right)$

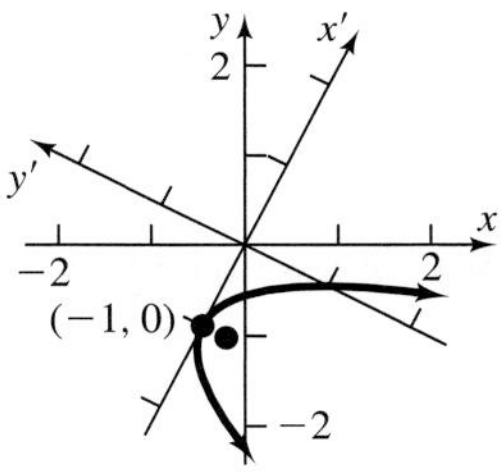

39. $\theta \approx 34°$ (see Problem 29)
$\frac{(x' - 2)^2}{4} + y'^2 = 1$
Ellipse
Center at $(2, 0)$
Major axis is the x'-axis.
Vertices at $(4, 0)$ and $(0, 0)$

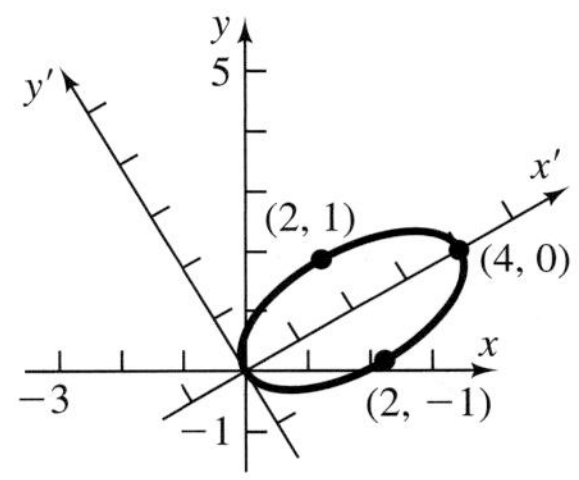

40. $\theta \approx 37°$ (see Problem 30)
$x'^2 + 2y'^2 = 1$
Ellipse
Center at $(0, 0)$
Major axis is the x'-axis.
Vertices at $(\pm 1, 0)$

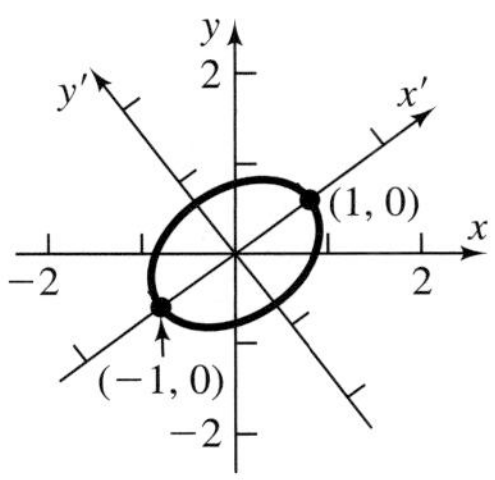

41. $\cot(2\theta) = \frac{7}{24}$;
$\theta = \sin^{-1}\left(\frac{3}{5}\right) \approx 37°$
$(x' - 1)^2 = -6\left(y' - \frac{1}{6}\right)$
Parabola
Vertex at $\left(1, \frac{1}{6}\right)$
Focus at $\left(1, -\frac{4}{3}\right)$

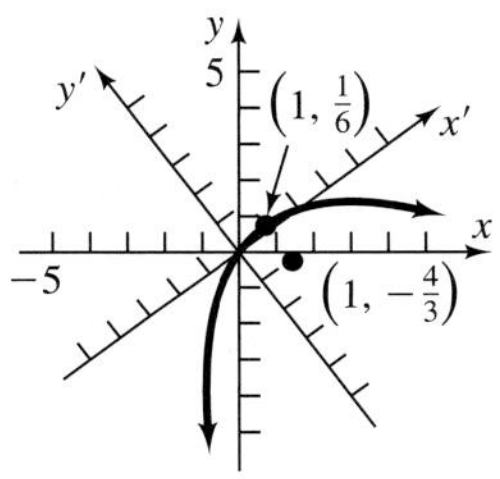

42. $\cot(2\theta) = \frac{7}{24}$;
$\theta = \sin^{-1}\left(\frac{3}{5}\right) \approx 37°$
$x'^2 = -4y'$
Parabola
Vertex at $(0, 0)$
Focus at $(0, -1)$

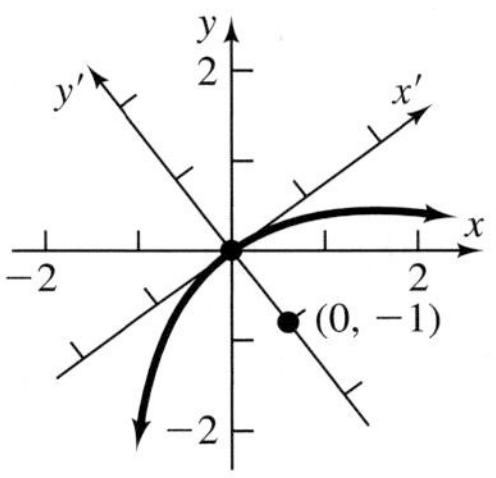

43. Hyperbola **44.** Ellipse **45.** Hyperbola **46.** Ellipse **47.** Parabola **48.** Ellipse **49.** Ellipse **50.** Parabola **51.** Ellipse **52.** Ellipse

53. Refer to equation (6): $A' = A\cos^2\theta + B\sin\theta\cos\theta + C\sin^2\theta$

$$B' = B(\cos^2\theta - \sin^2\theta) + 2(C - A)(\sin\theta\cos\theta)$$
$$C' = A\sin^2\theta - B\sin\theta\cos\theta + C\cos^2\theta$$
$$D' = D\cos\theta + E\sin\theta$$
$$E' = -D\sin\theta + E\cos\theta$$
$$F' = F$$

54. From equation (6), $A' + C' = (A\cos^2\theta + B\sin\theta\cos\theta + C\sin^2\theta) + (A\sin^2\theta - B\sin\theta\cos\theta + C\cos^2\theta)$ $= A(\cos^2\theta + \sin^2\theta) + C(\sin^2\theta + \cos^2\theta) = A + C$

55. Use Problem 53 to find $B'^2 - 4A'C'$. After much cancellation, $B'^2 - 4A'C' = B^2 - 4AC$.

56. **(a)** $B^2 - 4AC$ is invariant under axes rotation (see Problem 55). But with an appropriate rotation, $B = 0$, so $B^2 - 4AC = 0$ only if, after rotation, $A = 0$ or $C = 0$. In either case, the equation becomes quadratic in one variable and linear in the other.
(b) After axes rotation to make $B = 0$, $B^2 - 4AC < 0$ means that A and C have the same sign.
(c) After axes rotation to make $B = 0$, $B^2 - 4AC > 0$ means that A and C have opposite signs.

57. The distance between P_1 and P_2 in the $x'y'$-plane equals $\sqrt{(x_2' - x_1')^2 + (y_2' - y_1')^2}$.
Assuming that $x' = x\cos\theta - y\sin\theta$ and $y' = x\sin\theta + y\cos\theta$, then $(x_2' - x_1')^2 = (x_2\cos\theta - y_2\sin\theta - x_1\cos\theta + y_1\sin\theta)^2$ $= \cos^2\theta(x_2 - x_1)^2 - 2\sin\theta\cos\theta(x_2 - x_1)(y_2 - y_1) + \sin^2\theta(y_2 - y_1)^2$, and
$(y_2' - y_1')^2 = (x_2\sin\theta + y_2\cos\theta - x_1\sin\theta - y_1\cos\theta)^2 = \sin^2\theta(x_2 - x_1)^2 + 2\sin\theta\cos\theta(x_2 - x_1)(y_2 - y_1) + \cos^2\theta(y_2 - y_1)^2$.
Therefore, $(x_2' - x_1')^2 + (y_2' - y_1')^2 = \cos^2\theta(x_2 - x_1)^2 + \sin^2\theta(x_2 - x_1)^2 + \sin^2\theta(y_2 - y_1)^2 + \cos^2\theta(y_2 - y_1)^2$
$= (x_2 - x_1)^2(\cos^2\theta + \sin^2\theta) + (y_2 - y_1)^2(\sin^2\theta + \cos^2\theta) = (x_2 - x_1)^2 + (y_2 - y_1)^2$.

58. With a 45° rotation, $x^{1/2} + y^{1/2} = a^{1/2}$ becomes $\left[\frac{\sqrt{2}}{2}(x' - y')\right]^{1/2} + \left[\frac{\sqrt{2}}{2}(x' + y')\right]^{1/2} = a^{1/2}$, which is equivalent to $\sqrt{x' - y'} + \sqrt{x' + y'} = K$ with K a positive constant, which implies that $(x' - y') + 2\sqrt{(x' - y')(x' + y')} + (x' + y') = K^2$ $\Rightarrow 2\sqrt{x'^2 - y'^2} = K^2 - 2x' \Rightarrow 4x'^2 - 4y'^2 = 4x'^2 - 4Kx' + K^4 \Rightarrow y'^2 = Kx' - \frac{K^4}{4}$, which is a parabolic equation.

9.6 Concepts and Vocabulary *(page 663)*

3. $\frac{1}{2}$; ellipse; parallel; 4; below **4.** 1; < 1; > 1 **5.** True **6.** True

9.6 Exercises *(page 663)*

7. Parabola; directrix is perpendicular to the polar axis 1 unit to the right of the pole. **8.** Parabola; directrix is parallel to the polar axis 3 units below the pole. **9.** Hyperbola; directrix is parallel to the polar axis $\frac{4}{3}$ units below the pole.
10. Hyperbola; directrix is perpendicular to the polar axis 1 unit to the right of the pole. **11.** Ellipse; directrix is perpendicular to the polar axis $\frac{3}{2}$ units to the left of the pole. **12.** Ellipse, directrix is parallel to the polar axis 3 units above the pole.

13. Parabola; directrix is perpendicular to the polar axis 1 unit to the right of the pole; vertex is at $\left(\frac{1}{2}, 0\right)$.

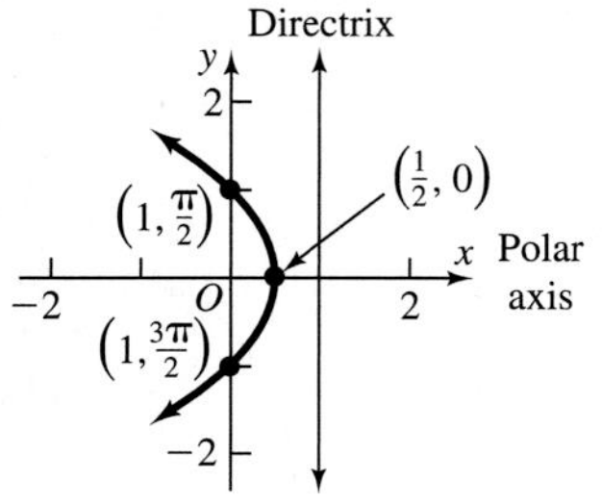

14. Parabola; directrix is parallel to the polar axis 3 units below the pole; vertex is at $\left(\frac{3}{2}, \frac{3\pi}{2}\right)$.

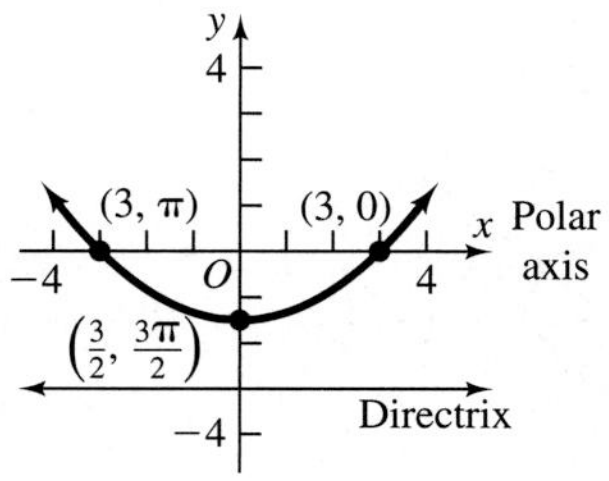

15. Ellipse; directrix is parallel to the polar axis $\frac{8}{3}$ units above the pole; vertices are at $\left(\frac{8}{7}, \frac{\pi}{2}\right)$ and $\left(8, \frac{3\pi}{2}\right)$.

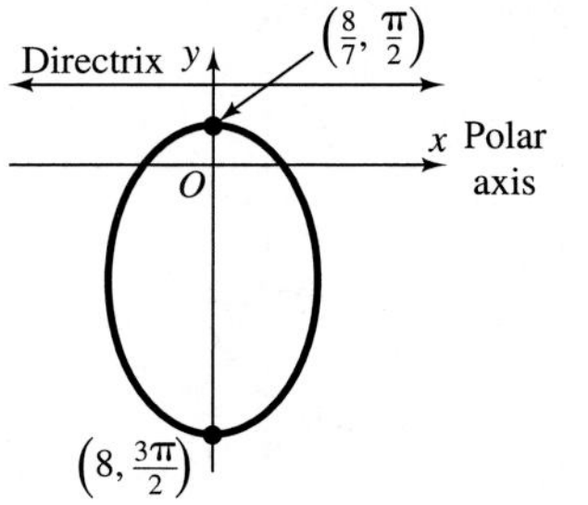

16. Ellipse; directrix is perpendicular to the polar axis $\frac{5}{2}$ units to the right of the pole; vertices are at $\left(\frac{10}{9}, 0\right)$ and $(10, \pi)$.

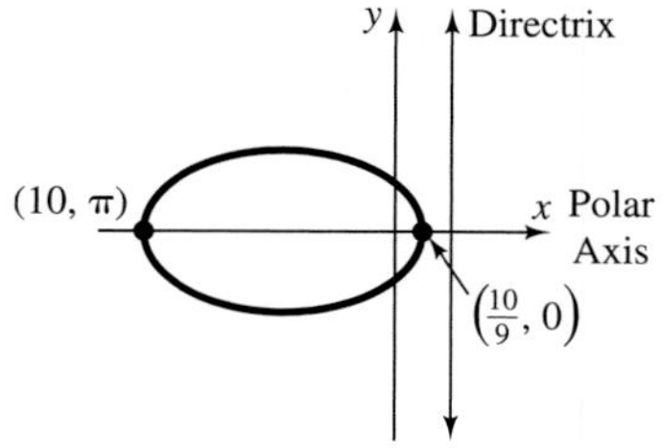

17. Hyperbola; directrix is perpendicular to the polar axis $\frac{3}{2}$ units to the left of the pole; vertices are at $(-3, 0)$ and $(1, \pi)$.

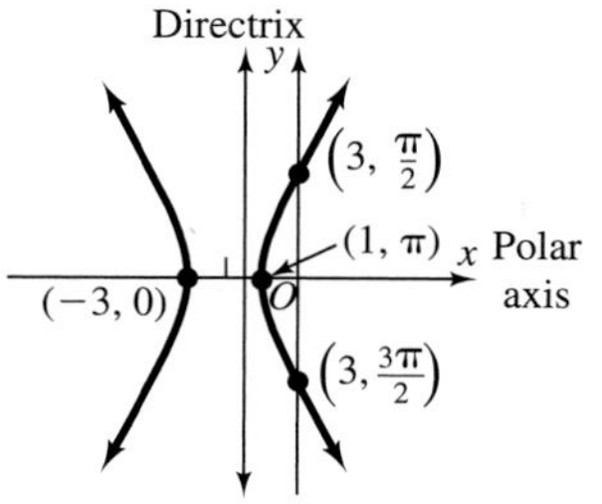

18. Hyperbola; directrix is parallel to the polar axis $\frac{3}{2}$ units above the pole; vertices are at $\left(-3, \frac{3\pi}{2}\right)$ and $\left(1, \frac{\pi}{2}\right)$.

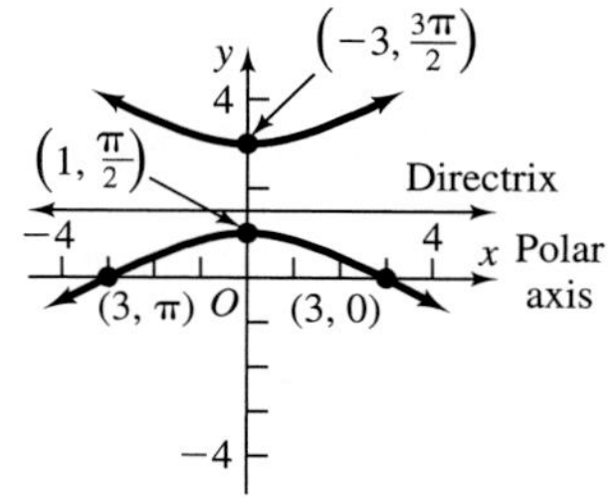

19. Ellipse; directrix is parallel to the polar axis 8 units below the pole; vertices are at $\left(8, \frac{\pi}{2}\right)$ and $\left(\frac{8}{3}, \frac{3\pi}{2}\right)$.

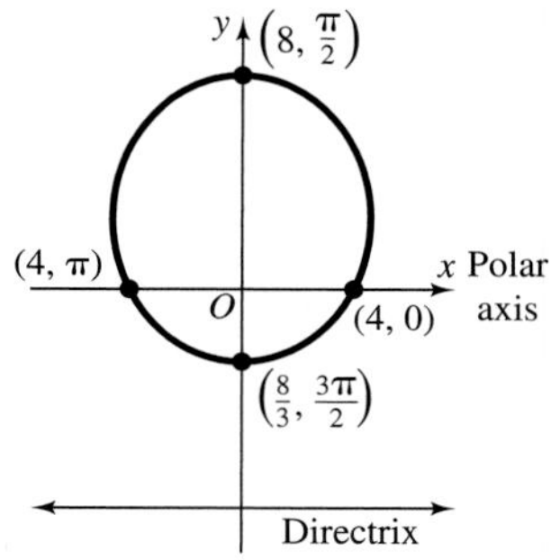

20. Hyperbola; directrix is perpendicular to the polar axis 2 units to the right of the pole; vertices are at $\left(\frac{4}{3}, 0\right)$ and $(-4, \pi)$.

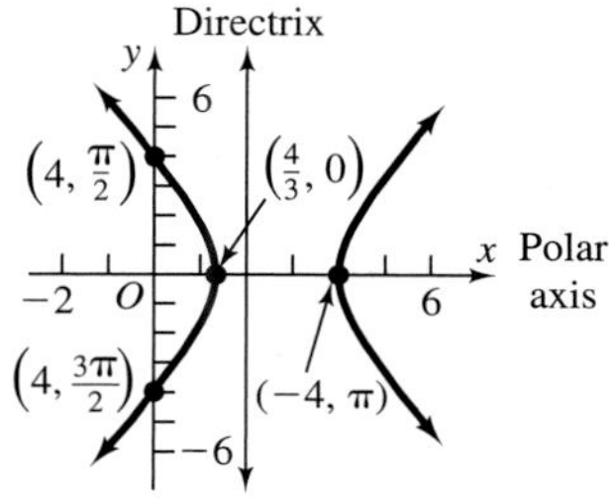

21. Ellipse; directrix is parallel to the polar axis 3 units below the pole; vertices are at $\left(6, \frac{\pi}{2}\right)$ and $\left(\frac{6}{5}, \frac{3\pi}{2}\right)$.

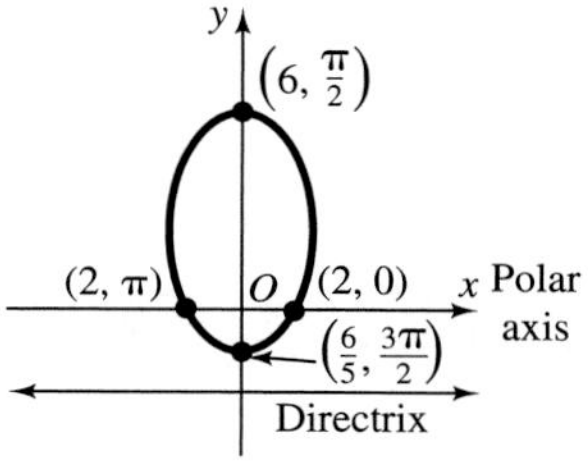

22. Ellipse; directrix is perpendicular to the polar axis 2 units to the left of the pole; vertices are at $\left(\frac{2}{3}, \pi\right)$ and $(2, 0)$.

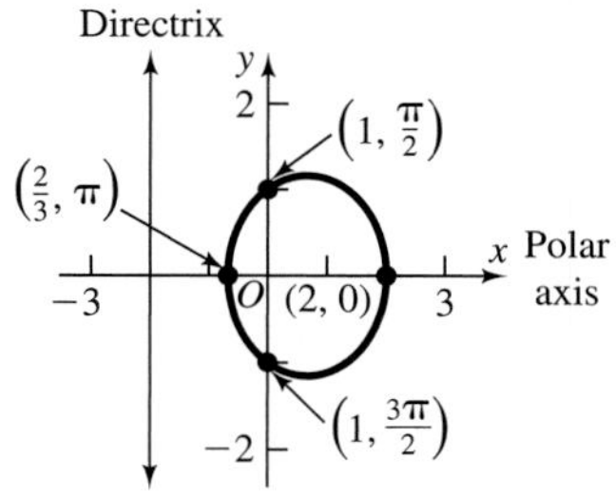

23. Ellipse; directrix is perpendicular to the polar axis 6 units to the left of the pole; vertices are at $(6, 0)$ and $(2, \pi)$.

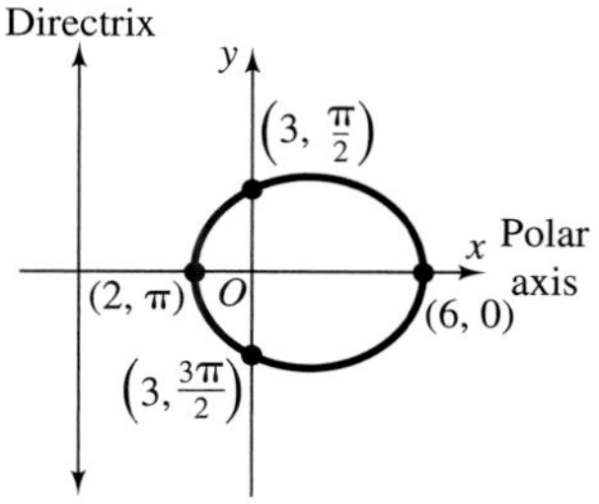

24. Parabola; directrix is parallel to the polar axis 3 units below the pole; vertex is at $\left(\frac{3}{2}, \frac{3\pi}{2}\right)$.

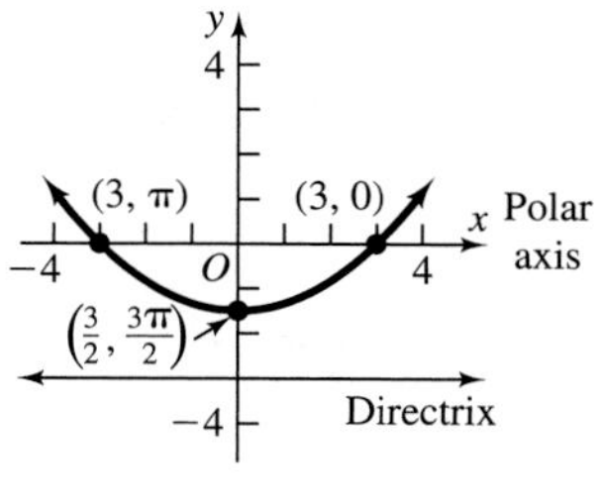

25. $y^2 + 2x - 1 = 0$ **26.** $x^2 - 6y - 9 = 0$ **27.** $16x^2 + 7y^2 + 48y - 64 = 0$ **28.** $9x^2 + 25y^2 + 80x - 100 = 0$ **29.** $3x^2 - y^2 + 12x + 9 = 0$ **30.** $x^2 - 3y^2 + 12y - 9 = 0$ **31.** $4x^2 + 3y^2 - 16y - 64 = 0$ **32.** $3x^2 - y^2 - 16x + 16 = 0$ **33.** $9x^2 + 5y^2 - 24y - 36 = 0$ **34.** $3x^2 + 4y^2 - 4x - 4 = 0$ **35.** $3x^2 + 4y^2 - 12x - 36 = 0$ **36.** $x^2 - 6y - 9 = 0$ **37.** $r = \frac{1}{1 + \sin\theta}$ **38.** $r = \frac{2}{1 - \sin\theta}$ **39.** $r = \frac{12}{5 - 4\cos\theta}$ **40.** $r = \frac{6}{3 + 2\sin\theta}$ **41.** $r = \frac{12}{1 - 6\sin\theta}$ **42.** $r = \frac{25}{1 + 5\cos\theta}$

43. Use $d(D, P) = p - r\cos\theta$ in the derivation of equation (a) in Table 5.

44. Use $d(D, P) = p - r \sin \theta$ in the derivation of equation (a) in Table 5.
45. Use $d(D, P) = p + r \sin \theta$ in the derivation of equation (a) in Table 5.
46. Aphelion: 4.335×10^7 mi; perihelion: 2.854×10^7 mi

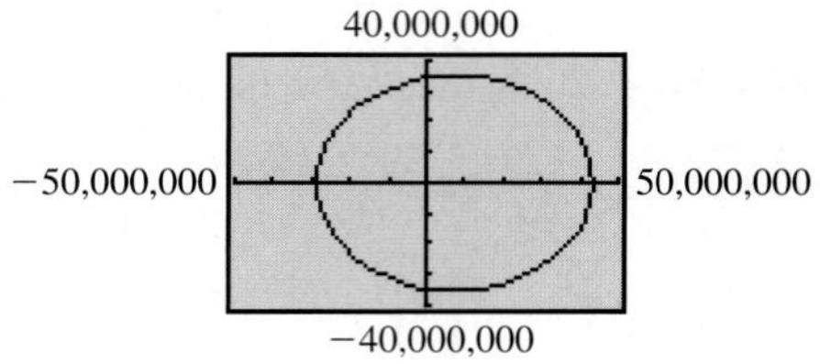

9.7 Concepts and Vocabulary *(page 674)*

2. plane curve; parameter **3.** ellipse **4.** cycloid **5.** False **6.** True

9.7 Exercises *(page 675)*

7.

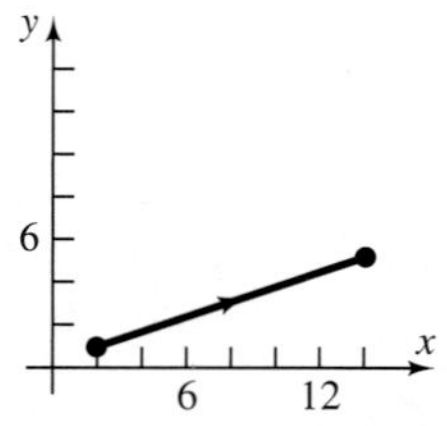

$x - 3y + 1 = 0$

8.

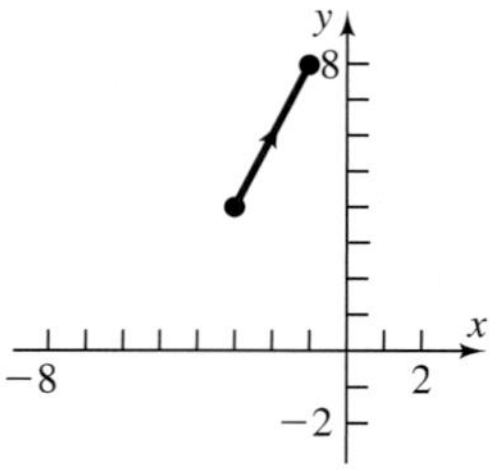

$2x - y + 10 = 0$

9.

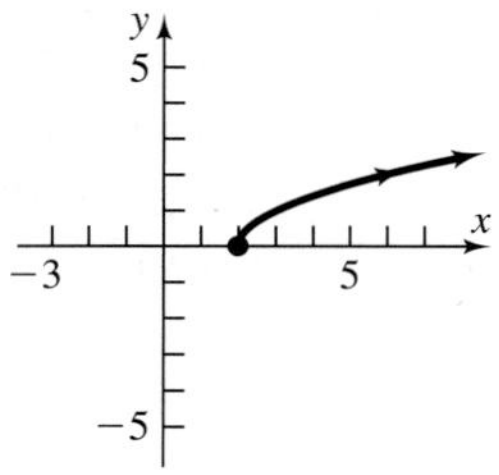

$y = \sqrt{x - 2}$

10.

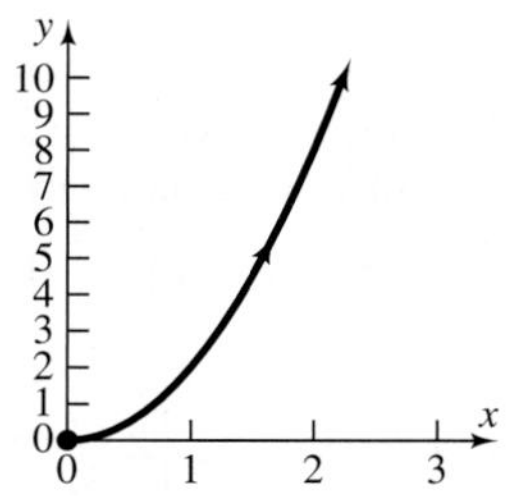

$x = \sqrt{\dfrac{y}{2}}$

11.

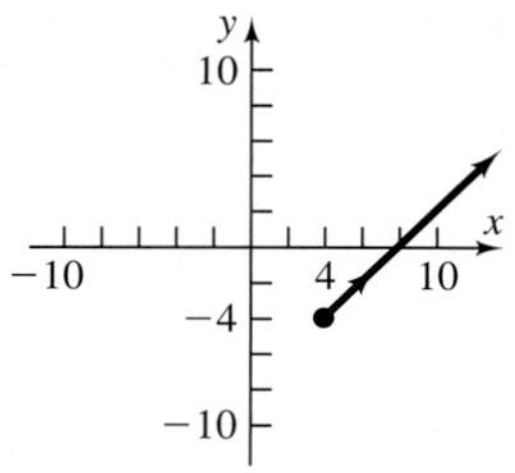

$x = y + 8$

12.

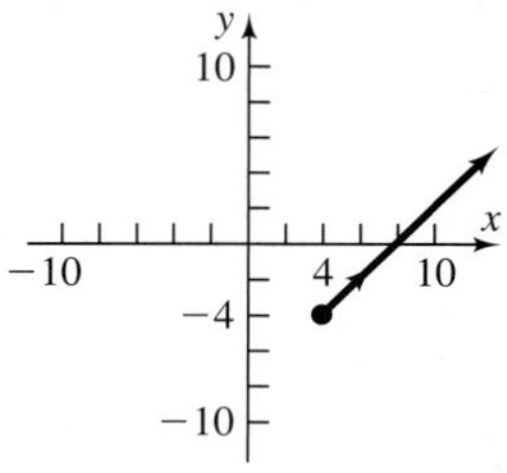

$y = x - 8$

13.

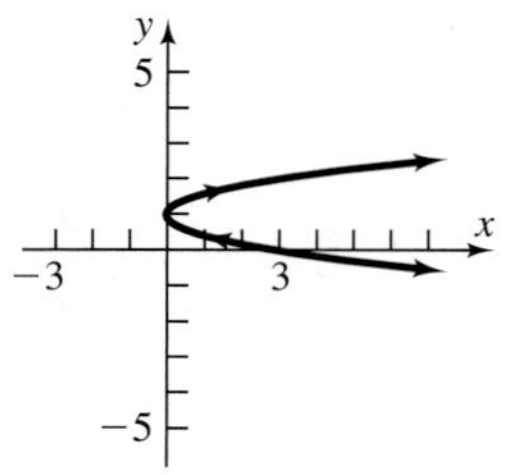

$x = 3(y - 1)^2$

14.

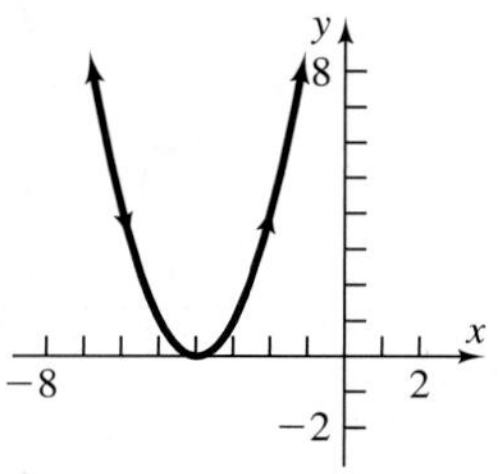

$y = (x + 4)^2$

15.

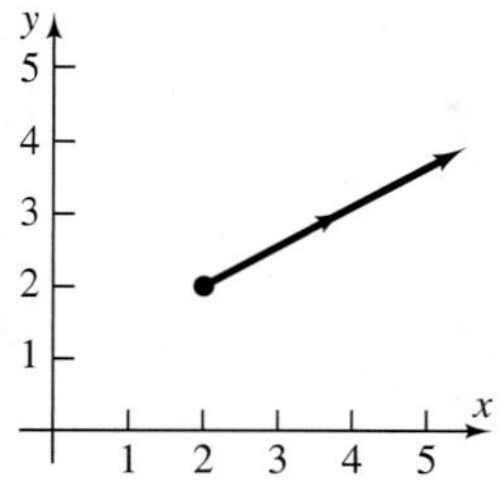

$2y = 2 + x$

16.

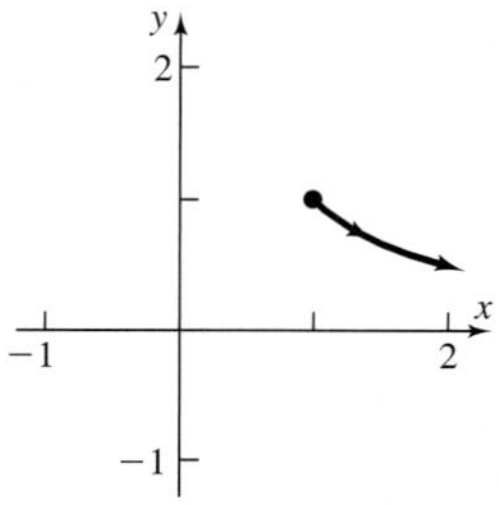

$xy = 1$

17.

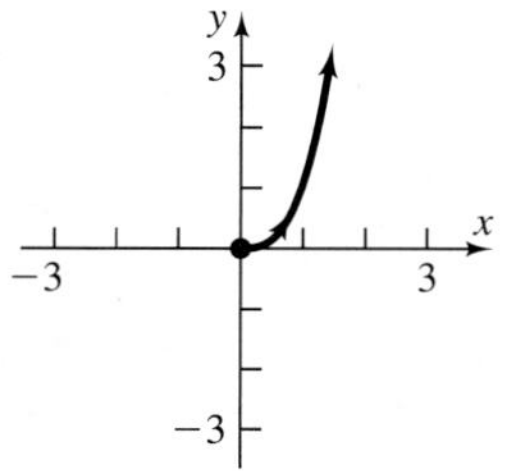

$y = x^3$

18.

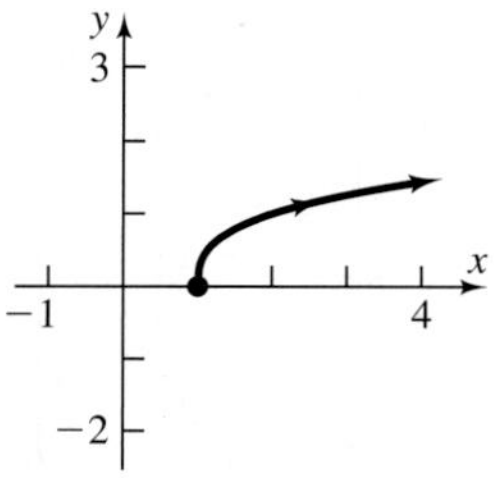

$x = y^3 + 1$

19.

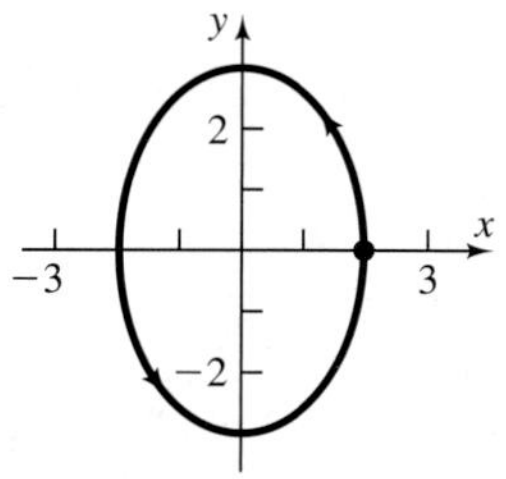

$\dfrac{x^2}{4} + \dfrac{y^2}{9} = 1$

20.

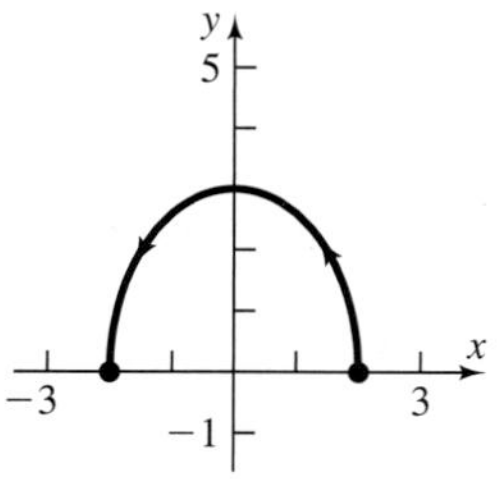

$\dfrac{x^2}{4} + \dfrac{y^2}{9} = 1$

21.

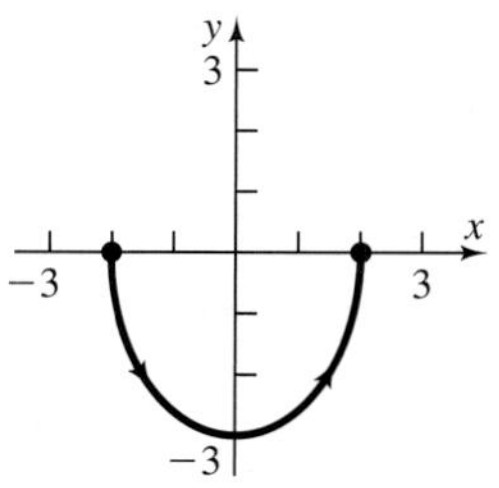

$\dfrac{x^2}{4} + \dfrac{y^2}{9} = 1$

22.

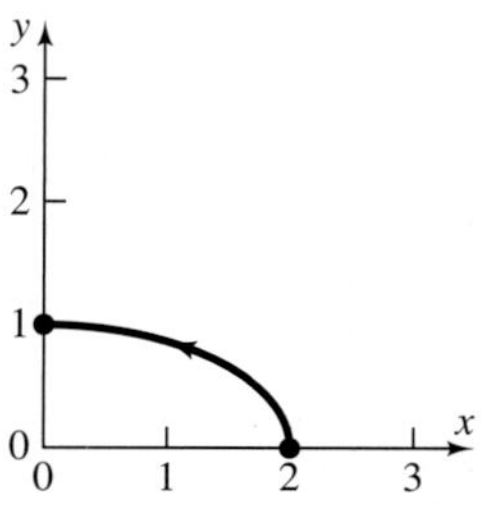

$\dfrac{x^2}{4} + y^2 = 1$

23.

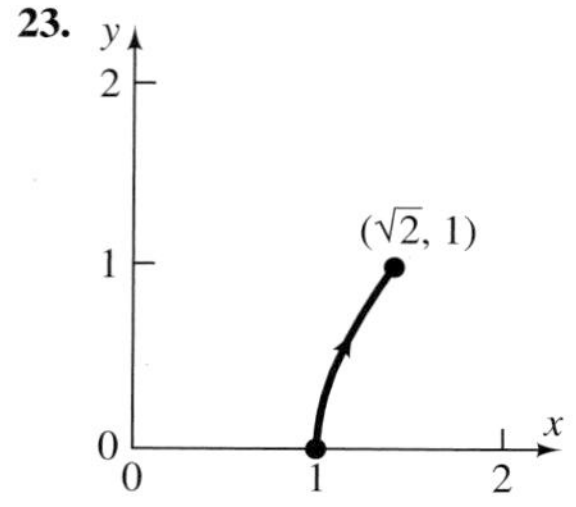

$x^2 - y^2 = 1$

24.

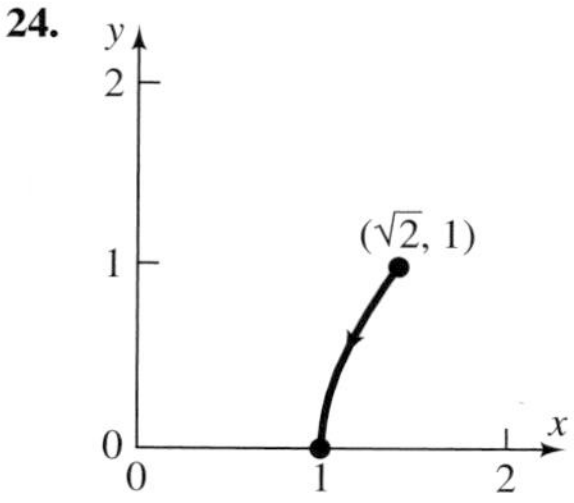

$x^2 - y^2 = 1$

25.

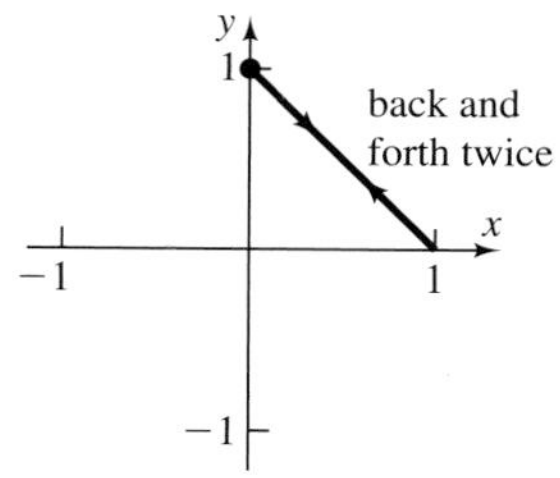

$x + y = 1$

26.

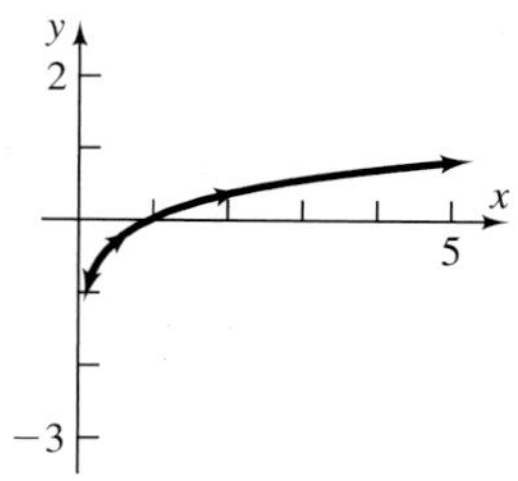

$y = \frac{1}{2} \ln x$

27. (a) $x = 3$
$y = -16t^2 + 50t + 6$
(b) 3.24 sec
(c) 1.5625 sec; 45.0625 ft
(d)

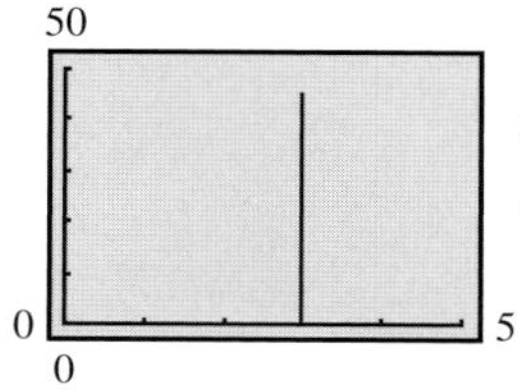

28. (a) $x = 3$
$y = -16t^2 + 40t + 5$
(b) 2.62 sec
(c) 1.25 sec; 30 ft
(d)

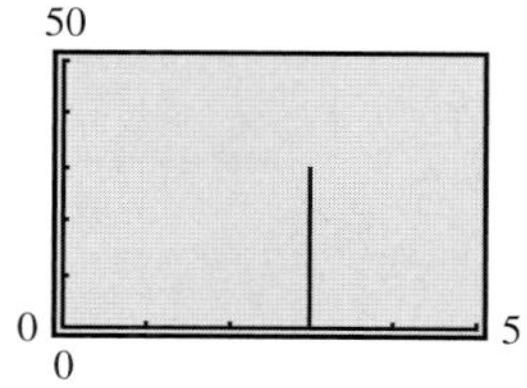

29. (a) Train: $x_1 = t^2, y_1 = 1$;
Bill: $x_2 = 5(t - 5), y_2 = 3$
(b) Bill won't catch the train.
(c)

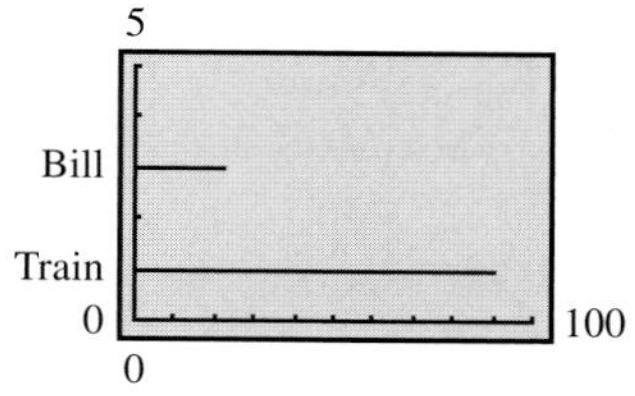

30. (a) Bus: $x_1 = \frac{3}{2}t^2$ Jodi: $x_2 = 5(t - 2)$
$y_1 = 1$ $y_2 = 3$
(b) Jodi won't catch the bus.
(c)

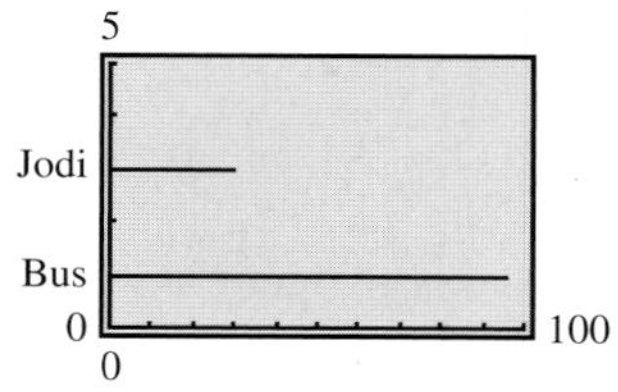

31. (a) $x = (145 \cos 20°)t$
$y = -16t^2 + (145 \sin 20°)t + 5$
(b) 3.197 sec
(c) 1.55 sec; 43.43 ft
(d) 435.61 ft
(e)

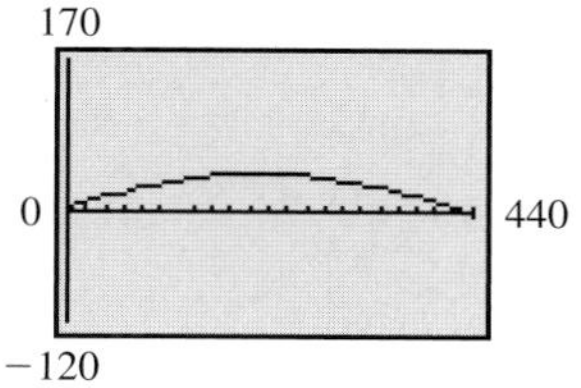

32. (a) $x = (180 \cos 40°)t$
$y = -16t^2 + (180 \sin 40°)t + 3$
(b) 7.257 sec
(c) 3.62 sec; 212.17 ft
(d) 1000.68 ft
(e)

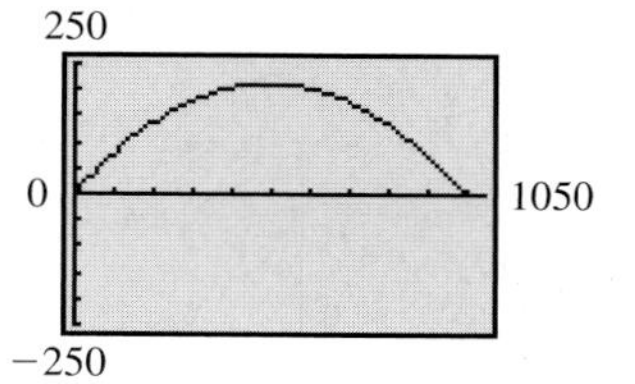

33. (a) $x = (40 \cos 45°)t$
$y = -4.9t^2 + (40 \sin 45°)t + 300$
(b) 11.226 sec
(c) 2.886 sec; 340.8 m
(d) 317.5 m
(e)

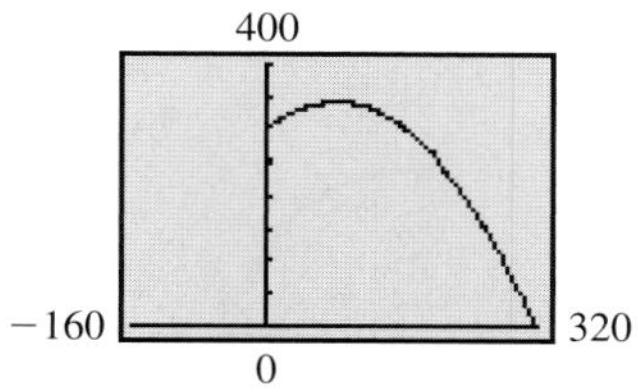

34. (a) $x = (40 \cos 45°)t$
$y = -0.8167t^2 + (40 \sin 45°)t + 300$
(b) 43.146 sec
(c) 17.317 sec; 544.888 m
(d) 1220.353 m
(e)

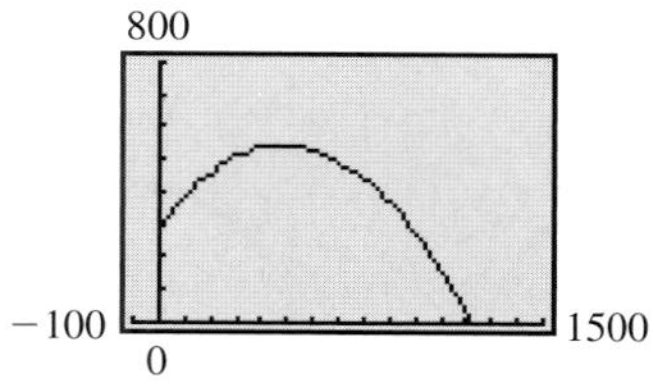

35. (a) Paseo: $x = 40t - 5, y = 0$; Bonneville: $x = 0, y = 30t - 4$ **(b)** $d = \sqrt{(40t - 5)^2 + (30t - 4)^2}$
(c)

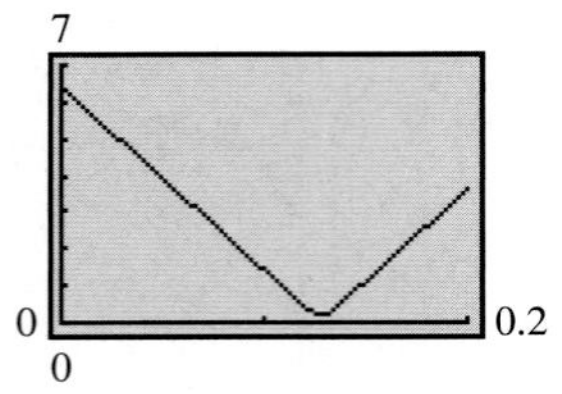

(d) 0.2 mi; 7.68 min **(e)** Turn axes off to see the graph:

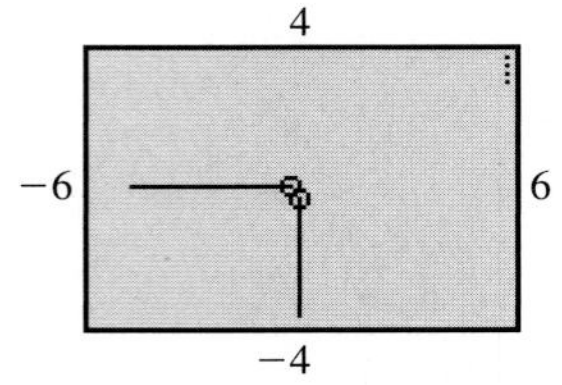

36. (a) Cessna: $x = 0, y = 100 - 120t$; 747: $x = 550 - 600t, y = 0$ **(b)** $d = \sqrt{(100 - 120t)^2 + (550t - 600)^2}$
(c)

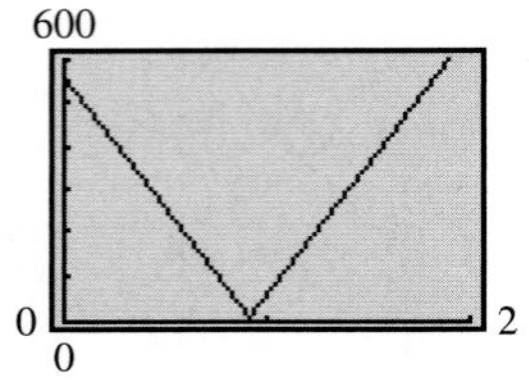

(d) 9.8 mi; 54.81 min **(e)** Turn axes off to see the graph:

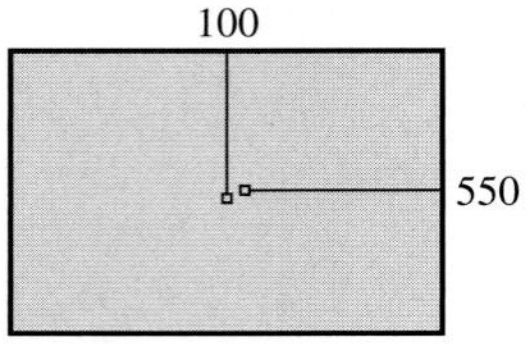

37. $x = t$, $y = 4t - 1$ or $x = \dfrac{t+1}{4}$, $y = t$

38. $x = t$, $y = -8t + 3$ or $x = \dfrac{3-t}{8}$, $y = t$

39. $x = t$, $y = t^2 + 1$ or $x = t^3$, $y = t^6 + 1$

40. $x = t$, $y = -2t^2 + 1$ or $x = t^2$, $y = -2t^4 + 1$

41. $x = t$, $y = t^3$ or $x = \sqrt[3]{t}$, $y = t$

42. $x = t$, $y = t^4 + 1$ or $x = \sqrt[4]{t - 1}$, $y = t$

43. $x = t$, $y = t^{2/3}$ or $x = t^3$, $y = t^2, t \geq 0$

44. $x = t, y = t^2, t \geq 0; x = \sqrt{t}, y = t$

45. $x = t + 2, y = t, 0 \leq t \leq 5$

46. $x = t - 1, y = -t + 2, 0 \leq t \leq 4$ **47.** $x = 3\cos t, y = 2\sin t, 0 \leq t \leq 2\pi$ **48.** $x = \cos\left(\dfrac{\pi}{2}(t - 1)\right), y = 4\sin\left(\dfrac{\pi}{2}(t - 1)\right), 0 \leq t \leq 2$

49. $x = 2\cos(\pi t), y = -3\sin(\pi t), 0 \leq t \leq 2$ **50.** $x = -2\sin(2\pi t), y = 3\cos(2\pi t), 0 \leq t \leq 1$

51. $x = 2\sin(2\pi t), y = 3\cos(2\pi t), 0 \leq t \leq 1$ **52.** $x = 2\cos\left(\dfrac{2}{3}\pi t\right), y = 3\sin\left(\dfrac{2}{3}\pi t\right), 0 \leq t \leq 3$

53.

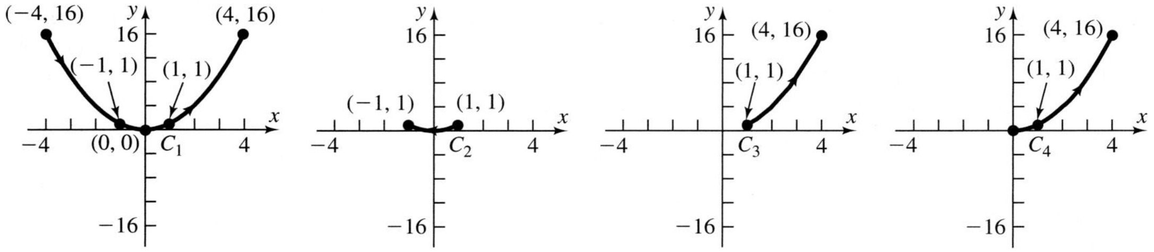

54.

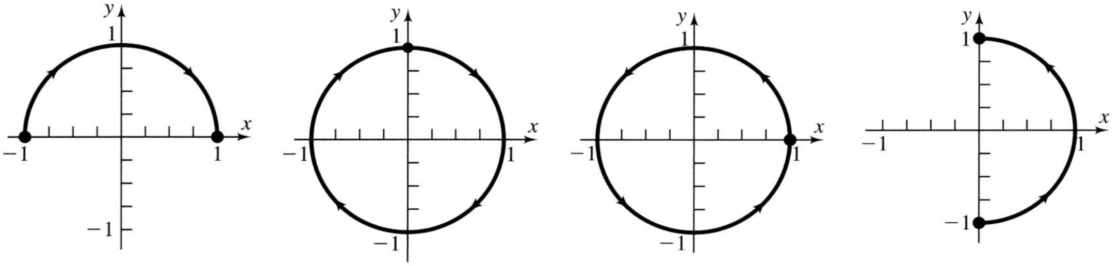

55. The orientation is from (x_1, y_1) to (x_2, y_2).

56. (a) $y = -\left(\dfrac{16}{v_0^2 \cos^2 \theta}\right)x^2 + (\tan \theta)x$; parabola

(b) $0 = (v_0 \sin \theta)t - 16t^2 \Rightarrow t = 0$ or $t = \dfrac{1}{16}v_0 \sin \theta$

(c) $\dfrac{1}{16}v_0^2 \sin \theta \cos \theta$

(d) $t = 0$ or $t = \dfrac{1}{16}v_0(\sin \theta - \cos \theta)$

$x = y = \dfrac{1}{16}v_0^2 \cos \theta(\sin \theta - \cos \theta)$

$\sqrt{x^2 + y^2} = \dfrac{\sqrt{2}}{16}v_0^2 \cos \theta(\sin \theta - \cos \theta)$

57.

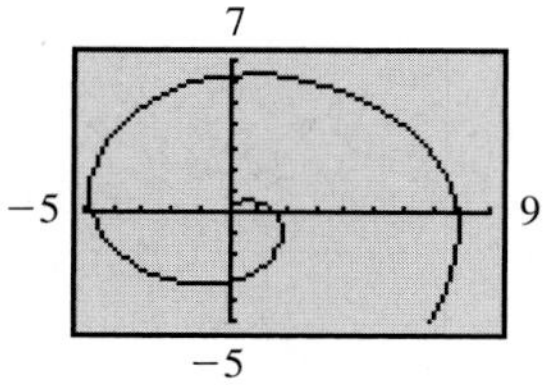

58.

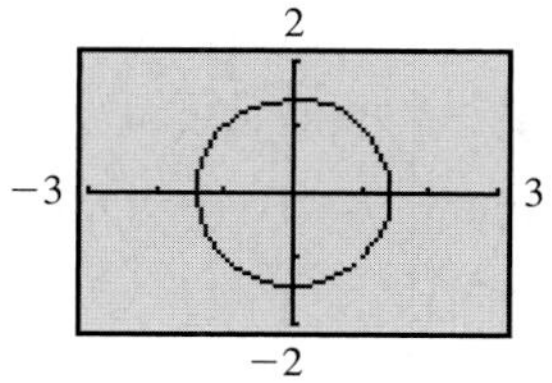

59.

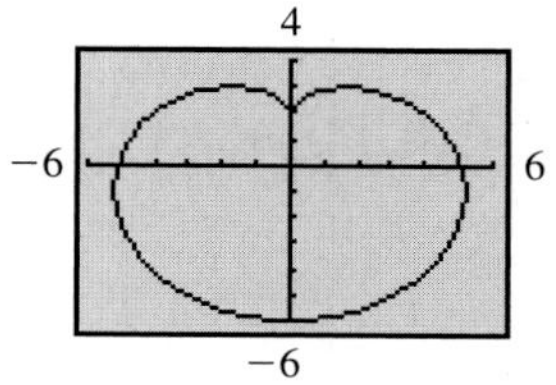

60.

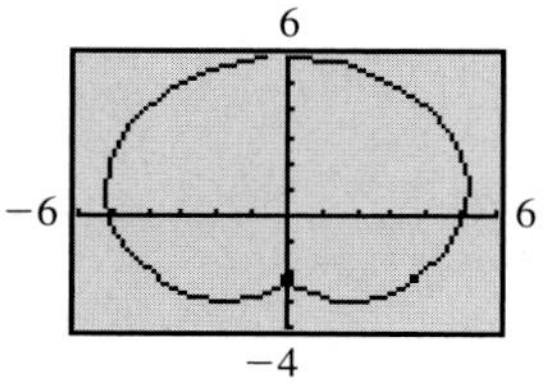

61. (a)

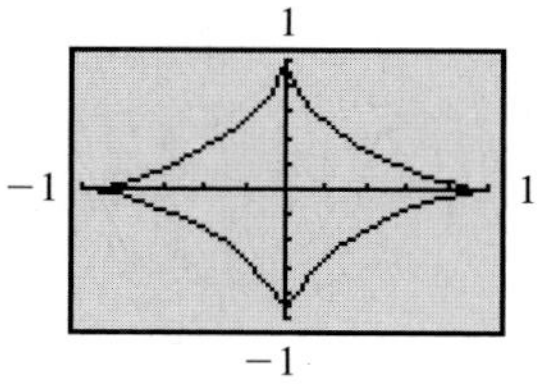

(b) $x^{2/3} + y^{2/3} = 1$

Review Exercises *(page 679)*

1. Parabola; vertex $(0, 0)$, focus $(-4, 0)$, directrix $x = 4$ **2.** Parabola; vertex $(0, 0)$, focus $\left(0, \dfrac{1}{64}\right)$, directrix $y = -\dfrac{1}{64}$

3. Hyperbola; center $(0, 0)$, vertices $(5, 0)$ and $(-5, 0)$, foci $(\sqrt{26}, 0)$ and $(-\sqrt{26}, 0)$, asymptotes $y = \dfrac{1}{5}x$ and $y = -\dfrac{1}{5}x$

4. Hyperbola; center $(0, 0)$, vertices $(0, 5)$ and $(0, -5)$, foci $(0, \sqrt{26})$ and $(0, -\sqrt{26})$, asymptotes $y = 5x$ and $y = -5x$

5. Ellipse; center $(0, 0)$, vertices $(0, 5)$ and $(0, -5)$, foci $(0, 3)$ and $(0, -3)$ **6.** Ellipse; center $(0, 0)$, vertices $(0, 4)$ and $(0, -4)$, foci $(0, \sqrt{7})$ and $(0, -\sqrt{7})$ **7.** $x^2 = -4(y - 1)$: Parabola; vertex $(0, 1)$, focus $(0, 0)$, directrix $y = 2$

8. $\dfrac{y^2}{3} - \dfrac{x^2}{9} = 1$: Hyperbola; center $(0, 0)$, vertices $(0, \sqrt{3})$ and $(0, -\sqrt{3})$, foci $(0, 2\sqrt{3})$ and $(0, -2\sqrt{3})$, asymptotes $y = \dfrac{\sqrt{3}}{3}x$, $y = -\dfrac{\sqrt{3}}{3}x$

9. $\dfrac{x^2}{2} - \dfrac{y^2}{8} = 1$: Hyperbola; center $(0, 0)$, vertices $(\sqrt{2}, 0)$ and $(-\sqrt{2}, 0)$, foci $(\sqrt{10}, 0)$ and $(-\sqrt{10}, 0)$, asymptotes $y = 2x$ and $y = -2x$

10. $\dfrac{x^2}{4} + \dfrac{y^2}{9} = 1$: Ellipse; center $(0, 0)$, vertices $(0, 3)$ and $(0, -3)$, foci $(0, \sqrt{5})$ and $(0, -\sqrt{5})$

11. $(x - 2)^2 = 2(y + 2)$: Parabola; vertex $(2, -2)$, focus $\left(2, -\dfrac{3}{2}\right)$, directrix $y = -\dfrac{5}{2}$

12. $(y - 1)^2 = \dfrac{1}{2}x$: Parabola; vertex $(0, 1)$, focus $\left(\dfrac{1}{8}, 1\right)$, directrix $x = -\dfrac{1}{8}$

13. $\dfrac{(y - 2)^2}{4} - (x - 1)^2 = 1$: Hyperbola; center $(1, 2)$, vertices $(1, 4)$ and $(1, 0)$, foci $(1, 2 + \sqrt{5})$ and $(1, 2 - \sqrt{5})$, asymptotes $y - 2 = \pm 2(x - 1)$

14. $(x + 1)^2 + \dfrac{(y - 2)^2}{4} = 1$: Ellipse; center $(-1, 2)$, vertices $(-1, 0)$ and $(-1, 4)$, foci $(-1, 2 + \sqrt{3})$ and $(-1, 2 - \sqrt{3})$

15. $\dfrac{(x - 2)^2}{9} + \dfrac{(y - 1)^2}{4} = 1$: Ellipse; center $(2, 1)$, vertices $(5, 1)$ and $(-1, 1)$, foci $(2 + \sqrt{5}, 1)$ and $(2 - \sqrt{5}, 1)$

16. $\dfrac{(x - 2)^2}{9} + \dfrac{(y + 1)^2}{4} = 1$: Ellipse; center $(2, -1)$, vertices $(5, -1)$ and $(-1, -1)$, foci $(2 + \sqrt{5}, -1)$ and $(2 - \sqrt{5}, -1)$

17. $(x - 2)^2 = -4(y + 1)$: Parabola; vertex $(2, -1)$, focus $(2, -2)$, directrix $y = 0$

18. $(y - 2)^2 = -\dfrac{3}{4}(x + 1)$: Parabola; vertex $(-1, 2)$, focus $\left(-\dfrac{19}{16}, 2\right)$, directrix $x = -\dfrac{13}{16}$

19. $\dfrac{(x-1)^2}{4} + \dfrac{(y+1)^2}{9} = 1$: Ellipse; center $(1, -1)$, vertices $(1, 2)$ and $(1, -4)$, foci $(1, -1 + \sqrt{5})$ and $(1, -1 - \sqrt{5})$

20. $(x-1)^2 - (y+1)^2 = 1$: Hyperbola; center $(1, -1)$, vertices $(0, -1)$ and $(2, -1)$, foci $(1 + \sqrt{2}, -1)$ and $(1 - \sqrt{2}, -1)$, asymptotes $y + 1 = x - 1$ and $y + 1 = -(x - 1)$

21. $y^2 = -8x$

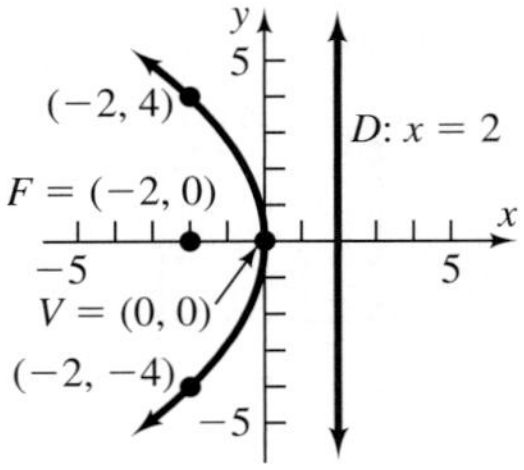

22. $\dfrac{x^2}{16} + \dfrac{y^2}{25} = 1$

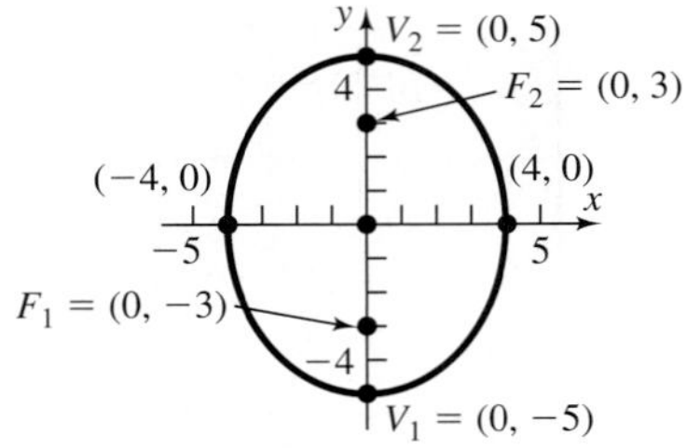

23. $\dfrac{y^2}{4} - \dfrac{x^2}{12} = 1$

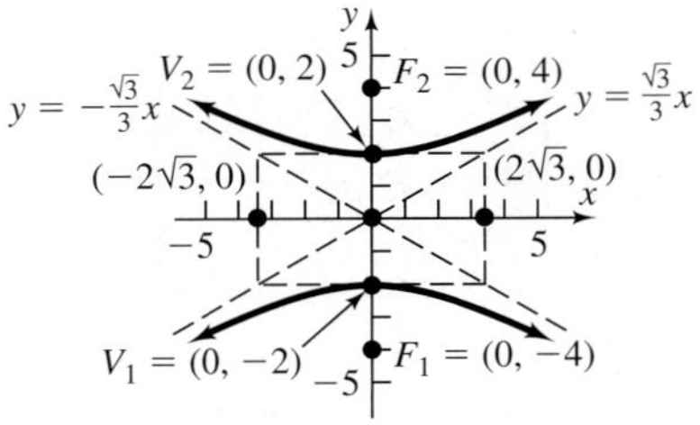

24. $x^2 = 12y$

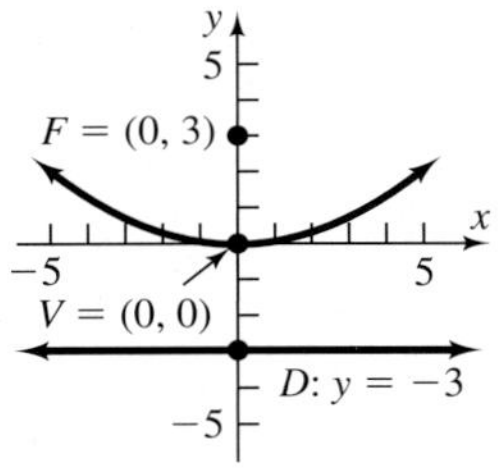

25. $\dfrac{x^2}{16} + \dfrac{y^2}{7} = 1$

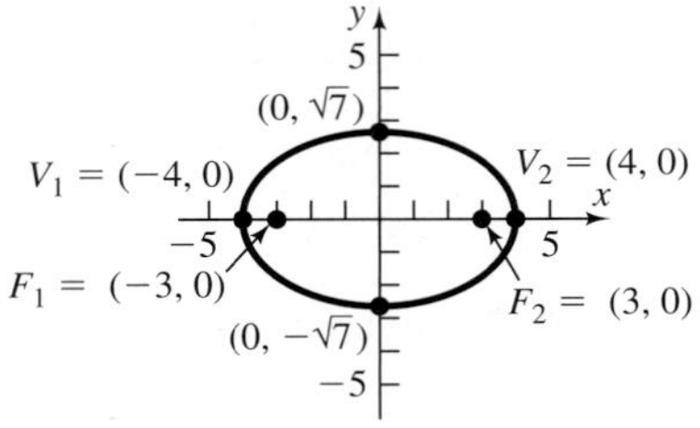

26. $\dfrac{x^2}{4} - \dfrac{y^2}{12} = 1$

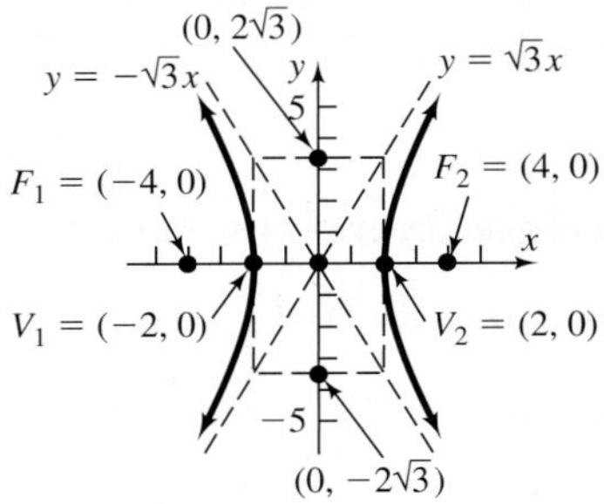

27. $(x-2)^2 = -4(y+3)$

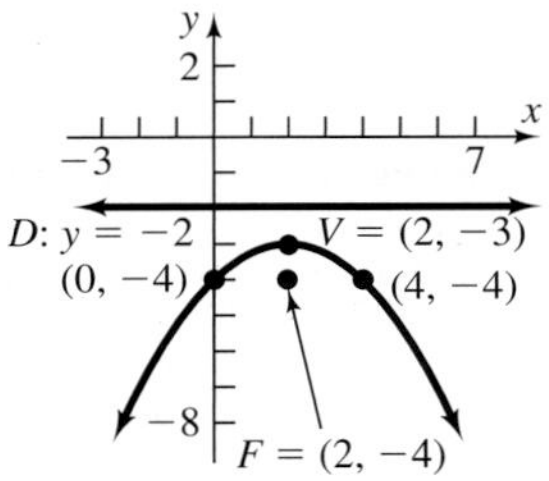

28. $\dfrac{(x+1)^2}{9} + \dfrac{(y-2)^2}{8} = 1$

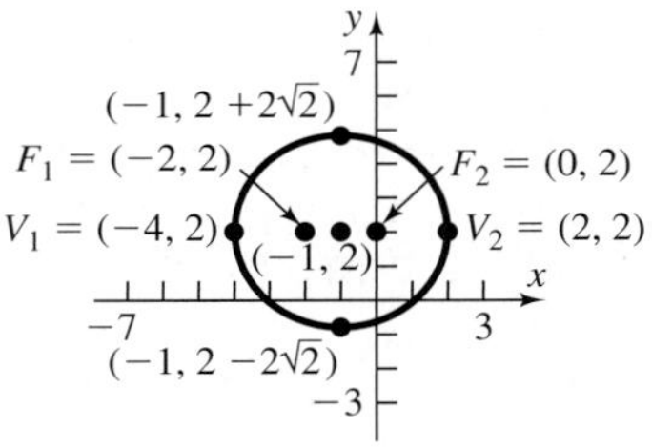

29. $(x+2)^2 - \dfrac{(y+3)^2}{3} = 1$

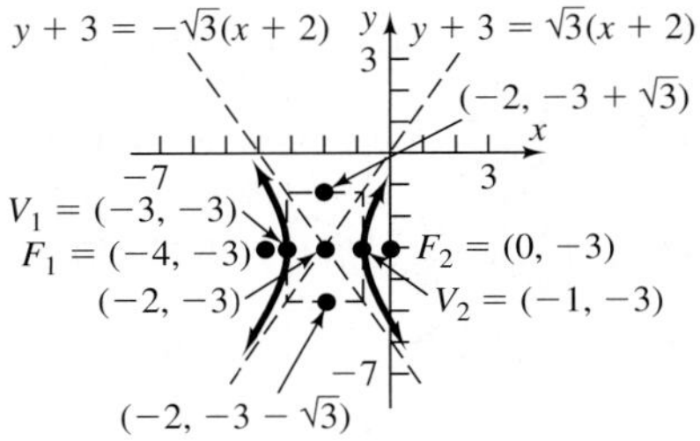

30. $(x-3)^2 = -4(y-7)$

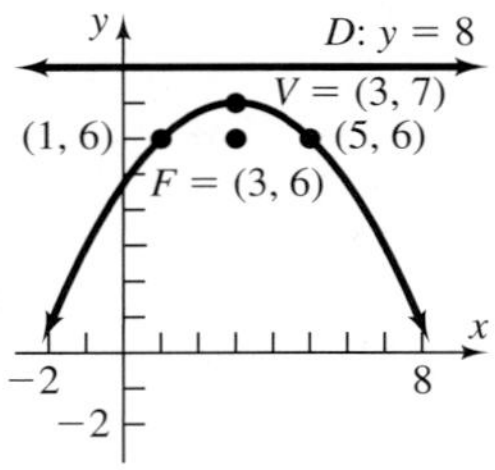

31. $\dfrac{(x+4)^2}{16} + \dfrac{(y-5)^2}{25} = 1$

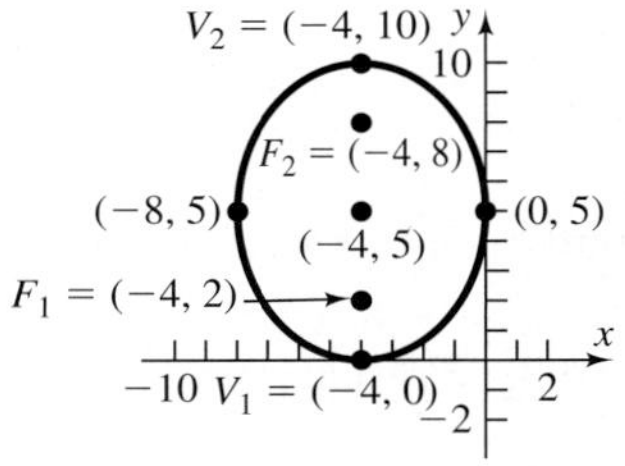

32. $\dfrac{(x-1)^2}{16} - \dfrac{(y-3)^2}{20} = 1$

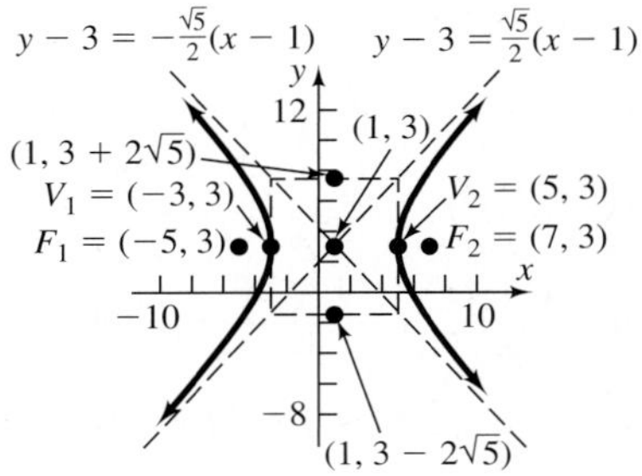

33. $\dfrac{(x+1)^2}{9} - \dfrac{(y-2)^2}{7} = 1$

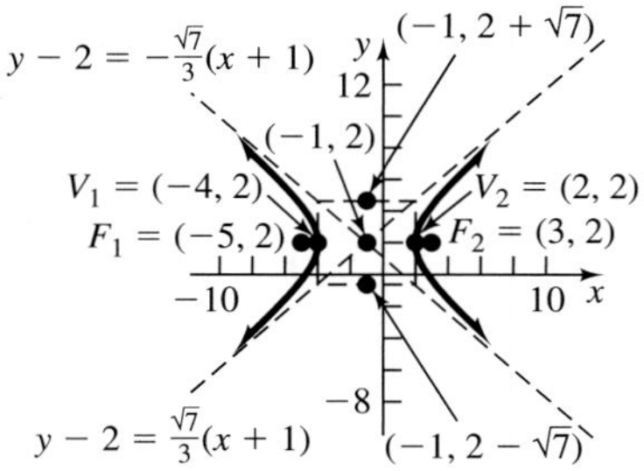

34. $(y+2)^2 - \dfrac{(x-4)^2}{15} = 1$

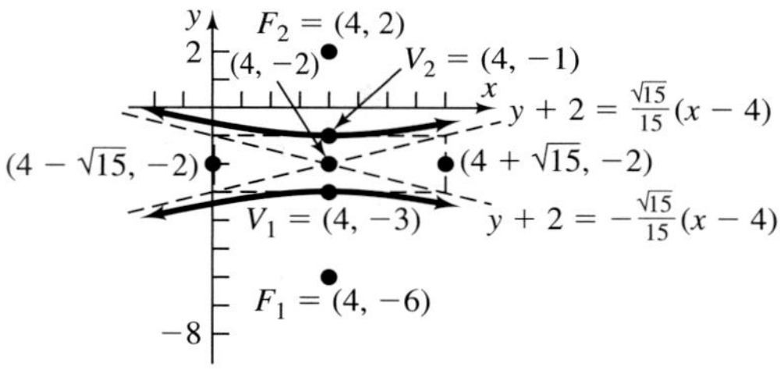

35. $\dfrac{(x-3)^2}{9} - \dfrac{(y-1)^2}{4} = 1$

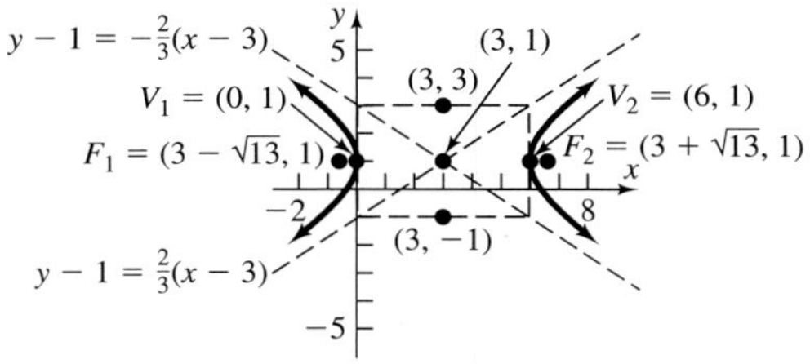

36. $\dfrac{(y-2)^2}{4} - \dfrac{(x-4)^2}{1} = 1$

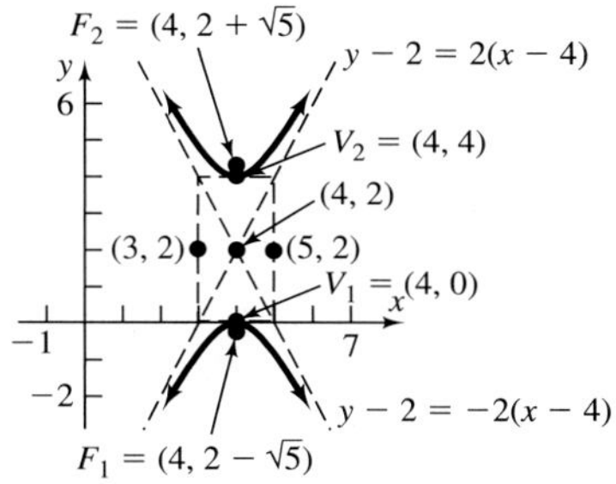

37. Parabola **38.** Parabola
39. Ellipse **40.** Hyperbola
41. Parabola **42.** Parabola
43. Hyperbola **44.** Hyperbola
45. Ellipse **46.** Hyperbola

47. $x'^2 - \dfrac{y'^2}{9} = 1$

Hyperbola
Center at the origin
Transverse axis the x'-axis
Vertices at $(\pm 1, 0)$

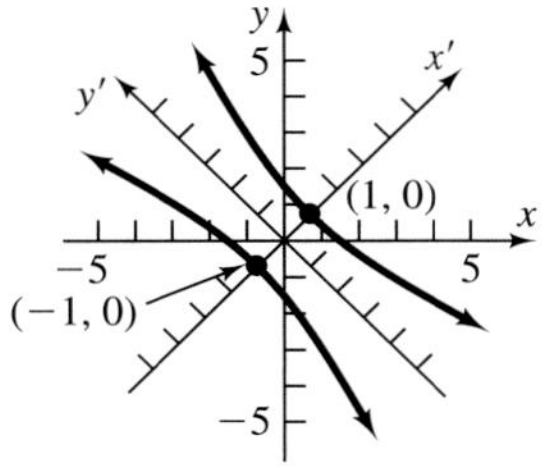

48. $y'^2 - \dfrac{x'^2}{9} = 1$

Hyperbola
Center at the origin
Transverse axis the y'-axis
Vertices at $(0, \pm 1)$

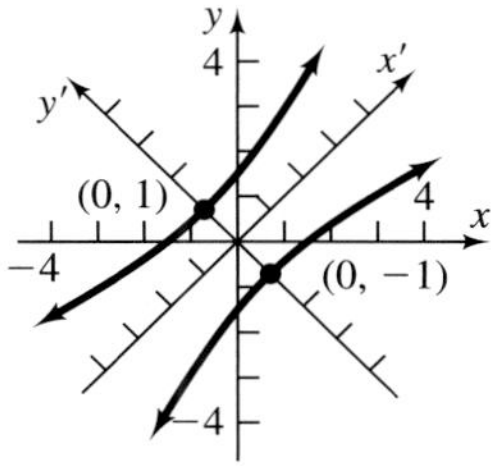

49. $\dfrac{x'^2}{2} + \dfrac{y'^2}{4} = 1$

Ellipse
Center at origin
Major axis the y'-axis
Vertices at $(0, \pm 2)$

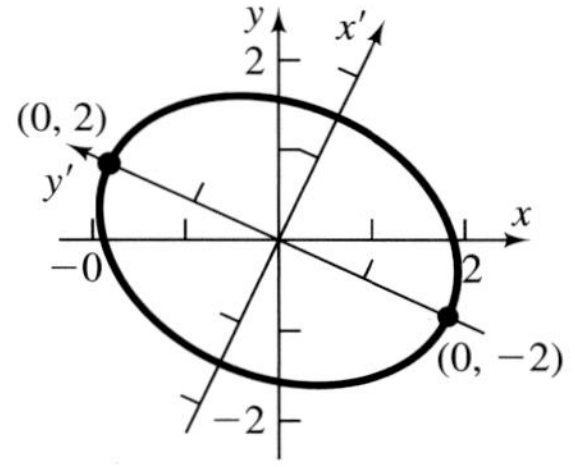

50. $x'^2 = 8y'$

Parabola
Vertex at the origin
Focus on the y'-axis at $(0, 2)$

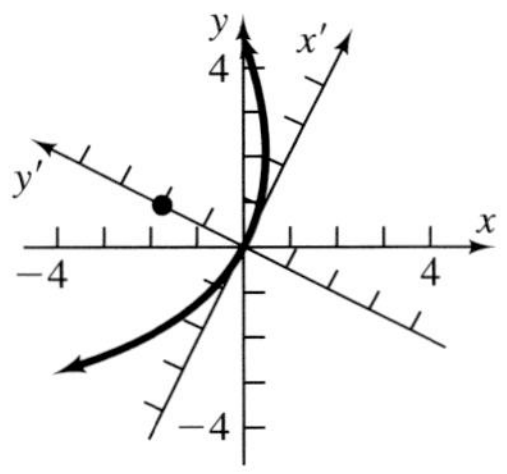

51. $y'^2 = -\dfrac{4\sqrt{13}}{13}x'$

Parabola
Vertex at the origin
Focus on the x'-axis at $\left(-\dfrac{\sqrt{13}}{13}, 0\right)$

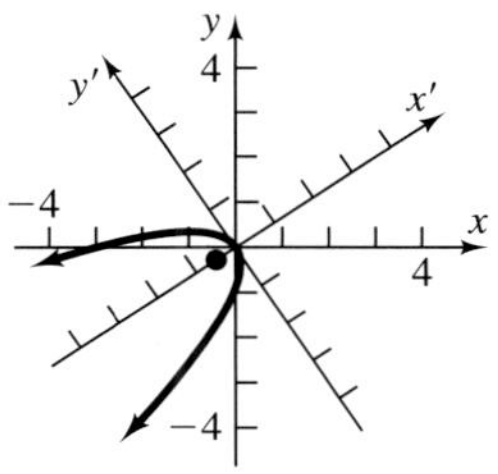

52. $y'^2 = -4x'$

Parabola
Vertex at the origin
Focus on the x'-axis at $(-1, 0)$

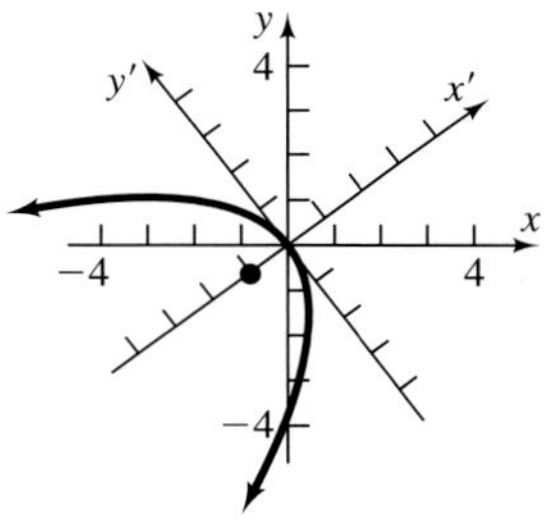

53. Parabola; directrix is perpendicular to the polar axis 4 units to the left of the pole; vertex is $(2, \pi)$.

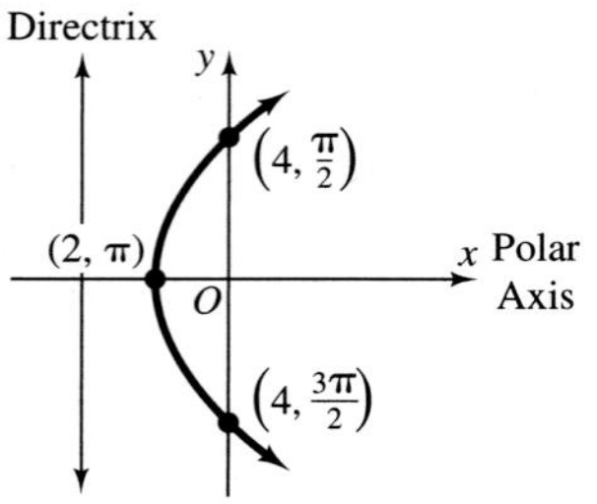

54. Parabola; directrix is parallel to the polar axis 6 units above the pole; vertex is $\left(3, \frac{\pi}{2}\right)$.

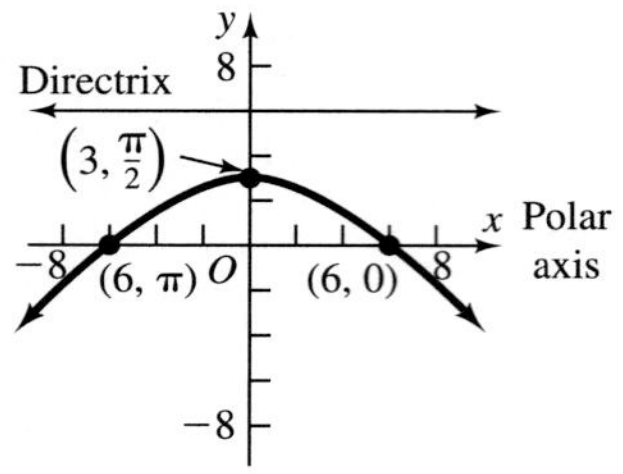

55. Ellipse; directrix is parallel to the polar axis 6 units below the pole; vertices are $\left(6, \frac{\pi}{2}\right)$ and $\left(2, \frac{3\pi}{2}\right)$.

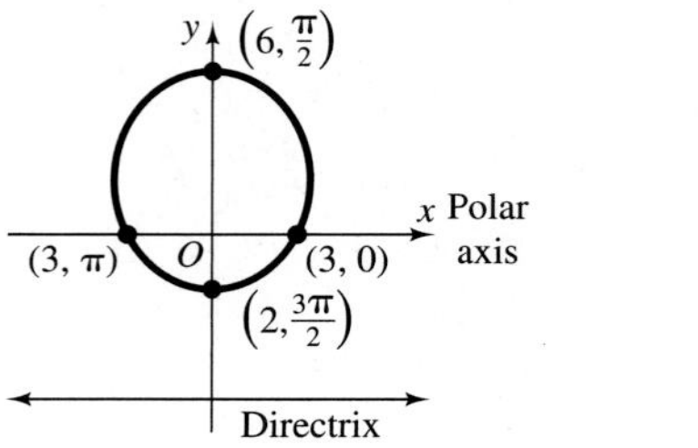

56. Ellipse; directrix is perpendicular to the polar axis 1 unit to the right of the pole; vertices are $\left(\frac{2}{5}, 0\right)$ and $(2, \pi)$.

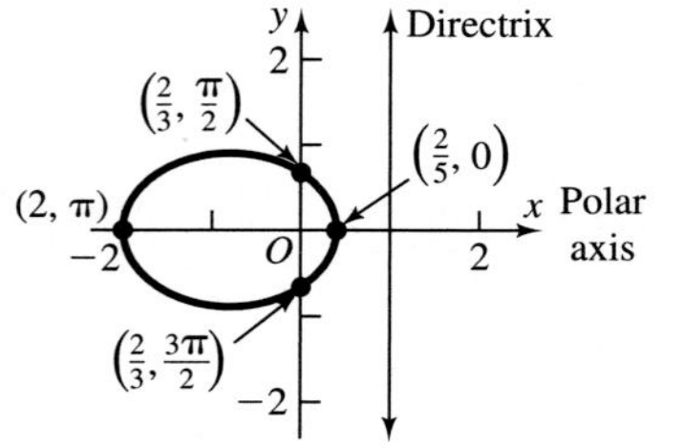

57. Hyperbola; directrix is perpendicular to the polar axis 1 unit to the right of the pole; vertices are $\left(\frac{2}{3}, 0\right)$ and $(-2, \pi)$.

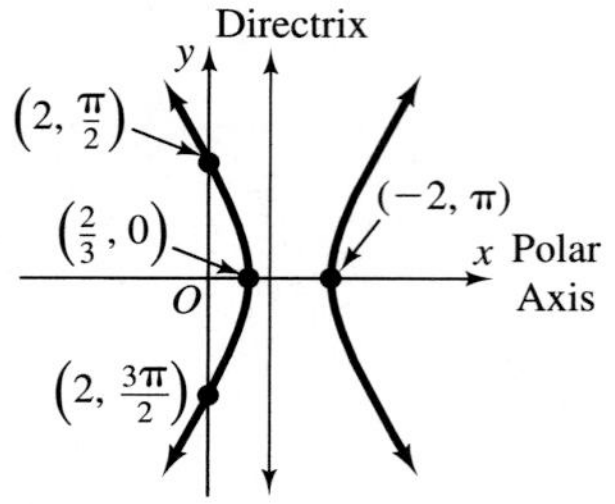

58. Hyperbola; directrix is parallel to the polar axis $\frac{1}{2}$ unit above the pole; vertices are $\left(-\frac{2}{3}, \frac{3\pi}{2}\right)$ and $\left(\frac{2}{5}, \frac{\pi}{2}\right)$.

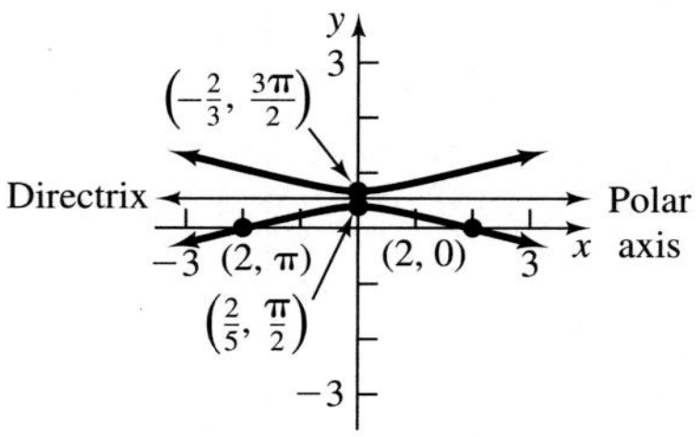

59. $y^2 - 8x - 16 = 0$ **60.** $4x^2 + 3y^2 - 12y - 36 = 0$ **61.** $3x^2 - y^2 - 8x + 4 = 0$ **62.** $5x^2 + 9y^2 + 8x - 4 = 0$

63.

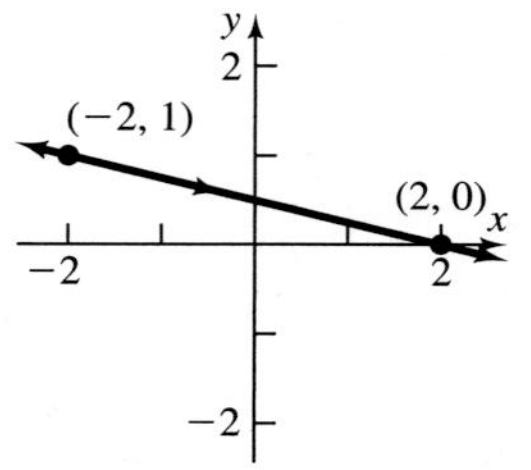

$x + 4y = 2$

64.

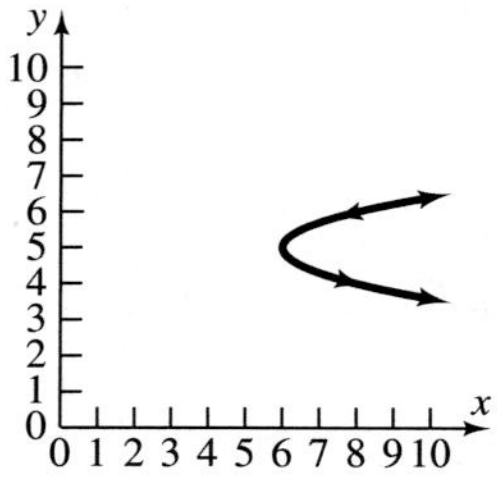

$x = 2(5 - y)^2 + 6$

65.

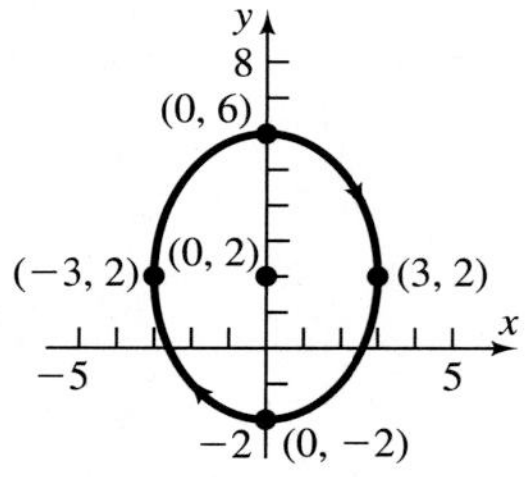

$\frac{x^2}{9} + \frac{(y - 2)^2}{16} = 1$

66.

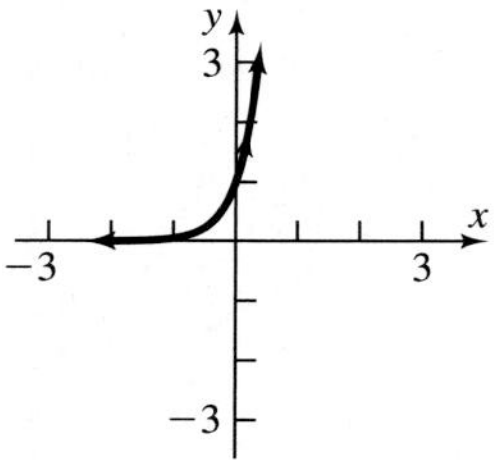

$x = \frac{1}{3} \ln y$

67.

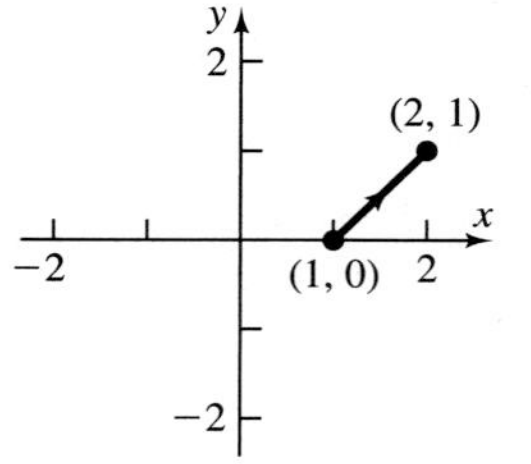

$1 + y = x$

68.

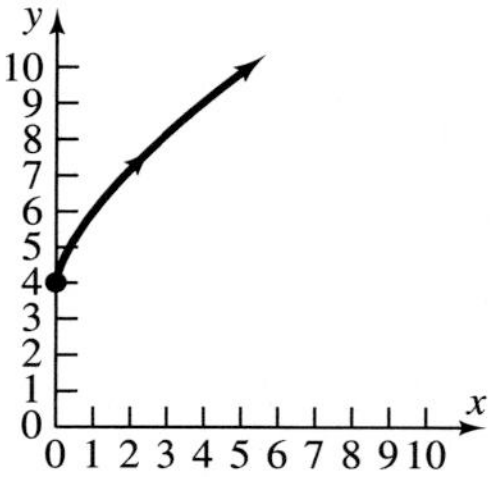

$y = 2x^{2/3} + 4$

69. $x = t,\ y = -2t + 4,\ -\infty < t < \infty$
$x = \frac{t - 4}{-2},\ y = t,\ -\infty < t < \infty$

70. $x = t,\ y = 2t^2 - 8,\ -\infty < t < \infty$
$x = 2t,\ y = 8t^2 - 8,\ -\infty < t < \infty$

71. $x = 4 \cos\left(\frac{\pi}{2} t\right),\ y = 3 \sin\left(\frac{\pi}{2} t\right),\ 0 \le t \le 4$

72. $x = 4 \sin\left(\frac{2\pi}{5} t\right),\ y = 3 \cos\left(\frac{2\pi}{5} t\right),\ 0 \le t \le 5$

73. $\frac{x^2}{5} - \frac{y^2}{4} = 1$ **74.** $\frac{x^2}{20} + \frac{y^2}{4} = 1$ **75.** The ellipse $\frac{x^2}{16} + \frac{y^2}{7} = 1$ **76.** The hyperbola $\frac{x^2}{16} - \frac{y^2}{9} = 1$ **77.** $\frac{1}{4}$ ft or 3 in.

78. 19.44 ft, 17.78 ft, 11.11 ft **79.** 19.72 ft, 18.86 ft, 14.91 ft **80.** The foci are located 8.78 ft in from each wall.

81. (a) 45.24 mi from the Master Station **(b)** 0.000645 sec **(c)** (66, 20)

82. **(a)** Train: $x_1 = \frac{3}{2}t^2$; Mary: $x_2 = 6(t - 2)$

$y_1 = 1$ $\quad y_2 = 3$

(b) Mary won't catch the train.

(c)

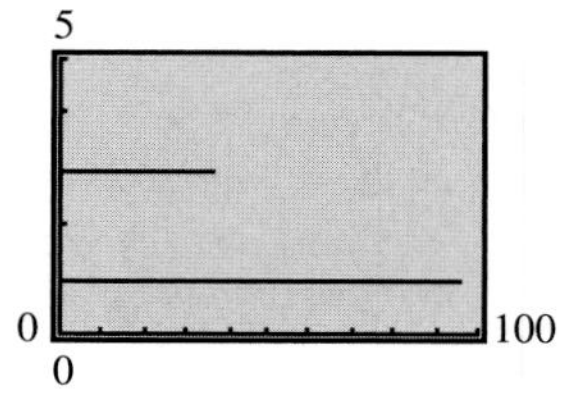

83. **(a)** $x = (100 \cos 35°)t$

$y = -16t^2 + (100 \sin 35°)t + 6$

(b) 3.6866 sec **(c)** 1.7924 sec; 57.4 ft

(d) 302 ft **(e)**

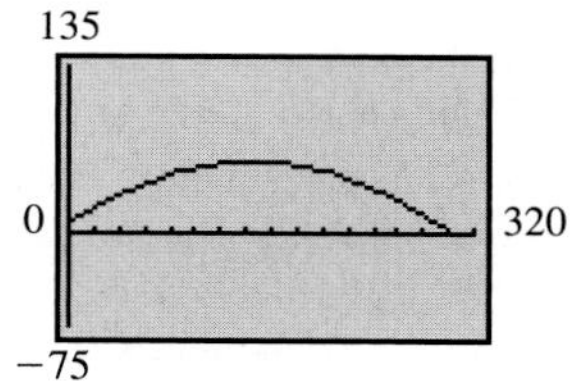

Cumulative Review *(page 682)*

1. $\theta = \frac{\pi}{12} \pm \pi k$, k is any integer; $\theta = \frac{5\pi}{12} \pm \pi k$, k is any integer **2.** $\theta = \frac{\pi}{6}$

3. $r = 8 \sin \theta$

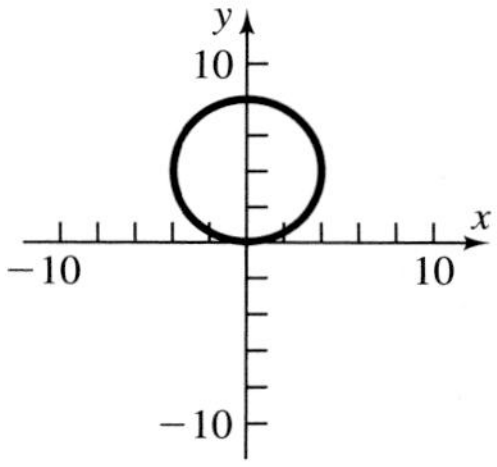

4. $\left\{x \,\middle|\, x \neq \frac{3\pi}{4} \pm \pi k, k \text{ is an integer}\right\}$ **5.** $-6x + 5 - 3h$

6. **(a)** Domain: $(-\infty, \infty)$; Range: $(2, \infty)$

(b) $y = \log_3(x - 2)$; Domain: $(2, \infty)$; Range: $(-\infty, \infty)$

7. $x = 2$ or $x = -\frac{1}{3}$ or $x = -5$

8. $-3 \leq x \leq 2$ or $[-3, 2]$ **9.** $\theta = 22.5°$

10. **(a)** $y = 2x - 2$ **(b)** $(x - 2)^2 + y^2 = 4$

(c) $\frac{x^2}{9} + \frac{y^2}{4} = 1$ **(d)** $y = 2(x - 1)^2$ **(e)** $\frac{y^2}{1} - \frac{x^2}{3} = 1$ **(f)** $y = 4^x$

11. **(a)** 18 **(b)** $(2, 18]$

CHAPTER 10 Systems of Equations and Inequalities

10.1 Concepts and Vocabulary *(page 696)*

3. inconsistent **4.** consistent **5.** False **6.** True

10.1 Exercises *(page 696)*

7. $\begin{cases} 2(2) - (-1) = 5 \\ 5(2) + 2(-1) = 8 \end{cases}$ **8.** $\begin{cases} 3(-2) + 2(4) = 2 \\ (-2) - 7(4) = -30 \end{cases}$ **9.** $\begin{cases} 3(2) - 4\left(\frac{1}{2}\right) = 4 \\ \frac{1}{2}(2) - 3\left(\frac{1}{2}\right) = -\frac{1}{2} \end{cases}$ **10.** $\begin{cases} 2\left(-\frac{1}{2}\right) + \frac{1}{2}(2) = 0 \\ 3\left(-\frac{1}{2}\right) - 4(2) = -\frac{19}{2} \end{cases}$ **11.** $\begin{cases} 4 - 1 = 3 \\ \frac{1}{2}(4) + 1 = 3 \end{cases}$

12. $\begin{cases} (-2) - (-5) = 3 \\ -3(-2) + (-5) = 1 \end{cases}$ **13.** $\begin{cases} 3(1) + 3(-1) + 2(2) = 4 \\ 1 - (-1) - 2 = 0 \\ 2(-1) - 3(2) = -8 \end{cases}$ **14.** $\begin{cases} 4(2) - 1 = 7 \\ 8(2) + 5(-3) - 1 = 0 \\ -2 - (-3) + 5(1) = 6 \end{cases}$ **15.** $\begin{cases} 3(2) + 3(-2) + 2(2) = 4 \\ 2 - 3(-2) + 2 = 10 \\ 5(2) - 2(-2) - 3(2) = 8 \end{cases}$

16. $\begin{cases} 4(4) - 5(2) = 6 \\ 5(-3) - 2 = -17 \\ -4 - 6(-3) + 5(2) = 24 \end{cases}$ **17.** $x = 6, y = 2$ **18.** $x = 1, y = 2$ **19.** $x = 3, y = 2$ **20.** $x = -1, y = 2$ **21.** $x = 8, y = -4$

22. $x = -\frac{13}{4}, y = 2$ **23.** $x = \frac{1}{3}, y = -\frac{1}{6}$ **24.** $x = -\frac{5}{3}, y = 1$ **25.** Inconsistent **26.** Inconsistent **27.** $x = 1, y = 2$

28. $x = 1, y = -\frac{4}{3}$ **29.** $x = 4 - 2y$, y is any real number **30.** $y = 3x - 7$, x is any real number **31.** $x = 1, y = 1$

32. $x = \frac{1}{5}, y = \frac{3}{10}$ **33.** $x = \frac{3}{2}, y = 1$ **34.** $x = 2, y = -3$ **35.** $x = 4, y = 3$ **36.** $x = 12, y = 6$ **37.** $x = \frac{4}{3}, y = \frac{1}{5}$

38. $x = \frac{1}{2}, y = 2$ **39.** $x = \frac{1}{5}, y = \frac{1}{3}$ **40.** $x = 4, y = 3$ **41.** $x = 8, y = 2, z = 0$ **42.** $x = -3, y = 2, z = 1$

43. $x = 2, y = -1, z = 1$ **44.** $x = \frac{56}{13}, y = -\frac{7}{13}, z = \frac{35}{13}$ **45.** Inconsistent **46.** Inconsistent

47. $x = 5z - 2,\ y = 4z - 3$, where z is any real number **48.** $x = \frac{4}{7}y + \frac{2}{7}$ and $z = -\frac{13}{7}y + \frac{4}{7}$, where y is any real number

49. Inconsistent **50.** Inconsistent **51.** $x = 1,\ y = 3,\ z = -2$ **52.** $x = 1,\ y = 3,\ z = -2$

53. $x = -3,\ y = \frac{1}{2},\ z = 1$ **54.** $x = 3,\ y = -\frac{8}{3},\ z = \frac{1}{9}$ **55.** Length 30 feet; width 15 feet **56.** 775 meters by 725 meters

57. Cheeseburger \$1.55; shake \$0.85 **58.** 110 adults; 215 seniors **59.** 22.5 pounds

60. **(a)** \$90,000 in AA bonds, \$60,000 in Bank Certificate **(b)** \$130,000 in AA bonds, \$20,000 in Bank Certificate

61. Average wind speed 25 mph; average airspeed 175 mph **62.** 30 miles per hour **63.** 80 \$25 sets and 120 \$45 sets

64. A single hot dog: \$1; a single soft drink: \$0.50 **65.** \$5.56

66. Pamela's average speed: 4 miles per hour; the speed of the current: 1 mile per hour

67. Mix 50 mg of first compound with 75 mg of second. **68.** First powder: 30 units; second powder: 15 units

69. $a = \frac{4}{3},\ b = -\frac{5}{3},\ c = 1$ **70.** $a = 3,\ b = -1,\ c = -6$ **71.** $I_1 = \frac{10}{71},\ I_2 = \frac{65}{71},\ I_3 = \frac{55}{71}$

72. $I_1 = \frac{11}{13},\ I_2 = \frac{6}{13},\ I_3 = \frac{17}{13}$ **73.** 100 orchestra, 210 main, and 190 balcony seats **74.** 110 adults; 220 children; 75 senior citizens

75. 1.5 chicken, 1 corn, 2 milk **76.** She should place \$8000 in Treasury bills, \$7000 in Treasury bonds, and \$5000 in corporate bonds.

77. If x = price of hamburgers, y = price of fries, z = price of colas, then $x = 2.75 - z,\ y = \frac{41}{60} + \frac{1}{3}z$, $\$0.60 \le z \le \$0.90$.

There is not sufficient information:

x	\$2.13	\$2.01	\$1.86
y	\$0.89	\$0.93	\$0.98
z	\$0.62	\$0.74	\$0.89

78. Yes; hamburgers are \$1.95, fries are \$0.95, cola is \$0.80. **79.** It will take Beth 30 hr, Bill 24 hr, and Edie 40 hr.

10.2 Concepts and Vocabulary *(page 713)*

1. matrix **2.** augmented **3.** True **4.** False

10.2 Exercises *(page 713)*

5. $\left[\begin{array}{rr|r} 1 & -5 & 5 \\ 4 & 3 & 6 \end{array}\right]$ **6.** $\left[\begin{array}{rr|r} 3 & 4 & 7 \\ 4 & -2 & 5 \end{array}\right]$ **7.** $\left[\begin{array}{rr|r} 2 & 3 & 6 \\ 4 & -6 & -2 \end{array}\right]$ **8.** $\left[\begin{array}{rr|r} 9 & -1 & 0 \\ 3 & -1 & 4 \end{array}\right]$ **9.** $\left[\begin{array}{rr|r} 0.01 & -0.03 & 0.06 \\ 0.13 & 0.10 & 0.20 \end{array}\right]$ **10.** $\left[\begin{array}{rr|r} \frac{4}{3} & -\frac{3}{2} & \frac{3}{4} \\ -\frac{1}{4} & \frac{1}{3} & \frac{2}{3} \end{array}\right]$

11. $\left[\begin{array}{rrr|r} 1 & -1 & 1 & 10 \\ 3 & 3 & 0 & 5 \\ 1 & 1 & 2 & 2 \end{array}\right]$ **12.** $\left[\begin{array}{rrr|r} 5 & -1 & -1 & 0 \\ 1 & 1 & 0 & 5 \\ 2 & 0 & -3 & 2 \end{array}\right]$ **13.** $\left[\begin{array}{rrr|r} 1 & 1 & -1 & 2 \\ 3 & -2 & 0 & 2 \\ 5 & 3 & -1 & 1 \end{array}\right]$ **14.** $\left[\begin{array}{rrr|r} 2 & 3 & -4 & 0 \\ 1 & 0 & -5 & -2 \\ 1 & 2 & -3 & -2 \end{array}\right]$ **15.** $\left[\begin{array}{rrr|r} 1 & -1 & -1 & 10 \\ 2 & 1 & 2 & -1 \\ -3 & 4 & 0 & 5 \\ 4 & -5 & 1 & 0 \end{array}\right]$

16. $\left[\begin{array}{rrrr|r} 1 & -1 & 2 & -1 & 5 \\ 1 & 3 & -4 & 2 & 2 \\ 3 & -1 & -5 & -1 & -1 \end{array}\right]$ **17.** $\left[\begin{array}{rr|r} 1 & -3 & -2 \\ 0 & 1 & 9 \end{array}\right]$ **18.** $\left[\begin{array}{rr|r} 1 & -3 & -3 \\ 0 & 1 & 2 \end{array}\right]$ **19.** **(a)** $\left[\begin{array}{rrr|r} 1 & -3 & 4 & 3 \\ 0 & 1 & -2 & 0 \\ -3 & 3 & 4 & 6 \end{array}\right]$ **(b)** $\left[\begin{array}{rrr|r} 1 & -3 & 4 & 3 \\ 2 & -5 & 6 & 6 \\ 0 & -6 & 16 & 15 \end{array}\right]$

20. **(a)** $\left[\begin{array}{rrr|r} 1 & -3 & 3 & -5 \\ 0 & 1 & -9 & 5 \\ -3 & -2 & 4 & 6 \end{array}\right]$ **(b)** $\left[\begin{array}{rrr|r} 1 & -3 & 3 & -5 \\ 2 & -5 & -3 & -5 \\ 0 & -11 & 13 & -9 \end{array}\right]$ **21.** **(a)** $\left[\begin{array}{rrr|r} 1 & -3 & 2 & -6 \\ 0 & 1 & -1 & 8 \\ -3 & -6 & 4 & 6 \end{array}\right]$ **(b)** $\left[\begin{array}{rrr|r} 1 & -3 & 2 & -6 \\ 2 & -5 & 3 & -4 \\ 0 & -15 & 10 & -12 \end{array}\right]$

22. **(a)** $\left[\begin{array}{rrr|r} 1 & -3 & -4 & -6 \\ 0 & 1 & 14 & 6 \\ -3 & 1 & 4 & 6 \end{array}\right]$ **(b)** $\left[\begin{array}{rrr|r} 1 & -3 & -4 & -6 \\ 2 & -5 & 6 & -6 \\ 0 & -8 & -8 & -12 \end{array}\right]$ **23.** **(a)** $\left[\begin{array}{rrr|r} 1 & -3 & 1 & -2 \\ 0 & 1 & 4 & 2 \\ -3 & 1 & 4 & 6 \end{array}\right]$ **(b)** $\left[\begin{array}{rrr|r} 1 & -3 & 1 & -2 \\ 2 & -5 & 6 & -2 \\ 0 & -8 & 7 & 0 \end{array}\right]$

24. **(a)** $\left[\begin{array}{rrr|r} 1 & -3 & -1 & 2 \\ 0 & 1 & 4 & 2 \\ -3 & -6 & 4 & 6 \end{array}\right]$ **(b)** $\left[\begin{array}{rrr|r} 1 & -3 & -1 & 2 \\ 2 & -5 & 2 & 6 \\ 0 & -15 & 1 & 12 \end{array}\right]$

25. $\begin{cases} x = 5 \\ y = -1 \end{cases}$ consistent; $x = 5,\ y = -1$

26. $\begin{cases} x = -4 \\ y = 0 \end{cases}$ consistent; $x = -4,\ y = 0$

27. $\begin{cases} x = 1 \\ y = 2 \\ 0 = 3 \end{cases}$ inconsistent

28. $\begin{cases} x = 0 \\ y = 0 \\ 0 = 2 \end{cases}$ inconsistent

29. $\begin{cases} x + 2z = -1 \\ y - 4z = -2 \\ 0 = 0 \end{cases}$ consistent; $x = -1 - 2z$, $y = -2 + 4z$, z is any real number

30. $\begin{cases} x + 4z = 4 \\ y + 3z = 2 \\ 0 = 0 \end{cases}$ consistent; $x = 4 - 4z$, $y = 2 - 3z$, z is any real number

31. $\begin{cases} x_1 = 1 \\ x_2 + x_4 = 2 \\ x_3 + 2x_4 = 3 \end{cases}$ consistent; $x_1 = 1, x_2 = 2 - x_4$, $x_3 = 3 - 2x_4$, x_4 is any real number

32. $\begin{cases} x_1 = 1 \\ x_2 + 2x_4 = 2 \\ x_3 + 3x_4 = 0 \end{cases}$ consistent; $x_1 = 1$, $x_2 = 2 - 2x_4$, $x_3 = -3x_4$, x_4 is any real number

33. $\begin{cases} x_1 + 4x_4 = 2 \\ x_2 + x_3 + 3x_4 = 3 \\ 0 = 0 \end{cases}$ consistent; $x_1 = 2 - 4x_4$, $x_2 = 3 - x_3 - 3x_4$, x_3, x_4 are any real numbers

34. $\begin{cases} x_1 = 1 \\ x_2 = 2 \\ x_3 + 2x_4 = 3 \end{cases}$ consistent; $x_1 = 1$, $x_2 = 2$, $x_3 = 3 - 2x_4$, x_4 is any real number

35. $\begin{cases} x_1 + x_4 = -2 \\ x_2 + 2x_4 = 2 \\ x_3 - x_4 = 0 \end{cases}$ consistent; $x_1 = -2 - x_4$, $x_2 = 2 - 2x_4$, $x_3 = x_4$, x_4 is any real number

36. $\begin{cases} x_1 = 1 \\ x_2 = 2 \\ x_3 = 3 \\ x_4 = 0 \end{cases}$ consistent; $x_1 = 1$, $x_2 = 2$, $x_3 = 3$, $x_4 = 0$

37. $x = 6, y = 2$ **38.** $x = 1, y = 2$ **39.** $x = \frac{1}{2}, y = \frac{3}{4}$ **40.** $x = \frac{1}{3}, y = \frac{2}{3}$ **41.** $x = 4 - 2y$, y is any real number
42. $y = 3x - 7$, x is any real number **43.** $x = \frac{3}{2}, y = 1$ **44.** $x = 2, y = -3$ **45.** $x = \frac{4}{3}, y = \frac{1}{5}$ **46.** $x = \frac{1}{2}, y = 2$
47. $x = 8, y = 2, z = 0$ **48.** $x = -3, y = 2, z = 1$ **49.** $x = 2, y = -1, z = 1$ **50.** $x = \frac{56}{13}, y = -\frac{7}{13}, z = \frac{35}{13}$ **51.** Inconsistent
52. Inconsistent **53.** $x = 5z - 2, y = 4z - 3$, where z is any real number
54. $x = -\frac{4}{13}z + \frac{6}{13}, y = -\frac{7}{13}z + \frac{4}{13}$, where z is any real number **55.** Inconsistent **56.** Inconsistent **57.** $x = 1, y = 3, z = -2$
58. $x = 1, y = 3, z = -2$ **59.** $x = -3, y = \frac{1}{2}, z = 1$ **60.** $x = 3, y = -\frac{8}{3}, z = \frac{1}{9}$ **61.** $x = \frac{1}{3}, y = \frac{2}{3}, z = 1$ **62.** $x = \frac{1}{3}, y = \frac{2}{3}, z = 1$
63. $x = 1, y = 2, z = 0, w = 1$ **64.** $x = 2, y = 1, z = 0, w = 1$ **65.** $y = 0, z = 1 - x$, x is any real number **66.** Inconsistent
67. $x = 2, y = z - 3$, z is any real number **68.** $x = 1 + \frac{4}{3}z, y = 2 - \frac{5}{3}z$, where z is any real number **69.** $x = \frac{13}{9}, y = \frac{7}{18}, z = \frac{19}{18}$
70. Inconsistent **71.** $x = \frac{7}{5} - \frac{3}{5}z - \frac{2}{5}w, y = -\frac{8}{5} + \frac{7}{5}z + \frac{13}{5}w$, where z and w are any real numbers
72. $x = -3 - w, y = -7 - 4w, z = 4 - w$, where w is any real number **73.** $y = -2x^2 + x + 3$ **74.** $y = x^2 - 4x + 2$
75. $f(x) = 3x^3 - 4x^2 + 5$ **76.** $f(x) = x^3 - 2x^2 + 6$ **77.** 1.5 salmon steak, 2 baked eggs, 1 acorn squash
78. 1 serving of pork chop; 2 servings of corn on the cob; 2 servings of 2% milk
79. $4000 in Treasury bills, $4000 in Treasury bonds, $2000 in corporate bonds **80.** $12,000 in Treasury bills, $2000 in Treasury bonds, $6000 in corporate bonds **81.** 8 Deltas, 5 Betas, 10 Sigmas **82.** 6 cases of orange juice, 20 cases of grapefruit juice, 12 cases of tomato juice
83. $I_1 = \frac{44}{23}, I_2 = 2, I_3 = \frac{16}{23}, I_4 = \frac{28}{23}$ **84.** $I_1 = 3.5, I_2 = 2.5, I_3 = 1$

85. (a) Amount Invested At

7%	9%	11%
0	10,000	10,000
1000	8000	11,000
2000	6000	12,000
3000	4000	13,000
4000	2000	14,000
5000	0	15,000

(b) Amount Invested At

7%	9%	11%
12,500	12,500	0
14,500	8500	2000
16,500	4500	4000
18,750	0	6250

(c) All the money invested at 7% provides $2100, more than what is required.

86. (a) Investing all the money at 7% yields more than $1500

(b)

7%	9%	11%
12,500	12,500	0
15,500	6500	3000
18,750	0	6250

(c)

7%	9%	11%
0	12,500	12,500
1000	10,500	13,500
6250	0	18,750

87.

First Liquid	Second Liquid	Third Liquid
50 mg	75 mg	0 mg
36 mg	76 mg	8 mg
22 mg	77 mg	16 mg
8 mg	78 mg	24 mg

88.

First Powder	Second Powder	Third Powder
30 units	15 units	0 units
20 units	14 units	8 units
10 units	13 units	16 units
0 units	12 units	24 units

10.3 Concepts and Vocabulary *(page 725)*

1. determinants **2.** $ad - bc$ **3.** False **4.** False

10.3 Exercises *(page 725)*

5. 2 **6.** 7 **7.** 22 **8.** 28 **9.** −2 **10.** −2 **11.** 10 **12.** −169 **13.** −26 **14.** −119 **15.** $x = 6, y = 2$ **16.** $x = \frac{11}{3}, y = \frac{2}{3}$ **17.** $x = 3, y = 2$ **18.** $x = -1, y = 2$ **19.** $x = 8, y = -4$ **20.** $x = -\frac{13}{4}, y = 2$ **21.** $x = 4, y = -2$ **22.** $x = 2, y = 3$ **23.** Not applicable **24.** Not applicable **25.** $x = \frac{1}{2}, y = \frac{3}{4}$ **26.** $x = \frac{1}{3}, y = \frac{2}{3}$ **27.** $x = \frac{1}{10}, y = \frac{2}{5}$ **28.** $x = \frac{1}{5}, y = \frac{3}{10}$ **29.** $x = \frac{3}{2}, y = 1$ **30.** $x = 2, y = -3$ **31.** $x = \frac{4}{3}, y = \frac{1}{5}$ **32.** $x = \frac{1}{2}, y = 2$ **33.** $x = 1, y = 3, z = -2$ **34.** $x = 1, y = 3, z = -2$ **35.** $x = -3, y = \frac{1}{2}, z = 1$ **36.** $x = 3, y = \frac{-8}{3}, z = \frac{1}{9}$ **37.** Not applicable **38.** Not applicable **39.** $x = 0, y = 0, z = 0$ **40.** $x = 0, y = 0, z = 0$ **41.** Not applicable **42.** Not applicable **43.** −5 **44.** 1 or −1 **45.** $\frac{13}{11}$ **46.** $-\frac{7}{6}$ **47.** 0 or −9 **48.** 0 or $-\frac{1}{2}$ **49.** −4 **50.** 8 **51.** 12 **52.** 4 **53.** 8 **54.** 4 **55.** 8 **56.** 12

57.
$$\begin{aligned}(y_1 - y_2)x - (x_1 - x_2)y + (x_1y_2 - x_2y_1) &= 0\\ (y_1 - y_2)x + (x_2 - x_1)y &= x_2y_1 - x_1y_2\\ (x_2 - x_1)y - (x_2 - x_1)y_1 &= (y_2 - y_1)x + x_2y_1 - x_1y_2 - (x_2 - x_1)y_1\\ (x_2 - x_1)(y - y_1) &= (y_2 - y_1)x - (y_2 - y_1)x_1\\ y - y_1 &= \frac{y_2 - y_1}{x_2 - x_1}(x - x_1)\end{aligned}$$

58. From Problem 57 and substitution we have $y_1 - y_2 = \frac{y_3 - y_2}{x_3 - x_2}(x_1 - x_2)$. The line containing (x_2, y_2) and (x_3, y_3) has the form $y - y_2 = \frac{y_3 - y_2}{x_3 - x_2}(x - x_2)$. Any other point on this line can be substituted for (x, y) to get a true statement, and any point that makes this a true statement is on the line.

59.
$$\begin{vmatrix} x^2 & x & 1\\ y^2 & y & 1\\ z^2 & z & 1\end{vmatrix} = x^2\begin{vmatrix} y & 1\\ z & 1\end{vmatrix} - x\begin{vmatrix} y^2 & 1\\ z^2 & 1\end{vmatrix} + \begin{vmatrix} y^2 & y\\ z^2 & z\end{vmatrix} = x^2(y - z) - x(y^2 - z^2) + yz(y - z)$$
$$= (y - z)[x^2 - x(y + z) + yz] = (y - z)[(x^2 - xy) - (xz - yz)] = (y - z)[x(x - y) - z(x - y)]$$
$$= (y - z)(x - y)(x - z)$$

60. If $a = 0$, we have
$$by = s$$
$$cx + dy = t$$
Thus, $y = \frac{s}{b}$ and $x = \frac{t - dy}{c} = \frac{tb - ds}{bc}$

Using Cramer's Rule we get
$$x = \frac{sd - tb}{-bc} = \frac{tb - sd}{bc}$$
$$y = \frac{-sc}{-bc} = \frac{s}{b}$$

If $b = 0$, we have
$$ax = s$$
$$cx + dy = t$$
Since $D = ad \neq 0$, then $a \neq 0$ and $d \neq 0$.

Thus, $x = \frac{s}{a}$ and $y = \frac{t - cx}{d} = \frac{ta - cs}{ad}$

Using Cramer's Rule we get
$$x = \frac{sd}{ad} = \frac{s}{a}$$
$$y = \frac{at - cs}{ad}$$

If $c = 0$, we have
$$ax + by = s$$
$$dy = t$$
Since $D = ad \neq 0$, then $a \neq 0$ and $d \neq 0$.

Thus, $y = \frac{t}{d}$ and $x = \frac{s - by}{a} = \frac{sd - bt}{ad}$

Using Cramer's Rule we get
$$x = \frac{sd - bt}{ad}$$
$$y = \frac{at}{ad} = \frac{t}{d}$$

If $d = 0$, we have
$$ax + by = s$$
$$cx = t$$
Since $D = -bc \neq 0$, then $b \neq 0$ and $c \neq 0$.

Thus, $x = \frac{t}{c}$ and $y = \frac{s - ax}{b} = \frac{sc - at}{bc}$

Using Cramer's Rule we get
$$x = \frac{-tb}{-bc} = \frac{t}{c}$$
$$y = \frac{at - sc}{-bc} = \frac{sc - at}{bc}$$

61. $\begin{vmatrix} a_{13} & a_{12} & a_{11} \\ a_{23} & a_{22} & a_{21} \\ a_{33} & a_{32} & a_{31} \end{vmatrix} = a_{13}(a_{22}a_{31} - a_{32}a_{21}) - a_{12}(a_{23}a_{31} - a_{33}a_{21}) + a_{11}(a_{23}a_{32} - a_{33}a_{22})$

$= -[a_{11}(a_{22}a_{33} - a_{32}a_{23}) - a_{12}(a_{21}a_{33} - a_{31}a_{23}) + a_{13}(a_{21}a_{32} - a_{31}a_{22})] = -\begin{vmatrix} a_{11} & a_{12} & a_{13} \\ a_{21} & a_{22} & a_{23} \\ a_{31} & a_{32} & a_{33} \end{vmatrix}$

62. $\begin{vmatrix} a_{11} & a_{12} & a_{13} \\ ka_{21} & ka_{22} & ka_{23} \\ a_{31} & a_{32} & a_{33} \end{vmatrix} = -ka_{21}(a_{12}a_{33} - a_{32}a_{13}) + ka_{22}(a_{11}a_{33} - a_{31}a_{13}) - ka_{23}(a_{11}a_{32} - a_{31}a_{12})$

$= k[-a_{21}(a_{12}a_{33} - a_{32}a_{13}) + a_{22}(a_{11}a_{33} - a_{31}a_{13}) - a_{23}(a_{11}a_{32} - a_{31}a_{12})]$

$= k\begin{vmatrix} a_{11} & a_{12} & a_{13} \\ a_{21} & a_{22} & a_{23} \\ a_{31} & a_{32} & a_{33} \end{vmatrix}$

63. $\begin{vmatrix} a_{11} & a_{12} & a_{11} \\ a_{21} & a_{22} & a_{21} \\ a_{31} & a_{32} & a_{31} \end{vmatrix} = a_{11}(a_{22}a_{31} - a_{32}a_{21}) - a_{12}(a_{21}a_{31} - a_{31}a_{21}) + a_{11}(a_{21}a_{32} - a_{31}a_{22})$

$= a_{11}a_{22}a_{31} - a_{11}a_{32}a_{21} - a_{12}(0) + a_{11}a_{21}a_{32} - a_{11}a_{31}a_{22} = 0$

64. $\begin{vmatrix} a_{11} + ka_{21} & a_{12} + ka_{22} & a_{13} + ka_{23} \\ a_{21} & a_{22} & a_{23} \\ a_{31} & a_{32} & a_{33} \end{vmatrix} = (a_{11} + ka_{21})(a_{22}a_{33} - a_{32}a_{23}) - (a_{12} + ka_{22})(a_{21}a_{33} - a_{31}a_{23}) + (a_{13} + ka_{23})(a_{21}a_{32} - a_{31}a_{22})$

$= a_{11}a_{22}a_{33} - a_{11}a_{32}a_{23} + ka_{21}a_{22}a_{33} - ka_{21}a_{32}a_{23} - a_{12}a_{21}a_{33} + a_{12}a_{31}a_{23}$
$- ka_{22}a_{21}a_{33} + ka_{22}a_{31}a_{23} + a_{13}a_{21}a_{32} - a_{13}a_{31}a_{22} + ka_{23}a_{21}a_{32} - ka_{23}a_{31}a_{22}$

$= a_{11}a_{22}a_{33} - a_{11}a_{32}a_{23} - a_{12}a_{21}a_{33} + a_{12}a_{31}a_{23} + a_{13}a_{21}a_{32} - a_{13}a_{31}a_{22}$

$= a_{11}(a_{22}a_{33} - a_{32}a_{23}) - a_{12}(a_{21}a_{33} - a_{31}a_{23}) + a_{13}(a_{21}a_{32} - a_{31}a_{22})$

$= \begin{vmatrix} a_{11} & a_{12} & a_{13} \\ a_{21} & a_{22} & a_{23} \\ a_{31} & a_{32} & a_{33} \end{vmatrix}$

Historical Problems *(page 740)*

1. (a) $2 - 5i \longleftrightarrow \begin{bmatrix} 2 & -5 \\ 5 & 2 \end{bmatrix}, 1 + 3i \longleftrightarrow \begin{bmatrix} 1 & 3 \\ -3 & 1 \end{bmatrix}$ **(b)** $\begin{bmatrix} 2 & -5 \\ 5 & 2 \end{bmatrix}\begin{bmatrix} 1 & 3 \\ -3 & 1 \end{bmatrix} = \begin{bmatrix} 17 & 1 \\ -1 & 17 \end{bmatrix}$ **(c)** $17 + i$ **(d)** $17 + i$

2. (a) $x = k(ar + bs) + l(cr + ds) = r(ka + lc) + s(kb + ld)$ **(b)** $A = \begin{bmatrix} ka + lc & kb + ld \\ ma + nc & mb + nd \end{bmatrix}$

$y = m(ar + bs) + n(cr + ds) = r(ma + nc) + s(mb + nd)$

10.4 Concepts and Vocabulary *(page 741)*

1. inverse **2.** square **3.** identity **4.** False **5.** False **6.** False

10.4 Exercises *(page 741)*

7. $\begin{bmatrix} 4 & 4 & -5 \\ -1 & 5 & 4 \end{bmatrix}$ **8.** $\begin{bmatrix} -4 & 2 & -5 \\ 3 & -1 & 8 \end{bmatrix}$ **9.** $\begin{bmatrix} 0 & 12 & -20 \\ 4 & 8 & 24 \end{bmatrix}$ **10.** $\begin{bmatrix} -12 & -3 & 0 \\ 6 & -9 & 6 \end{bmatrix}$ **11.** $\begin{bmatrix} -8 & 7 & -15 \\ 7 & 0 & 22 \end{bmatrix}$ **12.** $\begin{bmatrix} 16 & 10 & -10 \\ -6 & 16 & 4 \end{bmatrix}$

13. $\begin{bmatrix} 28 & -9 \\ 4 & 23 \end{bmatrix}$ **14.** $\begin{bmatrix} 22 & 6 \\ 14 & -2 \end{bmatrix}$ **15.** $\begin{bmatrix} 1 & 14 & -14 \\ 2 & 22 & -18 \\ 3 & 0 & 28 \end{bmatrix}$ **16.** $\begin{bmatrix} 14 & 7 & -2 \\ 20 & 12 & -4 \\ -14 & 7 & -6 \end{bmatrix}$ **17.** $\begin{bmatrix} 15 & 21 & -16 \\ 22 & 34 & -22 \\ -11 & 7 & 22 \end{bmatrix}$ **18.** $\begin{bmatrix} 50 & -3 \\ 18 & 21 \end{bmatrix}$ **19.** $\begin{bmatrix} 25 & -9 \\ 4 & 20 \end{bmatrix}$

20. $\begin{bmatrix} 6 & 14 & -14 \\ 2 & 27 & -18 \\ 3 & 0 & 33 \end{bmatrix}$ **21.** $\begin{bmatrix} -13 & 7 & -12 \\ -18 & 10 & -14 \\ 17 & -7 & 34 \end{bmatrix}$ **22.** $\begin{bmatrix} 50 & -3 \\ 18 & 21 \end{bmatrix}$ **23.** $\begin{bmatrix} -2 & 4 & 2 & 8 \\ 2 & 1 & 4 & 6 \end{bmatrix}$ **24.** $\begin{bmatrix} -22 & 29 & 8 & -1 \\ -10 & 17 & 6 & -1 \end{bmatrix}$ **25.** $\begin{bmatrix} 5 & 14 \\ 9 & 16 \end{bmatrix}$

26. $\begin{bmatrix} 1 & 2 & -1 \\ 0 & -12 & 3 \\ -15 & 30 & 0 \end{bmatrix}$ **27.** $\begin{bmatrix} 9 & 2 \\ 34 & 13 \\ 47 & 20 \end{bmatrix}$ **28.** $\begin{bmatrix} 6 & 32 \\ 1 & 3 \\ -2 & -4 \end{bmatrix}$ **29.** $\begin{bmatrix} 1 & -1 \\ -1 & 2 \end{bmatrix}$ **30.** $\begin{bmatrix} 1 & 1 \\ 2 & 3 \end{bmatrix}$ **31.** $\begin{bmatrix} 1 & -\frac{5}{2} \\ -1 & 3 \end{bmatrix}$ **32.** $\begin{bmatrix} -1 & -\frac{1}{2} \\ -3 & -2 \end{bmatrix}$

33. $\begin{bmatrix} 1 & \frac{-1}{a} \\ -1 & \frac{2}{a} \end{bmatrix}$ **34.** $\begin{bmatrix} -\frac{2}{b} & \frac{3}{b} \\ 1 & -1 \end{bmatrix}$ **35.** $\begin{bmatrix} 3 & -3 & 1 \\ -2 & 2 & -1 \\ -4 & 5 & -2 \end{bmatrix}$ **36.** $\begin{bmatrix} 3 & -2 & -4 \\ 3 & -2 & -5 \\ -1 & 1 & 2 \end{bmatrix}$ **37.** $\begin{bmatrix} -\frac{5}{7} & \frac{1}{7} & \frac{3}{7} \\ \frac{9}{7} & \frac{1}{7} & -\frac{4}{7} \\ \frac{3}{7} & -\frac{2}{7} & \frac{1}{7} \end{bmatrix}$ **38.** $\begin{bmatrix} \frac{3}{7} & \frac{-4}{7} & \frac{1}{7} \\ \frac{1}{7} & \frac{1}{7} & \frac{-2}{7} \\ \frac{-5}{7} & \frac{9}{7} & \frac{3}{7} \end{bmatrix}$

39. $x = 3, y = 2$ **40.** $x = 12, y = 28$ **41.** $x = -5, y = 10$ **42.** $x = 9, y = 23$ **43.** $x = 2, y = -1$ **44.** $x = -7, y = -28$
45. $x = \frac{1}{2}, y = 2$ **46.** $x = -\frac{1}{2}, y = 3$ **47.** $x = -2, y = 1$ **48.** $x = 2, y = 1$ **49.** $x = \frac{2}{a}, y = \frac{3}{a}$ **50.** $x = \frac{2}{b}, y = 4$

51. $x = -2, y = 3, z = 5$ **52.** $x = 4, y = -2, z = 1$ **53.** $x = \frac{1}{2}, y = -\frac{1}{2}, z = 1$ **54.** $x = 1, y = -1, z = \frac{1}{2}$
55. $x = -\frac{34}{7}, y = \frac{85}{7}, z = \frac{12}{7}$ **56.** $x = \frac{8}{7}, y = \frac{5}{7}, z = \frac{17}{7}$ **57.** $x = \frac{1}{3}, y = 1, z = \frac{2}{3}$ **58.** $x = 1, y = -1, z = 1$

59. $\left[\begin{array}{cc|cc} 4 & 2 & 1 & 0 \\ 2 & 1 & 0 & 1 \end{array}\right] \to \left[\begin{array}{cc|cc} 1 & \frac{1}{2} & \frac{1}{4} & 0 \\ 2 & 1 & 0 & 1 \end{array}\right] \to \left[\begin{array}{cc|cc} 1 & \frac{1}{2} & \frac{1}{4} & 0 \\ 0 & 0 & -\frac{1}{2} & 1 \end{array}\right]$ **60.** $\left[\begin{array}{cc|cc} -3 & \frac{1}{2} & 1 & 0 \\ 6 & -1 & 0 & 1 \end{array}\right] \to \left[\begin{array}{cc|cc} 1 & -\frac{1}{6} & -\frac{1}{3} & 0 \\ 6 & -1 & 0 & 1 \end{array}\right] \to \left[\begin{array}{cc|cc} 1 & -\frac{1}{6} & -\frac{1}{3} & 0 \\ 0 & 0 & 2 & 1 \end{array}\right]$

61. $\left[\begin{array}{cc|cc} 15 & 3 & 1 & 0 \\ 10 & 2 & 0 & 1 \end{array}\right] \to \left[\begin{array}{cc|cc} 1 & \frac{1}{5} & \frac{1}{15} & 0 \\ 10 & 2 & 0 & 1 \end{array}\right] \to \left[\begin{array}{cc|cc} 1 & \frac{1}{5} & \frac{1}{15} & 0 \\ 0 & 0 & -\frac{2}{3} & 1 \end{array}\right]$ **62.** $\left[\begin{array}{cc|cc} -3 & 0 & 1 & 0 \\ 4 & 0 & 0 & 1 \end{array}\right] \to \left[\begin{array}{cc|cc} 1 & 0 & -\frac{1}{3} & 0 \\ 4 & 0 & 0 & 1 \end{array}\right] \to \left[\begin{array}{cc|cc} 1 & 0 & -\frac{1}{3} & 0 \\ 0 & 0 & \frac{4}{3} & 1 \end{array}\right]$

63. $\left[\begin{array}{ccc|ccc} -3 & 1 & -1 & 1 & 0 & 0 \\ 1 & -4 & -7 & 0 & 1 & 0 \\ 1 & 2 & 5 & 0 & 0 & 1 \end{array}\right] \to \left[\begin{array}{ccc|ccc} 1 & 2 & 5 & 0 & 0 & 1 \\ 1 & -4 & -7 & 0 & 1 & 0 \\ -3 & 1 & -1 & 1 & 0 & 0 \end{array}\right] \to \left[\begin{array}{ccc|ccc} 1 & 2 & 5 & 0 & 0 & 1 \\ 0 & -6 & -12 & 0 & 1 & -1 \\ 0 & 7 & 14 & 1 & 0 & 3 \end{array}\right] \to \left[\begin{array}{ccc|ccc} 1 & 2 & 5 & 0 & 0 & 1 \\ 0 & 1 & 2 & 0 & -\frac{1}{6} & \frac{1}{6} \\ 0 & 1 & 2 & \frac{1}{7} & 0 & \frac{3}{7} \end{array}\right]$
$\to \left[\begin{array}{ccc|ccc} 1 & 2 & 5 & 0 & 0 & 1 \\ 0 & 1 & 2 & 0 & -\frac{1}{6} & \frac{1}{6} \\ 0 & 0 & 0 & \frac{1}{7} & \frac{1}{6} & \frac{11}{42} \end{array}\right]$

64. $\left[\begin{array}{ccc|ccc} 1 & 1 & -3 & 1 & 0 & 0 \\ 2 & -4 & 1 & 0 & 1 & 0 \\ -5 & 7 & 1 & 0 & 0 & 1 \end{array}\right] \to \left[\begin{array}{ccc|ccc} 1 & 1 & -3 & 1 & 0 & 0 \\ 0 & -6 & 7 & -2 & 1 & 0 \\ -5 & 7 & 1 & 0 & 0 & 1 \end{array}\right] \to \left[\begin{array}{ccc|ccc} 1 & 1 & -3 & 1 & 0 & 0 \\ 0 & 1 & -\frac{7}{6} & \frac{1}{3} & -\frac{1}{6} & 0 \\ -5 & 7 & 1 & 0 & 0 & 1 \end{array}\right] \to \left[\begin{array}{ccc|ccc} 1 & 1 & -3 & 1 & 0 & 0 \\ 0 & 1 & -\frac{7}{6} & \frac{1}{3} & -\frac{1}{6} & 0 \\ 0 & 12 & -14 & 5 & 0 & 1 \end{array}\right]$
$\to \left[\begin{array}{ccc|ccc} 1 & 1 & -3 & 1 & 0 & 0 \\ 0 & 1 & -\frac{7}{6} & \frac{1}{3} & -\frac{1}{6} & 0 \\ 0 & 0 & 0 & 1 & 2 & 1 \end{array}\right]$

65. $\begin{bmatrix} 0.01 & 0.05 & -0.01 \\ 0.01 & -0.02 & 0.01 \\ -0.02 & 0.01 & 0.03 \end{bmatrix}$ **66.** $\begin{bmatrix} 0.26 & -0.29 & -0.2 \\ -1.21 & 1.63 & 1.2 \\ -1.84 & 2.53 & 1.8 \end{bmatrix}$ **67.** $\begin{bmatrix} 0.02 & -0.04 & -0.01 & 0.01 \\ -0.02 & 0.05 & 0.03 & -0.03 \\ 0.02 & 0.01 & -0.04 & 0.00 \\ -0.02 & 0.06 & 0.07 & 0.06 \end{bmatrix}$ **68.** $\begin{bmatrix} 0.01 & 0.04 & 0.00 & 0.03 \\ 0.02 & -0.02 & 0.01 & 0.01 \\ -0.04 & 0.02 & 0.04 & 0.06 \\ 0.05 & -0.02 & 0.00 & -0.09 \end{bmatrix}$
69. $x = 4.57, y = -6.44, z = -24.07$ **70.** $x = 4.56, y = -6.06, z = -22.55$ **71.** $x = -1.19, y = 2.46, z = 8.27$
72. $x = -2.05, y = 3.88, z = 13.36$
73. (a) $\begin{bmatrix} 500 & 350 & 400 \\ 700 & 500 & 850 \end{bmatrix}$; $\begin{bmatrix} 500 & 700 \\ 350 & 500 \\ 400 & 850 \end{bmatrix}$ **(b)** $\begin{bmatrix} 15 \\ 8 \\ 3 \end{bmatrix}$ **(c)** $\begin{bmatrix} 11{,}500 \\ 17{,}050 \end{bmatrix}$ **(d)** $[0.10 \quad 0.05]$ **(e)** \$2002.50

74. (a) January sales $\begin{bmatrix} 400 & 250 & 50 \\ 450 & 200 & 140 \end{bmatrix}$; February sales $\begin{bmatrix} 350 & 100 & 30 \\ 350 & 300 & 100 \end{bmatrix}$ **(b)** $\begin{bmatrix} 750 & 350 & 80 \\ 800 & 500 & 240 \end{bmatrix}$ **(c)** $\begin{bmatrix} 100 \\ 150 \\ 200 \end{bmatrix}$ **(d)** $\begin{bmatrix} 143{,}500 \\ 203{,}000 \end{bmatrix}$

75. If $D = ad - bc \neq 0$, then $a \neq 0$ and $d \neq 0$, or $b \neq 0$ and $c \neq 0$. Assuming the former then,

$$\left[\begin{array}{cc|cc} a & b & 1 & 0 \\ c & d & 0 & 1 \end{array}\right] \rightarrow \left[\begin{array}{cc|cc} 1 & \frac{b}{a} & \frac{1}{a} & 0 \\ c & d & 0 & 1 \end{array}\right] \rightarrow \left[\begin{array}{cc|cc} 1 & \frac{b}{a} & \frac{1}{a} & 0 \\ 0 & \frac{D}{a} & -\frac{c}{a} & 1 \end{array}\right] \rightarrow \left[\begin{array}{cc|cc} 1 & \frac{b}{a} & \frac{1}{a} & 0 \\ 0 & 1 & -\frac{c}{D} & \frac{a}{D} \end{array}\right] \rightarrow \frac{1}{D}\left[\begin{array}{cc|cc} 1 & 0 & d & -b \\ 0 & 1 & -c & a \end{array}\right]$$

$$R_1 = \frac{1}{a}r_1 \qquad R_2 = -cr_1 + r_2 \qquad R_2 = \frac{a}{D}r_2 \qquad R_1 = -\frac{b}{a}r_2 + r_1$$

10.5 Exercises *(page 750)*

5. Proper **6.** Proper **7.** Improper; $1 + \dfrac{9}{x^2 - 4}$ **8.** Improper; $3 + \dfrac{1}{x^2 - 1}$ **9.** Improper; $5x + \dfrac{22x - 1}{x^2 - 4}$

10. Improper; $3x + \dfrac{x^2 - 24x - 2}{x^3 + 8}$ **11.** Improper; $1 + \dfrac{-2(x - 6)}{(x + 4)(x - 3)}$ **12.** Improper; $2x + \dfrac{6x}{x^2 + 1}$ **13.** $\dfrac{-4}{x} + \dfrac{4}{x - 1}$

14. $\dfrac{2}{x + 2} + \dfrac{1}{x - 1}$ **15.** $\dfrac{1}{x} + \dfrac{-x}{x^2 + 1}$ **16.** $\dfrac{\frac{1}{5}}{x + 1} + \dfrac{-\frac{1}{5}x + \frac{1}{5}}{x^2 + 4}$ **17.** $\dfrac{-1}{x - 1} + \dfrac{2}{x - 2}$ **18.** $\dfrac{1}{x + 2} + \dfrac{2}{x - 4}$

19. $\dfrac{\frac{1}{4}}{x + 1} + \dfrac{\frac{3}{4}}{x - 1} + \dfrac{\frac{1}{2}}{(x - 1)^2}$ **20.** $\dfrac{-\frac{3}{4}}{x} + \dfrac{\frac{-1}{2}}{x^2} + \dfrac{\frac{3}{4}}{x - 2}$ **21.** $\dfrac{\frac{1}{12}}{x - 2} + \dfrac{-\frac{1}{12}(x + 4)}{x^2 + 2x + 4}$ **22.** $\dfrac{2}{x - 1} + \dfrac{-2x - 2}{x^2 + x + 1}$

23. $\dfrac{\frac{1}{4}}{x - 1} + \dfrac{\frac{1}{4}}{(x - 1)^2} + \dfrac{-\frac{1}{4}}{x + 1} + \dfrac{\frac{1}{4}}{(x + 1)^2}$ **24.** $\dfrac{\frac{1}{2}}{x} + \dfrac{\frac{1}{4}}{x^2} + \dfrac{\frac{3}{4}}{(x - 2)^2} - \dfrac{\frac{1}{2}}{(x - 2)}$ **25.** $\dfrac{-5}{x + 2} + \dfrac{5}{x + 1} + \dfrac{-4}{(x + 1)^2}$

26. $\dfrac{\frac{2}{9}}{x + 2} + \dfrac{\frac{7}{9}}{x - 1} + \dfrac{\frac{2}{3}}{(x - 1)^2}$ **27.** $\dfrac{\frac{1}{4}}{x} + \dfrac{1}{x^2} + \dfrac{-\frac{1}{4}(x + 4)}{x^2 + 4}$ **28.** $\dfrac{\frac{14}{3}}{x - 1} + \dfrac{4}{(x - 1)^2} + \dfrac{\frac{-14}{3}x + \frac{4}{3}}{x^2 + 2}$ **29.** $\dfrac{\frac{2}{3}}{x + 1} + \dfrac{\frac{1}{3}(x + 1)}{x^2 + 2x + 4}$

30. $\dfrac{-6}{x} + \dfrac{7x + 7}{x^2 + 3x + 3}$ **31.** $\dfrac{\frac{2}{7}}{3x - 2} + \dfrac{\frac{1}{7}}{2x + 1}$ **32.** $\dfrac{\frac{-1}{7}}{2x + 3} + \dfrac{\frac{2}{7}}{4x - 1}$ **33.** $\dfrac{\frac{3}{4}}{x + 3} + \dfrac{\frac{1}{4}}{x - 1}$ **34.** $\dfrac{-3}{x + 1} + \dfrac{2}{x + 2} + \dfrac{2}{x + 3}$

35. $\dfrac{1}{x^2 + 4} + \dfrac{2x - 1}{(x^2 + 4)^2}$ **36.** $\dfrac{x}{x^2 + 16} + \dfrac{-16x + 1}{(x^2 + 16)^2}$ **37.** $\dfrac{-1}{x} + \dfrac{2}{x - 3} + \dfrac{-1}{x + 1}$ **38.** $\dfrac{-1}{x^2} + \dfrac{-1}{x^4} + \dfrac{1}{x - 1}$

39. $\dfrac{4}{x - 2} + \dfrac{-3}{x - 1} + \dfrac{-1}{(x - 1)^2}$ **40.** $\dfrac{\frac{3}{8}}{x - 1} + \dfrac{\frac{1}{2}}{(x - 1)^2} + \dfrac{\frac{5}{8}}{x + 3}$ **41.** $\dfrac{x}{(x^2 + 16)^2} + \dfrac{-16x}{(x^2 + 16)^3}$ **42.** $\dfrac{1}{(x^2 + 4)^2} + \dfrac{-4}{(x^2 + 4)^3}$

43. $\dfrac{-\frac{8}{7}}{2x + 1} + \dfrac{\frac{4}{7}}{x - 3}$ **44.** $\dfrac{\frac{4}{5}}{2x - 1} + \dfrac{\frac{8}{5}}{x + 2}$ **45.** $\dfrac{-\frac{2}{9}}{x} + \dfrac{-\frac{1}{3}}{x^2} + \dfrac{\frac{1}{6}}{x - 3} + \dfrac{\frac{1}{18}}{x + 3}$ **46.** $\dfrac{\frac{13}{24}}{x - 2} - \dfrac{\frac{13}{24}}{x + 2} - \dfrac{\frac{7}{6}}{x^2 + 2}$

Historical Problem *(page 756)*

$x = 6$ units, $y = 8$ units

10.6 Exercises *(page 757)*

5.

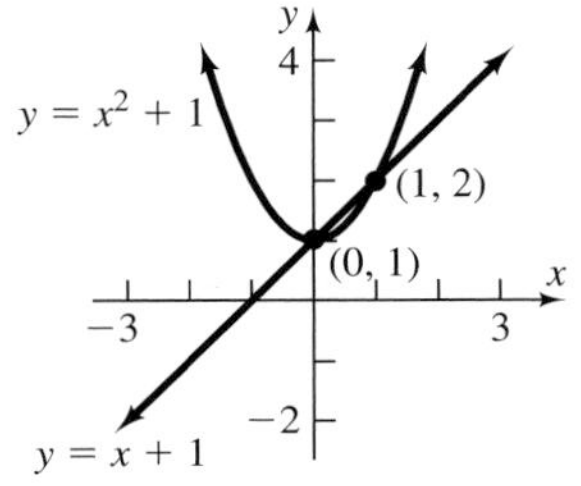

6.

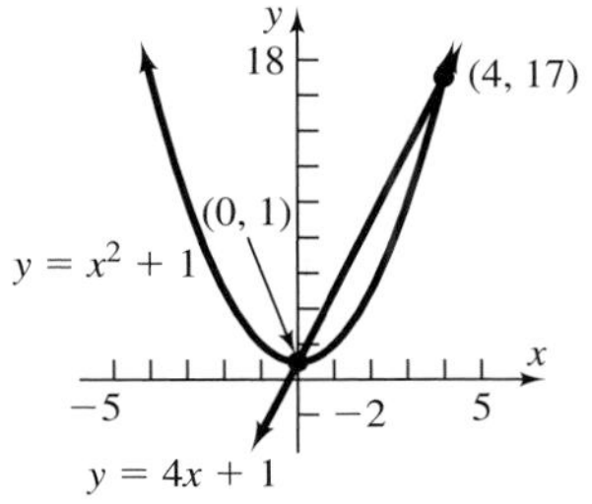

7.

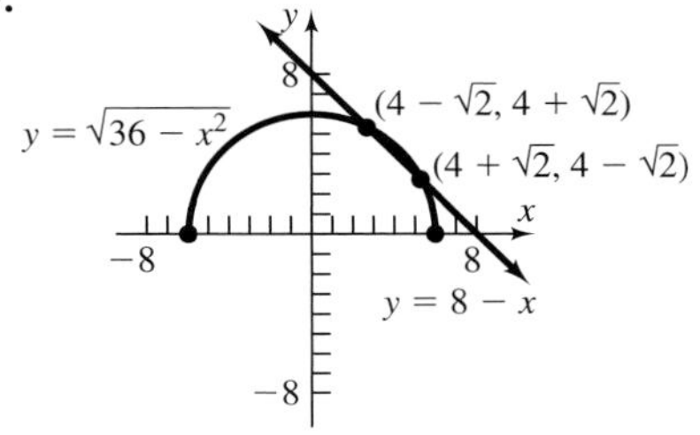

8.
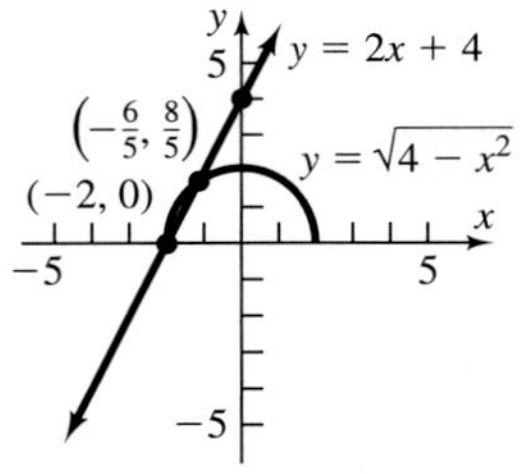

9.
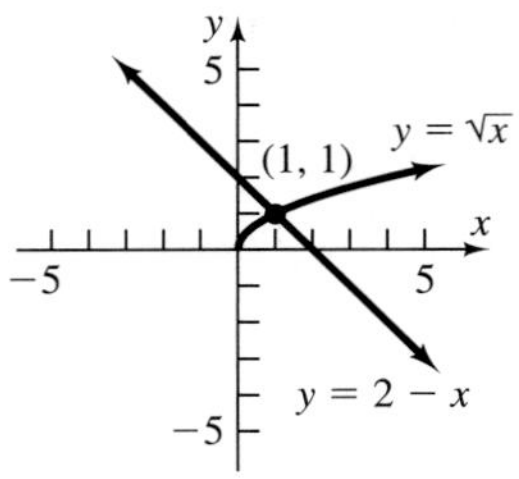

10.
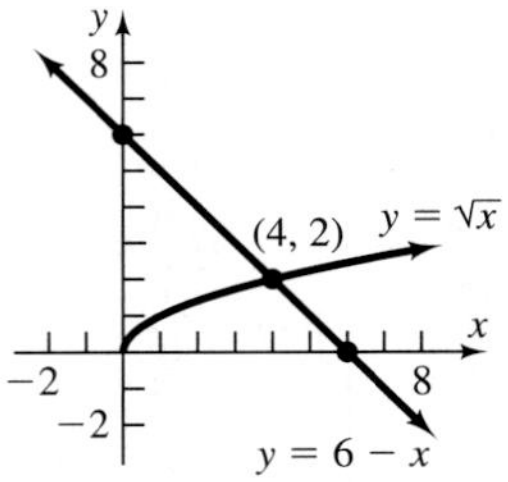

11.
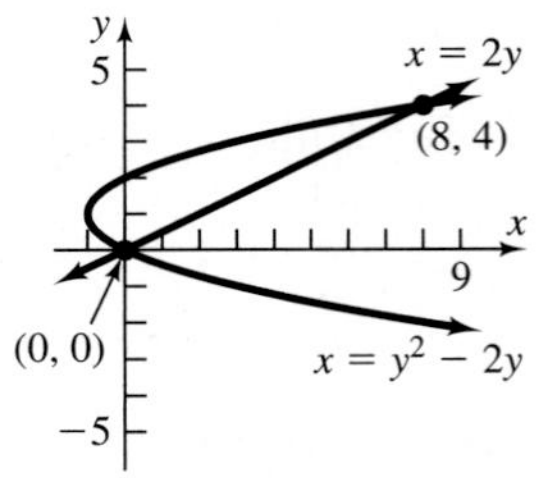

12.
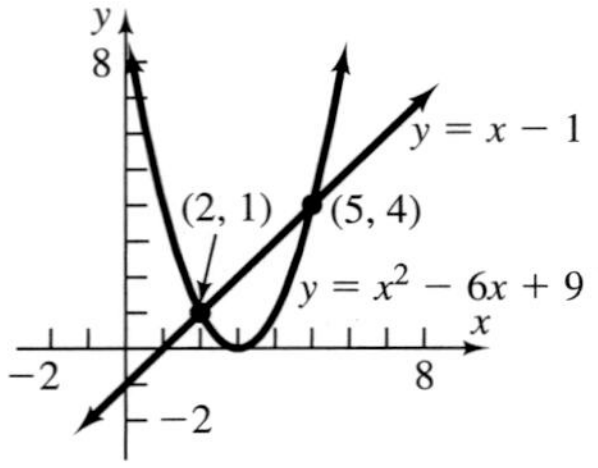

13.
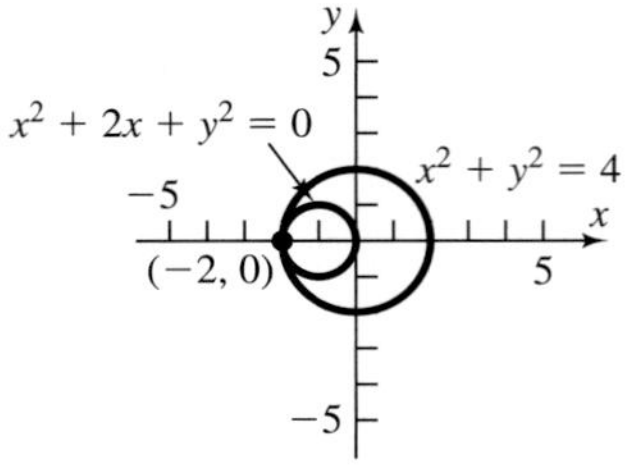

14.
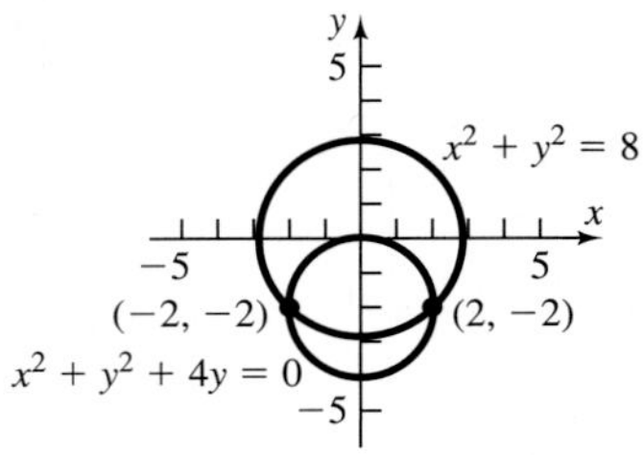

15.
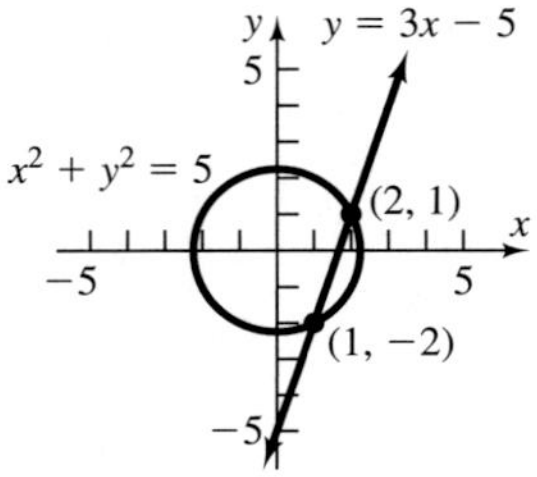

16.
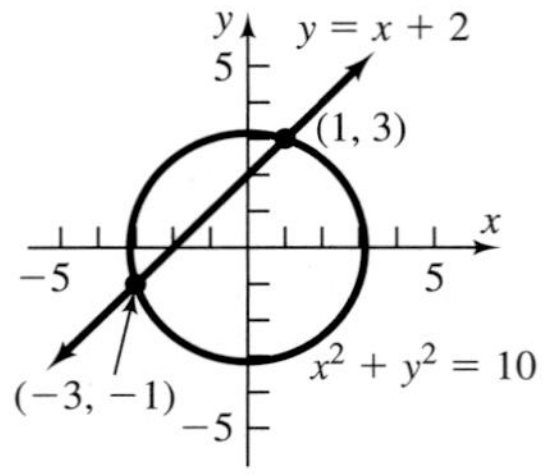

17.
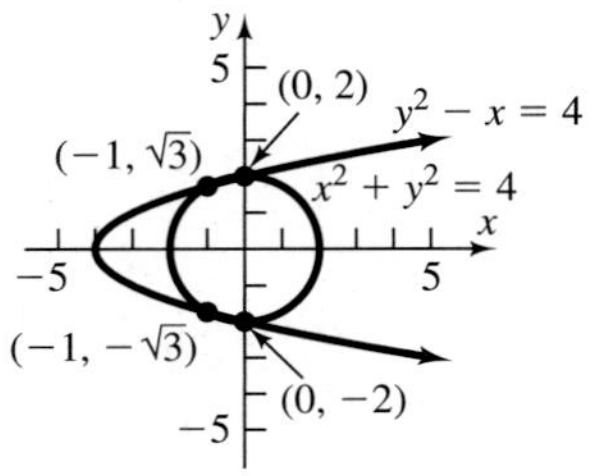

18.
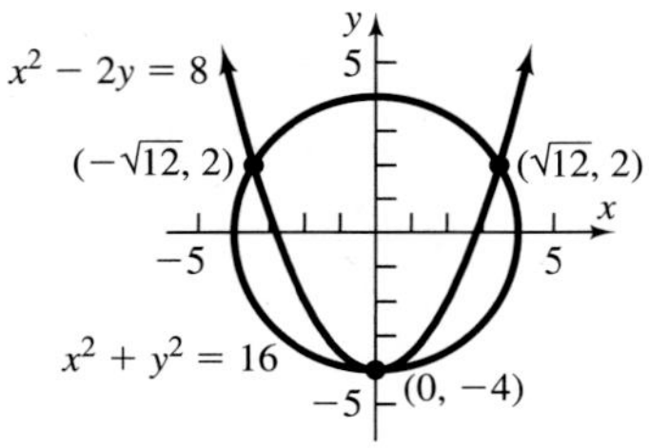

19.
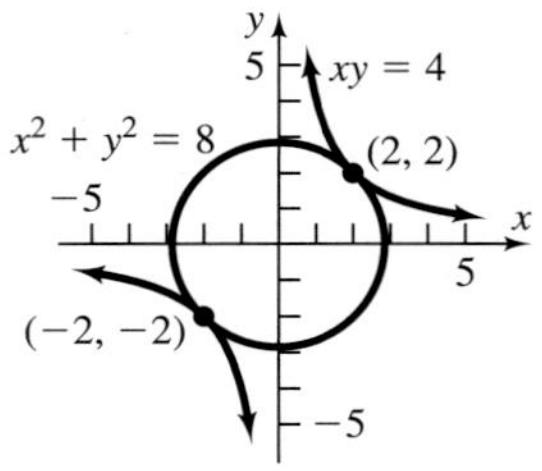

20.
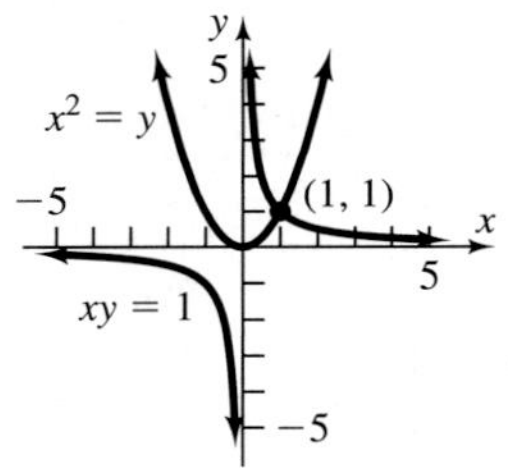

21. No points of intersection
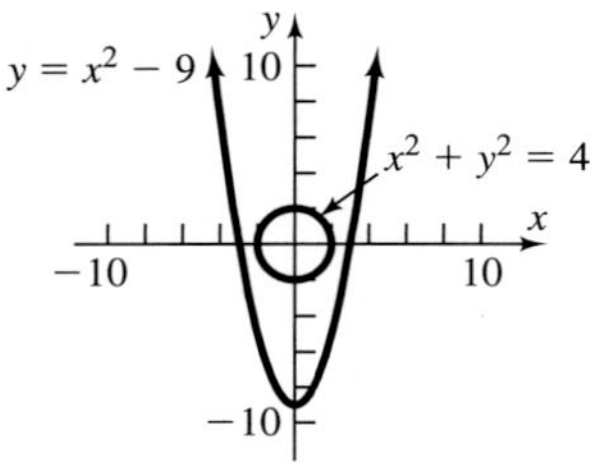

22.
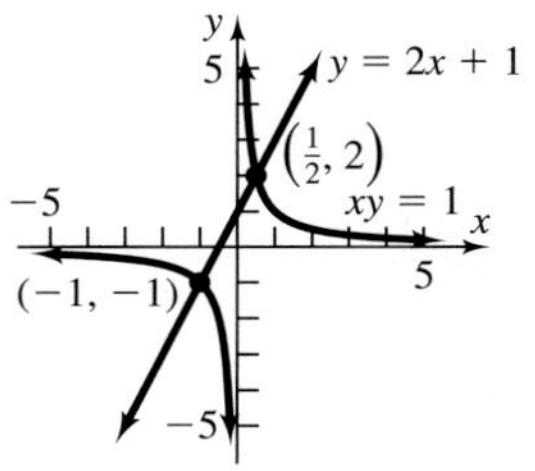

23.
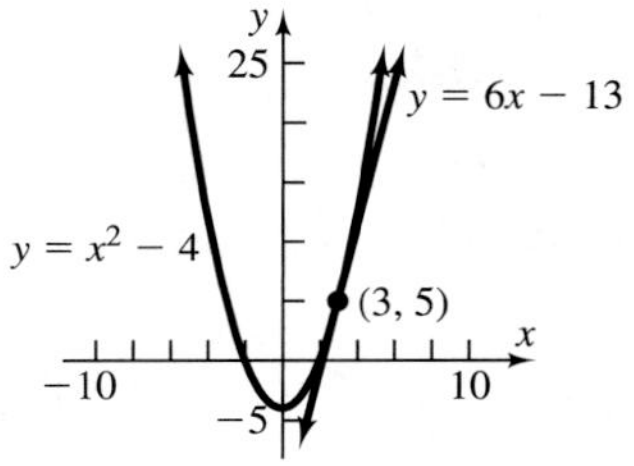

24.
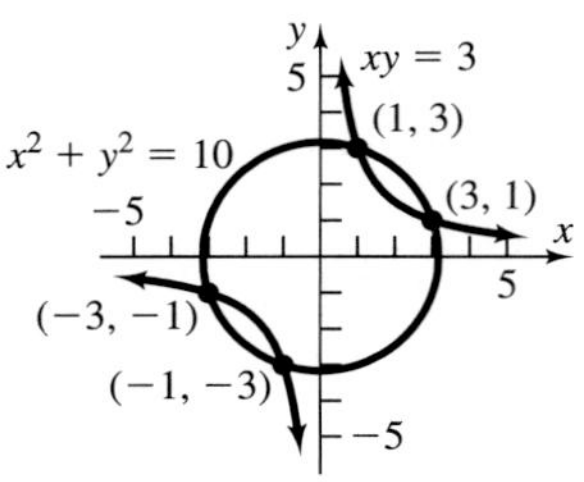

25. $x = 1, y = 4; x = -1, y = -4; x = 2\sqrt{2}, y = \sqrt{2}; x = -2\sqrt{2}, y = -\sqrt{2}$ **26.** $x = 5, y = 2$
27. $x = 0, y = 1; x = -\frac{2}{3}, y = -\frac{1}{3}$ **28.** $x = -5, y = -\frac{3}{2}$ **29.** $x = 0, y = -1; x = \frac{5}{2}, y = -\frac{7}{2}$
30. $x = -2, y = -2; x = -\sqrt{2}, y = -2\sqrt{2}; x = \sqrt{2}, y = 2\sqrt{2}; x = 2, y = 2$
31. $x = 2, y = \frac{1}{3}; x = \frac{1}{2}, y = \frac{4}{3}$ **32.** $x = -2, y = 0; x = 7, y = 6$
33. $x = 3, y = 2; x = 3, y = -2; x = -3, y = 2; x = -3, y = -2$
34. $x = -1, y = -2; x = -1, y = 2; x = 1, y = -2; x = 1, y = 2$

35. $x = \frac{1}{2}, y = \frac{3}{2}; x = \frac{1}{2}, y = -\frac{3}{2}; x = -\frac{1}{2}, y = \frac{3}{2}; x = -\frac{1}{2}, y = -\frac{3}{2}$ **36.** $x = -2\sqrt{2}, y = \sqrt{3}; x = 2\sqrt{2}, y = \sqrt{3}; x = 2\sqrt{2}, y = -\sqrt{3}; x = -2\sqrt{2}, y = -\sqrt{3}$ **37.** $x = \sqrt{2}, y = 2\sqrt{2}; x = -\sqrt{2}, y = -2\sqrt{2}$ **38.** $x = -5\sqrt{3}, y = \sqrt{3}; x = 5\sqrt{3}, y = -\sqrt{3}$

39. No real solution exists. **40.** No real solution exists. **41.** $x = \frac{8}{3}, y = \frac{2\sqrt{10}}{3}$; $x = -\frac{8}{3}, y = \frac{2\sqrt{10}}{3}$; $x = \frac{8}{3}, y = -\frac{2\sqrt{10}}{3}$; $x = -\frac{8}{3}, y = -\frac{2\sqrt{10}}{3}$ **42.** $x = \frac{-\sqrt{2}}{2}, y = \frac{-\sqrt{6}}{3}$; $x = \frac{-\sqrt{2}}{2}, y = \frac{\sqrt{6}}{3}$; $x = \frac{\sqrt{2}}{2}, y = \frac{-\sqrt{6}}{3}$; $x = \frac{\sqrt{2}}{2}, y = \frac{\sqrt{6}}{3}$ **43.** $x = 1, y = \frac{1}{2}$; $x = -1, y = \frac{1}{2}$; $x = 1, y = -\frac{1}{2}$; $x = -1, y = -\frac{1}{2}$ **44.** $x = -2, y = -\sqrt{2}$; $x = -2, y = \sqrt{2}$; $x = 2, y = -\sqrt{2}$; $x = 2, y = \sqrt{2}$

45. No real solution exists. **46.** $x = -\sqrt[4]{\frac{2}{5}}, y = -\sqrt[4]{\frac{2}{3}}$; $x = -\sqrt[4]{\frac{2}{5}}, y = \sqrt[4]{\frac{2}{3}}$; $x = \sqrt[4]{\frac{2}{5}}, y = -\sqrt[4]{\frac{2}{3}}$; $x = \sqrt[4]{\frac{2}{5}}, y = \sqrt[4]{\frac{2}{3}}$

47. $x = \sqrt{3}, y = \sqrt{3}$; $x = -\sqrt{3}, y = -\sqrt{3}$; $x = 2, y = 1$; $x = -2, y = -1$ **48.** $x = -2, y = 2$; $x = 3, y = -3$ **49.** $x = 0, y = -2$; $x = 0, y = 1$; $x = 2, y = -1$ **50.** $x = 2, y = 1$ **51.** $x = 2, y = 8$ **52.** $x = \frac{1}{2}, y = \frac{1}{16}$ **53.** $x = 81, y = 3$ **54.** $x = 32, y = 2$

55.

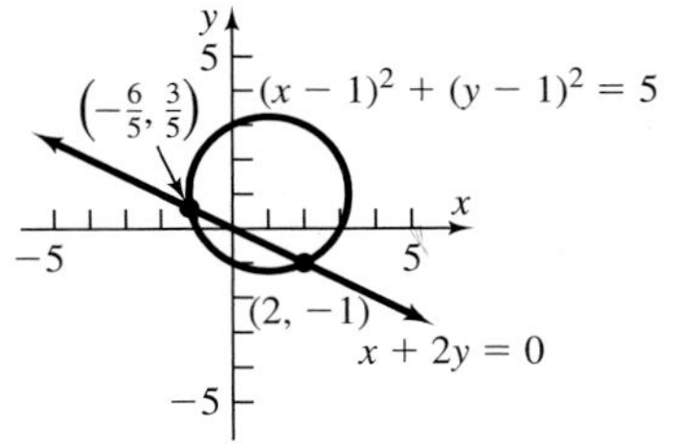

56.

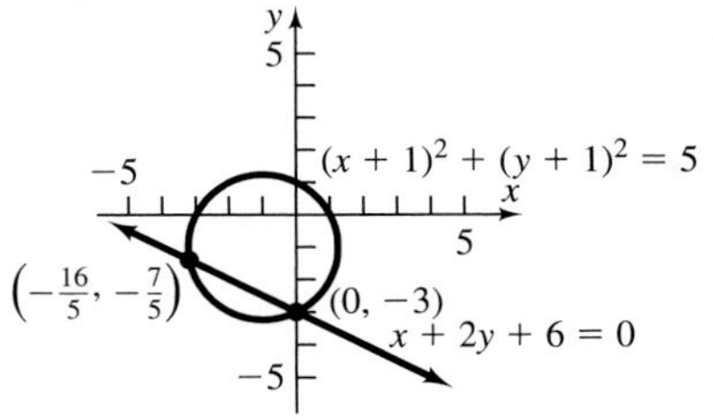

57.

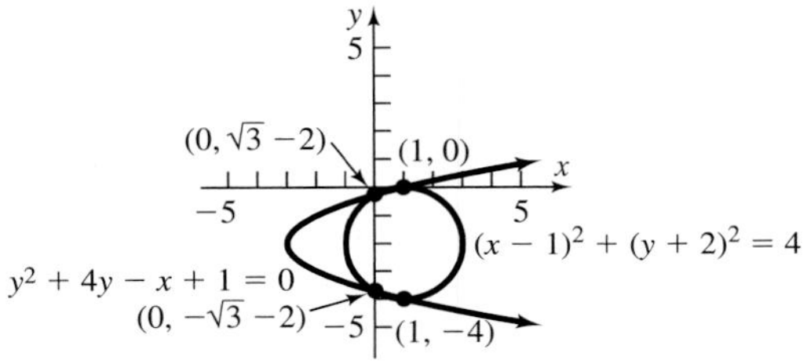

58.

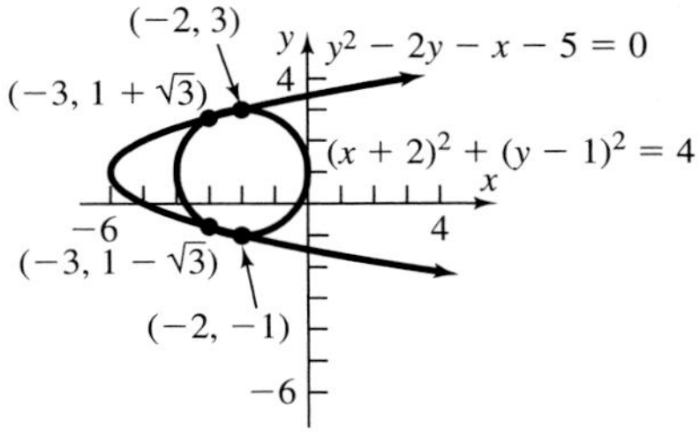

59.

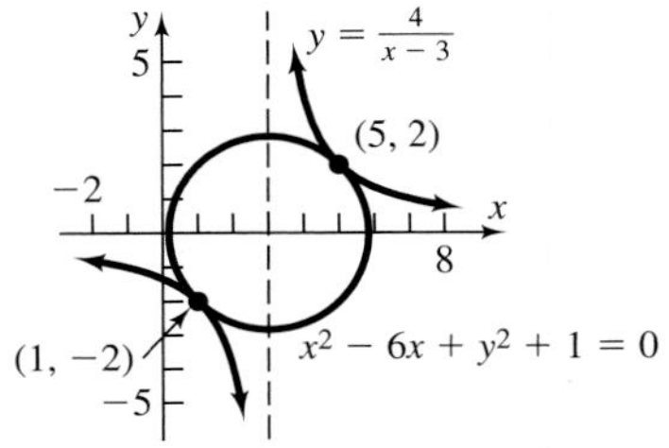

60.

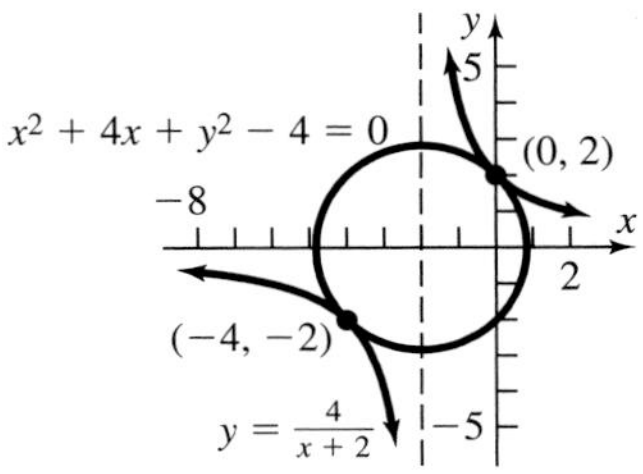

61. $x = 0.48, y = 0.62$ **62.** $x = 0.65, y = 0.52$ **63.** $x = -1.65, y = -0.89$ **64.** $x = -1.37, y = 2.14$
65. $x = 0.58, y = 1.86$; $x = 1.81, y = 1.05$; $x = 0.58, y = -1.86$; $x = 1.81, y = -1.05$
66. $x = 0.64, y = 1.55$; $x = -0.64, y = -1.55$; $x = 1.55, y = 0.64$; $x = -1.55, y = -0.64$ **67.** $x = 2.35, y = 0.85$
68. $x = 1.90, y = 0.64$; $x = 0.14, y = -2.00$ **69.** 3 and 1; −3 and −1 **70.** 2 and 5 **71.** 2 and 2; −2 and −2 **72.** 2 and 5; −2 and −5
73. $\frac{1}{2}$ and $\frac{1}{3}$ **74.** $\frac{1}{2}$ and −1 **75.** 5 **76.** $\frac{1}{7}$ **77.** 5 in. by 3 in. **78.** 4 ft, 6 ft **79.** 2 cm and 4 cm **80.** 8 cm

81. tortoise: 7 m/hr, hare: $7\frac{1}{2}$ m/hr **82.** 10.02 ft **83.** 12 cm by 18 cm **84.** 16.57 cm by 13.03 cm **85.** $x = 60$ ft; $y = 30$ ft

86. It is not possible. **87.** $l = \frac{P + \sqrt{P^2 - 16A}}{4}$; $w = \frac{P - \sqrt{P^2 - 16A}}{4}$ **88.** $b = P - \frac{4h^2 + P^2}{2P}$, $l = \frac{4h^2 + P^2}{4P}$ **89.** $y = 4x - 4$

90. $y = -\frac{1}{3}x + \frac{10}{3}$ **91.** $y = 2x + 1$ **92.** $y = 4x + 9$ **93.** $y = -\frac{1}{3}x + \frac{7}{3}$ **94.** $y = \frac{3}{2}x + \frac{7}{2}$ **95.** $y = 2x - 3$ **96.** $y = \frac{1}{3}x + \frac{7}{3}$

97. $r_1 = \frac{-b + \sqrt{b^2 - 4ac}}{2a}$; $r_2 = \frac{-b - \sqrt{b^2 - 4ac}}{2a}$ **99.** **(a)** 4.274 ft by 4.274 ft or 0.093 ft by 0.093 ft **(b)** yes

10.7 Concepts and Vocabulary *(page 766)*

7. satisfied **8.** half-plane **9.** False **10.** True

10.7 Exercises *(page 766)*

11.
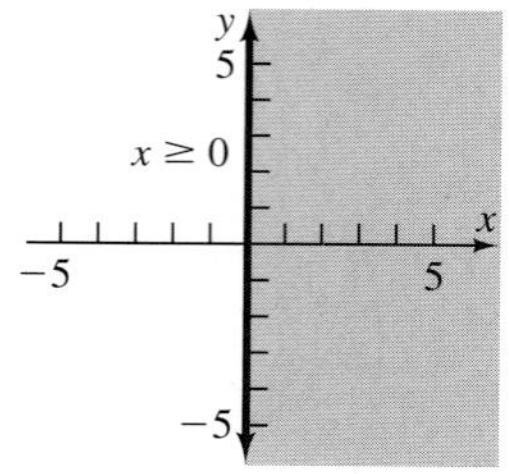

12.
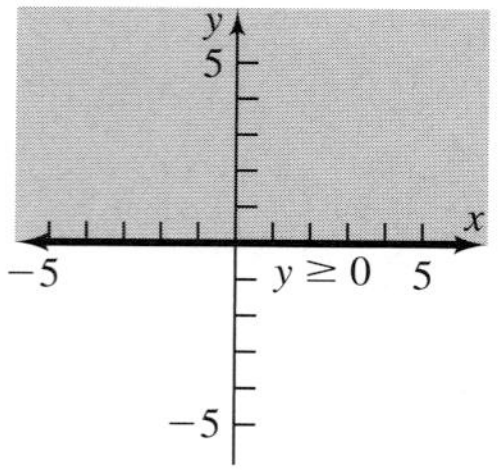

13.
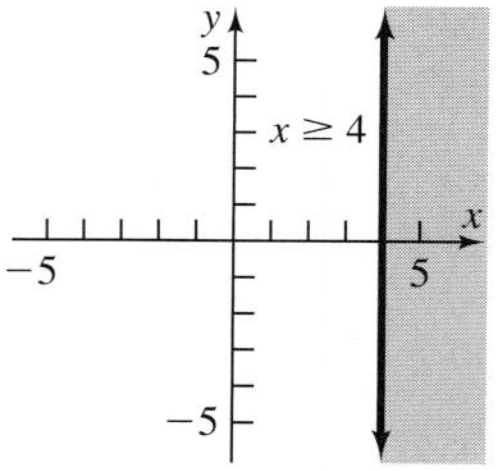

14.
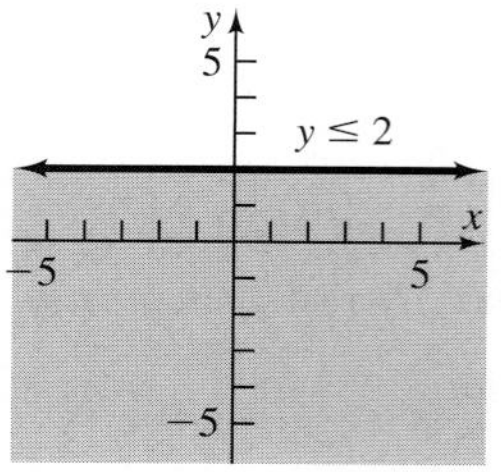

15.
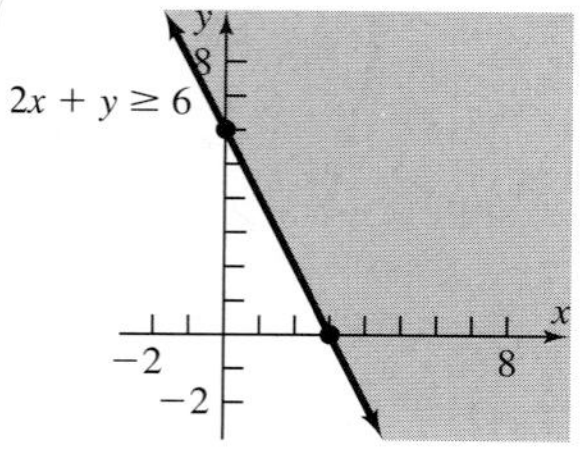

16.
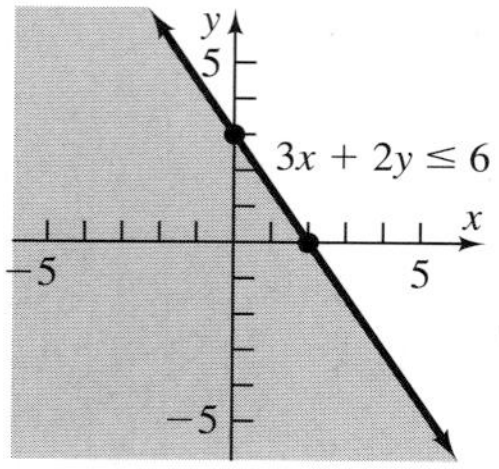

17.
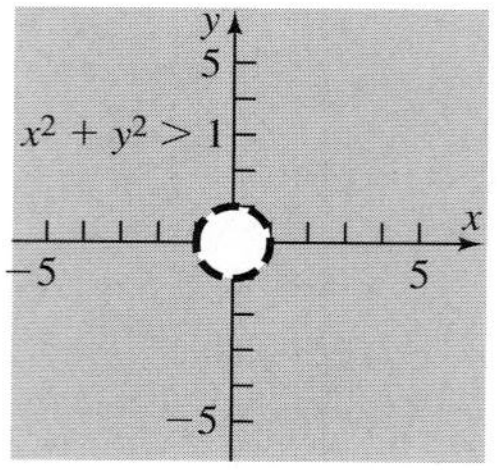

18.
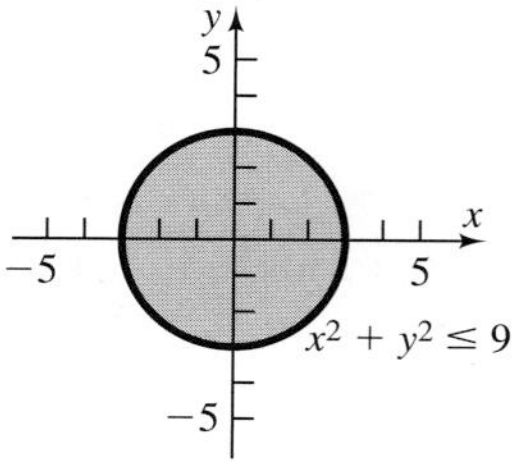

19.
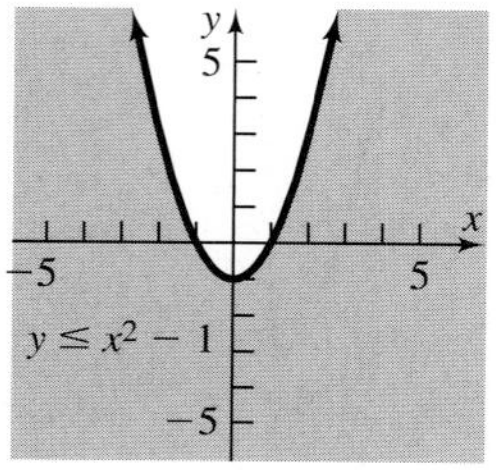

20.
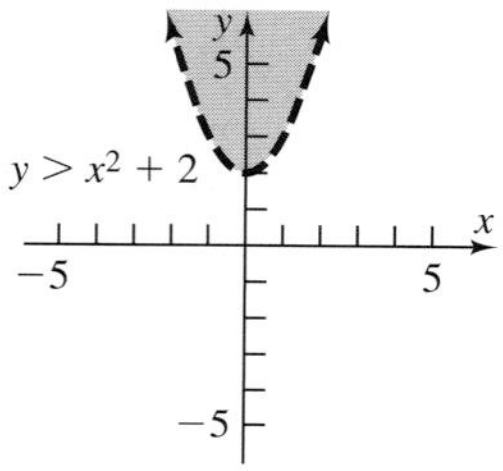

21.
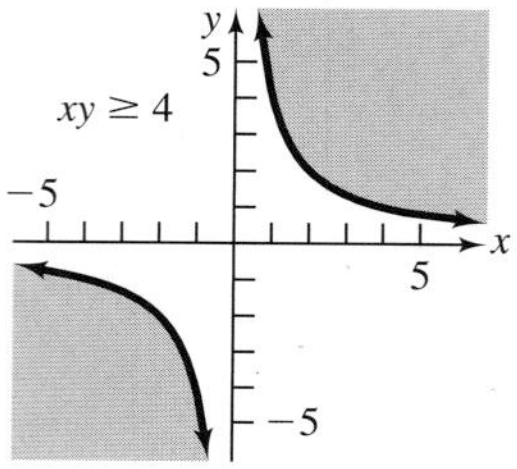

22.
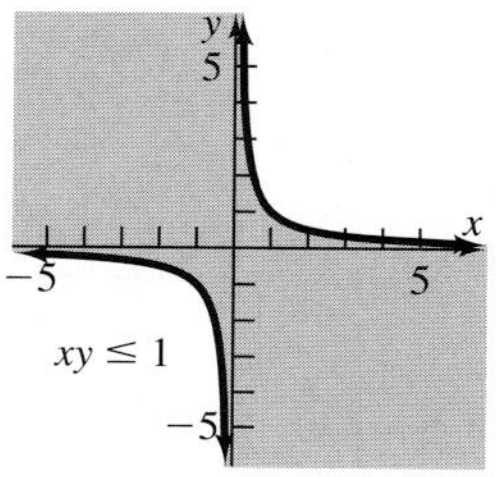

23.
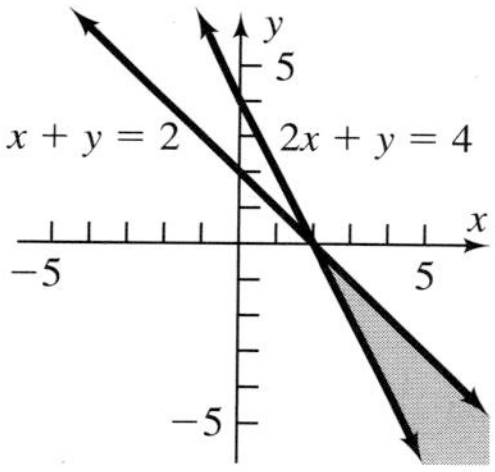

24.
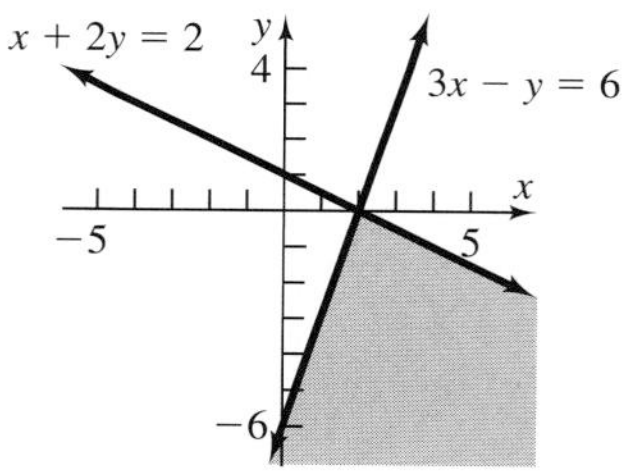

25.
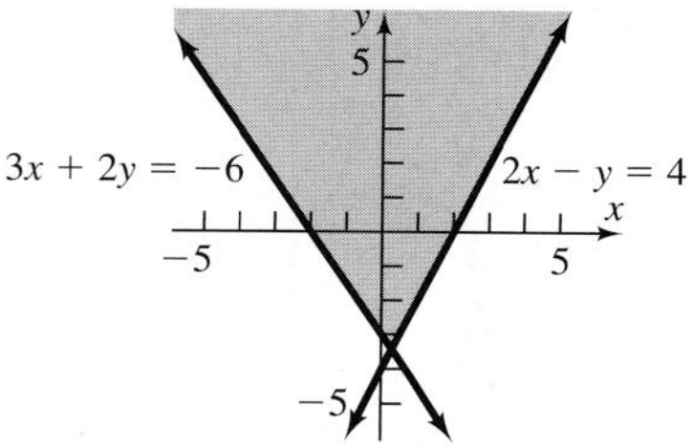

26.
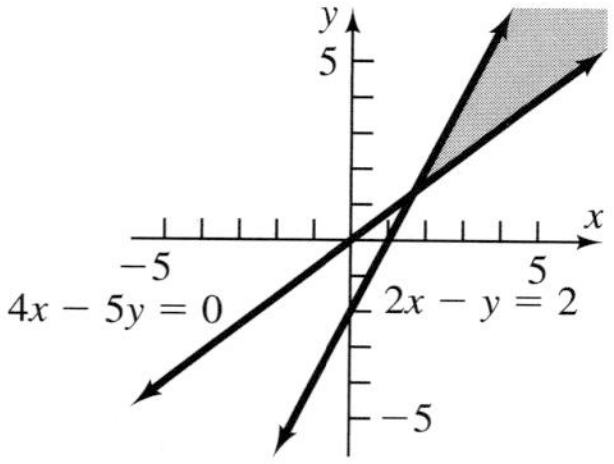

27.
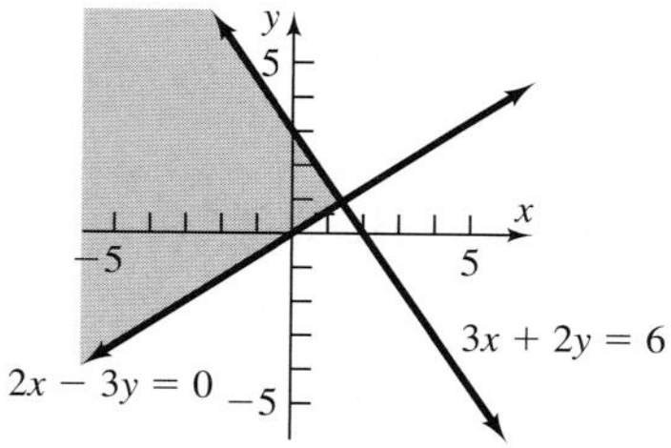

28.
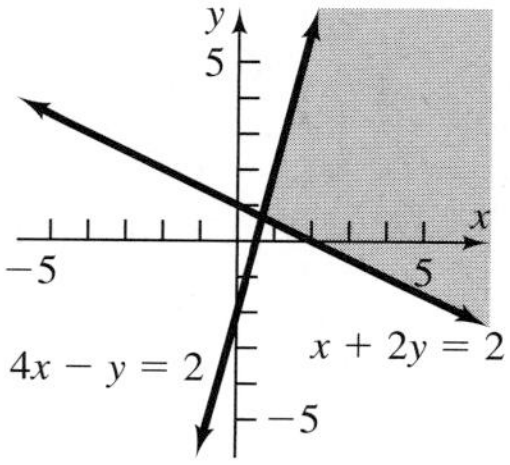

29.
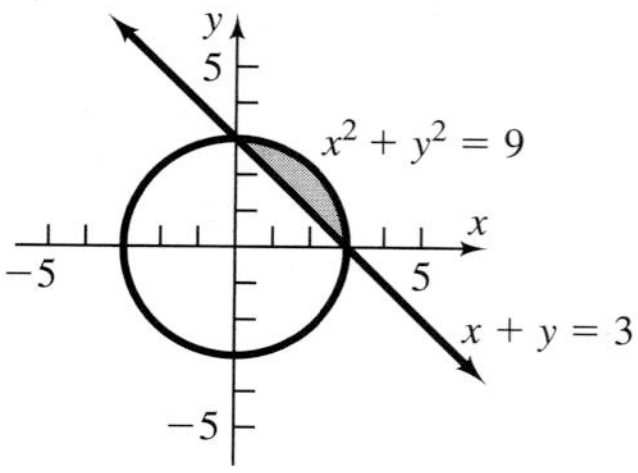

30.
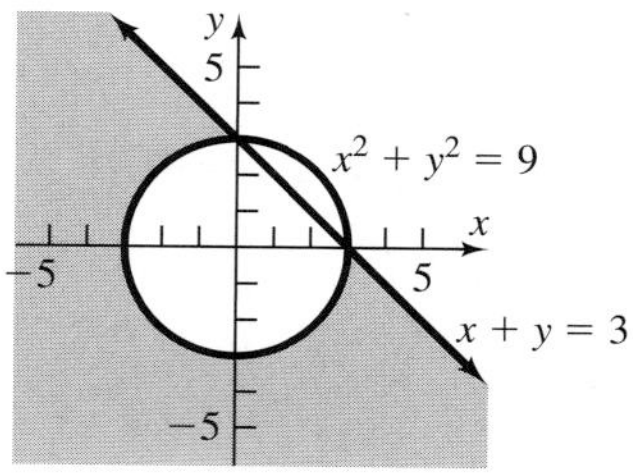

31.
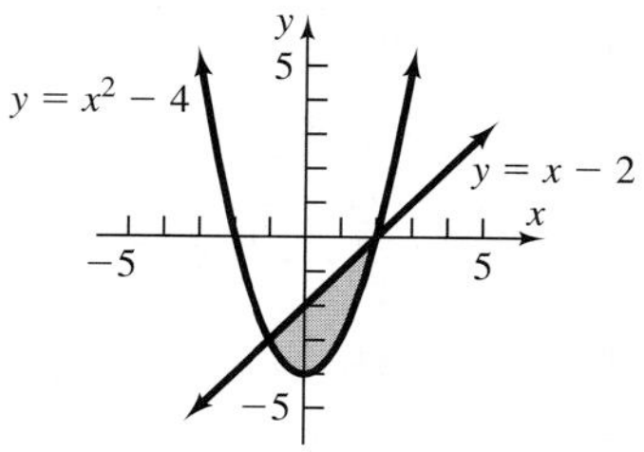

32.

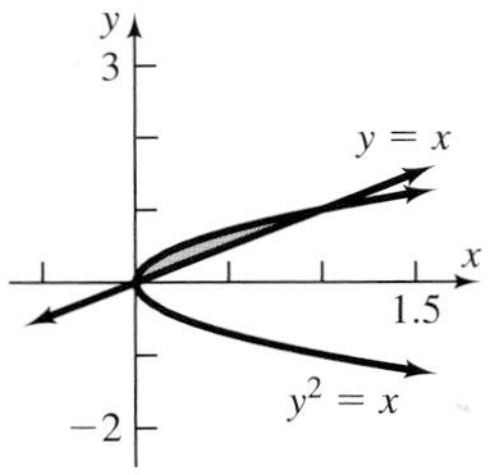

33.

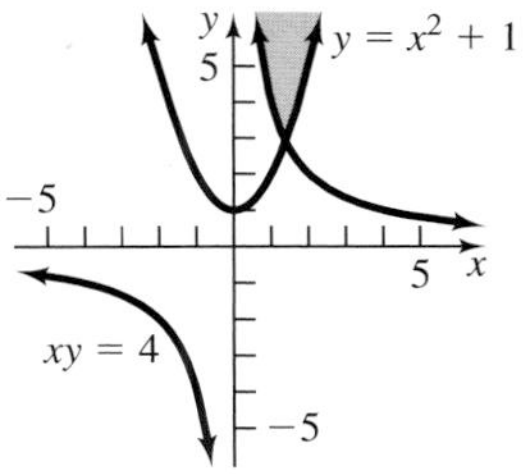

34.

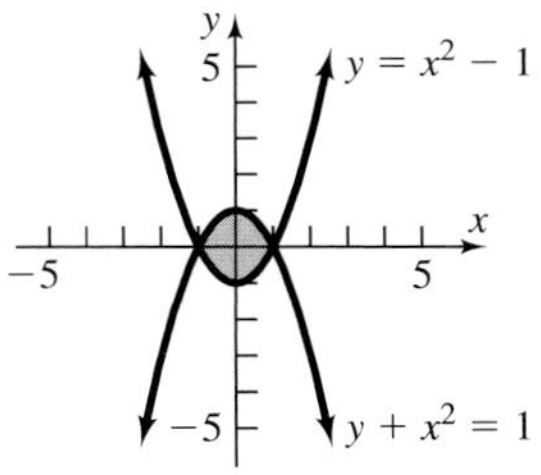

35.

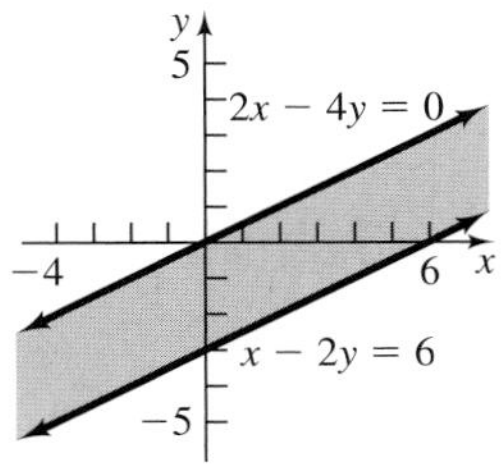

36.

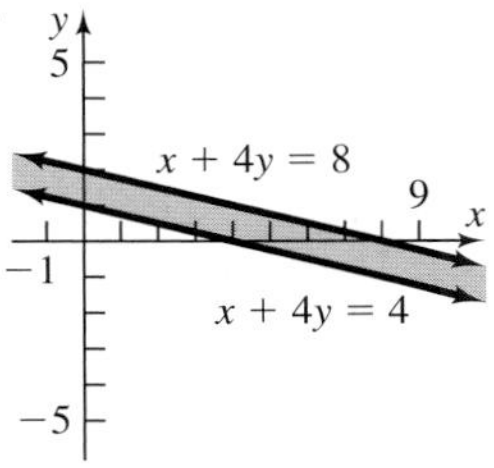

37.

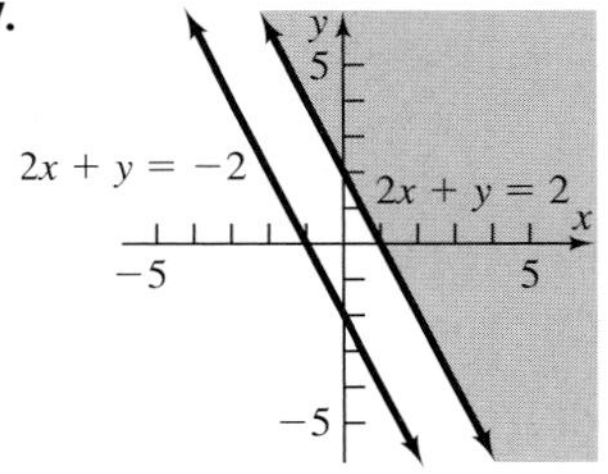

38.

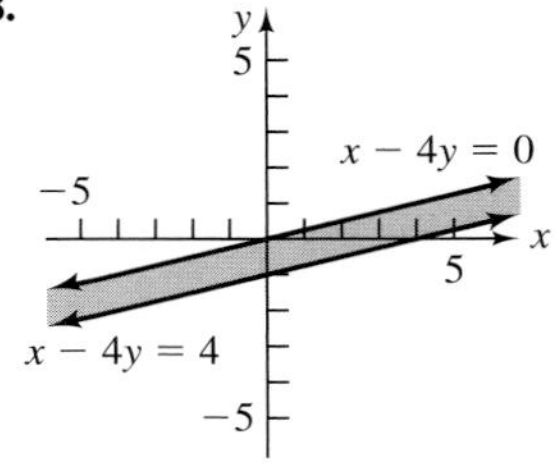

39. No solution

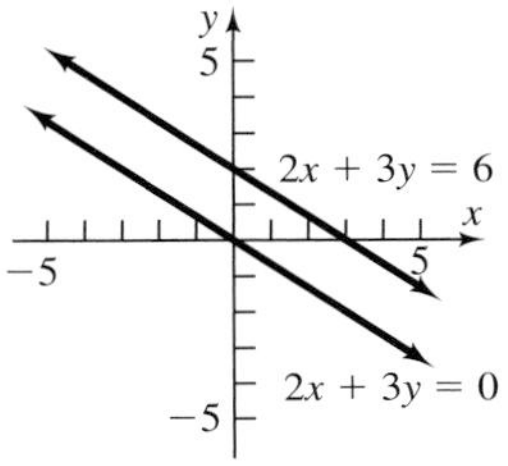

40.

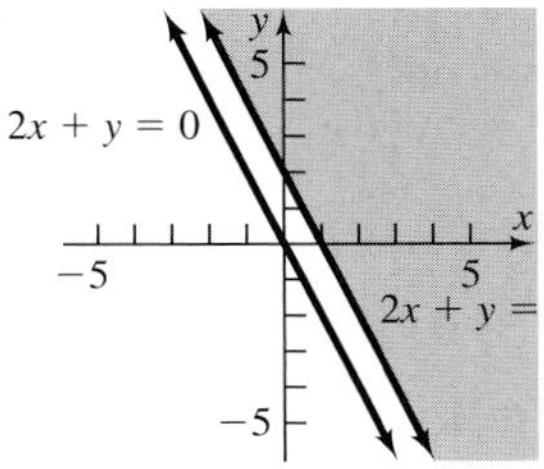

41. Bounded; corner points (0, 0), (3, 0), (2, 2), (0, 3)

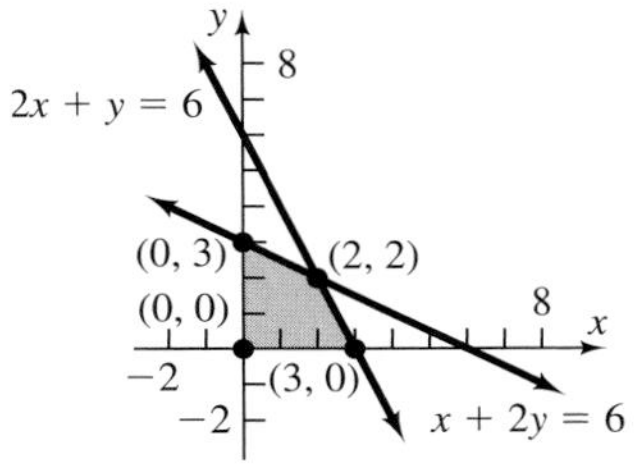

42. Unbounded; corner points (0, 4), (4, 0)

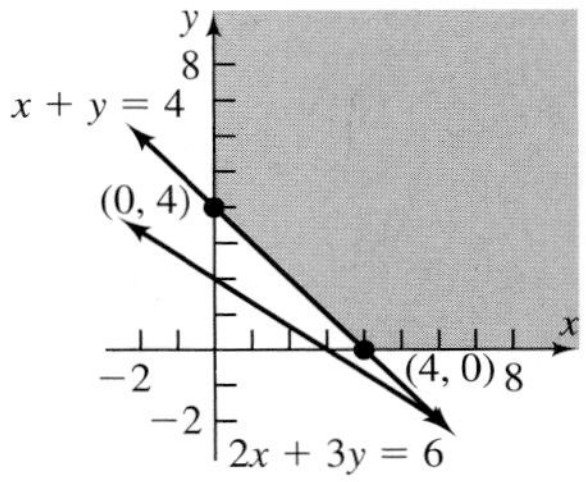

43. Unbounded; corner points (2, 0), (0, 4)

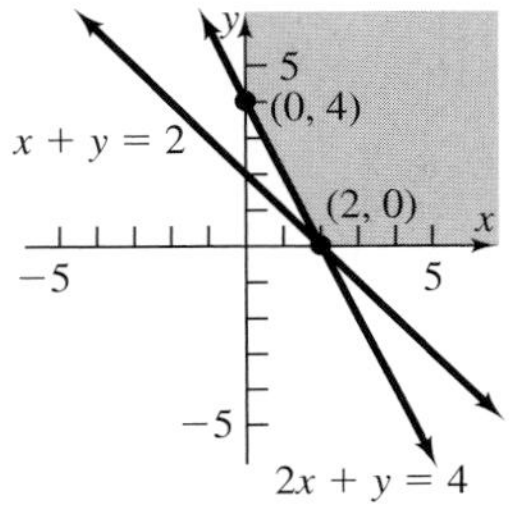

44. Bounded; corner points (0, 0), (0, 2), (1, 0)

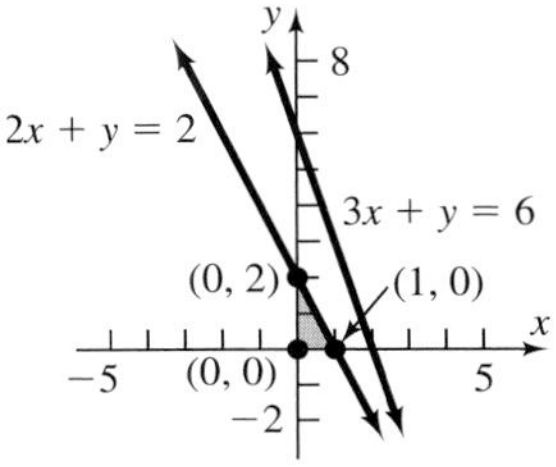

45. Bounded; corner points (2, 0), (4, 0), $\left(\frac{24}{7}, \frac{12}{7}\right)$, (0, 4), (0, 2)

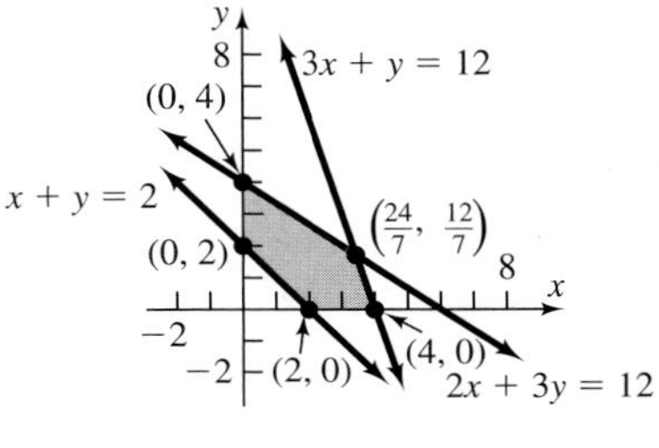

46. Bounded; corner points (0, 2), (0, 3), (1, 1)

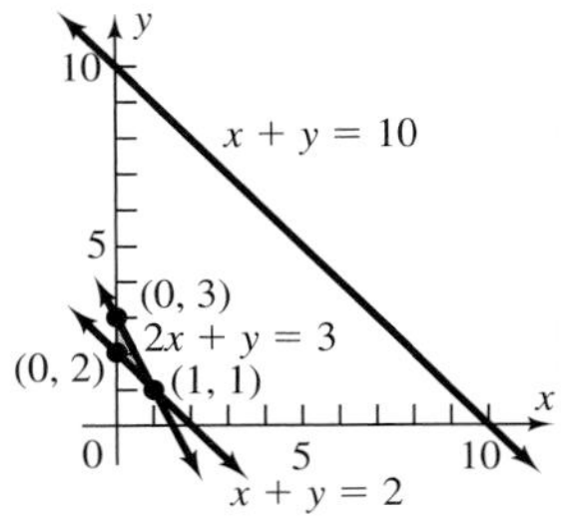

47. Bounded; corner points (2, 0), (5, 0), (2, 6), (0, 8), (0, 2)

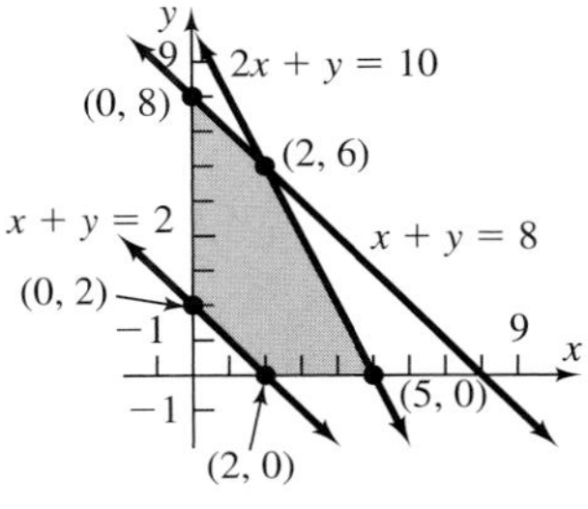

48. Bounded; corner points (2, 0), (0, 2), (0, 8), (8, 0)

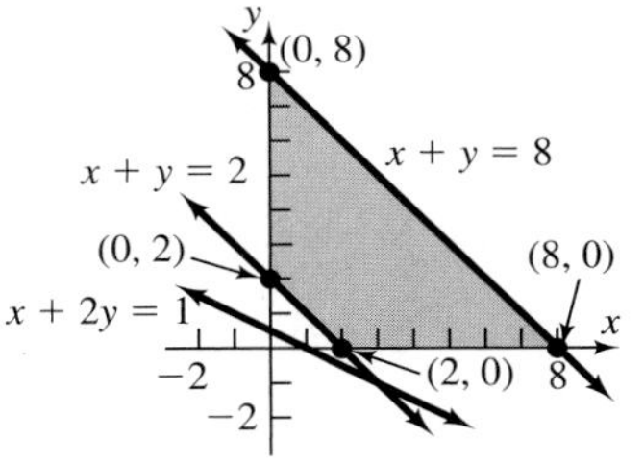

49. Bounded; corner points

$(1, 0), (10, 0), (0, 5), \left(0, \frac{1}{2}\right)$

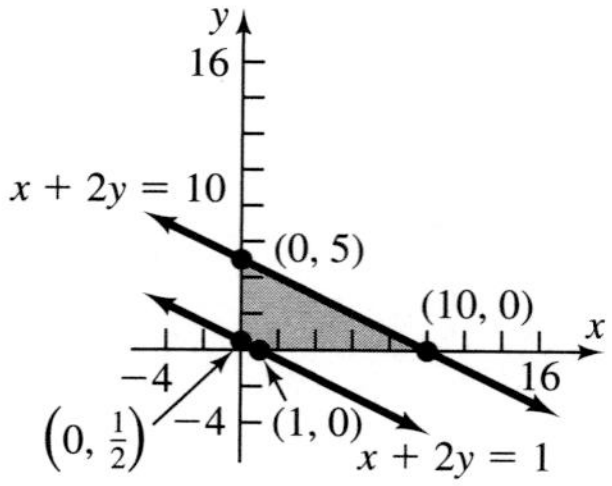

50. Bounded, corner points

$(0, 2), (0, 5), (6, 2), (8, 0), (2, 0)$

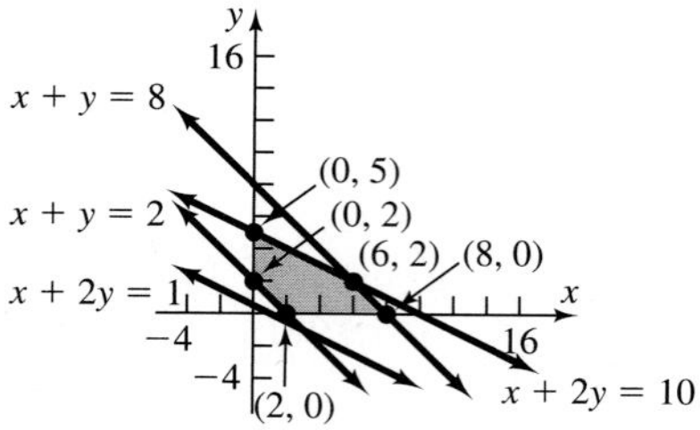

51. $\begin{cases} x \le 4 \\ x + y \le 6 \\ x \ge 0 \\ y \ge 0 \end{cases}$

52. $\begin{cases} y \le 5 \\ x + y \ge 2 \\ x \le 6 \\ x \ge 0, y \ge 0 \end{cases}$

53. $\begin{cases} x \le 20 \\ y \ge 15 \\ x + y \le 50 \\ x - y \le 0 \\ x \ge 0 \end{cases}$

54. $\begin{cases} y \le 6 \\ x \le 5 \\ 3x + 4y \ge 12 \\ y \ge 2x - 8 \\ x \ge 0, y \ge 0 \end{cases}$

55.

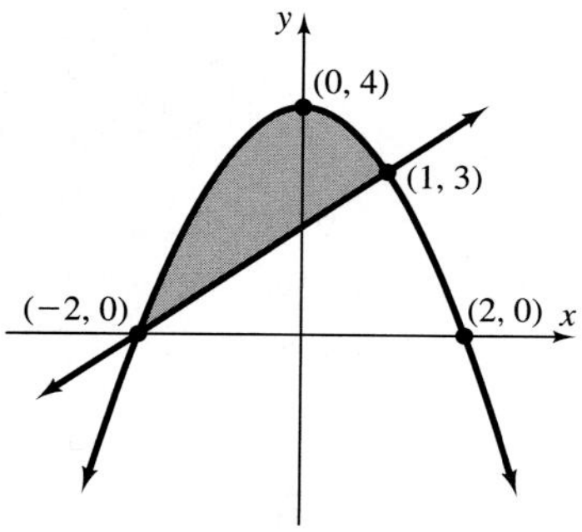

56.

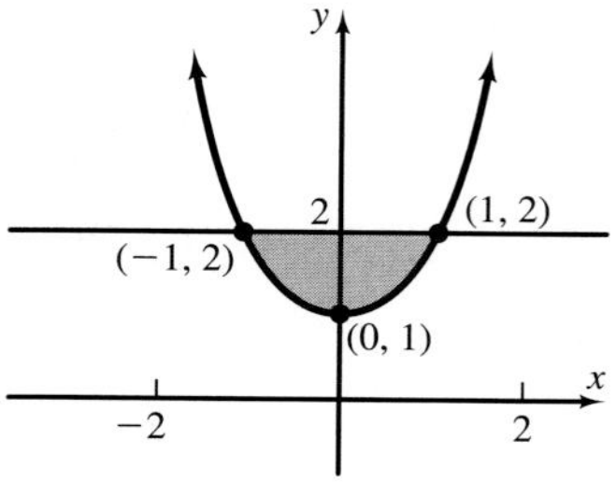

57.

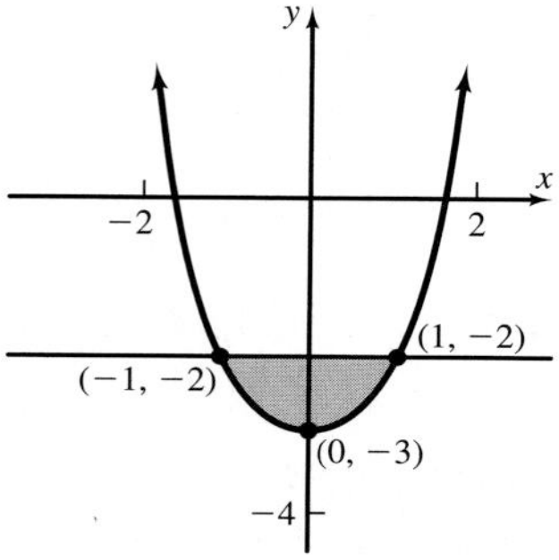

58.

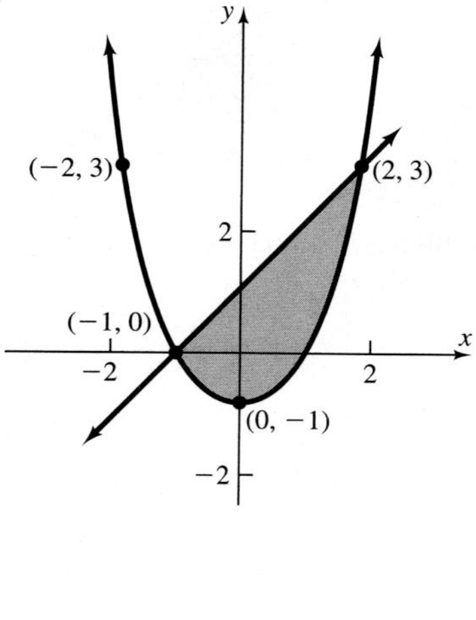

59.

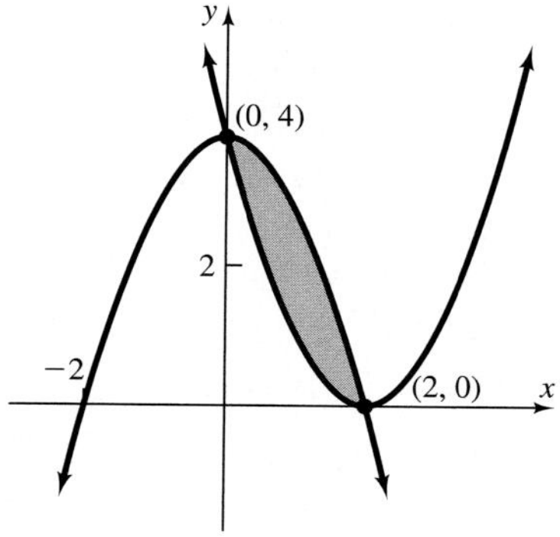

60.

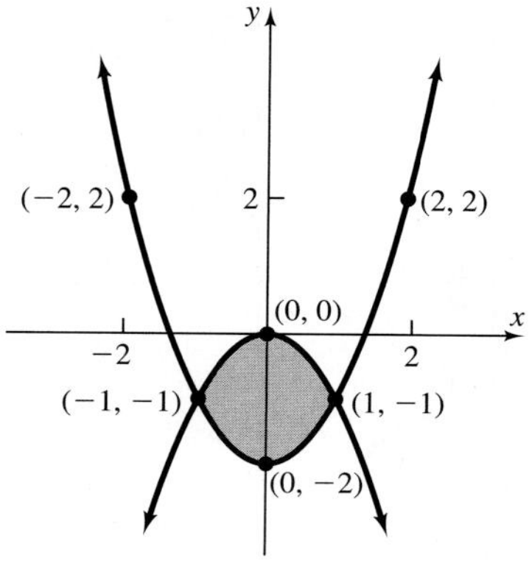

61. (a) $\begin{cases} x + y \le 50{,}000 \\ x \ge 35{,}000 \\ y \le 10{,}000 \\ x \ge 0 \\ y \ge 0 \end{cases}$

(b)

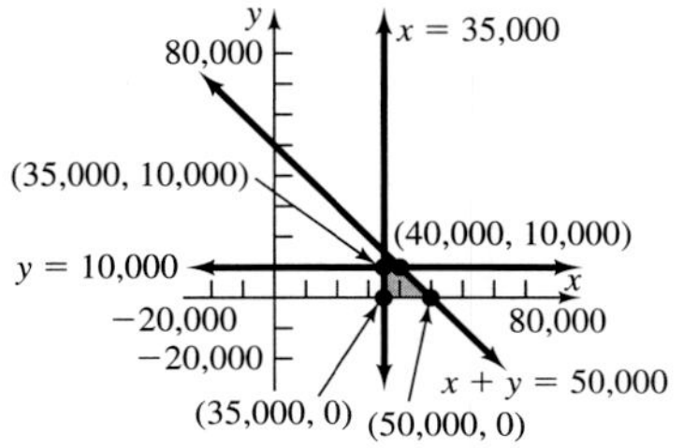

62. (a) $\begin{cases} x \ge 0 \\ y \ge 0 \\ 2x + 3y \le 80 \\ 3x + 4y \le 120 \end{cases}$

(b)

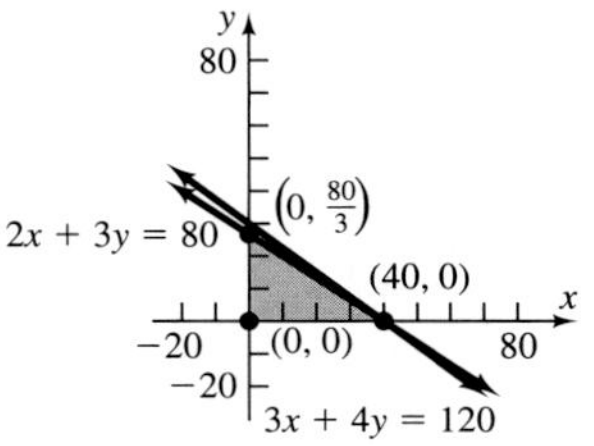

63. (a) $\begin{cases} x \ge 0 \\ y \ge 0 \\ x + 2y \le 300 \\ 3x + 2y \le 480 \end{cases}$

(b)

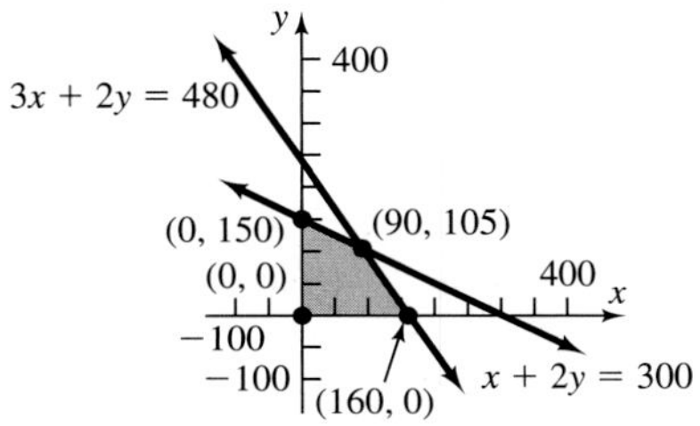

64. (a) $\begin{cases} 4x + 3y \le 960 \\ 2x + 3y \le 720 \\ x \ge 0 \\ y \ge 0 \end{cases}$

(b)

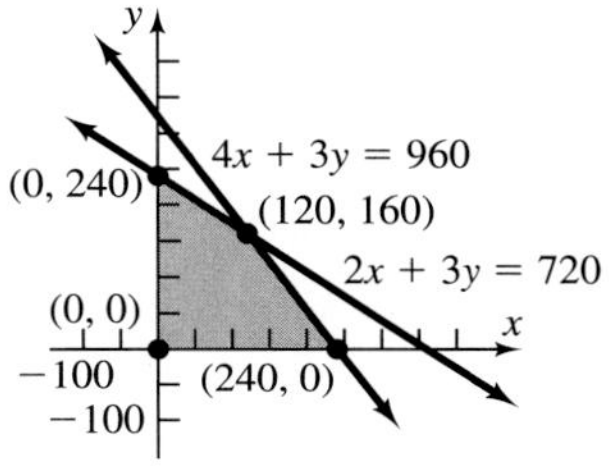

65. (a) $\begin{cases} 3x + 2y \le 160 \\ 2x + 3y \le 150 \\ x \ge 0 \\ y \ge 0 \end{cases}$

(b)

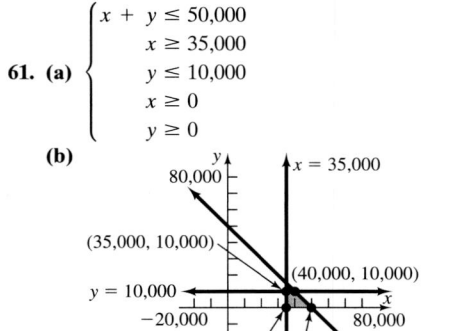

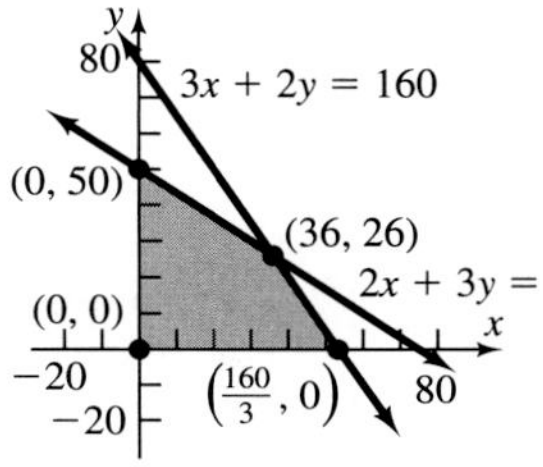

10.8 Concepts and Vocabulary *(page 774)*

1. objective function **2.** True

10.8 Exercises *(page 774)*

3. Maximum value is 11; minimum value is 3. **4.** Maximum value is 28; minimum value is 8. **5.** Maximum value is 65; minimum value is 4. **6.** Maximum value is 56; minimum value is 3. **7.** Maximum value is 67; minimum value is 20. **8.** Maximum value is 65; minimum value is 15. **9.** The maximum value of z is 12, and it occurs at the point $(6, 0)$. **10.** Maximum value is 26, and it occurs at the point $(5, 7)$. **11.** The minimum value of z is 4, and it occurs at the point $(2, 0)$. **12.** The minimum value of z is 8, and it occurs at the point $(0, 2)$. **13.** The maximum value of z is 20, and it occurs at the point $(0, 4)$. **14.** The maximum value of z is 28, and it occurs at the point $(2, 6)$. **15.** The minimum value of z is 8, and it occurs at the point $(0, 2)$. **16.** The minimum value of z is $\frac{15}{2}$, and it occurs at the point $\left(\frac{3}{2}, \frac{3}{2}\right)$. **17.** The maximum value of z is 50, and it occurs at the point $(10, 0)$. **18.** The maximum value of z is 36, and it occurs at the point $(0, 9)$. **19.** 8 downhill, 24 cross-country; \$1760; \$1920 **20.** 24 acres of soybeans and 12 acres of wheat for a maximum profit of \$5520; \$7200 **21.** 30 acres of soybeans and 10 acres of corn **22.** 15 oz of supplement A and 0 oz of supplement B or 7.5 oz of supplement A and 11.25 oz of supplement B, minimum cost \$22.50 **23.** $\frac{1}{2}$ hr on machine 1; $5\frac{1}{4}$ hr on machine 2 **24.** 15 newer trees and 10 older trees; minimum cost \$425 **25.** 100 lb of ground beef and 50 lb of pork **26. (a)** \$10,000 in a junk bond and \$10,000 in Treasury bills **(b)** \$12,000 in junk bonds and \$8,000 in Treasury bills **27.** 10 racing skates, 15 figure skates **28.** \$20,000 in a AAA bond and \$30,000 in a CD **29.** 2 metal samples, 4 plastic samples; \$34 **30.** 15 cans of "Gourmet Dog" and 25 cans of "Chow Hound" **31. (a)** 10 first class, 120 coach **(b)** 15 first class, 120 coach **32.** 5 oz of supplement A and 1 oz of supplement B; minimum cost \$0.38

Review Exercises *(page 778)*

1. $x = 2, y = -1$ **2.** $x = \frac{11}{23}, y = \frac{8}{23}$ **3.** $x = 2, y = \frac{1}{2}$ **4.** $x = -\frac{1}{2}, y = 1$ **5.** $x = 2, y = -1$ **6.** $x = 0, y = \frac{5}{3}$ **7.** $x = \frac{11}{5}, y = -\frac{3}{5}$ **8.** $x = -\frac{1}{2}, y = -\frac{1}{2}$ **9.** Inconsistent **10.** Inconsistent **11.** $x = 2, y = 3$ **12.** $x = 84, y = -63$

13. Inconsistent **14.** $y = -\frac{2}{5}x + 2$, x is any real number **15.** $x = -1, y = 2, z = -3$ **16.** $x = -1, y = 2, z = 7$

17. $x = \frac{7}{4}z + \frac{39}{4}, y = \frac{9}{8}z + \frac{69}{8}$, z is any real number. **18.** Inconsistent **19.** $\begin{cases} 3x + 2y = 8 \\ x + 4y = -1 \end{cases}$ **20.** $\begin{cases} x + 2y + 5z = -2 \\ 5x - 3z = 8 \\ 2x - y = 0 \end{cases}$

21. $\begin{bmatrix} 4 & -4 \\ 3 & 9 \\ 4 & 4 \end{bmatrix}$ **22.** $\begin{bmatrix} -2 & 4 \\ 1 & -1 \\ -6 & 0 \end{bmatrix}$ **23.** $\begin{bmatrix} 6 & 0 \\ 12 & 24 \\ -6 & 12 \end{bmatrix}$ **24.** $\begin{bmatrix} -16 & 12 & 0 \\ -4 & -4 & 8 \end{bmatrix}$ **25.** $\begin{bmatrix} 4 & -3 & 0 \\ 12 & -2 & -8 \\ -2 & 5 & -4 \end{bmatrix}$ **26.** $\begin{bmatrix} -2 & -12 \\ 5 & 0 \end{bmatrix}$ **27.** $\begin{bmatrix} 8 & -13 & 8 \\ 9 & 2 & -10 \\ 22 & -13 & -4 \end{bmatrix}$

28. $\begin{bmatrix} 9 & -31 \\ -6 & -3 \end{bmatrix}$ **29.** $\begin{bmatrix} \frac{1}{2} & -1 \\ -\frac{1}{6} & \frac{2}{3} \end{bmatrix}$ **30.** $\begin{bmatrix} -\frac{1}{2} & -\frac{1}{2} \\ -\frac{1}{4} & -\frac{3}{4} \end{bmatrix}$ **31.** $\begin{bmatrix} -\frac{5}{7} & \frac{9}{7} & \frac{3}{7} \\ \frac{1}{7} & \frac{1}{7} & -\frac{2}{7} \\ \frac{3}{7} & -\frac{4}{7} & \frac{1}{7} \end{bmatrix}$ **32.** $\begin{bmatrix} \frac{3}{7} & \frac{1}{7} & -\frac{5}{7} \\ -\frac{4}{7} & \frac{1}{7} & \frac{9}{7} \\ \frac{1}{7} & -\frac{2}{7} & \frac{3}{7} \end{bmatrix}$ **33.** Singular **34.** Singular

35. $x = \frac{2}{5}, y = \frac{1}{10}$ **36.** $x = 1, y = \frac{3}{2}$ **37.** $x = \frac{1}{2}, y = \frac{2}{3}, z = \frac{1}{6}$ **38.** Inconsistent **39.** $x = -\frac{1}{2}, y = -\frac{2}{3}, z = -\frac{3}{4}$

40. $x = \frac{1}{3}, y = \frac{4}{5}, z = -\frac{1}{15}$ **41.** $z = -1, x = y + 1$, y is any real number. **42.** $y = z, x = -\frac{1}{2}z$, z is any real number.

43. $x = 4, y = 2, z = 3, t = -2$ **44.** $x = -\frac{17}{15}, y = -\frac{1}{5}, z = \frac{22}{15}, t = -\frac{2}{15}$ **45.** 5 **46.** −12 **47.** 108 **48.** −19 **49.** −100 **50.** 11

51. $x = 2, y = -1$ **52.** $x = 0, y = \frac{5}{3}$ **53.** $x = 2, y = 3$ **54.** $x = 0, y = -3$ **55.** $x = -1, y = 2, z = -3$

56. $x = \frac{37}{9}, y = -\frac{19}{6}, z = \frac{13}{18}$ **57.** 16 **58.** −8 **59.** $\dfrac{-\frac{3}{2}}{x} + \dfrac{\frac{3}{2}}{x - 4}$ **60.** $\dfrac{\frac{2}{5}}{x + 2} + \dfrac{\frac{3}{5}}{x - 3}$ **61.** $\dfrac{-3}{x - 1} + \dfrac{3}{x} + \dfrac{4}{x^2}$

62. $\dfrac{4}{x - 2} + \dfrac{-2}{(x - 2)^2} + \dfrac{-4}{x - 1}$ **63.** $\dfrac{-\frac{1}{10}}{x + 1} + \dfrac{\frac{1}{10}x + \frac{9}{10}}{x^2 + 9}$ **64.** $\dfrac{\frac{6}{5}}{x - 2} + \dfrac{-\frac{6}{5}x + \frac{3}{5}}{x^2 + 1}$ **65.** $\dfrac{x}{x^2 + 4} + \dfrac{-4x}{(x^2 + 4)^2}$

66. $\dfrac{x}{x^2 + 16} + \dfrac{-16x + 1}{(x^2 + 16)^2}$ **67.** $\dfrac{\frac{1}{2}}{x^2 + 1} + \dfrac{\frac{1}{4}}{x - 1} + \dfrac{-\frac{1}{4}}{x + 1}$ **68.** $\dfrac{-\frac{4}{5}}{x^2 + 4} + \dfrac{-\frac{2}{5}}{x + 1} + \dfrac{\frac{2}{5}}{x - 1}$ **69.** $x = -\frac{2}{5}, y = -\frac{11}{5}$; $x = -2, y = 1$

70. $x = 2, y = -2\sqrt{3}$; $x = 2, y = 2\sqrt{3}$; $x = -4, y = 0$ **71.** $x = 2\sqrt{2}, y = \sqrt{2}$; $x = -2\sqrt{2}, y = -\sqrt{2}$

72. $x = -\sqrt{3}, y = -2\sqrt{2}$; $x = -\sqrt{3}, y = 2\sqrt{2}$; $x = \sqrt{3}, y = -2\sqrt{2}$; $x = \sqrt{3}, y = 2\sqrt{2}$

73. $x = 0, y = 0$; $x = -3, y = 3$; $x = 3, y = 3$ **74.** $x = 0, y = -3$; $x = 0, y = 3$ **75.** $x = \sqrt{2}, y = -\sqrt{2}$; $x = -\sqrt{2}, y = \sqrt{2}$; $x = \frac{4}{3}\sqrt{2}, y = -\frac{2}{3}\sqrt{2}$; $x = -\frac{4}{3}\sqrt{2}, y = \frac{2}{3}\sqrt{2}$ **76.** $x = -\frac{5}{21}\sqrt{42}, y = -\frac{2}{7}\sqrt{42}$; $x = \frac{5}{21}\sqrt{42}, y = \frac{2}{7}\sqrt{42}$; $x = -2, y = 1$; $x = 2, y = -1$

77. $x = 1, y = -1$ **78.** $x = -2, y = 0$; $x = 1, y = 0$; $x = -1, y = 2$

79.

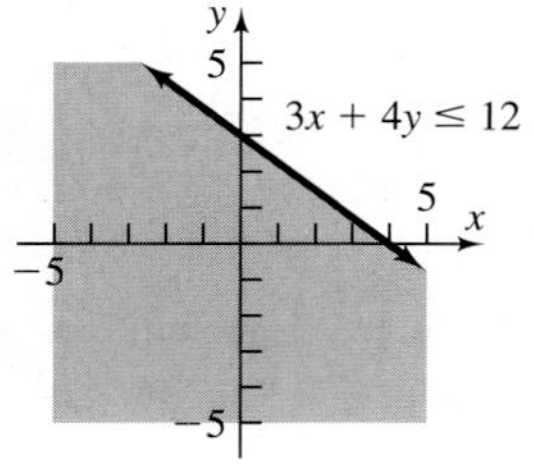

80.

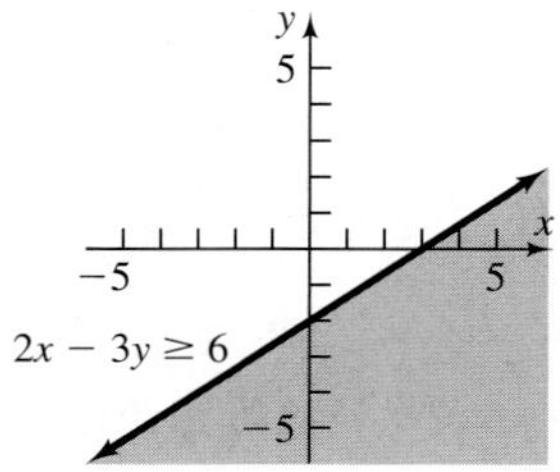

81. Unbounded

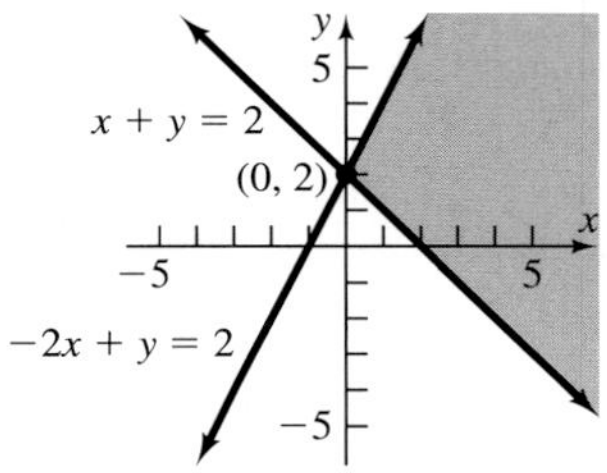

82. Unbounded

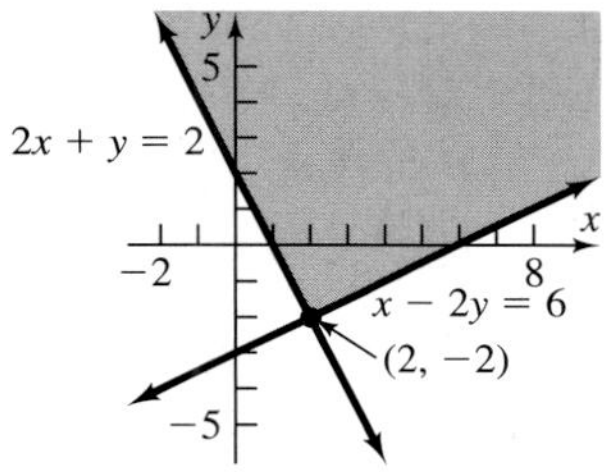

83. Bounded

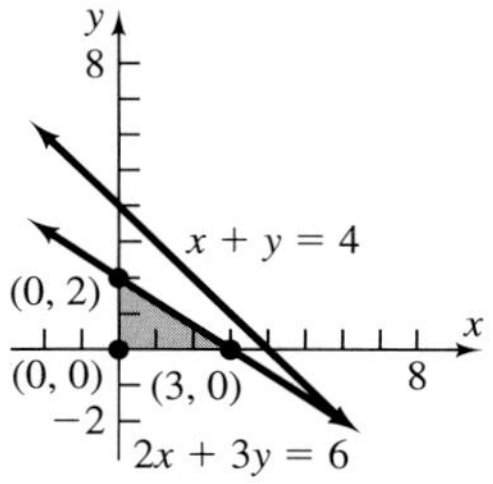

84. Unbounded

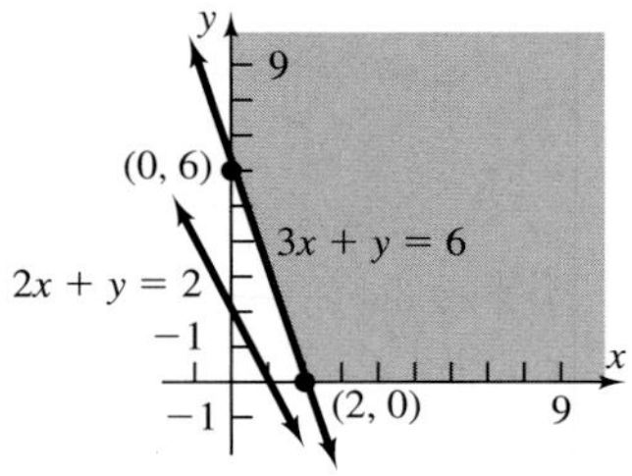

85. Bounded

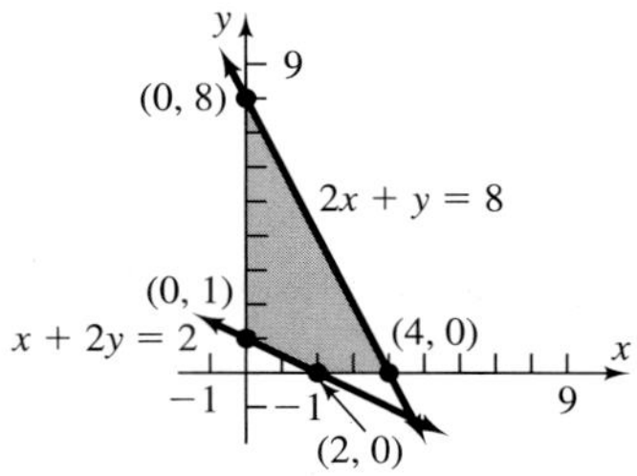

86. Bounded

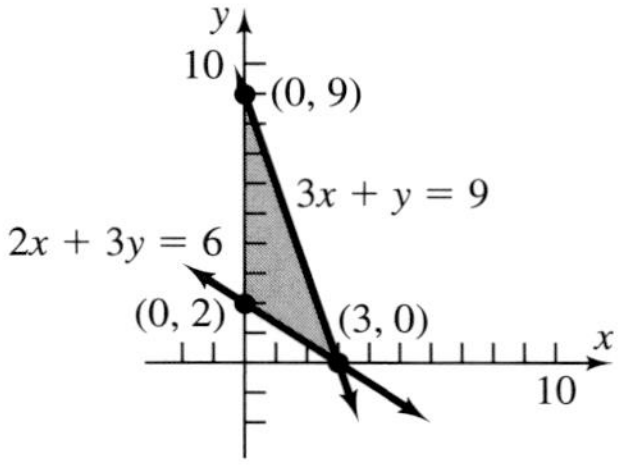

87.

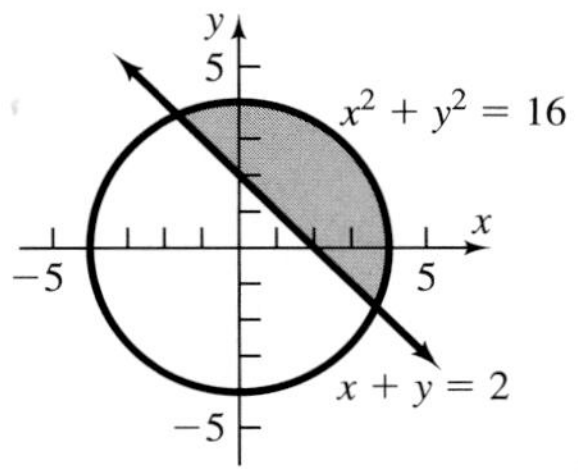

88.

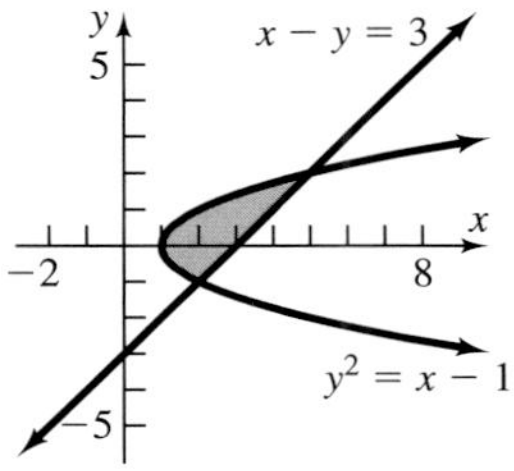

89.

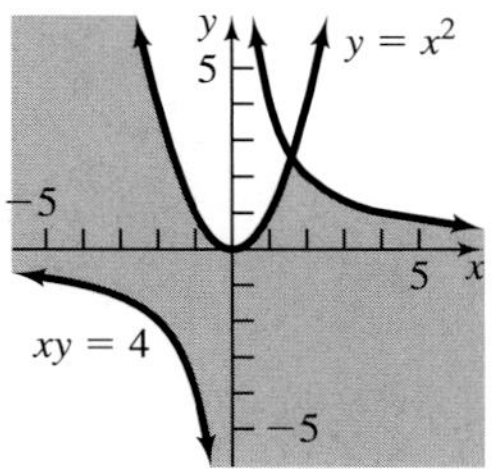

90.

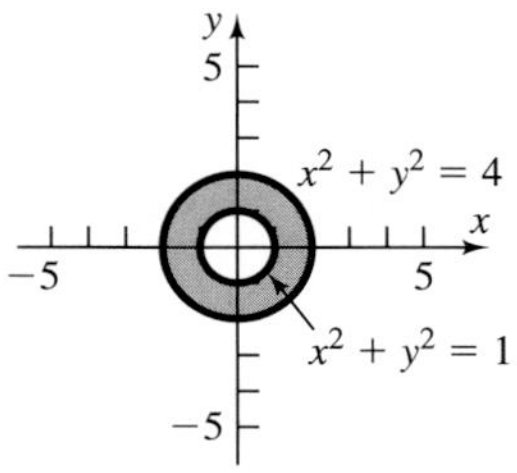

91. The maximum value is 32 when $x = 0$ and $y = 8$. **92.** The maximum value is 20 when $x = 2$ and $y = 4$.
93. The minimum value is 3 when $x = 1$ and $y = 0$. **94.** The minimum value is 4 when $x = 0$ and $y = 4$.

95. 10 **96.** A is any real number, $A \neq 10$. **97.** $y = -\frac{1}{3}x^2 - \frac{2}{3}x + 1$ **98.** $x^2 + 2x + y^2 + 2y - 3 = 0$

99. 70 pounds of \$3 coffee and 30 pounds of \$6 coffee **100.** 466.25 acres for corn and 533.75 acres for soy beans
101. 1 small, 5 medium, 2 large

102. (a) $x \geq 0$
$y \geq 0$
$4x + 3y \leq 960$
$2x + 3y \leq 576$

(b)

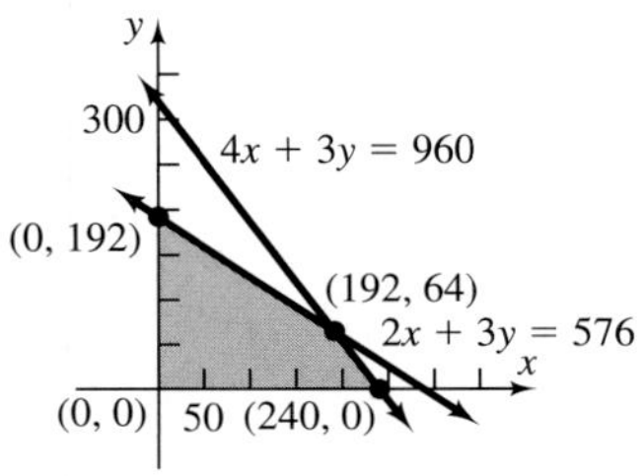

103. Speedboat: 36.67 km/hr; Aguarico River: 3.33 km/hr
104. 29.69 miles per hour
105. Bruce: 4 hours; Bryce: 2 hours; Marty: 8 hours
106. 12 mermaids and 12 dancing girls to have a profit of \$660; molding has 18 excess hours assigned to it.
107. 35 gasoline engines, 15 diesel engines; 15 gasoline engines, 0 diesel engines

Cumulative Review *(page 782)*

1. $\left\{0, \frac{1}{2}\right\}$ **2.** $\{5\}$ **3.** $\left\{-1, -\frac{1}{2}, 3\right\}$ **4.** $\{-2\}$ **5.** $\left\{\frac{5}{2}\right\}$ **6.** $\left\{\frac{1}{\ln 3}\right\}$ **7.** Odd; symmetric with respect to origin

8. Center: $(1, -2)$; radius $= 4$

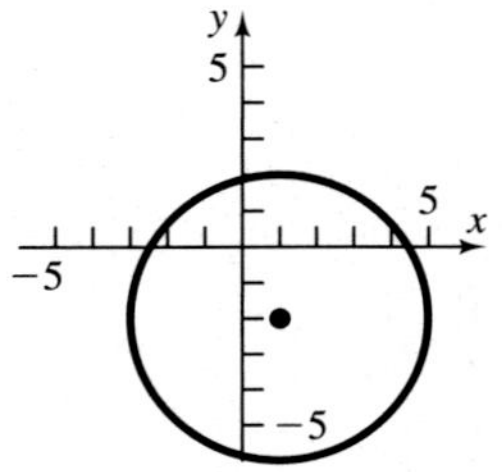

9. Domain: all real numbers
Range: $\{y \mid y > 1\}$
Horizontal asymptote: $y = 1$

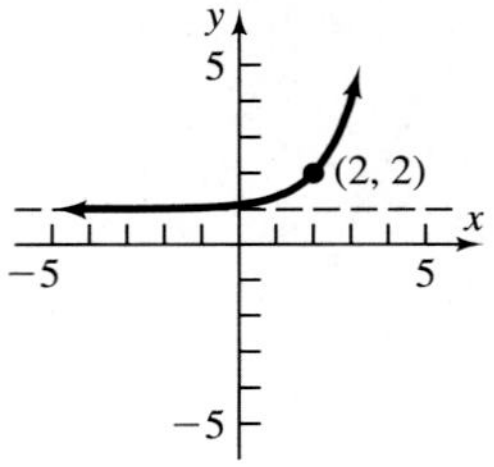

10. $f^{-1}(x) = \dfrac{5}{x} - 2$

Domain of f: $\{x \mid x \neq -2\}$
Range of f: $\{y \mid y \neq 0\}$
Domain of f^{-1}: $\{x \mid x \neq 0\}$
Range of f^{-1}: $\{y \mid y \neq -2\}$

11. (a)

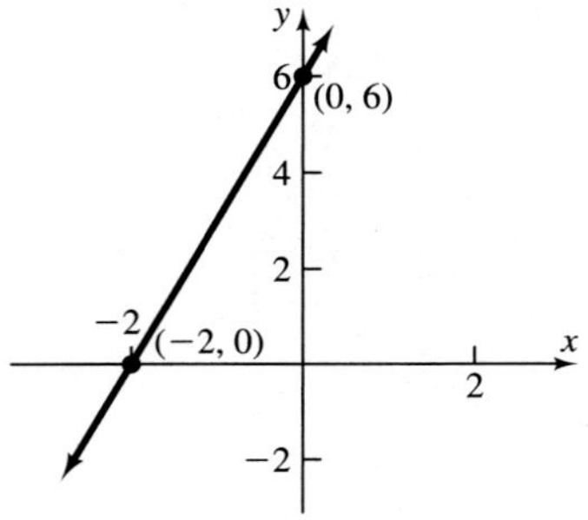

(b)

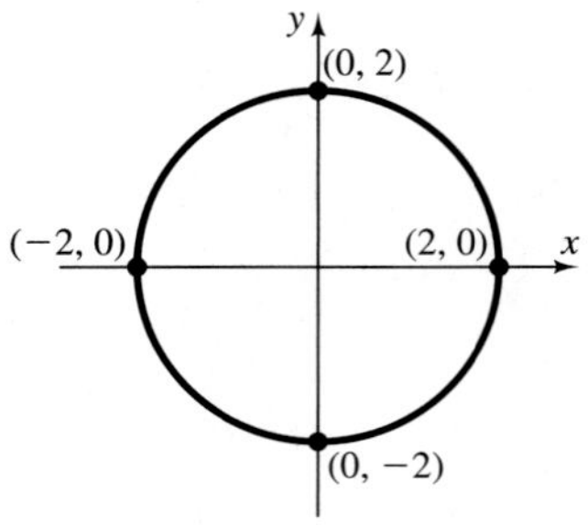

(c)

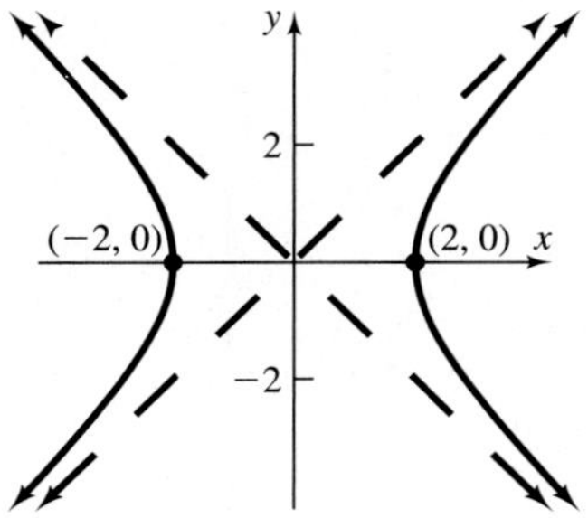

(d)

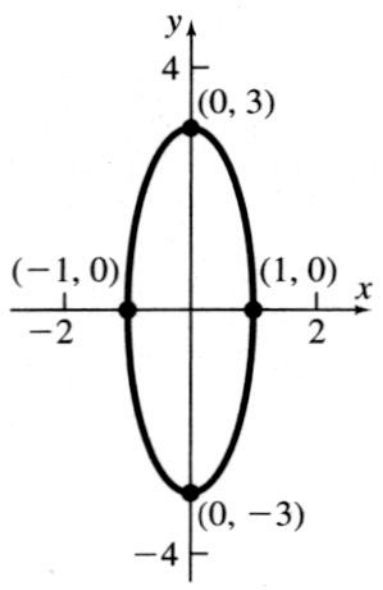

(e)

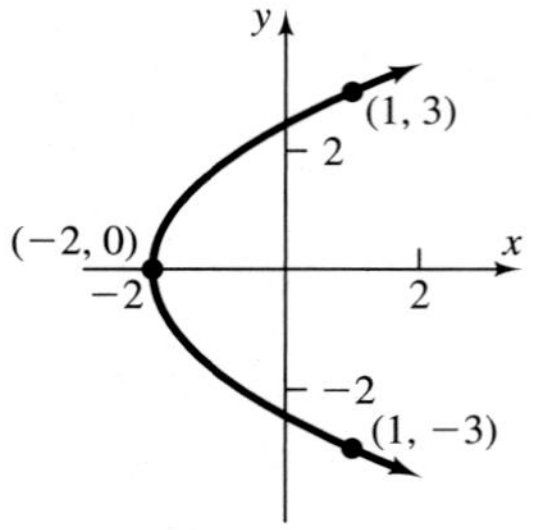

(f)

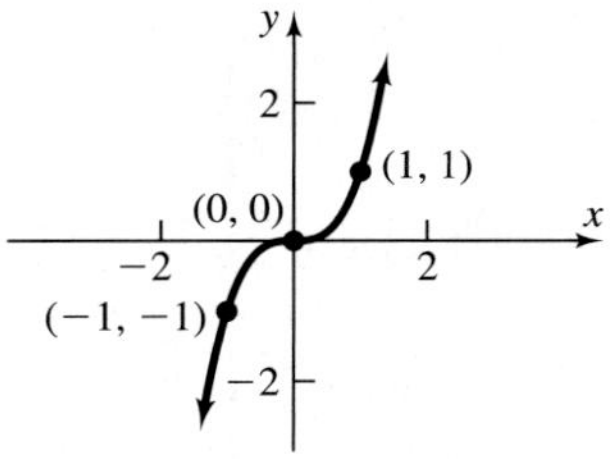

(g)

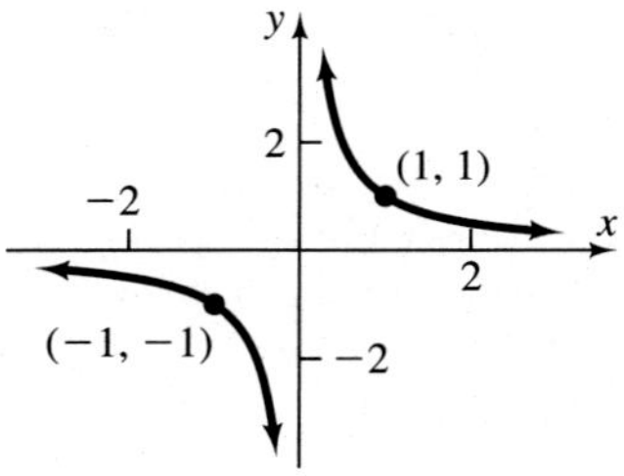

(h)

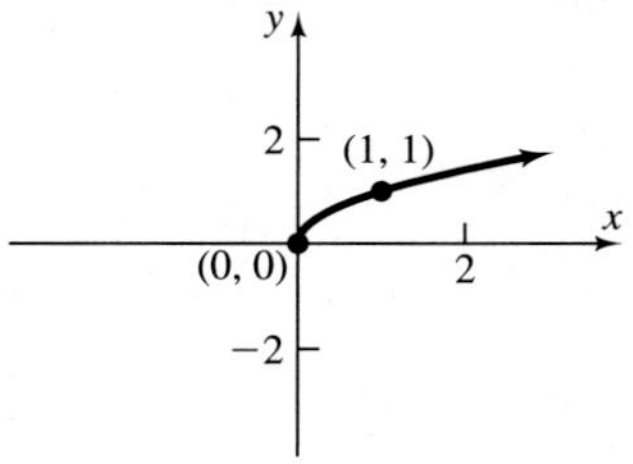

(i)

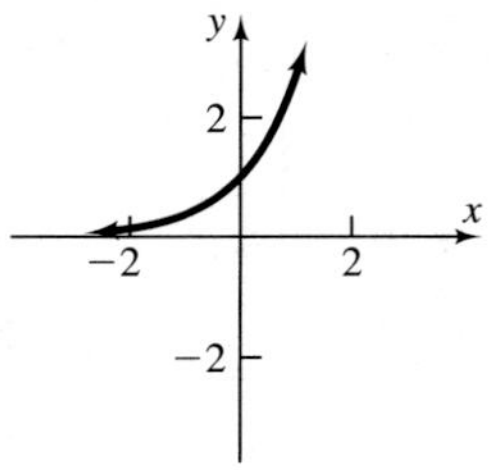

(j)

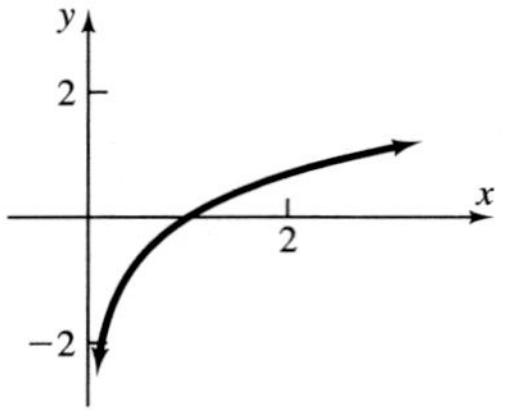

(k)

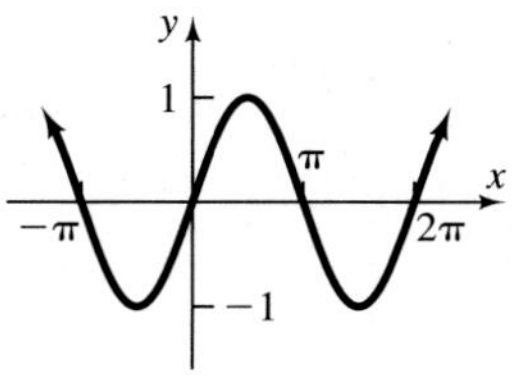

(l)

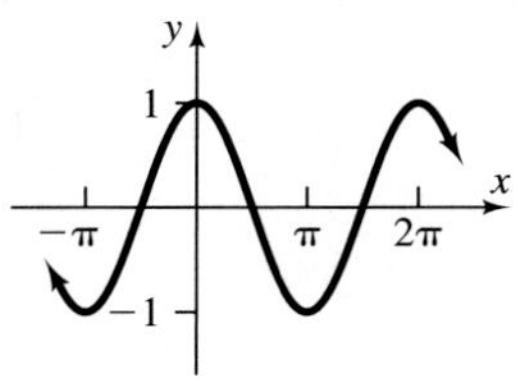

12. (a) $\left\{\dfrac{\pi}{6}, \dfrac{5\pi}{6}\right\}$ **(b)** $\left\{\dfrac{\pi}{9}, \dfrac{5\pi}{9}\right\}$

CHAPTER 11 Sequences; Induction; The Binomial Theorem

11.1 Concepts and Vocabulary *(page 791)*

3. sequence **4.** 3; 15 **5.** 20 **6.** True **7.** True **8.** True

11.1 Exercises *(page 791)*

9. 1, 2, 3, 4, 5 **10.** 2, 5, 10, 17, 26 **11.** $\frac{1}{3}, \frac{1}{2}, \frac{3}{5}, \frac{2}{3}, \frac{5}{7}$ **12.** $\frac{3}{2}, \frac{5}{4}, \frac{7}{6}, \frac{9}{8}, \frac{11}{10}$ **13.** 1, −4, 9, −16, 25 **14.** $1, -\frac{2}{3}, \frac{3}{5}, -\frac{4}{7}, \frac{5}{9}$ **15.** $\frac{1}{2}, \frac{2}{5}, \frac{2}{7}, \frac{8}{41}, \frac{8}{61}$

16. $\frac{4}{3}, \frac{16}{9}, \frac{64}{27}, \frac{256}{81}, \frac{1024}{243}$ **17.** $-\frac{1}{6}, \frac{1}{12}, -\frac{1}{20}, \frac{1}{30}, -\frac{1}{42}$ **18.** $3, \frac{9}{2}, 9, \frac{81}{4}, \frac{243}{5}$ **19.** $\frac{1}{e}, \frac{2}{e^2}, \frac{3}{e^3}, \frac{4}{e^4}, \frac{5}{e^5}$ **20.** $\frac{1}{2}, 1, \frac{9}{8}, 1, \frac{25}{32}$ **21.** $a_n = \frac{n}{n+1}$

22. $a_n = \frac{1}{n(n+1)}$ **23.** $a_n = \frac{1}{2^{n-1}}$ **24.** $a_n = \left(\frac{2}{3}\right)^n$ **25.** $a_n = (-1)^{n+1}$ **26.** $a_n = \begin{cases} n & \text{if } n \text{ is odd} \\ \frac{1}{n} & \text{if } n \text{ is even} \end{cases}$ or $a_n = n^{(-1)^{n-1}}$ **27.** $a_n = (-1)^{n+1}n$

28. $a_n = (-1)^{n-1} \cdot 2n$ **29.** $a_1 = 2, a_2 = 5, a_3 = 8, a_4 = 11, a_5 = 14$ **30.** $a_1 = 3, a_2 = 1, a_3 = 3, a_4 = 1, a_5 = 3$

31. $a_1 = -2, a_2 = 0, a_3 = 3, a_4 = 7, a_5 = 12$ **32.** $a_1 = 1, a_2 = 1, a_3 = 2, a_4 = 2, a_5 = 3$ **33.** $a_1 = 5, a_2 = 10, a_3 = 20, a_4 = 40, a_5 = 80$

34. $a_1 = 2, a_2 = -2, a_3 = 2, a_4 = -2, a_5 = 2$ **35.** $a_1 = 3, a_2 = \frac{3}{2}, a_3 = \frac{1}{2}, a_4 = \frac{1}{8}, a_5 = \frac{1}{40}$

36. $a_1 = -2, a_2 = -4, a_3 = -9, a_4 = -23, a_5 = -64$ **37.** $a_1 = 1, a_2 = 2, a_3 = 2, a_4 = 4, a_5 = 8$

38. $a_1 = -1, a_2 = 1, a_3 = 2, a_4 = 9, a_5 = 47$ **39.** $a_1 = A, a_2 = A + d, a_3 = A + 2d, a_4 = A + 3d, a_5 = A + 4d$

40. $a_1 = A, a_2 = rA, a_3 = r^2A, a_4 = r^3A, a_5 = r^4A$

41. $a_1 = \sqrt{2}, a_2 = \sqrt{2 + \sqrt{2}}, a_3 = \sqrt{2 + \sqrt{2 + \sqrt{2}}}, a_4 = \sqrt{2 + \sqrt{2 + \sqrt{2 + \sqrt{2}}}}, a_5 = \sqrt{2 + \sqrt{2 + \sqrt{2 + \sqrt{2 + \sqrt{2}}}}}$

42. $a_1 = \sqrt{2}, a_2 = \sqrt{\frac{\sqrt{2}}{2}}, a_3 = \sqrt{\frac{\sqrt{\frac{\sqrt{2}}{2}}}{2}}, a_4 = \sqrt{\frac{\sqrt{\frac{\sqrt{\frac{\sqrt{2}}{2}}}{2}}}{2}}, a_5 = \sqrt{\frac{\sqrt{\frac{\sqrt{\frac{\sqrt{\frac{\sqrt{2}}{2}}}{2}}}{2}}}{2}}$

43. 50 **44.** 160 **45.** 21 **46.** −10 **47.** 90 **48.** 21 **49.** 26 **50.** 10 **51.** 42 **52.** 60 **53.** 96 **54.** 44 **55.** $3 + 4 + \cdots + (n + 2)$

56. $3 + 5 + 7 + \cdots + (2n + 1)$ **57.** $\frac{1}{2} + 2 + \frac{9}{2} + \cdots + \frac{n^2}{2}$ **58.** $4 + 9 + 16 + \cdots + (n + 1)^2$ **59.** $1 + \frac{1}{3} + \frac{1}{9} + \cdots + \frac{1}{3^n}$

60. $1 + \frac{3}{2} + \frac{9}{4} + \cdots + \left(\frac{3}{2}\right)^n$ **61.** $\frac{1}{3} + \frac{1}{9} + \cdots + \frac{1}{3^n}$ **62.** $1 + 3 + 5 + \cdots + (2n - 1)$ **63.** $\ln 2 - \ln 3 + \ln 4 - \cdots + (-1)^n \ln n$

64. $8 - 16 + 32 - \cdots + (-1)^{n+1} 2^n$ **65.** $\sum_{k=1}^{20} k$ **66.** $\sum_{k=1}^{8} k^3$ **67.** $\sum_{k=1}^{13} \frac{k}{k+1}$ **68.** $\sum_{k=1}^{12} (2k - 1)$ **69.** $\sum_{k=0}^{6} (-1)^k \left(\frac{1}{3^k}\right)$ **70.** $\sum_{k=1}^{11} (-1)^{k+1} \left(\frac{2}{3}\right)^k$

71. $\sum_{k=1}^{n} \frac{3^k}{k}$ **72.** $\sum_{k=1}^{n} \frac{k}{e^k}$ **73.** $\sum_{k=0}^{n} (a + kd)$ or $\sum_{k=1}^{n+1} [a + (k - 1)d]$ **74.** $\sum_{k=1}^{n-1} ar^k$ or $\sum_{k=1}^{n} ar^{k-1}$

75. \$2930 **76.** \$18,058.03 **77.** 2162 **78.** 231 tons **79.** 21

80. (a) $u_1 = \frac{(1 + \sqrt{5}) - (1 - \sqrt{5})}{2\sqrt{5}} = \frac{2\sqrt{5}}{2\sqrt{5}} = 1;$

$$u_2 = \frac{(1 + \sqrt{5})^2 - (1 - \sqrt{5})^2}{2^2\sqrt{5}} = \frac{(1 + 2\sqrt{5} + 5) - (1 - 2\sqrt{5} + 5)}{4\sqrt{5}} = \frac{4\sqrt{5}}{4\sqrt{5}} = 1$$

(b) Set $A = 1 + \sqrt{5}, B = 1 - \sqrt{5}$.

$$\begin{aligned} u_{n+2} &= \frac{A^{n+2} - B^{n+2}}{2^{n+2}\sqrt{5}} \\ &= \frac{A^2 \cdot A^n - B^2 \cdot B^n}{2^{n+2}\sqrt{5}} \\ &= \frac{(1 + \sqrt{5})^2 A^n - (1 - \sqrt{5})^2 B^n}{2^{n+2}\sqrt{5}} \\ &= \frac{(6 + 2\sqrt{5}) A^n - (6 - 2\sqrt{5}) B^n}{2^{n+2}\sqrt{5}} \\ &= \frac{(3 + \sqrt{5}) A^n - (3 - \sqrt{5}) B^n}{2^{n+1}\sqrt{5}} \\ &= \frac{(1 + \sqrt{5}) A^n - (1 - \sqrt{5}) B^n + 2A^n - 2B^n}{2^{n+1}\sqrt{5}} \\ &= \frac{A^{n+1} - B^{n+1}}{2^{n+1}\sqrt{5}} + \frac{A^n - B^n}{2^n\sqrt{5}} \\ &= u_{n+1} + u_n \end{aligned}$$

(c) This is a Fibonacci sequence since $u_1 = 1 = u_2$ and $u_{n+2} = u_{n+1} + u_n$.

81. A Fibonacci sequence **82.** **(a)** 1, 1, 2, 3, 5, 8, 13, 21, 34, 55 **(b)** $1, 2, \frac{3}{2}, \frac{5}{3}, \frac{8}{5}, \frac{13}{8}, \frac{21}{13}, \frac{34}{21}, \frac{55}{34}, \frac{89}{55}$ **(c)** $\frac{1+\sqrt{5}}{2}$ **(d)** $1, \frac{1}{2}, \frac{2}{3}, \frac{3}{5}, \frac{5}{8}, \frac{8}{13}, \frac{13}{21}, \frac{21}{34}, \frac{34}{55}, \frac{55}{89}$ **(e)** $\frac{2}{1+\sqrt{5}}$ **83.** $2S = \underbrace{(1+n)+(1+n)+\cdots+(n+1)}_{n \text{ terms}} = n(n+1);\ S = \frac{1}{2}n(n+1)$

11.2 Concepts and Vocabulary *(page 798)*

1. arithmetic **2.** False

11.2 Exercises *(page 798)*

3. $d = 1; 5, 6, 7, 8$ **4.** $d = 1; -4, -3, -2, -1$ **5.** $d = 2; -3, -1, 1, 3$ **6.** $d = 3; 4, 7, 10, 13$ **7.** $d = -2; 4, 2, 0, -2$ **8.** $d = -2; 2, 0, -2, -4$ **9.** $d = -\frac{1}{3}; \frac{1}{6}, -\frac{1}{6}, -\frac{1}{2}, -\frac{5}{6}$ **10.** $d = \frac{1}{4}; \frac{11}{12}, \frac{7}{6}, \frac{17}{12}, \frac{5}{3}$ **11.** $d = \ln 3; \ln 3, 2\ln 3, 3\ln 3, 4\ln 3$ **12.** $d = 1; 1, 2, 3, 4$ **13.** $a_n = 3n - 1; a_5 = 14$ **14.** $a_n = 4n - 6; a_5 = 14$ **15.** $a_n = 8 - 3n; a_5 = -7$ **16.** $a_n = 8 - 2n; a_5 = -2$ **17.** $a_n = \frac{1}{2}(n-1); a_5 = 2$ **18.** $a_n = \frac{4}{3} - \frac{1}{3}n; a_5 = -\frac{1}{3}$ **19.** $a_n = \sqrt{2}n; a_5 = 5\sqrt{2}$ **20.** $a_n = -\pi + \pi n; a_5 = 4\pi$ **21.** $a_{12} = 24$ **22.** $a_8 = 13$ **23.** $a_{10} = -26$ **24.** $a_9 = -35$ **25.** $a_8 = a + 7b$ **26.** $a_7 = 14\sqrt{5}$ **27.** $a_1 = -13; d = 3; a_n = a_{n-1} + 3$ **28.** $a_1 = -3; d = 2; a_n = a_{n-1} + 2$ **29.** $a_1 = -53; d = 6; a_n = a_{n-1} + 6$ **30.** $a_1 = 74; d = -10; a_n = a_{n-1} - 10$ **31.** $a_1 = 28; d = -2; a_n = a_{n-1} - 2$ **32.** $a_1 = -18; d = 4; a_n = a_{n-1} + 4$ **33.** $a_1 = 25; d = -2; a_n = a_{n-1} - 2$ **34.** $a_1 = -40; d = 4; a_n = a_{n-1} + 4$ **35.** n^2 **36.** $n(n+1)$ **37.** $\frac{n}{2}(9 + 5n)$ **38.** $n(2n-3)$ **39.** 1260 **40.** 900 **41.** 324 **42.** 301

43.

44.

45.

46.
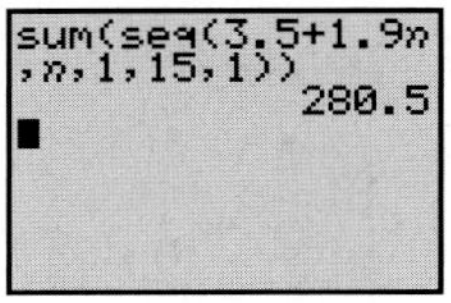

47.

48.

49. $-\frac{3}{2}$ **50.** 1 **51.** 1185 seats **52.** 2160 seats **53.** 210 of beige and 190 blue **54.** **(a)** 42 bricks **(b)** 2130 bricks **55.** 30 rows **56.** 8 yr

Historical Problems *(page 808)*

1. $1\frac{2}{3}$ loaves, $10\frac{5}{6}$ loaves, 20 loaves, $29\frac{1}{6}$ loaves, $38\frac{1}{3}$ loaves **2.** **(a)** 1 **(b)** 2401 **(c)** 2800

11.3 Concepts and Vocabulary *(page 808)*

3. geometric **4.** $\frac{a}{1-r}$ **5.** annuity **6.** True **7.** False **8.** True

11.3 Exercises *(page 808)*

9. $r = 3; 3, 9, 27, 81$ **10.** $r = -5; -5, 25, -125, 625$ **11.** $r = \frac{1}{2}; -\frac{3}{2}, -\frac{3}{4}, -\frac{3}{8}, -\frac{3}{16}$ **12.** $r = \frac{5}{2}; \frac{5}{2}, \frac{25}{4}, \frac{125}{8}, \frac{625}{16}$ **13.** $r = 2; \frac{1}{4}, \frac{1}{2}, 1, 2$ **14.** $r = 3; \frac{1}{3}, 1, 3, 9$ **15.** $r = 2^{1/3}; 2^{1/3}, 2^{2/3}, 2, 2^{4/3}$ **16.** $r = 9; 9, 81, 729, 6561$ **17.** $r = \frac{3}{2}; \frac{1}{2}, \frac{3}{4}, \frac{9}{8}, \frac{27}{16}$ **18.** $r = \frac{2}{3}; 2, \frac{4}{3}, \frac{8}{9}, \frac{16}{27}$ **19.** Arithmetic; $d = 1$ **20.** Arithmetic; $d = 2$ **21.** Neither **22.** Neither **23.** Arithmetic; $d = -\frac{2}{3}$ **24.** Arithmetic; $d = -\frac{3}{4}$ **25.** Neither **26.** Arithmetic; $d = 2$ **27.** Geometric; $r = \frac{2}{3}$ **28.** Geometric; $r = \frac{5}{4}$ **29.** Geometric; $r = 2$ **30.** Neither **31.** Geometric; $r = 3^{1/2}$ **32.** Geometric; $r = -1$ **33.** $a_5 = 162; a_n = 2 \cdot 3^{n-1}$ **34.** $a_5 = -512; a_n = -2 \cdot 4^{n-1}$ **35.** $a_5 = 5; a_n = 5 \cdot (-1)^{n-1}$ **36.** $a_5 = 96; a_n = 6 \cdot (-2)^{n-1}$ **37.** $a_5 = 0; a_n = 0$ **38.** $a_5 = \frac{1}{81}; a_n = \left(-\frac{1}{3}\right)^{n-1}$ **39.** $a_5 = 4\sqrt{2}; a_n = (\sqrt{2})^n$ **40.** $a_5 = 0; a_n = 0$ **41.** $a_7 = \frac{1}{64}$ **42.** $a_8 = 2187$ **43.** $a_9 = 1$ **44.** $a_{10} = 512$ **45.** $a_8 = 0.00000004$

46. $a_7 = 100{,}000$ **47.** $-\frac{1}{4}(1 - 2^n)$ **48.** $-\frac{1}{6}(1 - 3^n)$ **49.** $2\left[1 - \left(\frac{2}{3}\right)^n\right]$ **50.** $-2(1 - 3^n)$ **51.** $1 - 2^n$ **52.** $5\left[1 - \left(\frac{3}{5}\right)^n\right]$

53. 

54. 

55. 

56. 

57. 

58. 

59. $\frac{3}{2}$ **60.** 6 **61.** 16 **62.** 9 **63.** $\frac{8}{5}$ **64.** $\frac{4}{7}$ **65.** $\frac{20}{3}$ **66.** 12 **67.** $\frac{18}{5}$ **68.** $\frac{8}{3}$ **69.** -4 **70.** 2 **71.** \$21,879.11 **72.** \$6655.58

73. **(a)** 0.775 ft **(b)** 8th **(c)** 15.88 ft **(d)** 20 ft

74. **(a)** 15.36 ft **(b)** $30(0.8)^n$ **(c)** 19 times **(d)** 270 ft

75. \$349,496.41 **76.** \$16,712.73 **77.** \$96,885.98 **78.** \$66,438.85 **79.** \$305.10 **80.** \$312.44 **81.** Option *A* results in a higher salary in year 5 (\$25,250 versus \$24,761); Option *B* results in a higher 5-year total (\$116,801 versus \$ 112,742). **82.** Option *B* results in more money (\$524,287 versus \$500,500). **83.** Option 2 results in the most: \$16,038, 304; Option 1 results in the least: \$14,700,000.

84. December 20, 2001 (111 days); \$9999.92 **85.** 1.845×10^{19} **86.** $\frac{1}{3}$ **87.** 10 **88.** 20 **89.** \$72.67 per share **90.** \$39.64

91. Yes. A constant sequence is both arithmetic and geometric. For example, 3, 3, 3, ... is an arithmetic sequence with $a_1 = 3$ and $d = 0$ and is a geometric sequence with $a_1 = 3$ and $r = 1$.

11.4 Exercises *(page 814)*

1. (I) $n = 1: 2(1) = 2$ and $1(1 + 1) = 2$

(II) If $2 + 4 + 6 + \cdots + 2k = k(k + 1)$, then $2 + 4 + 6 + \cdots + 2k + 2(k + 1) = (2 + 4 + 6 + \cdots + 2k) + 2(k + 1)$ $= k(k + 1) + 2(k + 1) = k^2 + 3k + 2 = (k + 1)(k + 2) = (k + 1)[(k + 1) + 1]$.

2. (I) $n = 1: 4(1) - 3 = 1$ and $1(2 - 1) = 1$

(II) If $1 + 5 + 9 + \cdots + (4k - 3) = k(2k - 1)$, then $1 + 5 + 9 + \cdots + (4k - 3) + [4(k + 1) - 3]$ $= [1 + 5 + 9 + \cdots + (4k - 3)] + (4k + 1) = k(2k - 1) + (4k + 1) = 2k^2 + 3k + 1 = (k + 1)(2k + 1)$ $= (k + 1)[2(k + 1) - 1]$.

3. (I) $n = 1: 1 + 2 = 3$ and $\frac{1}{2}(1)(1 + 5) = \frac{1}{2}(6) = 3$

(II) If $3 + 4 + 5 + \cdots + (k + 2) = \frac{1}{2}k(k + 5)$, then $3 + 4 + 5 + \cdots + (k + 2) + [(k + 1) + 2]$

$= [3 + 4 + 5 + \cdots + (k + 2)] + (k + 3) = \frac{1}{2}k(k + 5) + k + 3 = \frac{1}{2}(k^2 + 7k + 6) = \frac{1}{2}(k + 1)(k + 6)$

$= \frac{1}{2}(k + 1)[(k + 1) + 5]$.

4. (I) $n = 1: 2(1) + 1 = 3$ and $1(1 + 2) = 3$

(II) If $3 + 5 + 7 + \cdots + (2k + 1) = k(k + 2)$, then $3 + 5 + 7 + \cdots + (2k + 1) + [2(k + 1) + 1]$ $= [3 + 5 + 7 + \cdots + (2k + 1)] + (2k + 3) = k(k + 2) + (2k + 3) = k^2 + 4k + 3 = (k + 1)(k + 3)$ $= (k + 1)[(k + 1) + 2]$.

5. (I) $n = 1: 3(1) - 1 = 2$ and $\frac{1}{2}(1)[3(1) + 1] = \frac{1}{2}(4) = 2$

(II) If $2 + 5 + 8 + \cdots + (3k - 1) = \frac{1}{2}k(3k + 1)$, then $2 + 5 + 8 + \cdots + (3k - 1) + [3(k + 1) - 1]$

$= [2 + 5 + 8 + \cdots + (3k - 1)] + (3k + 2) = \frac{1}{2}k(3k + 1) + (3k + 2) = \frac{1}{2}(3k^2 + 7k + 4) = \frac{1}{2}(k + 1)(3k + 4)$

$= \frac{1}{2}(k + 1)[3(k + 1) + 1]$.

6. (I) $n = 1: 3(1) - 2 = 1$ and $\frac{1}{2}(1)[3(1) - 1] = 1$

(II) If $1 + 4 + 7 + \cdots + (3k - 2) = \frac{1}{2}k(3k - 1)$, then $1 + 4 + 7 + \cdots + (3k - 2) + [3(k + 1) - 2]$

$= [1 + 4 + 7 + \cdots + (3k - 2)] + (3k + 1) = \frac{1}{2}k(3k - 1) + (3k + 1) = \frac{1}{2}(3k^2 + 5k + 2) = \frac{1}{2}(k + 1)(3k + 2)$

$= \frac{1}{2}(k + 1)[3(k + 1) - 1]$.

7. (I) $n = 1: 2^{1-1} = 1$ and $2^1 - 1 = 1$
(II) If $1 + 2 + 2^2 + \cdots + 2^{k-1} = 2^k - 1$, then $1 + 2 + 2^2 + \cdots + 2^{k-1} + 2^{(k+1)-1} = (1 + 2 + 2^2 + \cdots + 2^{k-1}) + 2^k$
$= 2^k - 1 + 2^k = 2(2^k) - 1 = 2^{k+1} - 1$.

8. (I) $n = 1: 3^{1-1} = 1$ and $\frac{1}{2}(3^1 - 1) = 1$

(II) If $1 + 3 + 3^2 + \cdots + 3^{k-1} = \frac{1}{2}(3^k - 1)$, then $1 + 3 + 3^2 + \cdots + 3^{k-1} + 3^{(k+1)-1} = (1 + 3 + 3^2 + \cdots + 3^{k-1}) + 3^k$

$= \frac{1}{2}(3^k - 1) + 3^k = \frac{1}{2}[3^k - 1 + 2(3^k)] = \frac{1}{2}[3(3^k) - 1] = \frac{1}{2}(3^{k+1} - 1)$.

9. (I) $n = 1: 4^{1-1} = 1$ and $\frac{1}{3}(4^1 - 1) = \frac{1}{3}(3) = 1$

(II) If $1 + 4 + 4^2 + \cdots + 4^{k-1} = \frac{1}{3}(4^k - 1)$, then $1 + 4 + 4^2 + \cdots + 4^{k-1} + 4^{(k+1)-1} = (1 + 4 + 4^2 + \cdots + 4^{k-1}) + 4^k$

$= \frac{1}{3}(4^k - 1) + 4^k = \frac{1}{3}[4^k - 1 + 3(4^k)] = \frac{1}{3}[4(4^k) - 1] = \frac{1}{3}(4^{k+1} - 1)$.

10. (I) $n = 1: 5^{1-1} = 1$ and $\frac{1}{4}(5^1 - 1) = \frac{1}{4}(4) = 1$

(II) If $1 + 5 + 5^2 + \cdots + 5^{k-1} = \frac{1}{4}(5^k - 1)$, then $1 + 5 + 5^2 + \cdots + 5^{k-1} + 5^{(k+1)-1} = (1 + 5 + 5^2 + \cdots + 5^{k-1}) + 5^k$

$= \frac{1}{4}(5^k - 1) + 5^k = \frac{1}{4}[5^k - 1 + 4(5^k)] = \frac{1}{4}[5(5^k) - 1] = \frac{1}{4}(5^{k+1} - 1)$.

11. (I) $n = 1: \frac{1}{1 \cdot 2} = \frac{1}{2}$ and $\frac{1}{1 + 1} = \frac{1}{2}$

(II) If $\frac{1}{1 \cdot 2} + \frac{1}{2 \cdot 3} + \frac{1}{3 \cdot 4} + \cdots + \frac{1}{k(k + 1)} = \frac{k}{k + 1}$, then $\frac{1}{1 \cdot 2} + \frac{1}{2 \cdot 3} + \frac{1}{3 \cdot 4} + \cdots + \frac{1}{k(k + 1)} + \frac{1}{(k + 1)[(k + 1) + 1]}$
$= \left[\frac{1}{1 \cdot 2} + \frac{1}{2 \cdot 3} + \frac{1}{3 \cdot 4} + \cdots + \frac{1}{k(k + 1)}\right] + \frac{1}{(k + 1)(k + 2)} = \frac{k}{k + 1} + \frac{1}{(k + 1)(k + 2)} = \frac{k(k + 2) + 1}{(k + 1)(k + 2)}$
$= \frac{k^2 + 2k + 1}{(k + 1)(k + 2)} = \frac{(k + 1)^2}{(k + 1)(k + 2)} = \frac{k + 1}{k + 2} = \frac{k + 1}{(k + 1) + 1}$.

12. (I) $n = 1: \frac{1}{(2 \cdot 1 - 1)(2 \cdot 1 + 1)} = \frac{1}{3}$ and $\frac{1}{2 \cdot 1 + 1} = \frac{1}{3}$

(II) If $\frac{1}{1 \cdot 3} + \frac{1}{3 \cdot 5} + \frac{1}{5 \cdot 7} + \cdots + \frac{1}{(2k - 1)(2k + 1)} = \frac{k}{2k + 1}$, then $\frac{1}{1 \cdot 3} + \frac{1}{3 \cdot 5} + \frac{1}{5 \cdot 7} + \cdots + \frac{1}{(2k - 1)(2k + 1)}$
$+ \frac{1}{[2(k + 1) - 1][2(k + 1) + 1]} = \left[\frac{1}{1 \cdot 3} + \frac{1}{3 \cdot 5} + \frac{1}{5 \cdot 7} + \cdots + \frac{1}{(2k - 1)(2k + 1)}\right] + \frac{1}{(2k + 1)(2k + 3)}$
$= \frac{k}{2k + 1} + \frac{1}{(2k + 1)(2k + 3)} = \frac{k(2k + 3) + 1}{(2k + 1)(2k + 3)} = \frac{2k^2 + 3k + 1}{(2k + 1)(2k + 3)} = \frac{(2k + 1)(k + 1)}{(2k + 1)(2k + 3)} = \frac{k + 1}{2k + 3} = \frac{k + 1}{2(k + 1) + 1}$.

13. (I) $n = 1: 1^2 = 1$ and $\frac{1}{6} \cdot 1 \cdot 2 \cdot 3 = 1$

(II) If $1^2 + 2^2 + 3^2 + \cdots + k^2 = \frac{1}{6}k(k + 1)(2k + 1)$, then $1^2 + 2^2 + 3^2 + \cdots + k^2 + (k + 1)^2$

$= (1^2 + 2^2 + 3^2 + \cdots + k^2) + (k + 1)^2 = \frac{1}{6}k(k + 1)(2k + 1) + (k + 1)^2 = \frac{1}{6}(2k^3 + 9k^2 + 13k + 6)$

$= \frac{1}{6}(k + 1)(k + 2)(2k + 3) = \frac{1}{6}(k + 1)[(k + 1) + 1][2(k + 1) + 1]$.

14. (I) $n = 1: 1^3 = 1$ and $\frac{1}{4} \cdot 1^2(1 + 1)^2 = 1$

(II) If $1^3 + 2^3 + 3^3 + \cdots + k^3 = \frac{1}{4}k^2(k + 1)^2$, then $1^3 + 2^3 + 3^3 + \cdots + k^3 + (k + 1)^3 = (1^3 + 2^3 + 3^3 + \cdots + k^3) + (k + 1)^3$

$= \frac{1}{4}k^2(k + 1)^2 + (k + 1)^3 = (k + 1)^2\left(\frac{1}{4}k^2 + k + 1\right) = \frac{1}{4}(k + 1)^2(k^2 + 4k + 4) = \frac{1}{4}(k + 1)^2(k + 2)^2$

$= \frac{1}{4}(k + 1)^2[(k + 1) + 1]^2$.

15. (I) $n = 1: 5 - 1 = 4$ and $\frac{1}{2}(1)(9 - 1) = \frac{1}{2} \cdot 8 = 4$

(II) If $4 + 3 + 2 + \cdots + (5 - k) = \frac{1}{2}k(9 - k)$, then $4 + 3 + 2 + \cdots + (5 - k) + [5 - (k + 1)]$

$= [4 + 3 + 2 + \cdots + (5 - k)] + 4 - k = \frac{1}{2}k(9 - k) + 4 - k = \frac{1}{2}(9k - k^2 + 8 - 2k) = \frac{1}{2}(-k^2 + 7k + 8)$

$= \frac{1}{2}(k + 1)(8 - k) = \frac{1}{2}(k + 1)[9 - (k + 1)]$.

16. (I) $n = 1: -(1 + 1) = -2$ and $-\frac{1}{2}(1)(1 + 3) = -2$

(II) If $-2 - 3 - 4 - \cdots - (k + 1) = -\frac{1}{2}k(k + 3)$, then $-2 - 3 - 4 - \cdots - (k + 1) - [(k + 1) + 1]$

$= [-2 - 3 - 4 - \cdots - (k + 1)] - (k + 2) = -\frac{1}{2}k(k + 3) - (k + 2) = -\frac{1}{2}(k^2 + 3k + 2k + 4) = -\frac{1}{2}(k^2 + 5k + 4)$

$= -\frac{1}{2}(k + 1)(k + 4) = -\frac{1}{2}(k + 1)[(k + 1) + 3]$.

17. (I) $n = 1: 1 \cdot (1 + 1) = 2$ and $\frac{1}{3} \cdot 1 \cdot 2 \cdot 3 = 2$

(II) If $1 \cdot 2 + 2 \cdot 3 + 3 \cdot 4 + \cdots + k(k + 1) = \frac{1}{3}k(k + 1)(k + 2)$, then $1 \cdot 2 + 2 \cdot 3 + 3 \cdot 4 + \cdots + k(k + 1)$

$+ (k + 1)[(k + 1) + 1] = [1 \cdot 2 + 2 \cdot 3 + 3 \cdot 4 + \cdots + k(k + 1)] + (k + 1)(k + 2)$

$= \frac{1}{3}k(k + 1)(k + 2) + \frac{1}{3} \cdot 3\,(k + 1)(k + 2) = \frac{1}{3}(k + 1)(k + 2)(k + 3) = \frac{1}{3}(k + 1)[(k + 1) + 1][(k + 1) + 2]$.

18. (I) $n = 1: (2 \cdot 1 - 1)(2 \cdot 1) = 2$ and $\frac{1}{3} \cdot 1 \cdot (1 + 1)(4 \cdot 1 - 1) = 2$

(II) If $1 \cdot 2 + 3 \cdot 4 + 5 \cdot 6 + \cdots + (2k - 1)(2k) = \frac{1}{3}k(k + 1)(4k - 1)$, then

$1 \cdot 2 + 3 \cdot 4 + 5 \cdot 6 + \cdots + (2k - 1)(2k) + [2(k + 1) - 1][2(k + 1)] = [1 \cdot 2 + 3 \cdot 4 + 5 \cdot 6 + \cdots + (2k - 1)(2k)]$

$+ (2k + 1)(2k + 2) = \frac{1}{3}k(k + 1)(4k - 1) + (2k + 1)(2k + 2) = \frac{1}{3}(4k^3 + 3k^2 - k) + (4k^2 + 6k + 2)$

$= \frac{1}{3}(4k^3 + 3k^2 - k + 12k^2 + 18k + 6) = \frac{1}{3}(4k^3 + 15k^2 + 17k + 6) = \frac{1}{3}(k^2 + 3k + 2)(4k + 3)$

$= \frac{1}{3}(k + 1)(k + 2)(4k + 3) = \frac{1}{3}(k + 1)[(k + 1) + 1][4(k + 1) - 1]$.

19. (I) $n = 1: 1^2 + 1 = 2$, which is divisible by 2.
(II) If $k^2 + k$ is divisible by 2, then $(k + 1)^2 + (k + 1) = k^2 + 2k + 1 + k + 1 = (k^2 + k) + 2k + 2$. Since $k^2 + k$ is divisible by 2 and $2k + 2$ is divisible by 2, $(k + 1)^2 + (k + 1)$ is divisible by 2.

20. (I) $n = 1: 1^3 + 2 \cdot 1 = 3$, which is divisible by 3.
(II) If $k^3 + 2k$ is divisible by 3, then $(k + 1)^3 + 2(k + 1) = k^3 + 3k^2 + 3k + 1 + 2k + 2 = (k^3 + 2k) + 3k^2 + 3k + 3$. Since $(k^3 + 2k)$ is divisible by 3 and $3k^2 + 3k + 3$ is divisible by 3, $(k + 1)^3 + 2(k + 1)$ is divisible by 3.

21. (I) $n = 1: 1^2 - 1 + 2 = 2$ which is divisible by 2.
(II) If $k^2 - k + 2$ is divisible by 2, then $(k + 1)^2 - (k + 1) + 2 = k^2 + 2k + 1 - k - 1 + 2 = (k^2 - k + 2) + 2k$. Since $k^2 - k + 2$ is divisible by 2 and $2k$ is divisible by 2, $(k + 1)^2 - (k + 1) + 2$ is divisible by 2.

22. (I) $n = 1: 1(1 + 1)(1 + 2) = 6$, which is divisible by 6.
(II) If $k(k + 1)(k + 2)$ is divisible by 6, then $(k + 1)[(k + 1) + 1][(k + 1) + 2] = (k + 1)(k + 2)(k + 3)$ $= (k + 1)(k + 2)k + (k + 1)(k + 2)3 = k(k + 1)(k + 2) + 3(k + 1)(k + 2)$. Now, $k(k + 1)(k + 2)$ is divisible by 6, and since $k + 1$ and $k + 2$ are consecutive integers, either $k + 1$ or $k + 2$ is even (divisible by 2). In either case, $3(k + 1)(k + 2)$ is divisible by 6, and so $(k + 1)[(k + 1) + 1][(k + 1) + 2]$ is divisible by 6.

23. (I) $n = 1$: If $x > 1$, then $x^1 = x > 1$.
(II) Assume, for an arbitrary natural number k, that if $x > 1$ then $x^k > 1$. Multiply both sides of the inequality $x^k > 1$ by x. If $x > 1$, then $x^{k+1} > x > 1$.

24. (I) $n = 1$: If $0 < x < 1$, then $x^1 = x$ and $0 < x^1 < 1$.
(II) Assume, for an arbitrary natural number k, that if $0 < x < 1$, then $0 < x^k < 1$. Multiply all sides of the inequalities $0 < x^k < 1$ by x. If $0 < x < 1$ then $0 < x^k x < x < 1$ and $0 < x^{k+1} < 1$.

25. (I) $n = 1$: $a - b$ is a factor of $a^1 - b^1 = a - b$.
(II) If $a - b$ is a factor of $a^k - b^k$, then $a^{k+1} - b^{k+1} = a(a^k - b^k) + b^k(a - b)$.
Since $a - b$ is a factor of $a^k - b^k$ and $a - b$ is a factor of $a - b$, then $a - b$ is a factor of $a^{k+1} - b^{k+1}$.

26. (I) $n = 1: a^{2\cdot 1+1} + b^{2\cdot 1+1} = a^3 + b^3 = (a+b)(a^2 - ab + b^2)$ and $a + b$ is a factor of $a^3 + b^3$.
(II) If $a + b$ is a factor of $a^{2k+1} + b^{2k+1}$, then $a^{2(k+1)+1} + b^{2(k+1)} = a^{2k+3} + b^{2k+3} = a^{2k+1}a^2 + b^{2k+1}b^2$
$= a^{2k+1}a^2 + b^{2k+1}a^2 - b^{2k+1}a^2 + b^{2k+1}b^2 = a^2(a^{2k+1} + b^{2k+1}) - b^{2k+1}(a^2 - b^2) = a^2(a^{2k+1} + b^{2k+1}) - b^{2k+1}(a+b)(a-b)$.
Since $a + b$ is a factor of $a^{2k+1} + b^{2k+1}$ and $a + b$ is a factor of $b^{2k+1}(a+b)(a-b)$, then $a + b$ is a factor of $a^{2(k+1)+1} + b^{2(k+1)+1}$.

27. $n = 1: 1^2 - 1 + 41 = 41$ which is a prime number.
$n = 41: 41^2 - 41 + 41 = 1681 = 41^2$, which is not prime.

28. If $2 + 4 + 6 + \cdots + 2k = k^2 + k + 2$, then $2 + 4 + 6 + \cdots + 2k + 2(k+1)$
$= (2 + 4 + 6 + \cdots + 2k) + 2k + 2 = k^2 + k + 2 + 2k + 2 = k^2 + 3k + 4 = (k^2 + 2k + 1) + (k+1) + 2$
$= (k+1)^2 + (k+1) + 2$.
But $2 \cdot 1 = 2$ and $1^2 + 1 + 2 = 4$. The fact is, $2 + 4 + 6 + \cdots + 2n = n^2 + n$, not $n^2 + n + 2$ (Problem 1)

29. (I) $n = 1: ar^{1-1} = a \cdot 1 = a$ and $a \cdot \dfrac{1 - r^1}{1 - r} = a$, because $r \neq 1$.

(II) If $a + ar + ar^2 + \cdots + ar^{k-1} = a \cdot \dfrac{1 - r^k}{1 - r}$, then $a + ar + ar^2 + \cdots + ar^{k-1} + ar^{(k+1)-1} = (a + ar + ar^2 + \cdots + ar^{k-1}) + ar^k$

$= a \cdot \dfrac{1 - r^k}{1 - r} + ar^k = \dfrac{a(1 - r^k) + ar^k(1 - r)}{1 - r} = \dfrac{a - ar^k + ar^k - ar^{k+1}}{1 - r} = a \cdot \dfrac{1 - r^{k+1}}{1 - r}$.

30. (I) $n = 1: [a + (1 - 1)d] = a$ and $1 \cdot a + d\dfrac{1 \cdot (1 - 1)}{2} = a$

(II) If $a + (a + d) + (a + 2d) + \cdots + [a + (k - 1)d] = ka + d\dfrac{k(k - 1)}{2}$, then

$a + (a + d) + (a + 2d) + \cdots + [a + (k - 1)d] + [a + ((k + 1) - 1)d] = ka + d\dfrac{k(k - 1)}{2} + a + kd$

$= (k + 1)a + d\dfrac{k(k - 1) + 2k}{2} = (k + 1)a + d\dfrac{(k + 1)(k)}{2} = (k + 1)a + d\dfrac{(k + 1)[(k + 1) - 1]}{2}$.

31. (I) $n = 4$: The number of diagonals in a convex polygon of 4 sides is 2 and $\dfrac{1}{2} \cdot 4 \cdot (4 - 3) = 2$

(II) If the number of diagonals in a convex polygon of k sides is $\dfrac{1}{2}k(k - 3)$ then that of $(k + 1)$ sides is increased by

$(k + 1) - 2 = k - 1$. Thus, the number of diagonals in a convex polygon of $(k + 1)$ sides is $\dfrac{1}{2}k(k - 3) + (k - 1)$

$= \dfrac{1}{2}[k^2 - 3k + 2k - 2] = \dfrac{1}{2}[k^2 - k - 2] = \dfrac{1}{2}(k + 1)(k - 2) = \dfrac{1}{2}(k + 1)[(k + 1) - 3]$.

32. (I) $n = 3$: The sum of the angles of a triangle is $(3 - 2) \cdot 180° = 180°$.
(II) Assume for some $k \geq 3$ that the sum of the angles of a convex polygon of k sides is $(k - 2) \cdot 180°$. A convex polygon of $k + 1$ sides consists of a convex polygon of k sides plus a triangle (see the illustration). The sum of the angles is
$(k - 2) \cdot 180° + 180° = (k - 1) \cdot 180° = [(k + 1) - 2] \cdot 180°$.

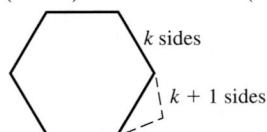

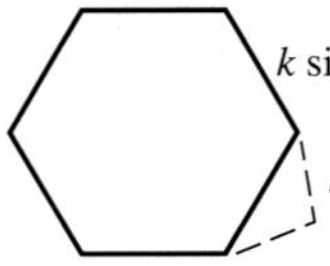

11.5 Concepts and Vocabulary *(page 821)*

1. Pascal triangle **2.** 15 **3.** False **4.** Binomial Theorem

11.5 Exercises *(page 821)*

5. 10 **6.** 35 **7.** 21 **8.** 36 **9.** 50 **10.** 4950 **11.** 1 **12.** 1 **13.** 1.8664×10^{15} **14.** 4.1918×10^{15} **15.** 1.4834×10^{13} **16.** 1.7673×10^{10}
17. $x^5 + 5x^4 + 10x^3 + 10x^2 + 5x + 1$ **18.** $x^5 - 5x^4 + 10x^3 - 10x^2 + 5x - 1$
19. $x^6 - 12x^5 + 60x^4 - 160x^3 + 240x^2 - 192x + 64$ **20.** $x^5 + 15x^4 + 90x^3 + 270x^2 + 405x + 243$
21. $81x^4 + 108x^3 + 54x^2 + 12x + 1$ **22.** $32x^5 + 240x^4 + 720x^3 + 1080x^2 + 810x + 243$
23. $x^{10} + 5x^8y^2 + 10x^6y^4 + 10x^4y^6 + 5x^2y^8 + y^{10}$ **24.** $x^{12} - 6x^{10}y^2 + 15x^8y^4 - 20x^6y^6 + 15x^4y^8 - 6x^2y^{10} + y^{12}$
25. $x^3 + 6\sqrt{2}x^{5/2} + 30x^2 + 40\sqrt{2}x^{3/2} + 60x + 24\sqrt{2}x^{1/2} + 8$ **26.** $x^2 - 4\sqrt{3}x^{3/2} + 18x - 12\sqrt{3}x^{1/2} + 9$
27. $a^5x^5 + 5a^4bx^4y + 10a^3b^2x^3y^2 + 10a^2b^3x^2y^3 + 5ab^4xy^4 + b^5y^5$ **28.** $a^4x^4 - 4a^3bx^3y + 6a^2b^2x^2y^2 - 4ab^3xy^3 + b^4y^4$
29. 17,010 **30.** −262,440 **31.** −101,376 **32.** 1760 **33.** 41,472 **34.** −314,928 **35.** $2835x^3$ **36.** $189x^5$ **37.** $314,928x^7$ **38.** $48,384x^3$
39. 495 **40.** −84 **41.** 3360 **42.** 252 **43.** 1.00501 **44.** 0.98806

45. $\binom{n}{n-1} = \frac{n!}{(n-1)![n-(n-1)]!} = \frac{n!}{(n-1)!1!} = \frac{n\cdot(n-1)!}{(n-1)!} = n; \binom{n}{n} = \frac{n!}{n!(n-n)!} = \frac{n!}{n!0!} = \frac{n!}{n!} = 1$

46. $\binom{n}{n-j} = \frac{n!}{(n-j)![n-(n-j)]!} = \frac{n!}{(n-j)!j!} = \frac{n!}{j!(n-j)!} = \binom{n}{j}$

47. $2^n = (1+1)^n = \binom{n}{0}1^n + \binom{n}{1}(1)^{n-1}(1) + \cdots + \binom{n}{n}1^n = \binom{n}{0} + \binom{n}{1} + \cdots + \binom{n}{n}$

48. $0 = 0^n = (1-1)^n = \binom{n}{0}1^n + \binom{n}{1}(1)^{n-1}(-1) + \binom{n}{2}(1)^{n-2}(-1)^2 + \cdots + \binom{n}{n}(-1)^n = \binom{n}{0} - \binom{n}{1} + \binom{n}{2} - \cdots + (-1)^n\binom{n}{n}$

49. 1 **50.** $12! = 479{,}001{,}600 \approx 479{,}013{,}972.4$
$20! = 2.432902008 \times 10^{18} \approx 2.43292403 \times 10^{18}$
$25! = 1.551121004 \times 10^{25} \approx 1.551129917 \times 10^{25}$

Review Exercises *(page 823)*

1. $-\frac{4}{3}, \frac{5}{4}, -\frac{6}{5}, \frac{7}{6}, -\frac{8}{7}$ **2.** $5, -7, 9, -11, 13$ **3.** $2, 1, \frac{8}{9}, 1, \frac{32}{25}$ **4.** $e, \frac{e^2}{2}, \frac{e^3}{3}, \frac{e^4}{4}, \frac{e^5}{5}$ **5.** $3, 2, \frac{4}{3}, \frac{8}{9}, \frac{16}{27}$ **6.** $4, -1, \frac{1}{4}, -\frac{1}{16}, \frac{1}{64}$ **7.** $2, 0, 2, 0, 2$ **8.** $-3, 1, 5, 9, 13$ **9.** $6 + 10 + 14 + 18 = 48$ **10.** $2 + (-1) + (-6) = -5$ **11.** $\sum_{k=1}^{13} (-1)^{k+1}\frac{1}{k}$ **12.** $\sum_{k=0}^{n} \frac{2^{k+1}}{3^k}$ **13.** Arithmetic; $d = 1$; $S_n = \frac{n}{2}(n + 11)$ **14.** Arithmetic; $d = 4$; $S_n = n(2n + 5)$ **15.** Neither **16.** Neither **17.** Geometric; $r = 8$; $S_n = \frac{8}{7}(8^n - 1)$ **18.** Geometric; $r = 9$; $S_n = \frac{9}{8}(9^n - 1)$ **19.** Arithmetic; $d = 4$; $S_n = 2n(n - 1)$ **20.** Arithmetic; $d = -4$; $S_n = n(3 - 2n)$

21. Geometric; $r = \frac{1}{2}$; $S_n = 6\left[1 - \left(\frac{1}{2}\right)^n\right]$ **22.** Geometric; $r = -\frac{1}{3}$; $S_n = \frac{15}{4}\left[1 - \left(-\frac{1}{3}\right)^n\right]$ **23.** Neither **24.** Neither **25.** 115

26. 50 **27.** 75 **28.** -18 **29.** $\frac{1093}{2187} \approx 0.49977$ **30.** 682 **31.** 35 **32.** -13 **33.** $\frac{1}{10^{10}}$ **34.** 1024 **35.** $9\sqrt{2}$ **36.** $16\sqrt{2}$

37. $a_n = 5n - 4$ **38.** $a_n = 4 - 3n$ **39.** $a_n = n - 10$ **40.** $a_n = 2n + 6$ **41.** $\frac{9}{2}$ **42.** 4 **43.** $\frac{4}{3}$ **44.** $\frac{18}{5}$ **45.** 8 **46.** $\frac{12}{7}$

47. (I) $n = 1: 3 \cdot 1 = 3$ and $\frac{3\cdot 1}{2}(1 + 1) = 3$

(II) If $3 + 6 + 9 + \cdots + 3k = \frac{3k}{2}(k + 1)$, then $3 + 6 + 9 + \cdots + 3k + 3(k + 1) = (3 + 6 + 9 + \cdots + 3k) + (3k + 3)$
$= \frac{3k}{2}(k + 1) + (3k + 3) = \frac{3k^2}{2} + \frac{3k}{2} + \frac{6k}{2} + \frac{6}{2} = \frac{3}{2}(k^2 + 3k + 2) = \frac{3}{2}(k + 1)(k + 2) = \frac{3(k + 1)}{2}[(k + 1) + 1]$

48. (I) $n = 1: (4 \cdot 1 - 2) = 2$ and $2 \cdot 1^2 = 2$

(II) If $2 + 6 + 10 + \cdots + (4k - 2) = 2k^2$, then $2 + 6 + 10 + \cdots + (4k - 2) + [4(k + 1) - 2]$
$= [2 + 6 + 10 + \cdots + (4k - 2)] + (4k + 2) = 2k^2 + (4k + 2) = 2(k + 1)^2$

49. (I) $n = 1: 2 \cdot 3^{1-1} = 2$ and $3^1 - 1 = 2$

(II) If $2 + 6 + 18 + \cdots + 2 \cdot 3^{k-1} = 3^k - 1$, then $2 + 6 + 18 + \cdots + 2 \cdot 3^{k-1} + 2 \cdot 3^{(k+1)-1} = (2 + 6 + 18 + \cdots + 2 \cdot 3^{k-1}) + 2 \cdot 3^k$
$= 3^k - 1 + 2 \cdot 3^k = 3 \cdot 3^k - 1 = 3^{k+1} - 1.$

50. (I) $n = 1: 3 \cdot 2^{1-1} = 3$ and $3(2^1 - 1) = 3$

(II) If $3 + 6 + 12 + \cdots + 3 \cdot 2^{k-1} = 3(2^k - 1)$, then $3 + 6 + 12 + \cdots + 3 \cdot 2^{k-1} + 3 \cdot 2^{(k+1)-1}$
$= (3 + 6 + 12 + \cdots + 3 \cdot 2^{k-1}) + 3 \cdot 2^k = 3(2^k - 1) + 3 \cdot 2^k = 3(2 \cdot 2^k - 1) = 3(2^{k+1} - 1).$

51. (I) $n = 1: (3 \cdot 1 - 2)^2 = 1$ and $\frac{1}{2} \cdot 1 \cdot [6(1)^2 - 3(1) - 1] = 1$

(II) If $1^2 + 4^2 + 7^2 + \cdots + (3k - 2)^2 = \frac{1}{2}k(6k^2 - 3k - 1)$, then $1^2 + 4^2 + 7^2 + \cdots + (3k - 2)^2 + [3(k + 1) - 2]^2$
$= [1^2 + 4^2 + 7^2 + \cdots + (3k - 2)^2] + (3k + 1)^2 = \frac{1}{2}k(6k^2 - 3k - 1) + (3k + 1)^2 = \frac{1}{2}(6k^3 - 3k^2 - k) + (9k^2 + 6k + 1)$
$= \frac{1}{2}(6k^3 + 15k^2 + 11k + 2) = \frac{1}{2}(k + 1)(6k^2 + 9k + 2) = \frac{1}{2}(k + 1)[6(k + 1)^2 - 3(k + 1) - 1].$

52. (I) $n = 1: 1 \cdot (1 + 2) = 3$ and

(II) If $1 \cdot 3 + 2 \cdot 4 + 3 \cdot 5 + \cdots + k(k + 2) = \frac{k}{6}(k + 1)(2k + 7)$, then $1 \cdot 3 + 2 \cdot 4 + 3 \cdot 5 + \cdots + k(k + 2)$
$+ (k + 1)[(k + 1) + 2] = [1 \cdot 3 + 2 \cdot 4 + 3 \cdot 5 + \cdots + k(k + 2)] + (k + 1)(k + 3)$
$= \frac{k}{6}(k + 1)(2k + 7) + (k + 1)(k + 3) = (k + 1)\left[\frac{k}{6}(2k + 7) + k + 3\right] = \frac{1}{6}(k + 1)(2k^2 + 13k + 18)$
$= \frac{1}{6}(k + 1)(k + 2)(2k + 9) = \frac{k + 1}{6}[(k + 1) + 1][2(k + 1) + 7].$

53. 10 **54.** 28 **55.** $x^5 + 10x^4 + 40x^3 + 80x^2 + 80x + 32$ **56.** $x^4 - 12x^3 + 54x^2 - 108x + 81$
57. $32x^5 + 240x^4 + 720x^3 + 1080x^2 + 810x + 243$ **58.** $81x^4 - 432x^3 + 864x^2 - 768x + 256$ **59.** 144 **60.** −13,608 **61.** 84
62. 1792 **63. (a)** 8 **(b)** 1100 **64.** 360 **65. (a)** $20\left(\frac{3}{4}\right)^3 = \frac{135}{16}$ ft **(b)** $20\left(\frac{3}{4}\right)^n$ ft **(c)** 13 times **(d)** 140 ft
66. \$151,873.77 **67.** \$244,129.08 **68.** \$23,397.17

Cumulative Review *(page 826)*

1. $-3, 3, -3i, 3i$

2. (a)

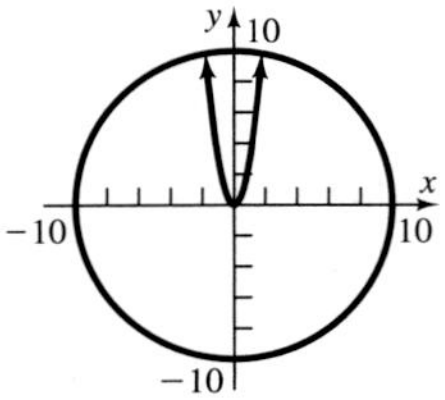

(b) $\left(\sqrt{\frac{-1 + \sqrt{3601}}{18}}, \frac{-1 + \sqrt{3601}}{6}\right), \left(-\sqrt{\frac{-1 + \sqrt{3601}}{18}}, \frac{-1 + \sqrt{3601}}{6}\right)$

(c) The circle and the parabola intersect at $\left(\sqrt{\frac{-1 + \sqrt{3601}}{18}}, \frac{-1 + \sqrt{3601}}{6}\right), \left(-\sqrt{\frac{-1 + \sqrt{3601}}{18}}, \frac{-1 + \sqrt{3601}}{6}\right)$.

3. $\left\{\ln\left(\frac{5}{2}\right)\right\}$ **4.** $y = 5x - 10$ **5.** $x^2 + y^2 + 2x - 4y - 20 = 0$ **6. (a)** 5 **(b)** 13 **(c)** $\frac{6x + 3}{2x - 1}$ **(d)** $\left\{x \middle| x \neq \frac{1}{2}\right\}$ **(e)** $\frac{7x - 2}{x - 2}$
(f) $\{x|x \neq 2\}$ **(g)** $g^{-1}(x) = \frac{1}{2}(x - 1)$; all reals **(h)** $f^{-1}(x) = \frac{2x}{x - 3}$; $\{x|x \neq 3\}$ **7.** $\frac{x^2}{7} + \frac{y^2}{16} = 1$ **8.** $(x + 1)^2 = 4(y - 2)$
9. $r = 8 \sin \theta$; $x^2 + (y - 4)^2 = 16$ **10.** $\left\{\frac{3\pi}{2}\right\}$ **11.** $\frac{2\pi}{3}$ **12. (a)** $-\frac{\sqrt{15}}{4}$ **(b)** $-\frac{\sqrt{15}}{15}$ **(c)** $-\frac{\sqrt{15}}{8}$ **(d)** $\frac{7}{8}$ **(e)** $\sqrt{\frac{1 + \frac{\sqrt{15}}{4}}{2}}$

CHAPTER 12 Counting and Probability

12.1 Concepts and Vocabulary *(page 833)*

1. union **2.** intersection **3.** True **4.** True

12.1 Exercises *(page 833)*

5. {1, 3, 5, 6, 7, 9} **6.** {1, 2, 3, 4, 5, 6, 7, 8, 9} **7.** {1, 5, 7} **8.** {1, 9} **9.** {1, 6, 9} **10.** {1, 6, 9} **11.** {1, 2, 4, 5, 6, 7, 8, 9}
12. {1, 2, 3, 4, 5, 6, 7, 8, 9} **13.** {1, 2, 4, 5, 6, 7, 8, 9} **14.** {1} **15.** {0, 2, 6, 7, 8} **16.** {0, 2, 5, 7, 8, 9} **17.** {0, 1, 2, 3, 5, 6, 7, 8, 9}
18. {0, 5, 9} **19.** {0, 1, 2, 3, 5, 6, 7, 8, 9} **20.** {0, 5, 9} **21.** {0, 1, 2, 3, 4, 6, 7, 8} **22.** {2, 7, 8} **23.** {0} **24.** {0, 1, 2, 3, 5, 6, 7, 8, 9}
25. ∅, {*a*}, {*b*}, {*c*}, {*d*}, {*a*, *b*}, {*a*, *c*}, {*a*, *d*}, {*b*, *c*}, {*b*, *d*}, {*c*, *d*}, {*a*, *b*, *c*}, {*b*, *c*, *d*}, {*a*, *c*, *d*}, {*a*, *b*, *d*}, {*a*, *b*, *c*, *d*}
26. ∅, {*a*}, {*b*}, {*c*}, {*d*}, {*e*}, {*a*, *b*}, {*a*, *c*}, {*a*, *d*}, {*a*, *e*}, {*b*, *c*}, {*b*, *d*}, {*b*, *e*}, {*c*, *d*}, {*c*, *e*}, {*d*, *e*}, {*a*, *b*, *c*}, {*a*, *b*, *d*}, {*a*, *b*, *e*}, {*a*, *c*, *d*}, {*a*, *c*, *e*}, {*a*, *d*, *e*}, {*b*, *c*, *d*}, {*b*, *c*, *e*}, {*b*, *d*, *e*}, {*c*, *d*, *e*}, {*a*, *b*, *c*, *d*}, {*a*, *b*, *c*, *e*}, {*a*, *b*, *d*, *e*}, {*a*, *c*, *d*, *e*}, {*b*, *c*, *d*, *e*}, {*a*, *b*, *c*, *d*, *e*}
27. 25 **28.** 25 **29.** 40 **30.** 50 **31.** 25 **32.** 20 **33.** 37 **34.** 8 **35.** 18 **36.** 31 **37.** 5 **38.** 52 **39.** 175; 125 **40.** 550
41. (a) 15 **(b)** 15 **(c)** 15 **(d)** 25 **(e)** 40 **42.** 8 **43. (a)** 57,886 thousand **(b)** 10,894 thousand **(c)** 14,126 thousand
44. (a) 58,748 thousand **(b)** 22,163 thousand **(c)** 26,285 thousand

12.2 Concepts and Vocabulary *(page 842)*

3. permutation **4.** combination **5.** True **6.** True

12.2 Exercises *(page 843)*

7. 30 **8.** 42 **9.** 24 **10.** 40,320 **11.** 1 **12.** 1 **13.** 1680 **14.** 336 **15.** 28 **16.** 28 **17.** 35 **18.** 15 **19.** 1 **20.** 18 **21.** 10,400,600
22. 48,620 **23.** {*abc*, *abd*, *abe*, *acb*, *acd*, *ace*, *adb*, *adc*, *ade*, *aeb*, *aec*, *aed*, *bac*, *bad*, *bae*, *bca*, *bcd*, *bce*, *bda*, *bdc*, *bde*, *bea*, *bec*, *bed*, *cab*, *cad*, *cae*, *cba*, *cbd*, *cbe*, *cda*, *cdb*, *cde*, *cea*, *ceb*, *ced*, *dab*, *dac*, *dae*, *dba*, *dbc*, *dbe*, *dca*, *dcb*, *dce*, *dea*, *deb*, *dec*, *eab*, *eac*, *ead*, *eba*, *ebc*, *ebd*, *eca*, *ecb*, *ecd*, *eda*, *edb*, *edc*}; 60 **24.** {*ab*, *ba*, *ac*, *ca*, *ad*, *da*, *ae*, *ea*, *bc*, *cb*, *bd*, *db*, *be*, *eb*, *cd*, *dc*, *ce*, *ec*, *de*, *ed*}; 20
25. {123, 124, 132, 134, 142, 143, 213, 214, 231, 234, 241, 243, 312, 314, 321, 324, 341, 342, 412, 413, 421, 423, 431, 432}; 24
26. {123, 132, 213, 231, 312, 321, 124, 142, 214, 241, 412, 421, 125, 152, 215, 251, 512, 521, 126, 162, 216, 261, 612, 621, 134, 143, 314, 341, 413, 431, 135, 153, 315, 351, 513, 531, 136, 163, 316, 361, 613, 631, 145, 154, 415, 451, 514, 541, 146, 164, 416, 461, 614, 641, 156, 165, 516, 561, 615, 651, 234, 243, 324, 342, 423, 432, 235, 253, 325, 352, 523, 532, 236, 263, 326, 362, 623, 632, 245, 254, 425, 452, 524, 542, 246, 264, 426, 462, 624, 642, 256, 265, 526, 562, 625, 652, 345, 354, 435, 453, 534, 543, 346, 364, 436, 463, 634, 643, 356, 365, 536, 563, 635, 653, 456, 465, 546, 564, 645, 654}; 120
27. {*abc*, *abd*, *abe*, *acd*, *ace*, *ade*, *bcd*, *bce*, *bde*, *cde*}; 10 **28.** {*ab*, *ac*, *ad*, *ae*, *bc*, *bd*, *be*, *cd*, *ce*, *de*}; 10 **29.** {123, 124, 134, 234}; 4
30. {123, 124, 125, 126, 134, 135, 136, 145, 146, 156, 234, 235, 236, 245, 246, 256, 345, 346, 356, 456}; 20 **31.** 15 **32.** 15 **33.** 16 **34.** 25
35. 8 **36.** 1000 (allowing numbers with initial zeros such as 012) **37.** 24 **38.** 120 **39.** 60 **40.** 360 **41.** 18,278 **42.** 12,338,352
43. 35 **44.** 56 **45.** 1024 **46.** 1024 **47.** 9000 **48.** 80,000 **49.** 120 **50. (a)** 6,760,000 **(b)** 3,407,040 **(c)** 3,276,000 **51.** 480

52. 117,600 (assuming a combination has no repeated numbers) **53.** 132,860 **54.** 6,302,555,018,760 **55.** 336 **56.** 5,209,344 **57.** 90,720 **58.** 4,989,600 **59. (a)** 63 **(b)** 35 **(c)** 1 **60. (a)** 3003 **(b)** 20,475 **(c)** 16,653 **61.** 1.157×10^{76} **62.** 302,400 **63.** 362,880 **64.** 40,320 **65.** 660 **66.** 70 **67.** 15 **68.** 126

Historical Problems *(page 853)*

1. (a) $\{AAAA, AAAB, AABA, AABB, ABAA, ABAB, ABBA, ABBB, BAAA, BAAB, BABA, BABB, BBAA, BBAB, BBBA, BBBB\}$

(b) $P(A \text{ wins}) = \dfrac{C(4,2) + C(4,3) + C(4,4)}{2^4} = \dfrac{6+4+1}{16} = \dfrac{11}{16}$; $P(B \text{ wins}) = \dfrac{C(4,3) + C(4,4)}{2^4} = \dfrac{4+1}{16} = \dfrac{5}{16}$

2. (a) $\$\dfrac{3}{2} = \$1.50$ **(b)** $\$\dfrac{1}{2} = \$0.50$

12.3 Concepts and Vocabulary *(page 853)*

1. equally likely **2.** complement **3.** False **4.** True

12.3 Exercises *(page 854)*

5. 0, 0.01, 0.35, 1 **6.** $\dfrac{1}{2}, \dfrac{3}{4}, \dfrac{2}{3}, 0$ **7.** Probability model **8.** Probability model **9.** Not a probability model **10.** Not a probability model

11. $S = \{\text{HH, HT, TH, TT}\}$; $P(\text{HH}) = \dfrac{1}{4}$, $P(\text{HT}) = \dfrac{1}{4}$, $P(\text{TH}) = \dfrac{1}{4}$, $P(\text{TT}) = \dfrac{1}{4}$

12. $S = \{\text{HH, HT, TH, TT}\}$; $P(\text{HH}) = \dfrac{1}{4}$, $P(\text{HT}) = \dfrac{1}{4}$, $P(\text{TH}) = \dfrac{1}{4}$, $P(\text{TT}) = \dfrac{1}{4}$

13. $S = \{$HH1, HH2, HH3, HH4, HH5, HH6, HT1, HT2, HT3, HT4, HT5, HT6, TH1, TH2, TH3, TH4, TH5, TH6, TT1, TT2, TT3, TT4, TT5, TT6$\}$; each outcome has the probability of $\dfrac{1}{24}$.

14. $S = \{$H1H, H2H, H3H, H4H, H5H, H6H, H1T, H2T, H3T, H4T, H5T, H6T, T1H, T2H, T3H, T4H, T5H, T6H, T1T, T2T, T3T, T4T, T5T, T6T$\}$; each outcome has the probability of $\dfrac{1}{24}$.

15. $S = \{$HHH, HHT, HTH, HTT, THH, THT, TTH, TTT$\}$; each outcome has the probability of $\dfrac{1}{8}$.

16. $S = \{$HHH, HTH, HHT, HTT, THH, TTH, THT, TTT$\}$; each outcome has the probability of $\dfrac{1}{8}$.

17. $S = \{$1 Yellow, 1 Red, 1 Green, 2 Yellow, 2 Red, 2 Green, 3 Yellow, 3 Red, 3 Green, 4 Yellow, 4 Red, 4 Green$\}$; each outcome has the probability of $\dfrac{1}{12}$; thus, $P(\text{2 Red}) + P(\text{4 Red}) = \dfrac{1}{12} + \dfrac{1}{12} = \dfrac{1}{6}$.

18. $S = \{$Forward Yellow, Forward Green, Forward Red, Backward Yellow, Backward Green, Backward Red$\}$; each outcome has the probability of $\dfrac{1}{6}$; thus, $P(\text{Forward Yellow}) + P(\text{Forward Green}) = \dfrac{1}{6} + \dfrac{1}{6} = \dfrac{1}{3}$.

19. $S = \{$1 Yellow Forward, 1 Yellow Backward, 1 Red Forward, 1 Red Backward, 1 Green Forward, 1 Green Backward, 2 Yellow Forward, 2 Yellow Backward, 2 Red Forward, 2 Red Backward, 2 Green Forward, 2 Green Backward, 3 Yellow Forward, 3 Yellow Backward, 3 Red Forward, 3 Red Backward, 3 Green Forward, 3 Green Backward, 4 Yellow Forward, 4 Yellow Backward, 4 Red Forward, 4 Red Backward, 4 Green Forward, 4 Green Backward$\}$; each outcome has the probability of $\dfrac{1}{24}$; thus, $P(\text{1 Red Backward}) + P(\text{1 Green Backward}) = \dfrac{1}{24} + \dfrac{1}{24} = \dfrac{1}{12}$.

20. $S = \{$Yellow 1 Forward, Yellow 1 Backward, Yellow 2 Forward, Yellow 2 Backward, Yellow 3 Forward, Yellow 3 Backward, Yellow 4 Forward, Yellow 4 Backward, Green 1 Forward, Green 1 Backward, Green 2 Forward, Green 2 Backward, Green 3 Forward, Green 3 Backward, Green 4 Forward, Green 4 Backward, Red 1 Forward, Red 1 Backward, Red 2 Forward, Red 2 Backward, Red 3 Forward, Red 3 Backward, Red 4 Forward, Red 4 Backward$\}$; each outcome has the probability of $\dfrac{1}{24}$; thus, $P(\text{Yellow 2 Forward}) + P(\text{Yellow 4 Forward}) = \dfrac{1}{24} + \dfrac{1}{24} = \dfrac{1}{12}$.

21. $S = \{$11 Red, 11 Yellow, 11 Green, 12 Red, 12 Yellow, 12 Green, 13 Red, 13 Yellow, 13 Green, 14 Red, 14 Yellow, 14 Green, 21 Red, 21 Yellow, 21 Green, 22 Red, 22 Yellow, 22 Green, 23 Red, 23 Yellow, 23 Green, 24 Red, 24 Yellow, 24 Green, 31 Red, 31 Yellow, 31 Green, 32 Red, 32 Yellow, 32 Green, 33 Red, 33 Yellow, 33 Green, 34 Red, 34 Yellow, 34 Green, 41 Red, 41 Yellow, 41 Green, 42 Red, 42 Yellow, 42 Green, 43 Red, 43 Yellow, 43 Green, 44 Red, 44 Yellow, 44 Green$\}$; each outcome has the probability of $\dfrac{1}{48}$; thus, $E = \{$22 Red, 22 Green, 24 Red, 24 Green$\}$; $P(E) = \dfrac{n(E)}{n(S)} = \dfrac{4}{48} = \dfrac{1}{12}$.

22. $S = \{$Forward 11, Forward 12, Forward 13, Forward 14, Forward 21, Forward 22, Forward 23, Forward 24, Forward 31, Forward 32, Forward 33, Forward 34, Forward 41, Forward 42, Forward 43, Forward 44, Backward 11, Backward 12, Backward 13, Backward 14, Backward 21, Backward 22, Backward 23, Backward 24, Backward 31, Backward 32, Backward 33, Backward 34, Backward 41, Backward 42, Backward 43, Backward 44$\}$; each outcome has the probability of $\frac{1}{32}$;

thus, $P(\text{Forward } 12) + P(\text{Forward } 14) + P(\text{Forward } 32) + P(\text{Forward } 34) = \frac{1}{32} + \frac{1}{32} + \frac{1}{32} + \frac{1}{32} = \frac{1}{8}$.

23. A, B, C, F **24.** A **25.** B **26.** F **27.** $P(H) = \frac{4}{5}; P(T) = \frac{1}{5}$ **28.** $P(H) = \frac{1}{3}; P(T) = \frac{2}{3}$

29. $P(1) = P(3) = P(5) = \frac{2}{9}; P(2) = P(4) = P(6) = \frac{1}{9}$ **30.** $P(1) = P(2) = P(3) = P(4) = P(5) = \frac{1}{5}; P(6) = 0$ **31.** $\frac{3}{10}$ **32.** $\frac{2}{5}$

33. $\frac{1}{2}$ **34.** $\frac{1}{2}$ **35.** $\frac{1}{6}$ **36.** $\frac{7}{30}$ **37.** $\frac{1}{8}$ **38.** $\frac{1}{8}$ **39.** $\frac{1}{4}$ **40.** $\frac{3}{8}$ **41.** $\frac{1}{6}$ **42.** $\frac{1}{18}$ **43.** $\frac{1}{18}$ **44.** $\frac{1}{36}$ **45.** 0.55 **46.** 0.10 **47.** 0.70 **48.** 0

49. 0.30 **50.** 0.5 **51.** 0.735 **52.** 0.961 **53.** 0.7 **54.** 0.96 **55.** $\frac{17}{20}$ **56.** $\frac{3}{5}$ **57.** $\frac{11}{20}$ **58.** $\frac{3}{5}$ **59.** $\frac{1}{2}$ **60.** $\frac{1}{10}$ **61.** $\frac{3}{10}$ **62.** $\frac{13}{20}$ **63.** $\frac{2}{5}$

64. $\frac{3}{5}$ **65. (a)** 0.57 **(b)** 0.95 **(c)** 0.83 **(d)** 0.38 **(e)** 0.29 **(f)** 0.05 **(g)** 0.78 **(h)** 0.71 **66. (a)** 0.45 **(b)** 0.75 **(c)** 0.90

67. (a) $\frac{25}{33}$ **(b)** $\frac{25}{33}$ **68. (a)** $\frac{7}{13}$ **(b)** $\frac{11}{13}$ **69.** 0.167 **70.** 0.814 **71.** 0.000033069

Review Exercises *(page 858)*

1. $\varnothing$, {Dave}, {Joanne}, {Erica}, {Dave, Joanne}, {Dave, Erica}, {Joanne, Erica}, {Dave, Joanne, Erica}

2. $\varnothing$, {Green}, {Blue}, {Red}, {Green, Blue}, {Green, Red}, {Blue, Red}, {Green, Blue, Red}

3. $\{1, 3, 5, 6, 7, 8\}$ **4.** $\{2, 3, 5, 6, 7, 8, 9\}$ **5.** $\{3, 7\}$ **6.** $\{3, 5, 7\}$ **7.** $\{1, 2, 4, 6, 8, 9\}$ **8.** $\{1, 4\}$ **9.** $\{1, 2, 4, 5, 6, 9\}$ **10.** $\{2, 4, 9\}$

11. 17 **12.** 24 **13.** 29 **14.** 34 **15.** 7 **16.** 45 **17.** 25 **18.** 7 **19.** 336 **20.** 210 **21.** 56 **22.** 35 **23.** 60 **24.** 120 **25.** 128 **26.** 64

27. 3024 **28.** 24 **29.** 70 **30.** 120 **31.** 91 **32. (a)** 14,400 **(b)** 14,400 **33.** 1,600,000 **34.** 60 **35.** 216,000

36. 256 (allowing numbers with initial zeros, such as 011) **37.** 1260 **38.** 12,600 **39. (a)** 381,024 **(b)** 1260 **40. (a)** 280 **(b)** 280 **(c)** 640 **41. (a)** $8.634628387 \times 10^{45}$ **(b)** 0.6531 **(c)** 0.3469 **42. (a)** 0.29 **(b)** 0.71 **43. (a)** 0.058 **(b)** 0.942 **44.** 0.15; 0.45

45. $\frac{4}{9}$ **46.** $\frac{1}{24}$ **47.** 0.2; 0.26 **48. (a)** 0.68 **(b)** 0.58 **(c)** 0.32

Cumulative Review *(page 860)*

1. $\left\{\frac{1}{3} - \frac{\sqrt{2}}{3}i, \frac{1}{3} + \frac{\sqrt{2}}{3}i\right\}$

2.

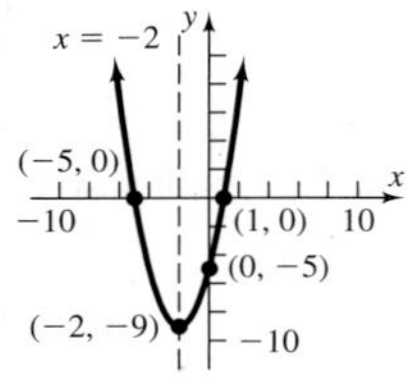

3.

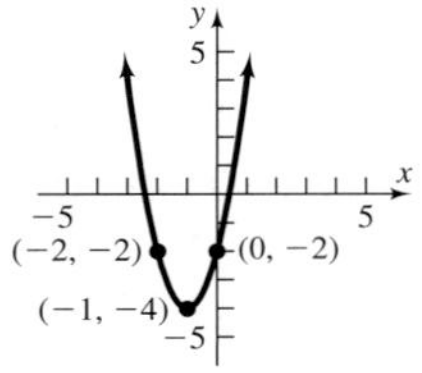

4. $\{x \mid 3.99 \le x \le 4.01\}$ or $[3.99, 4.01]$

5. $-\frac{1}{5}, 3, -\frac{1}{2} - \frac{\sqrt{7}}{2}i, -\frac{1}{2} + \frac{\sqrt{7}}{2}i$

6.

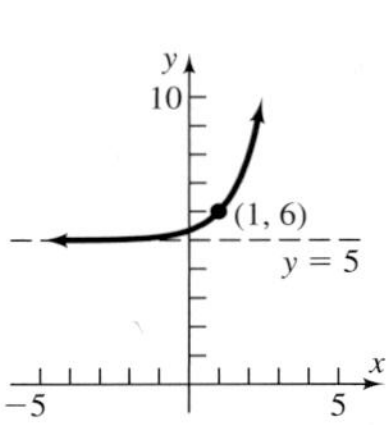

Domain: all real numbers

Range: $\{y \mid y > 5\}$

Horizontal asymptote: $y = 5$

7. 2 **8.** $\left\{\frac{8}{3}\right\}$

9. $x = 2, y = -5, z = 3$

10. 125; 700

11.

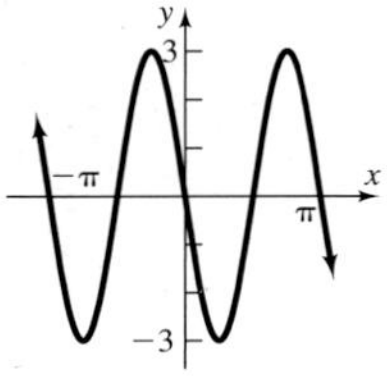

12.

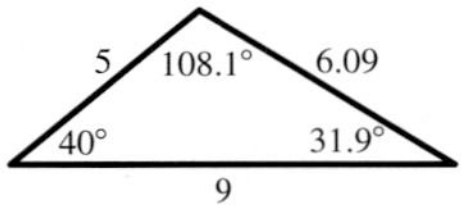

Area ≈ 14.46 square units

CHAPTER 13 A Preview of Calculus: The Limit, Derivative, and Integral of a Function

13.1 Concepts and Vocabulary *(page 866)*

3. $\lim_{x \to c} f(x)$ **4.** does not exist **5.** True **6.** False

13.1 Exercises *(page 866)*

7. 32 **8.** 19 **9.** 1 **10.** $\frac{1}{2}$ **11.** 4 **12.** 2 **13.** 2 **14.** 0 **15.** 0 **16.** 1 **17.** 3 **18.** −3 **19.** 4 **20.** 2 **21.** Does not exist

22. Does not exist

23.

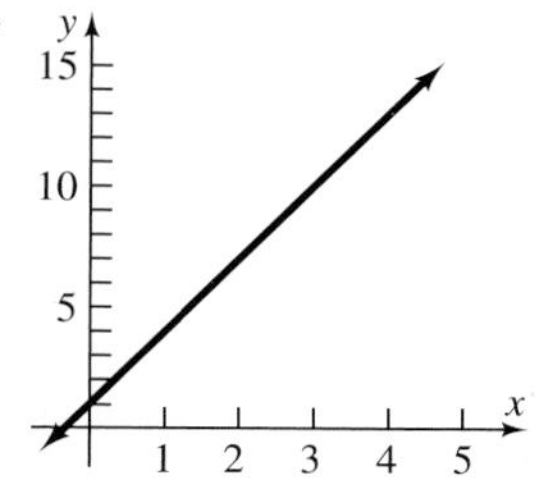

$\lim_{x \to 4} f(x) = 13$

24.

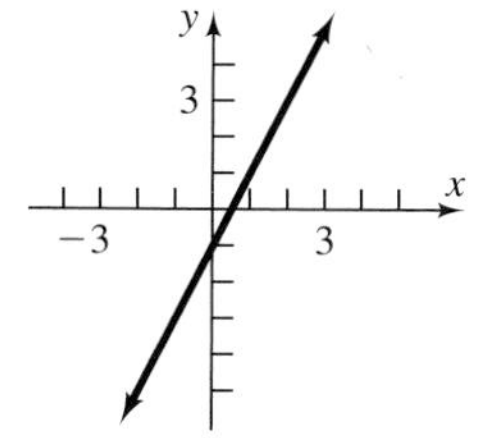

$\lim_{x \to -1} f(x) = -3$

25.

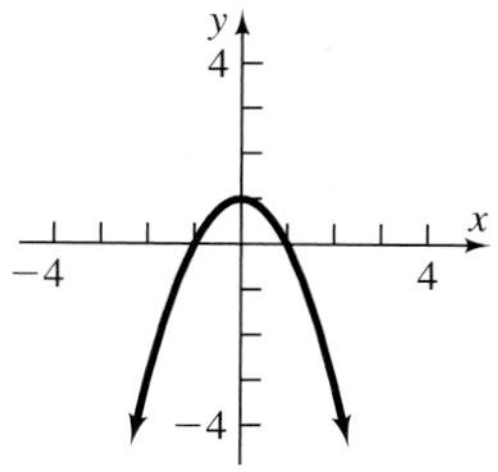

$\lim_{x \to 2} f(x) = -3$

26.

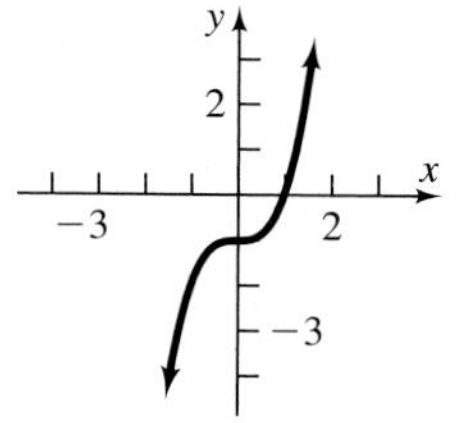

$\lim_{x \to -1} f(x) = -2$

27.

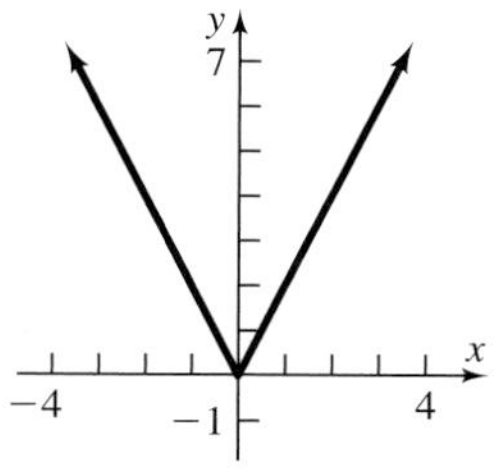

$\lim_{x \to -3} f(x) = 6$

28.

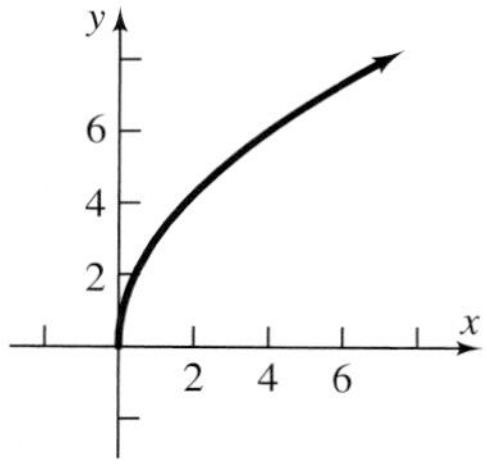

$\lim_{x \to 4} f(x) = 6$

29.

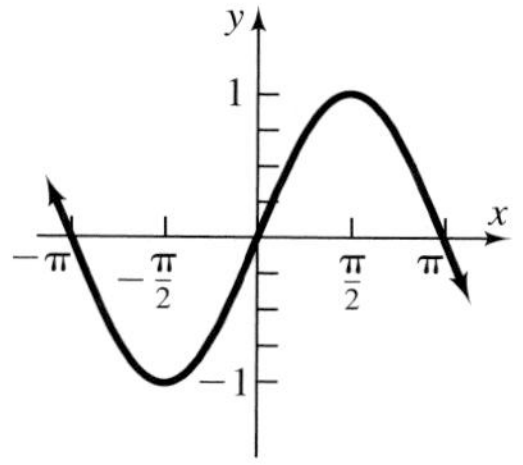

$\lim_{x \to \pi/2} f(x) = 1$

30.

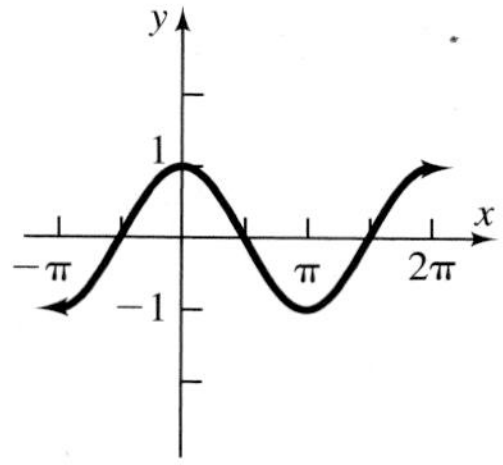

$\lim_{x \to \pi} f(x) = -1$

31.

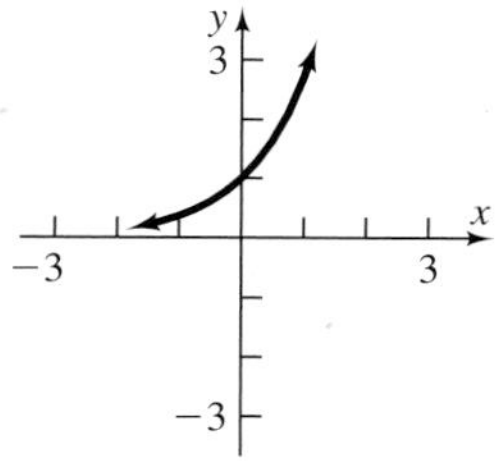

$\lim_{x \to 0} f(x) = 1$

32.

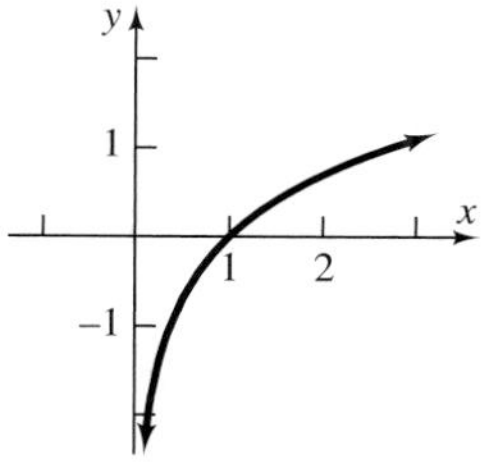

$\lim_{x \to 1} f(x) = 0$

33.

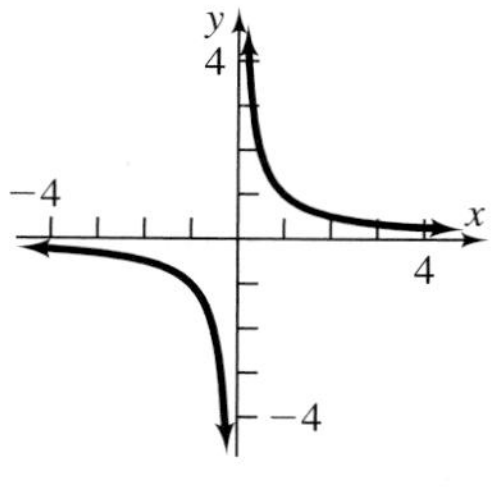

$\lim_{x \to -1} f(x) = -1$

34.

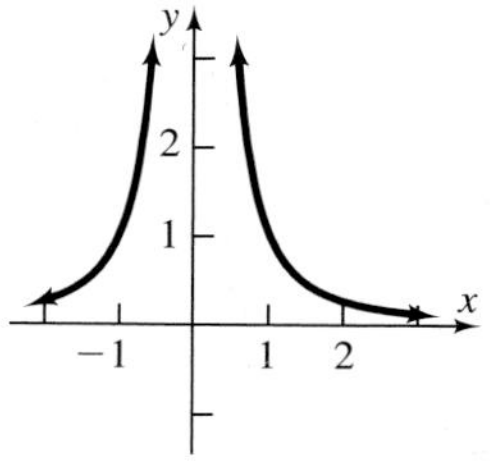

$\lim_{x \to 2} f(x) = \frac{1}{4}$

35.

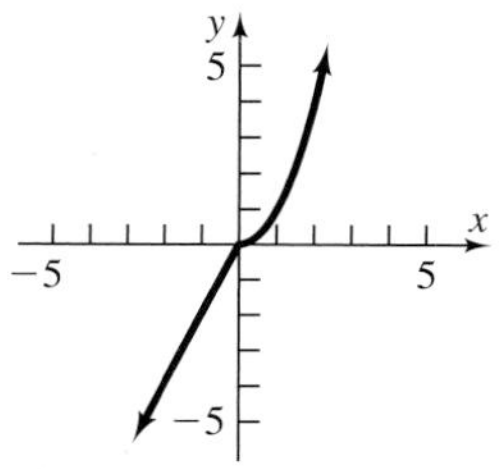

$\lim_{x \to 0} f(x) = 0$

36.

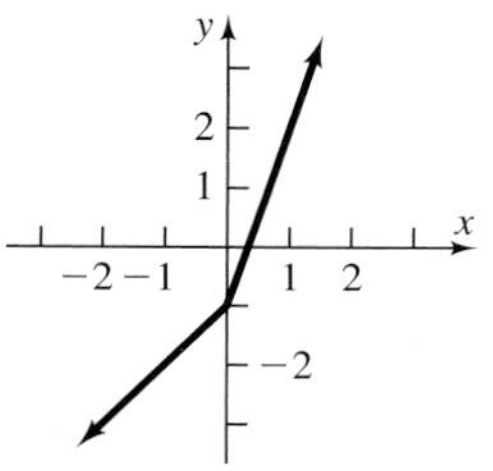

$\lim_{x \to 0} f(x) = -1$

37.

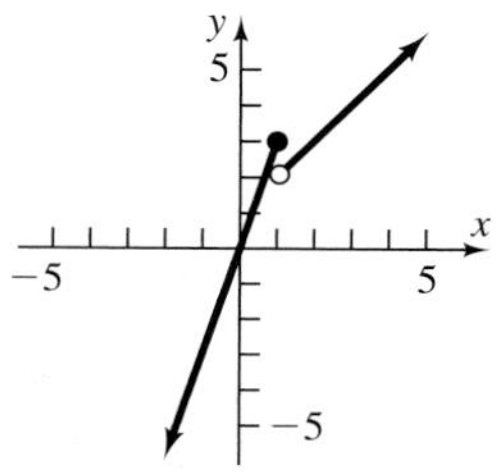

$\lim_{x \to 1} f(x)$ does not exist

38.

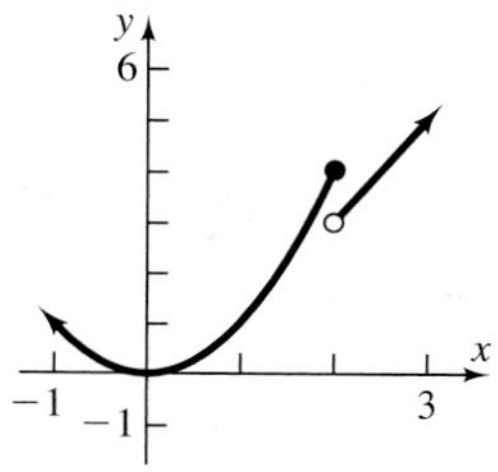

$\lim_{x \to 2} f(x)$ does not exist

39.

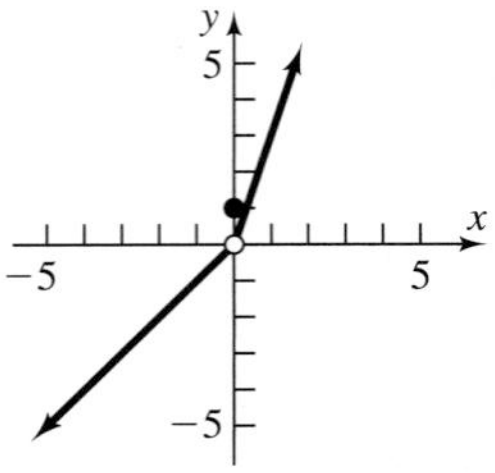

$\lim_{x \to 0} f(x) = 0$

40.

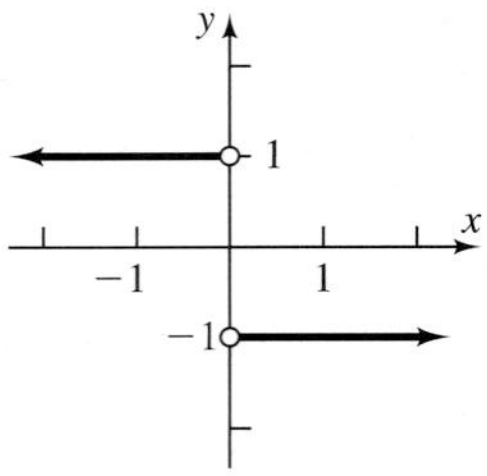

$\lim_{x \to 0} f(x)$ does not exist

41.

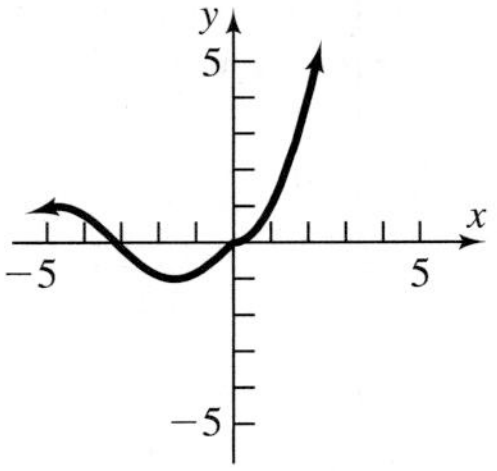

$\lim_{x \to 0} f(x) = 0$

42.

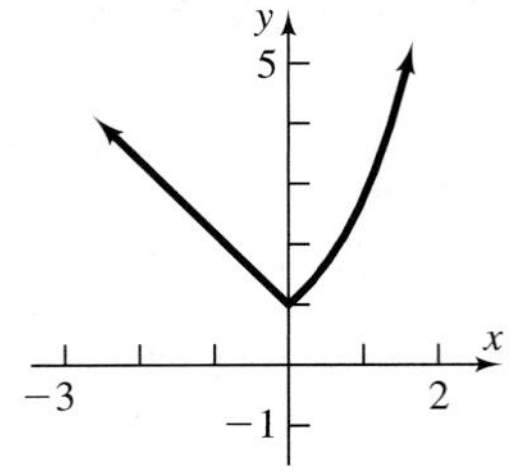

$\lim_{x \to 0} f(x) = 1$

43. 0.67 **44.** 4 **45.** 1.6 **46.** 0.8 **47.** 0 **48.** 0.46

13.2 Concepts and Vocabulary *(page 874)*

1. product **2.** 5 **3.** 1 **4.** True **5.** False **6.** False

13.2 Exercises *(page 874)*

7. 5 **8.** −3 **9.** 4 **10.** −3 **11.** 8 **12.** −13 **13.** 8 **14.** 28 **15.** −1 **16.** 2 **17.** 8 **18.** 4 **19.** 3 **20.** 1 **21.** −1 **22.** $\frac{5}{3}$ **23.** 32 **24.** −1 **25.** 2 **26.** $\frac{1}{2}$ **27.** $\frac{7}{6}$ **28.** $\frac{5}{4}$ **29.** 3 **30.** 4 **31.** 0 **32.** 3 **33.** $\frac{2}{3}$ **34.** 4 **35.** $\frac{8}{5}$ **36.** $\frac{4}{5}$ **37.** 0 **38.** $\frac{13}{28}$ **39.** 5 **40.** −3 **41.** 6 **42.** 27 **43.** 0 **44.** −7 **45.** 0 **46.** −5 **47.** −1 **48.** −2 **49.** 1 **50.** 2 **51.** $\frac{3}{4}$ **52.** 1

13.3 Concepts and Vocabulary *(page 881)*

7. one-sided **8.** $\lim_{x \to c^+} f(x) = R$ **9.** continuous; c **10.** False **11.** True **12.** True

13.3 Exercises *(page 881)*

13. $\{x \mid -8 \le x < -6 \text{ or } -6 < x < 4 \text{ or } 4 < x \le 6\}$ **14.** All real numbers **15.** −8, −5, −3 **16.** 3 **17.** $f(-8) = 0$; $f(-4) = 2$ **18.** $f(2) = 3$; $f(6) = 2$ **19.** ∞ **20.** $-\infty$ **21.** 2 **22.** −2 **23.** 1 **24.** −4 **25.** Limit exists; 0 **26.** Limit exists; 3 **27.** No **28.** No **29.** Yes **30.** No **31.** No **32.** Yes **33.** 5 **34.** 0 **35.** 7 **36.** 4 **37.** 1 **38.** −3 **39.** 4 **40.** 2 **41.** $-\frac{2}{3}$ **42.** −1 **43.** $\frac{3}{2}$ **44.** $\frac{7}{4}$ **45.** Continuous **46.** Continuous **47.** Continuous **48.** Continuous **49.** Not continuous **50.** Not continuous **51.** Not continuous **52.** Not continuous **53.** Not continuous **54.** Not continuous **55.** Continuous **56.** Continuous **57.** Not continuous **58.** Not continuous **59.** Continuous **60.** Continuous **61.** Continuous for all real numbers **62.** Continuous for all real numbers **63.** Continuous for all real numbers **64.** Continuous for all real numbers **65.** Continuous for all real numbers **66.** Continuous for all real numbers **67.** Continuous for all real numbers except $x = \frac{k\pi}{2}$, where k is an odd integer **68.** Continuous for all real numbers except $x = k\pi$, where k is an integer **69.** Continuous for all real numbers except $x = -2$ and $x = 2$ **70.** Continuous for all real numbers except $x = 3$ and $x = -3$ **71.** Continuous for all positive real numbers except $x = 1$ **72.** Continuous for all positive real numbers except $x = 3$

73. Discontinuous at $x = -1$ and $x = 1$; $\lim_{x \to 1} R(x) = \frac{1}{2}$: Hole at $\left(1, \frac{1}{2}\right)$ $\lim_{x \to -1^-} R(x) = -\infty$; $\lim_{x \to -1^+} R(x) = \infty$; Vertical Asymptote at $x = -1$

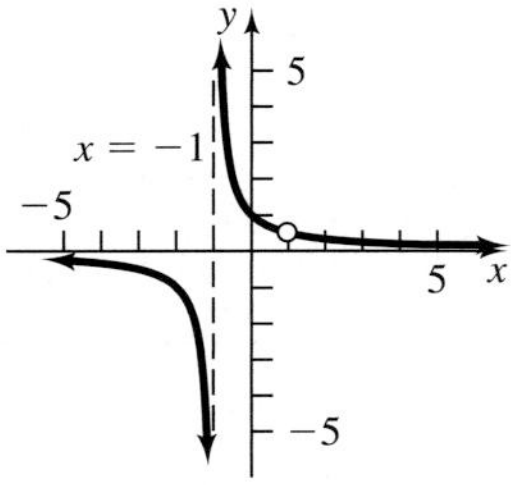

74. Discontinuous at $x = -2$ and $x = 2$; $\lim_{x \to -2} R(x) = -\frac{3}{4}$: Hole at $\left(-2, -\frac{3}{4}\right)$ $\lim_{x \to 2^-} R(x) = -\infty$; $\lim_{x \to 2^+} R(x) = \infty$; Vertical Asymptote at $x = 2$

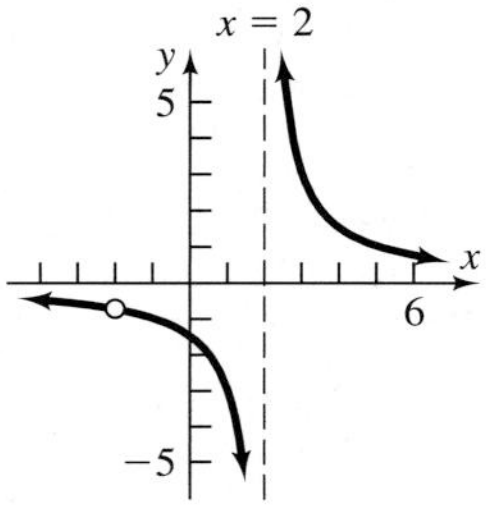

75. Discontinuous at $x = -1$ and $x = 1$; $\lim_{x \to -1} R(x) = \frac{1}{2}$: Hole at $\left(-1, \frac{1}{2}\right)$ $\lim_{x \to 1^-} R(x) = -\infty$; $\lim_{x \to 1^+} R(x) = \infty$; Vertical Asymptote at $x = 1$

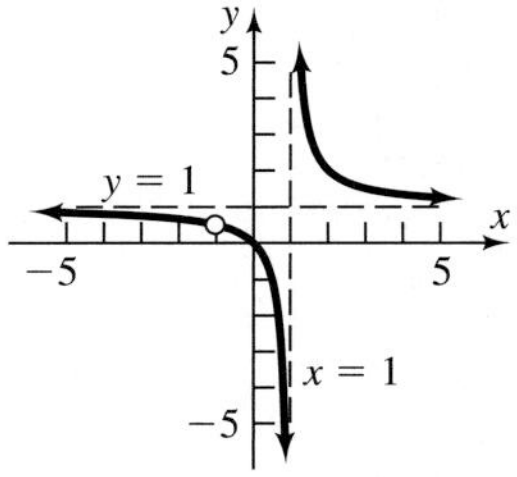

76. Discontinuous at $x = -4$ and $x = 4$;
$\lim_{x \to -4} R(x) = \frac{1}{2}$: Hole at $\left(-4, \frac{1}{2}\right)$
$\lim_{x \to 4^-} R(x) = -\infty$; $\lim_{x \to 4^+} R(x) = \infty$:
Vertical Asymptote at $x = 4$

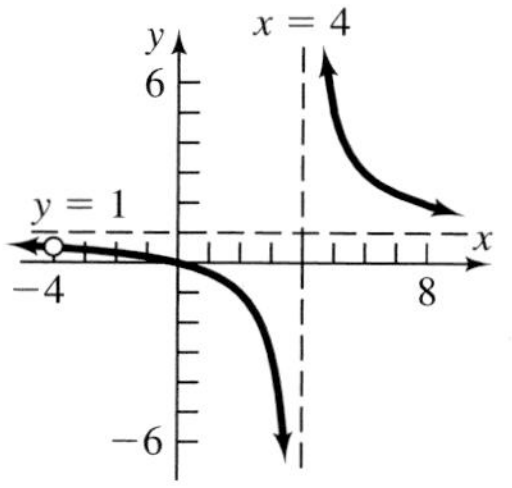

77. $x = -\sqrt[3]{2}$: Asymptote; $x = 1$: Hole

78. $x = -\sqrt[3]{2}$: Asymptote; $x = -1$: Hole

79. $x = -3$: Asymptote; $x = 2$: Hole

80. $x = -4$: Asymptote; $x = 1$: Hole

81. $x = -\sqrt[3]{2}$: Asymptote; $x = -1$: Hole

82. $x = -1$: Asymptote; $x = 3$: Hole

83.

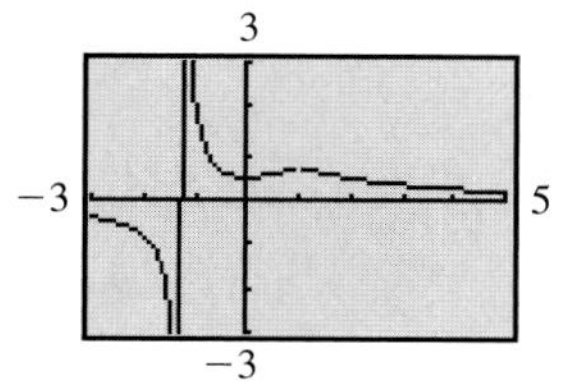

84.

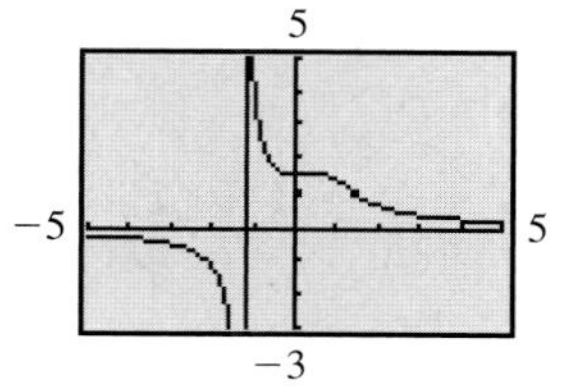

85.

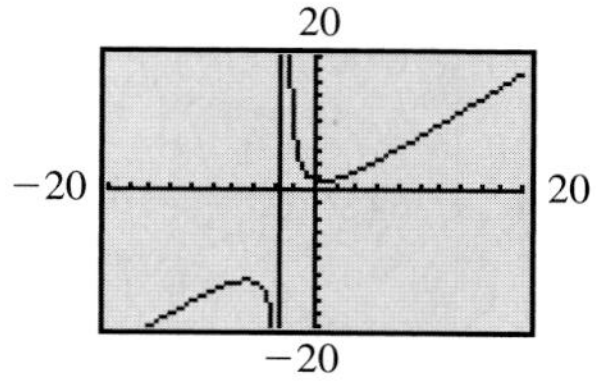

86.

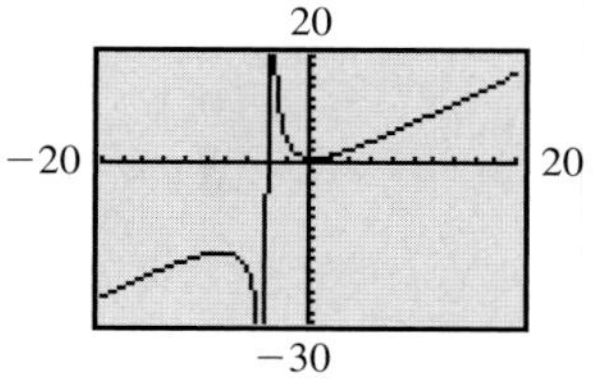

87.

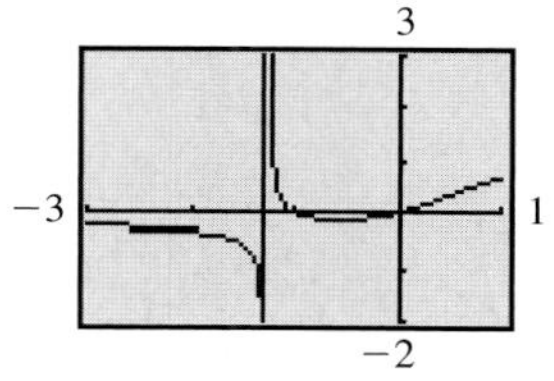

88.

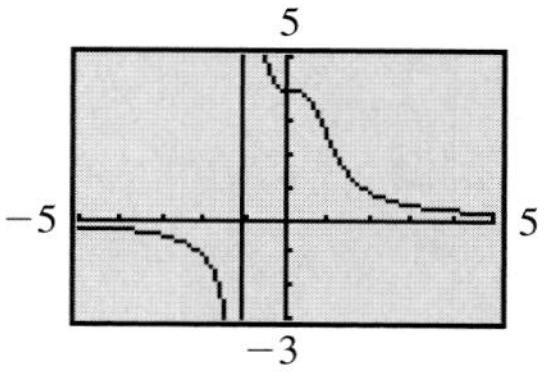

13.4 Concepts and Vocabulary *(page 889)*

3. tangent line **4.** derivative **5.** velocity **6.** True **7.** True **8.** True

13.4 Exercises *(page 889)*

9. $m_{\tan} = 3$

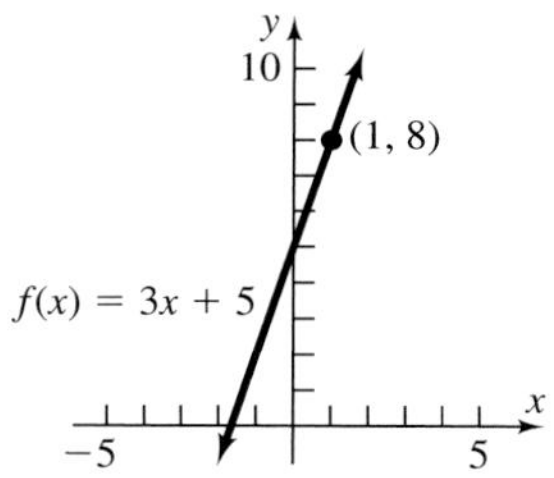

10. $m_{\tan} = -2$

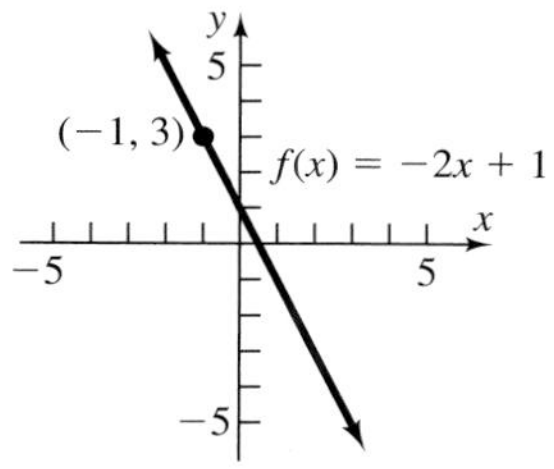

11. $m_{\tan} = -2$

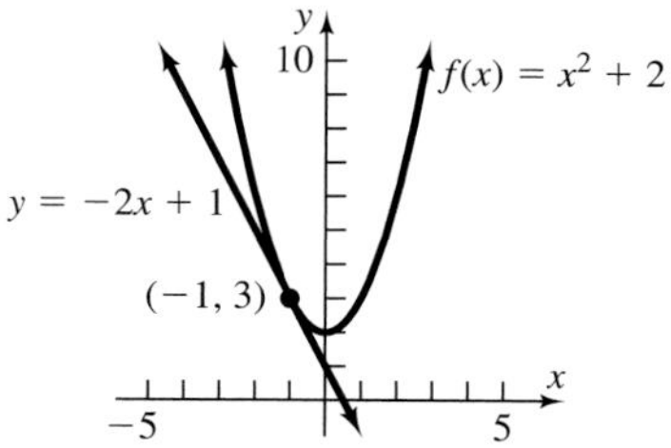

12. $m_{\tan} = -2$

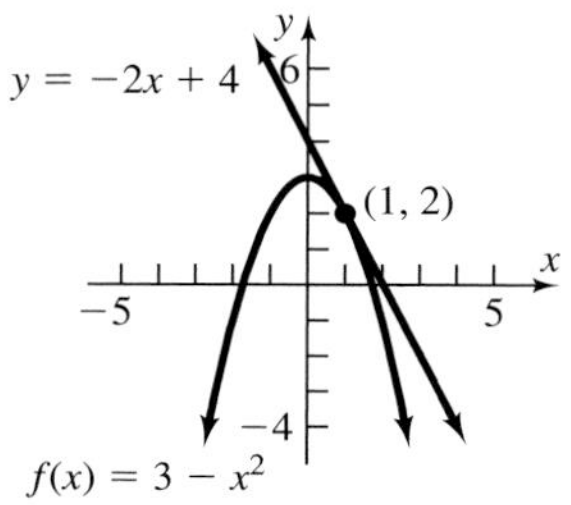

13. $m_{\tan} = 12$

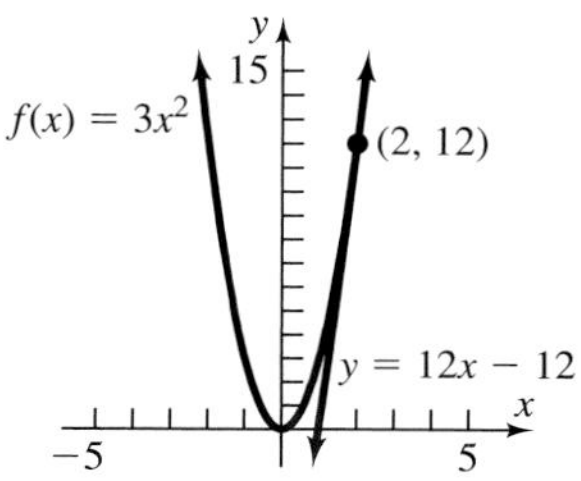

14. $m_{\tan} = 16$

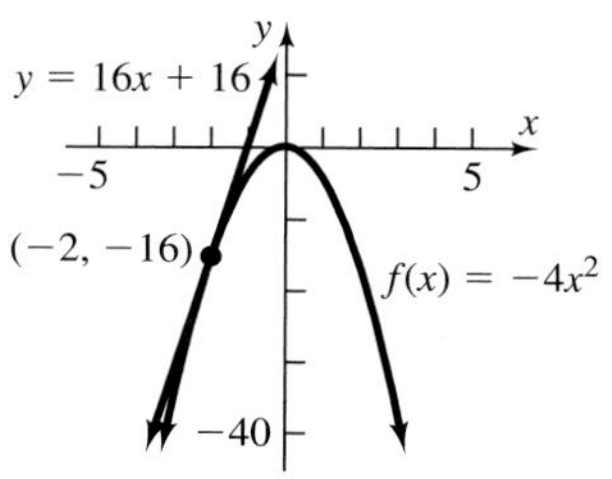

15. $m_{\tan} = 5$

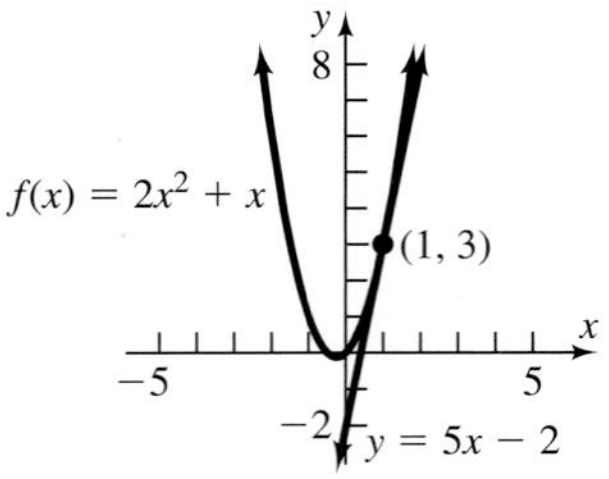

16. $m_{\tan} = -1$

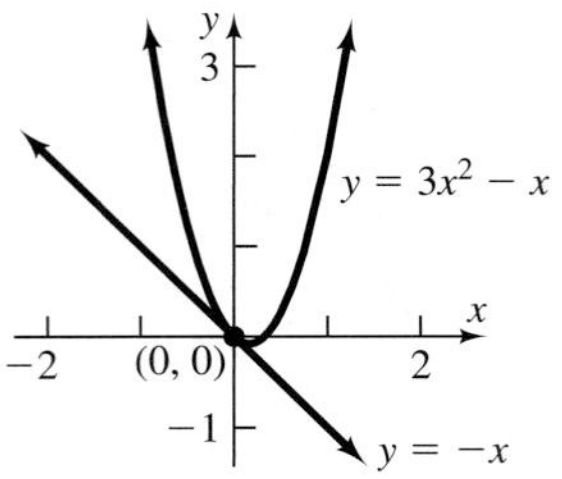

17. $m_{\tan} = -4$

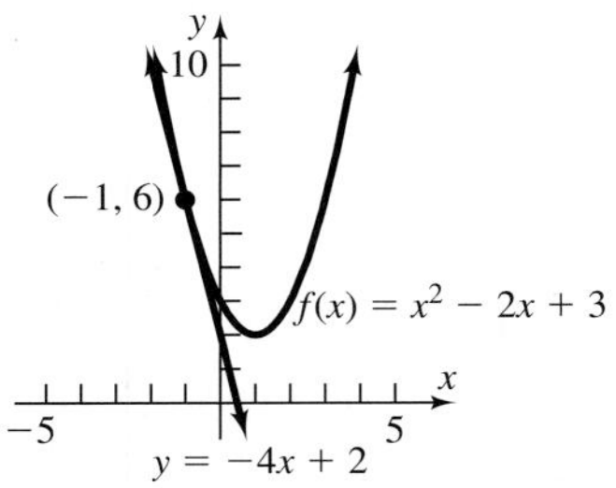

18. $m_{\tan} = -3$

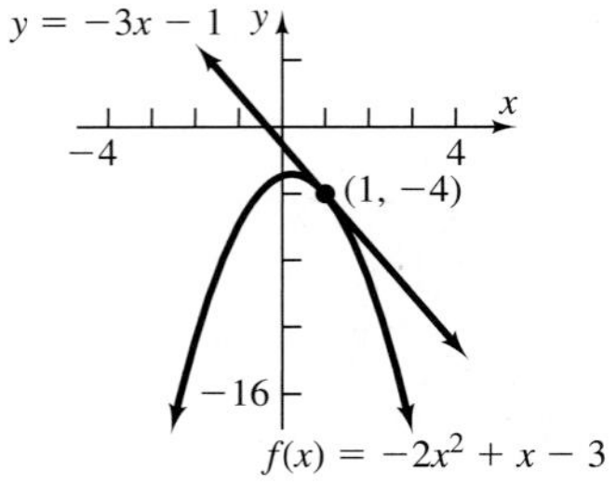

19. $m_{\tan} = 13$

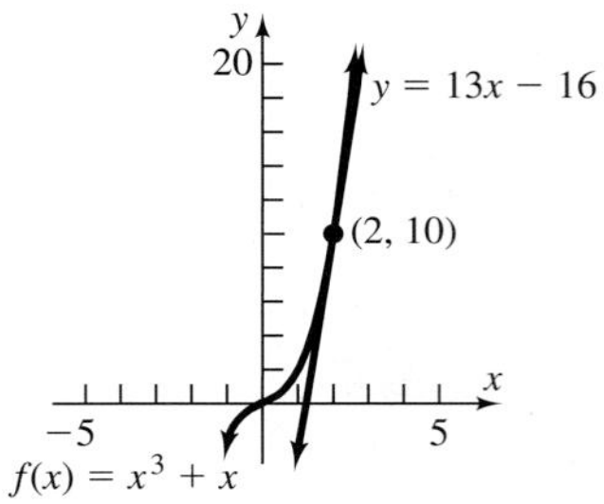

20. $m_{\tan} = 1$

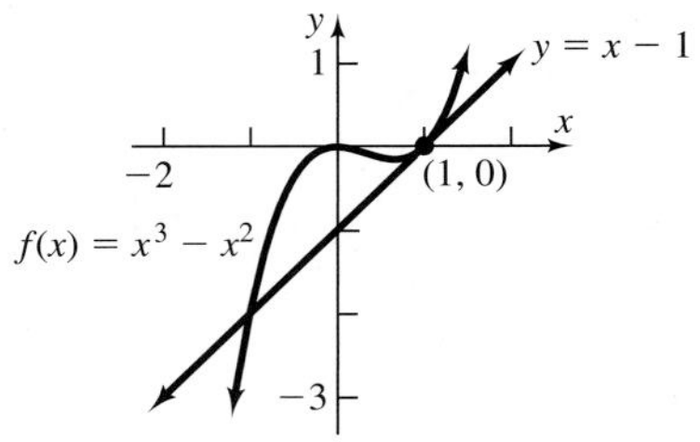

21. -4 **22.** 3 **23.** 0 **24.** -4 **25.** 7 **26.** 8 **27.** 7 **28.** 20 **29.** 3 **30.** 8 **31.** 1 **32.** 0 **33.** 60 **34.** -2440 **35.** -0.8587776956 **36.** 36.81251395 **37.** 1.389623659 **38.** 1.262466702 **39.** 2.362110222 **40.** 1.546899826 **41.** 3.643914112 **42.** -0.1793793526 **43.** 18π ft^3/ft **44.** 16π ft^2/ft **45.** 16π ft^3/ft **46.** 27 m^3/m **47.** **(a)** 6 sec **(b)** 64 ft/sec **(c)** $(-32t + 96)$ ft/sec **(d)** 32 ft/sec **(e)** 3 sec **(f)** 144 ft **(g)** -96 ft/sec **48.** **(a)** 2 sec **(b)** -64 ft/sec **(c)** $(-32t - 48)$ ft/sec **(d)** -80 ft/sec **(e)** -112 ft/sec

49. **(a)** $-23\frac{1}{3}$ ft/sec **(b)** -21 ft/sec **(c)** -18 ft/sec **(d)** $s(t) = -2.631t^2 - 10.269t + 999.933$ **(e)** -15.531 ft/sec

50. **(a)** \$249.28 per bicycle **(b)** Approximately \$329.87 per bicycle **(c)** Approximately \$485.71 per bicycle **(d)** $R(x) = -1.522x^2 + 597.791x + 7944.450$ **(e)** \$521.69 per bicycle

13.5 Concepts and Vocabulary *(page 896)*

3. $\int_a^b f(x)dx$ **4.** $\int_a^b f(x)dx$

13.5 Exercises *(page 896)*

5. 3 **6.** 6 **7.** 56 **8.** 38

9. **(a)**

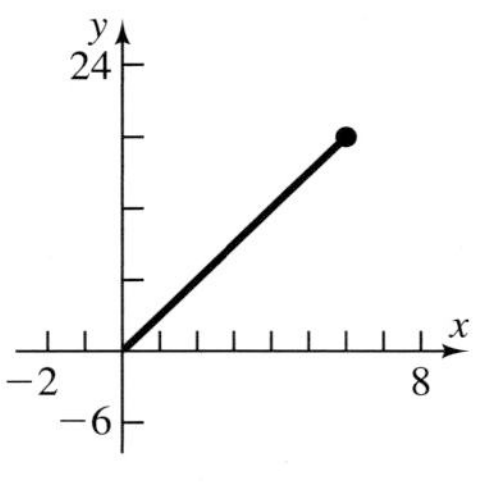

(b) 36 **(c)** 72

(d) 45 **(e)** 63 **(f)** 54

10. **(a)**

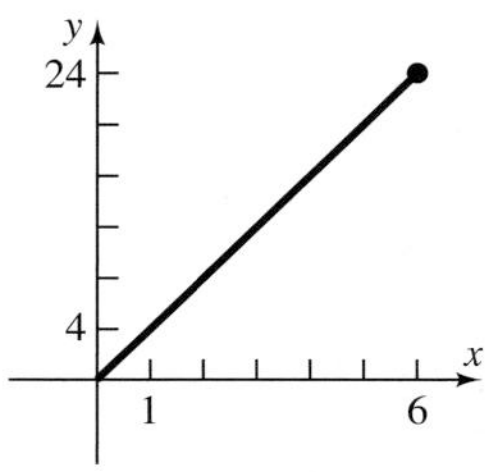

(b) 48 **(c)** 96

(d) 60 **(e)** 84 **(f)** 72

11. **(a)**

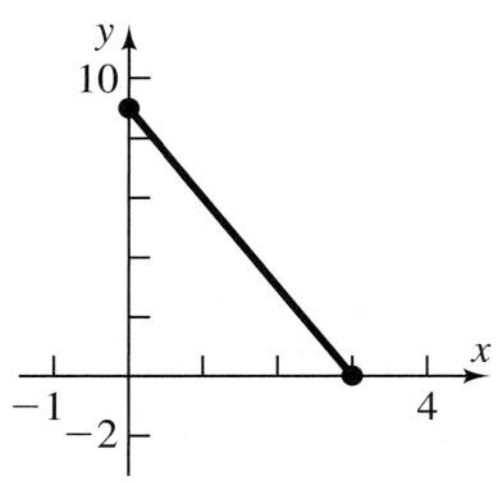

(b) 18 **(c)** 9

(d) $\frac{63}{4}$ **(e)** $\frac{45}{4}$ **(f)** $\frac{27}{2}$

12. (a)

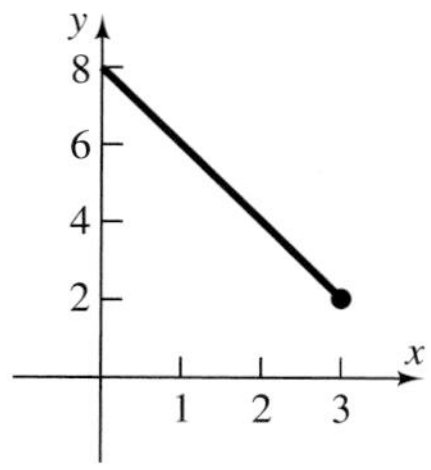

(b) 18 **(c)** 12

(d) $\frac{33}{2}$ **(e)** $\frac{27}{2}$ **(f)** 15

13. (a)

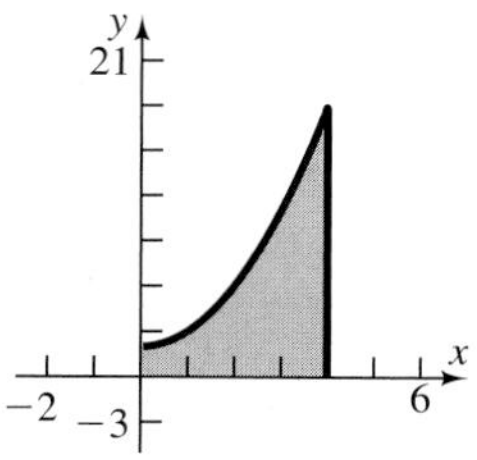

(b) 22 **(c)** $\frac{51}{2}$

(d) $\int_0^4 (x^2 + 2)dx$ **(e)** $\frac{88}{3}$

14. (a)

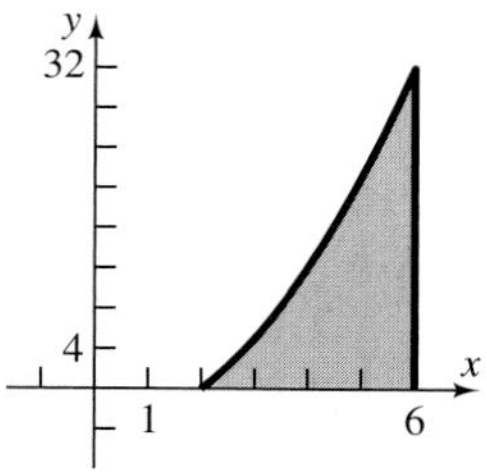

(b) 38 **(c)** $\frac{91}{2}$

(d) $\int_2^6 (x^2 - 4)dx$ **(e)** $\frac{160}{3}$

15. (a)

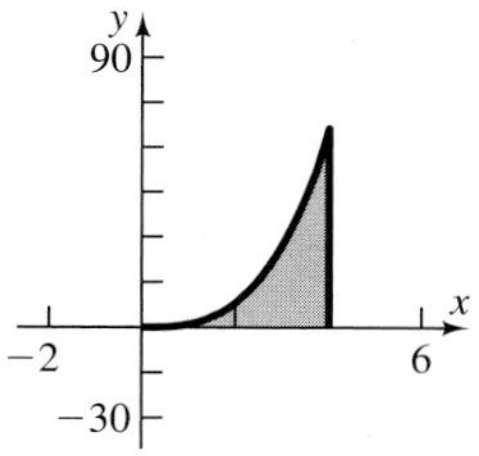

(b) 36 **(c)** 49

(d) $\int_0^4 x^3\,dx$ **(e)** 64

16. (a)

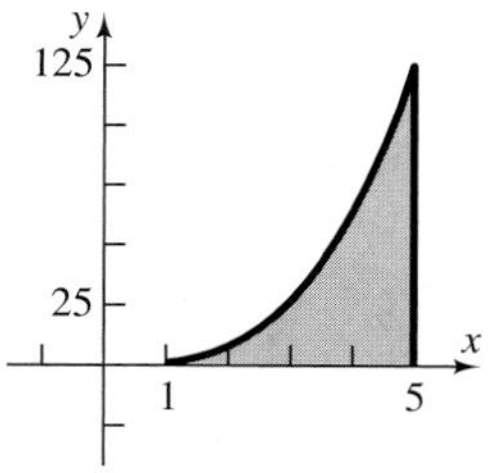

(b) 100 **(c)** $\frac{253}{2}$

(d) $\int_1^5 x^3\,dx$ **(e)** 156

17. (a)

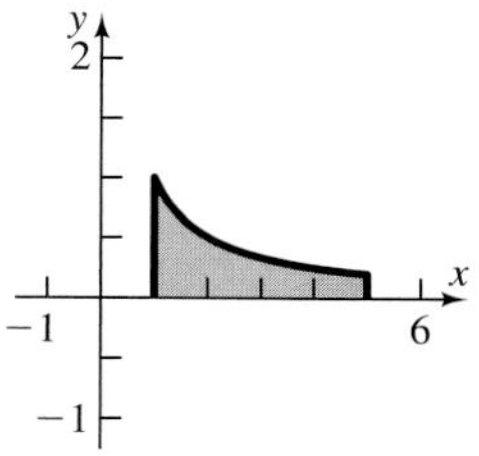

(b) $\frac{25}{12}$ **(c)** $\frac{4609}{2520}$

(d) $\int_1^5 \frac{1}{x}\,dx$ **(e)** 1.609

18. (a)

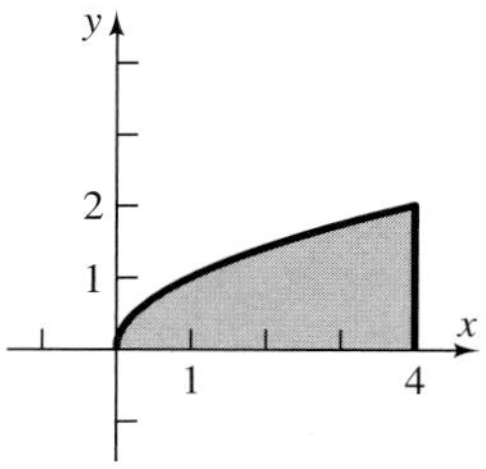

(b) 4.146 **(c)** 4.765

(d) $\int_0^4 \sqrt{x}\,dx$ **(e)** $\frac{16}{3}$

19. (a)

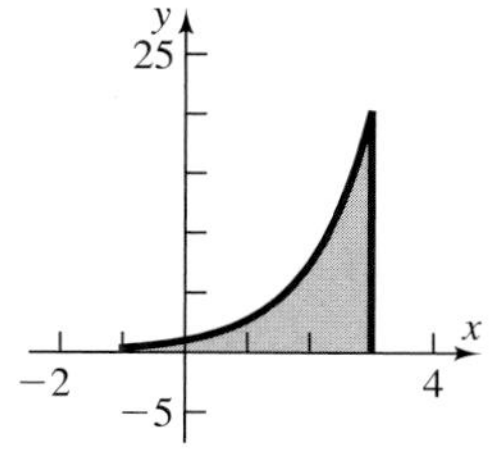

(b) 11.475 **(c)** 15.197

(d) $\int_{-1}^3 e^x\,dx$ **(e)** 19.718

20. (a)

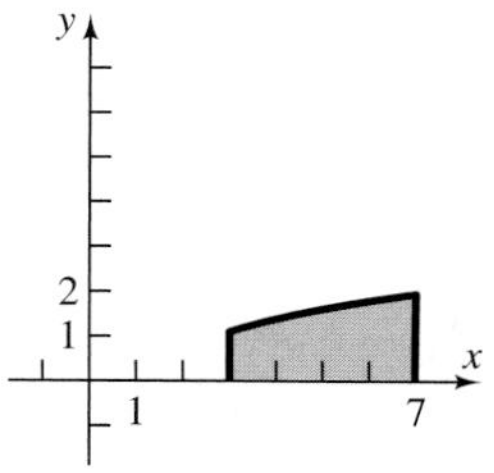

(b) 5.886 **(c)** 6.110

(d) $\int_3^7 \ln x\,dx$ **(e)** 6.326

21. (a)

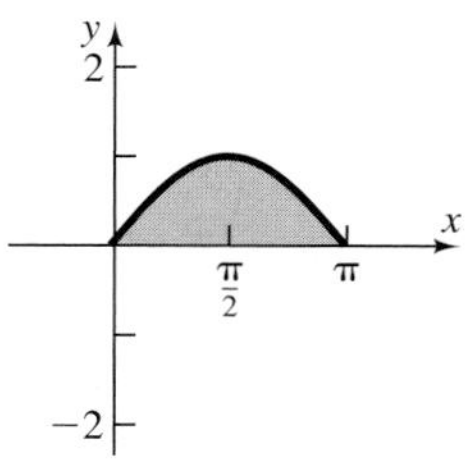

(b) 1.896 **(c)** 1.974

(d) $\int_0^{\pi} \sin x\,dx$ **(e)** 2

22. (a)

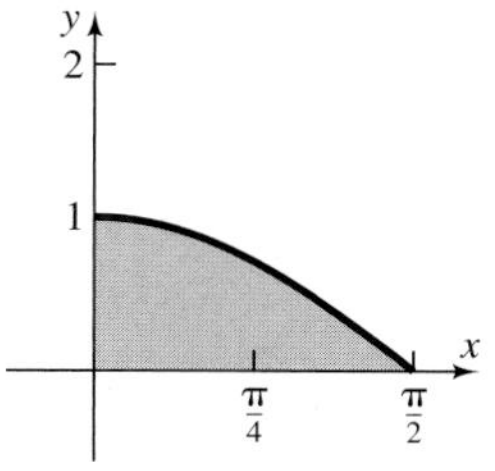

(b) 1.183 **(c)** 1.095

(d) $\int_0^{\pi/2} \cos x\,dx$ **(e)** 1

23. (a) Area under the graph of $f(x) = 3x + 1$ from 0 to 4

(b)

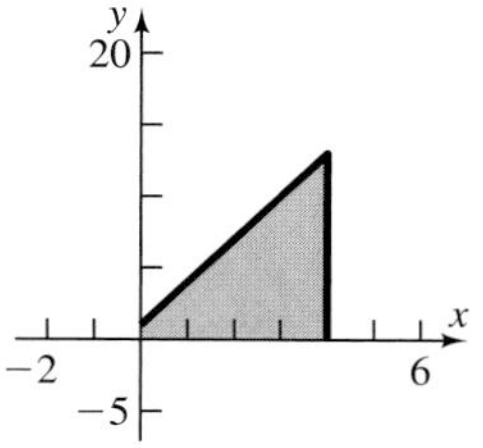

(c) 28

24. **(a)** Area under the graph of $f(x) = -2x + 7$ from 1 to 3

(b)

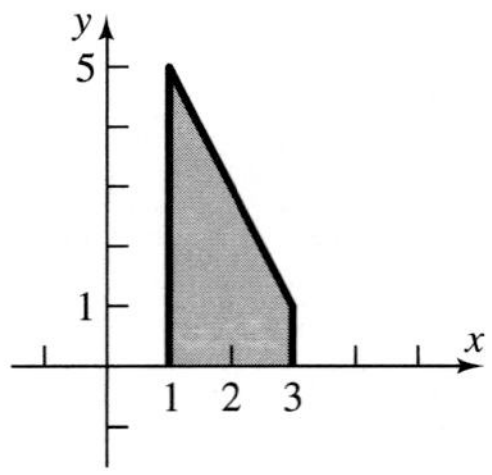

(c) 6

25. **(a)** Area under the graph of $f(x) = x^2 - 1$ from 2 to 5

(b)

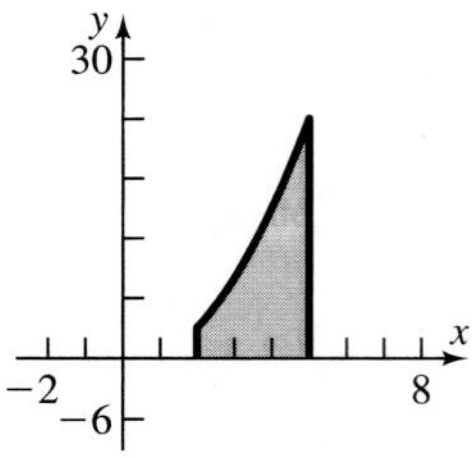

(c) 36

26. **(a)** Area under the graph of $f(x) = 16 - x^2$ from 0 to 4

(b)

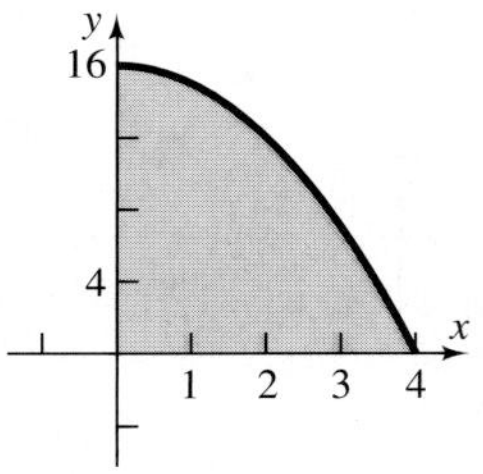

(c) $\frac{128}{3}$

27. **(a)** Area under the graph of $f(x) = \sin x$ from 0 to $\frac{\pi}{2}$

(b)

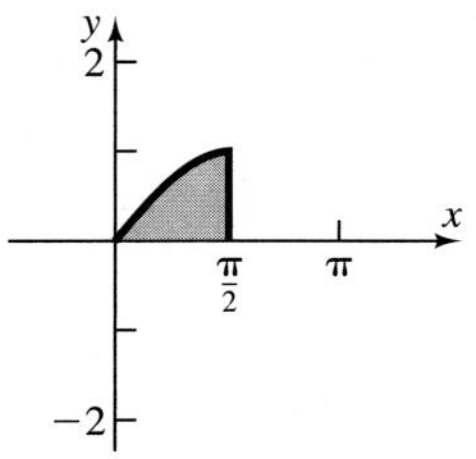

(c) 1

28. **(a)** Area under the graph of $f(x) = \cos x$ from $-\frac{\pi}{4}$ to $\frac{\pi}{4}$

(b)

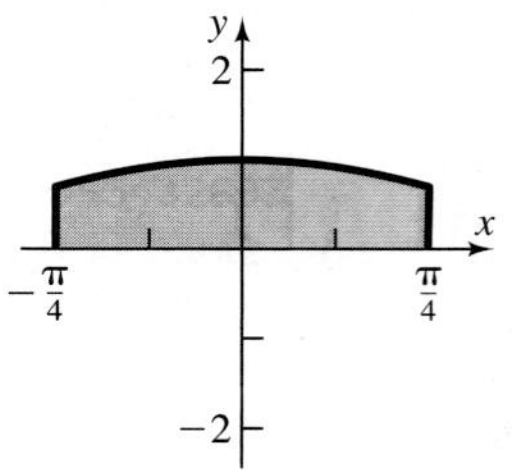

(c) 1.414

29. **(a)** Area under the graph of $f(x) = e^x$ from 0 to 2

(b)

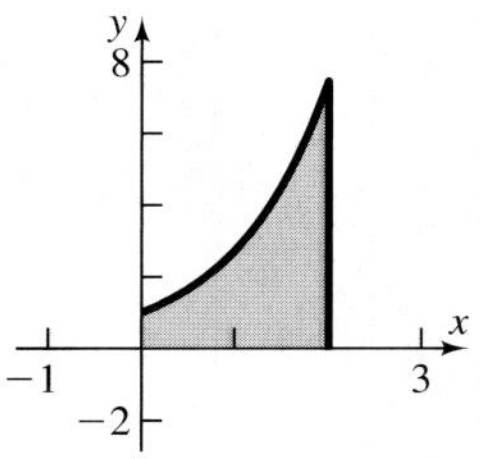

(c) 6.389

30. **(a)** Area under the graph of $f(x) = \ln x$ from e to $2e$

(b)

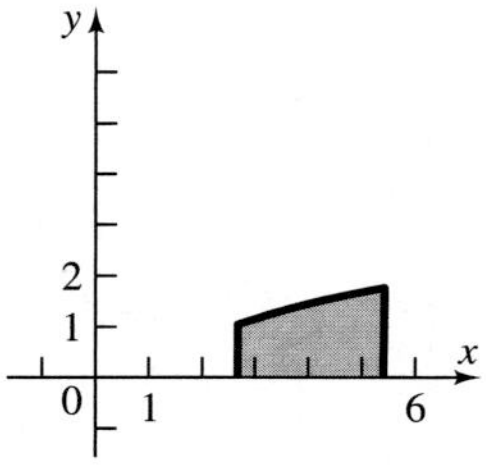

(c) 3.768

31. Using left endpoints: $n = 2: 0 + 0.5 = 0.5$; $n = 4: 0 + 0.125 + 0.25 + 0.375 = 0.75$;
$n = 10: 0 + 0.02 + 0.04 + 0.06 + \cdots + 0.18 = \frac{10}{2}(0 + 0.18) = 0.9$;
$n = 100: 0 + 0.0002 + 0.0004 + 0.0006 + \cdots + 0.0198 = \frac{100}{2}(0 + 0.0198) = 0.99$;
Using right endpoints: $n = 2: 0.5 + 1 = 1.5$; $n = 4: 0.125 + 0.25 + 0.375 + 0.5 = 1.25$;
$n = 10: 0.02 + 0.04 + 0.06 + \cdots + 0.20 = \frac{10}{2}(0.02 + 0.20) = 1.1$;
$n = 100: 0.0002 + 0.0004 + 0.0006 + \cdots + 0.02 = \frac{100}{2}(0.0002 + 0.02) = 1.01$

32. **(a)**

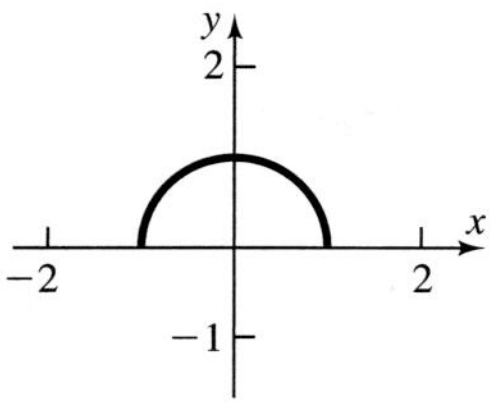

(b) 1.424 **(c)** 1.519 **(d)** $\int_{-1}^{1} \sqrt{1 - x^2}\, dx$ **(e)** 1.571 **(f)** $\frac{\pi}{2}$

Review Exercises *(page 899)*

1. 9 **2.** 3 **3.** 25 **4.** 49 **5.** 4 **6.** 2 **7.** 0 **8.** 2 **9.** 64 **10.** $\frac{\sqrt{6}}{288}$ **11.** $-\frac{1}{4}$ **12.** $\frac{13}{10}$ **13.** $\frac{1}{3}$ **14.** 2 **15.** $\frac{6}{7}$ **16.** $\frac{2}{3}$ **17.** 0 **18.** $\frac{1}{3}$ **19.** $\frac{3}{2}$ **20.** $\frac{3}{4}$ **21.** $\frac{28}{11}$ **22.** 1 **23.** Continuous **24.** Continuous **25.** Not continuous **26.** Not continuous **27.** Not continuous **28.** Not continuous **29.** Continuous **30.** Continuous **31.** $\{x \mid -6 \le x < 2 \text{ or } 2 < x < 5 \text{ or } 5 < x \le 6\}$ **32.** All real numbers **33.** 1, 6 **34.** 4 **35.** $f(-6) = 2; f(-4) = 1$ **36.** $f(-2) = 2; f(6) = 0$ **37.** 4 **38.** −2 **39.** −2 **40.** 2 **41.** $-\infty$ **42.** ∞ **43.** Does not exist **44.** Does not exist **45.** No **46.** No **47.** No **48.** No **49.** Yes **50.** No

51. R is discontinuous at $x = -4$ and $x = 4$.

$\lim_{x \to -4} R(x) = -\frac{1}{8}$: Hole at $\left(-4, -\frac{1}{8}\right)$

$\lim_{x \to 4^-} R(x) = -\infty$; $\lim_{x \to 4^+} R(x) = \infty$:

The graph of R has a vertical asymptote at $x = 4$.

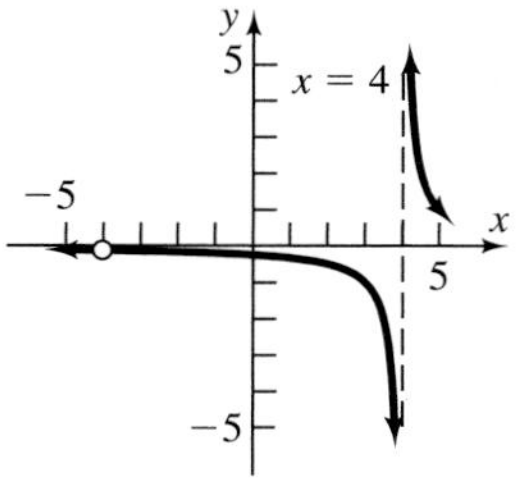

52. R is discontinuous at $x = -2$ and $x = 2$.

$\lim_{x \to -2} R(x) = \frac{3}{2}$: Hole at $\left(-2, \frac{3}{2}\right)$

$\lim_{x \to 2^-} R(x) = -\infty$; $\lim_{x \to 2^+} R(x) = \infty$:

The graph of R has a vertical asymptote at $x = 2$.

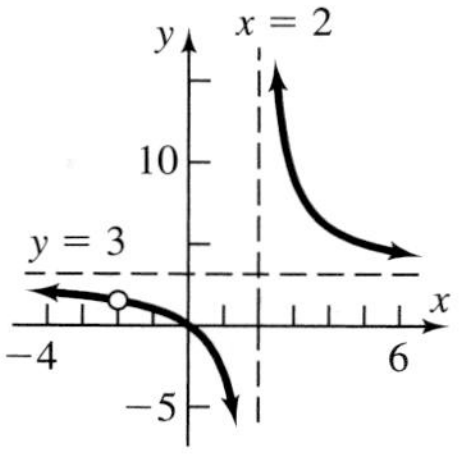

53. Undefined at $x = 2$ and $x = 9$; R has a hole at $x = 2$ and a vertical asymptote at $x = 9$.

54. Undefined at $x = -3$ and $x = 2$; R has a hole at $x = -3$ and a vertical asymptote at $x = 2$.

55. $m_{\tan} = 12$

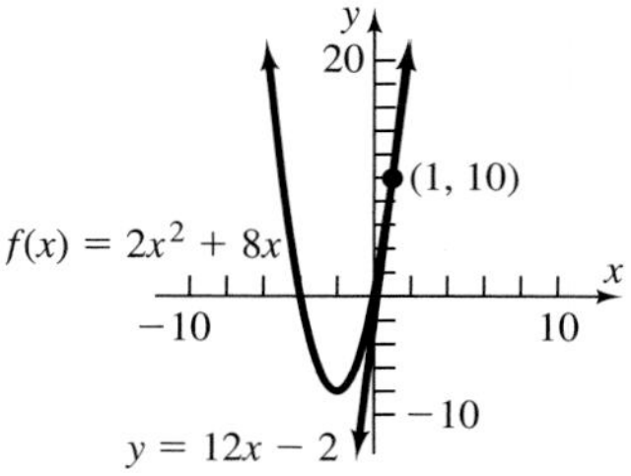

56. $m_{\tan} = -6$

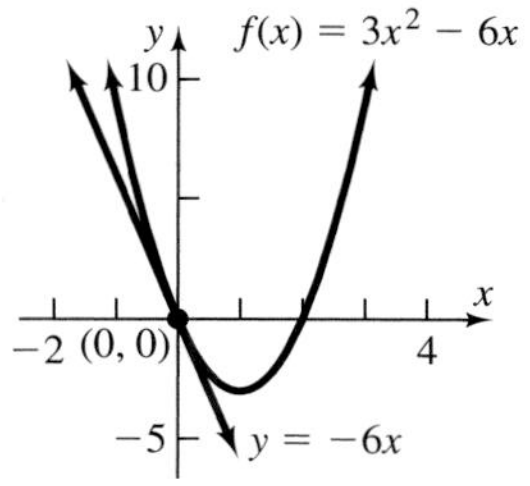

57. $m_{\tan} = 0$

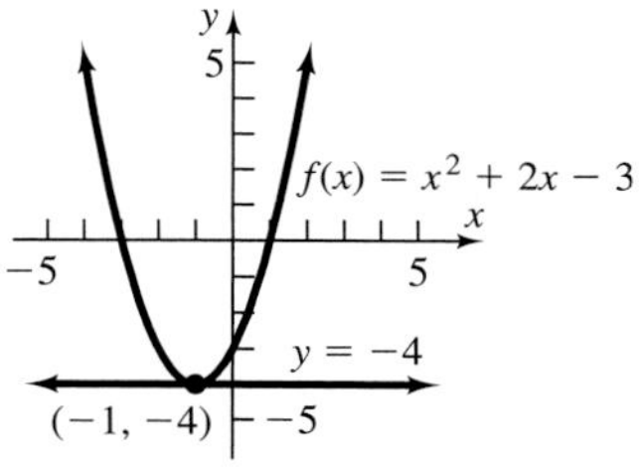

58. $m_{\tan} = 9$

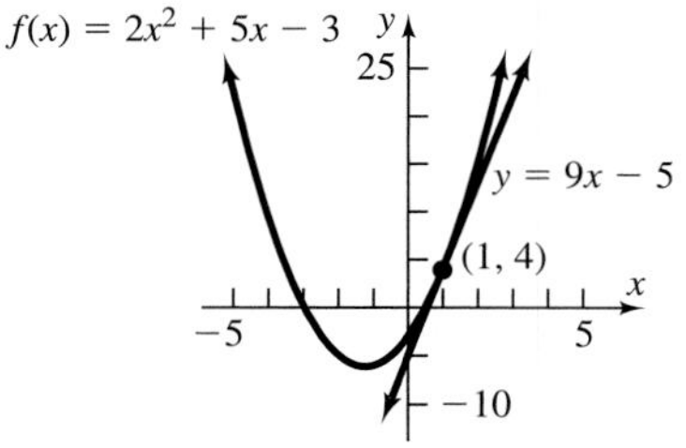

59. $m_{\tan} = 16$

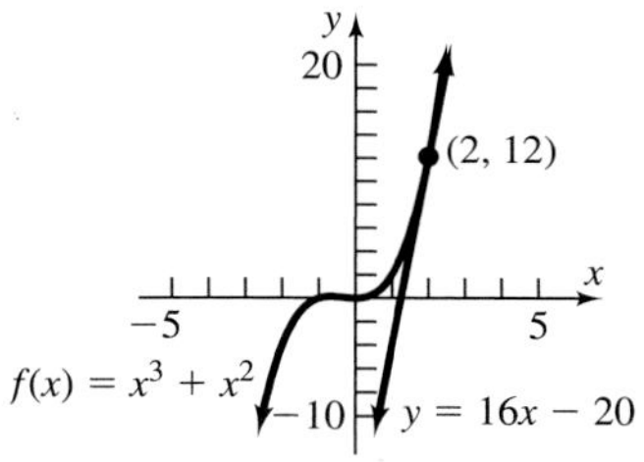

60. $m_{\tan} = 1$

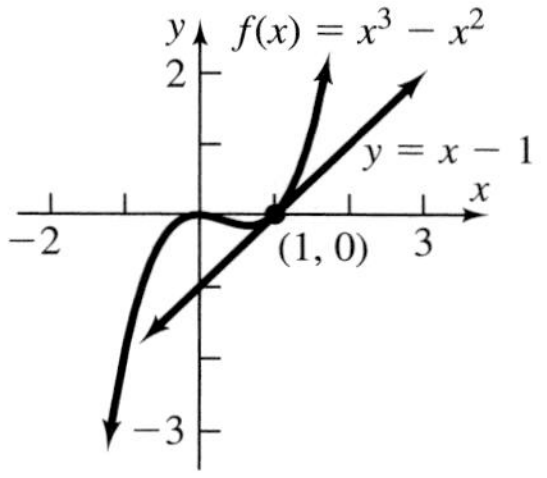

61. −24 **62.** 6 **63.** −3 **64.** 0 **65.** 7 **66.** 8 **67.** −158 **68.** −0.7663278737 **69.** 0.6662517653 **70.** 1.503767758
71. **(a)** 7 sec **(b)** 6 sec **(c)** 64 ft/sec **(d)** $(-32t + 96)$ ft/sec **(e)** 32 ft/sec **(f)** At $t = 3$ sec **(g)** −96 ft/sec **(h)** −128 ft/sec
72. 4π ft²/ft; 5π ft²/ft; 4.5π ft²/ft; 4.1π ft²/ft **73.** **(a)** \$61.29/watch **(b)** \$71.31/watch **(c)** \$81.40/watch
(d) $R(x) = -0.25x^2 + 100.01x - 1.24$ **(e)** \$87.51/watch **74.** **(a)** 80 ft/sec **(b)** 64 ft/sec **(c)** 48 ft/sec **(d)** $y = 16x^2$ **(e)** 32 ft/sec

75. **(a)**

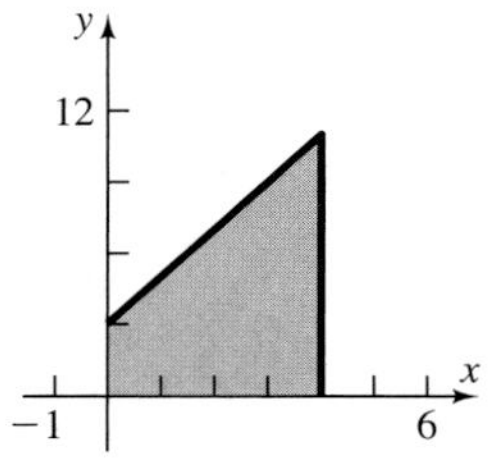

(b) 24 **(c)** 32

(d) 26 **(e)** 30 **(f)** 28

76. **(a)**

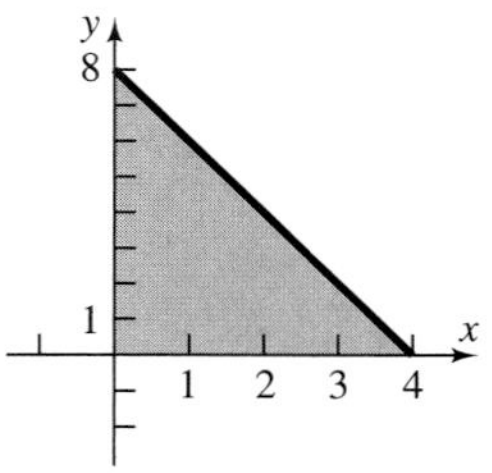

(b) 20 **(c)** 12

(d) 18 **(e)** 14 **(f)** 16

77. **(a)**

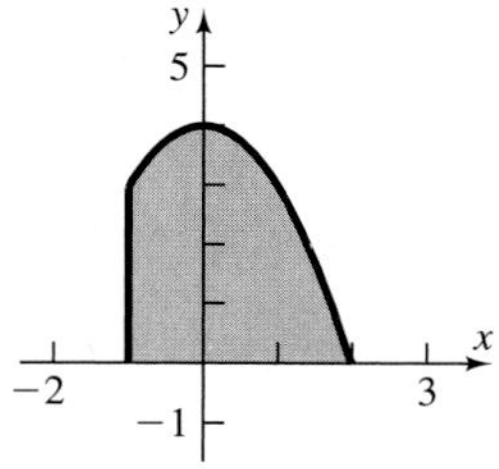

(b) 10 **(c)** $\frac{77}{8}$

(d) $\int_{-1}^{2} (4 - x^2)\, dx$ **(e)** 9

78. (a)

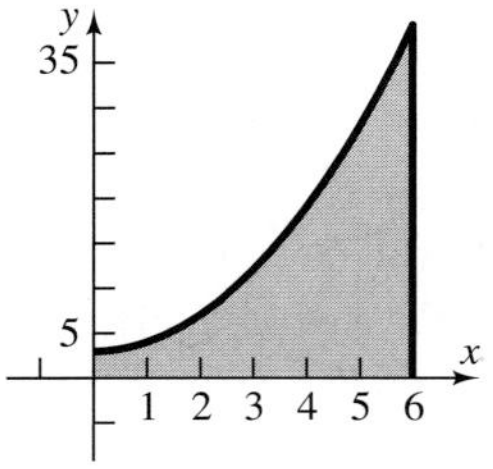

(b) 58 **(c)** 73

(d) $\int_0^6 (x^2 + 3)dx$ **(e)** 90

79. (a)

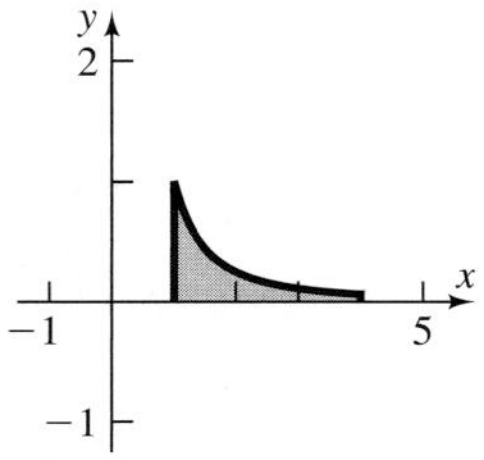

(b) $\frac{49}{36} \approx 1.36$ **(c)** 1.02

(d) $\int_1^4 \frac{1}{x^2}\,dx$ **(e)** 0.75

80. (a)

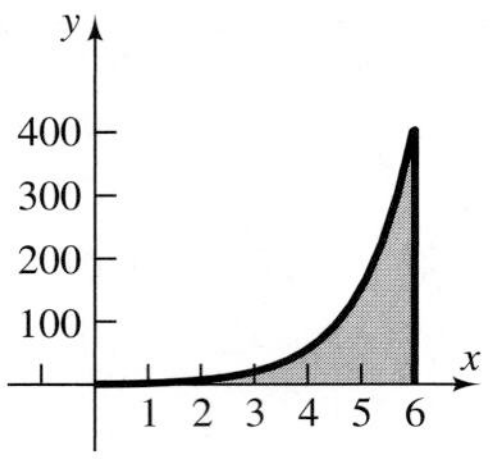

(b) 125.97 **(c)** 234.20

(d) $\int_0^6 e^x\,dx$ **(e)** 402.43

81. (a) Area under the graph of $f(x) = 9 - x^2$ from -1 to 3

(b)

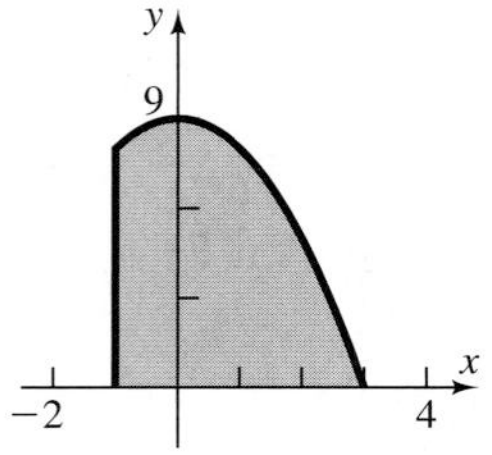

(c) $\frac{80}{3}$

82. (a) Area under the graph of $f(x) = \sqrt{x}$ from 1 to 4

(b)

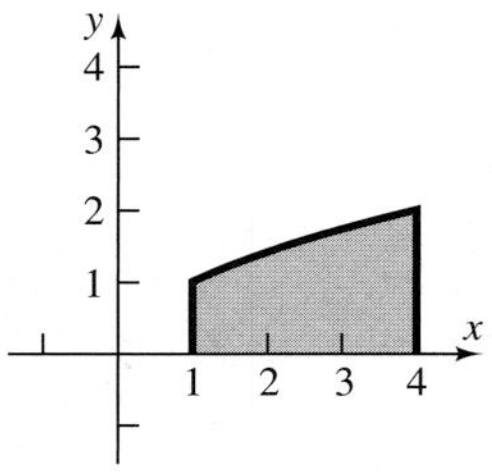

(c) $\frac{14}{3}$

83. (a) Area under the graph of $f(x) = e^x$ from -1 to 1

(b)

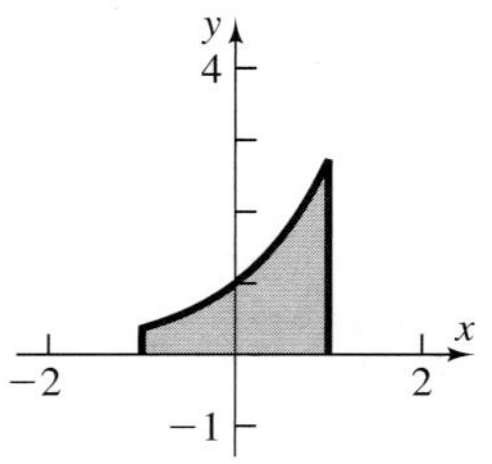

(c) 2.35

84. (a) Area under the graph of $f(x) = \sin x$ from $\frac{\pi}{3}$ to $\frac{2\pi}{3}$

(b)

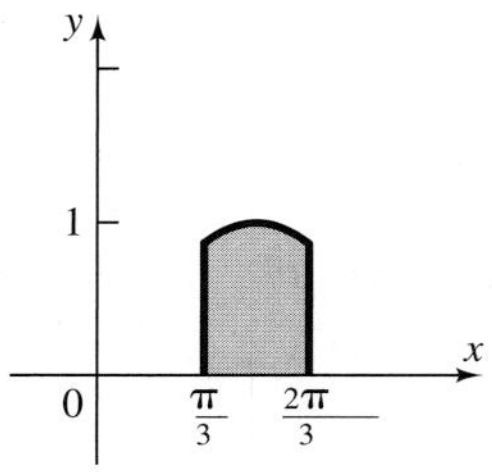

(c) 1

APPENDIX A Review

A.1 Concepts and Vocabulary *(page 911)*

1. variable **2.** origin **3.** strict **4.** base; exponent or power **5.** 5 **6.** False **7.** False **8.** False

A.1 Exercises *(page 911)*

9. 4 **10.** -3 **11.** -28 **12.** -2 **13.** $\frac{4}{5}$ **14.** $-\frac{1}{5}$ **15.** 0 **16.** $-\frac{7}{3}$ **17.** $x = 0$ **18.** $x = 0$ **19.** $x = 3$ **20.** None **21.** None **22.** $x = 1, x = -1$ **23.** $x = 1, x = 0, x = -1$ **24.** $x = 0$ **25.** $\{x \mid x \neq 5\}$ **26.** $\{x \mid x \neq -4\}$ **27.** $\{x \mid x \neq -4\}$ **28.** $\{x \mid x \neq 6\}$

29.

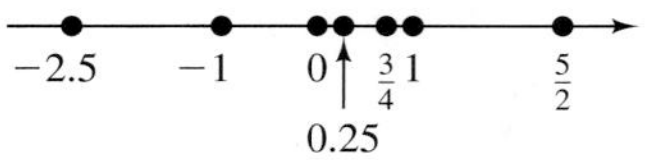

30.

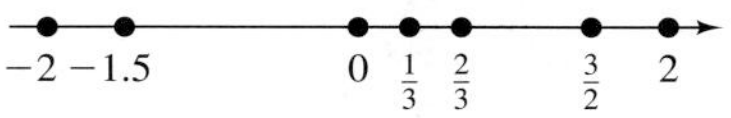

31. $>$ **32.** $<$ **33.** $>$ **34.** $<$ **35.** $>$ **36.** $>$ **37.** $=$ **38.** $>$ **39.** $<$ **40.** $=$ **41.** $x > 0$ **42.** $z < 0$ **43.** $x < 2$ **44.** $y > -5$ **45.** $x \leq 1$ **46.** $x \geq 2$

47.

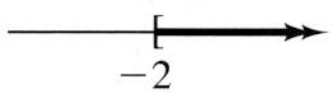

48.

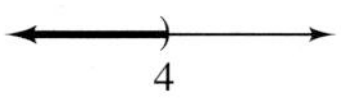

49.

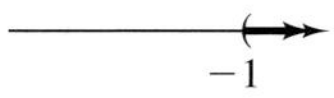

50.

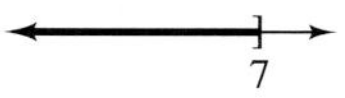

51. 1 **52.** 5 **53.** 5 **54.** 1 **55.** 1 **56.** -1 **57.** 22 **58.** 5

59. 2 **60.** 13 **61.** 1 **62.** 3 **63.** 2 **64.** 3 **65.** 6 **66.** 2 **67.** 16 **68.** −16 **69.** $\frac{1}{16}$ **70.** $-\frac{1}{16}$ **71.** $\frac{1}{9}$ **72.** 4 **73.** 9 **74.** 8 **75.** 5 **76.** 6 **77.** 4 **78.** 3 **79.** $\frac{1}{64x^6}$ **80.** $-\frac{1}{4x^2}$ **81.** $\frac{x^4}{y^2}$ **82.** $\frac{y^3}{x^3}$ **83.** $\frac{1}{x^3y}$ **84.** $\frac{1}{x^3y}$ **85.** $-\frac{8x^3z}{9y}$ **86.** $\frac{1}{2x^6y^2z}$ **87.** $\frac{16x^2}{9y^2}$ **88.** $\frac{216x^6}{125y^6}$ **89.** $A = lw$ **90.** $P = 2(l + w)$ **91.** $C = \pi d$ **92.** $A = \frac{1}{2}bh$ **93.** $A = \frac{\sqrt{3}}{4}x^2$ **94.** $P = 3x$ **95.** $V = \frac{4}{3}\pi r^3$ **96.** $S = 4\pi r^2$ **97.** $V = x^3$ **98.** $S = 6x^2$ **99. (a)** $2 \leq 5$ **(b)** $6 > 5$ **100. (a)** $6 \leq 8$ **(b)** $11 > 8$ **101. (a)** Yes **(b)** No **102. (a)** $1.6 \geq 1.5$ **(b)** $1.4 < 1.5$ **103.** No; $\frac{1}{3}$ is larger; 0.000333... **104.** No; $\frac{2}{3}$ is larger, 0.000666... **105.** No **106.** 3.15 or 3.16

A.2 Concepts and Vocabulary *(page 916)*

1. right; hypotenuse **2.** $A = \frac{1}{2}bh$ **3.** $C = 2\pi r$ **4.** True **5.** True **6.** False

A.2 Exercises *(page 916)*

7. 13 **8.** 10 **9.** 26 **10.** 5 **11.** 25 **12.** 50 **13.** Right triangle; 5 **14.** Right triangle; 10 **15.** Not a right triangle **16.** Not a right triangle **17.** Right triangle; 25 **18.** Right triangle; 26 **19.** Not a right triangle **20.** Not a right triangle **21.** 8 in^2 **22.** 36 cm^2 **23.** 4 in^2 **24.** 18 cm^2 **25.** $A = 25\pi$ m^2; $C = 10\pi$ m **26.** $A = 4\pi$ ft^2; $C = 4\pi$ ft **27.** $V = 224$ ft^3; $S = 232$ ft^2 **28.** $V = 288$ in^3; $S = 280$ in^2 **29.** $V = \frac{256}{3}\pi$ cm^3; $S = 64\pi$ cm^2 **30.** $V = 36\pi$ ft^3; $S = 36\pi$ ft^2 **31.** $V = 648\pi$ in^3; $S = 306\pi$ in^2 **32.** $V = 576\pi$ in^3; $S = 272\pi$ in^2 **33.** π square units **34.** $4 - \pi$ square units **35.** 2π square units **36.** $2\pi - 4$ square units **37.** About 16.8 ft **38.** About 1.6 revolutions **39.** 64 ft^2 **40.** 12 ft^2 **41.** $24 + 2\pi \approx 30.28$ ft^2; $16 + 2\pi \approx 22.28$ ft^2 **42.** $69\pi \approx 216.77$ ft^2; $26\pi \approx 81.68$ ft **43.** About 5.477 mi **44.** About 3.000 mi **45.** From 100 ft: 12.247 mi; From 150 ft: 15.000 mi

A.3 Concepts and Vocabulary *(page 927)*

1. 4; 3 **2.** $x^4 - 16$ **3.** $x^3 - 8$ **4.** False **5.** True **6.** False **7.** $3x(x - 2)(x + 2)$ **8.** prime **9.** True **10.** False **11.** in lowest terms **12.** least common multiple **13.** True **14.** False

A.3 Exercises *(page 927)*

15. $10x^5 + 3x^3 - 10x^2 + 6$ **16.** $9x^2 - 7x - 5$ **17.** $2ax + a^2$ **18.** $-2ax + a^2$ **19.** $2x^2 + 17x + 8$ **20.** $2x^2 + 3x - 2$ **21.** $x^4 - x^2 + 2x - 1$ **22.** $x^4 - x^3 - x^2 + 5x + 4$ **23.** $6x^2 + 2$ **24.** $-3x^2 - 9x - 7$ **25.** $(x - 6)(x + 6)$ **26.** $(x - 3)(x + 3)$ **27.** $(1 - 2x)(1 + 2x)$ **28.** $(1 - 3x)(1 + 3x)$ **29.** $(x + 2)(x + 5)$ **30.** $(x + 1)(x + 4)$ **31.** Prime **32.** Prime **33.** Prime **34.** $(x + 6)^2$ **35.** $(5 - x)(3 + x)$ **36.** Prime **37.** $3(x + 2)(x - 6)$ **38.** $x(x - 2)(x + 10)$ **39.** $y^2(y + 5)(y + 6)$ **40.** $3y(y - 8)(y + 2)$ **41.** $(2x + 3)^2$ **42.** $(3x - 2)^2$ **43.** $(3x + 1)(x + 1)$ **44.** $(4x - 1)(x + 1)$ **45.** $(x^2 + 9)(x - 3)(x + 3)$ **46.** $(x^2 + 1)(x - 1)(x + 1)$ **47.** $(x - 1)^2(x^2 + x + 1)^2$ **48.** $(x + 1)^2(x^2 - x + 1)^2$ **49.** $x^5(x - 1)(x + 1)$ **50.** $x^5(x - 1)(x^2 + x + 1)$ **51.** $(5 - 4x)(1 + 4x)$ **52.** $(5 + 16x)(1 - x)$ **53.** $(2y - 3)(2y - 5)$ **54.** $(3y - 1)(3y + 4)$ **55.** $(1 + x^2)(1 - 3x)(1 + 3x)$ **56.** $2(2 + x^2)(1 - 2x)(1 + 2x)$ **57.** $(x - 6)(x + 3)$ **58.** $(5 + x)(3x - 7)$ **59.** $(x + 2)(x - 3)$ **60.** $(x - 1)(x - 3)$ **61.** $3x(2 - x)^3(4 - 5x)$ **62.** $6x(1 - x)^3(1 + x)^3(1 - 5x^2)$ **63.** $(x + 2)(x - 1)(x + 1)$ **64.** $(x - 3)(x - 1)(x + 1)$ **65.** $(x - 1)(x + 1)(x^2 - x + 1)$ **66.** $(x + 1)^2(x^2 - x + 1)$ **67.** $\frac{3(x - 3)}{5x}$ **68.** $-\frac{(3x + 5)(1 + x)}{4}$ **69.** $\frac{x(2x - 1)}{x + 4}$ **70.** $\frac{6(x + 1)}{x(2x - 1)}$ **71.** $\frac{5x}{(x - 6)(x - 1)(x + 4)}$ **72.** $\frac{x^2 - 7x - 1}{(x - 3)(x + 8)}$ **73.** $\frac{2(x + 4)}{(x - 2)(x + 2)(x + 3)}$ **74.** $\frac{2x + 1}{(x - 1)^2}$ **75.** $\frac{x^3 - 2x^2 + 4x + 3}{x^2(x + 1)(x - 1)}$ **76.** $\frac{3x^3 - 5x^2 + 2x + 1}{x^2(x - 1)^2}$ **77.** $\frac{-1}{x(x + h)}$ **78.** $\frac{-(2x + h)}{x^2(x + h)^2}$ **79.** $2(3x + 4)(9x + 13)$ **80.** $(2x + 1)(30x - 19)$ **81.** $2x(3x + 5)$ **83.** $x^2(32x - 9)$ **83.** $5(x + 3)(x - 2)^2(x + 1)$ **84.** $6(x + 5)^3(x - 1)(x + 1)$ **85.** $3(4x - 3)(4x - 1)$ **86.** $3x^2(3x + 4)(5x + 4)$ **87.** $6(3x - 5)(2x + 1)^2(5x - 4)$ **88.** $2(4x + 5)^2(5x + 1)(50x + 31)$ **89.** $\frac{19}{(3x - 5)^2}$ **90.** $\frac{13}{(5x - 2)^2}$ **91.** $\frac{x^2 - 1}{(x^2 + 1)^2}$ **92.** $\frac{x^2 + 4}{(x^2 - 4)^2}$ **93.** $\frac{x(3x + 2)}{(3x + 1)^2}$ **94.** $\frac{x^2(4x - 15)}{(2x - 5)^2}$ **95.** $-\frac{3x^2 + 8x - 3}{(x^2 + 1)^2}$ **96.** $-\frac{2(x^2 - 5x - 9)}{(x^2 + 9)^2}$

A.4 Concepts and Vocabulary *(page 934)*

1. quotient; divisor; remainder **2.** $-3\overline{)2 \ 0 \ -5 \ 1}$ **3.** True **4.** True

A.4 Exercises *(page 935)*

5. $4x^2 - 11x + 23$; remainder -45 **6.** $3x^2 - 7x + 15$; remainder -32 **7.** $4x - 3$; remainder $x + 1$ **8.** $3x - 1$; remainder $x - 2$ **9.** $5x^2 - 13$; remainder $x + 27$ **10.** $5x^2 - 11$; remainder $x + 20$ **11.** $2x^2$; remainder $-x^2 + x + 1$ **12.** x^2; remainder $x - 2$ **13.** $x^2 - 2x + \frac{1}{2}$; remainder $\frac{5}{2}x + \frac{1}{2}$ **14.** $x^2 - \frac{2}{3}x - \frac{1}{9}$; remainder $\frac{16}{9}x - \frac{17}{9}$ **15.** $-4x^2 - 3x - 3$; remainder -7 **16.** $-3x^3 - 3x^2 - 3x - 5$; remainder -6 **17.** $x^2 - x - 1$; remainder $2x + 2$ **18.** $x^2 + x - 1$; remainder $-2x + 2$ **19.** $x^2 + ax + a^2$; remainder 0 **20.** $x^4 + ax^3 + a^2x^2 + a^3x + a^4$; remainder 0 **21.** $x^2 + x + 4$; remainder 12 **22.** $x^2 + x - 4$; remainder 5 **23.** $3x^2 + 11x + 32$; remainder 99 **24.** $-4x^2 + 10x - 21$; remainder 43 **25.** $x^4 - 3x^3 + 5x^2 - 15x + 46$; remainder -138 **26.** $x^3 + 2x^2 + 5x + 10$; remainder 22 **27.** $4x^5 + 4x^4 + x^3 + x^2 + 2x + 2$; remainder 7 **28.** $x^4 - x^3 + 6x^2 - 6x + 6$; remainder -16 **29.** $0.1x^2 - 0.11x + 0.321$; remainder -0.3531 **30.** $0.1x - 0.21$; remainder 0.241 **31.** $x^4 + x^3 + x^2 + x + 1$; remainder 0 **32.** $x^4 - x^3 + x^2 - x + 1$; remainder 0 **33.** No **34.** No **35.** Yes **36.** Yes **37.** Yes **38.** Yes **39.** No **40.** Yes **41.** Yes **42.** No **43.** $a = 1, b = -4, c = 11, d = -17; a + b + c + d = -9$

A.5 Concepts and Vocabulary *(page 945)*

5. equivalent equations **6.** identity **7.** False **8.** True **9.** add; $\frac{25}{4}$ **10.** discriminant; negative **11.** False **12.** False

A.5 Exercises *(page 945)*

13. 7 **14.** -8 **15.** -3 **16.** -6 **17.** 4 **18.** -4 **19.** $\frac{5}{4}$ **20.** $\frac{27}{4}$ **21.** -1 **22.** 1 **23.** -18 **24.** $\frac{7}{5}$ **25.** -3 **26.** -2 **27.** -16 **28.** -8 **29.** 0.5 **30.** -10 **31.** 2 **32.** $\frac{3}{10}$ **33.** 2 **34.** 3 **35.** 3 **36.** 2 **37.** $\{0, 9\}$ **38.** $\{0, 1\}$ **39.** $\{0, 9\}$ **40.** $\{0, 2\}$ **41.** 21 **42.** -10 **43.** $\{-2, 2\}$ **44.** $\{2, 5\}$ **45.** 6 **46.** $-\frac{1}{3}$ **47.** $\{-3, 3\}$ **48.** $\{-4, 4\}$ **49.** $\{-4, 1\}$ **50.** $\left\{-\frac{1}{3}, 1\right\}$ **51.** $\left\{-1, \frac{3}{2}\right\}$ **52.** $\{-1, 2\}$ **53.** $\{-4, 4\}$ **54.** $\{-1, 1\}$ **55.** 2 **56.** 3 **57.** No real solution **58.** No real solution **59.** $\{-2, 2\}$ **60.** $\{-3, 3\}$ **61.** $\{-1, 3\}$ **62.** $\{-4, 3\}$ **63.** $\{-2, -1, 0, 1\}$ **64.** $\{-4, -3, 0, 1\}$ **65.** $\{0, 4\}$ **66.** $\{-8, 0\}$ **67.** $\{-6, 2\}$ **68.** $\{-3, -4\}$ **69.** $\left\{-\frac{1}{2}, 3\right\}$ **70.** $\left\{-1, -\frac{2}{3}\right\}$ **71.** $\{3, 4\}$ **72.** $\{-4, 3\}$ **73.** $\frac{3}{2}$ **74.** $\frac{4}{5}$ **75.** $\left\{-\frac{2}{3}, \frac{3}{2}\right\}$ **76.** $\{3, 4\}$ **77.** $\left\{-\frac{3}{4}, 2\right\}$ **78.** $\left\{-\frac{5}{2}, 1\right\}$ **79.** $\{-5, 5\}$ **80.** $\{-6, 6\}$ **81.** $\{-1, 3\}$ **82.** $\{-3, -1\}$ **83.** $\{-3, 0\}$ **84.** $\left\{0, \frac{4}{3}\right\}$ **85.** 16 **86.** 4 **87.** $\frac{1}{16}$ **88.** $\frac{1}{36}$ **89.** $\frac{1}{9}$ **90.** $\frac{1}{25}$ **91.** $\{-7, 3\}$ **92.** $\{3 - \sqrt{22}, 3 + \sqrt{22}\}$ **93.** $\left\{-\frac{1}{4}, \frac{3}{4}\right\}$ **94.** $\left\{-1, \frac{1}{3}\right\}$ **95.** $\left\{\frac{-1 - \sqrt{7}}{6}, \frac{-1 + \sqrt{7}}{6}\right\}$ **96.** $\left\{\frac{3 - \sqrt{17}}{4}, \frac{3 + \sqrt{17}}{4}\right\}$ **97.** $\{2 - \sqrt{2}, 2 + \sqrt{2}\}$ **98.** $\{-2 - \sqrt{2}, -2 + \sqrt{2}\}$ **99.** $\left\{\frac{5 - \sqrt{29}}{2}, \frac{5 + \sqrt{29}}{2}\right\}$ **100.** $\left\{\frac{-5 - \sqrt{13}}{2}, \frac{-5 + \sqrt{13}}{2}\right\}$ **101.** $\left\{1, \frac{3}{2}\right\}$ **102.** $\left\{-\frac{3}{2}, -1\right\}$ **103.** No real solution **104.** No real solution **105.** $\left\{\frac{-1 - \sqrt{5}}{4}, \frac{-1 + \sqrt{5}}{4}\right\}$ **106.** $\left\{\frac{-1 - \sqrt{3}}{2}, \frac{-1 + \sqrt{3}}{2}\right\}$ **107.** $\left\{\frac{-\sqrt{3} - \sqrt{15}}{2}, \frac{-\sqrt{3} + \sqrt{15}}{2}\right\}$ **108.** $\left\{\frac{-\sqrt{2} - \sqrt{10}}{2}, \frac{-\sqrt{2} + \sqrt{10}}{2}\right\}$ **109.** No real solution **110.** No real solution **111.** Repeated real solution **112.** Repeated real solution **113.** Two unequal real solutions **114.** Two unequal real solutions **115.** $x = \frac{b + c}{a}$ **116.** $x = \frac{1 - b}{a}$ **117.** $x = \frac{abc}{a + b}$ **118.** $x = \frac{a + b}{c}$ **119.** $x = a^2$ **120.** $x = \frac{ab}{c}$ **121.** $R = \frac{R_1R_2}{R_1 + R_2}$ **122.** $r = \frac{A - P}{Pt}$ **123.** $R = \frac{mv^2}{F}$ **124.** $T = \frac{PV}{nR}$ **125.** $r = \frac{S - a}{S}$ **126.** $t = \frac{v_0 - v}{g}$ **127.** $\frac{-b + \sqrt{b^2 - 4ac}}{2a} + \frac{-b - \sqrt{b^2 - 4ac}}{2a} = \frac{-2b}{2a} = \frac{-b}{a}$ **128.** $\frac{-b + \sqrt{b^2 - 4ac}}{2a} \cdot \frac{-b - \sqrt{b^2 - 4ac}}{2a} = \frac{b^2 - (b^2 - 4ac)}{4a^2} = \frac{b^2 - b^2 + 4ac}{4a^2} = \frac{4ac}{4a^2} = \frac{c}{a}$ **129.** $k = -\frac{1}{2}$ or $\frac{1}{2}$ **130.** $k = -4$ or 4 **131.** The solutions of $ax^2 - bx + c = 0$ are $\frac{b + \sqrt{b^2 - 4ac}}{2a}$ and $\frac{b - \sqrt{b^2 - 4ac}}{2a}$.

132. The solution of $ax^2 + bx + c = 0$ is $\left\{\dfrac{-b + \sqrt{b^2 - 4ac}}{2a}, \dfrac{-b - \sqrt{b^2 - 4ac}}{2a}\right\}$. **133.** (b)

The solution of $cx^2 + bx + a = 0$ is $\left\{\dfrac{-b + \sqrt{b^2 - 4ca}}{2c}, \dfrac{-b - \sqrt{b^2 - 4ca}}{2c}\right\}$

$$\frac{-b + \sqrt{b^2 - 4ac}}{2a} \cdot \frac{-b - \sqrt{b^2 - 4ca}}{2c} = \frac{b^2 - (b^2 - 4ac)}{4ac} = \frac{b^2 - b^2 + 4ac}{4ac} = \frac{4ac}{4ac} = 1$$

$$\frac{-b - \sqrt{b^2 - 4ac}}{2a} \cdot \frac{-b + \sqrt{b^2 - 4ca}}{2c} = \frac{b^2 - (b^2 - 4ac)}{4ac} = \frac{b^2 - b^2 + 4ac}{4ac} = \frac{4ac}{4ac} = 1$$

Since the product of each solution of $ax^2 + bx + c = 0$ with a corresponding solution of $cx^2 + bx + a = 0$ is 1, the solutions of the two equations are reciprocals.

A.6 Concepts and Vocabulary *(page 954)*

1. False **2.** 5 **3.** False **4.** real; imaginary; imaginary unit **5.** $\{-2i, 2i\}$ **6.** False **7.** True **8.** False

A.6 Exercises *(page 955)*

9. $8 + 5i$ **10.** $-4 + 7i$ **11.** $-7 + 6i$ **12.** 6 **13.** $-6 - 11i$ **14.** $-10 + 6i$ **15.** $6 - 18i$ **16.** $-8 - 32i$ **17.** $6 + 4i$ **18.** $-12 - 9i$ **19.** $10 - 5i$ **20.** $13 + i$ **21.** 37 **22.** -10 **23.** $\frac{6}{5} + \frac{8}{5}i$ **24.** $\frac{5}{13} + \frac{12}{13}i$ **25.** $1 - 2i$ **26.** $\frac{1}{2} + i$ **27.** $\frac{5}{2} - \frac{7}{2}i$ **28.** $-\frac{1}{2} + \frac{5}{2}i$ **29.** $-\frac{1}{2} + \frac{\sqrt{3}}{2}i$ **30.** $\frac{1}{2} - \frac{\sqrt{3}}{2}i$ **31.** $2i$ **32.** $-2i$ **33.** $-i$ **34.** -1 **35.** i **36.** i **37.** -6 **38.** $4 - i$ **39.** $-10i$ **40.** $3 - 4i$ **41.** $-2 + 2i$ **42.** 82 **43.** 0 **44.** 0 **45.** 0 **46.** 0 **47.** $2i$ **48.** $3i$ **49.** $5i$ **50.** $8i$ **51.** $5i$ **52.** $5i$ **53.** $\{-2i, 2i\}$ **54.** $\{-2, 2\}$ **55.** $\{-4, 4\}$ **56.** $\{-5i, 5i\}$ **57.** $\{3 - 2i, 3 + 2i\}$ **58.** $\{-2 - 2i, -2 + 2i\}$ **59.** $\{3 - i, 3 + i\}$ **60.** $\{1 - 2i, 1 + 2i\}$ **61.** $\left\{\frac{1}{4} - \frac{1}{4}i, \frac{1}{4} + \frac{1}{4}i\right\}$ **62.** $\left\{-\frac{3}{10} - \frac{1}{10}i, -\frac{3}{10} + \frac{1}{10}i\right\}$ **63.** $\left\{\frac{1}{5} - \frac{2}{5}i, \frac{1}{5} + \frac{2}{5}i\right\}$ **64.** $\left\{\frac{3}{13} - \frac{2}{13}i, \frac{3}{13} + \frac{2}{13}i\right\}$ **65.** $\left\{-\frac{1}{2} - \frac{\sqrt{3}}{2}i, -\frac{1}{2} + \frac{\sqrt{3}}{2}i\right\}$ **66.** $\left\{\frac{1}{2} - \frac{\sqrt{3}}{2}i, \frac{1}{2} + \frac{\sqrt{3}}{2}i\right\}$ **67.** $\{2, -1 - \sqrt{3}i, -1 + \sqrt{3}i\}$ **68.** $\left\{-3, \frac{3}{2} - \frac{3\sqrt{3}}{2}i, \frac{3}{2} + \frac{3\sqrt{3}}{2}i\right\}$ **69.** $\{-2, 2, -2i, 2i\}$ **70.** $\{-1, 1 -i, i\}$ **71.** $\{-3i, -2i, 2i, 3i\}$ **72.** $\{-1, 1, -2i, 2i\}$ **73.** Two complex solutions that are conjugates of each other **74.** Two unequal real solutions **75.** Two unequal real solutions **76.** Two complex solutions that are conjugates of each other **77.** A repeated real solution **78.** A repeated real solution **79.** $2(3x + 4)(9x + 13)$ **80.** $(2x + 1)(30x - 19)$ **81.** $2x(3x + 5)$ **82.** $x^2(32x - 9)$ **83.** $5(x + 3)(x - 2)^2(x + 1)$ **84.** $6(x + 5)^3(x - 1)(x + 1)$ **85.** $3(4x - 3)(4x - 1)$ **86.** $3x^2(3x + 4)(5x + 4)$ **87.** $6(3x - 5)(2x + 1)^2(5x - 4)$ **88.** $2(4x + 5)^2(5x + 1)(50x + 31)$

A.7 Concepts and Vocabulary *(page 963)*

1. mathematical modeling **2.** interest **3.** uniform motion **4.** True **5.** True **6.** $100 - x$

A.7 Exercises *(page 964)*

7. $A = \pi r^2$; r = radius, A = area **8.** $C = 2\pi r$; r = radius, C = circumference **9.** $A = s^2$; A = area, s = length of a side **10.** $P = 4s$; s = length of a side, P = perimeter **11.** $F = ma$; F = force, m = mass, a = acceleration **12.** $P = \frac{F}{A}$; P = pressure, F = force, A = area **13.** $W = Fd$; W = work, F = force, d = distance **14.** $K = \frac{1}{2}mv^2$; K = kinetic energy, m = mass, v = velocity **15.** $C = 150x$; C = total variable cost, x = number of dishwashers **16.** $R = 250x$; R = total revenue, x = number of dishwashers **17.** $11,500 will be invested in bonds and $8500 in CDs. **18.** Sean will receive $6500 and George $3500. **19.** Brooke needs a score of 85. **20.** For a B, Mike needs 78; for an A, he needs 93. **21.** The length is 19 ft; the width is 11 ft. **22.** The length is 14 m; the width is 7 m. **23.** Invest $31,250 in bonds and $18,750 in CDs. **24.** Invest $43,750 in bonds and $6250 in CDs. **25.** $11,600 was loaned out at 8%. **26.** She can lend $333,333.33 at 16%. **27.** Mix 75 lb of Earl Grey tea with 25 lb of Orange Pekoe tea. **28.** 49 lb of coffee I should be mixed with 51 lb of Coffee II. **29.** Mix 40 lb of cashews with the peanuts. **30.** Each box should contain 20 caramels and 10 cremes. **31.** The speed of the current is 2.286 mi/hr. **32.** The speed of the boat is 9 mi/hr. **33.** The Metra commuter averages 30 mi/hr; the Amtrak averages 80 mi/hr. **34.** The average speed of the slower car is 60 mi/hr; the average speed of the faster car is 70 mi/hr. Each traveled 210 mi. **35.** Working together, it takes 12 min. **36.** April would take 15 hr. **37.** The dimensions should be 4 ft by 4 ft. **38.** The dimensions should be 6 ft by 3 ft. **39.** The speed of the current is 5 mph. **40.** The dimensions are 36 ft by 18 ft. **41.** The dimensions are 11 ft by 13 ft. **42.** The dimensions are 18 cm by 17 cm. **43.** The dimensions are 5 m by 8 m. **44.** The radius is 25 ft. **45. (a)** The ball strikes the ground after 6 sec. **(b)** The ball passes the top of the building on its way down after 5 sec.

46. **(a)** The object will be 15 m above the ground at about 0.99 sec and 3.09 sec. **(b)** The object will strike the ground at about 4.08 sec. **(c)** No **(d)** The maximum height is about 20.41 m. **47.** **(a)** The dimensions are 10 ft by 5 ft. **(b)** The area is 50 sq ft. **(c)** The dimensions would be 7.5 ft by 7.5 ft. **(d)** The area would be 56.25 sq ft. **48.** **(a)** Its dimensions are 19 ft by 19 ft. **(b)** Its dimensions are 28.5 ft by 9.5 ft. **(c)** The diameter is approximately 25.83 ft. **(d)** The circular one has the most area. **49.** The border will be 2.71 ft wide. **50.** The border will be about 2.13 ft wide. **51.** The border will be 2.56 ft wide. **52.** Its radius is 3 in. **53.** Add $\frac{2}{3}$ gal of water. **54.** 5 l should be drained. **55.** Evaporate 10.67 oz of water. **56.** Evaporate 96 oz of water. **57.** 40 g of 12 karat gold should be mixed with 20 g of pure gold. **58.** There are 11 atoms of oxygen and 22 atoms of hydrogen. **59.** Mike passes Dan $\frac{1}{3}$ mi from the start, 2 min from the time Mike started to race. **60.** It can fly 675 mi. **61.** Start the auxiliary pump at 9:45 AM. **62.** 5 lb of pure cement must be added. **63.** The tub will fill in 1 hr. **64.** It will take 1 hr and 45 min more for the 5 hp pump to empty the pool. **65.** Lewis would beat Burke by 16.75 m. **66.** The defensive back catches up to the tight end at the tight end's 45 yd line. **67.** Therese should be allowed 20,000 mi as a business expense. **68.** 36 **69.** 13 sides; No **70.** The average speed is 49.5 mph. **71.** The tail wind was 137 knots. **72.** Set the original price at \$40. At 50% off, there will be no profit.

A.8 Concepts and Vocabulary *(page 975)*

5. negative **6.** closed interval **7.** $-5, 5$ **8.** $-5 < x < 5$ **9.** True **10.** True

A.8 Exercises *(page 975)*

11. $[0, 2]; 0 \le x \le 2$ **12.** $[2, \infty); x \ge 2$ **13.** $(-1, 2); -1 < x < 2$ **14.** $(-\infty, 0]; x \le 0$ **15.** $[0, 3); 0 \le x < 3$ **16.** $(-1, 1]; -1 < x \le 1$

17. $[0, 4]$
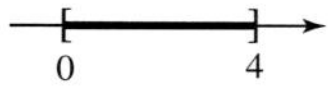

18. $(-1, 5)$
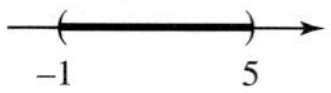

19. $[4, 6)$
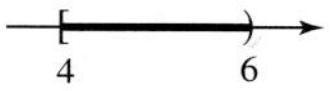

20. $(-2, 0)$
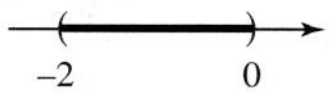

21. $[4, \infty)$
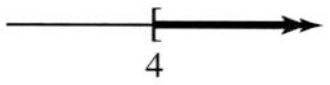

22. $(-\infty, 5]$
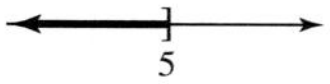

23. $(-\infty, -4)$
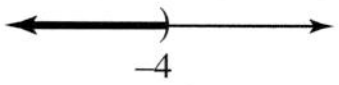

24. $(1, \infty)$
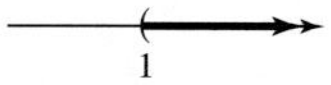

25. $2 \le x \le 5$
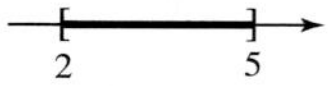

26. $1 < x < 2$
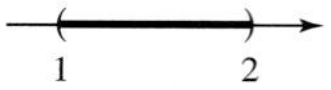

27. $-3 < x < -2$
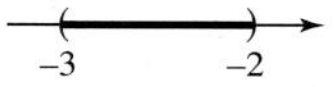

28. $0 \le x < 1$
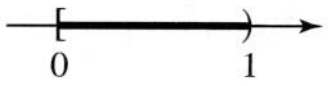

29. $x \ge 4$
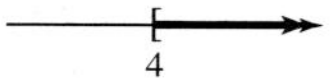

30. $x \le 2$
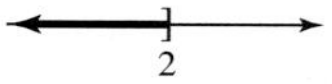

31. $x < -3$
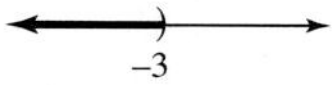

32. $x > -8$
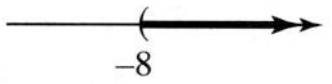

33. **(a)** $6 < 8$ **(b)** $-2 < 0$ **(c)** $9 < 15$ **(d)** $-6 > -10$ **34.** **(a)** $5 > 4$ **(b)** $-3 > -4$ **(c)** $6 > 3$ **(d)** $-4 < -2$
35. **(a)** $7 > 0$ **(b)** $-1 > -8$ **(c)** $12 > -9$ **(d)** $-8 < 6$ **36.** **(a)** $0 > -2$ **(b)** $-8 > -10$ **(c)** $-9 > -15$ **(d)** $6 < 10$
37. **(a)** $2x + 4 < 5$ **(b)** $2x - 4 < -3$ **(c)** $6x + 3 < 6$ **(d)** $-4x - 2 > -4$
38. **(a)** $4 - 2x > 8$ **(b)** $-4 - 2x > 0$ **(c)** $3 - 6x > 15$ **(d)** $-2 + 4x < -10$
39. $<$ **40.** $<$ **41.** $>$ **42.** $>$ **43.** $\ge$ **44.** $\le$ **45.** $<$ **46.** $<$ **47.** $\le$ **48.** $\ge$ **49.** $>$ **50.** $\le$ **51.** $\ge$ **52.** $<$

53. $\{x \mid x < 4\}; (-\infty, 4)$
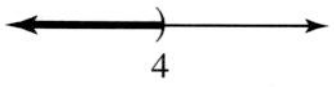

54. $\{x \mid x < 7\}; (-\infty, 7)$
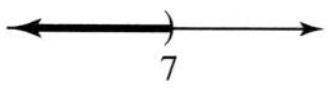

55. $\{x \mid x \ge -1\}; [-1, \infty)$
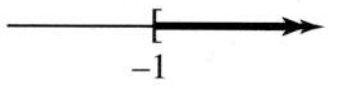

56. $\{x \mid x \ge -1\}; [-1, \infty)$
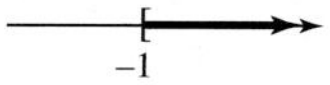

57. $\{x \mid x > 3\}; (3, \infty)$
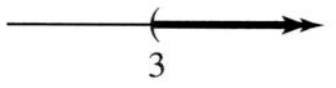

58. $\{x \mid x > -2\}; (-2, \infty)$
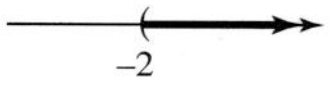

59. $\{x \mid x \ge 2\}; [2, \infty)$
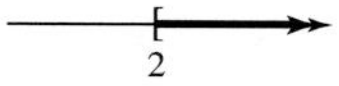

60. $\{x \mid x \ge 5\}; [5, \infty)$
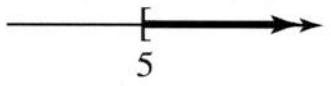

61. $\{x \mid x > -7\}; (-7, \infty)$
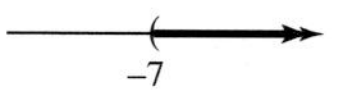

62. $\{x \mid x < 5\}; (-\infty, 5)$
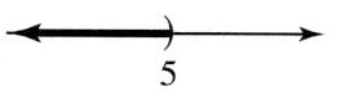

63. $\left\{x \,\middle|\, x \le \frac{2}{3}\right\}; \left(-\infty, \frac{2}{3}\right]$
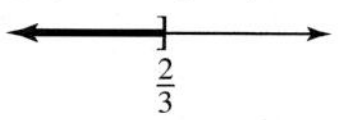

64. $\{x \mid x \le 0\}; (-\infty, 0]$
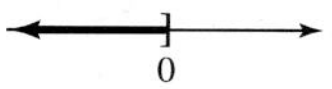

65. $\{x \mid x < -20\}; (-\infty, -20)$
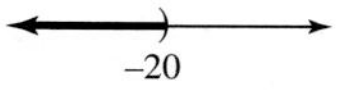

66. $\left\{x \,\middle|\, x > -\frac{7}{4}\right\}; \left(-\frac{7}{4}, \infty\right)$
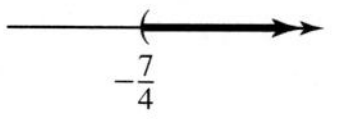

67. $\left\{x \,\middle|\, x \ge \frac{4}{3}\right\}; \left[\frac{4}{3}, \infty\right)$
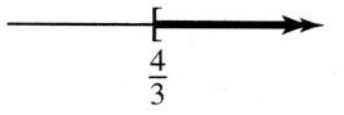

68. $\{x \mid x \ge 12\}; [12, \infty)$
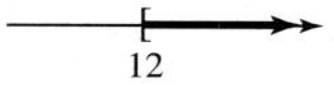

69. $\{x \mid 3 \le x \le 5\}; [3, 5]$
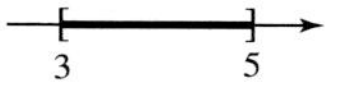

70. $\{x \mid 1 \le x \le 4\}; [1, 4]$
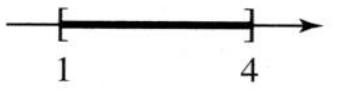

71. $\left\{x \,\middle|\, \frac{2}{3} \le x \le 3\right\}; \left[\frac{2}{3}, 3\right]$
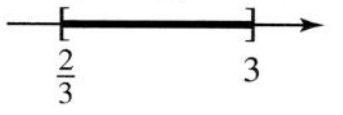

72. $\{x \mid -3 \le x \le 3\}; [-3, 3]$
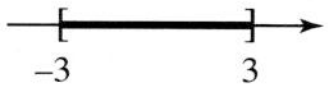

73. $\left\{x \middle| -\frac{11}{2} < x < \frac{1}{2}\right\}; \left(-\frac{11}{2}, \frac{1}{2}\right)$

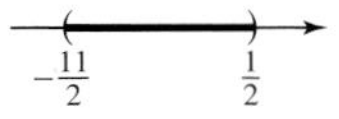

74. $\left\{x \middle| -\frac{2}{3} < x < 2\right\}; \left(-\frac{2}{3}, 2\right)$

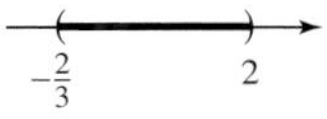

75. $\{x|-6 < x < 0\}; (-6, 0)$

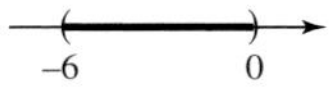

76. $\{x|0 < x < 3\}; (0, 3)$

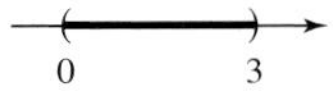

77. $\left\{x \middle| x < -\frac{1}{2}\right\}; \left(-\infty, -\frac{1}{2}\right)$

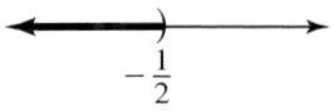

78. $\left\{x \middle| x > \frac{1}{2}\right\}; \left(\frac{1}{2}, \infty\right)$

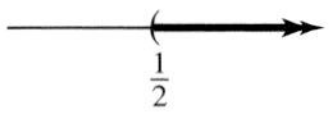

79. $\left\{x \middle| x > \frac{10}{3}\right\}; \left(\frac{10}{3}, \infty\right)$

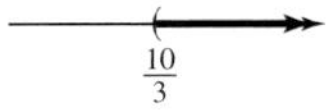

80. $\{x|x > 6\}; (6, \infty)$

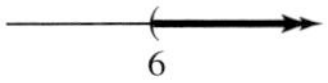

81. $\{x|-4 < x < 4\}; (-4, 4)$

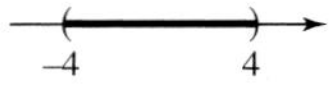

82. $\{x|-4 < x < 4\}; (-4, 4)$

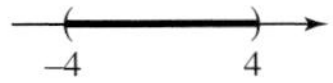

83. $\{x|x < -4 \text{ or } x > 4\}; (-\infty, -4) \text{ or } (4, \infty)$

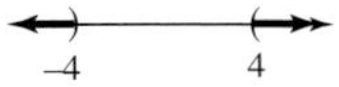

84. $\{x|x < -3 \text{ or } x > 3\}; (-\infty, -3) \text{ or } (3, \infty)$

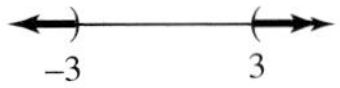

85. $\{x|0 \le x \le 1\}; [0, 1]$

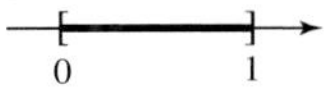

86. $\{x|-6 \le x \le 1\}; [-6, 1]$

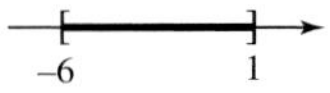

87. $\{x|x < -1 \text{ or } x > 2\}; (-\infty, -1) \text{ or } (2, \infty)$

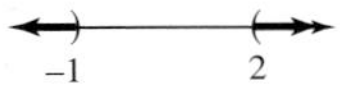

88. $\left\{x \middle| x < \frac{1}{3} \text{ or } x > 1\right\}; \left(-\infty, \frac{1}{3}\right) \text{ or } (1, \infty)$

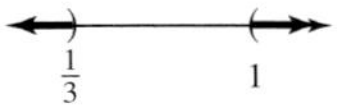

89. $\{x|-1 \le x \le 1\}; [-1, 1]$

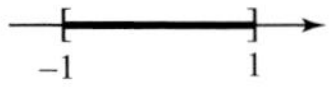

90. $\{x|-6 \le x \le 6\}; [-6, 6]$

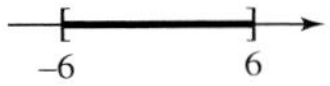

91. $\{x|x \le -2 \text{ or } x \ge 2\}; (-\infty, -2] \text{ or } [2, \infty)$

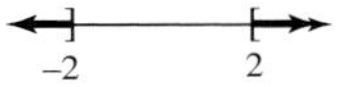

92. $\{x|x \le -3 \text{ or } x \ge -1\}; (-\infty, -3] \text{ or } [-1, \infty)$

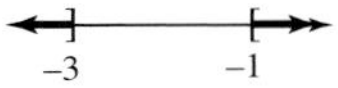

93. $|x - 2| < \frac{1}{2}; \left\{x \middle| \frac{3}{2} < x < \frac{5}{2}\right\}$ **94.** $|x + 1| < 1; \{x|-2 < x < 0\}$ **95.** $|x + 3| > 2; \{x|x < -5 \text{ or } x > -1\}$
96. $|x - 2| > 3; \{x|x < -1 \text{ or } x > 5\}$ **97.** $21 < \text{age} < 30$ **98.** $40 \le \text{age} < 60$ **99.** $|x - 98.6| \ge 1.5; \{x|x \le 97.1 \text{ or } x \ge 100.1\}$
100. $|x - 115| \le 5; \{x|110 \le x \le 120\}$ **101. (a)** Male ≥ 73.4 **(b)** Female ≥ 79.7 **(c)** A female can expect to live 6.3 yr longer.
102. Volume ranges from 1600 to 2400 cm^3, inclusive. **103.** The agent's commission ranges from \$45,000 to \$95,000, inclusive. As a percent of selling price, the commission ranges from 5% to 8.6%, inclusive. **104.** Commissions are at least \$53 and at most \$145.
105. The amount withheld varies from \$76.35 to \$126.35, inclusive. **106.** The amount withheld varies from \$126.35 to \$176.35, inclusive.
107. The usage varied from 675.41 kW hr to 2500.91 kW hr, inclusive. **108.** The water usage ranged from 16,000 to 37,968 gal.
109. The dealer's cost varied from \$7457.63 to \$7857.14, inclusive. **110.** Test score > 123.52 or $(123.52, \infty)$
111. You need at least a 74 on the fifth test. **112.** You need at least a 77 on the fifth test.

113. $\frac{a + b}{2} - a = \frac{a + b - 2a}{2} = \frac{b - a}{2} > 0$; therefore, $a < \frac{a + b}{2}$

$b - \frac{a + b}{2} = \frac{2b - a - b}{2} = \frac{b - a}{2} > 0$; therefore, $b > \frac{a + b}{2}$

114. $d\left(a, \frac{a + b}{2}\right) = \left|a - \frac{a + b}{2}\right| = \frac{b - a}{2}$ since $b > a$

$d\left(\frac{a + b}{2}, b\right) = \left|b - \frac{a + b}{2}\right| = \frac{b - a}{2}$ since $b > a$; thus $\frac{a + b}{2}$ is same distance from a and b.

115. $(\sqrt{ab})^2 - a^2 = ab - a^2 = a(b - a) > 0$; thus, $(\sqrt{ab})^2 > a^2$ and $\sqrt{ab} > a$

$b^2 - (\sqrt{ab})^2 = b^2 - ab = b(b - a) > 0$; thus $b^2 > (\sqrt{ab})^2$ and $b > \sqrt{ab}$

116. Since the square of any nonzero number is positive $(\sqrt{a} - \sqrt{b}) = a - 2\sqrt{ab} + b > 0$.

So, $a + b > 2\sqrt{ab}$, or equivalently, $\frac{a + b}{2} > \sqrt{ab}$

117. $h - a = \frac{2ab}{a + b} - a = \frac{ab - a^2}{a + b} = \frac{a(b - a)}{a + b} > 0$; thus, $h > a$

$b - h = b - \frac{2ab}{a + b} = \frac{b^2 - ab}{a + b} = \frac{b(b - a)}{a + b} > 0$; thus $h < b$

118. $\frac{1}{h} = \frac{1}{2}\left(\frac{1}{a} + \frac{1}{b}\right) = \frac{1}{2}\left(\frac{b + a}{ab}\right) = \frac{b + a}{2ab}$; thus $h = \frac{2ab}{b + a} = \frac{(\sqrt{ab})^2}{\left(\frac{b + a}{2}\right)}$

A.9 Concepts and Vocabulary *(page 983)*

3. index **4.** cube root **5.** True **6.** False

A.9 Exercises *(page 983)*

7. 3 **8.** 2 **9.** -2 **10.** -1 **11.** $2\sqrt{2}$ **12.** $3\sqrt[3]{2}$ **13.** $-2x\sqrt[3]{x}$ **14.** $2x\sqrt[4]{3x}$ **15.** x^3y^2 **16.** x^2y **17.** x^2y **18.** $\frac{1}{3x}$ **19.** $6\sqrt{x}$ **20.** $3x^2\sqrt{x}$ **21.** $6x\sqrt{x}$ **22.** $10x^2$ **23.** $15\sqrt[3]{3}$ **24.** $300\sqrt[3]{3}$ **25.** $12\sqrt{3}$ **26.** $-30\sqrt{6}$ **27.** $2\sqrt{3}$ **28.** $-1 + \sqrt{5}$ **29.** $x - 2\sqrt{x} + 1$ **30.** $x + 2\sqrt{5x} + 5$ **31.** $-5\sqrt{2}$ **32.** $(2 - x)\sqrt[3]{x}$ **33.** $(2x - 1)\sqrt[3]{2x}$ **34.** $(2 + x)\sqrt[4]{2x}$ **35.** $\frac{\sqrt{2}}{2}$ **36.** $3\sqrt[3]{2}$ **37.** $-\frac{\sqrt{15}}{5}$ **38.** $-\frac{\sqrt[3]{3}\sqrt{2}}{4}$ **39.** $\frac{\sqrt{3}(5 + \sqrt{2})}{23}$ **40.** $\frac{\sqrt{14} - 2\sqrt{2}}{3}$ **41.** $\frac{-19 + 8\sqrt{5}}{41}$ **42.** $\frac{9 - 5\sqrt{3}}{3}$ **43.** $\frac{5\sqrt[3]{4}}{2}$ **44.** $-\frac{2\sqrt[3]{3}}{3}$ **45.** $\frac{2x + h - 2\sqrt{x(x + h)}}{h}$ **46.** $\frac{x + \sqrt{x^2 - h^2}}{h}$ **47.** $\frac{9}{2}$ **48.** -3 **49.** 3 **50.** 3 **51.** 4 **52.** 8 **53.** -3 **54.** 8 **55.** 64 **56.** 512 **57.** $\frac{1}{27}$ **58.** $\frac{1}{3125}$ **59.** $\frac{27\sqrt{2}}{32}$ **60.** $\frac{9}{4}$ **61.** $\frac{27\sqrt{2}}{32}$ **62.** $\frac{9}{4}$ **63.** $x^{7/12}$ **64.** $x^{11/12}$ **65.** xy^2 **66.** x^3y^6 **67.** $x^{4/3}y^{5/3}$ **68.** $x^{5/4}y^{5/4}$ **69.** $\frac{8x^{3/2}}{y^{1/4}}$ **70.** $\frac{8y^{1/2}}{x^{3/2}}$ **71.** $\frac{3x + 2}{(1 + x)^{1/2}}$ **72.** $\frac{3x + 1}{2x^{1/2}}$ **73.** $\frac{x(3x^2 + 2)}{(x^2 + 1)^{1/2}}$ **74.** $\frac{4x + 3}{3(x + 1)^{2/3}}$ **75.** $\frac{22x + 5}{10\sqrt{x - 5}\sqrt{4x + 3}}$ **76.** $\frac{65x + 6}{24\sqrt[3]{(x - 2)^2}\sqrt[3]{(8x + 1)^2}}$ **77.** $\frac{2 + x}{2(1 + x)^{3/2}}$ **78.** $\frac{1}{(x^2 + 1)^{3/2}}$ **79.** $\frac{4 - x}{(x + 4)^{3/2}}$ **80.** $\frac{9}{(9 - x^2)^{3/2}}$ **81.** $\frac{1}{x^2(x^2 - 1)^{1/2}}$ **82.** $\frac{4}{(x^2 + 4)^{3/2}}$ **83.** $\frac{1 - 3x^2}{2\sqrt{x}(1 + x^2)^2}$ **84.** $\frac{6x - 4x^3}{3(1 - x^2)^{4/3}}$ **85.** $\frac{1}{2}(5x + 2)(x + 1)^{1/2}$ **86.** $\frac{1}{3}(x^2 + 4)^{1/3}(11x^2 + 12)$ **87.** $2x^{1/2}(3x - 4)(x + 1)$ **88.** $2x^{1/2}(10x + 9)$ **89.** $(x^2 + 4)^{1/3}(11x^2 + 12)$ **90.** $2x(3x + 4)^{1/3}(5x + 4)$ **91.** $(3x + 5)^{1/3}(2x + 3)^{1/2}(17x + 27)$ **92.** $12(6x + 1)^{1/3}(4x - 3)^{1/2}(5x - 1)$ **93.** $\frac{3(x + 2)}{2x^{1/2}}$ **94.** $\frac{4(2x - 1)}{x^{2/3}}$ **95.** $\frac{2(2 - x)(2 + x)}{(8 - x^2)^{1/2}}$ **96.** $x(1 - x)^{1/2}(1 + x)^{1/2}(2 - 5x^2)$

APPENDIX B Graphing Utilities

B.1 Exercises *(page 987)*

1. $(-1, 4)$; II **2.** $(3, 4)$; I **3.** $(3, 1)$; I **4.** $(-6, -4)$; III **5.** $X\min = -6$, $X\max = 6$, $X\text{scl} = 2$, $Y\min = -4$, $Y\max = 4$, $Y\text{scl} = 2$ **6.** $X\min = -3$, $X\max = 3$, $X\text{scl} = 1$, $Y\min = -2$, $Y\max = 2$, $Y\text{scl} = 1$ **7.** $X\min = -6$, $X\max = 6$, $X\text{scl} = 2$, $Y\min = -1$, $Y\max = 3$, $Y\text{scl} = 1$ **8.** $X\min = -9$, $X\max = 9$, $X\text{scl} = 3$, $Y\min = -12$, $Y\max = 4$, $Y\text{scl} = 4$ **9.** $X\min = 3$, $X\max = 9$, $X\text{scl} = 1$, $Y\min = 2$, $Y\max = 10$, $Y\text{scl} = 2$ **10.** $X\min = -22$, $X\max = -10$, $X\text{scl} = 2$, $Y\min = 4$, $Y\max = 8$, $Y\text{scl} = 1$ **11.** $X\min = -11$, $X\max = 5$, $X\text{scl} = 1$, $Y\min = -3$, $Y\max = 6$, $Y\text{scl} = 1$ **12.** $X\min = -3$, $X\max = 7$, $X\text{scl} = 1$, $Y\min = -4$, $Y\max = 9$, $Y\text{scl} = 1$ **13.** $X\min = -30$, $X\max = 50$, $X\text{scl} = 10$, $Y\min = -90$, $Y\max = 50$, $Y\text{scl} = 10$ **14.** $X\min = -90$, $X\max = 30$, $X\text{scl} = 10$, $Y\min = -50$, $Y\max = 70$, $Y\text{scl} = 10$ **15.** $X\min = -10$, $X\max = 110$, $X\text{scl} = 10$, $Y\min = -10$, $Y\max = 160$, $Y\text{scl} = 10$ **16.** $X\min = -20$, $X\max = 110$, $X\text{scl} = 10$, $Y\min = -10$, $Y\max = 60$, $Y\text{scl} = 10$ **17.** $4\sqrt{10}$ **18.** $\sqrt{149}$ **19.** $2\sqrt{65}$ **20.** $\sqrt{205}$

B.2 Exercises *(page 990)*

1. **(a)**

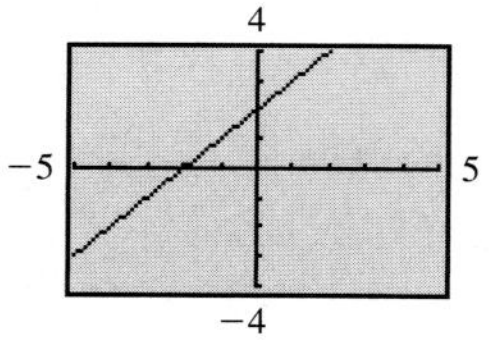

(b)

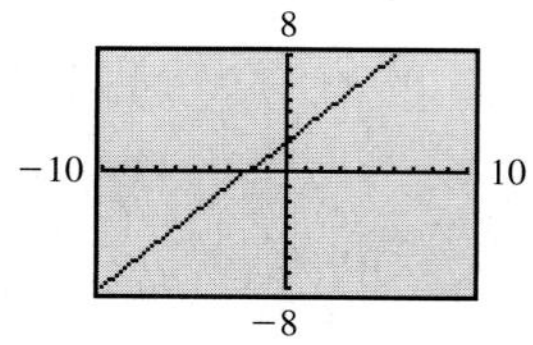

(c)

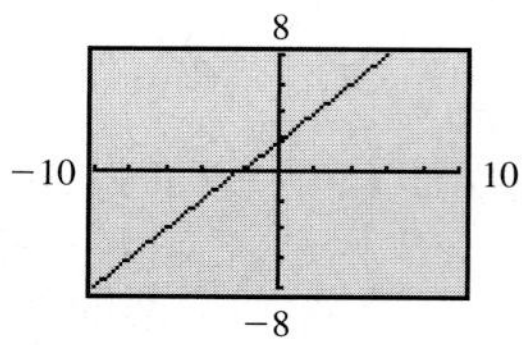

(d)

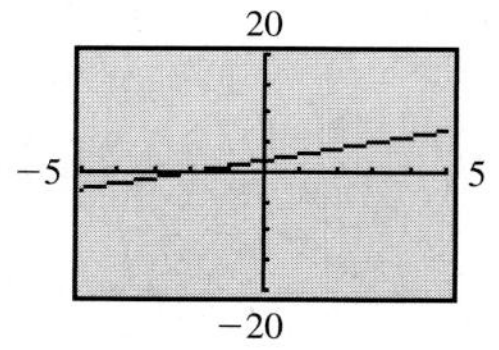

2. (a)
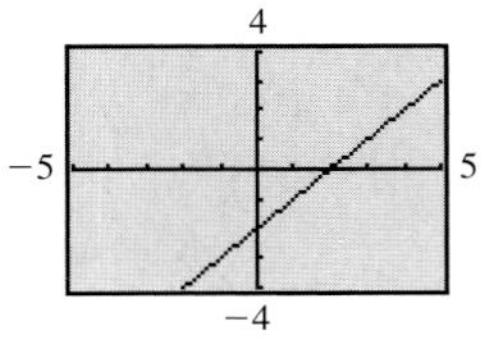

(b)
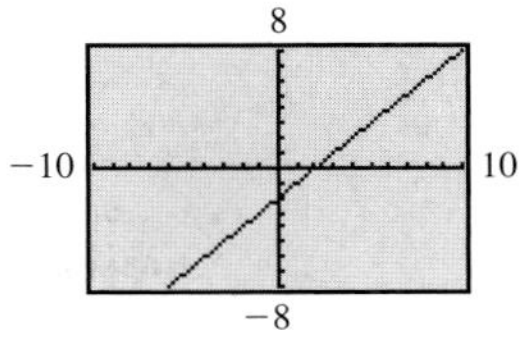

(c)
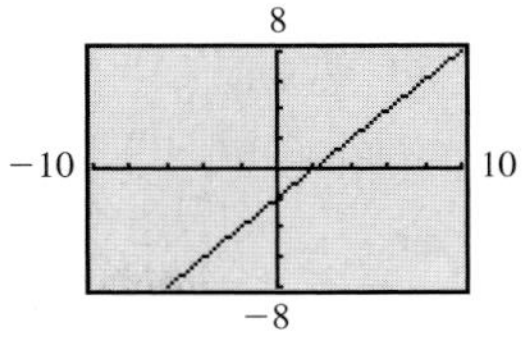

(d)
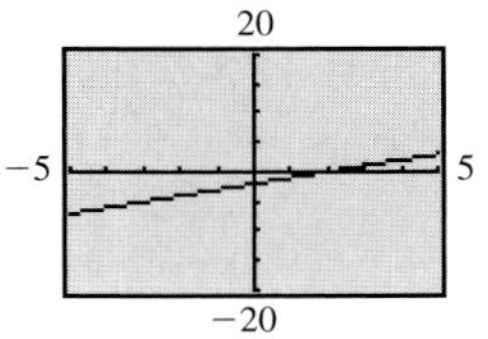

3. (a)
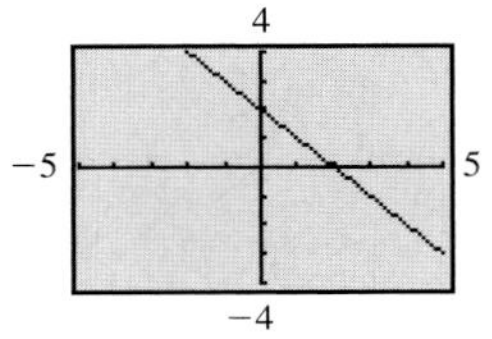

(b)
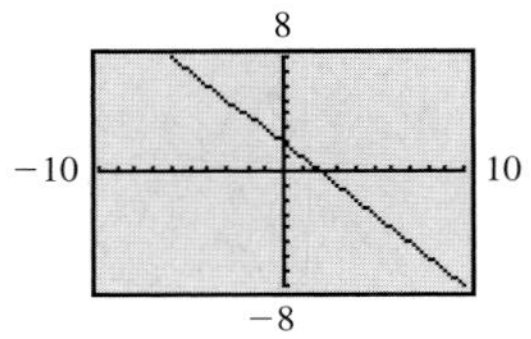

(c)
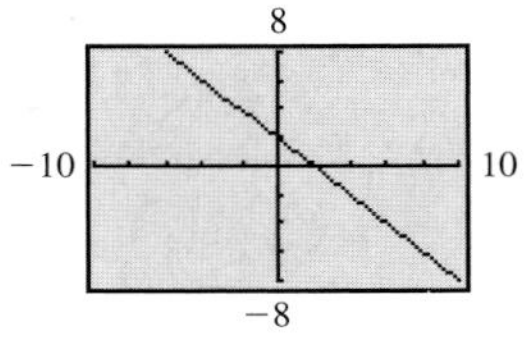

(d)
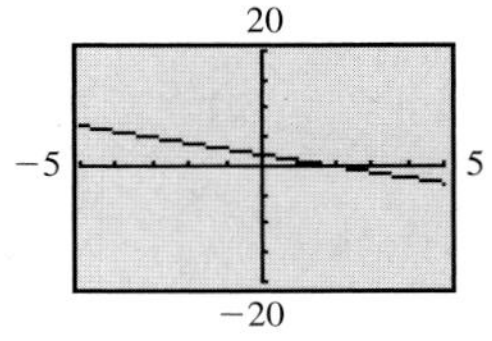

4. (a)
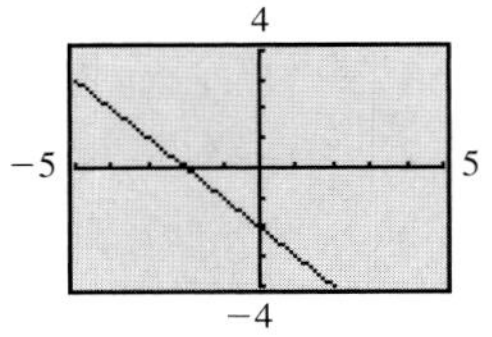

(b)
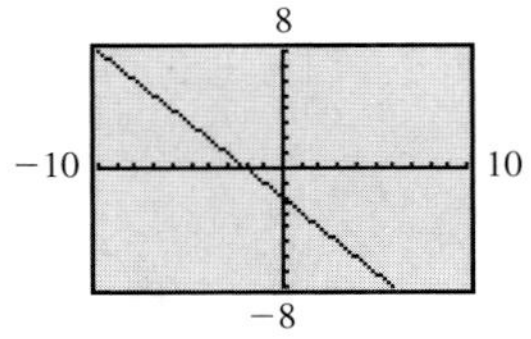

(c)
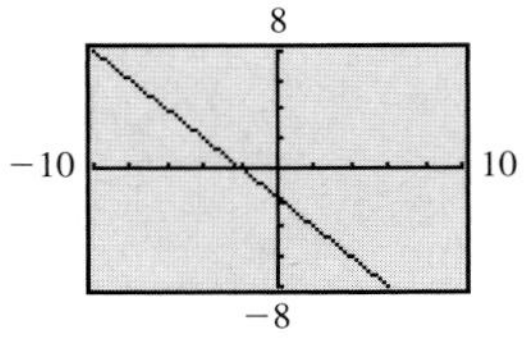

(d)
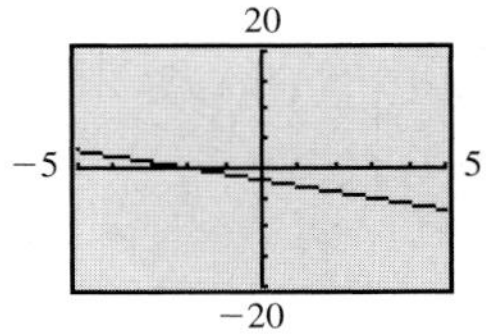

5. (a)
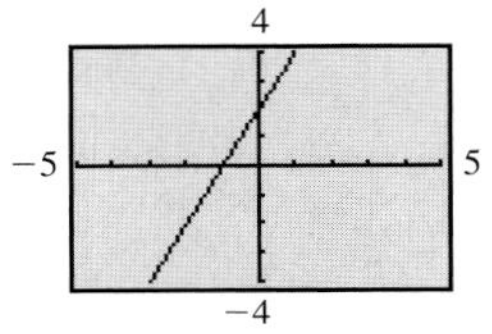

(b)
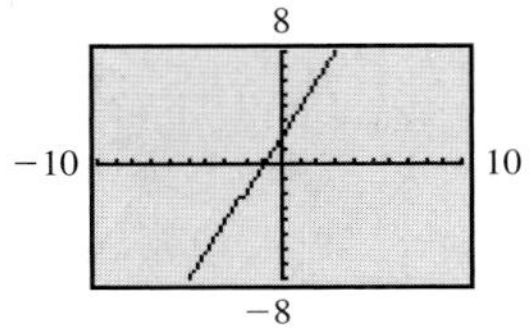

(c)
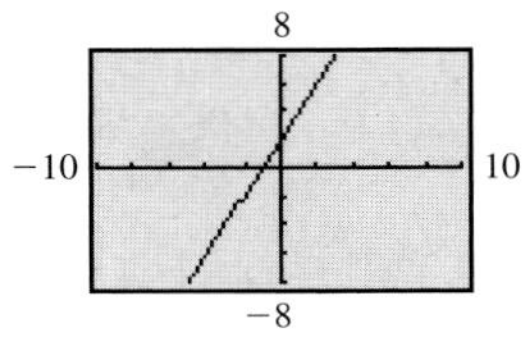

(d)
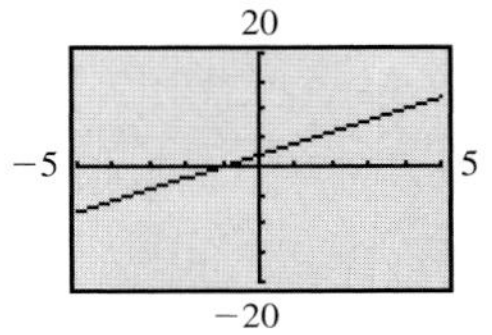

6. (a)
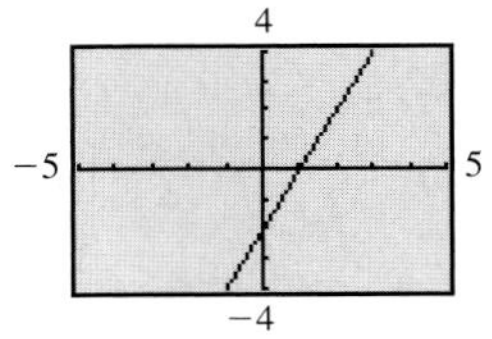

(b)
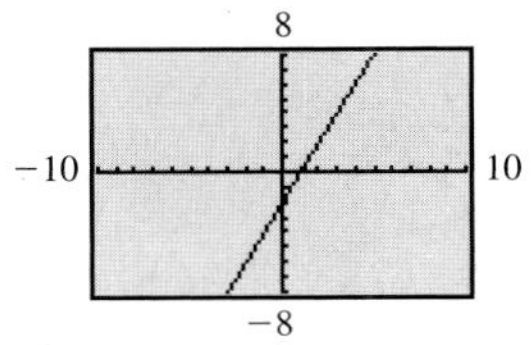

(c)
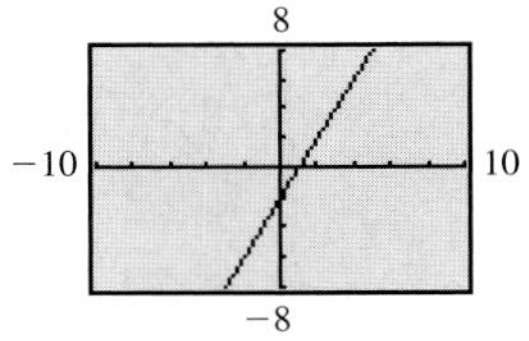

(d)
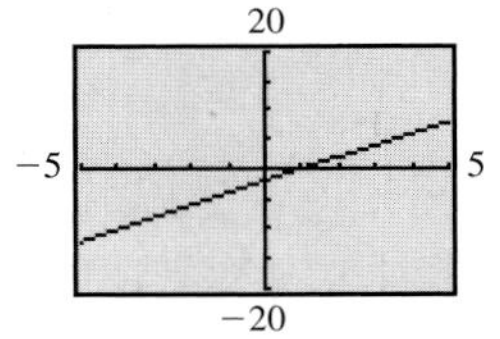

7. (a)
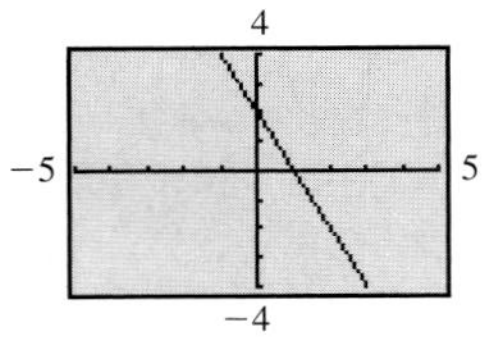

(b)
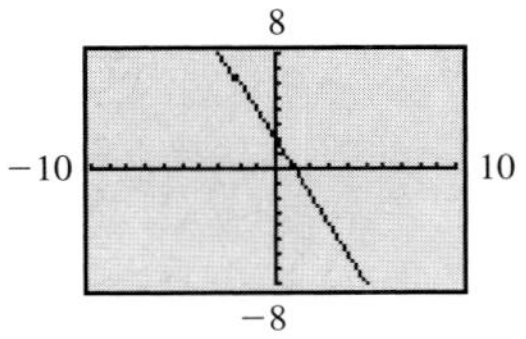

(c)
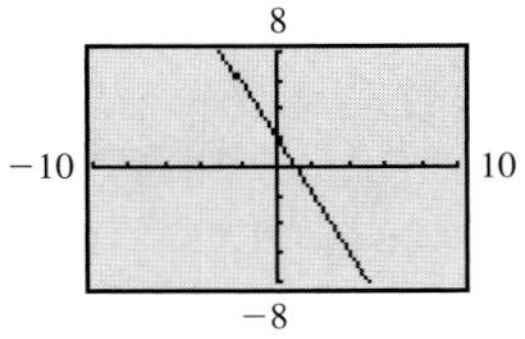

(d)
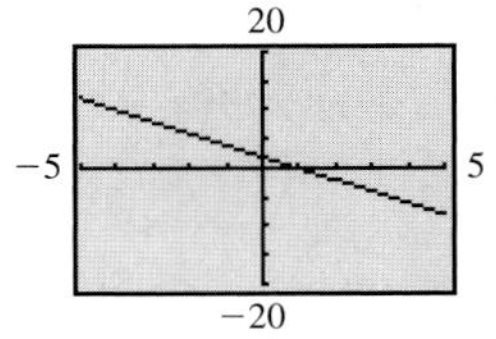

8. (a)
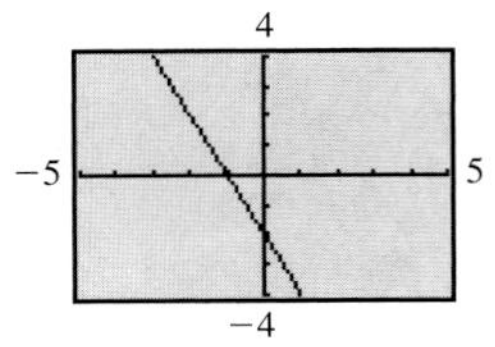

(b)
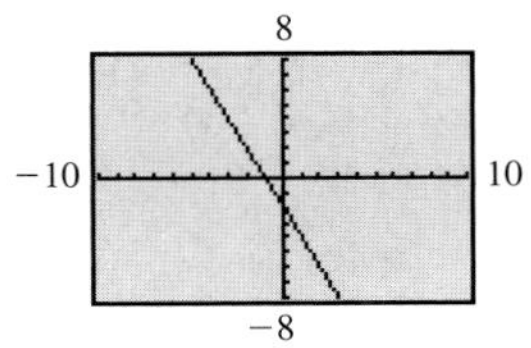

(c)
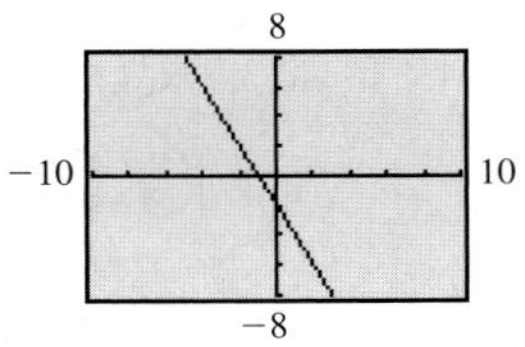

(d)
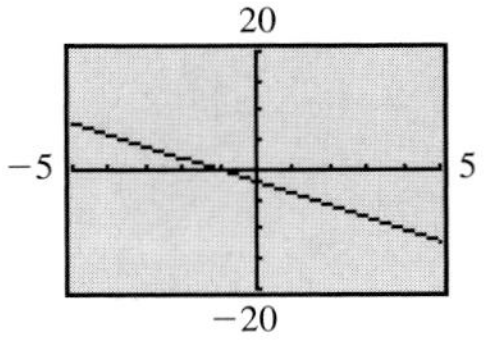

9. (a)
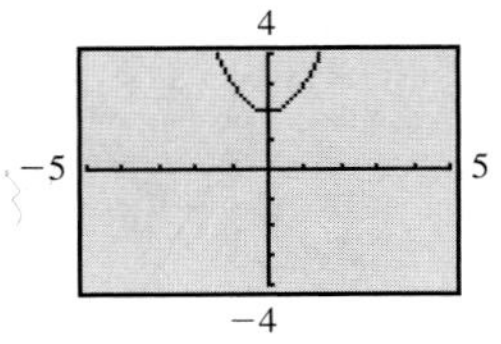

(b)
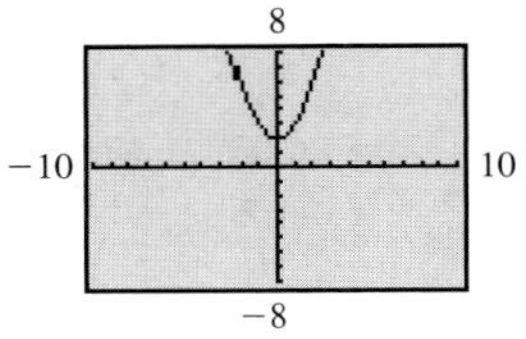

(c)
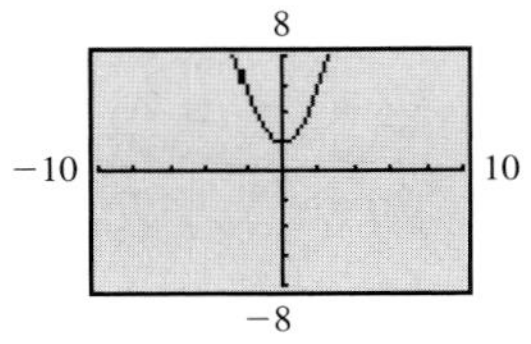

(d)
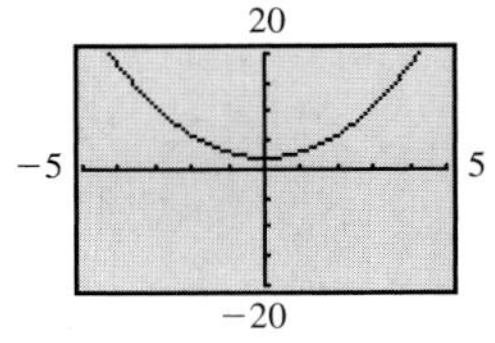

10. (a)
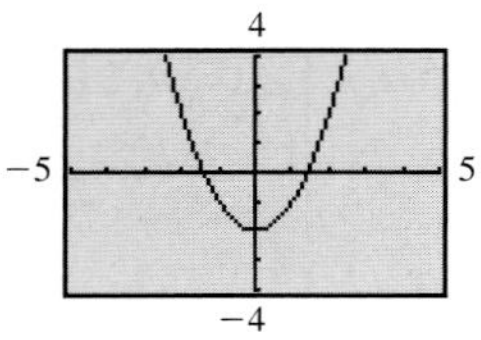

(b)
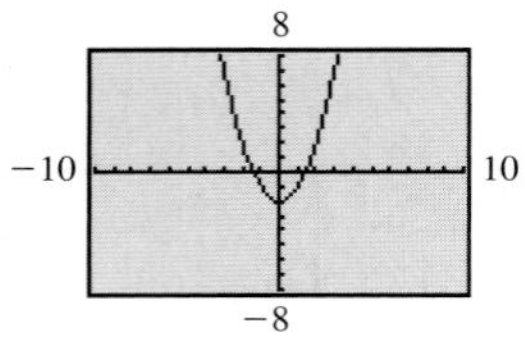

(c)
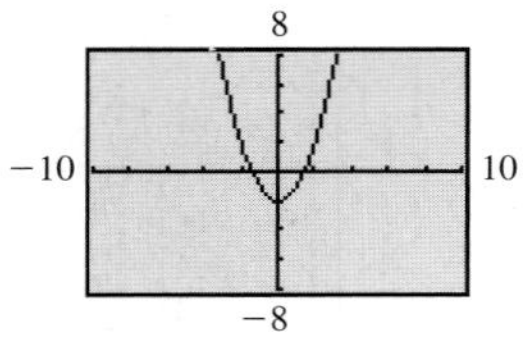

(d)
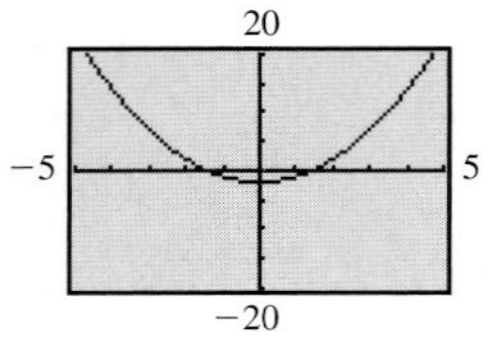

11. (a)
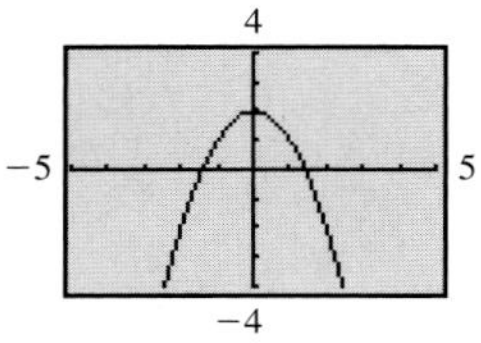

(b)
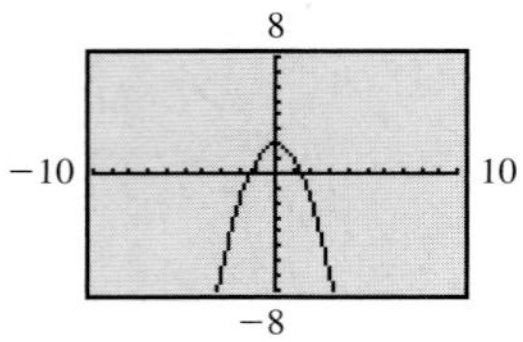

(c)
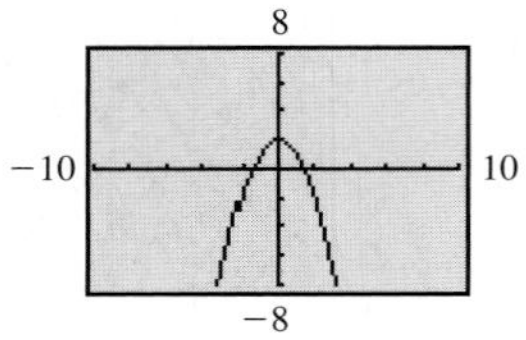

(d)
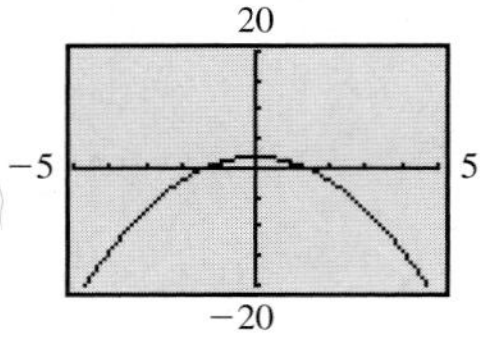

12. (a)
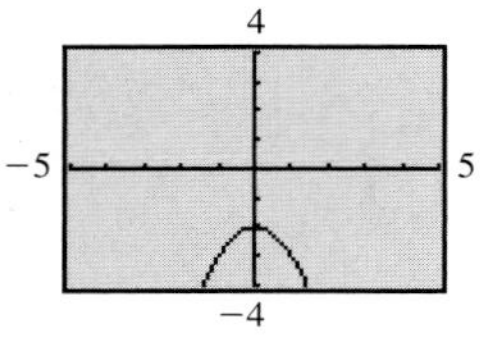

(b)
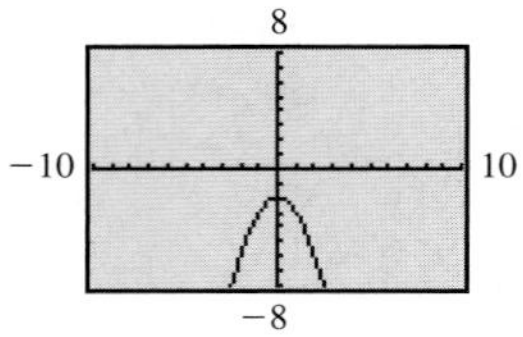

(c)
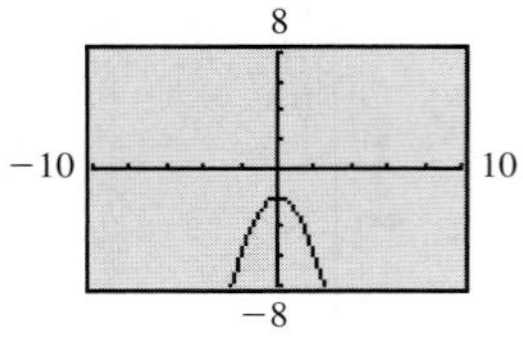

(d)
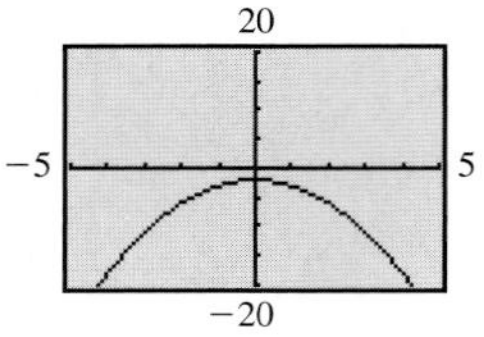

13. (a)
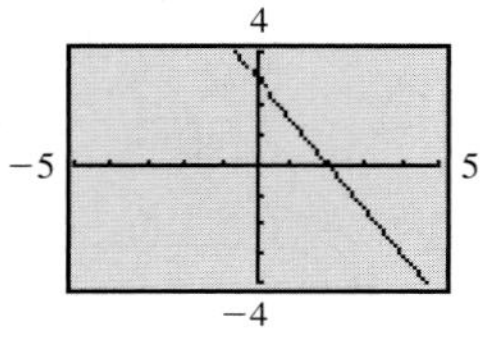

(b)
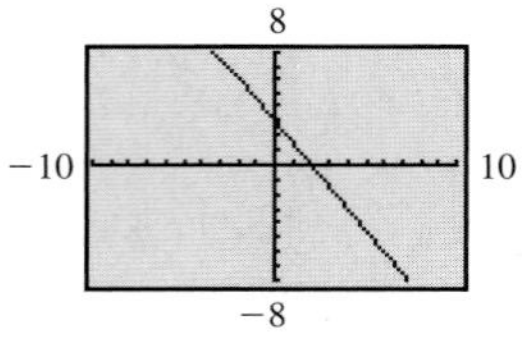

(c)
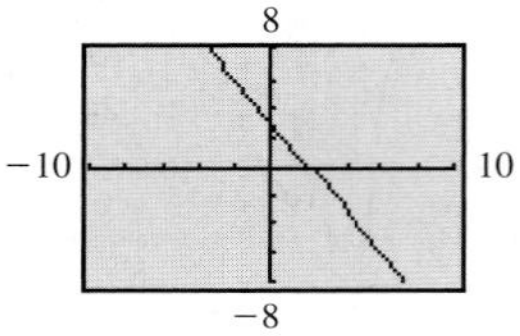

(d)
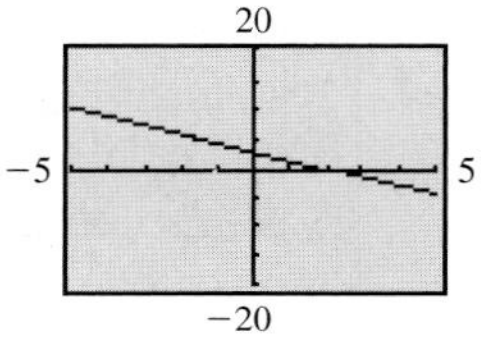

14. (a)
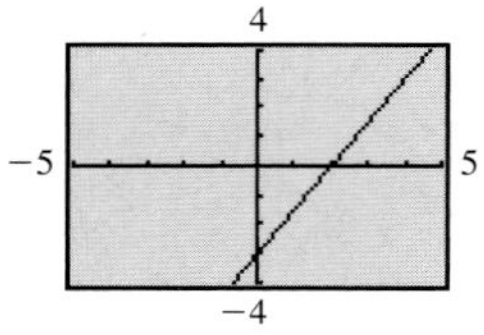

(b)
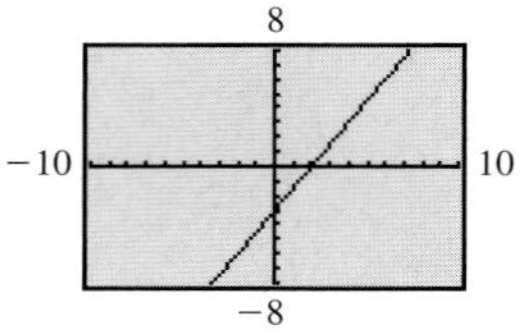

(c)
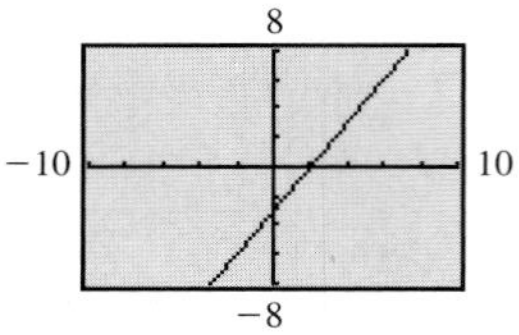

(d)
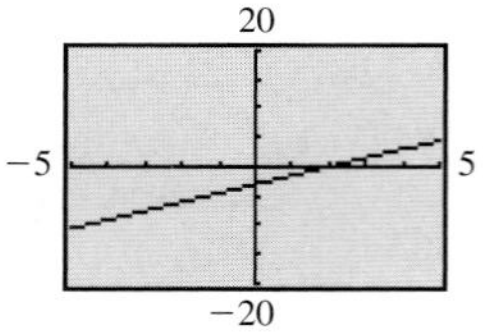

15. (a)
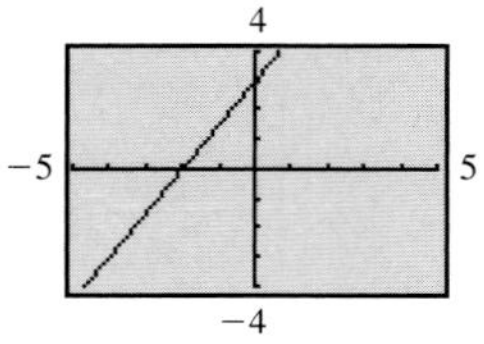

(b)
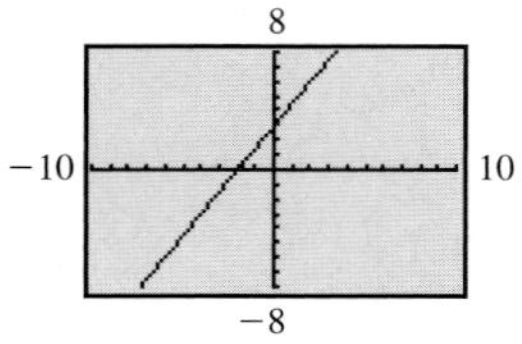

(c)
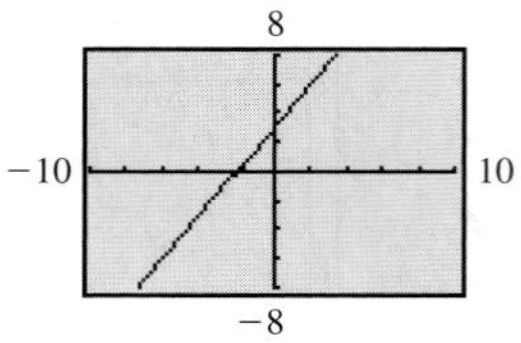

(d)
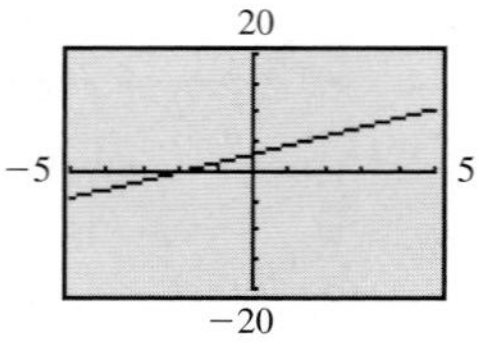

16. (a)
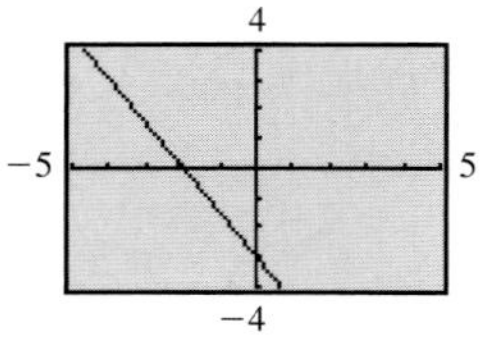

(b)
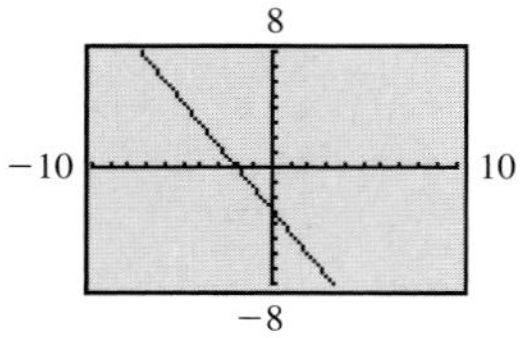

(c)
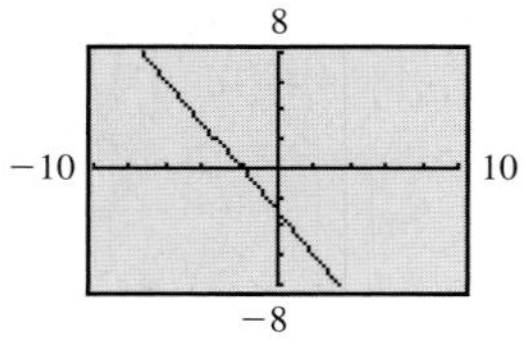

(d)
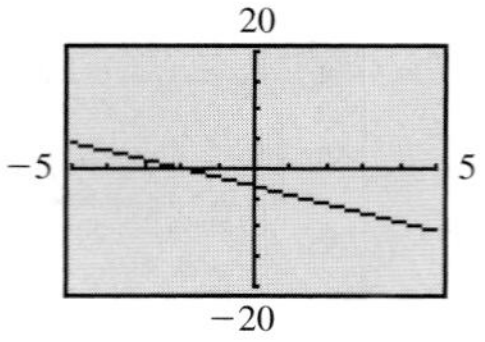

17.

X	Y1	
-3	-1	
-2	0	
-1	1	
0	2	
1	3	
2	4	
3	5	

Y1=X+2

18.

X	Y1	
-3	-5	
-2	-4	
-1	-3	
0	-2	
1	-1	
2	0	
3	1	

Y1=X−2

19.

X	Y1	
-3	5	
-2	4	
-1	3	
0	2	
1	1	
2	0	
3	-1	

Y1=-X+2

20.

X	Y1	
-3	1	
-2	0	
-1	-1	
0	-2	
1	-3	
2	-4	
3	-5	

Y1=-X−2

21. 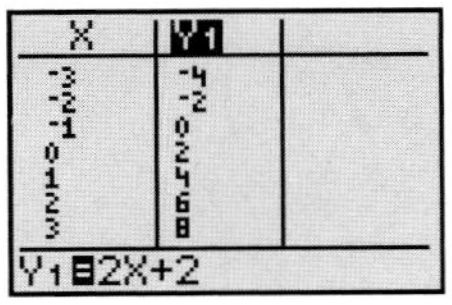

22. 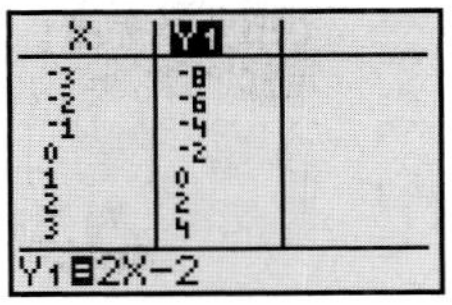

23. 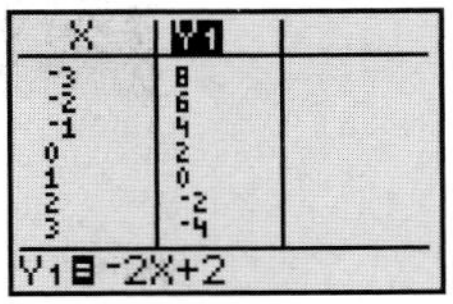

24. 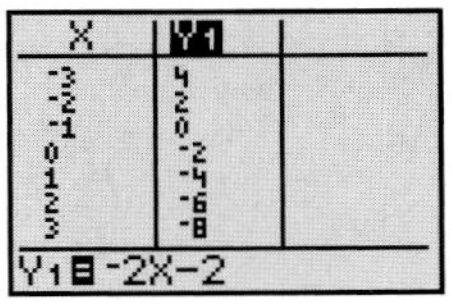

25. 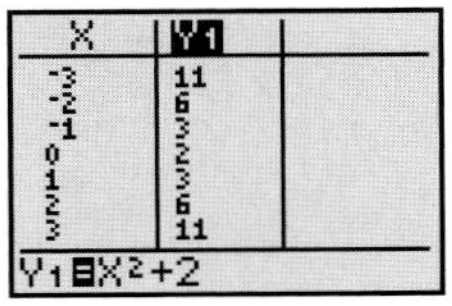

26. 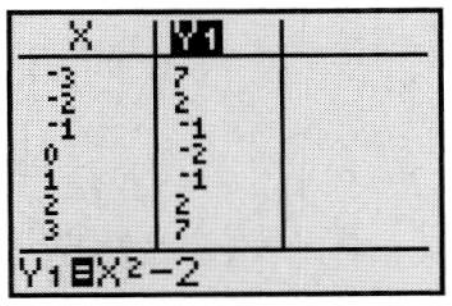

27. 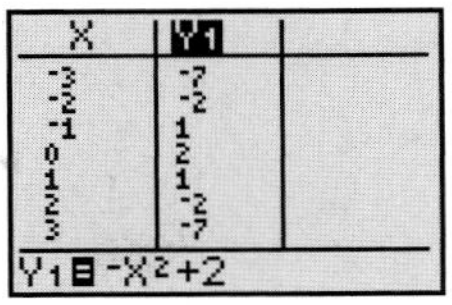

28. 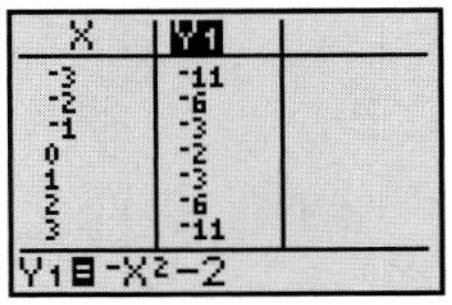

29. 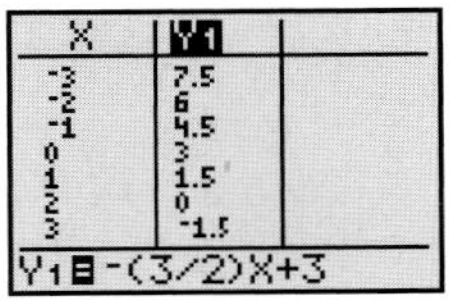

30. 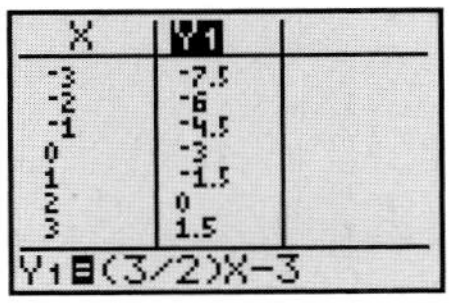

31. 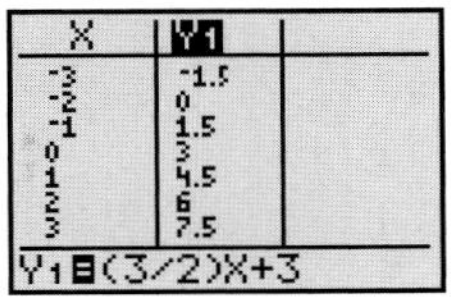

32. 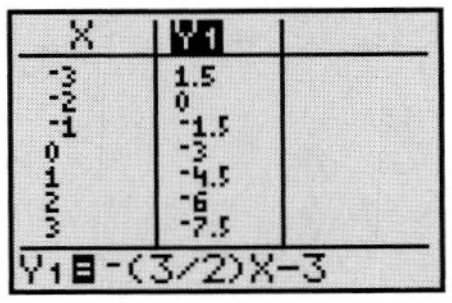

B.3 Exercises *(page 993)*

1. −3.41 **2.** −4.65 **3.** −1.71 **4.** −1.43 **5.** −0.28 **6.** −0.22 **7.** 3.00 **8.** 2.00 **9.** 4.50 **10.** 1.70 **11.** 0.32, 12.30 **12.** 0.64, 13.60 **13.** 1.00, 23.00 **14.** 1.07

B.5 Exercises *(page 996)*

1. Yes **2.** No **3.** Yes **4.** Yes **5.** No **6.** Yes **7.** Yes **8.** Yes

9. Answers may vary. A possible answer is $Y\text{min} = 4$, $Y\text{max} = 12$, and $Y\text{scl} = 1$.

10. Answers may vary. A possible answer is $Y\text{min} = -2$, $Y\text{max} = 10$, and $Y\text{scl} = 2$.

Index

CONICS

Parabola

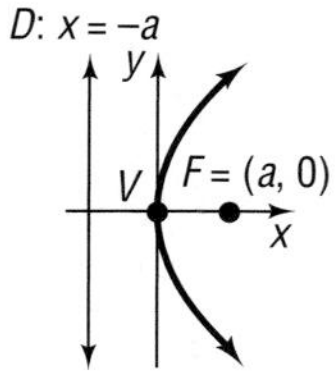

$y^2 = 4ax$

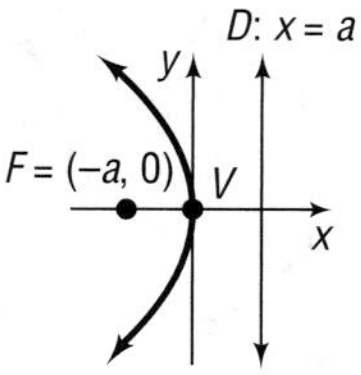

$y^2 = -4ax$

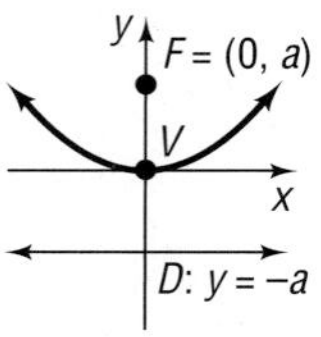

$x^2 = 4ay$

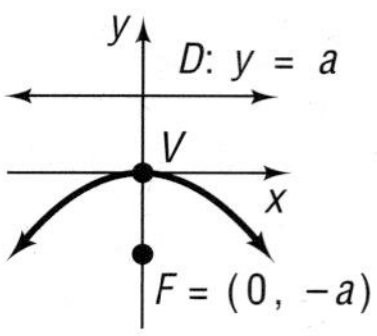

$x^2 = -4ay$

Ellipse

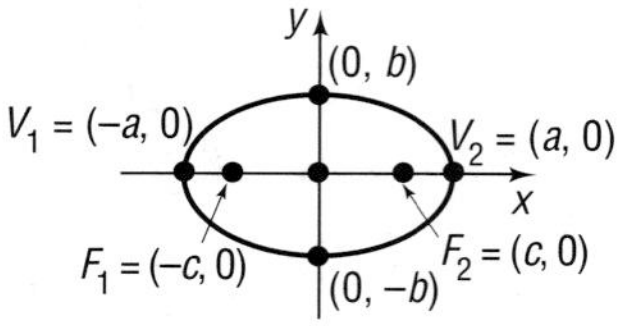

$$\frac{x^2}{a^2} + \frac{y^2}{b^2} = 1, \quad a > b, \quad c^2 = a^2 - b^2$$

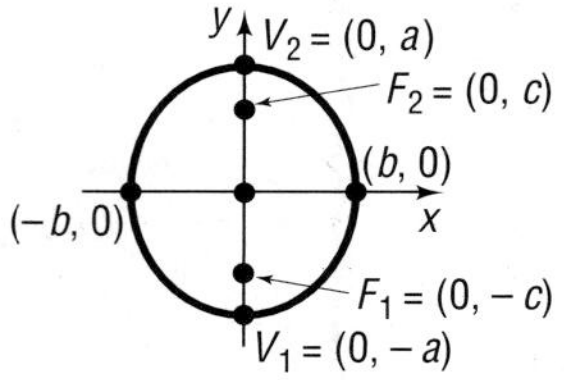

$$\frac{x^2}{b^2} + \frac{y^2}{a^2} = 1, \quad a > b, \quad c^2 = a^2 - b^2$$

Hyperbola

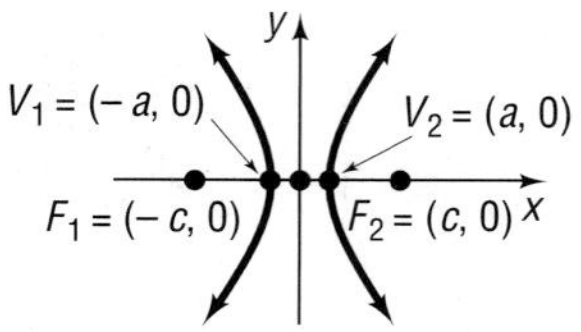

$$\frac{x^2}{a^2} - \frac{y^2}{b^2} = 1, \quad c^2 = a^2 + b^2$$

Asymptotes: $y = \frac{b}{a}x, \quad y = -\frac{b}{a}x$

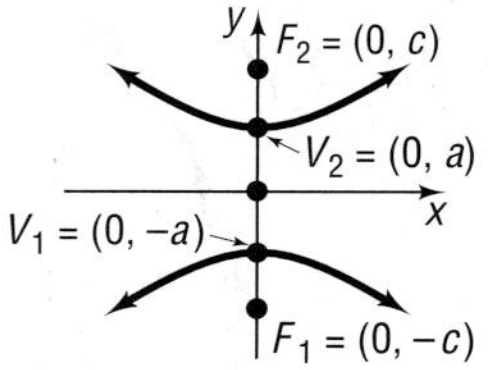

$$\frac{y^2}{a^2} - \frac{x^2}{b^2} = 1, \quad c^2 = a^2 + b^2$$

Asymptotes: $y = \frac{a}{b}x, \quad y = -\frac{a}{b}x$

PROPERTIES OF LOGARITHMS

$\log_a(MN) = \log_a M + \log_a N$

$\log_a\left(\frac{M}{N}\right) = \log_a M - \log_a N$

$\log_a M^r = r\log_a M$

$\log_a M = \frac{\log M}{\log a} = \frac{\ln M}{\ln a}$

PERMUTATIONS/COMBINATIONS

$0! = 1 \qquad 1! = 1$

$n! = n(n-1)\cdot \ldots \cdot (3)(2)(1)$

$P(n, r) = \frac{n!}{(n-r)!}$

$C(n, r) = \binom{n}{r} = \frac{n!}{(n-r)!r!}$

BINOMIAL THEOREM

$$(a + b)^n = a^n + \binom{n}{1}ba^{n-1} + \binom{n}{2}b^2a^{n-2} + \cdots + \binom{n}{n-1}b^{n-1}a + b^n$$

ARITHMETIC SEQUENCE

$$a + (a + d) + (a + 2d) + \cdots + [a + (n-1)d] = na + \frac{n(n-1)}{2}d$$

GEOMETRIC SEQUENCE

$$a + ar + ar^2 + \cdots + ar^{n-1} = a\frac{1 - r^n}{1 - r}$$

GEOMETRIC SERIES

If $|r| < 1$, $a + ar + ar^2 + \cdots = \sum_{k=1}^{\infty} ar^{k-1} = \frac{a}{1 - r}$

TRIGONOMETRIC FUNCTIONS

Let t be a real number and let $P = (x, y)$ be the point on the unit circle that corresponds to t.

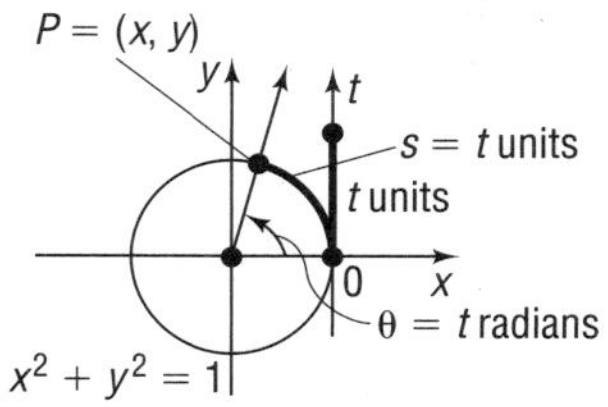

$\sin t = y$ $\qquad \cos t = x$ $\qquad \tan t = \dfrac{y}{x}, \quad x \neq 0$

$\csc t = \dfrac{1}{y}, \quad y \neq 0$ $\qquad \sec t = \dfrac{1}{x}, \quad x \neq 0$ $\qquad \cot t = \dfrac{x}{y}, \quad y \neq 0$

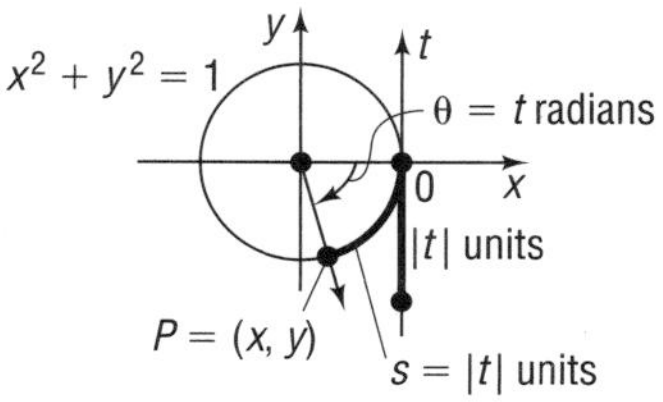

TRIGONOMETRIC IDENTITIES

Fundamental Identities

$$\tan\theta = \frac{\sin\theta}{\cos\theta} \qquad \cot\theta = \frac{\cos\theta}{\sin\theta}$$

$$\csc\theta = \frac{1}{\sin\theta} \qquad \sec\theta = \frac{1}{\cos\theta} \qquad \cot\theta = \frac{1}{\tan\theta}$$

$$\sin^2\theta + \cos^2\theta = 1$$
$$\tan^2\theta + 1 = \sec^2\theta$$
$$\cot^2\theta + 1 = \csc^2\theta$$

Half-Angle Formulas

$$\sin\frac{\theta}{2} = \pm\sqrt{\frac{1 - \cos\theta}{2}}$$
$$\cos\frac{\theta}{2} = \pm\sqrt{\frac{1 + \cos\theta}{2}}$$
$$\tan\frac{\theta}{2} = \frac{1 - \cos\theta}{\sin\theta}$$

Double-Angle Formulas

$$\sin(2\theta) = 2\sin\theta\cos\theta$$
$$\cos(2\theta) = \cos^2\theta - \sin^2\theta$$
$$\cos(2\theta) = 2\cos^2\theta - 1$$
$$\cos(2\theta) = 1 - 2\sin^2\theta$$
$$\tan(2\theta) = \frac{2\tan\theta}{1 - \tan^2\theta}$$

Even-Odd Identities

$$\sin(-\theta) = -\sin\theta \qquad \csc(-\theta) = -\csc\theta$$
$$\cos(-\theta) = \cos\theta \qquad \sec(-\theta) = \sec\theta$$
$$\tan(-\theta) = -\tan\theta \qquad \cot(-\theta) = -\cot\theta$$

Product-to-Sum Formulas

$$\sin\alpha\sin\beta = \tfrac{1}{2}[\cos(\alpha - \beta) - \cos(\alpha + \beta)]$$
$$\cos\alpha\cos\beta = \tfrac{1}{2}[\cos(\alpha - \beta) + \cos(\alpha + \beta)]$$
$$\sin\alpha\cos\beta = \tfrac{1}{2}[\sin(\alpha + \beta) + \sin(\alpha - \beta)]$$

Sum and Difference Formulas

$$\sin(\alpha + \beta) = \sin\alpha\cos\beta + \cos\alpha\sin\beta$$
$$\sin(\alpha - \beta) = \sin\alpha\cos\beta - \cos\alpha\sin\beta$$
$$\cos(\alpha + \beta) = \cos\alpha\cos\beta - \sin\alpha\sin\beta$$
$$\cos(\alpha - \beta) = \cos\alpha\cos\beta + \sin\alpha\sin\beta$$
$$\tan(\alpha + \beta) = \frac{\tan\alpha + \tan\beta}{1 - \tan\alpha\tan\beta}$$
$$\tan(\alpha - \beta) = \frac{\tan\alpha - \tan\beta}{1 + \tan\alpha\tan\beta}$$

Sum-to-Product Formulas

$$\sin\alpha + \sin\beta = 2\sin\frac{\alpha + \beta}{2}\cos\frac{\alpha - \beta}{2}$$
$$\sin\alpha - \sin\beta = 2\sin\frac{\alpha - \beta}{2}\cos\frac{\alpha + \beta}{2}$$
$$\cos\alpha + \cos\beta = 2\cos\frac{\alpha + \beta}{2}\cos\frac{\alpha - \beta}{2}$$
$$\cos\alpha - \cos\beta = -2\sin\frac{\alpha + \beta}{2}\sin\frac{\alpha - \beta}{2}$$

SOLVING TRIANGLES

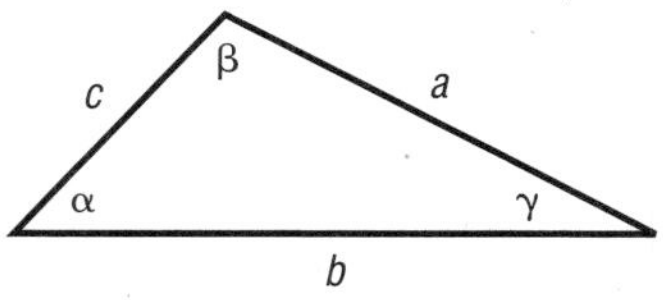

Law of Sines

$$\frac{\sin\alpha}{a} = \frac{\sin\beta}{b} = \frac{\sin\gamma}{c}$$

Law of Cosines

$$c^2 = a^2 + b^2 - 2ab\cos\gamma$$

FORMULAS/EQUATIONS

Distance Formula

If $P_1 = (x_1, y_1)$ and $P_2 = (x_2, y_2)$, the distance from P_1 to P_2 is

$$d(P_1, P_2) = \sqrt{(x_2 - x_1)^2 + (y_2 - y_1)^2}$$

Standard Equation of a Circle

The standard equation of a circle of radius r with center at (h, k) is

$$(x - h)^2 + (y - k)^2 = r^2$$

Slope Formula

The slope m of the line containing the points $P_1 = (x_1, y_1)$ and $P_2 = (x_2, y_2)$ is

$$m = \frac{y_2 - y_1}{x_2 - x_1} \qquad \text{if } x_1 \neq x_2$$

$$m \text{ is undefined} \qquad \text{if } x_1 = x_2$$

Point–Slope Equation of a Line

The equation of a line with slope m containing the point (x_1, y_1) is

$$y - y_1 = m(x - x_1)$$

Slope–Intercept Equation of a Line

The equation of a line with slope m and y-intercept b is

$$y = mx + b$$

Quadratic Formula

The solutions of the equation $ax^2 + bx + c = 0, a \neq 0$, are

$$x = \frac{-b \pm \sqrt{b^2 - 4ac}}{2a}$$

If $b^2 - 4ac > 0$, there are two unequal real solutions.
If $b^2 - 4ac = 0$, there is a repeated real solution.
If $b^2 - 4ac < 0$, there are two complex solutions that are not real.

GEOMETRY FORMULAS

Circle

r = Radius, A = Area, C = Circumference
$A = \pi r^2 \qquad C = 2\pi r$

Triangle

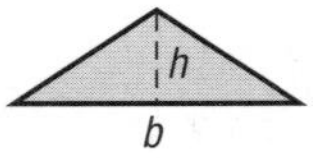

b = Base, h = Altitude (Height), A = area
$A = \frac{1}{2}bh$

Rectangle

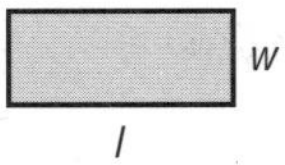

l = Length, w = Width, A = area, P = perimeter
$A = lw \qquad P = 2l + 2w$

Rectangular Box

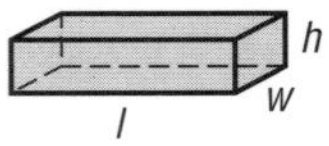

l = Length, w = Width, h = Height, V = Volume, S = Surface area
$V = lwh \qquad S = 2lw + 2lh + 2wh$

Sphere

r = Radius, V = Volume, S = Surface area
$V = \frac{4}{3}\pi r^3 \qquad S = 4\pi r^2$

Right Circular Cylinder

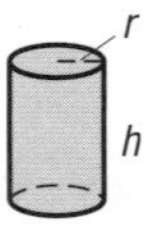

r = Radius, h = Height, V = Volume, S = Surface area
$V = \pi r^2 h \qquad S = 2\pi r^2 + 2\pi rh$